P9-CLI-226

Springer Handbook of Nanotechnology

Springer Handbook provides a concise compilation of approved key information on methods of research, general principles, and functional relationships in physics and engineering. The world's leading experts in the fields of physics and engineering will be assigned by one or several renowned editors to write the chapters comprising each volume. The content is selected by these experts from Springer sources (books, journals, online content) and other systematic and approved recent publications of physical and technical information.

The volumes will be designed to be useful as readable desk reference book to give a fast and comprehensive overview and easy retrieval of essential reliable key information, including tables, graphs, and bibliographies. References to extensive sources are provided.

Springer

Berlin
Heidelberg
New York
Hong Kong
London
Milan
Paris
Tokyo

Springer Handbook of Nanotechnology

Bharat Bhushan (Ed.)

With 972 Figures and 71 Tables

Professor Bharat Bhushan
Nanotribology Laboratory
for Information Storage and MEMS/NEMS
The Ohio State University
206 W. 18th Avenue
Columbus, Ohio 43210-1107
USA

Library of Congress Cataloging-in-Publication Data
Springer handbook of nanotechnology / Bharat Bhushan (ed.)
p. cm.
Includes bibliographical references and index
ISBN 3-540-01218-4 (alk. paper)
1. Nanotechnology--Handbooks, manuals, etc. I. Bhushan, Bharat; 1949-

T174.7S67 2003
620′.5--dc22 2003064953

ISBN 3-540-01218-4
Spinger-Verlag Berlin Heidelberg New York

Springer-Verlag is a part of Springer Science+Business Media

springeronline.com

Printed in Germany

Production and typesetting: LE-TeX GbR, Leipzig
Handbook coordinator: Dr. W. Skolaut, Heidelberg
Typography, layout and illustrations: schreiberVIS, Seeheim
Cover design: eStudio Calamar Steinen, Barcelona
Cover production: *design&production* GmbH, Heidelberg
Printing and binding: Stürtz AG, Würzburg

Printed on acid-free paper
SPIN 10890790 62/3141/YL 5 4 3 2 1

Foreword by Neal Lane

In a January 2000 speech at the California Institute of Technology, former President W. J. Clinton talked about the exciting promise of "nanotechnology" and the importance of expanding research in nanoscale science and engineering and in the physical sciences, more broadly. Later that month, he announced in his State of the Union Address an ambitious $497 million federal, multi-agency national nanotechnology initiative (NNI) in the fiscal year 2001 budget; and he made the NNI a top science and technology priority within a budget that emphasized increased investment in U.S. scientific research. With strong bipartisan support in Congress, most of this request was appropriated, and the NNI was born.

Nanotechnology is the ability to manipulate individual atoms and molecules to produce nanostructured materials and sub-micron objects that have applications in the real world. Nanotechnology involves the production and application of physical, chemical and biological systems at scales ranging from individual atoms or molecules to about 100 nanometers, as well as the integration of the resulting nanostructures into larger systems. Nanotechnology is likely to have a profound impact on our economy and society in the early 21st century, perhaps comparable to that of information technology or advances in cellular and molecular biology. Science and engineering research in nanotechnology promises breakthroughs in areas such as materials and manufacturing, electronics, medicine and healthcare, energy and the environment, biotechnology, information technology and national security. It is widely felt that nanotechnology will lead to the next industrial revolution.

Nanometer-scale features are built up from their elemental constituents. Micro- and nanosystems components are fabricated using batch-processing techniques that are compatible with integrated circuits and range in size from micro- to nanometers. Micro- and nanosystems include Micro/NanoElectroMechanical Systems (MEMS/NEMS), micromechatronics, optoelectronics, microfluidics and systems integration. These systems can sense, control, and activate on the micro/nanoscale and can function individually or in arrays to generate effects on the macroscale. Due to the enabling nature of these systems and the significant impact they can have on both the commercial and defense applications, industry as well as the federal government have taken special interest in seeing growth nurtured in this field. Micro- and nanosystems are the next logical step in the "silicon revolution".

Prof. Neal Lane

University Professor Department of Physics and Astronomy and James A. Baker III Institute for Public Policy Rice University Houston, Texas USA

Served in the Clinton Administration as Assistant to the President for Science and Technology and Director of the White House Office of Science and Technology Policy (1998–2001) and, prior to that, as Director of the National Science Foundation (1993–1998). While at the White House, he was instrumental in creating NNI.

The discovery of novel materials, processes, and phenomena at the nanoscale and the development of new experimental and theoretical techniques for research provide fresh opportunities for the development of innovative nanosystems and nanostructured materials. There is an increasing need for a multidisciplinary, systems-oriented approach to manufacturing micro/nanodevices which function reliably. This can only be achieved through the cross-fertilization of ideas from different disciplines and the systematic flow of information and people among research groups.

Nanotechnology is a broad, highly interdisciplinary, and still evolving field. Covering even the most important aspects of nanotechnology in a single book that reaches readers ranging from students to active researchers in academia and industry is an enormous challenge. To prepare such a wide-ranging book on nanotechnology, Professor Bhushan has harnessed his own knowledge and experience, gained in several industries and universities, and has assembled about 90 internationally recognized authors from three continents to write 38 chapters. The authors come from both academia and industry.

Professor Bharat Bhushan's comprehensive book is intended to serve both as a textbook for university courses as well as a reference for researchers. It is a timely addition to the literature on nanotechnology, which I anticipate will stimulate further interest in this important new field and serve as an invaluable resource to members of the international scientific and industrial community.

The Editor-in-Chief and his team are to be warmly congratulated for bringing together this exclusive, timely, and useful Nanotechnology Handbook.

Foreword by James R. Heath

Nanotechnology has become an increasingly popular buzzword over the past five years or so, a trend that has been fueled by a global set of publicly funded nanotechnology initiatives. Even as researchers have been struggling to demonstrate some of the most fundamental and simple aspects of this field, the term nanotechnology has entered into the public consciousness through articles in the popular press and popular fiction. As a consequence, the expectations of the public are high for nanotechnology, even while the actual public definition of nanotechnology remains a bit fuzzy.

Why shouldn't those expectations be high? The late 1990's witnessed a major information technology (IT) revolution and a minor biotechnology revolution. The IT revolution impacted virtually every aspect of life in the western world. I am sitting on an airplane at 30,000 feet at the moment, working on my laptop, as are about half of the other passengers on this plane. The plane itself is riddled with computational and communications equipment. As soon as we land, many of us will pull out cell phones, others will check email via wireless modem, some will do both. This picture would be the same if I was landing in Los Angeles, Beijing, or Capetown. I will probably never actually print this text, but will instead submit it electronically. All of this was unthinkable a dozen years ago. It is therefore no wonder that the public expects marvelous things to happen quickly. However, the science that laid the groundwork for the IT revolution dates back 60 years or more, with its origins in the fundamental solid state physics.

By contrast, the biotech revolution was relatively minor and, at least to date, not particularly effective. The major diseases that plagued mankind a quarter century ago are still here. In some third world countries, the average lifespan of individuals has actually decreased from where it was a full century ago. While the costs of electronics technologies have plummeted, health care costs have continued to rise. The biotech revolution may have a profound impact, but the task at hand is substantially more difficult to what was required for the IT revolution. In effect, the IT revolution was based on the advanced engineering of two-dimensional digital circuits constructed from relatively simple components – extended solids. The biotech revolution is really dependent upon the ability to reverse engineer three-dimensional analog systems constructed from quite complex components – proteins. Given that the basic science behind biotech is substantially younger than the science that has supported IT, it is perhaps not surprising that the biotech revolution has not really been a proper revolution yet, and it likely needs at least another decade or so to come to fruition.

Where does nanotechnology fit into this picture? In many ways, nanotechnology depends upon the ability to engineer two- and three-dimensional systems constructed from complex components such as macromolecules, biomolecules, nanostructured solids, etc. Furthermore, in terms of patents, publications, and other metrics that can be used to gauge the birth and evolution of a field, nanotech lags some 15–20 years behind biotech. Thus, now is the time that the fundamental science behind nanotechnology is being explored and developed. Nevertheless, progress with that science is moving forward at a dramatic pace. If the scientific community can keep up this pace and if the public sector will continue to support this science, then it is possible, and perhaps even likely, that in 20 years from now we may be speaking of the nanotech revolution.

The Nanotechnology Handbook is timely in assembling chapters in the broad field of nanotechnology with an emphasis on reliability. The handbook should be a valuable reference for experienced researchers as well as for a novice in the field.

Prof. James R. Heath

Department of Chemistry
Mail Code: 127-72
California Institute of Technology
Pasadena, CA 91125, USA

Worked in the group of Nobel Laureate Richard E. Smalley at Rice University (1984–88) and co-invented Fullerene molecules which led to a revolution in Chemistry including the realization of nanotubes. The work on Fullerene molecules was cited for the 1996 Nobel Prize in Chemistry. Later he joined the University of California at Los Angeles (1994–2002), and co-founded and served as a Scientific Director of The California Nanosystems Institute.

Preface

On December 29, 1959 at the California Institute of Technology, Nobel Laureate Richard P. Feynman gave a talk at the Annual meeting of the American Physical Society that has become one classic science lecture of the 20th century, titled "There's Plenty of Room at the Bottom." He presented a technological vision of extreme miniaturization in 1959, several years before the word "chip" became part of the lexicon. He talked about the problem of manipulating and controlling things on a small scale. Extrapolating from known physical laws, Feynman envisioned a technology using the ultimate toolbox of nature, building nanoobjects atom by atom or molecule by molecule. Since the 1980s, many inventions and discoveries in fabrication of nanoobjects have been a testament to his vision. In recognition of this reality, in a January 2000 speech at the same institute, former President W. J. Clinton talked about the exciting promise of "nanotechnology" and the importance of expanding research in nanoscale science and engineering. Later that month, he announced in his State of the Union Address an ambitious $ 497 million federal, multi-agency national nanotechnology initiative (NNI) in the fiscal year 2001 budget, and made the NNI a top science and technology priority. Nanotechnology literally means any technology done on a nanoscale that has applications in the real world. Nanotechnology encompasses production and application of physical, chemical and biological systems at size scales, ranging from individual atoms or molecules to submicron dimensions as well as the integration of the resulting nanostructures into larger systems. Nanofabrication methods include the manipulation or self-assembly of individual atoms, molecules, or molecular structures to produce nanostructured materials and sub-micron devices. Micro- and nanosystems components are fabricated using top-down lithographic and nonlithographic fabrication techniques. Nanotechnology will have a profound impact on our economy and society in the early 21st century, comparable to that of semiconductor technology, information technology, or advances in cellular and molecular biology. The research and development in nanotechnology will lead to potential breakthroughs in areas such as materials and manufacturing, nanoelectronics, medicine and healthcare, energy, biotechnology, information technology and national security. It is widely felt that nanotechnology will lead to the next industrial revolution.

Reliability is a critical technology for many micro- and nanosystems and nanostructured materials. No book exists on this emerging field. A broad based handbook is needed. The purpose of this handbook is to present an overview of nanomaterial synthesis, micro/nanofabrication, micro- and nanocomponents and systems, reliability issues (including nanotribology and nanomechanics) for nanotechnology, and industrial applications. The chapters have been written by internationally recognized experts in the field, from academia, national research labs and industry from all over the world.

The handbook integrates knowledge from the fabrication, mechanics, materials science and reliability points of view. This book is intended for three types of readers: graduate students of nanotechnology, researchers in academia and industry who are active or intend to become active in this field, and practicing engineers and scientists who have encountered a problem and hope to solve it as expeditiously as possible. The handbook should serve as an excellent text for one or two semester graduate courses in nanotechnology in mechanical engineering, materials science, applied physics, or applied chemistry.

We embarked on this project in February 2002, and we worked very hard to get all the chapters to the publisher in a record time of about 1 year. I wish to sincerely thank the authors for offering to write comprehensive chapters on a tight schedule. This is generally an added responsibility in the hectic work schedules of researchers today. I depended on a large number of reviewers who provided critical reviews. I would like to thank Dr. Phillip J. Bond, Chief of Staff and Under Secretary for Technology, US Department of Commerce, Washington, D.C. for suggestions for chapters as well as authors in the handbook. I would also like to thank my colleague, Dr. Huiwen Liu, whose efforts during the preparation of this handbook were very useful.

I hope that this handbook will stimulate further interest in this important new field, and the readers of this handbook will find it useful.

September 2003

Bharat Bhushan
Editor

Editors Vita

Dr. Bharat Bhushan received an M.S. in mechanical engineering from the Massachusetts Institute of Technology in 1971, an M.S. in mechanics and a Ph.D. in mechanical engineering from the University of Colorado at Boulder in 1973 and 1976, respectively, an MBA from Rensselaer Polytechnic Institute at Troy, NY in 1980, Doctor Technicae from the University of Trondheim at Trondheim, Norway in 1990, a Doctor of Technical Sciences from the Warsaw University of Technology at Warsaw, Poland in 1996, and Doctor Honouris Causa from the Metal-Polymer Research Institute of National Academy of Sciences at Gomel, Belarus in 2000. He is a registered professional engineer (mechanical). He is presently an Ohio Eminent Scholar and The Howard D. Winbigler Professor in the Department of Mechanical Engineering, Graduate Research Faculty Advisor in the Department of Materials Science and Engineering, and the Director of the Nanotribology Laboratory for Information Storage & MEMS/NEMS (NLIM) at the Ohio State University, Columbus, Ohio. He is an internationally recognized expert of tribology on the macro- to nanoscales, and is one of the most prolific authors in the field. He is considered by some a pioneer of the tribology and mechanics of magnetic storage devices and a leading researcher in the fields of nanotribology and nanomechanics using scanning probe microscopy and applications to micro/nanotechnology. He has authored 5 technical books, 45 handbook chapters, more than 450 technical papers in referred journals, and more than 60 technical reports, edited more than 25 books, and holds 14 U.S. patents. He is founding editor-in-chief of World Scientific Advances in Information Storage Systems Series, CRC Press Mechanics and Materials Science Series, and Microsystem Technologies – Micro- & Nanosystems and Information Storage & Processing Systems (formerly called Journal of Information Storage and Processing Systems). He has given more than 250 invited presentations on five continents and more than 60 keynote/plenary addresses at major international conferences.

Dr. Bhushan is an accomplished organizer. He organized the first symposium on Tribology and Mechanics of Magnetic Storage Systems in 1984 and the first international symposium on Advances in Information Storage Systems in 1990, both of which are now held annually. He is the founder of an ASME Information Storage and Processing Systems Division founded in 1993 and served as the founding chair during 1993–1998. His biography has been listed in over two dozen Who's Who books including Who's Who in the World and has received more than a dozen awards for his contributions to science and technology from professional societies, industry, and U.S. government agencies. He is also the recipient of various international fellowships including the Alexander von Humboldt Research Prize for Senior Scientists, Max Planck Foundation Research Award for Outstanding Foreign Scientists, and the Fulbright Senior Scholar Award. He is a foreign member of the International Academy of Engineering (Russia), Belorussian Academy of Engineering and Technology and the Academy of Triboengineering of Ukraine, an honorary member of the Society of Tribologists of Belarus, a fellow of ASME, IEEE, and the New York Academy of Sciences, and a member of STLE, ASEE, Sigma Xi and Tau Beta Pi.

Dr. Bhushan has previously worked for the R & D Division of Mechanical Technology Inc., Latham, NY; the Technology Services Division of SKF Industries Inc., King of Prussia, PA; the General Products Division Laboratory of IBM Corporation, Tucson, AZ; and the Almaden Research Center of IBM Corporation, San Jose, CA.

List of Authors

Chong H. Ahn
University of Cincinnati
Department of Electrical and Computer
Engineering and Computer Science
814 Rhodes Hall
Cincinnati, OH 45221-0030, USA
e-mail: *chong.ahn@uc.edu*

Boris Anczykowski
nanoAnalytics GmbH
Gievenbecker Weg 11
48149 Münster, Germany
e-mail: *anczykowski@nanoanalytics.com*

Massood Z. Atashbar
Western Michigan University
Department of Electrical
and Computer Engineering
200 Union Street SE
Kalamazoo, MI 49008-5329, USA
e-mail: *massood.atashbar@wmich.edu*

Wolfgang Bacsa
Université Paul Sabatier
Laboratoire de Physique des Solides (LPST)
118 Route de Narbonne
31062 Toulouse Cedex 4, France
e-mail: *bacsa@lpst.ups-tlse.fr*

William Sims Bainbridge
National Science Foundation
Division of Information and Intelligent Systems
4201 Wilson Boulevard
Arlington, VA 22230, USA
e-mail: *wbainbri@nsf.gov*

Antonio Baldi
Institut de Microelectronica de Barcelona (IMB)
Centro National Microelectrónica (CNM-CSIC)
Campus Universitat Autonoma de Barcelona
08193 Barcelona, Spain
e-mail: *Antoni.baldi@cnm.es*

Philip D. Barnes
Ohio State University
Biomedical Engineering Center
1080 Carmack Road
Columbus, OH 43210, USA
e-mail: *d_skill@yahoo.com*

James D. Batteas
National Institute of Standards and Technology
Surface and Microanalysis Science Division
100 Bureau Drive Mailstop 8372
Gaithersburg, MD 20899-8372, USA
e-mail: *james.batteas@nist.gov*

Roland Bennewitz
McGill University
Physics Department
3600 rue University
Montreal, QC H3A 2T8, Canada
e-mail: *roland@physics.mcgill.ca*

Alan D. Berman
Monitor Venture Enterprises
241 S. Figueroa St. Suite 300
Los Angeles, CA 90012, USA
e-mail: *alan.berman.2001@anderson.ucla.edu*

Bharat Bhushan
The Ohio State University
Nanotribology Laboratory for Information Storage
and MEMS/NEMS
206 W. 18th Avenue
Columbus, OH 43210-1107, USA
e-mail: *bhushan.2@osu.edu*

Gerd K. Binnig
IBM Zurich Research Laboratory
Micro-/Nanomechanics
Säumerstraße 4
8803 Rüschlikon, Switzerland
e-mail: *gbi@zurich.ibm.com*

Marcie R. Black
Massachusetts Institute of Technology
Department of Electrical Engineering
and Computer Science
77 Massachusetts Avenue
Cambridge, MA 02139, USA
e-mail: *marcie@alum.mit.edu*

Jean-Marc Broto
University Toulouse III
Laboratoire National
des Champs Magnétiques Pulsés (LNCMP)
143 Avenue de Rangueil
31432 Toulouse Cedex 4, France
e-mail: *broto@insa-tlse.fr*

Robert W. Carpick
University of Wisconsin-Madison
Department of Engineering Physics
1500 Engineering Drive
Madison, WI 53706-1687, USA
e-mail: *carpick@engr.wisc.edu*

Tsung-Lin Chen
National Chiao Tung University
Department of Mechanical Engineering
30050 Shin Chu, Taiwan
e-mail: *tsunglin@mail.nctu.edu.tw*

Yu-Ting Cheng
National Chiao Tung University
Department of Electronics Engineering
& Institute of Electronics
1001, Ta-Hsueh Road
300 HsinChu, Taiwan
e-mail: *ytcheng@faculty.nctu.edu.tw*

Giovanni Cherubini
IBM Zurich Research Laboratory
Storage Technologies
Säumerstraße 4
8803 Rüschlikon, Switzerland
e-mail: *cbi@zurich.ibm.com*

Jin-Woo Choi
Louisiana State University
Department of Electrical
and Computer Engineering
102 South Campus Drive
Baton Rouge, LA 70803-5901, USA
e-mail: *choi@ece.lsu.edu*

Shawn J. Cunningham
WiSpry, Inc.
Colorado Springs Design Center
7150 Campus Drive, Suite 255
Colorado Springs, CO 80920, USA
e-mail: *shawn.cunningham@wispry.com*

Michel Despont
IBM Zurich Research Laboratory
Micro-/Nanomechanics
Säumerstraße 4
8803 Rüschlikon, Switzerland
e-mail: *dpt@zurich.ibm.com*

Gene Dresselhaus
Massachusetts Institute of Technology
Francis Bitter Magnet Laboratory
77 Massachusetts Avenue
Cambridge, MA 02139, USA
e-mail: *gene@mgm.mit.edu*

Mildred S. Dresselhaus
Massachusetts Institute of Technology
Department of Electrical Engineering
and Computer Science and Department of Physics
77 Massachusetts Avenue
Cambridge, MA 02139, USA
e-mail: *millie@mgm.mit.edu*

Martin L. Dunn
University of Colorado at Boulder
Department of Mechanical Engineering
Campus Box 427
Boulder, CO 80309, USA
e-mail: *martin.dunn@colorado.edu*

Urs T. Dürig
IBM Zurich Research Laboratory
Micro-/Nanomechanics
Säumerstraße 4
8803 Rüschlikon, Switzerland
e-mail: *drg@zurich.ibm.com*

Evangelos Eleftheriou
IBM Zurich Research Laboratory
Storage Technologies
Säumerstraße 4
8803 Rüschlikon, Switzerland
e-mail: *ele@zurich.ibm.com*

Mauro Ferrari
Ohio State University
Biomedical Engineering Center
1080 Carmack Road
Columbus, OH 43210-1002, USA
e-mail: *Ferrari.5@osu.edu*

Emmanuel Flahaut
Université Paul Sabatier
CIRIMAT (Centre Interuniversitaire de Recherche et d'Ingénierie des Matériaux)
118 Route de Narbonne
31062 Toulouse Cedex 04, France
e-mail: *flahaut@chimie.ups-tlse.fr*

Lásló Forró
Swiss Federal Institute of Technology (EPFL)
Institute of Physics of Complex Matter
Ecublens
1015 Lausanne, Switzerland
e-mail: *laszlo.forro@epfl.ch*

Jane Frommer
IBM Almaden Research Center
Department of Science and Technology
650 Harry Road
San Jose, CA 95120, USA
e-mail: *frommer@Almaden.ibm.com*

Harald Fuchs
Universität Münster
Physikalisches Institut
Wilhelm-Klemm-Straße 10
48149 Münster, Germany
e-mail: *fuchsh@uni-muenster.de*

Franz J. Giessibl
Universität Augsburg
Lehrstuhl für Experimentalphysik VI
Universitätsstraße 1
86135 Augsburg, Germany
e-mail: *franz.giessibl@physik.uni-augsburg.de*

Enrico Gnecco
University of Basel
Department of Physics
Klingelbergstraße 82
4056 Basel, Switzerland
e-mail: *Enrico.Gnecco@unibas.ch*

Gérard Gremaud
Swiss Federal Institute of Technology (EPFL)
Institute of Physics of Complex Matter
Ecublens
1015 Lausanne, Switzerland
e-mail: *gremaud@epfl.ch*

Jason H. Hafner
Rice University
Department of Physics & Astronomy
PO BOX 1892
Houston, TX 77251-1892, USA
e-mail: *hafner@rice.edu*

Stefan Hengsberger
University of Applied Science of Fribourg
Bd de Pérolles
1705 Fribourg, Switzerland
e-mail: *stefan.hengsberger@eif.ch*

Peter Hinterdorfer
Johannes Kepler University of Linz
Institute for Biophysics
Altenbergerstraße 69
4040 Linz, Austria
e-mail: *peter.hinterdorfer@jku.at*

Roberto Horowitz
University of California at Berkeley
Department of Mechanical Engineering
5121 Etcheverry Hall
Berkeley, CA 94720-1742, USA
e-mail: *horowitz@me.berkeley.edu*

Hirotaka Hosoi
Japan Science and Technology Corporation
Innovation Plaza, Hokkaido
060-0819 Sapporo, Japan
e-mail: *hosoi@sapporo.jst-plaza.jp*

Jacob N. Israelachvili
University of California
Department of Chemical Engineering
and Materials Department
Santa Barbara, CA 93106, USA
e-mail: *Jacob@engineering.ucsb.edu*

Ghassan E. Jabbour
University of Arizona
Optical Sciences Center
1630 East University Boulevard
Tucson, AZ 85721, USA
e-mail: *gej@optics.arizona.edu*

Harold Kahn
Case Western Reserve University
Department of Materials Science and Engineering
10900 Euclid Avenue
Cleveland, OH 44106-7204, USA
e-mail: *kahn@cwru.edu*

András Kis
Swiss Federal Institute of Technology (EPFL)
Institute of Physics of Complex Matter
Ecublens
1015 Lausanne, Switzerland
e-mail: *andras@igahpse.epfl.ch*

Jané Kondev
Brandeis University
Physics Department
Waltham, MA 02454, USA
e-mail: *kondev@brandeis.edu*

Andrzej J. Kulik
Swiss Federal Institute of Technology (EPFL)
Institute of Physics of Complex Matter
1015 Lausanne, Switzerland
e-mail: *andrzej.kulik@epfl.ch*

Christophe Laurent
Université Paul Sabatier
CIRIMAT (Centre Interuniversitaire de Recherche
et d'Ingénierie des Matériaux)
118 Route de Narbonne
31062 Toulouse Cedex 04, France
e-mail: *laurent@chimie.ups-tlse.fr*

Stephen C. Lee
Ohio State University
Biomedical Engineering Center
1080 Carmack Road
Columbus, OH 43210-1002, USA
e-mail: *Lee@bme.ohio-state.edu*

Yunfeng Li
University of California at Berkeley
Department of Mechanical Engineering
5121 Etcheverry Hall
Berkeley, CA 94720-1740, USA
e-mail: *yunfeng@me.Berkeley.edu*

Liwei Lin
University of California at Berkeley
Mechanical Engineering Department
5126 Etcheverry
Berkeley, CA 94720-1740, USA
e-mail: *lwlin@me.berkeley.edu*

Yu-Ming Lin
Massachusetts Institute of Technology
Department of Electrical Engineering
and Computer Science
77 Massachusetts Avenue
Cambridge, MA 02139, USA
e-mail: *yming@mgm.mit.edu*

Huiwen Liu
Ohio State University
Nanotribology Laboratory for Information Storage and MEMS/NEMS
3070 St John Ct 7
Columbus, OH 43210-1107, USA
e-mail: *liu.403@osu.edu*

Adrian B. Mann
Rutgers University
Department of Ceramics and Materials Engineering
607 Taylor Road
Piscataway, NJ 08854, USA
e-mail: *abmann@rci.rutgers.edu*

Othmar Marti
University of Ulm
Department of Experimental Physics
Albert-Einstein-Allee 11
89069 Ulm, Germany
e-mail: *Othmar.Marti@physik.uni-ulm.de*

Jack Martin
Analog Devices, Inc.
Micromachined Products Division
21 Osborn Street
Cambridge, MA 02139, USA
e-mail: *jack.martin@analog.com*

Brendan McCarthy
University of Arizona
Optical Sciences Center
1630 East University Boulevard
Tucson, AZ 85721, USA
e-mail: *bmccarthy@optics.arizona.edu*

Mehran Mehregany
Case Western Reserve University
Department of Electrical Engineering and Computer Science
188 Bingham Building
Cleveland, OH 44106, USA
e-mail: *mxm31@cwru.edu*

Ernst Meyer
University of Basel
Institute of Physics
Klingelbergstraße 82
4056 Basel, Switzerland
e-mail: *Ernst.Meyer@unibas.ch*

Marc Monthioux
UPR A-8011 CNRS
Centre d'Elaboration des Matériaux et d'Etudes Structurales (CEMES)
29 Rue Jeanne Marvig
31055 Toulouse Cedex 4, France
e-mail: *monthiou@cemes.fr*

Markus Morgenstern
University of Hamburg
Institute of Applied Physics
Jungiusstraße 11
20355 Hamburg, Germany
e-mail: *mmorgens@physnet.uni-hamburg.de*

Seizo Morita
Osaka University
Department of Electronic Engineering
Yamada-Oka 2-1
565-0871 Suita-Citiy, Osaka, Japan
e-mail: *smorita@ele.eng.osaka-u.ac.jp*

Koichi Mukasa
Hokkaido University
Nanoelectronics Laboratory
Nishi-8, Kita-13, Kita-ku
060-8628 Sapporo, Japan
e-mail: *mukasa@nano.eng.hokudai.ac.jp*

Martin H. Müser
University of Western Ontario
Department of Applied Mathematics
WSC 139, Faculty of Science
London, Ontario N6A 5B7, Canada
e-mail: *mmuser@uwo.ca*

Kenn Oldham
University of California at Berkeley
Department of Mechanical Engineering
5121 Etcheverry Hall
Berkeley, CA 94720-1740, USA
e-mail: *oldham@newton.berkeley.edu*

Hiroshi Onishi
Kanagawa Academy of Science and Technology
Surface Chemistry Laboratory
KSP East 404, 3-2-1 Sakado, Takatsu-ku,
Kawasaki-shi
213-0012 Kanagawa, Japan
e-mail: *oni@net.ksp.or.jp*

René M. Overney
University of Washington
Department of Chemical Engineering
Seattle, WA 98195-1750, USA
e-mail: *roverney@u.Washington.edu*

Alain Peigney
Université Paul Sabatier
CIRIMAT (Centre Inter-universitaire de Recherches
et d'Ingénierie des Matériaux) – UMR CNRS 5085
118 Route de Narbonne
31062 Toulouse Cedex 4, France
e-mail: *peigney@chimie.ups-tlse.fr*

Oliver Pfeiffer
University of Basel
Institute of Physics
Klingelbergstraß 82
4056 Basel, Switzerland
e-mail: *Oliver.Pfeiffer@stud.unibas.ch*

Rob Phillips
California Institute of Technology
Mechanical Engineering and Applied Physics
1200 California Boulevard
Pasadena, CA 91125, USA
e-mail: *phillips@aero.caltech.edu*

Haralampos Pozidis
IBM Zurich Research Laboratory
Storage Technologies
Säumerstraße 4
8803 Rüschlikon, Switzerland
e-mail: *hap@zurich.ibm.com*

Prashant K. Purohit
California Institute of Technology
Mechanical Engineering
1200 California Boulevard
Pasadena, CA 91125, USA
e-mail: *prashant@caltech.edu*

Oded Rabin
Massachusetts Institute of Technology
Department of Chemistry
77 Massachusetts Avenue
Cambridge, MA 02139, USA
e-mail: *oded@mgm.mit.edu*

Françisco M. Raymo
University of Miami
Department of Chemistry
1301 Memorial Drive
Coral Gables, FL 33146-0431, USA
e-mail: *fraymo@miami.edu*

Manitra Razafinimanana
Université Paul Sabatier
Centre de Physique des Plasmas
et leurs Applications (CPPAT)
118 Route de Narbonne
31062 Toulouse Cedex, France
e-mail: *razafinimanana@cpat.ups-tlse.fr*

Mark O. Robbins
Johns Hopkins University
Department of Physics and Astronomy
3400 North Charles Street
Baltimore, MD 21218, USA
e-mail: *mr@jhu.edu*

John A. Rogers
University of Illinois
Department of Materials Science and Engineering
1304 W. Green Street
Urbana, IL 61801, USA
e-mail: *jrogers@uiuc.edu*

Mark Ruegsegger
Ohio State University
Biomedical Engineering Center
1080 Carmack Road
Columbus, OH 43210, USA
e-mail: *mark@bme.ohio-state.edu*

Marina Ruths
Åbo Akademi University
Department of Physical Chemistry
Porthansgatan 3–5
20500 Åbo, Finland
e-mail: *mruths@abo.fi*

Dror Sarid
University of Arizona
Optical Sciences Center
1630 East University Boulevard
Tucson, AZ 85721, USA
e-mail: *sarid@optics.arizona.edu*

Akira Sasahara
Kanagawa Academy of Science and Technology
Surface Chemistry Laboratory
KSP East 404, 3-2-1 Sakado, Takatsu-ku,
Kawasaki-shi
213-0012 Kanagawa, Japan
e-mail: *ryo@net.ksp.or.jp*

André Schirmeisen
University of Münster
Institute of Physics
Whilhem-Klemm-Straße 10
48149 Münster, Germany
e-mail: *schira@uni-muenster.de*

Alexander Schwarz
University of Hamburg
Institute of Applied Physics
Jungiusstraße 11
20355 Hamburg, Germany
e-mail: *aschwarz@physnet.uni-hamburg.de*

Udo D. Schwarz
Yale University
Department of Mechanical Engineering
15 Prospect Street
New Haven, CT 06510, USA
e-mail: *udo.schwarz@yale.edu*

Philippe Serp
Ecole Nationale Supèrieure d'Ingénieurs
en Arts Chimiques et Technologiques
Laboratoire de Catalyse, Chimie Fine et Polymères
118 Route de Narbonne
31077 Toulouse, France
e-mail: *Philippe.Serp@ensiacet.fr*

Bryan R. Smith
Ohio State University
Biomedical Engineering Center
1080 Carmack Road
Columbus, OH 43210, USA
e-mail: *bryan@bme.ohio-state.edu*

Anisoara Socoliuc
University of Basel
Institute of Physics
Klingelbergstraße 82
4056 Basel, Switzerland
e-mail: *A.Socoliuc@unibas.ch*

Yasuhiro Sugawara
Osaka University
Department of Applied Physics
Yamada-Oka 2-1
565-0871 Suita, Japan
e-mail: *sugawara@ap.eng.osaka-u.ac.jp*

George W. Tyndall
IBM Almaden Research Center
Science and Technology
6369 Didion Ct.
San Jose, CA 95123, USA
e-mail: *tyndallgw@netscape.net*

Peter Vettiger
IBM Zurich Research Laboratory
Manager Micro-/Nanomechanics
Säumerstraße 4
8803 Rüschlikon, Switzerland
e-mail: *pv@zurich.ibm.com*

Darrin J. Young
Case Western Reserve University
Electrical Engineering and Computer Science
10900 Euclid Avenue
Cleveland, OH 44106, USA
e-mail: *djy@po.cwru.edu*

Babak Ziaie
University of Minnesota
Department of Electrical
and Computer Engineering
200 Union Street SE
Minneapolis, MN 55455, USA
e-mail: *ziaie@ece.umn.edu*

Christian A. Zorman
Case Western Reserve University
Department of Electrical Engineering
and Computer Science
719 Glennan Building
Cleveland, OH 44106, USA
e-mail: *caz@po.cwru.edu*

Philippe K. Zysset
Technische Universität Wien
Institut für Leichtbau und Flugzeugbau (ILFB)
Gußhausstraße 27–29
1040 Wien, Austria
e-mail: *philippe.zysset@epfl.ch*

Contents

List of Tables ... XXIX
List of Abbreviations ... XXXIII

1 Introduction to Nanotechnology ... 1
1.1 Background and Definition of Nanotechnology ... 1
1.2 Why Nano? ... 2
1.3 Lessons from Nature ... 2
1.4 Applications in Different Fields ... 3
1.5 Reliability Issues of MEMS/NEMS ... 4
1.6 Organization of the Handbook ... 5
References ... 5

Part A Nanostructures, Micro/Nanofabrication, and Micro/Nanodevices

2 Nanomaterials Synthesis and Applications: Molecule-Based Devices ... 9
2.1 Chemical Approaches to Nanostructured Materials ... 10
2.2 Molecular Switches and Logic Gates ... 14
2.3 Solid State Devices ... 22
2.4 Conclusions and Outlook ... 35
References ... 36

3 Introduction to Carbon Nanotubes ... 39
3.1 Structure of Carbon Nanotubes ... 40
3.2 Synthesis of Carbon Nanotubes ... 45
3.3 Growth Mechanisms of Carbon Nanotubes ... 59
3.4 Properties of Carbon Nanotubes ... 63
3.5 Carbon Nanotube-Based Nano-Objects ... 68
3.6 Applications of Carbon Nanotubes ... 73
References ... 86

4 Nanowires ... 99
4.1 Synthesis ... 100
4.2 Characterization and Physical Properties of Nanowires ... 110
4.3 Applications ... 131
4.4 Concluding Remarks ... 138
References ... 138

5 Introduction to Micro/Nanofabrication ... 147
5.1 Basic Microfabrication Techniques ... 148
5.2 MEMS Fabrication Techniques ... 159

5.3 Nanofabrication Techniques 170
References 180

6 **Stamping Techniques for Micro and Nanofabrication: Methods and Applications** 185
6.1 High Resolution Stamps 186
6.2 Microcontact Printing 187
6.3 Nanotransfer Printing 190
6.4 Applications 193
6.5 Conclusions 200
References 200

7 **Materials Aspects of Micro- and Nanoelectromechanical Systems** 203
7.1 Silicon 203
7.2 Germanium-Based Materials 210
7.3 Metals 211
7.4 Harsh Environment Semiconductors 212
7.5 GaAs, InP, and Related III-V Materials 217
7.6 Ferroelectric Materials 218
7.7 Polymer Materials 219
7.8 Future Trends 220
References 221

8 **MEMS/NEMS Devices and Applications** 225
8.1 MEMS Devices and Applications 227
8.2 NEMS Devices and Applications 246
8.3 Current Challenges and Future Trends 249
References 250

9 **Microfluidics and Their Applications to Lab-on-a-Chip** 253
9.1 Materials for Microfluidic Devices and Micro/Nano Fabrication Techniques 254
9.2 Active Microfluidic Devices 257
9.3 Smart Passive Microfluidic Devices 262
9.4 Lab-on-a-Chip for Biochemical Analysis 270
References 276

10 **Therapeutic Nanodevices** 279
10.1 Definitions and Scope of Discussion 280
10.2 Synthetic Approaches: "top-down" versus "bottom-up" Approaches for Nanotherapeutic Device Components 285
10.3 Technological and Biological Opportunities 288
10.4 Applications for Nanotherapeutic Devices 307
10.5 Concluding Remarks: Barriers to Practice and Prospects 315
References 317

Part B Scanning Probe Microscopy

11 **Scanning Probe Microscopy – Principle of Operation, Instrumentation, and Probes** 325
11.1 Scanning Tunneling Microscope 327
11.2 Atomic Force Microscope 331
11.3 AFM Instrumentation and Analyses 347
References 364

12 **Probes in Scanning Microscopies** 371
12.1 Atomic Force Microscopy 372
12.2 Scanning Tunneling Microscopy 382
References 383

13 **Noncontact Atomic Force Microscopy and Its Related Topics** 385
13.1 Principles of Noncontact Atomic Force Microscope (NC-AFM) 386
13.2 Applications to Semiconductors 391
13.3 Applications to Insulators 397
13.4 Applications to Molecules 404
References 407

14 **Low Temperature Scanning Probe Microscopy** 413
14.1 Microscope Operation at Low Temperatures 414
14.2 Instrumentation 415
14.3 Scanning Tunneling Microscopy and Spectroscopy 419
14.4 Scanning Force Microscopy and Spectroscopy 433
References 442

15 **Dynamic Force Microscopy** 449
15.1 Motivation: Measurement of a Single Atomic Bond 450
15.2 Harmonic Oscillator: A Model System for Dynamic AFM 454
15.3 Dynamic AFM Operational Modes 455
15.4 *Q*-Control 464
15.5 Dissipation Processes Measured with Dynamic AFM 468
15.6 Conclusion 471
References 471

16 **Molecular Recognition Force Microscopy** 475
16.1 Ligand Tip Chemistry 476
16.2 Fixation of Receptors to Probe Surfaces 478
16.3 Single-Molecule Recognition Force Detection 479
16.4 Principles of Molecular Recognition Force Spectroscopy 482
16.5 Recognition Force Spectroscopy: From Isolated Molecules to Biological Membranes 484
16.6 Recognition Imaging 490
16.7 Concluding Remarks 492
References 493

Part C Nanotribology and Nanomechanics

17 **Micro/Nanotribology and Materials Characterization Studies Using Scanning Probe Microscopy** 497
17.1 Description of AFM/FFM and Various Measurement Techniques 499
17.2 Friction and Adhesion 507
17.3 Scratching, Wear, Local Deformation, and Fabrication/Machining ... 518
17.4 Indentation 526
17.5 Boundary Lubrication 530
17.6 Closure 538
References 539

18 **Surface Forces and Nanorheology of Molecularly Thin Films** 543
18.1 Introduction: Types of Surface Forces 544
18.2 Methods Used to Study Surface Forces 546
18.3 Normal Forces Between Dry (Unlubricated) Surfaces 550
18.4 Normal Forces Between Surfaces in Liquids 554
18.5 Adhesion and Capillary Forces 564
18.6 Introduction: Different Modes of Friction and the Limits of Continuum Models 569
18.7 Relationship Between Adhesion and Friction Between Dry (Unlubricated and Solid Boundary Lubricated) Surfaces 571
18.8 Liquid Lubricated Surfaces 580
18.9 Role of Molecular Shape and Surface Structure in Friction 591
References 594

19 **Scanning Probe Studies of Nanoscale Adhesion Between Solids in the Presence of Liquids and Monolayer Films** 605
19.1 The Importance of Adhesion at the Nanoscale 605
19.2 Techniques for Measuring Adhesion 606
19.3 Calibration of Forces, Displacements, and Tips 610
19.4 The Effect of Liquid Capillaries on Adhesion 612
19.5 Self-Assembled Monolayers 618
19.6 Concluding Remarks 624
References 624

20 **Friction and Wear on the Atomic Scale** 631
20.1 Friction Force Microscopy in Ultra-High Vacuum 632
20.2 The Tomlinson Model 636
20.3 Friction Experiments on Atomic Scale 638
20.4 Thermal Effects on Atomic Friction 642
20.5 Geometry Effects in Nanocontacts 646
20.6 Wear on the Atomic Scale 649
20.7 Molecular Dynamics Simulations of Atomic Friction and Wear 651
20.8 Energy Dissipation in Noncontact Atomic Force Microscopy 654
20.9 Conclusion 656
References 657

21 **Nanoscale Mechanical Properties – Measuring Techniques and Applications** 661
21.1 Local Mechanical Spectroscopy by Contact AFM 662
21.2 Static Methods – Mesoscopic Samples 667
21.3 Scanning Nanoindentation: An Application to Bone Tissue 674
21.4 Conclusions and Perspectives 682
References 682

22 **Nanomechanical Properties of Solid Surfaces and Thin Films** 687
22.1 Instrumentation 688
22.2 Data Analysis 694
22.3 Modes of Deformation 702
22.4 Thin Films and Multilayers 707
22.5 Developing Areas 711
References 712

23 **Atomistic Computer Simulations of Nanotribology** 717
23.1 Molecular Dynamics 718
23.2 Friction Mechanisms at the Atomic Scale 723
23.3 Stick-Slip Dynamics 732
23.4 Conclusions 734
References 735

24 **Mechanics of Biological Nanotechnology** 739
24.1 Science at the Biology–Nanotechnology Interface 740
24.2 Scales at the Bio-Nano Interface 746
24.3 Modeling at the Nano-Bio Interface 752
24.4 Nature's Nanotechnology Revealed: Viruses as a Case Study 755
24.5 Concluding Remarks 760
References 761

25 **Mechanical Properties of Nanostructures** 763
25.1 Experimental Techniques for Measurement of Mechanical Properties of Nanostructures 765
25.2 Experimental Results and Discussion 770
25.3 Finite Element Analysis of Nanostructures with Roughness and Scratches 778
25.4 Closure 785
References 786

Part D Molecularly Thick Films for Lubrication

26 **Nanotribology of Ultrathin and Hard Amorphous Carbon Films** 791
26.1 Description of Commonly Used Deposition Techniques 795
26.2 Chemical Characterization and Effect of Deposition Conditions on Chemical Characteristics and Physical Properties 800

26.3 Micromechanical and Tribological Characterizations of Coatings Deposited by Various Techniques ... 805
References ... 827

27 **Self-Assembled Monolayers for Controlling Adhesion, Friction and Wear** ... 831
27.1 A Primer to Organic Chemistry ... 834
27.2 Self-Assembled Monolayers: Substrates, Head Groups, Spacer Chains, and End Groups ... 839
27.3 Tribological Properties of SAMs ... 841
27.4 Closure ... 856
References ... 857

28 **Nanoscale Boundary Lubrication Studies** ... 861
28.1 Lubricants Details ... 862
28.2 Nanodeformation, Molecular Conformation, and Lubricant Spreading ... 864
28.3 Boundary Lubrication Studies ... 866
28.4 Closure ... 880
References ... 881

29 **Kinetics and Energetics in Nanolubrication** ... 883
29.1 Background: From Bulk to Molecular Lubrication ... 885
29.2 Thermal Activation Model of Lubricated Friction ... 887
29.3 Functional Behavior of Lubricated Friction ... 888
29.4 Thermodynamical Models Based on Small and Nonconforming Contacts ... 890
29.5 Limitation of the Gaussian Statistics – The Fractal Space ... 891
29.6 Fractal Mobility in Reactive Lubrication ... 892
29.7 Metastable Lubricant Systems in Large Conforming Contacts ... 894
29.8 Conclusion ... 895
References ... 895

Part E Industrial Applications and Microdevice Reliability

30 **Nanotechnology for Data Storage Applications** ... 899
30.1 Current Status of Commercial Data Storage Devices ... 901
30.2 Opportunities Offered by Nanotechnology for Data Storage ... 907
30.3 Conclusion ... 918
References ... 919

31 **The "Millipede" – A Nanotechnology-Based AFM Data-Storage System** ... 921
31.1 The Millipede Concept ... 923
31.2 Thermomechanical AFM Data Storage ... 924
31.3 Array Design, Technology, and Fabrication ... 926

31.4 Array Characterization ... 927
31.5 *x*/*y*/*z* Medium Microscanner ... 929
31.6 First Write/Read Results with the 32×32 Array Chip ... 931
31.7 Polymer Medium ... 932
31.8 Read Channel Model ... 939
31.9 System Aspects ... 943
31.10 Conclusions ... 948
References ... 948

32 **Microactuators for Dual-Stage Servo Systems in Magnetic Disk Files** ... 951
32.1 Design of the Electrostatic Microactuator ... 952
32.2 Fabrication ... 962
32.3 Servo Control Design of MEMS Microactuator Dual-Stage Servo Systems ... 968
32.4 Conclusions and Outlook ... 978
References ... 979

33 **Micro/Nanotribology of MEMS/NEMS Materials and Devices** ... 983
33.1 Introduction to MEMS ... 985
33.2 Introduction to NEMS ... 988
33.3 Tribological Issues in MEMS/NEMS ... 989
33.4 Tribological Studies of Silicon and Related Materials ... 995
33.5 Lubrication Studies for MEMS/NEMS ... 1003
33.6 Component-Level Studies ... 1009
References ... 1017

34 **Mechanical Properties of Micromachined Structures** ... 1023
34.1 Measuring Mechanical Properties of Films on Substrates ... 1023
34.2 Micromachined Structures for Measuring Mechanical Properties ... 1024
34.3 Measurements of Mechanical Properties ... 1034
References ... 1037

35 **Thermo- and Electromechanics of Thin-Film Microstructures** ... 1039
35.1 Thermomechanics of Multilayer Thin-Film Microstructures ... 1041
35.2 Electromechanics of Thin-Film Microstructures ... 1061
35.3 Summary and Mention of Topics not Covered ... 1078
References ... 1078

36 **High Volume Manufacturing and Field Stability of MEMS Products** ... 1083
36.1 Manufacturing Strategy ... 1086
36.2 Robust Manufacturing ... 1087
36.3 Stable Field Performance ... 1102
References ... 1106

37 **MEMS Packaging and Thermal Issues in Reliability** 1111
37.1 MEMS Packaging 1111
37.2 Hermetic and Vacuum Packaging and Applications 1116
37.3 Thermal Issues and Packaging Reliability 1122
37.4 Future Trends and Summary 1128
References 1129

Part F Social and Ethical Implication

38 **Social and Ethical Implications of Nanotechnology** 1135
38.1 Applications and Societal Impacts 1136
38.2 Technological Convergence 1139
38.3 Major Socio-technical Trends 1141
38.4 Sources of Ethical Behavior 1143
38.5 Public Opinion 1145
38.6 A Research Agenda 1148
References 1149

Acknowledgements 1153
About the Authors 1155
Detailed Contents 1171
Subject Index 1189

List of Tables

Part A Nanostructures, Micro/Nanofabrication, and Micro/Nanodevices

3 **Introduction to Carbon Nanotubes**

Table 3.1 Different carbon morphologies 50

Table 3.2 Carbon nanofilament morphologies and some basic synthesis conditions 62

Table 3.3 Adsorption properties and sites of SWNTs and MWNTs 65

Table 3.4 Preparation and catalytic performances of some nanotube-supported catalysts 77

Table 3.5 Applications for nanotube-based multifunctional materials 84

4 **Nanowires**

Table 4.1 Selected syntheses of nanowires by material 101

5 **Introduction to Micro/Nanofabrication**

Table 5.1 Typical dry etch chemistries 157

Table 5.2 Surface micromachined sacrificial/structural combinations 165

7 **Materials Aspects of Micro- and Nanoelectromechanical Systems**

Table 7.1 Selected materials for MEMS and NEMS 204

9 **Microfluidics and Their Applications to Lab-on-a-Chip**

Table 9.1 Overview of the different polymer micro/nano fabrication techniques 257

10 **Therapeutic Nanodevices**

Table 10.1 Ideal characteristics of nanodevices 282

Table 10.2 Orthogonal conjugation chemistries 295

Part B Scanning Probe Microscopy

11 **Scanning Probe Microscopy – Principle of Operation, Instrumentation, and Probes**

Table 11.1 Comparison of Various Conventional Microscopes with SPMs 326

Table 11.2 Relevant properties of materials used for cantilevers 339

Table 11.3 Measured vertical spring constants and natural frequencies of triangular (V-shaped) cantilevers made of PECVD Si_3N_4 341

Table 11.4 Vertical and torsional spring constants of rectangular cantilevers made of Si and PECVD 341

Table 11.5 Noise in Interferometers 353

13 **Noncontact Atomic Force Microscopy and Its Related Topics**
Table 13.1 Comparison between the adatom heights observed in an AFM image and the variety of properties for inequivalent adatoms 393

Part C Nanotribology and Nanomechanics

17 **Micro/Nanotribology and Materials Characterization Studies Using Scanning Probe Microscopy**
Table 17.1 Comparison of typical operating parameters in SFA, STM, and AFM/FFM used for micro/nanotribological studies 498
Table 17.2 Surface roughness and coefficients of friction of various samples in air 518

18 **Surface Forces and Nanorheology of Molecularly Thin Films**
Table 18.1 Types of surface forces in vacuum vs. in liquid (colloidal forces) 545
Table 18.2 Van der Waals interaction energy and force between macroscopic bodies of different geometries 551
Table 18.3 Electrical "double-layer" interaction energy and force between macroscopic bodies 556
Table 18.4 The three main tribological regimes characterizing the changing properties of liquids 570
Table 18.5 Effect of molecular shape and short-range forces on tribological properties 592

19 **Scanning Probe Studies of Nanoscale Adhesion Between Solids in the Presence of Liquids and Monolayer Films**
Table 19.1 Interactions evaluated by atomic force microscopy adhesion measurements 620

21 **Nanoscale Mechanical Properties – Measuring Techniques and Applications**
Table 21.1 Summary of the mechanical properties of carbon nanotubes.. 672
Table 21.2 Absolute and relative changes of the indentation modulus with respect to the initial values under wet conditions 680

22 **Nanomechanical Properties of Solid Surfaces and Thin Films**
Table 22.1 Results for some experimental studies of multilayer hardness 710

24 **Mechanics of Biological Nanotechnology**
Table 24.1 Sizes of entities in comparison to the size of *E. coli* 748
Table 24.2 Time scales of various events in biological nanotechnology 750

25 **Mechanical Properties of Nanostructures**
Table 25.1 Hardness, elastic modulus, fracture toughness, and critical load results of the bulk single-crystal Si(100) and thin films of undoped polysilicon, SiO_2, SiC, Ni-P, and Au 770

Table 25.2 Summary of measured parameters from quasi-static bending tests 775
Table 25.3 Stresses and displacements for materials that are elastic, elastic-plastic, or elastic-perfectly plastic 784

Part D Molecularly Thick Films for Lubrication

26 **Nanotribology of Ultrathin and Hard Amorphous Carbon Films**

Table 26.1 Summary of the most commonly used deposition techniques. 796
Table 26.2 Experimental results from EELS and Raman spectroscopy 801
Table 26.3 Experimental results of FRS analysis 802
Table 26.4 Experimental results of physical properties 802
Table 26.5 Hardness, elastic modulus, fracture toughness, fatigue life, critical load during scratch, coefficient of friction of various DLC coatings 809

27 **Self-Assembled Monolayers for Controlling Adhesion, Friction and Wear**

Table 27.1 Relative electronegativity of selected elements 835
Table 27.2 Names and formulas of selected hydrocarbons 836
Table 27.3 Names and formulas of selected alcohols, ethers, phenols, and thiols 836
Table 27.4 Names and formulas of selected aldehydes and ketones 837
Table 27.5 Names and formulas of selected carboxylic acids and esters ... 838
Table 27.6 Names and formulas of selected organic nitrogen compounds 838
Table 27.7 Some examples of polar and nonpolar groups 839
Table 27.8 Organic groups listed in the increasing order of polarity 839
Table 27.9 Selected substrates and precursors used for formation of SAMs 840
Table 27.10 The roughness, thickness, tilt angles, and spacer chain lengths of SAMs 846
Table 27.11 Melting point of typical organic compounds similar to HDT and BPT SAMs 851
Table 27.12 Calculated and measured relative heights of HDT and MHA self-assembled monolayers 855
Table 27.13 Calculated and measured residual film thickness for SAMs under critical load 855
Table 27.14 Bond strengths of the chemical bonds in SAMs 856

28 **Nanoscale Boundary Lubrication Studies**

Table 28.1 Typical properties of Z-15 and Z-DOL 863

Part E Industrial Applications and Microdevice Reliability

30 **Nanotechnology for Data Storage Applications**

Table 30.1 A comparison of the parameters of HDD and ODD 903

Table 30.2 Perceived technical limits for hard disk drive technology........ 904
Table 30.3 A table of probe storage operating parameters 909
Table 30.4 Probe storage projected performance 909

31 The "Millipede" – A Nanotechnology-Based AFM Data-Storage System
Table 31.1 Areal density and storage capacity 947

32 Microactuators for Dual-Stage Servo Systems in Magnetic Disk Files
Table 32.1 Parameters of the electrostatic microactuator 960

33 Micro/Nanotribology of MEMS/NEMS Materials and Devices
Table 33.1 Dimensions and masses in perspective 985
Table 33.2 Selected bulk properties of SiC and Si(100)............................ 996
Table 33.3 Surface roughness and micro- and macroscale coefficients of friction of selected samples.. 997
Table 33.4 RMS, microfriction, microscratching/microwear, and nanoindentation hardness data for various virgin, coated, and treated silicon samples 999
Table 33.5 Summary of micro/nanotribological properties of the sample materials 1001
Table 33.6 Surface roughness parameters and microscale coefficient of friction 1009
Table 33.7 Coefficient of static friction measurements of MEMS devices and structures 1015

35 Thermo- and Electromechanics of Thin-Film Microstructures
Table 35.1 Stresses, curvature, and strains in a film/substrate system...... 1045
Table 35.2 Predicted and measured curvature per unit negative temperature change for gold/polysilicon microstructures as a function of the polysilicon thickness. 1053

37 MEMS Packaging and Thermal Issues in Reliability
Table 37.1 Summary of bonding mechanisms... 1116
Table 37.2 The Maximum Likelihood Estimation for Mean Time To Failure (MTTF) 1127

List of Abbreviations

μCP	microcontact printing
μTAS	micro-total analysis systems
2-DEG	two-dimensional electron gas

A

A	adenine
ABS	air-bearing surface
ADEPTS	antibody directed enzyme-prodrug therapy
AFAM	atomic force acoustic microscopy
AFM	atomic force microscope/microscopy
AIDCN	2-amino-4,5-imidazoledicarbonitrile
AM	amplitude modulation
APCVD	atmospheric pressure chemical vapor deposition
ASA	anti-stiction agent
ATP	adenosine triphosphat

B

BE	boundary element
BioMEMS	biological or biomedical microelectromechanical systems
BP	bit pitch
BPI	bits per inch
bpsi	bits per square inch
BSA	bovine serum albumin

C

C	cytosine
CA	constant amplitude
CBA	cantilever beam array
CCVD	catalytic chemical vapor deposition
CDS	correlated double sampling
CDW	charge density wave
CE	capillary electrophoresis
CE	constant excitation mode
CFM	chemical force microscopy
CG	controlled geometry
CNT	carbon nanotube
COC	cyclic olefin copolymers
COF	chip-on-flex
COGs	cost of goods
CSM	continuous stiffness measurement
CTE	coefficient of thermal expansion
CTLs	cytoxic T-lymphocytes
CVD	chemical vapor deposition

D

DAS	dimer adatom stacking
DBR	distributed Bragg reflector
DCs	dendritic cells
DFB	distributed feedback
DFM	dynamic force microscopy
DFT	density functional theory
DLC	diamond-like carbon
DLP	digital light processing
DLVO	Derjaguin–Landau–Verwey–Overbeek
DMD	digital micromirror device
DMT	Derjaguin–Muller–Toporov
DOS	density of states
DPN	dip-pen nanolithography
DRIE	deep reactive ion etching
DSC	differential scanning calorimetry
DSP	digital signal processor
DT	diphteria toxin
DWNTs	double-wall nanotubes

E

EAM	embedded atom method
EBD	electron beam deposition
ECR-CVD	electron cyclotron resonance chemical vapor deposition
EDC	1-ethyl-3-(3-diamethylaminoprophyl) carbodiimide
EDP	ethylene diamine pyrocatechol
EDS	energy dispersive X-ray spectrometer
EELS	electron energy loss spectrometer/spectroscopy
EFC	electrostatic force constant
EFM	electric field gradient microscopy
EHD	electrohydrodynamic
EL	electro-luminiscence
EO	electro-osmosis
EOF	electro-osmotic flow
EPR	enhanced permeability and retention effect
ESD	electrostatic discharge

F

FAD	flavin adenine dinucleotide
FC	flip chip technique
FCA	filtered cathodic arc
FCP	force calibration plot
FD	finite difference
FE	finite element
FEM	finite element method/modeling

FET	field-effect transistor
FFM	friction force microscope/microscopy
FIB	focused ion beam
FID	free induction decay
FIM	field-ion microscope/microscopy
FKT	Frenkel-Kontorova-Tomlinson
FM	frequency modulation
FM-AFM	frequency modulation AFM
FMEA	failure mode effect analysis
FMM	force modulation mode
FM-SFM	frequency-modulation SFM
FS	force spectroscopy

G

G	guanine
GIO	grazing impact oscillator
GMR	giant magnetoresistance
GOD	glucose oxidase
Gox	flavoenzyme glucose oxidase

H

HARMEMS	high-aspect-ratio MEMS
HDD	hard disk drive
HF	hydrofluoric acid
HOP	highly oriented pyrolytic
HOPG	highly oriented pyrolytic graphite
HPMA	hydroxyl polymethacrylamide
HtBDC	hexa-tert-butyl-decacyclene
HTCS	high temperature superconductivity

I

IBD	ion beam deposition
IC	integrated circuit
ICAM-1	intercellular adhesion molecule-1
IFM	interfacial force microscope
ISE	indentation size effect
ITO	indium tin oxide

J

JKR	Johnson–Kendall–Roberts

K

KPFM	Kelvin probe force microscopy

L

LB	Langmuir–Blodgett
LCC	leadless chip carrier
LDOS	local density of states
LEDs	light-emitting diodes
LFA-1	leukocyte function-associated antigen-1
LFM	lateral force microscope
LFT	linear fractional transformation
LN	liquid nitrogen
LPCVD	low pressure chemical vapor deposition
LQG	linear quadratic Gaussian
LTR	loop-transfer recovery
LTSPM	low-temperature SPM
LTSTS	low-temperature scanning tunneling spectroscopy
LVDT	linear variable differential transformers

M

MAP	manifold absolute pressure
MD	molecular dynamics
MDS	molecular dynamics simulation
ME	metal evaporated
MEMS	microelectromechanical systems
MFM	magnetic field microscope/microscopy
MHA	16-mercaptohexadecanoic acid thiol
MHC	major histocompatibility complex
MHD	magnetohydrodynamic
MIM	metal/insulator/metal
MLE	maximum likelihood estimator
MOCVD	metalorganic CVD
MP	metal particle
MPTMS	mercaptopropyltrimethoxysilane
MRAM	magnetoresistive RAM
MRFM	magnetic resonance force microscopy
MRFM	molecular recognition force microscopy
MRI	magnetic resonance imaging
MTTF	mean time to failure
MWCNT	multiwall carbon nanotube

N

NA	nucleic acid
NBMN	3-nitrobenzal malonitrile
NC-AFM	noncontact atomic force microscopy
NCS	neocarzinostatin
NEMS	nanoelectromechanical systems
NMP	no moving part
NNI	National Nanotechnology Initiative
NSOM	near-field scanning optical microscope/microscopy
NTA	nitrilotriacetate
nTP	nanotransfer printing
NVRAM	nonvolatile random access memories

O

ODD	optical disk drives
OMVPE	organometallic vapor phase epitaxy
OT	optical tweezers
OTE	octadecyltrimethoxysilane
OTS	octadecyltrichlorosilane

OUM	Ovonyx unified memory

P

P/W	power to weight
PA	plasminogen activator
PAMAM	poly(amido) amine
PAPP	p-aminophenyl phosphate
PC	ploycarbonate
PC-RAM	phase change RAM
pDA	1,4-phenylenediamine
PDMS	polydimethylsiloxane
PDP	2-pyridyldithiopropionyl
PE	polyethylene
PECVD	plasma enhanced CVD
PEG	poly(ethylene glycol)
PES	photoemission spectroscopy
PES	position error signal
PET	poly(ethylene terephthalate)
PFDA	perfluorodecanoic acid
PFPE	perfluoropolyether
PL	photoluminescence
PMMA	poly(methylmethacrylate)
PS	polystyrene
PSG	phosphorus-doped glass
PSGL-1	P-selection glycoprotein ligand-1
PTFE	polytetrafluoroethylene
PZT	lead zirconate titanate

Q

QCM	quartz-crystal microbalance

R

RES	reticuloendothelial system
RF	radiofrequency
RH	relative humidity
RICM	reflection interface contrast microscopy
RIE	reactive ion etching
RLS	recursive least square algorithm
RPES	relative position error signal
RTP	rapid thermal processing

S

SACA	static advancing contact angle
SAED	selected area electron diffraction
SAM	self-assembling monolayer
SAM	scanning acoustic microscopy
SCM	scanning capitance microscopy
SCPM	scanning chemical potential
SEcM	scanning electrochemical microscopy
SEFM	scanning electrostatic force microscopy
SEM	scanning electron microscope/microscopy
SFA	surface forces apparatus
SFAM	scanning force acoustic microscopy
SFD	shear flow detachment
SFM	scanning force microscopy
SFS	scanning force spectroscopy
SICM	scanning ion conductance microscopy
SIMO	single-input–multi-output
SISO	single-input–single-output
SKPM	scanning Kelvin probe microscopy
SLAM	scanning local-acceleration microscopy
SMA	shape memory alloy
SMANCS	S-Methacryl-neocarzinostatin
SMM	scanning magnetic microscopy
SN	scanning nanoindentation
SNOM	scanning near-field optical microscopy
SPM	scanning probe microscopy
sPROMS	structurally programmable microfluidic system
SPS	spark plasma sintering
SRAM	static random access memory
ssDNA	single stranded DNA
SSNA	single-stranded nucleic acid molecule
SThM	scanning thermal microscopy
STM	scanning tunneling microscope/microscopy
SWCNT	single-wall carbon nanotubes
SWNT	single-wall nanotubes

T

T	thymine
TAAs	tumor associated antigens
TEM	transmission electron microscopy
TESP	tapping-mode etched silicon probe
TGA	thermo-gravimetric analysis
TIRM	total internal reflection microscopy
TMAH	tetramethyl ammonium hydroxide
TMAH	tetramethyl-aluminium hydroxide
TP	track pitch
TPI	tracks per inch
TRM/TMR	track mis-registration
T-SLAM	variable-temperature SLAM
TTF	tetrathiofulvane

U

UHV	ultrahigh vacuum

V

VCM	voice coil motor
VCO	voltage-controlled oscillator
VLS	vapor-liquid-solid
VLSI	very large-scale integration

X

XRD	X-ray diffraction

1. Introduction to Nanotechnology

A biological system can be exceedingly small. Many of the cells are very tiny, but they are very active; they manufacture various substances; they walk around; they wiggle; and they do all kinds of marvelous things – all on a very small scale. Also, they store information. Consider the possibility that we too can make a thing very small which does what we want – that we can manufacture an object that maneuvers at that level.

(From the talk "There's Plenty of Room at the Bottom", delivered by Richard P. Feynman at the annual meeting of the American Physical Society at the California Institute of Technology, Pasadena, CA, on December 29, 1959.)

by B. Bhushan

1.1 Background and Definition of Nanotechnology 1
1.2 Why Nano? 2
1.3 Lessons from Nature 2
1.4 Applications in Different Fields 3
1.5 Reliability Issues of MEMS/NEMS 4
1.6 Organization of the Handbook 5
References 5

1.1 Background and Definition of Nanotechnology

On Dec. 29, 1959, at the California Institute of Technology, Nobel Laureate Richard P. Feynman gave a talk at the annual meeting of the American Physical Society that has become one of the twentieth century's classic science lectures, titled "There's Plenty of Room at the Bottom" [1.1]. He presented a technological vision of extreme miniaturization several years before the word "chip" became part of the lexicon. He talked about the problem of manipulating and controlling things on a small scale. Extrapolating from known physical laws, Feynman envisioned a technology using the ultimate toolbox of nature, building nanoobjects atom by atom or molecule by molecule. Since the 1980s, many inventions and discoveries in the fabrication of nanoobjects have become a testament to his vision. In recognition of this reality, the National Science and Technology Council (NSTC) of the White House created the Interagency Working Group on Nanoscience, Engineering and Technology (IWGN) in 1998. In a January 2000 speech at the same institute, former President William J. Clinton talked about the exciting promise of nanotechnology and, more generally, the importance of expanding research in nanoscale science and technology. Later that month, he announced in his State of the Union Address an ambitious $497 million federal, multi-agency National Nanotechnology Initiative (NNI) in the fiscal year 2001 budget, and made it a top science and technology priority [1.2, 3]. The objective of this initiative was to form a broad-based coalition in which academe, the private sector, and local, state, and federal governments would work together to push the envelope of nanoscience and nanoengineering to reap nanotechnology's potential social and economic benefits.

Nanotechnology literally means any technology performed on a nanoscale that has applications in the real world. Nanotechnology encompasses the production and application of physical, chemical, and biological systems at scales ranging from individual atoms or molecules to submicron dimensions, as well as the integration of the resulting nanostructures into larger systems. Nanotechnology is likely to have a profound impact on our economy and society in the early twenty-first century, comparable to that of semiconductor technology, information technology, or cellular and molecular biology. Science and technology research in nanotechnology promises breakthroughs in such areas as materials and manufacturing, nanoelectronics, medicine and healthcare, energy, biotechnology, information technology, and national security. It is widely felt that nanotechnology will be the next industrial revolution.

Nanometer-scale features are mainly built up from their elemental constituents. Chemical synthesis – the spontaneous self-assembly of molecular clusters (molecular self-assembly) from simple reagents in solution – or biological molecules (e.g., DNA) are used as building blocks for the production of three-dimensional nanostructures, including quantum dots (nanocrystals) of arbitrary diameter (about 10 to 10^5 atoms). A variety of vacuum deposition and nonequilibrium plasma chemistry techniques are used to produce layered nanocomposites and nanotubes. Atomically controlled structures are produced using molecular beam epitaxy and organo-metallic vapor phase epitaxy. Micro- and nanosystem components are fabricated using top-down lithographic and nonlithographic fabrication techniques and range in size from micro- to nanometers. Continued improvements in lithography for use in the production of nanocomponents have resulted in line widths as small as 10 nanometers in experimental prototypes. The nanotechnology field, in addition to the fabrication of nanosystems, provides the impetus to development of experimental and computational tools.

The micro- and nanosystems include micro/nanoelectromechanical systems (MEMS/NEMS) (e.g., sensors, actuators, and miniaturized systems comprising sensing, processing, and/or actuating functions), micromechatronics, optoelectronics, microfluidics, and systems integration. These systems can sense, control, and activate on the micro/nanoscale and function individually or in arrays to generate effects on the macroscale. The microsystems market in 2000 was about $ 15 billion, and, with a projected 10–20 % annual growth rate, it is expected to increase to more than $ 100 billion by the end of this decade. The nanosystems market in 2001 was about $ 100 million and the integrated nanosystems market is expected to be more than $ 25 billion by the end of this decade. Due to the enabling nature of these systems, and because of the significant impact they can have on the commercial and defense applications, venture capitalists, industry, as well as the federal government have taken a special interest in nurturing growth in this field. Micro- and nanosystems are likely to be the next logical step in the "silicon revolution."

1.2 Why Nano?

The discovery of novel materials, processes, and phenomena at the nanoscale, as well as the development of new experimental and theoretical techniques for research provide fresh opportunities for the development of innovative nanosystems and nanostructured materials. Nanosystems are expected to find various unique applications. Nanostructured materials can be made with unique nanostructures and properties. This field is expected to open new venues in science and technology.

1.3 Lessons from Nature

Nanotechnology is a new word, but it is not an entirely new field. Nature has many objects and processes that function on a micro- to nanoscale [1.2, 4]. The understanding of these functions can guide us in imitating and producing nanodevices and nanomaterials.

Billions of years ago, molecules began organizing themselves into the complex structures that could support life. Photosynthesis harnesses solar energy to support plant life. Molecular ensembles are present in plants, which include light harvesting molecules, such as chlorophyll, arranged within the cells on the nanometer to micrometer scales. These structures capture light energy, and convert it into the chemical energy that drives the biochemical machinery of plant cells. Live organs use chemical energy in the body. The flagella, a type of bacteria, rotates at over 10,000 RPM [1.5]. This is an example of a biological molecular machine. The flagella motor is driven by the proton flow caused by the electrochemical potential differences across the membrane. The diameter of the bearing is about 20–30 nm, with an estimated clearance of about 1 nm.

In the context of tribology, some biological systems have anti-adhesion surfaces. First, many plant leaves (such as lotus leaf) are covered by a hydrophobic cuticle, which is composed of a mixture of large hydrocarbon molecules that have a strong hydrophobia. Second, the surface is made of a unique roughness distribution [1.6, 7]. It has been reported that for some leaf surfaces, the roughness of the hydrophobic leaf surface decreases wetness, which is reflected in a greater contact angle of water droplets on such surfaces.

1.4 Applications in Different Fields

Science and technology continue to move forward in making the fabrication of micro/nanodevices and systems possible for a variety of industrial, consumer, and biomedical applications. A range of MEMS devices have been produced, some of which are commercially used [1.4, 8–12]. A variety of sensors are used in industrial, consumer, and biomedical applications. Various microstructures or microcomponents are used in micro-instruments and other industrial applications, such as micromirror arrays. Two of the largest "killer" industrial applications are accelerometers (about 85 million units in 2002) and digital micromirror devices (about $400 million in sales in 2001). Integrated capacitive-type, silicon accelerometers have been used in airbag deployment in automobiles since 1991 [1.13, 14]. Accelerometer technology was about a billion-dollar-a-year industry in 2001, dominated by Analog Devices followed by Motorola and Bosch. Commercial digital light processing (DLP) equipment using digital micromirror devices (DMD) were launched in 1996 by Texas Instruments for digital projection displays in portable and home theater projectors, as well as table-top and projection TVs [1.15, 16]. More than 1.5 million projectors were sold before 2002. Other major industrial applications include pressure sensors, inkjet printer heads, and optical switches. Silicon-based piezoresistive pressure sensors for manifold absolute pressure sensing for engines were launched in 1991 by NovaSensor, and their annual sales were about 25 million units in 2002. Annual sales of inkjet printer heads with microscale functional components were about 400 million units in 2002. Capacitive pressure sensors for tire pressure measurements were launched by Motorola. Other applications of MEMS devices include chemical sensors; gas sensors; infrared detectors and focal plane arrays for earth observations; space science and missile defense applications; pico-satellites for space applications; and many hydraulic, pneumatic, and other consumer products. MEMS devices are also being pursued in magnetic storage systems [1.17], where they are being developed for super-compact and ultrahigh recording-density magnetic disk drives. Several integrated head/suspension microdevices have been fabricated for contact recording applications [1.18, 19]. High-bandwidth, servo-controlled microactuators have been fabricated for ultrahigh track-density applications, which serve as the fine-position control element of a two-stage, coarse/fine servo system, coupled with a conventional actuator [1.20–23]. Millimeter-sized wobble motors and actuators for tip-based recording schemes have also been fabricated [1.24].

BIOMEMS are increasingly used in commercial and defense applications (e.g., [1.4,25–28]). Applications of BIOMEMS include biofluidic chips (otherwise known as microfluidic chips, bioflips, or simply biochips) for chemical and biochemical analyses (biosensors) in medical diagnostics (e.g., DNA, RNA, proteins, cells, blood pressure and assays, and toxin identification) and implantable pharmaceutical drug delivery. The biosensors, also referred to as lab-on-a-chip, integrate sample handling, separation, detection, and data analysis onto one platform. Biosensors are designed to either detect a single or class of (bio)chemicals or system-level analytical capabilities for a broad range of (bio)chemical species known as micro total analysis systems (μTAS). The chips rely on microfluidics and involve the manipulation of tiny amounts of fluids in microchannels using microvalves for various analyses. The test fluid is pumped into the chip generally using an external pump for analyses. Some chips have been designed with an integrated electrostatically actuated diaphragm-type micropump. Silicon-based, disposable blood-pressure sensor chips were introduced in the early 1990s by NovaSensor for blood pressure monitoring (about 20 million units in 2002). A variety of biosensors are manufactured by various companies, including ACLARA, Agilent Technologies, Calipertech, and I-STAT.

After the tragedy of Sept. 11, 2001, concern over biological and chemical warfare has led to the development of handheld units with bio- and chemical sensors for the detection of biological germs, chemical or nerve agents, mustard agents, and chemical precursors to protect subways, airports, the water supply, and the population [1.29].

Other BIOMEMS applications include minimal invasive surgery, such as endoscopic surgery, laser angioplasty, and microscopic surgery. Implantable artificial organs can also be produced.

Micro-instruments and micro-manipulators are used to move, position, probe, pattern, and characterize nanoscale objects and nanoscale features. Miniaturized analytical equipment includes gas chromatography and mass spectrometry. Other instruments include micro-STM, where STM stands for scanning tunneling microscope.

Examples of NEMS include nanocomponents, nanodevices, nanosystems, and nanomaterials, such as microcantilever with integrated sharp nanotips for STM and atomic force microscopy (AFM), AFM array (millipede) for data storage, AFM tips for nanolithography, dip-pen nanolithography for printing molecules, biological (DNA) motors, molecular gears, molecularly thick films (e.g., in giant magneto-resistive or GMR heads and magnetic media), nanoparticles, (e.g., nanomagnetic particles in magnetic media), nanowires, carbon nanotubes, quantum wires (QWRs), quantum boxes (QBs), and quantum transistors [1.30–34]. BIONEMS include nanobiosensors – a microarray of silicon nanowires, roughly a few nm in size, to selectively bind and detect even a single biological molecule, such as DNA or protein, by using nanoelectronics to detect the slight electrical charge caused by such binding, or a microarray of carbon nanotubes to electrically detect glucose, implantable drug-delivery devices – e.g., micro/nanoparticles with drug molecules encapsulated in functionized shells for a site-specific targeting application, and a silicon capsule with a nanoporous membrane filled with drugs for long term delivery, nanodevices for sequencing single molecules of DNA in the Human Genome Project, cellular growth using carbon nanotubes for spinal cord repair, nanotubes for nanostructured materials for various applications, such as spinal fusion devices, organ growth, and growth of artificial tissues using nanofibers.

Nanoelectronics can be used to build computer memory, using individual molecules or nanotubes to store bits of information, as well as molecular switches, molecular or nanotube transistors, nanotube flat-panel displays, nanotube integrated circuits, fast logic gates, switches, nanoscopic lasers, and nanotubes as electrodes in fuel cells.

1.5 Reliability Issues of MEMS/NEMS

There is an increasing need for a multidisciplinary, system-oriented approach to manufacturing micro/nanodevices that function reliably. This can only be achieved through the cross-fertilization of ideas from different disciplines and the systematic flow of information and people among research groups. Common potential failure mechanisms for MEMS/NEMS that need to be addressed in order to increase reliability are: adhesion, friction, wear, fracture, fatigue, and contamination. Surface micro/nanomachined structures often include smooth and chemically active surfaces. Due to the large surface area to volume ratio in MEMS/NEMS, they are particularly prone to stiction (high static friction) as part of normal operation. Fracture occurs when the load on a microdevice is greater than the strength of the material. Fracture is a serious reliability concern, particularly for the brittle materials used in the construction of these components, since it can immediately, or eventually, lead to catastrophic failures. Additionally, debris can be formed from the fracturing of microstructures, leading to other failure processes. For less brittle materials, repeated loading over a long period causes fatigue that would also lead to the breaking and fracturing of the device. In principle, this failure mode is relatively easy to observe and simple to predict. However, the materials properties of thin films are often not known, making fatigue predictions prone to error.

Many MEMS/NEMS devices operate near their thermal dissipation limit. They may encounter hot spots that can cause failures, particularly in weak structures such as diaphragms or cantilevers. Thermal stressing and relaxation caused by thermal variations can create material delamination and fatigue in cantilevers. In large temperature changes, as experienced in the space environment, bimetallic beams will also experience warping due to mismatched coefficients of thermal expansion. Packaging has been a big problem. The contamination, which probably happens in packaging and during storage, can also strongly influence the reliability of MEMS/NEMS. For example, a particulate dust landed on one of the electrodes of a comb drive can cause catastrophic failure. There are no MEMS/NEMS fabrication standards, which makes it difficult to transfer fabrication steps in MEMS/NEMS between foundaries.

Obviously, studies of determination and suppression of active failure mechanisms affecting this new and promising technology are critical to the high reliability of MEMS/NEMS and are determining factors in successful practical application.

Mechanical properties are known to exhibit a dependence on specimen size. Mechanical property evaluation of nanometer-scaled structures is carried out to help design reliable systems, since good mechanical properties are of critical importance in such applications. Some of the properties of interest are: Young's modulus of elasticity, hardness, bending strength, fracture toughness, and fatigue life. Finite element modeling is carried out to study the effects of surface roughness and scratches on stresses in nanostructures. When nanostructures are smaller than a fundamental physical length scale, conventional theory may no longer apply, and new phenomena may emerge. Molecular mechanics is used to simulate the behavior of a nano-object.

1.6 Organization of the Handbook

The handbook integrates knowledge from the fabrication, mechanics, materials science, and reliability points of view. Organization of the book is straightforward. The handbook is divided into six parts. This first part introduces the nanotechnology field, including an introduction to nanostructures, micro/nanofabrication and, micro/nanodevices. The second part introduces scanning probe microscopy. The third part provides an overview of nanotribology and nanomechanics, which will prepare the reader to understand the tribology and mechanics of industrial applications. The fourth part provides an overview of molecularly thick films for lubrication. The fifth part focuses on industrial applications and microdevice reliability. And the last part focuses on the social and ethical implications of nanotechnology.

References

1.1 R. P. Feynmann: There's plenty of room at the bottom, Eng. Sci. **23** (1960) 22–36, and www.zyvex.com/nanotech/feynman.html (1959)

1.2 I. Amato: Nanotechnology, www.ostp.gov/nstc/html/iwgn/iwgn.public.brochure/welcome.htm or www.nsf.gov/home/crssprgm/nano/nsfnnireports.htm (2000)

1.3 Anonymous: National nanotechnology initiative, www.ostp.gov/nstc/html/iwgn.fy01budsuppl/nni.pdf or www.nsf.gov/home/crssprgm/nano/nsfnnireports.htm (2000)

1.4 I. Fujimasa: *Micromachines: A New Era in Mechanical Engineering* (Oxford Univ. Press, Oxford 1996)

1.5 C. J. Jones, S. Aizawa: The bacterial flagellum and flagellar motor: Structure, assembly, and functions, Adv. Microb. Physiol. **32** (1991) 109–172

1.6 V. Bergeron, D. Quere: Water droplets make an impact, Phys. World **14** (May 2001) 27–31

1.7 M. Scherge, S. Gorb: *Biological Micro- and Nanotribology* (Springer, Berlin, Heidelberg 2000)

1.8 B. Bhushan: *Tribology Issues and Opportunities in MEMS* (Kluwer, Dordrecht 1998)

1.9 G. T. A. Kovacs: *Micromachined Transducers Sourcebook* (WCB McGraw-Hill, Boston 1998)

1.10 S. D. Senturia: *Microsystem Design* (Kluwer, Boston 2001)

1.11 T. R. Hsu: *MEMS and Microsystems* (McGraw-Hill, Boston 2002)

1.12 M. Madou: *Fundamentals of Microfabrication: The Science of Miniaturization*, 2nd edn. (CRC, Boca Raton 2002)

1.13 T. A. Core, W. K. Tsang, S. J. Sherman: Fabrication technology for an integrated surface-micromachined sensor, Solid State Technol. **36** (October 1993) 39–47

1.14 J. Bryzek, K. Peterson, W. McCulley: Micromachines on the march, IEEE Spectrum (May 1994) 20–31

1.15 L. J. Hornbeck, W. E. Nelson: Bistable deformable mirror device, OSA Technical Digest **8** (1988) 107–110

1.16 L. J. Hornbeck: A digital light processing(tm) update – Status and future applications (invited), Proc. Soc. Photo-Opt. Eng. **3634** (1999) 158–170

1.17 B. Bhushan: *Tribology and Mechanics of Magnetic Storage Devices*, 2nd edn. (Springer, New York 1996)

1.18 H. Hamilton: Contact recording on perpendicular rigid media, J. Mag. Soc. Jpn. **15** (Suppl. S2) (1991) 481–483

1.19 T. Ohwe, Y. Mizoshita, S. Yonoeka: Development of integrated suspension system for a nanoslider with an MR head transducer, IEEE Trans. Magn. **29** (1993) 3924–3926

1.20 D. K. Miu, Y. C. Tai: Silicon micromachined scaled technology, IEEE Trans. Industr. Electron. **42** (1995) 234–239

1.21 L. S. Fan, H. H. Ottesen, T. C. Reiley, R. W. Wood: Magnetic recording head positioning at very high track densities using a microactuator-based, two-stage servo system, IEEE Trans. Ind. Electron. **42** (1995) 222–233

1.22 D. A. Horsley, M. B. Cohn, A. Singh, R. Horowitz, A. P. Pisano: Design and fabrication of an angular

microactuator for magnetic disk drives, J. Microelectromech. Syst. **7** (1998) 141–148
1.23 T. Hirano, L.S. Fan, D. Kercher, S. Pattanaik, T.S. Pan: HDD tracking microactuator and its integration issues, Proc. ASME Int. Mech. Eng. Congress, MEMS, New York 2000, ed. by A.P. Lee, J. Simon, F.K. Foster, R.S. Keynton (ASME, New York 2000) 449–452
1.24 L.S. Fan, S. Woodman: Batch fabrication of mechanical platforms for high-density data storage, 8th Int. Conf. Solid State Sensors and Actuators (Transducers '95)/Eurosensors IX, Stockholm (June, 1995) 434–437
1.25 P. Gravesen, J. Branebjerg, O.S. Jensen: Microfluidics – A review, J. Micromech. Microeng. **3** (1993) 168–182
1.26 C. Lai Poh San, E.P.H. Yap (Eds.): *Frontiers in Human Genetics* (World Scientific, Singapore 2001)
1.27 C.H. Mastrangelo, H. Becker (Eds.): *Microfluidics and BioMEMS*, Proc. SPIE **4560** (SPIE, Bellingham 2001)
1.28 H. Becker, L.E. Locascio: Polymer microfluidic devices, Talanta **56** (2002) 267–287
1.29 M. Scott: MEMS and MOEMS for national security applications, , Reliability, Testing, and Characterization of MEMS/MOEMS II, Proc. SPIE **4980** (SPIE, Bellingham 2003)
1.30 K.E. Drexler: *Nanosystems: Molecular Machinery, Manufacturing and Computation* (Wiley, New York 1992)
1.31 G. Timp (Ed.): *Nanotechnology* (Springer, Berlin, Heidelberg 1999)
1.32 E.A. Rietman: *Molecular Engineering of Nanosystems* (Springer, Berlin, Heidelberg 2001)
1.33 H.S. Nalwa (Ed.): *Nanostructured Materials and Nanotechnology* (Academic, San Diego 2002)
1.34 W.A. Goddard, D.W. Brenner, S.E. Lyshevski, G.J. Iafrate: *Handbook of Nanoscience, Engineering, and Technology* (CRC, Boca Raton 2003)

Part A

Nanostruc

Part A Nanostructures, Micro/Nanofabrication, and Micro/Nanodevices

2 **Nanomaterials Synthesis and Applications: Molecule-Based Devices**
Françisco M. Raymo, Coral Gables, USA

3 **Introduction to Carbon Nanotubes**
Marc Monthioux, Toulouse, France
Philippe Serp, Toulouse, France
Emmanuel Flahaut, Toulouse, France
Manitra Razafinimanana, Toulouse, France
Christophe Laurent, Toulouse, France
Alain Peigney, Toulouse, France
Wolfgang Bacsa, Toulouse, France
Jean-Marc Broto, Toulouse, France

4 **Nanowires**
Mildred S. Dresselhaus, Cambridge, USA
Yu-Ming Lin, Cambridge, USA
Oded Rabin, Cambridge, USA
Marcie R. Black, Cambridge, USA
Gene Dresselhaus, Cambridge, USA

5 **Introduction to Micro/Nanofabrication**
Babak Ziaie, Minneapolis, USA
Antonio Baldi, Barcelona, Spain
Massood Z. Atashbar, Kalamazoo, USA

6 **Stamping Techniques for Micro and Nanofabrication: Methods and Applications**
John A. Rogers, Urbana, USA

7 **Materials Aspects of Micro- and Nanoelectromechanical Systems**
Darrin J. Young, Cleveland, USA
Christian A. Zorman, Cleveland, USA
Mehran Mehregany, Cleveland, USA

8 **MEMS/NEMS Devices and Applications**
Darrin J. Young, Cleveland, USA
Christian A. Zorman, Cleveland, USA
Mehran Mehregany, Cleveland, USA

9 **Microfluidics and Their Applications to Lab-on-a-Chip**
Chong H. Ahn, Cincinnati, USA
Jin-Woo Choi, Baton Rouge, USA

10 **Therapeutic Nanodevices**
Stephen C. Lee, Columbus, USA
Mark Ruegsegger, Columbus, USA
Philip D. Barnes, Columbus, USA
Bryan R. Smith, Columbus, USA
Mauro Ferrari, Columbus, USA

2. Nanomaterials Synthesis and Applications: Molecule-Based Devices

The constituent components of conventional devices are carved out of larger materials relying on physical methods. This top-down approach to engineered building blocks becomes increasingly challenging as the dimensions of the target structures approach the nanoscale. Nature, on the other hand, relies on chemical strategies to assemble nanoscaled biomolecules. Small molecular building blocks are joined to produce nanostructures with defined geometries and specific functions. It is becoming apparent that nature's bottom-up approach to functional nanostructures can be mimicked to produce artificial molecules with nanoscaled dimensions and engineered properties. Indeed, examples of artificial nanohelices, nanotubes, and molecular motors are starting to be developed. Some of these fascinating chemical systems have intriguing electrochemical and photochemical properties that can be exploited to manipulate chemical, electrical, and optical signals at the molecular level. This tremendous opportunity has lead to the development of the molecular equivalent of conventional logic gates. Simple logic operations, for example, can be reproduced with collections of molecules operating in solution. Most of these chemical systems, however, rely on bulk addressing to execute combinational and sequential logic operations. It is essential to devise methods to reproduce these useful functions in solid-state configurations and, eventually, with single molecules. These challenging objectives are stimulating the design of clever devices that interface small assemblies of organic molecules with macroscaled and nanoscaled electrodes. These strategies have already produced rudimentary examples of diodes, switches, and transistors based on functional molecular components. The rapid and continuous progress of this exploratory research will, we hope, lead to an entire generation of molecule-based devices that might ultimately find useful applications in a variety of fields, ranging from biomedical research to information technology.

2.1 **Chemical Approaches to Nanostructured Materials** 10
- 2.1.1 From Molecular Building Blocks to Nanostructures 10
- 2.1.2 Nanoscaled Biomolecules: Nucleic Acids and Proteins 10
- 2.1.3 Chemical Synthesis of Artificial Nanostructures 12
- 2.1.4 From Structural Control to Designed Properties and Functions 12

2.2 **Molecular Switches and Logic Gates** 14
- 2.2.1 From Macroscopic to Molecular Switches 15
- 2.2.2 Digital Processing and Molecular Logic Gates 15
- 2.2.3 Molecular AND, NOT, and OR Gates .. 16
- 2.2.4 Combinational Logic at the Molecular Level 17
- 2.2.5 Intermolecular Communication 18

2.3 **Solid State Devices** 22
- 2.3.1 From Functional Solutions to Electroactive and Photoactive Solids 22
- 2.3.2 Langmuir–Blodgett Films 23
- 2.3.3 Self-Assembled Monolayers 27
- 2.3.4 Nanogaps and Nanowires 31

2.4 **Conclusions and Outlook** 35

References 36

2.1 Chemical Approaches to Nanostructured Materials

The fabrication of conventional devices relies on the assembly of macroscopic building blocks with specific configurations. The shapes of these components are carved out of larger materials by exploiting physical methods. This top-down approach to engineered building blocks is extremely powerful and can deliver effectively and reproducibly microscaled objects. This strategy becomes increasingly challenging, however, as the dimensions of the target structures approach the nanoscale. Indeed, the physical fabrication of nanosized features with subnanometer precision is a formidable technological challenge.

2.1.1 From Molecular Building Blocks to Nanostructures

Nature efficiently builds nanostructures by relying on chemical approaches. Tiny molecular building blocks are assembled with a remarkable degree of structural control in a variety of nanoscaled materials with defined shapes, properties, and functions. In contrast to the top-down physical methods, small components are connected to produce larger objects in these bottom-up chemical strategies. It is becoming apparent that the limitations of the top-down approach to artificial nanostructures can be overcome by mimicking nature's bottom-up processes. Indeed, we are starting to see emerge beautiful and ingenious examples of molecule-based strategies to fabricate chemically nanoscaled building blocks for functional materials and innovative devices.

2.1.2 Nanoscaled Biomolecules: Nucleic Acids and Proteins

Nanoscaled macromolecules play a fundamental role in biological processes [2.1]. Nucleic acids, for example, ensure the transmission and expression of genetic information. These particular biomolecules are linear polymers incorporating nucleotide repeating units (Fig. 2.1a). Each nucleotide has a phosphate bridge and a sugar residue. Chemical bonds between the phosphate of one nucleotide and the sugar of the next ensures the propagation of a polynucleotide strand from the 5′ to the 3′ end. Along the sequence of alternating sugar and phosphate fragments, an extended chain of robust covalent bonds involving carbon, oxygen, and phosphorous atoms forms the main backbone of the polymeric strand.

Every single nucleotide of a polynucleotide strand carries one of the four heterocyclic bases shown in Fig. 2.1b. For a strand incorporating 100 nucleotide repeating units, a total of 4^{100} unique polynucleotide sequences are possible. It follows that nature can fabricate a huge number of closely related nanostructures relying only on four building blocks. The heterocyclic bases appended to the main backbone of alternating phosphate and sugar units can sustain hydrogen bonding and $[\pi \cdots \pi]$ stacking interactions. Hydrogen bonds, formed between [N−H] donors and either N or O acceptors, encourage the pairing of adenine (A) with thymine (T) and of guanine (G) with cytosine (C). The stacking interactions involve attractive contacts between the extended π-surfaces of heterocyclic bases.

In the *B* conformation of deoxyribonucleic acid (DNA), the synergism of hydrogen bonds and $[\pi \cdots \pi]$ stacking glues pairs of complementary polynucleotide strands in fascinating double helical supermolecules (Fig. 2.1c) with precise structural control at the subnanometer level. The two polynucleotide strands wrap around a common axis to form a right-handed double helix with a diameter of ca. 2 nm. The hydrogen bonded and $[\pi \cdots \pi]$ stacked base pairs lie at the core of the helix with their π-planes perpendicular to the main axis of the helix. The alternating phosphate and sugar units define the outer surface of the double helix. In *B*-DNA, approximately ten base pairs define each helical turn corresponding to a rise per turn or helical pitch of ca. 3 nm. Considering that these molecules can incorporate up to approximately 10^{11} base pairs, extended end-to-end lengths spanning from only few nanometers to hundreds of meters are possible.

Nature's operating principles to fabricate nanostructures are not limited to nucleic acids. Proteins are also built joining simple molecular building blocks, the amino acids, by strong covalent bonds [2.1]. More precisely, nature relies on 20 amino acids differing in their side chains to assemble linear polymers, called polypeptides, incorporating an extended backbone of robust [C−N] and [C−C] bonds (Fig. 2.2a). For a single polymer strand of 100 repeating amino acid units, a total of 20^{100} unique combinations of polypeptide sequences are possible. Considering that proteins can incorporate more than one polypetide chain with over 4,000 amino acid residues each, it is obvious that nature can assemble an enormous number of different biomolecules relying on the same fabrication strategy and a relatively small pool of building blocks.

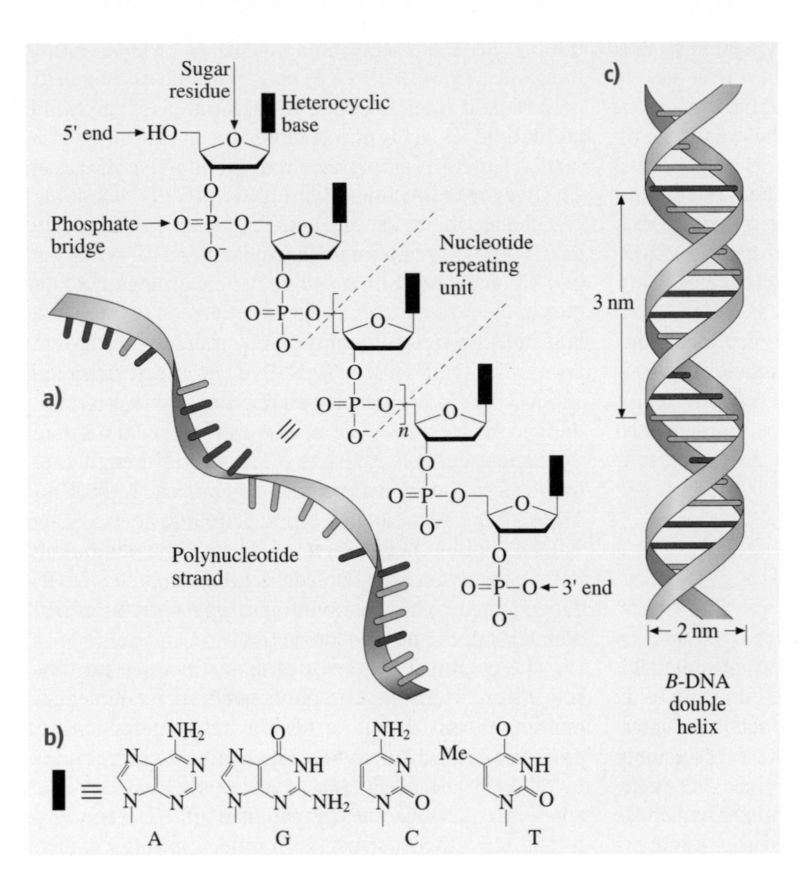

Fig. 2.1a–c A polynucleotide strand (**a**) incorporates alternating phosphate and sugar residues joined by covalent bonds. Each sugar carries one of four heterocyclic bases (**b**). Noncovalent interactions between complementary bases in two independent polynucleotide strands encourage the formation of nanoscaled double helixes (**c**)

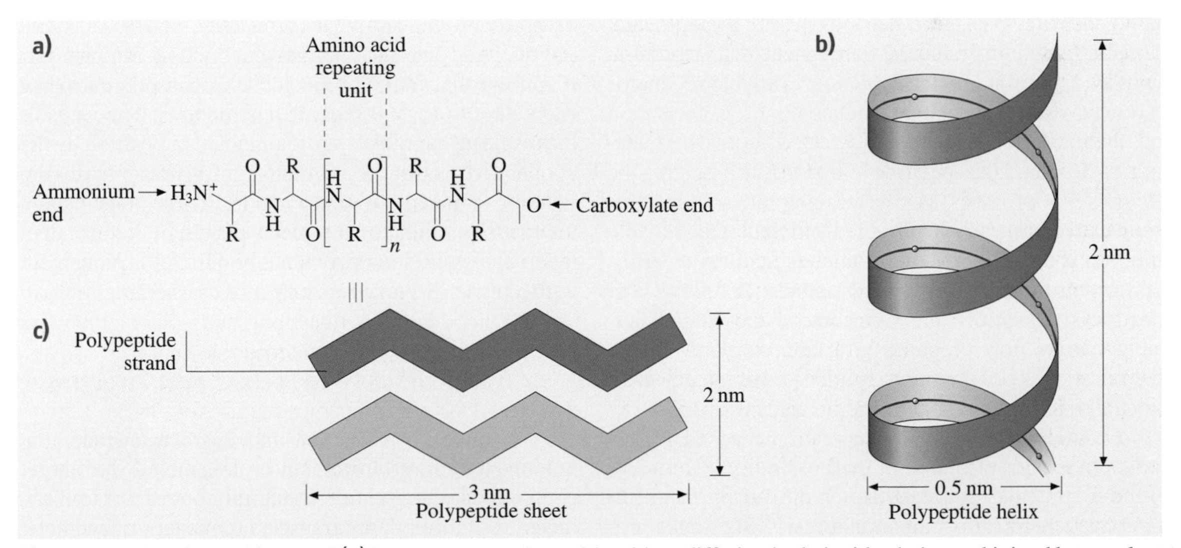

Fig. 2.2a–c A polypeptide strand (**a**) incorporates amino acid residues differing in their side chains and joined by covalent bonds. Hydrogen bonding interactions curl a single polypeptide strand into a helical arrangement (**b**) or lock pairs of strands into nanoscaled sheets (**c**)

The covalent backbones of the polypeptide strands form the main skeleton of a protein molecule. In addition, myriad secondary interactions, involving noncovalent contacts between portions of the amino acid residues, control the arrangement of the individual polypeptide chains. Intrastrand hydrogen bonds curl single polypeptide chains around a longitudinal axis in a helical fashion to form tubular nanostructures ca. 0.5 nm wide and ca. 2 nm long (Fig. 2.2b). Similarly, interstrand hydrogen bonds can align from 2 up to 15 parallel or antiparallel polypeptide chains to form nanoscaled sheets with average dimensions of 2×3 nm (Fig. 2.2c). Multiple nanohelices and/or nanosheets combine into a unique three-dimensional arrangement dictating the overall shape and dimensions of a protein.

2.1.3 Chemical Synthesis of Artificial Nanostructures

Nature fabricates complex nanostructures relying on simple criteria and a relatively small pool of molecular building blocks. Robust chemical bonds join the basic components into covalent scaffolds. Noncovalent interactions determine the three-dimensional arrangement and overall shape of the resulting assemblies. The multitude of unique combinations possible for long sequences of chemically connected building blocks provides access to huge libraries of nanoscaled biomolecules.

Modern chemical synthesis has evolved considerably over the past few decades [2.2]. Experimental procedures to join molecular components with structural control at the picometer level are available. A multitude of synthetic schemes to encourage the formation of chemical bonds between selected atoms in reacting molecules have been developed. Furthermore, the tremendous progress of crystallographic and spectroscopic techniques has provided efficient and reliable tools to probe directly the structural features of artificial inorganic and organic compounds. It follows that designed molecules with engineered shapes and dimensions can be now prepared in a laboratory relying on the many tricks of chemical synthesis and the power of crystallographic and spectroscopic analyses.

The high degree of sophistication reached in this research area translates into the possibility of mimicking the strategies successfully employed by nature to fabricate chemically nanostructures [2.3]. Small molecular building blocks can be synthesized and joined covalently following routine laboratory procedures. It is even possible to design the stereoelectronic properties of the assembling components in order to shape the geometry of the final product with the assistance of noncovalent interactions. For example, five bipyridine building blocks (Fig. 2.3) can be connected in five synthetic steps to produce an oligobipyridine strand [2.4]. The five repeating units are bridged by [C−O] bonds and can chelate metal cations in the bay regions defined by their two nitrogen atoms. The spontaneous assembly of two organic strands in a double helical arrangement occurs in the presence of inorganic cations. In the resulting helicate, the two oligobipyridine strands wrap around an axis defined by five Cu(I) centers. Each inorganic cation coordinates two bipyridine units with a tetrahedral geometry imposing a diameter of ca. 0.6 nm on the nanoscaled helicate [2.5]. The overall length from one end of the helicate to the other is ca. 3 nm [2.6]. The analogy between this artificial double helix and the *B*-DNA double helix shown in Fig. 2.1c is obvious. In both instances, a supramolecular glue combines two independent molecular strands into nanostructures with defined shapes and dimensions.

The chemical synthesis of nanostructures can borrow nature's design criteria as well as its molecular building blocks. Amino acids, the basic components of proteins, can be assembled into artificial macrocycles. In the example of Fig. 2.4, eight amino acid residues are joined through the formation of [C−N] bonds in multiple synthetic steps [2.7]. The resulting covalent backbone defines a circular cavity with a diameter of ca. 0.8 nm [2.8]. In analogy to the polypeptide chains of proteins, the amino acid residues of this artificial oligopeptide can sustain hydrogen bonding interactions. It follows that multiple macrocycles can pile on top of each other to form tubular nanostructures. The walls of the resulting nanotubes are maintained in position by the cooperative action of at least eight primary hydrogen bonding contacts per macrocycle. These noncovalent interactions maintain the mean planes of independent macrocycles in an approximately parallel arrangement with a plane-to-plane separation of ca. 0.5 nm.

2.1.4 From Structural Control to Designed Properties and Functions

The examples in Figs. 2.3 and 2.4 demonstrate that modular building blocks can be assembled into target compounds with precise structural control at the picometer level through programmed sequences of synthetic steps. Indeed, modern chemical synthesis offers access to complex molecules with nanoscaled dimensions and, thus, provides cost-effective strategies for the production and characterization of billions of engineered nanostructures in parallel. Furthermore, the high degree of structural control is accompanied by the possibility of designing specific properties into the target nanostructures. Electroactive and photoactive components can be integrated chemically into functional molecular machines [2.9]. Extensive electrochemical investigations have demonstrated that inorganic and organic compounds can exchange electrons with macroscopic electrodes [2.10]. These studies have unraveled the processes responsible for the oxidation and reduction of numerous functional groups and indicated viable design criteria to adjust the ability of molecules to accept or donate electrons [2.11]. Similarly, detailed photochemical and photophysical investigations have elucidated the mechanisms responsible for the absorption and emission of photons at the molecular level [2.12]. The vast knowledge established on the interactions between light and molecules offers the opportunity to engineer chromophoric and fluorophoric functional groups with defined absorption and emission properties [2.11, 13].

The power of chemical synthesis to deliver functional molecules is, perhaps, better illustrated by the molecular motor shown in Fig. 2.5. The preparation of this [2]rotaxane requires 12 synthetic steps starting from known precursors [2.14]. This complex molecule incorporates a Ru(II)-trisbipyridine stopper bridged to a linear tetracationic fragment by a rigid triaryl spacer. The other end of the tetracationic portion is terminated by a bulky tetraarylmethane stopper. The bipyridinium unit of this dumbbell-shaped compound is encircled by a macrocyclic polyether. No covalent bonds join the macrocyclic and linear components.

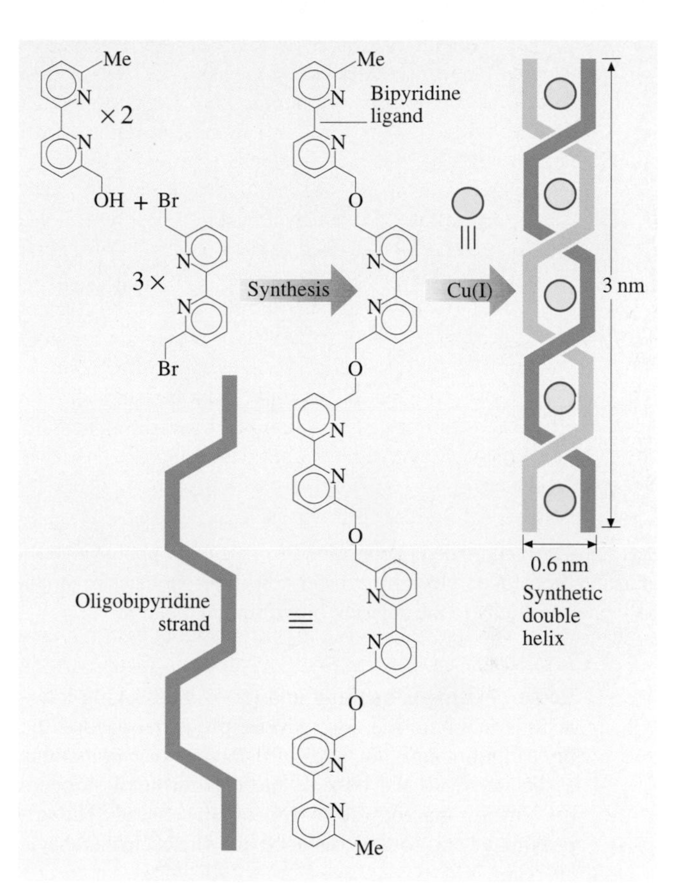

Fig. 2.3 An oligobipyridine strand can be synthesized joining five bipyridine subunits by covalent bonds. The tetrahedral coordination of pairs of bipyridine ligands by Cu(I) ions encourages the assembly two oligobipyridine strands into a double helical arrangement

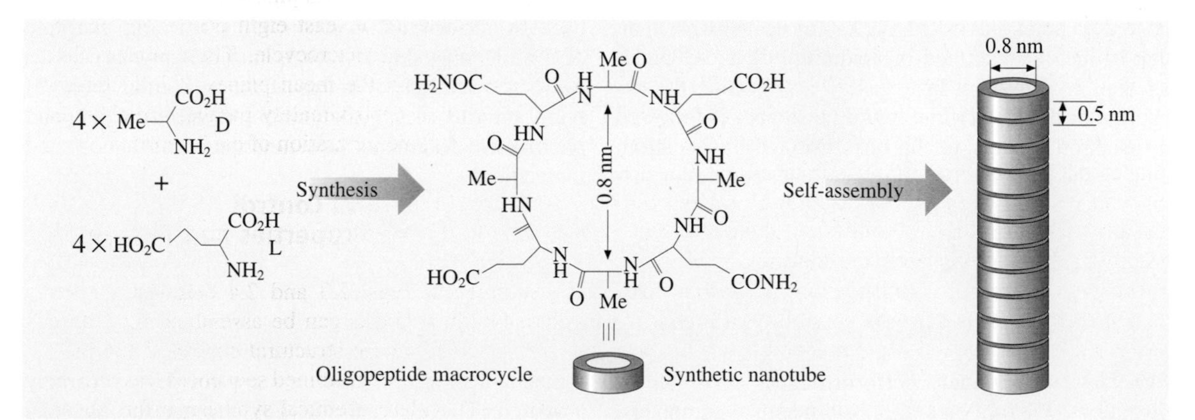

Fig. 2.4 Cyclic oligopeptides can be synthesized joining eight amino acid residues by covalent bonds. The resulting macrocycles self-assemble into nanoscaled tube-like arrays

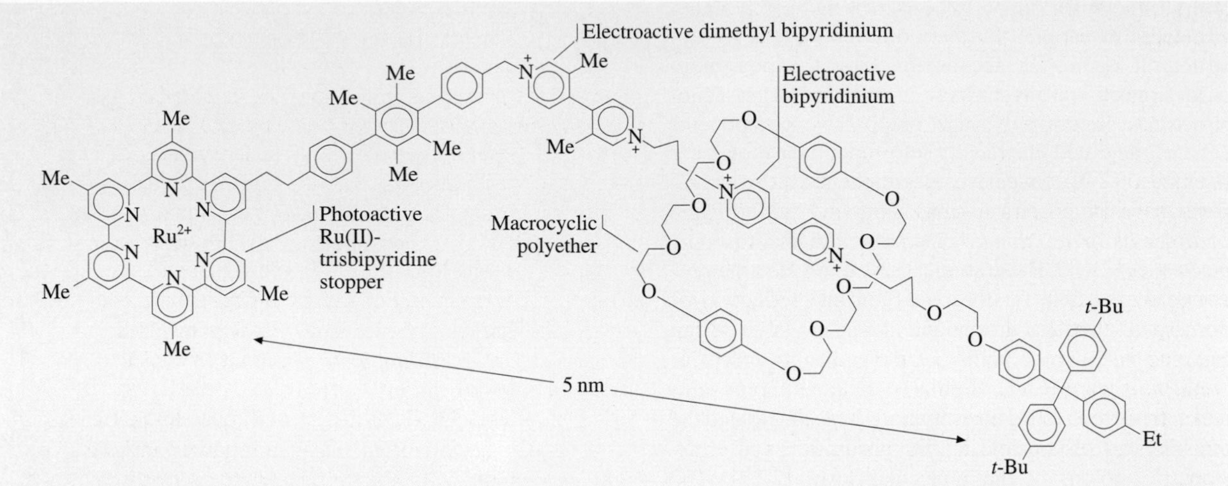

Fig. 2.5 This nanoscaled [2]rotaxane incorporates a photoactive Ru(II)-trisbipyridine stopper and two electroactive bipyridinium units. Photoinduced electron transfer from the photoactive stopper to the encircled electroactive unit forces the macrocyclic polyether to shuttle to the adjacent bipyridinium dication

Rather, hydrogen bonding and $[\pi \cdots \pi]$ stacking interactions maintain the macrocyclic polyether around the bipyridinium unit. In addition, mechanical constrains associated with the bulk of the two terminal stoppers prevent the macrocycle to slip off the thread. The approximate end-to-end distance for this [2]rotaxane is ca. 5 nm.

The bipyridinium and the 3,3′-dimethyl bipyridinium units within the dumbbell-shaped component undergo two consecutive and reversible monoelectronic reductions [2.14]. The two methyl substituents on the 3,3′-dimethyl bipyridinium dication make this electroactive unit more difficult to reduce. In acetonitrile, its redox potential is ca. 0.29 V more negative than that of the unsubstituted bipyridinium dication. Under irradiation at 436 nm in degassed acetonitrile, the excitation of the Ru(II)-trisbipyridine stopper is followed by electron transfer to the unsubstituted bipyridinium unit. In the presence of a sacrificial electron donor (triethanolamine) in solution, the photogenerated hole in the photoactive stopper is filled, and undesired back electron transfer is suppressed. The permanent and light-induced reduction of the dicationic bipyridinium unit to a radical cation depresses significantly the magnitude of the noncovalent interactions holding the macrocyclic polyether in position. As a result, the macrocycle shuttles from the reduced unit to the adjacent dicationic 3,3′-dimethyl bipyridinum. After the diffusion of molecular oxygen into the acetonitrile solution, oxidation occurs restoring the dicationic form of the bipyridinium unit and its ability to sustain strong noncovalent bonds. As a result, the macrocyclic polyether shuttles back to its original position. This amazing example of a molecular shuttle reveals that dynamic processes can be controlled reversibly at the molecular level relying on the clever integration of electroactive and photoactive fragments into functional and nanoscaled molecules.

2.2 Molecular Switches and Logic Gates

Everyday, we routinely perform dozens of switching operations. We turn on and off our personal computers, cellular phones, CD players, radios, or simple light bulbs at a click of a button. Every single time, our finger exerts a mechanical stimulation on a control device, namely a switch. The external stimulus changes the physical state of the switch closing or opening an electric circuit and enabling or preventing the passage of electrons.

Overall, the switch transduces a mechanical input into an electrical output.

2.2.1 From Macroscopic to Molecular Switches

The use of switching devices is certainly not limited to electric circuits. For example, a switch at the junction of a railroad can divert trains from one track to another. Similarly, a faucet in a lavatory pipe can block or release the flow of water. Of course, the nature of the control stimulations and the character of the final outcome vary significantly from case to case, but the operating principle behind each switching device is the same. In all cases, input stimulations reach the switch changing its physical state and producing a specific output.

The development of nanoscaled counterparts to conventional switches is expected to have fundamental scientific and technological implications. For instance, one can envisage practical applications for ultraminiaturized switches in areas ranging from biomedical research to information technology. The major challenge in the quest for nanoswitches, however, is the identification of reliable design criteria and operating principles for these innovative and fascinating devices. Chemical approaches to implement molecule-sized switches appear to be extremely promising. The intrinsically small dimensions of organic molecules coupled with the power of chemical synthesis are the main driving forces behind these exploratory investigations.

Certain organic molecules adjust their structural and electronic properties when stimulated with chemical, electrical, or optical inputs. Generally, the change is accompanied by an electrochemical or spectroscopic response. Overall, these nanostructures transduce input stimulations into detectable outputs and, appropriately, are called molecular switches [2.15, 16]. The chemical transformations associated with these switching processes are often reversible. The chemical system returns to the original state when the input signal is turned off. The interconverting states of a molecular switch can be isomers, an acid and its conjugated base, the oxidized and reduced forms of a redox active molecule, or even the complexed and uncomplexed forms of a receptor [2.9, 15, 16]. The output of a molecular switch can be a chemical, electrical, and/or optical signal that varies in intensity with the interconversion process. For example, changes in absorbance, fluorescence, pH, or redox potential can accompany the reversible transformation of a molecular switch.

2.2.2 Digital Processing and Molecular Logic Gates

In present computer networks, data are elaborated electronically by microprocessor systems [2.17] and are exchanged optically between remote locations [2.18]. Data processing and communication require the encoding of information in electrical and optical signals in the form of binary digits. Using arbitrary assumptions, logic thresholds can be established for each signal and, then, 0 and 1 digits can be encoded following simple conventions. Sequences of electronic devices manipulate the encoded bits executing logic functions as a result of basic switching operations.

The three basic AND, NOT, and OR operators combine binary inputs into binary outputs following precise logic protocols [2.17]. The NOT operator converts an input signal into an output signal. When the input is 0, the output is 1. When the input is 1, the output is 0. Because of the inverse relationship between the input and output values, the NOT gate is often called "inverter" [2.19]. The OR operator combines two input signals into a single output signal. When one or both inputs are 1, the output is 1. When both inputs are 0, the output is 0. The AND gate also combines two input signals into one output signal. In this instance, however, the output is 1 only when both inputs are 1. When at least one input is 0, the output is 0.

The output of one gate can be connected to one of the inputs of another operator. A NAND gate, for example, is assembled connecting the output of an AND operator to the input of a NOT gate. Now the two input signals are converted into the final output after two consecutive logic operations. In a similar fashion, a NOR gate can be assembled connecting the output of an OR operator to the input of a NOT gate. Once again, two consecutive logic operations determine the relation between two input signals and a single output. The NAND and NOR operations are termed universal functions because any conceivable logic operation can be implemented relying only on one of these two gates [2.17]. In fact, digital circuits are fabricated routinely interconnecting exclusively NAND or exclusively NOR operators [2.19].

The logic gates of conventional microprocessors are assembled interconnecting transistors, and their input and output signals are electrical [2.19]. But the concepts of binary logic can be extended to chemical, mechanical, optical, pneumatic, or any other type of signal. First it is necessary to design devices that can respond to these stimulations in the same way transistors respond to electrical signals. Molecular switches respond to a variety of input stimulations producing specific outputs and can, therefore, be exploited to implement logic functions [2.20, 21].

2.2.3 Molecular AND, NOT, and OR Gates

More than a decade ago, researchers proposed a potential strategy to execute logic operations at the molecular level [2.22]. Later, the analogy between molecular switches and logic gates was recognized in a seminal article [2.23], in which it was demonstrated that AND, NOT, and OR operations can be reproduced with fluorescent molecules. The pyrazole derivative **1** (Fig. 2.6) is a molecular NOT gate. It imposes an inverse relation between a chemical input (concentration of H^+) and an optical output (emission intensity). In a mixture of methanol and water, the fluorescence quantum yield of **1** is 0.13 in the presence of only 0.1 equivalents of H^+ [2.23]. The quantum yield drops to 0.003 when the equivalents of H^+ are 1,000. Photoinduced electron transfer from the central pyrazoline unit to the pendant benzoic acid quenches the fluorescence of the protonated form. Thus, a change in H^+ concentration (I) from a low to a high value switches the emission intensity (O) from a high to a low value. The inverse relationship between the chemical input I and the optical output O translates into the truth table of a NOT operation if a positive logic convention (low = 0, high = 1) is applied to both signals. The emission intensity is high (O = 1) when the concentration of H^+ is low (I = 0). The emission intensity is low (O = 0) when the concentration of H^+ is high (I = 1).

The anthracene derivative **2** (Fig. 2.6) is a molecular OR gate. It transduces two chemical inputs

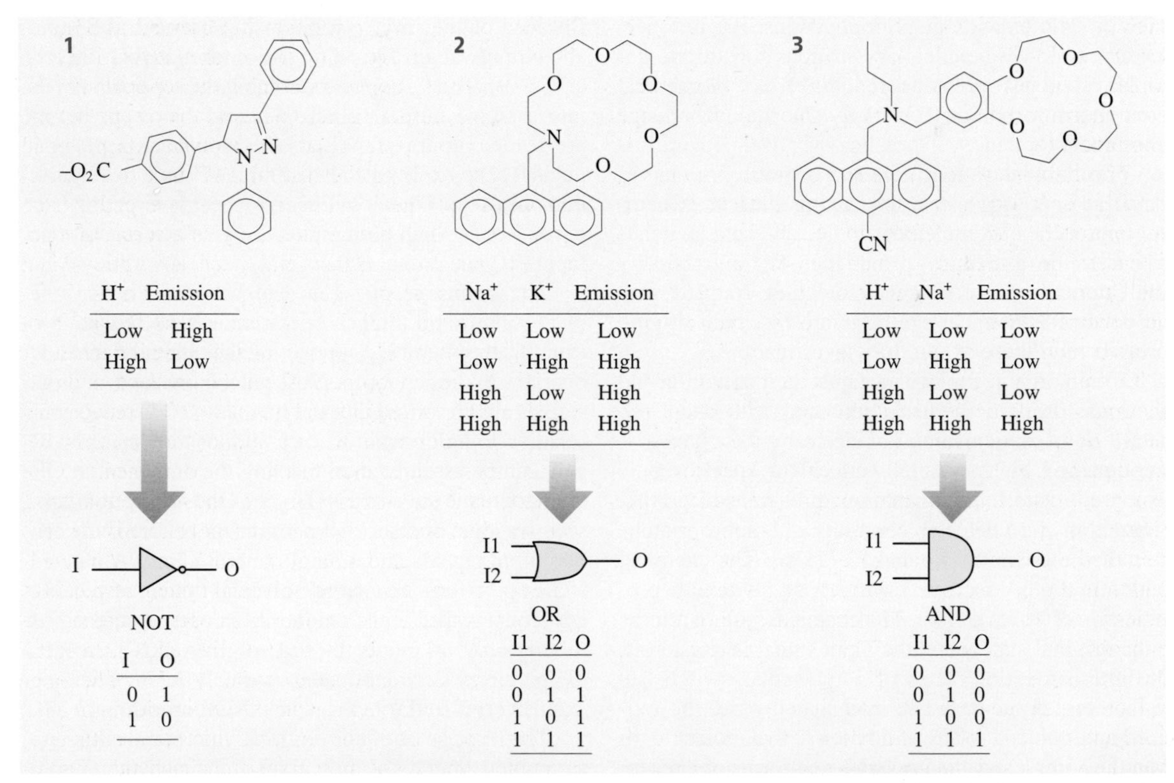

Fig. 2.6 The fluorescence intensity of the pyrazoline derivative **1** is high when the concentration of H^+ is low, and vice versa. The fluorescence intensity of the anthracene derivative **2** is high when the concentration of Na^+ and/or K^+ is high. The emission is low when both concentrations are low. The fluorescence intensity of the anthracene **3** is high only when the concentrations of H^+ and Na^+ are high. The emission is low in the other three cases. The signal transductions of the molecular switches **1**, **2**, and **3** translate into the truth tables of NOT, OR, and AND gates, respectively, if a positive logic convention is applied to all inputs and outputs (low = 0, high = 1)

(concentrations of Na^+ and K^+) into an optical output (emission intensity). In methanol, the fluorescence quantum yield is only 0.003 in the absence of metal cations [2.23]. Photoinduced electron transfer from the nitrogen atom of the azacrown fragment to the anthracene fluorophore quenches the emission. After the addition of 1,000 equivalents of either Na^+ or K^+, the quantum yield raises to 0.053 and 0.14, respectively. Similarly, the quantum yield is 0.14 when both metal cations are present in solution. The complexation of one of the two metal cations inside the azacrown receptor depresses the efficiency of the photoinduced electron transfer enhancing the fluorescence. Thus, changes in the concentrations of Na^+ (I1) and/or K^+ (I2) from low to high values switch the emission intensity (O) from a low to a high value. The relationship between the chemical inputs I1 and I2 and the optical output O translates into the truth table of an OR operation if a positive logic convention (low = 0, high = 1) is applied to all signals. The emission intensity is low (O = 0) only when the concentration of Na^+ and K^+ are low (I1 = 0, I2 = 0). The emission intensity is high (O = 1) for the other three input combinations.

The anthracene derivative **3** (Fig. 2.6) is a molecular AND gate. It transduces two chemical inputs (concentrations of H^+ and Na^+) into an optical output (emission intensity). In a mixture of methanol and *iso*-propanol, the fluorescence quantum yield is only 0.011 in the absence of H^+ or Na^+ [2.23]. Photoinduced electron transfer from either the tertiary amino group or the catechol fragment to the anthracene fluorophore quenches the emission. After the addition of either 100 equivalents of H^+ or 1,000 equivalents of Na^+, a modest change of the quantum yield to 0.020 and 0.011, respectively, is observed. Instead, the quantum yield increases to 0.068 when both species are present in solution. The protonation of the amino group and the insertion of the metal cation in the benzocrown ether receptor depress the efficiency of the photoinduced electron transfer processes enhancing the fluorescence. Thus, changes in the concentrations of H^+ (I1) and Na^+ (I2) from low to high values switch the emission intensity (O) from a low to a high value. The relationship between the chemical inputs I1 and I2 and the optical output O translates into the truth table of an AND operation if a positive logic convention (low = 0, high = 1) is applied to all signals. The emission intensity is high (O = 1) only when the concentration of H^+ and Na^+ are high (I1 = 1, I2 = 1). The emission intensity is low (O = 0) for the other three input combinations.

2.2.4 Combinational Logic at the Molecular Level

The fascinating molecular AND, NOT, and OR gates illustrated in Fig. 2.6 have stimulated the design of related chemical systems able to execute the three basic logic operations and simple combinations of them [2.20, 21]. Most of these molecular switches convert chemical inputs into optical outputs. But the implementation of logic operations at the molecular level is not limited to the use of chemical inputs. For example, electrical signals and reversible redox processes can be exploited to modulate the output of a molecular switch [2.24]. The supramolecular assembly **4** (Fig. 2.7) executes a XNOR function relying on these operating principles. The π-electron rich tetrathiafulvalene (TTF) guest threads the cavity of a π-electron deficient bipyridinium (BIPY) host. In acetonitrile, an absorption band associated with the charge-transfer interactions between the complementary π-surfaces is observed at 830 nm. Electrical stimulations alter the redox state of either the TTF or the BIPY units encouraging the separation of the two components of the complex and the disappearance of the charge-transfer band. Electrolysis at a potential of +0.5 V oxidizes the neutral TTF unit to a monocationic state. The now cationic guest is expelled from the cavity of the tetracationic host as a result of electrostatic repulsion. Consistently, the absorption band at 830 nm disappears. The charge-transfer band, however, is restored after the exhaustive back reduction of the TTF unit at a potential of 0 V. Similar changes in the absorption properties can be induced addressing the BIPY units. Electrolysis at −0.3 V reduces the dicationic BIPY units to their monocationic forms encouraging the separation of the two components of the complex and the disappearance of the absorption band. The original absorption spectrum is restored after the exhaustive back oxidation of the BIPY units at a potential of 0 V. Thus, this supramolecular system responds to electrical stimulations producing an optical output. One of the electrical inputs (I1) controls the redox state of the TTF unit switching between 0 and +0.5 V. The other (I2) determines the redox state of the bipyridinium units switching between −0.3 and 0 V. The optical output (O) is the absorbance of the charge-transfer band. A positive logic convention (low = 0, high = 1) can be applied to the input I1 and output O. A negative logic convention (low = 1, high = 0) can be applied to the input I2. The resulting truth table corresponds to that of a XNOR circuit (Fig. 2.7). The charge-transfer absorbance is high (O = 1) only when one voltage input is low and the other

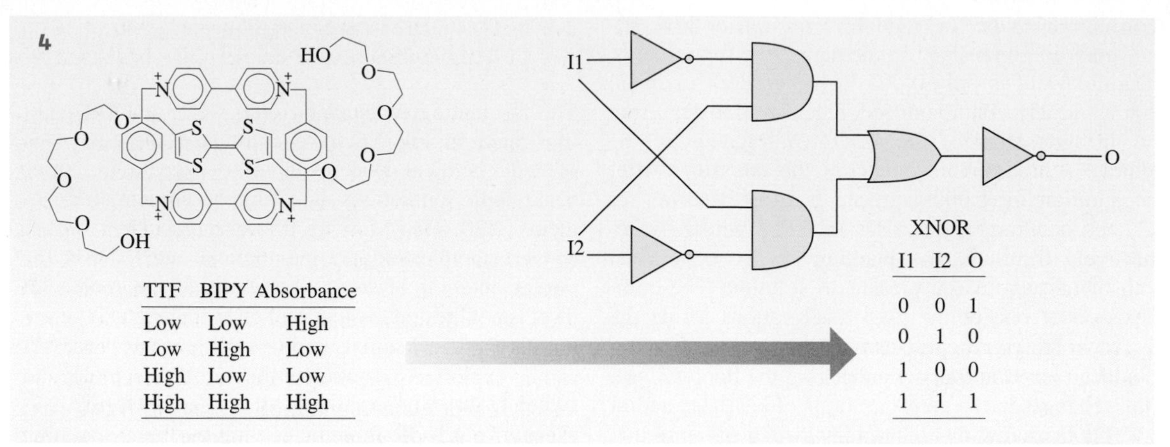

Fig. 2.7 The charge-transfer absorbance of the complex **4** is high when the voltage input addressing the tetrathiafulvalene (TTF) unit is low and that stimulating the bipyridinium (BIPY) units is high and vice versa. If a positive logic convention is applied to the TTF input and to the absorbance output (low = 0, high = 1) while a negative logic convention is applied to the BIPY input (low = 0, high = 1), the signal transduction of **4** translates into the truth table of a XNOR circuit

is high (I1 = 0, I2 = 0) or vice versa (I1 = 1, I2 = 1). It is important to note that the input string with both I1 and I2 equal to 1 implies that input potentials of +0.5 and −0.3 V are applied simultaneously to a solution containing the supramolecular assembly **4** and not to an individual complex. Of course, the concomitant oxidation of the TTF guest and reduction of the BIPY units in the very same complex would be unrealistic. In bulk solution, instead, some complexes are oxidized while others are reduced, leaving the average solution composition unaffected. Thus, the XNOR operation executed by this supramolecular system is a consequence of bulk properties and not a result of unimolecular signal transduction.

Optical inputs can be employed to operate the three-state molecular switch of Fig. 2.8 in acetonitrile solution [2.25]. This chemical system responds to three inputs producing two outputs. The three input stimulations are ultraviolet light (I1), visible light (I2), and the concentration of H^+ (I3). One of the two optical outputs is the absorbance at 401 nm (O1), which is high when the molecular switch is in the yellow-green state **6** and low in the other two cases. The other optical output is the absorbance at 563 nm (O2), which is high when the molecular switch is in the purple state **7** and low in the other two cases. The colorless spiropyran state **5** switches to the merocyanine form **7** upon irradiation with ultraviolet light. It switches to the protonated merocyanine from **6** when treated with H^+. The colored state **7** isomerizes back to **5** in the dark or upon irradiation with visible light. Alternatively, **7** switches to **6** when treated with H^+. The colored state **6** switches to **5**, when irradiated with visible light, and to **7**, after the removal of H^+. In summary, this three-state molecular switch responds to two optical inputs (I1 and I2) and one chemical input (I3) producing two optical outputs (O1 and O2). Binary digits can be encoded on each signal applying positive logic conventions (low = 0, high = 1). It follows that the three-state molecular switch converts input strings of three binary digits into output strings of two binary digits. The corresponding truth table (Fig. 2.8) reveals that the optical output O1 is high (O1 = 1) when only the input I3 is applied (I1 = 0, I2 = 0, I3 = 1), when only the input I2 is not applied (I1 = 1, I2 = 0, I3 = 0), or when all three inputs are applied (I1 = 1, I2 = 0, I3 = 0). The optical output O2 is high (O2 = 1) when only the input I1 is applied (I1 = 1, I2 = 0, I3 = 0) or when only the input I3 is not applied (I1 = 1, I2 = 0, I3 = 0). The combinational logic circuit (Fig. 2.8) equivalent to this truth table shows that all three inputs determine the output O1, while only I1 and I3 control the value of O2.

2.2.5 Intermolecular Communication

The combinational logic circuits in Figs. 2.7 and 2.8 are arrays of interconnected AND, NOT, and OR operators. The digital communication between these basic logic elements ensures the execution of a sequence of simple logic operations that results in the complex logic function processed by the entire circuit. It follows that

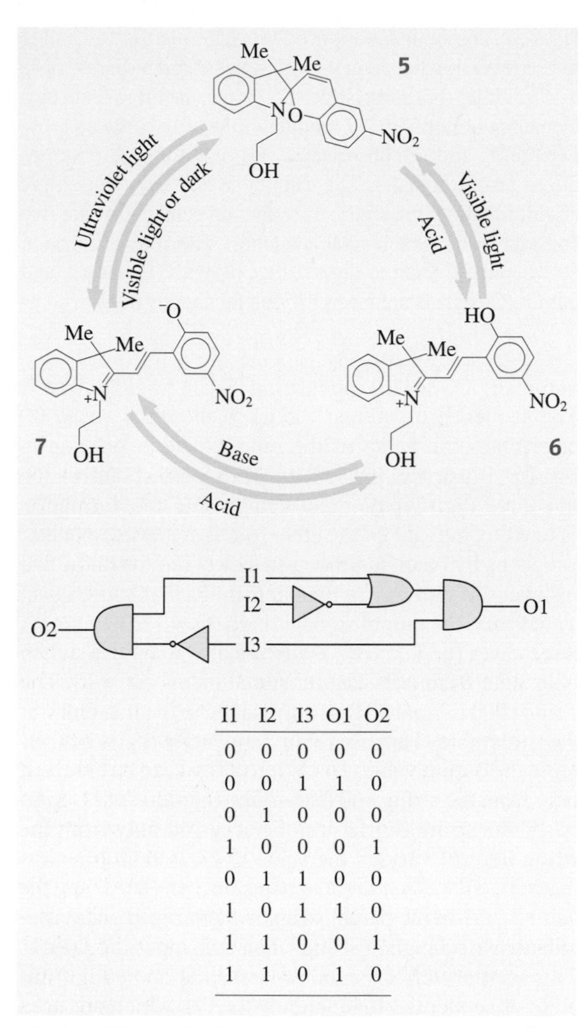

I1	I2	I3	O1	O2
0	0	0	0	0
0	0	1	1	0
0	1	0	0	0
1	0	0	0	1
0	1	1	0	0
1	0	1	1	0
1	1	0	0	1
1	1	0	1	0

Fig. 2.8 Ultraviolet light (I1), visible light (I2), and H^+ (I3) inputs induce the interconversion between the three states **5**, **6**, and **7**. The colorless state **5** does not absorb in the visible region. The yellow-green state **6** absorbs at 401 nm (O1). The purple state **7** absorbs at 563 nm (O2). The truth table illustrates the conversion of input strings of three binary digits (I1, I2, and I3) into output strings of two binary digits (O1 and O2) operated by this three-state molecular switch. A combinational logic circuit incorporating nine AND, NOT, and OR operators correspond to this particular truth table

the logic function of a given circuit can be adjusted altering the number and type of basic gates and their interconnection protocol [2.17]. This modular approach to combinational logic circuits is extremely powerful. Any logic function can be implemented connecting the appropriate combination of simple AND, NOT, and OR gates.

The strategies followed so far to implement complex logic functions with molecular switches are based on the careful design of the chemical system and on the judicious choice of the inputs and outputs [2.20, 21]. A specific sequence of AND, NOT, and OR operations is programmed in a single molecular switch. No digital communication between distinct gates is needed since they are built in the same molecular entity. Though extremely elegant, this strategy does not have the same versatility of a modular approach. A different molecule has to be designed, synthesized, and analyzed every single time a different logic function has to be realized. In addition, the degree of complexity that can be achieved with only one molecular switch is fairly limited. The connection of the input and output terminals of independent molecular AND, NOT, and OR operators, instead, would offer the possibility of assembling any combinational logic circuit from three basic building blocks.

In digital electronics, the communication between two logic gates can be realized connecting their terminals with a wire [2.19]. Methods to transmit binary data between distinct molecular switches are not so obvious and must be identified. Recently we developed two strategies to communicate signals between compatible molecular components. In one instance, a chemical signal is communicated between two distinct molecular switches [2.26]. They are the three-state switch illustrated in Fig. 2.8 and the two-state switch of Fig. 2.9. The merocyanine form **7** is a photogenerated base. Its *p*-nitrophenolate fragment, produced upon irradiation of the colorless state **5** with ultraviolet light, can abstract a proton from an acid present in the same solution. The resulting protonated form **6** is a photoacid. It releases a proton upon irradiation with visible light and can protonate a base co-dissolved in the same medium. The orange azopyridine **8** switches to the red-purple azopyridinium **9** upon protonation. This process is reversible, and the addition of a base restores the orange state **8**. It follows that photoinduced proton transfer can be exploited to communicate a chemical signal from **6** to **8** and from **9** to **7**. The two colored states **8** and **9** have different absorption properties in the visible region. In acetonitrile, the orange state **8** absorbs at 422 nm, and the red-purple state **9** absorbs at 556 nm. The changes in absorbance of these two bands can be exploited to monitor the photoinduced exchange of protons between the two communicating molecular switches.

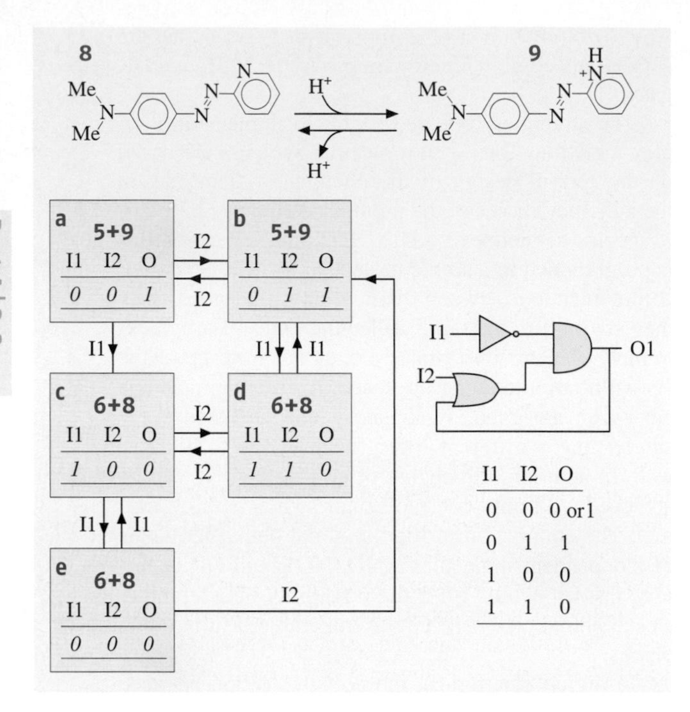

I1	I2	O
0	0	0 or 1
0	1	1
1	0	0
1	1	0

Fig. 2.9 The concentration of H^+ controls the reversible interconversion between the two states **8** and **9**. In response to ultraviolet (I1) and visible (I2) inputs, the three-state molecular switch in Fig. 2.7 modulates the ratio between these two forms and the absorbance (O) of **9** through photoinduced proton transfer. The truth table and sequential logic circuit illustrate the signal transduction behavior of the two communicating molecular switches. The interconversion between the five three-digit strings of input (I1 and I2) and output (O) data is achieved varying the input values in steps

The three-state molecular switch and the two-state molecular switch can be operated sequentially when dissolved in the same acetonitrile solution. In the presence of one equivalent of H^+, the two-state molecular switch is in state **9** and the absorbance at 556 nm is high (O = 1). Upon irradiation with ultraviolet light (I1 = 0), **5** switches to **7**. The photogenerated base deprotonates **9** producing **8** and **6**. As a result, the absorbance at 556 nm decreases (O = 0). Upon irradiation with visible light (I2 = 1), **6** switches to **5** releasing H^+. The result is the protonation of **8** to form **9** and restore the high absorbance at 556 nm (O = 1). In summary, the three-state molecular switch transduces two optical inputs (I1 = ultraviolet light, I2 = visible light) into a chemical signal (proton transfer) that is communicated to the two-state molecular switch and converted into a final optical output (O = absorbance at 556 nm).

The logic behavior of the two communicating molecular switches is significantly different from those of the chemical systems illustrated in Figs. 2.6, 2.7, and 2.8 [2.26]. The truth table in Fig. 2.9 lists the four possible combinations of two-digit input strings and the corresponding one-digit output. The output digit O for the input strings 01, 10, and 11 can take only one value. In fact, the input string 01 is transduced into a 1, and the input strings 10 and 11 are converted into 0. Instead, the output digit O for the input string 00 can be either 0 or 1. The sequence of events leading to the input string 00 determines the value of the output. The boxes ***a–e*** in Fig. 2.9 illustrates this effect. They correspond to the five three-digit input/output strings. The transformation of one box into any of the other four is achieved in one or two steps by changing the values of I1 and/or I2. In two instances (***a*** and ***b***), the two-state molecular switch is in state **9**, and the output signal is high (O = 1). In the other three cases (***c***, ***d***, and ***e***), the two-state molecular switch is in state **8**, and the output signal is low (O = 0). The strings 000 (***e***) and 001 (***a***) correspond to the first entry of the truth table. They share the same input digits but differ in the output value. The string 000 (***e***) can be obtained only from the string 100 (***c***) varying the value of I1. Similarly, the string 001 (***a***) can be accessed only from the string 011 (***b***) varying the value of I2. In both transformations, the output digit remains unchanged. Thus, the value of O1 in the parent string is memorized and maintained in the daughter string when both inputs become 0. This memory effect is the fundamental operating principle of sequential logic circuits [2.17], which are used extensively to assemble the memory elements of modern microprocessors. The sequential logic circuit equivalent to the truth table of the two communicating molecular switches is also shown in Fig. 2.9. In this circuit, the input data I1 and I2 are combined through NOT, OR, and AND operators. The output of the AND gate O is also an input of the OR gate and controls, together with I1 and I2, the signal transduction behavior.

The other strategy for digital transmission between molecules is based on the communication of optical signals between the three-state molecular switch (Fig. 2.8) and fluorescent compounds [2.27,28]. In the optical network of Fig. 2.10, three optical signals travel from an excitation source to a detector after passing through two quartz cells. The first cell contains an equimolar acetonitrile solution of naphthalene, anthracene, and tetracene. The second cell contains an acetonitrile solu-

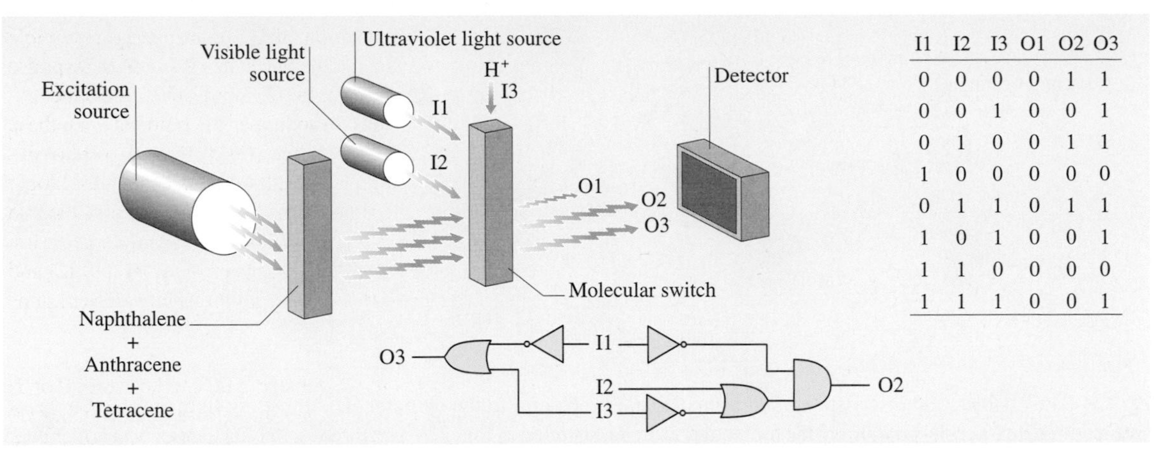

I1	I2	I3	O1	O2	O3
0	0	0	0	1	1
0	0	1	0	0	1
0	1	0	0	1	1
1	0	0	0	0	0
0	1	1	0	1	1
1	0	1	0	0	1
1	1	0	0	0	0
1	1	1	0	0	1

Fig. 2.10 The excitation source sends three monochromatic light beams (275, 357, and 441 nm) to a quartz cell containing an equimolar acetonitrile solution of naphthalene, anthracene and tetracene. The three fluorophores absorb the exciting beams and reemit at 305, 401, and 544 nm, respectively. The light emitted in the direction perpendicular to the exciting beams passes through another quartz cell containing an acetonitrile solution of the three-state molecular switch shown in Fig. 2.7. Ultraviolet (I1), visible (I2), and H^+ (I3) inputs control the interconversion between the three states of the molecular switch. They determine the intensity of the optical outputs reaching the detector and correspond to the naphthalene (O1), anthracene (O2), and tetracene (O3) emissions. The truth table and equivalent combinational logic illustrate the relation between the three inputs and the three outputs. The output O1 is always 0, and it is not influenced by the three inputs. Only two inputs determine the value of O3, while all of them control the output O2

tion of the three-state molecular switch. The excitation source sends three consecutive monochromatic light beams to the first cell stimulating the emission of the three fluorophores. The light emitted in the direction perpendicular to the exciting beam reaches the second cell. When the molecular switch is in state **5**, the naphthalene emission at 335 nm is absorbed and a low intensity output (O1) reaches the detector. Instead, the anthracene and tetracene emissions at 401 and 544 nm, respectively, pass unaffected and high intensity outputs (O2 and O3) reach the detector. When the molecular switch is in state **6**, the naphthalene and anthracene emissions are absorbed and only the tetracene emission reaches the detector (O1 = 0, O2 = 0, O3 = 1). When the molecular switch is state **7**, the emission of all three fluorophores is absorbed (O1 = 0, O2 = 0, O3 = 0). The interconversion of the molecular switch between the three states is induced addressing the second cell with ultraviolet (I1), visible (I2) and H^+ (I3) inputs. Thus, three independent optical outputs (O1, O2 and O3) can be modulated stimulating the molecular switch with two optical and one chemical input. The truth table in Fig. 2.10 illustrates the relation between the three inputs (I1, I2 and I3) and the three outputs (O1, O2 and O3), when positive logic conventions are applied to all signals. The equivalent logic circuit shows that all three inputs control the anthracene channel O2, but only I1 and I3 influence the tetracene channel O3. Instead, the intensity of the naphthalene channel O1 is always low, and it is not affected by the three inputs.

The operating principles of the optical network in Fig. 2.10 can be simplified to implement all-optical logic gates. The chemical input inducing the formation of the protonated form **6** of the molecular switch can be eliminated. The interconversion between the remaining two states **5** and **7** can be controlled relying exclusively on ultraviolet inputs. Indeed, ultraviolet irradiation induces the isomerization of the colorless form **5** to the colored species **7**, which reisomerizes to the original state in the dark. Thus, a single ultraviolet source is sufficient to control the switching from **5** to **7** and vice versa. On the basis of these considerations, all-optical NAND, NOR, and NOT gates can be implemented operating sequentially or in parallel from one to three independent switching elements [2.28]. For example, the all-optical network illustrated in Fig. 2.11 is a three-input NOR gate. A monochromatic optical signal travels from a visible source to a detector. Three switching elements are aligned along the path of the traveling light. They are quartz cells containing an acetonitrile solution of the mo-

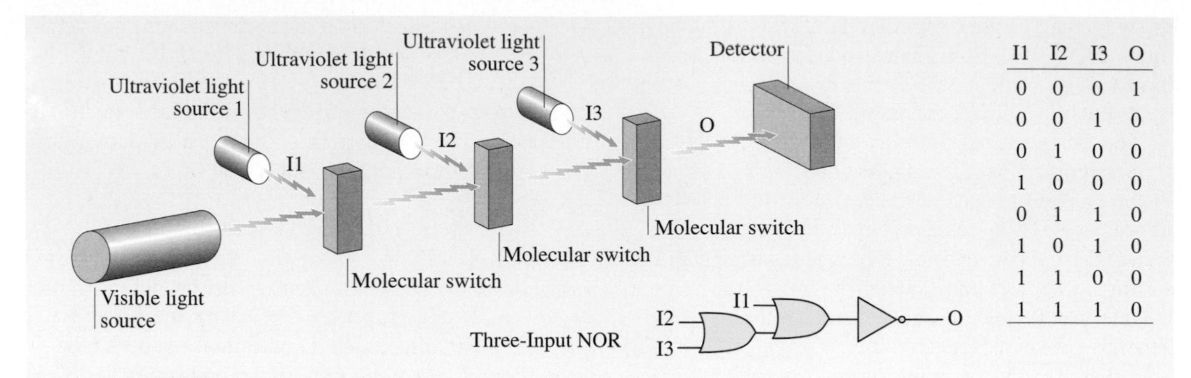

I1	I2	I3	O
0	0	0	1
0	0	1	0
0	1	0	0
1	0	0	0
0	1	1	0
1	0	1	0
1	1	0	0
1	1	1	0

Fig. 2.11 The visible source sends a monochromatic beam (563 nm) to the detector. The traveling light is forced to pass through three quartz cells containing the molecular switch illustrated in Fig. 2.7. The three switching elements are operated by independent ultraviolet inputs. When at least one of them is on, the associated molecular switch is in the purple form **7**, which can absorb and block the traveling light. The truth table and equivalent logic circuit illustrate the relation between the three inputs I1, I2, and I3 and the optical output O

lecular switch shown in Fig. 2.8. The interconversion of the colorless form **5** into the purple isomer **7** is induced stimulating the cell with an ultraviolet input. The reisomerization from **7** to **5** occurs spontaneously, as the ultraviolet sources is turned off. Using three distinct ultraviolet sources, the three switching elements can be controlled independently.

The colorless form **5** does not absorb in the visible region, while the purple isomer **7** has a strong absorption band at 563 nm. Thus, a 563 nm optical signal leaving the visible source can reach the detector unaffected only if all three switching elements are in the nonabsorbing state **5**. If one of the three ultraviolet inputs I1, I2, or I3 is turned on, the intensity of the optical output O drops to 3–4% of its original value. If two or three ultraviolet inputs are turned on simultaneously, the optical output drops to 0%. Indeed, the photogenerated state **7** absorbs and blocks the traveling light. Applying positive logic conventions to all signals, binary digits can be encoded in the three optical inputs and in the optical output. The resulting truth table is illustrated in Fig. 2.11. The output O is 1 only if all three inputs I1, I2, or I3 are 0. The output O is 0 if at least one of the three inputs I1, I2, or I3 is 1. This signal transduction corresponds to that executed by a three-input NOR gate, which is a combination of one NOT and two OR operators.

2.3 Solid State Devices

The fascinating chemical systems illustrated in Figs. 2.6–2.11 demonstrate that logic functions can be implemented relying on the interplay between designed molecules and chemical, electrical and/or optical signals [2.20, 21].

2.3.1 From Functional Solutions to Electroactive and Photoactive Solids

These molecular switches, however, are operated exclusively in solution and remain far from potential applications in information technology at this stage. The integration of liquid components and volatile organic solvents in practical digital devices is hard to envisage. Furthermore, the logic operations executed by these chemical systems rely on bulk addressing. Although the individual molecular components have nanoscaled dimensions, macroscopic collections of them are employed for digital processing. In some instances, the operating principles cannot even be scaled down to the unimolecular level. Often bulk properties are responsible for signal transduction. For example, a single fluorescent compound **2** cannot execute an OR operation. Its azacrown appendage can accommodate only one metal cation. As a result, an individual molecular switch can respond to only one of the two chemical inputs. It is a collection of numerous molecular switches dissolved in an organic solvent that responds to both inputs enabling an OR operation.

The development of miniaturized molecule-based devices requires the identification of methods to transfer the switching mechanisms developed in solution to the solid state [2.29]. Borrowing designs and fabrication strategies from conventional electronics, researchers are starting to explore the integration of molecular components into functional circuits and devices [2.30–33]. Generally, these strategies combine lithography and surface chemistry to assemble nanometer-thick organic films on the surfaces of microscaled or nanoscaled electrodes. Two main approaches for the deposition of organized molecular arrays on inorganic supports have emerged so far. In one instance, amphiphilic molecular building blocks are compressed into organized monolayers at air/water interfaces. The resulting films can be transferred on supporting solids employing the Langmuir–Blodgett technique [2.34]. Alternatively, certain molecules can be designed to adsorb spontaneously on the surfaces of compatible solids from liquid or vapor phases. The result is the self-assembly of organic layers on inorganic supports [2.35].

2.3.2 Langmuir–Blodgett Films

Films of amphiphilic molecules can be deposited on a variety of solid supports employing the Langmuir–Blodgett technique [2.34]. This method can be extended to electroactive compounds incorporating hydrophilic and hydrophobic groups. For example, the amphiphile **10** (Fig. 2.12) has a hydrophobic hexadecyl tail attached to a hydrophilic bipyridinium dication [2.36, 37]. This compound dissolves in mixtures of chloroform and methanol, but it is not soluble in moderately concentrated aqueous solutions of sodium perchlorate.

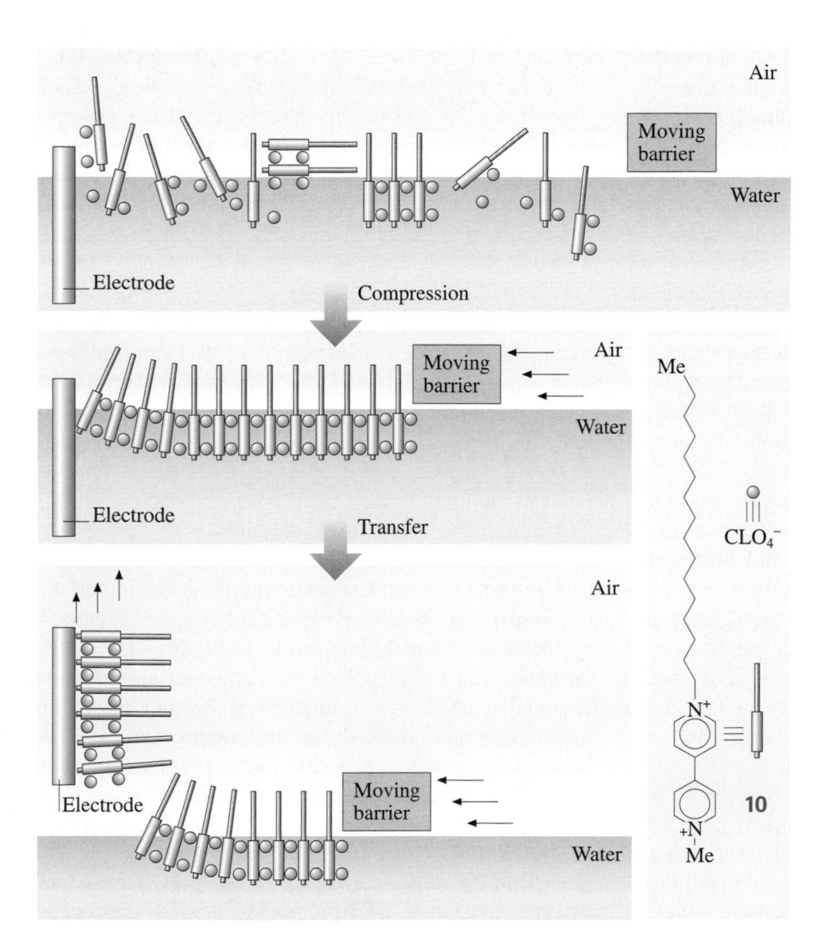

Fig. 2.12 The compression of the amphiphilic dication **10** with a moving barrier results in the formation of a packed monolayer at the air/water interface. The lifting of an electrode pre-immersed in the aqueous subphase encourages the transfer of part of the monolayer on the solid support

Thus the spreading of an organic solution of **10** on an aqueous sodium perchlorate subphase affords a collection of disorganized amphiphiles floating on the water surface (Fig. 2.12), after the organic solvent has evaporated. The molecular building blocks can be compressed into a monolayer with the aid of a moving barrier. The hydrophobic tails align away from the aqueous phase. The hydrophilic dicationic heads and the accompanying perchlorate counterions pack to form an organized monolayer at the air/water interface. The compression process can be monitored recording the surface pressure (π)-area per molecule (A) isotherm, which indicates a limiting molecular area of ca. 50 Å^2. This value is larger than the projected area of an oligomethylene chain. It correlates reasonably, however, with the overall area of a bipyridinium dication plus two perchlorate anions.

The monolayer prepared at the air/water interface (Fig. 2.12) can be transferred on the surface of a indium-tin oxide electrode pre-immersed in the aqueous phase. The slow lifting of the solid support drags the monolayer away from the aqueous subphase. The final result is the coating of the electrode with an organic film containing electroactive bipyridinium building blocks. The modified electrode can be integrated in a conventional electrochemical cell to probe the redox response of the electroactive layer. The resulting cyclic voltammograms reveal the characteristic waves for the first reduction process of the bipyridinium dications, confirming the successful transfer of the electroactive amphiphiles from the air/water interface to the electrode surface. The integration of the redox waves indicates a surface coverage of ca. 4×10^{10} mol cm^{-2}. This value corresponds to a molecular area of ca. 40 Å^2 and is in excellent agreement with the limiting molecular area of the π–A isotherm.

These seminal experiments demonstrate that electroactive amphiphiles can be organized into uniform monolayers at the air/water interface and then transferred efficiently on the surface of appropriate substrates to produce electrode/monolayer junctions. The resulting electroactive materials can become the functional components of molecule-based devices. For example, bipyridinium-based photodiodes can be fabricated following this approach [2.38, 39]. Their operating principles rely on photoinduced electron transfer from chromophoric units to bipyridinium acceptors. The electroactive and photoactive amphiphile **11** (Fig. 2.13) incorporates hydrophobic ferrocene and pyrene tails and a hydrophilic bipyridinium head. Chloroform solutions of **11** containing ten equivalents of arachidic acid can be spread on an aqueous calcium chloride subphase in a Langmuir trough. The amphiphiles can be compressed into a mixed monolayer, after the evaporation of the organic solvent. Pronounced steps in the corresponding π–A isotherm suggest that the bulky ferrocene and pyrene groups are squeezed away from the water surface. In the final arrangement, both photoactive groups align above the hydrophobic dication.

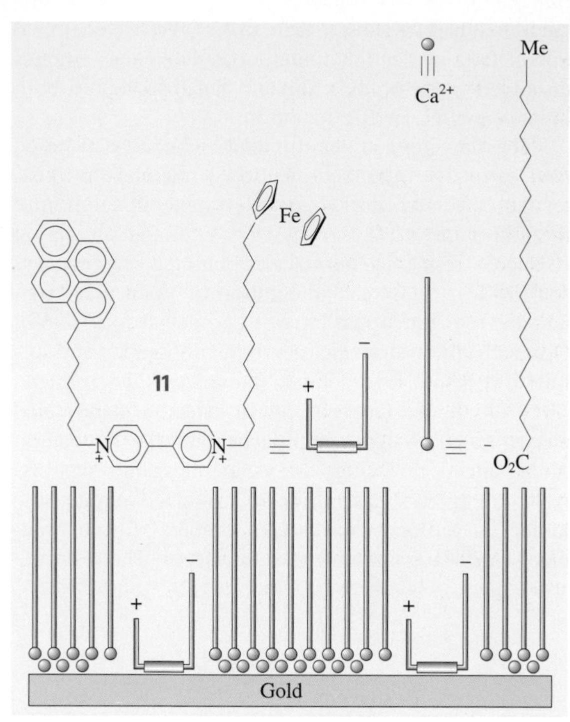

Fig. 2.13 Mixed monolayers of the amphiphile **11** and arachidic acid can be transferred from the air/water interface to the surface of an electrode to generate a molecule-based photodiode

A mixed monolayer of **11** and arachidic acid can be transferred from the air/water interface to the surface of a transparent gold electrode following the methodology illustrated for the system in Fig. 2.12. The coated electrode can be integrated in a conventional electrochemical cell. Upon irradiation at 330 nm under an inert atmosphere, an anodic photocurrent of ca. 2 nA develops at a potential of 0 V relative to a saturated calomel electrode. Indeed, the illumination of the electroactive monolayer induces the electron transfer from the pyrene appendage to the bipyridinium acceptor and then from the reduced acceptor to the electrode. A second intramolecular electron transfer from the ferrocene donor to the oxidized pyrene fills its photogenerated hole. Overall, a unidirectional flow of electrons across the monolayer/electrode junction is established under the influence of light.

The ability to transfer electroactive monolayers from air/water interfaces to electrode surfaces can be exploited to fabricate molecule-based electronic devices. In particular, arrays of interconnected electrode/monolayer/electrode tunneling junctions can be assembled combining the Langmuir–Blodgett technique with electron beam evaporation [2.33]. Fig. 2.14 illustrates a schematic representation of the resulting devices. Initially, parallel fingers are patterned on a silicon wafer with a silicon dioxide overlayer by electron beam evaporation. The bottom electrodes deposited on the support can be either aluminum wires covered by an aluminum oxide or n-doped silicon lines with silicon dioxide overlayers. Their widths are ca. 6 or 7 μm, respectively. The patterned silicon chip is immersed in the aqueous subphase of a Langmuir trough prior to monolayer formation. After the compression of electroactive amphiphiles at the air/water interface, the substrate is pulled out of the aqueous phase to encourage the transfer of the molecular layer on the parallel bottom electrodes as well as on the gaps between them. Then, a second set of electrodes orthogonal to the first is deposited through a mask by electron beam evaporation. They consist of a titanium underlayer plus an aluminum overlayer. Their thicknesses are ca. 0.05 and 1 μm, respectively, and their width is ca. 10 μm. In the final assembly, portions of the molecular layer become sandwiched between the bottom and top electrodes. The active areas of these electrode/monolayer/electrode junctions are ca. 60–70 μm^2 and correspond to ca. 10^6 molecules.

The [2]rotaxane **12** (Fig. 2.14) incorporates a macrocyclic polyether threaded onto a bipyridinium-based backbone [2.40, 41]. The two bipyridinium dications are bridged by a m-phenylene spacer and terminated by tetraarylmethane appendages. These two bulky groups trap mechanically the macrocycle preventing its dis-

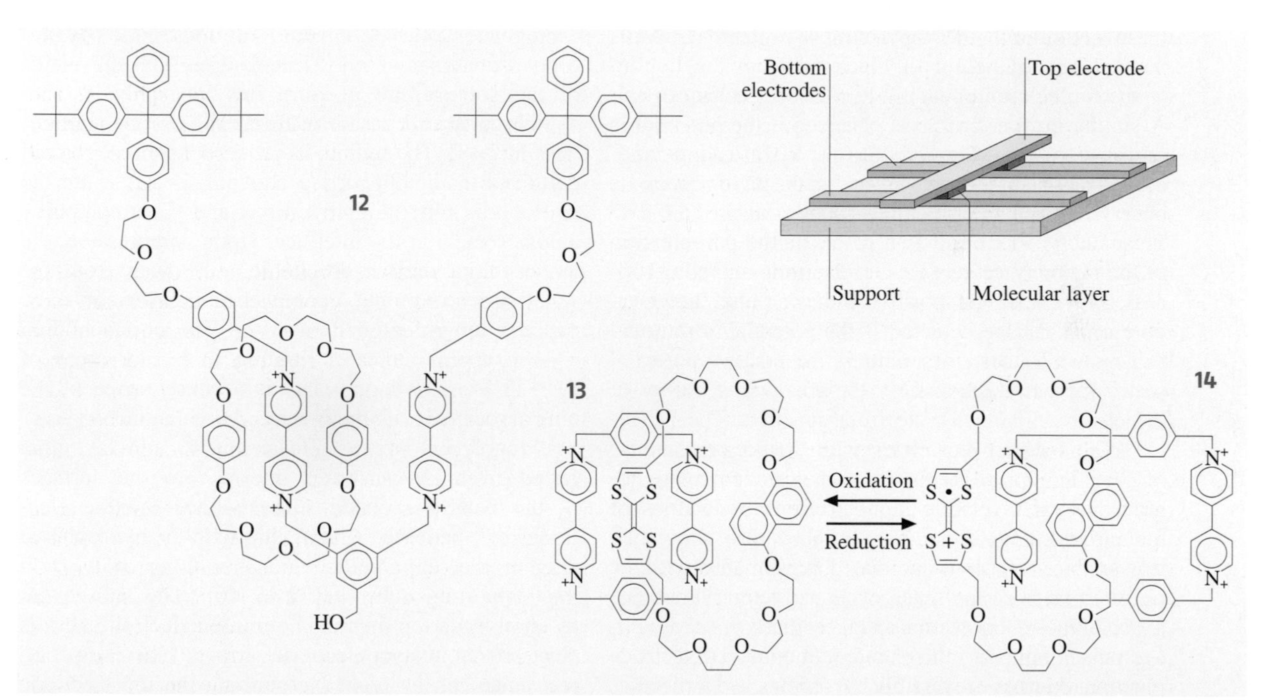

Fig. 2.14 The [2]rotaxane **12** and the [2]catenane **13** can be compressed into organized monolayers at air/water interfaces. The resulting monolayers can be transferred on the bottom electrodes of a patterned silicon support. After the deposition of a top electrode, electrode/monolayer/electrode junctions can be assembled. Note that only the portion of the monolayer sandwiched between the top and bottom electrodes is shown in the diagram. The oxidation of the tetrathiafulvalene unit of the [2]catenane **13** is followed by the circumrotation of the macrocyclic polyether to afford the [2]catenane **14**. The process is reversible, and the reduction of the cationic tetrathiafulvalene unit restores the original state

sociation from the tetracationic backbone. In addition, their hydrophobicity complements the hydrophilicity of the two bipyridinium dications imposing amphiphilic character on the overall molecular assembly. This compound does not dissolve in aqueous solutions and can be compressed into organized monolayers at air/water interfaces. The corresponding π–A isotherm reveals a limiting molecular area of ca. 130 Å^2. This large value is a consequence of the bulk associated with the hydrophobic tetraarylmethane tails and the macrocycle encircling the tetracationic backbone.

Monolayers of the [2]rotaxane **12** can be transferred from the air/water interface to the surfaces of the bottom aluminum/aluminum oxide electrodes of a patterned silicon chip with the hydrophobic tetraarylmethane groups pointing away from the supporting substrate. The subsequent assembly of a top titanium/aluminum electrode affords electrode/monolayer/electrode junctions. Their current/voltage signature can be recorded grounding the top electrode and scanning the potential of the bottom electrode. A pronounced increase in current is observed when the potential is lowered below -0.7 V. Under these conditions, the bipyridinium-centered LUMOs mediate the tunneling of electrons from the bottom to the top electrode leading to a current enhancement. A similar current profile is observed if the potential is returned to 0 and then back to -2 V. Instead, a modest increase in current in the opposite direction is observed when the potential is raised above $+0.7$ V. Presumably, this trend is a result of the participation of the phenoxy-centered HOMOs in the tunneling process. After a single positive voltage pulse, however, no current can be detected if the potential is returned to negative values. In summary, the positive potential scan suppresses irreversibly the conducting ability of the electrode/molecule/electrode junction. The behavior of this device correlates with the redox response of the [2]rotaxane **12** in solution. Cyclic voltammograms reveal reversible monoelectronic reductions of the bipyridinium dications. But they also show two irreversible oxidations associated, presumably, with the phenoxy rings of the macrocycle and tetraarylmethane groups. These observations suggest that a positive voltage pulse applied to the electrode/monolayer/electrode junction oxidizes irreversibly the sandwiched molecules suppressing their ability to mediate the transfer of electrons from the bottom to the top electrode under a negative bias.

The device incorporating the [2]rotaxane **13** can be exploited to implement simple logic operations [2.40]. The two bottom electrodes can be stimulated with voltage inputs (I1 and I2) while measuring a current output (O) at the common top electrode. When at least one of the two inputs is high (0 V), the output is low (< 0.7 nA). When both inputs are low (-2 V), the output is high (ca. 4 nA). If a negative logic convention is applied to the voltage inputs (low = 1, high = 0) and a positive logic convention is applied to the current output (low = 0, high = 1), the signal transduction behavior translates into the truth table of an AND gate. The output O is 1 only when both inputs are 1. Instead, an OR operation can be executed if the logarithm of the current is considered as the output. The logarithm of the current is -12 when both voltage inputs are 0 V. It raises to ca. -9 when one or both voltage inputs are lowered to -2 V. This signal transduction behavior translates into the truth table of an OR gate if a negative logic convention is applied to the voltage inputs (low = 1, high = 0) and a positive logic convention is applied to the current output (low = 0, high = 1). The output O is 1 when at least one of the two inputs is 1.

The [2]catenane **13** (Fig. 2.14) incorporates a macrocyclic polyether interlocked with a tetracationic cyclophane [2.42, 43]. Organic solutions of the hexafluorophosphate salt of this [2]catenane and six equivalents of the sodium salt of dimyristoylphosphatidic acid can be co-spread on the water surface of a Langmuir trough [2.44]. The sodium hexafluorophosphate formed dissolves in the supporting aqueous phase, while the hydrophilic bipyridinium cations and the amphiphilic anions remain at the interface. Upon compression, the anions align their hydrophobic tails away from the water surface forming a compact monolayer above the cationic bipyridinium derivatives. The corresponding π–A isotherm indicates limiting molecular areas of ca. 125 Å^2. This large value is a consequence of the bulk associated with the two interlocking macrocycles.

Monolayers of the [2]catenane **13** can be transferred from the air/water interface to the surfaces of the bottom n-doped silicon/silicon dioxide electrodes of a patterned silicon chip with the hydrophobic tails of the amphiphilic anions pointing away from the supporting substrate [2.45, 46]. The subsequent assembly of a top titanium/aluminum electrode affords electrode/monolayer/electrode arrays. Their junction resistance can be probed grounding the top electrode and maintaining the potential of the bottom electrode at $+0.1$ V. If a voltage pulse of $+2$ V is applied to the bottom electrode before the measurement, the junction resistance probed is ca. 0.7 GΩ. After a pulse of -2 V applied to the bottom electrode, the junction resistance probed at $+0.1$ V drops ca. 0.3 GΩ. Thus, alternating positive and negative voltage pulses can switch reversibly the junction resistance between high and low values. This intriguing behavior is a result of the redox and dynamic properties of the [2]catenane **13**.

Extensive spectroscopic and crystallographic studies [2.42, 43] demonstrated that the tetrathiafulvalene unit resides preferentially inside the cavity of the tetracationic cyclophane of the [2]catenane **13** (Fig. 2.14). Attractive $[\pi \cdots \pi]$ stacking interactions between the neutral tetrathiafulvalene and the bipyridinium dications are responsible for this co-conformation. Oxidation of the tetrathiafulvalene generates a cationic form that is expelled from the cavity of the tetracationic cyclophane. After the circumrotation of the macrocyclic polyether, the oxidized tetrathiafulvalene is exchanged with the neutral 1,5-dioxynaphthalene producing the [2]catenane **14** (Fig. 2.14). The reduction of the tetrathiafulvalene back to its neutral state is followed by the circumrotation of the macrocyclic polyether, which restores the original state **14**. The voltage pulses applied to the bottom electrode of the electrode/monolayer/junction oxidize and reduce the tetrathiafulvalene unit inducing the interconversion between the forms **13** and **14**. The difference in the stereoelectronic properties of these two states translates into distinct current/voltage signatures. Indeed, their ability to mediate the tunneling of electrons across the junction differs significantly. As a result, the junction resistance probed at a low voltage after an oxidizing pulse is significantly different from that determined under the same conditions after a reducing pulse.

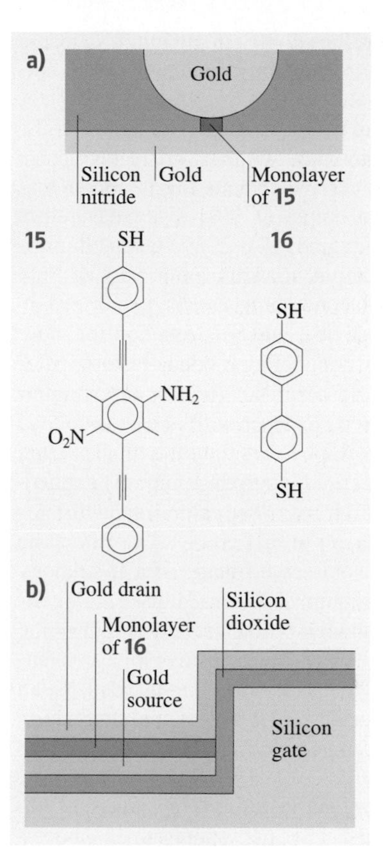

Fig. 2.15 (**a**) A monolayer of the thiol **15** is embedded between two gold electrodes maintained in position by a silicon nitride support. (**b**) A monolayer of the bisthiol **16** is located between two gold electrodes and stimulated by a silicon gate

2.3.3 Self-Assembled Monolayers

In the examples illustrated in Figs. 2.12–2.15, monolayers of amphiphilic and electroactive derivatives are assembled at air/water interfaces and then transferred on the surfaces of appropriate substrates. An alternative strategy to coat electrodes with molecular layers relies on the ability of certain compounds to adsorb spontaneously on solid supports from liquid or vapor phases [2.35]. In particular, the affinity of certain sulfurated functional groups for gold can be exploited to encourage the self-assembly of organic molecules on microscaled and nanoscaled electrodes.

The electrode/monolayer/electrode junction in Fig. 2.15a incorporates a molecular layer between two gold electrodes mounted on a silicon nitride support. This device can be fabricated combining chemical vapor deposition, lithography, anisotropic etching, and self-assembly [2.47]. Initially, a silicon wafer is coated with a 50 nm thick layer of silicon nitride by low pressure chemical vapor deposition. Then, a square of $400 \times 400\,\mu$m is patterned on one side of the coated wafer by optical lithography and reactive ion etching. Anisotropic etching of the exposed silicon up to the other side of the wafer leaves a suspended silicon nitride membrane of $40 \times 40\,\mu$m. Electron beam lithography and reactive ion etching can be used to carve a bowl-shaped hole (diameter = 30–50 nm) in the membrane. Evaporation of gold on the membrane fills the pore producing a bowl-shaped electrode. Immersion of the substrate in a solution of the thiol **15** results in the self-assembly of a molecular layer on the narrow part of the bowl-shaped electrode. The subsequent evaporation of a gold film on the organic monolayer produces an electrode/monolayer/electrode junction (Fig. 2.15a) with a contact area of less than 2,000 nm^2 and approximately 1,000 molecules.

Under the influence of voltage pulses applied to one of the two gold electrodes in Fig. 2.15a, the conductivity of the sandwiched monolayer switches reversibly between low and high values [2.48]. In the initial state, the monolayer is in a low conducting mode. A current output of only 30 pA is detected, when a probing voltage of +0.25 V is applied to the bowl-shaped electrode. If the same electrode is stimulated with a short voltage pulse of +5 V, the monolayer switches to a high conducting mode. Now a current output of 150 pA is measured at the same probing voltage of +0.25 V. Repeated probing of the current output at various intervals of time indicates that the high conducting state is memorized by the molecule-based device, and it is retained for more than 15 min. The low conducting mode is restored after either a relatively long period of time or the stimulation of the bowl-shaped electrode with a reverse voltage pulse of −5 V. Thus the current output switches from a low to a high value, if a high voltage input is applied. It switches from a high to a low value, under the influence of a low voltage pulse. This behavior offers the opportunity to store and erase binary data in analogy to a conventional random access memory [2.17]. Binary digits can be encoded on the current output of the molecule-based device applying a positive logic convention (low = 0, high = 1). It follows that a binary 1 can be stored in the molecule-based device applying a high voltage input, and it can be erased applying a low voltage input [2.48].

The electrode/monolayer/electrode junctions in Fig. 2.14 and in Fig. 2.15a are two-terminal devices. A voltage input is applied to one terminal, and a current output is measured at the other. The implementation of molecule-based transistors, however, requires three terminals. A third electrode is needed to gate the current flowing between the source and drain terminals. These types of devices are much more versatile than their two-terminal counterparts for the implementation of logic functions. But the incorporation of molecular components in a three-terminal configuration is considerably more challenging. A remarkable example is the composite assembly in Fig. 2.15b [2.49]. This field-effect transistor integrates a monolayer of the bisthiol **16** and three electrodes. Conventional lithography and anisotropic etching are employed to define a vertical step in a doped silicon substrate. The resulting structure is covered with an insulating layer of silicon dioxide with a thickness of ca. 30 nm. Then gold is evaporated on the bottom part of the step to allow the self-assembly the bisthiol **16**. Under the influence of the gate voltage, the molecular components embedded in this device modulate the current flowing between the source and drain electrodes. The intensity of the drain current increases by five orders of magnitude if the gate voltage is lowered below −0.2 V. It is important to note, however, that only the molecules within 5 nm from the monolayer/insulator interface are responsible for conduction. Thus the current/voltage behavior of this three-terminal device is determined only by few thousand molecules.

Conventional transistors are the basic building blocks of digital circuits [2.19]. They are routinely interconnected with specific configurations to execute designed logic functions. The molecule-based transistor in Fig. 2.15b is no exception. It can be used to assemble real electronic circuits incorporating molecular components. The connection of the drain terminals of two molecule-based transistors affords an inverter [2.49]. In this basic circuit, an output voltage switches from a high (0 V) to a low value (−2 V) when an input voltage varies from a low (−2 V) to a high value (0 V) and vice versa. This signal transduction behavior translates into the truth table of a NOT gate, if a positive logic convention (low = 0, high = 1) is applied to the input and output voltages.

The ability of thiols to self-assemble on the surface of gold can be exploited to fabricate nanocomposite materials integrating organic and inorganic components. For example, the bisthiol **17** forms monolayers (Fig. 2.16a) on gold electrodes with surface coverages of ca. 4.1×10^{10} mol cm^2 [2.50, 51]. The formation of a thiolate–gold bond at one of the two thiol ends of **17** is responsible for adsorption. The remaining thiol group points away from the supporting surface and can be exploited for further functionalization. Gold nanoparticles adsorb on the molecular layer (Fig. 2.16b), once again, as a result of thiolate–gold bond formation. The immersion of the resulting material in a methanol solution of **17** encourages the adsorption of an additional organic layer (Fig. 2.16c) on the composite material. Following these procedures, up to ten alternating organic and inorganic layers can be deposited on the electrode surface. The resulting assembly can mediate the unidirectional electron transfer from the supporting electrode to redox active species in solution. For example, the cyclic voltammogram of the $[Ru(NH_3)_6]^{3+/2+}$ couple recorded with a bare gold electrode reveals a reversible reduction process. In the presence of ten alternating molecular and nanoparticle layers on the electrode surface, the reduction potential shifts by ca. −0.2 V and the back oxidation wave disappears. The pronounced potential shift indicates that $[Ru(NH_3)_6]^{3+}$ accepts electrons only after the surface-confined bipyridinium dications have been reduced. The lack of reversibility indicates that the back oxidation to the bipyridinium dications inhibits the transfer of electrons from the $[Ru(NH_3)_6]^{2+}$ to the electrode. Thus the electroactive multilayer allows the flow of electrons in one direction only in analogy to conventional diodes.

The current/voltage behavior of individual nanoparticles in Fig. 2.16b can be probed by scanning tunneling spectroscopy in an aqueous electrolyte under an inert atmosphere [2.52]. The platinum-iridium tip of a scanning tunneling microscope is positioned above one of the gold particles. The voltage of the gold substrate relative to the tip is maintained at -0.2 V while that relative to a reference electrode immersed in the same electrolyte is varied to control the redox state of the electroactive units. Indeed, the bipyridinium dications in the molecular layer can be reduced reversibly to a monocationic state. The resulting monocations can be reduced further and, once again, reversibly to a neutral form. Finally, the current flowing from the gold support to the tip of the scanning tunneling microscope is monitored as the tip–particle distance increases. From the distance dependence of the current, inverse length decays of ca. 16 and 7 nm^{-1} for the dicationic and monocationic states, respectively, of the molecular spacer can be determined. The dramatic decrease indicates that the reduction of the electroactive unit facilitates the tunneling of electrons through the gold/molecule/nanoparticle/tip junction. In summary, a change in the redox state of the bipyridinium components can be exploited to gate reversibly the current flowing through this nanoscaled device.

Similar nanostructured materials, combining molecular and nanoparticles layers, can be prepared on layers on indium-tin oxide electrodes following multistep procedures [2.53]. The hydroxylated surfaces of indium-tin oxide supports can be functionalized with 3-ammoniumpropylysilyl groups and then exposed to gold nanoparticles having a diameter of ca. 13 nm [2.54, 55]. Electrostatic interactions promote the adsorption of the nanoparticles on the organic layer (Fig. 2.17a). The treatment of the composite film with the bipyridinium cyclophane **18** produces an organic layer on the gold nanoparticles (Fig. 2.18b). Following this approach, alternating layers of inorganic nanoparticles and organic building blocks can be assembled on the indium-tin oxide support. Cyclic voltammograms of the resulting materials show the oxidation of the gold nanoparticles and the reduction of the bipyridinium units. The peak current for both processes increases with the number of alternating layers. Comparison of these values indicates that the ratio between the number of tetracationic cyclophanes and that of the nanoparticles is ca. 100 : 1.

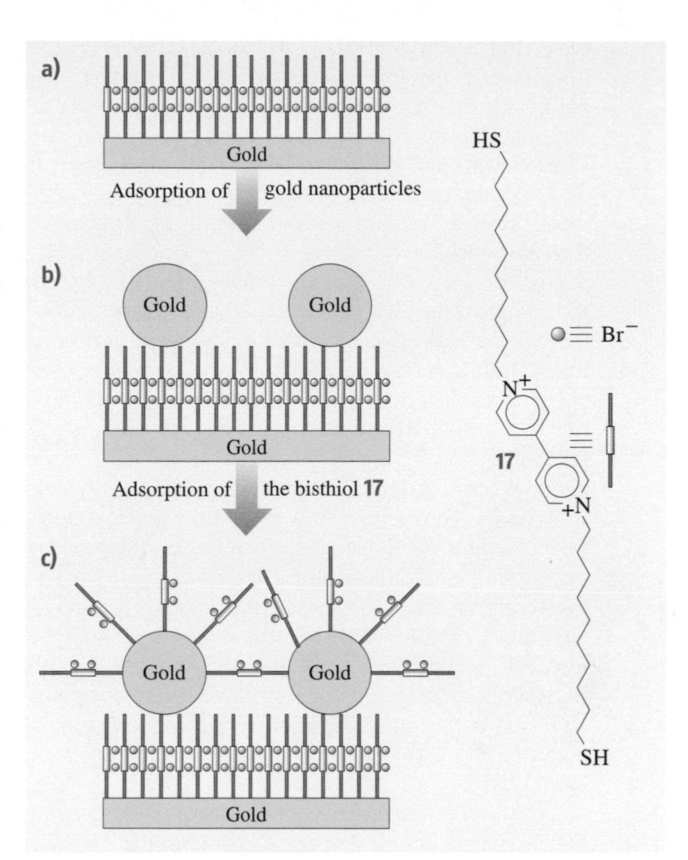

Fig. 2.16 (**a**) The bisthiol **18** self-assembles on gold electrodes as a result of thiolate–gold bond formation. (**b**) Gold nanoparticles adsorb spontaneously on the molecular layer. (**c**) Exposure of the composite assembly to a solution of **18** results in the formation of an additional molecular layer on the surface of the gold nanoparticles

The tetracationic cyclophane **18** binds dioxyarenes in solution [2.56, 57]. Attractive supramolecular forces between the electron deficient bipyridinium units and the electron rich guests are responsible for complexation. This recognition motif can be exploited to probe the ability of the composite films in Fig. 2.17b,c to sense electron rich analytes. In particular, hydroquinone is expected to enter the electron deficient cavities of the surface-confined cyclophanes. Cyclic voltammograms consistently reveal the redox waves associated with the reversible oxidation of hydroquinone even when very small amounts of the guest (ca. 1×10^5 M) are added to the electrolyte solution [2.54, 55]. No redox response can be detected with a bare indium–tin oxide electrode under otherwise identical conditions. The supramolecular association of the guest and the surface confined

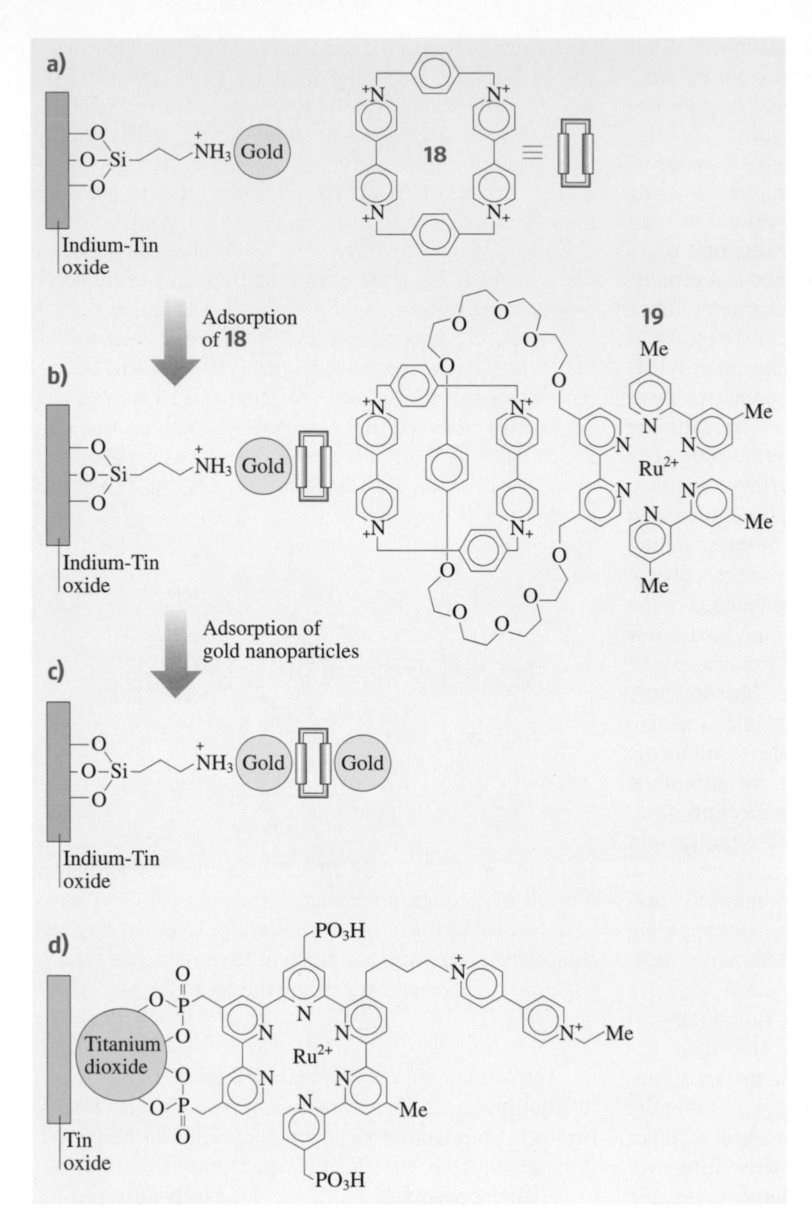

Fig. 2.17 **(a)** Gold nanoparticles assemble spontaneously on prefunctionalized indium-tin oxide electrodes. **(b)** Electrostatic interactions encourage the adsorption of the tetracationic cyclophane **18** on the surface-confined nanoparticles. **(c)** An additional layer of nanoparticles assembles on the cationic organic coating. Similar composite films can be prepared using the tetracationic [2]catenane **19** instead of the cyclophane **18**. **(d)** Phosphonate groups can be used to anchor molecular building blocks to titanium dioxide nanoparticles

cyclophanes increases the local concentration of hydroquinone at the electrode/solution interface enabling its electrochemical detection.

Following a related strategy, the [2]catenane **19** (Fig. 2.17) can be incorporated into similar composite arrays [2.58,59]. This interlocked molecule incorporates a Ru(II)/trisbipyridine sensitizer and two bipyridinium acceptors. Upon irradiation of the composite material at 440 nm, photoinduced electron transfer from the sensitizer to the appended acceptors occurs. The photogenerated hole in the sensitizer is filled after the transfer of an electron from a sacrificial electron donor present in the electrolyte solution. Under a positive voltage bias applied to the supporting electrode, an electron flow from the bipyridinium acceptors to the indium-tin oxide support is established. The resulting current switches between high and low values as the light source is turned on and off.

Another photoresponsive device, assembled combining inorganic nanoparticles with molecular building blocks, is illustrated in Fig. 2.17d. Phosphonate groups can be used to anchor a Ru(II)/trisbipyridine complex with an appended bipyridinium dication to titanium dioxide nanoparticles deposited on a doped tin oxide electrode [2.60, 61]. The resulting composite array can be integrated in a conventional electrochemical cell filled with an aqueous electrolyte containing triethanolamine. Under a bias voltage of -0.45 V and irradiation at 532 nm, 95% of the excited ruthenium centers transfer electrons to the titanium dioxide nanoparticles. The other 5% donate electrons to the bipyridinium dications. All the electrons transferred to the bipyridinium acceptors return to the ruthenium centers, while only 80% of those accepted by the nanoparticles return to the transition metal complexes. The remaining 15% reach the bipyridinium acceptors, while electron transfer from sacrificial triethanolamine donors fills the photogenerated holes left in the ruthenium sensitizers. The photoinduced reduction of the bipyridinium dication is accompanied by the appearance of the characteristic band of the radical cation in the absorption spectrum. This band persists for hours under open circuit conditions. But it fades in ca. 15 s under a voltage bias of $+1$ V, as the radical cation is oxidized back to the dicationic form. In summary, an optical stimulation accompanied by a negative voltage bias reduces the bipyridinium building block. The state of the photogenerated form can be read optically, recording the absorption spectrum in the visible region, and erased electrically, applying a positive voltage pulse.

2.3.4 Nanogaps and Nanowires

The operating principles of the electroactive and photoactive devices illustrated in Figs. 2.12–2.17 exploit the ability of small collections of molecular components to manipulate electrons and photons. Designed molecules are deposited on relatively large electrodes and can be addressed electrically and/or optically by controlling the voltage of the support and/or illuminating its surface. The transition from devices relying on collections of molecules to unimolecular devices requires the identification of practical methods to contact single molecules. This fascinating objective demands the rather challenging miniaturization of contacting electrodes to the nanoscale.

A promising approach to unimolecular devices relies on the fabrication of nanometer-sized gaps in metallic features followed by the insertion of individual molecules between the terminals of the gap. This strategy permits the assembly of nanoscaled three-terminal devices equivalent to conventional transistors [2.62–64]. A remarkable example is illustrated in Fig. 2.18a [2.62]. It incorporates a single molecule in the nanogap generated between two gold electrodes. Initially electron beam lithography is used to pattern a gold wire on a doped silicon wafer covered by an insulating silicon dioxide layer. Then the gold feature is broken by electromigration to generate the nanogap. The lateral size of the separated electrodes is ca. 100 nm and their thickness is ca. 15 nm. Scanning electron microcopy indicates that the facing surfaces of the separated electrodes are not uniform and that tiny gaps between their protrusions are formed. Current/voltage measurements suggest that the size of the smallest nanogap is ca. 1 nm. When the

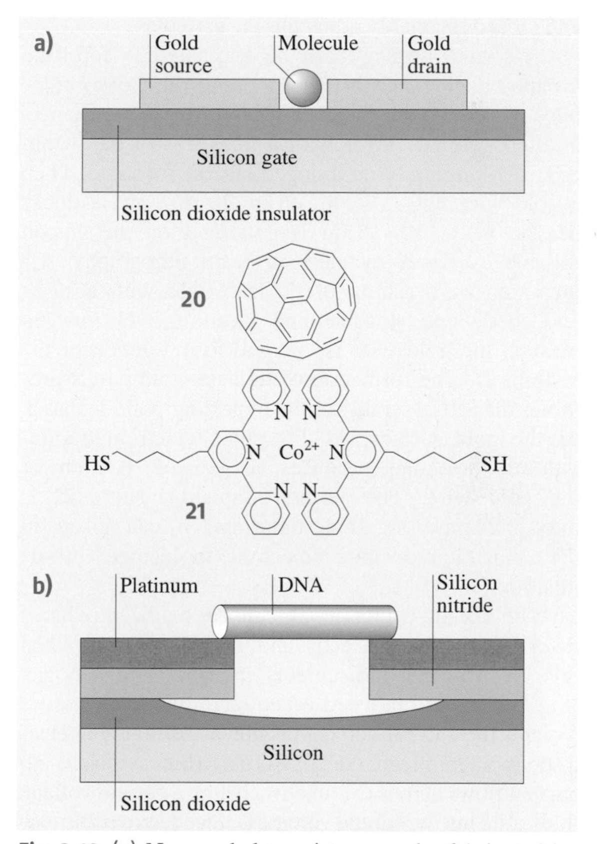

Fig. 2.18 (a) Nanoscaled transistors can be fabricated inserting a single molecule (**20** or **21**) between source and drain electrodes mounted on a silicon/silicon dioxide support. (b) A DNA nanowire can bridge nanoelectrodes suspended above a silicon dioxide support

breakage of the gold feature is preceded by the deposition of a dilute toluene solution of C_{60} (**20**), junctions with enhanced conduction are obtained. This particular molecule has a diameter of ca. 0.7 nm and can insert in the nanogap facilitating the flow of electrons across the junction.

The unique configuration of the molecule-based device in Fig. 2.18a can reproduce the functions of a conventional transistor [2.19] at the nanoscale. The two gold terminals of the junction are the drain and source of this nanotransistor, and the underlying silicon wafer is the gate. At a temperature of 1.5 K, the junction conductance is very small, when the gate bias is low, and increases in steps at higher voltages [2.62]. The conductance gap is a consequence of the finite energy required to oxidize/reduce the single C_{60} positioned in the junction. It is interesting that the zero-conductance window also changes with the gate voltage and can be opened and closed reversibly adjusting the gate bias.

A similar strategy can be employed to fabricate a nanoscaled transistor incorporating the Co(II) complex **21** shown in Fig. 2.18 [2.64]. In this instance, a silicon dioxide layer with a thickness of ca. 30 nm is grown thermally on a doped silicon substrate. Then a gold wire with a width of ca. 200 nm and a thicknesses of ca. 10–15 nm is patterned on the silicon dioxide overlayer by electron beam lithography. After extensive washing of the substrate with acetone and methylene chloride and cleaning with oxygen plasma, the gold wire is exposed to a solution of the bisthiol **21**. The formation of thiolate–gold bonds promotes the self-assembly of the molecular building block on the gold surface. At this point, electromigration-induced breakage produces a gap of 1–2 nm in the gold wire. The surface-confined bisthiol **21** is only 0.24 nm long and, therefore, it can insert in the nanogap producing an electrode/molecule/electrode junction.

The cobalt center in **21** can be oxidized/reduced reversibly between Co(II) and Co(III) [2.64]. When this electroactive molecule is inserted in a nanogap (Fig. 2.18a), its ability to accept and donate electrodes dictates the current/voltage profile of the resulting electrode/molecule/electrode junction. More precisely, no current flows across the junction below a certain voltage threshold. As the source voltage is raised above this particular value, the drain current increases in steps. The threshold associated with the source voltage varies in magnitude with the gate voltage. This intriguing behavior is a consequence of the finite energy necessary to oxidize/reduce the cobalt center and of a change in the relative stabilities of the oxidized and reduced forms Co(II) and Co(III) with the gate voltage. In summary, the conduction of the electrode/molecule/electrode junction can be tuned adjusting the voltage of the silicon support. The behavior of this molecule-based nanoelectronic device is equivalent to that of a conventional transistor [2.19]. In both instances, the gate voltage regulates the current flowing from the source to the drain.

The electromigration-induced breakage of preformed metallic features successfully produces nanogaps by moving apart two fragments of the same wire. Alternatively, nanogaps can be fabricated reducing the separation of the two terminals of much larger gaps. For example, gold electrodes separated by a distance of 20–80 nm can be patterned on a silicon/silicon dioxide substrate by electron beam lithography [2.65]. The relatively large gap between them can be reduced significantly by the electrochemical deposition of gold on the surfaces of both electrodes. The final result is the fabrication of two nanoelectrodes separated by ca. 1 nm and with a radius of curvature of 5–15 nm. The two terminals of this nanogap can be "contacted" by organic nanowires grown between them [2.66]. In particular, the electropolymerization of aniline produces polyaniline bridges between the gold nanoelectrodes. The conductance of the resulting junction can be probed immersing the overall assembly in an electrolyte solution. Employing a bipotentiostat, the bias voltage of the two terminals of the junction can be maintained at 20 mV, while their potentials are scanned relative to that of a silver/silver chloride reference electrode. Below ca. 0.15 V, the polymer wire is in an insulating state and the current flowing across the junction is less than 0.05 nA. At this voltage threshold, however, the current raises abruptly to ca. 30 nA. This value corresponds to a conductivity for the polymer nanojunction of 10–100 $\mathrm{S\,cm^{-1}}$. When the potential is lowered again below the threshold, the current returns back to very low values. The abrupt decrease in current in the backward scan is observed at a potential that is slightly more negative than that causing the abrupt current increase in the forward scan. In summary, the conductance of this nanoscaled junction switches on and off as a potential input is switched above and below a voltage threshold.

It is interesting to note that the influence of organic bridges on the junction conductance can be exploited for chemical sensing. Nanogaps fabricated following a similar strategy but lacking the polyaniline bridge alter their conduction after exposure to dilute solutions of small organic molecules [2.67]. Indeed, the organic analytes dock into the nanogaps producing a marked decrease in the junction conductance. The magnitude of the conductance drop happens to be proportional to the analyte–nanoelectrode binding strength. Thus the presence of the analyte in solution can be detected probing the current/voltage characteristics of the nanogaps.

Nanogaps between electrodes patterned on silicon/silicon dioxide supports can be bridged also by DNA double strands [2.68, 69]. The device in Fig. 2.18b has a 10.4 nm long poly(G)–poly(C) DNA oligomer suspended between two nanoelectrodes. It can be fabricated patterning a 30 nm wide slit in a silicon nitride overlayer covering a silicon/silicon dioxide support by electron beam evaporation. Underetching the silicon dioxide layer leaves a silicon nitride finger, which can be sputtered with a platinum layer and chopped to leave a nanogap of 8 nm. At this point, a microdroplet of a dilute solution of DNA is deposited on the device and a bias of 5 V is applied between the two electrodes. Electrostatic forces encourage the deposition of a single DNA wire on top of the nanogap. As soon as the nanowire is in position, current starts to flow across the junction. The current/voltage signature of the electrode/DNA/electrode junction shows currents below 1 pA at low voltage biases. Under these conditions, the DNA nanowire is an insulator. Above a certain voltage threshold, however, the nanowire becomes conducting and currents up to 100 nA can flow across the junction through a single nanowire. Assuming that direct tunneling from electrode to electrode is extremely unlikely for a relatively large gap of 8 nm, the intriguing current/voltage behavior has to be a consequence of the participation of the molecular states in the electron transport process. Two possible mechanisms can be envisaged. Sequential hopping of the electrons between states localized in the DNA base pairs can allow the current flow above a certain voltage threshold. But this mechanism would presumably result in a Coulomb blockade voltage gap that is not observed experimentally. More likely, electronic states delocalized across the entire length of the DNA nanowire are producing a molecular conduction band. The off-set between the molecular conduction band and the Fermi levels of the electrodes is responsible for the insulating behavior at low biases. Above a certain voltage threshold, the molecular band and one of the Fermi levels align facilitating the passage of electrons across the junction.

Carbon nanotubes are extremely versatile building blocks for the assembly of nanoscaled electronic devices. They can be used to bridge nanogaps [2.70–73] and assemble nanoscaled cross junctions [2.74–76]. In Fig. 2.19a, a single-wall carbon nanotube crosses over another one in an orthogonal arrangement [2.74]. Both nanotubes have electrical contacts at their ends.

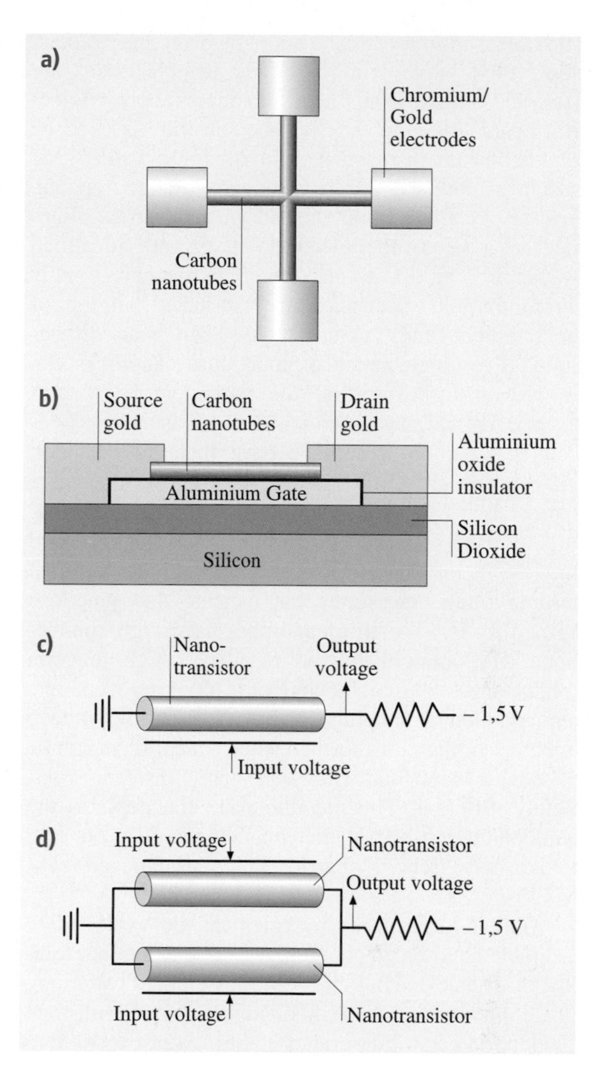

Fig. 2.19 (a) Nanoscaled junctions can be assembled on silicon/silicon dioxide supports crossing pairs of orthogonally arranged single-wall carbon nanotubes with chromium/gold electrical contacts at their ends. (b) Nanotransistors can be fabricated contacting the two ends of a single-wall carbon nanotube deposited on an aluminum/aluminum oxide gate with gold sources and drain. One or two nanotube transistors can be integrated into nanoscaled NOT (c) and NOR (d) logic gates

The fabrication of this device involves three main steps. First, alignment marks for the electrodes are patterned on a silicon/silicon dioxide support by electron beam lithography. Then the substrate is exposed to a dichloromethane suspension of single-wall SWNT carbon nanotubes. After washing with isopropanol, crosses of carbon nanotubes in an appropriate alignment relative to the electrode marks are identified by tapping mode atomic force microscopy. Finally chromium/gold electrodes are fabricated on top of the nanotube ends, again, by electron beam lithography. The conductance of individual nanotubes can be probed by exploiting the two electric contacts at their ends. These two-terminal measurements reveal that certain nanotubes have metallic behavior, while others are semiconducting. It follows that three distinct types of cross junctions differing in the nature of their constituent nanotubes can be identified on the silicon/silicon dioxide support. Four terminal current/voltage measurements indicate that junctions formed by two metallic nanotubes have high conductance and ohmic behavior. Similarly, high junction conductance and ohmic behavior is observed when two semiconducting nanotubes cross. The current/voltage signature of junctions formed when a metallic nanotube crosses a semiconducting one are, instead, completely different. The metallic nanotube depletes the semiconducting one at the junction region producing a nanoscaled Schottky barrier with a pronounced rectifying behavior.

Similar fabrication strategies can be exploited to assemble nanoscaled counterparts of conventional transistors. The device in Fig. 2.19b is assembled patterning an aluminum finger on a silicon/silicon dioxide substrate by electron beam lithography [2.76]. After exposure to air, an insulating aluminum oxide layer forms on the aluminum finger. Then a dichloromethane suspension of single-wall carbon nanotubes is deposited on the resulting substrate. Atomic force microscopy can be used to select carbon nanotubes with a diameter of ca. 1 nm positioned on the aluminum finger. After registering their coordinates relative to alignment markers, gold contacts can be evaporated on their ends by electron beam lithography. The final assembly is a nanoscaled three-terminal device equivalent to a conventional field effect transistor [2.19]. The two gold contacts are the source and drain terminals, while the underlying aluminum finger reproduces the function of the gate. At a source to drain bias of ca. -1.3 V, the drain current jumps from ca. 0 to ca. 50 nA when the gate voltage is lowered from -1.0 to -1.3 V. Thus moderate changes in the gate voltage vary significantly the current flowing through the nanotube-based device in analogy to a conventional enhancement mode *p*-type field effect transistor [2.19].

The nanoscaled transistor in Fig. 2.18a has a microscaled silicon gate that extends under the entire chip [2.62, 64]. The configuration in Fig. 2.19b, instead, has nanoscaled aluminum gates for every single carbon nanotube transistor fabricated on the same support [2.76]. It follows that multiple nanoscaled transistors can be fabricated on the same chip and operated independently following this strategy. This unique feature offers the possibility of fabricating nanoscaled digital circuits by interconnecting the terminals of independent nanotube transistors. The examples in Fig. 2.19c,d illustrate the configurations of nanoscaled NOT and NOR gates implemented using one or two nanotube transistors. In Fig. 2.19c, an off-chip bias resistor is connected to the drain terminal of a single transistor while the source is grounded. A voltage input applied to the gate modulates the nanotube conductance altering the voltage output probed at the drain terminal. In particular, a voltage input of -1.5 V lowers the nanotube resistance (26 MΩ) below that of the bias resistor (100 MΩ). As a result, the voltage output drops to 0 V. When the voltage input is raised to 0 V, the nanotube resistance increases above that of the bias resistor and the voltage output becomes -1.5 V. Thus the output of this nanoelectronic device switches from a high (0 V) and to a low (-1.5 V) level as the input shifts from a low (-1.5 V) to a high (0 V) value. The inverse relation between input and output translates into a NOT operation if a negative logic convention (low $=1$, high $=0$) is applied to both signals.

In Fig. 2.15d, the source terminals of two independent nanotube transistors fabricated on the same chip are connected by a gold wire and grounded [2.76]. Similarly, the two drain terminals are connected by another gold wire and contacted to an off-chip bias resistors. The gate of each nanotube can be stimulated with a voltage input and the voltage output of the device can be probed at their interconnected drain terminals. When the resistance of at least one of the two nanotubes is below that of the resistor, the output is 0 V. When both nanotubes are in a nonconducting mode, the output voltage is -1.5 V. Thus if a low voltage input -1.5 V is applied to one or both transistors, the output is high (0 V). When both voltage inputs are high (0 V), the output is low (-1.5 V). If a negative logic convention (low $=1$, high $=0$) is applied to all signals, the signal transduction behavior translates in to a NOR operation.

2.4 Conclusions and Outlook

Nature builds nanostructured biomolecules relying on a highly modular approach [2.1]. Small building blocks are connected by robust chemical bonds to generate long strands of repeating units. The synergism of a multitude of attractive supramolecular forces determines the three-dimensional arrangement of the resulting polymeric chains and controls the association of independent strands into single and well-defined entities. Nucleic acids and proteins are two representative classes of biomolecules assembled with subnanometer precision through the subtle interplay of covalent and noncovalent bonds starting from a relatively small pool of nucleotide and amino acid building blocks.

The power of chemical synthesis [2.2] offers the opportunity of mimicking nature's modular approach to nanostructured materials. Following established experimental protocols, small molecular building blocks can be joined together relying on the controlled formation of covalent bonds between designed functional groups. Thus artificial molecules with nanoscaled dimensions can be assembled piece by piece with high structural control. Indeed, helical, tubular, interlocked, and highly branched nanostructures have been all prepared already exploiting this general strategy and the synergism of covalent and noncovalent bonds [2.3].

The chemical construction of nanoscaled molecules from modular building blocks also offers the opportunity for engineering specific properties in the resulting assemblies. In particular, electroactive and photoactive fragments can be integrated into single molecules. The ability of these functional subunits to accept/donate electrons and photons can be exploited to design nanoscaled electronic and photonic devices. Indeed, molecules that respond to electrical and optical stimulations producing detectable outputs have been designed already [2.16]. These chemical systems can be employed to control the interplay of input and output signals at the molecular level. Their conceptual analogy with the signal transduction operated by conventional logic gates in digital circuits is evident. In fact, electroactive and photoactive molecules able to reproduce AND, NOT, and OR operations as well as simple combinational of these basic logic functions are already a reality [2.20, 21].

Most of the molecular switches for digital processing developed so far rely on bulk addressing. In general, relatively large collections of functional molecules are addressed simultaneously in solution. The realization of molecule-based devices with reduced dimensions as well as practical limitations associated with liquid phases in potential applications are encouraging a transition from the solution to the solid state. The general strategy followed so far relies on the deposition of functional molecules on the surfaces of appropriate electrodes following either the Langmuir–Blodgett methodology [2.34] or self-assembly processes [2.35]. The combination of these techniques with the nanofabrication of insulating, metallic, and semiconducting features on appropriate supports has already allowed the realization of fascinating molecule-based devices [2.30–33, 53]. The resulting assemblies integrate inorganic and organic components and, in some instances, even biomolecules to execute specific functions. They can convert optical stimulations into electrical signals. They can execute irreversible and reversible switching operations. They can sense qualitatively and quantitatively specific analytes. They can reproduce the functions of conventional rectifiers and transistors. They can be integrated within functioning nanoelectronic devices capable of simple logic operations.

The remarkable examples of molecule-based materials and devices now available demonstrate the great potential and promise for this research area. At this stage, the only limit left to the design of functional molecules is the imagination of the synthetic chemist. All sort of molecular building blocks with tailored dimensions, shapes, and properties are more or less accessible with the assistance of modern chemical synthesis. Now, the major challenges are (1) to master the operating principles of the molecule-based devices that have been and continue to be assembled and (2) to expand and improve the fabrication strategies available to incorporate molecules into reliable device architectures. As we continue to gather further insights in these directions, design criteria for a wide diversity of molecule-based devices will emerge. It is not unrealistic to foresee the evolution of an entire generation of nanoscaled devices, based on engineered molecular components, that will find applications in a variety of fields ranging from biomedical research to information technology. Perhaps nature can once again illuminate our path, teaching us not only how to synthesize nanostructured molecules but also how to use them. After all, nature is replete with examples of extremely sophisticated molecule-based devices. From tiny bacteria to higher animals, we are all a collection of molecule-based devices.

References

2.1 D. Voet, J. G. Voet: *Biochemistry* (Wiley, New York 1995)

2.2 K. C. Nicolau, E. C. Sorensen: *Classics in Total Synthesis* (VCH, Weinheim 1996)

2.3 J.-M. Lehn: *Supramolecular Chemistry: Concepts and Perspectives* (VCH, Weinheim 1995)

2.4 M. M. Harding, U. Koert, J.-M. Lehn, A. Marquis-Rigault, C. Piguet, J. Siegel: Synthesis of unsubstituted and 4,4'-substituted oligobipyridines as ligand strands for helicate self-assembly, Helv. Chim. Acta **74** (1991) 594–610

2.5 J.-M. Lehn, A. Rigault, J. Siegel, B. Harrowfield, B. Chevrier, D. Moras: Spontaneous assembly of double-stranded helicates from oligobipyridine ligands and copper(I) cations: Structure of an inorganic double helix, Proc. Natl. Acad. Sci. USA **84** (1987) 2565–2569

2.6 J.-M. Lehn, A. Rigault: Helicates: Tetra- and pentanuclear double helix complexes of Cu(I) and poly(bipyridine) strands, Angew. Chem. Int. Ed. Engl. **27** (1988) 1095–1097

2.7 J. D. Hartgerink, J. R. Granja, R. A. Milligan, M. R. Ghadiri: Self-assembling peptide nanotubes, J. Am. Chem. Soc. **118** (1996) 43–50

2.8 M. R. Ghadiri, J. R. Granja, R. A. Milligan, D. E. McRee, N. Khazanovich: Self-assembling organic nanotubes based on a cyclic peptide architecture, Nature **366** (1993) 324–327

2.9 V. Balzani, A. Credi, F. M. Raymo, J. F. Stoddart: Artificial molecular machines, Angew. Chem. Int. Ed. **39** (2000) 3348–3391

2.10 A. J. Bard, L. R. Faulkner: *Electrochemical Methods: Fundamentals and Applications* (Wiley, New York 2000)

2.11 V. Balzani (Ed.): *Electron Transfer in Chemistry* (Wiley-VCH, Weinheim 2001)

2.12 J. D. Coyle: *Principles and Applications of Photochemistry* (Wiley, New York 1988)

2.13 J. F. Stoddart (Ed.): Molecular machines, Acc. Chem. Res. **34** (2001) 409–522

2.14 P. R. Ashton, R. Ballardini, V. Balzani, A. Credi, K. R. Dress, E. Ishow, C. J. Kleverlaan, O. Kocian, J. A. Preece, N. Spencer, J. F. Stoddart, M. Venturi, S. Wenger: A photochemically driven molecular-level abacus, Chem. Eur. J. **6** (2000) 3558–3574

2.15 B. L Feringa (Ed.): Photochromism: memories and switches, Chem. Rev. **100** (2000) 1683–1890

2.16 B. L. Feringa (Ed.): *Molecular Switches* (Wiley-VCH, Weinheim 2001)

2.17 R. J. Mitchell: *Microprocessor Systems: An Introduction* (Macmillan, London 1995)

2.18 D. R. Smith: *Digital Transmission Systems* (Van Nostrand Reinhold, New York 1993)

2.19 S. Madhu: *Electronics: Circuits and Systems* (SAMS, Indianapolis 1985)

2.20 A. P. de Silva, N. D. McClenaghan, C. P. McCoy: Logic Gates. In: *Electron Transfer in Chemistry*, ed. by V. Balzani (Wiley-VCH, Weinheim 2001) pp. 156–185

2.21 F. M. Raymo: Digital processing and communication with molecular switches, Adv. Mater. **14** (2002) 401–414

2.22 A. Aviram: Molecules for memory, logic and amplification, J. Am. Chem. Soc. **110** (1988) 5687–5692

2.23 A. P. de Silva, H. Q. N. Gunaratne, C. P. McCoy: A molecular photoionic AND gate based on fluorescent signaling, Nature **364** (1993) 42–44

2.24 M. Asakawa, P. R. Ashton, V. Balzani, A. Credi, G. Mattersteig, O. A. Matthews, M. Montalti, N. Spencer, J. F. Stoddart, M. Venturi: Electrochemically induced molecular motions in pseudorotaxanes: A case of dual-mode (oxidative and reductive) dethreading, Chem. Eur. J. **3** (1997) 1992–1996

2.25 F. M. Raymo, S. Giordani: Signal processing at the molecular level, J. Am. Chem. Soc. **123** (2001) 4651–4652

2.26 F. M. Raymo, S. Giordani: Signal communication between molecular switches, Org. Lett. **3** (2001) 3475–3478

2.27 F. M. Raymo, S. Giordani: Digital communication through intermolecular fluorescence modulation, Org. Lett. **3** (2001) 1833–1836

2.28 F. M. Raymo, S. Giordani: All-optical processing with molecular switches, Proc. Natl. Acad. Sci. USA **99** (2002) 4941–4944

2.29 A. J. Bard: *Integrated Chemical Systems: A Chemical Approach to Nanotechnology* (Wiley, New York 1994)

2.30 R. M. Metzger: Electrical rectification by a molecule: The advent of unimolecular electronic devices, Acc. Chem. Res. **32** (1999) 950–957

2.31 C. Joachim, J. K. Gimzewski, A. Aviram: Electronics using hybrid-molecular and mono-molecular devices, Nature **408** (2000) 541–548

2.32 J. M. Tour: Molecular electronics. Synthesis and testing of components, Acc. Chem. Res. **33** (2000) 791–804

2.33 A. R. Pease, J. O. Jeppesen, J. F. Stoddart, Y. Luo, C. P. Collier, J. R. Heath: Switching devices based on interlocked molecules, Acc. Chem. Res. **34** (2001) 433–444

2.34 M. C. Petty: *Langmuir–Blodgett Films: An Introduction* (Cambridge Univ. Press, Cambridge 1996)

2.35 A. Ulman: *An Introduction to Ultrathin Organic Films* (Academic, Boston 1991)

2.36 C. Lee, A. J. Bard: Comparative electrochemical studies of *N*-Methyl-*N'*-hexadecyl viologen monomolecular films formed by irreversible adsorption and the Langmuir–Blodgett method, J. Electroanal. Chem. **239** (1988) 441–446

2.37 C. Lee, A. J. Bard: Cyclic voltammetry and Langmuir film isotherms of mixed monolayers of *N*-docosoyl-

N′-methyl viologen with arachidic acid, Chem. Phys. Lett. **170** (1990) 57–60

2.38 M. Fujihira, K. Nishiyama, H. Yamada: Photoelectrochemical responses of optically transparent electrodes modified with Langmuir–Blodgett films consisting of surfactant derivatives of electron donor, acceptor and sensitizer molecules, Thin Solid Films **132** (1985) 77–82

2.39 M. Fujihira: Photoelectric conversion with Langmuir–Blodgett films. In: *Nanostructures Based on Molecular Materials*, ed. by W. Göpel, C. Ziegler (VCH, Weinheim 1992) pp. 27–46

2.40 C. P. Collier, E. W. Wong, M. Belohradsky, F. M. Raymo, J. F. Stoddart, P. J. Kuekes, R. S. Williams, J. R. Heath: Electronically configurable molecular-based logic gates, Science **285** (1999) 391–394

2.41 E. W. Wong, C. P. Collier, M. Belohradsky, F. M. Raymo, J. F. Stoddart, J. R. Heath: Fabrication and transport properties of single-molecule-thick electrochemical junctions, J. Am. Chem. Soc. **122** (2000) 5831–5840

2.42 M. Asakawa, P. R. Ashton, V. Balzani, A. Credi, C. Hamers, G. Mattersteig, M. Montalti, A. N. Shipway, N. Spencer, J. F. Stoddart, M. S. Tolley, M. Venturi, A. J. P. White, D. J. Williams: A chemically and electrochemically switchable [2]catenane incorporating a tetrathiafulvalene unit, Angew. Chem. Int. Ed. **37** (1998) 333–337

2.43 V. Balzani, A. Credi, G. Mattersteig, O. A. Matthews, F. M. Raymo, J. F. Stoddart, M. Venturi, A. J. P. White, D. J. Williams: Switching of pseudorotaxanes and catenanes incorporating a tetrathiafulvalene unit by redox and chemical inputs, J. Org. Chem. **65** (2000) 1924–1936

2.44 M. Asakawa, M. Higuchi, G. Mattersteig, T. Nakamura, A. R. Pease, F. M. Raymo, T. Shimizu, J. F. Stoddart: Current/Voltage characteristics of monolayers of redox-switchable [2]catenanes on gold, Adv. Mater. **12** (2000) 1099–1102

2.45 C. P. Collier, G. Mattersteig, E. W. Wong, Y. Luo, K. Beverly, J. Sampaio, F. M. Raymo, J. F. Stoddart, J. R. Heath: A [2]catenane based solid-state electronically reconfigurable switch, Science **289** (2000) 1172–1175

2.46 C. P. Collier, J. O. Jeppesen, Y. Luo, J. Perkins, E. W. Wong, J. R. Heath, J. F. Stoddart: Molecular-based electronically switchable tunnel junction devices, J. Am. Chem. Soc. **123** (2001) 12632–12641

2.47 J. Chen, M. A. Reed, A. M. Rawlett, J. M. Tour: Large on-off ratios and negative differential resistance in a molecular electronic device, Science **286** (1999) 1550–1552

2.48 M. A. Reed, J. Chen, A. M. Rawlett, D. W. Price, J. M. Tour: Molecular random access memory cell, Appl. Phys. Lett. **78** (2001) 3735–3737

2.49 J. H. Schön, H. Meng, Z. Bao: Self-assembled monolayer organic field-effect transistors, Nature **413** (2001) 713–716

2.50 D. I. Gittins, D. Bethell, R. J. Nichols, D. J. Schiffrin: Redox-controlled multilayers of discrete gold particles: A novel electroactive nanomaterial, Adv. Mater. **9** (1999) 737–740

2.51 D. I. Gittins, D. Bethell, R. J. Nichols, D. J. Schiffrin: Diode-like electron transfer across nanostructured films containing a redox ligand, J. Mater. Chem. **10** (2000) 79–83

2.52 D. I. Gittins, D. Bethell, D. J. Schiffrin, R. J. Nichols: A nanometer-scale electronic switch consisting of a metal cluster and redox-addressable groups, Nature **408** (2000) 67–69

2.53 A. N. Shipway, M. Lahav, I. Willner: Nanostructured gold colloid electrodes, Adv. Mater. **12** (2000) 993–998

2.54 A. N. Shipway, M. Lahav, R. Blonder, I. Willner: Bis-bipyridinium cyclophane receptor–Au nanoparticle superstructure for electrochemical sensing applications, Chem. Mater. **11** (1999) 13–15

2.55 M. Lahav, A. N. Shipway, I. Willner, M. B. Nielsen, J. F. Stoddart: An enlarged bis-bipyridinum cyclophane–Au nanoparticle superstructure for selective electrochemical sensing applications, J. Electroanal. Chem. **482** (2000) 217–221

2.56 R. E. Gillard, F. M. Raymo, J. F. Stoddart: Controlling self-assembly, Chem. Eur. J. **3** (1997) 1933–1940

2.57 F. M. Raymo, J. F. Stoddart: From supramolecular complexes to interlocked molecular compounds, Chemtracts – Organic Chemistry **11** (1998) 491–511

2.58 M. Lahav, T. Gabriel, A. N. Shipway, I. Willner: Assembly of a Zn(II)-porphyrin-bipyridinium dyad and Au-nanoparticle superstructures on conductive surfaces, J. Am. Chem. Soc. **121** (1999) 258–259

2.59 M. Lahav, V. Heleg-Shabtai, J. Wasserman, E. Katz, I. Willner, H. Durr, Y. Hu, S. H. Bossmann: Photoelectrochemistry with integrated photosensitizer-electron acceptor Au-nanoparticle arrays, J. Am. Chem. Soc. **122** (2000) 11480–11487

2.60 G. Will, S. N. Rao, D. Fitzmaurice: Heterosupramolecular optical write-read-erase device, J. Mater. Chem. **9** (1999) 2297–2299

2.61 A. Merrins, C. Kleverlann, G. Will, S. N. Rao, F. Scandola, D. Fitzmaurice: Time-resolved optical spectroscopy of heterosupramolecular assemblies based on nanostructured TiO_2 films modified by chemisorption of covalently linked ruthenium and viologen complex components, J. Phys. Chem. B **105** (2001) 2998–3004

2.62 H. Park, J. Park, A. K. L. Lim, E. H. Anderson, A. P. Alivisatos, P. L. McEuen: Nanomechanical oscillations in a single C_{60} transistor, Nature **407** (2000) 57–60

2.63 W. Liang, M. P. Shores, M. Bockrath, J. R. Long, H. Park: Kondo resonance in a single-molecule transistor, Nature **417** (2002) 725–729

2.64 J. Park, A. N. Pasupathy, J. I. Goldsmith, C. Chang, Y. Yaish, J. R. Petta, M. Rinkoski, J. P. Sethna,

H. D. Abruna, P. L. McEuen, D. C. Ralph: Coulomb blockade and the Kondo effect in single-atom transistors, Nature **417** (2002) 722–725
2.65 C. Z. Li, H. X. He, N. J. Tao: Quantized tunneling current in the metallic nanogaps formed by electrodeposition and etching, Appl. Phys. Lett. **77** (2000) 3995–3997
2.66 H. He, J. Zhu, N. J. Tao, L. A. Nagahara, I. Amlani, R. Tsui: A conducting polymer nanojunction switch, J. Am. Chem. Soc. **123** (2001) 7730–7731
2.67 A. Bogozi, O. Lam, H. He, C. Li, N. J. Tao, L. A. Nagahara, I. Amlani, R. Tsui: Molecular adsorption onto metallic quantum wires, J. Am. Chem. Soc. **123** (2001) 4585–4590
2.68 A. Bezryadin, C. N. Lau, M. Tinkham: Quantum suppression of superconductivity in ultrathin nanowires, Nature **404** (2000) 971–974
2.69 D. Porath, A. Bezryadin, S. de Vries, C. Dekker: Direct measurement of electrical transport through DNA molecules, Nature **403** (2000) 635–638
2.70 S. J. Tans, M. H. Devoret, H. Dai, A. Thess, E. E. Smalley, L. J. Geerligs, C. Dekker: Individual single-wall carbon nanotubes as quantum wires, Nature **386** (1997) 474–477
2.71 A. F. Morpurgo, J. Kong, C. M. Marcus, H. Dai: Gate-controlled superconducting proximity effect in carbon nanotubes, Nature **286** (1999) 263–265
2.72 J. Nygård, D. H. Cobden, P. E. Lindelof: Kondo physics in carbon nanotubes, Nature **408** (2000) 342–346
2.73 W. Liang, M. Bockrath, D. Bozovic, J. H. Hafner, M. Tinkham, H. Park: Fabry–Perot interference in a nanotube electron waveguide, Nature **411** (2001) 665–669
2.74 M. S. Fuhrer, J. Nygård, L. Shih, M. Forero, Y.-G. Yoon, M. S. C. Mazzoni, H. J. Choi, J. Ihm, S. G. Louie, A. Zettl, P. L. McEuen: Crossed nanotube junctions, Science **288** (2000) 494–497
2.75 T. Rueckes, K. Kim, E. Joselevich, G. Y. Tseng, C.-L. Cheung, C. M. Lieber: Carbon nanotube-based nonvolatile random access memory for molecular computing, Science **289** (2000) 94–97
2.76 A. Bachtold, P. Hadley, T. Nakanishi, C. Dekker: Logic circuits with carbon nanotube transistors, Science **294** (2001) 1317–1320

3. Introduction to Carbon Nanotubes

Carbon nanotubes are among the amazing objects that science sometimes creates by accident, without meaning to, but that will likely revolutionize the technological landscape of the century ahead. Our society stands to be significantly influenced by carbon nanotubes, shaped by nanotube applications in every aspect, just as silicon-based technology still shapes society today. The world already dreams of space-elevators tethered by the strongest of cables, hydrogen-powered vehicles, artificial muscles, and so on – feasts that would be made possible by the emerging carbon nanotube science.

Of course, nothing is set in stone yet. We are still at the stage of possibilities and potential. The recent example of fullerenes – molecules closely related to nanotubes, whose importance was so anticipated that their discovery in 1985 brought a Nobel Prize to their finders in 1996 although few related applications have actually yet reached the market – teaches us to play the game of enthusiastic predictions with some caution. But in the case of carbon nanotubes, expectations are high. Taking again the example of electronics, the miniaturization of chips is about to reach its lowest limits. Are we going to accept that our video camera, computers, and cellular phones no longer decrease in size and increase in memory every six months? Surely not. Always going deeper, farther, smaller, higher is a characteristic unique to humankind, and which helps to explain its domination of the living world on earth. Carbon nanotubes can help us fulfill our expectation of constant technological progress as a source of better living.

In this chapter, after the structure, synthesis methods, growth mechanisms, and properties of carbon nanotubes will be described, an entire section will be devoted to nanotube-related nano-objects. Indeed, should pristine nanotubes reach any limitation in some area, their ready and close association to foreign atoms, molecules, and compounds offers the prospect of an even magnified set of properties. Finally, we will describe carbon nanotube applications supporting the idea that the future for the science and technology of carbon nanotubes looks very promising.

3.1 **Structure of Carbon Nanotubes** ... 40
3.1.1 Single-Wall Nanotubes ... 40
3.1.2 Multiwall Nanotubes ... 43

3.2 **Synthesis of Carbon Nanotubes** ... 45
3.2.1 Solid Carbon Source-Based Production Techniques for Carbon Nanotubes ... 45
3.2.2 Gaseous Carbon Source-Based Production Techniques for Carbon Nanotubes ... 52
3.2.3 Miscellaneous Techniques ... 57
3.2.4 Synthesis of Aligned Carbon Nanotubes ... 58

3.3 **Growth Mechanisms of Carbon Nanotubes** 59
3.3.1 Catalyst-Free Growth ... 59
3.3.2 Catalytically Activated Growth ... 60

3.4 **Properties of Carbon Nanotubes** ... 63
3.4.1 Variability of Carbon Nanotube Properties ... 63
3.4.2 General Properties ... 63
3.4.3 SWNT Adsorption Properties ... 63
3.4.4 Transport Properties ... 65
3.4.5 Mechanical Properties ... 67
3.4.6 Reactivity ... 67

3.5 **Carbon Nanotube-Based Nano-Objects** ... 68
3.5.1 Hetero-Nanotubes ... 68
3.5.2 Hybrid Carbon Nanotubes ... 68
3.5.3 Functionalized Nanotubes ... 71

3.6 **Applications of Carbon Nanotubes** ... 73
3.6.1 Current Applications ... 73
3.6.2 Expected Applications Related to Adsorption ... 76

References ... 86

Carbon nanotubes have been synthesized for a long time as products from the action of a catalyst over the gaseous species originating from the thermal decomposition of hydrocarbons (see Sect. 3.2). Some of the first evidence that the nanofilaments thus produced were actually nanotubes – i. e., exhibiting an inner cavity – can be found in the transmission electron microscope micrographs published by *Hillert* et al. in 1958 [3.1]. This was of course related to and made possible by the progress in transmission electron microscopy. It is then likely that the carbon filaments prepared by *Hughes* and *Chambers* in 1889 [3.2], reported by *Maruyama* et al. [3.3] as probably the first patent ever deposited in the field, and whose preparation method was also based on the catalytically enhanced thermal cracking of hydrocarbons, were already carbon nanotube-related morphologies. The preparation of vapor-grown carbon fibers had actually been reported as early as more than one century ago [3.4, 5]. Since then, the interest in carbon nanofilaments/nanotubes has been recurrent, though within a scientific area almost limited to the carbon material scientist community. The reader is invited to consult the review published by *Baker* et al. [3.6] regarding the early works. The worldwide enthusiasm came unexpectedly in 1991, after the catalyst-free formation of nearly perfect concentric multiwall carbon nanotubes (c-MWNTs, see Sect. 3.1) was reported [3.7] as by-products of the formation of fullerenes by the electric-arc technique. But the real breakthrough occurred two years later, when attempts in situ to fill the nanotubes with various metals (see Sect. 3.5) led to the discovery – again unexpected – of single-wall carbon nanotubes (SWNTs) simultaneously by *Iijima* et al. [3.8] and *Bethune* et al. [3.9]. Single-wall carbon nanotubes were really new nano-objects many of whose properties and behaviors are quite specific (see Sect. 3.4). They are also beautiful objects for fundamental physics as well as unique molecules for experimental chemistry, still keeping some mystery since their formation mechanisms are a subject of controversy and are still debated (see Sect. 3.3). Potential applications seem countless, although few have reached a marketable status so far (see Sect. 3.6). Consequently, about five papers a day with carbon nanotubes as the main topic are currently published by research teams from around the world, an illustration of how extraordinarily active – and highly competitive – is this field of research. It is a quite unusual situation, similar to that of fullerenes, which, by the way, are again carbon nano-objects structurally closely related to nanotubes.

This is not, however, only about scientific exaltation. Economical aspects are leading the game to a greater and greater extent. According to experts, the world market is predicted to be more than 430 M $ in 2004 and estimated to grow to several b $ before 2009. That is serious business, and it will be closely related to how scientists and engineers will be able to deal with the many challenges found on the path from the beautiful, ideal molecule to the reliable – and it is hoped, cheap – manufactured product.

3.1 Structure of Carbon Nanotubes

It is simple to imagine a single-wall carbon nanotube (SWNT). Ideally, it is enough to consider a perfect graphene sheet (a graphene being the same polyaromatic mono-atomic layer made of an hexagonal display of sp^2 hybridized carbon atoms that genuine graphite is built up with), to roll it into a cylinder (Fig. 3.1) paying attention that the hexagonal rings put in contact join coherently, then to close the tips by two caps, each cap being a hemi-fullerene with the appropriate diameter (Fig. 3.2a–c).

3.1.1 Single-Wall Nanotubes

Geometrically, there is no restriction regarding the tube diameter. But calculations have shown that collapsing the single-wall tube into a flattened two-layer ribbon is energetically more favorable than maintaining the tubular morphology beyond a diameter value of ~ 2.5 nm [3.10]. On the other hand, it intuitively comes to mind that the shorter the radius of curvature, the higher the stress and the energetic cost, although SWNTs with diameters as low as 0.4 nm have been successfully synthesized [3.11]. A suitable energetic compromise is thus reached for ~ 1.4 nm, the most frequent diameter encountered regardless of the synthesis techniques (at least those based on solid carbon source) when conditions for high SWNT yields are used. There is no such restriction for nanotube length, which only depends on limitations brought by the preparation method and the specific conditions used for the synthesis (thermal gradients, res-

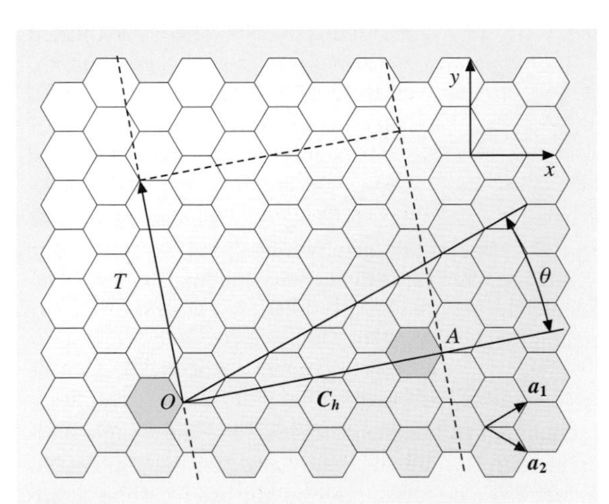

Fig. 3.1 Sketch of the way to make a single-wall carbon nanotube, starting from a graphene sheet (adapted from [3.12])

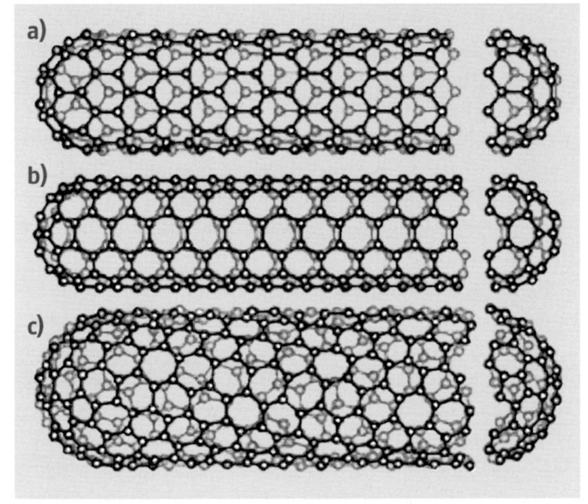

Fig. 3.2a–c Sketch of three different SWNT structures as examples for (**a**) a zig-zag-type nanotube, (**b**) an armchair-type nanotube, (**c**) a helical nanotube (adapted from [3.13])

idence time, etc.). Experimental data are consistent with these statements, since SWNTs wider than 2.5 nm are scarcely reported in literature, whatever the preparation method, while SWNT length can be in the micrometer or the millimeter range. These features make single-wall carbon nanotubes a unique example of single molecules with huge aspect ratios.

Two important consequences derive from the SWNT structure as described above:

1. All carbon atoms are involved in hexagonal aromatic rings only and are therefore in equivalent position, except at the nanotube tips where $6 \times 5 = 30$ atoms at each tip are involved in pentagonal rings (considering that adjacent pentagons are unlikely) – though not more, not less, as a consequence of the Euler's rule that also governs the fullerene structure. In case SWNTs are ideally perfect, their chemical reactivity will therefore be highly favored at the tube tips, at the very location of the pentagonal rings.
2. Though carbon atoms are involved in aromatic rings, the C=C bond angles are no longer planar as they should ideally be. This means that the hybridization of carbon atoms are no longer pure sp^2 but get some percentage of the sp^3 character, in a proportion that increases as the tube radius of curvature decreases. For example, the effect is the same as for the C_{60} fullerene molecules, whose radius of curvature is 0.35 nm, and the subsequent sp^3 character proportion about 30%. On the one hand, this is supposed to make the SWNT surface (though consisting of aromatic ring faces) a bit more reactive than regular, planar graphene, relatively speaking. On the other hand, this somehow induces a variable overlapping of the bands of density of states, thereby inducing a unique versatile electronic behavior (see Sect. 3.4).

As illustrated by Fig. 3.2, there are many ways to roll a graphene into a single-wall nanotube, some of the resulting nanotubes enabling symmetry mirrors both parallel and perpendicular to the nanotube axis (such as the SWNTs from Fig. 3.2a and 3.2b), some others not (such as the SWNT from Fig. 3.2c). By correspondence with the terms used for molecules, the latter are commonly called "chiral" nanotubes. "Helical" should be preferred, however, in order to respect the definition of chirality, which makes all chiral molecules unable to be superimposed on their own image in a mirror. The various ways to roll graphene into tubes are therefore mathematically defined by the vector of helicity $\boldsymbol{C}_\mathrm{h}$, and the angle of helicity θ, as follows (referring to Fig. 3.1):

$$\boldsymbol{OA} = \boldsymbol{C}_\mathrm{h} = n\boldsymbol{a}_1 + m\boldsymbol{a}_2$$

with

$$\boldsymbol{a}_1 = \frac{a\sqrt{3}}{2}\boldsymbol{x} + \frac{a}{2}\boldsymbol{y} \quad \text{and} \quad \boldsymbol{a}_2 = \frac{a\sqrt{3}}{2}\boldsymbol{x} - \frac{a}{2}\boldsymbol{y}$$

$$\text{where} \quad a = 2.46\,\text{Å}$$

and

$$\cos\theta = \frac{2n+m}{2\sqrt{n^2+m^2+nm}}$$

where n and m are the integers of the vector $\boldsymbol{OA}$ considering the unit vectors $\boldsymbol{a}_1$ and $\boldsymbol{a}_2$.

The vector of helicity $\boldsymbol{C}_\text{h}(=\boldsymbol{OA})$ is perpendicular to the tube axis, while the angle of helicity θ is taken with respect to the so-called zig-zag axis, i. e., the vector of helicity that makes nanotubes of the "zig-zag" type (see below). The diameter D of the corresponding nanotube is related to $\boldsymbol{C}_\text{h}$ by the relation:

$$D = \frac{|\boldsymbol{C}_\text{h}|}{\pi} = \frac{a_\text{CC}\sqrt{3(n^2+m^2+nm)}}{\pi}\,,$$

where

$$\underset{\text{(graphite)}}{1.41\,\text{Å}} \le a_{\text{C=C}} \le \underset{(\text{C}_{60})}{1.44\,\text{Å}}\,.$$

The C−C bond length is actually elongated by the curvature imposed, taking the average value for the C_{60} fullerene molecule as a reasonable upper limit, and the value for flat graphenes in genuine graphite as the lower limit (corresponding to an infinite radius of curvature). Since $\boldsymbol{C}_\text{h}$, θ, and D are all expressed as a function of the integers n and m, they are sufficient to define any SWNT specifically, by noting them (n, m). Obtaining the values of n and m for a given SWNT is simple by counting the number of hexagons that separate the extremities of the $\boldsymbol{C}_\text{h}$ vector following the unit vector $\boldsymbol{a}_1$ first, then $\boldsymbol{a}_2$ [3.12]. In the example of Fig. 3.1, the SWNT that will be obtained by rolling the graphene so that the two shaded aromatic cycles superimpose exactly is a (4,2) chiral nanotube. Similarly, SWNTs from Fig. 3.2a to 3.2c are (9,0), (5,5), and (10,5) nanotubes respectively, thereby providing examples of zig-zag–type SWNT (with an angle of helicity = 0°), armchair-type SWNT (with an angle of helicity of 30°) and a chiral SWNT, respectively. This also illustrates why the term "chiral" is sometimes inappropriate and should preferably be replaced by "helical". Armchair (n, n) nanotubes, though definitely achiral from a standpoint of symmetry, exhibit a "chiral angle" different from 0. "Zig-zag" and "armchair" qualifications for achiral nanotubes refer to the way carbon atoms are displayed at the edge of the nanotube cross section (Fig. 3.2a and 3.2b). Generally speaking, it is clear from Figs. 3.1 and 3.2a that having the vector of helicity perpendicular to any of the three overall C=C bond directions will provide zig-zag–type SWNTs, noted $(n,0)$, while having the vector of helicity parallel to one of the three C=C bond directions will provide armchair-type SWNTs, noted (n, n). On the other hand, because of the sixfold symmetry of the graphene sheet, the angle of helicity θ for the chiral (n, m) nanotubes is such as $0 < \theta < 30°$. Figure 3.3 provides two examples of what chiral SWNTs look like, as seen by means of atomic force microscopy.

Fig. 3.3 Image of two neighboring chiral SWNTs within a SWNT bundle as seen by high resolution scanning tunneling microscopy (by courtesy of Prof. Yazdani, University of Illinois at Urbana, USA)

Planar graphenes in graphite have π electrons, which are satisfied by the stacking of graphenes that allows van der Waals forces to develop. Similar reasons make fullerenes gather and order into fullerite crystals and SWNTs into SWNT ropes (Fig. 3.4a). Spontaneously, SWNTs in ropes tend to arrange into an hexagonal array, which corresponds to the highest compactness achievable (Fig. 3.4b). This feature brings new periodicities

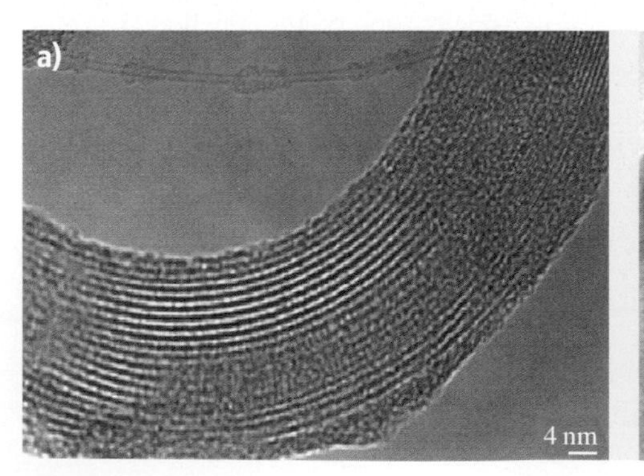

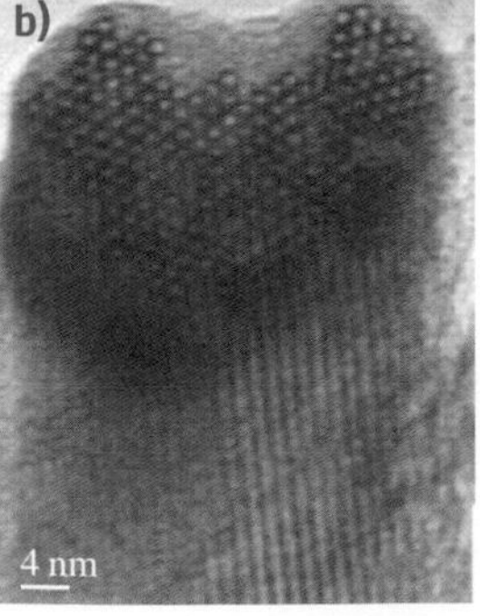

Fig. 3.4a,b High resolution transmission electron microscopy images of a SWNT rope **(a)** longitudinal view. At the top of the image an isolated single SWNT also appears. **(b)** cross section view (from [3.14])

with respect to graphite or turbostratic polyaromatic carbon crystals. Turbostatic structure corresponds to graphenes which are stacked with random rotations or translations instead of being piled up following sequential ABAB positions, as in graphite structure. This implies that no lattice atom plane exists anymore other than the graphene planes themselves (corresponding to the (001) atom plane family). These new periodicities make diffraction patterns specific and quite different from that of other sp^2-carbon-based crystals, although *hk* reflections, which account for the hexagonal symmetry of graphene plane, are still present. On the other hand, 00*l* reflections, which account for the stacking sequence of graphenes in regular, "multilayered" polyaromatic crystals that does not exist in SWNT ropes, are absent. Such a hexagonal packing of SWNTs within the ropes requires that SWNTs exhibit similar diameters, which is the common case for SWNTs prepared by the electric arc or the laser vaporization processes. SWNTs prepared these ways are actually about 1.35 nm wide (diameter value for a (10,10) tube, among others), for reasons still unclear but related to the growth mechanisms specific to the conditions provided by these techniques (see Sect. 3.3).

3.1.2 Multiwall Nanotubes

Building multiwall carbon nanotubes is a little bit more complex, since it has to deal with the various ways graphenes can be displayed and mutually arranged within a filament morphology. Considering the usual textural versatility of polyaromatic solids, a similar versatility can be expected. Likewise, diffraction patterns no longer differentiate from that of anisotropic polyaromatic solids. The easiest MWNT to imagine is the concentric type (c-MWNT), in which SWNTs with regularly increasing diameters are coaxially displayed according to a Russian-doll model into a multiwall nanotube (Fig. 3.5). Such nanotubes are generally formed either by the electric-arc technique (without need of any catalyst) or by catalyst-enhanced thermal cracking of gaseous hydrocarbons or CO disproportionation (see Sect. 3.2). The number of walls (or number of coaxial tubes) can be anything, starting from two, with no upper limit. The intertube distance is approximately that of the intergraphene distance in turbostratic, polyaromatic solids, i. e. 0.34 nm (as opposed to 0.335 nm in genuine graphite), since the increasing radius of curvature imposed on the concentric graphenes prevents the carbon atoms from being displayed as in graphite, i. e., each of the carbon atoms from a graphene facing alternatively either a ring center or a carbon atom from the neighboring graphene. But two cases allow such a nanotube to reach – totally or partially – the 3-D crystal periodicity of graphite. One is to consider a high number of concentric graphenes, i. e. concentric graphenes with long radius of curvature. In that case, the shift of the relative positions of carbon atoms from superimposed graphenes is so small with respect to that in graphite that some commensurability is possible. It may result in MWNTs with the association of both structures, i. e. turbostratic in the core and graphitic in the outer part [3.15]. The other case occurs for c-MWNTs exhibiting faceted morphologies, originating either from the synthesis process or more likely from

Fig. 3.5 High resolution transmission electron microscopy image (longitudinal view) of a concentric multiwall carbon nanotube (c-MWNT) prepared by electric arc. In *insert*, sketch of the Russian-doll-like display of graphenes

subsequent heat-treatments at high temperatures (e.g., 2,500 °C) in inert atmospheres. Obviously, facets allow graphenes to get back a flat arrangement of atoms – except at the junction between neighboring facets – in which the specific stacking sequence of graphite can develop.

Another frequent inner texture for multiwall carbon nanotubes is the so-called herringbone texture (h-MWNTs), in which graphenes make an angle with respect to the nanotube axis (Fig. 3.6). The angle value varies upon the processing conditions (e.g., the catalyst morphology or the atmosphere composition), from 0 (in which case the texture turns into that of a c-MWNT) to 90° (in that case, the filament is no longer a tube, see below), and the inner diameter varies so that the tubular feature can be lost [3.19], justifying that the latter can be called nanofibers rather than nanotubes. h-MWNTs are exclusively obtained by processes involving catalysts, generally catalyst-enhanced thermal cracking of hydrocarbons or CO disproportionation.

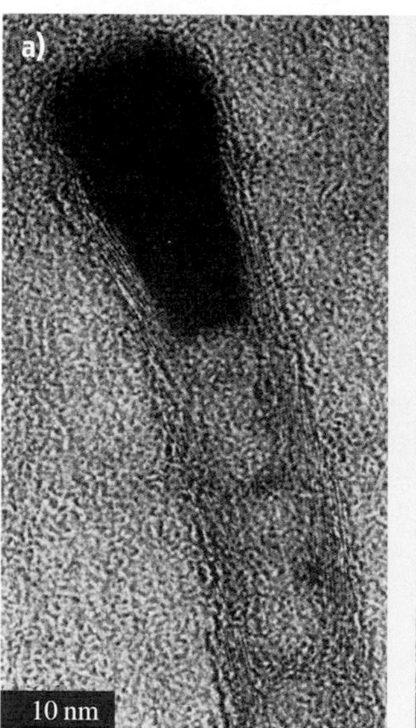

Fig. 3.6a,b One of the earliest high resolution transmission electron microscopy image of a herringbone (and bamboo) multi-wall nanotube (bh-MWNT, longitudinal view) prepared by CO disproportionation on Fe-Co catalyst. **(a)** as-grown. The nanotube surface is made of free graphene edges. **(b)** after 2,900 °C heat-treatment. Both the herringbone and the bamboo textures have become obvious. Graphene edges from the surface have buckled with neighbors (*arrow*), closing the access to the intergraphene spacing (adapted from [3.16])

One unresolved question is whether the herringbone texture, which actually describes the texture projection rather than the overall three-dimensional texture, originates from the scroll-like spiral arrangement of a single graphene ribbon or from the stacking of independent truncated-cone-like graphenes. Although the former is more likely for energetic reasons since providing a minimal and constant amount of transitorily unsatisfied bonds during the nanotube growth (similar to the well-known growth mechanism from a screw dislocation), the question is still debated.

Another common feature is the occurrence, at a variable frequency, of a limited amount of graphenes oriented perpendicular to the nanotube axis, thus forming the so-called "bamboo" texture. It cannot be a texture by its own but affects in a variable extent either c-MWNT (bc-MWNT) or h-MWNT (bh-MWNT) textures (Figs. 3.6 and 3.7). The question may be addressed whether such filaments, though hollowed, should still be called nanotubes, since the inner cavity is no longer opened all along the filament as it is for a genuine tube. They therefore are sometimes referred as "nanofibers" in literature.

Nanofilaments that definitely cannot be called nanotubes are those built from graphenes oriented perpendicular to the filament axis and stacked as piled-up plates. Although they actually correspond to h-MWNTs with the graphene/MWNT axis angle = 90°, the occurrence of inner cavity is no longer possible, and such filaments are therefore most often referred to as "platelet-nanofibers" in literature [3.19].

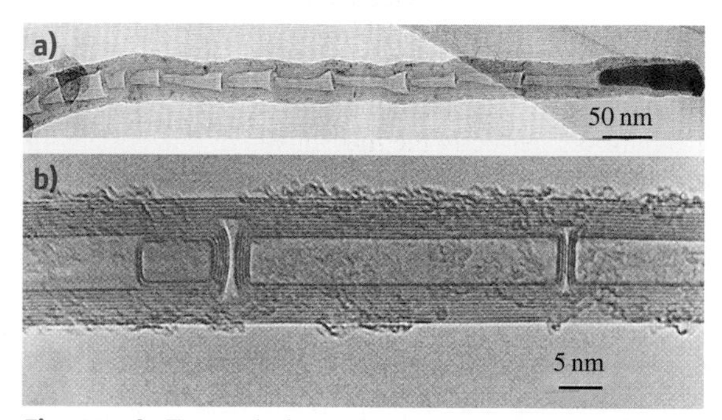

Fig. 3.7a,b Transmission electron microscopy images from bamboo-multi-wall nanotubes (longitudinal views). **(a)** low magnification of a bamboo-herringbone multi-wall nanotube (bh-MWNT) showing the nearly periodic feature of the texture, which is very frequent. (from [3.17]); **(b)** high resolution image of a bamboo-concentric multi-wall nanotube (bc-MWNT) (modified from [3.18])

As opposed to SWNTs, whose aspect ratio is so high that acceding the tube tips is almost impossible, aspect ratio for MWNTs (and carbon nanofibers) are generally lower and often allow one to image tube ends by transmission electron microscopy. Except for c-MWNTs from electric arc (see Fig. 3.5) that grow following a catalyst-free process, nanotube tips are frequently found associated with the catalyst crystal from which they have been formed.

MWNT properties (see Sect. 3.4) will obviously depend on the perfection and orientation of graphenes in the tube more than any other feature (e.g., the respective spiral angle of the constituting nanotubes for c-MWNTs has little importance). Graphene orientation is a matter of texture, as described above. Graphene perfection is a matter of nanotexture, which is commonly used to describe other polyaromatic carbon materials, and which is quantified by several parameters preferably obtained from high resolution transmission electron microscopy (Fig. 3.8). Both texture and nanotexture depend on the processing conditions. While the texture type is a permanent, intrinsic feature only able to go toward complete alteration upon severe degradation treatments (e.g., oxidation), however, nanotexture can be improved by subsequent thermal treatments at high temperatures (e.g., > 2,000 °C) or, reversibly, possibly degraded by chemical treatments (e.g., slight oxidation conditions).

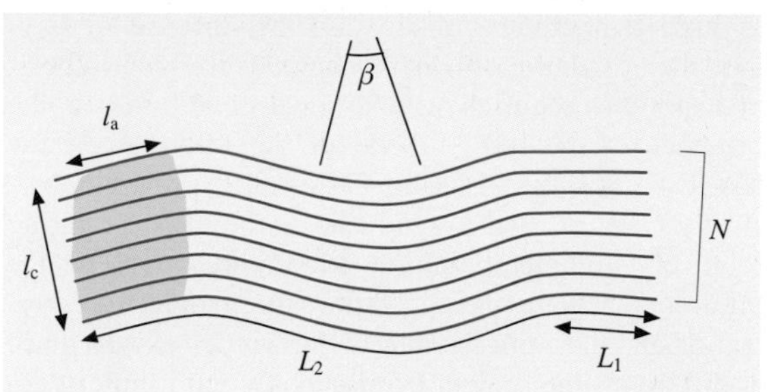

Fig. 3.8 Sketch explaining the various parameters obtained from high resolution (lattice fringe mode) transmission electron microscopy for quantifying nanotexture: L_1 is the average length of perfect (distortion-free) graphenes of coherent areas; N is the number of piled-up graphenes in the coherent (distortion-free) areas; L_2 is the average length of continuous though distorted graphenes within graphene stacks; β is the average distortion angle. L_1 and N are more or less related to L_a and L_c obtained from X-ray diffraction

3.2 Synthesis of Carbon Nanotubes

Producing carbon nanotubes so that the currently planned applications become marketable requires solving some problems that are more or less restrictive depending on the cases. One example is to specifically control the configuration (chirality), the purity, or the structural quality of SWNTs, with the production capacity adapted to the application. One condition would be to understand perfectly the mechanism of nanotube nucleation and growth, which remains the object of a lot of controversy in spite of an intense, worldwide experimental effort. This problem is partly explained by the lack of knowledge regarding several parameters controlling the synthesis conditions. For instance, neither the temperatures of nanotube condensation and formation nor the manner in which they are influenced by the synthesis parameters are known. Equally often unknown is the exact and accurate role of the catalysts in nanotube growth. Given the large number of experimental parameters and considering the large range of conditions that the several synthesis techniques correspond to, it is quite legitimate to think of more than one mechanism intervening in the nanotube formation.

3.2.1 Solid Carbon Source-Based Production Techniques for Carbon Nanotubes

Among the different SWNT production techniques, the three processes (laser ablation, solar energy, and electric arc) presented in this section have at least two common points: a high temperature ($1{,}000\,\mathrm{K} < T < 6{,}000\,\mathrm{K}$) medium and the fact that the carbon source originates from the erosion of solid graphite. In spite of these common points, both the morphologies of the carbon nanostructures and the SWNT yields can differ notably with respect to the experimental conditions.

Before being utilized for carbon nanotube synthesis, these techniques permitted the production of fullerenes. Laser vaporization of graphite was actually the very first method to demonstrate the existence of fullerenes, including the most popular one (because the most stable and therefore the most abundant), C_{60} [3.20]. On the other hand, the electric arc technique was (and still is) the first method to produce fullerenes in relatively large quantities [3.21]. As opposed to the fullerene formation, which requires the presence of carbon atoms in high temperature media and the absence of oxygen, the utilization of these techniques for the synthesis of nanotubes requires an additional condition, i. e. the presence of catalysts in one of the electrode or the target.

In case of these synthesis techniques requiring relatively high temperatures to sublime graphite, the different mechanisms such as the carbon molecule dissociation and the atom recombination processes take place at different time scales, from nanosecond to microsecond and even millisecond. The formation of nanotubes and other graphene-based products appears afterward with a relatively long delay.

The methods of laser ablation, solar energy, and electric arc are based on an essential mechanism, i. e. the energy transfer resulting from the interaction between either the target material and an external radiation (laser beam or radiation emanating from solar energy) or the electrode and the plasma (in case of electric arc). This interaction is at the origin of the target or anode erosion leading to the formation of a plasma, i. e. an electrically neutral ionized gas, composed of neutral atoms, charged particles (molecules and ionized species), and electrons. The ionization degree of this plasma, defined by the ratio $(n_e/n_e + n_o)$, where n_e and n_o are the density of electrons and that of neutral atoms respectively, highlights the importance of energy transfer between the plasma and the material. The characteristics of this plasma and notably the fields of temperature and of concentration of the various species present in the plasma thereby depend not only on the nature and composition of the target or that of the electrode but also on the energy transferred.

One of the advantages of these different synthesis techniques is the possibility of varying a large number of parameters, which allow the composition of the high temperature medium to be modified and, consequently, the most relevant parameters to be determined, in order to define the optimal conditions for the control of carbon nanotube formation. But a major drawback of these techniques – as for any other technique for the production of SWNTs – is that SWNTs never come pure, i. e. they are associated with other carbon phases and catalyst remnants. Although purification processes are proposed in literature and by some commercial companies to rid these undesirable phases, they are all based on oxidation (e.g. acidic) processes that are likely to affect deeply the SWNT structure [3.14]. Subsequent thermal treatments at $\sim 1{,}200\,°C$ under inert atmosphere, however, succeed somewhat in recovering the former structure quality [3.24].

Laser Ablation

After the first laser was built in 1960, physicists immediately made use of it as a means of concentrating a large quantity of energy inside a very small volume within a relatively short time. The consequences of this energy input naturally depends on the characteristics of the device employed. During the interaction between the laser beam and the material, numerous phenomena superimpose and/or follow each other within the time range, each of these processes being sensitive to such different parameters as the laser beam characteristics, the incoming power density (also termed "fluence"), the nature of the target, and the environment surrounding it. For instance, the solid target can merely heat up, or melt, or vaporize depending on the power provided.

While this technique was used successfully to synthesize fullerene-related structures for the very first time [3.20], the synthesis of SWNTs by laser ablation came only ten years later [3.22].

Laser Ablation – Experimental Devices

Two types of laser devices are utilized nowadays for carbon nanotube production: lasers operating in pulsed mode on the one hand and lasers operating in continuous mode on the other hand, the latter generally providing a smaller influence.

An example of the layout indicating the principle of a laser ablation device is given in Fig. 3.9. A graphite pellet containing the catalyst is put in the middle of an inert gas-filled quartz tube placed in an oven maintained at a temperature of $1{,}200\,°C$ [3.22, 23]. The energy of the laser beam focused on the pellet permits it to vaporize and sublime the graphite by uniformly bombarding its surface. The carbon species swept by a flow of neutral gas are thereafter deposited as soot in different regions: on the conical water-cooled copper collector, on the quartz tube walls, and on the backside of the pellet.

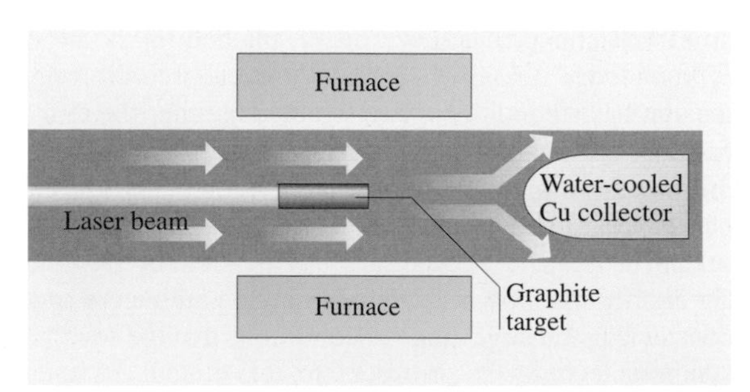

Fig. 3.9 Sketch of an early laser vaporization apparatus (adapted from [3.22, 23])

Various improvements have been made to this device in order to increase the production efficiency. For example, *Thess* et al. [3.25] employed a second pulsed laser that follows the initial impulsion but at a different frequency in order to ensure a more complete and efficient irradiation of the pellet. This second impulsion has the role of vaporizing the coarse aggregates issued from the first ablation and thereby making them participate in the active carbon feedstock involved in the nanotube growth. Other modifications were brought by *Rinzler* et al. [3.24], who inserted a second quartz tube of a smaller diameter coaxially disposed inside to the first one. This second tube has the role of reducing the vaporization zone and thereby permitting an increase in the quantity of sublimed carbon. They also arranged to place the graphite pellet on a revolving system so that the laser beam uniformly scans its whole surface.

Other groups have realized that, as far as the target contains both the catalyst and the graphite, the latter evaporates in priority and the pellet surface becomes more and more metal rich, resulting in a decrease of the efficiency in nanotube formation in the course of the process. To solve this problem, *Yudasaka* et al. [3.27] utilized two pellets facing each other, one entirely made from the graphite powder and the other from an alloy of transition metals (catalysts), and irradiated simultaneously.

A sketch of a synthesis reactor based on the vaporization of a target at a fixed temperature by a continuous CO_2 laser beam ($\lambda = 10.6\,\mu m$) is shown in Fig. 3.10 [3.26]. The power can be varied from 100 W to 1,600 W. The temperature of the target is measured with an optical pyrometer, and these measurements are used to regulate the laser power to maintain a constant vaporization temperature. The gas, heated by the contact with the target, acts as a local furnace and creates an extended hot zone, making an external furnace unnecessary. The gas is extracted through a silica pipe, and the solid products formed are carried away by the gas flow through the pipe and then collected on a filter. The synthesis yield is controlled by three parameters: the cooling rate of the medium where the active, secondary catalyst particles are formed, the residence time, and the temperature (in the 1,000–2,100 K range) at which SWNTs nucleate and grow [3.28].

But devices equipped with facilities to gather in situ data such as the target temperature are few and, generally speaking, among the numerous parameters of the laser ablation synthesis technique. The most studied are the nature of the target, the nature and concentration of the catalysts, the nature of the neutral gas flow, and the temperature of the outer oven (when any).

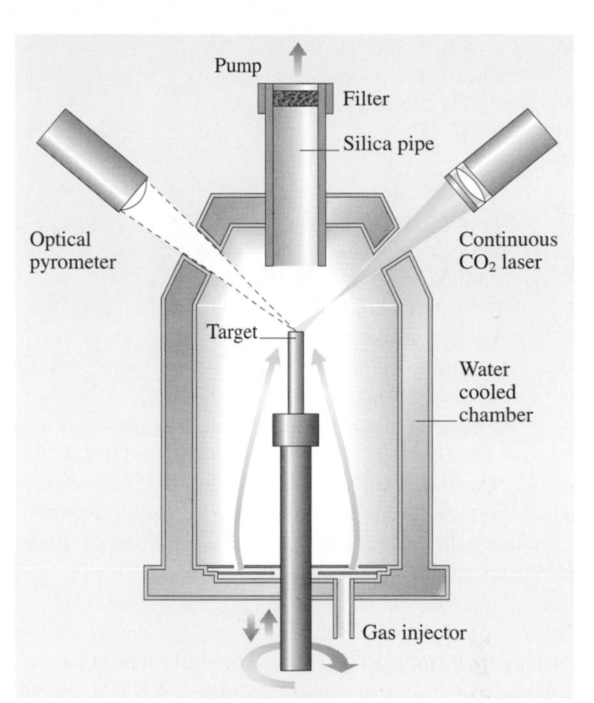

Fig. 3.10 Sketch of a synthesis reactor with a continuous CO_2 laser device (adapted from [3.26])

Laser Ablation – Results

In the absence of catalysts in the target, the soot collected mainly contains multiwall nanotubes (c-MWNTs). Their lengths can reach 300 nm. Their quantity and structure quality are dependent on the oven temperature. The best quality is obtained for an oven temperature set at 1,200 °C. At lower oven temperatures, the structure quality decreases, and the nanotubes start presenting many defects [3.23]. As soon as small quantities (few percents or less) of transition metals (Ni, Co) playing the role of catalysts are incorporated into the graphite pellet, products yielded undergo significant modifications, and SWNTs are formed instead of MWNTs. The yield of SWNTs strongly depends on the type of metal catalyst used and is seen to increase with furnace temperature, among other factors. The SWNTs have remarkably uniform diameter and they self-organize into rope-like crystallites 5–20 nm in diameter and tens to hundreds of micrometers in length (Fig. 3.11). The ends of all SWNTs appear to be perfectly closed with hemispherical end caps showing no evidence of any as-

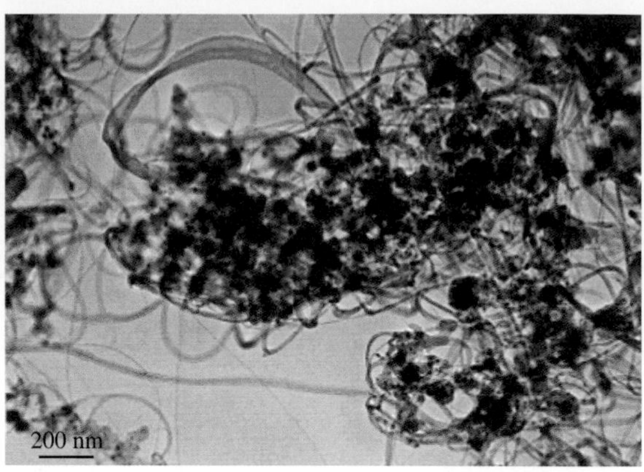

Fig. 3.11 Low magnification TEM images of a typical raw SWNT material obtained from the laser vaporization technique. Fibrous morphologies are SWNT bundles, and dark particles are catalyst remnants. Raw SWNT materials from electric arc exhibit a similar aspect (from [3.14])

sociated metal catalyst particle, although, as pointed out in Sect. 3.1, finding the two tips of a SWNT is rather challenging, considering the huge aspect ratio of the nanotube and their entanglement. Another feature of SWNTs produced with this technique is that they are supposedly "cleaner" than those produced employing other techniques, i. e., associated with a lower amount of an amorphous carbon phase, either coating the SWNTs or gathered into nano-particles. Such an advantage, however, stands only for synthesis conditions specifically set to ensure high quality SWNTs. It can no longer be true when conditions are such that high yields are preferred, and SWNTs from electric arc may appear cleaner than SWNTs from laser vaporization [3.14].

The laser vaporization technique is one of the three methods currently used to prepare SWNTs as commercial products. SWNTs prepared that way were first marketed by Carbon Nanotechnologies Inc. (Texas, USA), with prices as high as $1,000/g (raw materials) until December 2002. As a probable consequence of the impossibility to lower the amount of impurities in the raw materials, they have recently decided to focus on fabricating SWNTs by the HiPCo technique (see Sect. 3.2.2). Though laser-based method are generally considered not competitive in the long term for the low-cost production of SWNTs compared to CCVD-based methods (see Sect. 3.2.2), prediction of prices as low as $0.03 per gram of raw high concentration SWNT soot are expected in the near future (Acolt S.A., Switzerland).

Electric-Arc Method

Electric arcs between carbon electrodes have been studied as light sources and radiation standards for a very long time. Lately they have received renewed attention for their use in producing new fullerenes-related molecular carbon nanostructures such as genuine fullerenes or nanotubes. This technique was first brought to light by *Krätschmer* et al. [3.21] who utilized it to achieve the production of fullerenes in macroscopic quantities. In the course of investigating the other carbon nanostructures formed along with the fullerenes and more particularly the solid carbon deposit forming onto the cathode, Iijima [3.7] discovered the catalyst-free formation of perfect c-MWNT-type carbon nanotubes. Then, as mentioned in the introduction of this chapter, the catalyst-promoted formation of SWNTs was incidentally discovered after some amounts of transition metals were introduced into the anode in an attempt to fill the c-MWNTs with metals while they grow [3.8, 9]. Since then, a lot of work has been carried out by many groups using this technique to understand the mechanisms of nanotube growth as well as the role played by the catalysts (when any) for the synthesis of MWNTs and/or SWNTs [3.29–41].

Electric-Art Method – Experimental Devices

The principle of this technique is to vaporize carbon in the presence of catalysts (iron, nickel, cobalt, yttrium, boron, gadolinium, and so forth) under reduced atmosphere of inert gas (argon or helium). After the triggering of the arc between two electrodes, a plasma is formed consisting of the mixture of carbon vapor, the rare gas (helium or argon), and the vapors of catalysts. The vaporization is the consequence of the energy transfer from the arc to the anode made of graphite doped with catalysts. The anode erosion rate is more or less important depending on the power of the arc and also on the other experimental conditions. It is noteworthy that a high anode erosion does not necessarily lead to a high carbon nanotube production.

An example of a reactor layout is shown in Fig. 3.12. It consists of a cylinder of about 30 cm in diameter and about 1 m in height, equipped with diametrically opposed sapphire windows located so that they face the plasma zone in view of observing the arc. The reactor possesses two valves, one for carrying out the primary evacuation (0.1 Pa) of the chamber, the other permitting it to fill with a rare gas up to the desired working pressure.

Contrary to the solar energy technique, SWNTs are deposited (provided appropriate catalysts are used) in

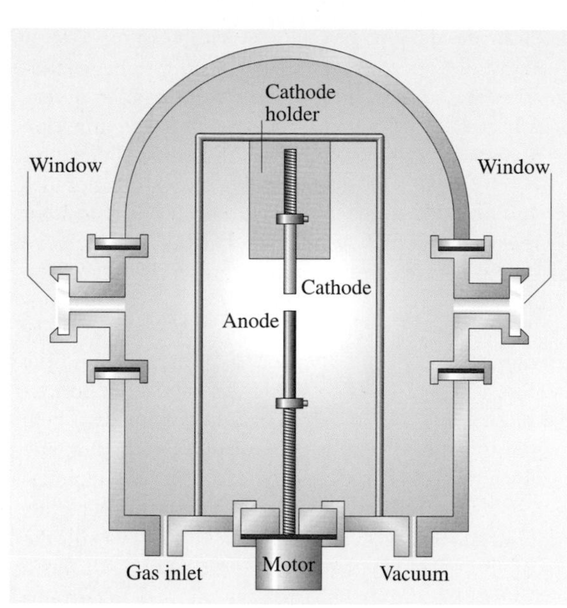

Fig. 3.12 Sketch of an electric arc reactor

different regions of the reactor: (1) the collaret, which forms around the cathode; (2) the web-like deposits found above the cathode; (3) the soot deposited all around the reactor walls and bottom. On the other hand, MWNTs are formed embedded in a hard deposit adherent to the cathode, whenever catalysts are used or not. The cathode deposits formed under the cathode. The formation of collaret and web is not systematic and depends on the experimental conditions as indicated in Table 3.1, as opposed to cathode deposit and soot that are systematically found.

Two graphite rods of few millimeters in diameter constitute the electrodes between which a potential difference is applied. The dimensions of these electrodes vary according to the authors. In certain cases, the cathode has a greater diameter than the anode in order to facilitate their alignment [3.32, 42]. Other authors utilize electrodes of the same diameter [3.41]. The whole device can be designed horizontally [3.33, 41] or vertically [3.34, 36–38]. The advantage of the latter is the symmetry brought by the verticality with respect to gravity, which facilitates computer modeling (regarding convection flows, for instance).

Two types of anodes can be utilized as soon as catalysts need to be introduced: (1) graphite anodes in which a coaxial hole is drilled several centimeters in length and in which catalyst and graphite powders are mixed; (2) graphite anodes in which the catalysts are homogeneously dispersed [3.43]. The former are by far the most popular, due to their ease of fabrication.

The optimization of the process regarding the nanotube yield and quality is attempted by studying the role of various parameters such as the type of doped anode (homogeneous or heterogeneous catalyst dispersion), the nature as well as the concentration of catalysts, the nature of the plasmagenic gas, the buffer gas pressure, the arc current intensity, and the distance between electrodes. Investigating the result of varying these parameters on the type and amount of carbon nanostructures formed is, of course, the preliminary work that has been done. Though electric arc reactors equipped with the related facilities are few (see Fig. 3.12), investigating the missing link, i. e. the effect of varying the parameters on the plasma characteristics (fields of concentration species and temperature), is likely to provide a more comprehensive understanding of the phenomena involved in nanotube formation. This has been recently developed by atomic and molecular optical emission spectroscopy [3.34, 36–39, 41].

Finally, mention has to be made of attempts to make the electric arc within liquid media such as liquid nitrogen [3.44] or water [3.45, 46], the goal being easier processing since they do not require pumping devices or a closed volume and therefore are likely to allow continuous synthesis. This adaptation has not, however, reached the state of mass production.

Electric-Arc Method – Results

In view of the numerous results obtained with this electric-arc technique, it appears clearly that both the nanotube morphology and the nanotube production efficiency strongly depend on the experimental conditions and, in particular, on the nature of the catalysts. It is worth noting that the products obtained do not consist solely in carbon nanotubes but also in nontubular forms of carbon such as nanoparticles, fullerene-like structures including C_{60}, poorly organized polyaromatic carbons, nearly amorphous nanofibers, multiwall shells, single-wall nanocapsules, and amorphous carbon as reported in Table 3.1 [3.35, 37, 38]. In addition, catalyst remnants are found all over the place, i. e. in the soot, collaret, web, and cathode deposit in various concentration. Generally, at pressure value of about 600 mbar of helium, for 80 A arc current and 1 mm electrode gap, the use of Ni/Y as coupled catalysts favors more particularly the synthesis of SWNTs [3.8, 33, 47]. For such conditions providing high SWNT yields, SWNT concentrations are higher in the collaret (in the range of 50–70 %), then in the web ($\sim$ 50% or less), then in the soot. On the

Table 3.1 Different carbon morphologies obtained by changing the type of anode, the type of catalysts, and pressure value, in a series of arc-discharge experiments (electrode gap = 1 mm)

Catalyst (atom%) Arc conditions	0.6Ni + 0.6Co (homogeneous anode) $P \sim 60$ kPa $I \sim 80$ A	0.6Ni + 0.6Co (homogeneous anode) $P \sim 40$ kPa $I \sim 80$ A	0.5Ni + 0.5Co $P \sim 60$ kPa $I \sim 80$ A	4.2Ni + 1Y $P \sim 60$ kPa $I \sim 80$ A
Soot	• **MWNT** + **MWS** + **POPAC** or **Cn** ± catalysts $\phi \sim 3-35$ nm • NANF + catalysts • AC particles + catalysts • [*DWNT*], [*SWNT*], ropes or isolated, + POPAC	• **POPAC** and **AC** particles + **catalysts** $\phi \sim 2-20$ nm • NANF + catalysts $\phi \sim 5-20$ nm + MWS • [*SWNT*] $\phi \sim 1-1.4$ nm, distorted or damaged, isolated or ropes + Cn	• **AC** and POPAC particles + catalysts $\phi \sim 3-35$ nm • **NANF** + catalysts $\phi \sim 4-15$ nm • [*SWNT*] $\phi \sim 1.2$ nm, isolated or ropes	• **POPAC** and **AC** + particles + catalysts $\phi \leq 30$ nm • SWNT $\phi \sim 1.4$ nm, clean + Cn, short with tips, [*damaged*], isolated or ropes $\phi \leq 25$ nm • [*SWNC*] particles
Web	• [*MWNT*], **DWNT**, $\phi\, 2.7 - \mathbf{4} - 5.7$ nm SWNT $\phi\, 1.2-1.8$ nm, isolated or **ropes** $\phi < 15$ nm, + POPAC ± **Cn** • AC particles + catalysts $\phi \sim 3-40$ nm + MWS • [*NANF*]	None	None	• **SWNT**, $\phi \sim 1.4$ nm, isolated or **ropes** $\phi \leq 20$ nm, + **AC** • POPAC and AC particles + catalysts $\phi \sim 3 - \mathbf{10} - 40$ nm + MWS
Collaret	• **POPAC** and **SWNC** particles • Catalysts $\phi \sim 3-250$ nm, $< \mathbf{50}$ nm + MWS • SWNT $\phi\, 1-1.2$ nm, [*opened*], **distorted**, isolated or **ropes** $\phi < 15$ nm, + **Cn** • [*AC*] particles	• **AC** and **POPAC** particles + catalysts $\phi \sim 3-25$ nm • SWNT $\phi \sim 1-1.4$ nm clean + Cn, [*isolated*] or ropes $\phi < 25$ nm • Catalysts $\phi \sim 5-50$ nm + MWS, • [SWNC]	• **Catalysts** $\phi \sim 3-170$ nm + MWS • AC or POPAC particles + **catalysts** $\phi \sim 3-50$ nm • SWNT $\phi \sim 1.4$ nm clean + Cn isolated or ropes $\phi < 20$ nm	• **SWNT** $\phi \sim 1.4-2.5$ nm, clean + Cn, [*damaged*], isolated or **ropes** $\phi < 30$ nm • POPAC or AC particles + catalysts $\phi \sim 3-30$ nm • [*MWS*] + catalysts or catalyst-free.
Cathode deposit	• **POPAC** and SWNC particles • **Catalysts** $\phi \sim$ 5–300 nm MWS • MWNT $\phi < 50$ nm • [*SWNT*] $\phi \sim 1.6$ nm clean + Cn, isolated or ropes	• **POPAC** and **SWNC** particles + Cn • Catalysts $\phi \sim 20-100$ nm + MWS	• **MWS**, catalyst-free • **MWNT** $\phi < 35$ nm • POPAC and PSWNC particles • [*SWNT*], isolated or ropes • [*Catalysts*] $\phi \sim 3-30$ nm	• **SWNT** $\phi \sim 1.4-4.1$ nm, clean + **Cn**, short with tips, isolated or ropes $\phi \leq 20$ nm. • **POPAC** or **AC** particles + catalysts $\phi \sim 3-30$ nm • MWS + catalysts $\phi < 40$ nm or catalyst-free • [*MWNT*]

Abundant – Present – [*Rare*]

Glossary: AC: amorphous carbon; POPAC: poorly organised polyaromatic carbon; Cn: fullerene-like structure, including C_{60}; NANF: nearly amorphous nanofiber; MWS: multiwall shell; SWNT: single-wall nanotube; DWNT: double-wall nanotube, MWNT: multiwall nanotube; SWNC: single-wall nanocapsule.

other hand, c-MWNTs are found in the cathode deposit. SWNT lengths are micrometric and, typically, outer diameters range around 1.4 nm. Taking the latter conditions as a reference experiment (Table 3.1, column 4), Table 3.1 illustrates the consequence of changing the parameters. For instance (Table 3.1, column 3), using Ni/Co instead of Ni/Y as catalysts prevents the formation of SWNTs. But when the Ni/Co catalysts are homogeneously dispersed in the anode (Table 3.1, column 1), the formation of nanotubes is promoted again, but MWNTs with two or three walls prevail over SWNTs, among which DWNTs (double-wall nanotubes) make a majority. But it is enough to decrease the ambient pressure from 60 to 40 kPa (Table 3.1, column 2) for getting back to conditions where the nanotube formation is impeached.

Recent works have attempted to replace graphite powder (*sp*^{2} hybridized carbon) by diamond powder (*sp*^{3} hybridized carbon) to mix with the catalyst powder and fill hollowed-type graphite anodes. The result was an unexpected but quite significant increase (up to +230%) in the SWNT yield [3.39,40]. Such experiments reveal, as for the comparison between the results from using homogeneous instead of heterogeneous anodes that the physical phenomena (charge and heat transfers) that occurred in the anode during the arc are of utmost importance, which was neglected until now.

It is clear that while the use of rare earth element (such as Y) as a single catalyst does not provide the conditions to grow SWNTs, associating it with a transition metal (such as Ni/Y) seems to correspond to the most appropriate combinations, leading to the highest SWNT yields [3.42]. On the other hand, using a single rare earth element may lead to unexpected results, such as the anticipated closure of graphene edges from a c-MWNT wall with the neighboring graphene edges from the same wall side, leading to the preferred formation of telescope-like and open c-MWNTs able to contain nested Gd crystals [3.36,38]. Such a need for bimetallic catalysts has just begun to be understood as a possible requirement to promote the transitory formation of nickel particles coated with yttrium carbide, whose lattice constants are somewhat commensurable with that of graphenes [3.48].

Figure 3.13 illustrates the kind of information provided by the analysis of the plasma by means of emission spectroscopy, i. e. radial temperature (Fig. 3.13a) or C_2 species concentration (Fig. 3.13b) profiles in the plasma. A common feature is that a huge vertical gradient (∼ 500 K/mm) rapidly establishes (at ∼ 0.5 mm from the center in the radial direction) from the bottom to the top of the plasma, as a probable consequence of convection phenomena (Fig. 3.13a). The zone of actual SWNT formation is beyond the limit of the volume analyzable in the radial direction, corresponding to colder areas. The C_2 concentration increases dramatically from the anode to the cathode and decreases dramatically in the radial direction (Fig. 3.13b). This demonstrates that C_2 moieties are secondary products resulting from the recombination of primary species formed from the anode. It also suggests that C_2 moieties may be the building blocks for MWNTs (formed at the cathode) but not for SWNTs [3.38,40].

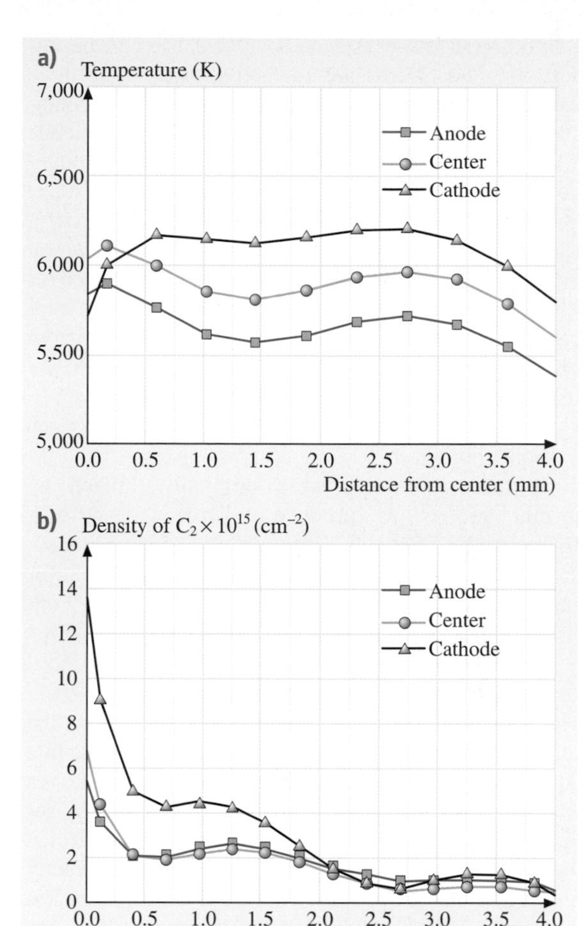

Fig. 3.13a,b Typical temperature (**a**) and C_2 concentration (**b**) profiles in the plasma at the anode surface (*square*), at the plasma center (*dot*), and at the cathode surface (*triangle*) for "standard" conditions (see text). Gradients are similar whatever the catalyst, although absolute value may vary

Although many aspects of it still need to be understood, the electric-arc method is one of the three methods currently used to produce SWNTs as commercial products. At the French company Nanoledge S.A. (Montpellier, France), for instance, current production (for 2003) reaches ~ 20 kg/year (raw SWNTs, i. e. not purified), with a marketed price of ~ 90 Euros/g, which is much cheaper than any other production method so far. A decrease to 2–5 Euros/g is expected for sometime before 2007, with a SWNT production capacity of 4–5 tons/year. Amazingly, raw SWNTs from electric arc proposed by Bucky USA (Texas, USA) are still proposed at a marketed price of $ 1,000/g.

Solar Furnace

Solar furnace devices were originally utilized by several groups to produce fullerenes [3.49–51]. *Heben* et al. [3.52] and *Laplaze* et al. [3.53] later modified their former devices in order to use them for carbon nanotube production. This modification consisted mainly in using more powerful ovens [3.54, 55].

Solar Furnace – Experimental Devices

The principle of this technique is again based on the sublimation in an inert gas of a mixture of graphite powder and catalysts formerly placed in a crucible. An example of such a device is shown in Fig. 3.14. The solar rays are collected by a plain mirror and reflected toward a parabolic mirror that focuses them directly onto a graphite pellet under controlled atmosphere. The high temperature of about 4,000 K permits both the carbon and the catalysts to vaporize. The vapors are then dragged by the neutral gas and condense onto the cold walls of the thermal screen. The reactor consists of a brass support cooled by water circulation, on which Pyrex® chambers with various shapes can be fixed (Fig. 3.14b). This support contains a watertight passage permitting the introduction of the neutral gas and a copper rod on which the target is mounted. The target is surrounded by a graphite tube that plays both the role of a thermal screen to reduce radiation losses (very important in the case of graphite) and the role of a duct to lead carbon vapors to a filter in order to avoid soot deposits on the Pyrex® chamber wall. A graphite crucible filled with powdered graphite (for fullerene synthesis) or a mixture of graphite and catalysts (for nanotube synthesis) is utilized in order to reduce the conduction losses.

These studies primarily investigated the target composition, the type and concentration of catalysts, the flow-rate, the composition and pressure of the plasmagenic gas inside the chamber, and the oven power. The objectives are similar to that of works carried out with the other solid-carbon-source-based processes. When possible, specific in situ diagnostics (pyrometry, optical emission spectroscopy, etc.) are also performed in order to understand better the role of the various parameters (temperature measurements at the crucible surface, along the graphite tube acting as thermal screen, C_2 radical concentration in the immediate vicinity of the crucible).

Solar Furnace – Results

Some results obtained by different groups concerning the influence of the catalysts can be summarized as follows. With Ni/Co and at low pressure, the sample collected contains mainly MWNTs with bamboo texture, carbon shells, and some bundles of SWNTs [3.54]. At higher pressures, only bundles of SWNTs are obtained with fewer carbon shells. With Ni/Y and at a high pressure, relatively long bundles of SWNTs are observed. With Co, SWNT bundles are obtained in the soot with SWNT diameters ranging from 1 to 2 nm. Laplaze et al. [3.54] observed very few nanotubes but a large quantity of carbon shells.

In order to proceed to a large-scale synthesis of single-wall carbon nanotubes, which is still a challenge for chemical engineers, *Flamant* et al. [3.56] recently demonstrated that solar energy–based synthesis is a versatile method to obtain SWNTs and can be scaled from 0.1–0.5 g/h to 10 g/h and then to 100 g/h productivity using existing solar furnaces. Experiments at medium scale (10-g/h, 50 kW solar power) have proven the feasibility of designing and building such a reactor and of the scaling-up method. Numerical simulation was meanwhile performed in order to improve the selectivity of the synthesis, in particular by controlling the carbon vapor cooling rate.

3.2.2 Gaseous Carbon Source-Based Production Techniques for Carbon Nanotubes

As mentioned in the introduction of this chapter, the catalysis-enhanced thermal cracking of gaseous carbon source (hydrocarbons, CO) – commonly referred to as catalytic chemical vapor deposition (CCVD) – has been known to produce carbon nanofilaments for a long time [3.4], so that reporting all the works published in the field since the beginning of the century is almost impossible. Until the 90s, however, carbon nanofilaments were mainly produced to act as a core substrate for the subsequent growth of larger (micrometric) carbon fibers – so-called vapor-grown carbon fibers – by means of thickening through mere catalyst-free CVD processes [3.57, 58]. We therefore are going to focus instead on the more recent attempts made to prepare genuine carbon nanotubes.

The synthesis of carbon nanotubes (either single- or multiwalled) by CCVD methods involves the catalytic decomposition of a carbon containing source on small metallic particles or clusters. This technique involves either an heterogeneous process if a solid substrate has a role or an homogeneous process if everything takes place in the gas phase. The metals generally used for these reactions are transition metals, such as Fe, Co, and Ni. It is a rather "low temperature" process compared to arc-discharge and laser-ablation methods, with the formation of carbon nanotubes typically occurring between 600 °C and 1,000 °C. Because of the low temperature, the selectivity of the CCVD method is generally better for the production of MWNTs with respect to graphitic particles and amorphous-like carbon, which remain an important part of the raw arc-discharge SWNT samples, for example. Both homogeneous and heterogeneous processes appear very sensitive to the nature and the structure of the catalyst used, as well as to the operating conditions. Carbon nanotubes prepared by CCVD methods are generally much longer (few tens to hundreds of micrometers) than those obtained by arc discharge (few micrometers). Depending on the experimental conditions, it is possible to grow dense arrays of nanotubes. It is a general statement that MWNTs from CCVD contain more structural defects (i. e., exhibit a lower nanotexture) than MWNTs from arc, due to the lower temperature reaction, which does not allow any structural rearrangements. These defects can be removed by subsequently applying heat treatments in vacuum or inert atmosphere to the products. Whether such a discrepancy is also true for SWNTs remains questionable. CCVD SWNTs are generally gathered into bundles that are generally of smaller diameter (few tens of nm) than their arc-discharge and laser-ablation counterparts (around 100 nm in diameter). CCVD provides reasonably good perspectives of large-scale and low-cost processes for the mass production of carbon nanotubes, a key point for their applications at the industrial scale.

A last word concerns the nomenclature. Because work in the field started more than one century ago, denomination of the carbon objects prepared by this method has changed with time with regard to the authors, research areas, and fashions. The same objects are found to be called vapor grown carbon fibers, nanofilaments, nanofibers, and from now on, nanotubes. Specifically regarding multilayered fibrous morphologies (since single-layered fibrous morphologies cannot be otherwise than SWNT anyway), the exact name should be vapor-grown carbon nanofilaments (VGCNF). Whether or not the filaments are then tubular is a matter of textural description, which should go with other textural features such as bamboo, herringbone, concentric, etc. (see Sect. 3.1.2). But the usage has decided otherwise. In the following, we will therefore use MWNTs for any hollowed nanofilament, whether they contain transversally oriented graphene walls or not. Any other nanofilament will be named "nanofiber."

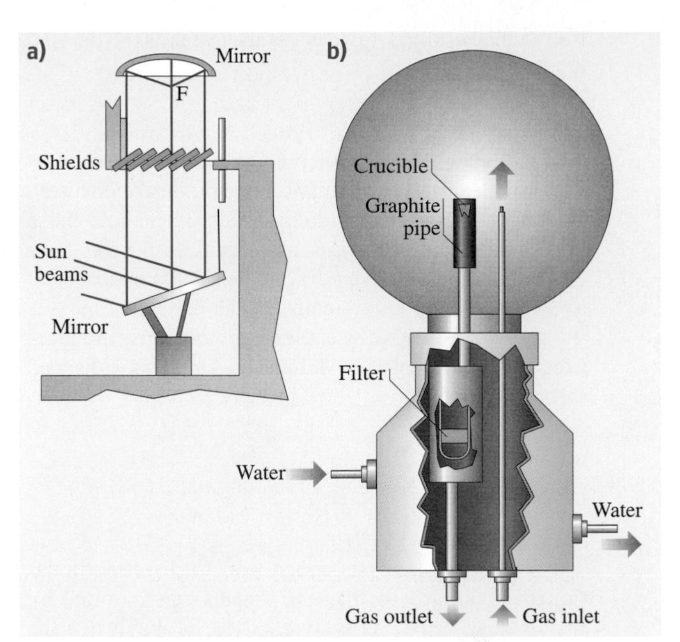

Fig. 3.14a,b Sketch of a solar energy reactor in use in Odeilho (France). (**a**) Gathering of sun rays, focused in F; (**b**) Example of Pyrex® chamber placed in (**a**) so that the graphite crucible is at the point F (adapted from [3.56])

Heterogeneous Processes

Heterogeneous CCVD processes are basically simple and consist, within a furnace heated to the desired temperature, in passing a gaseous flow containing a given proportion of a hydrocarbon (mainly CH_4, C_2H_2, C_2H_4, or C_6H_6, usually as a mixture with either H_2 or an inert gas such as Ar) over small transition metal particles (Fe, Co, Ni). The latter are previously deposited onto an inert substrate, for instance, by spraying a suspension of the metal particles on it, or any other method. The reaction is chemically defined as catalysis-enhanced thermal cracking

$$C_xH_y \rightarrow x\,C + y/2\,H_2 \,.$$

Catalysis-enhanced thermal cracking was used as early as the late 19th century (see beginning of chapter). Further extensive works before the 90s include those by Baker et al. [3.6, 59], or *Endo* et al. [3.60, 61]. Several review papers have been published since then, such as [3.62], in addition to many regular papers.

CO can be used instead of hydrocarbons, the reaction is then chemically defined as catalysis-enhanced disproportionation

$$2\,CO \rightarrow C + CO_2 \,.$$

Heterogeneous Processes – Experimental Devices

The ability of catalysis-enhanced CO disproportionation in making carbon nanofilaments was reported by *Davis* et al. [3.63] as early as 1953, probably for the first time. Extensive following works were performed by *Boehm* [3.64], *Audier* et al. [3.16, 65–67], and *Gadelle* et al. [3.68–71].

Although formation mechanisms for SWNTs and MWNTs can be quite different (see Sect. 3.3, or refer to a review article such as [3.72]), many of the parameters of the catalytic processes have a similar and important role on the type of nanotubes formed: the temperature, the duration of the treatment, the gas composition and flow rate, and of course the catalyst nature and size. At a given temperature, depending mainly on the nature of both the catalyst and the carbon-containing gas, the catalytic decomposition will take place at the surface of the metal particles, followed by a mass-transport of the freshly produced carbon either by surface or volume diffusion until the carbon concentration reaches the solubility limit, and the precipitation starts.

It is now agreed that the formation of CCVD carbon nanotubes occurs on metal particles of a very small size, typically in the nanometer-size range [3.72]. These catalytic metal particles are prepared mainly by reduction of transition metals compounds (salts, oxides) by H_2, prior to the nanotube formation step (where the carbon containing gas is required). It is possible, however, to produce these catalytic metal particles in situ in presence of the carbon source, allowing for a one-step process [3.73]. Because the control of the metal particle size is the key point (they have to be kept at a nanometric size), their coalescence is generally avoided by supporting them on an inert support such as an oxide (Al_2O_3, SiO_2, zeolites, $MgAl_2O_4$, MgO, etc.), or more rarely on graphite. A low concentration of the catalytic metal precursor is required to limit the coalescence of the metal particles that may happen during the reduction step.

There are two main ways for the preparation of the catalyst: (a) the impregnation of a substrate with a solution of a salt of the desired transition metal catalyst, and (b) the preparation of a solid solution of an oxide of the chosen catalytic metal in a chemically inert and thermally stable host oxide. The catalyst is then reduced to form the metal particles on which the catalytic decomposition of the carbon source will lead to carbon nanotube growth. In most cases, the nanotubes can then be separated from the catalyst (Fig. 3.15).

Heterogeneous Processes – Results with CCVD Involving Impregnated Catalysts

A lot of work was already done in this area even before the discovery of fullerenes and carbon nanotubes, but although the formation of tubular carbon structures by catalytic processes involving small metal particles was clearly identified, the authors did not focus on the preparation of SWNTs or MWNTs with respect to the other carbon species. Some examples will be given here to illustrate the most striking improvements obtained.

With the impregnation method, the process generally involves four different and successive steps: (1) the impregnation of the support by a solution of a salt (nitrate, chloride) of the chosen metal catalyst; (2) the drying and calcination of the supported catalyst to get the oxide of the catalytic metal; (3) the reduction in a H_2-containing atmosphere to make the catalytic metal particles and at last (4) the decomposition of a carbon-containing gas over the freshly prepared metal particles that will lead to the nanotube growth. For example, *Ivanov* et al. [3.74] have prepared nanotubes by the decomposition of C_2H_2 (pure or in mixture with H_2) on well-dispersed transition metal particles (Fe, Co, Ni, Cu) supported on graphite or SiO_2. Co-SiO_2 was found to be the best catalyst/support combination for the preparation of MWNTs, but most of the other combinations led to carbon filaments, sometimes covered with amorphous-like carbon. The same authors have developed a precipitation-ion-exchange method that provides a better dispersion of metals on silica compared to the classical impregnation technique. The same group has then proposed the use of a zeolite-supported Co catalyst [3.75, 76], resulting in very finely dispersed metal particles (from 1 to 50 nm in diameter). Only on this catalyst could they observe MWNTs with a diameter around 4 nm and only two or three walls. *Dai* et al. [3.77] have prepared SWNTs by CO disproportionation on nano-sized Mo particles. The diameters of the nanotubes obtained are closely related to that of the original particles and range from 1 to 5 nm. The nanotubes obtained by this method are free of amorphous carbon coating. It is also found that a synergetic effect occurs in the case of an alloy instead of the components alone, and one of the most striking examples is the addition of Mo to Fe [3.78] or Co [3.79].

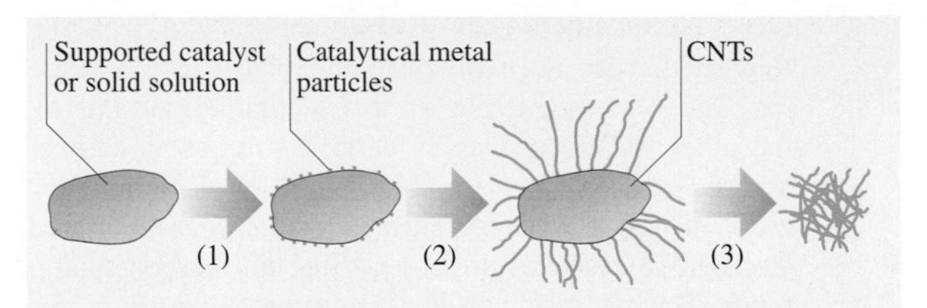

Fig. 3.15 Formation of nanotubes by the CCVD-based impregnation technique. (1) Formation of the catalytic metal particles by reduction of a precursor; (2) Catalytic decomposition of a carbon-containing gas, leading to the growth of carbon nanotubes (CNTs); (3) Removal of the catalyst to recover the CNTs (from [3.80])

Heterogeneous Processes – Results with CCVD Involving Solid-Solutions-Based Catalysts

A solid solution of two metal oxides is formed when some ions of one of the metals are found in substitution of the other metal ions. For example, Fe_2O_3 can be prepared in solid solution into Al_2O_3 to give a $Al_{2-2x}Fe_{2x}O_3$ solid solution. The use of a solid solution allows a perfect homogeneity of the dispersion of each oxide one in the other. These solid solutions can be prepared by different ways but the co-precipitation of mixed-oxalates and the combustion synthesis have been used mainly for the preparation of nanotubes. The synthesis of nanotubes by catalytic decomposition of CH_4 over $Al_{2-2x}Fe_{2x}O_3$ solid solutions was originated by *Peigney* et al. [3.73] and then studied extensively by the same group using different oxides such as spinel-based solid solutions ($Mg_{1-x}M_xAl_2O_4$ with M = Fe, Co, Ni, or a binary alloy [3.80, 81] or magnesia-based solid solutions [3.80, 82] ($Mg_{1-x}M_xO$, with M = Fe, Co or Ni)). Because of the very homogeneous dispersion of the catalytic oxide, it is possible to produce very small catalytic metal particles, at the high temperature required for the decomposition of CH_4 (which was chosen for its greater thermal stability compared to other hydrocarbons). The method proposed by these authors involves the heating of the solid solution from room temperature to a temperature between 850 °C and 1,050 °C in a mixture of H_2 and CH_4, typically containing 18 mol.% of CH_4. The nanotubes obtained depend clearly on the nature of both the transition metal (or alloy) used and the inert oxide (matrix), the latter because the Lewis acidity seems to play an important role [3.83]. For example, in the case of solid solutions containing around 10 wt% of Fe, the amount of carbon nanotubes obtained is decreasing in the following order depending of the matrixoxide: $MgO > Al_2O_3 > MgAl_2O_4$ [3.80]. In the case of MgO-based solid solutions the nanotubes can be very easily separated from the catalyst by dissolving it, in diluted HCl, for example [3.82]. The nanotubes obtained are typically gathered into small diameter bundles (less than 15 nm) with lengths up to 100 μm. The nanotubes are mainly SWNTs and DWNTs, with diameters ranging between 1 and 3 nm.

Obtaining pure nanotubes by the CCVD method requires, as for all the other techniques, the removal of the catalyst. When a catalyst supported (impregnation) in a solid solution is used, the supporting – and catalytically inactive – oxide is the main impurity, both in weight and volume. When oxides such as Al_2O_3 or SiO_2 (or even combinations) are used, aggressive treatments involving hot caustic solutions (KOH, NaOH) for Al_2O_3 or the use of HF for SiO_2 are required. These treatments have no effect, however, on the other impurities such as other forms of carbon (amorphous-like carbon, graphitized carbon particles and shells, and so on). Oxidizing treatments (air oxidation, use of strong oxidants such as HNO_3, $KMnO_4$, H_2O_2) are thus required and allow for the removal of most of the unwanted forms of carbon but are resulting in a low yield of remaining carbon nanotubes, which are often quite damaged. *Flahaut* et al. [3.82] were the first to use a MgCoO solid solution to prepare SWNTs and DWNTs that could be easily separated without any damage by a fast and safe washing with an aqueous HCl solution.

In most cases, only very small quantities of catalyst (typically less than 500 mg) have been used, and most claims for "high yield" productions of nanotubes are based on laboratory experimental data, without taking into account all the technical problems related to the scaling up of a laboratory-scale CCVD reactor. At the present time, although the production of MWNTs is possible at an industrial scale, the production of SWNTs at an affordable cost is still a challenge.

Homogeneous Processes

The homogenous route, also called "floating catalyst method," differs from the other CCVD-based methods because it uses only gaseous species and does not require the presence of any solid phase in the reactor. The basic principle of this technique, similar to the other CCVD processes, is to decompose a carbon source (ethylene, xylene, benzene, carbon monoxide, etc.) on nanometric transition metallic (generally Fe, Co, or Ni) particles in order to obtain carbon nanotubes. The catalytic particles are formed directly in the reactor, however, and are not introduced before the reaction, as it happens in the supported CCVD, for instance.

Homogeneous Processes – Experimental Devices

The typical reactor used in this technique is a quartz tube placed in an oven to which the gaseous feedstock, containing the metal precursor, the carbon source, some hydrogen, and a vector gas (N_2, Ar, or He), is sent. The first zone of the reactor is kept at a lower temperature, and the second zone, where the formation of tubes occurs, is heated to 700–1,200 °C. The metal precursor is generally a metal-organic compound, such as a zero-valent carbonyl compound like $Fe(CO)_5$ [3.84], or a metallocene [3.85–87], for instance ferrocene, nickelocene or cobaltocene. It may be advantageous to make the reactor vertical, in order for the effect of gravity to be symmetrically dispatched on the gaseous volume inside the furnace and to help in maintaining for a while the solid products in fluidized bed.

Homogeneous Processes – Results

The metal-organic compound decomposes in the first zone of the reactor to generate the nanometric metallic particles that can catalyze the nanotubes formation. In the second part of the reactor, the carbon source is decomposed to atomic carbon which then is responsible for the formation of nanotubes.

This technique is quite flexible and both single-walled nanotubes [3.88] and multiwalled nanotubes [3.89] have been obtained depending on the carbon feedstock gas; it has also been exploited for some years in the production of vapor grown carbon nanofibers [3.90].

The main drawback of this type of process is again, as for heterogeneous processes, the difficulty to control the size of the metal nanoparticles, and thus the nanotube formation is often accompanied by the production of undesired carbon forms (amorphous carbon or polyaromatic carbon phases found as various phases or as coatings). In particular, encapsulated forms have been often found, as the result of the creation of metallic particles that are too big to be active for growing nanotubes (but are still effective for catalytically decomposing the carbon sourcc) and be totally recovered by graphene layers.

The same kind of parameters as for heterogeneous processes have to be controlled in order to finely tune this process and obtain selectively the desired morphology and structure of the nanotubes formed, such as: the choice of the carbon source; the reaction temperature; the residence time; the composition of the incoming gaseous feedstock with a particular attention paid to the role played by hydrogen proportion, which can control the orientation of graphenes with respect to the nanotube axis thus switching from c-MWNT to h-MWNT [3.69]; and the ratio of the metallorganic precursor to the carbon source (for lower values, SWNTs are obtained [3.85]). As recently demonstrated, the overall process can be improved by adding other compounds such as ammonia or sulfur-containing species to the reactive gas phase. The former allows aligned nanotubes and mixed C-N nanotubes [3.91] to be obtained, the latter results in a significant increase in productivity [3.90].

It should be emphasized that only small productions have been achieved so far, and the scale up toward industrial exploitation seems quite difficult because of the large number of parameters that have to be considered. A critical one is to be able to increase the quantity of metallorganic compound that has to be sent in the reactor, as a requirement to increase the production, without obtaining too big particles. This problem has not yet been solved. An additional problem inherent in the process is the possibility of clogging the reactor due to the deposition of metallic nanoparticles on the reactor walls followed by carbon deposition.

A significant breakthrough concerning this technique could be the process developed at Rice University, the so-called HiPCo™ process to produce SWNTs of very high purity [3.92]. This gas phase catalytic reaction uses carbon monoxide to produce, from [$Fe(CO)_5$], a SWNT material claimed to be relatively free of by-products. The temperature and pressure conditions required are applicable to industrial plants. The company Carbon Nanotechnologies Inc. (Houston, TX, USA) actually sells raw SWNT materials prepared that way, at a marketed price of $375/g, doubled if purified (2003 data). Other companies are more specialized in MWNTs such as Applied Sciences Inc. (Cedarville, Ohio, USA), which currently has a production facility of ~ 40 tons/year of ~ 100 nm large MWNTs (Pyrograf-III), or Hyperion Catalysis (Cambridge, MA, USA), which makes MWNT-based materials.

Templating

Another technique interesting to describe briefly, though definitely not suitable for mass production, is the templating technique. It is the second method only (the first being the electric-arc technique, when considering the formation of MWNTs on the cathode) able to synthesize carbon nanotubes without any catalyst. Any other work reporting the catalyst-free formation of nanotubes is likely to have been fooled by metallic impurities present in the reactor or by some other factors having brought a chemical gradient to the system. Another original aspect is that it allows aligned nanotubes to be obtained naturally, without the help of any subsequent alignment procedure. But recovering the nanotubes only requires the template to be removed (dissolved), which means the former alignment of the nanotubes is lost.

Templating – Experimental Devices

The principle of the technique is to deposit a solid carbon coating obtained from CVD method onto the walls of an appropriate porous substrate whose pores are displayed as parallel channels. The feedstock is again a hydrocarbon, as a common carbon source. The substrate can be alumina or zeolite for instance, however, which present natural channel pores, while the whole is heated to a temperature able to crack the hydrocarbon selected as carbon source (Fig. 3.16).

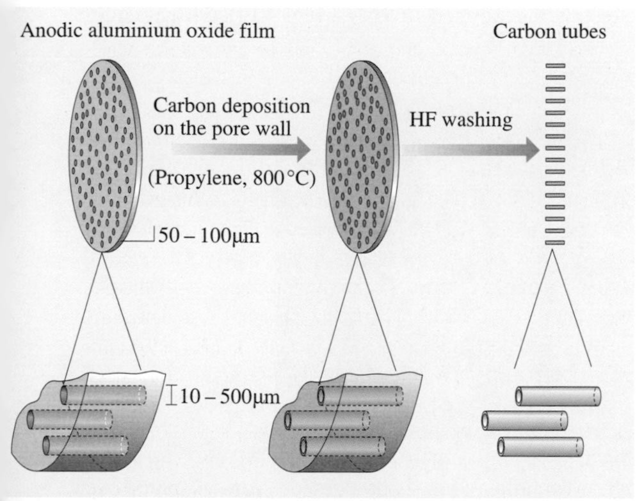

Fig. 3.16 Principle of the templating technique for the catalyst-free formation of single-walled or concentric-type multi-walled carbon nanotubes (from [3.93])

Templating – Results

Provided the chemical vapor deposition mechanism (which is actually better described as a chemical vapor infiltration mechanism) is well controlled, the synthesis results in the coating of the channel pore walls by a variable number of graphenes. Both MWNTs (exclusively concentric type) or SWNTs can be obtained. The smallest SWNTs (diameters ~ 0.4 nm) ever obtained mentioned in Sect. 3.1 have actually been synthesized by this technique [3.11]. Lengths are directly determined by channel lengths, i. e. by the thickness of the substrate plate. A main advantage is the purity of the tubes (no catalyst remnants, and little other carbon phases). On the other hand, the nanotube structure is not closed at both ends, which can be an advantage or a drawback, depending on the application. For instance, recovering the tubes requires the porous matrix to dissolve using one of the chemical treatment previously cited. The fact that tubes are open makes them even more sensitive to the acid attack.

3.2.3 Miscellaneous Techniques

In addition to the major techniques described in Sects. 3.2.1 and 3.2.2, many attempts can be found in literature to produce nanotubes by various ways, with a generally specific goal, such as looking for a low-cost or a catalyst-free production process. None has been sufficiently convincing so far to be presented as a serious alternative to the major processes described previously. Some examples are provided in the following.

Hsu et al. [3.94] have succeeded in preparing MWNTs (including coiled MWNTs, a peculiar morphology resembling a spring) by a catalyst-free (although Li was present) electrolytic method, by running a 3–5 A current between two graphite electrodes (a graphite crucible as the anode and a graphite rod as the cathode). The graphite crucible was filled with lithium chloride, while the whole was heated in air or argon at $\sim 600\,°C$. As with many other techniques, by-products such as encapsulated metal particles, carbon shells, amorphous carbon, and so on, are formed.

Cho et al. [3.95] have proposed a pure chemistry route, by the polyesterification of citric acid onto ethylen glycol at 50 °C, followed by a polymerization at 135 °C, then carbonized at 300 °C under argon, then oxidized at 400 °C in air. Despite the latter oxidation step, the solid product surprisingly contains short MWNTs, although obviously with a poor nanotexture. By-products such as carbon shells and amorphous carbon are also formed.

Li et al. [3.96] have also obtained short MWNTs by a catalyst-free (although Si is present) pyrolytic method which involves heating silicon carbonitride nanograins in a BN crucible to 1,200–1,900 °C in nitrogen within a graphite furnace. No details are given about the possible occurrence of by-products, but they are likely considering the complexity of the chemical system (Si-C-B-N) and the high temperatures involved.

Terranova et al. [3.98] have investigated the catalyzed reaction between a solid carbon source and atomic hydrogen. Graphite nanoparticles ($\sim$ 20 nm) are sent with a stream of H_2 onto a Ta filament heated at 2,200 °C. The species produced, whatever they are, then hit a Si polished plate warmed to 900 °C and supporting transition metal particles. The whole chamber is under a dynamic vacuum of 40 torr. SWNTs are supposed to form according to the authors, although images are poorly convincing. One major drawback of the method, besides its complexity compared to others, is the difficulty of recovering the "nanotubes" from the Si substrates onto which they seem to be firmly bonded.

A last example is an attempt to prepare nanotubes by diffusion flame synthesis [3.99]. A regular gaseous hydrocarbon source (ethylene, ...) added with ferrocene vapor is sent within a laminar diffusion flame made from air and CH_4 whose temperature range is 500–1,200 °C. SWNTs are actually formed, together with encapsulated metal particles, soot, and so on. In addition to a low yield, the SWNT structure is quite poor.

3.2.4 Synthesis of Aligned Carbon Nanotubes

Several applications (such as field-emission-based display, see Sect. 3.6) require that carbon nanotubes grow as highly aligned bunches, in highly ordered arrays, or located at specific positions. In that case, the purpose of the process is not mass production but controlled growth and purity, with subsequent control of nanotube morphology, texture, and structure. Generally speaking, the more promising methods for the synthesis of aligned nanotubes are based on CCVD processes, which involve molecular precursors as carbon source, and method of thermal cracking assisted by the catalytic activity of transition metal (Co, Ni, Fe) nanoparticles deposited onto solid supports. Few attempts have been made so far to obtain such materials with SWNTs. But the catalyst-free templating methods related to that described in Sect. 3.2.2 are not considered here, due to the lack of support after the template removal, which does not allow the former alignment to be maintained.

During the CCVD-growth, nanotubes can self-assemble into nanotube bunches aligned perpendicular to the substrate if the catalyst film on the substrate has a critical thickness [3.100], the driving forces for this alignment are the van der Waals interactions between the nanotubes, which allow them to grow perpendicularly to the substrates. If the catalyst nanoparticles are deposited onto a mesoporous substrate, the mesoscopic pores may also cause a certain effect on the alignment when the growth starts, thus controlling the growth direction of the nanotubes. Two kinds of substrates have been used so far in this purpose: mesoporous silica [3.101, 102] and anodic alumina [3.103].

Different methods have been reported in the literature for metal particles deposition onto substrates: (i) deposition of a thin film on alumina substrates from metallic-salt precursor impregnation followed by oxidation/reduction steps [3.104], (ii) embedding catalyst particles in mesoporous silica by sol-gel processes [3.101], (iii) thermal evaporation of Fe, Co, Ni, or Co-Ni metal alloys on SiO_2 or quartz substrates under high vacuum [3.105, 106], (iv) photolithographic patterning of metal-containing photoresist polymer with the aid of conventional black and white films as a mask [3.107] and, electrochemical deposition at the bottom of the pores in anodic aluminum oxide templates [3.103].

Depositing the catalyst nanoparticles onto a previously patterned substrate allows one to control the frequency of local occurrence and the display of the MWNT bunches formed. The as-produced materials mainly consist in arrayed, densely packed, freestanding, aligned MWNTs (Fig. 3.17), which is quite

Fig. 3.17 Example of a free-standing MWNT array obtained from the pyrolysis of a gaseous carbon source over catalyst nanoparticles previously deposited onto a patterned substrate. Each square-base rod is a bunch of MWNTs aligned perpendicular to the surface (from [3.97])

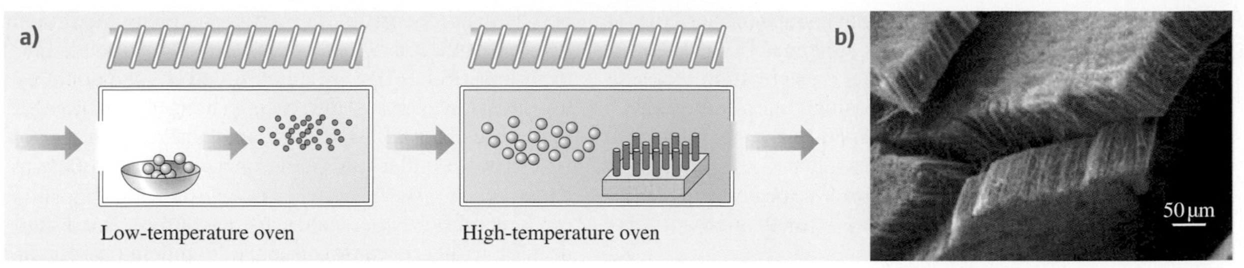

Fig. 3.18a–c Sketch of a double-furnace CCVD device for the organo-metallic/hydrocarbon co-pyrolysis process. In (**a**): sublimation of the precursor; in (**b**): decomposition of the precursor and MWNTs growth onto the substrate. (**c**) Example of the densely packed and aligned MWNT material obtained (from [3.108])

suitable for field emission-based applications for instance [3.97].

When a densely packed coating of vertically aligned MWNTs is desired (Fig. 3.18b), another route is the pyrolysis of hydrocarbons in the presence of organometallic precursor molecules like ferrocene or iron pentacarbonyl, operating in a dual furnace system (Fig. 3.18a). The organo-metallic precursor (e.g., ferrocene) is first sublimed at low temperature in the first furnace or is injected as a solution along with the hydrocarbon feedstock, then the whole components are pyrolyzed at higher temperature in the second furnace [3.88, 108–111]. The parameters that must be taken into account are the heating or feeding rate of ferrocene, the flow rate of the vector gas (Ar or N_2) and of the gaseous hydrocarbon, and the temperature of pyrolysis (650–1,050 °C). Generally speaking, the co-deposition process using $[Fe(CO)_5]$ as catalyst source implies its thermal decomposition at elevated temperatures to produce atomic iron that deposits on the substrates in the hot zone of the reactor. As the nanotube growth occurs simultaneously with the introduction of $[Fe(CO)_5]$, the temperatures chosen for the growth depend on the carbon feedstock utilized, e.g., they can vary from 750 °C for acetylene to 1,100 °C for methane. Mixtures of $[FeCp_2]$ and xylene or $[FeCp_2]$ and acetylene have also been used successfully for the production of freestanding MWNTs.

The nanotube yield and quality are directly linked with the amount and size of the catalyst particles, and since the planar substrates used do not exhibit high surface area, the dispersion of the metal can be a key step in the process. It has been observed that an etching pretreatment of the surface of the deposited catalyst thin film with NH_3 may be critical for an efficient growth of nanotubes by bringing the appropriate metal particle size distribution. It may also favor the alignment of MWNTs and prevent the formation of amorphous carbon from thermal cracking of acetylene [3.112].

3.3 Growth Mechanisms of Carbon Nanotubes

Growth mechanisms of carbon nanotubes are still debated. But researchers have been impressively imaginative and have made a number of hypotheses. One reason is that conditions allowing carbon nanofilaments to grow are very diverse, which means that related growing mechanisms are many. For given conditions, the truth is even probably a combination or a compromise between some of the proposals. Another reason is that phenomena are quite fast and difficult to observe in situ. It is generally agreed, however, that growth should occur so that the number of dangling bonds is limited, for energetic reasons.

3.3.1 Catalyst-Free Growth

As already mentioned, in addition to the templating technique which merely relates to chemical vapor infiltration mechanism for pyrolytic carbon, the growth of c-MWNT as a deposit onto the cathode in the electric-arc method is a rare example of catalyst-free carbon nanofilament growth. The driving forces are obviously related to the electric field, i. e. related to charge transfers from an electrode to the other via the particles contained in the plasma. How the MWNT nucleus forms is not clear, but once it has started, it may include the direct incorporation of C_2 species to the primary graphene structure as it was formerly proposed for fullerenes [3.113]. This is supported by recent C_2 radical concentration measurements that revealed an increasing concentration of C_2 from the anode being consumed to the growing cathode (Fig. 3.13), indicating, first, that C_2 are secondary species only, and second, that C_2 species could actually participate actively in the growth mechanism of c-MWNTs in the arc method.

3.3.2 Catalytically Activated Growth

Growth mechanisms involving catalysts are more difficult to ascertain, since they are more diverse. Although a more or less extensive contribution of a VLS (Vapor-Liquid-Solid [3.114]) mechanism is rather well admitted, it is quite difficult to find comprehensive and plausible explanations able to account for both the various conditions used and the various morphologies observed. What follows is an attempt to provide overall explanations of most of the phenomena, while remaining consistent with the experimental data. We did not consider hypotheses for which there are a lack of experimental evidence, such as the moving nano-catalyst mechanism, which proposed that dangling bonds from a growing SWNT may be temporarily stabilized by a nano-sized catalyst located at the SWNT tip [3.23], or the scooter mechanism, which proposed that C dangling bonds are temporarily stabilized by a single catalyst atom, moving all around the SWNT cross section edge, allowing subsequent C atom addition [3.115].

From the various results, it appears that the most important parameters are probably the thermodynamic conditions (only temperature will be considered here), the catalyst particle size, and the presence of a substrate. Temperature is critical and basically corresponds to the discrepancy between CCVD methods and solid-carbon-source based method.

Low Temperature Conditions

Low temperature conditions are typical from CCVD conditions, since nanotubes are frequently found to grow far below 1,000 °C. If conditions are such that the catalyst is a crystallized solid, the nanofilament is probably formed according to a mechanism close to a VLS mechanism, in which three steps are defined: (i) adsorption then decomposition of C-containing gaseous moieties at the catalyst surface; (ii) dissolution then diffusion of the C-species through the catalyst, thus forming a solid solution; (iii) back precipitation of solid carbon as nanotube walls. The texture is then determined by the orientation of the crystal faces relative to the filament axis (Fig. 3.19), as demonstrated beyond doubt by transmission electron microscopy images such as that in *Rodriguez* et al. [3.19]. This mechanism can therefore provide either c-MWNT, or h-MWNT, or platelet nanofiber.

If conditions are such that the catalyst is a liquid droplet, either because of higher temperature conditions or because of lower temperature melting catalyst, a mechanism similar to that above can still occur, really VLS (vapor = gaseous C species, liquid = molten catalyst, S = graphenes), but there are obviously no more crystal faces to orientate preferentially the rejected graphenes. Energy minimization requirements will therefore tend to make them concentric and parallel to the filament axis.

With large catalyst particles (or in the absence of any substrate), the mechanisms above will generally follow a "tip growth" scheme, i. e., the catalyst will move forward while the rejected carbon form the nanotube behind, whether there is a substrate or not. In that case,

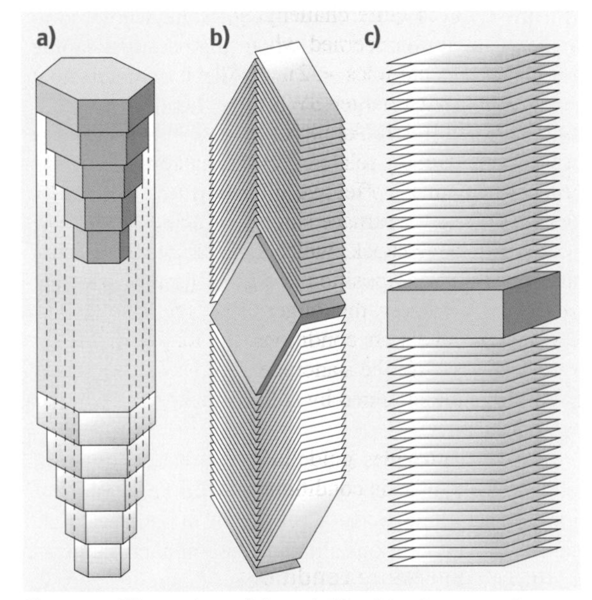

Fig. 3.19 Illustration of the relationships between the catalyst crystal outer morphology and the subsequent carbon nanofilament inner texture. Crystal are drawn from their images in the projected plane perpendicular to the electron beam in a transmission electron microscope, i. e. crystal morphology in the out-of-plane dimension is not ascertained (adapted from [3.19])

chances are high that one end is open. On the contrary, when catalyst particles deposited onto a substrate are small enough (nanoparticles) to be refrained to move by the interaction forces with the substrate, the growth mechanism will follow a "base growth" scheme, i. e. the carbon nanofilament grows away from the substrate leaving the catalyst nanoparticle attached to the substrate (Fig. 3.20).

The bamboo texture that affects both the herringbone and the concentric texture may reveal a specificity of the dissolution-rejection mechanism, i. e. a periodic, discontinuous dynamics of the phenomenon. Once the catalyst has reached the saturation threshold regarding its content in carbon, it expulses it quite suddenly. Then it becomes again able to incorporate a given amount of carbon without having any catalytic activity for a little while, then over-saturation is reached again, etc. Factors controlling such a behavior, however, are not determined.

It is then clear that, in any of the mechanisms above, 1 catalyst particle = 1 nanofilament. This explains why, although making SWNT by CCVD methods is possible, controlling the catalyst particle size is critical, since it thereby controls the subsequent nanofilament to be grown from it. Reaching a really narrow size distribution in CCVD is quite challenging, specifically as far as nanosizes are concerned when growing SWNTs is the goal. Only particles < 2 nm will be able to make it (Fig. 3.20), since larger SWNTs would energetically not be favored [3.10]. Another specificity of the CCVD method and from the related growth mechanisms is that the process can be effective all along the isothermal zone of the reactor furnace since continuously fed with a carbon-rich feedstock, which is generally in excess, with a constant composition at a given time of flight of the species. Coarsely, the longer the isothermal zone in gaseous carbon excess conditions, the longer the nanotubes. This is why the nanotube lengths can be much longer than that obtained by solid-carbon-source-based methods.

Table 3.2 provides guidelines for the relationships between the synthesis conditions and the type of nanotube formed.

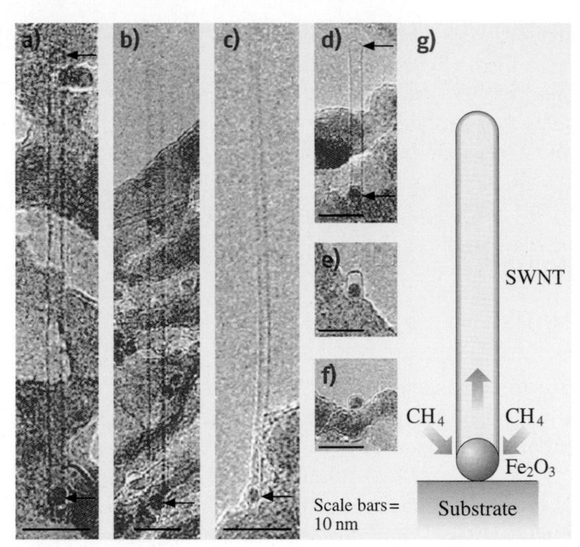

Fig. 3.20 High resolution transmission electron microscopy images of several SWNTs grown from iron-based nanoparticles by CCVD method, showing that particle sizes determine SWNT diameters in that case (adapted from [3.116])

High Temperature Conditions

High temperature conditions are typical from solid-carbon-source based methods such as electric arc, laser vaporization, or solar furnace (see Sect. 3.2). The huge temperatures involved (several thousands °C) atomize both the carbon source and the catalyst. Of course, catalyst-based SWNTs do not form in the area of highest temperatures (contrary to c-MWNT in the electric arc method), but the medium is made of atoms and radicals all mixed, some of which are likely to condense into the same droplet (since it will be liquid at the beginning). At some distance from the atomization zone, the medium is therefore made of carbon metal alloy droplets and of secondary carbon species that range from C_2 to higher order molecules such as corannulene which is made of a central pentagon surrounded by 5 hexagons. (The preferred formation of such a molecule can be explained by the former association of carbon atoms into a pentagon, because it is the fastest way to limit dangling bonds at low energetic cost, thereby providing a fixation site for other carbon atoms (or C_2) which also will tend to close into a ring, again to limit dangling bonds. Since adjacent pentagons are not energetically favoured, these cycles will be hexagons. Such a molecule is thought to be a probable precursor for fullerenes. Fullerenes are actually always produced, even in conditions which produce SWNTs.) The same event of saturation in C of the carbon-metal alloy droplet occurs as described in Sect. 3.3.2, resulting in the precipitation of excess C outside the particle upon the effect of the decreasing thermal gradient in the reactor, which decreases the solubility threshold of C in the metal [3.117]. Once the "inner" carbon atoms reach the catalyst particle surface, they meet the "outer" carbon species, including corannulene, that will con-

Table 3.2 Guidelines indicating the relationships between possible carbon nanofilament morphologies and some basic synthesis conditions

		Increasing temperature … → … and physical state of catalyst			Substrate		Thermal gradient	
		Solid (crystallized)	Liquid from melting	Liquid from clusters	Yes	No	Low	High
Catalyst particle size	<∼ **3** nm	SWNT	SWNT	?	base-growth	tip-growth	long length	short length
	>∼ **3** nm	MWNT (c,h,b) platelet nanofiber	c-MWNT	SWNT	tip-growth			
Nanotube diameter		(heterogeneous related to catalyst particle size)		homogeneous (independent) from particle size)				
Nanotube/particle		one nanotube/particle		several SWNTs/particle				

tribute to cap the merging nanotubes. Once formed and capped, nanotubes can grow both from the inner carbon atoms (Fig. 3.21a), according to the VLS mechanism proposed by *Saito* et al. [3.117], and from the outer carbon atoms, according the adatom mechanism proposed by *Bernholc* et al. [3.118]. In the latter, carbon atoms from the surrounding medium in the reactor are attracted then stabilized by the carbon/catalyst interface at the nanotube/catalyst surface contact, promoting their subsequent incorporation at the tube base. The growth mechanism therefore mainly follows the base growth scheme. But once the nanotubes are capped, C_2 species still remaining in the medium and which meet the growing nanotubes far from the nanotube/catalyst interface, may still incorporate the nanotubes from both the side wall or the tip, thereby giving rise to some proportion of Stone–Wales defects [3.40]. The occurrence of a nanometer-thick surface layer of yttrium carbide (onto the main Ni-containing catalyst core), whose some lattice distances are commensurable with that of the C−C distance in graphene, as recently revealed by *Gavillet* et al. [3.48], could possibly play a beneficial role in the stabilization effect at the nanotube/catalyst interface and explain why SWNT yield is enhanced by such bimetallic alloys (as opposed to single metal catalysts).

A major difference with the low temperature mechanisms described for CCVD methods is that many nanotubes are formed from a single, relatively large (∼ 10–50 nm) catalyst particle (Fig. 3.21b), whose size distribution is therefore not as critical as for the low

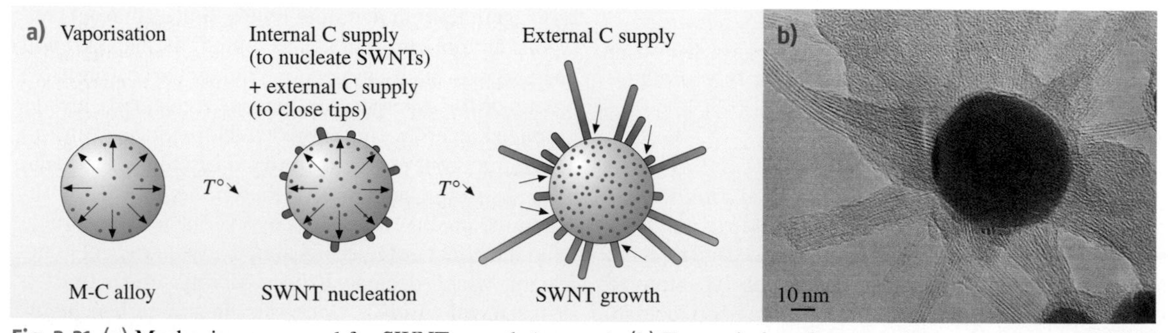

Fig. 3.21 (a) Mechanism proposed for SWNT growth (see text). (b) Transmission electron microscopy image of SWNT growing radial to a large Ni catalyst particle surface in the electric arc experiment. (modified from [3.17])

temperature mechanisms (particles that are too large, however, would induce polyaromatic shells instead of nanotubes). That is why SWNT diameters are much more homogeneous than for CCVD methods. Why the most frequent diameter is ~ 1.4 nm is again a matter of energy balance. Single-wall nanotubes larger than ~ 2.5 nm are not stable [3.10]. On the other hand, strain to the C−C angles increases as the radius of curvature decreases. Such diameter (1.4 nm) should therefore correspond to the best energetic compromise. Another difference with the low temperature mechanisms for CCVD is that temperature gradients are huge, and the gas phase composition surrounding the catalysts droplets is also subjected to rapid changes (as opposed to what could happen in a laminar flow of a gaseous feedstock whose carbon source is in excess). This explains why nanotubes from arc are generally shorter than nanotubes from CCVD, and why mass production by CCVD should be favored. In the latter, the catalytic role of the metallic particle can be recurrent as long as the conditions are maintained. In the former, surrounding conditions change continuously, and the window for efficient catalysis effect can be very narrow. Lowering temperature gradients in solid carbon source-based SWNT producing methods such as the electric-arc reactor should therefore be a way to increase the SWNT yield and length [3.119]. Amazingly, this is in opposition to fullerene production from arc.

3.4 Properties of Carbon Nanotubes

Carbon is a unique light atom that can form one-, two-, or threefold strong chemical bonds. The planar threefold configuration forms graphene planes that can, under certain growth conditions (see Sect. 3.3), adopt a tubular geometry. Properties of the so-called carbon nanotubes may change drastically depending on whether SWNTs or MWNTs are considered (see Sect. 3.1).

3.4.1 Variability of Carbon Nanotube Properties

Properties of MWNTs are generally not much different from that of regular polyaromatic solids (which may exhibit graphitic, or turbostratic, or intermediate crystallographic structure), and variations are then mainly driven by the textural type of the MWNTs considered (concentric, herringbone, bamboo) and the quality of the nanotexture (see Sect. 3.1), both of which control the extent of anisotropy. Actually, for polyaromatic solids, whose structural entities are built with stacked graphenes, bond strength is quite different depending on whether the in-plane direction is considered (which includes only very strong – covalent – and therefore very short – 0.142 nm bonds) or the direction perpendicular to it (which includes only very weak – van der Waals – and therefore very loose – ~ 0.34 nm bonds). Such heterogeneity is not found in single (isolated) SWNTs. But the heterogeneity is back along with the related consequences when SWNTs associate into bundles. Therefore, properties – and applicability – for SWNTs may also change dramatically depending on whether single SWNT or SWNT ropes are involved.

In the following, we will emphasize SWNT properties, as far as their original structure often leads to original properties with respect to that of regular polyaromatic solids. But we will also sometimes cite properties of MWNTs for comparison.

3.4.2 General Properties

SWNT-type carbon nanotube diameters fall in the nanometer range and can be hundreds of micrometers long. SWNTs are narrower in diameter than the thinnest line able to be obtained by electron beam lithography. SWNTs are stable up to 750 °C in air (but start being damaged earlier through oxidation mechanisms, as demonstrated by their subsequent ability to be filled with molecules, see Sect. 3.5) and up to $\sim 1{,}500$–$1{,}800$ °C in inert atmosphere beyond which they transform into regular, polyaromatic solids (i. e., phases built with stacked graphenes instead of single graphenes) [3.120]. They have half the mass density of aluminum. While the length of SWNTs can be macroscopic, the diameter has a molecular dimension. As a molecule, properties are closely influenced by the way atoms are displayed along the molecule direction. The physical and chemical behavior of SWNTs are therefore related to their unique structural features [3.121].

3.4.3 SWNT Adsorption Properties

An interesting feature of SWNTs is their very high surface area, the highest ever due to the fact that a single graphene sheet is probably the unique example of a material energetically stable in normal conditions while consisting of a single layer of atoms. Ideally, i. e. not considering SWNTs in bundles but isolated SWNTs, and provided the SWNTs have one end opened (by oxidation treatment for instance), the real surface area can be equal to that of a single, flat graphene, i. e. $\sim 2{,}700\,\mathrm{m^2/g}$ (accounting for both sides).

But practically, nanotubes – specifically SWNTs – are more often associated with many other nanotubes to form bundles, then fibers, films, papers, etc. than as a single entity. Such architectures develop various porosity ranges important for determining adsorption properties (the latter aspects are also developed in Sect. 3.6.2, focused on applications). It is thus most appropriate to discuss adsorption on SWNT-based materials in term of adsorption on the outer or inner surfaces of such bundles. Furthermore, theoretical calculations have predicted that the molecule adsorption on the surface or inside of the nanotube bundle is stronger than that on an individual tube. A similar situation exists for MWNTs where adsorption could occur either on or inside the tubes or between aggregated MWNTs. Additionally, it has been shown that the curvature of the graphene sheets constituting the nanotube walls can result in a lower heat for adsorption with respect to that of a planar graphene (see Sect. 3.1.1).

Accessible SWNT Surface Area

Various studies dealing with the adsorption of nitrogen on MWNTs [3.122, 123] and SWNTs [3.124] have highlighted the porous nature of these two materials. Pores in MWNTs can be divided mainly into inner hollow cavities of small diameter (narrowly distributed, mainly 3–10 nm) and aggregated pores (widely distributed, 20–40 nm), formed by interaction of isolated MWNTs. It is also worth noting that the ultra-strong nitrogen capillarity in the aggregated pores contributes to the major part of the total adsorption, showing that the aggregated pores are much more important than the inner cavities of the MWNTs for adsorption. Adsorption of N_2 has been studied on as-prepared and acid-treated SWNTs, and the results obtained point out the microporous nature of SWNT materials, as opposed to the mesoporous MWNT materials. Also, as opposed to isolated SWNTs (see above), surface areas well above $400\,\mathrm{m^2g^{-1}}$ have been measured for SWNT-bundle-containing materials, with internal surface areas of $300\,\mathrm{m^2g^{-1}}$ or higher.

The theoretical specific surface area of carbon nanotubes ranges over a broad scale, from 50 to $1{,}315\,\mathrm{m^2g^{-1}}$ depending on the number of walls, the diameter, or the number of nanotubes in a bundle of SWNTs [3.125]. Experimentally, the specific surface area of SWNTs is often larger than that of MWNTs. Typically, the total surface area of as-grown SWNTs ranges between 400 and $900\,\mathrm{m^2g^{-1}}$ (micropore volume, $0.15–0.3\,\mathrm{cm^3g^{-1}}$), whereas values ranging between 200 and $400\,\mathrm{m^2g^{-1}}$ for as-produced MWNTs are often reported. In the case of SWNTs, the diameter of the tubes and the number of tubes in the bundle will affect mainly the BET value. It is worth noting that opening/closing of the central canal noticeably contributes to the adsorption properties of nanotubes. In the case of MWNTs, chemical treatments such as KOH activation are efficient to develop a microporosity, and surface areas as high as $1{,}050\,\mathrm{m^2g^{-1}}$ have been reported [3.126]. Thus it appears that opening or cutting as well as chemical treatments (purification step for example) of carbon nanotubes can considerably affect their surface area and pore structure.

Adsorption Sites and Binding Energy of the Adsorbates

An important problem to solve when considering adsorption on nanotubes is the identification of the adsorption sites. Adsorption of gases in a SWNT bundle can occur inside the tubes (pore), in the interstitial triangular channels between the tubes, on the outer surface of the bundle, or in a groove formed at the contact between adjacent tubes on the outside of the bundle (Fig. 3.22). Starting from closed-end SWNTs, simple molecules can adsorb on the walls of the outer nanotubes of the bundle and preferably on the external grooves. For the more attractive sites, corresponding to the first adsorption stages, it seems that adsorption or condensation in the interstitial channel of SWNT bundles depends on the size of the molecule (or on the SWNT diameters) and on their interaction energies [3.127]. Then, adapted treatments to open the tubes will favour the gas adsorption (e.g., O_2, N_2 in the inner walls [3.128, 129]). For hydro-

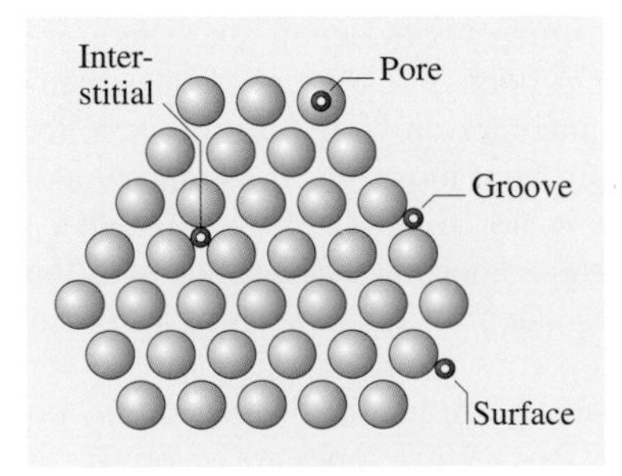

Fig. 3.22 Sketch of a SWNT bundle illustrating the four different adsorption sites

gen and other small molecules, computational methods have shown that, for open SWNTs, the pore, interstitial, and groove sites are energetically more favourable than the surface site, with respect to the increasing number of carbon nanotubes interacting with the adsorbed molecules [3.130].

For MWNTs, adsorption can occur in the aggregated pores, inside the tube or on the external walls. In the latter case, the presence of defects, as incomplete graphene layers, has to be taken into consideration. Although adsorption between the graphenes (intercalation) has been proposed in the case of hydrogen adsorption in h-MWNTs or platelet nanofibers [3.132], it is unlikely for many molecules due to steric effect and should not prevail for small molecules due to long diffusion path.

Few studies deal with adsorption sites in MWNTs, however it has been shown that butane adsorbs more in the case of MWNTs with smaller outside diameters, which is consistent with another statement that the strain brought to curved graphene surfaces affects sorption. Most of the butane adsorbs to the external surface of the MWNTs while only a small fraction of the gas condensed in the pores [3.133]. Comparative adsorption of krypton or of ethylene on MWNTs or on graphite has allowed scientists to determine the dependence of the adsorption and wetting properties on the specific morphology of the nanotubes. Higher condensation pressure and lower heat of adsorption were found on nanotubes with respect to graphite [3.134]. These differences result mainly from a decrease in the lateral interactions between the adsorbed molecules, related to the curvature of the graphene sheets.

A limited number of theoretical as well as experimental works exist on the values of the binding energies for gases on carbon nanotubes. If most of these studies report low binding energies on SWNTs, consistent with a physisorption, some experimental results, in particular in the case of hydrogen, are still controversial (see Sect. 3.6.2). In the case of platelet nanofibers, the initial dissociation of hydrogen on graphite edge sites, which constitute the majority of the nanofiber surface has been proposed [3.135]. For carbon nanotubes, a mechanism that involves H_2 dissociation on residual metal catalyst followed by H spillover and adsorption on the most reactive nanotube sites could be envisaged [3.136]. Doping nanotubes with alkali may enhance hydrogen adsorption, due to the charge transfer from the alkali metal to the nanotube, which polarizes the H_2 molecule and induces dipole interaction [3.137].

Generally speaking, the adsorbates can be either charge donors or acceptors to the nanotubes. The observed trends in the binding energies of gases with different van der Waals radii suggest that the groove sites of SWNTs are the preferred low coverage adsorption sites due to their higher binding energies. Finally, several studies have shown that, at low coverage, the binding energy of the adsorbate on SWNT is between 25% and 75% higher than the binding energy on a single graphene. This discrepancy can be attributed to an increase of the effective coordination in the binding sites, such as the groove sites, in SWNTs bundles [3.138,139]. Representative results concerning the adsorption properties of SWNTs and MWNTs are summarized on Table 3.3.

Table 3.3 Adsorption properties and sites of SWNTs and MWNTs. Letters in column 5 refer to Fig. 3.22. Data of two last columns are from [3.131]

Type of nanotube	Porosity (cm^3g^{-1})	Surface area (m^2/g)	Binding energy of the adsorbate	Adsorption sites	Attractive potential per site (eV)	Surface area per site (m^2/g)
SWNT (bundle)	Microporous V_{micro}: 0.15–0.3	400–900	Low, mainly physisorption 25–75 % > graphite	Surface (A)	0.049	483
				Groove (B)	0.089	22
				Pore (C)	0.062	783
				Interstitial (D)	0.119	45
MWNT	Mesoporous	200–400	Physisorption	Surface Pore Aggregated pores	–	–

3.4.4 Transport Properties

The narrow diameter of SWNTs has a strong influence on its electronic excitations due to its small size compared to the characteristic length scale of low energy electronic excitations. Combined with the particular shape of the electronic band structure of graphene, carbon nanotubes are ideal quantum wires. The conduction and the valence band of the graphene touch on the 6 corners (K points) of the Brillouin zone [3.140]. By rolling the graphene to form a tube, new periodic boundary conditions for the electronic wave functions superimpose, which give rise to one-dimensional sub-bands: $C_n K = 2q$ where q is an integer and C_n the vector $na_1 + ma_2$ characteristic of the diameter and the helicity of the tube (see Sect. 3.1). If one of these sub-bands passes through one of the K points, the nanotube will be metallic, otherwise it will be semiconductor with a gap inversely proportional with the diameter. For a tube with a diameter of 3 nm, the gap falls in the range of the thermal energy at room temperature. As pointed out in Sect. 3.1 already, knowing (n, m) then allows one to predict whether or not the tube exhibits a metallic behavior. To summarize, we can say that the electronic structure of SWNTs depends both on the orientation of the honeycomb lattice with respect to the tube axis and the radius of curvature imposed to the bent graphene sheet. This explains why properties of MWNTs quickly become no different from regular, polyaromatic solids as the number of walls (graphenes) increases, since the radius of curvature increases meanwhile. In addition, the conduction occurs essentially through the external tube (for c-MWNTs), but the interactions with the internal coaxial tubes may make the electronic properties vary. For small diameter MWNTs exhibiting a low number of walls, typically two (DWNTs), the relative lattice orientation of the two superimposed graphenes remains determinant, and the resulting overall electronic behavior can either be metallic or semiconductor upon structural considerations [3.141]. Another consequence, valid for any nano-sized carbon nanotube whatever the number of walls, is that applying stresses to the tube is also likely to change its electronic behavior, as a consequence of the variation brought to the position of the sub-bands with respect to the K points of the Brillouin zone [3.142].

The details of electronic properties of SWNTs are not well understood yet, and numerous theoretical and experimental works have revealed fascinating effects [3.143]. An assembly of agglomerated tubes (ropes) increases the complexity. Moreover, investigating the electron properties of a single SWNT requires one to characterize and control the properties of the electrical contacts. Likewise, the interaction with the substrate influence the bulk transport properties a lot when SWNTs embedded in a medium (e.g., a composite) are considered. The conduction of charge carriers in metallic SWNTs is thought to be ballistic (i. e., independent from length) due to the expected small number of defects [3.144–146] with a predicted electric conductance twice the fundamental conductance unit $G_0 = 4e/h$ because of the existence of two propagating modes. Because of the reduced scattering, metallic SWNTs can transport huge current densities (max $10^9\ \mathrm{A/cm^2}$) without being damaged, i. e. about three orders of magnitude higher than in Cu. On the other hand, in a one-dimensional system, electrons often localize due to structural defects or disorder. In case the contact resistance is too high, a tunnel barrier is formed, and the charge transport is dominated by tunneling effects. Several transport phenomena were reported in the literature of this regime: weak localization due to quantum interferences of the electronic wave functions that leads to an increase of the resistivity [3.147–149], Coulomb blockade at low temperature that characterizes by conductance oscillations as the gate voltage increases [3.148], superconductivity induced by superconductor contacts [3.150], or spin polarization effects induced by ferromagnetic contacts [3.151]. Furthermore, due to enhanced Coulomb interactions, one-dimensional metals are expected to show drastic changes in the density of states at the Fermi level and are described by a Luttinger liquid [3.148]. This behavior is expected to result in the variation of conductance vs. temperature that follows a power law, with a zero conductance at low temperature.

As mentioned previously, SWNTs are model systems to study one-dimensional charge transport phenomena. Due to the of mesoscopic dimension of the morphology, the one-dimensional structure of the nanotube has strong singularities in the electronic density of states (DOS) that fall in the energy range of visible light. As a consequence, visible light is strongly absorbed. It has been observed that flash illumination with a broadband light can lead to spontaneous burning of a macroscopic sample of agglomerated (i. e., ropes) carbon nanotubes in air and room temperature [3.152].

As a probable consequence of both the small number of defects (at least the kind of defects that oppose phonon transport) and the cylindrical topography, SWNTs exhibit a large phonon mean free path, which results in a high thermal conductivity. The thermal conductivity of SWNTs is comparable to that of a single, isolated graphene layer or high purity diamond [3.153] or possibly higher (∼ 6,000 W/mK).

Finally, carbon nanotubes may also exhibit a positive or negative magneto-resistance depending on the current, the temperature, and the field, whether they are MWNTs [3.154] or SWNTs [3.149].

3.4.5 Mechanical Properties

While tubular nano-morphology is also observed for many two-dimensional solids, carbon nanotubes are unique through the particularly strong threefolded bonding (sp^2 hybridization of the atomic orbitals) of the curved graphene sheet, which is stronger than in diamond (sp^3 hybridization) as revealed by their difference in C−C bond length (0.142 vs. 0.154 nm for graphene and diamond respectively). This makes carbon nanotubes – SWNTs or c-MWNTs – particularly stable against deformations. The tensile strength of SWNTs can be 20 times that of steel [3.155] and has actually been measured equal to ∼ 45 GPa [3.156]. Very high tensile strength values are also expected for ideal (defect-free) c-MWNTs, since combining perfect tubes concentrically is not supposed to be detrimental to the overall tube strength, provided the tube ends are well capped (otherwise, concentric tubes could glide relative to each other, inducing high strain). Tensile strength values as high as ∼ 150 GPa have actually been measured for perfect MWNTs from electric arc [3.157], although the reason for such a high value compared to that measured for SWNTs is not clear. It probably reveals the difficulty in carrying out such measurements in a reliable manner. At least, flexural modulus, for perfect MWNTs should logically exhibit higher values than SWNTs [3.155], with a flexibility that decreases as the number of walls increases. On the other hand, measurements performed on defective MWNTs obtained from CCVD gave a range of 3–30 GPa [3.158]. Tensile modulus, also, reaches the highest values ever, 1 TPa for MWNTs [3.159], and possibly even higher for SWNTs, up to 1.3 TPa [3.160,161]. Figure 3.23 illustrates how spectacularly defect-free carbon nanotubes could revolutionize the current panel of high performance fibrous materials.

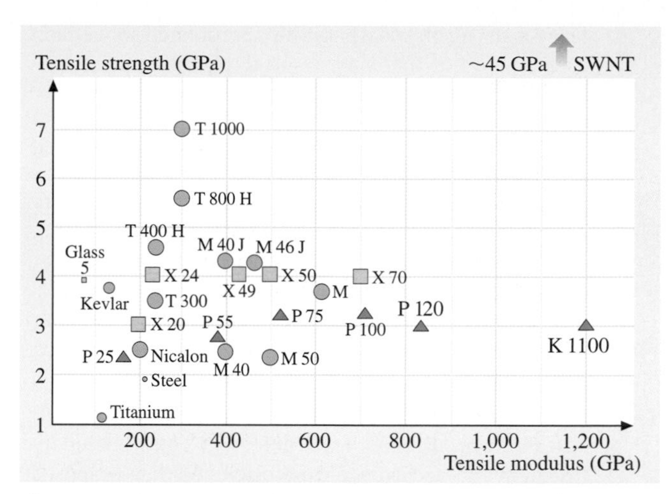

Fig. 3.23 Plot of the tensile strength versus tensile modulus for current fibrous materials, as compared to SWNTs. *Large circles* are PAN-based carbon fibers, which include the highest tensile strength fiber available on the market (T1000 from Torayca); *Triangles* are pitch-based carbon fibers, which include the highest tensile modulus fiber on the market (K1100 from Amoco)

3.4.6 Reactivity

The chemical reactivity of graphite, fullerenes, and carbon nanotubes exhibits some common features. Like any small object, carbon nanotubes have a large surface with which they can interact with their environment (see Sect. 3.4.1). It is worth noting, however, that the chemistry of nanotubes chemistry differs from that of regular polyaromatic carbon materials because of their unique shape, small diameter, and structural properties. Compared to graphite, perfect SWNTs have no chemically active dangling bonds (the reactivity of polyaromatic solids is known to occur mainly through graphene edges). Compared to fullerenes, the ratio of "weak" sites (C−C bonds involved in heterocycles) over strong sites (C−C bonds between regular hexagons) is only slightly different from 0 for ideally perfect tubes (as opposed to 1 in C_{60} fullerenes – C_{60} molecules have 12 pentagons (therefore accounting for $5 \times 12 = 30$ C−C bonds) and 20 hexagons, each of them having three C−C bonds not involved in an adjacent pentagon but shared with a neighboring hexagon (therefore accounting for $20 \times 3 \times 1/2 = 30$ C−C bonds involved in hexagons only). Although graphene faces are chemically relatively inert, the radius of curvature imposed on the graphene in nanotubes enforces the three formerly planar C−C bonds characteristic of the genuine sp^2 hybridization to accept distortions in bond angles that make them closer and closer to the situation of three of the four C−C bonds in diamond (characteristic of the genuine sp^3 hybridization) as the radius of curvature decreases. Even though it is not enough to bring a genuine chemical reactivity to the carbon atoms, a consequence is to consider that either nesting sites are created at the concave surface, or strong physisorption sites are created above each carbon atom of the convex surface, both with a bonding efficiency that increases as the nanotube diameter decreases.

As already pointed out in Sect. 3.1, the chemical reactivity of SWNTs (and c-MNWTs) is supposed to come mainly from the caps, since they contain six pentagons each, as opposed to the tube body, which supposedly contains hexagons only. Indeed, oxidizing treatments applied to carbon nanotubes (air oxidation, wet-chemistry oxidation) are known to open the nanotube tips rather selectively [3.162]. But the ability of SWNTs to be opened by oxidation methods then filled with foreign molecules such as fullerenes (see Sect. 3.5) reveals the occurrence of side defects [3.14], whose identification and occurrence were discussed then proposed to be an average of one Stone–Wales defect every 5 nm along the tube length, involving an amount of about 2% of the carbon atoms for the example of a regular (10,10) SWNT [3.163] – a Stone–Wales defect is made of four adjacent heterocycles, two pentagons and two heptagons, displayed by pairs in opposition. Such a defect actually allows localized double bonds to be formed between the carbon atoms involved in the defect (instead of the electrons participating to the delocalized electron cloud above the graphene as usual, enhancing the chemical reactivity, e.g. toward chlorocarbenes [3.163]). This means that the overall chemical reactivity of carbon nanotubesshould strongly depend on the way they are synthesized. For example, SWNTs prepared by the arc-discharge method are supposed to contain fewer structural defects compared to CCVD-synthesized SWNTs, which have a higher chemical reactivity. Of course, the reactivity of h-MWNT-type nanotubes is intrinsically higher, due to the occurrence of accessible graphene edges at the nanotube surface.

3.5 Carbon Nanotube-Based Nano-Objects

3.5.1 Hetero-Nanotubes

Hetero-nanotubes are defined here as carbon nanotubes whose carbon atoms, part or all of them, are substituted with another element while the overall honeycomb lattice-based graphene structure is maintained. Elements involved are, therefore, boron and/or nitrogen. Benefits can be new behaviors (e.g., BN nanotubes are electrical insulators), improved properties (e.g., regarding the resistance to oxidation), or better control of the properties. One current challenge is actually to control the processing so that the desired SWNT structure is selectively synthesized (metallic or semiconductor). In this regard, it was demonstrated that substituting some C atoms by N or B atoms lead to SWNTs whose electrical behavior is systematically metallic [3.164, 165].

Some examples of hetero-nanotubes, though mainly MWNTs, are found in literature, whose hetero-atom involved is generally nitrogen due to the ease of introducing some gaseous or solid nitrogen and/or boron containing species (e.g., N_2, NH_3, BN, HfB_2) in existing MWNT synthesis devices [3.164, 166] until complete substitution of carbon [3.167, 168]. An amazing result of such attempts to synthesize hetero-MWNTs was the subsequent formation of "multilayered" c-MWNTs, i. e. MWNTs whose constituting coaxial tubes were alternatively made of carbon graphenes and boron nitride graphenes [3.169]. On the other hand, examples of hetero-SWNTs are few. But the synthesis of B- or N-containing SWNTs has been recently reported [3.165, 170].

3.5.2 Hybrid Carbon Nanotubes

Hybrid carbon nanotubes are here defined as carbon nanotubes, SWNTs or MWNTs, whose inner cavity has been filled, partially or entirely, by foreign atoms, molecules, compounds, or crystals. Such hybrids are thereby noted as X@SWNT (or X@MWNT, if appropriate, where X is the atom, molecule, etc. involved) [3.171].

Motivation

Why filling carbon nanotubes? The very small inner cavity of nanotubes is an amazing tool to prepare and study the properties of confined nanostructures of any nature, such as salts, metals, oxides, gases, or even discrete molecules like C_{60}, for example. Because of the almost one-dimensional structure of carbon nanotubes (specifically considering SWNTs), different physical and/or chemical properties can be expected for the encapsulated foreign materials or possibly for the overall hybrid materials. Indeed, when the volume available inside a carbon nanotube is small enough, the foreign material can be made mainly of "surface atoms" of reduced coordinence. The overall former motivation was then to tentatively form metal nanowires likely to be interesting for electronic applications (quantum wires). Metals were the preferred fillers, and nanotubes were mostly considered merely as nano-molds, likely to be removed afterward. But it is likely that the removal of the SWNT "container" to liberate the confined one-dimensional structure may destroy or at least transform it because of the stabilizing effect brought by the interactions with the nanotube wall.

Filling nanotubes while they grow (in situ filling) was one of the pioneering methods. In most cases, however, the filling step is separated from the step of synthesis of the nanotubes. Three methods can then be distinguished: (a) wet chemistry procedures, and physical procedures, involving (b) capillarity filling by a molten material or (c) filling by a sublimated material.

Generally speaking, the estimation of the filling rate is problematic and most of the time is taken merely from the TEM observation, without any statistics on the number of tubes observed. Moreover, as far as SWNTs are concerned, the fact that the nanotubes are gathered into bundles makes difficult the observation of the exact number of filled tubes, as well as the estimation of the filled length for each tube.

Filling carbon nanotubes with materials that could not have been introduced directly is also possible. This can be accomplished by first filling the nanotubes with an appropriate precursor (i. e., able to sublime, or melt, or solubilize) that will later be transformed by chemical reaction or by a physical interaction such as electron beam irradiation, for example [3.172]. In the case of secondary chemical transformation, the reduction by H_2 is the most frequently used to obtain nanotubes filled with metals [3.173]. Sulfides can be obtained as well by using H_2S as reducing agent [3.173].

Because the inner diameter of SWNTs is generally smaller than that of MWNTs, it is more difficult to fill them, and the driving forces involved in the phenomena are not yet totally understood (see the review paper by [3.163]). Consequently, more than one hundred articles dealing with filled MWNTs were already published by the time this chapter was written. We, therefore, did not attempt to review all the works. We chose to cite the pioneering works, then to focus on the more recent works dealing with the more challenging topic of filling SWNTs.

In Situ Filling Method

Most of the first hybrid carbon nanotubes ever synthesized were directly obtained in the course of their processing. This concerns only MWNTs, prepared by the arc-discharge method in its early hours, for which the filling materials were easily introduced in the system by drilling a central hole in the anode and filling it with the heteroelements. First attempts were all reported the same year [3.174–178], with elements such as Pb, Bi, YC_2, TiC, etc. Later on, *Loiseau* et al. [3.179] showed that MWNTs could also be filled to several μm in length by elements such as Se, Sb, S, and Ge, but only as nanoparticles with elements such as Bi, B, Al and Te. Sulfur was suggested to play an important role during the in situ formation of filled MWNTs by arc-discharge [3.180]. This technique is no longer the preferred one because of the difficulty in controling the filling ratio and yield and in achieving mass production.

Wet Chemistry Filling Method

The wet chemistry method requires the opening of the nanotubes tips by chemical oxidation prior to the filling step. This is generally obtained by refluxing the nanotubes in diluted nitric acid [3.181–183], although other oxidizing liquid media may work as well, e.g. [$HCl + CrO_3$] [3.184] or chlorocarbenes formed from the photolytic dissociation of $CHCl_3$ [3.163] as a rare example of a nonacidic liquid route for opening SWNTs. If a dissolved form (salt, oxide, etc.) of the desired metal is meanwhile introduced during the opening step, some of it will get inside the nanotubes. An annealing treatment (after washing and drying of the treated nanotubes) may then lead to the oxide or to the metal, depending on the annealing atmosphere [3.162]. Although the wet chemistry method looked promising because a wide variety of materials can be introduced within nanotubes this way, and because it operates at temperature not much different from room temperature, attention has to be paid to the oxidation method that has to be carried out as the first steps. The damages brought to the nanotubes by severe treatments (e.g., nitric acid) make them improper for use with SWNTs. Moreover, the filling yield is not very important, as a probable consequence of the presence of the solvent molecules that also enter the tube cavity, and the subsequent filled lengths rarely exceed 100 nm. Recent results have been obtained, however, using wet chemistry filling by *Mittal* et al. [3.184] who have filled SWNTs ropes by CrO_3 with an average yield of $\sim 20\%$.

Molten State Filling Method

The physical method involving a liquid (molten) phase is more restrictive, first, because some materials may start decomposing when they melt, and second because the melting point has to be compatible with the nanotubes, i. e. thermal treatment temperature should remain below the temperature of transformation or damaging of the nanotubes. Because the filling is occurring by capillarity, the surface tension threshold of the molten material was found to be 100–200 N/cm^2 [3.185]. But this threshold was proposed for MWNTs, whose inner diameters (5–10 nm) are generally larger than those of SWNTs (1–2 nm). In a typical filling experiment, the MWNTs are closely mixed with the desired amount of filler by gentle grinding, and the mixture is then vacuum-sealed in a silica ampoule. The ampoule is then slowly heated to a temperature above the melting point of the filler, then is slowly cooled. The use of this method does not require any opening of the nanotubes prior to the heat treatment. The mechanism of nanotube opening is not yet established clearly, but it is certainly related to the chemical aggressivity of the molten materials toward carbon and more precisely toward graphene defects in the tube structure (see Sect. 3.4.4).

Most of the works on SWNTs involving this method were performed by Sloan and coworkers at Oxford University [3.187], although other groups followed the same procedure [3.86, 181, 183]. Precursors to fill the nanotubes were mainly metals halides. Although little is known yet about the physical properties of halides crystallized within carbon nanotubes, the crystallization itself of molten salts within small diameter SWNTs has been studied in detail, revealing a strong interaction between the one-dimensional crystals and the surrounding graphene wall. For example, *Sloan* et al. [3.188] have described two-layer 4 : 4 coordinated KI crystals formed within SWNTs of around 1.4 nm diameter. These two-layer crystals are "all surface" and have no "internal" atoms. Significant lattice distortions occur compared to the bulk structure of KI, where the normal coordinence is 6 : 6 (meaning that each ion is surrounded by six identical closer neighbors). Indeed, the distance between two ions across the SWNT capillary is 1.4 times the same distance along the tube axis. This denotes an accommodation of the KI crystal to the confined space provided by the inner nanotube cavity in the constrained crystal direction (across the tube axis). This implies that the interactions between the ions and the surrounding carbon atoms are strong. The volume dimension available within the nanotubes thus somehow controls the crystal structure of inserted materials. For instance, the structure and orientation of encapsulated PbI_2 crystals inside their capillaries were found to arrange in different ways inside SWNTs and DWNTs, with respect to the diameter of the confining nanotubes [3.186]. In the case of SWNTs, most of the obtained encapsulated one-dimensional PbI_2 crystals exhibit a strong preferred orientation with their (110) planes aligning at an angle of ca. 60° to the SWNT axes as shown in Fig. 3.24a and 3.24b. Due to the extremely small diameter of the nanotube capillaries, individual crystallites are often only a few polyhedral layers thick, as outlined in Fig. 3.24d to 3.24h. As a result of lattice terminations enforced by capillary confinement, the edging polyhedra must be of reduced coordination, as indicated in Fig. 3.24g and 3.24h. In the case of PbI_2 formed inside DWNTs, similar crystal growth behavior was generally observed to occur in narrow nanotubes with diameters comparable to that of SWNTs. As the diameter of the encapsulating capillaries increases, however, different preferred orientations are frequently observed (Fig. 3.25). In this example, the PbI_2 crystal is oriented with the [121] direction parallel to the direction of the electron beam (Fig. 3.25a to 3.25d). If the PbI_2@DWNT hybrid is viewed "side-on" (as indicated by the arrow in Fig. 3.25e) polyhedral slabs are seen to arrange along the capillary, oriented at an angle of ca. 45° with respect to the tubule axis.

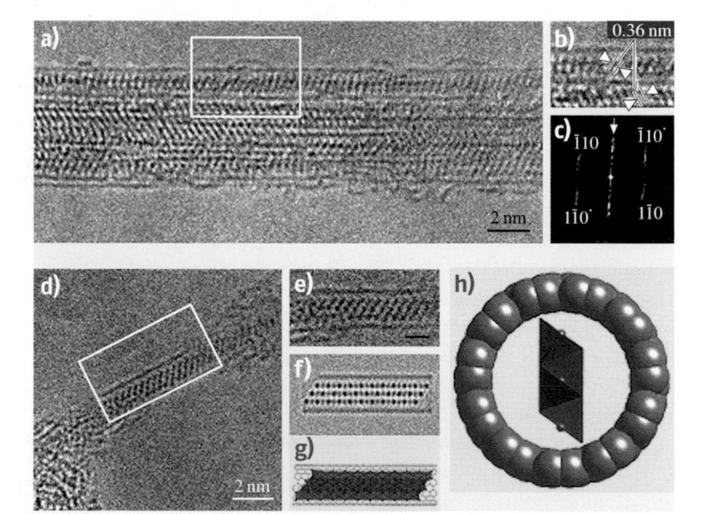

Fig. 3.24 HRTEM images and corresponding structural model for PbI_2 filled SWNTs (from [3.186])

Sublimation Filling Method

This method is even more restrictive than the previous one, since it is applicable only to a very limited number of compounds due to the requirement for the filling materials to sublimate within the temperature range of thermal stability of the nanotubes. Examples are therefore few. Actually, except for an attempt to fill SWNTs with $ZrCl_4$ [3.189], the only example published so far is the formation of C_{60}@SWNT (whose popular nickname is "peapods"), reported for the first time in 1998 [3.190], where regular ∼ 1.4 nm large SWNTs are filled with C_{60} fullerene molecule chains (Fig. 3.26a). Of course, the process requires a pre-opening of the SWNTs by some method, as discussed previously, typically either acid attack [3.191] or heat-treatment in air [3.192]. Then the opened SWNTs are put into a sealed glass tube together with fullerene powder and placed into a furnace heated above the sublimation temperature for fullerite (>∼ 350 °C). Since there is no limitation related to Laplace's law or the presence of solvent since only gaseous molecules are involved, filling efficiency may actually reach ∼ 100% [3.192].

Remarkable behaviors have been revealed since then, such as the ability of the C_{60} molecules to move freely within the SWNT cavity (Fig. 3.26b and 3.26c) upon random ionization effects from electron irradiation [3.193], or to coalesce into 0.7 nm large inner elongated capsules upon electron irradiation [3.194], or into a 0.7 nm large nanotube upon subsequent thermal treatment above 1,200 °C [3.193, 195]. Annealing of C_{60}@SWNT materials may thus appear as an efficient way to produce DWNTs with nearly constant inner (∼ 0.7 nm) and outer (∼ 1.4 nm) diameters. DWNTs produced this way are actually the smallest MWNTs synthesized so far.

Based on the successful synthesis of so-called endofullerenes previously achieved [3.13], further development of the process has recently led to the synthesis of more complex nanotube-based hybrid materials such as La_2@C_{80}@SWNTs [3.196], Gd@C_{82}@SWNTs [3.197], or $Er_xSc_{3-x}N$@C_{80}@SWNT [3.198] among other examples. These provide even more extended perspective regarding the potential applications of peapods, which are still being investigated.

Finally the last example to cite is the successful attempt to produce peapods by a related method, by using accelerated fullerene ions (instead of neutral gaseous molecules) to enforce the fullerenes to enter the SWNT structure [3.199].

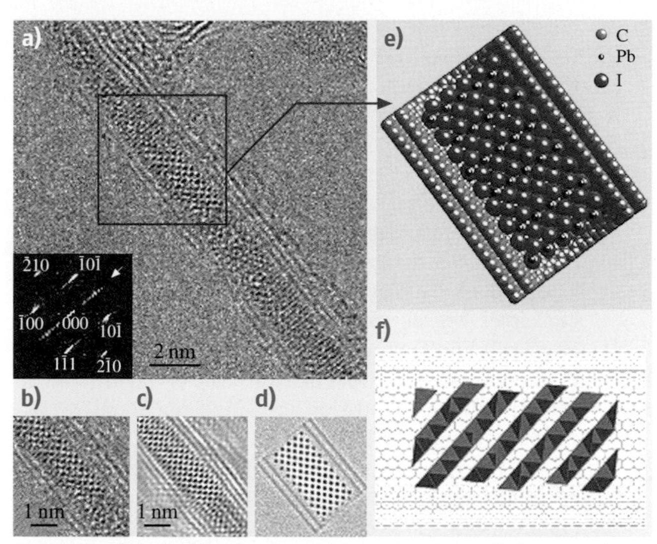

Fig. 3.25 HRTEM images (experimental and simulations), and corresponding structural model for a PbI_2 filled double-wall carbon nanotube (from [3.186])

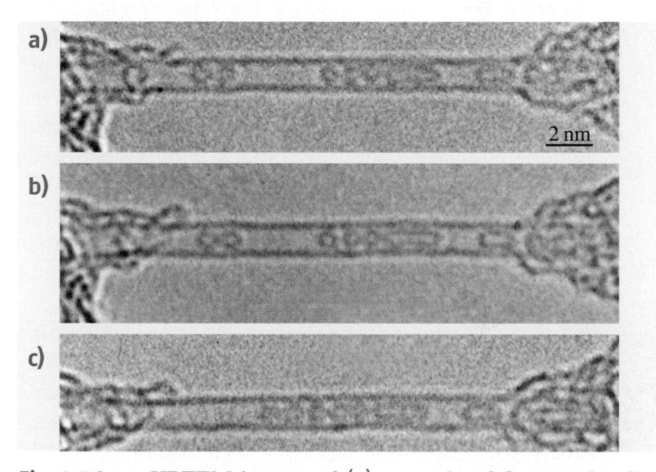

Fig. 3.26a–c HRTEM images of (a) example of five regular C_{60} molecules encapsulated together with two higher fullerenes (C_{120} and C_{180}) as distorted capsules (*on the right*) within a regular 1.4 nm diameter SWNT. (a)–(c) Example of the diffusion of the C_{60}-molecules along the SWNT cavity. The time between each image of the sequence is about 10 s. The fact that nothing occurs between (a) and (b) reveals the randomness of the ionisation events generated by the microscope electron beam and assumed to be responsible for the molecule displacement

3.5.3 Functionalized Nanotubes

Based on the reactivity of carbon nanotubes as discussed in Sect. 3.4.6, the functionalization reactions can be divided into two main groups. One is based on the chemical oxidation of the nanotubes (tips, structural defects) leading to carboxylic, carbonyl, and/or hydroxyl functions. These functions are then used for additional reactions, to attach oligomeric or polymeric functional entities. The second group is based on direct addition to the graphitic-like surface of the nanotubes (i. e. without any intermediate step). Examples of the latter reactions include oxidation or fluorination (an important first step for further functionalization with other organic groups). Properties and applications of functionalized nanotubes have been reviewed recently [3.200].

Oxidation of Carbon Nanotubes

Carbon nanotubes are often oxidized and therefore opened before chemical functionalization to increase their chemical reactivity (creation of dangling bonds). Chemical oxidation of nanotubes is mainly achieved using either wet chemistry or gaseous oxidants such as oxygen (typically air) or CO_2. Depending on the synthesis process, the oxidation resistance of the nanotubes can vary. When oxidation is carried out through gas phases, thermo-gravimetric analysis (TGA) is of great use to determine at which temperature the treatment should be applied. It is important to know that TGA accuracy increases as the heating rate diminishes, while the literature often provides TGA analyses obtained in nonoptimized conditions, leading to overestimated oxidation temperatures. The presence of catalyst remnants (metals, more rarely oxides), the type of nanotubes (SWNTs, c-MWNTs, h-MWNTs), and the oxidizing agent (air, O_2 is an inert gas, CO_2, etc.) as well as the flow rate used make difficult the comparison of published results. It is generally agreed, however, that amorphous carbon burns first, followed by SWNTs and then multiwall materials (shells, MWNTs), even if TGA is often unfortunately not able to clearly separate the different oxidation steps. The more common method used for nanotube oxidation is that of aqueous solutions of oxidizing reagents. The main one is nitric acid, either concentrated or diluted (around 3 moles per liter in most cases), but oxidants such as potassium dichromate ($K_2Cr_2O_7$), hydrogen peroxide (H_2O_2), or potassium permanganate ($KMnO_4$) are very often used as well. HCl, like HF, do not damage nanotubes because these acids are not oxidizing.

Functionalization of Oxidized Carbon Nanotubes

Carboxylic groups located at the nanotube tips can be coupled with different chemical groups. Oxidized nanotubes are usually reacted with thionyl chloride ($SOCl_2$) to generate the acyl chloride, even if a direct reaction is theoretically possible with alcohols or amines, for example. Reaction of SWNTs with octadecylamine (ODA) was reported by *Chen* et al. [3.201] after reaction of oxidized SWNTs with $SOCl_2$. The functionalized SWNTs have substantial solubility in chloroform ($CHCl_3$), dichloromethane (CH_2Cl_2), aromatic solvents, and carbon bisulfide (CS_2). Many other reactions between functionalized nanotubes (after reaction with $SOCl_2$) and amines have been reported in the literature and will not be reviewed here. Noncovalent reactions between the carboxylic groups of oxidized nanotubes and octadecylammonium ions are possible [3.202], providing solubility in tetrahydrofuran (THF) and CH_2Cl_2. Functionalization by glucosamine using similar procedures [3.203] allowed the water solubilization of SWNTs, which is of special interest when considering biological applications of functionalized nanotubes. Functionalization with lipophilic and hydrophilic dendra (with long alkyl chains and oligomeric poly(ethyleneglycol) groups) has been achieved by amination and esterification reactions [3.204], leading to solubility of the functionalized nanotubes in hexane, chloroform, and water. It is interesting to note that in the latter case, the functional groups could be removed just by modifying the pH of the solution (base- and acid-catalyzed hydrolysis reaction conditions, [3.205]). A last example is the possible interconnection of nanotubes via chemical functionalization. This has been recently achieved by *Chiu* et al. [3.206] using the already detailed acyl chloride way and a bifunctionalized amine to link the nanotubes by the formation of amide bonds.

Sidewall Functionalization of Carbon Nanotubes

The covalent functionalization of the nanotube walls is possible by means of fluorination reactions. It was first reported by *Mickelson* et al. [3.207], with F_2 gas (the nanotubes can then be defluorinated, if required, with anhydrous hydrazine). As recently reviewed by *Khabashesku* et al. [3.208], it is then possible to use these fluorinated nanotubes to carry out subsequent derivatization reactions. Thus, sidewall-alkylated nanotubes can be prepared by nucleophilic substitution (Grignard synthesis or reaction with alkyllithium precursors [3.209]). These alkyl sidewall groups can be removed by air oxidation. Electrochemical addition of aryl radicals (from the reduction of aryl diazonium salts) to nanotubes has also been reported by *Bahr* et al. [3.210]. Functionalization of the nanotube external wall by cycloaddition of nitrenes, addition of nuclephilic carbenes, or addition of radicals has been described by *Holzinger* et al. [3.211]. Electrophilic addition of dichlorocarbene to SWNTs occurs by reaction with the deactivated double bonds in the nanotube wall [3.212]. Silanization reactions are another way to functionalize nanotubes, although only tested with MWNTs. *Velasco-Santos* et al. [3.213] have reacted oxidized MWNTs with an organosilane ($RsiR'_3$, with R being an organo-functional group attached to silicon) and obtained organo-functional groups attached to the nanotubes by silanol groups.

Noncovalent sidewall functionalization of nanotubes is important because the covalent bonds are associated with changes from sp^2 to sp^3 in the hybridization of the carbon atoms involved, which corresponds to a loss of the graphite-like character. As a consequence, physical properties of functionalized nanotubes, specifically SWNTs, may be modified. One possibility for achieving noncovalent functionalization of nanotubes is to wrap the nanotubes in a polymer [3.214], which allows their solubilization (thus improving the processing possibilities) while preserving their physical properties.

Finally, it is worth keeping in mind that all these chemical reactions are not specific to nanotubes and may occur to most of the carbonaceous impurities present in the raw materials. This makes difficult the characterization of the functionalized samples and thus requires one to perform the experiments starting with very pure carbon nanotube samples, which is unfortunately not always the case for the results reported in the literature. On the other hand, purifying the nanotubes to start with may also bias the functionalization experiments since purification requires chemical treatments. But the demand for such products does exist already, and purified then fluorinated SWNTs can be found on the market at $ 900/g (Carbon Nanotechnologies Inc., 2003).

3.6 Applications of Carbon Nanotubes

Carbon nanotubes can be inert and can have a high aspect ratio, high tensile strength, low mass density, high heat conductivity, large surface area, and versatile electronic behavior including high electron conductivity. While these are the main characteristics of the properties for individual nanotubes, a large number of them can form secondary structures such as ropes, fibers, papers, thin films with aligned tubes, etc., with their own specific properties. The combination of these properties makes them ideal candidates for a large number of applications provided their cost is sufficiently low. The cost of carbon nanotubes depends strongly on both the quality and the production process. High quality single shell carbon nanotubes can cost 50–100 times more than gold. But synthesis of carbon nanotubes is constantly improving, and sale prizes fall by 50% per year. Application of carbon nanotubes is therefore a very fast moving area with new potential applications found every year, even several times per year. Making an exhaustive list is again a challenge we will not take. Below are listed applications considered "current" (Sect. 3.6.1) either because already marketed, or because up-scaling synthesis is being processed, or because prototypes are currently developed by profit-based companies. Other applications are said to be "expected" (Sect. 3.6.2).

3.6.1 Current Applications

High mechanical strength of carbon nanotubes makes them a near to ideal force sensor in scanning probe microscopy (SPM) with a higher durability and the ability to image surfaces with a high lateral resolution, the latter being a typical limitation of conventional force sensors (ceramic tips). The idea was first proposed and experimented by Dai et al. [3.77] using c-MWNTs. It was extended to SWNTs by *Hafner* et al. [3.215], since small diameters of SWNTs were supposed to bring higher resolution than MWNTs due to the extremely short radius of curvature of the tube end. But commercial nanotube-based tips (e.g., Piezomax, Middleton, WI, USA) use MWNTs for processing convenience. It is also likely that the flexural modulus of SWNTs is too low and therefore induces artifacts affecting the lateral resolution when scanning a rough surface. On the other hand, the flexural modulus of c-MWNTs is supposed to increase with the number of walls, but the radius of curvature of the tip meanwhile increases.

Near-Field Microscope Probes

SWNT or MWNT, such nanotube-based SPM tips also offer the perspective of being functionalized, in the prospect of making selective images based on chemical discrimination by "chemical force microscopy" (CFM). Chemical function imaging using functionalized nanotubes represents a huge step forward in CFM because the tip can be functionalized very accurately (ideally at the very nanotube tip only, where the reactivity is the highest), increasing the spatial resolution. The interaction between chemical species present at the end of the nanotube tip and a surface containing chemical functions can be recorded with great sensitivity, allowing the chemical mapping of molecules [3.216, 217].

Today the cost of nanotube-based SPM tips is in the range of $ 200/tip. Such a high cost is explained both

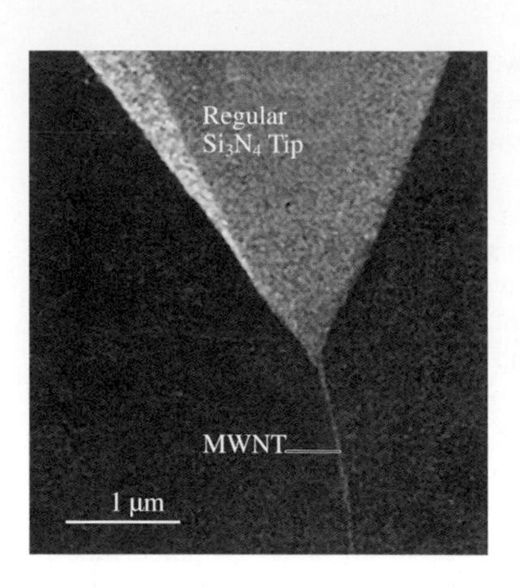

Fig. 3.27 Scanning electron microscopy image of a carbon nanotube (MWNT) mounted onto a regular ceramic tip as a probe for atomic force microscopy

by the processing difficulty, which ideally requires one to grow or mount a single MWNT in the appropriate direction at the tip of a regular SPM probe (Fig. 3.27), and by the need to individually control the tip quality. Such a market has been estimated at ~ $20 M/year.

Field Emission-Based Devices

Based on a pioneering work by *de Heer* et al. [3.218], carbon nanotubes have been demonstrated to be efficient field emitters and are currently being incorporated in several applications, including flat panel display for television sets or computers (whose a first prototype was exhibited by Samsung in 1999) or any devices requiring an electron producing cathode, such as X-ray sources. The principle of a field-emission-based screen is demonstrated in Fig. 3.28a. Briefly, a potential difference is brought between emitting tips and an extraction grid so that electrons are taking out from the tips and sent onto an electron sensitive screen layer. Replacing the glass support and protection of the screen by some polymer-based material will even allow the develope of flexible screens. As opposed to regular (metallic) electron emitting tips, the structural perfection of carbon nanotubes allows higher electron emission stability, higher mechanical resistance, and longer life time. First of all, it allows energy savings since it needs lower (or no) heating temperature of the tips and requires much lower threshold voltage. As an illustration for the latter, producing a current density of 1 mA/cm^2 is possible for a threshold voltage of 3 V/µm with nanotubes, while it requires 20 V/µm for graphite powder and 100 V/µm for regular Mo or Si tips. The subsequent reductions in cost and energy consumption are estimated at 1/3 and 1/10 respectively. Generally speaking, the maximum current density obtainable ranges between 10^6 A/cm^2 and 10^8 A/cm^2 depending on the nanotubes involved (e.g., SWNT or MWNT, opened or capped, aligned or not) [3.219–221]. Although nanotube side-walls seem to emit as well as nanotube tips, many works have dealt (and are still dealing) with growing nanotubes perpendicular to the substrate surface as regular arrays (Fig. 3.28b). Besides, using SWNTs instead of MWNTs does not appear necessary for many of these applications when they are used as bunches. On the other hand, when considering single, isolated nanotubes, SWNTs are generally less preferable since they allow much lower electron doses than do MWNTs, although they are likely meanwhile to provide an even higher coherent source (an useful feature for devices such as electron microscopes or X-ray generators).

The market associated with this application is huge. With such major companies involved as Motorola, NEC, NKK, Samsung, Thales, Toshiba, etc., the first commercial flat TV sets and computers using nanotube-based screens are about to be seen in stores, and companies such as Oxford Instruments and Medirad are about to commercialize miniature X-ray generators for medical applications on the basis of nanotube-based cold cathodes developed by Applied Nanotech Inc.

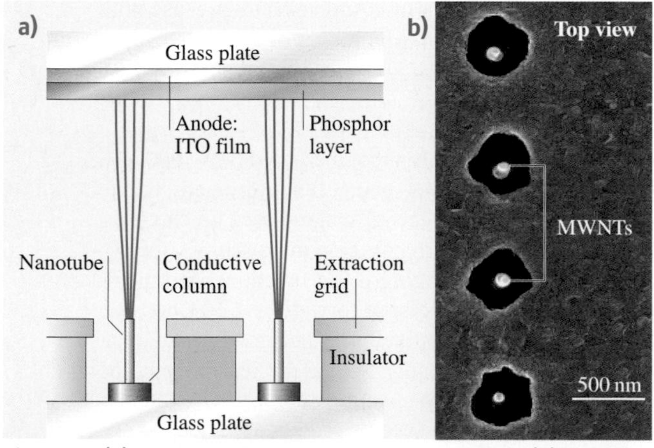

Fig. 3.28 (**a**) Principle of a field-emitter-based screen. (**b**) Scanning electron microscope image of a nanotube-based emitter system (*top view*). *Round dots* are MWNT tips seen through the holes corresponding to the extraction grid. By courtesy of *Legagneux* (Thales Research & Technology, Orsay, USA)

Chemical Sensors

The electrical conductance of semiconductor SWNTs was recently demonstrated to be highly sensitive to the change in the chemical composition of the surrounding atmosphere at room temperature, due to the charges transfer between the nanotubes and the molecules from the gases adsorbed onto the SWNT surface. It has also been shown that there is a linear dependence between the concentration of the adsorbed gas and the difference in electrical properties, and that the adsorption is reversible. First tries involved NO_2 or NH_3 [3.222] and O_2 [3.223]. SWNT-based chemical NO_2 and NH_3 sensors are characterized by extremely short response time (Fig. 3.29), thus being different from conventionally used sensors [3.222, 224]. High sensitivity toward water or ammonia vapors has been measured on SWNT-SiO_2 composite [3.225]. This study indicated the presence of p-type SWNTs dispersed among the predominant metallic SWNTs, and that chemisorption of gases on the surface of the semiconductor SWNTs is responsible for the sensing action. The determination of CO_2 and O_2 concentrations on SWNT-SiO_2 composite has also been reported [3.226]. By doping nanotubes with palladium nanoparticles, *Kong* et al. [3.227] have also shown that

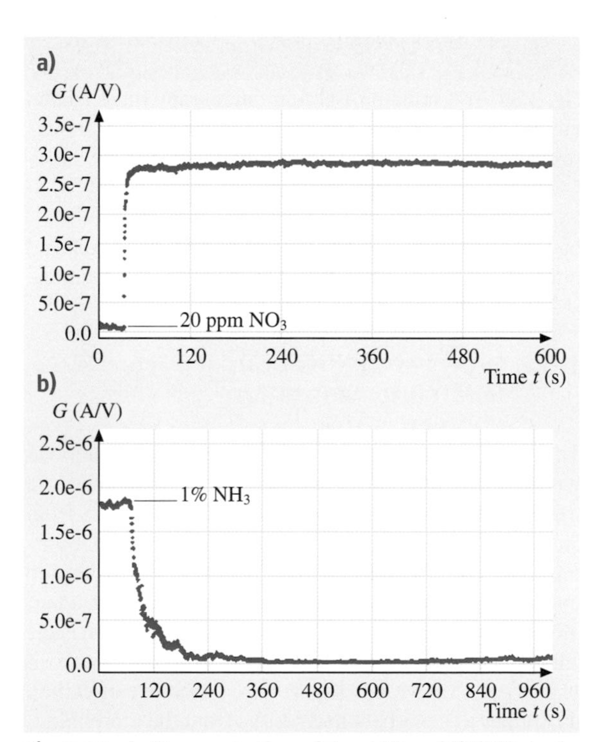

Fig. 3.29a,b Demonstration of the ability of SWNTs in detecting molecule traces in inert gases. (**a**) Increase in a single SWNT conductance when 20 ppm of NO_2 are added to an argon gas flow. (**b**) Same with 1% NH_3 added to the argon gas flow (from [3.222])

the modified material could reveal the presence of hydrogen up to 400 ppm whereas the as-grown material was totally ineffective.

Generally speaking, the sensitivity of the new nanotube-based sensors is three orders of magnitude higher than that of standard solid state devices. In addition, the interest in using nanotubes as opposed to current sensors is the simplicity and the very small size of the system in which they can be placed, the fact that they can operate at room temperature, and their selectivity, which allows a limited number of sensor device architectures to be built for a variety of industrial purposes, while the current technology requires a large diversity of devices based on mixed metal oxide, opto-mechanics, catalytic beads, electrochemistry, etc. The market opportunity is expected to be $ 1.6 billion by 2006, including both the biological area and the chemical industrial area. Nanotube-based sensors are currently developed in both large and small companies, such as Nanomix (USA), for example.

Catalyst Support

Carbon materials are attractive supports in heterogeneous catalytic processes due to their ability to be tailored to specific needs: indeed, activated carbons are already currently employed as catalyst supports because of their high surface area, their stability at high temperatures (under not oxidizing atmosphere), and the possibility of controlling both their porous structure and the chemical nature of their surface [3.228, 229]. The attention has been brought to nano-sized fibrous morphologies of carbon that appeared during the last decade, showing novel interesting potentialities as supports [3.230]. Carbon nanofibers (also improperly called graphite nanofibers) and carbon nanotubes have been used successfully and shown to present, as catalyst-supporting materials, properties superior to that of such other regular catalyst-supports as activated carbon, soot, or graphite [3.231]. The use of graphite nanofibers as a direct catalyst for oxidative dehydrogenation has also been reported [3.232].

The morphology and size of carbon nanotubes, especially since they do present huge lengths vs. diameters (= aspect ratio), can play a significant role in catalytic applications due to their ability to disperse catalytically active metal particles. Their electronic properties are also of primary importance [3.233] since this conductive support may be exerting electronic perturbation as well as geometric constraint on the dispersed metal particles. A recent comparison between the interaction of transition metal atoms with carbon nanotube walls and that with graphite indicates major differences in bonding sites, magnetic moments, and charge-transfer direction [3.234]. Thus the possibility of strong metal-support interaction has to be taken into account. Their mechanical strength is also important and makes them resistant to attrition in view of recycling. Their external or internal surface presents strong hydrophobicity and strong adsorption ability toward organic molecules. Finally, for MWNT-based catalyst-supports, the relatively high surface area and the absence of microporosity (i. e., pores < 2 nm) associated with a high meso- and macropore volume (see Sect. 3.4.3) result in significant improvements of the catalytic activity, in the case of liquid-phase reactions when compared to catalysts supported on activated carbon. With nanotube supports, the mass transfer limitation of the reactants to the active sites is actually prevented due to the absence of microporosity, and the apparent contact time of the products with the catalyst is diminished, leading to more active and more selective catalytic effects.

The most widely used technique to prepare carbon nanotube-supported catalysts is the incipient wetness impregnation, in which the purified support is impregnated with a solution of the metal precursor, then dried, calcinated and/or reduced in order to obtain metal particles dispersed on the support. The chemical treatment and/or the modification of the carbon nanotube surface was found to be a useful tool to control its hydrophobic or hydrophilic character [3.236]. A strong metal/support interaction can thus be expected from the occurrence of functionalized groups created by the oxidation of the support surface, resulting in smaller particle sizes [3.237]. A more sophisticated technique to achieve the grafting of metal particles onto carbon nanotubes consists of functionalizing the outer surface of the tubes and then performing a chemical reaction with a metal complex, thus resulting in a good dispersion of the metallic particles (Fig. 3.30) [3.235].

Fig. 3.30 Transmission electron microscopy image showing rhodium nanoparticles supported on MWNT surface (from [3.235])

Selected examples of some carbon nanotube-based catalysts together with the related preparation routes and catalytic activities are listed in Table 3.4.

The market is huge for this application since it most often concerns the "heavy" chemical industry. It implies and requires mass production of low-cost nanotubes, processed by methods other than solid-carbon-source-based methods (see Sect. 3.2.1). Such an application also requires some surface reactivity, making the h-MWNT type nanotubes with a poor nanotexture (see Sect. 3.1.2) the most suitable starting material for preparing such catalyst supports. Catalysis-enhanced thermal cracking of gaseous carbon precursors is therefore preferred, and pilot plants are already being built by major chemical industrial companies (e.g., Ato-Fina, France).

3.6.2 Expected Applications Related to Adsorption

The adsorption and interaction of gases, liquids, or metal with carbon nanotubes has attracted much attention in the past several years. The applications resulting from the adsorption properties of carbon nanotubes can be arbitrarily divided into two groups. The first group is based on the consequences of molecule adsorption on nanotube electronic properties, whose main application is the chemical sensors (see Sect. 3.6.1). The second group includes gas storage, gas separation, or the use of carbon nanotubes as adsorbent and results from the morphological characteristics of carbon nanotubes (surface area, aspect ratio, and so forth). Among the three latter potential applications, the possibility of storing gases, and more particularly hydrogen, on carbon nanotubes has received most of the attention.

Table 3.4 Preparation and catalytic performances of some nanotube-supported catalysts

Catalyst	Preparation route	Catalytic reaction	Comments
Ru/MWNT+SWNT [3.230]	Liquid phase impregnation, no pretreatment of the tubes	Liquid phase cinnamaldehyde hydrogenation	Different kind of metal support interaction compared to activated carbon
Pt/MWNT electrodes [3.149]	Electrode-less plating with pre-functionalization of MWNT	Oxygen reduction for fuel cells applications	High electrocatalytic activity
Rh/MWNT [3.235]	Surface mediated organometallic synthesis, pre-functionalization of MWNT	Liquid phase hydroformylation and hydrogenation	Higher activity of Rh/MWNT compared to Rh/activated carbon
Ru-alkali/MWNT [3.238]	Liquid phase impregnation, no pretreatment of the tubes	Ammonia synthesis, gas phase reaction	Higher activity with MWNT than with graphite
Rh-phosphine/MWNT [3.239]	Liquid phase grafting from [RhH(CO)(PPh$_3$)$_3$]	Liquid phase hydroformylation	Highly active and regioselective catalyst

Gas Storage – Hydrogen

The development of lightweight and safe system for hydrogen storage is necessary for a wide use of highly efficient H_2-air fuel cells in transportation vehicles. The U.S. Department of Energy Hydrogen Plan has set a standard by providing a commercially significant benchmark for the amount of reversible hydrogen adsorption. This benchmark requires a system-weight efficiency (the ratio of H_2 weight to system weight) of 6.5 wt% hydrogen and a volumetric density of 63 kgH_2/m^3.

From now on, the failure to produce a practical storage system for hydrogen has prevented hydrogen from coming to the commercial forefront as a transportation fuel. The ideal hydrogen storage system needs to be light, compact, relatively inexpensive, safe, easy to use, and reusable without the need for regeneration. While research and development are continuing on such technologies as liquid hydrogen systems, compressed hydrogen systems, metal hydride systems, and superactivated carbon systems, all have serious disadvantages. For example, liquid hydrogen systems are very expensive, primarily because the hydrogen must be cooled to about −252 °C. For example, a liquid hydrogen system will cost about four times more than an equivalent amount of gasoline. Further, liquid hydrogen must be kept cooled to prevent it from boiling away, even when the vehicle is parked.

Compressed hydrogen is much cheaper than liquid hydrogen but also much bulkier.

Metal hydride systems store hydrogen as a solid in combination with other materials. For example, metal hydrides are produced by bathing a metal, such as palladium or magnesium, in hydrogen. The metal splits the dihydrogen gas molecules and binds the hydrogen atoms to the metal until released by heating. The disadvantages of a metal hydride system are its weight (typically about eight times more than an equivalent amount of liquid hydrogen or an equivalent amount of gasoline) and the need to warm it up to release the hydrogen.

Superactivated carbon is the basis of another system for storing hydrogen that initially showed commercial potential. Superactivated carbon is a material similar to the highly porous activated carbon used in water filters but can gently hold hydrogen molecules by physisorption at sub-zero temperatures. The colder the carbon, the less heat is needed to disturb the weak forces holding the carbon and hydrogen together. Again, a major disadvantage of such a system is that preventing the hydrogen from escaping requires it to constantly remain at very low temperatures, even when the vehicle is parked.

Consequently there still remains a great need for a material that can store hydrogen and is light, compact, relatively inexpensive, safe, easy to use, and reusable without regeneration. Recently some articles or patents concerning the very high, reversible adsorption of hydrogen in carbon nanotubes or platelet nanofibers have aroused tremendous interest in the research community, stimulating much experimental and theoretical works. Most of the works on hydrogen adsorption on carbon nanotubes have been recently reviewed [3.240–243].

A group from Northeastern University [3.132, 244] was the first to report the supposedly successful hydrogen storage in carbon layered nanostructures possessing some crystallinity. The authors claimed that, in platelet nanofibers (3–50 nm in width), hydrogen can be stored up to 75 wt%, meaning a C/H ratio of 1/9. Complete hydrogen desorption occurs only at very high temperature. The same authors have recently pointed out the dramatic role of the treatment that has to be applied to the carbon nanofibers prior to H_2 adsorption [3.244]. This thermal treatment, though simple (1,000 °C in an inert gas), is a critical procedure in order to remove chemisorbed gases from the edge and step regions of the carbon structures. Failure to achieve this condition results in a dramatic decrease in the hydrogen adsorption performance of the materials. In this work, the authors have performed the experiments at room temperature, and 100 bar of H_2, hydrogen uptake between 20 and 40 wt% were recorded. The authors have also demonstrated by XRD measurements that structural perturbation of the material was produced following hydrogen treatment, and that it was manifested by an expansion of the graphitic interplane distance (from 0.34 before H_2 adsorption to 0.347 after H_2 uptake).

Hydrogen adsorption and desorption measurements on nanofibers similar to that used by *Rodriguez* et al. [3.19] were also performed by *Ahn* et al. [3.245]. The authors have measured a hydrogen uptake of 2.4 wt%. In fact, in their study, none of the graphite nanofiber material tested has indicated hydrogen storage abilities that exceed the values already reported for activated carbons.

Adsorption of H_2 on platelet nanofibers was computed by Monte Carlo simulations [3.246]. The graphene spacing was optimized to maximize the weight fraction of H_2 adsorbed. Given the results of their calculations, the authors concluded that no physically realistic graphite-hydrogen potential can account for the tremendous adsorption ability reported by the group from Penn State.

Rzepka et al. [3.247] have also used theoretical Monte Carlo calculation to calculate the amount of physically adsorbed hydrogen molecules in carbon slit pores of dimensions similar to that used by the Penn State group (i. e., 0.34 nm as the interplanar distance between graphenes in the nanofibers). They reached the conclusion that no hydrogen can be adsorbed at all. Even if the interplanar distance is assumed to expand during the adsorption process, the maximum calculated adsorption is 1 wt% for $d = 0.7$ nm. These authors concluded that the Penn State experimental results could not be attributed to some "abnormal capillary condensation effect" as initially claimed. Since then, nobody has been able to duplicate it experimentally.

Meanwhile, hydrogen storage experiments were also attempted on SWNTs. The first work was reported by *Dillon* et al. [3.248] who used temperature-programmed desorption measurements. A major drawback of their experimental procedure was that they used only 1 mg of *unpurified* soot containing 0.1% of SWNTs. They attributed the reversible hydrogen capacity of 0.01 wt% observed to the SWNTs only, therefore leading to a presumable capacity of 5–10 wt% for a pure nanotubes sample. TPD experiments shown H_2 desorption between −170 and +100 °C. Despite the weakness of the demonstration, the paper, which was published not much before that of Chambers et al. [3.132], has induced a worldwide excitement for the field.

Hydrogen adsorption measurements were also performed on SWNT samples of high purity [3.249]. At −193 °C, 160 bar of H_2 were admitted and the hydrogen adsorption was found to exceed 8 wt%. In this case, the authors proposed that hydrogen is first adsorbed on the outer surfaces of the crystalline material. But no data concerning H_2 desorption are given.

In recent reports, Dresselhaus and coworkers [3.240, 250] reported hydrogen storage in SWNTs (1.85 nm of average diameter) at room temperature. A hydrogen storage capacity of 4.2 wt% was achieved reproducibly under 100 bar for a SWNT-containing material of about 0.5 g that was previously soaked in hydrochloric acid and then heat-treated (500 °C) in vacuum. The purity of the sample was estimated from TGA and TEM observations to be about 50 to 60%. Moreover, 78% of the adsorbed hydrogen could be released under ambient pressure at room temperature, while the release of the residual stored hydrogen required some heating at 200 °C.

A valuable input to the topic was brought by *Hirscher* et al. [3.251], who demonstrated that several of the supposedly successful experiments regarding the storage of H_2 in SWNTs were actually misled by the hydrogen storage capacity of Ti nanoparticles originating from the sonoprobe frequently used at one step or another of the procedure, specifically when the SWNT material is previously purified.

Density functional theory (DFT) was used [3.252] to estimate H_2 adsorption in SWNTs. From their calculations, within the regime of operating conditions where adsorptive storage seems attractive, the storage properties of H_2 in a SWNT system appear to fall far short of the DOE target. In their model, rolling graphite into a cylindrical sheet does not significantly alter the nature of the carbon–H_2 interaction, in contradiction with the latest calculations, which indicate that the possibilities of physisorption increase as the radius of curvature of SWNTs decreases [3.253].

More recently, however, *Lee* et al. [3.254, 255] have also used DFT calculations to search for hydrogen adsorption sites and predict maximum storage capacity in SWNTs. They have found two chemisorption sites, at the exterior and the interior of the tube wall. Thus they predict a maximum hydrogen storage capacity that can exceed 14 wt% (160 kgH_2/m^3). The authors have also considered H_2 adsorption in multiwall carbon nanotubes and have predicted lower storage capacities for the latter.

Calculations are constrained, however, by the starting hypotheses. While considering the same (10,10) SWNT, although calculations based on DFT predict a 14.3% storage [3.254], calculations based on a geometrical model predict 3.3% [3.240], and calculations based on quantum mechanical molecular dynamics model predict 0.47% [3.256].

In conclusion, neither the experimental results, obviously biased by some problem in the procedures, nor theoretical results are yet able to demonstrate that an efficient storage of H_2 is achievable for carbon nanotubes, whatever the type. But the definitive failure statement cannot yet be claimed. Further efforts have to be made to enhance H_2 adsorption of these materials, in particular (i) by adjusting the surface properties that can be modified by chemical or mechanical treatments, and (ii) by adjusting the material structure (pore size and curvature). Whether the best carbon material will then be nanotube-based is another story.

Gas Storage – Gases Other than Hydrogen

Encouraged by the potential applications related to hydrogen adsorption, several research groups have tried to use carbon nanotubes as a stocking and transporting mean for other gases such as: oxygen, nitrogen, noble gases (argon and xenon), and hydrocarbons (methane, ethane, and ethylene). These studies have shown that carbon nanotubes could become the world's smallest gas cylinders combining low weight, easy transportability, and safe use with acceptable adsorbed quantities. Thanks to their nano-sizes, nanotubes might also be used in medicine where physically confining special gases (like ^{133}Xe for instance) prior to injection would be extremely useful.

Kusnetzova et al. [3.257] have conducted experiments with xenon and found that a very significant enhancement (280 times more, up to a molar ratio $N_{Xe}/N_C = 0.045$) can be achieved by opening the SWNT bundles by thermal activation at 800 °C. With this treatment the gas can be adsorbed inside the nanotubes and the rates of adsorption are also increased.

The possibility of storing argon in carbon nanotubes has been studied with encouraging results by *Gadd* et al. [3.258]: their experiments show that large amounts of argon can be trapped into catalytically grown MWNTs (20–150 nm) by hot isostatic pressing (HIP-ing) for 48 hours at 650 °C under an argon pressure of 1,700 bar. Energy dispersive X-ray spectroscopy was used to determine that the gas was located inside the tubes and not on their walls. Further studies determined the argon pressure inside the tubes at room temperature. The authors estimated this feature to be around 600 bars, indicating that the equilibrium pressure was attained in the tubes during the HIP-ing and that MWNTs would be a convenient material for storing that gas.

Gas Separation

As SWNTs or MWNTs have regular geometries that can, to some extent, be controllable, they could be used to develop precise separation tools. If the sorption mechanisms are known, it should be possible to control sorption of various gases through combinations of temperature, pressure, and nanotube morphology. Since the large-scale production of nanotubes is now constantly progressing and may result in low costs, accurate separation methods based on carbon nanotubes have started to be investigated.

A theoretical study has aimed to determine the effect of different factors such as the diameter of the tubes, the density and the type of the gas used on the flow of molecules inside the nanotubes. An atomistic simulation with methane, ethane, and ethylene [3.259] has shown that the molecule mobility decreases with the decrease of the diameter of the tube for each of the three gases. Ethane and ethylene have a smaller mobility due to the stronger interaction they seem to have with the nanotube walls. In another theoretical study on the possibility of hydrocarbon mixture separation on SWNT bundles, the authors conclude on the possibility of using carbon nanotubes to separate methane/n-butane and methane/isobutene mixtures [3.260], with an efficiency that increases as the average tube diameter decreases. An experimental work was also performed by the same group on the sorption of butane on MWNTs [3.133].

Simulation by grand canonical Monte Carlo for separation of hydrogen and carbon monoxide by adsorption on SWNT has also been reported [3.261]. In most of the situations studied, SWNTs are found to adsorb more CO than H_2, and an excellent separation effect could probably be obtained, again, by varying the SWNT average diameter.

Adsorbents

Carbon nanotubes were recently found to be able to adsorb some toxic gases such as dioxins [3.262], fluoride [3.263], lead [3.264], or alcohols [3.265] better than the materials used so far, such as activated carbon. These pioneering works opened a new field of applications as cleaning filters for many industrial processes having hazardous by-products. Adsorption of dioxins, which are very common and persistent carcinogenic by-products of many industrial processes, is a good example of the interest of nanotubes in this field. Ecological consciousness has imposed emission limits on dioxin-generating sources in many countries, but finding materials that can act as effective filters, even at extremely low concentrations, is a big issue. *Long* et al. [3.262] have found that nanotubes can attract and trap more dioxins than activated carbons or other polyaromatic materials that are currently used as filters. This improvement is probably due to the stronger interaction forces that exist between dioxin molecules and the nanotube curved surfaces compared to those for flat graphene sheets.

The adsorption capacity of Al_2O_3/MWNT for fluoride from water has been reported to be 13.5 times that of activated carbon and 4 times that of Al_2O_3 [3.263]. The same group has also reported an adsorption capacity of lead from water by MWNTs higher than that by activated carbon [3.264]. The possibility to using graphite nanofibers to purify water from alcohols has also been explored [3.265]. These experimental results suggest that carbon nanotubes may be promising adsorbents for the removal of polluting agents from water.

Bio-Sensors

Attaching molecules of biological interest to carbon nanotubes is an ideal way to realize nanometer-sized biosensors. Indeed, the electrical conductivity of such functionalized nanotubes would depend on modifications of the interaction of the probe with the studied media, because of chemical changes or as result of their interaction with target species. The science of attaching biomolecules to nanotubes is rather recent and was inspired by similar research in the fullerene area. Some results have already been patented, and what was looking like a dream a couple of years ago may become reality in the near future. The use of the internal cavity of nanotubes for drug delivery would be another amazing application, but little work has been carried out so far to investigate the harmfulness of nanotubes in the human body. MWNTs have been used by *Mattson* et al. [3.266] as a substrate for neuronal growth. They have compared the activity of untreated MWNTs with that of MWNTs coated with a bioactive molecule (4-hydroxynonenal) and observed that on these latter functionalized nanotubes, neurons elaborated multiple neurites. This is an important result illustrating the feasibility of using nanotubes as substrate for nerve cell growth.

Davis et al. [3.267] have immobilized different proteins (metallothionein, cytochrome c and c_3, β-lactamase I) in MWNTs and checked that these molecules were still catalytically active compared to the nonimmobilized ones. They have shown that confining a protein within a nanotube provides some protection toward the external environment. Protein immobilization via noncovalent sidewall functionalization was proposed by *Chen* et al. [3.268] by using a bifunctional molecule (1-pyrenebutanoic acid, succinimidyl ester). This molecule is maintained to the nanotube wall by the pyrenyl group, and amine groups or biological molecules can react with the ester function to form amid bonds. This method was also used to immobilize ferritin and streptavidin onto SWNTs. It has the main advantage of not modifying the SWNT wall and keeping unperturbed the sp^2 structure, maintaining the physical properties of the nanotubes. *Shim* et al. [3.269] have functionalized SWNTs with biotin and observed specific binding with streptavidin, illustrating biomolecular recognition possibilities. *Dwyer* et al. [3.270] have functionalized SWNTs by covalently coupling DNA strands to them using EDC(1-ethyl-3-(3-dimethylaminopropyl) carbodiimide hydrochloride) but did not test biomolecular recognition; other proteins such as bovine serum albumin (BSA) [3.271] have been attached to nanotubes using the same process (diimide-activated amidation with EDC) and most of the attached proteins remained bioactive. Instead of working with individual nanotubes (or more likely nanotube bundles in the case of SWNTs), *Nguyen* et al. [3.272] have functionalized nanotubes arrayed with a nucleic acid, still using EDC as the coupling agent, in order to facilitate the realization of biosensors based on protein-functionalized nanotubes. Recently *Azamian* et al. [3.273] have immobilized a series of biomolecules (cytochrome c, ferritin, and glucose oxydase) on SWNTs and observed that the use of EDC was not always necessary, indicating that the binding was predominantly noncovalent. In the case of glucose oxydase, they have tested the catalytic activity of functionalized nanotubes immobilized on a glassy carbon electrode and observed a tenfold greater catalytic response compared to that in the absence of modified SWNTs.

Functionalization of nanotubes with biomolecules is still in its infancy, and their use as biosensors may lead to practical applications earlier than expected. For example, functionalized nanotubes can be used as AFM tips (see Sect. 3.6.1) allowing to perform "chemical force microscopy" (CFM), allowing single-molecule measurements. Even in the case of nonfunctionalized CNT-based tips, important improvements have been obtained for the characterization of biomolecules (see the review by [3.215]).

Composites

Because of their exceptional morphological, electrical, thermal, and mechanical characteristics, carbon nanotubes are particularly promising materials as reinforcement in composite materials with metal, ceramic, or polymer matrix. Key basic issues include the good dispersion of the nanotubes, the control of the nanotube/matrix bonding, the densification of bulk composites and thin films, and the possibility of aligning the nanotubes. In addition, the nanotube type (SWNT, c-MWNT, h-MWNT, etc.) and origin (arc, laser, CCVD, etc.) is also an important variable since determining the structural perfection, surface reactivity, and aspect ratio of the reinforcement.

Considering the major breakthrough that carbon nanotubes are expected to make in the field, the following will give an overview of the current work on metal-, ceramic- and polymer-matrix composites reinforced with nanotubes. We will not consider nanotubes merely coated with another material will not be considered here. We discussed filled nanotubes in Sect. 3.5.2. Applications involving the incorporation of nanotubes to materials for purposes other than structural will be mentioned in Sect. 3.6.2.

Metal Matrix Composites

Nanotube-metal matrix composites are still rarely studied. The materials are generally prepared by standard powder metallurgy techniques, but the dispersion of the nanotubes is not optimal. Thermal stability and electrical and mechanical properties of the composites are investigated. The room temperature electrical resistivity of hot-pressed CCVD MWNTs-Al composites increases slightly by increasing the MWNT volume fraction [3.274]. The tensile strength and elongation of nonpurified arc-discharge MWNT-Al composites are only slightly affected by annealing at 873 K in contrast to that of pure Al [3.275]. The Young's modulus of nonpurified arc-discharge MWNTs-Ti composites is about 1.7 times that of pure Ti [3.276]. The formation of TiC, probably resulting from the reaction between amorphous carbon and the matrix, was observed, but the MWNTs themselves were not damaged. An increase of the Vickers hardness by a factor 5.5 over that of pure Ti was associated with the suppression of coarsening of the Ti grains, the TiC formation, and the addition of MWNTs with an extremely high Young's modulus. CCVD MWNTs-Cu composites [3.277] also show a higher hardness and a lower friction coefficient and wear loss. A deformation of 50–60 % of the composites was observed. A MWNTs-metallic glass Fe82P18 composite [3.278] with a good dispersion of the MWNTS was prepared by the rapid solidification technique.

Ceramic Matrix Composites

Carbon nanotube-containing ceramic-matrix composites are a bit more frequently studied, most efforts made to obtain tougher ceramics [3.279]. Composites can be processed following the regular processing route, in which the nanotubes are usually mechanically mixed with the matrix (or a matrix precursor) then densified using hot-pressing sintering of the slurry or powders. *Zhan* et al. [3.280] ball-milled SWNT bundles (from the HiPCo technique) and nanometric alumina powders, producing a fairly homogeneous dispersion, supposedly without damaging the SWNTs. On the other hand, interesting results were obtained using the spark plasma sintering (SPS) technique, which was able to prepare fully dense composites without damage to the SWNTs [3.281]. Other original composite processing include sol-gel route, by which bulk and thin film composites were prepared [3.282], and in situ SWNT growth in ceramic foams, using procedures closely related to that described in Sect. 3.2.2 [3.283].

Ma et al. [3.284] prepared MWNT-SiC composites by hot-pressing mixtures of CCVD MWNTs and nano-SiC powders. They claimed that the presence of the MWNTs provides an increase of about 10% of both the bending strength and fracture toughness, but the dispersion of the MWNTs looks poor. Several detailed studies [3.82, 285–288] have dealt with the preparation of nanotube-Fe-Al_2O_3, nanotube-Co-MgO, and nanotube-Fe/Co-$MgAl_2O_4$ composites by hot-pressing of the corresponding composite powders. The nanotubes were grown in situ in the starting powders by a CCVD method [3.73, 80, 289–293] and thus are very homogeneously dispersed between the metal-oxide grains, in a way that could be impossible to achieve by mechanical mixing. The nanotubes (mostly SWNTs and DWNTs) gather in long, branched bundles smaller than 50 nm in diameter, which appear to be very flexible. Depending on the matrix and hot-pressing temperature, a fraction of the nanotubes seems to be destroyed during hot-pressing in a primary vacuum. The increase of the quantity of nanotubes (ca. 2–12 wt%) leads to a refinement of the microstructure but also to a strong decrease in relative density (70–93 %). Probably because of this, the fracture strength and the toughness of the nanotube-containing composites are generally lower than those of the nanotube-free metal-oxide composites and only marginally higher than those of the corresponding ceramics. SEM observations revealed both the trapping of some of the nanotubes within the matrix grains or at grain boundaries and a relatively good wetting in the case of alumina. Most nanotubes are cut near the fracture surface after some pull-out and could contribute to a mechanical reinforcement.

Siegel et al. [3.294] claimed that the fracture toughness of 10 vol.% MWNTs-alumina hot-pressed composites is increased by 24% (to $4.2\,\mathrm{MPa\,m^{1/2}}$) over that of pure alumina. But the density and grain size are not reported, so it is difficult to consider this result as an evidence for the beneficial role of the MWNTs themselves. A confirmation could be found in the work by *Sun* et al. [3.281] who developed a colloidal route to coat the nanotubes with alumina particles prior to another mixing step with alumina particles then densification by SPS. A very large gain in fracture toughness (calculated from Vickers indentation) is reported (from $3.7\,\mathrm{MPa\,m^{1/2}}$ for pure Al_2O_3 to $4.9\,\mathrm{MPa\,m^{1/2}}$ for an addition of 0.1 wt% of SWNT nanotubes, and to $9.7\,\mathrm{MPa\,m^{1/2}}$ for an addition of 10 wt%). But it is possible that the beneficial input of the nanotubes is due to the previous coating of the nanotubes before sintering, resulting in an improved bonding with the matrix.

Bulk properties other than mechanical are also worth being investigated. Interestingly, the presence of well dispersed nanotubes confers an electrical conductivity to the otherwise insulating ceramic-matrix composites. The percolation threshold is very low – in the range 0.2–0.6 wt% of carbon – due to the very high aspect ratio of nanotubes. Varying the nanotube quantity in the starting powders allows the electrical conductivity to be controlled in the range $0.01–10\,\mathrm{S\,cm^{-1}}$. Moreover, the extrusion at high temperatures in vacuum allows the nanotubes to be aligned in the matrix, thus inducing an anisotropy in the electrical conductivity [3.295]. One of the early works involved MWNTs in a superconducting $Bi_2Sr_2CaCu_2O_{8+\delta}$ matrix [3.296]. The application of SWNTs for thermal management in partially stabilized zirconia [3.297] and of MWNTs-lanthanum cobaltate composites for application as an oxygen electrode in zinc/air [3.298] were reported more recently. The friction coefficient was also shown to increase and the wear loss to decrease upon the increase in nanotube volume fraction in carbon/carbon composites [3.299].

The latter is one of the few examples in which carbon has been considered as a matrix. Another one is the work reported by *Andrews* et al. [3.300], who introduced 5 wt% of SWNTs into a pitch material before spinning the whole as a SWNT-reinforced pitch-fiber. Subsequent carbonization transformed it into a SWNT-reinforced carbon fiber. Due to the contribution of the SWNTs, which were aligned by the spinning stresses, the resulting gain in tensile strength, modulus, and electrical conductivity with respect to the nonreinforced carbon fiber was claimed to be 90, 150, and 340%, respectively.

Polymer Matrix Composites

Nanotube-polymer composites, first reported by *Ajayan* et al. [3.301] are now intensively studied, notably epoxy- and polymethylmethacrylate (PMMA)-matrix composites. Regarding the mechanical characteristics, the three key issues affecting the performance of a fiber-polymer composite are the strength and toughness of the fibrous reinforcement, its orientation, and a good interfacial bonding crucial for the load transfer to occur [3.302]. The ability of the polymer to form large-diameter helices around individual nanotubes favors the formation of a strong bond with the matrix [3.302]. Isolated SWNTs may be more desirable than MWNTs or bundles for dispersion in a matrix because of the weak frictional interactions between layers of MWNTs and between SWNTs in bundles [3.302]. The main mechanisms of load transfer are micromechanical interlocking, chemical bonding, and van der Waals bonding between the nanotubes and the matrix. A high interfacial shear stress between the fiber and the matrix will transfer the applied load to the fiber over a short distance [3.303]. SWNTs longer than $10–100\,\mu\mathrm{m}$ would be needed for significant load-bearing ability in the case of nonbonded SWNT-matrix interactions, whereas the critical length for SWNTs cross-linked to the matrix is only $1\,\mu\mathrm{m}$ [3.304]. Defects are likely to limit the working length of SWNTs [3.305], however.

The load transfer to MWNTs dispersed in an epoxy resin was much higher in compression than in tension [3.303]. It was proposed that all the walls of the MWNTs are stressed in compression, whereas only the outer walls are stressed in tension because all the inner tubes are sliding within the outer. Mechanical tests performed on 5 wt% SWNT-epoxy composites [3.306] showed that SWNTs bundles were pulled out of the matrix during the deformation of the material. The influence of the nanotube/matrix interfacial interaction was evidenced by *Gong* et al. [3.307]. It was also reported that coating regular carbon fiber with MWNTs prior to their dispersion into an epoxy matrix improves the interfacial load transfer, possibly by local stiffening of the matrix near interface [3.308].

As for ceramic matrix composites, the electrical characteristics of SWNT- and MWNT-epoxy composites are described with the percolation theory. Very low percolation threshold (below 1 wt%) is often reported [3.309–311]. Industrial epoxy loaded with 1 wt% unpurified CCVD-prepared SWNTs showed an increase in thermal conductivity of 70% and 125% at 40 K and at room temperature, respectively [3.309]. Also, the Vickers hardness rose with the SWNT loading up to a factor of 3.5 at 2 wt%. Thus both the thermal and mechanical properties of such composites are improved without the need to chemically functionalize the SWNTs.

Likewise, the conductivity of laser-prepared SWNT-PMMA composites increases with the load in SWNTs (1–8 wt% SWNTs) [3.302]. Thermogravimetric analysis shows that, compared to pure PMMA, the thermal degradation of PMMA films occurs at a slightly higher temperature when 26 wt% MWNTs are added [3.312]. Improving the wetting between the MWNTs and the PMMA by coating the MWNTs with poly(vinylidene fluoride) prior to melt-blending with PMMA resulted in an increased storage modulus [3.313]. The impact strength in aligned SWNT-PMMA composites was significantly increased with only 0.1 wt% of SWNTs, possibly because of a weak interfacial adhesion, and/or of the high flexibility of the SWNTs, and/or the pullout and sliding effects of individual SWNTs within bundles [3.314]. The transport properties of arc-discharge SWNTs-PMMA composite films (10 μm thick) were studied in great detail [3.315, 316]. The electrical conductivity increases by nine orders of magnitude from 0.1 to 8 wt% SWNTs. The room temperature conductivity is again well described with the standard percolation theory, confirming the good dispersion of the SWNTs in the matrix.

Polymer composites with other matrices include CCVD-prepared MWNT-polyvinyl alcohol [3.317], arc-prepared MWNT-polyhydroxyaminoether [3.318], arc-prepared MWNT-polyurethane acrylate [3.319, 320], SWNT-polyurethane acrylate [3.321], SWNT-polycarbonate [3.322], MWNT-polyaniline [3.323], MWNT-polystyrene [3.324], SWNT-polyethylene [3.325], CCVD-prepared MWNT-polyacrylonitrile. [3.326], MWNT-oxotitanium phthalocyanine [3.327], arc-prepared MWNT-poly(3-octylthiophene) [3.328], SWNT-poly(3-octylthiophene) [3.329], and CCVD MWNTs-poly(3-hexylthiophene) [3.330]. These works deal mainly with films 100–200 micrometer thick and aim to study the glass transition of the polymer, mechanical, and electrical characteristics as well as photoconductivity.

A great deal of work is also devoted to nano--tube-polymer composites for applications as materials for molecular optoelectronics, using primarily poly(m-phenylenevinylene-co-2,5-dioctoxy-p-phenylenevinylene) (PmPV) as the matrix. This conjugated polymer tends to coil, forming a helical structure. The electrical conductivity of the composite films (4–36 wt% MWNTs) is increased by eight orders of magnitude compared to that of PmPV [3.331]. Using the MWNT-PmPV composites as the electron-transport-layer in light-emitting diodes results in a significant increase in brightness [3.332]. The SWNTs act as a hole-trapping material blocking the holes in the composites, this being probably induced through long-range interactions within the matrix [3.333]. Similar investigations were carried out on arc-discharge SWNTs-polyethylene dioxythiophene (PEDOT) composite layers [3.334] and MWNTs-polyphenylenevinylene composites [3.335].

To conclude, two critical issues have to be considered for the application of nanotubes as reinforcement for advanced composites. One is to chose between SWNTs and MWNTs. The former seem more beneficial for the purpose of mechanical strengthening, provided they are isolated or arranged into cohesive yarns so that the load is able to be conveniently transferred from a SWNT to another. Unfortunately, despite recent advances [3.336–339], this is still a technical challenge. The other issue is to tailor the nanotube/matrix interface with respect to the matrix. This is a current topic for many laboratories involved in the field.

Multifunctional Materials

One of the major benefits expected from incorporating carbon nanotubes in other solid or liquid materials is bringing some electrical conductivity to them while not affecting the other properties or behaviors of these materials. As already mentioned in the previous section, the percolation threshold is reached at very low load with nanotubes. Tailoring the electrical conductivity of a bulk material is then achievable by adjusting the nanotube volume fraction in the formerly insulating material while not making this fraction too large anyway. As demonstrated by Maruyama [3.3], there are three regimes of interest regarding the electrical conductivity: (i) electrostatic discharge, e.g., to prevent fire or explosion hazard in combustible environments or perturbations in electronics, which requires electrical resistivity less than $10^{12}\,\Omega$ cm; (ii) electrostatic painting, which requires the material to be painted to be electrical conductive enough (electrical resistivity below $10^6\ \Omega$ cm) to prevent the charged paint droplets to be repelled; (iii) electromagnetic interference shielding, which is achieved for electrical resistivity lower than $10\ \Omega$ cm.

Materials are often asked to be multifunctional, e.g. having both high electrical conductivity and high toughness, or high thermal conductivity and high thermal stability, etc. The association of several materials, each of them bringing one of the desired features, therefore generally meets this need. Exceptional features and properties of carbon nanotubes make them likely to be a perfect multifunctional material in many cases. For instance, materials for satellites are often required to be electrical conductive, mechanically self-supporting, able to transport the excess heat away, sometimes be protected against electromagnetic interferences, etc., while exhibiting minimal weight and volume. All these properties should be possible with a single nanotube-containing composite material instead of complex multi-materials combining layers of polymers, aluminum, copper, etc. Table 3.5 provides an overview of various fields in which nanotube-based multifunctional materials should find an application.

Nano-Electronics

As reported in Sects. 3.1.1 and 3.4.4, SWNT nanotubes can be either metallic (with an electrical conductivity higher than that of copper), or semiconductor. This has inspired the design of several components for nanoelectronics. First, metallic SWNTs can be used as mere ballistic conductors. Moreover, as early as 1995,

Table 3.5 Applications for nanotube-based multifunctional materials (from [3.3]), by courtesy of B. Maruyama (WPAFB, Dayton, Ohio)

		Mechanical			Electrical			Thermal		Thermo-mechanical	
Fiber fraction	Applications system	Strength/stiffness	Specific strength	through-thickness strength	Static dissipation	Surface Conduction[a]	EMI shielding	Service[b] temp.	conduction/dissipation[c]	Dimensional Stability[d]	CTE reduction[e]
Low Volume fraction (fillers)											
Elastomers	Tires	×			×				×		
Thermo Plastics	Chip package				×				×		
	Electronics/Housing	×					×	×	×		
Thermosets	Epoxy products	×	×	×		×				×	×
	Composites			×						×	
High Volume Fraction											
Structural composites	Space/aircraft components		×	×							
High conduction composites	Radiators	×							×	×	
	Heat exchangers	×						×	×		×
	EMI shield	×					×				

[a] For electrostatic painting, to mitigate lightning strikes on aircraft, etc.
[b] To increase service temperature rating of product
[c] To reduce operating temperatures of electronic packages
[d] Reduces warping
[e] Reduces microcracking damage in composites

realizing a rectifying diode by joining one metallic SWNT to one semiconductor SWNT (hetero-junction) was proposed by *Lambin* et al. [3.340], then later by *Chico* et al. [3.341] and *Yao* et al. [3.342]. Also, field-effect transistors (FET) can be built by the attachment of a semiconductor SWNT across two electrodes (source and drain) deposited on an insulating substrate that serves as a gate electrode [3.343, 344]. Recently the association of two such SWNT-based FETs has made a voltage inverter [3.345].

All the latter developments are fascinating and provide promising perspectives for nanotube-based electronics. But progress is obviously needed before making SWNT-based integrated circuits on a routine basis. A key point is the need to be able to prepare selectively either metallic or semiconductor nanotubes. Although a way has been proposed to destroy selectively metallic SWNTs from bundles of undifferentiated SWNTs [3.347], the method is not scalable and selective synthesis would be preferable. Also, defect-free nanotubes are required. Generally speaking, it relates to another major challenge, which is to be able to fabricate at industrial scale integrated circuits including nanometer-size components that only sophisticated imaging methods (AFM) are able to visualize.

Nano-Tools, Nano-Devices, Nano-Systems

Due to the ability of graphene to expand slightly when electrically charged, nanotubes have been found to act conveniently as actuators. *Kim* et al. [3.346] demonstrated it by designing "nano"-tweezers able to grab, manipulate, release nano-objects (the "nano"-bead having been handled for the demonstration was actually closer to micrometer than nanometer), and measure their electrical properties. This was made possible quite simply by depositing two non-interconnected gold coatings onto a pulled glass micropipette (Fig. 3.31), then attaching two MWNTs (or two SWNT-bundles) ~ 20–50 nm in diameter to each of the gold electrodes. Applying a voltage (0–8.5 V) between the two electrodes then makes the tube tips to open and close reversibly in a controlled manner.

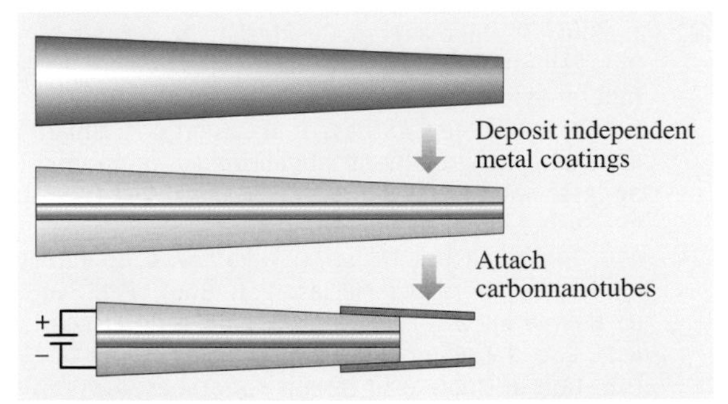

Fig. 3.31 Sketch explaining how the first nano-tweezers were designed. First is a glass micropipette (dark cone, *top*). Then two Au coatings (in grey, *middle*) are deposited so that they are not in contact. Then a voltage is applied to the electrodes (from [3.346])

A similar experiment, again rather simple, was proposed by *Baughman* et al. the same year (1999) [3.348], consisting in mounting two SWNT-based paper strips ("bucky-paper") on both sides of an insulating double-side tape. The two bucky-paper strips were previously loaded with Na^+ and Cl^-, respectively. When 1 V was applied between the two paper strips, both expand, but the strip loaded with Na^+ expands a bit more, forcing the whole system to bend. Though performed in a liquid environment, such a behavior has inspired the authors to predict a future for their system as "artificial muscles."

Another example of amazing nano-tools is the nano-thermometer proposed by *Gao* et al. [3.349]. A single MWNT was used, in that case, partially filled with liquid gallium. Upon the effect of temperature variations in the range 50–500 °C, the gallium goes up and down reversibly within the nanotube cavity at reproducible level with respect to the values of the temperature applied.

Of course, nano-tools such as nano-tweezers or nano-thermometers will hardly reach a commercial development so to justify industrial investments. But such experiments are more than amazing laboratory curiosities. They definitely demonstrate the ability of carbon nanotubes as parts for future nano-devices, including nanomechanics-based systems.

Supercapacitors

Supercapacitors include two electrodes immersed in an electrolyte (e.g., 6 M KOH), separated by an insulating ion-permeable membrane. Charging the capacitors is achieved by applying a potential between the two electrodes, making the cations and the anions moving toward the electrode oppositely charged. Suitable electrodes should exhibit a high electrical conductivity and a high surface area since the capacitance is proportional to it. Actually, the surface area should originate mainly from the appropriate combination of mesopores (to allow the electrolyte components to circulate well, which is related to the charging speed) and micropores (whose walls are the surface of attraction and fixation sites for the ions). Based on early works by

Niu et al. [3.350], such a combination was found to be reached with the specific architecture offered by packed and entangled h-MWNTs with a poor nanotexture (see Sect. 3.1.2). But, activation pre-treatments are necessary. For instance, a capacitor made from nanotubes with surface area $= 220\,m^2/g$ exhibited a capacitance of $20\,F/g$, which increased up to $100\,F/g$ after an activation treatment was applied to the nanotubes so that their surface area reaches $880\,m^2/g$ [3.126]. Alternatively, again due to their specific architecture induced by their huge aspect ratio, nanotubes can also be used as supports for conductive polymer coatings, e.g., polypyrrole or polyaniline [3.351], which otherwise would make a too dense phase (i. e. not allowing an easy circulation and penetration of ions). Supercapacitors built from such composites can overcome more than 2,000 charging cycles, with current density as high as $350\,mA/g$ [3.352]. Capacitors including nanotubes already have capacitance as high as $180–200\,F/g$, i. e. equivalent to that obtained with electrodes built from regular carbon materials, with the advantage of faster charging [3.126]. Current works will certainly lead to further optimization of both the nanotube material architecture and the nanotube-supported conductive polymers, giving reasonable perspectives for the commercial use of nanotubes as components for supercapacitors.

References

3.1 M. Hillert, N. Lange: The structure of graphite filaments, Zeitschr. Kristall. **111** (1958) 24–34

3.2 T. V. Hughes, C. R. Chambers: US Patent 405,480 (1889)

3.3 B. Maruyama, K. Alam: Carbon nanotubes and nanofibers in composite materials, SAMPE J. **38** (2002) 59–70

3.4 P. Schützenberger, L. Schützenberger: Sur quelques faits relatifs à l'histoire du carbone, C. R. Acad. Sci. Paris **111** (1890) 774–778

3.5 C. Pélabon, H. Pélabon: Sur une variété de carbone filamenteux, C. R. Acad. Sci. (Paris) **137** (1903) 706–708

3.6 R. T. K. Baker, P. S. Harris: The formation of filamentous carbon. In: *Chemistry and Physics of Carbon*, Vol. 14, ed. by P. L. Walker Jr., P. A. Thrower (Dekker, New York 1978) pp. 83–165

3.7 S. Iijima: Helical microtubules of graphite carbon, Nature **354** (1991) 56–58

3.8 S. Iijima, T. Ichihashi: Single-shell carbon nanotubes of 1-nm diameter, Nature **363** (1993) 603–605

3.9 D. S. Bethune, C. H. Kiang, M. S. de Vries, G. Gorman, R. Savoy, J. Vazquez, R. Bayers: Cobalt-catalysed growth of carbon nanotubes with single-atomic-layer walls, Nature **363** (1993) 605–607

3.10 J. Tersoff, R. S. Ruoff: Structural properties of a carbon-nanotube crystal, Phys. Rev. Lett. **73** (1994) 676–679

3.11 N. Wang, Z. K. Tang, G. D. Li, J. S. Chen: Single-walled 4 Å carbon nanotube arrays, Nature **408** (2000) 50–51

3.12 N. Hamada, S. I. Sawada, A. Oshiyama: New one-dimensional conductors, graphitemicrotubules, Phys. Rev. Lett. **68** (1992) 1579–1581

3.13 M. S. Dresselhaus, G. Dresselhaus, P. C. Eklund: *Science of Fullerenes and Carbon Nanotubes* (Academic, San Diego 1995)

3.14 M. Monthioux, B. W. Smith, B. Burteaux, A. Claye, J. Fisher, D. E. Luzzi: Sensitivity of single-wall nanotubes to chemical processing: An electron microscopy investigation, Carbon **39** (2001) 1261–1272

3.15 H. Allouche, M. Monthioux: Chemical vapor deposition of pyrolytic carbon onto carbon nanotubes II – Structure and texture, Carbon (2003) accepted

3.16 M. Audier, A. Oberlin, M. Oberlin, M. Coulon, L. Bonnetain: Morphology and crystalline order in catalytic carbons, Carbon **19** (1981) 217–224

3.17 Y. Saito: Nanoparticles and filled nanocapsules, Carbon **33** (1995) 979–988

3.18 P. J. F. Harris: *Carbon Nanotubes and Related Structures* (Cambridge Univ. Press, Cambridge 1999)

3.19 N. M. Rodriguez, A. Chambers, R. T. Baker: Catalytic engineering of carbon nanostructures, Langmuir **11** (1995) 3862–3866

3.20 H. W. Kroto, J. R. Heath, S.C. O'Brien, R. F. Curl, R. E. Smalley: C_{60} Buckminsterfullerene, Nature **318** (1985) 162–163

3.21 W. Krätschmer, L. D. Lamb, K. Fostiropoulos, D. R. Huffman: Solid C_{60}: A new form of carbon, Nature **347** (1990) 354–358

3.22 T. Guo, P. Nikolaev, A. G. Rinzler, D. Tomanek, D. T. Colbert, R. E. Smalley: Self-assembly of tubular fullerenes, J. Phys. Chem. **99** (1995) 10694–10697

3.23 T. Guo, P. Nikolaev, A. Thess, D. T. Colbert, R. E. Smalley: Catalytic growth of single-walled nanotubes by laser vaporisation, Chem. Phys. Lett. **243** (1995) 49–54

3.24 A. G. Rinzler, J. Liu, H. Dai, P. Nikolaev, C. B. Huffman, F.J. Rodriguez-Macias, P.J. Boul, A. H. Lu, D. Heymann, D.T. Colbert, R.S. Lee, J.E. Fischer, A. M. Rao, P. C. Eklund, R. E. Smalley: Large scale purification of single wall carbon nanotubes: Process, product and characterization, Appl. Phys. A **67** (1998) 29–37

3.25 A. Thess, R. Lee, P. Nikolaev, H. Dai, P. Petit, J. Robert, C. Xu, Y. H. Lee, S. G. Kim, D. T. Colbert, G. Scuseria, D. Tomanek, J. E. Fischer, R. E. Smalley: Crystalline ropes of metallic carbon nanotubes, Science **273** (1996) 487–493

3.26 L. M. de la Chapelle, J. Gavillet, J. L. Cochon, M. Ory, S. Lefrant, A. Loiseau, D. Pigache: A continuous wave CO_2 laser reactor for nanotube synthesis, Proc. Electronic Properties Novel Materials-XVI Int. Winterschool – AIP Conf. Proc., Melville 1999, ed. by H. Kuzmany, J. Fink, M. Mehring, S. Roth (Springer, Berlin, Heidelberg 1999) **486** 237– 240

3.27 M. Yudasaka, T. Komatsu, T. Ichihashi, S. Iijima: Single wall carbon nanotube formation by laser ablation using double targets of carbon and metal, Chem. Phys. Lett. **278** (1997) 102–106

3.28 M. Castignolles, A. Foutel-Richard, A. Mavel, J. L. Cochon, D. Pigache, A. Loiseau, P. Bernier: Combined experimental and numerical study of the parameters controlling the C-SWNT synthesis via laser vaporization, Proc. Electronic Properties of Novel Materials-XVI Int. Winterschool – AIP Conf. Proc., Melville 2002, ed. by H. Kuzmany, J. Fink, M. Mehring, S. Roth (Springer, Berlin, Heidelberg 2002) **633** 385–389

3.29 T. W. Ebbesen, P. M. Ajayan: Large-scale synthesis of carbon nanotubes, Nature **358** (1992) 220–221

3.30 D. Ugarte: Morphology and structure of graphitic soot particles generated in arc-discharge C_{60} production, Chem. Phys. Lett. **198** (1992) 596–602

3.31 T. W. Ebbesen: Carbon nanotubes, Ann. Rev. Mater. Sci. **24** (1994) 235–264

3.32 T. Beltz, J. Find, D. Herein, N. Pfänder, T. Rühle, H. Werner, M. Wohlers, R. Schlögl: On the production of different carbon forms by electric arc graphite evaporation, Ber. Bunsenges. Phys. Chem. **101** (1997) 712–725

3.33 C. Journet, W. K. Maser, P. Bernier, A. Loiseau, L. M. de la Chapelle, S. Lefrant, P. Deniard, R. Lee, J. E. Fischer: Large-scale production of single-walled carbon nanotubes by the electric-arc technique, Nature **388** (1997) 756–758

3.34 K. Saïdane, M. Razafinimanana, H. Lange, M. Baltas, A. Gleizes, J. J. Gonzalez: Influence of the carbon arc current intensity on fullerene synthesis, Proc. 24th Int. Conf. on Phenomena in Ionized Gases, Warsaw 1999, ed. by P. Pisarczyk, T. Pisarczyk, J. Wotowski, 203–204

3.35 H. Allouche, M. Monthioux, M. Pacheco, M. Razafinimanana, H. Lange, A. Huczko, T. P. Teulet, A. Gleizes, T. Sogabe: Physical characteristics of the graphite-electrode electric-arc as parameters for the formation of single-wall carbon nanotubes, Proc. Eurocarbon (Deutsche Keram. Ges., Berlin 2000) 1053–1054

3.36 M. Razafinimanana, M. Pacheco, M. Monthioux, H. Allouche, H. Lange, A. Huczko, A. Gleizes: Spectroscopic study of an electric arc with Gd and Fe doped anodes for the carbon nanotube formation, Proc. 25th Int. Conf. on Phenomena in Ionized Gases, Nagoya 2001, ed. by E. Goto (Nagoya Univ., Nagoya 2001) Extend. Abstr. **3** 297–298

3.37 M. Razafinimanana, M. Pacheco, M. Monthioux, H. Allouche, H. Lange, A. Huczko, P. Teulet, A. Gleizes, C. Goze, P. Bernier, T. Sogabe: Influence of doped graphite electrode in electric arc for the formation of single wall carbon nanotubes, Proc. 6th Eur. Conf. on Thermal Plasma Processes – Progress in Plasma Processing of Materials, New York 2000, ed. by P. Fauchais (Begell House, New York 2001) 649–654

3.38 M. Pacheco, H. Allouche, M. Monthioux, A. Razafinimanana, A. Gleizes: Correlation between the plasma characteristics and the morphology and structure of the carbon phases synthesised by electric arc discharge, Proc. 25th Biennial Conf. on Carbon, Lexington 2001 (Univ. of Kentucky, Lexington 2001) Extend. Abstr.(CD-Rom), Novel/14.1

3.39 M. Pacheco, M. Monthioux, M. Razafinimanana, L. Donadieu, H. Allouche, N. Caprais, A. Gleizes: New factors controlling the formation of single-wall carbon nanotubes by arc plasma, Proc. Carbon 2002 Int. Conf., Beijing 2002 (Tsinghua Univ., Beijing 2002) (CD-Rom/Oral/I014)

3.40 M. Monthioux, M. Pacheco, H. Allouche, M. Razafinimanana, N. Caprais, L. Donnadieu, A. Gleizes: New data about the formation of SWNTs by the electric arc method, Electronic Properties of Molecular Nanostructures, AIP Conf. Proc., Melville 2002, ed. by H. Kuzmany, J. Fink, M. Mehring, S. Roth (Springer, Berlin, Heidelberg 2002) 182–185

3.41 H. Lange, A. Huczko, M. Sioda, M. Pacheco, M. Razafinimanana, A. Gleizes: Influence of gadolinium on carbon arc plasma and formation of fullerenes and nanotubes, Plasma Chem. Plasma Process **22** (2002) 523–536

3.42 C. Journet: La production de nanotubes de carbone. Ph.D. Thesis (University of Montpellier II, Montpellier 1998)

3.43 T. Sogabe, T. Masuda, K. Kuroda, Y. Hirohaya, T. Hino, T. Ymashina: Preparation of B_4C-mixed graphite by pressureless sintering and its air oxidation behavior, Carbon **33** (1995) 1783–1788

3.44 M. Ishigami, J. Cumings, A. Zettl, S. Chen: A simple method for the continuous production of carbon nanotubes, Chem. Phys. Lett. **319** (2000) 457–459

3.45 Y. L. Hsin, K. C. Hwang, F. R. Chen, J. J. Kai: Production and in-situ metal filling of carbon nanotube in water, Adv. Mater. **13** (2001) 830–833

3.46 H. W. Zhu, X. S. Li, B. Jiang, C. L. Xu, C. L. Zhu, Y. F. Zhu, D. H. Wu, X. H. Chen: Formation of carbon nanotubes in water by the electric arc technique, Chem. Phys. Lett. **366** (2002) 664–669

3.47 W. K. Maser, P. Bernier, J. M. Lambert, O. Stephan, P. M. Ajayan, C. Colliex, V. Brotons, J. M. Planeix, B. Coq, P. Molinie, S. Lefrant: Elaboration and characterization of various carbon nanostructures, Synth. Met. **81** (1996) 243–250

3.48 J. Gavillet, A. Loiseau, J. Thibault, A. Maigné, O. Stéphan, P. Bernier: TEM study of the influence of the catalyst composition on the formation and growth of SWNT, Proc. Electronic Properties Novel Materials-XVI Int. Winterschool – AIP Conf. Proc., Melville 2002, ed. by H. Kuzmany, J. Fink, M. Mehring, S. Roth (Springer, Berlin, Heidelberg 2002) **633** 202–206
3.49 L. P. F. Chibante, A. Thess, J. M. Alford, M. D. Diener, R. E. Smalley: Solar generation of the fullerenes, J. Phys. Chem. **97** (1993) 8696–8700
3.50 C. L. Fields, J. R. Pitts, M. J. Hale, C. Bingham, A. Lewandowski, D. E. King: Formation of fullerenes in highly concentrated solar flux, J. Phys. Chem. **97** (1993) 8701–8702
3.51 P. Bernier, D. Laplaze, J. Auriol, L. Barbedette, G. Flamant, M. Lebrun, A. Brunelle, S. Della-Negra: Production of fullerenes from solar energy, Synth. Met. **70** (1995) 1455–1456
3.52 M. J. Heben, T. A. Bekkedhal, D. L. Schultz, K. M. Jones, A. C. Dillon, C. J. Curtis, C. Bingham, J. R. Pitts, A. Lewandowski, C. L. Fields: Production of single wall carbon nanotubes using concentrated sunlight, Proc. Symp. Recent Adv. Chem. Phys. Fullerenes Rel. Mater., Pennington 1996, ed. by K. M. Kadish, R. S. Ruoff (Electrochemical Society, Pennington 1996) INC**3** 803–811
3.53 D. Laplaze, P. Bernier, C. Journet, G. Vié, G. Flamant, E. Philippot, M. Lebrun: Evaporation of graphite using a solar furnace, Proc. 8th Int. Symp. Solar Concentrating Technol., Köln 1996, ed. by M. Becker, M. Balmer (C. F. Müller Verlag, Heidelberg 1997)
3.54 D. Laplaze, P. Bernier, W. K. Maser, G. Flamant, T. Guillard, A. Loiseau: Carbon nanotubes : The solar approach, Carbon **36** (1998) 685–688
3.55 T. Guillard, S. Cetout, L. Alvarez, J. L. Sauvajol, E. Anglaret, P. Bernier, G. Flamant, D. Laplaze: Production of carbon nanotubes by the solar route, Eur. Phys. J. **5** (1999) 251–256
3.56 G. Flamant, D. Luxembourg, C. Mas, D. Laplaze: Carbon nanotubes from solar energy; a chemical engineering approach of the scale up, Proc. Am. Inst. Chem. Eng. Ann. Meeting, Indianapolis 2002, in print
3.57 G. G. Tibbets, M. Endo, C. P. Beetz: Carbon fibers grown from the vapor phase: A novel material, SAMPE J. **22** (1989) 30
3.58 R. T. K. Baker: Catalytic growth of carbon filaments, Carbon **27** (1989) 315–323
3.59 R. T. K. Baker, P. S. Harris, R. B. Thomas, R. J. Waite: Formation of filamentous carbon from iron, cobalt, and chromium catalyzed decomposition of acetylene, J. Catal. **30** (1973) 86–95
3.60 T. Koyama, M. Endo, Y. Oyuma: Carbon fibers obtained by thermal decomposition of vaporized hydrocarbon, Jap. J. Appl. Phys. **11** (1972) 445–449
3.61 M. Endo, A. Oberlin, T. Koyama: High resolution electron microscopy of graphitizable carbon fiber prepared by benzene decomposition, Jap. J. Appl. Phys. **16** (1977) 1519–1523
3.62 N. M. Rodriguez: A review of catalytically grown carbon nanofibers, J. Mater. Res. **8** (1993) 3233–3250
3.63 W. R. Davis, R. J. Slawson, G. R. Rigby: An unusual form of carbon, Nature **171** (1953) 756
3.64 H. P. Boehm: Carbon from carbon monoxide disproportionation on nickel and iron catalysts; morphological studies and possible growth mechanisms, Carbon **11** (1973) 583–590
3.65 M. Audier, A. Oberlin, M. Coulon: Crystallographic orientations of catalytic particles in filamentous carbon; case of simple conical particles, J. Cryst. Growth **55** (1981) 546–549
3.66 M. Audier, M. Coulon: Kinetic and microscopic aspects of catalytic carbon growth, Carbon **23** (1985) 317–323
3.67 M. Audier, A. Oberlin, M. Coulon: Study of biconic microcrystals in the middle of carbon tubes obtained by catalytic disproportionation of CO, J. Cryst. Growth **57** (1981) 524–534
3.68 A. Thaib, G. A. Martin, P. Pinheiro, M. C. Schouler, P. Gadelle: Formation of carbon nanotubes from the carbon monoxide disproportionation reaction over CO/Al$_2$O′ and Co/SiO′ catalysts, Catal. Lett. **63** (1999) 135–141
3.69 P. Pinheiro, M. C. Schouler, P. Gadelle, M. Mermoux, E. Dooryhée: Effect of hydrogen on the orientation of carbon layers in deposits from the carbon monoxide disproportionation reaction over Co/Al$_2$O$_3$ catalysts, Carbon **38** (2000) 1469–1479
3.70 P. Pinheiro, P. Gadelle: Chemical state of a supported iron-cobalt catalyst during CO disproportionation. I. Thermodynamic study, J. Phys. Chem. Solids **62** (2001) 1015–1021
3.71 P. Pinheiro, P. Gadelle, C. Jeandey, J. L. Oddou: Chemical state of a supported iron-cobalt catalyst during CO disproportionation. II. Experimental study, J. Phys. Chem. Solids **62** (2001) 1023–1037
3.72 C. Laurent, E. Flahaut, A. Peigney, A. Rousset: Metal nanoparticles for the catalytic synthesis of carbon nanotubes, New J. Chem. **22** (1998) 1229–1237
3.73 A. Peigney, C. Laurent, F. Dobigeon, A. Rousset: Carbon nanotubes grown in situ by a novel catalytic method, J. Mater. Res. **12** (1997) 613–615
3.74 V. Ivanov, J. B. Nagy, P. Lambin, A. Lucas, X. B. Zhang, X. F. Zhang, D. Bernaerts, G. Van Tendeloo, S. Amelinckx, J. Van Landuyt: The study of nanotubules produced by catalytic method, Chem. Phys. Lett. **223** (1994) 329–335
3.75 V. Ivanov, A. Fonseca, J. B. Nagy, A. Lucas, P. Lambin, D. Bernaerts, X. B. Zhang: Catalytic production and purification of nanotubules having fullerene-scale diameters, Carbon **33** (1995) 1727–1738
3.76 K. Hernadi, A. Fonseca, J. B. Nagy, D. Bernaerts, A. Fudala, A. Lucas: Catalytic synthesis of carbon nanotubes using zeolite support, Zeolites **17** (1996) 416–423

3.77 H. Dai, A. G. Rinzler, P. Nikolaev, A. Thess, D. T. Colbert, R. E. Smalley: Single-wall nanotubes produced by metal-catalysed disproportionation of carbon monoxide, Chem. Phys. Lett. **260** (1996) 471–475

3.78 A. M. Cassel, J. A. Raymakers, J. Kong, H. Dai: Large scale CVD synthesis of single-walled carbon nanotubes, J. Phys. Chem. B **109** (1999) 6484–6492

3.79 B. Kitiyanan, W. E. Alvarez, J. H. Harwell, D. E. Resasco: Controlled production of single-wall carbon nanotubes by catalytic decomposition of CO on bimetallic Co-Mo catalysts, Chem. Phys. Lett. **317** (2000) 497–503

3.80 E. Flahaut: Synthèse par voir catalytique et caractérisation de composites nanotubes de carbone-metal-oxyde Poudres et matériaux denses. Ph.D. Thesis (Univers. Paul Sabatier, Toulouse 1999)

3.81 A. Govindaraj, E. Flahaut, C. Laurent, A. Peigney, A. Rousset, C. N. R. Rao: An investigation of carbon nanotubes obtained from the decomposition of methane over reduced $Mg_{1-x}M_xAl_2O_4$ spinel catalysts, J. Mater. Res. **14** (1999) 2567–2576

3.82 E. Flahaut, A. Peigney, C. Laurent, A. Rousset: Synthesis of single-walled carbon nanotube-Co-MgO composite powders and extraction of the nanotubes, J. Mater. Chem. **10** (2000) 249–252

3.83 J. Kong, A. M. Cassel, H. Dai: Chemical vapor deposition of methane for single-walled carbon nanotubes, Chem. Phys. Lett. **292** (1998) 567–574

3.84 R. Marangoni, P. Serp, R. Feurrer, Y. Kihn, P. Kalck, C. Vahlas: Carbon nanotubes produced by substrate free metalorganic chemical vapor deposition of iron catalyst and ethylene, Carbon **39** (2001) 443–449

3.85 R. Sen, A. Govindaraj, C. N. R. Rao: Carbon nanotubes by the metallocene route, Chem. Phys. Lett. **267** (1997) 276–280

3.86 Y. Y. Fan, H. M. Cheng, Y. L. Wei, G. Su, S. H. Shen: The influence of preparation parameters on the mass production of vapor grown carbon nanofibers, Carbon **38** (2000) 789–795

3.87 L. Ci, J. Wei, B. Wei, J. Liang, C. Xu, D. Wu: Carbon nanofibers and single-walled carbon nanotubes prepared by the floating catalyst method, Carbon **39** (2001) 329–335

3.88 O. A. Nerushev, M. Sveningsson, L. K. L. Falk, F. Rohmund: Carbon nanotube films obtained by thermal vapour deposition, J. Mater. Chem. **11** (2001) 1122–1132

3.89 F. Rohmund, L. K. L. Falk, F. E. B. Campbell: A simple method for the production of large arrays of aligned carbon nanotubes, Chem. Phys. Lett. **328** (2000) 369–373

3.90 G. G. Tibbets, C. A. Bernardo, D. W. Gorkiewicz, R. L. Alig: Role of sulfur in the production of carbon fibers in the vapor phase, Carbon **32** (1994) 569–576

3.91 W. Q. Han, P. Kholer-Riedlich, T. Seeger, F. Ernst, M. Ruhle, N. Grobert, W. K. Hsu, B. H. Chang, Y. Q. Zhu, H. W. Kroto, M. Terrones, H. Terrones: Aligned CN_x nanotubes by pyrolysis of ferrocene under NH_3 atmosphere, Appl. Phys. Lett. **77** (2000) 1807–1809

3.92 R. E. Smalley, J. H. Hafner, D. T. Colbert, K. Smith: Catalytic growth of single-wall carbon nanotubes from metal particles, US patent US19980601010903 (1998)

3.93 T. Kyotani, L. F. Tsai, A. Tomita: Preparation of ultrafine carbon tubes in nanochannels of an anodic aluminum oxide film, Chem. Mater. **8** (1996) 2109–2113

3.94 W. K. Hsu, J. P. Hare, M. Terrones, H. W. Kroto, D. R. M. Walton, P. J. F. Harris: Condensed-phase nanotubes, Nature **377** (1995) 687

3.95 W. S. Cho, E. Hamada, Y. Kondo, K. Takayanagi: Synthesis of carbon nanotubes from bulk polymer, Appl. Phys. Lett. **69** (1996) 278–279

3.96 Y. L. Li, Y. D. Yu, Y. Liang: A novel method for synthesis of carbon nanotubes: Low temperature solid pyrolysis, J. Mater. Res. **12** (1997) 1678–1680

3.97 S. Fan, M. Chapline, N. Franklin, T. Tombler, A. M. Cassel, H. Dai: Self-oriented regular arrays of carbon nanotubes and their field emission properties, Science **283** (1999) 512–514

3.98 M. L. Terranova, S. Piccirillo, V. Sessa, P. Sbornicchia, M. Rossi, S. Botti, D. Manno: Growth of single-walled carbon nanotubes by a novel technique using nanosized graphite as carbon source, Chem. Phys. Lett. **327** (2000) 284–290

3.99 R. L. Vander Wal, T. Ticich, V. E. Curtis: Diffusion flame synthesis of single-walled carbon nanotubes, Chem. Phys. Lett. **323** (2000) 217–223

3.100 Y. Y. Wei, G. Eres, V. I. Merkulov, D. H. Lowdens: Effect of film thickness on carbon nanotube growth by selective area chemical vapor deposition, Appl. Phys. Lett. **78** (2001) 1394–1396

3.101 W. Z. Li, S. S. Xie, L. X. Qian, B. H. Chang, B. S. Zou, W. Y. Zhou, R. A. Zha, G. Wang: Large scale synthesis of aligned carbon nanotubes, Science **274** (1996) 1701–1703

3.102 F. Zheng, L. Liang, Y. Gao, J. H. Sukamto, L. Aardahl: Carbon nanotubes synthesis using mesoporous silica templates, Nanolett. **2** (2002) 729–732

3.103 S. H. Jeong, O.-K. Lee, K. H. Lee, S. H. Oh, C. G. Park: Preparation of aligned carbon nanotubes with prescribed dimension: Template synthesis and sonication cutting approach, Chem. Mater. **14** (2002) 1859–1862

3.104 N. S. Kim, Y. T. Lee, J. Park, H. Ryu, H. J. Lee, S. Y. Choi, J. Choo: Dependence of vertically aligned growth of carbon nanotubes on catalyst, J. Phys. Chem. B **106** (2002) 9286–9290

3.105 C. J. Lee, D. W. Kim, T. J. Lee, Y. C. Choi, Y. S. Park, Y. H. Lee, W. B. Choi, N. S. Lee, G.-S. Park, J. M. Kim: Synthesis of aligned carbon nanotubes using thermal chemical vapor deposition, Chem. Phys. Lett. **312** (1999) 461–468

3.106 W. D. Zhang, Y. Wen, S. M. Liu, W. C. Tjiu, G. Q. Xu, L. M. Gan: Synthesis of vertically aligned carbon

nanotubes on metal deposited quartz plates, Carbon **40** (2002) 1981–1989
3.107 S. Huang, L. Dai, A. W. H. Mau: Controlled fabrication of large scale aligned carbon nanofiber/nanotube patterns by photolithography, Adv. Mater. **14** (2002) 1140–1143
3.108 R. Andrews, D. Jacques, A. M. Rao, F. Derbyshire, D. Qian, X. Fan, E. C. Dickey, J. Chen: Continous production of aligned carbon nanotubes: A step closer to commercial realization, Chem. Phys. Lett. **303** (1999) 467–474
3.109 C. N. R. Rao, R. Sen, B. C. Satishkumar, A. Govindaraj: Large aligned carbon nanotubes bundles from ferrocene pyrolysis, Chem. Commun. **15** (1998) 1525–1526
3.110 X. Zhang, A. Cao, B. Wei, Y. Li, J. Wei, C. Xu, D. Wu: Rapid growth of well-aligned carbon nanotube arrays, Chem. Phys. Lett. **362** (2002) 285–290
3.111 X. Zhang, A. Cao, Y. Li, C. Xu, J. Liang, D. Wu, B. Wei: Self-organized arrays of carbon nanotube ropes, Chem. Phys. Lett. **351** (2002) 183–188
3.112 K. S. Choi, Y. S. Cho, S. Y. Hong, J. B. Park, D. J. Kim: Effects of ammonia on the alignment of carbon nanotubes in metal-assisted chemical vapor deposition, J. Eur. Cer. Soc. **21** (2001) 2095–2098
3.113 M. Endo, H. W. Kroto: Formation of carbon nanofibers, J. Phys. Chem. **96** (1992) 6941–6944
3.114 R. S. Wagner: VLS mechanisms of crystal growth. In: *Whisker Technology*, ed. by P. Levit A. (Wiley, New York 1970) pp. 47–72
3.115 Y. H. Lee, S. G. Kim, D. Tomanek: Catalytic growth of single-wall carbon nanotubes: An ab initio study, Phys. Rev. Lett. **78** (1997) 2393–2396
3.116 H. Dai: Carbon Nanotubes: Synthesis, integration, and properties, Acc. Chem. Res. **35** (2002) 1035–1044
3.117 Y. Saito, M. Okuda, N. Fujimoto, T. Yoshikawa, M. Tomita, T. Hayashi: Single-wall carbon nanotubes growing radially from Ni fine particles formed by arc evaporation, Jpn. J. Appl. Phys. **33** (1994) L526–L529
3.118 J. Bernholc, C. Brabec, M. Buongiorno Nardelli, A. Malti, C. Roland, B. J. Yakobson: Theory of growth and mechanical properties of nanotubes, Appl. Phys. A **67** (1998) 39–46
3.119 M. Pacheco: Synthèse des nanotubes de carbone par arc electrique. Ph.D. Thesis (Université Toulouse III, Toulouse 2003)
3.120 K. Méténier, S. Bonnamy, F. Béguin, C. Journet, P. Bernier, L. M. de la Chapelle, O. Chauvet, S. Lefrant: Coalescence of single walled nanotubes and formation of multi-walled carbon nanotubes under high temperature treatments, Carbon **40** (2002) 1765–1773
3.121 P. G. Collins, P. Avouris: Nanotubes for electronics, Sci. Am. **283** (2000) 38–45
3.122 Q.-H. Yang, P. X. Hou, S. Bai, M. Z. Wang, H. M. Cheng: Adsorption and capillarity of nitrogen in aggregated multi-walled carbon nanotubes, Chem. Phys. Lett. **345** (2001) 18–24
3.123 S. Inoue, N. Ichikuni, T. Suzuki, T. Uematsu, K. Kaneko: Capillary condensation of N_2 on multiwall carbon nanotubes, J. Phys. Chem. **102** (1998) 4689–4692
3.124 M. Eswaramoorthy, R. Sen, C. N. R. Rao: A study of micropores in single-walled carbon nanotubes by the adsorption of gases and vapors, Chem. Phys. Lett. **304** (1999) 207–210
3.125 A. Peigney, Ch. Laurent, E. Flahaut, R. R. Bacsa, A. Rousset: Specific surface area of carbon nanotubes and bundles of carbon nanotubes, Carbon **39** (2001) 507–514
3.126 E. Frackowiak, S. Delpeux, K. Jurewicz, K. Szostak, D. Cazorla-Amoros, F. Béguin: Enhanced capacitance of carbon nanotubes through chemical activation, Chem. Phys. Lett. **336** (2002) 35–41
3.127 M. Muris, N. Dupont-Pavlosky, M. Bienfait, P. Zeppenfeld: Where are the molecules adsorbed on single-walled nanotubes?, Surf. Sci. **492** (2001) 67–74
3.128 A. Fujiwara, K. Ishii, H. Suematsu, H. Kataura, Y. Maniwa, S. Suzuki, Y. Achiba: Gas adsorption in the inside and outside of single-walled carbon nanotubes, Chem. Phys. Lett. **336** (2001) 205–211
3.129 C. M. Yang, H. Kanoh, K. Kaneko, M. Yudasaka, S. Iijima: Adsorption behaviors of HiPco single-walled carbon nanotubes aggregates for alcohol vapors, J. Phys. Chem. **106** (2002) 8994–8999
3.130 J. Zhao, A. Buldum, J. Han, J. P. Lu: Gas molecule adsorption in carbon nanotubes and nanotube bundles, Nanotechnology **13** (2002) 195–200
3.131 K. A. Williams, P. C. Eklund: Monte Carlo simulation of H_2 physisorption in finite diameter carbon nanotube ropes, Chem. Phys. Lett. **320** (2000) 352–358
3.132 A. Chambers, C. Park, R. T. K. Baker, N. Rodriguez: Hydrogen storage in graphite nanofibers, J. Phys. Chem. B. **102** (1998) 4253–4256
3.133 J. Hilding, E. A. Grulke, S. B. Sinnott, D. Qian, R. Andrews, M. Jagtoyen: Sorption of butane on carbon multiwall nanotubes at room temperature, Langmuir **17** (2001) 7540–7544
3.134 K. Masenelli-Varlot, E. McRae, N. Dupont-Pavlosky: Comparative adsorption of simple molecules on carbon nanotubes. Dependence of the adsorption properties on the nanotube morphology, Appl. Surf. Sci. **196** (2002) 209–215
3.135 D. J. Browning, M. L. Gerrard, J. B. Lakeman, I. M. Mellor, R. J. Mortimer, M. C. Turpin: Studies into the storage of hydrogen in carbon nanofibers: Proposal of a possible mechanism, Nanolett. **2** (2002) 201–205
3.136 F. H. Yang, R. T. Yang: Ab initio molecular orbital study of adsorption of atomic hydrogen on graphite: insight into hydrogen storage in carbon nanotubes, Carbon **40** (2002) 437–444
3.137 G. E. Froudakis: Why alkali-metal-doped carbon nanotubes possess high hydrogen uptake, Nanolett. **1** (2001) 531–533

3.138 H. Ulbricht, G. Moos, T. Hertel: Physisorption of molecular oxygen on single-wall carbon nanotube bundles and graphite, Phys. Rev. B **66** (2002) 075404-1–075404-7

3.139 H. Ulbricht, J. Kriebel, G. Moos, T. Hertel: Desorption kinetics and interaction of Xe with single-wall carbon nanotube bundles, Chem. Phys. Lett. **363** (2002) 252–260

3.140 R. Saito, G. Dresselhaus, M.S. Dresselhaus: *Physical Properties of Carbon Nanotubes* (Imperial College Press, London 1998)

3.141 A. Charlier, E. McRae, R. Heyd, M.F. Charlier, D. Moretti: Classification for double-walled carbon nanotubes, Carbon **37** (1999) 1779–1783

3.142 A. Charlier, E. McRae, R. Heyd, M.F. Charlier: Metal semi-conductor transitions under uniaxial stress for single- and double-walled carbon nanotubes, J. Phys. Chem. Solids **62** (2001) 439–444

3.143 H. Ajiki, T. Ando: Electronic states of carbon nanotubes, J. Phys. Soc. Jap. **62** (1993) 1255–1266

3.144 C.T. White, T.N. Todorov: Carbon nanotubes as long ballistic conductors, Nature **393** (1998) 240–242

3.145 S. Frank, P. Poncharal, Z.L. Wang, W.A. de Heer: Carbon nanotube quantum resistors, Science **280** (1998) 1744–1746

3.146 W. Liang, M. Bockrath, D. Bozovic, J.H. Hafner, M. Tinkham, H. Park: Fabry–Perot interference in a nanotube electron waveguide, Nature **411** (2001) 665–669

3.147 L. Langer, V. Bayot, E. Grivei, J.-P. Issi, J.-P. Heremans, C.H. Olk, L. Stockman, C. van Haesendonck, Y. Buynseraeder: Quantum transport in a multiwalled carbon nanotube, Phys. Rev. Lett. **76** (1996) 479–482

3.148 M. Bockrath, D.H. Cobden, J. Lu, A.G. Rinzler, R.E. Smalley, L. Balents, P.L. McEuen: Luttinger-liquid behaviour in carbon nanotubes, Nature **397** (1999) 598–601

3.149 K. Liu, S. Roth, G.S. Duesberg, G.T. Kim, D. Popa, K. Mukhopadhyay, R. Doome, J. B'Nagy: Antilocalization in multiwalled carbon nanotubes, Phys. Rev. B **61** (2000) 2375–2379

3.150 Y.A. Kasumov, R. Deblock, M. Kociak, B. Reulet, H. Bouchiat, I.I. Khodos, Y.B. Gorbatov, V.T. Volkov, C. Journet, M. Burghard: Supercurrents through single-walled carabon nanotubes, Science **284** (1999) 1508–1511

3.151 B.W. Alphenaar, K. Tsukagoshi, M. Wagner: Magnetoresistance of ferromagnetically contacted carbon nanotubes, Phys. Eng. **10** (2001) 499–504

3.152 P.M. Ajayan, M. Terrrones, A. de la Guardia, V. Hue, N. Grobert, B.Q. Wei, H. Lezec, G. Ramanath, T.W. Ebbesen: Nanotubes in a flash – Ignition and reconstruction, Science **296** (2002) 705

3.153 S. Berber, Y. Kwon, D. Tomanek: Unusually high thermal conductivity of carbon nanotubes, Phys. Rev. Lett. **84** (2000) 4613–4616

3.154 S.N. Song, X.K. Wang, R.P.H. Chang, J.B. Ketterson: Electronic properties of graphite nanotubules from galvanomagnetic effects, Phys. Rev. Lett. **72** (1994) 697–700

3.155 M.-F. Yu, O. Lourie, M.J. Dyer, K. Moloni, T.F. Kelley, R.S. Ruoff: Strength and breaking mechanism of multiwalled carbon nanotubes under tensile load, Science **287** (2000) 637–640

3.156 D.A. Walters, L.M. Ericson, M.J. Casavant, J. Liu, D.T. Colbert, K.A. Smith, R.E. Smalley: Elastic strain of freely suspended single-wall carbon nanotube ropes, Appl. Phys. Lett. **74** (1999) 3803–3805

3.157 B.G. Demczyk, Y.M. Wang, J. Cumingd, M. Hetamn, W. Han, A. Zettl, R.O. Ritchie: Direct mechanical measurement of the tensile strength and eleastic modulus of multiwalled carbon nanotubes, Mater. Sci. Eng. A **334** (2002) 173–178

3.158 R.P. Gao, Z.L. Wang, Z.G. Bai, W.A. De Heer, L.M. Dai, M. Gao: Nanomechanics of inidividual carbon nanotubes from pyrolytically grown arrays, Phys. Rev. Lett. **85** (2000) 622–625

3.159 M.M.J. Treacy, T.W. Ebbesen, J.M. Gibson: Exceptionally high Young's modulus observed for individual carbon nanotubes, Nature **381** (1996) 678–680

3.160 N. Yao, V. Lordie: Young's modulus of single-wall carbon nanotubes, J. Appl. Phys. **84** (1998) 1939–1943

3.161 O. Lourie, H.D. Wagner: Transmission electron microscopy observations of fracture of single-wall carbon nanotubes under axial tension, Appl. Phys. Lett. **73** (1998) 3527–3529

3.162 S.C. Tsang, Y.K. Chen, P.J.F. Harris, M.L.H. Green: A simple chemical method of opening and filling carbon nanotubes, Nature **372** (1994) 159–162

3.163 M. Montioux: Filling single-wall carbon nanotubes, Carbon **40** (2002) 1809–1823

3.164 W.K. Hsu, S.Y. Chu, E. Munoz-Picone, J.L. Boldu, Firth S., P. Franchi, B.P. Roberts, A. Shilder, H. Terrones, N. Grobert, Y.Q. Zhu, M. Terrones, M.E. McHenry, H.W. Kroto, D.R.M. Walton: Metallic behaviour of boron-containing carbon nanotubes, Chem. Phys. Lett. **323** (2000) 572–579

3.165 R. Czerw, M. Terrones, J.C. Charlier, X. Blasé, B. Foley, R. Kamalakaran, N. Grobert, H. Terrones, D. Tekleab, P.M. Ajayan, W. Blau, M. Rühle, D.L. Caroll: Identification of electron donor states, in N-doped carbon nanotubes, Nanolett. **1** (2001) 457–460

3.166 O. Stephan, P.M. Ajayan, C. Colliex, P. Redlich, J.M. Lambert, P. Bernier, P. Lefin: Doping graphitic and carbon nanotube structures with boron and nitrogen, Science **266** (1994) 1683–1685

3.167 A. Loiseau, F. Willaime, N. Demoncy, N. Schramchenko, G. Hug, C. Colliex, H. Pascard: Boron nitride nanotubes, Carbon **36** (1998) 743–752

3.168 C.C. Tang, L.M. de la Chapelle, P. Li, Y.M. Liu, H.Y. Dang, S.S. Fan: Catalytic growth of nanotube

and nanobamboo structures of boron nitride, Chem. Phys. Lett. **342** (2001) 492–496
3.169 K. Suenaga, C. Colliex, N. Demoncy, A. Loiseau, H. Pascard, F. Willaime: Synthesis of nanoparticles and nanotubes with well separated layers of boron-nitride and carbon, Science **278** (1997) 653–655
3.170 D. Golberg, Y. Bando, L. Bourgeois, K. Kurashima, T. Sato: Large-scale synthesis and HRTEM analysis of single-walled B-and N-doped carbon nanotube bundles, Carbon **38** (2000) 2017–2027
3.171 B. Burteaux, A. Claye, B. W. Smith, M. Monthioux, D. E. Luzzi, J. E. Fischer: Abundance of encapsulated C_{60} in single-wall carbon nanotubes, Chem. Phys. Lett. **310** (1999) 21–24
3.172 D. Ugarte, A. Châtelain, W. A. de Heer: Nanocapillarity and chemistry in carbon nanotubes, Science **274** (1996) 1897–1899
3.173 J. Cook, J. Sloan, M. L. H. Green: Opening and filling carbon nanotubes, Fuller. Sci. Technol. **5** (1997) 695–704
3.174 P. M. Ajayan, S. Iijima: Capillarity-induced filling of carbon nanotubes, Nature **361** (1993) 333–334
3.175 P. M. Ajayan, T. W. Ebbesen, T. Ichihashi, S. Iijima, K. Tanigaki, H. Hiura: Opening carbon nanotubes with oxygen and implications for filling, Nature **362** (1993) 522–525
3.176 S. Seraphin, D. Zhou, J. Jiao, J. C. Withers, R. Loufty: Yttrium carbide in nanotubes, Nature **362** (1993) 503
3.177 S. Seraphin, D. Zhou, J. Jiao, J. C. Withers, R. Loufty: Selective encapsulation of the carbides of yttrium and titanium into carbon nanoclusters, Appl. Phys. Lett. **63** (1993) 2073–2075
3.178 R. S. Ruoff, D. C. Lorents, B. Chan, R. Malhotra, S. Subramoney: Single-crystal metals encapsulated in carbon nanoparticles, Science **259** (1993) 346–348
3.179 A. Loiseau, H. Pascard: Synthesis of long carbon nanotubes filled with Se, S, Sb, and Ge by the arc method, Chem. Phys. Lett. **256** (1996) 246–252
3.180 N. Demoncy, O. Stephan, N. Brun, C. Colliex, A. Loiseau, H. Pascard: Filling carbon nanotubes with metals by the arc discharge method: The key role of sulfur, Eur. Phys. J. B **4** (1998) 147–157
3.181 C. H. Kiang, J. S. Choi, T. T. Tran, A. D. Bacher: Molecular nanowires of 1 nm diameter from capillary filling of single-walled carbon nanotubes, J. Phys. Chem. B **103** (1999) 7449–7551
3.182 Z. L. Zhang, B. Li, Z. J. Shi, Z. N. Gu, Z. Q. Xue, L. M. Peng: Filling of single-walled carbon nanotubes with silver, J. Mater. Res. **15** (2000) 2658–2661
3.183 A. Govindaraj, B. C. Satishkumar, M. Nath, C. N. R. Tao: Metal nanowires and intercalated metal layers in single-walled carbon nanotubes bundles, Chem. Mater. **12** (2000) 202–205
3.184 J. Mittal, M. Monthioux, H. Allouche: Synthesis of SWNT-based hybrid nanomaterials from photolysis-enhanced chemical processes, Chem. Phys. Lett. **339** (2001) 311–318
3.185 E. Dujardin, T. W. Ebbesen, H. Hiura, K. Tanigaki: Capillarity and wetting of carbon nanotubes, Science **265** (1994) 1850–1852
3.186 E. Flahaut, J. Sloan, K. S. Coleman, V. C. Williams, S. Friedrichs, N. Hanson, M. L. H. Green: 1D p-block halide crystals confined into single walled carbon nanotubes, Proc. Mater. Res. Soc. Symp. **633** (2001) A13.15.1–A13.15.6
3.187 J. Sloan, A. I. Kirkland, J. L. Hutchison, M. L. H. Green: Integral atomic layer architectures of 1D crystals inserted into single walled carbon nanotubes, Chem. Commun. (2002) 1319–1332
3.188 J. Sloan, M. C. Novotny, S. R. Bailey, G. Brown, C. Xu, V. C. Williams, S. Friedrichs, E. Flahaut, R. L. Callender, A. P. E. York, K. S. Coleman, M. L. H. Green, R. E. Dunin-Borkowski, J. L. Hutchison: Two layer 4 : 4 co-ordinated KI crystals grown within single walled carbon nanotubes, Chem. Phys. Lett. **329** (2000) 61–65
3.189 G. Brown, S. R. Bailey, J. Sloan, C. Xu, S. Friedriechs, E. Flahaut, K. S. Coleman, J. L. Hutchinson, R. E. Dunin-Borkowski, M. L. H. Green: Electron beam induced in situ clusterisation of 1D $ZrCl_4$ chains within single-walled carbon nanotubes, Chem. Commun. **9** (2001) 845–846
3.190 B. W. Smith, M. Monthioux, D. E. Luzzi: Encapsulated C_{60} in carbon nanotubes, Nature **396** (1998) 323–324
3.191 B. W. Smith, D. E. Luzzi: Formation mechanism of fullerene peapods and coaxial tubes: A path to large scale synthesis, Chem. Phys. Lett. **321** (2000) 169–174
3.192 K. Hirahara, K. Suenaga, S. Bandow, H. Kato, T. Okazaki, H. Shinohara, S. Iijima: One-dimensional metallo-fullerene crystal generated inside single-walled carbon nanotubes, Phys. Rev. Lett. **85** (2000) 5384–5387
3.193 B. W. Smith, M. Monthioux, D. E. Luzzi: Carbon nanotube encapsulated fullerenes: A unique class of hybrid material, Chem. Phys. Lett. **315** (1999) 31–36
3.194 D. E. Luzzi, B. W. Smith: Carbon cage structures in single wall carbon nanotubes: A new class of materials, Carbon **38** (2000) 1751–1756
3.195 S. Bandow, M. Takisawa, K. Hirahara, M. Yudasoka, S. Iijima: Raman scattering study of double-wall carbon nanotubes derived from the chains of fullerenes in single-wall carbon nanotubes, Chem. Phys. Lett. **337** (2001) 48–54
3.196 B. W. Smith, D. E. Luzzi, Y. Achiba: Tumbling atoms and evidence for charge transfer in $La_2@C_{80}$@SWNT, Chem. Phys. Lett. **331** (2000) 137–142
3.197 K. Suenaga, M. Tence, C. Mory, C. Colliex, H. Kato, T. Okazaki, H. Shinohara, K. Hirahara, S. Bandow, S. Iijima: Element-selective single atom imaging, Science **290** (2000) 2280–2282
3.198 D. E. Luzzi, B. W. Smith, R. Russo, B. C. Satishkumar, F. Stercel, N. R. C. Nemes: Encapsulation of metallofullerenes and metallocenes in carbon nanotubes, Proc. Electronic Properties of Novel Materials-XVI Int. Winterschool – AIP Conf. Proc., Melville 2001, ed. by

H. Kuzmany, J. Fink, M. Mehring, S. Roth (Springer, Berlin, Heidelberg 2001) **591** 622–626
3.199 G.H. Jeong, R. Hatakeyama, T. Hirata, K. Tohji, K. Motomiya, N. Sato, Y. Kawazoe: Structural deformation of single-walled carbon nanotubes and fullerene encapsulation due to magnetized plasma ion irradiation, Appl. Phys. Lett. **79** (2001) 4213–4215
3.200 Y.P. Sun, K. Fu, Y. Lin, W. Huang: Functionalized carbon nanotubes: Properties and applications, Acc. Chem. Res. **35** (2002) 1095–1104
3.201 J. Chen, M.A. Hamon, M. Hui, C. Yongsheng, A.M. Rao, P.C. Eklund, R.C. Haddon: Solution properties of single-walled carbon nanotubes, Science **282** (1998) 95–98
3.202 J. Chen, A.M. Rao, S. Lyuksyutov, M.E. Itkis, M.A. Hamon, H. Hu, R.W. Cohn, P.C. Eklund, D.T. Colbert, R.E. Smalley, R.C. Haddon: Dissolution of full-length single-walled carbon nanotubes, J. Phys. Chem. B **105** (2001) 2525–2528
3.203 F. Pompeo, D.E. Resasco: Water solubilization of single-walled carbon nanotubes by functionalization with glucosamine, Nanolett. **2** (2002) 369–373
3.204 Y.P. Sun, W. Huang, Y. Lin, K. Fu, A. Kitaygorodskiy, L.A. Riddle, Y.J. Yu, D.L. Carroll: Soluble dendron-functionalized carbon nanotubes: Preparation, characterization, and properties, Chem. Mater. **13** (2001) 2864–2869
3.205 K. Fu, W. Huang, Y. Lin, L.A. Riddle, D.L. Carroll, Y.P. Sun: Defunctionalization of functionalized carbon nanotubes, Nanolett. **1** (2001) 439–441
3.206 P.W. Chiu, G.S. Duesberg, U. Dettlaff-Weglikowska, S. Roth: Interconnection of carbon nanotubes by chemical functionalization, Appl. Phys. Lett. **80** (2002) 3811–3813
3.207 E.T. Mickelson, C.B. Huffman, A.G. Rinzler, R.E. Smalley, R.H. Hauge, J.L. Margrave: Fluorination of single-wall carbon nanotubes, Chem. Phys. Lett. **296** (1998) 188–194
3.208 V.N. Khabashesku, W.E. Billups, J.L. Margrave: Fluorination of single-wall carbon nanotubes and subsequent derivatization reactions, Acc. Chem. Res. **35** (2002) 1087–1095
3.209 P.J. Boul, J. Liu, E.T. Michelson, C.B. Huffman, L.M. Ericson, I.W. Chiang, K.A. Smith, D.T. Colbert, R.H. Hauge, J.L. Margrave, R.E. Smalley: Reversible side-wall functionalization of buckytubes, Chem. Phys. Lett. **310** (1999) 367–372
3.210 J.L. Bahr, J. Yang, D.V. Kosynkin, M.J. Bronikowski, R.E. Smalley, J.M. Tour: Functionalization of carbon nanotubes by electrochemical reduction of aryl diazonium salts: A bucky paper electrode, J. Am. Chem. Soc. **123** (2001) 6536–6542
3.211 M. Holzinger, O. Vostrowsky, A. Hirsch, F. Hennrich, M. Kappes, R. Weiss, F. Jellen: Sidewall functionalization of carbon nanotubes, Angew. Chem. Int. Ed. **40** (2001) 4002–4005
3.212 Y. Chen, R.C. Haddon, S. Fang, A.M. Rao, P.C. Eklund, W.H. Lee, E.C. Dickey, E.A. Grulke, J.C. Pendergrass, A. Chavan, B.E. Haley, R.E. Smalley: Chemical attachment of organic functional groups to single-walled carbon nanotube material, J. Mater. Res. **13** (1998) 2423–2431
3.213 C. Velasco-Santos, A.L. Martinez-Hernandez, M. Lozada-Cassou, A. Alvarez-Castillo, V.M. Castano: Chemical functionalization of carbon nanotubes through an organosilane, Nanotechnology **13** (2002) 495–498
3.214 A. Star, J.F. Stoddart, D. Steuerman, M. Diehl, A. Boukai, E.W. Wong, X. Yang, S.W. Chung, H. Choi, J.R. Heath: Preparation and properties of polymer-wrapped single-walled carbon nanotubes, Angew. Chem. Int. Ed. **41** (2002) 1721–1725
3.215 J.H. Hafner, C.L. Cheung, A.T. Wooley, C.M. Lieber: Structural and functional imaging with carbon nanotube AFM probes, Progr. Biophys. Molec. Biol. **77** (2001) 73–110
3.216 S.S. Wong, E. Joselevich, A.T. Woodley, C.L. Cheung, C.M. Lieber: Covalently functionalized nanotubes as nanometre-size probes in chemistry and biology, Nature **394** (1998) 52–55
3.217 C.L. Cheung, J.H. Hafner, C.M. Lieber: Carbon nanotube atomic force microscopy tips: Direct growth by chemical vapor deposition and application to high-resolution imaging, Proc. Nat. Acad. Sci. (2000) 973809–973813
3.218 W.A. de Heer, A. Châtelain, D. Ugarte: A carbon nanotube field-emission electron source, Science **270** (1995) 1179–1180
3.219 J.M. Bonard, J.P. Salvetat, T. Stockli, W.A. de Heer, L. Forro, A. Chatelain: Field emission from single-wall carbon nanotube films, Appl. Phys. Lett. **73** (1998) 918–920
3.220 W. Zhu, C. Bower, O. Zhou, G. Kochanski, S. Jin: Large curent density from carbon nanotube field emitters, Appl. Phys. Lett. **75** (1999) 873–875
3.221 Y. Saito, R. Mizushima, T. Tanaka, K. Tohji, K. Uchida, M. Yumura, S. Uemura: Synthesis, structure, and field emission of carbon nanotubes, Fuller. Sci. Technol. **7** (1999) 653–664
3.222 J. Kong, N.R. Franklin, C. Zhou, M.G. Chapline, S. Peng, K. Cho, H. Dai: Nanotube molecular wire as chemical sensors, Science **287** (2000) 622–625
3.223 P.G. Collins, K. Bradley, M. Ishigami, A. Zettl: Extreme oxygen sensitivity of electronic properties of carbon nanotubes, Science **287** (2000) 1801–1804
3.224 H. Chang, J.D. Lee, S.M. Lee, Y.H. Lee: Adsorption of NH_3 and NO_2 molecules on carbon nanotubes, Appl. Phys. Lett. **79** (2001) 3863–3865
3.225 O.K. Varghese, P.D. Kichambre, D. Gong, K.G. Ong, E.C. Dickey, C.A. Grimes: Gas sensing characteristics of multi-wall carbon nanotubes, Sens. Actuators B **81** (2001) 32–41
3.226 K.G. Ong, K. Zeng, C.A. Grimes: A wireless, passive carbon nanotube-based gas sensor, IEEE Sens. J. **2/2** (2002) 82–88

3.227 J. Kong, M. G. Chapline, H. Dai: Functionalized carbon nanotubes for molecular hydrogen sensors, Adv. Mater. **13** (2001) 1384–1386
3.228 F. Rodriguez-Reinoso: The role of carbon materials in heterogeneous catalysis, Carbon **36** (1998) 159–175
3.229 E. Auer, A. Freund, J. Pietsch, T. Tacke: Carbon as support for industrial precious metal catalysts, Appl. Catal. A **173** (1998) 259–271
3.230 J. M. Planeix, N. Coustel, B. Coq, B. Botrons, P. S. Kumbhar, R. Dutartre, P. Geneste, P. Bernier, P. M. Ajayan: Application of carbon nanotubes as supports in heterogeneous catalysis, J. Am. Chem. Soc. **116** (1994) 7935–7936
3.231 B. Coq, J. M. Planeix, V. Brotons: Fullerene-based materials as new support media in heterogeneous catalysis by metals, Appl. Catal. A **173** (1998) 175–183
3.232 G. Mestl, N. I. Maksimova, N. Keller, V. V. Roddatis, R. Schlögl: Carbon nanofilaments in heterogeneous catalysis: An industrial application for new carbon materials?, Angew. Chem. Int. Ed. Engl. **40** (2001) 2066–2068
3.233 J. E. Fischer, A. T. Johnson: Electronic properties of carbon nanotubes, Current Opinion Solid State Mater. Sci. **4** (1999) 28–33
3.234 M. Menon, A. N. Andriotis, G. E. Froudakis: Curvature dependence of the metal catalyst atom interaction with carbon nanotubes walls, Chem. Phys. Lett. **320** (2000) 425–434
3.235 R. Giordano, Ph. Serp, Ph. Kalck, Y. Kihn, J. Schreiber, C. Marhic, J.-L. Duvail: Preparation of Rhodium supported on carbon canotubes catalysts via surface mediated organometallic reaction, Eur. J. Inorg. Chem. (2003) 610–617
3.236 T. Kyotani, S. Nakazaki, W.-H. Xu, A. Tomita: Chemical modification of the inner walls of carbon nanotubes by HNO_3 oxidation, Carbon **39** (2001) 782–785
3.237 Z. J. Liu, Z. Y. Yuan, W. Zhou, L. M. Peng, Z. Xu: Co/carbon nanotubes monometallic system: The effects of oxidation by nitric acid, Phys. Chem. Chem. Phys. **3** (2001) 2518–2521
3.238 H.-B. Chen, J. D. Lin, Y. Cai, X. Y. Wang, J. Yi, J. Wang, G. Wei, Y. Z. Lin, D. W. Liao: Novel multiwalled nanotubes-supported and alkali-promoted Ru catalysts for ammonia synthesis under atmospheric pressure, Appl. Surf. Sci. **180** (2001) 328–335
3.239 Y. Zhang, H. B. Zhang, G. D. Lin, P. Chen, Y. Z. Yuan, K. R. Tsai: Preparation, characterization and catalytic hydroformylation properties of carbon nanotubes-supported Rh-phosphine catalyst, Appl. Catal. A **187** (1999) 213–224
3.240 M. S. Dresselhaus, K. A. Williams, P. C. Eklund: Hydrogen adsorption in carbon materials, Mater. Res. Soc. Bull. (November 1999) 45–50
3.241 H.-M. Cheng, Q.-H. Yang, C. Liu: Hydrogen storage in carbon nanotubes, Carbon **39** (2001) 1447–1454
3.242 G. G. Tibbetts, G. P. Meisner, C. H. Olk: Hydrogen storage capacity of carbon nanotubes, filaments, and vapor-grown fibers, Carbon **39** (2001) 2291–2301
3.243 F. L. Darkrim, P. Malbrunot, G. P. Tartaglia: Review of hydrogen storage adsorption in carbon nanotubes, Int. J. Hydrogen Energy **27** (2002) 193–202
3.244 C. Park, P. E. Anderson, C. D. Tan, R. Hidalgo, N. Rodriguez: Further studies of the interaction of hydrogen with graphite nanofibers, J. Phys. Chem. B **103** (1999) 10572–1058
3.245 C. C. Ahn, Y. Ye, B. V. Ratnakumar, C. Witham, R. C. Bowman, B. Fultz: Hydrogen adsorption measurements on graphite nanofibers, Appl. Phys. Lett. **73** (1998) 3378–3380
3.246 Q. Wang, J. K. Johnson: Computer simulations of hydrogen adsorption on graphite nanofibers, J. Phys. Chem. B **103** (1999) 277–281
3.247 M. Rzepka, P. Lamp, M. A. de la Casa-Lillo: Physisorption of hydrogen on microporous carbon and carbon nanotubes, J. Phys. Chem. B **102** (1998) 10894–10898
3.248 A. C. Dillon, K. M. Jones, T. A. Bekkedahl, C. H. Kiang, D. S. Bethune, M. J. Heben: Storage of hydrogen in single-walled carbon nanotubes, Nature **386** (1997) 377–379
3.249 Y. Ye, C. C. Ahn, C. Witham, R. C. Bowman, B. Fultz, J. Liu, A. G. Rinzler, D. Colbert, K. A. Smith, R. E. Smalley: Hydrogen adsorption and cohesive energy of single-walled carbon nanotubes, Appl. Phys. Lett. **74** (1999) 2307–2309
3.250 C. Liu, Y. Y. Fan, M. Liu, H. T. Cong, H. M. Cheng, M. S. Dresselhaus: Hydrogen storage in single-walled carbon nanotubes at room temperature, Science **286** (1999) 1127–1129
3.251 M. Hirscher, M. Becher, M. Haluska, U. Dettlaff-Weglikowska, A. Quintel, G. S. Duesberg, Y. M. Choi, P. Dwones, M. Hulman, S. Roth, I. Stepanek, P. Bernier: Hydrogen storage in sonicated carbon materials, Appl. Phys. A **72** (2001) 129–132
3.252 P. A. Gordon, R. B. Saeger: Molecular modeling of adsorptive energy storage: Hydrogen storage in single-walled carbon nanotubes, Ind. Eng. Chem. Res. **38** (1999) 4647–4655
3.253 P. Marinelli, R. Pellenq, J. Conard: H stocké dans les carbones un site légèrement métastable, matériaux, Tours 2002 AF-14-020
3.254 S. M. Lee, H. Y. Lee: Hydrogen storage in single-walled carbon nanotubes, Appl. Phys. Lett. **76** (2000) 2877–2879
3.255 S. M. Lee, K. S. Park, Y. C. Choi, Y. S. Park, J. M. Bok, D. J. Bae, K. S. Nahm, Y. G. Choi, C. S. Yu, N. Kim, T. Frauenheim, Y. H. Lee: Hydrogen adsorption in carbon nanotubes, Synth. Met. **113** (2000) 209–216
3.256 H. Cheng, G. P. Pez, A. C. Cooper: Mechanism of hydrogen sorption in single-walled carbon nanotubes, J. Am. Chem. Soc. **123** (2001) 5845–5846
3.257 A. Kusnetzova, D. B. Mawhinney, V. Naumenko, J. T. Yates, J. Liu, R. E. Smalley: Enhancement

of adsorption inside of single-walled nanotubes: Opening the entry ports, Chem. Phys. Lett. **321** (2000) 292–296

3.258 G. E. Gadd, M. Blackford, S. Moricca, N. Webb, P. J. Evans, A. M. Smith, G. Jacobsen, S. Leung, A. Day, Q. Hua: The world's smallest gas cylinders?, Science **277** (1997) 933–936

3.259 Z. Mao, S. B. Sinnott: A computational study of molecular diffusion and dynamic flow through carbon nanotubes, J. Phys. Chem. B **104** (2000) 4618–4624

3.260 Z. Mao, S. B. Sinnott: Separation of organic molecular mixtures in carbon nanotubes and bundles: Molecular dynamics simulations, J. Phys. Chem. B **105** (2001) 6916–6924

3.261 C. Gu, G.-H. Gao, Y. X. Yu, T. Nitta: Simulation for separation of hydrogen and carbon monoxide by adsorption on single-walled carbon nanotubes, Fluid Phase Equilibria **194/197** (2002) 297–307

3.262 R. Q. Long, R. T. Yang: Carbon nanotubes as superior sorbent for dioxine removal, J. Am. Chem. Soc. **123** (2001) 2058–2059

3.263 Y. H. Li, S. Wang, A. Cao, D. Zhao, X. Zhang, C. Xu, Z. Luan, D. Ruan, J. Liang, D. Wu, B. Wei: Adsorption of fluoride from water by amorphous alumina supported on carbon nanotubes, Chem. Phys. Lett. **350** (2001) 412–416

3.264 Y. H. Li, S. Wang, J. Wei, X. Zhang, C. Xu, Z. Luan, D. Wu, B. Wei: Lead adsorption on carbon nanotubes, Chem. Phys. Lett. **357** (2002) 263–266

3.265 C. Park, E. S. Engel, A. Crowe, T. R. Gilbert, N. M. Rodriguez: Use of carbon nanofibers in the removal of organic solvents from water, Langmuir **16** (2000) 8050–8056

3.266 M. P. Mattson, R. C. Haddon, A. M. Rao: Molecular functionalization of carbon nanotubes and use as substrates for neuronal growth, J. Molec. Neurosci. **14** (2000) 175–182

3.267 J. J. Davis, M. L. H. Green, H. A. O. Hill, Y. C. Leung, P. J. Sadler, J. Sloan, A. V. Xavier, S. C. Tsang: The immobilization of proteins in carbon nanotubes, Inorg. Chim. Acta **272** (1998) 261–266

3.268 R. J. Chen, Y. Zhang, D. Wang, H. Dai: Noncovalent sidewall functionalization of single-walled carbon nanotubes for protein immobilization, J. Am. Chem. Soc. **123** (2001) 3838–3839

3.269 M. Shim, N. W. S. Kam, R. J. Chen, Y. Li, H. Dai: Functionalization of carbon nanotubes for biocompatibility and biomolecular recognition, Nanolett. **2** (2002) 285–288

3.270 C. Dwyer, M. Guthold, M. Falvo, S. Washburn, R. Superfine, D. Erie: DNA-functionalized single-walled carbon nanotubes, Nanotechnology **13** (2002) 601–604

3.271 H. Huang, S. Taylor, K. Fu, Y. Lin, D. Zhang, T. W. Hanks, A. M. Rao, Y. Sun: Attaching proteins to carbon nanotubes via diimide-activated amidation, Nanolett. **2** (2002) 311–314

3.272 C. V. Nguyen, L. Delzeit, A. M. Cassell, J. Li, J. Han, M. Meyyappan: Preparation of nucleic acid functionalized carbon nanotube arrays, Nanolett. **2** (2002) 1079–1081

3.273 B. R. Azamian, J. J. Davis, K. S. Coleman, C. B. Bagshaw, M. L. H. Green: Bioelectrochemical single-walled carbon nanotubes, J. Am. Chem. Soc. **124** (2002) 12664–12665

3.274 C. L. Xu, B. Q. Wei, R. Z. Ma, J. Liang, X. K. Ma, D. H. Wu: Fabrication of aluminum-carbon nanotube composites and their electrical properties, Carbon **37** (1999) 855–858

3.275 T. Kuzumaki, K. Miyazawa, H. Ichinose, K. Ito: Processing of carbon nanotube reinforced aluminum composite, J. Mater. Res. **13** (1998) 2445–2449

3.276 T. Kuzumaki, O. Ujiie, H. Ichinose, K. Ito: Mechanical characteristics and preparation of carbon nanotube fiber-reinforced Ti composite, Adv. Eng. Mater. **2** (2000) 416–418

3.277 S. R. Dong, J. P. Tu, X. B. Zhang: An investigation of the sliding wear behavior of Cu-matrix composite reinforced by carbon nanotubes, Mater. Sci. Eng. A **313** (2001) 83–87

3.278 Y. B. Li, Q. Ya, B. Q. Wei, J. Liang, D. H. Wu: Processing of a carbon nanotubes-Fe82P18 metallic glass composite, J. Mater. Sci. Lett. **17** (1998) 607–609

3.279 A. Peigney: Tougher ceramics with carbon nanotubes, Nature Mater. **2** (2003) 15–16

3.280 G. D. Zhan, J. D. Kuntz, J. Wan, A. K. Mukherjee: Single-wall carbon nanotubes as attractive toughening agents in alumina-based composites, Nature Mater. **2** (2003) 38–42

3.281 J. Sun, L. Gao, W. Li: Colloidal processing of carbon nanotube/alumina composites, Chem. Mater. **14** (2002) 5169–5172

3.282 V. G. Gavalas, R. Andrews, D. Bhattacharyya, L. G. Bachas: Carbon nanotube sol-gel composite materials, Nanolett. **1** (2001) 719–721

3.283 S. Rul, Ch. Laurent, A. Peigney, A. Rousset: Carbon nanotubes prepared in-situ in a cellular ceramic by the gelcasting-foam method, J. Eur. Ceram. Soc. **23** (2003) 1233–1241

3.284 R. Z. Ma, J. Wu, B. Q. Wei, J. Liang, D. H. Wu: Processing and properties of carbon nanotube/nano-SiC ceramic, J. Mater. Sci. **33** (1998) 5243–5246

3.285 C. Laurent, A. Peigney, O. Dumortier, A. Rousset: Carbon nanotubes-Fe-alumina nanocomposites. Part II: Microstructure and mechanical properties of the hot-pressed composites, J. Eur. Ceram. Soc. **18** (1998) 2005–2013

3.286 A. Peigney, C. Laurent, A. Rousset: Synthesis and characterization of alumina matrix nanocomposites containing carbon nanotubes, Key Eng. Mater. **132-136** (1997) 743–746

3.287 A. Peigney, C. Laurent, E. Flahaut, A. Rousset: Carbon nanotubes in novel ceramic matrix nanocomposites, Ceram. Intern. **26** (2000) 677–683

3.288 S. Rul: Synthèse de composites nanotubes de carbone-métal-oxyde. Ph.D. Thesis (Université Toulouse III, Toulouse 2002)
3.289 E. Flahaut, A. Peigney, C. Laurent, C. Marliere, F. Chastel, A. Rousset: Carbon nanotube-metal-oxide nanocomposites: Microstructure, electrical conductivity and mechanical properties, Acta Mater. **48** (2000) 3803–3812
3.290 R.R. Bacsa, C. Laurent, A. Peigney, W.S. Bacsa, T. Vaugien, A. Rousset: High specific surface area carbon nanotubes from catalytic chemical vapor deposition process, Chem. Phys. Lett. **323** (2000) 566–571
3.291 P. Coquay, A. Peigney, E. De Grave, R.E. Vandenberghe, C. Laurent: Carbon nanotubes by a CVD method. Part II: Formation of nanotubes from (Mg,Fe)O catalysts, J. Phys. Chem. B **106** (2002) 13199–13210
3.292 E. Flahaut, Ch. Laurent, A. Peigney: Double-walled carbon nanotubes in composite powders, J. Nanosci. Nanotech. **3** (2003) 151–158
3.293 A. Peigney, P. Coquay, E. Flahaut, R.E. Vandenberghe, E. De Grave, C. Laurent: A study of the formation of single- and double-walled carbon nanotubes by a CVD method, J Phys. Chem. B **105** (2001) 9699–9710
3.294 R.W. Siegel, S.K. Chang, B.J. Ash, J. Stone, P.M. Ajayan, R.W. Doremus, L.S. Schadler: Mechanical behavior of polymer and ceramic matrix nanocomposites, Scr. Mater. **44** (2001) 2061–2064
3.295 A. Peigney, E. Flahaut, C. Laurent, F. Chastel, A. Rousset: Aligned carbon nanotubes in ceramic-matrix nanocomposites prepared by high-temperature extrusion, Chem. Phys. Lett. **352** (2002) 20–25
3.296 S.L. Huang, M.R. Koblischka, K. Fossheim, T.W. Ebbesen, T.H. Johansen: Microstructure and flux distribution in both pure and carbon-nanotube-embedded $Bi_2Sr_2CaCu_2O_{8+\delta}$ superconductors, Physica C **311** (1999) 172–186
3.297 L. L. Yowell: Thermal management in ceramics: Synthesis and characterization of a zirconia-carbon nanotube composite, Proc Mater. Res. Soc. Symp. **633** (2002) A17.4.1–A17.4.6
3.298 A. Weidenkaff, S.G. Ebbinghaus, T. Lippert: $Ln_{1-x}A_xCoO_3$ (Ln = Er, La; A = Ca, Sr)/carbon nanotube composite materials applied for rechargeable Zn/Air batteries, Chem. Mater. **14** (2002) 1797–1805
3.299 D.S. Lim, J.W. An, H.J. Lee: Effect of carbon nanotube addition on the tribological behavior of carbon/carbon composites, Wear **252** (2002) 512–517
3.300 R. Andrews, D. Jacques, A.M. Rao, T. Rantell, F. Derbyshire, Y. Chen, J. Chen, R.C. Haddon: Nanotube composite carbon fibers, Appl. Phys. Lett. **75** (1999) 1329–1331
3.301 P.M. Ajayan, O. Stephan, C. Colliex, D. Trauth: Aligned carbon nanotube arrays formed by cutting a polymer resin-nanotube composite, Science **265** (1994) 1212–14
3.302 R. Haggenmueller, H.H. Gommans, A.G. Rinzler, J.E. Fischer, K.I. Winey: Aligned single-wall carbon nanotubes in composites by melt processing methods, Chem. Phys. Lett. **330** (2000) 219–225
3.303 L.S. Schadler, S.C. Giannaris, P.M. Ajayan: Load transfer in carbon nanotube epoxy composites, Appl. Phys. Lett. **73** (1998) 3842–3844
3.304 S.J.V. Frankland, A. Caglar, D.W. Brenner, M. Griebel: Molecular simulation of the influence of chemical cross-links on the shear strength of carbon nanotube-polymer interfaces, J. Phys. Chem. B **106** (2002) 3046–3048
3.305 H. D. Wagner: Nanotube-polymer adhesion: A mechanics approach, Chem. Phys. Lett. **361** (2002) 57–61
3.306 P.M. Ajayan, L.S. Schadler, C. Giannaris, A. Rubio: Single-walled carbon nanotube-polymer composites: Strength and weakness, Adv. Mater. **12** (2000) 750–753
3.307 X. Gong, J. Liu, S. Baskaran, R.D. Voise, J.S. Young: Surfactant-assisted processing of carbon nanotube/polymer composites, Chem. Mater. **12** (2000) 1049–1052
3.308 E.T. Thostenson, W.Z. Li, D.Z. Wang, Z.F. Ren, T.W. Chou: Carbon nanotube/carbon fiber hybrid multiscale composites, J. Appl. Phys. **91** (2002) 6034–6037
3.309 M.J. Biercuk, M.C. Llaguno, M. Radosavljevic, J.K. Hyun, A.T. Johnson, J.E. Fischer: Carbon nanotube composites for thermal management, Appl. Phys. Lett. **80** (2002) 2767–2769
3.310 S. Barrau, P. Demont, A. Peigney, C. Laurent, C. Lacabanne: DC and AC conductivity of carbon nanotube – Epoxy composites macromolecules, Macromol. **36** (2003) 5187–5194
3.311 J. Sandler, M.S.P. Shaffer, T. Prasse, W. Bauhofer, K. Schulte, A.H. Windle: Development of a dispersion process for carbon nanotubes in an epoxy matrix and the resulting electrical properties, Polymer **40** (1999) 5967–5971
3.312 Z. Jin, K.P. Pramoda, G. Xu, S.H. Goh: Dynamic mechanical behavior of melt-processed multi-walled carbon nanotube/poly(methyl methacrylate) composites, Chem. Phys. Lett. **337** (2001) 43–47
3.313 Z. Jin, K.P. Pramoda, S.H. Goh, G. Xu: Poly(vinylidene fluoride)-assisted melt-blending of multi-walled carbon nanotube/poly(methyl methacrylate) composites, Mat. Res. Bull. **37** (2002) 271–278
3.314 C.A. Cooper, D. Ravich, D. Lips, J. Mayer, H.D. Wagner: Distribution and alignment of carbon nanotubes and nanofibrils in a polymer matrix, Compos. Sci. Technol. **62** (2002) 1105–1112
3.315 J.M. Benoit, B. Corraze, S. Lefrant, W.J. Blau, P. Bernier, O. Chauvet: Transport properties of PMMA-carbon nanotubes composites, Synth. Met. **121** (2001) 1215–1216

3.316 J. M. Benoit, B. Corraze, O. Chauvet: Localization, Coulomb interactions, and electrical heating in single-wall carbon nanotubes/polymer composites, Phys. Rev. B: Cond. Matter. Mater. Phys. **65** (2002) 241405/1–241405/4

3.317 M. S. P. Shaffer, A. H. Windle: Fabrication and characterization of carbon nanotube/poly(vinyl alcohol) composites, Adv. Mater. **11** (1999) 937–941

3.318 L. Jin, C. Bower, O. Zhou: Alignment of carbon nanotubes in a polymer matrix by mechanical stretching, Appl. Phys. Lett. **73** (1998) 1197–1199

3.319 H. D. Wagner, O. Lourie, Y. Feldman, R. Tenne: Stress-induced fragmentation of multiwall carbon nanotubes in a polymer matrix, Appl. Phys. Lett. **72** (1998) 188–190

3.320 H. D. Wagner, O. Lourie, X. F. Zhou: Macrofragmentation and microfragmentation phenomena in composite materials, Composites Part A **30A** (1998) 59–66

3.321 J. R. Wood, Q. Zhao, H. D. Wagner: Orientation of carbon nanotubes in polymers and its detection by Raman spectroscopy, Composites Part A **32A** (2001) 391–399

3.322 Q. Zhao, J. R. Wood, H. D. Wagner: Using carbon nanotubes to detect polymer transitions, J. Polym. Sci., Part B: Polym. Phys. **39** (2001) 1492–1495

3.323 M. Cochet, W. K. Maser, A. M. Benito, M. A. Callejas, M. T. Martinesz, J. M. Benoit, J. Schreiber, O. Chauvet: Synthesis of a new polyaniline/nanotube composite: In-situ polymerisation and charge transfer through site-selective interaction, Chem. Commun. **16** (2001) 1450–1451

3.324 D. Qian, E. C. Dickey, R. Andrews, T. Rantell: Load transfer and deformation mechanisms in carbon nanotube-polystyrene composites, Appl. Phys. Lett. **76** (2000) 2868–2870

3.325 C. Wei, D. Srivastava, K. Cho: Thermal expansion and diffusion coefficients of carbon nanotube-polymer composites, Los Alamos Nat. Lab., Preprint Archive, Condensed Matter (archiv:cond-mat/0203349) (2002) 1–11

3.326 C. Pirlot, I. Willems, A. Fonseca, J. B. Nagy, J. Delhalle: Preparation and characterization of carbon nanotube/polyacrylonitrile composites, Adv. Eng. Mater. **4** (2002) 109–114

3.327 L. Cao, H. Chen, M. Wang, J. Sun, X. Zhang, F. Kong: Photoconductivity study of modified carbon nanotube/oxotitanium phthalocyanine composites, J. Phys. Chem. B **106** (2002) 8971–8975

3.328 I. Musa, M. Baxendale, G. A. J. Amaratunga, W. Eccleston: Properties of regular poly(3-octylthiophene)/multi-wall carbon nanotube composites, Synth. Met. **102** (1999) 1250

3.329 E. Kymakis, I. Alexandou, G. A. J. Amaratunga: Single-walled carbon nanotube-polymer composites: Electrical, optical and structural investigation, Synth. Met. **127** (2002) 59–62

3.330 K. Yoshino, H. Kajii, H. Araki, T. Sonoda, H. Take, S. Lee: Electrical and optical properties of conducting polymer-fullerene and conducting polymer-carbon nanotube composites, Fuller. Sci. Technol. **7** (1999) 695–711

3.331 S. A. Curran, P. M. Ajayan, W. J. Blau, D. L. Carroll, J. N. Coleman, A. B. Dalton, A. P. Davey, A. Drury, B. McCarthy, S. Maier, A. Strevens: A composite from poly(m-phenylenevinylene-co-2,5-dioctoxy-p-phenylenevinylene) and carbon nanotubes. A novel material for molecular optoelectronics, Adv. Mater. **10** (1998) 1091–1093

3.332 P. Fournet, D.F. O'Brien, J. N. Coleman, H. H. Horhold, W. J. Blau: A carbon nanotube composite as an electron transport layer for M3EH-PPV based light-emitting diodes, Synth. Met. **121** (2001) 1683–1684

3.333 H. S. Woo, R. Czerw, S. Webster, D. L. Carroll, J. Ballato, A. E. Strevens, D. O'Brien, W. J. Blau: Hole blocking in carbon nanotube-polymer composite organic light-emitting diodes based on poly (m-phenylene vinylene-co-2, 5-dioctoxy-p-phenylene vinylene), Appl. Phys. Lett. **77** (2000) 1393–1395

3.334 H. S. Woo, R. Czerw, S. Webster, D. L. Carroll, J. W. Park, J. H. Lee: Organic light emitting diodes fabricated with single wall carbon nanotubes dispersed in a hole conducting buffer: The role of carbon nanotubes in a hole conducting polymer, Synth. Met. **116** (2001) 369–372

3.335 H. Ago, K. Petritsch, M. S. P. Shaffer, A. H. Windle, R. H. Friend: Composites of carbon nanotubes and conjugated polymers for photovoltaic devices, Adv. Mater. **11** (1999) 1281–1285

3.336 B. Vigolo, A. Pénicaud, C. Coulon, C. Sauder, R. Pailler, C. Journet, P. Bernier, P. Poulin: Macroscopic fibers and ribbons of oriented carbon nanotubes, Science **290** (2000) 1331–1334

3.337 B. Vigolo, P. Poulin, M. Lucas, P. Launois, P. Bernier: Improved structure and properties of single-wall carbon nanotube spun fibers, Appl. Phys. Lett. **11** (2002) 1210–1212

3.338 P. Poulin, B. Vigolo, P. Launois: Films and fibers of oriented single wall nanotubes, Carbon **40** (2002) 1741–1749

3.339 K. Jiang, Q. Li, S. Fan: Spinning continuous carbon nanotube yarn, Nature **419** (2002) 801

3.340 P. Lambin, A. Fonseca, J. P. Vigneron, J. B'Nagy, A. A. Lucas: Structural and electronic properties of bent carbon nanotubes, Chem. Phys. Lett. **245** (1995) 85–89

3.341 L. Chico, V. H. Crespi, L. X. Benedict, S. G. Louie, M. L. Cohen: Pure carbon nanoscale devices: Nanotube heterojunctions, Phys. Rev. Lett. **76** (1996) 971–974

3.342 Z. Yao, H. W. C. Postma, L. Balents, C. Dekker: Carbon nanotube intramolecular junctions, Nature **402** (1999) 273–276

3.343 S.J. Tans, A.R.M. Verschueren, C. Dekker: Room temperature transistor based on single carbon nanotube, Nature **393** (1998) 49–52

3.344 R. Martel, T. Schmidt, H.R. Shea, T. Hertel, P. Avouris: Single and multi-wall carbon nanotube field effect transistors, Appl. Phys. Lett. **73** (1998) 2447–2449

3.345 V. Derycke, R. Martel, J. Appenzeller, P. Avouris: Carbon nanotube inter- and intramolecular logic gates, Nanolett. **1** (2001) 453–456

3.346 P. Kim, C.M. Lieber: Nanotube nanotweezers, Science **286** (1999) 2148–2150

3.347 P.G. Collins, M.S. Arnold, P. Avouris: Engineering carbon nanotubes using electrical breakdown, Science **292** (2001) 706–709

3.348 R.H. Baughman, C. Changxing, A.A. Zakhidov, Z. Iqbal, J.N. Barisci, G.M. Spinks, G.G. Wallace, A. Mazzoldi, D. de Rossi, A.G. Rinzler, O. Jaschinki S. Roth, M. Kertesz: Carbon nanotubes actuators, Science **284** (1999) 1340–1344

3.349 Y. Gao, Y. Bando: Carbon nanothermometer containing gallium, Nature **415** (2002) 599

3.350 C. Niu, E.K. Sichel, R. Hoch, D. Moy, H. Tennent: High power electro-chemical capacitors based on carbon nanotube electrodes, Appl. Phys. Lett. **70** (1997) 1480–1482

3.351 E. Frackowiak, F. Béguin: Electrochemical storage of energy in carbon nanotubes and nanostructured carbons, Carbon **40** (2002) 1775–1787

3.352 E. Frackowiak, K. Jurewicz, K. Szostak, S. Delpeux, F. Béguin: Nanotubular materials as electrodes for supercapacitors, Fuel Process. Technol. **77** (2002) 213–219

4. Nanowires

Nanowires are especially attractive for nanoscience studies as well as for nanotechnology applications. Nanowires, compared to other low dimensional systems, have two quantum confined directions while still leaving one unconfined direction for electrical conduction. This allows them to be used in applications which require electrical conduction, rather than tunneling transport. Because of their unique density of electronic states, nanowires in the limit of small diameters are expected to exhibit significantly different optical, electrical, and magnetic properties from their bulk 3-D crystalline counterparts. Increased surface area, very high density of electronic states and joint density of states near the energies of their van Hove singularities, enhanced exciton binding energy, diameter-dependent bandgap, and increased surface scattering for electrons and phonons are just some of the ways in which nanowires differ from their corresponding bulk materials. Yet the sizes of nanowires are typically large enough (>1 nm in the quantum confined direction) to have local crystal structures closely related to their parent materials, thereby allowing theoretical predictions about their properties to be made on the basis of an extensive literature relevant to their bulk properties.

Not only do nanowires exhibit many properties that are similar to, and others that are distinctly different from those of their bulk counterparts, nanowires have the advantage from an applications standpoint in that some of the materials parameters critical for certain properties can be independently controlled in nanowires but not in their bulk counterparts. Certain properties can also be enhanced nonlinearly in small diameter nanowires by exploiting the singular aspects of the 1-D electronic density of states. Furthermore, nanowires have been shown to provide a promising framework for applying the "bottom-up" approach [4.1] for the design of nanostructures for nanoscience investigations and for potential nanotechnology applications. Driven by: (1) these new research and development opportunities, (2) the smaller and smaller length scales now being used in the semiconductor, opto-electronics, and magnetics industries, and (3) the dramatic development of the biotechnology industry where the action is also at the nanoscale, the nanowire research field has developed with exceptional speed in the last few years. A review of the current status of nanowire research is therefore of significant broad interest at the present time. This review aims to focus on nanowire properties that differ from those of their parent crystalline bulk materials, with an eye toward possible applications that might emerge from the unique properties of nanowires and from future discoveries in this field.

4.1 **Synthesis** ... 100
- 4.1.1 Template-Assisted Synthesis ... 100
- 4.1.2 VLS Method for Nanowire Synthesis ... 105
- 4.1.3 Other Synthesis Methods ... 107
- 4.1.4 Hierarchical Arrangement and Superstructures of Nanowires ... 108

4.2 **Characterization and Physical Properties of Nanowires** ... 110
- 4.2.1 Structural Characterization ... 110
- 4.2.2 Transport Properties ... 115
- 4.2.3 Optical Properties ... 126

4.3 **Applications** ... 131
- 4.3.1 Electrical Applications ... 131
- 4.3.2 Thermoelectric Applications ... 133
- 4.3.3 Optical Applications ... 134
- 4.3.4 Chemical and Biochemical Sensing Devices ... 137
- 4.3.5 Magnetic Applications ... 137

4.4 **Concluding Remarks** ... 138

References ... 138

For quick reference, examples of typical nanowires that have been synthesized and studied are listed in Table 4.1. Also of use to the reader are review articles that focus on a comparison between nanowire and nanotube properties [4.2] and the many reviews that have been written about carbon nanotubes [4.3–5], which can be considered as a model one-dimensional system.

4.1 Synthesis

In this section we survey the most common synthetic approaches that have successfully afforded high quality nanowires of a large variety of materials (see Table 4.1). In Sect. 4.1.1, we discuss methods that make use of various templates with nanochannels to confine the nanowire growth to two dimensions. In Sect. 4.1.2, we present the synthesis of nanowires by the vapor-liquid-solid mechanism and its many variations. In Sect. 4.1.3, we present examples of other synthetic methods of general applicability. The last part of this section (Sect. 4.1.4) features several approaches that have been developed to organize nanowires into simple architectures.

4.1.1 Template-Assisted Synthesis

The template-assisted synthesis of nanowires is a conceptually simple and intuitive way to fabricate nanostructures [4.7–10]. These templates contain very small cylindrical pores or voids within the host material, and the empty spaces are filled with the chosen material, which adopts the pore morphology, to form nanowires. In this section, we describe the templates first and then describe strategies for filling them to make nanowires.

Template Synthesis

In template-assisted synthesis of nanostructures, the chemical stability and mechanical properties of the template, as well as the diameter, uniformity, and density of the pores are important characteristics to consider. Templates frequently used for nanowire synthesis include anodic alumina (Al_2O_3), nano-channel glass, ion track-etched polymers, and mica films.

Porous anodic alumina templates are produced by anodizing pure Al films in various acids [4.11–13]. Under carefully chosen anodization conditions, the resulting oxide film possesses a regular hexagonal array of parallel and nearly cylindrical channels, as shown in Fig. 4.1a. The self-organization of the pore structure in an anodic alumina template involves two coupled processes: pore formation with uniform diameters and pore ordering. The pores form with uniform diameters because of a delicate balance between electric-field-enhanced diffusion, which determines the growth rate of the alumina, and dissolution of the alumina into the acidic electrolyte [4.12]. The pores are believed to self-order because of mechanical stress at the aluminum-alumina interface due to volume expansion during the anodization. This stress produces a repulsive force between the pores, causing them to arrange in a hexagonal lattice [4.14]. Depending on the anodization conditions, the pore diameter can be systematically varied from < 10 nm up to 200 nm with a pore density in the range of 10^9–10^{11} pores/cm^2 [4.11, 12, 15, 16]. Many groups have shown that the pore size distribution and the pore ordering of the anodic alumina templates can be significantly improved by a two-step anodization technique [4.6, 17, 18] in which the aluminum oxide layer is dissolved after the first anodization in an acidic solution followed by a second anodization under the same conditions.

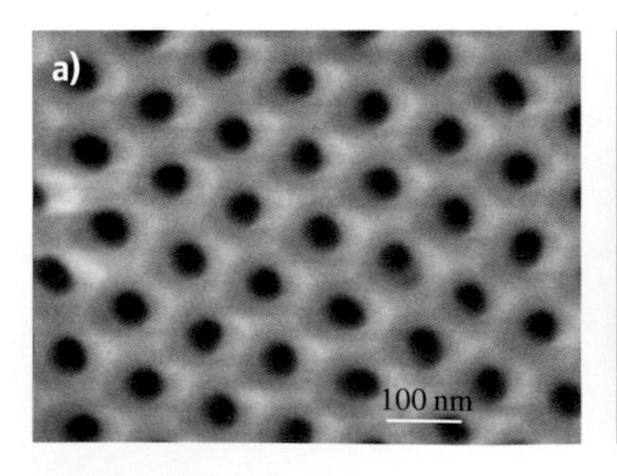

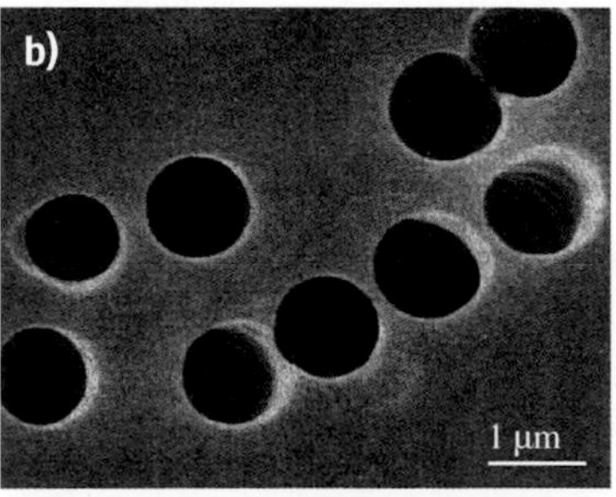

Fig. 4.1 (**a**) SEM images of the top surfaces of porous anodic alumina templates anodized with an average pore diameter of 44 nm [4.6]. (**b**) SEM image of the particle track-etched polycarbonate membrane, with a pore diameter of 1 μm [4.7]

Table 4.1 Selected syntheses of nanowires by material

Material	Growth Technique	Reference
Ag	DNA-template, redox	[4.23]
	template, pulsed ECD[a]	[4.24]
Au	template, ECD[a]	[4.25, 26]
Bi	stress-induced	[4.27]
	template, vapor-phase	[4.28]
	template, ECD[a]	[4.29], [4.30], [4.31]
	template, pressure-injection	[4.32], [4.16, 33]
Bi_2Te_3	template, dc ECD[a]	[4.34]
CdS	liquid-phase (surfactant), recrystallization	[4.35]
	template, ac ECD[a]	[4.36], [4.37]
CdSe	liquid-phase (surfactant), redox	[4.38]
	template, ac ECD[a]	[4.39], [4.40]
Cu	vapor deposition	[4.41]
	template, ECD[a]	[4.42]
Fe	template, ECD[c]	[4.15], [4.43]
	shadow deposition	[4.44]
GaN	template, CVD[c]	[4.45]
	VLS[b]	[4.46, 47]
GaAs	template, liquid/vapor OMCVD[d]	[4.48]
Ge	high-T, high-P liquid-phase, redox	[4.49]
	VLS[b]	[4.50]
	oxide-assisted	[4.51]
InAs	template, liquid/vapor OMCVD[d]	[4.48]
InP	VLS[b]	[4.52]
Mo	step decoration, ECD[a] + redox	[4.53]
Ni	template, ECD[a]	[4.22], [4.31, 54]
PbSe	liquid phase	[4.55]
Pd	step decoration, ECD[a]	[4.56]
Se	liquid-phase, recrystallization	[4.57]
	template, pressure injection	[4.58]
Si	VLS[b]	[4.59]
	laser-ablation VLS[b]	[4.60]
	oxide-assisted	[4.61]
	low-T VLS[b]	[4.62]
Zn	template, vapor-phase	[4.63]
	template, ECD[a]	[4.64]
ZnO	VLS[b]	[4.65]
	template, ECD[a]	[4.64, 66]

[a] Electrochemical deposition
[b] Vapor-liquid-solid growth
[c] Chemical vapor deposition
[d] Organometallic chemical vapor deposition

Another type of porous template commonly used for nanowire synthesis is the template type fabricated by chemically etching particle tracks originating from ion bombardment [4.19], such as track-etched polycarbonate membranes (Fig. 4.1b) [4.7,20,21] and also mica films [4.22].

Other porous materials can be used as host templates for nanowire growth, as discussed by *Ozin* [4.8]. Nanochannel glass (NCG), for example, contains a regular hexagonal array of capillaries similar to the pore structure in anodic alumina with a packing density as high as 3×10^{10} pores/cm^2 [4.9]. Porous Vycor glass that contains an interconnected network of pores less than 10 nm was also employed for the early study of nanostructures [4.68]. Mesoporous molecular sieves [4.69], termed MCM-41, possess hexagonally packed pores with very small channel diameters that can be varied between 2 nm and 10 nm. Conducting organic filaments have been fabricated in the nanochannels of MCM-41 [4.70]. Recently the DNA molecule has also been used as a template for growing nanometer-sized wires [4.23].

Diblock copolymers, polymers that consist of two chain segments with different properties, have also been utilized as templates for nanowire growth. When the two segments are immiscible in each other, phase segregation occurs, and depending on their volume ratio, spheres, cylinders, and lamellae may self-assemble. To form self-assembled arrays of nanopores, copolymers composed of polystyrene and polymethylmethacrylate [P(S-b-MMA)] [4.71] were used. By applying an electric field while the copolymer was heated above the glass transition temperature of the two constituent polymers, the self-assembled cylinders of PMMA could be aligned with their main axis perpendicular to the film. Selective removal of the PMMA component afforded the preparation of 14-nm-diameter ordered pore arrays with a packing density of 1.9×10^{11} cm^{-3}.

Nanowire Template Assisted Growth by Pressure Injection

The pressure injection technique is often employed when fabricating highly crystalline nanowires from a low-melting point material and when using porous templates with robust mechanical strength. In the high-pressure injection method, the nanowires are formed by pressure injecting the desired material in liquid form into the evacuated pores of the template. Due to the heating and the pressurization processes, the templates used for the pressure injection method must be chemically stable and be able to maintain their structural integrity at high temperatures and high pressures. Anodic aluminum oxide films and nano-channel glass are two typical materials used as templates in conjunction with the pressure injection filling technique. Metal nanowires (Bi, In, Sn, and Al) and semiconductor nanowires (Se, Te, GaSb, and Bi_2Te_3) have been fabricated in anodic aluminum oxide templates using this method [4.32,58,67].

The pressure P required to overcome the surface tension for the liquid material to fill the pores with a diameter d_W is determined by the Washburn equation [4.72]:

$$d_W = -4\gamma\cos\theta/P\,, \tag{4.1}$$

where γ is the surface tension of the liquid, and θ is the contact angle between the liquid and the template. To reduce the required pressure and to maximize the filling factor, some surfactants are used to decrease the surface tension and the contact angle. For example, the introduction of Cu in the Bi melt can facilitate filling the

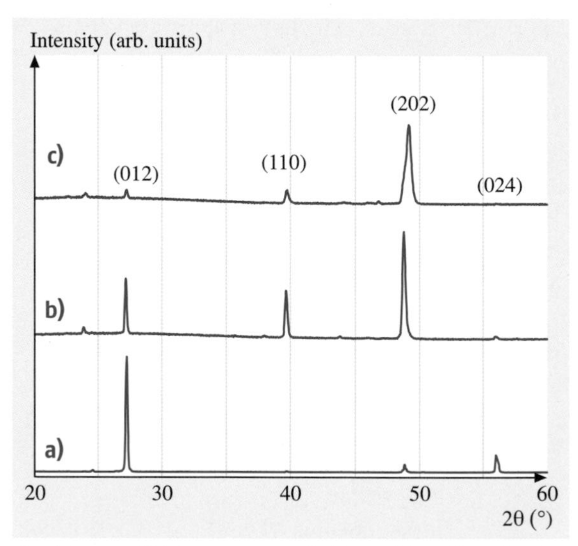

Fig. 4.2a–c XRD patterns of bismuth/anodic alumina nano-composites with average bismuth wire diameters of **(a)** 40 nm, **(b)** 52 nm, and **(c)** 95 nm [4.67]. The Miller indices corresponding to the lattice planes of bulk Bi are indicated above the individual peaks. The majority of the Bi nanowires are oriented along the $[10\bar{1}1]$ and $[01\bar{1}2]$ directions for $d_W \geq 60$ nm and $d_W \leq 50$ nm, respectively [4.16,67]. The existence of more than one dominant orientation in the 52-nm Bi nanowires is attributed to the transitional behavior of *intermediate*-diameter nanowires as the preferential growth orientation is shifted from $[10\bar{1}1]$ to $[01\bar{1}2]$ with decreasing d_W

pores in the anodic alumina template with liquid Bi and can increase the number of nanowires prepared [4.16]. But some of the surfactants might cause contamination problems and therefore should be avoided. Nanowires produced by the pressure injection technique usually possess high crystallinity and a preferred crystal orientation along the wire axis. For example, Fig. 4.2 shows the X-ray diffraction (XRD) patterns of Bi nanowire arrays of three different wire diameters with an injection pressure of $\sim 5{,}000$ psi [4.67], demonstrating that the major ($> 80\%$) crystal orientation of the wire axes in the 95-nm and 40-nm diameter Bi nanowire arrays are, respectively, normal to the (202) and (012) lattice planes. These are denoted by $[10\bar{1}1]$ and $[01\bar{1}2]$ when using a hexagonal unit cell, suggesting a wire-diameter-dependent crystal growth direction. On the other hand, 30 nm Bi nanowires produced using a much higher pressure of $> 20{,}000$ psi show a different crystal orientation of (001) along the wire axis [4.33], indicating that the preferred crystal orientation may also depend on the applied pressure, with the most dense packing direction along the wire axis for the highest applied pressure.

Electrochemical Deposition

The electrochemical deposition technique has attracted increasing attention as an alternative method for fabricating nanowires. Traditionally, electrochemistry has been used to grow thin films on conducting surfaces. Since electrochemical growth is usually controllable in the direction normal to the substrate surface, this method can be readily extended to fabricate 1-D or 0-D nanostructures, if the deposition is confined within the pores of an appropriate template. In the electrochemical methods, a thin conducting metal film is first coated on one side of the porous membrane to serve as the cathode for electroplating. The length of the deposited nanowires can be controlled by varying the duration of the electroplating process. This method has been used to synthesize a wide variety of nanowires e.g., metals (Bi [4.21, 29]; Co [4.73, 74]; Fe [4.15, 75]; Cu [4.20, 76]; Ni [4.22, 73]; Ag [4.24, 77]; Au [4.25, 26]); conducting polymers [4.7, 29]; superconductors (Pb [4.78]); semiconductors (CdS [4.37]); and even superlattice nanowires with A/B constituents (such as Cu/Co [4.20, 76]) have been synthesized electrochemically (see Table 4.1).

In the electrochemical deposition process, the chosen template has to be chemically stable in the electrolyte during the electrolysis process. Cracks and defects in the templates are detrimental to the nanowire growth, since the deposition processes primarily occur in the more accessible cracks, leaving most of the nanopores unfilled. Particle track-etched mica films or polymer membranes are typical templates used in the simple dc electrolysis. To use anodic aluminum oxide films in the dc electrochemical deposition, the insulating barrier layer that separates the pores from the bottom aluminum substrate has to be removed, and a metal film is then evaporated onto the back of the template membrane [4.79]. Compound nanowire arrays, such as Bi_2Te_3, have been fabricated in alumina templates with a high filling factor using the dc electrochemical deposition [4.34]. Figures 4.3a and b, respectively, show the top view and cross-sectional SEM images of a Bi_2Te_3 nanowire array [4.34]. The light areas are associated with Bi_2Te_3 nanowires, the dark regions denote empty pores, and the surrounding gray matrix is alumina.

Surfactants are also used with electrochemical deposition when necessary. For example, when using templates derived from PMMA/PS diblock copolymers (see above), methanol is used as a surfactant is used to facilitate pore filling [4.71], thereby achieving $\sim 100\%$ filling factor.

It is also possible to employ an ac electrodeposition method in anodic alumina templates without the removal of the barrier layer, by utilizing the rectifying

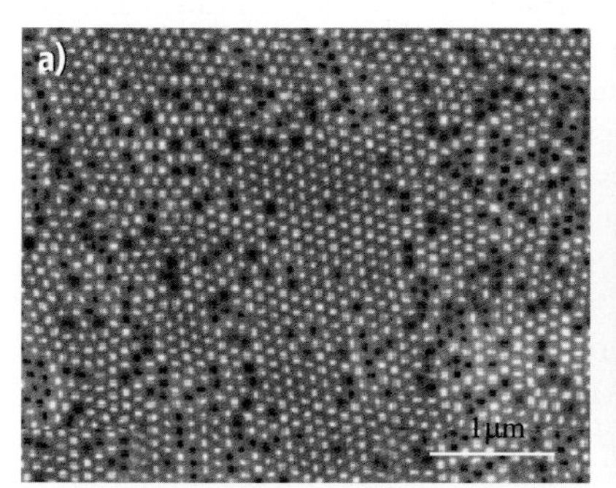

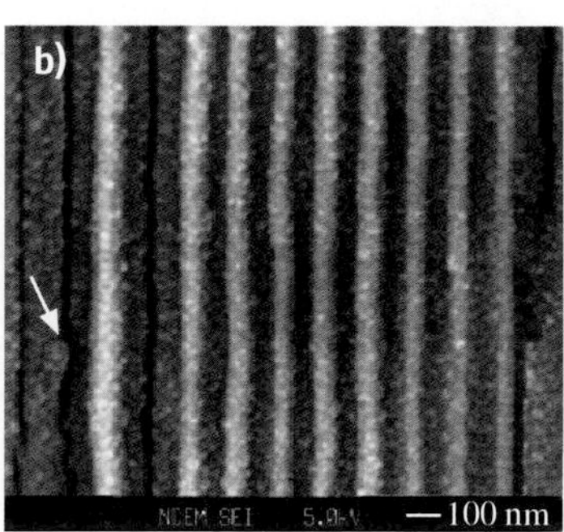

Fig. 4.3 (**a**) SEM image of a Bi_2Te_3 nanowire array in cross section showing a relatively high pore filling factor. (**b**) SEM image of a Bi_2Te_3 nanowire array composite along the wire axis [4.34]

properties of the oxide barrier. In ac electrochemical deposition, although the applied voltage is sinusoidal and symmetric, the current is greater during the cathodic half-cycles, making deposition dominant over the stripping, which occurs in the subsequent anodic half-cycles. Since no rectification occurs at defect sites, the deposition and stripping rates are equal, and no material is deposited. Hence, the difficulties associated with cracks are avoided. In this fashion, metals, such as Co [4.74] and Fe [4.15, 75], and semiconductors, such as CdS [4.37], have been deposited into the pores of anodic aluminum oxide templates without removing the barrier layer.

In contrast to nanowires synthesized by the pressure injection method, nanowires fabricated by the electrochemical process are usually polycrystalline, with no preferred crystal orientations, as observed by XRD studies. Some exceptions exist, however. For example, polycrystalline CdS nanowires, fabricated by an ac electrodeposition method in anodic alumina templates [4.37], possibly have a preferred wire growth orientation along the *c*-axis. In addition, *Xu* et al. have prepared a number of single-crystal II-VI semiconductor nanowires, including CdS, CdSe, and CdTe, by dc electrochemical deposition in anodic alumina templates with a nonaqueous electrolyte [4.36, 40]. Furthermore, single-crystal Pb nanowires can be formed by pulse electrodeposition under over-potential conditions, but no specific crystal orientation along the wire axis was observed [4.78]. The use of pulse currents is believed to be advantageous for the growth of crystalline wires because the metal ions in the solution can be regenerated between the electrical pulses and, therefore, uniform deposition conditions can be produced for each deposition pulse. Similarly, single crystal Ag nanowires were fabricated by pulsed electro-deposition [4.24].

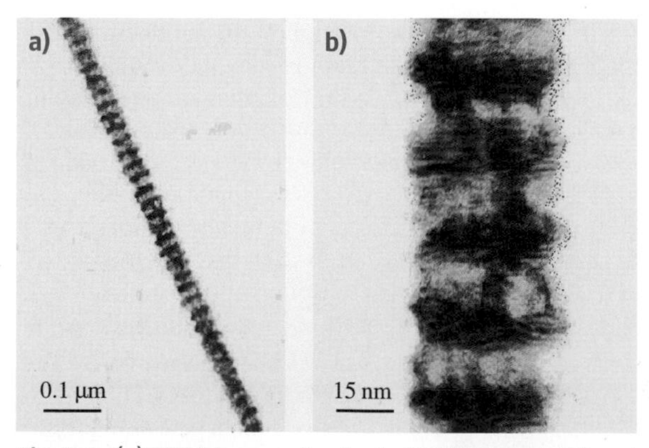

Fig. 4.4 (a) TEM image of a single Co(10 nm)/Cu(10 nm) multilayered nanowire. (b) A selected region of the sample at high magnification [4.76]

One advantage of the electrochemical deposition technique is the possibility of fabricating multilayered structures within nanowires. By varying the cathodic potentials in the electrolyte that contains two different kinds of ions, different metal layers can be controllably deposited. In this fashion, Co/Cu multilayered nanowires have been synthesized [4.20, 76]. Figure 4.4 shows TEM images of a single Co/Cu nanowire of about 40 nm in diameter [4.76]. The light bands represent Co-rich regions, and the dark bands represent Cu-rich layers. This electrodeposition method provides a low-cost approach to preparing multilayered 1-D nanostructures.

Vapor Deposition

Vapor deposition of nanowires includes physical vapor deposition (PVD) [4.28], chemical vapor deposition (CVD) [4.80], and metallorganic chemical vapor deposition (MOCVD) [4.48]. Like electrochemical deposition, vapor deposition is usually capable of preparing smaller-diameter ($\leq$ 20 nm) nanowires than pressure injection methods, since it does not rely on the high pressure and surface tension involved to insert the material into the pores.

In the physical vapor deposition technique, the material to be filled is first heated to produce a vapor, which is then introduced through the pores of the template and cooled to solidify. Using an especially designed experimental setup [4.28], nearly single-crystal Bi nanowires in anodic aluminum templates with pore diameters as small as 7 nm have been synthesized, and these Bi nanowires were found to possess a preferred crystal growth orientation along the wire axis, similar to the Bi nanowires prepared by pressure injection [4.16, 28].

Compound materials that result from two reacting gases have also been prepared by the chemical vapor deposition (CVD) technique. For example, single-crystal GaN nanowires have been synthesized in anodic alumina templates through a gas reaction of Ga_2O vapor with a flowing ammonia atmosphere [4.45, 80]. A different liquid/gas phase approach has been used to prepare polycrystalline GaAs and InAs nanowires in a nanochannel glass array [4.48]. In this method, the nanochannels are filled with one liquid precursor (e.g., Me_3Ga or Et_3In) via a capillary effect, and the nanowires are formed within the template by reactions between the liquid precursor and the other gas reactant (e.g., AsH_3).

Nanotube Synthesis with Templates and as Templates

Recently carbon nanotubes, an important class of 1-D nanostructures, have been fabricated within the pores of anodic alumina templates using a chemical vapor deposition technique [4.81–84] to form highly ordered two-dimensional carbon nanotube arrays. A small amount of metal catalyst (e.g., Co) is first electrochemically deposited on the bottom of the pores. The templates are then placed in a furnace and heated to $\sim 700-800\,°C$ with a flowing gas consisting of a mixture of N_2 and acetylene (C_2H_2) or ethylene (C_2H_4). The hydrocarbon molecules are pyrolyzed to form carbon nanotubes in the pores of the template with the help of the metal catalysts. Well-aligned nanotube arrays have stimulated much interest because of their great potential in various applications, such as cold-cathode flat panel displays. Of particular interest is the use of zeolite templates with very narrow pores (< 1 nm diameter) that permit the growth of carbon nanotubes with diameters of 0.42 nm, having only 10 carbon atoms around the circumference [4.85].

The hollow cores of carbon nanotubes have also been used to synthesize a variety of nanowires of very small diameter [4.86]. These very small diameter nanowires have been extensively studied by high resolution TEM but have not yet been characterized to the same degree regarding their physical properties.

4.1.2 VLS Method for Nanowire Synthesis

Some of the recent successful syntheses of semiconductor nanowires are based on the so-called vapor-liquid-solid (VLS) mechanism of anisotropic crystal growth. This mechanism was first proposed for the growth of single crystal silicon whiskers 100 nm to hundreds of microns in diameter [4.87]. The proposed growth mechanism (see Fig. 4.5) involves the absorption of source material from the gas phase into a liquid droplet of catalyst (a molten particle of gold on a silicon substrate in the original work [4.87]). Upon supersaturation of the liquid alloy, a nucleation event generates a solid precipitate of the source material. This seed serves as a preferred site for further deposition of material at the interface of the liquid droplet, promoting the elongation of the seed into a nanowire or a whisker, and suppressing further nucleation events on the same catalyst. Since the liquid droplet catalyzes the incorporation of material from the gas source to the growing crystal, the deposit grows anisotropically as a whisker whose diameter is dictated by the diameter of the liquid alloy droplet. The nanowires thus obtained are of high purity, except for the end containing the solidified catalyst as an alloy particle (see Figs. 4.5, 4.6a). Real-time observations of the alloying, nucleation, and elongation steps in the growth of germanium nanowires from gold nanoclusters by the VLS method were recorded by in situ TEM [4.88].

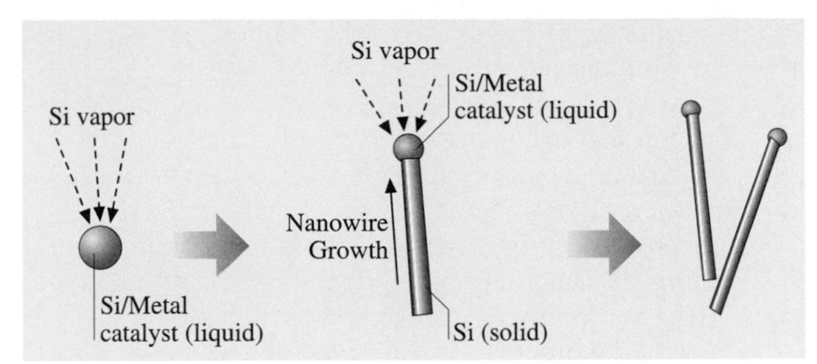

Fig. 4.5 Schematic diagram illustrating the growth of silicon nanowires by the VLS mechanism

Reduction of the average wire diameter to the nanometer scale requires the generation of nanosized catalyst droplets. Due to the balance between the liquid-vapor surface free energy and the free energy of condensation, however, the size of a liquid droplet, in equilibrium with its vapor, is usually limited to the micrometer range. This obstacle was overcome in recent years by several new methodologies: 1) Advances in the synthesis of metal nanoclusters have made monodispersed nanoparticles commercially available. These can be dispersed on a solid substrate in high dilution so that when the temperature is raised above the melting point, the liquid clusters do not aggregate [4.59]. 2) Alternatively, metal islands of nanoscale sizes can self-form when a strained thin layer is grown or heat treated on a non-epitaxial substrate [4.50]. 3) Laser-assisted catalytic VLS growth is a method used to generate nanowires under nonequilibrium conditions. By laser ablation of a target containing both the catalyst and the source materials, a plasma is generated from which catalyst nanoclusters nucleate as the plasma cools down. Single crystal nanowires grow as long as the particle remains liquid [4.60]. 4) Interestingly, by optimization of the material properties of the catalyst-nanowire system, conditions can be achieved for which nanocrystals nucleate in a liquid catalyst pool supersaturated with the nanowire material, migrate to the surface due to a large surface tension, and continue growing as nanowires perpendicular to the liquid surface [4.62]. In this case, supersaturated nanodroplets are sustained on the outer

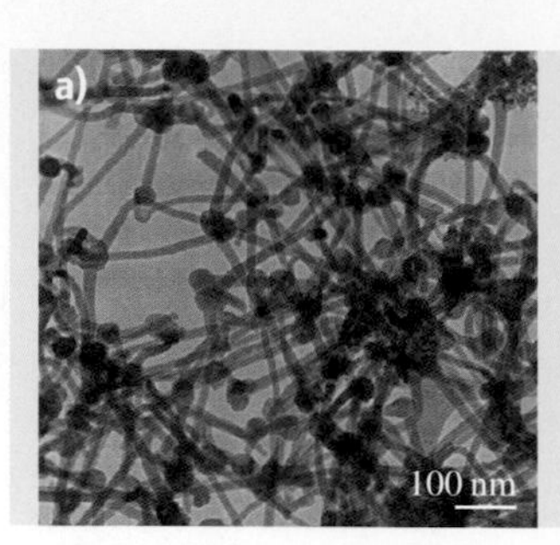

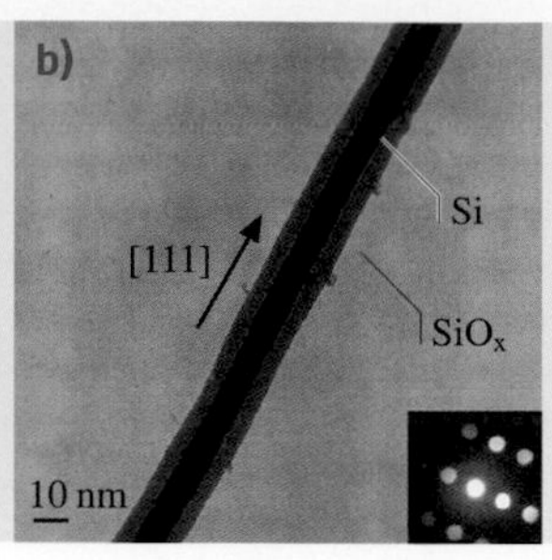

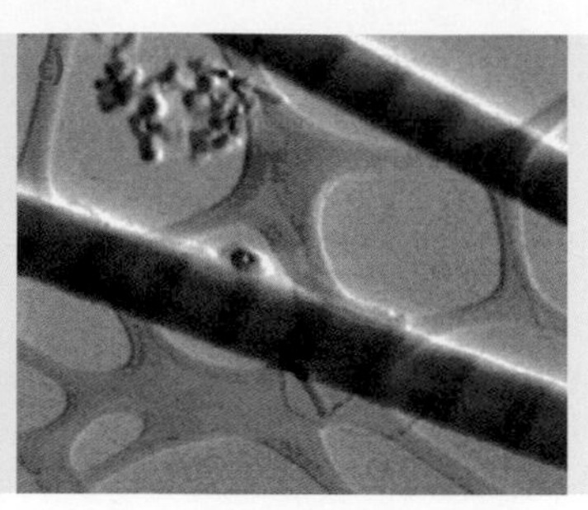

Fig. 4.6 (a) TEM images of Si nanowires produced after laser ablating a $Si_{0.9}Fe_{0.1}$ target. The dark spheres with a slightly larger diameter than the wires are solidified catalyst clusters [4.60]. (b) Diffraction contrast TEM image of a Si nanowire. The crystalline Si core appears darker than the amorphous oxide surface layer. The inset shows the convergent beam electron diffraction pattern recorded perpendicular to the wire axis, confirming the nanowire crystallinity [4.60]. (c) STEM image of $Si/Si_{1-x}Ge_x$ superlattice nanowires in the bright field mode. The scale bar is 500 nm [4.90]

end of the nanowire due to the low solubility of the nanowire material in the liquid [4.89].

A wide variety of elemental, binary, and compound semiconductor nanowires has been synthesized by the VLS method, and relatively good control over the nanowire diameter and diameter distribution has been achieved. Researchers are currently focusing attention on the controlled variation of the materials properties along the nanowire axis. To this context, researchers have modified the VLS synthesis apparatus to generate compositionally modulated nanowires. GaAs/GaP modulated nanowires have been synthesized by alternately ablating targets of the corresponding materials in the presence of gold nanoparticles [4.92]. p-Si/n-Si nanowires were grown by chemical vapor deposition from alternating gaseous mixtures containing the appropriate dopant [4.92]. $Si/Si_{1-x}Ge_x$ nanowires were grown by combining silicon from a gaseous source with germanium from a periodically ablated target (see Fig. 4.6c) [4.90]. Finally, using an ultra-high vacuum chamber and molecular beams, InAs/InP nanowires with atomically sharp interfaces were obtained [4.93]. These compositionally modulated nanowires are expected to exhibit exciting electronic, photonic, and thermoelectric properties.

Silicon and germanium nanowires grown by the VLS method consist of a crystalline core coated by a relatively thick amorphous oxide layer (2–3 nm) (see Fig. 4.6b). These layers are too thick to be the result of ambient oxidation, and it has been shown that these oxides play an important role in the nanowire growth process [4.61, 91]. Silicon oxides were found to serve as a special and highly selective catalyst, that significantly enhances the yield of Si nanowires, without the need for metal catalyst particles [4.61, 91, 94]. A similar yield enhancement was also found in the synthesis of Ge nanowires from the laser ablation of Ge powder mixed with GeO_2 [4.51]. The Si and Ge nanowires produced from these metal-free targets generally grow along the [112] crystal direction [4.95] and have the benefit that no catalyst clusters are found on either ends of the nanowires. Based on these observations

Fig. 4.7 TEM image showing the two major morphologies of Si nanowires prepared by the oxide assisted growth method [4.91]. Notice the absence of metal particles when compared to Fig. 4.6a. The *arrow* points at an oxide-linked chain of Si nanoparticles

and other TEM studies, an oxide-enhanced nanowire growth mechanism different from the classical VLS mechanism was proposed, in which no metal catalyst is required during the laser ablation-assisted synthesis [4.91]. It is postulated that the nanowire growth is dependent on the presence of SiO (or GeO) vapor. This decomposes in the nanowire tip region into Si (or Ge), which is incorporated into the crystalline phase, and SiO_2 (or GeO_2), which contributes to the outer coating. The initial nucleation events generate oxide-coated spherical nanocrystals. The [112] crystal faces have the fastest growth rate, and therefore the nanocrystals soon begin elongating along this direction to form one-dimensional structures. The Si_mO or Ge_mO ($m > 1$) layer on the nanowire tips may be at temperatures near their molten states, catalyzing the incorporation of gas molecules in a directional fashion [4.95]. Besides nanowires with smooth walls, a second morphology of chains of non-oriented nanocrystals linked by oxide necks is frequently observed (indicated by an arrow in Fig. 4.7). In addition, it was found by STM studies that about 1% of the wires consist of a regular array of two alternating segments, 10 nm and 5 nm in length, respectively [4.96]. The segments, whose junctions form an angle of 30°, are probably a result of alternating growth along different crystallographic orientations (see Sect. 4.2.1).

4.1.3 Other Synthesis Methods

In this section we review several other general procedures available for the synthesis of a variety of nanowires. We focus on "bottom-up" approaches, which afford many kinds of nanowires in large numbers and do not require highly sophisticated equipment (such as scanning microscopy or lithography based methods). We exclude cases for which the nanowires are not self sustained (such as in the case of atomic rows on the surface of crystals).

Gates et al. have demonstrated a solution-phase synthesis of nanowires with controllable diameters [4.57, 97], without the use of templates, catalysts, or surfactants. They make use of the anisotropy of the crystal structure of trigonal selenium and tellurium, which can be viewed as rows of 1-D helical atomic chains. They base their approach on the mass transfer of atoms during an aging step from a high free-energy solid phase (e.g., amorphous selenium) to a seed (e.g., trigonal selenium nanocrystal) that grows preferentially along one crystallographic axis. The lateral dimension of the seed, which dictates the diameter of the nanowire, can be controlled by the temperature of the nucleation step. Furthermore, Se/Te alloy nanowires were synthesized by this method, and Ag_2Se compound nanowires were obtained by treating selenium nanowires with $AgNO_3$ [4.98–100].

More often, however, the use of surfactants is necessary to promote the anisotropic 1-D growth of nanocrystals. Solution phase synthetic routes have been optimized to produce monodispersed quantum dots, i. e., zero-dimensional isotropic nanocrystals [4.101]. Surfactants are necessary in this case to stabilize the interfaces of the nanoparticles and retard oxidation and aggregation processes. Detailed studies on the effect of growth conditions revealed that they can be manipulated to induce a directional growth of the nanocrystals, usually generating nanorods (aspect ratio of ≈ 10), and in favorable cases, nanowires of high-aspect ratios. *Heath* and *LeGoues* [4.49] synthesized germanium nanowires by reducing a mixture of $GeCl_4$ and phenyl-$GeCl_3$ at high temperature and high pressure. The phenyl ligand was essential for the formation of high-aspect ratio nanowires [4.49]. In growing CdSe nanorods [4.38], Alivisatos et al. used a mixture of two surfactants, whose concentration ratio influenced the structure of the nanocrystal. It is believed that different surfactants have different affinities and different absorption rates for the different crystal faces of CdSe, thereby regulating the growth rate of these faces. A coordinating alkyl-diamine solvent was used to grow polycrystalline PbSe nanowires at low temperatures [4.55]. Here, the surfactant-induced directional growth is believed to occur, through the formation of organometallic complexes in which the bidentate ligand assumes the equatorial positions, thus hindering the ions from approaching each other in this plane. Additionally, the alkyl-amine molecules coat the external surface of the wire, preventing lateral growth. The aspect ratio of the wires increased as the temperature was lowered in the range $10 < T < 117$ °C. Ethylenediamine was used to grow CdS nanowires and tetrapods by a solvo-thermal recrystallization process starting with CdS nanocrystals or amorphous particles [4.35]. While the coordinating solvent was crucial for the nanowire growth, the researchers did not clarify its role in the shape and phase control.

Stress-induced crystalline bismuth nanowires have been grown from sputtered films of layers of Bi and CrN. The nanowires presumably grow from defects and cleavage fractures in the film and are up to several millimeter in lengths with diameters ranging from 30 to 200 nm [4.27]. While the exploration of this technique has only begun, stress-induced unidirectional growth should be applicable to a variety of composite films.

Selective electrodeposition along the step edges in highly oriented pyrolytic graphite (HOPG) was used to obtain MoO_2 nanowires as shown in Fig. 4.8. The site-selectivity was achieved by applying a low overpotential to the electrochemical cell in which the HOPG served as cathode, thus minimizing the nucleation events on less favorable sites (i. e., plateaus). While these nanowires cannot be removed from the substrate, they can be reduced to metallic molybdenum nanowires, which can then be released as free-standing nanowires. Other metallic nanowires were also obtained by this method [4.53, 102]. In contrast to the template synthesis approaches described above, in this method the substrate defines only the position and orientation of the nanowire, not its diameter. In this context, other surface morphologies, such as self-assembled grooves in etched crystal planes, have been used to generate nanowire arrays via gas-phase shadow deposition (for example: Fe nanowires on (110)NaCl [4.44]). The cross-section of artificially prepared superlattice structures has also been used for site-selective electrodeposition of parallel and closely spaced nanowires [4.103]. Nanowires prepared on the above-mentioned substrates would have semicircular, rectangular, or other unconventional cross-sectional shapes.

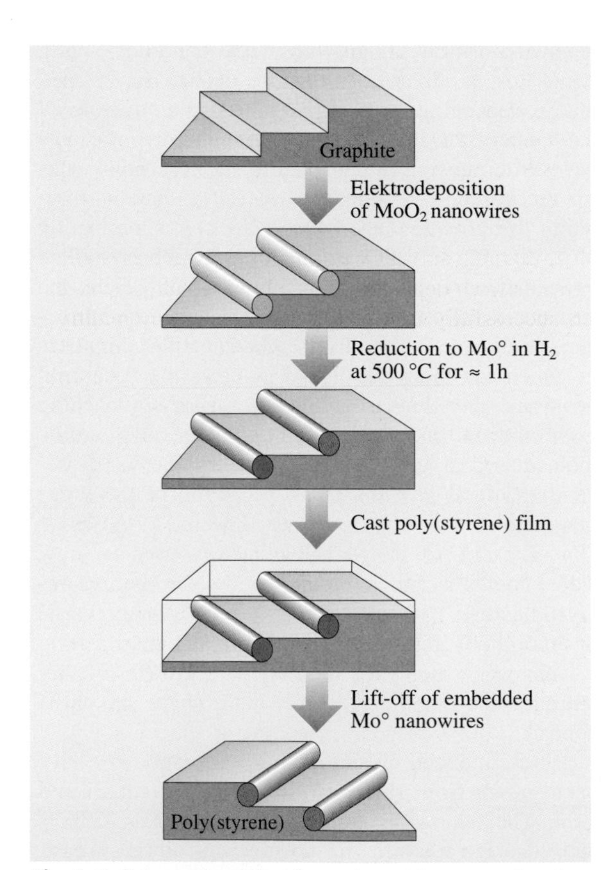

Fig. 4.8 Schematic of the electrodeposition step edge decoration of HOPG (highly oriented pyrolytic graphite) for the synthesis of molybdenum nanowires [4.53, 102]

4.1.4 Hierarchical Arrangement and Superstructures of Nanowires

Ordering nanowires into useful structures is another challenge to address in order to harness the full potential of nanowires for applications. We will first review examples of nanowires having a nontrivial structure and then proceed to describe methods to create assemblies of nanowires of a predetermined structure.

We have already mentioned in Sect. 4.1.2 that the preparation of nanowires with a graded composition or with a superlattice structure along their main axis was demonstrated by controlling the gas-phase chemistry as a function of time during the growth of the nanowires by the VLS method. Control of the composition along the axial dimension was also demonstrated by a template-assisted method, for example by the consecutive electrochemical deposition of different metals in the pores of an alumina template [4.104]. Alternatively, the composition can be varied along the radial dimension of the nanowire, for example, by first growing a nanowire by the VLS method and then switching the synthesis conditions to grow a different material on the surface of the nanowire by CVD. This technique was demonstrated for the synthesis of Si/Ge and Ge/Si coaxial (or core-shell) nanowires [4.105], and it was shown that by a thermal annealing process the outer shell can be formed epitaxially on the inner core. A different approach was adopted by *Wang* et al. who generated a mixture of coaxial and biaxial SiC-SiO_x nanowires by the catalyst-free high-temperature reaction of amorphous silica and a carbon/graphite mixture [4.106].

A different category of nontrivial nanowires is that of nanowires having a nonlinear structure, resulting from multiple one-dimensional growth steps. Members of this category are tetrapods, which were mentioned in the context of liquid-phase synthesis (Sect. 4.1.3). In this process, a tetrahedral quantum-dot core is first grown, and then the conditions are modified to induce a one-dimensional growth of a nanowire from each one of the facets of the tetrahedron. A similar process

Fig. 4.9a–d SEM images of (**a**) 6-fold (**b**) 4-fold and (**c**) 2-fold symmetry nanobrushes made of an In_2O_3 core and ZnO nanowire brushes [4.107], and of (**d**) ZnO nanonails [4.108]

produced high-symmetry In_2O_3/ZnO hierarchical nanostructures. From a mixture of heat-treated In_2O_3, ZnO and graphite powders, faceted In_2O_3 nanowires were first obtained, on which oriented shorter ZnO nanowires crystallized [4.107]. Brush-like structures were obtained as a mixture of 11 structures of different symmetries. For example, two, four, or six rows of ZnO nanorods could be found on different core nanowires, depending on the crystallographic orientation of the main axis of the core nanowire, as shown in Fig. 4.9. Comb-like structures entirely made of ZnO were also reported [4.65].

Control of the position of a nanowire in the growth process is important for preparing devices or test structures containing nanowires, especially when involving a large array of nanowires. Post-synthesis methods to align and position nanowires include microfluidic channels [4.110], Langmuir–Blodgett assemblies [4.109], and electric-field assisted assembly [4.111]. The first method involves the orientation of the nanowires by the liquid flow direction when a nanowire solution is injected into a microfluidic channel assembly and by the interaction of the nanowires with the side walls of the channel. The second method involves the alignment of nanowires at a liquid-gas or liquid-liquid interface by the application of compressive forces on the interface (Fig. 4.10). The third technique is based on dielectrophoretic forces that pull polarizable nanowires toward regions of high field strength. The nanowires align between two isolated electrodes that are capacitatively coupled to a pair of buried electrodes biased with an AC voltage. Once a nanowire shorts the electrodes, the electric field is eliminated, preventing more nanowires from depositing. The above techniques have been successfully used to prepare electronic circuitry and optical devices out of nanowires (see Sects. 4.3.1 and 4.3.3). Alternatively, alignment and positioning of the nanowires can be specified and controlled during their growth by the proper design of the synthesis. For example, ZnO nanowires prepared by the VLS method were grown into an array in which both their position on the substrate and their growth direction and orientation were controlled [4.65]. The nanowire growth region was defined by patterning the gold film, which serves as a catalyst for the ZnO nanowire growth, employing soft-lithography, e-beam lithography, or photolithography. The orientation of the nanowires was achieved by selecting a substrate with a lattice structure matching that of the nanowire material to facilitate the epitaxial growth. These conditions result in an array of nanowire posts at predetermined positions, all vertically aligned with the same crystal growth orientation (see Fig. 4.11). A similar structure could be obtained by the template-mediated electrochemical synthesis of nanowires (see Sect. 4.1.1), particularly if anodic alumina with its parallel and ordered channels is used. The control over the

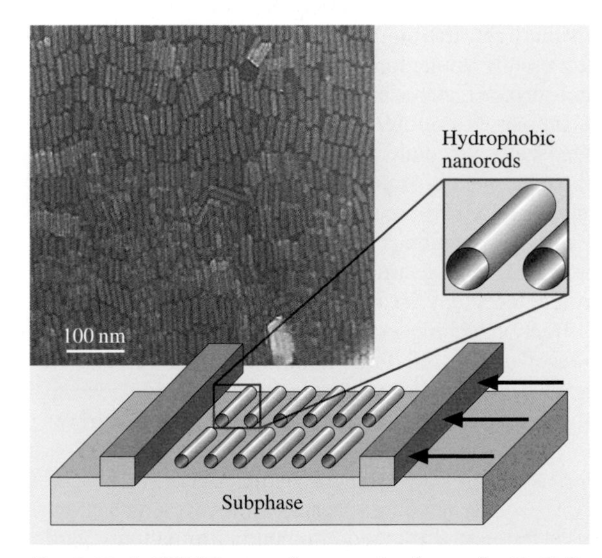

Fig. 4.10 A TEM image of a smectic phase of a $BaCrO_4$ nanorod film (*left inset*) achieved by the Langmuir–Blodgett technique, as depicted by the illustration [4.109]

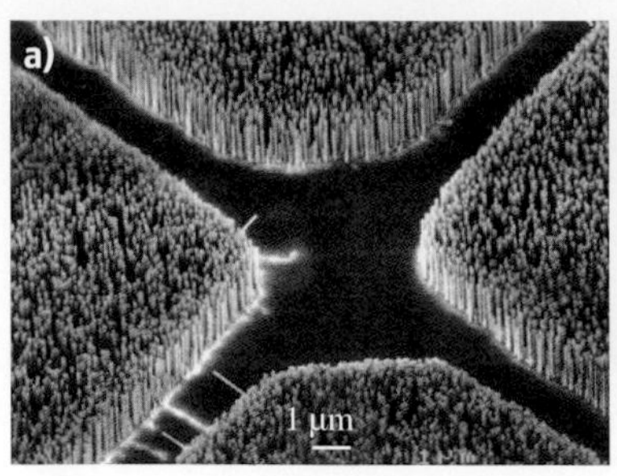

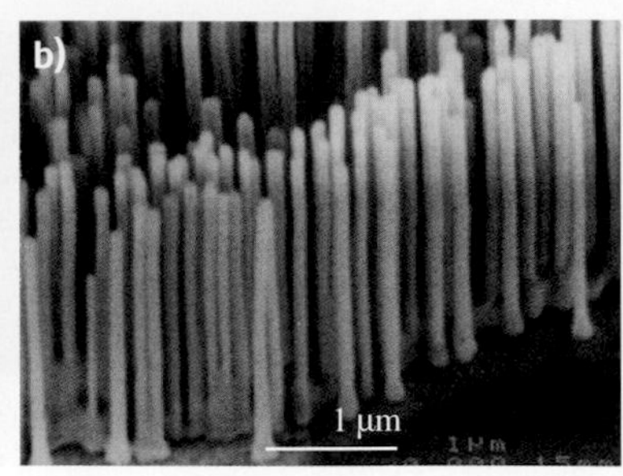

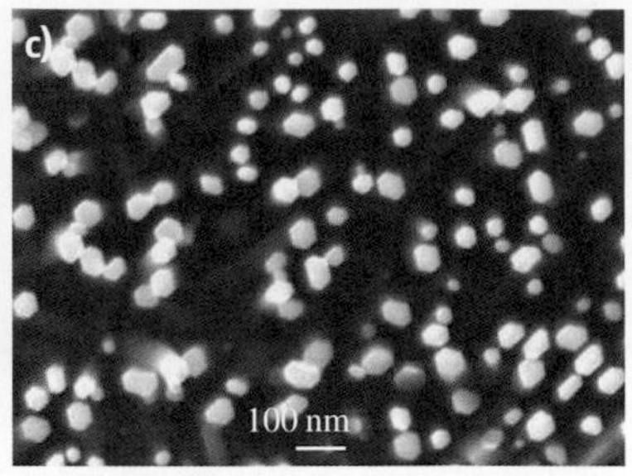

Fig. 4.11a–c SEM images of ZnO nanowire arrays grown on a sapphire substrate, where (**a**) shows patterned growth, (**b**) shows a higher resolution image of the parallel alignment of the nanowires, and (**c**) shows the faceted side-walls and the hexagonal cross section of the nanowires. For nanowire growth, the sapphire substrates were coated with a 1.0 to 3.5 nm thick patterned layer of Au as the catalyst, using a TEM grid as the shadow mask. These nanowires have been used for nanowire laser applications [4.115]

location of the nucleation of nanowires in the electrochemical deposition is determined by the pore positions and the back-electrode geometry. The pore positions can be precisely controlled by imprint lithography [4.112]. By growing the template on a patterned conductive substrate that serves as a back-electrode [4.113,114], different materials can be deposited in the pores at different regions of the template.

4.2 Characterization and Physical Properties of Nanowires

In this section we review the structure and properties of nanowires, and the interrelationship between the two. The discovery and investigation of nanostructures were galvanized by advances in various characterization and microscopy techniques that enable materials characterization to take place at smaller and smaller length scales, reaching down to individual atoms. For applications, characterization of the nanowire structural properties is especially important so that a reproducible relationship between their desired functionality and their geometrical and structural characteristics can be established. Due to the enhanced surface-to-volume ratio in nanowires, their properties may depend sensitively on their surface condition and geometrical configuration. Even nanowires made of the same material may possess dissimilar properties due to differences in their crystal phase, crystalline size, surface conditions, and aspect ratios, which depend on the synthesis methods and conditions used in their preparation.

4.2.1 Structural Characterization

Structural and geometric factors play an important role in determining the various attributes of nanowires, such as their electrical, optical and magnetic properties. Various novel tools, therefore, have been developed and employed to obtain this important structural information at the nanoscale. At the micron scale, optical techniques are extensively used for imaging structural features. Since the sizes of nanowires are usually comparable to or, in most cases, much smaller than the wavelength of visible light, traditional optical microscopy techniques are usually limited in characterizing the morphology and surface features of nanowires. Electron microscopy techniques, therefore, play a more dominant role at the nanoscale. Since electrons interact more strongly than photons, electron microscopy is particularly sensitive relative to X-rays for the analysis of tiny samples.

In this section we review and give examples of how scanning electron microscopy, transmission electron microscopy, scanning probe spectroscopies, and diffraction techniques are used to characterize the structure of nanowires. To provide the necessary basis for developing reliable structure-property relations, we apply multiple characterization tools to the same samples.

Scanning Electron Microscopy (SEM)

SEM usually produces images down to length scales of ∼ 10 nm and provides valuable information regarding the structural arrangement, spatial distribution, wire density, and geometrical features of the nanowires. Examples of SEM micrographs shown in Figs. 4.1 and

4.3 indicate that structural features at the 10 nm to 10 μm length scales can be probed, providing information on the size, size distribution, shapes, spatial distributions, density, nanowire alignment, filling factors, granularity etc. As another example, Fig. 4.11a shows an SEM image of ZnO nanowire arrays grown on a sapphire substrate [4.115], which provides evidence for the nonuniform spatial distribution of the nanowires on the substrate. This distribution was attained by patterning the catalyst film to define high density growth regions and nanowire-free regions. Figure 4.11b, showing a higher magnification of the same system, indicates that these ZnO nanowires grow perpendicular to the substrate, are well aligned with approximately equal wire lengths, and have wire diameters in the range of $20 < d_W < 150$ nm. The SEM micrograph in Fig. 4.11c provides further information about the surface of the nanowires, showing it to be well-faceted and forming a hexagonal cross-section, indicative of nanowire growth along the ⟨0001⟩ direction. The uniformity of the nanowire size, its alignment perpendicular to the substrate, and its uniform growth direction, as suggested by the SEM data, are linked to the good epitaxial interface between the (0001) plane of the ZnO nanowire and the (110) plane of the sapphire substrate. (The crystal structures of ZnO and sapphire are essentially incommensurate, with the exception that the *a* axis of ZnO and the *c* axis of sapphire are related almost exactly by a factor of 4, with a mismatch of less than 0.08% at room temperature [4.115].) The well-faceted nature of these nanowires has important implications for their lasing action (see Sect. 4.3.2). Figure 4.12 shows an SEM image of GaN nanowires synthesized by a laser-assisted catalytic growth method [4.46], indicating a random spatial orientation of the nanowire axes and a wide diameter distribution for these nanowires, in contrast to the ZnO wires in Fig. 4.11 and to arrays of well-aligned nanowires prepared by template-assisted growth (see Fig. 4.3).

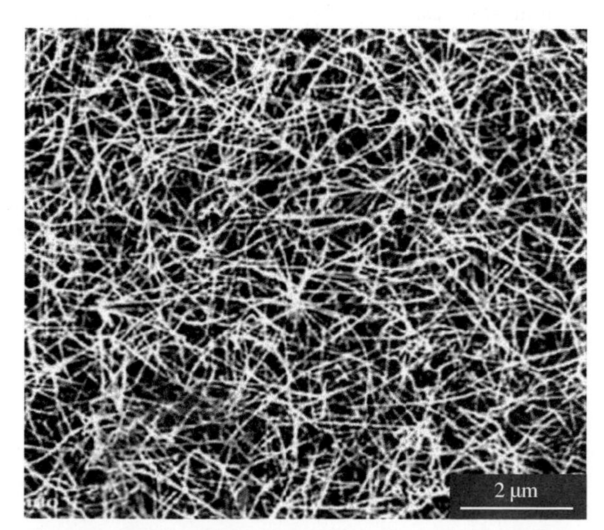

Fig. 4.12 SEM image of GaN nanowires in a mat arrangement synthesized by laser-assisted catalytic growth. The nanowires have diameters and lengths on the order of 10 nm and 10 μm, respectively [4.46]

Transmission Electron Microscopy (TEM)

TEM and high resolution transmission electron microscopy (HRTEM) are powerful imaging tools to study nanowires at the atomic scale, and they usually provide more detailed geometrical features than are seen in SEM images. TEM studies also yield information regarding the crystal structure, crystal quality, grain size, and crystal orientation of the nanowire axis. When operating in the diffraction mode, selected area electron diffraction (SAED) patterns can be made to determine the crystal structure of nanowires. As an example, the TEM images in Fig. 4.13 show four different morphologies for Si nanowires prepared by the laser ablation of a Si target [4.116]: (a) spring-shaped; (b) fishbone-shaped (indicated by solid arrow) and frog-egg-shaped (indicated by the hollow arrow), (c) pearl-shaped, while (d) shows the poly-sites of nanowire nucleation. The crystal quality of nanowires is revealed from high resolution TEM images with atomic resolution, along with selected area electron diffraction (SAED) patterns. For example, Fig. 4.14 shows a TEM image of one of the GaN nanowires from Fig. 4.12, indicating single crystallinity and showing (100) lattice planes, thus demonstrating the growth direction of the nanowire. This information is supplemented by the corresponding electron diffraction pattern in the upper right.

The high resolution of the TEM also allows for the investigation of the surface structure of the nanowires. In many cases the nanowires are sheathed with a native oxide layer or an amorphous oxide layer that forms during the growth process. This can be seen in Fig. 4.6b for silicon nanowires and in Fig. 4.15 for germanium nanowires [4.51], showing a mass-thickness contrast TEM image and a selected-area electron diffraction pattern of a Ge nanowire. The main TEM image shows that these Ge nanowires possess an amorphous GeO_2 sheath with a crystalline Ge core oriented in the [211] direction.

Dynamical processes of the surface layer of nanowires can be studied by using an in situ environ-

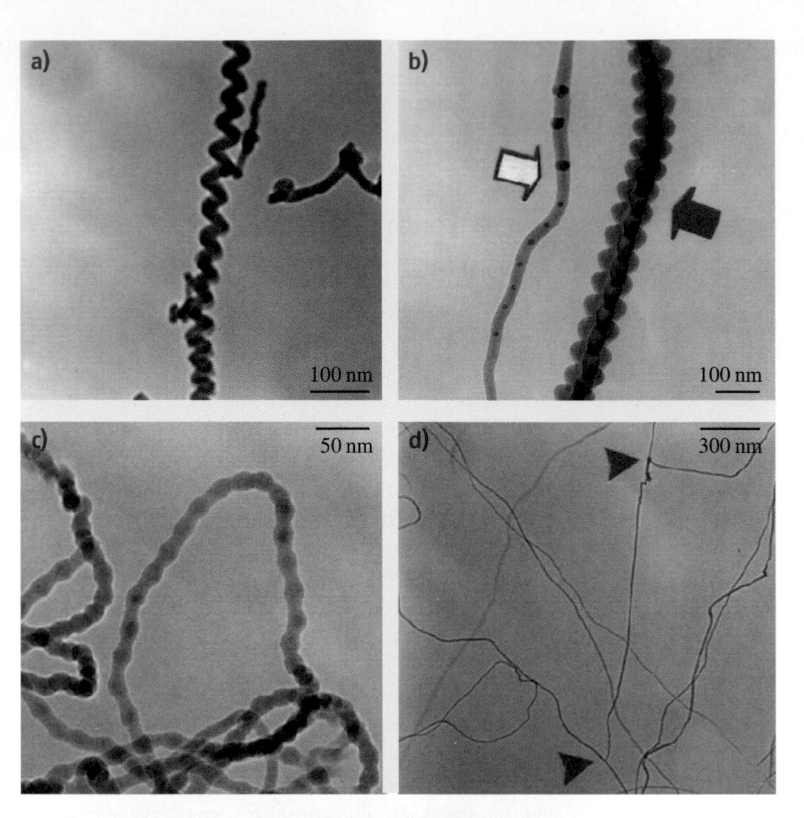

Fig. 4.13a–d TEM morphologies of four special forms of Si nanowires synthesized by the laser ablation of a Si powder target. (**a**) A spring-shaped Si nanowire; (**b**) fishbone-shaped (indicated by a *solid arrow*) and frog-egg-shaped (indicated by a *hollow arrow*) Si nanowires; and (**c**) pearl-shaped nanowires, while (**d**) shows poly-sites for the nucleation of silicon nanowires (indicated by *arrows*) [4.116]

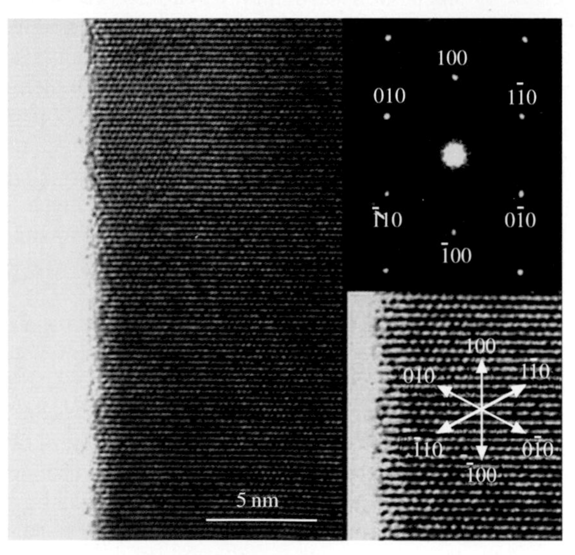

Fig. 4.14 Lattice resolved high resolution TEM image of one GaN nanowire (*left*) showing that (100) lattice planes are visible perpendicular to the wire axis. The electron diffraction pattern (*top right*) was recorded along the ⟨001⟩ zone axis. A lattice-resolved TEM image (*lower right*) highlights the continuity of the lattice up to the nanowire edge, where a thin native oxide layer is found. The directions of various crystallographic planes are indicated in the *lower right* figure [4.46]

mental TEM chamber, which allows TEM observations to be made while different gases are introduced or as the sample is heat treated at various temperatures, as illustrated in Fig. 4.16. The figure shows high resolution TEM images of a Bi nanowire with an oxide coating and the effect of a dynamic oxide removal process carried out within the environmental chamber of the TEM [4.117]. The amorphous bismuth-oxide layer

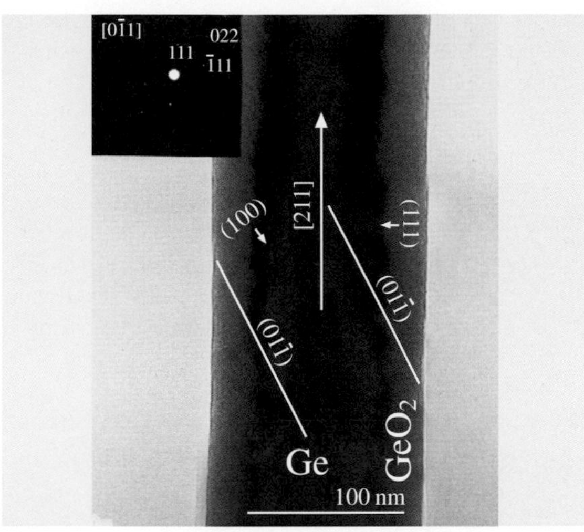

Fig. 4.15 A mass-thickness contrast TEM image of a Ge nanowire taken along the [0$\bar{1}$1] zone axis and a selected-area electron diffraction pattern (*upper left inset*) [4.51]. The Ge nanowires were synthesized by laser ablation of a mixture of Ge and GeO_2 powder. The core of the Ge nanowire is crystalline, while the surface GeO_2 is amorphous

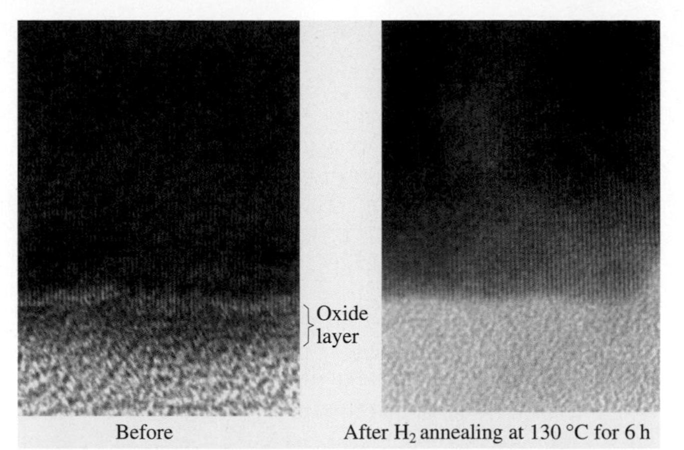

Fig. 4.16 High resolution transmission electron microscope (HRTEM) image of a Bi nanowire (*left*) before and (*right*) after annealing in hydrogen gas at 130 °C for 6 hours within the environmental chamber of the HRTEM instrument to remove the oxide surface layer [4.117]

coating the nanowire (Fig. 4.16a) is removed by exposure to hydrogen gas within the environmental chamber of the TEM, as indicated in Fig. 4.16b.

By coupling the powerful imaging capabilities of TEM with other characterization tools, such as an electron energy loss spectrometer (EELS) or an energy dispersive X-ray spectrometer (EDS) within the TEM instrument, additional properties of the nanowires can be probed with high spatial resolution. With the EELS technique, the energy and momentum of the incident and scattered electrons are measured in an inelastic electron scattering process to provide information on the energy and momentum of the excitations in the nanowire sample. Fig. 4.17 shows the dependence on nanowire diameter of the electron energy loss spectra of Bi nanowires. The spectra were taken from the center of the nanowire, and the shift in the energy of the peak position (Fig. 4.17) indicates the effect of the nanowire diameter on the plasmon frequency in the nanowires. The results show changes in the electronic structure of Bi nanowires as the wire diameter decreases [4.118]. Such changes in electronic structure as a function of nanowire diameter are also observed in their transport (Sect. 4.2.2) and optical (Sect. 4.2.3) properties and are related to quantum confinement effects.

EDS measures the energy and intensity distribution of X-rays generated by the impact of the electron beam on the surface of the sample. The elemental composition within the probed area can be determined to a high degree of precision. The technique was particularly useful for the compositional characterization of superlattice

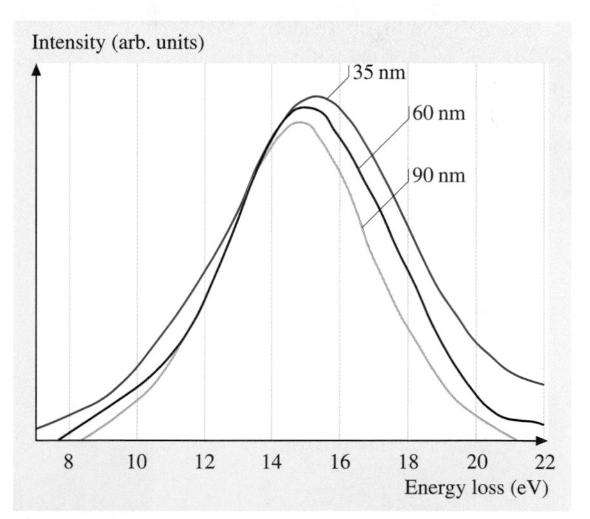

Fig. 4.17 Electron energy loss spectra (EELS) taken from the center of bismuth nanowires with diameters of 35, 60, and 90 nm. The shift in the volume plasmon peaks is due to the wire diameter effects on the electronic structure [4.118]

nanowires [4.90] and core-sheath nanowires [4.105] (see Sect. 4.1.2).

Scanning Tunneling Probes

Several scanning probe techniques [4.119], such as scanning tunneling microscopy (STM), electric field gradient microscopy (EFM) [4.16], magnetic field microscopy (MFM) [4.54], and scanning thermal microscopy (SThM) [4.120], combined with atomic force microscopy (AFM), have been employed to study the structural, electronic, magnetic, and thermal properties of nanowires. A scanning tunneling microscope can be employed to reveal both topographical structural information, such as that illustrated in Fig. 4.18, as well as information on the local electronic density of states of a nanowire, when used in the STS (scanning tunneling spectroscopy) mode. Figure 4.18 shows STM height images (taken in the constant current STM mode) of MoSe molecular wires deposited from a methanol or acetonitrile solution of $Li_2Mo_6Se_6$ on to Au substrates. The STM image of a single MoSe wire (Fig. 4.18a) exhibits a 0.45 nm lattice repeat distance in a MoSe molecular wire. When both STM and STS measurements are made on the same sample, the electronic and structural properties can be correlated, for example, as in the joint STM/STS studies on Si nanowires [4.96], showing alternating segments of a single nanowire identified with growth along [110] and [112] directions, and different I–V characteristics measured for the [110] segments as compared with the [112] segments.

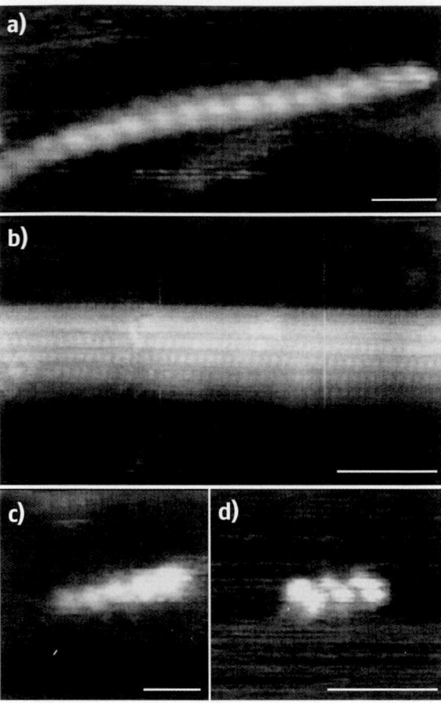

Fig. 4.18a–d STM height images, obtained in the constant current mode, of MoSe chains deposited on an Au(111) substrate. (**a**) A single chain image, and (**b**) a MoSe wire bundle. (**c**) and (**d**) Images of MoSe wire fragments containing 5 and 3 unit cells, respectively [4.119]. The scale bars are all 1 nm

Magnetic field microscopy (MFM) has been employed to study magnetic polarization of magnetic nanowires embedded in an insulating template, such as an anodic alumina template. For example, Fig. 4.19a shows the topographic image of an anodic alumina template filled with Ni nanowires, and Fig. 4.19b demonstrates the corresponding magnetic polarization of each nanowire in the template. This micrograph shows that a magnetic field microscopy probe can distinguish between spin-up and spin-down nanowires in the nanowire

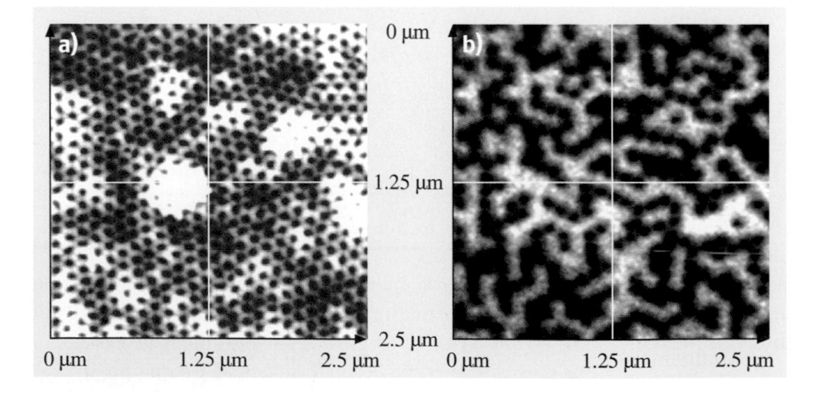

Fig. 4.19 (**a**) Topographic image of a highly ordered porous alumina template with a period of 100 nm filled with 35 nm diameter nickel nanowires. (**b**) The corresponding MFM (magnetic force microscope) image of the nano-magnet array, showing that the pillars are magnetized alternately "up" (*white*) and "down" (*black*) [4.54]

array, thereby providing a method for measuring interwire magnetic dipolar interactions [4.54].

X-Ray Analysis

Other characterization techniques commonly used to study the crystal structure and chemical composition of nanowires include X-ray diffraction and X-ray energy dispersion analysis (EDAX). The peak positions in the X-ray diffraction pattern can be used to determine the chemical composition and the crystal phase structure of the nanowires. For example, Fig. 4.2 shows that Bi nanowires have the same crystal structure and lattice constants as bulk bismuth. Both the X-ray diffraction pattern (XRD) for an array of aligned Bi nanowires (Fig. 4.2) and the SAED pattern for individual Bi nanowires [4.16] suggest that the nanowires have a common axis of crystal orientation.

As another example of an XRD pattern for an array of aligned nanowires, Fig. 4.20 shows the X-ray diffraction pattern of the ZnO nanowires displayed in Fig. 4.11. Only (00ℓ) diffraction peaks are observed for these aligned ZnO nanowires, indicating that their preferred growth direction is (001) along the wire axis. A similar preferred growth orientation was also observed for Bi nanowires produced by high pressure injection (see Sect. 4.1.1), while the vapor phase and low pressure injection filling techniques give preferred alignment along lower symmetry axes (see Fig. 4.2).

EDAX has been used to determine the chemical composition, stoichiometry of compound nanowires, or the impurity content in the nanowires. But the results from EDAX analysis should be interpreted carefully to avoid systematic errors.

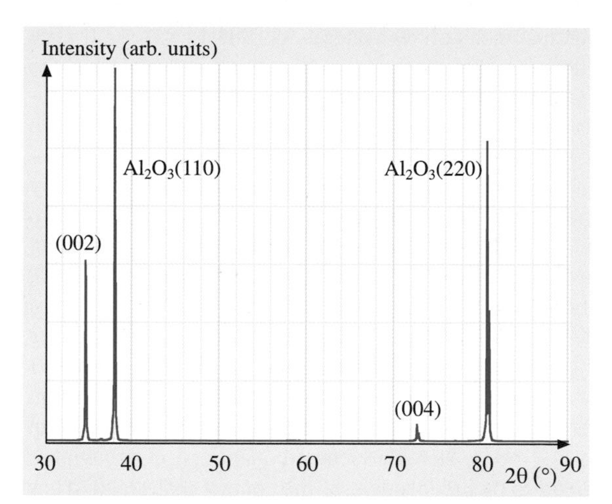

Fig. 4.20 X-ray diffraction pattern of aligned ZnO nanowires (see Fig. 4.11) grown on a sapphire substrate. Only $[00\ell]$ diffraction peaks are observed for the nanowires, owing to their well-oriented growth orientation. Strong diffraction peaks for the sapphire substrate are also found [4.115]

4.2.2 Transport Properties

The study of nanowire electrical transport properties is important for nanowire characterization, electronic device applications, and the investigation of unusual transport phenomena arising from one-dimensional quantum effects. Important factors that determine the transport properties of nanowires include the wire diameter (important for both classical and quantum size effects), material composition, surface conditions, crystal quality, and the crystallographic orientation along the wire axis, which is important for materials with anisotropic materials parameters, such as the effective mass tensor, the Fermi surface, or the carrier mobility.

Electronic transport phenomena in low-dimensional systems can be roughly divided into two categories: ballistic and diffusive transport. Ballistic transport phenomena occur when electrons travel across the nanowire without any scattering. In this case, the conduction is mainly determined by the contacts between the nanowire and the external circuit, and the conductance is quantized into an integral number of universal conductance units $G_0 = 2e^2/h$ [4.121, 122]. Ballistic transport phenomena are usually observed in very short quantum wires, such as those produced by using mechanically controlled break junctions (MCBJ) [4.123, 124] where the electron mean free path is much longer than the wire length, and the conduction is a pure quantum phenomenon. To observe ballistic transport, the thermal energy must also obey the relation $k_B T \ll \varepsilon_j - \varepsilon_{j-1}$, where $\varepsilon_j - \varepsilon_{j-1}$ is the energy separation between subband levels j and $j-1$. On the other hand, for nanowires with lengths much larger than the carrier mean free path, the electrons (or holes) undergo numerous scattering events when they travel along the wire. In this case, the transport is in the diffusive regime, and the conduction is dominated by carrier scattering within the wires due to phonons (lattice vibrations), boundary scattering, lattice and other structural defects, and impurity atoms.

Conductance Quantization in Metallic Nanowires

The ballistic transport of 1-D systems has been extensively studied since the discovery of quantized conductance in 1-D systems in 1988 [4.121, 122]. The phenomena of conductance quantization occur when the diameter of the nanowire is comparable to the electron Fermi wavelength, which is on the order of 0.5 nm for most metals [4.125]. Most conductance quantization experiments up to the present were performed by joining and separating two metal electrodes. As the two metal electrodes are slowly separated, a nano-contact is formed before it breaks completely (see Fig. 4.21a), and conductance in integral multiple values of G_0 is observed through these nano-contacts. Fig. 4.21b shows the conductance histogram built with 18,000 contact breakage curves between two gold electrodes at room temperature [4.126], with the electrode separation up to $\sim$ 1.8 nm. The conductance quantization behavior is found to be independent of the contact material and has been observed in various metals, such as Au [4.126], Ag, Na, Cu [4.127], and Hg [4.128]. For semimetals such as Bi, conductance quantization has also been observed for an electrode separation of as long as 100 nm at 4 K because of the long Fermi wavelength ($\sim$ 26 nm) [4.125], indicating that the conductance quantization may be due to the existence of well-defined quantum states localized at a constriction instead of resulting from the atom rearrangement as the electrodes separate. Since the conductance quantization is observed only in breaking contacts, or for very narrow and very short nanowires, most nanowires of practical interest (possessing lengths of several microns) lie in the diffusive transport regime, where the carrier scattering is significant and should be considered.

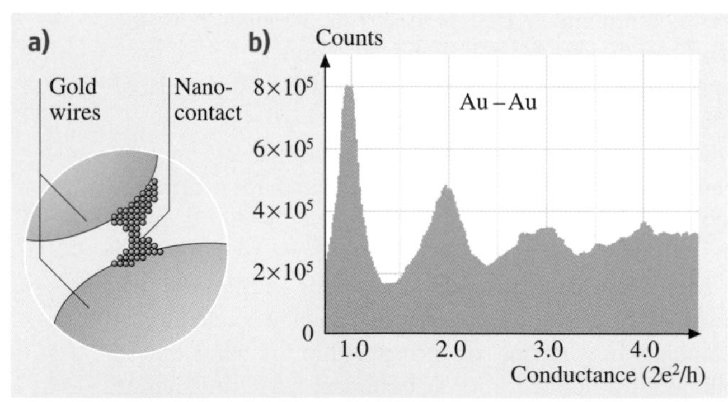

Fig. 4.21 **(a)** Schematic representation of the last stages of the contact breakage process [4.126]. **(b)** Histogram of conductance values built with 18,000 gold contact breakage experiments in air at room temperature, showing conductance peaks at integral values of G_0. In this experiment the gold electrodes approach and separate at 89,000 Å/s [4.126]

I–V Characterization of Semiconducting Nanowires

The electronic transport behavior of nanowires may be categorized based on the relative magnitude of three length scales: carrier mean free path ℓ_W, the de Broglie wavelength of electrons λ_e, and the wire diameter d_W. For wire diameters much larger than the carrier mean free path ($d_W \gg \ell_W$), the nanowires exhibit transport properties similar to bulk materials, which are independent of the wire diameter, since the scattering due to the wire boundary is negligible compared to other scattering mechanisms. For wire diameters comparable or smaller than the carrier mean free path ($d_W \sim \ell_W$ or $d_W < \ell_W$) but still much larger than the de Broglie wavelength of the electrons ($d_W \gg \lambda_e$), the transport in nanowires is in the classical finite size regime, in which the band structure of the nanowire is still similar to that of bulk while the scattering events at the wire boundary alter their transport behavior. For wire diameters comparable to the electronic wavelength $d_W \sim \lambda_e$, the electronic density of states is altered dramatically and quantum subbands are formed due to the quantum confinement effect at the wire boundary. In this regime, the transport properties are further influenced by the change in the band structure. Transport properties for nanowires in the classical finite size and quantum size regimes, therefore, are highly diameter-dependent.

Researchers have investigated the transport properties of various semiconducting nanowires and have demonstrated their potential for diverse electronic devices, such as for p-n diodes [4.129, 130], field effect transistors [4.129], memory cells, and switches [4.131] (see Sect. 4.3.1). The nanowires studied so far in this context have usually been made from conventional semiconducting materials, such as group IV and III-V compound semiconductors via the VLS growth method (see Sect. 4.1.2), and their nanowire properties have been compared to their well-established bulk properties. Interestingly, the physical principles for describing bulk semiconductor devices also hold for devices based on these semiconducting nanowires with wire diameters of tens of nanometers. For example, Fig. 4.22 shows the current–voltage (I–V) behavior of a 4-by-1 crossed p-Si/n-GaN junction array at room temperature [4.129]. The long horizontal wire in the figure is a p-Si nanowire (10–25 nm in diameter) and the four short vertical wires are n-GaN nanowires (10–30 nm in diameter). Each of the four nanoscale cross points independently forms a p-n junction with current rectification behavior, as shown by the I–V curves in Fig. 4.22, and the junction behavior (e.g., the turn-on voltage) can

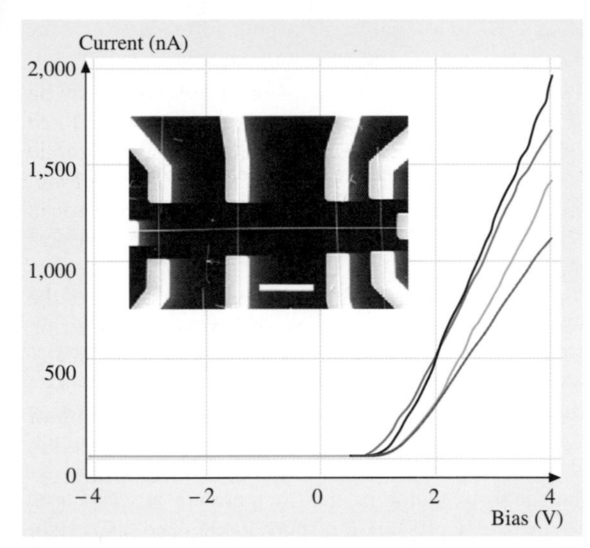

Fig. 4.22 I–V behavior for a 4(p) by 1(n) crossed p-Si/n-GaN junction array shown schematically in the inset. The four curves represent the I–V response for each of the four junctions, showing similar current rectifying characteristics in each case. The length scale bar between the two middle junctions is 2 μm [4.129]. The p-Si and n-GaN nanowires are 10–25 nm and 10–30 nm in diameter, respectively

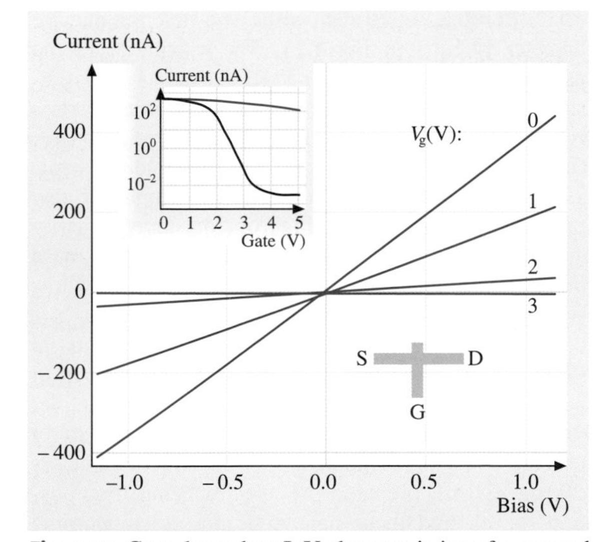

Fig. 4.23 Gate-dependent I–V characteristics of a crossed nanowire field-effect transistor (FET). The n-GaN nanowire is used as the nano-gate, with the gate voltage indicated (0, 1, 2, and 3 V). The inset shows the current vs. V_{gate} for a nanowire gate (*lower curve*) and for a global back-gate (*top curve*) when the bias voltage is set at 1 V [4.129]

be controlled by varying the oxide coating on these nanowires [4.129].

Huang et al. have demonstrated nanowire junction diodes with a high turn-on voltage ($\sim$ 5 V) by increasing the oxide thickness at the junctions. The high turn-on voltage enables the use of the junction in a nanoscale FET as shown in Fig. 4.23 [4.129], where I–V data for a p-Si nanowire are presented, for which the n-GaN nanowire with a thick oxide coating is used as a nano-gate. By varying the nano-gate voltage, the conductance of the p-Si nanowire can be changed by more than a factor of 10^5 (lower curve in the inset), whereas the conductance changes by only a factor of ten when a global back-gate is used (top curve in the inset of Fig. 4.23). This behavior may be due to the thin gate dielectric between the crossed nanowires and the better control of the local carrier density through a nano-gate. Based on the gate-dependent I–V data of these p-Si nanowires, it is found that the mobility for holes in the p-Si nanowires may be higher than that for bulk p-Si, although further investigation is required for a complete understanding.

Because of the enhanced surface-to-volume ratio of nanowires, their transport behavior may be modified by changing their surface conditions. For example, researchers have found that by coating n-InP nanowires with a layer of redox molecules such as cobalt phthalocyanine, the conductance of the InP nanowires may change by orders of magnitude by altering the charge state of the redox molecules to provide bistable nanoscale switches [4.131]. The resistance (or conductance) of some nanowires (e.g., Pd nanowires) is also very sensitive to the presence of certain gases (e.g., H_2) [4.132, 133], and this property may be utilized for sensor applications to provide improved sensitivity compared to conventional sensors based on bulk material (see Sect. 4.3.4).

Although it remains unclear how the size may influence the transport properties and device performance of semiconducting nanowires, many of the larger diameter semiconducting nanowires are expected to be described by classical physics, since their quantization energies $\hbar^2/(2m_e d_W^2)$ are usually smaller than the thermal energy $k_B T$. By comparing the quantization energy with the thermal energy, the critical wire diameter below which quantum confinement effects become significant is estimated to be 1 nm for Si nanowires at room temperature, which is much smaller than the size of many of the semiconducting nanowires that have been investigated so far. By using material systems with much smaller effective carrier masses m_e (such as bismuth), the critical diameter for which such quantum effects can be observed is increased, thereby facilitating the study of quantum confinement effects. It is for this reason that the bismuth nanowire system has been studied so extensively. Furthermore, since the crystal structure and lattice constants of bismuth nanowires are the same as for 3-D crystalline bismuth, it is possible to carry out detailed model calculations to guide and to interpret transport and optical experiments on bismuth nanowires. For these reasons, bismuth can be considered as a model system for studying 1-D effects in nanowires.

Temperature-Dependent Resistance Measurements

Although nanowires with electronic properties similar to their bulk counterparts are promising for constructing nano-devices based on well-established knowledge of their bulk counterparts, it is expected that quantum size effects in nanowires will likely be utilized to generate new phenomena absent in bulk materials and thus provide enhanced performance and novel functionality for certain applications. In this context, the transport properties of bismuth (Bi) nanowires have been extensively studied, both theoretically [4.134] and experimentally [4.28, 30, 67, 135–138] because of their promise for enhanced thermoelectric performance. Transport studies of ferromagnetic nanowire arrays, such as Ni or Fe, have also received much attention because of their potential for high-density magnetic storage applications.

The very small electron effective mass components and the long carrier mean free paths in Bi facilitate the study of quantum size effects in the transport properties of nanowires. Quantum size effects are expected to become significant in bismuth nanowires with diameters smaller than 50 nm [4.134], and the fabrication of crystalline nanowires of this diameter range is relatively easy.

Figure 4.24a shows the T dependence of the resistance $R(T)$ for Bi nanowires ($7 \leq d_W < 200$ nm) synthesized by vapor deposition and pressure injection [4.28], illustrating the quantum effects in their temperature-dependent resistance. In Fig. 4.24a, the $R(T)$ behavior of Bi nanowires is dramatically different from that of bulk Bi and is highly sensitive to the wire diameter. The $R(T)$ curves in Fig. 4.24a show a non-monotonic trend for larger-diameter (70 and 200 nm) nanowires, although $R(T)$ becomes monotonic with T for small-diameter (≤ 48 nm) nanowires. This dramatic change in the behavior of $R(T)$ as a function of d_W is attributed to a unique semimetal-semiconductor transition phenomena in Bi [4.139], induced by quantum size effects. Bi is a semimetal in bulk form, in which the T-point valence band overlaps with the L-point conduction band by 38 meV at 77 K. As the wire diameter decreases, the lowest conduction subband increases in energy, and the highest valence subband decreases in energy. Model calculations predict that the band overlap should vanish in Bi nanowires (with their wire axes along the trigonal direction) at a wire diameter ~ 50 nm [4.134].

The resistance of Bi nanowires is determined by two competing factors: the carrier density that increases with T, and the carrier mobility that decreases with T. The non-monotonic $R(T)$ for large-diameter Bi nanowires is due to a smaller carrier concentration variation at low temperature (≤ 100 K) in semimetals, so that the electrical resistance is dominated by the mobility factor in this temperature range. Based on the semiclassical transport model and the established band structure of Bi nanowires, the calculated $R(T)/R(300\,\text{K})$ for 36-nm and 70-nm Bi nanowires is shown by the solid curves in Fig. 4.24c to illustrate different $R(T)$ trends for semiconducting and semimetallic nanowires, respectively [4.67]. The curves in Fig. 4.24c exhibit trends consistent with experimental results. The condition for the semimetal–semiconductor transition in Bi nanowires can be experimentally determined, as shown by the measured resistance ratio $R(10\,\text{K})/R(100\,\text{K})$ of Bi nanowires as a function of wire diameter [4.140] in Fig. 4.25. The maximum in the resistance ratio $R(10\,\text{K})/R(100\,\text{K})$ at $d_W \sim 48$ nm indicates the wire diameter for the transition of Bi nanowires from a semimetallic phase to a semiconducting phase. The semimetal–semiconductor transition and the semiconducting phase in Bi nanowires are examples of new transport phenomena, resulting from low dimensionality and absent in the bulk 3-D phase, which further increase the possible benefits from the properties of nanowires for desired applications (see Sect. 4.3.2).

It should be noted that good crystal quality is essential for observing the quantum size effect in nanowires, as shown by the $R(T)$ plots in Fig. 4.24a. For example, Fig. 4.24b shows the normalized $R(T)$ measurements of Bi nanowires with larger diameters (200 nm–2 μm) prepared by electrochemical deposition [4.30], and these nanowires possess monotonic $R(T)$ behaviors, quite different from those of the corresponding nanowire diameters shown in Fig. 4.24a. The absence of the resistance maximum in Fig. 4.24b is due to the lower crystalline quality for nanowires prepared by electrochemical deposition, which tends to produce polycrystalline nanowires with a much lower

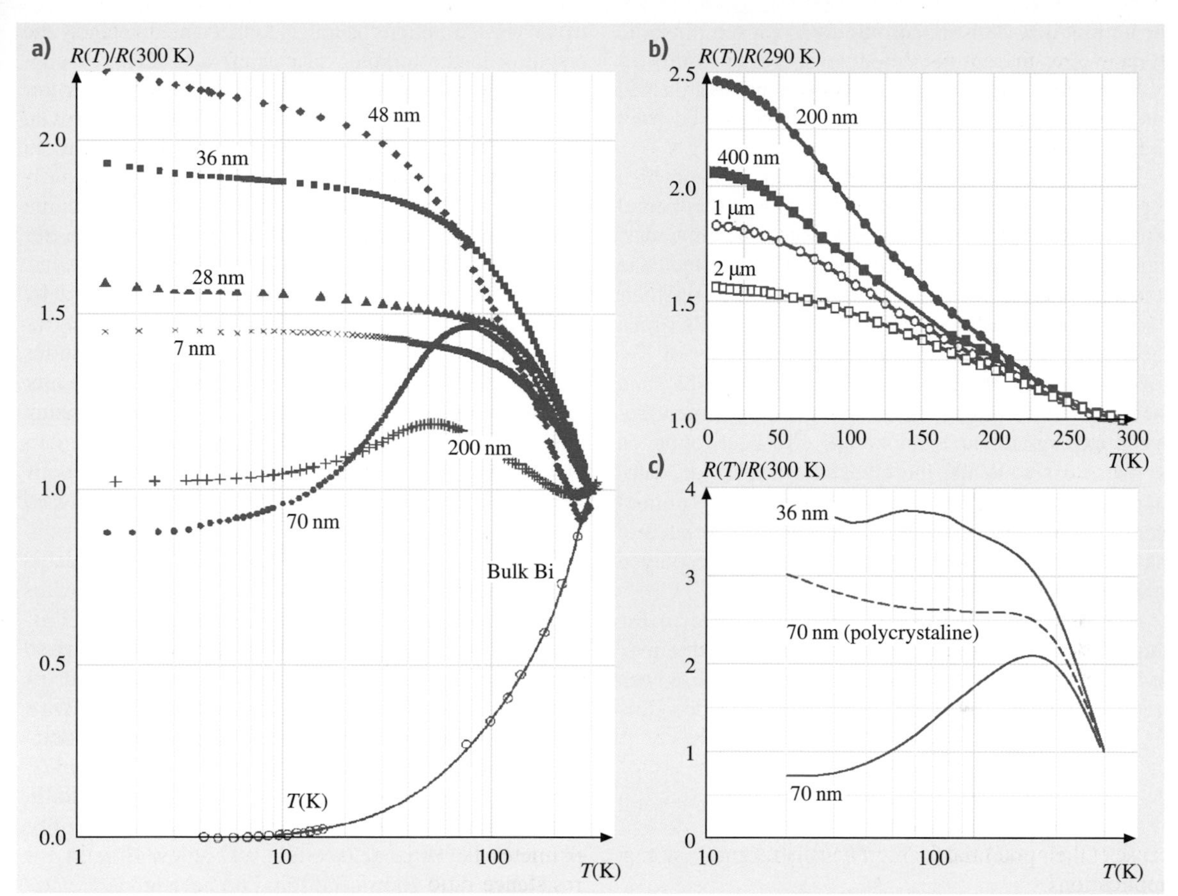

Fig. 4.24 (a) Measured temperature dependence of the resistance $R(T)$ normalized to the room temperature (300 K) resistance for bismuth nanowire arrays of various wire diameters d_W [4.28]. (b) $R(T)/R(290\,K)$ for bismuth wires of larger d_W and lower mobility [4.30]. (c) Calculated $R(T)/R(300\,K)$ of 36-nm and 70-nm bismuth nanowires. The *dashed curve* refers to a 70-nm polycrystalline wire with increased boundary scattering [4.139]

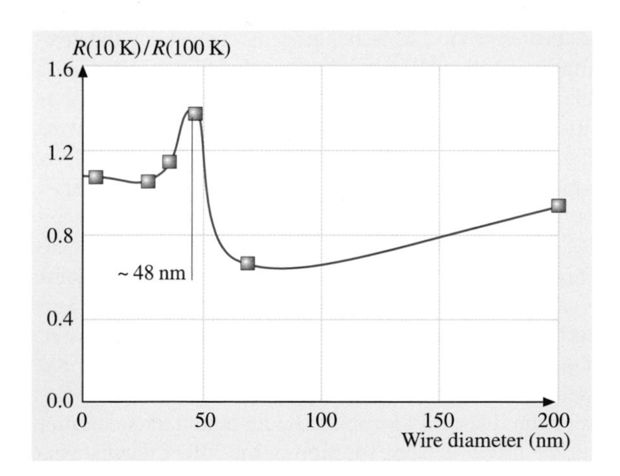

Fig. 4.25 Measured resistance ratio $R(10\,K)/R(100\,K)$ of Bi nanowires as a function of diameter. The peak indicates the transition from a semimetallic phase to a semiconducting phase as the wire diameter decreases [4.141]

carrier mobility. This monotonic $R(T)$ for semimetallic Bi nanowires at a higher defect level is also confirmed by theoretical calculations, as shown by the dashed curve in Fig. 4.24c for 70-nm wires with increased grain boundary scattering [4.139].

The theoretical model developed for Bi nanowires not only provides good agreement with experimental results, but it also plays an essential role in understanding the influence of the quantum size effect, the boundary scattering, and the crystal quality on their electrical properties. The transport model has also been generalized to predict the transport properties of Te-doped Bi nanowires [4.67], Sb nanowires [4.142], and BiSb alloy nanowires [4.143], and good agreement between experiment and theory has also been obtained for these cases. While the electronic density of states may be significantly altered due to quantum confinement effects, various scattering mechanisms related to the transport properties of nanowires can be accounted for by Matthiessen's rule.

For nanowires with diameters comparable to the phase-breaking length, their transport properties may be further influenced by localization effects. It has been predicted that in disordered systems, the extended electronic wave functions become localized near defect sites, resulting in the trapping of carriers and giving rise to a different transport behavior. Localization effects are also expected to be more pronounced as dimensionality and sample size are reduced. Localization effects on the transport properties of nanowire systems have been studied on Bi nanowires [4.144] and, more recently, on Zn nanowires [4.63]. Figure 4.26 shows the measured $R(T)/R(300\,\mathrm{K})$ of Zn nanowires fabricated by vapor deposition in porous silica or alumina [4.63]. While 15 nm Zn nanowires exhibit an $R(T)$ behavior with a T^1 dependence as expected for a metallic wire, the $R(T)$ of 9 nm and 4 nm Zn nanowires exhibits a temperature dependence of $T^{-1/2}$ at low temperature, consistent with 1-D localization theory. Thus, due to this localization effect, the use of nanowires with very small diameters for transport applications may be limited.

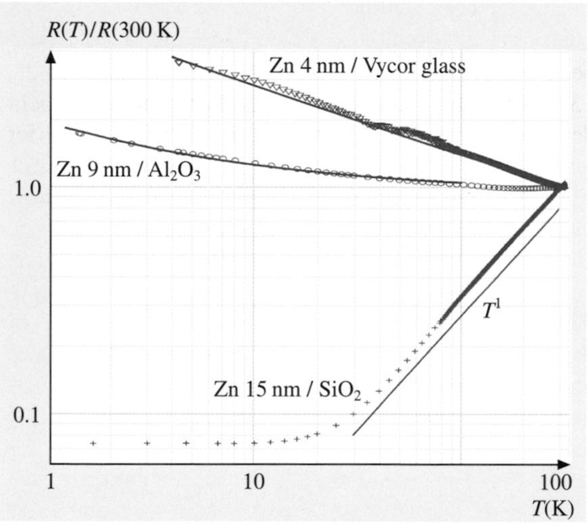

Fig. 4.26 Temperature dependence of the resistance of Zn nanowires synthesized by vapor deposition in various porous templates [4.63]. The data are given as points, the *full line* arc fits to a T^1 law for 15 nm diameter Zn nanowires in an SiO_2 template, denoted by Zn/SiO_2. Fits to a combined T^1 and $T^{-1/2}$ law were made for the smaller nanowire diameter composite denoted by 9 nm Zn/Al_2O_3 and 4 nm Zn/Vycor glass

Magnetoresistance

Magnetoresistance (MR) measurements provide an informative technique for characterizing nanowires because these measurements yield a great deal of information about the electron scattering with wire boundaries, the effects of doping and annealing on scattering, and localization effects in the nanowires [4.137]. For example, at low fields the MR data show a quadratic dependence on the B field from which carrier mobility estimates can be made (see Fig. 4.27 at low B field).

Figure 4.27 shows the longitudinal magnetoresistance ($\boldsymbol{B}$ parallel to the wire axis) for 65 nm and 109 nm Bi nanowire samples (before thermal annealing) at 2 K. The MR maxima in Fig. 4.27a are due to the classical size effect, where the wire boundary scattering is reduced as the cyclotron radius becomes smaller than the wire radius in the high field limit, resulting in a decrease in the resistivity. This behavior is typical for the longitudinal MR of Bi nanowires in the diameter range of 45 nm to 200 nm [4.28, 136, 137, 145], and the peak position B_{m} moves to lower B field values as the wire diameter increases, as shown in Fig. 4.27c [4.145], in which B_{m} varies linearly with $1/d_{\mathrm{W}}$. The condition for the occurrance of B_{m} is approximately given by $B_{\mathrm{m}} \sim 2c\hbar k_{\mathrm{F}}/ed_{\mathrm{W}}$ where k_{F} is the wave vector at the Fermi energy. The peak position, B_{m}, is found to increase linearly with temperature in the range of 2 to 100 K, as shown in Fig. 4.27b [4.145]. As T is increased, phonon scattering becomes increasingly important, and therefore a higher magnetic field is required to reduce the resistivity associated with boundary scattering sufficiently to change the sign of the MR. Likewise, in-

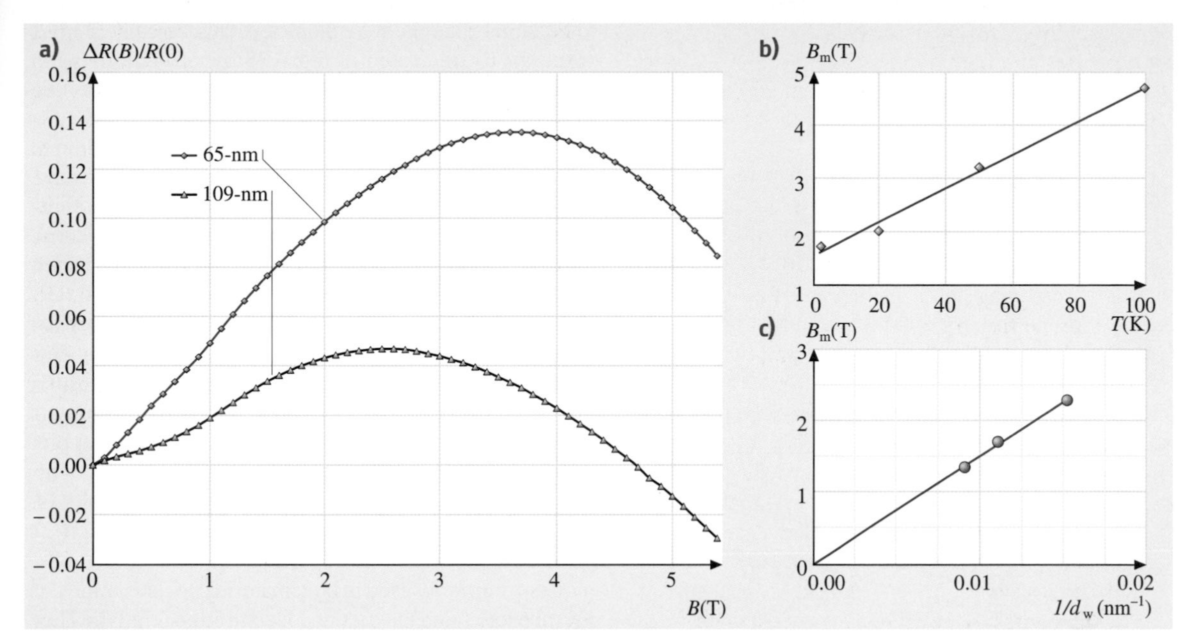

Fig. 4.27 (**a**) Longitudinal magnetoresistance, $\Delta R(B)/R(0)$, at 2 K as a function of B for Bi nanowire arrays with diameters 65 and 109 nm before thermal annealing. (**b**) The peak position B_m as a function of temperature for the 109 nm diameter Bi nanowire array after thermal annealing. (**c**) The peak position B_m of the longitudinal MR (after thermal annealing) at 2 K as a function of $1/d_W$, the reciprocal of the nanowire diameter [4.145]

creasing the grain boundary scattering is also expected to increase the value of B_m at a given T and wire diameter.

The presence of the peak in the longitudinal MR of nanowires requires a high crystal quality with long carrier mean free paths along the nanowire axis, so that most scattering events occur at the wire boundary instead of at a grain boundary, at impurity sites, or at defect sites within the nanowire. *Liu* et al. have investigated the MR of 400-nm Bi nanowires synthesized by electrochemical deposition [4.21], and no peak in the longitudinal MR is observed. The absence of a magnetoresistance peak may be attributed to a higher defect level in the nanowires produced electrochemically and to a large wire diameter, much longer than the carrier mean free path. The negative MR observed for the Bi nanowire arrays above B_m (see Fig. 4.27) shows that wire boundary scattering is a dominant scattering process for the longitudinal magnetoresistance, thereby establishing that the mean free path is larger than the wire diameter and that a ballistic transport behavior is indeed observed in the high field regime.

In addition to the longitudinal magnetoresistance measurements, transverse magnetoresistance measurements (***B*** perpendicular to the wire axis) have also been performed on Bi nanowires array samples [4.28, 137, 145], where a monotonically increasing B^2 dependence over the entire range $0 \leq B \leq 5.5$ T is found for all Bi nanowires studied thus far. This is as expected, since the wire boundary scattering cannot be reduced by a magnetic field perpendicular to the wire axis. The transverse magnetoresistance is also found always to be larger than the longitudinal magnetoresistance in nanowire arrays.

By applying a magnetic field to nanowires at very low temperatures (≤ 5 K), one can induce a transition from a 1-D confined system at low magnetic fields to a 3-D confined system as the field strength increases, as shown in Fig. 4.28 for the longitudinal MR of Bi nanowire arrays of various nanowire diameters (28–70 nm) for $T < 5$ K [4.137]. In these curves, a subtle step-like feature is seen at low magnetic fields, which is found to depend only on the wire diameter, and is independent of temperature, the orientation of the magnetic field, and even on the nanowire material (e.g., in Sb nanowires [4.142]). The lack of a dependence of the magnetic field at which the step appears on temperature, field orientation, and material type indicates that

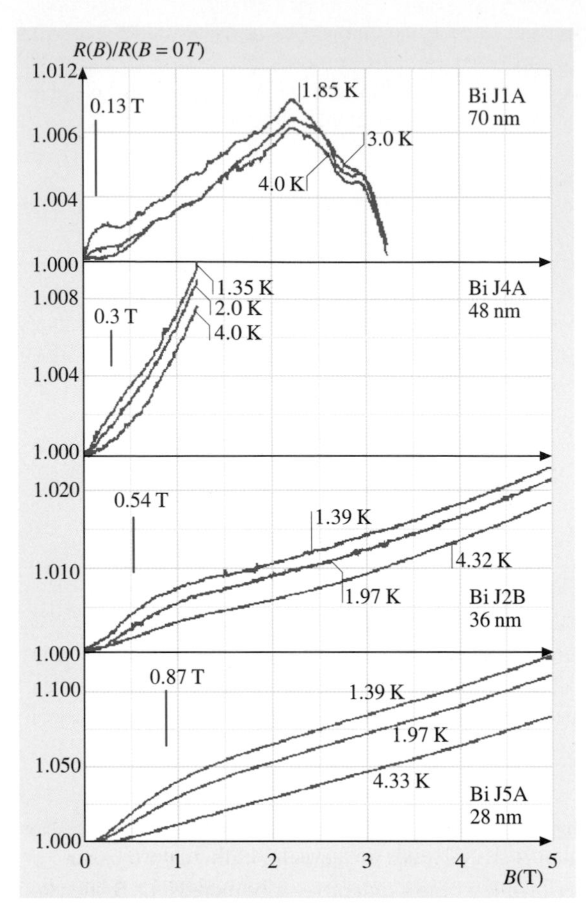

Fig. 4.28 Longitudinal magnetoresistance as a function of magnetic field for Bi nanowires of the diameters indicated. The vertical bars indicate the critical magnetic field B_c at which the magnetic length equals the nanowire diameter [4.137]

the phenomenon is related to the magnetic field length, $L_H = (\hbar/eB)^{1/2}$. The characteristic length L_H is the spatial extent of the wave function of electrons in the lowest Landau level, and L_H is independent of the carrier effective masses. Setting $L_H(B_c)$ equal to the diameter d_W of the nanowire defines a critical magnetic field strength, B_c, below which the wave function is confined by the nanowire boundary (the 1-D regime), and above which the wave function is confined by the magnetic field (the 3-D regime). The physical basis for this phenomenon is associated with confinement of a single magnetic flux quantum within the nanowire cross section [4.137]. This phenomenon, though independent of temperature, is observed for $T \leq 5$ K, since the phase breaking length has to be larger than the wire diameter. This calculated field strength, B_c, indicated in Fig. 4.28 by vertical lines for the appropriate nanowire diameters, provides a good fit to the step-like features in these MR curves.

The Shubnikov–de Haas (SdH) quantum oscillatory effect, which results from the passage of the quantized Landau levels through the Fermi energy as the field strength varies, should, in principle, provide the most direct measurement of the Fermi energy and carrier density. For example, *Heremans* et al. have demonstrated that SdH oscillations can be observed in Bi nanowire samples with diameters down to 200 nm [4.146] and have demonstrated that Te doping can be used to raise the Fermi energy in Bi nanowires. Such information on the Fermi energy is important because, for certain applications based on nanowires, it is necessary to place the Fermi energy near a subband edge where the density of states has a sharp feature. But due to the unusual 1-D geometry for nanowires, other characterization techniques commonly used in bulk materials to determine the Fermi energy and the carrier concentration (e.g., the Hall measurement) cannot be applied for nanowire systems. The observation of the SdH oscillatory effect requires very high crystal quality samples that allow carriers to execute a complete cyclotron orbit in the nanowire before they are scattered. For small nanowire diameters, large magnetic fields are required to produce cyclotron radii smaller than the wire radius. For some nanowire systems, all Landau levels may have passed through the Fermi level at such a high field strength, and in such a case, no oscillations can be observed. The localization effect may also prevent the observation of SdH oscillations for very small diameter (≤ 10 nm) nanowires. Observing SdH oscillations in highly doped samples (as may be required for certain applications) may be difficult because impurity scattering reduces the mean free path, requiring high $\boldsymbol{B}$ fields to satisfy the requirement that carriers complete a cyclotron orbit prior to scattering. Although SdH oscillations provide the most direct method of measuring the Fermi energy and carrier density of nanowire samples, this technique may not work, however, for smaller diameter nanowires, or for heavily doped nanowires.

Thermoelectric Properties

Nanowires are predicted to be promising for thermoelectric applications [4.134, 147], due to their novel band structure compared to their bulk counterparts and the expected reduction in thermal conductivity associated with enhanced boundary scattering (see Sect. 4.2.2). Due to the sharp density of states at the 1-D subband edges (where the van Hove singularities occur), nanowires are expected to exhibit enhanced Seebeck coefficients compared to their bulk counterparts. Since the Seebeck coefficient measurement is independent of the number of nanowires contributing to the signal, the measurements on nanowire arrays of uniform wire diameter are, in principle, as informative as single-wire measurements. The major challenge in measuring the Seebeck coefficient of nanowires lies in the design of tiny temperature probes to determine accurately the temperature difference across the nanowire. Figure 4.29a shows the schematic experimental setup for the Seebeck coefficient measurement of nanowire arrays [4.148], where two thermocouples are placed on both faces of a nanowire array and a heater is attached on one face of the array to generate a temperature gradient along the nanowire axis. Ideally the size of the thermocouples should be much smaller than

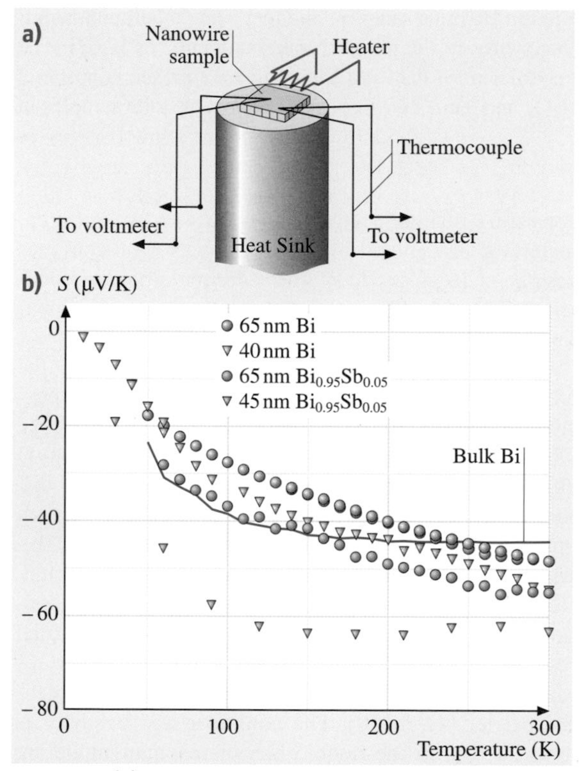

Fig. 4.29 **(a)** Experimental setup for the measurement of the Seebeck coefficient in nanowire arrays [4.148]. **(b)** Measured Seebeck coefficient as a function of temperature for Bi (●,▽) and $Bi_{0.95}Sb_{0.05}$ (●,▽) nanowires with different diameters. The *solid curve* denotes the Seebeck coefficient for bulk Bi [4.139]

the thickness of the nanowire array template (i. e., the nanowire length) to minimize error. However, due to the thinness of most templates ($\leq 50\,\mu m$) and the large size of commercially available thermocouples ($\sim 12\,\mu m$), the measured Seebeck coefficient values are usually underestimated.

The thermoelectric properties of Bi nanowire systems have been investigated extensively because of their potential as good thermoelectric materials. Figure 4.29b shows the measured Seebeck coefficient $S(T)$ as a function of temperature for nanowire arrays with diameters of 40 and 65 nm and different isoelectronic Sb alloy concentrations [4.139]; $S(T)$ results for bulk Bi are shown (*solid curve*) for comparison. Enhancement in the thermopower is observed in Fig. 4.29b as the wire diameter decreases and as the Sb content increases, which is attributed to the semimetal–semiconductor transition induced by quantum confinement and to Sb alloying effects in $Bi_{1-x}Sb_x$ nanowires. *Heremans* et al. have observed a substantial increase in the thermopower of Bi nanowires as the wire diameter further decreases, as shown in Fig. 4.30a for 15 nm Bi/silica and 9 nm Bi/alumina nanocomposites [4.63]. The enhancement is due to the sharp density of states near the Fermi energy in a 1-D system. Although the samples in Fig. 4.30a also possess very high electrical resistance ($\sim$ GΩ), the results for the 9 nm Bi/alumina samples show that the Seebeck coefficient can be enhanced by almost 1,000 times relative to bulk material. But for Bi nanowires with very small diameters (~ 4 nm), the localization effect becomes dominant, which compromises the thermopower enhancement. Therefore, for Bi nanowires, the optimal wire diameter range for the largest thermopower enhancement is found to be between 4 to 15 nm [4.63].

The effect of the nanowire diameter on the thermopower of nanowires has also been observed in Zn nanowires [4.63]. Figure 4.30b shows the Seebeck coefficient of 9 nm Zn/alumina and 4 nm Zn/Vycor glass nanocomposites, exhibiting enhanced thermopower as the wire diameter decreases. It is found that while 9 nm Zn nanowires still exhibit metallic behavior, the thermopower of 4 nm Zn nanowires shows a different temperature dependence, which may be due to the 1-D localization effect, although further investigation is required for a definitive identification of the conduction mechanism in such small nanowires.

Quantum Wire Superlattices

The studies on superlattice nanowires, which possess a periodic modulation in their materials composition

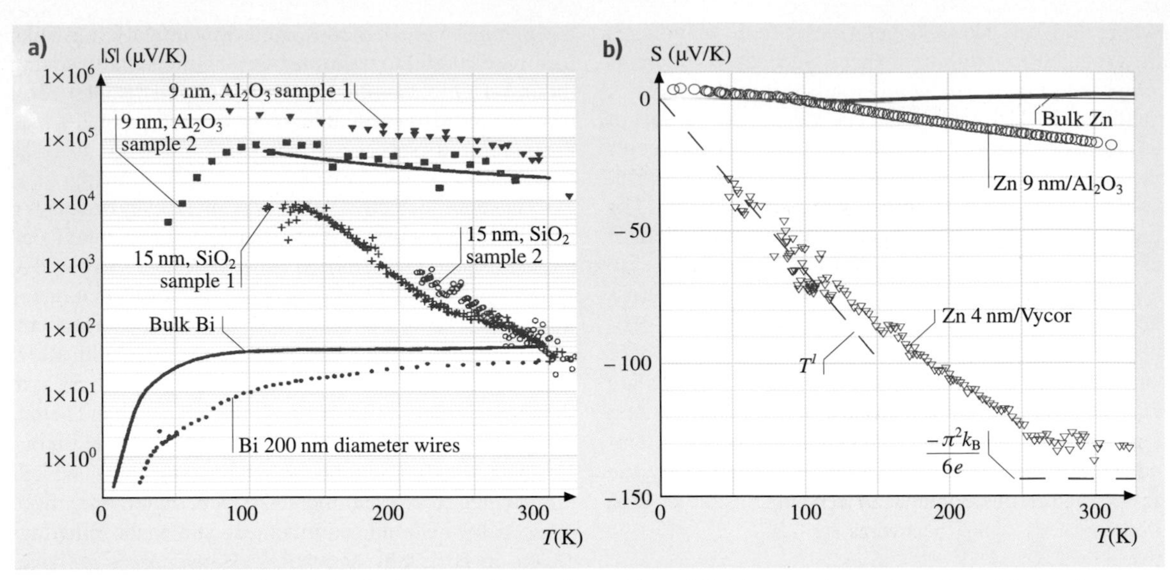

Fig. 4.30 **(a)** Absolute value of the Seebeck coefficient of two 15 nm Bi/silica samples, and two 9 nm Bi/alumina nanocomposite samples, in comparison to bulk Bi and 200 nm Bi nanowires in the pores of alumina templates [4.63]. The *full line* on top of the figure is a fit to a T^{-1} law. The Seebeck coefficient of the 9 nm Bi/alumina composite is positive; the rest are negative. **(b)** The Seebeck coefficient of 9 nm Zn/Al_2O_3 and 4 nm Zn/Vycor glass nanocomposite samples in comparison to bulk Zn [4.63]

along the wire axis, have attracted much attention recently because of their promise in such applications as thermoelectrics (see Sect. 4.3.2) [4.149], nanobarcodes (see Sect. 4.3.3) [4.90], nanolasers (see Sect. 4.3.3) [4.92], one-dimensional waveguides, and resonant tunneling diodes [4.93, 150]. Figure 4.31a shows a schematic structure of a superlattice nanowire consisting of interlaced quantum dots of two different materials, as denoted by A and B. Various techniques have been developed to synthesize superlattice nanowire structures with different interface conditions, as mentioned in Sect. 4.1.1 and Sect. 4.1.2.

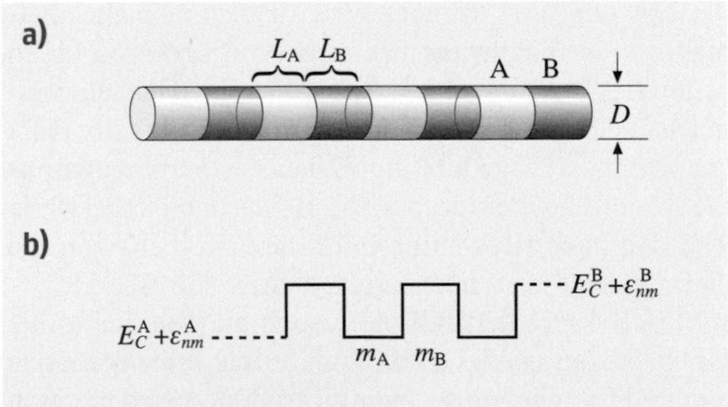

Fig. 4.31 **(a)** Schematic diagram of superlattice (segmented) nanowires consisting of interlaced nanodots A and B of the indicated length and wire diameter. **(b)** Schematic potential profile of the subbands in the superlattice nanowire [4.149]

In this superlattice (SL) nanowire structure, the electronic transport along the wire axis is made possible by the tunneling between adjacent quantum dots, while the uniqueness of each quantum dot and its 0-D characteristic behavior is maintained by the energy difference of the conduction or valence bands between quantum dots of different materials (see Fig. 4.31b), which provides some amount of quantum confinement. Recently *Björk* et al. have observed interesting nonlinear I–V characteristics with a negative differential resistance in one-dimensional heterogeneous structures made of InAs and InP, where InP serves as the potential barrier [4.93, 150]. The nonlinear I–V behavior is associated with the double barrier resonant tunneling in one-dimensional structures, demonstrating the capability of transport phenomena in superlattice nanowires via tunneling and the possibility of controlling the electronic band structure of the SL nanowires by carefully selecting the constituent materials. This kind of new structure is especially attractive for thermoelectric ap-

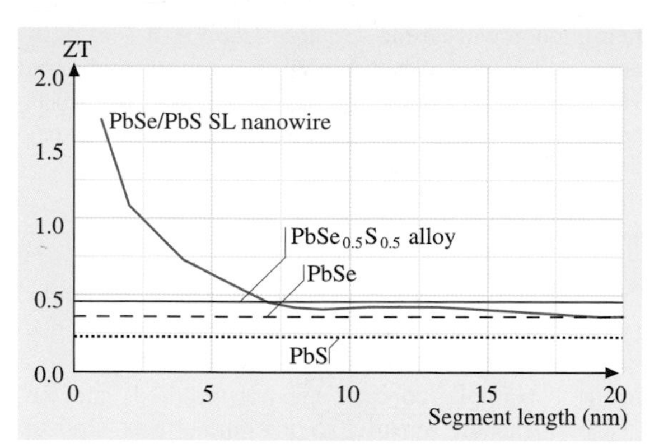

Fig. 4.32 Optimal ZT calculated as a function of segment length for 10-nm diameter PbSe/PbS nanowires at 77 K, where "optimal" refers to the placement of the Fermi level to optimize ZT. The optimal ZT for 10-nm diameter PbSe, PbS, and $PbSe_{0.5}S_{0.5}$ nanowires are 0.33, 0.22, and 0.48, respectively [4.141]

plications, because the interfaces between the nanodots can reduce the lattice thermal conductivity by blocking the phonon conduction along the wire axis, while electrical conduction may be sustained and even benefit from the unusual electronic band structures due to the periodic potential perturbation. For example, Fig. 4.32 shows the calculated dimensionless thermoelectric figure of merit $ZT = S^2\sigma T/\kappa$ (see Sect. 4.2.2) where κ is the total thermal conductivity (including both the lattice and electronic contribution) of 10-nm diameter PbS/PbSe superlattice nanowires as a function of the segment length. A higher thermoelectric performance than for $PbSe_{0.5}S_{0.5}$ alloy nanowires can be achieved for a 10 nm diameter superlattice nanowire with segment lengths ≤ 7 nm. But the localization effect, which may become important for very short segment lengths, may jeopardize this enhancement in the ZT of superlattice nanowires [4.141].

Thermal Conductivity of Nanowires

Experimental measurements of the temperature dependence of the thermal conductivity $\kappa(T)$ of individual suspended nanowires have been carried out to study the dependence of $\kappa(T)$ on the wire diameter. In this context, measurements have been made on nanowires down to only 22 nm in diameter [4.151]. Such measurements are very challenging and are now possible because of technological developments in fabricating and using nanometer size thermal scanning probes [4.120, 152, 153]. The experiments show that the thermal conductivity of small homogeneous nanowires may be more than one order of magnitude smaller than in the bulk, arising mainly from strong boundary scattering effects [4.154]. Phonon confinement effects may eventually become important at still smaller diameter nanowires (see Sect. 4.2.3). Measurements on mats of nanowires (see, for example, Fig. 4.12) do not generally give reliable results because the contact thermal resistance between adjacent nanowires tends to be high, which is in part due to the thin surface oxide coating that most nanowires have. This surface oxide coating may also be important for thermal conductivity measurements on individual suspended nanowires because of the relative importance of phonon scattering at the lateral walls of the nanowire.

The most extensive experimental thermal conductivity measurements have been done on Si nanowires [4.151] where $\kappa(T)$ measurements have been made on nanowires in the diameter range of $22 \leq d_W \leq 115$ nm. The results show a large decrease in the peak of $\kappa(T)$ associated with umklapp processes as d_W decreases, indicating a growing importance of boundary scattering and a corresponding decreasing importance of phonon–phonon scattering. At the smallest wire diameter of 22 nm, a linear $\kappa(T)$ dependence is found experimentally, consistent with a linear T dependence of the specific heat for a 1-D system and a temperature independent mean free path and velocity of sound.

Model calculations for $\kappa(T)$ based on a radiative heat transfer model have been carried out for Si nanowires [4.155]. These results show that the predicted $\kappa(T)$ behavior for Si nanowires is similar to that observed experimentally in the range of $37 \leq d_W \leq 115$ nm regarding both the functional form of $\kappa(T)$ and the magnitude of the relative decrease in the maximum thermal conductivity κ_{max} as a function of d_W. But the model calculations predict a substantially larger magnitude for $\kappa(T)$ (by 50% or more) than is observed experimentally. Furthermore, the model calculations (see Fig. 4.33) do not reproduce the experimentally observed linear T dependence for the 22 nm nanowires, but rather predict a 3-D behavior for both the density of states and the specific heat in 22 nm nanowires [4.155, 156].

Thermal conductance measurements on GaAs nanowires below 6 K show a power law dependence, but the T dependence becomes somewhat less pronounced below ~ 2.5 K [4.152]. This deviation from the power law temperature dependence led to a more detailed study of the quantum limit for the thermal conductance. To carry out these more detailed experiments, a mesoscopic phonon resonator and waveguide device were

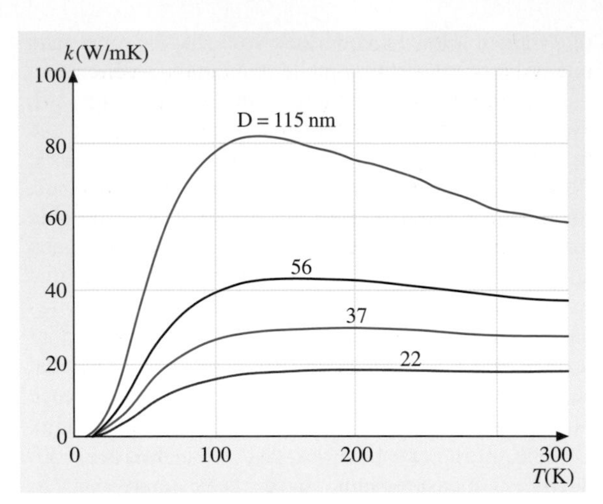

Fig. 4.33 Predicted thermal conductivity of Si nanowires of various diameters [4.155]

constructed that included four ∼ 200 nm wide and 85 nm thick silicon nitride nanowire-like nano-constrictions (see Fig. 4.34a) to establish the quantized thermal conductance limit of $g_0 = \pi^2 k_B^2 T/3h$ (see Fig. 4.34b) for ballistic phonon transport [4.157, 158]. For temperatures above 0.8 K, the thermal conductance in Fig. 4.34b follows a T^3 law, but as T is further reduced, a transition to a linear T dependence is observed, consistent with a phonon mean free path of ∼ 1 μm, and a thermal conductance value approaching $16g_0$, corresponding to four massless phonon modes per channel and four channels in their phonon waveguide structure (see Fig. 4.34a). Ballistic phonon transport occurs when the thermal phonon wavelength (380 nm for the experimental structure) is somewhat greater than the width of the phonon waveguide at its constriction.

4.2.3 Optical Properties

Optical methods provide an easy and sensitive tool for measuring the electronic structure of nanowires since optical measurements require minimal sample preparation (for example, contacts are not required) and the measurements are sensitive to quantum effects. Optical spectra of 1-D systems, such as carbon nanotubes, often show intense features at specific energies near singularities in the joint density of states formed under strong quantum confinement conditions. A variety of optical techniques have shown that the properties of nanowires are different from those of their bulk counterparts, and this section of the review focuses on these differences.

Although optical properties have been shown to provide an extremely important tool for the characterization of nanowires, the interpretation of these measurements is not always straightforward. The wavelength of light used to probe the sample is usually smaller than the wire length but larger than the wire diameter. Hence, the probe light used in an optical measurement cannot be focused solely onto the wire, and the wire and the substrate on which the wire rests (or host material, if the wires are embedded in a template) are simultaneously probed. For measurements, such as photo-luminescence

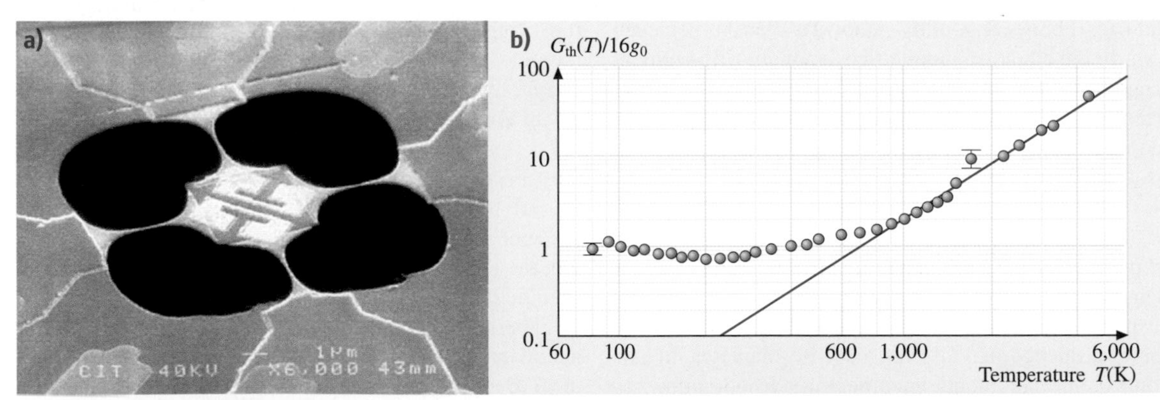

Fig. 4.34 (a) Suspended mesoscopic phonon device used to measure ballistic phonon transport. The device consists of an 4×4 μm "phonon cavity" (*center*) connected to four Si_3N_4 membranes, 60 nm thick and less than 200 nm wide. The two bright "C" shaped objects on the phonon cavity are thin film heating and sensing Cr/Au resistors, whereas the *dark regions* are empty space. (b) Log–log plot of the temperature dependence of the thermal conductance G_0 of the structure in (a) normalized to $16g_0$ (see text) [4.158]

(PL), if the substrate does not luminescence or absorb in the frequency range of the measurements, PL directly measures the luminescence of the nanowires and the substrate can be ignored. In reflection and transmission measurements, however, even a nonabsorbing substrate can modify the measured spectra of nanowires.

In this section, we discuss the determination of the dielectric function for nanowires in the context of effective medium theories. We then discuss various optical techniques with appropriate examples that sensitively differentiate nanowire properties from those also found in the parent bulk material, giving particular emphasis to electronic quantum confinement effects. Finally, we review phonon confinement effects.

The Dielectric Function

In this subsection we review the use of effective medium theory as a method to handle the optical properties of nanowires whose diameters are typically smaller than the wavelength of light, noting that observable optical properties of materials can be related to the complex dielectric function [4.159, 160]. Effective medium theories [4.161, 162] can be applied to model the nanowire and substrate as one continuous composite with a single complex dielectric function ($\epsilon_1 + \mathrm{i}\epsilon_2$), where the real and imaginary parts of the dielectric function ϵ_1 and ϵ_2 are related to the index of refraction (n) and the absorption coefficient (K) by the relation $\epsilon_1 + \mathrm{i}\epsilon_2 = (n + \mathrm{i}K)^2$. Since photons at visible or infrared wavelengths "see" a dielectric function for the composite nanowire array/substrate system that is different from that of the nanowire itself, the optical transmission and reflection are different from what they would be if the light were focused only on the nanowire. One commonly observed consequence of effective medium theory is the shift in the plasma frequency in accordance with the percentage of nanowire material contained in the composite [4.163]. The plasma resonance occurs when $\epsilon_1(\omega)$ becomes zero, and the plasma frequency of the nanowire composite will shift to lower (higher) energies when the magnitude of the dielectric function of the host materials is larger (smaller) than that of the nanowire.

Although reflection and transmission measurements probe both the nanowire and the substrate, the optical properties of the nanowires can be independently determined. One technique for separating the dielectric function of the nanowires from the host is to use an effective medium theory in reverse. Since the dielectric function of the host material is often known, and since the dielectric function of the composite material can be measured by the standard method of using reflection and transmission measurements in combination with either the Kramer–Kronig relations or Maxwell's equations, the complex dielectric function of the nanowires can be deduced. This approach has been used successfully, for example, in determining the frequency dependence of the real and imaginary parts of the dielectric function $\epsilon_1(\omega)$ and $\epsilon_2(\omega)$ for a parallel array of bismuth nanowires filling the pores of an alumina template [4.164].

Optical Properties Characteristic of Nanowires

A wide range of optical techniques are available for the characterization of nanowires to distinguish their properties from those of their parent bulk materials. Some differences in properties are geometric, such as the small diameter size and the large length-to-diameter ratio (also called the aspect ratio), while other differences focus on quantum confinement issues.

Probably the most basic optical technique is to measure the reflection and/or transmission of a nanowire to determine the frequency-dependent real and imaginary parts of the dielectric function. This technique has been used, for example, to study the band gap and its temperature dependence in gallium nitride nanowires in the 10–50 nm range in comparison to bulk values [4.165]. The plasma frequency, free carrier density, and donor impurity concentration as a function of temperature were also determined from the infrared spectra, which is especially useful for nanowire research since Hall effect measurements cannot be made on nanowires.

Photo-luminescence (PL) or fluorescence spectroscopy is a common method to study nanowires. Emission techniques probe the nanowires directly, and the effect of the host material does not have to be considered. This characterization method has been used to study many properties of nanowires, such as the optical gap behavior, oxygen vacancies in ZnO nanowires [4.66], strain in Si nanowires [4.166], and quantum confinement effects in InP nanowires [4.92]. Figure 4.35 shows the photo-luminescence of InP nanowires as a function of wire diameter, thereby providing direct information on the effective band gap. As the wire diameter of an InP nanowire is decreased so that it becomes smaller than the bulk exciton diameter of 19 nm, quantum confinement effects set in, and the band gap is increased. This results in an increase in the PL peak energy because of the stronger electron–hole Coulomb binding energy within the quantum-confined nanowires as the wire radius gets smaller than the effective Bohr radius for the exciton for bulk InP. The smaller the effective mass, the larger the quantum con-

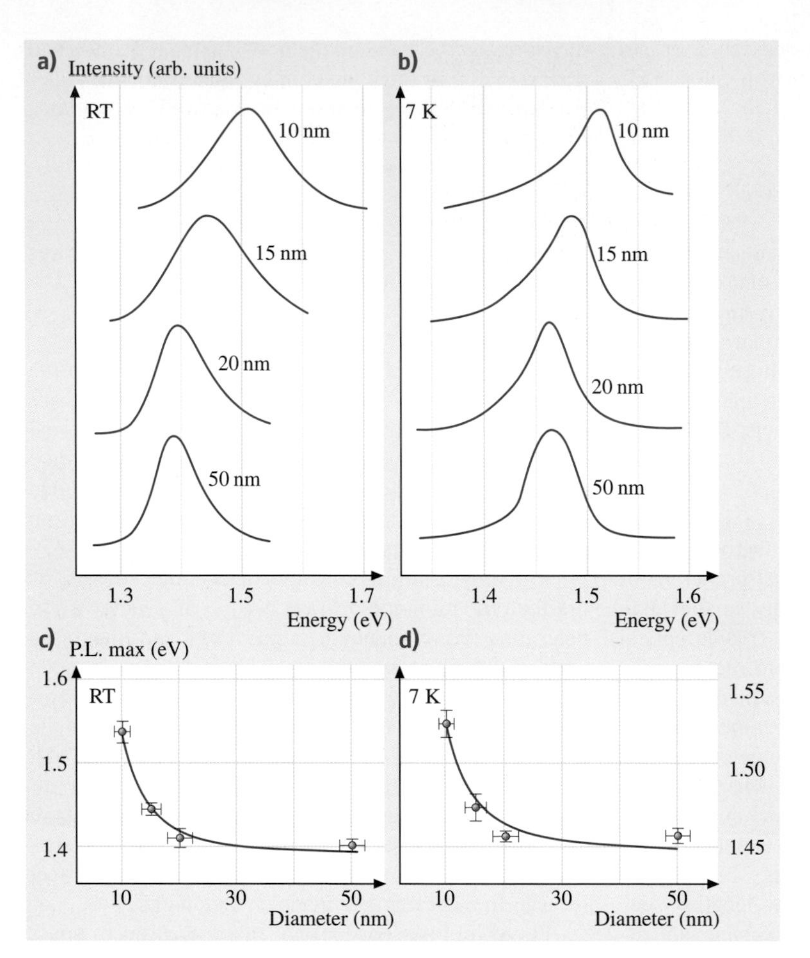

Fig. 4.35a–d Photo-luminescence of InP nanowires of varying diameters at 7 K ((**b**) and (**d**)) and room temperature ((**a**) and (**c**)), showing quantum confinement effects of the exciton for wire diameters less than 20 nm [4.92]

finement effects. When the shift in the peak energy as a function of nanowire diameter (Fig. 4.35a) is analyzed using an effective mass model, the reduced effective mass of the exciton is deduced to be $0.052m_0$, which agrees quite well with the literature value of $0.065m_0$ for bulk InP. Although the line widths of the PL peak for the small diameter nanowires (10 nm) are smaller at low temperature (7 K), the observation of strong quantum confinement and band gap tunability effects at room temperature are significant for photonics applications of nanowires (see Sect. 4.3.3).

The resolution of photo-luminescence (PL) optical imaging of a nanowire is, in general, limited by the wavelength of light. But when a sample is placed very close to the detector, the light is not given a chance to diffract, and so samples much smaller than the wavelength of light can be resolved. This technique is known as near-field scanning optical microscopy (NSOM) and has been used successfully [4.167] to image nanowires. For example, Fig. 4.36 shows the topographical (a) and PL (b) NSOM images of a single ZnO nanowire.

Magneto-optics can be used to measure the electronic band structure of nanowires. For example, magneto-optics in conjunction with photo-conductance has been proposed [4.168] as a tool to determine band parameters for nanowires, such as the Fermi energy, electron effective masses, and the number of subbands to be considered [4.168]. Since different nanowire subbands have different electrical transmission properties, the electrical conductivity changes when light is used to excite electrons to higher subbands, thereby providing a method for studying the electronic structure of nanowires optically. Magneto-optics can also be used to study the magnetic properties of nanowires in relation to bulk properties [4.44, 169]. For example, the surface magneto-optical Kerr effect has been used to measure

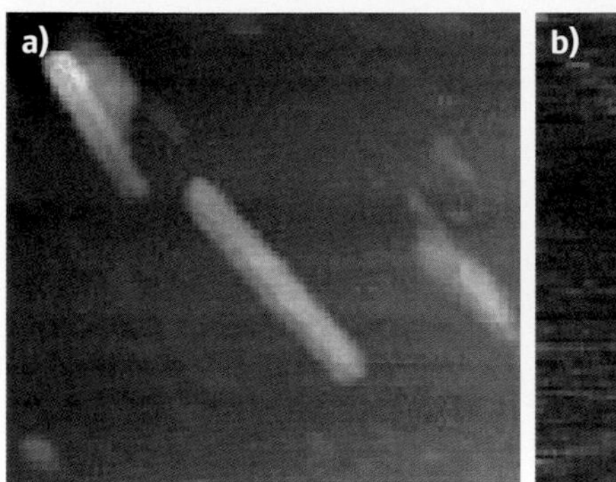

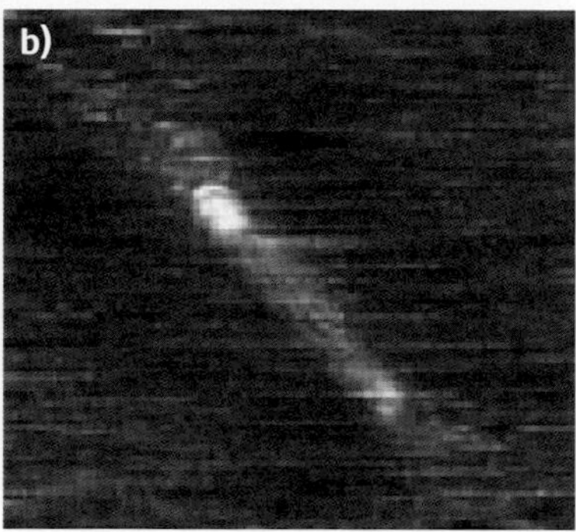

Fig. 4.36 (a) Topographical and (b) photoluminescence (PL) near-field scanning optical microscopy (NSOM) images of a single ZnO nanowire waveguide [4.167]

the dependence of the magnetic ordering temperature of Fe-Co alloy nanowires on the relative concentration of Fe and Co [4.169], and it was used to find that, unlike the case of bulk Fe-Co alloys, cobalt in nanowires inhibits magnetic ordering.

Nonlinear optical properties of nanowires have received particular attention since the nonlinear behavior is often enhanced over that in bulk materials and since the nonlinear effects can be utilized for many applications. One such study measured the second harmonic generation (SHG) and third harmonic generation (THG) in a single nanowire by using near-field optical microscopy [4.170]. ZnO nanowires were shown to have strong SHG and THG effects that are highly polarization-sensitive, and this polarization sensitivity can be explained on the basis of optical and geometrical considerations. Some components of the second harmonic polarization tensor are found to be enhanced in nanowires while others are suppressed as the wire diameter is decreased, and such effects could be of interest for device applications. The authors also showed that the second-order nonlinearities are mostly wavelength independent for $\lambda < 400\,\text{nm}$, which is in the transparent regime for ZnO, below the onset of band gap absorption; this observation is also of interest for device applications.

Reflectivity and transmission measurements have also been used to study the effects of quantum confinement and surface effects on the low energy indirect transition in bismuth nanowires [4.171]. *Black* et al. [4.171] investigated an intense and sharp absorption peak in bismuth nanowires, which is not observed in bulk bismuth. The energy position E_p of this strong absorption peak increases with decreasing diameter. But the rate of increase in energy with decreasing diameter $|\partial E_\text{p}/\partial d_\text{W}|$ is an order of magnitude less than that predicted for either a direct interband transition or for intersubband transitions in bismuth nanowires. On the other hand, the magnitude of $|\partial E_\text{p}/\partial d_\text{W}|$ agrees well with that predicted for an indirect L-point valence to T-point valence band transition (see Fig. 4.37). Since both the initial and final states for the indirect L-T point valence band transition downshift in energy as the wire diameter d_W is decreased, the shift in the absorption peak results from a *difference* between the effective masses and not from the actual value of either of the masses. Hence the diameter dependence of the absorption peak energy is an order of magnitude less for a valence to valence band indirect transition than for a direct interband L-point transition. Furthermore, the band-tracking effect for the indirect transition gives rise to a large value for the joint density of states, thus accounting for the high intensity of this feature. The enhanced coupling of this indirect transition to an applied optical field arises through the gradient of the dielectric function, which is large at the bismuth-air or bismuth-alumina interfaces. It should be noted that, in contrast to the surface effect for bulk samples, the whole nanowire contributes to the optical absorption due to the spatial variation in the dielectric function, since the penetration depth is larger than or comparable to the wire diameter. In addition, the intensity can be quite significant because abundant initial state electrons, final state holes, and appropriate phonons exist for making an indirect L-T point valence band transition at room temperature. Interestingly, the polarization dependence of this absorption peak is such that the strong absorption is present when the electric field is perpendicular to the wire axis but is absent when the electric field is parallel to the wire axis; this is contrary to a traditional polarizer, such as a carbon nanotube in which the optical E field is polarized by the nanotube itself to be aligned along the carbon nanotube axis. The observed polarization dependence for bismuth nanowires is consistent

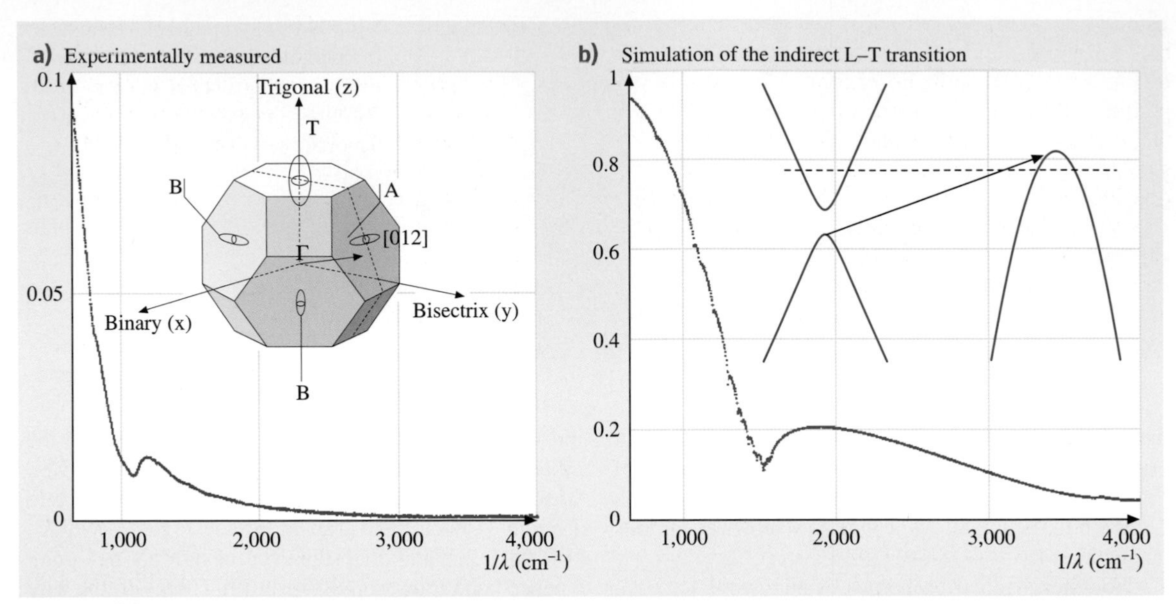

Fig. 4.37 **(a)** The measured optical transmission spectra as a function of wavenumber ($1/\lambda$) of a ~ 45 nm diameter bismuth nanowire array. **(b)** The simulated optical transmission spectrum resulting from an indirect transition of an L point electron to a T point valence subband state. The insert in **(a)** shows the bismuth Brillouin zone, and the location of the T-point hole and of the three L-point electron pockets, including the nondegenerate A, and the doubly degenerate B pockets. The insert in **(b)** shows the indirect L to T point electronic transition induced by a photon with an energy equal to the energy difference between the initial and final states minus the phonon energy (about $100\,\mathrm{cm}^{-1}$) needed to satisfy conservation of energy in a Stokes process [4.171]

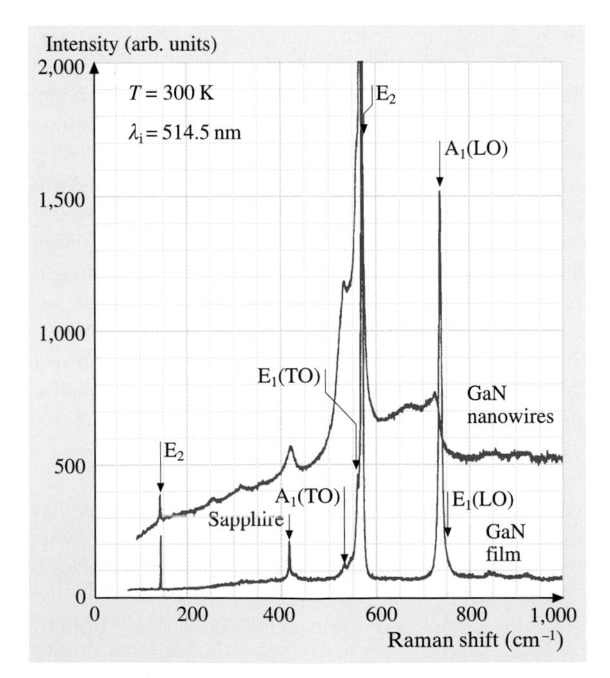

Fig. 4.38 Room temperature Raman-scattering spectra of GaN nanowires and of a 5 μm thick GaN epilayer film with green (514.5 nm) laser excitation. The Raman-scattering response was obtained by dividing the measured spectra by the Bose–Einstein thermal factor [4.172]

with a surface-induced effect that increases the coupling between the L-point and T-point bands throughout the full volume of the nanowire. Figure 4.37 shows the experimentally observed transmission spectrum in bismuth nanowires of ~ 45 nm diameter (a), and for comparison, the simulated optical transmission from an indirect transition in bismuth nanowires of ~ 45 nm diameter is also shown in (b). The indirect L to T point valence band transition mechanism [4.171] is also consistent with observations of the effect on the optical spectra of a decrease in the nanowire diameter and of n-type doping of bismuth nanowires with Te.

Phonon Confinement Effects

Phonons in nanowires are spatially confined by the nanowire cross-sectional area, crystalline boundaries and surface disorder. These finite size effects give rise to phonon confinement, causing an uncertainty in the phonon wave vector, which typically gives rise to a frequency shift and a lineshape broadening. Since zone center phonons tend to correspond to maxima in the phonon dispersion curves, the inclusion of contributions from a broader range of phonon wave vectors results in both a downshift in frequency and an asymmetric broadening of the Raman line that developes a low frequency tail. These phonon confinement effects have been theoretically predicted [4.173,174] and experimentally observed in GaN [4.172], as shown in Fig. 4.38 for GaN nanowires with diameters in the range 10–50 nm. The application of these theoretical models indicates that broadening effects should be noticeable as the wire diameter in GaN nanowires decreases to ~ 20 nm. When the wire diameter further decreases to ~ 10 nm, the frequency downshift and asymmetric Raman line broadening effects should become observable in the Raman spectra for the GaN nanowires but are not found in the corresponding spectra for bulk GaN.

The experimental spectra in Fig. 4.38 show the four $A_1 + E_1 + 2E_2$ modes expected from symmetry considerations for bulk GaN crystals. Two types of quantum confinement effects are observed. The first type is the observation of the downshift and asymmetric broadening effects discussed above. Observations of such downshifts and asymmetric broadening have also been recently reported in 7 nm diameter Si nanowires [4.175]. A second type of confinement effect found in Fig. 4.38 for GaN nanowires is the appearance of additional Raman features not found in the corresponding bulk spectra and associated with combination modes and a zone boundary mode. Resonant enhancement effects were also observed for the A_1(LO) phonon at 728 cm^{-1} (see Fig. 4.38) at higher laser excitation energies [4.172].

4.3 Applications

In the preceding sections we have reviewed many of the central characteristics that make nanowires in some cases similar and in some cases very different from their parent materials. We have also shown that some properties are diameter dependent and are therefore tunable during synthesis. Thus it is of great interest to find applications for nanowires that could benefit in unprecedented ways from both the unique and tunable properties of nanowires and the small size of these nanostructures, for use in the miniaturization of conventional devices. As the synthetic methods for the production of nanowires are maturing (Sect. 4.1), and nanowires can be made in reproducible and cost-effective ways, it is only a matter of time before applications will be explored seriously. This is a timely development, as the semiconductor industry will soon be reaching what seems to be its limit in feature-size reduction and approaching a classical-to-quantum size transition. At the same time the field of biotechnology is expanding through the availability of tremendous genome information and innovative screening assays. Since nanowires are the size of the shrinking electronic components and of cellular biomolecules, it is only natural for nanowires to be good candidates for applications in these fields. Commercialization of nanowire devices, however, will require reliable mass-production, effective assembly techniques, and quality-control methods.

In this section, we discuss applications of nanowires to electronics (Sect. 4.3.1), thermoelectrics (Sect. 4.3.2), optics (Sect. 4.3.3), chemical and biological sensing (Sect. 4.3.4), and magnetic media (Sect. 4.3.5).

4.3.1 Electrical Applications

The microelectronics industry continues to face technology (e.g., lithography) and economic challenges as the device feature size is decreased, especially below 100 nm. The self-assembly of nanowires might present a way to construct unconventional devices that do not rely on improvements in photo-lithography and, therefore, do not necessarily imply increasing fabrication costs. Devices made from nanowires have several advantages over those made by photolithography. A variety of approaches has been devised to organize nanowires via self-assembly (see Sect. 4.1.4), thus eliminating the need for the expensive lithographic techniques normally required to produce devices the size of typical nanowires, which we discuss earlier (see Sect. 4.1). In addition, unlike traditional silicon processing, different semiconductors can be simultaneously used in nanowire devices to produce diverse functionalities. Not only can wires of different materials be combined, but a single wire can be made of different materials. For example, junctions of GaAs and GaP show rectifying behavior [4.92], thus demonstrating that good electronic interfaces between two different semiconductors can be achieved in the synthesis of multicomponent nanowires. Transistors made from nanowires could also have advantages because of their unique morphology. For example, in bulk field effect transistors (FETs), the depletion layer formed below the source and drain region results in a source-drain capacitance that limits the operation speed. In nanowires, however, the conductor is surrounded by an oxide and thus the depletion layer cannot be formed. Depending on the device design, the source-drain capacitance in nanowires could be greatly minimized and possibly eliminated.

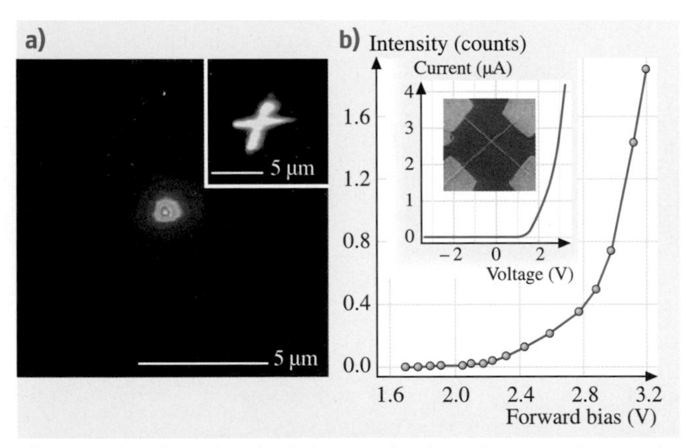

Fig. 4.39a,b Optoelectrical characterization of a crossed nanowire junction formed between 65-nm n-type and 68-nm p-type InP nanowires. **(a)** Electroluminescence (EL) image of the light emitted from a forward-biased nanowire p-n junction at 2.5 V. *Inset*, photoluminescence (PL) image of the junction. **(b)** EL intensity as a function of operation voltage. *Inset*, the SEM image and the I–V characteristics of the junction [4.52]. The scale bar in the *inset* is 5 µm

Device functionalities common in conventional semiconductor technologies, such the p-n junction diodes [4.129], field-effect transistors [4.131], logic gates [4.129], and light-emitting diodes [4.52, 92], have been recently demonstrated in nanowires, showing their promise as the building blocks for the construction of complex integrated circuits by employing the "bottom-up" paradigm. Several approaches have been investigated to form nanowire diodes (see Sect. 4.2.2). For example, Schottky diodes can be formed by contacting a GaN nanowire with Al electrodes [4.130]. Furthermore, p-n junction diodes can be formed at the crossing of two nanowires, such as the crossing of n and p-type InP nanowires doped by Te and Zn, respectively [4.52], or Si nanowires doped by phosphorus (n-type) and boron (p-type) [4.176]. In addition to the crossing of two distinctive nanowires, heterogeneous junctions have also been constructed inside a single wire, either along the wire axis in the form of a nanowire superlattice [4.92] or perpendicular to the wire axis by forming a core-shell structure of silicon and germanium [4.105]. These various nanowire junctions not only possess similar current rectifying properties (see Fig. 4.22) as expected for bulk semiconductor devices, but they also exhibit electro-luminescence (EL) that may be interesting for optoelectronic applications, as shown in Fig. 4.39 for the electroluminescence of a crossed junction of n and p-type InP nanowires [4.52] (see Sect. 4.3.3).

In addition to the two-terminal nanowire devices, such as the p-n junctions described above, it is found that the conductance of a semiconductor nanowire can be significantly modified by applying voltage at a third gate terminal, implying the utilization of nanowires as a field-effect transistor (FET). This gate terminal can either be the substrate [4.46, 47], a separate metal contact located close to the nanowire [4.177], or another nanowire with a thick oxide coating in the crossed nanowire junction configuration [4.129]. We discuss the operation principles of these nanowire-based FETs in Sect. 4.2.2. Various logic devices performing basic logic functions have been demonstrated using nanowire junctions [4.129], as shown in Fig. 4.40 for the OR and AND logic gates constructed from 2-by-1 and 1-by-3 nanowire p-n junctions, respectively. By functionalizing nanowires with redox active molecules to store charges, the nanowire FETs can exhibit bi-stable logic *on* or *off* states [4.131], which may be used for nonvolatile memory or as switches.

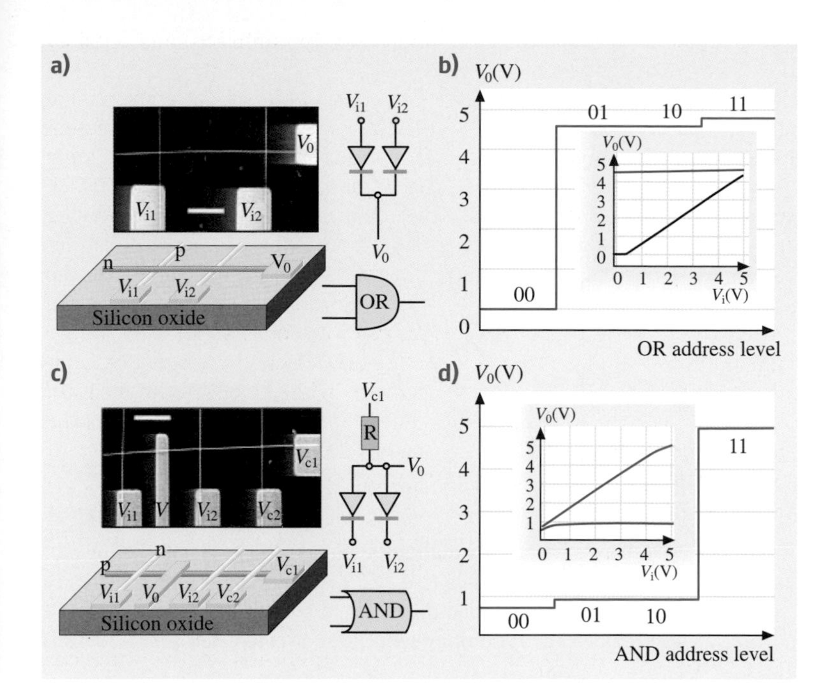

Fig. 4.40a–d Nanowire logic gates: (**a**) Schematic of logic OR gate constructed from a 2(p-Si) by 1(n-GaN) crossed nanowire junction. The *inset* shows the SEM image (*bar:* 1 μm) of an assembled OR gate and the symbolic electronic circuit. (**b**) The output voltage of the circuit in (**a**) versus the four possible logic address level inputs: (0,0); (0,1); (1,0); (1,1), where logic 0 input is 0 V and logic 1 is 5 V (same for below). (**c**) Schematic of logic AND gate constructed from a 1(p-Si) by 3(n-GaN) crossed nanowire junction. The *inset* shows the SEM image (*bar:* 1 μm) of an assembled AND gate and the symbolic electronic circuit. (**d**) The output voltage of the circuit in (**c**) versus the four possible logic address level inputs [4.129]

Nanowires have also been proposed for applications associated with electron field emission [4.178], such as flat panel displays, because of their small diameter and large curvature at the nanowire tip, which may reduce the threshold voltage for the electron emission [4.179]. In this connection, the demonstration of very high field emission currents from the sharp tip (~ 10 nm radius) of a Si cone [4.178] and from carbon nanotubes [4.180] stimulates interest in this potential application opportunity for nanowires.

The concept of constructing electronic devices based on nanowires has already been demonstrated, and the next step for electronic applications would be to devise a feasible method for integration and mass production. We expect that in order to maintain the growing rate of device density and functionality in the existing electronics industry, new kinds of complementary electronic devices will emerge from this "bottom-up" scheme for nanowire electronics, different from what has been produced by the traditional "top-down" approach pursued by conventional electronics.

4.3.2 Thermoelectric Applications

One proposed application for nanowires is for thermoelectric cooling and for the conversion between thermal and electrical energy [4.156, 181]. The efficiency of a thermoelectric device is measured in terms of a dimensionless figure of merit ZT, where Z is defined as

$$Z = \frac{\sigma S^2}{\kappa}, \tag{4.2}$$

in which σ is the electrical conductivity, S is the Seebeck coefficient, κ is the thermal conductivity, and T is the temperature. In order to achieve a high ZT and therefore efficient thermoelectric performance, a high electrical conductivity, a high Seebeck coefficient, and a low thermal conductivity are required. In 3-D systems, the electronic contribution to κ is proportional to σ in accordance with the Wiedemann–Franz law, and normally materials with high S have a low σ. Hence an increase in the electrical conductivity (e.g., by electron donor doping) results in an adverse variation in both the Seebeck coefficient (decreasing) and the thermal conductivity (increasing). These two trade-offs set the upper limit for increasing ZT in bulk materials, with the maximum ZT remaining at ~ 1 at room temperature for the 1960–1995 time frame.

The high electronic density of states in quantum-confined structures is proposed as a promising possibility to bypass the Seebeck/electrical conductivity trade-off and to control each thermoelectric-related variable independently, thereby allowing for an increased electrical conductivity, a relatively low thermal

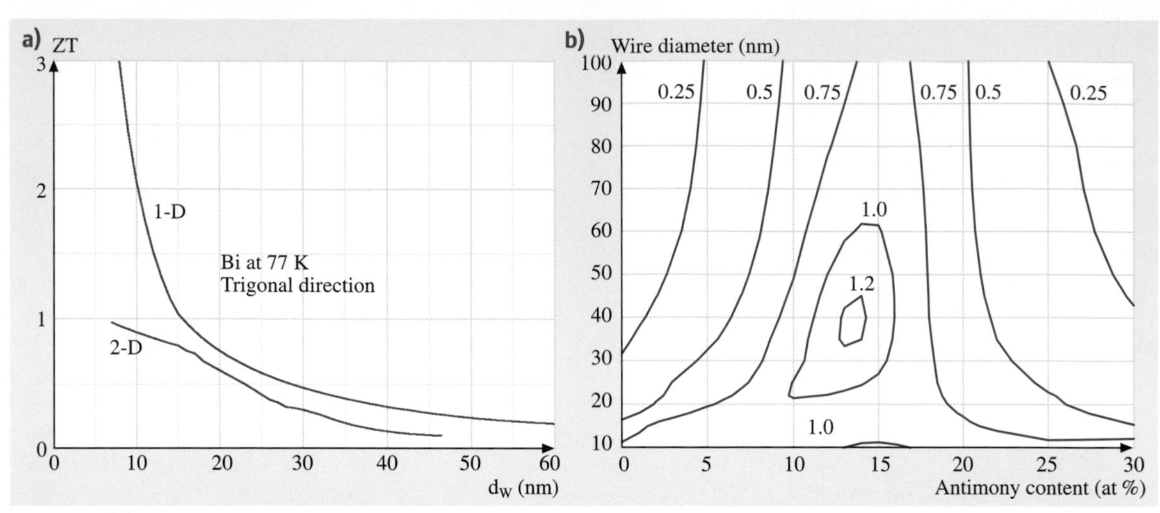

Fig. 4.41 **(a)** Calculated ZT of 1-D(nanowire) and 2-D(quantum well) bismuth systems at 77 K as a function of d_W, denoting the wire diameter or film thickness. The thermoelectric performance (i. e., ZT) is expected to improve greatly when the wire diameter is small enough so that the nanowire become a one-dimensional system. **(b)** Contour plot of optimal ZT values for p-type $Bi_{1-x}Sb_x$ nanowires vs. wire diameter and antimony concentration calculated at 77 K [4.183]

conductivity, and a large Seebeck coefficient simultaneously [4.182]. For example, Figs. 4.29b and 4.30a in Sect. 4.2.2 show an enhancement in S for bismuth an dbismuth-antimony nanowires as the wire diameter decreases. In addition to alleviating the undesired connections between σ, S, and the electronic contribution to the thermal conductivity, nanowires also have the advantage that the phonon contribution to the thermal conductivity is greatly reduced because of boundary scattering (see Sect. 4.2.2), thereby achieving a high ZT. Figure 4.41a shows the theoretical values for ZT vs. sample size for both bismuth thin films (2-D) and nanowires (1-D) in the quantum confined regime, exhibiting a rapidly increasing ZT as the quantum size effect becomes more and more important [4.182]. In addition, the quantum size effect in nanowires can be combined with other parameters to tailor the band structure and electronic transport behavior (e.g., Sb alloying in Bi) to further optimize ZT. For example, Fig. 4.41b shows the predicted ZT for p-type $Bi_{1-x}Sb_x$ alloy nanowires as a function of wire diameter and Sb content x [4.183]. The occurrence of a local ZT maxima in the vicinity of $x \sim 0.13$ and $d_W \sim 45$ nm is due to the coalesce of ten valence bands in the nanowire and the resulting unusual high density of states for holes, which is a phenomenon absent in bulk $Bi_{1-x}Sb_x$ alloys. For nanowires with very small diameters, it is speculated that localization effects will eventually limit the enhancement of ZT. But in bismuth nanowires, localization effects are not significant for wires with diameters larger than 9 nm [4.63]. In addition to 1-D nanowires, ZT values as high as ~ 2 have also been experimentally demonstrated in macroscopic samples containing PbSe quantum dots (0-D) [4.184] and stacked 2-D films [4.154].

Although the application of nanowires to thermoelectrics seems very promising, these materials are still in the research phase of the development cycle and quite far from being commercialized. One challenge for thermoelectric devices based on nanowires lies in finding a suitable host material that will not reduce ZT too much due to the unwanted heat conduction through the host material. The host material should, therefore, have a low thermal conductivity and occupy as low a volume percentage in the composite material as possible while still providing the quantum confinement and the support for the nanowires.

4.3.3 Optical Applications

Nanowires also hold promise for optical applications. One-dimensional systems exhibit a singularity in their joint density of states, allowing quantum effects in nanowires to be optically observable, sometimes, even at room temperature. Since the density of states of a nanowire in the quantum limit (small wire diameter) is highly localized in energy, the available states quickly fill up with electrons as the intensity of the incident light is increased. This filling up of the subbands, as well as other effects unique to low-dimensional materials, lead to strong optical nonlinearities in quantum wires. Quantum wires may thus yield optical switches with a lower switching energy and increased switching speed compared to currently available optical switches.

Light emission from nanowires can be achieved by photo-luminescence (PL) or electro-luminescence (EL), distinguished by whether the electronic excitation is achieved by optical illumination or by electrical stimulation across a p-n junction, respectively. PL is often used for optical properties characterization, as described in Sect. 4.2.3, but from the applications point of view, EL is a more convenient excitation method. Light emitting diodes (LEDs) have been achieved in junctions between a p-type and an n-type nanowire (Fig. 4.39) [4.52] and in superlattice nanowires with p-type and n-type segments [4.92]. The light emission was localized to the junction area and was polarized in the superlattice nanowire.

Light emission from quantum wire p-n junctions is especially interesting for laser applications, because quantum wires can form lasers with lower excitation thresholds compared to their bulk counterparts, and they also exhibit a decreased temperature sensitivity in their performance [4.185]. Furthermore, the emission wavelength can be tuned for a given material composition by only altering the geometry of the wire.

Lasing action has been reported in ZnO nanowires with wire diameters much smaller than the wavelength of the light emitted ($\lambda = 385$ nm) [4.115] (see Fig. 4.42). Since the edge and lateral surface of ZnO nanowires are faceted (see Sect. 4.2.1), they form optical cavities that sustain desired cavity modes. Compared to conventional semiconductor lasers, the exciton laser action employed in zinc oxide nanowire lasers exhibits a lower lasing threshold ($\sim 40\,kW/cm^2$) than their 3-D counterparts ($\sim 300\,kW/cm^2$). In order to utilize exciton confinement effects in the lasing action, the exciton binding energy (~ 60 meV in ZnO) must be greater than the thermal energy (~ 26 meV at 300 K). Decreasing the wire diameter increases the excitation binding energy and lowers the threshold for lasing. PL NSOM imaging confirmed the waveguiding properties of the anisotropic and well-faceted structure of ZnO nanowires, limiting the emission to the tips of the ZnO nanowires [4.167].

Laser action has been also observed in GaN nanowires [4.186]. Unlike ZnO, GaN has a small exciton binding energy, of only ~ 25 meV. Furthermore, since the wire radii used in this study (15–75 nm) [4.186] are larger than the Bohr radius of excitons in GaN (11 nm), the exciton binding energy is not expected to increase in these GaN wires and quantum confinement effects such as shown in Fig. 4.35 for InP are not expected. Some tunability of the center of the spectral intensity, however, was achieved by increasing the intensity of the pump power, causing a redshift in the laser emission, which is explained as a band gap renormalization as a result of the formation of an electron-hole plasma. Heating effects were excluded as the source of the spectral shift.

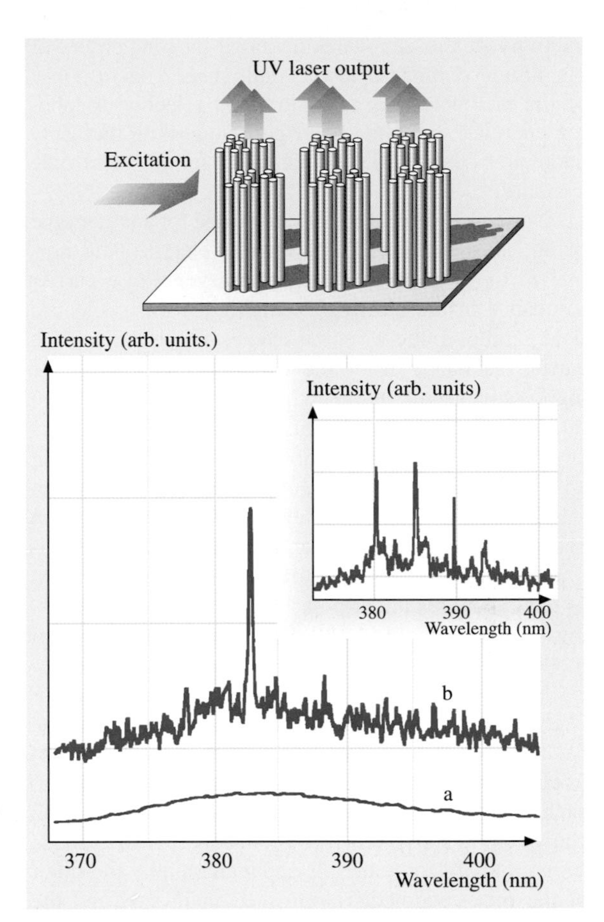

Fig. 4.42 A schematic of lasing in ZnO nanowires and the PL spectra of ZnO nanowires at two excitation intensities. One PL spectra is taken below the lasing threshold, and the other above it [4.115]

Nanowire photodetectors were also proposed. ZnO nanowires were found to display a strong photocurrent response to UV light irradiation [4.187]. The conductivity of the nanowire increased by four orders of magnitude compared to the dark state. The response of the nanowire was reversible and selective to photon energies above the band gap, suggesting that ZnO nanowires could be a good candidate for optoelectronic switches.

Nanowires have also been proposed for another type of optical switching. Light with its electric field normal to the wire axis excites a transverse free carrier resonance inside the wire, while light with its electric field parallel to the wire axis excites a longitudinal free carrier resonance inside the wire. Since nanowires are highly anisotropic, these two resonances occur at two different wavelengths and thus result in absorption peaks at two different energies. Gold nanowires dispersed in an aqueous solution align along the electric field when a DC voltage is applied. The energy of the absorption peak can be toggled between the transverse and longitudinal resonance energies by changing the alignment of the nanowires under polarized light illumination using an electric field [4.188, 189]. Thus, electro-optical modulation is achieved.

Nanowires may also be used as barcode tags for optical read out. Nanowires containing gold, silver, nickel, palladium, and platinum were fabricated [4.104] by electrochemical filling of porous anodic alumina, so that each nanowire consisted of segments of various metal constituents. Thus many types of nanowires can be made from a handful of materials and identified by the order of the metal segments along their main axis and the length of each segment. Barcode read out is possible by reflectance optical microscopy. The segment length is limited by the Rayleigh diffraction limit and not by synthesis limitations; thus it can be as small as 145 nm. Figure 4.43a shows an optical image of many Au-Ag-Au-Ag bar-coded wires, where the silver segments show higher reflectivity. Figure 4.43b is a backscattering mode FE-SEM image of a single nanowire, highlighting the composition and segment length variations along the nanowire.

Both the large surface area and the high conductivity along the length of nanowires are favorable for their use in inorganic-organic solar cells [4.190], which offer promise from a manufacture-ability and cost effective standpoint. In a hybrid nanocrystal-organic solar cell, the incident light forms bound electron–hole pairs (excitons) in both the inorganic nanocrystal and in the surrounding organic medium. These excitons diffuse to the inorganic–organic interface and disassociate to form an electron and a hole. Since conjugated polymers usually have poor electron mobilities, the inorganic phase is chosen to have a higher electron affinity than the organic phase so that the organic phase carries the holes and the semiconductor carries the electrons. The separated electrons and holes drift to the external electrodes through the inorganic and organic materials, respectively. But only those excitons formed within an exciton diffusion length from an interface can disassociate before recombining, and therefore the distance between the dissociation sites limits the efficiency of a solar cell. A solar cell prepared from a composite of CdSe nanorods inside poly (3-ethylthiophene) [4.190] yielded 6.9% monochromatic power efficiencies and 1.7% power conversion efficiencies under A.M. 1.5 illumination (equal to solar irradiance through 1.5 times the air mass of the earth at direct normal incidence). The nanorods provide a large surface area with good chemical bonding to the polymer for efficient charge transfer and exciton dissociation. Furthermore, they provide a good conduction path for the electrons to reach the electrode. Their enhanced absorption coefficient and their tunable band gap are also characteristics that can be used to enhance the energy conversion efficiency of solar cells.

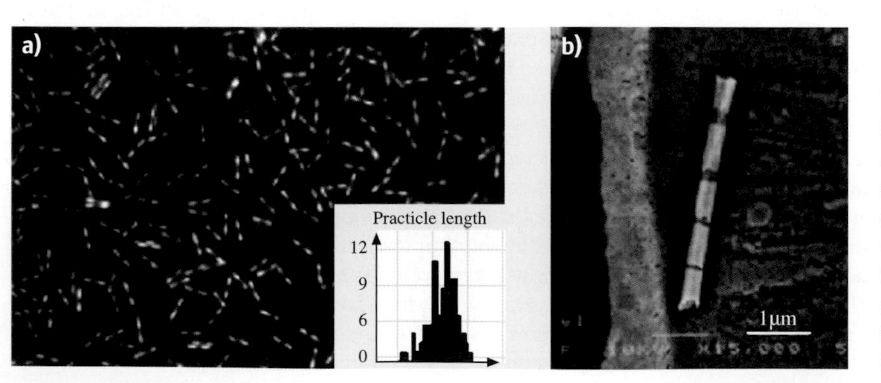

Fig. 4.43 (a) An optical image of many short Au-Ag-Au-Au bar-coded wires and (b) an FE-SEM image of an Au/Ag bar-coded wire with multiple strips of varying length. The insert in (a) shows a histogram of the particle lengths for 106 particles in this image [4.104]

4.3.4 Chemical and Biochemical Sensing Devices

Sensors for chemical and biochemical substances with nanowires as the sensing probe are a very attractive application area. Nanowire sensors will potentially be smaller, more sensitive, demand less power, and react faster than their macroscopic counterparts. Arrays of nanowire sensors could, in principle, achieve nanometer scale spatial resolution and therefore provide accurate real-time information regarding not only the concentration of a specific analyte but also its spatial distribution. Such arrays, for example, could be very useful for dynamic studies on the effects of chemical gradients on biological cells. The operation of sensors made with nanowires, nanotubes, or nano-contacts is based mostly on the reversible change in the conductance of the nanostructure upon absorption of the agent to be detected, but other detection methods, such as mechanical and optical detection, are conceptually plausible. The increased sensitivity and faster response time of nanowires are a result of the large surface-to-volume ratio and the small cross-section available for conduction channels. In the bulk, on the other hand, the abundance of charges can effectively shield external fields, and the abundance of material can afford many alternative conduction channels. Therefore a stronger chemical stimulus and longer response time are necessary to observe changes in the physical properties of a 3-D sensor in comparison to a nanowire.

Cui et al. placed silicon nanowires made by the VLS method (Sect. 4.1.2) between two metal electrodes and modified the silicon oxide coating of the wire by the addition of molecules sensitive to the analyte to be detected [4.191]. For example, a pH sensor was made by covalently linking an amine containing silane to the surface of the nanowire. Variations in the pH of the solution into which the nanowire was immersed caused protonation and deprotonation of the $-NH_2$ and the $-SiOH$ groups on the surface of the nanowire. The variation in surface charge densities regulates the conductance of the nanowire; due to the p-type characteristics of a silicon wire, the conductance increases with the addition of negative surface charge. The combined acid and base behavior of the surface groups results in an approximately linear dependence of the conductance on pH in the pH range 2 to 9, thus leading to a direct readout pH meter. This same type of approach was used for the detection of the binding of biomolecules, such as streptavidin using biotin-modified nanowires (see Fig. 4.44). This nanowire-based device has high sensitivity and could detect streptavidin binding down to a 10 pM (10^{-12} mole) concentration. The chemical detection devices were made in a field-effect transistor geometry, so that the back-gate potential could be used to regulate the conductance in conjugation with the chemical detection and to provide a real time direct read out [4.191]. The extension of this device to detect multiple analytes, using multiple nanowires each sensitized to a different analyte, could provide for fast, sensitive and in-situ screening procedures.

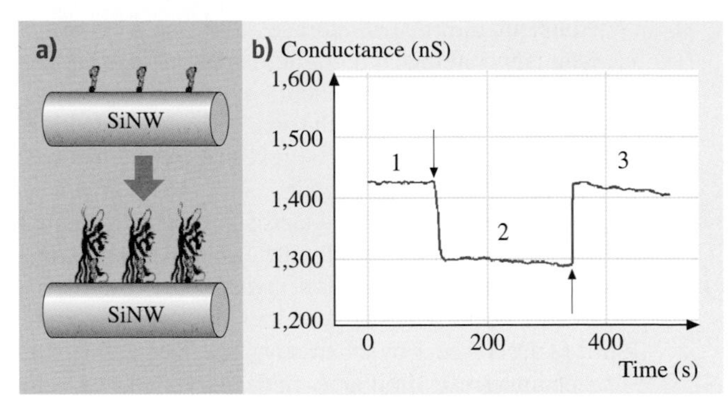

Fig. 4.44 (a) Streptavidin molecules bind to a silicon nanowire functionalized with biotin. The binding of streptavidin to biotin causes the nanowire to change its resistance. (b) The conductance of a biotin-modified silicon nanowire exposed to streptavidin in a buffer solution (regions 1 and 3) and with the introduction of a solution of antibiotin monoclonal antibody (region 2) [4.191]

Favier et al. used a similar approach, making a nanosensor for the detection of hydrogen out of an array of palladium nanowires between two metal contacts [4.56]. They demonstrated that nanogaps were present in their nanowire structure, and upon absorption of H_2 and formation of Pd hydride, the nanogap structure would close and improve the electrical contact, thereby increasing the conductance of the nanowire array. The response time of these sensors was 75 msec, and they could operate in the range of 0.5 to 5% H_2 before saturation occurred.

4.3.5 Magnetic Applications

It has been demonstrated that arrays of single domain magnetic nanowires can be prepared with controlled nanowire diameter and length, aligned along a common direction and arranged in a close-packed ordered array (see Sect. 4.1.1), and that the magnetic properties (coercivity, remanence, and dipolar magnetic interwire interaction) can be controlled to achieve a variety of magnetic applications [4.54, 71].

Magnetic information storage is the most interesting of these applications. The large nanowire aspect ratio (length/diameter) prevents the onset of the "superparamagnetic" limit at which the magnetization direction in the magnetic grains can be reversed by the thermal energy $k_B T$, thereby resulting in loss of recorded data in the magnetic recording medium. The magnetic energy in a grain can be increased by increasing either the volume or the anisotropy of the grain. If the volume is increased, the particle size increases, and thus, the resolution is decreased. For spherical magnetized grains, the superparamagnetic limit at room temperature is reached at 70 Gbit/in^2. In nanowires, the anisotropy is very large and yet the wire diameters are small, so the magnetostatic switching energy can easily be above the thermal energy while the spatial resolution is large. For magnetic data storage applications, a large aspect ratio is needed for the nanowires to maintain a high coercivity, and a sufficient separation between nanowires is needed to suppress inter-wire magnetic dipolar coupling (see Sect. 4.2.1). Thus nanowires can form stable and highly dense magnetic memory arrays with packing densities in excess of 10^{11} wires/cm^2.

The onset of superparamagnetism can be prevented in the single domain magnetic nanowire arrays that have already been fabricated either using porous alumina templates to make Ni nanowires with 35 nm diameters [4.54] or diblock copolymer templates [4.71] to make Co nanowires, with mean diameters of 14 nm and 100% filling of the template pores (see Sect. 4.1.1). Ordered magnetic nanowire arrays that have already been demonstrated offer the exciting promise for achieving systems permitting 10^{12} bits/in^2 data storage.

4.4 Concluding Remarks

A review has been presented of synthesis, characterization, and physical properties of nanowires, with particular reference to nanowire properties that differ from those of their bulk counterparts and to potential applications that might result from the special structure and properties of nanowires. We have shown in this chapter that the emerging field of nanowire research has developed rapidly over the past few years, driven by the development of a variety of complementary nanowire synthesis methods and of effective tools to measure both their structure and properties, as reviewed, respectively, in Sect. 4.1 and Sect. 4.2. At present, much of the progress is at the demonstration-of-concept level, with many gaps in knowledge remaining to be elucidated, theoretical models to be developed, and new nanowire systems to be explored. Having demonstrated that many of the most interesting discoveries to date relate to nanowire properties not present in their bulk material counterparts, we can expect future research emphasis to be focused increasingly on smaller diameter nanowires, where new unexplored physical phenomena related to quantum confinement effects are more likely to be found. We can also expect the development of applications to soon follow. Many promising applications are now at the early demonstration stage (see Sect. 4.3), but are moving ahead rapidly because of their promise for new functionality, not previously available, to the fields of electronics, opto-electronics, biotechnology, magnetics, and energy conversion and generation, among others. Many exciting challenges remain in advancing both the nanoscience and the nano-technological promise already demonstrated by the nanowire research described in this review.

References

4.1 R. P. Feynman: There's plenty of room at the bottom, Eng. Sci. (February 1960) 22

4.2 M. S. Dresselhaus, Y.-M. Lin, O. Rabin, A. Jorio, A. G. Souza Filho, M. A. Pimenta, R. Saito, G. G. Samsonidze, G. Dresselhaus: Nanowires and nanotubes, Mater. Sci. & Eng. C **1008** (2002) 1–12

4.3 R. Saito, G. Dresselhaus, M. S. Dresselhaus: *Physical Properties of Carbon Nanotubes* (Imperial College Press, London 1998)

4.4 M. S. Dresselhaus, G. Dresselhaus, P. Avouris (Eds.): *Carbon Nanotubes: Synthesis Structure, Properties and Applications* (Springer, Berlin, Heidelberg 2001)

4.5 R. C. Haddon: Special issue on carbon nanotubes, Accounts Chem. Res. **35** (2002) 997–1113

4.6 Y.-M. Lin, X. Sun, S. Cronin, Z. Zhang, J. Y. Ying, M. S. Dresselhaus: Fabrication and transport properties of Te-doped bismuth nanowire arrays, Molecular Electronics: MRS Symp. Proc., Boston 1999, ed. by S. T. Pantelides, M. A. Reed, J. Murday, A. Aviran (Materials Research Society Press, Pittsburgh 2000) **582** H10.3 (1–6)

4.7 C. R. Martin: Nanomaterials: A membrane-based synthetic approach, Science **266** (1994) 1961–1966

4.8 G. A. Ozin: Nanochemistry: synthesis in diminishing dimensions, Adv. Mater. **4** (1992) 612–649

4.9 R. J. Tonucci, B. L. Justus, A. J. Campillo, C. E. Ford: Nanochannel array glass, Science **258** (1992) 783–785

4.10 J. Y. Ying: Nanoporous systems and templates, Sci. Spectra **18** (1999) 56–63

4.11 J. W. Diggle, T. C. Downie, C. W. Goulding: Anodic oxide films on aluminum, Chem. Rev. **69** (1969) 365–405

4.12 J. P. O'Sullivan, G. C. Wood: The morphology and mechanism of formation of porous anodic films on aluminum, Proc. Roy. Soc. London A **317** (1970) 511–543

4.13 A. P. Li, F. Müller, A. Birner, K. Neilsch, U. Gösele: Hexagonal pore arrays with a 50–420 nm interpore distance formed by self-organization in anodic alumina, J. Appl. Phys. **84** (1998) 6023–6026

4.14 O. Jessensky, F. Müller, U. H. Gösele: Self-organized formation of hexagonal pore arrays in anaodic alumina, Appl. Phys. Lett. **72** (1998) 1173–1175

4.15 D. AlMawlawi, N. Coombs, M. Moskovits: Magnetic-properties of Fe deposited into anodic aluminum-oxide pores as a function of particle-size, J. Appl. Phys. **70** (1991) 4421–4425

4.16 Z. Zhang, D. Gekhtman, M. S. Dresselhaus, J. Y. Ying: Processing and characterization of single-crystalline ultrafine bismuth nanowires, Chem. Mater. **11** (1999) 1659–1665

4.17 F. Li, L. Zhang, R. M. Metzger: On the growth of highly ordered pores in anodized aluminum oxide, Chem. Mater. **10** (1998) 2470–2480

4.18 H. Masuda, M. Satoh: Fabrication of gold nanodot array using anodic porous alumina as an evaporation mask, Jpn. J. Appl. Phys. **35** (1996) L126–L129

4.19 E. Ferain, R. Legras: Track-etched membrane – dynamics of pore formation, Nucl. Instrum. Methods B **84** (1993) 331–336

4.20 A. Blondel, J. P. Meier, B. Doudin, J.-P. Ansermet: Giant magnetoresistance of nanowires of multilayers, Appl. Phys. Lett. **65** (1994) 3019–3021

4.21 K. Liu, C. L. Chien, P. C. Searson, Y. Z. Kui: Structural and magneto-transport properties of electrodeposited bismuth nanowires, Appl. Phys. Lett. **73** (1998) 1436–1438

4.22 L. Sun, P. C. Searson, C. L. Chien: Electrochemical deposition of nickel nanowire arrays in single-crystal mica films, Appl. Phys. Lett. **74** (1999) 2803–2805

4.23 E. Braun, Y. Eichen, U. Sivan, G. Ben-Yoseph: DNA-templated assembly and electrode attachment of a conducting silver wire, Nature **391** (1998) 775–778

4.24 G. Sauer, G. Brehm, S. Schneider, K. Nielsch, R. B. Wehrspohn, J. Choi, H. Hofmeister, U. Gösele: Highly ordered monocrystalline silver nanowire arrays, J. Appl. Phys. **91** (2002) 3243–3247

4.25 G. L. Hornyak, C. J. Patrissi, C. M. Martin: Fabrication, characterization, and optical properties of gold nanoparticle/porous alumina composites: the nonscattering Maxwell-Garnett limit, J. Phys. Chem. B **101** (1997) 1548–1555

4.26 X. Y. Zhang, L. D. Zhang, Y. Lei, L. X. Zhao, Y. Q. Mao: Fabrication and characterization of highly ordered Au nanowire arrays, J. Mater. Chem. **11** (2001) 1732–1734

4.27 Y.-T. Cheng, A. M. Weiner, C. A. Wong, M. P. Balogh, M. J. Lukitsch: Stress-induced growth of bismuth nanowires, Appl. Phys. Lett. **81** (2002) 3248–3250

4.28 J. Heremans, C. M. Thrush, Y.-M. Lin, S. Cronin, Z. Zhang, M. S. Dresselhaus, J. F. Mansfield: Bismuth nanowire arrays: synthesis and galvanomagnetic properties, Phys. Rev. B **61** (2000) 2921–2930

4.29 L. Piraux, S. Dubois, J. L. Duvail, A. Radulescu, S. Demoustier-Champagne, E. Ferain, R. Legras: Fabrication and properties of organic and metal nanocylinders in nanoporous membranes, J. Mater. Res. **14** (1999) 3042–3050

4.30 K. Hong, F. Y. Yang, K. Liu, D. H. Reich, P. C. Searson, C. L. Chien, F. F. Balakirev, G. S. Boebinger: Giant positive magnetoresistance of Bi nanowire arrays in high magnetic fields, J. Appl. Phys. **85** (1999) 6184–6186

4.31 A. J. Yin, J. Li, W. Jian, A. J. Bennett, J. M. Xu: Fabrication of highly ordered metallic nanowire arrays by electrodeposition, Appl. Phys. Lett. **79** (2001) 1039–1041

4.32 Z. Zhang, J. Y. Ying, M. S. Dresselhaus: Bismuth quantum-wire arrays fabricated by a vacuum melting and pressure injection process, J. Mater. Res. **13** (1998) 1745–1748

4.33 T. E. Huber, M. J. Graf, C. A. Foss Jr., P. Constant: Processing and characterization of high-conductance bismuth wire array composites, J. Mater. Res. **15** (2000) 1816–1821

4.34 M. S. Sander, A. L. Prieto, R. Gronsky, T. Sands, A. M. Stacy: Fabrication of high-density, high aspect ratio, large-area bismuth telluride nanowire arrays by electrodeposition into porous anodic alumina templates, Adv. Mater. **14** (2002) 665–667

4.35 M. Chen, Y. Xie, J. Lu, Y. J. Xiong, S. Y. Zhang, Y. T. Qian, X. M. Liu: Synthesis of rod-, twinrod-, and tetrapod-shaped CdS nanocrystals using a highly oriented solvothermal recrystallization technique, J. Mat. Chem. **12** (2002) 748–753

4.36 D. Xu, Y. Xu, D. Chen, G. Guo, L. Gui, Y. Tang: Preparation of CdS single-crystal nanowires by electrochemically induced deposition, Adv. Mater. **12** (2000) 520–522

4.37 D. Routkevitch, T. Bigioni, M. Moskovits, J. M. Xu: Electrochemical fabrication of CdS nanowire arrays in porous anodic aluminum oxide templates, J. Phys. Chem. **100** (1996) 14037–14047

4.38 L. Manna, E. C. Scher, A. P. Alivisatos: Synthesis of soluble and processable rod-, arrow-, teardrop-, and tetrapod-shaped CdSe nanocrystals, J. Am. Chem. Soc. **122** (2000) 12700–12706

4.39 D. Routkevitch, A. A. Tager, J. Haruyama, D. Al-Mawlawi, M. Moskovits, J. M. Xu: Nonlithographic nano-wire arrays: fabrication, physics, and device applications, IEEE Trans. Electron. Dev. **43** (1996) 1646–1658

4.40 D. S. Xu, D. P. Chen, Y. J. Xu, X. S. Shi, G. L. Guo, L. L. Gui, Y. Q. Tang: Preparation of II–VI group semiconductor nanowire arrays by dc electrochemical deposition in porous aluminum oxide templates, Pure Appl. Chem. **72** (2000) 127–135

4.41 R. Adelung, F. Ernst, A. Scott, M. Tabib-Azar, L. Kipp, M. Skibowski, S. Hollensteiner, E. Spiecker, W. Jäger, S. Gunst, A. Klein, W. Jägermann, V. Zaporojtchenko, F. Faupel: Self-assembled nanowire networks by deposition of copper onto layered-crystal surfaces, Adv. Mater. **14** (2002) 1056–1061

4.42 T. Gao, G. W. Meng, J. Zhang, Y. W. Wang, C. H. Liang, J. C. Fan, L. D. Zhang: Template synthesis of single-crystal Cu nanowire arrays by electrodeposition, Appl. Phys. A **73** (2001) 251–254

4.43 F. Li, R. M. Metzger: Activation volume of α-Fe particles in alumite films, J. Appl. Phys. **81** (1997) 3806–3808

4.44 A. Sugawara, T. Coyle, G. G. Hembree, M. R. Scheinfein: Self-organized Fe nanowire arrays prepared by shadow deposition on NaCl(110) templates, Appl. Phys. Lett. **70** (1997) 1043–1045

4.45 G. S. Cheng, L. D. Zhang, Y. Zhu, G. T. Fei, L. Li, C. M. Mo, Y. Q. Mao: Large-scale synthesis of single crystalline gallium nitride nanowires, Appl. Phys. Lett. **75** (1999) 2455–2457

4.46 Y. Huang, X. Duan, Y. Cui, C. M. Lieber: Gallium nitride nanowire nanodevices, Nano. Lett. **2** (2002) 101–104

4.47 Y. Cui, X. Duan, J. Hu, C. M. Lieber: Doping and electrical transport in silicon nanowires, J. Phys. Chem. B **104** (2000) 5213–5216

4.48 A. D. Berry, R. J. Tonucci, M. Fatemi: Fabrication of GaAs and InAs wires in nanochannel glass, Appl. Phys. Lett. **69** (1996) 2846–2848

4.49 J. R. Heath, F. K. LeGoues: A liquid solution synthesis of single-crystal germanium quantum wires, Chem. Phys. Lett. **208** (1993) 263–268

4.50 Y. Wu, P. Yang: Germanium nanowire growth via simple vapor transport, Chem. Mater. **12** (2000) 605–607

4.51 Y. F. Zhang, Y. H. Tang, N. Wang, C. S. Lee, I. Bello, S. T. Lee: Germanium nanowires sheathed with an oxide layer, Phys. Rev. B **61** (2000) 4518–4521

4.52 X. Duan, Y. Huang, Y. Cui, J. Wang, C. M. Lieber: Indium phosphide nanowires as building blocks for nanoscale electronic and optoelectronic devices, Nature **409** (2001) 66–69

4.53 M. P. Zach, K. H. Ng, R. M. Penner: Molybdenum nanowires by electrodeposition, Science **290** (2000) 2120–2123

4.54 K. Nielsch, R. Wehrspohn, S. F. H. Kronmuller, J. Barthel, J. Kirschner, U. Gosele: Magnetic properties of 100 nm nickel nanowire arrays obtained from ordered porous alumina templates, MRS Symp. Proc. **636** (2001) D19 (1–6)

4.55 E. Lifshitz, M. Bashouti, V. Kloper, A. Kigel, M. S. Eisen, S. Berger: Synthesis and characterization of PbSe quantum wires, multipods, quantum rods, and cubes, Nano. Lett. **3** (2003) 857–862

4.56 F. Favier, E. C. Walter, M. P. Zach, T. Benter, R. M. Penner: Hydrogen sensors and switches from electrodeposited palladium mesowire arrays, Science **293** (2001) 2227–2231

4.57 B. Gates, B. Mayers, B. Cattle, Y. Xia: Synthesis and characterization of uniform nanowires of trigonal selenium, Adv. Funct. Mat. **12** (2002) 219–227

4.58 C. A. Huber, T. E. Huber, M. Sadoqi, J. A. Lubin, S. Manalis, C. B. Prater: Nanowire array composites, Science **263** (1994) 800–802

4.59 Y. Cui, L. J. Lauhon, M. S. Gudiksen, J. Wang, C. M. Lieber: Diameter-controlled synthesis of single crystal silicon nanowires, Appl. Phys. Lett. **78** (2001) 2214–2216

4.60 A. M. Morales, C. M. Lieber: A laser ablation method for the synthesis of crystalline semiconductor nanowires, Science **279** (1998) 208–211

4.61 N. Wang, Y. F. Zhang, Y. H. Tang, C. S. Lee, S. T. Lee: SiO_2-enhanced synthesis of Si nanowires by laser ablation, Appl. Phys. Lett. **73** (1998) 3902–3904

4.62 M. K. Sunkara, S. Sharma, R. Miranda, G. Lian, E. C. Dickey: Bulk synthesis of silicon nanowires using a low-temperature vapor-liquid-solid method, Appl. Phys. Lett. **79** (2001) 1546–1548

4.63 J. P. Heremans, C. M. Thrush, D. T. Morelli, M.-C. Wu: Thermoelectric power of bismuth nanocomposites, Phys. Rev. Lett. **88** (2002) 216801(1–4)

4.64 Y. Li, G. S. Cheng, L. D. Zhang: Fabrication of highly ordered ZnO nanowire arrays in anodic alumina membranes, J. Mater. Res. **15** (2000) 2305–2308

4.65 P. Yang, H. Yan, S. Mao, R. Russo, J. Johnson, R. Saykally, N. Morris, J. Pham, R. He, H.-J. Choi: Controlled growth of ZnO nanowires and their optical properties, Adv. Func. Mat. **12** (2002) 323–331

4.66 M. J. Zheng, L. D. Zhang, G. H. Li, W. Z. Chen: Fabrication and optical properties of large-scale uniform zinc oxide nanowire arrays by one-step electrochemical deposition technique, Chem. Phys. Lett. **363** (2002) 123–128

4.67 Y.-M. Lin, S. B. Cronin, J. Y. Ying, M. S. Dresselhaus, J. P. Heremans: Transport properties of Bi nanowire arrays, Appl. Phys. Lett. **76** (2000) 3944–3946

4.68 C. A. Huber, T. E. Huber: A novel microstructure: semiconductor-impregnated porous Vycor glass, J. Appl. Phys. **64** (1988) 6588–6590

4.69 J. S. Beck, J. C. Vartuli, W. J. Roth, M. E. Leonowicz, C. T. Kresge, K. D. Schmitt, C. T.-W. Chu, D. H. Olson, E. W. Sheppard, S. B. McCullen, J. B. Higgins, J. L. Schlenker: A new family of mesoporous molecular sieves prepared with liquid crystal templates, J. Am. Chem. Soc. **114** (1992) 10834–10843

4.70 C.-G. Wu, T. Bein: Conducting polyaniline filaments in a mesoporous channel host, Science **264** (1994) 1757–1759

4.71 T. Thurn-Albrecht, J. Schotter, G. A. Kästle, N. Emley, T. Shibauchi, L. Krusin-Elbaum, K. Guarini, C. T. Black, M. T. Tuominen, T. P. Russell: Ultrahigh-density nanowire arrays grown in self-assembled diblock copolymer templates, Science **290** (2000) 2126–2129

4.72 A. W. Adamson: *Physical Chemistry of Surfaces* (Wiley, New York 1982) p. 338

4.73 R. Ferré, K. Ounadjela, J. M. George, L. Piraux, S. Dubois: Magnetization processes in nickel and cobalt electrodeposited nanowires, Phys. Rev. B **56** (1997) 14066–14075

4.74 H. Zeng, M. Zheng, R. Skomski, D. J. Sellmyer, Y. Liu, L. Menon, S. Bandyopadhyay: Magnetic properties of self-assembled Co nanowires of varying length and diameter, J. Appl. Phys. **87** (2000) 4718–4720

4.75 Y. Peng, H.-L. Zhang, S.-L. Pan, H.-L. Li: Magnetic properties and magnetization reversal of α-Fe nanowires deposited on alumina film, J. Appl. Phys. **87** (2000) 7405–7408

4.76 L. Piraux, J. M. George, J. F. Despres, C. Leroy, E. Ferain, R. Legras, K. Ounadjela, A. Fert: Giant magnetoresistance in magnetic multilayered nanowires, Appl. Phys. Lett. **65** (1994) 2484–2486

4.77 S. Bhattacharrya, S. K. Saha, D. Chakravorty: Nanowire formation in a polymeric film, Appl. Phys. Lett. **76** (2000) 3896–3898

4.78 G. Yi, W. Schwarzacher: Single crystal superconductor nanowires by electrodeposition, Appl. Phys. Lett. **74** (1999) 1746–1748

4.79 D. AlMawlawi, C. Z. Liu, M. Moskovits: Nanowires formed in anodic oxide nanotemplates, J. Mater. Res. **9** (1994) 1014–1018

4.80 G. S. Cheng, L. D. Zhang, S. H. Chen, Y. Li, L. Li, X. G. Zhu, Y. Zhu, G. T. Fei, Y. Q. Mao: Ordered nanostructure of single-crystalline GaN nanowires in a honeycomb structure of anodic alumina, J. Mater. Res. **15** (2000) 347–350

4.81 D. N. Davydov, P. A. Sattari, D. AlMawlawi, A. Osika, T. L. Haslett, M. Moskovits: Field emitters based on porous aluminum oxide templates, J. Appl. Phys. **86** (1999) 3983–3987

4.82 J. Li, C. Papadopoulos, J. M. Xu, M. Moskovits: Highly-ordered carbon nanotube arrays for electronic applications, Appl. Phys. Lett. **75** (1999) 367–369

4.83 T. Iwasaki, T. Motoi, T. Den: Multiwalled carbon nanotubes growth in anodic alumina nanoholes, Appl. Phys. Lett. **75** (1999) 2044–2046

4.84 J. S. Suh, J. S. Lee: Highly ordered two-dimensional carbon nanotube arrays, Appl. Phys. Lett. **75** (1999) 2047–2049

4.85 Z. M. Li, Z. K. Tang, H. J. Liu, N. Wang, C. T. Chan, R. Saito, S. Okada, G. D. Li, J. S. Chen, N. Nagasawa, S. Tsuda: Polarized absorption spectra of single-walled 4 Å carbon nanotubes aligned in channels of an $AlPO_4$-5 single crystal, Phys. Rev. Lett. **87** (2001) 127401 (1–4)

4.86 J. Sloan, A. I. Kirkland, J. L. Hutchison, M. L. H. Green: Structural characterization of atomically regulated nanocrystals formed within single-walled carbon nanotubes using electron microscopy, Acc. Chem. Res. **35** (2002) 1054–1062

4.87 R. S. Wagner, W. C. Ellis: Vapor-liquid-solid mechanism of single crystal growth, Appl. Phys. Lett. **4** (1964) 89–90

4.88 Y. Wu, P. Yang: Direct observation of vapor-liquid-solid nanowire growth, J. Am. Chem. Soc. **123** (2001) 3165–3166

4.89 S. Sharma, M. K. Sunkara, R. Miranda, G. Lian, E. C. Dickey: A novel low temperature synthesis method for semiconductor nanowires Synthesis, Functional Properties and Applications of Nanostructures: Mat. Res. Soc. Symp. Proc., San Francisco 2001, ed. by H. W. Hahn, D. L. Feldheim, C. P. Kubiak, R. Tannenbaum, R. W. Siegel (Materials Research Society Press, Pittsburgh 2001) **676** Y1.6

4.90 Y. Wu, R. Fan, P. Yang: Block-by-block growth of single-crystalline Si/SiGe superlattice nanowires, Nano. Lett. **2** (2002) 83–86

4.91 N. Wang, Y. H. Tang, Y. F. Zhang, C. S. Lee, S. T. Lee: Nucleation and growth of Si nanowires from silicon oxide, Phys. Rev. B **58** (1998) R16024–R16026

4.92 M. S. Gudiksen, L. J. Lauhon, J. Wang, D. C. Smith, C. M. Lieber: Growth of nanowire superlattice structures for nanoscale photonics and electronics, Nature **415** (2002) 617–620

4.93 M. T. Björk, B. J. Ohlsson, T. Sass, A. I. Persson, C. Thelander, M. H. Magnusson, K. Deppert, L. R. Wallenberg, L. Samuelson: One-dimensional steeplechase for electrons realized, Nano. Lett. **2** (2002) 87–89

4.94 Y. F. Zhang, Y. H. Tang, N. Wang, C. S. Lee, I. Bello, S. T. Lee: One-dimensional growth mechanism of crystalline silicon nanowires, J. Cryst. Growth **197** (1999) 136–140

4.95 S. T. Lee, Y. F. Zhang, N. Wang, Y. H. Tang, I. Bello, C. S. Lee, Y. W. Chung: Semiconductor nanowires from oxides, J. Mater. Res. **14** (1999) 4503–4507

4.96 D. D. D. Ma, C. S. Lee, Y. Lifshitz, S. T. Lee: Periodic array of intramolecular junctions of silicon nanowires, Appl. Phys. Lett. **81** (2002) 3233–3235

4.97 B. Gates, Y. Yin, Y. Xia: A solution-phase approach to the synthesis of uniform nanowires of crystalline selenium with lateral dimensions in the range of 10–30 nm, J. Am. Chem. Soc. **122** (2000) 12582–12583

4.98 B. Mayers, B. Gates, Y. Yin, Y. Xia: Large-scale synthesis of monodisperse nanorods of Se/Te alloys through a homogeneous nucleation and solution growth process, Adv. Mat. **13** (2001) 1380–1384
4.99 B. Gates, Y. Wu, Y. Yin, P. Yang, Y. Xia: Single-crystalline nanowires of Ag_2Se can be synthesized by templating against nanowires of trigonal Se. J. Am. Chem. Soc., **123** (2001) 11500–11501
4.100 B. Gates, B. Mayers, Y. Wu, Y. Sun, B. Cattle, P. Yang, Y. Xia: Synthesis and characterization of crystalline Ag_2Se nanowires through a template-engaged reaction at room temperature, Adv. Funct. Mat. **12** (2002) 679–686
4.101 X. Peng, J. Wickham, A. P. Alivisatos: Kinetics of II-VI and III-V colloidal semiconductor nanocrystal growth: 'focusing' of size distributions, J. Am. Chem. Soc. **120** (1998) 5343–5344
4.102 M. P. Zach, K. Inazu, K. H. Ng, J. C. Hemminger, R. M. Penner: Synthesis of molybdenum nanowires with millimeter-scale lengths using electrochemical step edge decoration, Chem. Mater. **14** (2002) 3206–3216
4.103 N. A. Melosh, A. Boukai, F. Diana, B. Gerardot, A. Badolato, P. M. Petroff, J. R. Heath: Ultrahigh-density nanowire lattices and circuits, Science **300** (2003) 112–115
4.104 S. R. Nicewarner-Peña, R. G. Freeman, B. D. Reis, L. He, D. J. Peña, I. D. Walton, R. Cromer, C. D. Keating, M. J. Natan: Submicrometer metallic barcodes, Science **294** (2001) 137–141
4.105 L. J. Lauhon, M. S. Gudiksen, D. Wang, C. M. Lieber: Epitaxial core-shell and core-multishell nanowire heterostructures, Nature **420** (2002) 57–61
4.106 Z. L. Wang, Z. R. Dai, R. P. Gao, Z. G. Bai, J. L. Gole: Side-by-side silicon carbide-silica biaxial nanowires: Synthesis, structure, and mechanical properties, Appl. Phys. Lett. **77** (2000) 3349–3351
4.107 J. Y. Lao, J. G. Wen, Z. F. Ren: Hierarchical ZnO nanostructures, Nano. Lett. **2** (2002) 1287–1291
4.108 J. Y. Lao, J. Y. Huang, D. Z. Wang, Z. F. Ren: ZnO nanobridges and nanonails, Nano. Lett. **3** (2003) 235–238
4.109 P. Yang, F. Kim: Langmuir–Blodgett assembly of one-dimensional nanostructures, CHEMPHYSCHEM **3** (2002) 503–506
4.110 B. Messer, J. H. Song, P. Yang: Microchannel networks for nanowire patterning, J. Am. Chem. Soc. **122** (2000) 10232–10233
4.111 P. A. Smith, C. D. Nordquist, T. N. Jackson, T. S. Mayer, B. R. Martin, J. Mbindyo, T. E. Mallouk: Electric-field assisted assembly and alignment of metallic nanowires, Appl. Phys. Lett. **77** (2000) 1399–1401
4.112 H. Masuda, H. Yamada, M. Satoh, H. Asoh, M. Nakao, T. Tamamura: Highly ordered nanochannel-array architecture in anodic alumina, Appl. Phys. Lett. **71** (1997) 2770–2772
4.113 O. Rabin, P. R. Herz, S. B. Cronin, Y.-M. Lin, A. I. Akinwande, M. S. Dresselhaus: Nanofabrication using self-assembled alumina templates, Nonlithographic and Lithographic Methods for Nanofabrication: MRS Symposium Proceedings, Boston 2000, ed. by J. A. Rogers, A. Karim, L. Merhari, D. Norris, Y. Xia (Materials Research Society Press, Pittsburgh 2001) **636** D4.7 (1–6)
4.114 A. Rabin, P. R. Herz, Y.-M. Lin, A. I. Akinwande, S. B. Cronin, M. S. Dresselhaus: Formation of thick porous anodic alumina films and nanowire arrays on silicon wafers and glass, Adv. Funct. Mater. **13** (2003) 631–638
4.115 M. H. Huang, S. Mao, H. Feick, H. Yan, Y. Wu, H. Kind, E. Weber, R. Russo, P. Yang: Room-temperature ultraviolet nanowire nanolasers, Science **292** (2001) 1897–1899
4.116 Y. H. Tang, Y. F. Zhang, N. Wang, C. S. Lee, X. D. Han, I. Bello, S. T. Lee: Morphology of Si nanowires synthesized by high-temperature laser ablation, J. Appl. Phys. **85** (1999) 7981–7983
4.117 S. B. Cronin, Y.-M. Lin, O. Rabin, M. R. Black, G. Dresselhaus, M. S. Dresselhaus, P. L. Gai: Bismuth nanowires for potential applications in nanoscale electronics technology, Microsc. Microanal. **8** (2002) 58–63
4.118 M. S. Sander, R. Gronsky, Y.-M. Lin, M. S. Dresselhaus: Plasmon excitation modes in nanowire arrays, J. Appl. Phys. **89** (2001) 2733–2736
4.119 L. Venkataraman, C. M. Lieber: Molybdenum selenide molecular wires as one-dimensional conductors, Phys. Rev. Lett. **83** (1999) 5334–5337
4.120 A. Majumdar: Scanning thermal microscopy, Annual Rev. Mater. Sci. **29** (1999) 505–585
4.121 D. A. Wharam, T. J. Thornton, R. Newbury, M. Pepper, H. Ahmed, J. E. F. Frost, D. G. Hasko, D. C. Peacock, D. A. Ritchie, G. A. C. Jones: One-dimensional transport and the quantization of the ballistic resistance, J. Phys. C: Solid State Phys. **21** (1988) L209–L214
4.122 B. J. van Wees, H. van Houten, C. W. J. Beenakker, J. G. Williamson, L. P. Kouvenhoven, D. van der Marel, C. T. Foxon: Quantized conductance of point contacts in a two-dimensional electron gas, Phys. Rev. Lett. **60** (1988) 848–850
4.123 C. J. Muller, J. M. van Ruitenbeek, L. J. de Jongh: Conductance and supercurrent discontinuities in atomic-scale metallic constrictions of variable width, Phys. Rev. Lett. **69** (1992) 140–143
4.124 C. J. Muller, J. M. Krans, T. N. Todorov, M. A. Reed: Quantization effects in the conductance of metallic contacts at room temperature, Phys. Rev. B **53** (1996) 1022–1025
4.125 J. L. Costa-Krämer, N. Garcia, H. Olin: Conductance quantization in bismuth nanowires at 4 K, Phys. Rev. Lett. **78** (1997) 4990–4993
4.126 J. L. Costa-Krämer, N. Garcia, H. Olin: Conductance quantization histograms of gold nanowires at 4 K, Phys. Rev. B **55** (1997) 12910–12913
4.127 C. Z. Li, H. X. He, A. Bogozi, J. S. Bunch, N. J. Tao: Molecular detection based on conductance quan-

tization of nanowires, Appl. Phys. Lett. **76** (2000) 1333–1335

4.128 J. L. Costa-Krämer, N. Garcia, P. Garcia-Mochales, P. A. Serena, M. I. Marques, A. Correia: Conductance quantization in nanowires formed between micro and macroscopic metallic electrodes, Phys. Rev. B **55** (1997) 5416–5424

4.129 Y. Huang, X. Duan, Y. Cui, L. J. Lauhon, K.-H. Kim, C. Lieber: Logic gates and computation from assembled nanowire building blocks, Science **294** (2001) 1313–1317

4.130 J.-R. Kim, H. Oh, H. M. So, J.-J. Kim, J. Kim, C. J. Lee, S. C. Lyu: Schottky diodes based on a single GaN nanowire, Nanotechnology **13** (2002) 701–704

4.131 X. Duan, Y. Huang, C. M. Lieber: Nonvolatile memory and programmable logic from molecule-gated nanowires, Nano. Lett. **2** (2002) 487–490

4.132 E. C. Walter, R. M. Penner, H. Liu, K. H. Ng, M. P. Zach, F. Favier: Sensors from electrodeposited metal nanowires, Surf. Inter. Anal. **34** (2002) 409–412

4.133 E. C. Walter, K. H. Ng, M. P. Zach, R. M. Penner, F. Favier: Electronic devices from electrodeposited metal nanowires, Microelectron. Eng. **61–62** (2002) 555–561

4.134 Y.-M. Lin, X. Sun, M. S. Dresselhaus: Theoretical investigation of thermoelectric transport properties of cylindrical Bi nanowires, Phys. Rev. B **62** (2000) 4610–4623

4.135 K. Liu, C. L. Chien, P. C. Searson: Finite-size effects in bismuth nanowires, Phys. Rev. B **58** (1998) R14681–R14684

4.136 Z. Zhang, X. Sun, M. S. Dresselhaus, J. Y. Ying, J. Heremans: Magnetotransport investigations of ultrafine single-crystalline bismuth nanowire arrays, Appl. Phys. Lett. **73** (1998) 1589–1591

4.137 J. Heremans, C. M. Thrush, Z. Zhang, X. Sun, M. S. Dresselhaus, J. Y. Ying, D. T. Morelli: Magnetoresistance of bismuth nanowire arrays: A possible transition from one-dimensional to three dimensional localization, Phys. Rev. B **58** (1998) R10091–R10095

4.138 L. Sun, P. C. Searson, C. L. Chien: Finite-size effects in nickel nanowire arrays, Phys. Rev. B **61** (2000) R6463–R6466

4.139 Y.-M. Lin, O. Rabin, S. B. Cronin, J. Y. Ying, M. S. Dresselhaus: Semimetal-semiconductor transition in $Bi_{1-x}Sb_x$ alloy nanowires and their thermoelectric properties, Appl. Phys. Lett. **81** (2002) 2403–2405

4.140 Y.-M. Lin, S. B. Cronin, O. Rabin, J. Y. Ying, M. S. Dresselhaus: Transport properties and observation of semimetal-semiconductor transition in bi-based nanowires, Nanocrystalline Semiconductor Materials and Devices: MRS Symp. Proc., Boston 2002, ed. by J. M. Buriak, D. D. M. Wayner, L. Tsyeskov, F. Priolo, B. E. White, Boston 2003) **737** F3.14

4.141 Y.-M. Lin, M. S. Dresselhaus: Transport properties of superlattice nanowires and their potential for thermoelectric applications, Nanocrystalline Semiconductor Materials and Devices: MRS Symp. Proc., Boston 2002, ed. by J. M. Buriak, D. D. M. Wayner, L. Tsyeskov, F. Priolo, B. E. White (Materials Research Society Press, Boston 2003) **737** F8.18

4.142 J. Heremans, C. M. Thrush, Y.-M. Lin, S. B. Cronin, M. S. Dresselhaus: Transport properties of antimony nanowires, Phys. Rev. B **63** (2001) 085406(1–8)

4.143 Y.-M. Lin, S. B. Cronin, O. Rabin, J. Y. Ying, M. S. Dresselhaus: Transport properties of $Bi_{1-x}Sb_x$ alloy nanowires synthesized by pressure injection, Appl. Phys. Lett. **79** (2001) 677–679

4.144 D. E. Beutler, N. Giordano: Localization and electron–electron interaction effects in thin Bi wires and films, Phys. Rev. B **38** (1988) 8–19

4.145 Z. Zhang, X. Sun, M. S. Dresselhaus, J. Y. Ying, J. Heremans: Electronic transport properties of single crystal bismuth nanowire arrays, Phys. Rev. B **61** (2000) 4850–4861

4.146 J. Heremans, C. M. Thrush: Thermoelectric power of bismuth nanowires, Phys. Rev. B **59** (1999) 12579–12583

4.147 L. D. Hicks, M. S. Dresselhaus: Thermoelectric figure of merit of a one-dimensional conductor, Phys. Rev. B **47** (1993) 16631–16634

4.148 Y.-M. Lin, S. B. Cronin, O. Rabin, J. Heremans, M. S. Dresselhaus, J. Y. Ying: Transport properties of Bi-related nanowire systems, Anisotropic Nanoparticles: Synthesis, Characterization and Applications: MRS Symp. Proc., Boston 2000, ed. by S. Stranick, P. C. Searson, L. A. Lyon, C. Keating (Materials Research Society Press, Pittsburgh 2001) **635** C4301–C4306

4.149 Y.-M. Lin, M. S. Dresselhaus: Thermoelectric properties of superlattice nanowires, Phys. Rev. B **68** (2003) 075304 (1–14)

4.150 M. T. Björk, B. J. Ohlsson, C. Thelander, A. I. Persson, K. Deppert, L. R. Wallenberg, L. Samuelson: Nanowire resonant tunneling diodes, Appl. Phys. Lett. **81** (2002) 4458–4460

4.151 D. Li: Thermal transport in individual nanowires and nanotubes. Ph.D. Thesis (Univ. California, Berkeley 2002)

4.152 T. S. Tighe, J. M. Worlock, M. L. Roukes: Direct thermal conductance measurements on suspended monocrystalline nanostructures, Appl. Phys. Lett. **70** (1997) 2687–2689

4.153 S. T. Huxtable, A. R. Abramson, C.-L. Tien, A. Majumdar, C. LaBounty, X. Fan, G. Zeng, J. E. Bowers, A. Shakouri, E. T. Croke: Thermal conductivity of Si/SiGe and SiGe/SiGe superlattices, Appl. Phys. Lett. **80** (2002) 1737–1739

4.154 R. Venkatasubramanian, E. Siivola, T. Colpitts, B. O'Quinn: Thin-film thermoelectric devices with high room-temperature figures of merit, Nature **413** (2001) 597–602

4.155 C. Dames, G. Chen: Modeling the thermal conductivity of a SiGe segmented nanowire, 21st Int. Conf. Thermoelectrics: Proc. ICT Symposium, Long Beach 2002 (IEEE, Piscataway 2002) 317–320
4.156 G. Chen, M. S. Dresselhaus, G. Dresselhaus, J.-P. Fluerial, T. Caillat: Recent developments in thermoelectric materials, Int. Mater. Rev. **48** (2003) 45–66
4.157 K. Schwab, E. A. Henriksen, J. M. Worlock, M. L. Roukes: Measurement of the quantum of thermal conductance, Nature **404** (2000) 974–977
4.158 K. Schwab, J. L. Arlett, J. M. Worlock, M. L. Roukes: Thermal conductance through discrete quantum channels, Physica E **9** (2001) 60–68
4.159 M. Cardona: *Light scattering in solids* (Springer, Berlin, Heidelberg 1982)
4.160 P. Y. Yu, M. Cardona: *Fundamentals of Semiconductors* (Springer, Berlin, Heidelberg 1995) Chap. 7
4.161 J. C. M. Garnett: Colours in metal glasses, in metallic films, and in metallic solutions, Philos. Trans. Roy. Soc. London A **205** (1906) 237–288
4.162 D. E. Aspnes: Optical properties of thin films, Thin Solid Films **89** (1982) 249–262
4.163 U. Kreibig, L. Genzel: Optical absorption of small metallic particles, Surf. Sci. **156** (1985) 678–700
4.164 M. R. Black, Y.-M. Lin, S. B. Cronin, O. Rabin, M. S. Dresselhaus: Infrared absorption in bismuth nanowires resulting from quantum confinement, Phys. Rev. B **65** (2002) 195417(1–9)
4.165 M. W. Lee, H. Z. Twu, C.-C. Chen, C.-H. Chen: Optical characterization of wurtzite gallium nitride nanowires, Appl. Phys. Lett. **79** (2001) 3693–3695
4.166 D. M. Lyons, K. M. Ryan, M. A. Morris, J. D. Holmes: Tailoring the optical properties of silicon nanowire arrays through strain, Nano. Lett. **2** (2002) 811–816
4.167 J. C. Johnson, H. Yan, R. D. Schaller, L. H. Haber, R. J. Saykally, P. Yang: Single nanowire lasers, J. Phys. Chem. B **105** (2001) 11387–11390
4.168 S. Blom, L. Y. Gorelik, M. Jonson, R. I. Shekhter, A. G. Scherbakov, E. N. Bogachek, U. Landman: Magneto-optics of electronic transport in nanowires, Phys. Rev. B **58** (1998) 16305–16314
4.169 J. P. Pierce, E. W. Plummer, J. Shen: Ferromagnetism in cobalt-iron alloy nanowire arrays on W(110), Appl. Phys. Lett. **81** (2002) 1890–1892
4.170 J. C. Johnson, H. Yan, R. D. Schaller, P. B. Petersen, P. Yang, R. J. Saykally: Near-field imaging of nonlinear optical mixing in single zinc oxide nanowires, Nano. Lett. **2** (2002) 279–283
4.171 M. R. Black, Y.-M. Lin, S. B. Cronin, M. S. Dresselhaus: Using optical measurements to improve electronic models of bismuth nanowires, 21st Int. Conf. Thermoelectrics: Proc. ICT Symposium, Long Beach 2002 (IEEE, Piscataway 2002) 253–256
4.172 H.-L. Liu, C.-C. Chen, C.-T. Chia, C.-C. Yeh, C.-H. Chen, M.-Y. Yu, S. Keller, S. P. DenBaars: Infrared and Raman-scattering studies in single-crystalline GaN nanowires, Chem. Phys. Letts. **345** (2001) 245–251
4.173 H. Richter, Z. P. Wang, L. Ley: The one phonon Raman-spectrum in microcrystalline silicon, Solid State Commun. **39** (1981) 625–629
4.174 I. H. Campbell, P. M. Fauchet: The effects of microcrystal size and shape on the one phonon Raman-spectra of crystalline semiconductors, Solid State Commun. **58** (1986) 739–741
4.175 R. Gupta, Q. Xiong, C. K. Adu, U. J. Kim, P. C. Eklund and: Laser-induced Fanoresonance scattering in silicon nanowires, Nano. Lett. **3** (2003) 627–631
4.176 Y. Cui, C. M. Lieber: Functional nanoscale electronic devices assembled using silicon nanowire building blocks, Science **291** (2001) 851–853
4.177 S.-W. Chung, J.-Y. Yu, J. R. Heath: Silicon nanowire devices, Appl. Phys. Lett. **76** (2000) 2068–2070
4.178 M. Ding, H. Kim, A. I. Akinwande: Observation of valence band electron emission from n-type silicon field emitter arrays, Appl. Phys. Lett. **75** (1999) 823–825
4.179 F. C. K. Au, K. W. Wong, Y. H. Tang, Y. F. Zhang, I. Bello, S. T. Lee: Electron field emission from silicon nanowires, Appl. Phys. Lett. **75** (1999) 1700–1702
4.180 P. M. Ajayan, O. Z. Zhou: Applications of carbon nanotubes. In: *Carbon Nanotubes: Synthesis, Structure, Properties and Applications*, Vol. 80, ed. by M. S. Dresselhaus, G. Dresselhaus, P. Avouris (Springer, Berlin, Heidelberg 2001) pp. 391–425
4.181 G. Dresselhaus, M. S. Dresselhaus, Z. Zhang, X. Sun, J. Ying, G. Chen: Modeling thermoelectric behavior in Bi nano-wires, Seventeenth Int. Conf. Thermoelectrics: Proc. ICT'98, Nagoya 1998, ed. by K. Koumoto (Institute of Electrical and Electronics Engineers Inc, Piscataway 1998) 43–46
4.182 L. D. Hicks, M. S. Dresselhaus: The effect of quantum well structures on the thermoelectric figure of merit, Phys. Rev. B **47** (1993) 12727–12731
4.183 O. Rabin, Y.-M. Lin, M. S. Dresselhaus: Anomalously high thermoelectric figure of merit in $Bi_{1-x}Sb_x$ nanowires by carrier pocket alignment, Appl. Phys. Lett. **79** (2001) 81–83
4.184 T. C. Harman, P. J. Taylor, M. P. Walsh, B. E. LaForge: Quantum dot supertattice thermoelectric materials and devices, Science **297** (2002) 2229–2232
4.185 V. Dneprovskii, E. Zhukov, V. Karavanskii, V. Poborchii, I. Salamatini: Nonlinear optical properties of semiconductor quantum wires, Superlattices Microstruct. **23** (1998) 1217–1221
4.186 J. C. Johnson, H.-J. Choi, K. P. Knutsen, R. D. Schaller, P. Yang, R. J. Saykally: Single gallium nitride nanowire lasers, Nature Mater. **1** (2002) 106–110
4.187 H. Kind, H. Yan, B. Messer, M. Law, P. Yang: Nanowire ultraviolet photodetectors and optical switches, Adv. Mat. **14** (2002) 158–160

4.188 B. M. I. van der Zande, M. Böhmer, L. G. J. Fokkink, C. Schönenberger: Colloidal dispersions of gold rods: Synthesis and optical properties, Langmuir **16** (2000) 451–458
4.189 B. M. I. van der Zande, G. J. M. Koper, H. N. W. Lekkerkerker: Alignment of rod-shaped gold particles by electric fields, J. Phys. Chem. B **103** (1999) 5754–5760
4.190 W. U. Huynh, J. J. Dittmer, A. P. Alivisatos: Hybrid nanorod-polymer solar cells, Science **295** (2002) 2425–2427
4.191 Y. Cui, Q. Wei, H. Park, C. Lieber: Nanowire nanosensors for highly sensitive and selective detection of biological and chemical species, Science **293** (2001) 1289–1292

5. Introduction to Micro/Nanofabrication

In this chapter, we discuss various *micro/nanofabrication* techniques used to manufacture structures in a wide range of dimensions (mm–nm). Starting with some of the most common microfabrication techniques (lithography, deposition, and etching), we present an array of *micromachining* and *MEMS* technologies that can be used to fabricate microstructures down to ~1 μm. These techniques have attained an adequate level of maturity to allow for a variety of MEMS-based commercial products (pressure sensors, accelerometers, gyroscopes, etc.). More recently, nanometer-sized structures have attracted an enormous amount of interest. This is mainly due to their unique electrical, magnetic, optical, thermal, and mechanical properties. These could lead to a variety of electronic, photonic, and sensing devices with a superior performance compared to their macro counterparts. Subsequent to our discussion on MEMS and micromachining, we present several important *nanofabrication* techniques currently under intense investigation. Although e-beam and other high resolution lithographies can be used to fabricate nanometer-sized structures, their serial nature and/or cost preclude their widespread application. This has forced investigators to explore alternative and potentially superior techniques such as *strain engineering*, *self-assembly*, and *nano-imprint* lithography. Among these, self-assembly is the most promising method, due to its low cost and the ability to produce nanostructures at different length scales.

5.1 **Basic Microfabrication Techniques** 148
5.1.1 Lithography 148
5.1.2 Thin Film Deposition and Doping ... 149
5.1.3 Etching and Substrate Removal 153
5.1.4 Substrate Bonding 157

5.2 **MEMS Fabrication Techniques** 159
5.2.1 Bulk Micromachining 159
5.2.2 Surface Micromachining 163
5.2.3 High-Aspect-Ratio Micromachining 166

5.3 **Nanofabrication Techniques** 170
5.3.1 E-Beam and Nano-Imprint Fabrication 171
5.3.2 Epitaxy and Strain Engineering 172
5.3.3 Scanned Probe Techniques 173
5.3.4 Self-Assembly and Template Manufacturing 176

References 180

Recent innovations in the area of *micro/nanofabrication* have created a unique opportunity for manufacturing structures in the nanometer–millimeter range. The available six orders of magnitude dimensional span can be used to fabricate novel electronic, optical, magnetic, mechanical, and chemical/biological devices with applications ranging from sensors to computation and control. In this chapter, we will introduce major micro/nanofabrication techniques currently used to fabricate structures in the nanometer to several hundred micrometer range. We will focus mainly on the most important and widely used techniques and will not discuss specialized methods. After a brief introduction to basic microfabrication, we will discuss *MEMS*-fabrication techniques used to build microstructures down to about 1 μm in dimensions. Following this, we will discuss several major top-down and bottom-up nanofabrication methods that have shown great promise in manufacturing *nanostructures* (dimensions $< 1\,\mu m$).

5.1 Basic Microfabrication Techniques

Most micro/nanofabrication techniques have their roots in the standard fabrication methods developed for the semiconductor industry [5.1–3]. Therefore, a clear understanding of these techniques is necessary for anyone embarking on a research and development path in the micro/nano area. In this section, we will discuss major microfabrication techniques used most frequently in the manufacturing of micro/nanostructures. Some of these techniques such as thin-film deposition and etching are common between the micro/nano and VLSI microchip fabrication disciplines. However, several other techniques that are more specific to the micro/nanofabrication area will also be discussed in this section.

5.1.1 Lithography

Lithography is the technique used to transfer a computer generated pattern onto a substrate (silicon, glass, GaAs, etc.). This pattern is subsequently used to etch an underlying thin film (oxide, nitride, etc.) for various purposes (doping, etching, etc.). Although photolithography, i. e., the lithography using a UV light source, is by far the most widely used lithography technique in microelectronic fabrication, electron-beam (e-beam) and X-ray lithography are two other alternatives that have attracted considerable attention in the MEMS and nanofabrication areas. We will discuss photolithography in this section and postpone the discussion of e-beam and X-ray techniques to the subsequent sections dealing with MEMS and nanofabrication.

The starting point following the creation of the computer layout for a specific fabrication sequence is the generation of a photomask. This involves a sequence of photographic processes (using optical or e-beam pattern generators) that results in a glass plate having the desired pattern in the form of a thin ($\sim$ 100 nm) chromium layer. Following the generation of photomask, the lithography process can proceed as shown in Fig. 5.1. This sequence demonstrates the pattern transfer onto a substrate coated with silicon dioxide. However, the same technique is applicable to other materials. After depositing the desired material on the substrate, the photolithography process starts with spin coating the substrate with a photoresist. This is a polymeric photosensitive material that can be spun onto the wafer in liquid form (usually an adhesion promoter such as hexamethyldisilazane HMDS is used prior to the application of the resist). The spinning speed and photoresist viscosity will determine the final resist thickness, which is typically between 0.5–2.5 μm. Two different kinds of photoresist are available: positive and negative. In the positive resist, the UV-exposed areas will be dissolved in the subsequent development stage, whereas in the negative photoresist, the exposed areas will remain intact after the development. Due to its better performance with regard to the process control in small geometries, the positive resist is the most extensively used photoresist in the VLSI processes. After spinning the photoresist on the wafer, the substrate is soft baked (5–30 min at 60–100 °C) in order to remove the solvents from the resist and improve the adhesion. Subsequently, the mask is aligned to the wafer and the photoresist is exposed to a UV source.

Depending on the separation between the mask and the wafer, three different exposure systems are available: 1) contact, 2) proximity, and 3) projection. Although contact printing gives better resolution compared to the proximity technique, the constant contact of the mask with photoresist reduces the process yield and can damage the mask. Projection printing uses a dual-lens optical system to project the mask image onto the wafer. Since only one die at a time can be exposed, this requires a step-and-repeat system to totally cover the wafer area. Projection printing is by far the most widely used system in microfabrication and can yield superior resolutions compared to contact and proximity methods. The exposure source for photolithography depends on the resolution. Above 0.25 μm minimum line width, high pressure mercury lamp is adequate (436 nm g-line and 365 nm i-line). However, between 0.25 and 0.13 μm, deep UV sources such as excimer lasers (248 nm KrF and 193 nm ArF) are required. Although there has been extensive competition for the below 0.13 μm regime (including e-beam and X-ray), extreme UV (EUV) with

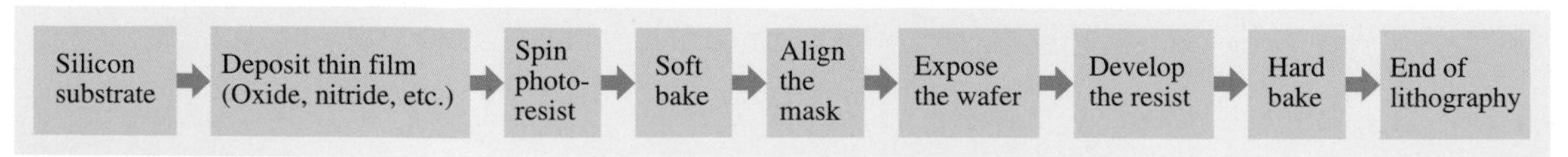

Fig. 5.1 Lithography process flow

a wavelength of 10–14 nm seems to be the preferred technique [5.4].

After exposure, the photoresist is developed in a process similar to the development of photographic film. The resist is subsequently hard baked (20–30 min at 120–180 °C) in order to further improve the adhesion. The hard bake step concludes the photolithography sequence by creating the desired pattern on the wafer. Next, the underlying thin film is etched, and the photoresist is stripped in acetone or other organic removal solvents. Figure 5.2 shows schematic drawing of photolithography steps with a positive photoresist.

5.1.2 Thin Film Deposition and Doping

Thin-film deposition and doping are used extensively in micro/nanofabrication technologies. Most of the fabricated micro/nanostructures contain materials other than that of the substrate, which are obtained by various deposition techniques, or by modification of the substrate. The following is a list of a few typical applications for the deposited and/or doped materials used in micro/nanofabrication that gives an idea of the required properties:

- Mechanical structure
- Electrical isolation
- Electrical connection
- Sensing or actuating
- Mask for etching and doping
- Support or mold during deposition of other materials (sacrificial materials)
- Passivation

Most of the deposited thin films have different properties than those of their corresponding "bulk" forms (for example, metals show higher resistivities as thin films). In addition, the techniques utilized to deposit these materials have a great impact on their final properties. For instance, internal stress (compressive or tensile) in a film is strongly process-dependent. Excessive stress may crack or detach the film from the substrate and therefore must be minimized, although it may also be useful for certain applications. Adhesion is another important issue that needs to be taken into account when depositing thin films. In some cases such as the deposition of noble metals (e.g., gold) an intermediate layer (chromium or titanium) may be needed to improve the adhesion. Finally, step coverage and conformality are two properties that can also influence the choice of deposition technique. Figure 5.3 illustrates these concepts.

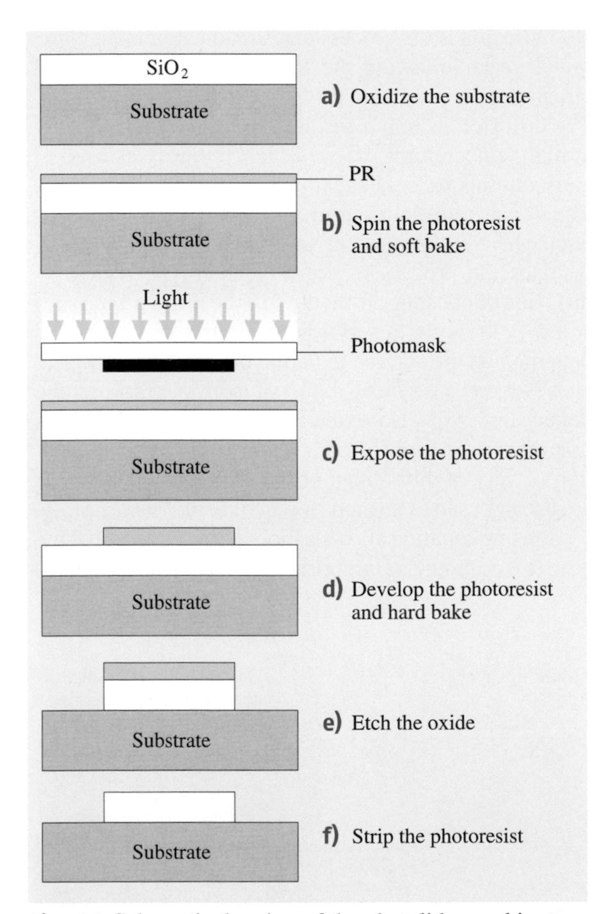

Fig. 5.2 Schematic drawing of the photolithographic steps with a positive photoresist (PR)

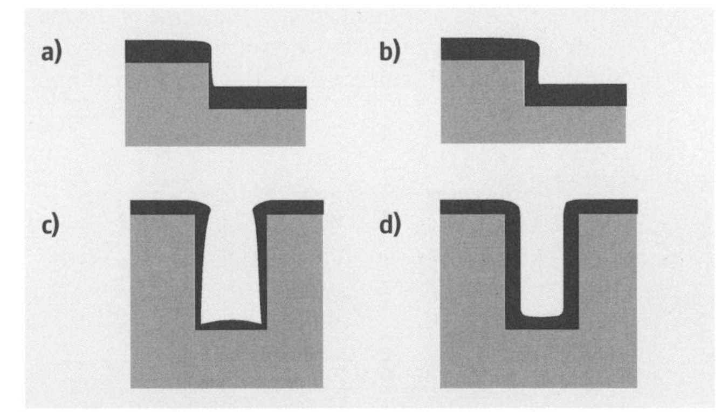

Fig. 5.3a–d Step coverage and conformality: (a) poor step coverage, (b) good step coverage, (c) nonconformal layer, and (d) conformal layer

Oxidation

Oxidation of silicon is a process used to obtain a thin film of SiO_2 of excellent quality (very low density of defects) and thickness homogeneity. Although it is not properly a deposition, the result is the same, i. e., a thin layer of a new material covering the surface is produced. The oxidation process is typically carried out at temperatures in the range of 900 °C to 1,200 °C in the presence of O_2 (dry oxidation) or H_2O (wet oxidation). The reactions for oxide formation are:

$$\mathrm{Si}_{(\mathrm{solid})} + \mathrm{O}_{2(\mathrm{gas})} \Rightarrow \mathrm{SiO}_{2(\mathrm{solid})}$$

and

$$\mathrm{Si}_{(\mathrm{solid})} + 2\mathrm{H}_2\mathrm{O}_{(\mathrm{steam})} \Rightarrow \mathrm{SiO}_{2(\mathrm{solid})} + 2\mathrm{H}_{2(\mathrm{gas})} \ .$$

Although the rate of the oxide growth is higher for wet oxidation, it is achieved at the expense of a lower oxide quality (density). Since silicon atoms from the substrate participate in the reaction, the substrate is consumed as the oxide grows ($\sim 44\%$ of the total thickness lies above the line of the original silicon surface). The oxidation of silicon also occurs at room temperature, however, a layer of about 20 Å (native oxide) is enough to passivate the surface and prevent further oxidation. To grow thicker oxides, wafers are introduced into an electric resistance furnace such as the one represented in Fig. 5.4. Tens of wafers can be processed in a single batch in such equipment. By strictly controlling the timing, temperature, and gas flow entering the quartz tube, the desired thickness can be achieved with a high accuracy. Thicknesses ranging from a few tens of Angstroms to 2 μm can be obtained in reasonable times. Despite the good quality of the SiO_2 obtained by silicon oxidation (also called *thermal oxide*), the use of this process is often limited to the early stages of fabrication, since some of the materials added during the formation of structures may not withstand the high temperatures. The contamination of the furnace when the substrates have been previously in contact with certain etchants such as KOH, or when materials such as metals have been deposited also pose limitations in most of the cases.

Doping

The introduction of certain impurities in a semiconductor can change its electrical, chemical, and even mechanical properties. Typical impurities, or *dopants*, used in silicon include boron (to form p-type regions) and phosphorous or arsenic (to form n-type regions). Doping is the main process used in the microelectronic industry to fabricate major components such as diodes and transistors. In micro/nanofabrication technologies doping has additional applications such as the formation of piezoresistors for mechanical transducers and the creation of etch stop layers. Two different techniques are used to introduce the impurities into a semiconductor substrate: diffusion and ion implantation.

Diffusion is the process that became dominant in the initial years following the invention of the integrated circuit to form n- and p-type regions in the silicon. The diffusion of impurities into the silicon occurs only at high temperatures (above 800 °C). Furnaces used to carry out this process are similar to the oxidation ones. The dopants are introduced into the furnace's gaseous atmosphere from liquid or solid sources. Figure 5.5 illustrates the process of creating an n-type region by diffusion of phosphor from the surface into a p-type substrate. A masking material is previously deposited and patterned on the surface to define the areas to be doped. However, because diffusion is an isotropic process, the doped area will also extend underneath the mask. In microfabrication, diffusion is mainly used for the formation of very highly doped boron regions (p^{++}), which are usually used as an etch stop in bulk *micromachining*.

Ion implantation allows a more precise control of the dose (total amount of impurities introduced per area unit)

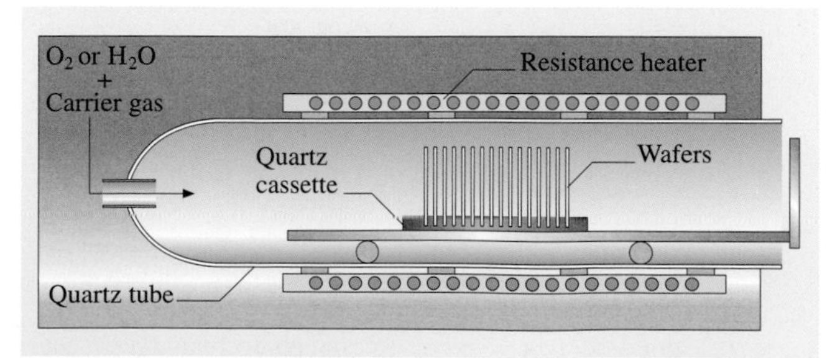

Fig. 5.4 Schematic representation of a typical oxidation furnace

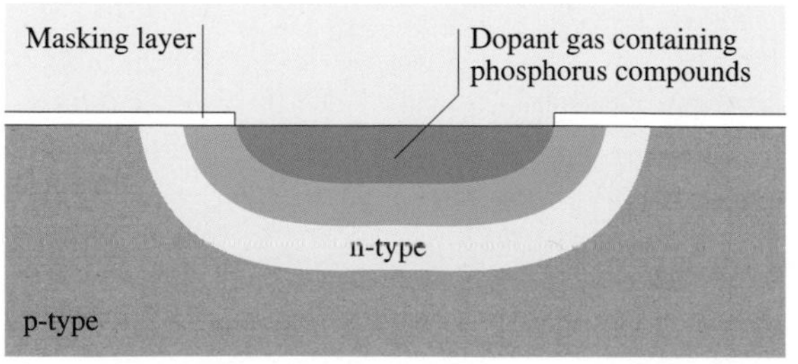

Fig. 5.5 Formation of an n-type region on a p-type silicon substrate by diffusion of phosphorous

and the impurity profile (concentration versus depth). In ion implantation, the impurities are ionized and accelerated toward the semiconductor surface. The penetration of impurities in the material follows a Gaussian distribution. After implantation, an annealing process is needed to activate the impurities and repair the damage in the crystal structure produced by the ion collisions. A *drive-in* process to redistribute the impurities performed in a standard furnace such as the ones used for oxidation or diffusion may be required as well.

Chemical Vapor Deposition and Epitaxy

As its name suggests, chemical vapor deposition (CVD) includes all the deposition techniques using the reaction of chemicals in a gas phase to form the deposited thin film. The energy needed for the chemical reaction to occur is usually supplied by maintaining the substrate at elevated temperatures. Other alternative energy sources such as plasma or optical excitation are also used, with the advantage of requiring a lower temperature at the substrate. The most common CVD processes in microfabrication are LPCVD (low pressure CVD) and PECVD (plasma enhanced CVD).

The LPCVD process is typically carried out in electrically heated tubes, similar to the oxidation tubes, equipped with pumping capabilities to achieve the needed low pressures (0.1 to 1.0 torr). A large number of wafers can be processed simultaneously, and the material is deposited in both sides of the wafers. The process temperatures depend on the material to be deposited, but generally are in the range of 550 to 900 °C. As in the oxidation, high temperatures and contamination issues can restrict the type of processes used prior to the LPCVD. Typical materials deposited by LPCVD include silicon oxide (e.g., $SiCl_2H_2 + 2N_2O \Rightarrow SiO_2 + 2N_2 + 2HCl$ at 900 °C), silicon nitride (e.g., $3SiH_4 + 4NH_3 \Rightarrow Si_3N_4 + 12H_2$ at 700–900 °C), and polysilicon (e.g., $SiH_4 \Rightarrow Si + 2H_2$ at 600 °C). Due to its faster etch rate in HF, in situ phosphorous-doped LPCVD oxide (phosphosilicate glass or PSG) is used extensively as the sacrificial layer in surface micromachining. Conformality in this process is excellent, even for very high-aspect-ratio structures. Mechanical properties of LPCVD materials are good compared to others such as PECVD, and LPCVD materials often used as structural materials in microfabricated devices. Stress in the deposited layers depends on the material, deposition conditions, and subsequent thermal history (e.g., post-deposition anneal). Typical values are: 100–300 MPa (compressive) for oxide, ∼ 1 GPa (tensile) for stoichiometric nitride, and ∼ 200–300 MPa (tensile) for polysilicon. The stress in nitride layers can be reduced to nearly zero by using a silicon-rich composition. Since the stress values can vary over a wide range, one has to measure and characterize the internal stress of deposited thin films for any specific equipment and deposition conditions.

The PECVD process is performed in plasma systems such as the one represented in Fig. 5.6. The use of RF energy to create highly reactive species in the plasma allows for the use of lower temperatures at the substrates (150 to 350 °C). Parallel-plate plasma reactors normally used in microfabrication can only process a limited number of wafers per batch. The wafers are positioned horizontally on top of the lower electrode, so only one side gets deposited. Typical materials deposited with PECVD include silicon oxide, nitride, and amorphous silicon. Conformality is good for low-aspect-ratio structures, but becomes very poor for deep trenches (20% of the surface thickness inside through-wafer holes with aspect ratio of ten). Stress depends on deposition parameters and can be either compressive or tensile. PECVD nitrides are typically non-stoichiometric (Si_xN_y) and are much less resistant to etchants in masking applications.

Another interesting type of CVD is epitaxial growth. In this process, a single crystalline material is grown as an extension of the crystal structure of the substrate. It is possible to grow dissimilar materials if the crystal structures are somehow similar (lattice-matched). Silicon-on-sapphire (SOS) substrates and some heterostructures are fabricated in this way. However, most

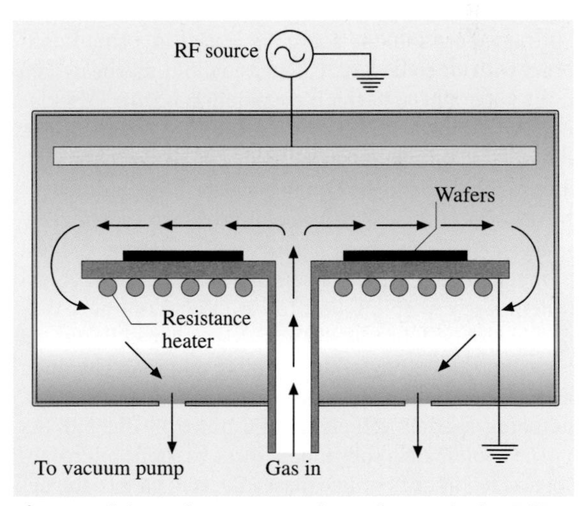

Fig. 5.6 Schematic representation of a typical PECVD system

common in microfabrication is the growth of silicon on another silicon substrate. Of particular interest for the formation of microstructures is the selective epitaxial growth. In this process the silicon crystal is allowed to grow only in windows patterned on a masking material. Many CVD techniques have been used to produce epitaxial growth. The most common for silicon is thermal chemical vapor deposition or *vapor phase epitaxy* (VPE). *Metallorganic chemical vapor deposition* (MOCVD) and *molecular beam epitaxy* (MBE) are the most common for growing high quality III-V compound layers with nearly atomic abrupt interfaces. The former uses vapors of organic compounds with group III atoms such as trimethylgallium [$Ga(CH_3)_3$] and group V hydrides such as AsH_3 in a CVD chamber with fast gas switching capabilities. The latter typically uses molecular beams from thermally evaporated elemental sources aiming at the substrate in an ultrahigh vacuum chamber. In this case, rapid on/off control of the beams is achieved by using shutters in front of the sources. Finally, it should be mentioned that many metals (molybdenum, tantalum, titanium, and tungsten) can also be deposited using LPCVD. These are attractive for their low resistivities and their ability to form silicides with silicon. Due to its application in the new interconnect technologies, copper CVD is an active area of research.

Physical Vapor Deposition (Evaporation and Sputtering)

In physical deposition systems the material to be deposited is transported from a source to the wafers, both being in the same chamber. Two physical principles are used to do so: evaporation and sputtering.

In evaporation, the source is placed in a small container with tapered walls, called a crucible, and heated up to a temperature at which evaporation occurs. Various techniques are utilized to reach the high temperatures needed, including the induction of high currents with coils wound around the crucible and the bombardment of the material surface with an electron beam (e-beam evaporators). This process is mainly used to deposit metals, although dielectrics can also be evaporated. In a typical system, the crucible is located at the bottom of a vacuum chamber, whereas the wafers are placed lining the dome-shaped ceiling of the chamber, Fig. 5.7. The main characteristic of this process is very poor step coverage, including shadow effects, as illustrated in Fig. 5.8. As will be explained in subsequent sections, some microfabrication techniques utilize these effects to pattern the deposited layer. One way to improve the step coverage is by rotating and/or heating the wafers during the deposition.

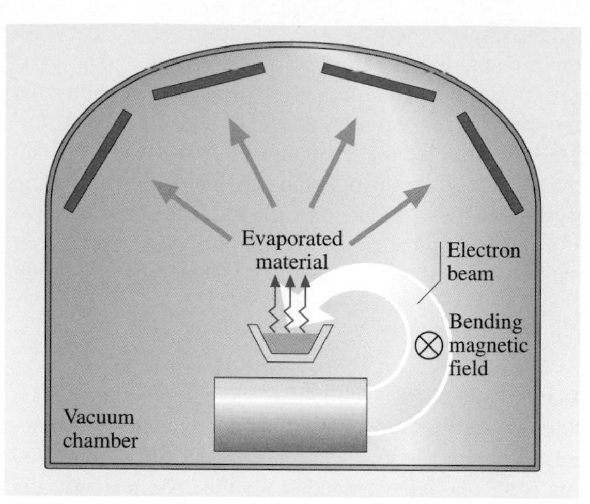

Fig. 5.7 Schematic representation of an electron-beam deposition system

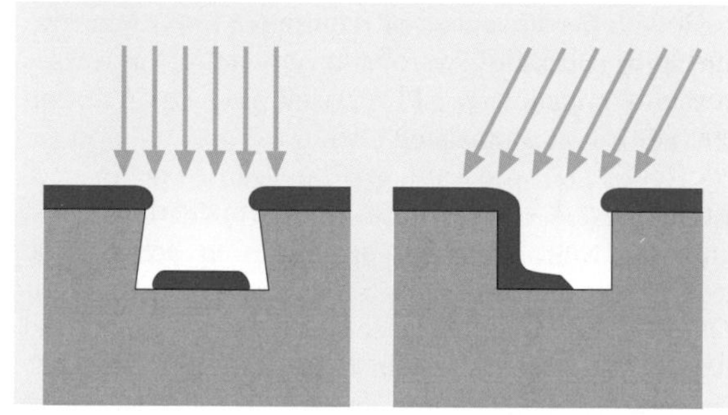

Fig. 5.8 Shadow effects observed in evaporated films. *Arrows* show the trajectory of the material atoms being deposited

In sputtering, a target of the material to be deposited is bombarded with high-energy inert ions (usually argon). The outcome of the bombardment is that individual atoms or clusters are removed from the surface and ejected toward the wafer. The physical nature of this process allows its use with virtually any existing material. Examples of interesting materials for microfabrication that are frequently sputtered include metals, dielectrics, alloys (such as shape memory alloys), and all kinds of compounds (for example, piezoelectric PZT). The inert ions bombarding the target are produced in DC or RF plasma. In a simple parallel-plate system, the top electrode is the target and the wafers are placed horizontally on top of the bottom electrode. In spite of its lower deposition rate, step coverage in sputtering is much better than in evaporation. Yet, the films obtained

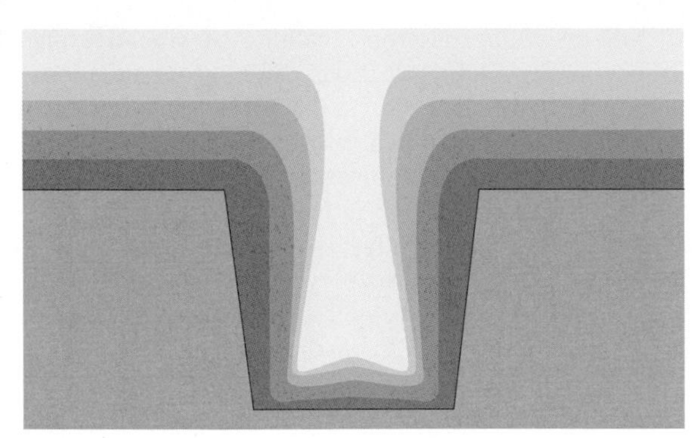

Fig. 5.9 Typical cross section evolution of a trench while being filled with sputter deposition

with this deposition process are nonconformal. Figure 5.9 illustrates successive sputtering profiles in a trench.

Both evaporation and sputtering systems are often capable of depositing more than one material simultaneously or sequentially. This capability is very useful in obtaining alloys and multilayers (e.g., multilayer magnetic recording heads are sputtered). For certain low reactivity metals such as Au and Pt, the previous deposition of a thin layer of another metal is needed to improve the adhesion. Ti and Cr are two frequently used adhesion promoters. Stress in evaporated or sputtered layers is typically tensile. The deposition rates are much higher than most CVD techniques. However, due to stress accumulation and cracking, a thickness beyond 2 μm is rarely deposited with these processes. For thicker depositions a technique described in the next section is sometimes used.

Electroplating

Electroplating (or electrodeposition) is a process typically used to obtain thick (tens of micrometers) metal structures. The sample to be electroplated is introduced in a solution containing a reducible form of the ion of the desired metal and is maintained at a negative potential (cathode) relative to a counter electrode (anode). The ions are reduced at the sample surface and the insoluble metal atoms are incorporated into the surface. As an example, copper electrodeposition is frequently done in copper sulfide-based solutions. The reaction taking place at the surface is $Cu^{2+} + 2e^- \rightarrow Cu_{(s)}$. Recommended current densities for electrodeposition processes are on the order of 5 to 100 mA/cm^2.

As can be deduced from the process mechanism, the surface to be electroplated has to be electrically conductive and preferably of the same material as the deposited one if a good adhesion is desired. In order to electrodeposit metals on top of an insulator (the most common case) a thin film of the same metal, called the seed layer, is previously deposited on the surface. Masking of the seed layer with a resist permits selective electroplatingat the patterned areas. Figure 5.10 illustrates a typical sequence of the steps required to obtain isolated metal structures.

5.1.3 Etching and Substrate Removal

Thin film and bulk substrate etching is another fabrication step that is of fundamental importance to both VLSI processes and micro/nanofabrication. In the VLSI area, various conducting and dielectric thin films deposited for passivation or masking purposes must be removed at some point or another. In micro/nanofabrication, in addition to thin film etching, very often the substrate (silicon, glass, GaAs, etc.) also needs to be removed in order to create various mechanical micro-/nanostructures (beams, plates, etc.). Two important figures of merit for any etching process are selectivity and directionality. Selectivity is the degree to which the etchant can differentiate between the masking layer and the layer to be

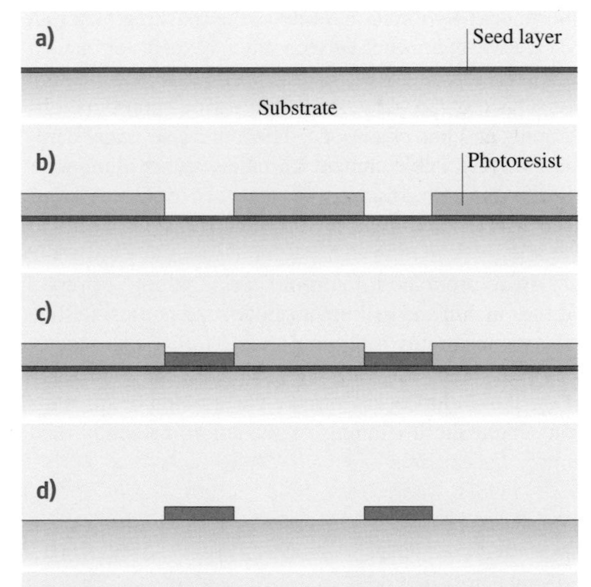

Fig. 5.10a–d Formation of isolated metal structures by electroplating through a mask: (**a**) seed layer deposition, (**b**) photoresist spinning and patterning, (**c**) electroplating, (**d**) photoresist and seed layer stripping

etched. Directionality has to do with the etch profile under the mask. In an isotropic etch, the etchant attacks the material in all directions at the same rate, creating a semicircular profile under the mask, Fig. 5.11a. In an anisotropic etch, the dissolution rate depends on specific directions, and one can obtain straight sidewalls or other noncircular profiles, Fig. 5.11b. One can also divide the various etching techniques into wet and dry categories. In this chapter, we will use this classification and discuss different wet etchants, followed by the dry etching techniques used most often in micro/nanofabrication.

Wet Etching

Historically, wet etching techniques preceded the dry ones. These still constitute an important group of etchants for micro/nanofabrication, in spite of their less frequent application in the VLSI processes. Wet etchants are by and large isotropic and show superior selectivity to the masking layer compared to various dry techniques. In addition, due to the lateral undercut, the minimum feature achievable with wet etchants is limited to $> 3\,\mu m$. Silicon dioxide is commonly etched in a dilute (6:1, 10:1, or 20:1 by volume) or buffered HF (BHF, $HF + NH_4F$) solutions (etch rate of $\sim 1{,}000\,Å/min$ in BHF). Photoresist and silicon nitride are the two most common masking materials for the wet oxide etch. The wet etchant for silicon nitride is hot (140–200 °C) phosphoric acid with silicon oxide as the masking material. Nitride wet etch is not very common (except for blanket etch), due to the masking difficulty and unrepeatable etch rates. Metals can be etched using various combinations of acids and base solutions. There are also many commercially available etchant formulations for aluminum, chromium, and gold that can be used easily. A comprehensive table of various metal etchants can be found in [5.5].

Anisotropic and isotropic wet etching of crystalline (silicon and gallium arsenide) and non-crystalline (glass) substrates are important topics in micro/nanofabrication [5.6–10]. In particular, the realization of the possibility of anisotropic wet etching of silicon is considered the beginning of the micromachining and MEMS discipline. Isotropic etching of silicon using $HF/HNO_3/CH_3COOH$ (various different formulations have been used) dates back to the 1950s and is still frequently used to thin down the silicon wafer. The etch mechanism for this combination has been elucidated and is as follows: HNO_3 is used to oxidize the silicon, which is subsequently dissolved away in the HF. The acetic acid is used to prevent the dissociation of HNO_3 (the etch works as well without the acetic acid). For short etch times, silicon dioxide can be used as the masking material However, one needs to use silicon nitride if a longer etch time is desired. This etch also shows dopant selectivity with the etch rate dropping at lower doping concentrations ($< 10^{17}\,cm^{-3}$ n- or p-type). Although this effect can potentially be used as an etch stop mechanism in order to fabricate microstructures, the difficulty in masking has prevented a widespread application of this approach. Glass can also be isotropically etched using the HF/HNO_3 combination with the etch surfaces showing considerable roughness. This has been used extensively in fabricating microfluidic components (mainly channels). Although Cr/Au is usually used as the masking layer, long etch times require a more robust mask (bonded silicon has been used for this purpose).

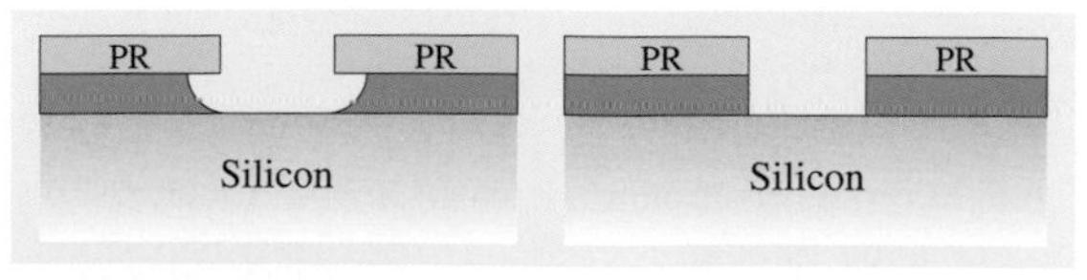

Fig. 5.11a,b Profile for isotropic (**a**) and anisotropic (**b**) etch through a photoresist (PR) mask

Silicon anisotropic wet etch constitutes an important technique in bulk micromachining. The three most important silicon etchants in this category are potassium hydroxide (KOH), ethylene diamine pyrochatechol (EDP), and tetramethyl ammonium hydroxide (TMAH). These are all anisotropic etchants that attack silicon along preferred crystallographic directions. In addition, they all show a marked reduction in the etch rate in heavily doped ($> 5\times10^{19}\,cm^{-3}$) boron ($p^{++}$) regions. The chemistry behind the action of these etchants is not yet very clear, but it seems that silicon atom oxidation at the surface and its reaction with hydroxyl ions (OH^-) are responsible for the formation of a soluble silicon complex ($SiO_2(OH)^{2-}$). The etch rate depends on the concentration and temperature and is usually around $1\,\mu m/min$ at temperatures of 85–115 °C. Common masking materials for anisotropic wet etchants are silicon dioxide and nitride, the latter being superior for longer etch times. The crystallographic plane which shows the slowest etch rate is the (111) plane. Although the lower atomic concentration along these planes have been speculated as the reason for this phenomenon, the evidence is inconclusive, and other factors must be included to account for this remarkable etch stop property. The anisotropic behavior of these etchants with respect to the (111) planes have been used extensively to create

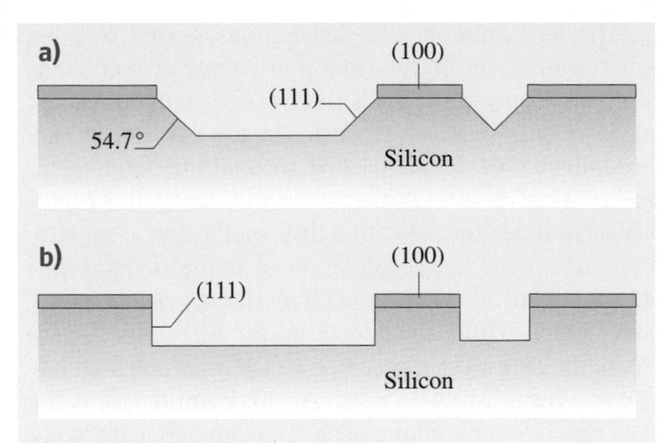

Fig. 5.12a,b Anisotropic etch profiles for: (a) (100) and (b) (110) silicon wafers

beams, membranes, and other mechanical and structural components. Figure 5.12 shows the typical cross sections of (100) and (110) silicon wafers etched with an anisotropic wet etchant. As can be seen, the (111) slow planes are exposed in both situations, one creating 54.7° sloped sidewalls in the (100) wafer and the other creating vertical sidewalls in the (110) wafer. Depending on the dimensions of the mask opening, a V-groove or a trapezoidal trench is formed in the (100) wafer. A large enough opening will allow the silicon to be etched all the way through the wafer, thus creating a thin dielectric membrane of the other side. It should be mentioned that exposed convex corners have a higher etch rate than the concave ones, resulting in an undercut that can be used to create dielectric (e.g., nitride) cantilever beams. Figure 5.13 shows a cantilever beam fabricated using the convex corner undercut on a (100) wafer.

The three above mentioned etchants show different directional and dopant selectivities. KOH has the best (111) selectivity (400/1) followed by TMAH and EDP. However, EDP has the highest selectivity with respect to the deep boron diffusion regions. Safety and CMOS compatibility are other important criteria for choosing a particular anisotropic etchant. Among the three mentioned etchants TMAH is the most benign, whereas EDP is extremely corrosive and carcinogenic. Silicon can be dissolved in TMAH in order to improve its selectivity with respect to aluminum. This property has made TMAH very appealing for post-CMOS micromachining for which aluminum lines have to be protected. Finally, it should be mentioned that one can modulate the etch rate using a reverse biased p–n junction (electrochemical etch stop). Figure 5.14 shows the setup commonly used to perform electrochemical etching. The silicon wafer under etch consists of an n-epi region on a p-type substrate. Upon the application of a reverse bias voltage to the structure (p-substrate is in contact with the solution and n-epi is protected using a water tight fixture), the p-substrate is etched away. When the n-epi regions are exposed to the solution, an oxide passivation layer is formed and etch is stopped. This technique can be used to fabricate single-crystalline silicon membranes for pressure sensors and other mechanical transducers.

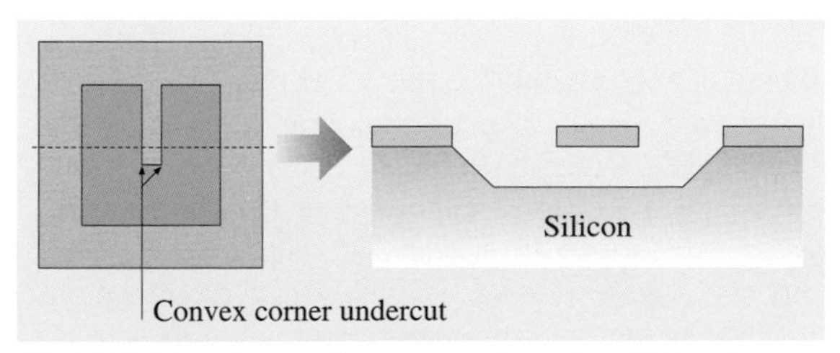

Fig. 5.13 Top view and cross section of a dielectric cantilever beam fabricated using convex corner undercut

Dry Etching

Most dry etching techniques are plasma-based. They have several advantages compared with wet etching. These include smaller undercut (allowing smaller lines to be patterned) and higher anisotropicity (allowing high-aspect-ratio vertical structures). However, the selectivity of dry etching techniques is lower than the wet etchants, and one must take into account the finite etch rate of the masking materials. The three basic dry etching techniques, namely, high-pressure plasma etching, reactive ion etching (RIE), and ion milling utilize different mechanisms to obtain directionality.

Ion milling is a purely physical process that utilizes accelerated inert ions (e.g., Ar^+) striking perpendic-

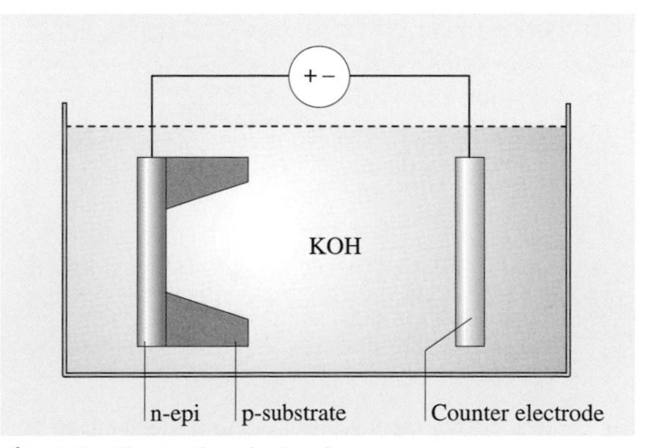

Fig. 5.14 Electrochemical etch setup

ular to the surface to remove the material (pressure $\sim 10^{-4}$–10^{-3} torr), Fig. 5.15a. The main characteristics of this technique are very low etch rates (on the order of a few nanometers per minute) and poor selectivity (close to 1:1 for most materials); hence it is generally used to etch very thin layers. In high-pressure (10^{-1}–5 torr) plasma etchers, highly reactive species are created that react with the material to be etched. The products of the reaction are volatile, so that they diffuse away and new material is exposed to the reactive species. Directionality can be achieved, if desired, with the sidewall passivation technique (Fig. 5.15b). In this technique, nonvolatile species produced in the chamber deposit and passivate the surfaces. The deposit can only be removed by physical collision with incident ions. Because the movement of the ions has a vertical directionality, the deposit is removed mainly at the horizontal surfaces, while the vertical walls remain passivated. In this fashion, the vertical etch rate becomes much higher than the lateral one.

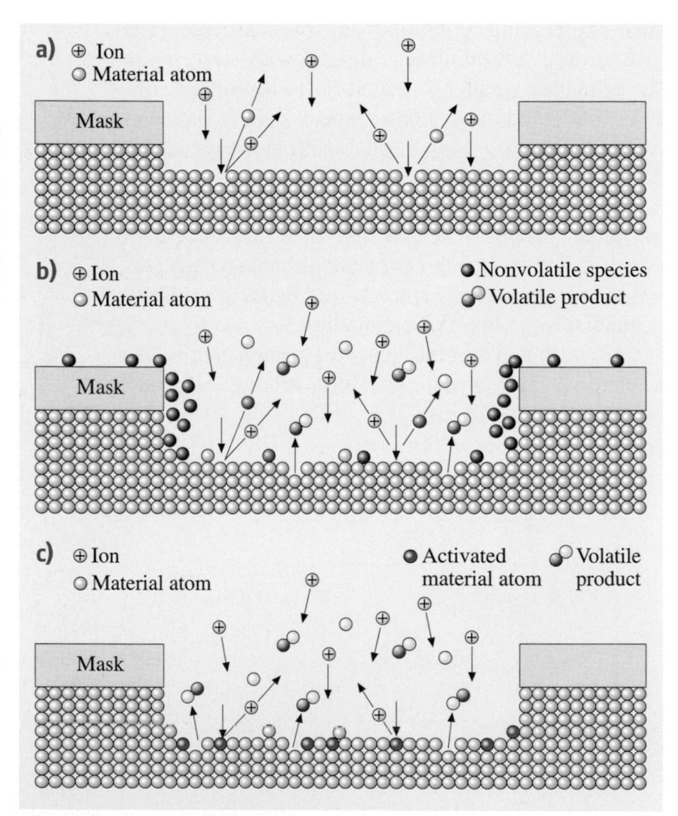

Fig. 5.15a–c Simplified representation of etching mechanisms for (**a**) ion milling, (**b**) high-pressure plasma etching, and (**c**) RIE

The RIE etching, also called ion-assisted etching, is a combination of physical and chemical processes. In this technique, the reactive species react with the material only when the surfaces are "activated" by the collision of incident ions from the plasma (e.g., by breaking bonds at the surface). As in the previous technique, the directionality of the ion's velocity produces many more collisions in the horizontal surfaces than in the walls, thus generating faster etching rates in the vertical direction (Fig. 5.15c). To further increase the etch anisotropy, sidewall passivation methods in some cases, are also used. An interesting case is the *deep reactive ion etching* (DRIE) technique, capable of achieving aspect ratios of 30:1 and silicon etching rates of 2 to 3 μm/min (through-wafer etch is possible). In this technique, the passivation deposition and etching steps are performed sequentially in a two-step cycle, as shown in Fig. 5.16. In commercial silicon DRIE etchers, SF_6/Ar is typically used for the etching step and a combination of Ar and a fluoropolymer ($n CF_2$) for the passivation step. A Teflon-like polymer about 50 nm thick is deposited during the latter step, covering only the sidewalls (Ar^+ ion bombardment removes the Teflon on the horizontal surfaces). Due to the cyclic nature of this process, the sidewalls of the etched features show a periodic "wave-shape" roughness in the range of 50 to 400 nm.

Dry etching can also be performed in non-plasma equipment if the etching gases are reactive enough. The so-called vapor phase etching (VPE) processes can be carried out in a simple chamber with gas feeding and pumping capabilities. Two examples of VPE are xenon difluoride (XeF_2) etching of silicon and HF-vapor etching of silicon dioxide. Due to its isotropic nature, these processes are typically used for etching sacrificial layers and releasing structures while avoiding stiction problems (see Sects. 5.2.1 and 5.2.2).

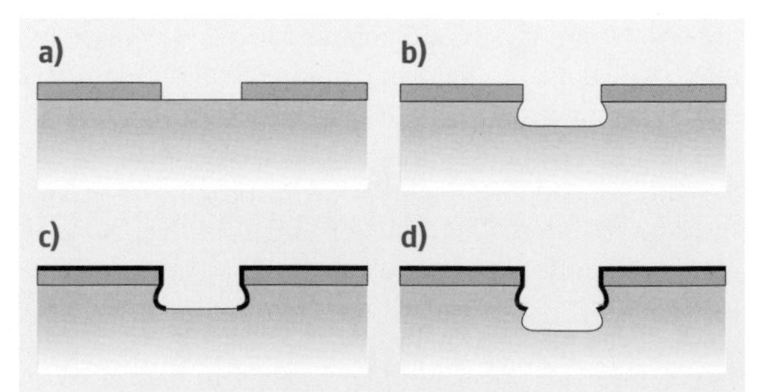

Fig. 5.16a–d DRIE cyclic process: (**a**) photoresist patterning, (**b**) etch step, (**c**) passivation step, and (**d**) etch step

Table 5.1 Typical dry etch chemistries

Si	CF_4/O_2, CF_2Cl_2, CF_3Cl, $SF_6/O_2/Cl_2$, $Cl_2/H_2/C_2F_6/CCl_4$, C_2ClF_5/O_2, Br_2, SiF_4/O_2, NF_3, ClF_3, CCl_4, CCl_3F_5, C_2ClF_5/SF_6, C_2F_6/CF_3Cl, CF_3Cl/Br_2
SiO_2	CF_4/H_2, C_2F_6, C_3F_8, CHF_3/O_2
Si_3N_4	$CF_4/O_2/H_2$, C_2F_6, C_3F_8, CHF_3
Organics	O_2, CF_4/O_2, SF_6/O_2
Al	BCl_3, BCl_3/Cl_2, $CCl_4/Cl_2/BCl_3$, $SiCl_4/Cl_2$
Silicides	CF_4/O_2, NF_3, SF_6/Cl_2, CF_4/Cl_2
Refractories	CF_4/O_2, NF_3/H_2, SF_6/O_2
GaAs	BCl_3/Ar, $Cl_2/O_2/H_2$, $CClH_3/H_2$ H_2, CH_4/H_2, $CCl_2F_2/O_2/Ar/He$
InP	CH_4/H_2, C_2H_6/H_2, Cl_2/Ar
Au	$C_2Cl_2F_4$, Cl_2, $CClF_3$

Most importantly, materials can be etched with the aforementioned techniques, and for each material a variety of chemistries is available. Table 5.1 lists some of the most common materials along with selected etch recipes [5.11]. For each chemistry, the etch rate, directionality, and selectivity with respect to the mask materials depend on parameters such as the flow/rates of the gases entering the chamber, the working pressure, and the RF power applied to the plasma.

5.1.4 Substrate Bonding

Substrate (wafer) bonding (silicon-silicon, silicon-glass, and glass-glass) is among the most important fabrication techniques in microsystem technology [5.12, 13]. It is frequently used to fabricate complex 3-D structures both as a functional unit and as part of the final microsystem package and encapsulation. The two most important bonding techniques are silicon-silicon fusion (or silicon direct bonding) and silicon-glass electrostatic (or anodic) bonding. In addition to these techniques, several other alternative methods that utilize an intermediate layer (eutectic, adhesive, and glass frit) have also been investigated. All these techniques can be used to bond the substrates at the wafer level. In this chapter, we will only discuss wafer-level techniques and will not treat the device level bonding methods (e.g., e-beam and laser welding).

Silicon Direct Bonding

Direct silicon, or fusion bonding, is used in the fabrication of micromechanical devices and silicon-on-insulator (SOI) substrates. Although it is mostly used to bond two silicon wafers with or without an oxide layer, it has also been used to bond different semiconductors such as GaAs and InP [5.13]. One main requirement for a successful bond is sufficient planarity (< 10 Å surface roughness and < 5 μm bow across a 4" wafer) and cleanliness of the surfaces. In addition, thermal expansion mismatch also needs to be considered if bonding of two dissimilar materials is contemplated. The bonding procedure is as follows: The silicon- or oxide-coated silicon wafers are first thoroughly cleaned. Subsequently, the surfaces are hydrated (activated) in HF or boiling nitric acid (RCA clean also works). This renders the surfaces hydrophilic by creating an abundance of hydroxyl ions. Then the substrates are brought together in close proximity (starting from the center to avoid the void formation). The close approximation of the bonding surfaces allows the short-range attractive van der Waals forces to bring the surfaces into intimate contact on an atomic scale. Following this step, a hydrogen bond between the two hydroxyl-coated silicon wafers bonds the substrates together. These steps can be performed at room temperatures; however, in order to increase the bond strength a high temperature (800–1,200 °C) anneal is usually required. A big advantage of silicon fusion bonding is the thermal matching of the substrates.

Anodic Bonding

Silicon-glass anodic (electrostatic) bonding is another major substrate joining technique that has been used extensively for microsensor packaging and device fabrication. The main advantage of this technique is its lower bonding temperature, which is around 300–400 °C. Figure 5.17 shows the bonding setup. A glass wafer (usually Pyrex 7740 for its thermal expansion match to silicon)

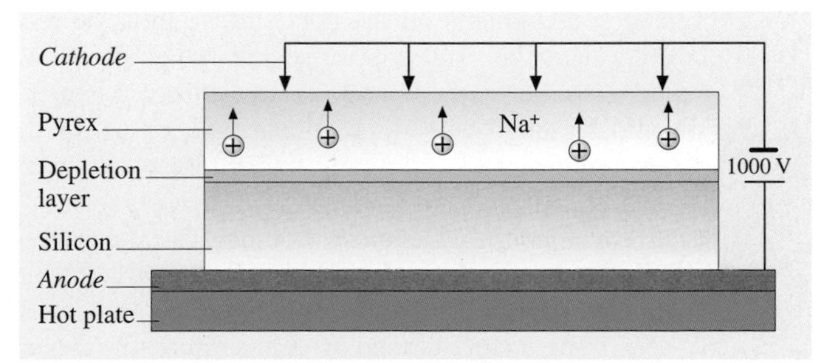

Fig. 5.17 Glass-to-silicon anodic bonding setup

is placed on top of a silicon wafer, and the sandwich is heated to 300–400 °C. Subsequently, a voltage of ~ 1,000 V is applied to the glass-silicon sandwich with the glass connected to the cathode. The bond starts immediately after the application of the voltage and spreads outward from the cathode contact point. The bond can be visually observed as a dark greyish front that expands throughout the wafer.

The bonding mechanism is as follows: During the heating period, glass sodium ions move toward the cathode and create a depletion layer at the silicon-glass interface. A strong electrostatic force is therefore created at the interface that pulls the substrates into intimate contact. The exact chemical reaction responsible for anodic bonding is not yet clear, but covalent silicon oxygen bond at the interface seems to be responsible for the bond. Silicon-to-silicon anodic bonding using sputtered or evaporated glass interlayer is also possible.

Bonding with Intermediate Layers

Various other wafer bonding techniques utilizing an intermediate layer have also been investigated [5.13]. Among the most important ones are adhesive, eutectic, and glass frit bonds. Adhesive bond using a polymer (e.g., polyimides, epoxies, thermoplast adhesives, and photoresists) in between the wafers has been used to join different wafer substrates [5.14]. Complete curing (in an oven or using dielectric heating) of the polymer before or during the bonding process prevents subsequent solvent outgas and void formation. Although reasonably high bonding strengths can be obtained, these bonds are non-hermetic and unstable over time.

In the eutectic bonding process, gold-coated silicon wafers are bonded together at temperatures greater than the silicon-gold eutectic point (363 °C, 2.85% silicon and 97.1% Au) [5.15]. This process can achieve high bonding strength and good stability at relatively low temperatures. For good bond uniformity, silicon dioxide must be removed from the silicon surface prior to the deposition of the gold. In addition, all organic contaminants on the gold surface must be removed (using UV light) prior to the bond. Pressure should also be applied in order to achieve a better contact. Although eutectic bond can be accomplished at low temperatures, achieving uniformity over large areas has proven to be challenging. More recently, localized silicon-gold eutectic bonding has been reported. In this technique, a silicon cap is bonded to another silicon wafer having a gold micro-heater (0.27 Å current for a bonding temperature of about 800 °C), Fig. 5.18 [5.16].

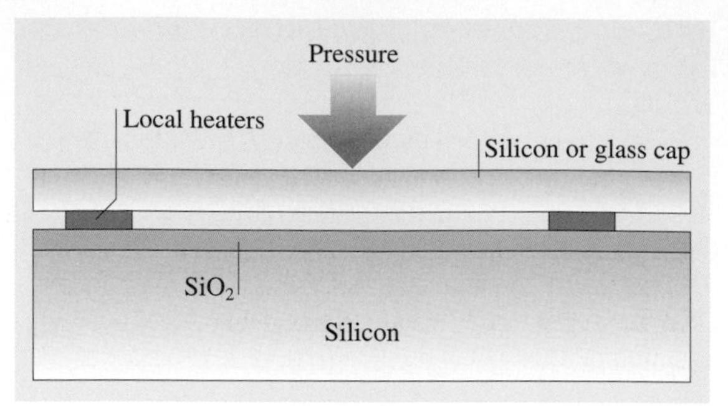

Fig. 5.18 Eutectic bonding with localized heating

Glass frit can also be used as an interlayer in substrate bonding. In this technique, first a thin layer of glass is deposited and pre-glazed. The glass-coated substrates are then brought into contact, and the sandwich is heated to above the glass melting temperature (typically < 600 °C). Like in the eutectic process, pressure must be applied for adequate contact [5.17].

Bond Characterization

Two major characterization techniques for substrate bonding evaluation are: 1) visual inspection and 2) blade test [5.12]. Once the bond has been completed, the bonding surfaces can be visually inspected for the presence of voids, cracks, and nonuniformity. The silicon-glass anodic bond interface does not require any particular imaging instrument and can be inspected with an unaided eye or under a light microscope. However, infrared imaging, X-ray photography, and ultrasonic methods are necessary for opaque substrates (mainly silicon-silicon). Although infrared transmission has some resolution limits due to its wavelength, it

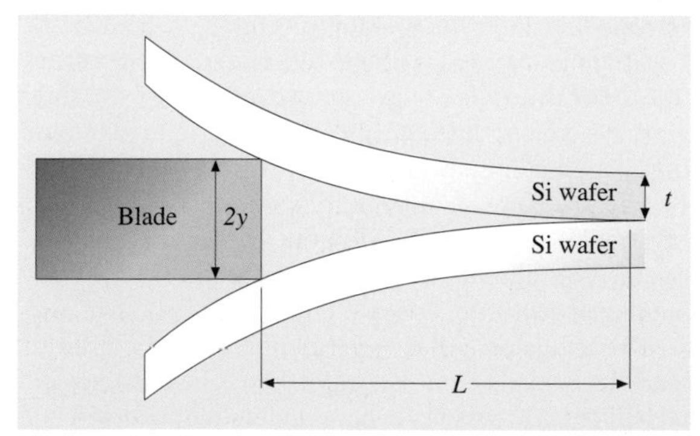

Fig. 5.19 Blade test technique for the measurement of bond strength

is the most common tool in bond inspection and offers a quality technique with a simple IR source and a CCD camera.

The blade test is often performed in order to estimate the specific surface energy (and hence bond strength), Fig. 5.19. In this test, a blade is inserted between the bonded substrates and a crack is induced. The length of the crack is then measured and the specific surface energy is calculated using the following equation [5.18]:

$$g = \frac{3Et^3y^2}{8L^4} ,$$

where g is the surface energy (erg/cm^2), E is the Young's modulus of substrate (dyne/cm^2), t is the thickness of the wafer, y is the half-thickness of the blade, and L is the crack length. The crack length is usually measured using the interference fringes parallel to theblade edge under an IR light source. A semi-quantitative technique for evaluation of the bond strength is the pull test. This technique is based on using a test fixture (or tensometer) for measuring the force required to pull two bonded substrates apart. The pull test is more effective on small samples and can be used to compare different bonding techniques without the need to accurately measure the surface energy.

5.2 MEMS Fabrication Techniques

In this section, we will discuss various important MEMS fabrication techniques commonly used to build various microdevices (microsensors and microactuators) [5.6–10]. The dimensional spectrum of the microstructures that can be fabricated using these techniques span from 1 mm to 1 μm. As mentioned in the introduction, we will mostly emphasize the more important techniques and will not discuss specialized methods.

5.2.1 Bulk Micromachining

Bulk micromachining is the oldest MEMS technology and hence probably one of the more mature ones [5.8, 9]. It is currently by far the most commercially successful one, helping manufacture devices such as pressure sensors and ink-jet print heads. Although there are many different variations, the basic concept behind bulk micromachining is selective removal of the substrate (silicon, glass, GaAs, etc.). This allows the creation of various micromechanical components such as beams, plates, and membranes that can be used to fabricate a variety of sensors and actuators. The most important microfabrication techniques used in bulk micromachining are wet and dry etch and substrate bonding. Although one can use different criteria in dividing the bulk micromachining techniques into separate categories, we will use a historical time line for this purpose. Starting with the more traditional wet etching techniques, we will proceed to the more recent ones that use deep RIE and wafer bonding.

Bulk Micromachining Using Wet Etch and Wafer Bonding

The use of anisotropic wet etchants to remove silicon can be regarded as the beginning of the micromachining era. Back side etch was used to create movable structures such as beams, membranes, and plates, Fig. 5.20. Initially, the etching was timed to create a specified thickness. However, this technique proved inadequate in creating thin structures ($< 20\,\mu$m). Subsequent use of various etch stop techniques allowed the creation of thinner membranes in a more controlled fashion. As mentioned in Sect. 5.1.3, heavily doped boron regions and electrochemical bias can be used to drastically slow down the etch process and hence create controllable thickness microstructures. Figure 5.21a,b show the cross section of two piezoresistive pressure sensors fabricated using electrochemical and P^{++} etch stop techniques. The use of the P^{++} method requires the epitaxial growth of a lightly doped region on top of a P^{++} etch stop layer.

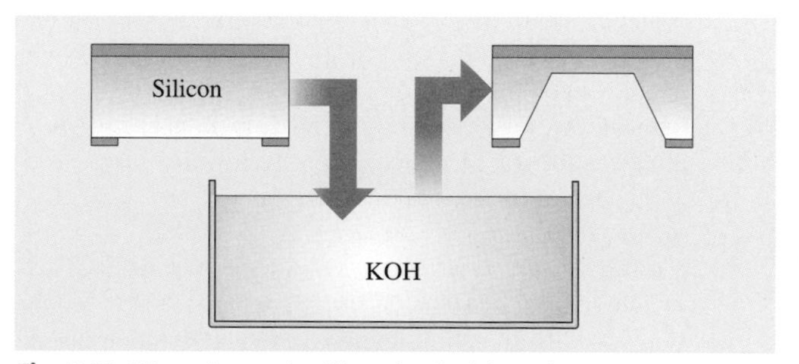

Fig. 5.20 Wet anisotropic silicon back side etch

This layer is subsequently used for the placement of piezeoresistors. However, if no active component is required, one can simply use the P^{++} region to create a thin membrane, Fig. 5.21c.

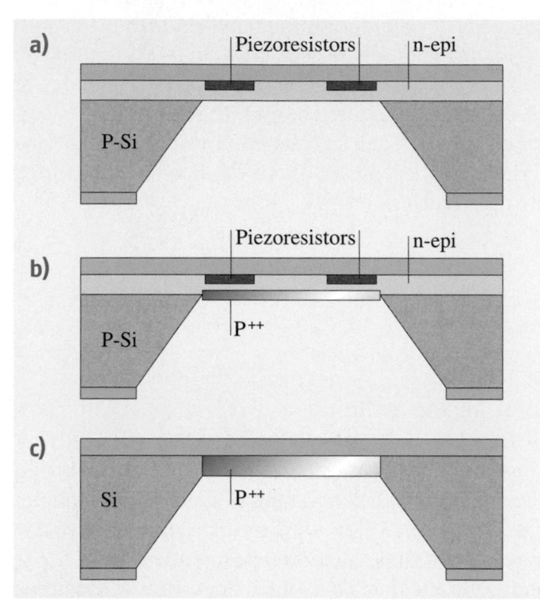

Fig. 5.21a–c Wet micromachining etch stop techniques: (a) electrochemical with n-epi on p substrate, (b) P^{++} etch stop with n-epi, and (c) P^{++} etch stop without n-epi

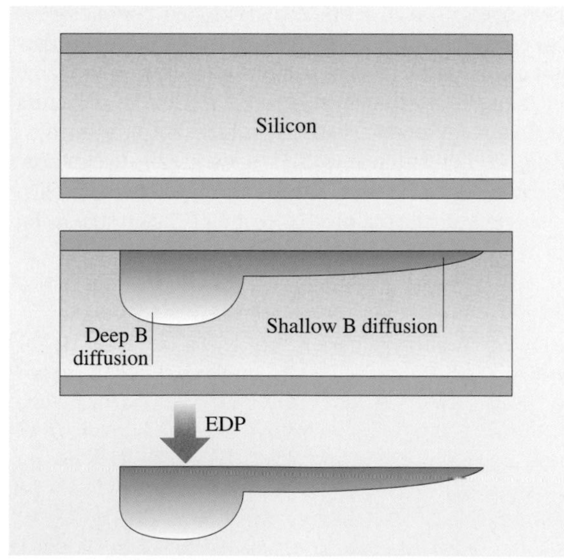

Fig. 5.22 Freestanding microstructure fabrication using deep and shallow boron diffusions and EDP release

The P^{++} etch stop technique can also be used to create isolated thin silicon structures through the dissolution of the entire lightly doped region [5.19]. This technique was successfully used to fabricate silicon recording and stimulate electrodes for biomedical applications. Figure 5.22 shows the cross section of such a process that relies on deep (15–20 μm) and shallow boron (2–5 μm) diffusion steps to create microelectrodes with flexible connecting ribbon cables. An extension of this process that uses a combination of P^{++} etch stop layers and silicon-glass anodic bonding has also been developed. This process is commonly known as the dissolved wafer process and has been used to fabricate a variety of microsensors and microactuators [5.20]. Figure 5.23 shows the cross section of this process. Figure 5.24 shows an SEM photograph of a micro-accelerometer fabricated using the dissolved wafer process.

It is also possible to merge wet bulk micromachining and microelectronics fabrication processes to build mi-

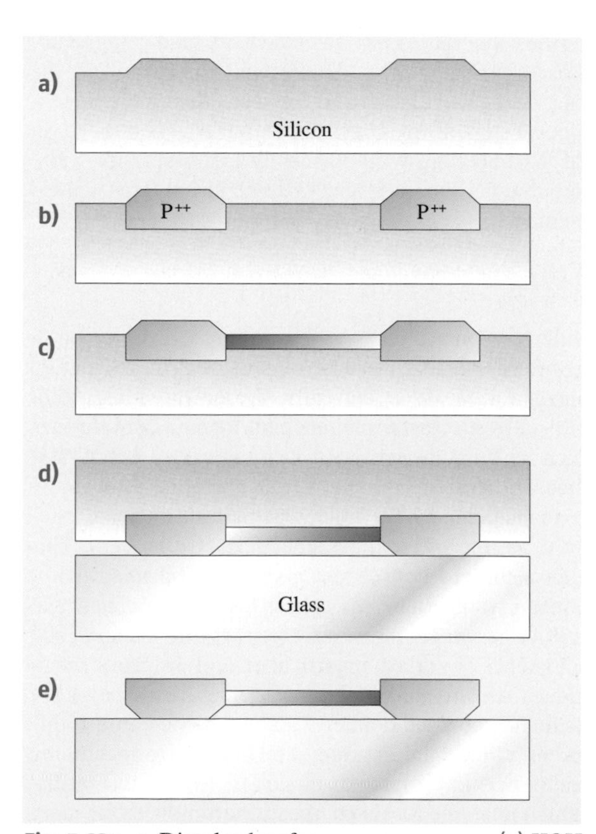

Fig. 5.23a–e Dissolved wafer process sequence, (a) KOH etch, (b) deep B diffusion, (c) shallow B diffusion, (d) silicon-glass anodic bond, and (e) release in EDP

Fig. 5.24 SEM photograph of a micro-accelerometer fabricated using the dissolved wafer process [5.20]

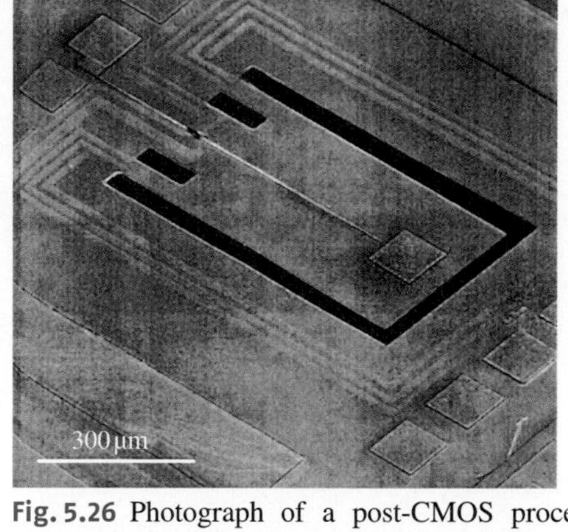

Fig. 5.26 Photograph of a post-CMOS processed cantilever beam resonator for chemical sensing [5.21]

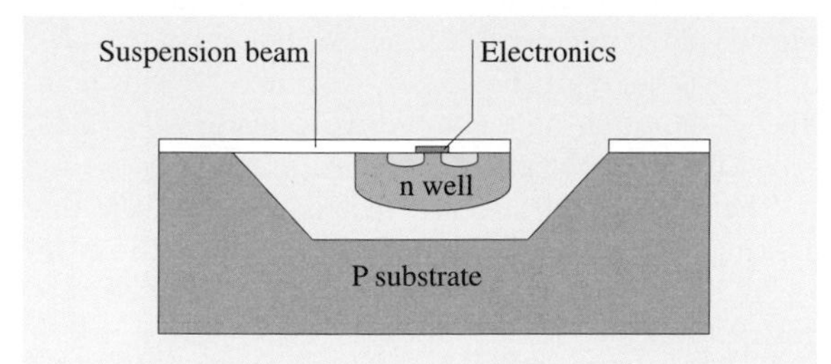

Fig. 5.25 Suspended island created on a prefabricated CMOS chip using front-side wet etch and electrochemical etch stop

cromechanical components on the same substrate as the integrated circuits (CMOS, Bipolar, or BiCMOS) [5.21]. This is very appealing since it allows the integration of interface and signal processing circuitry with MEMS structures on a single chip. However, important fabrication issues such as process compatibility and yield have to be carefully considered. Among the most popular techniques in this category is the post-processing of the CMOS integrated circuits by front side etching in TMAH solutions. As mentioned previously, silicon-rich TMAH does not attack aluminum and therefore can be used to undercut microstructures in an already processed CMOS chip. Figure 5.25 shows a schematic of such a process in which front side wet etch and electrochemical etch stop have been used to produce suspended beams. This technique has been used extensively to fabricate a variety of microsensors (e.g., humidity, gas, chemical, and pressure). Figure 5.26 shows a photograph of a post-CMOS processed chemical sensor.

Bulk Micromachining Using Dry Etch

Bulk silicon micromachining using dry etch is a very attractive alternative to the wet techniques described in the previous section. These techniques were developed in the mid-1990s, following the successful efforts to develop processes for anisotropic dry silicon etch. More recent advances in deep silicon RIE and the availability of SOI wafers with a thick top silicon layer have increased the application of these techniques. These techniques allow the fabrication of high-aspect-ratio vertical structures in isolation or along with on-chip electronics. Process compatibility with active microelectronics is less of a concern in dry methods, since many of them do not damage the circuit or its interconnect.

The most simple dry bulk micromachining technique relies on the front side undercut of microstructures using XeF_2 vapor phase etch [5.22]. As mentioned before, this, however, is an isotropic etch and, therefore, has a limited application. A combination of isotropic/anisotropic dry etch is more useful and can be used to create a variety of interesting structures. Two successful techniques using this combination are single-crystal reactive etching and metallization (SCREAM) [5.23] and post-CMOS dry release using aluminum/silicon dioxide laminate [5.24]. The first technique relies on the combination of isotropic/anisotropic dry etch to create single crystalline suspended structures. Figure 5.27 shows the cross section of the process. It starts with an anisotropic (Cl_2/BCl_3) silicon etch using an oxide mask (Fig. 5.27b). This is followed by a conformal PECVD oxide deposition (Fig. 5.27c). Subsequently, an anisotropic oxide etch is used to remove the oxide at the bottom of the

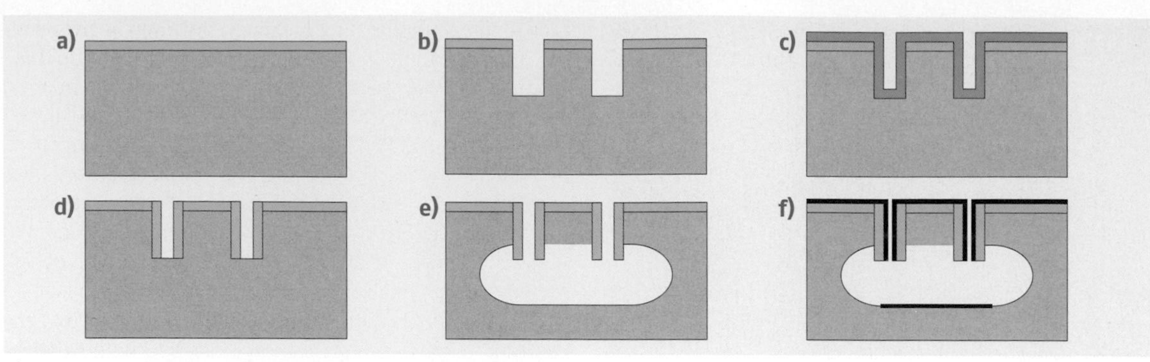

Fig. 5.27 Cross section of the SCREAM process

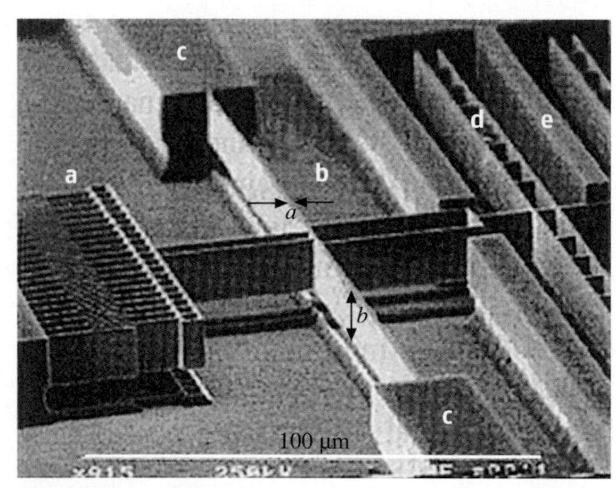

Fig. 5.28a–e SEM photograph of a structure fabricated using SCREAM process: (a) comb-drive actuator, (b) suspended spring, (c) spring support, (d) moving suspended capacitor plate, and (e) fixed capacitor plate [5.25]

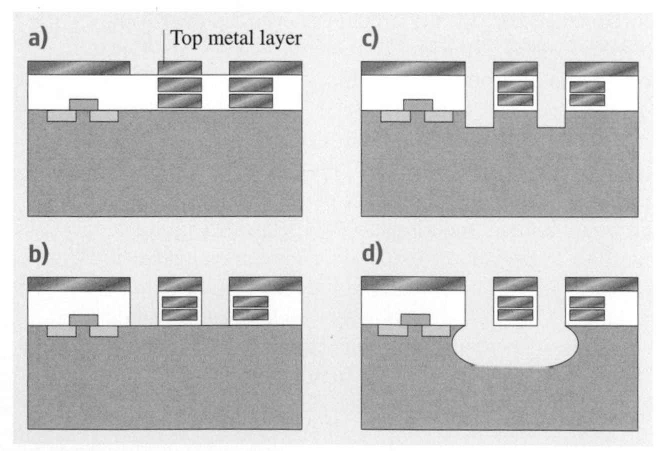

Fig. 5.29 Cross section of the process flow for the post-CMOS dry microstructure fabrication

trenches leaving the sidewall oxide intact (Fig. 5.27d). At this stage, an isotropic silicon etch (SF_6) is performed that results in undercut and release of the silicon structures (Fig. 5.27e). Finally, if electrostatic actuation is desired, a metal can be sputtered to cover the top and sidewall of the microstructure and the bottom of the cavity formed below it (Fig. 5.27f). Figure 5.28 shows an SEM photograph of a comb-drive actuator fabricated using SCREAM technology.

The second dry release technique relies on the masking capability of aluminum interconnect lines in a CMOS integrated circuit to create suspended microstructures. Figure 5.29 shows the cross section of this process. As can be seen, the third level Al of a prefabricated CMOS chip is used as a mask to anisotropically etch the underlying oxide layers all the way to the silicon (CHF_3/O_2), Fig. 5.29b. This is followed by an anisotropic silicon etch to create a recess in the silicon, which will be used in the final step to facilitate the undercut and release, Fig. 5.29c. Finally, an isotropic silicon etch is used to undercut and release the structures, Fig. 5.29d. Figure 5.30 shows an SEM photograph of a comb-drive actuator fabricated using this technology.

In addition to the methods described above, recent advancements in the development of deep reactive ion etching of silicon (DRIE, see Sect. 5.1.3) have created new opportunities for dry bulk micromachining techniques (see Sect. 5.2.3). One of the most important ones uses thick silicon SOI wafers that are commercially available in various top silicon thicknesses [5.26]. Figure 5.31 shows the cross section of a typical process using DRIE and SOI wafers. The top silicon layer is patterned and etched all the way to the buried oxide, Fig. 5.31b. The oxide is subsequently removed in HF, releasing suspended single crystalline microstructures, Fig. 5.31c. In a modification of this process, the substrate can also be removed from the back side, allowing

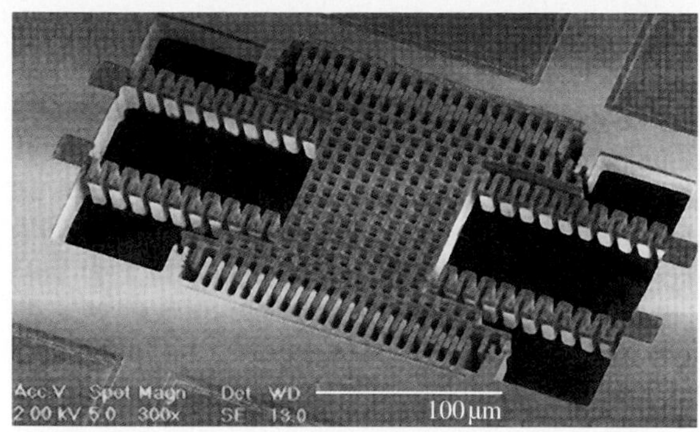

Fig. 5.30 SEM photograph of a comb-drive actuator fabricated using aluminum mask post-CMOS dry release [5.27]

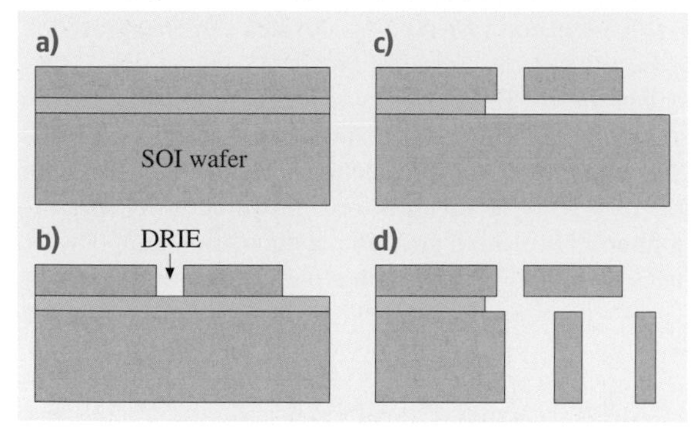

Fig. 5.31 DRIE processes using SOI wafers

easy access from both sides (this allows an easier release and prevents stiction), Fig. 5.31d.

5.2.2 Surface Micromachining

Surface micromachining is another important MEMS microfabrication technique that can be used to create movable microstructures on top of a silicon substrate [5.28]. This technique relies on the deposition of structural thin films on a sacrificial layer that is subsequently etched away, resulting in movable micromechanical structures (beams, membranes, plates, etc.). The main advantage of surface micromachining is that extremely small sizes can be obtained. In addition, it is relatively easy to integrate the micromachined structures with on-chip electronics for increased functionality. However, due to the increased surface nonplanarity with any additional layer, there is a limit to the number of layers that can be deposited. Although one of the earliest reported MEMS structures was a surface micromachined resonant gate transistor [5.29], material-related difficulties resulted in the termination of efforts in this area. In the mid-1980s, improvements in the field of thin film deposition rekindled interest in surface micromachining [5.30]. Later in the same decade, polysilicon surface micromachining was introduced, which opened the door to the fabrication of a variety of microsensors (accelerometers, gyroscopes, etc.) and microactuators (micromirrors, RF switches, etc.). In this section, we will concentrate on the key process steps involved in surface micromachining fabrication and the various materials used in the surface micromachining process. In addition, monolithic integration of CMOS with MEMS structures and 3-D surface micromachining are also discussed.

Basic Surface Micromachining Processes

The basic surface micromachining process is illustrated in Fig. 5.32. The process begins with a silicon substrate on top of which a sacrificial layer is grown and patterned, Fig. 5.32a. Subsequently, the structural material is deposited and patterned, Fig. 5.32b. As can be seen, the structural material is anchored to the substrate through the openings created in the sacrificial layer during the previous step. Finally, the sacrificial layer is removed, resulting in the release of the microstructures Fig. 5.32c. In wide structures, it is usually necessary to provide access holes in the structural layer for a fast sacrificial layer removal. It is also possible to seal microcavities cre-

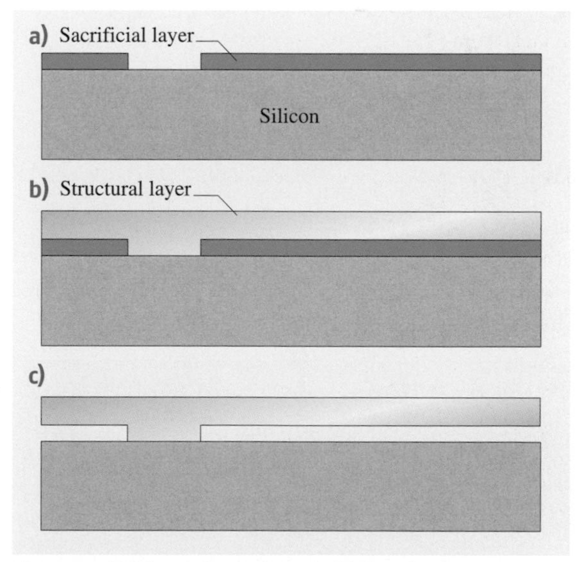

Fig. 5.32 Basic surface micromachining fabrication process

ated by the surface micromachining technique [5.8, 9]. This can be done at the wafer level and is a big advantage in applications such as pressure sensors that require a sealed cavity. Figure 5.33 shows two different techniques that can be used for this purpose. In the first, following the etching of the sacrificial layer, a LPCVD dielectric layer (oxide or nitride) is deposited to cover and seal the etch holes in the structural material, Fig. 5.33a. Since the LPCVD deposition is performed at reduced pressures, a sub-atmospheric pill-box microcavity can be created. In the second technique, also called reactive sealing, the polysilicon structural material is oxidized following the sacrificial layer removal, Fig. 5.33b. If access holes are small enough, the grown oxide can seal the cavity. Due to the consumption of the oxygen during the growth process, the cavity is sub-atmospheric.

The most common sacrificial and structural materials are phosphosilicate glass (PSG) and polysilicon, respectively (low temperature oxide, or LTO, is also frequently used as the sacrificial layer). However, there are several other sacrificial/structural combinations that have been used to create a variety of surface micromachined structures. Important design issues related to the choice of the sacrificial layer are: 1) quality (pinholes, etc.), 2) ease of deposition, 3) deposition rate, 4) deposition temperature, and 5) etch difficulty and selectivity (sacrificial layer etchant should not attack the structural layer). The particular choice of material for structural layers depends on the desired properties and the specific application. Several important requirements are: 1) ease of deposition, 2) deposition rate, 3) step coverage, 4) mechanical properties (internal stress, stress gradient, Young's modulus, fracture strength, and internal damping), 5) etch selectivity, 6) thermal budget and history, 7) electrical conductivity, and 8) optical reflectivity. Two examples from the commercially available surface micromachined devices illustrate various successful sacrificial/structural combinations. Texas Instruments's deformable mirror display (DMD) spatial light modulator uses aluminum as the structural material (good optical reflectivity) and photoresist as the sacrificial layer (easy dry etch and low processing temperatures, allowing easy post-IC integration with CMOS) [5.32], Fig. 5.34. In contrast, Analog Devices's microgyroscope uses polysilicon structural material and a PSG sacrificial layer, Fig. 5.35. Two recent additions to the collection of available structural layers are poly-silicon-germanium and poly-germanium [5.33, 34]. These are intended to be a substitute for polysilicon in applications where the high polysilicon deposition temperature (around 600 °C) is forbidding (e.g., CMOS integration).

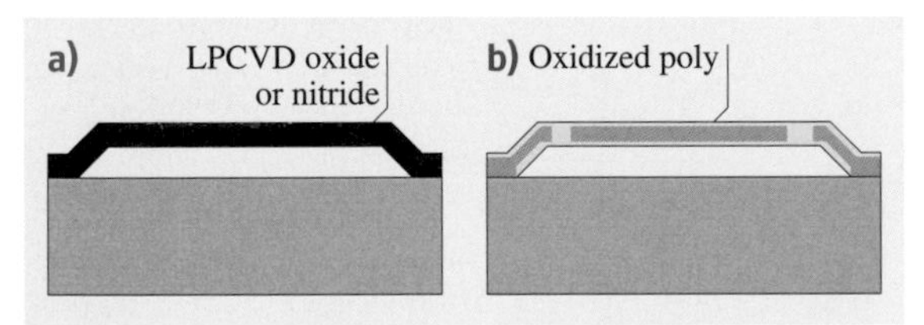

Fig. 5.33 Two sealing techniques for surface micromachined created cavities

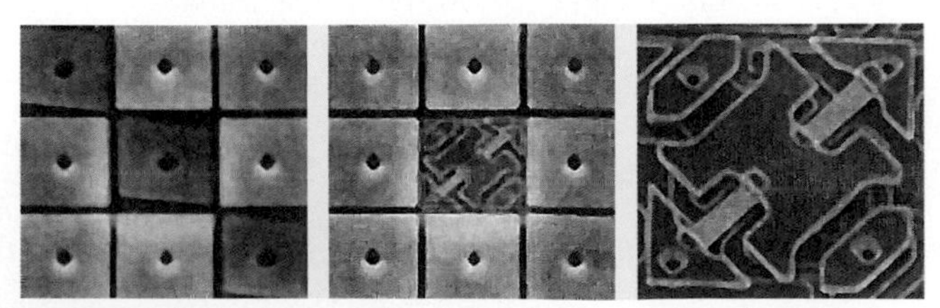

Fig. 5.34 SEM photographs of Texas Instrument micromirror array [5.28]

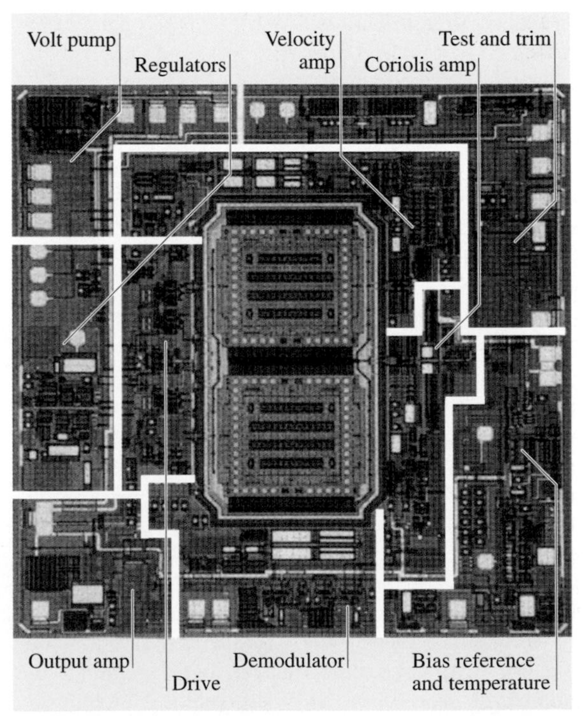

Fig. 5.35 SEM photograph of Analog Devices gyroscope [5.31]

Table 5.2 Several important surface micromachined sacrificial/structural combinations

System	Sacrificial layer	Structural layer	Structural layer etchant	Sacrificial layer etchant
1	PSG or LTO	Poly-Si	RIE	Wet or vapor HF
2	Photoresist, polyimide	Metals (Al, Ni, Co, Ni-Fe)	Various metal etchants	Organic solvents, plasma O_2
3	Poly-Si	Nitride	RIE	KOH
4	PSG or LTO	Poly-Ge	H_2O_2 or RCA1	Wet or vapor HF
5	PSG or LTO	Poly-Si-Ge	H_2O_2 or RCA1	Wet or vapor HF

Unlike LPCVD polysilicon, poly-germanium (poly-Ge) and poly-silicon-germanium (poly-$Si_{1-x}Ge_x$) can be deposited at temperatures as low as $350\,°C$ (poly-Ge deposition temperature is usually lower than poly-SiGe). Table 5.2 summarizes important surface micromachined sacrificial/structural combinations.

An important consideration in the design and processing of surface micromachined structures is the issue of stiction [5.8, 9, 35, 36]. This can happen during the release step if a wet etchant is used to remove the sacrificial layer, or during the device lifetime. The reason for stiction during release is the surface tension of the liquid etchant, which can hold the microstructure down and cause stiction. This usually happens when the structure is compliant and does not possess enough spring constant to overcome the surface tension force of the rinsing liquid (i. e., water). There are several ways one can alleviate the release-related stiction problem. These include: 1) the use of dry or vapor phase etchant, 2) the use of solvents with lower surface tension, 3) geometrical modifications, 4) CO_2 critical drying, 5) freeze drying, and 6) *self-assembled monolayer* (SAM) or organic thin-film surface modification. The first technique prevents stiction by not using a wet etchant, although in the case of vapor phase release, the condensation possibility is existent and can cause some stiction. The second method uses rinsing solvents such as methanol with a lower surface tension than water. This is usually followed by a rapid evaporation of the solvent on a hot plate. However, this technique is not optimum, and many structures still stick. The third technique is geometrical, which provides dimples in the structural layer in order to reduce the contact surface area and hence reduce the attractive force. The fourth and fifth techniques rely on the phase change (in one case CO_2 and the other butyl-alcohol), which avoids the liquid phase altogether by directly going to the gas phase. The last technique uses self-assembled monolayers or organic thin films to coat the surfaces with a hydrophobic layer. The stiction that occurs during the operating lifetime of the device (in-use stiction) is due to the condensation of moisture on the surfaces, electrostatic charge accumulation, or direct chemical bonding. Surface passivation using self-assembled monolayers or organic thin films can be used to reduce the surface energy and reduce or eliminate the capillary forces and direct chemical bonding. These organic coatings also reduce electrostatic forces if a thin layer is applied directly to the semiconductor (without the intervening oxide layer). Commonly used organic coatings include fluorinated fatty acids (Texas Instruments aluminum micromirrors), silicone polymeric layers (Analog Devices accelerometers), and siloxane self-assembled monolayers.

Surface Micromachining Integration with Active Electronics

Integration of surface micromachined structures with on-chip circuitry can increase performance and simplify packaging. However, issues related to process compatibility and yield must be carefully considered. The two most common techniques are MEMS-first and MEMS-last techniques. In the MEMS-last technique, the integrated circuit is first fabricated and surface micromachined structures are subsequently built on top of the silicon wafer. An aluminum structural layer with a photoresist sacrificial layer is an attractive combination due to the low thermal budget of the process (Texeas Instruments micromirror array). However, in applications where mechanical properties of Al are not adequate, polysilicon structural material with LTO or PSG sacrificial layer must be used. Due to the rather high deposition temperature of polysilicon, this combination requires special attention with regard to the thermal budget. For example, aluminum metallization must be avoided and substituted with refractory metals such as tungsten. This can only be achieved at a greater process complexity and lower transistor performance.

MEMS-first technique alleviates these difficulties by fabricating the microstructures at the very beginning of the process. But if the microstructures are processed first, they have to be buried in a sealed trench to eliminate the interference of microstructures with subsequent CMOS processes. Figure 5.36 shows a cross section of a MEMS-first fabrication process developed at the Sandia National Laboratory [5.37]. The process starts with a shallow anisotropic etching of trenches in a silicon substrate to accommodate the height of the polysilicon structures fabricated later on. A silicon nitride layer is then deposited to provide isolation at the bottom of the trenches. Next, several layers of polysilicon and sacrificial oxide are deposited and patterned in a standard surface micromachining process. Subsequently, the trenches are completely filled with sacrificial oxide and the wafers are planarized with chemical-mechanical polishing (this avoids complication in the following lithographic steps). After an annealing step, the trenches are sealed with a nitride cap. At this point, a standard CMOS fabrication process is performed. At the end of the CMOS process, the nitride cap is etched and the buried structures released by etching the sacrificial oxide.

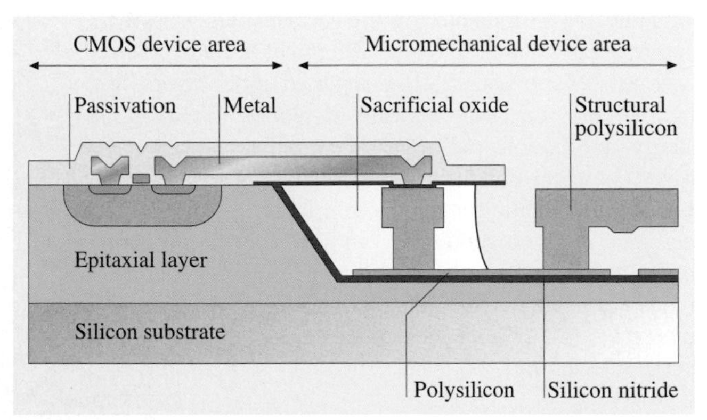

Fig. 5.36 Cross section of the Sandia MEMS-first integrated fabrication process

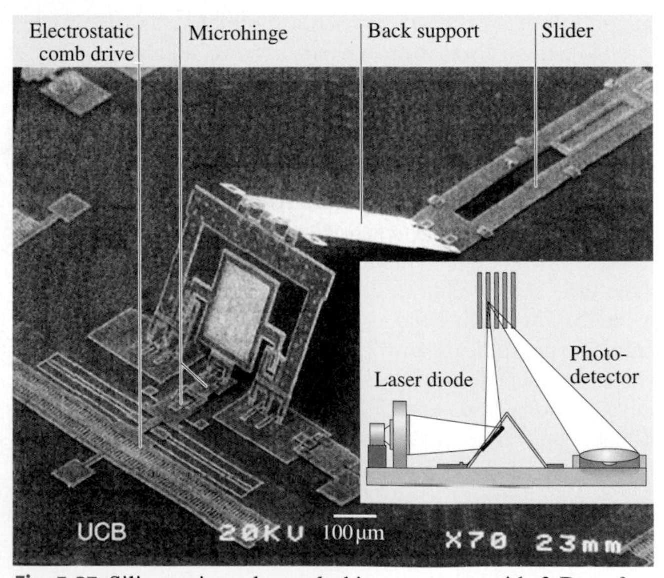

Fig. 5.37 Silicon pin-and-sample hinge scanner with 3-D surface micromachined structures [5.38]

3-D Microstructures Using Surface Micromachining

Three-dimensional surface microstructures can be fabricated using surface micromachining. The fabrication of hinges for the vertical assembly of MEMS was a major advance toward achieving 3-D microstructures [5.39]. Optical microsytems have greatly benefited from surface micromachined 3-D structures. These microstructures are used as passive or active components (micromirror, Fresnel lens, optical cavity, etc.) on a silicon optical bench (silicon microphotonics). An example is the Fresnel lens that has been surface micromachined in polysilicon and then erected using hinge structures and locked in place using micromachined tabs, thus liberating the structure from the horizontal plane of the wafer [5.38, 40]. Various microactuators (e.g., comb drive and vibromotors) have been used to move these structures out of the silicon plane and into position. Figure 5.37 shows an SEM photograph of a bar-code micro-scanner using a silicon optical microbench with 3-D surface micromachined structures.

5.2.3 High-Aspect-Ratio Micromachining

The bulk and surface micromachining technologies presented in the previous sections fulfill the requirements of a large group of applications. Certain applications, however, require the fabrication of high-aspect-ratio structures, which is not possible with the aforementioned technologies. In this section, we describe three technologies, LIGA, HEXSIL, and HARPSS, capable of producing structures with vertical dimensions much larger than the lateral dimensions by means of X-ray lithography (LIGA) and DRIE etching (HEXSIL and HARPSS).

LIGA

LIGA is a high-aspect-ratio micromachining process that relies on X-ray lithography and electroplating (in German: *LI*thographie *G*lvanoformung *A*bformung) [5.41, 42]. We already introduced the concept of the plating-through-mask technique in Sect. 5.1.2 (see Fig. 5.10). With standard UV photolithography and photoresists, the maximum thickness achievable is on the order of a few tens of microns and the resulting metal structures show tapered walls. LIGA is a technology based on the same plating-through-mask idea, but can be used to fabricate metal structures of thicknesses ranging from a few microns to a few millimeters with almost vertical sidewalls. This is achieved using X-ray lithography and special photoresists. Due to their short wavelength, X-rays are capable of penetrating a thick photoresist layer with no scattering and defining features with lateral dimensions down to 0.2 μm (aspect ratio $> 100:1$).

The photoresists used in LIGA should comply with certain requirements, including sensitivity to X-rays, resistance to electroplating chemicals, and good adhesion to the substrate. Based on such requirements Poly-(methylmethacrylate) (PMMA) is considered an optimal choice for the LIGA process. Application of the thick photoresist on top of the substrate can be performed by various techniques such as multiple spin coating, pre-cast PMMA sheets, and plasma polymerization coating. The mask structure and materials for X-ray lithography must also comply with certain requirements. The traditional masks based on

Fig. 5.38 SEM of assembled LIGA-fabricated nickel structures [5.41]

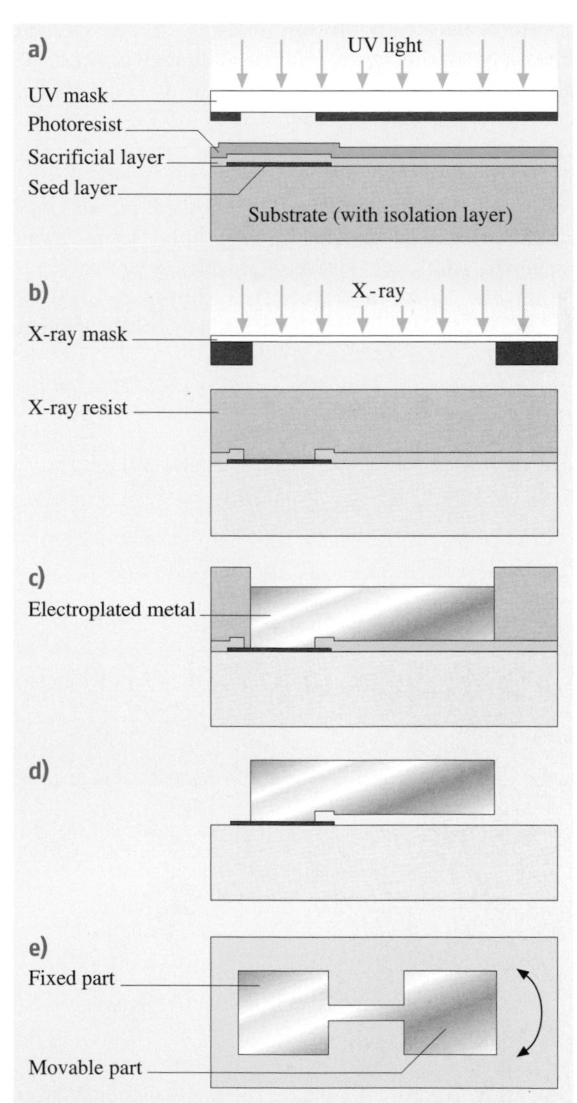

Fig. 5.39a–e Sacrificial LIGA process: (a) UV lithography for sacrificial layer patterning, (b) X-ray lithography, (c) electroplating, (d) structure releasing, (e) top view of the movable structure

glass plates with a patterned chrome thin layer are not suitable, because X-rays are not absorbed in the chromium layer and the glass plate is not transparent enough. Instead, X-ray lithography uses a silicon nitride mask with gold as the absorber material (typically formed by electroplating gold to a thickness of 10–20 μm). The nitride membrane is supported by a silicon frame, which can be fabricated using bulk micromachining techniques. Once the photoresist is exposed to the X-rays and developed, the process proceeds with the electroplating of the desired metal. Ni is the most common, although other metals and metallic compounds such as Cu, Au, NiFe, and NiW are also electroplated in LIGA processes. A good agitation of the plating solution is the key in obtaining a uniform and repeatable result during this step. A paddle plating cell, based on a windshield wiper-like device moving only 1 mm away from the substrate surface, provides extremely reproducible agitation. Figure 5.38 shows an SEM of a LIGA microstructure fabricated by electroplating nickel.

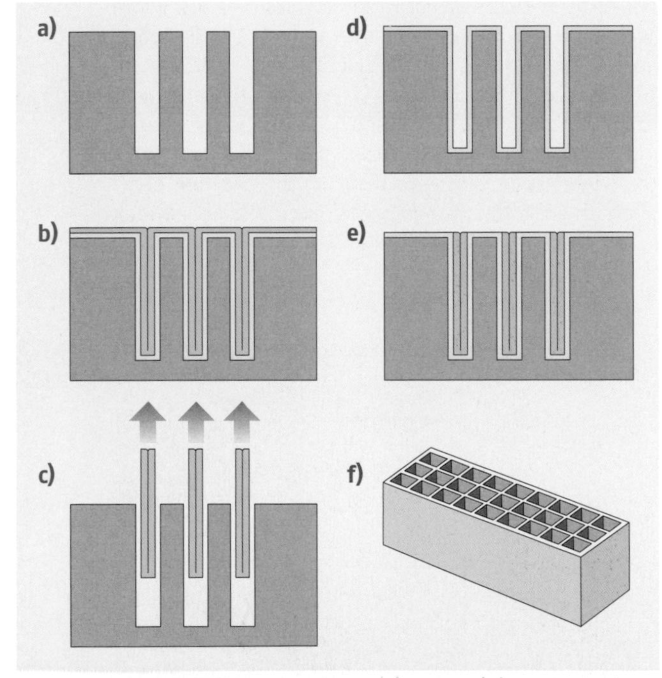

Fig. 5.40a–f HEXSIL process flow: (**a**) DRIE, (**b**) sacrificial layer deposition, (**c**) structural material deposition and trench filling, (**d**) etch structural layer from the surface, (**e**) etch sacrificial layer and pulling out of the structure, (**f**) example of a HEXSIL fabricated structure

Part A | 5.2

Due to the high cost of the X-ray sources (synchrotron radiation), the LIGA technology was initially intended for the fabrication of molds that could be used many times in hot embossing or injection molding processes. However, it has been also used in many applications to directly form high-aspect-ratio metal structures on top of a substrate. A cheaper alternative to the LIGA process (with somewhat poorer qualities) called UV-LIGA, or "poor man's LIGA", has been proposed [5.43, 44]. This process uses SU-8 negative photoresists (available for spin coating at various thicknesses ranging from 1 to 500 μm) and standard contact lithography equipment. Using this technique, aspect ratios larger than 20:1 have been demonstrated. A major problem of this alternative is the removal of the SU-8 photoresist after plating. Various methods with different degrees of success have been proposed. These include: wet etching with special solvents, burning at high temperatures (600 °C), dry etching, use of a release layer, and high-pressure water jet etching.

A variation of the basic LIGA process, shown in Fig. 5.39, permits the fabrication of electrically isolated movable structures, and thus opens more possibilities for sensor and actuator design using this technology [5.45]. The so-called sacrificial LIGA (SLIGA) starts with the patterning of the seed layer. Subsequently, a sacrificial layer (e.g., titanium) is deposited and patterned. The process then proceeds as usual in standard LIGA until the last step when the sacrificial layer is removed. The electroplated structures that overlap with the sacrificial layer are released in this step.

HEXSIL

The second method for fabricating high-aspect-ratio structures, which is based on a template replication technology, is *HEXSIL* (*HEX*agonal honeycomb poly*SIL*icon) [5.46]. Figure 5.40 shows a simplified process flow. A high-aspect-ratio template is first formed in a silicon substrate using DRIE. Next, a sacrificial multilayer is deposited to allow the final release of the structures. The multilayer is composed of one or more PSG nonconformal layers to provide fast etch release (∼ 20 μm/min in 49% HF) alternated with conformal layers of either oxide or nitride to provide enough thickness for proper release of the structures. The total thickness of the sacrificial layer has to be larger than the shrinkage or elongation of the structures caused by the relaxation of the internal stress (compressive or tensile) during the release step. Otherwise, the structures will clamp themselves to the walls of the template and their retrieval will not be possible. Any material

Fig. 5.41 SEM micrograph of an angular microactuator fabricated using HEXSIL [5.48]

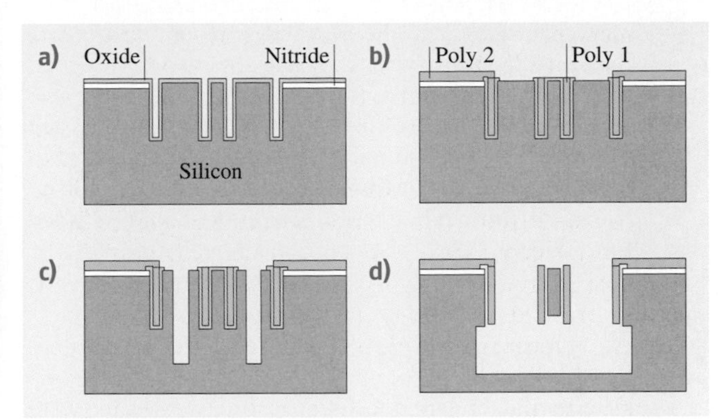

Fig. 5.42a–d HARPSS process flow. (a) Nitride deposition and patterning, DRIE etching and oxide deposition, (b) poly 1 deposition and etch back, oxide patterning and poly 2 deposition and patterning, (c) DRIE etching, (d) silicon isotropic etching

that can be conformally deposited and yet not damaged during the HF release step is suitable for the structural layer. Structures made of polysilicon, nitride, and electroless nickel [5.47] have been reported. Nickel can only be deposited in combination with polysilicon since a conductive surface is needed for the deposition to occur. After deposition of structural materials, a blanket etch (poly or nitride) or a mechanical lapping (nickel) is performed to remove the excess materials from the surface. Finally, a 49% HF with surfactant is used to dissolve the sacrificial layers. The process can be repeated many times using the same template, thus considerably lowering the fabrication costs. Figure 5.41 shows an SEM photograph of a microactuator fabricated using the HEXSIL process.

HARPSS

The *h*igh *a*spect *r*atio combined with *p*oly and *s*ingle-crystal *s*ilicon (*HARPSS*) technology is another technique capable of producing high-aspect-ratio, electrically isolated polycrystalline and single-crystal silicon microstructures with capacitive air gaps ranging from submicrometer to tens of micrometers [5.50]. The structures, tens to hundreds of micrometers thick, are defined by trenches etched with DRIE and filled with oxide and poly layers. The release of the microstructures is done at the end by means of a directional silicon etch followed by an isotropic etch. The small, vertical gaps and thick structures possible with this technology can be applied to the fabrication of a variety of MEMS devices, particularly inertial sensors [5.51] and RF beam resonators [5.52]. Figure 5.42 shows the process flow at the cross section of a single-crystal silicon beam resonator. The HARPSS process starts with the deposition and patterning of a silicon nitride layer that will be used to isolate the poly structure's connection pads from the substrate. High-aspect-ratio trenches ($\sim 5\,\mu m$ wide) are then etched into the substrate using a DRIE etcher. Next, a conformal oxide layer (LPCVD) is deposited. This layer has two functions: 1) to protect the structures during the dry etch release and 2) to define the submicrometer gap between silicon and polysilicon structures. Following the oxide deposition, the trenches are completely filled with LPCVD polysilicon. The polysilicon

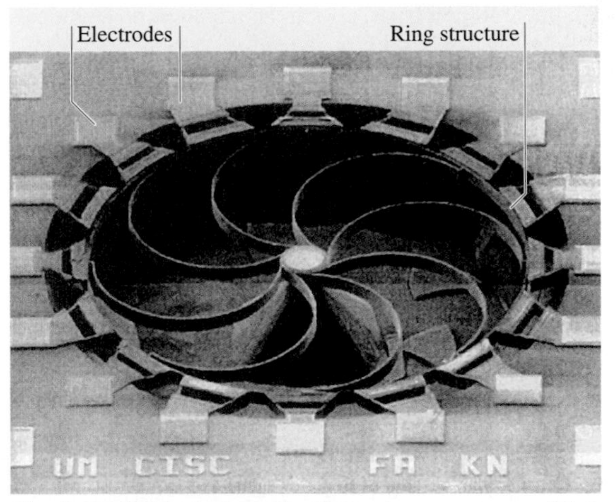

Fig. 5.43 SEM photograph of a micro-gyroscope fabricated using HARPSS process [5.49]

is etched back, and the underneath oxide is patterned to provide anchor points for the structures. A second layer of polysilicon is then deposited and patterned. Finally, the structures are released using a DRIE step followed by an isotropic silicon etch through a photoresist mask that exposes only the areas of silicon substrate surrounding the structures. It should be noted that single-crystal silicon structures are not protected at the bottom during the isotropic etch. This causes the single-crystal silicon structures to be etched vertically from the bottom, and thus be shorter than the polysilicon structures. Figure 5.43 shows an SEM photograph of a micro-gyroscope fabricated using the HARPSS process.

5.3 Nanofabrication Techniques

The microfabrication techniques discussed so far were mostly geared toward fabricating devices in the 1 mm to 1 μm dimensional range (submicron dimensions being possible in certain techniques such as *HARPSS* using a dielectric sacrificial layer). Recent years have witnessed a tremendous surge of interest in fabricating sub-micro (1 μm–100 nm) and nanostructures (1–100 nm range) [5.53]. This interest arises from both practical and fundamental view points. At the more scientific and fundamental level, nanostructures provide an interesting tool in studying electrical, magnetic, optical, thermal, and mechanical properties of matter at the nanometer scale. These include important quantum mechanical phenomena (e.g., conductance quantization, band-gap modification, coulomb blockade, etc.) arising from the confinement of charged carriers in structures such as quantum wells, wires, and dots, Fig. 5.44. On the practical side, nanostructures can provide significant improvements in the performance of electronic/optical devices and sensors. In the device area, investigators have been mostly interested in fabricating nanometer-sized transistors in anticipation of technical difficulties forecasted in extending Moore's law beyond 100 nm resolution. In addition, optical sources and detectors having nanometer-sized dimensions exhibit improved characteristics unachievable in larger devices (e.g., lower threshold current, improved dynamic behavior, and improved emission line-width in quantum dot lasers). These improvements create novel possibilities for next-generation computation and communication devices. In the sensors area, shrinking dimensions beyond conventional optical lithography can provide major improvements in sensitivity and selectivity.

One can broadly divide various nanofabrication techniques into top-down and bottom-up categories. The first approach starts with a bulk or thin film material and removes selective regions to fabricate nanostructures (similar to micromachining techniques). The second method relies on molecular recognition and self-assembly to fabricate nanostructures out of smaller building blocks (molecules, colloids, and clusters). The top-down approach is an offshoot of standard lithography and micromachining techniques. On the other hand, the bottom-up approach has more of a chemical engineering and material science flavor and relies on fundamentally different principles. In this chapter, we will discuss four major nano-fabrication techniques. These include: i) e-beam and *nano-imprint* fabrication, ii) *epitaxy* and *strain engineering*, iii) scanned probe techniques, and iv) *self-assembly* and *template manufacturing*.

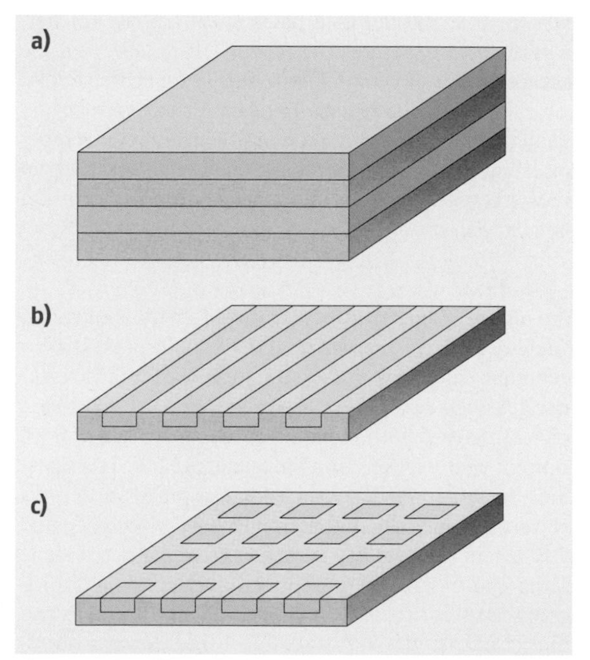

Fig. 5.44a–c Several important quantum confinement structures, (**a**) quantum well, (**b**) quantum wire, and (**c**) quantum dot

5.3.1 E-Beam and Nano-Imprint Fabrication

In previous sections, we discussed several important lithography techniques used commonly in MEMS and microfabrication. These include various forms of UV (regular, deep, and extreme) and X-ray lithographies. However, due to the lack of resolution (in the case of the UV), or the difficulty in manufacturing mask and radiation sources (X-ray), these techniques are not suitable for nanometer-scale fabrication. E-beam lithography is an attractive alternative technique for fabricating nanostructures [5.54]. It uses an electron beam to expose an electron-sensitive resist such as polymethyl methacrylate (PMMA) dissolved in trichlorobenzene (positive) or poly chloromethylstyrene (negative) [5.55]. The e-beam gun is usually part of a scanning electron microscope (SEM), although transmission electron microscopes (TEM) can also be used. Although electron wavelengths on the order of 1 Å can be easily achieved, electron scattering in the resist limits the attainable resolutions to $> 10\,\mathrm{nm}$. The beam control and pattern generation are achieved through a computer interface. E-beam lithography is serial and, hence, it has a low throughput. Although this is not a major concern in fabricating devices used in studying fundamental microphysics, it severely limits large scale nanofabrication. E-beam lithography, in conjunction with such processes as lift-off, etching, and electro-deposition, can be used to fabricate various nanostructures.

An interesting new technique that circumvents the serial and low throughput limitations of the e-beam lithography for fabricating nanostructures is the nano-imprint technology [5.56]. This technique uses an e-beam fabricated hard material master (or mold) to stamp and deform a polymeric resist. This is usually followed by a reactive ion etching step to transfer the stamped pattern to the substrate. This technique is economically superior, since a single stamp can be used repeatedly to fabricate a large number of nanostructures. Figure 5.45 shows a schematic illustration of nano-imprint fabrication. First, a hard material (e.g., silicon or SiO_2) stamp is created using e-beam lithography and reactive ion etching. Then, a resist-coated substrate is stamped, and, finally, an anisotropic RIE is performed to remove the resist residue in the stamped area. At this stage, the process is complete and one can either etch the substrate or, if metallic nanostructures are desired, evaporate the metal and perform a lift off. The resist used in nano-imprint technology can be thermal plastic, UV-curable, or thermal-curable polymer. For a thermal plastic resist (e.g., PMMA), the substrate is heated to above the glass transition temperature (T_g) of the polymer before stamping and is cooled to below T_g before the stamp is removed. Similarly, the UV and thermal-curable resists are fully cured before the stamp is separated. Nano-imprint technology resolution is limited by the mold and polymer strengths and can be as small as $10\,\mathrm{nm}$. More recently, the nano-imprint technique has been used to stamp a silicon substrate in less than $250\,\mathrm{ns}$ using a XeCl excimer laser ($308\,\mathrm{nm}$) and a quartz mask (laser assisted direct imprint, LADI), Fig. 5.46 [5.57].

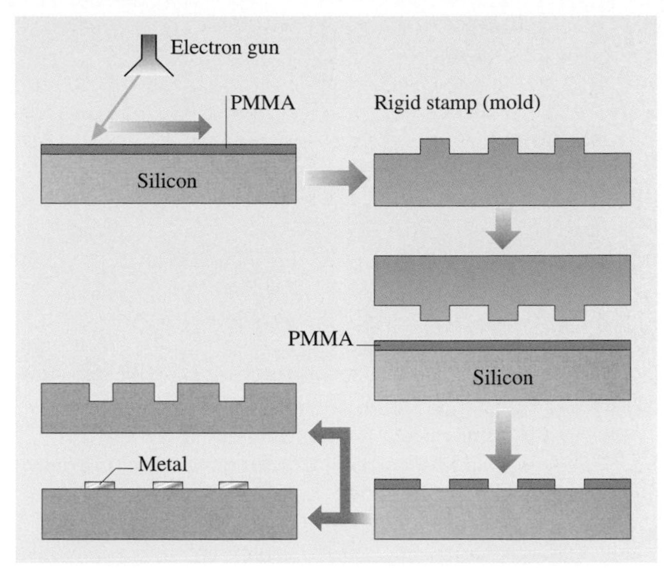

Fig. 5.45 Schematic illustration of nano-imprint fabrication

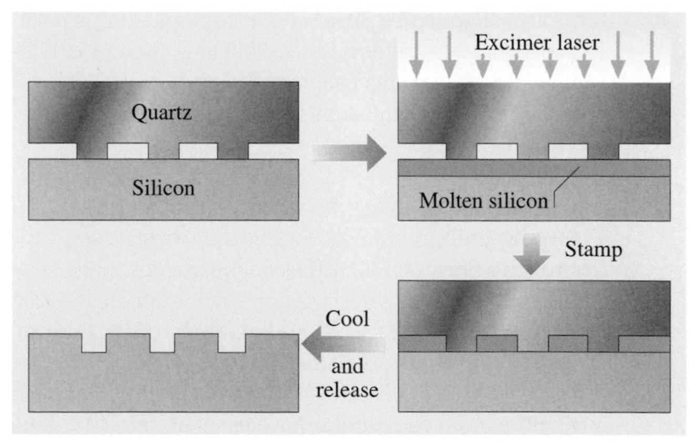

Fig. 5.46 Ultrafast silicon nano-imprinting using an excimer laser

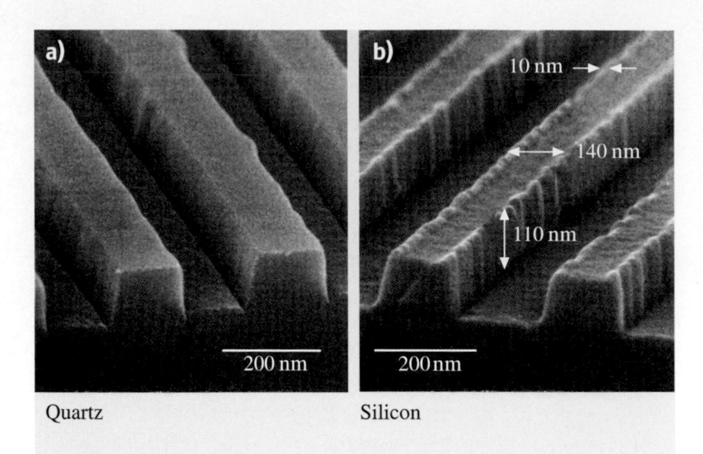

Fig. 5.47a,b SEM micrographs of (a) quartz mold and (b) imprinted silicon surface using LADI [5.57]

Figure 5.47 shows SEM micrographs of the quartz mold and imprinted silicon substrate with 140-nm lines using LADI.

5.3.2 Epitaxy and Strain Engineering

Atomic precision deposition techniques such as molecular beam epitaxy (MBE) and metallo-organic chemical vapor deposition (MOCVD) have proven to be effective tools in fabricating a variety of quantum confinement structures and devices (quantum well lasers, photodetectors, resonant tunneling diodes, etc.) [5.58–60]. Although quantum wells and superlattices are the structures that lend themselves most easily to these techniques (see Fig. 5.44a), quantum wires and dots have also been fabricated by adding subsequent steps such as etching and selective growth. Fabrication of quantum well and superlattice structures using epitaxial growth is a mature and well developed field and, therefore, will not be discussed in this chapter. Instead, we will concentrate on quantum wire and dot nanostructure fabrication using basic epitaxial techniques [5.61, 62].

Quantum Structure Nanofabrication Using Epitaxy on Patterned Substrates

There have been several different approaches to the fabrication of quantum wires and dots using epitaxial layers. The most straight forward technique involves e-beam lithography and etching of an epitaxial grown layer (e.g., InGaAs on GaAs substrate) [5.63]. However, due to the damage and/or contamination during lithography, this method is not very suitable for active device fabrication (e.g., quantum dot lasers). Several other methods involving regrowth of epitaxial layers over nonplanar surfaces such as step-edge, cleaved-edge, and patterned substrate have been used to fabricate quantum wires and dots without the need for lithography and etching of the quantum confined structure [5.62, 64]. These nonplanar surface templates can be fabricated in a variety of ways such as etching through a mask or cleavage along crystallographic planes. Subsequent epitaxial growth on top of these structures results in a set of planes with different growth rates depending on the geometry or surface diffusion and adsorption effects. These effects can significantly enhance or limit the growth rate on certain planes, resulting in lateral patterning and confinement of deposited epitaxial layers and formation of quantum wires (in V grooves) and dots (in inverted pyramids). Figure 5.48a shows a schematic cross section of an InGaAs quantum wire fabricated in a V-groove InP. As can be seen, the growth rate on the sidewalls is lower than that of the top and bottom surfaces. Therefore, the thicker InGaAs layer at the bottom of the V-groove forms a quantum wire confined from the sides by a thinner layer with a wider bandgap. Figure 5.48b shows a quantum wire formed using epitaxial growth over a dielectric pat-

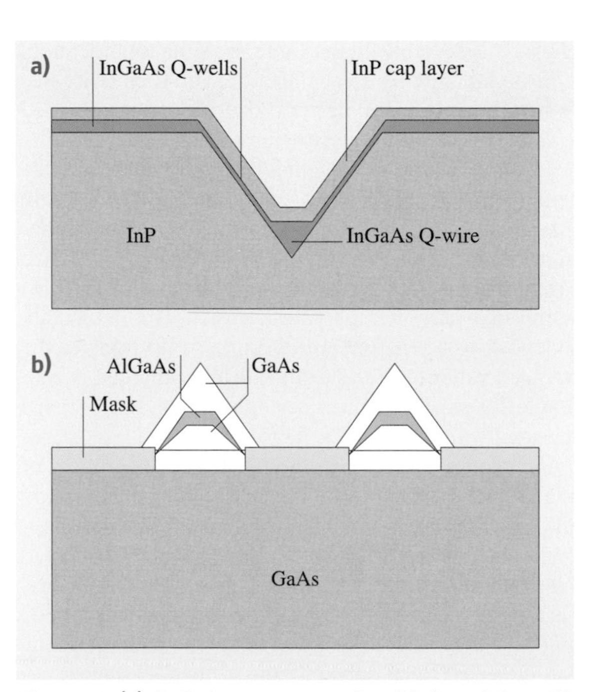

Fig. 5.48 (a) InGaAs quantum wire fabricated in a V-groove InP and (b) AlGaAs quantum wire fabricated through epitaxial growth on a masked GaAs substrate

terned planar substrate. In both of these techniques, it is relatively easy to create quantum wells. However, in order to create quantum wires and dots one still needs e-beam lithography to pattern the groove and window templates.

Quantum Structure Nanofabrication Using Strain-Induced Self-Assembly

A more recent technique for fabricating quantum wires and dots involves strain-induced *self-assembly* [5.62, 65]. The term *self-assembly* represents a process whereby a strained 2-D system reduces its energy by a transition into a 3-D morphology. The most commonly used material combination for this technique is the $In_xGa_{1-x}As/GaAs$ system, which offers a large lattice mismatch (7.2% between InAs and GaAs) [5.66,67], although recently Ge dots on Si substrate have also attracted considerable attention [5.68]. This method relies on lattice mismatch between an epitaxial grown layer and its substrate to form an array of quantum dots or wires. Figure 5.49 shows a schematic of the strain-induced self-assembly process. When the lattice constants of the substrate and the epitaxial layer differ considerably, only the first few deposited monolayers crystallize in the form of an epitaxial strained layer in which the lattice constants are equal. When a critical thickness is exceeded, a significant strain in the layer leads to the breakdown of this ordered structure and the spontaneous formation of randomly distributed islets of regular shape and similar size (usually < 30 nm in diameter). This mode of growth is usually referred to as the Stranski-Krastanow mode. The quantum dot size, separation, and height depend on the deposition parameters (i. e., total deposited material, growth rate, and temperature) and material combinations. As can be seen, this is a very convenient method to grow perfect crystalline nanostructures over a large area without any lithography and etching. One major drawback of this technique is the randomness of the quantum dot distribution. It should be mentioned that this technique can also be used to fabricate quantum wires by strain relaxation bunching at the step edges.

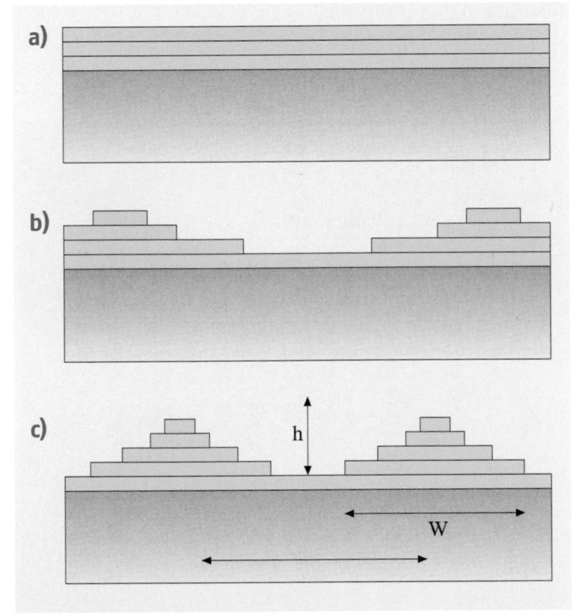

Fig. 5.49a–c Stranski-Krastanow growth mode, (a) 2-D wetting layer, (b) growth front roughening and breakup, and (c) coherent 3-D self-assembly

5.3.3 Scanned Probe Techniques

The invention of scanned probe microscopes in the 1980s revolutionized atomic-scale imaging and spectroscopy. In particular, scanning tunneling and atomic force microscopes (STM and AFM) have found widespread applications in physics, chemistry, material science, and biology. The possibility of atomic-scale manipulation, lithography, and nanomachining using such probes was considered from the beginning and has matured considerably over the past decade. In this section, after a brief introduction to scanned probe microscopes, we will discuss several important nano-lithography and machining techniques that have been used to create nanometer-sized structures.

The scanning probe microscopy (SPM) systems are capable of controlling the movement of an atomically sharp tip in close proximity to or in contact with a surface with subnanometer accuracy. Piezoelectric positioners are typically used to achieve such accuracy. High resolution images can be acquired by raster scanning the tips over a surface while simultaneously monitoring the interaction of the tip with the surface. In scanning tunneling microscope systems, a bias voltage is applied to the sample and the tip is positioned close enough to the surface, so that a tunneling current develops through the gap (Fig. 5.50a). Because this current is extremely sensitive to the distance between the tip and the surface, scanning the tip in the x-y plane while recording the tunnel current permits the mapping of the surface topography with resolution on the atomic scale. In a more common mode of operation, the amplified current signal is connected to the z-axis piezoelectric positioner through

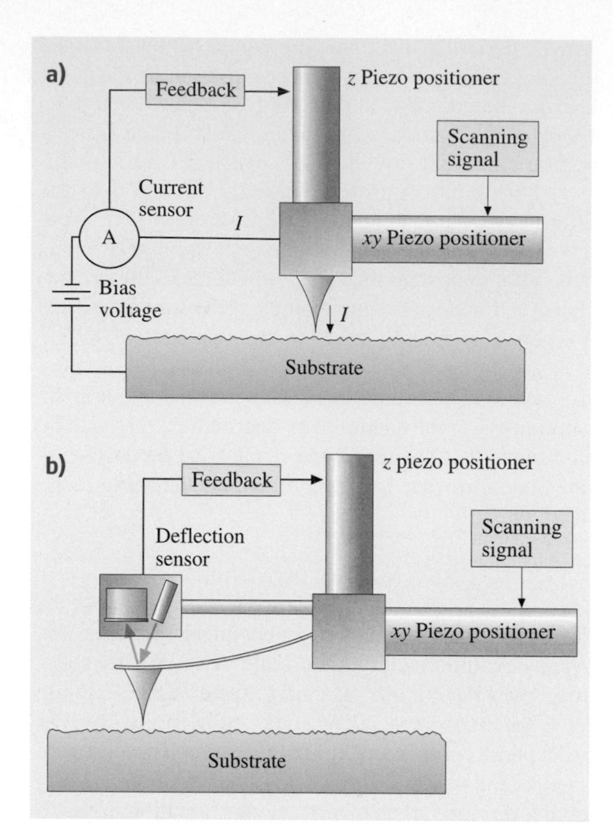

Fig. 5.50a,b Scanning probe systems: **(a)** STM and **(b)** AFM

a feedback loop, so that the current and, therefore, the distance are kept constant throughout the scanning. In this configuration the picture of the surface topography is obtained by recording the vertical position of the tip at each x-y position.

The STM system only works for conductive surfaces because of the need to establish a tunneling current. The atomic force microscopy was developed as an alternative for imaging either conducting or nonconducting surfaces. In AFM, the tip is attached to a flexible cantilever and is brought in contact with the surface (Fig. 5.50b). The force between the tip and the surface is detected by sensing the cantilever deflection. A topographic image of the surface is obtained by plotting the deflection as a function of the x-y position. In a more common mode of operation, a feedback loop is used to maintain a constant deflection, while the topographic information is obtained from the cantilever vertical displacement. Some scanning probe systems use a combination of the AFM and STM modes, i. e., the tip is mounted in a cantilever with an electrical connection, so both the surface forces and the tunneling currents can be controlled or monitored. STM systems can be operated in ultrahigh vacuum (UHV STM) or in air, whereas *AFM* systems are typically operated in air. When a scanning probe system is operated in air, water adsorbed onto the sample surface accumulates underneath the tip, forming a meniscus between the tip and the surface. This water meniscus plays an important role in some of the scanning probe techniques described below.

Scanning Probe Induced Oxidation

Nanometer-scale local oxidation of various materials can be achieved using scanning probes operated in air and biased at a sufficiently high voltage, Fig. 5.51. Tip bias of -2 to -10 V is normally used with writing speeds of 0.1–100 μm/s in an ambient humidity of 20–40 %. It is believed that the water meniscus formed at the contact point serves as an electrolyte such that the biased tip anodically oxidizes a small region of the surface [5.70]. The most common application of this principle is the oxidation of hydrogen-passivated silicon. A dip in HF solution is typically used to passivate the silicon surfaces with hydrogen atoms. Patterns of oxide "written" on a silicon surface can be used as a mask for wet or dry etching. Ten-nm line width patterns have been successfully transferred to a silicon substrate in this fashion [5.71]. Various metals have also been locally anodized using this approach such as alu-

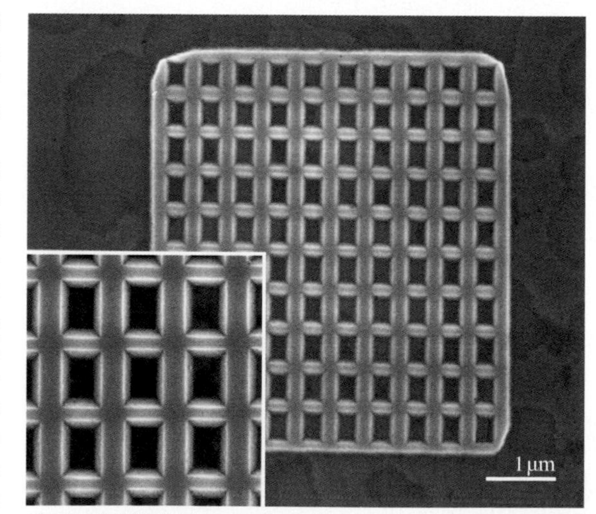

Fig. 5.51 SEM image of an inverted, truncated pyramid array fabricated on a silicon SOI wafer by SPM oxidation and subsequent etch in TMAH (pitch is 500 nm) [5.69]

minum or titanium [5.72]. An interesting variation of this process is the anodization of deposited amorphous silicon [5.73]. Amorphous silicon can be deposited at low temperature on top of many materials. The deposited silicon layer can be patterned and used as, for example, the gate of a 0.1 μm CMOS transistor [5.74], or it can be used as a mask to pattern an underlying film. The major drawback of this technique is poor reproducibility due to tip wear during the anodization. However, using AFM in noncontact mode has overcome this problem [5.70].

Scanning Probe Resist Exposure and Lithography

Electrons emitted from a biased SPM tip can be used to expose a resist the same way e-beam lithography does (Fig. 5.52) [5.74]. Various systems have been used for this lithographic technique. These include constant current STM, noncontact AFM, and AFM with constant tip-resist force and constant current. The systems using AFM cantilevers have the advantage of performing imaging and alignment tasks without exposing the resist. Resists well characterized for e-beam lithography (e.g., PMMA or SAL601) have been used with scanning probe lithography to achieve reliable sub-100-nm lithography. The procedure for this process is as follows. The wafers are cleaned and the native oxide (in the case of silicon or poly) is removed with a HF dip. Subsequently, 35–100 nm-thick resist is spin-coated on top of the surface. The exposure is done by moving the SPM tip over the surface while applying a bias voltage high enough to produce an emission of electrons from the tip (a few tens of volts). Development of the resist is performed in standard solutions following the exposure. Features below 50 nm in width have been achieved with this procedure.

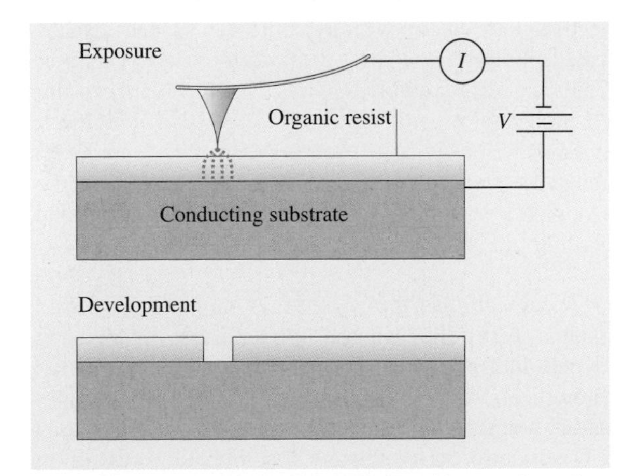

Fig. 5.52 Scanning probe lithography with organic resist

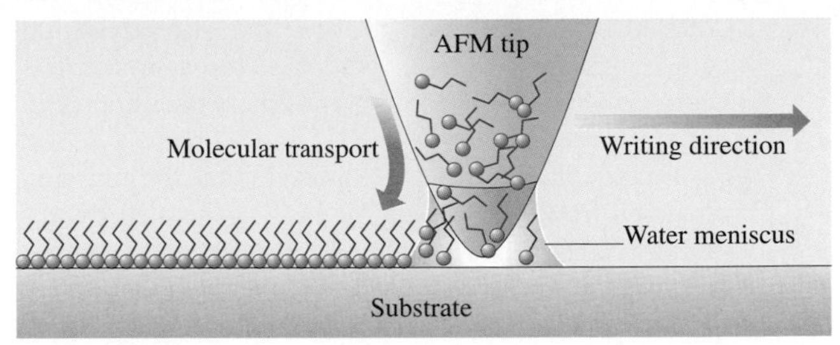

Fig. 5.53 Schematic representation of the dip-pen nanolithography working principle

Dip-Pen Nanolithography

In dip-pen nanolithography (DPN), the tip of an AFM operated in air is "inked" with a chemical of interest and brought into contact with a surface. The ink molecules flow from the tip onto the surface as with a fountain pen. The water meniscus that naturally forms between the tip and the surface enables the diffusion and transport of the molecules, as shown in Fig. 5.53. Inking can be done by dipping the tip in a solution containing a small concentration of the molecules followed by a drying step (e.g., blowing dry with compressed difluoroethane). Line widths down to 12 nm with spatial resolution of 5 nm have been demonstrated with this technique [5.75]. Species patterned with DPN include conducting polymers, gold, dendrimers, DNA, organic dyes, antibodies, and alkanethiols. Alkanethiols have also been used as an organic monolayer mask to etch a gold layer and subsequently etch the exposed silicon substrate.

Other Scanning Probe Nanofabrication Techniques

A great variety of nanofabrication techniques using scanning probe systems have been demonstrated. Some of these are proof of concept demonstrations, and their utility as a viable and repeatable fabrication process has yet to be evaluated. For example, a substrate can be mechanically machined using STM/AFM tips acting as plows or engraving tools [5.76]. This can be used to directly create structures in the substrate, although it is more commonly used to pattern a resist for a subsequent etch, liftoff, or electro-deposition step. Mechanical nanomachining with SPM probes can be facilitated by heating the tip above the glass transition of a polymeric substrate material. This approach has been applied to SPM-based high-density data storage in polycarbonate substrates [5.77].

Electric fields strong enough to induce the emission of atoms from the tip can be easily generated by applying voltage pulses above 3 V. This phenomenon has been used to transfer material from the tip to the surface and vice versa. Ten- to 20-nm mounds of metals such as Au, Ag, or Pt have been deposited or removed from a surface in this fashion [5.78]. The same approach has been used to extract single atoms from a semiconductor surface and re-deposit them elsewhere [5.79]. Manipulation of nanoparticles, molecules, and single atoms on top of a surface has also been achieved by simply pushing or sliding them with the SPM tip [5.80]. Metals can also be locally deposited by the STM chemical vapor deposition technique [5.81]. In this technique, a precursor organometallic gas is introduced in the STM chamber. A voltage pulse applied between the tip and the surface dissociates the precursor gas into a thin layer of metal. Local electro-chemical etch [5.82] and electrodeposition [5.83] are also possible using SPM systems. A droplet of suitable solution is first placed on the substrate. Then the STM tip is immersed into the droplet and a voltage is applied. In order to reduce Faradaic currents the tip is coated with wax such that only the very end is exposed to the solution. Sub-100-nm feature size has been achieved using this technique.

Using a single tip to serially produce the desired modification in a surface leads to very slow fabrication processes that are impractical for mass production. Many of the scanning probe techniques developed thus far, however, could also be performed by an array of tips, which would increase their throughput and make them more competitive compared with other parallel nanofabrication processes. This approach has been demonstrated for imaging, lithography [5.84], and data storage [5.85] using both 1-D and 2-D arrays of scanning probes. With the development of larger arrays with individual control of force, vertical position, and current advances, we might see these techniques become standard fabrication processes in the industry.

5.3.4 Self-Assembly and Template Manufacturing

Self-assembly is a nanofabrication technique that involves aggregation of colloidal nano-particles into the final desired structure [5.86]. This aggregation can be either spontaneous (entropic) and due to the thermodynamic minima (energy minimization) constraints, or chemical and due to the complementary binding of organic molecules and supramolecules (molecular self-assembly) [5.87]. Molecular self-assembly is one of the most important techniques used in biology for the development of complex functional structures. Since these techniques require that the target structures be thermodynamically stable, they tend to produce structures that are relatively defect-free and self-healing. Self-assembly is by no means limited to molecules or the nano-domain and can be carried out on just about any scale, making it a powerful bottom-up assembly and manufacturing method (multiscale ordering). Another attractive feature of this technique relates to the possibility of combining self-assembly properties of organic molecules with the electronic, magnetic, and photonic properties of inorganic components. *Template manufacturing* is another bottom-up technique that utilizes material deposition (electroplating, CVD, etc.) in nano-templates in order to fabricate nanostructures. The nano-templates used in this technique are usually prepared using self-assembly techniques. In the following sections, we will discuss various important self-assembly and template manufacturing techniques currently under heavy investigation.

Physical and Chemical Self-Assembly

The central theme behind the self-assembly process is spontaneous (physical) or chemical aggregation of colloidal nano-particles [5.88]. Spontaneous self-assembly exploits the tendency of mono-dispersed nano- or sub-micro colloidal spheres to organize into a face-centered cubic (FCC) lattice. The force driving this process is the desire of the system to achieve a thermodynamically stable state (minimum free energy). In addition to spontaneous thermal self-assembly, gravitational, convective, and electrohydrodynamic forces can also be used to induce aggregation into complex 3-D structures. Chemical self-assembly requires the attachment of a single molecular organic layer (*self-assembled monolayer*, or *SAM*) to the colloidal particles (organic or inorganic) and subsequent self-assembly of these components into a complex structure using molecular recognition and binding.

Physical Self-Assembly. This is an entropic-driven method that relies on the spontaneous organization of colloidal particles into a relatively stable structure through non-covalent interactions. For example, colloidal polystyrene spheres can be assembled into a 3-D structure on a substrate that is held vertically in the colloidal solution, Fig. 5.54 [5.89, 90]. Upon the

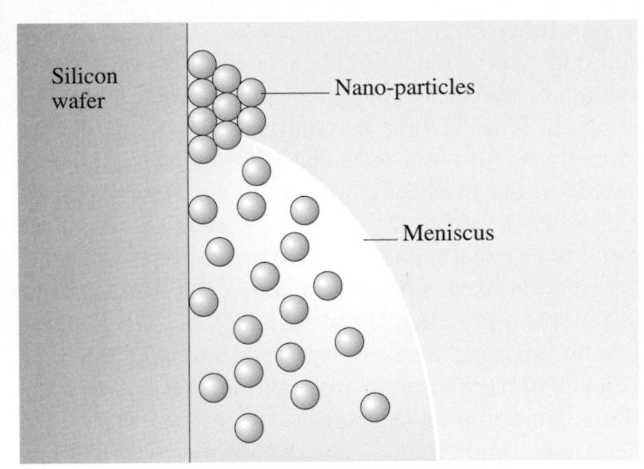

Fig. 5.54 Colloidal particle self-assembly onto solid substrates upon drying in vertical position

Fig. 5.55 Cross-sectional SEM image of a thin planar opal silica template (spheres 855 nm in diameter) assembled directly on a Si wafer [5.89]

evaporation of the solvent, the spheres aggregate into a hexagonal close packed (HCP) structure. The interstitial pore size and density are determined by the polymer sphere size. The polymer spheres can be etched into smaller sizes after forming the HCP arrays, thereby altering the template pore separations [5.91]. This technique can fabricate large patterned areas in a quick, simple, and cost-effective way. A classic example is the natural assembly of on-chip silicon photonic bandgap crystals [5.89], which are capable of reflecting the light arriving from any direction in a certain wavelength range [5.92]. In this method, a thin layer of silica colloidal spheres is assembled on a silicon substrate. This is achieved by placing a silicon wafer vertically in a vial containing an ethanolic suspension of silica spheres. A temperature gradient across the vial aids the flow of silica spheres. Figure 5.55 shows the cross-sectional SEM image of a thin planar opal template assembled directly on a Si wafer from 855 nm spheres. Once such a template is prepared, LPCVD can be used to fill the interstitial spaces with Si, so that the high refractive index of silicon provides the necessary bandgap.

One can also deposit colloidal particles into a patterned substrate (template-assisted self-assembly, TASA) [5.93, 94]. This method is based on the principle that when aqueous dispersion of colloidal particles is allowed to dewet from a solid surface that is already patterned, the colloidal particles are trapped by the recessed regions and assembled into aggregates of shapes and sizes determined by the geometric confinement provided by the template. The patterned arrays of templates can be fabricated using conventional contact-mode photolithography, which gives control over the shape and dimensions of the templates, thereby allowing the assembly of complex structures from colloidal particles. The cross-sectional view of a fluidic cell used in TASA is shown in Fig. 5.56. The fluidic cell has two parallel glass substrates to confine the aqueous dispersion of the colloidal particles. The surface of the bottom substrate is patterned with a 2-D array of templates. When the aqueous dispersion is allowed to slowly dewet across the cell, the capillary force exerted on the liquid pushes the colloidal spheres across the surface of the bottom substrate until they are physically trapped by the templates. If the concentration of the colloidal dispersion is high enough, the template will be filled by the maximum number of colloidal particles determined by the geometrical confinement. This method can be used to fabricate a variety of polygonal and polyhedral aggregates that are difficult to generate [5.95].

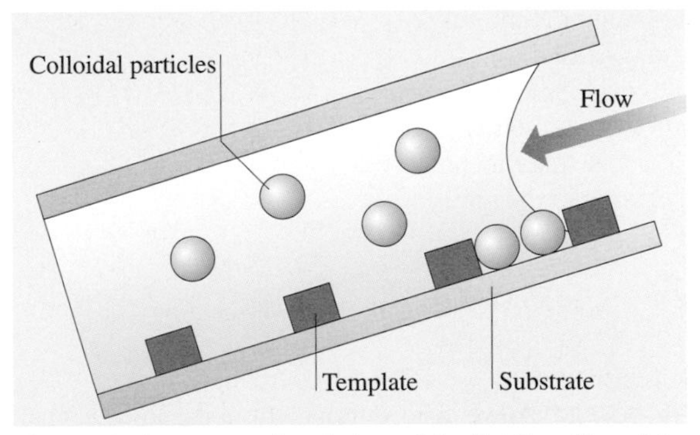

Fig. 5.56 A cross-sectional view of the fluidic cell used for template-assisted self-assembly

Chemical Self-Assembly. Organic and supramolecular SAMs play a critical role in colloidal particle self-assembly. SAMs are robust organic molecules that are chemically adsorbed onto solid substrates [5.96]. Most often, they have a hydrophilic (polar) head that can be bonded to various solid surfaces and a long, hydrophobic (nonpolar) tail that extends outward. SAMs are formed by the immersion of a substrate in a dilute solution of the molecule in an organic solvent. The resulting film is a dense organization of molecules arranged to expose the end group. The durability of a SAM is highly dependent on the effectiveness of the anchoring to the surface of the substrate. SAMs have been widely studied, because the end group can be functionalized to form precisely arranged molecular arrays for various applications ranging from simple, ultrathin insulators and lubricants to complex biological sensors. Chemical self-assembly uses organic or supramolecular SAMs as the binding and recognition sites for fabricating complex 3-D structures from colloidal nano-particles. Most commonly used organic monolayers include: 1) organosilicon compounds on glass and native surface oxide layer of silicon, 2) alkanethiols, dialkyl disulfides, and dialkyl sulfides on gold, 3) fatty acids on alumina and other metal oxides, and 4) DNA.

Octadecyltrichlorosilane (OTS) is the most common organosilane used in the formation of SAMs, mainly because of the fact that it is simple, readily available, and forms good, dense layers [5.97, 98]. Alkyltrichlorosilane monolayers can be prepared on clean silicon wafers whose surface is SiO_2 (with almost 5×10^{14} SiOHgroups/cm^2). Figure 5.57 shows the schematic representation of the formation of alkylsiloxane monolayers by adsorption of alkyltrichlorosilanes from solution onto Si/SiO_2 substrates. Since the silicon-chloride bond is susceptible to hydrolysis, a limited amount of water has to be present in the system in order to obtain good quality monolayers. Monolayers made of methyl- and vinyl-terminated alkylsilanes are autophobic to the hydrocarbon solution and hence emerge uniformly dry from the solution, whereas monolayers made of ester-terminated alkylsilanes emerge wet from the solution used in their formation. The disadvantage of this method is that if the alkyltrichlorosilane in the solvent adhering to the substrate is exposed to water, a cloudy film is deposited on the surface due to formation of a gel of polymeric siloxane.

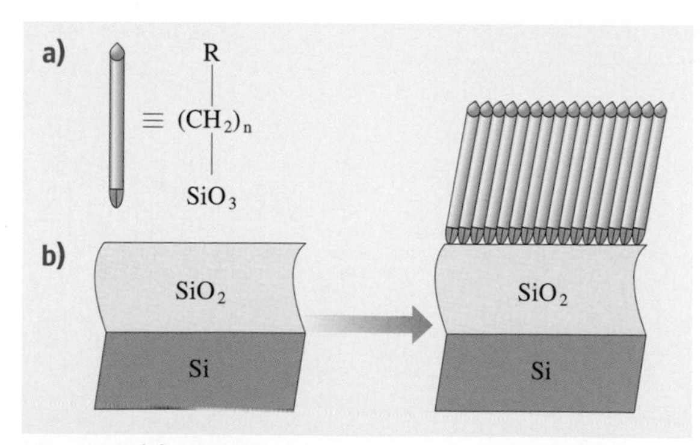

Fig. 5.57 **(a)** Alkylsiloxane formed from the adsorption of alkyltrichlorosilane on Si/SiO_2 substrates. **(b)** Schematic representation of the process

Another important organic SAM system is the alkanethiols ($X(CH_2)_nSH$, where X is the endgroup) on gold [5.96, 99–101]. A major advantage of using gold as the substrate material is that it does not have a stable oxide, and thus it can be handled in ambient conditions. When a fresh, clean, hydrophilic gold substrate is immersed (several minutes to several hours) into a dilute solution (10^{-3} M) of the organic sulfur compound (alkanethiols) in an inorganic solvent, a close-packed, oriented monolayer can be obtained. Sulfur is used as the head group, because of its strong interaction with gold substrate (44 kcal/mol), resulting in the formation of a close-packed, ordered monolayer. The end group of alkanethiol can be modified to render hydrophobic or hydrophilic properties to the adsorbed layer. Another method for depositing alkanethiol SAM is soft-lithography. This technique is based on inking a PDMS stamp with alkanethiol and its subsequent transfer to planar and nonplanar substrates. Alkanethiol functionalized surfaces (planar, nonplanar, spherical) can also be used to self-assemble a variety of intricate 3-D structures [5.102].

Carboxylic acid derivatives self-assemble on surfaces (e.g., glass, Al_2O_3, and Ag_2O) through an acid-base reaction, giving rise to monolayers of fatty acids [5.103]. The time required for the formation of a complete monolayer increases with decreased concentration. A higher concentration of carboxylic acid is required to form a monolayer on gold compared to Al_2O_3. This is due to differences in the affinity of the COOH groups (more affinity to Al_2O_3 and glass than gold) and also the surface concentration of the salt-forming oxides in the two substrates. In the case of amorphous metal oxide surfaces, the chemisorption of alkanoic acids is not unique. For example, on Ag_2O the carboxylate two oxygen atoms bind to the substrate in a nearly symmetrical manner, resulting in ordered monolayers with a chain tilt angle from the surface normally of 15° to 25°. But on CuO and Al_2O_3, the oxygen atoms bind themselves symmetrically and the chain tilt angle is close to 0°. The structure of the monolayers is thus a balance of the various interactions taking place in the polymer chains.

Deoxyribonucleic acid (DNA), the framework on which all life is built, can be used to self-assemble nanomaterials into useful macroscopic aggregates that display a number of desired physical properties [5.104]. DNA consists of two strands, which are coiled around each other to form a double helix. When the two strands are uncoiled, singular strands of nucleotides are left. These nucleotides consist of a sugar (pentose ring) a phosphate (PO_4), and a nitrogenous base. The order and architecture of these components are essential for the proper structure of a nucleotide. There are typically four nucleotides found in DNA: Adenine (A), Guanine (G), Cytosine (C), and Thyamine (T). A key property of the DNA structure is that the described nucleotides bind specifically to another nucleotide when arranged in the two-strand double helix (A to T and C to G). This specific bonding capability can be used to assemble nanophase material and nano-structures [5.105]. For example, nucleotide functionalized nano-gold particles have been assembled into complex 3-D structures by attaching DNA strands to the gold via an enabler or linker [5.106]. In a separate work, DNA was used to assemble nanoparticles into macroscopic materials. This method uses alkane dithiol as the linker molecule to connect the DNA template to the nanoparticle. The thiol groups at each end of the linker molecule covalently attach themselves to the colloidal particles to form aggregate structures [5.107].

Template Manufacturing

Template manufacturing refers to a set of techniques that can be used to fabricate 3-D organic or inorganic structures from a nano-template. These templates differ in material, pattern, feature size, overall template size, and periodicity. Although nano-templates can be fabricated using e-beam lithography, the serial nature of this technique prohibits its widespread application. Self-assembly is the preferred technique that can produce large-area nano-templates in a massively parallel fashion. Several nano-templates have been investigated for use in template manufacturing. These include polymer colloidal spheres, alumina membranes, and nuclear-track etched membranes. Colloidal spheres can be deposited in a regular 3-D array using the techniques described in the previous section (see Figs. 5.54–5.56). Porous aluminum oxide membranes can be fabricated by the anodic oxidation of aluminum [5.108]. The oxidized film consists of columnar arrays of hexagonal, close-packed pores with a separation comparable to the pore size. By controlling the electrolyte species, temperature, anodizing voltage and time, different pore sizes, density and height can be obtained. The pore size and depth can further be adjusted by etching the oxide in an appropriate acid. Templates of porous polycarbonate or mica membranes can be fabricated by nuclear-track etched membranes [5.109]. This technique is based on the passage of high-energy decay fragments from a radioactive source through a dielectric material. The particles leave behind chemically active damaged tracks that subsequently can be etched to create pores throughout the thickness of the membrane [5.110, 111]. Unlike the other methods, the pore separation and, hence, the pore density are independent of the pore size. The pore density is only determined by the irradiation process.

Following template fabrication, the interstitial spaces (in the case of colloidal spheres) or pores (in the case of alumina and polycarbonate membranes) in the template are filled with the desired material [5.91, 112]. This can be done using a variety of deposition techniques such as electroplating and CVD. The final structure can be a composite of nano-template and deposited mater-

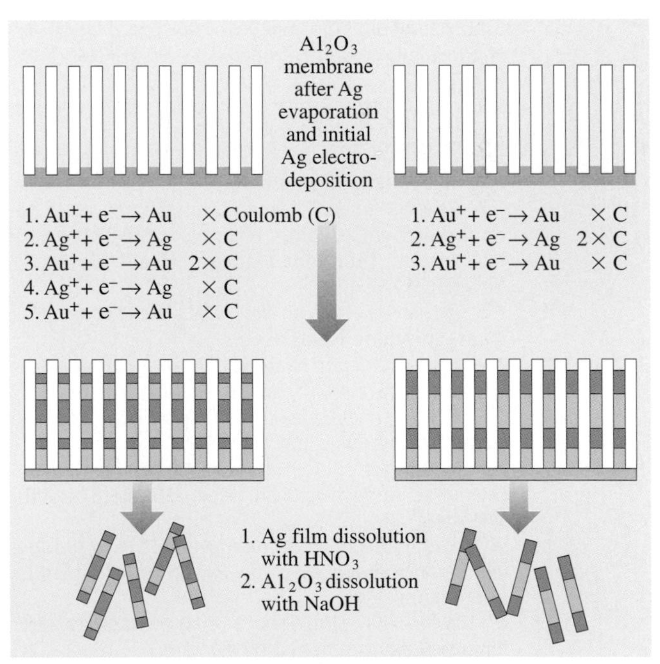

Fig. 5.58 Synthesis of nano-bar-code particles

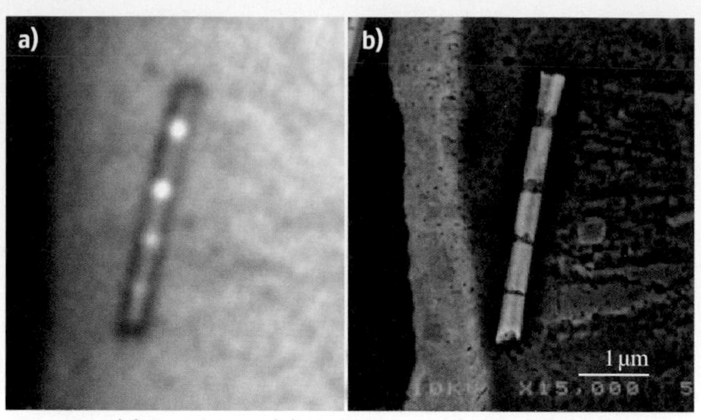

Fig. 5.59 (a) Optical and (b) FE-SEM images of Au-Ag multistripe particle 5.116

ial, or the template can be selectively etched, resulting in an air-filled 3-D complex structure. For example, nickel [5.111], iron [5.113], and cobalt [5.114] nanowires have been electrochemically grown into porous template matrices. Three-dimensional photonic crystals have been fabricated by electrochemical deposition of CdSe and silicon into polystyrene and silica colloidal assembly templates [5.89, 115]. An interesting example of template-assisted manufacturing is the synthesis of nanometer metallic bar codes [5.116]. These nano-bar codes are prepared by the electrochemical reduction of metallic ions into the pores of an aluminum oxide membrane followed by their release through etching of the template [5.117–119]. This procedure is schematically illustrated in Fig. 5.58. A back-side silver film is used as the working electrode for the reduction of metallic ions (silver and gold in this case) from solution. Up to seven different metallic segments as short as 10 nm and as long as several micrometers with 13 distinguishable stripes have been fabricated using this technique. Optical reflectivity is used to read out the striping pattern encoded in the metal particles [5.116]. Figure 5.59 shows optical and field emission scanning electron microscope images of a Au-Ag multistripe nano-bar code (Ag stripes ranging in length from 60 to 240 nm separated by Au segments of 550 nm). These coded nano-particles can be used in fluorescence and mass-spectrometry-based assays, enabling a wide variety of bioanalytical measurements.

References

5.1 S. A. Campbell: *The Science and Engineering of Microelectronic Fabrication* (Oxford Univ. Press, New York 2001)

5.2 C. J. Jaeger: *Introduction to Microelectronic Fabrication* (Prentice Hall, New Jersey 2002)

5.3 J. D. Plummer, M. D. Deal, P. B. Griffin: *Silicon VLSI Technology* (Prentice Hall, New Jersey 2000)

5.4 J. E. Bjorkholm: EUV lithography: the successor to optical lithography, Intel Technol. J. **2** (1998) 1–8

5.5 J. L. Vossen: *Thin Film Processes* (Academic Press, New York 1976)

5.6 M. Gad-el-Hak (Ed.): *The MEMS Handbook* (CRC Press, Boca Raton 2002)

5.7 T.-R. Hsu: *MEMS and Microsystems Design and Manufacture* (McGraw-Hill, New York 2002)

5.8 G. T. A. Kovacs: *Micromachined Transducers Sourcebook* (McGraw-Hill, New York 1998)

5.9 G. T. A. Kovacs, N. I. Maluf, K. A. Petersen: Bulk micromachining of silicon, Proc. IEEE **86**(8) (1998) 1536–1551

5.10 P. Rai-Choudhury (Ed.): *Handbook of Microlithography, Micromachining and Microfabrication* (SPIE, Bellingham 1997)

5.11 T. J. Cotler, M. E. Elta: Plasma-etch technology, IEEE Circuits & Devices Mag. **6** (1990) 38–43

5.12 U. Gosele, Q. Y. Tong: Semiconductor wafer bonding, Annu. Rev. Mater. Sci. **28** (1998) 215–241

5.13 Q. Y. Tong, U. Gosele: *Semiconductor Wafer Bonding: Science and Technology* (Wiley, New York 1999)

5.14 F. Niklaus, P. Enoksson, E. Kalveston, G. Stemme: Void-free full-wafer adhesive bonding, J. Micromech. Microeng. **11** (2000) 100–107

5.15 C. A. Harper: *Electronic Packaging and Interconnection Handbook* (McGraw-Hill, New York 2000)

5.16 Y. T. Cheng, L. Lin, K. Najafi: Localized silicon fusion and eutectic bonding for MEMS fabrication and packaging, J. Microelectromech. Syst. **9**(1) (2000) 3–8

5.17 W. H. Ko, J. T. Suminto, G. J. Yeh: Bonding techniques for microsensors. In: *Micromachining and Micropackaging for Transducers* (Elsevier, Amsterdam 1985)

5.18 W. P. Maszara, G. Goetz, A. Caviglia, J. B. McKitterick: Bonding of silicon wafers for silicon-on-insulator, J. Appl. Phys. **64**(10) (1988) 4943–4950

5.19 K. Najafi, K. D. Wise, T. Mochizuki: A high-yield IC-compatible multichannel recording array, IEEE Trans. Electron Devices **32** (1985) 1206–1211

5.20 A. Selvakumar, K. Najafi: A high-sensitivity *z* axis capacitive silicon microaccelerometer with a tortional suspension, J. Microelectromech. Syst. **7** (1998) 192–200

5.21 H. Baltes, O. Paul, O. Brand: Micromachined thermally based CMOS microsensors, Proc. IEEE **86**(8) (1998) 1660–1678

5.22 B. Eyre, K. S. J. Pister, W. Gekelman: Multi-axis microcoil sensors in standard CMOS, Proc. SPIE Conf. Micromachined Devices and Components, Austin 1995, 183–191

5.23 K. A. Shaw, Z. L. Zhang, N. C. MacDonnald: SCREAM: a single mask single-crystal silicon process for microelectromechanical structures, Proc. IEEE Workshop Micro Electro Mechanical Systems, Fort Lauderdale 1993, 155–160

5.24 G. K. Fedder, S. Santhanam, M. L. Reed, S. C. Eagle, D. F. Guillo, M. S. C. Lu, L. R. Carley: Laminated high-aspect-ratio microstructures in a conventional CMOS process, Proc. IEEE Workshop Micro Electro Mechanical Systems, San Diego 1996, 13–18

5.25 N. C. MacDonald: SCREAM MicroElectroMechanical Systems, Microelectron. Eng. **32** (1996) 51–55

5.26 B. P. Van Drieenhuizen, N. I. Maluf, I. E. Opris, G. T. A. Kovacs: Force-balanced accelerometer with mG resolution fabricated using silicon fusion bonding and deep reactive ion etching, Proc. Int. Conf. Solid-State Sensors and Actuators, Chicago 1997, 1229–1230

5.27 X. Huikai, L. Erdmann, Z. Xu, K. J. Gabriel, G. K. Fedder: Post-CMOS processing for high-aspect-ratio integrated silicon microstructures, J. Microelectromech. Syst. **11** (2002) 93–101

5.28 J. M. Bustillo, R. S. Muller: Surface micromachining for microelectromechanical systems, Proc. IEEE **86**(8) (1998) 1552–1574

5.29 H. C. Nathanson, W. E. Newell, R. A. Wickstrom, J. R. Davis: The resonant gate transistor, IEEE Trans. Electron Devices **14** (1967) 117–133

5.30 R. T. Howe, R. S. Muller: Polycrystalline silicon micromechanical beams, Proc. Electrochem. Soc. Spring Meeting, Montreal 1982, 184–185

5.31 J. A. Geen, S. J. Sherman, J. F. Chang, S. R. Lewis: Single-chip surface-micromachined integrated gyroscope with 50 degrees/hour root Allan variance, IEEE J. Solid-State Circuits **37** (2002) 1860–1866

5.32 P. F. Van Kessel, L. J. Hornbeck, R. E. Meier, M. R. Douglass: A MEMS-based projection display, Proc. IEEE **86**(8) (1998) 1687–1704

5.33 A. E. Franke, D. Bilic, D. T. Chang, P. T. Jones, R. T. Howe, G. C. Johnson: Post-CMOS integration of germanium microstructures, Proc. Micro Electro Mechanical Systems, Orlando 1999, 630–637

5.34 S. Sedky, P. Fiorini, M. Caymax, S. Loreti, K. Baert, L. Hermans, R. Mertens: Structural and mechanical properties of polycrystalline silicon germanium for micromachining applications, J. Microelectromech. Syst. **7** (1998) 365–372

5.35 N. Tas, T. Sonnenberg, H. Jansen, R. Legtenberg, M. Elwenspoek: Stiction in surface micromachining, J. Micromech. Microeng. **6** (1996) 385–397

5.36 R. Maboudian, R. T. Howe: Critical review: adhesion in surface micromechanical structures, J. Vac. Sci. Technol. B **15**(1) (1997) 1–20

5.37 J. H. Smith, S. Montague, J. J. Sniegowski, J. R. Murray, P. J. McWhorter: Embedded micromechanical devices for the monolithic integration of MEMS with CMOS, Proc. Int. Electron Devices Meeting, Washington 1995, 609–612

5.38 R. S. Muller, K. Y. Lau: Surface-micromachined microoptical elements and systems, Proc. IEEE **86**(8) (1998) 1705–1720

5.39 K. S. J. Pister, M. W. Judy, S. R. Burgett, R. S. Fearing: Microfabricated hinges: 1 mm vertical features with surface micromachining, Proc. 6th Int. Conf. Solid-State Sensors and Actuators, San Francisco 1991, 647–650

5.40 L. Y. Lin, S. S. Lee, M. C. Wu, K. S. J. Pister: Micromachined integrated optics for free space interconnection, Proc. IEEE MicroElectroMechanical Systems Workshop, Amsterdam 1995, 77–82

5.41 H. Guckel: High-aspect-ratio micromachining via deep X-ray lithography, Proc. IEEE **86**(8) (1998) 1586–1593

5.42 E. W. Becker, W. Ehrfeld, P. Hagmann, A. Maner, D. Munchmeyer: Fabrication of microstructures with high aspect ratios and great structural heights by synchrotron radiation lithography, galvanoforming, and plastic molding (LIGA process), Microelectron. Eng. **4** (1986) 35–56

5.43 K. Y. Lee, N. LaBianca, S. A. Rishton, S. Zolgharnain, J. D. Gelorme, J. Shaw, T. H. P. Chang: Micromachining applications of a high resolution ultra-thick photoresist, J. Vac. Sci. & Technol. B **13** (1995) 3012–3016

5.44 K. Roberts, F. Williamson, G. Cibuzar, L. Thomas: The fabrication of an array of microcavities utilizing SU-8 photoresist as an alternative 'LIGA' technology, Proc. Thirteenth Biennial University/Government/Industry Microelectronics Symposium, Minnesota 1999 (IEEE, Piscataway 1999) 139–141

5.45 C. Burbaum, J. Mohr, P. Bley, W. Ehrfeld: Fabrication of capacitive acceleration sensors by the LIGA technique, Sensors & Actuators A **27** (1991) 559–563

5.46 C. G. Keller, R. T. Howe: Hexsil bimorphs for vertical actuation, Digest of Technical Papers 8th Int. Conf. Solid-State Sensors and Actuators and Eurosensors IX, Stockholm 1995, 99–102

5.47 C. G. Keller, R. T. Howe: Nickel-filled hexsil thermally actuated tweezers, Digest of Technical Papers 8th Int. Conf. Solid-State Sensors and Actuators and Eurosensors IX, Stockholm 1995, 376–379

5.48 D. A. Horsley, M. B. Cohn, A. Singh, R. Horowitz, A. P. Pisano: Design and fabrication of an angular microactuator for magnetic disk drives, J. Microelectromech. Syst. **7** (1998) 141–148

5.49 N. Yazdi, F. Ayazi, K. Najafi: Micromachined inertial sensors, Proc. IEEE **86** (1998) 1640–1659
5.50 F. Ayazi, K. Najafi: High aspect-ratio combined poly and single-crystal silicon (HARPSS) MEMS technology, J. Microelectromech. Syst. **9** (2000) 288–294
5.51 F. Ayazi, K. Najafi: A HARPSS polysilicon vibrating ring gyroscope, J. Microelectromech. Syst. **10** (2001) 169–179
5.52 Y.S. No, F. Ayazi: The HARPSS process for fabrication of nano-precision silicon electromechanical resonators, Proc. 2001 1st IEEE Conference on Nanotechnology, 2001, 489–494
5.53 G. Timp: *Nanotechnology* (Springer, New York 1998)
5.54 P. Rai-Choudhury (Ed.): *Handbook of Microlithography, Micromachining and Microfabrication* (SPIE, Bellingham 1997)
5.55 L. Ming, C. Bao-qin, Y. Tian-Chun, Q. He, X. Qiuxia: The sub-micron fabrication technology, Proc. 6th Int. Conf. Solid-State and Integrated-Circuit Technol. (IEEE, 2001) 452–455
5.56 S.Y. Chou: Nano-imprint lithography and lithographically induced self-assembly, MRS Bulletin **26** (2001) 512–517
5.57 S.Y. Chou, C. Keimel, J. Gu: Ultrafast and direct imprint of nanostructures in silicon, Nature **417** (2002) 835–837
5.58 M.A. Herman: *Molecular Beam Epitaxy: Fundamentals and Current Status* (Springer, New York 1996)
5.59 J.S. Frood, G.J. Davis, W.T. Tsang: *Chemical Beam Epitaxy and Related Techniques* (Wiley, New York 1997)
5.60 S. Mahajan, K.S. Sree Harsha: *Principles of Growth and Processing of Semiconductors* (McGraw-Hill, New York 1999)
5.61 S. Kim, M. Razegi: Advances in quantum dot structures. In: *Processing and Properties of Compound Semiconductors*, ed. by R. Willardson, H.S. Navawa (Academic Press, New York 2001)
5.62 D. Bimberg, M. Grundmann, N.N. Ledentsov: *Quantum Dot Heterostructures* (Wiley, New York 1999)
5.63 G. Seebohm, H.G. Craighead: Lithography and patterning for nanostructure fabrication. In: *Quantum Semiconductor Devices and Technologies*, ed. by T.P. Pearsall (Kluwer, Boston 2000)
5.64 E. Kapon: Lateral patterning of quantum well heterostructures by growth on nonplanar substrates. In: *Epitaxial Microstructures*, ed. by A.C. Gossard (Academic Press, New York 1994)
5.65 F. Guffarth, R. Heitz, A. Schliwa, O. Stier, N.N. Ledentsov, A.R. Kovsh, V.M. Ustinov, D. Bimberg: Strain engineering of self-organized InAs quantum dots, Phys. Rev. B **64** (2001) 085305(1)–085305(7)
5.66 M. Sugawara: *Self-Assembled InGaAs/GaAs Quantum Dots* (Academic Press, New York 1999)
5.67 B.C. Lee, S.D. Lin, C.P. Lee, H.M. Lee, J.C. Wu, K.W. Sun: Selective growth of single InAs quantum dots using strain engineering, Appl. Phys. Lett. **80** (2002) 326–328
5.68 K. Brunner: Si/Ge nanostructures, Rep. Prog. Phys. **65** (2002) 27–72
5.69 F.S.S. Chien, W.F. Hsieh, S. Gwo, A.E. Vladar, J.A. Dagata: Silicon nanostructures fabricated by scanning probe oxidation and tetra-methyl ammonium hydroxide etching, J. Appl. Phys. **91** (2002) 10044–10050
5.70 M. Calleja, J. Anguita, R. Garcia, K. Birkelund, F. Perez-Murano, J.A. Dagata: Nanometer-scale oxidation of silicon surfaces by dynamic force microscopy: reproducibility, kinetics and nanofabrication, Nanotechnology **10** (1999) 34–38
5.71 E.S. Snow, P.M. Campbell, F.K. Perkins: Nanofabrication with proximal probes, Proc. IEEE **85** (1997) 601–611
5.72 H. Sugimura, T. Uchida, N. Kitamura, H. Masuhara: Tip-induced anodization of titanium surfaces by scanning tunneling microscopy: a humidity effect on nanolithography, Appl. Phys. Lett. **63** (1993) 1288–1290
5.73 N. Kramer, J. Jorritsma, H. Birk, C. Schonenberger: Nanometer lithography on silicon and hydrogenated amorphous silicon with low energy electrons, J. Vac. Sci. & Technol. B **13** (1995) 805–811
5.74 H.T. Soh, K.W. Guarini, C.F. Quate: *Scanning Probe Lithography* (Kluwer, Boston 2001)
5.75 C.A. Mirkin: Dip-pen nanolithography: automated fabrication of custom multicomponent, sub-100 nanometer surface architectures, MRS Bulletin **26** (2001) 535–538
5.76 L.L. Sohn, R.L. Willett: Fabrication of nanostructures using atomic-force-microscope-based lithography, Appl. Phys. Lett. **67** (1995) 1552–1554
5.77 H.J. Mamin, B.D. Terris, L.S. Fan, S. Hoen, R.C. Barrett, D. Rugar: High-density data storage using proximal probe techniques, IBM J. Res. & Dev. **39** (1995) 681–699
5.78 K. Bessho, S. Hashimoto: Fabricating nanoscale structures on Au surface with scanning tunneling microscope, Appl. Phys. Lett. **65** (1994) 2142–2144
5.79 I.W. Lyo, P. Avouris: Field-induced nanometer-to atomic-scale manipulation of silicon surfaces with the STM, Science **253** (1991) 173–176
5.80 M.F. Crommie, C.P. Lutz, D.M. Eigler: Confinement of electrons to quantum corrals on a metal surface, Science **262** (1993) 218–220
5.81 A. de Lozanne: Pattern generation below 0.1 micron by localized chemical vapor deposition with the scanning tunneling microscope,, Japan. J. Appl. Physic **33** (1994) 7090–7093
5.82 L.A. Nagahara, T. Thundat, S.M. Lindsay: Nanolithography on semiconductor surfaces under an etching solution, Appl. Phys. Lett. **57** (1990) 270–272
5.83 T. Thundat, L.A. Nagahara, S.M. Lindsay: Scanning tunneling microscopy studies of semiconductor

electrochemistry, J. Vac. Sci. & Technol. A **8** (1990) 539–543
5.84 S. C. Minne, S. R. Manalis, A. Atalar, C. F. Quate: Independent parallel lithography using the atomic force microscope, J. Vac. Sci. & Technol. B **14** (1996) 2456–2461
5.85 M. Lutwyche, C. Andreoli, G. Binnig, J. Brugger, U. Drechsler, W. Haeberle, H. Rohrer, H. Rothuizen, P. Vettiger: Microfabrication and parallel operation of 5 × 5 2D AFM cantilever arrays for data storage and imaging, Proc. MEMS '98 (1998), 8–11
5.86 G. M. Whitesides, B. Grzybowski: Self-assembly at all scales, Science **295** (2002) 2418–2421
5.87 P. Kazmaier, N. Chopra: Bridging size scales with self-assembling supramolecular materials, MRS Bulletin **25** (2000) 30–35
5.88 R. Plass, J. A. Last, N. C. Bartelt, G. L. Kellogg: Self-assembled domain patterns, Nature **412** (2001) 875
5.89 Y. A. Vlasov, X.-Z. Bo, J. G. Sturm, D. J. Norris: On-chip natural self-assembly of silicon photonic bandgap crystals, Nature **414** (2001) 289–293
5.90 C. Gigault, D.-K. Veress, J. R. Dutcher: Changes in the morphology of self-assembled polystyrene microsphere monolayers produced by annealing, J. Colloid Interface Sci. **243** (2001) 143–155
5.91 J. C. Hulteen, P. Van Duyne: Nanosphere lithography: a materials general fabrication process for periodic particle array surfaces, J. Vac. Sci. Technol. A **13** (1995) 1553–1558
5.92 J. D. Joannopoulos, P. R. Villeneuve, S. Fan: Photonic crystals: putting a new twist on light, Nature **386** (1997) 143–149
5.93 T. D. Clark, R. Ferrigno, J. Tien, K. E. Paul, G. M. Whitesides: Template-directed self-assembly of 10-μm-sized hexagonal plates, J. Am. Chem. Soc. **124** (2002) 5419–5426
5.94 S. A. Sapp, D. T. Mitchell, C. R. Martin: Using template-synthesized micro-and nanowires as building blocks for self-assembly of supramolecular architectures, Chem. Mater. **11** (1999) 1183–1185
5.95 Y. Yin, Y. Lu, B. Gates, Y. Xia: Template assisted self-assembly: a practical route to complex aggregates of monodispersed colloids with well-defined sizes, shapes and structures, J. Am. Chem. Soc. **123** (2001) 8718–8729
5.96 J. L. Wilbur, G. M. Whitesides: Self-assembly and self-assembles monolayers in micro and nanofabrication. In: *Nanotechnology*, ed. by G. Timp (Springer, New York 1999)
5.97 S. R. Wasserman, Y. T. Tao, G. M. Whitesides: Structure and reactivity of alkylsiloxane monolayers formed by reaction of alkyltrichlorosilanes on silicon substrates, Langmuir **5** (1989) 1074–1087
5.98 C. P. Tripp, M. L. Hair: An infrared study of the reaction of octadecyltrichlorosilane with silica, Langmuir **8** (1992) 1120–1126
5.99 D. R. Walt: Nanomaterials: top-to-bottom functional design, Nature **1** (2002) 17–18
5.100 J. Noh, T. Murase, K. Nakajima, H. Lee, M. Hara: Nanoscopic investigation of the self-assembly processes of dialkyl disulfides and dialkyl sulfides on Au(111), J. Phys. Chem. B **104** (2000) 7411–7416
5.101 M. Himmelhaus, F. Eisert, M. Buck, M. Grunze: Self-assembly of n-alkanethiol monolayers: a study by IR-visible sum frequency spectroscopy (SFG), J. Phys. Chem. **104** (2000) 576–584
5.102 A. K. Boal, F. Ilhan, J. E. DeRouchey, T. Thurn-Albrecht, T. P. Russell, V. M. Rotello: Self-assembly of nanoparticles into structures spherical and network aggregates, Nature **404** (2000) 746–748
5.103 A. Ulman: *An Introduction to Ultrathin Organic Films: From Langmuir–Blodgett to Self-Assembly* (Academic Press, New York 1991)
5.104 E. Winfree, F. Liu, L. A. Wenzler, N. C. Seeman: Design and self-assembly of two-dimensional DNA crystals, Nature **394** (1998) 539–544
5.105 J. H. Reif, T. H. LaBean, N. C. Seeman: Programmable assembly at the molecular scale: self-assembly of DNA lattices, Proc. 2001 IEEE Int. Conf. Robotics and Automation (IEEE, 2001) 966–971
5.106 A. P. Alivisatos, K. P. Johnsson, X. Peng, T. E. Wilson, C. J. Loweth, M. P. Bruchez Jr, P. G. Schultz: Organization of 'nanocrystal molecules' using DNA, Nature **382** (1996) 609–611
5.107 C. Y. Cao, R. Jin, C. A. Mirkin: Nanoparticles with Raman spectroscopic fingerprints for DNA and RNA detection, Science **297** (2002) 1536–1540
5.108 H. Masuda, H. Yamada, M. Satoh, H. Asoh: Highly ordered nanochannel-array architecture in anodic alumina, Appl. Phys. Lett. **71** (1997) 2770–2772
5.109 R. L. Fleischer: *Nuclear Tracks in Solids: Principles and Applications* (Univ. of California Press, Berkeley 1976)
5.110 R. E. Packard, J. P. Pekola, P. B. Price, R. N. R. Spohr, K. H. Westmacott, Y. Q. Zhu: Manufacture observation and test of membranes with locatable single pores, Rev. Sci. Instrum. **57** (1986) 1654–1660
5.111 L. Sun, P. C. Searson, C. L. Chien: Electrochemical deposition of nickel nanowire arrays in single-crystal mica films, Appl. Phys. Lett. **74** (1999) 2803–2805
5.112 Y. Du, W. L. Cai, C. M. Mo, J. Chen, L. D. Zhang, X. G. Zhu: Preparation and photoluminescence of alumina membranes with ordered pore arrays, Appl. Phys. Lett. **74** (1999) 2951–2953
5.113 S. Yang, H. Zhu, D. Yu, Z. Jin, S. Tang, Y. Du: Preparation and magnetic property of Fe nanowire array, J. Magn. Mater. **222** (2000) 97–100
5.114 M. Sun, G. Zangari, R. M. Metzger: Cobalt island arrays with in-plane anisotropy electrodeposited in highly ordered alumina, IEEE Trans. Magn. **36** (2000) 3005–3008
5.115 P. V. Braun, P. Wiltzius: Electrochemically grown photonic crystals, Nature **402** (1999) 603–604

5.116 S. R. Nicewarner-Pena, R. G. Freeman, B. D. Reiss, L. He, D. J. Pena, I. D. Walton, R. Cromer, C. D. Keating, M. J. Natan: Submicrometer metallic barcodes, Science **294** (2001) 137–141
5.117 D. Almalawi, C. Z. Ziu, M. Moskovits: Nanowires formed in anodic oxide nanotemplates, J. Mater. Res. **9** (1996) 1014
5.118 J. C. Hulteen, C. R. Martin: A general template-based method for the preparation of nanomaterials, J. Mater. Chem. **7** (1997) 1075–1087
5.119 B. R. Martin, D. J. Dermody, B. D. Reiss, M. Fang, L. A. Lyon, M. J. Natan, T. E. Mallouk: Orthogonal self-assembly on colloidal gold-platinum nanorods, Adv. Mater. **11** (1999) 1021–1025

6. Stamping Techniques for Micro and Nanofabrication: Methods and Applications

This chapter highlights some recent advances in high resolution printing methods, in which a "stamp" forms a pattern of "ink" on the surface it contacts. It focuses on two approaches whose capabilities, level of development, and demonstrated applications indicate a strong potential for widespread use, especially in areas where conventional methods are unsuitable. The first of these, known as microcontact printing, uses a high resolution rubber stamp to print patterns of chemical inks, mainly those that lead to the formation of organic self-assembled monolayers (SAMs). These printed SAMs can be used either as resists in selective wet etching, or as templates in selective deposition to form structures of a variety of materials. The other approach, referred to as nanotransfer printing, uses similar high resolution stamps, but ones inked with solid thin film materials. In this case, SAMs, or other types of surface chemistries, bond these films to a substrate that the stamp contacts. The material transfer that results upon removal of the stamp forms a pattern in the geometry of the relief features, in a purely additive fashion. In addition to providing detailed descriptions of these micro/nanoprinting techniques, this chapter illustrates their use in some areas where these methods may provide attractive alternatives to more established lithographic methods.

6.1 **High Resolution Stamps** 186

6.2 **Microcontact Printing** 187

6.3 **Nanotransfer Printing** 190

6.4 **Applications** 193
6.4.1 Unconventional Electronic Systems 193
6.4.2 Lasers and Waveguide Structures 198

6.5 **Conclusions** 200

References 200

The demonstrator applications span fields as diverse as biotechnology (intravascular stents), fiber optics (tunable fiber devices), nanoanalytical chemistry (high resolution nuclear magnetic resonance), plastic electronics (paper-like displays), and integrated optics (distributed feedback lasers). The growing interest in nanoscience and nanotechnology motivates research and the development of new methods that can be used for nanofabricating the relevant test structures or devices. The attractive capabilities of the techniques described here, together with the interesting and subtle materials science, chemistry, and physics associated with them, make this a promising area for basic and applied study.

There is considerable interest in methods for building structures that have micron or nanometer dimensions. Historically, research and development in this area has been driven mainly by the needs of the microelectronics industry. The spectacularly successful techniques that have emerged from those efforts – photolithography, electron beam lithography, etc. – are extremely well suited to the tasks for which they were principally designed: forming structures of radiation sensitive materials (e.g., photoresists or electron beam resists) on ultra-flat glass or semiconductor surfaces. Significant challenges exist in adapting these methods for new emerging applications and areas of research that require patterning of unusual systems and materials (e.g., those in biotechnology, plastic electronics, etc.), structures with nanometer dimensions (i. e., below 50–100 nm), large areas in a single step (i. e., larger than a few square centimeters), or non-planar (i. e., rough or curved) surfaces. These established techniques also have the disadvantage of high capital and operational costs. As a result, some of the oldest and conceptually simplest forms of lithography – embossing, molding, stamping, writing, etc. – are now being re-examined for their potential to serve as the basis for nanofabrication techniques that can avoid these limitations [6.1]. Considerable progress has been made in the last few years, mainly by combining these approaches or variants of them with new materials, chemistries, and processing techniques. This chapter highlights some recent advances in high resolution printing methods, in which a "stamp" forms a pattern of "ink" on a surface that it contacts. It focuses on approaches whose capabilities, level of development, and demonstrated applications indicate a strong potential for widespread use, especially in areas where conventional methods are unsuitable.

Contact printing involves the use of an element with surface relief (i. e., the stamp) for transferring material applied to its surface (i. e. the ink) to locations on a substrate that it contacts. The printing press, one of the earliest manufacturable implementations of this approach, was introduced by Gutenberg in the fifteenth century. The general approach has since been used almost exclusively for producing printed text or images with features that are one hundred microns or larger in their minimum dimension. The resolution is determined by the nature of the ink and its interaction with the stamp and/or substrate, the resolution of the stamp, and the processing conditions that are used for printing or converting the pattern of ink into a pattern of functional material. This chapter focuses on (1) printing techniques that are capable of micron and nanometer resolution, and (2) their use for fabricating key elements of active electronic or optical devices and subsystems. It begins with an overview of some methods for fabricating high resolution stamps and then illustrates two different ways that these stamps can be used to print patterns of functional materials. Applications that highlight the capabilities of these techniques and the performance of systems that are constructed with them are also presented.

6.1 High Resolution Stamps

The printing process can be separated into two parts: fabrication of the stamp and the use of this stamp to pattern features defined by the relief on its surface. These two processes are typically quite different, although it is possible in some cases to use patterns generated by a stamp to produce a replica of that stamp. The structure from which the stamp is derived, which is known as the "master", can be fabricated with any technique that is capable of producing well-defined structures of relief on a surface. This master can then be used directly as the stamp, or to produce stamps via molding or printing procedures. It is important to note that the technique for producing the master does not need to be fast or low in cost. It also does not need to possess many other characteristics that might be desirable for a given patterning task: It is used just once to produce a master, which is directly or indirectly used to fabricate stamps. Each one of these stamps can then be used many times for printing.

In a common approach for the high resolution techniques that are the focus of this chapter, an established lithographic technique, such as one of those developed for the microelectronics industry, defines the master. Figure 6.1 schematically illustrates typical processes. Here, photolithography patterns a thin layer of resist on a silicon wafer. Stamps are generated from this structure in one of two ways: By casting against this master, or by etching the substrate with the patterned resist as a mask. In the first approach, the master itself can be used multiple times to produce many stamps, typically using a light or heat-curable prepolymer. In the second, the etched substrate serves as the stamp. Additional stamps can be generated either by repeating the lithography and etching, or by using the original stamp to print replica stamps. For minimum lateral feature sizes greater than $\sim 1-2$ microns, contact or proximity mode photolithography with a mask produced by direct write photolithography represents a convenient method to fabricate the master. For features smaller than ~ 2 microns, several different techniques can be used [6.2], including: (1) projection mode photolithography [6.3], (2) direct write electron beam (or focused ion beam) lithography [6.4, 5], (3) scanning probe lithography [6.6–9] or (4) laser interference lithography [6.10]. The first approach requires a photomask generated by some other method, such as direct write photolithography or electron beam lithography. The reduction (typically 4×) provided by the projection optics relaxes the resolution requirements on the mask and enables features as small as ~ 90 nm when deep ultraviolet radiation and phase shifting masks are used. The costs for these systems are, however, very high and their availability for general research purposes is limited. The second method is flexible in the geometry of patterns that can be produced, and the writing sys-

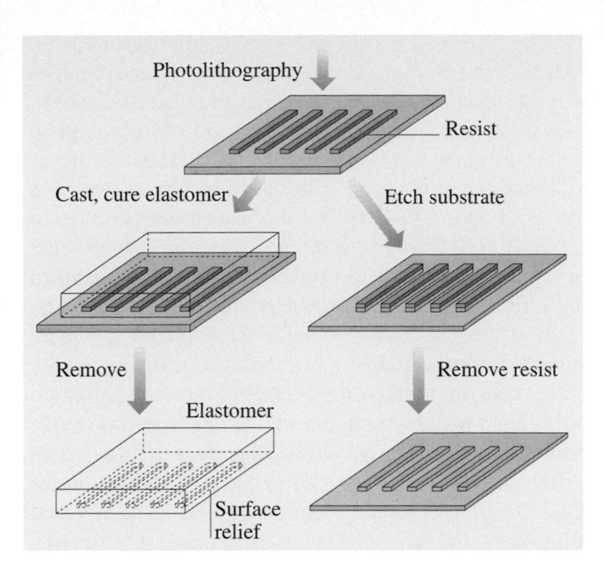

Fig. 6.1 Schematic illustration of two methods for producing high resolution stamps. The first step involves patterning a thin layer of some radiation sensitive material, known as the resist, on a flat substrate, such as a silicon wafer. It is convenient to use an established technique, such as photolithography or electron beam lithography, for this purpose. This structure, known as the master, is converted to a stamp either by etching or molding. In the first case, the resist acts as a mask for etching the underlying substrate. Removing the resist yields a stamp. This structure can be used directly as a stamp to print patterns, or to produce additional stamps. In the molding approach, a prepolymer is cast against the relief structure formed by the patterned resist on the substrate. Curing (thermally or optically) and then peeling the resulting polymer away from the substrate yields a stamp. In this approach, many stamps can be made with a single master, and each stamp can be used many times

tems are highly developed: 30–50 nm features can be achieved with commercial systems [6.11], and < 10 nm features are possible with research tools, as first demonstrated more than 25 years ago by *Broers* [6.12]. The main drawbacks of this method are that it is relatively slow and it is difficult to pattern large areas. Like projection mode photolithography, it can be expensive. The third method, scanning probe lithography, is quite powerful in principle, but the tools are not as well established as those for other approaches. This technique has atomic resolution, but its writing speed can be lower and the areas that can be patterned are smaller than electron beam systems. Interference lithography provides a powerful, low cost tool for generating periodic arrays of features with dimensions down to 100–200 nm; smaller sizes demand ultraviolet lasers, and patterns with aperiodic or non-regular features are difficult to produce.

6.2 Microcontact Printing

Microcontact printing (μCP) [6.13] is one of several soft lithographic techniques – replica molding, micromolding in capillaries, microtransfer molding, near-field conformal photolithography using an elastomeric phase-shifting mask, etc. – that have been developed as alternatives to established methods for micro- and nanofabrication [6.14–18]. μCP uses an elastomeric element (usually polydimethylsiloxane – PDMS) with high resolution features of relief as a stamp to print patterns of chemical inks. It was mainly developed for use with inks that form self-assembled monolayers (SAMs) of alkanethiolates on gold and silver. The procedure for carrying out μCP in these systems is remarkably simple: A stamp, inked with a solution of alkanethiol, is brought into contact with the surface of a substrate to transfer ink molecules to regions where the stamp and substrate contact. The resolution and effectiveness of μCP rely on conformal contact between the stamp and the surface of the substrate, rapid formation of highly ordered monolayers [6.19], and autophobicity of the SAM, which effectively blocks the reactive spreading of the ink across the surface [6.20]. It can pattern SAMs over relatively large areas ($\sim$ up to 0.25 ft^2 have been demonstrated in prototype electronic devices) in a single impression [6.21]. The edge resolution of SAMs printed onto thermally evaporated gold films is on the order of 50 nm, as determined by lateral force microscopy [6.22]. Microcontact printing has been used with a range of different SAMs on various substrates [6.14]. Of these, alkanethiolates on gold, silver, and palladium [6.23] presently give the highest resolution. In many cases, the mechanical properties of the stamp limit the sizes of the smallest features that can be achieved: The most commonly used elastomer (Sylgard 184, Dow Corning) has a low modulus, which can lead to mechanical collapse or sagging for features of relief with aspect

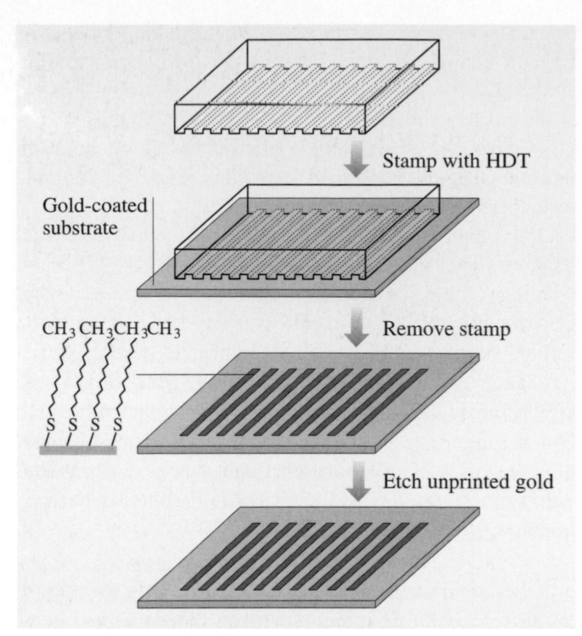

Fig. 6.2 Schematic illustration of microcontact printing. The first step involves inking a stamp with a solution of a material that is capable of forming a self-assembled monolayer (SAM) on a substrate that will be printed. In the case illustrated here, the ink is a millimolar concentration of hexadecanethiol (HDT) in ethanol. Directly applying the ink to the surface of the stamp with a pipette prepares the stamp for printing. Blowing the surface of the stamp dry and placing it on a substrate delivers the ink to areas where the stamp contacts the substrate. The substrate consists of a thin layer of Au on a flat support. Removing the stamp after a few seconds of contact leaves a patterned SAM of HDT on the surface of the Au film. The printed SAM can act as a resist for the aqueous-based wet etching of the exposed regions of the Au. The resulting pattern of conducting gold can be used to build devices of various types

ratios greater than ~ 2 or less than ~ 0.05. Stamps fabricated with high modulus elastomers avoid some of these problems [6.24, 25]. Conventional stamps are also susceptible to in-plane mechanical strains that can cause distortions in the printed patterns. Composite stamps that use thin elastomer layers on stiff supports are effective at minimizing this source of distortion [6.26]. Methods for printing that avoid direct mechanical manipulation of the stamp can reduce distortions with conventional and composite stamps [6.21]. This approach has proven effective in large area flexible circuit applications that require accurate multilevel registration.

The patterned SAM can be used either as a resist in selective wet etching or as a template in selective deposition to form structures of a variety of materials: metals, silicon, liquids, organic polymers, and even biological species. Figure 6.2 schematically illustrates the use of μCP and wet etching to pattern a thin film of Au. Figure 6.3 shows SEM images of nanostructures of gold (20 nm thick, thermally evaporated with a 2.5 nm layer of Ti as an adhesion promoter) and silver (~ 100 nm thick formed by electroless deposition using commercially available plating baths) [6.27] that were fabricated using this approach. In the first and second examples, the masters for the stamps consisted of photoresist patterned on silicon wafers with projection and contact mode photolithography, respectively. Placing these masters in a desiccator for ~ 1 h with a few drops of tridecafluoro-1,1,2,2-tetrahydrooctyl-1-trichlorosilane forms a silane monolayer on the exposed native oxide of the silicon. This monolayer prevents adhesion of the master to PDMS (Sylgard 184), which is cast and cured from a 10:1 mixture of prepolymer and curing agent. Placing a few drops of a ~ 1 mM solution of hexadecanethiol (HDT) in ethanol on the surface of the stamps, and then blowing them dry with a stream of nitrogen prepares them for printing. Contacting the metal film for a few seconds with the stamp produces a patterned self-assembled monolayer (SAM) of HDT. An aqueous etchant (1 mM $K_4Fe(CN)_6$, 10 mM $K_3Fe(CN)_6$, and 0.1 M $Na_2S_2O_3$) removes the unprinted regions of the silver [6.28]. A similar solution (1 mM $K_4Fe(CN)_6$, 10 mM $K_3Fe(CN)_6$, 1.0 M KOH, and 0.1 M $Na_2S_2O_3$)

Fig. 6.3 Scanning electron micrographs of typical structures formed by microcontact printing a self-assembled monolayer ink of hexadecanethiol onto a thin metal film followed by etching of the unprinted areas of the film. The *left frame* shows an array of Au (20 nm-thick) dots with ~ 500 nm diameters. The *right frame* shows a printed structure of Ag (100 nm thick) in the geometry of interdigitated source/drain electrodes for a transistor in a simple inverter circuit. The edge resolution of patterns that can be easily achieved with microcontact printing is ~ 50–100 nm

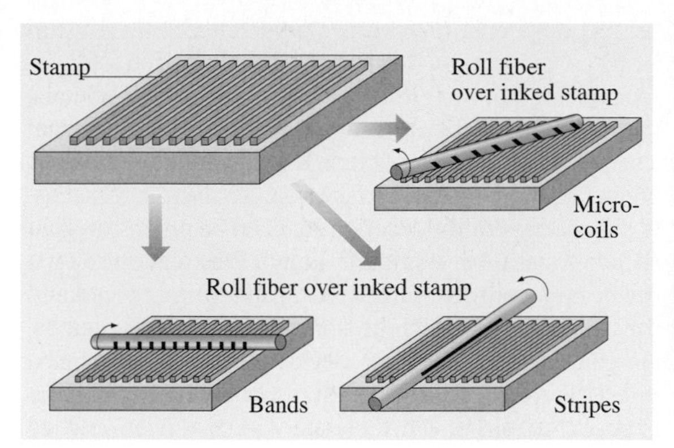

Fig. 6.4 Schematic illustration of a simple method to print lines on the surfaces of optical fibers. Rolling a fiber over the inked stamp prints a pattern onto the fiber surface. Depending on the orientation of the fiber axis with the line stamp illustrated here, it is possible in a single rotation of the fiber to produce a continuous microcoil, or arrays of bands or stripes

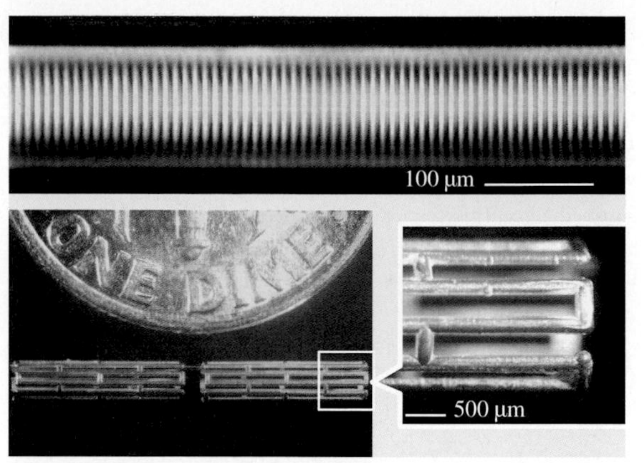

Fig. 6.5 Optical micrographs of some three-dimensional microstructures formed by microcontact printing on curved surfaces. The *top frame* shows an array of 3-micron lines of Au (20 nm)/Ti (1.5 nm) printed onto the surface of an optical fiber. This type of structure can be used as an integrated photomask for producing mode-coupling gratings in the core of the fiber. The *bottom frames* show a freestanding metallic microstructure formed by (1) microcontact printing and etching a thin (100 nm-thick) film of Ag on the surface of a glass microcapillary tube, (2) electroplating the Ag to increase its thickness (to tens of microns) and (3) etching away the glass microcapillary with concentrated hydrofluoric acid. The structure shown here has the geometry and mechanical properties of an intravascular stent, which is a biomedical device commonly used in balloon angioplasty

can be used for etching the bare gold [6.29]. The results of Fig. 6.3 show that the roughness on the edges of the patterns is $\sim 50-100$ nm. The resolution is determined by the grain size of the metal films, the isotropic etching process, slight reactive spreading of the inks, and edge disorder in the patterned SAMs.

The structures of Fig. 6.3 were formed on the flat surfaces of silicon wafers (left image) and glass slides (right image). An attractive feature of μCP and certain other contact printing techniques is their ability to pattern features with high resolution on highly curved or rough surfaces [6.18, 30, 31]. This type of patterning task is difficult or impossible to accomplish with photolithography due to its limited depth of focus and the difficulty of casting uniform films of photoresist on non-flat surfaces. Figure 6.4 shows, as an example, a straightforward approach for high resolution printing on the highly curved surfaces of optical fibers. Here, simply rolling the fiber over an inked stamp prints a pattern on the entire outer surface of the fiber. Simple staging systems allow alignment of features to the fiber axis; they also ensure registration of the pattern from one side of the fiber to the other [6.16]. Figure 6.5 shows 3 micron-wide lines and spaces printed onto the surface of a single mode optical fiber (diameter 125 μm). The bottom frame shows a freestanding metallic structure with the geometry and mechanical properties of an intravascular stent, which is a biomedical device that is commonly used in balloon angioplasty procedures. In this latter case μCP followed by electroplating generated the Ag microstructure on a sacrificial glass cylinder that was subsequently etched away with concentrated hydrofluoric acid [6.32]. Other examples of microcontact printing on non-flat surfaces (i. e., low cost plastic sheets and optical ridge waveguides) appear in the applications section of this chapter.

6.3 Nanotransfer Printing

Nanotransfer printing (nTP) is a more recent high resolution printing technique, which uses surface chemistries as interfacial "glues" and "release" layers (rather than "inks", as in μCP) to control the transfer of solid material layers from relief features on a stamp to a substrate [6.33–35]. This approach is purely additive (i. e., material is only deposited in locations where it is needed), and it can generate complex patterns of single or multiple layers of materials with nanometer resolution over large areas in a single process step. It does not suffer from surface diffusion or edge disorder in the patterned inks of μCP, nor does it require post-printing etching or deposition steps to produce structures of functional materials. The method involves four components: (1) a stamp (rigid, flexible, or elastomeric) with relief features in the geometry of the desired pattern, (2) a method for depositing a thin layer of solid material onto the raised features of this stamp, (3) a means of bringing the stamp into intimate physical contact with a substrate, and (4) surface chemistries that prevent adhesion of the deposited material to the stamp and promote its strong adhesion to the substrate. nTP has been demonstrated with SAMs and other surface chemistries for printing onto flexible and rigid substrates with hard inorganic and soft polymer stamps. Figure 6.6 presents a set of procedures for using nTP to pattern a thin metal bilayer of Au/Ti with a surface transfer chemistry that relies on a dehydration reaction [6.33]. The process begins with fabrication of a suitable stamp. Elastomeric stamps can be built using the same casting and curing procedures described for μCP. Rigid stamps can be fabricated by (1) patterning resist (e.g., electron beam resist or photoresist) on a substrate (e.g., Si or GaAs), (2) etching the exposed regions of the substrate with an anisotropic reactive ion etch, and (3) removing the resist, as illustrated in Fig. 6.1. For both types of stamps, careful control of the lithography and the etching steps yields features of relief with nearly vertical or slightly re-entrant sidewalls. The stamps typically have depths of relief $> 0.2\,\mu m$ for patterning metal films with thicknesses $< 50\,nm$.

Electron beam evaporation of Au (20 nm; 1 nm/s) and Ti (5 nm; 0.3 nm/s) generates uniform metal bilayers on the surfaces of the stamp. A vertical, collimated flux of metal from the source ensures uniform deposition only on the raised and recessed regions of relief. The gold adheres poorly to the surfaces of stamps made of GaAs, PDMS, glass, or Si. In the process of Fig. 6.6, a fluorinated silane monolayer acts to reduce further the adhesion when a Si stamp (with native oxide) is used. The Ti layer serves two purposes: (1) it promotes adhesion between the Au layer and the substrate after pattern

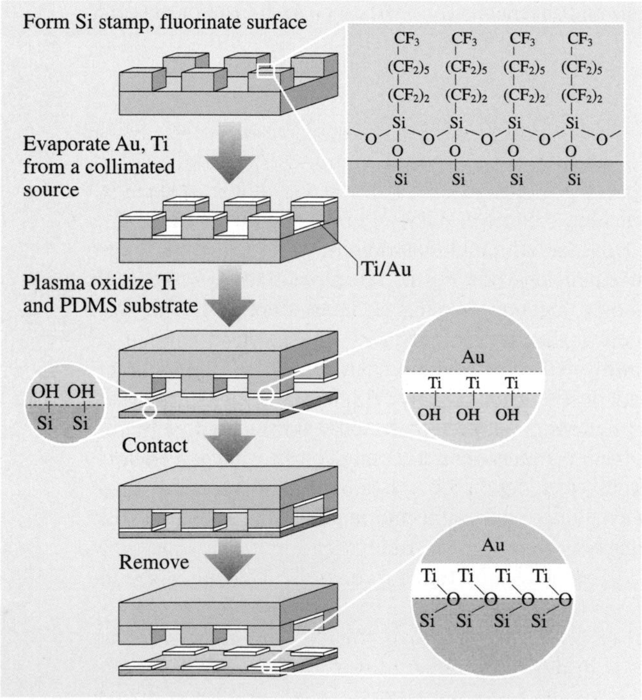

Fig. 6.6 Schematic illustration of nanotransfer printing procedures. Here, interfacial dehydration chemistries control the transfer of a thin metal film from a hard inorganic stamp to a conformable elastomeric substrate (thin film of polydimethylsiloxane (PDMS) on a plastic sheet). The process begins with the fabrication of a silicon stamp (by conventional lithography and etching) followed by surface functionalization of the native oxide with a fluorinated silane monolayer. This layer ensures poor adhesion between the stamp and a bilayer metal film (Au and Ti) deposited by electron beam evaporation. A collimated flux of metal oriented perpendicular to the surface of the stamp avoids deposition on the sidewalls of relief. Exposing the surface Ti layer to an oxygen plasma produces titanol groups. A similar exposure for the PDMS produces silanol groups. Contacting the metal-coated stamp to the PDMS results in a dehydration reaction that links the metal to the PDMS. Removing the stamp leaves a pattern of metal in the geometry of the relief features

transfer, and (2) it readily forms a ~ 3 nm oxide layer at ambient conditions, which provides a surface where the dehydration reaction can take place. Exposing the titanium oxide (TiO_x) surface to an oxygen plasma breaks bridging oxygen bonds, thus creating defect sites where water molecules can adsorb. The result is a titanium oxide surface with some fractional coverage of hydroxyl (–OH) groups (titanol).

In the case of Fig. 6.6, the substrate is a thin film of PDMS (10–50 μm thick) cast onto a sheet of poly(ethylene terephthalate) (PET; 175 μm thick). Exposing the PDMS to an oxygen plasma produces surface (–OH) groups (silanol). Placing the plasma oxidized, Au/Ti-coated stamp on top of these substrates leads to intimate, conformal contact between the raised regions of the stamp and the substrate, without the application of any external pressure. (The soft, conformable PDMS is important in this regard.) It is likely that a dehydration reaction takes place at the (–OH)-bearing interfaces during contact; this reaction results in permanent Ti–O–Si bonds that produce strong adhesion between the two surfaces. Peeling the substrate and stamp apart transfers the Au/Ti bilayer from the raised regions of the stamp (to which the metal has extremely poor adhesion) to the substrate. Complete pattern transfer from an elastomeric stamp to a thin elastomeric substrate occurs readily at room temperature in open air with contact times of less than 15 seconds. When a rigid stamp is employed, slight heating is needed to induce transfer. While the origin of this difference is unclear, it may reflect the comparatively poor contact when rigid stamps are used; similar differences are also observed in cold welding of gold films [6.36].

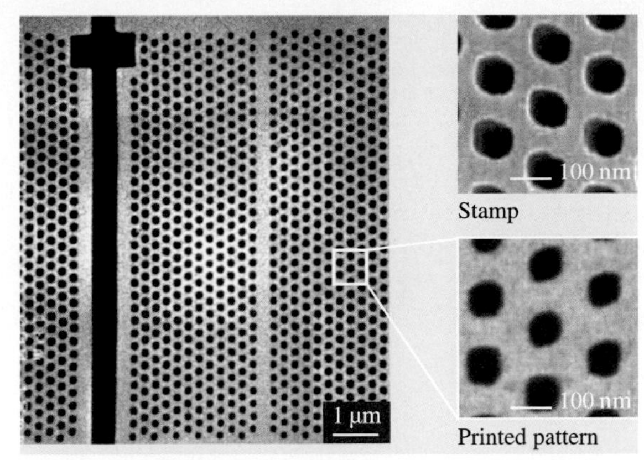

Fig. 6.7 Scanning electron micrograph (SEM) of a pattern produced by nanotransfer printing. The structure consists of a bilayer of Au (20 nm)/Ti (1 nm) (*white*) in the geometry of a photonic bandgap waveguide printed onto a thin layer of polydimethylsiloxane on a sheet of plastic (*black*). Electron beam lithography and etching of a GaAs wafer produced the stamp that was used in this case. The transfer chemistry relied on condensation reactions between titanol groups on the surface of the Ti and silanol groups on the surface of the PDMS. The *frames on the right* show SEMs of the Au/Ti-coated stamp (*top*) before printing and on the substrate (*bottom*) after printing. The electron beam lithography and etching used to fabricate the stamp limit the minimum feature size (~ 70 nm) and the edge resolution (~ 5–10 nm) of this pattern

Figure 6.7 shows scanning electron micrographs of a pattern produced using a GaAs stamp generated by electron beam lithography and etching. The frames on the right show images of the metal-coated stamp before printing (top) and the transferred pattern (bottom). The resolution appears to be limited only by the resolution of the stamp itself, and perhaps by the grain size of the metal films. Although the accuracy in multilevel registration that is possible with nTP has not yet been quantified, its performance is likely similar to that of embossing techniques when rigid stamps are used [6.37].

A wide range of surface chemistries can be used for the transfer. SAMs are particularly attractive due to their chemical flexibility. Figure 6.8 illustrates the use of a thiol terminated SAM and nTP for forming patterns of Au on a silicon wafer [6.34]. Here, the vapor phase co-condensation of the methoxy groups of molecules of 3-mercaptopropyltrimethoxysilane (MPTMS) with the –OH terminated surface of the wafer produces a SAM of MPTMS with exposed thiol (–SH) groups. PDMS stamps can be prepared for printing on this surface by coating them with a thin film (~ 15 nm) of Au using conditions (thermal evaporation 1.0 nm/s; ~ 10^{-7} torr base pressure) that yield optically smooth, uniform films without the buckling that has been observed in the past with similar systems [6.38]. Nanocracking that sometimes occurs in the films deposited in this way can be reduced or eliminated by evaporating a small amount of Ti onto the PDMS before Au deposition and/or by exposing the PDMS surface briefly to an oxygen plasma. Bringing this coated stamp into contact with the MPTMS SAM leads to the formation of sulfur–gold bonds in the regions of contact. Removing the stamp after a few seconds efficiently transfers the gold from the raised regions of the stamp (Au does not adhere to the PDMS) to the substrate. Covalent bonding of the SAM glue to both the substrate and the gold leads to good adhesion of the printed patterns: They easily pass Scotch

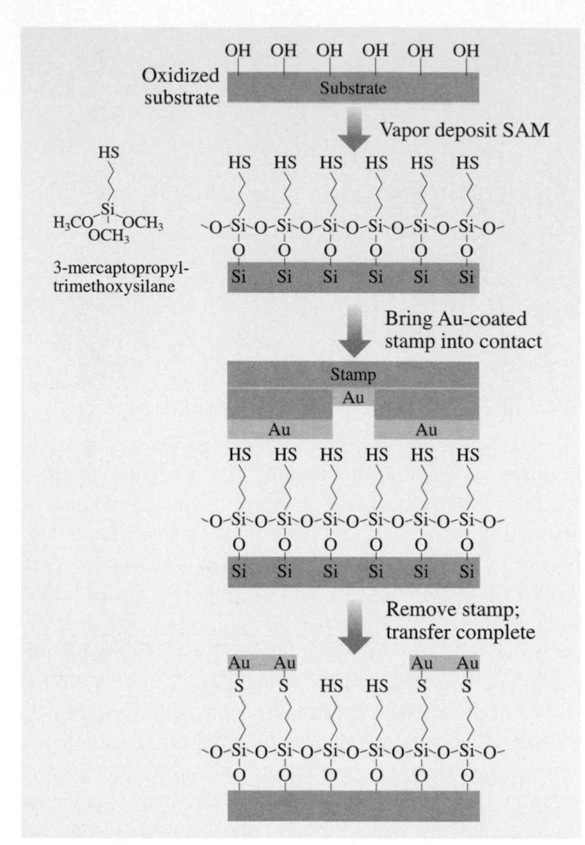

Fig. 6.8 Schematic illustration of steps for nanotransfer printing a pattern of a thin layer of Au onto a silicon wafer using a self-assembled monolayer (SAM) surface chemistry. Plasma oxidizing the surface of the wafer generates $-OH$ groups. Solution or vapor phase exposure of the wafer to 3-mercaptopropyl-trimethoxysilane yields a SAM with exposed thiol groups. Contacting an Au-coated stamp to this surface produces thiol linkages that bond the gold to the substrate. Removing the stamp completes the transfer printing process

tape adhesion tests. Similar results can be obtained with other substrates containing surface $-OH$ groups. For example, Au patterns can be printed onto $\sim 250\,\mu m$-thick sheets of poly(ethylene terephthalate) (PET) by first spin casting and curing (130 °C for 24 h) a thin film of an organosilsesquioxane on the PET. Exposing the cured film ($\sim 1\,\mu m$ thick) to an oxygen plasma and then to air produces the necessary surface ($-OH$) groups. Figure 6.9 shows some optical micrographs of typical printed patterns in this case [6.34].

Similar surface chemistries can guide transfer to other substrates. Alkanedithiols, for example, are useful

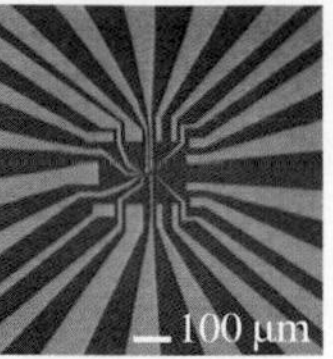

Fig. 6.9 Optical micrographs of patterns of Au (15 nm thick) formed on plastic (*left frame*) and silicon (*right frame*) substrates with nanotransfer printing. The transfer chemistries in both cases rely on self-assembled monolayers with exposed thiol groups. The minimum feature sizes and the edge resolution are both limited by the photolithography used to fabricate the stamps

for printing Au onto GaAs wafers [6.35]. Immersing these substrates (freshly etched with 37% HCl for ~ 2 min to remove the surface oxide) in a 0.05 M solution of 1,8-octanedithiol in ethanol for 3 h produces a monolayer of dithiol on the surface. Although the chemistry of this system is not completely clear, it is generally believed that the thiol end groups bond chemically to the surface. Surface spectroscopy suggests the formation of Ga$-$S and As$-$S bonds. Contacting an Au-coated PDMS stamp with the treated substrate causes the exposed thiol endgroups to react with Au in the regions of contact. This reaction produces permanent Au$-$S bonds at the stamp/substrate interface (see insets in Fig. 6.2 for idealized chemical reaction schemes). Figure 6.10 illustrates a representative optical micrograph of an Au pattern generated by nTP on GaAs.

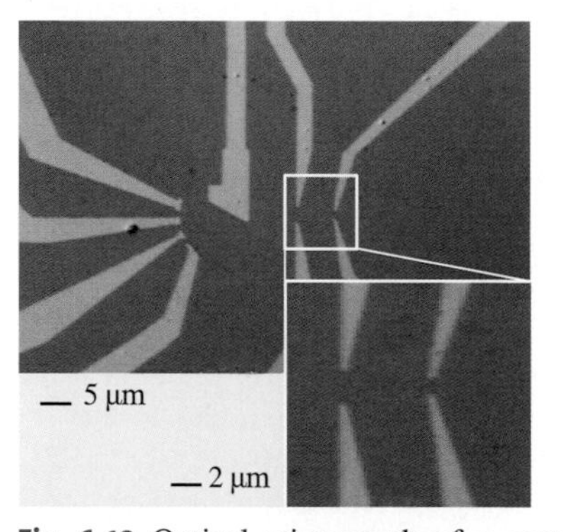

Fig. 6.10 Optical micrograph of a pattern of Au (15 nm thick) formed on a GaAs substrate using nanotransfer printing. The transfer chemistry in this case relies on a dithiol layer formed on the substrate. The minimum feature sizes and the edge resolution are both limited by the photolithography used to fabricate the stamps

6.4 Applications

Promising potential applications for the types of printing techniques described in the previous sections include those where the patterning requirements cannot be met with existing tools or those where the costs of these conventional methods are prohibitively high.We have focused our efforts partly on unusual electronic systems such as flexible plastic circuits and devices that rely on electrodes patterned on curved objects such as microcapillaries and optical fibers. We have also explored photonic systems such as distributed feedback structures for lasers and other integrated optical elements that demand sub-micron features.The sections below highlight several examples in each of these areas.

6.4.1 Unconventional Electronic Systems

A relatively new direction in electronics research seeks to establish low cost plastic materials, substrates, and printing techniques for large area flexible electronic devices, such as paper-like displays. These types of novel devices can complement those (e.g., high density memories, high speed microprocessors, etc.) that are well suited to existing inorganic (e.g., silicon) electronics technologies. High resolution patterning methods for defining the separation between the source and drain electrodes (i. e., channel length) of transistors in these plastic circuits are particularly important because this dimension determines current output and other important characteristics [6.39].

Although conventional patterning techniques, such as photolithography or electron beam lithography, have the required resolution, they are not appropriate because they are expensive and generally require multiple processing steps with resists, solvents, and developers, which can be difficult to use with organic active materials and plastic substrates. Microcontact and nanotransfer printing are both particularly well suited for this application. They can be combined and matched with other techniques, such as ink jet or screen printing, to form a complete system for patterning all layers in practical plastic electronic devices [6.40].

Figure 6.11 illustrates schematically a cross-sectional view of a typical organic transistor. The frame on the right shows the electrical switching characteristics of a device that uses source/drain electrodes of Au patterned by μCP, a dielectric layer of an organosilsesquioxane, a gate of indium tin oxide (ITO), and a PET substrate. The effective semiconductor mobility extracted from these data is comparable to those measured in devices that use the same semiconductor (pentacene in this case) with inorganic substrates and dielectrics, and gold source/drain electrodes defined by photolithography. Our recent work [6.1,27,41]with μCP in the area of plastic electronics demonstrates: (1) methods for using cylindrical "roller" stamps mounted on fixed axles for printing in a continuous reel-to-reel fashion, high resolution source/drain electrodes in ultrathin gold and silver deposited from solution at room temperature using electroless deposition, (2) techniques for performing registration and alignment of the printed features with other elements of a circuit over large areas, (3) strategies for achieving densities of defects that are as good as those observed with photolithography when the

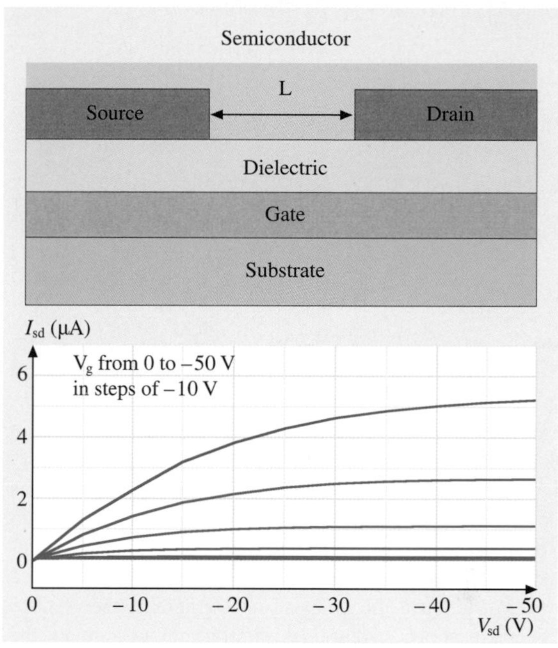

Fig. 6.11 Schematic cross-sectional view (*top*) and electrical performance (*bottom*) of an organic thin film transistor with microcontact printed source and drain electrodes. The structure consists of a substrate (PET), a gate electrode (indium tin oxide), a gate dielectric (spin cast layer of organosilsesquioxane), source and drain electrodes (20 nm Au and 1.5 nm Ti), and a layer of the organic semiconductor pentacene. The electrical properties of this device are comparable to or better than those that use pentacene with photolithographically defined source/drain electrodes and inorganic dielectrics, gates, and substrates

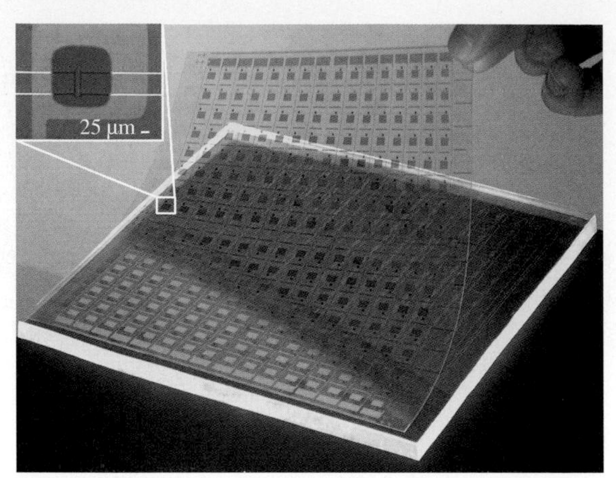

Fig. 6.12 Image of a flexible plastic active matrix backplane circuit whose finest features (transistor source/drain electrodes, and related interconnects) are patterned by microcontact printing. The circuit rests partly on the elastomeric stamp that was used for printing. The circuit consists of a square array of interconnected organic transistors, each of which acts locally as a voltage-controlled switch to control the color of an element in the display. The *inset* shows an optical micrograph of one of the transistors

patterning is performed outside of clean room facilities, (4) methods for removing the printed SAMs to allow good electrical contact of the electrodes with organic semiconductors deposited on top of them, and (5) materials and fabrication sequences that can efficiently exploit these printed electrodes for working organic TFTs in large scale circuits.

Figure 6.12 provides an image of a large area plastic circuit with critical features defined by µCP. This circuit is a flexible active matrix backplane for a display. It consists of a square array of interconnected transistors, each of which serves as a switching element that controls the color of a display pixel [6.21,42]. The transistors themselves have the layout illustrated in Fig. 6.11, and they use similar materials. The semiconductor in this image is blue (pentacene), the source/drain level is Au, the ITO appears green in the optical micrograph in the inset. Part of the circuit rests on the stamp that was used for µCP. The smallest features are the source and drain electrodes ($\sim 15\,\mu m$ lines), the interconnecting lines ($\sim 15\,\mu m$ lines), and the channel length of the transistor ($\sim 15\,\mu m$). This circuit incorporates five layers of material patterned with good registration of the source/drain, gate, and semiconductor levels. The simple printing approach is illustrated in Fig. 6.13 [6.21]. Just before use, the surface of the stamp is cleaned using a conventional adhesive roller lint remover; this procedure removes dust from the stamp in a manner that does

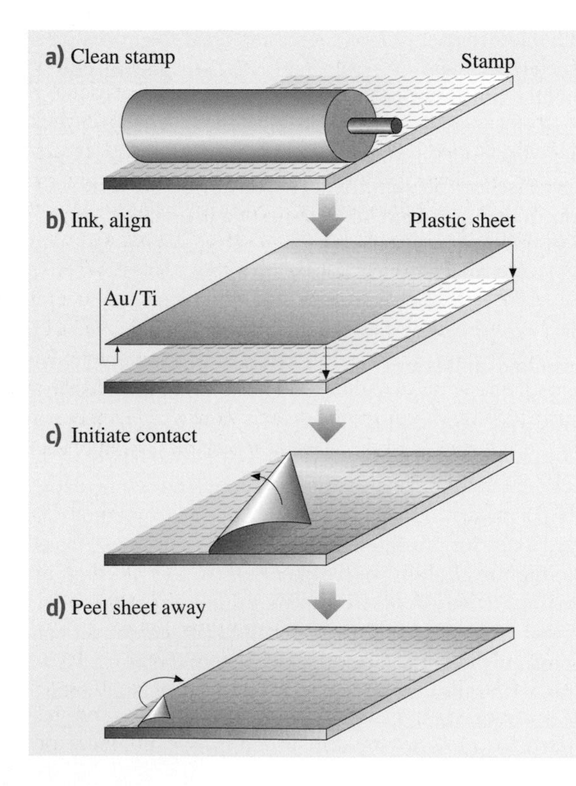

Fig. 6.13 Schematic illustration of fabrication steps for microcontact printing onto large areas of plastic sheets. The process begins with cleaning the stamp using a conventional adhesive roller lint remover. This procedure effectively removes dust particles. To minimize distortions, the stamp rests face up on a flat surface and is not manipulated directly during the printing. Alignment and registration are achieved with alignment marks on one side of the substrate and the stamp. By bending the plastic sheet, contact is initiated on one side of the stamp. The contact line is then allowed to progress gradually across the stamp. This approach avoids the formation of air bubbles that can frustrate good contact. After the substrate is in contact with the stamp for a few seconds, the plastic substrate is separated from the stamp by peeling it away, beginning in one corner. Good registration (maximum cumulative distortions less than 50 microns over an area of 0.25 square feet) and low density of defects can be achieved with this simple approach. It is also well suited for use with rigid composite stamps designed to even further reduce the level of distortion

not contaminate or damage its surface. Inking the stamp and placing it face up on a flat surface prepares it for printing. Matching the cross hair alignment marks on the corners of one edge of the stamp with those patterned in the ITO brings the substrate into registration with the stamp. During this alignment, features on the stamp are viewed directly through the semitransparent substrate. By bending the PET sheet, contact with the stamp is initiated on the edge of the substrate that contains the cross hair marks. Gradually unbending the sheet allows contact to progress across the rest of the surface. This printing procedure is attractive because it avoids distortions that can arise when directly manipulating the flexible rubber stamp. It also minimizes the number and size of trapped air pockets that can form between the stamp and substrate. Careful measurements performed after etching the unprinted areas of the gold show that over the entire 6" × 6" area of the circuit, (1) the overall alignment accuracy for positioning the stamp relative to the substrate (i. e., the offset of the center of the distribution of registration errors) is $\sim 50{-}100\,\mu m$, even with the simple approach used here, and (2) the distortion in the positions of features in the source/drain level, when referenced to the gate level, can be as small as $\sim 50\,\mu m$ (i. e., the full width at half maximum of the distribution of registration errors). These distortions represent the cumulative effects of deformations in the stamp and distortions in the gate and column electrodes that may arise during the patterning and processing of the flexible PET sheet. The density of defects in the printed patterns is comparable to (or smaller than) that in resist patterned by contact mode photolithography when both procedures are performed outside of a cleanroom facility (i. e., when dust is the dominant source of defects).

Figure 6.14 shows an "exploded" view of a paper-like display that consists of a printed flexible plastic backplane circuit, like the one illustrated in Fig. 6.13, laminated against a thin layer of "electronic ink" [6.21, 43]. The electronic ink is composed of a monolayer of transparent polymer microcapsules that contain a suspension of charged white pigment particles suspended in a black liquid. The printed transistors in the backplane circuit act as local switches, which control electric fields that drive the pigments to the front or back of the display. When the particles flow to the front of a microcapsule, it appears white; when they flow to the back, it appears black. Figure 6.15 shows a working sheet of active matrix electronic paper that uses this design. This prototype display has several hundred pixels and an optical contrast that is both independent of the viewing angle and significantly better than newsprint. The device is ~ 1 mm thick, is mechanically flexible, and weighs $\sim 80\%$ less than a conventional liquid crystal display of similar size. Although these displays have only a relatively coarse resolution, all of the processing techniques, the μCP method, the materials, and the electronic inks are suitable for the large numbers of

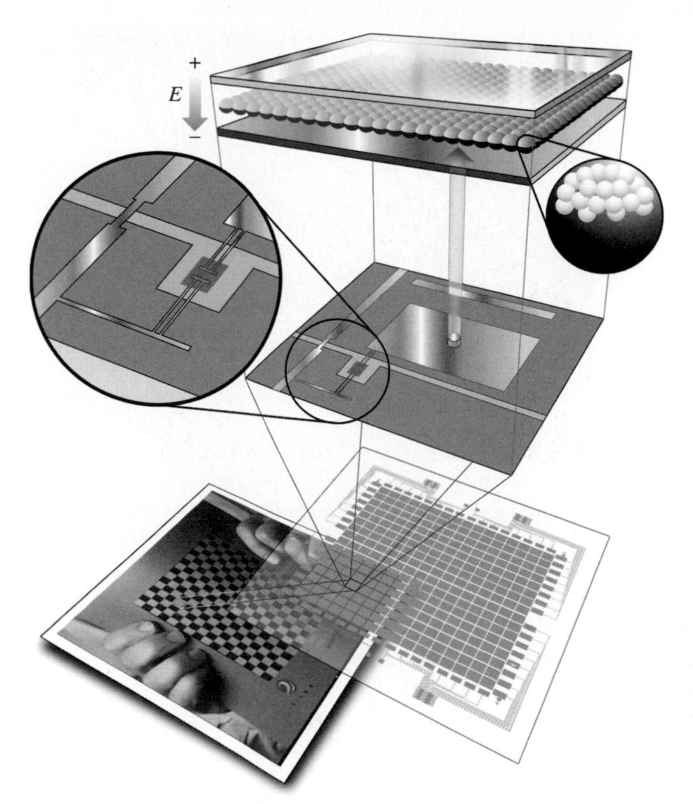

Fig. 6.14 Schematic exploded view of the components of a pixel in an electronic paper-like display (*bottom frame*) that uses a microcontact printed flexible active matrix backplane circuit (*illustration near the bottom frame*). The circuit is laminated against an unpatterned thin sheet of electronic ink (*top frame*) that consists of a monolayer of transparent polymer microcapsules (diameter ~ 100 microns). These capsules contain a heavily dyed black fluid and a suspension of charged white pigment particles (see *right inset*). When one of the transistors turns on, electric fields develop between an unpatterned transparent frontplane electrode (indium tin oxide) and a backplane electrode that connects to the transistor. Electrophoretic flow drives the pigment particles to the front or the back of the display, depending on the polarity of the field. This flow changes the color of the pixel, as viewed from the front of the display, from black to white or vice versa

Fig. 6.15 Electronic paper-like display showing two different images. The device consists of several hundred pixels controlled by a flexible active matrix backplane circuit formed by microcontact printing. The relatively coarse resolution of the display is not limited by material properties or by the printing techniques. Instead, it is set by practical considerations for achieving high pixel yields in the relatively uncontrolled environment of the chemistry laboratory in which the circuits were fabricated

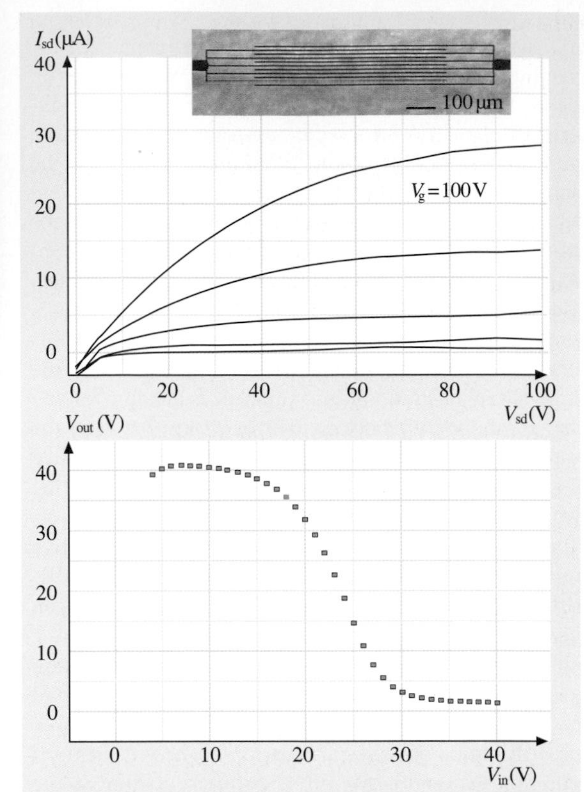

Fig. 6.16 The *top frame* shows current-voltage characteristics of an n-channel transistor formed with electrodes patterned by nanotransfer printing that are laminated against a substrate that supports an organic semiconductor, a gate dielectric, and a gate. The *inset* shows an optical micrograph of the interdigitated electrodes. The *bottom frame* on the right shows the transfer characteristics of a simple CMOS inverter circuit that uses this device and a similar one for the p-channel transistor

pixels required for high information content electronic newspapers and other systems.

Like μCP, nTP is well suited to forming high resolution source/drain electrodes for plastic electronics. nTP of Au/Ti features in the geometry of the drain and source level of organic transistors and with appropriate interconnects on a thin layer of PDMS on PET yields a substrate that can be used in an unusual but powerful way for building circuits: Soft, room temperature lamination of such a structure against a plastic substrate that supports the semiconductor, gate dielectric, and gate levels yields a high performance circuit embedded between two plastic sheets [6.33,44]. (Details of this lamination procedure are presented elsewhere.) The left frame of Fig. 6.16 shows the current-voltage characteristics of a laminated n-channel transistor that uses the organic semiconductor copper hexadecafluorophthalocyanine (n-type) and source/drain electrodes patterned with nTP. The inset shows an optical micrograph of the printed interdigitated source/drain electrodes of this device. The bottom frame of Fig. 6.16 shows the transfer characteristics of a laminated complementary organic inverter circuit whose electrodes and connecting lines are defined by nTP. The p-channel transistor in this circuit used pentacene for the semiconductor [6.33].

In addition to high resolution source/drain electrodes, it is possible to use nTP to form complex multilayer devices with electrical functionality on plastic substrates [6.34]. Figure 6.17 shows a metal/insulator/metal (MIM) structure of Au (50 nm), SiN_x (100 nm; by plasma enhanced vapor deposition, PECVD), Ti (5 nm), and Au (50 nm) formed by transfer printing with a silicon stamp that is coated sequentially with these layers. In this case, a short reactive ion etch (with CF_4), after the second Au deposition, removes the SiN_x from the sidewalls of the stamp. nTP transfers these layers in a patterned geometry to a substrate of Au(15 nm)/Ti(1 nm)-coated PDMS(50 μm)/PET(250 μm). Interfacial cold welding between the Au on the surfaces of the stamp and substrate bonds the multilayers to the substrate. Figure 6.3 illustrates the procedures, the structures (lateral dimensions of 250 μm × 250 μm, for ease of electrical probing), and their electrical characteristics. These MIM capacitors have performance similar to devices fabricated on silicon wafers by photolithography and lift-off. This example illustrates the ability of nTP to print patterns of materials whose growth conditions (high temperature SiN_x by PECVD, in this case) prevent their direct deposition or processing on the substrate of interest (PET, in this case). The cold welding transfer approach has also been exploited in other ways for patterning components for plastic electronics [6.45,46].

Another class of unusual electronic/optoelectronic devices relies on circuits or circuit elements on curved surfaces. This emerging area of research was stimulated primarily by the ability of μCP to print high resolution features on fibers and cylinders. Figure 6.18 shows a conducting microcoil printed with μCP on a microcapillary tube using the approach illustrated in Fig. 6.4. The coil serves as the excitation and detection element for high resolution proton nuclear magnetic resonance of nanoliter volumes of fluid that are housed in the bore of the microcapillary [6.47]. The high fill factor and other considerations lead to extremely high sensitivity with such printed coils. The bottom frame of Fig. 6.18 shows the spectrum of an ∼ 8 nL volume of ethylbenzene. The narrow lines demonstrate the high resolution that is possible with this approach. Similar coils can be used as magnets [6.48], springs [6.32], and electrical transformers [6.49]. Figure 6.19 shows an optical micrograph and the electrical measurements from a concentric cylindrical microtransformer that uses a microcoil printed on a microcapillary tube with a ferromagnetic wire threaded through its core. Inserting this

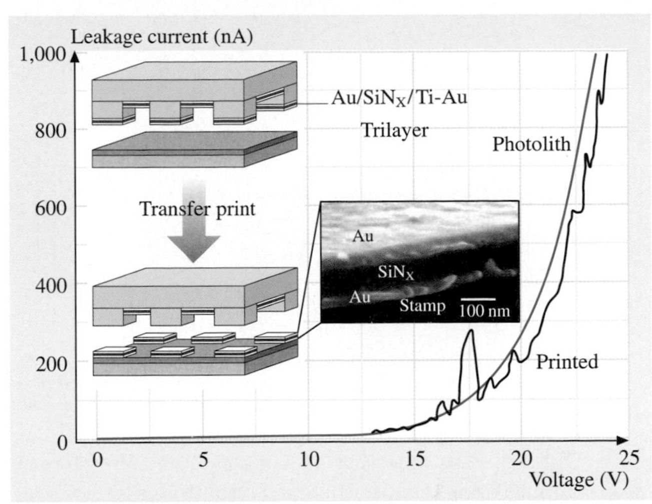

Fig. 6.17 Multilayer thin film capacitor structure printed in a single step onto a plastic substrate using the nanotransfer printing technique. A multilayer of Au/SiN_x/Ti/Au was first deposited onto a silicon stamp formed by photolithography and etching. Contacting this stamp to a substrate of Au/PDMS/PET forms a cold weld that bonds the exposed Au on the stamp to the Au coating on the substrate. Removing the stamp produces arrays of square (250 μm × 250 μm) metal/insulator/metal capacitors on the plastic support. The *black line* shows the measured current–voltage characteristics of one of these printed capacitors. The *brown line* corresponds to a similar structure formed on a rigid glass substrate using conventional photolithographic procedures. The characteristics are the same for these two cases. The slightly higher level of noise in the printed devices results, at least partly, from the difficulty in making good electrical contacts to structures on the flexible plastic substrate

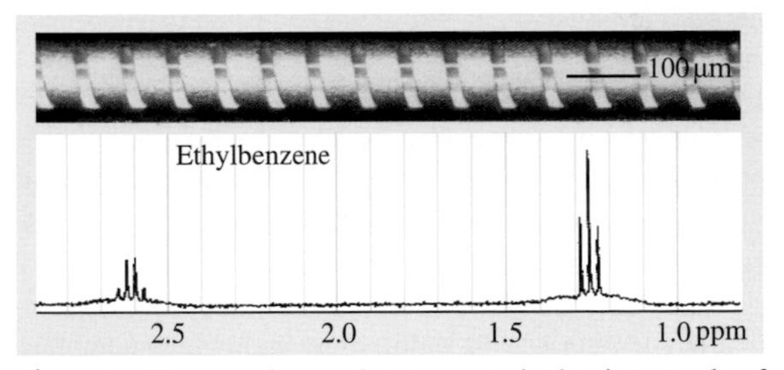

Fig. 6.18 The *top frame* shows an optical micrograph of a continuous conducting microcoil formed by microcontact printing onto a microcapillary tube. This type of printed microcoil is well suited for excitation and detection of nuclear magnetic resonance spectra from nanoliter volumes of fluid housed in the bore of the microcapillary. The *bottom frame* shows a spectra trace collected from an ∼ 8 nL volume of ethyl benzene using a structure similar to the one shown in the top frame

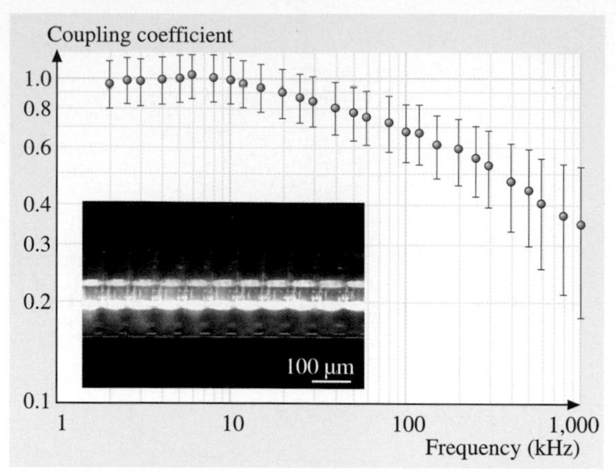

Fig. 6.19 The *inset* shows a concentric microtransformer formed using microcoils printed onto two different microcapillary tubes. The smaller of the tubes (outer diameter 135 microns) has a ferromagnetic wire threaded through its core. The larger one (outer diameter 350 microns) has the smaller tube threaded through its core. The resulting structure is a microtransformer that shows good coupling coefficients at frequencies up to ~ 1 MHz. The graph shows its performance

structure into the core of a larger microcapillary that also supports a printed microcoil completes the transformer [6.49]. This type of device shows good coupling coefficients up to relatively high frequencies. Examples of other optoelectronic components appear in fiber optics where microfabricated on-fiber structures serve as integrated photomasks [6.16] and distributed thermal actuators [6.18].

6.4.2 Lasers and Waveguide Structures

In addition to integral components of unconventional electronic systems, useful structures for integrated optics can be built by using μCP and nTP to print sacrificial resist layers for etching glass waveguides. These printing techniques offer the most significant potential value for this area when they are used to pattern fea-

Fig. 6.20 ▶ Schematic illustration of the use of microcontact printing (μCP) for fabricating high resolution gratings that can be incorporated into distributed planar laser structures or other components for integrated optics. The geometries illustrated here are suitable for 3rd order distributed feedback (DFB) lasers that operate in the red

Fig. 6.21 ▶ Schematic illustrations and lasing spectra of plastic lasers that use microcontact printed resonators based on surface relief distributed Bragg reflectors (DBRs) and distributed feedback gratings (DFBs) on glass substrates. The grating periods are ~ 600 nm in both cases. The lasers use thin film plastic gain media deposited onto the printed gratings. The laser shows emission in a narrow wavelength range, with a width that is limited by the resolution of the spectrometer used to characterize the output. In both cases, the emission profiles, the lasing thresholds and other characteristics of the devices are comparable to similar lasers that use resonators formed by high resolution projection mode photolithography

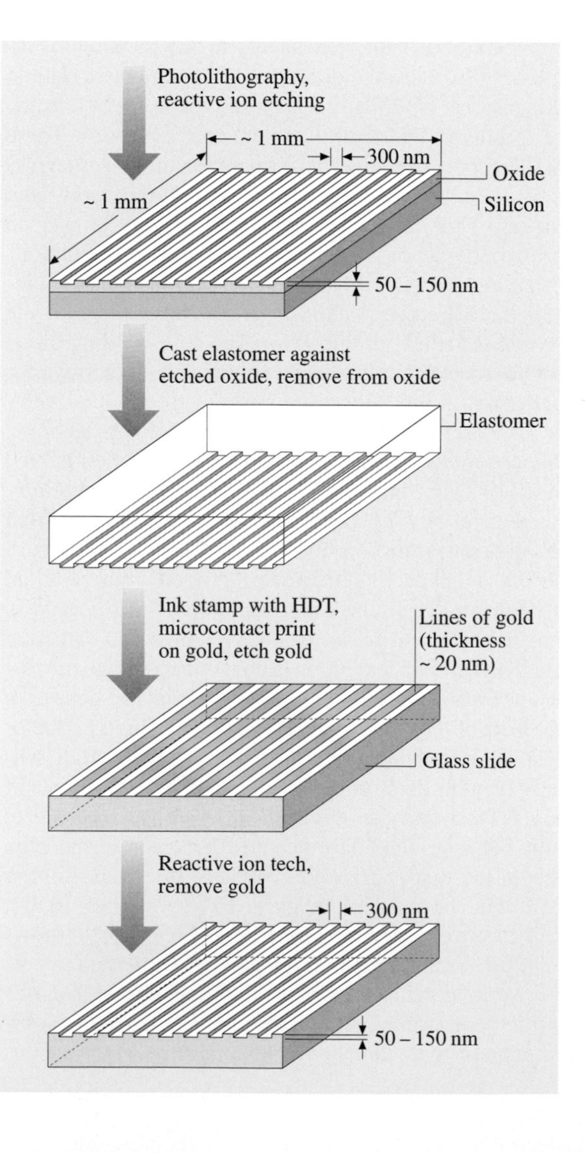

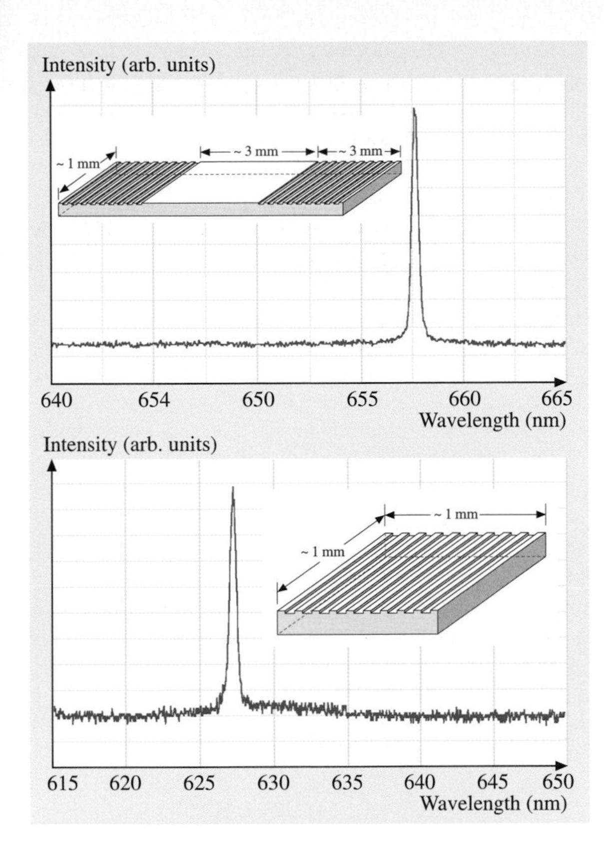

tures that are smaller than those that can be achieved with contact mode photolithography. Mode coupling gratings and distributed laser resonators are two such classes of structures. We have demonstrated μCP for forming distributed feedback (DFB) and distributed Bragg reflector (DBR) lasers that have narrow emission line widths [6.50]. This challenging fabrication

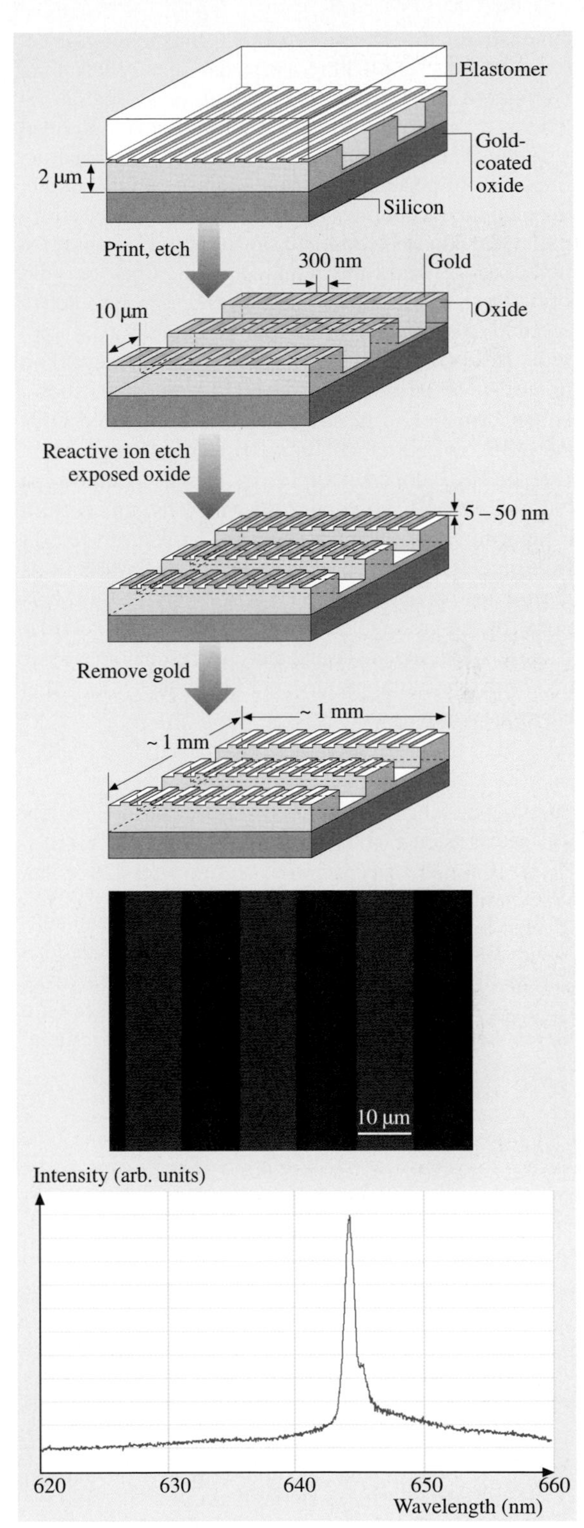

Fig. 6.22 The *top frames* gives a schematic illustration of steps for microcontact printing high resolution gratings directly onto the top surfaces of ridge waveguides. The printing defines a sacrificial etch mask of gold that is subsequently removed. Producing this type of structure with photolithography is difficult because of severe thickness nonuniformities that appear in photoresist spin cast on this type of non-planar substrate. The *upper bottom frame* shows a top view optical micrograph of printed gold lines on the ridge waveguides. The *lower bottom frame* shows the emission output of a plastic photopumped laser that uses the printed structure and a thin evaporated layer of gain media

demonstrates the suitability of μCP for building structures that have (1) feature sizes significantly less than one micron (∼ 300 nm), *and* (2) long-range spatial coherence (∼ 1 mm). The lasers employ optically pumped gain material deposited onto DFB or DBR resonators formed from periodic relief on a transparent substrate. The gain media confines light to the surface of the structure; its thickness is chosen to support a single transverse mode. To generate the required relief, lines of gold formed by μCP on a glass slide act as resists for reactive ion etching of the glass. Removing the gold leaves a periodic pattern of relief (600 nm period, 50 nm depth) on the surface of the glass (see Fig. 6.20). Figure 6.21 shows the performance of plastic lasers that use printed DFB and DBR resonators with gain media consisting of thin films of PBD doped with 1% by weight of coumarin 490 and DCMII, photopumped with 2 ns pulses from a nitrogen laser with intensities $> 5\,\mathrm{kW/cm^2}$ [6.50]. Multimode lasing at resolution-limited line widths was observed at wavelengths corresponding to the third harmonic of the gratings. These characteristics are similar to those observed in lasers that use resonators generated with photolithography and better than those that use imprinted polymers [6.51].

Contact printing not only provides a route to low cost equivalents of gratings fabricated with other approaches, but also allows the fabrication of structures that would be difficult or impossible to generate with photolithography. For example, μCP can be used to form DFB resonators directly on the top surfaces of ridge waveguides [6.52]. Figure 6.22 illustrates the procedures. The bottom left frame shows an optical micrograph of the printed gold lines. Sublimation of a ∼ 200 nm film of tris(8-hydroxyquinoline) aluminum (Alq) doped with 0.5–5.0 weight percent of the laser dye DCMII onto the resonators produces waveguide DFB lasers. The layer of gain material itself provides a planar waveguide with air and polymer as the cladding layers. The relief waveguide provides lateral confinement of the light. Photopumping these devices with the output of a pulsed nitrogen laser (∼ 2 ns, 337 nm) causes lasing due to Bragg reflections induced by the DFB structures on the top surfaces of the ridge waveguides. Some of the laser emission scatters out of the plane of the waveguide at an angle allowed by phase matching conditions. In this way, the grating also functions as an output coupler and offers a convenient way to characterize the laser emission. The bottom right frame of Fig. 6.22 shows the emission profile.

6.5 Conclusions

This chapter provides an overview of two contact printing techniques that are capable of micron and submicron resolution. It also illustrates some applications of these methods that may provide attractive alternatives to more established lithographic methods. The growing interest in nanoscience and technology makes crucial the development of new methods for fabricating the relevant test structures and devices. The simplicity of the techniques described here, together with the interesting and subtle materials science, chemistry, and physics associated with them, make this a promising area for basic and applied study.

References

6.1 C. A. Mirkin, J. A. Rogers: Emerging methods for micro- and nanofabrication, MRS Bull. **26** (2001) 506–507

6.2 H. I. Smith, H. G. Craighead: Nanofabrication, Phys. Today **43** (February 1990) 24–43

6.3 W. M. Moreau (Ed.): *Semiconductor Lithography: Principles and Materials* (Plenum, New York 1988)

6.4 S. Matsui, Y. Ochiai: Focused ion beam applications to solid state devices, Nanotechnology **7** (1996) 247–258

6.5 J. M. Gibson: Reading and writing with electron beams, Phys. Today **50** (1997) 56–61

6.6 L. L. Sohn, R. L. Willett: Fabrication of nanostructures using atomic-force microscope-based lithography, Appl. Phys. Lett. **67** (1995) 1552–1554

6.7 E. Betzig, K. Trautman: Near-Field optics – microscopy, spectroscopy, and surface modification beyond the diffraction limit, Science **257** (1992) 189–195

6.8 A. J. Bard, G. Denault, C. Lee, D. Mandler, D. O. Wipf: Scanning electrochemical microscopy: A new technique for the characterization and modification of surfaces, Acc. Chem. Res. **23** (1990) 357

6.9 J. A. Stroscio, D. M. Eigler: Atomic and molecular manipulation with the scanning tunneling microscope, Science **254** (1991) 1319–1326
6.10 J. Nole: Holographic lithography needs no mask, Laser Focus World **33** (1997) 209–212
6.11 A. N. Broers, A. C. F. Hoole, J. M. Ryan: Electron beam lithography – resolution limits, Microelectron. Eng. **32** (1996) 131–142
6.12 A. N. Broers, W. Molzen, J. Cuomo, N. Wittels: Electron-beam fabrication of 80 A metal structures, Appl. Phys. Lett. **29** (1976) 596
6.13 A. Kumar, G. M. Whitesides: Features of gold having micrometer to centimeter dimensions can be formed through a combination of stamping with an elastomeric stamp and an alkanethiol ink followed by chemical etching, Appl. Phys. Lett. **63** (1993) 2002–2004
6.14 Y. Xia, G. M. Whitesides: Soft lithography, Angew. Chem. Int. Ed. **37** (1998) 550–575
6.15 Y. Xia, J. A. Rogers, K. E. Paul, G. M. Whitesides: Unconventional methods for fabricating and patterning nanostructures, Chem. Rev. **99** (1999) 1823–1848
6.16 J. A. Rogers, R. J. Jackman, J. L. Wagener, A. M. Vengsarkar, G. M. Whitesides: Using microcontact printing to generate photomasks on the surface of optical fibers: A new method for producing in-fiber gratings, Appl. Phys. Lett. **70** (1997) 7–9
6.17 B. Michel, A. Bernard, A. Bietsch, E. Delamarche, M. Geissler, D. Juncker, H. Kind, J. P. Renault, H. Rothuizen, H. Schmid, P. Schmidt-Winkel, R. Stutz, H. Wolf: Printing meets lithography: Soft approaches to high-resolution printing, IBM J. Res. Dev. **45** (2001) 697–719
6.18 J. A. Rogers: Rubber stamping for plastic electronics and fiber optics, MRS Bull. **26** (2001) 530–534
6.19 N. B. Larsen, H. Biebuyck, E. Delamarche, B. Michel: Order in microcontact printed self-assembled monolayers, J. Am. Chem. Soc. **119** (1997) 3017–3026
6.20 H. A. Biebuyck, G. M. Whitesides: Self-organization of organic liquids on patterned self-assembled monolayers of alkanethiolates on gold, Langmuir **10** (1994) 2790–2793
6.21 J. A. Rogers, Z. Bao, K. Baldwin, A. Dodabalapur, B. Crone, V. R. Raju, V. Kuck, H. Katz, K. Amundson, J. Ewing, P. Drzaic: Paper-like electronic displays: Large area, rubber stamped plastic sheets of electronics and electrophoretic inks, Proc. Nat. Acad. Sci. **98** (2001) 4835–4840
6.22 J. L. Wilbur, H. A. Biebuyck, J. C. MacDonald, G. M. Whitesides: Scanning force microscopies can image patterned self-assembled monolayers, Langmuir **11** (1995) 825–831
6.23 J. C. Love, D. B. Wolfe, M. L. Chabinyc, K. E. Paul, G. M. Whitesides: Self-assembled monolayers of alkanethiolates on palladium are good etch resists, J. Am. Chem. Soc. **124** (2002) 1576–1577
6.24 H. Schmid, B. Michel: Siloxane polymers for high-resolution, high-accuracy soft lithography, Macromolecules **33** (2000) 3042–3049
6.25 K. Choi, J. A. Rogers: A photocurable poly(dimethylsiloxane) chemistry for soft lithography in the nanometer regime, J. Am. Chem. Soc. **125** (2003) 4060–4061
6.26 J. A. Rogers K. E. Paul, G. M. Whitesides: Quantifying distortions in soft lithography, J. Vac. Sci. Technol. B **16** (1998) 88–97
6.27 J. Tate, J. A. Rogers, C. D. W. Jones, W. Li, Z. Bao, D. W. Murphy, R. E. Slusher, A. Dodabalapur, H. E. Katz, A. J. Lovinger: Anodization and microcontact printing on electroless silver: Solution-based fabrication procedures for low voltage organic electronic systems, Langmuir **16** (2000) 6054–6060
6.28 Y. Xia, E. Kim, G. M. Whitesides: Microcontact printing of alkanethiols on silver and its application to microfabrication, J. Electrochem. Soc. **143** (1996) 1070–1079
6.29 Y. N. Xia, X. M. Zhao, E. Kim, G. M. Whitesides: A selective etching solution for use with patterned self-assembled monolayers of alkanethiolates on gold, Chem. Mater. **7** (1995) 2332–2337
6.30 R. J. Jackman, J. Wilbur, G. M. Whitesides: Fabrication of submicrometer features on curved substrates by microcontact printing, Science **269** (1995) 664–666
6.31 R. J. Jackman, S. T. Brittain, A. Adams, M. G. Prentiss, G. M. Whitesides: Design and fabrication of topologically complex, three-dimensional microstructures, Science **280** (1998) 2089–2091
6.32 J. A. Rogers, R. J. Jackman, G. M. Whitesides: Microcontact printing and electroplating on curved substrates: a new means for producing free-standing three-dimensional microstructures with possible applications ranging from micro-coil springs to coronary stents, Adv. Mater. **9** (1997) 475–477
6.33 Y.-L. Loo, R. W. Willett, K. Baldwin, J. A. Rogers: Additive, nanoscale patterning of metal films with a stamp and a surface chemistry mediated transfer process: Applications in plastic electronics, Appl. Phys. Lett. **81** (2002) 562–564
6.34 Y.-L. Loo, R. W. Willett, K. Baldwin, J. A. Rogers: Interfacial chemistries for nanoscale transfer printing, J. Am. Chem. Soc. **124** (2002) 7654–7655
6.35 Y.-L. Loo, J. W. P. Hsu, R. L. Willett, K. W. Baldwin, K. W. West, J. A. Rogers: High-resolution transfer printing on GaAs surfaces using alkane dithiol self-assembled monolayers, J. Vac. Sci. Technol. B. **20** (2002) 2853–2856
6.36 G. S. Ferguson, M. K. Chaudhury, G. B. Sigal, G. M. Whitesides: Contact adhesion of thin gold-films on elastomeric supports – cold welding under ambient conditions, Science **253** (1991) 776–778
6.37 W. Zhang, S. Y. Chou: Multilevel nanoimprint lithography with submicron alignment over 4 in Si wafers, Appl. Phys. Lett. **79** (2001) 845–847

6.38 N. Bowden, S. Brittain, A. G. Evans, J. W. Hutchinson, G. M. Whitesides: Spontaneous formation of ordered structures in thin films of metals supported on an elastomeric polymer, Nature **393** (1998) 146–149

6.39 J. A. Rogers, Z. Bao, A. Dodabalapur, A. Makhija: Organic smart pixels and complementary inverter circuits formed on plastic substrates by casting, printing and molding, IEEE Electron Dev. Lett. **21** (2000) 100–103

6.40 Z. Bao, J. A. Rogers, H. E. Katz: Printable organic and polymeric semiconducting materials and devices, J. Mater. Chem. **9** (1999) 1895–1904

6.41 J. A. Rogers, Z. Bao, A. Makhija: Non-photolithographic fabrication sequence suitable for reel-to-reel production of high performance organic transistors and circuits that incorporate them, Adv. Mater. **11** (1999) 741–745

6.42 P. Mach, S. Rodriguez, R. Nortrup, P. Wiltzius, J. A. Rogers: Active matrix displays that use printed organic transistors and polymer dispersed liquid crystals on flexible substrates, Appl. Phys. Lett. **78** (2001) 3592–3594

6.43 J. A. Rogers: Toward paperlike displays, Science **291** (2001) 1502–1503

6.44 Y.-L. Loo, T. Someya, K. W. Baldwin, P. Ho, Z. Bao, A. Dodabalapur, H. E. Katz, J. A. Rogers: Soft, conformable electrical contacts for organic transistors: High resolution circuits by lamination, Proc. Nat. Acad. Sci. USA **99** (2002) 10252–10256

6.45 C. Kim, P. E. Burrows, S. R. Forrest: Micropatterning of organic electronic devices by cold-welding, Science **288** (2000) 831–833

6.46 C. Kim, M. Shtein, S. R. Forrest: Nanolithography based on patterned metal transfer and its application to organic electronic devices, Appl. Phys. Lett. **80** (2002) 4051–4053

6.47 J. A. Rogers, R. J. Jackman G. M. Whitesides, D. L. Olson, J. V. Sweedler: Using microcontact printing to fabricate microcoils on capillaries for high resolution ^{1}H-NMR on nanoliter volumes, Appl. Phys. Lett. **70** (1997) 2464–2466

6.48 J. A. Rogers, R. J. Jackman, G. M. Whitesides: Constructing single and multiple helical microcoils and characterizing their performance as components of microinductors and microelectromagnets, J. Microelectromech. Syst. (JMEMS) **6** (1997) 184–192

6.49 R. J. Jackman, J. A. Rogers, G. M. Whitesides: Fabrication and characterization of a concentric, cylindrical microtransformer, IEEE Trans. Magn. **33** (1997) 2501–2503

6.50 J. A. Rogers, M. Meier, A. Dodabalapur: Using stamping and molding techniques to produce distributed feedback and Bragg reflector resonators for plastic lasers, Appl. Phys. Lett. **73** (1998) 1766–1768

6.51 M. Berggren, A. Dodabalapur, R. E. Slusher, A. Timko, O. Nalamasu: Organic solid-state lasers with imprinted gratings on plastic substrates, Appl. Phys. Lett. **72** (1998) 410–411

6.52 J. A. Rogers, M. Meier, A. Dodabalapur: Distributed feedback ridge waveguide lasers fabricated by nanoscale printing and molding on non-planar substrates, Appl. Phys. Lett. **74** (1999) 3257–3259

7. Materials Aspects of Micro- and Nanoelectromechanical Systems

Two of the more significant technological achievements during the last 20 years have been the development of MEMS and its new offshoot, NEMS. These developments were made possible in large measure by significant advancements in the materials and processes used in the fabrication of MEMS and NEMS devices. And while initial developments capitalized on a mature Si infrastructure, recent advances have used materials and processes not associated with IC fabrication, a trend that is likely to continue as new application areas are identified.

A well-rounded understanding of MEMS and NEMS requires a basic knowledge of the materials used to construct the devices, since material properties often govern device performance. An understanding of the materials used in MEMS and NEMS is really an understanding of material systems. Devices are rarely constructed of a single material, but rather a collection of materials, each providing a critical function and often working in conjunction with each other. It is from this perspective that the following chapter is constructed. A preview of the materials selected for inclusion is presented in Table 7.1. From this table it is easy to see that this chapter is not a summary of all materials used in MEMS and NEMS, as such a work would itself constitute a text of significant size. It does, however, present a selection of some of the more important material systems, including examples that illustrate the importance of viewing MEMS and NEMS in terms of material systems.

7.1 **Silicon** ... 203
7.1.1 Single Crystal Silicon ... 204
7.1.2 Polysilicon ... 205
7.1.3 Porous Silicon ... 208
7.1.4 Silicon Dioxide ... 208
7.1.5 Silicon Nitride ... 209

7.2 **Germanium-Based Materials** ... 210
7.2.1 Polycrystalline Ge ... 210
7.2.2 Polycrystalline SiGe ... 210

7.3 **Metals** ... 211

7.4 **Harsh Environment Semiconductors** ... 212
7.4.1 Silicon Carbide ... 212
7.4.2 Diamond ... 215

7.5 **GaAs, InP, and Related III-V Materials** ... 217

7.6 **Ferroelectric Materials** ... 218

7.7 **Polymer Materials** ... 219
7.7.1 Polyimide ... 219
7.7.2 SU-8 ... 220
7.7.3 Parylene ... 220

7.8 **Future Trends** ... 220

References ... 221

7.1 Silicon

Use of silicon (Si) as a material for microfabricated sensors dates back to the 1950s when *C. S. Smith* published a paper describing the piezoresistive effect in germanium (Ge) and Si [7.1]. *Smith* found that the piezoresistive coefficients of Si were significantly higher than those associated with conventional metal strain gauges, a finding that led to the development of Si strain gauges in the early 1960s. Throughout the 1960s and early 1970s, various Si bulk micromachining techniques were developed, which ultimately led to the commercial production of piezoresistive Si pressure sensors during this period. The subsequent development of surface micromachining techniques for Si, along with the recognition that micromachined Si structures could potentially be integrated with Si IC devices, marked the advent of MEMS and positioned Si as the primary MEMS material.

Table 7.1 Distinguishing characteristics and application examples of selected materials for MEMS and NEMS

Material	Distinguishing Characteristics	Application Examples
Single crystal Silicon (Si)	High quality electronic material, selective anisotropic etching	Bulk micromachining, piezoresistive sensing
Polycrystalline Si (polysilicon)	Doped Si films on sacrificial layers	Surface micromachining, electrostatic actuation
Silicon Dioxide (SiO_2)	Insulating, etched by HF, compatible with polysilicon	Sacrificial layer in polysilicon surface micromachining, passivation layer for devices
Silicon Nitride (Si_3N_4, Si_xN_y)	Insulating, chemically resistant, mechanically durable	Isolation layer for electrostatic devices, membrane and bridge material
Polycrystalline Germanium (poly Ge), Polycrystalline Silicon-Germanium (poly Si-Ge)	Deposited at low temperatures	Integrated surface micromachined MEMS
Gold (Au), Aluminum (Al)	Conductive thin films, flexible deposition techniques	Innerconnect layers, masking layers, electromechanical switches
Nickel-Iron (NiFe)	Magnetic alloy	Magnetic actuation
Titanium-Nickel (TiNi)	Shape-memory alloy	Thermal actuation
Silicon Carbide (SiC) Diamond	Electrically and mechanically stable at high temperatures, chemically inert, high Young's modulus to density ratio	Harsh environment MEMS, high frequency MEMS/NEMS
Gallium Arsenide (GaAs), Indium Phosphide (InP), Indium Arsenide (InAs), and related materials	Wide bandgap, epitaxial growth on related ternary compounds	RF MEMS, optoelectronic devices, single crystal bulk and surface micromachining
Lead Zirconate Titanate (PZT)	Piezoelectric material	Mechanical sensors and actuators
Polyimide	Chemically resistant, high temperature polymer	Mechanically flexible MEMS, bioMEMS
SU-8	Thick, photo-definable resist	Micromolding, high-aspect-ratio structures
Parylene	Biocompatible polymer, deposited at room temperature by CVD	Protective coatings, molded polymer structures

7.1.1 Single Crystal Silicon

For MEMS applications, single crystal Si serves several key functions. It is one of the most versatile materials for bulk micromachining, due to the availability of anisotropic etching processes in conjunction with good mechanical properties. Having a Young's modulus of about 190 GPa, Si compares favorably, from a mechanical perspective, with steel, which has a Young's modulus of about 210 GPa. Favorable mechanical properties enable Si to be used as a material for membranes, beams, and other such structures. For surface micromachining applications, single crystal Si substrates are used primarily as mechanical platforms on which Si and non-Si device structures can be fabricated. Use of high-quality single crystal wafers enables the fabrication of integrated MEMS devices, at least for materials and processes that are compatible with Si ICs.

Bulk micromachining is a process whereby etching techniques are used in conjunction with etch masks and etch stops to selectively sculpt micromechanical structures from a bulk substrate. From the materials perspective, single crystal Si is a relatively easy material to bulk micromachine, due to the availability of anisotropic etchants, such as potassium hydroxide (KOH) and tetramethyl-aluminum hydroxide (TMAH), that attack the (100) and (110) Si crystal planes significantly faster than the (111) crystal planes. For example, the etching rate ratio of (100) to (111) planes in Si is about 400:1 for a typical KOH/water etch solution. Silicon dioxide (SiO_2), silicon nitride (Si_3N_4), and some metallic thin films (e.g., Cr, Au, etc.) provide good etch masks for most Si anisotropic etchants.

In contrast to anisotropic etching, isotropic etching exhibits no selectivity to the various crystal planes. From a processing perspective, isotropic etching of Si is commonly used for removal of work-damaged surfaces, creation of structures in single-crystal slices, and patterning single-crystal or polycrystalline films. The most commonly used isotropic Si etchants are mixtures of hydrofluoric (HF) and nitric (HNO_3) acid in water or acetic acid (CH_3COOH), with the etch rate dependent on the ratio of HF to HNO_3.

In terms of etch stops, boron-doped Si is effective for some liquid reagents. Boron-doped etch stops are often less than 10 μm thick, since the boron concentration in Si must exceed $7 \times 10^{19}\,cm^3$ for the etch stop to be effective, and the doping is done by diffusion. It is possible to create a boron-doped etch stop below an undoped Si layer using ion implantation; however, the practical implant depth is limited to a few microns.

Well-established dry etching processes are available to pattern single crystal Si. The process spectrum ranges from physical mechanisms such as sputtering and ion milling to chemical mechanisms such as plasma etching. Reactive ion etching (RIE) is the most commonly used dry etching technique to etch Si. By combining both physical and chemical processes, RIE is a highly effective anisotropic Si etching technique that is independent of crystalline orientation. Fluorinated compounds such as CF_4, SF_6, and NF_3, or chlorinated compounds such as CCl_4 or Cl_2, sometimes mixed with He, O_2, or H_2, are commonly used in Si RIE. The RIE process is highly directional, which enables direct pattern transfer from an overlying masking material to the etched Si surface. SiO_2 thin films are often used as a masking material, owing to its chemical durability and ease in patterning. Process limitations (i. e., etch rates) restrict the etch depths of Si RIE to less than 10 microns. However, a process called deep reactive ion etching (DRIE) has extended the use of anisotropic dry etching to depths well beyond several hundred microns.

Using the aforementioned processes and techniques, a wide variety of microfabricated devices have been made from single crystal Si, such as piezoresistive pressure sensors, accelerometers, and mechanical resonators, to name a few. Using nearly the same approaches but on a smaller scale, top-down nanomachining techniques have been used to fabricate nanoelectromechanical devices from single crystal Si. Single crystal Si is particularly well-suited for nanofabrication, because high crystal quality substrates with very smooth surfaces are readily available. By coupling electron-beam (e-beam) lithographic techniques with conventional Si etching, device structures with submicron dimensions have been fabricated. Submicron, single crystal Si nanomechanical structures have been successfully micromachined from bulk Si wafers [7.2] and silicon-on-insulator (SOI) wafers [7.3]. In the former case, an isotropic Si etch was performed to release the device structures, whereas in the latter, the 50 nm to 200 nm structures were released by dissolving the underlying oxide layer in HF. An example of nanoelectromechanical beam structures fabricated from a single crystal Si substrate is shown in Fig. 7.1.

Fig. 7.1 A collection of Si nanoelectromechanical beam resonators fabricated from a single-crystal Si substrate (courtesy M. Roukes, Caltech)

7.1.2 Polysilicon

Surface micromachining is a process whereby a sequence of thin films, often of different materials, is deposited and selectively etched to form the desired micromechanical (or microelectromechanical) structure. In contrast to bulk micromachining, the substrate serves primarily as a platform to support the device. For Si-based surface micromachined MEMS, polycrystalline Si (polysilicon) is most often used as the structural material, silicon dioxide (SiO_2) as the sacrificial material, silicon nitride (Si_3N_4) for electrical isolation of device structures, and single crystal Si as the substrate. Like single crystal Si, polysilicon can be doped during or after film deposition. SiO_2 can be thermally grown or deposited on Si over a broad temperature range (e.g., 200 °C to 1,150 °C) to meet various process and material

Fig. 7.2 SEM micrograph of a surface-micromachined polysilicon micromotor fabricated using a SiO_2 sacrificial layer

requirements. SiO_2 is readily dissolvable in hydrofluoric acid (HF), which does not etch polysilicon and, thus, can be used to dissolve SiO_2 sacrificial layers. Si_3N_4 is an insulating film that is highly resistant to oxide etchants. The polysilicon micromotor shown in Fig. 7.2 was surface micromachined using a process that included these materials.

For MEMS and IC applications, polysilicon films are commonly deposited using a process known as low-pressure chemical vapor deposition (LPCVD). The typical polysilicon LPCVD reactor is based on a hot wall, resistance-heated furnace. Typical processes are performed at temperatures ranging from 580 °C to 650 °C and pressures from 100 to 400 mtorr. The most commonly used source gas is silane (SiH_4).

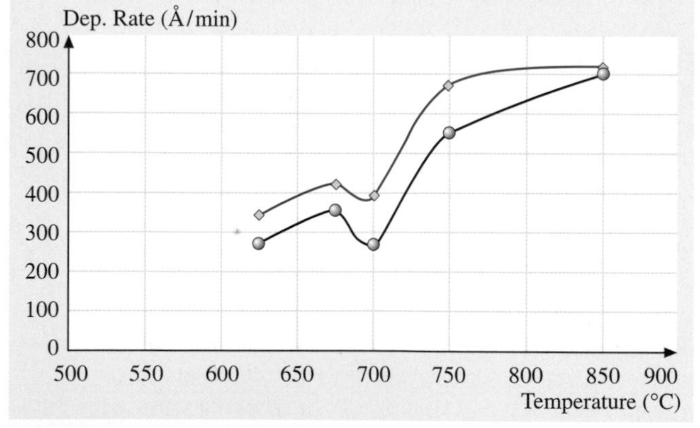

Fig. 7.3 Deposition rate versus substrate temperature for in situ boron-doped (◇) and undoped (●) polysilicon films grown by atmospheric pressure chemical vapor deposition [7.4]

Polysilicon thin films consist of a collection of small single crystal grains whose microstructure and orientation is a function of the deposition conditions [7.5]. For typical LPCVD processes (e.g., 200 mtorr), the amorphous-to-polycrystalline transition temperature is about 570 °C, with amorphous films deposited below the transition temperature. At 600 °C, the grains are small and equiaxed, while at 625 °C, the grains are large and columnar [7.5]. The crystal orientation is predominantly (110) Si for temperatures between 600 °C and 650 °C, while the (100) orientation is dominant for temperatures between 650 °C and 700 °C.

In terms of chemical resistance, polysilicon is very much like single crystal Si. The oxidation rate of undoped polysilicon is typically between that of (100)- and (111)-oriented single crystal Si. For temperatures below 1,000 °C, heavily phosphorus-doped polysilicon oxidizes at a rate significantly higher than undoped polysilicon.

The resistivity of polysilicon can be modified using the doping methods developed for single crystal Si. Diffusion is an effective method for doping polysilicon films, especially for heavy doping of thick films. Phosphorous, which is the most commonly used dopant in polysilicon MEMS, diffuses significantly faster in polysilicon than in single crystal Si, due primarily to enhanced diffusion rates along grain boundaries. The diffusivity in polysilicon thin films (i. e., small equiaxed grains) is about $1 \times 10^{12}\,\mathrm{cm^2/s}$.

Ion implantation is also used to dope polysilicon films. A high-temperature annealing step is usually required to electrically activate the implanted dopants, as well as to repair implant-related damage in the polysilicon films. In general, the conductivity of implanted polysilicon films is not as high as films doped by diffusion.

In situ doping of polysilicon is performed by simply including a dopant gas, usually diborane (B_2H_6) or phosphine (PH_3), in the CVD process. The addition of dopants during the deposition process not only modifies the conductivity, but also affects the deposition rate of the polysilicon films. As shown in Fig. 7.3, the inclusion of boron generally increases the deposition rate of polysilicon relative to undoped films [7.4], while phosphorus (not shown) reduces the rate. In situ doping can be used to produce conductive films with uniform doping profiles without requiring the high-temperature steps usually associated with diffusion or ion implantation. Although commonly used to produce doped polysilicon for electrostatic devices, *Cao* et al. [7.6] have used in situ phosphorus-doped polysilicon films in piezoresistive strain gauges, achieving gauge factors as high as 15 for a single strip sensor.

The thermal conductivity of polysilicon is a strong function of its microstructure and, therefore, the conditions used during deposition [7.5]. For fine-grained films, the thermal conductivity is about 25% of the value of single crystal Si. For thick films with large grains, the thermal conductivity ranges between 50% and 85% of the single crystal value.

Like the electrical and thermal properties of polysilicon, the as-deposited residual stress in polysilicon films depends on microstructure. For films deposited under typical conditions (200 mtorr, 625 °C), the as-deposited polysilicon films have compressive residual stresses. The highest compressive stresses are found in amorphous Si films and polysilicon films with a strong, columnar (110) texture. For films with fine-grained microstructures, the stress tends to be tensile. Annealing can be used to reduce the compressive stress in as-deposited polysilicon films. For instance, compressive residual stresses on the order of 500 MPa can be reduced to less than 10 MPa by annealing the as-deposited films at 1,000 °C in a N_2 ambient [7.7, 8]. Recent advances in the area of rapid thermal annealing (RTA) indicate that RTA provides an effective method of stress reduction in polysilicon films. *Zhang* et al. [7.9] reported that a 10 s anneal at 1,100 °C was sufficient to completely relieve the stress in films that originally had a compressive stress of about 340 MPa. RTA is particularly attractive in situations where the process parameters require a low thermal budget.

As an alternative to high-temperature annealing, *Yang* et al. [7.10] have developed an approach that actually utilizes the residual stress characteristics of polysilicon deposited under various conditions to construct polysilicon multilayers that have the desired thickness and stress values. The multilayers are comprised of alternating tensile and compressive polysilicon layers that are deposited in a sequential manner. The tensile layers consist of fine-grained polysilicon grown at a temperature of 570 °C, while the compressive layers are made up of columnar polysilicon deposited at 615 °C. The overall stress in the composite film depends on the number of alternating layers and the thickness of each layer. With the proper set of parameters, a composite polysilicon multilayer can be deposited with near zero residual stress and no stress gradient. The process achieves stress reduction without high-temperature annealing, a considerable advantage for integrated MEMS processes.

Many device designs require thick polysilicon structural layers that are not readily achievable using conventional LPCVD polysilicon, due to process-related limitations. For these designs, epitaxial Si reactors can be used to grow polysilicon films. Unlike conventional LPCVD processes with deposition rates of less than 100 Å/min, epitaxial processes have deposition rates on the order of 1 micron/min [7.11]. The high deposition rates result from the much higher substrate temperatures ($>$ 1,000 °C) and deposition pressures ($>$ 50 torr) used in these processes. The polysilicon films are usually deposited on SiO_2 sacrificial layers to enable surface micromachining. An LPCVD polysilicon seed layer is sometimes used to control nucleation, grain size, and surface roughness. As with conventional polysilicon, the microstructure and residual stress of the epi-poly films, as they are known, is related to deposition conditions. Compressive films generally have a mixture of [110] and [311] grains [7.12, 13], while tensile films have a random mix of [110], [100], [111], and [311] grains [7.12]. The Young's modulus of epi-poly measured from micromachined test structures is comparable to LPCVD polysilicon [7.13]. Mechanical properties test structures [7.11–13], thermal actuators [7.11], electrostatically actuated accelerometers [7.11], and gryoscopes [7.14] have been fabricated from these films.

All of the aforementioned polysilicon deposition processes utilize substrate temperatures in excess of 570 °C either during film deposition, or in post-deposition processing steps. As a low temperature alternative to LPCVD polysilicon, *Abe* et al. [7.15] and *Honer* et al. [7.16] have developed sputtering processes for polysilicon. Early work [7.15] emphasized the ability to deposit very smooth (25 Å) polysilicon films on thermally oxidized wafers at reasonable deposition rates (191 Å/min) and with low residual compressive stresses. The process involved DC magnitron sputtering from a Si target using an Ar sputtering gas, a chamber pressure of 5 mtorr, and a power of 100 W. The authors reported that a post-deposition anneal at 700 °C in N_2 for two hours was needed to crystallize the deposited film and perhaps lower the stress. *Honer* et al. [7.16] sought to develop a polymer-friendly, Si-based surface micromachining process based on polysilicon sputtered onto polyimide and PSG sacrificial layers. To improve the conductivity of the micromachined Si structures, the sputtered Si films were sandwiched between two TiW cladding layers. The device structures on polyimide were released using oxygen plasma etching. The processing step with the highest temperature was, in fact, the polyimide cure at 350 °C. To test the robustness of the process, sputter-deposited Si microstructures were fabricated on substrates containing CMOS devices. As expected from thermal budget considerations, the authors reported no measurable degradation of device performance.

7.1.3 Porous Silicon

Porous Si is produced by room temperature electrochemical etching of Si in HF. If configured as an electrode in an HF-based electrochemical circuit, positive charge carriers (holes) at the Si surface facilitate the exchange of F atoms with H atoms, that terminate the Si surface. The exchange continues in the subsurface region, leading to the eventual removal of the fluorinated Si. The quality of the etched surface is related to the density of holes at the surface, which is controlled by the applied current density. For high current densities, the density of holes is high and the etched surface is smooth. For low current densities, the hole density is low and clustered in highly localized regions associated with surface defects. Surface defects become enlarged by etching, which leads to the formation of pores. Pore size and density are related to the type of Si used and the conditions of the electrochemical cell. Both single crystal and polycrystalline Si can be converted to porous Si.

The large surface-to-volume ratios make porous Si attractive for gaseous and liquid applications, including filter membranes and absorbing layers for chemical and mass sensing [7.17]. When single crystal substrates are used, the unetched porous layer remains single crystalline and is suitable for epitaxial Si growth. It has been shown that CVD coatings do not generally penetrate the porous regions, but rather overcoat the pores at the surface of the substrate [7.18]. The formation of localized Si-on-insulator structures is, therefore, possible by simply combining pore formation with epitaxial growth, followed by dry etching to create access holes to the porous region, and thermal oxidation of the underlying porous region. A third application uses porous Si as a sacrificial layer for polysilicon and single crystalline Si surface micromachining. As shown by *Lang* et al. [7.18], the process involves the electrical isolation of the solid structural Si layer by either pn-junction formation through selective doping, or use of electrically insulating thin films, since the formation of pores only occurs on electrically charged surfaces. A weak Si etchant will aggressively attack the porous regions with little damage to the structural Si layers and can be used to release the devices.

7.1.4 Silicon Dioxide

Silicon dioxide (SiO_2) is one of the most widely used materials in the fabrication of MEMS. In polysilicon surface micromachining, SiO_2 is used as a sacrificial material, since it can be easily dissolved using etchants that do not attack polysilicon. SiO_2 is widely used as etch mask for dry etching of thick polysilicon films, since it is chemically resistant to dry etching processes for polysilicon. SiO_2 films are also used as passivation layers on the surfaces of environmentally sensitive devices.

The most common processes used to produce SiO_2 films for polysilicon surface micromachining are thermal oxidation and LPCVD. Thermal oxidation of Si is performed at temperatures of 900 °C to 1,200 °C in the presence of oxygen or steam. Since thermal oxidation is a self-limiting process, the maximum practical film thickness that can be obtained is about 2 μm, which is sufficient for many sacrificial applications. As noted by its name, thermal oxidation of Si can only be performed on Si surfaces.

SiO_2 films can be deposited on a wide variety of substrate materials by LPCVD. In general, LPCVD provides a means for depositing thick (> 2 μm) SiO_2 films at temperatures much lower than thermal oxidation. Known as low-temperature oxides, or LTO for short, these films have a higher etch rate in HF than thermal oxides, which translates to significantly faster release times when LTO films are used as sacrificial layers. Phosphosilicate glass (PSG) can be formed using nearly the same deposition process as LTO by adding a phosphorus-containing gas to the precursor flows. PSG films are useful as sacrificial layers, since they generally have higher etching rates in HF than LTO films.

PSG and LTO films are deposited in hot-wall, low-pressure, fused silica furnaces in systems similar to those described previously for polysilicon. Precursor gases include SiH_4 as a Si source, O_2 as an oxygen source, and, in the case of PSG, PH_3 as a source of phosphorus. LTO and PSG films are typically deposited at temperatures of 425 °C to 450 °C and pressures ranging from 200 mtorr to 400 mtorr. The low deposition temperatures result in LTO and PSG films that are slightly less dense than thermal oxides, due to the incorporation of hydrogen in the films. LTO films can, however, be densified by an annealing step at high temperature (1,000 °C). The low density of LTO and PSG films is partially responsible for the increased etch rate in HF.

Thermal SiO_2 and LTO are electrical insulators used in numerous MEMS applications. The dielectric constants of thermal oxide and LTO are 3.9 and 4.3, respectively. The dielectric strength of thermal SiO_2 is 1.1×10^6 V/cm, and for LTO it is about 80% of that value [7.19]. The stress in thermal SiO_2 is compressive with a magnitude of about 300 MPa [7.19]. For LTO, however, the as-deposited residual stress is tensile, with a magnitude of about 100 MPa to 400 MPa [7.19]. The addition of phosphorous to LTO decreases the tensile residual stress to about 10 MPa for phosphorus concentrations of 8% [7.20]. As with polysilicon, the properties of LTO and PSG are dependent on processing conditions.

Plasma-enhanced chemical vapor deposition (PECVD) is another common method to produce oxides of silicon. Using a plasma to dissociate the gaseous precursors, the deposition temperatures needed to deposit PECVD oxide films is lower than for LPCVD films. For this reason, PECVD oxides are quite commonly used as masking, passivation, and protective layers, especially on devices that have been coated with metals.

Quartz is the crystalline form of SiO_2 and has interesting properties for MEMS. Quartz is optically transparent, piezoelectric, and electrically insulating. Like single crystal Si, quartz substrates are available as high quality, large area wafers that can be bulk micromachined using anisotropic etchants. A short review of the basics of quartz etching by *Danel* et al. [7.21] is recommended for those interested in the subject. Quartz has recently become a popular substrate material for microfluidic devices due to its optical, electronic, and chemical properties.

Another SiO_2-related material that has recently found uses in MEMS is spin-on-glass (SOG). SOG is a polymeric material with a viscosity suitable for spin coating. Two recent publications illustrate the potential for SOG in MEMS fabrication. In the first example, *Yasseen* et al. [7.22] detailed the development of SOG as a thick-film sacrificial molding material for thick polysilicon films. The authors reported a process to deposit, polish, and etch SOG films that were 20 microns thick. The thick SOG films were patterned into molds and filled with 10 micron-thick LPCVD polysilicon films, planarized by selective CMP, and subsequently dissolved in a wet etchant containing HCl, HF, and H_2O to reveal the patterned polysilicon structures. The cured SOG films were completely compatible with the polysilicon deposition process. In the second example, *Liu* et al. [7.23] fabricated high-aspect ratio channel plate microstructures from SOG. Electroplated nickel (Ni) was used as a molding material, with Ni channel plate molds fabricated using a conventional LIGA process. The Ni molds were then filled with SOG, and the sacrificial Ni molds were removed in a reverse electroplating process. In this case, the fabricated SOG structures (over 100 microns tall) were micromachined glass structures fabricated using a molding material more commonly used for structural components.

7.1.5 Silicon Nitride

Silicon nitride (Si_3N_4) is widely used in MEMS for electrical isolation, surface passivation, etch masking, and as a mechanical material. Two deposition methods are commonly used to deposit Si_3N_4 thin films: LPCVD and PECVD. PECVD silicon nitride is generally non-stoichiometric (sometimes denoted as $Si_xN_y:H$) and may contain significant concentrations of hydrogen. Use of PECVD silicon nitride in micromachining applications is somewhat limited because it has a high etch rate in HF (e.g., often higher than that of thermally grown SiO_2). However, PECVD offers the ability to deposit nearly stress-free silicon nitride films, an attractive property for encapsulation and packaging.

Unlike its PECVD counterpart, LPCVD Si_3N_4 is extremely resistant to chemical attack, making it the material of choice for many Si bulk and surface micromachining applications. LPCVD Si_3N_4 is commonly used as an insulating layer because it has a resistivity of $10^{16}\,\Omega \times$cm and field breakdown limit of 10^7 V/cm. LPCVD Si_3N_4 films are deposited in horizontal furnaces similar to those used for polysilicon deposition. Typical deposition temperatures and pressures range between 700 °C to 900 °C and 200 mtorr to 500 mtorr, respectively. The standard source gases are dichlorosilane (SiH_2Cl_2) and ammonia (NH_3). To produce stoichiometric Si_3N_4, a NH_3 to SiH_2Cl_2 ratio of 10:1 is commonly used. The microstructure of films deposited under these conditions is amorphous.

The residual stress in stoichiometric Si_3N_4 is large and tensile, with a magnitude of about 1 GPa. Such a large residual stress causes films thicker than a few thousand angstroms to crack. Nonetheless, thin stoichiometric Si_3N_4 films have been used as mechanical support structures and electrical insulating layers in piezoresistive pressure sensors [7.24]. To enable the use of Si_3N_4 films for applications that require micron-thick, durable, and chemically resistant membranes, Si_xN_y films can be deposited by LPCVD. These films, often referred to as Si-rich or low-stress nitride, are intentionally deposited with an excess of Si by simply decreasing the ratio of NH_3 to SiH_2Cl_2 during deposition. Nearly stress-free films can be deposited using a NH_3 to SiH_2Cl_2 ratio of 1 : 6, a deposition temperature of 850 °C, and a pressure of 500 mtorr [7.25]. The increase in Si content not only leads to a reduction in tensile stress, but also a decrease in the etch rate in HF. Such properties have enabled the development of fabrication techniques that would otherwise not be feasible with stoichiometric Si_3N_4. For example, low stress silicon nitride has been bulk micromachined using silicon as the sacrificial material. In this case, an anisotropic Si etchant (TMAH) was used to dissolve the sacrificial silicon. *French* et al. [7.26] used PSG as a sacrificial layer to surface micromachine low-stress nitride, capitalizing on the HF resistance of the nitride films.

7.2 Germanium-Based Materials

Like Si, Ge has a long history as a semiconductor device material, dating back to the development of the earliest transistors and semiconductor strain gauges. Issues related to the water solubility of germanium oxide, however, stymied the development of Ge for microelectronic devices. Nonetheless, there is a renewed interest in using Ge in micromachined devices due to the relatively low processing temperatures required to deposit the material.

7.2.1 Polycrystalline Ge

Thin polycrystalline Ge (poly-Ge) films can be deposited by LPCVD at temperatures as low as 325 °C on Si, Ge, and SiGe substrates [7.27]. Ge does not nucleate on SiO_2 surfaces, which prohibits the use of thermal oxides and LTO films as sacrificial layers, but enables the use of these films as sacrificial molds. Residual stress in poly-Ge films deposited on Si substrates can be reduced to nearly zero after short anneals at modest temperatures (30 s at 600 °C). Poly-Ge is essentially impervious to KOH, TMAH, and BOE, enabling the fabrication of Ge membranes on Si substrates [7.27]. The mechanical properties of poly-Ge are comparable to polysilicon, having a Young's modulus of 132 GPa and a fracture stress between 1.5 GPa and 3.0 GPa [7.28]. Mixtures of HNO_3, H_2O, and HCl and H_2O, H_2O_2, and HCl, as well as the RCA SC-1 cleaning solution isotropically etch Ge. Since these mixtures do not etch Si, SiO_2, Si_3N_4, and SiN, poly-Ge can be used as a sacrificial substrate layer in polysilicon surface micromachining. Using these techniques, devices such as poly-Ge-based thermistors and Si_3N_4 membrane-based pressure sensors, made using poly-Ge sacrificial layers, have been fabricated [7.27]. *Franke* et al. [7.28] found no performance degradation in Si CMOS devices following the fabrication of surface micromachined poly-Ge structures, thus demonstrating the potential for on-chip integration of Ge electromechanical devices with Si circuitry.

7.2.2 Polycrystalline SiGe

Like poly-Ge, polycrystalline SiGe (poly-SiGe) is a material that can be deposited at temperatures lower than those required for polysilicon. Deposition processes include LPCVD, APCVD, and RTCVD (rapid thermal CVD) using SiH_4 and GeH_4 as precursor gases. Deposition temperatures range from 450 °C for LPCVD [7.29] to 625 °C by rapid thermal CVD (RTCVD) [7.30]. In general, the deposition temperature is related to the concentration of Ge in the films, with higher Ge concentrations resulting in lower deposition temperatures. Like polysilicon, poly-SiGe can be doped with boron and phosphorus to modify its conductivity. In situ boron doping can be performed at temperatures as low as 450 °C [7.29]. *Sedky* et al. [7.30] showed that the deposition temperature of conductive films doped with boron could be further reduced to 400 °C if the Ge content was kept at or above 70%.

Unlike poly-Ge, poly-SiGe can be deposited on a number of sacrificial substrates, including SiO_2 [7.30], PSG [7.28], and poly-Ge [7.28]. For Ge rich films, a thin polysilicon seed layer is sometimes used on SiO_2 surfaces since Ge does not readily nucleate on oxide surfaces. Like many compound materials, variations in film composition can change the physical properties of the material. For instance, etching of poly-SiGe by H_2O_2 becomes significant for Ge concentrations over 70%. *Sedky* et al. [7.31] have shown that the microstructure, film conductivity, residual-stress, and residual stress gradient are related to the concentration of Ge in the material. With respect to residual stress, *Franke* et al. [7.29] produced in situ boron-doped films with residual compressive stresses as low as 10 MPa.

The poly-SiGe, poly-Ge material system is particularly attractive for surface micromachining since H_2O_2 can be used as a release agent. It has been reported that poly-Ge etches at a rate of 0.4 microns/min in H_2O_2, while poly-SiGe with Ge concentrations below 80% have no observable etch rate after 40 hrs [7.32]. The ability to use H_2O_2 as a sacrificial etchant makes the combination of poly-SiGe and poly-Ge extremely attractive for surface micromachining from the processing, safety, and materials compatibility points of view. Due to the conformal nature of LPCVD processing, poly-SiGe structural elements, such as gimbal-based microactuator structures, have been made by high-aspect ratio micromolding [7.32]. Capitalizing on the low deposition temperatures, an integrated MEMS fabrication process with Si ICs has been demonstrated [7.29]. In this process, CMOS structures are first fabricated on Si wafers. Poly-SiGe mechanical structures are then surface micromachined using a poly-Ge sacrificial layer. A significant advantage of this design is that the MEMS structure is positioned directly above the CMOS structure, thus reducing the parasitic capacitance and contact resistance characteristic of interconnects associated with side-by-side integration schemes. Use of H_2O_2 as the sacrificial etchant means that no special protective layers are required to protect the underlying CMOS layer during release. In addition to its utility as a material for integrated MEMS devices, poly-SiGe has been identified as a material well suited for micromachined thermopiles [7.33] due to its lower thermal conductivity relative to Si.

7.3 Metals

It can be argued that of all the material categories associated with MEMS, metals may be among the most enabling, since metallic thin films are used in many different capacities ranging from etch masks used in device fabrication to interconnects and structural elements in microsensors and microactuators. Metallic thin films can be deposited using a wide range of techniques, including evaporation, sputtering, CVD, and electroplating. Since a complete review of the metals used in MEMS is far beyond the scope of this chapter, the examples presented in this section were selected to represent a broad cross section of where metals have found uses in MEMS.

Aluminum (Al) and gold (Au) are among the most widely employed metals in microfabricated electronic and electromechanical devices, as a result of their use as innerconnect and packaging materials. In addition to these critical electrical functions, Al and Au are also desirable as electromechanical materials. One such example is the use of Au micromechanical switches for RF MEMS. For conventional RF applications, chip level switching is currently performed using FET- and PIN diode-based solid state devices fabricated from gallium arsenide (GaAs) substrates. Unfortunately, these devices suffer from insertion losses and poor electrical isolation. In an effort to develop replacements for GaAs-based solid state switches, *Hyman* et al. [7.34] reported the development of an electrostatically actuated, cantilever-based micromechanical switch fabricated on GaAs substrates. The device consisted of a silicon nitride-encased Au cantilever constructed on a sacrificial silicon dioxide layer. The silicon nitride and silicon dioxide layers were deposited by PECVD, and the Au beam was electroplated from a sodium sulfite solution inside a photoresist mold. A thin multilayer of Ti and Au was sputter deposited in the mold prior to electroplating. The trilayer cantilever structure was chosen to minimize the deleterious effects of thermal and process-related stress gradients in order to produce unbent and thermally stable beams. After deposition and pattering, the cantilevers were released in HF. The processing steps proved to be completely compatible with GaAs substrates. The released cantilevers demonstrated switching speeds better than 50 μs at 25 V with contact lifetimes exceeding 10^9 cycles.

In a second example from RF MEMS, *Chang* et al. [7.35] reported the fabrication of an Al-based micromachined switch as an alternative to GaAs FETs and PIN diodes. In contrast to the work by *Hyman* et al. [7.34], this switch utilizes the differences in the residual stresses in Al and Cr thin films to create bent cantilever switches that capitalize on the stress differences in the materials. Each switch is comprised of a series of linked bimorph cantilevers designed in such a way that the resulting structure bends significantly out of the plane of the wafer due to the stress differences in the bimorph. The switch is drawn closed by electrostatic attraction. The bimorph consists of metals that can easily be processed with GaAs wafers, thus making integration with GaAs devices possible. The released switches were relatively slow, at 10 ms, but an actuation voltage of only 26 V was needed to close the switch.

Thin-film metallic alloys that exhibit the shape-memory effect are of particular interest to the MEMS community for their potential in microactuators. The shape-memory effect relies on the reversible transformation from a ductile martensite phase to a stiff austenite phase in the material with the application of heat. The reversible phase change allows the shape-memory effect to be used as an actuation mechanism, since the material changes shape during the transition. It has been found that high forces and strains can be generated from shape-memory thin films at reasonable power inputs, thus enabling shape-memory actuation to be used in MEMS-based microfluidic devices, such as microvalves and micropumps. Titanium-nickel (TiNi) is among the most popular of the shape-memory alloys, owing to its high actuation work density ($50\,MJ/m^3$) and large bandwidth (up to 0.1 kHz) [7.36]. TiNi is also attractive because conventional sputtering techniques can be employed to deposit thin films, as detailed in a recent report by *Shih* et al. [7.36]. In this study, TiNi films were deposited by co-sputtering elemental Ti and Ni targets, and a co-sputtering TiNi alloy and elemental Ti targets. It was reported that co-sputtering from TiNi and Ti targets produced better films due to process variations related to the roughening of the Ni target in the case of Ti and Ni co-sputtering. The TiNi/Ti co-sputtering process has been used to produce shape-memory material for a silicon spring-based microvalve [7.37].

Use of thin-film metal alloys in magnetic actuator systems is another example of the versatility of metallic materials in MEMS. Magnetic actuation in microdevices generally requires the magnetic layers to be relatively thick (tens to hundreds of microns) to generate magnetic fields of sufficient strength to generate the desired actuation. To this end, magnetic materials are often deposited by thick film methods, such as electroplating. The thickness of these layers exceeds what can feasibly be patterned by etching, so plating is often performed in microfabricated molds made from materials such as polymethylmethacrylate (PMMA). The PMMA mold thickness can exceed several hundred microns, so X-rays are used as the exposure source during the patterning steps. When necessary, a metallic thin film seed layer is deposited prior to plating. After plating, the mold is dissolved, which frees the metallic component. Known as LIGA (short for Lithography, Galvanoforming and Abformung), this process has been used to produce a wide variety of high-aspect ratio structures from plateable materials, such as nickel-iron (NiFe) magnetic alloys [7.38] and Ni [7.39].

In addition to elemental metals and simple compound alloys, more complex metallic alloys commonly used in commerical macroscopic applications are finding their way into MEMS applications. One such example is an alloy of titanium known as Ti-6Al-4V. Composed of 88% titanium, 6% aluminum and 4% vanadium, this alloy is widely used in commercial aviation due to its weight, strength, and temperature tolerance. *Pornsin-sirirak* et al. [7.40] have explored the use of this alloy in the manufacture of MEMS-based winged structures for micro aerial vehicles. The authors considered this alloy not only because of its weight and strength, but also because of its ductility and its etching rate at room temperature. The designs for the wing prototype were modelled after the wings of bats and various flying insects. For this application, Ti-alloy structures patterned from bulk (250 μm-thick) material by an $HF/HNO_3/H_2O$ etching solution were used, rather than thin films. Parylene-C (detailed in Sect. 7.7.3) was deposited on the patterned alloy to serve as the wing membrane. The miniature micromachined wings were integrated into a test setup, and several prototypes actually demonstrated short duration flight.

7.4 Harsh Environment Semiconductors

Silicon Carbide (SiC) has long been recognized as the leading semiconductor for use in high temperature and high power electronics and is currently receiving attention as a material for harsh environment MEMS.

7.4.1 Silicon Carbide

SiC is a polymorphic material that exists in cubic, hexagonal, and rhombehedral polytypes. The cubic polytype, called 3C-SiC, has an electronic bandgap of 2.3 eV, which is over twice that of Si. Numerous hexagonal and rhombehedral polytypes have been identified, the two most common being 4H-SiC and 6H-SiC. The electronic bandgap of 4H- and 6H-SiC is even higher than 3C-SiC – 2.9 and 3.2 eV, respectively. SiC films can be doped to create n-type and p-type material. The stiffness of SiC is quite large relative to Si, with a Young's modulus ranging from 300 GPa to 450 GPa, depending on the microstructure. SiC is not etched in any wet Si etchants or by XeF_2, a popular dry Si etchant used for releasing device structures [7.41]. SiC is a material that does not melt, but rather sublimes at temperatures in excess of 1,800 °C. Single crystal 4H- and 6H-SiC wafers are commercially available, but smaller in diameter (3 inches) and much more expensive than Si.

SiC thin films can be grown or deposited using a number of different techniques. For high-quality single crystal films, APCVD and LPCVD processes are most commonly employed. Homoepitaxial growth of 4H- and 6H-SiC yields high quality films suitable for microelectronic applications, but only on substrates of the same polytype. These processes usually employ dual precursors, such as SiH_4 and C_3H_8, and are performed at temperatures ranging from 1,500 °C to 1,700 °C. Epitaxial films with p-type or n-type conductivity can be grown using aluminum and boron for p-type films and nitrogen and phosphorus for n-type films. Nitrogen is so effective at modifying the conductivity of SiC that growth of undoped SiC is extremely challenging, because the concentrations of residual nitrogen in typical deposition systems are sufficient for n-type doping.

APCVD and LPCVD can also be used to deposit 3C-SiC on Si substrates. Heteroepitaxy is possible, despite a 20% lattice mismatch, because 3C-SiC and Si have similar lattice structures. The growth process involves two key steps. The first step, called carbonization, converts the near surface region of the Si substrate to 3C-SiC by simply exposing it to a hydrocarbon/hydrogen mixture at high substrate temperatures (> 1,200 °C). The carbonized layer forms a crystalline template on which a 3C-SiC film can be grown by adding a silicon-containing gas to the hydrogen/hydrocarbon mix. The lattice mismatch between Si and 3C-SiC results in the formation of crystalline defects in the 3C-SiC film, with the density being highest in the carbonization layer and decreasing with increasing thickness. The crystal quality of 3C-SiC films is nowhere near that of epitaxially grown 4H- and 6H-SiC films; however, the fact that 3C-SiC can be grown on Si substrates enables the use of Si bulk micromachining techniques to fabricate a host of 3C-SiC-based mechanical devices. These include microfabricated pressure sensors [7.42] and nanoelectromechanical resonant structures [7.43].

Polycrystalline SiC (poly-SiC) has proven to be a very versatile material for SiC MEMS. Unlike single crystal versions of SiC, poly-SiC can be deposited on a variety of substrate types, including such common surface micromachining materials as polysilicon, SiO_2, and Si_3N_4, using a much wider range of processes than epitaxial films. Commonly used deposition techniques include LPCVD [7.41, 44, 45] and APCVD [7.46, 47]. The deposition of poly-SiC requires much lower substrate temperatures than epitaxial films, ranging from roughly 700 °C to 1,200 °C. Amorphous SiC can be deposited at even lower temperatures (25 °C to 400 °C) by PECVD [7.48] and sputtering [7.49]. The microstructure of poly-SiC films is temperature, substrate, and process dependent. For amorphous substrates, such as SiO_2 and Si_3N_4, APCVD poly-SiC films deposited from silane and propane are oriented randomly with equiaxed grains [7.47], whereas for oriented substrates, such as polysilicon, the texture of the poly-SiC film matches that of the substrate [7.46]. By comparison, poly-SiC films deposited by LPCVD from dichlorosilane and acetylene are highly textured (111) films with a columnar microstructure [7.44], while films deposited from disilabutane have a distribution of orientations [7.41]. This variation suggests that device performance can be tailored by selecting the proper substrate and deposition conditions.

Direct bulk micromachining of SiC is very difficult due to its chemical inertness. Although conventional wet chemical techniques are not effective, several electrochemical etch processes have been demonstrated and used in the fabrication of 6H-SiC pressure sensors [7.50]. The etching processes are selective to the conductivity of the material, so dimensional control of the etched structures depends on the ability to form doped layers, which can only be formed by in situ or ion implantation processes, since solid source diffusion is not possible at reasonable processing temperatures. This constraint somewhat limits the geometrical complexity of the patterned structures, as compared with conventional plasma-based etching. To fabricate thick (hundreds of microns), 3-D, high-aspect ratio SiC structures, a molding technique has been developed [7.39]. The molds are fabricated from Si substrates using deep reactive ion etching and then filled with SiC using a combination of thin epitaxial and thick polycrystalline film CVD processes. The thin film process is used to protect the mold from pitting during the more aggressive mold filling SiC growth step. The mold filling process coats all surfaces of the mold with a SiC film as thick as the mold is deep. To release the SiC structure, the substrate is first mechanically polished to expose sections of the Si mold, then the substrate is immersed in a Si etchant to completely dissolve the mold. This process has been used to fabricate solid SiC fuel atomizers [7.39], and a variant has been used to fabricate SiC structures for micropower systems [7.51].

In addition to CVD processes, bulk micromachined SiC structures can be fabricated using sintered SiC powders. *Tanaka* et al. [7.52] describe a process in which SiC components, such as micro gas turbine engine rotors, can be fabricated from SiC powders using a micro-reaction-sintering process. The molds are microfabricated from Si using DRIE and filled with SiC and graphite powders mixed with a phenol resin. The molds are then reaction-sintered using a hot isostatic pressing technique. The SiC components are then released from the Si mold by wet chemical etching. The authors reported that the component shrinkage was less than 3%. The bending strength and Vickers hardness of the micro-reaction-sintered material was roughly 70 to 80% of commercially available reaction-sintered SiC, the difference being attributed to the presence of unreacted Si in the micro-scale components.

In a related process, *Liew* et al. [7.54] detail a technique to create silicon carbon nitride (SiCN) MEMS structures by molding injectable polymer precursors. Unlike the aforementioned processes, this technique uses SU-8 photoresists for the molds. SU-8, detailed later in this chapter, is a versatile photo-definable polymer in which thick films (hundreds of microns) can be patterned using conventional UV photolithographic techniques. After patterning, the molds are filled with the SiCN-containing polymer precursor, lightly polished, and then subjected to a multistep heat-treating process. During the thermal processing steps, the SU-8 mold decomposes, and the SiCN structure is released. The resulting SiCN structures retain many of the same properties of stoichiometric SiC.

Although SiC cannot be etched using conventional wet etch techniques, thin SiC films can be patterned using conventional dry etching techniques. RIE processes using fluorinated compounds, such as CHF_3 and SF_6 combined with O_2, an inert gas, or H_2. The high oxygen content in these plasmas generally prohibits the use of photoresist as a masking material, therefore, hard masks made of Al, Ni, and ITO are often used. RIE-based SiC surface micromachining processes with polysilicon and SiO_2 sacrificial layers have been developed for single layer devices [7.55,56]. Multilayered structures are very difficult to fabricate by direct RIE because the etch rates of the sacrificial layers are much higher than the SiC structural layers, making critical dimensional control very difficult and RIE-based SiC multilayer processes impractical.

To address the issues related to RIE of multilayer SiC structures, a micromolding process for patterning SiC films on sacrificial layer substrates has been developed [7.57]. In essence, the micromolding technique is the thin film analog to the molding-based, bulk micromachining technique presented earlier. The micromolding process utilizes polysilicon and SiO_2 films as both molds and sacrificial substrate layers, with SiO_2 molds used with polysilicon sacrificial layers and vica versa. These films are deposited and patterned using conventional methods, thus leveraging the well characterized and highly selective processes developed for polysilicon MEMS. Poly-SiC films are simply deposited into the micromolds, and mechanical polishing is used to remove poly-SiC from the top of the molds. Appropriate etchants are then used to dissolve the molds and sacrificial layers. The micromolding method utilizes the differences in chemical properties of the three materials in this system in a way that bypasses the difficulties associated with chemical etching of SiC. This technique has been extended to multilayer processes that have been used to fabricate SiC micromotors [7.57] and the lateral resonant structure shown in Fig. 7.4 [7.53].

Yang et al. [7.43] have recently shown that the chemical inertness of SiC facilitates the fabrication of NEMS devices. In this work, the authors present a fabrication method to realize SiC mechanical resonators with submicron thickness and width dimensions. The resonators were fabricated from ~ 260 nm-thick 3C-SiC films epitaxially grown on (100) Si wafers. The films were patterned into 150 nm-wide beams ranging in length from 2 to 8 μm. The beams were etched in a $NF_3/O_2/Ar$

V2-ANN 10.0kV x200 150 μm

Fig. 7.4 SEM micrograph of a poly-SiC lateral resonant structure fabricated using a multilayer, micromolding-based micromachining process [7.53]

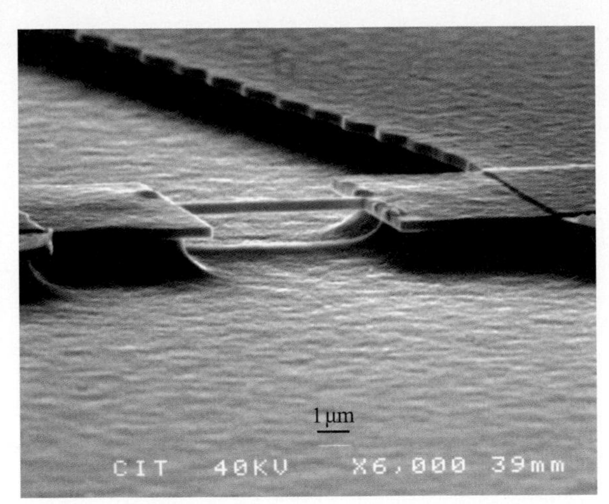

Fig. 7.5 SEM micrograph of a 3C-SiC nanomechanical beam resonator fabricated by electron-beam lithography and dry etching processes (courtesy M. Roukes, Caltech)

plasma using an evaporated Cr etch mask. After patterning, the beams were released by etching isotropically the underlying Si using a NF_3/Ar plasma. The inertness of the SiC film to the Si etchant enables the dry release of the nanomechanical beams. An example of a 3C-SiC nanomechanical beam is shown in Fig. 7.5.

7.4.2 Diamond

Diamond is commonly known as nature's hardest material, making it ideal for high wear environments. Diamond has a very large electronic bandgap (5.5 eV), which makes it attractive for high temperature electronics. Undoped diamond is a high quality insulator with a dielectric constant of 5.5; however it can be relatively easily doped with boron to create p-type conductivity. Diamond has a very high Young's modulus (1,035 GPa), making it suitable for high frequency micromachined resonators, and it is among nature's most chemically inert materials, making it well suited for harsh chemical environments.

Unlike SiC, fabrication of diamond MEMS is currently restricted to polycrystalline and amorphous material, since epitaxial-quality single-crystal diamond wafers are not yet commercially available. Polycrystalline diamond films can be deposited on Si and SiO_2 substrates, but the surfaces often must be seeded by diamond powders, or biased with a negative charge to initiate growth. In general, diamond nucleates much more readily on Si surfaces than on SiO_2 surfaces, an effect that has been used to selectively pattern diamond films into micromachined AFM cantilever probes using SiO_2 molding masks [7.58].

Bulk micromachining of diamond is very difficult given its extreme chemical inertness. Diamond structures have nevertheless been fabricated using bulk micromachined Si molds to pattern the structures [7.59]. The Si molds were fabricated using conventional micromachining techniques and filled with polycrystalline diamond deposited by hot filament chemical vapor deposition (HFCVD). The HFCVD process uses hydrogen as a carrier gas and methane as the carbon source. The process was performed at a substrate temperature of 850 °C to 900 °C and a pressure of 50 mtorr. The Si substrate was seeded prior to deposition using a diamond particle/ethanol solution. After deposition, the top surface of the structure was polished using a hot iron plate. After polishing, the Si mold was removed in a Si etchant, leaving behind the micromachined diamond structure. This process was used to produce high-aspect-ratio capillary channels for microfluidic applications [7.60] and components for diffractive optics, laser-to-fiber alignment, and power device cooling structures [7.61].

Due to the nucleation processes associated with diamond film growth, surface micromachining of polycrystalline diamond thin films requires modifications to conventional micromachining to facilitate film growth on sacrificial substrates. Conventional RIE methods are generally ineffective, so initial work focused on developing selective deposition techniques. One early method used selective seeding to form patterned templates for diamond nucleation. The selective seeding process employed the lithographic patterning of photoresist that contained diamond powders [7.62]. The diamond-loaded photoresist was deposited and patterned onto a Cr-coated Si wafer. During the onset of diamond growth, the patterned photoresist rapidly evaporates, leaving behind the diamond seed particles in the desired locations. A patterned diamond film is then selectively grown on these locations.

A second process utilizes selective deposition directly on sacrificial substrate layers. This process combines conventional diamond seeding with photolithographic patterning and etching to fabricate micromachined diamond structures on SiO_2 sacrificial layers [7.63]. The process can be performed in one of two ways. The first approach begins with the seeding of an oxidized Si wafer. The wafer is coated with photoresist and photolithographically patterned. Unmasked regions of the seeded SiO_2 film are then partially etched, forming a surface unfavorable for diamond growth. The photoresist is then removed, and a diamond film is deposited on the seeded regions. The second approach also begins with an oxidized Si wafer. The wafer is coated with photoresist, photolithographically patterned, and then seeded with diamond particles. The photoresist is removed, leaving behind a patterned seed layer suitable for selective growth. These techniques have been successfully used to fabricate cantilever beams and bridge structures.

A third method to surface micromachine polycrystalline diamond films follows the conventional approach of film deposition, dry etching, and release. The chemical inertness of diamond renders most conventional plasma chemistries useless; however, oxygen-based ion beam plasmas can be used to etch diamond thin films [7.64]. A simple surface micromachining process begins with the deposition of a polysilicon sacrificial layer on a Si_3N_4-coated Si wafer. The polysilicon layer is seeded using diamond slurry, and a diamond film is deposited by HFCVD. Since photoresist is not resistant to O_2 plasmas, an Al masking film is deposited and patterned. The diamond films are then etched in the O_2 ion beam plasma, and the structures are released by etching the polysilicon with KOH. This process has been used to create lateral resonant structures, but a significant stress gradient in the films rendered the devices nonoperational.

Fig. 7.6 SEM micrograph of the folded beam truss of a diamond lateral resonator. The diamond film was deposited using a seeding-based hot filament CVD process. The micrograph illustrates the challenges currently facing diamond MEMS, namely, roughened surfaces and residual stress gradients

In general, conventional HFCVD requires that the substrate be pretreated with a seeding layer prior to diamond film growth. However, a method called biased enhanced nucleation (BEN) has been developed that enables the growth of diamond on unseeded Si surfaces. *Wang* et al. [7.65] have shown that if Si substrates are masked with patterned SiO_2 films, selective diamond growth will occur primarily on the exposed Si surfaces, and a slight HF etch is sufficient to remove the adventitious diamond from the SiO_2 mask. This group was able to use this method to fabricate diamond micromotor rotors and stators on Si surfaces.

Diamond is a difficult but not impossible material to etch using RIE techniques. It is well-known that diamond will be etched in oxygen plasmas, but these plasmas are not generally well suited for device fabrication, because the etches tend to be isotropic. A recent development, however, suggests that RIE processes for diamond are close at hand. *Wang* et al. [7.65] describe a process to fabricate a vertically actuated, doubly clamped micromechanical diamond beam resonator using RIE. The process outlined here addresses two key issues related to diamond surface micromachining, namely, residual stress gradients in the diamond films and diamond patterning techniques. A microwave plasma CVD reactor was used to grow the diamond films on sacrificial SiO_2 layers pretreated with a nanocrystalline diamond powder, resulting in a uniform nucleation density at the diamond/SiO_2 interface. The diamond films were etched in a CF_4/O_2 plasma using Al as a hard mask. Reasonably straight sidewalls were created, with roughness attributable to the surface roughness of the faceted diamond film. An Au/Cr drive electrode beneath the sacrificial oxide remained covered throughout the diamond patterning steps and, thus, was undamaged during the diamond etching process.

In conjunction with recent advances in RIE and micromachining techniques, work is being performed to develop diamond deposition processes specifically for MEMS applications. Diamond films grown using conventional techniques, especially processes that require pre-growth seeding, tend to have high residual stress gradients and roughened surface morphologies as a result of the highly faceted, large grain polycrystalline films that are produced by these methods (see Fig. 7.6). The rough surface morphology degrades the patterning process, resulting in roughened sidewalls in etched structures and roughened surfaces of films deposited over these layers. Unlike polysilicon and SiC, a post-deposition polishing process is not technically feasible for diamond due to its extreme hardness. For the fabrication of multilayer diamond devices, methods to reduce the surface roughness of the as-deposited films are highly desirable. Along these lines, *Krauss* et al. [7.66] reported the development of an ultrananocrystalline diamond film that exhibits a much smoother surface morphology than comparable diamond films grown using conventional methods. Unlike conventional CVD diamond films that are grown using a mixture of H_2 and CH_4, the ultrananocrystalline diamond films are grown from mixtures of Ar, H_2, and C_{60} or Ar, H_2, and CH_4. Although the processes are still under development, *Krauss* et al. [7.66] have demonstrated the viability of the ultrananocrystalline diamond films as conformal coatings on Si surfaces, as well as their utility in several micromachining processes.

7.5 GaAs, InP, and Related III-V Materials

Gallium Arsenide (GaAs), Indium Phosphide (InP), and related III-V compounds have favorable piezoelectric and optoelectric properties, high piezoresistive constants, and wide electronic bandgaps relative to Si, making them attractive for various sensor and optoelectronic applications. Like Si, significant research in bulk crystal growth has led to the development of GaAs and InP substrates, which are commercially available as high quality, single crystal wafers. Unlike compound semiconductors, such as SiC, III-V materials can be deposited as ternary and quaternary alloys with lattice constants that closely match the binary compounds from which they are derived (i. e., $Al_xGa_{1-x}As$ and GaAs), thus permitting the fabrication of a wide variety of heterostructures that facilitate device performance.

Crystalline GaAs has a zinc blend crystal structure with an electronic bandgap of 1.4 eV, enabling GaAs electronic devices to function at temperatures as high as 350 °C [7.67]. High quality, single crystal wafers are commercially available, as are well developed metalorganic chemical vapor deposition (MOCVD) and molecular beam epitaxy (MBE) growth processes for epitaxial layers of GaAs and its alloys. GaAs does not outperform Si in terms of mechanical properties, however, its stiffness and fracture toughness are still suitable for micromechanical devices.

Micromachining of GaAs is relatively straightforward, since many of its lattice-matched ternary and quaternary alloys have sufficiently different chemical properties to allow their use as sacrificial layers [7.68]. For example, the most common ternary alloy for GaAs is $Al_xGa_{1-x}As$. For values of x less than or equal to 0.5, etchants containing mixtures of HF and H_2O will etch $Al_xGa_{1-x}As$ without attacking GaAs, while etchants containing NH_4OH and H_2O_2 attack GaAs isotropically, but do not etch $Al_xGa_{1-x}As$. Such selectivity enables the micromachining of GaAs wafers using lattice-matched etch stops and sacrificial layers. Devices fabricated using these methods include comb drive lateral resonant structures [7.68], pressure sensors [7.69, 70], thermopile sensors [7.71], Fabry–Perot detectors [7.72], and cantilever-based sensors and actuators [7.73, 74]. In addition, nanoelectromechanical devices, such as suspended micromechanical resonators [7.75] and tethered membranes [7.76], have been fabricated using these techniques. An example of a nanoelectromechanical beam structure fabricated from GaAs is shown in Fig. 7.7.

In addition to using epitaxial layers as etch stops, ion implantation methods can also be used to produce etch stops in GaAs layers. *Miao* et al. [7.77] describe a process that uses electrochemical etching to selectively remove n-type GaAs layers. The process relies on the creation of a highly resistive near surface GaAs layer on an n-type GaAs substrate by low dose nitrogen implantation in the MeV energy range. A pulsed electrochemical etch method using a H_2PtCl_6, H_3PO_4, H_2SO_4 platinum electrolytic solution at 40 °C with 17 V, 100 ms pulses is sufficient to selectively remove n-type GaAs at about 3 μm/min. Using this method, stress-free, tethered membranes could readily be fabricated from the high resistive GaAs layer. The high implant energies enable the fabrication of membranes several microns in thickness. Moreover, the authors demonstrated that if the GaAs wafer were etched in such a way as to create an undulating surface prior to ion implantation, corrugated membranes could be fabricated. These structures can sustain much higher deflection amplitudes than flat structures.

Micromachining of InP closely resembles the techniques used for GaAs. Many of the properties of InP are similar to GaAs in terms of crystal structure, mechanical stiffness, and hardness. However, the optical properties of InP make it particularly attractive for micro-optomechanical devices to be used in the 1.3 to 1.55 μm wavelength range [7.78]. Like GaAs, single crystal wafers of InP are readily available, and

Fig. 7.7 SEM micrograph of a GaAs nanomechanical beam resonator fabricated by epitaxial growth, electron-beam lithography, and selective etching (courtesy of M. Roukes, Caltech)

ternary and quaternary lattice-matched alloys, such as InGaAs, InAlAs, InGaAsP, and InGaAlAs, can be used as either etch stop and/or sacrificial layers, depending on the etch chemistry [7.68]. For instance, InP structural layers deposited on $In_{0.53}Al_{0.47}As$ sacrificial layers can be released using etchants containing $C_6H_8O_7$, H_2O_2, and H_2O. In addition, InP films and substrates can be etched in solutions containing HCl and H_2O using $In_{0.53}Ga_{0.47}As$ films as etch stops. Using InP-based micromachining techniques, multi-air gap filters [7.79], bridge structures [7.78], and torsional membranes [7.72] have been fabricated from InP and its related alloys.

In addition to GaAs and InP, materials such as indium arsenide (InAs) can be micromachined into device structures. Despite a 7% lattice mismatch between InAs and (111) GaAs, high quality epitaxial layers can be grown on GaAs substrates. As described by *Yamaguchi* et al. [7.80], the surface Fermi level of InAs/GaAs structures is pinned in the conduction band, enabling the fabrication of very thin conductive membranes. In fact, the authors have successfully fabricated freestanding InAs structures that range in thickness from 30 to 300 nm. The thin InAs films were grown directly on GaAs substrates by MBE and etched using a solution containing H_2O, H_2O_2, and H_2SO_4. The structures, mainly doubly clamped cantilevers, were released by etching the GaAs substrate using a $H_2O/H_2O_2/NH_4OH$ solution.

7.6 Ferroelectric Materials

Piezoelectric materials play an important role in MEMS technology for sensing and mechanical actuation applications. In a piezoelectric material, mechanical stress produces a polarization, and, conversely, a voltage-induced polarization produces a mechanical stress. Many asymmetric materials, such as quartz, GaAs, and zinc oxide (ZnO), exhibit some piezoelectric behavior. Recent work in MEMS has focussed on the development of ferroelectric compounds, such as lead zirconate titanate, $Pb(Zr_xTi_{1-x})O_3$, or PZT for short, because such compounds have high piezoelectric constants that result in high mechanical transduction. It is relatively straightforward to fabricate a PZT structure on top of a thin freestanding structural layer (i. e., cantilever, diaphragm). Such a capability enables the piezoelectric material to be used in sensor applications or actuator applications for which piezoelectric materials are particularly well suited. Like Si, PZT films can be patterned using dry etch techniques based on chlorine chemistries, such as Cl_2/CCl_4, as well as ion beam milling using inert gases like Ar.

PZT has been successfully deposited in thin film form using co-sputtering, CVD, and sol-gel processing. Sol-gel processing is particularly attractive because the composition and homogeneity of the deposited material over large surface areas can be readily controlled. The sol-gel process outlined by *Lee* et al. [7.81] uses PZT solutions made from liquid precursors containing Pb, Ti, Zr, and O. The solution is deposited by spin coating on a Si wafer that has been coated with a $Pt/Ti/SiO_2$ thin film multilayer. The process is executed to produce a PZT film in layers, with each layer consisting of a spin-coated layer that is dried at 110 °C for 5 min and then heat treated at 600 °C for 20 min. After building up the PZT layer to the desired thickness, the multilayer was heated at 600 °C for up to 6 hrs. Prior to this anneal, a PbO top layer was deposited on the PZT surface. An Au/Cr electrode was then sputter-deposited on the surface of the piezoelectric stack. This process was used to fabricate a PZT-based force sensor. *Xu* et al. [7.82] describe a similar sol-gel process to produce 12 μm-thick, crack-free PZT films on Pt-coated Si wafers and 5 μm-thick films on insulating ZrO_2 layers to produce micromachined MHz-range, two-dimensional transducer arrays for acoustic imaging.

Thick film printing techniques for PZT have been developed to produce thick films in excess of 100 μm. Such thicknesses are desired for applications that require actuation forces that cannot be achieved with the much thinner sol-gel films. *Beeby* et al. [7.83] describe a thick film printing process whereby a PZT paste is made from a mixture of 95% PZT powder, 5% lead borosilicate powder, and an organic carrier. The paste was then printed through a stainless steel screen using a thick film printer. Printing was performed on an oxidized Si substrate that is capped with a Pt electrode. After printing, the paste was dried and then fired at 850 °C to 950 °C. Printing could be repeated to achieve the desired thickness. The top electrode consisted of an evaporated Al film. The authors found that it was possible to perform plasma-based processing on the printed substrates, but that the porous nature of the printed PZT films made them unsuitable for wet chemical processing.

7.7 Polymer Materials

Polyimides comprise an important class of durable polymers that are well suited for many of the techniques used in conventional MEMS processing. In general, polyimides can be acquired in bulk, or deposited as thin films by spin coating. Polyimides can be patterned using conventional dry etching techniques and processed at relatively high temperatures. These attributes make polyimides an attractive group of polymers for MEMS that require polymer structural and/or substrate layers, such as microfabricated biomedical devices, for which inertness and flexibility are important parameters.

7.7.1 Polyimide

Shearwood et al. [7.84] explored the use of polyimide as a robust mechanical material for microfabricated audio membranes. The authors fabricated 7 μm-thick, 8 mm-diameter membranes on GaAs substrates by bulk micromachining the GaAs substrate using a NH_3/H_2O_2 solution. They realized 100% yield and, despite a low Young's modulus ($\sim$ 3 GPa), observed flat membranes to within 1 nm after fabrication.

Jiang et al. [7.85] capitalized on the strength and flexibility of polyimide to fabricate a flexible sheer-stress sensor array based on Si sensors. The sensor array consisted of a collection of Si islands linked by two polyimide layers. Each Si sensor island was 250×250 μm^2 in area and 80 μm in thickness. Al was used as an electrical innerconnect layer. The two polyimide layers served as highly flexible hinges, giving the sensor array the ability to be mounted on curved surfaces. The sensor array was successful in profiling the shear-stress distribution along the leading edge of a rounded delta wing.

The chemical and temperature durability of polyimides enables their use as a sacrificial layer for a number of commonly used materials, such as evaporated or sputter-deposited metals. *Memmi* et al. [7.86] developed a fabrication process for capacitive micromechanical ultransonic transducers using polyimide as a sacrificial layer. The authors showed that the polyimide could withstand the conditions used to deposit silicon monoxide by evaporation and silicon nitride by PECVD at 400 °C. Recent work by *Bagolini* et al. [7.87] has shown that polyimides can even be used as sacrificial layers for PECVD SiC.

In the area of microfabricated biomedical devices, polyimide is receiving attention as a substrate material for implantable devices, owing to its potential biocompatiblity and mechanical flexibility. *Stieglitz* [7.88] reported on the fabrication of multichannel microelectrodes on polyimide substrates. Instead of using polyimide sheets as starting substrates, Si carrier wafers coated with a 5 μm-thick polyimide film were used. Pt microelectrodes were then fabricated on these substrates using conventional techniques. Thin polyimide layers were deposited between various metal layers to serve as insulating layers. A capping polyimide layer was then deposited on the top of the substrates, and the entire polyimide/metal structure was peeled off the Si carrier wafers. Backside processing was then performed on the freestanding polyimide structures to create devices that have exposed electrodes on both surfaces. In a later paper, *Stieglitz* et al. [7.89] describe a variation of this process for neural prostheses.

7.7.2 SU-8

SU-8 is a negative-tone, epoxy-like photoresist that is receiving much attention for its versatility in MEMS processing. It is a high-aspect-ratio, UV-sensitive resist designed for applications requiring single-coat resists with thicknesses on the order of 500 microns [7.90]. SU-8 has favorable chemical properties that enable it to be used as a molding material for high-aspect-ratio electroplated structures (as an alternative to LIGA), and as a structural material for microfluidics [7.90]. In terms of mechanical properties, *Lorenz* et al. [7.91] reported that SU-8 has a modulus of elasticity of 4.02 GPa, which compares favorably with a commonly used polyamid (3.4 GPa).

In addition to the aforementioned conventional uses for SU-8, several interesting alternative uses are beginning to appear in the literature. *Conradie* et al. [7.92] have used SU-8 to trim the mass of silicon paddle oscillators as a means to adjust the resonant frequency of the beams. The trimming process involves the patterning of SU-8 posts on Si paddles. The process capitalizes on the relative chemical stability of the SU-8 resin, in conjunction with the relatively large masses that can be patterned using standard UV exposure processes.

SU-8 is also of interest as a bonding layer material for wafer bonding processes using patterned bonding layers. *Pan* et al. [7.93] compared several UV photo-definable polymeric materials and found that SU-8 exhibited the highest bonding strength (20.6 MPa) for layer thicknesses up to 100 μm.

7.7.3 Parylene

Parylene (poly-paraxylylene) is another emerging polymeric MEMS material due in large part to its biocompatibility. It is particularly attractive from the fabrication perspective, because it can be deposited by CVD at room temperature. Moreover, the deposition process is conformal, which enables parylene coatings to be applied to prefabricated structures, such as Si microneedles [7.94], low stress silicon nitride membrane particle filters [7.95], and micromachined polyimide/Au optical scanners [7.96]. In the first two cases, the parylene coating served to strengthen the microfabricated structures, while in the last case, it served to protect the structure from condensing water vapor.

In addition to its function as a protective coating, parylene can actually be micromachined into freestanding components. *Noh* et al. [7.97] demonstrated a method to create bulk micromachined parylene microcolumns for miniature gas chromatographs. The structure is fabricated using a micromolding technique by which Si molds are fabricated by DRIE and coated with parylene to form three sides of the microcolumn. A second wafer is coated with parylene, and the two are bonded together via a fusion bonding process. After bonding, the structure is released from the Si mold by KOH etching. In a second example, *Yao* et al. [7.98] describe a dry release process for parylene surface micromachining. In this process, sputtered Si is used as a sacrificial layer onto which a thick sacrificial photoresist is deposited. Parylene is then deposited on the photoresist and patterned into the desired structural shape. The release procedure is a two step process. First, the photoresist is dissolved in acetone. This results in the parylene structure sticking to the sputtered Si. Next, a dry BrF_3 etch is performed, which dissolves the Si and releases the parylene structures. Parylene beams that were 1 mm long and 4.5 μm thick were successfully fabricated using this technique.

7.8 Future Trends

The rapid expansion of MEMS in recent years is due in large part to the inclusion of new materials that have expanded the functionality of microfabricated devices beyond what is achievable in silicon. This trend will certainly continue as new application areas for micro- and nanofabricated devices are identified. Many of these applications will likely require both new materials and new processes to fabricate the micro- and nanomachined devices for these yet to be identified applications. Currently, conventional micromachining techniques employ a top-down approach that begins with either bulk substrates or thin films. Future MEMS and NEMS will likely incorporate materials that are created using a bottom-up approach. A significant challenge facing device design and fabrication engineers alike will be how to marry top-down and bottom-up approaches to create devices and systems that cannot be made using either process alone.

References

7.1 C. S. Smith: Piezoresistive effect in germanium and silicon, Phys. Rev. **94** (1954) 1–10
7.2 A. N. Cleland, M. L. Roukes: Fabrication of high frequency nanometer scale mechanical resonators from bulk Si crystals, Appl. Phys. Lett. **69** (1996) 2653–2655
7.3 D. W. Carr, H. G. Craighead: Fabrication of nanoelectromechanical systems in single crystal silicon using silicon on insulator substrates and electron beam lithography, J. Vac. Sci. Technol. B **15** (1997) 2760–2763
7.4 J. J. McMahon, J. M. Melzak, C. A. Zorman, J. Chung, M. Mehregany: Deposition and characterization of in situ boron doped polycristalline silicon films for microelectromechanical systems applications, Mater. Res. Soc. Proc. **605** (2000) 31–36
7.5 T. Kamins: *Polycrystalline Silicon for Integrated Circuits and Displays*, 2nd edn. (Kluwer, Boston 1998)
7.6 L. Cao, T. S. Kin, S. C. Mantell, D. Polla: Simulation and fabrication of piezoresistive membrane type MEMS strain sensors, Sens. Actuators **80** (2000) 273–279
7.7 H. Guckel, T. Randazzo, D. W. Burns: A simple technique for the determination of mechanical strain in thin films with application to polysilicon, J. Appl. Phys. **57** (1985) 1671–1675
7.8 R. T. Howe, R. S. Muller: Stress in polysilicon and amorphous silicon thin films, J. Appl. Phys. **54** (1983) 4674–4675
7.9 X. Zhang, T. Y. Zhang, M. Wong, Y. Zohar: Rapid thermal annealing of polysilicon thin films, J. Microelectromech. Syst. **7** (1998) 356–364
7.10 J. Yang, H. Kahn, A.-Q. He, S. M. Phillips, A. H. Heuer: A new technique for producing large-area as-deposited zero-stress LPCVD polysilicon films: The multipoly process, J. Microelectromech. Syst. **9** (2000) 485–494
7.11 P. Gennissen, M. Bartek, P. J. French, P. M. Sarro: Bipolar-compatible epitaxial poly for smart sensors: Stress minimization and applications, Sens. Actuators A **62** (1997) 636–645
7.12 P. Lange, M. Kirsten, W. Riethmuller, B. Wenk, G. Zwicker, J. R. Morante, F. Ericson, J. A. Schweitz: Thick polycrystalline silicon for surface-micromechanical applications: Deposition, structuring, and mechanical characterization, Sens. Actuators A **54** (1996) 674–678
7.13 S. Greek, F. Ericson, S. Johansson, M. Furtsch, A. Rump: Mechanical characterization of thick polysilicon films: Young's modulus and fracture strength evaluated with microstructures, **9** (1999) 245–251
7.14 K. Funk, H. Emmerich, A. Schilp, M. Offenberg, R. Neul, F. Larmer: A surface micromachined silicon gyroscope using a thick polysilicon layer, Proc. of the 12th Int. Conf. Microelectromech. Systems, Orlando 1999, ed. by K. J. Gabriel, K. Najafi (IEEE, Piscataway 1999) 57–60
7.15 T. Abe, M. L. Reed: Low strain sputtered polysilicon for micromechanical structures, Proc. 9th Int. Workshop Microelectromech. Systems, San Diego 1996, ed. by M. G. Allen, M. L. Reed (IEEE, Piscataway 1996) 258–262
7.16 K. Honer, G. T. A. Kovacs: Integration of sputtered silicon microstructures with pre-fabricated CMOS circuitry, Sens. Actuators A **91** (2001) 392–403
7.17 R. Anderson, R. S. Muller, C. W. Tobias: Porous polycrystalline silicon: A new material for MEMS, J. Microelectromech. Syst. **3** (1994) 10–18
7.18 W. Lang, P. Steiner, H. Sandmaier: Porous silicon: A novel material for microsystems, Sens. Actuators A **51** (1995) 31–36
7.19 S. K. Ghandhi: *VLSI Fabrication Principles – Silicon and Gallium Arsenide* (Wiley, New York 1983)
7.20 W. A. Pilskin: Comparison of properties of dielectric films deposited by various methods, J. Vac. Sci. Technol. **21** (1977) 1064–1081
7.21 J. S. Danel, F. Michel, G. Delapierre: Micromachining of quartz and its application to an acceleration sensor, Sens. Actuators A **21–23** (1990) 971–977
7.22 A. Yasseen, J. D. Cawley, M. Mehregany: Thick glass film technology for polysilicon surface micromachining, J. Microelectromech. Syst. **8** (1999) 172–179
7.23 R. Liu, M. J. Vasile, D. J. Beebe: The fabrication of nonplanar spin-on glass microstructures, J. Microelectromech. Syst. **8** (1999) 146–151
7.24 B. Folkmer, P. Steiner, W. Lang: Silicon nitride membrane sensors with monocrystalline transducers, Sens. Actuators A **51** (1995) 71–75
7.25 M. Sekimoto, H. Yoshihara, T. Ohkubo: Silicon nitride single-layer X-ray mask, J. Vac. Sci. Technol. **21** (1982) 1017–1021
7.26 P. J. French, P. M. Sarro, R. Mallee, E. J. M. Fakkeldij, R. F. Wolffenbuttel: Optimization of a low-stress silicon nitride process for surface micromachining applications, Sens. Actuators A **58** (1997) 149–157
7.27 B. Li, B. Xiong, L. Jiang, Y. Zohar, M. Wong: Germanium as a versatile material for low-temperature micromachining, J. Microelectromech. Syst. **8** (1999) 366–372
7.28 A. Franke, D. Bilic, D. T. Chang, P. T. Jones, T. J. King, R. T. Howe, C. G. Johnson: Post-CMOS integration of germanium microstructures, Proc. 12th Int. Conf. Microelectromech. Syst., Orlando 1999, ed. by K. J. Gabriel, K. Najafi (IEEE, Piscataway 1999) 630–637
7.29 A. E. Franke, Y. Jiao, M. T. Wu, T. J. King, R. T. Howe: Post-CMOS modular integration of poly-SiGe microstructures using poly-Ge sacrificial layers, Technical Digest – Solid State Sens. Actuator Workshop,

Hilton Head 2000, ed. by L. Bousse (Transducers Research Foundation, Cleveland Heights Ohio 2000) 18–21

7.30 S. Sedky, P. Fiorini, M. Caymax, S. Loreti, K. Baert, L. Hermans, R. Mertens: Structural and mechanical properties of polycrystalline silicon germanium for micromachining applications, J. Microelectromech. Syst. **7** (1998) 365–372

7.31 S. Sedky, A. Witvrouw, K. Baert: Poly SiGe, a promising material for MEMS monolithic integration with the driving electronics, Sens. Actuators A **97-98** (2002) 503–511

7.32 J.M. Heck, C.G. Keller, A.E. Franke, L. Muller, T.-J. King, R.T. Howe: High aspect ratio polysilicon-germanium microstructures, Proc. 10th Int. Conf. Solid State Sens. Actuators, Sendai 1999, ed. by M. Esashi, T. Morizumi (IEE of Japan, Tokyo 1999) 328–334

7.33 P. Van Gerwen, T. Slater, J.B. Chevrier, K. Baert, R. Mertens: Thin-film boron-doped polycrystalline $silicon_{70\%}$-$germanium_{30\%}$ for thermopiles, Sens. Actuators A **53** (1996) 325–329

7.34 D. Hyman, J. Lam, B. Warneke, A. Schmitz, T.Y. Hsu, J. Brown, J. Schaffner, A. Walson, R.Y. Loo, M. Mehregany, J. Lee: Surface micromachined RF MEMS switches on GaAs substrates, Int. J. Radio Frequency Microwave Commun. Eng. **9** (1999) 348–361

7.35 C. Chang, P. Chang: Innovative micromachined microwave switch with very low insertion loss, Sens. Actuators **79** (2000) 71–75

7.36 C.L. Shih, B.K. Lai, H. Kahn, S.M. Phillips, A.H. Heuer: A robust co-sputtering fabrication procedure for TiNi shape memory alloys for MEMS, J. Microelectromech. Syst. **10** (2001) 69–79

7.37 G. Hahm, H. Kahn, S.M. Phillips, A.H. Heuer: Fully microfabricated silicon spring biased shape memory actuated microvalve, Technical Digest – Solid State Sens. Actuator Workshop, Hilton Head Island 2000, ed. by L. Bousse (Transducers Research Foundation, Cleveland Heights Ohio 2000) 230–233

7.38 S.D. Leith, D.T. Schwartz: High-rate through-mold electrodeposition of thick (>200 micron) NiFe MEMS components with uniform composition, J. Microelectromech. Syst. **8** (1999) 384–392

7.39 N. Rajan, M. Mehregany, C.A. Zorman, S. Stefanescu, T. Kicher: Fabrication and testing of micromachined silicon carbide and nickel fuel atomizers for gas turbine engines, J. Microelectromech. Syst. **8** (1999) 251–257

7.40 T. Pornsin-sirirak, Y.C. Tai, H. Nassef, C.M. Ho: Titanium-alloy MEMS wing technology for a microaerial vehicle application, Sens. Actuators A **89** (2001) 95–103

7.41 C.R. Stoldt, C. Carraro, W.R. Ashurst, D. Gao, R.T. Howe, R. Maboudian: A low temperature CVD process for silicon carbide MEMS, Sens. Actuators A **97-98** (2002) 410–415

7.42 M. Eickhoff, H. Moller, G. Kroetz, J. von Berg, R. Ziermann: A high temperature pressure sensor prepared by selective deposition of cubic silicon carbide on SOI substrates, Sens. Actuators **74** (1999) 56–59

7.43 Y.T. Yang, K.L. Ekinci, X.M.H. Huang, L.M. Schiavone, M.L. Roukes, C.A. Zorman, M. Mehregany: Monocrystalline silicon carbide nanoelectromechanical systems, Appl. Phys. Lett. **78** (2001) 162–164

7.44 C.A. Zorman, S. Rajgolpal, X.A. Fu, R. Jezeski, J. Melzak, M. Mehregany: Deposition of polycrystalline 3C-SiC films on 100 mm-diameter (100) Si wafers in a large-volume LPCVD furnace, Electrochem. Solid State Lett. **5** (2002) G99–G101

7.45 I. Behrens, E. Peiner, A.S. Bakin, A. Schlachetzski: Micromachining of silicon carbide on silicon fabricated by low-pressure chemical vapor deposition, J. Micromech. Microeng. **12** (2002) 380–384

7.46 C.A. Zorman, S. Roy, C.H. Wu, A.J. Fleischman, M. Mehregany: Characterization of polycrystalline silicon carbide films grown by atmospheric pressure chemical vapor deposition on polycrystalline silicon, J. Mater. Res. **13** (1996) 406–412

7.47 C.H. Wu, C.H. Zorman, M. Mehregany: Growth of polycrystalline SiC films on SiO_2 and Si_3N_4 by APCVD, Thin Solid Films **355-356** (1999) 179–183

7.48 P. Sarro: Silicon carbide as a new MEMS technology, Sens. Actuators **82** (2000) 210–218

7.49 N. Ledermann, J. Baborowski, P. Muralt, N. Xantopoulos, J.M. Tellenbach: Sputtered silicon carbide thin films as protective coatings for MEMS applications, Surf. Coat. Technol. **125** (2000) 246–250

7.50 R.S. Okojie, A.A. Ned, A.D. Kurtz: Operation of a 6H-SiC pressure sensor at 500 °C, Sens. Actuators A **66** (1998) 200–204

7.51 K. Lohner, K.S. Chen, A.A. Ayon, M.S. Spearing: Microfabricated silicon carbide microengine structures, Mater. Res. Soc. Symp. Proc. **546** (1999) 85–90

7.52 S. Tanaka, S. Sugimoto, J.-F. Li, R. Watanabe, M. Esashi: Silicon carbide micro-reaction-sintering using micromachined silicon molds, J. Microelectromech. Syst. **10** (2001) 55–61

7.53 X. Song, S. Rajgolpal, J.M. Melzak, C.A. Zorman, M. Mehregany: Development of a multilayer SiC surface micromachining process with capabilities and design rules comparable with conventional polysilicon surface micromachining, Mater. Sci. Forum **389-393** (2001) 755–758

7.54 L.A. Liew, W. Zhang, V.M. Bright, A. Linan, M.L. Dunn, R. Raj: Fabrication of SiCN ceramic MEMS using injectable polymer-precursor technique, Sens. Actuators A **89** (2001) 64–70

7.55 A.J. Fleischman, S. Roy, C.A. Zorman, M. Mehregany: Polycrystalline silicon carbide for surface micromachining, Proc.9th Int. Workshop Microelectromech. Systems, San Diego 1996, ed. by M.G. Allen, M.L. Reid (IEEE, Piscataway 1996) 234–238

7.56 A. J. Fleischman, X. Wei, C. A. Zorman, M. Mehregany: Surface micromachining of polycrystalline SiC deposited on SiO_2 by APCVD, Mater. Sci. Forum **264–268** (1998) 885–888
7.57 A. Yasseen, C. H. Wu, C. A. Zorman, M. Mehregany: Fabrication and testing of surface micromachined polycrystalline SiC micromotors, Electron Device Lett. **21** (2000) 164–166
7.58 T. Shibata, Y. Kitamoto, K. Unno, E. Makino: Micromachining of diamond film for MEMS applications, J. Microelectromech. Syst. **9** (2000) 47–51
7.59 H. Bjorkman, P. Rangsten, P. Hollman, K. Hjort: Diamond replicas from microstructured silicon masters, Sens. Actuators **73** (1999) 24–29
7.60 P. Rangsten, H. Bjorkman, K. Hjort: Microfluidic components in diamond, Proc. 10th Int. Conf. Solid State Sens. Actuators, Sendai 1999, ed. by M. Esashi, T. Morizumi (IEE of Japan, Tokyo 1999) 190–193
7.61 H. Bjorkman, P. Rangsten, K. Hjort: Diamond microstructures for optical microelectromechanical systems, Sens. Actuators **78** (1999) 41–47
7.62 M. Aslam, D. Schulz: Technology of diamond microelectromechanical systems, Proc. 8th Int. Conf. Solid State Sens. Actuators, Stockholm 1995, ed. by I. Lundstrum (Foundation for Sensor and Actuator Technology, Stockholm 1995) 222–224
7.63 R. Ramesham: Fabrication of diamond microstructures for microelectromechanical systems (MEMS) by a surface micromachining process, Thin Solid Films **340** (1999) 1–6
7.64 Y. Yang, X. Wang, C. Ren, J. Xie, P. Lu, W. Wang: Diamond surface micromachining technology, Diamond Rel. Mater. **8** (1999) 1834–1837
7.65 X. D. Wang, G. D. Hong, J. Zhang, B. L. Lin, H. Q. Gong, W. Y. Wang: Precise patterning of diamond films for MEMS application, J. Mater. Processing Technol. **127** (2002) 230–233
7.66 A. R. Krauss, O. Auciello, D. M. Gruen, A. Jayatissa, A. Sumant, J. Tucek, D. C. Mancini, N. Moldovan, A. Erdemire, D. Ersoy, M. N. Gardos, H. G. Busmann, E. M. Meyer, M. Q. Ding: Ultrananocrystalline diamond thin films for MEMS and moving mechanical assembly devices, Diamond Rel. Mater. **10** (2001) 1952–1961
7.67 K. Hjort, J. Soderkvist, J.-A. Schweitz: Galium arsenide as a mechanical material, J. Micromech. Microeng. **4** (1994) 1–13
7.68 K. Hjort: Sacrificial etching of III-V compounds for micromechanical devices, J. Micromech. Microeng. **6** (1996) 370–365
7.69 K. Fobelets, R. Vounckx, G. Borghs: A GaAs pressure sensor based on resonant tunnelling diodes, J. Micromech. Microeng. **4** (1994) 123–128
7.70 A. Dehe, K. Fricke, H. L. Hartnagel: Infrared thermopile sensor based on AlGaAs-GaAs micromachining, Sens. Actuators A **46–47** (1995) 432–436
7.71 A. Dehe, K. Fricke, K. Mutamba, H. L. Hartnagel: A piezoresistive GaAs pressure sensor with GaAs/AlGaAs membrane technology, J. Micromech. Microeng. **5** (1995) 139–142
7.72 A. Dehe, J. Peerlings, J. Pfeiffer, R. Riemenschneider, A. Vogt, K. Streubel, H. Kunzel P. Meissner, H. L. Hartnagel: III-V compound semiconductor micromachined actuators for long resonator tunable Fabry–Perot detectors, Sens. Actuators A **68** (1998) 365–371
7.73 T. Lalinsky, S. Hascik, Z. Mozolova, E. Burian, M. Drzik: The improved performance of GaAs micromachined power sensor microsystem, Sens. Actuators **76** (1999) 241–246
7.74 T. Lalinsky, E. Burian, M. Drzik, S. Hascik, Z. Mozolova, J. Kuzmik, Z. Hatzopoulos: Performance of GaAs micromachined microactuator, Sens. Actuators **85** (2000) 365–370
7.75 H. X. Tang, X. M. H. Huang, M. L. Roukes, M. Bichler, W. Wegscheider: Two-dimensional electron-gas actuation and transduction for GaAs nanoelectromechanical systems, Appl. Phys. Lett. **81** (2002) 3879–3881
7.76 T. S. Tighe, J. M. Worlock, M. L. Roukes: Direct thermal conductance measurements on suspended monocrystalline nanostructures, Appl. Phys. Lett. **70** (1997) 2687–2689
7.77 J. Miao, B. L. Weiss, H. L. Hartnagel: Micromachining of three-dimensional GaAs membrane structures using high-energy nitrogen implantation, J. Micromech. Microeng. **13** (2003) 35–39
7.78 C. Seassal, J. L. Leclercq, P. Viktorovitch: Fabrication of InP-based freestanding microstructures by selective surface micromachining, J. Micromech. Microeng. **6** (1996) 261–265
7.79 J. Leclerq, R. P. Ribas, J. M. Karam, P. Viktorovitch: III-V micromachined devices for microsystems, Microelectron. J. **29** (1998) 613–619
7.80 H. Yamaguchi, R. Dreyfus, S. Miyashita, Y. Hirayama: Fabrication and elastic properties of InAs freestanding structures based on InAs/GaAs(111) A heteroepitaxial systems, Physica E **13** (2002) 1163–1167
7.81 J. B. Lee, J. English, C. H. Ahn, M. G. Allen: Planarization techniques for vertically integrated metallic MEMS on silicon foundry circuits, J. Micromech. Microeng. **7** (1997) 44–54
7.82 B. Xu, L. E. Cross, J. J. Bernstein: Ferroelectric and antiferroelectric films for microelectromechanical systems applications, Thin Solid Films **377–378** (2000) 712–718
7.83 S. P. Beeby, A. Blackburn, N. M. White: Processing of PZT piezoelectric thick films on silicon for microelectromechanical systems, J. Micromech. Microeng. **9** (1999) 218–229
7.84 C. Shearwood, M. A. Harradine, T. S. Birch, J. C. Stevens: Applications of polyimide membranes to MEMS technology, Microelectron. Eng. **30** (1996) 547–550

7.85 F. Jiang, G.B. Lee, Y.C. Tai, C.M. Ho: A flexible micromachine-based shear-stress sensor array and its application to separation-point detection, Sens. Actuators **79** (2000) 194–203

7.86 D. Memmi, V. Foglietti, E. Cianci, G. Caliano, M. Pappalardo: Fabrication of capacitive micromechanical ultrasonic transducers by low-temperature process, Sens. Actuators A **99** (2002) 85–91

7.87 A. Bagolini, L. Pakula, T.L.M. Scholtes, H.T.M. Pham, P.J. French, P.M. Sarro: Polyimide sacrificial layer and novel materials for post-processing surface micromachining, J. Micromech. Microeng. **12** (2002) 385–389

7.88 T. Stieglitz: Flexible biomedical microdevices with double-sided electrode arrangements for neural applications, Sens. Actuators A **90** (2001) 203–211

7.89 T. Stieglitz, G. Matthias: Flexible BioMEMS with electrode arrangements on front and back side as key component in neural prostheses and biohybrid systems, Sens. Actuators B **83** (2002) 8–14

7.90 H. Lorenz, M. Despont, N. Fahrni, J. Brugger, P. Vettiger, P. Renaud: High-aspect-ratio, ultrathick, negative-tone-near-UV photoresist and its applications in MEMS, Sens. Actuators A **64** (1998) 33–39

7.91 H. Lorenz, M. Despont, N. Fahrni, N. LaBianca, P. Renaud, P. Vettiger: SU-8: A low-cost negative resist for MEMS, J. Micromech. Microeng. **7** (1997) 121–124

7.92 E.H. Conradie, D.F. Moore: SU-8 thick photoresist processing as a functional material for MEMS applications, J. Micromech. Microeng. **12** (2002) 368–374

7.93 C.T. Pan, H. Yang, S.C. Shen, M.C. Chou, H.P. Chou: A low-temperature wafer bonding technique using patternable materials, J. Micromech. Microeng. **12** (2002) 611–615

7.94 P.A. Stupar, A.P. Pisano: Silicon parylene, and silicon/parylene micro-needles for strength and toughness, Technical Digest 11th Int. Conf. Solid State Sens. Actuators, Munich 2001, ed. by E. Obermeier (Springer, Berlin, Heidelberg 2001) 1368–1389

7.95 X. Yang, J.M. Yang, Y.C. Tai, C.M. Ho: Micromachined membrane particle filters, Sens. Actuators **73** (1999) 184–191

7.96 J.M. Zara, S.W. Smith: Optical scanner using a MEMS actuator, Sens. Actuators A **102** (2002) 176–184

7.97 H.S. Noh, P.J. Hesketh, G.C. Frye-Mason: Parylene gas chromatographic column for rapid thermal cycling, J. Microelectromech. Syst. **11** (2002) 718–725

7.98 T.J. Yao, X. Yang, Y.C. Tai: BrF_3 dry release technology for large freestanding parylene microstructures and electrostatic actuators, Sens. Actuators A **97–98** (2002) 771–775

8. MEMS/NEMS Devices and Applications

Microelectromechanical Systems (MEMS) have played key roles in many important areas, for example transportation, communication, automated manufacturing, environmental monitoring, health care, defense systems, and a wide range of consumer products. MEMS are inherently small, thus offering attractive characteristics such as reduced size, weight, and power dissipation and improved speed and precision compared to their macroscopic counterparts. Integrated circuits (IC) fabrication technology has been the primary enabling technology for MEMS besides a few special etching, bonding and assembly techniques. Microfabrication provides a powerful tool for batch processing and miniaturization of electromechanical devices and systems into a dimensional scale, which is not achievable by conventional machining techniques. As IC fabrication technology continues to scale toward deep sub-micron and nano-meter feature sizes, a variety of nanoelectromechanical systems (NEMS) can be envisioned in the foreseeable future. Nano-scale mechanical devices and systems integrated with nanoelectronics will open a vast number of new exploratory research areas in science and engineering. NEMS will most likely serve as an enabling technology merging engineering with the life sciences in ways that are not currently feasible with the micro-scale tools and technologies. MEMS has been applied to a wide range of fields. Over hundreds of micro-devices have been developed for specific applications. It is thus difficult to provide an overview covering every aspect of the topic. In this chapter, key aspects of MEMS technology and application impacts are illustrated through selecting a few demonstrative device examples, which consist of pressure sensors, inertial sensors, optical and wireless communication devices. Microstructure examples with dimensions on the order of sub-micron are presented with fabrication technologies for future NEMS applications. Although MEMS has experienced significant growth over the past decade, many challenges still remain. In broad terms, these challenges can be grouped into three general categories; (1) fabrication challenges, (2) packaging challenges, and (3) application challenges. Challenges in these areas will, in large measure, determine the commercial success of a particular MEMS device both in technical and economic terms. This chapter presents a brief discussion of some of these challenges as well as possible approaches to address them.

8.1 **MEMS Devices and Applications** 227
8.1.1 Pressure Sensor 227
8.1.2 Inertial Sensor 229
8.1.3 Optical MEMS 233
8.1.4 RF MEMS 239

8.2 **NEMS Devices and Applications** 246

8.3 **Current Challenges and Future Trends** 249

References 250

Microelectromechanical systems, generally referred to as MEMS, has had a history of research and development spanning a few decades. Besides the traditional microfabricated sensors and actuators, the field covers micromechanical components and systems integrated or micro-assembled with electronics on the same substrate or package, achieving high performance functional systems. These devices and systems have played key roles in many important areas, such as transportation, communication, automated manufacturing, environmental monitoring, health care, defense systems, and a wide range of consumer products. MEMS are inherently small, thus offering such attractive characteristics as reduced size, weight, and power dissipation, as well as improved speed and precision compared to their macroscopic counterparts. The development of MEMS requires appropriate fabrication technologies that enable the definition of small geometries, precise dimension

Fig. 8.1 SEM micrograph of a polysilicon microelectromechanical motor [8.1]

control, design flexibility, interfacing with microelectronics, repeatability, reliability, high yield, and low cost. Integrated circuits (IC) fabrication technology meets all of the above criteria and has been the primary enabling fabrication technology for MEMS, besides a few special etching, bonding and assembly techniques. Microfabrication provides a powerful tool for batch processing and miniaturization of electromechanical devices and systems into a dimensional scale, which is not achievable by conventional machining techniques. Most MEMS devices exhibit a length or width ranging from micrometers to several hundreds of micrometers with a thickness from submicrometer up to tens of micrometers, depending upon the fabrication technique employed. A physical displacement of a sensor or an actuator is typically on the same order of magnitude. Figure 8.1 shows an SEM micrograph of a microelectromechanical motor developed in the late 1980s [8.1]. Polycrystalline silicon (polysilicon) surface micromachining technology was used to fabricate the micromotor, achieving a diameter of 150 μm and a minimum vertical feature size on the order of a micrometer. A probe tip is also shown in the micrograph for a size comparison. This device example, and similar others [8.2], demonstrated what MEMS technology could accomplish in microscale machining at the time and served as a strong technology indicator for continued MEMS development. The field has expanded greatly in recent years along with rapid technology advances. Figure 8.2, for example, shows a photo of micro-gears fabricated in the mid-1990s using a five-level polysilicon surface micromachining technology [8.3]. This device represents one of the most advanced surface micromachining fabrication processes developed to date. One can imagine that a wide range of sophisticated microelectromechanical devices and systems can be realized through applying such technology in the future. As IC fabrication technology continues to scale toward deep submicron and nanometer feature sizes, a variety of nanoelectromechanical systems (NEMS) can be envisioned in the near future. Nanoscale mechanical devices and systems integrated with nanoelectronics will open a vast number of new exploratory research areas in science and engineering. NEMS will most likely serve as an enabling technology, merging engineering with the life sciences in ways that are not currently feasible with microscale tools and technologies.

This chapter will provide a general overview of MEMS and NEMS devices and their applications. MEMS technology has been applied to a wide range of fields. Over hundreds of micro-devices have been developed for specific applications. Therefore, it is difficult to provide an overview that covers every aspect of the topic. It is the authors' intent to illustrate key aspects of MEMS technology, and its impact on specific applications by selecting a few demonstrative device examples. For a wide-ranging discussion of nearly all types of micromachined sensors and actuators, books by *Kovacs* [8.4] and *Senturia* [8.5] are recommended.

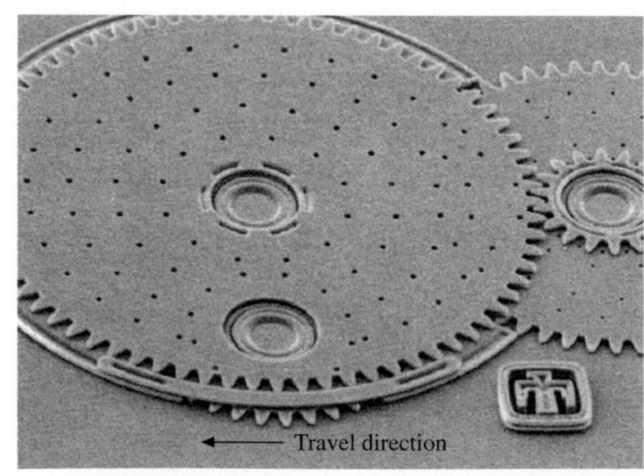

Fig. 8.2 SEM micrograph of polysilicon micro-gears [8.5]

8.1 MEMS Devices and Applications

MEMS devices have played key roles in many areas of development. Microfabricated sensors, actuators, and electronics are the most critical components in implementing a complete system for a specific function. Microsensors and actuators can be fabricated by various micromachining processing technologies. In this section, a number of selected MEMS devices are presented to illustrate the basic device operating principles, as well as demonstrate key aspects of microfabrication technology and its application impact.

8.1.1 Pressure Sensor

Pressure sensors are one of the early devices realized by silicon micromachining technologies and have become successful commercial products. The devices have been widely used in various industrial and biomedical applications. The sensors can be based on piezoelectric, piezoresistive, capacitive, and resonant sensing mechanisms. Silicon bulk and surface micromachining techniques have been used for sensor batch fabrication, thus achieving size miniaturization and low cost. Two types of pressure sensors – piezoresistive and capacitive – are described here for illustration purposes.

Piezoresistive Sensor

The piezoresistive effect in silicon has been widely used for implementing pressure sensors. A pressure-induced strain deforms the silicon band structure, thus changing the resistivity of the material. The piezoresistive effect is typically crystal-orientation dependent and is also affected by doping and temperature. A practical piezoresistive pressure sensor can be implemented by fabricating four sensing resistors along the edges of a thin silicon diaphragm, which acts as a mechanical amplifier to increase the stress and strain at the sensor site. The four sensing elements are connected in a bridge configuration with push-pull signals to increase the sensitivity. The measurable pressure range for such a sensor can be from 10^{-3} to 10^{6} torr, depending on the design. An example of a piezoresistive pressure sensor is shown in Fig. 8.3. The device consists of a silicon diaphragm suspended over a reference vacuum cavity to form an absolute pressure sensor. An external pressure applied over the diaphragm introduces a stress on the sensing resistors, resulting in a resistance value change corresponding to the pressure. The fabrication sequence is outlined as follows. The piezoresistors are typically first formed through a boron diffusion process, followed by a high temperature annealing step in order to achieve a resistance value on the order of a few kilo-ohms. The wafer is then passivated with a silicon dioxide layer, and contact windows are opened for metallization. At this point, the wafer is patterned on the backside, followed by a timed silicon wet etch to form the diaphragm, typically having a thickness around a few tens of micrometers. The diaphragm can have a length of several hundreds of micrometers. A second silicon wafer is then bonded to the device wafer in a vacuum to form a reference vacuum cavity, thus completing the fabrication process. The second wafer can also be further etched through to form an inlet port, implementing a gauge pressure sensor [8.6]. The piezoresistive sensors are simple to fabricate and can be readily interfaced with electronic systems. However, the resistors are temperature dependent and consume DC power. Long-term characteristic drift and resistor thermal noise ultimately limit the sensor resolution.

Capacitive Sensor

Capacitive pressure sensors are attractive because they are virtually temperature independent and consume zero DC power. The devices do not exhibit initial turn-on drift and are stable over time. Furthermore, CMOS microelectronic circuits can be readily interfaced with the sensors to provide advanced signal conditioning and processing, thus improving overall system performance. An example of a capacitive pressure sensor is shown in Fig. 8.4. The device consists of an edge-clamped silicon diaphragm suspended over a vacuum cavity. The diaphragm can be square or circular with a typical thickness of a few micrometers and a length or radius of a few hundred micrometers, respectively. The vacuum

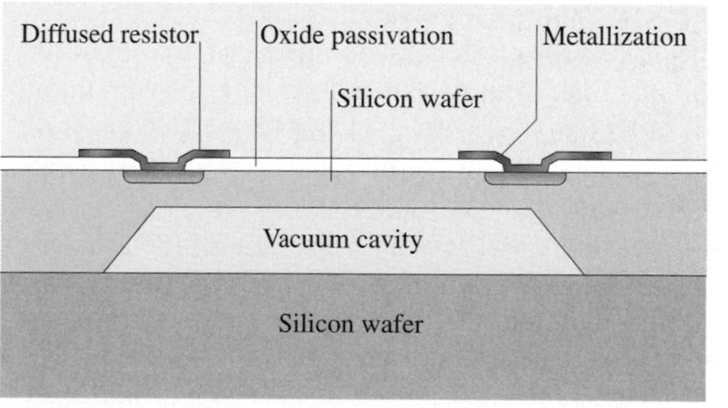

Fig. 8.3 Cross-sectional schematic of a piezoresistive pressure sensor

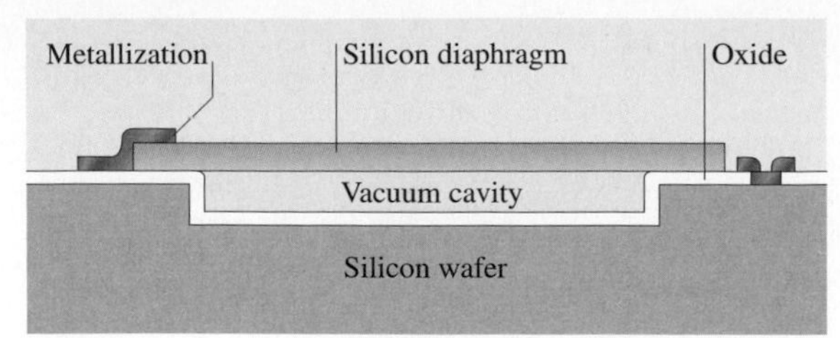

Fig. 8.4 Cross-sectional schematic of a capacitive pressure sensor

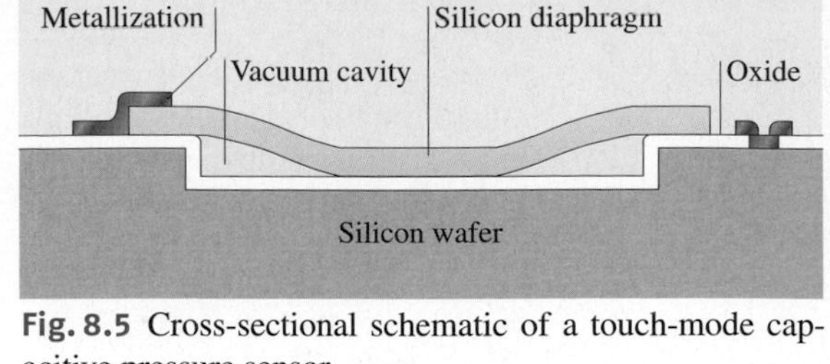

Fig. 8.5 Cross-sectional schematic of a touch-mode capacitive pressure sensor

cavity typically has a depth of a few micrometers. The diaphragm and substrate form a pressure dependent air-gap variable capacitor. An increased external pressure causes the diaphragm to deflect towards the substrate, resulting in an increase in the capacitance value. A simplified fabrication process can be outlined as follows. A silicon wafer is first patterned and etched to form the cavity. The wafer is then oxidized and bonded to a second silicon wafer with a heavily doped boron layer, which defines the diaphragm thickness, at the surface. The bonding process can be performed in a vacuum to realize the vacuum cavity. If the vacuum bonding is not performed at this stage, a low-pressure sealing process can be used to form the vacuum cavity after patterning the sensor diaphragm, provided that sealing channels are available. The silicon substrate above the boron layer is then removed through a wet etching process, followed by patterning to form the sensor diaphragm, which serves as the device top electrode. Contact pads are formed by metallization and patterning. This type of pressure sensor exhibits a nonlinear characteristic and a limited dynamic range. These phenomena, however, can be alleviated by applying an electrostatic-force-balanced feedback architecture. A common practice is to introduce another electrode above the sensing diaphragm through wafer bonding [8.7], forming two capacitors in series, with the diaphragm being the middle electrode. The capacitors are interfaced with electronic circuits, that convert the sensor capacitance value to an output voltage corresponding to the diaphragm position. This voltage is further processed to generate a feedback signal to the top electrode, thereby introducing an electrostatic pull up force to maintain the deflectable diaphragm at its nominal position. This negative feedback loop would substantially minimize the device nonlinearity and also extend the sensor dynamic range.

A capacitive pressure sensor achieving an inherent linear characteristic response and a wide dynamic range can be implemented by employing a touch mode architecture [8.8]. Figure 8.5 shows the cross-sectional view of a touch-mode pressure sensor. The device consists of an edge-clamped silicon diaphragm suspended over a vacuum cavity. The diaphragm deflects under an increasing external pressure and touches the substrate, causing a linear increase in the sensor capacitance value beyond the touch point pressure. Figure 8.6 shows a typical device characteristic curve. The touch point pressure can be designed by engineering the sensor geometric parameters, such as diaphragm size, thickness, cavity depth, etc., for various application requirements. The device can be fabricated using a process flow similar to the flow outlined for the basic capacitive pressure sensor. Figure 8.7 presents a photo of a fabricated touch-mode sensor employing a circular diaphragm with a diameter of 800 μm and a thickness of 5 μm suspended over a 2.5 μm vacuum cavity. The device achieves a touch point pressure of 8 psi and exhibits a linear capacitance range of 33 pF at 10 psi to 40 pF at 32 psi (absolute pressures).

The above processes use bulk silicon materials for machining and are usually referred to as bulk

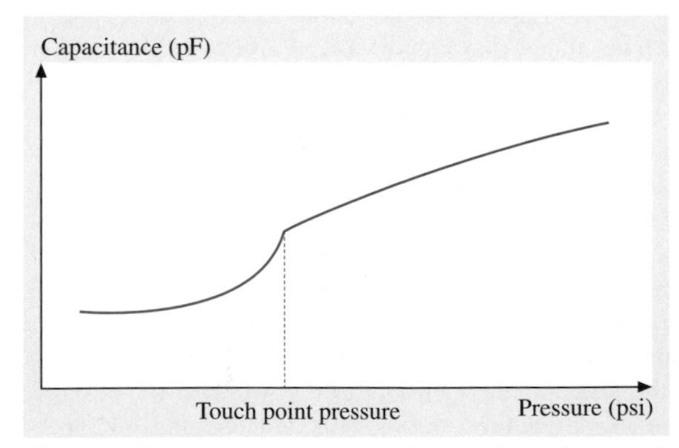

Fig. 8.6 Characteristic response of a touch-mode capacitive pressure

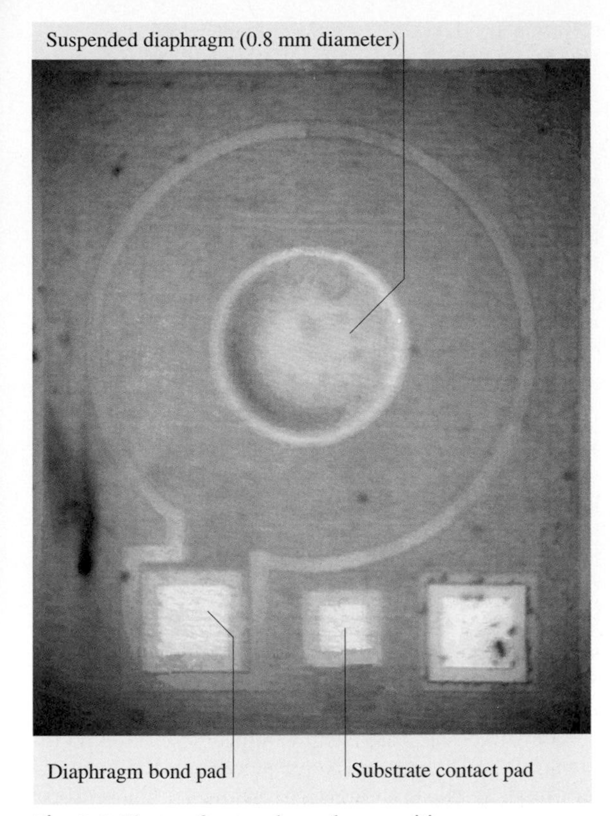

Fig. 8.7 Photo of a touch-mode capacitive pressure sensor [8.12]

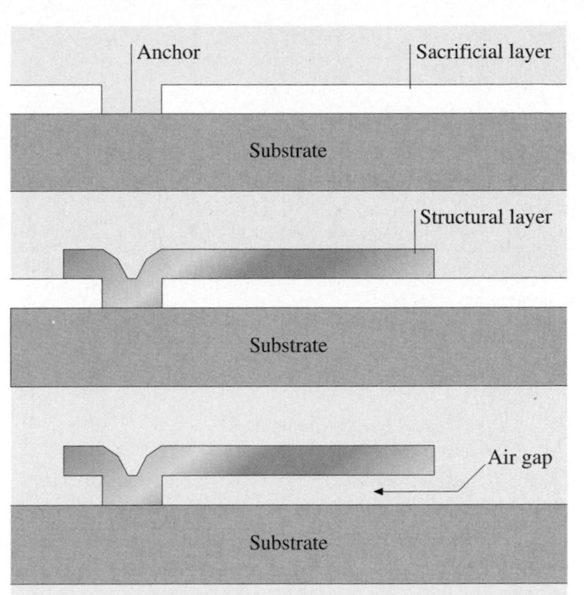

Fig. 8.8 Simplified fabrication sequence of surface micromachining technology

micromachining. The same devices can also be fabricated using so-called surface micromachining. Surface micromachining technology is attractive for integrating MEMS sensors with on-chip electronic circuits. As a result, advanced signal processing capabilities, such as data conversion, offset and noise cancellation, digital calibration, temperature compensation, etc., can be implemented adjacent to microsensors on the same substrate, providing a complete, high performance microsystem solution. The single chip approach also eliminates external wiring, which is critical for minimizing noise pickup and enhancing system performance. Surface micromachining, simply stated, is a method of fabricating MEMS through depositing, patterning, and etching a sequence of thin films with thicknesses on the order of a micrometer. Figure 8.8 illustrates a typical surface micromachining process flow [8.9]. The process starts by depositing a layer of sacrificial material, such as silicon dioxide, over a wafer, followed by anchor formation. A structural layer, typically a polysilicon film, is deposited and patterned. The underlying sacrificial layer is then removed to release the suspended microstructure and complete the fabrication sequence. The processing materials and steps are compatible with the standard integrated circuit process and, thus, can be readily incorporated as an add-on module to an IC process [8.9–11]. A similar surface micromachining technology has been developed to produce monolithic pressure sensor systems [8.12]. Figure 8.9 shows an SEM micrograph of an array of MEMS capacitive pressure sensors fabricated with BiCMOS electronics on the same substrate. Each sensor consists of a 0.8 μm-thick circular polysilicon membrane with a diameter on the order of 20 μm suspended over a 0.3 μm-deep vacuum cavity. The devices operate using the same principle as the sensor shown in Fig. 8.4. A close view of the sensor cross section is shown in Fig. 8.10, which shows the suspended membrane and underneath air gap. These sensors have demonstrated operations in pressure ranges up to 400 bar with an accuracy of 1.5%.

8.1.2 Inertial Sensor

Micromachined inertial sensors consist of accelerometers and gyroscopes. These devices are one of the important types of silicon-based MEMS sensors that have been successfully commercialized. MEMS ac-

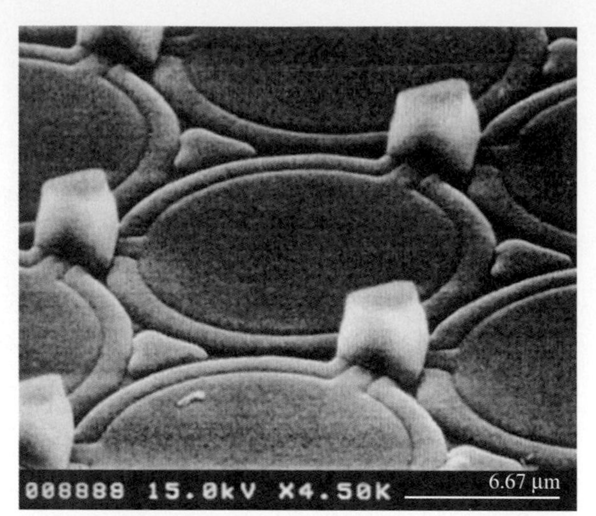

Fig. 8.9 SEM micrograph of polysilicon surface-micromachined capacitive pressure sensors [8.12]

Fig. 8.10 SEM micrograph of a close-up view of a polysilicon surface-micromachined capacitive pressure sensor [8.12]

celerometers alone have the second largest sales volume after pressure sensors. Gyroscopes are expected to reach a comparable sales volume in the foreseeable future. Accelerometers have been used in a wide range of applications, including automotive application for safety systems, active suspension and stability control, biomedical application for activity monitoring, and numerous consumer products, such as head-mount displays, camcorders, three-dimensional mouse, etc. High sensitivity accelerometers are crucial for implementing self-contained navigation and guidance systems. A gyroscope is another type of inertial sensor that measures rate or angle of rotation. The devices can be used along with accelerometers to provide heading information in an inertial navigation system. Gyroscopes are also useful in such applications as automotive ride stabilization and rollover detection, camcorder stabilization, virtual reality, etc. Inertial sensors fabricated by micromachining technology can achieve reduced size, weight, and cost, all which are critical for consumer applications. More importantly, these sensors can be integrated with microelectronic circuits to achieve a functional microsystem with high performance.

Accelerometer

An accelerometer generally consists of a proof mass suspended by compliant mechanical suspensions anchored to a fixed frame. An external acceleration displaces the support frame relative to the proof mass. The displacement can result in an internal stress change in the suspension, which can be detected by piezoresistive sensors as a measure of the external acceleration. The displacement can also be detected as a capacitance change in capacitive accelerometers. Capacitive sensors are attractive for various applications because they exhibit high sensitivity and low temperature dependence, turn-on drift, power dissipation, and noise. The sensors can also be readily integrated with CMOS electronics to perform advanced signal processing for high system performance. Capacitive accelerometers may be divided into two categories: vertical and lateral sensors. Figure 8.11 shows sensor structures for the two versions. In a vertical device, the proof mass is suspended above the substrate electrode by a small gap typically on the order of a micrometer, forming a parallel-plate sense capacitance. The proof mass moves in the direction perpendicular to the substrate (z-axis) upon a vertical input acceleration, thus changing the gap and the capacitance value. The lateral accelerometer consists of a number of movable fingers attached to the proof mass, forming a sense capacitance with an array of fixed parallel fingers. The sensor proof mass moves in a plane parallel to the substrate when subjected to a lateral input acceleration, thus changing the overlap area of these fingers and, hence, the capacitance value. Figure 8.12 shows an SEM top view of a surface micromachined polysilicon z-axis accelerometer [8.13]. The device consists of a $400\,\mu\text{m} \times 400\,\mu\text{m}$ proof mass with a thickness of $2\,\mu\text{m}$ suspended above the substrate electrode by four folded beam suspensions with an air gap around $2\,\mu\text{m}$, thus achieving a sense capacitance of approximately 500 fF.

The visible holes are used to ensure complete removal of the sacrificial oxide underneath the proof mass at the end of the fabrication process. The sensor can be interfaced with a microelectronic charge amplifier, converting the capacitance value to an output voltage for further signal processing and analysis. Force feedback architecture can be applied to stabilize the proof mass position. The combs around the periphery of the proof mass can exert an electrostatic levitation force on the proof mass to achieve the position control, thus improving the system frequency response and linearity performance [8.13].

Surface micromachined accelerometers typically suffer from severe mechanical thermal vibration, commonly referred to as Brownian motion [8.14], due to the small proof mass, resulting in a high mechanical noise floor that ultimately limits the sensor resolution. Vacuum packaging can be employed to minimize this adverse effect, but with a penalty of increasing system complexity and cost. Accelerometers using large proof masses fabricated by bulk micromachining, or a combination of surface and bulk micromachining techniques are attractive for circumventing this problem. Figure 8.13 shows an SEM micrograph of an all-silicon z-axis accelerometer fabricated through a single silicon wafer by using a combined surface and bulk micromachining process to obtain a large proof mass with dimensions of approximately $2\,\mathrm{mm} \times 1\,\mathrm{mm} \times 450\,\mu\mathrm{m}$ [8.15]. The large mass suppresses the Brownian motion effect, achieving a high performance with a resolution on the order of several μg.

A surface micromachined lateral accelerometer developed by Analog Devices Inc. is shown in Fig. 8.14. The sensor consists of a center proof mass supported by folded beam suspensions with arrays of attached movable fingers, forming a sense capacitance with the fixed parallel fingers. The device is fabricated using a $6\,\mu$m-thick polysilicon structural layer with a small air gap on the order of a micrometer to increase the sensor capacitance value, thus improving the device resolution. Figure 8.15 shows a close-up view of the finger structure for a typical lateral accelerometer. Each movable finger forms differential capacitances with two adjacent fixed fingers. This sensing capacitance configuration is attractive for interfacing with differential electronic detection circuits to suppress common-mode noise and other undesirable signal coupling. Monolithic accelerometers with a three-axis sensing capability integrated with on-chip electronic detection circuits have been realized using surfacing micromachining and CMOS microelectronics fabrication technologies [8.16]. Figure 8.16 shows

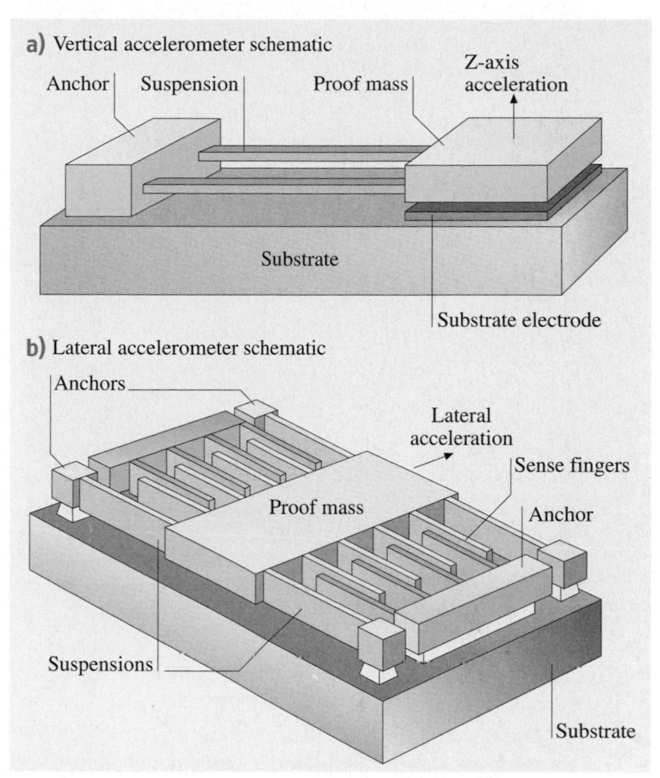

Fig. 8.11a,b Schematics of vertical (a) and lateral (b) accelerometers

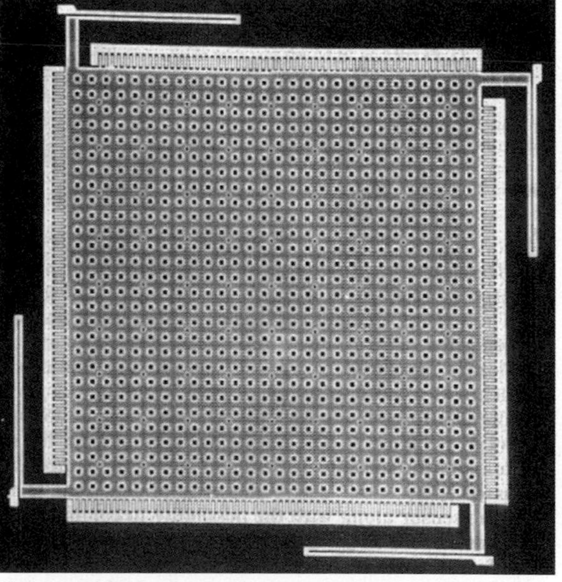

Fig. 8.12 SEM micrograph of a polysilicon surface-micromachined z-axis accelerometer [8.13]

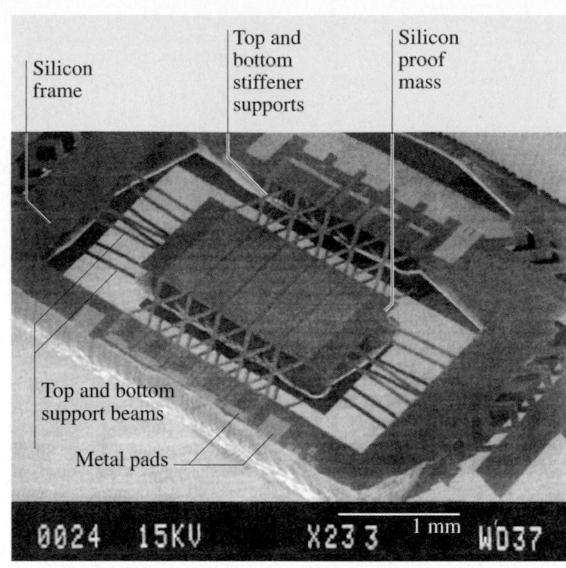

Fig. 8.13 SEM micrograph of a MEMS z-axis accelerometer fabricated using a combined surface and bulk micromachining technology [8.15]

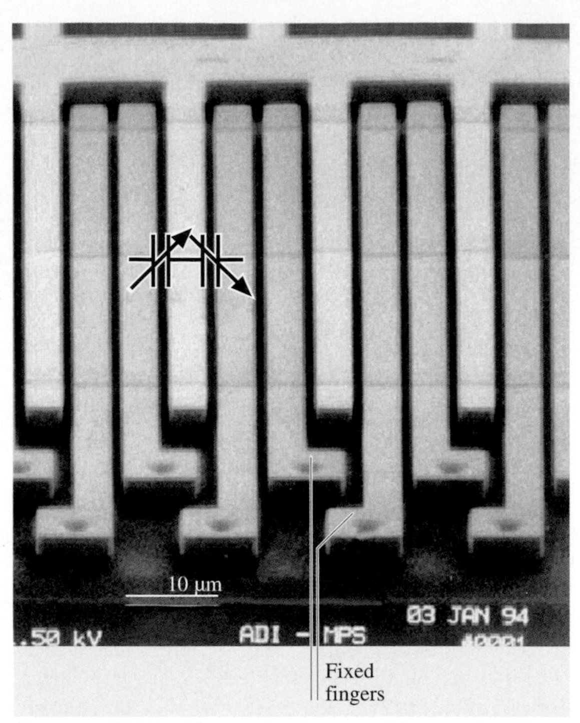

Fig. 8.15 SEM micrograph of a capacitive sensing finger structure

a photo of one of these microsystem chips, which has an area of 4 mm × 4 mm. One vertical accelerometer and two lateral accelerometers are placed at the chip center with corresponding detection electronics along the periphery. A z-axis reference device, which is not movable, is used with the vertical sensor for electronic interfacing. The prototype system achieves a sensing resolution on the order of 1 mG with a 100 Hz bandwidth along each axis. The level of performance is adequate for automobile safety activation systems, vehicle stability, active suspension control, and various consumer products.

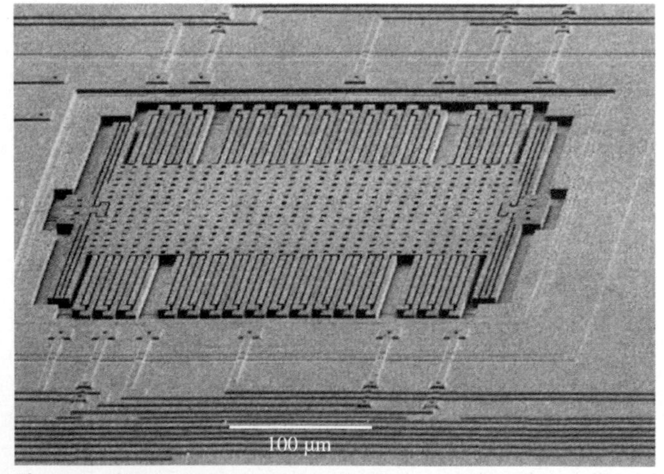

Fig. 8.14 SEM micrograph of a polysilicon surface-micromachined lateral accelerometer (courtesy of Analog Devices Inc.)

Gyroscope

Most micromachined gyroscopes employ vibrating mechanical elements to sense rotations. The sensors rely on the energy transfer between two vibration modes of a structure caused by Coriolis acceleration. Figure 8.17 presents a schematic of a z-axis vibratory rate gyroscope. The device consists of an oscillating mass electrostatically driven into resonance along the drive-mode axis using comb fingers. An angular rotation along the vertical axis (z-axis) introduces a Coriolis acceleration, which results in a structure deflection along the sense-mode axis, shown in the figure. The deflection changes the differential sense capacitance value, which can be detected as a measure of input angular rotation. A z-axis vibratory rate gyroscope operating on this principle is fabricated using surface micromachining technology and integrated together with electronic detection circuits, as illustrated in Fig. 8.18 [8.17]. The micromachined sensor is fabricated using polysilicon structural material with a thickness around 2 μm and occupies an area of 1 mm × 1 mm. The sensor achieves a resolution of approximately $1°/\text{s}/\sqrt{\text{Hz}}$ under a vacuum pressure around 50 mtorr.

A dual-axis gyroscope based on a rotational disk at its resonance can be used to sense angular rotation along two lateral axes (x-axis and y-axis). Figure 8.19 shows a device schematic demonstrating the operating principle. A rotor disk supported by four mechanical suspensions can be driven into angular resonance along the z-axis. An input angular rotation along the x-axis will generate a Coriolis acceleration, causing the disk to rotate along the y-axis and vice versa. This Coriolis acceleration-induced rotation will change the sensor capacitance values between the disk and the different sensing electrodes underneath. The capacitance change can be detected and processed by electronic interface circuits. Angular rotations along the two lateral axes can be measured simultaneously using this device architecture. Figure 8.20 shows a photo of a dual-axis gyroscope fabricated using a 2 μm-thick polysilicon surface micromachining technology [8.18]. As shown in the figure, curved electrostatic drive combs are positioned along the circumference of the rotor dick to drive it into resonance along the vertical axis. The gyroscope exhibits a low random walk of $1°/\sqrt{\text{h}}$ under a vacuum pressure around 60 mtorr. With accelerometers and gyroscopes, each capable of three-axis sensing, a micromachine-based inertial measurement system providing a six-degree-of-freedom sensing capability can be realized. Figure 8.21 presents a photo of such a system containing a dual-axis gyroscope, a z-axis gyroscope, and a three-axis accelerometer chip integrated with microelectronic circuitry. Due to the precision in device layout and fabrication, the system can measure angular rotation and acceleration without the need to align individual sensors.

8.1.3 Optical MEMS

Surface micromachining has served as a key enabling technology to realize microeletromechanical optical devices for various applications ranging from sophisticated visual information displays and fiber-optic telecommunication to bar-code reading. Most of the existing optical systems are implemented using conventional optical components, which suffer from bulky size, high cost, large power consumption, and poor efficiency and reliability issues. MEMS technology is promising for producing miniaturized, reliable, inexpensive

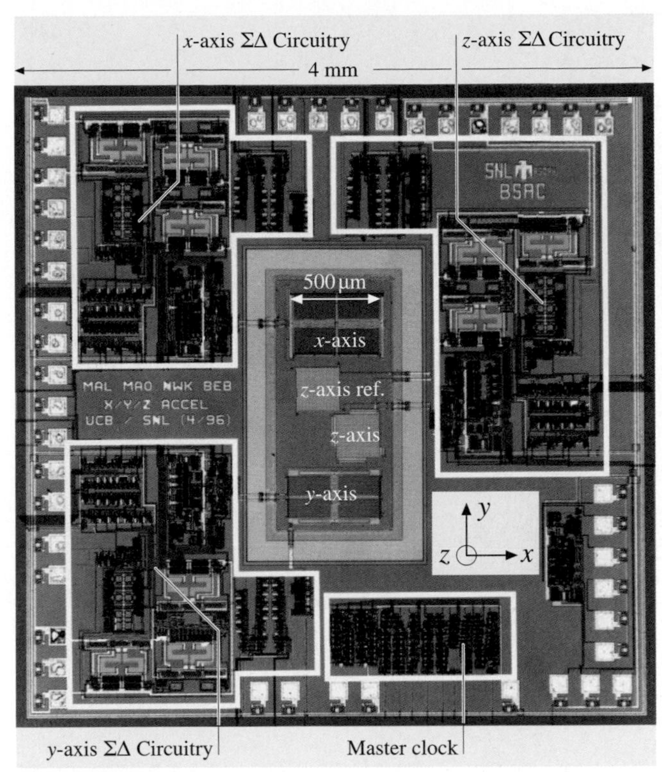

Fig. 8.16 Photo of a monolithic three-axis polysilicon surface-micromachined accelerometer with integrated interface and control electronics [8.16]

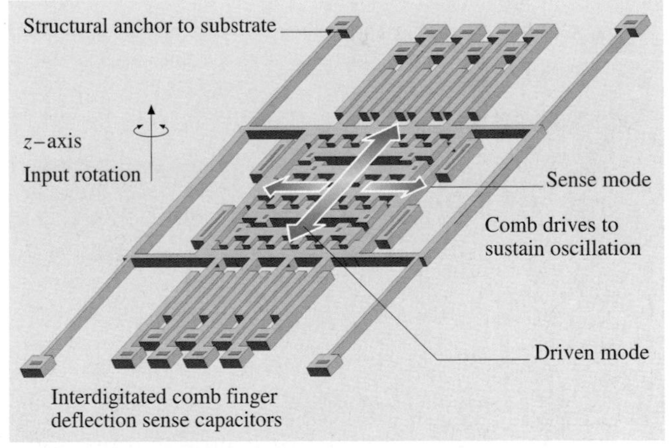

Fig. 8.17 Schematic of a vibratory rate gyroscope

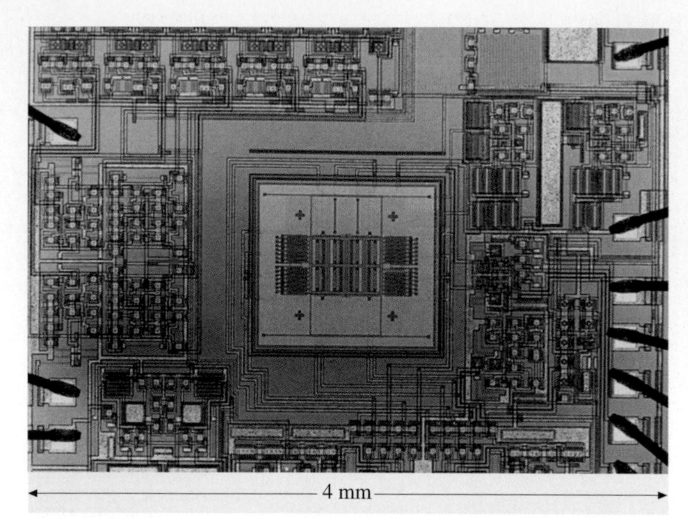

Fig. 8.18 Photo of a monolithic polysilicon surface-micromachined z-axis vibratory gyroscope with integrated interface and control electronics [8.17]

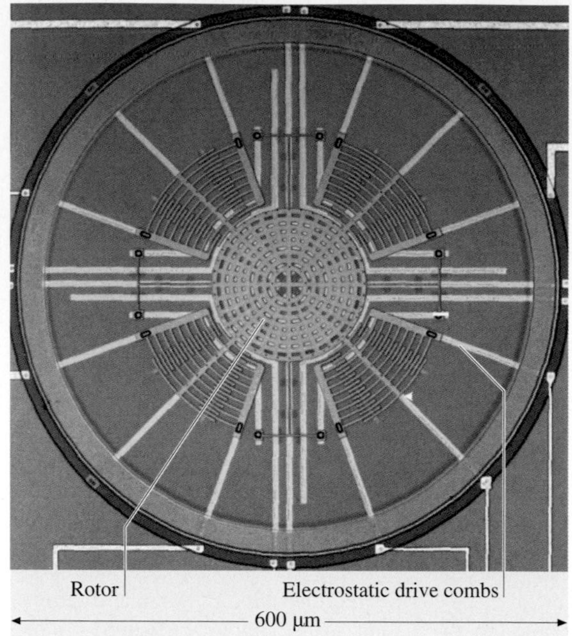
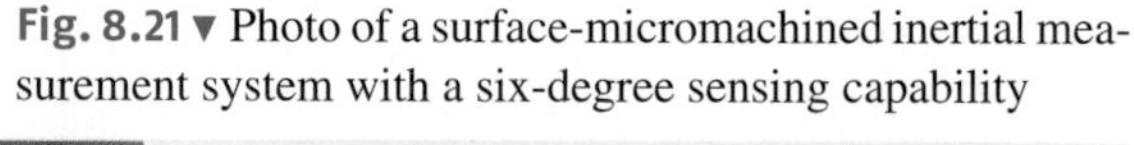

Fig. 8.20 Photo of a polysilicon surface-micromachined dual-axis gyroscope [8.18]

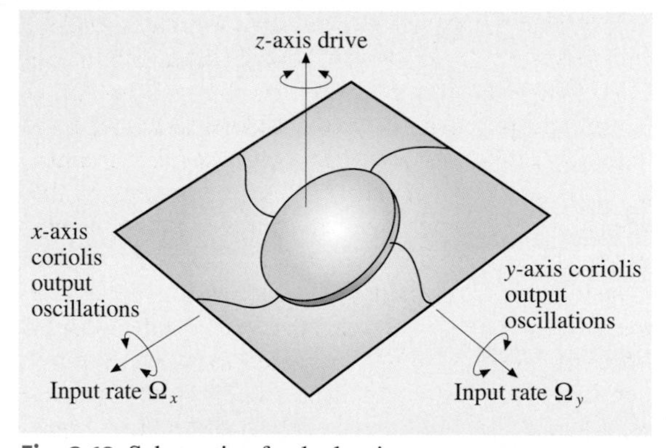

Fig. 8.19 Schematic of a dual-axis gyroscope

Fig. 8.21 ▼ Photo of a surface-micromachined inertial measurement system with a six-degree sensing capability

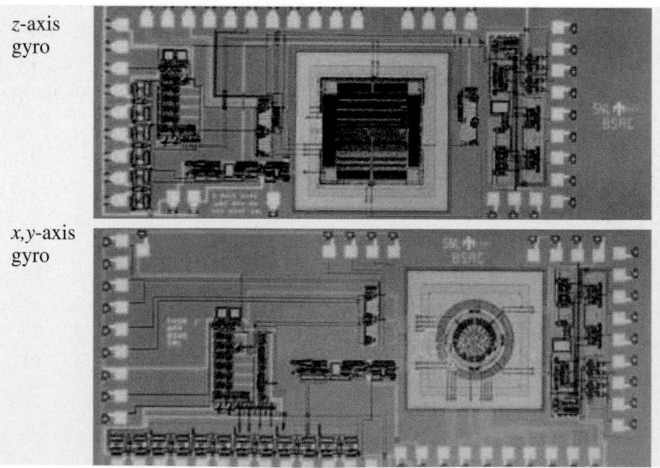

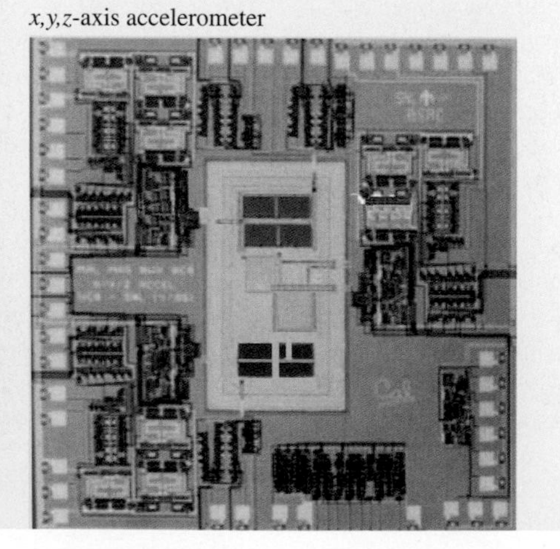

optical components to revolutionize conventional optical systems [8.19]. In this section of the chapter, a few selected MEMS optical devices will be presented to illustrate their impact on the fields of visual display, precision optical platform, and data switching for optical communication.

Visual Display

An early MEMS device successfully used for various display applications is the Texas Instruments Digital Micromirror Device (DMD). The DMD technology can achieve higher performance in terms of resolution, brightness, contrast ratio, and convergence than the conventional cathode ray tube and is critical for digital high definition television applications. A DMD consists of a large array of small mirrors with a typical area of $16\,\mu\text{m}\times 16\,\mu\text{m}$, as illustrated in Fig. 8.22. A probe tip is shown in the figure for a size comparison. Figure 8.23 shows an SEM micrograph of a close-up view of a DMD pixel array [8.20]. Each mirror is capable of rotating by ± 10 degrees, corresponding to either the "on" or "off" position, due to an electrostatic actuation force. Light reflected from any on-mirrors passes through a projection lens and creates images on a large screen. Light from the remaining off-mirrors is reflected away from the projection lens to an absorber. The amount of time during each video frame that a mirror remains in the on-state determines shades of gray, from black for zero on-time to white for a hundred percent on-time. Color can be added by a color wheel or a three-DMD chip setup. The three-DMD chips are used for projecting red, green, and blue colors. Each DMD pixel consists of a mirror connected by a mirror support post to an underlying yoke. The yoke, in turn, is connected by torsion hinges to the support posts, as shown in Fig. 8.24 [8.21]. The support post and hinges are hidden under the mirror to avoid light diffraction and, thus, improve contrast ratio and optical efficiency. There are two gaps on the order of a micrometer, one between the mirror, the underlying hinges and the address electrodes, and a second between the coplanar address electrodes, the hinges and an underlying metal layer from the CMOS static random access memory (SRAM) structure. The yoke is tilted over the second gap by an electrostatic actuation force, thereby rotating the mirror plate. The SRAM determines which angle the mirror needs to be tilted by applying proper actuation voltages to the mirror and address electrodes. The DMD is fabricated using an aluminum-based surface micromachining technology. Three layers of aluminum thin film are deposited and patterned to form the mirror and its suspension system. Polymer material is used as the sacrificial layer and is removed by a plasma etch at the end of the process to freely release the micromirror structure. The micromachining process is compatible with standard CMOS fabrication, allowing the DMD to be monolithically integrated with a mature CMOS address circuit technology, thus achieving high yield and low cost. Figure 8.25 shows an SEM micrograph of a fabricated DMD pixel revealing its cross section after an ion milling. A close-up view of the yoke and hinge support under the mirror is shown in Fig. 8.26.

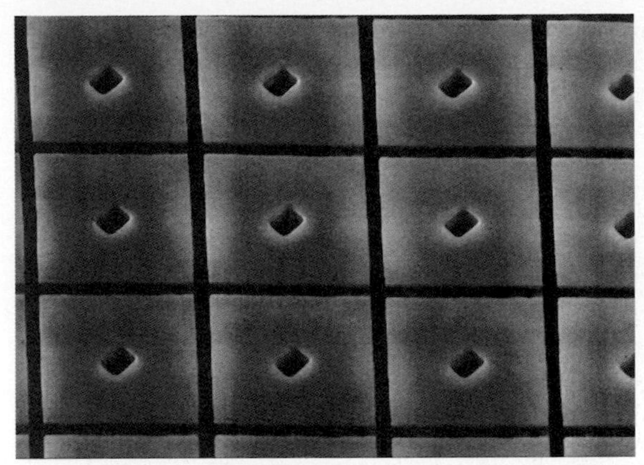
Fig. 8.23 SEM micrograph of a close-up view of a DMD pixel array [8.20]

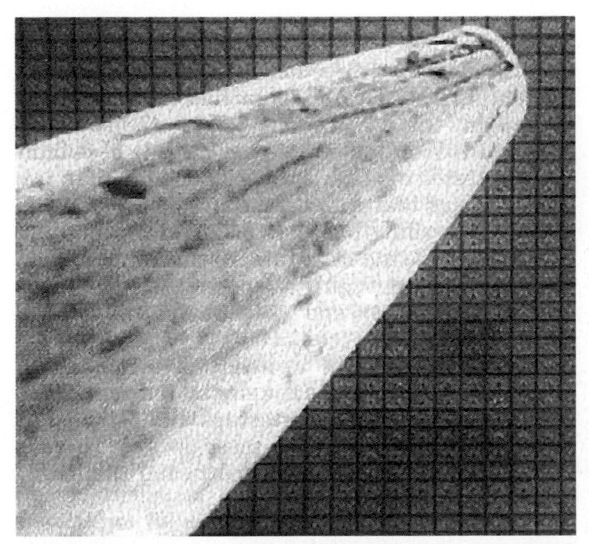
Fig. 8.22 Photo of a digital micromirror device (DMD) array (courtesy of Texas Instruments)

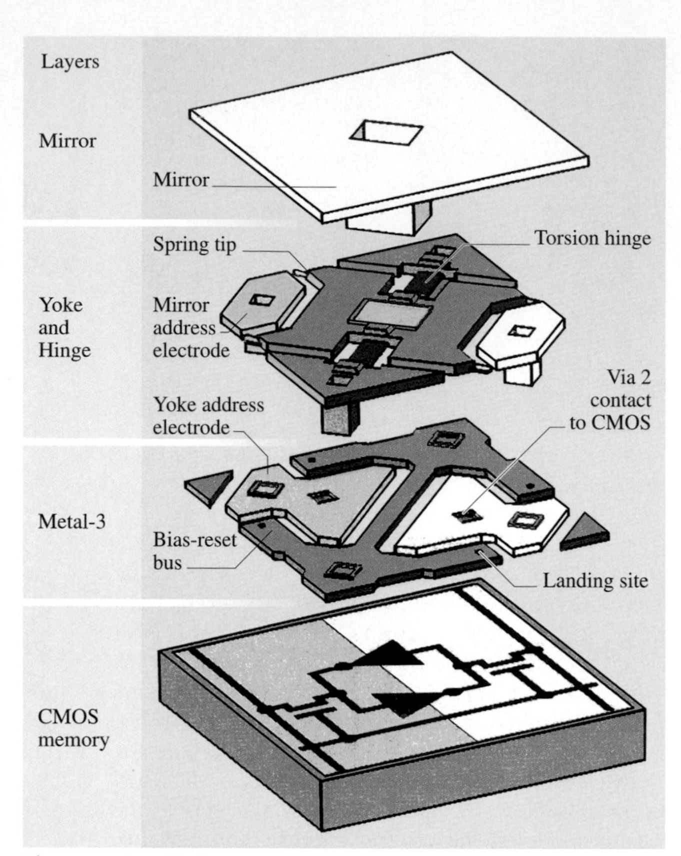

Fig. 8.24 Detailed structure layout of a DMD pixel [8.21]

Fig. 8.25 SEM micrograph of a DMD pixel after removing half of the mirror plate using ion milling (courtesy of Texas Instruments)

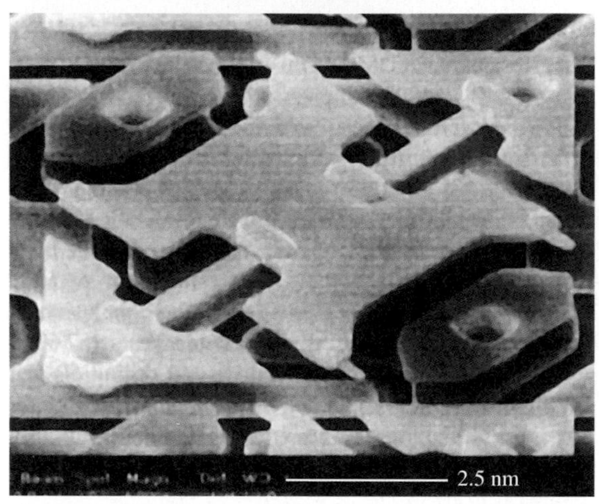

Fig. 8.26 SEM micrograph of a close view of a DMD yoke and hinges [8.21]

Precision Optical Platform

The growing optical communication and measurement industry requires low cost, high performance optoelectronic modules, such as laser-to-fiber couplers, scanners, interferometers, etc. A precision alignment and the ability to actuate optical components, such as mirrors, gratings, and lenses, with sufficient accuracy are critical for high performance optical applications. Conventional hybrid optical integration approaches, such as the silicon optical bench, suffers from a limited alignment tolerance of $\pm 1\ \mu$m and also lacks component actuation capability [8.22, 23]. As a result, only simple optical systems can be constructed with no more than a few components, thus severely limiting the performance. Micromachining, however, provides a critical enabling technology, allowing movable optical components to be fabricated on a silicon substrate. High-precision component movement can be achieved through electrostatic actuation. By combining micromachined movable optical components with lasers, lenses, and fibers on the same substrate, an on-chip, complex, self-aligning optical system can be realized. Figure 8.27a shows an SEM micrograph of a surface micromachined, electrostatically actuated microreflector for laser-to-fiber coupling and external cavity laser applications [8.24]. The device consists of a polysilicon mirror plate hinged to a support beam. The mirror and the support, in turn, are hinged to a vibromotor-actuator slider. The micro-hinge technology [8.25] allows the joints to rotate out of the substrate plane to achieve large aspect ratios. Common-mode actuation of the sliders results in a translational motion, while differential slider motion produces an out-of-plane mirror rotation. These motions permit the microreflector to redirect an optical beam in a desirable location. Each of the two slides is actuated with an integrated microvibromotor, shown in Fig. 8.27b. The vibromotor consists of four electrostatic comb resonators with attached impact arms driving a slider through oblique

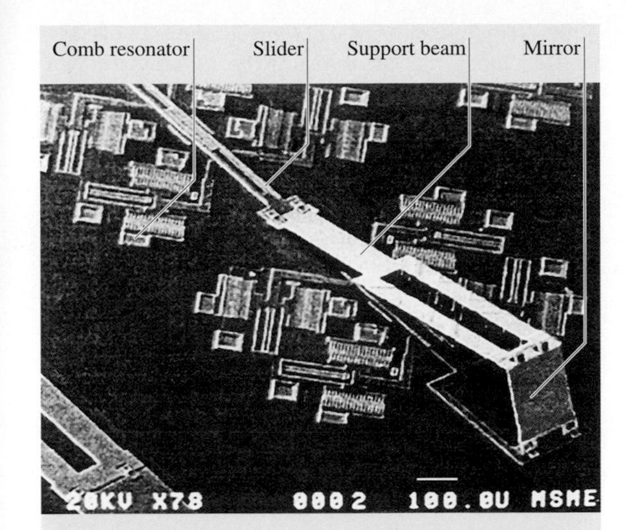

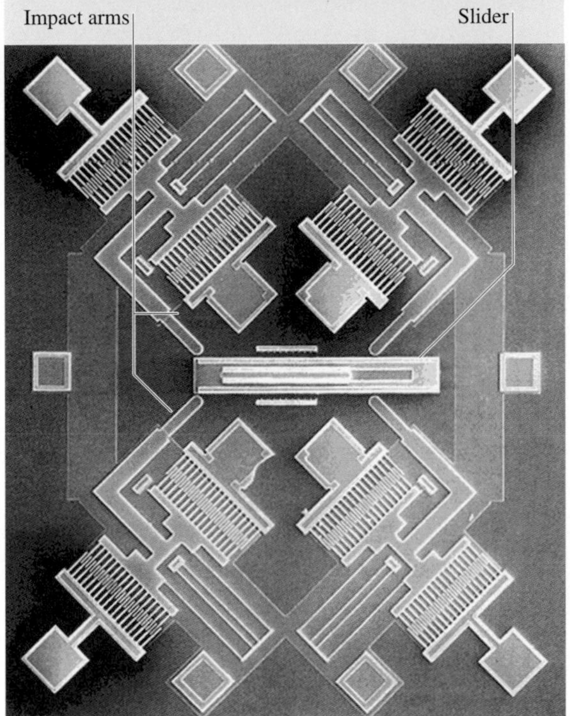

Fig. 8.27 (**a**) SEM micrograph of a surface-micromachined, electrostatically actuated microreflector; (**b**) SEM micrograph of a surface-micromachined vibromotor [8.24]

impact. The two opposing impacters are used for each travel direction to balance the forces. The resonator is a capacitively driven mass anchored to the substrate through a folded beam flexure. The flexure compliance determines the resonant frequency and travel range of the resonator. When the comb structures are driven at their resonant frequency (around 8 KHz), the slider exhibits a maximum velocity of over 1 mm/s. Characterization of the vibromotor also shows that a slider step resolution of less than 0.3 μm can be achieved [8.26], making it attractive for precision alignment of various optical components. The prototype microreflector can obtain an angular travel range over 90 degrees and a translational travel range of 60 μm. By using this device, beam steering, fiber coupling, and optical scanning have been demonstrated.

Optical Data Switching

High-speed communication infrastructures are highly desirable for transferring and processing real-time voice and video information. Optical fiber communication technology has been identified as the critical backbone to support such systems. A high performance optical data switching network, which routes various optical signals from sources to destinations, is one of the key building blocks for system implementation. At present, optical signal switching is performed by using hybrid optical-electronic-optical (O-E-O) switches. These devices first convert incoming light from input fibers to electrical signals and then route the electrical signals to the proper output ports after signal analyses. At the output ports, the electrical signals are converted back to streams of photons or optical signals for further transmission over the fibers. The O-E-O switches are expensive to build, integrate, and maintain. Furthermore, they consume a substantial amount of power and introduce additional latency. It is, therefore, highly desirable to develop an all-optical switching network in which optical signals can be routed without intermediate conversion into electrical form, thus minimizing power dissipation and system delay. While a number of approaches are being considered for building all-optical switches, MEMS technology is attractive because it can provide arrays of tiny movable mirrors that can redirect incoming beams from input fibers to corresponding output fibers. As described in the previous sections, these micromirrors can be batch fabricated using silicon micromachining technologies, achieving an integrated solution with the potential for low cost. A significant reduction in power dissipation is also expected.

Figure 8.28 shows an architecture of a two-dimensional micromirror array forming a switching matrix with rows of input fibers and columns of output fibers (or vice versa). An optical beam from an input fiber can be directed to an output fiber by activating the cor-

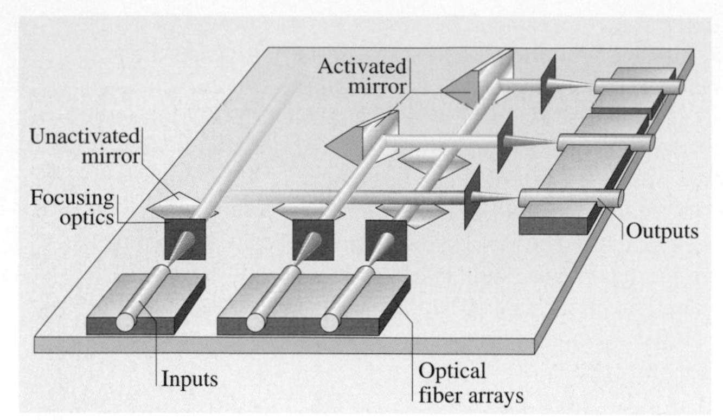

Fig. 8.28 Schematic of a two-dimensional micromirror-based fiber optic switching matrix

responding reflecting micromirror. Switches with eight inputs and eight outputs can be readily implemented using this technique, which can be further extended to a 64×64 matrix. The micromirrors are moved between two fixed stops by digital control, eliminating the need for precision motion control. Figure 8.29 presents an SEM micrograph of a simple 2×2 MEMS fiber optic switching network prototype for an illustration purpose [8.27]. The network includes a mirror chip passively integrated with a silicon submount, which contains optical fibers and ball lenses. The mirror chip consists of four surface-micromachined vertical torsion mirrors. The four mirrors are arranged so that in the "reflection" mode, the input beams are reflected by two 45-degree vertical torsion mirrors and coupled with the output fibers located on the same side of the chip. In the "transmission" mode, the vertical torsion mirrors are rotated out of the optical paths, allowing the input beams to be coupled with the opposing output fibers. Figure 8.30 shows an SEM micrograph of a polysilicon vertical torsion mirror. The device consists of a mirror plate attached to a vertical supporting frame by torsion beams and a vertical back electrode plate. The mirror plate is approximately 200 μm wide, 160 μm long, and 1.5 μm thick. The mirror surface is coated with a thin layer of gold to improve the optical reflectivity. The back plate is used to electrostatically actuate the mirror plate, so that the mirror can be rotated out of the optical path in the "transmission" mode. Surface micromachining with micro-hinge technology is used to realize the overall structure. The back electrode plate is integrated with a scratch drive actuator array [8.28] for self-assembly. The self-assembly approach is critical when multiple vertical torsion mirrors are used to implement more advanced functions.

A more sophisticated optical switching network with a large scaling potential can be implemented by using a three-dimensional (3-D) switching architecture, as shown in Fig. 8.31. The network consists of arrays of two-axis mirrors to steer optical beams from input fibers to output fibers. A precision analog closed-loop mirror position control is required to accurately direct a beam along two angles, so that one input fiber can be optically

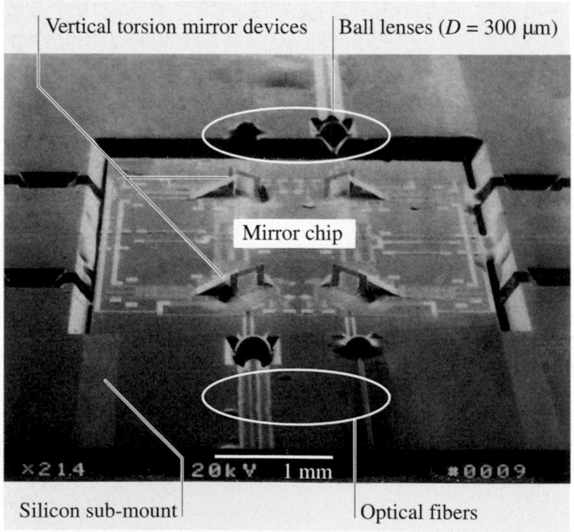

Fig. 8.29 SEM micrograph of a 2×2 MEMS fiber optic switching network [8.27]

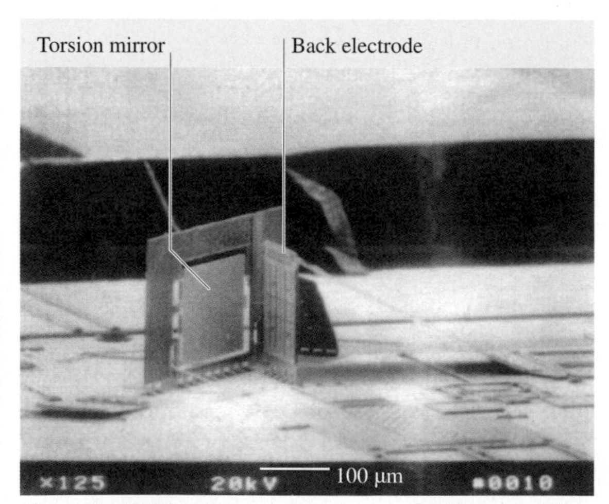

Fig. 8.30 SEM micrograph of a polysilicon surface-micromachined vertical torsion mirror [8.27]

connected to any output fiber. The optical length depends little on which sets of fibers are connected, thus achieving a more uniform switching characteristic, which is critical for implementing large scale networks. Two-axis mirrors are the crucial components for implementing the 3-D architecture. Figure 8.32 shows an SEM micrograph of a surface micromachined two-axis beam-steering mirror positioned by using self-assembly technique [8.29]. The self-assembly is accomplished during the final release step of the mirror processing sequence. Mechanical energy is stored in a special high-stress layer during the deposition, which is put on top of the four assembly arms. Immediately after the assembly arms are released, the tensile stress in this layer causes the arms to bend upward, pushing the mirror frame and lifting it above the silicon substrate. All mirrors used in the switching network can be fabricated simultaneously without any human intervention or external power supply.

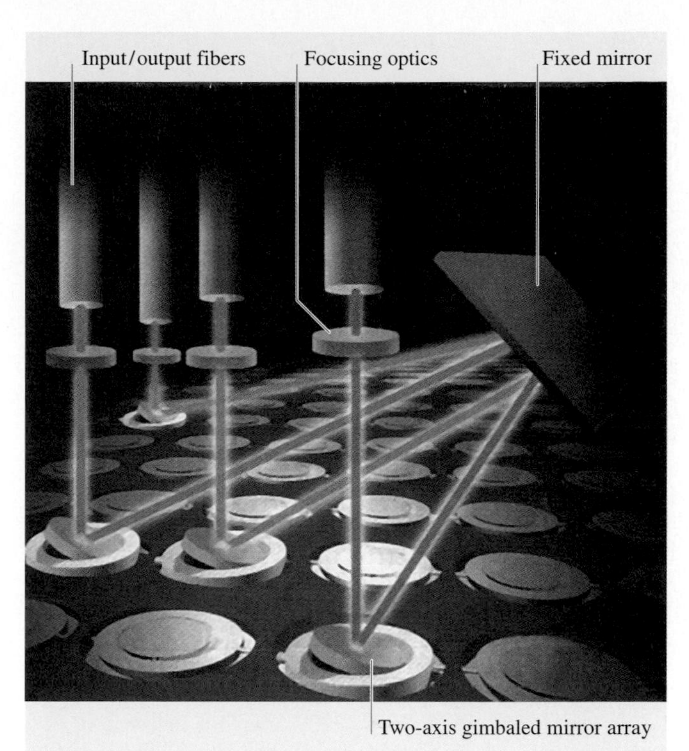

Fig. 8.31 Schematic of a three-dimensional micromirror-based fiber optic switching matrix

Fig. 8.32 SEM micrograph of a surface micromachined two-axis beam-steering micromirror positioned using self-assembly technique [8.29]

8.1.4 RF MEMS

The increasing demand for wireless communication applications, such as cellular and cordless telephony, wireless data networks, two-way paging, global positioning systems, etc., feeds a growing interest in building miniaturized wireless transceivers with multi-standard capabilities. Such transceivers will greatly enhance the convenience and accessibility of various wireless services independent of geographic location. Miniaturizing current single-standard transceivers through a high level of integration is a critical step toward building transceivers that are compatible with multiple standards. Highly integrated transceivers will also result in reduced package complexity, power consumption, and cost. At present, most radio transceivers rely on a large number of discrete frequency-selection components, such as radio frequency (RF) and intermediate frequency (IF) band-pass filters, RF voltage-controlled oscillators (VCOs), quartz crystal oscillators, solid-state switches, etc., to perform the necessary analog signal processing. Figure 8.33 shows a schematic of a super-heterodyne radio architecture, in which discrete components are shaded. These off-chip devices occupy the majority of the system area, thus severely hindering transceiver miniaturization. MEMS technology, however, offers a potential solution to integrate these discrete components onto silicon substrates with microelectronics, achieving a size reduction of a few orders of magnitude. It is, therefore, expected to become a technology that will ultimately enable the miniaturization of radio transceivers for future wireless communications.

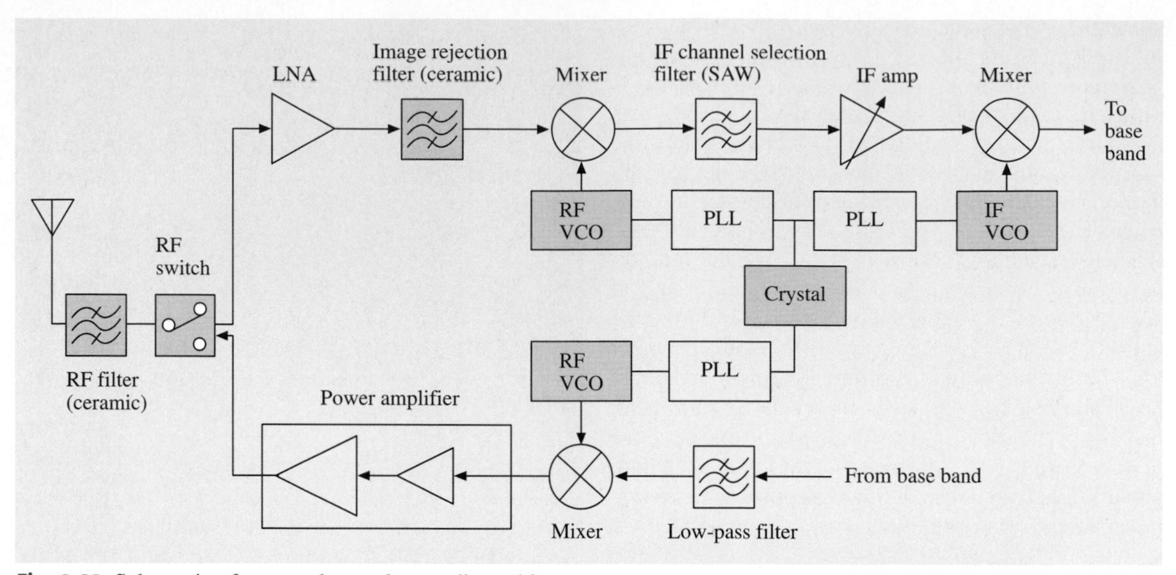

Fig. 8.33 Schematic of a super-heterodyne radio architecture

MEMS Variable Capacitors

Integrated high performance variable capacitors are critical for low noise VCOs, antenna tuning, tunable matching networks, etc. Capacitors with high quality factors (Q), large tuning range, and linear characteristics are crucial for achieving system performance requirements. On-chip silicon PN junction and MOS-based variable capacitors suffer from low quality factors (below 10 at 1 GHz), limited tuning range, and poor linearity, and are, therefore, inadequate for building high performance transceivers. MEMS technology has demonstrated monolithic variable capacitors, achieving stringent performance requirements. These devices typically rely on an electrostatic actuation method to vary the air gap between a set of parallel plates [8.30–33], or vary the capacitance area between a set of conductors [8.34], or mechanically displace a dielectric layer in an air-gap capacitor [8.35]. Improved tuning ranges have been achieved with various device configurations. Figure 8.34 shows SEM micrographs of an aluminum micromachined variable capacitor fabricated on a silicon substrate [8.30]. The device consists of a 200 μm × 200 μm aluminum plate with a thickness of

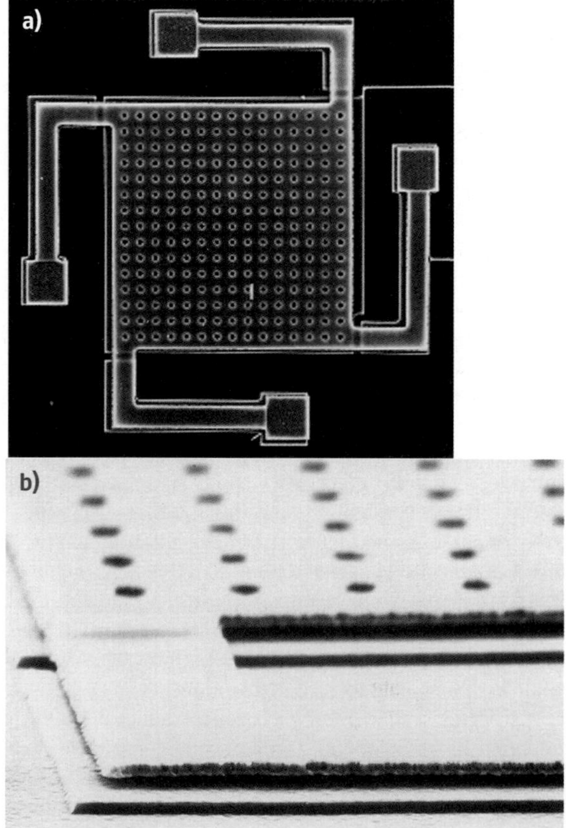

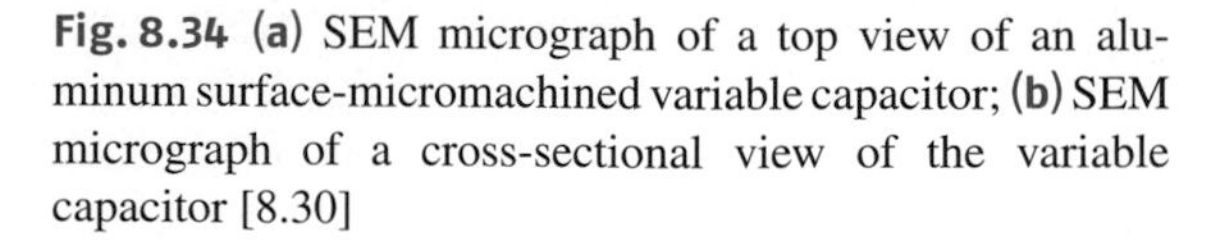

Fig. 8.34 **(a)** SEM micrograph of a top view of an aluminum surface-micromachined variable capacitor; **(b)** SEM micrograph of a cross-sectional view of the variable capacitor [8.30]

1 μm suspended above the bottom electrode by an air gap of 1.5 μm. Aluminum is selected as the structural material due to its low resistivity, which is critical for achieving a high quality factor at high frequencies. A DC voltage applied across the top and bottom electrodes introduces an electrostatic pull-down force, which pulls the top plate towards the bottom electrode, thus changing the device capacitance value. The capacitors are fabricated using aluminum-based surface micromachining technology. Sputtered aluminum is used for building the capacitor top and bottom electrodes. Photoresist serves as the sacrificial layer, which is then removed through an oxygen-based plasma dry etch to release the microstructure. The processing technology requires a low thermal budget, allowing the variable capacitors to be fabricated on top of wafers with completed electronic circuits without degrading the performance of active devices. Figure 8.35 presents an SEM micrograph of four MEMS tunable capacitors connected in parallel. This device achieves a nominal capacitance value of 2 pF and a tuning range of 15% with 3 V. A quality factor of 62 has been demonstrated at 1 GHz, which matches or exceeds that of discrete varactor diodes and is at least an order of magnitude larger than that of a typical junction capacitor implemented in a standard IC process.

MEMS tunable capacitors, based on the varying capacitance area between a set of conductors, have been demonstrated. Figure 8.36 shows an SEM micrograph of such a device [8.34]. The capacitor is composed of arrays of interdigitated electrodes, which can be electrostatically actuated to vary the electrode overlap area. A close-up view of the electrodes is shown in Fig. 8.37. The capacitor is fabricated using a silicon-on-insulator (SOI) substrate with a top silicon layer thickness of about 20 μm to obtain a high-aspect ratio for the electrodes, which is critical for achieving a large capacitance density and reduced tuning voltage. The silicon layer is etched to form the device structure, followed by the removal of the oxide underneath to release the capacitor. A thin aluminum layer is then sputtered over the capacitor to reduce the series resistive loss. The device exhibits a quality factor of 34 at 500 MHz and can be tuned between 2.48 pF and 5.19 pF with an actuation voltage under 5 V, corresponding to a tuning range of 100%. Figure 8.38 shows the variation of electrode overlap under different tuning voltages.

Tunable capacitors relying on a movable dielectric layer have been fabricated using MEMS technology. Figure 8.39 presents an SEM micrograph of a copper-based micromachined tunable capacitor [8.35]. The device consists of an array of copper top electrodes suspended above a bottom copper plate with an air gap of approximately 1 μm. A thin nitride layer is deposited, patterned, and suspended between the two copper layers by lateral mechanical spring suspensions after sacrificial release. A DC voltage applied across the copper layers introduces a lateral electrostatic pull-in force on the nitride, resulting in a movement that changes the overlapping area between each copper electrode and the bottom plate and, hence, the device capacitance. The

Fig. 8.35 SEM micrograph of four MEMS aluminum variable capacitors connected in parallel [8.30]

Fig. 8.36 SEM micrograph of a silicon tunable capacitor using a comb drive actuator [8.34]

Fig. 8.37 SEM micrograph of a close view of tunable capacitor comb fingers [8.34]

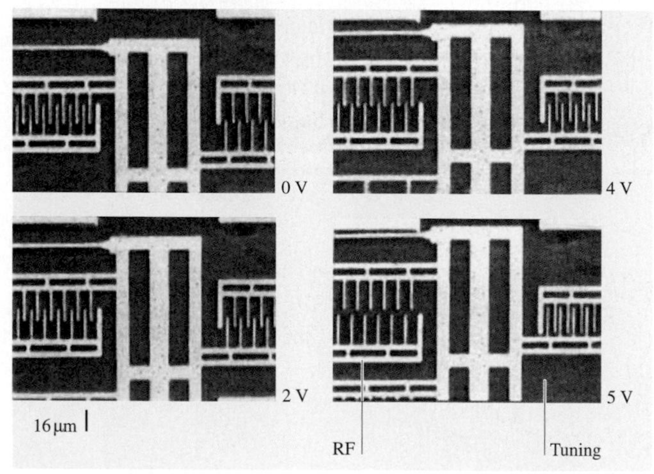

Fig. 8.38 Photos of comb fingers at different actuation voltages [8.34]

tunable capacitor achieves a quality factor of over 200 at 1 GHz with 1 pF capacitance, due to the highly conductive copper layers, and a tuning range around 8% with 10 V.

Micromachined Inductors

Integrated inductors with high quality factors are as critical as the tunable capacitors for high performance RF system implementation. They are the key components

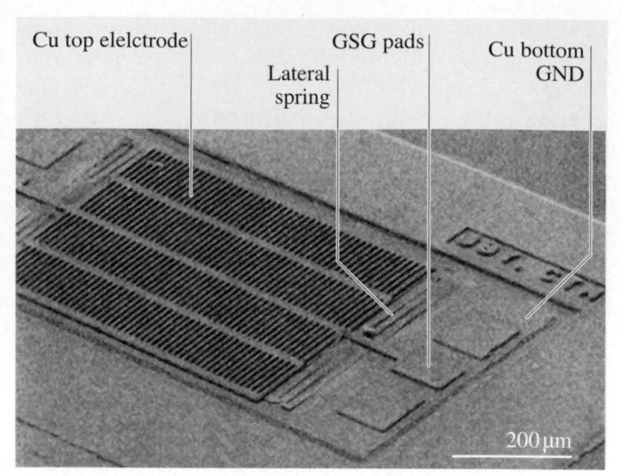

Fig. 8.39 SEM micrograph of a copper surface-micromachined tunable capacitor with a movable dielectric layer [8.35]

for building low noise oscillators, low loss matching networks, etc. Conventional on-chip spiral inductors suffer from limited quality factors of around 5 at 1 GHz, an order of magnitude lower than the required values from discrete counterparts. The poor performance is mainly due to substrate loss and metal resistive loss at high frequencies. Micromachining technology provides an attractive solution to minimizing these loss contributions, hence, enhancing the device quality factors. Figure 8.40 shows an SEM micrograph of a 3-D coil inductor fabricated on a silicon substrate [8.36]. The device consists of four-turn 5 μm-thick copper traces electroplated around an insulating Alumina core with a 650 μm × 500 μm cross section. Compared to spiral inductors, this geometry minimizes the coil area, which is in close proximity to the substrate, and, hence, the eddy-current loss, resulting in a maximized Q-factor and device self-resonant frequency. Copper is selected as the interconnect metal because of its low sheet resistance, critical for achieving a high Q-factor. The inductor achieves a 14 nH inductance value with a quality factor of 16 at 1 GHz. A single-turn 3-D device exhibits a Q-factor of 30 at 1 GHz, which matches the performance of discrete counterparts. The high-Q 3-D inductor and MEMS tunable capacitors, shown in Fig. 8.35, have been employed to implement a RF CMOS VCO achieving a low phase noise performance suitable for typical wireless communication applications, such as GMS cellular telephony [8.37].

Other 3-D inductor structures, such as the levitated spiral inductors, have been demonstrated using micromachining fabrication technology. Figure 8.41 shows an SEM micrograph of a levitated copper inductor,

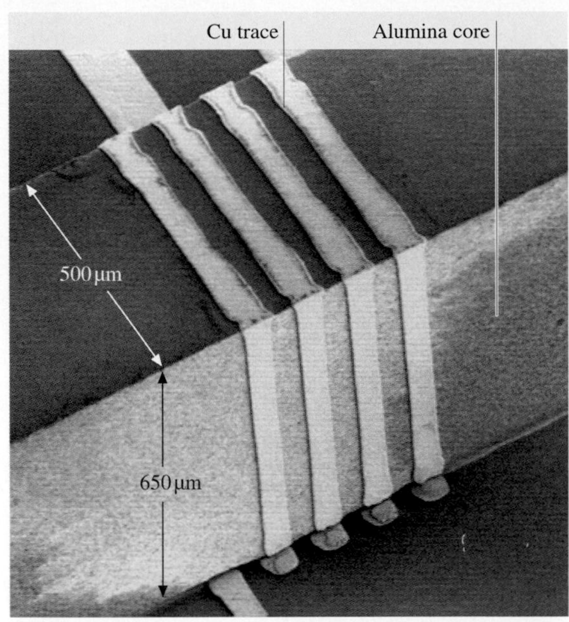

Fig. 8.40 SEM micrograph of a 3-D coil inductor fabricated on a silicon substrate [8.36]

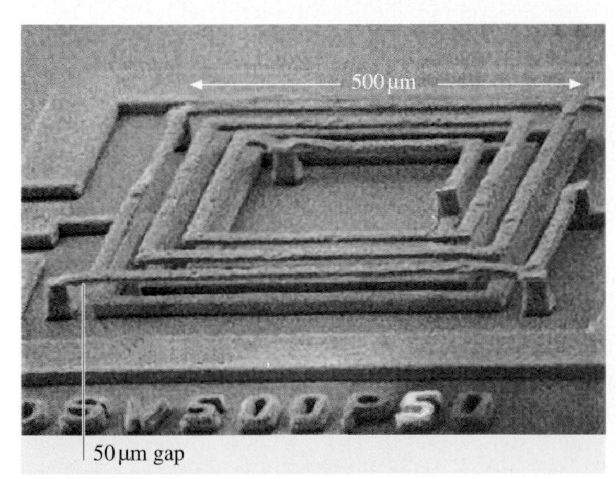

Fig. 8.41 SEM micrograph of a levitated spiral inductor fabricated on a glass substrate [8.38]

which is suspended above the substrate by supporting posts [8.38]. The levitated geometry can minimize the substrate loss, thus achieving an improved quality factor. The inductor shown in the figure achieves a 14 nH inductance value with a Q-factor of 38 at 1.8 GHz using a glass substrate.

A self-assembled out-of-plane coil has been fabricated using micromachining technology. The inductor winding traces are made of refractory metals with controlled built-in stress such that the traces can curl out of the substrate surface upon release and interlock into each other to form coil windings. Figure 8.42 shows an SEM micrograph of a self-assembled out-of-plane coil inductor [8.39]. A close-up view of an interlocking trace is shown in Fig. 8.43. Copper is plated on the interlocked traces to form highly conductive windings at the end of the processing sequence. The inductor shown in Fig. 8.42 achieves a quality factor around 40 at 1 GHz.

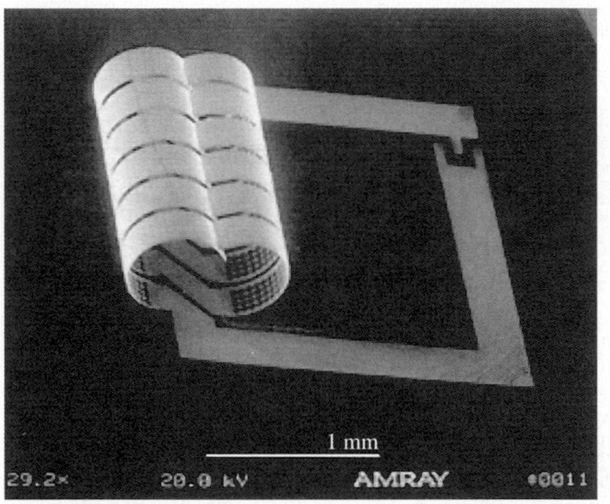

Fig. 8.42 SEM micrograph of a self-assembled out-of-plane coil inductor [8.39]

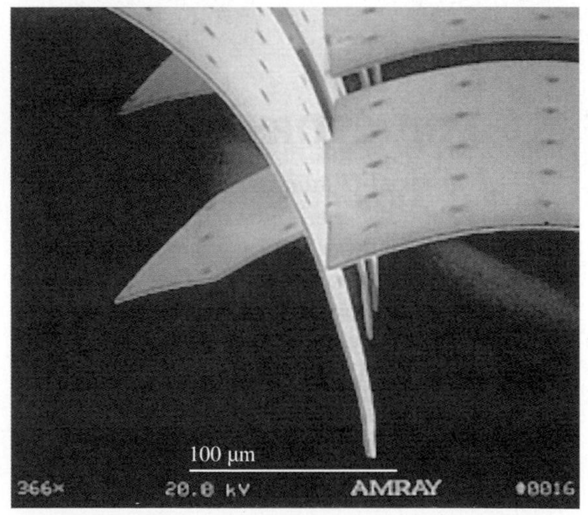

Fig. 8.43 SEM micrograph of an interlocking trace from a self-assembled out-of-plane oil inductor [8.39]

MEMS Switches

The microelectromechanical switch is another potentially attractive miniaturized component enabled by micromachining technologies. These switches offer superior electrical performance in terms of insertion loss, isolation, linearity, etc. and are intended to replace off-chip solid state counterparts, which provide switching between the receiver and transmitter signal paths. They are also critical for building phase shifters, tunable antennas, and filters. The MEMS switches can be broken into two categories: capacitive and metal-to-metal contact types. Figure 8.44 presents a cross-sectional schematic of a RF MEMS capacitive switch. The device consists of a conductive membrane, typically made of aluminum or gold alloy, suspended above a coplanar electrode by an air gap of a few micrometers. For RF or microwave applications, actual metal-to-metal contact is not necessary; rather, a step change in the plate-to-plate capacitance realizes the switching function. A thin silicon nitride layer with a thickness on the order of 1,000 Å is typically deposited above the bottom electrode. When the switch is in the on-state, the membrane is high, resulting in a small plate-to-plate capacitance and, hence, a minimum high-frequency signal coupling (high isolation) between the two electrodes. In the off-state (with a large enough applied DC voltage), the switch provides a large capacitance due to the thin dielectric layer, thus causing a strong signal coupling (low insertion loss). The capacitive switch consumes near zero power, which is attractive for low power portable applications. Switching cycles numbering in the millions for this type of device have been demonstrated. Figure 8.45 shows a top view photo of a fabricated MEMS capacitive switch [8.40]. Surface micromachining technology, using metal for the electrodes and polymer as the sacrificial layer, is used to fabricate the device. The switch can be actuated with a DC voltage on the order of 50 V and exhibits a low insertion loss of approximately −0.28 dB at 35 GHz and a high isolation of −35 dB at the same frequency.

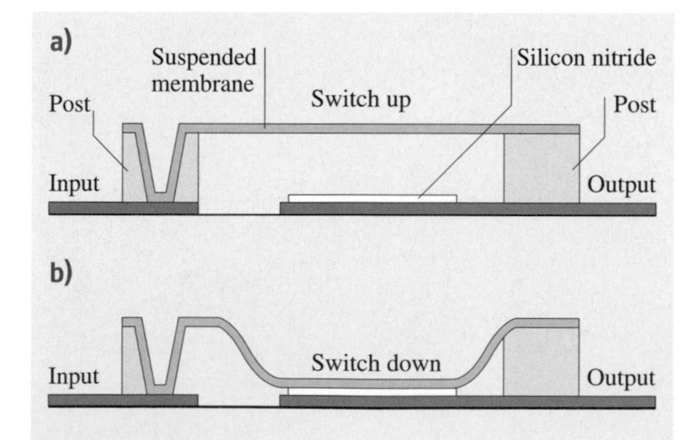

Fig. 8.44 Cross-sectional schematics of an RF MEMS capacitive switch

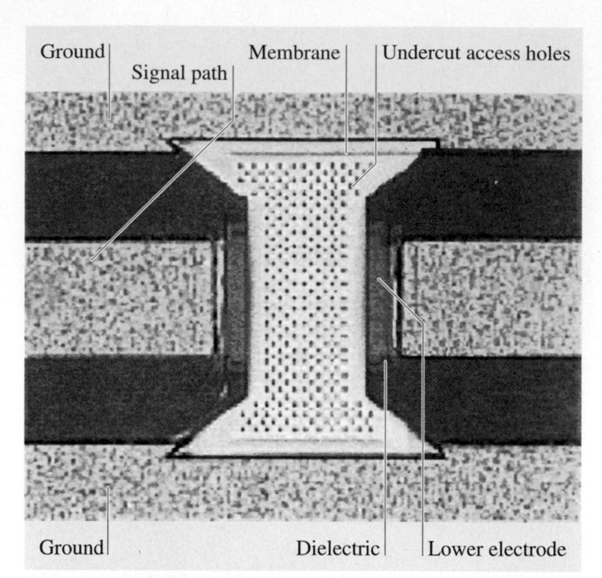

Fig. 8.45 Top view photo of a fabricated RF MEMS capacitive switch [8.40]

Metal-to-metal contact switches are important for interfacing large bandwidth signals, including DC. This type of device typically consists of a cantilever beam, or clamped-clamped bridge with a metallic contact pad positioned at the beam tip or underneath the bridge center. Through an electrostatic actuation, a contact can be formed between the suspended contact pad and an underlying electrode on the substrate [8.41–43]. Figure 8.46 shows a cross-sectional schematic of a metal-to-metal contact switch [8.43]. The top view of the fabricated device is presented in Fig. 8.47. The switch exhibits an actuation voltage of 30 V, a response time of 20 μs, and the mechanical strength to withstand 10^9 actuations. An isolation greater than 50 dB below 2 GHz and insertion loss less than 0.2 dB from DC through 40 GHz have been demonstrated.

MEMS Resonators

Microelectromechanical resonators based on polysilicon comb-drive structures, suspended beams, and center-pivoted disk configurations have been demonstrated for performing analog signal processing [8.44–48]. These micro-resonators can be excited into mechanical reso-

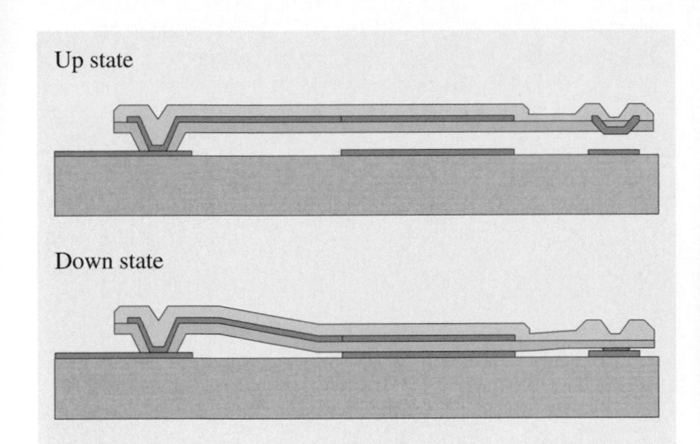

Fig. 8.46 Cross-sectional schematic of a metal-to-metal contact switch [8.43]

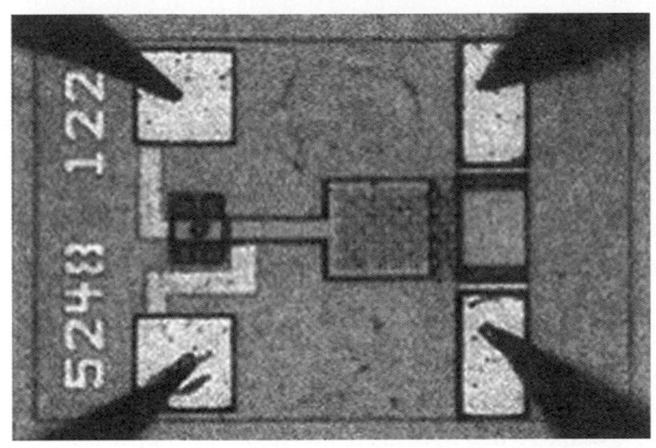

Fig. 8.47 Top view photo of a fabricated metal-to-metal contact switch [8.43]

nance through an electrostatic drive. The mechanical motion causes a change of device capacitance, resulting in an output electrical current when a proper DC bias voltage is used. The output current exhibits the same frequency as the mechanical resonance, thus achieving an electrical filtering function through the electro mechanical coupling. Micromachined polysilicon flexural-mode mechanical resonators have demonstrated a quality factor greater than 80,000 in a 50 μtorr vacuum [8.49]. This level of performance is comparable to a typical quartz crystal and is, therefore, attractive for implementing monolithic low noise and low drift reference signal sources. Figure 8.48 shows an SEM micrograph of a surface micromachined comb-drive resonator integrated with CMOS sustaining electronics on the same substrate to form a monolithic high-Q MEMS resonator-based oscillator [8.44]. The oscillator achieves an operating frequency of 16.5 KHz with a clean spectral purity.

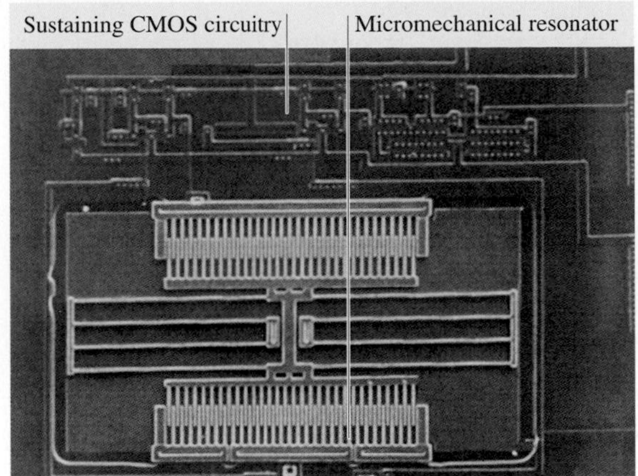

Fig. 8.48 SEM micrograph of a surface micromachined comb drive resonator integrated with CMOS sustaining electronics [8.44]

A chip area of approximately 420 μm × 230 μm is consumed for fabricating the overall system, representing a size reduction by orders of magnitude compared to conventional quartz crystal oscillators.

Micromachined high-Q resonators can be coupled to implement low-loss frequency selection filters. Figure 8.49 shows an SEM micrograph of a surface micromachined polysilicon two-resonator, spring-coupled bandpass micromechanical filter [8.46]. The filter consists of two silicon micromechanical clamped-clamped beam resonators, coupled mechanically by a soft spring, all suspended 0.1 μm above the substrate. Polysilicon strip lines underlie the central regions of each resonator and serve as capacitive transducer electrodes positioned

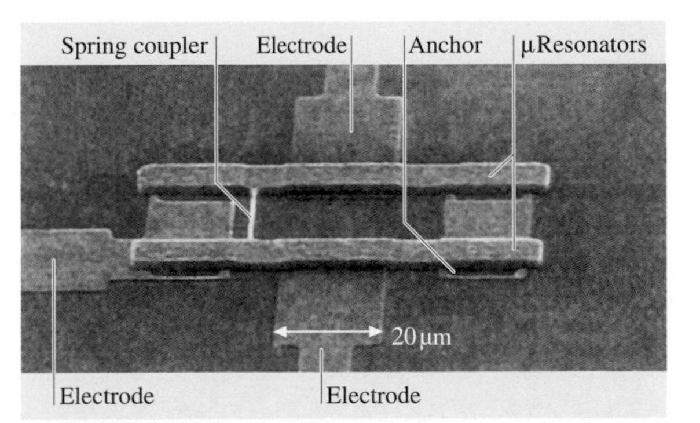

Fig. 8.49 SEM micrograph of a polysilicon surface micromachined two-resonator spring-coupled bandpass micromechanical filter [8.46]

to induce resonator vibration in a direction perpendicular to the substrate. Under normal operation, the device is excited capacitively by a signal voltage applied to the input electrode. The output is taken at the other end of the structure, also by capacitive transduction. The filter achieves a center frequency of 7.81 MHz, a bandwidth of 0.23%, and an insertion loss less than 2 dB. The achieved performance is attractive for implementing filters in the low MHz range.

To obtain a higher mechanical resonant frequency with low losses, a surface micromachined contour-mode disk resonator has been proposed, as shown in Fig. 8.50 [8.48]. The resonator consists of a polysilicon disk suspended 5,000 Å above the substrate with a single anchor at its center. Plated metal input electrodes surround the perimeter of the disk with a narrow separation of around 1,000 Å, which defines the capacitive, electromechanical transducer of the device. To operate the device, a DC bias voltage is applied to the structure with an AC input signal applied to the electrodes, resulting in a time varying electrostatic force acting radially on the disk. When the input signal matches the device resonant frequency, the resulting electrostatic force is amplified by the Q factor of the resonator, resulting in the expansion and contraction of the disk along its radius. This motion, in turn, produces a time varying output current at the same frequency, thus achieving the desirable filtering. The prototype resonator demonstrates an operating frequency of 156 MHz with a Q factor of 9,400 in vacuum. The increased resonant frequency is comparable to the first intermediate frequency used in a typical wireless transceiver design and is thus suitable for implementing IF bandpass filters. A higher operating frequency is needed in order to potentially replace the RF frequency selection filters with microelectromechanical resonators.

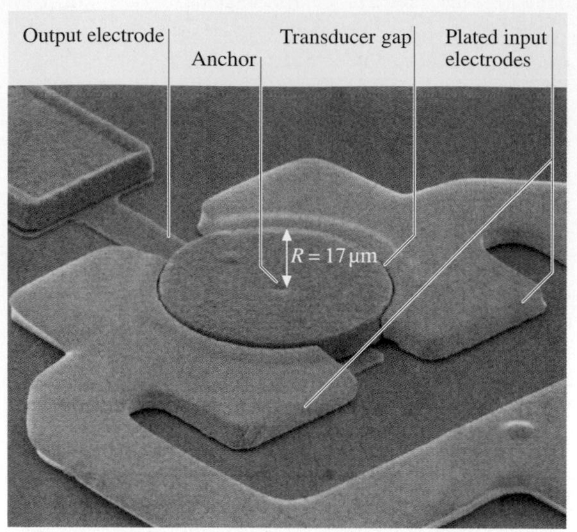

Fig. 8.50 SEM micrograph of a polysilicon surface-micromachined contour-mode disk resonator [8.48]

8.2 NEMS Devices and Applications

Unlike their microscale counterparts, nanoelectromechanical systems (NEMS) are made of electromechanical devices that have critical structural dimensions on the submicron scale. These devices are attractive for applications where structures of very small mass provide essential functionality, such as force sensors, chemical sensors, biological sensors, and ultrahigh frequency resonators, to name a few. NEMS fabrication processes can be categorized according to the two approaches used to create the structures. Top-down approaches utilize submicron lithographic techniques to fabricate device structures from bulk material, either thin films or thick substrates. Bottom-up approaches involve the fabrication of nanoscale devices in much the same way that nature constructs objects, by sequential assembly using atomic and molecular building blocks. While many contend that bottom-up approaches will eventually dominate the fabrication of NEMS, most of the functional NEMS devices are currently created utilizing top-down techniques that combine existing process technologies, such as submicron electron-beam lithography, conventional film growth, and chemical etching. Top-down approaches make integration with microscale packaging relatively straightforward since the only significant difference between the nanoscale and microscale processing steps is the method used to pattern the various features.

In large measure, NEMS has followed a developmental path similar to that of MEMS in that both have leveraged existing processing techniques. For instance, the electron-beam lithographic techniques used in top-down NEMS fabrication are the same techniques that have become standard in the fabrication of submicron ICs. Furthermore, the materials used in many of the first generation, top-down NEMS devices (Si, GaAs, Si_3N_4, SiC) were first used in ICs and then in MEMS. Like the first MEMS devices, the first generation NEMS structures consist of freestanding nanomechanical beams, paddle oscillators, and tethered plates made using simple bulk and single layer surface "nano" machining processes. As an introduction to this new and exciting class of devices, the following section reviews the development of some of these first generation NEMS devices.

Like Si MEMS, Si NEMS capitalizes on well developed processing techniques for Si and the availability of high quality substrates. *Cleland* et al. [8.50] reported a relatively simple process to fabricate nanomechanical clamped-clamped beams directly from single crystal (100) Si substrates. As illustrated in Fig. 8.51, the process begins with the thermal oxidation of a Si substrate (Fig. 8.51a). Large Ni contact pads were then fabricated using optical lithography and lift-off. A polymethyl methacrylate (PMMA) lift-off mold was then deposited and patterned, using electron-beam lithography, into the shape of nanomechanical beams (Fig. 8.51b). Ni was then deposited and patterned by lifting off the PMMA (Fig. 8.51c). Next, the underlying oxide film was patterned by RIE using the Ni film as an etch mask. After oxide etching, nanomechanical beams were patterned by etching the Si substrate using RIE, as shown in Fig. 8.51d. Following Si RIE, the Ni etch mask was removed, and the sidewalls of the Si nanomechanical beams were lightly oxidized in order to protect them during the release step (Fig. 8.51e). After performing an anisotropic SiO_2 etch to clear any oxide from the field areas, the Si beams were released using an isotropic Si RIE step, as shown in Fig. 8.51f. After release, the protective SiO_2 film was removed by wet etching in HF (Fig. 8.51g). Using this process, the authors reported the successful fabrication of nanomechanical Si beams with micron-scale lengths ($\sim 8\,\mu$m) and submicron widths (330 nm) and heights (800 nm).

The advent of silicon-on-insulator (SOI) substrates with high quality, submicron-thick silicon top layers enables the fabrication of nanomechanical Si beams with fewer processing steps than the aforementioned technique, since the buried oxide layer makes these device structures relatively easy to pattern and release. Additionally, the buried SiO_2 layer electrically isolates the beams from the substrate. *Carr* et al. [8.51] detail a process that uses SOI substrates to fabricate submicron clamped-clamped mechanical beams and

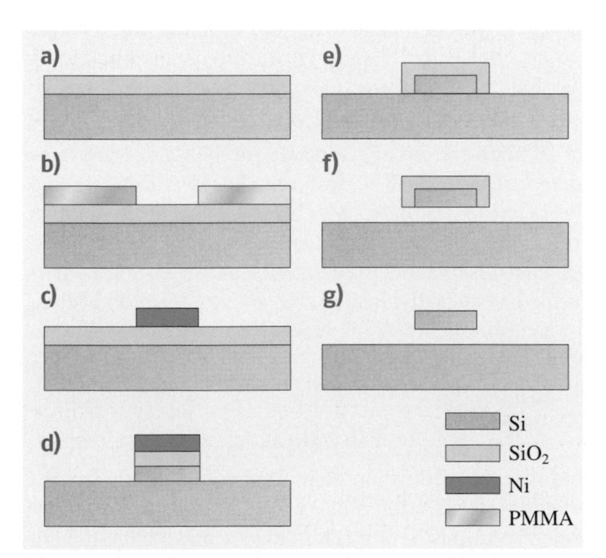

Fig. 8.51a–g Cross-sectional schematics of a process to fabricate nanomechanical beams directly from a silicon substrate: (**a**) after thermal oxidation; (**b**) after PMMA coating and electron beam lithography; (**c**) after Ni deposition and lift-off; (**d**) after anisotropic SiO_2 and Si etching; (**e**) after Ni etching, thermal oxidation, and anisotropic SiO_2 etching; (**f**) after release by isotropic Si etching; (**g**) after oxide removal

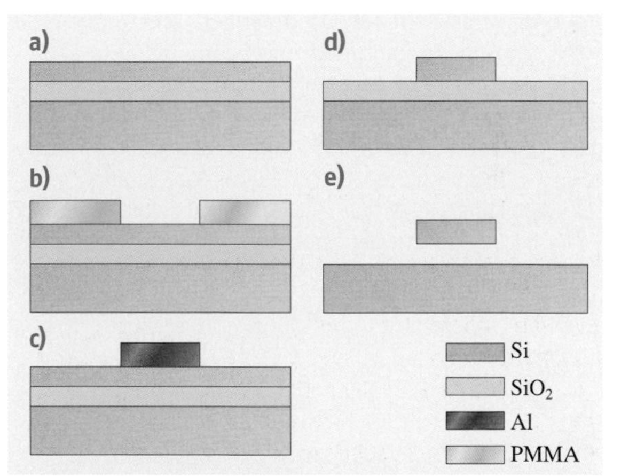

Fig. 8.52a–e Cross-sectional schematics of a process to fabricate nanomechanical structures using silicon-on-insulator substrates: (**a**) starting SOI substrate; (**b**) after coating with PMMA and electron beam lithography; (**c**) after Al deposition and lift-off; (**d**) after anisotropic Si etching; (**e**) after release by SiO_2 etching

suspended plates with submicron tethers. The process, presented in Fig. 8.52, begins with the deposition of PMMA on an SOI substrate. The SOI substrate has a top Si layer that is either 50 nm or 200 nm in thickness. The PMMA is patterned into a metal lift-off mask by electron beam lithography (Fig. 8.52b). An aluminum film is then deposited and patterned by lift-off into a Si etch mask, as shown in Fig. 8.52c. The nanomechanical beams are then patterned by Si RIE and released by etching the underlying SiO_2 in a buffered hydrofluoric acid solution, as shown in Fig. 8.52d and Fig. 8.52e, respectively. Using this process, nanomechanical beams that were 7 to 16 μm in length, 120 to 200 nm in width, and 50 or 200 nm in thickness were successfully fabricated.

Fabrication of NEMS structures is not limited to Si. In fact, III-V compounds, such as gallium arsenide (GaAs), make particularly good NEMS materials from a fabrication perspective because thin epitaxial GaAs films can be grown on lattice-matched materials that can be used as sacrificial release layers. A collection of clamped-clamped nanomechanical GaAs beams fabricated on lattice-matched sacrificial layers having micron-scale lengths and submicron widths and thicknesses is shown in Fig. 8.53. *Tighe* et al. [8.52] reported on the fabrication of GaAs plates suspended with nanomechanical tethers. The structures were made from single crystal GaAs films that were epitaxially grown on aluminum arsenide (AlAs) sacrificial layers. Ni etch masks were fabricated using electron beam lithography and lift-off, as described previously. The GaAs films were patterned into beams using a chemically assisted ion beam etching process and released using a highly selective AlAs etchant. In a second example, *Tang* et al. [8.53] capitalized on the ability to grow high quality GaAs layers on ternary compounds, such as $Al_xGa_{1-x}As$, to fabricate complex GaAs-based structures, such as submicron clamped-clamped beams from GaAs/AlGaAs quantum well heterostructures. As with the process described by *Tighe* et al. [8.52], this process exploits a lattice matched sacrificial layer, in this case $Al_{0.8}Ga_{0.2}As$, which can be selectively etched to release the heterostructure layers.

Several unique approaches have been developed to actuate and sense the motion of NEMS devices. Electrostatic actuation can be used to actuate beams [8.54], tethered meshes [8.55], and paddle oscillators [8.56]. *Sekaric* et al. [8.57] have recently shown that low power lasers can be used to drive paddle oscillators into self-oscillation by induced thermal effects on the structures.

Fig. 8.53 SEM micrograph of a set of nanomechanical GaAs clamped-clamped beam resonators (courtesy of M. Roukes, Caltech)

In these examples, an optical detection scheme based on the modulation of incident laser light by a vibrating beam is used to detect the motion of the beams. *Cleland* et al. [8.50] describe a magnetomotive transduction technique that capitalizes on a time-varying Lorentz force created by an alternating current in the presence of a strong magnetic field. In this case, the nanomechanical beam is positioned in the magnetic field so that an AC current passing through the beam is transverse to the field lines. The resulting Lorentz force causes the beam to oscillate, which creates an electromotive force along the beam that can be detected as a voltage. Thus, in this method, the excitation and detection are performed electrically. In all of the above-mentioned cases, the measurements were performed in vacuum, presumably to minimize the effects of squeeze-film damping, as well as mass loading due to adsorbates from the environment.

For the most part, NEMS technology is still in the initial stage of development. Technological challenges related to fabrication and packaging will require innovative solutions before such devices make a significant commercial impact. Nevertheless, NEMS devices have already been used for precision measurements [8.58], enabling researchers to probe the properties of matter [8.59, 60] on a nanoscopic level. Whether or not NEMS makes an impact in the commercial sector, they will undoubtedly prove to be useful platforms for a host of experiments and scientific discoveries in fields ranging from physics to biology.

8.3 Current Challenges and Future Trends

Although the field of MEMS has experienced significant growth over the past decade, many challenges still remain. In broad terms, these challenges can be grouped into three general categories: (1) fabrication, (2) packaging, and (3) application. Challenges in these areas will, in large measure, determine the commercial success of a particular MEMS device, both in technical and economic terms. The following presents a brief discussion of some of these challenges, as well as possible approaches to address them.

In terms of fabrication, MEMS is currently dominated by planar processing techniques, which find their roots in silicon IC fabrication. The planar approach and the strong dependence on silicon worked well in the early years, since many of the processing tools and methodologies that are commonplace in IC fabrication could be directly utilized in the fabrication of MEMS devices. This approach lends itself to the integration of MEMS with silicon ICs. Therefore, it is still popular for various applications. However, modular process integration of micromachining with standard IC fabrication is not straightforward and represents a great challenge in terms of processing material compatibility, thermal budget requirements, etc. Furthermore, planar processing places significant geometric restrictions on device designs, especially for complex mechanical components requiring high-aspect ratio three-dimensional geometries, which are certain to increase as the application areas for MEMS continue to grow. Along the same lines, new applications will likely demand materials other than silicon that may not be compatible with the conventional microfabrication approach, posing a significant challenge if integration with silicon microelectronics is required. Micro-assembly technique can become an attractive solution to alleviate these issues. Multifunctional microsystems can be implemented by assembling various MEMS devices and electronic building blocks fabricated through disparate processing technologies. Microsystem on a common substrate will likely become the ultimate solution. Development of sophisticated modeling programs for device design and performance will become increasingly important as fabrication processes and device designs become more complex. In terms of NEMS, the most significant challenge is likely the integration of nano- and microfabrication techniques into a unified process, since NEMS devices are likely to consist of both nanoscale and microscale structures. Integration will be particularly challenging for nanoscale devices fabricated using a bottom-up approach, since no analog is found in microfabrication. Nevertheless, hybrid systems consisting of nanoscale and microscale components will become increasingly common as the field continues to expand.

Fabrication issues notwithstanding, packaging is and will continue to be a significant challenge to the implementation of MEMS. MEMS is unlike IC packaging, which benefits from a high degree of standardization. MEMS devices inherently require interaction with the environment, and since each application has in some way a unique environment, standardization of packaging becomes extremely difficult. This lack of standardization tends to drive up the costs associated with packaging, making MEMS less competitive with alternative approaches. In addition, packaging tends to negate the effects of miniaturization based upon microfabrication, especially for MEMS devices requiring protection from certain environmental conditions. Moreover, packaging can cause performance degradation of MEMS devices, especially in situations where the environment exerts mechanical stresses on the package, which, in turn, results in a long-term device performance drift. To address many of these issues, wafer-level packaging schemes that are customized to the device of interest will likely become more common. In essence, packaging of MEMS will move away from the conventional IC methods, which utilize independently manufactured packages, toward custom packages, which are created specifically for the device as part of the batch fabrication process.

Without question, the advancement of MEMS will open many new potential application areas to the technology. In most cases, MEMS will be one of several alternatives available for implementation. For cost-sensitive applications, the trade off between technical capabilities and cost will challenge those who desire to commercialize the technology. The biggest challenge to the field will be to identify application areas that are well suited for MEMS/NEMS technology and have no serious challengers. As MEMS technology moves away from the component level and towards microsystems solutions, it is likely that such application areas will come to the fore.

References

8.1 M. Mehregany, S. F. Bart, L. S. Tavrow, J. H. Lang, S. D. Senturia: Principles in design and microfabrication of variable-capacitance side-drive motors, J. Vac. Sci. Tech. A **8** (1990) 3614–3624

8.2 Y.-C. Tai, R. S. Muller: IC-processed electrostatic synchronous micromotors, Sens. Actuators **20** (1989) 49–55

8.3 M. S. Rodgers, J. J. Sniegowski: 5-level polysilicon surface micromachine technology: Application to complex mechanical systems, Technical Digest, Solid-State Sens. Actuator Workshop, Hilton Head Island 1998 (Transducers Research Foundation, Cleveland 1998) 144–149

8.4 G. T. A. Kovacs: *Micromachined Transducer Sourcebook* (McGraw-Hill, Boston 1998)

8.5 S. D. Senturia: *Microsystem Design* (Kluwer, Dordrecht 1998)

8.6 J. E. Gragg, W.E. McCulley, W. B. Newton, C. E. Derrington: Compensation and calibration of a monolithic four terminal silicon pressure transducer, Technical Digest, IEEE Solid-State Sens. Actuators Workshop, Hilton Head Island 1984 (IEEE, Piscataway 1984) 21–27

8.7 Y. Wang, M. Esashi: A novel electrostatic servo capacitive vacuum sensor, Technical Digest, IEEE Int. Conf. Solid-State Sens. Actuators, Chicago 1997 (Transducers Conference, Chicago 1997) 1457–1460

8.8 W. H. Ko, Q. Wang: Touch mode capacitive pressure sensors, Sens. Actuators **75** (1999) 242–251

8.9 J. M. Bustillo, R. T. Howe, R. S. Muller: Surface micromachining for microelectromechanical systems, Proc. IEEE **86** (1998) 1552–1574

8.10 J. H. Smith, S. Montague, J. J. Sniegowski, J. R. Murray, P. J. McWhorter: Embedded micromechanical devices for the monolithic integration of MEMS with CMOS, Technical Digest, IEEE Int. Electron Devices Meeting, Washington D.C. 1993 (IEEE, Piscataway 1993) 609–612

8.11 T. A. Core, W. K. Tsang, S. J. Sherman: Fabrication technology for an integrated surface-micromachined sensor, Solid State Technol. **36**(10) (1993) 39–40, 42, 44, 46–47

8.12 H. Kapels, R. Aigner, C. Kolle: Monolithic surface-micromachined sensor system for high pressure applications, Technical Digest, Int. Conf. Solid-State Sens. Actuators, Munich 2001 (Transducers Conference, Chicago 2001) 56–59

8.13 C. Lu, M. Lemkin, B. E. Boser: A monolithic surface micromachined accelerometer with digital output, Technical Digest, IEEE Int.Solid-State Circuits Conferernce, San Francisco 1995 (IEEE, Piscataway 1995) 160–161

8.14 T. B. Gabrielson: Mechanical-thermal noise in micromachined acoustic and vibration sensors, IEEE Trans. Electron Devices **40** (1993) 903–909

8.15 N. Yazdi, K. Najafi: An all-silicon single-wafer fabrication technology for precision microaccelerometers, Technical Digest, IEEE Int. Conf. Solid-State Sens. Actuators, Chicago 1997 (Transducers Conference, Chicago 1997) 1181–1184

8.16 M. Lemkin, M. A. Ortiz, N. Wongkomet, B. E. Boser, J. H. Smith: A 3-axis surface micromachined $\Sigma\Delta$ accelerometer, Technical Digest, IEEE Int. Solid-State Circuits Conference, San Francisco 1997 (IEEE, Piscataway 1997) 202–203

8.17 W. A. Clark, R. T. Howe: Surface micromachined *Z*-axis vibratory rate gyroscope, Technical Digest, IEEE Solid-State Sens. Actuators Workshop, Hilton Head Island 1996 (Transducers Research Foundation, Inc., Cleveland 1996) 283–287

8.18 T. Juneau, A. P. Pisano: Micromachined dual input axis angular rate sensor, Technical Digest, IEEE Solid-State Sens. Actuators Workshop, Hilton Head Island 1996 (Transducers Research Foundation, Inc., Cleveland 1996) 299–302

8.19 R. S. Muller, K. Y. Lau: Surface-micromachined microoptical elements and systems, Proc. IEEE **86** (1998) 1705–1720

8.20 L. J. Hornbeck: Current status of the digital micromirror device (DMD) for projection television applications, Technical Digest, IEEE Int. Electron Devices Meeting, Washington D.C. 1993 (IEEE, Piscataway 1993) 381–384

8.21 P. F. Van Kessel, L. J. Hornbeck, R. E. Meier, M. R. Douglass: A MEMS-based projection display, Proc. IEEE **86** (1998) 1687–1704

8.22 M. S. Cohen, M. F. Cina, E. Bassous, M. M. Opyrsko, J. L. Speidell, F. J. Canora, M. J. DeFranza: Packaging of high density fiber/laser modules using passive alignment techniques, IEEE Trans. Comp., Hybrids, Manufact. Technol. **15** (1992) 944–954

8.23 M. J. Wale, C. Edge: Self-aligned flip-chip assembly of photonic devices with electrical and optical connections, IEEE Trans. Comp., Hybrids, Manufact. Technol. **13** (1990) 780–786

8.24 M. J. Daneman, N. C. Tien, O. Solgaard, K. Y. Lau, R. S. Muller: Linear vibromotor-actuated micromachined microreflector for integrated optical systems, Technical Digest, IEEE Solid-State Sens. Actuators Workshop, Hilton Head Island 1996 (Transducers Research Foundation, Inc., Cleveland 1996) 109–112

8.25 K. S. J. Pister, M. W. Judy, S. R. Burgett, R. S. Fearing: Microfabricated hinges, Sens. Actuators A **33** (1992) 249–256

8.26 O. Solgaard, M. Daneman, N. C. Tien, A. Friedberger, R. S. Muller, K. Y. Lau: Optoelectronic packaging using silicon surface-micromachined alignment mirrors, IEEE Photon. Technol. Lett. **7** (1995) 41–43

8.27 S.S. Lee, L.S. Huang, C.J. Kim, M.C. Wu: 2×2 MEMS fiber optic switches with silicon sub-mount for low-cost packaging, Technical Digest, IEEE Solid-State Sens. Actuators Workshop, Hilton Head Island 1998 (Transducers Research Foundation, Inc., Cleveland 1998) 281–284
8.28 T. Akiyama, H. Fujita: A quantitative analysis of scratch drive actuator using buckling motion, Technical Digest, 8th IEEE Int. MEMS Workshop, Amsterdam 1995 (IEEE, Piscataway 1995) 310–315
8.29 V.A. Aksyuk, M.E. Simon, F. Pardo, S. Arney: Optical MEMS design for telecommunications applications, Technical Digest, Solid-State Sensors and Actuators Workshop, Hilton Head Island 2002 (Transducers Research Foundation, Cleveland 2002) 1–6
8.30 D.J. Young, B.E. Boser: A micromachined variable capacitor for monolithic low-noise VCOs, Technical Digest, Solid-State Sens. Actuator Workshop, Hilton Head Island 1996 (Transducers Research Foundation, Cleveland 1996) 86–89
8.31 A. Dec, K. Suyama: Micromachined electro-mechanically tunable capacitors and their applications to RF IC's, IEEE Trans. Microwave Theory Techniques **46** (1998) 2587–2596
8.32 Z. Li, N.C. Tien: A high tuning-ratio silicon-micromachined variable capacitor with low driving voltage, Technical Digest, Solid-State Sens. Actuator Workshop, Hilton Head Island 2002 (Transducers Research Foundation, Inc., Cleveland 2002) 239–242
8.33 Z. Xiao, W. Peng, R.F. Wolffenbuttel, K.R. Farmer: Micromachined variable capacitor with wide tuning range, Technical Digest, Solid-State Sens. Actuator Workshop, Hilton Head Island, South Carolina 2002 (Transducers Research Foundation, Inc., Cleveland 2002) 346–349
8.34 J.J. Yao, S.T. Park, J. DeNatale: High tuning-ratio MEMS-based tunable capacitors for RF communications applications, Technical Digest, Solid-State Sens. Actuator Workshop, Hilton Head Island 1998 (Transducers Research Foundation, Inc., Cleveland 1998) 124–127
8.35 J.B. Yoon, C.T.-C. Nguyen: A high-Q tunable micromechanical capacitor with movable dielectric for RF applications, Technical Digest, IEEE Int. Electron Devices Meeting, San Francisco 2000 (IEEE, Piscataway 2000) 489–492
8.36 D.J. Young, V. Malba, J.J. Ou, A.F. Bernhardt, B.E. Boser: Monolithic high-performance three-dimensional coil inductors for wireless communication applications, Technical Digest, IEEE Int. Electron Devices Meeting, Washington D.C. 1997 (IEEE, Piscataway 1997) 67–70
8.37 D.J. Young, B.E. Boser, V. Malba, A.F. Bernhardt: A micromachined RF low phase noise voltage-controlled oscillator for wireless communication, Int. J. RF Microwave Computer-Aided Eng. **11** (2001) 285–300
8.38 J.B. Yoon, C.H. Han, E. Yoon, K. Lee, C.K. Kim: Monolithic High-Q Overhang Inductors Fabricated on Silicon and Glass Substrates, Technical Digest, IEEE Int. Electron Devices Meeting, Washington D.C. 1999 (IEEE, Piscataway 1999) 753–756
8.39 C.L. Chua, D.K. Fork, K.V. Schuylenbergh, J.P. Lu: Self-assembled out-of-plane high Q inductors, Technical Digest, Solid-State Sens. Actuator Workshop, Hilton Head Island 2002 (Transducers Research Foundation, Inc., Cleveland 2002) 372–373
8.40 C.L. Goldsmith, Z. Yao, S. Eshelman, D. Denniston: Performance of low-loss RF MEMS capacitive switches, IEEE Microwave Guided Wave Lett. **8** (1998) 269–271
8.41 J.J. Yao, M.F. Chang: A surface micromachined miniature switch for telecommunication applications with signal frequencies from DC up to 40 GHz, Technical Digest, 8th Int. Conf. Solid-State Sens. Actuators, Stockholm 1995 (Transducers Conference, Chicago 1995) 384–387
8.42 P.M. Zavracky, N.E. McGruer, R.H. Morrison, D. Potter: Microswitches and microrelays with a view toward microwave applications, Int. J. RF Microwave Computer-Aided Eng. **9** (1999) 338–347
8.43 D. Hyman, J. Lam, B. Warneke, A. Schmitz, T.Y. Hsu, J. Brown, J. Schaffner, A. Walston, R.Y. Loo, M. Mehregany, J. Lee: Surface-micromachined RF MEMs switches on GaAs substrates, Int. J. RF Microwave Computer-Aided Eng. **9** (1999) 348–361
8.44 C.T.C. Nguyen, R.T. Howe: CMOS microelectromechanical resonator oscillator, Technical Digest, IEEE Int. Electron Devices Meeting, Washington D.C. 1993 (IEEE, Piscataway 1993) 199–202
8.45 L. Lin, R.T. Howe, A.P. Pisano: Microelectromechanical filters for signal processing, IEEE J. Microelectromechan. Syst. **7** (1998) 286–294
8.46 F.D. Bannon III, J.R. Clark, C.T.C. Nguyen: High frequency micromechanical filter, IEEE J. Solid-State Circuits **35** (2000) 512–526
8.47 K. Wang, Y. Yu, A.C. Wong, C.T.C. Nguyen: VHF free-free beam high-Q micromechanical resonators, Technical Digest, 12th IEEE Int. Conf. Micro Electro Mechanical Systems, San Francisco 1999 (IEEE, Piscataway 1999) 453–458
8.48 J.R. Clark, W.T. Hsu, C.T.C. Nguyen: High-Q VHF micromechanical contour-mode disk resonators, Technical Digest, IEEE Int. Electron Devices Meeting, San Francisco 2000 (IEEE, Piscataway 2000) 493–496
8.49 C.T.C. Nguyen, R.T. Howe: Quality factor control for micromechanical resonator, Technical Digest, IEEE Int. Electron Devices Meeting, San Francisco 1992 (IEEE, Piscataway 1992) 505–508
8.50 A.N. Cleland, M.L. Roukes: Fabrication of high frequency nanometer scale mechanical resonators from bulk Si crystals, Appl. Phys. Lett. **69** (1996) 2653–2655

8.51 D.W. Carr, H.G. Craighead: Fabrication of nanoelectromechanical systems in single crystal silicon using silicon on insulator substrates and electron beam lithography, J. Vac. Sci. Technol. B **15** (1997) 2760–2763

8.52 T.S. Tighe, J.M. Worlock, M.L. Roukes: Direct thermal conductance measurements on suspended monocrystalline nanostructures, Appl. Phys. Lett. **70** (1997) 2687–2689

8.53 H.X. Tang, X.M.H. Huang, M.L. Roukes, M. Bichler, W. Wegsheider: Two-dimensional electron-gas actuation and transduction for GaAs nanoelectromechanical systems, Appl. Phys. Lett. **81** (2002) 3879–3881

8.54 D.W. Carr, S. Evoy, L. Sekaric, H.G. Craighead, J.M. Parpia: Measurement of mechanical resonance and losses in nanometer scale silicon wires, Appl. Phys. Lett. **75** (1999) 920–922

8.55 D.W. Carr, L. Sekaric, H.G. Craighead: Measurement of nanomechanical resonant structures in single-crystal silicon, J. Vac. Sci. Technol. B **16** (1998) 3821–3824

8.56 S. Evoy, D.W. Carr, L. Sekaric, A. Olkhovets, J.M. Parpia, H.G. Craighead: Nanofabrication and electrostatic operation of single-crystal silicon paddle oscillators, J. Appl. Phys. **86** (1999) 6072–6077

8.57 L. Sekaric, M. Zalalutdinov, S.W. Turner, A.T. Zehnder, J.M. Parpia, H.G. Craighead: Nanomechanical resonant structures as tunable passive modulators, Appl. Phys. Lett. **80** (2002) 3617–3619

8.58 A.N. Cleland, M.L. Roukes: A nanometre-scale mechanical electrometer, Nature **392** (1998) 160–162

8.59 K. Schwab, E.A. Henriksen, J.M. Worlock, M.L. Roukes: Measurement of the quantum of thermal conductance, Nature **404** (2000) 974–977

8.60 S. Evoy, A. Olkhovets, L. Sekaric, J.M. Parpia, H.G. Craighead, D.W. Carr: Temperature-dependent internal friction in silicon nanoelectromechanical systems, Appl. Phys. Lett. **77** (2000) 2397–2399

9. Microfluidics and Their Applications to Lab-on-a-Chip

Various microfluidic components and their characteristics, along with the demonstration of two recent achievements of lab-on-a-chip systems have been reviewed and discussed. Many microfluidic devices and components have been developed during the past few decades, as introduced earlier for various applications. The design and development of microfluidic devices still depend on the specific purposes of the devices (actuation or sensing) due to a wide variety of application areas, which encourages researchers to develop novel, purpose-specific microfluidic devices and systems. Microfluidics is the real multidisciplinary research field that requires basic knowledge in fluidics, micromachining, electromagnetics, materials, and chemistry for better applications.

Among the various application areas of microfluidics, one of the most important application areas is the lab-on-a-chip system. Lab-on-a-chip is becoming a revolutionary tool for many different applications in chemical and biological analyses due to its fascinating advantages (fast and low cost) over conventional chemical or biological laboratories. Furthermore, the simplicity of lab-on-a-chip systems will enable self-testing capability for patients or health consumers overcoming space limitation.

9.1 **Materials for Microfluidic Devices and Micro/Nano Fabrication Techniques** ... 254
9.1.1 Silicon ... 254
9.1.2 Glass ... 254
9.1.3 Polymer ... 255

9.2 **Active Microfluidic Devices** ... 257
9.2.1 Microvalves ... 258
9.2.2 Micropumps ... 260

9.3 **Smart Passive Microfluidic Devices** ... 262
9.3.1 Passive Microvalves ... 262
9.3.2 Passive Micromixers ... 265
9.3.3 Passive Microdispensers ... 266
9.3.4 Microfluidic Multiplexer Integrated with Passive Microdispenser ... 267
9.3.5 Passive Micropumps ... 269
9.3.6 Advantages and Disadvantages of the Passive Microfluidic Approach ... 269

9.4 **Lab-on-a-Chip for Biochemical Analysis** ... 270
9.4.1 Magnetic Micro/Nano Bead-Based Biochemical Detection System ... 270
9.4.2 Disposable Smart Lab-on-a-Chip for Blood Analysis ... 273

References ... 276

Microfluidics covers the science of fluidic behaviors on the micro/nanoscales and the engineering of design, simulation, and fabrication of the fluidic devices for the transport, delivery, and handling of fluids on the order of microliters or smaller volumes. It is the backbone of the BioMEMS (Biological or Biomedical Microelectromechanical Systems) and lab-on-a-chip concept, as most biological analyses involve fluid transport and reaction. Biological or chemical reactions on the micro/nanoscale are usually rapid since small amounts of samples and reagents are used, which offers a quick and low cost analysis.

A fluidic volume of 1 nanoliter (1 nl) can be understood as the volume in a cube surrounded by 100 μm in each direction. It is much smaller than the size of a grain of table salt! Microfluidic devices and systems handle sample fluids in this range for various applications, including inkjet printing, blood analysis, biochemical detection, chemical synthesis, drug screening/delivery, protein analysis, DNA sequencing, and so on.

Microfluidic systems consist of microfluidic platforms or devices for fluidic sampling, control, monitoring, transport, mixing, reaction, incubation, and analysis. To construct microfluidic systems, or lab-on-a-chips, microfluidic devices must be functionally integrated on a microfluidic platform using proper micro/nano fabrication techniques. In this chapter, the basics of microfluidicdevices and their applications to lab-on-a-chip are briefly reviewed and summarized. Basic materials and fabrication techniques for microfluidic devices will be introducedfirst and various active and passive microfluidic components will be described. Then, their applications to lab-on-a-chip, or biochemical analysis, will be discussed.

9.1 Materials for Microfluidic Devices and Micro/Nano Fabrication Techniques

Various materials are being used for the fabrication of microfluidic devices and systems. Silicon is one of the most popular materials in micro/nano fabrication because its micromachining has been well established over decades. In general, the advantages to using silicon as a substrate or structural material include good mechanical properties, excellent chemical resistance, well characterized processing techniques, and the capability for integration of control/sensing circuitry as a semiconductor. Other materials such as glass, quartz, ceramics, metals, and polymers are also being used for substrates and structures in micro/nano fabrication, depending on the application. Among these materials, polymers or plastics have recently become one of the more promising materials for lab-on-a-chip applications, due to their excellent material properties for biochemical fluids and their low cost manufacturability. Main issues in the fabrication techniques of microfluidic devices and systems usually lie in forming microfluidic channels, key micro/nano structures of lab-on-a-chip. In this section, the basic micro/nano fabrication techniques for silicon, glass, and polymers are described.

9.1.1 Silicon

Microfluidic channels on silicon substrates are usually formed either by wet (chemical) etching or by dry (plasma) etching. Crystalline silicon has a preferential etch direction, depending on which crystalline plane is exposed to an etchant. Etch rate is slowest in ⟨111⟩ crystalline direction – approximately 100:1 etch rate anisotropy compared with ⟨100⟩:⟨111⟩ or ⟨110⟩:⟨111⟩. Potassium hydroxide (KOH), tetra-methyl ammonium hydroxide (TMAH), and ethylene diamine pyrocatechol (EDP) are commonly used silicon anisotropic etchants. In most cases, silicon dioxide (SiO_2) or silicon nitride (Si_3N_4) is used as a masking material during the etching process. Anisotropic etchants and basic etching mechanisms are summarized by *Ristic* et al. [9.1]. There is also an isotropic wet etching process available using a mixture of fluoric acid (HF), nitric acid (HNO_3), and acetic acid (CH_3COOH), so-called "HNA". HNA etches in all directions with almost the same etch rate regardless of crystalline directions. Figure 9.1 illustrates wet anisotropic and isotropic etching profiles.

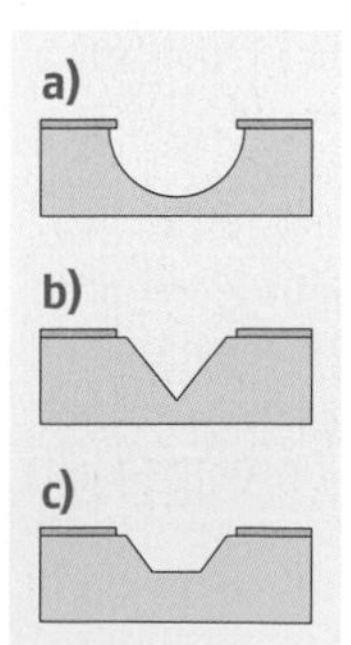

Fig. 9.1a–c Wet etching of silicon substrate for anisotropic and isotropic etching: (**a**) isotropic etching profile, (**b**) anisotropic etching profile (long term), and (**c**) anisotropic etching profile (short term)

Reactive ion etching (RIE) is also one of the most commonly used dry etching processes to generate microfluidic channels or deep trench structures on silicon substrate. In this dry etching technique, radio frequency (RF) energy is used to excite ions in a gas to an energetic state. The energized ions supply the necessary energy to generate physical and chemical reactions on the exposed area of the substrate, which starts the etching process. RIE can generate strong anisotropic, as well as isotropic profiles, depending on the gases used, the condition of plasma, and the applied power. Further information on reactive ion etching process, including deep reactive ion etching (DRIE), on silicon substrate can be found in the literature [9.2–5].

Many microfluidic devices were realized using silicon as a substrate material, including microvalves and micropumps, which are covered in the next section.

9.1.2 Glass

Glass substrate has been widely used for the fabrication of microfluidic systems and lab-on-a-chip due to its excellent optical transparency and ease of electro-osmotic flow (EOF). Chemical wet etching and thermal fusion bonding are the common fabrication techniques for glass substrate. Chemical wet etching and the bonding technique have also been widely reported [9.6, 7]. Most commonly used etchants are hydrofluoric acid (HF), buffered hydrofluoric acid, and a mixture of hydrofluoric acid, nitric acid, and deionized water (HF, HNO_3, H_2O). Gold with an adhesion layer of chrome is most often used as etch mask for wet etching of glass substrate. Since glass has no crystalline structure, only isotropic etch profiles are obtained as forming a hemispherical-shaped channel. Often the problem is that stresses within the surface layers of the glass cause preferential etching, and scratches created by polishing or handling errors cause spikes etched in the channels. Pre-etching is one method to release stress, which causes defects in channels after etching. Another way to improve the channel etching process is to anneal the glass wafers before etching. Annealing can be done nearly at glass transition temperature for at least a couple of hours. Figure 9.2 shows two examples of poorly etched channels without the pre-etching and annealing steps compared with an etched channel without defects. Other fabrication techniques for glass substrate include photoimageable glass, as *Dietrich* et al. [9.8] reported. Anisotropy is introduced in glass by making the glass photosensitive using lithium/aluminium/silicates composition.

One of the most successful examples of using glass as a substrate material in lab-on-a-chip applications is the capillary electrophoresis (CE) chip, fabricated using the glass etching and fusion bonding technique, since the most advantageous property of glass is an excellent optical transparency that is required for the most lab-on-a-chip using optical detection including capillary electrophoresis microchips [9.6, 7, 9].

9.1.3 Polymer

Among several substrates available for lab-on-a-chip, polymers, or plastics, have recently become one of the most popular and promising substrates due to their low cost, ease-of-fabrication, and favorable biochemical reliability and compatibility. Plastic substrates, such as polyimide, PMMA, PDMS, polyethylene or polycarbonate, offer a wide range of physical and chemical material parameters for the applications of lab-on-a-chip, generally at low cost using replication approaches. Polymers and plastics are promising materials in microfluidic and lab-on-a-chip applications because they can be used for mass production using casting, hot embossing, and injection molding techniques. This mass production capability allows the successful commercialization of lab-on-a-chip technology, including disposable lab-on-a-chip. While several fabrication methods have recently been developed, three fabrication techniques – casting, hot embossing, and injection molding – are major techniques of great interest. Figure 9.3 shows schematic illustrations of these polymer microfabrication techniques.

For polymer or plastic micro/nano fabrication, a mold master is essential for replication. Mold masters are fabricated using photolithography, silicon/glass bulk etching, and metal electroplating, depending on the application. Figure 9.4 summarizes mold masters from different fabrication techniques.

Photolithography, including UV-LIGA [9.10, 11] and LIGA [9.12], is used to fabricate mold masters for casting or soft lithography replication, while silicon-based and electroplated mold masters are used for hot embossing replication. For injection molding, electroplated metallic mold masters are preferable.

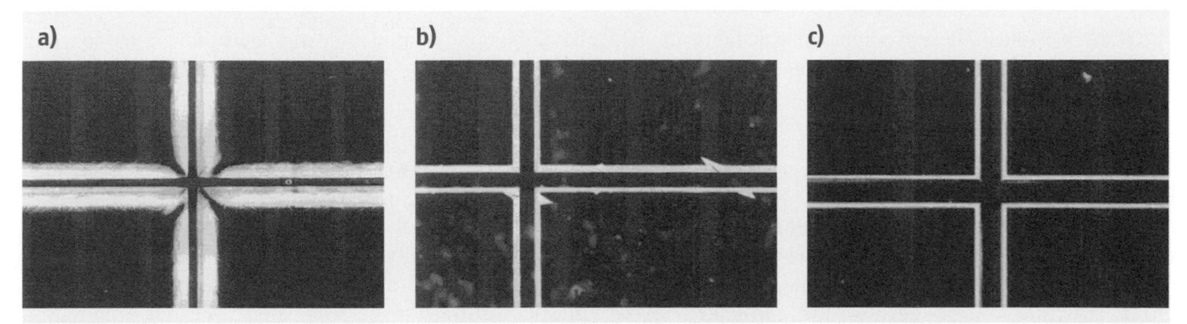

Fig. 9.2a–c Isotropically etched microfluidic channels on glass substrate: (**a**) poorly etched channel with large under etching, (**b**) poorly etched channel with spikes, and (**c**) well etched channel without any defects

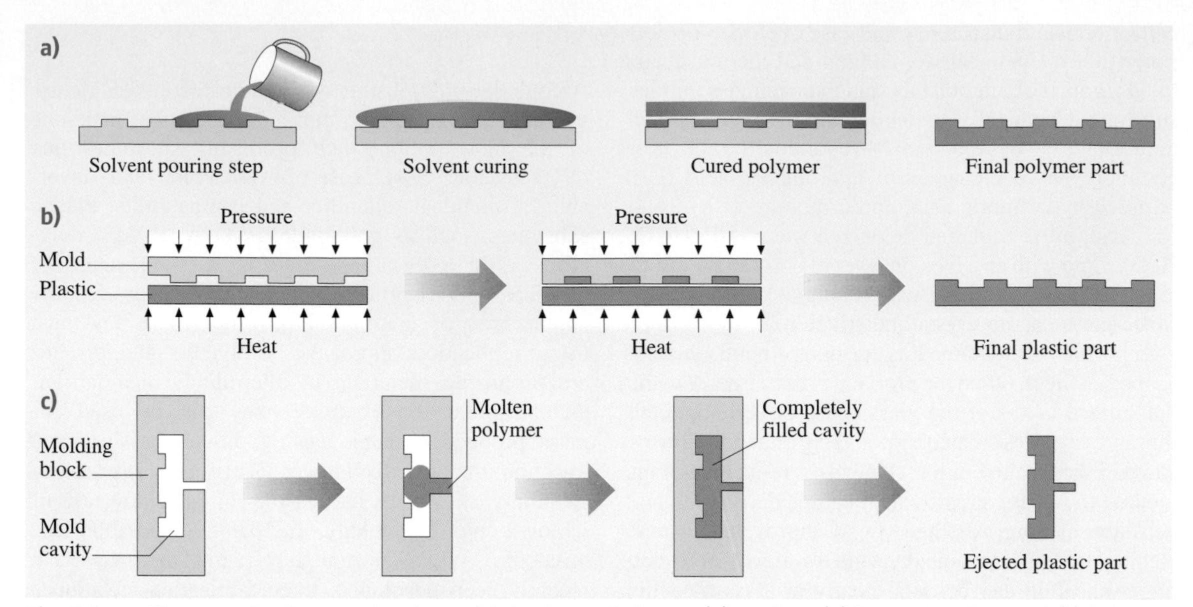

Fig. 9.3a–c Concept of polymer micro/nano fabrication techniques: (a) casting, (b) hot embossing, and (c) injection molding

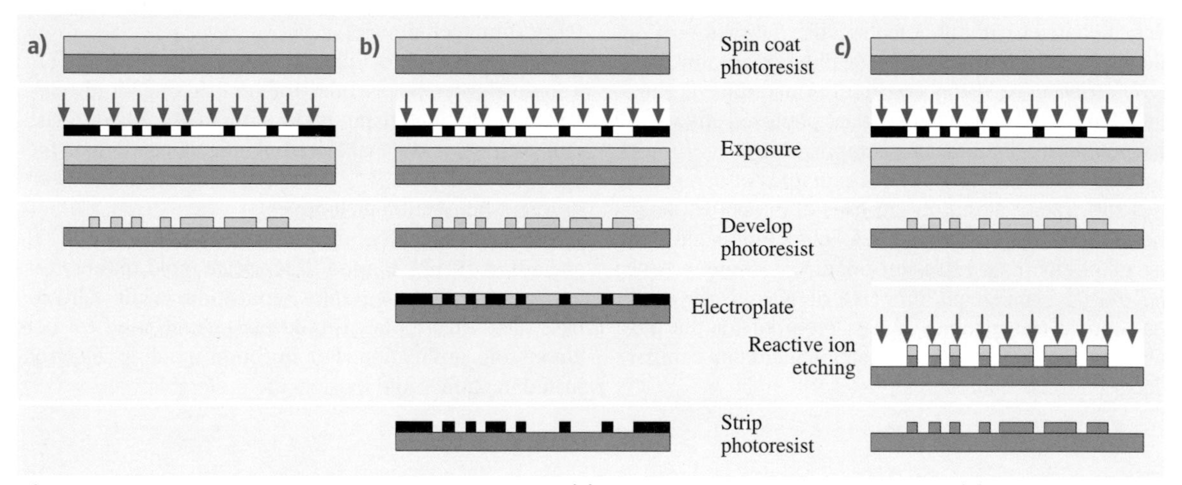

Fig. 9.4a–c Mold masters in polymer/plastic fabrication: (a) photolithography-based mold master, (b) silicon-based mold master, and (c) mold master by electroplating

Casting or soft lithography [9.13, 14] usually offers flexible access to microfluidic structures using mostly poly(dimethylsiloxane) (PDMS) as a casting material. A mixture of the elastomer precursor and curing agent is poured over a master mold structure. After curing, the replicated elastomer is released from the mold master, having transferred a reverse structure of the mold master. Patterns of a few nm can be achieved using this technique.

While casting can be carried out in room temperature, hot embossing requires a slightly higher temperature – up to glass transition temperature of the plastic substrate to be replicated. The hot embossing technique has been developed by several research groups [9.15–17]. A mold master is placed in the chamber

Table 9.1 Overview of the different polymer micro/nano fabrication techniques

Fabrication type	Casting	Hot embossing	Injection molding
Investment	Low	Moderate	High
Manufacturability	Low	Moderate	High
Cycle time	8–10 hrs	1 hr	1 min
Polymer choices	Low	Moderate	Moderate
Mold replication	Good	Good	Good
Reusability of mold	No (photolithography-based molds)	Yes	Yes

of a hot embossing system with the plastic substrate, then heated plates press both the plastic substrate and the mold master, as illustrated in Fig. 9.3b. After a certain amount of time (typically 5–20 minutes, depending on the plastic substrate), the plates are cooled down to release the replicated plastic substrate. Hot embossing offers mass production of polymer microstructures, as its cycle time is less than an hour.

Injection molding is a technique to fabricate polymer microstructures at low cost and high volumes [9.18]. A micromachined mold master is placed in the molding block of the injection molding machine, as illustrated in Fig. 9.3c. The plastic in granular form is melted and then injected into the cavity of a closed mold block, where the mold master is located. The molten plastic continues to flow into and fill the mold cavity until the plastic cools down to a highly viscous melt, and a cooled plastic part is ejected. In order to ensure good flow properties during the injection molding process, thermoplastics with low or medium viscosity are desirable. So the filling of the mold cavity, and, subsequently, the microstructures depend on the viscosity of the plastic melt, injection speed, molding block temperature, and nozzle temperature of the injection unit. This technique allows very rapid replication and high volume mass production. Typical cycle time is several seconds for most applications. However, due to high shear force on the mold master inside of the mold cavity during injection molding, metallic mold masters are highly recommended. Polymethylmethacrylate (PMMA), polyethylene (PE), polystyrene (PS), polycarbonate (PC), and cyclic olefin copolymers (COC) are common polymer/plastic materials for both hot embossing and injection mold replication.

All of these polymer/plastic replication techniques are summarized in Table 9.1. Since each technique differs from the others, fabrication techniques and materials have to be selected according to the application.

9.2 Active Microfluidic Devices

Microfluidics devices are essential for the development of lab-on-a-chip or μTAS (Micro-Total Analysis Systems). A number of different microfluidic devices have been developed with basic structures analogous to the macroscale fluidic devices. Such devices include microfluidic valves, microfluidic pumps, microfluidic mixers, etc. The devices listed above have been developed both as active and passive devices. While passive microfluidic devices are generally easy to fabricate, they do not offer the same functional diversity that the active microfluidic devices provide. For example, passive microvalves based on surface tension effect [9.19] can operate a few times to hold the liquid, which is acceptable for a disposable format. Once the air-liquid interface passes over the valve mechanism, the operation characteristics of the valve will differ, due to the change of surface energy over the channel. Similarly, passive check valves [9.20–22] are dependent on the pressure of the fluid for operation. Since the active microvalves can be triggered on/off depending on an external signal regardless of the fluid system status, there have been considerable research efforts to develop active microfluidic devices. However, the active devices usually require its expense as desired functional and fabrication complexity.

This section reviews some of the active microfluidic devices, such as microvalves, micropumps, and active microfluidic mixers. Passive counterparts of these microfluidic devices will be discussed in the next section.

9.2.1 Microvalves

Active microvalves have been an area of intense research over the past decade, and a number of novel design and actuation schemes have been developed. This makes the categorization of active microvalves a confusing enterprise. Classification schemes for active microvalves include:

1. Fluidics handled – liquid/gas/liquid and gas
2. Materials used for the structure – silicon/polysilicon/glass/polymer
3. Actuation mechanisms – electrostatic/pneumatic/thermopneumatic, etc.
4. Physical actuating microstructures – membrane type, flap type, ball valve, etc.

All of the classification schemes listed above are valid, but the most commonly used method [9.23, 24] is the classification based on the actuation mechanisms.

In this section, the various microvalves are discussed in terms of their actuation mechanisms and their relevance to the valve mechanism, as well as fluid handled and special design criteria.

Pneumatically/Thermopneumatically Actuated Microvalves

Pneumatic actuation uses an external air line (or pneumatic source) to actuate a flexible diaphragm. Pneumatic actuation offers such attractive features as high force, high displacement, and rapid response time. Figure 9.5 illustrates a schematic concept of pneumatically actuated microvalves. *Schomburg* et al. [9.25] demonstrated pneumatically actuated microvalves.

Pneumatically actuated microvalves have also been demonstrated using polymeric substrate. *Hosokawa* et al. [9.26] demonstrated a pneumatically actuated three-way microvalve system using a PDMS platform. The microfluidic lines and the pneumatic lines are fabricated on separate layers.

Thermopneumatic actuation is typically performed by heating a fluid (usually gas) in a confined cavity, as illustrated in Fig. 9.6. The increase in temperature leads to a rise in pressure of the gas, and this pressure is used for deflecting a membrane for valve operation. Thermopneumatic actuation is an inherently slow technique but offers very high forces when compared to other techniques [9.23]. Thermopneumatically actuated microvalves have been realized by many researchers using various substrates and diaphragm materials [9.27–29].

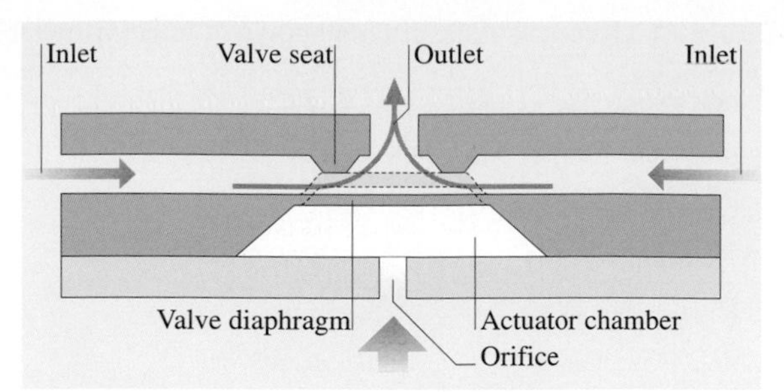

Fig. 9.5 Schematic concept of pneumatically actuated microvalve

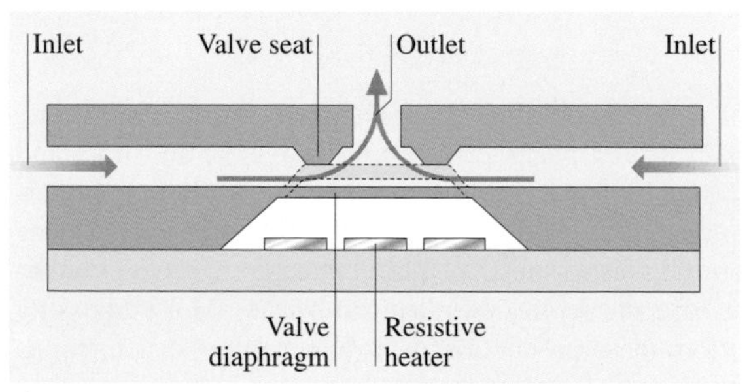

Fig. 9.6 Schematic concept of thermopneumatically actuated microvalve

Electrostatically Actuated Microvalves

Electrostatic actuation has been widely explored for a number of applications, including pressure sensors, comb drives, active mirror arrays, etc. Electrostatically actuated devices typically have a fairly simple structure and are easy to fabricate. A number of fabrication issues, such as stiction and release problems of membranes and valve flaps, need to be addressed for realizing practical electrostatic microvalves. *Sato* et al. [9.30] have developed a novel membrane design in which the deflection "propagates" through the membrane, rather than deforming it entirely. Figure 9.7 shows a schematic sketch of the actuation mechanism and valve design. The use of this "S-shaped" design allows them to have relatively large gaps across the two surfaces, as the electrostatic force must only be concentrated at the edges of the S-shape where the membrane is deflected. *Robertson* et al.

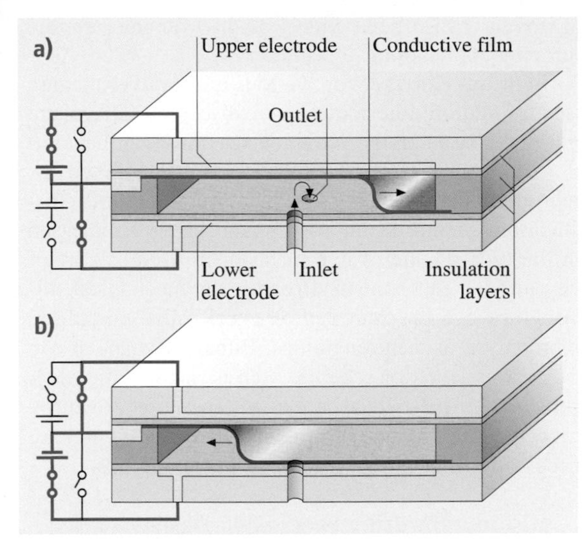

Fig. 9.7a,b Electrostatically actuated microvalve with S-Shaped film element: propagation of bend in the film as (a) open and (b) closed (adapted from [9.30]). Reproduced by permission of IOP Publishing Limited

[9.31] have developed an array of electrostatically actuated valves using a flap design (rather than a membrane) for sealing fluid flow. The demonstrated system is suitable for very low pressure gas control systems such as the ones needed in a clean room environment.

Wijngaart et al. [9.33] have developed a high-stroke, high pressure electrostatic actuator for valve applications. This reference provides a good overview of the theoretical design parameters used to design and analyze an electrostatically actuated microvalve.

Piezoelectrically Actuated Microvalve

Piezoelectric actuation schemes offer a significant advantage in terms of operating speed; they are typically the fastest actuation scheme at the expense of a reduced actuator stroke. Also, piezoelectric materials are more challenging to incorporate into fully integrated MEMS devices. *Watanabe* et al. [9.34] and *Stehr* et al. [9.35] demonstrated piezoelectric actuators for valve applications. A film of piezoelectric material is deposited on the movable membrane, and upon application of an electric potential, a small deformation occurs in the piezo film that is transmitted to the valve membrane.

Electromagnetically Actuated Microvalves

Electromagnetic actuators are typically capable of delivering high force and range of motion. A significant advantage of electromagnetic microvalves is that they are relatively insensitive to external interference. However, electromagnetic actuators involve a fairly complex fabrication process. Usually, a soft electromagnetic material like Ni-Fe (nickel-iron permalloy) is used as a membrane layer, and an external electromagnet is used to actuate this layer. *Sadler* et al. [9.32] have developed a microvalve using the electromagnetic actuation scheme shown in Fig. 9.8.

They have demonstrated a fully integrated magnetic actuator with magnetic interconnection vias to guide the magnetic flux. The valve seat design, also shown in Fig. 9.8, allows for very intimate contact between the Ni-Fe valve membrane and the valve seat, hence, achieving an ultralow leak rate when the valve is closed.

Jackson et al. [9.36] demonstrate an electromagnetic microvalve using "magnetic" PDMS as the membrane material. For this application, the PDMS pre-polymer is loaded with soft magnetic particles and then cured to form the valve membrane. The PDMS membrane

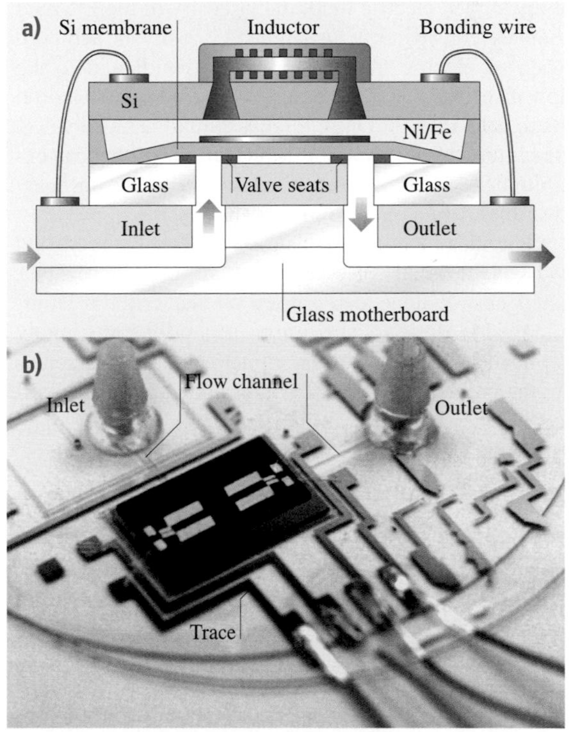

Fig. 9.8a,b Electromagnetically actuated microvalves: (a) schematic illustration and (b) photograph of the electromagnetically actuated microvalve as a part of lab-on-a-chip (after [9.32])

is then assembled over the valve body, and miniature electromagnets are used to actuate the membrane.

Other Microvalve Actuation Schemes

Microvalves have most commonly been implemented with one of the actuation schemes listed above. However, these are not the only actuation schemes that are used for microvalves. Some others include the use of SMA (shape memory alloys) [9.37], electrochemical actuation [9.38], etc. SMA actuation schemes offer the advantage of generating very large forces when the SMA material is heated to its original state. *Neagu* et al. [9.38] present an electrochemically actuated microvalve. In their device, an electrolysis reaction is used to generate oxygen in a confined chamber. This chamber is sealed by a deformable membrane that deflects due to increased pressure. The reported microvalve has relatively fast actuation time and can generate very high pressures. *Yoshida* et al. [9.39] present a novel approach to the microvalve design: a micro-electro-rheological valve (ER valve). An electro-rheological fluid is loaded into the microchannel, and, depending on the strength of an applied electric field, the viscosity of the ER fluid changes considerably. A higher viscosity is achieved when an electric field is applied perpendicular to the flow direction. This increased viscosity leads to a drop in the flow rate, allowing the ER fluid to act as a valve. Of course, this technique is limited to fluids that can exhibit such properties; nevertheless, it provides a novel idea to generate on-chip microvalves.

The ideal characteristics for a microvalve are listed by Kovacs [9.24]. However, of all the microvalves listed above, none can satisfy all the criteria. Thus, microvalve design, fabrication, and utility are highly application-specific and most microvalves try to generate the performance characteristics that are most useful for the intended application.

9.2.2 Micropumps

One of the most challenging tasks in developing a fully integrated microfluidic system has been the development of efficient and reliable micropumps. In macroscale, a number of pumping techniques exist such as peristaltic pumps, vacuum-driven pumps, venturi-effect pumps, etc. However, in microscale, most mechanical pumps rely on pressurizing the working fluid and forcing it to flow through the system. Practical vacuum pumps are not available in microscale. There are also some effects such as electro-osmotic pumping, that are only possible on the microscale. Consequently, electrokinetic driving mechanisms have also been widely studied for microfluidic pumping applications.

In the previous section, various microvalves that are used for microfluidic control were discussed. They were presented based on the actuation schemes that they employ. A similar classification can also be adopted for micropumps. However, rather than using the same classification scheme, the micropumps are categorized based on the type of microvalve mechanism used as part of the pumping mechanism. Broadly, the mechanical micropumps can be classified as check-valve controlled microvalves or diffuser pumps. Either mechanism can use various actuation schemes such as electrostatic, electromagnetic, piezoelectric, etc. Micropumps driven by direct electrical control form separate category and are discussed following the mechanical micropumps.

Micropumps Using Check Valve Design

Figure 9.9 shows a typical mechanical micropump with check valves. This pump consists of an inlet and an outlet check valve with a pumping chamber in between. A membrane is deflected upwards, and a low pressure zone is created in the pumping chamber. This forces the inlet check valve open, and fluid is sucked into the pumping chamber. As the membrane returns to its original state and continues to travel downwards, a positive pressure builds up, which seals the inlet valve while simultaneously opening the outlet valve. The fluid is then ejected and the pump is ready for another cycle.

A number of other techniques have been used to realize mechanical micropumps with check valves and a pumping chamber. *Jeon* et al. [9.40] present a micropump that uses PDMS flap valves to control the pumping mechanism, *Koch* et al. [9.41], *Cao* et al. [9.42], *Park* et al. [9.43], and *Koch* et al. [9.44] present piezoelectrically driven micropumps. *Xu* et al. [9.45] and *Makino*

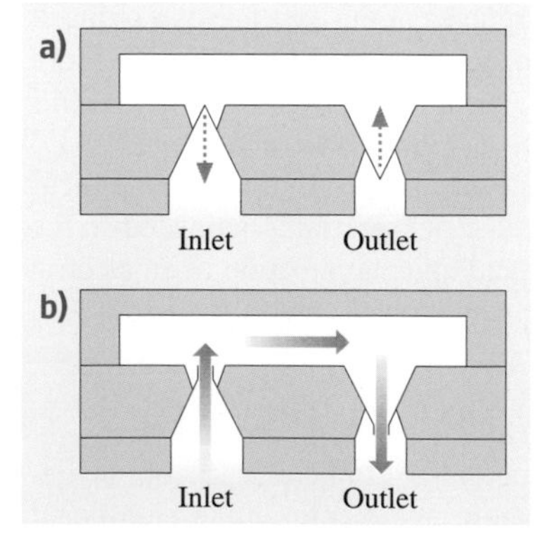

Fig. 9.9a,b Operation of micropump using check valve design: **(a)** check valves are closed and **(b)** check valves are open

et al. [9.46] present SMA driven micropumps. *Chou* et al. [9.47] present a novel rotary pump using a soft lithography approach. As explained in the active valve section, valves can be created on a PDMS layer using separate liquid and air layers. When a pressure is applied to the air lines, the membranes deflect to seal the fluidic path. *Chou* et al. [9.47] have implemented a series of such valves in a loop. When they are deflected in a set sequence, the liquid within the ring is pumped by peristaltic motion.

An interesting approach toward the development of a bidirectional micropump has been used by *Zengerle* et al. [9.48]. Their design has two flap valves at the inlet and outlet of the micropump, which work in the forward mode at low actuation frequencies and in the reverse direction at higher frequencies. *Zengerle* et al. [9.48] attribute the change in pumping direction to the phase shift between the response of the valves and the pressure difference that drives the fluid. *Carrozza* et al. [9.49] use a different approach to generate the check valves. Rather than using the conventional membrane or flap type valves, they use ball valves by employing a stereolithographic approach. The developed pump is actuated using a piezoelectric actuation scheme.

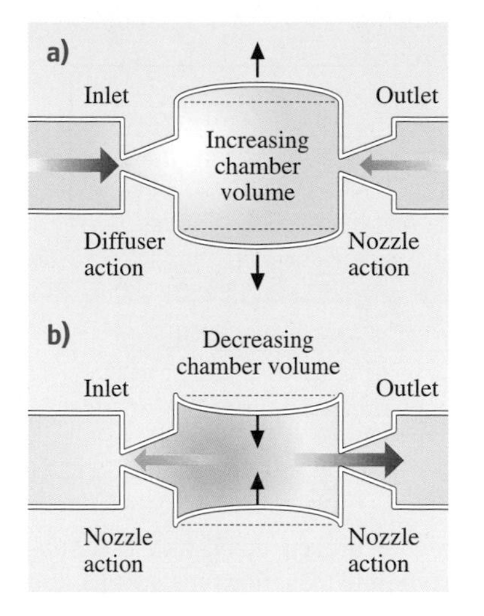

Fig. 9.10a,b Operating principle of a diffuser micropump: (**a**) suction mode and (**b**) pumping mode

Diffuser Micropumps

The use of nozzle-diffuser sections, or pumps with fixed valves, or pumps with dynamic valves has been extensively researched. The basic principle of these pumps is based on the idea that the geometrical structure used as an inlet valve has a preferential flow direction toward the pumping chamber, and the outlet valve structure has a preferential flow direction away from the pumping chamber. An illustrative example of this concept is shown in Fig. 9.10. As can be readily seen, these pumps are designed to work with liquids only. When the pump is in suction mode, the flow in the inlet diffuser structure is primarily directed toward the pump, and a slight backflow occurs from the outlet diffuser (acting as a nozzle) section. When the pump is in pressure mode, the outlet diffuser section allows most of the flow out of the pump, whereas the inlet diffuser section allows a slight backflow. The net effect is that liquid pumping occurs from left to right.

Diffuser micropumps are simple and valveless structures that improve pumping reliability [9.50], but they cannot eliminate backflow problems.

These micropumps can also be described as flow rectifiers, analogous to diodes in electrical systems. *Forster* et al. [9.51] have used the Tesla valve geometry, instead of the diffuser section, and the reference presents a detailed discussion of the design parameters and operational characteristics of their fixed-valve micropumps.

Electric/Magnetic Field Driven Micropumps

An electric field can be used to directly pump liquids in microchannels using such techniques as electro-omosis (EO), electrohydrodynamic (EHD) pumping, magnetohydrodynamic (MHD) pumping, etc. These pumping techniques rely on creating an attractive force for some of the ions in the liquid, and the remaining liquid is dragged along to form a bulk flow.

When a liquid is introduced into a microchannel, a double-layer charge exists at the interface of the liquid and the microchannel wall. The magnitude of this charge is governed by the zeta potential of the channel-liquid pair. Figure 9.11 shows a schematic view of the electro-osmotic transport phenomenon.

As shown in Fig. 9.11, the channel wall is negatively charged, which attracts the positive ions in solution. When a strong electric field is applied along the length of the microchannel, the ions at the interface experience an attractive force toward the cathode. As the positive ions move toward the cathode, they exert a drag force on the bulk fluid, and net fluid transport occurs from the anode to the cathode. It is interesting to note that the flow profile of the liquid plug is significantly different from the pressure-driven flow. Unlike the parabolic flow profile of pressure-driven flow, electro-osmotic transport leads to an almost vertical flow profile. The electro-osmotic transport phenomenon is only effective across very narrow channels.

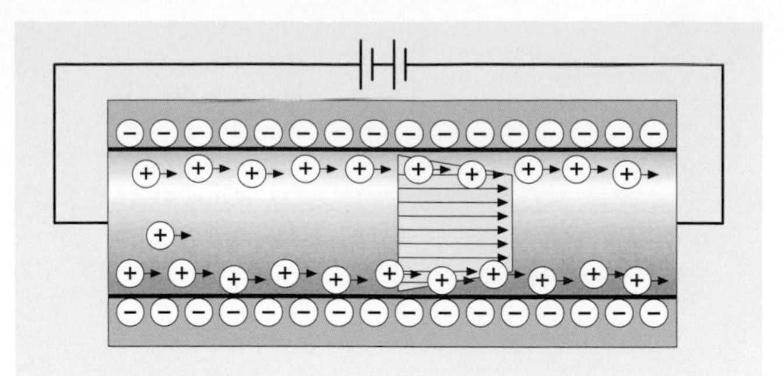

Fig. 9.11 Schematic sketch explaining the principle of electro-osmotic fluid transport

EHD can be broadly broken down into two subcategories: injection type and induction type. In injection type EHD, a strong electric field ($\sim 100\,\mathrm{kV/cm}$) is applied across a dielectric liquid. This induces charge formation in the liquid, and these induced (or injected) charges are then acted upon by the electrical field for pumping. Induction type EHD relies on generating a gradient/discontinuity in the conductivity and/or permitivity of the liquid. *Fuhr* et al. [9.52] explain various techniques that can be used to generate gradients for non-injection type EHD pumping.

MHD pumps rely on creating a Lorentz force on the liquid particles in the presence of an externally applied electric field. MHD has been demonstrated using both DC [9.53] and AC [9.54] excitations. MHD, EHD, and electro-osmotic transport share one feature in common that makes them very appealing for microfluidic systems; namely, none requires microvalves to regulate the flow. This makes these pumping techniques very reliable, as there is no concern of wear and tear of the microvalves, or any other moving parts of the micropump. However, it has been difficult to implement these actuation schemes on a complete microscale, owing to the high voltages/electromagnets, etc. required for these actuation schemes.

9.3 Smart Passive Microfluidic Devices

Passive microfluidics is a powerful technique for the rapidly evolving discipline of BioMEMS. It is a fluid control topology in which the physical configuration of the microfabricated system primarily determines the functional characteristics of the device/system. Typically, passive microfluidic devices do not require an external power source, and the control exerted by the devices is based, in part, on energy drawn from the working fluid, or based purely on surface effects, such as surface tension, selective hydrophobic/hydrophilic control, etc. Most passive microfluidic devices exploit various physical properties such as shape, contact angle, and flow characteristics to achieve the desired function. Passive microfluidic systems are usually easier to implement and allow for a simple microfluidic system with little or no control circuitry. A further list of advantages and disadvantages of passive microfluidic systems (or devices) is considered toward the end of this section.

Passive microfluidic devices can be categorized based on:

- Function – microvalves, micromixers, filters, reactors, etc.
- Fluidic medium – gas or liquid
- Application – biological, chemical, or other
- Substrate material – silicon, glass, polysilicon, polymer, or others

In this section, we will study various passive microfluidic devices that are categorized based on their function. Passive microfluidic devices include, but are not limited to, microvalves, micromixers, filters, dispensers, etc. [9.24].

9.3.1 Passive Microvalves

Passive microvalves have been a subject of great interest ever since the inception of the lab-on-a-chip concept. Microvalves are a key component of any microfluidic system and are essential for fluidic sequencing operations. Since most chemical and biochemical reactions require about five to six reaction steps, passive microvalves with limited functionality are ideally suited for such simple tasks. Passive microvalves can be broadly categorized as follows:

- Silicon/polysilicon or polymer-based check valves
- Passive valves based on surface tension effects
- Hydrogel-based biomimetic valves

Passive Check Valves

Shoji et al. [9.23] provide an excellent review of check-type passive microvalves. Some of the valves, shown in Fig. 9.12, illustrate the various techniques that can be used to fabricate check valves.

Figure 9.12a shows a microvalve fabricated using silicon bulk etching techniques. A through hole (pyramidal cavity) is etched through a silicon wafer that is sandwiched between two glass wafers. The normally closed valve is held in position by the spring effect of the silicon membrane. Upon applying pressure at the lower fluidic port, the membrane deflects upwards, allowing fluid flow through the check valve. The same working principle is employed by the microvalve shown in Fig. 9.12b. However, instead of using a membrane supported on all sides, a cantilever structure is used for the flap. This reduces the burst pressure, i.e., the minimum pressure required to open the microvalve. Figure 9.12c shows that the membrane structure can also be realized using a polysilicon layer deposited on a bulk-etched silicon wafer. Poly-silicon processes typically allow tighter control over dimensions and, consequently, offer more repeatable operating characteristics. Figure 9.12d shows the simplest type of check valve where the V-groove etched in a bulk silicon substrate acts as a check valve. However, the low contact area between the flaps of the microvalve leads to nontrivial leakage rates in the forward direction. Figures 9.12e and f show check valve designs that are realized using polymer/metal films, in addition to the traditional glass/silicon platform. This technique offers a significant advantage in terms of biocompatibility characteristic and controllable operating characteristics. The surface properties of polymers can be easily tailored using a wide variety of techniques such as plasma treatment and surface adsorption [9.55]. Thus, in applications for which the biocompatibility requirements are very stringent, it is preferable to have polymers as the fluid contacting material. Furthermore, polymer properties such as stiffness, can also be controlled in some cases based on the composition and/or processing conditions. Thus, it may be possible to fabricate microvalves with different burst pressures by using different processing conditions for the same polymer.

In addition to the passive check valves reviewed by *Shoji* et al. [9.23], other designs include check valves using composite titanium/polyimide membranes [9.20,21], polymeric membranes such as Mylar or KAPTON [9.56], or PDMS [9.40], and metallic membranes such as [9.22].

Terray et al. [9.57] present an interesting approach to fabricating ultra-small passive valve structures. They have demonstrated a technique to polymerize colloidal particles into linear structures using an optical trap to form microscale particulate valves.

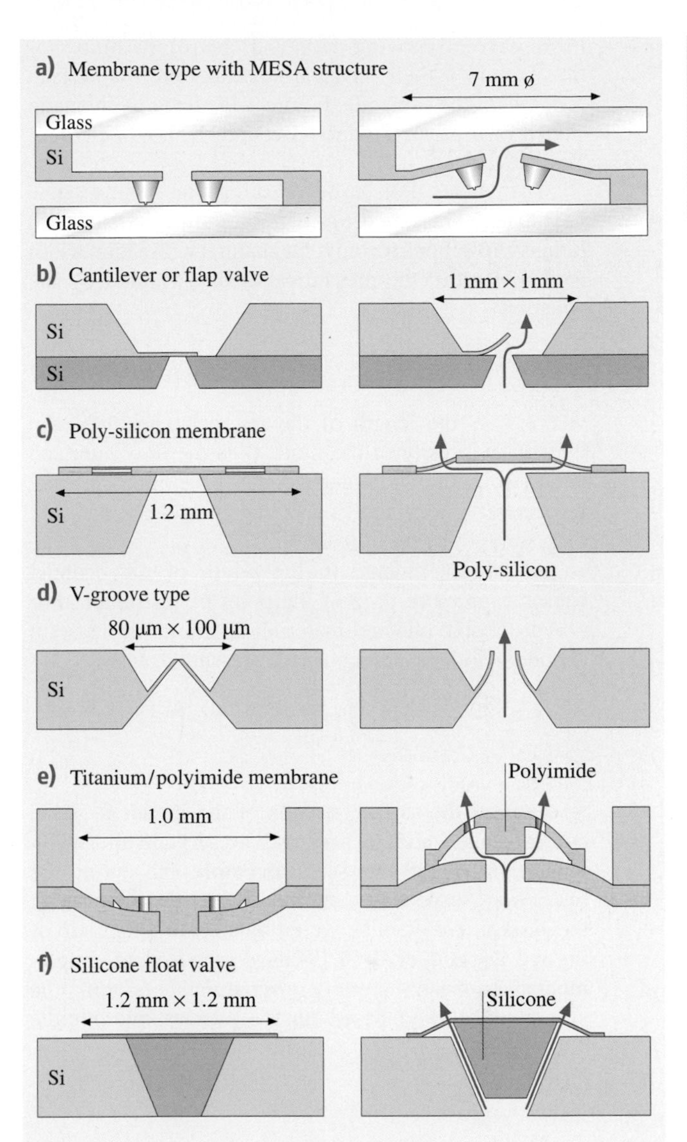

Fig. 9.12a–f Various types of passive microvalve designs: (**a**) membrane type with MESA structure; (**b**) cantilever or flap valve; (**c**) polysilicon membrane; (**d**) V-groove type; (**e**) titanium/polyimide membrane; and (**f**) silicone float valve. Adapted from [9.23]. Reproduced by permissions of IOP Publishing Limited

Passive Valves Based on Surface Tension Effects

The passive valves listed in the previous section use the forced motion of the membrane or flap to control the flow of fluids. These valves are prone to such problems as clogging and mechanical wear and tear. Passive valves based on surface tension effects, on the other hand, have no moving parts and control the fluid motion based on their physical structure and the surface property of the substrate. Figure 9.13 shows a schematic sketch of a passive microvalve on a hydrophobic substrate [9.19].

The Hagen–Poiseuille equation for laminar flow governs the pressure drops in microfluidic systems with laminar flow. For a rectangular channel with a high width to height ratio, the pressure drop is governed by the equation

$$\Delta P = \frac{12L\mu.Q}{wh^3} , \quad (9.1)$$

where L is the length of the microchannel, μ is the dynamic viscosity of the fluid, Q is the flow rate, and w and h are the width and height of the microchannel, respectively. Varying L or Q can control the pressure drop for a given set of w and h.

An abrupt change in the width of the channel causes a pressure drop at the point of restriction. For a hydrophobic channel material, an abrupt decrease in channel width causes a positive pressure drop:

$$\Delta P_2 = 2\sigma_1 \cos(\theta_c) \left[\left(\frac{1}{w_1} + \frac{1}{h_1} \right) - \left(\frac{1}{w_2} + \frac{1}{h_2} \right) \right] , \quad (9.2)$$

where σ_l is the surface tension of the liquid, θ_c is the contact angle, and w_1, h_1 and w_2, h_2 are the width and height of the two sections before and during the restriction, respectively. Setting h as constant through the system, ΔP_2 can be varied by adjusting the ratio of w_1 and w_2. *Ahn* et al. [9.19] have proposed and implemented a novel structurally programmable microfluidic system (sPROMs) based on the passive microfluidic approach. In short, the sPROMs system consists of a network of microchannels with passive valves of the type shown in the passive valve section. If the pressure drop of the passive valves is set to be significantly higher than the pressure drop of the microchannel network, then the position of the liquid in the microchannel network can be controlled accurately. By applying sequentially higher pressure pulses, the liquid is forced to move from one passive valve to another. Thus, the movement of the fluid within the microfluidic channels is "programmed" using the physical structure of the microfluidic system, and this forms the basic idea behind sPROMs.

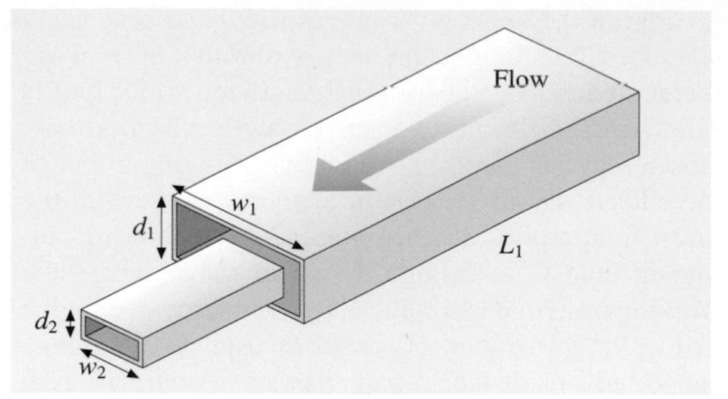

Fig. 9.13 Structure of a passive microvalve based on surface tension effects (adapted from [9.19])

The abrupt transition from a wide channel to a narrow channel can also be affected along the height of the microchannel. Furthermore, the passive valve geometry shown in Fig. 9.13 is not exclusive. *Puntambekar* et al. [9.58] have demonstrated different geometries of passive valves, as shown in Fig. 9.14a. As shown in Fig. 9.14b,

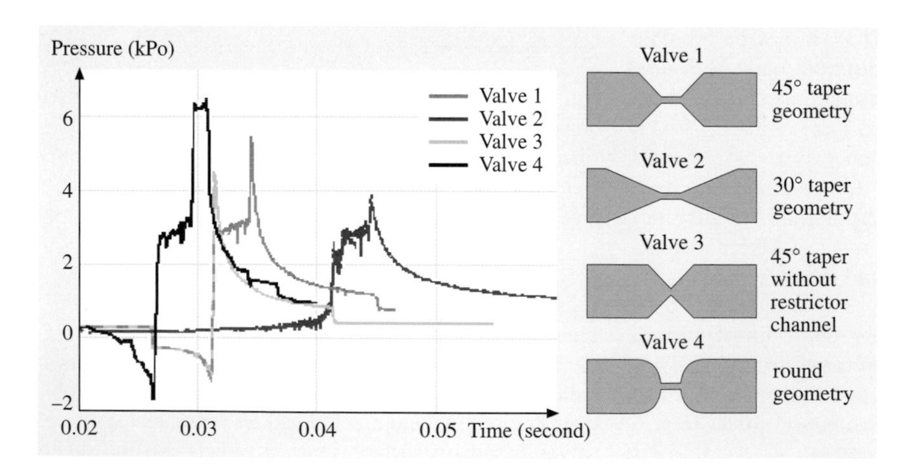

Fig. 9.14 Analysis of different geometries of passive valves. After [9.58], reproduced by permission of The Royal Society of Chemistry

the various geometries of the passive valves in Fig. 9.14a can act as effective passive valves without having an abrupt transition. This is important in order to avoid the dead volume that is commonly encountered across an abrupt step junction.

The use of surface tension to control the operation of passive valves is not limited to hydrophobic substrates. *Madou* et al. [9.59] demonstrate a capillarity-driven stop valve on a hydrophilic substrate. On a hydrophilic substrate, the fluid can easily "wick through" in the narrow region. However, at the abrupt transition to a larger channel section, the surface tension effects will not allow the fluid to leave the narrow channel. Thus, in this case, the fluid is held at the transition from the narrow capillary to the wide outlet channel.

Another mechanism to implement passive valves is the use of hydrophobic patches on a normally hydrophilic channel. *Handique* et al. [9.60] demonstrate the use of this technique to implement passive valves for a DNA analysis system. The fluid is sucked into the microfluidic channels via capillary suction force. The hydrophobic patch exerts a negative capillary pressure that stops the further flow of fluid. The use of hydrophobic patches as passive valves is reported by *Andersson* et al. [9.61].

Other Passive Microvalves

A novel approach to realizing passive valves that are responsive to their surrounding environment is demonstrated by *Low* et al. [9.62] and *Yu* et al. [9.63]. *Yu* et al. [9.63] have developed customized polymer "cocktails" that are polymerized in situ around prefabricated posts. The specialized polymer is selectively responsive to stimuli such as pH, temperature, electric fields, light, carbohydrates, and antigens.

Another interesting approach in developing passive valves has been adopted by *Foster* et al. [9.51]. They have developed the so-called no moving part (NMP) valves, which are based on a physical configuration that allows a higher flow rate along a direction compared to the reverse direction.

9.3.2 Passive Micromixers

The successful implementation of microfluidic systems for many lab-on-a-chip systems is partly owing to the significant reduction in volumes handled by such systems. This reduction in volume is made possible by the use of microfabricated features and channel dimensions ranging from a few μm to several hundred μm. Despite the advantage offered by the micron-sized channels, one of the significant challenges has been the implementation of effective microfluidic mixers on the microscale. Mixing on a macroscale is a turbulent flow regime process. However, on the microscale, because of the low Reynolds numbers as a result of the small channel dimensions, most flow streams are laminar in nature, which does not allow for efficient mixing. On the other hand, diffusion is an important factor in mixing because of the short diffusion lengths.

There have been numerous attempts to realize micromixers using both active and passive techniques. Active micromixers rely on creating localized turbulence to enhance the mixing process, whereas passive micromixers usually enhance the diffusion process. The diffusion process can be modeled by the following equation:

$$\tau = \frac{d^2}{D}\,, \qquad (9.3)$$

where τ is the mixing time, d is the distance, and D is the diffusion coefficient. Equation (9.3) illustrates the diffusion-dominated mixing at the microscale. Because of the small diffusion lengths (d), the mixing times can be made very short. The simplest category of micromixers is illustrated in Fig. 9.15. In these mixers, creating a convoluted path increases the path length that the two fluids share, leading to higher diffusion and more complete mixing. However, these mixers only exhibit good mixing performance at low flow rates, in the range of a few μl/min.

Mitchell et al. [9.64] have demonstrated three-dimensional micromixers that can achieve better

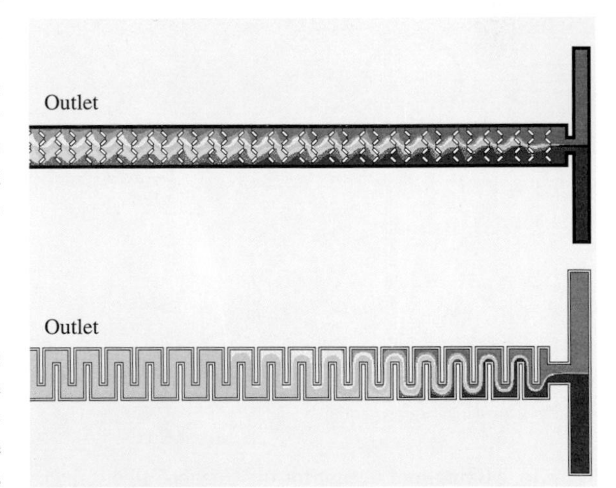

Fig. 9.15 Diffusion-enhanced mixers based on long, convoluted flow path

performance by alternately laminating the two fluid streams to be mixed. *Beebe* et al. [9.65] have created a chaotic mixer that has the convoluted channel along three dimensions. This micromixer works on the principle of forced advection resulting from repeated turns in three dimensions. Furthermore, at each turn, eddies are generated because of the difference in flow velocities along the inner and outer radii, which enhances the mixing.

Stroock et al. [9.68] demonstrate a passive micromixer that uses chaotic mixing by superimposing a transverse flow component on the axial flow. Ridges are fabricated at the bottom of microchannels. The flow resistance is lower along the ridges (peak/valley) and higher in the axial direction. This generates a helical flow pattern that is superimposed on the laminar flow. The demonstrated mixer shows good mixing performance over a wide range of flow velocities.

Hong et al. [9.66] have demonstrated a passive micromixer based on the Koanda effect. Their design uses the effects of diffusion mixing at low flow velocities; at high flow velocities, a convective component is added perpendicular to the flow direction, allowing for rapid mixing. This mixer shows excellent mixing performances across a wide range of flow rates because of the dual mixing effects. Figure 9.16 shows a schematic sketch of the mixer structure.

The mixer works on the principle of superimposing a parabolic flow profile in a direction perpendicular to

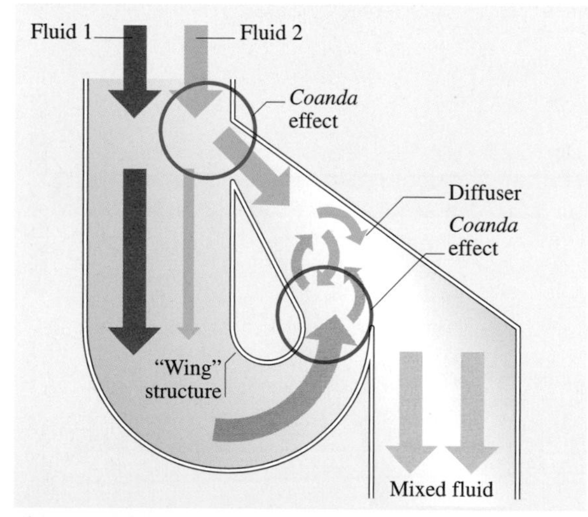

Fig. 9.16 Mixing unit design for the Coanda effect mixer. The actual mixer has mixing unit pairs in series (after [9.66])

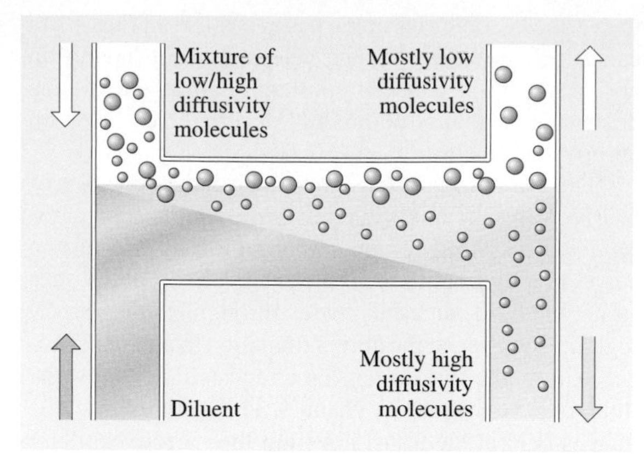

Fig. 9.17 Conceptual illustration of the H-Sensor. Adapted from [9.67]

the flow direction. The parabolic profile creates a Taylor dispersion pattern across the cross section of the flow path. The dispersion is directly proportional to the flow velocity, and higher flow rates generate more dispersion mixing.

Brody et al. [9.67] have used the laminar flow characteristics in a microchannel to develop a diffusion-based extractor. When two fluid streams, where fluid 1 is loaded with particles of different diffusivity and fluid 2 is a diluent, are forced to flow together in a microchannel, they form two laminar streams with little mixing. If the length that the two streams are in contact is carefully adjusted, only particles with high diffusivity (usually small molecules) can diffuse across into the diluent stream, as shown in Fig. 9.17. The same idea can be extended to a T-filter. *Weigl* et al. [9.69] have demonstrated a rapid diffusion immunoassay using the T-filter.

9.3.3 Passive Microdispensers

The principle of the structurally programmable microfluidic systems (sPROMs) was introduced earlier in the passive valve section. One of the key components of the sPROMs system is the microdispenser, which is designed to accurately and repeatedly dispense fluidic volumes in the micro- to nanoliter range. This would allow the dispensing of a controlled amount of the analyte into the system that could be used for further biochemical analysis. Figure 9.18 shows a schematic sketch of the microdispenser design [9.58].

The microdispenser works on the principle of graduated volume measurement. The fluid fills up the exact fixed volume of the reservoir, and the second passive valve at the other end of the microdispenser stops further

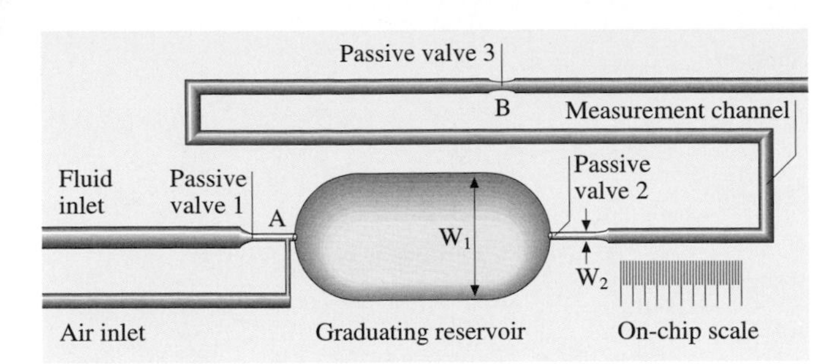

Fig. 9.18 Schematic sketch of the microdispenser. Adapted from [9.58], reproduced by permission of The Royal Society of Chemistry

motion. When the reservoir is filled with fluid, the fluidic actuation is stopped, and, simultaneously, pneumatic actuation from the air line causes a split in the fluid column at point A (Fig. 9.18). Thus, the accuracy of the reservoir decides the accuracy of the dispensed volume. Since the device is manufactured using UV-LIGA lithography techniques, highly accurate and reproducible volumes can be defined. The precisely measured volume of fluid is expelled to the right from the reservoir. The expelled fluid then starts to fill up the measuring channel. When the fluid reaches point B (Fig. 9.18), the third microvalve holds the fluid column.

Figure 9.19 shows an actual operation sequence of the microdispenser. Figure 9.19f shows that the dispensed volume is held by passive valve 3, and at this stage, the length (and hence volume) can be calculated using the on-chip scale. In experiments only, the region in the immediate vicinity of the scale was viewed using a stereomicroscope for measuring the length of the fluid column. The microdispenser demonstrated above is reported to have dispensing variation of less than 1% between multiple dispensing cycles.

9.3.4 Microfluidic Multiplexer Integrated with Passive Microdispenser

Ahn et al. [9.19] have demonstrated the sPROMs technology to be an innovative method of controlling liquid

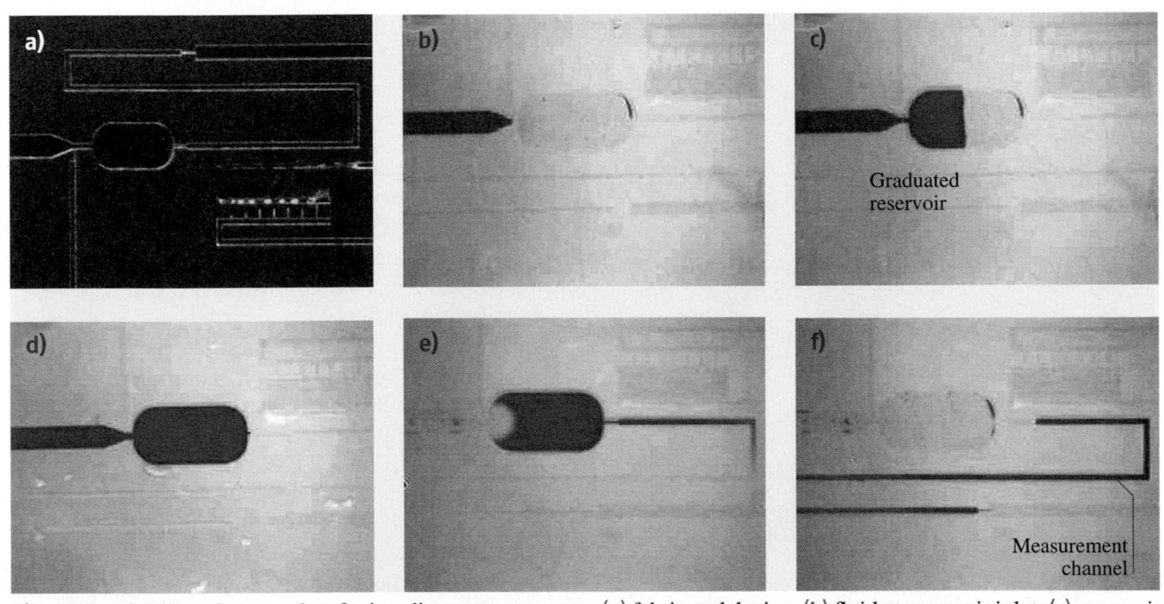

Fig. 9.19a–f Microphotographs of microdispenser sequence: (**a**) fabricated device; (**b**) fluid at reservoir inlet; (**c**) reservoir filling; (**d**) reservoir filled; (**e**) split in liquid column due to pneumatic actuation; and (**f**) fluid ejected to measurement channel and locked in by passive valve. After [9.58], reproduced by permission of The Royal Society of Chemistry

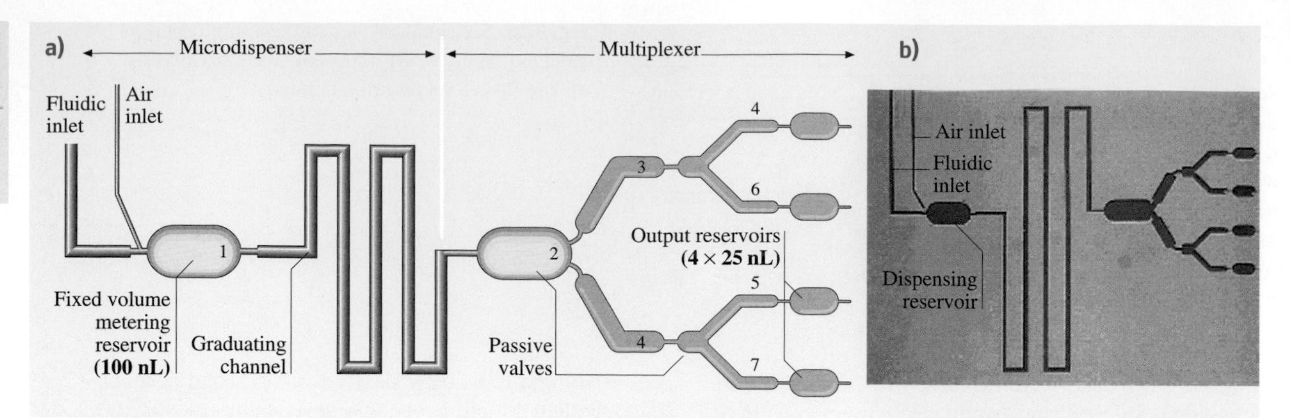

Fig. 9.20a,b Microfluidic multiplexer with integrated dispenser: (**a**) schematic sketch and (**b**) fabricated device filled with dye (after [9.58])

movement in a programmed fashion in a microfluidic network. By integrating this technique with the microdispensers, a more functionally useful microfluidic system can be realized. Figure 9.20a shows a schematic illustration of the microfluidic multiplexer with the integrated dispenser, and Fig. 9.20b shows an actual device fabricated using rapid prototyping techniques [9.70].

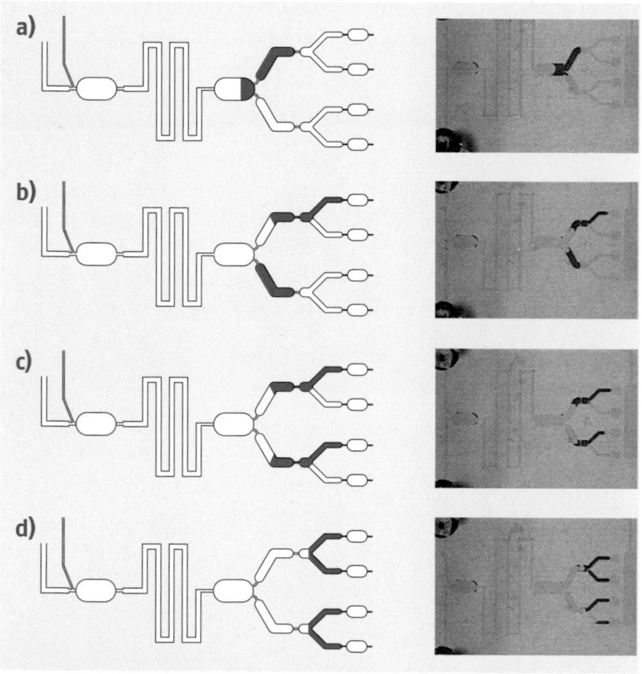

The operation of the microdispenser has been explained earlier in this section. Briefly, the fluid is loaded in the fixed-volume metering reservoir via a syringe pump. The fluid is locked in the reservoir by the passive valve at the outlet of the reservoir. When a higher pressure is applied via the air inlet line, the liquid column is split and the fluid is dispensed into the graduating channel.

The microfluidic multiplexer is designed to have a programmed delivery sequence, as shown in Fig. 9.20a, where the numbers on each branch of the multiplexer indicates the filling sequence. This sequential filling is achieved by using different ratios of passive valves along the multiplexer section. For instance, at the first branching point, the passive valve at the upper branch offers less resistance than the passive valve at the beginning of the lower branch. Thus, the dispensed fluid will first fill the top branch. After filling the top branch, the liquid encounters another passive valve at the end of the top branch. The pressure needed to push beyond this valve is higher than the pressure needed to push liquid into the lower branch of the first split. The liquid will then fill the lower branch. This sequence of nonsymmetrical passive valves continues along all the branches of the multiplexer, as shown in Fig. 9.21.

The ability to sequentially divide and deliver liquid volumes was demonstrated for the first time using a passive microfluidic system. This approach has the potential to deliver very simple microfluidic control systems that are capable of a number of sequential microfluidic manipulation steps required in a biochemical analysis system.

a)

b)

c)

d)

Fig. 9.21a–d Microphotographs showing operation of sequential multiplexer: (**a**) first level division; (**b**) second level division; (**c**) continued multiplexing; and (**d**) end of sequential multiplexing sequence (after [9.58])

9.3.5 Passive Micropumps

A passive system is defined inherently as one that does not require an external energy source. Thus, the term passive micropump might seem misnomer upon initial inspection. However, there have been some efforts dedicated to realizing a passive micropump that essentially does not draw energy from an external source, but stores the required actuation energy in some form and converts it to mechanical energy on demand.

Passive Micropump Based on Osmotic Pressure

Nagakura et al. [9.71] have demonstrated a mesoscale osmotic actuator that converts chemical energy to mechanical displacement. Osmosis is a well-known phenomenon by which liquid is transported across a semipermeable membrane to achieve a uniform concentration distribution across the membrane. If the membrane is flexible, such as the one used by *Nagakura* et al. [9.71], then the transfer of liquid would cause the membrane to deform and act as an actuator. The inherent drawback of using osmosis as an actuation mechanism is that it is a very slow process: Typical response times (on a macroscale) are on the order of several hours. However, osmotic transport scales favorably to the microscale, and it is expected that these devices will have response times on the order of several minutes, rather than hours. Based on this idea, *Nagakura* et al. [9.71] are developing a miniature insulin pump. *Su* et al. [9.72] have demonstrated a microscale osmotic actuator that is capable of developing pressures as high as 35 MPa. This is still a relatively unexplored realm in BioMEMS actuation, and it has good potential for applications such as sustained drug delivery.

Passive Pumping Based on Surface Tension

Glenn et al. have demonstrated pumping action using the difference between the surface tension pressure at the inlet and outlet of a microfluidic channel. In the simplest case, a small drop of a fluid is placed at one end of a straight microchannel, and a much larger drop of fluid is placed at the opposite end of the microchannel. The pressure within the small drop is significantly higher than the pressure within the large drop, due the difference in the surface tension effects across the two drops. Consequently, the liquid will flow from the small drop and add to the larger drop. The flow rate can be varied by changing various parameters such as the volume of the pumping drop, the surface free energy of the liquid, or the resistance of the microchannel, etc. This pumping scheme is very easy to realize and can be used for a wide variety of fluids.

Evaporation-Based Continuous Micropump

Effenhauser et al. [9.73] have demonstrated a continuous flow micropump based on a controlled evaporation approach. Their concept is based on the controlled evaporation of a liquid through a membrane into a gas reservoir. The reservoir contains a suitable adsorption agent that draws out the liquid vapors and maintains a low vapor pressure conducive to further evaporation. If the liquid being pumped is replenished from a reservoir, capillary forces will ensure that the fluid is continuously pumped through the microchannels as it evaporates on the other end into the adsorption reservoir. Though the pump suffers from inherent disadvantages such as strong temperature dependence and operation only in suction mode, it offers a very simple technique for fluidic transport.

9.3.6 Advantages and Disadvantages of the Passive Microfluidic Approach

This chapter has covered a number of different passive microfluidic devices and systems. Passive microfluidic devices have only recently been a subject of considerable research effort. One of the reasons for this interest is the long list of advantages that passive microfluidic devices offer. However, since most microfluidic devices are very application-specific (and even more so for passive microfluidic devices), the advantages are not to be considered universally applicable for all the devices/systems. Some of the advantages that are commonly found are:

- avoid the need for an "active" control system;
- usually very easy to fabricate;
- the passive microfluidic systems with no moving parts are inherently more reliable because of the lack of mechanical wear and tear;
- offer very repeatable performance once the underlying phenomena are well understood and characterized;
- highly suited for BioMEMS applications – can easily handle a limited number of microfluidic manipulation sequences;
- well suited for low cost, mass production approach;

- low cost offers the possibility of having disposable microfluidic systems for specific applications, such as working with blood;
- can offer other interesting possibilities, such as biomimetic response.

However, like all MEMS devices, passive microfluidic devices or systems are not the solution to the microfluidic handling problem. Usually they are very application-specific – they cannot be reconfigured for another application easily. Other disadvantages are listed below:

- suited for well understood, niche applications for which the fluidic sequencing steps are well decided,
- strongly dependent on variances in the fabrication process,
- usually not very suitable for a wide range of fluidic mediums.

9.4 Lab-on-a-Chip for Biochemical Analysis

Recent development in MEMS (microelectromechanical systems) has brought a new and revolutionary tool in biological or chemical applications: "lab-on-a-chip". New terminology, such as micro total analysis systems and lab-on-a-chip, was introduced in the last decade, and several prototype systems have been reported.

The idea of lab-on-a-chip is basically to reduce biological or chemical laboratories to a microscale system, hand-held size or smaller. Lab-on-a-chip systems can be made out of silicon, glass, and polymeric materials, and the typical microfluidic channel dimensions are in the range of several tens to hundreds of μm. Liquid samples or reagents can be transported through the microchannels from reservoirs to reactors using electrokinetic, magnetic, or hydrodynamic pumping methods. Fluidic motions or biochemical reactions can also be monitored using various sensors, which are often used for biochemical detection of products.

There are many advantages to using lab-on-a-chip over conventional chemical or biological laboratories. One of the important advantages lies in its low cost. Many reagents and chemicals used in biological and chemical reactions are expensive, so the prospect of using very small amounts (in micro- to nanoliter range) of reagents and chemicals for an application is very appealing. Another advantage is that lab-on-a-chip requires very small amounts of reagents/chemicals (which enables rapid mixing and reaction) because biochemical reaction is mainly involved in the diffusion of two chemical or biological reagents, and microscale fluidics reduces diffusion time as it increases reaction probabilities. In practical terms, reaction products can be produced in a matter of seconds/minutes, whereas laboratory scale can take hours, or even days. In addition, lab-on-a-chip systems minimize harmful by-products since their volume is so small. Complex reactions of many reagents could happen on lab-on-a-chip that have ultimate potential in DNA analysis, biochemical warfare agent detection, biological cell/molecule sorting, blood analysis, drug screening/development, combinatorial chemistry, and protein analysis. In this section, two recent developments of microfluidic systems for lab-on-a-chip applications will be introduced: (a) magnetic micro/nano bead-based biochemical detection system and (b) disposable smart lab-on-a-chip for blood analysis.

9.4.1 Magnetic Micro/Nano Bead-Based Biochemical Detection System

In the past few years, a large number of microfluidic prototype devices and systems have been developed, specifically for biochemical warfare detection systems and portable diagnostic applications. The BioMEMS team at the University of Cincinnati has been working on the development of a remotely accessible generic microfluidic system for biochemical detection and biomedical analysis, based on the concepts of surface-mountable microfluidic motherboards, sandwich immunoassays, and electrochemical detection techniques [9.74, 75]. The limited goal of this work is to develop a generic MEMS-based microfluidic system and to apply the fluidic system to detect bio-molecules, such as specific proteins and/or antigens, in liquid samples. Figure 9.22 illustrates the schematic diagram of a generic microfluidic system for biochemical detection using a magnetic bead approach for both sampling and manipulating the target bio-molecules [9.76, 77].

The analytical concept is based on sandwich immunoassay and electrochemical detection [9.78], as illustrated in Fig. 9.23. Magnetic beads are used as both substrates of antibodies and carriers of target antigens.

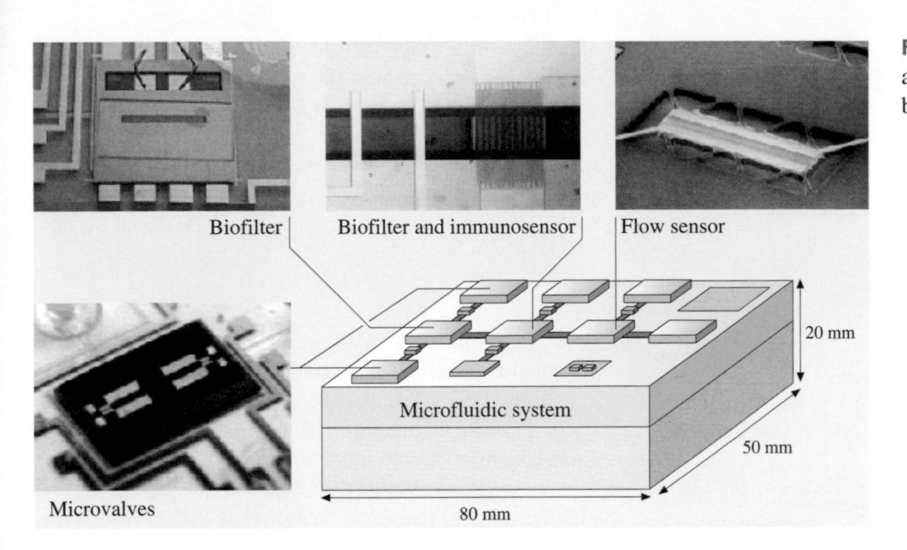

Fig. 9.22 Schematic diagram of a generic microfluidic system for biochemical detection (after [9.75])

A simple concept of magnetic bead-based bio-sampling with electromagnet for the case of sandwich immunoassay is shown in Fig. 9.24.

Antibody-coated beads are introduced on the electromagnet and separated by applying magnetic fields. While holding the antibody-coated beads, antigens are injected into the channel. Only target antigens are immobilized and, thus, separated on the magnetic bead surface due to antibody/antigen reaction. Other antigens get washed out with the flow. Next, enzyme-labeled secondary antibodies are introduced and incubated, along with the immobilized antigens. The chamber is then rinsed to remove all unbound secondary antibodies. Substrate solution, which will react with enzyme, is injected into the channel, and the electrochemical detection is performed. Finally, the magnetic beads are released to the waste chamber, and the bio-separator is ready for another immunoassay. Alkaline phosphatase (AP) and p-aminophenyl phosphate (PAPP) were chosen as enzyme and electrochemical substrate, respectively. Alkaline phosphatase makes PAPP turn into its electrochemical product, p-aminophenol (PAP). By applying potential, PAP gives electrons and turns into 4-quinoneimine (4QI), which is the oxidant form of PAP.

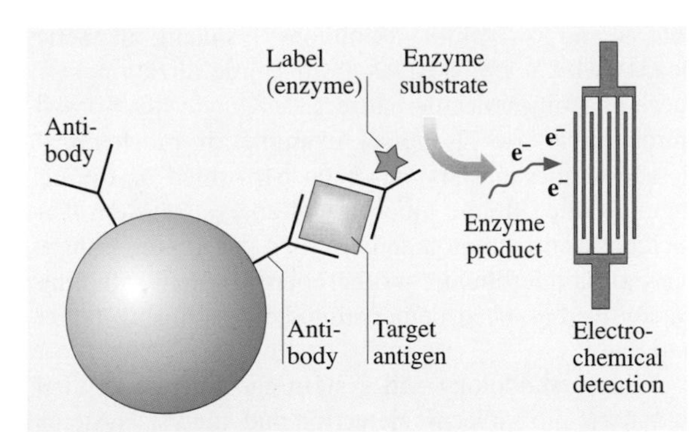

Fig. 9.23 Analytical concept based on sandwich immunoassay and electrochemical detection. After [9.77], reproduced by permission of The Royal Society of Chemistry

For the successful immunoassay, the biofilter [9.76] and the immunosensor were fabricated separately and integrated together. The integrated biofilter and immunosensor were surface-mounted using a fluoropolymer bonding technique [9.79] on a microfluidic motherboard, which contains microchannels fabricated by glass etching and glass-to-glass direct bonding technique. Each inlet and outlet were connected to sample reservoirs through custom-designed microvalves. Figure 9.25 shows the integrated microfluidic biochemical detection system for magnetic bead-based immunoassay.

After a fluidic sequencing test, full immunoassays were performed in the integrated microfluidic system to prove magnetic bead-based biochemical detection and sampling function. Magnetic beads (Dynabeads® M-280, Dynal Biotech Inc.) coated with biotinylated sheep anti-mouse immunoglobulin G (IgG) were injected into the reaction chamber and separated on the surface of the biofilter by applying magnetic fields. While holding the magnetic beads, antigen (mouse IgG) was injected into the chamber and incubated. Then

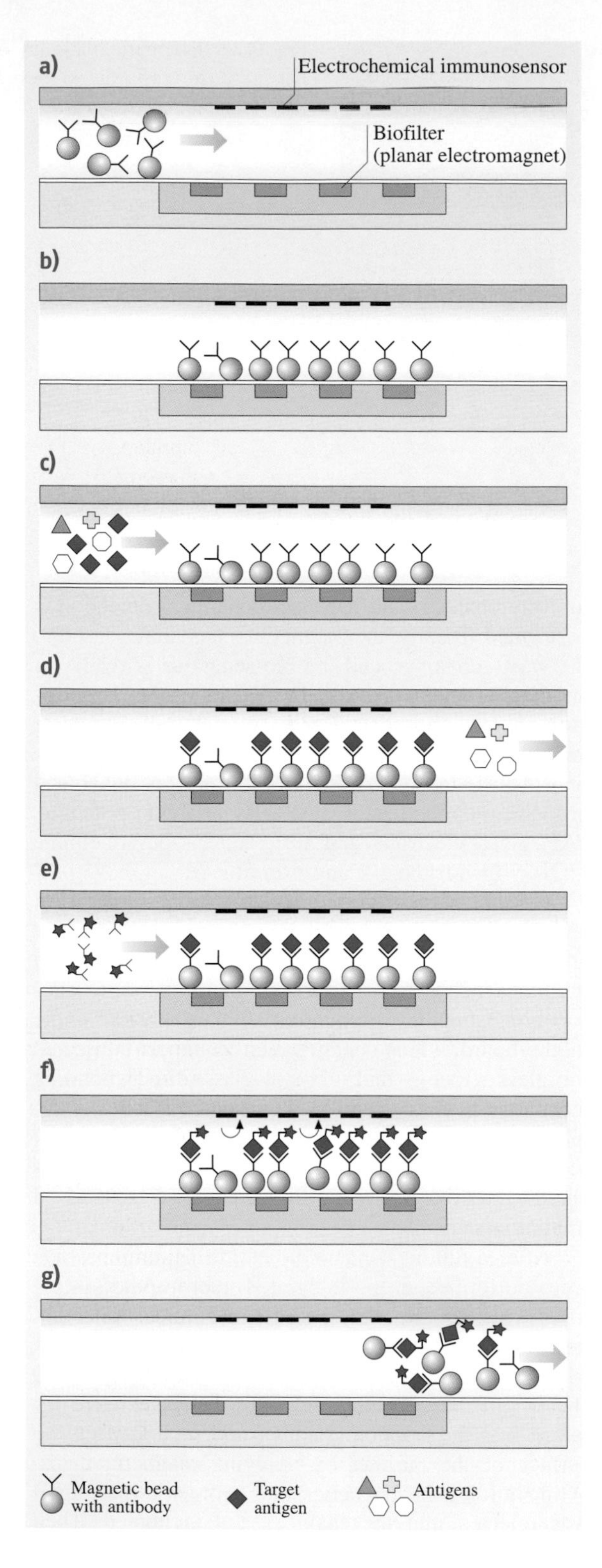

Fig. 9.24a–g Conceptual illustration of bio-sampling and immuno-assay procedure: (**a**) injection of magnetic beads; (**b**) separation and holding of beads; (**c**) flowing samples; (**d**) immobilization of target antigen; (**e**) flowing labeled antibody; (**f**) electrochemical detection; and (**g**) washing out magnetic beads and ready for another immunoassay (after [9.77], reproduced by permission of The Royal Society of Chemistry)

secondary antibody with label (rat anti-mouse IgG conjugated alkaline phosphatase) and electrochemical substrate (PAPP) to alkaline phosphatase was sequentially injected and incubated to ensure production of PAP. Electrochemical detection using an amperometric time-based detection method was performed during incubation. After detection, magnetic beads with all the reagents were washed away, and the system was ready for another immunoassay. This sequence was repeated for every new immunoassay. The flow rate was set to $20\,\mu\text{l/min}$ in every step. After calibration of the electrochemical immunosensor, full immunoassays were performed following the sequence stated above for different antigen concentrations: 50, 75, 100, 250, and $500\,\text{ng/ml}$. Concentration of primary antibody-coated magnetic beads and conjugated secondary antibody was 1.02×10^{7} beads/ml and $0.7\,\mu\text{g/ml}$, respectively. Immunoassay results for different antigen concentrations are shown in Fig. 9.26.

Immunoreactant consumed during one immunoassay was $10\,\mu\text{l}$ ($20\,\mu\text{l/min}\times30\,\text{s}$), and total assay time was less than 20 min, including all incubation and detection steps.

The integrated microfluidic biochemical detection system has been successfully developed and fully tested for fast and low volume immunoassays using magnetic beads, which are used as both immobilization surfaces and bio-molecule carriers. Magnetic bead-based immunoassay, as a typical example of biochemical detection and analysis, has been performed on the integrated microfluidic biochemical analysis system that includes a surface-mounted biofilter and immunosensor on a glass microfluidic motherboard. Protein sampling capability has been demonstrated by capturing target antigens.

The methodology and system can also be applied to generic bio-molecule detection and analysis systems by replacing the antibody/antigen with appropriate bio receptors/reagents, such as DNA fragments or oligonucleotides, for the application to DNA analysis and/or high throughput protein analysis.

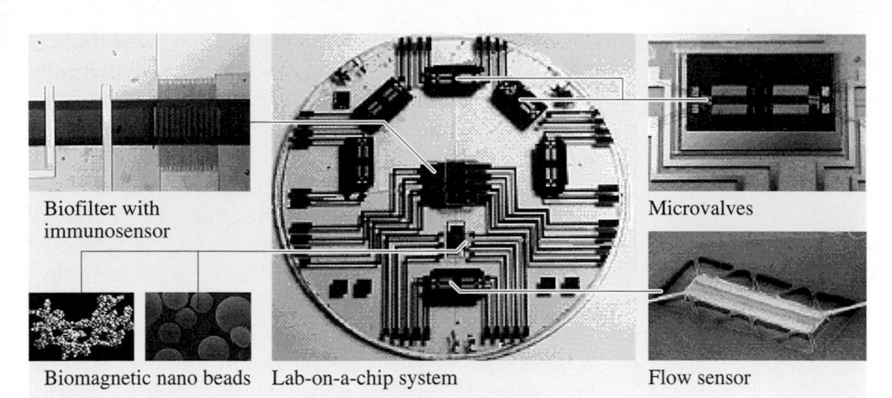

Fig. 9.25 Photograph of the fabricated lab-on-a-chip for magnetic bead-based immunoassay

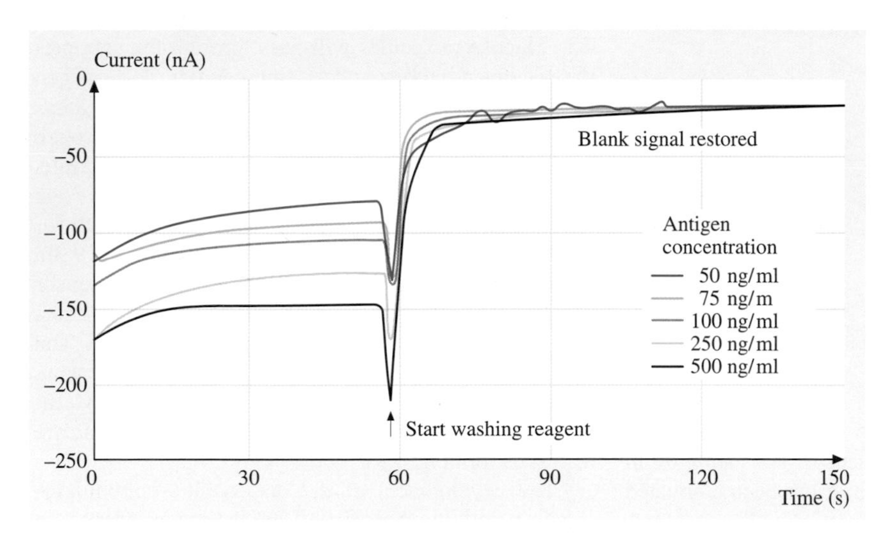

Fig. 9.26 Immunoassay results measured by amperometric time-based detection method (after [9.77], reproduced by permission of The Royal Society of Chemistry)

9.4.2 Disposable Smart Lab-on-a-Chip for Blood Analysis

One of several substrates available for bio-fluidic chips, plastics have recently become one of the most popular and promising substrates due to their low cost, ease of fabrication, and favorable biochemical reliability and compatibility. Plastic substrates, such as polyimide, PMMA, PDMS, polyethylene, or polycarbonate, offer a wide range of physical and chemical material parameters for the applications of bio-fluidic chips, generally at low cost using replication approaches. The disposable smart plastic biochip is composed of integrated modules of plastic fluidic chips for fluid regulation, chemical and biological sensors, and electronic controllers. As a demonstration vehicle, the biochip has the specific goal of detecting and identifying three metabolic parameters such as PO_2 (partial pressure of oxygen), lactate, and glucose from a blood sample. The schematic concept of the cartridge-type disposable lab-on-a-chip for blood analysis is illustrated in Fig. 9.27. The disposable lab-on-a-chip cartridge has been fabricated using plastic micro-injection molding and plastic-to-plastic direct bonding techniques. The biochip cartridge consists of a fixed volume microdispenser based on the sPROMs (structurally programmable microfluidic system) technique [9.70], an air-bursting, on-chip pressure source [9.80], and electrochemical biosensors [9.81].

A passive microfluidic dispenser measures exact amounts of sample to be analyzed, and then the air-bursting on-chip power source is detonated to push the graduated sample fluid from the dispenser reservoir. Upon air bursting, the graduated sample fluid travels through the microfluidic channel into sensing reservoirs, under which the biosensor array is located, as shown in Fig. 9.28.

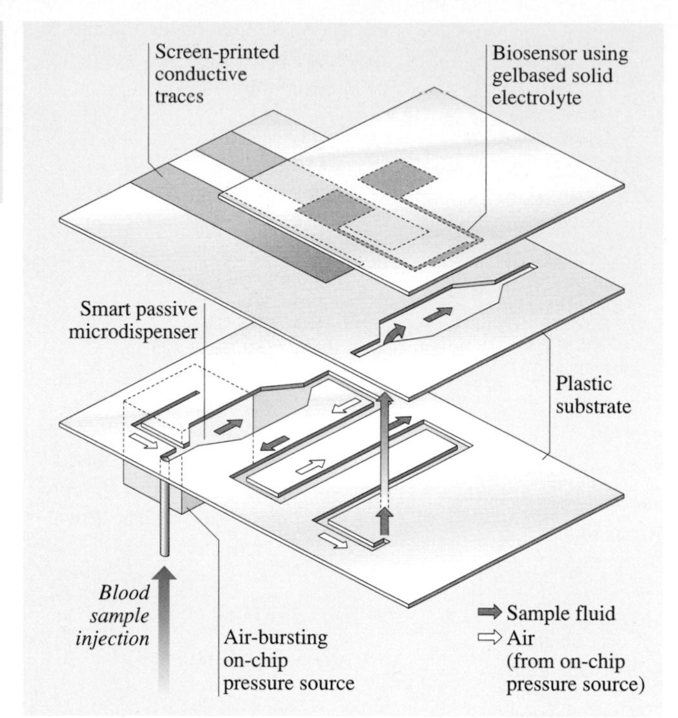

Fig. 9.27 Schematic illustration of smart and disposable plastic lab-on-a-chip by *Ahn* et al. [9.82]

An array of disposable biosensors consisting of an oxygen sensor and a glucose sensor has been fabricated using screen-printing technology [9.81]. Measurements from the developed biosensor array can be done based on tiny amounts of sample (as low as 100 nl). The glucose sensor is designed by adding two additional layers above the oxygen sensor (Fig. 9.29), which are glucose semipermeable membrane and glucose oxidase (GOD) layers.

The principle of the oxygen sensor is based on amperometric detection that when the oxygen diffusion profile from the sample to electrode surface is saturated, which means a constant oxygen gradient distribution profile is generated, the current is proportional only to the oxygen concentration in the solution. A silicone layer is spin-coated and utilized as an oxygen semipermeable membrane because of its high permeability and low signal-to-noise ratio. Water molecules pass through the silicone membrane and reconstitute the gel-based electrolyte so the Cl^- ions can move close to the anode to coalesce with Ag^+. The number of electrons in this reaction is counted by the measuring system.

For the glucose sensor, the additional layers – glucose semipermeable membrane (polyurethane) and the immobilized GOD (in a polyacrylamide gel) – allow direct conversion of the oxygen sensor to a glucose sensor. The glucose molecules will pass through the semipermeable layer and be oxidized immediately. The oxygen sensor will measure hydrogen peroxide, which is a by-product of glucose oxidation. The level of hydrogen peroxide is proportional to the glucose level in sample. The fabricated disposable plastic lab-on-a-chip cartridge was inserted into a hand-held type biochip analyzer for analysis of human blood sample, as shown in Fig. 9.30. The prototype biochip analyzer consists of biosensor detection circuitry, timing/sequence circuitry for the air-bursting, on-chip power source, and a display unit. The battery-operated biochip analyzer initiated the sensing sequence and displayed readings in one minute. Measured glucose and PO_2 levels in human blood sample are also shown in Fig. 9.31.

The development of the disposable smart microfluidic-based biochips is of immediate relevance to several patient-monitoring systems, specifically for point-of-care health monitors. Since the developed biochip is a low cost, plastic-based system, we envision a disposable application for monitoring clinically significant parameters such as PO_2, glucose, lactate, hematocrit, and pH. These health indicators provide an early warning system for the detection of patient status and can also serve as markers for disease and toxicity monitoring.

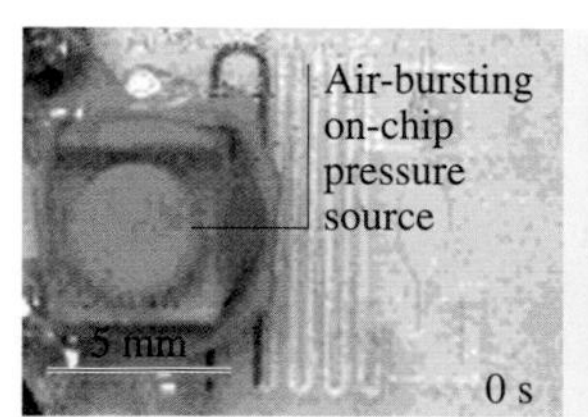

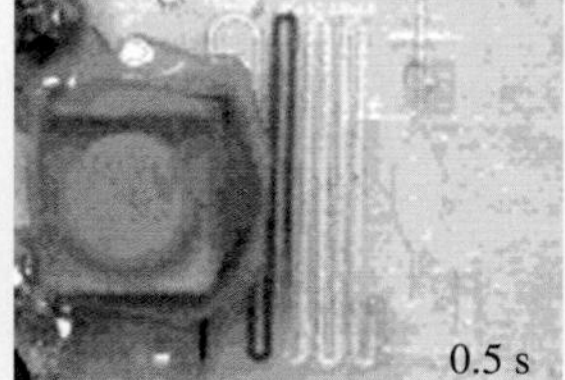

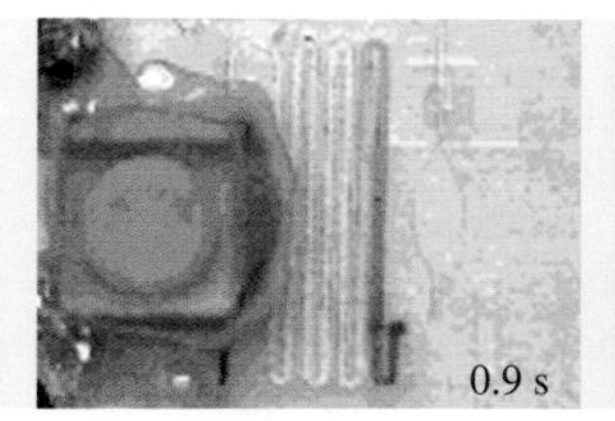

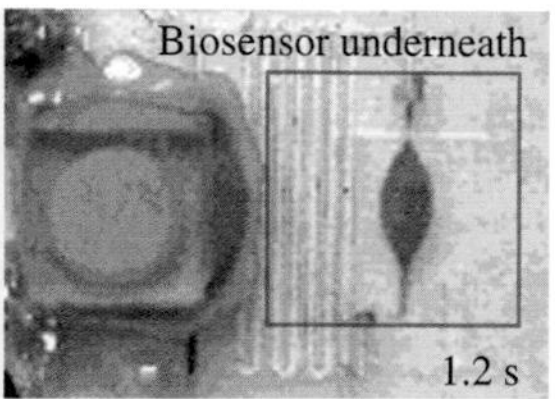

Fig. 9.28 Upon air-bursting, sample fluid travels through the microchannel into biosensor detection chamber. Biosensor was detached for clear micro-photographs. Dotted line indicates where biosensor is supposed to be attached (after [9.82])

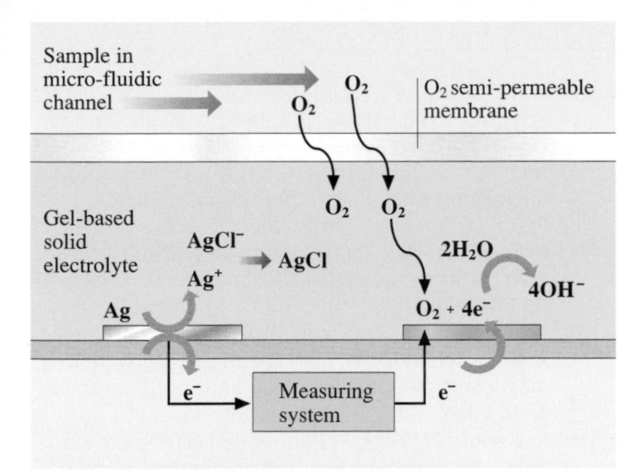

Fig. 9.29 Electrochemical and analytical principle of the developed disposable biosensor for partial oxygen concentration sensing (after, © 2002 IEEE [9.81])

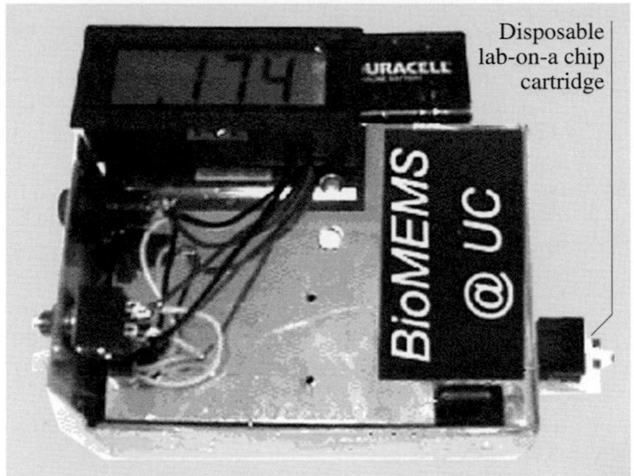

Fig. 9.30 Hand-held type disposable smart lab-on-a-chip analyzer (after [9.82])

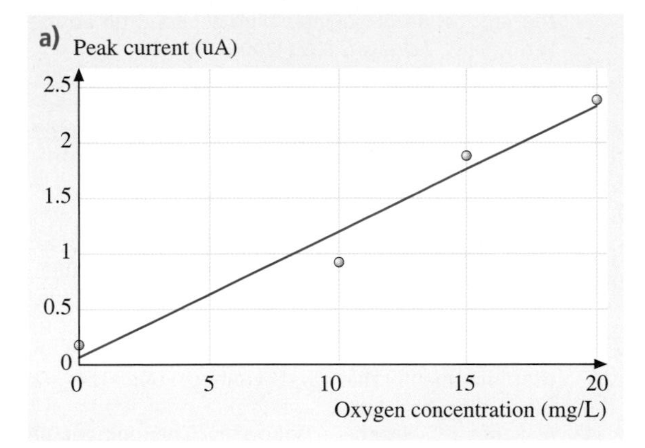

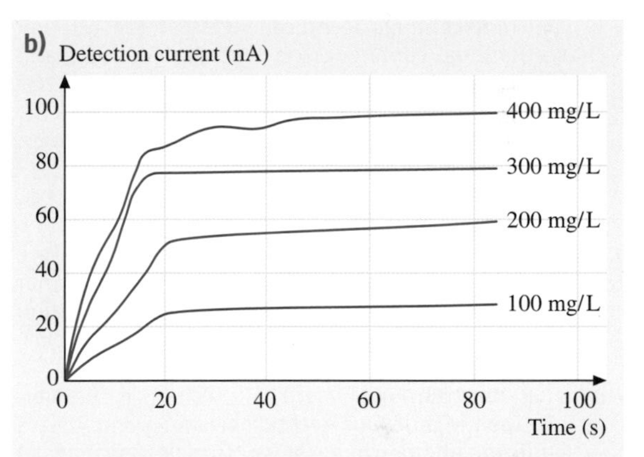

Fig. 9.31a,b Measurement results from the biochip cartridge and analyzer: (**a**) PO_2 calibration curve and (**b**)glucose level (after [9.81])

References

9.1 L. Ristic, H. Hughes, F. Shemansky: Bulk micromachining technology. In: *Sensor Technology and Devices*, ed. by L. Ristic (Artech House, Norwood 1994) pp. 49–93

9.2 M. Esashi, M. Takinami, Y. Wakabayashi, K. Minami: High-rate directional deep dry etching for bulk silicon micromachining, J. Micromech. Microeng. **5** (1995) 5–10

9.3 H. Jansen, M. de Boer, R. Legtenberg, M. Elwenspoek: The black silicon method: A universal method for determining the parameter setting of a fluorine-based reactive ion etcher in deep silicon trench etching with profile control, J. Micromech. Microeng. **5** (1995) 115–120

9.4 Z. L. Zhang, N. C. MacDonald: A RIE process for submicron silicon electromechanical structures, J. Micromech. Microeng. **2** (1992) 31–38

9.5 C. Linder, T. Tschan, N. F. de Rooij: Deep dry etching techniques as a new IC compatible tool for silicon micromachining, Technical Digest of the 6th International Conference on Solid-State Sensors and Actuators (Transducers '91), San Francisco June 24–27, 1991 (IEEE, Piscataway 1991) 524–527

9.6 D. J. Harrison, A. Manz, Z. Fan, H. Ludi, H. M. Widmer: Capillary electrophoresis and sample injection systems integrated on a planar glass chip, Anal. Chem. **64** (1992) 1926–1932

9.7 A. Manz, D. J. Harrison, E. M. J. Verpoorte, J. C. Fettinger, A. Paulus, H. Ludi, H. M. Widmer: Planar chips technology for miniaturization and integration of separation technique into monitoring systems, J. Chromatography **593** (1992) 253–258

9.8 T. R. Dietrich, W. Ehrfeld, M. Lacher, M. Krämer, B. Speit: Fabrication technologies for microsystems utilizing photoetchable glass, Microelectron. Eng. **30** (1996) 497–504

9.9 S. C. Jacobson, R. Hergenroder, L. B. Koutny, J. M. Ramsey: Open channel electrochromatography on a microchip, Anal. Chem **66** (1994) 2369–2373

9.10 M. G. Allen: Polyimide-based process for the fabrication of thick electroplated microstructures, Proc. of the 7th International Conference on Solid-State Sensors and Actuators (Transducers '93), Yokohama June 7–10, 1993 (IEE of Jpn., Tokyo 1993) 60–65

9.11 C. H. Ahn, M. G. Allen: A fully integrated surface micromachined magnetic microactuator with a multilevel meander magnetic core, IEEE/ASME J. Microelectromech. Syst. (MEMS) **2** (1993) 15–22

9.12 E. W. Becker, W. Ehrfeld, P. Hagmann, A. Maner, D. Munchmeyer: Fabrication of microstructures with high aspect ratios and great structural heights by synchrotron radiation lithography, galvanofoming, and plastic moulding (LIGA Process), Microelectronic Eng. **4** (1986) 35–36

9.13 C. S. Effenhauser, G. J. M. Bruin, A. Paulus, M. Ehrat: Integrated capillary electrophoresis on flexible silicone microdevices: Analysis of DNA restriction fragments and detection of single DNA molecules on microchips, Anal. Chem. **69** (1997) 3451–3457

9.14 D. C. Duffy, J. C. McDonald, O. J. A. Schueller, G. M. Whitesides: Rapid prototyping of microfluidic systems in poly(dimethylsiloxane), Anal. Chem. **70** (1998) 4974–4984

9.15 L. Martynova, L. Locascio, M. Gaitan, G. W. Kramer, R. G. Christensen, W. A. MacCrehan: Fabrication of plastic microfluid channels by imprinting methods, Anal. Chem. **69** (1997) 4783–4789

9.16 H. Becker, W. Dietz, P. Dannberg: Microfluidic manifolds by polymer hot embossing for micro-TAS applications, Proc. of Micro-Total Analysis Systems '98, Banff, Canada 1998, ed. by D. J. Harrison, A. van den Berg (Kluwer, Dordrecht 1998) 253–256

9.17 H. Becker, U. Heim: Hot embossing as a method for the fabrication of polymer high aspect ratio structures, Sens. Actuators A **83** (2000) 130–135

9.18 R. M. McCormick, R. J. Nelson, M. G. Alonso-Amigo, J. Benvegnu, H. H. Hooper: Microchannel electrophoretic separations of DNA in injection-molded plastic substrates, Anal. Chem. **69** (1997) 2626–2630

9.19 C. H. Ahn, A. Puntambekar, S. M. Lee, H. J. Cho, C.-C. Hong: Structurally programmable microfluidic systems, Proc. of Micro-Total Analysis systems 2000, Enschede 2000, ed. by A. van den Berg, W. Olthius, P. Bergveld (Kluwer, Dordrecht 2000) 205–208

9.20 W. Schomburg, B. Scherrer: 3.5 mm thin valves in titanium membranes, J. Micromech. Microeng. **2** (1992) 184–186

9.21 A. Ilzofer, B. Ritter, C. Tsakmasis: Development of passive microvalves by the finite element method, J. Micromech. Microeng. **5** (1995) 226–230

9.22 B. Paul, T. Terharr: Comparison of two passive microvalve designs for microlamination architectures, J. Micromech. Microeng. **10** (2000) 15–20

9.23 S. Shoji, M. Esashi: Microflow device and systems, J. Micromech. Microeng. **4** (1994) 157–171

9.24 G. T. A. Kovacs: *Micromachined Transducers Sourcebook* (McGraw-Hill, New York 1998)

9.25 W. Schomburg, J. Fahrenberg, D. Maas, R. Rapp: Active valves and pumps for microfluidics, J. Micromech. Microeng. **3** (1993) 216–218

9.26 K. Hosokawa, R. Maeda: A pneumatically-actuated three way microvalve fabricated with polydimethylsiloxane using the membrane transfer technique, J. Micromech. Microeng. **10** (2000) 415–420

9.27 J. Fahrenberg, W. Bier, D. Maas, W. Menz, R. Ruprecht, W. Schomburg: A microvalve system fabricated by thermoplastic molding, J. Micromech. Microeng. **5** (1995) 169–171

9.28 C. Goll, W. Bacher, B. Bustgens, D. Maas, W. Menz, W. Schomburg: Microvalves with bistable buckled diaphragms, J. Micromech. Microeng **6** (1996) 77–79

9.29 A. Ruzzu, J. Fahrenberg, M. Heckele, T. Schaller: Multi-functional valve components fabricated by combination of LIGA process and high precision mechanical engineering, Microsystem Technol. **4** (1998) 128–131

9.30 K. Sato, M. Shikida: An electrostatically actuated gas valve with S-shaped film element, J. Micromech. Microeng. **4** (1994) 205–209

9.31 J. Robertson, K. D. Wise: A low pressure micromachined flow modulator, Sens. Actuators A **71** (1998) 98–106

9.32 D.J. Sadler, K.W. Oh, C.H. Ahn, S. Bhansali, H.T. Henderson: A new magnetically actuated microvalve for liquid and gas control applications, Proc. of the 6th International Conference on Solid-State Sensors and Actuators (Transducers '99), Sendai 1999 (IEE of Jpn., Tokyo 1999) 1812–1815

9.33 W. Wijngaart, H. Ask, P. Enoksson, G. Stemme: A high-stroke, high-pressure electrostatic actuator for valve applications, Sens. Actuators A **100** (2002) 264–271

9.34 T. Watanabe, H. Kuwano: A microvalve matrix using piezoelectric actuators, Microsystem Technol. **3** (1997) 107–111

9.35 M. Stehr, S. Messner, H. Sandmaier, R. Zangerle: The VAMP – A new device for handling liquids and gases, Sens. Actuators A **57** (1996) 153–157

9.36 W. Jackson, H. Tran, M. O'Brien, E. Rabinovich, G. Lopez: Rapid prototyping of active microfluidic components based on magnetically modified elastomeric materials, J. Vac. Sci. Technol. B **19** (2001) 596

9.37 M. Kohl, K. Skrobanek, S. Miyazaki: Development of stress-optimized shape memory microvalves, Sens. Actuators A **72** (1999) 243–250

9.38 C. Neagu, J. Gardeniers, M. Elwenspoek, J. Kelly: An electrochemical active valve, Electrochemica Acta **42** (1997) 3367–3373

9.39 K. Yoshida, M. Kikuchi, J. Park, S. Yokota: Fabrication of micro-electro-rheological valves (ER valves) by micromachining and experiments, Sens. Actuators A **95** (2002) 227–233

9.40 N. Jeon, D. Chiu, C. Wargo, H. Wu, I. Choi, J. Anderson, G.M. Whitesides: Design and fabrication of integrated passive valves and pumps for flexible polymer 3-dimensional microfluidic systems, Biomedical Microdevices **4** (2002) 117–121

9.41 M. Koch, N. Harris, A. Evans, N. White, A. Brunnschweiler: A novel micropump design with thick-film piezoelectric actuation, Meas. Sci. Technol. **8** (1997) 49–57

9.42 L. Cao, S. Mantell, D. Polla: Design and simulation of an implantable medical drug delivery system using microelectromechanical systems technology, Sens. Actuators A **94** (2001) 117–125

9.43 J. Park, K. Yoshida, S. Yokota: Resonantly driven piezoelectric micropump fabrication of a micropump having high power density, Mechatronics **9** (1999) 687–702

9.44 M. Koch, A. Evans, A. Brunnschweiler: The dynamic micropump driven with a screen printed PZT actuator, J. Micromech. Microeng. **8** (1998) 119–122

9.45 D. Xu, L. Wang, G. Ding, Y. Zhou, A. Yu, B. Cai: Characteristics and fabrication of NiTi/Si diaphragm micropump, Sens. Actuators A **93** (2001) 87–92

9.46 E. Makino, T. Mitsuya, T. Shibata: Fabrication of TiNi shape memory micropump, Sens. Actuators A **88** (2001) 256–262

9.47 H. Chou, M. Unger, S. Quake: A microfabricated rotary pump, Biomedical Microdevices **3** (2001) 323–330

9.48 R. Zengerle, J. Ulrich, S. Kulge, M. Richter, A. Richter: A bidirectional silicon micropump, Sens. Actuators A **50** (1995) 81–86

9.49 M. Carrozza, N. Croce, B. Magnani, P. Dario: A piezoelectric driven stereolithography-fabricated micropump, J. Micromech. Microeng. **5** (1995) 177–179

9.50 H. Andersson, W. Wijngaart, P. Nilsson, P. Enoksson, G. Stemme: A valveless diffuser micropump for microfluidic analytical systems, Sens. Actuators B **72** (2001) 259–265

9.51 F. Forster, R. Bardell, M. Afromowitz, N. Sharma: Design, fabrication and testing of fixed-valve micropumps, Proc. ASME Fluids Engineering Division, ASME Int. Mechanical Engineering Congress and Exposition **234** (1995) 39–44

9.52 G. Fuhr, T. Schnelle, B. Wagner: Travelling wave driven microfabricated electrohydrodynamic pumps for liquids, J. Micromech. Microeng. **4** (1994) 217–226

9.53 J. Jang, S. Lee: Theoretical and experimental study of MHD micropump, Sens. Actuators A **80** (2000) 84–89

9.54 A. Lemoff, A.P. Lee: An AC magnetohydrodynamic micropump, Sens. Actuators B **63** (2000) 178–185

9.55 J. DeSimone, G. York, J. McGrath: Synthesis and bulk, surface and microlithographic characterization of poly(1-butene sulfone)-g-poly(dimethylsiloxane), Macromolecules **24** (1994) 5330–5339

9.56 A. Wego, L. Pagel: A self-filling micropump based on PCB technology, Sens. Actuators A **88** (2001) 220–226

9.57 A. Terray, J. Oakey, D. Marr: Fabrication of colloidal structures for microfluidic applications, Appl. Phys. Lett. **81** (2002) 1555–1557

9.58 A. Puntambekar, J.-W. Choi, C.H. Ahn, S. Kim, V.B. Makhijani: Fixed-volume metering microdispenser module, Lab on a Chip **2** (2002) 213–218

9.59 M. Madou, Y. Lu, S. Lai, C. Koh, L. Lee, B. Wenner: A novel design on a CD disc for 2-point calibration measurement, Sens. Actuators A **91** (2001) 301–306

9.60 K. Handique, D. Burke, C. Mastrangelo, C. Burns: On-chip thermopneumatic pressure for discrete drop pumping, Anal. Chem. **73** (2001) 1831–1838

9.61 H. Andersson, W. Wijngaart, P. Griss, F. Niklaus, G. Stemme: Hydrophobic valves of plasma deposited octafluorocyclobutane in DRIE channels, Sens. Actuators B **75** (2001) 136–141

9.62 L. Low, S. Seetharaman, K. He, M. Madou: Microactuators toward microvalves for responsive controlled drug delivery, Sens. Actuators B **67** (2000) 149–160

9.63 Q. Yu, J. Bauer, J. Moore, D. Beebe: Responsive biomimetic hydrogel valve for microfluidics, Appl. Phys. Lett. **78** (2001) 2589–2591

9.64 M. Mitchell, V. Spikmans, F. Bessoth, A. Manz, A. de Mello: Towards organic synthesis in microfluidic devices: Multicomponent reactions for the construction of compound libraries, Proc. of Micro-Total Analysis systems 2000, Enschede 2000, ed. by A. van den Berg, W. Olthius, P. Bergveld (Kluwer, Dordrecht 2000) 463–465

9.65 D. Beebe, R. Adrian, M. Olsen, M. Stremler, H. Aref, B. Jo: Passive mixing in microchannels: Fabrication and flow experiments, Mecanique & Industries, **2** (2001) 343–348

9.66 C.-C. Hong, J.-W. Choi, C. H. Ahn: A novel in-plane passive micromixer using Coanda effect, Proc. of the μTAS 2001 Symposium, Monterey 2001, ed. by J. M. Ramsey, A. van den Berg (Kluwer, Dordrecht 2001) 31–33

9.67 J. Brody, P. Yager: Diffusion-based extraction in a microfabricated device, Sens. Actuators A **54** (1996) 704–708

9.68 A. Stroock, S. Dertinger, A. Ajdari, I. Mezic, H. Stone, G. M. Whitesides: Chaotic mixer for microchannels, Science **295** (2002) 647–651

9.69 B. Weigl, P. Yager: Microfluidic diffusion-based separation and detection, Science **283** (1999) 346–347

9.70 A. Puntambekar, J.-W. Choi, C. H. Ahn, S. Kim, S. Bayyuk, V. B. Makhijani: An air-driven fluidic multiplexer integrated with microdispensers, Proc. of the μTAS 2001 Symposium, Monterey 2001, ed. by J. M. Ramsey, A. van den Berg (Kluwer, Dordrecht 2001) 78–80

9.71 T. Nagakura, K. Ishihara, T. Furukawa, K. Masuda, T. Tsuda: Auto-regulated osmotic pump for insulin therapy by sensing glucose concentration without energy supply, Sens. Actuators B **34** (1996) 229–233

9.72 Y. Su, L. Lin, A. Pisano: Water-powered osmotic microactuator, Proc. of the 14th IEEE MEMS Workshop (MEMS 2001), Interlaken 2001 (IEEE, Piscataway 393–396

9.73 C. Effenhauser, H. Harttig, P. Kramer: An evaporation-based disposable micropump concept for continuous monitoring applications, Biomedical Devices **4** (2002) 27–32

9.74 C. H. Ahn, H. T. Henderson, W. R. Heineman, H. B. Halsall: Development of a generic microfluidic system for electrochemical immunoassay-based remote bio/chemical sensors, Proc. of Micro-Total Analysis Systems '98, Banff, Canada 1998, ed. by D. J. Harrison, A. van den Berg (Kluwer, Dordrecht 1998) 225–230

9.75 J.-W. Choi, K. W. Oh, A. Han, N. Okulan, C. A. Wijayawardhana, C. Lannes, S. Bhansali, K. T. Schlueter, W. R. Heineman, H. B. Halsall, J. H. Nevin, A. J. Helmicki, H. T. Henderson, C. H. Ahn: Development and characterization of microfluidic devices and systems for magnetic bead-based biochemical detection, Biomedical Microdevices **3** (2001) 191–200

9.76 J.-W. Choi, C. H. Ahn, S. Bhansali, H. T. Henderson: A new magnetic bead-based, filterless bio-separator with planar electromagnet surfaces for integrated bio-detection systems, Sens. Actuators B **68** (2000) 34–39

9.77 J.-W. Choi, K. W. Oh, J. H. Thomas, W. R. Heineman, H. B. Halsall, J. H. Nevin, A. J. Helmicki, H. T. Henderson, C. H. Ahn: An integrated microfluidic biochemical detection system for protein analysis with magnetic bead-based sampling capabilities, Lab on a Chip **2** (2002) 27–30

9.78 O. Niwa, Y. Xu, H. B. Halsall, W. R. Heineman: Small-volume voltammetric detection of 4-aminophenol with interdigitated array electrodes and its application to electrochemical enzyme immunoassay, Anal. Chem. **65** (1993) 1559–1563

9.79 K. W. Oh, A. Han, S. Bhansali, H. T. Henderson, C. H. Ahn: A low-temperature bonding technique using spin-on fluorocarbon polymers to assemble microsystems, J. Micromech. Microeng. **12** (2002) 187–191

9.80 C.-C. Hong, J.-W. Choi, C. H. Ahn: A disposable on-chip air detonator for driving fluids on point-of-care systems, Proc. of the 6th International Conference on Micro Total Analysis Systems (μ-TAS), Nara 2002, ed. by Y. Baba, S. Shoji, A. van den Berg (Kluwer, Dordrecht 2002)

9.81 C. Gao, J.-W. Choi, M. Dutta, S. Chilukuru, J. H. Nevin, J. Y. Lee, M. G. Bissell, C. H. Ahn: A fully integrated biosensor array for measurement of metabolic parameters in human blood, Proc. of the 2nd Second Annual International IEEE-EMBS Special Topic Conference on Microtechnologies in Medicine and Biology, Madison May 2–4, 2002, ed. by A. Dittmar, D. Beebe (IEEE, New York 2002) 223–226

9.82 C. H. Ahn, J.-W. Choi, A. Puntambekar, C.-C. Hong, X. Zhu, C. Gao, R. Trichur, S. Chilukuru, M. Dutta, S. Murugesan, S. Kim, Y.-S. Sohn, J. H. Nevin, G. Beaucage, J.-B. Lee, J. Y. Lee, M. G. Bissell: Disposable biochip cartridge for clinical diagnostics toward point-of-care systems, Proc. of the 6th International Conference on Micro Total Analysis Systems (μ-TAS), Nara 2002, ed. by Y. Baba, S. Shoji, A. van den Berg (Kluwer, Dordrecht 2002) 187–189

10. Therapeutic Nanodevices

Therapeutic nanotechnology offers minimally invasive therapies with high densities of function concentrated in small volumes, features that may reduce patient morbidity and mortality. Unlike other areas of nanotechnology, novel physical properties associated with nanoscale dimensionality are not the raison d'etre of therapeutic nanotechnology, whereas the aggregation of multiple biochemical (or comparably precise) functions into controlled nanoarchitectures is. Multifunctionality is a hallmark of emerging nanotherapeutic devices, and multifunctionality can allow nanotherapeutic devices to perform multi-step work processes, with each functional component contributing to one or more nanodevice subroutine such that, in aggregate, subroutines sum to a cogent work process. Canonical nanotherapeutic subroutines include tethering (targeting) to sites of disease, dispensing measured doses of drug (or bioactive compound), detection of residual disease after therapy and communication with an external clinician/operator. Emerging nanotherapeutics thus blur the boundaries between medical devices and traditional pharmaceuticals. Assembly of therapeutic nanodevices generally exploits either (bio)material self assembly properties or chemoselective bioconjugation techniques, or both. Given the complexity, composition, and the necessity for their tight chemical and structural definition inherent in the nature of nanotherapeutics, their cost of goods (COGs) might exceed that of (already expensive) biologics. Early therapeutic nanodevices will likely be applied to disease states which exhibit significant unmet patient need (cancer and cardiovascular disease), while application to other disease states well-served by conventional therapy may await perfection of nanotherapeutic design and assembly protocols.

10.1 **Definitions and Scope of Discussion** 280
10.1.1 Design Issues 281
10.1.2 Utility and Scope of Therapeutic Nanodevices 285

10.2 **Synthetic Approaches: "top-down" versus "bottom-up" Approaches for Nanotherapeutic Device Components** 285
10.2.1 Production of Nanoporous Membranes by Microfabrication Methods: A top-down Approach 285
10.2.2 Synthesis of Poly(amido) Amine (PAMAM) Dendrimers: A bottom-up Approach 286
10.2.3 The Limits of top-down and bottom-up Distinctions with Respect to Nanomaterials and Nanodevices 287

10.3 **Technological and Biological Opportunities** 288
10.3.1 Assembly Approaches 288
10.3.2 Targeting: Delimiting Nanotherapeutic Action in Three-Dimensional Space 296
10.3.3 Triggering: Delimiting Nanotherapeutic Action in Space and Time 298
10.3.4 Sensing Modalities 302
10.3.5 Imaging Using Nanotherapeutic Contrast Agents 304

10.4 **Applications for Nanotherapeutic Devices** 307
10.4.1 Nanotherapeutic Devices in Oncology 307
10.4.2 Cardiovascular Applications of Nanotherapeutics 310
10.4.3 Nanotherapeutics and Specific Host Immune Responses 311

10.5 **Concluding Remarks: Barriers to Practice and Prospects** 315
10.5.1 Complexity in Biology 315
10.5.2 Dissemination of Biological Information 315
10.5.3 Cultural Differences Between Technologists and Biologists 316

References 317

10.1 Definitions and Scope of Discussion

Nanotechnology is a field in rapid flux and development, as cursory examination of this volume shows, and definition of its meets and bounds, as well as identification of sub-disciplines embraced by it, can be elusive. The word means many things to many people, and aspects of multiple disciplines, from physics to information technology to biotechnology, legitimately fall into the intersection of the Venn diagram of disciplines that defines nanotechnology. The breadth of the field allows almost any interested party to contribute to it, but the same ambiguity can render the field diffuse and amorphous. If nanotechnology embraces everything, what then is it? The ambiguity fuels cognitive dissonance that can result in frustrating interactions between investigators and funders, authors and editors, and entrepreneurs and investors. Some consideration of the scope of the field therefore is useful.

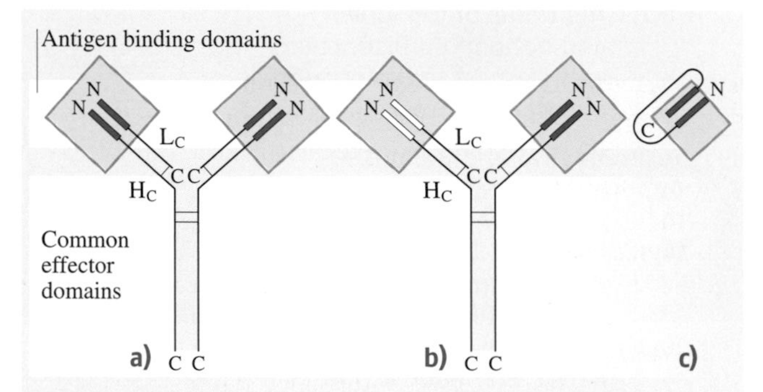

Fig. 10.1a–c Antibodies resemble purpose-built devices with distinct functional domains [10.1]. Native antibodies are composed of four polypeptide chains: two heavy chains (H_C) and two light chains (L_C), joined by interchain disulfide linkages (lines between H_C and L_C moieties). Amino and carboxy termini of individual polypeptide chains are indicated (by N and C). Antigen binding domains are responsible for specific antigen recognition, vary from antibody to antibody, and are indicated by the thicker lines. Common effector functions (F_C receptor binding, complement fixation, etc.) are delimited to domains of the antibodies that are constant from molecule to molecule. **(a)** A native IgG antibody is monospecific but bivalent in its antigen binding capacity. **(b)** An engineered, bispecific, bivalent antibody capable of recognizing two distinct antigens. **(c)** An engineered antibody fragment (single chain Fv or SCFv) that is monospecific and monovalent can recognize only one antigenic determinant and is engineered to lack common effector functions. This construct is translated as a single, continuous polypeptide chain (hence the name SCFv) because a peptide linker (indicated by the connecting line in the figure) is incorporated to connect the carboxy end of the H_C fragment and the amino end of the L_C fragment

To frame the discussion, we will define nanotechnology as the discipline that aims to satisfy desired objectives using materials and devices whose valuable properties are based on a specific nanometer-scale element of their structures. The field is unabashedly application-oriented, so its raison d'être is fulfillment of tasks of interest: technical information is important primarily to the extent that it bears on device design, function, or application.

The meaning of "therapeutic" is largely self-explanatory and refers here to intervention in *human* disease processes (although many of the approaches discussed are equally applicable to veterinary medicine). Our discussion will be confined primarily to therapeutics used in vivo, because such applications clearly benefit from the low invasiveness that ultra-small, but multipotent, nanotherapeutics potentially offer. It is debatable whether imaging, diagnostic, or sensing devices can be considered therapeutic in this context, though, as we will see, sensing/diagnostic functionalities are often inextricable elements of therapeutic nanodevices, and it is difficult to consider so-called smart nanotherapeutics without discussion of their sensing capabilities.

Our definition of nanotechnology projects several corollaries. First, it embraces macroscale structures whose useful properties derive from their nanoscale aspects. Second, the modifier "specific" (as in, "specific nanometer-scale elements") is intended to exclude materials whose utility derives solely from properties inherent in being finely divided (high surface-to-volume ratios, for instance), or other bulk chemical and physical properties. We made the exclusion based on our assessment that therapeutic nanodevices are more intriguing than nanomaterials per se (see below), though we will engage these attributes where they are germane to specific devices or therapeutic applications. Third, our definition implies that limited nanotechnology has been available since the 1970s in the form of biotechnology. Based on their nanoscale structures, individual biological macromolecules (such as proteins) often exhibit the coordinated, modular multifunctionality that is characteristic of purpose-built devices (Fig. 10.1). An analogous, but perhaps less persuasive, argument can be made that organic chemistry is an early form of nanotechnology. Compared to organic small molecules, protein functional capabilities and properties are generally more complex and extremely dependent on their conformation in three-dimensional space at nanometer scale. The nanotechnology sobriquet, therefore, may be more appropriate to biotechnology than organic chemistry.

Biological macromolecules rely on the deployment of specific chemical functionalities to specific relative distributions in space with nanometer (and greater) resolution for their function, so the inclusion of molecular engineering aspects of biotechnology practice under the nanotechnology rubric is legitimate, despite the discomfort it may cause traditionally trained engineers. As we will see, intervention in human disease often requires inclusion of biomolecules in therapeutic devices: frequently, no functional synthetic analogue of active proteins and nucleic acids is available.

As described above, this chapter focuses primarily on nanoscale therapeutic devices as opposed to therapeutic nanomaterials. Devices are integrated functional structures and not mixtures of materials. Devices exhibit desirable emergent properties inherent in their design: the properties emerge as the result of the spatial and/or temporal organization, and coordination and regulation of action of individual components. The organization of components in devices allows them to perform multi-step, cogent work processes that can't be mimicked by simple admixtures of individual components. In fact, if device functions can be mimicked well by simple mixtures of components, the labor involved in configuring and constructing a nanoscale device is not warranted. Our device definition thus excludes nanomaterials used as drug formulation excipients (pharmacologically inert materials included in formulations that improve pharmacophore uptake, biodistribution, pharmacokinetic, handling, storage, or other properties), but embraces those same materials as integral components of drug delivery or other clinical devices.

10.1.1 Design Issues

The biotechnology industry historically has focused on production of individual soluble protein and nucleic acid molecules for pharmaceutical use, with only limited attention paid to functional supramolecular structures [10.2–7]. This bias toward free molecules flies in the face of the obvious importance of integrated supramolecular structures in biology and, to the casual observer, may seem an odd gap in attention and emphasis on the part of practicing biotechnologists. The bias toward single molecule, protein therapeutics, however, follows from the fact that biotechnology is an industrial activity, governed by market considerations. Of the myriad potential therapeutics that might be realized from biotechnology, single protein therapeutics are among the easiest to realize from both technical and regulatory perspectives and so warrant extensive industrial attention. This is changing, however, and more complex entities (actual supramolecular therapeutic devices) have and will appear with increasing frequency in the twenty-first century.

New top-down and bottom-up materials derived from micro/nanotechnology provide the opportunity to complement the traditional limits of biotechnology by providing scaffolds that can support higher level organization of multiple biomolecules to perform work activities they could not perform as free, soluble molecules. Such supramolecular structures have been called nanobiotechnological devices [10.8], nanobiological devices [10.2–7], or semi-synthetic nanodevices and figure prominently in therapeutic nanotechnology.

Incorporation of / Interaction with Biomolecules

In general, design of nanodevices is similar to design of other engineered structures, providing that the special properties of the materials (relating to their nanoscale aspects such as quantum, electrical, mechanical, biological properties, etc.), as well as their impact in therapy, are considered. Therapeutics can interact with patients on multiple levels, ranging from organismal to molecular, but it is reasonable to expect that most nanotherapeutics will interface with patients at the nanoscale at least to some extent [10.2–5, 9–14]. Typically, this means interaction between therapeutics and biological macromolecules, supramolecular structures and organelles, which, in turn, often dictates the incorporation of biological macromolecules (and other biostructures) into nanodevices [10.2, 5, 13–15]. Incorporating biological structures into (nanobiological) devices presents special challenges that do not occur in other aspects of engineering practice.

Unlike fully synthetic devices, semi-biological nanodevices must incorporate pre-fabricated biological components (or derivatives thereof), and therefore the intact nanodevices are seldom made entirely de novo. As a corollary, knowledge of properties of biological device components is often incomplete (as they were not made by human design), and therefore the range of activities inherent in any nanobiological device design may be much less obvious and less well-defined than it is for fully synthetic devices. Further complicating the issue, the activities of biological molecules are often multifaceted (many genes and proteins exhibit plieotropic activities), and the full range of functionality of individual biological molecules in interactions with other biological systems (as in nanotherapeutics) is often not known. This makes design and prototyping of biological nanodevices an empirically intensive, iterative process [10.3–5, 14].

Biological macromolecules have properties, particularly those relating to their stability, that can limit their use in device contexts. In general, proteins, nucleic acids, lipids, and other biomolecules are more labile to physical insult than are synthetic materials. With the possible exceptions of topical agents or oral delivery and endosomal uptake of nanotherapeutics (both involving exposure to low pH), patients can tolerate conditions encountered by nanobiological therapeutics in vivo, and device lability in the face of *physical* insult is generally a major consideration only in ex vivo settings (relating to storage, sterilization, ex vivo cell culture, etc.). Living organisms remodel themselves constantly in response to stress, development, pathology, and external stimuli. For instance, epithelial tissues and blood components are constantly eliminated and regenerated, and bone and vasculature are continuously remodeled. The metabolic facilities responsible (circulating and tissue-bound proteases and other enzymes, various clearance organs, the immune system, etc.) can potentially process biological components of nanobiological therapeutic devices as well as endogenous materials, leading to partial or complete degradation of nanotherapeutic structure, function, or both. Furthermore, the host immune and wound responses protect the host against pathogenic organism incursions by mechanisms that involve sequestering and degrading the pathogens. Nanobiological therapeutics are subject to the actions of these host defense systems and to normal remodeling processes. As we will discuss, various strategies to stabilize biomolecules and structures in heterologous in vivo environments are applicable to nanobiological therapeutics [10.3, 16–20]. Conversely, instability of active biocomponents can offer a valuable and simple way to delimit the activity of nanotherapeutics containing biomolecules.

Nanotherapeutic Design Paradigms

Several early attempts to codify the canonical properties of ideal nanobiological devices, and therapeutic nanodevices in particular, have been made [10.5, 13, 15, 21] and are summarized in Table 10.1. In general, nanobiological devices contain biological components that retain their function in new (device) contexts. In other words, one must abstract enough of a functional biological unit from its native context to allow it to perform the function for which it was selected. If one wishes, for example, to appropriate the specific antigen recognition property of an antibody for a device function (say, in targeting, discussed later), it is not necessary to incorporate the entire 150,000 atomic mass unit (AMU) antibody, the bulk of which is devoted to functions other than antigen recognition (see Fig. 10.1 [10.1]), but it *is* critical to incorporate the approximately 20,000 AMU of the antibody essential for specific antibody–antigen

Table 10.1 Some ideal characteristics of nanodevices. (**A**) Characteristics of all nanobiological devices [10.2, 5, 13–15]. (**B**) Desirable characteristics of therapeutic platforms [10.2, 5, 13–15]

A
Biological molecules must retain function.
Device function is the result of the summed activities of device components.
The relative organization of device components drives device function.
Device functions can be unprecedented in the biological world.
B
Therapeutics should be minimally invasive.
Therapeutics should have the capacity to target sites of disease.
Therapeutics should be able to sense disease states in order to:
– report conditions at the disease site to clinicians.
– administer metered therapeutic interventions.
Therapeutic functions should be segregated into standardized modules.
Modules should be interchangeable to tune therapeutic function.

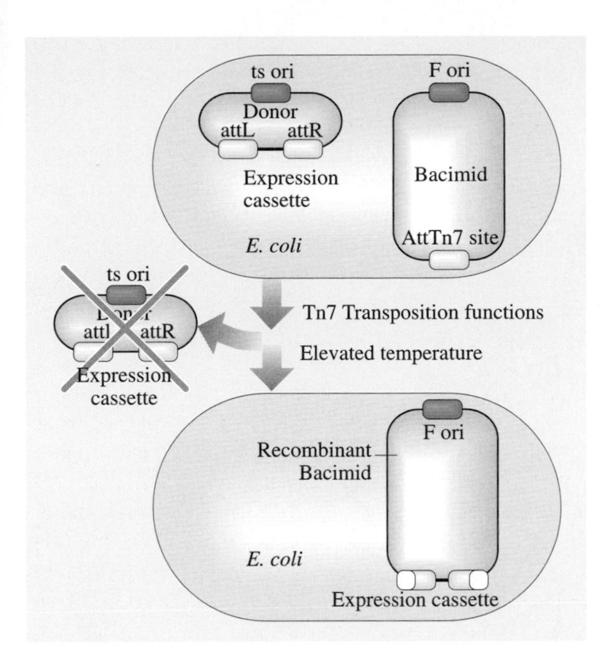

Fig. 10.2 The bacmid molecular cloning system is a molecular device designed to allow efficient production of recombinant insect viruses (baculovirus) in *Escherichia coli* [10.22–24]. Baculovirus is replicated in *E. coli* by the F plasmid origin of replication (F ori), and as such, is called a bacmid. The bacmid also includes an engineered transposable DNA element 7 (Tn7) attachment site isolated from the chromosome of an enteric bacteria (AttTn7). AttTn7 can receive Tn7 elements transposed from other cellular locations. A donor plasmid (donor) is replicated by a temperature-sensitive plasmid pSC101 origin of replication (ts ori). The donor also incorporates an expression cassette containing both the gene of interest for ultimate expression in insect cells and a selectable genetic marker operable in *E. coli*. The expression cassette is flanked by DNA sequences (attL and attR) that are recognized by the Tn7 transposition machinery. Tn7 transposition machinery resides elsewhere in the same *E. coli* cell. When donor plasmid is introduced into *E. coli* containing bacmid, Tn7 transposition machinery causes the physical relocation of expression cassettes from donor plasmid to bacmid. Unreacted donor plasmid is conveniently removed by elevating the incubation temperature, causing the ts pSC101 replicon to cease to function, in turn, causing the donor to be lost. If selection for the genetic markers within the expression cassette is applied at this point, the only *E. coli* that survive are those containing recombinant bacmid (i. e., those that have received the gene for insect cell expression by transposition from the donor). Recombinant bacmid are conveniently isolated from *E. coli* and introduced into insect cell culture, where expression of the gene of interest occurs

binding. Device function is the result of the summed and various activities of biological and synthetic device components, as well, though functional biological components generally exist in the context of higher order systems that support the organisms of which they are a part by the control the nanobiological device designer can exert on the relative organization of biological device components allows biomolecules abstracted from their native context and incorporated in nanobiotechnological devices to contribute to functions entirely different from those they performed in their organismal contexts. All of these features are illustrated in the bacmid, or Bac-to-Bac, system, a commercially available molecular cloning device ([10.22–24] and Fig. 10.2). This system configures prokaryotic genetic elements from multiple sources into a device for producing recombinant eukaryotic viruses, a function that is unprecedented in nature. The system is feasible because of the modularity of the genetic elements involved and because of the strict control of the relative arrangement of genetic elements allowed by recombinant DNA technology. Analogous devices based on bacterial and eukaryotic regulatory elements to pre-program the micro- and nanoscale architectural properties and physiological behavior of living things are now being realized [10.25, 26].

Bacmid provides an example of a nontherapeutic nanobiological device and illustrates some specific design approaches for building functional devices with biocomponents. Hypothetical properties of nanoscale devices specifically for therapeutic purposes have been codified (Table 10.1) and bear examination [10.13, 15].

In nanotherapeutic applications, devices should be noninvasive and target therapeutic payloads to sites of disease to maximize therapeutic benefit while minimizing undesired side effects. This of course implies the existence of therapeutic effector functions in these nanodevices, to give devices the ability to remediate a physiologically undesirable condition. Beyond that, several attributes relate to sensing of biomolecules, cells, or physical conditions (sensing disease itself, identification of residual disease, and, potentially, targeting capacity, responding to intrinsic or externally supplied triggers for payload release). Other properties relate to communication between device subunits (for instance, between sensor and effector domains of the device)

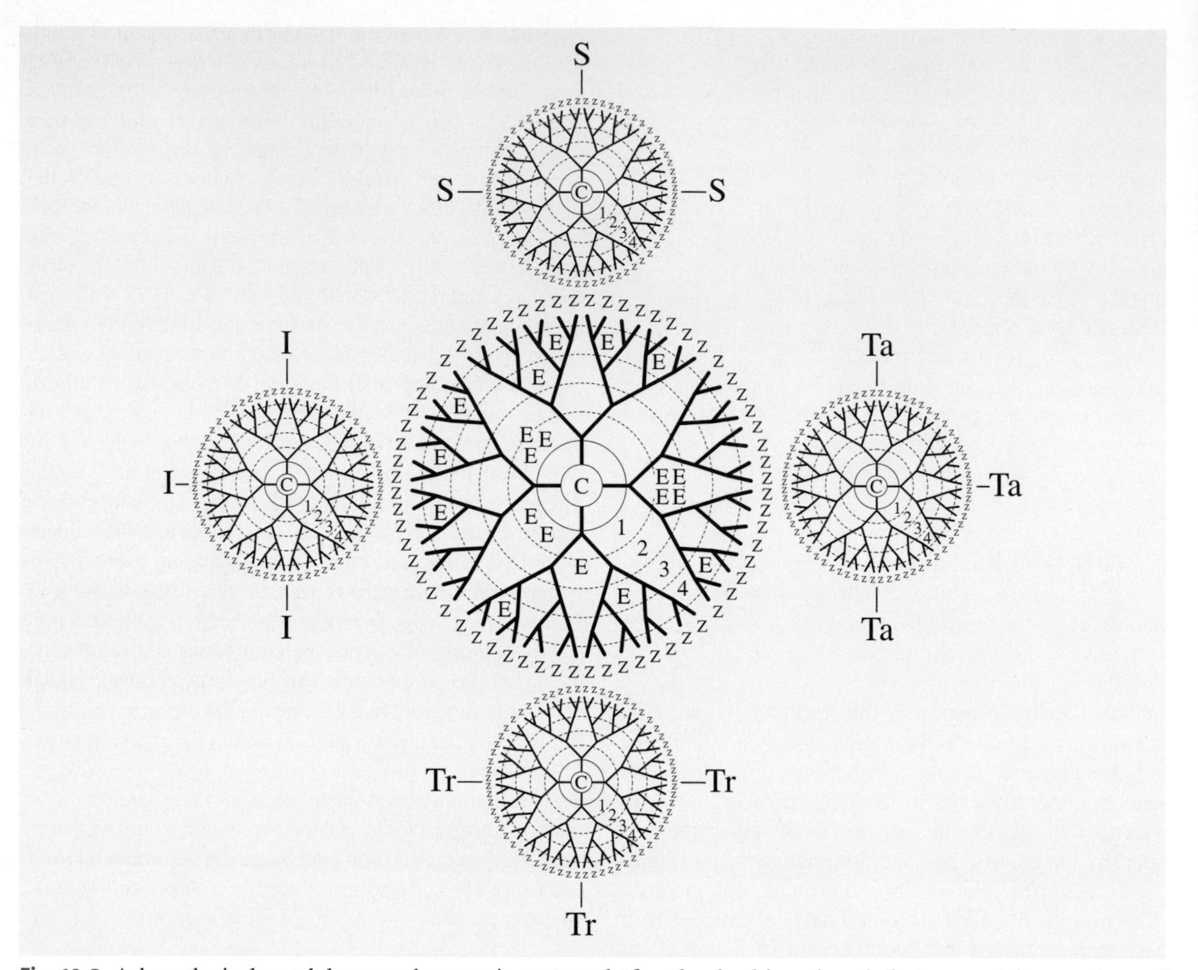

Fig. 10.3 A hypothetical, modular nanotherapeutic patterned after the dendrimer-based cluster agent for oncology of Baker [10.15]. As described in the text, each dendrimer subunit is grown from an initiator core (C), and the tunable surface groups of the dendrimers are represented by Z. Each dendrimer subunit has a specific, dedicated function in the device: the central dendrimer encapsulates small molecule therapeutics (E), whereas other functional components are segregated to other dendrimer components. These include biochemical targeting/tethering functions (Ta), therapeutic triggering functions to allow activation of prodrug portions of the device by an external operator (Tr), metal or other constituents for imaging (I), and sensing functions (S) to mediate intrinsically controlled activation/release of therapeutic. This design constitutes a therapeutic platform [10.13, 15] because of its modular design. The depicted device is only one possible configuration of an almost infinite number of analogous therapeutics that can be tuned to fit particular therapeutic needs by interchanging functional modules

or between the device and an external operator (external triggering and data documentation capability). With appropriate design, device functions can be modular [10.13, 15]. As discussed below, this approach allows construction of nanotherapeutic platforms (such as the dendrimer-based therapeutic of Fig. 10.3, dendrimer synthesis and assembly are discussed later), as opposed to single one-off devices that are capable of only one therapeutic task.

The vision of nanoscale therapeutic platforms arose from collaboration between the National Cancer Institute (NCI) and the National Aeronautics and Space Agency (NASA). NASA is concerned with minimal mass therapeutics: therapeutics, along with almost everything else used by astronauts, must be launched from Earth. NCI is interested in early detection of disease to improve prognosis. Since this requires screening a population predominantly of healthy patients, the screening mode must be minimally invasive. The considerations of both agencies might be met with ultra-small (micro- or nanoscale) multipotent therapeutic devices. Furthermore, the proposed therapeutic platforms should not only remediate undesired physiological conditions but also have the capacity to recognize them and report them. Extensive capability for molecular recognition and communication with external clinicians/operators is integral to the requirements of NASA and NCI. These capacities would allow drugs or other therapeutic interventions to be provided in a controlled fashion, to maximize benefit and minimize side effects, and is the essence of "smart" therapeutics (see the discussion of targeting and triggering below).

Though substantial progress has been made in device design and realization, no fully realized multipotent nanoscale therapeutic platform has yet been commercialized, but the therapeutic platform paradigm, in which devices are modular, with functional tasks segregated into individual modules, has potential to be extremely powerful. Classes of broadly similar devices could be tailored to specific disease states by interchanging modules (targeting modules, drugs dispensers, etc.) as appropriate to the disease state or therapeutic course. Thus, the hypothetical device represents a possible therapeutic platform composed of functional modules that could be for use in particular indications or in specific individuals.

10.1.2 Utility and Scope of Therapeutic Nanodevices

Therapeutic nanotechnology will be useful, of course, when the underlying biology of the disease states involved is amenable to intervention at the nanoscale. As we will discuss, while several disease states and physiological conditions (cancer, vaccination, cardiovascular disease, etc.) are particularly accessible to nanoscale interventions, some nanotechnological approaches may be applicable more broadly. Much as was the case with the introduction of recombinant protein therapeutics over the last 20 years, nanotherapeutics may present regulatory and pharmacoeconomic challenges related to their novelty and their cost of goods (COGs).

10.2 Synthetic Approaches: "top-down" versus "bottom-up" Approaches for Nanotherapeutic Device Components

Synthesis of nanomaterials is commonly thought of in terms of "top-down" or "bottom-up" processes. Top-down approaches begin with larger starting materials and, in a more or less controlled fashion (depending on the technique), remove material until the desired structure is achieved. Most microfabrication techniques for inorganic materials (lithography and milling techniques, etc.) fit this description. In contrast, bottom-up approaches begin with smaller subunits that are assembled, again with varying levels of control, depending on technique, into the final product. Key examples of materials made by bottom-up approaches include some inorganic structures. This includes "handmade" structures created by using direct atomic or molecular placement by force microscopy (discussed below), structures built using various deposition or growth methods, as well as all polymerization methods of synthesis. Thus, almost all biologic macromolecules and most biogenic structures, including mineralized biomaterials, are made by bottom-up methods. To make the distinction between top-down and bottom-up approaches more concrete, we will consider an example of each.

10.2.1 Production of Nanoporous Membranes by Microfabrication Methods: A top-down Approach

Lithography can be summarized by three basic steps: 1) Pattern Design (generation of masks), 2) Pattern Definition (exposure), and 3) Pattern Transfer (etching/liftoff). Optical lithography uses masks to form patterns on resist/substrate surfaces to produce features. The technique's power lies in its reproducibility and its capacity to manufacture via highly parallel processes. The key limitation of photolithography lies in the fact that resolution of features is diffraction-limited by the wavelength of light used. To address this limitation, short-wave radiation (i. e., X-rays with wavelengths of about 1nm wavelength) can be generated by synchrotron or other sources (from X-ray tubes, discharge plasma, or laser plasma), controlled, and focused for use in X-ray lithographic techniques [10.27]. The process is identical conceptually to optical lithography but requires special masks and resists amenable to the high-frequency radiation used. Combinations of filters and mirrors can produce resolutions in feature size of less than 100 nm, with fabrication throughputs congruent with those of other optical lithography processes. Limited access to synchrotron sources in turn limits the wide application of the method, however. The relatively long wavelengths used in conventional photolithography are generally unsuitable for formation of nanoscale features unless some clever technical expedient, like the use of a sacrificial layer (described below for generating nanoporous membranes) is employed.

As we will discuss in consideration of applications of nanotherapeutic devices, there are numerous potential applications for tunable nanopore membranes [10.16–18]. For certain applications, such as immunoisolation (see below), the distribution of pore sizes must be tight and nearly perfect. Until the late 1990s, this was an unattainable objective.

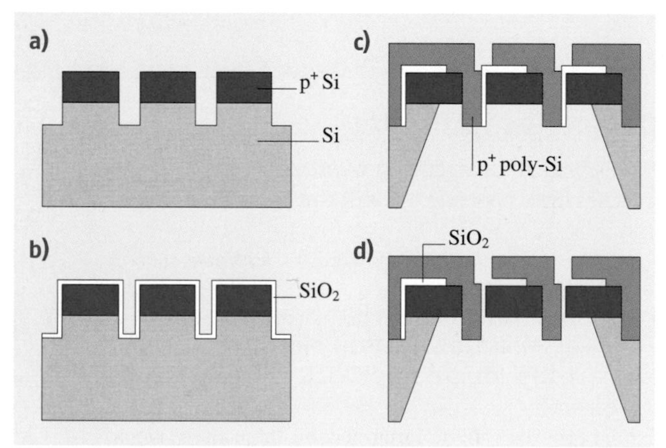

Fig. 10.4a–d Nanoporous silicon membranes are fabricated by a top-down approach [10.16–18]. **(a)** Pore exit holes are plasma-etched through p+ silicon (*dark brown*) and boronated silicon (*grey*). **(b)** Controlled depth silicon oxide (*white*) layer is grown on the structure. **(c)** A thick polysilicon layer (*light brown*) is deposited over the oxide, and pore exits are established by back-side plasma etch. **(d)** Pores are opened by removal of the sacrificial layer with hydrofluoric acid. The membranes themselves are macroscale objects that fit under the nanotechnology rubric because they derive their useful properties from controlled nanoscale architecture

The key technical innovation facilitating the microfabrication of highly defined nanoporous membranes was an approach featuring a sacrificial oxide layer sandwiched between two structural layers that ultimately is etched away to define the pore pathway and diameter [10.16–18]. The sacrificial layer (SiO_2) is sandwiched between a silicon wafer and a polysilicon layer: pore channel diameter is determined by the SiO_2 thickness. In the process (Fig. 10.4), the top of a silicon wafer is doped with boron to increase its mechanical robustness, and p+ silicon is overlaid. Pore exit holes are plasma etched through the p+ and doped material. In the pore dimension-determining step, SiO_2 is grown on the wafer by dry thermal oxidation, a process allowing control of oxide layer thickness to within 1 nm. This oxide constitutes the critical sacrificial layer. A thick polysilicon layer (again, boron-doped for mechanical strength) is then deposited over the oxide. Pore entry holes are etched by plasma etch, offset from the exit pores. Ultimately, the offset will require the diffusion pathway of the finished structure to pass through a "bottleneck" whose diameter is determined by the thickness of the sacrificial layer. The backside of the wafer is anisotropically etched to expose the doped layer of the wafer (now the bottom of the membrane structure). Pores are opened by removal of exposed sacrificial layer using concentrated hydrofluoric acid.

10.2.2 Synthesis of Poly(amido) Amine (PAMAM) Dendrimers: A bottom-up Approach

PAMAM dendrimers are remarkably defined synthetic molecules made using polymer chemistry (Fig. 10.5). Their unique structural attribute is their fractal geometry, and their unique physical property is their high monodispersity. Clever use of orthogonal conjugation strategies in their synthesis drives their monodispersity [10.28–30].

Orthogonal conjugation involves the mutually exclusive reactivity of the chemical specificities present in the reaction: when the orthogonal reactants are present in a vessel, and the reaction is carried at appropriate stoichiometry, only a monodisperse single product is formed. PAMAM dendrimers are among the most monodisperse synthetic materials available and, within limits, their sizes, surface chemistries, and shapes can be controlled at the synthetic level. To a greater or lesser extent, these desirable material properties of dendrimers are caused by an orthogonal synthetic strategy.

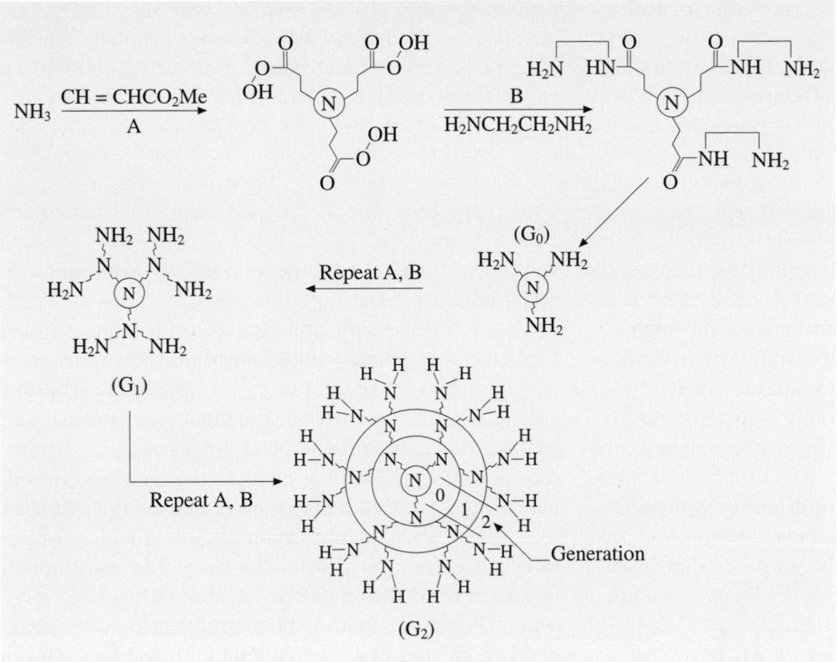

Fig. 10.5 PAMAM dendrimers are grown bottom-up using sequential orthogonal synthetic steps [10.28–30]. Dendrimers are grown from amine cores, in this case ammonia, by sequential addition of methyl acrylate, followed by an independent reaction addition reaction involving ethylene diamine. Together, these two half-steps constitute a full generation of dendrimer growth. PAMAM dendrimers are classified by their generation of growth as being G_0, G_1, G_2, etc., as indicated in the figure

PAMAM dendrimers are usually built from an initiator core (an amine) in sequential shells, called generations (i. e., G_0, G_1, G_2, etc.). Each generation comprises two synthetic half-steps, each of which is self-limiting in that the reagents performing each addition step are reactive only with the distinct chemical functionalities added to the growing dendrimer in the *previous* addition reaction. Reagent is added until the chemical functionalities from the previous step are fully consumed (Fig. 10.5). In the first synthetic half-step, methyl acrylate is added to the amine precusor, adding a carboxy-terminated functionality to all amine branch points. Polymer intermediates terminating in such carboxy functionalities are called "half-generation." In the second half-step, the generation is completed. Ethylene diamine is added, resulting in a branched amine-terminated adduct for each carboxylate group of the half-generation precursor. In general, before proceeding to downstream synthetic operations, products of individual reactions are purified so that only functionalities incorporated into growing dendrimer are available for subsequent reactions. Under ideal conditions, only a single chemical structure can result from each generation of growth, and with adequate purification between addition reactions; in the absence of limitations on the completeness of each reaction (such as steric limitations which become manifest in dendrimers of generation 6 and above), dendrimers of each generation are perfectly identical to each other.

10.2.3 The Limits of top-down and bottom-up Distinctions with Respect to Nanomaterials and Nanodevices

Some materials can be produced by alternate means, some of which are bottom-up, some of which are top-down approaches. For instance, carbon nanotubes can be synthesized in an arguably top-down approach from graphite sheets in an arc oven or can be grown bottom-up, by a metal-catalyzed polymerization method [10.31]. Additionally, not all finished materials can be classified as either top-down or bottom-up: synthetic protocols can contain both steps. For instance, while proteins are synthesized from lower molecular weight amino acid precursors by chemically or biologically mediated polymerization (bottom-up), they are often made as precursor molecules that are processed to a final product by chemical or enzymatic cleavage (top-down, see [10.32] for an overview). While the question of whether any given material is made top-down or bottom-up can be ambiguous, the synthetic provenance of multicomponent nanodevices can be even more so. In analogy to biogenic materials, synthetic polymers are bottom-up materials per se, but fabrication of raw polymeric materials into final device architectures often involves top-down steps [10.33]. As we will see, therapeutic nanodevices frequently contain both synthetic and biologic components, and strict top-down and bottom-up categorization of these devices is often not applicable.

10.3 Technological and Biological Opportunities

This section considers selected enablers for therapeutic nanodevices. Some are purely technological: nanomaterial self-assembly properties, bioconjugation methods, engineered polymers for conditional release of therapeutics, external triggering strategies, and so forth. Others relate to disease-state tissue or cell-specific biology that can be exploited by nanotherapeutics, such as the emerging vascular address system and intrinsic triggering approaches. In association with those applications, we will consider additional biological opportunities for nanoscale approaches specific to particular disease states.

10.3.1 Assembly Approaches

Assembly of components into devices is amenable to multiple approaches. In the case of devices comprising a single molecule or processed from a single crystal (some microfabricated structures, single polymers, or grafted polymeric structures) assembly may not be an issue. Integration of multiple, separately microfabricated components may sometimes be necessary (as in the immunoisolation capsule discussed below) and may sometimes drive the need for assembly, even for silicon devices. Furthermore, many therapeutic nanodevices contain multiple, chemically diverse components that must be assembled precisely to support their harmonious contribution to device function.

"One off" Nanostructures and Low Throughput Construction Methods

Direct-write technologies can obtain high (nanometer scale) resolution. For instance, electron-beam (e-beam) lithography is a technique requiring no mask, and that can yield resolutions on the order of tens of nanometers, depending on the resist materials used [10.34]. Resolution in e-beam lithography ultimately is limited by electron scattering in the resist and electron optics, and like most direct-write approaches, e-beam lithography is limited in its throughput. Parallel approaches involving simultaneous writing with up to 1,000 shaped e-beams are under development [10.34] and may mitigate limitations in manufacturing rate.

Force microscopy approaches utilize an ultrafine cantilever tip (typically with point diameters of 50 nm or less, Fig. 10.6) in contact with, or tapping, a surface or a stage. The technique can be used to image molecules, to analyze molecular biochemical properties (like ligand-receptor affinity [10.35]), or to manipulate materials at nanoscale. In the latter mode, force microscopy has been used to manipulate atoms to build individual nanostructures since the mid-1980s. This has led to construction of structures that are precise to atomic levels of resolution (Fig. 10.6), though the manufacturing throughput of "manual" placement of atoms by force microscopy is limited.

Dip-pen nanolithography (DPN) is a force microscopy methodology that can achieve high resolution features (features of 100 nm or less) in a single step. In DPN, the AFM tip is coated with molecules to be deployed on a surface, and the molecules are transferred from the AFM tip to the surface as the coated tip contacts it. DPN also can be used to functionalize surfaces with two or more constituents and is well suited for deployment of functional biomolecules on synthetic surfaces with nanoscale precision [10.36, 37]. DPN suffers the limitations of synthetic throughput typical of AFM construction strategies.

Much as multibeam strategies might improve throughput in e-beam lithography [10.34], multiple tandem probes may increase assembly throughput for construction methods that depend on force microscopy significantly, but probably not sufficiently to allow manufacture of bulk quantities of nanostructures, as will likely be needed for consumer nanotherapeutic devices. As standard of care evolves increasingly toward tailored courses of therapy [10.38] and individual therapeutics become increasingly multicapable and powerful, however, relatively low throughput synthesis/assembly methods may become more desirable. For the moment, though, ideal manufacturing approaches for nanotherapeutic devices resemble either industrial polymer chemistry, occurring in bulk, in convenient buffer

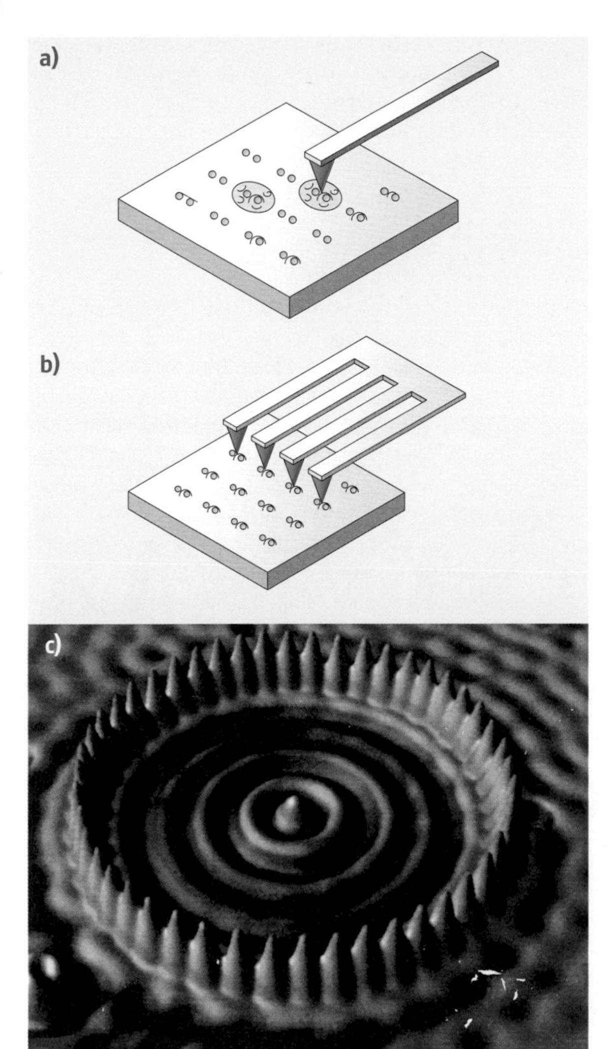

Fig. 10.6 (a) A schematic depiction of an atomic force microscope cantilever and tip interacting with materials on a surface. Tips typically have points of 50 nm or less in diameter [10.31, 39]. (b) Schematic of multiplexed AFM tips performing multiple operations in parallel [10.31, 39]. (c) AFM image of a quantum corral, a structure built using AFM manipulation of individual atoms (from the IBM Image Gallery)

systems, or in massively parallel industrial microfabrication approaches. In any case, therapy for a single patient may involve billions of billions of individual nanotherapeutic units, so each individual nanotherapeutic structure must require only minimal input from a human synthesis/manufacturing technician.

Self-Assembly of Nanostructures

Self-assembly has been long recognized as a potentially critical labor-saving approach to construction of nanostructures [10.40], and many organic and inorganic materials have self-assembly properties that can be exploited to build structures with controlled configurations. Self-assembly processes are driven by thermodynamic forces and generally result in structures that are not covalently linked. Intra/intermolecular forces driving assembly can be electrostatic or hydrophobic interactions, hydrogen bonds, and van der Waals interactions between and within subunits of the self-assembling structures and the assembly environment. Thus, final configurations are limited by the ability to "tune" the properties of the subunits and control the assembly environment to generate particular structures.

Self-Assembly of Carbon Nanostructures

Carbon nanotubes (Fig. 10.7) spontaneously assemble into higher order [10.31] structures (nanoropes) as the result of hydrophobic interactions between individual tubes. Multi-wall carbon nanotubes (MWCNT) are well-known structures that can be viewed as self-assembled, nested structures of nanotubes with tube diameters decreasing serially from the outermost to innermost tubes. The striking resemblance that MWCNT have to macroscale bearings has been noted and exploited [10.41]. MWCNT linear bearings can be actuated by application of mechanical force to the inner nanotubes of the MWCNT assembly. Actuation causes the assembly to undergo a reversible telescoping motion. Interestingly, these linear MWCNT bearings exhibit essentially no wear as the result of friction between bearing components.

C_{60} fullerenes and single wall carbon nanotubes (SWCNT) also spontaneously assemble (Fig. 10.7) into higher order nanostructures called "peapods" [10.42] in which fullerene molecules are encapsulated in nanotubes. The fullerenes of peapods modulate the local electronic properties of the SWCNT in which they are encapsulated and may allow tuning of carbon nanotube electrical properties. The potentially fine-level control of nanotube properties may prove useful in nanotube-containing electrical devices, particularly in cases wherein nanotubes are serving as molecular wires. In this capacity, carbon nanotubes have been incorporated into FETs (field effect transistors, discussed below under sensing architectures [10.43]), and other molecular electronic structures. Ultimately, these architectures may result in powerful, ultra-small computers to provide the intelligence of "smart," indwelling nanotherapeutic

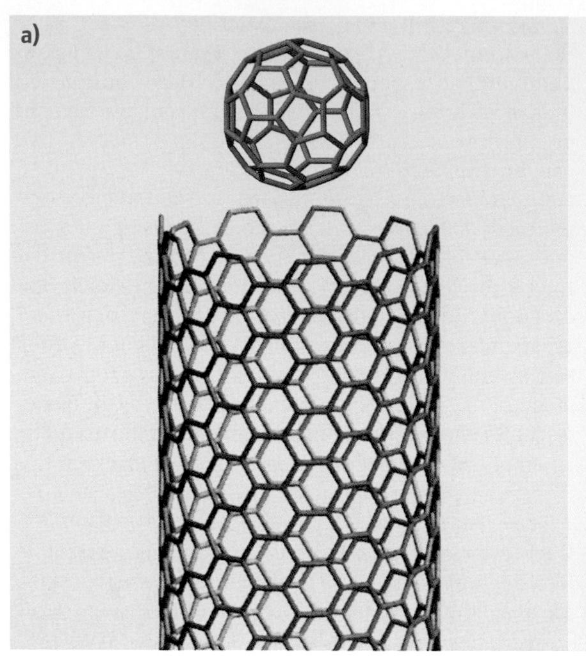

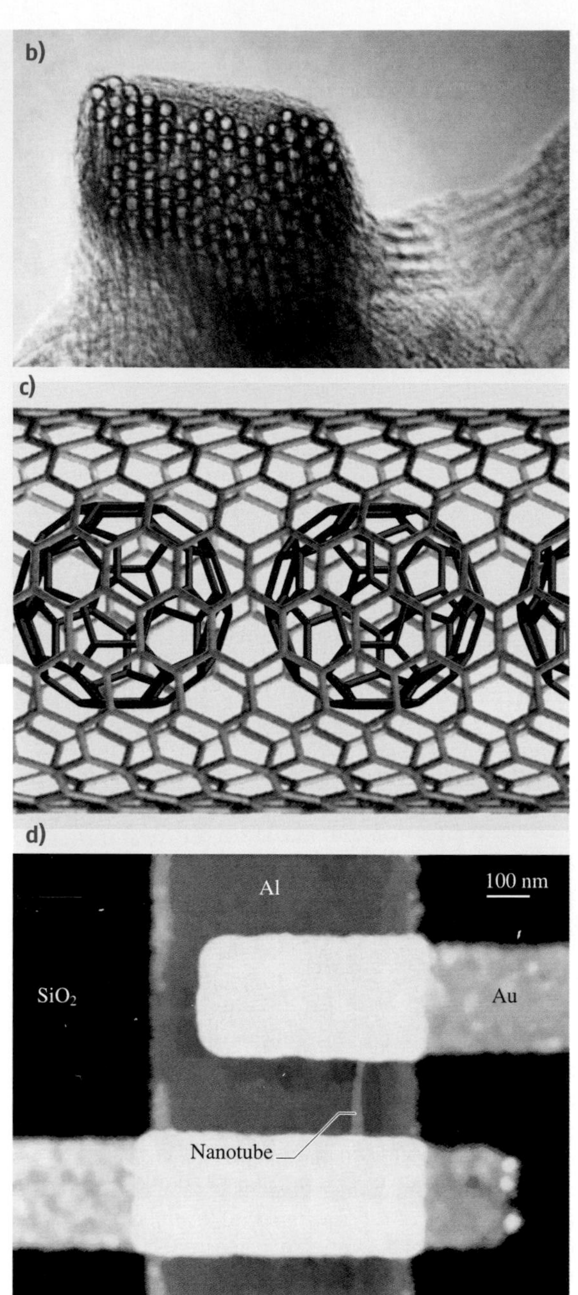

Fig. 10.7a–d (**a**) Shown to scale are two highly defined carbon nanostructures: a C_{60} fullerene and (10, 10) single wall carbon nanotube (SWCNT). (**b**) A self-assembled nanorope composed of carbon nanotubes that assemble by virtue of hydrophobic interactions [10.31,39]. (**c**) Schematic depiction of another self-assembled carbon nanostructure (a peapod) consisting of the fullerenes and SWCNT of 7a, and wherein the fullerenes are encapsulated in the SWCNT [10.42]. Fullerene encapsulation in the peapods modulates local electronic properties of the SWCNT. (**d**) A nanotube field effect transistor (FET) consisting of gold source and drain electrodes on an aluminum stage with a carbon nanotube serving as the FET channel [10.43]

devices. In general, though, fullerenes, nanotubes carbon nanoropes, nanotube bearings, and peapods have yet to find extensive biological application, in part because of their extreme hydrophobicity, presumably poor biocompatibility, and high chemical stability [10.31, 39]. But controlled derivatization of nanotubes may be possible through a number of approaches [10.31, 39] including controlled introduction of bond strain to render individual carbons of the tubes selectively chemically reactive (so-called mechanosynthesis). Carbon nanotubes and fullerenes, however, owe many of their remarkable properties (chemical stability, mechanical robustness, some electrical properties) to the fact that all the valences of the constituent carbon atoms (except those at the ends of nanotubes that are "open") are satisfied. Thus, to derivatize carbon nanostructures is to degrade them, a fact one must consider when the design purposes that drove incorporation of carbon structures into a therapeutic depends on the chemical perfection of the material.

Self-Assembly of Materials Made by Traditional Polymer Chemistry

In the realms of drug delivery and biomedical micro and nanodevices, the most familiar self-assembled structures are micelles [10.44–46]. These structures are formed from the association of block co-polymer subunits (Fig. 10.8), each individual subunit containing hydrophobic and hydrophilic domains. Micelles spontaneously form when the concentration of their subunits exceeds the critical micelle concentration (cmc) in a solvent in which one of the polymeric domains is immiscible (Fig. 10.8). The cmc is determined by the immiscible polymeric domain and can be adjusted by control of the chemistry and length of the immiscible domain, as well as by control of solvent conditions. Micelles formed at low concentrations from low-cmc polymers are stable at high dilution. Micelles formed from polymer monomers with high cmcs can dissociate upon dilution, a phenomenon that might be exploited to control release of therapeutic cargos. If desired, micelles can be stabilized by covalent cross-linking to generate shell-stabilized structures [10.44–47].

The size dispersity and other properties of micelles can be manipulated by control of solvent conditions, incorporation of excipients (to modulate polymer packing properties), temperature, and agitation. From the standpoint of size, reasonably monodisperse preparations (polydispersity of 1–5%) of nanoscale micellar structures can be prepared [10.44–47]. The immense versatility of industrial polymer chemistry allows micellar structures to be tuned chemically to suit the task at hand. They can be modified for targeting or to support higher order assembly properties. They can be made to imbibe therapeutic or other molecules for delivery and caused to dissociate or disgorge themselves of payloads at desired times or bodily sites under the influence of local physical/chemical conditions. The tunability of these and other properties at the level of monomeric polymer subunits as well as the level of assembled higher order structures make micelles potentially powerful nanoscale drug delivery and imaging vehicles.

Fractal materials, such as dendritic polymers, whose synthesis and structure were discussed previously, exhibit packing properties that can be exploited to assemble higher order aggregate structures called "tecto(dendrimers)" [10.48]. In fact, these self-assembly properties are being exploited in oncological nanotherapeutics as Fig. 10.3 shows [10.15, 49]. In principle, these self-assembling therapeutic complexes need not be pre-formed prior to administration. Individual functional modules of the therapeutic assembly might be administered sequentially, potentially to tailor therapies more precisely to individual patient responses.

Stoichiometric Control and Self-Assembly

As the preceding examples demonstrate, self-assembly approaches sometimes do not feature precise control of subunit identity and stoichiometry in the assembled complexes. This can be a limitation when the stoichiometry and relative arrangement of differentiable individual subunits is critical to device function. Stoichiometry is less an issue when the self-assembling components are identical and functionally fungible, as in the synthetic, peptidyl anti-infective illustrated in Fig. 10.9 [10.50,51].

In the anti-infective architecture, individual peptide components are flat, circular molecules. The planar character of the toroidal subunits is a consequence of the alternating chirality of alternating D-L amino acids (aas) in the primary sequence of the peptide rings. Alternating D and L aas is not possible in proteins made by ribosomal synthesis [10.32]. Ribosomes recognize and incorporate into nascent polypeptides only L amino acids, and so, as the result of aa chirality and bond strain, peptides made by ribosomes cannot be made flat, closed toroids like those of the peptidyl anti-infectives. Much as in α-helical domains of ribosomally synthesized proteins, however, the aa R-groups (which are of varying hydrophobic or hydrophilic chemical specificities [10.32]) are arranged in the plane of the closed D, L rings extending out from the center of the rings. Hydrogen bonds between individual rings govern self-assembly of the toroids into rod-like stacks, while the R-groups dominate interactions between multiple stacks of toroids and other macromolecules and structures (see below).

The planar toroidal subunits can be administered as monomers and self-assemble into multi-toroid rods at the desired site of action (in biological membranes). But the peptide toroids R-groups are chemically tuned so that the rod structures into which they spontaneously assemble intercalate preferentially in specific lipid bilayers (i. e., in pathogen vs. host membranes). Moreover, the assembled rods may undergo an additional level of self-assembly into multi-rod structures, spanning pathogen membranes [10.50, 51]. Whether as single rod or multiple rod assemblies, membrane intercalation by stacked toroids reduces the integrity of pathogen membranes selectively, and therefore particular toroid species exhibit selective toxicity to specific pathogens.

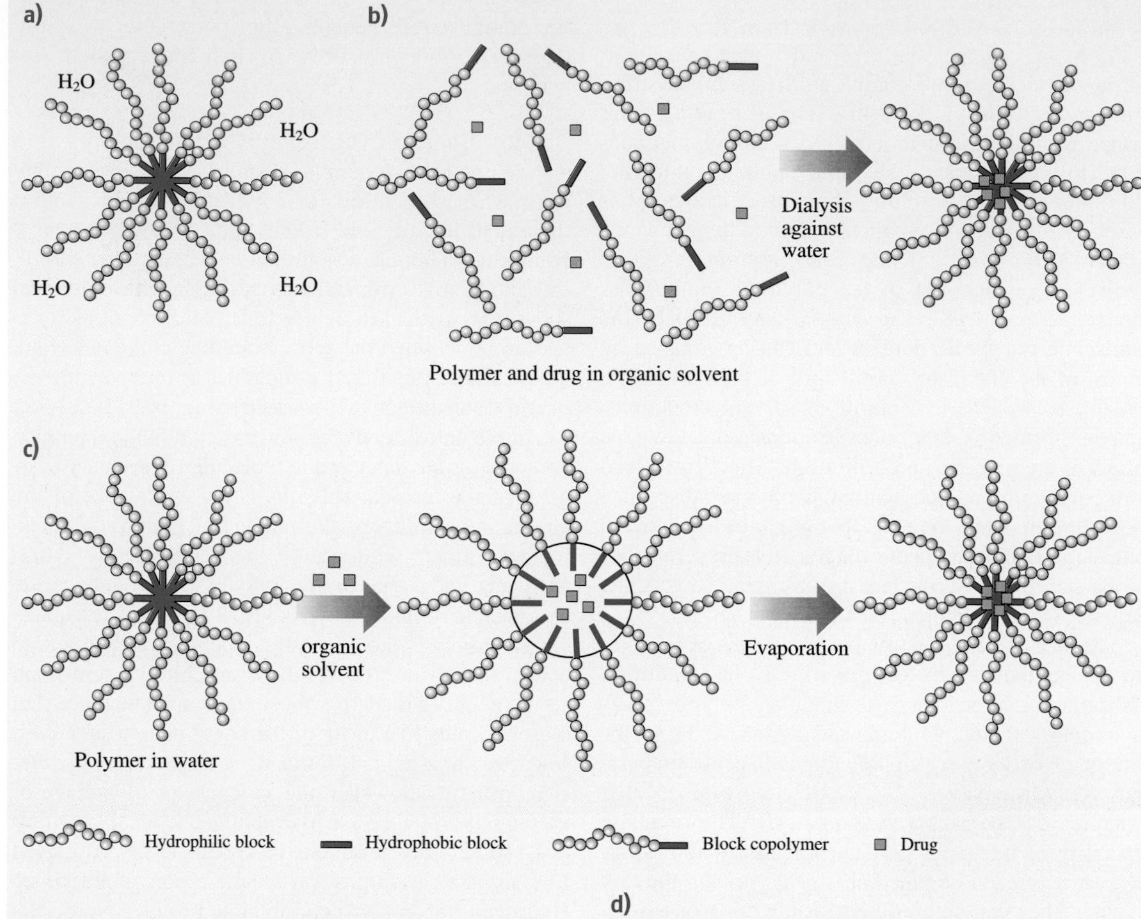

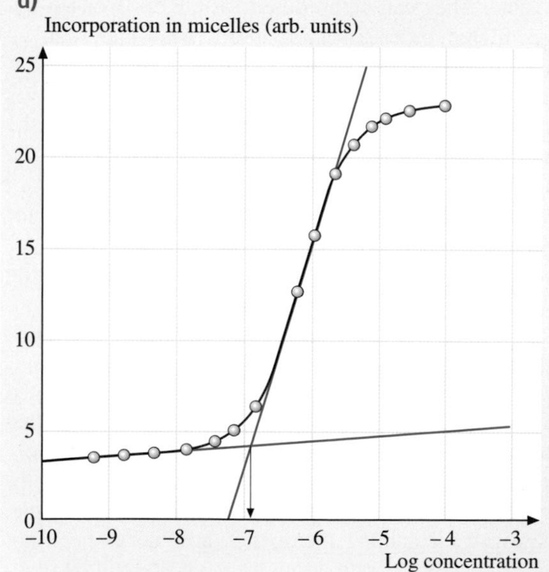

Fig. 10.8a–d Micellar drug delivery vehicles and their self-assembly from block copolymers [10.44–46]. (**a**) Morphology of a micelle in aqueous buffer. Hydrophobic and hydrophilic polymer blocks, copolymers containing the blocks, micelles generated from the block copolymers, and (hydrophobic) drugs for encapsulation in the micelles are indicated. (**b**) Micelle self-assembly and charging with drug occuring simultaneously when the drug-polymer formulation is transitioned from organic to aqueous solvent by dialysis. (**c**) Pre-formed micelles can be passively imbibed with drugs in organic solvent. Organic solvent is then removed by evaporation, resulting in compression of the (now) drug-bearing hydrophobic core of the micelle. (**d**) An illustration of concentration-driven micelle formation. At and above the critical micelle concentration (cmc), block copolymer monomers assemble into micelles, rather than exist as free block copolymer molecules. The *arrow* indicates the cmc for this system

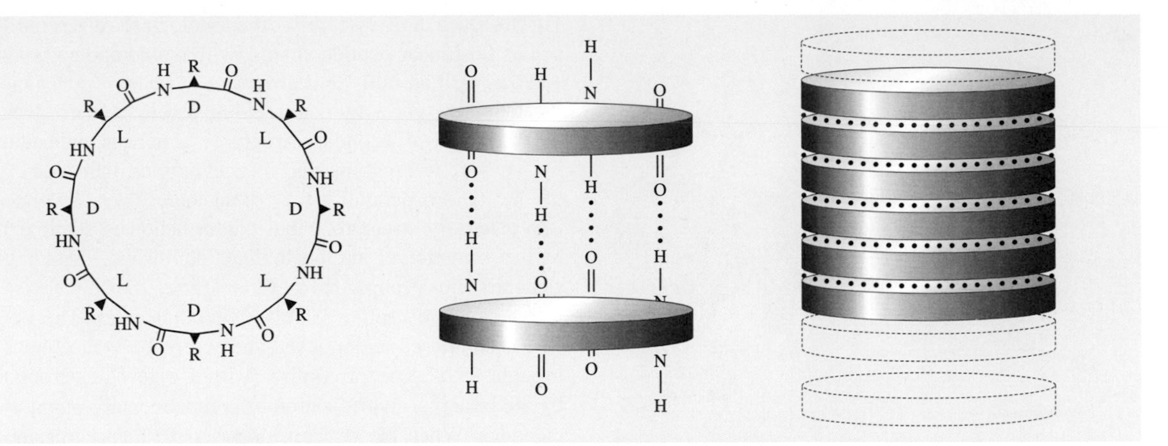

Fig. 10.9 A self-assembling peptide antibiotic nanostructure [10.50, 51]. Peptide linkages and the α-carbons and their pendant R groups are indicated. The synthetic peptide rings are planar as the result of the alternating chirality (D or L) of their amino acid (aa) constituents. R groups of aas radiate out from the center of the toroid structure. Individual toroids self-assemble (stack) as the result of hydrogen bonding interaction between amine and carboxy groups of the peptide backbones of adjacent toroids. The surface chemistry of multi-toroid stacks is tuned at the level of the aa sequence and, therefore, R group content of the synthetic peptide rings. The chemical properties of the stacked toroid surfaces allow them to intercalate into the membranes of pathogenic organisms with lethal consequences. The specific membrane preferences for intercalation of the compound are tuned by control of R group content of torroids

These toroidal, synthetic antibiotics, and other nanoscale antimicrobials represent critically needed, novel antibacterial agents. Resistance to traditional, microbially derived antibiotics often is tied to detoxifying functions associated with secondary metabolite synthesis; these detoxifying functions are essential for the viability of many antibiotic producing organisms (for instance, see [10.52]). The genes encoding such detoxifying functions are rapidly disseminated to other microorganisms, accounting for the rapid evolution of drug resistant organisms that has bedeviled antimicrobial chemotherapy for the last 25 years. Synthetic nanoscale antibiotics, like the peptide toroids [10.50, 51] and the N8N antimicrobial nanoemulsion [10.53], act by mechanisms entirely distinct from those of traditional secondary metabolite antibiotics, and no native detoxifying gene exists. Therefore, novel nanoscale antimicrobials may not be subject to the unfortunately rapid rise in resistant organisms associated with most secondary metabolite antibiotics, though this remains to be seen. As bacterial infection continues to re-emerge as a major cause of morbidity and mortality in the developed world, consequence of increasing antibiotic-resistant pathogens, novel nanoscale antibiotics will become more important.

Biomolecules in Therapeutic Nanodevices: Self-Assembly and Orthogonal Conjugation

Biological macromolecules undergo self-assembly at multiple levels, and like all instances of such construction, biological self-assembly processes are driven by thermodynamic forces. Some biomolecules undergo intramolecular self-assembly (as in protein folding from linear peptide sequences, Fig. 10.10). Higher order structures are, in turn, built by self-assembly of smaller self-assembled subunits (for instance, structures assembled by hybridization of multiple oligonucleotides, enzyme complexes, fluid mosaic membranes, ribosomes, organelles, cells, tissues, etc.).

Proteins are nonrandom copolymers of 20 chemically distinct amino acid (aa) subunits [10.32]. The precise order of aas (i. e., via interactions between aa side chains) drives the linear polypeptide chains to form specific secondary structures (the α helices and β sheet structures seen in Fig. 10.10). The secondary structures have their own preferences for association, which, in turn, leads to the formation of the tertiary and quartenary structures that constitute the folded protein structures. In its entirety, this process produces consistent structures that derive their biological functions from strict control of the deployment of chemical

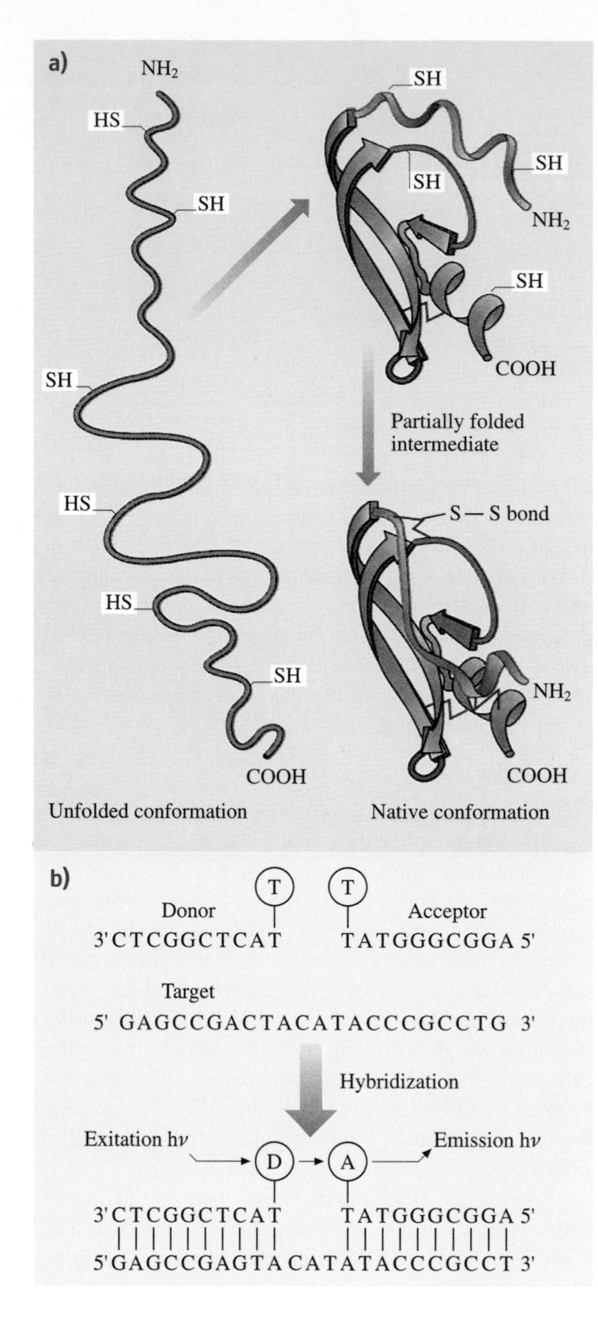

Fig. 10.10a,b Self-assembly and biological macromolecules (**a**) Linear peptide chains (with amino and carboxy ends, as well as sulfhydral groups of cysteine residues indicated) undergoe a multistep folding process that involves the formation of secondary structures (α-helices, indicated by heavy helical regions, and β-sheet regions, indicated by the heavy anti-parallel arrows) that themselves associate into a tertiary structure. Final conformation is stabilized by the formation of intrachain disulfide linkages involving cysteine thiol groups. (**b**) A fluorescence transfer device that depends on self-assembly of biomolecules. The device is composed of donor (D) and acceptor (A) molecules brought into close proximity (within a few angstroms) by the base pair hybridization of complementary oligonucleotides. When the structure is assembled, acceptor and donor are energetically coupled, and fluorescence transfer can occur [10.54–57]

specificities (the aa side chains) in three-dimensional space.

Biomolecules can be used to drive assembly of nanostructures, either as free molecules or conjugated to heterologous nanomaterials. For instance, three-dimensional nanostructures can be made by DNA hybridization [10.54–57]. Such DNA nanostructures can exhibit tightly controlled topographies but limited integrity in terms of geometry [10.57], due to the flexibility of DNA strands. Oligonucleotides, antibodies and other specific biological affinity reagents can also be used to assemble nanostructures (Fig. 10.10, see also targeting and triggering discussions below). Often the domains of biomolecules responsible for assembly and recognition are small, continuous, and discrete enough that they can be abstracted from their native context as modules and appended to other nanomaterials of interest to direct formation of controlled nanoscale architectures (for instance, see the SCFv antibody fragement of Fig. 10.1).

Several orthogonal bioconjugate approaches have arisen from the field of protein semisynthesis [10.20, 58, 59]. These protein synthetic chemistries allow site-specific conjugation of polypeptides to heterologous materials in bulk, as the result of conjugation between exclusively, mutually reactive electrophile-nucleophile pairs (analogous to dendrimer synthesis discussed above, see Table 10.2). They have been applied to the synthesis of multiple therapeutic nanodevices [10.2–5, 7, 14, 58, 59].

As described above, proteins are profoundly dependent on their three-dimensional shapes: chemical derivatization at critical aa sites can profoundly impact protein bioactivity. Because conjugation can be directed to pre-selected sites via orthogonal approaches, and since the sites of conjugation in the protein can be chosen because the proteins involved tolerate adducts at those positions, proteins coupled to nanomaterials by such orthogonal methodologies often retain their

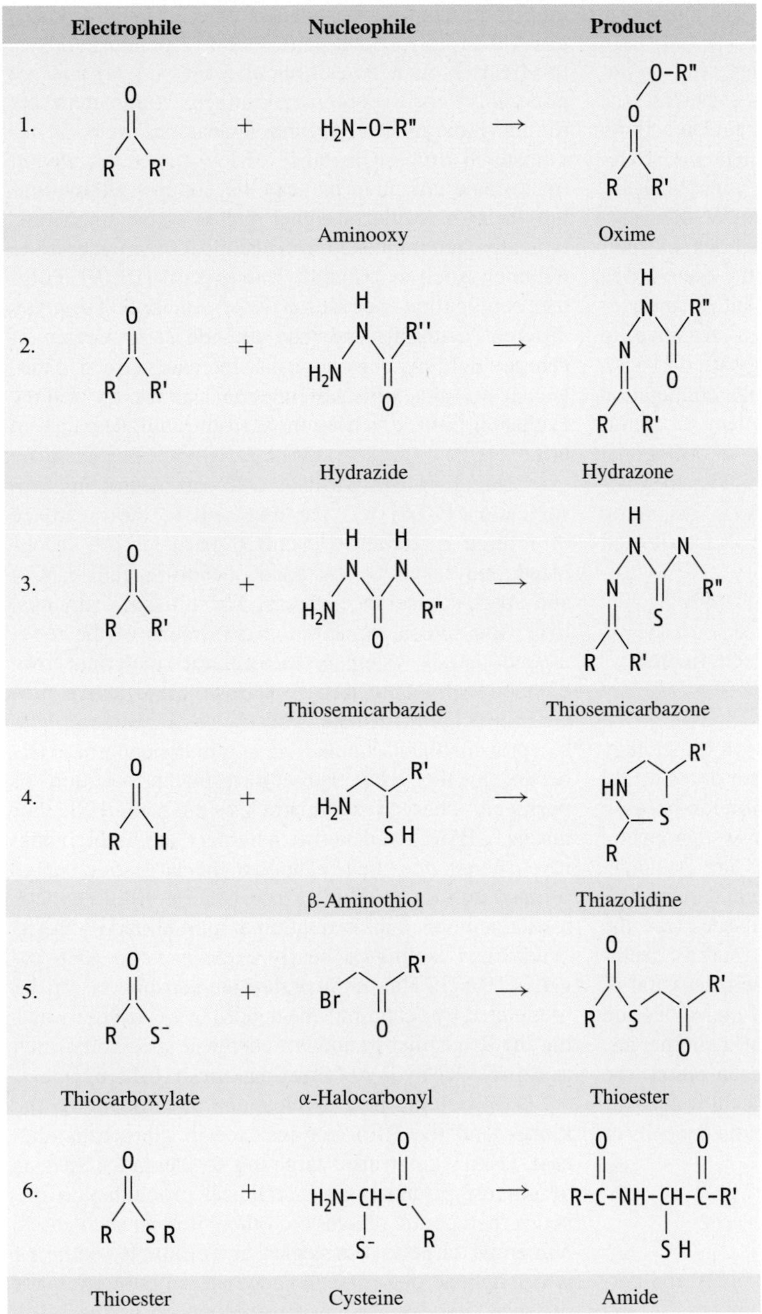

Table 10.2 Orthogonal conjugation chemistries originally derived for protein semi-synthesis but applicable to conjugation of proteins to nanomaterials [10.20, 58, 59] (R' = H: Aldehyde, R = alkyl: Ketone)

	Electrophile		Nucleophile		Product
1.	$R-CO-R'$	+	H_2N-O-R'' Aminooxy	→	Oxime
2.	$R-CO-R'$	+	Hydrazide	→	Hydrazone
3.	$R-CO-R'$	+	Thiosemicarbazide	→	Thiosemicarbazone
4.	$R-CO-H$	+	β-Aminothiol	→	Thiazolidine
5.	Thiocarboxylate	+	α-Halocarbonyl	→	Thioester
6.	Thioester	+	$H_2N-CH(S^-)-CO-R$ Cysteine	→	$R-C(O)-NH-CH(SH)-C(O)-R'$ Amide

biological activity. In contrast, protein bioactivity in conjugates generally is lost or profoundly impaired when proteins are coupled to nanomaterials using promiscuous chemistries. For instance, proteins such as cytokines (and other protein hormones) elicit their effects by interacting with a receptor, and a large fraction of their surface (20% or more) is involved in receptor binding, directly or indirectly. Promiscuous chemistries (1-ethyl-3-(3-diamethylaminopropyl) carbodiimide or EDC, conjugation, see [10.60]) used to conjugate cytokines to nanoparticles tend to inactivate hIL-3 and other cytokines whereas the same protein/particle bioconjugates retain bioactivity if judiciously chosen orthogonal conjugation strategies are used (Lee & Parthasarathy, unpublished). Proteins for which only a small portion of their surfaces contribute to the interesting portions of their bioactivities (from the standpoint of the nanodevice designer), such as some enzymes or intact antibodies (Fig. 10.1) may be somewhat less sensitive to promiscuity of the bioconjugate strategy used [10.20, 58, 59], but the benefits of orthogonal conjugation strategies can also apply to these protein bioconjugates [10.20, 58, 59]. The potential utility of orthogonal conjugation for incorporation of active biological structures into semi-synthetic nanodevices is becoming more fully recognized and cannot be overestimated [10.58, 59].

10.3.2 Targeting: Delimiting Nanotherapeutic Action in Three-Dimensional Space

Delivery of therapeutics to sites of action is a key strategy to enhance clinical benefit, particularly for drugs useful within only narrow windows of concentration because of their toxicity (i. e., drugs with narrow therapeutic windows). Diverse targeting approaches are available, ranging from methods exploiting differential extravasation limits of vasculature of different tissues (see the discussion of oncology below), sizes, and surface chemistry preferences for cellular uptake (see discussion of vaccines below); preferential partition of molecules and particles into specific tissues by virtue of their charges, sizes, surface chemistry, or extent of opsonization (see below); or the affinity of biological molecules decorating the nanodevice for counter-receptors on the cells or tissues of interest.

The Reticuloendothelial System and Clearance of Foreign Materials

Physical properties such as surface chemistry and particle size can drive targeting of nanomaterials (and presumably nanodevices containing them) to some tissues. For instance, the pharmacokinetic (P_k) and biodistribution (B_d) properties [10.32] of many drugs and nanomaterials are driven by their clearance in urine, which is in turn governed by the filtration preferences of the kidney. Most molecules making transit into urine have masses of less than 25 to 50 kilodaltons (kDa; 25–50 kDa particles corresponding *loosely* to effective diameters of about 5 nm or less) and are preferably positively charged; these parameters are routinely modulated to control clearance rates of administered drugs. Clearance of low molecular weight (nano)materials in urine can be suppressed by tuning their molecular weights and effective diameters, typically accomplished by chemical conjugation, to polymers such as poly(ethylene glycol) [10.19]. Polymer conjugation (pegylation) has been applied to many different materials and may provide some degree of charge shielding. Pegylation also increases effective molecular weights of small materials above the kidney exclusion limit, diverting them from rapid clearance in urine.

Coating foreign particles with serum proteins (opsonization [10.61]) is the first step in the clearance of foreign materials. Opsonized particles are recognized and taken up by tissue dendritic cells (DCs) and specific clearance organs. These tissues (thymus, liver, and spleen, constituting the organs of the reticuloendothelial system or RES) extract materials from circulation by both passive diffusion and active processes (receptor-mediated endocytosis). Charge-driven, receptor-mediated uptake of synthetic nanomaterials occurs in the RES and can result in partition of positively charged nanoparticles into the RES. For instance, PAMAM dendritic polymers exhibit high positive charge densities related to the large number of primary amines on their surfaces [10.28–30]. In experimental animals, biodistribution of unmodified PAMAM dendrimers is limited nearly exclusively to RES organs [10.62]. This unfavorable biodistribution can be modulated by "capping" the dendrimers (i. e., derivatizing the dendrimer to another chemical specificity, such as carboxy or hydroxyl functionalities [10.15, 62]).

Despite legitimate applications of targeting to the kidney and the RES (for instance in glomerular disease [10.63]), intrinsic targeting to clearance sites is of interest primarily as a technical problem that impedes therapeutic delivery to other sites. In such cases, numerous targeting strategies are available, some of which depend on synthetic nanomaterial properties (for instance, see discussion of the enhanced permeability and retention or EPR effect in the context of oncology, below) to minimize uptake of nanotherapeutic devices by clearance systems and maximize delivery to desired sites. Targeting via biological affinity reagents decorating the surfaces of therapeutic nanodevices may be the most direct approach.

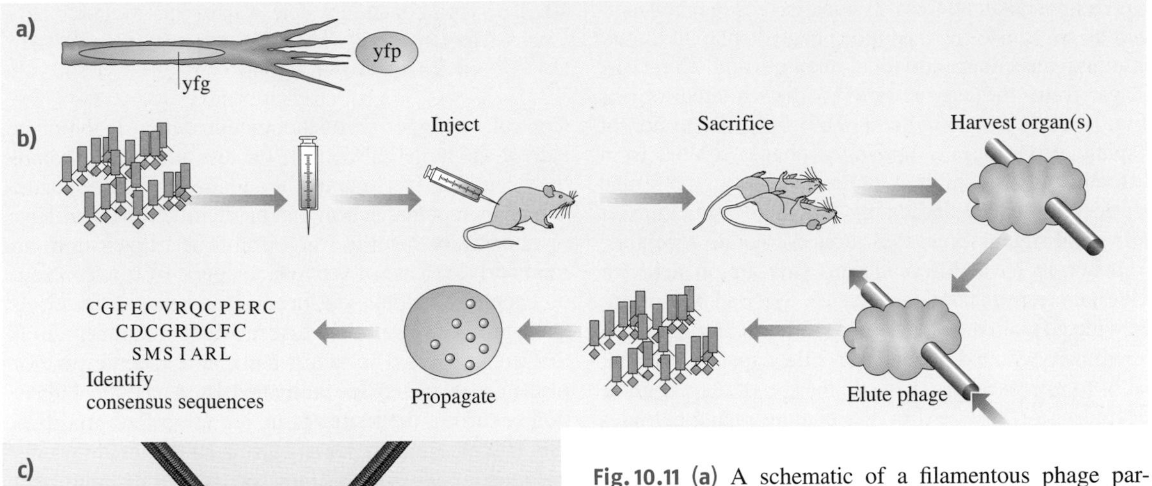

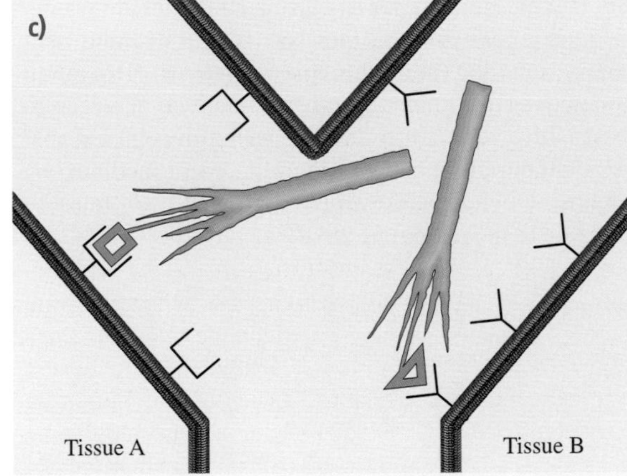

Fig. 10.11 (**a**) A schematic of a filamentous phage particle encapsulating DNA encoding a gene of interest (your favorite gene, or yfg) and presenting the corresponding protein (your favorite protein or yfp) on its surface. The linkage between yfp and yfg provided by the phage allows simultaneous affinity isolation and recovery of proteins of interest along with replicable genetic elements encoding them. When variant protein libraries are built by this method, they can be sorted for their ability to bind receptors by affinity. (**b**) A library of filamentous phage presenting random peptides is shown schematically. The library is injected into the vasculature of an animal (in this case, a mouse), where individual phage bind specific receptors present in the vasculature of different tissues. The animals are sacrificed, and organs of interest are harvested. Phage particles bound to receptors of vasculature of the organ at hand can be eluted by any of several methods (low pH elution is schematically shown here). *E. coli* is infected with the eluted phage particles and clonally propagated, allowing identification of the peptides encoded by the eluted phage, which mediated binding to the vascular receptors. Ultimately, this allows the identification of consensus peptide sequences that recognize specific receptors on the vasculature of the organ or tissue involved. These receptors constitute the molecular addresses for these organs, and the peptides isolated from phage eluted from them are biological affinity reagents that can be used to direct nanotherapeutic devices to vasculature of particular organs. (**c**) Specific recognition of vascular beds of two distinct tissues (A and B) via recognition of organ-specific vascular addresses (the Y and box-shaped receptors shown on vascular endothelium) by specific peptides (the *triangle* and the *square*) presented on the phage surface is depicted [10.64–66]

Nanotherapeutic Targeting Exploiting Biological Affinity Properties

Tissue-specific delivery by biological affinity requires the presence of tissue-specific surface features, most commonly proteins or glycoproteins (tissue-specific antigens). Historically the search for tissue-specific antigens for drug targeting, whether associated with tumors or other cells, organs or tissues, has been arduous and not entirely gratifying. The primary problems are specificity (few antigens are uniquely present in any single tissue), availability (some tissues may not have their own unique antigenic signature or marker), and therapeutic extravasation or directed migration in tissue spaces (markers in tissue may not be accessible from vasculature). An exciting recent development in biochemical targeting is the discovery of a vascular address system [10.64–66].

The vascular address system has been characterized by administering a peptide phage display to library animals, resecting individual organs, and extracting phage from the vasculature of the isolated organs (Fig. 10.11 [10.64–66]; see [10.67] for a discussion of display technology). Amazingly, phage isolated from different organs exhibited distinct consensus presented peptide sequences, indicating that the vasculature of individual organs presented unique cognate receptors, each bound by a different short (ten amino acids or fewer) consensus peptide sequence that had been affinity selected from the phage display library. Furthermore, the affinity-selected peptides have the capacity to tether nano- to microscale particles to the site of their cognate receptors (as illustrated by the binding phage particles presenting the peptides to specific vascular locations). Site-specific drug delivery using the vascular address system has already been demonstrated [10.68, 69]: it has further been used to target an apoptotic (cytocidal) agent to the prostate and to direct destruction of the organ in an animal model [10.68].

Mapping of the vascular address system is currently underway [10.64] and holds the promise of specific delivery of therapeutic agents to vasculature of specific organs. It remains to be seen whether each organ has a single molecular marker constituting its address that is amenable to binding a single peptide sequence; organs may instead have unique constellations of antigenic markers. If so, specific targeting may be possible using multiple peptides, each peptide binding its cognate receptor on target organ vasculature very weakly. Peptides used in such a multivalent targeting strategy would be chosen to reflect the unique constellation of address markers present in the target tissue. Affinities of cognates for linear peptides are often very low [10.67], though their aggregate affinity may be substantially higher than that of any peptide-vascular address cognate alone. Under ideal conditions, the affinity of such a multipeptide, multicognate complex should be the equivalent of the products of the affinities of each constituent peptide for its individual constituent cognate. Such multivalent interaction avidities can be extremely high (and the corresponding effective affinity constants are also high) but seldom fully realize their theoretical maximums (Lee & Parthasarathy, unpublished; see [10.32] for a review of receptor biochemistry).

It should be noted that most of the vascular addresses identified to date deliver materials to the organ vasculature: extravasation and access of organ tissue spaces by nanotherapeutics remains a separate issue.

10.3.3 Triggering: Delimiting Nanotherapeutic Action in Space and Time

Controlled triggering of therapeutic action is the other side of the targeting coin. If the site and time of nanotherapeutic delivery cannot be adequately controlled, the site of therapeutic action can be delimited by spatially- or temporally-specific triggering. The triggering event might drive release of active therapeutic from a reservoir, or chemical or physical processing of drug materials from an inert to an active form (inert administrations that are converted to active form at a specific time or place drugs which are activated by a chemical reaction occurring their sites of action are called prodrugs, Fig. 10.12). Three major triggering strategies are widely used: external stimuli, intrinsic triggering, and secondary signaling (multicomponent systems). Triggering strategies require nanotherapeutic delivery devices to be sensitive to a controlled triggering event, or a spatially/temporally intrinsic triggering event mediated by the host. Obviously, the triggering event itself must be tolerable to the patient.

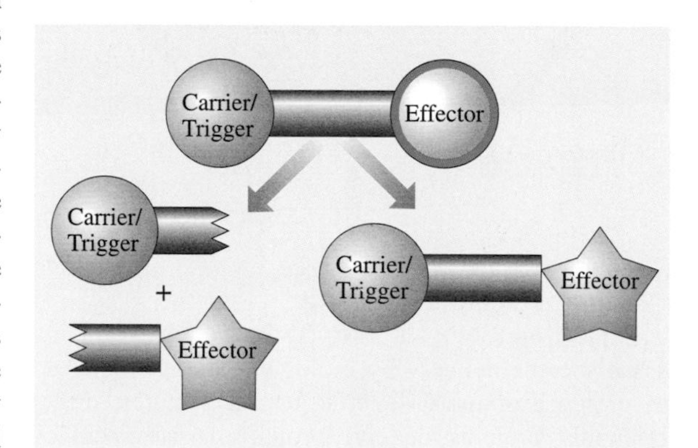

Fig. 10.12 One possible configuration of a prodrug, in which carrier/trigger and effector functions are separate functional domains of the therapeutic. The key feature of prodrugs is that they are therapeutically inert (as indicated by the colorless effector domain) until an activation event occurs (mediated here through the carrier/trigger domain, with effector activation indicated by its change to a star shape). Activation events can involve cleavage of inhibitory carrier/trigger domains from effectors. Other activation strategies involve a chemical change or shift in conformation of the effector, mediated through the carrier/trigger domain and in response to an environmental condition

Nanotherapeutic Triggering Using External Stimuli

External stimuli are provided by an external nanodevice operator/clinician, usually in the form of a site-specific energy input, typically light, ultrasound, or magnetic or electrical fields. Organic polymeric structures are very amenable to interaction with these energy sources. For instance, micellar structures can be reversibly dissociated with ultrasound, in which case they disgorge their contents or expose their internal spaces to the environment during ultrasound pulses (20 to 90 kilohertz range). The process has been used to control release of cytotoxin (doxorubicin) from micelles (Fig. 10.13), and short ultrasound transients might be used for pulsatile or intermittent exposure of patients to therapeutics [10.70].

Light is another popular external triggering modality. Bioactive materials can be covalently associated to a nanoscale-delivery vehicle by photo-labile linkages [10.15, 71, 72], or micelles can be constructed so that their permeability is altered as the result of exposure to light [10.72]. In the latter case, light input can cause photopolymerization resulting in micelle compaction that drives release of therapeutic cargo or photo-oxidation, which causes loss of micellar integrity to release encapsulated materials. The ability of light to penetrate dense tissues is a clear limitation to this approach. This concern can be accommodated by polymer systems responsive to wavelengths that penetrate tissue efficiently (usually in the red-infrared region of the spectrum), or by use of systems such as fiber optics to deliver light to deep tissues [10.15, 71, 72].

Externally applied magnetic fields can also be used to control nanotherapeutic activity. For instance, eddy currents induced by alternating magnetic fields can heat nanometallic particles and their immediate vicinity, an approach that has been successfully applied to control the bioactivity of individual biological macromolecules [10.73]. Colloidal gold particles are covalently conjugated to biomolecules (nucleic acid, or NA, duplexes or proteins), and alternating magnetic

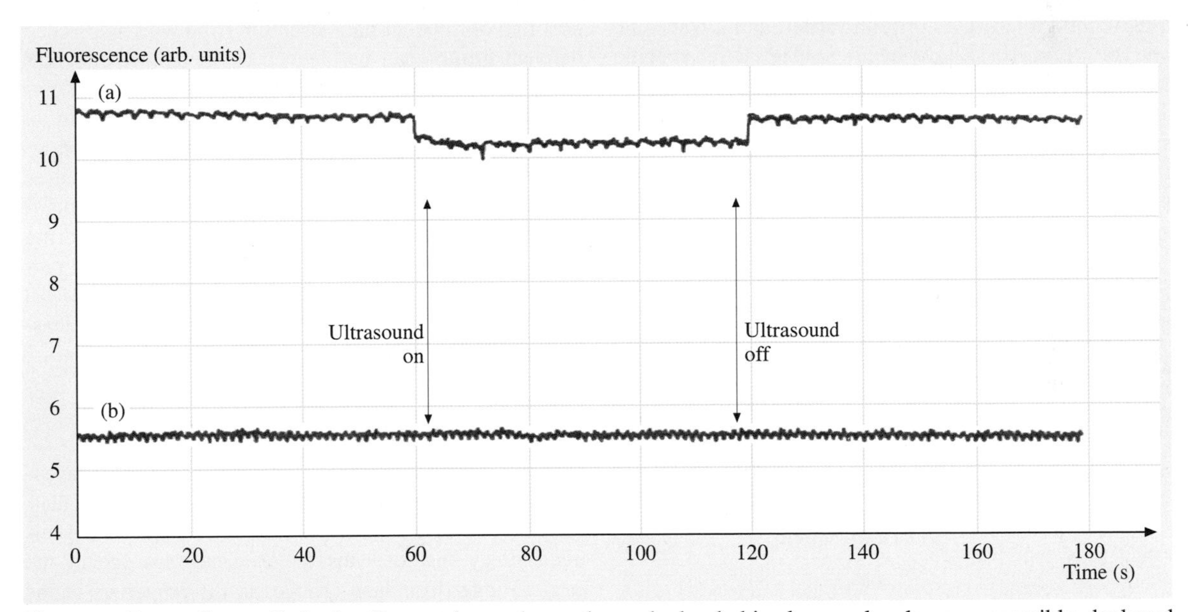

Fig. 10.13 Externally applied stimuli can trigger drug release: hydrophobic drug molecules are reversibly deployed from micelles in response to acoustic stimuli [10.70]. Intact pluoronic micelles maintain the hydrophobic cytotoxin doxorubicin in a hydrophobic environment (the core of the micelle, as in Fig. 10.4). Doxorubicin fluorescence is quenched in aqueous environments, and hence changes in integrity of pluoronic micelles carrying doxorubicin can be monitored by doxorubicin fluorescence. Here, an ultrasound pulse is applied to suspensions of such micelles or to doxorubicin in solution. Application of ultrasound triggers exposure of micelle-encapsulated doxorubicin to the aqueous solvent (i. e., drug release), as demonstrated by the reduction in fluorescence. After cessation of the acoustic pulse, doxorubicin is repackaged into the micelles, as evidenced by the increase in fluorescence after the ultrasound administration (*Line* a). Ultrasound has no impact on fluorescence of doxorubicin in solution, as expected (*Line* b)

fields are used to induce heating sufficient to cause dissociation of hybridized NA structures or denaturation of protein three-dimensional structures. Within certain parameters, the process is reversible and so allows the construction of semi-biological assembly or release switches. The modality has clear application to temporally specific triggered release, but the current inability to direct magnetic fields to pre-selected tissue locations may limit its use for spatially specific triggering.

Nanotherapeutic Triggering by Intrinsic Physiological Conditions

Intrinsic triggering is a prodrug strategy that depends on the conditions at the desired site of action to trigger the activity of the nanotherapeutic. Much as the three-dimensional conformation of, and therefore the activity of, proteins can be controlled in external triggering strategies [10.73], proteins can be engineered to make their conformations sensitive to intrinsic conditions at their desired sites of activity (see [10.3] for a discussion of protein engineering for use in nanobiotechnological devices). For instance, noncovalent complexes of a diphtheria toxin (DT) protein variant and a bispecific antibody (Fig. 10.1) have been engineered for specific toxicity to cells that can take up the complexes [10.75]. The bi-specific antibody recognizes both a cell-specific surface receptor as well as the DT in an inactive (nontoxic) configuration. Antibody binding to the receptor results in receptor-mediated endocytosis of the complex in target cells. As a result of the lower pH of the endosomal compartment (reaching pH 4–5 in endosomes, as opposed to the constant pH 7–8 in circulation) into which the bound complex is taken up, the DT variant protein undergoes a three-dimensional conformational shift to its toxic form. This conformation of DT is not recognized by the bispecific antibody, so the toxin dissociates from the complex and kills the cell. Hypoxia-triggered prodrugs have also been developed that are activated either by chemical reduction or by enzymatic activities induced in hypoxic tissues. Extreme hypoxia [10.76] is a unique feature of neoplastic tissue, so these strategies have clear application in oncology. Clinical manipulation of the extent of tumor oxygenation is possible [10.77] and may also be possible in other tissues. If so, redox state-dependent triggering approaches may have applications beyond cancer (Fig. 10.14a).

Enzyme-activated delivery (EAD [10.74], Fig.10.14) is another prodrug-like triggering strategy in which the properties of a nanoscale drug-delivery vehicle are altered at the site of action by an enzymatic activity endogenous to that site. Most typically, this involves a liposomal or micellar nanostructure from which designed pendent groups can be cleaved by metabolic enzymes (alkaline phosphatase, phospholipases, proteases, glycosidic enzymes, etc.) that are highly expressed at the site of therapy. Nanostructures are designed so that enzymatic cleavage of the pendent group causes a conformational or electrostatic change in the polymeric components of the delivery device, rendering the micellar structure fusogenic, leaky, or causing partial or complete dissociation of the structure [10.12] (Fig. 10.14c). This has the result of delivering or releasing therapeutic payloads at pre-selected sites.

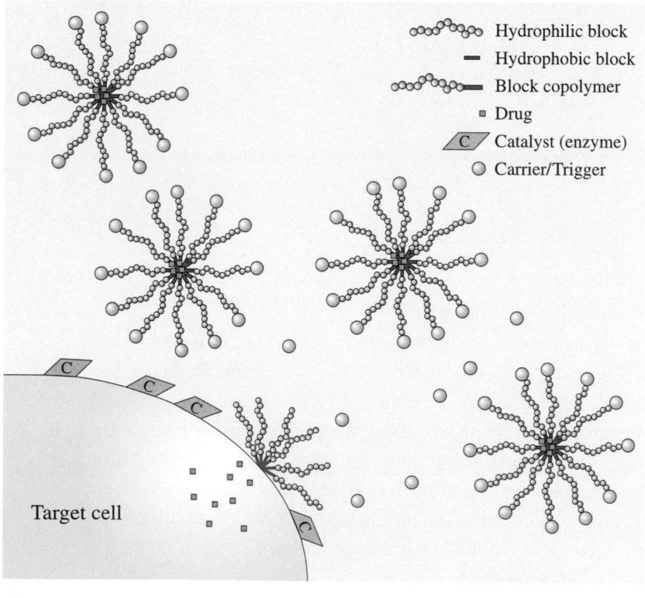

Hydrophilic block
Hydrophobic block
Block copolymer
Drug
C Catalyst (enzyme)
Carrier/Trigger
C
C
C
C
Target cell

Fig. 10.14 Intrinsic conditions at desired sites of therapeutic action can trigger release of active drug molecules. Enzyme activated delivery (EAD) is an intrinsic triggering strategy that depends on chemical cleavage of the carrier/trigger to induce changes in the properties of the delivery vehicle to allow delivery of drug [10.74]. Here a surface functionality is cleaved from a micellar drug delivery vehicle, rendering the micelle fusogenic and allowing it to fuse with the cell membrane. As a result, the therapeutic payload of the micelle is delivered to the cytosol of target cells. Tuning of micellar surface functionalities and their linkages to the block copolymers constituting micelles theoretically can be used as target delivery to a variety of intracellular compartments

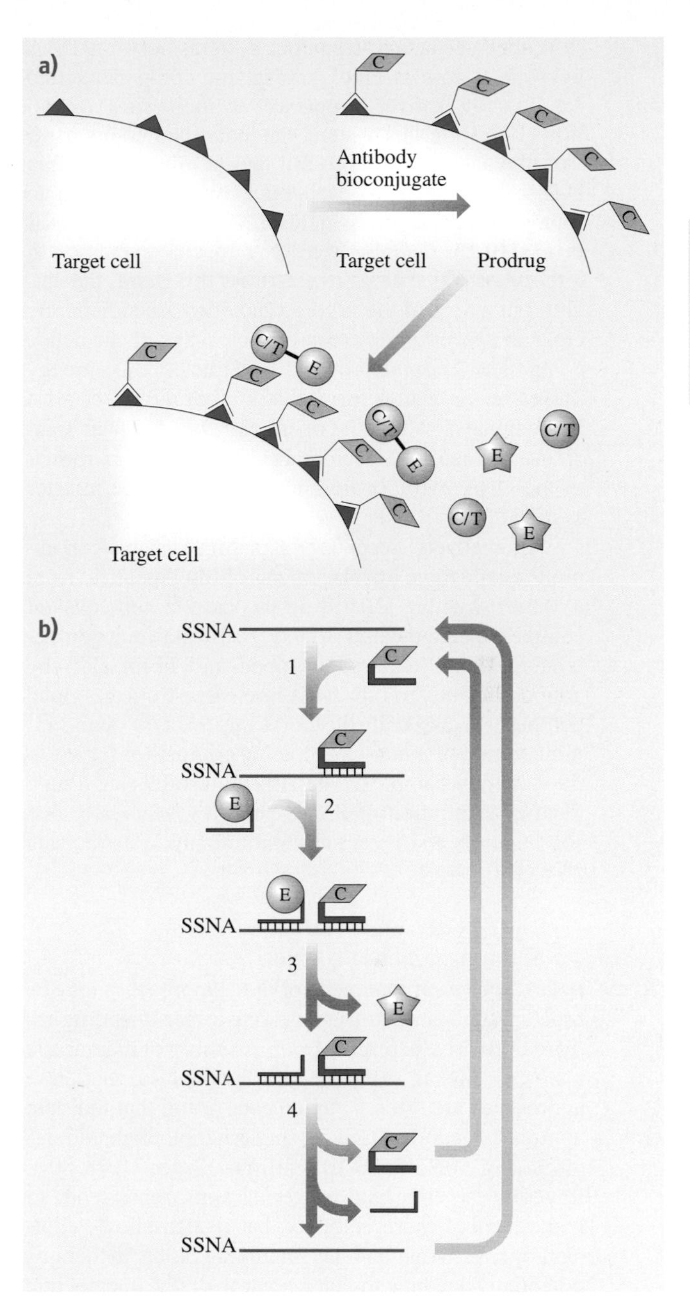

Fig. 10.15a,b Secondary biochemical signals can trigger release of active therapeutics. **(a)** Antibody directed enzyme-prodrug therapy (ADEPT) exploits enxyme-monoclonal antibody (mAB) bioconjugates to activate prodrugs [10.78]. Here, a mAB (like the native IgG of Fig. 10.1) is covalently liked to a catalyst (C in rhombus, usually an enzyme). The binding specificity of the monoclonal antibody causes the bioconjugate to bind to a desired target site (possibly by recognition of a vascular address, as shown in Fig. 10.11). The catalyst (enzyme) specificity of the bioconjugate is capable of triggering activation of a prodrug, much as in Fig. 10.12. The corresponding prodrug is then administered (represented as in Fig. 10.12). The enzyme localized at the desired target site of therapeutic action then activates the prodrug (represented here by cleavage of carrier/trigger moiety and change in shape of the effector to a star, as in Fig. 10.12). **(b)** Another secondary signaling triggering strategy [10.55]. Here, single stranded nucleic acid (SSNA) drives assembly of an activating complex for a prodrug. A catalyst (indicated again by C) is conjugated to an oligonucleotide that hybridizes to a single stranded nucleic acid (SSNA). A prodrug, containing an effector domain (E) and a second oligonucleotide, in this case homologous to a region of the same SSNA, and proximal to the site of hybridization of the first oligonucleotide, is also administered. Catalyst specificity is tuned to correspond to linkage between the prodrug effector and oligonucleotide. The catalyst in the assembled structure is sufficiently close to the prodrug to cleave the linkage between effector and oligonucleotide, releasing active drug (star). Potentially, prodrug and catalyst oligonucleotides might dissociate from the SSNA and the cycle could be repeated. Note the similarity to Fig. 10.10

Nanotherapeutic Triggering Using Secondary Signaling

Secondary signaling or multicomponent delivery systems are more complex delivery strategies (Fig. 10.15). These systems can feature site-specific and systemic delivery of one or more components. One class of such systems is the so-called ADEPTS devices (antibody directed enzyme-prodrug therapy, Fig. 10.15 [10.78]). ADEPTS systems feature affinity-based targeting of an enzymatic prodrug activator to a desired site, followed by systemic administration of a prodrug. ADEPTS is used primarily for oncology, and so the prodrug involved is generally one that can be activated to cytotoxicity. Since prodrug dosing is systemic, prodrug toxicity must be de minimus in the inactive form, and the activating catalyst must not be present at locations where cytotoxicity would be deleterious. To address this issue, the nanotherapeutic designer can chose to engineer an activation mechanism that has no physiological analogue in the host. In addition, tuned, high lability of the active drug can potentially delimit the site and extent of cytotoxic effect mediated by ADEPTS systems.

Other secondary signaling systems rely on DNA hybridization to assemble therapeutic components into catalytically active complexes at their sites of action [10.55]: in this system, the individual components lack therapeutic activity until they are brought within a few nanometers of each other by hybridization to a target single-stranded nucleic acid molecule (SSNA in Fig. 10.15, potentially a RNA species expressed in a tissue or cell type of interest). At this point, the catalytic moiety and the prodrug moieties are sufficiently close in space to allow catalytic cleavage of the active component from the prodrug. As depicted, this strategy makes no provision for delivery of drug and catalyst bioconjugates to the site of interest, nor for their transit across biological membranes. The system is similar to the DNA hybridization-driven fluorescence transfer system of Fig. 10.10.

Alternatively, secondary signaling systems can exploit competitive displacement of therapeutics from a carrier structure [10.79]. In this case, a non-covalent complex of engineered antibody and plasminogen activator (PA) is tethered to blood and fibrin clots by antibody affinity for fibrin. PA is released from the bound complex (to dissociate the clot) using bolus systemic administration of a nontoxic binding competitor for PA to the antibody complex. The strategy establishes a high local concentration of PA at clot sites, efficiently dissolving clots and potentially minimizing systemic side effects.

Layering Strategies for Fine Control of Nanotherapeutic Action

It should be clear that many of these approaches may be broadly applicable to trigger events other than drug release or to drive assembly or disassembly of therapeutic nanostructures in situ. It should also be clear that these approaches are often complementary, and that multiple approaches can be used in single nanotherapeutic devices (Fig. 10.3, Table 10.1 [10.11–13, 15, 71, 78, 80]). Layering targeting and triggering approaches tends to make devices more complex, but it also allows clinician/operators to intervene at multiple points in therapy, potentially leading to finer control of the therapeutic process and better clinical outcomes.

10.3.4 Sensing Modalities

The need for "smart" therapy is a key theme of therapeutic nanotechnology and pharmacology as a whole. Drugs with narrow therapeutic windows should be delivered only to their desired site of action and be pharmacologically active only when that activity is needed. These strategies can limit undesired secondary effects of therapy, some of which can be debilitating or life threatening, as discussed elsewhere in this chapter. One possible approach to this issue is the incorporation of sensing capability (specifically, the capacity to recognize appropriate contexts for therapeutic activity) into nanotherapeutic devices. Sensing capability may allow self-regulation of a therapeutic device, reporting to an external clinician/device operator (though this begins to touch on imaging applications, see below), or both. In the context of our discussion of therapeutic nano- and microscale devices, we will consider primarily electrical and electrochemical sensor systems, particularly microfabricated (Field Effect Transistor or FET, and cantilever) and conducting polymer sensors.

Sensor Systems

Sensing is predominantly a higher-order device functionality, depending on multiple device components, though one could argue that some targeting/triggering strategies, particularly targeting by bioaffinity and intrinsic triggering strategies, must, a priori, incorporate at least limited sensing capability. But biosensors, as they are typically considered, are multifunctional, multicomponent devices [10.81]. Usually a biosensor system is composed of signal transducer, sensor interface, biological detection (bioaffinity) agent, and an associated assay methodology, with each system component governed by its inherent operational considerations.

The transducer component determines the physical size and portability of the biosensor system. Signal transducers are moieties that are sensitive to a physical-chemical change in their environment and that undergo some detectable change in chemistry, structure, or state as the result of analyte (the thing to be sensed) recognition. Analytes for nanotherapeutic application could be biomolecules, like proteins, small molecules (organic or inorganic), ions (salts or hydrogen ions), or physical conditions (such as redox state, temperature). Interfaces are the sensor components that interact directly with the analyte. For sensor use in nanotherapeutic devices, immobilized or otherwise captured biological molecules (proteins, nucleic acids) often constitute the sensor interface. Whatever the chemical nature of the interface, it determines the selectivity, sensitivity, and stability of the sensing system and also is a dominant determinant of sensor operational limits. Assay methodology determines the need (or lack thereof) for analyte tracers, the number of analytical reagents, and the complexity and rapidity of the sensing process. Nanotherapeutics are of interest at least in part because they can be minimally invasive, low complexity, yet robust and accurate; convenient assay methods are therefore highly desirable.

Cantilever Biosensors

Micromechanical cantilevers (discussed above in conjunction with force microscopy, Fig. 10.6) transduce sensed events by mechanical means [10.82]. Both changes in the resonant frequency and deflection of cantilevers resulting from analyte binding or dissociation can be conveniently and sensitively detected. These changes in cantilever state can be conveniently detected by optical, capacitive, interferometry, or piezoresistive/piezoelectric methods, among others. Microfabricated cantilever dimensions range from micron to sub-micron range, with potential for further dimensional optimization (by carbon nanotubes appended to them, for instance [10.82]). They are operationally versatile and can be used in air, vacuum, or liquid, although they suffer some degradation in performance in liquid media. Like most micromachined structures, they can be batch fabricated and conveniently multiplexed.

Cantilevers used in atomic force microscopy (AFM, see also discussion of one-off nanostructures above) approaches can be used to study individual biomolecular interactions [10.35, 81]. In this approach, cantilever tips are derivatized with biomolecules (effectively, one member of a receptor/counter-receptor pair), and the tip-bound biomolecule is allowed to bind its counter-receptor (itself bound on a surface). Under non-equilibrium conditions, (i. e., conditions that result in thermodynamically irreversible changes in analyte molecular structures), the force required to disrupt single molecular interactions can be measured and related to classical biochemical parameters of receptor binding. The method has been applied to interactions between hormones and their receptors, sugars and lectins, as well as hybridizing DNA strands.

Cantilever systems sensitively detect changes in mass at their surfaces: changes as small as a mass density of $0.67\,\mathrm{ng/cm^{-3}}$ are theoretically detectable. This can allow detection of binding of extremely small objects to the cantilever and has been applied to detection and enumeration of prokaryotic and eukaryotic cells, as well as small numbers of macromolecules [10.82]. The incorporation of biological receptors or affinity reagents on the cantilever surface can drive specific binding events for particular sensing tasks. Microcantilevers are also highly sensitive to temperature, detecting changes as low as $10^{-5}\,\mathrm{K}$; they can also detect small changes in pH.

Field Effect Transistor Biosensors

Field effect transistor (FET) architectures are another sensing architecture that can be conveniently produced by micro-nanofabrication. FETs consist of a current source, a current drain, a conductive path (sensing channel) between them, and a sensing gate to which a bias can be applied. Analyte binding to the sensing channel induces a charge transfer resulting in a dipole between the surface and the underlying depletion region of the semiconductor: current that passes between the source and drain of a semiconductor FET is quite sensitive to the charge state and potential of the surface in the connecting channel region. Moving a standard silicon FET from depletion to strong inversion (i. e., shifting the surface potential by $>\sim 0.5\,\mathrm{eV}$) requires less than $\sim 10^{-7}\,\mathrm{C/cm^2}$ or $\sim 6\times10^{12}\,\mathrm{charges/cm^2}$, corresponding to transfer of $6.25\times10^{11}\,\mathrm{e/cm^2}$. With FETs of 2,000 square micrometers, detection of biological analytes in sub-nanomolar concentrations is easily feasible. Specificity for binding of macromolecular analytes of interest can be provided by deployment of biological affinity reagents in the FET sensing channel. Submicron FETs are routinely manufactured; use of carbon nanotubes in FETS will offer still greater miniaturization [10.43, 83].

Carbon nanotubes also have excellent mechanical properties and chemical stability in addition to potentially tunable electrical properties, making them highly desirable electrode/nanoelectrical materials for any number of nanoelectrical applications [10.84]. Biomolecules can be bound to carbon nanotubes, particularly in FET and nanoelectrode applications. Most biomolecules bound to carbon nanotubes are not covalently bound (as discussed above) and do not exhibit direct electrical communication with the nanotube, though redox enzymes bound to nanotubes and other conductive nanomaterials may [10.84, 85]. Flavin adenine dinucleotide (FAD) and flavoenyzme glucose oxidase (Gox) both display quasi-reversible one electron transfer when absorbed onto unannealed carbon nanotubes in glassy carbon electrodes. Gox so immobilized retains its substrate-specific (glucose) oxidative activity, leading to applications in sensing circulating glucose for diabetes and, perhaps, to a strategy of harvesting electrical power from metabolic energy.

Conducting Polymers and Sensor Biocompatibility

Biocompatibility of most metallic structures (as might be used in the bioelectrical sensors described) is limited at best; metal structures rapidly foul with serum proteins (i. e., become opsonized), undergo electrochemical degradation, or have other problematic properties. Polymer chemistry, however, has the capacity to tune composition to enhance biocompatibility properties and is commonly used to make synthetic surfaces more biologically tolerable. Electrically conductive polymers are potentially attractive in this context.

Polymeric materials are available with intrinsic conductivities comparable to that of metals (up to $1.5 \times 10^{7}\ (\Omega\,\mathrm{m})^{-1}$, which, by weight, is about twice that of copper [10.86]). Significant conductivity has been documented for a dozen or so polymers, including polyacetylene, polyparaphenylene, polypyrrole, and polyaniline, doped with various impurities. Careful control of doping can tune electrical conductivity properties over several orders of magnitude [10.87].

Conductive polymeric materials may be extremely well suited to biosensing applications. The polymeric materials themselves are often compatible with proteins and other biomolecules in solution. Furthermore, polymers are amenable to very simple assembly of sensor transducer-interface components by deposition of the polymer and trapped protein (sensor interface) directly on a metallic micro- or nanoelectrode surface [10.87,88]. Combined with a facile nanoelectrode array microfabrication method [10.89], simultaneous conducting polymer-interface protein deposition may offer an extremely simple way to fabricate multiplexed sensor arrays.

These electrochemical and electrophysical sensing modalities are of little use in and of themselves: they must communicate with either other device components (such as drug-dispensing effector components) or with external observers or operators. Sensor coupling in autonomously operating devices produced by microfabrication can be done directly, and the coupling linkages incorporated into the fabrication protocol. Coupling with biological device components can also be direct, as when conductive materials are conjugated to biomolecules, and can directly modulate their activity [10.73].

Sensor-device coupling can also be indirect, through electrochemically produced mediator molecules [10.85]. Sensors might be independent of nanotherapeutics and report conditions at the site of therapeutic action to an external operator, who would use any one of the external triggering strategies discussed above to engage therapy when and where appropriate. Communication/sensor interrogation might be accomplished most crudely by direct electrical wiring of sensors to an external observation station. Alternatively, if the event to be detected is transduced optically, as in colorimetric smart polymers that change optical properties in the presence of analyte [10.90], sensors might be interrogated by fiber optics. These modalities are most applicable to sensor arrays delivered to the site of interest by a catheter. The use of a catheter may be justified in some therapeutic applications, but it is an invasive procedure and is not optimal.

We have already seen multiple examples of prodrugs that are activated by cleavage of an inhibitory domain from the complex. One noninvasive approach to communication with external operators could exploit this phenomenon by detecting the cleaved fragment in bodily fluids. If the cleaved moiety cleared through urine, the extent of drug activation could be monitored noninvasively via urinalysis. There is no a priori need to connect the cleavage event to drug activation. For instance, an operator might administer a catalyst that cleaves a detectable material from a nanotherapeutic. If this secondary signaling moeity was independently targeted to the desired site of therapy, presence of the detectable cleaved product would provide information regarding the bodily location of the therapeutic nanodevice. The cleaved product would not be detectable unless the therapeutic and secondary signaling moiety co-localized at a single site. Other sophisticated communication involves ultrasound or electro-magnetic radiation to carry information. Application of these sorts of modalities currently occurs in in vivo imaging approaches.

10.3.5 Imaging Using Nanotherapeutic Contrast Agents

Imaging is a minimally invasive procedure that allows visualization of organs and tissues following the administration of a detectable moiety (contrast agent). The contrast agent is then exposed to some condition that interacts with the contrast agent so as to produce an emission or response detectable to an external monitoring device. Nano-sized particles (5–100 nm in diameter) have found application as contrast-enhancing agents for medical imaging modalities such as magnetic resonance imaging (MRI, our primary topic, reviewed in [10.91, 92]). MRI currently provides cross-sectional and volumetric images with high spatial resolution (< 1 mm) and is potentially applicable to many clinical purposes, though enhanced imaging capabilities are desirable. For instance, blood flow measurements of healthy and diseased arteries can be quantified better by the aid of improved contrast agents that highlight blood flow at the vessel wall. Similarly, some tissues (certain tumors) do not exhibit strong contrast within the MRI field and cannot be readily identified or characterized at present. It is of great interest, therefore, to develop nanoscale particles that provide enhanced contrast for many applications.

Magnetic Resonance Imaging (MRI): the Basics

Objects to be imaged are exposed to a strong magnetic field and a well-defined radio frequency pulse. The external magnetic field (B_0) serves to loosely align protons either with (lower energy level) or against (high energy level) the field, the difference between the two energy levels being proportional to B_0. Once the protons are separated into these two populations, a short multi-wavelength burst (or pulse) of radio frequency energy is applied. Any particular proton will absorb only the frequency that matches its particular energy (the Larmor frequency). This *resonance* absorption is followed by the excitation of protons from the low to high energy level and of equivalent protons moving from high to low energy levels. After the radio frequency pulse, protons rapidly return to their original equilibrium energy levels. This process is called relaxation and involves the release of absorbed energy. Once equilibrium is again established, another pulse can be applied.

Data is collected by positioning a receiver perpendicular to the transmitter: relaxation energy release induces a detectable, quantifiable signal (i. e., in amplitude, phase, and frequency) at the receiver coil. Since multiple protons in multiple chemical environments are involved, the signal at the receiver includes many frequencies. The received signal consists of multiple, superimposed signals (called the free induction decay or FID) signal, resulting from the relaxation of multiple, chemically distinct protons. FID is converted from the time domain to frequency domain (by Fourier transformation), within which individual proton types can be identified. Iterations of this procedure in two or three dimensions can create a high-resolution image of anatomical cross section or volume.

Intrinsic factors affecting image quality include the proton density of the tissues, local blood flow, and two relaxation time constants [10.93]: longitudinal relaxation time (T_1) and transverse relaxation time (T_2). Control of T_1 and T_2 relaxation effects are most critical for high-resolution MR images. T_1 relaxation measures energy transfer from an excited proton to its environment. In tissues, protons of fats and cholesterol molecules (relatively movement-constrained macromolecules) relax efficiently after a pulse and exhibit a short T_1 time. Water in solutions (a small molecule tumbling relatively freely in solution) has a much longer T_1 time. T_2 relaxation measures the duration of coherency between resonating protons after a pulse, prior to their return to equilibrium. Tightly packed, solid tissues with closely interacting hydrogen nuclei relax more quickly than loosely structured liquids, so tissues such as skeletal muscle have a short T_2s, while cerebrospinal fluid has a very long T_2. Intrinsic factors are manipulated through extrinsic factors, such as the external magnetic field strength, the specific pulse sequence, etc., allowing the collection of meaningful MR images.

Nanoparticle Contrast Agents

Signal intensity in tissue is influenced linearly by proton density, while changes in T_1 or T_2 result in exponential changes in signal intensity. T_1 and T_2, therefore, are manipulated to enhance imaging by administration of exogenous contrast-enhancing agents. MRI contrast agents are divided into paramagnetic, ferromagnetic, or superparamagnetic materials. Metal ion toxicity is an unfortunate consequence of physiologic administration of contrast agents but can be mitigated somewhat by complexation of the metals with organic molecules.

Paramagnetic metals used for enhanced MRI contrast (gadolinium, Gd, iron, Fe, chromium, Cr, and manganese, Mn) have permanent magnetic fields, though the magnetic moments of individual domains are unaligned [10.94]. Upon exposure to an external magnetic field, individual domain moments become aligned, generating a strong local field (up to 10^4 gauss [10.95]). Paramagnetic metal ions interact with water molecules, causing an enhanced relaxation of the water molecules via tumbling of the water-metal complex, dramatically decreasing the T_1 value for the water molecules and enhancing the proton signal [10.94]. Contrast enhancement by paramagnetics is thus due to the indirect effect the contrast agent has on water and its magnetic resonance properties.

Ferromagnetic and superparamagnetic materials both contain iron (Fe) clusters, which generate magnetic moments 10 to 1,000 times greater than do individual iron ions. Clusters greater than 30 nm in diameter are ferromagnetic, whereas smaller particles are superparamagnetic [10.93]. Ferromagnetic materials maintain their magnetic moment after the external field is removed, but superparamagnetic materials lose their magnetic field after the field is removed, as do paramagnetics. Both ferromagnetic and superparamagnetic substances minimize the proton signal by shortening T_2 [10.96], resulting in negative contrast (i. e., darkening of the image [10.97]).

Nanobiotechnological Contrast Agent Design

First generation contrast agents often contained a single metal ion/complex, whereas emerging agents incorporate nanoscale metal clusters, crystals, or aggregates, sometimes encapsulated within a synthetic or biopolymer matrix or shell [10.98–100]. These metal cluster agents improve contrast effects and, hence, output MR images profoundly. Furthermore, surface chemical groups (from the matrix or shell) can be derivatized to improve biocompatibility or allow targeting to a tissue or site of interest.

Typically particles are prepared from colloidal suspension where metallic cores are thoroughly mixed with the matrix material before being aggregated out of solution with a non-solvent. Dextran (a polymer of 1,6-β-D-glucose) is a typical matrix material used in commercial imaging reagents: Combidex is coated with 10,000 molecular weight dextran [10.99, 101, 102], Feridex has an incomplete, variable dextran coating [10.102], and Resovist is coated with carboxydextran [10.97]. Other polymeric materials are also used (oxidized starch [10.98, 100]), and matrix-less particles are also produced [10.103].

Nanoparticle contrast agents must be purified under tightly controlled conditions (generally by centrifugation or high pressure liquid chromatography [10.104]) and accurately characterized (for size dispersity by light scattering, chromatography, photon correlation, or electron microscopy [10.100, 104]) to assure the reproducibility of imaging agent production lots. Elemental analysis, X-ray powder diffraction, and Mössbauer spectroscopy have also been used to characterize metallic cores [10.103, 105]. Tight definition of the finished particles ensures accurate correlation between structural properties of imaged materials and prior in vitro and in vivo studies, allowing collection and interpretation of meaningful images. Metallic cores generally range from 4 to 20 nm in diameter [10.106], while coated particles can be up to 100 nm or more in diameter [10.105]. Compared to larger cores, however, nanoparticles less than 20 nm in diameter exhibit considerably longer blood half-life and improved T_1 and T_2 relaxivity effects [10.107].

As previously discussed, clearance of nanoparticles via the RES is a critical problem usually approached by surface modifications to mitigate nonspecific adsorption of proteins to the biomaterial surface (i. e., the opsonization of synthetic materials). Neutral, hydrophilic surfaces tend to adsorb less serum protein than hydrophobic or charged surfaces. Bisphosphonate and phosphorylcholine-derived thin film coatings have been applied to nanoparticles to stabilize iron oxide particles against pH, opsonization, and aggregation [10.108]. Such thin films do not fully eliminate protein adsorption, and dense layers or thick brushes of polysaccharides or hydrophilic polymers may more effectively avoid opsonization [10.109].

Other, as yet incompletely understood, biological factors influence the use of nanoparticles in vivo. For instance, a direct correlation between the circulating half-life of nanoparticles (i. e., their $T_{1/2}$s) and age of animals used has been reported [10.106]. The observed increase in $T_{1/2}$ may be correlated to age-related changes in phagocytic activity [10.106]. Local environments also influence particle stability: iron oxide particles are degraded at a pH of 4.5 or less [10.100], a condition sometimes attained in some intracellular vesicles.

First generation contrast agents were primarily blood-pool agents that moved freely through the entire vasculature. Targeting contrast-enhancing nanoparticles to sites of interest can reduce heavy metal toxicity associated with the commonly used agents by diminishing the dose required to obtain an acceptable image. Contrast agent targeting can also provide enhanced diagnostic information. For instance, nanoparticles that bind to molecular fibrin at a clot site on a vessel wall have been developed [10.110], potentially allowing differentiation between vulnerable and stable atherosclerotic plaques. Similarly prognostically valuable data regarding disruptions of the blood–brain barrier (BBB) have been visualized with MRI as the result of delivery of contrast-enhancing nanoparticles to the affected site [10.102].

10.4 Applications for Nanotherapeutic Devices

As discussed above, nanotherapeutic devices are novel, emerging therapeutics with properties not fully understood or predictable. Nanotherapeutics, therefore, must be justifiable on at least two levels. As we have seen, the nature of the therapeutic task and the state of current nanoscale-materials technology make the incorporation of biological macromolecules unavoidable for many nanotherapeutics. Proteins, for example, typically are substantially more expensive than small molecule therapeutics, and precise nanostructures containing proteins will be more costly still. Nanotherapeutics must justify their high COGs. Secondly, as new therapeutic modalities, nanotherapeutics may carry significantly larger risks than those associated with more conventional therapies. Expensive, novel moieties, such as nanobiotechnological therapeutic devices, are therefore most likely to be accepted for treatment of conditions that not only are accessible to intervention at the nanoscale but also for which existing therapeutic modalities have acknowledged shortcomings in patient morbidity or mortality. We have selected two disease states sufficiently grave and sufficiently unserved to warrant nanotherapy: cancer and cardiovascular disease. Modulation of immune responses and vaccination is our third application area.

10.4.1 Nanotherapeutic Devices in Oncology

The economic burden imposed by cancer is immense, measuring in the billions of dollars annually in the United States alone. Existing therapies such as surgical resection, radiotherapy, and chemotherapy have profoundly limited efficacy and frequently provide unfavorable outcomes as the result of catastrophic therapeutic side effects. Additionally, the biology, chemistry, and physics of cancer, in general, and solid tumors in particular, provide therapeutic avenues accessible only by nanoscale therapeutics. Oncology is thus an ideal arena for emerging nanotechnological therapies.

Tumor Architecture and Properties

Tumors as tissues are relatively chaotic structures exhibiting vast structural heterogeneity as a function of both time and space [10.9, 77, 111, 112]. In healthy tissues, vasculature resembles a regular mesh in which the mean distance of tissue spaces to the nearest vessel is tightly controlled and highly uniform. On the other hand, the vasculature of tumors resembles a percolation network containing regions experiencing vastly different levels of perfusion. Tumors of 1 mm^3 or larger typically contain measurably hypoxic domains, with pO2 values as much as two- to threefold lower than in normal tissue. High levels of hypoxia are characteristic of enhanced metastatic potential and tumor progression. Also due to insufficiency of perfusion, tumors frequently contain necrotic domains.

The average torturosity of vascular flow paths in tumors is also much greater than in healthy tissue, and transient thrombotic events lead to enhanced resistance to flow and ongoing vascular remodeling. Aside from its plasticity, tumor vasculature itself is highly irregular, may be incompletely lined with endothelial cells, and often exhibits significantly higher extravasation limits (the highest molecular weight of materials that can leave the vasculature and diffuse into the interstitial spaces) than normal vasculature. Tumor tissue is also poorly drained by lymphatics, so extravasated materials tend to remain in situ in tumors and are not cleared efficiently.

These biological phenomena all differentiate tumor tissue from normal tissue and can be exploited in therapy. For instance, the special vascular integrity and lymphatic drainage properties of tumors constitute the Enhanced Permeability and Retention effect (EPR) [10.11, 12, 80, 113]. EPR presents an obvious opportunity for intervention with nanoscale therapeutics (Fig. 10.3). Extravasation limits for normal tissues are variable, but structures larger than a few nanometers in diameter do not leave circulation efficiently in most tissues, whereas tumor vasculature frequently allows egress of materials in the tens to hundreds of nanometer range. Further, nanomaterials, once extravasated, are not cleared by lymphatic drainage. EPR provides tumor targeting that does not depend on biological affinity reagents: tumor tissue provides preferential depot sites for extravasated drug delivery devices. Targeting to a desired site of action is highly desirable for cytotoxic therapeutics, and EPR provides the basis for a growing class of polymer therapeutics [10.11, 12, 80, 113]. That said, EPR and targeting by biological affinity are not necessarily mutually exclusive. A number of antigens more or less specifically related to tumors are known (tumor associated antigens or TAAs), and as discussed above tumor-/organ-specific vascular addresses might be exploited for delivery of therapeutics [10.65, 66]. For instance, various biological reagents that recognize TAAs (i. e., antibody to carcinoembryonic antigen, transferrin) have been conjugated to nanoscale contrast agents to enhance contrast agent localization to tumoral sites [10.104].

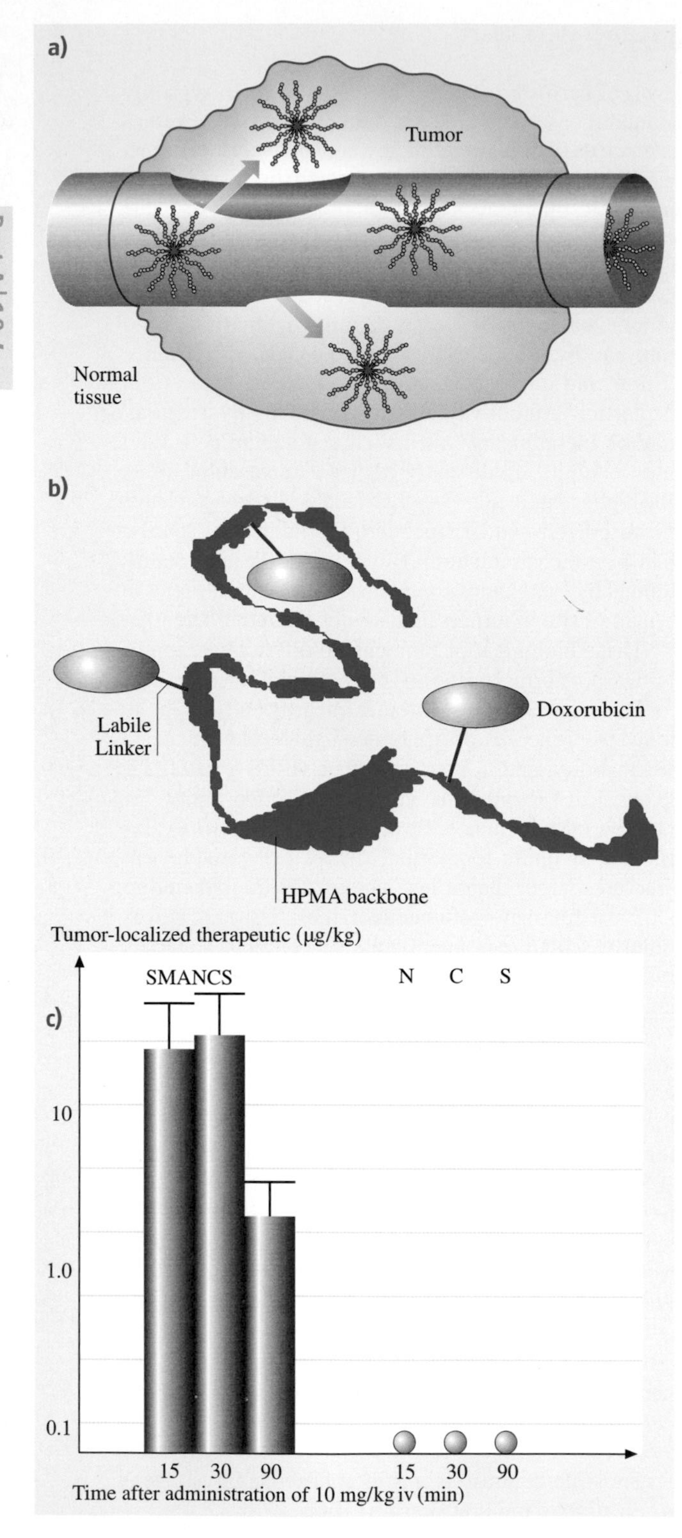

Fig. 10.16a–c Therapeutic nanodevices can be specifically delivered to tumors by virtue of their size [10.11, 12, 80, 113]. (**a**) An illustration of the relatively low integrity of tumor vasculature. Therapeutic nanodevices (in this case, micelles carrying cytotoxins) cannot leave (i. e., extravasate from) the vasculature of normal tissue, but the vasculature of tumors is sufficiently leaky to allow specific delivery to tumoral interstitial spaces. This is the basis of size-dependent targeting to tumors by the enhanced permeability and retention, or EPR effect. (**b**) A polymer nanotherapeutic (PK1, [10.80]) that exploits EPR effect to deliver a cytotoxin (doxorubicin) to tumors. The overall mass of the complex is tuned by control of the mass of the N-2-(hydroxypropyl) methacrylimide backbone, so as to be above extravasation limits for normal tissues but below both tumor extravasation limits and the size limit for excretion through the kidney. A final feature of the device that favors delimiting doxorubicin cytotoxicity to tumoral sites is the fact that the device is a prodrug. The cytotoxicity of doxorubicin is severely curtailed when the doxorubicin is covalently linked to the polymer therapeutic, though it manifests its full toxicity when it is released from the complex by cleavage of the labile (peptidic) linker. The labile linker contains a cleavage site recognized by a protease known to be over-expressed in the target tumor type. Thus, the device produces a depot of inactive doxorubicin in tumors that is activated preferentially to full toxicity by conditions prevalent in the tumor. (**c**) EPR can drive localization of appropriately sized therapeutics to tumoral sites [10.113]. The polymer therapeutic SMANCS accumulates rapidly in murine tumors, whereas the NCS, the parent therapeutic (not engineered for EPR delivery), does not

Tumor (EPR) Properties and Nanotherapeutics

PK1 (Fig. 10.16) is an early example of a growing and increasingly sophisticated class of polymer therapeutics [10.11, 12, 80, 113]. It is a nanoscale molecular therapeutic device wherein the therapeutic moiety (doxorubicin) is covalently linked to the polymeric backbone. As we will see, each component of the polymer therapeutic fulfills a discrete function, so the device rubric is warranted [10.6].

PK1 consists of a polymeric backbone (HPMA, hydroxyl polymethacrylamide) which, as discussed below, targets desired sites by virtue of its specific nanoscale size (by EPR effect) and a cytotoxin (doxorubicin), covalently linked together by a peptide. PK1 exhibits multiple design features that allow it to preferentially deliver doxorubicin to tumors. The size of the complex is tuned to be too large to extravasate in healthy tissue but still small enough for renal clearance. Effectively, tumors and the excretory system compete for PK1. This aspect of the relationship between therapeutic size and systemic tolerability is illustrated in Fig. 10.16. The therapeutic device has to be sufficiently large to have limited access to healthy tissues but must have relatively unimpaired access to tumors and the excretory system via the kidneys.

A second engineered feature of PK1 relates to tumor-specific release of cytotoxin. Doxorubicin is toxic as a free molecule but is nontoxic as a part of a polymer complex. In PK1, doxorubicin is linked covalently to the polymeric backbone by a peptide chosen specifically because it is the substrate for a highly expressed protease in the target tumor. Thus PK1 deposits cytotoxin to tumors in an inactive form that is processed to an active form preferentially by tumoral metabolic activity. In all, the design features of PK1 act to minimize systemic exposure to cytotoxin and augment tumoral exposure, thereby enhancing therapeutic benefit and diminishing undesired side effects.

Therapeutics need not be covalently linked complexes to be delivered by EPR. In some cases, bioactives can be engineered to interact with blood constituents non-covalently so as to bring the apparent size of the complex into a range that allows preferential tumor deposition. This is the mechanism of delivery for the oldest EPR-exploiting therapeutic in clinical use, SMANCS (S-Methacryl-neocarzinostatin) [10.113]. Neocarzinostatin (NCS) is a small (12 kDa, a few nm in diameter), cytotoxic protein that elicits its toxicity by generating intracellular superoxide radicals enzymatically, which in turn damage DNA. Unmodified NCS is cleared rapidly and exhibits dose-limiting bone-marrow toxicity. Both problems are mitigated by the site-specific conjugation of two small (about 2 kDa and less than 1 nm in length) styrene-comaleic acid polymer chains, resulting in SMANCS. The $T_{1/2}$ of SMANCS is about ten-fold higher than that of NCS, though SMANCS itself is too small to accumulate preferentially in tumors by EPR. This is mitigated by the fact that SMANCS (but not NCS) is efficiently bound by serum albumin. The resulting non-covalent complex is sufficiently large not to extravasate into normal tissue, though it still partitions efficiently into many tumors. The EPR-driven partitioning of the non-covalent SMANCS-albumin complexes into tumors suggests that other non-covalent complexes (such as the multi-dendrimer device of Fig. 10.3, were it pre-assembled prior to administration) might also target to tumors via EPR.

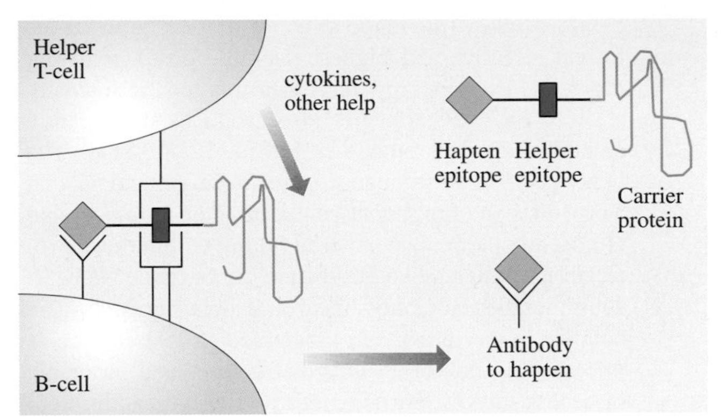

Fig. 10.17 Antibody responses [10.1] are efficiently triggered by antigens that contain both B-cell epitopes (diamond) and helper T-cell epitopes (rectangle). The B-cell epitope binds a receptor on the B-cell surface (the Y shape shown interacting with the B-cell epitope) that is essentially a membrane-bound form of the antibody species that the particular B-cell can produce. The peptidic T-helper epitope engages another receptor complex on the B-cell surface (class II MHC, the two part receptor schematically depicted as interacting with the T-helper epitope, see text). When presented in the context of the class II MHC complex of the B-cell, the T-helper epitope can be recognized by a cognate T-cell receptor on the helper T-cell surface (shown as a receptor on the T-cell surface interacting with the T-helper epitope/class II MHC complex). In response to recognition of its appropriately presented, cognate T-helper antigen, the helper T-cell provides multiple stimulatory signals that cause the B-cell to proliferate, secrete antibody, and differentiate. This is a classical hapten-carrier configuration of an immunogen [10.1], in which a synthetic B-cell epitope is the hapten (in our cases, a PAMAM dendrimer or SWCNT epitope), and the protein covalently linked to the hapten is the carrier, providing the T-helper epitope. Note that the antibody produced recognizes *only* the B-cell epitope and *not* the T-helper epitope. The T-helper epitope also need not be present for antibody recognition of carrier epitopes, nor is its presence necessary to maintain the ongoing antibody response once antibody production is triggered. Note that whether the T-helper epitope remains covalently linked to the B-cell epitope during antigen presentation (here, the intact carrier protein is shown linked to the B-cell epitope) is debatable. Proximity of the B-cell epitope/receptor complex and T-helper epitope/class II MHC complex on the B-cell surface may be sufficient to trigger a robust antibody response

Delivery of antitumor agents using EPR can be highly efficacious. As shown in Fig. 10.17, conjugation of therapeutics to produce controlled-size delivery complexes can enhance tumoral accumulation by multiple orders of magnitude relative to free therapeutics [10.11, 12, 80, 113]. In effect, EPR targeting broadens the therapeutic window (the range of concentration between the lowest effective and highest tolerable dose), allowing more intensive therapy. Increased therapeutic intensity allowed by EPR strategies also produces significantly enhanced clinical outcomes [10.11, 12, 80, 113]. The benefits of EPR strategies seem to be general in that a broad range of materials can be targeted to tumors and almost any particulate material falling within the appropriate size range will partition into tumor tissue via EPR. This includes not only linear and dendritic polymers but also higher-order, self-assembled structures such as micelles or shell cross-linked structures and inorganic materials such as ferromagnetic particles. In addition to EPR tumor targeting, each of these structures has intrinsic properties that might be exploited in therapeutic nanodevices (see the discussion of imaging above). Targeted contrast agents may be particularly important for early tumor detection.

10.4.2 Cardiovascular Applications of Nanotherapeutics

The cardiovascular system is one of the most dynamic and vital systems of the body. In principle, every cell in the body is accessible to the bloodstream via the vasculature. If controlled navigation of, and extravasation from, the vasculature can be accomplished, therefore, the potential of nanotherapeutic devices is virtually unlimited. But as mentioned throughout this chapter, opsonization and immune clearance, as well as targeting and triggering of therapeutics, remain key issues to address in order to facilitate wide application of nanotherapeutic devices to cardiovascular disease. In this section we provide a brief overview of categories and examples of nanotherapeutic devices used in blood-contacting applications.

Cardiovascular Tissue Engineering

A usable, immunologically compatible artery for coronary bypass is not always available, so synthetic or semi-synthetic artery substitutes are highly valuable. Tissue engineering approaches might mitigate the lack of acceptable homologous artery, as long as an acceptable artery substitute can be produced. Current tissue engineering approaches involve synthesis of three-dimensional, porous scaffolds that allow adhesion, growth, and proliferation of seeded cells to generate a functional vessel. Cellular organization and growth in synthetic scaffolds are often chaotic and random, and sufficient blood supply is seldom achieved within the scaffold. These conditions often result in death of the seeded cells and compromise the integrity of the vascular prosthesis.

Microfabrication techniques have allowed miniaturization of the tissue engineering scaffolds. These techniques allow the molecular-level control of cellular adhesion and propagation by control of surface topography, surface chemistry, and arrangement of morphogenic and proliferative signaling ligands and cytokines. Microfabrication schemes are being developed for the elucidation of parameters necessary for cellular attachment and orientation on well-defined silicon- and polymer-based surfaces. One particularly powerful application of controlled cellular growth is for the development of artificial capillaries to remediate vascular disease [10.114] (Moldovan, unpublished). Micro-machined silicon or polymer channels of the dimensions of capillaries have shown promise as effective conduits to direct endothelial cells to form tubes through which a fluid (blood) could eventually flow.

MEMS technology and nanoscale control of molecular events and interactions has also been applied to the development of cardiovascular sensors. For instance, blood cell counters have successfully detected red and white blood cells and show promise for quick, reliable blood cell analysis [10.115]. Additionally, force transducers have been developed to measure the contractile force of a single cardiac myocyte [10.116]. Basic research into cellular function and response to stimuli such as these will eventually support more efficacious and robust cardiovascular tissue engineering approaches.

Nanoparticulate Carriers for Therapy and Imaging

Two major cardiovascular applications of nanoparticles are targeted contrast agents for imaging (MRI, CT, X-ray, etc.) and targeted drug delivery vehicles to address vascular lesions (atherosclerotic plaques) and other vascular disease states. In either application, nanoparticle surfaces must be appropriately modified to avoid RES clearance and to drive interaction with the desired effector cells. Decades of research in the area of molecular and cell biology have provided researchers with many tools to make the appropriate protein component selection for nanotherapeutic contrast agents. For instance, knowing that the procoagulant protein tissue factor is expressed only at the surface of injured endothelial cells provides a means to differentiate healthy from diseased vasculature. It also provides a method to target disease sites. Similar sorts of information regarding disease and normal vasculature at specific tissue sites can be critically important for targeting specialized vascular beds, such as that which occurs at the BBB. Several groups have begun efforts to develop nanoparticles that selectively pass across the BBB to deliver chemotherapeutic or other drugs to the brain [10.117, 118], and this may improve outcomes from stroke.

With regard to MRI approaches, particles used for targeted contrast enhancement in specific tissues or vascular beds may be quite powerful for detection of unstable atherosclerotic plaques and quantitative analysis of blood flow. For instance, nonspecific accumulation of nanoparticles within human atherosclerotic plaques has been reported [10.119, 120] and may ultimately provide a way to distinguish vulnerable and stable plaques. Such data might be used to direct therapy, with the intent, as always, to improve therapeutic outcome. Furthermore, nanotherapeutic devices have been designed to include both contrast and drug delivery functions (Fig. 10.3). Single-platform, multifunctional particles for cardiovascular therapy represent a cutting edge in the cardiovascular field. These delivery vehicles are being designed and improved continual for longer circulation time, more efficient targeting [10.66, 121], degradability, and biocompatibility [10.122].

10.4.3 Nanotherapeutics and Specific Host Immune Responses

Immune responses directed to immunogens (molecules or structures capable of eliciting specific immune responses, also called antigens) are a pervasive aspect of the responses of vertebrates to exposure to foreign macromolecules, proteins, particles, and organisms (for overviews, see [10.1, 123, 124]). Undesired antibody responses, in particular, are the single most important hurdle for clinical use of recombinant proteins. On the other hand, undesired cellular immune responses are key mediators of frequently devastating disorders for patients, whether the responses are directed to self (as in autoimmune disease) or to foreign antigens (as in transplant rejection). As we shall see, host immune responses to nanotherapeutic devices are a critical determinant of their efficacy. Any strategy (including the biotechnological strategies) that allows controlled manipulation of immune responses either to augment desirable responses, as in vaccines, or to mitigate deleterious responses holds substantial potential benefit.

Basic Immunology

Specific immune responses are directed to individual macromolecules or assemblages of them and can be categorized into cellular responses, mediated by cytotoxic T-lymphocytes (CTLs), and humeral responses, mediated by soluble proteins (found in blood serum and other biological fluids) secreted by B-cells (antibodies). Regardless of the effector (CTLs or antibodies), specific responses are directed to individual molecular features of antigens called epitopes. The three types of epitopes are: cytotoxic T-cell (CTL) epitopes, helper T-cell epitopes, and B-cell epitopes. The first two types are peptides that must be presented on cell surfaces in the context of the class I (in the case of CTL epitopes for CTL responses) or class II (in the case of T-helper epitomes, involved in both antibody and CTL responses) major histocompatibility complex (MHC) molecules. B-cell epitopes are a chemically and structurally heterogeneous group of epitopes and include synthetic molecules as well as peptides. In addition to CTL and B-cell epitopes, which are recognized by immune effectors, T-helper epitopes are recognized by regulatory components of the immune system (helper T-cells). Vibrant CTL and antibody responses require immunogens containing T-helper epitopes along with either B-cell or CTL epitopes.

While epitope-presenting MHC class I and class II complexes must be present on cell surfaces to trigger immune responses, the machinery for antigen processing and loading of MHC class I or class II proteins for antigen presentation reside in distinct membrane-bound compartments in the cell [10.1]. For instance, processing for class I presentation (to trigger CTL responses) occurs through the cytosol, but charging of MHC class I complexes is accomplished in the endoplasmic reticulum. On the other hand, processing for class II presentation (to augment antibody and CTL responses), as well as the loading of class II, occur in endosomal compartments. Although these are membrane bounded compartments in the cytosol, topographically the interiors of endosomal compartments are part of the extracellular environment and are not inside the cell at all.

In the case of both class I and class II, the antigen presenting complexes are moved to the cell surface after charging. The moving of molecules between different cellular locations (called trafficking) is tightly regulated by and driven by specific cellular proteins, systems, and structures. Proper trafficking is essential to assure that all cellular components assume their correct position in the cellular organization, which in this case, is to assure antigens are directed to the compartments that will (most often) produce most protective immune responses. These complex antigen processing and presentation systems can be perturbed and exploited to offer novel opportunities in vaccine design.

Nanotherapeutic Vaccines: Eliciting Desired Host Immune Responses

Both therapeutic and prophylactic vaccines are naturals for nanobiological approaches. Therapeutics that modulate immune responses (antibody, CTL responses, or both), vaccines typically produce desired immunity to disease or pathogens. Ideally they are chemically well defined, safe, and induce long term (preferably, multiyear) protective immune responses [10.125]. Recent developments in immunobiology have opened opportunities for nanobiological vaccine design [10.126, 127].

Particulate Vaccines

The immune system, through dendritic cells (DCs), other antigen presenting cells (APCs), and immune effectors, performs ongoing surveillance for non-self antigens using the MHC antigen presentation system described above [10.1]. It has long been known that APCs sample serum for antigens that they process and present using the MHC class II pathway for antibody production [10.1]. Research of the last few years has shown that DCs, macrophages, and other APCs also take up particulate materials and process and present the constituent peptide epitopes via the MHC class I pathway to trigger specific CTL responses. Previous dogma held that only endogenous antigens (such as those produced in virally infected cells) could be presented through the class I MHC pathway. Both soluble and particulate antigens can be presented via the class I pathway, but particulate antigens are orders of magnitude more potent than the same immunogens formulated in soluble form [10.125–128].

APCs are not particularly fastidious regarding chemical composition of the particles they take up: latex beads [10.126, 128] as well as iron beads [10.126] have been used. Self-assembled nanoparticles made of lipidated epitope peptides also deliver orders of magnitude of better immunization than do soluble formulations of the same peptides [10.129]. The specific size for optimal CTL responses is not fully defined but clearly is in the nanometer to micrometer size range [10.125, 126, 128, 129]. One could imagine a systematic exploration of optimum size and compositional preferences for uptake by APCs using such chemically and morphologically tunable polymeric nanomaterials as dendrimers and tectodendrimers.

Nanobiological Design of Nanotherapeutic Vaccines

As mentioned earlier, antigen trafficking determines which specific MHC complex (class I or class II) epitopes are presented on and, therefore, which immune effectors are produced in the subsequent response. The process, while tightly regulated, is amenable to intervention using nanotherapeutic vaccines. Nanobiological design strategies (discussed above) can take advantage of the numerous proteins and peptides now known to mediate the trafficking of carried materials to specific sites. In the device context, these trafficking moieties might assure delivery of antigens to the appropriate cellular compartments to trigger desired class I- or class II-mediated responses.

Vaccine polypeptides, for instance, might be fused or formulated with pathogen polypeptides that will deliver them across epithelium from outside the host for use in mucosal vaccination [10.130]. Still other peptides are known to carry injected materials across the cell envelope into the cytosol [10.131]. Alternatively, vaccine entities taken up by receptor-mediated endocytosis (into endosomal vesicles, the traditional site of uptake for antigens destined for MHC class II presentation) could be shuttled into the cytosol (and hence to the class I MHC pathway) by the incorporation of peptide domains of bacterial toxins that mediate endosomal escape to cytoplasm [10.132]. Potentially, nanobiological therapeutic design strategies [10.5, 13, 15] can combine knowledge of polymer chemistry, nanomaterials, chemistry, cell, molecular, and immunobiology to generate highly effective nanotherapeutic vaccines.

Nanotechnology and Modulation of Immune Responses

On the other hand, protein immunogenicity represents a critical limitation to the therapeutic use of recombinant proteins, particularly when specific antibody responses to the therapeutics are engendered [10.32]. The consequences of antibody responses to proteins are variable, ranging from no effect to decidedly negative consequences to the host. Antibodies can neutralize therapeutic bioactivity, rendering drug entities inactive in repeated rounds of therapy, distort drug P_k and B_d properties, and, in the worst case, trigger autoimmune responses with cross-reactive host antigens (host antigens that share some identity with the immunizing epitopes).

Antibody Responses to Nanotherapeutic Devices

Antibody responses to protein components are a major challenge for nanobiological therapeutic devices in vivo. We, and others [10.6, 14, 21, 133–136], have demonstrated that protein components of bioconjugates to diverse nanomaterials can not only retain their intrinsic immunogenicity but, worse, can render the otherwise non-immunogenic synthetic nanomaterials to which they are linked capable of inducing antibody responses [10.6, 7, 14, 21, 126, 133–135]. The mechanism of immunogenicity seems to be haptenization, a well-known phenomenon in immunology. Haptenization occurs when antigens containing only B-cell epitopes are covalently linked to proteins containing T-helper epitopes, thereby creating hybrid antigens that are fully immunogenic with respect to antibody responses. The process is illustrated in Fig. 10.17.

As a result of the chemically repetitive structure of many synthetic nanomaterials, the antibodies directed to them have interesting properties. Most notably, antibodies generated to nanostructures are often cross-reactive to other chemically similar nanostructures. For example, antibodies raised using haptenized fullerenes also recognize carbon nanotubes [10.133], and antibodies raised to haptenized generation 0 (G_0) PAMAM dendrimers also recognize higher generation PAMAM dendrimers (G_1, G_2, G_3, etc. [10.14, 21, 133–136]. Presumably, cross-reactivity results from the presence of common B-cell epitopes in the immunogen used and in structurally similar cross-reactive antigens. Thus, patients immunized with any therapeutic containing one haptenized nanostructure could potentially raise antibodies that would recognize any nanotherapeutic containing a material structurally similar to the haptenized nanomaterial.

This is a potentially serious problem, in light of the documented ability of antibody responses to interfere with therapy by neutralizing drug or distorting P_k and B_d properties. As discussed before, molecular weights of serum antibodies are around 150,000 AMU with an effective diameter in excess of 10 nm [10.1, 123, 124]. Many of the desirable properties of nanotherapeutics derive from their precisely controlled sizes, and size-dependent properties like EPR targeting could be jeopardized if the therapeutic to be delivered were non-covalently complexed with antibodies directed to its synthetic components. Due to long-term persistence of memory immune responses, exposure to haptenized nanotherapeutics could endanger patient responsiveness to structurally similar materials for the duration of their lives. Obviously, immunogenicity is a critical challenge to the nanobiological therapeutic paradigm that has yet to be satisfactorily addressed.

Polymer Conjugation to Mitigate Immune Responses

There are multiple approaches to mitigating immunogenicity, some of which are applicable to nanobiotechnological therapeutic devices. Polymer conjugation, particularly conjugation of proteins to poly (ethylene glycol) (PEG, a flexible, biologically well-tolerated hydrophilic polymer) has been used to control immunogenicity of a broad class of proteins and synthetic molecules since the mid-1970s [10.19, 137]. Typically, pegylation involves covalent linkage (often by promiscuous conjugation chemistry) of variable numbers (usually two to six polymer molecules) of size-polydisperse (usually 3,000 to 10,000 AMU) linear or branched PEG chains to the potential immunogen. This approach results in polydisperse bioconjugates, which flies in the face of the molecular level-structural precision often considered essential for nanotechnology. Still, pegylation may be useful in some limited nanobiological applications.

The mechanism of molecule-specific immunosuppression remains obscure, despite almost 30 years of research. The flexibility of the PEG strands allows them to wrap around the molecules they are conjugated to, perhaps delimiting access of immune effectors and immune processing machinery to the potential immunogens. Pegylated molecules exhibit dramatically enhanced apparent diameters and vastly enhanced solubility in aqueous buffer. Both of these phenomena result in substantially diminished clearance of pegylated molecules by the RES and may account for the retention of in vivo bioactivity despite the apparent steric hindrance that may result from wrapping PEG moieties around their protein bioconjugate partners. Pegylation-related, bioconjugate-specific immunosuppressive phenomena cannot be fully accounted for by sterics, though. Immunological experiments in syngeneic mice show that the protein-specific immunosuppression associated with pegylated proteins is transmissible from one animal to another by surgically harvested splenocytes. This finding is more consistent with a mechanism of suppression involving immunologically induced tolerance of foreign antigens as opposed to one caused by simple steric considerations [10.137].

Synthetic Nanoporous Membranes for Immunoisolation

A second approach to avoiding immune responses to nanodevices or biocomponents is immunoisolation by encapsulation [10.16, 17]. In this strategy, potentially immunogenic structures (nanodevices, cells, macromolecules) are sequestered from high molecular weight, large host immune effectors, and immunosurveillance, while small molecules, such as glucose and small proteins, diffuse freely (Fig. 10.18). While conceptually simple, immunoisolation is technically challenging in that the number of pores must be high to maximize

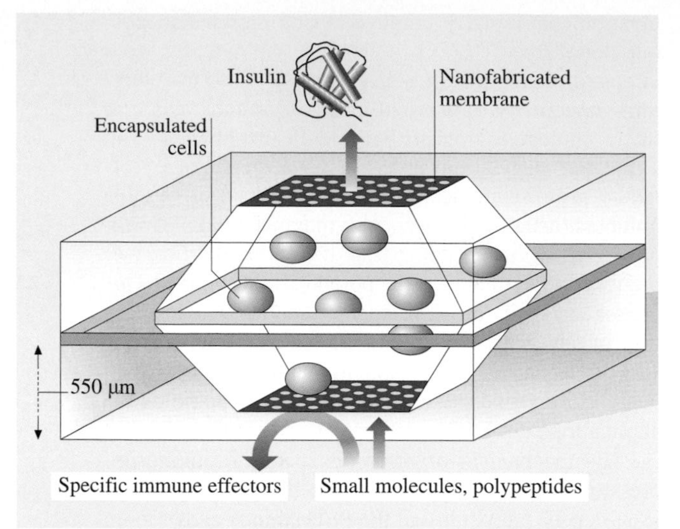

Fig. 10.18 Biological components can be protected from unfavorable environments by encapsulation [10.16–18]. Shown is an immunoisolation capsule containing heterologous (immunogenic) cells producing insulin (a four helical bundle protein, see text). The cells are protected from immune responses by membranes with 10 nm pores that allow free diffusion of small molecules (oxygen, nutrients) into the capsule and small proteins (insulin, shown here not to scale) out of it. However, immune effectors with diameters greater than pore size (immune cells like CTLs and APCs, antibodies, other effectors) don't access the capsule interior, protecting the xenograft from immune rejection. The structure integrates both micro- and nanoscale components (including the nanoporous membranes of Fig. 10.4) into a macroscale construct

flux, and the pores must be highly uniform in dimension.

Because of the exquisite sensitivity of the immune system, immunoisolation must be nearly perfect if it is to work at all. Particularly in the case of immunoisolation of cells for xenografts, if only a tiny fraction (as few as 1%) of the pores of an immunoisolation capsule are large enough to admit immune effectors (for instance, antibodies and complement components, exclusion of which requires pores of 50 nm or less), or the viability of the graft will be compromised [10.16, 17, 138]. Thus, the strategy requires fabrication of membranes with pores that have very tightly controlled dispersities in terms of their maximum diameters. Until recently, immunoisolation for xenografts has failed for lack of such membranes.

The technological innovations that made the synthesis of these high-quality nanoporous membranes were described above and in Fig. 10.4. The nanoporous membranes can be used to form immunoisolation capsules when either two such membranes are joined so that they bind an enclosed space or that a nanoporous membrane is used to enclose a compartment of some sort in a therapeutic structure. The enclosed spaces can be charges with functional nanodevices or, for that matter, living cells. Such immunoisolation capsules charged with heterologous pancreatic islet cells have demonstrated immense promise for regulating normoglycemia in diabetic animal models [10.16–18].

Biogenic, Nanoporous, Immunoisolation Membranes

The novel microfabrication strategy has made it possible to make these highly defined membranes in an industrial process, and they are applicable to numerous biomedical applications beyond immunoisolation [10.16, 17]. Still the fabrication process is rigorous, and an alternative biogenic method to derive selectively porous silicon capsules has been proposed [10.139, 140]. Diatoms are encapsulated in biogenic amorphous silicon shells (called frustules), which the living cells secrete. Frustules must necessarily be porous to allow the cells to take on nutrients and excrete wastes, and the porosity, pore dimensions, and other micro- and nanoscale morphological frustule features are tightly controlled at genetic and physiological levels. Using techniques well known to microbiologists (manipulation of culture conditions), geneticists (mutagenesis), and cell biologists (optically driven cell sorting in flow systems), it is theoretically possible to select and propagate diatoms whose frustule morphology, size, and pore characteristics fall within pre-selected limits [10.139, 140]. Frustules can be isolated intact from diatom cultures, so biogenic production of silicon membranes with porosity characteristics similar to the microfabricated membranes discussed above may be feasible. Though they have yet to be realized, immunoisolation capsules made from genetically/physiologically tuned frustules have been proposed [10.140]: one wonders what other biomineralized structures eventually might also be put to nanobiotechnological use.

10.5 Concluding Remarks: Barriers to Practice and Prospects

The effort to produce nanoscale therapeutic devices is clearly highly interdisciplinary. As we have seen, it touches on numerous established disciplines, encompassing elements of physiology, biotechnology, bioconjugate chemistry, electrical engineering, and materials science, to name just a few of the fields involved. Obviously, this broad sweep of knowledge is difficult for any one investigator to master fully. The breadth of the effort constitutes just one of the major barriers to entry in the field. Other challenges include the raw complexity of biology, the fashion in which biologists hold and distribute information, and cultural differences between engineers and biological scientists.

10.5.1 Complexity in Biology

Biology is characterized by particularity: nuances of biological systems are often unique to the system at hand and highly idiosyncratic. This follows from the fact that biological systems are not purpose-built, as are designed devices (although they can give that appearance, as seen in Fig. 10.1), but rather arose as the consequence of evolutionary processes. Evolution is a highly chaotic business, with the outcome of any given evolutionary process highly sensitive to initial conditions (populations of organisms subjected to selection, other organisms in the environment, biology of all organisms involved, resource availability, other environmental factors, etc.). Moreover, many of the variables affecting natural selection processes are nonstatic and change, as a function of time or space or both, *while* selection is exerted against a population of organisms. Conditions leading to one evolutionary adaptation or another are therefore seldom duplicated exactly, so that individual adaptive features are idiosyncratic, with elements relating not only to their biological functions but also to their evolutionary history. Individual systems and adaptive features thereof can be almost baroquely complex because the structures themselves arose under unique conditions, under unique selective pressures, and from unique initial biological systems. The extent of the complexity of biological systems is apparent even in individual macromolecular constituents of biological systems and can be made clear by comparison of synthetic nanostructures to biological nanostructures with similar dimensional aspects.

SWCNT (single wall carbon nanotubes) have a minimal diameter of about 1.3 nm, and many proteins likewise have diameters of few nms. SWCNTs are regular and homogeneous polymeric structures composed entirely of carbon atoms (Fig. 10.6). Proteins, on the other hand, are not polymers of one repeating subunit, but rather are nonrandom copolymers of 20 chemically distinct amino acids (aas). The order of the aas drives specific folding events that produce three-dimensional structures that, in turn, present specific aas and their side chains at specific positions in space (Fig. 10.10). This control of aa position in space drives protein activity: small changes in aa sequence can perturb structure and function profoundly. Therefore, though they fall within the same broad size regime, proteins are structurally and functionally much more complex than SWCNTs. The extent of biological complexity can be glimpsed when one considers that individual living things are ordered aggregates of multiple macromolecules and supramolecular structures (potentially, billions and billions of them), belonging to distinct chemical classes (lipids, proteins, nucleic acids, etc.), with each individual macromolecule being at least as complex in structure and function as the protein of Fig. 10.10. The nearly irreducible complexity of biological systems is a central fact of practice in the biological sciences and drives the way biologists gather and disseminate information. It is the key informant of the culture of biologists.

10.5.2 Dissemination of Biological Information

Biologists typically consider themselves primarily as scientists, as opposed to being engineers or technology developers. The product biologists generate, then, is information rather than devices or structures. Their interest in technological application of the knowledge they produce is typically secondary to their desire to develop the information itself. Additionally, as we have just seen, biology is a ferociously complex discipline.

As a practical matter, biological data itself is often much more rich (and often more ambiguous) than data from harder scientific endeavors. The data can be so very rich that biologists must make choices as to what data is relevant to a given phenomena and, therefore, what data they will publish. With absolutely no intent to conceal or mislead, biologists are often driven to publish only a small fraction of the data they gather.

Since they see themselves primarily as scientists, biologists tend to publish data they believe to be of broad scientific significance. Typically, the chosen information does not include data that might be critical to technology development, frequently making the biological literature an inadequate resource for engineers. The omitted information thus becomes lore. That is to say, the infomation is critical for the practice or development of a given technology but usually is not accessible to persons outside the field (as many nanotherapeutic device producers may be). This can be illustrated by recent experiences around phage display-mediated affinity maturation of four helical bundle proteins [10.141–143].

Four helical bundle proteins are loosely related small proteins consisting of four α helices arranged in a specific configuration (see the dynamite stick representation of insulin that occurs in Fig. 10.18). They include insulin, growth hormone, most cytokines, and various other molecules involved in cell to cell signaling. They have a wide range of therapeutically valuable properties that might be made even more desirable if their potencies could be enhanced. There is thus significant interest in variant four-helical bundle proteins with improved bioactivities.

Phage display [10.67] is a method that can be used to sort protein variant libraries for variants with enhanced affinity for a target receptor. This is usually accomplished by iterative rounds of affinity selection on the receptor followed by propagation of selected phage. As rounds of selection and propagation proceed, the mean affinity of the selected variants for the receptor increases (so-called affinity maturation). A phage display method is depicted in Fig. 10.11.

Affinity maturation by phage display can be used to identify variant proteins from libraries that have enhanced biological activities, providing that affinity for receptor is limiting in the overall activity of the parent protein [10.134]. Phage display affinity maturation has been successfully used to select some enhanced activity variants of four helical bundle proteins [10.144, 145] but is *not* applicable to engineering increased activity for *all* four helical bundle proteins [10.134]. The reason for the inapplicability of display methods to engineer high-activity human growth hormone (hGH) variants became apparent in a recent publication from Genentech [10.142].

In the case of hGH, the affinity of the parent molecule for receptor is in excess of that required to drive maximal biological responses [10.142]. This fact was inferred after a multiyear effort (beginning in the late 1980s and culminating in 1999) to derive more active hGH proteins, involving literally thousands of hGH variants. Many of the variants did indeed exhibit enhanced affinity for the hGH binding protein (receptor), but none exhibited significantly enhanced biological activity. Each data point (i. e., the biological activity of each individual variant) essentially constituted a failure of a technology to provide a desired result and was therefore seen, not unreasonably, as negative (and therefore uninteresting) data by the investigators. Consequently, the information was not broadly disseminated until a significant conclusion could be presented (i. e., that biological responses to hGH are not limited by ligand-receptor affinity). Thus, between the late 1980s and the 1999 publication, this information was known primarily and unambiguously only to the investigators involved, even though it would have been critical to any technologist planning to affinity mature hGH by phage display for enhanced activity [10.134].

10.5.3 Cultural Differences Between Technologists and Biologists

Biology is the realm of particularity, whereas more physical sciences are realms within which general rules can be derived and broadly applied. The vast complexity of biology usually drives biologists to be highly specialized, perhaps making a career from the study of particular macromolecules (specific enzymes, growth factors, genetic elements, etc.) or biological systems. Biologists are, therefore, neither trained nor encouraged by funding agencies to see themselves as fungible, and frequently they are very reluctant to step away from their primary research focuses into new areas. Doing so is a major undertaking, and a substantial risk, for most biologists.

Engineers, on the other hand, often see themselves as operating from first principles and are more (if not entirely) willing to step into new areas. After all, crossing disciplines is feasible for engineers working in the physical sciences, whereas it is much more difficult, not only for reasons of training but also because of the biological realities discussed above, for biologists. Also by reason of their training and experience, engineers often vastly underestimate the complexity of biological systems and sometimes exhibit what biologists perceive as naïveté in their approach to biological systems. Unfortunately, or fortunately, depending on your perspective, therapeutic nanodevices require both biological and engineering expertise. An often bruising debate as to how best to prepare students for practice in this extremely challenging field is currently playing out in biomedical engineering and other related academic departments across the country.

The traditional approach for training biomedical engineers has been to train conventional engineers with a smattering of biological knowledge. This "primarily engineers" model of biomedical engineering has been reasonably successful in multiple endeavors (orthopedics, imaging technologies, etc.), but it seems unlikely in the extreme that it will suffice for robust practice in nanotherapeutic devices. As this chapter shows, the role of biology is pivotal to success in the field, and sophisticated nanotherapeutics are virtually impossible to design without deep knowledge of the biology involved. A reasonable argument can be made that successful engineering of therapeutic nanostructures will demand intense focus on biology and a smattering of engineering knowledge (a "primarily biologists" approach, as opposed to the historically prevalent "primarily engineers" training model). This view is often strongly and passionately resisted in colleges of engineering, which are the usual homes of biomedical engineering and related efforts. The outcome of the debate is not finalized, but it is critically important: it will determine whether or not a ready cadre of researchers is prepared for a career in the area. If not, the field will progress much as it has, very slowly bringing the benefits of the research to patients. It seems apparent to us, and it also seems that experience to date shows, that a substantial biological component of training for the discipline is simply unavoidable if we are committed to realizing the vast potential of nanotechnology in human therapy.

References

10.1 C. A. Janeway, P. Travers, M. Walport, J. D. Capra: *Immunobiology* (Elsevier, London 1999)

10.2 S. C. Lee: Biotechnology for nanotechnology, Trends Biotechnol. **16** (1998) 239–240

10.3 S. C. Lee: Engineering the protein components of nanobiological devices. In: *Biological Molecules in Nanotechnology: The Convergence of Biotechnology, Polymer Chemistry and Materials Science*, ed. by S. C. Lee, L. Savage (IBC, Southborough 1998) pp. 67–74

10.4 S. C. Lee: How a molecular biologist can wind up organizing nanotechnology meetings. In: *Biological Molecules in Nanotechnology: The Convergence of Biotechnology, Polymer Chemistry and Materials Science*, ed. by S. C. Lee, L. Savage (IBC, Southborough 1998)

10.5 S. C. Lee: The nanobiological strategy for construction of nanodevices. In: *Biological Molecules in Nanotechnology: The Convergence of Biotechnology, Polymer Chemistry and Materials Science*, ed. by S. C. Lee, L. Savage (IBC, Southborough 1998) pp. 3–14

10.6 S. C. Lee: A biological nanodevice for drug delivery. In: *National Science and Technology Council. IWGN Workshop Report: Nanotechnology Research Directions. International Technology Research Institute, World Technology Division* (Kluwer, Baltimore 1999) pp. 91–92

10.7 S. C. Lee, R. Parthasarathy, K. Botwin: Protein-polymer conjugates: Synthesis of simple nanobiotechnological devices, Polymer Preprints **40** (1999) 449–450

10.8 L. Jelinski: Biologically related aspects of nanoparticles, nanostructured materials and nanodevices. In: *Nanostructure Science and Technology*, ed. by R. W. Siegel, E. Hu, M. C. Roco (Kluwer, Dordrecht 1999) pp. 113–130

10.9 J. Baish, Y. Gazit, D. Berk, M. Nozue, L. T. Baxter: Role of tumor vascular architecture in nutrient and drug delivery: An invasion percolation-based network model, Microvasc. Res. **51** (1996) 327–346

10.10 J. R. Baker Jr.: Therapeutic nanodevices. In: *Biological Molecules in Nanotechnology: The Convergence of Biotechnology, Polymer Chemistry and Materials Science*, ed. by S. C. Lee, L. Savage (IBC, Southborough 1998) pp. 173–183

10.11 R. Duncan: Drug targeting: Where are we now and where are we heading?, J. Drug Targeting **5** (1997) 1–4

10.12 R. Duncan, S. Gac-Breton, R. Keane, Y. N. Sat, R. Satchi, F. Searle: Polymer-drug conjugates, PDEPT and PELT: Basic principles for design and transfer from the laboratory to clinic, J. Cont. Release **74** (2001) 135–146

10.13 D. S. Goldin, C. A. Dahl, K. L. Olsen, L. H. Ostrach, R. D. Klausner: Biomedicine. The NASA-NCI collaboration on biomolecular sensors, Science **292** (2001) 443–444

10.14 S. C. Lee: Dendrimers in nanobiological devices. In: *Dendrimers and other Dendritic Polymers*, ed. by D. Tomalia, J. Frechet (Wiley, London 2001) pp. 548–557

10.15 J. R. Baker Jr., A. Quintana, L. Piehler, M. Banazak-Holl, D. Tomalia, E. Racka: The synthesis and testing of anti-cancer therapeutic nanodevices, Biomed. Microdevices **3** (2001) 61–69

10.16 T. Desai, D. Hansford, L. Kulinsky, A. Nashat, G. Rasi, J. Tu, Y. Wang, M. Zhang, M. Ferrari: Nanopore technology for biomedical applications, Biomed. Microdevices **21** (1999) 11–40

10.17 T.A. Desai, W.H. Chu, J.K. Tu, G.M. Beattie, A. Hayek, M. Ferrari: Microfabricated immunoisolating biocapsules, Biotechnol. Bioeng. **57** (1998) 118–120

10.18 M. Ferrari, W.H. Chu, T.A, Desai, J. Tu: *Microfabricated silicon biocapsule for immunoisolation of pancreatic islets*, Advanced Manufacturing Systems, CISM Courses and Lectures, Vol. 372, ed. by E. Kuljanic (Springer, Wien 1996) pp. 559–568

10.19 J.M. Harris, N.E. Martin, M. Modi: Pegylation: A novel process for modifying pharmacokinetics, Clin. Pharmacokin. **40** (2001) 539–551

10.20 S.B.H. Kent: Building proteins through chemistry: Total chemical synthesis of protein molecules by chemical ligation of unprotected protein segments. In: *Biological Molecules in Nanotechnology: The Convergence of Biotechnology, Polymer Chemistry and Materials Science*, ed. by S.C. Lee, L. Savage (IBC, Southborough 1998) pp. 75–92

10.21 S.C. Lee, R. Parthasarathy, T. Duffin, K. Botwin, T. Beck, G. Lange, J. Zobel, D. Jansson, D. Kunneman, E. Rowold, C.F. Voliva: Antibodies to PAMAM dendrimers: Reagents for immune detection assembly and patterning of dendrimers. In: *Dendrimers and other Dendritic Polymers*, ed. by D. Tomalia, J. Frechet (Wiley, London 2001) pp. 559–566

10.22 S.C. Lee, M.S. Leusch, V.A. Luckow, P. Olins: Method of producing recombinant viruses in bacteria, (1993) US Patent 5,348,886

10.23 M.S. Leusch, S.C. Lee, P.O. Olins: A novel host-vector system for direct selection of recombinant baculoviruses (bacmids) in *Escherichia coli*, Gene. **160** (1995) 191–194

10.24 V.A. Luckow, S.C. Lee, G.F. Barry, P.O. Olins: Efficient generation of infectious recombinant baculoviruses by site-specific, transposon-mediated insertion of foreign DNA into a baculovirus genome propagated in *E. coli*, J. Virol. **67** (1993) 4566–4579

10.25 T. Gardner, C.R. Cantor, J.J. Collins: Construction of a genetic toggle switch in *Escherichia coli*, Nature **403** (2000) 339–342

10.26 J. Hasty, F. Isaacs, M. Dolnik, D. McMillen, J.J. Collins: Designer gene networks: Towards fundamental cellular control, Chaos **11** (2001) 107–220

10.27 E. Di Fabrisio, A. Nucara, M. Gentili, R. Cingolani: Design of a beamline for soft and deep lithography on third generation synchrotron radiation source, Rev. Sci. Instrum. **70** (1999) 1605–1613

10.28 G.M. Dykes: Dendrimers: A review of their appeal and applications, J. Chem. Tech. Biotech. **76** (2001) 903–918

10.29 R. Spindler: PAMAM starburst dendrimers: Designed nanoscopic reagents for biological applications. In: *Biological Molecules in Nanotechnology: The Convergence of Biotechnology, Polymer Chemistry and Materials Science*, ed. by S.C. Lee, L. Savage (IBC, Southborough 1998) pp. 15–32

10.30 D. Tomalia, H.M. Brothers II: Regiospecific conjugation to dendritic polymers to produce nanodevices. In: *Biological Molecules in Nanotechnology: The Convergence of Biotechnology, Polymer Chemistry and Materials Science*, ed. by S.C. Lee, L. Savage (IBC, Southborough 1998) pp. 107–120

10.31 B.I. Yacobson, R.E. Smalley: Fullerene nanotubes: C1000000 and beyond, Amer. Scient. **85** (1997) 324–337

10.32 D.J.A. Crommelin, R.D. Sindelar: *Pharmaceutical Biotechnology* (Harwood Academic Publishers, Amsterdam 1997)

10.33 L.J. Lee: BioMEMS and micro-/nano-processing of polymers – An overview, Chin. J. Chem. Eng. **34** (2003) 25–46

10.34 T.R. Groves, D. Pickard, B. Rafferty, N. Crosland, D. Adam, G. Schubert: Maskless electron beam lithography: Propects, progress and challenges, Microelectron. Eng. **61** (2002) 285–293

10.35 M. Guthold, R. Superfine, R. Taylor: The rules are changing: Force measurements on single molecules and how they relate to bulk reaction kinetics and energies, Biomed. Microdevices **3** (2001) 9–18

10.36 L.M. Demers, D.S. Ginger, S.-J. Park, Z. Li, S.-W. Chung, C.A. Mirkin: Direct patterning of modified oligonucleotides on metals and insulatos by dip-pen nanolithography, Science **296** (2002) 1836–1838

10.37 K.-B. Lee, S.-J. Park, C.A. Mirkin, J.C. Smith, M. Mrksich: Protein nanoarrays generated by dip-pen nanolithography, Science **295** (2002) 1702–1705

10.38 M. Ferrari, J. Liu: The engineered course of treatment, Mech. Eng. **123** (2001) 44–47

10.39 H.W. Rohrs, R.S. Ruoff: The use of carbon nanotubes in hybrid nanometer-scale devices. In: *Biological Molecules in Nanotechnology: The Convergence of Biotechnology, Polymer Chemistry and Materials Science*, ed. by S.C. Lee, L. Savage (IBC, Southborough 1998) pp. 33–38

10.40 K.E. Drexler: *Engines of Creation: The Coming Era of Nanotechnology* (Anchor Books, New York 1986)

10.41 J. Cumings, A. Zetti: Low-friction nanoscale linear bearing realized from multiwall carbon nanotubes, Science **289** (2000) 602–604

10.42 D.J. Hornbaker, S.-J. Kahng, S. Mirsa, B.W. Smith, A.T. Johnson, E.J. Mele, D.E. Luzzi, A. Yazdoni: Mapping the one-dimensional electronic states of nanotube peapod structures, Science **295** (2002) 828–831

10.43 C. Dekker: Carbon nanotubes as molecular quantum wires, Phys. Today **28** (1999) 22–28

10.44 M.-C. Jones, J.-C. Leroux: Polymeric micelles-a new generation of colloidal drug carriers, Eur. J. Pharma. Biopharma. **48** (1999) 101–111

10.45 I. Uchegbu: Parenteral drug delivery: 1, Pharma. J. **263** (1999) 309–318

10.46 I. Uchegbu: Parenteral drug delivery: 2, Pharma. J. **263** (1999) 355–359

10.47 K.B. Thurmond II, H. Huang, K.L. Wooley: Stabilized micellar structures in nanodevices. In: *Biological Molecules in Nanotechnology: The Convergence of Biotechnology, Polymer Chemistry and Materials Science*, ed. by S.C. Lee, L. Savage (IBC, Southborough 1998) pp. 39–43

10.48 S. Uppuluri, D.R. Swanson, L.T. Peihler, J. Li, G. Hagnauer, D.A. Tomalia: Core shell tecto(dendrimers). I. Synthesis and characterization of saturated shell models, Adv. Mater. **12** (2000) 796–800

10.49 A.K. Patri, I.J. Majoros, J.R. Baker Jr.: Dendritic polymer macromolecular carriers for drug delivery, Curr. Opin. Chem. Biol. **6** (2002) 466–471

10.50 S. Fernandez-Lopez, H.-S. Kim, E.C. Choi, M. Delgado, J.R. Granja, A. Khasanov, K. Kraehenbuehl, G. Long, D.A. Weinberger, K.M. Wilcoxen, M. Ghardiri: Antibacterial agents based on the cyclic D,L-alpha-peptide architecture, Nature **412** (2001) 452–455

10.51 A. Saghatelian, Y. Yokobayashi, K. Soltani, M.R. Ghadiri: A chiroselective peptide replicator, Nature **409** (2001) 777–778

10.52 J. Davies: Aminoglycoside-aminocyclitol antibiotics and their modifying enzymes. In: *Antibiotics in Laboratory Medicine*, ed. by V. Lorian (Williams and Wilkins, Baltimore 1984) pp. 474–489

10.53 T. Hamouda, A. Myc, B. Donovan, A.Y. Shih, J.D. Reuter, J.R. Baker Jr.: A novel surfactant nanoemulsion with a unique non-irritant topical antimicrobial activity against bacteria, enveloped viruses and fungi, Microbiol. Res. **156** (2001) 1–7

10.54 M.J. Heller: Utilization of synthetic DNA for molecular electronic and photonic-based device applications. In: *Biological Molecules in Nanotechnology: The Convergence of Biotechnology, Polymer Chemistry and Materials Science*, ed. by S.C. Lee, L. Savage (IBC, Southborough 1998) pp. 59–66

10.55 Z. Ma, S. Taylor: Nucleic acid triggered catalytic drug release, Proc. Nat. Acad. Sci. USA **97** (2000) 11159–11163

10.56 R.C. Merkle: Biotechnology as a route to nanotechnology, Trends Biotechnol. **17** (1999) 271–274

10.57 N.C. Seeman, J. Chen, Z. Zhang, B. Lu, H. Qiu, T.-J. Fu, Y. Wang, X. Li, J. Qi, F. Liu, L.A. Wenzler, S. Du, J.E. Mueller, H. Wang, C. Mao, W. Sun, Z. Shen, M.H. Wong, R. Sha: A bottom-up approach to nanotechnology using DNA. In: *Biological Molecules in Nanotechnology: The Convergence of Biotechnology, Polymer Chemistry and Materials Science*, ed. by S.C. Lee, L. Savage (IBC, Southborough 1998) pp. 45–58

10.58 G. Lemieux, C. Bertozzi: Chemoselective ligation reactions with proteins, oligosaccharides and cells, Trends Biotechnol. **16** (1998) 506–512

10.59 R. Offord, K. Rose: Multicomponent synthetic constructs. In: *Biological Molecules in Nanotechnology: The Convergence of Biotechnology, Polymer Chemistry and Materials Science*, ed. by S.C. Lee, L. Savage (IBC, Southborough 1998) pp. 93–105

10.60 G.T. Hermanson: *Bioconjugate Chemistry* (Academic, San Diego 1996)

10.61 S.S. Davis: Biomedical applications of nanotechnology-implications for drug targeting and gene therapy, Trends Biotechnol. **15** (1997) 217–224

10.62 J.C. Roberts, M.K. Bhalgat, R.T. Zera: Preliminary biological evaluation of polyamidoamine (PAMAM) starburst dendrimers, J. Biomed. Mater. Res. **30** (1996) 53–65

10.63 N.S. Nahman, T. Drost, U. Bhatt, T. Sferra, A. Johnson, P. Gamboa, G. Hinkle, A. Haynam, V. Bergdall, C. Hickey, J.D. Bonagura, L. Brannon-Pappas, J. Ellison, A. Mansfield, S. Shiwe, N. Shen: Biodegradable microparticles for in vivo glomerular targeting: Implications for gene therapy of glomerular disease, Biomed. Microdevices **4** (2002) 189–196

10.64 W. Arap, M. Kolonin, M. Trepel, J. Lahdenranta, M. Cardo-Vila, R. Giordano, P.J. Mintz, P. Ardelt, V. Yao, C. Vidal, L. Chen, A. Flamm, H. Valtanen, L.M. Weavind, M.E. Hicks, R. Pollock, G.H. Botz, C.D. Bucana, E. Koivunen, D. Cahil, P. Troncosco, K.A. Baggerly, R.D. Pentz, K.-A. Do, C. Logothetis, R. Pasqualini: Steps towards mapping the human vasculature by phage display, Nature Med. **8** (2002) 121–127

10.65 M. Kolonin, R. Pasqualini, W. Arap: Molecular addresses in blood vessels as targets for therapy, Curr. Opin. Chem. Biol. **5** (2001) 308–313

10.66 E. Ruoslahti: Special delivery of drugs by targeting to tissue-specific receptors in vasculature, Pharmaceutical News **7** (2000) 35–40

10.67 B.K. Kay, J. Winter, J. McCofferty: *Phage Display of Peptides and Proteins* (Academic, San Diego 1996)

10.68 W. Arap, W. Haedicke, M. Bernasconi, R. Kain, D. Rajotte, S. Krajewski, M. Ellerby, R. Pasqualini, E. Ruoslahti: Targeting the prostate for destruction through a vascular address, Proc. Nat. Acad. Sci. USA **99** (2002) 1527–1531

10.69 M. Essier, E. Ruoslahti: Molecular specialization of breast vasculature: A breast homing phage displayed peptide binds to aminopeptidase P in breast vasculature, Proc. Nat. Acad. Sci. USA **99** (2002) 2252–2257

10.70 G.A. Husseini, G.D. Myrup, W.G. Pitt, D. Christensen, N.Y. Rapoport: Factors affecting acoustically triggered release of drugs from polymeric micelles, J. Cont. Release **69** (2000) 43–52

10.71 A. Quintana, E. Raczka, L. Piehler, I. Lee, A. Myc, I. Majoros, A. K. Patri, T. Thomas, J. Mule, J. R. Baker Jr.: Design and function of a dendrimer-based therapeutic nanodevice targeted to tumor cells through the folate receptor, Pharma. Res. **19** (2002) 1310–1316

10.72 P. Shum, J.-M. Kim, D. H. Thompson: Phototriggering of liposomal delivery systems, Adv. Drug Deliv. Rev. **53** (2001) 273–284

10.73 K. Hamad-Schifferli, J. J. Schwartz, A. T. Santos, S. Zhang, J. M. Jacobson: Remote electronic control of DNA hybridization through inductive coupling to an attached metal nanocrystal antenna, Nature **415** (2002) 152–155

10.74 P. Meers: Enzyme-activated targeting of liposomes, Adv. Drug Deliv. Rev. **53** (2001) 265–272

10.75 V. Raso, M. Brown, J. McGrath: Intracellular triggering with low pH-triggered bispecific antibodies, J. Biol. Chem. **272** (1997) 27623–27628

10.76 W. A. Denny: The role of hypoxia activated prodrugs in cancer therapy, The Lancet Oncol. **1** (2000) 25–29

10.77 P. Vaupel, D. K. Kelleher, O. Thews: Modulation of tumor oxygenation, Int. J. Radiation Oncology Biol. Phys. **42** (1998) 843–848

10.78 P. D. Senter, C. J. Springer: Selective activation of anticancer prodrugs by monoclonal antibody-enzyme conjugates, Adv. Drug Deliv. Rev. **53** (2001) 247–264

10.79 H. Wang, H. Song, V. C. Yang: A recombinant prodrug type approach for triggered delivery of streptokinase, J. Cont. Release **59** (1999) 119–122

10.80 R. Duncan: Polymer therapeutics for tumor specific delivery, Chem. Industry **7** (1997) 262–264

10.81 K. Rogers: Principles of affinity-based biosensors, Mol. Biotechnol. **14** (2000) 109–129

10.82 R. Raiteri, M. Grattarola, H.-J. Butt, P. Skladl: Micromechanical cantilever-based biosensors, Sensors Actuat. B **79** (2001) 115–126

10.83 R. Martel, T. Schmidt, H. R. Shea, T. Hertel, P. Avouris: Single and multiwall carbon nanotube field-effect transistors, Appl. Phys. Lett. **73** (1998) 2447–2449

10.84 A. Guiseppi-Elie, C. Lei, R. H. Baughman: Direct electron transfer of glucose oxidase on carbon nanotubes, Nanotechnology **13** (2002) 559–564

10.85 C. N. Campbell: How far are we from detecting single bioconjugation events?. In: *Biological Molecules in Nanotechnology: The Convergence of Biotechnology Polymer Chemistry and Materials Science*, ed. by S. C. Lee, L. Savage (IBC, Southborough 1998) pp. 163–171

10.86 W. D. Callister Jr.: *Material Science and Engineering: an Introduction* (Wiley, New York 1997)

10.87 M. Gerard, A. Chaubey, B. D. Malhotra: Applications of conducting polymers to biosensors, Biosens. Bioelectron. **17** (2002) 345–349

10.88 P. N. Bartlett, Y. Astier: Microelectrochemical enzyme transistors, Chem. Commun. **2** (2000) 105–112

10.89 I. Kleps, A. Angelscu, R. Valisco, D. Dascalu: New micro and nanoelectrode arrays for biomedical applications, Biomed. Microdevices **3** (2001) 29–33

10.90 J. Song, Q. Cheng, S. Zhu, R. C. Stevens: "smart" materials for biosensing devices: cell-mimicking supramolecular assemblies and colorometric detection of pathogenic agents, Biomed. Microdevices **4** (2002) 213–222

10.91 M. Brown, R. Semelka: *MRI: Basic Principles and Applications* (Wiley, New York 1999)

10.92 A. Elster, S. Handel, A. Goldman: *Magnetic Resonance Imaging: A Reference Guide and Atlas* (Lippincott, Philadelphia 1997)

10.93 H. Paajanen, M. Kormano: Contrast agents in magnetic resonance imaging. In: *Radiographic Contrast Agents*, ed. by J. Skucas (Aspen, Rockville 1989) pp. 377–406

10.94 L. Thunus, R. Lejeune: Overview of transition metal and lanthanide complexes as diagnostic tools, Coord. Chem. Rev. **184** (1999) 125–155

10.95 M. Mendoca-Dias, E. Gaggelli, P. Lauterbur: Paramagnetic contrast agents in nuclear magnetic resonance medical imaging, Sem. Nuclear Med. **13** (1983) 364–376

10.96 M. Ollsen, B. Persson, L. Salford: Ferromagnetic particles as contrast agent in T2 NMR imaging, Magn. Reson. Imag. **4** (1986) 437–440

10.97 D. Kehagias, A. Gouliamos, V. Smyrniotis, L. Vlahos: Diagnostic efficacy and safety of MRI of the liver with superparamagnetic iron oxide particles (SH U 555 A), J. Magn. Reson. Im. **14** (2001) 595–601

10.98 C. Nolte-Ernsting, G. Adam, A. Bucker, S. Berges, A. Bjornerud, R. Gunther: Abdominal MR angiography performed using blood pool contrast agents, Am. J. Roentgenol. **171** (1998) 107–113

10.99 F. Rety, O. Clement, N. Siauve, C.-A. Cuenod, F. Carnot, M. Sich, A. Buisine, G. Frija: MR lymphography using iron oxide nanoparticles in rats: pharmacokinetics in the lymphatic system after intravenous injection, J. Magn. Reson. Im. **12** (2000) 734–739

10.100 T. Skotland, P. Sontum, I. Oulie: In vitro stability analyses as a model for metabolism of ferromagnetic particles (Clariscan^{TM}), a contrast agent for magnetic resonance imaging, J. Pharm. Biomed. Anal. **28** (2002) 323–329

10.101 D. Hogemann, L. Josephson, R. Weissleder, J. Basilion: Improvement of MRI probes to allow efficient detection of gene expression, Bioconjugate Chem. **11** (2000) 941–946

10.102 L. L. Muldoon, M. A. Pagel, R. A. Kroll, S. Roman-Goldstein, R. S. Jones, E. A. Neuwelt: A physiological barrier distal to the anatomic blood-brain barrier in a model of transvascular delivery, AJNR Am. J. Neuroradial. **20** (1999) 217–222

10.103 G. Biddlecombe, Y. Gun'ko, J. Kelly, S. Pillai, J. Coey, M. Ventatesan, A. Douvalis: Preparation of magnetic nanoparticles and their assemblies using a new Fe(II) alkoxide precursor, J. Mater. Chem. **11** (2001) 2937–2939

10.104 L. Tiefenauer, G. Kuhne, R. Andres: Antibody-magnetite nanoparticles: in vitro characterization of a potential tumor-specific contrast agent for magnetic resonance imaging, Bioconjugate Chem. **4** (1993) 347–352

10.105 C. Liu, Z. Zhang: Size-dependent superparamagnetic properties of Mn spinel ferrite nanoparticles synthesized from reverse micelles, Chem. Mater. **13** (2001) 2092–2096

10.106 J. Schnorr, M. Taupitz, S. Wagner, H. Pilgrimm, J. Hansel, B. Hamm: Age-related blood half-life of particulate contrast materials: Experimental results with a USPIO in rats, J. Magn. Reson. Im. **12** (2000) 740–744

10.107 R. Weissleder, G. Elizondo, J. Wittenberg, C. Rabito, H. Bengele, L. Josephson: Ultrasmall superparamagnetic iron oxide: Characterization of a new class of contrast agents for MR imaging, Radiology **175** (1990) 489–493

10.108 D. Portet, B. Denizot, E. Rump, J.-J. Lejeune, P. Jallet: Nonpolymeric coatings of iron oxide colloids for biological use as magnetic resonance imaging contrast agents, J. Coll. Inter. Sci. **238** (2001) 37–42

10.109 M. Ruegsegger, R. Marchant: Reduced protein adsorption and platelet adhesion by controlled variation of oligomaltose surfactant polymer coatings, J. Biomed. Mater. Res. **56** (2001) 159–167

10.110 S. Flack, S. Fischer, M. Scott, R. Fuhrhop, J. Allen, M. McLean, P. Winter, G. Sicard, P. Gaffney, S. Wickline, G. Lanza: Novel MRI contrast agent for molecular fibrin, Circulation **104** (2001) 1280–1285

10.111 D.F. Baban, L.W. Seymour: Control of tumor vascular permeability, Adv. Drug Deliv. Rev. **34** (1998) 109–119

10.112 R.K. Jain: Delivery of molecular and cellular medicine to tumors, J. Cont. Release **53** (1998) 49–67

10.113 H. Maeda, T. Sawa, T. Konno: Mechanism of tumor-targeted delivery of macromolecular drugs, including the EPR effect in solid tumor and clinical overview of the prototype polymeric drug SMANCS, J. Cont. Release **74** (2001) 47–61

10.114 J. Borenstein, H. Terai, K. King, E. Weinberg, M. Kaazempour-Mofrad, J. Vacanti: Microfabrication technology for vascularized tissue engineering, Biomed. Microdevices **4** (2002) 167–175

10.115 D. Satake, H. Ebi, N. Oku, K. Matsuda, H. Takao, M. Ashiki, M. Ishida: A sensor for blood cell counters using MEMS technology, Sensors Actuat. B – Chem. **83** (2002) 77–81

10.116 G. Lin, R. Palmer, K. Pister, K. Roos: Miniature heart cell force transducer system implemented in MEMS technology, IEEE Trans. Biomed. Engin. **48** (2001) 996–1006

10.117 J. Kreuter, D. Shamenkov, V. Petrov, P. Ramge, K. Cychutek, C. Koch-Brandt, R. Alyautdin: Apolipoprotein-mediated transport of nanoparticle-bound drugs across the blood-brain barrier, J. Drug Target **10** (2002) 317–325

10.118 P. Lockman, R. Mumper, M. Khan, D. Allen: Nanoparticle technology for drug delivery across the blood-brain barrier, Drug Devel. Indust. Pharm. **28** (2002) 1–13

10.119 H. Quick, J. Debatin, M. Ladd: MR imaging of the vessel wall, Eur. Radiol. **12** (2002) 889–900

10.120 S. Schmitz, M. Taupitz, S. Wagner, K.-J. Wolf, D. Beyersdorff, D. Hamm: Magnetic resonance imaging of atherosclerotic plaques using supermagnetic iron oxide particles, J. Magn. Reson. Im. **14** (2001) 355–361

10.121 D. Ranney: Biomimetic transport and rational drug delivery, Biochem. Pharm. **59** (2000) 105–114

10.122 K. Soppimath, T. Aminabhavi, A. Kulkarni, W. Rudzinski: Biodegradable polymeric nanoparticles as drug delivery devices, J. Cont. Release **70** (2001) 1–20

10.123 F. Breitling, S. Dubel: *Recombinant Antibodies* (Wiley, London 1998)

10.124 E. Harlow, D. Lane: *Antibodies: A Laboratory Manual* (Cold Spring Harbor Press, Cold Spring Harbor 1988)

10.125 S. Raychanduri, K.L. Rock: Fully mobilizing host defense: Building better vaccines, Nature Biotech. **16** (1998) 1025–1031

10.126 M. Kovasovics-Bankowski, K. Clark, B. Benacerraf, K.L. Rock: Efficient major histocompatibility complex class I presentation of exogenous antigen upon phagocytosis by macrophages, Proc. Nat. Acad. Sci. USA **90** (1993) 4942–4946

10.127 K. Rock, S. Gamble, L. Rothstein: Presentation of exogenous antigen with class I major histocompatibility complex molecules, Science **249** (1990) 918–921

10.128 C.V. Harding, R. Song: Phagocytic processing of exogenous particulate antigens by macrophages for presentation by class I MHC molecules, J. Immunol. **152** (1994) 4925–4933

10.129 C. Oseroff, A. Sette, P. Wentworth, E. Celis, A. Maewal, C. Dahlberg, J. Fikes, R.T. Kubo, R.W. Chestnut, H.M. Grey, J. Alexander: Pools of lapidated HTL-CTL constructs prime for multiple HBV and HCV CTL epitope responses, Vaccine **16** (1998) 823–833

10.130 R. Mrsny, A.L. Daughtery, M. Mckee, D. Fitzgerald: Bacterial toxins as tools for mucosal vaccination, Drug Disc. Today **7** (2000) 247–257

10.131 S. Fawell, J. Seery, Y. Daikh, C. Moore, L.L. Chen, B. Pepinsky, J. Barsoum: Tat-mediated delivery of heterologous proteins to cells, Proc. Nat. Acad. Sci. USA **91** (1994) 664–668

10.132 T.J. Golentz, K. Klimpel, S. Leppla, J.M. Keith, J.A. Berzofsky: Delivery of antigens to the MHC class I pathway using bacterial toxins, Hum. Immunol. **54** (1997) 129–136

10.133 B.-X. Chen, S.R. Wilson, M. Das, D.J. Coughlin, B. F. Erlanger: Antigenicity of fullerenes: antibodies specific for fullerenes and their characteristics, Proc. Nat. Acad. Sci. USA **95** (1998) 10809–10813

10.134 S.C. Lee, R. Ibdah, C.D. van Valkenburgh, E. Rowold, A. Donelly, A. Abegg, J. Klover, S. Merlin, J. McKearn: Phage display mutagenesis of the chimeric dual cytokine receptor agonist myelopoietin, Leukemia **15** (2001) 1277–1285

10.135 S.C. Lee, R. Parthasarathy, K. Botwin, D. Kunneman, E. Rowold, G. Lange, J. Zobel, T. Beck, T. Miller, C.F. Voliva: Humeral immune responses to polymeric nanomaterials. In: *Functional Condensation Polymers*, ed. by C. Carraher, G. Swift (Kluwer, New York 2002) pp. 31–41

10.136 S.C. Lee, R. Parthasarathy, T. Duffin, K. Botwin, T. Beck, G. Lange, J. Zobel, D. Kunneman, E. Rowold, C.F. Voliva: Recognition properties of antibodies to PAMAM dendrimers and their use in immune detection of dendrimers, Biomed. Microdevices **3** (2001) 51–57

10.137 A.H. Sehon: Suppression of antibody responses by conjugates of antigens and monomethoxypoly(ethylene glycol). In: *Poly(ethylene glycol) Chemistry*, ed. by J.M. Harris (Plenum, New York 1992) pp. 139–151

10.138 R.P. Lanza, S.J. Sullivan, W.L. Chick: Perspectives in diabetes. Islet transplantation with immunoisolation, Diabetes **41** (1992) 1503–1510

10.139 R. Gordon: Computer controlled evolution of diatoms: Design for a compustat, Nova Hedwigia **112** (1996) 215–219

10.140 J. Parkinson, R. Gordon: Beyond micromachining: The potential of diatoms, Trends Biotechnol. **17** (1999) 190–196

10.141 S.C. Lee: Antibody responses to nanomaterials. In: *Nanospace 2001: Exploring Interdisciplinary Frontiers*, ed. by T. Nicodemus (Institute for Advanced Interdisciplinary Research, Houston 2002)

10.142 K. Pearce, B. Cunningham, G. Fuh, T. Teeri, J.A. Wells: Growth hormone affinity for its receptor surpasses the requirements for cellular activity, Biochem. **38** (1999) 81–89

10.143 T.L. Ciardelli: Reengineering growth factors through the looking glass, Nat. Biotechnol. **14** (1996) 1652

10.144 P.J. Buchli, Z. Wu, T.L. Ciardelli: Functional display of interleukin-2 on filamentous phage, Arch. Biochem. Biophys. **339** (1997) 79–84

10.145 I. Saggio, I. Gloaguen, R. Laufer: Functional phage display of cilliary neurotropic factor, Gene **152** (1995) 35–39

Part B

Part B Scanning Probe Microscopy

11 Scanning Probe Microscopy – Principle of Operation, Instrumentation, and Probes
Bharat Bhushan, Columbus, USA
Othmar Marti, Ulm, Germany

12 Probes in Scanning Microscopies
Jason H. Hafner, Houston, USA

13 Noncontact Atomic Force Microscopy and Its Related Topics
Seizo Morita, Suita-Citiy, Osaka, Japan
Franz J. Giessibl, Augsburg, Germany
Yasuhiro Sugawara, Suita, Japan
Hirotaka Hosoi, Sapporo, Japan
Koichi Mukasa, Sapporo, Japan
Akira Sasahara, Kanagawa, Japan
Hiroshi Onishi, Kanagawa, Japan

14 Low Temperature Scanning Probe Microscopy
Markus Morgenstern, Hamburg, Germany
Alexander Schwarz, Hamburg, Germany
Udo D. Schwarz, New Haven, USA

15 Dynamic Force Microscopy
A. Schirmeisen, Münster, Germany
B. Anczykowski, Münster, Germany
Harald Fuchs, Münster, Germany

16 Molecular Recognition Force Microscopy
Peter Hinterdorfer, Linz, Austria

11. Scanning Probe Microscopy – Principle of Operation, Instrumentation, and Probes

Since the introduction of the STM in 1981 and AFM in 1985, many variations of probe based microscopies, referred to as SPMs, have been developed. While the pure imaging capabilities of SPM techniques is dominated by the application of these methods at their early development stages, the physics of probe–sample interactions and the quantitative analyses of tribological, electronic, magnetic, biological, and chemical surfaces have now become of increasing interest. Nanoscale science and technology are strongly driven by SPMs which allow investigation and manipulation of surfaces down to the atomic scale. With growing understanding of the underlying interaction mechanisms, SPMs have found applications in many fields outside basic research fields. In addition, various derivatives of all these methods have been developed for special applications, some of them targeted far beyond microscopy.

This chapter presents an overview of STM and AFM and various probes (tips) used in these instruments, followed by details on AFM instrumentation and analyses.

11.1 **Scanning Tunneling Microscope** 327
11.1.1 Binnig et al.'s Design.................. 327
11.1.2 Commercial STMs......................... 328
11.1.3 STM Probe Construction 330

11.2 **Atomic Force Microscope** 331
11.2.1 Binnig et al.'s Design.................. 333
11.2.2 Commercial AFM.......................... 333
11.2.3 AFM Probe Construction 338
11.2.4 Friction Measurement Methods...... 342
11.2.5 Normal Force and Friction Force Calibrations of Cantilever Beams..................... 346

11.3 **AFM Instrumentation and Analyses** 347
11.3.1 The Mechanics of Cantilevers 347
11.3.2 Instrumentation and Analyses of Detection Systems for Cantilever Deflections 350
11.3.3 Combinations for 3-D-Force Measurements 358
11.3.4 Scanning and Control Systems 359

References 364

Part B | 11

The Scanning Tunneling Microscope (STM) developed by *Dr. Gerd Binnig* and his colleagues in 1981 at the IBM Zurich Research Laboratory, Rueschlikon, Switzerland, is the first instrument capable of directly obtaining three-dimensional (3-D) images of solid surfaces with atomic resolution [11.1]. *Binnig* and *Rohrer* received a Nobel Prize in Physics in 1986 for their discovery. STMs can only be used to study surfaces which are electrically conductive to some degree. Based on their design of the STM, in 1985, *Binnig* et al. developed an Atomic Force Microscope (AFM) to measure ultrasmall forces (less than 1 μN) present between the AFM tip surface and the sample surface [11.2] (also see [11.3]). AFMs can be used for measurement of all engineering surfaces which may be either electrically conductive or insulating. The AFM has become a popular surface profiler for topographic and normal force measurements on the micro to nanoscale [11.4]. AFMs modified in order to measure both normal and lateral forces, are called Lateral Force Microscopes (LFMs) or Friction Force Microscopes (FFMs) [11.5–11]. FFMs have been further modified to measure lateral forces in two orthogonal directions [11.12–16]. A number of researchers have continued to improve the AFM/FFM designs and used them to measure adhesion and friction of solid and liquid surfaces on micro- and nanoscales [11.4, 17–30]. AFMs have been used for scratching, wear, and measurement of elastic/plastic mechanical properties (such as indentation hardness and the modulus of elasticity) [11.4,10,11,21,23,26–29,31–36]. AFMs have been used for manipulation of individual atoms of Xenon [11.37], molecules [11.38], silicon surfaces [11.39], and polymer surfaces [11.40]. STMs have been used for formation of nanofeatures by localized heating or by inducing chemical reactions under the STM tip [11.41–43] and nanomachining [11.44]. AFMs

Table 11.1 Comparison of Various Conventional Microscopes with SPMs

	Optical	SEM/TEM	Confocal	SPM
Magnification	10^3	10^7	10^4	10^9
Instrument Price (U.S. $)	$10 k	$250 k	$30 k	$100 k
Technology Age	200 yrs	40 yrs	20 yrs	20 yrs
Applications	Ubiquitous	Science and technology	New and unfolding	Cutting edge
Market 1993	$800 M	$400 M	$80 M	$100 M
Growth Rate	10%	10%	30%	70%

have been used for nanofabrication [11.4,10,45–47] and nanomachining [11.48].

STMs and AFMs are used at extreme magnifications ranging from 10^3 to 10^9 in x, y and z directions for imaging macro to atomic dimensions with high resolution information and for spectroscopy. These instruments can be used in any environment such as ambient air [11.2,49], various gases [11.17], liquid [11.50–52], vacuum [11.1,53], at low temperatures (lower than about 100 K) [11.54–58] and high temperatures [11.59,60]. Imaging in liquid allows the study of live biological samples and it also eliminates water capillary forces present in ambient air present at the tip–sample interface. Low temperature (liquid helium temperatures) imaging is useful for the study of biological and organic materials and the study of low-temperature phenomena such as superconductivity or charge-density waves. Low-temperature operation is also advantageous for high-sensitivity force mapping due to the reduction in thermal vibration. They also have been used to image liquids such as liquid crystals and lubricant molecules on graphite surfaces [11.61–64]. While the pure imaging capabilities of SPM techniques dominated the application of these methods at their early development stages, the physics and chemistry of probe–sample interactions and the quantitative analyses of tribological, electronic, magnetic, biological, and chemical surfaces have now become of increasing interest. Nanoscale science and technology are strongly driven by SPMs which allow investigation and manipulation of surfaces down to the atomic scale. With growing understanding of the underlying interaction mechanisms, SPMs have found applications in many fields outside basic research fields. In addition, various derivatives of all these methods have been developed for special applications, some of them targeting far beyond microscopy.

Families of instruments based on STMs and AFMs, called Scanning Probe Microscopes (SPMs), have been developed for various applications of scientific and industrial interest. These include – STM, AFM, FFM (or LFM), scanning electrostatic force microscopy (SEFM) [11.65,66], scanning force acoustic microscopy (SFAM) (or atomic force acoustic microscopy (AFAM)) [11.21, 22, 36, 67–69], scanning magnetic microscopy (SMM) (or magnetic force microscopy (MFM)) [11.70–73], scanning near field optical microscopy (SNOM) [11.74–77], scanning thermal microscopy (SThM) [11.78–80], scanning electrochemical microscopy (SEcM) [11.81], scanning Kelvin Probe microscopy (SKPM) [11.82–86], scanning chemical potential microscopy (SCPM) [11.79], scanning ion conductance microscopy (SICM) [11.87,88], and scanning capacitance microscopy (SCM) [11.82, 89–91]. Families of instruments which measure forces (e.g., AFM, FFM, SEFM, SFAM, and SMM) are also referred to as scanning force microscopy (SFM). Although these instruments offer atomic resolution and are ideal for basic research, they are used for cutting edge industrial applications which do not require atomic resolution. Commercial production of SPMs started with the STM in 1987 and the AFM in 1989 by Digital Instruments Inc. For comparisons of SPMs with other microscopes, see Table 11.1 (Veeco Instruments, Inc.). Numbers of these instruments are equally divided between the U.S., Japan and Europe, with the following industry/university and Government labs. splits: 50/50, 70/30, and 30/70, respectively. It is clear that research and industrial applications of SPMs are rapidly expanding.

11.1 Scanning Tunneling Microscope

The principle of electron tunneling was proposed by *Giaever* [11.92]. He envisioned that if a potential difference is applied to two metals separated by a thin insulating film, a current will flow because of the ability of electrons to penetrate a potential barrier. To be able to measure a tunneling current, the two metals must be spaced no more than 10 nm apart. *Binnig* et al. [11.1] introduced vacuum tunneling combined with lateral scanning. The vacuum provides the ideal barrier for tunneling. The lateral scanning allows one to image surfaces with exquisite resolution, lateral-less than 1 nm and vertical-less than 0.1 nm, sufficient to define the position of single atoms. The very high vertical resolution of the STM is obtained because the tunnel current varies exponentially with the distance between the two electrodes, that is, the metal tip and the scanned surface. Typically, tunneling current decreases by a factor of 2 as the separation is increased by 0.2 nm. Very high lateral resolution depends upon sharp tips. *Binnig* et al. overcame two key obstacles for damping external vibrations and for moving the tunneling probe in close proximity to the sample. Their instrument is called the scanning tunneling microscope (STM). Today's STMs can be used in the ambient environment for atomic-scale image of surfaces. Excellent reviews on this subject are presented by *Hansma* and *Tersoff* [11.93], *Sarid* and *Elings* [11.94], *Durig* et al. [11.95]; *Frommer* [11.96], *Güntherodt* and *Wiesendanger* [11.97], *Wiesendanger* and *Güntherodt* [11.98], *Bonnell* [11.99], *Marti* and *Amrein* [11.100], *Stroscio* and *Kaiser* [11.101], and *Güntherodt* et al. [11.102].

The principle of the STM is straightforward. A sharp metal tip (one electrode of the tunnel junction) is brought close enough (0.3–1 nm) to the surface to be investigated (the second electrode) that, at a convenient operating voltage (10 mV–1 V), the tunneling current varies from 0.2 to 10 nA which is measurable. The tip is scanned over a surface at a distance of 0.3–1 nm, while the tunneling current between it and the surface is measured. The STM can be operated in either the constant current mode or the constant height mode, Fig. 11.1. The left-hand column of Fig. 11.1 shows the basic constant current mode of operation. A feedback network changes the height of the tip z to keep the current constant. The displacement of the tip given by the voltage applied to the piezoelectric drives then yields a topographic map of the surface. Alternatively, in the constant height mode, a metal tip can be scanned across a surface at nearly constant height and constant voltage while the current is monitored, as shown in the right-hand column of Fig. 11.1. In this case, the feedback network responds only rapidly enough to keep the average current constant. A current mode is generally used for atomic-scale images. This mode is not practical for rough surfaces. A three-dimensional picture $[z(x, y)]$ of a surface consists of multiple scans $[z(x)]$ displayed laterally from each other in the y direction. It should be noted that if different atomic species are present in a sample, the different atomic species within a sample may produce different tunneling currents for a given bias voltage. Thus the height data may not be a direct representation of the topography of the surface of the sample.

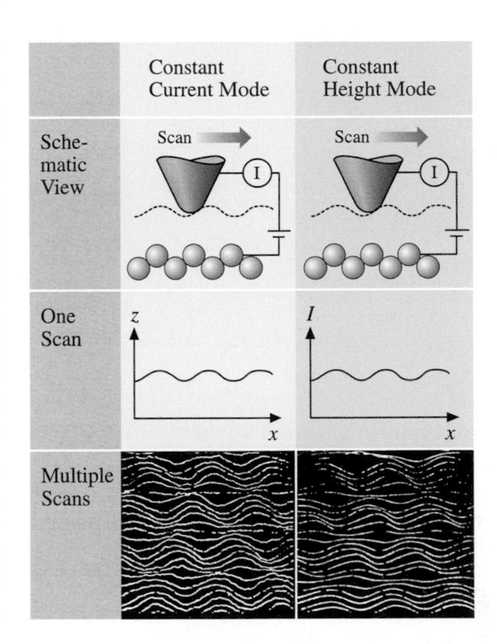

Fig. 11.1 STM can be operated in either the constant-current or the constant-height mode. The images are of graphite in air [11.93]

11.1.1 Binnig et al.'s Design

Figure 11.2 shows a schematic of one of Binnig and Rohrer's designs for operation in ultrahigh vacuum [11.1, 103]. The metal tip was fixed to rectangular piezodrives P_x, P_y, and P_z made out of commercial piezoceramic material for scanning. The sample is mounted via either superconducting magnetic levitation or a two-stage spring system to achieve a stability of the gap width of about 0.02 nm. The tunnel current J_T is a sensitive function of the gap width d that is $J_T \propto V_T \exp(-A\phi^{1/2}d)$, where V_T is the bias voltage, ϕ is the average barrier height (work function) and the constant $A = 1.025\,\text{eV}^{-1/2}\text{Å}^{-1}$. With a work function of a few eV, J_T changes by an order of magnitude for

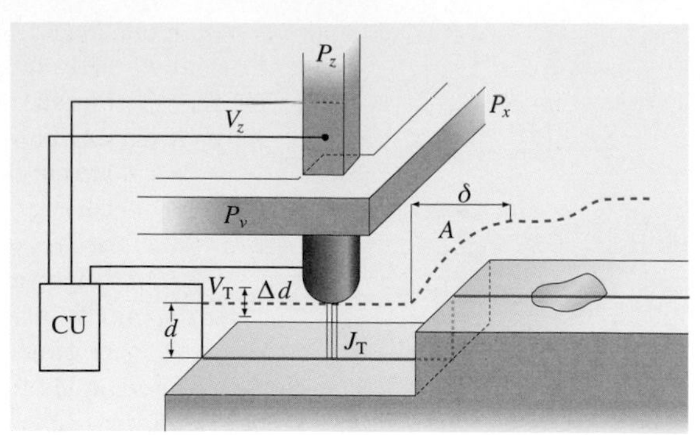

Fig. 11.2 Principle of operation of the STM made by *Binnig* and *Rohrer* [11.103]

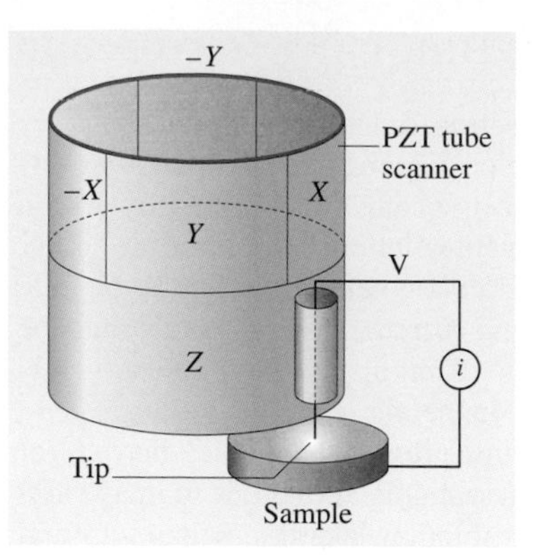

Fig. 11.3 Principle of operation of a commercial STM, a sharp tip attached to a piezoelectric tube scanner is scanned on a sample

every angstrom change of d. If the current is kept constant to within, for example, 2%, then the gap d remains constant to within 1 pm. For operation in the constant current mode, the control unit CU applies a voltage V_z to the piezo P_z such that J_T remains constant when scanning the tip with P_y and P_x over the surface. At constant work function ϕ, $V_z(V_x, V_y)$ yields the roughness of the surface $z(x, y)$ directly, as illustrated at a surface step at A. Smearing the step, δ (lateral resolution) is on the order of $(R)^{1/2}$, where R is the radius of the curvature of the tip. Thus, a lateral resolution of about 2 nm requires tip radii on the order of 10 nm. A 1-mm-diameter solid rod ground at one end at roughly 90° yields overall tip radii of only a few hundred nm, but with closest protrusion of rather sharp microtips on the relatively dull end yields a lateral resolution of about 2 nm. In-situ sharpening of the tips by gently touching the surface brings the resolution down to the 1-nm range; by applying high fields (on the order of 10^8 V/cm) during, for example, half an hour, resolutions considerably below 1 nm could be reached. Most experiments were done with tungsten wires either ground or etched to a radius typically in the range of 0.1–10 μm. In some cases, in-situ processing of the tips was done for further reduction of tip radii.

11.1.2 Commercial STMs

There are a number of commercial STMs available on the market. Digital Instruments, Inc. located in Santa Barbara, CA introduced the first commercial STM, the Nanoscope I, in 1987. In a recent Nanoscope IV STM for operation in ambient air, the sample is held in position while a piezoelectric crystal in the form of a cylindrical tube (referred to as PZT tube scanner) scans the sharp metallic probe over the surface in a raster pattern while sensing and outputting the tunneling current to the control station (Fig. 11.3). The digital signal processor (DSP) calculates the desired separation of the tip from the sample by sensing the tunneling current flowing between the sample and the tip. The bias voltage applied between the sample and the tip encourages the tunneling current to flow. The DSP completes the digital feedback loop by outputting the desired voltage to the piezoelectric tube. The STM operates in both the "constant height" and "constant current" modes depending on a parameter selection in the control panel. In the constant current mode, the feedback gains are set high, the tunneling tip closely tracks the sample surface, and the variation in the tip height required to maintain constant tunneling current is measured by the change in the voltage applied to the piezo tube. In the constant height mode, the feedback gains are set low, the tip remains at a nearly constant height as it sweeps over the sample surface, and the tunneling current is imaged.

Physically, the Nanoscope STM consists of three main parts: the head which houses the piezoelectric tube scanner for three dimensional motion of the tip and the preamplifier circuit (FET input amplifier) mounted on top of the head for the tunneling current, the base on which the sample is mounted, and the base support, which supports the base and head [11.4]. The base accommodates samples up to 10 mm by 20 mm and 10 mm in thickness. Scan sizes available for the STM are 0.7 μm (for atomic resolution), 12 μm, 75 μm and 125 μm square.

The scanning head controls the three dimensional motion of the tip. The removable head consists of a piezo tube scanner, about 12.7 mm in diameter, mounted into an invar shell used to minimize vertical thermal drifts because of a good thermal match between the piezo tube and the Invar. The piezo tube has separate electrodes for X, Y and Z which are driven by separate drive circuits.

The electrode configuration (Fig. 11.3) provides x and y motions which are perpendicular to each other, minimizes horizontal and vertical coupling, and provides good sensitivity. The vertical motion of the tube is controlled by the Z electrode which is driven by the feedback loop. The x and y scanning motions are each controlled by two electrodes which are driven by voltages of the same magnitude, but opposite signs. These electrodes are called $-Y$, $-X$, $+Y$, and $+X$. Applying complimentary voltages allows a short, stiff tube to provide a good scan range without large voltages. The motion of the tip due to external vibrations is proportional to the square of the ratio of vibration frequency to the resonant frequency of the tube. Therefore, to minimize the tip vibrations, the resonant frequencies of the tube are high at about 60 kHz in the vertical direction and about 40 kHz in the horizontal direction. The tip holder is a stainless steel tube with a 300 μm inner diameter for 250 μm diameter tips, mounted in ceramic in order to keep the mass on the end of the tube low. The tip is mounted either on the front edge of the tube (to keep mounting mass low and resonant frequency high) (Fig. 11.3) or the center of the tube for large range scanners, namely 75 and 125 μm (to preserve the symmetry of the scanning). This commercial STM accepts any tip with a 250 μm diameter shaft. The piezotube requires X-Y calibration which is carried out by imaging an appropriate calibration standard. Cleaved graphite is used for the small-scan length head while two dimensional grids (a gold plated ruling) can be used for longer range heads.

The Invar base holds the sample in position, supports the head, and provides coarse X-Y motion for the sample. A spring-steel sample clip with two thumb screws holds the sample in place. An x-y translation stage built into the base allows the sample to be repositioned under the tip. Three precision screws arranged in a triangular pattern support the head and provide coarse and fine adjustment of the tip height. The base support consists of the base support ring and the motor housing. The stepper motor enclosed in the motor housing allows the tip to be engaged and withdrawn from the surface automatically.

Samples to be imaged with the STM must be conductive enough to allow a few nanoamperes of current to flow from the bias voltage source to the area to be scanned. In many cases, nonconductive samples can be coated with a thin layer of a conductive material to facilitate imaging. The bias voltage and the tunneling current depend on the sample. Usually they are set at a standard value for engagement and fine tuned to enhance the quality of the image. The scan size depends on the sample and the features of interest. A maximum scan rate of 122 Hz can be used. The maximum scan rate is usually related to the scan size. Scan rates above 10 Hz are used for small scans (typically 60 Hz for atomic-scale imaging with a 0.7 μm scanner). The scan rate should be lowered for large scans, especially if the sample surfaces are rough or contain large steps. Moving the tip quickly along the sample surface at high scan rates with large scan sizes will usually lead to a tip crash. Essentially, the scan rate should be inversely proportional to the scan size (typically 2–4 Hz for 1 μm, 0.5–1 Hz for 12 μm, and 0.2 Hz for 125 μm scan sizes). Scan rate in length/time, is equal to scan length divided by the scan rate in Hz. For example, for 10 μm × 10 μm scan size scanned at 0.5 Hz, the scan rate is 10 μm/s. Typically, 256 × 256 data formats are most commonly used. The lateral resolution at larger scans is approximately equal to scan length divided by 256.

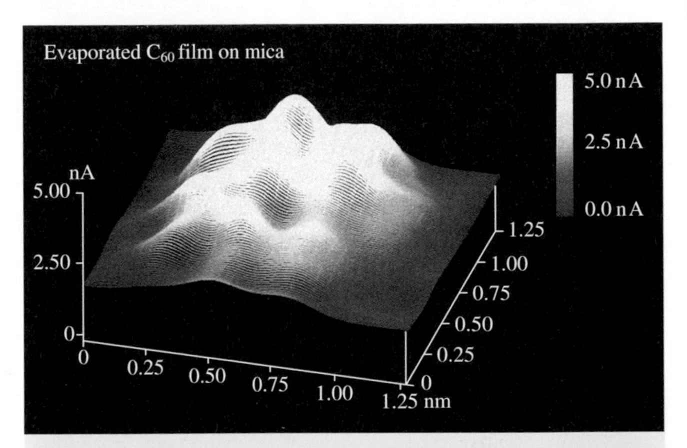

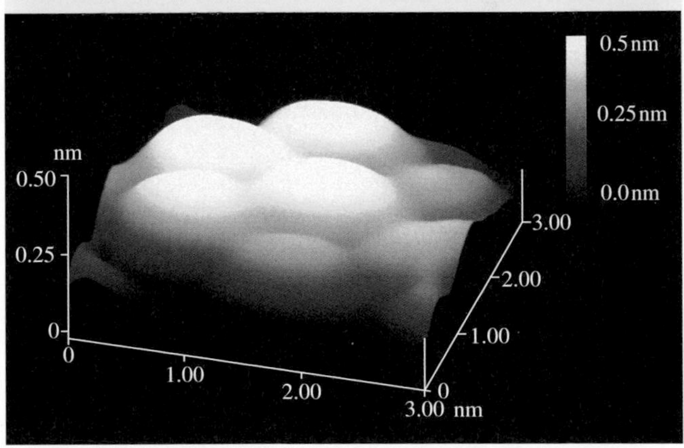

Fig. 11.4 STM images of evaporated C_{60} film on a gold-coated freshly-cleaved mica using a mechanically sheared Pt-Ir (80-20) tip in constant height mode [11.104]

Figure 11.4 shows an example of STM images of an evaporated C_{60} film on a gold-coated freshly-cleaved mica taken at room temperature and ambient pressure [11.104]. Images with atomic resolution at two scan sizes are obtained. Next we describe STM designs which are available for special applications.

Electrochemical STM
The electrochemical STM is used to perform and monitor the electrochemical reactions inside the STM. It includes a microscope base with an integral potentiostat, a short head with a 0.7 μm scan range and a differential preamp and the software required to operate the potentiostat and display the result of the electrochemical reaction.

Standalone STM
The stand alone STMs are available to scan large samples which rest directly on the sample. From Digital instruments, it is available in 12 and 75 μm scan ranges. It is similar to the standard STM except the sample base has been eliminated.

11.1.3 STM Probe Construction

The STM probe should have a cantilever integrated with a sharp metal tip with a low aspect ratio (tip length/tip shank) to minimize flexural vibrations. Ideally, the tip should be atomically sharp, but in practice most tip preparation methods produce a tip with a rather ragged profile that consists of several asperities with the one closest to the surface responsible for tunneling. STM cantilevers with sharp tips are typically fabricated from metal wires of tungsten (W), platinum-iridium (Pt-Ir), or gold (Au) and sharpened by grinding, cutting with a wire cutter or razor blade, field emission/evaporator, ion milling, fracture, or electrochemical polishing/etching [11.105, 106]. The two most commonly used tips are made from either a Pt-Ir (80/20) alloy or tungsten wire. Iridium is used to provide stiffness. The Pt-Ir tips are generally mechanically formed and are readily available. The tungsten tips are etched from tungsten wire with an electrochemical process, for example by using 1 molar KOH solution with a platinum electrode in a electrochemical cell at about 30 V. In general, Pt-Ir tips provide better atomic resolution than tungsten tips, probably due to the lower reactivity of Pt. But tungsten tips are more uniformly shaped and may perform better on samples with steeply sloped features. The tungsten wire diameter used for the cantilever is typically 250 μm with the radius of curvature ranging from 20 to 100 nm and a cone angle ranging from 10 to 60° (Fig. 11.5). The wire can be bent in an *L* shape, if so required, for use in the instrument. For calculations of normal spring constant and natural frequency of round cantilevers, see *Sarid* and *Elings* [11.94].

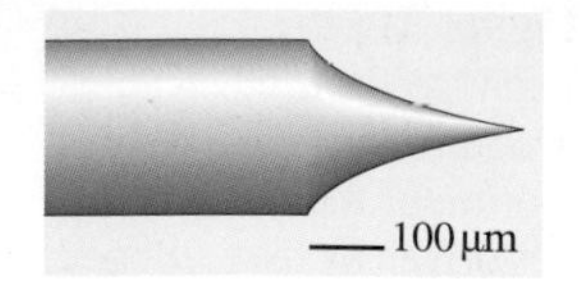

Fig. 11.5 Schematic of a typical tungsten cantilever with a sharp tip produced by electrochemical etching

For imaging of deep trenches, high-aspect-ratio, controlled geometry (CG) Pt-Ir probes are commercially available (Fig. 11.6). These probes are electrochemically etched from Pt-Ir (80/20) wire and polished to a specific shape which is consistent from tip to tip. Probes have a full cone angle of approximately 15°, and a tip radius of less than 50 nm. For imaging of very deep trenches (> 0.25 μm) and nanofeatures, focused ion beam (FIB) milled CG probes with an extremely sharp tip (radius < 5 nm) are used. For electrochemistry, Pt/Ir probes are coated with a nonconducting film (not shown in the figure). These probes are available from Materials Analytical Services, Raleigh, North Carolina.

Pt alloy and W tips are very sharp and have high resolution, but are fragile and sometimes break when contacting a surface. Diamond tips have been used by *Kaneko* and *Oguchi* [11.107]. The diamond tip made conductive by boron ion implantation is found to be chip resistant.

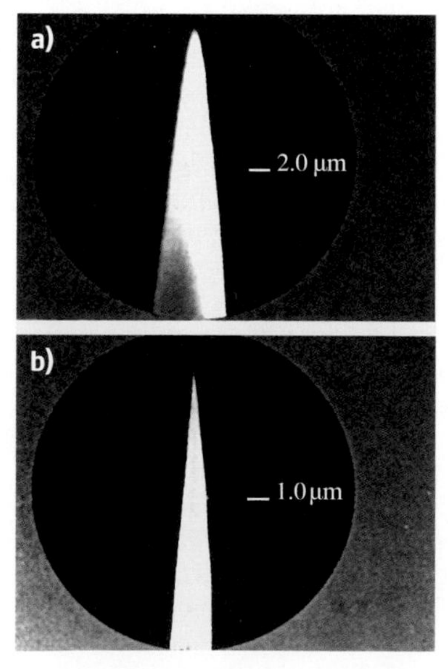

Fig. 11.6a,b Schematics of (**a**) CG Pt-Ir probe, and (**b**) CG Pt-Ir FIB milled probe

11.2 Atomic Force Microscope

Like the STM, the AFM relies on a scanning technique to produce very high resolution, 3-D images of sample surfaces. The AFM measures ultrasmall forces (less than 1 nN) present between the AFM tip surface and a sample surface. These small forces are measured by measuring the motion of a very flexible cantilever beam having an ultrasmall mass. While STMs require that the surface to be measured be electrically conductive, AFMs are capable of investigating surfaces of both conductors and insulators on an atomic scale if suitable techniques for measurement of cantilever motion are used. In the operation of a high resolution AFM, the sample is generally scanned instead of the tip as with the STM, because the AFM measures the relative displacement between the cantilever surface and reference surface and any cantilever movement would add vibrations. For measurements of large samples, AFMs are available where the tip is scanned and the sample is stationary. As long as the AFM is operated in the so-called contact mode, little if any vibration is introduced.

The AFM combines the principles of the STM and the stylus profiler (Fig. 11.7). In an AFM, the force between the sample and tip is detected, rather than the tunneling current, to sense the proximity of the tip to the sample. The AFM can be used either in a static or dynamic mode. In the static mode, also referred to as repulsive mode or contact mode [11.2], a sharp tip at the end of a cantilever is brought in contact with a sample surface. During initial contact, the atoms at the end of the tip experience a very weak repulsive force due to electronic orbital overlap with the atoms in the sample surface. The force acting on the tip causes a cantilever deflection which is measured by tunneling, capacitive, or optical detectors. The deflection can be measured to within 0.02 nm, so for typical cantilever spring constant of 10 N/m a force as low as 0.2 nN (corresponding normal pressure ~ 200 MPa for a Si_3N_4 tip with radius of about 50 nm against single-crystal silicon) can be detected. (To put these number in perspective, individual atoms and human hair are typically a fraction of a nanometer and about 75 μm in diameter, respectively, and a drop of water and an eyelash have a mass of about 10 μN and 100 nN, respectively.) In the dynamic mode of operation for the AFM, also referred to as attractive force imaging or noncontact imaging mode, the tip is brought in close proximity (within a few nm) to, and not in contact with the sample. The cantilever is deliberately vibrated either in amplitude modulation (AM) mode [11.65] or frequency modulation (FM) mode [11.65, 94, 108, 109]. Very weak van der Waals attractive forces are present at the tip–sample interface. Although in this technique, the normal pressure exerted at the interface is zero (desirable to avoid any surface deformation), it is slow, and is difficult to use, and is rarely used outside research environments. In the two modes, surface topography is measured by laterally scanning the sample under the tip while simultaneously measuring the separation-dependent force or force gradient (derivative) between the tip and the surface (Fig. 11.7). In the contact (static) mode, the interaction force between tip and sample is measured by measuring the cantilever deflection. In the noncontact (or dynamic) mode, the force gradient is obtained by vibrating the cantilever and measuring the shift of resonant frequency of the cantilever. To obtain topographic information, the interaction force is either recorded directly, or used as a control parameter for a feedback circuit that maintains the force or force derivative at a constant value. With an AFM operated in the contact mode, topographic images with a vertical resolution of less than 0.1 nm (as low as 0.01 nm) and a lateral resolution of about 0.2 nm have been obtained [11.3, 50, 110–114]. With a 0.01 nm displacement sensitivity, 10 nN to 1 pN forces are measurable. These forces are comparable to the forces associated with chemical bonding, e.g., 0.1 μN for an ionic bond and 10 pN for a hydrogen bond [11.2]. For further reading, see [11.94–96, 100, 102, 115–119].

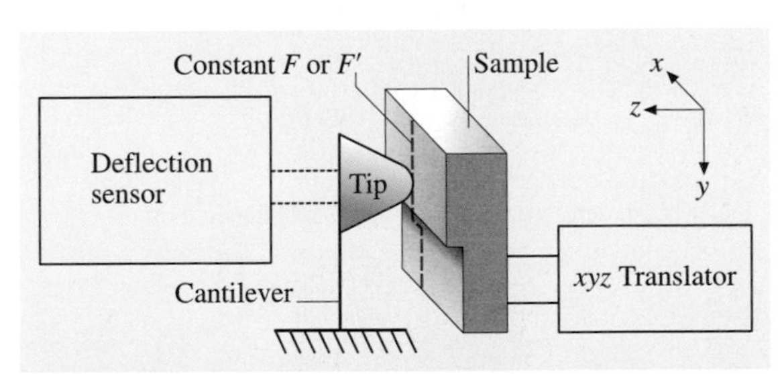

Fig. 11.7 Principle of operation of the AFM. Sample mounted on a piezoelectric tube scanner is scanned against a short tip and the cantilever deflection is measured, mostly, using a laser deflection technique. Force (contact mode) or force gradient (noncontact mode) is measured during scanning

Lateral forces being applied at the tip during scanning in the contact mode, affect roughness measurements [11.120]. To minimize the effects of friction and other lateral forces in the topography measurements in the contact-mode, and to measure the topography of soft surfaces, AFMs can be operated in the so called tapping mode or force modulation mode [11.32, 121].

The STM is ideal for atomic-scale imaging. To obtain atomic resolution with the AFM, the spring constant of the cantilever should be weaker than the equivalent spring between atoms. For example, the vibration frequencies ω of atoms bound in a molecule or in a crystalline solid are typically 10^{13} Hz or higher. Combining this with the mass of the atoms m, on the order of 10^{-25} kg, gives an interatomic spring constants k, given by $\omega^2 m$, on the order of 10 N/m [11.115]. (For comparison, the spring constant of a piece of household aluminum foil that is 4 mm long and 1 mm wide is about 1 N/m.) Therefore, a cantilever beam with a spring constant of about 1 N/m or lower is desirable. Tips have to be as sharp as possible. Tips with a radius ranging from 5 to 50 nm are commonly available.

Atomic resolution cannot be achieved with these tips at the normal load in the nN range. Atomic structures at these loads have been obtained from lattice imaging or by imaging of the crystal periodicity. Reported data show either perfectly ordered periodic atomic structures or defects on a larger lateral scale, but no well-defined, laterally resolved atomic-scale defects like those seen in images routinely obtained with a STM. Interatomic forces with one or several atoms in contact are 20–40 or 50–100 pN, respectively. Thus, atomic resolution with an AFM is only possible with a sharp tip on a flexible cantilever at a net repulsive force of 100 pN or lower [11.122]. Upon increasing the force from 10 pN, *Ohnesorge* and *Binnig* [11.122] observed that monoatomic steplines were slowly wiped away and a perfectly ordered structure was left. This observation explains why mostly defect-free atomic resolution has been observed with AFM. Note that for atomic-resolution measurements, the cantilever should not be too soft to avoid jumps. Further note that measurements in the noncontact imaging mode may be desirable for imaging with atomic resolution.

The key component in an AFM is the sensor for measuring the force on the tip due to its interaction with the sample. A cantilever (with a sharp tip) with extremely low spring constant is required for high vertical and lateral resolutions at small forces (0.1 nN or lower) but at the same time a high resonant frequency is desirable (about 10 to 100 kHz) in order to minimize the sensitivity to vibration noise from the building which is near 100 Hz. This requires a spring with extremely low vertical spring constant (typically 0.05 to 1 N/m) as well as low mass (on the order of 1 ng). Today, the most advanced AFM cantilevers are microfabricated from silicon or silicon nitride using photolithographic techniques. (For further details on cantilevers, see a later section). Typical lateral dimensions are on the order of 100 μm, with the thicknesses on the order of 1 μm. The force on the tip due to its interaction with the sample is sensed by detecting the deflection of the compliant lever with a known spring constant. This cantilever deflection (displacement smaller than 0.1 nm) has been measured by detecting a tunneling current similar to that used in the STM in the pioneering work of *Binnig* et al. [11.2] and later used by *Giessibl* et al. [11.56], by capacitance detection [11.123, 124], piezoresistive detection [11.125, 126], and by four optical techniques namely (1) by optical interferometry [11.5, 6, 127, 128] with the use of optical fibers [11.57, 129] (2) by optical polarization detection [11.72, 130], (3) by laser diode feedback [11.131] and (4) by optical (laser) beam deflection [11.7, 8, 53, 111, 112]. Schematics of the four more commonly used detection systems are shown in Fig. 11.8. The tunneling method originally used by *Binnig* et al. [11.2] in the first version of the AFM, uses a second tip to monitor the deflection of the cantilever with its force sensing tip. Tunneling is rather sensitive to contaminants and the interaction between the tunneling tip and the rear side of the cantilever can become comparable to the interaction between the tip and sample. Tunneling is rarely used and is mentioned first for historical purposes. *Giessibl* et al. [11.56] have used it for a low temperature AFM/STM design. In contrast to tunneling, other deflection sensors are far

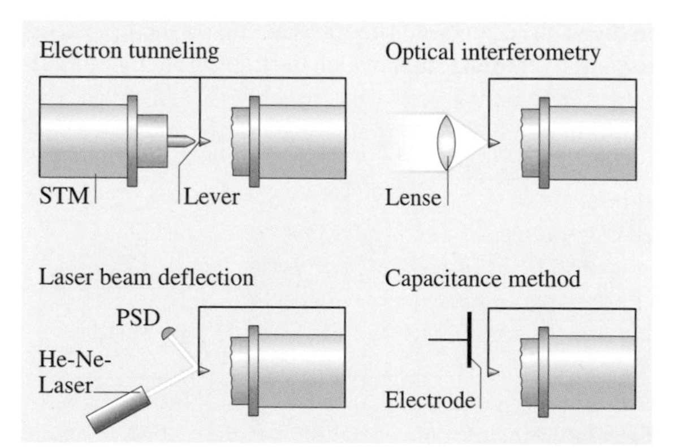

Fig. 11.8 Schematics of the four more commonly used detection systems for measurement of cantilever deflection. In each set up, the sample mounted on piezoelectric body is shown on the right, the cantilever in the middle, and the corresponding deflection sensor on the left [11.118]

away from the cantilever at distances of microns to tens of mm. The optical techniques are believed to be more sensitive, reliable and easily implemented detection methods than others [11.94, 118]. The optical beam deflection method has the largest working distance, is insensitive to distance changes and is capable of measuring angular changes (friction forces), therefore, it is most commonly used in the commercial SPMs.

Almost all SPMs use piezo translators to scan the sample, or alternatively, to scan the tip. An electric field applied across a piezoelectric material causes a change in the crystal structure, with expansion in some directions and contraction in others. A net change in volume also occurs [11.132]. The first STM used a piezo tripod for scanning [11.1]. The piezo tripod is one way to generate three-dimensional movement of a tip attached to its center. However, the tripod needs to be fairly large ($\sim 50\,\mathrm{mm}$) to get a suitable range. Its size and asymmetric shape makes it susceptible to thermal drift. The tube scanners are widely used in AFMs [11.133]. These provide ample scanning range with a small size. Control electronics systems for AFMs can use either analog or digital feedback. Digital feedback circuits are better suited for ultra low noise operation.

Images from the AFMs need to be processed. An ideal AFM is a noise free device that images a sample with perfect tips of known shape and has a perfectly linear scanning piezo. In reality, scanning devices are affected by distortions and these distortions must be corrected for. The distortions can be linear and nonlinear. Linear distortions mainly result from imperfections in the machining of the piezo translators causing cross talk between the Z-piezo to the X- and Y-piezos, and vice versa. Nonlinear distortions mainly result because of the presence of a hysteresis loop in piezoelectric ceramics. These may also result if the scan frequency approaches the upper frequency limit of the X- and Y-drive amplifiers or the upper frequency limit of the feedback loop (z-component). In addition, electronic noise may be present in the system. The noise is removed by digital filtering in real space [11.134] or in the spatial frequency domain (Fourier space) [11.135].

Processed data consists of many tens of thousand of points per plane (or data set). The output of the first STM and AFM images were recorded on an X-Y chart recorder, with z-value plotted against the tip position in the fast scan direction. Chart recorders have slow response so computers are used to display the data. The data are displayed as a wire mesh display or gray scale display (with at least 64 shades of gray).

11.2.1 Binnig et al.'s Design

In the first AFM design developed by *Binnig* et al. [11.2], AFM images were obtained by measurement of the force on a sharp tip created by the proximity to the surface of the sample mounted on a 3-D piezoelectric scanner. Tunneling current between STM tip and the backside of the cantilever beam with attached tip was measured to obtain the normal force. This force was kept at a constant level with a feedback mechanism. The STM tip was also mounted on a piezoelectric element to maintain the tunneling current at a constant level.

11.2.2 Commercial AFM

A review of early designs of AFMs is presented by *Bhushan* [11.4]. There are a number of commercial AFMs available on the market. Major manufacturers of AFMs for use in an ambient environment are: Digital Instruments Inc., a subsidiary of Veeco Instruments, Inc., Santa Barbara, California; Topometrix Corp., a subsidiary of Veeco Instruments, Inc., Santa Clara, California; other subsidiaries of Veeco Instruments Inc., Woodbury, New York; Molecular Imaging Corp., Phoenix, Arizona; Quesant Instrument Corp., Agoura Hills, California; Nanoscience Instruments Inc., Phoenix, Arizona; Seiko Instruments, Japan; and Olympus, Japan. AFM/STMs for use in UHV environments are manufactured by Omicron Vakuumphysik GMBH, Taunusstein, Germany.

We describe here two commercial AFMs – small sample and large sample AFMs – for operation in the contact mode, produced by Digital Instruments, Inc., Santa Barbara, CA, with scanning lengths ranging from about 0.7 μm (for atomic resolution) to about 125 μm [11.9,111,114,136]. The original design of these AFMs comes from *Meyer* and *Amer* [11.53]. Basically the AFM scans the sample in a raster pattern while outputting the cantilever deflection error signal to the control station. The cantilever deflection (or the force) is measured using laser deflection technique (Fig. 11.9). The DSP in the workstation controls the z position of the piezo based on the cantilever deflection error signal. The AFM operates in both the "constant height" and "constant force" modes. The DSP always adjusts the height of the sample under the tip based on the cantilever deflection error signal, but if the feedback gains are low the piezo remains at a nearly "constant height" and the cantilever deflection data is collected. With the high gains the piezo height changes to keep the cantilever deflection nearly constant (therefore the force is

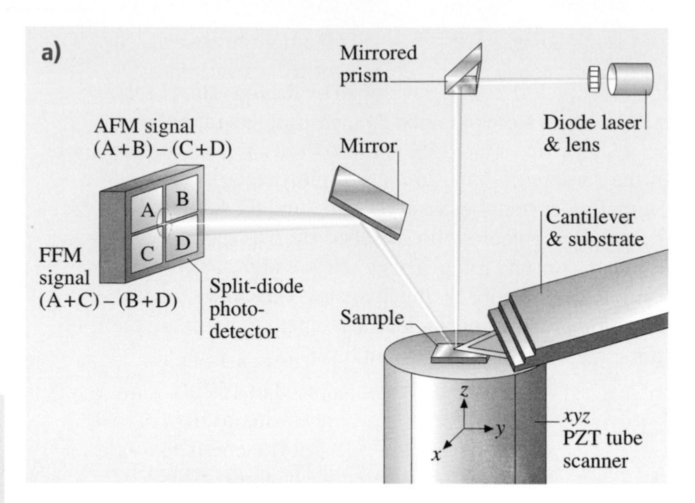

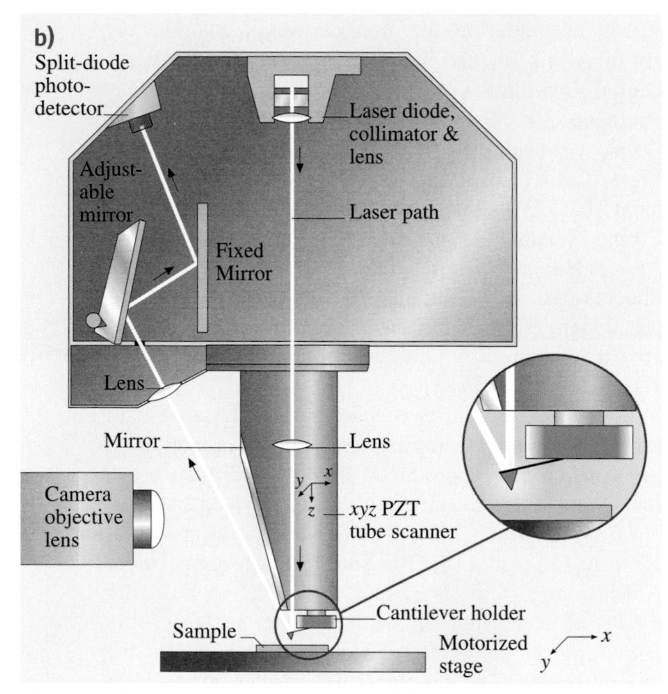

Fig. 11.9a,b Principles of operation of (**a**) a commercial small sample AFM/FFM, and (**b**) a large sample AFM/FFM

constant) and the change in piezo height is collected by the system.

To further describe the principle of operation of the commercial small sample AFM shown in Fig. 11.9a, the sample, generally no larger than 10 mm × 10 mm, is mounted on a PZT tube scanner which consists of separate electrodes to scan precisely the sample in the x-y plane in a raster pattern and to move the sample in the vertical (z) direction. A sharp tip at the free end of a flexible cantilever is brought in contact with the sample. Features on the sample surface cause the cantilever to deflect in the vertical and lateral directions as the sample moves under the tip. A laser beam from a diode laser (5 mW max peak output at 670 nm) is directed by a prism onto the back of a cantilever near its free end, tilted downward at about 10° with respect to the horizontal plane. The reflected beam from the vertex of the cantilever is directed through a mirror onto a quad photodetector (split photodetector with four quadrants) (commonly called position-sensitive detector or PSD, produced by Silicon Detector Corp., Camarillo, California). The differential signal from the top and bottom photodiodes provides the AFM signal which is a sensitive measure of the cantilever vertical deflection. Topographic features of the sample cause the tip to deflect in the vertical direction as the sample is scanned under the tip. This tip deflection will change the direction of the reflected laser beam, changing the intensity difference between the top and bottom sets of photodetectors (AFM signal). In the AFM operating mode called the height mode, for topographic imaging or for any other operation in which the applied normal force is to be kept a constant, a feedback circuit is used to modulate the voltage applied to the PZT scanner to adjust the height of the PZT, so that the cantilever vertical deflection (given by the intensity difference between the top and bottom detector) will remain constant during scanning. The PZT height variation is thus a direct measure of the surface roughness of the sample.

In a large sample AFM, both force sensors using optical deflection methods and scanning unit are mounted on the microscope head (Fig. 11.9b). Because of vibrations added by cantilever movement, lateral resolution of this design is somewhat poorer than the design in Fig. 11.9a in which the sample is scanned instead of the cantilever beam. The advantage of the large sample AFM is that large samples can be measured readily.

Most AFMs can be used for topography measurements in the so-called tapping mode (intermittent contact mode), also referred to as dynamic force microscopy. In the tapping mode, during scanning over the surface, the cantilever/tip assembly is sinusoidally vibrated by a piezo mounted above it, and the oscillating tip slightly taps the surface at the resonant frequency of the cantilever (70–400 Hz) with a constant (20–100 nm) oscillating amplitude introduced in the vertical direction with a feedback loop keeping the average normal force constant (Fig. 11.10). The oscillating amplitude is kept

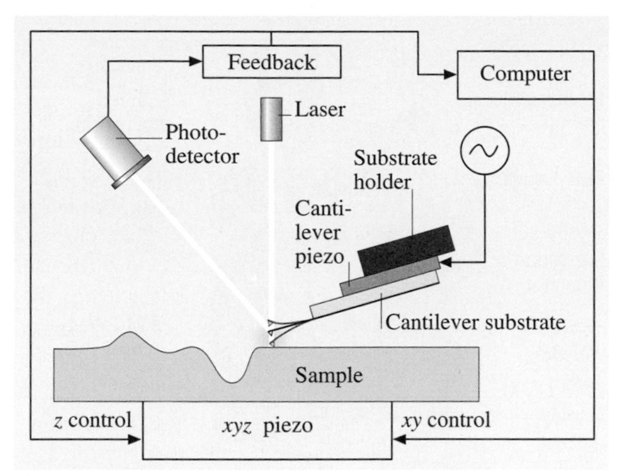

Fig. 11.10 Schematic of tapping mode used for surface roughness measurements

large enough so that the tip does not get stuck to the sample because of adhesive attractions. The tapping mode is used in topography measurements to minimize effects of friction and other lateral forces to measure the topography of soft surfaces.

Topographic measurements are made at any scanning angle. At a first instance, scanning angle may not appear to be an important parameter. However, the friction force between the tip and the sample will affect the topographic measurements in a parallel scan (scanning along the long axis of the cantilever). Therefore a perpendicular scan may be more desirable. Generally, one picks a scanning angle which gives the same topographic data in both directions; this angle may be slightly different than that for the perpendicular scan.

For measurement of the friction force being applied at the tip surface during sliding, the left hand and right hand sets of quadrants of the photodetector are used. In the so-called friction mode, the sample is scanned back and forth in a direction orthogonal to the long axis of the cantilever beam. A friction force between the sample and the tip will produce a twisting of the cantilever. As a result, the laser beam will be reflected out of the plane defined by the incident beam and the beam reflected vertically from an untwisted cantilever. This produces an intensity difference of the laser beam received in the left hand and right hand sets of quadrants of the photodetector. The intensity difference between the two sets of detectors (FFM signal) is directly related to the degree of twisting and hence to the magnitude of the friction force. This method provides three-dimensional maps of the friction force. One problem associated with this method is that any misalignment between the laser beam and the photodetector axis would introduce error in the measurement. However, by following the procedures developed by *Ruan* and *Bhushan* [11.136], in which the average FFM signal for the sample scanned in two opposite directions is subtracted from the friction profiles of each of the two scans, the misalignment effect is eliminated. By following the friction force calibration procedures developed by *Ruan* and *Bhushan* [11.136], voltages corresponding to friction forces can be converted to force units. The coefficient of friction is obtained from the slope of friction force data measured as a function of normal loads typically ranging from 10 to 150 nN. This approach eliminates any contributions due to the adhesive forces [11.10]. For calculation of the coefficient of friction based on a single point measurement, friction force should be divided by the sum of applied normal load and intrinsic adhesive force. Furthermore, it should be pointed out that for a single asperity contact, the coefficient of friction is not independent of load. For further details, refer to a later section.

The tip is scanned in such a way that its trajectory on the sample forms a triangular pattern (Fig. 11.11). Scanning speeds in the fast and slow scan directions depend on the scan area and scan frequency. Scan sizes ranging from less than 1 nm × 1 nm to 125 μm × 125 μm and scan rates from less than 0.5 to 122 Hz typically can be used. Higher scan rates are used for smaller scan lengths. For example, scan rates in the fast and slow scan directions for an area of 10 μm × 10 μm scanned at 0.5 Hz are 10 μm/s and 20 nm/s, respectively.

We now describe the construction of a small sample AFM in more detail. It consists of three main parts: the optical head which senses the cantilever deflection, a PZT tube scanner which controls the scanning motion of the sample mounted on its one end, and the base

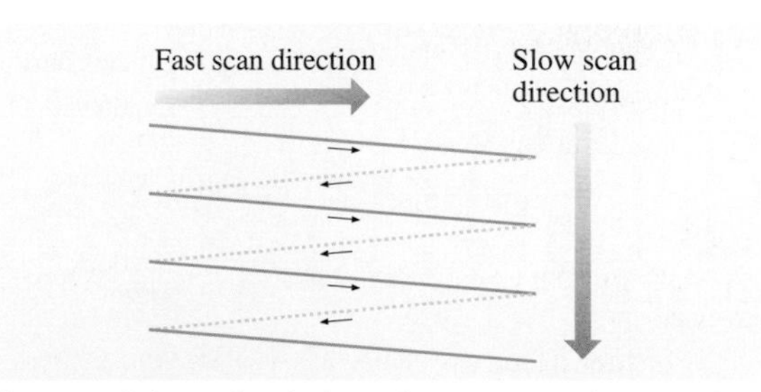

Fig. 11.11 Schematic of triangular pattern trajectory of the AFM tip as the sample is scanned in two dimensions. During imaging, data are recorded only during scans along the solid scan lines

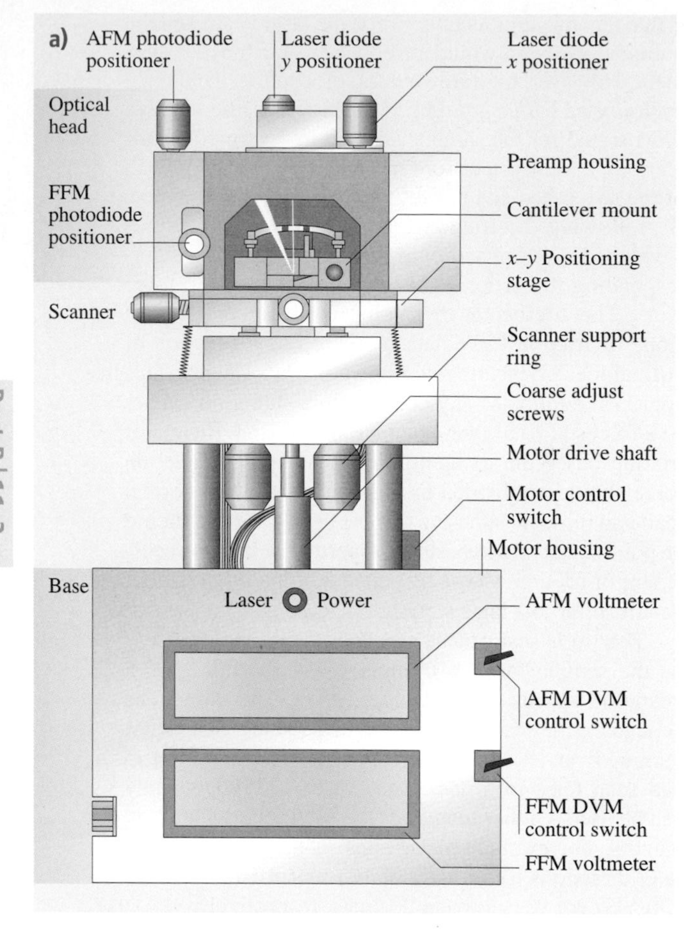

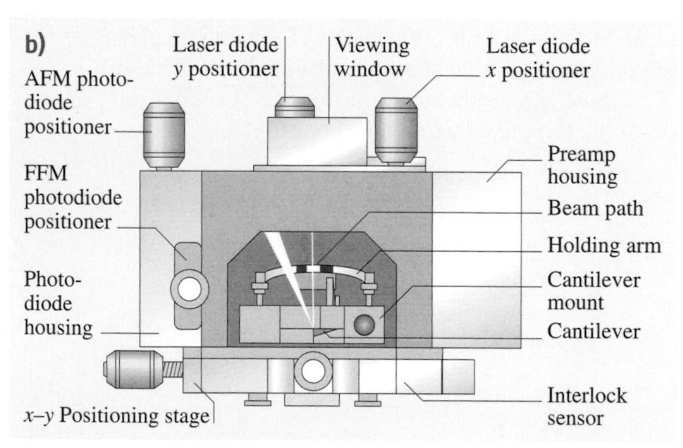

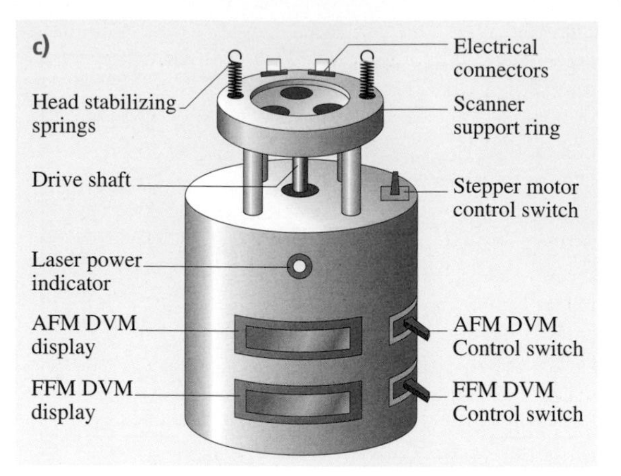

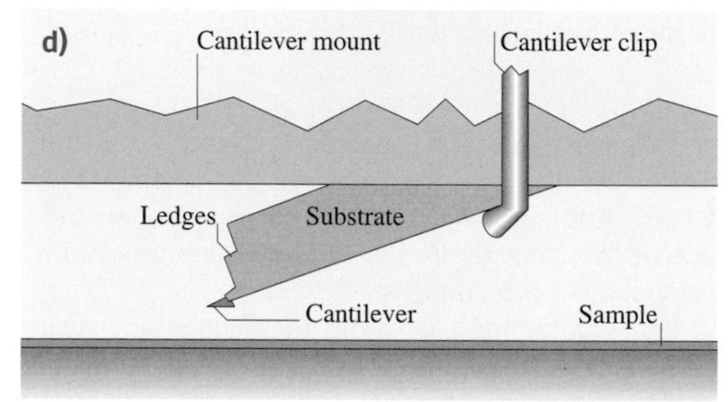

Fig. 11.12a–d Schematics of a commercial AFM/FFM made by Digital Instruments Inc. (**a**) front view, (**b**) optical head, (**c**) base, and (**d**) cantilever substrate mounted on cantilever mount (not to scale)

which supports the scanner and head and includes circuits for the deflection signal (Fig. 11.12a). The AFM connects directly to a control system. The optical head consists of laser diode stage, photodiode stage preamp board, cantilever mount and its holding arm, and deflection beam reflecting mirror (Fig. 11.12b). The laser diode stage is a tilt stage used to adjust the position of the laser beam relative to the cantilever. It consists of the laser diode, collimator, focusing lens, baseplate, and the X and Y laser diode positioners. The positioners are used to place the laser spot on the end of the cantilever. The photodiode stage is an adjustable stage used to position the photodiode elements relative to the reflected laser beam. It consists of the split photodiode, the base plate, and the photodiode positioners. The deflection beam reflecting mirror is mounted on the upper left in the interior of the head which reflects the deflected beam toward the photodiode. The cantilever mount is a metal (for operation in air) or glass (for operation in water) block which holds the cantilever firmly at the proper angle (Fig. 11.12d). Next, the tube scanner consists of an Invar cylinder holding a single tube made of piezoelectric crystal which provides the necessary three-dimensional motion to the sample. Mounted on top of the tube is a magnetic cap on which the steel sample puck is placed. The tube is rigidly held at one end with the sample mounted on the other end of the tube. The scanner also contains three fine-pitched screws which form the mount for the optical head. The optical head rests on the tips of the screws which are used to adjust the position of the head relative to the sample. The scanner fits into the scanner support ring mounted on the base of the microscope (Fig. 11.12c). The stepper motor is controlled manually with the switch on the upper surface of the base and automatically by the computer during the tip engage and tip-withdraw processes.

The scan sizes available for these instruments are 0.7 μm, 12 μm and 125 μm. The scan rate must be decreased as the scan size is increased. A maximum scan rate of 122 Hz can be used. Scan rates of about 60 Hz should be used for small scan lengths (0.7 μm). Scan rates of 0.5 to 2.5 Hz should be used for large scans on samples with tall features. High scan rates help reduce drift, but they can only be used on flat samples with small scan sizes. Scan rate, or scanning speed in length/time in the fast scan direction, is equal to twice the scan length times the scan rate in Hz, and in the slow direction, it is equal to scan length times the scan rate in Hz divided by number of data points in the transverse direction. For example, for 10 μm × 10 μm scan size scanned at 0.5 Hz, the scan rates in the fast and slow scan directions are 10 μm/s and 20 nm/s, respectively. Normally 256 × 256 data points are taken for each image. The lateral resolution at larger scans is approximately equal to scan length divided by 256. The piezo tube requires x-y calibration which is carried out by imaging an appropriate calibration standard. Cleaved graphite is used for small scan heads while two-dimensional grids (a gold plating ruling) can be used for longer range heads.

Examples of AFM images of freshly-cleaved highly-oriented pyrolytic (HOP) graphite and mica surfaces are shown in Fig. 11.13 [11.50, 110, 114]. Images with near atomic resolution are obtained.

The force calibration mode is used to study interactions between the cantilever and the sample surface. In the force calibration mode, the X and Y voltages applied to the piezo tube are held at zero and a sawtooth voltage is applied to the Z electrode of the piezo tube, Fig. 11.14a. The force measurement starts with the sample far away and the cantilever in its rest position. As a result of the applied voltage, the sample is moved up and down relative to the stationary cantilever tip. As the

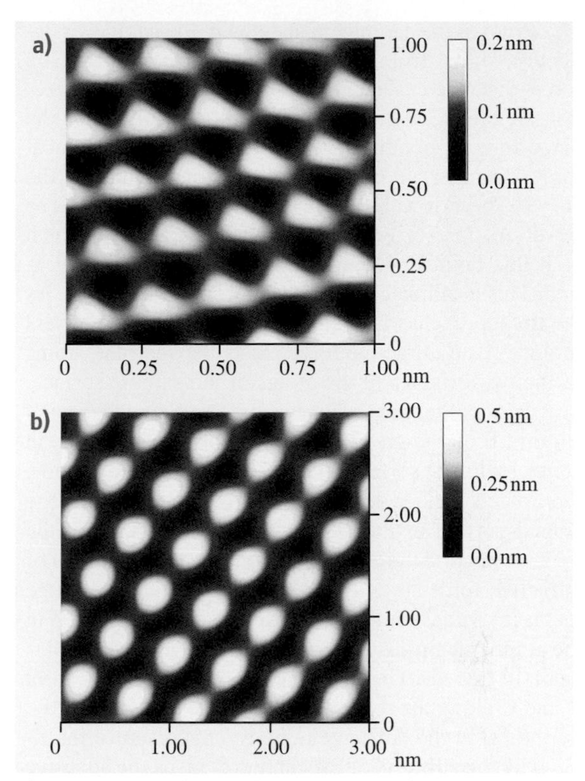

Fig. 11.13a,b Typical AFM images of freshly-cleaved (**a**) highly-oriented pyrolytic graphite and (**b**) mica surfaces taken using a square pyramidal Si_3N_4 tip

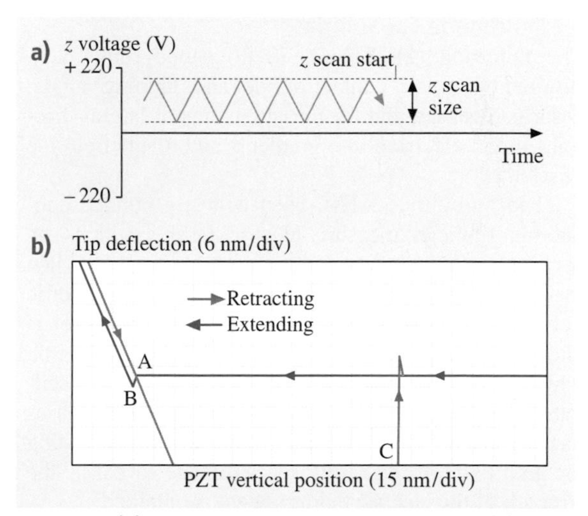

Fig. 11.14 (**a**) Force calibration Z waveform, and (**b**) a typical force-distance curve for a tip in contact with a sample. Contact occurs at point B; tip breaks free of adhesive forces at point C as the samples moves away from the tip

piezo moves the sample up and down, the cantilever deflection signal from the photodiode is monitored. The force–distance curve, a plot of the cantilever tip deflection signal as a function of the voltage applied to the piezo tube, is obtained. Figure 11.14b shows a typical force–distance curve showing the various features of the curve. The arrow heads reveal the direction of piezo travel. As the piezo extends, it approaches the tip, which is at this point in free air and hence shows no deflection. This is indicated by the flat portion of the curve. As the tip approaches the sample within a few nanometers (point A), an attractive force exists between the atoms of the tip surface and the atoms of the sample surface. The tip is pulled towards the sample and contact occurs at point B on the graph. From this point on, the tip is in contact with the surface and as the piezo further extends, the tip gets further deflected. This is represented by the sloped portion of the curve. As the piezo retracts, the tip goes beyond the zero deflection (flat) line because of attractive forces (van der Waals forces and long range meniscus forces), into the adhesive regime. At point C in the graph, the tip snaps free of the adhesive forces, and is again in free air. The horizontal distance between point B and C along the retrace line gives the distance moved by the tip in the adhesive regime. This distance multiplied by the stiffness of the cantilever gives the adhesive force. Incidentally, the horizontal shift between the loading and unloading curves results from the hysteresis in the PZT tube [11.4].

Multimode Capabilities

The multimode AFM can be used for topography measurements in the contact mode and tapping mode, described earlier, and for measurements of lateral (friction) force, electric force gradients and magnetic force gradients.

The multimode AFM, used with a grounded conducting tip, can measure electric field gradients by oscillating the tip near its resonant frequency. When the lever encounters a force gradient from the electric field, the effective spring constant of the cantilever is altered, changing its resonant frequency. Depending on which side of the resonance curve is chosen, the oscillation amplitude of the cantilever increases or decreases due to the shift in the resonant frequency. By recording the amplitude of the cantilever, an image revealing the strength of the electric field gradient is obtained.

In the magnetic force microscope (MFM) used with a magnetically-coated tip, static cantilever deflection is detected which occurs when a magnetic field exerts a force on the tip and the MFM images of magnetic materials can be produced. MFM sensitivity can be enhanced by oscillating the cantilever near its resonant frequency. When the tip encounters a magnetic force gradient, the effective spring constant, and hence the resonant frequency, is shifted. By driving the cantilever above or below the resonant frequency, the oscillation amplitude varies as the resonance shifts. An image of magnetic field gradients is obtained by recording the oscillation amplitude as the tip is scanned over the sample.

Topographic information is separated from the electric field gradients and magnetic field images by using a so-called lift mode. Measurements in lift mode are taken in two passes over each scan line. On the first pass, topographical information is recorded in the standard tapping mode where the oscillating cantilever lightly taps the surface. On the second pass, the tip is lifted to a user-selected separation (typically 20–200 nm) between the tip and local surface topography. By using the stored topographical data instead of the standard feedback, the separation remains constant without sensing the surface. At this height, cantilever amplitudes are sensitive to electric field force gradients or relatively weak but long-range magnetic forces without being influenced by topographic features. Two-pass measurements are taken for every scan line, producing separate topographic and magnetic force images.

Electrochemical AFM

This option allows one to perform electrochemical reactions on the AFM. It includes a potentiostat, a fluid cell with a transparent cantilever holder and electrodes, and the software required to operate the potentiostat and display the results of the electrochemical reaction.

11.2.3 AFM Probe Construction

Various probes (cantilevers and tips) are used for AFM studies. The cantilever stylus used in the AFM should meet the following criteria: (1) low normal spring constant (stiffness), (2) a high resonant frequency, (3) a high quality factor of the cantilever Q, (4) high lateral spring constant (stiffness), (5) short cantilever length, (6) incorporation of components (such as mirror) for deflection sensing, and (7) a sharp protruding tip [11.137]. In order to register a measurable deflection with small forces, the cantilever must flex with a relativly low force (on the order of few nN) requiring vertical spring constants of 10^{-2} to 10^{2} N/m for atomic resolution in the contact profiling mode. The data rate or imaging rate in the AFM is limited by the mechanical resonant frequency

Table 11.2 Relevant properties of materials used for cantilevers

Property	Young's Modulus (E) (GPa)	Density (ρg) (kg/m^3)	Microhardness (GPa)	Speed of sound ($\sqrt{E/\rho}$) (m/s)
Diamond	900–1,050	3,515	78.4–102	17,000
Si_3N_4	310	3,180	19.6	9,900
Si	130–188	2,330	9–10	8,200
W	350	19,310	3.2	4,250
Ir	530	–	~ 3	5,300

of the cantilever. To achieve a large imaging bandwidth, the AFM cantilever should have a resonant frequency greater than about 10 kHz (30–100 kHz is preferable) in order to make the cantilever the least sensitive part of the system. Fast imaging rates are not just a matter of convenience, since the effects of thermal drifts are more pronounced with slow-scanning speeds. The combined requirements of a low spring constant and a high resonant frequency is met by reducing the mass of the cantilever. The quality factor Q ($= \omega_R/(c/m)$, where ω_R is the resonant frequency of the damped oscillator and c is the damping constant and m is the mass of the oscillator) should have a high value for some applications. For example, resonance curve detection is a sensitive modulation technique for measuring small force gradients in noncontact imaging. Increasing the Q increases the sensitivity of the measurements. Mechanical Q values of 100–1,000 are typical. In contact modes, the Q is of less importance. High lateral spring constant in the cantilever is desirable to reduce the effect of lateral forces in the AFM as frictional forces can cause appreciable lateral bending of the cantilever. Lateral bending results in error in the topography measurements. For friction measurements, cantilevers with less lateral rigidity is preferred. A sharp protruding tip must be formed at the end of the cantilever to provide a well defined interaction with sample over a small area. The tip radius should be much smaller than the radii of corrugations in the sample in order for these to be measured accurately. The lateral spring constant depends critically on the tip length. Additionally, the tip should be centered at the free end.

In the past, cantilevers have been cut by hand from thin metal foils or formed from fine wires. Tips for these cantilevers were prepared by attaching diamond fragments to the ends of the cantilevers by hand, or in the case of wire cantilevers, electrochemically etching the wire to a sharp point. Several cantilever geometries for wire cantilevers have been used. The simplest geometry is the L-shaped cantilever, usually made by bending a wire at a 90° angle. Other geometries include single-V and double-V geometries with a sharp tip attached at the apex of V, and double-X configuration with a sharp tip attached at the intersection [11.31, 138]. These cantilevers can be constructed with high vertical spring constants. For example, double-cross cantilever with an effective spring constant of 250 N/m was used by *Burnham* and *Colton* [11.31]. The small size and low mass needed in the AFM make hand fabrication of the cantilever a difficult process with poor reproducibility. Conventional microfabrication techniques are ideal for constructing planar thin-film structures which have submicron lateral dimensions. The triangular (V-shaped) cantilevers have improved (higher) lateral spring constant in comparison to rectangular cantilevers. In terms of spring constants, the triangular cantilevers are approximately equivalent to two rectangular cantilevers in parallel [11.137]. Although the macroscopic radius of a photolithographically patterned corner is seldom much less than about 50 nm, microscopic asperities on the etched surface provide tips with near atomic dimensions.

Cantilevers have been used from a whole range of materials. Most commonly are cantilevers made of Si_3N_4, Si, and diamond. Young's modulus and the density are the material parameters which determine the resonant frequency, besides the geometry. Table 11.2 shows the relevant properties and the speed of sound, indicative of the resonant frequency for a given shape. Hardness is important to judge the durability of the cantilevers, and is also listed in the table. Materials used for STM cantilevers are also included.

Silicon nitride cantilevers are less expensive than those made of other materials. They are very rugged and well suited to imaging in almost all environments. They are especially compatible to organic and biological materials. Microfabricated silicon nitride triangular beams with integrated square pyramidal tips made of plasma-enhanced chemical vapor deposition (PECVD) are most commonly used [11.137]. Four cantilevers with different

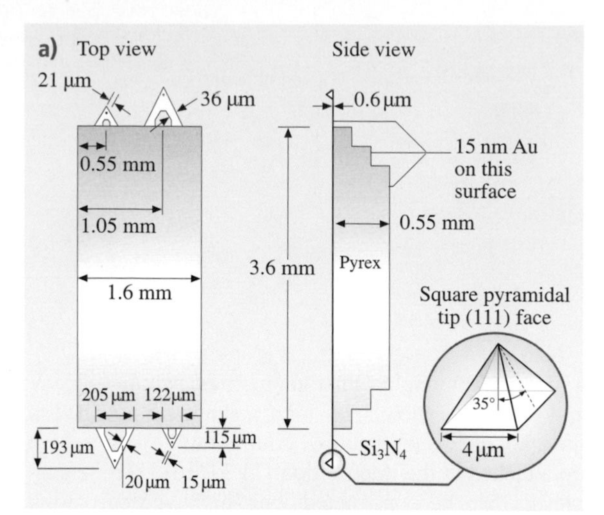

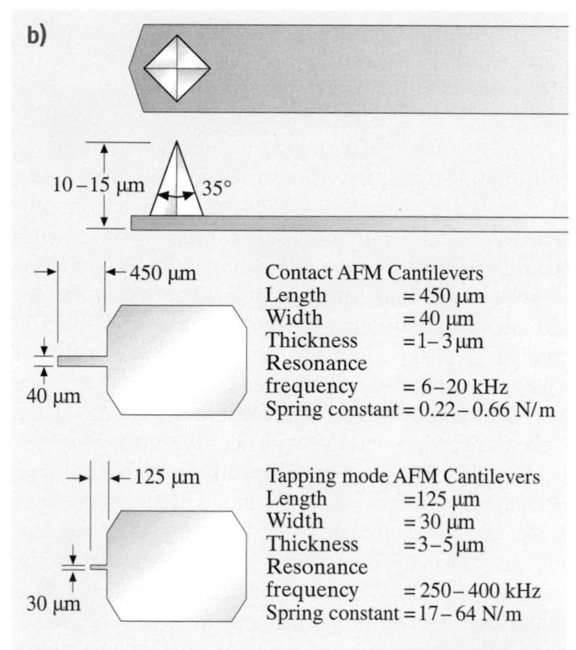

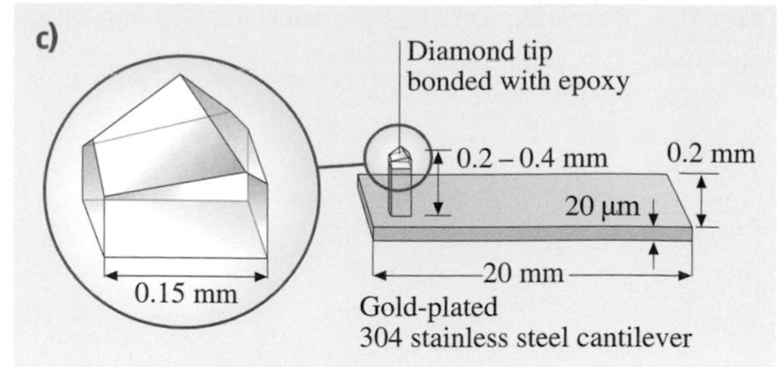

Fig. 11.15a–c Schematics of (**a**) triangular cantilever beam with square pyramidal tips made of PECVD Si_3N_4, (**b**) rectangular cantilever beams with square pyramidal tips made of etched single-crystal silicon, and (**c**) rectangular cantilever stainless steel beam with three-sided pyramidal natural diamond tip

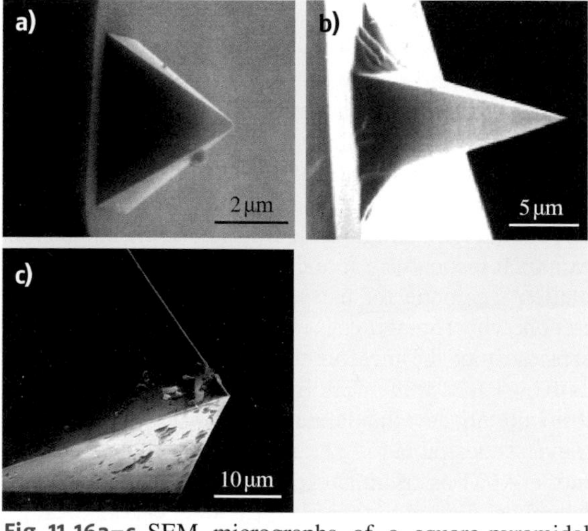

Fig. 11.16a–c SEM micrographs of a square-pyramidal PECVD Si_3N_4 tip (**a**), a square pyramidal etched single-crystal silicon tip (**b**), and a three-sided pyramidal natural diamond tip (**c**)

sizes and spring constants on each cantilever substrate made of boron silicate glass (Pyrex), marketed by Digital Instruments, are shown in Figs. 11.15a and 11.16. Two pairs of the cantilevers on each substrate measure about 115 and 193 µm from the substrate to the apex of the triangular cantilever with base widths of 122 and 205 µm, respectively. Both cantilever legs, with the same thickness (0.6 µm) of all the cantilevers, are available with wide and narrow legs. Only one cantilever is selected and used from each substrate. Calculated spring constant and measured natural frequencies for each of the configurations are listed in Table 11.3. The most commonly used cantilever beam is the 115-µm long, wide-legged cantilever (vertical spring constant = 0.58 N/m). Cantilevers with smaller spring constants should be used on softer samples. The pyramidal tips are highly symmetric with its end having a radius of about 20–50 nm. The

Table 11.3 Measured vertical spring constants and natural frequencies of triangular (V-shaped) cantilevers made of PECVD Si_3N_4 (Data provided by Digital Instruments, Inc.)

Cantilever dimension	Spring constant (k_z) (N/m)	Natural frequency (ω_0) (kHz)
115-μm long, narrow leg	0.38	40
115-μm long, wide leg	0.58	40
193-μm long, narrow leg	0.06	13–22
193-μm long, wide leg	0.12	13–22

Table 11.4 Vertical (k_z), lateral (k_y), and torsional (k_{yT}) spring constants of rectangular cantilevers made of Si (IBM) and PECVD Si_3N_4 (Veeco Instruments, Inc.)

Dimensions/stiffness	Si cantilever	Si_3N_4 cantilever
Length (L) (μm)	100	100
Width (b) (μm)	10	20
Thickness (h) (μm)	1	0.6
Tip length (ℓ) (μm)	5	3
k_z (N/m)	0.4	0.15
k_y (N/m)	40	175
k_{yT} (N/m)	120	116
ω_0 (kHz)	~ 90	~ 65

Note: $k_z = Ebh^3/4L^3$, $k_y = Eb^3h/4\ell^3$, $k_{yT} = Gbh^3/3L\ell^2$, and $\omega_0 = [k_z/(m_c + 0.24bhL\rho)]^{1/2}$, where E is Young's modulus, G is the modulus of rigidity [$= E/2(1+\nu)$, ν is Poisson's ratio], ρ is the mass density of the cantilever, and m_c is the concentrated mass of the tip (~ 4 ng) [11.94]. For Si, $E = 130$ GPa, $\rho g = 2{,}300\,kg/m^3$, and $\nu = 0.3$. For Si_3N_4, $E = 150$ GPa, $\rho g = 3{,}100\,kg/m^3$, and $\nu = 0.3$

tip side walls have a slope of 35 deg and the length of the edges of the tip at the cantilever base is about 4 μm.

An alternative to silicon nitride cantilevers with integrated tips are microfabricated single-crystal silicon cantilevers with integrated tips. Si tips are sharper than Si_3N_4 tips because they are directly formed by the anisotropic etch in single-crystal Si rather than using an etch pit as a mask for deposited materials [11.139]. Etched single-crystal n-type silicon rectangular cantilevers with square pyramidal tips with a lower radius of less than 10 nm for contact and tapping mode (tapping-mode etched silicon probe or TESP) AFMs are commercially available from Digital Instruments and Nanosensors GmbH, Aidlingen, Germany, Figs. 11.15b and 11.16. Spring constants and resonant frequencies are also presented in the Fig. 11.15b.

Commercial triangular Si_3N_4 cantilevers have a typical width-thickness ratio of 10 to 30 which results in 100 to 1000 times stiffer spring constants in the lateral direction compared to the normal direction. Therefore these cantilevers are not well suited for torsion. For friction measurements, the torsional spring constant should be minimized in order to be sensitive to the lateral forces. Rather long cantilevers with small thickness and large tip length are most suitable. Rectangular beams have lower torsional spring constants in comparison to the triangular (V-shaped) cantilevers. Table 11.4 lists the spring constants (with full length of the beam used) in three directions of the typical rectangular beams. We note that lateral and torsional spring constants are about two orders of magnitude larger than the normal spring constants. A cantilever beam required for the tapping mode is quite stiff and may not be sensitive enough for friction measurements. *Meyer* et al. [11.140] used a specially designed rectangular silicon cantilever with length = 200 μm, width = 21 μm, thickness = 0.4 μm, tip length = 12.5 μm, and shear modulus = 50 GPa, giving a normal spring constant of 0.007 N/m and torsional spring constant of 0.72 N/m which gives a lateral force sensitivity of 10 pN and an angle of resolution of 10^{-7} rad. With this particular geometry, sensitivity to lateral forces could be improved by about a factor of 100 compared with commercial V-shaped Si_3N_4 or rectangular Si or Si_3N_4 cantilevers used by *Meyer* and *Amer* [11.8] with torsional spring constant of ~ 100 N/m. *Ruan* and *Bhushan* [11.136] and *Bhushan* and *Ruan* [11.9] used 115-μm long, wide-legged V-shaped cantilevers made of Si_3N_4 for friction measurements.

For scratching, wear, and indentation studies, single-crystal natural diamond tips ground to the shape of a three-sided pyramid with an apex angle of either 60° or 80° whose point is sharpened to a radius of about 100 nm are commonly used [11.4, 10] (Figs. 11.15c and 11.16). The tips are bonded with conductive epoxy to a gold-plated 304 stainless steel spring sheet (length = 20 mm, width = 0.2 mm, thickness = 20 to 60 μm) which acts as a cantilever. The free length of the spring is varied to change the beam stiffness. The normal spring con-

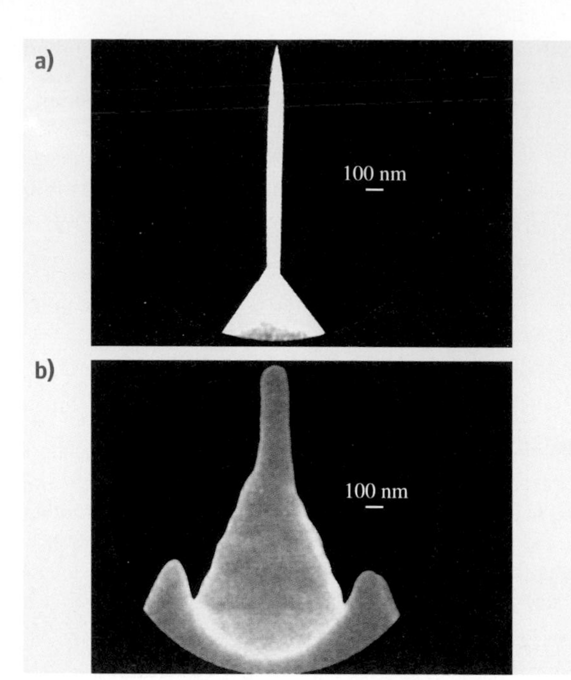

Fig. 11.17a,b Schematics of (a) HART Si_3N_4 probe, and (b) FIB milled Si_3N_4 probe

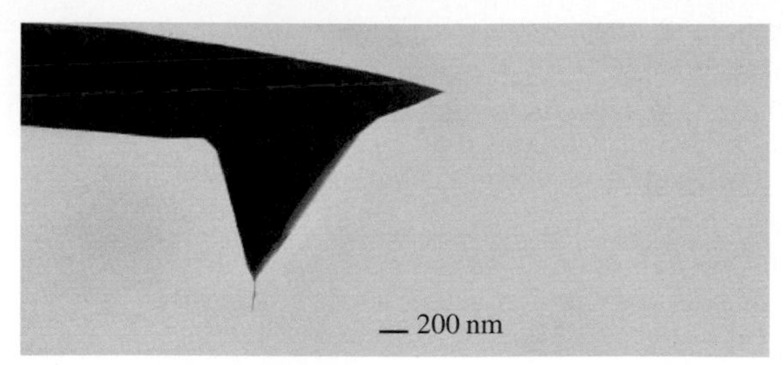

Fig. 11.18 SEM micrograph of a multi-walled carbon nanotube (MWNT) tip physically attached on the single-crystal silicon, square-pyramidal tip (Courtesy Piezomax Technologies, Inc.)

stant of the beam ranges from about 5 to 600 N/m for a 20 μm thick beam. The tips are produced by R-DEC Co., Tsukuba, Japan.

For imaging within trenches by an AFM, high aspect ratio tips are used. Examples of the two probes are shown in Fig. 11.17. The high-aspect ratio tip (Hart) probes are produced by starting with a conventional Si_3N_4 pyramidal probe. Through a combination of focused ion beam (FIB) and high resolution scanning electron microscopy (SEM) techniques, a thin filament is grown at the apex of the pyramid. The probe filament is approximately 1 μm long and 0.1 μm in diameter. It tapers to an extremely sharp point (radius better than the resolution of most SEMs). The long thin shape and sharp radius make it ideal for imaging within "vias" of microstructures and trenches (> 0.25 μm). Because of flexing of the probe, it is unsuitable for imaging structures at the atomic level since the flexing of the probe can create image artifacts. For atomic-scale imaging, a FIB-milled probe is used which is relatively stiff yet allows for closely spaced topography. These probes start out as conventional Si_3N_4 pyramidal probes but the pyramid is FIB milled until a small cone shape is formed which has a high aspect ratio with 0.2–0.3 μm in length. The milled probes allow nanostructure resolution without sacrificing rigidity. These types of probes are manufactured by various manufacturers including Materials Analytical Services, Raleigh, North Carolina.

Carbon nanotube tips having small diameter and high aspect ratio are used for high resolution imaging of surfaces and of deep trenches, in the tapping mode or noncontact mode. Single-walled carbon nanotubes (SWNT) are microscopic graphitic cylinders that are 0.7 to 3 nm in diameter and up to many microns in length. Larger structures called multi-walled carbon nanotubes (MWNT) consist of nested, concentrically arranged SWNT and have diameters ranging from 3 to 50 nm. MWNT carbon nanotube AFM tips are produced by manual assembly [11.141], chemical vapor deposition (CVD) synthesis, and hybrid fabrication process [11.142]. Figure 11.18 shows TEM micrograph of a carbon nanotube tip, ProbeMax^{TM}, commercially produced by mechanical assembly by Piezomax Technologies, Inc., Middleton, Wisconsin. For production of these tips, MWNT nanotubes are produced by carbon arc. They are physically attached on the single-crystal silicon, square-pyramidal tips in the SEM using a manipulator and the SEM stage to control the nanotubes and the tip independently. Once the nanotube is attached to the tip, it is usually too long to image with. It is shortened by using an AFM and applying voltage between the tip and the sample. Nanotube tips are also commercially produced by CVD synthesis by NanoDevices, Santa Barbara, California.

11.2.4 Friction Measurement Methods

Based on the work by *Ruan* and *Bhushan* [11.136], the two methods for friction measurements are now described in more detail. (Also see [11.8].) A scanning angle is defined as the angle relative to the y-axis in Fig. 11.19a. This is also the long axis of the cantilever.

A zero degree scanning angle corresponds to the sample scanning in the y direction, and a 90 degree scanning angle corresponds to the sample scanning perpendicular to this axis in the xy plane (along x axis). If the scanning direction is in both the y and $-y$ directions, we call this "parallel scan". Similarly, a "perpendicular scan" means the scanning direction is in the x and $-x$ directions. The sample traveling direction for each of these two methods is illustrated in Fig. 11.19b.

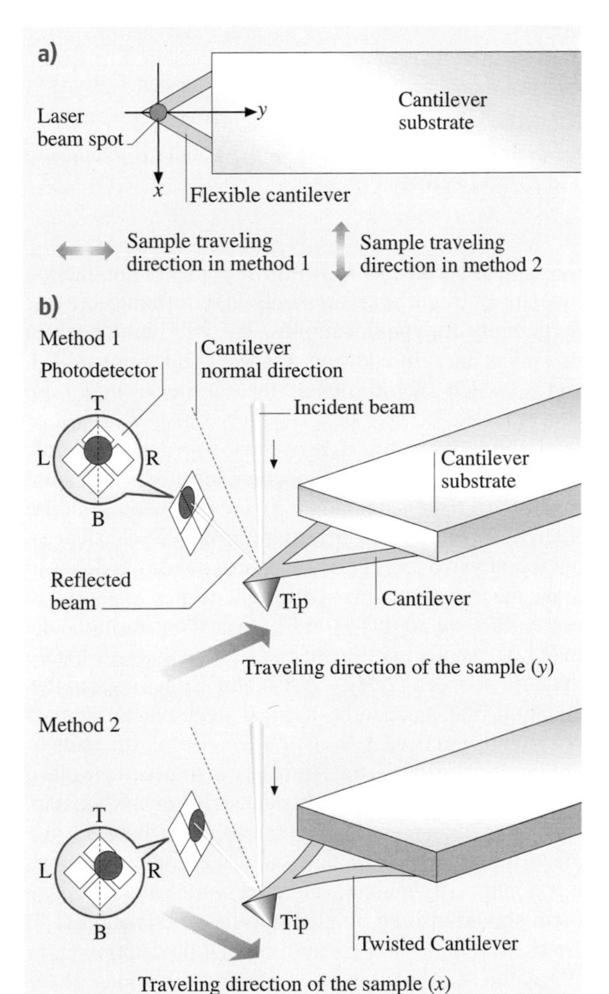

Fig. 11.19 **(a)** Schematic defining the x- and y-directions relative to the cantilever, and showing the sample traveling direction in two different measurement methods discussed in the text, **(b)** schematic of deformation of the tip and cantilever shown as a result of sliding in the x- and y-directions. A twist is introduced to the cantilever if the scanning is in the x-direction ((**b**), *lower part*) [11.136]

In method 1 (using "height" mode with parallel scans) in addition to topographic imaging, it is also possible to measure friction force when the scanning direction of the sample is parallel to the y direction (parallel scan). If there were no friction force between the tip and the moving sample, the topographic feature would be the only factor which cause the cantilever to be deflected vertically. However, friction force does exist on all contact surfaces where one object is moving relative to another. The friction force between the sample and the tip will also cause a cantilever deflection. We assume that the normal force between the sample and the tip is W_0 when the sample is stationary (W_0 is typically in the range of 10 nN to 200 nN), and the friction force between the sample and the tip is W_f as the sample scans against the tip. The direction of friction force (W_f) is reversed as the scanning direction of the sample is reversed from positive (y) to negative ($-y$) directions ($\vec{W}_{f(y)} = -\vec{W}_{f(-y)}$).

When the vertical cantilever deflection is set at a constant level, it is the total force (normal force and friction force) applied to the cantilever that keeps the cantilever deflection at this level. Since the friction force is in opposite directions as the traveling direction of the sample is reversed, the normal force will have to be adjusted accordingly when the sample reverses its traveling direction, so that the total deflection of the cantilever will remain the same. We can calculate the difference of the normal force between the two traveling directions for a given friction force W_f. First, by means of a constant deflection, the total moment applied to the cantilever is constant. If we take the reference point to be the point where the cantilever joins the cantilever holder (substrate), point P in Fig. 11.20, we have the following relationship:

$$(W_0 - \Delta W_1)L + W_f \ell = (W_0 + \Delta W_2)L - W_f \ell \qquad (11.1)$$

or

$$(\Delta W_1 + \Delta W_2)L = 2W_f \ell \,. \qquad (11.2)$$

Thus

$$W_f = (\Delta W_1 + \Delta W_2)L/(2\ell)\,, \qquad (11.3)$$

where ΔW_1 and ΔW_2 are the absolute value of the changes of normal force when the sample is traveling in $-y$ and y directions, respectively, as shown in Fig. 11.20; L is the length of the cantilever; ℓ is vertical distance between the end of the tip and point P. The coefficient of

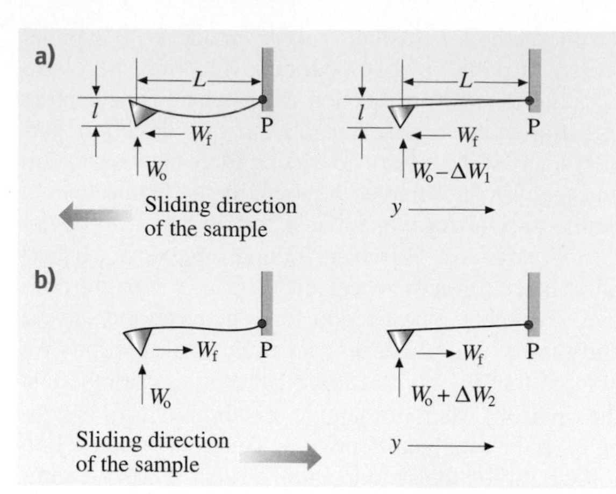

Fig. 11.20 **(a)** Schematic showing an additional bending of the cantilever – due to friction force when the sample is scanned in the y- or $-y$-direction (*left*). **(b)** This effect will be canceled by adjusting the piezo height by a feedback circuit (*right*) [11.136]

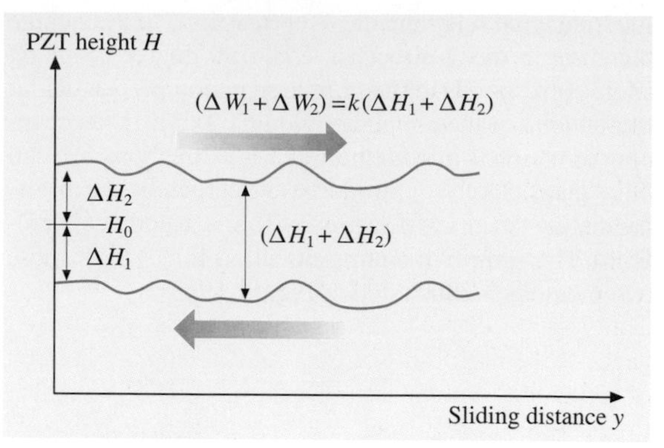

Fig. 11.21 Schematic illustration of the height difference of the piezoelectric tube scanner as the sample is scanned in y and $-y$ directions

friction (μ) between the tip and the sample is then given as

$$\mu = \frac{W_f}{W_0} = \left(\frac{(\Delta W_1 + \Delta W_2)}{W_0}\right)\left(\frac{L}{2\ell}\right) . \tag{11.4}$$

In all circumstances, there are adhesive and interatomic attractive forces between the cantilever tip and the sample. The adhesive force can be due to water from the capillary condensation and other contaminants present at the surface which form meniscus bridges [11.4, 143, 144] and the interatomic attractive force includes van der Waals attraction [11.18]. If these forces (and indentation effect as well, which is usually small for rigid samples) can be neglected, the normal force W_0 is then equal to the initial cantilever deflection H_0 multiplied by the spring constant of the cantilever. $(\Delta W_1 + \Delta W_2)$ can be measured by multiplying the same spring constant by the height difference of the piezo tube between the two traveling directions (y and $-y$ directions) of the sample. This height difference is denoted as $(\Delta H_1 + \Delta H_2)$, shown schematically in Fig. 11.21. Thus, (11.4) can be rewritten as

$$\mu = \frac{W_f}{W_0} = \left(\frac{(\Delta H_1 + \Delta H_2)}{H_0}\right)\left(\frac{L}{2\ell}\right) . \tag{11.5}$$

Since the piezo tube vertical position is affected by the surface topographic profile of the sample in addition to the friction force being applied at the tip, this difference has to be taken point by point at the same location on the sample surface as shown in Fig. 11.21. Subtraction of point by point measurements may introduce errors, particularly for rough samples. We will come back to this point later. In addition, precise measurement of L and ℓ (which should include the cantilever angle) are also required.

If the adhesive forces between the tip and the sample are large enough that it can not be neglected, one should include it in the calculation. However, there could be a large uncertainty in determining this force, thus an uncertainty in using (11.5). An alternative approach is to make the measurements at different normal loads and to use $\Delta(H_0)$ and $\Delta(\Delta H_1 + \Delta H_2)$ from the measurements in (11.5). Another comment on (11.5) is that, since only the ratio between $(\Delta H_1 + \Delta H_2)$ and H_0 comes into this equation, the piezo tube vertical position H_0 and its position difference $(\Delta H_1 + \Delta H_2)$ can be in the units of volts as long as the vertical traveling distance of the piezo tube and the voltage applied to it has a linear relationship. However, if there is a large nonlinearity between the piezo tube traveling distance and the applied voltage, this nonlinearity must be included in the calculation.

It should also be pointed out that (11.4) and (11.5) are derived under the assumption that the friction force W_f is the same for the two scanning directions of the sample. This is an approximation since the normal force is slightly different for the two scans and there may also be a directionality effect in friction. However, this difference is much smaller than W_0 itself. We can ignore the second order correction.

Method 2 ("aux" mode with perpendicular scan) to measure friction was suggested by *Meyer* and *Amer* [11.8]. The sample is scanned perpendicular to the long axis of the cantilever beam (i. e., to scan along the x or $-x$ direction in Fig. 11.19a) and the output of the horizontal two quadrants of the photodiode-detector is measured. In this arrangement, as the sample moves under the tip, the friction force will cause the cantilever to twist. Therefore the light intensity between the left and right (L and R in Fig. 11.19b, right) detectors will be different. The differential signal between the left and right detectors is denoted as FFM signal $[(\mathrm{L}-\mathrm{R})/(\mathrm{L}+\mathrm{R})]$. This signal can be related to the degree of twisting, hence to the magnitude of friction force. Again, because of a possible error in determining normal force due to the presence of an adhesive force at the tip–sample interface, the slope of the friction data (FFM signal vs. normal load) needs to be taken for an accurate value of the coefficient of friction.

While friction force contributes to the FFM signal, friction force may not be the only contributing factor in commercial FFM instruments (for example, NanoScope IV). One can notice this fact by simply engaging the cantilever tip with the sample. Before engaging, the left and right detectors can be balanced by adjusting the position of the detectors so that the intensity difference between these two detectors is zero (FFM signal is zero). Once the tip is engaged with the sample, this signal is no longer zero even if the sample is not moving in the xy plane with no friction force applied. This would be a detrimental effect. It has to be understood and eliminated from the data acquisition before any quantitative measurement of friction force becomes possible.

One of the fundamental reasons for this observation is the following. The detectors may not have been properly aligned with respect to the laser beam. To be precise, the vertical axis of the detector assembly (the line joining T-B in Fig. 11.22) is not in the plane defined by the incident laser beam and the beam reflected from an untwisted cantilever (we call this plane the "beam plane"). When the cantilever vertical deflection changes due to a change of applied normal force (without having the sample scanned in the xy plane), the laser beam will be reflected up and down and form a projected trajectory on the detector. (Note that this trajectory is in the defined beam plane.) If this trajectory is not coincident with the vertical axis of the detector, the laser beam will not evenly bisect the left and right quadrants of the detectors, even under the condition of no torsional motion of the cantilever, see Fig. 11.22. Thus when the laser beam is reflected up and down due a change of the normal force, the intensity difference between the left and right detectors will also change. In other words, the FFM signal will change as the normal force applied to the tip is changed, even if the tip is not experiencing any friction force. This (FFM) signal is unrelated to friction force or to the actual twisting of the cantilever. We will call this part of FFM signal "$\mathrm{FFM_F}$", and the part which is truly related to friction force "$\mathrm{FFM_T}$".

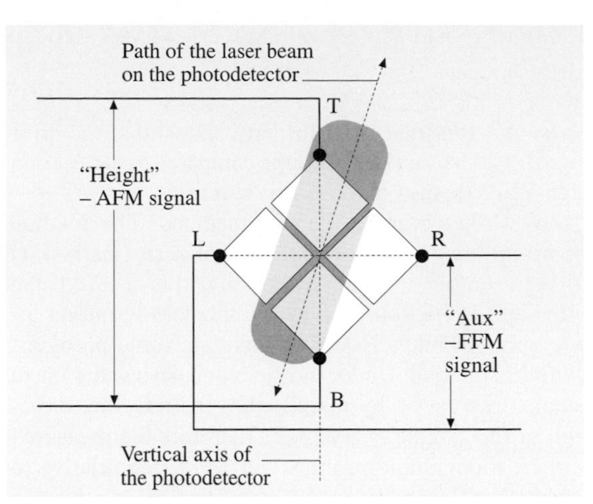

Fig. 11.22 The trajectory of the laser beam on the photodetectors in as the cantilever is vertically deflected (with no torsional motion) for a misaligned photodetector with respect to the laser beam. For a change of normal force (vertical deflection of the cantilever), the laser beam is projected at a different position on the detector. Due to a misalignment, the projected trajectory of the laser beam on the detector is not parallel with the detector vertical axis (the line joint T-B) [11.136]

The $\mathrm{FFM_F}$ signal can be eliminated. One way of doing this is as follows. First the sample is scanned in both x and $-x$ directions and the FFM signal for scans in each direction is recorded. Since friction force reverses its directions when the scanning direction is reversed from x to $-x$ direction, the $\mathrm{FFM_T}$ signal will have opposite signs as the scanning direction of the sample is reversed ($\mathrm{FFM_T}(x) = -\mathrm{FFM_T}(-x)$). Hence the $\mathrm{FFM_T}$ signal will be canceled out if we take the sum of the FFM signals for the two scans. The average value of the two scans will be related to $\mathrm{FFM_F}$ due to the misalignment,

$$\mathrm{FFM}(x) + \mathrm{FFM}(-x) = 2\mathrm{FFM_F}\,. \tag{11.6}$$

This value can therefore be subtracted from the original FFM signals of each of these two scans to obtain the true FFM signal ($\mathrm{FFM_T}$). Or, alternately, by taking the difference of the two FFM signals, one directly gets the $\mathrm{FFM_T}$ value

$$\begin{aligned}\mathrm{FFM}(x) - \mathrm{FFM}(-x) &= \mathrm{FFM_T}(x) - \mathrm{FFM_T}(-x)\\ &= 2\mathrm{FFM_T}(x)\,.\end{aligned} \quad (11.7)$$

Ruan and *Bhushan* [11.136] have shown that the error signal ($\mathrm{FFM_F}$) can be very large compared to the friction signal $\mathrm{FFM_T}$, thus correction is required.

Now we compare the two methods. The method of using "height" mode and parallel scan (method 1) is very simple to use. Technically, this method can provide 3-D friction profiles and the corresponding topographic profiles. However, there are some problems with this method. Under most circumstances, the piezo scanner displays a hysteresis when the traveling direction of the sample is reversed. Therefore the measured surface topographic profiles will be shifted relative to each other along the y-axis for the two opposite (y and $-y$) scans. This would make it difficult to measure the local height difference of the piezo tube for the two scans. However, the average height difference between the two scans and hence the average friction can still be measured. The measurement of average friction can serve as an internal means of friction force calibration. Method 2 is a more desirable approach. The subtraction of $\mathrm{FFM_F}$ signal from FFM for the two scans does not introduce error to local friction force data. An ideal approach in using this method would be to add the average value of the two profiles in order to get the error component ($\mathrm{FFM_F}$) and then subtract this component from either profiles to get true friction profiles in either directions. By making measurements at various loads, we can get the average value of the coefficient of friction which then can be used to convert the friction profile to the coefficient of friction profile. Thus any directionality and local variations in friction can be easily measured. In this method, since topography data are not affected by friction, accurate topography data can be measured simultaneously with friction data and a better localized relationship between the two can be established.

11.2.5 Normal Force and Friction Force Calibrations of Cantilever Beams

Based on *Ruan* and *Bhushan* [11.136], we now discuss normal force and friction force calibrations. In order to calculate the absolute value of normal and friction forces in Newtons using the measured AFM and $\mathrm{FFM_T}$ voltage signals, it is necessary to first have an accurate value of the spring constant of the cantilever (k_c). The spring constant can be calculated using the geometry and the physical properties of the cantilever material [11.8,94,137]. However, the properties of the PECVD Si_3N_4 (used in fabricating cantilevers) could be different from those of the bulk material. For example, by using an ultrasonic measurement, we found the Young's modulus of the cantilever beam to be about 238 ± 18 GPa which is less than that of bulk Si_3N_4 (310 GPa). Furthermore the thickness of the beam is nonuniform and difficult to measure precisely. Since the stiffness of a beam goes as the cube of thickness, minor errors in precise measurements of thickness can introduce substantial stiffness errors. Thus one should experimentally measure the spring constant of the cantilever. *Cleveland* et al. [11.145] measured the normal spring constant by measuring resonant frequencies of the beams.

For normal spring constant measurement, *Ruan* and *Bhushan* [11.136] used a stainless steel spring sheet of known stiffness (width = 1.35 mm, thickness = 15 μm, free hanging length = 5.2 mm). One end of the spring was attached to the sample holder and the other end was made to contact with the cantilever tip during the measurement, see Fig. 11.23. They measured the piezo traveling distance for a given cantilever deflection. For a rigid sample (such as diamond), the piezo traveling distance Z_t (measured from the point where the tip touches the sample) should equal the cantilever deflection. To keep the cantilever deflection at the same level using a flexible spring sheet, the new piezo traveling distance $Z_{t'}$ would be different from Z_t. The difference between $Z_{t'}$ and Z_t corresponds to the deflection of the spring

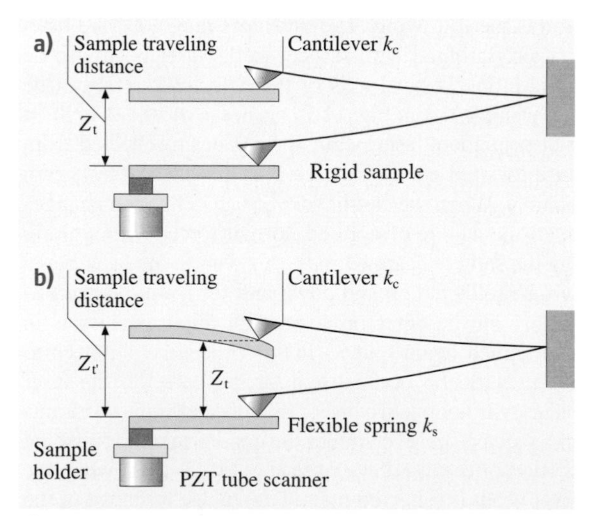

Fig. 11.23a,b Illustration showing the deflection of cantilever as it is pushed by **(a)** a rigid sample or by **(b)** a flexible spring sheet [11.136]

sheet. If the spring constant of the spring sheet is k_s, the spring constant of the cantilever k_c can be calculated by

$$(Z_{t'} - Z_t)k_s = Z_t k_c$$

or

$$k_c = k_s(Z_{t'} - Z_t)/Z_t . \qquad (11.8)$$

The spring constant of the spring sheet (k_s) used in this study is calculated to be 1.54 N/m. For a wide-legged cantilever used in our study (length = 115 μm, base width = 122 μm, leg width = 21 μm, and thickness = 0.6 μm), k_c was measured to be 0.40 N/m instead of 0.58 N/m reported by its manufacturer – Digital Instruments Inc. To relate photodiode detector output to the cantilever deflection in nm, they used the same rigid sample to push against the AFM tip. Since for a rigid sample the cantilever vertical deflection equals the sample traveling distance measured from the point where the tip touches the sample, the photodiode output as the tip is pushed by the sample can be converted directly to cantilever deflection. For these measurements, they found the conversion factor to be 20 nm/V.

The normal force applied to the tip can be calculated by multiplying the cantilever vertical deflection by the cantilever spring constant for samples which have very small adhesive force with the tip. If the adhesive force between the sample and the tip is large, it should be included in the normal force calculation. This is particularly important in atomic-scale force measurement because in this region, the typical normal force that is measured is in the range of a few hundreds of nN to a few mN. The adhesive force could be comparable to the applied force.

The conversion of friction signal (from FFM_T to friction force) is not as straightforward. For example, one can calculate the degree of twisting for a given friction force using the geometry and the physical properties of the cantilever [11.53, 144]. One would need the information on the detectors such as the quantum efficiency of the detector, the laser power, the instrument's gain, etc. in order to be able convert the signal into the degree of twisting. Generally speaking, this procedure can not be accomplished without having some detailed information about the instrument. This information is not usually provided by the manufactures. Even if this information is readily available, error may still occur in using this approach because there will always be variations as a result of the instrumental set up. For example, it has been noticed that the measured FFM_T signal could be different for the same sample when different AFM microscopes of the same kind are used. The essence is that, one can not calibrate the instrument experimentally using this calculation. *O'Shea* et al. [11.144] did perform a calibration procedure in which the torsional signal was measured as the sample is displaced a known distance laterally while ensuring that the tip does not slide over the surface. However, it is difficult to verify that tip sliding does not occur.

Apparently, a new method of calibration is required. There is a more direct and simpler way of doing this. The first method described (method 1) to measure friction can directly provide an absolute value of the coefficient of friction. It can therefore be used just as an internal means of calibration for the data obtained using method 2. Or for a polished sample which introduces least error in friction measurement using method 1, method 1 can be used to obtain calibration for friction force for method 2. Then this calibration can be used for measurement on all samples using method 2. In method 1, the length of the cantilever required can be measured using an optical microscope; the length of the tip can be measured using a scanning electron microscope. The relative angle between the cantilever and the horizontal sample surface can be measured directly. Thus the coefficient of friction can be measured with few unknown parameters. The friction force can then be calculated by multiplying the coefficient of friction by the normal load. The FFM_T signal obtained using method 2 can then be converted into friction force. For their instrument, they found the conversion to be 8.6 nN/V.

11.3 AFM Instrumentation and Analyses

The performance of AFMs and the quality of AFM images greatly depend on the instrument available and the probes (cantilever and tips) in use. This section describes the mechanics of cantilevers, instrumentation and analysis of force detection systems for cantilever deflections, and scanning and control systems.

11.3.1 The Mechanics of Cantilevers

Stiffness and Resonances of Lumped Mass Systems

Any one of the building blocks of an AFM, be it the body of the microscope itself or the force measuring cantilevers, are mechanical resonators. These resonances can be excited either by the surroundings or by the rapid movement of the tip or the sample. To avoid problems due to building or air induced oscillations, it is of paramount importance to optimize the design of AFMs for high resonant frequencies. This usually means decreasing the size of the microscope [11.146]. By using cube-like or sphere-like structures for the microscope, one can considerably increase the lowest eigenfrequency. The fundamental natural frequency, ω_0, of any spring is given by

$$\omega_0 = \frac{1}{2\pi}\sqrt{\frac{k}{m_{\text{eff}}}}\,, \tag{11.9}$$

where k is the spring constant (stiffness) in the normal direction and m_{eff} is the effective mass. The spring constant k of a cantilever beam with uniform cross section (Fig. 11.24) is given by [11.147]

$$k = \frac{3EI}{L^3}\,, \tag{11.10}$$

where E is the Young's modulus of the material, L is the length of the beam and I is the moment of inertia of the cross section. For a rectangular cross section with a width b (perpendicular to the deflection) and a height h one obtains an expression for I

$$I = \frac{bh^3}{12}\,. \tag{11.11}$$

Combining (11.9), (11.10) and (11.11) we get an expression for ω_0

$$\omega_0 = \sqrt{\frac{Ebh^3}{4L^3 m_{\text{eff}}}}\,. \tag{11.12}$$

The effective mass can be calculated using Raleigh's method. The general formula using Raleigh's method for the kinetic energy T of a bar is

$$T = \frac{1}{2}\int_0^L \frac{m}{L}\left(\frac{\partial z(x)}{\partial t}\right)^2 \mathrm{d}x\,. \tag{11.13}$$

For the case of a uniform beam with a constant cross section and length L one obtains for the deflection $z(x) = z_{\text{max}}\left[1-(3x/2L)+(x^3/2L^3)\right]$. Inserting z_{max} into (11.13) and solving the integral gives

$$T = \frac{1}{2}\int_0^L \frac{m}{L}\left[\frac{\partial z_{\text{max}}(x)}{\partial t}\left(1-\frac{3x}{2L}\right)+\left(\frac{x^3}{L^3}\right)\right]^2 \mathrm{d}x$$
$$= \frac{1}{2}m_{\text{eff}}(z_{\text{max}}t)^2\,,$$

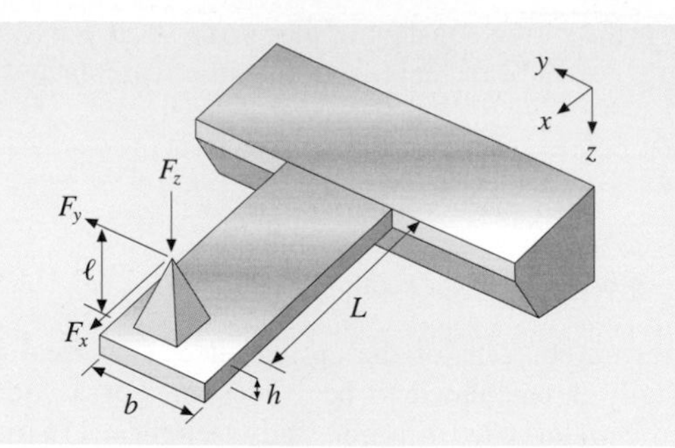

Fig. 11.24 A typical AFM cantilever with length L, width b, and height h. The height of the tip is ℓ. The material is characterized by Young's modulus E, the shear modulus G and a mass density ρ. Normal (F_z), axial (F_x), and lateral (F_y) forces exist at the end of the tip

which gives

$$m_{\text{eff}} = \frac{9}{20}m\,. \tag{11.14}$$

Substituting (11.14) into (11.12) and noting that $m = \rho Lbh$, where ρ is the mass density, one obtains the following expression

$$\omega_0 = \left(\frac{\sqrt{5}}{3}\sqrt{\frac{\mathrm{E}}{\rho}}\right)\frac{\mathrm{h}}{L^2}\,. \tag{11.15}$$

It is evident from (11.15), that one way to increase the natural frequency is to choose a material with a high ratio E/ρ; see Table 11.2 for typical values of $\sqrt{E/\rho}$ of various commonly used materials. Another way to increase the lowest eigenfrequency is also evident in (11.15). By optimizing the ratio h/L^2, one can increase the resonant frequency. However it does not help to make the length of the structure smaller than the width or height. Their roles will just be interchanged. Hence the optimum structure is a cube. This leads to the design rule, that long, thin structures like sheet metal should be avoided. For a given resonant frequency, the quality factor Q should be as low as possible. This means that an inelastic medium such as rubber should be in contact with the structure to convert kinetic energy into heat.

Stiffness and Resonances of Cantilevers

Cantilevers are mechanical devices specially shaped to measure tiny forces. The analysis given in the previous section is applicable. However, to understand better the intricacies of force detection systems we will discuss the example of a cantilever beam with uniform cross section, Fig. 11.24. The bending of a beam due to a normal load on the beam is governed by the Euler equation [11.147]

$$M = EI(x)\frac{d^2 z}{dx^2}, \quad (11.16)$$

where M is the bending moment acting on the beam cross section. $I(x)$ the moment of inertia of the cross section with respect to the neutral axis defined by

$$I(x) = \int_z \int_y z^2 \, dy \, dz . \quad (11.17)$$

For a normal force F_z acting at the tip,

$$M(x) = (L - x) F_z \quad (11.18)$$

since the moment must vanish at the endpoint of the cantilever. Integrating (11.16) for a normal force F_z acting at the tip and observing that EI is a constant for beams with a uniform cross section, one gets

$$z(x) = \frac{L^3}{6EI}\left(\frac{x}{L}\right)^2\left(3 - \frac{x}{L}\right) F_z . \quad (11.19)$$

The slope of the of the beam is

$$z'(x) = \frac{Lx}{2EI}\left(2 - \frac{x}{L}\right) F_z . \quad (11.20)$$

From (11.19) and (11.20), at the end of the cantilever, i. e. for $x = L$, for a rectangular beam, and by using an expression for I in (11.11), one gets,

$$z(L) = \frac{4}{Eb}\left(\frac{L}{h}\right)^3 F_z , \quad (11.21)$$

$$z'(L) = \frac{3}{2}\left(\frac{z}{L}\right) . \quad (11.22)$$

Now the stiffness in the normal (z) direction, k_z, is

$$k_z = \frac{F_z}{z(L)} = \frac{Eb}{4}\left(\frac{h}{L}\right)^3 . \quad (11.23)$$

and a change in angular orientation of the end of cantilever beam is

$$\Delta\alpha = \frac{3}{2}\frac{z}{L} = \frac{6}{Ebh}\left(\frac{L}{h}\right)^2 F_z . \quad (11.24)$$

Now we ask what will, to first order, happen if we apply a lateral force F_y to the end of the tip (Fig. 11.24). The cantilever will bend sideways and it will twist. The stiffness in the lateral (y) direction, k_y, can be calculated with (11.23) by exchanging b and h

$$k_y = \frac{Eh}{4}\left(\frac{b}{L}\right)^3 . \quad (11.25)$$

Therefore the bending stiffness in lateral direction is larger than the stiffness for bending in the normal direction by $(b/h)^2$. The twisting or torsion on the other hand is more complicated to handle. For a wide, thin cantilever ($b \gg h$) we obtain torsional stiffness along y axis, k_{yT}

$$k_{yT} = \frac{Gbh^3}{3L\ell^2}, \quad (11.26)$$

where G is the modulus of rigidity ($= E/2(1+\nu)$, where ν is the Poisson's ratio). The ratio of the torsional stiffness to the lateral bending stiffness is

$$\frac{k_{yT}}{k_y} = \frac{1}{2}\left(\frac{\ell b}{hL}\right)^2 , \quad (11.27)$$

where we assume $\nu = 0.333$. We see that thin, wide cantilevers with long tips favor torsion while cantilevers with square cross sections and short tips favor bending. Finally we calculate the ratio between the torsional stiffness and the normal bending stiffness,

$$\frac{k_{yT}}{k_z} = 2\left(\frac{L}{\ell}\right)^2 . \quad (11.28)$$

Equations (11.26) to (11.28) hold in the case where the cantilever tip is exactly in the middle axis of the cantilever. Triangular cantilevers and cantilevers with tips not on the middle axis can be dealt with by finite element methods.

The third possible deflection mode is the one from the force on the end of the tip along the cantilever axis, F_x (Fig. 11.24). The bending moment at the free end of the cantilever is equal to the $F_x\ell$. This leads to the following modification of (11.18) for the case of forces F_z and F_x

$$M(x) = (L - x) F_z + F_x \ell . \quad (11.29)$$

Integration of (11.16) now leads to

$$z(x) = \frac{1}{2EI}\left[Lx^2\left(1 - \frac{x}{3L}\right) F_z + \ell x^2 F_x\right] \quad (11.30)$$

and

$$z'(x) = \frac{1}{EI}\left[\frac{Lx}{2}\left(2 - \frac{x}{L}\right) F_z + \ell x F_x\right] . \quad (11.31)$$

Evaluating (11.30) and (11.31) at the end of the cantilever, we get the deflection and the tilt

$$z(L) = \frac{L^2}{EI}\left(\frac{L}{3}F_z - \frac{\ell}{2}F_x\right),$$

$$z'(L) = \frac{L}{EI}\left(\frac{L}{2}F_z + \ell F_x\right) . \quad (11.32)$$

From these equations, one gets

$$F_z = \frac{12EI}{L^3}\left[z(L) - \frac{Lz'(L)}{2}\right],$$
$$F_x = \frac{2EI}{\ell L^2}\left[2Lz'(L) - 3z(L)\right]. \qquad (11.33)$$

A second class of interesting properties of cantilevers is their resonance behavior. For cantilever beams one can calculate the resonant frequencies [11.147, 148]

$$\omega_n^{\text{free}} = \frac{\lambda_n^2}{2\sqrt{3}}\frac{h}{L^2}\sqrt{\frac{E}{\rho}} \qquad (11.34)$$

with $\lambda_0 = (0.596864\ldots)\pi$, $\lambda_1 = (1.494175\ldots)\pi$, $\lambda_n \to (n+1/2)\pi$. The subscript n represents the order of the frequency, e.g., fundamental, second mode, and the nth mode.

A similar equation to (11.34) holds for cantilevers in rigid contact with the surface. Since there is an additional restriction on the movement of the cantilever, namely the location of its end point, the resonant frequency increases. Only the λ_n's terms change to [11.148]

$$\lambda_0' = (1.2498763\ldots)\pi, \quad \lambda_1' = (2.2499997\ldots)\pi,$$
$$\lambda_n' \to (n+1/4)\pi. \qquad (11.35)$$

The ratio of the fundamental resonant frequency in contact to the fundamental resonant frequency not in contact is 4.3851.

For the torsional mode we can calculate the resonant frequencies as

$$\omega_0^{\text{tors}} = 2\pi\frac{h}{Lb}\sqrt{\frac{G}{\rho}}. \qquad (11.36)$$

For cantilevers in rigid contact with the surface, we obtain the expression for the fundamental resonant frequency [11.148]

$$\omega_0^{\text{tors, contact}} = \frac{\omega_0^{\text{tors}}}{\sqrt{1+3(2L/b)^2}}. \qquad (11.37)$$

The amplitude of the thermally induced vibration can be calculated from the resonant frequency using

$$\Delta z_{\text{therm}} = \sqrt{\frac{k_B T}{k}}, \qquad (11.38)$$

where k_B is Boltzmann's constant and T is the absolute temperature. Since AFM cantilevers are resonant structures, sometimes with rather high Q, the thermal noise is not evenly distributed as (11.38) suggests. The spectral noise density below the peak of the response curve is [11.148]

$$z_0 = \sqrt{\frac{4k_B T}{k\omega_0 Q}} \quad (\text{in m}/\sqrt{\text{Hz}}), \qquad (11.39)$$

where Q is the quality factor of the cantilever, described earlier.

11.3.2 Instrumentation and Analyses of Detection Systems for Cantilever Deflections

A summary of selected detection systems was provided in Fig. 11.8. Here we discuss in detail pros and cons of various systems.

Optical Interferometer Detection Systems

Soon after the first papers on the AFM [11.2] appeared, which used a tunneling sensor, an instrument based on an interferometer was published [11.149]. The sensitivity of the interferometer depends on the wavelength of the light employed in the apparatus. Figure 11.25 shows the principle of such an interferometeric design. The light incident from the left is focused by a lens on the cantilever. The reflected light is collimated by the same lens and interferes with the light reflected at the flat. To separate the reflected light from the incident light a $\lambda/4$ plate converts the linear polarized incident light to circular polarization. The reflected light is made linear polarized again by the $\lambda/4$-plate, but with a polarization orthogonal to that of the incident light. The polarizing

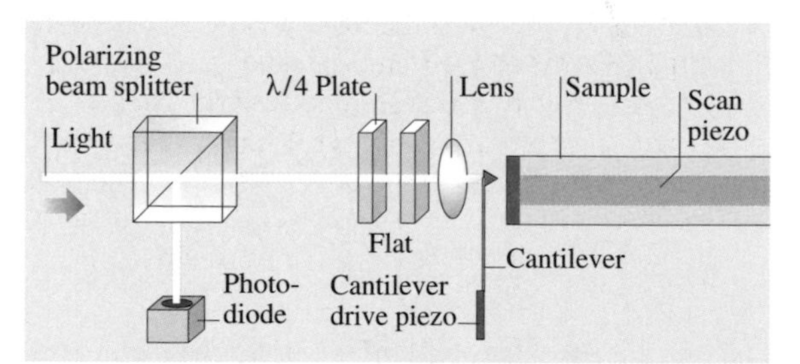

Fig. 11.25 Principle of an interferometric AFM. The light of the laser light source is polarized by the polarizing beam splitter and focused on the back of the cantilever. The light passes twice through a quarter wave plate and is hence orthogonally polarized to the incident light. The second arm of the interferometer is formed by the flat. The interference pattern is modulated by the oscillating cantilever

beam splitter then deflects the reflected light to the photo diode.

Homodyne Interferometer. To improve the signal to noise ratio of the interferometer the cantilever is driven by a piezo near its resonant frequency. The amplitude Δz of the cantilever as a function of driving frequency Ω is

$$\Delta z\,(\Omega)=\Delta z_0\frac{\Omega_0^2}{\sqrt{\left(\Omega^2-\Omega_0^2\right)^2+\frac{\Omega^2\Omega_0^2}{Q^2}}}\,, \tag{11.40}$$

where Δz_0 is the constant drive amplitude and Ω_0 the resonant frequency of the cantilever. The resonant frequency of the cantilever is given by the effective potential

$$\Omega_0=\sqrt{\left(k+\frac{\partial^2 U}{\partial z^2}\right)\frac{1}{m_{\text{eff}}}}\,, \tag{11.41}$$

where U is the interaction potential between the tip and the sample. Equation (11.41) shows that an attractive potential decreases Ω_0. The change in Ω_0 in turn results in a change of the Δz (see (11.40)). The movement of the cantilever changes the path difference in the interferometer. The light reflected from the cantilever with the amplitude $A_{\ell,0}$ and the reference light with the amplitude $A_{r,0}$ interfere on the detector. The detected intensity $I(t)=[A_\ell(t)+A_r(t)]^2$ consists of two constant terms and a fluctuating term

$$2A_\ell\,(t)\,A_r\,(t)=\\ A_{\ell,0}A_{r,0}\sin\left[\omega t+\frac{4\pi\delta}{\lambda}+\frac{4\pi\,\Delta z}{\lambda}\sin\,(\Omega t)\right]\sin\,(\omega t)\,. \tag{11.42}$$

Here ω is the frequency of the light, λ is the wavelength of the light, δ is the path difference in the interferometer, and Δz is the instantaneous amplitude of the cantilever, given according to (11.40) and (11.41) as a function of Ω, k, and U. The time average of (11.42) then becomes

$$\begin{aligned}\langle 2A_\ell\,(t)\,A_r\,(t)\rangle_T &\propto \cos\left[\frac{4\pi\delta}{\lambda}+\frac{4\pi\,\Delta z}{\lambda}\sin\,(\Omega t)\right]\\ &\approx \cos\left(\frac{4\pi\delta}{\lambda}\right)-\sin\left[\frac{4\pi\,\Delta z}{\lambda}\sin\,(\Omega t)\right]\\ &\approx \cos\left(\frac{4\pi\delta}{\lambda}\right)-\frac{4\pi\,\Delta z}{\lambda}\sin\,(\Omega t)\,.\end{aligned} \tag{11.43}$$

Here all small quantities have been omitted and functions with small arguments have been linearized. The amplitude of Δz can be recovered with a lock-in technique. However, (11.43) shows that the measured amplitude is also a function of the path difference δ in the interferometer. Hence this path difference δ must be very stable. The best sensitivity is obtained when $\sin(4\delta/\lambda)\approx 0$.

Heterodyne Interferometer. This influence is not present in the heterodyne detection scheme shown in Fig. 11.26. Light incident from the left with a frequency ω is split in a reference path (upper path in Fig. 11.26) and a measurement path. Light in the measurement path is shifted in frequency to $\omega_1=\omega+\Delta\omega$ and focused on the cantilever. The cantilever oscillates at the frequency Ω, as in the homodyne detection scheme. The reflected light $A_\ell(t)$ is collimated by the same lens and interferes on the photo diode with the reference light $A_r(t)$. The fluctuating term of the intensity is given by

$$2A_\ell\,(t)\,A_r\,(t)=A_{\ell,0}A_{r,0}\sin\left[(\omega+\Delta\omega)\,t+\frac{4\pi\delta}{\lambda}\right.\\ \left.+\frac{4\pi\,\Delta z}{\lambda}\sin\,(\Omega t)\right]\sin\,(\omega t)\,, \tag{11.44}$$

where the variables are defined as in (11.42). Setting the path difference $\sin(4\pi\delta/\lambda)\approx 0$ and taking the time average, omitting small quantities and linearizing functions with small arguments we get

$$\langle 2A_\ell\,(t)\,A_r\,(t)\rangle_T\\ \propto \cos\left[\Delta\omega t+\frac{4\pi\delta}{\lambda}+\frac{4\pi\,\Delta z}{\lambda}\sin\,(\Omega t)\right]$$

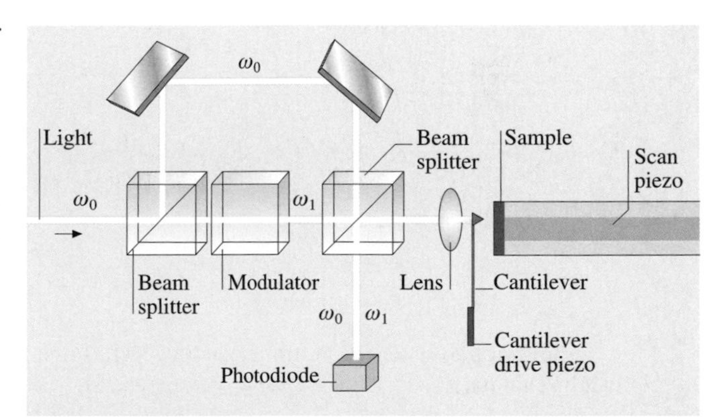

Fig. 11.26 Principle of a heterodyne interferometric AFM. Light with the frequency ω_0 is split into a reference path (upper path) and a measurement path. The light in the measurement path is frequency shifted to ω_1 by an acousto-optical modulator (or an electro-optical modulator) The light reflected from the oscillating cantilever interferes with the reference beam on the detector

$$
\begin{aligned}
&= \cos\left(\Delta\omega t + \frac{4\pi\delta}{\lambda}\right)\cos\left[\frac{4\pi\,\Delta z}{\lambda}\sin(\Omega t)\right] \\
&\quad - \sin\left(\Delta\omega t + \frac{4\pi\delta}{\lambda}\right)\sin\left[\frac{4\pi\,\Delta z}{\lambda}\sin(\Omega t)\right] \\
&\approx \cos\left(\frac{4\pi\delta}{\lambda}\right) - \sin\left[\frac{4\pi\,\Delta z}{\lambda}\sin(\Omega t)\right] \\
&\approx \cos\left(\Delta\omega t + \frac{4\pi\delta}{\lambda}\right)\left[1 - \frac{8\pi^2\Delta z^2}{\lambda^2}\sin(\Omega t)\right] \\
&\quad - \frac{4\pi\,\Delta z}{\lambda}\sin\left(\Delta\omega t + \frac{4\pi\delta}{\lambda}\right)\sin(\Omega t) \\
&= \cos\left(\Delta\omega t + \frac{4\pi\delta}{\lambda}\right) - \frac{8\pi^2\Delta z^2}{\lambda^2}\cos\left(\Delta\omega t + \frac{4\pi\delta}{\lambda}\right) \\
&\quad \times\sin(\Omega t) - \frac{4\pi\,\Delta z}{\lambda}\sin\left(\Delta\omega t + \frac{4\pi\delta}{\lambda}\right)\sin(\Omega t) \\
&= \cos\left(\Delta\omega t + \frac{4\pi\delta}{\lambda}\right) - \frac{4\pi^2\Delta z^2}{\lambda^2}\cos\left(\Delta\omega t + \frac{4\pi\delta}{\lambda}\right) \\
&\quad + \frac{4\pi^2\Delta z^2}{\lambda^2}\cos\left(\Delta\omega t + \frac{4\pi\delta}{\lambda}\right)\cos(2\Omega t) \\
&\quad - \frac{4\pi\,\Delta z}{\lambda}\sin\left(\Delta\omega t + \frac{4\pi\delta}{\lambda}\right)\sin(\Omega t) \\
&= \cos\left(\Delta\omega t + \frac{4\pi\delta}{\lambda}\right)\left(1 - \frac{4\pi^2\Delta z^2}{\lambda^2}\right) \\
&\quad + \frac{2\pi^2\Delta z^2}{\lambda^2}\left\{\cos\left[(\Delta\omega + 2\Omega)\,t + \frac{4\pi\delta}{\lambda}\right]\right. \\
&\quad \left. + \cos\left[(\Delta\omega - 2\Omega)\,t + \frac{4\pi\delta}{\lambda}\right]\right\} \\
&\quad + \frac{2\pi\,\Delta z}{\lambda}\left\{\cos\left[(\Delta\omega + \Omega)\,t + \frac{4\pi\delta}{\lambda}\right]\right. \\
&\quad \left. + \cos\left[(\Delta\omega - \Omega)\,t + \frac{4\pi\delta}{\lambda}\right]\right\} .
\end{aligned}
\tag{11.45}
$$

Multiplying electronically the components oscillating at $\Delta\omega$ and $\Delta\omega + \Omega$ and rejecting any product except the one oscillating at Ω we obtain

$$
\begin{aligned}
A &= \frac{2\Delta z}{\lambda}\left(1 - \frac{4\pi^2\Delta z^2}{\lambda^2}\right)\cos\left[(\Delta\omega + 2\Omega)\,t + \frac{4\pi\delta}{\lambda}\right] \\
&\quad \times\cos\left(\Delta\omega t + \frac{4\pi\delta}{\lambda}\right) \\
&= \frac{\Delta z}{\lambda}\left(1 - \frac{4\pi^2\Delta z^2}{\lambda^2}\right)\left\{\cos\left[(2\Delta\omega + \Omega)\,t + \frac{8\pi\delta}{\lambda}\right]\right. \\
&\quad \left. + \cos(\Omega t)\right\} \\
&\approx \frac{\pi\,\Delta z}{\lambda}\cos(\Omega t) .
\end{aligned}
\tag{11.46}
$$

Unlike in the homodyne detection scheme the recovered signal is independent from the path difference δ of the interferometer. Furthermore a lock-in amplifier with the reference set $\sin(\Delta\omega t)$ can measure the path difference δ independent of the cantilever oscillation. If necessary, a feedback circuit can keep $\delta = 0$.

Fiber-optical Interferometer. The fiber-optical interferometer [11.129] is one of the simplest interferometers to build and use. Its principle is sketched in Fig. 11.27. The light of a laser is fed into an optical fiber. Laser diodes with integrated fiber pigtails are convenient light sources. The light is split in a fiber-optic beam splitter into two fibers. One fiber is terminated by index matching oil to avoid any reflections back into the fiber. The end of the other fiber is brought close to the cantilever in the AFM. The emerging light is partially reflected back into the fiber by the cantilever. Most of the light, however, is lost. This is not a big problem since only 4% of the light is reflected at the end of the fiber, at the glass-air interface. The two reflected light waves interfere with each other. The product is guided back into the fiber coupler and again split into two parts. One half is analyzed by the photodiode. The other half is fed back into the laser. Communications grade laser diodes are sufficiently resistant against feedback to be operated in this environment. They have, however, a bad coherence length, which in this case does not matter, since the optical path difference is in any case no larger than 5 μm. Again the end of the fiber has to be positioned on a piezo drive to set the distance between the fiber and the cantilever to $\lambda(n + 1/4)$.

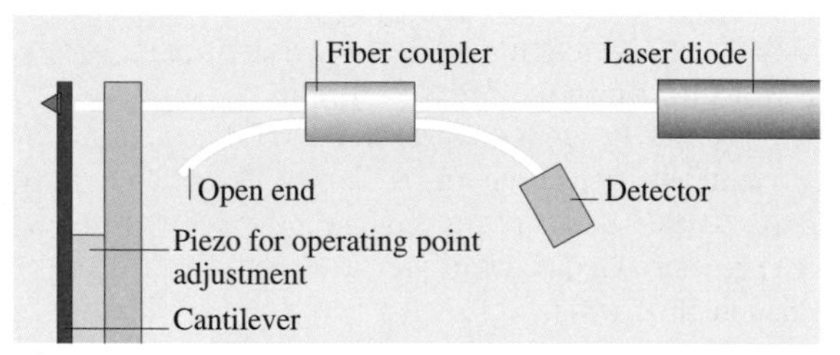

Fig. 11.27 A typical setup for a fiber optic interferometer readout

Nomarski-Interferometer. Another solution to minimize the optical path difference is to use the Nomarski interferometer [11.130]. Figure 11.28 shows a schematic of the microscope. The light of a laser is focused on the cantilever by lens. A birefringent crystal (for instance calcite) between the cantilever and the lens with its optical axis 45° off the polarization direction of the light

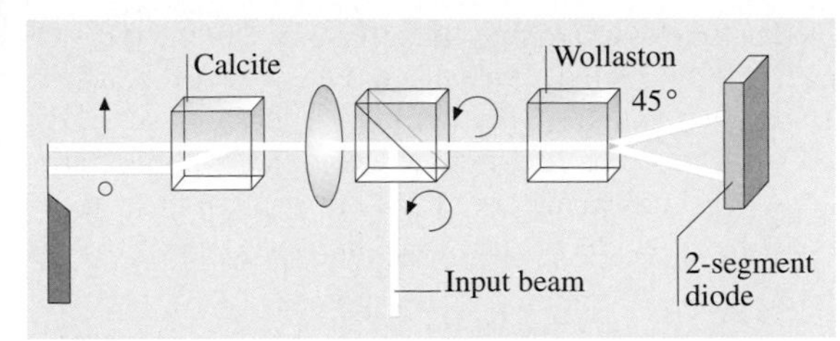

Fig. 11.28 Principle of Nomarski AFM. The circular polarized input beam is deflected to the left by a non-polarizing beam splitter. The light is focused onto a cantilever. The calcite crystal between the lens and the cantilever splits the circular polarized light into two spatially separated beams with orthogonal polarizations. The two light beams reflected from the lever are superimposed by the calcite crystal and collected by the lens. The resulting beam is again circular polarized. A Wollaston prism produces two interfering beams with $\pi/2$ phase shift between them. The minimal path difference accounts for the excellent stability of this microscope

splits the light beam into two paths, offset by a distance given by the length of the crystal. Birefringent crystals have varying indexes of refraction. In calcite, one crystal axis has a lower index than the other two. This means, that certain light rays will propagate at a different speed through the crystal than the others. By choosing a correct polarization, one can select the ordinary ray, the extraordinary ray or one can get any distribution of the intensity amongst those two rays. A detailed description of birefringence can be found in textbooks (e.g., [11.150]). A calcite crystal deflects the extraordinary ray at an angle of 6° within the crystal. By choosing a suitable length of the calcite crystal, any separation can be set.

The focus of one light ray is positioned near the free end of the cantilever while the other is placed close to the clamped end. Both arms of the interferometer pass through the same space, except for the distance between the calcite crystal and the lever. The closer the calcite crystal is placed to the lever, the less influence disturbances like air currents have.

Sarid [11.116] has given values for the sensitivity of the different interferometeric detection systems. Table 11.5 presents a summary of his results.

Optical Lever

The most common cantilever deflection detection system is the optical lever [11.53, 111]. This method, depicted in Fig. 11.29, employs the same technique as light beam deflection galvanometers. A fairly well collimated light beam is reflected off a mirror and projected to a receiving target. Any change in the angular position of the mirror will change the position, where the light ray hits the target. Galvanometers use optical path lengths of several meters and scales projected to the target wall as a read-out help.

For the AFM using the optical lever method a photodiode segmented into two (or four) closely spaced devices detects the orientation of the end of the cantilever. Initially, the light ray is set to hit the photodiodes in the middle of the two sub-diodes. Any deflection of

Table 11.5 Noise in Interferometers. F is the finesse of the cavity in the homodyne interferometer, P_i the incident power, P_d is the power on the detector, η is the sensitivity of the photodetector and RIN is the relative intensity noise of the laser. P_R and P_S are the power in the reference and sample beam in the heterodyne interferometer. P is the power in the Nomarski interferometer, $\delta\theta$ is the phase difference between the reference and the probe beam in the Nomarski interferometer. B is the bandwidth, e is the electron charge, λ is the wavelength of the laser, k the cantilever stiffness, ω_0 is the resonant frequency of the cantilever, Q is the quality factor of the cantilever, T is the temperature, and δi is the variation of current i

	Homodyne interferometer, fiber optic interferometer	**Heterodyne interferometer**	**Nomarski interferometer**
Laser noise $\left\langle \delta i^2 \right\rangle_\mathrm{L}$	$\frac{1}{4}\eta^2 F^2 P_i^2$ RIN	$\eta^2\left(P_R^2+P_S^2\right)$ RIN	$\frac{1}{16}\eta^2 P^2 \delta\theta$
Thermal noise $\left\langle \delta i^2 \right\rangle_\mathrm{T}$	$\frac{16\pi^2}{\lambda^2}\eta^2 F^2 P_i^2 \frac{4k_\mathrm{B}TBQ}{\omega_0 k}$	$\frac{4\pi^2}{\lambda^2}\eta^2 P_d^2 \frac{4k_\mathrm{B}TBQ}{\omega_0 k}$	$\frac{\pi^2}{\lambda^2}\eta^2 P^2 \frac{4k_\mathrm{B}TBQ}{\omega_0 k}$
Shot Noise $\left\langle \delta i^2 \right\rangle_\mathrm{S}$	$4e\eta P_\mathrm{d} B$	$2e\eta\left(P_\mathrm{R}+P_\mathrm{S}\right) B$	$\frac{1}{2}e\eta PB$

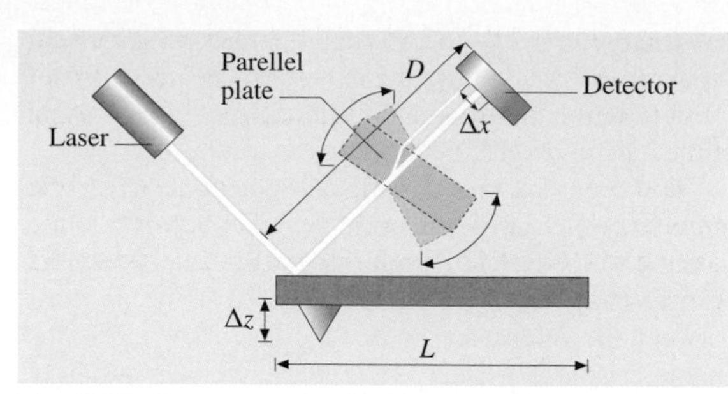

Fig. 11.29 The setup of optical lever detection microscope

the cantilever will cause an imbalance of the number of photons reaching the two halves. Hence the electrical currents in the photodiodes will be unbalanced too. The difference signal is further amplified and is the input signal to the feedback loop. Unlike the interferometeric AFMs, where often a modulation technique is necessary to get a sufficient signal to noise ratio, most AFMs employing the optical lever method are operated in a static mode. AFMs based on the optical lever method are universally used. It is the simplest method for constructing an optical readout and it can be confined in volumes smaller than 5 cm on the side.

The optical lever detection system is a simple yet elegant way to detect normal and lateral force signals simultaneously [11.7, 8, 53, 111]. It has the additional advantage that it is the fact that it is a remote detection system.

Implementations. Light from a laser diode or from a super luminescent diode is focused on the end of the cantilever. The reflected light is directed onto a quadrant diode that measures the direction of the light beam. A Gaussian light beam far from its waist is characterized by an opening angle β. The deflection of the light beam by the cantilever surface tilted by an angle α is 2α. The intensity on the detector then shifts to the side by the product of 2α and the separation between the detector and the cantilever. The readout electronics calculates the difference of the photocurrents. The photocurrents in turn are proportional to the intensity incident on the diode.

The output signal is hence proportional to the change in intensity on the segments

$$I_{\text{sig}} \propto 4\frac{\alpha}{\beta} I_{\text{tot}} . \tag{11.47}$$

For the sake of simplicity, we assume that the light beam is of uniform intensity with its cross section increasing proportional to the distance between the cantilever and the quadrant detector. The movement of the center of the light beam is then given by

$$\Delta x_{\text{Det}} = \Delta z \frac{D}{L} . \tag{11.48}$$

The photocurrent generated in a photodiode is proportional to the number of incoming photons hitting it. If the light beam contains a total number of N_0 photons then the change in difference current becomes

$$\Delta\,(I_{\text{R}} - I_{\text{L}}) = \Delta I = \text{const}\ \Delta z\ D\ N_0 . \tag{11.49}$$

Combining (11.48) and (11.49), one obtains that the difference current ΔI is independent of the separation of the quadrant detector and the cantilever. This relation is true, if the light spot is smaller than the quadrant detector. If it is greater, the difference current ΔI becomes smaller with increasing distance. In reality the light beam has a Gaussian intensity profile. For small movements Δx (compared to the diameter of the light spot at the quadrant detector), (11.49) still holds. Larger movements Δx, however, will introduce a nonlinear response. If the AFM is operated in a constant force mode, only small movements Δx of the light spot will occur. The feedback loop will cancel out all other movements.

The scanning of a sample with an AFM can twist the microfabricated cantilevers because of lateral forces [11.5, 7, 8] and affect the images [11.120]. When the tip is subjected to lateral forces, it will twist the cantilever and the light beam reflected from the end of the cantilever will be deflected perpendicular to the ordinary deflection direction. For many investigations this influence of lateral forces is unwanted. The design of the triangular cantilevers stems from the desire, to minimize the torsion effects. However, lateral forces open up a new dimension in force measurements. They allow, for instance, a distinction of two materials because of the different friction coefficient, or the determination of adhesion energies. To measure lateral forces the original optical lever AFM has to be modified. The only modification compared with Fig. 11.29 is the use of a quadrant detector photodiode instead of a two-segment photodiode and the necessary readout electronics, see Fig. 11.9a. The electronics calculates the following signals:

$$\begin{aligned} U_{\text{Normal Force}} &= \alpha \left[\left(I_{\text{Upper Left}} + I_{\text{Upper Right}} \right) \right. \\ &\quad \left. - \left(I_{\text{Lower Left}} + I_{\text{Lower Right}} \right) \right] , \\ U_{\text{Lateral Force}} &= \beta \left[\left(I_{\text{Upper Left}} + I_{\text{Lower Left}} \right) \right. \\ &\quad \left. - \left(I_{\text{Upper Right}} + I_{\text{Lower Right}} \right) \right] . \end{aligned} \tag{11.50}$$

The calculation of the lateral force as a function of the deflection angle does not have a simple solution for cross-sections other than circles. An approximate formula for the angle of twist for rectangular beams is [11.151]

$$\theta = \frac{M_t L}{\beta G b^3 h}\,, \tag{11.51}$$

where $M_t = F_y \ell$ is the external twisting moment due to lateral force, F_y, and β a constant determined by the value of h/b. For the equation to hold, h has to be larger than b.

Inserting the values for a typical microfabricated cantilever with integrated tips

$$\begin{aligned} b &= 6\times 10^{-7}\,\text{m}\,,\\ h &= 10^{-5}\,\text{m}\,,\\ L &= 10^{-4}\,\text{m}\,,\\ \ell &= 3.3\times 10^{-6}\,\text{m}\,,\\ G &= 5\times 10^{10}\,\text{Pa}\,,\\ \beta &= 0.333 \end{aligned} \tag{11.52}$$

into (11.51) we obtain the relation

$$F_y = 1.1\times 10^{-4}\,\text{N}\times\theta\,. \tag{11.53}$$

Typical lateral forces are of order 10^{-10} N.

Sensitivity. The sensitivity of this setup has been calculated in various papers [11.116, 148, 152]. Assuming a Gaussian beam the resulting output signal as a function of the deflection angle is dispersion like. Equation (11.47) shows that the sensitivity can be increased by increasing the intensity of the light beam I_{tot} or by decreasing the divergence of the laser beam. The upper bound of the intensity of the light I_{tot} is given by saturation effects on the photodiode. If we decrease the divergence of a laser beam we automatically increase the beam waist. If the beam waist becomes larger than the width of the cantilever we start to get diffraction. Diffraction sets a lower bound on the divergence angle. Hence one can calculate the optimal beam waist w_{opt} and the optimal divergence angle β [11.148, 152]

$$\begin{aligned} w_{opt} &\approx 0.36b\,,\\ \theta_{opt} &\approx 0.89\frac{\lambda}{b}\,. \end{aligned} \tag{11.54}$$

The optimal sensitivity of the optical lever then becomes

$$\varepsilon\,[\text{mW/rad}] = 1.8\frac{b}{\lambda} I_{tot}\ [\text{mW}]\,. \tag{11.55}$$

The angular sensitivity optical lever can be measured by introducing a parallel plate into the beam. A tilt of the parallel plate results in a displacement of the beam, mimicking an angular deflection.

Additional noise sources can be considered. Of little importance is the quantum mechanical uncertainty of the position [11.148, 152], which is, for typical cantilevers at room temperature

$$\Delta z = \sqrt{\frac{\hbar}{2m\omega_0}} = 0.05\,\text{fm}\,, \tag{11.56}$$

where $\hbar$ is the Planck constant ($= 6.626\times 10^{-34}$ J s). At very low temperatures and for high frequency cantilevers this could become the dominant noise source. A second noise source is the shot noise of the light. The shot noise is related to the particle number. We can calculate the number of photons incident on the detector

$$n = \frac{I\tau}{\hbar\omega} = \frac{I\lambda}{2\pi B\hbar c} = 1.8\times 10^{9}\,\frac{I[\text{W}]}{B[\text{Hz}]}\,, \tag{11.57}$$

where I is the intensity of the light, τ the measurement time, $B = 1/\tau$ the bandwidth, and c the speed of light. The shot noise is proportional to the square root of the number of particles. Equating the shot noise signal with the signal resulting for the deflection of the cantilever one obtains

$$\Delta z_{shot} = 68\frac{L}{w}\sqrt{\frac{B\,[\text{kHz}]}{I\,[\text{mW}]}}\ [\text{fm}]\,, \tag{11.58}$$

where w is the diameter of the focal spot. Typical AFM setups have a shot noise of 2 pm. The thermal noise can be calculated from the equipartition principle. The amplitude at the resonant frequency is

$$\Delta z_{therm} = 129\sqrt{\frac{B}{k\,[\text{N/m}]\,\omega_0 Q}}\ [\text{pm}]\,. \tag{11.59}$$

A typical value is 16 pm. Upon touching the surface, the cantilever increases its resonant frequency by a factor of 4.39. This results in a new thermal noise amplitude of 3.2 pm for the cantilever in contact with the sample.

Piezoresistive Detection

Implementation. An alternative detection system which is not as widely used as the optical detection schemes are piezoresistive cantilevers [11.125, 126, 132]. These cantilevers are based on the fact that the resistivity of certain materials, in particular of Si, changes with the applied stress. Figure 11.30 shows a typical implementation of a piezo-resistive cantilever. Four

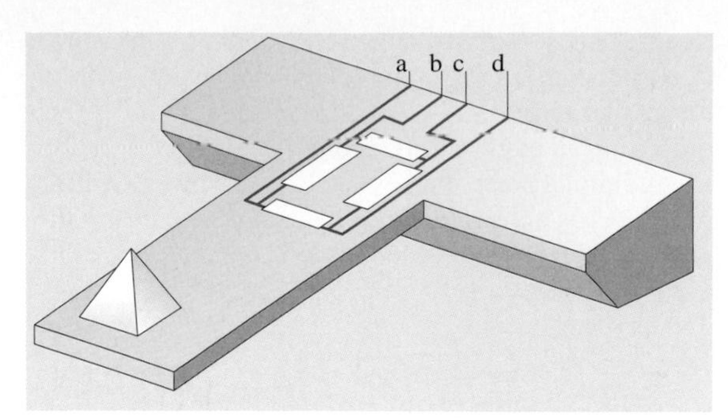

Fig. 11.30 A typical setup for a piezoresistive readout

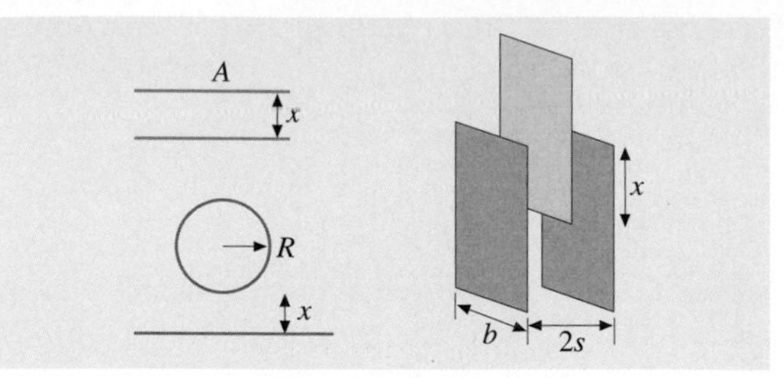

Fig. 11.31 Three possible arrangements of a capacitive readout. The *upper left* shows the cross section through a parallel plate capacitor. The *lower left* shows the geometry of sphere versus plane. The *right side* shows the more complicated, but linear capacitive readout

resistances are integrated on the chip, forming a Wheatstone bridge. Two of the resistors are in unstrained parts of the cantilever, the other two are measuring the bending at the point of the maximal deflection. For instance when an AC voltage is applied between terminals a and c one can measure the detuning of the bridge between terminals b and d. With such a connection the output signal varies only due to bending, but not due to changing of the ambient temperature and thus the coefficient of the piezoresistance.

Sensitivity. The resistance change is [11.126]

$$\frac{\Delta R}{R_0} = \Pi\delta\,, \tag{11.60}$$

where Π is the tensor element of the piezo-resistive coefficients, δ the mechanical stress tensor element and R_0 the equilibrium resistance. For a single resistor they separate the mechanical stress and the tensor element in longitudinal and transverse components

$$\frac{\Delta R}{R_0} = \Pi_t\delta_t + \Pi_l\delta_l\,. \tag{11.61}$$

The maximum value of the stress components are $\Pi_t = -64.0\times10^{-11}\,\mathrm{m^2/N}$ and $\Pi_l = -71.4\times10^{-11}\,\mathrm{m^2/N}$ for a resistor oriented along the (110) direction in silicon [11.126]. In the resistor arrangement of Fig. 11.30 two of the resistors are subject to the longitudinal piezo-resistive effect and two of them are subject to the transversal piezo-resistive effect. The sensitivity of that setup is about four times that of a single resistor, with the advantage that temperature effects cancel to first order. The resistance change is then calculated as

$$\frac{\Delta R}{R_0} = \Pi\frac{3Eh}{2L^2}\Delta z = \Pi\frac{6L}{bh^2}F_z\,, \tag{11.62}$$

where $\Pi = 67.7\times10^{-11}\,\mathrm{m^2/N}$ is the averaged piezo-resistive coefficient. Plugging in typical values for the dimensions (Fig. 11.24) ($L = 100\,\mu\mathrm{m}$, $b = 10\,\mu\mathrm{m}$, $h = 1\,\mu\mathrm{m}$) one obtains

$$\frac{\Delta R}{R_0} = \frac{4\times10^{-5}}{\mathrm{nN}}F_z\,. \tag{11.63}$$

The sensitivity can be tailored by optimizing the dimensions of the cantilever.

Capacitance Detection

The capacitance of an arrangement of conductors depends on the geometry. Generally speaking, the capacitance increases for decreasing separations. Two parallel plates form a simple capacitor (see Fig. 11.31, upper left), with the capacitance

$$C = \frac{\varepsilon\varepsilon_0 A}{x}\,, \tag{11.64}$$

where A is the area of the plates, assumed equal, and x is the separation. Alternatively one can consider a sphere versus an infinite plane (see Fig. 11.31, lower left). Here the capacitance is [11.116]

$$C = 4\pi\varepsilon_0 R\sum_{n=2}^{\infty}\frac{\sinh(\alpha)}{\sinh(n\alpha)} \tag{11.65}$$

where R is the radius of the sphere, and α is defined by

$$\alpha = \ln\left(1 + \frac{z}{R} + \sqrt{\frac{z^2}{R^2} + 2\frac{z}{R}}\right)\,. \tag{11.66}$$

One has to keep in mind that capacitance of a parallel plate capacitor is a nonlinear function of the separation. Using a voltage divider one can circumvent this problem.

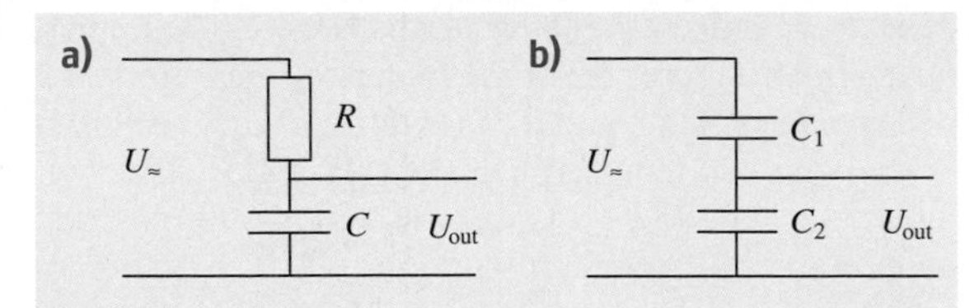

Fig. 11.32a,b Measuring the capacitance. The *left side*, (**a**), shows a low pass filter, the *right side*, (**b**), shows a capacitive divider. C (*left*) or C_2 (*right*) are the capacitances under test

Figure 11.32a shows a low pass filter. The output voltage is given by

$$U_{out} = U_\approx \frac{\frac{1}{j\omega C}}{R + \frac{1}{j\omega C}} = U_\approx \frac{1}{j\omega CR + 1} \cong \frac{U_\approx}{j\omega CR} . \qquad (11.67)$$

Here C is given by (11.64), ω is the excitation frequency and j is the imaginary unit. The approximate relation in the end is true when $\omega CR \gg 1$. This is equivalent to the statement that C is fed by a current source, since R must be large in this setup. Plugging (11.64) into (11.67) and neglecting the phase information one obtains

$$U_{out} = \frac{U_\approx x}{\omega R \varepsilon \varepsilon_0 A} , \qquad (11.68)$$

which is linear in the displacement x.

Figure 11.32b shows a capacitive divider. Again the output voltage U_{out} is given by

$$U_{out} = U_\approx \frac{C_1}{C_2 + C_1} = U_\approx \frac{C_1}{\frac{\varepsilon\varepsilon_0 A}{x} + C_1} . \qquad (11.69)$$

If there is a stray capacitance C_s then (11.69) is modified as

$$U_{out} = U_\approx \frac{C_1}{\frac{\varepsilon\varepsilon_0 A}{x} + C_s + C_1} . \qquad (11.70)$$

Provided $C_s + C_1 \ll C_2$ one has a system which is linear in x. The driving voltage $U_\approx$ has to be large (more than 100 V) to have the output voltage in the range of 1 V. The linearity of the readout depends on the capacitance C_1 (Fig. 11.33).

Another idea is to keep the distance constant and to change the relative overlap of the plates (see Fig. 11.31, right side). The capacitance of the moving center plate versus the stationary outer plates becomes

$$C = C_s + 2\frac{\varepsilon\varepsilon_0 bx}{s} , \qquad (11.71)$$

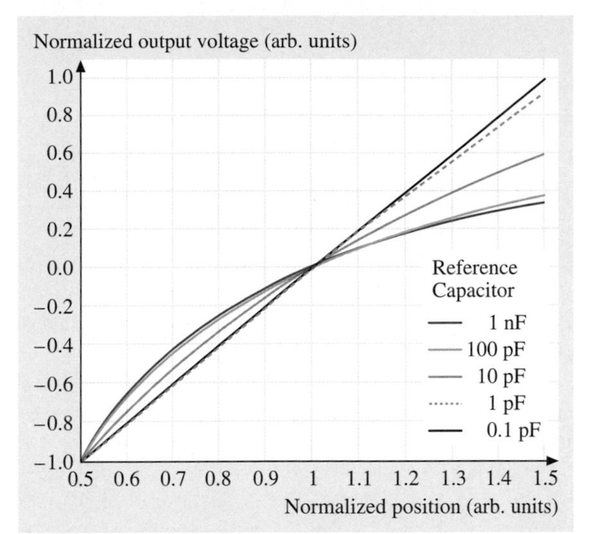

Fig. 11.33 Linearity of the capacitance readout as a function of the reference capacitor

where the variables are defined in Fig. 11.31. The stray capacitance comprises all effects, including the capacitance of the fringe fields. When the length x is comparable to the width b of the plates one can safely assume that the stray capacitance is constant and independent of x. The main disadvantage of this setup is that it is not as easily incorporated in a microfabricated device as the others.

Sensitivity. The capacitance itself is not a measure of the sensitivity, but its derivative is indicative of the signals one can expect. Using the situation described in Fig. 11.31, upper left, and in (11.64) one obtains for the parallel plate capacitor

$$\frac{dC}{dx} = -\frac{\varepsilon\varepsilon_0 A}{x^2} . \qquad (11.72)$$

Assuming a plate area A of 20 μm by 40 μm and a separation of 1 μm one obtains a capacitance of 31 fF (neglecting stray capacitance and the capacitance of the connection leads) and a dC/dx of 3.1×10^{-8} F/m $= 31$ fF/μm. Hence it is of paramount importance to maximize the area between the two contacts and to minimize the distance x. The latter however is far from being trivial. One has to go to the limits of microfabrication to achieve a decent sensitivity.

If the capacitance is measured by the circuit shown in Fig. 11.32 one obtains for the sensitivity

$$\frac{\mathrm{d}U_{\mathrm{out}}}{U_{\approx}} = \frac{\mathrm{d}x}{\omega R \varepsilon \varepsilon_0 A} . \quad (11.73)$$

Using the same value for A as above, setting the reference frequency to 100 kHz, and selecting $R = 1\,\mathrm{G\Omega}$, we get the relative change of the output voltage U_{out} to

$$\frac{\mathrm{d}U_{\mathrm{out}}}{U_{\approx}} = \frac{22.5 \times 10^{-6}}{\text{Å}} \times \mathrm{d}x . \quad (11.74)$$

A driving voltage of 45 V then translates to a sensitivity of 1 mV/Å. A problem in this setup is the stray capacitances. They are in parallel to the original capacitance and decrease the sensitivity considerably.

Alternatively one could build an oscillator with this capacitance and measure the frequency. *RC*-oscillators typically have an oscillation frequency of

$$f_{\mathrm{res}} \propto \frac{1}{RC} = \frac{x}{R \varepsilon \varepsilon_0 A} . \quad (11.75)$$

Again the resistance R must be of the order of $1\,\mathrm{G\Omega}$, when stray capacitances C_{s} are neglected. However C_{s} is of the order of 1 pF. Therefore one gets $R = 10\,\mathrm{M\Omega}$. Using these values the sensitivity becomes

$$\mathrm{d}f_{\mathrm{res}} = \frac{C\,\mathrm{d}x}{R\,(C + C_{\mathrm{s}})^2\,x} \approx \frac{0.1\,\mathrm{Hz}}{\text{Å}}\,\mathrm{d}x . \quad (11.76)$$

The bad thing is that the stray capacitances have made the signal nonlinear again. The linearized setup in Fig. 11.31 has a sensitivity of

$$\frac{\mathrm{d}C}{\mathrm{d}x} = 2\frac{\varepsilon \varepsilon_0 b}{s} . \quad (11.77)$$

Substituting typical values, $b = 10\,\mathrm{\mu m}$, $s = 1\,\mathrm{\mu m}$ one gets $\mathrm{d}C/\mathrm{d}x = 1.8 \times 10^{-10}\,\mathrm{F/m}$. It is noteworthy that the sensitivity remains constant for scaled devices.

Implementations. The readout of the capacitance can be done in different ways [11.123, 124]. All include an alternating current or voltage with frequencies in the 100 kHz to the 100 MHz range. One possibility is to build a tuned circuit with the capacitance of the cantilever determining the frequency. The resonance frequency of a high quality Q tuned circuit is

$$\omega_0 = (LC)^{-1/2} . \quad (11.78)$$

where L is the inductance of the circuit. The capacitance C includes not only the sensor capacitance but also the capacitance of the leads. The precision of a frequency measurement is mainly determined by the ratio of L and C

$$Q = \left(\frac{L}{C}\right)^{1/2} \frac{1}{R} . \quad (11.79)$$

Here R symbolizes the losses in the circuit. The higher the quality the more precise the frequency measurement. For instance a frequency of 100 MHz and a capacitance of 1 pF gives an inductance of 250 μH. The quality becomes then 2.5×10^8. This value is an upper limit, since losses are usually too high.

Using a value of $\mathrm{d}C/\mathrm{d}x = 31\,\mathrm{fF/\mu m}$ one gets $\Delta C/\text{Å} = 3.1\,\mathrm{aF}/\text{Å}$. With a capacitance of 1 pF one gets

$$\frac{\Delta\omega}{\omega} = \frac{1}{2}\frac{\Delta C}{C} ,$$

$$\Delta\omega = 100\,\mathrm{MHz} \times \frac{1}{2}\frac{3.1\mathrm{aF}}{1\,\mathrm{pF}} = 155\,\mathrm{Hz} . \quad (11.80)$$

This is the frequency shift for 1 Å deflection. The calculation shows, that this is a measurable quantity. The quality also indicates that there is no physical reason why this scheme should not work.

11.3.3 Combinations for 3-D-Force Measurements

Three dimensional force measurements are essential if one wants to know all the details of the interaction between the tip and the cantilever. The straightforward attempt to measure three forces is complicated, since force sensors such as interferometers or capacitive sensors need a minimal detection volume, which often is too large. The second problem is that the force-sensing tip has to be held by some means. This implies that one of the three Cartesian axes is stiffer than the others.

However by the combination of different sensors one can achieve this goal. Straight cantilevers are employed for these measurements, because they can be handled analytically. The key observation is, that the optical lever method does not determine the position of the end of the cantilever. It measures the orientation. In the previous sections, one has always made use of the fact, that for a force along one of the orthogonal symmetry directions at the end of the cantilever (normal force, lateral force, force along the cantilever beam axis) there is a one to one correspondence of the tilt angle and the deflection. The problem is, that the force along the cantilever beam axis and the normal force create a deflection in the same direction. Hence what is called the normal force component is actually a mixture of two forces. The deflection of the cantilever is the third quantity, which is not considered in most of the AFMs. A fiber optic interferometer in parallel to the optical lever measures the deflection. Three measured quantities then allow the separation of the three orthonormal force directions, as is evident from (11.27) and (11.33) [11.12–16].

Alternatively one can put the fast scanning direction along the axis of the cantilever. Forward and backward scans then exert opposite forces F_x. If the piezo movement is linearized, both force components in AFM based on the optical lever detection can be determined. In this case, the normal force is simply the average of the forces in the forward and backward direction. The force, F_x, is the difference of the forces measured in forward and backward direction.

11.3.4 Scanning and Control Systems

Almost all SPMs use piezo translators to scan the tip or the sample. Even the first STM [11.1, 103] and some of the predecessor instruments [11.153, 154] used them. Other materials or setups for nano-positioning have been proposed, but were not successful [11.155, 156].

Piezo Tubes

A popular solution is tube scanners (Fig. 11.34). They are now widely used in SPMs due to their simplicity and their small size [11.133, 157]. The outer electrode is segmented in four equal sectors of 90 degrees. Opposite sectors are driven by signals of the same magnitude, but opposite sign. This gives, through bending, a two-dimensional movement on, approximately, a sphere. The inner electrode is normally driven by the z signal. It is possible, however, to use only the outer electrodes for scanning and for the z-movement. The main drawback of applying the z-signal to the outer electrodes is, that the applied voltage is the sum of both the x- or y-movement and the z-movement. Hence a larger scan size effectively reduces the available range for the z-control.

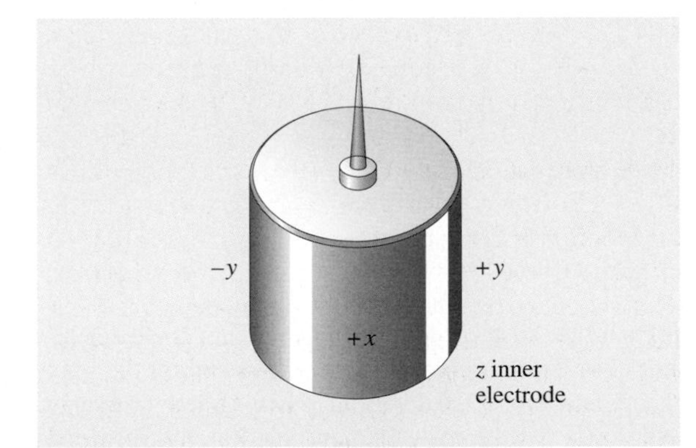

Fig. 11.34 Schematic drawing of a piezoelectric tube scanner. The piezo ceramic is molded into a tube form. The outer electrode is separated into four segments and connected to the scanning voltages. The z-voltage is applied to the inner electrode

Piezo Effect

An electric field applied across a piezoelectric material causes a change in the crystal structure, with expansion in some directions and contraction in others. Also, a net volume change occurs [11.132]. Many SPMs use the transverse piezo electric effect, where the applied electric field $\boldsymbol{E}$ is perpendicular to the expansion/contraction direction.

$$\Delta L = L\,(\boldsymbol{E}\cdot\boldsymbol{n})\,d_{31} = L\frac{V}{t}d_{31}\,, \tag{11.81}$$

where d_{31} is the transverse piezoelectric constant, V is the applied voltage, t is the thickness of the piezo slab or the distance between the electrodes where the voltage is applied, L is the free length of the piezo slab, and $\boldsymbol{n}$ is the direction of polarization. Piezo translators based on the transverse piezoelectric effect have a wide range of sensitivities, limited mainly by mechanical stability and breakdown voltage.

Scan Range

The calculation of the scanning range of a piezotube is difficult [11.157–159]. The bending of the tube depends on the electric fields and the nonuniform strain induced. A finite element calculation where the piezo tube was divided into 218 identical elements was used [11.158] to calculate the deflection. On each node the mechanical stress, stiffness, strain and piezoelectric stress was calculated when a voltage was applied on one electrode. The results were found to be linear on the first iteration and higher-order corrections were very small even for large electrode voltages. It was found that to first order the x- and z-movement of the tube could be reasonably well approximated by assuming that the piezo tube is a segment of a torus. Using this model one obtains

$$\mathrm{d}x = (V_+ - V_-)\,|d_{31}|\,\frac{L^2}{2t\,\mathrm{d}}\,, \tag{11.82}$$

$$\mathrm{d}z = (V_+ + V_- - 2V_z)\,|d_{31}|\,\frac{L}{2t}\,, \tag{11.83}$$

where $|d_{31}|$ is the coefficient of the transversal piezoelectric effect, L is tube's free length, t is tube's wall thickness, d is tube's diameter, V_+ is voltage on positive outer electrode while V_- is voltage of the opposite quadrant negative electrode, and V_z is voltage of the inner electrode.

The cantilever or sample mounted on the piezotube has an additional lateral movement because the point of measurement is not in the end plane of the piezotube. The additional lateral displacement of the end of the tip is $\ell \sin\varphi \approx \ell\varphi$, where ℓ is the tip length and φ is the deflection angle of the end surface. Assuming that the sample or cantilever are always perpendicular to the end of the walls of the tube and calculating with the torus model one gets for the angle

$$\varphi = \frac{L}{R} = \frac{2\,\mathrm{d}x}{L}\,, \tag{11.84}$$

where R is the radius of curvature the piezo tube. Using the result of (11.84) one obtains for the additional x-movement

$$\begin{aligned}\mathrm{d}x_{\mathrm{add}} &= \ell\varphi = \frac{2\,\mathrm{d}x\ell}{L} \\ &= (V_+ - V_-)\,|d_{31}|\,\frac{\ell\,L}{td}\end{aligned} \tag{11.85}$$

and for the additional z-movement due to the x-movement

$$\begin{aligned}\mathrm{d}z_{\mathrm{add}} &= \ell - \ell\cos\varphi = \frac{\ell\varphi^2}{2} = \frac{2\ell\,(\mathrm{d}x)^2}{L^2} \\ &= (V_+ - V_-)^2\,|d_{31}|^2\,\frac{\ell L^2}{2t^2d^2}\,.\end{aligned} \tag{11.86}$$

Carr [11.158] assumed for his finite element calculations that the top of the tube was completely free to move and, as a consequence, the top surface was distorted, leading to a deflection angle about half that of the geometrical model. Depending on the attachment of the sample or the cantilever this distortion may be smaller, leading to a deflection angle in-between that of the geometrical model and the one of the finite element calculation.

Nonlinearities and Creep

Piezo materials with a high conversion ratio, i. e. a large d_{31} or small electrode separations, with large scanning ranges are hampered by substantial hysteresis resulting in a deviation from linearity by more than 10%. The sensitivity of the piezo ceramic material (mechanical displacement divided by driving voltage) decreases with reduced scanning range, whereas the hysteresis is reduced. A careful selection of the material for the piezo scanners, the design of the scanners, and of the operating conditions is necessary to get optimum performance.

Passive Linearization: Calculation. The analysis of images affected by piezo nonlinearities [11.160–163] shows that the dominant term is

$$x = AV + BV^2\,, \tag{11.87}$$

where x is the excursion of the piezo, V the applied voltage and A and B two coefficients describing the sensitivity of the material. Equation (11.87) holds for scanning from $V = 0$ to large V. For the reverse direction the equation becomes

$$x = \tilde{A}V - \tilde{B}\,(V - V_{\mathrm{max}})^2\,, \tag{11.88}$$

where $\tilde{A}$ and $\tilde{B}$ are the coefficients for the back scan and V_{max} is the applied voltage at the turning point. Both equations demonstrate that the true x-travel is small at the beginning of the scan and becomes larger towards the end. Therefore images are stretched at the beginning and compressed at the end.

Similar equations hold for the slow scan direction. The coefficients, however, are different. The combined action causes a greatly distorted image. This distortion can be calculated. The data acquisition systems record the signal as a function of V. However the data is measured as a function of x. Therefore we have to distribute the x-values evenly across the image this can be done by inverting an approximation of (11.87). First we write

$$x = AV\left(1 - \frac{B}{A}V\right)\,. \tag{11.89}$$

For $B \ll A$ we can approximate

$$V = \frac{x}{A}\,. \tag{11.90}$$

We now substitute (11.90) into the nonlinear term of (11.89). This gives

$$\begin{aligned}x &= AV\left(1 + \frac{Bx}{A^2}\right)\,,\\ V &= \frac{x}{A}\,\frac{1}{(1 + Bx/A^2)} \approx \frac{x}{A}\left(1 - \frac{Bx}{A^2}\right)\,.\end{aligned} \tag{11.91}$$

Hence an equation of the type

$$\begin{aligned}&x_{\mathrm{true}} = x\,(\alpha - \beta x/x_{\mathrm{max}})\\ &\text{with}\quad 1 = \alpha - \beta\end{aligned} \tag{11.92}$$

takes out the distortion of an image. α and β are dependent on the scan range, the scan speed and on the scan history and have to be determined with exactly the same settings as for the measurement. x_{max} is the maximal scanning range. The condition for α and β guarantees that the image is transformed onto itself.

Similar equations as the empirical one shown above (11.92) can be derived by analyzing the movements of domain walls in piezo ceramics.

Passive Linearization: Measuring the Position. An alternative strategy is to measure the position of the piezo translators. Several possibilities exist.

1. The interferometers described above can be used to measure the elongation of the piezo elongation. The fiber optic interferometer is especially easy to implement. The coherence length of the laser only limits the measurement range. However the signal is of a periodic nature. Hence a direct use of the signal in a feedback circuit for the position is not possible. However as a measurement tool and, especially, as a calibration tool the interferometer is without competition. The wavelength of the light, for instance in a HeNe laser, is so well defined that the precision of the other components determines the error of the calibration or measurement.
2. The movement of the light spot on the quadrant detector can be used to measure the position of a piezo [11.164]. The output current changes by $0.5\,\mathrm{A/cm} \times P(\mathrm{W})/R(\mathrm{cm})$. Typical values ($P = 1\,\mathrm{mW}$, $R = 0.001\,\mathrm{cm}$) give $0.5\,\mathrm{A/cm}$. The noise limit is typically $0.15\,\mathrm{nm} \times \sqrt{\Delta f(\mathrm{Hz})/H(\mathrm{W/cm^2})}$. Again this means that the laser beam above would have a 0.1 nm noise limitation for a bandwidth of 21 Hz. The advantage of this method is that, in principle, one can linearize two axes with only one detector.
3. A knife-edge blocking part of a light beam incident on a photodiode can be used to measure the position of the piezo. This technique, commonly used in optical shear force detection [11.75, 165], has a sensitivity of better than 0.1 nm.
4. The capacitive detection [11.166, 167] of the cantilever deflection can be applied to the measurement of the piezo elongation. Equations (11.64) to (11.79) apply to the problem. This technique is used in some commercial instruments. The difficulties lie in the avoidance of fringe effects at the borders of the two plates. While conceptually simple, one needs the latest technology in surface preparation to get a decent linearity. The electronic circuits used for the readout are often proprietary.
5. Linear Variable Differential Transformers (LVDT) are a convenient means to measure positions down to 1 nm. They can be used together with a solid state joint setup, as often used for large scan range stages. Unlike the capacitive detection there are few difficulties in implementation. The sensors and the detection circuits LVDTs are available commercially.
6. A popular measurement technique is the use of strain gauges. They are especially sensitive when mounted on a solid state joint where the curvature is maximal. The resolution depends mainly on the induced curvature. A precision of 1 nm is attainable. The signals are low – a Wheatstone bridge is needed for the readout.

Active Linearization. Active linearization is done with feedback systems. Sensors need to be monotonic. Hence all the systems described above, with the exception of the interferometers are suitable. The most common solutions include the strain gauge approach, the capacitance measurement or the LVDT, which are all electronic solutions. Optical detection systems have the disadvantage that the intensity enters into the calibration.

Alternative Scanning Systems

The first STMs were based on piezo tripods [11.1]. The piezo tripod (Fig. 11.35) is an intuitive way to generate the three dimensional movement of a tip attached to its center. However, to get a suitable stability and scanning range, the tripod needs to be fairly large (about 50 mm). Some instruments use piezo stacks instead of monolithic piezoactuators. They are arranged in the tripod arrangement. Piezo stacks are thin layers of piezoactive materials glued together to form a device with up to 200 μm of actuation range. Preloading with a suitable metal casing reduces the nonlinearity.

If one tries to construct a homebuilt scanning system, the use of linearized scanning tables is recommended. They are built around solid state joints and actuated by piezo stacks. The joints guarantee that the movement is parallel with little deviation from the predefined scanning plane. Due to the construction it is easy to add measurement devices such as capacitive sensors, LVDTs

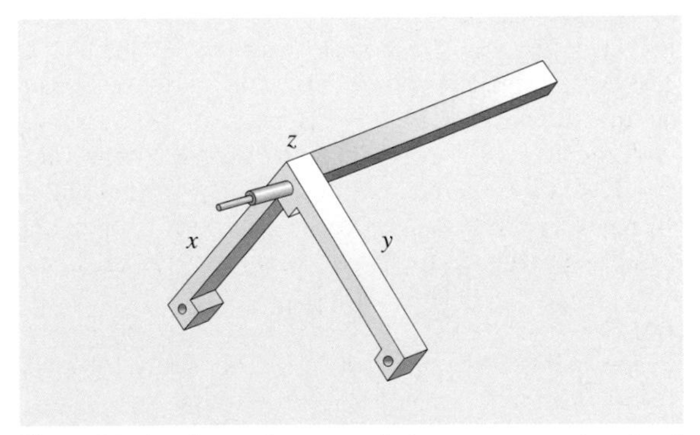

Fig. 11.35 An alternative type of piezo scanners: the tripod

or strain gauges which are essential for a closed loop linearization. Two-dimensional tables can be bought from several manufacturers. They have a linearity of better than 0.1% and a noise level of 10^{-4} to 10^{-5} of the maximal scanning range.

Control Systems

Basics. The electronics and software play an important role in the optimal performance of an SPM. Control electronics and software are supplied with commercial SPMs. Control electronic systems can use either analog or digital feedback. While digital feedback offers greater flexibility and the ease of configuration, analog feedback circuits might be better suited for ultralow noise operation. We will describe here the basic setups for AFMs.

Figure 11.36 shows a block schematic of a typical AFM feedback loop. The signal from the force transducer is fed into the feedback loop consisting mainly of a subtraction stage to get an error signal and an integrator. The gain of the integrator (high gain corresponds to short integration times) is set as high as possible without generating more than 1% overshoot. High gain minimizes the error margin of the current and forces the tip to follow the contours of constant density of states as well as possible. This operating mode is known as Constant Force Mode. A high voltage amplifier amplifies the outputs of the integrator. As AFMs using piezotubes usually require ±150 V at the output, the output of the integrator needs to be amplified by a high voltage amplifier.

In order to scan the sample, additional voltages at high tension are required to drive the piezo. For example, with a tube scanner, four scanning voltages are required, namely $+V_x$, $-V_x$, $+V_y$ and $-V_y$. The x- and y-scanning voltages are generated in a scan generator (analog or computer controlled). Both voltages are input to the two respective power amplifiers. Two inverting amplifiers generate the input voltages for the other two power amplifiers. The topography of the sample surface is determined by recording the input-voltage to the high voltage amplifier for the z-channel as a function of x and y (Constant Force Mode).

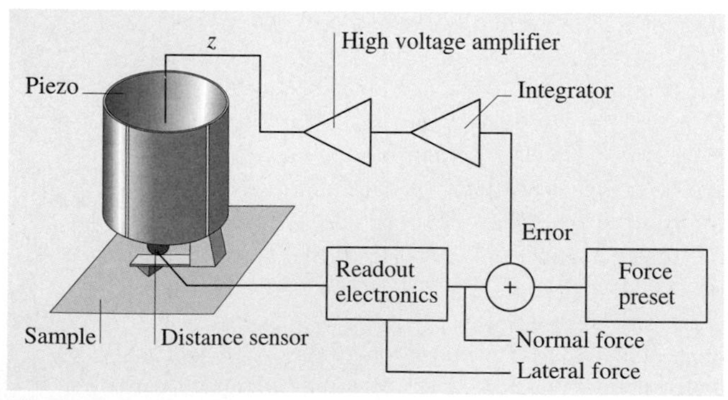

Fig. 11.36 Block schematic of the feedback control loop of an AFM

Another operating mode is the Variable Force Mode. The gain in the feedback loop is lowered and the scanning speed increased such that the force on the cantilever is no longer constant. Here the force is recorded as a function of x and y.

Force Spectroscopy. Four modes of spectroscopic imaging are in common use with force microscopes: measuring lateral forces, $\partial F/\partial z$, $\partial F/\partial x$ spatially resolved, and measuring force versus distance curves. Lateral forces can be measured by detecting the deflection of a cantilever in a direction orthogonal to the normal direction. The optical lever deflection method most easily does this. Lateral force measurements give indications of adhesion forces between the tip and the sample.

$\partial F/\partial z$ measurements probe the local elasticity of the sample surface. In many cases the measured quantity originates from a volume of a few cubic nanometers. The $\partial F/\partial z$ or local stiffness signal is proportional to Young's modulus, as far as one can define this quantity. Local stiffness is measured by vibrating the cantilever by a small amount in the z-direction. The expected signal for very stiff samples is zero: for very soft samples one gets, independent of the stiffness, also a constant signal. This signal is again zero for the optical lever deflection and equal to the driving amplitude for interferometeric measurements. The best sensitivity is obtained when the compliance of the cantilever matches the stiffness of the sample.

A third spectroscopic quantity is the lateral stiffness. It is measured by applying a small modulation in the x-direction on the cantilever. The signal is again optimal when the lateral compliance of the cantilever matches the lateral stiffness of the sample. The lateral stiffness is, in turn, related to the shear modulus of the sample.

Detailed information on the interaction of the tip and the sample can be gained by measuring force versus distance curves. It is necessary to have cantilevers with high enough compliance to avoid instabilities due to the attractive forces on the sample.

Using the Control Electronics as a Two-Dimensional Measurement Tool. Usually the control electronics of an AFM is used to control the x- and y-piezo signals while several data acquisition channels record the position dependent signals. The control electronics can be used in another way: it can be viewed as a two-dimensional function generator. What is normally the x- and y-signal can be used to control two independent variables of an experiment. The control logic of the AFM then ensures that the available parameter space is systematically probed at equally spaced points. An example is friction force curves measured along a line across a step on graphite.

Figure 11.37 shows the connections. The z-piezo is connected as usual, like the x-piezo. However the y-output is used to command the desired input parameter. The offset of the y-channel determines the position of the tip on the sample surface, together with the x-channel.

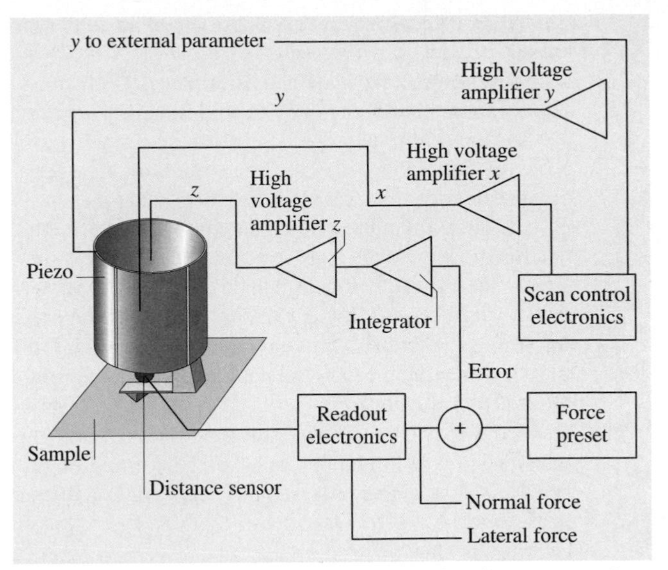

Fig. 11.37 Wiring of an AFM to measure friction force curves along a line

Some Imaging Processing Methods

The visualization and interpretation of images from AFMs is intimately connected to the processing of these images. An ideal AFM is a noise-free device that images a sample with perfect tips of known shape and has perfect linear scanning piezos. In reality, AFMs are not that ideal. The scanning device in AFMs is affected by distortions. The distortions are both linear and nonlinear. Linear distortions mainly result from imperfections in the machining of the piezotranslators causing crosstalk from the Z-piezo to the X- and Y-piezos, and vice versa. Among the linear distortions, there are two kinds which are very important. First, scanning piezos invariably have different sensitivities along the different scan axes due to the variation of the piezo material and uneven sizes of the electrode areas. Second, the same reasons might cause the scanning axes not to be orthogonal. Furthermore, the plane in which the piezoscanner moves for constant height z is hardly ever coincident with the sample plane. Hence, a linear ramp is added to the sample data. This ramp is especially bothersome when the height z is displayed as an intensity map.

The nonlinear distortions are harder to deal with. They can affect AFM data for a variety of reasons. First, piezoelectric ceramics do have a hysteresis loop, much like ferromagnetic materials. The deviations of piezoceramic materials from linearity increase with increasing amplitude of the driving voltage. The mechanical position for one voltage depends on the previously applied voltages to the piezo. Hence, to get the best positional accuracy, one should always approach a point on the sample from the same direction. Another type of nonlinear distortion of the images occurs when the scan frequency approaches the upper frequency limit of the X- and Y-drive amplifiers or the upper frequency limit of the feedback loop (z-component). This distortion, due to feedback loop, can only be minimized by reducing the scan frequency. On the other hand, there is a simple way to reduce distortions due to the X- and Y-piezo drive amplifiers. To keep the system as simple as possible, one normally uses a triangular waveform for driving the scanning piezos. However, triangular waves contain frequency components as multiples of the scan frequency. If the cutoff frequency of the X- and Y-drive electronics or of the feedback loop is too close to the scanning frequency (two or three times the scanning frequency), the triangular drive voltage is rounded off at the turning points. This rounding error causes, first, a distortion of the scan linearity and, second, through phase lags, the projection of part of the backward scan onto the forward scan. This type of distortion can be minimized by carefully selecting the scanning frequency and by using driving voltages for the X- and Y-piezos with waveforms like trapezoidal waves, which are closer to a sine wave. The values measured for X, Y, or Z piezos are affected by noise. The origin of this noise can be either electronic, some disturbances, or a property of the sample surface due to adsorbates. In addition to this incoherent noise, interference with main and other equipment nearby might be present. Depending on the type of noise, one can filter it in the real space or in Fourier space. The most important part of image processing is to visualize the measured data. Typical AFM data sets can consist of many thousands to over a million points per plane. There may be more than one image plane present.

The AFM data represents a topography in various data spaces.

Most commercial data acquisition systems use implicitly some kind of data processing. Since the original data is commonly subject to slopes on the surface, most programs use some kind of slope correction. The least disturbing way is to substrate a plane $z(x, y) = Ax + By + C$ from the data. The coefficients are determined by fitting $z(x, y)$ to the data. Another operation is to subtract a second order function such as $z(x, y) = Ax^2 + By^2 + Cxy + Dx + Ey + F$. Again, the parameters are determined with a fit. This function is appropriate for almost planar data, where the nonlinearity of the piezos caused such a distortion.

In the image processing software from Digital Instruments, up to three operations are performed on the raw data. First, a zero-order flatten is applied. The flatten operation is used to eliminate image bow in the slow scan direction (caused by a physical bow in the instrument itself), slope in the slow scan direction, bands in the image (caused by differences in the scan height from one scan line to the next). The flattening operation takes each scan line and subtracts the average value of the height along each scan line from each point in that scan line. This brings each scan line to the same height. Next, a first-order plane-fit is applied in the fast scan direction. The plane-fit operation is used to eliminate bow and slope in the fast scan direction. The plane-fit operation calculated a best-fit plane for the image and subtracts it from the image. This plane has a constant non-zero slope in the fast scan direction. In some cases, a higher-order polynomial "plane" may be required. Depending upon the quality of the raw data, the flattening operation and/or the planefit operation may not be required at all.

References

11.1 G. Binnig, H. Rohrer, Ch. Gerber, E. Weibel: Surface studies by scanning tunneling microscopy, Phys. Rev. Lett. **49** (1982) 57–61

11.2 G. Binnig, C. F. Quate, Ch. Gerber: Atomic force microscope, Phys. Rev. Lett. **56** (1986) 930–933

11.3 G. Binnig, Ch. Gerber, E. Stoll, T. R. Albrecht, C. F. Quate: Atomic resolution with atomic force microscope, Europhys. Lett. **3** (1987) 1281–1286

11.4 B. Bhushan: *Handbook of Micro/Nanotribology*, 2nd edn. (CRC, Boca Raton 1999)

11.5 C. M. Mate, G. M. McClelland, R. Erlandsson, S. Chiang: Atomic-scale friction of a tungsten tip on a graphite surface, Phys. Rev. Lett. **59** (1987) 1942–1945

11.6 R. Erlandsson, G. M. McClelland, C. M. Mate, S. Chiang: Atomic force microscopy using optical interferometry, J. Vac. Sci. Technol. A **6** (1988) 266–270

11.7 O. Marti, J. Colchero, J. Mlynek: Combined scanning force and friction microscopy of mica, Nanotechnol. **1** (1990) 141–144

11.8 G. Meyer, N. M. Amer: Simultaneous measurement of lateral and normal forces with an optical-beam-deflection atomic force microscope, Appl. Phys. Lett. **57** (1990) 2089–2091

11.9 B. Bhushan, J. Ruan: Atomic-scale friction measurements using friction force microscopy: Part II – Application to magnetic media, ASME J. Tribol. **116** (1994) 389–396

11.10 B. Bhushan, V. N. Koinkar, J. Ruan: Microtribology of magnetic media, Proc. Inst. Mech. Eng., Part J: J. Eng. Tribol. **208** (1994) 17–29

11.11 B. Bhushan, J. N. Israelachvili, U. Landman: Nanotribology: Friction, wear, and lubrication at the atomic scale, Nature **374** (1995) 607–616

11.12 S. Fujisawa, M. Ohta, T. Konishi, Y. Sugawara, S. Morita: Difference between the forces measured by an optical lever deflection and by an optical interferometer in an atomic force microscope, Rev. Sci. Instrum. **65** (1994) 644–647

11.13 S. Fujisawa, E. Kishi, Y. Sugawara, S. Morita: Fluctuation in 2-dimensional stick-slip phenomenon observed with 2-dimensional frictional force microscope, Jpn. J. Appl. Phys. **33** (1994) 3752–3755

11.14 S. Grafstrom, J. Ackermann, T. Hagen, R. Neumann, O. Probst: Analysis of lateral force effects on the topography in scanning force microscopy, J. Vac. Sci. Technol. B **12** (1994) 1559–1564

11.15 R. M. Overney, H. Takano, M. Fujihira, W. Paulus, H. Ringsdorf: Anisotropy in friction and molecular stick-slip motion, Phys. Rev. Lett. **72** (1994) 3546–3549

11.16 R. J. Warmack, X. Y. Zheng, T. Thundat, D. P. Allison: Friction effects in the deflection of atomic force microscope cantilevers, Rev. Sci. Instrum. **65** (1994) 394–399

11.17 N. A. Burnham, D. D. Domiguez, R. L. Mowery, R. J. Colton: Probing the surface forces of monolayer films with an atomic force microscope, Phys. Rev. Lett. **64** (1990) 1931–1934

11.18 N. A. Burham, R. J. Colton, H. M. Pollock: Interpretation issues in force microscopy, J. Vac. Sci. Technol. A **9** (1991) 2548–2556

11.19 C. D. Frisbie, L. F. Rozsnyai, A. Noy, M. S. Wrighton, C. M. Lieber: Functional group imaging by chemical force microscopy, Science **265** (1994) 2071–2074
11.20 V. N. Koinkar, B. Bhushan: Microtribological studies of unlubricated and lubricated surfaces using atomic force/friction force microscopy, J. Vac. Sci. Technol. A **14** (1996) 2378–2391
11.21 V. Scherer, B. Bhushan, U. Rabe, W. Arnold: Local elasticity and lubrication measurements using atomic force and friction force microscopy at ultrasonic frequencies, IEEE Trans. Mag. **33** (1997) 4077–4079
11.22 V. Scherer, W. Arnold, B. Bhushan: Lateral force microscopy using acoustic friction force microscopy, Surf. Interface Anal. **27** (1999) 578–587
11.23 B. Bhushan, S. Sundararajan: Micro/Nanoscale friction and wear mechanisms of thin films using atomic force and friction force microscopy, Acta Mater. **46** (1998) 3793–3804
11.24 U. Krotil, T. Stifter, H. Waschipky, K. Weishaupt, S. Hild, O. Marti: Pulse force mode: A new method for the investigation of surface properties, Surf. Interface Anal. **27** (1999) 336–340
11.25 B. Bhushan, C. Dandavate: Thin-film friction and adhesion studies using atomic force microscopy, J. Appl. Phys. **87** (2000) 1201–1210
11.26 B. Bhushan: *Micro/Nanotribology and its Applications* (Kluwer, Dordrecht 1997)
11.27 B. Bhushan: *Principles and Applications of Tribology* (Wiley, New York 1999)
11.28 B. Bhushan: *Modern Tribology Handbook Vol. 1: Principles of Tribology* (CRC, Boca Raton 2001)
11.29 B. Bhushan: *Introduction to Tribology* (Wiley, New York 2002)
11.30 M. Reinstaedtler, U. Rabe, V. Scherer, U. Hartmann, A. Goldade, B. Bhushan, W. Arnold: On the nanoscale measurement of friction using atomic force microscope cantilever torsional resonances, Appl. Phys. Lett. **82** (2003) 2604–2606
11.31 N. A. Burnham, R. J. Colton: Measuring the nanomechanical properties and surface forces of materials using an atomic force microscope, J. Vac. Sci. Technol. A **7** (1989) 2906–2913
11.32 P. Maivald, H. J. Butt, S. A. C. Gould, C. B. Prater, B. Drake, J. A. Gurley, V. B. Elings, P. K. Hansma: Using force modulation to image surface elasticities with the atomic force microscope, Nanotechnol. **2** (1991) 103–106
11.33 B. Bhushan, A. V. Kulkarni, W. Bonin, J. T. Wyrobek: Nano/Picoindentation measurements using capacitive transducer in atomic force microscopy, Philos. Mag. A **74** (1996) 1117–1128
11.34 B. Bhushan, V. N. Koinkar: Nanoindentation hardness measurements using atomic force microscopy, Appl. Phys. Lett. **75** (1994) 5741–5746
11.35 D. DeVecchio, B. Bhushan: Localized surface elasticity measurements using an atomic force microscope, Rev. Sci. Instrum. **68** (1997) 4498–4505
11.36 S. Amelio, A. V. Goldade, U. Rabe, V. Scherer, B. Bhushan, W. Arnold: Measurements of mechanical properties of ultra-thin diamond-like carbon coatings using atomic force acoustic microscopy, Thin Solid Films **392** (2001) 75–84
11.37 D. M. Eigler, E. K. Schweizer: Positioning single atoms with a scanning tunnelling microscope, Nature **344** (1990) 524–528
11.38 A. L. Weisenhorn, J. E. MacDougall, J. A. C. Gould, S. D. Cox, W. S. Wise, J. Massie, P. Maivald, V. B. Elings, G. D. Stucky, P. K. Hansma: Imaging and manipulating of molecules on a zeolite surface with an atomic force microscope, Science **247** (1990) 1330–1333
11.39 I. W. Lyo, Ph. Avouris: Field-induced nanometer-to-atomic-scale manipulation of silicon surfaces with the STM, Science **253** (1991) 173–176
11.40 O. M. Leung, M. C. Goh: Orientation ordering of polymers by atomic force microscope tip-surface interactions, Science **225** (1992) 64–66
11.41 D. W. Abraham, H. J. Mamin, E. Ganz, J. Clark: Surface modification with the scanning tunneling microscope, IBM J. Res. Dev. **30** (1986) 492–499
11.42 R. M. Silver, E. E. Ehrichs, A. L. de Lozanne: Direct writing of submicron metallic features with a scanning tunnelling microscope, Appl. Phys. Lett. **51** (1987) 247–249
11.43 A. Kobayashi, F. Grey, R. S. Williams, M. Ano: Formation of nanometer-scale grooves in silicon with a scanning tunneling microscope, Science **259** (1993) 1724–1726
11.44 B. Parkinson: Layer-by-layer nanometer scale etching of two-dimensional substrates using the scanning tunneling microscopy, J. Am. Chem. Soc. **112** (1990) 7498–7502
11.45 A. Majumdar, P. I. Oden, J. P. Carrejo, L. A. Nagahara, J. J. Graham, J. Alexander: Nanometer-scale lithography using the atomic force microscope, Appl. Phys. Lett. **61** (1992) 2293–2295
11.46 B. Bhushan: Micro/Nanotribology and its applications to magnetic storage devices and MEMS, Tribol. Int. **28** (1995) 85–96
11.47 L. Tsau, D. Wang, K. L. Wang: Nanometer scale patterning of silicon(100) surface by an atomic force microscope operating in air, Appl. Phys. Lett. **64** (1994) 2133–2135
11.48 E. Delawski, B. A. Parkinson: Layer-by-layer etching of two-dimensional metal chalcogenides with the atomic force microscope, J. Am. Chem. Soc. **114** (1992) 1661–1667
11.49 B. Bhushan, G. S. Blackman: Atomic force microscopy of magnetic rigid disks and sliders and its applications to tribology, ASME J. Tribol. **113** (1991) 452–458
11.50 O. Marti, B. Drake, P. K. Hansma: Atomic force microscopy of liquid-covered surfaces: atomic resolution images, Appl. Phys. Lett. **51** (1987) 484–486

11.51 B. Drake, C.B. Prater, A.L. Weisenhorn, S.A.C. Gould, T.R. Albrecht, C.F. Quate, D.S. Cannell, H.G. Hansma, P.K. Hansma: Imaging crystals, polymers and processes in water with the atomic force microscope, Science **243** (1989) 1586–1589
11.52 M. Binggeli, R. Christoph, H.E. Hintermann, J. Colchero, O. Marti: Friction force measurements on potential controlled graphite in an electrolytic environment, Nanotechnol. **4** (1993) 59–63
11.53 G. Meyer, N.M. Amer: Novel optical approach to atomic force microscopy, Appl. Phys. Lett. **53** (1988) 1045–1047
11.54 J.H. Coombs, J.B. Pethica: Properties of vacuum tunneling currents: Anomalous barrier heights, IBM J. Res. Dev. **30** (1986) 455–459
11.55 M.D. Kirk, T. Albrecht, C.F. Quate: Low-temperature atomic force microscopy, Rev. Sci. Instrum. **59** (1988) 833–835
11.56 F.J. Giessibl, Ch. Gerber, G. Binnig: A low-temperature atomic force/scanning tunneling microscope for ultrahigh vacuum, J. Vac. Sci. Technol. B **9** (1991) 984–988
11.57 T.R. Albrecht, P. Grutter, D. Rugar, D.P.E. Smith: Low temperature force microscope with all-fiber interferometer, Ultramicroscopy **42–44** (1992) 1638–1646
11.58 H.J. Hug, A. Moser, Th. Jung, O. Fritz, A. Wadas, I. Parashikor, H.J. Güntherodt: Low temperature magnetic force microscopy, Rev. Sci. Instrum. **64** (1993) 2920–2925
11.59 C. Basire, D.A. Ivanov: Evolution of the lamellar structure during crystallization of a semicrystalline-amorphous polymer blend: Time-resolved hot-stage SPM study, Phys. Rev. Lett. **85** (2000) 5587–5590
11.60 H. Liu, B. Bhushan: Investigation of nanotribological properties of self-assembled monolayers with alkyl and biphenyl spacer chains, Ultramicroscopy **91** (2002) 185–202
11.61 J. Foster, J. Frommer: Imaging of liquid crystal using a tunneling microscope, Nature **333** (1988) 542–547
11.62 D. Smith, H. Horber, C. Gerber, G. Binnig: Smectic liquid crystal monolayers on graphite observed by scanning tunneling microscopy, Science **245** (1989) 43–45
11.63 D. Smith, J. Horber, G. Binnig, H. Nejoh: Structure, registry and imaging mechanism of alkylcyanobiphenyl molecules by tunnelling microscopy, Nature **344** (1990) 641–644
11.64 Y. Andoh, S. Oguchi, R. Kaneko, T. Miyamoto: Evaluation of very thin lubricant films, J. Phys. D **25** (1992) A71–A75
11.65 Y. Martin, C.C. Williams, H.K. Wickramasinghe: Atomic force microscope-force mapping and profiling on a sub 100-A scale, J. Appl. Phys. **61** (1987) 4723–4729
11.66 J.E. Stern, B.D. Terris, H.J. Mamin, D. Rugar: Deposition and imaging of localized charge on insulator surfaces using a force microscope, Appl. Phys. Lett. **53** (1988) 2717–2719
11.67 K. Yamanaka, H. Ogisco, O. Kolosov: Ultrasonic force microscopy for nanometer resolution subsurface imaging, Appl. Phys. Lett. **64** (1994) 178–180
11.68 K. Yamanaka, E. Tomita: Lateral force modulation atomic force microscope for selective imaging of friction forces, Jpn. J. Appl. Phys. **34** (1995) 2879–2882
11.69 U. Rabe, K. Janser, W. Arnold: Vibrations of free and surface-coupled atomic force microscope: Theory and experiment, Rev. Sci. Instrum. **67** (1996) 3281–3293
11.70 Y. Martin, H.K. Wickramasinghe: Magnetic imaging by force microscopy with 1000 Å resolution, Appl. Phys. Lett. **50** (1987) 1455–1457
11.71 D. Rugar, H.J. Mamin, P. Guethner, S.E. Lambert, J.E. Stern, I. McFadyen, T. Yogi: Magnetic force microscopy – General principles and application to longitudinal recording media, J. Appl. Phys. **63** (1990) 1169–1183
11.72 C. Schoenenberger, S.F. Alvarado: Understanding magnetic force microscopy, Z. Phys. B **80** (1990) 373–383
11.73 U. Hartmann: Magnetic force microscopy, Annu. Rev. Mater. Sci. **29** (1999) 53–87
11.74 D.W. Pohl, W. Denk, M. Lanz: Optical stethoscopy-image recording with resolution lambda/20, Appl. Phys. Lett. **44** (1984) 651–653
11.75 E. Betzig, J.K. Troutman, T.D. Harris, J.S. Weiner, R.L. Kostelak: Breaking the diffraction barrier – optical microscopy on a nanometric scale, Science **251** (1991) 1468–1470
11.76 E. Betzig, P.L. Finn, J.S. Weiner: Combined shear force and near-field scanning optical microscopy, Appl. Phys. Lett. **60** (1992) 2484
11.77 P.F. Barbara, D.M. Adams, D.B. O'Connor: Characterization of organic thin film materials with near-field scanning optical microscopy (NSOM), Annu. Rev. Mater. Sci. **29** (1999) 433–469
11.78 C.C. Williams, H.K. Wickramasinghe: Scanning thermal profiler, Appl. Phys. Lett. **49** (1986) 1587–1589
11.79 C.C. Williams, H.K. Wickramasinghe: Microscopy of chemical-potential variations on an atomic scale, Nature **344** (1990) 317–319
11.80 A. Majumdar: Scanning thermal microscopy, Annu. Rev. Mater. Sci. **29** (1999) 505–585
11.81 O.E. Husser, D.H. Craston, A.J. Bard: Scanning electrochemical microscopy – high resolution deposition and etching of materials, J. Electrochem. Soc. **136** (1989) 3222–3229
11.82 Y. Martin, D.W. Abraham, H.K. Wickramasinghe: High-resolution capacitance measurement and potentiometry by force microscopy, Appl. Phys. Lett. **52** (1988) 1103–1105

11.83 M. Nonnenmacher, M. P. O'Boyle, H. K. Wickramasinghe: Kelvin probe force microscopy, Appl. Phys. Lett. **58** (1991) 2921–2923
11.84 J. M. R. Weaver, D. W. Abraham: High resolution atomic force microscopy potentiometry, J. Vac. Sci. Technol. B **9** (1991) 1559–1561
11.85 D. DeVecchio, B. Bhushan: Use of a nanoscale Kelvin probe for detecting wear precursors, Rev. Sci. Instrum. **69** (1998) 3618–3624
11.86 B. Bhushan, A. V. Goldade: Measurements and analysis of surface potential change during wear of single-crystal silicon (100) at ultralow loads using Kelvin probe microscopy, Appl. Surf. Sci. **157** (2000) 373–381
11.87 P. K. Hansma, B. Drake, O. Marti, S. A. C. Gould, C. B. Prater: The scanning ion-conductance microscope, Science **243** (1989) 641–643
11.88 C. B. Prater, P. K. Hansma, M. Tortonese, C. F. Quate: Improved scanning ion-conductance microscope using microfabricated probes, Rev. Sci. Instrum. **62** (1991) 2634–2638
11.89 J. Matey, J. Blanc: Scanning capacitance microscopy, J. Appl. Phys. **57** (1985) 1437–1444
11.90 C. C. Williams: Two-dimensional dopant profiling by scanning capacitance microscopy, Annu. Rev. Mater. Sci. **29** (1999) 471–504
11.91 D. T. Lee, J. P. Pelz, B. Bhushan: Instrumentation for direct, low frequency scanning capacitance microscopy, and analysis of position dependent stray capacitance, Rev. Sci. Instrum. **73** (2002) 3523–3533
11.92 I. Giaever: Energy gap in superconductors measured by electron tunneling, Phys. Rev. Lett. **5** (1960) 147–148
11.93 P. K. Hansma, J. Tersoff: Scanning tunneling microscopy, J. Appl. Phys. **61** (1987) R1–R23
11.94 D. Sarid, V. Elings: Review of scanning force microscopy, J. Vac. Sci. Technol. B **9** (1991) 431–437
11.95 U. Durig, O. Zuger, A. Stalder: Interaction force detection in scanning probe microscopy: Methods and applications, J. Appl. Phys. **72** (1992) 1778–1797
11.96 J. Frommer: Scanning tunneling microscopy and atomic force microscopy in organic chemistry, Angew. Chem. Int. Ed. Engl. **31** (1992) 1298–1328
11.97 H. J. Güntherodt, R. Wiesendanger (Eds.): *Scanning Tunneling Microscopy I: General Principles and Applications to Clean and Adsorbate-Covered Surfaces* (Springer, Berlin, Heidelberg 1992)
11.98 R. Wiesendanger, H. J. Güntherodt (Eds.): *Scanning Tunneling Microscopy, II: Further Applications and Related Scanning Techniques* (Springer, Berlin, Heidelberg 1992)
11.99 D. A. Bonnell (Ed.): *Scanning Tunneling Microscopy and Spectroscopy – Theory, Techniques, and Applications* (VCH, New York 1993)
11.100 O. Marti, M. Amrein (Eds.): *STM and SFM in Biology* (Academic, San Diego 1993)
11.101 J. A. Stroscio, W. J. Kaiser (Eds.): *Scanning Tunneling Microscopy* (Academic, Boston 1993)
11.102 H. J. Güntherodt, D. Anselmetti, E. Meyer (Eds.): *Forces in Scanning Probe Methods* (Kluwer, Dordrecht 1995)
11.103 G. Binnig, H. Rohrer: Scanning tunnelling microscopy, Surf. Sci. **126** (1983) 236–244
11.104 B. Bhushan, J. Ruan, B. K. Gupta: A scanning tunnelling microscopy study of fullerene films, J. Phys. D **26** (1993) 1319–1322
11.105 R. L. Nicolaides, W. E. Yong, W. F. Packard, H. A. Zhou: Scanning tunneling microscope tip structures, J. Vac. Sci. Technol. A **6** (1988) 445–447
11.106 J. P. Ibe, P. P. Bey, S. L. Brandon, R. A. Brizzolara, N. A. Burnham, D. P. DiLella, K. P. Lee, C. R. K. Marrian, R. J. Colton: On the electrochemical etching of tips for scanning tunneling microscopy, J. Vac. Sci. Technol. A **8** (1990) 3570–3575
11.107 R. Kaneko, S. Oguchi: Ion-implanted diamond tip for a scanning tunneling microscope, Jpn. J. Appl. Phys. **28** (1990) 1854–1855
11.108 F. J. Giessibl: Atomic resolution of the silicon(111)-(7×7) surface by atomic force microscopy, Science **267** (1995) 68–71
11.109 B. Anczykowski, D. Krueger, K. L. Babcock, H. Fuchs: Basic properties of dynamic force spectroscopy with the scanning force microscope in experiment and simulation, Ultramicroscopy **66** (1996) 251–259
11.110 T. R. Albrecht and C. F. Quate: Atomic resolution imaging of a nonconductor by atomic force microscopy, J. Appl. Phys. **62** (1987) 2599–2602
11.111 S. Alexander, L. Hellemans, O. Marti, J. Schneir, V. Elings, P. K. Hansma: An atomic-resolution atomic-force microscope implemented using an optical lever, J. Appl. Phys. **65** (1989) 164–167
11.112 G. Meyer, N. M. Amer: Optical-beam-deflection atomic force microscopy: The NaCl(001) surface, Appl. Phys. Lett. **56** (1990) 2100–2101
11.113 A. L. Weisenhorn, M. Egger, F. Ohnesorge, S. A. C. Gould, S. P. Heyn, H. G. Hansma, R. L. Sinsheimer, H. E. Gaub, P. K. Hansma: Molecular resolution images of Langmuir–Blodgett films and DNA by atomic force microscopy, Langmuir **7** (1991) 8–12
11.114 J. Ruan, B. Bhushan: Atomic-scale and microscale friction of graphite and diamond using friction force microscopy, J. Appl. Phys. **76** (1994) 5022–5035
11.115 D. Rugar, P. K. Hansma: Atomic force microscopy, Phys. Today **43** (1990) 23–30
11.116 D. Sarid: *Scanning Force Microscopy* (Oxford Univ. Press, Oxford 1991)
11.117 G. Binnig: Force microscopy, Ultramicroscopy **42–44** (1992) 7–15
11.118 E. Meyer: Atomic force microscopy, Surf. Sci. **41** (1992) 3–49
11.119 H. K. Wickramasinghe: Progress in scanning probe microscopy, Acta Mater. **48** (2000) 347–358

11.120 A.J. den Boef: The influence of lateral forces in scanning force microscopy, Rev. Sci. Instrum. **62** (1991) 88–92
11.121 M. Radmacher, R.W. Tillman, M. Fritz, H.E. Gaub: From molecules to cells: Imaging soft samples with the atomic force microscope, Science **257** (1992) 1900–1905
11.122 F. Ohnesorge, G. Binnig: True atomic resolution by atomic force microscopy through repulsive and attractive forces, Science **260** (1993) 1451–1456
11.123 G. Neubauer, S.R. Coben, G.M. McClelland, D. Horne, C.M. Mate: Force microscopy with a bidirectional capacitance sensor, Rev. Sci. Instrum. **61** (1990) 2296–2308
11.124 T. Goddenhenrich, H. Lemke, U. Hartmann, C. Heiden: Force microscope with capacitive displacement detection, J. Vac. Sci. Technol. A **8** (1990) 383–387
11.125 U. Stahl, C.W. Yuan, A.L. Delozanne, M. Tortonese: Atomic force microscope using piezoresistive cantilevers and combined with a scanning electron microscope, Appl. Phys. Lett. **65** (1994) 2878–2880
11.126 R. Kassing, E. Oesterschulze: Sensors for scanning probe microscopy. In: *Micro/Nanotribology and Its Applications*, ed. by B. Bhushan (Kluwer, Dordrecht 1997) pp. 35–54
11.127 C.M. Mate: Atomic-force-microscope study of polymer lubricants on silicon surfaces, Phys. Rev. Lett. **68** (1992) 3323–3326
11.128 S.P. Jarvis, A. Oral, T.P. Weihs, J.B. Pethica: A novel force microscope and point contact probe, Rev. Sci. Instrum. **64** (1993) 3515–3520
11.129 D. Rugar, H.J. Mamin, P. Guethner: Improved fiber-optical interferometer for atomic force microscopy, Appl. Phys. Lett. **55** (1989) 2588–2590
11.130 C. Schoenenberger, S.F. Alvarado: A differential interferometer for force microscopy, Rev. Sci. Instrum. **60** (1989) 3131–3135
11.131 D. Sarid, D. Iams, V. Weissenberger, L.S. Bell: Compact scanning-force microscope using laser diode, Opt. Lett. **13** (1988) 1057–1059
11.132 N.W. Ashcroft, N.D. Mermin: *Solid State Physics* (Holt Reinhart and Winston, New York 1976)
11.133 G. Binnig, D.P.E. Smith: Single-tube three-dimensional scanner for scanning tunneling microscopy, Rev. Sci. Instrum. **57** (1986) 1688
11.134 S.I. Park, C.F. Quate: Digital filtering of STM images, J. Appl. Phys. **62** (1987) 312
11.135 J.W. Cooley, J.W. Tukey: An algorithm for machine calculation of complex fourier series, Math. Computation **19** (1965) 297
11.136 J. Ruan, B. Bhushan: Atomic-scale friction measurements using friction force microscopy: Part I – General principles and new measurement techniques, ASME J. Tribol. **116** (1994) 378–388
11.137 T.R. Albrecht, S. Akamine, T.E. Carver, C.F. Quate: Microfabrication of cantilever styli for the atomic force microscope, J. Vac. Sci. Technol. A **8** (1990) 3386–3396
11.138 O. Marti, S. Gould, P.K. Hansma: Control electronics for atomic force microscopy, Rev. Sci. Instrum. **59** (1988) 836–839
11.139 O. Wolter, T. Bayer, J. Greschner: Micromachined silicon sensors for scanning force microscopy, J. Vac. Sci. Technol. B **9** (1991) 1353–1357
11.140 E. Meyer, R. Overney, R. Luthi, D. Brodbeck: Friction force microscopy of mixed Langmuir–Blodgett films, Thin Solid Films **220** (1992) 132–137
11.141 H.J. Dai, J.H. Hafner, A.G. Rinzler, D.T. Colbert, R.E. Smalley: Nanotubes as nanoprobes in scanning probe microscopy, Nature **384** (1996) 147–150
11.142 J.H. Hafner, C.L. Cheung, A.T. Woolley, C.M. Lieber: Structural and functional imaging with carbon nanotube AFM probes, Prog. Biophys. Mol. Biol. **77** (2001) 73–110
11.143 G.S. Blackman, C.M. Mate, M.R. Philpott: Interaction forces of a sharp tungsten tip with molecular films on silicon surface, Phys. Rev. Lett. **65** (1990) 2270–2273
11.144 S.J. O'Shea, M.E. Welland, T. Rayment: Atomic force microscope study of boundary layer lubrication, Appl. Phys. Lett. **61** (1992) 2240–2242
11.145 J.P. Cleveland, S. Manne, D. Bocek, P.K. Hansma: A nondestructive method for determining the spring constant of cantilevers for scanning force microscopy, Rev. Sci. Instrum. **64** (1993) 403–405
11.146 D.W. Pohl: Some design criteria in STM, IBM J. Res. Dev. **30** (1986) 417
11.147 W.T. Thomson, M.D. Dahleh: *Theory of Vibration with Applications*, 5th edn. (Prentice Hall, Upper Saddle River 1998)
11.148 J. Colchero: Reibungskraftmikroskopie. Ph.D. Thesis (University of Konstanz, Konstanz 1993)
11.149 G.M. McClelland, R. Erlandsson, S. Chiang: Atomic force microscopy: General principles and a new implementation. In: *Review of Progress in Quantitative Nondestructive Evaluation*, Vol. 6B, ed. by D.O. Thompson, D.E. Chimenti (Plenum, New York 1987) pp. 1307–1314
11.150 Y.R. Shen: *The Principles of Nonlinear Optics* (Wiley, New York 1984)
11.151 T. Baumeister, S.L. Marks: *Standard Handbook for Mechanical Engineers*, 7th edn. (McGraw-Hill, New York 1967)
11.152 J. Colchero, O. Marti, H. Bielefeldt, J. Mlynek: Scanning force and friction microscopy, Phys. Stat. Sol. **131** (1991) 73–75
11.153 R. Young, J. Ward, F. Scire: Observation of metal-vacuum-metal tunneling, field emission, and the transition region, Phys. Rev. Lett. **27** (1971) 922
11.154 R. Young, J. Ward, F. Scire: The topographiner: An instrument for measuring surface microtopography, Rev. Sci. Instrum. **43** (1972) 999
11.155 C. Gerber, O. Marti: Magnetostrictive positioner, IBM Tech. Disclosure Bull. **27** (1985) 6373

11.156 R. Garcìa Cantù, M.A. Huerta Garnica: Long-scan imaging by STM, J. Vac. Sci. Technol. A **8** (1990) 354
11.157 C.J. Chen: In situ testing and calibration of tube piezoelectric scanners, Ultramicroscopy **42–44** (1992) 1653–1658
11.158 R.G. Carr: J. Microscopy **152** (1988) 379
11.159 C.J. Chen: Electromechanical deflections of piezoelectric tubes with quartered electrodes, Appl. Phys. Lett. **60** (1992) 132
11.160 N. Libioulle, A. Ronda, M. Taborelli, J.M. Gilles: Deformations and nonlinearity in scanning tunneling microscope images, J. Vac. Sci. Technol. B **9** (1991) 655–658
11.161 E.P. Stoll: Restoration of STM images distorted by time-dependent piezo driver aftereffects, Ultramicroscopy **42–44** (1991) 1585–1589
11.162 R. Durselen, U. Grunewald, W. Preuss: Calibration and applications of a high precision piezo scanner for nanometrology, Scanning **17** (1995) 91–96
11.163 J. Fu: In situ testing and calibrating of Z-piezo of an atomic force microscope, Rev. Sci. Instrum. **66** (1995) 3785–3788
11.164 R.C. Barrett, C.F. Quate: Optical scan-correction system applied to atomic force microscopy, Rev. Sci. Instrum. **62** (1991) 1393
11.165 R. Toledo-Crow, P.C. Yang, Y. Chen, M. Vaez-Iravani: Near-field differential scanning optical microscope with atomic force regulation, Appl. Phys. Lett. **60** (1992) 2957–2959
11.166 J.E. Griffith, G.L. Miller, C.A. Green: A scanning tunneling microscope with a capacitance-based position monitor, J. Vac. Sci. Technol. B **8** (1990) 2023–2027
11.167 A.E. Holman, C.D. Laman, P.M.L.O. Scholte, W.C. Heerens, F. Tuinstra: A calibrated scanning tunneling microscope equipped with capacitive sensors, Rev. Sci. Instrum. **67** (1996) 2274–2280

12. Probes in Scanning Microscopies

Scanning probe microscopy (SPM) provides nanometer-scale mapping of numerous sample properties in essentially any environment. This unique combination of high resolution and broad applicability has lead to the application of SPM to many areas of science and technology, especially those interested in the structure and properties of materials at the nanometer scale. SPM images are generated through measurements of a tip-sample interaction. A well-characterized tip is the key element to data interpretation and is typically the limiting factor.

Commercially available atomic force microscopy (AFM) tips, integrated with force sensing cantilevers, are microfabricated from silicon and silicon nitride by lithographic and anisotropic etching techniques. The performance of these tips can be characterized by imaging nanometer-scale standards of known dimension, and the resolution is found to roughly correspond to the tip radius of curvature, the tip aspect ratio, and the sample height. Although silicon and silicon nitride tips have a somewhat large radius of curvature, low aspect ratio, and limited lifetime due to wear, the widespread use of AFM today is due in large part to the broad availability of these tips. In some special cases, small asperities on the tip can provide resolution much higher than the tip radius of curvature for low-*Z* samples such as crystal surfaces and ordered protein arrays.

Several strategies have been developed to improve AFM tip performance. Oxide sharpening improves tip sharpness and enhances tip asperities. For high-aspect-ratio samples such as integrated circuits, silicon AFM tips can be modified by focused ion beam (FIB) milling. FIB tips reach three-degree cone angles over lengths of several microns and can be fabricated at arbitrary angles. Other high resolution and high-aspect-ratio tips are produced by electron beam deposition (EBD) in which a carbon spike is deposited onto the tip apex from the background gases in an electron microscope. Finally, carbon nanotubes have been employed as AFM tips. Their nanometer-scale diameter, long length, high stiffness, and elastic buckling properties make carbon nanotubes possibly the ultimate tip material for AFM. Nanotubes can be manually attached to silicon or silicon nitride AFM tips or "grown" onto tips by chemical vapor deposition (CVD), which should soon make them widely available. In scanning tunneling microscopy (STM), the electron tunneling signal decays exponentially with tip-sample separation, so that in principle only the last few atoms contribute to the signal. STM tips are, therefore, not as sensitive to the nanoscale tip geometry and can be made by simple mechanical cutting or electrochemical etching of metal wires. In choosing tip materials, one prefers hard, stiff metals that will not oxidize or corrode in the imaging environment.

12.1 **Atomic Force Microscopy** 372
12.1.1 Principles of Operation 372
12.1.2 Standard Probe Tips 373
12.1.3 Probe Tip Performance 374
12.1.4 Oxide-Sharpened Tips 375
12.1.5 FIB tips 376
12.1.6 EBD tips 376
12.1.7 Carbon Nanotube Tips 376

12.2 **Scanning Tunneling Microscopy** 382
12.2.1 Mechanically Cut STM Tips 382
12.2.2 Electrochemically Etched STM Tips 383

References 383

In scanning probe microscopy (SPM), an image is created by raster scanning a sharp probe tip over a sample and measuring some highly localized tip-sample interaction as a function of position. SPMs are based on several interactions, the major types including scanning tunneling microscopy (STM), which measures an electronic tunneling current; atomic force microscopy (AFM), which measures force interactions; and near-field scanning optical microscopy (NSOM), which measures local optical properties by exploiting near-field effects (Fig. 12.1). These methods allow the characterization of many properties (structural, mechanical, electronic, optical) on essentially any material (metals, semiconductors, insulators, biomolecules) and in essentially any environment (vacuum, liquid, or ambient air conditions). The unique combination of nanoscale resolution, previously the domain of electron microscopy, *and broad applicability* has led to the proliferation of SPM into virtually all areas of nanometer-scale science and technology.

Several enabling technologies have been developed for SPM, or borrowed from other techniques. Piezoelectric tube scanners allow accurate, sub-angstrom positioning of the tip or sample in three dimensions. Optical deflection systems and microfabricated cantilevers can detect forces in AFM down to the picoNewton range. Sensitive electronics can measure STM currents less than 1 picoamp. High transmission fiber optics and sensitive photodetectors can manipulate and detect small optical signals of NSOM. Environmental control has been developed to allow SPM imaging in UHV, cryogenic temperatures, at elevated temperatures, and in fluids. Vibration and drift have been controlled such that a probe tip can be held over a single molecule for hours of observation. Microfabrication techniques have been developed for the mass production of probe tips, making SPMs commercially available and allowing the development of many new SPM modes and combinations with other characterization methods. However, of all this SPM development over the past 20 years, what has received the least attention is perhaps the most important aspect: the probe tip.

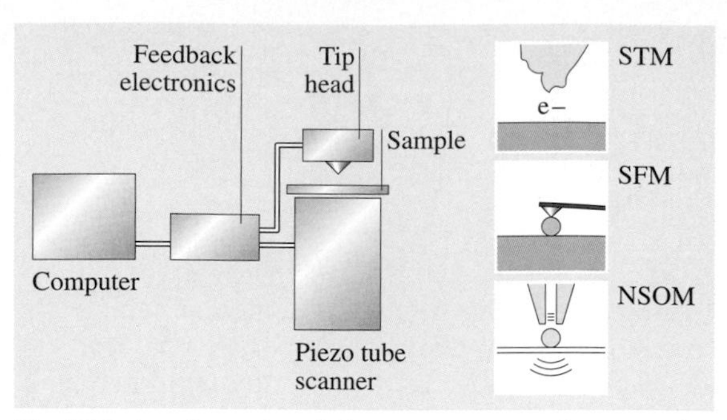

Fig. 12.1 A schematic of the components of a scanning probe microscope and the three types of signals observed: STM senses electron tunneling currents, AFM measures forces, and NSOM measures near-field optical properties via a sub-wavelength aperture

Interactions measured in SPMs occur at the tip-sample interface, which can range in size from a single atom to tens of nanometers. The size, shape, surface chemistry, electronic and mechanical properties of the tip apex will directly influence the data signal and the interpretation of the image. Clearly, the better characterized the tip the more useful the image information. In this chapter, the fabrication and performance of AFM and STM probes will be described.

12.1 Atomic Force Microscopy

AFM is the most widely used form of SPM, since it requires neither an electrically conductive sample, as in STM, nor an optically transparent sample or substrate, as in most NSOMs. Basic AFM modes measure the topography of a sample with the only requirement being that the sample is deposited on a flat surface and rigid enough to withstand imaging. Since AFM can measure a variety of forces, including van der Waals forces, electrostatic forces, magnetic forces, adhesion forces and friction forces, specialized modes of AFM can characterize the electrical, mechanical, and chemical properties of a sample in addition to its topography.

12.1.1 Principles of Operation

In AFM, a probe tip is integrated with a microfabricated force-sensing cantilever. A variety of silicon and silicon nitride cantilevers are commercially available with micron-scale dimensions, spring constants ranging from 0.01 to 100 N/m, and resonant frequencies ranging from 5 kHz to over 300 kHz. The cantilever deflection is detected by optical beam deflection, as illustrated in Fig. 12.2. A laser beam bounces off the back of the cantilever and is centered on a split photodiode. Cantilever deflections areproportional to the difference signal $V_A - V_B$. Sub-angstrom deflections can be deflected and, therefore, forces down to tens of picoNewtons can be measured. A more recently developed method of cantilever deflection measurement is through a piezoelectric layer on the cantilever that registers a voltage upon deflection [12.1].

A piezoelectric scanner rasters the sample under the tip while the forces are measured through deflections of the cantilever. To achieve more controlled imaging con-

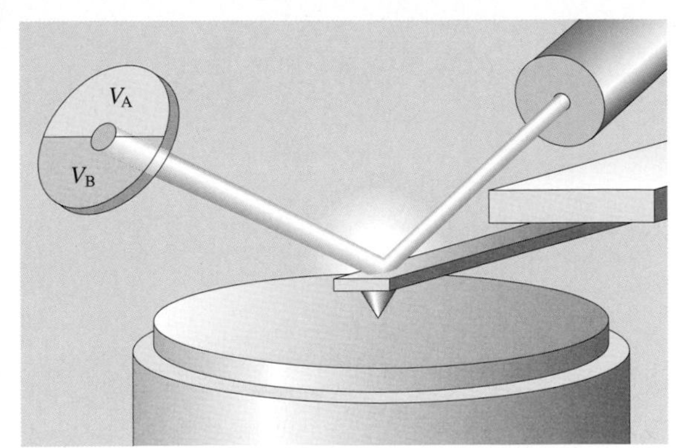

Fig. 12.2 An illustration of the optical beam deflection system that detects cantilever motion in the AFM. The voltage signal $V_A - V_B$ is proportional to the deflection

ditions, a feedback loop monitors the tip-sample force and adjusts the sample Z-position to hold the force constant. The topographic image of the sample is then taken from the sample Z-position data. The mode described is called contact mode, in which the tip is deflected by the sample due to repulsive forces, or "contact". It is generally only used for flat samples that can withstand lateral forces during scanning. To minimize lateral forces and sample damage, two AC modes have been developed. In these, the cantilever is driven into AC oscillation near its resonant frequency (tens to hundreds of kHz) with amplitudes of 5 to tens of s. When the tip approaches the sample, the oscillation is damped, and the reduced amplitude is the feedback signal, rather than the DC deflection. Again, topography is taken from the varying Z-position of the sample required to keep the tip oscillation amplitude constant. The two AC modes differ only in the nature of the interaction. In intermittent contact mode, also called tapping mode, the tip contacts the sample on each cycle, so the amplitude is reduced by ionic repulsion as in contact mode. In non-contact mode, long-range van der Waals forces reduce the amplitude by effectively shifting the spring constant experienced by the tip and changing its resonant frequency.

12.1.2 Standard Probe Tips

In early AFM work, cantilevers were made by hand from thin metal foils or small metal wires. Tips were created by gluing diamond fragments to the foil cantilevers or electrochemically etching the wires to a sharp point. Since these methods were labor intensive and not highly reproducible, they were not amenable to large-scale production. To address this problem, and the need for smaller cantilevers with higher resonant frequencies, batch fabrication techniques were developed (see Fig. 12.3). Building on existing methods to batch fabricate Si_3N_4 cantilevers, *Albrecht* et al. [12.2] etched an array of small square openings in an SiO_2 mask layer over a (100) silicon surface. The exposed square (100) regions were etched with KOH, an anisotropic etchant that terminates at the (111) planes, thus creating pyramidal etch pits in the silicon surface. The etch pit mask was then removed and another was applied to define the cantilever shapes with the pyramidal etch pits at the end. The Si wafer was then coated with a low stress Si_3N_4 layer by LPCVD. The Si_3N_4 fills the etch pit, using it as a mold to create a pyramidal tip. The silicon was later removed by etching to free the cantilevers and tips. Further steps resulting in the attachment of the cantilever to a macroscopic piece of glass are not described here. The resulting pyramidal tips were highly symmetric and had a tip radius of less than 30 nm, as determined by scanning electron microscopy (SEM). This procedure has likely not changed significantly, since commercially available Si_3N_4 tips are still specified to have a curvature radius of 30 nm.

Wolter et al. [12.3] developed methods to batch fabricate single-crystal Si cantilevers with integrated tips. Microfabricated Si cantilevers were first prepared using previously described methods, and a small mask was formed at the end of the cantilever. The Si around the mask was etched by KOH, so that the mask was under cut. This resulted in a pyramidal silicon tip under the mask, which was then removed. Again, this partial description of the full procedure only describes tip fabrication. With some refinements the silicon tips were made

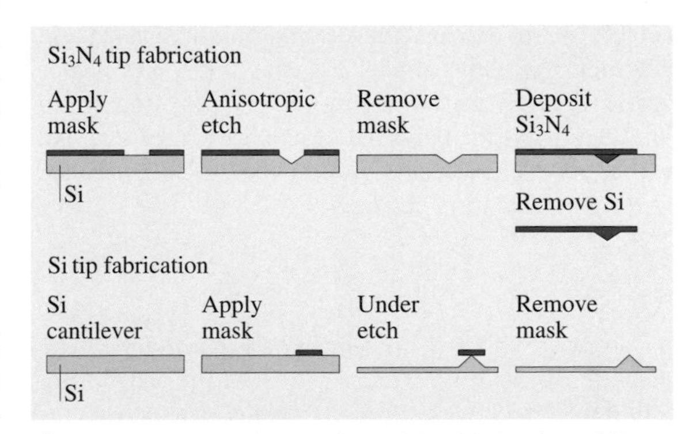

Fig. 12.3 A schematic overview of the fabrication of Si and Si_3N_4 tip fabrication as described in the text

in high yield with curvature radii of less than 10 nm. Si tips are sharper than Si_3N_4 tips, because they are directly formed by the anisotropic etch in single-crystal Si, rather than using an etch pit as a mask for deposited material. Commercially available silicon probes are made by similar refined techniques and provide a curvature typical radius of < 10 nm.

12.1.3 Probe Tip Performance

In atomic force microscopy the question of resolution can be a rather complicated issue. As an initial approximation, resolution is often considered strictly in geometrical terms that assume rigid tip-sample contact. The topographical image of a feature is broadened or narrowed by the size of the probe tip, so the resolution is approximately the width of the tip. Therefore, the resolution of AFM with standard commercially available tips is on the order of 5 to 10 nm. *Bustamante* and *Keller* [12.4] carried the geometrical model further by drawing an analogy to resolution in optical systems. Consider two sharp spikes separated by a distance d to be point objects imaged by AFM (see Fig. 12.4). Assume the tip has a parabolic shape with an end radius R. The tip-broadened image of these spikes will appear as inverted parabolas. There will be a small depression between the images of depth Δz. The two spikes are considered "resolved" if Δz is larger than the instrumental noise in the z direction. Defined in this manner, the resolution d, the minimum separation at which the spikes are resolved, is

$$d = 2\sqrt{2R(\Delta z)}\,, \quad (12.1)$$

where one must enter a minimal detectable depression for the instrument (Δz) to determine the resolution. So for a silicon tip with radius 5 nm and a minimum detectable Δz of 0.5 nm, the resolution is about 4.5 nm. However, the above model assumes the spikes are of equal height. *Bustamante* and *Keller* [12.4] went on to point out that if the height of the spikes is not equal, the resolution will be affected. Assuming a height difference of Δh, the resolution becomes:

$$d = \sqrt{2R}\left(\sqrt{\Delta z} + \sqrt{\Delta z + \Delta h}\right)\,. \quad (12.2)$$

For a pair of spikes with a 2 nm height difference, the resolution drops to 7.2 nm for a 5 nm tip and 0.5 nm minimum detectable Δz. While geometrical considerations are a good starting point for defining resolution, they ignore factors such as the possible compression and deformation of the tip and sample. *Vesenka* et al. [12.5] confirmed a similar geometrical resolution model by imaging monodisperse gold nanoparticles with tips characterized by transmission electron microscopy (TEM).

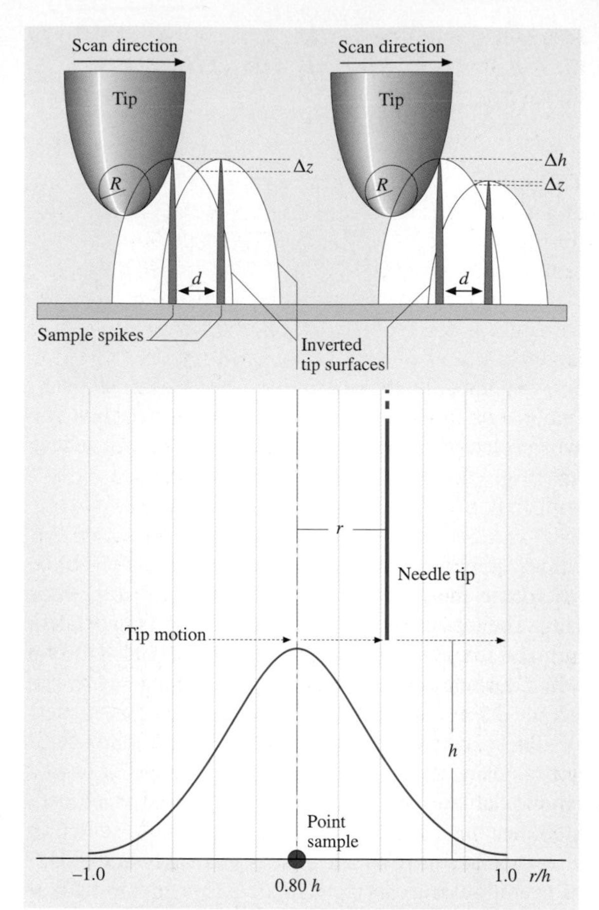

Fig. 12.4 The factors that determine AFM imaging resolution in contact mode (*top*) and noncontact mode (*bottom*), adapted from [12.4]

Noncontact AFM contrast is generated by long-range interactions such as van der Waals forces, so resolution will not simply be determined by geometry because the tip and sample are not in rigid contact. *Bustamante* and *Keller* [12.4] have derived an expression for the resolution in noncontact AFM for an idealized, infinitely thin "line" tip and a point particle as the sample (Fig. 12.4). Noncontact AFM is sensitive to the gradient of long-range forces, so the van der Waals force gradient was calculated as a function of position for the tip at height h above the surface. If the resolution d is defined as the full width at half maximum of this curve, the resolution is:

$$d = 0.8h\,. \qquad (12.3)$$

This shows that even for an ideal geometry, the resolution is fundamentally limited in noncontact mode by the tip-sample separation. Under UHV conditions, the tip-sample separation can be made very small, so atomic resolution is possible on flat, crystalline surfaces. Under ambient conditions, however, the separation must be larger to keep the tip from being trapped in the ambient water layer on the surface. This larger separation can lead to a point where further improvements in tip sharpness do not improve resolution. It has been found that imaging 5 nm gold nanoparticles in noncontact mode with carbon nanotube tips of 2 nm diameter leads to particle widths of 12 nm, larger than the 7 nm width one would expect assuming rigid contact [12.8]. However, in tapping mode operation, the geometrical definition of resolution is relevant, since the tip and sample come into rigid contact. When imaging 5 nm gold particles with 2 nm carbon nanotube tips in tapping mode, the expected 7 nm particle width is obtained [12.9].

The above descriptions of AFM resolution cannot explain the sub-nanometer resolution achieved on crystal surfaces [12.10] and ordered arrays of biomolecules [12.11] in contact mode with commercially available probe tips. Such tips have nominal radii of curvature ranging from 5 nm to 30 nm, an order of magnitude larger than the resolution achieved. A detailed model to explain the high resolution on ordered membrane proteins has been put forth by [12.6]. In this model, the larger part of the silicon nitride tip apex balances the tip-sample interaction through electrostatic forces, while a very small tip asperity interacts with the sample to provide contrast (see Fig. 12.5). This model is supported by measurements at varying salt concentrations to vary the electrostatic interaction strength and the observation of defects in the ordered samples. However, the existence of such asperities has never been confirmed by independent electron microscopy images of the tip. Another model, considered especially applicable to atomic resolution on crystal surfaces, assumes the tip is in contact with a region of the sample much larger than the resolution observed, and that force components matching the periodicity of the sample are transmitted to the tip, resulting in an "averaged" image of the periodic lattice. Regardless of the mechanism, the structures determined are accurate and make this a highly valuable method for membrane proteins. However, this level of resolution should not be expected for most biological systems.

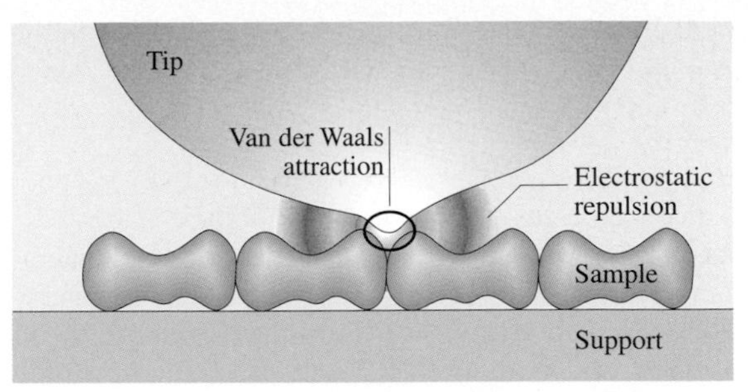

Fig. 12.5 A tip model to explain the high resolution obtained on ordered samples in contact mode, from [12.6]

12.1.4 Oxide-Sharpened Tips

Both Si and Si_3N_4 tips with increased aspect ratio and reduced tip radius can be fabricated through oxide sharpening of the tip. If a pyramidal or cone-shaped silicon tip is thermally oxidized to SiO_2 at low temperature ($< 1{,}050\,°C$), Si-SiO_2 stress formation reduces the oxidation rate at regions of high curvature. The result is a sharper, higher-aspect-ratio cone of silicon at the high curvature tip apex inside the outer pyramidal layer of SiO_2 (see Fig. 12.6). Etching the SiO_2 layer with HF then leaves tips with aspect ratios up to 10:1 and radii down to 1 nm [12.7], although 5–10 nm is the nominal specification for most commercially available tips. This oxide sharpening technique can also be applied to Si_3N_4 tips by oxidizing the silicon etch pits that are used as molds. As with tip fabrication, oxide sharpening is not quite as effective for Si_3N_4. Si_3N_4 tips were reported to have an

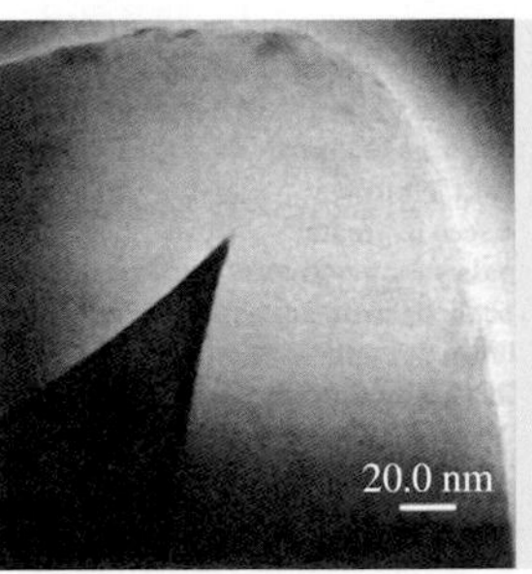

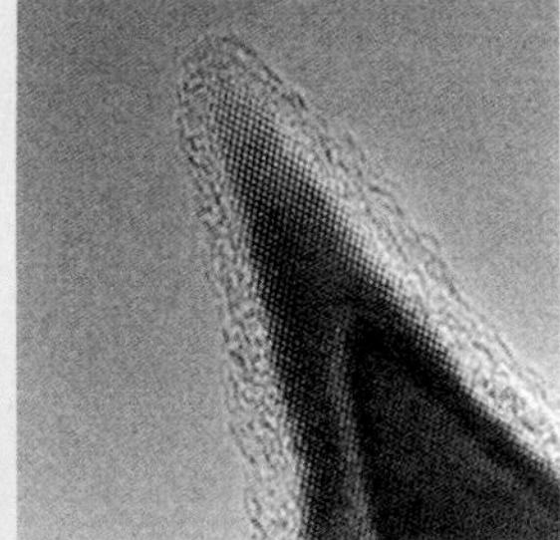

Fig. 12.6 Oxide sharpening of silicon tips. The *left image* shows a sharpened core of silicon in an outer layer of SiO_2. The *right image* is a higher magnification view of such a tip after the SiO_2 is removed. Adapted from [12.7]

11 nm radius of curvature [12.12], while commercially available oxide-sharpened Si_3N_4 tips have a nominal radius of < 20 nm.

12.1.5 FIB tips

A common AFM application in integrated circuit manufacture and MEMs is to image structures with very steep sidewalls such as trenches. To accurately image these features, one must consider the micron-scale tip structure, rather than the nanometer-scale structure of the tip apex. Since tip fabrication processes rely on anisotropic etchants, the cone half-angles of pyramidal tips are approximately 20 degrees. Images of deep trenches taken with such tips display slanted sidewalls and may not reach the bottom of the trench due to the tip broadening effects. To image such samples more faithfully, high-aspect-ratio tips are fabricated by focused ion beam (FIB) machining a Si tip to produce a sharp spike at the tip apex. Commercially available FIB tips have half cone angles of < 3 degrees over lengths of several microns, yielding aspect ratios of approximately 10:1. The radius of curvature at the tip end is similar to that of the tip before the FIB machining. Another consideration for high-aspect-ratio tips is the tip tilt. To ensure that the pyramidal tip is the lowest part of the tip-cantilever assembly, most AFM designs tilt the cantilever about 15 degrees from parallel. Therefore, even an ideal "line tip" will not give an accurate image of high steep sidewalls, but will produce an image that depends on the scan angle. Due to the versatility of the FIB machining, tips are available with the spikes at an angle to compensate for this effect.

12.1.6 EBD tips

Another method of producing high-aspect-ratio tips for AFM is called electron beam deposition (EBD). First developed for STM tips [12.13, 14], EBD tips were introduced for AFM by focusing an SEM onto the apex of a pyramidal tip arranged so that it pointed along the electron beam axis (see Fig. 12.7). Carbon material was deposited by the dissociation of background gases in the SEM vacuum chamber. *Schiffmann* [12.15] systematically studied the following parameters and how they affected EBD tip geometry:

Deposition time :	0.5 to 8 min
Beam current :	3–300 pA
Beam energy :	1–30 keV
Working distance :	8–48 mm .

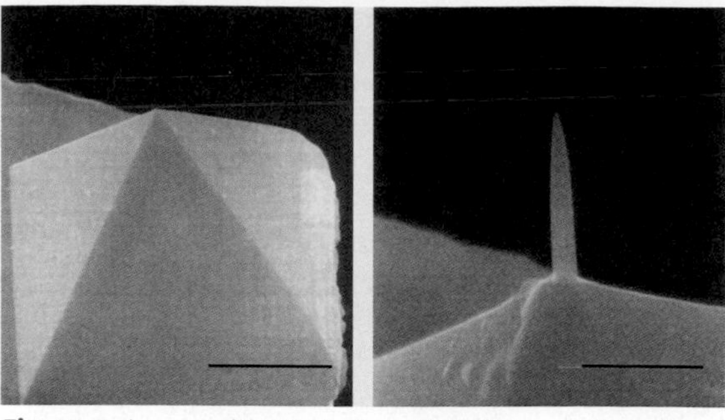

Fig. 12.7 A pyramidal tip before (*left*, 2-μm-scale bar) and after (*right*, 1-μm-scale bar) electron beam deposition, adapted from [12.13]

EBD tips were cylindrical with end radii of 20–40 nm, lengths of 1 to 5 μm, and diameters of 100 to 200 nm. Like FIB tips, EBD tips were found to achieve improved imaging of steep features. By controlling the position of the focused beam, the tip geometry can be further controlled. Tips were fabricated with lengths over 5 μm and aspect ratios greater than 100:1, yet these were too fragile to use as a tip in AFM [12.13].

12.1.7 Carbon Nanotube Tips

Carbon nanotubes are microscopic graphitic cylinders that are nanometers in diameter, yet many microns in length. Single-walled carbon nanotubes (SWNT) consist of single sp^2 hybridized carbon sheets rolled into seamless tubes and have diameters ranging from 0.7 to 3 nm.

Carbon Nanotube Structure

Larger structures called multiwalled carbon nanotubes (MWNT) consist of nested, concentrically arranged SWNT and have diameters ranging from 3 to 50 nm. Figure 12.8 shows a model of nanotube structure, as well as TEM images of a SWNT and a MWNT. The small diameter and high aspect ratio of carbon nanotubes suggests their application as high resolution, high-aspect-ratio AFM probes.

Carbon Nanotube Mechanical Properties

Carbon nanotubes possess exceptional mechanical properties that impact their use as probes. Their lateral stiffness can be approximated from that of a solid elastic rod:

$$k_{\text{lat}} = \frac{3\pi Y r^4}{4l^3} , \qquad (12.4)$$

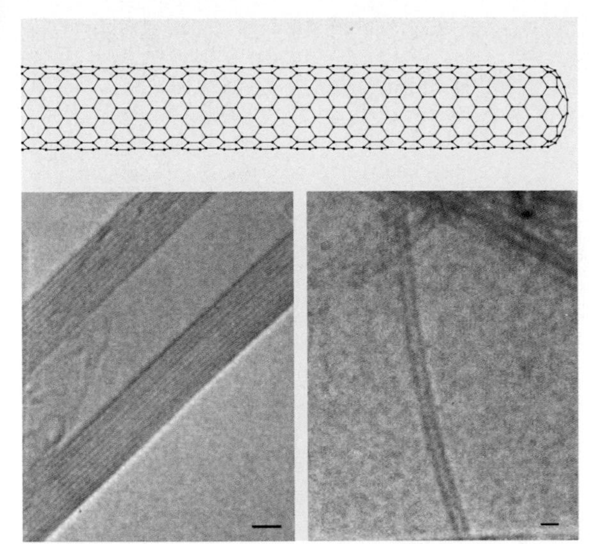

Fig. 12.8 The structure of carbon nanotubes, including TEM images of a MWNT (*left*) and a SWNT (*right*), from [12.16]

where the spring constant k_{lat} represents the restoring force per unit lateral displacement, r is the radius, l is the length, and Y is the Young's modulus (also called the elastic modulus) of the material. For the small diameters and extreme aspect ratios of carbon nanotube tips, the thermal vibrations of the probe tip at room temperature can become sufficient to degrade image resolution. These thermal vibrations can be approximated by equating $\frac{1}{2}k_B T$ of thermal energy to the energy of an oscillating nanotube:

$$\frac{1}{2}k_B T = \frac{1}{2}k_{lat}a^2 , \tag{12.5}$$

where k_B is Boltzmann's constant, T is the temperature, and a is the vibration amplitude. Substituting for k_{lat} from (12.4) yields:

$$a = \sqrt{\frac{4k_B T l^3}{3\pi Y r^4}} . \tag{12.6}$$

The strong dependence on radius and length reveals that one must carefully control the tip geometry at this size scale. Equation (12.6) implies that the stiffer the material, i. e., the higher its Young's modulus, the smaller the thermal vibrations and the longer and thinner a tip can be. The Young's moduli of carbon nanotubes have been determined by measurements of the thermal vibration amplitude by TEM [12.18, 19] and by directly measuring the forces required to deflect a pinned carbon nanotube in an AFM [12.20]. These experiments revealed that the Young's modulus of carbon nanotubes is 1–2 TPa, in agreement with theoretical predictions [12.21]. This makes carbon nanotubes the stiffest known material and, therefore, the best for fabricating thin, high-aspect-ratio tips. A more detailed and accurate derivation of the thermal vibration amplitudes was derived for the Young's modulus measurements [12.18, 19].

Carbon nanotubes elastically buckle under large loads, rather than fracture or plastically deform like most materials. Nanotubes were first observed in the buckled state by transmission electron microscopy [12.17], as shown in Fig. 12.9. The first experimental evidence that nanotube buckling is elastic came from the application of nanotubes as probe tips [12.22], described in detail below. A more direct experimental observation of elastic buckling was obtained by deflecting nanotubes pinned to a low friction surface with an AFM tip [12.20]. Both reports found that the buckling force could be approximated with the macroscopic Euler buckling formula for an elastic column:

$$F_{Euler} = \frac{\pi^3 Y r^4}{4l^2} . \tag{12.7}$$

The buckling force puts another constraint on the tip length: If the nanotube is too long the buckling force

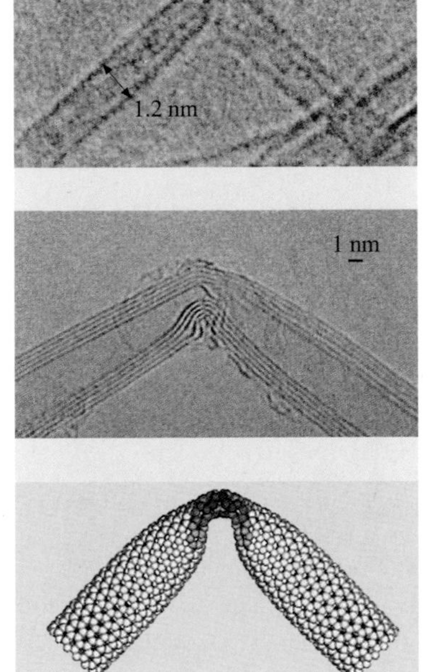

Fig. 12.9 TEM images and a model of a buckled nanotube, adapted from [12.17]

will be too low for stable imaging. The elastic buckling property of carbon nanotubes has significant implications for their use as AFM probes. If a large force is applied to the tip inadvertently, or if the tip encounters a large step in sample height, the nanotube can buckle to the side, then snap back without degraded imaging resolution when the force is removed, making these tips highly robust. No other tip material displays this buckling characteristic.

Manually Assembled Nanotube Probes

The first carbon nanotube AFM probes [12.22] were fabricated by techniques developed for assembling single-nanotube field emission tips [12.23]. This process, illustrated in Fig. 12.10, used purified MWNT material synthesized by the carbon arc procedure. The raw material, which must contain at least a few percent of long nanotubes ($> 10\,\mu$m) by weight, purified by oxidation to approximately 1% of its original mass. A torn edge of the purified material was attached to a micromanipulator by carbon tape and viewed under a high power optical microscope. Individual nanotubes and nanotube bundles were visible as filaments under dark field illumination. A commercially available AFM tip was attached to another micromanipulator opposing the nanotube material. Glue was applied to the tip apex from high vacuum carbon tape supporting the nanotube material. Nanotubes were then manually attached to the tip apex by micromanipulation. As assembled, MWNT tips were often too long for imaging due to thermal vibrations and low buckling forces described in Sect. 12.1.7. Nanotubes tips were shortened by applying 10 V pulses to the tip while it was near a sputtered niobium surface. This process etched ~ 100 nm lengths of nanotube per pulse.

The manually assembled MWNT tips demonstrated several important nanotube tip properties [12.22]. First, the high aspect ratio of the MWNT tips allowed the accurate imaging of trenches in silicon with steep sidewalls, similar to FIB and EBD tips. Second, elastic buckling was observed indirectly through force curves (see Fig. 12.11). Note that as the tip taps the sample, the amplitude drops to zero and a DC deflection is observed, because the nanotube is unbuckled and is essentially rigid. As the tip moves closer, the force on the nanotube eventually exceeds the buckling force. The nanotube buckles, allowing the vibration amplitude to partially recover, and the deflection remains constant. Numeric tip trajectory simulations could only reproduce these force curves if elastic buckling was included in the nanotube response. Finally, the nanotube tips were highly robust. Even after "tip crashes" or hundreds of controlled buckling cycles, the tip retained its resolution and high aspect ratio.

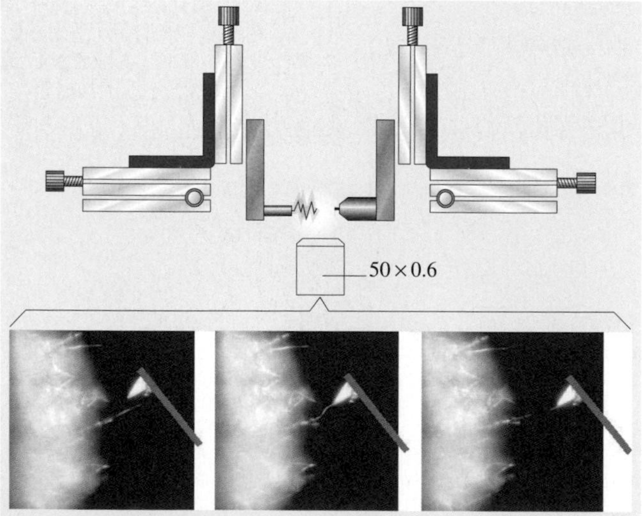

Fig. 12.10 A schematic drawing of the setup for manual assembly of carbon nanotube tips (*top*) and optical microscopy images of the assembly process (the cantilever was drawn in for clarity)

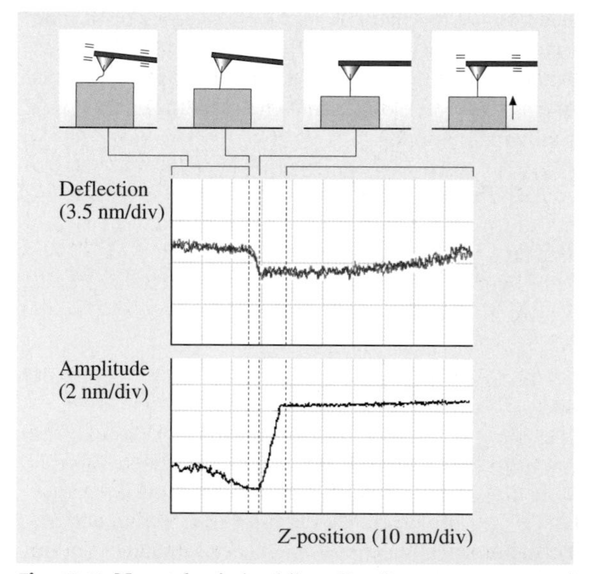

Fig. 12.11 Nanotube tip buckling. *Top* diagrams correspond to labeled regions of the force curves. As the nanotube tip buckles, the deflection remains constant and the amplitude increases, from [12.16]

Manual assembly of carbon nanotube probe tips is straightforward, but has several limitations. It is labor intensive and not amenable to mass production. Although MWNT tips have been made commercially available by this method, they are about ten times more expensive than silicon probes. The manual assembly method has also been carried out in an SEM, rather than an optical microscope [12.24]. This eliminates the need for pulse-etching, since short nanotubes can be attached to the tip, and the "glue" can be applied by EBD. But this is still not the key to mass production, since nanotube tips are made individually. MWNT tips provided a modest improvement in resolution on biological samples, but typical MWNT radii are similar to that of silicon tips, so they cannot provide the ultimate resolution possible with a SWNT tip. SWNT bundles can be attached to silicon probes by manual assembly. Pulse etching at times produces very high resolution tips that likely result from exposing a small number of nanotubes from the bundle, but this is not reproducible [12.25]. Even if a sample could be prepared that consisted of individual SWNT for manual assembly, such nanotubes would not be easily visible by optical microscopy or SEM.

CVD Nanotube Probe Synthesis

The problems of manual assembly of nanotube probes discussed above can largely be solved by directly growing nanotubes onto AFM tips by metal-catalyzed chemical vapor deposition (CVD). The key features of the nanotube CVD process are illustrated in Fig. 12.12. Nanometer-scale metal catalyst particles are heated in a gas mixture containing hydrocarbon or CO. The gas molecules dissociate on the metal surface, and carbon is adsorbed into the catalyst particle. When this carbon precipitates, it nucleates a nanotube of similar diameter to the catalyst particle. Therefore, CVD allows control over nanotube size and structure, including the production of SWNTs [12.26] with radii as low as 3.5 Angstrom [12.27].

Several key issues must be addressed to grow nanotube AFM tips by CVD: (1) the alignment of the nanotubes at the tip, (2) the number of nanotubes that grow at the tip, and (3) the length of the nanotube tip. *Li* et al. [12.28] found that nanotubes grow perpendicular to a porous surface containing embedded catalyst. This approach was exploited to fabricate nanotube tips by CVD [12.29] with the proper alignment, as illustrated in Fig. 12.13. A flattened area of approximately 1–5 μm^2 was created on Si tips by scanning in contact mode at high load (1 μN) on a hard, synthetic diamond surface. The tip was then anodized in HF to create 100 nm-diameter pores in this flat surface [12.30]. It is important to only anodize the last 20–40 μm of the cantilever, which includes the tip, so that the rest

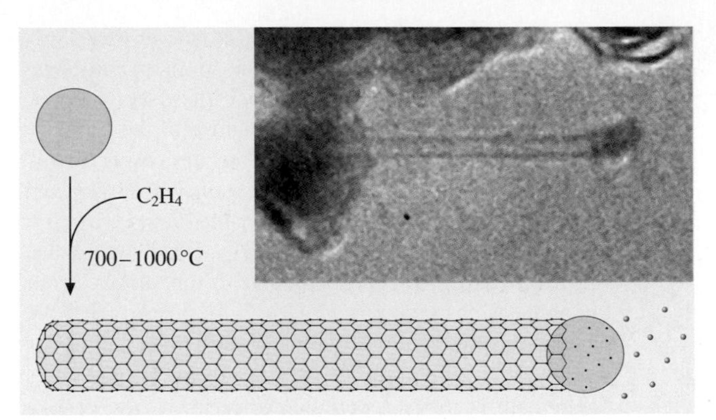

Fig. 12.12 CVD nanotube synthesis. Ethylene reacts with a nanometer-scale iron catalyst particle at high temperature to form a carbon nanotube. The *inset* in the upper right is a TEM image showing a catalyst particle at the end of a nanotube, from [12.16]

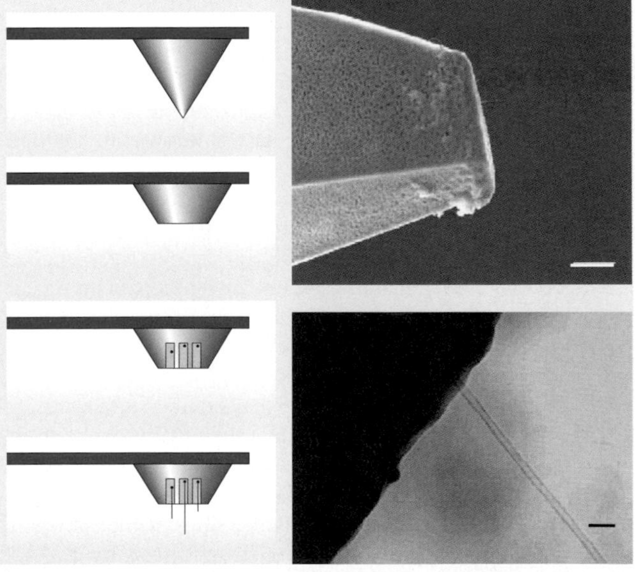

Fig. 12.13 Pore-growth CVD nanotube tip fabrication. The *left panel*, from top to bottom, shows the steps described in the text. The *upper right* is an SEM image of such a tip with a small nanotube protruding from the pores (scale bar is 1 μm). The *lower right* is a TEM of a nanotube protruding from the pores (scale bar is 20 nm), from [12.16]

of the cantilever is still reflective for use in the AFM. This was achieved by anodizing the tip in a small drop of HF under the view of an optical microscope. Next, iron was electrochemically deposited into the pores to form catalyst particles [12.31]. Tips prepared in this way were heated in low concentrations of ethylene at 800 °C, which is known to favor the growth of thin nanotubes [12.26]. When imaged by SEM, nanotubes were found to grow perpendicular to the surface from the pores as desired (Fig. 12.13). TEM revealed that the nanotubes were thin, individual, multiwalled nanotubes with typical radii ranging from 3–5 nm. If nanotubes did not grow in an acceptable orientation, the carbon could be removed by oxidation, and then CVD repeated to grow new nanotube tips.

These "pore-growth" CVD nanotube tips were typically several microns in length – too long for imaging – and were pulse-etched to a usable length of < 500 nm. The tips exhibited elastic buckling behavior and were very robust in imaging. In addition, the thin, individual nanotube tips enabled improved resolution [12.29] on isolated proteins. The pore-growth method demonstrated the potential of CVD to simplify the fabrication of nanotube tips, although there were still limitations. In particular, the porous layer was difficult to prepare and rather fragile.

An alternative approach for CVD fabrication of nanotube tips involves direct growth of SWNTs on the surface of a pyramidal AFM tip [12.32, 33]. In this "surface-growth" approach, an alumina/iron/molybdenum-powdered catalyst known to produce SWNT [12.26] was dispersed in ethanol at 1 mg/mL. Silicon tips were dipped in this solution and allowed to dry, leaving a sparse layer of ~ 100 nm catalyst clusters on the tip. When CVD conditions were applied, single-walled nanotubes grew along the silicon tip surface. At a pyramid edge, nanotubes can either bend to align with the edge, or protrude from the surface. If the energy required to bend the tube and follow the edge is less than the attractive nanotube-surface energy, then the nanotube will follow the pyramid edge to the apex. Therefore, nanotubes were effectively steered toward the tip apex by the pyramid edges. At the apex, the nanotube protruded from the tip, since the energetic cost of bending around the sharp silicon tip was too high. The high aspect ratio at the oxide-sharpened silicon tip apex was critical for good nanotube alignment. A schematic of this approach is shown in Fig. 12.14. Evidence for this model came from SEM investigations that show that a very high yield of tips contains nanotubes only at the apex, with very few protruding elsewhere

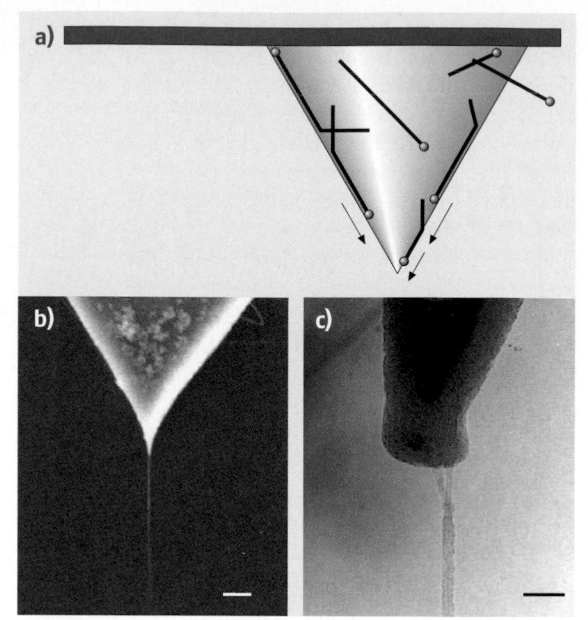

Fig. 12.14a–c Surface-growth nanotube tip fabrication. (a) Schematic represents the surface growth process in which nanotubes growing on the pyramidal tip are guided to the tip apex. (b),(c) Images show (b) SEM (200-nm-scale bar) and (c) TEM (20-nm-scale bar) images of a surface growth tip, from [12.16]

from the pyramid. TEM analysis demonstrated that the tips typically consist of small SWNT bundles that are formed by nanotubes coming together from different edges of the pyramid to join at the apex, supporting the surface growth model described above (Fig. 12.14). The "surface growth" nanotube tips exhibit a high aspect ratio and high resolution imaging, as well as elastic buckling.

The surface growth method has been expanded to include wafer-scale production of nanotube tips with yields of over 90% [12.34], yet one obstacle remains to the mass production of nanotube probe tips. Nanotubes protruding from the tip are several microns long, and since they are so thin, they must be etched to less than 100 nm. While the pulse-etching step is fairly reproducible, it must be carried out on nanotube tips in a serial fashion, so surface growth does not yet represent a true method of batch nanotube tip fabrication.

Hybrid Nanotube Tip Fabrication: Pick-up Tips

Another method of creating nanotube tips is something of a hybrid between assembly and CVD. The motivation was to create AFM probes that have an *individual* SWNT at the tip to achieve the ultimate imaging resolution. In order to synthesize isolated SWNT, they must be nucleated at sites separated farther than their typical length. The alumina-supported catalyst contains a high density of catalyst particles per 100 nm cluster, so nanotube bundles cannot be avoided. To fabricate completely isolated nanotubes, isolated catalyst particles were formed by dipping a silicon wafer in an isopropyl alcohol solution of $Fe(NO_3)_3$. This effectively left a submonolayer of iron on the wafer, so that when it was heated in a CVD furnace, the iron became mobile and aggregated to form isolated iron particles. During CVD conditions, these particles nucleated and grew SWNTs. By controlling the reaction time, the SWNT lengths were kept shorter than their typical separation, so that the nanotubes never had a chance to form bundles. AFM analysis of these samples revealed 1–3 nm-diameter SWNT and un-nucleated particles on the surface (Fig. 12.15). However, there were tall objects that were difficult to image at a density of about 1 per 50 μm^2. SEM analysis at an oblique angle demonstrated that these were SWNTs that had grown perpendicular to the surface (Fig. 12.15).

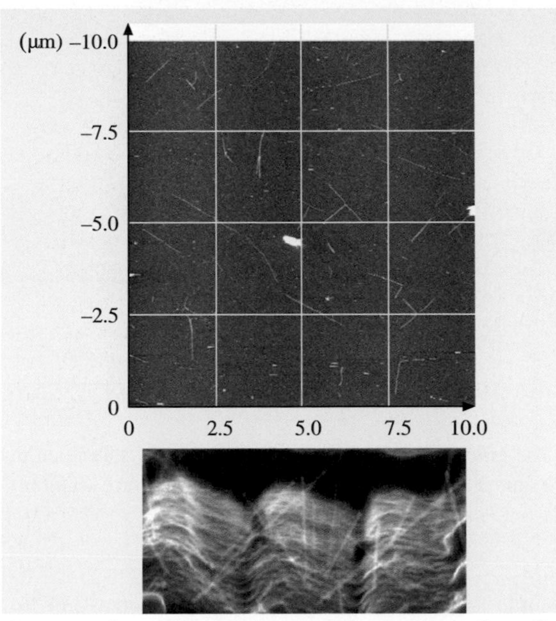

Fig. 12.15 Atomic force microscopy image (*top*) of a wafer with isolated nanotubes synthesized by CVD. The SEM view provides evidence that some of these nanotubes are arranged vertically

In the "pick-up tip" method, these isolated SWNT substrates were imaged by AFM with silicon tips in air [12.9]. When the tip encountered a vertical SWNT, the oscillation amplitude was damped, so the AFM pulled the sample away from the tip. This pulled the SWNT into contact with the tip along its length, so that it became attached to the tip. This assembly process happened automatically when imaging in tapping mode – no special tip manipulation was required. When imaging a wafer with the density shown in Fig. 12.15, one nanotube was attached per 8 μm × 8 μm scan at 512×512 and 2 Hz, so a nanotube tip could be made in about 5 min. Since the as-formed SWNT tip continued to image, there was usually no evidence in the topographic image that a nanotube had been attached. However, the pick-up event was identified when the Z-voltage suddenly stepped to larger tip-

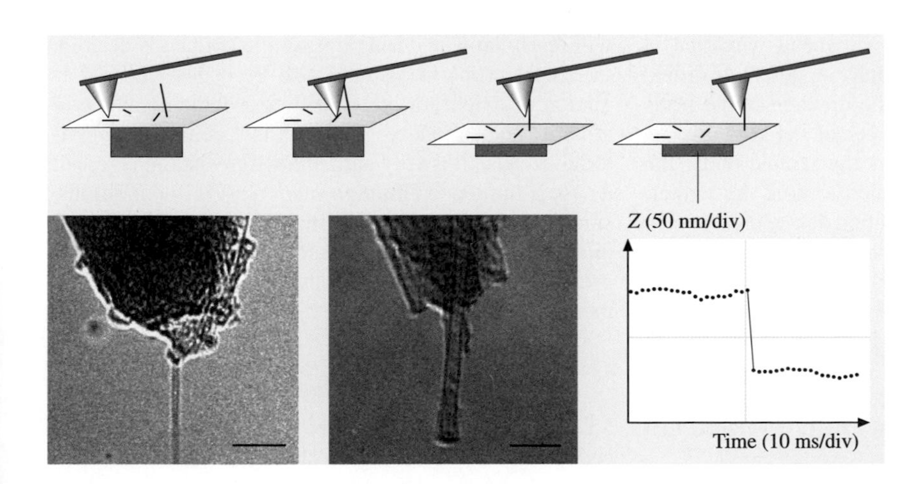

Fig. 12.16 Pick-up tip assembly of nanotube probes. The *top* illustrates the nanotube pick-up process that occurs while imaging vertical nanotubes in an AFM, including a trace of the Z-position during a pick-up event. The *lower left* TEM images show single nanotubes (diameters 0.9 nm and 2.8 nm) on an AFM tip fabricated by this method, adapted from [12.9]

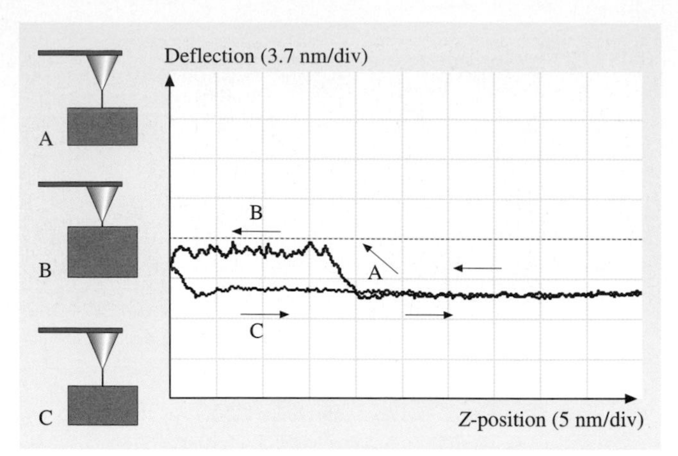

Fig. 12.17 The process by which nanotube tips can be shortened in AFM force curves. The hysteresis in the deflection trace (*bottom*) reveals that ∼ 17 nm were removed

sample separation due to the effective increase in tip length.

Individual SWNT tips must be quite short, typically less than 50 nm in length, for reasons outlined above. Pulse-etching, which removes 50–100 nm of nanotube length at a time, lacked the necessary precision for shortening pick-up tips, so the tips were shortened through force curves. A pick-up tip force curve is shown in Fig. 12.17. When the tip first interacted with the sample, the amplitude decreased to zero, and further approach generated a small deflection that ultimately saturated. However, this saturation was not due to buckling. Note that the amplitude did not recover, as in the buckling curves. This leveling was due to the nanotube sliding on the pyramidal tip, which was confirmed by the hysteresis in the amplitude and deflection curves. If the force curve was repeated, the tip showed no deflection or amplitude drop until further down, because the tip was essentially shorter.

Pick-up SWNT tips achieve the highest resolution of all nanotube tips, since they are always individuals rather than bundles. In tapping mode, they produce images of 5 nm gold particles that have a *full width* of ∼ 7 nm, the expected geometrical resolution for a 2 nm cylindrical probe [12.9]. Although the pick-up method is serial in nature, it may still be the key to the mass production of nanotube tips. Note that the original nanotube tip length can be measured electronically from the size of the *Z*-piezo step. The shortening can be electronically controlled through the hysteresis in the force curves. Therefore, the entire procedure (including tip exchange) can be automated by computer.

12.2 Scanning Tunneling Microscopy

Scanning tunneling microscopy (STM) was the original scanning probe microscopy and generally produces the highest resolution images, routinely achieving atomic resolution on flat, conductive surfaces. In STM, the probe tip consists of a sharpened metal wire that is held 0.3 to 1 nm from the sample. A potential difference of 0.1 V to 1 V between the tip and sample leads to tunneling currents on the order of 0.1 to 1 nA. As in AFM, a piezo-scanner rasters the sample under the tip, and the *Z*-position is adjusted to hold the tunneling current constant. The *Z*-position data represents the "topography", or in this case the surface, of constant electron density. As with other SPMs, the tip properties and performance greatly depend on the experiment being carried out. Although it is nearly impossible to prepare a tip with a known atomic structure, a number of factors are known to affect tip performance, and several preparation methods have been developed that produce good tips.

The nature of the sample being investigated and the scanning environment will affect the choice of the tip material and how the tip is fabricated. Factors to consider are mechanical properties – a hard material that will resist damage during tip-sample contact is desired. Chemical properties should also be considered – formation of oxides, or other insulating contaminants will affect tip performance. Tungsten is a common tip material because it is very hard and will resist damage, but its use is limited to ultrahigh vacuum (UHV) conditions, since it readily oxidizes. For imaging under ambient conditions an inert tip material such as platinum or gold is preferred. Platinum is typically alloyed with iridium to increase its stiffness.

12.2.1 Mechanically Cut STM Tips

STM tips can be fabricated by simple mechanical procedures such as grinding or cutting metal wires. Such tips are not formed with highly reproducible shapes and have a large opening angle and a large radius of curvature in the range of 0.1 to 1 μm (see Fig. 12.18a). They are not useful for imaging samples with surface roughness above a few nanometers. However, on atomically flat samples, mechanically cut tips can achieve atomic resolution due to the nature of the tunneling signal, which drops exponentially with tip-sample separation. Since mechanically cut tips contain many small asperities on the larger tip structure, atomic resolution is easily achieved as long as one atom of the tip is just a few angstroms lower than all of the others.

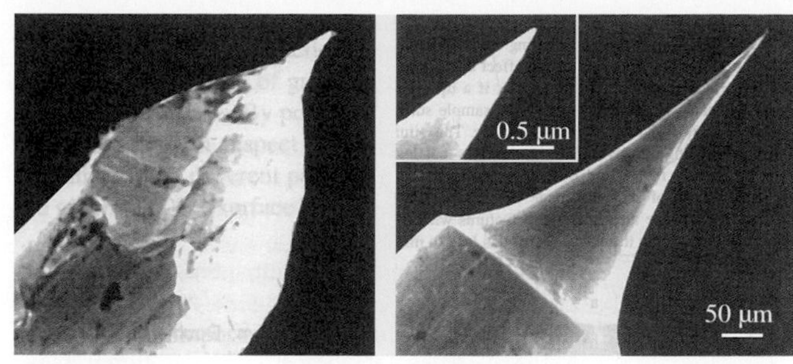

Fig. 12.18 A mechanically cut STM tip (*left*) and an electrochemically etched STM tip (*right*), from [12.35]

12.2.2 Electrochemically Etched STM Tips

For samples with more than a few nanometers of surface roughness, the tip structure in the nanometer-size range becomes an issue. Electrochemical etching can provide tips with reproducible and desirable shapes and sizes (Fig. 12.18), although the exact atomic structure of the tip apex is still not well controlled. The parameters of electrochemical etching depend greatly on the tip material and the desired tip shape. The following is an entirely general description. A fine metal wire (0.1–1 mm diameter) of the tip material is immersed in an appropriate electrochemical etchant solution. A voltage bias of 1–10 V is applied between the tip and a counterelectrode such that the tip is etched. Due to the enhanced etch rate at the electrolyte-air interface, a neck is formed in the wire. This neck is eventually etched thin enough so that it cannot support the weight of the part of the wire suspended in the solution, and it breaks to form a sharp tip. The widely varying parameters and methods will be not be covered in detail here, but many recipes are found in the literature for common tip materials [12.36–39].

References

12.1 R. Linnemann, T. Gotszalk, I. W. Rangelow, P. Dumania, E. Oesterschulze: Atomic force microscopy and lateral force microscopy using piezoresistive cantilevers, J. Vac. Sci. Technol. B **14**(2) (1996) 856–860

12.2 T. R. Albrecht, S. Akamine, T. E. Carver, C. F. Quate: Microfabrication of cantilever styli for the atomic force microscope, J. Vac. Sci. Technol. A **8**(4) (1990) 3386–3396

12.3 O. Wolter, T. Bayer, J. Greschner: Micromachined silicon sensors for scanning force microscopy, J. Vac. Sci. Technol. B **9**(2) (1991) 1353–1357

12.4 C. Bustamante, D. Keller: Scanning force microscopy in biology, Phys. Today **48**(12) (1995) 32–38

12.5 J. Vesenka, S. Manne, R. Giberson, T. Marsh, E. Henderson: Colloidal gold particles as an incompressible atomic force microscope imaging standard for assessing the compressibility of biomolecules, Biophys. J. **65** (1993) 992–997

12.6 D. J. Muller, D. Fotiadis, S. Scheuring, S. A. Muller, A. Engel: Electrostatically balanced subnanometer imaging of biological specimens by atomic force microscope, Biophys. J. **76**(2) (1999) 1101–1111

12.7 R. B. Marcus, T. S. Ravi, T. Gmitter, K. Chin, D. Liu, W. J. Orvis, D. R. Ciarlo, C. E. Hunt, J. Trujillo: Formation of silicon tips with <1 nm radius, Appl. Phys. Lett. **56**(3) (1990) 236–238

12.8 J. H. Hafner, C. L. Cheung, C. M. Lieber: unpublished results (2001)

12.9 J. H. Hafner, C. L. Cheung, T. H. Oosterkamp, C. M. Lieber: High-yield assembly of individual single-walled carbon nanotube tips for scanning probe microscopies, J. Phys. Chem. B **105**(4) (2001) 743–746

12.10 F. Ohnesorge, G. Binnig: True atomic resolution by atomic force microscopy through repulsive and attractive forces, Science **260** (1993) 1451–1456

12.11 D. J. Muller, D. Fotiadis, A. Engel: Mapping flexible protein domains at subnanometer resolution with the atomic force microscope, FEBS Lett. **430**(1–2 Special Issue SI) (1998) 105–111

12.12 S. Akamine, R. C. Barrett, C. F. Quate: Improved atomic force microscope images using microcantilevers with sharp tips, Appl. Phys. Lett. **57**(3) (1990) 316–318

12.13 D. J. Keller, C. Chih-Chung: Imaging steep, high structures by scanning force microscopy with electron beam deposited tips, Surf. Sci. **268** (1992) 333–339
12.14 T. Ichihashi, S. Matsui: In situ observation on electron beam induced chemical vapor deposition by transmission electron microscopy, J. Vac. Sci. Technol. B **6**(6) (1988) 1869–1872
12.15 K. I. Schiffmann: Investigation of fabrication parameters for the electron-beam-induced deposition of contamination tips used in atomic force microscopy, Nanotechnology **4** (1993) 163–169
12.16 J. H. Hafner, C. L. Cheung, A. T. Woolley, C. M. Lieber: Structural and functional imaging with carbon nanotube AFM probes, Prog. Biophys. Mol. Biol. **77**(1) (2001) 73–110
12.17 S. Iijima, C. Brabec, A. Maiti, J. Bernholc: Structural flexibility of carbon nanotubes, J. Chem. Phys. **104**(5) (1996) 2089–2092
12.18 M. M. J. Treacy, T. W. Ebbesen, J. M. Gibson: Exceptionally high Young's modulus observed for individual carbon nanotubes, Nature **381** (1996) 678–680
12.19 A. Krishnan, E. Dujardin, T. W. Ebbesen, P. N. Yianilos, M. M. J. Treacy: Young's modulus of single-walled nanotubes, Phys. Rev. B **58**(20) (1998) 14013–14019
12.20 E. W. Wong, P. E. Sheehan, C. M. Lieber: Nanobeam mechanics – elasticity, strength, and toughness of nanorods and nanotubes, Science **277**(5334) (1997) 1971–1975
12.21 J. P. Lu: Elastic properties of carbon nanotubes and nanoropes, Phys. Rev. Lett. **79**(7) (1997) 1297–1300
12.22 H. J. Dai, J. H. Hafner, A. G. Rinzler, D. T. Colbert, R. E. Smalley: Nanotubes as nanoprobes in scanning probe microscopy, Nature **384**(6605) (1996) 147–150
12.23 A. G. Rinzler, Y. H. Hafner, P. Nikolaev, L. Lou, S. G. Kim, D. Tomanek, D. T. Colbert, R. E. Smalley: Unraveling nanotubes: Field emission from atomic wire, Science **269** (1995) 1550
12.24 H. Nishijima, S. Kamo, S. Akita, Y. Nakayama, K. I. Hohmura, S. H. Yoshimura, K. Takeyasu: Carbon-nanotube tips for scanning probe microscopy: Preparation by a controlled process and observation of deoxyribonucleic acid, Appl. Phys. Lett. **74**(26) (1999) 4061–4063
12.25 S. S. Wong, A. T. Woolley, T. W. Odom, J. L. Huang, P. Kim, D. V. Vezenov , C. M. Lieber: Single-walled carbon nanotube probes for high-resolution nanostructure imaging, Appl. Phys. Lett. **73**(23) (1998) 3465–3467
12.26 J. H. Hafner, M. J. Bronikowski, B. R. Azamian, P. Nikolaev, A. G. Rinzler, D. T. Colbert, K. A. Smith, R. E. Smalley: Catalytic growth of single-wall carbon nanotubes from metal particles, Chem. Phys. Lett. **296**(1–2) (1998) 195–202
12.27 P. Nikolaev, M. J. Bronikowski, R. K. Bradley, F. Rohmund, D. T. Colbert, K. A. Smith, R. E. Smalley: Gas-phase catalytic growth of single-walled carbon nanotubes from carbon monoxide, Chem. Phys. Lett. **313**(1–2) (1999) 91–97
12.28 W. Z. Li, S. S. Xie, L. X. Qian, B. H. Chang, B. S. Zou, W. Y. Zhou, R. A. Zhao, G. Wang: Large-scale synthesis of aligned carbon nanotubes, Science **274**(5293) (1996) 1701–1703
12.29 J. H. Hafner, C. L. Cheung, C. M. Lieber: Growth of nanotubes for probe microscopy tips, Nature **398**(6730) (1999) 761–762
12.30 V. Lehmann: The physics of macroporous silicon formation, Thin Solid Films **255** (1995) 1–4
12.31 F. Ronkel, J. W. Schultze, R. Arensfischer: Electrical contact to porous silicon by electrodeposition of iron, Thin Solid Films **276**(1–2) (1996) 40–43
12.32 J. H. Hafner, C. L. Cheung, C. M. Lieber: Direct growth of single-walled carbon nanotube scanning probe microscopy tips, J. Am. Chem. Soc. **121**(41) (1999) 9750–9751
12.33 E. B. Cooper, S. R. Manalis, H. Fang, H. Dai, K. Matsumoto, S. C. Minne, T. Hunt, C. F. Quate: Terabit-per-square-inch data storage with the atomic force microscope, Appl. Phys. Lett. **75**(22) (1999) 3566–3568
12.34 E. Yenilmez, Q. Wang, R. J. Chen, D. Wang, H. Dai: Wafer scale production of carbon nanotube scanning probe tips for atomic force microscopy, Appl. Phys. Lett. **80**(12) (2002) 2225–2227
12.35 A. Stemmer, A. Hefti, U. Aebi, A. Engel: Scanning tunneling and transmission electron microscopy on identical areas of biological specimens, Ultramicroscopy **30**(3) (1989) 263
12.36 R. Nicolaides, L. Yong, W. E. Packard, W. F. Zhou, H. A. Blackstead, K. K. Chin, J. D. Dow, J. K. Furdyna, M. H. Wei, R. C. Jaklevic, W. J. Kaiser, A. R. Pelton, M. V. Zeller, J. J. Bellina: Scanning tunneling microscope tip structures, J. Vac. Sci. Technol. A **6**(2) (1988) 445–447
12.37 J. P. Ibe, P. P. Bey, S. L. Brandow, R. A. Brizzolara, N. A. Burnham, D. P. DiLella, K. P. Lee, C. R. K. Marrian, R. J. Colton: On the electrochemical etching of tips for scanning tunneling microscopy, J. Vac. Sci. Technol. A **8** (1990) 3570–3575
12.38 L. Libioulle, Y. Houbion, J.-M. Gilles: Very sharp platinum tips for scanning tunneling microscopy, Rev. Sci. Instrum. **66**(1) (1995) 97–100
12.39 A. J. Nam, A. Teren, T. A. Lusby, A. J. Melmed: Benign making of sharp tips for STM and FIM: Pt, Ir, Au, Pd, and Rh, J. Vac. Sci. Technol. B **13**(4) (1995) 1556–1559

13. Noncontact Atomic Force Microscopy and Its Related Topics

The scanning probe microscopy (SPM) such as the STM and the NC-AFM is the basic technology for the nanotechnology and also for the future bottom-up process. In Sect. 13.1, the principles of AFM such as operating modes and the frequency modulation method of the NC-AFM are fully explained. Then, in Sect. 13.2, applications of NC-AFM to semiconductors that make clear its potentials such as spatial resolution and functions are introduced. Next, in Sect. 13.3, applications of NC-AFM to insulators such as alkali halides, fluorides and transition metal oxides are introduced. At last, in Sect. 13.4, applications of NC-AFM to molecules such as carboxylate ($RCOO^-$) with $R = H$, CH_3, $C(CH_3)_3$ and CF_3 are introduced. Thus, the NC-AFM can observe atoms and molecules on various kinds of surfaces such as semiconductor, insulator and metal oxide with atomic/molecular resolutions. These sections are essential to understand the status of the art and the future possibility of NC-AFM that is the second generation of atom/molecule technology.

13.1 **Principles of Noncontact Atomic Force Microscope (NC-AFM)** ... 386
13.1.1 Imaging Signal in AFM ... 386
13.1.2 Experimental Measurement and Noise ... 387
13.1.3 Static AFM Operating Mode ... 387
13.1.4 Dynamic AFM Operating Mode ... 388
13.1.5 The Four Additional Challenges Faced by AFM ... 388
13.1.6 Frequency-Modulation AFM (FM-AFM) ... 389
13.1.7 Relation Between Frequency Shift and Forces ... 390
13.1.8 Noise in Frequency-Modulation AFM – Generic Calculation ... 391
13.1.9 Conclusion ... 391

13.2 **Applications to Semiconductors** ... 391
13.2.1 Si(111)7×7 Surface ... 392
13.2.2 Si(100)2×1 and Si(100)2×1:H Monohydride Surfaces ... 393
13.2.3 Metal-Deposited Si Surface ... 395

13.3 **Applications to Insulators** ... 397
13.3.1 Alkali Halides, Fluorides, and Metal Oxides ... 397
13.3.2 Atomically Resolved Imaging of a NiO(001) Surface ... 402
13.3.3 Atomically Resolved Imaging Using Noncoated and Fe-Coated Si Tips ... 402

13.4 **Applications to Molecules** ... 404
13.4.1 Why Molecules and What Molecules? ... 404
13.4.2 Mechanism of Molecular Imaging ... 404
13.4.3 Perspectives ... 407

References ... 407

The scanning tunneling microscope (STM) is an atomic tool based on an electric method that measures the tunneling current between a conductive tip and a conductive surface. It can electrically observe individual atoms/molecules. It can characterize or analyze the electronic nature around surface atoms/molecules. In addition, it can manipulate individual atoms/molecules. Hence, the STM is the first generation of atom/molecule technology. On the other hand, the atomic force microscope (AFM) is a unique atomic tool based on a mechanical method that can deal with even the insulator surface. Since the invention of noncontactAFM (NC-AFM) in 1995, the NC-AFM and the NC-AFM-based method have rapidly developed into a powerful surface tool on atomic/molecular scales, because the NC-AFM has the following characteristics: (1) it has true atomic resolution, (2) it can measure atomic force (so-called atomic force spectroscopy), (3) it can observe even insulators, and (4) it can measure mechanical responses such as elastic deformation. Thus, the NC-AFM is the second generation of atom/molecule technology. The scanning probe microscopy (SPM) such as the STM and the NC-AFM is the basic technology for nanotechnology and also for the future bottom-up process.

In Sect. 13.1, principles of NC-AFM will be fully introduced. Then, in Sect. 13.2, applications to semiconductors will be presented. Next, in Sect. 13.3, applications to insulators will be described. And, in Sect. 13.4, applications to molecules will be introduced. These sections are essential to understanding the status of the art and the future possibility of the NC-AFM.

13.1 Principles of Noncontact Atomic Force Microscope (NC-AFM)

The atomic force microscope (AFM), invented by *Binnig* [13.1] and introduced in 1986 by *Binnig*, *Quate*, and *Gerber* [13.2], is an offspring of the scanning tunneling microscope (STM) [13.3]. The STM is covered in books and review articles, e.g., [13.4–9]. Early on in the development of STM it became evident that relatively strong forces act between a tip in close proximity to a sample. It was found that these forces could be put to good use in the atomic force microscope (AFM). Detailed information about the noncontact AFM can be found in [13.10–12].

13.1.1 Imaging Signal in AFM

Figure 13.1 shows a sharp tip close to a sample. The potential energy between the tip and sample V_{ts} causes a z component of the tip-sample force $F_{ts} = -\partial V_{ts}/\partial z$. Depending on the mode of operation, the AFM uses F_{ts} or some entity derived from F_{ts} as the imaging signal.

Unlike the tunneling current, which has a very strong distance dependence, F_{ts} has long- and short-range contributions. We can classify the contributions by their range and strength. In vacuum, there are van der Waals, electrostatic, and magnetic forces with long-range (up to 100 nm) and short-range chemical forces (fractions of a nm).

The van der Waals interaction is caused by fluctuations in the electric dipole moment of atoms and their mutual polarization. For a spherical tip with radius R next to a flat surface (z is the distance between the plane connecting the centers of the surface atoms and the center of the closest tip atom), the van der Waals potential is given by [13.13]:

$$V_{vdW} = -\frac{A_H}{6z} . \qquad (13.1)$$

The "Hamaker constant", A_H, depends on the type of materials (atomic polarizability and density) of the tip and sample and is on the order of 1 eV for most solids [13.13].

When the tip and sample are both conductive and have an electrostatic potential difference, $U \neq 0$, electrostatic forces are important. For a spherical tip with radius R, the force is given by [13.14]:

$$F_{electrostatic} = -\frac{\pi \varepsilon_0 R U^2}{z} . \qquad (13.2)$$

Chemical forces are more complicated. Empirical model potentials for chemical bonds are the Morse potential (see e.g., [13.13]).

$$V_{Morse} = -E_{bond}\left(2\,e^{-\kappa(z-\sigma)} - e^{-2\kappa(z-\sigma)}\right) \qquad (13.3)$$

and the Lennard-Jones potential [13.13]:

$$V_{Lennard\text{-}Jones} = -E_{bond}\left(2\frac{\sigma^6}{z^6} - \frac{\sigma^{12}}{z^{12}}\right) . \qquad (13.4)$$

These potentials describe a chemical bond with bonding energy E_{bond} and equilibrium distance σ. The Morse potential has an additional parameter – a decay length κ.

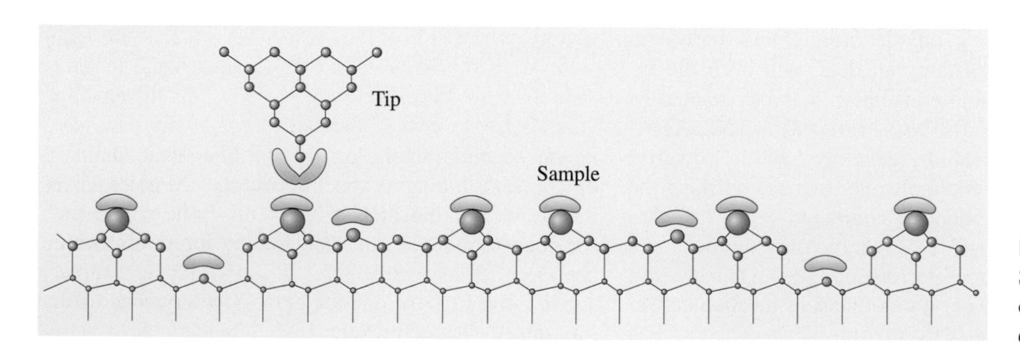

Fig. 13.1 Schematic view of an AFM tip close to a sample

13.1.2 Experimental Measurement and Noise

Forces between the tip and sample are typically measured by recording the deflection of a cantilever beam that has a tip mounted to its end (see Fig. 13.2). Today's microfabricated silicon cantilevers were first created by the *Calvin F. Quate* group [13.15–17] and *Wolter* et al. at IBM [13.18].

The cantilever is characterized by its spring constant k, eigenfrequency f_0, and quality factor Q.

For a rectangular cantilever with dimensions w, t, and L (see Fig. 13.2), the spring constant k is given by [13.6]:

$$k = \frac{E_Y w t^3}{4L^3}, \tag{13.5}$$

where E_Y is Young's modulus. The eigenfrequency f_0 is given by [13.6]:

$$f_0 = 0.162 \frac{t}{L^2} \sqrt{E_Y/\rho}, \tag{13.6}$$

where ρ is the mass density of the cantilever material. The Q-factor depends on the damping mechanisms present in the cantilever. For micromachined cantilevers operated in air, Q is typically a few hundred but can reach hundreds of thousands in vacuum.

In the first AFM, the deflection of the cantilever was measured with an STM – the backside of the cantilever was metalized, and a tunneling tip was brought close to it to measure the deflection [13.2]. Today's designs use optical (interferometer, beam-bounce) or electrical methods (piezoresistive, piezoelectric) for measuring the cantilever deflection. A discussion of the various techniques can be found in [13.19], descriptions of piezoresistive detection schemes are found in [13.17, 20], and piezoelectric methods are explained in [13.21–24].

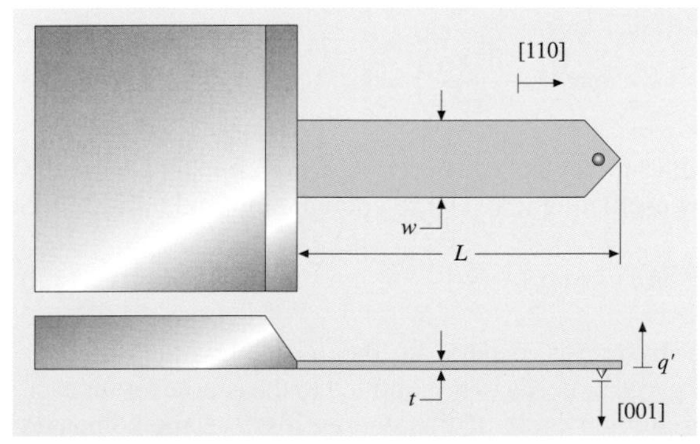

Fig. 13.2 Top view and side view of a microfabricated silicon cantilever (schematic)

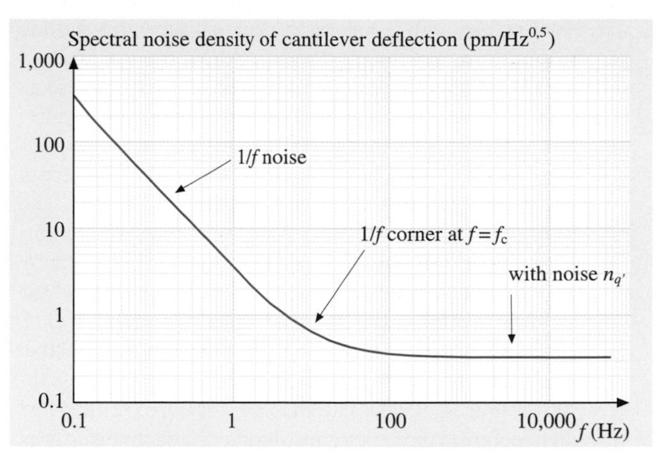

Fig. 13.3 Schematic view of $1/f$ noise apparent in force detectors. Static AFMs operate in a frequency range from 0.01 Hz to a few hundred Hz, while dynamic AFMs operate at frequencies around 10 kHz to a few hundred kHz. The noise of the cantilever deflection sensor is characterized by the $1/f$ corner frequency f_c and the constant deflection noise density $n_{q'}$ for the frequency range where white noise is dominating

The quality of the cantilever deflection measurement can be expressed in a schematic plot of the deflection noise density versus frequency, as in Fig. 13.3.

The noise density has a $1/f$ dependence for low frequency and merges into a constant noise density ("white noise") above the "$1/f$ corner frequency".

13.1.3 Static AFM Operating Mode

In the static mode of operation, the force translates into a deflection $q' = F_{ts}/k$ of the cantilever, yielding images as maps $z(x, y, F_{ts} = \text{const.})$. The noise level of the force measurement is then given by the cantilever's spring constant k times the noise level of the deflection measurement. In this respect, a small value for k increases force sensitivity. On the other hand, instabilities are more likely to occur with soft cantilevers (see Sect. 13.1.5). Because the deflection of the cantilever should be significantly larger than the deformation of the tip and sample, the cantilever should be much softer than the bonds between the bulk atoms in the tip and sample. Inter-atomic force constants in solids are in a range from 10 N/m to about 100 N/m – in biological samples, they can be as small as 0.1 N/m. Thus, typical values for k in the static mode are 0.01–5 N/m.

Even though it has been demonstrated that atomic resolution is possible with static AFM, the method can only be applied in certain cases. The detrimental effects of $1/f$ noise can be limited by working at low temperatures [13.25], where the coefficients of thermal expansion are very small, or by building the AFM of a material with a low thermal expansion coefficient [13.26]. The long-range attractive forces have to be cancelled by immersing the tip and sample in a liquid [13.26], or by partly compensating the attractive force by pulling at the cantilever after jump-to-contact has occurred [13.27]. *Jarvis* et al. have cancelled the long-range attractive force with an electromagnetic force applied to the cantilever [13.28]. Even with these restrictions, static AFM does not produce atomic resolution on reactive surfaces like silicon, as the chemical bonding of the AFM tip and sample pose an unsurmountable problem [13.29, 30].

13.1.4 Dynamic AFM Operating Mode

In the dynamic operation modes, the cantilever is deliberately vibrated. There are two basic methods of dynamic operation: amplitude modulation (AM) and frequency modulation (FM) operation. In AM-AFM [13.31], the actuator is driven by a fixed amplitude A_{drive} at a fixed frequency f_{drive}, where f_{drive} is close to f_0. When the tip approaches the sample, elastic and inelastic interactions cause a change in both the amplitude and the phase (relative to the driving signal) of the cantilever. These changes are used as the feedback signal. While the AM mode was initially used in a noncontact mode, it was later implemented very successfully at a closer distance range in ambient conditions involving repulsive tip-sample interactions.

The change in amplitude in AM mode does not occur instantaneously with a change in the tip-sample interaction, but on a time scale of $\tau_{\text{AM}} \approx 2Q/f_0$, and the AM mode is slow with high-Q cantilevers. However, the use of high Q-factors reduces noise. *Albrecht* et al. found a way to combine the benefits of high Q and high speed by introducing the frequency modulation (FM) mode [13.32], where the change in the eigenfrequency settles on a time scale of $\tau_{\text{FM}} \approx 1/f_0$.

Using the FM mode, the resolution was improved dramatically, and finally atomic resolution [13.33, 34] was obtained by reducing the tip-sample distance and working in vacuum. For atomic studies in vacuum, the FM mode (see Sect. 13.1.6) is now the preferred AFM technique. However, atomic resolution in vacuum can also be obtained with the AM mode, as demonstrated by *Erlandsson* et al. [13.35].

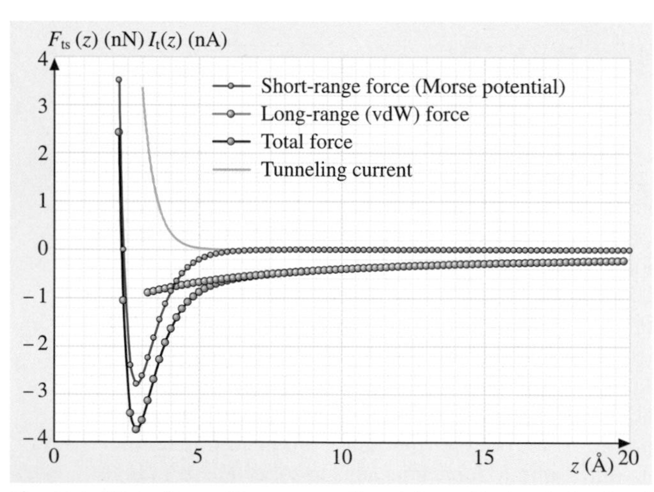

Fig. 13.4 Plot of tunneling current I_t and force F_{ts} (typical values) as a function of distance z between front atom and surface atom layer

13.1.5 The Four Additional Challenges Faced by AFM

Some of the inherent AFM challenges are apparent by comparing the tunneling current and tip-sample force as a function of distance (Fig. 13.4).

The tunneling current is a monotonic function of the tip-sample distance and has a very sharp distance dependence. In contrast, the tip-sample force has long- and short-range components and is not monotonic.

Jump-to-Contact and Other Instabilities

If the tip is mounted on a soft cantilever, the initially attractive tip-sample forces can cause a sudden "jump-to-contact" when the tip approaches the sample. This instability occurs in the quasi-static mode if [13.36, 37]

$$k < \max\left(-\frac{\partial^2 V_{\text{ts}}}{\partial z^2}\right) = k_{\text{ts}}^{\max} . \quad (13.7)$$

Jump-to-contact can be avoided even for soft cantilevers by oscillating it at a large enough amplitude A [13.38]:

$$kA > \max(-F_{\text{ts}}) . \quad (13.8)$$

If hysteresis occurs in the $F_{\text{ts}}(z)$-relation, the energy ΔE_{ts} needs to be supplied to the cantilever for each oscillation cycle. If this energy loss is large compared to the intrinsic energy loss of the cantilever, amplitude control can become difficult. An additional approximate criterion for k and A is then

$$\frac{kA^2}{2} \geq \frac{\Delta E_{\text{ts}} Q}{2\pi} . \tag{13.9}$$

Contribution of Long-Range Forces

The force between tip and sample is composed of many contributions: electrostatic, magnetic, van der Waals, and chemical forces in vacuum. All of these force types except for the chemical forces have strong long-range components that conceal the atomic-force components. For imaging by AFM with atomic resolution, it is desirable to filter out the long-range force contributions and only measure the force components that vary at the atomic scale. While there is no way to discriminate between long- and short-range forces in static AFM, it is possible to enhance the short-range contributions in dynamic AFM through the proper choice of the oscillation amplitude A of the cantilever.

Noise in the Imaging Signal

Measuring the cantilever deflection is subject to noise, especially at low frequencies ($1/f$ noise). In static AFM, this noise is particularly problematic because of the approximate $1/f$ dependence. In dynamic AFM, the low-frequency noise is easily discriminated when using a bandpass filter with a center frequency around f_0.

Non-monotonic Imaging Signal

The tip-sample force is not monotonic. In general, the force is attractive for large distances, and upon decreasing the distance between tip and sample, the force turns repulsive (see Fig. 13.4). Stable feedback is only possible on a monotonic subbranch of the force curve.

Frequency-modulation AFM (FM-AFM) helps to overcome challenges. The non-monotonic imaging signal in AFM is a remaining complication for FM-AFM.

13.1.6 Frequency-Modulation AFM (FM-AFM)

In FM-AFM, a cantilever with eigenfrequency f_0 and spring constant k is subject to controlled positive feedback such that it oscillates with a constant amplitude A [13.32], as shown in Fig. 13.5.

Experimental Setup

The deflection signal is phase shifted, routed through an automatic gain control circuit and fed back to the actuator. The frequency f is a function of f_0, its quality factor Q, and the phase shift ϕ between the mechanical excitation generated at the actuator and the deflection of the cantilever. If $\phi = \pi/2$, the loop oscillates at $f = f_0$.

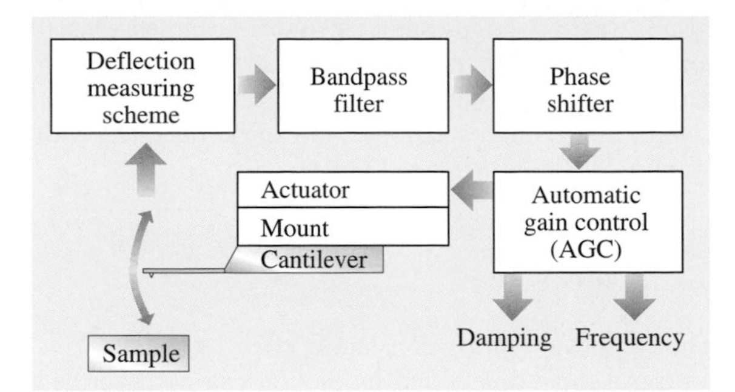

Fig. 13.5 Block diagram of a frequency-modulation force sensor

Three physical observables can be recorded: 1) a change in resonance frequency, Δf, 2) the control signal of the automatic gain control unit as a measure of the tip-sample energy dissipation and 3) an average tunneling current (for conducting cantilevers and tips).

Applications

FM-AFM was introduced by *Albrecht* and coworkers in magnetic force microscopy [13.32]. The noise level and imaging speed were enhanced significantly compared to amplitude modulation techniques. Achieving atomic resolution on the Si(111)-(7×7) surface has been an important step in the development of the STM [13.39], and in 1994, this surface was imaged by AFM with true atomic resolution for the first time [13.33] (see Fig. 13.6).

The initial parameters that provided *true atomic resolution* (see caption of Fig. 13.6) were found empirically. Surprisingly, the amplitude necessary for obtaining good

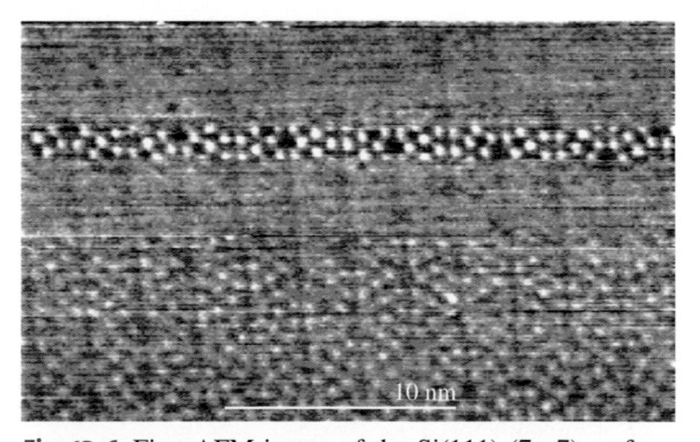

Fig. 13.6 First AFM image of the Si(111)-(7×7) surface. Parameters $k = 17\,\text{Nm}$, $f_0 = 114\,\text{kHz}$, $Q = 28{,}000$, $A = 34\,\text{nm}$, $\Delta f = -70\,\text{Hz}$, $V_t = 0\,\text{V}$

results was very large compared to atomic dimensions. It turned out later that the amplitudes had to be that large in order to fulfill the stability criteria listed in Sect. 13.1.5. Cantilevers with $k \approx 2{,}000\,\mathrm{N/m}$ can be operated with amplitudes in the Å-range [13.24].

13.1.7 Relation Between Frequency Shift and Forces

The cantilever (spring constant k, effective mass m^*) is a macroscopic object and its motion can be described by classical mechanics. Figure 13.7 shows the deflection $q'(t)$ of the tip of the cantilever: It oscillates with an amplitude A at a distance $q(t)$ to a sample.

Generic Calculation

The Hamiltonian of the cantilever is:

$$H = \frac{p^2}{2m^*} + \frac{kq'^2}{2} + V_{\mathrm{ts}}(q)\,, \tag{13.10}$$

where $p = m^*\,\mathrm{d}q'/\mathrm{d}t$. The unperturbed motion is given by:

$$q'(t) = A\cos(2\Box f_0 t)\,, \tag{13.11}$$

and the frequency is:

$$f_0 = \frac{1}{2\pi}\sqrt{\frac{k}{m^*}}\,. \tag{13.12}$$

If the force gradient $k_{\mathrm{ts}} = -\partial F_{\mathrm{ts}}/\partial z = \partial^2 V_{\mathrm{ts}}/\partial z^2$ is constant during the oscillation cycle, the calculation of the frequency shift is trivial:

$$\Delta f = \frac{f_0}{2k} k_{\mathrm{ts}}\,. \tag{13.13}$$

However, in classic FM-AFM, k_{ts} varies in orders of magnitude during one oscillation cycle, and a perturbation approach, as shown below, has to be employed for the calculation of the frequency shift.

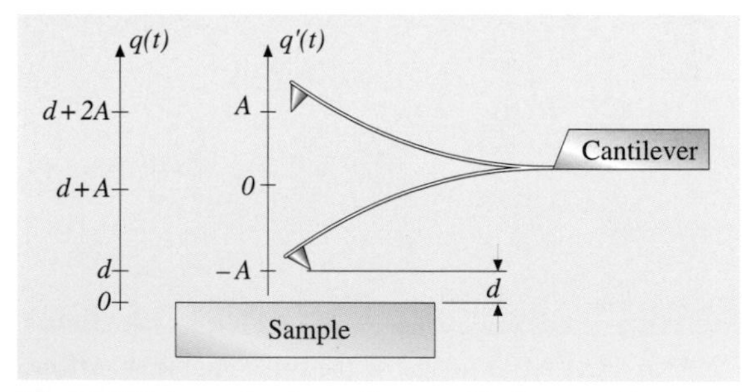

Fig. 13.7 Schematic view of an oscillating cantilever and definition of geometric terms

Hamilton-Jacobi Method

The first derivation of the frequency shift in FM-AFM was achieved in 1997 [13.38] using canonical perturbation theory [13.40]. The result of this calculation is:

$$\Delta f = -\frac{f_0}{kA^2}\langle F_{\mathrm{ts}} q'\rangle = -\frac{f_0}{kA^2}\int_0^{1/f_0} F_{\mathrm{ts}}(d + A + q'(t))q'(t)\,\mathrm{d}t\,. \tag{13.14}$$

The applicability of first-order perturbation theory is justified, because in FM-AFM, E is typically in the range of several keVs, while V_{ts} is on the order of a few eVs. *Dürig* [13.41] has found a generalized algorithm that even allows the reconstruction of the tip-sample potential if not only the frequency shift, but the higher harmonics of the cantilever oscillation are known.

A Descriptive Expression for Frequency Shifts as a Function of the Tip-Sample Forces

With integration by parts, the complicated expression (13.14) transforms into a very simple expression that resembles (13.13) [13.42].

$$\Delta f = \frac{f_0}{2k}\int_{-A}^{A} k_{\mathrm{ts}}(z - q')\frac{\sqrt{A^2 - q'^2}}{\frac{\pi}{2}kA^2}\,\mathrm{d}q'\,. \tag{13.15}$$

This expression is closely related to (13.13): The constant k_{ts} is replaced by a weighted average, where the weight function $w(q', A)$ is a semicircle with radius A divided by the area of the semicircle $\pi A^2/2$ (see Fig. 13.8). For $A \to 0$, $w(q', A)$ is a representation of Dirac's delta function and the trivial zero-amplitude result of (13.13) is immediately recovered. The frequency shift results from a convolution between the tip-sample force gradient and the weight function. This convolution can easily be reversed with a linear transformation, and the tip-sample force can be recovered from the frequency shift versus distance curve [13.42].

The dependence of the frequency shift with amplitude confirms an empirical conjecture: Small amplitudes increase the sensitivity to short-range forces! Adjusting the amplitude in FM-AFM resembles tuning an optical spectrometer to a passing wavelength. When short-range interactions are to be probed, the amplitude should be in the range of the short-range forces. While using amplitudes in the Å-range has been elusive with conventional cantilevers, because of the instability problems described in Sect. 13.1.5, cantilevers with a stiffness on the order of $1{,}000\,\mathrm{N/m}$ like the ones introduced in [13.23] are well suited for small-amplitude operation.

13.1.8 Noise in Frequency-Modulation AFM – Generic Calculation

The vertical noise in FM-AFM is given by the ratio between the noise in the imaging signal and the slope of the imaging signal with respect to z:

$$\delta z = \frac{\delta \Delta f}{\left|\frac{\partial \Delta f}{\partial z}\right|} . \tag{13.16}$$

Figure 13.9 shows a typical frequency shift versus distance curve. Because the distance between the tip and sample is measured indirectly through the frequency shift, it is clearly evident from Fig. 13.9 that the noise in the frequency measurement $\delta\Delta f$ translates into vertical noise δz and is given by the ratio between $\delta\Delta f$ and the slope of the frequency shift curve $\Delta f(z)$ (13.16). Low vertical noise is obtained for a low-noise frequency measurement and a steep slope of the frequency shift curve.

The frequency noise $\delta\Delta f$ is typically inversely proportional to the cantilever amplitude A [13.32,43]. The derivative of the frequency shift versus distance is constant for $A \ll \lambda$, where λ is the range of the tip-sample interaction and proportional to $A^{-1.5}$ for $A \gg \lambda$ [13.38]. Thus, minimal noise occurs if [13.44]

$$A_{\text{optimal}} \approx \lambda , \tag{13.17}$$

for chemical forces, $\lambda \approx 1$ Å. However, for stability reasons (Sect. 13.1.5), extremely stiff cantilevers are needed for small amplitude operation. The excellent noise performance of the stiff cantilever and small amplitude technique has been verified experimentally [13.24].

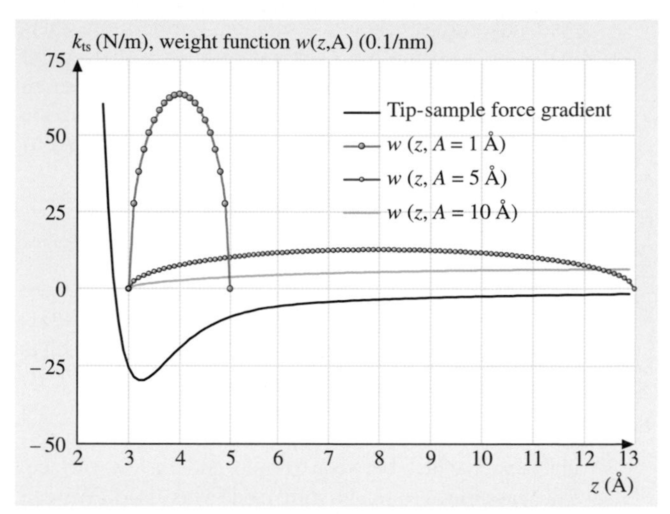

Fig. 13.8 Tip-sample force gradient k_{ts} and weight function for the calculation of the frequency shift

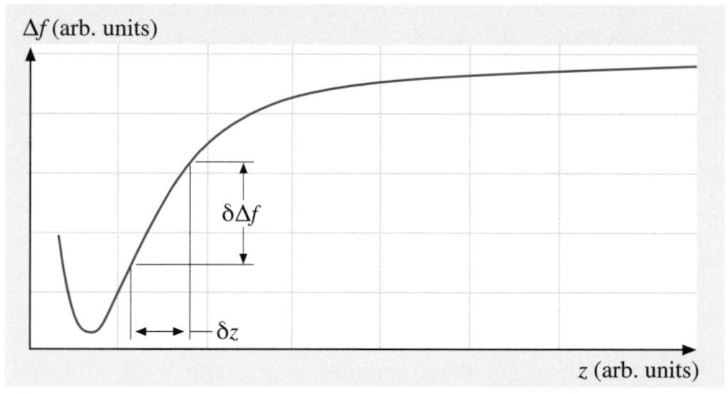

Fig. 13.9 Plot of frequency shift Δf as a function of tip-sample distance z. The noise in the tip-sample distance measurement is given by the noise of the frequency measurement $\delta\Delta f$ divided by the slope of the frequency shift curve

13.1.9 Conclusion

Dynamic force microscopy, in particular frequency-modulation atomic force microscopy, has matured into a viable technique that allows true atomic resolution of conducting and insulating surfaces and spectroscopic measurements on individual atoms [13.10, 45]. Even true atomic resolution in lateral force microscopy is now possible [13.46]. Challenges remain in the chemical composition and structural arrangement of the AFM tip.

13.2 Applications to Semiconductors

For the first time, corner holes and adatoms on Si(111)7×7 surface have been observed at a very local area by *Giessibl* using the pure noncontact AFM in ultrahigh vacuum (UHV) [13.33]. It became the breakthrough for the true atomic resolution imaging on a well-defined clean surface using noncontact AFM. Then, Si(111)7×7 [13.34, 35, 45, 47], InP(110) [13.48], and Si(100)2×1 [13.34] surfaces were successively resolved with true atomic resolution. Furthermore, thermally induced motion of atoms or atomic-scale point defects on InP(110) surface have been observed at room temperature [13.48]. In this section, we will describe typical results of atomically resolved noncontact AFM imaging of semiconductor surfaces.

13.2.1 Si(111)7×7 Surface

Figure 13.10 shows the atomic resolution images of the Si(111)7×7 surface [13.49]. Here, Fig. 13.10a (TYPE I) was obtained using the Si tip without dangling, which is covered with an inert oxide layer. Fig. 13.10b (TYPE II) was obtained using the Si tip with dangling bond on which the Si atoms were deposited, due to the mechanical soft contact between the tip and the Si surface. The variable frequency shift mode was used. We can see not only adatoms and corner holes, but also missing adatoms described by the dimer-adatom-stacking fault (DAS) model. We can see that the image contrast in Fig. 13.10b is clearly stronger than that in Fig. 13.10a.

Interestingly, by using the Si tip with dangling bond, we observed contrast between inequivalent halves and between inequivalent adatoms of the 7×7 unit cell. Namely, as shown in Fig. 13.11a, the faulted halves surrounded with a solid line show brighter than the unfaulted halves surrounded with a broken line. Here, the positions of the faulted and unfaulted halves were decided from the step direction. From the cross-sectional profile along the long diagonal of the 7×7 unit cell in Fig. 13.11b, the heights of the corner adatoms are slightly higher than those of the adjacent center adatoms in the faulted and unfaulted halves of the unit cell. The measured corrugations are in the following decreasing order:

$$\text{Co-F} > \text{Ce-F} > \text{Co-U} > \text{Ce-U} ,$$

where Co-F and Ce-F indicate the corner and center adatoms in faulted halves, and Co-U and Ce-U indicate the corner and center adatoms in unfaulted halves, respectively. As averaging over several units, the corrugation height differences are estimated to be 0.25 Å, 0.15 Å, and 0.05 Å for Co-F, Ce-F, and Co-U, respectively, referring to Ce-U. This tendency, that the heights of the corner adatoms are higher than those of the center adatoms, is consistent with the experimental results using a silicon tip [13.47], although they could not determine faulted and unfaulted halves of the unit cell in the measured AFM images. However, this tendency is completely contrary to the experimental results using a tungsten tip [13.35]. This difference may originate from the difference in the materials of the tip, which seems to affect the interaction between the tip and the reactive sample surface. Another possibility is that the tip is in contact with the surface during the small fraction of the oscillating cycle in their experiments [13.35].

We consider that contrast between inequivalent adatoms is not caused by tip artifacts for the following reasons: 1) Each adatom, corner hole and defect were clearly observed; 2) the apparent heights of adatoms are the same whether they are located adjacent to defects or not; 3) the same contrast in several images for the different tips has been observed.

It should be noted that the corrugation amplitude of adatoms ∼ 1.4 Å in Fig. 13.11b is higher than that of 0.8–1.0 Å obtained with the STM, although the depth of corner holes obtained with noncontact AFM is almost the same as that observed with STM. Besides, in noncontact mode AFM images, the corrugation ampli-

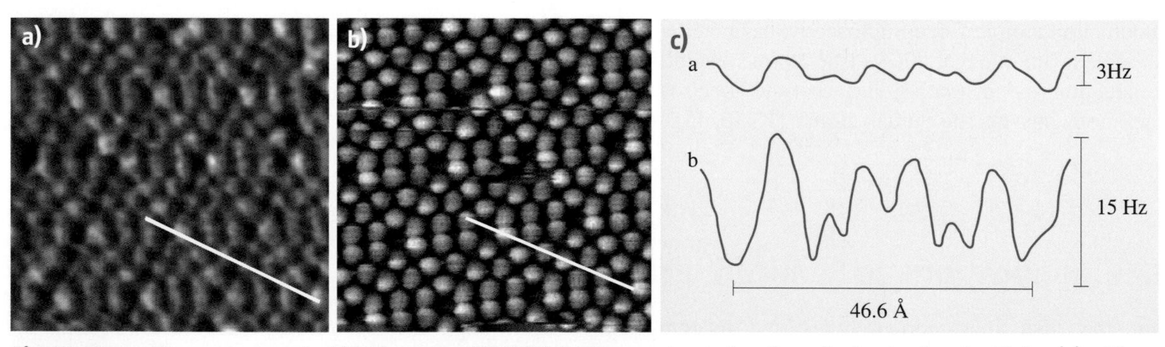

Fig. 13.10a–c Noncontact mode AFM images of Si(111) 7×7 reconstructed surface obtained using the Si tips (**a**) without and (**b**) with dangling bond. The scan area is 99 Å×99 Å. (**c**) The cross sectional profiles along long diagonal of the 7×7 unit cell indicated by *white lines* in (**a**) and (**b**)

tude of adatoms was frequently larger than the depth of the corner holes. The origin of such a large corrugation of adatoms may be due to the effect of the chemical interaction, but that is yet unclear.

The atom positions, surface energies, dynamic properties, and chemical reactivities on the Si(111)7×7 reconstructed surface have been extensively investigated theoretically and experimentally. From these investigations, the possible origins of the contrast between inequivalent adatoms in AFM images are the following: the true atomic heights that correspond to the adatom core positions, the stiffness (spring constant) of inter-atomic bonding with the adatoms that corresponds to the frequencies of the surface mode, the amount of charge of adatom, and the chemical reactivity of adatoms. Table 13.1 summarizes the decreasing orders of the inequivalent adatoms for individual property. From Table 13.1, we can see that the calculated adatom heights and the stiffness of inter-atomic bonding cannot explain the AFM data, while the amount of charge of adatom and the chemical reactivity of adatoms can explain our data. The contrast due to the amount of charge of adatom means that the AFM image originated from the difference of the vdW, or the electrostatic physical interactions between the tip and the valence electrons at the adatoms. The contrast due to the chemical reactivity of adatoms means that the AFM image originated from the difference of covalent bonding chemical interaction between the atoms at the tip apex and the dangling bond of adatoms. Thus, we can see there are two possible interactions that explain the strong contrast between inequivalent adatoms of the 7×7 unit cell observed using the Si tip with dangling bond.

A weak contrast image in Fig. 13.10a is due to vdW and/or electrostatic force interactions. On the other hand, strong contrast images in Figs. 13.10b and 13.11a are due to a covalent bonding formation between the AFM tip with Si atoms and Si adatoms. These results signify the capability of the noncontact mode AFM to image the variation in chemical reactivity of Si adatoms. In the future, by controlling an atomic species at the tip apex, the study of chemical reactivity on an atomic scale will be possible using the noncontact AFM.

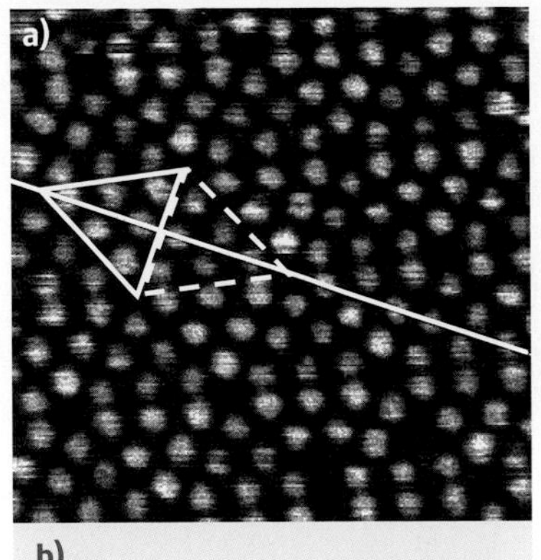

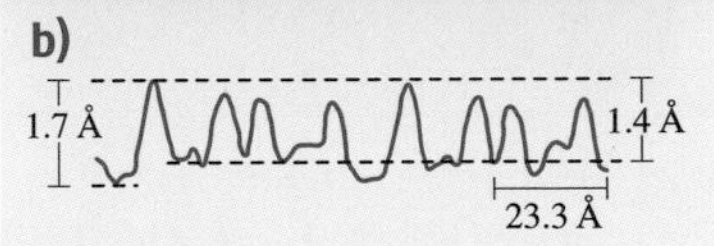

Fig. 13.11 (a) Noncontact mode AFM image with contrast of unequivalent adatoms and (b) a cross-sectional profile indicated by *white line*. The halves of the 7×7 unit cell surrounded with *solid line* and *broken line* correspond to faulted and unfaulted halves, respectively. The scan area is 89 Å×89 Å

13.2.2 Si(100)2×1 and Si(100)2×1:H Monohydride Surfaces

In order to investigate the imaging mechanism of the noncontact AFM, a comparative study between a reactive surface and an insensitive surface using the same tip is very useful. Si(100)2×1:H monohydride surface is a Si(100)2×1 reconstructed surface that is terminated by a hydrogen atom. It does not reconstruct as metal is deposited on the semiconductor surface. The surface structure hardly changes. Thus, Si(100)2×1:H monohydride surface is one of most useful surfaces for a model system to investigate the imaging mechanism, experimentally and theoretically. Furthermore, whether the interaction between a very small atom such as hy-

Table 13.1 Comparison between the adatom heights observed in an AFM image and the variety of properties for inequivalent adatoms

	Decreasing order	Agreement
AFM image	Co–F > Ce–F > Co–U > Ce–U	–
Calculated height	Co–F > Co–U > Ce–F > Ce–U	×
Stiffness of inter-atomic bonding	Ce–U > Co–U > Ce–F > Co–F	×
Amount of charge of adatom	Co–F > Ce–F > Co–U > Ce–U	○
Calculated chemical reactivity	Faulted > Unfaulted	○
Experimental chemical reactivity	Co–F > Ce–F > Co–U > Ce–U	○

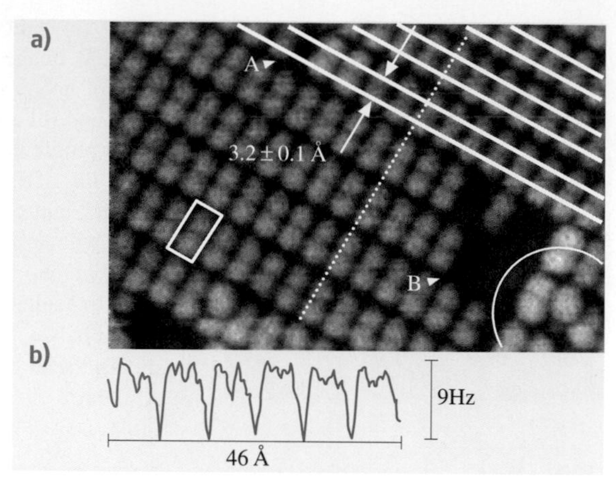

Fig. 13.12 (a) Noncontact AFM image of a Si(001)2×1 reconstructed surface. The scan area was 69×46 Å. One 2×1 unit cell is outlined with a *box*. *White rows* are superimposed to show the bright spots arrangement. The distance between bright spots on dimer row is 3.2±0.1 Å. Into *white arc*, the alternative bright spots are shown. (b) Cross-sectional profile indicated by a *white dotted line*

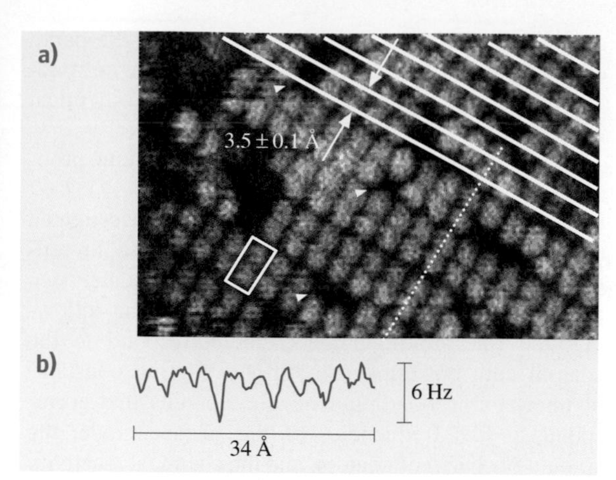

Fig. 13.13 (a) Noncontact AFM image of Si(001)2×1-H surface. The scan area was 69×46 Å. One 2×1 unit cell is outlined with a *box*. *White rows* are superimposed to show the bright spots arrangement. The distance between bright spots on dimer row is 3.5±0.1 Å. (b) Cross-sectional profile indicated by a *white dotted line*

drogen and a tip apex is observable with noncontact AFM is interesting. Here, we show noncontact AFM images measured on a Si(100)2×1 reconstructed surface with a dangling bond and a Si(100)2×1:H monohydride surface with a dangling bond that is terminated by a hydrogen atom [13.50].

Figure 13.12a shows the atomic resolution image of a Si(100)2×1 reconstructed surface. The paired bright spots arranged like rows with a 2×1 symmetry was observed with clear contrast. Missing pairs of bright spots were observed, as indicated by arrows. Further, pairs of bright spots are shown by the white dashed arc and appear to be the stabilize-buckled asymmetric dimer structure. The distance between paired bright spots is 3.2±0.1 Å.

Figure 13.13a shows the atomic resolution image of a Si(100)2×1:H monohydride surface. Paired bright spots arranged like rows were observed. Missing paired bright spots, as well as those paired in rows and single bright spots were observed, as indicated by arrows. The distance between paired bright spots is 3.5±0.1 Å. This distance is 0.2 Å larger than that of a Si(100)2×1 reconstructed surface. It was found that the distance between bright spots increases in size due to the hydrogen termination.

The bright spots in Fig. 13.12 do not merely image the silicon atom site, because the distance between bright spots forming the dimer structure of Fig. 13.12a (3.2±0.1 Å) is larger than the distance between silicon atoms of every dimer structure model (2.9 Å is the maximum distance between upper silicons on asymmetric dimer structures). This seems to be due to the contribution to imaging of the chemical bonding interaction between the dangling bond out of the silicon tip apex and the dangling bond on the Si(100)2×1 reconstructed surface. The chemical bonding interaction works with strong direction dependence between the dangling bond out of the silicon dimer structure on the Si(100)2×1 reconstructed surface and the dangling bond out of the silicon tip apex, and the dimer structure is obtained with a larger distance than between silicons on the surface.

The bright spots of Fig. 13.13 seem to be located at hydrogen atom sites on a Si(100)2×1:H monohydride surface, because the distance between bright spots forming the dimer structures (3.5±0.1 Å) agrees approximately with the distance between the hydrogen, i. e., 3.52 Å. Thus, noncontact AFM atomically resolves the individual hydrogen atoms on the topmost layer. On this surface, the dangling bond is terminated by a hydrogen atom, and the hydrogen atom on the topmost layer does not have a chemical reactivity. Therefore, the interaction between the hydrogen atom on the top most layer and the silicon tip apex does not contribute to the chemical bonding interaction with strong direction dependence as in a silicon surface, and the bright spots of a noncontact AFM image correspond to the hydrogen atom sites on the topmost layer.

13.2.3 Metal-Deposited Si Surface

In this section, we will introduce the comparative study of force interaction between a Si tip and a metal deposited Si surface and between a metal adsorbed Si tip and a metal-deposited Si surface [13.51, 52]. As for the metal-deposited Si surface, a Si(111)$\sqrt{3}\times\sqrt{3}$-Ag (hereafter referred to as $\sqrt{3}$-Ag) surface was used.

For the $\sqrt{3}$-Ag surface, the honeycomb-chained trimer (HCT) model has been accepted as the appropriate model. As shown in Fig. 13.14, this structure contains Si trimer at the second layer, 0.75 Å below the Ag trimer at the topmost layer. The topmost Ag atoms and the lower Si atoms form covalent bonds. The inter-atomic distances between the nearest neighbor Ag atoms forming Ag trimer and between lower Si atoms forming Si trimer are 3.43 Å and 2.31 Å, respectively. The apexes of Si trimers and Ag trimers face the [11$\bar{2}$] direction and the direction tilted a little to the [$\bar{1}$12] direction, respectively.

In Fig. 13.15, we show the noncontact AFM images measured using a normal Si tip at the frequency shift of (a) −37 Hz, (b) −43 Hz, and (c) −51 Hz, respectively. These frequency shifts correspond to the tip-sample distance of about 0–3 Å. We defined the zero position of tip-sample distance, i. e., the contact point, where the vibration amplitude began to decrease. The rhombus indicates the $\sqrt{3}\times\sqrt{3}$ unit cell. When the tip approached the surface, the contrast of the noncontact AFM images became strong and the pattern remarkably changed. That is, by moving the tip toward the sample surface, the hexagonal pattern, the trefoil-like pattern composed of three dark lines, and the triangle pattern can be observed sequentially. In Fig. 13.15a, the distance between the bright spots is 3.9 ± 0.2 Å. In Fig. 13.15c, the distance

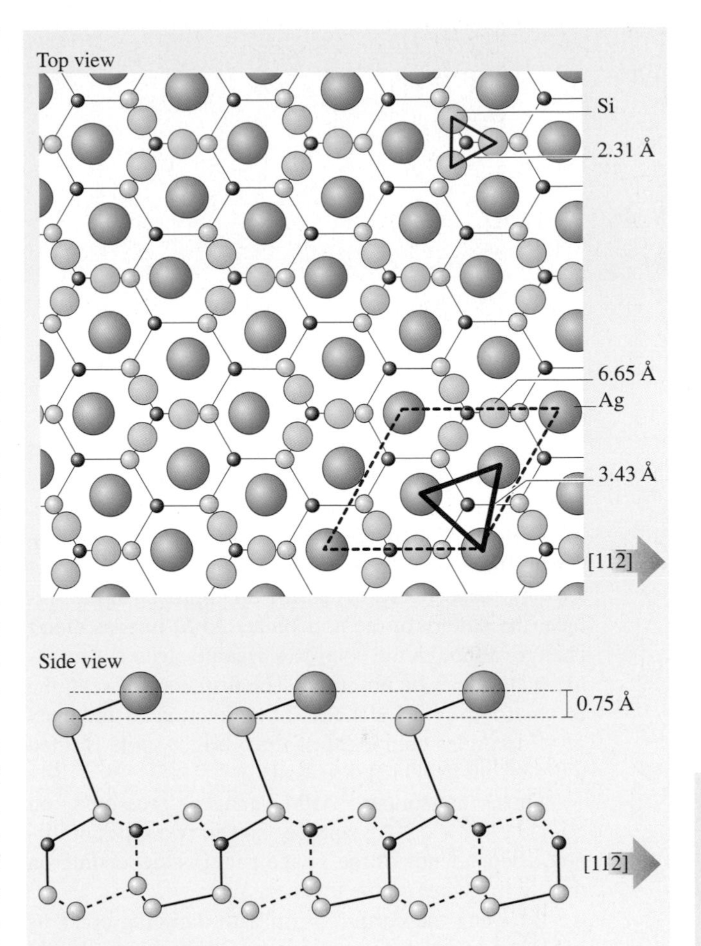

Fig. 13.14 HCT model for the structure of the Si(111)$\sqrt{3}\times\sqrt{3}$-Ag surface. *Black closed circle*, *gray closed circle*, *open circle*, and *closed circle* with *horizontal line* indicate Ag atom at the topmost layer, Si atom at the second layer, Si atom at the third layer, and Si atom at the fourth layer, respectively. *Rhombus* indicates $\sqrt{3}\times\sqrt{3}$ unit cell. *Thick*, *large solid triangle* indicates an Ag trimer. *Thin*, *small*, *solid triangle* indicates a Si trimer

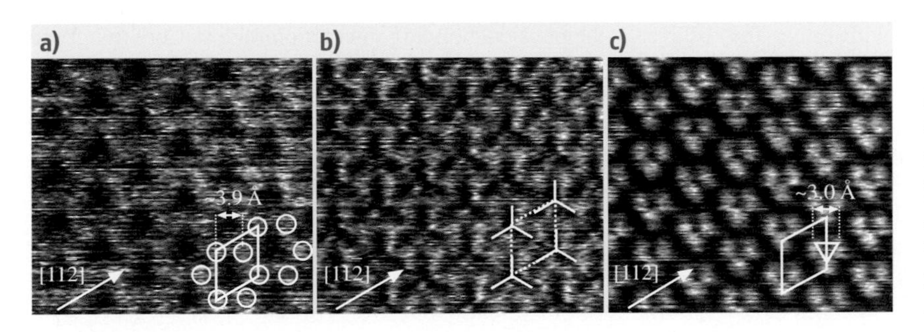

Fig. 13.15a–c Noncontact AFM images obtained at the frequency shifts of (a) −37 Hz, (b) −43 Hz, and (c) −51 Hz on a Si(111)$\sqrt{3}\times\sqrt{3}$-Ag surface. This distance dependence was obtained with the Si tip. Scan area is 38 Å × 34 Å. A *rhombus* indicates $\sqrt{3}\times\sqrt{3}$ unit cell

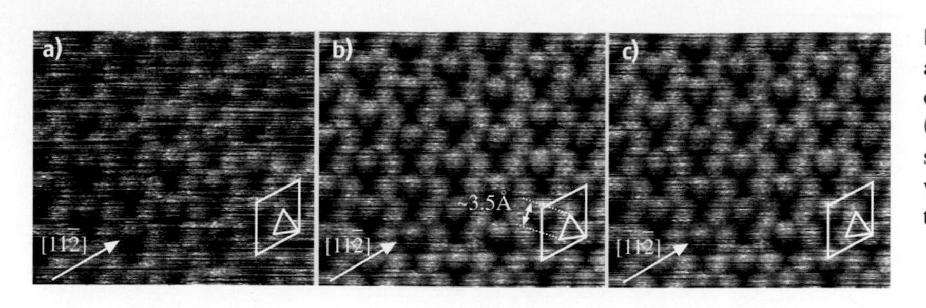

Fig. 13.16a–c Noncontact AFM images obtained at the frequency shifts of (a) −4.4 Hz, (b) −6.9 Hz, and (c) −9.4 Hz on a Si(111)$\sqrt{3}\times\sqrt{3}$-Ag surface. This distance dependence was obtained with the Ag adsorbed tip. Scan area is 38 Å × 34 Å

between the bright spots is 3.0 ± 0.2 Å, and the direction of the apex of all triangles composed of three bright spots is $[11\bar{2}]$.

In Fig. 13.16, we show the noncontact AFM images measured by using an Ag adsorbed tip at the frequency shift of (a) −4.4 Hz, (b) −6.9 Hz, and (c) −9.4 Hz, respectively. The tip-sample distances Z are roughly estimated to be $Z = 1.9$ Å, 0.6 Å, and ∼ 0 Å (in noncontact region), respectively. When the tip approached the surface, the pattern of the noncontact AFM images didn't change, although the contrasts became clearer. A triangle pattern can be observed. The distance between the bright spots is 3.5 ± 0.2 Å. The direction of the apex of all triangles composed of three bright spots is tilted a little from the $[\bar{1}\bar{1}2]$ direction.

Thus, noncontact AFM images measured on a Si(111)$\sqrt{3}\times\sqrt{3}$-Ag surface showed two types of distance dependence of the image patterns, depending on the atom species on the tip apex.

By using the normal Si tip with dangling bond, as in Fig. 13.15a, the measured distance between the bright spot of 3.9 ± 0.2 Å agrees with the distance of 3.84 Å between the center of the Ag trimer in the HCT model within the experimental error. Furthermore, the hexagonal pattern composed of six bright spots also agrees with the honeycomb structure of the Ag trimer in the HCT model. So the most appropriate site corresponding to the bright spots in Fig. 13.15a is the site of the center of the Ag trimers. In Fig. 13.15c, the measured distance of 3.0 ± 0.2 Å between the bright spots forming the triangle pattern does not agree with either the distance between the Si trimer of 2.31 Å or the distance between the Ag trimer of 3.43 Å in the HCT model, while the direction of the apex of triangles composed of three bright spots agrees with the $[11\bar{2}]$ direction of the apex of the Si trimer in the HCT model. So the most appropriate site corresponding to the bright spots in Fig. 13.15c is the intermediate site between Si atoms and Ag atoms. On the other hand, by using the Ag adsorbed tip, the measured distance of 3.5 ± 0.2 Å between the bright spots in Fig. 13.16 agrees with the distance of 3.43 Å between the nearest neighbor Ag atoms forming the Ag trimer at the topmost layer in the HCT model within the experimental error. In addition, the direction of the apex of the triangles composed of three bright spots also agrees with the direction of the apex of the Ag trimer, i.e., tilted $[\bar{1}\bar{1}2]$, in the HCT model. So the most appropriate site corresponding to the bright spots in Fig. 13.16 is the site of individual Ag atoms forming the Ag trimer at the topmost layer.

It should be noted that by using the noncontact AFM with an Ag adsorbed tip the individual Ag atom on the $\sqrt{3}$-Ag surface could be resolved in real space for the first time, although it could not be resolved by using the noncontact AFM with a Si tip. So far, the $\sqrt{3}$-Ag surface has been observed by the scanning tunneling microscope (STM) with atomic resolution. However, the STM can measure the local charge density of state near Fermi level on the surface. From the first principle calculation, it was proven that the unoccupied surface state is densely distributed around the center of the Ag trimer. As a result, bright contrast is obtained at the center of the Ag trimer with the STM.

Finally, we consider the origin of the atomic resolution imaging of the individual Ag atoms on the $\sqrt{3}$-Ag surface. Here, we discuss the difference of the force interactions between the Si and the Ag adsorbed tips. As shown in Fig. 13.17a, by using the Si tip, there is the dangling bond out of the topmost Si atom on the Si tip apex. As a result, the force interaction is dominated by the physical bonding interaction such as Coulomb force far from the surface and by chemical bonding interaction very close to the surface. In other words, if the reactive Si tip with dangling bond approaches the surface, at the distance far from the surface, the Coulomb force acts between the electron localized at the dangling bond out of the topmost Si atom on the tip apex and the positive charge distributed around the center of the Ag trimer. At the distance very close to the surface, the chemical bonding interaction will occur due to the onset of the or-

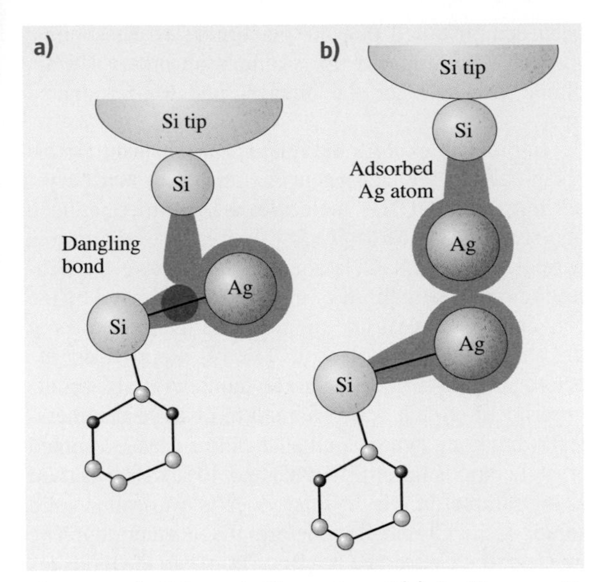

Fig. 13.17a,b Schematic illustration of (a) the Si atom with dangling bond and (b) the Ag adsorbed tip above the Si-Ag covalent bond on Si(111)$\sqrt{3}\times\sqrt{3}$-Ag surface

bital hybridization between the dangling bond out of the topmost Si atom on the Si tip apex and the Si-Ag covalent bond on the surface. Hence, the individual Ag atoms will not be resolved, the image pattern will change depending on the tip-sample distance. On the other hand, as shown in Fig. 13.17b, by using an Ag adsorbed tip, the dangling bond localized out of the topmost Si atom on the Si tip apex is terminated by the adsorbed Ag atom. As a result, even at very close tip-sample distance, the force interaction is dominated by the physical bonding interaction such as the vdW force interaction. If the Ag adsorbed tip approaches the surface, the vdW force acts between the Ag atom on the tip apex and Ag or Si atom on the surface. Ag atoms at the topmost layer on the $\sqrt{3}$-Ag surface are located higher than the Si atoms at the lower layer. Hence, the individual Ag atoms (or nearly true topography) will be resolved, and the image pattern will not change even at very close tip-sample distance. It should be emphasized that there is a possibility to identify or recognize atom species on a sample surface using the noncontact AFM if we can control an atomic species at the tip apex.

13.3 Applications to Insulators

Insulators such as alkali halides, fluorides, and metal oxides are key materials in many applications, including optics, microelectronics, catalysis, and so on. Surface properties are important in these technologies, but they are usually poorly understood. This is due to their low conductivity, which makes it difficult to investigate them using electron- and ion-based measurement techniques such as low-energy electron diffraction, ion scattering spectroscopy, and scanning tunneling microscopy (STM). Surface imaging by noncontact atomic force microscopy (NC-AFM) does not require the sample to exhibit high conductivity, because NC-AFM detects a force between the tip on the cantilever and the surface of the sample. Since the first report of atomically resolved NC-AFM on a Si(111) 7×7 surface [13.33], several groups have succeeded in obtaining "true" atomic resolution images of insulators, including defects, and it has been shown that NC-AFM is a powerful new tool for the atomic-scale surface investigation of insulators.

In this section, we will describe typical results of atomically resolved NC-AFM imaging of insulators such as alkali halides, fluorides, and metal oxides. For alkali halides and fluorides, we will focus on the contrast formation, which is the most important issue for interpreting atomically resolved images of binary compounds on the basis of experimental and theoretical results. For the metal oxides, typical examples of atomically resolved imaging will be shown. Also, simultaneous imaging using STM and NC-AFM will be described, and the difference between the STM and NC-AFM images will be demonstrated. Finally, we will describe the results obtained when we imaged an antiferromagnetic NiO(001) surface with a ferromagnetic Fe-coated tip in order to explore the possibility of detecting short-range magnetic interaction using the NC-AFM.

13.3.1 Alkali Halides, Fluorides, and Metal Oxides

The surfaces of alkali halides were the first insulating materials to be imaged by NC-AFM with true atomic resolution [13.53]. Up until now, there have been reports on atomically resolved images of (001) cleaved surfaces for single crystal NaF, RbBr, LiF, KI, NaCl [13.54], KBr [13.55], and thin films of NaCl(001) on Cu(111) [13.56]. In this section, we describe the contrast formation of alkali halides surface on the basis of experimental and theoretical results.

Alkali Halides

In experiments of alkali halides, the symmetry of the observed topographic images indicates that the protrusions exhibit only one type of ions, either positively or negatively charged ions. This leads to the conclusion that the atomic contrast is dominantly caused by electrostatic interactions between a charged atom at the tip apex and the surface ions, i. e., the long-range forces between the macroscopic tip and the sample such as van der Waals force is modulated by an alternating short-range electrostatic interaction of the surface ions. Theoretical work employing the atomistic simulation technique has revealed the mechanism for contrast formation on an ionic surface [13.57]. A significant part of the contrast is due to the displacement of ions in the force field, not only enhancing the atomic corrugations, but also contributing to the electrostatic potential by forming dipoles at the surface. The experimentally observed atomic corrugation height is determined by the interplay of the long- and short-range forces. In the case of NaCl, it has been experimentally demonstrated that a more blunt tip produces a larger corrugation when the tip-sample distance is shorter [13.54]. This result shows that the increased long-range forces induced by a blunt tip allow for more stable imaging closer to the surface. The stronger electrostatic short-range interaction and the larger ion displacement produce a more pronounced atomic corrugation.

At steps and kinks of an NaCl thin film on Cu(111), the amplitude of the atomic corrugations has been observed to increase by up to a factor of two, as large as that of atomically flat terraces [13.56]. This increase in the contrast is due to the low coordination of the step-edge and the kinked ions. The low coordination of the ions results in an enhancement of the electrostatic potential over the site and an increase in the displacement induced by the interaction with the tip. These results demonstrate that the short-range interaction dominates the atomic contrast and leads to a considerable displacement of the surface ions from the ideal position.

Theoretical study predicts that the image contrast depends on the chemical species of the tip apex. *Bennewitz* et al. [13.56] have performed the calculations using an MgO tip terminated by oxygen, an Mg ion, and an OH. The magnitude of the atomic contrast for the Mg-terminated tip shows a slight increase in comparison with an oxygen-terminated tip. The atomic contrast with the oxygen-terminated tip is dominated by the attractive electrostatic interaction between the oxygen on the tip apex and the Na ion, but the Mg-terminated tip attractively interacts with the Cl ion. Also, it has been demonstrated that the magnitude of the contrast for an OH-terminated tip is almost an order of magnitude less than for the oxygen and Mg-terminated tips.

In order to explore the imaging mechanism for the alkali halides, the heterogeneous surface of NaCl crystal containing $(OCN)^-$ molecules as impurities has been observed by NC-AFM [13.58]. In this case, the atomic periodicity of the NaCl lattice is imaged. However, high-resolution investigations of impurities are limited by the convolution between the tip apex and the molecules at the surface. *Bennewitz* et al. [13.59] reported on the heterogeneous surfaces of alkali halide crystals. In this experiment, which was designed to observe the chemically nonhomogeneous alkali halide surface, a mixed crystal composing of 60% KCl and 40% KBr was used as the sample. In this crystal, which is a so-called solid solution, the Cl and Br ions are mixed randomly. The image of the cleaved $KCl_{0.6}Br_{0.4}(001)$ surface indicates that only one type of ion is imaged as protrusions as if it were a pure alkali halide crystal. However, there is a significant difference in comparison with the surface of a pure crystal. The amplitude of atomic corrugation varies strongly between the positions of the ions imaged as depressions. This variation in the corrugations corresponds to the constituents of the crystal, i. e., the Cl and Br ions, and it is concluded that the tip apex is negatively charged. Also, it is demonstrated that a comparison between the distribution of the deep depressions and the composition of the mixed crystal assigns the deep depressions to Br ions. Although an evaluation of the effect of the surface buckling induced by a random distribution of anions is needed in order to understand the contrast mechanism, this result demonstrates that NC-AFM measurement exhibits chemical sensitivity on an atomic scale. For applications to alkali halide surfaces, real space imaging with NC-AFM presents the possibility for chemical discrimination at the surface.

Fluorides

Fluorides are important materials for the progress of atomic-scale resolution NC-AFM imaging on insulators. There are reports in the literature of surface images for single crystal BaF_2, SrF_2 [13.60], and CaF_2 [13.61–63]. Surfaces of fluorite-type crystals are prepared by cleaving along the (111) planes. Their structure is more complex than the structure of alkali halides with the NaCl structure. The complexity is of great interest for atomic-resolution imaging using an NC-AFM and also for theoretical predictions of the interpretation of atomic-scale contrast information.

The first atomically resolved images of a $CaF_2(111)$ surface have been obtained in topographic mode [13.62] in which the tip-sample distance is regulated by maintaining a constant frequency shift. In this imaging mode, the surface ions mostly appear as spherical caps. Similar measurements have been performed on the (111) surfaces of SrF_2 and BaF_2.

Barth et al. [13.64] have found that the $CaF_2(111)$ surface images obtained using the constant height mode, in which the frequency shift is recorded with a very low loop gain, can be categorized into two contrast patterns. In the first, the ions appear as triangles, and in the other, they have the appearance of circles, similar to the contrast obtained in a topographic image. These differences are revealed by comparing the cross sections along the [121] direction. Theoretical studies demonstrated that these two different contrast patterns could be explained as a result of imaging with tips of different polarity [13.64–66]. When imaging with a positively charged (cation-terminated) tip, the triangular pattern appears. In this case, the contrast is dominated by the strong short-range electrostatic attraction between the positive tip and the negative F ions. The cross section along the [121] direction of the triangular image shows two maxima: one is a larger peak over the F ions located in the topmost layer, and the other is a smaller peak at the position of the F ions in the third layer. The minima appear at the position of the Ca ions in the second layer. When imaging with a negatively charged (anion-terminated) tip, the spherical image contrast appears, and the main periodicity is created by the Ca ions in the second layer between the topmost and the third F ion layers. In the cross section along the [121] direction, the large maxima correspond to the Ca sites, because of the strong attraction of the negative tip, and the minima appear at the sites of maximum repulsion over the F ions in the first layer. At a position between two F ions, there are smaller maxima. This reflects the weaker repulsion over the third layer F ion sites compared to the protruding topmost F ion sites and a slower decay in the contrast on one side of the Ca ions. These results demonstrate that the image contrast of a $CaF_2(111)$ surface depends on the state of the tip apex.

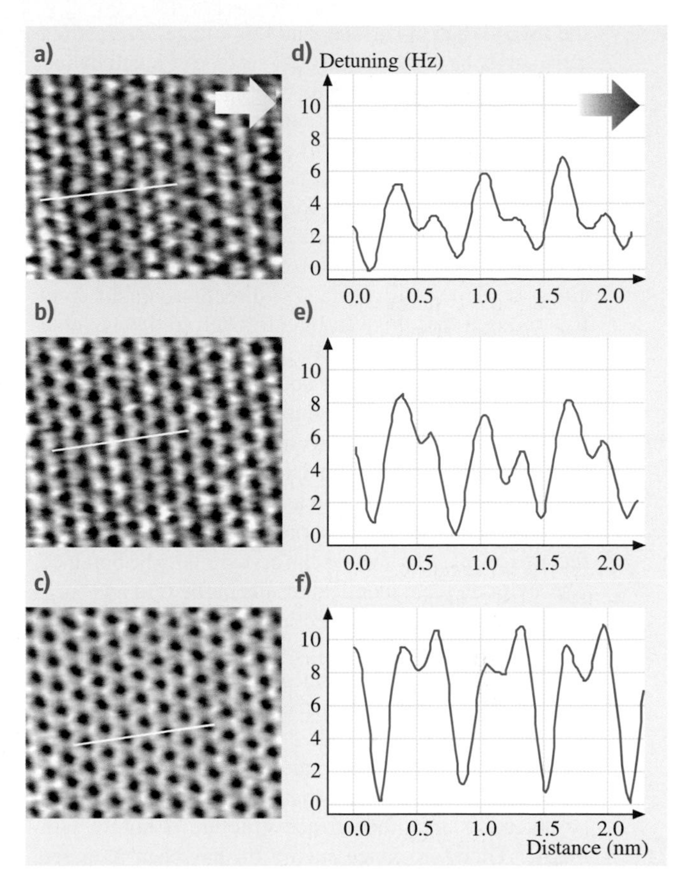

Fig. 13.18 (a)–(c) $CaF_2(111)$ surface images obtained by using the constant height mode. From (a) to (c) the frequency shift was lowered. The *white lines* represent the positions of the cross section. (d)–(f) The cross section extracted from the Fourier-filtered images of (a)–(c). The *white* and *black arrows* represent the scanning direction. The images and the cross sections are from [13.64]

The triangular pattern obtained with a positively charged tip appears at relatively large tip-sample distance, as shown in Fig. 13.18a. On the cross section along the [121] direction, experimental results and theoretical studies both demonstrate the large peak and small shoulder characteristic for the triangular pattern image (Fig. 13.18d). When the tip approaches the surface more closely, the triangular pattern of the experimental images is more vivid (Fig. 13.18b), just as predicted in the theoretical works. On the cross section along the [121] direction, as the tip approaches, the amplitude of the shoulder increases until it is equal to that of the main peak, and this feature gives rise to the honeycomb pattern image, as shown Fig. 13.18c. These results demonstrate that there is a more complex process of image contrast formation. Also, from detailed theoretical analysis of the electrostatic potential [13.67], it is suggested that the change in displacement of the ions due to the proximity of the tip plays an important role in the formation of the image contrast. Such a drastic change in image contrast, depending on both the polarity of the terminated tip atom and the tip-sample distance, is inherent in the fluoride (111) surface, and this image contrast feature cannot be seen on the (001) surface of alkali halides with a simple crystal structure.

These results have demonstrated that there is good agreement between experiment and theory. However, the results of careful experiments show another feature: The cross sections in the forward and backward scan directions do not yield identical results [13.64]. Also, it is shown that the structures of cross sections taken along the three equivalent [121] directions are different. It is thought that this can be attributed to the asymmetry of the nano-cluster at the tip apex, which leads to different interactions in the equivalent directions. A better understanding of the asymmetric image contrast may require a more complicated modeling of the tip structure. However, the most desirable solution would be the development of the suitable techniques for well-defined tip preparation. In fact, it should be mentioned that perfect tips on an atomic scale can occasionally be obtained. These tips do yield identical results in forward and backward scanning, and cross sections in the three equivalent directions taken with this tip are almost identical [13.68].

The fluoride (111) surface is an excellent standard surface for calibrating tips on an atomic scale. The polarity of the tip-terminated atom can be determined from the image contrast pattern (spherical or triangular pattern). The irregularities in the tip structure can be detected, since the surface structure is highly symmetric. Therefore, once such a tip has been prepared, it can be used as a calibrated tip for imaging unknown surfaces.

The states of the tip apex play an important role in interpreting NC-AFM images of alkali halide and fluoride surfaces. It is expected that the achievement of a good correlation between experimental and theoretical studies will help advance surface imaging for insulators by NC-AFM.

Metal Oxides

Most of the metal oxides that have attracted strong interest for their technological importance are insulative. Therefore, in the case of atomically resolved imaging of metal oxide surfaces by STM, efforts to increase the conductivity of the sample are needed, for example, the introduction of defects of anions or cations, doping with other atoms, and surface observations during heating of the sample. However, in principle, NC-AFM provides the possibility of observing nonconductive metal oxides without these efforts. In cases where the conductivity of the metal oxides is high enough for a tunneling current to flow, it should be noted that most of the surface images obtained by NC-AFM and STM are not identical.

Since the first report of atomically resolved images on a $TiO_2(110)$ surface with oxygen point defects [13.69], they have also been reported on $TiO_2(100)$ [13.70–72], $SnO_2(110)$ [13.73], NiO(001) [13.74, 75], $SrTiO_3(100)$ [13.76], and $CeO_2(111)$ [13.77] surfaces. Also, *Barth* et al. [13.78] have succeeded in obtaining atomically resolved NC-AFM images of a clean α-$Al_2O_3(0001)$ surface, which is impossible to investigate using STM. In this section, we describe typical results of the imaging of metal oxides by NC-AFM.

The α-$Al_2O_3(0001)$ surface exists in several ordered phases that can be reversibly transformed into each other by thermal treatments and oxygen exposure. It is known that the high temperature phase has a large $(\sqrt{31}\times\sqrt{31})R\pm 9°$ unit cell. However, the details of the atomic structure of this surface have not been revealed, and two models are proposed from the results of low-energy electron diffraction and X-ray diffraction studies. Recently, *Barth* et al. [13.78] have directly observed this reconstructed α-$Al_2O_3(0001)$ surface by NC-AFM. They confirmed that the dominant contrast of the low magnification image corresponds to a rhombic grid representing a unit cell of $(\sqrt{31}\times\sqrt{31})R+9°$, as shown in Fig. 13.19a. Also, more details of atomic structures are determined from the higher magnification image (Fig. 13.19b), which was taken at a reduced tip-sample distance. In this atomically resolved image, it is revealed that each side of the rhombus is intersected by 10 atomic rows, and that a hexagonal arrangement of atoms exists in the center of the rhombi (Fig. 13.19c). This feature agrees with the proposed surface structure that predicts order in the center of hexagonal surface domains and disorder at the domain boundaries. Their result is an excellent demonstration of the capabilities of the NC-AFM as a powerful tool for the atomic-scale surface investigation of insulators.

In addition, *Barth* et al. [13.78] have performed observations of the $Al_2O_3(0001)$ surface when it is exposed to water and hydrogen. It is suggested that the region of atomic disorder has an extremely high reactivity. Such investigations of the surface structure of adsorbed insulator substrates can only be provided by NC-AFM. The NC-AFM imaging of metal growth on oxides [13.79] has also been reported. A description of the adsorbed molecular arrangement on a TiO_2 surface can be found in the next section.

The atomic structure of $SrTiO_3(100)$-$(\sqrt{5}\times\sqrt{5})$ $R26.6°$ surface, as well as that of $Al_2O_3(0001)$-$(\sqrt{31}\times\sqrt{31})R\pm 9°$ can be determined on the basis of

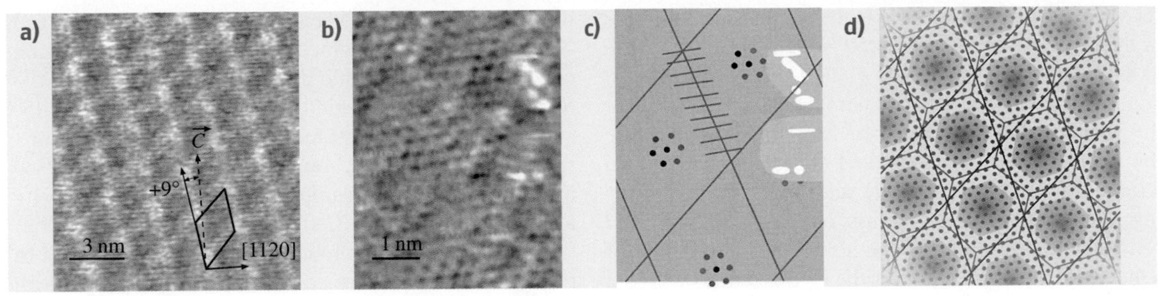

Fig. 13.19 (**a**) Image of the high temperature, reconstructed clean α-Al_2O_3 surface obtained by using the constant height mode. The *rhombus* represents the unit cell of the $(\sqrt{31}\times\sqrt{31})R+9°$ reconstructed surface. (**b**) Higher magnification image of (**a**). Imaging was performed at a reduced tip-sample distance. (**c**) Schematic representation of the indicating regions of hexagonal order in the center of reconstructed rhombi. (**d**) Superposition of the hexagonal domain with reconstruction rhombi found by NC-AFM imaging. Atoms in *gray shaded regions* are well-ordered. The images and the schematic representations are from [13.78]

the results of NC-AFM imaging [13.76]. $SrTiO_3$ is one of the perovskite oxides and its (100) surface exhibits the many different kinds of reconstructed structures. In the case of the $(\sqrt{5}\times\sqrt{5})R26.6°$ reconstruction, the oxygen vacancy-Ti^{3+}-oxygen model (where the terminated surface is TiO_2 and the observed spots are related to oxygen vacancies) was proposed from the results of STM imaging. As shown in Fig. 13.20, *Kubo* et al. [13.76] have performed measurements using both STM and NC-AFM and have found that the size of the bright spots, as observed by NC-AFM, is always smaller than that for STM measurement and that the dark spots, which are not observed by STM, are arranged along the [001] and [010] directions in the NC-AFM image. A theoretical simulation of the NC-AFM image using first-principles calculations shows that the bright and dark spots correspond to Sr and oxygen atoms, respectively. It has been proposed that the structural model of the reconstructed surface consists of an ordered Sr adatom located at the oxygen fourfold site on the TiO_2-terminated layer (Fig. 13.20c).

Because STM images are related to the spatial distribution of the wave functions near the Fermi level, atoms, which do not have a local density of states near the Fermi level, are generally invisible even on the conductive materials. On the other hand, the NC-AFM image reflects the strength of the tip-sample interaction force originating from chemical, electrostatic interactions, and so on. Therefore, even STM and NC-AFM images obtained using an identical tip and sample will generally differ.

Simultaneous imaging of a $TiO_2(110)$ surface with STM and NC-AFM shows a typical result that differs in terms of the atoms that are visualized [13.72]. For simultaneous imaging, the tip-sample distance is controlled by keeping a constant frequency shift, and the tunneling current between the oscillating cantilever and the surface is measured. The two images show different atomic contrasts, i. e., the atomic corrugation of the

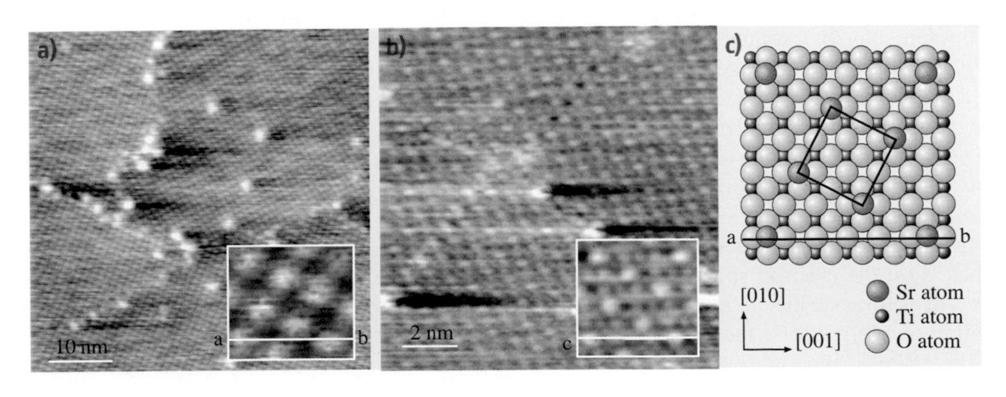

Fig. 13.20 (**a**) STM and (**b**) NC-AFM images of a $SrTiO_3(100)$ surface. (**c**) A proposed model of the $SrTiO_3(100)$-$(\sqrt{5}\times\sqrt{5})R26.6°$ surface reconstruction. The images and the schematic representations are from [13.76]

STM image is in antiphase in the (1×1) and in phase in the (1×2) with respect to the NC-AFM image. In the (1×1), the STM image shows that the dangling bond states at the tip apex overlap with the dangling bonds of the $3d$ states that are protruding from the Ti atom, while the NC-AFM primarily imaged the uppermost oxygen atom. The results of $SrTiO_3$ and TiO_2 surface imaging show that the information obtained by STM and NC-AFM differ, and that the simultaneous imaging of a metal oxide surface enables the investigation of a more detailed surface structure.

13.3.2 Atomically Resolved Imaging of a NiO(001) Surface

The transition metal oxides such as NiO, CoO, and FeO feature the simultaneous existence of an energy gap and unpaired electrons, which gives rise to magnetic properties. Such magnetic insulators are widely used for the exchange biasing in spin valve devices. The NC-AFM enables the direct surface imaging of magnetic insulators on an atomic scale. The forces detected by NC-AFM originate from several kinds of interaction between the surface and the tip, including magnetic interactions in some cases. Theoretical studies predict that short-range magnetic interactions such as the exchange interaction should enable the NC-AFM to image magnetic moments on an atomic scale. In this section, we will describe an imaging of the antiferromagnetic NiO(001) surface using a ferromagnetic Fe-coated tip in order to explore the detection of the short-range magnetic interaction. Also, theoretical studies of the exchange force interaction between a magnetic tip and a sample will be described.

Theoretical Studies of the Exchange Force

The NC-AFM measures various kinds of interaction between tip and sample such as van der Waals, and electrostatic and short-range chemical interactions. In the case of a magnetic tip and sample, the interaction detected by the NC-AFM includes the long-range magnetic dipole interaction and the short-range magnetic interaction. The short-range magnetic interaction between the atoms at the tip apex and the sample surface is governed by a combination of the Coulomb interaction and the Pauli exclusion principle and may be included among the short-range chemical interactions. The magnetic interaction energy depends on the electron spin state of the atoms, and the energy difference in spin alignments (parallel or anti-parallel) is referred to as the exchange interaction energy. Therefore, if the short-range magnetic interaction can be detected, an atomically resolved NC-AFM image, in which the short-range interaction contributes to the atomic protrusions, should reveal the local energy difference in the spin alignment.

In the past, extensive theoretical studies on the short-range magnetic interaction between a ferromagnetic tip and a ferromagnetic sample have been performed by a simple calculation [13.80], a tight-binding approximation [13.81], and first-principles calculations [13.82]. In the calculations performed by *Nakamura* et al. [13.82], three-atomic-layer Fe(001) films, which are separated by a vacuum gap, are used as a model for the tip and sample. The exchange force (F_{ex}) is defined as the difference of the forces (F_P, F_{AP}) in each spin configuration (parallel: P, anti-parallel: AP) of the tip and sample. The tip-sample distance dependency of the exchange force demonstrates that the amplitude of the exchange force is on the order of 10^{-10} N or above and is large enough to be detected by NC-AFM. Also, the site dependency of the exchange force indicates that the discrimination of the exchange force enables the direct imaging of the magnetic moments on an atomic scale. Recently, the interaction between a spin-polarized H atom and a Ni atom on a NiO(001) surface has been investigated theoretically [13.83]. *Foster* and *Shluger* demonstrated that the difference in magnitude of the exchange interaction between opposite spin Ni ions in a NiO surface could be sufficient to be measured in a low temperature NC-AFM experiment. Also, in order to investigate the effect of surface relaxation, they calculated the interaction of different metal tips with the MgO surface used as a model system. The results indicated that the interaction over the oxygen site is stronger than that over the Mg site, and it was pointed out that the tips should be coated in metals that interact weakly with the surface for the detection of the exchange interaction between the atom at the tip apex and the Ni atom on the NiO surface. However, we must take into account another possibility, that is, that a metal atom at the ferromagnetic tip apex may interact with a Ni atom on the second layer through a magnetic interaction mediated by the electrons in an oxygen atom on the surface.

13.3.3 Atomically Resolved Imaging Using Noncoated and Fe-Coated Si Tips

It is thought that the magnetic insulator NiO is one of the best candidates for detecting short-range magnetic interactions such as the exchange interaction, because an experimental configuration of a ferromagnetic tip and an antiferromagnetic sample minimizes the long-range magnetic dipole interaction that might hinder the detection of the short-range interaction.

Below the Néel temperature of 525 K, NiO is antiferromagnetic and has a rhombohedral structure contracted along the ⟨111⟩ axes. The direction of the spin on nickel atoms in the same {111} plane is parallel, a so-called ferromagnetic sheet. The direction of spin in the adjacent {111} planes is anti-parallel. As a result, the spins at the Ni atoms on the (001) surface are arranged in a checkerboard pattern, i. e., the spin alignment along the ⟨111⟩ direction is antiferromagnetic in order and that along the ⟨1 −1 0⟩ direction is ferromagnetic in order.

Figure 13.21a shows an atomically resolved image of a NiO(001) surface with a ferromagnetic Fe-coated tip [13.84]. The bright protrusions correspond to atoms spaced about 0.42 nm apart, consistent with the expected periodic arrangement of the NiO(001) surface.

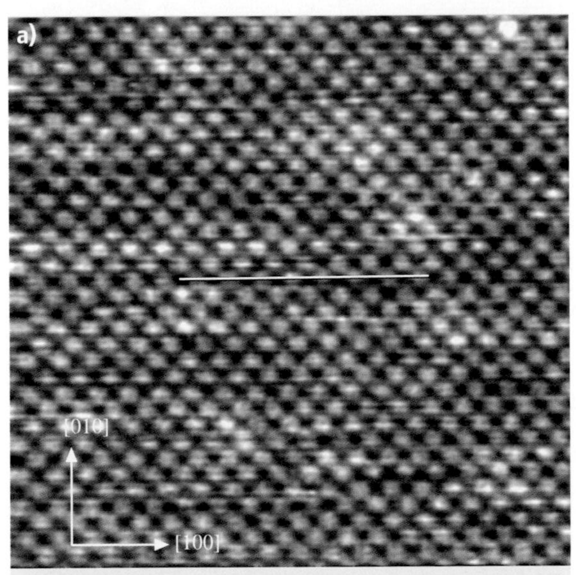

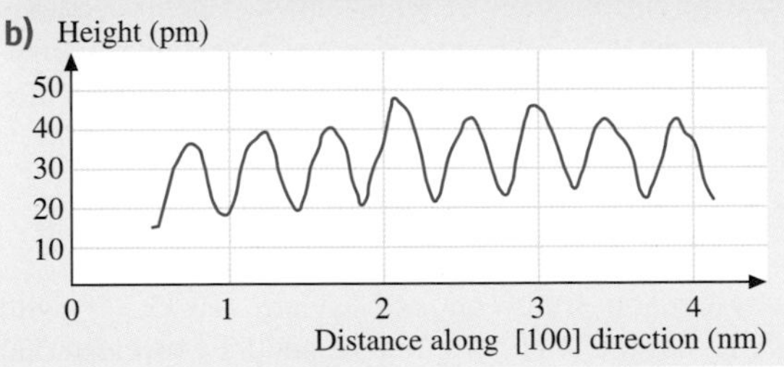

Fig. 13.21 (**a**) Atomically resolved image obtained with an Fe-coated tip. (**b**) shows the cross sections of the middle part in (**a**). Their corrugations are about 30 pm

The corrugation amplitude is typically 30 pm, which is comparable to the value previously reported [13.74, 75], as shown in Fig. 13.21b. Also, because of the symmetry of the image, it was concluded that only one type of atom is visualized by NC-AFM. However, from this image, we think that it is difficult to distinguish which of the atoms are observed, oxygen or nickel. The theoretical works indicate that a metal tip interacts strongly with the oxygen atoms on the MgO(001) surface [13.83]. From this result, it is presumed that the bright protrusions correspond to the oxygen atoms. However, it is still questionable which of the atoms are visible with a Fe-coated tip.

If the short-range magnetic interaction is included in the interactions detected by NC-AFM, the corrugation amplitude of the atoms should depend on the direction of the spin at the atom-site. The results of a comparison of the corrugation amplitudes of the same species of atoms with opposite spin can reveal the difference in the short-range magnetic interaction, because the nonmagnetic interactions peculiar to a particular atomic species should be identical. From the results of first-principles calculations [13.82], the contribution of the short-range magnetic interaction to the measured corrugation amplitude is expected to be about a few percent of the total interaction. Discrimination of such small perturbations is therefore needed. In order to reduce the noise, the corrugation amplitude was added on the basis of the periodicity of the NC-AFM image. In addition, the topographical asymmetry, which is the index characterizing the difference in atomic corrugation amplitude, has been defined [13.85]. The result shows that the value of the topographical asymmetry calculated from the image obtained with an Fe-coated Si tip depends on the direction of summing of the corrugation amplitude, and that the dependency corresponds to the antiferromagnetic spin ordering of the NiO(001) surface. With a paramagnetic Si tip, no significant indication of a directional dependency of the topographical asymmetry can be seen. Therefore, it is believed that the results obtained using a Fe-coated tip support the conclusion that the dependency of the topographical asymmetry originates in the short-range magnetic interaction.

The measurements presented here demonstrate the possibility of imaging magnetic structures on an atomic scale by NC-AFM. However, the origin of the detected short-range magnetic interaction is under discussion. Therefore, further experiments such as the measurement of force-distance curves and a theoretical study of the interaction between the ferromagnetic tip and an antiferromagnetic sample are needed.

13.4 Applications to Molecules

In the future, it is expected that electronic, chemical, and medical devices will be downsized to a nanoscale. To achieve this, visualizing and assembling individual molecular components is of fundamental importance. Topographic imaging of nonconductive materials is a challenge for atomic force microscopy (AFM). Nanometer-sized domains of surfactants terminated with different functional groups have been identified by lateral force microscopy (LFM) [13.88] and chemical force microscopy (CFM) [13.89] as extensions of AFM. At a higher resolution, a periodic array of molecules, Langmuir–Blodgett films [13.90] for example, was recognized by AFM. However, it remains difficult to visualize an isolated molecule, molecule vacancy, or the boundary of different periodic domains with a microscope with the tip in contact.

13.4.1 Why Molecules and What Molecules?

The access to individual molecules has not been a trivial task even for noncontact atomic force microscopy (NC-AFM). The force pulling the tip into the surface is less sensitive to the gap width (r), especially when chemically stable molecules cover the surface. The attractive potential between two stable molecules is shallow and exhibits r^{-6} decay [13.13].

High-resolution topography of formate ($HCOO^-$) [13.91] was first reported in 1997 as a molecular adsorbate. The number of imaged molecules is now increasing because of the technological importance of molecular interfaces. To date, the following studies on molecular topography have been published: C_{60} [13.86, 92], DNAs [13.87, 93], adenine and thymine [13.94], alkanethiols [13.94, 95], a perylene-derivative (PTCDA) [13.96], a metal porphyrin (Cu-TBPP) [13.97], glycine sulfate [13.98], polypropylene [13.99], vinylidene fluoride [13.100], and a series of carboxylates ($RCOO^-$) [13.101–107]. Two of these are presented in Figs. 13.22 and 13.23 to demonstrate the current stage of achievement. The proceedings of the annual NC-AFM conference represent a convenient opportunity for us to update the list of molecules imaged.

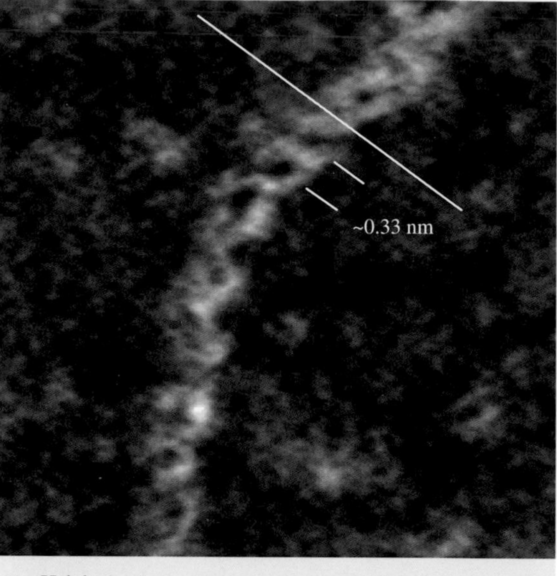

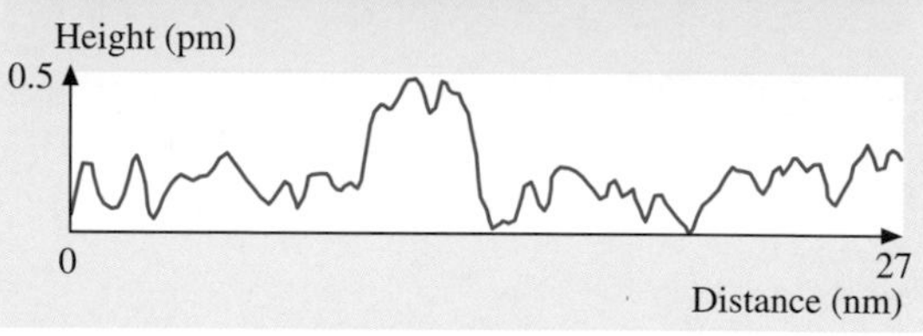

Fig. 13.23 The constant frequency-shift topography of a DNA helix on a mica surface based on [13.87]. Image size: 43 × 43 nm^2. The image revealed features with a spacing of 3.3 nm, consistent with the helix turn of B-DNA

Fig. 13.22 The constant frequency-shift topography of domain boundaries on a C_{60} multilayered film deposited on a Si(111) surface based on [13.86]. Image size: 35 × 35 nm^2

13.4.2 Mechanism of Molecular Imaging

A systematic study on carboxylate ($RCOO^-$) with R = H, CH_3, $C(CH_3)_3$, C ≡ CH, and CF_3 revealed that the van der Waals force is responsible for the molecule-dependent microscope topography despite its long-range (r^{-6}) nature. Carboxylates adsorbed on the (110) sur-

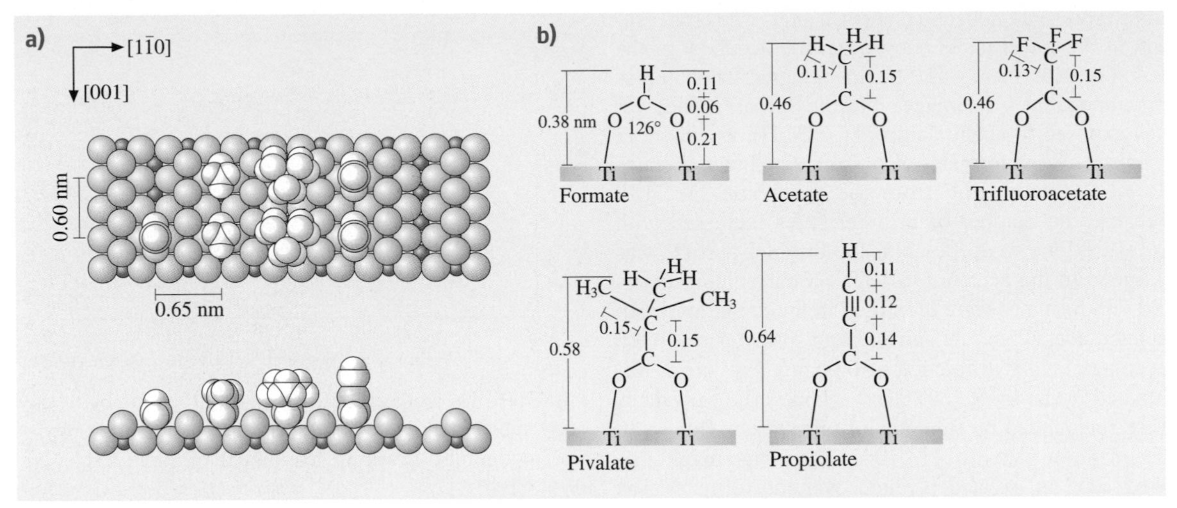

Fig. 13.24a,b The carboxylates and TiO_2 substrate. (**a**) Top and side view ball model. *Small shaded* and *large shaded balls* represent Ti and O atoms in the substrate. Protons yielded in the dissociation reaction are not shown. (**b**) Atom geometry of formate, acetate, pivalate, propiolate, and trifluoroacetate adsorbed on the $TiO_2(110)$ surface. The O-Ti distance and O – C – O angle of the formate were determined in the quantitative analysis of photoelectron diffraction [13.109]

face of rutile TiO_2 have been extensively studied as a prototype of organic materials interfaced with an inorganic metal oxide [13.108]. A carboxylic acid molecule (RCOOH) dissociates on this surface to a carboxylate ($RCOO^-$) and a proton (H^+) at room temperature, as illustrated in Fig. 13.24. The pair of negatively charged oxygen atoms in $RCOO^-$ coordinate two positively charged Ti atoms at the surface. The adsorbed carboxylates create a long-range ordered monolayer. The lateral distances of the adsorbates in the ordered monolayer are regulated at 0.65 and 0.59 nm along the [1$\bar{1}$0] and [001] directions. By scanning a mixed monolayer containing different carboxylates, the microscope topography of the terminal groups can be quantitatively compared while minimizing tip-dependent artifacts.

Figure 13.25 presents the observed constant frequency-shift topography of four carboxylates terminated by different alkyl groups. On the formate-covered surface of panel (a) individual formates (R = H) were resolved as protrusions of a uniform brightness. The dark

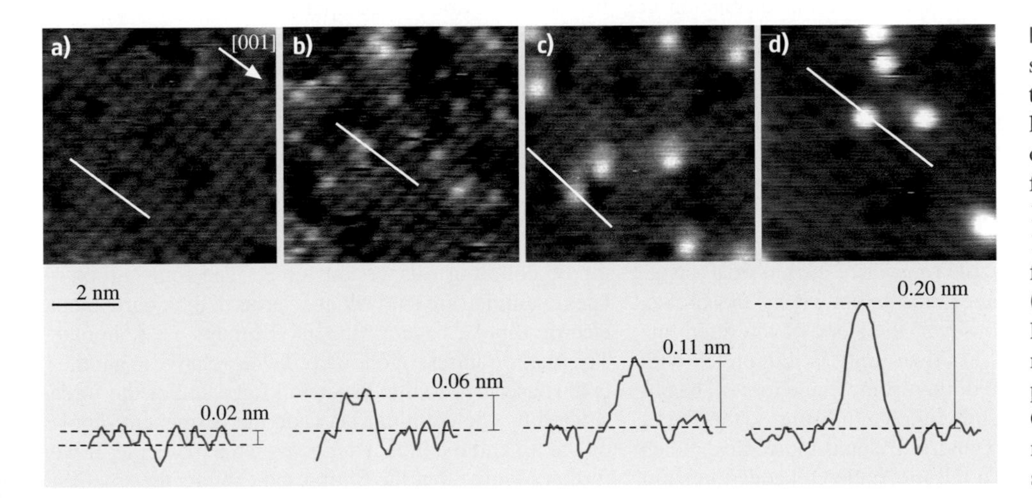

Fig. 13.25a–d The constant frequency-shift topography of carboxylate monolayers prepared on the $TiO_2(110)$ surface based on [13.102, 104, 106]. Image size: $10 \times 10\,nm^2$. (**a**) Pure formate monolayer; (**b**) formate-acetate mixed layer; (**c**) formate-pivalate mixed layer; (**d**) formate-propiolate mixed layer. Cross sections determined on the *lines* are shown in the lower panel

holes represent unoccupied surface sites. The cross section in the lower panel shows that the accuracy of the height measurement was 0.01 nm or better. Brighter particles appeared in the image when the formate monolayer was exposed to acetic acid (CH_3COOH), as shown in panel (b). Some formates were exchanged with acetates ($R = CH_3$) impinging from the gas phase [13.110]. Because the number of brighter spots increased with exposure time to acetic acid, the brighter particle was assigned to the acetate [13.102]. Twenty-nine acetates and 188 formates were identified in the topography. An isolated acetate and its surrounding formates exhibited an image height difference of 0.06 nm. Pivalate is terminated by bulky $R = C(CH_3)_3$. Nine bright pivalates were surrounded by formates of ordinary brightness in the image of panel (c) [13.104]. The image height difference of an isolated pivalate over the formates was 0.11 nm. Propiolate with C≡CH is a needle-like adsorbate of single-atom diameter. That molecule exhibited in panel (d) a microscope topography 0.20 nm higher than that of the formate [13.106].

The image topography of formate, acetate, pivalate, and propiolate followed the order of the size of the alkyl groups. Their physical topography can be assumed on the C−C and C−H bond lengths in the corresponding RCOOH molecules in the gas phase [13.111] and is illustrated in Fig. 13.24. The top hydrogen atom of the formate is located 0.38 nm above the surface plane containing the Ti atom pair, while three equivalent hydrogen atoms of the acetate are more elevated at 0.46 nm. The uppermost H atoms in the pivalate are raised by 0.58 nm relative to the Ti plane. The H atom terminating the triple-bonded carbon chain in the propiolate is at 0.64 nm. Figure 13.26 summarizes the observed image heights relative to the formate as a function of the physical height of the topmost H atoms given in the model. The straight line fitted the four observations [13.104]. When the horizontal axis was scaled with other properties (molecular weight, the number of atoms in a molecule, or the number of electrons in valence states), the correlation became poor.

On the other hand, if the tip apex traced the contour of a molecule composed of hard-sphere atoms, the image topography would reproduce the physical topography in a one-to-one ratio, as shown by the broken line in Fig. 13.26. However, the slope of the fitted line was 0.7. A slope of less than unity is interpreted with the long-range nature of the tip-molecule force. The observable frequency shift reflects the sum of the forces between the tip apex and individual molecules. When the tip passes above a tall molecule embedded in short

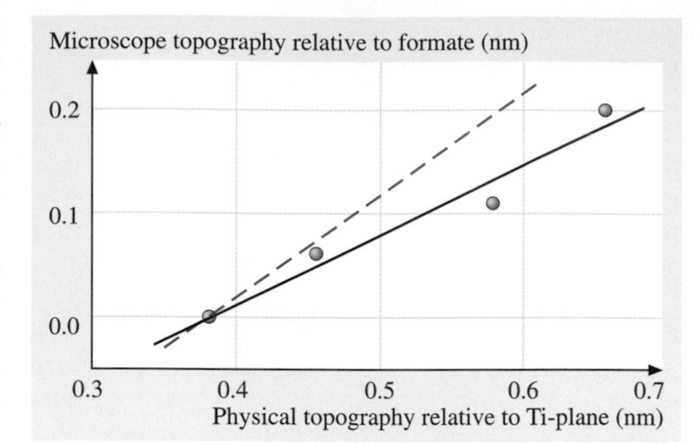

Fig. 13.26 The constant frequency-shift topography of the alkyl-substituted carboxylates as a function of their physical topography given in the model of Fig. 13.3 based on [13.104]

molecules, it is pulled up to compensate for the increased force originating from the tall molecule. Forces between the lifted tip and the short molecules are reduced due to the increased tip-surface distance. Feedback regulation pushes down the probe to restore the lost forces.

This picture predicts that microscope topography is sensitive to the lateral distribution of the molecules, and that was in fact the case. Two-dimensionally clustered acetates exhibited enhanced image height over an isolated acetate [13.102]. The tip-molecule force, therefore, remained non-zero at distances over the lateral separation of the carboxylates on this surface (0.59–0.65 nm). Chemical bond interactions cannot be important across such a wide tip-molecule gap, whereas atom-scale images of Si(111)-(7×7) are interpreted with the fractional formation of tip-surface chemical bonds [13.24, 45, 49]. Instead, the attractive component of the van der Waals force is probable for the observed molecule-dependent topography. The absence of the tip-surface chemical bond is reasonable on the carboxylate-covered surface terminated with stable C−H bonds.

The attractive component of the van der Waals force contains electrostatic terms caused by permanent-dipole/permanent-dipole coupling, permanent-dipole/induced-dipole coupling, and induced-dipole/induced-dipole coupling (dispersion force). The four carboxylates examined are equivalent in terms of their permanent electric dipole, because the alkyl groups are nonpolar. The image contrast of one carboxylate relative to another is thus ascribed to the dispersion force and/or the force created by the coupling between the permanent dipole on the tip and the induced dipole on the molecule. If we further assume that the Si tip used exhibits the smallest permanent dipole, the dispersion force remains dominant to create the NC-AFM topography dependent on the nonpolar groups of atoms. A numerical simulation on this assumption [13.106] successfully reproduced the propiolate topography of Fig. 13.25d. A calculation that does not include quantum chemical treatments is expected to work, unless the tip approaches the surface too closely, or the molecule possesses a dangling bond.

In addition to the contribution of the dispersion force, the permanent dipole moment of molecules may perturb the microscope topography through electrostatic coupling with the tip. Its possible role was demonstrated by imaging a fluorine-substituted acetate. The strongly polarized C−F bonds were expected to perturb the electrostatic field over the molecule. The constant frequency-shift topography of acetate ($R = CH_3$) and trifluoroacetate ($R = CF_3$) was indeed sensitive to the fluorine substitution. The acetate was observed 0.05 nm higher than the trifluoroacetate [13.103], although the F atoms in the trifluoroacetate, as well as the H atoms in the acetate were lifted by 0.46 nm from the surface plane, as illustrated in Fig. 13.24.

13.4.3 Perspectives

Experimental results summarized in this section prove the feasibility of NC-AFM in identifying individual molecules. A systematic study on the constant frequency-shift topography of carboxylates with $R = H$, CH_3, $C(CH_3)_3$, $C{\equiv}CH$, and CF_3 has revealed the mechanism behind the high-resolution imaging of the chemically stable molecules. The dispersion force is primarily responsible for the molecule-dependent topography. The permanent dipole moment of the imaged molecule, if it exists, perturbs the topography through the electrostatic coupling with the tip. A tiny calculation containing empirical force fields works in simulating the microscope topography.

These results make us optimistic about analyzing physical and chemical properties of nanoscale supramolecular assemblies constructed on a solid surface. If the accuracy of topographic measurement is developed by one more order of magnitude, which is not an unrealistic concept, it may be possible to identify structural isomers, chiral isomers, and conformational isomers of a molecule. Kelvin probe force microscopy (KPFM), an extension of NC-AFM, provides a nanoscale analysis of molecular electronic properties [13.99, 100]. Force spectroscopy with chemically modified tips seems promising for the detection of a selected chemical force. Operation in a liquid atmosphere [13.112] is demanded to observe biochemical materials in the raw.

References

13.1 G. Binnig: Atomic force microscope and method for imaging surfaces with atomic resolution, US Patent 4,724,318 (1986)

13.2 G. Binnig, C. F. Quate, C. Gerber: Atomic force microscope, Phys. Rev. Lett. **56** (1986) 930–933

13.3 G. Binnig, H. Rohrer, C. Gerber, E. Weibel: Surface studies by scanning tunneling microscopy, Phys. Rev. Lett. **49** (1982) 57–61

13.4 G. Binnig, H. Rohrer: The scanning tunneling microscope, Sci. Am. **253** (1985) 50–56

13.5 G. Binnig, H. Rohrer: In touch with atoms, Rev. Mod. Phys. **71** (1999) S324–S320

13.6 C. J. Chen: *Introduction to Scanning Tunneling Microscopy* (Oxford Univ. Press, Oxford 1993)

13.7 H.-J. Güntherodt, R. Wiesendanger (Eds.): *Scanning Tunneling Microscopy I–III* (Springer, Berlin, Heidelberg 1991)

13.8 J. A. Stroscio, W. J. Kaiser (Eds.): *Scanning Tunneling Microscopy* (Academic, Boston 1993)

13.9 R. Wiesendanger: *Scanning Probe Microscopy and Spectroscopy: Methods and Applications* (Cambridge Univ. Press, Cambridge 1994)

13.10 S. Morita, R. Wiesendanger, E. Meyer (Eds.): *Noncontact Atomic Force Microscopy* (Springer, Berlin, Heidelberg 2002)

13.11 R. Garcia, R. Perez: Dynamic atomic force microscopy methods, Surf. Sci. Rep. **47** (2002) 197–301

13.12 F. J. Giessibl: Advances in atomic force microscopy, Rev. Mod. Phys. **75** (2003) 949–983

13.13 J. Israelachvili: *Intermolecular and Surface Forces*, 2nd edn. (Academic, London 1992)

13.14 L. Olsson, N. Lin, V. Yakimov, R. Erlandsson: A method for *in situ* characterization of tip shape in ac-mode atomic force microscopy using electrostatic interaction, J. Appl. Phys. **84** (1998) 4060–4064

13.15 S. Akamine, R. C. Barrett, C. F. Quate: Improved atomic force microscopy images using cantilevers with sharp tips, Appl. Phys. Lett. **57** (1990) 316–318

13.16 T. R. Albrecht, S. Akamine, T. E. Carver, C. F. Quate: Microfabrication of cantilever styli for the atomic force microscope, J. Vac. Sci. Technol. A **8** (1990) 3386–3396

13.17 M. Tortonese, R.C. Barrett, C. Quate: Atomic resolution with an atomic force microscope using piezoresistive detection, Appl. Phys. Lett. **62** (1993) 834–836

13.18 O. Wolter, T. Bayer, J. Greschner: Micromachined silicon sensors for scanning force microscopy, J. Vac. Sci. Technol. **9** (1991) 1353–1357

13.19 D. Sarid: *Scanning Force Microscopy*, 2nd edn. (Oxford Univ. Press, New York 1994)

13.20 F.J. Giessibl, B.M. Trafas: Piezoresistive cantilevers utilized for scanning tunneling and scanning force microscope in ultrahigh vacuum, Rev. Sci. Instrum. **65** (1994) 1923–1929

13.21 P. Güthner, U.C. Fischer, K. Dransfeld: Scanning near-field acoustic microscopy, Appl. Phys. B **48** (1989) 89–92

13.22 K. Karrai, R.D. Grober: Piezoelectric tip-sample distance control for near field optical microscopes, Appl. Phys. Lett. **66** (1995) 1842–1844

13.23 F.J. Giessibl: High-speed force sensor for force microscopy and profilometry utilizing a quartz tuning fork, Appl. Phys. Lett. **73** (1998) 3956–3958

13.24 F.J. Giessibl, S. Hembacher, H. Bielefeldt, J. Mannhart: Subatomic features on the silicon (111)-(7×7) surface observed by atomic force microscopy, Science **289** (2000) 422–425

13.25 F. Giessibl, C. Gerber, G. Binnig: A low-temperature atomic force/scanning tunneling microscope for ultrahigh vacuum, J. Vac. Sci. Technol., B **9** (1991) 984–988

13.26 F. Ohnesorge, G. Binnig: True atomic resolution by atomic force microscopy through repulsive and attractive forces, Science **260** (1993) 1451–1456

13.27 F.J. Giessibl, G. Binnig: True atomic resolution on KBr with a low-temperature atomic force microscope in ultrahigh vacuum, Ultramicroscopy **42–44** (1992) 281–286

13.28 S.P. Jarvis, H. Yamada, H. Tokumoto, J.B. Pethica: Direct mechanical measurement of interatomic potentials, Nature **384** (1996) 247–249

13.29 L. Howald, R. Lüthi, E. Meyer, P. Guthner, H.-J. Güntherodt: Scanning force microscopy on the Si(111)7×7 surface reconstruction, Z. Phys. B **93** (1994) 267–268

13.30 L. Howald, R. Lüthi, E. Meyer, H.-J. Güntherodt: Atomic-force microscopy on the Si(111)7×7 surface, Phys. Rev. B **51** (1995) 5484–5487

13.31 Y. Martin, C.C. Williams, H.K. Wickramasinghe: Atomic force microscope – force mapping and profiling on a sub 100 Å scale, J. Appl. Phys. **61** (1987) 4723–4729

13.32 T.R. Albrecht, P. Grutter, H.K. Horne, D. Rugar: Frequency modulation detection using high-Q cantilevers for enhanced force microscope sensitivity, J. Appl. Phys. **69** (1991) 668–673

13.33 F.J. Giessibl: Atomic resolution of the silicon (111)-(7×7) surface by atomic force microscopy, Science **267** (1995) 68–71

13.34 S. Kitamura, M. Iwatsuki: Observation of silicon surfaces using ultrahigh-vacuum noncontact atomic force microscopy, Jpn. J. Appl. Phys. **35** (1995) L668–L671

13.35 R. Erlandsson, L. Olsson, P. Martensson: Inequivalent atoms and imaging mechanisms in ac-mode atomic-force microscopy of Si(111)7×7, Phys. Rev. B **54** (1996) R8309–R8312

13.36 N. Burnham, R.J. Colton: Measuring the nanomechanical and surface forces of materials using an atomic force microscope, J. Vac. Sci. Technol. A **7** (1989) 2906–2913

13.37 D. Tabor, R.H.S. Winterton: Direct measurement of normal and retarded van der Waals forces, Proc. R. Soc. London A **312** (1969) 435

13.38 F.J. Giessibl: Forces and frequency shifts in atomic resolution dynamic force microscopy, Phys. Rev. B **56** (1997) 16010–16015

13.39 G. Binnig, H. Rohrer, C. Gerber, E. Weibel: 7×7 reconstruction on Si(111) resolved in real space, Phys. Rev. Lett. **50** (1983) 120–123

13.40 H. Goldstein: *Classical Mechanics* (Addison Wesley, Reading 1980)

13.41 U. Dürig: Interaction sensing in dynamic force microscopy, New J. Phys. **2** (2000) 5.1–5.12

13.42 F.J. Giessibl: A direct method to calculate tip-sample forces from frequency shifts in frequency-modulation atomic force microscopy, Appl. Phys. Lett. **78** (2001) 123–125

13.43 U. Dürig, H.P. Steinauer, N. Blanc: Dynamic force microscopy by means of the phase-controlled oscillator method, J. Appl. Phys. **82** (1997) 3641–3651

13.44 F.J. Giessibl, H. Bielefeldt, S. Hembacher, J. Mannhart: Calculation of the optimal imaging parameters for frequency modulation atomic force microscopy, Appl. Surf. Sci. **140** (1999) 352–357

13.45 M.A. Lantz, H.J. Hug, R. Hoffmann, P.J.A. van Schendel, P. Kappenberger, S. Martin, A. Baratoff, H.-J. Güntherodt: Quantitative measurement of short-range chemical bonding forces, Science **291** (2001) 2580–2583

13.46 F.J. Giessibl, M. Herz, J. Mannhart: Friction traced to the single atom, Proc. Nat. Acad. Sci. USA **99** (2002) 12006–12010

13.47 N. Nakagiri, M. Suzuki, K. Oguchi, H. Sugimura: Site discrimination of adatoms in Si(111)-7×7 by noncontact atomic force microscopy, Surf. Sci. Lett. **373** (1997) L329–L332

13.48 Y. Sugawara, M. Ohta, H. Ueyama, S. Morita: Defect motion on an InP(110) surface observed with noncontact atomic force microscopy, Science **270** (1995) 1646–1648

13.49 T. Uchihashi, Y. Sugawara, T. Tsukamoto, M. Ohta, S. Morita: Role of a covalent bonding interaction in noncontact-mode atomic-force microscopy on Si(111)7×7, Phys. Rev. B **56** (1997) 9834–9840

13.50 K. Yokoyama, T. Ochi, A. Yoshimoto, Y. Sugawara, S. Morita: Atomic resolution imaging on

Si(100)2 × 1 and Si(100)2 × 1-H surfaces using a noncontact atomic force microscope, Jpn. J. Appl. Phys. **39** (2000) L113–L115
13.51 Y. Sugawara, T. Minobe, S. Orisaka, T. Uchihashi, T. Tsukamoto, S. Morita: Non-contact AFM images measured on Si(111)$\sqrt{3} \times \sqrt{3}$-Ag and Ag(111) surfaces, Surf. Interface Anal. **27** (1999) 456–461
13.52 K. Yokoyama, T. Ochi, Y. Sugawara, S. Morita: Atomically resolved Ag imaging on Si(111)$\sqrt{3} \times \sqrt{3}$-Ag surface with noncontact atomic force microscope, Phys. Rev. Lett. **83** (1999) 5023–5026
13.53 M. Bammerlin, R. Lüthi, E. Meyer, A. Baratoff, J. Lü, M. Guggisberg, Ch. Gerber, L. Howald, H.-J. Güntherodt: True atomic resolution on the surface of an insulator via ultrahigh vacuum dynamic force microscopy, Probe Microsc. **1** (1997) 3–7
13.54 M. Bammerlin, R. Lüthi, E. Meyer, A. Baratoff, J. Lü, M. Guggisberg, C. Loppacher, Ch. Gerber, H.-J. Güntherodt: Dynamic SFM with true atomic resolution on alkali halide surfaces, Appl. Phys. A **66** (1998) S293–S294
13.55 R. Hoffmann, M.A. Lantz, H.J. Hug, P.J.A. van Schendel, P. Kappenberger, S. Martin, A. Baratoff, H.-J. Güntherodt: Atomic resolution imaging and force versus distance measurements on KBr(001) using low temperature scanning force microscopy, Appl. Surf. Sci. **188** (2002) 238–244
13.56 R. Bennewitz, A.S. Foster, L.N. Kantorovich, M. Bammerlin, Ch. Loppacher, S. Schär, M. Guggisberg, E. Meyer, A.L. Shluger: Atomically resolved edges and kinks of NaCl islands on Cu(111): experiment and theory, Phys. Rev. B **62** (2000) 2074–2084
13.57 A.I. Livshits, A.L. Shluger, A.L. Rohl, A.S. Foster: Model of noncontact scanning force microscopy on ionic surfaces, Phys. Rev. B **59** (1999) 2436–2448
13.58 R. Bennewitz, M. Bammerlin, Ch. Loppacher, M. Guggisberg, L. Eng, E. Meyer, H.-J. Güntherodt, C.P. An, F. Luty: Molecular impurities at the NaCl(100) surface observed by scanning force microscopy, Rad. Effects Defects Solids **150** (1998) 321–326
13.59 R. Bennewitz, O. Pfeiffer, S. Schär, V. Barwich, E. Meyer, L.N. Kantorovich: Atomic corrugation in NC-AFM of alkali halides, Appl. Surf. Sci. **188** (2002) 232–237
13.60 C. Barth, M. Reichling: Resolving ions and vacancies at step edges on insulating surfaces, Surf. Sci. **470** (2000) L99–L103
13.61 R. Bennewitz, M. Reichling, E. Matthias: Force microscopy of cleaved and electron-irradiated CaF_2(111) surfaces in ultra-high vacuum, Surf. Sci. **387** (1997) 69–77
13.62 M. Reichling, C. Barth: Scanning force imaging of atomic size defects on the CaF_2(111) surface, Phys. Rev. Lett. **83** (1999) 768–771
13.63 M. Reichling, M. Huisinga, S. Gogoll, C. Barth: Degradation of the CaF_2(111) surface by air exposure, Surf. Sci. **439** (1999) 181–190
13.64 C. Barth, A.S. Foster, M. Reichling, A.L. Shluger: Contrast formation in atomic resolution scanning force microscopy of CaF_2(111): experiment and theory, J. Phys. Condens. Matter **13** (2001) 2061–2079
13.65 A.S. Foster, C. Barth, A.L. Shluger, M. Reichling: Unambiguous interpretation of atomically resolved force microscopy images of an insulator, Phys. Rev. Lett. **86** (2001) 2373–2376
13.66 A.S. Foster, A.L. Rohl, A.L. Shluger: Imaging problems on insulators: what can be learnt from NC-AFM modeling on CaF_2?, Appl. Phys. A **72** (2001) S31–S34
13.67 A.S. Foster, A.L. Shluger, R.M. Nieminen: Quantitative modeling in scanning force microscopy on insulators, Appl. Surf. Sci. **188** (2002) 306–318
13.68 M. Reichling, C. Barth: Atomically resoluiton imaging on fluorides. In: *Noncontact Atomic Force Microscopy*, ed. by S. Morita, R. Wiesendanger, E. Meyer (Springer, Berlin, Heidelberg 2002)
13.69 K. Fukui, H. Ohnishi, Y. Iwasawa: Atom-resolved image of the TiO_2(110) surface by noncontact atomic force microscopy, Phys. Rev. Lett. **79** (1997) 4202–4205
13.70 H. Raza, C.L. Pang, S.A. Haycock, G. Thornton: Non-contact atomic force microscopy imaging of TiO_2(100) surfaces, Appl. Surf. Sci. **140** (1999) 271–275
13.71 C.L. Pang, H. Raza, S.A. Haycock, G. Thornton: Imaging reconstructed TiO_2(100) surfaces with noncontact atomic force microscopy, Appl. Surf. Sci. **157** (2000) 223–238
13.72 M. Ashino, T. Uchihashi, K. Yokoyama, Y. Sugawara, S. Morita, M. Ishikawa: STM and atomic-resolution noncontact AFM of an oxygen-deficient TiO_2(110) surface, Phys. Rev. B **61** (2000) 13955–13959
13.73 C.L. Pang, S.A. Haycock, H. Raza, P.J. Møller, G. Thornton: Structures of the 4 × 1 and 1 × 2 reconstructions of SnO_2(110), Phys. Rev. B **62** (2000) R7775–R7778
13.74 H. Hosoi, K. Sueoka, K. Hayakawa, K. Mukasa: Atomic resolved imaging of cleaved NiO(100) surfaces by NC-AFM, Appl. Surf. Sci. **157** (2000) 218–221
13.75 W. Allers, S. Langkat, R. Wiesendanger: Dynamic low-temperature scanning force microscopy on nickel oxide (001), Appl. Phys. A **72** (2001) S27–S30
13.76 T. Kubo, H. Nozoye: Surface structure of $SrTiO_3$(100)-$(\sqrt{5} \times \sqrt{5}) - R26.6°$, Phys. Rev. Lett. **86** (2001) 1801–1804
13.77 K. Fukui, Y. Namai, Y. Iwasawa: Imaging of surface oxygen atoms and their defect structures on CeO_2(111) by noncontact atomic force microscopy, Appl. Surf. Sci. **188** (2002) 252–256
13.78 C. Barth, M. Reichling: Imaging the atomic arrangements on the high-temperature reconstructed α-Al_2O_3 surface, Nature **414** (2001) 54–57

13.79 C.L. Pang, H. Raza, S.A. Haycock, G. Thornton: Growth of copper and palladium α-Al_2O_3(0001), Surf. Sci. **460** (2000) L510–L514

13.80 K. Mukasa, H. Hasegawa, Y. Tazuke, K. Sueoka, M. Sasaki, K. Hayakawa: Exchange interaction between magnetic moments of ferromagnetic sample and tip: Possibility of atomic-resolution images of exchange interactions using exchange force microscopy, Jpn. J. Appl. Phys. **33** (1994) 2692–2695

13.81 H. Ness, F. Gautier: Theoretical study of the interaction between a magnetic nanotip and a magnetic surface, Phys. Rev. B **52** (1995) 7352–7362

13.82 K. Nakamura, H. Hasegawa, T. Ohuchi, K. Sueoka, K. Hayakawa, K. Mukasa: First-principles calculation of the exchange interaction and the exchange force between magnetic Fe films, Phys. Rev. B **56** (1997) 3218–3221

13.83 A.S. Foster, A.L. Shluger: Spin-contrast in noncontact SFM on oxide surfaces: theoretical modeling of NiO(001) surface, Surf. Sci. **490** (2001) 211–219

13.84 H. Hosoi, K. Kimura, K. Sueoka, K. Hayakawa, K. Mukasa: Non-contact atomic force microscopy of an antiferromagnetic NiO(100) surface using a ferromagnetic tip, Appl. Phys. A **72** (2001) S23–S26

13.85 H. Hosoi, K. Sueoka, K. Hayakawa, K. Mukasa: Atomically resolved imaging of a NiO(001) surface. In: *Noncontact Atomic Force Microscopy*, ed. by S. Morita, R. Wiesendanger, E. Meyer (Springer, Berlin, Heidelberg 2002) pp. 125–134

13.86 K. Kobayashi, H. Yamada, T. Horiuchi, K. Matsushige: Structures and electrical properties of fullerene thin films on Si(111)-7×7 surface investigated by noncontact atomic force microscopy, Jpn. J. Appl. Phys. **39** (2000) 3821–3829

13.87 T. Uchihashi, M. Tanigawa, M. Ashino, Y. Sugawara, K. Yokoyama, S. Morita, M. Ishikawa: Identification of B-form DNA in an ultrahigh vacuum by noncontact-mode atomic force microscopy, Langmuir **16** (2000) 1349–1353

13.88 R.M. Overney, E. Meyer, J. Frommer, D. Brodbeck, R. Lüthi, L. Howald, H.-J. Güntherodt, M. Fujihira, H. Takano, Y. Gotoh: Friction measurements on phase-separated thin films with a modified atomic force microscope, Nature **359** (1992) 133–135

13.89 D. Frisbie, L.F. Rozsnyai, A. Noy, M.S. Wrighton, C.M. Lieber: Functional group imaging by chemical force microscopy, Science **265** (1994) 2071–2074

13.90 E. Meyer, L. Howald, R.M. Overney, H. Heinzelmann, J. Frommer, H.-J. Güntherodt, T. Wagner, H. Schier, S. Roth: Molecular-resolution images of Langmuir–Blodgett films using atomic force microscopy, Nature **349** (1992) 398–400

13.91 K. Fukui, H. Onishi, Y. Iwasawa: Imaging of individual formate ions adsorbed on TiO_2(110) surface by non-contact atomic force microscopy, Chem. Phys. Lett. **280** (1997) 296–301

13.92 K. Kobayashi, H. Yamada, T. Horiuchi, K. Matsushige: Investigations of C_60 molecules deposited on Si(111) by noncontact atomic force microscopy, Appl. Surf. Sci. **140** (1999) 281–286

13.93 Y. Maeda, T. Matsumoto, T. Kawai: Observation of single- and double-strand DNA using non-contact atomic force microscopy, Appl. Surf. Sci. **140** (1999) 400–405

13.94 T. Uchihashi, T. Ishida, M. Komiyama, M. Ashino, Y. Sugawara, W. Mizutani, K. Yokoyama, S. Morita, H. Tokumoto, M. Ishikawa: High-resolution imaging of organic monolayers using noncontact AFM, Appl. Surf. Sci **157** (2000) 244–250

13.95 T. Fukuma, K. Kobayashi, T. Horiuchi, H. Yamada, K. Matsushige: Alkanethiol self-assembled monolayers on Au(111) surfaces investigated by non-contact AFM, Appl. Phys. A **72** (2001) S109–S112

13.96 B. Gotsmann, C. Schmidt, C. Seidel, H. Fuchs: Molecular resolution of an organic monolayer by dynamic AFM, Europ. Phys. J. B **4** (1998) 267–268

13.97 Ch. Loppacher, M. Bammerlin, M. Guggisberg, E. Meyer, H.-J. Güntherodt, R. Lüthi, R. Schlittler, J.K. Gimzewski: Forces with submolecular resolution between the probing tip and Cu-TBPP molecules on Cu(100) observed with a combined AFM/STM, Appl. Phys. A **72** (2001) S105–S108

13.98 L.M. Eng, M. Bammerlin, Ch. Loppacher, M. Guggisberg, R. Bennewitz, R. Lüthi, E. Meyer, H.-J. Güntherodt: Surface morphology, chemical contrast, and ferroelectric domains in TGS bulk single crystals differentiated with UHV non-contact force microscopy, Appl. Surf. Sci. **140** (1999) 253–258

13.99 S. Kitamura, K. Suzuki, M. Iwatsuki: High resolution imaging of contact potential difference using a novel ultrahigh vacuum non-contact atomic force microscope technique, Appl. Surf. Sci. **140** (1999) 265–270

13.100 H. Yamada, T. Fukuma, K. Umeda, K. Kobayashi, K. Matsushige: Local structures and electrical properties of organic molecular films investigated by non-contact atomic force microscopy, Appl. Surf. Sci **188** (2000) 391–398

13.101 K. Fukui, Y. Iwasawa: Fluctuation of acetate ions in the (2×1)-acetate overlayer on TiO_2(110)-(1×1) observed by noncontact atomic force microscopy, Surf. Sci. **464** (2000) L719–L726

13.102 A. Sasahara, H. Uetsuka, H. Onishi: Single-molecule analysis by non-contact atomic force microscopy, J. Phys. Chem. B **105** (2001) 1–4

13.103 A. Sasahara, H. Uetsuka, H. Onishi: NC-AFM topography of HCOO and CH_3COO molecules co-adsorbed on TiO_2(110), Appl. Phys. A **72** (2001) S101–S103

13.104 A. Sasahara, H. Uetsuka, H. Onishi: Image topography of alkyl-substituted carboxylates observed by noncontact atomic force microscopy, Surf. Sci. **481** (2001) L437–L442

13.105 A. Sasahara, H. Uetsuka, H. Onishi: Noncontact atomic force microscope topography dependent on permanent dipole of individual molecules, Phys. Rev. B **64** (2001) 121406(R)

13.106 A. Sasahara, H. Uetsuka, T. Ishibashi, H. Onishi: A needle-like organic molecule imaged by noncontact atomic force microscopy, Appl. Surf. Sci. **188** (2002) 265–271

13.107 H. Onishi, A. Sasahara, H. Uetsuka, T. Ishibashi: Molecule-dependent topography determined by noncontact atomic force microscopy: Carboxylates on TiO_2(110), Appl. Surf. Sci. **188** (2002) 257–264

13.108 H. Onishi: Carboxylates adsorbed on TiO_2(110). In: *Chemistry of Nano-molecular Systems*, ed. by T. Nakamura (Springer, Berlin, Heidelberg 2002) pp. 75–89

13.109 S. Thevuthasan, G. S. Herman, Y. J. Kim, S. A. Chambers, C. H. F. Peden, Z. Wang, R. X. Ynzunza, E. D. Tober, J. Morais, C. S. Fadley: The structure of formate on TiO_2(110) by scanned-energy and scanned-angle photoelectron diffraction, Surf. Sci. **401** (1998) 261–268

13.110 H. Uetsuka, A. Sasahara, A. Yamakata, H. Onishi: Microscopic identification of a bimolecular reaction intermediate, J. Phys. Chem. B **106** (2002) 11549–11552

13.111 D. R. Lide: *Handbook of Chemistry and Physics*, 81st edn. (CRC, Boca Raton 2000)

13.112 K. Kobayashi, H. Yamada, K. Matsushige: Dynamic force microscopy using FM detection in various environments, Appl. Surf. Sci. **188** (2002) 430–434

14. Low Temperature Scanning Probe Microscopy

This chapter is dedicated to scanning probe microscopy, one of the most important techniques in nanotechnology. In general, scanning probe techniques allow the measurement of physical properties down to the nanometer scale. Some techniques, such as the scanning tunneling microscope and the scanning force microscope even go down to the atomic scale. The properties that are accessible are various. Most importantly, one can image the arrangement of atoms on conducting surfaces by scanning tunneling microscopy and on insulating substrates by scanning force microscopy. But also the arrangement of electrons (scanning tunneling spectroscopy), the force interaction between different atoms (scanning force spectroscopy), magnetic domains (magnetic force microscopy), the local capacitance (scanning capacitance microscopy), the local temperature (scanning thermo microscopy), and local light-induced excitations (scanning near-field microscopy) can be measured with high spatial resolution. In addition, some techniques even allow the manipulation of atomic configurations.

Probably the most important advantage of the low-temperature operation of scanning probe techniques is that they lead to a significantly better signal-to-noise ratio than measuring at room temperature. This is why many researchers work below 100 K. However, there are also physical reasons to use low-temperature equipment. For example, the manipulation of atoms or scanning tunneling spectroscopy with high energy resolution can only be realized at low temperatures. Moreover, some physical effects such as superconductivity or the Kondo effect are restricted to low temperatures. Here, we describe the design criteria of low-temperature scanning probe equipment and summarize some of the most spectacular results achieved since the invention of the method about 20 years ago. We first focus on the scanning tunneling microscope, giving examples of atomic manipulation and the analysis of electronic properties in different material arrangements. Afterwards, we describe results obtained by scanning force microscopy, showing atomic-scale imaging on insulators, as well as force spectroscopy analysis. Finally, the magnetic force microscope, which images domain patterns in ferromagnets and vortex patterns in superconductors, is discussed. Although this list is far from complete, we feel that it gives an adequate impression of the fascinating possibilities of low-temperature scanning probe instruments.

In this chapter low temperatures are defined as lower than about 100 K and are normally achieved by cooling with liquid nitrogen or liquid helium. Applications in which SPMs are operated close to 0 °C are not covered in this chapter.

14.1 **Microscope Operation at Low Temperatures** ... 414
14.1.1 Drift ... 414
14.1.2 Noise ... 415
14.1.3 Stability ... 415
14.1.4 Piezo Relaxation and Hysteresis ... 415

14.2 **Instrumentation** ... 415
14.2.1 A Simple Design for a Variable Temperature STM ... 416
14.2.2 A Low Temperature SFM Based on a Bath Cryostat ... 417

14.3 **Scanning Tunneling Microscopy and Spectroscopy** ... 419
14.3.1 Atomic Manipulation ... 419
14.3.2 Imaging Atomic Motion ... 420
14.3.3 Detecting Light from Single Atoms and Molecules ... 421
14.3.4 High Resolution Spectroscopy ... 422
14.3.5 Imaging Electronic Wave Functions ... 427
14.3.6 Imaging Spin Polarization: Nanomagnetism ... 431

14.4 **Scanning Force Microscopy and Spectroscopy** ... 433
14.4.1 Atomic-Scale Imaging ... 434
14.4.2 Force Spectroscopy ... 436
14.4.3 Electrostatic Force Microscopy ... 438
14.4.4 Magnetic Force Microscopy ... 439

References ... 442

More than two decades ago, the first design of an experimental setup was presented where a sharp tip was systematically scanned over a sample surface in order to obtain local information on the tip-sample interaction down to the atomic scale. This original instrument used the tunneling current between a conducting tip and a conducting sample as a feedback signal and was named *scanning tunneling microscope* accordingly [14.1]. Soon after this historic breakthrough, it became widely recognized that virtually any type of tip-sample interaction can be used to obtain local information on the sample by applying the same general principle, provided that the selected interaction was reasonably short-ranged. Thus, a whole variety of new methods has been introduced, which are denoted collectively as *scanning probe methods*. An overview is given by *Wiesendanger* [14.2].

The various methods, especially the above mentioned scanning tunneling microscopy (STM) and scanning force microscopy (SFM) – which is often further classified into subdisciplines such as the topography-reflecting atomic force microscopy (AFM), the magnetic force microscopy (MFM), or the electrostatic force microscopy (EFM) – have been established as standard methods for surface characterization on the nanometer scale. The reason is that they feature extremely high resolution (often down to the atomic scale for STM and AFM), despite a principally simple, compact, and comparatively inexpensive design.

A side effect of the simple working principle and the compact design of many scanning probe microscopes (SPMs) is that they can be adapted to different environments such as air, all kinds of gaseous atmospheres, liquids, or vacuum with reasonable effort. Another advantage is their ability to work within a wide temperature range. A microscope operation at higher temperatures is chosen to study surface diffusion, surface reactivity, surface reconstructions that only manifest at elevated temperatures, high-temperature phase transitions, or to simulate conditions as they occur, e.g., in engines, catalytic converters, or reactors. Ultimately, the upper limit for the operation of an SPM is determined by the stability of the sample, but thermal drift, which limits the ability to move the tip in a controlled manner over the sample, as well as the depolarization temperature of the piezoelectric positioning elements might further restrict successful measurements.

On the other hand, low-temperature (LT) application of SPMs is much more widespread than operation at high temperatures. Essentially five reasons make researchers adapt their experimental setups to low-temperature compatibility. These are: (1) the reduced thermal drift, (2) lower noise levels, (3) enhanced stability of tip and sample, (4) the reduction in piezo hysteresis/creep, and (5) probably the most obvious, the fact that many physical effects are restricted to low temperature. Reasons (1) to (4) only apply unconditionally if the whole microscope body is kept at low temperature (typically in or attached to a bath cryostat, see Sect. 14.2). Setups in which only the sample is cooled may show considerably less favorable operating characteristics. As a result of (1) to (4), ultrahigh resolution and long-term stability can be achieved on a level that significantly exceeds what can be accomplished at room temperature even under the most favorable circumstances. Typical examples for (5) are superconductivity [14.3] and the Kondo effect [14.4].

14.1 Microscope Operation at Low Temperatures

Nevertheless, before we devote ourselves to a small overview of experimental LTSPM work, we will take a closer look at the specifics of microscope operation at low temperatures, including a discussion of the corresponding instrumentation.

14.1.1 Drift

Thermal drift originates from thermally activated movements of the individual atoms, which are reflected by the thermal expansion coefficient. At room temperature, typical values for solids are on the order of $(1-50)\times 10^{-6}\,\mathrm{K}^{-1}$. If the temperature could be kept precisely constant, any thermal drift would vanish, regardless of the absolute temperature of the system. The close coupling of the microscope to a large temperature bath that keeps a constant temperature ensures a significant reduction in thermal drift and allows for distortion-free, long-term measurements. Microscopes that are efficiently attached to sufficiently large bath cryostats, therefore, show a one- to two-order-of-magnitude increase in thermal stability compared with non-stabilized setups operated at room temperature.

A second effect also helps suppress thermally induced drift of the probing tip relative to a specific location on the sample surface. The thermal expansion coefficients are at liquid helium temperature two or more orders of magnitude smaller than at room temperature. Consequently, the thermal drift during low-temperature operation decreases accordingly.

For some specific scanning probe methods, there may be additional ways a change in temperature can affect the quality of the data. In *frequency-modulation SFM* (FM-SFM), for example, the measurement principle relies on the accurate determination of the eigenfrequency of the cantilever, which is determined by its spring constant and its effective mass. However, the spring constant changes with temperature due to both thermal expansion (i.e., the resulting change in the cantilever dimensions) and the variation of Young's modulus with temperature. Assuming drift rates of about 2 mK/min, as is typical for room temperature measurements, this effect might have a significant influence on the obtained data.

14.1.2 Noise

The theoretically achievable resolution in SPMs often increases with decreasing temperature due to a decrease in thermally induced noise. An example is the thermal noise in SFMs, which is proportional to the square root of the temperature [14.5, 6]. Lowering the temperature from $T = 300\,\mathrm{K}$ to $T = 10\,\mathrm{K}$ thus results in a reduction of the thermal frequency noise of more than a factor of five. Graphite, e.g., has been imaged atomically resolved only at low temperatures due to its extremely low corrugation, which was below the room temperature noise level [14.7, 8].

Another, even more striking, example is the spectroscopic resolution in *scanning tunneling spectroscopy* (STS). It depends linearly on the temperature [14.2] and is consequently reduced even more at LT than the thermal noise in AFM. This provides the opportunity to study structures or physical effects not accessible at room temperature such as spin and Landau levels in semiconductors [14.9].

Finally, it might be worth mentioning that the enhanced stiffness of most materials at low temperatures (increased Young's modulus) leads to a reduced coupling to external noise. Even though this effect is considered small [14.6], it should not be ignored.

14.1.3 Stability

There are two major stability issues that considerably improve at low temperature. First, low temperatures close to the temperature of liquid helium inhibit most of the thermally activated diffusion processes. As a consequence, the sample surfaces show a significantly increased long-term stability, since defect motion or adatom diffusion is massively suppressed. Most strikingly, even single xenon atoms deposited on suitable substrates can be successfully imaged [14.10, 11], or even manipulated [14.12]. In the same way, the low temperatures also stabilize the atomic configuration at the tip end by preventing sudden jumps of the most loosely bound foremost tip atom(s). Second, the large cryostat that usually surrounds the microscope acts as an effective cryo pump. Thus samples can be kept clean for several weeks, which is a multiple of the corresponding time at room temperature (about 3–4 h).

14.1.4 Piezo Relaxation and Hysteresis

The last important benefit from low-temperature operation of SPMs is that artifacts given by the response of the piezoelectric scanners are substantially reduced. After applying a voltage ramp to one electrode of a piezoelectric scanner, its immediate initial deflection, l_0, is followed by a much slower relaxation, Δl, with a logarithmic time dependence. This effect, known as piezo relaxation or "creep", diminishes substantially at low temperatures, typically by a factor of ten or more. As a consequence, piezo nonlinearities and piezo hysteresis decrease accordingly. Additional information is given by *Hug* et al. [14.13].

14.2 Instrumentation

The two main design criteria for all vacuum-based scanning probe microscope systems are (1) to provide an efficient decoupling of the microscope from the vacuum system and other sources of external vibrations, and (2) to avoid most internal noise sources through a high mechanical rigidity of the microscope body itself. In vacuum systems designed for low-temperature applications, a significant degree of complexity is added, since, on the one hand, close thermal contact of the SPM and cryogen is necessary to ensure the (approximately) drift-free conditions described above, while, on the other hand, a good vibration isolation (both from the outside world, as well as from the boiling or flowing cryogen) has to be maintained.

Plenty of microscope designs have been presented in the last ten to 15 years, predominantly in the field of STM. Due to the variety of the different approaches, we will, somewhat arbitrarily, give two examples on different levels of complexity that might serve as illustrative model designs.

14.2.1 A Simple Design for a Variable Temperature STM

A simple design for a variable temperature STM system is presented in Fig. 14.1 (Similar systems are also offered by Omicron (Germany) or Jeol (Japan)). It should give an impression of what the minimum requirements are, if samples are to be investigated successfully at low temperatures. It features a single ultrahigh vacuum (UHV) chamber that houses the microscope in its center. The general idea to keep the setup simple is that only the sample is cooled by means of a flow cryostat that ends in the small liquid nitrogen (LN) reservoir. This reservoir is connected to the sample holder with copper braids. The role of the copper braids is to thermally attach the LN reservoir to the sample located on the sample holder in an effective manner, while vibrations due to the flow of the cryogen should be blocked as much as possible. In this way, a sample temperature of about 100 K is reached. Alternatively, with liquid helium

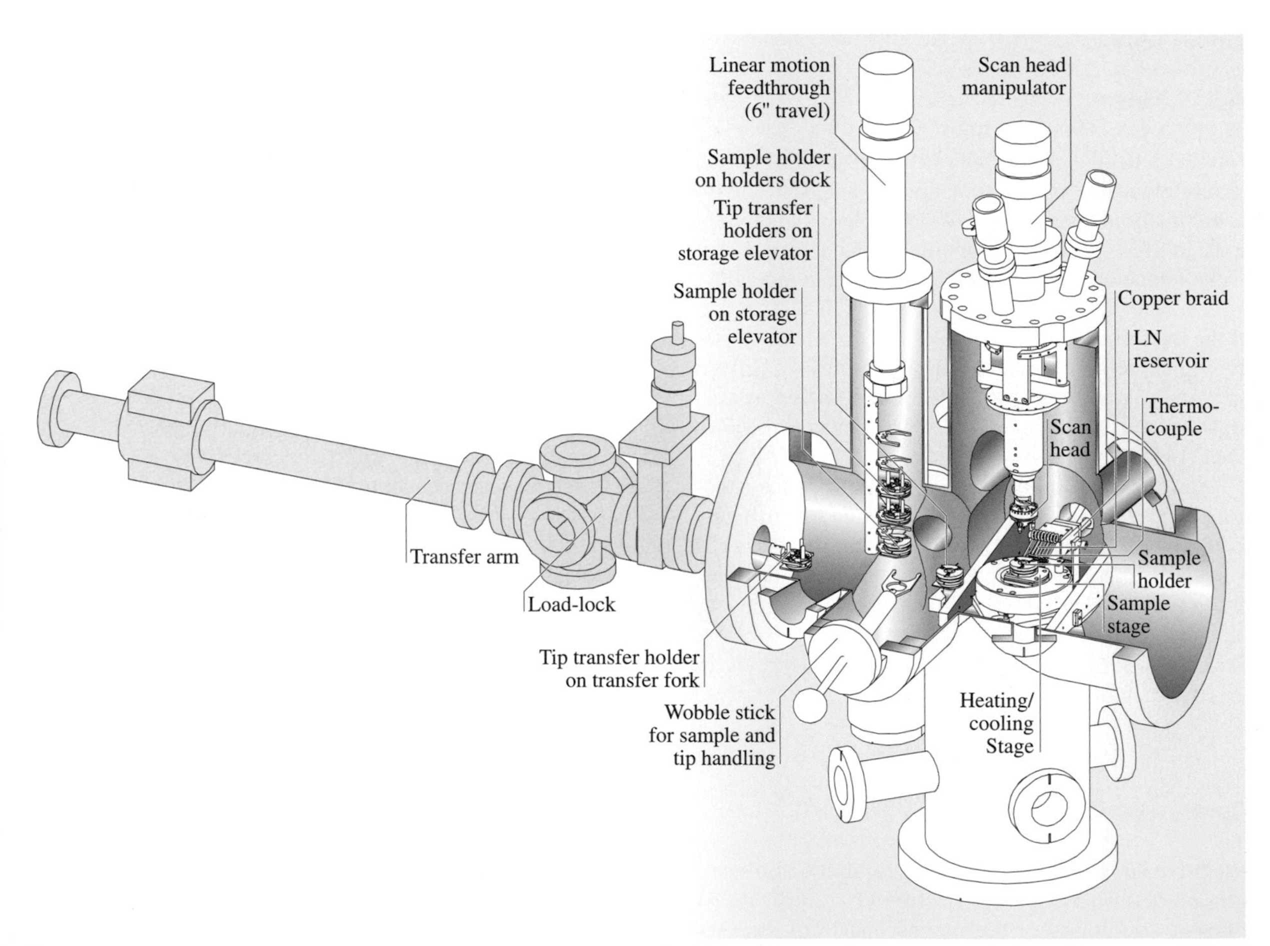

Fig. 14.1 One chamber UHV system with variable temperature STM based on a flow cryostat design. (Courtesy of RHK Technology, USA)

operation, a base temperature of below 30 K can be achieved, while a heater that is integrated into the sample stage enables high-temperature operation up to 1,000 K.

A typical experiment would run as follows: First, the sample is brought into the system by placing it in the so-called *load-lock*. This small part of the chamber can be separated from the rest of the system by a valve, so that the main part of the system can remain under vacuum at all times (i. e., even if the load-lock is opened for introducing the sample). After vacuum is re-established, the sample is transferred to the main chamber using the transfer arm. A linear motion feedthrough enables the storage of sample holders or, alternatively, specialized holders that carry replacement tips for the STM. Extending the transfer arm further, the sample can be placed on the sample stage and subsequently cooled down to the desired temperature. The scan head, which carries the STM tip, is then lowered with the scan head manipulator onto the sample holder (see Fig. 14.2). The special design of the scan head (see [14.14] for details) allows not only a flexible positioning of the tip on any desired location on the sample surface, but also compensates to a certain degree for thermal drift that inevitably occurs in such a design due to temperature gradients.

In fact, thermal drift is often much more prominent in LT-SPM designs, where only the sample is cooled, than in room temperature designs. Therefore, to fully benefit from the high stability conditions described in the introduction, it is mandatory to keep the whole microscope at the exact same temperature. This is mostly realized by using bath cryostats, which add a certain degree of complexity.

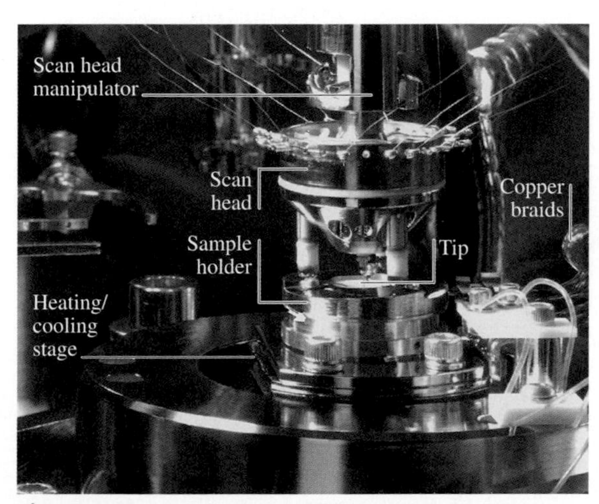

Fig. 14.2 Photograph of the STM located inside the system sketched in Fig. 14.1. After the scan head has been lowered onto the sample holder, it is fully decoupled from the scan head manipulator and can be moved laterally using the three piezo legs it stands on. (Courtesy of RHK Technology, USA)

14.2.2 A Low Temperature SFM Based on a Bath Cryostat

As an example for an LT-SPM setup based on a bath cryostat, let us take a closer look at the LT-SFM system sketched in Fig. 14.3, which has been used to acquire the images on graphite, xenon, NiO, and InAs presented in Sect. 14.4. The force microscope is

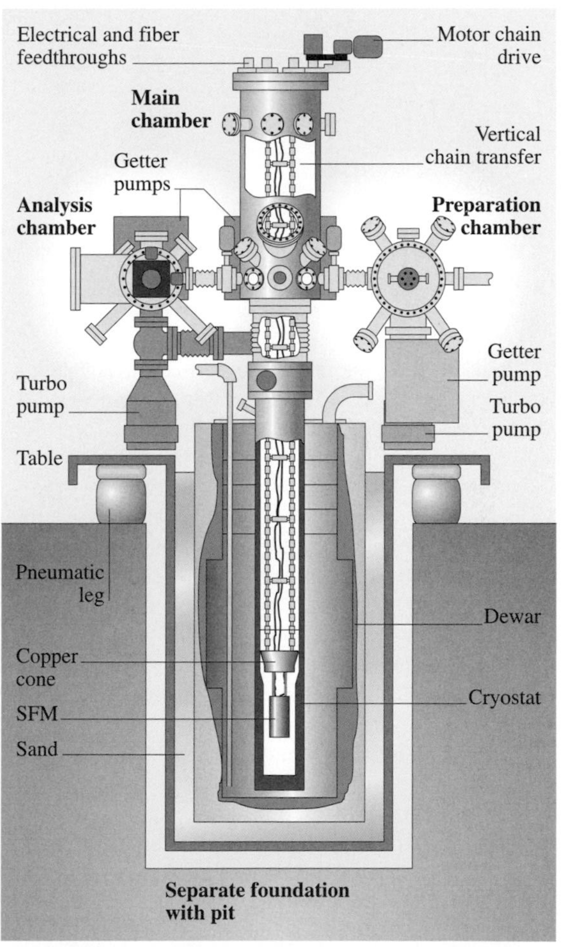

Fig. 14.3 Three-chamber UHV and bath cryostat system for scanning force microscopy, front view

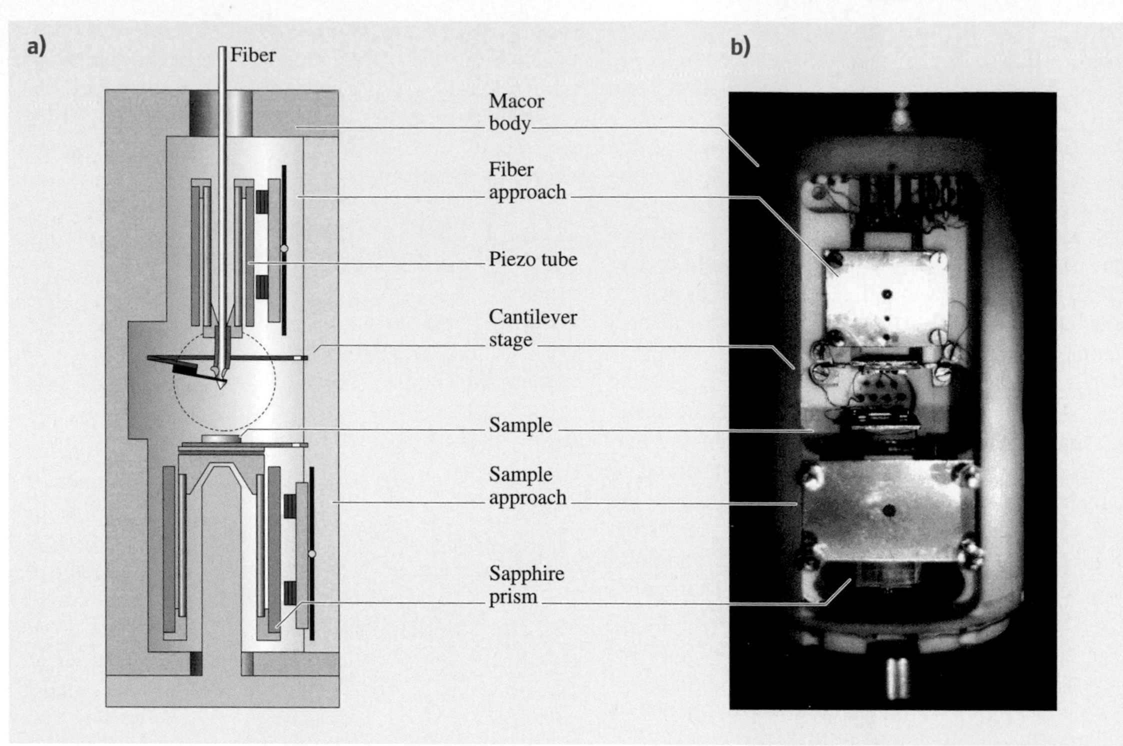

Fig. 14.4a,b The scanning force microscope incorporated into the system presented in Fig. 14.3. **(a)** Section along plane of symmetry. **(b)** Photo from the front

built into a UHV system that comprises three vacuum chambers: one for cantilever and sample preparation, which also serves as a transfer chamber, one for analysis purposes, and a main chamber that houses the microscope. A specially designed vertical transfer mechanism based on a double chain allows the lowering of the microscope into a UHV compatible bath cryostat attached underneath the main chamber. To damp the system, it is mounted on a table carried by pneumatic damping legs, which, in turn, stand on a separate foundation to decouple it from building vibrations. The cryostat and dewar are separated from the rest of the UHV system by a bellow. In addition, the dewar is surrounded by sand for acoustic isolation.

In this design, tip and sample are exchanged at room temperature in the main chamber. After the transfer into the cryostat, the SFM can be cooled by either liquid nitrogen or liquid helium, reaching temperatures down to 10 K. An all-fiber interferometer as the detection mechanism for the cantilever deflection ensures high resolution, while simultaneously allowing the construction of a comparatively small, rigid, and symmetric microscope.

Figure 14.4 highlights the layout of the SFM body itself. Along with the careful choice of materials, the symmetric design eliminates most of the problems with drift inside the microscope encountered when cooling or warming it up. The microscope body has an overall cylindrical shape with a 13 cm height and a 6 cm diameter and exact mirror symmetry along the cantilever axis. The main body is made of a single block of macor, a machinable glass ceramic, which ensures a rigid and stable design. For most of the metallic parts titanium was used, which has a temperature coefficient similar to macor. The controlled but stable accomplishment of movements, such as coarse approach and lateral positioning in other microscope designs, is a difficult task at low temperatures. The present design uses a special type of piezo motor that moves a sapphire prism (see the "fiber approach" and the "sample approach" signs in Fig. 14.3). It is described in detail in [14.15]. More information regarding this design is given in [14.16].

14.3 Scanning Tunneling Microscopy and Spectroscopy

In this section, we review some of the most important results achieved by LTSTM. After summarizing the results with emphasis on the necessity of LT equipment, we turn to details of the different experiments and the physical meaning of the obtained results.

As described in Sect. 14.1, the LT equipment has basically three advantages for scanning tunneling microscopy (STM) and spectroscopy (STS). First, the instruments are much more stable with respect to thermal drift and the coupling to external noise, allowing the establishment of new functionalities of the instrument. In particular, the LTSTM has been used to move atoms on a surface [14.12], cut molecules in pieces [14.17], reform bonds [14.18], and, consequently, establish new structures on the nanometer scale. Also, the detection of light resulting from tunneling into a particular atom [14.19] and the visualization of thermally induced atomic movements [14.20] partly require LT instrumentation.

Second, the spectroscopic resolution in STS depends linearly on temperature and is, therefore, considerably reduced at LT. This provides the opportunity to study physical effects unaccessible at room temperature. Obvious examples are the resolution of spin and Landau levels in semiconductors [14.9], or the investigation of lifetime broadening effects of particular electronic states on the nanometer scale [14.21]. More spectacular, electronic wave functions have been imaged for the first time in real space using an LTSTM [14.22], and vibrational levels giving rise to additional inelastic tunneling have been detected [14.23] and localized within particular molecules [14.24].

Third and most obvious, many physical effects, in particular, effects guided by electronic correlations, are restricted to low temperature. Typical examples are superconductivity [14.3], the Kondo effect [14.4], and many of the electron phases found in semiconductors [14.25]. Here, the LTSTM provides the possibility to study electronic effects on a local scale, and intensive work has been done in this field, the most elaborate with respect to high temperature superconductivity [14.26].

14.3.1 Atomic Manipulation

Although manipulation of surfaces on the atomic scale can be achieved at room temperature [14.27], only the use of LTSTMs allow the placement of individual atoms at desired atomic positions [14.28].

The usual technique to manipulate atoms is to increase the current above a certain atom, which reduces the tip-atom distance, then to move the tip with the atom to a desired position, and finally to reduce the current again in order to decouple atom and tip. The first demonstration of this technique was performed by Eigler and Schweizer (1990), who used Xe atoms on a Ni(110) surface to write the three letters "IBM" (their employer) on the atomic scale (Fig. 14.5a). Nowadays, many laboratories are able to move different kinds of atoms and molecules on different surfaces with high precision. An example featuring CO molecules on Cu(110) is shown in Fig. 14.5b–g. Basic modes of controlled motion, pushing, pulling, and sliding of the molecules have been established that depend on the tunneling current, i. e., the distance and the particular molecule-substrate combination [14.29]. It is believed that the electric field between tip and molecule is the strongest force moving the molecules, but other mechanisms such as electromigration caused by the high current density [14.28] or modifications of the surface potential due to the presence of the tip [14.30] have been put forth as important for some of the manipulation modes.

Meanwhile, other types of manipulation on the atomic scale have been developed. Some of them require inelastic tunneling into vibrational or rotational modes of molecules or atoms. They lead to controlled desorption [14.31], diffusion [14.32], pick-up of molecules by the tip [14.18], or rotation of individual entities [14.33, 34]. Also, dissociation of molecules by voltage pulses [14.17], conformational changes induced by dramatic change of tip-molecule distance [14.35], and association of pieces into larger molecules by reducing their lateral distance [14.18] have been shown. Fig. 14.5h–m shows the production of biphenyl from two iodobenzene molecules [14.36]. The iodine is abstracted by voltage pulses (Fig. 14.5i,j), then the iodine is moved to the terrace by the pulling mode (Fig. 14.5k,l), and finally the two phenyl parts are slid along the step edge until they are close enough to react (Fig. 14.5m). The chemical identification of the components is not deduced straightforwardly from STM images and partly requires detailed calculations of their apparent shape.

Low temperatures are not always required in these experiments, but they increase the reproducibility because of the higher stability of the instrument, as discussed in Sect. 14.1. Moreover, rotation or diffusion of entities could be excited at higher temperatures, making the intentionally produced configurations unstable.

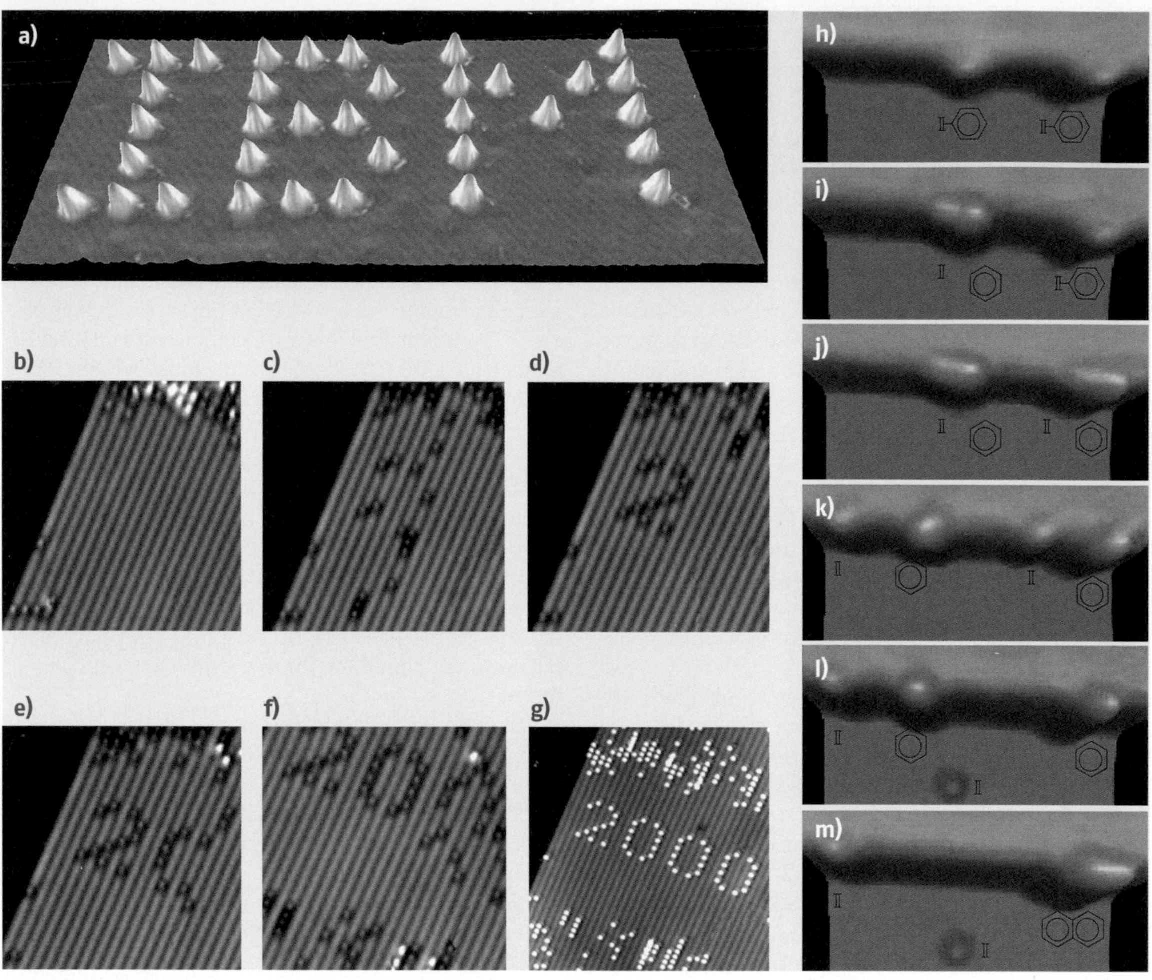

Fig. 14.5 (**a**) STM image of single Xe atoms positioned on a Ni(110) surface in order to realize the letters IBM on the atomic scale (courtesy of Eigler, IBM); (**b**)–(**f**) STM images recorded after different positioning processes of CO molecules on a Cu(110) surface; (**g**) final artwork greeting the new millennium on the atomic scale ((**b**)–(**g**) courtesy of Meyer, Berlin). (**h**)–(**m**) Synthesis of biphenyl from two iodobenzene molecules on Cu(111): First, iodine is abstracted from both molecules (**i**),(**j**), then the iodine between the two phenyl groups is removed from the step (**k**), and finally one of the phenyls is slid along the Cu-step (**l**) until it reacts with the other phenyl (**m**); the line drawings symbolize the actual status of the molecules ((**h**)–(**m**) courtesy of Hla and Rieder, Ohio)

14.3.2 Imaging Atomic Motion

Since individual manipulation processes last seconds to minutes, they probably cannot be used to manufacture large and repetitive structures. A possibility to construct such structures is self-assembled growth. It partly relies on the temperature dependence of different diffusion processes on the surface. A detailed knowledge of the diffusion parameters is required, which can be deduced from sequences of STM images measured at temperatures close to the onset of the process of interest [14.37]. Since many diffusion processes set on at LT, LT are partly required [14.20]. Consecutive images of so-called HtBDC molecules on Cu(110) recorded at $T = 194\,\mathrm{K}$ are shown in Fig. 14.6a–c [14.38]. As indicated by the arrows, the position of the molecules changes with time,

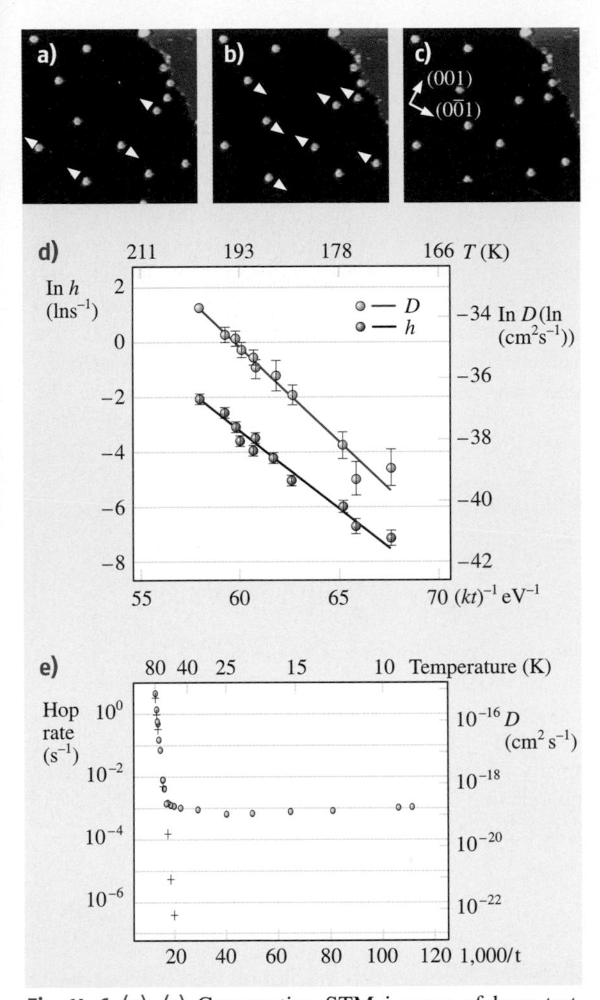

Fig. 14.6 (**a**)–(**c**) Consecutive STM images of hexa-tert-butyl decacyclene molecules on Cu(110) imaged at $T = 194$ K; *arrows* indicate the direction of motion of the molecules between two images. (**d**) Arrhenius plot of the hopping rate h determined from images like (**a**)–(**c**) as a function of inverse temperature (*grey symbols*); the *brown symbols* show the corresponding diffusion constant D; *lines* are fit results revealing an energy barrier of 570 meV for molecular diffusion ((**a**)–(**d**) courtesy of M. Schuhnack and F. Besenbacher, Aarhus). (**e**) Arrhenius plot for D (*crosses*) and H (*circles*) on Cu(001). The constant hopping rate of H below 65 K indicates a non-thermal diffusion process, probably tunneling (courtesy of Ho, Irvine)

implying diffusion. Diffusion parameters are obtained from Arrhenius plots of the determined hopping rate h, as shown in Fig. 14.6d. Of course, one must make sure that the diffusion process is not influenced by the presence of the tip, since it is known from manipulation experiments that the presence of the tip can move a molecule. However, particularly at low tunneling voltages these conditions can be fulfilled.

Besides the determination of diffusion parameters, studies of the diffusion of individual molecules showed the importance of mutual interactions in diffusion, which can lead to concerted motion of several molecules [14.20], or, very interestingly, the influence of quantum tunneling [14.39]. The latter is deduced from the Arrhenius plot of hopping rates of H and D on Cu(001), as shown in Fig. 14.6e. The hopping rate of H levels off at about 65 K, while the hopping rate of the heavier D atom goes down to nearly zero, as expected from thermally induced hopping.

Other diffusion processes such as the movement of surface vacancies [14.40] or of bulk interstitials close to the surface [14.41], and the Brownian motion of vacancy islands [14.42] have also been displayed.

14.3.3 Detecting Light from Single Atoms and Molecules

It had already been realized in 1988 that STM experiments are accompanied by light emission [14.43]. The fact that molecular resolution in the light intensity was achieved at LT (Fig. 14.7a,b) [14.19] raised the hope of performing quasi-optical experiments on the molecular scale. Meanwhile, it is clear that the basic emission process observed on metals is the decay of a local plasmon induced in the area around the tip by inelastic tunneling processes [14.44, 45]. Thus, the molecular resolution is basically a change in the plasmon environment, largely given by the increased height of the tip with respect to the surface above the molecule [14.46]. However, the electron can, in principle, also decay via single-particle excitations. Indeed, signatures of single-particle levels are observed. Figure 14.7c shows light spectra measured at different tunneling voltage V above a nearly complete Na monolayer on Cu(111) [14.47]. The plasmon mode peak energy (arrow) is found, as usual, to be proportional to V, but an additional peak that does not move with V appears at 1.6 eV (p). Plotting the light intensity as a function of photon energy and V (Fig. 14.7d) clearly shows that this additional peak is fixed in photon energy and corresponds to the separation of quantum well levels of the Na ($E_n - E_m$).

Light has also been detected from semiconductors [14.48], including heterostructures [14.49]. There, the light is mostly caused by single-particle relaxation

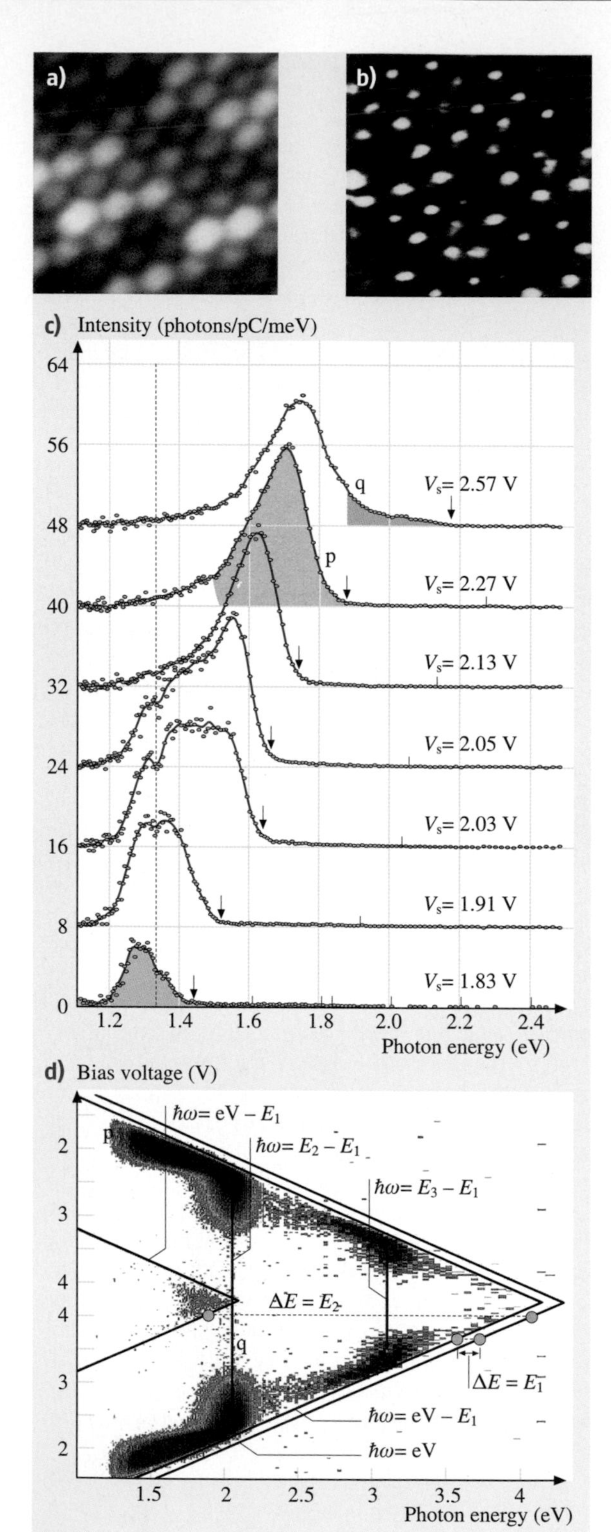

Fig. 14.7 (a) STM image of C_{60} molecules on Au(110) imaged at $T = 50$ K. (b) STM-induced photon intensity map of the same area; all photons from 1.5 eV to 2.8 eV contribute to the image, tunneling voltage $V = -2.8$ V ((a),(b) courtesy of Berndt, Kiel). (c) STM-induced photon spectrum measured on 0.6 monolayer of Na on Cu(111) at different tunneling voltages as indicated. Besides the shifting plasmon mode marked by an *arrow*, an energetically constant part named p is recognizable. (d) Greyscale map of photon intensity as a function of tunnelling voltage and photon energy measured on 2.0 monolayer Na on Cu(111). The energetically constant photons are identified with intersubband transitions of the Na quantum well, as marked by $E_n - E_m$ ((c),(d) courtesy of Hoffmann, Hamburg)

of the injected electrons, allowing a very local source of electron injection.

14.3.4 High Resolution Spectroscopy

One of the most important modes of LTSTM is STS. It detects the differential conductivity $\mathrm{d}I/\mathrm{d}V$ as a function of the applied voltage V and the position (x, y). The $\mathrm{d}I/\mathrm{d}V$-signal is basically proportional to the local density of states (LDOS) of the sample, the sum over squared single-particle wave functions Ψ_i [14.2]

$$\frac{\mathrm{d}I}{\mathrm{d}V}(V, x, y) \propto \mathrm{LDOS}(E, x, y) = \sum_{\Delta E} |\Psi_i(E, x, y)|^2 , \quad (14.1)$$

where ΔE is the energy resolution of the experiment. In simple terms, each state corresponds to a tunneling channel if it is located between the Fermi levels (E_F) of the tip and the sample. Thus, all states located in this energy interval contribute to I, while $\mathrm{d}I/\mathrm{d}V(V)$ detects only the states at the energy E corresponding to V. The local intensity of each channel depends further on the LDOS of the state at the corresponding surface position and its decay length into vacuum. For s-like tip states, *Tersoff* and *Hamann* have shown that it is simply proportional to the LDOS at the position of the tip [14.50]. Therefore, as long as the decay length is spatially constant, one measures the LDOS at the surface (14.1). Note that the contributing states are not only surface states, but also bulk states. However, surface states usually dominate if present. *Chen* has shown that higher orbital tip states lead to the so-called derivation rule [14.51]: p_z-type tip states detect $\mathrm{d(LDOS)}/\mathrm{d}z$, d_z^2-states detect $\mathrm{d^2(LDOS)}/\mathrm{d}z^2$, and so on. As long as the decay into vacuum is exponential and spatially constant, this leads only to an additional factor in dI/dV. Thus, it is still the LDOS that is measured (14.1). The requirement of a spatially constant decay is usually fulfilled on larger length scales, but not on the atomic scale [14.51]. There, states located close to the atoms show a stronger decay into vacuum than the less localized states in the interstitial region. This effect can lead to corrugations that are larger than the real LDOS corrugations [14.52].

The voltage dependence of dI/dV is sensitive to a changing decay length with V, which increases with V. Additionally, $dI/dV(V)$-curves might be influenced by possible structures in the DOS of the tip, which also contributes to the number of tunneling channels [14.53]. However, these structures can usually be identified, and only tips free of characteristic DOS structures are used for quantitative experiments.

Importantly, the energy resolution ΔE is largely determined by temperature. It is defined as the smallest energy distance of two δ-peaks in the LDOS that can still be resolved as two individual peaks in $dI/dV(V)$-curves and is $\Delta E = 3.3\,\mathrm{kT}$ [14.2]. The temperature dependence is nicely demonstrated in Fig. 14.8, where the tunneling gap of the superconductor Nb is measured at different temperatures [14.54]. The peaks at the rim of the gap get wider at temperatures well below the critical temperature of the superconductor ($T_c = 9.2\,\mathrm{K}$).

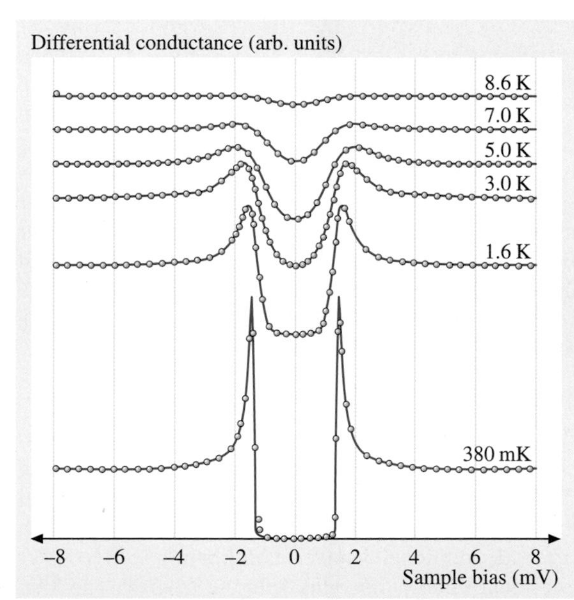

Fig. 14.8 Differential conductivity curve $dI/dV(V)$ measured on a Au surface by a Nb tip (*symbols*). Different temperatures are indicated; the lines are fits according to the superconducting gap of Nb folded with the temperature broadened Fermi distribution of the Au (courtesy of Pan, Houston)

Lifetime Broadening

Besides ΔE, intrinsic properties of the sample lead to a broadening of spectroscopic features. Basically, the finite lifetime of the electron or hole in the corresponding state broadens its energetic width. Any kind of interaction such as electron-electron interaction can be responsible. Lifetime broadening has usually been measured by photoemission spectroscopy (PES), but it turned out that lifetimes of surface states on noble metal surfaces determined by STS (Fig. 14.9a,b) are up to a factor of three larger than the ones measured by PES [14.55]. The reason is probably that defects broaden the PES spectrum. Defects are unavoidable in a spatially integrating technique such as PES, thus, STS has the advantage of choosing a particularly clean area for lifetime measurements. The STS results can be successfully compared to theory, highlighting the dominating influence of intraband transitions for the surface state lifetime on Au(111) and Cu(111), at least close to the onset of the surface band [14.21].

With respect to band electrons, the analysis of the width of the band onset in $dI/dV(V)$-curves has the disadvantage of being restricted to the onset energy. Another method circumvents this problem by measuring the decay of standing electron waves scattered from a step edge as a function of energy [14.56]. Figure 14.9c,d shows the resulting oscillating dI/dV-signal measured for two different energies. To deduce the coherence length L_Φ, which is inversely proportional to the lifetime τ_Φ, one has to consider that the finite ΔE in the experiment also leads to a decay of the standing wave away from the step edge. The dotted fit line using $L_\Phi = \infty$ indicates this effect and, more importantly, shows a discrepancy with the measured curve. Only including a finite coherence length of 6.2 nm results in good agreement, which, in turn, determines L_Φ and thus τ_Φ, as displayed in Fig. 14.9c. The found $1/E^2$-dependence of τ_Φ points to a dominating influence of electron-electron interactions at higher energies in the surface band.

Landau and Spin Levels

Moreover, the increased energy resolution at LT allows the resolution of electronic states that are not resolvable at RT. For example, Landau and spin quantization appearing in magnetic field B have been probed on InAs(110) [14.9, 57]. The corresponding quantization

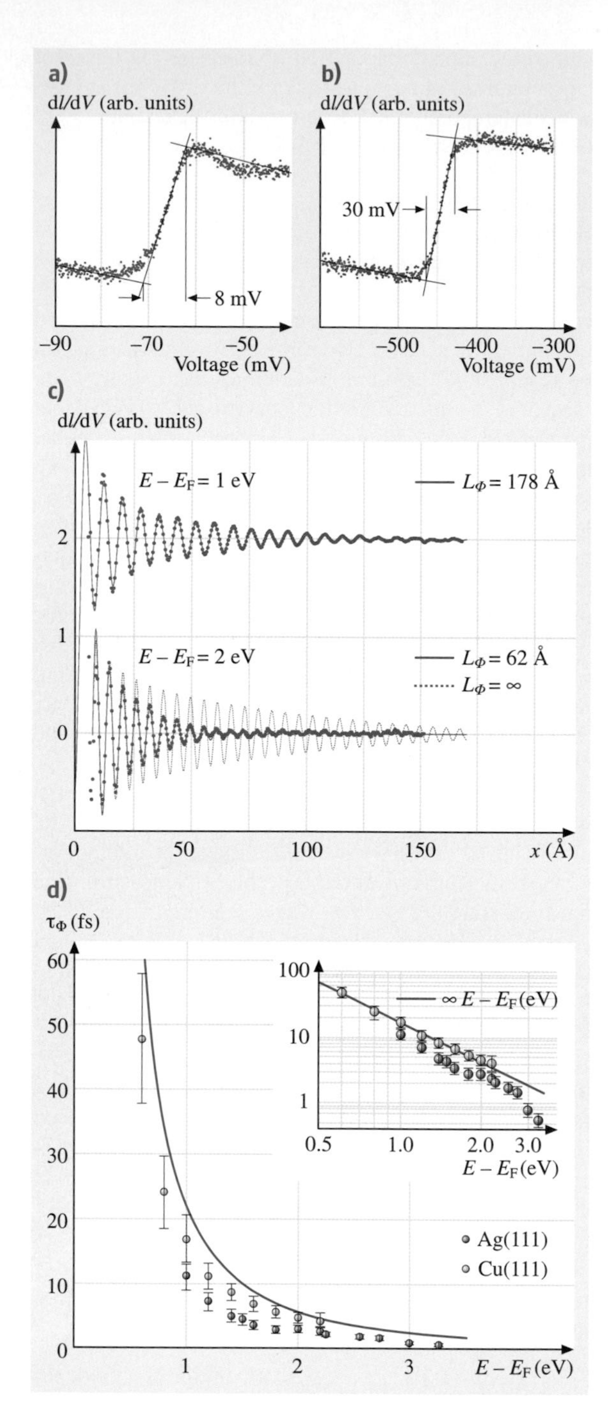

Fig. 14.9 **(a)**,**(b)** spatially averaged $dI/dV(V)$-curves of Ag(111) and Cu(111); both surfaces exhibit a surface state with parabolic dispersion starting at -65 meV and -430 meV, respectively. The lines are drawn to determine the energetic width of the onset of these surface bands (**(a)**,**(b)** courtesy of Berndt, Kiel); **(c)** dI/dV-intensity as a function of position away from a step edge of Cu(111) measured at the voltages ($E - E_F$), as indicated (*points*); the lines are fits assuming standing electron waves with a phase coherence length L_Φ as marked; **(d)** resulting phase coherence time as a function of energy for Ag(111) and Cu(111). *Inset* shows the same data on a double logarithmic scale evidencing the E^{-2}-dependence (*line*) (**(c)**,**(d)** courtesy of Brune, Lausanne)

energies are given by $E_{\text{Landau}} = \hbar eB/m_{\text{eff}}$ and $E_{\text{spin}} = g\mu B$. Thus InAs is a good choice, since it exhibits a low effective mass $m_{\text{eff}}/m_e = 0.023$ and a high g-factor of 14 in the bulk conduction band. The values in metals are $m_{\text{eff}}/m_e \approx 1$ and $g \approx 2$, resulting in energy splittings of only 1.25 meV and 1.2 meV at $B = 10$ T. This is obviously lower than the typical lifetime broadenings discussed in the previous section and also close to $\Delta E = 1.1$ meV achievable at $T = 4$ K.

Fortunately, the electron density in doped semiconductors is much lower, and thus the lifetime increases significantly. Figure 14.10a shows a set of spectroscopy curves obtained on InAs(110) in different magnetic fields [14.9]. Above E_F, oscillations with increasing intensity and energy distance are observed. They show the separation expected from Landau quantization. In turn, they can be used to deduce m_{eff} from the peak separation (Fig. 14.10b). An increase of m_{eff} with increasing E has been found as expected from theory. Also, at high fields spin quantization is observed (Fig. 14.10c). It is larger than expected from the bare g-factor due to contributions from exchange enhancement [14.58]. A detailed discussion of the peaks revealed that they belong to the so-called tip-induced quantum dot resulting from the work function difference between tip and sample.

Vibrational Levels

As discussed with respect to light emission in STM, inelastic tunneling processes contribute to the tunneling current. The coupling of electronic states to vibrational levels is one source of inelastic tunneling [14.23]. It provides additional channels contributing to $dI/dV(V)$ with final states at energies different from V. The final energy is simply shifted by the energy of the vibrational level. If only discrete vibrational energy levels couple to a smooth electronic DOS, one expects a peak in d^2I/dV^2 at the vibrational energy. This situation appears for molecules on noble metal surfaces. As usual, the isotope effect on the vibrational energy can be used

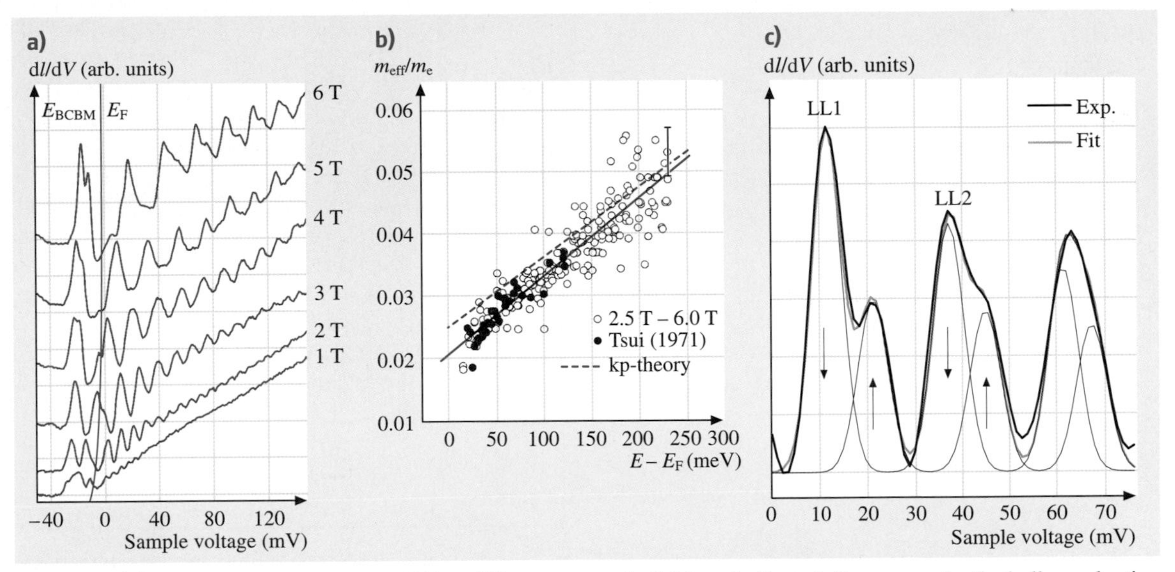

Fig. 14.10 **(a)** dI/dV-curves of n-InAs(110) at different magnetic fields as indicated; E_{BCBM} marks the bulk conduction band minimum; oscillations above E_{BCBM} are caused by Landau quantization; the double peaks at $B = 6\,T$ are caused by spin quantization. **(b)** Effective mass data deduced from the distance of adjacent Landau peaks ΔE according to $\Delta E = heB/m_{eff}$ (*open symbols*); *filled symbols* are data from planar tunnel junctions (Tsui), the *solid line* is a mean-sqare fit of the data and the *dashed line* is the expected effective mass of InAs according to kp-theory. **(c)** Magnification of a dI/dV-curve at $B = 6\,T$ exhibiting spin splitting; the Gaussian curves marked by *arrows* are the fitted spin levels

to verify the vibrational origin of the peak. First indications of vibrational levels have been found for H_2O and D_2O on TiO_2 [14.59], and completely convincing work has been performed for C_2H_2 and C_2D_2 on Cu(001) [14.23] (Fig. 14.11a). The technique can be used to identify individual molecules on the surface by their characteristic vibrational levels. In particular, surface reactions, as described in Fig. 14.5h–m, can be directly verified. Moreover, the orientation of complexes with respect to the surface can be determined to a certain extent, since the vibrational excitation depends on the position of the tunneling current within the molecule. Finally, the excitation of certain molecular levels can induce such corresponding motions as hopping [14.32], rotation [14.34] (Fig. 14.11b–e), or desorption [14.31], leading to additional possibilities for manipulation on the atomic scale.

Kondo Resonance

A rather intricate interaction effect is the Kondo effect. It results from a second order scattering process between itinerate states and a localized state [14.60]. The two states exchange some degree of freedom back and forth, leading to a divergence of the scattering probability at the Fermi level of the itinerate states. Due to the divergence, the effect strongly modifies sample properties. For example, it leads to an unexpected increase in resistance with decreasing temperature for metals containing magnetic impurities [14.4]. Here, the exchanged degree of freedom is the spin. A spectroscopic signature of the Kondo effect is a narrow peak in the DOS at the Fermi level disappearing above a characteristic temperature (Kondo temperature). STS provides the opportunity to study this effect on the local scale [14.61, 62].

Figure 14.12a–d shows an example of Co clusters deposited on a carbon nanotube [14.63]. While a small dip at the Fermi level, which is probably caused by curvature influences on the π-orbitals, is observed without Co (Fig. 14.12b) [14.64], a strong peak is found around a Co cluster deposited on top of the tube (Fig. 14.12a, arrow). The peak is slightly shifted with respect to $V = 0\,mV$ due to the so-called Fano resonance [14.65], which results from interference of the tunneling processes into the localized Co-level and the itinerant nanotube levels. The resonance disappears within a several nanometer distance from the cluster, as shown in Fig. 14.12d.

The Kondo effect has also been detected for different magnetic atoms deposited on noble metal sur-

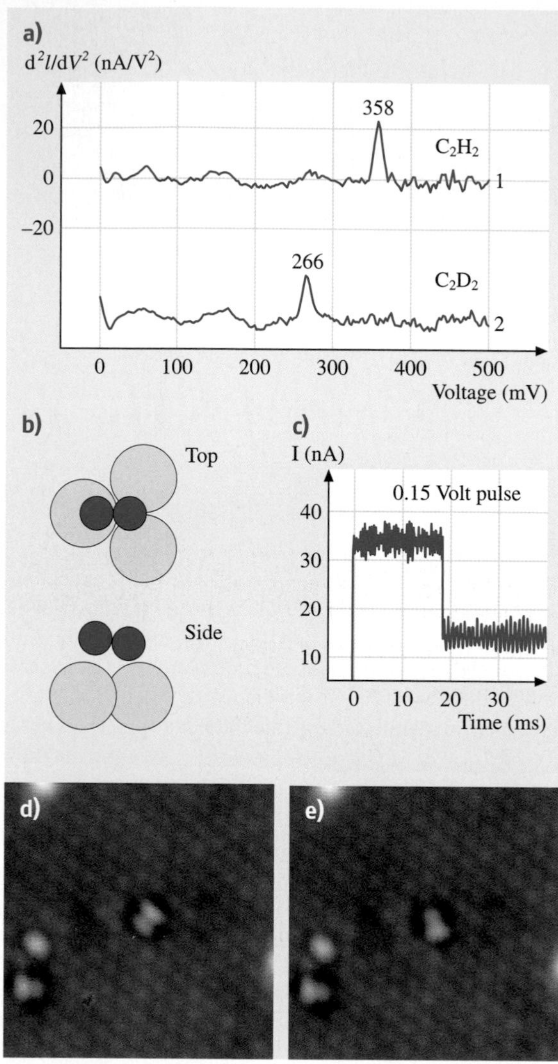

Fig. 14.11 (**a**) d^2I/dV^2-curves taken above a C_2H_2 and a C_2D_2 molecule on Cu(100); the peaks correspond to the C-H, respectively, C-D stretch mode energy of the molecule. (**b**) Sketch of O_2 molecule on Pt(111). (**c**) Tunnelling current above an O_2 molecule on Pt(111) during a voltage pulse of 0.15 V; the jump in current indicates rotation of the molecule. (**d**), (**e**) STM image of an O_2 molecule on Pt(111) ($V = 0.05$ V) prior and after rotation induced by a voltage pulse to 0.15 V ((**a**)–(**e**) courtesy of Ho, Irvine)

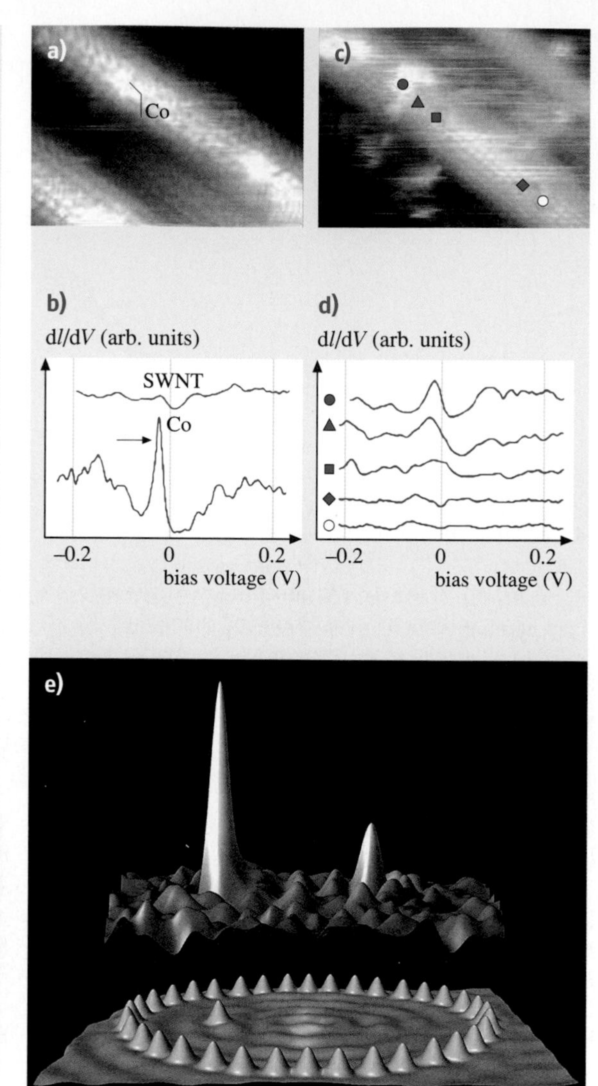

Fig. 14.12 (**a**) STM image of a Co cluster on a single-wall carbon nanotube (SWNT). (**b**) dI/dV-curves taken directly above the Co cluster (Co) and far away from the Co cluster (SWNT); the *arrow* marks the Kondo peak. (**c**) STM image of another Co cluster on a SWNT with *symbols* marking the positions where the dI/dV-curves displayed in (**d**) are taken. (**d**) dI/dV-curves taken at the positions marked in (**c**) ((**a**)–(**d**) courtesy of Lieber, Cambridge). (**e**) *Lower part:* STM image of a quantum corral of elliptic shape made from Co atoms on Cu(111); one Co atom is placed in one of the focii of the ellipse. *Upper part:* map of the strength of the Kondo signal in the corral; note that there is also a Kondo signal in the focus, which is not covered by a Co atom ((**e**) courtesy of Eigler, Almaden)

faces [14.61,62]. There, it disappears at about 1 nm away from the magnetic impurity, and the effect of the Fano resonance is more pronounced, contributing to dips in $dI/dV(V)$-curves instead of peaks.

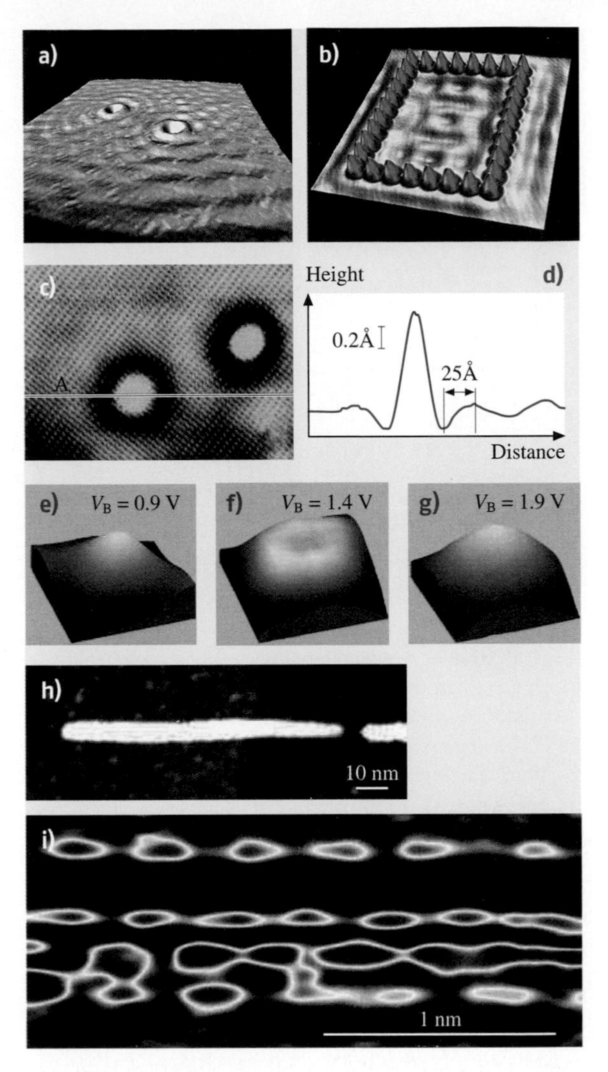

Fig. 14.13 (**a**) Low voltage STM image of Cu(111), including two defect atoms; the waves are electronic Bloch waves scattered at the defects. (**b**) Low voltage STM image of a rectangular quantum corral made from single atoms on Cu(111); the pattern inside the corral is the confined state of the corral close to E_F; ((**a**),(**b**) courtesy of Eigler, Almaden). (**c**) STM image of GaAs(110) around a Si donor, $V = -2.5$ V; the line scan along A shown in (**d**) exhibits an additional oscillation around the donor caused by a standing Bloch wave; the grid like pattern corresponds to the atomic corrugation of the Bloch wave ((**c**),(**d**) courtesy of van Kempen, Nijmegen). (**e**)–(**g**) STM images of an InAs/ZnSe-core/shell-nanocluster at different V. The image is measured in the so-called constant-height mode, i. e., the images display the tunneling current at constant height above the surface; the hill in (**e**) corresponds to the s-state of the cluster, the ring in (**f**) to the degenerate p_x- and p_y-state and the hill in (**g**) to the p_z-state ((**e**)–(**g**) courtesy of Millo, Jerusalem). (**h**) STM-image of a short-cut carbon nanotube. (**i**) Colour plot of the dI/dV intensity inside the short-cut nanotube as a function of position and tunneling voltage; four wavy patterns of different wavelength are visible in the voltage range from -0.1 to 0.15 V ((**h**),(**i**) courtesy of Dekker, Delft)

A fascinating experiment has been performed by *Manoharan* et al. [14.66], who used manipulation to form an elliptic cage for the surface states of Cu(111) (Fig. 14.12e, bottom). This cage was constructed to have a quantized level at E_F. Then, a cobalt atom was placed in one focus of the elliptic cage, producing a Kondo resonance. Surprisingly, the same resonance reappeared in the opposite focus, but not away from the focus (Fig. 14.12e, top). This shows amazingly that complex local effects such as the Kondo resonance can be guided to remote points.

Orbital scattering as a source of the Kondo resonance has also been found around a defect on Cr(001) [14.67]. Here, it is believed that itinerate sp-levels scatter at a localized d-level to produce the Kondo peak.

14.3.5 Imaging Electronic Wave Functions

Since STS measures the sum of squared wave functions (14.1), it is an obvious task to measure the local appearance of the most simple wave functions in solids, namely, the Bloch waves.

Bloch Waves

The atomically periodic part of the Bloch wave is always measured if atomic resolution is achieved (inset of Fig. 14.14a). However, the long-range, wavy part requires the presence of scatterers. The electron wave impinges on the scatterer and gets reflected, leading to self-interference. In other words, the phase of the Bloch wave gets fixed by the scatterer.

Such self-interference patterns were first found on Graphite(0001) [14.68] and later on noble metal surfaces, where adsorbates or step edges scatter the surface states (Fig. 14.13a) [14.22]. Fourier transforms of the real-space images reveal the k-space distribution of the corresponding states [14.69], which may include additional contributions besides the surface state [14.70]. Using particular geometries as the so-called quantum corrals to form a cage for the electron wave, the scatter-

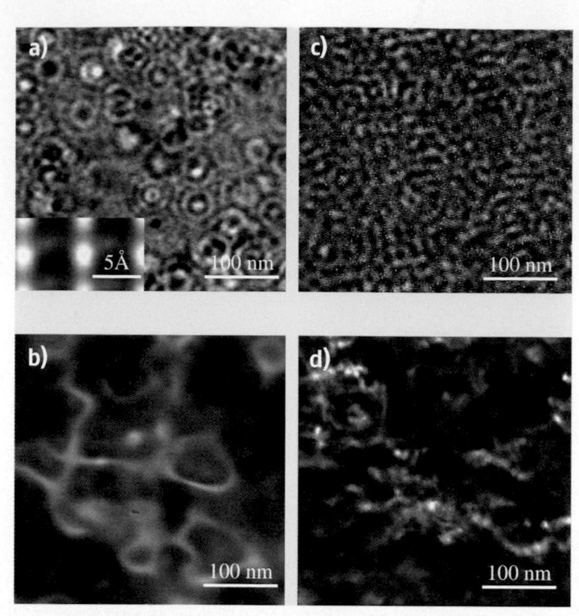

Fig. 14.14 (**a**) dI/dV-image of InAs(110) at $V = 50\,\text{mV}$, $B = 0\,\text{T}$; circular wave patterns corresponding to standing Bloch waves around each S donor are visible; inset shows a magnification revealing the atomically periodic part of the Bloch wave. (**b**) Same as (**a**), but at $B = 6\,\text{T}$; the stripe structures are drift states. (**c**) dI/dV-image of a 2-D electron system on InAs(110) induced by the deposition of Fe, $B = 0\,\text{T}$. (**d**) Same as (**c**) but at $B = 6\,\text{T}$; note that the contrast in (**a**) is increased by a factor of ten with respect to (**b**)–(**d**)

ing state can be rather complex (Fig. 14.13b). Anyway, it can usually be reproduced by simple calculations involving single-particle states [14.71].

Bloch waves in semiconductors scattered at charged dopants (Fig. 14.13c,d) [14.72], Bloch states confined in semiconductor quantum dots (Fig. 14.13e–g) [14.73], and Bloch waves confined in short-cut carbon nanotubes (Fig. 14.13h,i) [14.74, 75] have been visualized.

Drift States

More complex wave functions result from interactions. A nice playground to study such interactions are doped semiconductors. The reduced electron density with respect to metals increases the importance of electron interactions with potential disorder and other electrons. Applying a magnetic field quenches the kinetic energy, further enhancing the importance of interactions. A dramatic effect can be observed on InAs(110), where 3-D bulk states are displayed. While the usual scattering states around individual dopants are observed at $B = 0\,\text{T}$ (Fig. 14.14a) [14.76], stripe structures are found in high magnetic field (Fig. 14.14b) [14.77]. They run along equipotential lines of the disorder potential. This can be understood by recalling that the electron tries to move in a cyclotron orbit, which is accelerated and decelerated in electrostatic potential, leading to a drift motion along an equipotential line [14.78].

The same effect has been found in 2-D electron systems (2-DES) of the same substrate, where the scattering states at $B = 0\,\text{T}$ are, however, found to be more complex (Fig. 14.14c) [14.79]. The reason is the tendency of a 2-DES to exhibit closed scattering paths [14.80]. Consequently, the self-interference does not result from scattering at individual scatterers, but from complicated self-interference paths involving many scatterers. But drift states are also observed in the 2-DES at high magnetic fields (Fig. 14.14d) [14.81].

Charge Density Waves

Another interaction modifying the LDOS is the electron-phonon interaction. Phonons scatter electrons between different Fermi points. If the wave vectors connecting Fermi points exhibit a preferential orientation, a so-called Peierls instability occurs [14.82]. The corresponding phonon energy goes to zero, the atoms are slightly displaced with the periodicity of the corresponding wave vector, and a charge density wave (CDW) with the same periodicity appears. Essentially, the CDW increases the overlap of the electronic states with the phonon by phase fixing with respect to the atomic lattice. The Peierls transition naturally occurs in 1-D systems, where only two Fermi points are present, and, hence, preferential orientation is pathological. It can also occur in 2-D systems if large areas of the Fermi line run in parallel.

STS studies of CDWs are numerous (e.g., [14.83, 84]). Examples of a 1-D CDW on a quasi 1-D-like bulk material and of a 2-D CDW are shown in Fig. 14.15a–d and Fig. 14.15e, respectively [14.85, 86]. In contrast to usual scattering states where LDOS corrugations are only found close to the scatterer, the corrugations of CDWs are continous across the surface. Heating the substrate toward the transition temperature leads to a melting of the CDW lattice, as shown in Fig. 14.15f–h.

CDWs have also been found on monolayers of adsorbates such as a monolayer of Pb on Ge(111) [14.87]. These authors performed a nice temperature-dependent study revealing that the CDW is nucleated by scattering states around defects as one might expect [14.88]. 1-D systems have also been prepared on surfaces show-

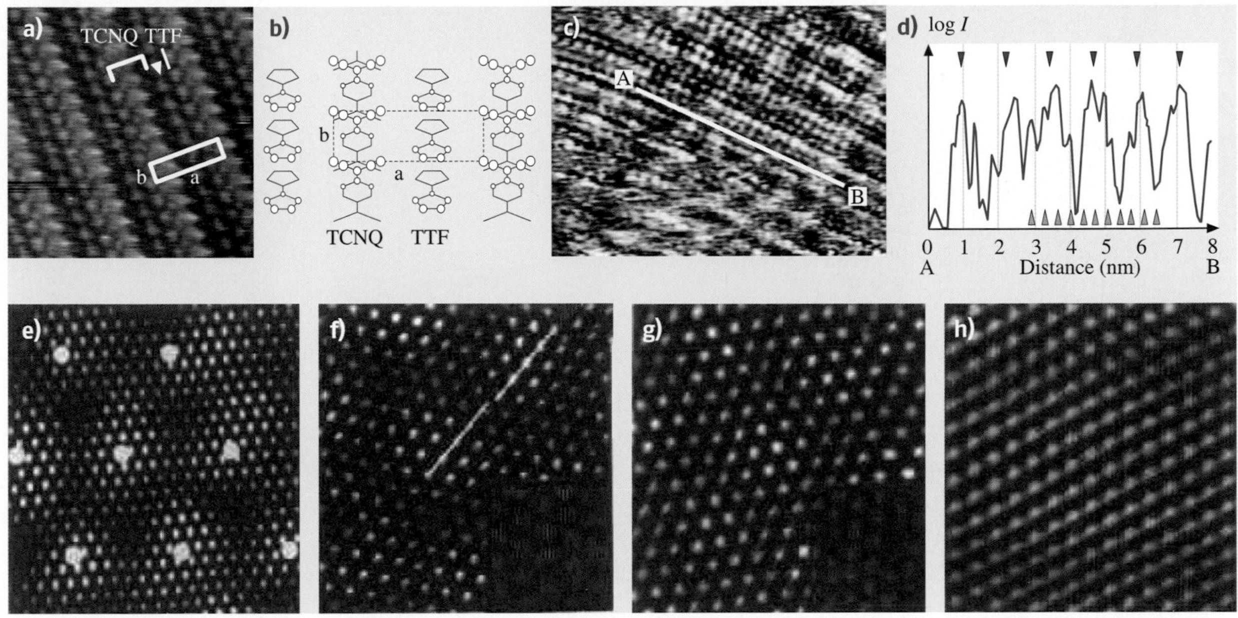

Fig. 14.15 (**a**) STM image of the ab-plane of the organic quasi 1-D conductor TTF-TCNQ, $T = 300\,\mathrm{K}$; while the TCNQ chains are conducting, the TTF chains are insulating. (**b**) Stick and ball model of the ab-plane of TTF-TCNQ. (**c**) STM image taken at $T = 61\,\mathrm{K}$, the additional modulation due to the Peierls-transition is visible in the profile along line A shown in (**d**); the *open triangles* mark the atomic periodicity and the *filled triangles* the expected CDW periodicity ((**a**)–(**d**) courtesy of Kageshima, Kanagawa). (**e**)–(**h**) Low voltage STM-images of the two-dimensional CDW-system 1 T-TaS_2 at $T = 242\,\mathrm{K}$ (**e**), 298 K (**f**), 349 K (**g**), 357 K (**h**). A long-range, hexagonal modulation is visible besides the atomic spots; its periodicity is highlighted by *big white dots* in (**e**); the additional modulation obviously weakens with increasing T, but is still apparent in (**f**) and (**g**), as evidenced in the lower magnification images in the insets ((**e**)–(**h**) courtesy of Lieber, Cambridge)

ing Peierls transitions [14.89,90]. Finally, the energy gap occurring at the transition has been studied by measuring $\mathrm{d}I/\mathrm{d}V(V)$-curves [14.91].

Superconductors

An intriguing effect resulting from electron-phonon interaction is superconductivity. Here, the attractive part of the electron-phonon interaction leads to the coupling of electronic states with opposite wave vector and mostly opposite spin [14.92]. Since the resulting Cooper pairs are bosons, they can condense at LT, forming a coherent many-particle phase, which can carry current without resistance. Interestingly, defect scattering does not influence the condensate if the coupling along the Fermi surface is homogeneous (s-wave superconductor). The reason is that the symmetry of the scattering of the two components of a Cooper pair effectively leads to a scattering from one Cooper pair state to another without affecting the condensate. This is different if the scatterer is magnetic, since the different spin components of the pair are scattered differently, leading to an effective pair breaking visible as a single-particle excitation within the superconducting gap. On a local scale, the effect has first been demonstrated by putting Mn, Gd, and Ag atoms on a Nb(110) surface [14.93]. While the nonmagnetic Ag does not modify the gap shown in Fig. 14.16a, it is modified in an asymmetric fashion close to Mn or Gd adsorbates, as shown in Fig. 14.16b. The asymmetry of the additional intensity is caused by the breaking of the particle-hole symmetry due to the exchange interaction between the localized Mn state and the itinerate Nb states.

Another important local effect is caused by the relatively large coherence length of the condensate. At a material interface, the condensate wave function cannot stop abruptly, but laps into the surrounding material (proximity effect). Consequently, a superconducting gap can be measured in areas of non-superconducting material. Several studies have shown this effect on the

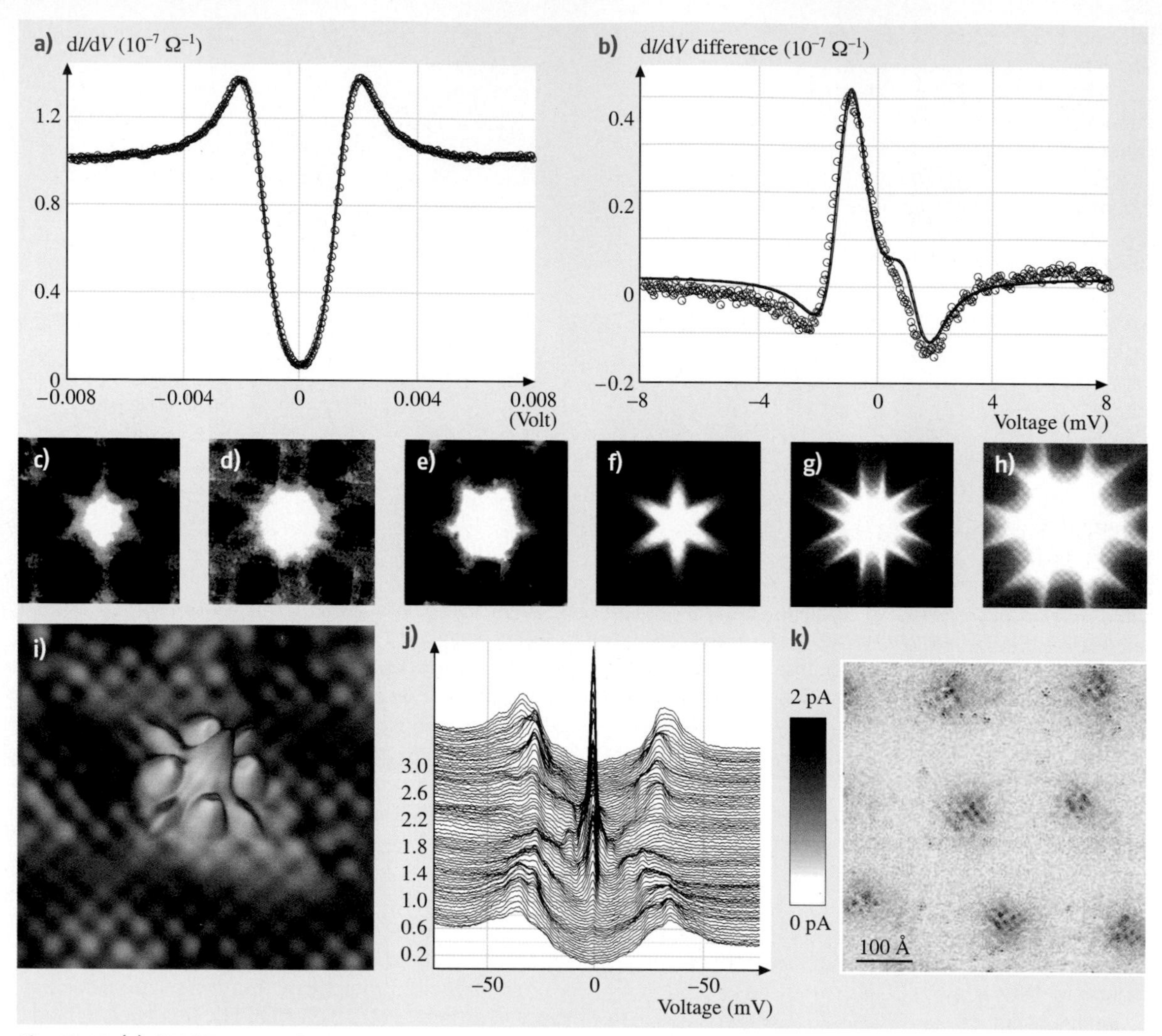

Fig. 14.16 **(a)** dI/dV-curve of Nb(110) at $T = 3.8$ K (*symbols*) in comparison with a BCS-fit of the superconducting gap of Nb (*line*). **(b)** Difference between the dI/dV-curve taken directly above a Mn-atom on Nb(110) and the dI/dV-curve taken above the clean Nb(110) (*symbols*) in comparison with a fit using the Bogulubov-de Gennes equations (*line*) (**(a)**,**(b)** courtesy of Eigler, Almaden). **(c)**–**(e)** dI/dV-images of a vortex core in the type-II superconductor 2H-$NbSe_2$ at 0 mV **(c)**, 0.24 mV **(d)**, and 0.48 mV **(e)** (**(c)**–**(e)** courtesy of H. F. Hess). **(f)**–**(h)** Corresponding calculated LDOS images within the Eilenberger framework (**(f)**–**(h)** courtesy of Machida, Okayama). **(i)** Overlap of an STM image at $V = -100$ mV (backround 2-D image) and a dI/dV-image at $V = 0$ mV (overlapped 3-D image) of optimally doped $Bi_2Sr_2CaCu_2O_{8+\delta}$ containing 0.6% Zn impurities. The STM image shows the atomic structure of the cleavage plane, while the dI/dV-image shows a bound state within the superconducting gap, which is located around a single Zn impurity. The fourfold symmetry of the bound state reflects the d-like symmetry of the superconducting pairing function; **(j)** dI/dV-curves taken at different positions across the Zn impurity; the bound state close to 0 mV is visible close to the Zn atom; **(k)** LDOS in the vortex core of slightly overdoped $Bi_2Sr_2CaCu_2O_{8+\delta}$, $B = 5$ T; the dI/dV-image taken at $B = 5$ T is integrated over $V = 1$–12 mV, and the corresponding dI/dV-image at $B = 0$ T is subtracted to highlight the LDOS induced by the magnetic field. The checkerboard pattern within the seven vortex cores exhibits a periodicity, which is fourfold with respect to the atomic lattice shown in **(i)** and is thus assumed to be a CDW (**(i)**–**(k)** courtesy of S. Davis, Cornell and S. Uchida, Tokyo)

local scale using metals and doped semiconductors as surrounding materials [14.94, 95].

While the classical type-I superconductors are ideal diamagnets, the so-called type-II superconductors can contain magnetic flux. The flux forms vortices, each containing one flux quantum. These vortices are accompanied by a disappearance of the superconducting gap and, therefore, can be probed by STS [14.96]. LDOS maps measured inside the gap lead to bright features in the area of the vortex core. Importantly, the length scale of these features is different from the length scale of the magnetic flux due to the difference between London's penetration depth and coherence length. Thus, STS probes another property of the vortex than the usual magnetic imaging techniques (see Sect. 14.4.4). Surprisingly, first measurements of the vortices on $NbSe_2$ revealed vortices shaped as a sixfold star [14.97] (Fig. 14.16c). With increasing voltage inside the gap, the orientation of the star rotates by 30° (Fig. 14.16d,e). The shape of these stars could finally be reproduced by theory, assuming an anisotropic pairing of electrons in the superconductor (Fig. 14.16f–h) [14.98]. Additionally, bound states inside the vortex core, which result from confinement by the surrounding superconducting material, are found [14.97]. Further experiments investigated the arrangement of the vortex lattice, including transitions between hexagonal and quadratic lattices [14.99], the influence of pinning centers [14.100], and the vortex motion induced by current [14.101].

An important topic is still the understanding of high temperature superconductivity (HTCS). An almost accepted property of HTCS is its d-wave pairing symmetry. In contrast to s-wave superconductors, scattering can lead to pair breaking, since the Cooper pair density vanishes in certain directions. Indeed, scattering states (bound states in the gap) around nonmagnetic Zn impurities have been observed in $Bi_2Sr_2CaCu_2O_{8+\delta}$ (BSCCO) (Fig. 14.16i,j) [14.26]. They reveal a d-like symmetry, but not the one expected from simple Cooper pair scattering. Other effects such as magnetic polarization in the environment probably have to be taken into account [14.102]. Moreover, it has been found that magnetic Ni impurities exhibit a weaker scattering structure than Zn impurities [14.103]. Thus, BSCCO shows exactly the opposite behavior of the Nb discussed above (Fig. 14.16a,b). An interesting topic is the importance of inhomogeneities in HTCS materials. Evidence for inhomogeneities has indeed been found in underdoped materials, where puddles of the superconducting phase are shown to be embedded in non-superconducting areas [14.104].

Of course, vortices have also been investigated in HTCS materials [14.105]. Bound states are found, but at energies that are in disagreement with simple models, assuming a BCS-like d-wave superconductor [14.106, 107]. Theory predicts, instead, that the bound states are magnetic-field-induced spin-density waves stressing the competition between anti-ferromagnetic order and superconductivity in HTCS materials [14.108]. Since the spin density wave is accompanied by a charge density wave of half wavelength, it can be probed by STS [14.109]. Indeed, a checkerboard pattern of the right periodicity has been found in and around vortex cores in BSCCO (Fig. 14.16k). It exceeds the width of an individual vortex core, implying that the superconducting coherence length is different from the anti-ferromagnetic one.

Complex Systems (Manganites)

Complex phase diagrams are not restricted to HTCS materials (cuprates). They exist with similar complexity for other doped oxides such as manganites. Only a few studies of these materials have been performed by STS, mainly showing the inhomogeneous evolution of metallic and insulating phases [14.110, 111]. Similarities to the granular case of an underdoped HTCS material are obvious. Since inhomogeneities seem to be crucial in many of these materials, a local method such as STS might continue to be important for the understanding of their complex properties.

14.3.6 Imaging Spin Polarization: Nanomagnetism

Conventional STS couples to the LDOS, i. e., the charge distribution of the electronic states. Since electrons also have spin, it is desirable to also probe the spin distribution of the states. This can be achieved if the tunneling tip is covered by a ferromagnetic material [14.112]. The coating acts as a spin filter or, more precisely, the tunneling current depends on the relative angle α_{ij} between the spins of the tip and the sample according to $\cos(\alpha_{ij})$. In ferromagnets, the spins mostly have one preferential orientation along the so-called easy axis, i. e., a particular tip is not sensitive to spin orientations of the sample perpendicular to the spin orientation of the tip. Different tips have to be prepared to detect different spin orientations of the sample. Moreover, the magnetic stray field of the tip can perturb the spin orientation of the sample. To avoid this, a technique

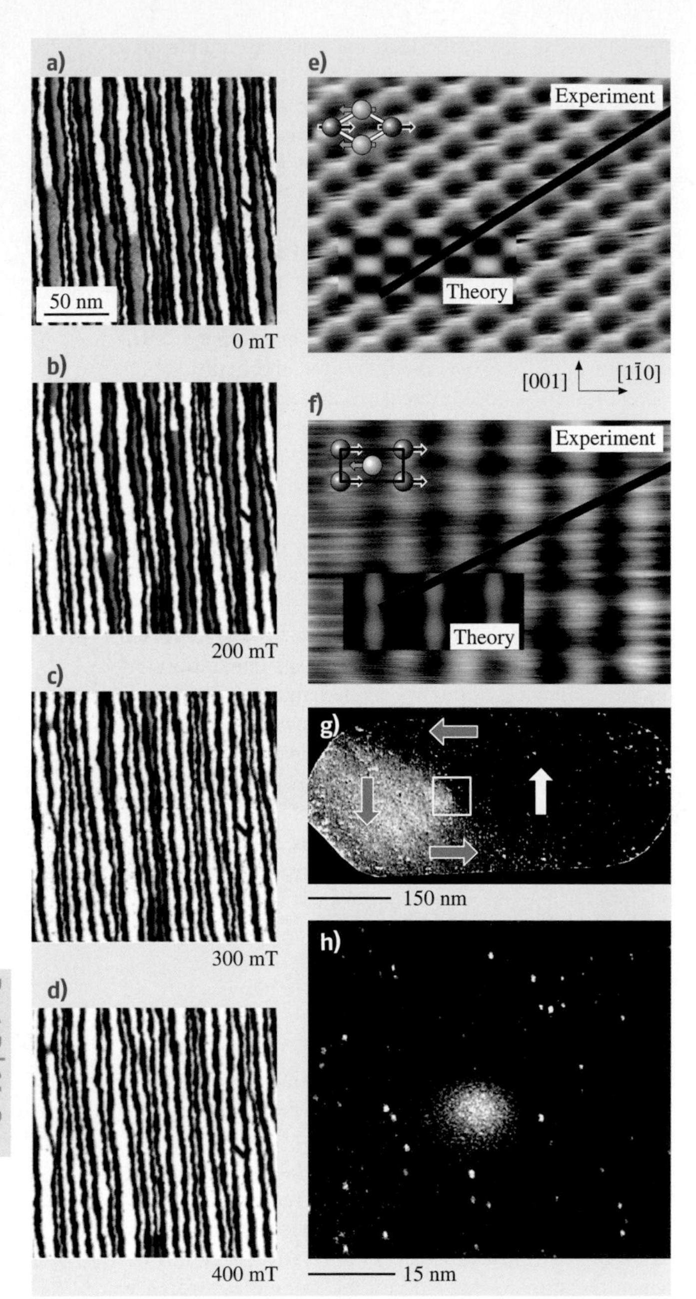

Fig. 14.17 (a)–(d) Spin-polarized STM images of 1.65 monolayer of Fe deposited on a stepped W(110) surface measured at different B-fields, as indicated. Double layer and monolayer Fe stripes are formed on the W substrate; only the double layer stripes exhibit magnetic contrast with an out-of-plane sensitive tip as used here. *White* and *grey areas* correspond to different domains. Note that more white areas appear with increasing field. (e) STM image of an antiferromagnetic Mn monolayer on W(110). (f) Spin-polarized STM-image of the same surface (in-plane tip). The insets in (e) and (f) show the calculated STM and spin-polarized STM images, respectively, and the stick and ball models symbolize the atomic and the magnetic unit cell ((a)–(f) courtesy of M. Bode, Hamburg). (g) Spin-polarized STM image of a 6-nm-high Fe island on W(110) (in-plane tip). Four different areas are identified as four different domains with domain orientations as indicated by the *arrows*. (h) Spin-polarized STM image of the central area of an island; the size of the area is indicated by the rectangle in (g); the measurement is performed with an out-of-plane sensitive tip showing that the magnetization turns out-of-plane in the center of the island

using anti-ferromagnetic Cr as a tip coating material has been developed [14.113]. This avoids stray fields, but still provides a preferential spin orientation of the few atoms at the tip apex that dominate the tunneling current. Depending on the thickness of the Cr coating, spin orientations perpendicular or parallel to the sample surface are prepared.

So far, the described technique has been used to image the evolution of magnetic domains with increasing B-field(Fig. 14.17a–d) [14.114], the antiferromagnetic order of a Mn monolayer on W(110) (Fig. 14.17e,f) [14.115], and the out-of-plane orientation predicted for a magnetic vortex core as it exists in the center of a Fe island exhibiting four domains in the flux closure configuration (Fig. 14.17g,h) [14.116].

Besides the obvious strong impact on nanomagnetism, the technique might also be used to investigate other electronic phases such as the proposed spin density wave around a HTCS vortex core.

14.4 Scanning Force Microscopy and Spectroscopy

SFSThe examples discussed in the previous section show the wide variety of physical questions that have been tackled with the help of LTSTM. Here, we turn to the other prominent scanning probe method that is applied at low temperatures, namely, SFM, which gives complementary information on sample properties on the atomic scale.

The ability to sensitively detect *forces* with spatial resolution down to the atomic scale is of great interest, since force is one of the most fundamental quantities in physics. Mechanical force probes usually consist of a cantilever with a tip at its free end that is brought close to the sample surface. The cantilever can be mounted parallel or perpendicular to the surface (general aspects of force probe designs are described in Chapt. 12). Basically, two methods exist to detect forces with cantilever-based probes – the *static* and the *dynamic* mode (see Chapt. 11). They can be used to generate a laterally resolved image (*microscopy* mode) or determine its distance dependence (*spectroscopy* mode). One can argue about the terminology, since spectroscopy is usually related to energies and not to distance dependences. Nevertheless, we will use it throughout the text, because it avoids lenghty paraphrases and is established in this sense throughout the literature.

In the static mode, a force that acts on the tip bends the cantilever. By measuring its deflection Δz the tip-sample force F_{ts} can be directly calculated with Hooke's law: $F_{ts} = c_z \cdot \Delta z$, where c_z denotes the spring constant of the cantilever. In the various dynamic modes, the cantilever is oscillated with amplitude A at or near its eigenfrequency f_0, but in some applications also off resonance. At ambient pressures or in liquids, amplitude modulation (AM-SFM) is used to detect amplitude changes or the phase shift between driving force and cantilever oscillation. In vacuum, the frequency shift Δf of the cantilever due to a tip-sample interaction is measured by frequency modulation technique (FM-SFM). The nomenclature is not standardized. Terms like tapping mode or intermittent contact mode are used instead of AM-SFM, and NC-AFM (noncontact atomic force microscopy) or DFM (dynamic force microscopy) instead of FM-SFM or FM-AFM. However, all these modes are *dynamic*, i. e., they involve an oscillating cantilever and can be used in the noncontact, as well as in the contact regime. Therefore, we believe that the best and most consistent way is to distinguish them by their different detection schemes. Converting the measured quantity (amplitude, phase, or frequency shift) into a physically meaningful quantity, e.g., the tip-sample interaction force F_{ts} or the force gradient $\partial F_{ts}/\partial z$, is not always straightforward and requires an analysis of the equation of motion of the oscillating tip (see Chaps. 15, 13).

Whatever method is used, the resolution of a cantilever-based force detection is fundamentally limited by its intrinsic *thermomechanical* noise. If the cantilever is in thermal equilibrium at a temperature T, the equipartition theorem predicts a thermally induced *root mean square* (rms) motion of the cantilever in z direction of $z_{rms} = (k_B T/c_{eff})^{1/2}$, where k_B is the Boltzmann constant and $c_{eff} = c_z + \partial F_{ts}/\partial z$. Note that usually $\mathrm{d}F_{ts}/\mathrm{d}z \gg c_z$ in contact and $\mathrm{d}F_{ts}/\mathrm{d}z < c_z$ in noncontact. Evidently, this fundamentally limits the force resolution in the static mode, particularly if operated in noncontact. Of course, the same is true for the different dynamic modes, because the thermal energy $k_B T$ excites the eigenfrequency f_0 of the cantilever. Thermal noise is *white* noise, i. e., its spectral density is flat. However, if the cantilever transfer function is taken into account, one can see that the thermal energy mainly excites f_0. This explains the term "thermo" in thermomechanical noise, but what is the "mechanical" part?

A more detailed analysis reveals that the thermally induced cantilever motion is given by

$$z_{rms} = \sqrt{\frac{2k_B TB}{\pi c_z f_0 Q}}\,, \tag{14.2}$$

where B is the measurement bandwidth and Q is the quality factor of the cantilever. Analog expressions can be obtained for all quantities measured in dynamic modes, because the deflection noise translates, e.g., into frequency noise [14.5]. Note that f_0 and c_z are correlated with each other via $2\pi f_0 = (c_z/m_{eff})^{1/2}$, where the effective mass m_{eff} depends on the geometry, density, and elasticity of the material. The Q-factor of the cantilever is related to the external damping of the cantilever motion in a medium and on the intrinsic damping within the material. This is the "mechanical" part of the fundamental cantilever noise.

It is possible to operate a low temperature force microscope directly immersed in the cryogen [14.117, 118] or in the cooling gas [14.119], whereby the cooling is simple and very effective. However, it is evident from (14.2) that the smallest fundamental noise is achievable in vacuum, where the Q-factors are more than 100 times larger than in air, and at low temperatures.

The best force resolution up to now, which is better than $1 \times 10^{-18}\,\mathrm{N/Hz^{1/2}}$, has been achieved by *Mamin* et al. [14.120] in vacuum at a temperature below 300 mK. Due to the reduced thermal noise and the lower thermal drift, which results in a higher stability of the tip-sample gap and a better signal-to-noise ratio, the highest resolution is possible at low temperatures in ultrahigh vacuum with FM-SFM. A vertical rms-noise below 2 pm [14.121, 122] and a force resolution below 1 aN [14.120] have been reported.

Besides the reduced noise, the application of force detection at low temperatures is motivated by the increased stability and the possibility to observe phenomena, which appear below a certain critical temperature T_c, as outlined at page 414. The experiments, which have been performed at low temperatures up until now, were motivated by at least one of these reasons and can be roughly divided into four groups: (i) atomic-scale imaging, (ii) force spectroscopy, (iii) investigation of quantum phenomena by measuring electrostatic forces, and (iv) utilizing magnetic probes to study ferromagnets, superconductors, and single spins. In the following, we describe some exemplary results.

14.4.1 Atomic-Scale Imaging

In a simplified picture, the dimensions of the tip end and its distance to the surface limit the lateral resolution of force microscopy, since it is a near-field technique. Consequently, atomic resolution requires a stable single atom at the tip apex that has to be brought within a distance of some tenths of a nanometer to an atomically flat surface. Both conditions can be fulfilled by FM-AFM, and nowadays *true* atomic resolution is routinely obtained (see Chapt. 13). However, the above statement says nothing about the nature and the underlying processes of imaging with atomic resolution. Si(111)-(7×7) was the first surface on which true atomic resolution was achieved [14.123], and several studies have been performed at low temperatures on this well-known material [14.124–126]. First principle simulations performed on semiconductors with a silicon tip revealed that *chemical* interactions, i. e., a significant charge redistribution between the dangling bonds of tip and sample, dominate the atomic-scale contrast [14.127–129]. On V-III semiconductors, it was found that only one atomic species, the group V atoms, is imaged as protrusions with a silicon tip [14.128, 129]. Furthermore, these simulations revealed that the sample, as well as the tip atoms are noticeably displaced from their equilibrium position due to the interaction forces. At low temperatures, both aspects could be observed with silicon tips on indium arsenide [14.121, 130].

Chemical Sensitivity of Force Microscopy

The (110) surface of the III-V semiconductor indium arsenide exhibits both atomic species in the top layer (see Fig. 14.18a). Therefore, this sample is well suited to study the chemical sensitivity of force microscopy [14.121]. In Fig. 14.18b, the usually observed atomic-scale contrast on InAs(110) is displayed. As predicted, the arsenic atoms, which are shifted by 80 pm above the indium layer due to the (1×1) relaxation, are imaged as protrusions. While this general appearance was similar for most tips, two other distinctively different contrasts were also observed: a second protru-

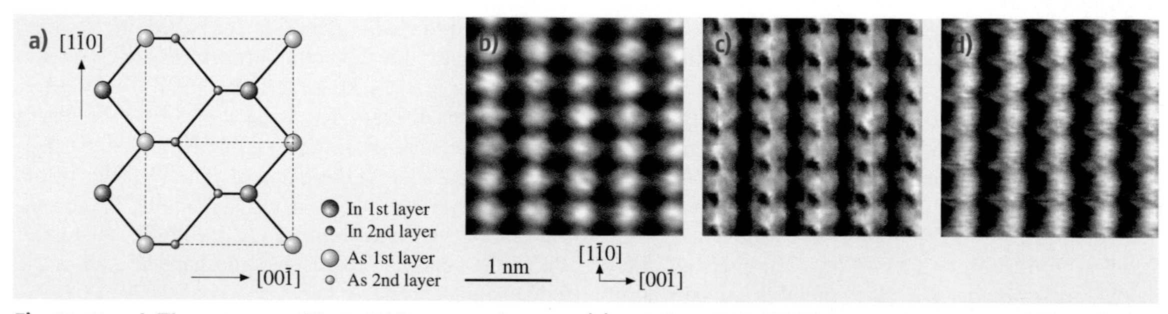

Fig. 14.18a–d The structure of InAs(110) as seen from top (a) and three FM-AFM images of this surface obtained with different tips at 14 K (b)–(d). In (b), only the arsenic atoms are imaged as protrusions, as predicted for a silicon tip. The two features in (c) and (d) corresponds to the zigzag arrangement of the indium and arsenic atoms. Since force microscopy is sensitive to short-range chemical forces, the appearance of the indium atoms can be associated with a chemically different tip apex

sion (c) and a sharp depression (d). The arrangement of these two features corresponds well to the zigzag configuration of the indium and arsenic atoms along the [1$\bar{1}$0]-direction. A sound explanation would be as follows: The contrast usually obtained with one feature per surface unit cell corresponds to a silicon-terminated tip, as predicted by simulations. A different atomic species at the tip apex, however, can result in a very different charge redistribution. Since the atomic-scale contrast is due to a chemical interaction, the two other contrasts would then correspond to a tip that has been accidentally contaminated with sample material (arsenic or indium-terminated tip apex). Nevertheless, this explanation has not yet been verified by simulations for this material.

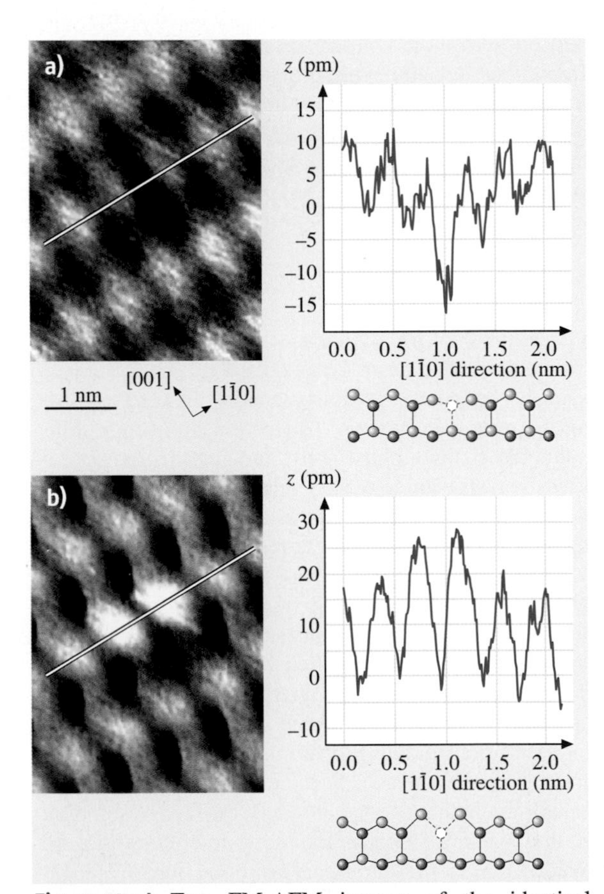

Fig. 14.19a,b Two FM-AFM images of the identical indium-site point defect (presumably an indium vacancy) recorded at 14 K. If the tip is relatively far away, the theoretically predicted inward relaxation of two arsenic atoms adjacent to an indium vacancy is visible (**a**). At a closer tip-sample distance (**b**), the two arsenic atoms are pulled farther toward the tip compared to the other arsenic atoms, since they have only two instead of three bonds

Tip-Induced Atomic Relaxation

Schwarz et al. [14.121] were able to directly visualize the predicted tip-induced relaxation during atomic-scale imaging near a point defect. Figure 14.19 shows two FM-AFM images of the same point defect recorded with different constant frequency shifts on InAs(110), i. e., the tip was closer to the surface in (b) compared to (a). The arsenic atoms are imaged as protrusions with the used silicon tip. From the symmetry of the defect, an indium-site defect can be inferred, since the distance-dependent contrast is consistent with what has to be expected for an indium vacancy. This expectation is based on calculations performed for the similar III-V semiconductor GaP(110), where the two surface gallium atoms around a P-vacancy were found to relax downward [14.131]. This corresponds to the situation in Fig. 14.19a, where the tip is relatively far away and an inward relaxation of the two arsenic atoms is observed. The considerably larger attractive force in Fig. 14.19b, however, pulls the two arsenic atoms toward the tip. All other arsenic atoms are also pulled, but they are less displaced, because they have three bonds to the bulk, while the two arsenic atoms in the neighborhood of an indium vacancy have only two bonds. This direct experimental proof of the presence of tip-induced relaxations is also relevant for STM measurements, because the tip-sample distances are similar during atomic resolution imaging. Moreover, the result demonstrates that FM-AFM can probe elastic properties on an atomic level.

Imaging of Weakly Interacting van der Waals Surfaces

For weakly interacting van der Waals surfaces like the (0001) surface of graphite, much smaller atomic corrugation amplitudes are expected compared to strongly interacting surfaces such as semiconductors. In graphite, a layered material, the carbon atoms are covalently bonded and arranged in a honeycomb structure within the (0001) plane. Individual layers of (0001) planes ($ABA\dots$ stacking) stick together only by weak van der Waals forces. Three distinctive sites exist on the surface: carbon atoms with (A-type) and without (B-type) a neighbor in the next graphite layer and the *hollow site* (H-site) in the hexagon center. Neither static contact force microscopy nor STM can resolve the three different sites. For both methods, the contrast is well understood and exhibits protrusions with a sixfold symmetry and a periodicity of 246 pm. In high resolution FM-AFM images acquired at low temperatures, a large maximum and two different minima have been resolved (see Fig. 14.20a). A simulation using the Lennard-Jones potential, in which the attractive part is given by the short-range interatomic van der Waals force, reproduced these three features very well. By comparison with the experimental data (cf. the solid and the dotted line in the section of Fig. 14.20a), the large maximum could be assigned to the hollow site, while the two different minima represent A- and B-type carbon atoms [14.132]. Note that the image contrast is inverted with respect the arrangement of the atoms, i. e., the minima correspond to the position of the carbon atoms. The carbon-carbon distance of only 142 pm is the smallest interatomic distance that has been resolved with FM-AFM so far. Moreover, the weak tip-sample interaction results in typical corrugation amplitudes around 10 pm peak-to-peak, i. e., just above the noise level [14.7].

While experiments on graphite basically gain advantages from the increased stability and signal-to-noise ratio at low temperatures, solid xenon (melting temperature $T_m = 161$ K) can only be observed at sufficiently low temperatures [14.8]. In addition, xenon is a pure van der Waals crystal, and since it is an insulator, FM-AFM is the only real space method available today that allows the study of solid xenon on the atomic scale.

Allers et al. [14.8] adsorbed a well-ordered xenon film on cold graphite(0001) ($T < 55$ K) and subsequently studied it at 22 K by FM-AFM (see Fig. 14.20b). The six-fold symmetry and the distance between the protrusions correspond well with the nearest neighbor distance in the closed, packed (111) plane of bulk xenon, which crystallizes in the face-centered cubic structure. A comparison between experiment and simulation confirmed that the protrusions correspond to the position of the xenon atoms [14.133]. However, the simulated corrugation amplitudes do not fit as well as for graphite (see sections in Fig. 14.20b). A possible reason is that tip-induced relaxations, which were not considered in the simulations, are more important for this pure van der Waals crystal xenon than they are for graphite, because in-plane graphite exhibits strong covalent bonds. Nevertheless, the results demonstrated that a weakly bonded van der Waals crystal could be imaged nondestructively on the atomic scale for the first time.

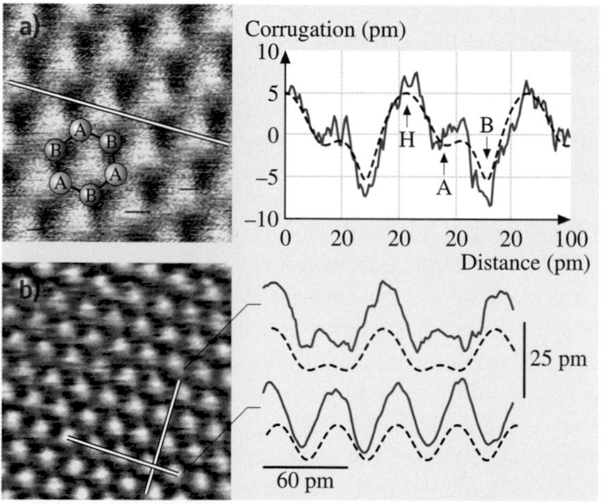

Fig. 14.20a,b FM-AFM images of graphite(0001) **(a)** and Xe(111) **(b)** recorded at 22 K. On the right side, line sections taken from the experimental data (*solid lines*) are compared to simulations (*dotted lines*). A- and B-type carbon atoms, as well as the hollow site (H-site) on graphite can be distinguished, but are imaged with inverted contrast, i. e., the carbon sites are displayed as minima. Such an inversion does not occur on Xe(111)

14.4.2 Force Spectroscopy

A wealth of information about the nature of the tip-sample interaction can be obtained by measuring its distance dependence. This is usually done by recording the measured quantity (deflection, frequency shift, amplitude change, phase shift) and applying an appropriate voltage ramp to the z-electrode of the scanner piezo, while the z-feedback is switched off. According to (14.2), low temperatures and high Q-factors (vacuum) considerably increase the force resolution. In the static mode, long-range forces and contact forces can be examined. Force measurements at small tip-sample distances are inhibited by the *jump-to-contact* phenomenon: If the force gradient $\partial F_{ts}/\partial z$ becomes larger than the spring constant c_z, the cantilever cannot resist the attractive tip-sample forces and the tip snaps onto the surface. Sufficiently large spring constants prevent this effect, but reduce the force resolution. In the dynamic modes, the jump-to-contact can be avoided due to the additional restoring force ($c_z A$) at the lower turnaround point. The highest sensitivity can be achieved in vacuum by using the FM technique, i. e., by recording $\Delta f(z)$-curves. An alternative FM spectroscopy method, the recording of $\Delta f(A)$-curves, has been suggested by *Hölscher* et al. [14.134]. Note that if the amplitude is much larger than the characteristic decay length of the tip-sample force, the frequency shift cannot simply be converted into force gradients by using $\partial F_{ts}/\partial z = 2c_z \cdot \Delta f/f_0$ [14.135]. Several methods have been published to convert $\Delta f(z)$ data into the tip-sample potential $V_{ts}(z)$ and tip-sample force $F_{ts}(z)$ (see, e.g., [14.136, 137]).

Measurement of Interatomic Forces at Specific Atomic Sites

FM force spectroscopy has been successfully used to measure and determine quantitatively the short-range chemical force between the foremost tip atom and specific surface atoms [14.109, 138, 139]. Figure 14.21 displays an example for the quantitative determination of the short-range force. Figure 14.21a shows two $\Delta f(z)$-curves measured with a silicon tip above a corner hole and above an adatom. Their position is indicated by arrows in the inset, which displays the atomically resolved Si(111)-(7×7) surface. The two curves differ from each other only for small tip-sample distances, because the long-range forces do not contribute to the atomic-scale contrast. The low, thermally induced lateral drift and the high stability at low temperatures were required to precisely address the two specific sites. To extract the short-range force, the long-range van der Waals and/or electrostatic forces can be subtracted from the total force. The grey curve in Fig. 14.21b has been reconstructed from the $\Delta f(z)$-curve recorded above an adatom and represents the total force. After removing the long-range contribution from the data, the much steeper black line is obtained, which corresponds to the short-range force between the adatom and the atom at the tip apex. The measured maximum attractive force (−2.1 nN) agrees well with first principle calculations (−2.25 nN).

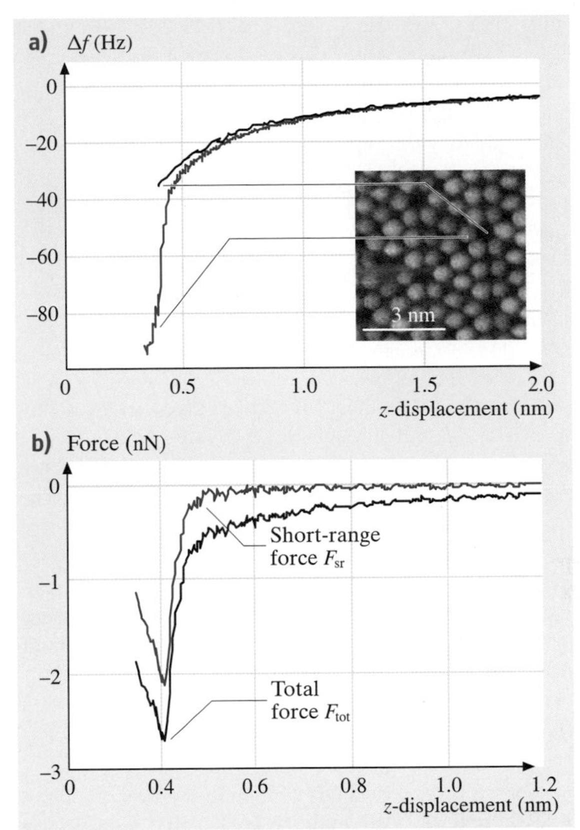

Fig. 14.21a,b FM force spectroscopy on specific atomic sites at 7.2 K. In (**a**), an FM-SFM image of the Si(111)-(7×7) surface is displayed together with two $\Delta f(z)$-curves, which have been recorded at the positions indicated by the *arrows*, i. e., above the cornerhole (*brown*) and above an adatom (*black*). In (**b**), the total force above an adatom (*brown line*) has been recovered from the $\Delta f(z)$-curve. After subtraction of the long-range part, the short-range force can be determined (*black line*) (courtesy of H. J. Hug; cf. [14.138])

Three-dimensional Force Field Spectroscopy

Further progress with the FM technique has been made by *Hölscher* et al. [14.140]. They acquired a complete 3-D force field on NiO(001) with atomic resolution (*3-D force field spectroscopy*). In Fig. 14.22a, the atomically resolved FM-AFM image of NiO(001) is shown together with the used coordinate system and the tip to illustrate the measurement principle. NiO(001) crystallizes in the rock salt structure. The distance between the protrusions corresponds to the lattice constant of 417 pm, i. e., only one type of atom (most likely the oxygen) is imaged as protrusion. In an area of 1 nm × 1 nm, 32×32 individual $\Delta f(z)$-curves have been recorded at every (x, y) image point and converted into $F_{ts}(z)$-curves. The $\Delta f(x, y, z)$ data set is thereby converted into the 3-D force field $F_{ts}(x, y, z)$. Figure 14.22b, where a specific x-z-plane is displayed, demonstrates that atomic resolution is achieved. It represents a 2-D cut $F_{ts}(x, y = \text{const}, z)$ along the [100]-direction (corresponding to the shaded slice marked in Fig. 14.22a). Since a large number of curves have been recorded, *Langkat* et al. [14.139] could evaluate the whole data set by standard statistical means to extract the long- and short-range forces. A possible future application of 3-D force field spectroscopy could be to map the short-range forces of complex molecules with functionalized tips in order to locally resolve their chemical reactivity.

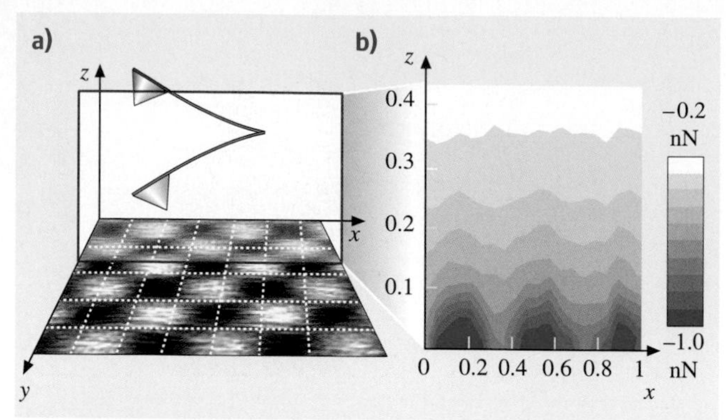

Fig. 14.22a,b Principle of the 3-D force filed spectroscopy method (**a**) and a 2-D cut through the 3-D force field $F_{ts}(x,y,z)$ recorded at 14 K (**b**). At all 32×32 image points of the 1 nm × 1 nm large scan area on NiO(001), a $\Delta f(z)$-curve has been recorded. The obtained $\Delta f(x, y, z)$ data set is then converted into the 3-D tip-sample force field $F_{ts}(x, y, z)$. The *shaded slice* $F_{ts}(x, y = \text{const}, z)$ in (**a**) corresponds to a cut along the [100]-direction and demonstrates that atomic resolution has been obtained, because the distance between the protrusions corresponds well to the lattice constant of nickel oxide

Noncontact Friction

Another approach to achieve small tip-sample distances in combination with high force sensitivity is to use soft springs in a perpendicular configuration. The much higher cantilever stiffness along the cantilever axis prevents the jump-to-contact, but the lateral resolution is limited by the magnitude of the oscillation amplitude. However, with such a setup at low temperatures, *Stipe* et al. [14.141] measured the distance dependence of the very small force due to noncontact friction between tip and sample in vacuum. The effect was attributed to electric charges, which are moved parallel to the surface by the oscillating tip. Since the topography was not recorded in situ, the influence of contaminants or surface steps remained unknown.

14.4.3 Electrostatic Force Microscopy

Electrostatic forces are readily detectable by a force microscope, because tip and sample can be regarded as two electrodes of a capacitor. If they are electrically connected via their back sides and have different work functions, electrons will flow between tip and sample until their Fermi levels are equalized. As a result, an electric field and, consequently, an attractive electrostatic force exists between them at zero bias. This *contact potential difference* can be balanced by applying an appropriate bias voltage. It has been demonstrated that individual doping atoms in semiconducting materials can be detected by electrostatic interactions due to the local variation of the surface potential around them [14.142, 143].

Detection of Edge Channels in the Quantum Hall Regime

At low temperatures, electrostatic force microscopy has been used to measure the electrostatic potential in the quantum Hall regime of a *two-dimensional electron gas* (2-DEG) buried in epitaxially grown GaAs/AlGaAs heterostructures [14.144–147]. In the 2-DEG, electrons can move freely in the x-y-plane, but they cannot move in z-direction. Electrical transport properties of a 2-DEG are very different compared to normal metallic conduction. Particularly, the Hall resistance $R_H = h/ne^2$ (where h represents Planck's constant, e is the electron charge, and $n = 1, 2, \ldots$) is quantized in the quantum Hall regime, i.e., at sufficiently low temperatures ($T < 4$ K) and high magnetic fields (up to 20 T). Under these conditions, theoretical calculations predict the existence of *edge channels* in a Hall bar. A Hall bar is a strip conductor that is contacted in a specific way to allow longitudinal and transversal transport measurements in a perpendicular magnetic field. The current is not evenly distributed over the cross section of the bar, but passes mainly along rather thin paths close to the edges. This prediction has been verified by measuring profiles of the electrostatic potential across a Hall bar in different perpendicular external magnetic fields [14.144–146].

Figure 14.23a shows the experimental setup used to observe these edge channels on top of a Hall bar with a force microscope. The tip is positioned above the surface of a Hall bar under which the 2-DEG is buried. The direction of the magnetic field is oriented perpendicular to the 2-DEG. Note that although the 2-DEG is located several tens of nanometers below the surface, its influence on the electrostatic surface potential can be detected. In Fig. 14.23b, the results of scans perpendicular to the Hall bar are plotted against the magnitude of the external magnetic field. The value of the electrostatic potential is grey-coded in arbitrary units. In certain field ranges, the potential changes linearly across the Hall bar, while in other field ranges the potential drop is confined to the edges of the Hall bar. The predicted edge channels can explain this behavior. The periodicity of the phenomenon is related to the filling factor ν, i.e., the number of Landau levels that are filled with electrons (see also Sect. 14.3.4). Its value depends on $1/B$

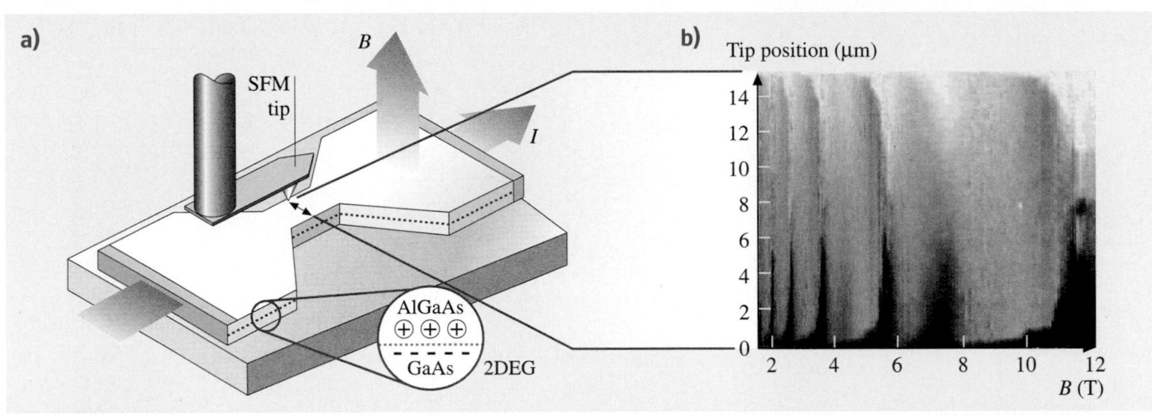

Fig. 14.23a,b Configuration of the Hall bar within a low temperature ($T < 1$ K) force microscope (**a**) and profiles (y-axis) at different magnetic field (x-axis) of the electrostatic potential across a 14-μm-wide Hall bar in the quantum Hall regime (**b**). The external magnetic field is oriented perpendicular to the 2-DEG, which is buried below the surface. *Bright* and *dark regions* reflect the characteristic changes of the electrostatic potential across the Hall bar at different magnetic fields and can be explained by the existence of the theoretically predicted edge channels (courtesy of E. Ahlswede; cf. [14.146])

and is proportional to the electron concentration n_e in the 2-DEG ($\nu = n_e h/eB$, where h represents Planck's constant and e the electron charge).

14.4.4 Magnetic Force Microscopy

To detect magnetostatic tip-sample interactions with magnetic force microscopy (MFM), a ferromagnetic probe has to be used. Such probes are readily prepared by evaporating a thin magnetic layer, e.g., 10 nm iron, onto the tip. Due to the in-plane shape anisotropy of thin films, the magnetization of such tips lies predominantly along the tip axis, i. e., perpendicular to the surface. Since magnetostatic interactions are long-range, they can be separated from the topography by scanning in a certain constant height (typically around 20 nm) above the surface, where the z-component of the sample stray field is probed (see Fig. 14.24a). Therefore, MFM is always operated in noncontact. The signal from the cantilever is directly recorded while the z-feedback is switched off. MFM can be operated in the static mode or in the dynamic modes (AM-MFM at ambient pressures and FM-MFM in vacuum). A lateral resolution below 50 nm can be routinely obtained.

Observation of Domain Patterns

MFM is widely used to visualize domain patterns of ferromagnetic materials. At low temperatures, *Moloni* et al. [14.148] observed the domain structure of magnetite below its Verwey transition temperature ($T_V = 122$ K), but most of the work concentrated on thin films of $La_{1-x}Ca_xMnO_3$ [14.149–151]. Below T_V, the conductivity decreases by two orders of magnitude and a small structural distortion is observed. The domain structure of this mixed valence manganite is of great interest, because its resistivity strongly depends on the external magnetic field, i. e., it exhibits a large colossal magneto resistive effect. To investigate the field dependence of domain patterns in ambient conditions, electromagnets have to be used. They can cause severe thermal drift problems due to Joule-heating of the coils by large currents. Field on the order of 100 mT can be achieved. In contrast, much larger fields (more than 10 T) can be rather easily produced by implementing a superconducting magnet in low temperature setups. With such a design, *Liebmann* et al. [14.151] recorded the domain structure along the major hysteresis loop of $La_{0.7}Ca_{0.3}MnO_3$ epitaxially grown on $LaAlO_3$ (see Fig. 14.24b–f). The film geometry (thickness is 100 nm) favors an in-plane magnetization, but the lattice mismatch with the substrate induces an out-of-plane anisotropy. Thereby, an irregular pattern of strip domains appears in zero field. If the external magnetic field is increased, the domains with anti-parallel orientation shrink and finally disappear in saturation (see Fig. 14.24b,c). The residual contrast in saturation (d) reflects topographic features. If the field is decreased after saturation (see Fig. 14.24e,f), cylindrical domains first nucleate and then start to grow. At zero field, the maze-type domain pattern has been evolved again.

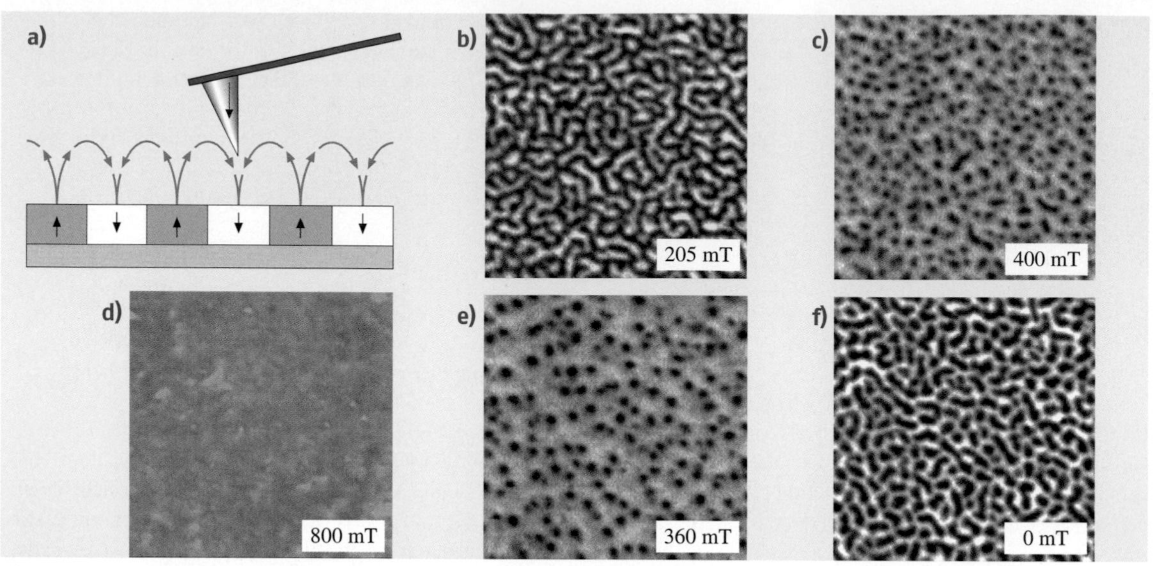

Fig. 14.24a–f Principle of MFM operation (**a**) and field-dependent domain structure of a ferromagnetic thin film (**b**)–(**f**) recorded at 5.2 K with FM-MFM. All images were recorded on the same 4 μm×4 μm large scan area. The $La_{0.7}Ca_{0.3}MnO_3/LaAlO_3$ system exhibits a substrate-induced out-of-plane anisotropy. *Bright* and *dark areas* are visible and correspond to attractive and repulsive magnetostatic interactions, respectively. The series shows how the domain pattern evolves along the major hysteresis loop from, i. e., zero field to saturation at 600 mT and back to zero field

Such data sets can be used to analyze domain nucleation and the domain growth mode. Moreover, due to the negligible drift, domain structure and surface morphology can be directly compared, because every MFM can be used as a regular topography-imaging force microscope.

Detection of Individual Vortices in Superconductors

Numerous low temperature MFM experiments have been performed on superconductors [14.152–159]. Some basic features of superconductors have been mentioned already in Sect. 14.3.5. The main difference of STM/STS compared to MFM is the high sensitivity to the electronic properties of the surface. Therefore, careful sample preparation is a prerequisite. This is not so important for MFM experiments, since the tip is scanned at a certain distance above the surface.

Superconductors can be divided into two classes with respect to their behavior in an external magnetic field. For type-I superconductors, any magnetic flux is entirely excluded below their critical temperature T_c (Meissner effect), while for type-II superconductors, cylindrical inclusions (*vortices*) of normal material exist in a superconducting matrix (*vortex* state). The radius of the vortex *core*, where the Cooper pair density decreases to zero, is on the order of the coherence length ξ. Since the superconducting gap vanishes in the core, they can be detected by STS (see Sect. 14.3.5). Additionally, each vortex contains one magnetic quantum flux $\Phi = h/2e$ (where h represents Planck's constant and e the electron charge). Circular supercurrents around the core screen the magnetic field associated with a vortex; their radius is given by the London penetration depth λ of the material. This magnetic field of the vortices can be detected by MFM. Investigations have been performed on the two most popular copper oxide high-T_c superconductors, $YBa_2Cu_3O_7$ [14.152, 153, 155] and $Bi_2Sr_2CaCu_2O_8$ [14.153, 159], on the only elemental conventional type-II superconductor Nb [14.156, 157], and on the layered compound crystal $NbSe_2$ [14.154, 156].

Most often, vortices have been generated by cooling the sample from the normal state to below T_c in an external magnetic field. After such a *field cooling* procedure, the most energetically favorable vortex arrangement is a regular triangular Abrikosov lattice. *Volodin* et al. [14.154] were able to observe such an Abrikosov lattice on $NbSe_2$. The intervortex distance d is related to the external field during B cool down

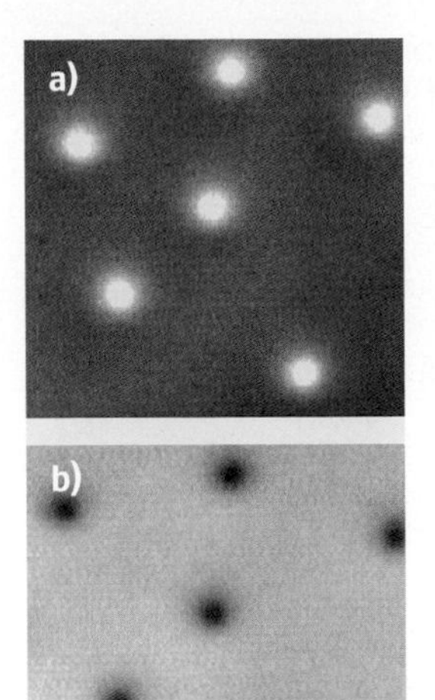

Fig. 14.25a,b Two $5\,\mu m \times 5\,\mu m$ FM-MFM images of vortices in a niobium thin film after field-cooling at 0.5 mT (**a**) and -0.5 mT (**b**), respectively. Since the external magnetic field was parallel in (**a**) and antiparallel in (**b**) with respect to the tip magnetization, the vortices exhibit reversed contrast. Strong pinning dominates the position of the vortices, since they appear at identical locations in (**a**) and (**b**) and are not arranged in a regular Abrikosov lattice (courtesy of P. Grütter; cf. [14.157])

via $d = (4/3)^{1/4}(\Phi/B)^{1/2}$. Another way to introduce vortices into a type-II superconductor is vortex penetration from the edge by applying a magnetic field at temperatures below T_c. According to the Bean model, a vortex density gradient exists under such conditions within the superconducting material. *Pi* et al. [14.159] slowly increased the external magnetic field until the vortex front approaching from the edge reached the scanning area.

If the vortex configuration is dominated by the *pinning* of vortices at randomly distributed structural defects, no Abrikosov lattice emerges. The influence of pinning centers can be studied easily by MFM, because every MFM can be used to scan the topography in its AFM mode. This has been done for natural growth defects by *Moser* et al. [14.155] on $YBa_2Cu_3O_7$ and for $YBa_2Cu_3O_7$ and niobium thin films, respectively, by *Volodin* et al. [14.158]. *Roseman* et al. [14.160] investigated the formation of vortices in the presence of an artificial structure on niobium films, while *Pi* et al. [14.159] produced columnar defects by heavy ion bombardment in a $Bi_2Sr_2CaCu_2O_8$ single crystal to study the strong pinning at these defects.

Figure 14.25 demonstrates that MFM is sensitive to the polarity of vortices. In Fig. 14.25a, six vortices have been produced in a niobium film by field cooling in $+0.5$ mT. External magnetic field and tip magnetization are parallel, and, therefore, the tip-vortex interaction is attractive (bright contrast). To remove the vortices, the niobium was heated above T_c (≈ 9 K). Thereafter, vortices of opposite polarity were produced by field cooling in -0.5 mT, which appear dark in Fig. 14.25b. The vortices are probably bound to strong pinning sites, because the vortex positions are identical in both images of Fig. 14.25. By imaging the vortices at different scanning heights, *Roseman* et al. [14.157] tried to extract values for the London penetration depth from the scan height dependence of their profiles. While good qualitative agreement with theoretical predictions has been found, the absolute values do not agree with published literature values. The disagreement was attributed to the convolution between tip and vortex stray field. Better values might be obtained with calibrated tips.

Toward Single Spin Detection

So far, only collective magnetic phenomena like ferromagnetic domains have been observed via magnetostatic tip-sample interactions detected by MFM. However, magnetic ordering exists due to the exchange interaction between the electron spins of neighboring atoms in a solid. The most energetically favorable situation can be either ferromagnetic (parallel orientation) or anti-ferromagnetic (anti-parallel orientation) ordering. It has been predicted that the exchange force between an individual spin of a magnetically ordered sample and the spin of the foremost atom of a magnetic tip can be detected at sufficiently small tip-sample distances [14.161, 162].

The experimental realization, however, is very difficult, because the exchange force is about a factor of ten weaker and of even shorter range than the chemical interactions that are responsible for the atomic-scale contrast. FM-AFM experiments with a ferromagnetic tip have been performed on the anti-ferromagnetic NiO(001) surface at room temperature [14.163] and with a considerable better signal-to-noise ratio at low temperatures [14.122]. Although it was possible to achieve atomic resolution, a periodic contrast that could be attributed to the anti-ferromagnetically ordered spins of the nickel atoms could not be observed.

Even more ambitious is the proposed detection of individual nuclear spins by magnetic resonance force microscopy (MRFM) using a magnetic tip [14.164, 165]. The interest in this technique is driven by the possibility of reaching subsurface true 3-D imaging with atomic resolution and chemical specificity. The idea of MRFM is to combine aspects of force microscopy (atomic resolution capability) with nuclear magnetic resonance (3-D imaging and elemental selectivity). Low temperatures are required, because the forces are extremely small.

Up until now, no individual spins have been detected by MRFM. However, electron spin resonance [14.166, 167], nuclear magnetic resonance [14.168], and ferromagnetic resonance [14.169] experiments of spin ensembles have been performed with micrometer resolution. On the way toward single spin detection, the design of ultrasensitive cantilevers made considerable progress, and the detection of forces below 1×10^{-18} N has been achieved [14.120].

References

14.1 G. Binnig, H. Rohrer, Ch. Gerber, E. Weibel: Surface studies by scanning tunneling microscopy, Phys. Rev. Lett. **49** (1982) 57–61
14.2 R. Wiesendanger: *Scanning Probe Microscopy and Spectroscopy* (Cambridge Univ. Press, Cambridge 1994)
14.3 M. Tinkham: *Introduction to Superconductivity* (McGraw-Hill, New York 1996)
14.4 J. Kondo: Theory of dilute magnetic alloys, Solid State Phys. **23** (1969) 183–281
14.5 T. R. Albrecht, P. Grütter, H. K. Horne, D. Rugar: Frequency modulation detection using high-Q cantilevers for enhanced force microscope sensitivity, J. Appl. Phys. **69** (1991) 668–673
14.6 F. J. Giessibl, H. Bielefeld, S. Hembacher, J. Mannhart: Calculation of the optimal imaging parameters for frequency modulation atomic force microscopy, Appl. Surf. Sci. **140** (1999) 352–357
14.7 W. Allers, A. Schwarz, U. D. Schwarz, R. Wiesendanger: Dynamic scanning force microscopy at low temperatures on a van der Waals surface: graphite(0001), Appl. Surf. Sci. **140** (1999) 247–252
14.8 W. Allers, A. Schwarz, U. D. Schwarz, R. Wiesendanger: Dynamic scanning force microscopy at low temperatures on a noble-gas crystal: atomic resolution on the xenon(111) surface, Europhys. Lett. **48** (1999) 276–279
14.9 M. Morgenstern, D. Haude, V. Gudmundsson, C. Wittneven, R. Dombrowski, R. Wiesendanger: Origin of Landau oscillations observed in scanning tunneling spectroscopy on *n*-InAs(110), Phys. Rev. B **62** (2000) 7257–7263
14.10 D. M. Eigler, P. S. Weiss, E. K. Schweizer, N. D. Lang: Imaging Xe with a low-temperature scanning tunneling microscope, Phys. Rev. Lett. **66** (1991) 1189–1192
14.11 P. S. Weiss, D. M. Eigler: Site dependence of the apparent shape of a molecule in scanning tunneling micoscope images: Benzene on Pt{111}, Phys. Rev. Lett. **71** (1992) 3139–3142
14.12 D. M. Eigler, E. K. Schweizer: Positioning single atoms with a scanning tunneling microscope, Nature **344** (1990) 524–526
14.13 H. Hug, B. Stiefel, P. J. A. van Schendel, A. Moser, S. Martin, H.-J. Güntherodt: A low temperature ultrahigh vacuum scanning force microscope, Rev. Sci. Instrum. **70** (1999) 3627–3640
14.14 S. Behler, M. K. Rose, D. F. Ogletree, F. Salmeron: Method to characterize the vibrational response of a beetle type scanning tunneling microscope, Rev. Sci. Instrum. **68** (1997) 124–128
14.15 C. Wittneven, R. Dombrowski, S. H. Pan, R. Wiesendanger: A low-temperature ultrahigh-vacuum scanning tunneling microscope with rotatable magnetic field, Rev. Sci. Instrum. **68** (1997) 3806–3810
14.16 W. Allers, A. Schwarz, U. D. Schwarz, R. Wiesendanger: A scanning force microscope with atomic resolution in ultrahigh vacuum and at low temperatures, Rev. Sci. Instrum. **69** (1998) 221–225
14.17 G. Dujardin, R. E. Walkup, Ph. Avouris: Dissociation of individual molecules with electrons from the tip of a scanning tunneling microscope, Science **255** (1992) 1232–1235
14.18 H. J. Lee, W. Ho: Single-bond formation and characterization with a scanning tunneling microscope, Science **286** (1999) 1719–1722
14.19 R. Berndt, R. Gaisch, J. K. Gimzewski, B. Reihl, R. R. Schlittler, W. D. Schneider, M. Tschudy: Photon emission at molecular resolution induced by a scanning tunneling microscope, Science **262** (1993) 1425–1427
14.20 B. G. Briner, M. Doering, H. P. Rust, A. M. Bradshaw: Microscopic diffusion enhanced by adsorbate interaction, Science **278** (1997) 257–260
14.21 J. Kliewer, R. Berndt, E. V. Chulkov, V. M. Silkin, P. M. Echenique, S. Crampin: Dimensionality effects in the lifetime of surface states, Science **288** (2000) 1399–1401
14.22 M. F. Crommie, C. P. Lutz, D. M. Eigler: Imaging standing waves in a two-dimensional electron gas, Nature **363** (1993) 524–527
14.23 B. C. Stipe, M. A. Rezaei, W. Ho: Single-molecule vibrational spectroscopy and microscopy, Science **280** (1998) 1732–1735
14.24 H. J. Lee, W. Ho: Structural determination by single-molecule vibrational spectroscopy and microscopy: Contrast between copper and iron carbonyls, Phys. Rev. B **61** (2000) R16347–R16350
14.25 C. W. J. Beenakker, H. van Houten: Quantum transport in semiconductor nanostructures, Solid State Phys. **44** (1991) 1–228
14.26 S. H. Pan, E. W. Hudson, K. M. Lang, H. Eisaki, S. Uchida, J. C. Davis: Imaging the effects of in-

dividual zinc impurity atoms on superconductivity in $Bi_2Sr_2CaCu_2O_{8+\delta}$, Nature **403** (2000) 746–750
14.27 R. S. Becker, J. A. Golovchenko, B. S. Swartzentruber: Atomic-scale surface modifications using a tunneling microscope, Nature **325** (1987) 419–42
14.28 J. A. Stroscio, D. M. Eigler: Atomic and molecular manipulation with the scanning tunneling microscope, Science **254** (1991) 1319–1326
14.29 L. Bartels, G. Meyer, K. H. Rieder: Basic steps of lateral manipulation of single atoms and diatomic clusters with a scanning tunneling microscope, Phys. Rev. Lett. **79** (1997) 697–700
14.30 J. J. Schulz, R. Koch, K. H. Rieder: New mechanism for single atom manipulation, Phys. Rev. Lett. **84** (2000) 4597–4600
14.31 T. C. Shen, C. Wang, G. C. Abeln, J. R. Tucker, J. W. Lyding, Ph. Avouris, R. E. Walkup: Atomic-scale desorption through electronic and vibrational excitation mechanisms, Science **268** (1995) 1590–1592
14.32 T. Komeda, Y. Kim, M. Kawai, B. N. J. Persson, H. Ueba: Lateral hopping of molecules induced by excitations of internal vibration mode, Science **295** (2002) 2055–2058
14.33 Y. W. Mo: Reversible rotation of antimony dimers on the silicon(001) surface with a scanning tunneling microscope, Science **261** (1993) 886–888
14.34 B. C. Stipe, M. A. Rezaei, W. Ho: Inducing and viewing the rotational motion of a single molecule, Science **279** (1998) 1907–1909
14.35 F. Moresco, G. Meyer, K. H. Rieder, H. Tang, A. Gourdon, C. Joachim: Conformational changes of single molecules by scanning tunneling microscopy manipulation: a route to molecular switching, Phys. Rev. Lett. **86** (2001) 672–675
14.36 S. W. Hla, L. Bartels, G. Meyer, K. H. Rieder: Inducing all steps of a chemical reaction with the scanning tunneling microscope tip: Towards single molecule engineering, Phys. Rev. Lett. **85** (2000) 2777–2780
14.37 E. Ganz, S. K. Theiss, I. S. Hwang, J. Golovchenko: Direct measurement of diffusion by hot tunneling microscopy: Activations energy, anisotropy, and long jumps, Phys. Rev. Lett. **68** (1992) 1567–1570
14.38 M. Schuhnack, T. R. Linderoth, F. Rosei, E. Laegsgaard, I. Stensgaard, F. Besenbacher: Long jumps in the surface diffusion of large molecules, Phys. Rev. Lett. **88** (2002) 156102, 1–4
14.39 L. J. Lauhon, W. Ho: Direct observation of the quantum tunneling of single hydrogen atoms with a scanning tunneling microscope, Phys. Rev. Lett. **85** (2000) 4566–4569
14.40 N. Kitamura, M. Lagally, M. B. Webb: Real-time observation of vacancy diffusion on Si(001)-(2 × 1) by scanning tunneling microscopy, Phys. Rev. Lett. **71** (1993) 2082–2085
14.41 M. Morgenstern, T. Michely, G. Comsa: Onset of interstitial diffusion determined by scanning tunneling microscopy, Phys. Rev. Lett. **79** (1997) 1305–1308
14.42 K. Morgenstern, G. Rosenfeld, B. Poelsema, G. Comsa: Brownian motion of vacancy islands on Ag(111), Phys. Rev. Lett. **74** (1995) 2058–2061
14.43 B. Reihl, J. H. Coombs, J. K. Gimzewski: Local inverse photoemission with the scanning tunneling microscope, Surf. Sci. **211–212** (1989) 156–164
14.44 R. Berndt, J. K. Gimzewski, P. Johansson: Inelastic tunneling excitation of tip-induced plasmon modes on noble-metal surfaces, Phys. Rev. Lett. **67** (1991) 3796–3799
14.45 P. Johansson, R. Monreal, P. Apell: Theory for light emission from a scanning tunneling microscope, Phys. Rev. B **42** (1990) 9210–9213
14.46 J. Aizpurua, G. Hoffmann, S. P. Apell, R. Berndt: Electromagnetic coupling on an atomic scale, Phys. Rev. Lett. **89** (2002) 156803, 1–4
14.47 G. Hoffmann, J. Kliewer, R. Berndt: Luminescence from metallic quantum wells in a scanning tunneling microscope, Phys. Rev. Lett. **78** (2001) 176803, 1–4
14.48 A. Downes, M. E. Welland: Photon emission from Si(111)-(7 × 7) induced by scanning tunneling microscopy: atomic scale and material contrast, Phys. Rev. Lett. **81** (1998) 1857–1860
14.49 M. Kemerink, K. Sauthoff, P. M. Koenraad, J. W. Geritsen, H. van Kempen, J. H. Wolter: Optical detection of ballistic electrons injected by a scanning-tunneling microscope, Phys. Rev. Lett. **86** (2001) 2404–2407
14.50 J. Tersoff, D. R. Hamann: Theory and application for the scanning tunneling microscope, Phys. Rev. Lett. **50** (1983) 1998–2001
14.51 C. J. Chen: *Introduction to Scanning Tunneling Microscopy* (Oxford Univ. Press, Oxford 1993)
14.52 J. Winterlin, J. Wiechers, H. Brune, T. Gritsch, H. Hofer, R. J. Behm: Atomic-resolution imaging of close-packed metal surfaces by scanning tunneling microscopy, Phys. Rev. Lett. **62** (1989) 59–62
14.53 A. L. Vazquez de Parga, O. S. Hernan, R. Miranda, A. Levy Yeyati, N. Mingo, A. Martin-Rodero, F. Flores: Electron resonances in sharp tips and their role in tunneling spectroscopy, Phys. Rev. Lett. **80** (1998) 357–360
14.54 S. H. Pan, E. W. Hudson, J. C. Davis: Vacuum tunneling of superconducting quasiparticles from atomically sharp scanning tunneling microscope tips, Appl. Phys. Lett. **73** (1998) 2992–2994
14.55 J. T. Li, W. D. Schneider, R. Berndt, O. R. Bryant, S. Crampin: Surface-state lifetime measured by scanning tunneling spectroscopy, Phys. Rev. Lett. **81** (1998) 4464–4467
14.56 L. Bürgi, O. Jeandupeux, H. Brune, K. Kern: Probing hot-electron dynamics with a cold scanning tunneling microscope, Phys. Rev. Lett. **82** (1999) 4516–4519

14.57 J. W. G. Wildoer, C. J. P. M. Harmans, H. van Kempen: Observation of Landau levels at the InAs(110) surface by scanning tunneling spectroscopy, Phys. Rev. B **55** (1997) R16013–R16016
14.58 M. Morgenstern, V. Gudmundsson, C. Wittneven, R. Dombrowski, R. Wiesendanger: Nonlocality of the exchange interaction probed by scanning tunneling spectroscopy, Phys. Rev. B **63** (2001) 201301(R), 1–4
14.59 M.V. Grishin, F.I. Dalidchik, S.A. Kovalevskii, N.N. Kolchenko, B.R. Shub: Isotope effect in the vibrational spectra of water measured in experiments with a scanning tunneling microscope, JETP Lett. **66** (1997) 37–40
14.60 A. Hewson: *From the Kondo Effect to Heavy Fermions* (Cambridge Univ. Press, Cambridge 1993)
14.61 V. Madhavan, W. Chen, T. Jamneala, M. F. Crommie, N. S. Wingreen: Tunneling into a single magnetic atom: Spectroscopic evidence of the Kondo resonance, Science **280** (1998) 567–569
14.62 J. Li, W. D. Schneider, R. Berndt, B. Delley: Kondo scattering observed at a single magnetic impurity, Phys. Rev. Lett. **80** (1998) 2893–2896
14.63 T. W. Odom, J. L. Huang, C. L. Cheung, C. M. Lieber: Magnetic clusters on single-walled carbon nanotubes: the Kondo effect in a one-dimensional host, Science **290** (2000) 1549–1552
14.64 M. Ouyang, J. L. Huang, C. L. Cheung, C. M. Lieber: Energy gaps in metallic single-walled carbon nanotubes, Science **292** (2001) 702–705
14.65 U. Fano: Effects of configuration interaction on intensities and phase shifts, Phys. Rev. **124** (1961) 1866–1878
14.66 H. C. Manoharan, C. P. Lutz, D. M. Eigler: Quantum mirages formed by coherent projection of electronic structure, Nature **403** (2000) 512–515
14.67 O. Y. Kolesnychenko, R. de Kort, M. I. Katsnelson, A. I. Lichtenstein, H. van Kempen: Real-space observation of an orbital Kondo resonance on the Cr(001) surface, Nature **415** (2002) 507–509
14.68 H. A. Mizes, J. S. Foster: Long-range electronic perturbations caused by defects using scanning tunneling microscopy, Science **244** (1989) 559–562
14.69 P.T. Sprunger, L. Petersen, E.W. Plummer, E. Laegsgaard, F. Besenbacher: Giant Friedel oscillations on beryllium (0001) surface, Science **275** (1997) 1764–1767
14.70 P. Hofmann, B. G. Briner, M. Doering, H. P. Rust, E. W. Plummer, A. M. Bradshaw: Anisotropic two-dimensional Friedel oscillations, Phys. Rev. Lett. **79** (1997) 265–268
14.71 E. J. Heller, M. F. Crommie, C. P. Lutz, D. M. Eigler: Scattering and adsorption of surface electron waves in quantum corrals, Nature **369** (1994) 464–466
14.72 M. C. M. M. van der Wielen, A. J. A. van Roij, H. van Kempen: Direct observation of Friedel oscillations around incorporated Si_{Ga} dopants in GaAs by low-temperature scanning tunneling microscopy, Phys. Rev. Lett. **76** (1996) 1075–1078
14.73 O. Millo, D. Katz, Y. W. Cao, U. Banin: Imaging and spectroscopy of artificial-atom states in core/shell nanocrystal quantum dots, Phys. Rev. Lett. **86** (2001) 5751–5754
14.74 L. C. Venema, J. W. G. Wildoer, J. W. Janssen, S. J. Tans, L. J. T. Tuinstra, L. P. Kouwenhoven, C. Dekker: Imaging electron wave functions of quantized energy levels in carbon nanotubes, Nature **283** (1999) 52–55
14.75 S. G. Lemay, J. W. Jannsen, M. van den Hout, M. Mooij, M. J. Bronikowski, P. A. Willis, R. E. Smalley, L. P. Kouwenhoven, C. Dekker: Two-dimensional imaging of electronic wavefunctions in carbon nanotubes, Nature **412** (2001) 617–620
14.76 C. Wittneven, R. Dombrowski, M. Morgenstern, R. Wiesendanger: Scattering states of ionized dopants probed by low temperature scanning tunneling spectroscopy, Phys. Rev. Lett. **81** (1998) 5616–5619
14.77 D. Haude, M. Morgenstern, I. Meinel, R. Wiesendanger: Local density of states of a three-dimensional conductor in the extreme quantum limit, Phys. Rev. Lett. **86** (2001) 1582–1585
14.78 R. Joynt, R. E. Prange: Conditions for the quantum Hall effect, Phys. Rev. B **29** (1984) 3303–3317
14.79 M. Morgenstern, J. Klijn, C. Meyer, M. Getzlaff, R. Adelung, R. A. Römer, K. Rossnagel, L. Kipp, M. Skibowski, R. Wiesendanger: Direct comparison between potential landscape and local density of states in a disordered two-dimensional electron system, Phys. Rev. Lett. **89** (2002) 136806, 1–4
14.80 E. Abrahams, P. W. Anderson, D. C. Licciardello, T. V. Ramakrishnan: Scaling theory of localization: absence of quantum diffusion in two dimensions, Phys. Rev. Lett. **42** (1979) 673–676
14.81 M. Morgenstern, J. Klijn, R. Wiesendanger: Real space observation of drift states in a two-dimensional electron system at high magnetic fields, Phys. Rev. Lett. **90** (2003) 056804, 1–4
14.82 R. E. Peierls: *Quantum Theory of Solids* (Clarendon, Oxford 1955)
14.83 C. G. Slough, W. W. McNairy, R. V. Coleman, B. Drake, P. K. Hansma: Charge-density waves studied with the use of a scanning tunneling microscope, Phys. Rev. B **34** (1986) 994–1005
14.84 X. L. Wu, C. M. Lieber: Hexagonal domain-like charge-density wave of TaS_2 determined by scanning tunneling microscopy, Science **243** (1989) 1703–1705
14.85 T. Nishiguchi, M. Kageshima, N. Ara-Kato, A. Kawazu: Behaviour of charge density waves in a one-dimensional organic conductor visualized by scanning tunneling microscopy, Phys. Rev. Lett. **81** (1998) 3187–3190
14.86 X. L. Wu, C. M. Lieber: Direct observation of growth and melting of the hexagonal-domain

charge-density-wave phase in 1T-TaS_2 by scanning tunneling microscopy, Phys. Rev. Lett. **64** (1990) 1150–1153
14.87 J.M. Carpinelli, H.H. Weitering, E.W. Plummer, R. Stumpf: Direct observation of a surface charge density wave, Nature **381** (1996) 398–400
14.88 H.H. Weitering, J.M. Carpinelli, A.V. Melechenko, J. Zhang, M. Bartkowiak, E.W. Plummer: Defect-mediated condensation of a charge density wave, Science **285** (1999) 2107–2110
14.89 H.W. Yeom, S. Takeda, E. Rotenberg, I. Matsuda, K. Horikoshi, J. Schäfer, C.M. Lee, S.D. Kevan, T. Ohta, T. Nagao, S. Hasegawa: Instability and charge density wave of metallic quantum chains on a silicon surface, Phys. Rev. Lett. **82** (1999) 4898–4901
14.90 K. Swamy, A. Menzel, R. Beer, E. Bertel: Charge-density waves in self-assembled halogen-bridged metal chains, Phys. Rev. Lett. **86** (2001) 1299–1302
14.91 J.J. Kim, W. Yamaguchi, T. Hasegawa, K. Kitazawa: Observation of Mott localization gap using low temperature scanning tunneling spectroscopy in commensurate 1T-$TaSe_2$, Phys. Rev. Lett. **73** (1994) 2103–2106
14.92 J. Bardeen, L.N. Cooper, J.R. Schrieffer: Theory of superconductivity, Phys. Rev. **108** (1957) 1175–1204
14.93 A. Yazdani, B.A. Jones, C.P. Lutz, M.F. Crommie, D.M. Eigler: Probing the local effects of magnetic impurities on superconductivity, Science **275** (1997) 1767–1770
14.94 S.H. Tessmer, M.B. Tarlie, D.J. van Harlingen, D.L. Maslov, P.M. Goldbart: Probing the superconducting proximity effect in $NbSe_2$ by scanning tunneling micrsocopy, Phys. Rev. Lett **77** (1996) 924–927
14.95 K. Inoue, H. Takayanagi: Local tunneling spectroscopy of Nb/InAs/Nb superconducting proximity system with a scanning tunneling microscope, Phys. Rev. B **43** (1991) 6214–6215
14.96 H.F. Hess, R.B. Robinson, R.C. Dynes, J.M. Valles, J.V. Waszczak: Scanning-tunneling-microscope observation of the Abrikosov flux lattice and the density of states near and inside a fluxoid, Phys. Rev. Lett. **62** (1989) 214–217
14.97 H.F. Hess, R.B. Robinson, J.V. Waszczak: Vortex-core structure observed with a scanning tunneling microscope, Phys. Rev. Lett. **64** (1990) 2711–2714
14.98 N. Hayashi, M. Ichioka, K. Machida: Star-shaped local density of states around vortices in a type-II superconductor, Phys. Rev. Lett. **77** (1996) 4074–4077
14.99 H. Sakata, M. Oosawa, K. Matsuba, N. Nishida: Imaging of vortex lattice transition in YNi_2B_2C by scanning tunneling spectroscopy, Phys. Rev. Lett. **84** (2000) 1583–1586
14.100 S. Behler, S.H. Pan, P. Jess, A. Baratoff, H.-J. Güntherodt, F. Levy, G. Wirth, J. Wiesner: Vortex pinning in ion-irrediated $NbSe_2$ studied by scanning tunneling microscopy, Phys. Rev. Lett. **72** (1994) 1750–1753
14.101 R. Berthe, U. Hartmann, C. Heiden: Influence of a transport current on the Abrikosov flux lattice observed with a low-temperature scanning tunneling microscope, Ultramicroscopy **42–44** (1992) 696–698
14.102 A. Polkovnikov, S. Sachdev, M. Vojta: Impurity in a d-wave superconductor: Kondo effect and STM spectra, Phys. Rev. Lett. **86** (2001) 296–299
14.103 E.W. Hudson, K.M. Lang, V. Madhavan, S.H. Pan, S. Uchida, J.C. Davis: Interplay of magnetism and high-T_c superconductivity at individual Ni impurity atoms in $Bi_2Sr_2CaCu_2O_{8+\delta}$, Nature **411** (2001) 920–924
14.104 K.M. Lang, V. Madhavan, J.E. Hoffman, E.W. Hudson, H. Eisaki, S. Uchida, J.C. Davis: Imaging the granular structure of high-T_c superconductivity in underdoped $Bi_2Sr_2CaCu_2O_{8+\delta}$, Nature **415** (2002) 412–416
14.105 I. Maggio-Aprile, C. Renner, E. Erb, E. Walker, Ø. Fischer: Direct vortex lattice imaging and tunneling spectroscopy of flux lines on $YBa_2Cu_3O_{7-\delta}$, Phys. Rev. Lett. **75** (1995) 2754–2757
14.106 C. Renner, B. Revaz, K. Kadowaki, I. Maggio-Aprile, Ø. Fischer: Observation of the low temperature pseudogap in the vortex cores of $Bi_2Sr_2CaCu_2O_{8+\delta}$, Phys. Rev. Lett. **80** (1998) 3606–3609
14.107 S.H. Pan, E.W. Hudson, A.K. Gupta, K.W. Ng, H. Eisaki, S. Uchida, J.C. Davis: STM studies of the electronic structure of vortex cores in $Bi_2Sr_2CaCu_2O_{8+\delta}$, Phys. Rev. Lett. **85** (2000) 1536–1539
14.108 D.P. Arovas, A.J. Berlinsky, C. Kallin, S.C. Zhang: Superconducting vortex with antiferromagnetic core, Phys. Rev. Lett. **79** (1997) 2871–2874
14.109 J.E. Hoffmann, E.W. Hudson, K.M. Lang, V. Madhavan, H. Eisaki, S. Uchida, J.C. Davis: A four unit cell periodic pattern of quasi-particle states surrounding vortex cores in $Bi_2Sr_2CaCu_2O_{8+\delta}$, Science **295** (2002) 466–469
14.110 M. Fäth, S. Freisem, A.A. Menovsky, Y. Tomioka, J. Aaarts, J.A. Mydosh: Spatially inhomogeneous metal–insulator transition in doped manganites, Science **285** (1999) 1540–1542
14.111 C. Renner, G. Aeppli, B.G. Kim, Y.A. Soh, S.W. Cheong: Atomic-scale images of charge ordering in a mixed-valence manganite, Nature **416** (2000) 518–521
14.112 M. Bode, M. Getzlaff, R. Wiesendanger: Spin-polarized vacuum tunneling into the exchange-split surface state of Gd(0001), Phys. Rev. Lett. **81** (1998) 4256–4259
14.113 A. Kubetzka, M. Bode, O. Pietzsch, R. Wiesendanger: Spin-polarized scanning tunneling microscopy with antiferromagnetic probe tips, Phys. Rev. Lett. **88** (2002) 057201, 1–4

14.114 O. Pietzsch, A. Kubetzka, M. Bode, R. Wiesendanger: Observation of magnetic hysteresis at the nanometer scale by spin-polarized scanning tunneling spectroscopy, Science **292** (2001) 2053–2056
14.115 S. Heinze, M. Bode, A. Kubetzka, O. Pietzsch, X. Xie, S. Blügel, R. Wiesendanger: Real-space imaging of two-dimensional antiferromagnetism on the atomic scale, Science **288** (2000) 1805–1808
14.116 A. Wachowiak, J. Wiebe, M. Bode, O. Pietzsch, M. Morgenstern, R. Wiesendanger: Internal spin-structure of magnetic vortex cores observed by spin-polarized scanning tunneling microscopy, Science **298** (2002) 577–580
14.117 M. D. Kirk, T. R. Albrecht, C. F. Quate: Low-temperature atomic force microscopy, Rev. Sci. Instrum. **59** (1988) 833–835
14.118 D. Pelekhov, J. Becker, J. G. Nunes: Atomic force microscope for operation in high magnetic fields at milliKelvin temperatures, Rev. Sci. Instrum. **70** (1999) 114–120
14.119 J. Mou, Y. Jie, Z. Shao: An optical detection low temperature atomic force microscope at ambient pressure for biological research, Rev. Sci. Instrum. **64** (1993) 1483–1488
14.120 H. J. Mamin, D. Rugar: Sub-attoNewton force detection at milliKelvin temperatures, Appl. Phys. Lett. **79** (2001) 3358–3360
14.121 A. Schwarz, W. Allers, U. D. Schwarz, R. Wiesendanger: Dynamic mode scanning force microscopy of *n*-InAs(110)-(1 × 1) at low temperatures, Phys. Rev. B **61** (2000) 2837–2845
14.122 W. Allers, S. Langkat, R. Wiesendanger: Dynamic low-temperature scanning force microscopy on nickel oxide(001), Appl. Phys. A **72** (2001) S27–S30
14.123 F. J. Giessibl: Atomic resolution of the silicon(111)-(7 × 7) surface by atomic force microscopy, Science **267** (1995) 68–71
14.124 M. A. Lantz, H. J. Hug, P. J. A. van Schendel, R. Hoffmann, S. Martin, A. Baratoff, A. Abdurixit, H.-J. Güntherodt: Low temperature scanning force microscopy of the Si(111)-(7 × 7) surface, Phys. Rev. Lett. **84** (2000) 2642–2465
14.125 K. Suzuki, H. Iwatsuki, S. Kitamura, C. B. Mooney: Development of low temperature ultrahigh vacuum force microscope/scanning tunneling microscope, Jpn. J. Appl. Phys. **39** (2000) 3750–3752
14.126 N. Suehira, Y. Sugawara, S. Morita: Artifact and fact of Si(111)-(7 × 7) surface images observed with a low temperature noncontact atomic force microscope (LT-NC-AFM), Jpn. J. Appl. Phys. **40** (2001) 292–294
14.127 R. Peréz, M. C. Payne, I. Štich, K. Terakura: Role of covalent tip-surface interactions in noncontact atomic force microscopy on reactive surfaces, Phys. Rev. Lett. **78** (1997) 678–681
14.128 S. H. Ke, T. Uda, R. Pérez, I. Štich, K. Terakura: First principles investigation of tip-surface interaction on GaAs(110): Implication for atomic force and tunneling microscopies, Phys. Rev. B **60** (1999) 11631–11638
14.129 J. Tobik, I. Štich, R. Peréz, K. Terakura: Simulation of tip-surface interactions in atomic force microscopy of an InP(110) surface with a Si tip, Phys. Rev. B **60** (1999) 11639–11644
14.130 A. Schwarz, W. Allers, U. D. Schwarz, R. Wiesendanger: Simultaneous imaging of the In and As sublattice on InAs(110)-(1 × 1) with dynamic scanning force microscopy, Appl. Surf. Sci. **140** (1999) 293–297
14.131 G. Schwarz, A. Kley, J. Neugebauer, M. Scheffler: Electronic and structural properties of vacancies on and below the GaP(110) surface, Phys. Rev. B **58** (1998) 1392–1499
14.132 H. Hölscher, W. Allers, U. D. Schwarz, A. Schwarz, R. Wiesendanger: Interpretation of 'true atomic resolution' images of graphite (0001) in noncontact atomic force microscopy, Phys. Rev. B **62** (2000) 6967–6970
14.133 H. Hölscher, W. Allers, U. D. Schwarz, A. Schwarz, R. Wiesendanger: Simulation of NC-AFM images of xenon(111), Appl. Phys. A **72** (2001) S35–S38
14.134 H. Hölscher, W. Allers, U. D. Schwarz, A. Schwarz, R. Wiesendanger: Determination of tip-sample interaction potentials by dynamic force spectroscopy, Phys. Rev. Lett. **83** (1999) 4780–4783
14.135 H. Hölscher, U. D. Schwarz, R. Wiesendanger: Calculation of the frequency shift in dynamic force microscopy, Appl. Surf. Sci. **140** (1999) 344–351
14.136 B. Gotsman, B. Anczykowski, C. Seidel, H. Fuchs: Determination of tip-sample interaction forces from measured dynamic force spectroscopy curves, Appl. Surf. Sci. **140** (1999) 314–319
14.137 U. Dürig: Extracting interaction forces and complementary observables in dynamic probe microscopy, Appl. Phys. Lett. **76** (2000) 1203–1205
14.138 M. A. Lantz, H. J. Hug, R. Hoffmann, P. J. A. van Schendel, P. Kappenberger, S. Martin, A. Baratoff, H.-J. Güntherodt: Quantitative measurement of short-range chemical bonding forces, Science **291** (2001) 2580–2583
14.139 S. M. Langkat, H. Hölscher, A. Schwarz, R. Wiesendanger: Determination of site specific forces between an iron coated tip and the NiO(001) surface by force field spectroscopy, Surf. Sci. (2002) in press
14.140 H. Hölscher, S. M. Langkat, A. Schwarz, R. Wiesendanger: Measurement of three-dimensional force fields with atomic resolution using dynamic force spectroscopy, Appl. Phys. Lett. (2002) in press
14.141 B. C. Stipe, H. J. Mamin, T. D. Stowe, T. W. Kenny, D. Rugar: Noncontact friction and force fluctuations between closely spaced bodies, Phys. Rev. Lett. **87** (2001)
14.142 C. Sommerhalter, T. W. Matthes, T. Glatzel, A. Jäger-Waldau, M. C. Lux-Steiner: High-sensitivity quantitative Kelvin probe microscopy by noncontact

ultra-high-vacuum atomic force microscopy, Appl. Phys. Lett. **75** (1999) 286–288
14.143 A. Schwarz, W. Allers, U. D. Schwarz, R. Wiesendanger: Dynamic mode scanning force microscopy of *n*-InAs(110)-(1 × 1) at low temperatures, Phys. Rev. B **62** (2000) 13617–13622
14.144 K. L. McCormick, M. T. Woodside, M. Huang, M. Wu, P. L. McEuen, C. Duruoz, J. S. Harris: Scanned potential microscopy of edge and bulk currents in the quantum Hall regime, Phys. Rev. B **59** (1999) 4656–4657
14.145 P. Weitz, E. Ahlswede, J. Weis, K. v. Klitzing, K. Eberl: Hall-potential investigations under quantum Hall conditions using scanning force microscopy, Physica E **6** (2000) 247–250
14.146 E. Ahlswede, P. Weitz, J. Weis, K. v. Klitzing, K. Eberl: Hall potential profiles in the quantum Hall regime measured by a scanning force microscope, Physica B **298** (2001) 562–566
14.147 M. T. Woodside, C. Vale, P. L. McEuen, C. Kadow, K. D. Maranowski, A. C. Gossard: Imaging interedge-state scattering centers in the quantum Hall regime, Phys. Rev. B **64** (2001) 041310-1–041310-4
14.148 K. Moloni, B. M. Moskowitz, E. D. Dahlberg: Domain structures in single crystal magnetite below the Verwey transition as observed with a low-temperature magnetic force microscope, Geophys. Res. Lett. **23** (1996) 2851–2854
14.149 Q. Lu, C. C. Chen, A. de Lozanne: Observation of magnetic domain behavior in colossal magnetoresistive materials with a magnetic force microscope, Science **276** (1997) 2006–2008
14.150 G. Xiao, J. H. Ross, A. Parasiris, K. D. D. Rathnayaka, D. G. Naugle: Low-temperature MFM studies of CMR manganites, Physica C **341–348** (2000) 769–770
14.151 M. Liebmann, U. Kaiser, A. Schwarz, R. Wiesendanger, U. H. Pi, T. W. Noh, Z. G. Khim, D. W. Kim: Domain nucleation and growth of $La_{07}Ca_{0.3}MnO_{3-\delta}$/ $LaAlO_3$ films studied by low temperature MFM, J. Appl. Phys. **93** (2003) 8319–8321
14.152 A. Moser, H. J. Hug, I. Parashikov, B. Stiefel, O. Fritz, H. Thomas, A. Baratoff, H. J. Güntherodt, P. Chaudhari: Observation of single vortices condensed into a vortex-glass phase by magnetic force microscopy, Phys. Rev. Lett. **74** (1995) 1847–1850
14.153 C. W. Yuan, Z. Zheng, A. L. de Lozanne, M. Tortonese, D. A. Rudman, J. N. Eckstein: Vortex images in thin films of $YBa_2Cu_3O_{7-x}$ and $Bi_2Sr_2Ca_1Cu_2O_{8-x}$ obtained by low-temperature magnetic force microscopy, J. Vac. Sci. Technol. B **14** (1996) 1210–1213
14.154 A. Volodin, K. Temst, C. van Haesendonck, Y. Bruynseraede: Observation of the Abrikosov vortex lattice in $NbSe_2$ with magnetic force microscopy, Appl. Phys. Lett. **73** (1998) 1134–1136
14.155 A. Moser, H. J. Hug, B. Stiefel, H. J. Güntherodt: Low temperature magnetic force microscopy on $YBa_2Cu_3O_{7-\delta}$ thin films, J. Magn. Magn. Mater. **190** (1998) 114–123
14.156 A. Volodin, K. Temst, C. van Haesendonck, Y. Bruynseraede: Imaging of vortices in conventional superconductors by magnetic force microscopy images, Physica C **332** (2000) 156–159
14.157 M. Roseman, P. Grütter: Estimating the magnetic penetration depth using constant-height magnetic force microscopy images of vortices, New J. Phys. **3** (2001) 24.1–24.8
14.158 A. Volodin, K. Temst, C. van Haesendonck, Y. Bruynseraede, M. I. Montero, I. K. Schuller: Magnetic force microscopy of vortices in thin niobium films: Correlation between the vortex distribution and the thickness-dependent film morphology, Europhys. Lett. **58** (2002) 582–588
14.159 U. H. Pi, T. W. Noh, Z. G. Khim, U. Kaiser, M. Liebmann, A. Schwarz, R. Wiesendanger: Vortex dynamics in $Bi_2Sr_2CaCu_2O_8$ single crystal with low density columnar defects studied by magnetic force microscopy, J. Low Temp. Phys. **131** (2003) 993–1002
14.160 M. Roseman, P. Grütter, A. Badia, V. Metlushko: Flux lattice imaging of a patterned niobium thin film, J. Appl. Phys. **89** (2001) 6787–6789
14.161 K. Nakamura, H. Hasegawa, T. Oguchi, K. Sueoka, K. Hayakawa, K. Mukasa: First-principles calculation of the exchange interaction and the exchange force between magnetic Fe films, Phys. Rev. B **56** (1997) 3218–3221
14.162 A. S. Foster, A. L. Shluger: Spin-contrast in non-contact AFM on oxide surfaces: Theoretical modeling of NiO(001) surface, Surf. Sci. **490** (2001) 211–219
14.163 H. Hoisoi, M. Kimura, K. Hayakawa, K. Sueoka, K. Mukasa: Non-contact atomic force microscopy of an antiferromagnetic NiO(100) surface using a ferromagnetic tip, Appl. Phys. A **72** (2001) S23–S26
14.164 J. A. Sidles, J. L. Garbini, G. P. Drobny: The theory of oscillator-coupled magnetic resonance with potential applications to molecular imaging, Rev. Sci. Instrum. **63** (1992) 3881–3899
14.165 J. A. Sidles, J. L. Garbini, K. J. Bruland, D. Rugar, O. Züger, S. Hoen, C. S. Yannoni: Magnetic resonance force microscopy, Rev. Mod. Phys. **67** (1995) 249–265
14.166 D. Rugar, C. S. Yannoni, J. A. Sidles: Mechanical detection of magnetic resonance, Nature **360** (1992) 563–566
14.167 K. Wago, D. Botkin, O. Züger, R. Kendrick, C. S. Yannoni, D. Rugar: Force-detected electron spin resonance: Adiabatic inversion, nutation and spin echo, Phys. Rev. B **57** (1998) 1108–1114
14.168 D. Rugar, O. Züger, S. Hoen, C. S. Yannoni, H. M. Vieth, R. D. Kendrick: Force detection of nuclear magnetic resonance, Science **264** (1994) 1560–1563
14.169 Z. Zhang, P. C. Hammel, P. E. Wigen: Observation of ferromagnetic resonance in a microscopic sample using magnetic resonance force microscopy, Appl. Phys. Lett. **68** (1996) 2005–2007

15. Dynamic Force Microscopy

This chapter presents an introduction to the concept of the dynamic operational mode of the atomic force microscope (dynamic AFM). While the static, or contact mode AFM is a widespread technique to obtain nanometer resolution images on a wide variety of surfaces, true atomic resolution imaging is routinely observed only in the dynamic mode. We will explain the jump-to-contact phenomenon encountered in static AFM and present the dynamic operational mode as a solution to overcome this effect. The dynamic force microscope is modeled as a harmonic oscillator to gain a basic understanding of the underlying physics in this mode.

Dynamic AFM comprises a whole family of operational modes. A systematic overview of the different modes typically encountered in force microscopy is presented, and special care is taken to explain the distinct features of each mode. Two modes of operation dominate the application of dynamic AFM. First, the amplitude modulation mode (also called tapping mode) is shown to exhibit an instability, which separates the purely attractive force interaction regime from the attractive–repulsive regime. Second, the self-excitation mode is derived and its experimental realization is outlined. While the first is primarily used for imaging in air and liquid, the second dominates imaging in UHV (ultrahigh vacuum) for atomic resolution imaging. In particular, we explain the influence of different forces on spectroscopy curves obtained in dynamic force microscopy. A quantitative link between the measurement values and the interaction forces is established.

Force microscopy in air suffers from small quality factors of the force sensor (i. e., the cantilever beam), which are shown to limit the achievable resolution. Also, the above mentioned instability in amplitude modulation mode often hinders imaging of soft and fragile samples. A combination of the amplitude modulation with the self-excitation mode is shown to increase the quality, or Q-factor, and extend the regime of stable operation, making the so-called Q-control module a valuable tool. Apart from the advantages of dynamic force microscopy as a nondestructive, high-resolution imaging method, it can also be used to obtain information about energy dissipation phenomena at the nanometer scale. This measurement channel can provide crucial information on electric and magnetic surface properties. Even atomic resolution imaging has been obtained in the dissipation mode. Therefore, in the last section, the quantitative relation between the experimental measurement values and the dissipated power is derived.

15.1 **Motivation: Measurement of a Single Atomic Bond** 450

15.2 **Harmonic Oscillator: A Model System for Dynamic AFM** 454

15.3 **Dynamic AFM Operational Modes** 455
15.3.1 Amplitude-Modulation/Tapping-Mode AFMs 456
15.3.2 Self-Excitation Modes 461

15.4 ***Q*-Control** 464

15.5 **Dissipation Processes Measured with Dynamic AFM** 468

15.6 **Conclusion** 471

References 471

15.1 Motivation: Measurement of a Single Atomic Bond

The direct measurement of the force interaction between two distinct molecules has been the motivation for scientists for many years now. The fundamental forces responsible for the solid state of matter can be directly investigated, ultimately between defined single molecules. But it has not been until very recently that the chemical forces could be quantitatively measured for a single atomic bond [15.1]. How can we reliably measure forces that may be as small as one billionth of 1 N? How can we identify one single pair of atoms as the source of the force interaction?

The same mechanical principle that is used to measure the gravitational force exerted by your body weight (e.g., with the scale in your bathroom) can be employed to measure the forces between single atoms. A spring with a defined elasticity is compressed by an arbitrary force (e.g., your weight). The compression Δz of the spring (with spring constant k) is a direct measure of the force F exerted, which in the regime of elastic deformation obeys Hooke's law:

$$F = k\Delta z\,. \qquad (15.1)$$

The only difference with regard to your bathroom scale is the sensitivity of the measurement. Typically springs with a stiffness of 0.1 N/m to 10 N/m are used, which will be deflected by 0.1 nm to 10 nm upon application of an interatomic force of some nN. Experimentally, a laser deflection technique is used to measure the movement of the spring. The spring is a bendable cantilever microfabricated from silicon wafers. If a sufficiently sharp tip, usually directly attached to the cantilever, is approached toward a surface within some nanometers, we can measure the interacting forces through changes in the deflected laser beam. This is a static measurement; hence it is called *static AFM*. Alternatively, the cantilever can be excited to vibrate at its resonant frequency. Under the influence of tip-sample forces the resonant frequency (and consequently also amplitude and phase) of the cantilever will change and serve as measurement parameters. This is called the *dynamic AFM*. Due to the multitude of possible operational modes, expressions like noncontact mode, intermittent-contact mode, tapping mode, FM-mode, AM-mode, self-excitation, constant-excitation, or constant-amplitude mode AFM are found in the literature, which will be systematically categorized in the following paragraphs.

In fact, the first AFMs were operated in the dynamic mode. In 1986, *Binnig*, *Quate* and *Gerber* presented the concept of the atomic force microscope [15.2]. The deflection of the cantilever with the tip was measured with sub-angstrom precision by an additional scanning tunneling microscope (STM). While the cantilever was externally oscillated close to its resonant frequency, the amplitude and phase of the oscillation were measured. If the tip is approached toward the surface, the oscillation parameters, amplitude and phase, are influenced by the tip-surface interaction, and can, therefore, be used as feedback channels. A certain set-point, for example, the amplitude, is given, and the feedback loop will adjust the tip-sample distance such that the amplitude remains constant. The controller parameter is recorded as a function of the lateral position of the tip with respect to the sample, and the scanned image essentially represents the surface topography.

What then is the difference between the static and the dynamic mode of operation for the AFM? The static deflection AFM directly gives the interaction force between tip and sample using (15.1). In the dynamic mode, we find that the resonant frequency, amplitude, and phase of the oscillation change as a consequence of the interaction forces (and also dissipative processes, as discussed in the last section).

In order to get a basic understanding of the underlying physics, it is instructive to consider a very simplified case. Assume that the vibration amplitude is small compared to the range of force interaction. Since van der Waals forces range over typical distances of 10 nm, the vibration amplitude should be less than 1 nm. Furthermore, we require that the force gradient $\partial F_{ts}/\partial z$ does not vary significantly over one oscillation cycle.

We can view the AFM setup as a coupling of two springs (see Fig. 15.1). Whereas the cantilever is represented by a spring with spring constant k, the force interaction between tip and surface can be modeled by

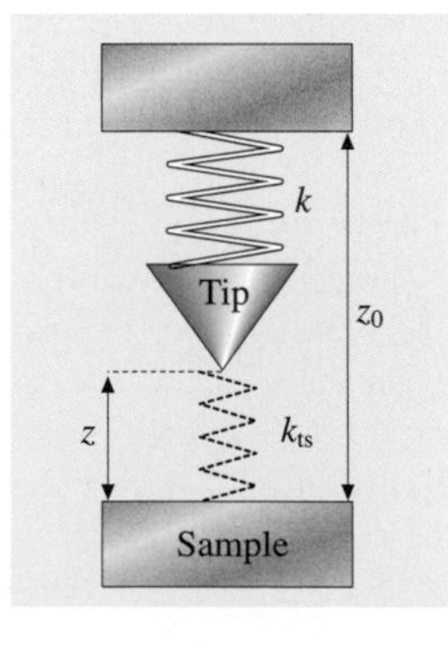

Fig. 15.1 Model of the AFM tip while experiencing tip-sample forces. The tip is attached to a cantilever with spring constant k, and the force interaction is modeled by a spring with a stiffness k_{ts} equal to the force gradient. Note that the force interaction spring is not constant, but depends on the tip-sample distance z

a second spring. The derivative of the force with respect to tip-sample distance is the force gradient and represents the spring constant k_{ts} of the interaction spring. This spring constant k_{ts} is constant only with respect to one oscillation cycle, but varies with the average tip-sample distance as the probe is approached to the sample. The two springs are coupled in series. Therefore, we can write for the total spring constant of the AFM system:

$$k_{\text{total}} = k + k_{\text{ts}} = k - \frac{\partial F_{\text{ts}}}{\partial z} . \quad (15.2)$$

From the simple harmonic oscillator (neglecting any damping effects) we find that the resonant frequency ω of the system is shifted by $\Delta\omega$ from the free resonant frequency ω_0 due to the force interaction:

$$\begin{aligned}\omega^2 &= (\omega_0 + \Delta\omega)^2 \\ &= k_{\text{total}}/m^* = \left(k + \frac{\partial F_{\text{ts}}}{\partial z}\right)/m^* .\end{aligned} \quad (15.3)$$

Here m^* represents the effective mass of the cantilever. A detailed analysis of how m^* is related to the geometry and total mass of the cantilever can be found in the literature [15.3]. In the approximation that $\Delta\omega$ is much smaller than ω_0, we can write:

$$\frac{\Delta\omega}{\omega_0} \cong -\frac{1}{2k}\frac{\partial F_{\text{ts}}}{\partial z} . \quad (15.4)$$

Therefore, we find that the frequency shift of the cantilever resonance is proportional to the force gradient of the tip-sample interaction.

Although the above consideration is based on a very simplified model, it shows qualitatively that in dynamic force microscopy we will find that the oscillation frequency depends on the force gradient, while static force microscopy measures the force itself. In principle, we can calculate the force curve from the force gradient and vice versa (neglecting a constant offset). It seems, therefore, that the two methods are equivalent, and our choice will depend on whether we can measure the beam deflection or the frequency shift with better precision at the cost of technical effort.

However, we have neglected one important issue for the operation of the AFM thus far: the mechanical stability of the measurement. In static AFM, the tip is slowly approached toward the surface. The force between the tip and the surface will always be counteracted by the restoring force of the cantilever. In Fig. 15.2, you can see a typical force-distance curve. Upon approach of the tip toward the sample, the negative attractive forces, representing, e.g., van der Waals or chemical interaction forces, increase until a maximum is reached. This turnaround point is due to the onset of repulsive forces caused by coulomb repulsion, which will start to dominate upon further approach. The spring constant of the cantilever is represented by the slope of the straight line. The position of the z-transducer (typically a piezo element), which moves the probe, is at the intersection of the line with the horizontal axis. The position of the tip, shifted from the probe's base due to the lever bending, can be found at the intersection of the cantilever line with the force curve. Hence, the total force is zero, i. e., the cantilever is in its equilibrium position (note that the spring constant line here shows attractive forces, although in reality the forces are repulsive, i. e., pulling back the tip from the surface). As soon as the position A in Fig. 15.2 is reached, we find two possible intersection points, and upon further approach there are even three force equilibrium points. However, between point A to B the tip is at a local energy minimum and, therefore, will still follow the force curve. But at

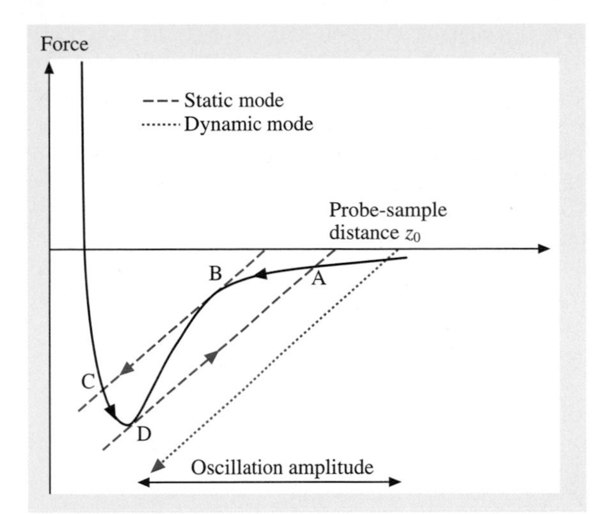

Fig. 15.2 Force-distance curve of a typical tip-sample interaction. In static mode AFM, the tip would follow the force curve until point B is reached. If the slope of the force curve becomes larger than the spring constant of the cantilever (*dashed line*), the tip will suddenly jump to position C. Upon retraction, a different path will be followed along D and A again. In dynamic AFM, the cantilever oscillates with amplitude A. Although the equilibrium position of the oscillation is far from the surface, the tip will experience the maximum attractive force at point D during some parts of the oscillation cycle. However, the total force is always pointing away from the surface, therefore, avoiding an instability

point B, when the adhesion force upon further approach would become larger than the spring restoring force, the tip will suddenly jump to point C. We can then probe the predominantly repulsive force interaction by further reducing the tip-sample distance. Then, while retracting the tip, we will pass point C, because the tip is still in a local energy minimum. At position D the tip will jump suddenly to point A again, since the restoring force now exceeds the adhesion. From Fig. 15.2 we can see that the sudden instability will happen at exactly the point where the slope of the adhesion force exceeds the slope of the spring constant. Therefore, if the negative force gradient of the tip-sample interaction will at any point exceed the spring constant, a mechanical instability occurs. Mathematically speaking, we demand that for a stable measurement:

$$\left.\frac{\partial F_{\mathrm{ts}}}{\partial z}\right|_z > k \quad \text{for all points } z\,. \tag{15.5}$$

The phenomenon of mechanical instability is often referred to as the "jump-to-contact".

Looking at Fig. 15.2, we realize that large parts of the force curve cannot be measured if the jump-to-contact phenomenon occurs. We will not be able to measure the point at which the attractive forces reach their maximum, representing the temporary chemical bonding of the tip and the surface atoms. Secondly, the sudden instability, the jump-to-contact, will often cause the tip to change the very last tip or surface atoms. A smooth, careful approach needed to measure the full force curve does not seem feasible. Our goal of measuring the chemical interaction forces of two single molecules may become impossible.

There are several solutions to the jump-to-contact problem: On the one hand, we can simply choose a sufficiently stiff spring, so that (15.5) is fulfilled at all points of the force curve. On the other hand, we can resort to a trick to enhance the counteracting force of the cantilever: We can oscillate the cantilever with large amplitude, thereby making it virtually stiffer at the point of strong force interaction.

Consider the first solution, which seems simpler at first glance. Chemical bonding forces extend over a distance range of about 0.1 nm. Typical binding energies of a couple of eV will lead to adhesion forces on the order of some nN. Force gradients will, therefore, reach values of some 10 N/m. A spring for stable force measurements will have to be as stiff as 100 N/m to ensure that no instability occurs (a safety factor of ten seems to be a minimum requirement, since usually one cannot be sure a priori that only one atom will dominate the interaction). In order to measure the nN interaction force, a static cantilever deflection of 0.01 nm has to be detected. With standard beam deflection AFM setups this becomes a challenging task.

This problem was solved [15.4, 5] using an in situ optical interferometer measuring the beam deflection at liquid nitrogen temperature in a UHV environment. In order to ensure that the force gradients are smaller than the lever spring constant (50 N/m), the tips were fabricated to terminate in only three atoms, therefore, minimizing the total force interaction. The only tool known to be able to engineer SPM tips down to atomic dimensions is the field ion microscope (FIM). This technique not only allows imaging of the tip apex with atomic precision, but also can be used to manipulate the tip atoms by field evaporation [15.6], as shown in Fig. 15.3. Atomic interaction forces were measured with sub-nanonewton precision, revealing force curves of only a few atoms interacting without mechanical hysteresis. However, the technical effort to achieve this type of measurement is considerable, and most researchers today have resorted to the second solution.

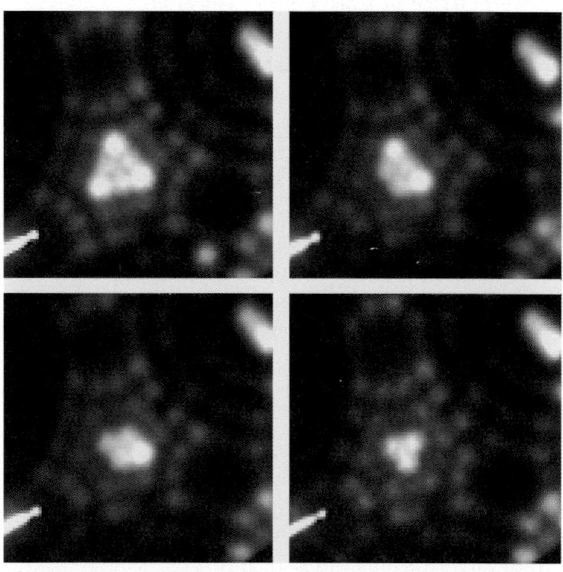

Fig. 15.3 Manipulation of the apex atoms of an AFM tip using field ion microscopy (FIM). Images are obtained at a tip bias of 4.5 kV. The last six atoms of the tip can be inspected in this example. Field evaporation to remove single atoms is performed by increasing the bias voltage for a short time to 5.2 kV. Each of the outer three atoms can be consecutively removed, eventually leaving a trimer tip apex

The alternative solution can be visualized in Fig. 15.2. The straight, dashed line now represents the force values of the oscillating cantilever, with amplitude A assuming Hooke's law is valid. This is the tensile force of the cantilever spring pulling the tip away from the sample. The restoring force of the cantilever is at all points stronger than the adhesion force. For example, the total force at point D is still pointing away from the sample, although the spring has the same stiffness as before. Mathematically speaking, the measurement is stable as long as the cantilever spring force $F_{\mathrm{cb}} = kA$ is larger than the attractive tip-sample force F_{ts} [15.7]. In the static mode we would already experience an instability at that point. However, in the dynamic mode, the spring is pre-loaded with a force stronger than the attractive tip-sample force. The equilibrium point of the oscillation is still far away from the point of closest contact of the tip and surface atoms. The total force curve can be probed by varying the equilibrium point of the oscillation, i. e., by adjusting the z-piezo.

The diagram also shows that the oscillation amplitude has to be quite large if fairly soft cantilevers are to be used. With lever spring constants of 10 N/m, the amplitude must be at least 1 nm to ensure forces of 1 nN can be reliably measured. In practical applications, amplitudes of 10–100 nm are used to stay on the safe side. This means that the oscillation amplitude is much larger than the force interaction range. The above simplification, that the force gradient remains constant within one oscillation cycle, does not hold anymore. Measurement stability is gained at the cost of a simple quantitative analysis of the experiments. In fact, dynamic AFM was first used to obtain atomic resolution images of clean surfaces [15.8], and it took another six years [15.1] before quantitative measurements of single bond forces were obtained.

The technical realization of dynamic mode AFMs is based on the same key components as a DC-AFM setup. The most common principle is the method of laser deflection sensing (see, e.g., Fig. 15.4). A laser beam is focused on the back side of a microfabricated cantilever. The reflected laser spot is detected with a positional sensitive diode (PSD). This photodiode is sectioned into two parts that are read-out separately (usually even a four-quadrant diode is used to detect torsional movements of the cantilever for lateral friction measurements). With the cantilever at equilibrium, the spot is adjusted such that the two sections show the same intensity. If the cantilever bends up or down, the spot moves, and the difference signal between the upper and lower sections is a measure of the bending.

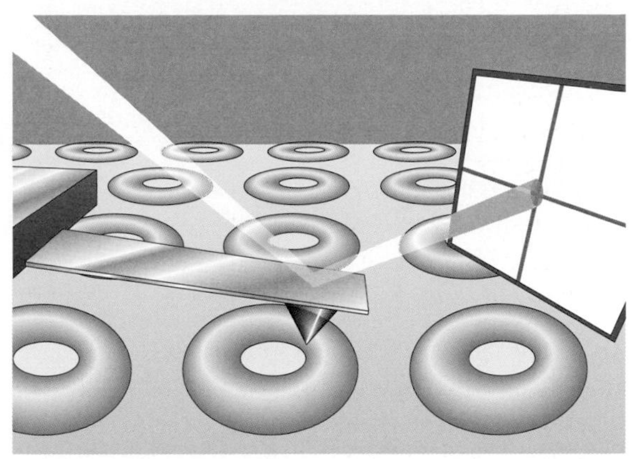

Fig. 15.4 Representation of an AFM setup with the laser beam deflection method. Cantilever and tip are microfabricated from silicon wafers. A laser beam is deflected from the back side of the cantilever and again focused on a photosensitive diode. The diode is segmented into four quadrants, which allows the measurement of vertical and torsional bending of the cantilever (artwork from *Jörg Heimel* rendered with POV-Ray 3.0)

In order to enhance sensitivity, several groups have adopted an interferometer system to measure the cantilever deflection. A thorough comparison of different measurement methods with analysis of sensitivity and noise level is given by [15.3].

The cantilever is mounted on a device that allows oscillation of the beam. Often a piezo-element serves this purpose. The reflected laser beam is analyzed for oscillation amplitude, frequency, and phase difference between excitation and vibration. Depending on the mode of operation, a feedback mechanism will adjust oscillation parameters and/or tip-sample distance during the scanning. The setup can be operated in air, UHV, and even in fluids. This allows measurement of a wide range of surface properties from atomic resolution imaging [15.8] up to studying biological processes in liquid [15.9, 10].

15.2 Harmonic Oscillator: A Model System for Dynamic AFM

The oscillating cantilever has three degrees of freedom: the amplitude, the frequency, and the phase difference between excitation and oscillation. Let us consider the damped driven harmonic oscillator. The cantilever is mounted on a piezoelectric element that is oscillating with amplitude A_d at frequency ω:

$$z_d(t) = A_d \cos(\omega t) . \quad (15.6)$$

We assume that the cantilever spring obeys Hooke's law. Secondly, we introduce a friction force that is proportional to the speed of the cantilever motion, whereas α denotes the damping coefficient (Amontons's law). With Newton's first law we find for the oscillating system the following equation of motion for the position $z(t)$ of the cantilever tip (see also Fig. 15.1):

$$m\ddot{z}(t) = -\alpha\dot{z}(t) - kz(t) - kz_d(t) . \quad (15.7)$$

We define $\omega_0^2 = k/m^*$, which turns out to be the resonant frequency of the free (undamped, i. e., $\alpha = 0$) oscillating beam. We further define the dimensionless quality factor $Q = m^*\omega_0/\alpha$, antiproportional to the damping coefficient. The quality factor describes the number of oscillation cycles, after which the damped oscillation amplitude decays to $1/\mathrm{e}$ of the initial amplitude with no external excitation ($A_d = 0$). After some basic math, this results in the following differential equation:

$$\ddot{z}(t) + \frac{\omega_0}{Q}\dot{z}(t) + \omega_0^2 z(t) = A_d\omega_0^2 \cos(\omega t) . \quad (15.8)$$

The solution is a linear combination of two regimes [15.11]. Starting from rest and switching on the piezo-excitation at $t = 0$, the amplitude will increase from zero to the final magnitude and reach a steady state, where amplitude, phase, and frequency of the oscillation stay constant over time. The steady-state solution $z_1(t)$ is reached after $2Q$ oscillation cycles and follows the external excitation with amplitude A_0 and phase difference φ:

$$z_1(t) = A_0 \cos(\omega t + \varphi) . \quad (15.9)$$

The oscillation amplitude in the transient regime during the first $2Q$ cycles follows:

$$z_2(t) = A_t \mathrm{e}^{-\omega_0 t/2Q} \sin(\omega_0 t + \varphi_t) . \quad (15.10)$$

We emphasize the important fact that the exponential term causes $z_2(t)$ to diminish exponentially with time constant τ:

$$\tau = 2Q/\omega_0 . \quad (15.11)$$

In vacuum conditions, only the internal dissipation due to bending of the cantilever is present, and Q reaches values of 10,000 at typical resonant frequencies of 100,000 Hz. This results in a relatively long transient regime of $\tau \cong 30$ ms, which limits the possible operational modes for dynamic AFM (detailed analysis by [15.11]). Changes in the measured amplitude, which reflect a change of atomic forces, will have a time lag of 30 ms, which is very slow considering one wants to scan a 200×200 point image within a few minutes. In air, however, viscous damping due to air friction dominates and Q goes down to less than 1,000, resulting in a time constant below the millisecond level. This response time is fast enough to use the amplitude as a measurement parameter.

If we evaluate the steady state solution $z_1(t)$ in the differential equation, we find the following well-known

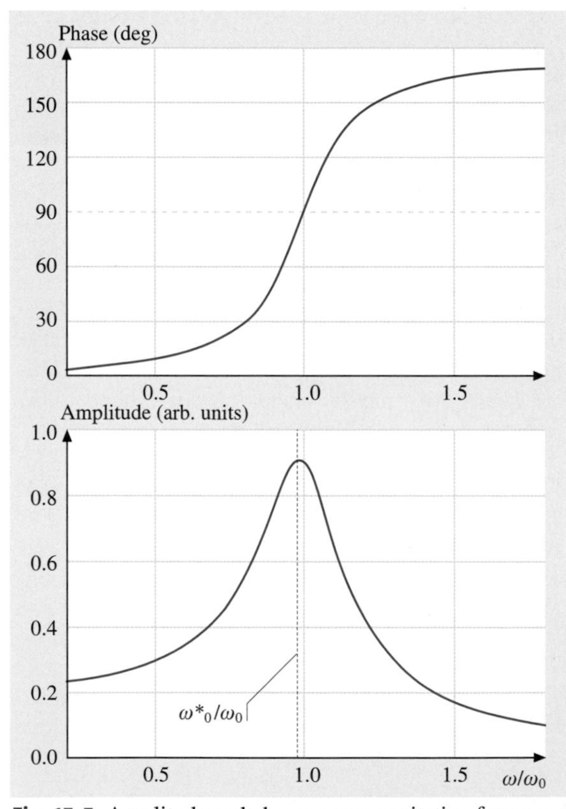

Fig. 15.5 Amplitude and phase versus excitation frequency curves for the damped harmonic oscillator with a quality factor of $Q = 4$

solution for amplitude and phase of the oscillation as a function of the excitation frequency ω:

$$A_0 = \frac{A_d Q \omega_0^2}{\sqrt{\omega^2\omega_0^2 + Q^2\left(\omega_0^2 - \omega^2\right)^2}} \qquad (15.12)$$

$$\varphi = \arctan\left(\frac{\omega\omega_0}{Q\left(\omega_0^2 - \omega^2\right)}\right) . \qquad (15.13)$$

Amplitude and phase diagrams are depicted in Fig. 15.5. As can be seen from (15.12), the amplitude will reach its maximum at a frequency different from ω_0, if Q has a finite value. The damping term of the harmonic oscillator causes the resonant frequency to shift from ω_0 to ω_0^*:

$$\omega_0^* = \omega_0\sqrt{1 - \frac{1}{2Q^2}} . \qquad (15.14)$$

The shift is negligible for Q-factors of 100 and above, which is the case for most applications in vacuum or air. However, for measurements in liquids, Q can be smaller than 10 and ω_0 differs significantly from ω_0^*. As we will discuss later, it is also possible to enhance Q by using a special excitation method.

In the case that the excitation frequency is equal to the resonant frequency of the undamped cantilever, $\omega = \omega_0$, we find the useful relation:

$$A_0 = QA_d \quad \text{for } \omega = \omega_0 . \qquad (15.15)$$

Since $\omega_0^* \approx \omega_0$ for most cases, we find that (15.15) holds true for exciting the cantilever at its resonance. From a similar argument, the phase becomes approximately 90 deg for the resonance case. We also see that in order to reach vibration amplitudes of some 10 nm, the excitation only has to be as small as 1 pm for typical cantilevers operated in vacuum.

So far, we have not considered an additional force term, describing the interaction between the probing tip and the sample. For typical, large vibration amplitudes of 10–100 nm no general solution for this analytical problem has been found yet. The cantilever tip will experience a whole range of force interactions during one single oscillation cycle, rather than one defined tip-sample force. Only in the special case of a self-excited cantilever oscillation has the problem been solved semi-analytically, as we will see later.

15.3 Dynamic AFM Operational Modes

While the quantitative interpretation of force curves in contact AFM is straightforward using (15.1), we explained in the previous paragraphs that its application to assess short-range attractive interatomic forces is rather limited. The dynamic mode of operation seems to open a viable direction toward achieving this task. However interpretation of the measurements generally appears to be more difficult. Different operational modes are employed in dynamic AFM, and the following paragraphs are intended to distinguish these modes and categorize them in a systematic way.

The oscillation trajectory of a dynamically driven cantilever is determined by three parameters: the amplitude, the phase, and the frequency. Tip-sample interactions can influence all three parameters, in the following, termed the internal parameters. The oscillation is driven externally, with excitation amplitude A_d and excitation frequency ω. These variables will be referred to as the external parameters. The external parameters are set by the experimentalist, whereas the internal parameters are measured and contain the crucial information about the force interaction. In scanning probe applications, it is common to control the probe-surface distance z_0 in order to keep an internal parameter constant (i. e., the tunneling current in STM or the beam deflection in contact AFM), which represents a certain tip-sample interaction. In z-spectroscopy mode, the distance is varied in a certain range, and the change of the internal parameters are measured as a fingerprint of the tip-sample interactions.

In dynamic AFM the situation is rather complex. Any of the internal parameters can be used for feedback of the tip-sample distance z_0. However, we already realized that, in general, the tip-sample forces could only be fully assessed by measuring all three parameters. This makes it difficult to obtain images where the distance z_0 is representative of the surface at one force set-point. A solution to this problem is to establish additional feedback loops, which keep the internal parameters constant by adjusting the external variables.

In the simplest setup, the excitation frequency is set to a predefined value, and the excitation amplitude remains constant by a feedback loop. This is called the AM mode (amplitude modulation), or tapping mode (TM). As stated before, in principle, any of the internal parameters can be used for feedback to the tip-sample distance – in AM mode the amplitude signal is used. A certain amplitude (smaller than the free oscillation amplitude) at a frequency close to the resonance of the cantilever is chosen, the tip is approached toward the surface under investigation, and the approach is stopped as soon as the set-point amplitude is reached. The oscillation-phase is usually recorded during the scan, however, the shift of the resonant frequency of the cantilever cannot be directly accessed, since this degree of freedom is blocked by the external excitation at a fixed frequency. It turns out that this mode is simple to operate from a technical perspective, but quantitative information about the tip-sample interaction forces has so far not been reliably extracted from AM-mode AFM. Still, it is one of the most commonly used modes in dynamic AFM operated in air, and even in liquid. The strength of this mode is the qualitative imaging of a large variety of surfaces.

It is interesting to discuss the AM mode in the extreme situation that the external excitation frequency is much lower than the resonant frequency [15.12,13]. This results in a quasi-static measurement, although a dynamic oscillation force is applied, and, therefore, this mode can be viewed as a hybrid between static and dynamic AFM. Unfortunately, it has the drawbacks of the static mode, namely, that stiff spring constants must be used and, therefore, the sensitivity of the deflection measurement must be very good, typically employing a high resolution interferometer. Still, it has the advantage of the static measurement in quantitative interpretation, since in the regime of small amplitudes (< 0.1 nm) a direct interpretation of the experiments is possible. In particular, the force gradient at tip-sample distance z_0 is given by the change of the amplitude A and the phase angle φ:

$$\left.\frac{\partial F_{\text{ts}}}{\partial z}\right|_{z_0} = k\left(1 - \frac{A_0}{A}\cos\varphi\right) . \tag{15.16}$$

In effect, the modulated AFM, in contrast to the purely static AFM, can enhance sensitivity due to the use of lock-in techniques, which allows the measurement of the amplitude and phase of the oscillation signal with high precision.

As stated before, the internal parameters can be fed back to the external excitation variables. One of the most useful applications in this direction is the self-excitation system. Here the resonant frequency of the cantilever is detected and selected again as the excitation frequency. In a typical setup, this is done with a phase shift of 90 deg by feeding back the detector signal to the excitation piezo, i. e., the cantilever is always excited in resonance. Influences of the tip-sample interaction forces on the resonant frequency do not change the two other parameters of the oscillation, i. e., amplitude and phase; only the oscillation frequency is shifted. Therefore, it is sufficient to measure the frequency shift between the free oscillation and the oscillation with tip-sample interaction. Since the phase remains at a fixed value, the oscillating system is much better defined than before, and the degrees of freedom for the oscillation are reduced. To even reduce the last degree of freedom, the oscillation amplitude, another feedback loop can be established to keep the oscillation amplitude A constant by varying the excitation amplitude A_{d}. Now, all internal parameters have a fixed relation to the external excitation variables, the system is well-defined, and all parameters can be assessed during the measurement. As it turns out, this mode is the only dynamic mode in which a quantitative relation between tip-sample forces and the change of the resonant frequency can be established.

In the following section we want to discuss the two most popular operational modes, tapping mode and self-excitation mode, in more detail.

15.3.1 Amplitude-Modulation/ Tapping-Mode AFMs

In tapping mode, or AM-AFM, the cantilever is excited externally at a constant frequency close to its resonance. Oscillation amplitude and phase during approach of tip and sample serve as the experimental observation channels. Figure 15.6 shows a diagram of a typical tapping-mode AFM setup. The oscillation amplitude and the phase (not shown in diagram) detected with the photodiode are analyzed with a lock-in amplifier. The amplitude is compared to the set-point, and the difference or error signal is used to adjust the z-piezo, i. e., the probe-sample distance. The external modulation unit supplies the signal for the excitation piezo, and, at the same time, the oscillation signal serves as the reference for the lock-in amplifier.

During one oscillation cycle with amplitudes of 10–100 nm, the tip-sample interaction will range over a wide distribution of forces, including attractive, as well as repulsive forces. We will, therefore, measure a convolution of the force-distance curve with the oscillation trajectory. This complicates the interpretation of AM-AFM measurements appreciably.

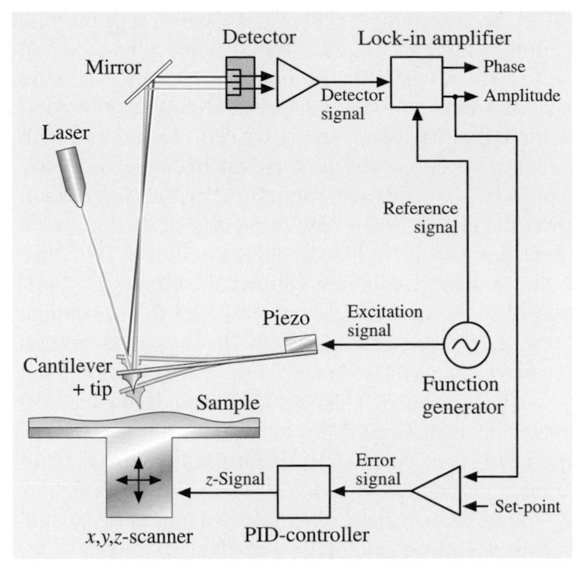

Fig. 15.6 Setup of a dynamic force microscope operated in the AM or tapping mode. A laser beam is deflected by the back side of the cantilever, and the deflection is detected by a split photodiode. The excitation frequency is chosen externally with a modulation unit, which drives the excitation piezo. A lock-in amplifier analyzes phase and amplitude of the cantilever oscillation. The amplitude is used as the feedback signal for the probe-sample distance control

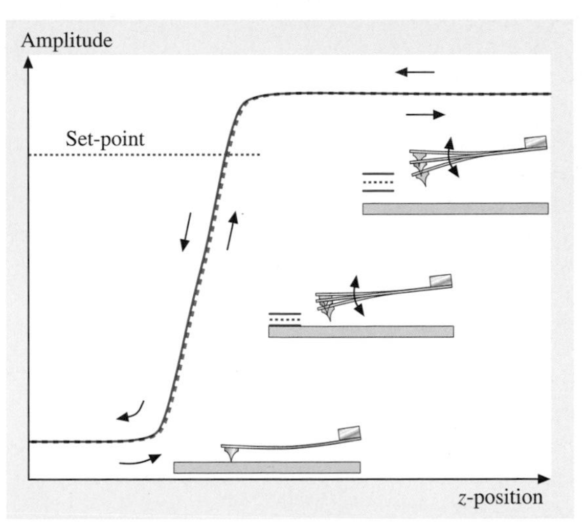

Fig. 15.7 Simplified model showing the oscillation amplitude in tapping-mode AFM for various probe-sample distances

At the same time, the resonant frequency of the cantilever will change due to the appearing force gradients, as could already be seen in the simplified model from (15.4). If the cantilever is excited exactly at its resonant frequency before, it will be excited off-resonance after interaction forces are encountered. This, in turn, changes amplitude and phase (15.12, 15.13), which serve as the measurement signals. Consequently, a different amplitude will cause a change in the encountered effective force. We can see already from this simple *gedanken*-experiment that the interpretation of force curves will be highly complicated. In fact, there is no quantitative theory for AM-AFM available, which allows the experimentalist to unambiguously convert the experimental data to a force-distance relationship.

The qualitative behavior for amplitude versus z_0-position curves is depicted in Fig. 15.7. At large distances, where the forces between tip and sample are negligible, the cantilever oscillates with its free oscillation amplitude. Upon approach of the probe toward the surface the interaction forces cause the amplitude to change, resulting typically in an amplitude getting smaller with continuously decreased tip-sample distance. This is expected, since the force-distance curve will eventually reach the repulsive part and the tip is hindered from indenting further into the sample, resulting in smaller oscillation amplitudes.

However, in order to gain some qualitative insight into the complex relationship between forces and oscillation parameters, we resort to numerical simulations. *Anczykowski* et al. [15.14, 15] have calculated the oscillation trajectory of the cantilever under the influence of a given force model. Van der Waals interactions were considered the only effective, attractive forces, and the total interaction resembled a Lennard-Jones-type potential. Mechanical relaxations of the tip and sample surface were treated in the limits of continuum theory with the numerical MYD/BHW [15.16, 17] approach, which allows the simulations to be compared to corresponding experiments. Figure 15.8 shows the force-distance curves for different tip radii underlying the dynamic AFM simulations.

The cantilever trajectory was analyzed by solving the differential equation (15.7) extended by the force-distance relations from Fig. 15.8 using the numerical Verlet algorithm [15.18, 19]. The results of the simulation for the amplitude and phase of the tip oscillation as a function of z-position of the probe are presented in Fig. 15.9. One has to keep in mind that the z-position

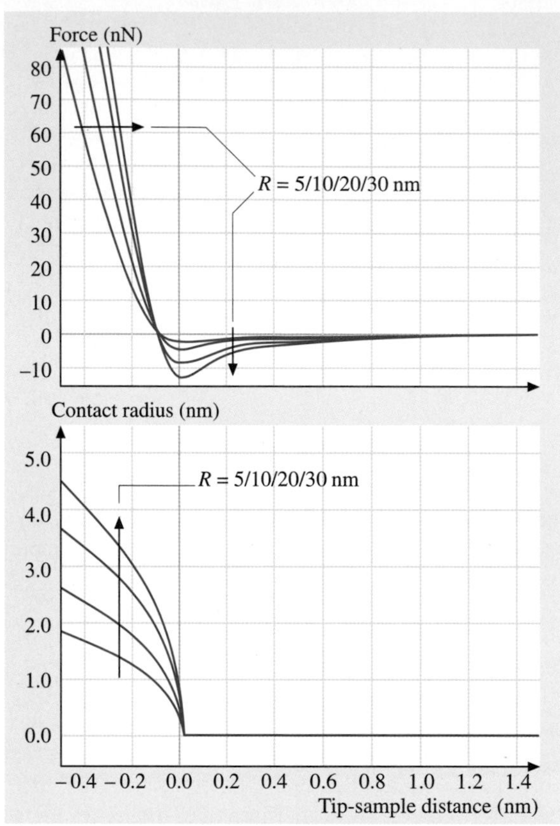

Fig. 15.8 Force curves and corresponding contact radius calculated with the MYD/BHW-model as a function of tip radius for a Si–Si contact. These force curves are used for the tapping-mode AFM simulations

of the probe is not equivalent to the real tip-sample distance at equilibrium position, since the cantilever might bend statically due to the interaction forces. The behavior of the cantilever can be subdivided into three different regimes. We distinguish the cases in which the beam is oscillated below its resonant frequency ω_0, exactly at ω_0, and above ω_0. In the following, we will refer to ω_0 as the resonant frequency, although the correct resonant frequency is ω_0^* if taking into account the finite Q-value.

Clearly, Fig. 15.9 exhibits more features than were anticipated from the initial, simple arguments. Amplitude and phase seem to change rather abruptly at certain points when the z_0-position is decreased. Besides, the amplitude or phase-distance curves don't resemble the force-distance curves from Fig. 15.8 in a simple, direct manner. Additionally, we find a hysteresis between approach and retraction.

As an example, let us start by discussing the discontinuous features in the AFM spectroscopy curves of the first case, where the excitation frequency is smaller than ω_0. Consider the oscillation amplitude as a function of excitation frequency in Fig. 15.5 in conjunction with a typical force curve, as depicted in Fig. 15.8. Upon approach of probe and sample, attractive forces will lower the effective resonant frequency of the oscillator. Therefore, the excitation frequency will now be closer to the resonant frequency, causing the vibration amplitude to increase. This, in turn, reduces the tip-sample distance, which again gives rise to a stronger attractive force. The system becomes unstable until the point $z_0 = d_{app}$ is reached, where repulsive forces stop the self-enhancing instability. This can be clearly observed in Fig. 15.9. Large parts of the force-distance curve cannot be measured due to this instability.

In the second case, where the excitation equals the free resonant frequency, only a small discontinuity is observed upon reduction of the z-position. Here, a shift of the resonant frequency toward smaller values, induced by the attractive force interaction, will reduce the oscillation amplitude. The distance between tip and sample is, therefore, reduced as well, and the self-amplifying effect with the sudden instability does not occur as long as repulsive forces are not encountered. However, at closer tip-sample distances, repulsive forces will cause the resonant frequency to shift again toward higher values, increasing the amplitude with decreasing tip-sample distance. Therefore, a self-enhancing instability will also occur in this case, but at the crossover from purely attractive forces to the regime where repulsive forces occur. Correspondingly, a small kink in the amplitude curve can be observed in Fig. 15.9. An even clearer indication of this effect is manifested by the sudden change in the phase signal at d_{app}.

In the last case, with $\omega > \omega_0$, the effect of amplitude reduction due to the resonant frequency shift is even larger. Again, we find no instability in the amplitude signal during approach in the attractive force regime. However, as soon as the repulsive force regime is reached, the instability occurs due to the induced positive frequency shift. Consequently, a large jump in the phase curve from values smaller than 90 deg to values larger than 90 deg is observed. The small change in the amplitude curve is not resolved in the simulated curves in Fig. 15.9, however, it can be clearly seen in the experimental curves in Fig. 15.10.

Figure 15.10 depicts the corresponding experimental amplitude and phase curves. The measurements were performed in air with a Si cantilever approaching

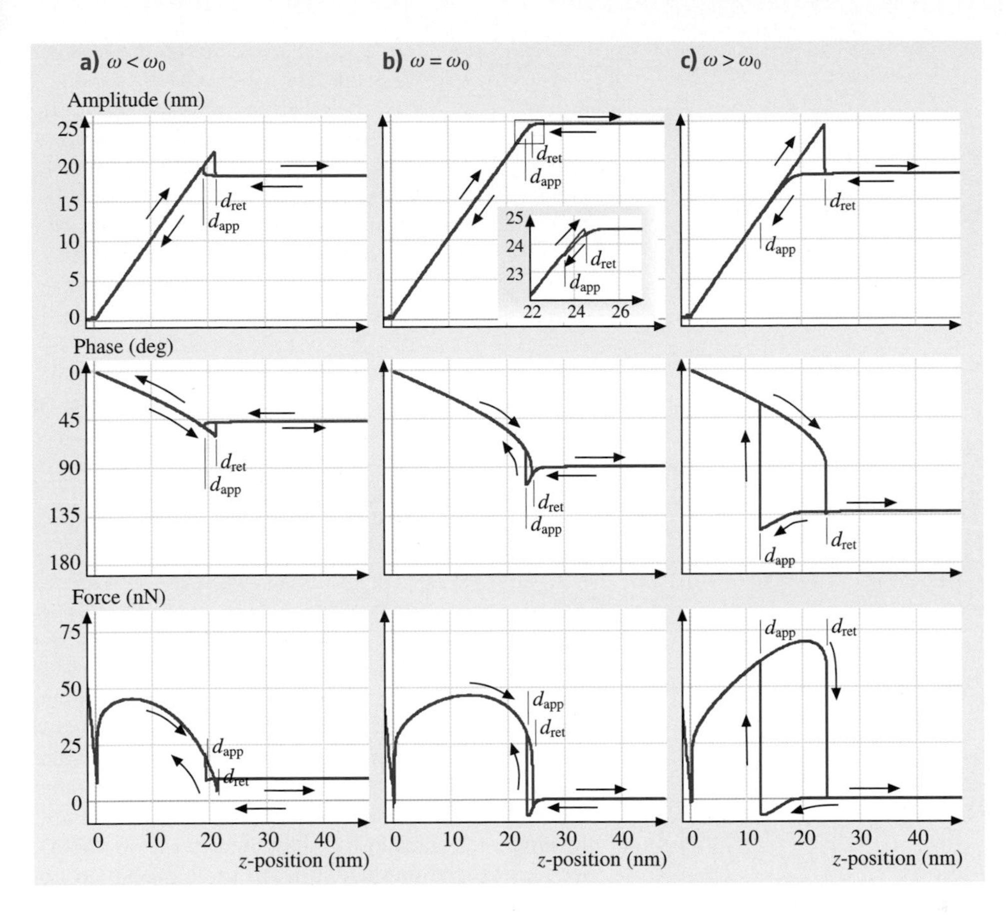

Fig. 15.9a–c Amplitude and phase diagrams with excitation frequency (**a**) below, (**b**) exactly at, and (**c**) above the resonant frequency for tapping-mode AFM from the numerical simulations. Additionally, the bottom diagrams show the interaction forces at the point of closest tip–sample distance, i. e., the lower turnaround point of the oscillation

a Si wafer, with a cantilever resonant frequency of 299.95 kHz. Qualitatively, all prominent features of the simulated curves can also be found in the experimental data sets. Hence, the above model seems to capture the important factors necessary for an appropriate description of the experimental situation.

But what is the reason for this unexpected behavior? We have to turn to the numerical simulations again, where we have access to all physical parameters, in order to understand the underlying processes. The lower part of Fig. 15.9 also shows the interaction force between the tip and the sample at the point of closest approach, i. e., the sample-sided turnaround point of the oscillation. We see that exactly at the points of the discontinuities the total interaction force changes from the net-attractive regime to the attractive-repulsive regime, also termed the intermittent contact regime. The term net-attractive is used to emphasize that the total force is attractive, despite the fact that some minor contributions might still originate from repulsive forces. As soon as a minimum distance is reached, the tip also starts to experience repulsive forces, which completely changes the oscillation behavior. In other words, the dynamic system switches between two oscillatory states.

Directly related to this fact is the second phenomenon: the hysteresis effect. We find separate curves for the approach of the probe toward the surface and the retraction. This seems to be somewhat counterintuitive, since the tip is constantly approaching and retracting from the surface, and the average values of amplitude and phase should be independent of the direction of the average tip-sample distance movement. A hysteresis between approach and retraction within one oscillation due to dissipative processes should directly influence amplitude and phase. However, no dissipation models were included in the simulation. In this case, the hysteresis in Fig. 15.9 is due to the fact that the oscillation jumps into different modes, the system exhibits bistability. This effect is often observed

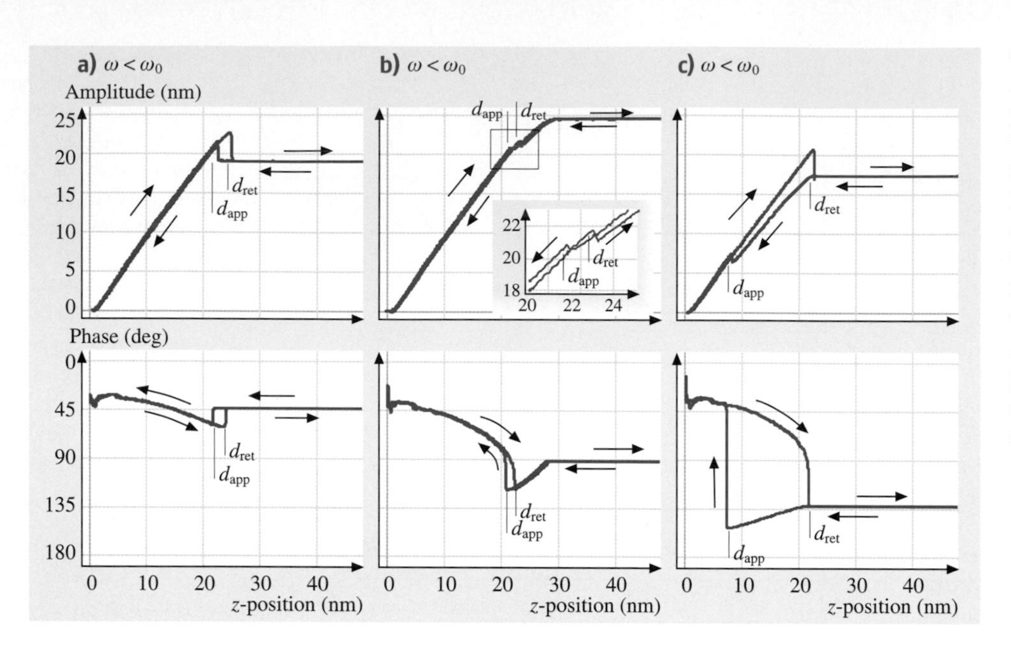

Fig. 15.10a–c Amplitude and phase diagrams with excitation frequency (**a**) below, (**b**) exactly at, and (**c**) above the resonant frequency for tapping-mode AFM from experiments with a Si cantilever on a Si wafer in air

in oscillators under the influence of nonlinear forces [15.20].

For the interpretation of these effects it is helpful to look at Fig. 15.11, which shows the behavior of the simulated tip vibration and force during one oscillation cycle over time. The data is recorded at the same z-position at a point where hysteresis is observed, while a) was taken during the approach and b) during the retraction. Excitation is in resonance, representing case 2, where the amplitude shows a small hysteresis. Also note that the amplitude is almost exactly the same in a) and b). We see that the oscillation at the same z-position exhibits two different modes: While in a) the experienced force is net-attractive, in b) the tip is exposed to attractive and repulsive interactions. Experimental and simulated data show that the change between the net-attractive and intermittent contact mode takes place at different z-positions (d_{app} and d_{ret}) for approach and retraction. Between d_{app} and d_{ret} the system is in a bistable mode. Depending on the history of the measurement, e.g., whether the position d_{app} during the approach (or d_{ret} during retraction) has been reached, the system flips to the other oscillation mode. While the amplitude might not be influenced strongly (case 2), the phase is a clear indicator of the mode switch. On the other hand, if point d_{app} is never reached during the approach, the system will stay in the net-attractive regime and no hysteresis is observed, i. e., the system remains stable.

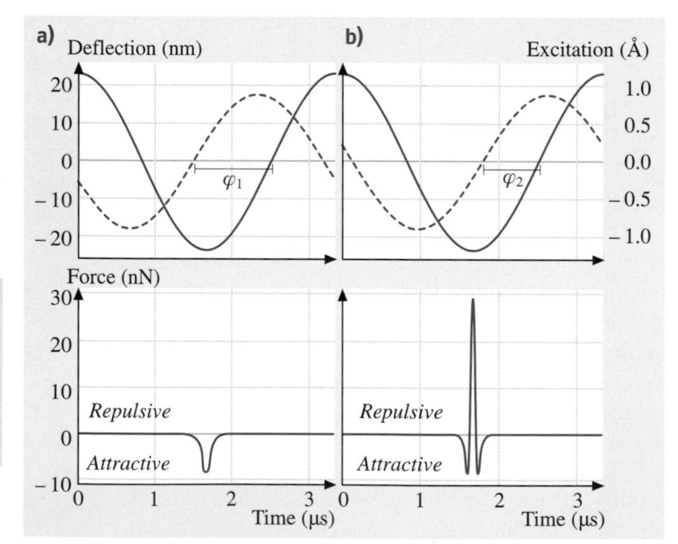

Fig. 15.11a,b Simulation of the tapping-mode cantilever oscillation in the (**a**) net-attractive and (**b**) the intermittent contact regime. The *dashed line* represents the excitation amplitude and the *solid line* is the oscillation amplitude

In conclusion, we find that although a qualitative interpretation of the interaction forces is possible, the AM- nor tapping-mode AFM is not suitable to gain direct quantitative knowledge of tip-sample force interactions. However, it is a very useful tool for imaging nanometer-sized structures in a wide variety of setups, in air or even in liquid. We find that two distinct modes exist for the externally excited oscillation – the net-attractive and the intermittent contact mode – which describe what kind of forces the tip-sample interaction is governed by. The phase can be used as an indicator of which mode the system is running in.

In particular, it can be easily seen that if the free resonant frequency of the cantilever is higher than the excitation frequency, the system cannot stay in the net-attractive regime due to a self-enhancing instability. Since in many applications involving soft and delicate biological samples strong repulsive forces should be avoided, the tapping-mode AFM should be operated at frequencies equal to or above the free resonant frequency [15.21]. Even then, statistical changes of tip-sample forces during the scan might induce a sudden jump into the intermittent contact mode, and the previously explained hysteresis will tend to keep the system in this mode. It is, therefore, of high importance to tune the oscillation parameters in such a way that the AFM stays in the net-attractive regime [15.22]. A concept that achieves this task is the Q-control system, which will be discussed in some detail in the forthcoming paragraphs.

A last word concerning the overlap of simulation and experimental data: While the qualitative agreement down to the detailed shape of hysteresis and instabilities is rather striking, we still find some quantitative discrepancies between the positions of the instabilities d_{app} and d_{ret}. This is probably due to the simplified force model, which only takes into account van der Waals and repulsive forces. Especially at ambient conditions, an omnipresent water meniscus between tip and sample will give rise to much stronger attractive and also dissipative forces than considered in the model. A very interesting feature is that the simulated phase curves in the intermittent contact regime tend to have a steeper slope in the simulation than in the experiments (see also Fig. 15.12). We will later show that this effect is a fingerprint of an effect that had not been included in the above simulation at all: dissipative processes during the oscillation, giving rise to an additional loss of oscillation energy.

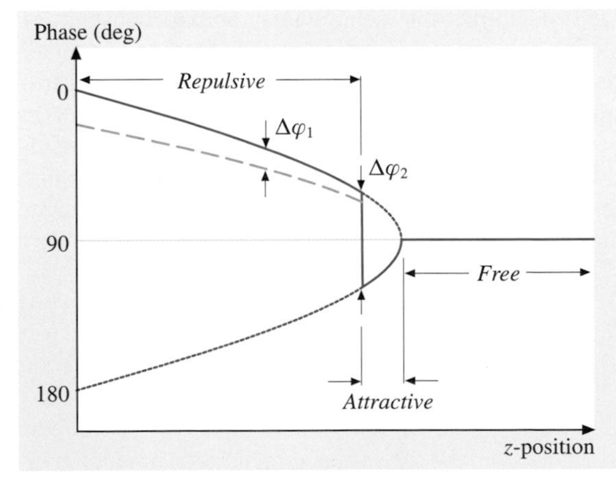

Fig. 15.12 Phase shift in tapping mode as a function of tip-sample distance

15.3.2 Self-Excitation Modes

Despite the wide range of technical applications of the AM or tapping mode of dynamic AFM, it has been found unsuitable for measurements in an environment extremely useful for scientific research: vacuum or ultrahigh vacuum (UHV) with pressures reaching 1×10^{-10} mbar. The STM has already shown how much insight can be gained from some highly defined experiments under those conditions. Consider (15.11) from the above section. The time constant τ for the amplitude to adjust to a different tip-sample force scales with $1/Q$. In vacuum applications, Q of the cantilever is on the order of 10,000, which means that τ is in the range of some 10 ms. This is clearly too long for a scan of at least (100×100) data points. The temperature-induced drift of the sample will render useful interpretation of images an impossible task. However, the resonant frequency of the system will react instantaneously to tip-sample forces. This has led *Albrecht* et al. [15.11] to use a modified excitation scheme.

The system is always oscillated at its resonant frequency. This is achieved by feeding back the oscillation signal from the cantilever into the excitation piezo-element. Figure 15.13 pictures the method in a block diagram. The signal from the PSD is phase shifted by 90 deg (and, therefore, always exciting in resonance) and used as the excitation signal of the cantilever. An additional feedback loop adjusts the excitation amplitude in such a way that the oscillation amplitude remains constant. This ensures that the tip-sample distance is not influenced by changes in the oscillation amplitude. The only degree of freedom of the oscillation system that can react to the tip-sample forces is the change of the resonant frequency. This shift of the frequency is detected and used as the set-point signal for surface scans. Therefore, this mode is also called the FM (frequency modulated) mode.

Let us take a look at the sensitivity of the dynamic AFM. If electronic noise, laser noise, and thermal drift

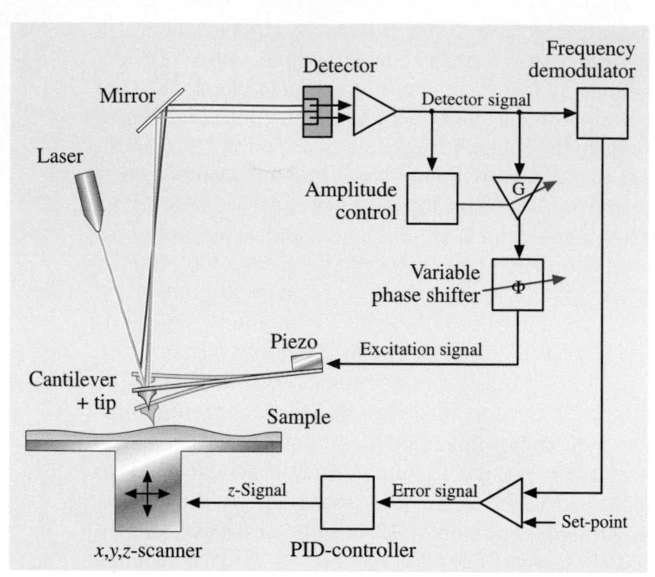

Fig. 15.13 Dynamic AFM operated in the self-excitation mode, where the oscillation signal is directly feedbacked to the excitation piezo. The detector signal is amplified with the variable gain G and phase shifted by phase ϕ. The frequency demodulator detects the frequency shift due to tip-sample interactions, which serves as the control signal for the probe-sample distance

can be neglected, the main noise contribution will come from thermal excitation of the cantilever. A detailed analysis of a dynamic system yields for the minimum detectable force gradient the following relation [15.11]:

$$\left.\frac{\partial F}{\partial z}\right|_{\mathrm{MIN}} = \sqrt{\frac{4k_{\mathrm{B}}TB \cdot k}{\omega_0 Q \langle z_{\mathrm{osc}}^2 \rangle}} \,. \tag{15.17}$$

Here B is the bandwidth of the measurement, T the temperature, and $\langle z_{\mathrm{osc}}^2 \rangle$ is the mean-square amplitude of the oscillation. Please note that this sensitivity limit was deliberately calculated for the FM mode. A similar analysis of the AM mode, however, yields virtually the same result [15.23]. We find that the minimum detectable force gradient, i. e., the measurement sensitivity, is inversely proportional to the square root of the Q-factor of the cantilever. This means that it should be possible to achieve very high-resolution imaging in vacuum conditions. In contrast, AM- or tapping-mode AFM cannot be usefully pursued with large Q. Only the FM mode makes it possible to take practical advantage of (15.17).

A breakthrough in high-resolution AFM application was the atomic resolution imaging of the Si(111)-(7×7) surface reconstruction by Giessibl [15.8] under UHV conditions. Today, atomic resolution imaging has become a standard feature guaranteed by industrially produced dynamic AFM systems. While STM has already proven to be an indispensable tool to gain detailed insight into surface structures of conductors with atomic resolution, the dynamic AFM has opened up the avenue into investigating nonconductive surfaces with equal precision (for example, on aluminium oxide by *Barth* et al. [15.24]).

However, we are concerned with measuring atomic force potentials of a single pair of molecules. Clearly, FM-mode AFM will allow us to identify single atoms, and with sufficient care we will be able to ensure that only one atom from the tip contributes to the total force interaction. Can we, therefore, fill in the last bit of information and find a quantitative relation between the oscillation parameters and the force?

Gotsmann et al. [15.25] investigated this relation by employing a numerical approach. During each oscillation cycle the tip experiences a whole range of forces. For each step during the approach the differential equation for the whole oscillation loop (including also the feedback system) was evaluated and the relation between force and frequency shift revealed. It was possible to determine the quantitative interaction forces of a metallic contact of nanometer dimensions.

However, it was shown that there also exists an analytical relationship if some approximations are accepted [15.7, 26, 27]. Here, we will follow the route indicated by *Dürig* [15.27], although different alternative ways have also proven successful. Consider the tip oscillation trajectory reaching over a large part of the force gradient curve in Fig. 15.2. We model the tip-sample interaction as a spring constant of stiffness $k_{\mathrm{ts}}(z) = \partial F/\partial z|_{z_0}$, as in Fig. 15.1. For small oscillation amplitudes, we already found that the frequency shift is proportional to the force gradient in (15.4). For large amplitudes, we can calculate an effective force gradient k_{eff} as a convolution of the force and the fraction of time the tip spends between the positions x and $x + \mathrm{d}x$:

$$k_{\mathrm{eff}}(z) = \frac{2}{\pi A^2} \int_{z}^{z+2A} F(x) g\left(\frac{x-z}{A} - 1\right) \mathrm{d}x$$
$$\text{with } g(u) = -\frac{u}{\sqrt{1-u^2}} \,. \tag{15.18}$$

In the approximation that the vibration amplitude is much larger than the range of the tip-sample forces, the above equation can be simplified to:

$$k_{\text{eff}}(z) = \frac{\sqrt{2}}{\pi} A^{3/2} \int_z^\infty \frac{F(x)}{\sqrt{x-z}}\,\mathrm{d}x\,. \tag{15.19}$$

This effective force gradient can now be used in (15.4), the relation between frequency shift and force gradient. We find:

$$\Delta f = \frac{f_0}{\sqrt{2}\pi k A^{3/2}} \int_z^\infty \frac{F(x)}{\sqrt{x-z}}\,\mathrm{d}x\,. \tag{15.20}$$

If we separate the integral from other parameters, we can define:

$$\Delta f = \frac{f_0}{kA^{3/2}}\gamma(z)$$
$$\text{with } \gamma(z) = \frac{1}{\sqrt{2}\pi} \int_z^\infty \frac{F(x)}{\sqrt{x-z}}\,\mathrm{d}x\,. \tag{15.21}$$

This means we can define $\gamma(z)$, which is only dependent on the shape of the force curve $F(z)$ but independent of the external parameters of the oscillation. The function $\gamma(z)$ is also referred to as the "normalized frequency shift" [15.7], a very useful parameter, which allows us to compare measurements independent of resonant frequency, amplitude and spring constant of the cantilever.

The dependence of the frequency shift on the vibration amplitude is an especially useful relation, since this parameter can be easily varied during one experiment. A nice example is depicted in Fig. 15.14, where frequency shift curves for different amplitudes were found to coincide very well in the $\gamma(z)$ diagrams [15.28].

This relationship has been nicely exploited for the calibration of the vibration amplitude by *Guggisberg* [15.29], which is a problem often encountered in dynamic AFM operation and worthwhile discussing. One approaches tip and sample and takes frequency shift curves versus distance, which show a reproducible shape. Then, the z-feedback is permanently turned off, and several curves with different amplitudes are taken. The amplitudes are typically chosen by adjusting the amplitude set-point in volts. One has to take care that drift in the z-direction is negligible. An analysis of the corresponding $\gamma(z)$-curves will show the same curves (as in Fig. 15.14), but the curves will be shifted in the horizontal axis. These shifts correspond to the change in amplitude, allowing one to correlate the voltage values with the z-distances.

For the oft-encountered force contributions from electrostatic, van der Waals, and chemical binding forces the frequency shift has been calculated from the force laws. In the approximation that the tip radius R is larger than the tip-sample distance z, an electrostatic potential V will yield a normalized frequency shift of (adapted from [15.30]):

$$\gamma(z) = \frac{\pi\varepsilon_0 R V^2}{\sqrt{2}} z^{-1/2}\,. \tag{15.22}$$

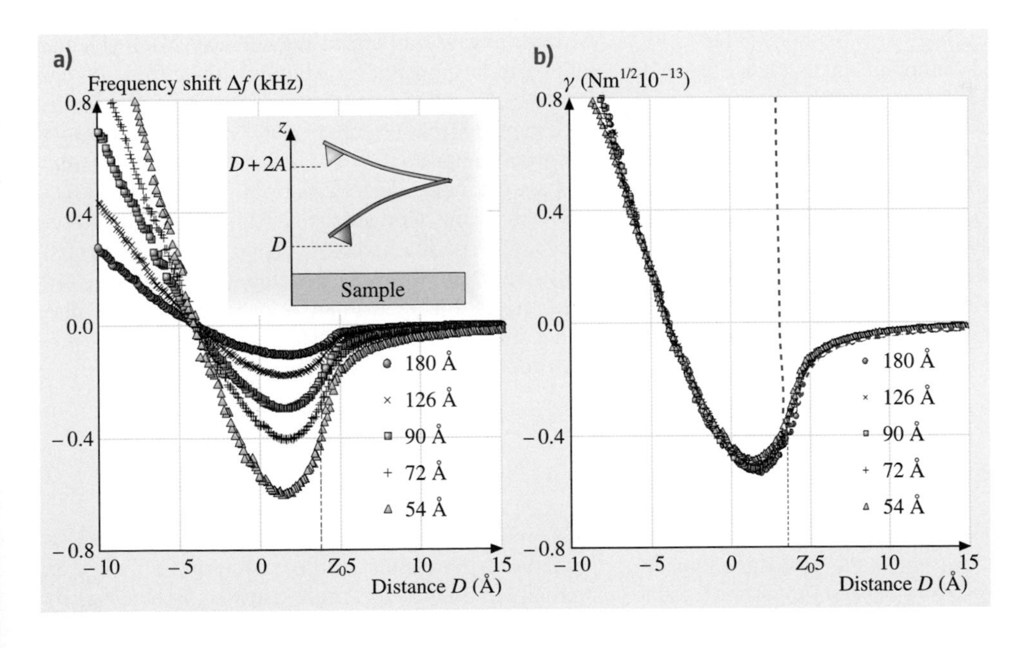

Fig. 15.14 (a) Frequency-shift curves for different oscillation amplitudes for a silicon tip on a graphite surface in UHV, (b) γ-curves calculated from the Δf-curves in (a). (After [15.28] with permission, Copyright (2000) by The American Physical Society)

For van der Waals forces with Hamaker constant H and with R larger than z, we find accordingly:

$$\gamma(z) = \frac{HR}{12\sqrt{2}} z^{-3/2} . \tag{15.23}$$

Finally, short-range chemical forces represented by the well-known Morse potential (with the parameters binding energy U_0, decay length λ, and equilibrium distance z_{equ}) yield:

$$\gamma(z) = \frac{U_0\sqrt{2}}{\sqrt{\pi\lambda}} \exp\left(-\frac{(z - z_{equ})}{\lambda}\right) . \tag{15.24}$$

These equations allow the experimentalist to directly interpret the spectroscopic measurements. For example, the contributions of the electrostatic and van der Waals forces can be easily distinguished by their slope in a log-log plot (for an example, see [15.30]).

Alternatively, if the force law is not known beforehand, the experimentalist wants to analyze the experimental frequency-shift data curves and extract the force or energy potential curves. We, therefore, have to invert the integral in (15.21) to find the tip-sample interaction potential V_{ts} from the $\gamma(z)$-curves [15.27]:

$$V_{ts}(z) = \sqrt{2} \int_z^\infty \frac{\gamma(x)}{\sqrt{x - z}}\, dx . \tag{15.25}$$

Using this method, quantitative force curves were extracted from Δf-spectroscopy measurements on different, atomically resolved sites of the Si(111)-(7×7) reconstruction [15.1]. Comparison to theoretical MD simulations showed good quantitative agreement with theory and confirmed the assumption that force interactions were governed by a single atom at the tip apex. Our initially formulated goal seems to be achieved: With FM-AFM we have found a powerful method that allows us to measure the chemical bond formation of single molecules. The last uncertainty, the exact shape and identity of the tip apex atom, can possibly be resolved by employing the FIM technique to characterize the tip surface in combination with FM-AFM.

All the above equations are strictly valid only in the approximation that the oscillation amplitudes are much larger than the distance range of the encountered forces. For amplitudes of 10 nm and long-range forces like electrostatic interactions this approximation may not be valid. An iterative approach has been presented by *Dürig* [15.31] to overcome this problem. The interaction force is calculated from the frequency shift curves [gradient of (15.25)]. This force curve again is used to calculate the frequency shift with the exact relation from (15.18). The difference between the experimental Δf-curve and the reconstructed Δf-curve can now be used to calculate a first-order correction term to the interaction force. This procedure can be iteratively followed until sufficient agreement is found.

In this context it is worthwhile to point out a slightly different dynamic AFM method. While in the typical FM-AFM setup the oscillation amplitude is controlled to stay constant by a dedicated feedback circuit, one could simply keep the excitation amplitude constant (this has been termed *CA = constant amplitude*, as opposed to the *CE = constant excitation* mode). It is expected that this mode is more gentle to the surface, because any dissipative interaction will reduce the amplitude and therefore increase the tip-sample distance. The tip is prevented from deeply indenting the surface. This mode has been employed to image soft biological molecules like DNA or thiols in UHV [15.32]. However, quantitative interpretation of the obtained frequency spectra is more complicated, since the amplitude and tip-sample distance are altered during the measurement. Until now, we have always associated the self-excitation scheme with vacuum applications. Although it is difficult to operate the FM-AFM in constant amplitude mode in air, since large dissipative effects make it difficult to ensure a constant amplitude, it is indeed possible to use the constant-excitation FM-AFM in air or even in liquid. Still, only a few applications of FM-AFM under ambient or liquid conditions have been reported so far. In fact, a low budget construction set (employing a tuning-fork force sensor) for a CE-mode dynamic AFM setup has been published on the Internet (*http://www.sxm4.uni-muenster.de*).

15.4 *Q*-Control

We have already discussed the virtues of a high Q value for high sensitivity measurements: The minimum detectable force gradient was inversely proportional to the square root of Q. In vacuum, Q mainly represents the internal dissipation of the cantilever during oscillation, an internal damping factor. Little damping is obtained by using high quality cantilevers, which are cut (or etched) from defect-free, single-crystal silicon wafers. Under ambient or liquid conditions, the quality factor is dominated by dissipative interactions between the cantilever and the surrounding medium, and Q values can be as low as 100 for air or even 5 in liquid. Still, we ask if it is somehow possible to compensate for the damping effect by exciting the cantilever in a sophisticated way?

It turns out that the shape of the resonance curves in Fig. 15.15 can be influenced toward higher (or lower) Q values by an amplitude feedback loop. In principle, there are several mechanisms to couple the amplitude signal back to the cantilever, by the photothermal effect [15.33] or capacitive forces [15.34]. Figure 15.16 shows a method in which the amplitude feedback is mediated directly by the excitation piezo [15.35]. This has the advantage that no additional mechanical setups are necessary.

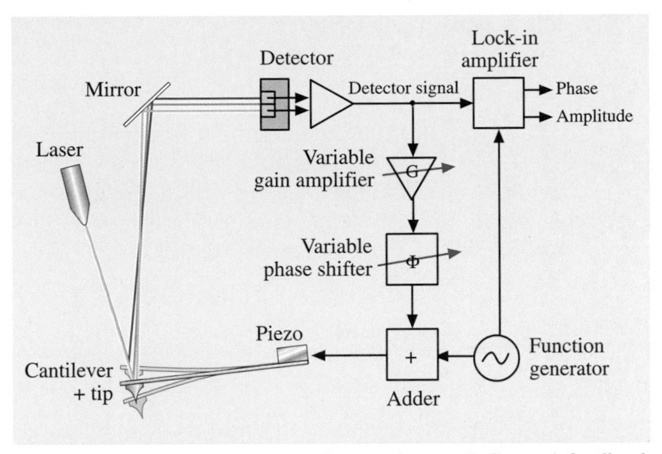

Fig. 15.16 Schematic diagram of operating a Q-Control feedback circuit with an externally driven dynamic AFM. The tapping mode setup is in effect extended by an additional feedback loop

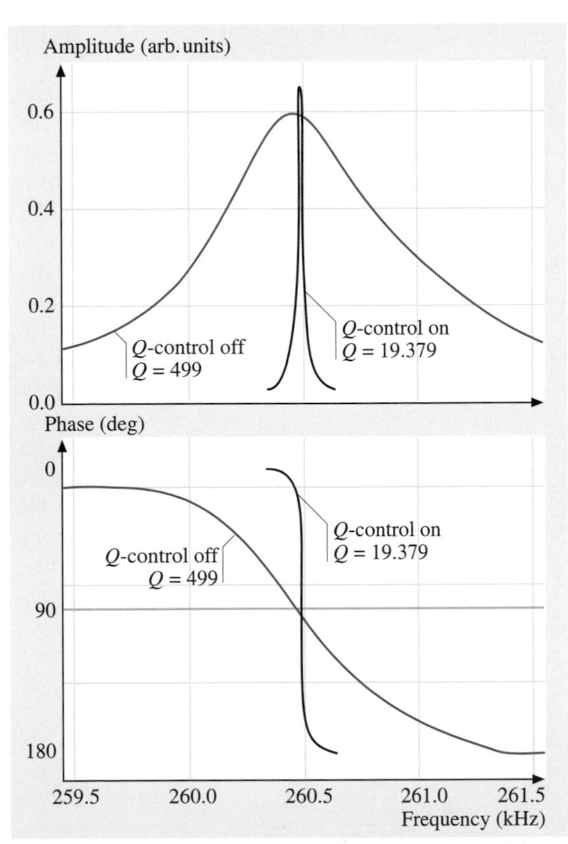

Fig. 15.15 Amplitude and phase diagrams measured in air with a Si cantilever far away from the sample. The quality factor can be increased from 450 to 20,000 by using the Q-Control feedback method

The working principle of the feedback loop can be understood by analyzing the equation of motion of the modified dynamic system:

$$m^*\ddot{z}(t) + \alpha\dot{z}(t) + kz(t) - F_{\text{ts}}\left(z_0 + z(t)\right) = F_{\text{ext}}\cos(\omega t) + G\,e^{i\phi}z(t)\,. \tag{15.26}$$

This ansatz takes into account the feedback of the detector signal through a phase shifter, amplifier, and adder as an additional force, which is linked to the cantilever deflection $z(t)$ through the gain G and the phase shift $e^{i\phi}$. To a good approximation, we assume that the oscillation can be described by a harmonic trajectory. With a phase

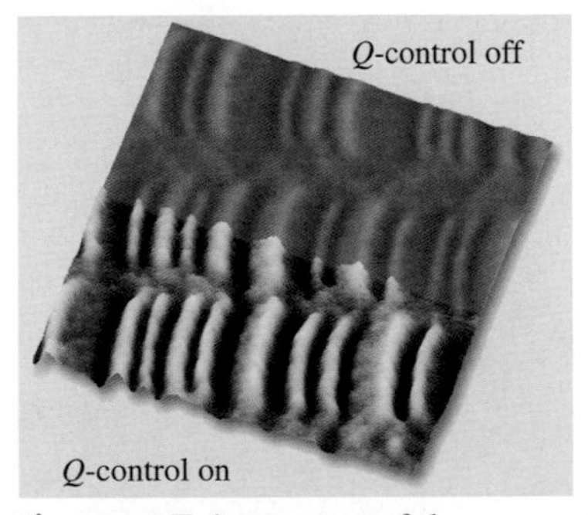

Fig. 15.17 Enhancement of the contrast in the phase channel due to Q-Control on a magnetic hard disk measured with a magnetic tip in tapping-mode AFM in air. Scan size $5\times5\,\mu$m, phase range 10 deg (www.nanoanalytics.com)

shift of $\phi = \pm\pi/2$ we find:

$$e^{\pm i\pi/2} z(t) = \pm\frac{1}{\omega}\dot{z}(t) . \quad (15.27)$$

This means that the additional feedback force signal $G e^{i\phi} z(t)$ is proportional to the velocity of the cantilever, just like the damping term in the equation of motion. We can define an effective damping constant α_{eff}, which combines the two terms:

$$m^*\ddot{z}(t) + \alpha_{eff}\dot{z}(t) + kz(t) - F_{ts}(z_0 + z(t)) = F_{ext}\cos(\omega t)$$
$$\text{with } \alpha_{eff} = \alpha \mp \frac{1}{\omega}G \quad \text{for } \phi = \pm\frac{\pi}{2} . \quad (15.28)$$

Equation (15.28) shows that the dampening of the oscillator can be enhanced or weakened by the choice of G, $\phi = +\pi/2$ or $\phi = -\pi/2$, respectively. The feedback loop, therefore, allows us to vary the effective quality factor $Q_{eff} = m\omega_0/\alpha_{eff}$ of the complete dynamic system. Hence, we term this system Q-Control. Figure 15.15 shows experimental data on the effect of Q-Control on the amplitude and phase as a function of the external excitation frequency [15.35]. While $Q = 449$, without Q-Control the quality factor of the system operated in air is enhanced by a factor of more than 40 by applying the feedback loop.

The effect of improved image contrast is demonstrated in Fig. 15.17. Here, a computer hard disk was analyzed with a magnetic tip in tapping mode, where the magnetic contrast is observed in the phase image. The upper part shows the recorded magnetic data structures in standard mode, whereas in the lower part of the image Q-Control feedback was activated, giving rise to an improved signal, i. e., magnetic contrast. A more detailed analysis of measurements on a magnetic tape shows that the signal amplitude (upper diagrams in Fig. 15.18), i. e.,

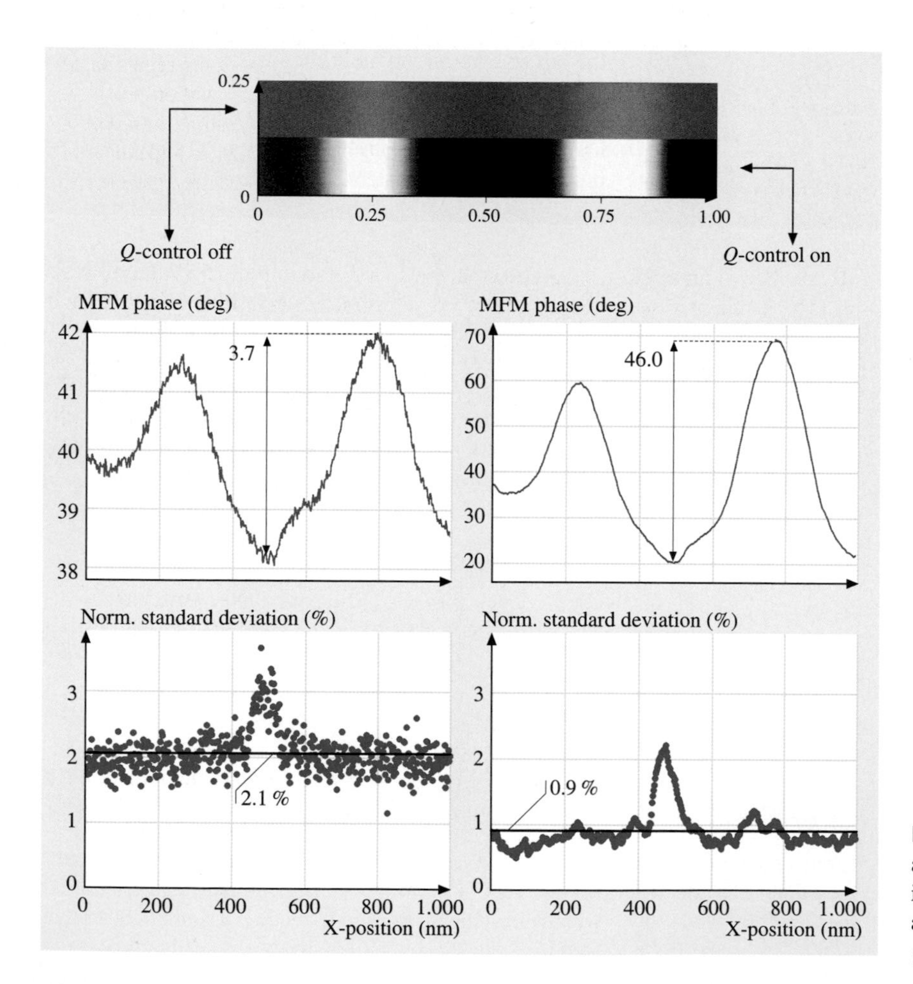

Fig. 15.18 Signal-to-noise analysis with a magnetic tip in tapping-mode AFM on a magnetic tape sample with Q-Control

the image contrast, was increased by a factor of 12.4 by the Q-Control feedback. The lower image shows a noise analysis of the signal, indicating an increase of the signal-to-noise ratio by a factor of 2.3.

Note that the diagrams represent measurements in air with an AFM operated in AM mode. Only then can we make a distinction between excitation and vibration frequency, since in the FM mode these two frequencies are equal by definition. Although the relation between sensitivity and Q-factor in (15.17) is the same for AM and FM mode, it must be critically investigated to see whether the enhanced quality factor by Q-Control can be inserted in the equation for FM-mode AFM. In vacuum applications, Q is already very high, which makes it oblique to operate an additional Q-Control module.

As stated before, we can also use Q-Control to enhance the dampening in the oscillating system. This would decrease the sensitivity of the system. But on the other hand, the response time of the amplitude change is decreased as well. For tapping-mode applications, where high speed scanning is the goal, Q-Control will be able to reduce the relaxation time [15.36].

A large quality factor Q does not only have the virtue of increasing the force sensitivity of the instrument. It also has the advantage of increasing the parameter space of stable AFM operation in tapping- or AM-mode AFM. Consider the resonance curve of Fig. 15.5. When approaching the tip toward the surface there are two competing mechanisms: On the one hand, we bring the tip closer to the sample, which results in an increase in attractive forces (see Fig. 15.2). On the other hand, for the case $\omega > \omega_0$, the resonant frequency of the cantilever is shifted toward smaller values due to the attractive forces, which causes the amplitude to become smaller, preventing a tip-sample contact. This is the desirable regime, where stable operation of the AFM is possible in the net-attractive regime. But as explained before, below a certain tip-sample separation d_{app}, the system switches suddenly into intermittent contact mode, where surface modifications are likely due to the onset of strong repulsive forces. The steeper the amplitude curve the larger the regime of stable, net-attractive AFM operation. Looking at Fig. 15.5 we find that the slope of the amplitude curve is governed by the quality factor Q. A high Q, therefore, facilitates stable operation of the AM-AFM in the net-attractive regime.

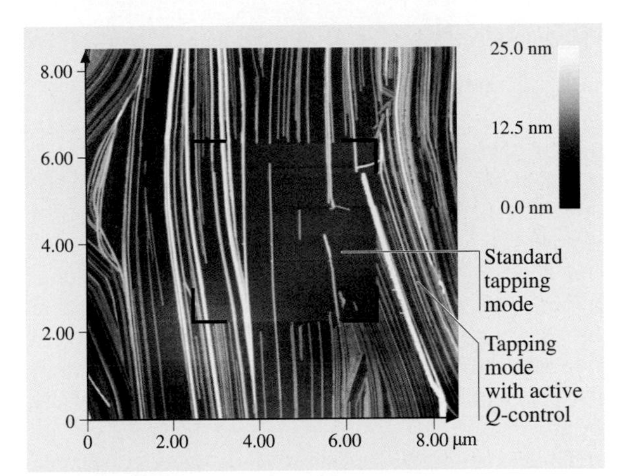

Fig. 15.19 Imaging of a delicate organic surface with Q-Control. Sample was a Langmuir-Blodgett film (ethyl-2,3-dihydroxyoctadecanoate) on a mica substrate. The topographical image clearly shows that the highly sensitive sample surface can only be imaged non-destructively with active Q-Control, whereas the periodic repulsive contact with the probe in standard operation without Q-Control leads to a significant modification or destruction of the surface structure. (Data courtesy of *Lifeng Chi* and coworkers, Westfälische Wilhelms-Universität, Münster, Germany)

An example can be found in Fig. 15.19. Here, a surface scan of an ultrathin organic film is acquired in tapping mode under ambient conditions. First, the inner square is scanned without the Q enhancement, and then a wider surface area was scanned with applied Q-Control. The high quality factor provides a larger parameter space for operating the AFM in the net-attractive regime, allowing good resolution of the delicate organic surface structure. Without the Q-Control the surface structures are deformed and even destroyed due to the strong repulsive tip-sample interactions [15.37–39]. This also allowed imaging of DNA structures without predominantly depressing the soft material during imaging. It was then possible to observe a DNA diameter close to the theoretical value with the Q-Control feedback [15.40].

In conclusion, we have shown that by applying an additional feedback circuit to the dynamic AFM system it is possible to influence the quality factor Q of the oscillator system. High-resolution, high-speed, or low-force scanning is then possible.

15.5 Dissipation Processes Measured with Dynamic AFM

Dynamic AFM methods have proven their great potential for imaging surface structures at the nanoscale, and we have also discussed methods that allow the assessment of forces between distinct single molecules. However, there is another physical mechanism that can be analyzed with the dynamic mode and has been mentioned in some previous paragraphs: energy dissipation. In Fig. 15.12, we have already shown an example, where the phase signal in tapping mode cannot be explained by conservative forces alone; dissipative processes must also play a role. In constant-amplitude FM mode, where the quantitative interpretation of experiments has proven to be less difficult, an intuitive distinction between conservative and dissipative tip-sample interaction is possible. While we have shown the correlation between forces and frequency shifts of the oscillating system, we have neglected one experimental input channel. The excitation amplitude, which is necessary to keep the oscillation amplitude constant, is a direct indication of the energy dissipated during one oscillation cycle. *Dürig* [15.41] has shown that in self-excitation mode (with an excitation-oscillation phase difference of 90 deg), conservative and dissipative interactions can be strictly separated. Part of this energy is dissipated in the cantilever itself, another part is due to external viscous forces in the surrounding medium. But more interestingly, some energy is dissipated at the tip-sample junction. This is the focus of the following paragraphs.

In contrast to conservative forces acting at the tip-sample junction, which at least in vacuum can be understood in terms of van der Waals, electrostatic, and chemical interactions, the dissipative processes are poorly understood. *Stowe* et al. [15.42] have shown that if a voltage potential is applied between tip and sample, charges are induced in the sample surface, which will follow the tip motion (here the oscillation is parallel to the surface). Due to the finite resistance of the sample material, energy will be dissipated during the charge movement. This effect has been exploited to image the doping level of semiconductors. Energy dissipation has also been observed in imaging magnetic materials. *Liu* et al. [15.43] found that energy dissipation due to magnetic interactions was enhanced at the boundaries of magnetic domains, which was attributed to domain wall oscillations. But also in the absence of external electromagnetic fields, energy dissipation was observed in close proximity of tip and sample, within 1 nm. Clearly, mechanical surface relaxations must give rise to energy losses. One could model the AFM tip as a small hammer, hitting the surface at high frequency, possibly resulting in phonon excitations. From a contiuum mechanics point of view, we assume that the mechanical relaxation of the surface is not only governed by elastic responses. Viscoelastic effects of soft surfaces will also render a significant contribution to energy dissipation. The whole area of phase imaging in tapping mode is concerned with those effects [15.44–47].

In the atomistic view, the last tip atom can be envisaged to change position while yielding to the tip-sample force field. A strictly reversible change of position would not result in a loss of energy. Still, it has been pointed out by *Sasaki* et al. [15.48] that a change in atom position would result in a change in the force interaction itself. Therefore, it is possible that the tip atom changes position at different tip-surface distances during approach and retraction, effectively causing an atomic-scale hysteresis to develop. In fact, *Hoffmann* et al. [15.13] have measured short-range energy dissipation for a tungsten tip on silicon in UHV. A theoretical model explaining this effect on the basis of a two-energy-state system was developed. However, a clear understanding of the underlying physical mechanism is still lacking.

Nonetheless, the dissipation channel has been used to image surfaces with atomic resolution [15.49]. Instead of feedbacking the distance on the frequency shift, the excitation amplitude in FM mode has been used as the control signal. The Si(111)-(7×7) reconstruction was successfully imaged in this mode. The step edges of monoatomic NaCl islands on single-crystalline copper have also rendered atomic resolution contrast in the dissipation channel [15.50]. The dissipation processes discussed so far are mostly in the configuration in which the tip is oscillated perpendicular to the surface. Friction is usually referred to as the energy loss due to lateral movement of solid bodies in contact. It is interesting to note in this context that *Israelachivili* [15.51] has pointed out a quantitative relationship between lateral and vertical (with respect to the surface) dissipation. He states that the hysteresis in vertical force-distance curves should equal the energy loss in lateral friction. An experimental confirmation of this conjecture at the molecular level is still missing.

Physical interpretation of energy dissipation processes at the atomic scale seems to be a daunting task at this point. Notwithstanding, we can find a quantitative relation between the energy loss per oscillation cycle and the experimental parameters in dynamic AFM, as will be shown in the following section.

In static AFM it was found that permanent changes of the sample surface by indentations can cause a hysteresis between approach and retraction (e.g., [15.5]). The area between the approach and retraction curves in a force-distance diagram represents the lost or dissipated energy caused by the irreversible change of the surface structure. In dynamic-mode AFM, the oscillation parameters like amplitude, frequency, and phase must contain the information about the dissipated energy per cycle. So far, we have resorted to a treatment of the equation of motion of the cantilever vibration in order to find a quantitative correlation between forces and the experimental parameters. For the dissipation it is useful to treat the system from the energy conservation point of view.

Assuming a dynamic system is in equilibrium, the average energy input must equal the average energy output or dissipation. Applying this rule to an AFM running in a dynamic mode means that the average power fed into the cantilever oscillation by an external driver, denoted $\bar{P}_{\text{in}}$, must equal the average power dissipated by the motion of the cantilever beam $\bar{P}_0$ and by tip-sample interaction $\bar{P}_{\text{tip}}$:

$$\bar{P}_{\text{in}} = \bar{P}_0 + \bar{P}_{\text{tip}} \,. \tag{15.29}$$

The term $\bar{P}_{\text{tip}}$ is what we are interested in, since it gives us a direct physical quantity to characterize the tip-sample interaction. Therefore, we have first to calculate and then measure the two other terms in (15.29) in order to determine the power dissipated when the tip periodically probes the sample surface. This requires an appropriate rheological model to describe the dynamic system. Although there are investigations in which the complete flexural motion of the cantilever beam has been considered [15.52], a simplified model, comprising a spring and two dashpots (Fig. 15.20), represents a good approximation in this case [15.53].

The spring, characterized by the constant k according to Hooke's law, represents the only channel through which power P_{in} can be delivered to the oscillating tip $z(t)$ by the external driver $z_{\text{d}}(t)$. Therefore, the instantaneous power fed into the dynamic system is equal to the force exerted by the driver times the velocity of the driver (the force that is necessary to move the base side of the dashpot can be neglected, since this power is directly dissipated and, therefore, does not contribute to the power delivered to the oscillating tip):

$$P_{\text{in}}(t) = F_{\text{d}}(t)\dot{z}_{\text{d}}(t) = k\,[z(t) - z_{\text{d}}(t)]\,\dot{z}_{\text{d}}(t) \,. \tag{15.30}$$

Assuming a sinusoidal steady state response and that the base of the cantilever is driven sinusoidally (see (15.6))

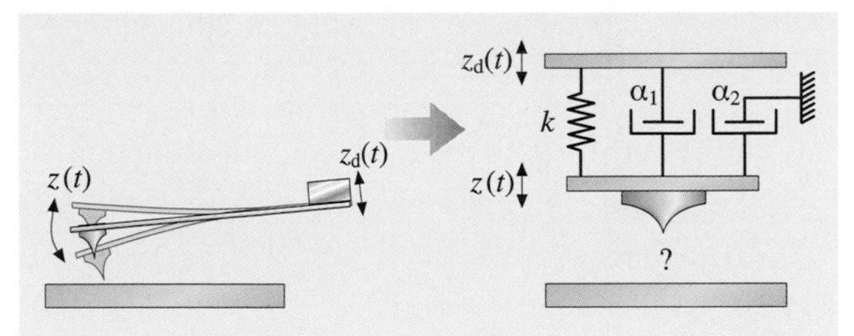

Fig. 15.20 Rheological model applied to describe the dynamic AFM system, comprising the oscillating cantilever and tip interacting with the sample surface. The movement of the cantilever base and the tip is denoted as $z_{\text{d}}(t)$ and $z(t)$, respectively. The cantilever is characterized by the spring constant k and the damping constant α. In a first approach, damping is broken into two pieces, α_1 and α_2. First, intrinsic damping caused by the movement of the cantilever's tip relative to its base. Second, damping related to the movement of the cantilever body in a surrounding medium, e.g., air damping

with amplitude A_{d} and frequency ω, the deflection from equilibrium of the end of the cantilever follows (15.9), where A and $0 \le \varphi \le \pi$ are the oscillation amplitude and phase shift, respectively. This allows us to calculate the average power input per oscillation cycle by integrating (15.30) over one period $T = 2\pi/\omega$:

$$\bar{P}_{\text{in}} = \frac{1}{T}\int_0^T P_{\text{in}}(t)\,\text{d}t = \frac{1}{2}k\omega A_{\text{d}}A\sin\varphi \,. \tag{15.31}$$

This contains the familiar result that the maximum power is delivered to an oscillator when the response is 90 deg out of phase with the drive.

The simplified rheological model, as it is depicted in Fig. 15.20, exhibits two major contributions to the damping term $\bar{P}_0$. Both are related to the motion of the cantilever body and assumed to be well modeled by viscous damping with coefficients α_1 and α_2. The dominant damping mechanism in UHV conditions is intrinsic damping caused by the deflection of the cantilever beam, i. e., the motion of the tip relative to the cantilever base. Therefore, the instantaneous power dissipated by such a mechanism is given by

$$P_{01}(t) = |F_{01}(t)\dot{z}(t)| = |\alpha_1\,[\dot{z}(t) - \dot{z}_{\text{d}}(t)]\,\dot{z}(t)| \,. \tag{15.32}$$

Note that the absolute value has to be calculated, since all dissipated power is "lost" and, therefore, cannot be returned to the dynamic system.

However, when running an AFM in ambient conditions an additional damping mechanism has to be considered. Damping due to the motion of the cantilever body in the surrounding medium, e.g., air damping, is in most cases the dominant effect. The corresponding instantaneous power dissipation is given by

$$P_{02}(t) = |F_{02}(t)\dot{z}(t)| = \alpha_2 \dot{z}^2(t) . \tag{15.33}$$

In order to calculate the average power dissipation, (15.32) and (15.33) have to be integrated over one complete oscillation cycle. This yields

$$\begin{aligned}\bar{P}_{01} &= \frac{1}{T}\int_0^T P_{01}(t)\,\mathrm{d}t \\ &= \frac{1}{\pi}\alpha_1\omega^2 A \left[(A - A_\mathrm{d}\cos\varphi) \right. \\ &\quad \times \arcsin\left(\frac{A - A_\mathrm{d}\cos\varphi}{\sqrt{A^2 + A_\mathrm{d}^2 - 2AA_\mathrm{d}\cos\varphi}}\right) \\ &\quad \left. + A_\mathrm{d}\sin\varphi \right] ,\end{aligned} \tag{15.34}$$

and

$$P_{02}(t) = \frac{1}{T}\int_0^T P_{02}(t)\,\mathrm{d}t = \frac{1}{2}\alpha_2\omega^2 A^2 . \tag{15.35}$$

Considering the fact that commonly used cantilevers exhibit a quality factor of at least several hundreds (in UHV several tens of thousands), we can assume that the oscillation amplitude is significantly larger than the drive amplitude when the dynamic system is driven at or near its resonance frequency: $A \gg A_\mathrm{d}$. Therefore, (15.34) can be simplified in first order approximation to an expression similar to (15.35). Combining the two equations yields the total average power dissipated by the oscillating cantilever

$$\bar{P}_0 = \frac{1}{2}\alpha\omega^2 A^2 \quad \text{with } \alpha = \alpha_1 + \alpha_2 , \tag{15.36}$$

where α denotes the overall effective damping constant.

We can now solve (15.29) for the power dissipation localized to the small interaction volume of the probing tip with the sample surface, represented by the question mark in Fig. 15.20. Furthermore, by expressing the damping constant α in terms of experimentally accessible quantities such as the spring constant k, the quality factor Q, and the natural resonant frequency ω_0 of the free oscillating cantilever, $\alpha = \frac{k}{Q\omega_0}$, we obtain:

$$\begin{aligned}\bar{P}_\mathrm{tip} &= \bar{P}_\mathrm{in} - \bar{P}_0 \\ &= \frac{1}{2}\frac{k\omega}{Q}\left(Q_\mathrm{cant}A_\mathrm{d}A\sin\varphi - A^2\frac{\omega}{\omega_0}\right) .\end{aligned} \tag{15.37}$$

Note that so far no assumptions have been made regarding how the AFM is operated, except that the motion of the oscillating cantilever has to remain sinusoidal to a good approximation. Therefore, (15.37) is applicable to a variety of different dynamic AFM modes.

For example, in FM-mode AFM the oscillation frequency ω changes due to tip-sample interaction, while at the same time the oscillation amplitude A is kept constant by adjusting the drive amplitude A_d. By measuring these quantities, one can apply (15.37) to determine the average power dissipation related to tip-sample interaction. In spectroscopy applications, usually $A_\mathrm{d}(d)$ is not measured directly, but a signal $G(d)$ proportional to $A_\mathrm{d}(d)$ is acquired representing the gain factor applied to the excitation piezo. With the help of (15.15) we can write:

$$A_\mathrm{d}(d) = \frac{A_0 G(d)}{Q G_0} , \tag{15.38}$$

while A_0 and G_0 are the amplitude and gain at large tip-sample distances, where the tip-sample interactions are negligible.

Now let us consider the tapping-mode, or AM-AFM. In this case, the cantilever is driven at a fixed frequency and with constant drive amplitude, while the oscillation amplitude and phase shift may change when the probing tip interacts with the sample surface. Assuming that the oscillation frequency is chosen to be ω_0, (15.37) can be further simplified by again employing (15.15) for the free oscillation amplitude A_0. This yields

$$\bar{P}_\mathrm{tip} = \frac{1}{2}\frac{k\omega_0}{Q}\left(A_0 A\sin\varphi - A^2\right) . \tag{15.39}$$

Equation (15.39) implies that if the oscillation amplitude A is kept constant by a feedback loop, like it is commonly done in tapping mode, simultaneously acquired phase data can be interpreted in terms of energy dissipation [15.45, 47, 54, 55]. When analyzing such phase images [15.56–58], one has also to consider the fact that the phase may also change due to the transition from net-attractive ($\varphi > 90°$) to intermittent contact ($\varphi < 90°$) interaction between the tip and the sample [15.15, 35, 59, 60]. For example, consider the phase shift in tapping mode as a function of z-position, Fig. 15.12. If phase measurements are performed close to the point where the oscillation switches from the net-attractive to the intermittent contact regime, a large contrast in the phase channel is observed. However, this contrast is not due to dissipative processes. Only a variation of the phase signal within the intermittent contact regime will give information of the tip-sample dissipative processes.

An example of a dissipation measurement is depicted in Fig. 15.21. The surface of a polymer blend was imaged in air, simultaneously acquiring the topography and dissipation. The dissipation on the softer polyurethane-matrix is significantly larger than on the embedded, mechanically stiffer polypropylene particles.

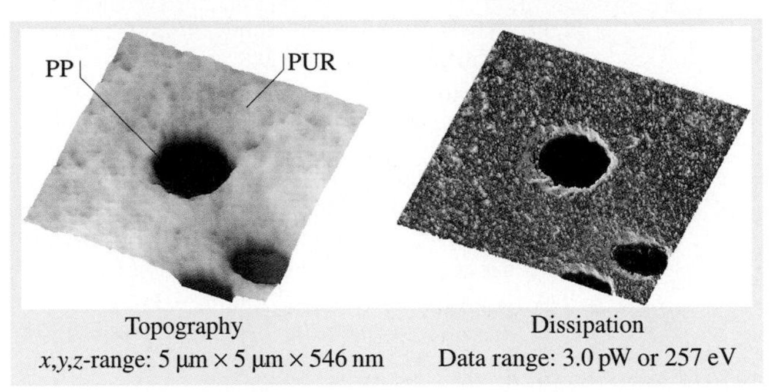

Fig. 15.21 Topography and phase image in tapping-mode AFM of a polymer blend composed of polypropylene (PP) particles embedded in a polyurethane (PUR) matrix. The dissipation image shows a strong contrast between the harder PP (little dissipation, *dark*) to the softer PUR (large dissipation, *bright*) surface

15.6 Conclusion

Dynamic force microscopy is a powerful tool that is capable of imaging surfaces with atomic precision. It also allows us to look at surface dynamics, and it can operate in vacuum, air, or even in liquid. However, the oscillating cantilever system introduces a level of complexity that disallows a straightforward interpretation of acquired images. An exception is the self-excitation mode, where tip-sample forces can be successfully extracted from spectroscopic experiments. However, not only conservative forces can be investigated with dynamic AFM; energy dissipation also influences the cantilever oscillation and can, therefore, serve as a new information channel.

Open questions still exist concerning the exact geometric and chemical identity of the probing tip, which significantly influences the imaging and spectroscopic results. Using predefined tips like single-walled nanotubes or atomic resolution techniques like field ion microscopy to image the tip itself are possible approaches to addressing this issue. Furthermore, little is known about the dissipative processes at the scale where only a few atoms between tip and sample interact. It is also desirable to learn more about the interpretation of images acquired in tapping mode.

References

15.1 M. A. Lantz, H. J. Hug, R. Hoffmann, P. J. A. van Schendel, P. Kappenberger, S. Martin, A. Baratoff, H.-J. Güntherodt: Quantitative measurement of short-range chemical bonding forces, Science **291** (2001) 2580–2583

15.2 G. Binnig, C. F. Quate, Ch. Gerber: Atomic force microscope, Phys. Rev. Lett. **56** (1986) 930–933

15.3 O. Marti: AFM instrumentation and tips. In: *Handbook of Micro/Nanotribology*, 2nd edn., ed. by B. Bhushan (CRC, Boca Raton 1999) pp. 81–144

15.4 G. Cross, A. Schirmeisen, A. Stalder, P. Grütter, M. Tschudy, U. Dürig: Adhesion interaction between atoically defined tip and sample, Phys. Rev. Lett. **80** (1998) 4685–4688

15.5 A. Schirmeisen, G. Cross, A. Stalder, P. Grütter, U. Dürig: Metallic adhesion and tunneling at the atomic scale, New J. Phys. **2** (2000) 29.1–29.10

15.6 A. Schirmeisen: Metallic adhesion and tunneling at the atomic scale. Ph.D. Thesis (McGill University, Montréal 1999) pp. 29–38

15.7 F. J. Giessibl: Forces and frequency shifts in atomic-resolution dynamic-force microscopy, Phys. Rev. B **56** (1997) 16010–16015

15.8 F. J. Giessibl: Atomic resolution of the silicon (111)-(7×7) surface by atomic force microscopy, Science **267** (1995) 68–71

15.9 M. Bezanilaa, B. Drake, E. Nudler, M. Kashlev, P. K. Hansma, H. G. Hansma: Motion and enzymatic

degradation of DNA in the atomic force microscope, Biophys. J. **67** (1994) 2454–2459
15.10 Y. Jiao, D. I. Cherny, G. Heim, T. M. Jovin, T. E. Schäffer: Dynamic interactions of p53 with DNA in solution by time-lapse atomic force microscopy, J. Mol. Biol. **314** (2001) 233–243
15.11 T. R. Albrecht, P. Grütter, D. Horne, D. Rugar: Frequency modulation detection using high-Q cantilevers for enhanced force microscopy sensitivity, J. Appl. Phys. **69** (1991) 668–673
15.12 S. P. Jarvis, M. A. Lantz, U. Dürig, H. Tokumoto: Off resonance AC mode force spectroscopy and imaging with an atomic force microscope, Appl. Surf. Sci. **140** (1999) 309–313
15.13 P. M. Hoffmann, S. Jeffery, J. B. Pethica, H. Ö. Özer, A. Oral: Energy dissipation in atomic force microscopy and atomic loss processes, Phys. Rev. Lett. **87** (2001) 265502–265505
15.14 B. Anczykowski, D. Krüger, H. Fuchs: Cantilever dynamics in quasinoncontact force microscopy: Spectroscopic aspects, Phys. Rev. B **53** (1996) 15485–15488
15.15 B. Anczykowski, D. Krüger, K. L. Babcock, H. Fuchs: Basic properties of dynamic force spectroscopy with the scanning force microscope in experiment and simulation, Ultramicroscopy **66** (1996) 251–259
15.16 V. M. Muller, V. S. Yushchenko, B. V. Derjaguin: On the influence of molecular forces on the deformation of an elastic sphere and its sticking to a rigid plane, J. Coll. Interf. Sci. **77** (1980) 91–101
15.17 B. D. Hughes, L. R. White: 'Soft' contact problems in linear elasticity, Quart. J. Mech. Appl. Math. **32** (1979) 445–471
15.18 L. Verlet: Computer "experiments" on classical fluids. I. Thermodynamical properties of Lennard–Jones molecules, Phys. Rev. **159** (1967) 98–103
15.19 L. Verlet: Computer "experiments" on classical fluids. II. Equilibrium correlation functions, Phys. Rev. **165** (1968) 201–214
15.20 P. Gleyzes, P. K. Kuo, A. C. Boccara: Bistable behavior of a vibrating tip near a solid surface, Appl. Phys. Lett. **58** (1991) 2989–2991
15.21 A. San Paulo, R. García: High-resolution imaging of antibodies by tapping-mode atomic force microscopy: Attractive and repulsive tip-sample interaction regimes, Biophys. J. **78** (2000) 1599–1605
15.22 D. Krüger, B. Anczykowski, H. Fuchs: Physical properties of dynamic force microscopies in contact and noncontact operation, Ann. Phys. **6** (1997) 341–363
15.23 Y. Martin, C. C. Williams, H. K. Wickramasinghe: Atomic force microscope – force mapping and profiling on a sub 100-Å scale, J. Appl. Phys. **61** (1987) 4723–4729
15.24 C. Barth, M. Reichling: Imaging the atomic arrangement on the high-temperature reconstructed α-Al_2O_3(0001) surface, Nature **414** (2001) 54–57
15.25 B. Gotsmann, H. Fuchs: Dynamic force spectroscopy of conservative and dissipative forces in an Al-Au(111) tip-sample system, Phys. Rev. Lett. **86** (2001) 2597–2600
15.26 H. Hölscher, W. Allers, U. D. Schwarz, A. Schwarz, R. Wiesendanger: Determination of tip-sample interaction potentials by dynamic force spectroscopy, Phys. Rev. Lett. **83** (1999) 4780–4783
15.27 U. Dürig: Relations between interaction force and frequency shift in large-amplitude dynamic force microscopy, Appl. Phys. Lett. **75** (1999) 433–435
15.28 H. Hölscher, A. Schwarz, W. Allers, U. D. Schwarz, R. Wiesendanger: Quantitative analysis of dynamic-force-spectroscopy data on graphite(0001) in the contact and noncontact regime, Phys. Rev. B **61** (2000) 12678–12681
15.29 M. Guggisberg: Lokale Messung von atomaren Kräften. Ph.D. Thesis (University Basel, Basel 2000) pp. 9–11
15.30 M. Guggisberg, M. Bammerlin, E. Meyer, H.-J. Güntherodt: Separation of interactions by noncontact force microscopy, Phys. Rev. B **61** (2000) 11151–11155
15.31 U. Dürig: Extracting interaction forces and complementary observables in dynamic probe microscopy, Appl. Phys. Lett. **76** (2000) 1203–1205
15.32 T. Uchihasi, T. Ishida, M. Komiyama, M. Ashino, Y. Sugawara, W. Mizutani, K. Yokoyama, S. Morita, H. Tokumoto, M. Ishikawa: High-resolution imaging of organic monolayers using noncontact AFM, Appl. Surf. Sci. **157** (2000) 244–250
15.33 J. Mertz, O. Marti, J. Mlynek: Regulation of a microcantilever response by force feedback, Appl. Phys. Lett. **62** (1993) 2344–2346
15.34 D. Rugar, P. Grütter: Mechanical parametric amplification and thermomechanical noise squeezing, Phys. Rev. Lett. **67** (1991) 699–702
15.35 B. Anczykowski , J. P. Cleveland, D. Krüger, V. B. Elings, H. Fuchs: Analysis of the interaction mechanisms in dynamic mode SFM by means of experimental data and computer simulation, Appl. Phys. A **66** (1998) 885–889
15.36 T. Sulchek, G. G. Yaralioglu, C. F. Quate, S. C. Minne: Characterization and optimisation of scan speed for tapping-mode atomic force microscopy, Rev. Sci. Instrum. **73** (2002) 2928–2936
15.37 L. F. Chi, S. Jacobi, B. Anczykowski, M. Overs, H.-J. Schäfer, H. Fuchs: Supermolecular periodic structures in monolayers, Adv. Mater. **12** (2000) 25–30
15.38 S. Gao, L. F. Chi, S. Lenhert, B. Anczykowski, C. Niemeyer, M. Adler, H. Fuchs: High-quality mapping of DNA-protein complexes by dynamic scanning force microscopy, ChemPhysChem **6** (2001) 384–388
15.39 B. Zou, M. Wang, D. Qiu, X. Zhang, L. F. Chi, H. Fuchs: Confined supramolecular nanostructures of mesogen-bearing amphiphiles, Chem. Commun. **9** (2002) 1008–1009

15.40 B. Pignataro, L. F. Chi, S. Gao, B. Anczykowski, C. Niemeyer, M. Adler, H. Fuchs: Dynamic scanning force microscopy study of self-assembled DNA-protein nanostructures, Appl. Phys. A **74** (2002) 447–452

15.41 U. Dürig: Interaction sensing in dynamic force microscopy, New J. Phys. **2** (2000) 5.1–5.2

15.42 T. D. Stowe, T. W. Kenny, D. J. Thomson, D. Rugar: Silicon dopant imaging by dissipation force microscopy, Appl. Phys. Lett. **75** (1999) 2785–2787

15.43 Y. Liu, P. Grütter: Magnetic dissipation force microscopy studies of magnetic materials, J. Appl. Phys. **83** (1998) 7333–7338

15.44 J. Tamayo, R. García: Effects of elastic and inelastic interactions on phase contrast images in tapping-mode scanning force microscopy, Appl. Phys. Lett. **71** (1997) 2394–2396

15.45 J. P. Cleveland, B. Anczykowski, A. E. Schmid, V. B. Elings: Energy dissipation in tapping-mode atomic force microscopy, Appl. Phys. Lett. **72** (1998) 2613–2615

15.46 R. García, J. Tamayo, M. Calleja, F. García: Phase contrast in tapping-mode scanning force microscopy, Appl. Phys. A **66** (1998) 309–312

15.47 B. Anczykowski, B. Gotsmann, H. Fuchs, J. P. Cleveland, V. B. Elings: How to measure energy dissipation in dynamic mode atomic force microscopy, Appl. Surf. Sci. **140** (1999) 376–382

15.48 N. Sasaki, M. Tsukada: Effect of microscopic nonconservative process on noncontact atomic force microscopy, Jpn. J. Appl. Phys. **39** (2000) L1334–L1337

15.49 R. Lüthi, E. Meyer, M. Bammerlin, A. Baratoff, L. Howald, C. Gerber, H.-J Güntherodt: Ultrahigh vacuum atomic force microscopy: true atomic resolution, Surf. Rev. Lett. **4** (1997) 1025–1029

15.50 R. Bennewitz, A. S. Foster, L. N. Kantorovich, M. Bammerlin, Ch. Loppacher, S. Schär, M. Guggisberg, E. Meyer, A. L. Shluger: Atomically resolved edges and kinks of NaCl islands on Cu(111): experiment and theory, Phys. Rev. B **62** (2000) 2074–2084

15.51 J. Israelachvili: *Intermolecular and Surface Forces* (Academic, London 1992)

15.52 U. Rabe, J. Turner, W. Arnold: Analysis of the high-frequency response of atomic force microscope cantilevers, Appl. Phys. A **66** (1998) 277–282

15.53 T. R. Rodríguez, R. García: Tip motion in amplitude modulation (tapping-mode) atomic-force microscopy: Comparison between continuous and point-mass models, Appl. Phys. Lett. **80** (2002) 1646–1648

15.54 J. Tamayo, R. García: Relationship between phase shift and energy dissipation in tapping-mode scanning force microscopy, Appl. Phys. Lett. **73** (1998) 2926–2928

15.55 R. García, J. Tamayo, A. San Paulo: Phase contrast and surface energy hysteresis in tapping mode scanning force microsopy, Surf. Interface Anal. **27** (1999) 312–316

15.56 S. N. Magonov, V. B. Elings, M. H. Whangbo: Phase imaging and stiffness in tapping-mode atomic force microscopy, Surf. Sci. **375** (1997) L385–L391

15.57 J. P. Pickering, G. J. Vancso: Apparent contrast reversal in tapping mode atomic force microscope images on films of polystyrene-b-polyisoprene-b-polystyrene, Polymer Bull. **40** (1998) 549–554

15.58 X. Chen, S. L. McGurk, M. C. Davies, C. J. Roberts, K. M. Shakesheff, S. J. B. Tendler, P. M. Williams, J. Davies, A. C. Dwakes, A. Domb: Chemical and morphological analysis of surface enrichment in a biodegradable polymer blend by phase-detection imaging atomic force microscopy, Macromolecules **31** (1998) 2278–2283

15.59 A. Kühle, A. H. Sørensen, J. Bohr: Role of attractive forces in tapping tip force microscopy, J. Appl. Phys. **81** (1997) 6562–6569

15.60 A. Kühle, A. H. Sørensen, J. B. Zandbergen, J. Bohr: Contrast artifacts in tapping tip atomic force microscopy, Appl. Phys. A **66** (1998) 329–332

16. Molecular Recognition Force Microscopy

Atomic force microscopy (AFM), developed in the late eighties to explore atomic details on hard material surfaces, has evolved to an imaging method capable of achieving fine structural details on biological samples. Its particular advantage in biology is that the measurements can be carried out in aqueous and physiological environment, which opens the possibility to study the dynamics of biological processes in vivo. The additional potential of the AFM to measure ultra-low forces at high lateral resolution has paved the way for measuring inter- and intra-molecular forces of bio-molecules on the single molecule level. Molecular recognition studies using AFM open the possibility to detect specific ligand–receptor interaction forces and to observe molecular recognition of a single ligand–receptor pair. Applications include biotin–avidin, antibody–antigen, NTA nitrilotriacetate–hexahistidine 6, and cellular proteins, either isolated or in cell membranes.

The general strategy is to bind ligands to AFM tips and receptors to probe surfaces (or vice versa), respectively. In a force–distance cycle, the tip is first approached towards the surface whereupon a single receptor–ligand complex is formed, due to the specific ligand receptor recognition. During subsequent tip–surface retraction a temporarily increasing force is exerted to the ligand–receptor connection thus reducing its lifetime until the interaction bond breaks at a critical force (unbinding force). Such experiments allow for estimation of affinity, rate constants, and structural data of the binding pocket. Comparing them with values obtained from ensemble-average techniques and binding energies is of particular interest. The dependences of unbinding force on the rate of load increase exerted to the receptor–ligand bond reveal details of the molecular dynamics of the recognition process and energy landscapes. Similar experimental strategies were also used for studying intra-molecular force properties of polymers and unfolding–refolding kinetics of filamentous proteins. Recognition imaging, developed by combing dynamic force microscopy with force spectroscopy, allows for localization of receptor sites on surfaces with nanometer positional accuracy.

16.1 **Ligand Tip Chemistry** 476
16.2 **Fixation of Receptors to Probe Surfaces** 478
16.3 **Single-Molecule Recognition Force Detection** 479
16.4 **Principles of Molecular Recognition Force Spectroscopy** 482
16.5 **Recognition Force Spectroscopy: From Isolated Molecules to Biological Membranes** 484
16.5.1 Forces, Energies, and Kinetic Rates 484
16.5.2 Complex Bonds and Energy Landscapes 486
16.5.3 Live Cells and Membranes 489
16.6 **Recognition Imaging** 489
16.7 **Concluding Remarks** 491
References 492

Macromolecular interactions play a key role in cellular regulation and biological function. Molecular recognition is one of the most important regulatory elements, as it is often the initiating step in reaction pathways and cascades. Since the immune system has evolved recognition molecules that are both highly specific and versatile, antibody-antigen interaction is often used as a paradigm for molecular recognition. However, molecular recognition also plays fundamental roles in the regulation of gene expression, cellular metabolism, and drug design.

Molecular recognition studies emphasize specific biological interactions between receptors and their cognitive ligands. Despite a growing literature on the structure and function of receptor-ligand complexes, it is still not possible to predict reaction kinetics or energetics for any given complex formation, even when the structures are known. Additional insights, in particular about the molecular dynamics within the complex during the association and dissociation process, are needed. The high-end strategy is to probe quantitatively the science that underlies a specific biological recognition in which the chemical structures can be defined at atomic resolution, and genetic or chemical means are available to vary the structure of the interacting partners.

Receptor-ligand complexes are usually formed by a few tens of parallel weak interactions of contacting chemical groups in complementary determining regions supported by framework residues providing structurally conserved scaffolding. Both the complementary determining regions and the framework have a considerable amount of plasticity and flexibility resulting in conformational changes upon association and dissociation (induced fit). In addition to what is known about structure, energies, and kinetic constants, information about conformational movements is required for the understanding of the recognition process. It is likely that insight into the temporal and spatial action of the many weak interactions, in particular the cooparativity of bond formation, is the key to understanding receptor-ligand recognition.

For this, experiments on the single-molecule level at time scales typical for receptor-ligand complex formation and dissociation appear to be required. The potential of the atomic force microscope (AFM) [16.1] to measure ultralow forces at high lateral resolution has paved the way for single-molecule recognition force microscopy studies. The particular advantage of AFM in biology is that the measurements can be carried out in aqueous and physiological environments, which opens the possibility for studying biological processes in vivo. The methodology for investigating the molecular dynamics of receptor-ligand interactions described in this chapter, *Molecular Recognition Force Microscopy* (MRFM) [16.2–4], was developed from scanning probe microscopy (SPM) [16.1]. A force is exerted on a receptor-ligand complex, and the dissociation process is followed over time. Dynamic aspects of recognition are addressed in force spectroscopy (FS) experiments, in which distinct force-time profiles are applied to give insight into the changes in conformations and states during receptor-ligand dissociation. It will be shown that MRFM is a versatile tool to explore kinetic and structural details of receptor-ligand recognition.

16.1 Ligand Tip Chemistry

In MRFM experiments, the binding of ligands on AFM tips to surface-bound receptors (or vice versa) is studied by applying a force to the receptor-ligand complex that reduces its lifetime until the bond breaks at a measurable unbinding force. This requires a careful AFM tip sensor design, including tight attachment of the ligands to the tip surface. In the first pioneering demonstrations of single-molecule recognition force measurements [16.2, 3], strong physical adsorption of bovine serum albumin (BSA) was used to directly coat the tip [16.3] or, alternatively, a glass bead glued to it [16.2]. This physisorbed protein layer may then serve as a functional matrix for the biochemical modification with chemically active ligands (Fig. 16.1). In spite of the large number of probe molecules on the tip ($10^3–10^4/\mu m^2$), the low fraction of properly oriented molecules, or internal blocks of most reactive sites (see Fig. 16.1), allowed the measurement of single receptor-ligand unbinding forces. Nevertheless, parallel breakage of multiple bonds was predominately observed with this configuration.

For measuring interactions between isolated receptor-ligand pairs, strictly defined conditions need to be fulfilled. Covalently coupling ligands to gold-coated tip surfaces via freely accessible SH-groups guarantees a sufficiently stable attachment, because these bonds are about ten times stronger than typical ligand-receptor interactions [16.5]. This chemistry has been used to detect the forces between complementary DNA strands [16.6] and single nucleotides [16.7]. Self-assembled monolayers of dithio-bis(succinimidylundecanoate) were formed to enable covalent coupling of biomolecules via amines [16.8], and were used to study the binding strength between cell adhesion proteoglycans [16.9] and between biotin-directed IgG antibodies and biotin [16.10]. A vectorial orientation of Fab molecules on gold tips was achieved by a firsthand site-directed chemical binding via their SH-groups [16.11], without the need of additional linkers. For this, antibodies were

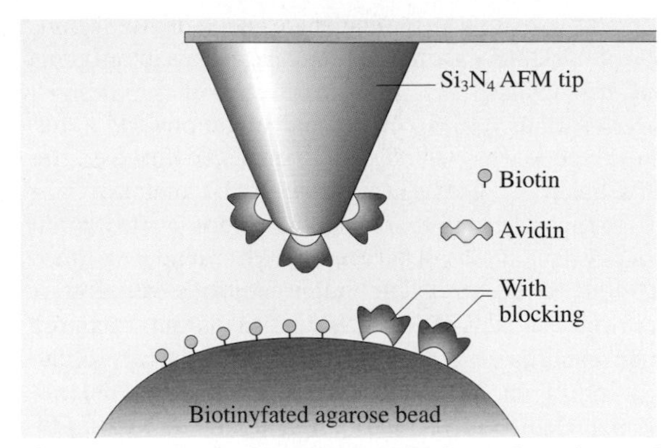

Fig. 16.1 Avidin-functionalized AFM tip. A dense layer of biotinylated BSA was adsorbed to the tip and subsequently saturated with avidin. The biotinylated agarose bead opposing the tip also contained a high surface density of reactive sites, which were, therefore, partly blocked with avidin to achieve single-molecule binding events (after [16.3])

digested with papain and subsequently purified to generate Fab fragments with freely accessible SH-groups in the hinge region.

Gold surfaces provide a unique and selective affinity for thiols, although the adhesion strength of the formed bond is weaker than that of other covalent bonds [16.5]. Since all commercially available AFM tips are etched of silicon nitride or silicon oxide material, deposition of a gold layer onto the tip surface is required prior to using this chemistry. Therefore, designing a sensor with covalent attachments of the biomolecules to the silicon surface may be more straightforward. Amine-functionalization procedures, a strategy widely used in surface bio-chemistry, were applied using ethanolamine [16.4, 12] and various silanization methods [16.13–15] as a first step in thoroughly developed surface anchoring protocols suitable for single molecule experiments. Since the amine density achieved in these procedures often determines the number of ligands on the tip that can specifically bind to the receptors on the surface, it has to be sufficiently low to guarantee single molecule detection [16.4, 12]. One molecule per tip apex ($\sim 5-20$ nm) corresponds to a macroscopic density of about 500 molecules per μm^2. A most striking example of a single ligand molecule tip was realized by gluing a single nanotube to the cantilever, thus minimizing the contact between tip and surface with a chemically well-defined and extremely small tip apex [16.16]. The ligand molecule was reacted to an activated chemical site at the open end of the nanotube.

In a number of laboratories, a distensible and flexible linker was used to space the ligand molecule from the tip surface (e.g., [16.4, 14]) (Fig. 16.2). At a given low number of spacer molecules per tip, the ligand can freely orient and diffuse within a certain volume provided by the length of the tether to achieve unconstrained binding to its receptor. The unbinding process occurs with little torque, and the ligand molecule escapes the danger of being squeezed between the tip and the surface. It also opens the possibility of site-directed coupling for a defined orientation of the ligand relative to the receptor at receptor-ligand unbinding. As a crosslinking element, poly(ethylene glycol) (PEG), a water soluble, nontoxic polymer with a wide range of applications in surface technology and clinical research, was often used [16.17]. PEG is known to prevent surface adsorption of proteins and lipid structures and therefore appears ideally suited for this purpose. Gluteraldehyde [16.13] and DNA [16.6] were also successfully applied in recognition force studies. Crosslinker lengths, ideally arriving at a good compromise between high tip molecule mobility and narrow lateral resolution of the target recognition site, varied from 2 to 100 nm.

For coupling to the tip surface and to the ligands, respectively, the crosslinker is mostly equipped with two different functional ends, e.g., an N-hydroxy-

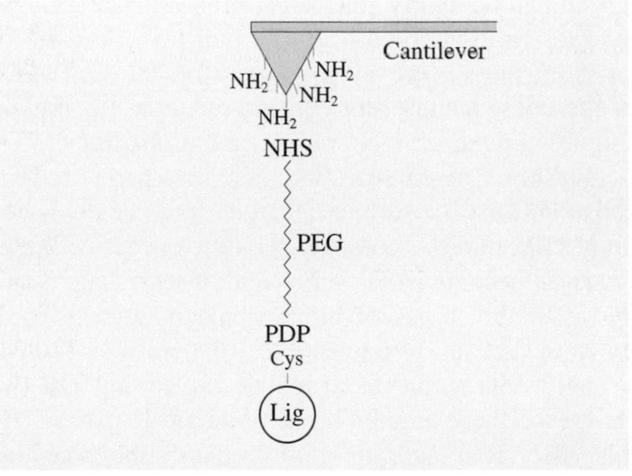

Fig. 16.2 Linkage of ligands to AFM tips. Ligands were covalently coupled to AFM tips via a heterobifunctional polyethylene glycol (PEG) derivative of 8 nm length. Silicon tips were first functionalized with ethanolamine ($NH_2-C_2H_4OH \cdot HCl$). Then, the NHS (N-hydroxysuccinimide) end of the PEG linker was covalently bound to amines on the tip surface before ligands were attached to the PDP (Pyridoyldithiopropionate) end via a free thiol or cysteine

succinimidyl (NHS) residue on the one end is reactive to amines on the tip and a 2-pyridyldithiopropionyl (PDP) [16.18] or a vinyl sulfon [16.19] residue on the other end, can be covalently bound to thiols of ligands (Fig. 16.2). This sulfur chemistry is highly advantageous, since it is very reactive and renders site-directed coupling possible. However, free thiols are hardly available on native ligands and must, therefore, be generated.

Different strategies were used to achieve this goal. Lysins of protein-ligands were derived from the short heterobifunctional linker N-succinnimidyl-3-(S-acethylthio)propionate (SATP) and subsequent deprotection with NH_2OH led to reactive SH groups [16.18]. Since it is very delicate to react distinct lysins with this method, the coupling to the crosslinker is often not specifically site directed. Several protocols are commercially available (Pierce, Rockford, IL) to generate active antibody fragments with free cysteines. Half-antibodies are produced by cleaving the two disulfides in the central region of the heavy chain using 2-mercaptoethylamine HCl [16.20] and Fab fragments are generated from digestion using papain [16.11]. The most elegant methods are to mutate a cysteine into the primary sequence of proteins and to append a thiol to the end of synthesized DNA strands [16.6], because they allow for a defined, sequence-specific coupling of the ligand to the crosslinker.

A nice alternative for a most common, strong but noncovalent site-directed anchor on the spacer has been introduced recently. The binding strength of the NTA (nitrilotriacetate) - His_6 (histidine 6) system, routinely used on chromatographic and biosensor matrices for the binding of recombinant proteins to which a His_6 tag is appended to the primary sequence, was found to be significantly larger than typical values of other ligand-receptor systems [16.21–23]. Therefore, a crosslinker containing an NTA residue is ideally suited for coupling a recombinant ligand carrying a His_6 in its sequence to the AFM tip. This general, site-directed, and oriented coupling strategy also allows rigid and fast control of the specificity of ligand-receptor recognition by using Ni^{++} as a molecular switch of the NTA-His_6 bond.

16.2 Fixation of Receptors to Probe Surfaces

For the recognition by ligands on the AFM tip, receptors should be tightly attached to probe surfaces. Loose receptor fixation could lead to a pull-off of the receptor from the surface by the ligand on the tip, which would consequently block ligand-receptor recognition and obscure the recognition force experiments.

Freshly cleaved muscovite mica is a perfectly pure and atomically flat surface and, therefore, ideally suited for MRFM studies. Moreover, the strong negative charge of mica accomplishes very tight electrostatic binding of certain types of biomolecules. Some receptor proteins such as lysozyme [16.20] or avidin [16.24] strongly adhere to mica due to the strong positive charge of these highly basic proteins at $pH < 8$. In this case, it is safe to purely adsorb the receptors from the solution, since the unspecific attachment to the surface is sufficiently strong for recognition force experiments. Nucleic acids are firmly bound to mica via Zn^{2+}, Ni^{2+}, or Mg^{2+} bridges [16.25]. Electrostatic interaction has also been used to adsorb the strongly acidic sarcoplasmic domain of the Ca^{2+} release channel via Ca^{2+}-bridges to probe the cytoplasmic surface [16.26]. Similarly, protein crystals and bacterial layers have been deposited onto mica in defined orientations with an appropriate choice of buffer conditions [16.27, 28].

In many cases, however, one cannot rely on electrostatic binding toward surfaces. Ideally, water-soluble receptors like globular antigenic proteins or extracellular protein chimeras are then anchored covalently. When glass, silicon, or mica is used as the probe surface, exactly the same surface chemistry described for the silicon AFM tips applies (see above). The number of reactive SiOH groups of the relatively chemically inert mica can be optionally increased by water plasma treatment [16.29]. Frequently, crosslinkers were also used on probe surfaces to provide the receptors with motional freedom, too [16.4]. Protein binding to photo-crosslinkers thereby enables rigid temporal control of the reaction [16.30].

A major limitation of the silicon chemistry is the impossibility of reaching a high surface density of functional sites ($< 1{,}000/\mu m^2$). By comparison, a monolayer of streptavidin would consist of 60,000 molecules per μm^2 and a phospholipid monolayer would consist of 1.7×10^6 molecules per μm^2. The latter density is also achieved by chemisorption of alkanethiols to gold. Tightly bound, ordered, self-assembled monolayers are formed on ultraflat gold surfaces and display excellent probes for AFM [16.10]. Alkanethiols can carry reactive groups at the free end that allow for covalent attachment of biomolecules [16.10, 31] (Fig. 16.3).

A recently developed anchoring strategy to gold uses dithio-phospholipids consisting of a propyldithio group at their hydrophobic end and either a phosphocholine head group (as host lipid) or a phosphethanolamine head group [16.32]. The latter head group is chemically reactive and was derived from a long chain biotin for the molecular recognition of streptavidin molecules in an initial study [16.32]. This phospholipid layer closely mimicks a cell surface and has been optimized by nature to afford little nonspecific adsorption. Additionally, it can be spread as an insoluble monolayer at an air-water interface. Thereby, the ratio of functionalized thio-lipids in host thio-lipids accurately defines the surface density of bio-reactive sites in the monolayer. Subsequent transfer onto gold substrates leads to covalent and, therefore, tight attachment of the monolayer, and thus it is suited for recognition force studies.

Immobilization of cells mainly depends on the type of cell. Cells with adherent growth are readily usable for MRFM, whereas cells that grow in solution without contact to a surface have to be adsorbed. Various protocols for a tight cell anchoring are available. The easiest way is to grow the cells directly on glass or other surfaces in their cell culture medium [16.33]. Various adhesive coatings like Cell-Tak [16.34], gelatin, and poly-lysin increase the strength of adhesion and/or display appropriate surfaces to immobilize spherical cells. Other hydrophic surfaces like gold or carbon are suitable matrices as well [16.35]. Covalent binding of cells to surfaces can be accomplished by using PEG crosslinkers as the one described for the tip chemistry, since they react with free thiols on the cell surface [16.34]. Alternatively, PEG crosslinkers carrying a fatty acid penetrate into the interior of the cell membrane, which guarantees a sufficiently strong fixation without interference with membrane proteins [16.34].

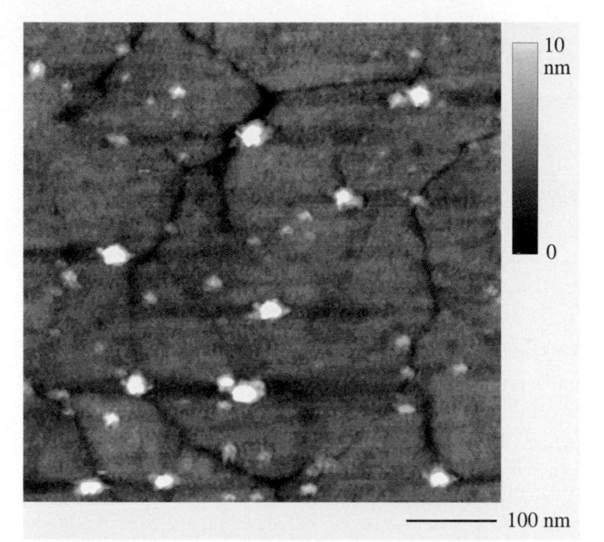

Fig. 16.3 AFM image of hisRNAP molecules specifically bound to nickel-NTA domains on a functionalized gold surface. Alkanethiols terminated with ethylene glycol groups to resist unspecific protein adsorption served as host matrices and were doped with 10% nickel-NTA alkenthiols. The sample was prepared to achieve full monolayer coverage. Ten individual hisRNAP molecules can be clearly seen bound to the surface. The more abundant, smaller, lower features are NTA islands with no bound molecules. The underlying morphology of the gold can also be distinguished (after [16.31])

16.3 Single-Molecule Recognition Force Detection

Quantification of recognition forces goes back to ensemble techniques such as shear flow detachment measurements (SFD) [16.36] and the surface force apparatus (SFA) [16.37]. In SFD, receptors are fixed to a surface to which beads or cells containing ligands attach via their specific bonds. A fluid shear stress is applied to the surface-bound particles by a flow of given velocity that disrupts the ligand-receptor-bonds at a critical force. However, the unbinding force per single bond can only be estimated, because the calculation of the net force on the particle is complicated by the stress distribution and the number of bonds per particle is assumed from the geometry of the contact area.

SFA directly measures the vertical forces between two macroscopic surfaces containing receptors and ligands, respectively, using a spring as a force prober and interferometry for detection. Although SFA does not have a great sensitivity in measuring absolute forces and requires assumptions to be made for down-scaling to

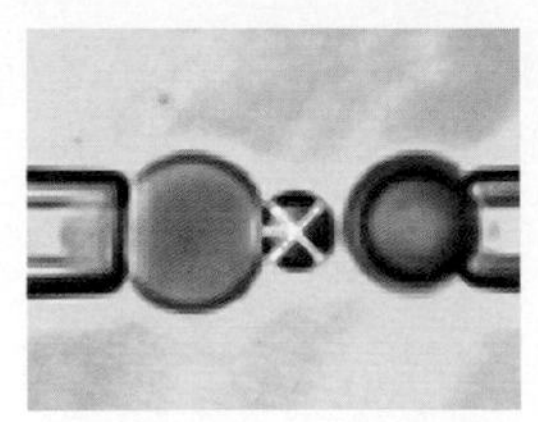

Fig. 16.4 Experimental setup of the biomembrane force probe (BFP). The spring in the BFP is a pressurized membrane capsule. Its spring constant is set by the membrane tension, which is controlled by micropipette suction. The BFP tip, a glass microbead of 1–2 μm diameter, was chemically glued to the membrane. Operated on the stage of an inverted microscope, the BFP (on the *left*) was kept stationary and the microbead test surface (on the *right*) was translated to/from contact with the BFP tip by precision piezo control (after [16.38])

single bonds, its z-resolution is highly accurate (better than 1 Å) over a long range. Adhesion and compression forces, as well as rapid transient effects can be followed in real time.

The bio force probe (BFP) technique (for a recent paper see [16.38]) (Fig. 16.4) uses soft membranes as the force transducer, rather than a mechanical spring. A single cell or giant vesicle is held by suction at the tip of a glass micropipette and its spring constants are adjusted by the aspiration pressure and can be varied over several orders of magnitude. Interaction forces with opposing cells, vesicles, or functionalized beads are obtained from the micromechanical analysis of the deformations of the membranes and/or via interference contrast microscopy. A force range from 0.1 pN to 1 nN is accessible with high force sensitivity down to the single-bond level.

In optical tweezers (OT) small particles are manipulated by optical traps [16.39]. Three-dimensional light intensity gradients of a focused laser beam are used to pull or push the particles against another one. Particle displacements can be followed with nanometer resolution. Movements of single molecular motors [16.40] and force-extension profiles of single extensible molecules such as titin [16.41] and DNA [16.42] were measured. Defined, force-controlled twisting of DNA using rotating, magnetically manipulated particles gave even further insights into DNA viscoelastic properties [16.43].

The atomic force microscope (AFM) is the force measuring method with the smallest-sized force sensor and therefore achieves the highest lateral resolution. Radii of commercially available AFM tips vary between 2 and 50 nm. In contrast, the sensing particles in SFD, BFP, and OT are in the 1 to 10 μm range, and the surfaces used in SFA exceed millimeter extensions. The small apex of the AFM tip allows for the probing of single biomolecules. Originally designed to explore the topography of rigorous surfaces with atomic resolution [16.1] in area scans, the AFM was used in biology to image the secondary structures of nucleotides [16.25], as well as polar lipid head groups and subunits of proteins embedded in membranes in liquid environments [16.27, 28].

The additional potential of the AFM to detect ultralow forces has been successfully applied to the detection of interaction forces of single receptor-ligand pairs [16.2–4]. These forces are measured in so called force-distance cycles using a ligand-carrying tip mounted to a cantilever and a probe surface with receptors firmly attached. A force-distance cycle, where a distensible tether was used to couple ligands to tips and receptors to surfaces is shown in Fig. 16.5 (from [16.4]). At a fixed lateral position the tip approaches the probe surface and is subsequently retracted, and the cantilever deflection Δx is measured in dependence of the tip-surface separation Δz. Cantilever deflections are detected by reflecting a laser beam off the back of the cantilever onto a split photodiode. Bending of the cantilever causes a change in the position where the laser beam strikes the photodiode and, therefore, a change in the output voltage of the photodiode.

The force F acting on the cantilever directly relates to the cantilever deflection Δx according to Hook's law $F = k\Delta x$, where k is the cantilever spring constant. During tip-surface approach (trace, and line 1 to 5), the cantilever deflection remains zero far away from the surface (1 to 4), because there is no detectable tip-surface interaction. Upon tip-surface contact (4), the cantilever bends upward (4 to 5), consistent with a repulsive force that increases linearly with z the harder the tip is pushed into the surface. Subsequent tip-surface retraction (retrace, 5 to 7) first leads to relaxation of the cantilever bending and the repulsive force drops back down to zero (5 to 4).

When ligand-receptor binding has occurred, the cantilever bends downward upon retraction, reflecting an attractive force (retrace, 4 to 7) that increases with increasing tip-surface separation. Since the cognitive molecules were tethered to the surfaces via flexible and distensible crosslinkers, the shape of the attractive force-distance profile is, in contrast to the repulsive force-distance profile of the contact region

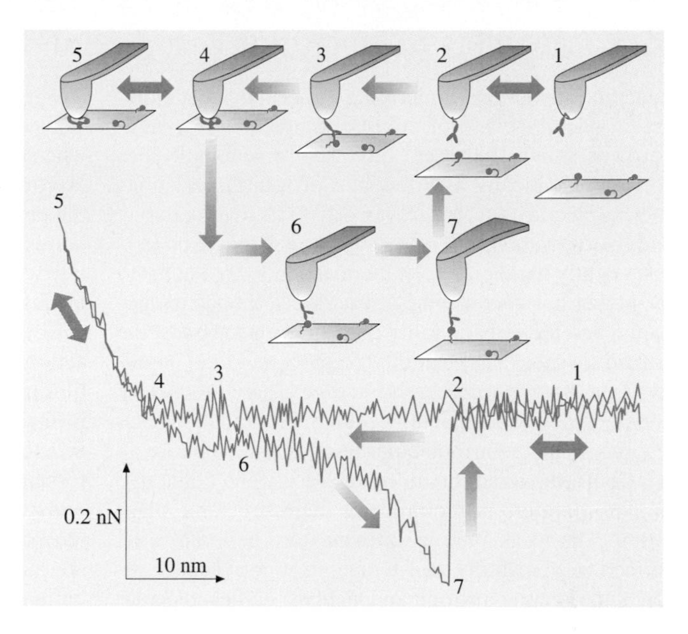

Fig. 16.5 Single-molecule recognition event detected with AFM. Raw data from a force-distance cycle with 100 nm z-amplitude at 1 Hz sweep frequency measured in PBS. Binding of the antibody on the tip to the antigen on the surface during approach (trace points 1 to 5) physically connects tip to probe. This causes a distinct force signal of distinct shape (points 6 to 7) during tip retraction, reflecting extension of the distensible crosslinker-antibody-antigen connection. The force increases until unbinding occurs at an unbinding force of 268 pN (points 7 to 2). (After [16.4])

(4 to 7), nonlinear. Its shape is determined by the elastic properties of the flexible PEG crosslinker [16.17,44] and shows parabolic-like characteristics, which reflects the increase of the spring constant of the crosslinker during extension. The force attractive profile contains the features of a single-molecule recognition event.

The physical connection between the tip and surface sustains the increasing force until the ligand-receptor complex dissociates at a certain critical force (unbinding force, 7), and the cantilever finally jumps back to the resting position (7 to 2). The quantitative force measure of the unbinding force f of a single receptor-ligand pair is directly given by the force at the moment of unbinding (7).

If the ligand on the tip does not form a specific bond with one of the receptors on the probe surface, the recognition event is missing and retrace looks like trace. Also, the specificity of the ligand-receptor binding is usually demonstrated in block experiments. Free ligands are injected in solution so as to block receptor sites on the surfaces. As a further consequence, recognition signals completely disappear, because the receptor sites on the surface are blocked by the ligand of the solution and, thus, prevent recognition by the ligand on the tip.

The force resolution of the AFM is limited by the thermal noise of the cantilever that is determined by its spring constant. According to the equipartition theorem, the cantilever has on average a thermal energy of $0.5\,k_\mathrm{B}T$, therefore, $0.5\,k\Delta z^2 = 0.5\,k_\mathrm{B}T$, where k_B is Boltzmann's constant and T the absolute temperature. Since $F = k\Delta x$, the force sensitivity is given by $\Delta F = (k_\mathrm{B}Tk)^{1/2}$ and is, therefore, better the softer the cantilever. The smallest force that can be detected with commercially available cantilevers is in the few picoNewton range. For thinner levers with low spring constants, the force resolution decreases by an order of magnitude [16.45]. Due to their smaller mass, they have a higher resonance frequency (30–100 kHz in liquid). When deflection is recorded at a typical bandwidth of 10 kHz, only a fraction of the thermal noise is included and fast movements of the tip can still be detected [16.45]. An alternative route for improving the force resolution was found by electronically increasing the apparent damping constant, which results in the reduction of thermal noise [16.46].

Besides the detection of intermolecular forces between receptors and ligands, the AFM also shows great potential in measuring intramolecular force profiles of single molecules. The molecule is clamped in between the tip and the probe surface and its visco-elastic properties are studied in force-distance cycles. The detailed force extension profile of the recognition event shown in Fig. 16.5 contains, for example, the elastic features of the crosslinker (PEG in this particular case) [16.44] with which the ligand is tethered to the tip of the AFM.

16.4 Principles of Molecular Recognition Force Spectroscopy

Specific interactions underlying molecular recognition are a distinct class of highly complementary, non-covalent bonds between biological macromolecules composed typically of a few tens of weak interactions such as electrostatic, polar, van der Waals, hydrophobic, and hydrogen bonds. The energy of each single bond is just slightly higher than the thermal energy $k_{\mathrm{B}}T$. Due to the power law dependence of each weak interaction potential and the directionality of the hydrogen bonds, the attractive forces between the receptors and their cognitive ligands are extremely short-range. Therefore, a close geometrical and chemical fit within the binding pocket is a prerequisite for molecular recognition.

Conformational flexibility of at least one of the two binding partners is required for their structural adaptation. The weak interaction bonds are believed to be formed in a spatially and temporarily correlated fashion, thus creating a strong and highly specific molecular interaction. Recognition binding sites are structurally and chemically unique and selective for distinct receptor/ligand combinations. The overall affinity or binding energy of a receptor-ligand bond of a few tens of $k_{\mathrm{B}}T$ is typically sufficient to effect chemical or physical changes of the binding partners, which often initiate biochemical reaction cascades or cellular metabolic pathways.

The binding energy E_{B}, given by the free energy difference between the bound and the free state, is the common parameter to describe the strength of a bond and can be determined in ensemble average calorimetric experiments. E_{B} determines the ratio of bound complexes, [RL], to the product of free reactants, [R] [L], at equilibrium in solution and is related to the equilibrium dissociation constant K_{D}, the affinity parameter, according to $E_{\mathrm{B}}\alpha - \log(K_{\mathrm{D}})$. At thermodynamic equilibrium, the number of complexes that form per unit of time, k_{on} [R] [L], equals the number of complexes that dissociate per time unit, k_{off} [RL]. The empirical kinetic rate constants, on-rate k_{on} and off-rate k_{off}, are related to the equilibrium dissociation constant K_{D} through $K_{\mathrm{D}} = k_{\mathrm{off}}/k_{\mathrm{on}}$.

In order to get an estimate for the interaction forces f from binding energies E_{B}, the dimension of the binding pocket may be used as the characteristic length dimension l; thus $f = E_{\mathrm{B}}/l$. Typical values of $E_{\mathrm{B}} = 20\,k_{\mathrm{B}}T$ and $l = 0.5\,\mathrm{nm}$ yields $f \sim 170\,\mathrm{pN}$ as an order of magnitude feeling for the strength of a single receptor-ligand bond. Classical mechanics describes the force required to separate interacting molecules as the gradient in energy along the interaction potential. The complexed molecules would, therefore, dissociate when force exceeds the steepest gradient in energy. However, activation barriers, temperature, time scales, and the detailed characteristics of the energy landscape, all known to be essential for the understanding of receptor-ligand dissociation, are not considered in this purely mechanical picture.

Ligand-receptor binding is generally a reversible reaction. Viewed on the single-molecule level, the average lifetime of a ligand-receptor bond, $\tau(0)$, is given by the inverse of the kinetic off-rate, k_{off}, so $\tau(0) = k_{\mathrm{off}}^{-1}$. Therefore, ligands will dissociate from receptors without any force applied to the bond at times larger than $\tau(0)$ (covering a range from milliseconds to days for different receptor-ligand combinations). In contrast, if molecules are pulled faster than $\tau(0)$, the bond will resist and require a force for detachment. The size of the unbinding force may vary over more than one order of magnitude, and even exceed the adiabatic limit given by the steepest energy gradient of the interaction potential if bond breakage occurs faster than diffusive relaxation (nanosecond range for biomolecules in viscous aqueous medium) and friction effects become dominant [16.47]. Therefore, unbinding forces do not resemble unitary values, and the dynamics of the experiment defines the physics underlying the unbinding process.

At the millisecond to second time scale of AFM experiments, thermal impulses govern the unbinding process. In the thermal activation model, the lifetime of a complex in solution is described by a Boltzmann ansatz, $\tau(0) = \tau_{\mathrm{osc}} \exp(E_{\mathrm{b}}/k_{\mathrm{B}}T)$ [16.48], where τ_{osc} is the inverse of the natural oscillation frequency and E_{b} the energy barrier for dissociation. Hence, due to the thermal energy there is a finite probability of overcoming the energy barrier E_{b}, which leads to the separation of the receptor-ligand complex. The mean complex lifetime in solution is, therefore, determined by the ratio of E_{b} over the thermal energy.

A force acting on a binding complex deforms the interaction energy landscape and lowers the activation energy barrier, very similar to the mode of operation of an enzyme (Fig. 16.6). Thus the input of thermal energy by forces acting in this time regime reduces the bond lifetime. The lifetime $\tau(f)$ of a bond loaded with a constant force f is given as $\tau(f) = \tau_{\mathrm{osc}} \exp[(E_{\mathrm{b}} - x_{\beta} f)/k_{\mathrm{B}}T]$ [16.48], x_{β} being interpreted as the distance of the energy barrier E_{b} from the energy minimum along the direction of the

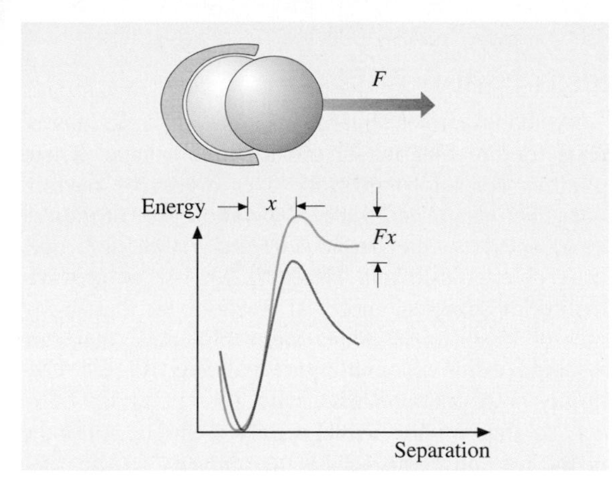

Fig. 16.6 Dissociation over a single sharp energy barrier. Under a constant force, the barrier is linearly decreased by the applied force F, giving rise to a characteristic length scale x that is interpreted as the distance of the energy barrier from the energy minimum along the projection of the force (after [16.50])

applied force. In this model, the energy barrier decreases linearly with the force applied. A detailed physical basis for Bell's theory has been derived in [16.49], where the strength of weak non-covalent bonds was studied in liquid using Kramer's theory for reaction kinetics in liquids under the influence of force.

The lifetime $\tau(f)$ under constant force f compares to the lifetime at zero force, $\tau(0)$, according to $\tau(f) = \tau(0)\exp(-x_\beta f/k_\mathrm{B}T)$ [16.4] for a single sharp energy barrier for which a mono-exponential dependence is characteristic. Using AFM, receptor-ligand unbinding is commonly measured in force-distance cycles (cf. previous chapter), during which the applied force to the complex does not remain constant. Rather, it increases in a complex fashion at a nonlinear rate determined by the pulling velocity, the spring constant of the cantilever, and the force-distance profile of the complexed biomolecules. The main contribution of the thermal activation, however, comes from the part of the force curve, which is close to unbinding. Therefore, an effective force increase or loading rate r can be deduced from $r = \mathrm{d}f/\mathrm{d}t$ = pulling velocity times effective spring constant at the end of the force curve just before unbinding occurs. The approximation of a constant loading rate using force-distance cycles with AFM is sound with a properly defined effective loading rate [16.51].

The dependence of the unbinding force on a linear loading rate in the thermally activated regime was first derived in [16.49] and further described in [16.50]. Force-induced dissociation of receptor-ligand complexes using AFM or BFP can be regarded as an irreversible process, because the two binding partners are further separated after dissociation has occurred. Therefore, rebinding and back-reactions are negligible. The unbinding process itself is of a stochastic nature, and the likelihood of bond survival is expressed in the master equation as time-dependent probability $N(t)$ to be in the bound state under a linearly increasing force $f = rt$, thus $\mathrm{d}N(t) = -k_\mathrm{off}(rt)N(t)$ [16.50].

Rewriting Bell's formula using $\tau(f) = k_\mathrm{off}^{-1}(f)$ relates the kinetic off-rate at a given force f, $k_\mathrm{off}(f)$, to the off-rate at zero force, $k_\mathrm{off}(0)$: $k_\mathrm{off}(f) = k_\mathrm{off}(0)\exp(l_\mathrm{r} f/k_\mathrm{B}T)$. The master equation combined with Bell's formula results in a spectrum of Gaussian-like distributions of unbinding forces (frequency of occurrence versus force) parameterized by the loading rate r [16.50]. Thus the maximum of each force distribution, $f^*(r)$, reflects the most probable force of unbinding for the respective loading rate r. f^* is related to r through $f^* = f_\beta \log_\mathrm{e}(rk_\mathrm{off}^{-1}/f_\beta)$, where the slope is governed by f_β, a force scale set by the ratio of the thermal energy $k_\mathrm{B}T$ to the length scale x_β, which marks the thermally averaged position of the energy barrier along the direction of the force [16.49, 50]. Apparently, the unbinding force f^* scales linearly with the logarithm of the loading rate. For a single barrier, this would give rise to a simple, linear dependence of the force versus log loading rate. In cases where more barriers are involved along the escape path, the curve will follow a sequence of linear regimes, each of which marks a particular barrier [16.38]. Hence, transition from one regime to the other is associated with an abrupt change of slope determined by the characteristic barrier length scale.

In force spectroscopy experiments, the dynamics of pulling on specific receptor-ligand bonds is varied, which leads to detailed structural and kinetic information of the bond breakage, as discussed above and shown in the following chapter on explicit examples. Length scales of energy barriers are obtained from the slope of the spectroscopy plot (force versus loading rate) and their relative heights may be gained from the force shifts [16.38]. Extrapolation to zero forces yields the kinetic off-rate constant for the dissociation of the complex in solution [16.52].

16.5 Recognition Force Spectroscopy: From Isolated Molecules to Biological Membranes

In the early pioneering AFM studies receptor-ligand bond strengths were only measured at one loading rate. Therefore, only a single point in the spectrum of forces, that depend on the dynamics of the experiment, was determined. Avidin–biotin is often regarded as the prototype of receptor-ligand interaction due to its enormously high affinity ($K_D = 10^{-13}$ M) and long bond lifetime ($\tau(0) = 80$ days). Hence, it may not be too surprising that the first realizations of single-molecule recognition force detections were made with biotin and its cognitive receptors, streptavidin [16.2], the succinilated form of avidin, and avidin [16.3]. Unbinding forces of 250–300 pN and 160 pN were found for streptavidin and avidin, respectively. It was also shown [16.53] that the forces vary with the spring constant of the cantilever using the same pulling velocity, which is consistent with the loading rate dependency of the unbinding force discussed above. The recognition forces of various biotin analogs with avidin and streptavidin [16.53], as well as with site-directed streptavidin mutants [16.54] were related to their energies and found to be correlated with the equilibrium binding enthalpy [16.53] and the enthalpic activation barrier [16.54] but independent of free energy parameters. It was suggested that internal energies of the bond breakage, rather than entropic changes were probed by the force measurements [16.54].

16.5.1 Forces, Energies, and Kinetic Rates

In first measurements of inter- and intramolecular forces of the DNA double helix [16.6], complementary single-strand DNA oligonucleotides covalently attached to the AFM tip and the probe surface, respectively, were probed. The multimodal force distributions obtained (three distinct, separated force peaks) were associated with interchain interactions of single pairs of 12, 16, and 20 base sequence length. A long strain single-strand DNA with cohesive ends was sandwiched in between tip and substrate and its intramolecular force pattern revealed elastic properties comparable to polymeric molecules [16.6]. Single molecule interactions between all possible 16 combinations of the four bases were also studied [16.7] and yielded specific binding only when complemetary bases were present on the tip and the probe surface, indicating that AFM is capable of following Watson-Crick base pairing.

Anibody-antigen interactions are of key importance for the function of the immune system. Their affinities are known to vary over orders of magnitudes. For single-molecule recognition AFM studies, the molecules were surface-coupled via flexible and distensible crosslinkers [16.4, 10, 13, 14] to provide them with sufficient motional freedom, so that problems of misorientation and steric hindrance that can obscure specific recognition are avoided. In [16.4, 12], the tips were functionalized with a low antibody density, so that on the average only a single antibody on the tip end could access the antigens on the surface. Hence, isolated single molecular antibody-antigen complexes could be examined. It was observed that the interaction sites of the two Fab fragments of the antibody (the antibody consists of two antigen-active Fab-fragments and one Fc-portion) are able to bind simultaneously and independently with equal binding probability. Single antibody-antigen binding events were also studied with tip-bound antigens [16.10] using an antibody biosensor probe surface [16.13], or single chain antibody-fragments attached to gold via freely accessible sulfhidryl groups [16.14]. In the latter study, the unbinding forces of two mutant single-chain antibodies, varying by an order of magnitude in their affinity, were compared. The mutant of ten times lower affinity (consistent with 12% lower free energy due to the logarithmic dependence of the affinity on free energy) showed 20% lower unbinding force.

All these studies describe detection of intermolecular forces between single receptors and their cognitive ligands. However, the AFM can also be used to characterize single-molecule intramolecular forces, e.g., interactions responsible for stabilizing the conformations and tertiary structures of polymers and proteins. The force-extension profile (cf. also stretching of PEG in Fig. 16.5) of polysaccharide dextran revealed multiple elastic regimes dominated by entropic effects, internal bond angle twistings, and conformational changes [16.55, 56]. The complex quantities of elastic and viscous force contributions upon stretching a single dextran molecule were resolved by resonant energy tracking using dynamic force spectroscopy [16.57]. Titin, a fibrillose muscle protein, contains a large number of repetitive IgG domains that shows a characteristic sawtooth pattern in the force profile, reflecting their sequential unfolding (Fig. 16.7) [16.58]. Unfolding and refolding kinetics and energetics of proteins

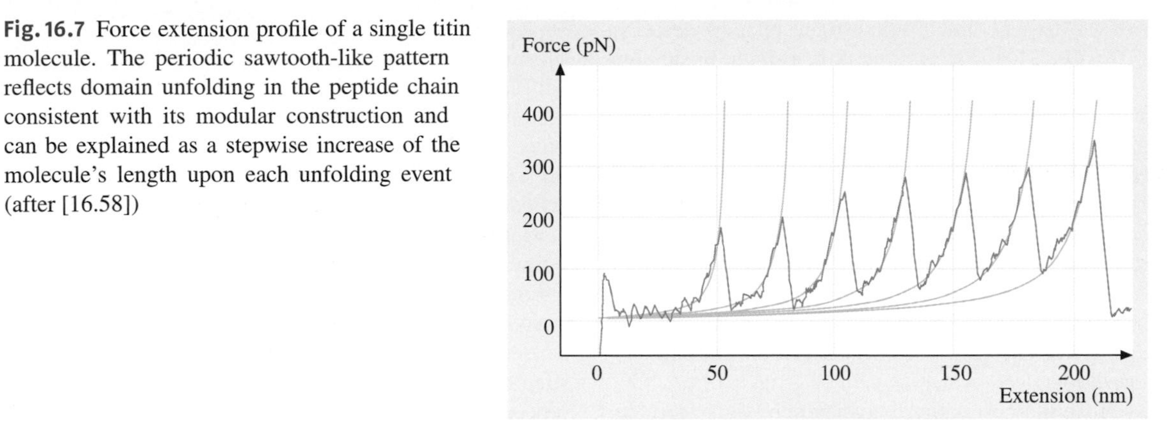

Fig. 16.7 Force extension profile of a single titin molecule. The periodic sawtooth-like pattern reflects domain unfolding in the peptide chain consistent with its modular construction and can be explained as a stepwise increase of the molecule's length upon each unfolding event (after [16.58])

were studied in pull-hold-release cycles [16.59] by carefully handling the molecule between tip and surface. These studies yield crucial information on the relation between structure and function of single molecules.

Besides the investigation of forces, single-molecule recognition force microscopy studies allow for the estimation of association and dissociation rates [16.4, 12, 22, 52, 60, 61], energies [16.38], and structural parameters of the binding pocket [16.4, 12, 15, 60, 61]. For single molecule studies, quantification of the on-rate constant k_{on} for the association of the ligand on the tip to a receptor on the surface requires determination of the interaction time $t_{0.5}$ needed for half-maximal probability of binding. This value is obtained from an experiment in which the encounter duration of the ligand-coated tip on the receptor-containing surface is varied until a range of the unbinding activity from zero to saturation is achieved [16.60]. With the knowledge of the effective ligand concentration c_{eff} on the tip available for receptor interaction, k_{on} is given by $k_{on} = t_{0.5}^{-1} c_{eff}^{-1}$. The effective concentration c_{eff} is described by the effective volume V_{eff} the tip-tethered ligand diffuses about the tip, which yields $c_{eff} = N_A^{-1} V_{eff}^{-1}$, where N_A is the Avogadro's number. V_{eff} is essentially a half-sphere with a radius of the effective tether length, and the kinetic on-rate can, therefore, be calculated from $k_{on} = t_{0.5}^{-1} N_A 2\pi/3r^3$. However, since k_{on} critically depends on the tether length r, only order of magnitude estimates are gained [16.60].

Additional information about the unbinding process is contained in the unbinding force distributions. Not only the maximum f^*, but also its intrinsic width σ increases with increasing loading rate [16.22, 38]. Apparently, at slower loading rates the systems adjust closer to equilibrium, which leads to smaller values of both the force f^* and its variation σ. This is another indication that dynamics is an important issue when studying recognition forces. An effective lifetime $\tau(f)$ of the bond under an applied force f was estimated by the time the cantilever spends in the force window spanned by the standard deviation σ of the f distribution [16.4]. The time the force increases from $f^* - \sigma$ to $f^* + \sigma$ is thus given by $\tau(f) \approx 2\sigma/\mathrm{d}f/\mathrm{d}t$ [16.4]. In an example of the Ni^{2+}-chelating ligand/receptor interaction nitrotrilacetate/hexa-histidin [16.22], the lifetime $\tau(f)$ decreased with increasing pulling force f from 17 ms at 150 pN to 2.5 ms at 194 pN. The data were fitted with Bell's formula, confirming the theoretically predicted exponential lifetime-force relation for the reduction of the lifetime $\tau(f)$ by the applied force f and, thus yielding a lifetime at zero force, $\tau_0 = 15$ s. Direct measurements of lifetimes are only possible by using a force clamp, where a constant adjustable force is applied to the complex via a feedback loop and the time duration of bond survival is detected. This configuration has been developed recently to study the force dependence of the unfolding probability of the filament muscle protein titin [16.62].

Theory predicts (cf. previous chapter) that the unbinding force of specific and reversible ligand-receptor bonds is dependent on the rate of the increasing force [16.47, 49, 63] during force-distance cycles. Indeed, in experiments unbinding forces were found to assume not a unitary value, but rather were dependent on both the pulling velocity [16.52, 60] and the cantilever spring constant [16.2]. The theoretical findings were confirmed by experimental studies and revealed a logarithmic dependence of the unbinding force on the loading rate [16.15, 22, 38, 52, 60, 61], consistent with the exponential lifetime-force relation described above. On a half-logarithmic plot, a single linear slope is characteristic for a single energy barrier probed in the thermally activated regime [16.38]. The same relation was also found for unfolding forces of proteins [16.55]. The slope of the force spectroscopy curve contains the length scale of the energy barrier, which may be connected to the length dimension of the interaction binding pocket [16.60]. The kinetic off-rate k_{off} at zero force is gained from the extrapolation to zero force [16.52, 60]. Combining k_{off} with k_{on} leads to a value for the equilibrium dissociation constant K_{D}, according to $K_{\text{D}} = k_{\text{off}}/k_{\text{on}}$.

A simple correlation between unbinding force and solution kinetics was found for nine different single-chain Fv fragments constructed from point mutations of three unrelated anti-fluorescein antibodies [16.61]. These molecules provide a good model system because they are monovalent and follow a 1:1 binding model without allosteric effects, and the ligand (fluoerscein) is inert and rigid. For each mutant, the affinity K_{D}, the kinetic rate constants k_{on} and k_{off}, and the temperature dependence of k_{off} yielding the activation barrier E_{b}, were determined in solution. In a series of single-molecule AFM recognition events, the loading rate (r) dependence of the unbinding forces (f) was measured. The maxima of the force distributions f^* scaled linearly with $\log r$. Therefore, unbinding was dictated by a single energy barrier along the reaction path (Fig. 16.8). Since the measured unbinding forces of all mutants precisely correlated with k_{off}, the unbinding path followed and the transition state crossed cannot be too dissimilar in spontaneous and forced unbinding. Interestingly, the distance of the energy barrier along the unbinding pathway determined from the force spectroscopy curve was found to be proportional to the barrier height and, thus, most likely includes elastic stretching of the antibody construct during the unbinding process.

16.5.2 Complex Bonds and Energy Landscapes

More complex force spectra were found from unbinding force measurements between complementary DNA strands of different lengths [16.15]. DNA duplexes with 10, 20, and 30 base pairs, were investigated to reveal unbinding forces from 20 to 50 pN of single duplexes, depending on the loading rate and sequence length. The unbinding force of the respective duplexes scaled

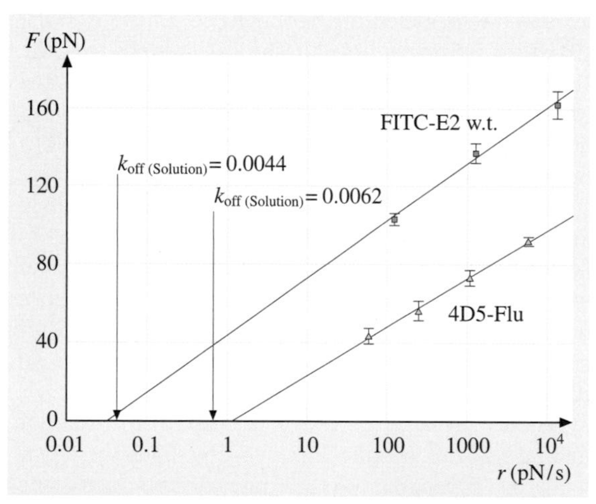

Fig. 16.8 Loading rate dependence of the unbinding force dependence of two anti-fluoresceins. For both FITC-E2 w.t. and 4D5-Flu, a strictly mono-logarithmic dependence was found in the range accessed, indicating that only a single energy barrier was probed. The same energy barrier dominates dissociation without forces applied, because extrapolation to zero force matches kinetic off-rates determined in solution (indicated by *arrow*) (after [16.61])

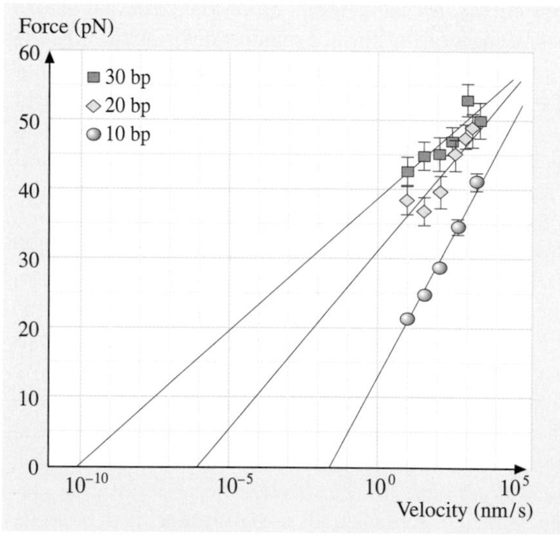

Fig. 16.9 Dependence of the unbinding force between DNA single-strand duplexes on the retract velocity. Besides the mono-logarithmic behavior, the unbinding force scales with the length of the strands, increasing from the 10- to 20- to 30-base-pair duplexes (after [16.15])

with the logarithm of the loading rate, however, each of them with a different slope (Fig. 16.9). The distance of the energy barrier along the separation path and the logarithm of the kinetic off-rate, both obtained from the force spectroscopy plot as parameters describing the energy landscape, were found to be proportional to the number of DNA complexes per duplex. This defined scaling leads to a model of fully cooperative bonds in a series for the unbinding of the base pairs in the duplex, described by one distinct length and time scale for each duplex length, where each single bond contributes a defined increment [16.15]. A temperature study revealed that entropic contributions also play a role in the unbinding between complementary DNA strands [16.64].

Energy landscapes of receptor-ligand bonds in biology can be rugged terrains with more than one prominent energy barrier. The inner barriers are undetectable in solution assays, but can become predominant for certain time scales in force-induced unbinding. Therefore, force spectroscopy appears to be a proper tool to discover hidden activation barriers and explore the energy landscape in greater detail. In initial AFM studies [16.53, 54], it was concluded that the interaction strength of biotin-streptavidin is in the range of 200–300 pN and the strength of biotin-avidin is somewhat lower (~ 150 pN). However, using BFP, the loading rate was varied over eight orders of magnitude (from 10^{-2} to 10^{6} pN/s) and gained a detailed picture of the force spectroscopy curve with forces increasing from 5 to 200 pN (Fig. 16.10) [16.38].

Distinct linear regimes that demonstrate the thermally activated nature of the bond breakage are visible, abrupt changes in slope imply a number of sharp barriers along the dissociation pathway. Above 85 pN, there is a common high strength regime for both biotin-streptavidin and biotin-avidin with a slope of $f_\beta = 34$ pN, locating a barrier deep in the binding pocket at $x_\beta = 0.12$ nm. Below 85 pN, the slopes for biotin-streptavidin and biotin-avidin are different. $f_\beta = 8$ pN for biotin-strepavidin marks a barrier at $x_\beta = 0.5$ nm, whereas the steeper slope $f_\beta = 13–14$ pN between 38 and 85 pN in the biotin-avidin spectrum indicates that its next barrier is located at 0.3 nm. Below 11 pN, the biotin-avidin spectrum exhibits a very low strength regime with a slope of $f_\beta = 1.4$ pN that maps to $x_\beta = 3$ nm. In addition to the map of barrier locations, the logarithmic intercepts found by extrapolation of each linear regime to zero force also yield estimates of the energy differences between the activation barriers (Fig. 16.10).

A meticulous picture of the biotin-avidin bond rupture process with amino acid resolution was obtained

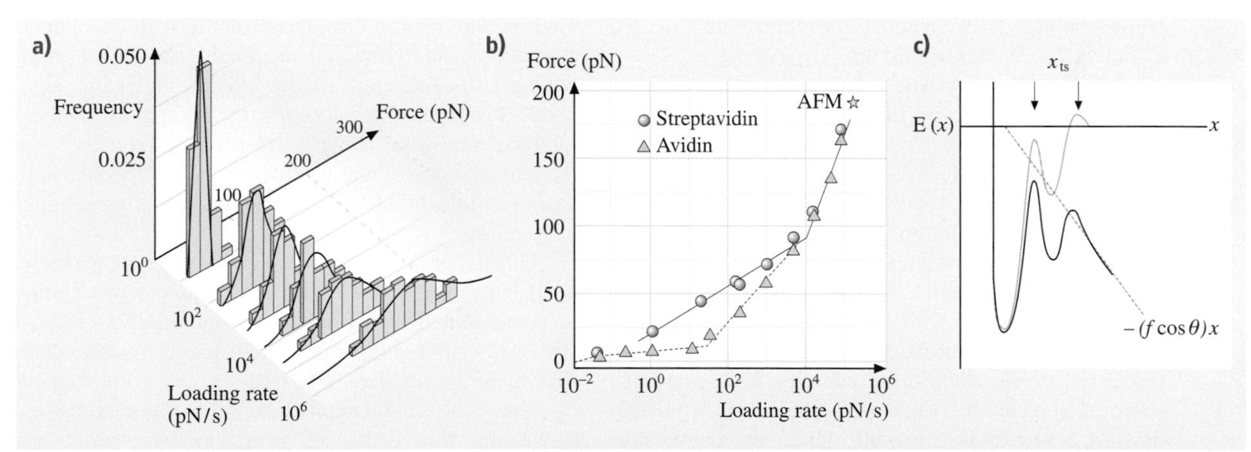

Fig. 16.10a–c Unbinding force distributions and energy landscape of a complex molecular bond. **(a)** Force histograms of single biotin-streptavidin bonds recorded at different loading rates. The shift in peak location and the increase in width with increasing loading rate is clearly demonstrated. **(b)** Dynamic force spectra for biotin-streptavidin (*circles*) and biotin-avidin (*triangles*). The slopes of the linear regimes mark distinct activation barriers along the direction of force. **(c)** Conceptual energy landscape traversed along a reaction coordinate under force under an angle θ to the molecular coordinate. The external force f adds a mechanical potential that tilts the energy landscape and lowers the barriers. The inner barrier starts to dominate when the outer has fallen below it due to the applied force (after [16.38])

with molecular dynamics simulation (MDS) [16.47,63]. Although the time scales of MDS for AFM pulling experiments are orders of magnitudes faster, forces in laboratory speeds were approximated. By reconstructing the interaction potential, the experimentally found transition states were readily identified. Simulations of antibody-antigen unbinding, however, revealed a large heterogeneity of enforced unbinding pathways and a correspondingly large flexibility of the binding pocket region, which exhibited significant induced fit motions [16.65].

The velocity dependence of rupture forces and adhesion probability of the receptor/ligand system P-selectin/P-selectin glycoprotein ligand-1 (PSGL-1) and their relation to kinetic and thermodynamic parameters obtained in ensemble average assays were analyzed in detail in [16.52]. The adhesive interactions of this system play a central immunogenic role in directing the PSGLT-1-containing leukocytes out of the blood stream and into sites of inflammation on cell layers via P-selectin. The high molecular elasticity of the P-selectin/PSGL-1 complex was determined from the force extension profile gained in force-distance cycles and described by the freely jointed chain, a polymer model that yields the Contour length (overall maximum extension), the persistence length (effective segment dimension), and the molecular spring constant. Rupture forces of single complex unbinding increased logarithmically with the loading rate (spring constant times pulling velocity), as expected from theory, and yielded $k_{off} = 0.02\,s^{-1}$ and $l = 0.25$ nm.

Another characteristic of the P-selectin/PSGL-1 relevant to its biological function was found by investigating the adhesion probability, defined as the ratio between the force-distance cycles showing an adhesion event and the total number of cycles, on the pulling velocity. Counterintuitively and in contrast to experiments with antibody-antigen [16.4], cell adhesion proteins [16.9] and cadherins [16.60], the adhesion probability increased with increasing velocities [16.52]. Since the adhesion probability approaches 1.0 at fast velocities, P-selectin/PSGL-1 complex formation must occur as soon as the tip reaches the surface, thus reflecting a fast kinetic on-rate. Upon the increase in force during pulling, the complexes only dissociate under a measurable force when pulled at fast velocities. No force will be detected at low velocities because the complex has already dissociated by the pure thermal energy and/or a force beneath the noise limit. A quantitative analysis revealed a fast kinetic off-rate of $k_{off} = 15 \pm 2\,s^{-1}$. The fast kinetic rates and high chain-like elasticity of the PSGL-1 system guarantee a rapid exchange of the leukocytes on endothelial cell surfaces and therefore provide the basis of the physiological leukocyte *roling* process [16.52].

PSGL-1 recognition to another member of the selectin family, L-selectin, was studied using BFP [16.66]. The loading rate was varied over three orders of magnitude, thereby gaining rupture forces from a few to 200 pN. Plotted on a logarithmic scale of loading rates, the rupture forces reveal two prominent energy barriers along the unbinding pathway. Strengths above 75 pN arise from rapid detachment (< 0.01 s) impeded by an inner barrier that requires Ca^{2+}, whereas strengths below 75 pN occur under slow detachment (> 0.01 s) impeded by an outer barrier, most likely constituted by an array of weak hydrogen bonds. It was speculated that a complex hierarchy of inner and outer barriers is significant in selectin-mediated function [16.66].

Single molecule recognition was also used to investigate the mechanism of the cadherin-mediated adhesion process of vascular endothelial cells. A monolayer of these cells constitutes the major barrier of the body that separates the blood compartment from the extracellular space of tissues. Homophilic cadherin adhesion acting within the cell junctions allows fast dynamic cellular remodeling to change the barrier properties of the cell layers. Recordings of force-distance cycles showed specific recognition events between tip- and surface-bound chimeric VE-cadherin dimers. Measuring the binding activity in dependence of the free Ca^{2+} concentration [16.60] revealed an apparent K_D of 1.15 mM with a Hill coefficient of $n_H = 5.04$, indicating high cooperativity and steep dependency. It might be of physiological relevance that a local drop of free Ca^{2+} in the narrow intercellular cleft weakens intercellular adhesion and is, therefore, involved in facilitating cellular remodeling.

The rather low trans-interaction force of a single bond between two cadherin strand dimers was found to be amplified by complexes of cumulative binding strength, visible as distinct force quanta in the adhesion force distributions. These associates were formed by time-controlled lateral and trans-oligomerization, suggesting that cadherins from opposing cells associate to form complexes and, thus, increase the intercellular binding strength. The rather low values for the kinetic rates and the adhesive binding activity determined as described above, make cadherins ideal candidates for adhesion regulation by cytoskeletal tethering, thereby setting the kinetic rate of lateral self-association [16.60].

16.5.3 Live Cells and Membranes

So far, there have been only a few attempts to apply recognition force microscopy to living cells. Recently, *Lehenkari* et al. [16.67] measured the binding forces between integrin receptors in intact cells and Arg-Gly-Asp (RGD) amino acid sequence containing extracellular matrix protein ligands. Ligands, which had a higher predicted affinity for the receptor, gave greater unbinding forces, and the amino acid sequence/pH/divalent cation dependency of the receptor-ligand interaction examined by AFM was similar to the one observed under bulk measurement conditions using other techniques. In another study [16.68], the AFM was used to detect the adhesive strength between concavalin A (con A) coupled to an AFM tip and Con A receptors on the surface of NIH3T3 fibroblast cells. Crosslinking of receptors led to an increase in adhesion that could be attributed to enhanced cooperativity among adhesion complexes, which has been proposed as a physiological mechanism for modulating cell adhesion.

The sidedness and accessability of protein epitopes of the Na^{2+}/D-glucose co-transporter in intact brush border membrane vesicles was sensed with an AFM tip containing an antibody directed against an a sequence close to the glucose binding site [16.35]. Orientations and conformations of the transporter molecule were found to change during glucose binding and transmembrane transport, as evident from the observation of three distinct states discernable in interaction strength and binding activity. Force spectroscopy experiments between leukocyte function-associated antigen-1 (LFA-1) on the tip and intercellular adhesion molecule-1 (ICAM-1) expressed on cell surfaces [16.69] revealed a bilinear behavior in the force vs. log loading rate curve, marking a steep inner activation barrier and a wide outer activation barrier. Since addition of Mg^{2+}, a co-factor that stabilizes the LFA-1/ICAM-1 interaction, elevated the unbinding force of the complex in the slow loading regime and EDTA suppressed the inner barrier, it was suggested that the equilibrium dissociation constant is regulated by the energetics of the outer barrier, whereas the inner barrier determines the ability to resist an external force.

Forces about ten times higher (3 nN) were achieved when whole cells were used to modify the tip instead of single molecules, as shown in measurements of inter-cell force detections between trophoplasts and uterine pithelium cells [16.70]. In a similar configuration but with defined control of the interaction, de-adhesion forces at the resolution of individual cell adhesion molecules in cell membranes were detected and the expression of a gene was quantitatively linked to the function of its product in cell adhesion [16.71].

16.6 Recognition Imaging

Besides studying ligand-receptor recognition processes, the identification and localization of receptor binding sites on bio-surfaces such as cells and membranes are of particular interest. For this, high resolution imaging must by combined with force detection, so that binding sites can be assigned to structures.

A method for imaging functional groups using chemical force microscopy was developed in [16.72]. Both tip and surface were chemically modified, and the adhesive interactions between tip and surface directly correlated with friction images of the patterned sample surfaces. Simultaneous information for topography and recognition forces on a μm-sized streptavidin pattern was obtained by lateral force mapping on a μm-sized streptavidin pattern with a biotinylated AFM tip. An approach-retract cycle was performed in every pixel of the image, and the information for topography was gained from the contact region, whereas a recognition image was constructed from the recognition forces of the force curve [16.72]. This method was also used to map functional receptors on living cells [16.73] and differentiate red blood cells of different blood groups within a mixed population using AFM tips conjugated with a blood group A specific lectin [16.74].

A first identification and localization of single-molecule antigenic sites was realized by laterally scanning an AFM tip containing antibodies in one dimension across a surface containing a low density of antigens during recording force-distance cycles. The binding probability was determined in dependence of the lateral position, resulting in Gaussian-like profiles [16.4, 12]. The position of the antigen could be determined from the position of the maximum, yielding a positional accuracy of 1.5 nm. The width of the binding profile of 6 nm reflects the dynamical reach of the antibody on the tip, provided by the crosslinker used for coupling the antibody to the tip, and can be considered the lateral resolution obtained by this technique [16.4]. With a similar configuration, height and adhesion force images were simultaneously obtained with resolution approaching the single-molecule level as well [16.75].

The strategies of force mapping, however, lack high lateral resolution [16.72] and/or are much slower in data acquisition [16.4,12,75] than topography imaging, since the frequency of the retract-approach cycles performed in every pixel is limited by hydrodynamic forces in the aqueous solution. In addition, obtaining the force image requires the ligand to be disrupted from the receptor in each retract-approach cycle. For this, the z-amplitude of the retract-approach cycle must be at least 50 nm and, therefore, the ligand on the tip is without access to the receptor on the surface for most of the experiment. The development of shorter cantilevers [16.45], however, increases the speed for force mapping, because the hydrodynamic forces are significantly reduced and the resonance frequency is higher than that of commercially available cantilevers. The short cantilevers were also recently applied to follow the association and dissociation of individual chaperonin proteins GroES to GroEL in real time using dynamic force microscopy topography imaging [16.76].

An imaging method for the mapping of antigenic sites on surfaces was recently developed [16.20] by combining molecular recognition force spectroscopy [16.4] with dynamic force microscopy (DFM) [16.25, 77]. In DFM, the AFM tip is oscillated across a surface and the amplitude reduction arising from tip-surface interactions is held constant by a feedback loop that lifts or lowers the tip according to the detected amplitude signal. Since the tip only intermittently contacts the surface, this technique provides very gentle tip-surface interactions, and the specific interaction of the antibody on the tip with the antigen on the surface can be used to localize antigenic sites for recording recognition images. The principles of this method are described in detail below and in Fig. 16.11. A magnetically coated tip is oscillated by an alternating magnetic field at an ampli-

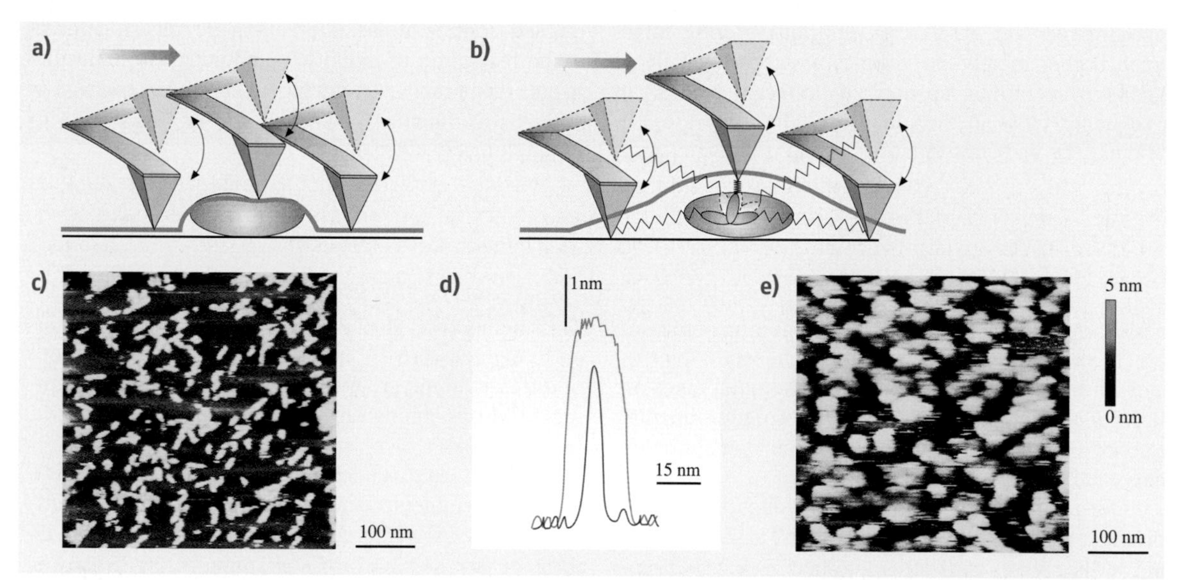

Fig. 16.11a–e Topography and recognition image. **(a)** AFM tip-lysozyme interaction during topography imaging. The *brown line* indicates the height profile obtained from a single lysozyme molecule with the AFM using a bare tip. **(b)** AFM tip-lysozyme interaction during recognition imaging. Half-antibodies are bound to the AFM tip via a flexible tether (*jagged line*) for the recognition of lysozyme on the surface. Imaging results in a height profile, as indicated (*brown line*). **(c)** Topography image. Single lysozyme molecules can be clearly resolved. Sometimes small lysozyme aggregates are observed. Image size was 500 nm. False color bar for heights from 0 (*dark*) to 5 (*bright*) nm. **(d)** Height profiles. Cross-sectional profiles of single lysozyme molecules obtained from the topography (*black line*) and the recognition (*brown line*) image. **(e)** Recognition image. The *bright dots* represent recognition profiles of single lysozyme molecules. Imaging was performed using an AFM tip carrying one half-antibody with access to the antigens on the surface. Conditions were exactly the same as in **(c)**. (After [16.20])

tude of 5 nm while being scanned along the surface. Since the tether has a length of 8 nm, the antibody on the tip always has a chance of recognition when passing an antigenic site, which increases the binding probability enormously. Since the oscillation frequency is more than 100 times faster than typical frequencies in conventional force mapping, the data acquisition rate is much higher.

Half-antibodies were used to provide a monovalent ligand on the tip. The antigen, lysozyme, was tightly adsorbed to mica in conditions that yielded a low surface coverage (for details see [16.20]). A topographic image of this preparation was first recorded in buffer using a bare AFM tip as a control (Fig. 16.11c). Single lysozyme molecules were clearly resolved (Fig. 16.11c). A cross-sectional analysis (Fig. 16.11d, profile in black) reveals that the molecules appear 8 to 12 nm in diameter and 2.0 to 2.5 nm in height. Imaging with a half-antibody tethered to the tip in conditions identical to those used to obtain the topographical image gave strikingly different images (Fig. 16.11e). They differ significantly both in height and diameter compared to the topography image (Fig. 16.11c). Cross-sectional analysis (Fig. 16.11d, trace in red) reveals a height of 3.0 to 3.5 nm and a diameter of 20 to 25 nm. Therefore, profiles obtained from the recognition image appear at least 1 nm higher and 10 nm broader than profiles from the topography image.

The antibody-antigen recognition process during imaging is depicted in Fig. 16.11b. Approaching the antigen in a lateral scan from the left, the antibody on the tip binds to the antigen about 10 nm before the tip end is above the antigenic site (Fig. 16.11b, left tip), due to the flexible tethering provided by the crosslinker. In the bound state, the z-oscillation of the cantilever is further reduced by the attractive force of the crosslinker-antibody-antigen connection, which is acting as a nonlinear spring. Since the AFM detects the z-projection of the force, the amount of the attractive force measured increases when the tip moves further to the right and reaches its maximum just above the position of the antigenic site (Fig. 16.11b, tip in middle). At lateral distances comparable to the length of the antibody-crosslinker connection, the antibody on the tip dissociates from the antibody on the surface and the attractive force goes to zero.

The diameter of cross-sectional profiles obtained from the recognition image (Fig. 16.11d, brown trace) corresponds to about twice the length of the crosslinker (6 nm) plus antibody (6 nm). Increased heights detected in comparison to profiles of the topography image (Fig. 16.11d, black trace) reflect the amplitude reduction owing to antibody-antigen recognition. The maxima of the cross-sectional profiles of the recognition image, as shown in Fig. 16.11d (brown trace), indicate the position of the antigenic site. The accuracy of maximum determination, which in turn reflects the positional accuracy of determining the position of the antigenic site, was 3 nm. The specific nature of the antibody-antigen interaction was tested by injecting free antibody into the liquid cell, so as to block the antigenic sites on the surface, and subsequent images showed a reduction in apparent height.

With this methodology, topography and recognition images can be obtained at the same time and distinct receptor sites in the recognition image can be assigned to structures from the topography image. The described methodology [16.20] is applicable to any ligand and, therefore, it should prove possible to recognize many types of proteins or protein layers and carry out epitope mapping on a nanometer scale on membranes and cells.

16.7 Concluding Remarks

Atomic force microscopy has evolved into the imaging method that yields the greatest structural details on live, biological samples like proteins, nucleotides, membranes and cells in their native, aqueous environment at ambient conditions. Due to its high lateral resolution and sensitive force detection capability, the exciting possibility of measuring inter- and intramolecular forces of biomolecules on the single-molecule level has become possible. The proof-of-principle stage of the pioneering experiments has already developed into a high-end analysis method for exploring kinetic and structural details of interactions underlying protein folding and molecular recognition by taking the advantages of single molecule studies. The information obtained from force spectroscopy includes physical parameters unachievable by other methods and opens new perspectives in exploring the regulation of the dynamics of biological processes such as molecular recognition. With ongoing instrumental developments to increase sensitivity and speed, exciting new fields may also open up in drug and cell screening technologies.

References

16.1 G. Binnig, C.F. Quate, Ch. Gerber: Atomic force microscope, Phys. Rev. Lett. **56** (1986) 930–933

16.2 G.U. Lee, D.A. Kidwell, R.J. Colton: Sensing discrete streptavidin-biotin interactions with atomic force microscopy, Langmuir **10** (1994) 354–357

16.3 E.L. Florin, V.T. Moy, H.E. Gaub: Adhesion forces between individual ligand receptor pairs, Science **264** (1994) 415–417

16.4 P. Hinterdorfer, W. Baumgartner, H.J. Gruber, K. Schilcher, H. Schindler: Detection and localization of individual antibody-antigen recognition events by atomic force microscopy, Proc. Nat. Acad. Sci. USA **93** (1996) 3477–3481

16.5 M. Grandbois, W. Dettmann, M. Benoit, H.E. Gaub: How strong is a covalent bond?, Science **283** (1999) 1727–1730

16.6 G.U. Lee, A.C. Chrisey, J.C. Colton: Direct measurement of the forces between complementary strands of DNA, Science **266** (1994) 771–773

16.7 T. Boland, B.D. Ratner: Direct measurement of hydrogen bonding in DNA nucleotide bases by atomic force microscopy, Proc. Nat. Acad. Sci. USA **92** (1995) 5297–5301

16.8 P. Wagner, M. Hegner, P. Kernen, F. Zaugg, G. Semenza: Covalent immobilization of native biomolecules onto Au(111) via N-hydroxysuccinimide ester functionalized self assembled monolayers for scanning probe microscopy, Biophys. J **70** (1996) 2052–2066

16.9 U. Dammer, O. Popescu, P. Wagner, D. Anselmetti, H.-J. Güntherodt, G.M. Misevic: Binding strength between cell adhesion proteoglycans measured by atomic force microscopy, Science **267** (1995) 1173–1175

16.10 U. Dammer, M. Hegner, D. Anselmetti, P. Wagner, M. Dreier, W. Huber, H.-J. Güntherodt: Specific antigen/antibody interactions measured by force microscopy, Biophys. J. **70** (1996) 2437–2441

16.11 Y. Harada, M. Kuroda, A. Ishida: Specific and quantized antibody-antigen interaction by atomic force microscopy, Langmuir **16** (2000) 708–715

16.12 P. Hinterdorfer, K. Schilcher, W. Baumgartner, H.J. Gruber, H. Schindler: A mechanistic study of the dissociation of individual antibody-antigen pairs by atomic force microscopy, Nanobiology **4** (1998) 39–50

16.13 S. Allen, X. Chen, J. Davies, M.C. Davies, A.C. Dawkes, J.C. Edwards, C.J. Roberts, J. Sefton, S.J.B. Tendler, P.M. Williams: Spatial mapping of specific molecular recognition sites by atomic force microscopy, Biochemistry **36** (1997) 7457–7463

16.14 R. Ros, F. Schwesinger, D. Anselmetti, M. Kubon, R. Schäfer, A. Plückthun, L. Tiefenauer: Antigen binding forces of individually addressed single-chain Fv antibody molecules, Proc. Nat. Acad. Sci. USA **95** (1998) 7402–7405

16.15 T. Strunz, K. Oroszlan, R. Schäfer, H.-J. Güntherodt: Dynamic force spectroscopy of single DNA molecules, Proc. Nat. Acad. Sci. USA **96** (1999) 11 277–11 282

16.16 S.S. Wong, E. Joselevich, A.T. Woolley, C.L. Cheung, C.M. Lieber: Covalently functionalyzed nanotubes as nanometre-sized probes in chemistry and biology, Nature **394** (1998) 52–55

16.17 P. Hinterdorfer, F. Kienberger, A. Raab, H.J. Gruber, W. Baumgartner, G. Kada, C. Riener, S. Wielert-Badt, C. Borken, H. Schindler: Poly-(ethylene glycol): an ideal spacer for molecular recognition force microscopy/spectroscopy, Single Mol. **1** (2000) 99–103

16.18 Th. Haselgrübler, A. Amerstorfer, H. Schindler, H.J. Gruber: Synthesis and applications of a new poly(ethylene glycol) derivative for the crosslinking of amines with thiols, Bioconj. Chem. **6** (1995) 242–248

16.19 C.K. Riener, G. Kada, C. Borken, F. Kienberger, P. Hinterdorfer, H. Schindler, G.J. Schütz, T. Schmidt, C.D. Hahn, H.J. Gruber: Bioconjugation for biospecific detection of single molecules in atomic force microscopy (AFM) and in single dye tracing (SDT), Recent Res. Devel. Bioconj. Chem. **1** (2002) 133–149

16.20 A. Raab, W. Han, D. Badt, S.J. Smith-Gill, S.M. Lindsay, H. Schindler, P. Hinterdorfer: Antibody recognition imaging by force microscopy, Nature Biotech. **17** (1999) 902–905

16.21 M. Conti, G. Falini, B. Samori: How strong is the coordination bond between a histidine tag and Ni-Nitriloacetate? An experiment of mechanochemistry on single molecules, Angew. Chem. **112** (2000) 221–224

16.22 F. Kienberger, G. Kada, H.J. Gruber, V.Ph. Pastushenko, C. Riener, M. Trieb, H.-G. Knaus, H. Schindler, P. Hinterdorfer: Recognition force spectroscopy studies of the NTA-His6 bond, Single Mol. **1** (2000) 59–65

16.23 L. Schmitt, M. Ludwig, H.E. Gaub, R. Tampe: A metal-chelating microscopy tip as a new toolbox for single-molecule experiments by atomic force microscopy, Biophys. J. **78** (2000) 3275–3285

16.24 C. Yuan, A. Chen, P. Kolb, V.T. Moy: Energy landscape of avidin-biotin complexes measured by atomic force microscopy, Biochemistry **39** (2000) 10 219–10 223

16.25 W. Han, S.M. Lindsay, M. Dlakic, R.E. Harrington: Kinked DNA, Nature **386** (1997) 563

16.26 G. Kada, L. Blaney, L.H. Jeyakumar, F. Kienberger, V.Ph. Pastushenko, S. Fleischer, H. Schindler,

F. A. Lai, P. Hinterdorfer: Recognition force microscopy/spectroscopy of ion channels: applications to the skeletal muscle Ca^{2+} release channel (RYR1), Ultramicroscopy **86** (2001) 129–137

16.27 D. J. Müller, W. Baumeister, A. Engel: Controlled unzipping of a bacterial surface layer atomic force microscopy, Proc. Nat. Acad. Sci. USA **96** (1999) 13170–13174

16.28 F. Oesterhelt, D. Oesterhelt, M. Pfeiffer, A. Engle, H. E. Gaub, D. J. Müller: Unfolding pathways of individual bacteriorhodopsins, Science **288** (2000) 143–146

16.29 E. Kiss, C.-G. Gölander: Chemical derivatization of muscovite mica surfaces, Coll. Surf. **49** (1990) 335–342

16.30 S. Karrasch, M. Dolder, F. Schabert, J. Ramsden, A. Engel: Covalent binding of biological samples to solid supports for scanning probe microscopy in buffer solution, Biophys. J. **65** (1993) 2437–2446

16.31 N. H. Thomson, B. L. Smith, N. Almquist, L. Schmitt, M. Kashlev, E. T. Kool, P. K. Hansma: Oriented, active *escherichia coli* RNA polymerase: an atomic force microscopy study, Biophys. J. **76** (1999) 1024–1033

16.32 G. Kada, C. K. Riener, P. Hinterdorfer, F. Kienberger, C. M. Stroh, H. J. Gruber: Dithio-phospholipids for biospecific immobilization of proteins on gold surfaces, Single Mol. **3** (2002) 119–125

16.33 C. Le Grimellec, E. Lesniewska, M. C. Giocondi, E. Finot, V. Vie, J. P. Goudonnet: Imaging of the surface of living cells by low-force contact-mode atomic force microscopy, Biophys. J. **75**(2) (1998) 695–703

16.34 K. Schilcher, P. Hinterdorfer, H. J. Gruber, H. Schindler: A non-invasive method for the tight anchoring of cells for scanning force microscopy, Cell. Biol. Int. **21** (1997) 769–778

16.35 S. Wielert-Badt, P. Hinterdorfer, H. J. Gruber, J.-T. Lin, D. Badt, H. Schindler, R. K.-H. Kinne: Single molecule recognition of protein binding epitopes in brush border membranes by force microscopy, Biophys. J. **82** (2002) 2767–2774

16.36 P. Bongrand, C. Capo, J.-L. Mege, A.-M. Benoliel: Use of hydrodynamic flows to study cell adhesion. In: *Physical Basis of Cell Adhesion*, ed. by P. Bongrand (CRC Press, Boca Raton 1988) pp. 125–156

16.37 J. N. Israelachvili: *Intermolecular and Surface Forces*, 2nd edn. (Academic Press, London 1991)

16.38 R. Merkel, P. Nassoy, A. Leung, K. Ritchie, E. Evans: Energy landscapes of receptor–ligand bonds explored with dynamic force spectroscopy, Nature **397** (1999) 50–53

16.39 A. Askin: Optical trapping and manipulation of neutral particles using lasers, Proc. Nat. Acad. Sci. USA **94** (1997) 4853–4860

16.40 K. Svoboda, C. F. Schmidt, B. J. Schnapp, S. M. Block: Direct observation of kinesin stepping by optical trapping interferometry, Nature **365** (1993) 721–727

16.41 M. S. Z. Kellermayer, S. B. Smith, H. L. Granzier, C. Bustamante: Folding-unfolding transitions in single titin molecules characterized with laser tweezers, Science **276** (1997) 1112–1216

16.42 S. Smith, Y. Cui, C. Bustamante: Overstretching B-DNA: the elastic response of individual double-stranded and single-stranded DNA molecules, Science **271** (1996) 795–799

16.43 T. R. Strick, J. F. Allemend, D. Bensimon, A. Bensimon, V. Croquette: The elasticity of a single supercoiled DNA molecule, Biophys. J. **271** (1996) 1835–1837

16.44 F. Kienberger, V. Ph. Pastushenko, G. Kada, H. J. Gruber, C. Riener, H. Schindler, P. Hinterdorfer: Static and dynamical properties of single poly(ethylene glycol) molecules investigated by force spectroscopy, Single Mol. **1** (2000) 123–128

16.45 B. V. Viani, T. E. Schäffer, A. Chand, M. Rief, H. E. Gaub, P. K. Hansma: Small cantilevers for force spectroscopy of single molecules, J. Appl. Phys. **86** (1999) 2258–2262

16.46 S. Liang, D. Medich, D. M. Czajkowsky, S. Sheng, J.-Y. Yuan, Z. Shao: Thermal noise reduction of mechanical oscillators by actively controlled external dissipative forces, Ultramicroscopy **84** (2000) 119–125

16.47 H. Grubmüller, B. Heymann, P. Tavan: Ligand binding: molecular mechanics calculation of the streptavidin-biotin rupture force, Science **271** (1996) 997–999

16.48 G. I. Bell: Models for the specific adhesion of cells to cells, Science **200** (1978) 618–627

16.49 E. Evans, K. Ritchie: Dynamic strength of molecular adhesion bonds, Biophys. J. **72** (1997) 1541–1555

16.50 T. Strunz, K. Oroszlan, I. Schumakovitch. H.-J. Güntherodt, M. Hegner: Model energy landscapes and the force-induced dissociation of ligand-receptor bonds, Biophys. J. **79** (2000) 1206–1212

16.51 E. Evans, K. Ritchie: Strength of weak bondconnecting flexible polymer chains, Biophys. J. **76** (1999) 2439–2447

16.52 J. Fritz, A. G. Katopidis, F. Kolbinger, D. Anselmetti: Force-mediated kinetics of single P-selectin/ligand complexes observed by atomic force microscopy, Proc. Nat. Acad. Sci. USA **95** (1998) 12 283–12 288

16.53 V. T. Moy, E.-L. Florin, H. E. Gaub: Adhesive forces between ligand and receptor measured by AFM, Science **266** (1994) 257–259

16.54 A. Chilkoti, T. Boland, B. Ratner, P. S. Stayton: The relationship between ligand-binding thermodynamics and protein-ligand interaction forces measured by atomic force microscopy, Biophys. J. **69** (1995) 2125–2130

16.55 M. Rief, F. Oesterhelt, B. Heymann, H. E. Gaub: Single molecule force spectroscopy on polysaccharides by atomic force microscopy, Science **275** (1997) 1295–1297
16.56 P. E. Marzsalek, A. F. Oberhauser, Y.-P. Pang, J. M. Fernandez: Polysaccharide elasticity governed by chair-boat transitions of the glucopyranose ring, Nature **396** (1998) 661–664
16.57 A. D. L. Humphries, J. Tamayo, M. J. Miles: Active quality factor control in liquids for force spectroscopy, Langmuir **16** (2000) 7891–7894
16.58 M. Rief, M. Gautel, F. Oesterhelt, J. M. Fernandez, H. E. Gaub: Reversible unfolding of individual titin immunoglobulin domains by AFM, Science **276** (1997) 1109–1112
16.59 A. F. Oberhauser, P. E. Marzsalek, H. P. Erickson, J. M. Fernandez: The molecular elasticity of the extracellular matrix tenascin, Nature **393** (1998) 181–185
16.60 W. Baumgartner, P. Hinterdorfer, W. Ness, A. Raab, D. Vestweber, H. Schindler, D. Drenckhahn: Cadherin interaction probed by atomic force microscopy, Proc. Nat. Acad. Sci. USA **97** (2000) 4005–4010
16.61 F. Schwesinger, R. Ros, T. Strunz, D. Anselmetti, H.-J. Güntherodt, A. Honegger, L. Jermutus, L. Tiefenauer, A. Plückthun: Unbinding forces of single antibody-antigen complexes correlate with their thermal dissociation rates, Proc. Nat. Acad. Sci. USA **29** (2000) 9972–9977
16.62 A. F. Oberhauser, P. K. Hansma, M. Carrion-Vazquez, J. M. Fernandez: Stepwise unfolding of titin under force-clamp atomic force microscopy, Proc. Nat. Acad. Sci. USA **98** (2001) 468–472
16.63 S. Izraelev, S. Stepaniants, M. Balsera, Y. Oono, K. Schulten: Molecular dynamics study of unbinding of the avidin-biotin complex, Biophys. J. **72** (1997) 1568–1581
16.64 I. Schumakovitch, W. Grange, T. Strunz, P. Bertoncini, H.-J. Güntherodt, M. Hegner: Temperature dependence of unbinding forces between complementary DNA strands, Biophys. J. **82** (2002) 517–521
16.65 B. Heymann, H. Grubmüller: Molecular dynamics force probe simulations of antibody/antigen unbinding: entropic control and non additivity of unbinding forces, Biophys. J. **81** (2001) 1295–1313
16.66 E. Evans, E. Leung, D. Hammer, S. Simon: Chemically distinct transition states govern rapid dissociation of single L-selectin bonds under force, Proc. Nat. Acad. Sci. USA **98** (2001) 3784–3789
16.67 P. P. Lehenkari, M. A. Horton: Single integrin molecule adhesion forces in intact cells measured by atomic force microscopy, Biochem. Biophys. Res. Com. **259** (1999) 645–650
16.68 A. Chen, V. T. Moy: Cross-linking of cell surface receptors enhances cooperativity of molecular adhesion, Biophys. J. **78** (2000) 2814–2820
16.69 X. Zhang, E. Woijcikiewicz, V. T. Moy: Force spectroscopy of the leukocyte function-associated antigen-1/intercellular adhesion molecule-1 interaction, Biophys. J. **83** (2002) 2270–2279
16.70 M. Thie, R. Rospel, W. Dettmann, M. Benoit, M. Ludwig, H. E. Gaub, H. W. Denker: Interactions between trophoblasts and uterine epithelium: monitoring of adhesive forces, Hum. Reprod. **13** (1998) 3211–3219
16.71 M. Benoit, D. Gabriel, G. Gerisch, H. E. Gaub: Discrete interactions in cell adhesion measured by single molecule force spectroscopy, Nature Cell Biol. **2** (2000) 313–317
16.72 M. Ludwig, W. Dettmann, H. E. Gaub: Atomic force microscopy imaging contrast based on molecular recognition, Biophys. J. **72** (1997) 445–448
16.73 P. P. Lehenkari, G. T. Charras, G. T. Nykänen, M. A. Horton: Adapting force microscopy for cell biology, Ultramicroscopy **82** (2000) 289–295
16.74 M. Grandbois, M. Beyer, M. Rief, H. Clausen-Schaumann, H. E. Gaub: Affinity imaging of red blood cells using an atomic force microscope, J. Histochem. Cytochem. **48** (2000) 719–724
16.75 O. H. Willemsen, M. M. E. Snel, K. O. van der Werf, B. G. de Grooth, J. Greve, P. Hinterdorfer, H. J. Gruber, H. Schindler, Y. van Kyook, C. G. Figdor: Simultaneous height and adhesion imaging of antibody antigen interactions by atomic force microscopy, Biophys. J. **57** (1998) 2220–2228
16.76 B. V. Viani, L. I. Pietrasanta, J. B. Thompson, A. Chand, I. C. Gebeshuber, J. H. Kindt, M. Richter, H. G. Hansma, P. K. Hansma: Probing protein-protein interactions in real time, Nature Struct. Biol. **7** (2000) 644–647
16.77 W. Han, S. M. Lindsay, T. Jing: A magnetically driven oscillating probe microscope for operation in liquid, Appl. Phys. Lett. **69** (1996) 1–3

Part C

Nanotribo

Part C Nanotribology and Nanomechanics

17 **Micro/Nanotribology and Materials Characterization Studies Using Scanning Probe Microscopy**
Bharat Bhushan, Columbus, USA

18 **Surface Forces and Nanorheology of Molecularly Thin Films**
Marina Ruths, Åbo, Finland
Alan D. Berman, Los Angeles, USA
Jacob N. Israelachvili, Santa Barbara, USA

19 **Scanning Probe Studies of Nanoscale Adhesion Between Solids in the Presence of Liquids and Monolayer Films**
Robert W. Carpick, Madison, USA
James D. Batteas, Gaithersburg, USA

20 **Friction and Wear on the Atomic Scale**
Enrico Gnecco, Basel, Switzerland
Roland Bennewitz, Montreal, Canada
Oliver Pfeiffer, Basel, Switzerland
Anisoara Socoliuc, Basel, Switzerland
Ernst Meyer, Basel, Switzerland

21 **Nanoscale Mechanical Properties – Measuring Techniques and Applications**
Andrzej J. Kulik, Lausanne, Switzerland
András Kis, Lausanne, Switzerland
Gérard Gremaud, Lausanne, Switzerland
Stefan Hengsberger, Fribourg, Switzerland
Philippe K. Zysset, Wien, Austria
Lásló Forró, Lausanne, Switzerland

22 **Nanomechanical Properties of Solid Surfaces and Thin Films**
Adrian B. Mann, Piscataway, USA

23 **Atomistic Computer Simulations of Nanotribology**
Martin H. Müser, London, Canada
Mark O. Robbins, Baltimore, USA

24 **Mechanics of Biological Nanotechnology**
Rob Phillips, Pasadena, USA
Prashant K. Purohit, Pasadena, USA
Jané Kondev, Waltham, USA

25 **Mechanical Properties of Nanostructures**
Bharat Bhushan, Columbus, USA

17. Micro/Nanotribology and Materials Characterization Studies Using Scanning Probe Microscopy

A sharp AFM/FFM tip sliding on a surface simulates just one asperity contact. However, asperities come in all shapes and sizes. The effect of radius of a single asperity (tip) on the friction/adhesion performance can be studied using tips of different radii. AFM/FFM are used to study the various tribological phenomena, which include surface/roughness, adhesion, friction, scratching, wear, indentation, detection of material transfer, and boundary lubrication.

Directionality in the friction is observed on both micro- and macroscales and results from the surface roughness and surface preparation. Microscale friction is generally found to be smaller than the macrofriction, as there is less plowing contribution in microscale measurements. The mechanism of material removal on the microscale is studied. Evolution of wear has also been studied using AFM. Wear is found to be initiated at nanoscratches. For a sliding interface requiring near-zero friction and wear, contact stresses should be below the hardness of the softer material to minimize plastic deformation, and surfaces should be free of nanoscratches. Wear precursors can be detected at early stages of wear by using surface potential measurements. Detection of material transfer on a nanoscale is possible with AFM. In situ surface characterization of local deformation of materials and thin coatings can be carried out using a tensile stage inside an AFM.

Boundary lubrication studies can be conducted using AFM. Chemically bonded lubricant films and self-assembled monolayers are superior in friction and wear resistance. For chemically bonded lubricant films, the adsorption of water, the formation of meniscus and its change during sliding, viscosity, and surface properties play an important role on the friction, adhesion, and durability of these films. For SAMs, their friction mechanism is explained by a so-called "molecular spring" model.

17.1 **Description of AFM/FFM and Various Measurement Techniques** 499
17.1.1 Surface Roughness and Friction Force Measurements ... 500
17.1.2 Adhesion Measurements ... 502
17.1.3 Scratching, Wear and Fabrication/Machining ... 503
17.1.4 Surface Potential Measurements 503
17.1.5 In Situ Characterization of Local Deformation Studies ... 504
17.1.6 Nanoindentation Measurements 504
17.1.7 Localized Surface Elasticity and Viscoelasticity Mapping ... 505
17.1.8 Boundary Lubrication Measurements ... 507

17.2 **Friction and Adhesion** ... 507
17.2.1 Atomic-Scale Friction ... 507
17.2.2 Microscale Friction ... 507
17.2.3 Directionality Effect on Microfriction 511
17.2.4 Velocity Dependence on Microfriction ... 513
17.2.5 Effect of Tip Radii and Humidity on Adhesion and Friction ... 515
17.2.6 Scale Dependence on Friction ... 518

17.3 **Scratching, Wear, Local Deformation, and Fabrication/Machining** ... 518
17.3.1 Nanoscale Wear ... 518
17.3.2 Microscale Scratching ... 519
17.3.3 Microscale Wear ... 520
17.3.4 In Situ Characterization of Local Deformation ... 524
17.3.5 Nanofabrication/Nanomachining ... 526

17.4 **Indentation** ... 526
17.4.1 Picoindentation ... 526
17.4.2 Nanoscale Indentation ... 527
17.4.3 Localized Surface Elasticity and Viscoelasticity Mapping ... 528

17.5 **Boundary Lubrication** ... 530
17.5.1 Perfluoropolyether Lubricants ... 530
17.5.2 Self-Assembled Monolayers ... 536
17.5.3 Liquid Film Thickness Measurements ... 537

17.6 **Closure** ... 538

References ... 539

The mechanisms and dynamics of the interactions of two contacting solids during relative motion ranging from atomic- to microscale need to be understood in order to develop a fundamental understanding of adhesion, friction, wear, indentation, and lubrication processes. At most solid-solid interfaces of technological relevance contact occurs at many asperities. Consequently, the importance of investigating single asperity contacts in studies of the fundamental micro/nanomechanical and micro/nanotribological properties of surfaces and interfaces has long been recognized. The recent emergence and proliferation of proximal probes, including scanning probe microscopies (the scanning tunneling microscope and the atomic force microscope), the surface force apparatus, and computational techniques for simulating tip-surface interactions and interfacial properties, have allowed systematic investigations of interfacial problems with high resolution, as well as ways and means of modifying and manipulating nanoscale structures. These advances have led to the appearance of the new field of micro/nanotribology, which pertains to experimental and theoretical investigations of interfacial processes on scales ranging from the atomic- and molecular- to the microscale, occurring during adhesion, friction, scratching, wear, indentation, and thin-film lubrication at sliding surfaces [17.1–12].

Micro/nanotribological studies are needed to develop a fundamental understanding of interfacial phenomena on a small scale and study interfacial phenomena in micro/nanostructures used in magnetic storage systems, micro/nanoelectromechanical systems (MEMS/NEMS), and other applications [17.3, 7–9, 13, 14]. Friction and wear of lightly loaded micro/nanocomponents are highly dependent on the surface interactions (few atomic layers). These structures are generally lubricated with molecularly thin films. Micro/nanotribological studies are also valuable in understanding interfacial phenomena in macrostructures and provide a bridge between science and engineering.

The surface force apparatus (SFA), the scanning tunneling microscopes (STM), atomic force and friction force microscopes (AFM and FFM) are widely used in micro/nanotribological studies. Typical operating parameters are compared in Table 17.1. The SFA was developed in 1968 and is commonly employed to study both static and dynamic properties of molecularly thin films sandwiched between two molecularly smooth surfaces. The STM, developed in 1981, allows imaging of electrically conducting surfaces with atomic resolution and has been used for the imaging of clean surfaces, as well as of lubricant molecules. The introduction of the atomic force microscope in 1985 provided a method for measuring ultra-small forces between a probe tip and an engineering (electrically conducting or insulating) surface and has been used for topographical measurements of surfaces on the nanoscale, as well as for adhesion and electrostatic force measurements. Subsequent modifications of the AFM led to the development of the friction

Table 17.1 Comparison of typical operating parameters in SFA, STM, and AFM/FFM used for micro/nanotribological studies

Operating parameter	SFA	STM[a]	AFM/FFM
Radius of mating surface/tip	$\sim$ 10 mm	5–100 nm	5–100 nm
Radius of contact area	10–40 μm	N/A	0.05–0.5 nm
Normal load	10–100 mN	N/A	< 0.1 nN – 500 nN
Sliding velocity	0.001–100 μm/s	0.02–200 μm/s (scan size $\sim$ 1 nm ×1 nm to 125 μm ×125 μm; scan rate < 1–122 Hz)	0.02–200 μm/s (scan size $\sim$ 1 nm ×1 nm to 125 μm ×125 μm; scan rate < 1–122 Hz)
Sample limitations	Typically atomically smooth, optically transparent mica; opaque ceramic, smooth surfaces can also be used	Electrically conducting samples	None

[a] Can only be used for atomic-scale imaging

force microscope (FFM), designed for atomic-scale and microscale studies of friction. This instrument measures forces in the scanning direction. The AFM is also used in investigations of scratching, wear, indentation, detection of transfer of material, boundary lubrication, and fabrication and machining. Meanwhile, significant progress in understanding the fundamental nature of bonding and interactions in materials, combined with advances in computer-based modeling and simulation methods, have allowed theoretical studies of complex interfacial phenomena with high resolution in space and time. Such simulations provide insights into atomic-scale energetics, structure, dynamics, thermodynamics, transport and rheological aspects of tribological processes.

The nature of interactions between two surfaces brought close together and those between two surfaces in contact as they are separated have been studied experimentally with the surface force apparatus. This has led to a basic understanding of the normal forces between surfaces and the way in which these are modified by the presence of a thin liquid or a polymer film. The frictional properties of such systems have been studied by moving the surfaces laterally, and such experiments have provided insights into the molecular-scale operation of lubricants such as thin liquid or polymer films. Complementary to these studies are those in which the AFM or FFM is used to provide a model asperity in contact with a solid or lubricated surface, Fig. 17.1. These experiments have demonstrated that the relationship between friction and surface roughness is not always simple or obvious. AFM studies have also revealed much about the nanoscale nature of intimate contact during wear and indentation.

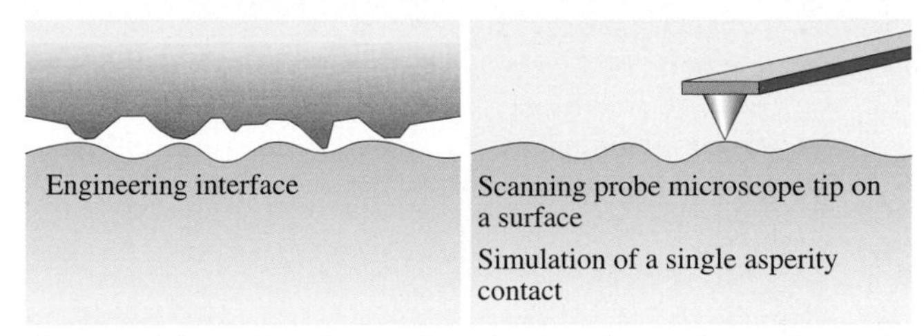

Fig. 17.1 Schematics of an engineering interface and scanning probe microscope tip in contact with an engineering interface

In this chapter, we present a review of significant aspects of micro/nanotribological studies conducted using AFM/FFM.

17.1 Description of AFM/FFM and Various Measurement Techniques

An atomic force microscope (AFM) was developed by *Binnig* et al. in 1985. It is capable of investigating surfaces of scientific and engineering interest on an atomic scale [17.15, 16]. The AFM relies on a scanning technique to produce very high resolution, 3-D images of sample surfaces. It measures ultrasmall forces (less than 1 nN) present between the AFM tip surface mounted on a flexible cantilever beam and a sample surface. These small forces are determined by measuring the motion of a very flexible cantilever beam with an ultra-small mass by a variety of measurement techniques including optical deflection, optical interference, capacitance, and tunneling current. The deflection can be measured to within 0.02 nm, so for a typical cantilever spring constant of 10 N/m, a force as low as 0.2 nN can be detected. To put these numbers in perspective, individual atoms and human hair are typically a fraction of a nanometer and about 75 μm in diameter, respectively, and a drop of water and an eyelash have a mass of about 10 μN and 100 nN, respectively. In the operation of high resolution AFM, the sample is generally scanned, rather than the tip, because any cantilever movement would add vibrations. AFMs are available for measurement of large samples where the tip is scanned and the sample is stationary. To obtain atomic resolution with the AFM, the spring constant of the cantilever should be weaker than the equivalent spring between atoms. A cantilever beam with a spring constant of about 1 N/m or lower is desirable. For high lateral resolution, tips should be as sharp as possible. Tips with a radius ranging from 10–100 nm are commonly available.

A modification to AFM, providing a sensor to measure the lateral force, led to the development of the friction force microscope (FFM), or the lateral force microscope (LFM), designed for atomic-scale and microscale studies of friction [17.3–5, 7, 8, 11, 17–25] and lubrication [17.26–30]. This instrument measures lateral or friction forces (in the plane of sample surface and in the scanning direction). By using a standard or a sharp diamond tip mounted on a stiff cantilever beam, AFM is also used in investigations of scratching and wear [17.6, 9, 11, 22, 31–33], indentation [17.6, 9, 11, 22, 34, 35], and fabrication/machining [17.4, 11, 22].

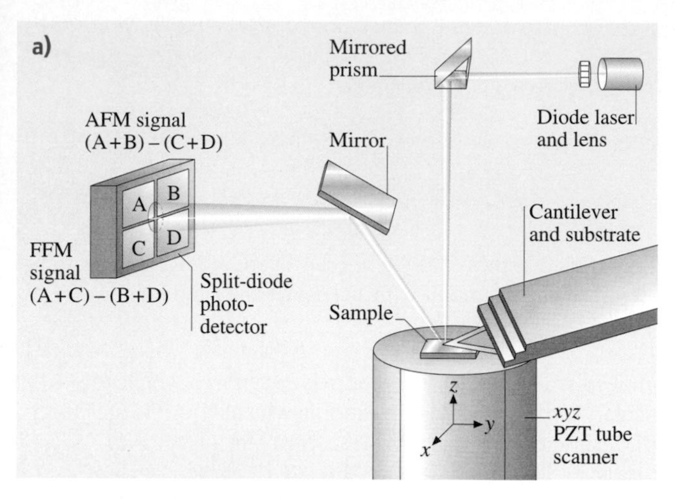

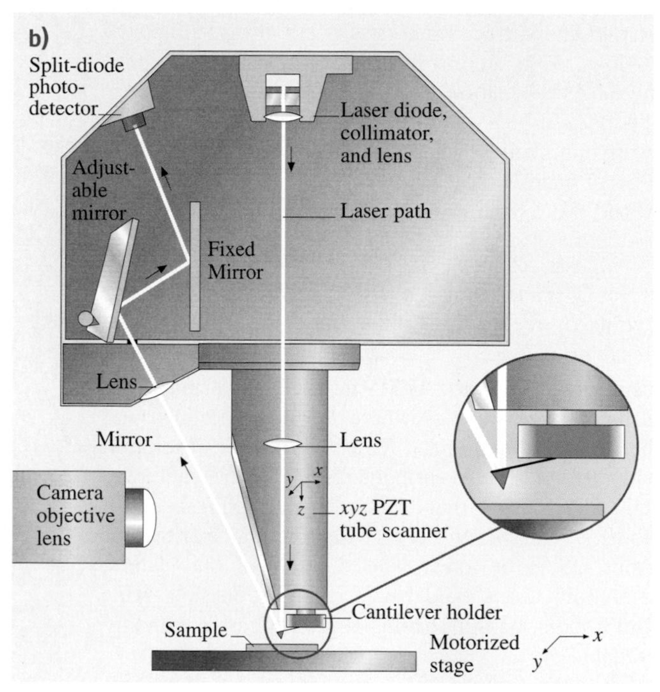

Fig. 17.2a,b Schematics (**a**) of a commercial small sample atomic force microscope/friction force microscope (AFM/FFM) and (**b**) a large sample AFM/FFM

Surface roughness, including atomic-scale imaging, is routinely measured using the AFM, Fig. 17.2. Adhesion, friction, wear, and boundary lubrication at the interface between two solids with and without liquid films are studied using AFM and FFM. Nanomechanical properties are also measured using an AFM. In situ surface characterization of local deformation of materials and thin coatings has been carried out by imaging the sample surfaces using an AFM during tensile deformation using a tensile stage.

17.1.1 Surface Roughness and Friction Force Measurements

Simultaneous measurements of surface roughness and friction force can be made with a commercial AFM/FFM. These instruments are available for the measurement of small samples and large samples. In a small sample AFM, shown in Fig. 17.2a, the sample, generally no larger than 10 mm × 10 mm, is mounted on a PZT tube scanner that consists of separate electrodes to scan precisely the sample in the x-y plane in a raster pattern and move it in the vertical (z) direction. A sharp tip at the free end of a flexible cantilever is brought in contact with the sample. Normal and frictional forces being applied at the tip-sample interface are measured using a laser beam deflection technique. A laser beam from a diode laser is directed by a prism onto the back of a cantilever near its free end, which is tilted downward at about 10° with respect to the horizontal plane. The reflected beam from the vertex of the cantilever is directed through a mirror onto a quad photodetector (split photodetector with four quadrants). The differential signal from the top and bottom photodiodes provides the AFM signal, which is a sensitive measure of the cantilever vertical deflection. Topographic features of the sample cause the tip to deflect in the vertical direction as the sample is scanned under the tip. This tip deflection will change the direction of the reflected laser beam, changing the intensity difference between the top and bottom sets of photodetectors (AFM signal). In the AFM operating mode, called the height mode, for topographic imaging, or for any other operation in which the applied normal force is to be kept a constant, a feedback circuit is used to modulate the voltage applied to the PZT scanner to adjust the height of the PZT, so that the cantilever vertical deflection (given by the intensity difference between the top and bottom detector) will remain constant during scanning. The PZT height variation is thus a direct measure of the surface roughness of the sample.

In a large sample AFM both force sensors using optical deflection method and scanning unit are mounted on the microscope head, Fig. 17.2b. Because of vibrations added by cantilever movement, lateral resolution of this design is somewhat poorer than the design in Fig. 17.2a in which the sample is scanned instead of cantilever beam. The advantage of the large sample AFM is that large samples can be measured readily.

Most AFMs can be used for topography measurements in the so-called tapping mode (intermittent contact mode), also referred to as dynamic force microscopy. In the tapping mode, during scanning over the surface, the cantilever/tip assembly is sinusoidally vibrated by a piezo mounted above it, and the oscillating tip slightly taps the surface at the resonant frequency of the cantilever (70–400 Hz) with a constant (20–100 nm) oscillating amplitude introduced in the vertical direction with a feedback loop keeping the average normal force constant, Fig. 17.3. The oscillating amplitude is kept large enough so that the tip does not get stuck to the sample because of adhesive attractions. The tapping mode is used in topography measurements to minimize the effects of friction and other lateral forces and to measure topography of soft surfaces.

For measurement of friction force being applied at the tip surface during sliding, left-hand and right-hand sets of quadrants of the photodetector are used. In the so-called friction mode, the sample is scanned back and forth in a direction orthogonal to the long axis of the cantilever beam. A friction force between the sample and the tip will produce a twisting of the cantilever. As a result, the laser beam will be reflected out of the plane defined by the incident beam and the beam reflected vertically from an untwisted cantilever. This produces an intensity difference of the laser beam received in the left-hand and right-hand sets of quadrants of the photodetector. The intensity difference between the two sets of detectors (FFM signal) is directly related to the degree of twisting and, hence, to the magnitude of the friction force. One problem associated with this method is that any misalignment between the laser beam and the photodetector axis would introduce error in the measurement. However, by following the procedures developed by *Ruan* and *Bhushan* [17.19], in which the average FFM signal for the sample scanned in two opposite directions is subtracted from the friction profiles of each of the two scans, the misalignment effect is eliminated. This method provides three-dimensional maps of friction force. By following the friction force calibration procedures developed by *Ruan* and *Bhushan* [17.19], voltages corresponding to friction forces can be converted to force units. The coefficient of friction is obtained from the slope of friction force data measured as a function of normal loads typically ranging from 10–150 nN. This approach eliminates any contributions due to the adhesive forces [17.22]. For calculation of the coefficient of friction based on a single point measurement, friction force should be divided by the sum of applied normal load and intrinsic adhesive force. Furthermore, it should be pointed out that for a single asperity contact, the coefficient of friction is not independent of load. This will be discussed in greater detail later in the chapter.

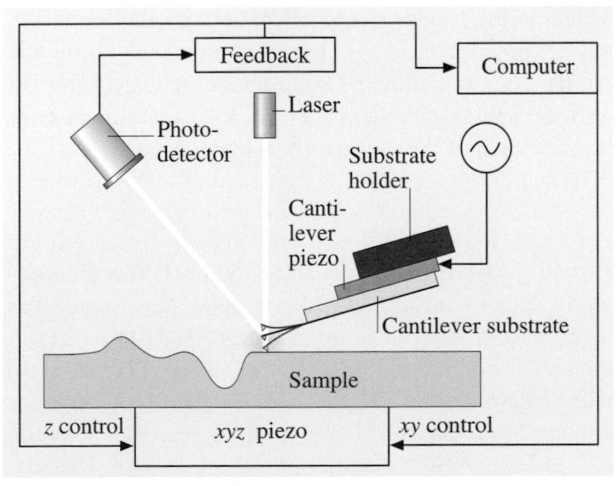

Fig. 17.3 Schematic of tapping mode used for surface topography measurements

Topographic measurements in the contact mode are typically made using a sharp, microfabricated square-pyramidal Si_3N_4 tip with a radius of 30–50 nm on a triangular cantilever beam (Fig. 17.4a) with normal stiffness on the order of 0.5 N/m at a normal load of about 10 nN, and friction measurements are carried out in the load range of 10–150 nN. Topography measurements in the tapping mode utilize a stiff cantilever with high resonant frequency; typically, a square-pyramidal, etched, single-crystal silicon tip with a tip radius of 5–10 nm mounted on a stiff rectangular silicon cantilever beam (Fig. 17.4a) with a normal stiffness on the order of 50 N/m is used. Carbon nanotube tips with small diameters (few nm) and high aspect ratios attached on the single-crystal silicon, square pyramidal tips are used for high resolution imaging of surfaces and deep trenches in the tapping mode or noncontact mode (Fig. 17.4b). To study the effect of the radius of a single asperity (tip) on adhesion and friction, microspheres of silica with radii ranging from about 4–15 μm are attached with epoxy at the ends of tips of Si_3N_4 cantilever beams. Optical micrographs of two of the microspheres mounted at the ends of triangular cantilever beams are shown in Fig. 17.4c.

The tip is scanned in such a way that its trajectory on the sample forms a triangular pattern, Fig. 17.5.

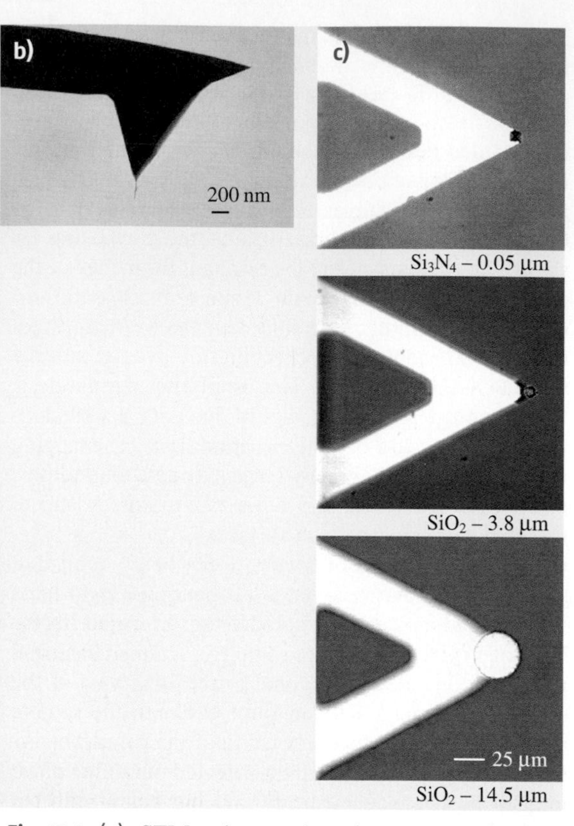

Fig. 17.4 (a) SEM micrographs of a square-pyramidal PECVD Si_3N_4 tip with a triangular cantilever beam (*top*), a square-pyramidal etched single-crystal silicon tip with a rectangular silicon cantilever beam (*middle*), and a three-sided pyramidal natural diamond tip with a square stainless steel cantilever beam (*bottom*), (b) SEM micrograph of a multiwalled carbon nanotube (MWNT) physically attached on the single-crystal silicon, square-pyramidal tip, and (c) optical micrographs of a commercial Si_3N_4 tip and two modified tips showing SiO_2 spheres mounted over the sharp tip and at the end of the triangular Si_3N_4 cantilever beams. (Radii of the tips are given in the figure)

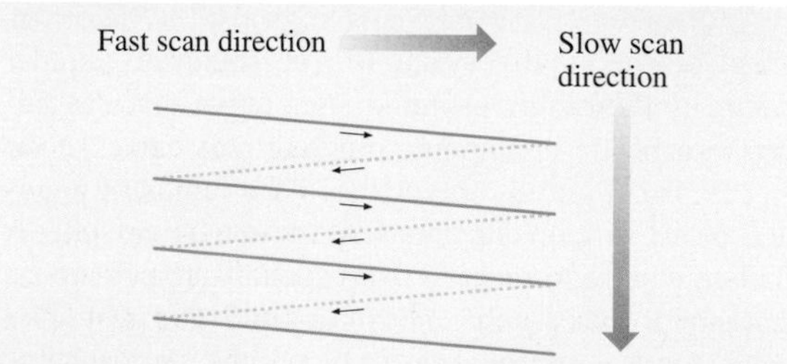

Fig. 17.5 Schematic of triangular pattern trajectory of the tip as the sample (or the tip) is scanned in two dimensions. During scanning, data are recorded only during scans along the solid scan lines

Scanning speeds in the fast and slow scan directions depend on the scan area and scan frequency. Scan sizes ranging from less than 1 nm × 1 nm to 125 µm × 125 µm and scan rates from less than 0.5 to 122 Hz typically can be used. Higher scan rates are used for smaller scan lengths. For example, scan rates in the fast and slow scan directions for an area of 10 µm × 10 µm scanned at 0.5 Hz are 10 µm/s and 20 nm/s, respectively.

17.1.2 Adhesion Measurements

Adhesive force measurements are performed in the so-called force calibration mode. In this mode, force-distance curves are obtained, for an example, see Fig. 17.6. The horizontal axis gives the distance the piezo (and hence the sample) travels, and the vertical axis gives the tip deflection. As the piezo extends, it approaches the tip, which is at this point in free air and hence shows no deflection. This is indicated by the flat portion of the curve. As the tip approaches within a few nanometers of the sample (point A), an attractive force exists between the atoms of the tip surface and the atoms of the sample surface. The tip is pulled toward the sample and contact occurs at point B on the graph. From this point on, the tip is in contact with the surface, and as the piezo extends farther, the tip gets deflected farther. This is represented by the sloped portion of the curve. As the piezo retracts, the tip goes beyond the zero deflection (flat) line, due to attractive forces (van der Waals forces and long-range meniscus forces), into the adhesive regime. At point C in the graph, the tip snaps free of the adhesive forces and is again in free air. The horizontal distance between points B and C along the retrace line gives the distance the tip moved in the adhesive regime. This distance multiplied by the stiffness of the cantilever gives the adhesive force. Incidentally, the horizontal shift between the loading and unloading curves results from the hysteresis in the PZT tube [17.4, 36].

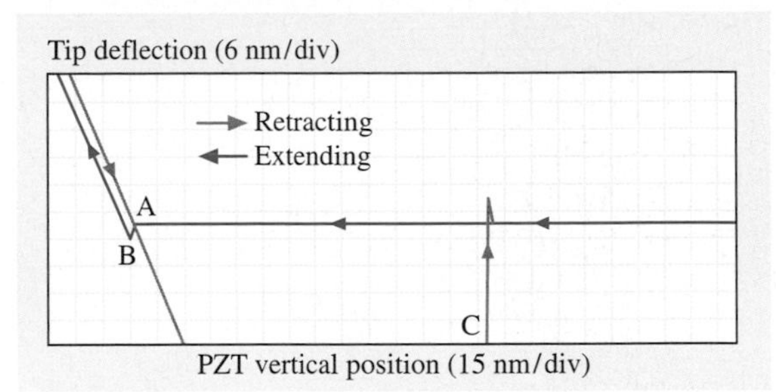

Fig. 17.6 Typical force-distance curve for a contact between Si_3N_4 tip and single-crystal silicon surface in measurements made in the ambient environment. Contact between the tip and silicon occurs at point B; tip breaks free of adhesive forces at point C as the sample moves away from the tip

17.1.3 Scratching, Wear and Fabrication/Machining

For microscale scratching, microscale wear, nanofabrication/nanomachining, and nanoindentation hardness measurements, an extremely hard tip is required. A three-sided pyramidal single-crystal natural diamond tip with an apex angle of 80° and a radius of about 100 nm mounted on a stainless steel cantilever beam with normal stiffness of about 25 N/m is used at relatively higher loads (1–150 μN), Fig. 17.4a. For scratching and wear studies, the sample is generally scanned in a direction orthogonal to the long axis of the cantilever beam (typically at a rate of 0.5 Hz), so that friction can be measured during scratching and wear. The tip is mounted on the cantilever such that one of its edges is orthogonal to the long axis of the beam. Therefore, wear during scanning along the beam axis is higher (about two to three times) than during scanning orthogonal to the beam axis. For wear studies, an area on the order of 2 μm × 2 μm is scanned at various normal loads (ranging from 1 to 100 μN) for a selected number of cycles [17.4, 22].

Scratching can also be performed at ramped loads and the coefficient of friction can be measured during scratching [17.33]. A linear increase in the normal load approximated by a large number of normal load increments of small magnitude is applied using a software interface (lithography module in Nanoscope III) that allows the user to generate controlled movement of the tip with respect to the sample. The friction signal is tapped out of the AFM and is recorded on a computer. A scratch length on the order of 25 μm and a velocity on the order of 0.5 μm/s are used and the number of loading steps is usually 50.

Nanofabrication/nanomachining is conducted by scratching the sample surface with a diamond tip at specified locations and scratching angles. The normal load used for scratching (writing) is on the order of 1–100 μN with a writing speed on the order of 0.1–200 μm/s [17.4, 5, 11, 22, 37].

17.1.4 Surface Potential Measurements

To detect wear precursors and study the early stages of localized wear, the multimode AFM can be used to measure the potential difference between the tip and the sample by applying a DC bias potential and an oscillating (AC) potential to a conducting tip over a grounded substrate in a Kelvin probe microscopy, or so-called "nano-Kelvin probe" technique [17.38–40].

Mapping of the surface potential is made in the so-called "lift mode" (Fig. 17.7). These measurements are made simultaneously with the topography scan in the

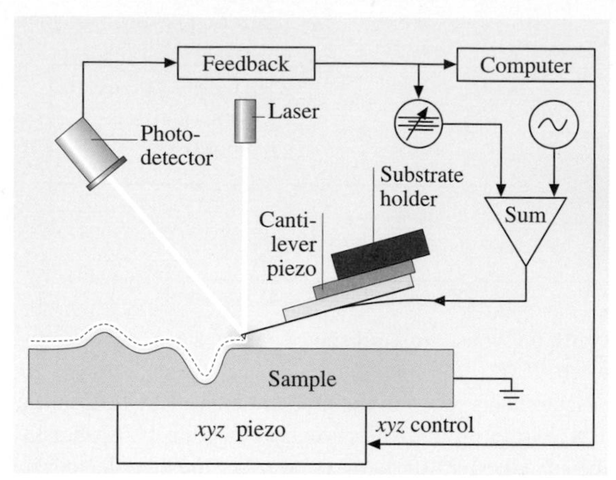

Fig. 17.7 Schematic of lift mode used to make surface potential measurement. The topography is collected in tapping mode in the primary scan. The cantilever piezo is deactivated. Using topography information of the primary scan, the cantilever is scanned across the surface at a constant height above the sample. An oscillating voltage at the resonant frequency is applied to the tip and a feedback loop adjusts the DC bias of the tip to maintain the cantilever amplitude at zero. The output of the feedback loop is recorded by the computer and becomes the surface potential map [17.4]

tapping mode using an electrically conducting (nickel-coated single-crystal silicon) tip. After each line of the topography scan is completed, the feedback loop controlling the vertical piezo is turned off, and the tip is lifted from the surface and traced over the same topography at a constant distance of 100 nm. During the lift mode, a DC bias potential and an oscillating potential (3–7 V) is applied to the tip. The frequency of oscillation is chosen to be equal to the resonant frequency of the cantilever (∼ 80 kHz). When a DC bias potential equal to the negative value of surface potential of the sample (on the order of ±2 V) is applied to the tip, it does not vibrate. During scanning, a difference between the DC bias potential applied to the tip and the potential of the surface will create DC electric fields that interact with the oscillating charges (as a result of the AC potential), causing the cantilever to oscillate at its resonant frequency, as in tapping mode. However, a feedback loop is used to adjust the DC bias on the tip to exactly nullify the electric field and, thus, the vibrations of the cantilever. The required bias voltage follows the localized potential of the surface. The surface potential was obtained by reversing the sign of the bias potential provided by the electronics [17.39, 40]. Surface and subsurface changes of structure and/or chemistry can cause changes in the measured potential of a surface. Thus, the mapping of the surface potential after sliding can be used for detecting wear precursors and studying the early stages of localized wear.

17.1.5 In Situ Characterization of Local Deformation Studies

In situ characterization of local deformation of materials can be carried out by performing tensile, bending, or compression experiments inside an AFM and by observing nanoscale changes during the deformation experiment [17.6]. In these experiments, small deformation stages are used to deform the samples inside an AFM. In tensile testing of the polymeric films carried out by *Bobji* and *Bhushan* [17.41, 42], a tensile stage was used (Fig. 17.8). The stage with a left-right combination lead screw (which helps to move the slider in the opposite direction) was used to stretch the sample and minimize the movement of the scanning area, which was kept close to the center of the tensile specimen. One end of the sample was mounted on the slider via a force sensor to monitor the tensile load. The samples were stretched for various strains using a stepper motor, and the same control area at different strains was imaged. In order to better locate the control area for imaging, a set of four markers was created at the corners of a 30 μm × 30 μm square at the center of the sample by scratching the sample with a sharp silicon tip. The scratching depth was controlled such that it did not affect cracking behavior of the coating. A minimum displacement of 1.6 μm could be obtained. This corresponded to a strain increment of 8×10^{-3}% for a sample length of 38 mm. The maximum travel was about 100 mm. The resolution of the force sensor was 10 mN with a capacity of 45 N. During stretching, a stress-strain curve was obtained during the experiment to study any correlation between the degree of plastic strain and propensity of cracks.

17.1.6 Nanoindentation Measurements

For nanoindentation hardness measurements, the scan size is set to zero, and a normal load is then applied to make the indents using the diamond tip. During this procedure, the tip is continuously pressed against the sample surface for about two seconds at various indentation loads. The sample surface is scanned before and after the scratching, wear, or indentation to obtain the

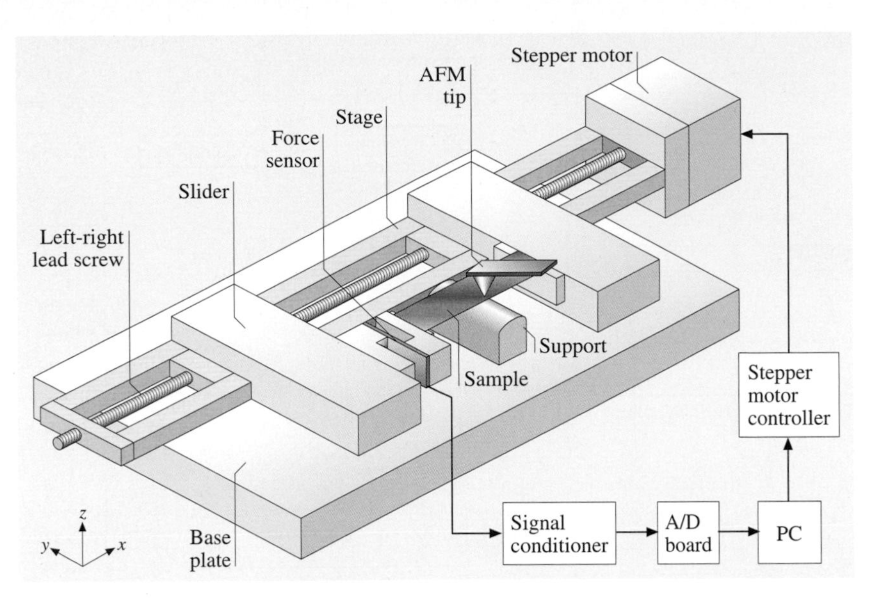

Fig. 17.8 Schematic of the tensile stage to conduct in situ tensile testing of the polymeric films in AFM [17.42]

initial and final surface topographies at a low normal load of about 0.3 μN using the same diamond tip. An area larger than the indentation region is scanned to observe the indentation marks. Nanohardness is calculated by dividing the indentation load by the projected residual area of the indents [17.34].

Direct imaging of the indent allows one to quantify piling up of ductile material around the indenter. However, it becomes difficult to identify the boundary of the indentation mark with great accuracy. This makes the direct measurement of contact area somewhat inaccurate. A technique with the dual capability of depth-sensing, as well as in situ imaging, which is most appropriate in nanomechanical property studies, is used for accurate measurement of hardness with shallow depths [17.4, 35]. This indentation system is used to make load-displacement measurement and, subsequently, carry out in situ imaging of the indent, if required. The indentation system consists of a three-plate transducer with electrostatic actuation hardware used for direct application of normal load and a capacitive sensor used for measurement of vertical displacement. The AFM head is replaced with this transducer assembly while the specimen is mounted on the PZT scanner, which remains stationary during indentation experiments. Indent area and, consequently, hardness value can be obtained from the load-displacement data. The Young's modulus of elasticity is obtained from the slope of the unloading curve.

17.1.7 Localized Surface Elasticity and Viscoelasticity Mapping

Indentation experiments provide a single-point measurement of the Young's modulus of elasticity calculated from the slope of the indentation curve during unloading. Localized surface elasticity maps can be obtained using dynamic force microscopy in which an oscillating tip is scanned over the sample surface in contact under steady and oscillating load. Lower frequency operation mode in the kHz range such as force modulation microscopy [17.43–45] and pulsed force microscopy [17.46] are well suited for soft samples such as polymers. However, if the tip-sample contact stiffness becomes significantly higher than the cantilever stiffness, the sensitivity of these techniques strongly decreases. In this case, sensitivity of the measurement of stiff materials can be improved by using high-frequency operation modes in the MHz range such as acoustic (ultrasonic) force microscopy [17.47]. We only describe here the force modulation technique.

In the force modulation technique, the oscillation is applied to the cantilever substrate with a cantilever piezo (bimorph), Fig. 17.9a. For measurements, an etched silicon tip is first brought in contact with a sample under a static load of 50–300 nN. In addition to the static load applied by the sample piezo, a small oscillating (modulating) load is applied by a bimorph, generally at a frequency (about 8 kHz) far below that of the natural resonance of the cantilever (70–400 kHz). When the

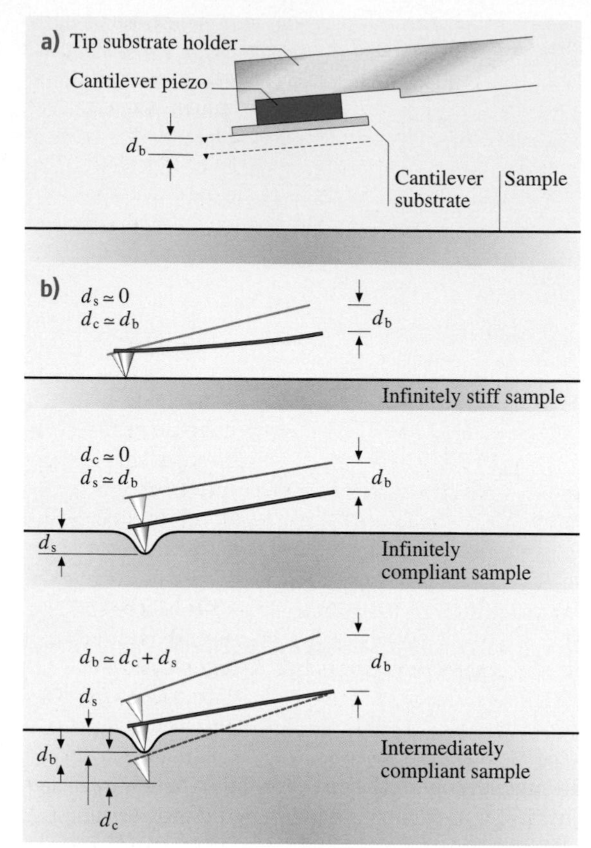

Fig. 17.9a,b Schematics (**a**) of the bimorph assembly and (**b**) of the motion of the cantilever and tip as a result of the oscillations of the bimorph for an infinitely stiff sample, an infinitely compliant sample, and an intermediate sample. The *thin line* represents the cantilever at the top cycle, and the *thick line* corresponds to the bottom of the cycle. The *dashed line* represents the position of the tip if the sample was not present or was infinitely compliant. d_c, d_s and d_b are the oscillating (AC) deflection amplitude of the cantilever, penetration depth, and oscillating (AC) amplitude of the bimorph, respectively [17.44]

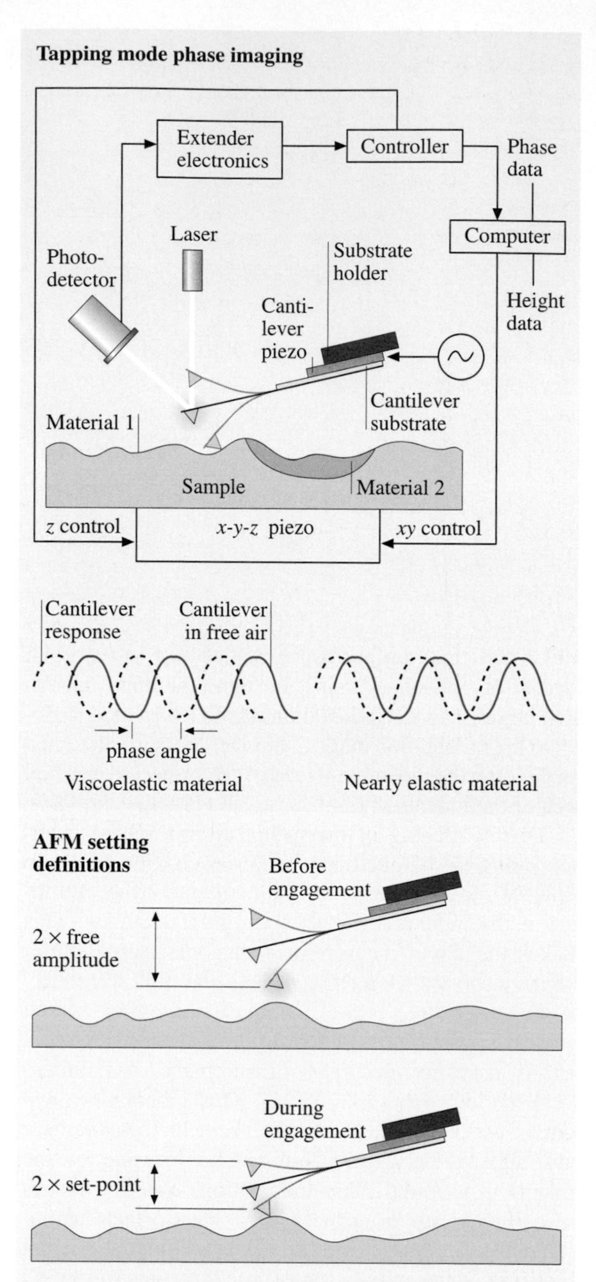

Fig. 17.10 Schematic of tapping mode used to obtain height and phase data and definitions of tapping amplitude and set point. During scanning the cantilever is vibrated at its resonant frequency and the sample x-y-z piezo is adjusted by feedback control in the z-direction to maintain a constant set point. The computer records height and phase angle (which is a function of the viscoelastic properties of the sample materials) data

tip is brought into contact with the sample, the surface resists the oscillations of the tip, and the cantilever deflects. Under the same applied load, a stiff area on the sample would deform less than a soft one; i. e., stiffer areas cause greater deflection amplitudes of the cantilever, Fig. 17.9b. The variations in the deflection amplitudes provide a measure of the relative stiffness of the surface. Contact analyses can be used to obtain a quantitative measure of localized elasticity of soft surfaces [17.44]. The elasticity data are collected simultaneously with the surface height data using a so-called negative lift mode technique. In this mode, each scan line of each topography image (obtained in tapping mode) is replaced with the tapping action disabled and the tip lowered into steady contact with the surface.

Another form of dynamic force microscopy, phase contrast microscopy, is used to detect the contrast in viscoelastic (viscous energy dissipation) properties of the different materials across the surface [17.48–52]. In this technique, both deflection amplitude and phase angle contrasts are measured, which are the measures of the relative stiffness and viscoelastic properties, respectively. Tapping mode is utilized. For tapping mode during scanning, the cantilever/tip assembly is sinusoidally vibrated at its resonant frequency, and the sample X-Y-Z piezo is adjusted using feedback control in the z-direction to maintain a constant set point. As shown in Fig. 17.10, the cantilever/tip assembly is vibrated at some amplitude, here referred to as the tapping amplitude, by a cantilever piezo before the tip engages the sample. The tip engages the sample at some set point, which may be thought of as the amplitude of the cantilever as influenced by contact with the sample. A lower set point gives a reduced amplitude and closer mean tip-to-sample distance. This is also depicted in Fig. 17.10. The feedback signal to the z-direction sample piezo (to keep the set point constant) is a measure of topography. This height data is recorded on a computer. The extender electronics is used to measure the phase angle lag between the cantilever piezo drive signal and the cantilever response during sample engagement. As illustrated in Fig. 17.10, the phase angle lag is (at least partially) a function of the viscoelastic properties of the sample material. A range of tapping amplitudes and set points can be used for measurements. A commercially etched single-crystal silicon tip (DI TESP) used for tapping mode, with a radius of 5–10 nm, a stiffness of 20–100 N/m, and a natural frequency of 350–400 kHz is normally used. Scanning is normally set to a rate of 1 Hz along the fast axis.

17.1.8 Boundary Lubrication Measurements

To study nanoscale boundary lubrication, adhesive forces are measured in the force calibration mode, as previously described. The adhesive forces are also calculated from the horizontal intercept of friction versus normal load curves at a zero value of friction force. For friction measurements, the samples are typically scanned using an Si_3N_4 tip over an area of $2 \times 2\,\mu m$ at the normal load ranging from 5 to 130 nN. The samples are generally scanned with a scan rate of 0.5 Hz resulting in a scanning speed of $2\,\mu m/s$. Velocity effects on friction are studied by changing the scan frequency from 0.1 to 60 Hz while the scan size is maintained at $2 \times 2\,\mu m$, which allows the velocity to vary from 0.4 to $240\,\mu m/s$. To study the durability properties, the friction force and coefficient of friction are monitored during scanning at a normal load of 70 nN and a scanning speed of $0.8\,\mu m/s$ for a desired number of cycles [17.27, 28, 30].

17.2 Friction and Adhesion

17.2.1 Atomic-Scale Friction

To study friction mechanisms on an atomic scale, a well characterized, freshly cleaved surface of highly oriented pyrolytic graphite (HOPG) has been used by *Mate* et al. [17.17] and *Ruan* and *Bhushan* [17.20].

The atomic-scale friction force of HOPG exhibited the same periodicity as that of the corresponding topography (Fig. 17.11a), but the peaks in friction and those in topography are displaced relative to each other (Fig. 17.11b). A Fourier expansion of the interatomic potential was used to calculate the conservative interatomic forces between atoms of the FFM tip and those of the graphite surface. Maxima in the interatomic forces in the normal and lateral directions do not occur at the same location, which explains the observed shift between the peaks in the lateral force and those in the corresponding topography. Furthermore, the observed local variations in friction force were explained by variation in the intrinsic lateral force between the sample and the FFM tip resulting from an atomic-scale stick-slip process [17.17, 20].

17.2.2 Microscale Friction

Local variations in the microscale friction of cleaved graphite have been observed (Fig. 17.12). These arise from structural changes that occur during the cleaving process [17.21]. The cleaved HOPG surface is largely atomically smooth, but exhibits line-shaped regions in which the coefficient of friction is more than an order of magnitude larger. Transmission electron microscopy indicates that the line-shaped regions consist of graphite planes of different orientation, as well as of amorphous

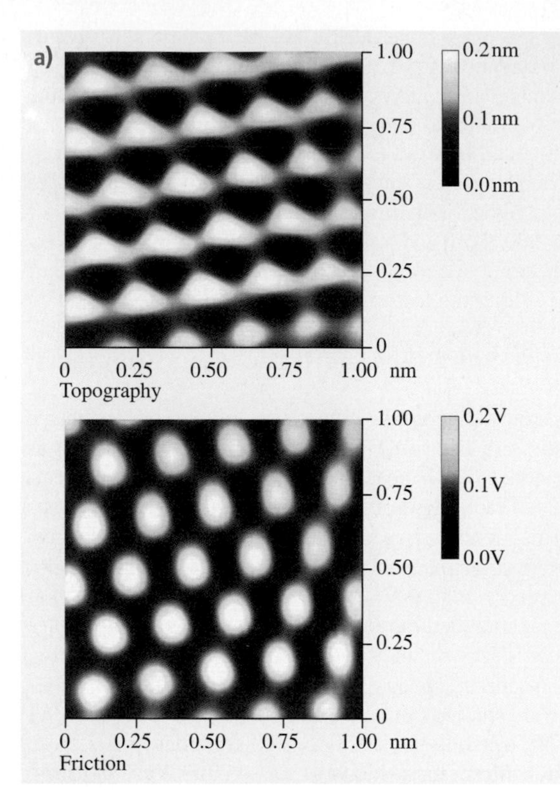

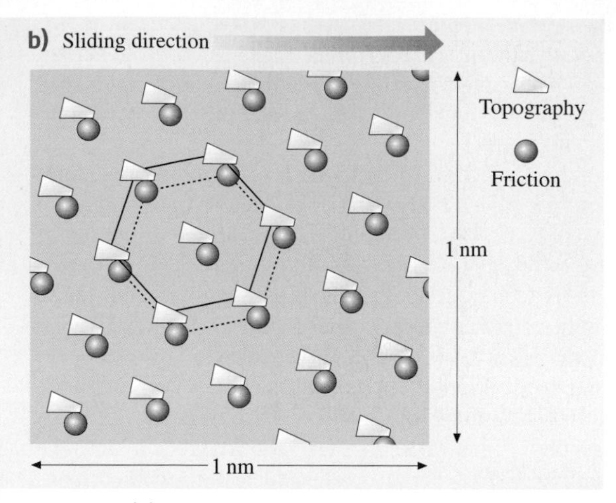

Fig. 17.11 (**a**) Gray-scale plots of surface topography and friction force maps of a 1 nm × 1 nm area of freshly cleaved HOPG showing the atomic-scale variation of topography and friction and (**b**) schematic of superimposed topography and friction maps from (**a**); the symbols correspond to maxima. Note the spatial shift between the two plots [17.20]

carbon. Differences in friction have also been observed for multiphase ceramic materials [17.53]. Figure 17.13 shows the surface roughness and friction force maps of Al_2O_3-TiC (70–30 wt%). TiC grains have a Knoop hardness of about 2,800 kg/mm^2. Therefore, they do not polish as much and result in a slightly higher elevation (about 2–3 nm higher than that of Al_2O_3 grains). TiC grains exhibit higher friction force than Al_2O_3 grains. The coefficients of friction of TiC and Al_2O_3 grains are 0.034 and 0.026, respectively, and the coefficient of friction of Al_2O_3-TiC composite is 0.03. Local variation in friction force also arises from the scratches present on the Al_2O_3-TiC surface. *Meyer* et al. [17.54] also used FFM to measure structural variations of organic mono- and multilayer films. All of these measurements suggest that the FFM can be used for structural mapping of the

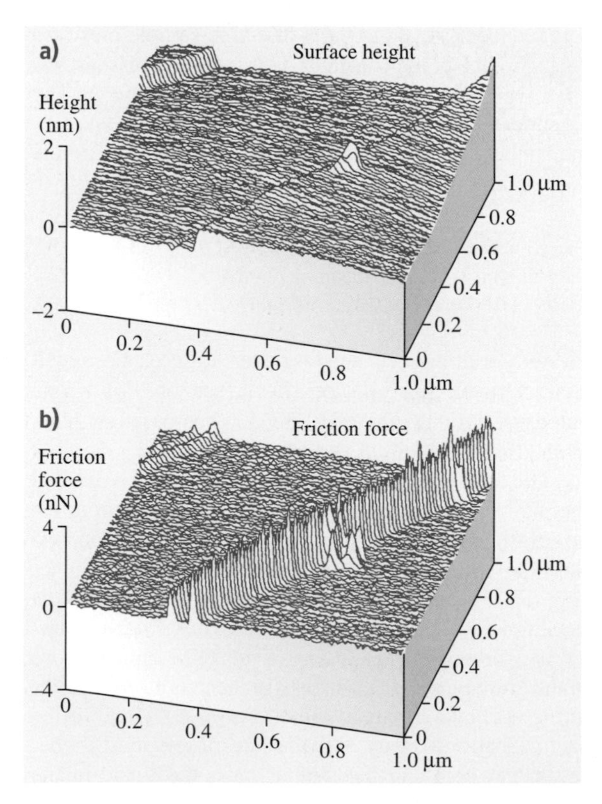

Fig. 17.12 (**a**) Surface roughness and (**b**) friction force maps at a normal load of 42 nN of freshly cleaved HOPG surface against an Si_3N_4 FFM tip. Friction in the *line-shaped region* is over an order of magnitude larger than the *smooth areas* [17.20]

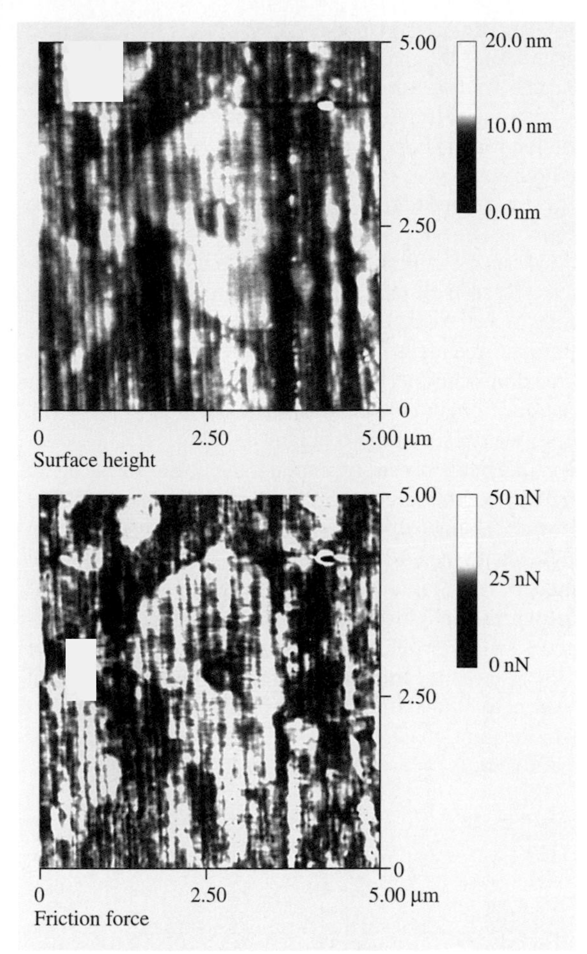

Fig. 17.13 Gray-scale surface roughness ($\sigma = 0.80$ nm) and friction force maps (mean = 7.0 nN, $\sigma = 0.90$ nN) for Al_2O_3-TiC (70–30 wt%) at a normal load of 138 nN [17.53]

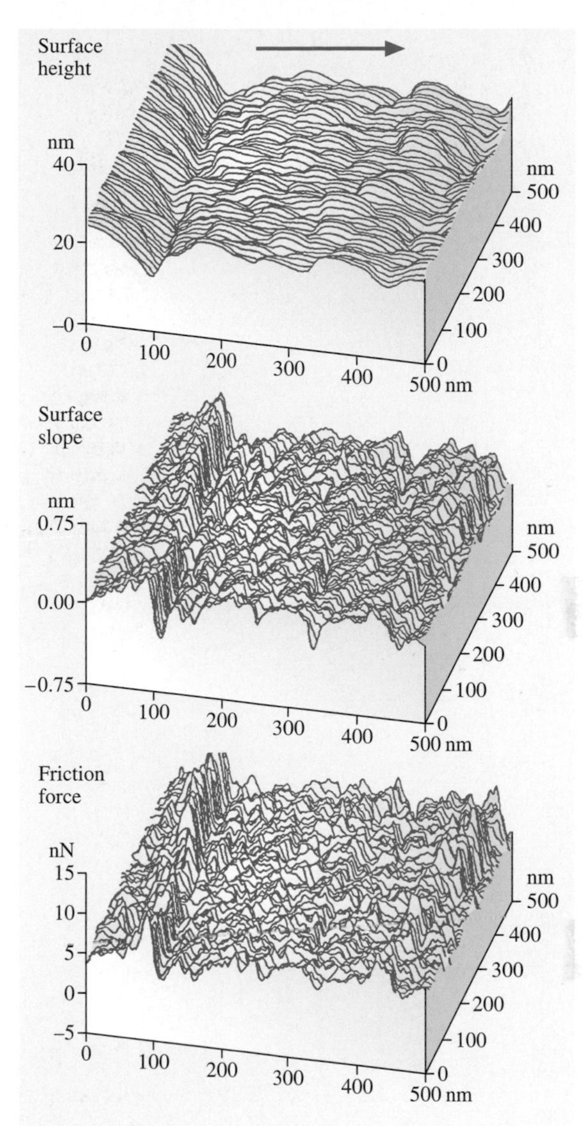

Fig. 17.14 Surface roughness map ($\sigma = 4.4$ nm); surface slope map taken in the sample sliding direction, the horizontal axis, (mean = 0.023, $\sigma = 0.197$); and friction force map (mean = 6.2 nN, $\sigma = 2.1$ nN) for a lubricated thin-film magnetic rigid disk for a normal load of 160 nN [17.22]

surfaces. FFM measurements can be used to map chemical variations, as indicated by the use of the FFM with a modified probe tip to map the spatial arrangement of chemical functional groups in mixed organic monolayer films [17.55]. Here, sample regions that had stronger interactions with the functionalized probe tip exhibited larger friction.

Local variations in the microscale friction of nominally rough, homogeneous surfaces can be significant and are seen to depend on the local surface slope, rather than the surface height distribution, Fig. 17.14. This dependence was first reported by *Bhushan* and *Ruan* [17.18], *Bhushan* et al. [17.22], and *Bhushan* [17.37] and later discussed in more detail in [17.56] and [17.57]. In order to show elegantly any correlation between local values of friction and surface roughness, surface roughness and friction force maps of a gold-coated ruling with somewhat rectangular grids and a silicon grid with square pits were obtained, see Fig. 17.15 [17.57]. Figures 17.14 and 17.15 show the

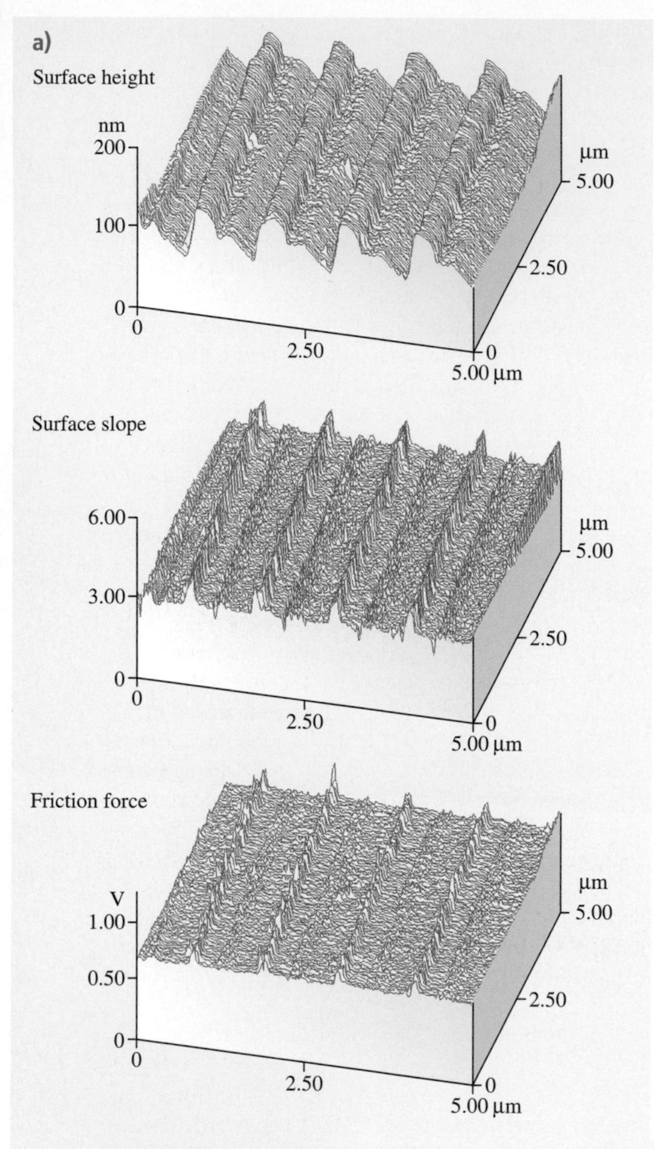

surface roughness map, the slopes of the roughness map taken along the sliding direction (surface slope map), and the friction force map for various samples. There is a strong correlation between the surface slopes and friction forces. For example, in Fig. 17.15, friction force is high locally at the edge of the grids and pits with a positive slope and is low at the edges with a negative slope.

We now examine the mechanism of microscale friction, which may explain the resemblance between the slope of surface roughness maps and the corresponding friction force maps [17.4, 5, 18, 20–22, 56, 57]. There are three dominant mechanisms of friction: adhesive, adhesive and roughness (ratchet), and plowing [17.10, 57]. First, we may assume these to be additive. The adhesive mechanism cannot explain the local variation in friction. Next, we consider the ratchet mechanism. We consider a small tip sliding over an asperity making an angle θ with the horizontal plane, Fig. 17.16. The normal force W (normal to the general surface) applied by the tip to the sample surface is constant. The friction force F on the sample would be a constant for a smooth surface if the friction mechanism does not change. For a rough surface, as shown in Fig. 17.16, if the adhesive mechanism does not change during sliding, the local value of coefficient of friction remains constant,

$$\mu_0 = S/N , \tag{17.1}$$

Fig. 17.15a,b Surface roughness map, surface slope map taken in the sample sliding direction (the horizontal axis), and friction force map for (**a**) a gold-coated ruling (with somewhat rectangular grids with a pitch of 1 μm and a ruling step height of about 70 nm) at a normal load of 25 nN and (**b**) a silicon grid (with 5 μm square pits of 180-nm depth and a pitch of 10 μm) [17.57]

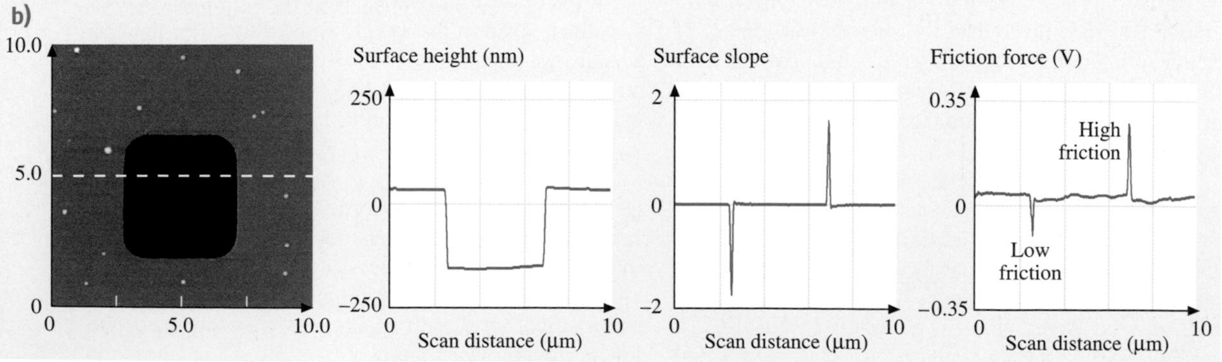

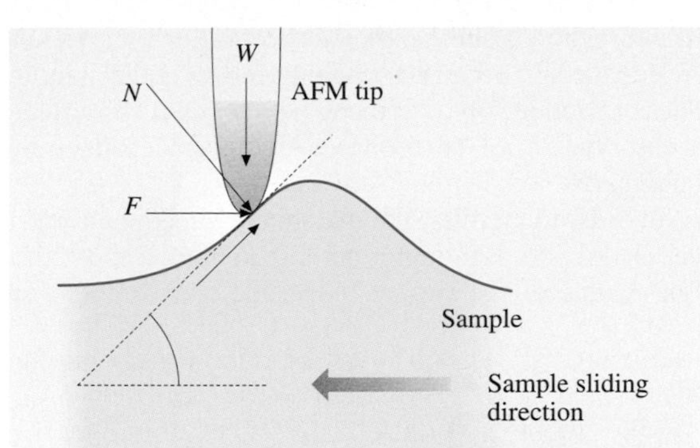

Fig. 17.16 Schematic illustration showing the effect of an asperity (making an angle θ with the horizontal plane) on the surface in contact with the tip on local friction in the presence of adhesive friction mechanism. W and F are the normal and friction forces, respectively, and S and N are the force components along and perpendicular to the local surface of the sample at the contact point, respectively

where S is the local friction force and N is the local normal force. However, the friction and normal forces are measured with respect to global horizontal and normal axes, respectively. The measured local coefficient of friction μ_1 in the ascending part is

$$\begin{aligned}\mu_1 &= F/W \\ &= (\mu_0 + \tan\theta)\,/\,(1 - \mu_0 \tan\theta) \sim \mu_0 + \tan\theta\,, \\ &\text{for small } \mu_0 \tan\theta\,,\end{aligned} \tag{17.2}$$

indicating that in ascending part of the asperity one may simply add the friction force and the asperity slope to one another. Similarly, on the right-hand side (descending part) of the asperity,

$$\begin{aligned}\mu_2 &= (\mu_0 - \tan\theta)\,/\,(1 + \mu_0 \tan\theta) \sim \mu_0 - \tan\theta\,, \\ &\text{for small } \mu_0 \tan\theta\,.\end{aligned} \tag{17.3}$$

For a symmetrical asperity, the average coefficient of friction experienced by the FFM tip traveling across the whole asperity is

$$\begin{aligned}\mu_{\text{ave}} &= (\mu_1 + \mu_2)\,/2 \\ &= \mu_0 \left(1 + \tan^2\theta\right) / \left(1 - \mu_0^2 \tan^2\theta\right) \\ &\sim \mu_0 \left(1 + \tan^2\theta\right)\,, \\ &\text{for small } \mu_0 \tan\theta\,.\end{aligned} \tag{17.4}$$

Finally, we consider the plowing component of friction with tip sliding in either direction, which is [17.10, 58]

$$\mu_{\text{p}} \sim \tan\theta\,. \tag{17.5}$$

Because we notice little damage of the sample surface in FFM measurements, the contribution by plowing is expected to be small and the ratchet mechanism is believed to be the dominant mechanism for the local variations in the friction force map. With the tip sliding over the leading (ascending) edge of an asperity, the surface slope is positive; it is negative during sliding over the trailing (descending) edge of an asperity. Thus, measured friction is high at the leading edge of asperities and low at the trailing edge. In addition to the slope effect, the collision of the tip when encountering an asperity with a positive slope produces additional torsion of the cantilever beam, leading to higher measured friction force. When encountering an asperity with the same negative slope, however, there is no collision effect and hence no effect on torsion. This effect also contributes to the difference in friction forces when the tip scans up and down on the same topography feature. The ratchet mechanism and the collision effects thus explain semiquantitatively the correlation between the slopes of the roughness maps and friction force maps observed in Figs. 17.14 and 17.15. We note that in the ratchet mechanism, the FFM tip is assumed to be small compared to the size of asperities. This is valid since the typical radius of curvature of the tips is about 10–50 nm. The radii of curvature of samples, the asperities measured here (the asperities that produce most of the friction variation) are found to typically be about 100–200 nm, which is larger than that of the FFM tip [17.59]. It is important to note that the measured local values of friction and normal forces are measured with respect to global (and not local) horizontal and vertical axes, which are believed to be relevant in applications.

17.2.3 Directionality Effect on Microfriction

During friction measurements, the friction force data from both the forward (trace) and backward (retrace) scans are useful in understanding the origins of the observed friction forces. Magnitudes of material-induced effects are independent of the scanning direction, whereas topography-induced effects are different between forward and backward scanning directions. Since the sign of the friction force changes as the scanning direction is reversed (because of the reversal of torque applied to the end of the tip), addition of the friction force data of the forward and backward scan eliminates the material-induced effects, while topography-induced effects still remain. Subtraction of the data between forward and backward scans does not eliminate either effect, see Fig. 17.17 [17.57].

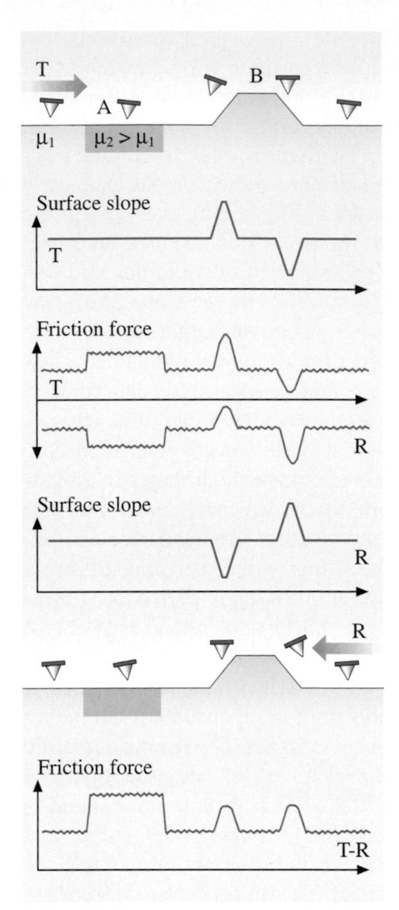

Fig. 17.17 Schematic of friction forces expected when a tip traverses a sample that is composed of different materials and sharp changes in topography. A schematic of surface slope is also shown (T = Trace, R = Retrace)

Owing to the reversal of the sign of the retrace (R) friction force with respect to the trace (T) data, the friction force variations due to topography are in the same direction (peaks in trace correspond to peaks in retrace). However, the magnitudes of the peaks in trace and retrace at a given location are different. An increase in the friction force experienced by the tip when scanning up a sharp change in topography is more than the decrease in the friction force experienced when scanning down the same topography change, partly due to the collision effects discussed earlier. Asperities on engineering surfaces are asymmetrical, which also affect the magnitude of friction force in the two directions. In addition, asymmetry in tip shape may have an effect on the directionality effect of friction. We will later note that the magnitude of surface slopes are virtually identical. Therefore, the tip shape asymmetry should not have much effect.

Figure 17.18 shows surface height and friction force data for gold ruler and a silicon grid in the trace and retrace directions. Subtraction of two friction data yields a residual peak, because of the differences in the magnitudes of friction forces in the two directions. This effect is observed at all locations of significant changes in topography.

In order to facilitate comparison of directionality effect on friction, it is important to take into account the sign change of the surface slope and friction force in

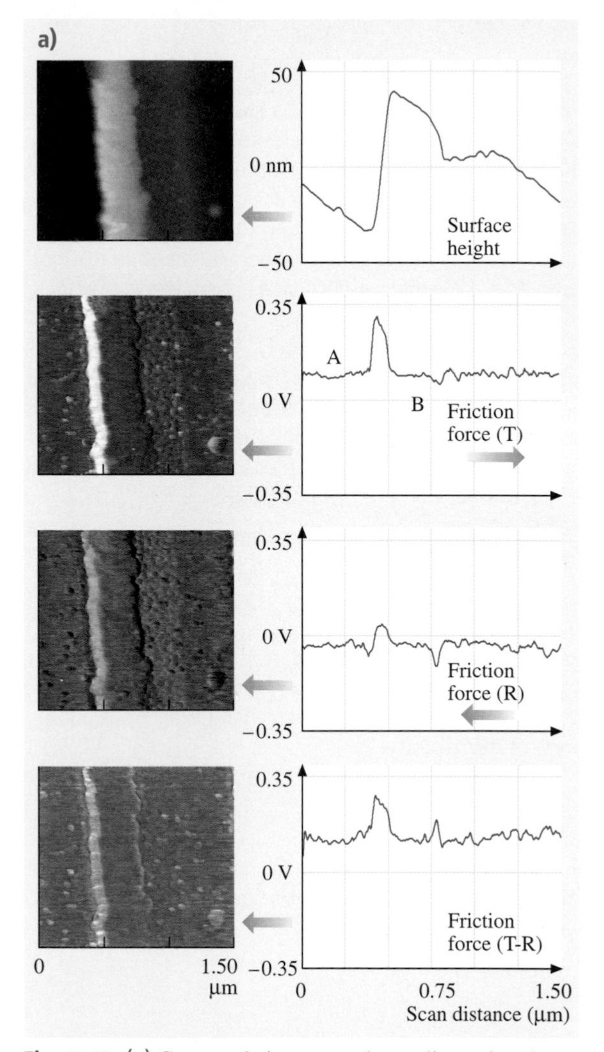

Fig. 17.18 (a) Gray-scale images and two-dimensional profiles of surface height and friction forces across a single ruling of the gold-coated ruling and (b) (*see next page*) two-dimensional profiles of surface height and friction forces across a silicon grid pit. Friction force data in trace (T) and retrace (R) directions and substrated force data are presented

the trace and retrace directions. Figure 17.19 shows surface height, surface slope, and friction force data for the two samples in the trace and retrace directions. The correlation between surface slope and friction forces is clear. The third column in the figures shows retrace slope and friction data with an inverted sign (−retrace). Now we can compare trace data with −retrace data. It is clear that the friction experienced by the tip is dependent upon the scanning direction because of surface topography. In addition to the effect of topographical changes discussed earlier, during surface-finishing processes, material can be transferred preferentially onto one side of the asperities, which also causes asymmetry and direction dependence. Reduction in local variations and in directionality of friction properties requires careful optimization of surface roughness distributions and of surface-finishing processes.

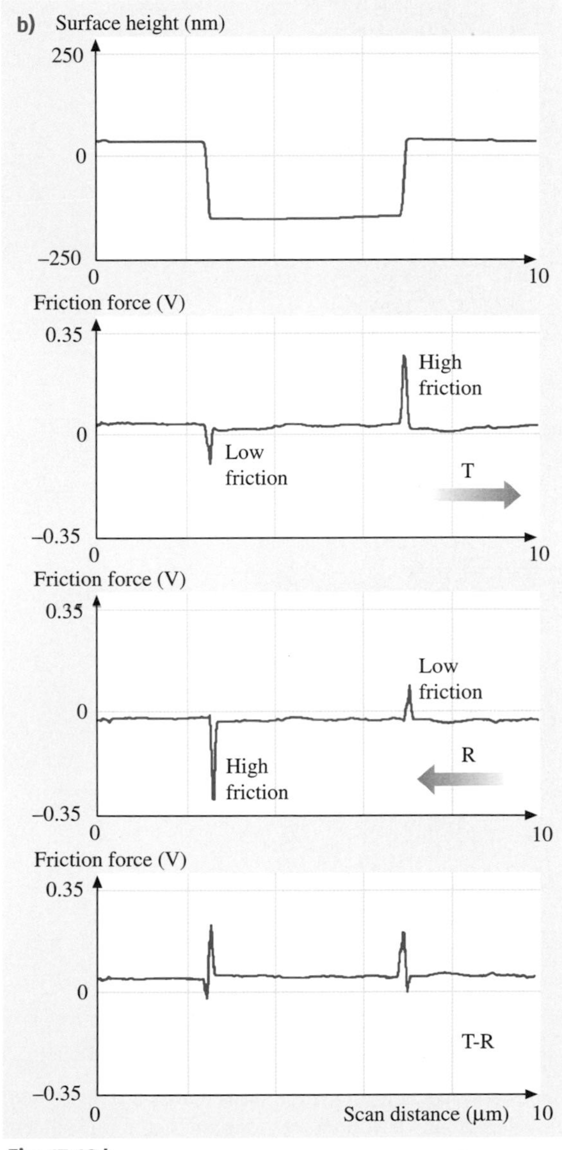

Fig. 17.18 b

The directionality as a result of surface asperities effect will also be manifested in macroscopic friction data, i. e., the coefficient of friction may be different in one sliding direction than in the other direction. Asymmetrical shape of asperities accentuates this effect. The frictional directionality can also exist in materials with particles with a preferred orientation. The directionality effect in friction on a macroscale is observed in some magnetic tapes. In a macroscale test, a 12.7-mm wide polymeric magnetic tape was wrapped over an aluminum drum and slid in a reciprocating motion with a normal load of 0.5 N and a sliding speed of about 60 mm/s [17.3]. The coefficient of friction as a function of sliding distance in either direction is shown in Fig. 17.20. We note that the coefficient of friction on a macroscale for this tape varies in different directions. Directionality in friction is sometimes observed on the macroscale; on the microscale this is the norm [17.4, 13]. On the macroscale, the effect of surface asperities is normally averaged out over a large number of contacting asperities.

17.2.4 Velocity Dependence on Microfriction

AFM/FFM experiments are generally conducted at relative velocities as high as a few tens of µm/s. To simulate applications, it is of interest to conduct friction experiments at higher velocities. High velocities can be achieved by mounting either the sample or the cantilever beam on a shear wave transducer (ultrasonic transducer) to produce surface oscillations at MHz frequencies [17.24, 25, 60]. The velocities on the order of a few mm/s can thus be achieved. The effect of in-plane and out-of-plane sample vibration amplitude on the coefficient of friction is shown in Fig. 17.21. Vibration of a sample at ultrasonic frequencies (> 20 kHz) can substantially reduce the coefficient of friction, known as ultrasonic lubrication or sonolubrication. When the surface is vibrated in-plane, classical hydrodynamic lubrication develops hydrodynamic pressure, which supports the tip and reduces friction. When the surface is vibrated out-of-plane, a lift-off caused by the squeeze-film lubrication (a form of hydrodynamic lubrication), reduces friction.

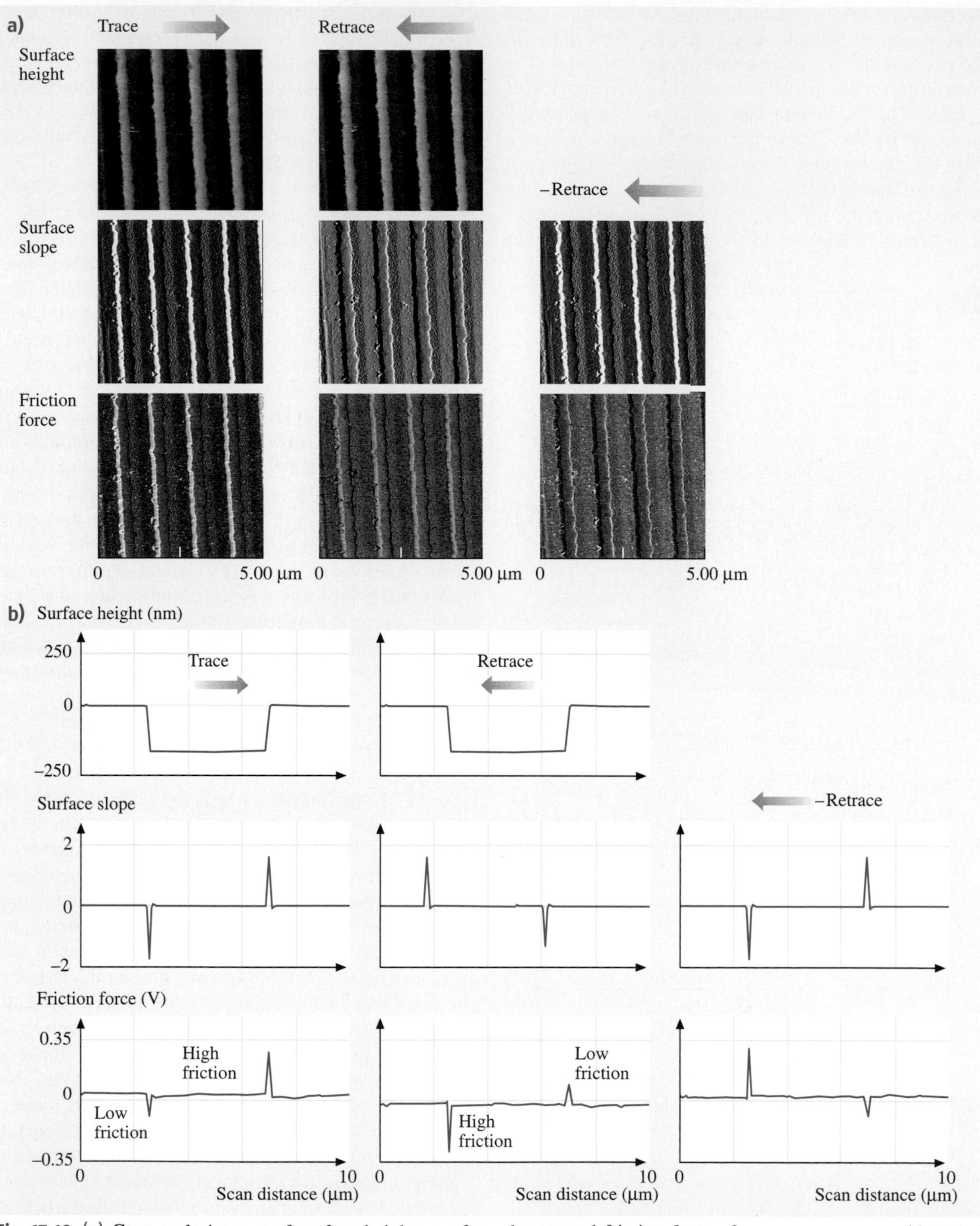

Fig. 17.19 (**a**) Gray-scale images of surface heights, surface slopes, and friction forces for scans across a gold-coated ruling, and (**b**) two-dimensional profiles of surface heights, surface slopes, and friction forces for scans across the silicon grid pit. *Arrows* indicate the tip sliding direction [17.57]

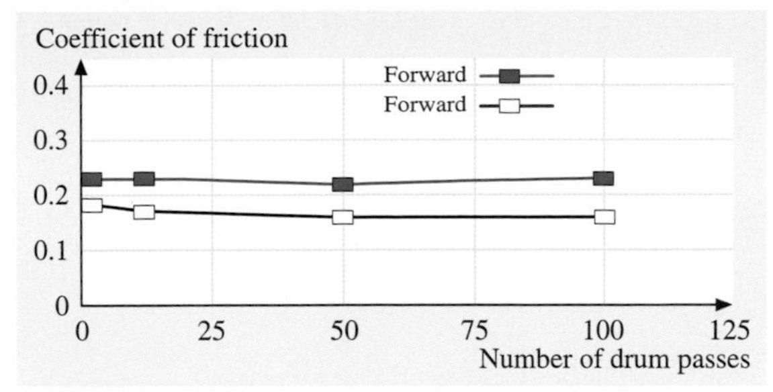

Fig. 17.20 Coefficient of macroscale friction as a function of drum passes for a polymeric magnetic tape sliding over an aluminum drum in a reciprocating mode in both directions. Normal load = 0.5 N over 12.7-mm wide tape, sliding speed = 60 mm/s [17.37]

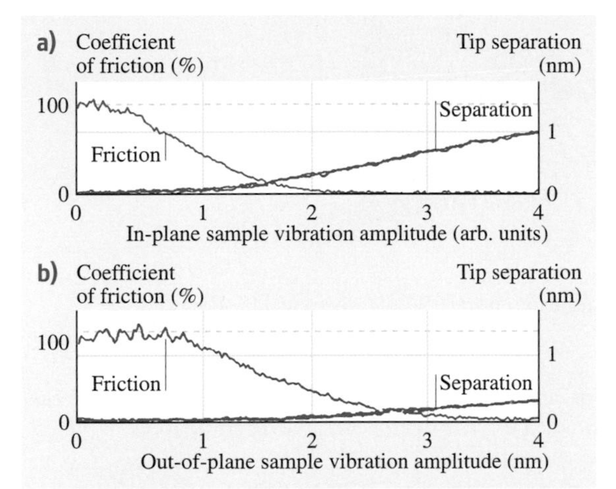

Fig. 17.21a,b Reduction of coefficient of friction measured at a normal load of 100 nN and average tip separation as a function of surface amplitude on a single-crystal silicon subjected to (**a**) in-plane and (**b**) out-of-plane vibrations at about 1 MHz against a silicon nitride tip [17.24]

17.2.5 Effect of Tip Radii and Humidity on Adhesion and Friction

The tip radius and relative humidity affect adhesion and friction for dry and lubricated surfaces [17.23, 36].

Experimental Observations

Figure 17.22 shows the variation of single-point adhesive-force measurements as a function of tip radius on a Si(100) sample for several humidities. The adhesive force data are also plotted as a function of relative humidity for several tip radii. The general trend at humidities up to the ambient is that a 50-nm radius Si_3N_4 tip exhibits a slightly lower adhesive force compared to the other microtips of larger radii; in the latter case, values are similar. Thus, for the microtips there is no appreciable variation in adhesive force with tip radius at a given humidity up to the ambient. The adhesive force increases as relative humidity increases for all tips.

Sources of adhesive force between a tip and a sample surface are van der Waals attraction and meniscus formation [17.10, 58]. Relative magnitudes of the forces from the two sources are dependent upon various factors, including the distance between the tip and the sample surface, their surface roughness, their hydrophobicity, and relative humidity [17.61]. For most rough surfaces, meniscus contribution dominates at moderate to high humidities, which arise from capillary condensation of water vapor from the environment. If enough liquid is present to form a meniscus bridge, the meniscus force should increase with an increase in tip radius (proportional to tip radius for a spherical tip). In addition, an increase in tip radius results in increased contact area leading to higher values of van der Waals forces. However, if nanoasperities on the tip and the sample are considered, then the number of contacting and near-contacting asperities forming meniscus bridges increases with an increase of humidity leading to an increase in meniscus forces. These explain the trends observed in Fig. 17.22. From the data, the tip radius has little effect on the adhesive forces at low humidities, but increases with tip radius at high humidity. Adhesive force also increases with an increase in humidity for all tips. This observation suggests that thickness of the liquid film at low humidities is insufficient to form continuous meniscus bridges to affect adhesive forces in the case of all tips.

Figure 17.22 also shows the variation in coefficient of friction as a function of tip radius at a given humidity and as a function of relative humidity for a given tip radius for Si(100). It can be observed that for 0% RH, the coefficient of friction is about the same for the tip radii except for the largest tip, which shows a higher value. At all other humidities, the trend consistently shows that the coefficient of friction increases with tip radius. An increase in friction with tip radius at low to moderate humidities arises from increased contact area (higher van der Waals forces) and higher values of shear forces required for larger contact area. At high humidities, similar to adhesive force data, an increase with tip radius occurs due to both contact area and meniscus effects. Although AFM/FFM measurements are able to measure the combined effect of the contribution of van der Waals

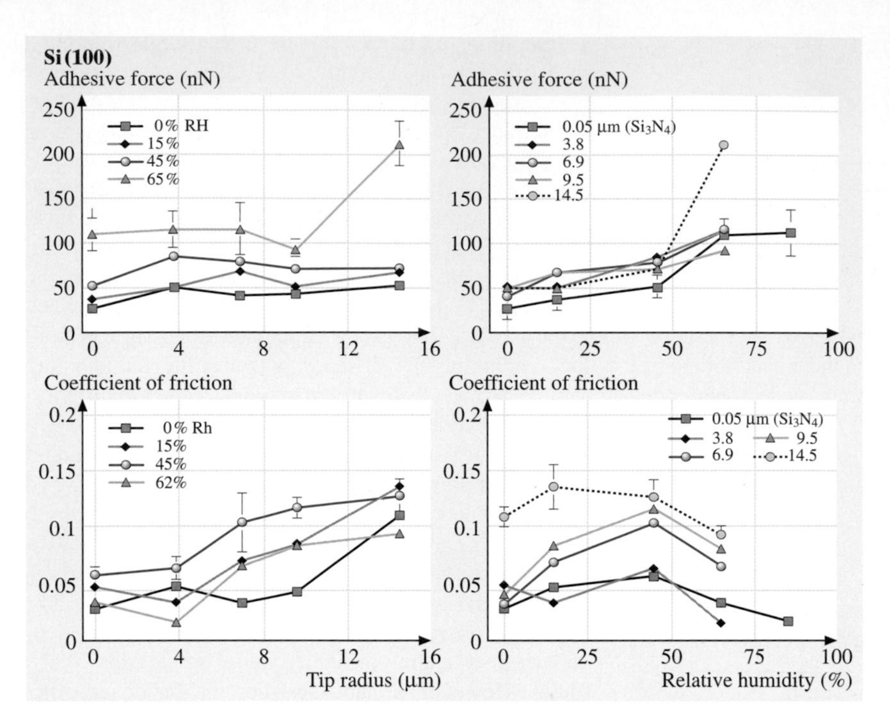

Fig. 17.22 Adhesive force and coefficient of friction as a function of tip radius at several humidities and as a function of relative humidity (RH) at several tip radii on Si(100) [17.23]

and meniscus forces toward friction force or adhesive force, it is difficult to measure their individual contributions separately. It can be seen that for all tips, the coefficient of friction increases with humidity to about ambient, beyond which it starts to decrease. The initial increase in the coefficient of friction with humidity arises from the fact that the thickness of the water film increases with an increase in the humidity, which results in a larger number of nanoasperities forming meniscus bridges and leads to higher friction (larger shear force). The same trend is expected with the microtips beyond 65% RH. This is attributed to the fact that at higher humidities, the adsorbed water film on the surface acts as a lubricant between the two surfaces. Thus the interface is changed at higher humidities, resulting in lower shear strength and, hence, lower friction force and coefficient of friction.

Adhesion and Friction Force Expressions for a Single-Asperity Contact

We now obtain the expressions for the adhesive force and coefficient of friction for a single-asperity contact with a meniscus formed at the interface. For a spherical asperity of radius R in contact with a flat and smooth surface with the composite modulus of elasticity E^* and with a concave meniscus, the attractive meniscus force (adhesive force) W_{ad} is given as [17.10, 58]

$$W_{ad} = 2\pi R\gamma_1 (\cos\theta_1 + \cos\theta_2) \; . \tag{17.6}$$

For an elastic contact for both extrinsic (W) and intrinsic (W_{ad}) normal load, the friction force is given as

$$F_e = \pi\tau \left(\frac{3(W + W_{ad})R}{4E^*} \right)^{2/3} , \tag{17.7}$$

where γ_1 is the surface tension of the liquid in air, θ_1 and θ_2 are the contact angles between the liquid and the two surfaces, W is the external load, and τ is the average shear strength of the contacts. (Surface energy effects are not considered here.) Note that adhesive force increases linearly with an increase in the tip radius, and the friction force increases with an increase in tip radius as $R^{2/3}$ and with normal load as $(W + W_{ad})^{2/3}$. The experimental data in support of $W^{2/3}$ dependence on the friction force can be found in various references (see, for example, [17.62]). The coefficient of friction μ_e is obtained from (17.7) as

$$\mu_e = \frac{F_e}{(W + W_{ad})} = \pi\tau \left(\frac{3R}{4E^*} \right)^{2/3} \frac{1}{(W + W_{ad})^{1/3}} \; . \tag{17.8}$$

In the plastic contact regime [17.58], the coefficient of friction μ_p is obtained as

$$\mu_p = \frac{F_p}{(W + W_{ad})} = \frac{\tau}{H_s}\,, \tag{17.9}$$

where H_s is the hardness of the softer material. Note that in the plastic contact regime, the coefficient of friction is independent of external load, adhesive contributions, and surface geometry.

In comparison, for multiple-asperity contacts in the elastic contact regime, the total adhesive force W_{ad} is the summation of adhesive forces at n individual contacts,

$$W_{ad} = \sum_{i=1}^{n} (W_{ad})_i \quad \text{and}$$

$$\mu_e \approx \frac{3.2\tau}{E^* \left(\sigma_p / R_p\right)^{1/2} + W_{ad}/W}\,, \tag{17.10}$$

where σ_p and R_p are the standard deviation of summit heights and average summit radius, respectively. Note that the coefficient of friction depends upon the surface roughness. In the plastic contact regime, the expression for μ_p in (17.9) does not change.

The source of the adhesive force, in a wet contact in the AFM experiments being performed in an ambient environment, includes mainly attractive meniscus force due to capillary condensation of water vapor from the environment. The meniscus force for a single contact increases with an increase in tip radius. A sharp AFM tip in contact with a smooth surface at low loads (on the order of a few nN) for most materials can be simulated as a single-asperity contact. At higher loads, for rough and soft surfaces, multiple contacts would occur. Furthermore, at low loads (nN range) for most materials, the local deformation would be primarily elastic. Assuming that shear strength of contacts does not change, the adhesive force for smooth and hard surfaces at low normal load (on the order of few nN) (for a single-asperity contact in the elastic contact regime) would increase with an increase in tip radius, and the coefficient of friction would decrease with an increase in total normal load as $(W + W_{ad})^{-1/3}$ and would increase with an increase of tip radius as $R^{2/3}$. In this case, the Amontons law of friction, which states that the coefficient of friction is independent of normal load and is independent of apparent area of contact, does not hold. For a single-asperity plastic contact and multiple-asperity plastic contacts, neither the normal load nor tip radius come into play in the calculation of the coefficient of friction. In the case of multiple-asperity contacts, the number of contacts increase with an increase of normal load. Therefore, adhesive force increases with an increase in load.

In the data presented earlier in this section, the effect of tip radius and humidity on the adhesive forces and coefficient of friction is investigated for experiments with Si(100) surface at loads in the range of 10–100 nN. The multiple-asperity elastic contact regime is relevant for this study. An increase in humidity generally results in an increase in the number of meniscus bridges that would increase the adhesive force. As suggested earlier, that increase in humidity may also decrease the shear strength of contacts. A combination of an increase in adhesive

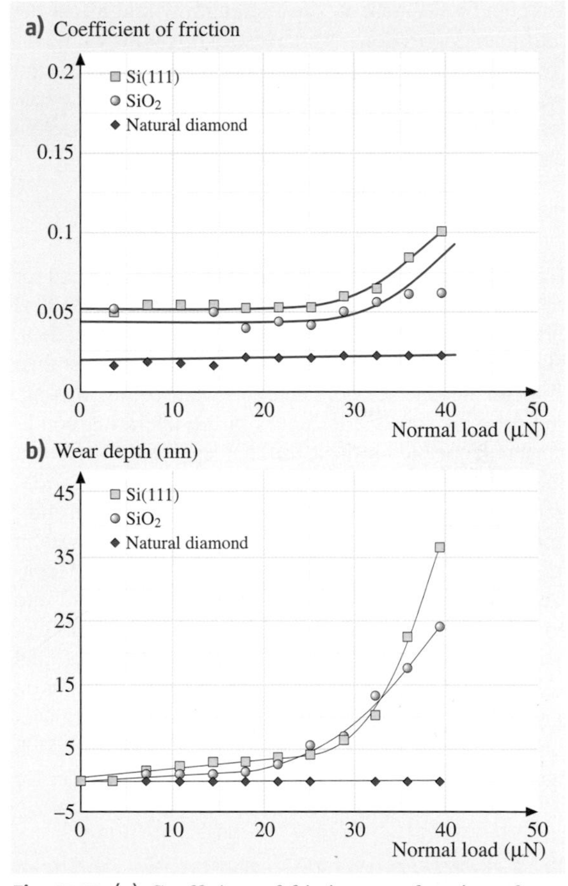

Fig. 17.23 **(a)** Coefficient of friction as a function of normal load and **(b)** corresponding wear depth as a function of normal load for silicon, SiO_2 coating, and natural diamond. Inflections in the curves for silicon and SiO_2 correspond to the contact stresses equal to the hardnesses of these materials [17.63]

Table 17.2 Surface roughness (standard deviation of surface heights or σ) and coefficients of friction on nano- and macroscales of various samples in air

Material	σ (nm)	Coefficient of nanoscale friction versus Si_3N_4 tip[a]	Coefficient of macroscale friction versus Si_3N_4 ball[b]
Graphite (HOPG)	0.09	0.006	0.1
Natural diamond	2.3	0.04	0.2
Si(100)	0.14	0.07	0.4

[a] Tip radius of about 50 nm in the load range of 10–150 nN (2.5–6.1 GPa) and a scanning speed of 0.5 nm/s and scan area of 1 nm × 1 nm for HOPG and a scanning speed of 4 μm/s and scan area of 1 μm × 1 μm for diamond and Si(100)

[b] Ball radius of 3 mm at a normal load of 1 N (0.6 GPa) and average sliding speed of 0.8 mm/s

force and a decrease in shear strength would affect the coefficient of friction. An increase in tip radius would increase the meniscus force (adhesive force). A substantial increase in the tip radius may also increase interatomic forces. These effects influence the coefficient of friction with an increase in the tip radius.

17.2.6 Scale Dependence on Friction

Table 17.2 shows the coefficient of friction measured for various materials on nano- and macroscales [17.19]. It is clearly observed that friction values are scale-dependent. The values on the nano/microscale are much lower than those on the macroscale. There can be the following four, and possibly more, differences in the operating conditions responsible for these differences. First, the contact stresses at AFM conditions, in spite of small tip radii, generally do not exceed the sample hardness that minimizes plastic deformation. Average contact stresses in macrocontacts are generally lower than in AFM contacts, however, a large number of asperities come into contact that go through some plastic deformation. Second, when measured for the small contact areas and very low loads used in microscale studies, indentation hardness is higher than at the macroscale, as will be discussed later [17.11, 34, 35]. Lack of plastic deformation and improved mechanical properties reduce the degree of wear and friction. Third, the small apparent area of contact reduces the number of particles trapped at the interface, and thus minimizes the third-body plowing contribution to the friction force [17.11]. As a fourth and final difference, we have seen in the previous section that coefficient of friction increases with an increase in the AFM tip radius. AFM data are taken with a sharp tip, whereas asperities coming in contact in macroscale tests range from nanoasperities to much larger asperities, which may be responsible for larger values of friction force on the macroscale.

To demonstrate the load dependence on the coefficient of friction, stiff cantilevers were used to conduct friction experiments at high loads, see Fig. 17.23 [17.63]. At higher loads (with contact stresses exceeding the hardness of the softer material), as anticipated, the coefficient of friction for microscale measurements increases toward values comparable to those obtained from macroscale measurements, and surface damage also increases. At high loads, plowing is a dominant contributor to the friction force. Based on these results, Amontons rule of friction, which states that the coefficient of friction is independent of apparent area and normal load, does not hold for microscale measurements. These findings also suggest that microcomponents sliding under lightly loaded conditions should experience ultralow friction and near-zero wear [17.11].

17.3 Scratching, Wear, Local Deformation, and Fabrication/Machining

17.3.1 Nanoscale Wear

Bhushan and *Ruan* [17.18] conducted nanoscale wear tests on polymeric magnetic tapes using conventional silicon nitride tips at two different loads of 10 and 100 nN (Fig. 17.24). For a low normal load of 10 nN, measurements were made twice. There was no discernible difference between consecutive measurements for this load. However, as the load was increased from 10 nN to 100 nN, topographical changes were ob-

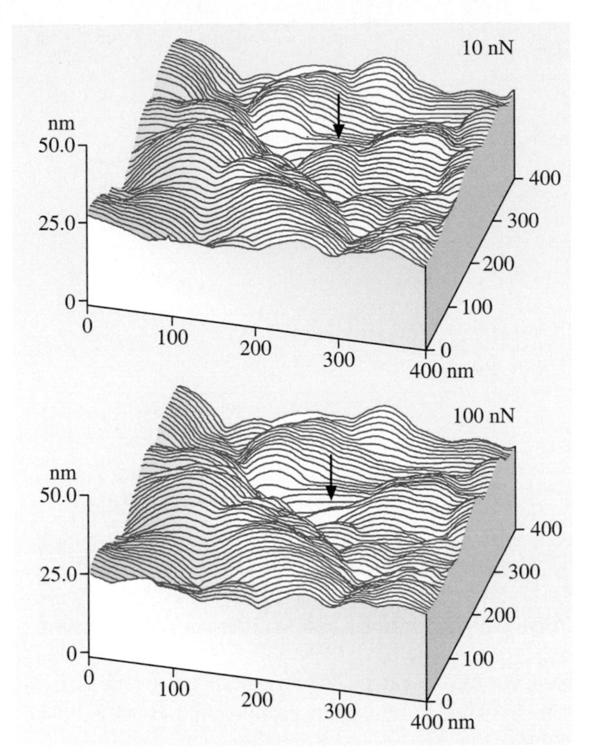

Fig. 17.24 Surface roughness maps of a polymeric magnetic tape at the applied normal load of 10 nN and 100 nN. Location of the change in surface topography as a result of nanowear is indicated by *arrows* [17.18]

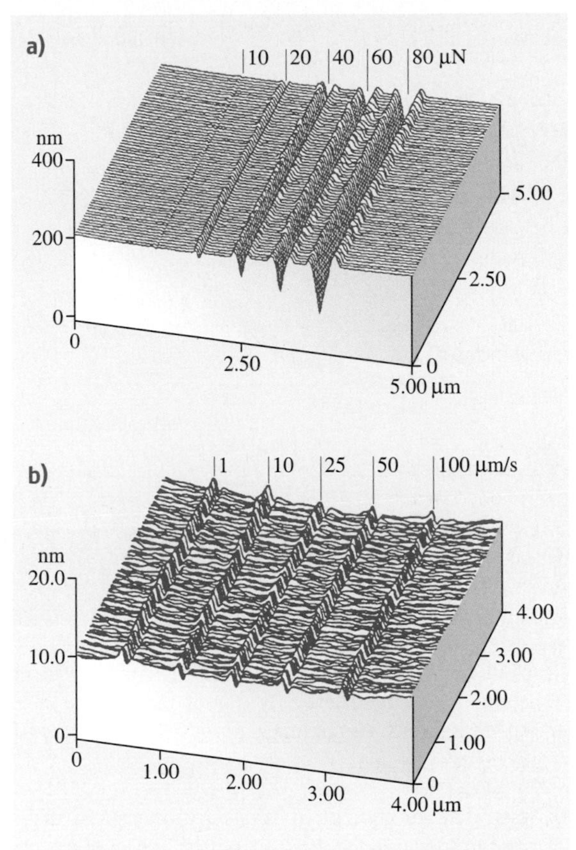

Fig. 17.25a,b Surface plots of (**a**) Si(111) scratched for 10 cycles at various loads and a scanning velocity of 2 μm/s. Note that x and y axes are in μm and z axis is in nm. (**b**) Si(100) scratched in one unidirectional scan cycle at a normal force of 80 μN and different scanning velocities

served during subsequent scanning at a normal load of 10 nN; material was pushed in the sliding direction of the AFM tip relative to the sample. The material movement is believed to occur as a result of plastic deformation of the tape surface. Thus, deformation and movement of the soft materials on a nanoscale can be observed.

17.3.2 Microscale Scratching

The AFM can be used to investigate how surface materials can be moved or removed on micro- to nanoscales, for example, in scratching and wear [17.4] (where these things are undesirable), and nanofabrication/nanomachining (where they are desirable). Figure 17.25a shows microscratches made on Si(111) at various loads and scanning velocity of 2 μm/s after 10 cycles [17.22]. As expected, the scratch depth increases linearly with load. Such microscratching measurements can be used to study failure mechanisms on the microscale and evaluate the mechanical integrity (scratch resistance) of ultrathin films at low loads.

To study the effect of scanning velocity, unidirectional scratches, 5 μm in length, were generated at scanning velocities ranging from 1–100 μm/s at various normal loads ranging from 40–140 μN. There is no effect of scanning velocity obtained at a given normal load. (For representative scratch profiles at 80 μN, see Fig. 17.25b.) This may be because of a small effect of frictional heating with the change in scanning velocity used here. Furthermore, for a small change in interface temperature, there is a large underlying volume to dissipate the heat generated during scratching.

Scratching can be performed under ramped loading to determine the scratch resistance of materials and

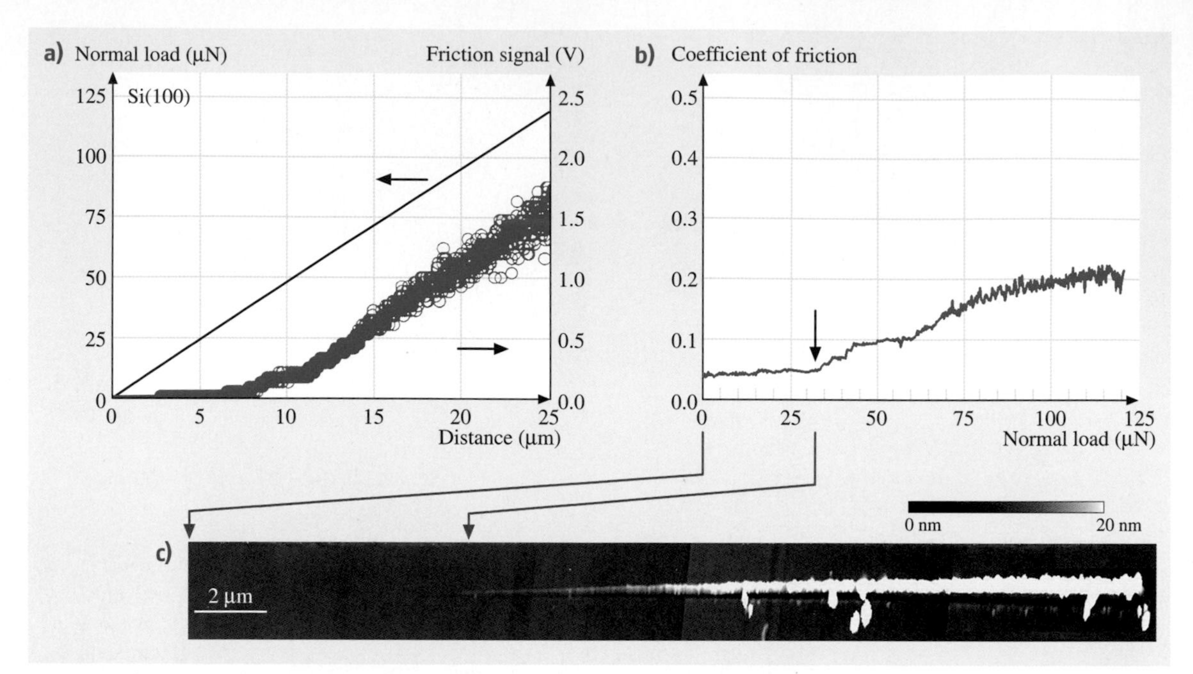

Fig. 17.26 (a) Applied normal load and friction signal measured during the microscratch experiment on Si(100) as a function of scratch distance, (b) friction data plotted in the form of coefficient of friction as a function of normal load, and (c) AFM surface height image of scratch obtained in tapping mode [17.33]

coatings. The coefficient of friction is measured during scratching, and the load at which the coefficient of friction increases rapidly is known as the "critical load", which is a measure of scratch resistance. In addition, post-scratch imaging can be performed in situ with the AFM in tapping mode to study failure mechanisms. Figure 17.26 shows data from a scratch test on Si(100) with a scratch length of 25 µm and a scratching velocity of 0.5 µm/s. At the beginning of the scratch, the coefficient of friction is 0.04, which indicates a typical value for silicon. At about 35 µN (indicated by the arrow in the figure), there is a sharp increase in the coefficient of friction, which indicates the critical load. Beyond the critical load, the coefficient of friction continues to increase steadily. In the post-scratch image, we note that at the critical load, a clear groove starts to form. This implies that Si(100) was damaged by plowing at the critical load, associated with the plastic flow of the material. At and after the critical load, small and uniform debris is observed and the amount of debris increases with increasing normal load. *Sundararajan* and *Bhushan* [17.33] have also used this technique to measure the scratch resistance of diamond-like carbon coatings ranging in thickness from 3.5–20 nm.

17.3.3 Microscale Wear

By scanning the sample in two dimensions with the AFM, wear scars are generated on the surface. Figure 17.27 shows the effect of normal load on wear depth. We note that wear depth is very small below 20 µN of normal load [17.64, 65]. A normal load of 20 µN corresponds to contact stresses comparable

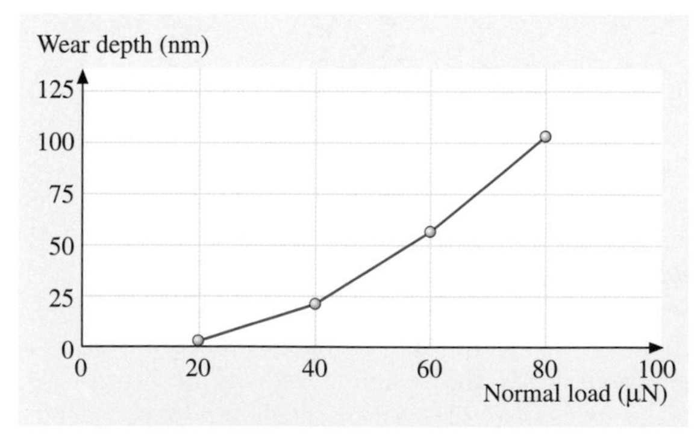

Fig. 17.27 Wear depth as a function of normal load for Si(100) after one cycle [17.65]

to the hardness of the silicon. Elastic deformation at loads below 20 μN is primarily responsible for low wear [17.63].

A typical wear mark, of the size 2 μm × 2 μm, generated at a normal load of 40 μN for one scan cycle and imaged using AFM with scan size of 4 μm × 4 μm at 300 nN load is shown in Fig. 17.28a [17.64]. The inverted map of wear marks shown in Fig. 17.28b indicates the uniform material removal at the bottom of the wear mark. An AFM image of the wear mark shows small debris at the edges, swiped during AFM scanning. Thus, the debris is loose (not sticky) and can be removed during the AFM scanning.

Next, we examine the mechanism of material removal on a microscale in AFM wear experiments [17.23, 64, 65]. Figure 17.29 shows a secondary electron image of the wear mark and associated wear particles. The specimen used for the scanning electron microscope (SEM) was not scanned with the AFM after initial wear, in order to retain wear debris in the wear region. Wear debris is clearly observed. In the SEM micrographs, the wear debris appear to be agglomerated because of high surface energy of the fine particles. Particles appear to be a mixture of rounded and so-called cutting type (feather-like or ribbon-like material). *Zhao* and *Bhushan* [17.65] reported an increase in the number and size of cutting-type particles with the normal load. The presence of cutting-type particles indicates that the material is removed primarily by plastic deformation.

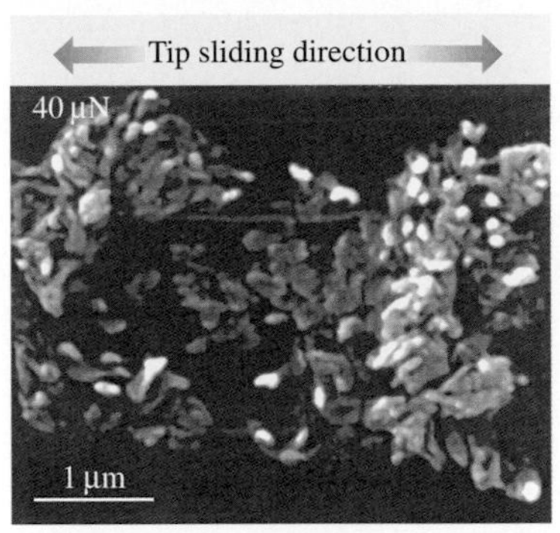

Fig. 17.29 Secondary electron image of wear mark and debris for Si(100) produced at a normal load of 40 μN and one scan cycle

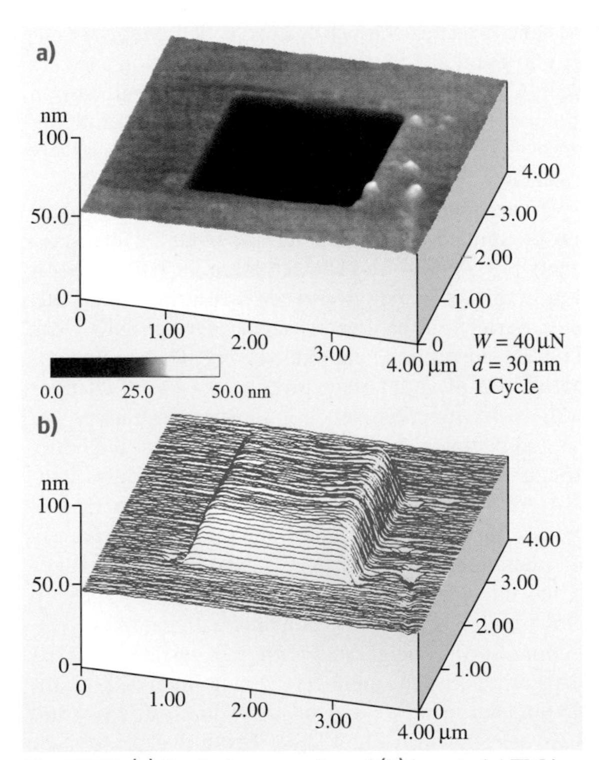

Fig. 17.28 **(a)** Typical gray-scale and **(b)** inverted AFM images of wear mark created using a diamond tip at a normal load of 40 μN and one scan cycle on Si(100) surface

To better understand the material removal mechanisms, transmission electron microscopy (TEM) has been used. The TEM micrograph of the worn region and associated diffraction pattern are shown in Fig. 17.30a,b. The bend contours are observed to pass through the wear mark in the micrograph. The bend contours around and inside the wear mark are indicative of a strain field, which, in the absence of applied stresses, can be interpreted as plastic deformation and/or elastic residual stresses. Often, localized plastic deformation during loading would lead to residual stresses during unloading; therefore, bend contours reflect a mix of elastic and plastic strains. The wear debris is observed outside the wear mark. The enlarged view of the wear debris in Fig. 17.30c shows that much of the debris is ribbon-like, indicating that material is removed by a cutting process via plastic deformation, which is consistent with the SEM observations. The diffraction pattern from inside the wear mark is similar to that of virgin silicon, showing no evidence of any phase transformation (amorphization) during wear. A selected area diffraction pattern of the wear debris shows some diffuse rings, which indicates the existence of amorphous

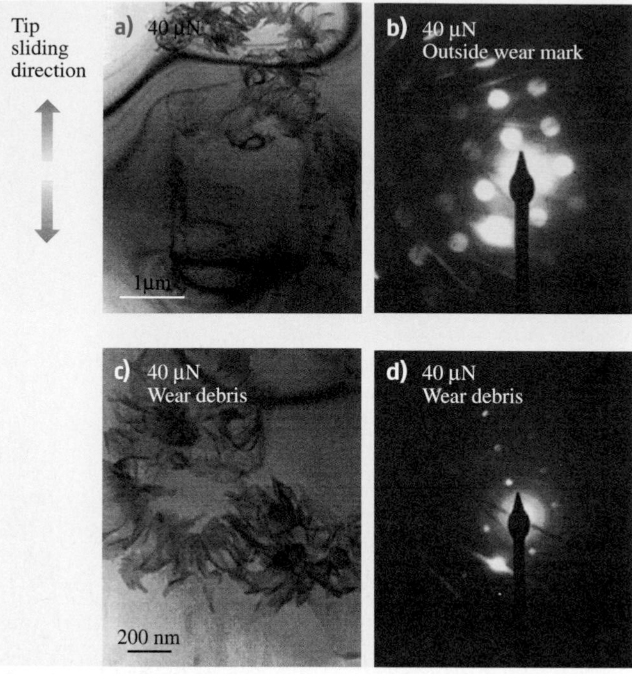

Fig. 17.30a–d Bright field TEM micrographs (*left*) and diffraction patterns (*right*) of wear mark (a),(b) and wear debris (c),(d) in Si(100) produced at a normal load of 40 µN and one scan cycle. Bend contours around and inside wear mark are observed

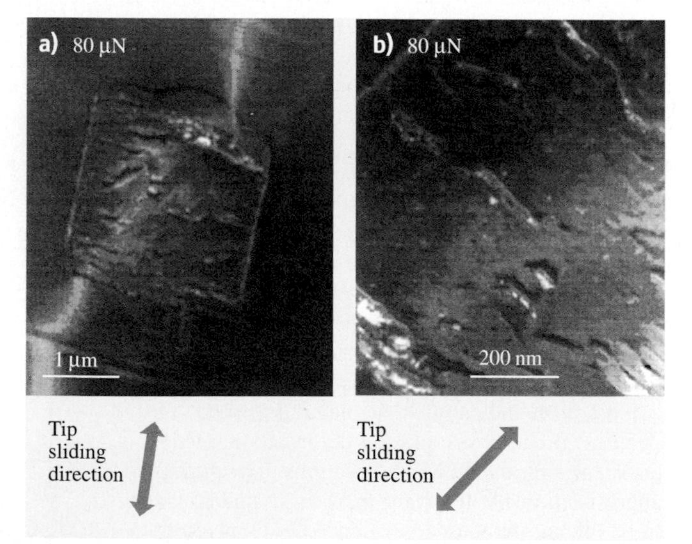

Fig. 17.31 (a) Bright field and (b) weak beam TEM micrographs of wear mark in Si(100) produced at a normal load of 80 µN and one scan cycle showing bend contours and dislocations [17.65]

material in the wear debris, confirmed as silicon oxide products from chemical analysis. It is known that plastic deformation occurs by generation and propagation of dislocations. No dislocation activity or cracking was observed at 40 µN. However, dislocation arrays could be observed at 80 µN. Figure 17.31 shows the TEM micrographs of the worn region at 80 µN; for better observation of the worn surface, wear debris was moved out of the wear mark using AFM with a large area scan at 300 nN after the wear test. The existence of dislocation arrays confirms that material removal occurs by plastic deformation. This corroborates the observations made in scratch tests at ramped load in the previous section. It is concluded that the material on microscale at high loads is removed by plastic deformation with a small contribution from elastic fracture [17.65].

To understand wear mechanisms, evolution of wear can be studied using AFM. Figure 17.32 shows evolution of wear marks of a DLC-coated disk sample. The data illustrate how the microwear profile for a load of 20 µN develops as a function of the number of scanning cycles [17.22]. Wear is not uniform, but is initiated at the nanoscratches. Surface defects (with high surface energy) present at nanoscratches act as initiation sites for wear. Coating deposition also may not be uniform on and near nanoscratches, which may lead to coating delamination. Thus, scratch-free surfaces will be relatively resistant to wear.

Wear precursors (precursors to measurable wear) can be studied by making surface potential measurements [17.38–40]. The contact potential difference, or simply surface potential between two surfaces, depends on a variety of parameters such as electronic work function, adsorption, and oxide layers. The surface potential map of an interface gives a measure of changes in the work function, which is sensitive to both physical and chemical conditions of the surfaces, including structural and chemical changes. Before material is actually removed in a wear process, the surface experiences stresses that result in surface and subsurface changes of structure and/or chemistry. These can cause changes in the measured potential of a surface. An AFM tip allows mapping of surface potential with nanoscale resolution. Surface height and change in surface potential maps of a polished single-crystal aluminum (100) sample abraded using a diamond tip at loads of 1 µN and 9 µN are shown in Fig. 17.33a. (Note that the sign of the change in surface potential is reversed here from that in [17.38].) It is evident that both abraded regions show a large potential contrast (∼ 0.17 V) with respect

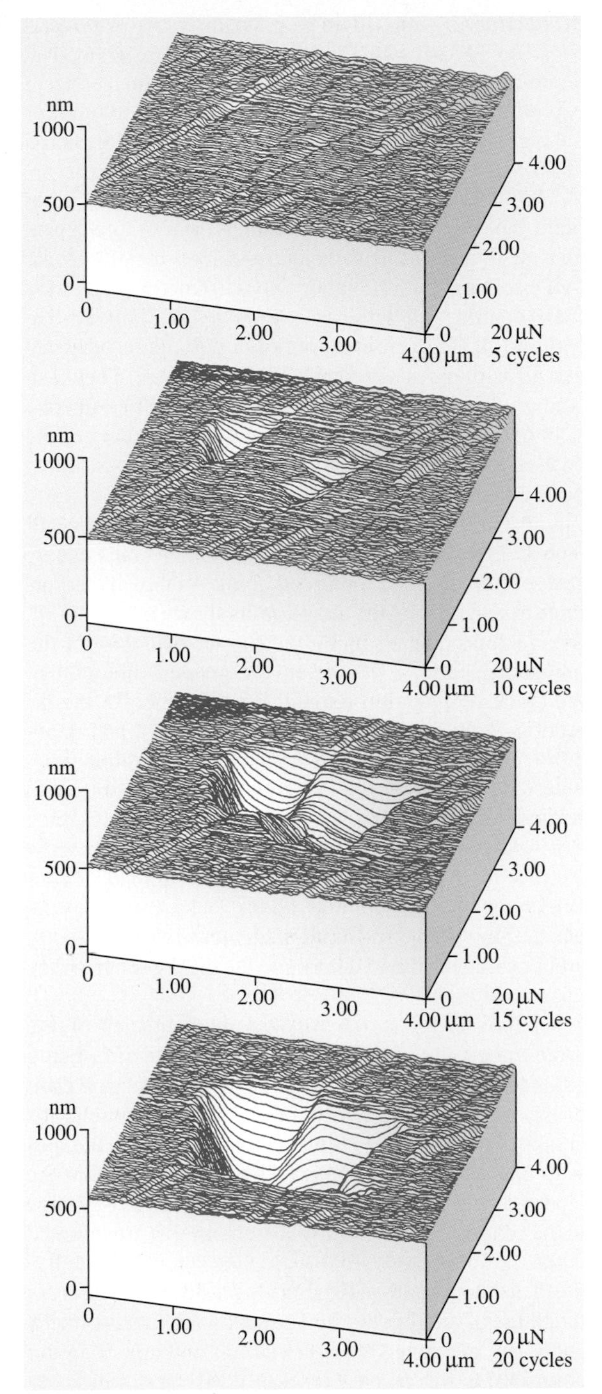

Fig. 17.32 Surface plots of diamond-like carbon-coated thin-film disk showing the worn region; the normal load and number of test cycles are indicated [17.22]

to the non-abraded area. The black region in the lower right-hand part of the topography scan shows a step that was created during the polishing phase. There is no potential contrast between the high region and the low region of the sample, indicating that the technique is independent of surface height. Figure 17.33b shows a close-up scan of the upper (low load) wear region in Fig. 17.33a. Notice that while there is no detectable change in the surface topography, there is, nonetheless, a large change in the potential of the surface in the worn region. Indeed, the wear mark of Fig. 17.33b might not have been visible at all in the topography map were it not for the noted absence of wear debris generated nearby and then swept off during the low load scan. Thus, even in the case of zero wear (no mea-

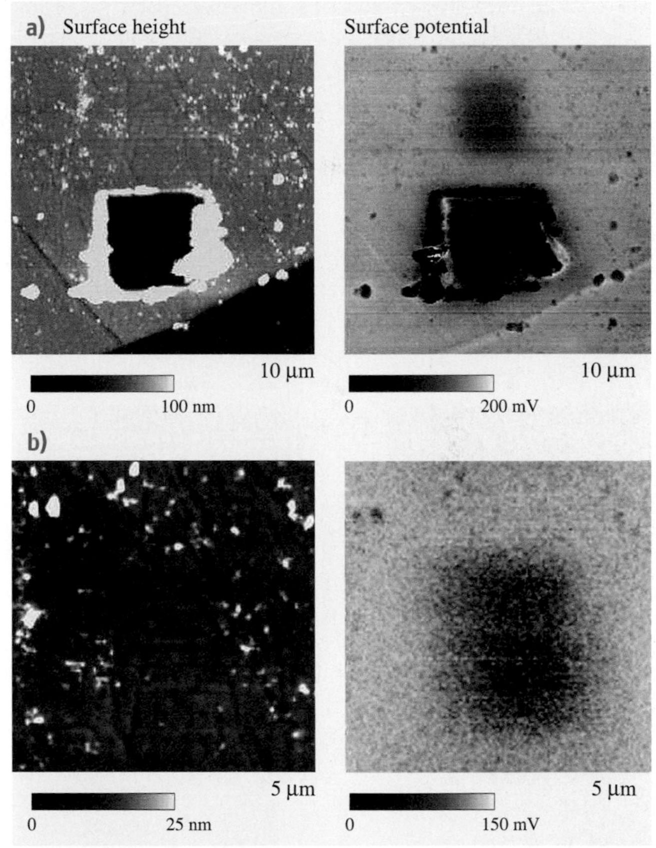

Fig. 17.33 (**a**) Surface height and change in surface potential maps of wear regions generated at 1 µN (*top*) and 9 µN (*bottom*) on a single-crystal aluminum sample showing bright contrast in the surface potential map on the worn regions. (**b**) Close up of upper (low load) wear region [17.38]

surable deformation of the surface using AFM), there can be a significant change in the surface potential inside the wear mark, which is useful for the study of wear precursors. It is believed that the removal of the thin contaminant layer including the natural oxide layer gives rise to the initial change in surface potential. The structural changes, which precede generation of wear debris and/or measurable wear scars, occur under ultralow loads in the top few nanometers of the sample and are primarily responsible for the subsequent changes in surface potential.

17.3.4 In Situ Characterization of Local Deformation

In situ surface characterization of local deformation of materials and thin films is carried out using a tensile stage inside an AFM. Failure mechanisms of polymeric thin films were studied by *Bobji* and *Bhushan* [17.41, 42]. The specimens were strained at a rate of 4×10^{-3}% per second, and AFM images were captured at different strains up to about 10% to monitor generation and propagation of cracks and deformation bands.

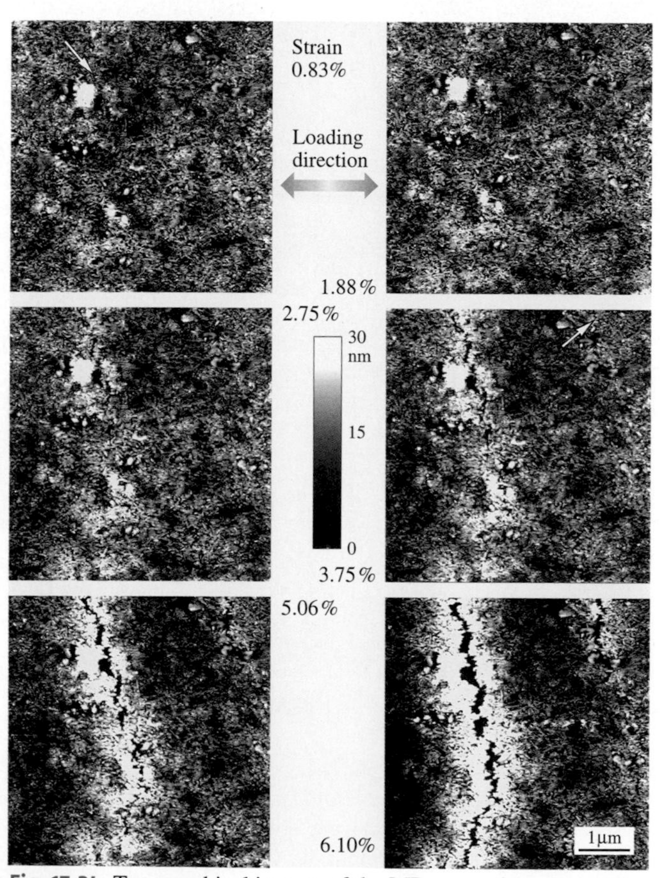

Fig. 17.34 Topographical images of the MP magnetic tape at different strains [17.41]

Bobji and *Bhushan* [17.41, 42] studied three magnetic tapes with thickness ranging from 7 to 8.5 μm. One of these was with acicular-shaped metal particle (MP) coating, and the other two with metal-evaporated (ME) coating and with and without a thin diamond-like carbon (DLC) overcoat, both on a polymeric substrate and all with particulate back coating [17.13]. They also studied the polyethylene terephthalate (PET) substrate with 6 μm thickness. They reported that cracking of the coatings started at about 1% strain for all tapes, much before the substrate starts to yield at about 2% strain. Figure 17.34 shows the topographical images of the MP tape at different strains. At 0.83% strain, a crack can be seen originating at the marked point. As the tape is further stretched along the direction, as shown in Fig. 17.34, the crack propagates along the shorter boundary of the ellipsoidal particle. However, the general direction of the crack propagation remains perpendicular to the direction of the stretching. The length, width, and depth of the cracks increase with strain, and at the same time, newer cracks keep on nucleating and propagating with reduced crack spacing. At 3.75% strain, another crack can be seen nucleating. This crack continues to grow parallel to the first. When the tape is unloaded after stretching up to a strain of about 2%, i. e., within the elastic limit of the substrate, the cracks rejoin perfectly and it is impossible to determine the difference from the unstrained tape.

Figure 17.35 shows topographical images of the three magnetic tapes and the PET substrate after being strained to 3.75%, which is well beyond the elastic limit of the substrate. MP tape develops short and numerous cracks perpendicular to the direction of loading. In tapes with metallic coating, the cracks extend throughout the tape width. In ME tape with DLC coating, there is a bulge in the coating around the primary cracks that are initiated when the substrate is still elastic, like crack A in the figure. The white band on the right-hand side of the figure is the bulge of another crack. The secondary cracks like B and C are generated at higher strains and are straighter compared to the primary cracks. In ME tape, which has a Co-O film on PET substrate, with a thickness ratio of 0.03, both with and without DLC coating, no difference is observed in the rate of growth between primary and secondary cracks. The failure is cohesive with no

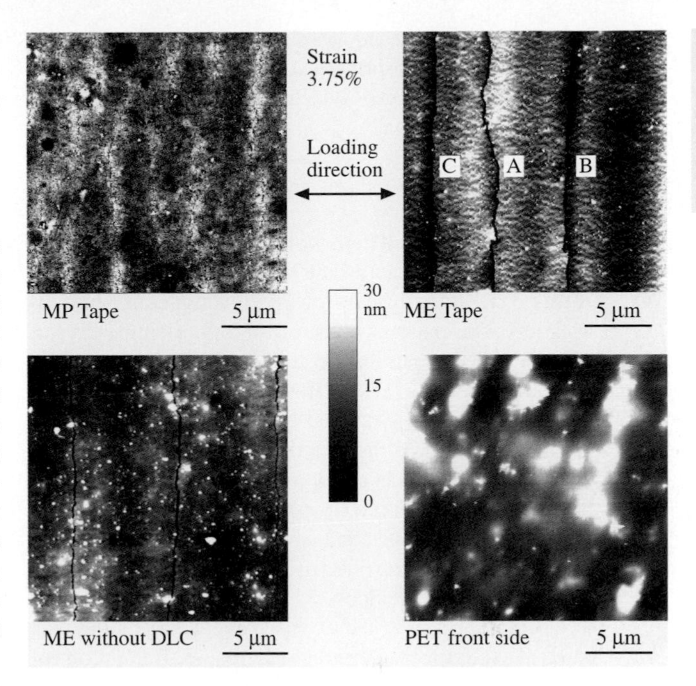

Fig. 17.35 Comparison of crack morphologies at 3.75% strain in three magnetic tapes and PET substrate. Cracks B and C, nucleated at higher strains, are more linear than crack A [17.42]

bulging of the coating. This seems to suggest that the DLC coating has residual stresses that relax when the coating cracks, causing delamination. Since the stresses are already relaxed, the secondary crack does not result in delamination. The presence of the residual stress is confirmed by the fact that a freestanding ME tape curls up (in a cylindrical form with its axis perpendicular to the tape length) with a radius of curvature of about 6 mm; the ME tape without the DLC does not curl. The magnetic coating side of PET substrate is much smoother at smaller scan lengths. However, in 20 μm scans it has a lot of bulging out, which appears as white spots in the figure. These spots change shape even while scanning the samples in tapping mode at very low contact forces.

The variation of average crack width and average crack spacing with strain is plotted in Fig. 17.36. The crack width is measured at a spot along a given crack over a distance of 1 μm in the 5 μm scan image at different strains. The crack spacing is obtained by averaging the inter-crack distance measured in five separate 50 μm scans at each strain. It can be seen that the cracks nucleate at a strain of about 0.7–1.0%, well within the elastic limit of the substrate. There is a definite change in the slope of the load-displacement curve at the strain where cracks nucleate, and the slope after that is closer to the slope of the elastic portion of the substrate. This would mean that most of the load is supported by the substrate once the coating fails by cracking.

In situ surface characterization of unstretched and stretched films has been used to measure Poisson's ratio of polymeric thin films [17.66]. Uniaxial tension is applied by the tensile stage. Surface height profiles obtained from the AFM images of unstretched and stretched samples are used to monitor simultaneously the changes in displacements of the polymer films in the longitudinal and lateral directions.

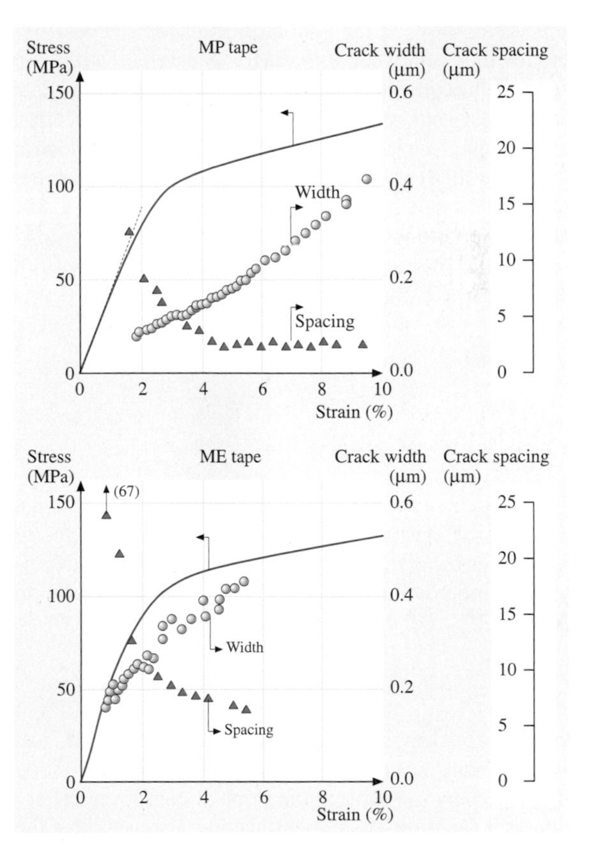

Fig. 17.36 Variation of stress, crack width, and crack spacing with strain in three magnetic tapes and PET substrate [17.41]

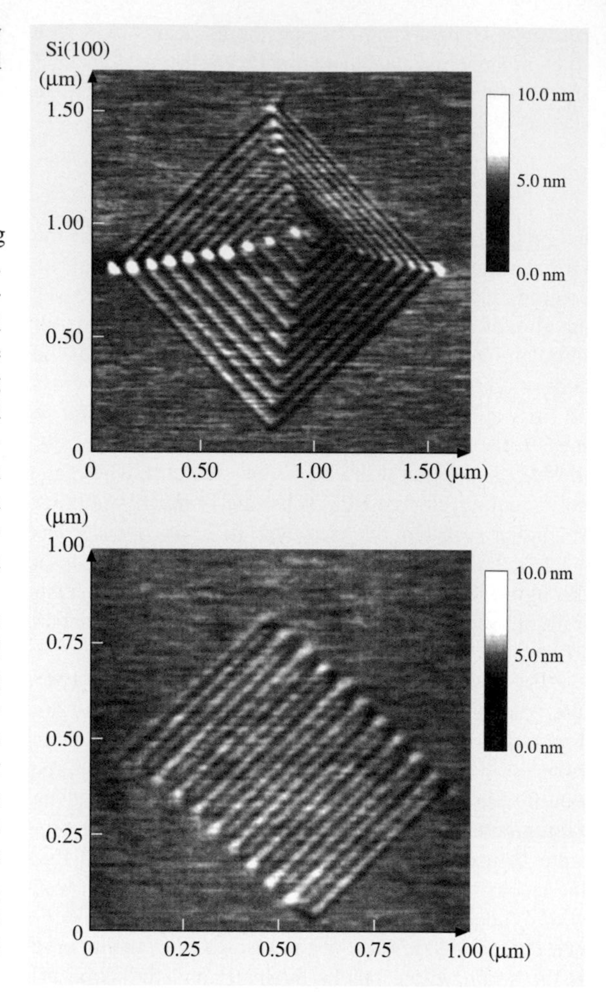

Fig. 17.37 (a) Trim and (b) spiral patterns generated by scratching a Si(100) surface using a diamond tip at a normal load of 15 µN and writing speed of 0.5 µm/s

17.3.5 Nanofabrication/Nanomachining

An AFM can be used for nanofabrication/nanomachining by extending the microscale scratching operation [17.4, 11,22,37]. Figure 17.37 shows two examples of nanofabrication. The patterns were created on a single-crystal silicon (100) wafer by scratching the sample surface with a diamond tip at specified locations and scratching angles. Each line is scribed manually at a normal load of 15 µN and a writing speed of 0.5 µm/s. The separation between lines is about 50 nm, and the variation in line width is due to the tip asymmetry. Nanofabrication parameters – normal load, scanning speed, and tip geometry – can be controlled precisely to control the depth and length of the devices.

Nanofabrication using mechanical scratching has several advantages over other techniques. Better control over the applied normal load, scan size, and scanning speed can be used for nanofabrication of devices. Using the technique, nanofabrication can be performed on any engineering surface. Use of chemical etching or reactions is not required, and this dry nanofabrication process can be used where use of chemicals and electric field is prohibited. One disadvantage of this technique is the formation of debris during scratching. At light loads, debris formation is not a problem compared to high-load scratching. However, debris can be removed easily from the scan area at light loads during scanning.

17.4 Indentation

Mechanical properties such as hardness and Young's modulus of elasticity can be determined on micro- to picoscales using the AFM [17.18,22,31,34] and a depth-sensing indentation system used in conjunction with an AFM [17.35,67–69].

17.4.1 Picoindentation

Indentability on the scale of subnanometers of soft samples can be studied in the force calibration mode (Fig. 17.6) by monitoring the slope of cantilever deflection as a function of sample traveling distance after the tip is engaged and the sample is pushed against the tip. For a rigid sample, cantilever deflection equals the sample traveling distance, but the former quantity is smaller if the tip indents the sample. In an example of a polymeric magnetic tape shown in Fig. 17.38, the line in the left portion of the figure is curved with a slope of less than 1 shortly after the sample touches the tip, which suggests that the tip has indented the sample [17.18]. Later, the slope is equal to 1, suggesting that the tip no longer indents the sample. This observation indicates that the tape surface is soft locally (polymer-rich) but hard (as a result of magnetic particles) underneath. Since

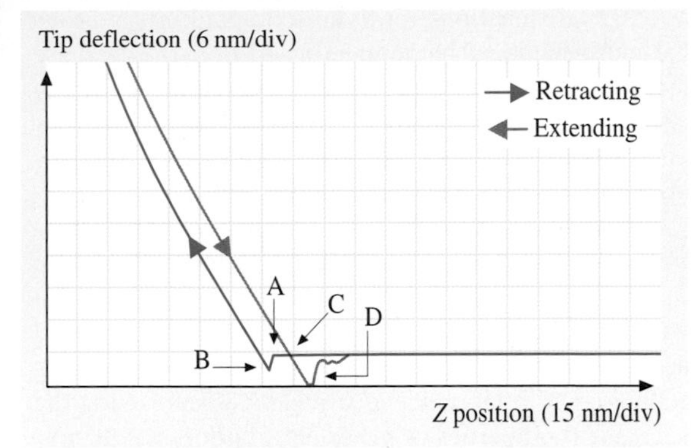

Fig. 17.38 Tip deflection (normal load) as a function of the Z (separation distance) curve for a polymeric magnetic tape [17.18]

the curves in extending and retracting modes are identical, the indentation is elastic up to a maximum load of about 22 nN used in the measurements.

Detection of transfer of material on a nanoscale is possible with the AFM. Indentation of C_{60}-rich fullerene films with an AFM tip has been shown [17.70] to result in the transfer of fullerene molecules to the AFM tip, as indicated by discontinuities in the cantilever deflection as a function of sample traveling distance in subsequent indentation studies.

17.4.2 Nanoscale Indentation

The indentation hardness of surface films with an indentation depth as small as about 1 nm can be measured using an AFM [17.11, 34, 35]. Figure 17.39 shows the gray-scale plots of indentation marks made on Si(111) at normal loads of 60, 65, 70, and 100 μN. Triangular indents can be clearly observed with very shallow depths. At a normal load of 60 μN, indents are observed and the depth of penetration is about 1 nm. As the normal load is increased, the indents become clearer and indentation depth increases. For the case of hardness measurements at shallow depths on the same order as variations in surface roughness, it is desirable to subtract the original (unindented) map from the indent map for accurate measurement of the indentation size and depth [17.22].

To make accurate measurements of hardness at shallow depths, a depth-sensing indentation system is used [17.35]. Figure 17.40 shows the load-displacement curves at different peak loads for Si(100). Loading/unloading curves often exhibit sharp discontinuities,

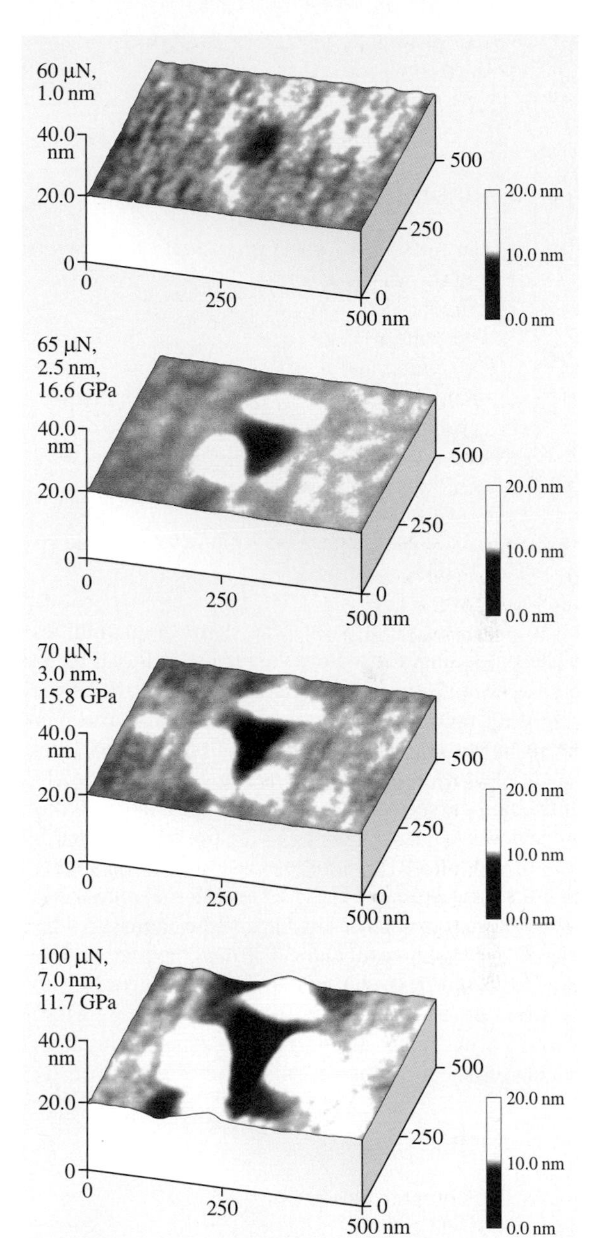

Fig. 17.39 Gray-scale plots of indentation marks on the Si(111) sample at various indentation loads. Loads, indentation depths, and hardness values are listed in the figure [17.34]

particularly at high loads. Discontinuities, referred to as pop-ins, during the initial loading part of the curve mark a sharp transition from pure elastic loading to a plastic deformation of the specimen surface, thus correspond-

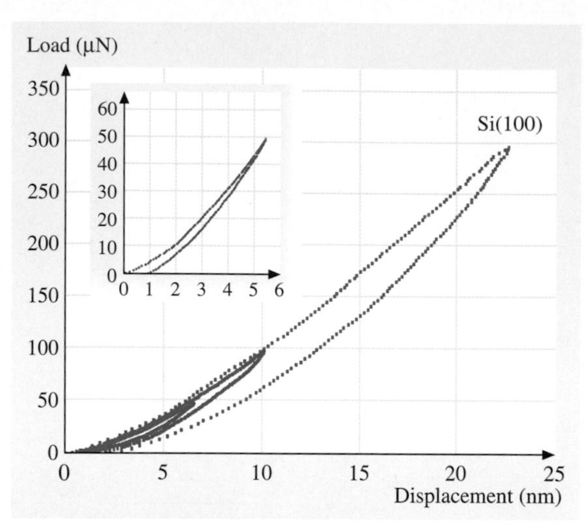

Fig. 17.40 Load-displacement curves at various peak loads for Si(100) [17.35]

ing to an initial yield point. The sharp discontinuities in the unloading part of the curves are believed to be due to the formation of lateral cracks that form at the base of the median crack, which results in the surface of the specimen being thrust upward. Load-displacement data at residual depths as low as about 1 nm can be obtained, and the indentation hardness of surface films has been measured for Si(100) [17.35, 67–69]. The hardness of silicon on a nanoscale is found to be higher than on a microscale (Fig. 17.41). Microhardness has also been reported to be higher than on the millimeter scale by several investigators. The data reported to date show that hardness exhibits size effect. According to the strain gradient plasticity theory advanced by *Fleck* et al. [17.71], large strain gradients inherent in small indentations lead to the accumulation of geometrically necessary dislocations for strain compatibility reasons that cause enhanced hardening. In addition, the decrease in hardness with an increase in indentation depth can possibly be rationalized on the basis that as the volume of deformed material increases, there is a higher probability of encountering material defects.

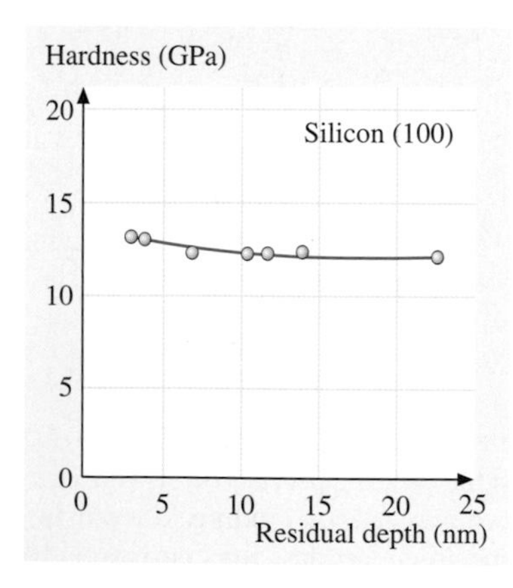

Fig. 17.41 Indentation hardness as a function of residual indentation depth for Si(100) [17.35]

Bhushan and *Koinkar* [17.31] have used AFM measurements to show that ion implantation of silicon surfaces increases their hardness and thus their wear resistance. Formation of surface alloy films with improved mechanical properties by ion implantation is of growing technological importance as a means of improving the mechanical properties of materials. Hardness of 20 nm-thick diamond-like carbon films has been measured by *Kulkarni* and *Bhushan* [17.69].

The creep and strain-rate effects (viscoelastic effects) of ceramics can be studied using a depth-sensing indentation system. *Bhushan* et al. [17.35] and *Kulkarni* and *Bhushan* [17.67–69] have reported that ceramics exhibit significant plasticity and creep on a nanoscale. Figure 17.42a shows the load-displacement curves for single-crystal silicon at various peak loads held at 180 s. To demonstrate the creep effects, the load-displacement curves for a 500 μN peak load held at 0 and 30 s are also shown as an inset. Note that significant creep occurs at room temperature. Nanoindenter experiments conducted by *Li* et al. [17.72] exhibited significant creep only at high temperatures (greater than or equal to 0.25 times the melting point of silicon). The mechanism of dislocation glide plasticity is believed to dominate the indentation creep process on the macroscale. To study the strain-rate sensitivity of silicon, data at two different (constant) rates of loading are presented in Fig. 17.42b. Note that a change in the loading rate by a factor of about five results in a significant change in the load-displacement data. The viscoelastic effects observed here for silicon at ambient temperature could arise from size effects mentioned earlier. Most likely, creep and strain rate experiments are being conducted on the hydrated films present on the silicon surface in ambient environment, and these films are expected to be viscoelastic.

17.4.3 Localized Surface Elasticity and Viscoelasticity Mapping

The Young's modulus of elasticity is calculated from the slope of the indentation curve during unloading. However, these measurements provide a single-point measurement. By using the force modulation technique, it is possible to get localized elasticity maps of soft and

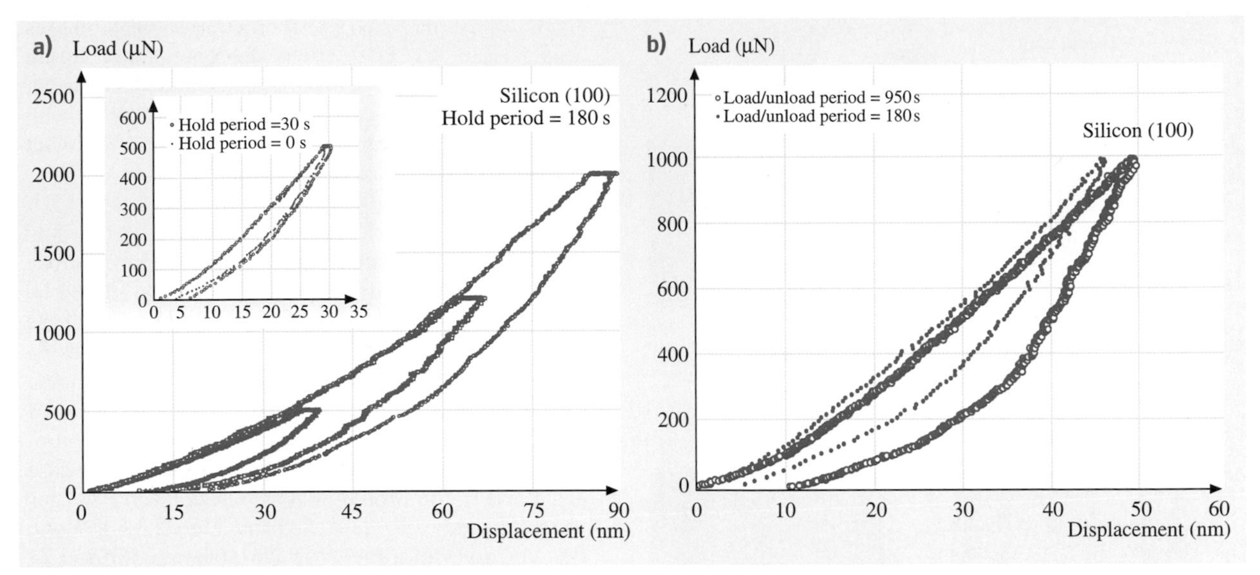

Fig. 17.42 (a) Creep behavior and (b) strain-rate sensitivity of Si(100) [17.35]

compliant materials with penetration depths of less than 100 nm. This technique has been successfully used for polymeric magnetic tapes, which consist of magnetic and nonmagnetic ceramic particles in a polymeric matrix. Elasticity maps of a tape can be used to identify relative distribution of hard magnetic and nonmagnetic ceramic particles on the tape surface, which has an effect on friction and stiction at the head-tape interface [17.13]. Figure 17.43 shows surface height and elasticity maps on a polymeric magnetic tape [17.44]. The elasticity image reveals sharp variations in surface elasticity due to the composite nature of the film. As can be clearly seen, regions of high elasticity do not always correspond to high or low topography. Based on a Hertzian elastic-contact analysis, the static indentation depth of these samples during the force modulation scan is estimated to be about 1 nm. We conclude that the contrast seen is influenced most strongly by material properties in the top few nanometers, independent of the composite structure beneath the surface layer.

By using phase contrast microscopy, it is possible to get phase contrast maps or the contrast in viscoelastic properties of near surface regions. This technique has been used successfully for polymeric films and magnetic tapes that consist of ceramic particles in a polymeric matrix [17.51, 52]. Figure 17.44 shows topography, phase angle, and friction force images for polyethylene terephthalate (PET) film. Three particles, which show up as approximately circular, lightly shaded (high points) regions, are seen in the topography image. In the phase angle images, the particles show up dark against the film background. A combination of high tapping amplitude and low set point was found to maximize contrast in phase images. This result may be explained by the degree to which the tip is able to penetrate and thus deform the sample. A high tapping amplitude and low set point should maximize the force with which the tip strikes the sample. This should give more penetration. So the mater-

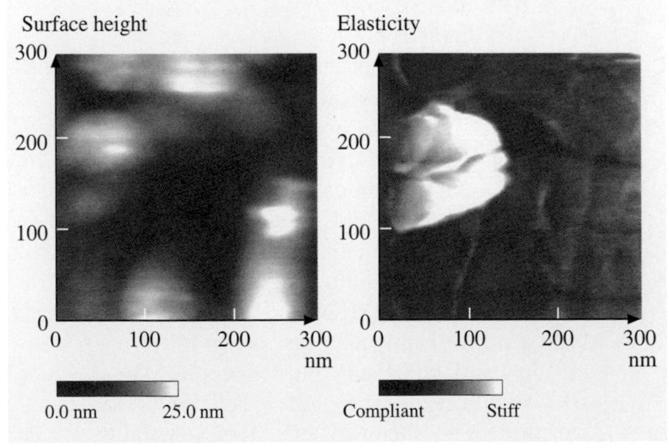

Fig. 17.43 Surface height and elasticity maps on a polymeric magnetic tape ($\sigma = 6.72$ nm and P-V = 31.7 nm; σ and P-V refer to standard deviation of surface heights and peak-to-valley distance, respectively). The gray scale on the elasticity map is arbitrary [17.44]

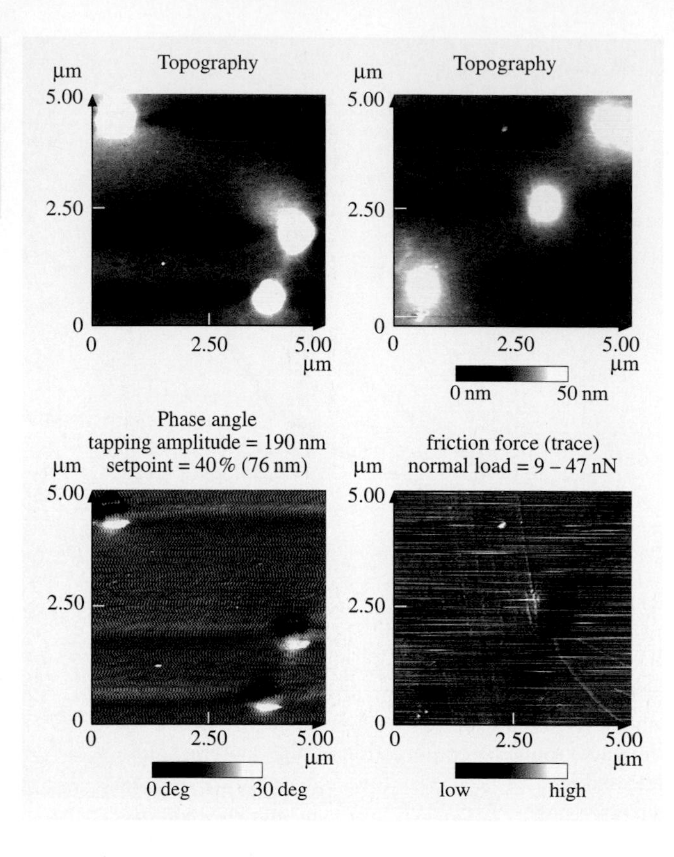

Fig. 17.44 Tapping mode height and phase angle images acquired using a TESP tip at a tapping amplitude of 190 nm and a setpoint of 40% (74 nm), and contact mode height and friction force (trace) images using an MESP tip with a normal load between 9 and 47 nN for PET film with embedded ceramic particles [17.51]

ial or viscoelastic contribution, as opposed to the surface force and adhesion hysteresis contributions, should be more dominant for this tapping condition. The phase angle image shown in Fig. 17.44 was acquired using tapping mode with a relatively high tapping amplitude of 190 nm and a relatively low set point of 40% (of the tapping amplitude) to emphasize viscoelastic properties. Very little correlation is found between the phase angle and friction force images. Friction force images were obtained using an etched single-crystal silicon tip (DI MESP) with a radius of 20–50 nm, a stiffness of 1–5 N/m, and a natural frequency of 60–70 kHz. A lack of correlation may be a further indication that something other than adhesion, which should correlate to friction force, causes the phase angle contrast; this may imply that viscoelastic properties may be dominant. These results also indicate that phase angle imaging yields information that cannot be obtained using other, more conventional AFM modes.

17.5 Boundary Lubrication

17.5.1 Perfluoropolyether Lubricants

The classical approach to lubrication uses freely supported multimolecular layers of liquid lubricants [17.10, 13, 58, 73]. The liquid lubricants are sometimes chemically bonded to improve their wear resistance [17.10, 13, 58]. Partially chemically bonded, molecularly thick perfluoropolyether (PFPE) films are used for lubrication of magnetic storage media [17.13]. These are considered potential candidate lubricants for micro/nanoelectromechanical systems (MEMS/NEMS). Molecularly thick PFPEs are well suited because of the following properties: low surface tension and low contact angle, which allow easy spreading on surfaces and provide hydrophobic properties; chemical and thermal stability, which minimize degradation under use; low vapor pressure, which provides low out-gassing; high adhesion to substrate via organic functional bonds; and good lubricity, which reduces contact surface wear.

For boundary lubrication studies, friction, adhesion, and durability experiments have been performed on virgin Si(100) surfaces and silicon surfaces lubricated with two PFPE lubricants – Z-15 (with $-CF_3$ nonpolar end groups) and Z-DOL (with $-OH$ polar end groups) [17.27, 28, 30]. Z-DOL film was thermally bonded at 150 °C for 30 minutes and an unbonded fraction was removed by a solvent (BW) [17.13]. The thicknesses of Z-15 and Z-DOL films were 2.8 nm and 2.3 nm, respectively.

The adhesive forces of Si(100), Z-15, and Z-DOL (BW) measured by force calibration plot and friction force versus normal load plot are summarized in Fig. 17.45. The results measured by these two methods are in good agreement. Figure 17.45 shows that the presence of mobile Z-15 lubricant film increases the adhesive force as compared to that of Si(100) by meniscus formation. The presence of solid phase Z-DOL (BW) film reduces the adhesive force as compared to that of

Si(100) because of the absence of mobile liquid. The schematic (bottom) in Fig. 17.45 shows relative size and sources of meniscus. It is well-known that the native oxide layer (SiO_2) on the top of a Si(100) wafer exhibits hydrophilic properties, and some water molecules can be adsorbed on this surface. The condensed water will form meniscus as the tip approaches the sample surface. The larger adhesive force in Z-15 is not only caused by the Z-15 meniscus; the nonpolarized Z-15 liquid does not have good wettability and strong bonding with Si(100). In the ambient environment, the condensed water molecules from the environment will permeate through the liquid Z-15 lubricant film and compete with the lubricant molecules present on the substrate. The interaction of the liquid lubricant with the substrate is weakened, and a boundary layer of the liquid lubricant forms puddles [17.27, 28]. This de-wetting allows water molecules to be adsorbed on the Si(100) surface as aggregates along with Z-15 molecules. And both of them can form meniscus while the tip approaches the surface. Thus, the de-wetting of liquid Z-15 film results in higher adhesive force and poorer lubrication performance. In addition, as the Z-15 film is pretty soft compared to the solid Si(100) surface, and penetration of the tip in the film occurs while pushing the tip down. This leads the large area of the tip involved to form the meniscus at the tip-liquid (mixture of Z-15 and water) interface. It should also be noted that Z-15 has a higher viscosity compared to water and, therefore, Z-15 film provides higher resistance to motion and coefficient of friction. In the case of Z-DOL (BW) film, both of the active groups of Z-DOL molecules are mostly bonded on Si(100) substrate. Thus, the Z-DOL (BW) film has low free surface energy and cannot be displaced readily by water molecules or readily adsorb water molecules. Therefore, the use of Z-DOL (BW) can reduce the adhesive force.

To study the velocity effect on friction and adhesion, the variation of friction force, adhesive force, and coefficient of friction of Si(100), Z-15 and Z-DOL (BW) as a function of velocity are summarized in Fig. 17.46. It indicates that for silicon wafer the friction force decreases logarithmically with increasing velocity. For Z-15, the friction force decreases with increasing velocity up to 10 μm/s, after which it remains almost constant. The velocity has very small effect on the friction force of Z-DOL (BW): It reduced slightly only at very high velocity. Figure 17.46 also indicates that the adhesive force of Si(100) increases when the velocity is higher than 10 μm/s. The adhesive force of Z-15 is reduced dramatically with a velocity increase up to 20 μm/s, after which it is reduced slightly. And the adhesive force of

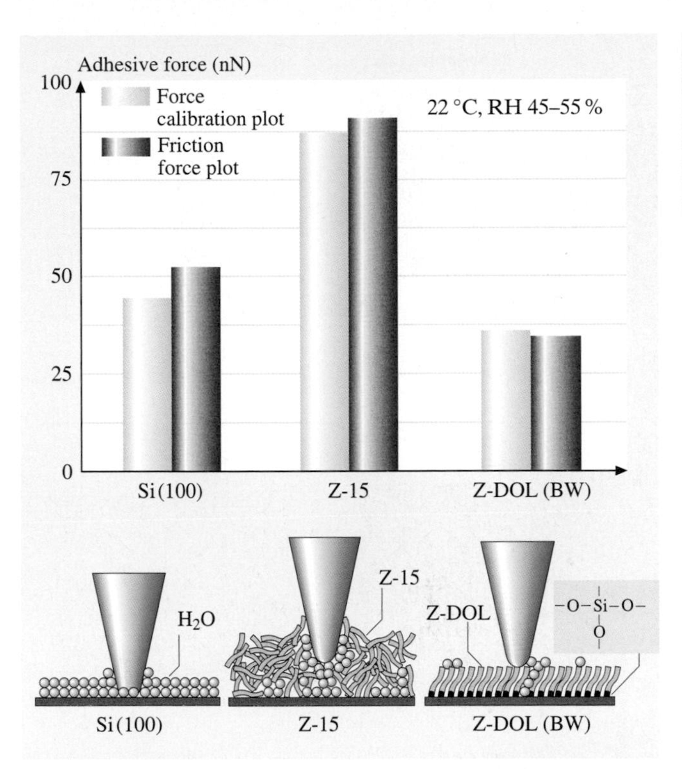

Fig. 17.45 Summary of the adhesive forces of Si(100) and Z-15 and Z-DOL (BW) films measured by force calibration plots and friction force versus normal load plots in ambient air. The schematic (*bottom*) showing the effect of meniscus formed between the AFM tip and surface sample on the adhesive and friction forces [17.30]

Z-DOL (BW) also decreases at high velocity. In the testing range of velocity, only the coefficient of friction of Si(100) decreases with velocity, but the coefficients of friction of Z-15 and Z-DOL (BW) almost remain constant. This implies that the friction mechanisms of Z-15 and Z-DOL (BW) do not change with the variation of velocity.

The mechanisms of the effect of velocity on the adhesion and friction are explained based on schematics shown in Fig. 17.46 (right). For Si(100), tribochemical reaction plays a major role. Although at high velocity the meniscus is broken and does not have enough time to rebuild, the contact stresses and high velocity lead to tribochemical reactions of Si(100) wafer (which has native oxide (SiO_2)) and a Si_3N_4 tip with water molecules and form $Si(OH)_4$. The $Si(OH)_4$ is removed and continuously replenished during sliding. The $Si(OH)_4$ layer between the tip and Si(100) surface is known to be of low shear strength and causes a decrease in friction force and

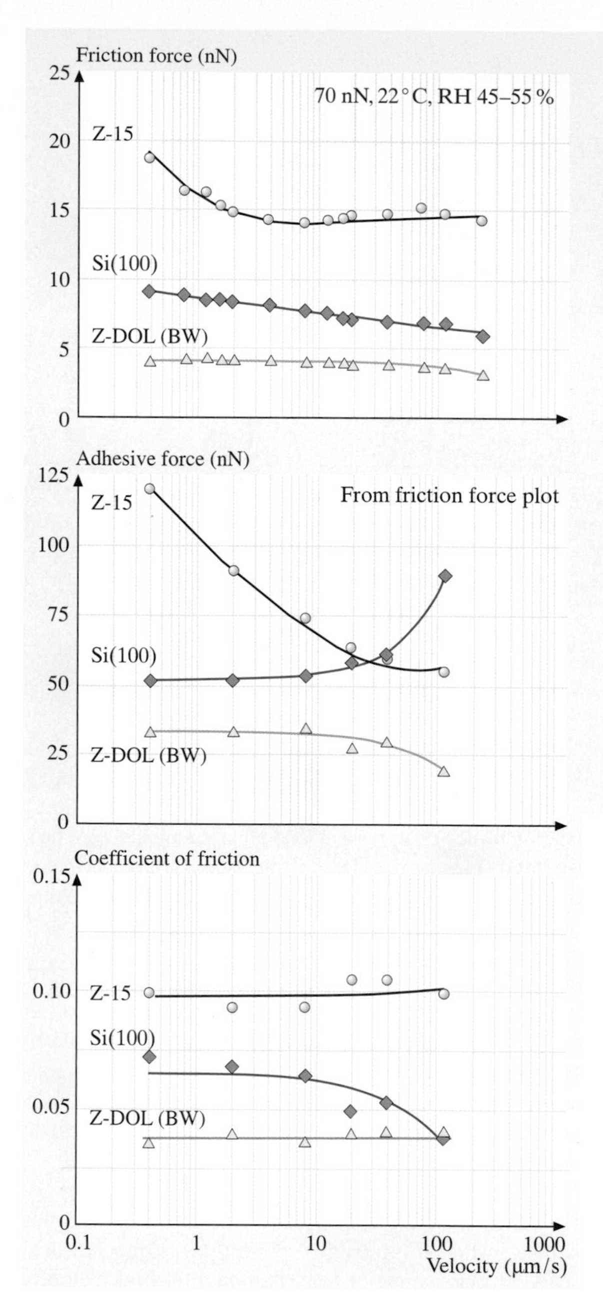

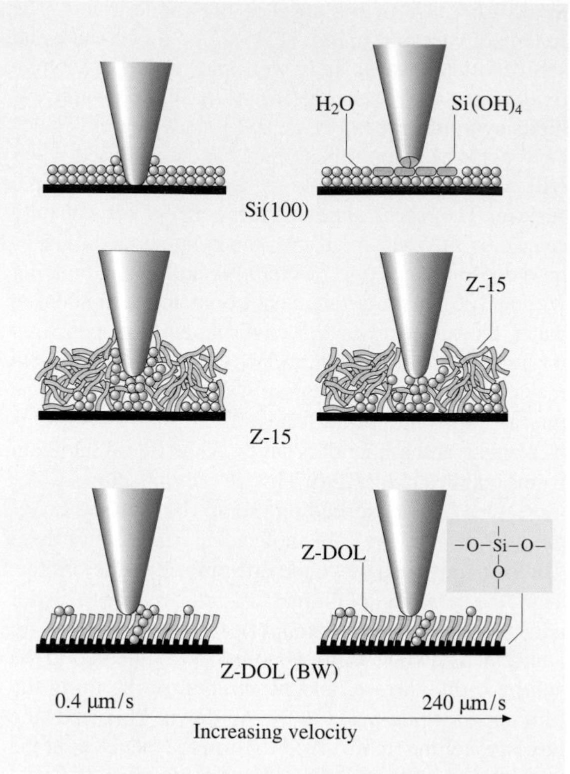

Fig. 17.46 The influence of velocity on the friction force, adhesive force, and coefficient of friction of Si(100) and Z-15 and Z-DOL (BW) films at 70 nN, in ambient air. The schematic (*right*) shows the change of surface composition (by tribochemical reaction) and formation of meniscus while increasing the velocity [17.30]

coefficient of friction [17.10,58]. The chemical bonds of Si–OH between the tip and Si(100) surface induce large adhesive force. For Z-15 film, at high velocity the meniscus formed by condensed water and Z-15 molecules is broken and does not have enough time to rebuild. Therefore, the adhesive force and, consequently, friction force are reduced. The friction mechanism for Z-15 film still is shearing the same viscous liquid even at high velocity range, thus, the coefficient of friction of Z-15 does not change with velocity. For Z-DOL (BW) film, the surface can adsorb few water molecules in ambient condition, and at high velocity these molecules are displaced, which causes a slight decrease in friction and adhesive forces. *Koinkar* and *Bhushan* [17.27, 28] have suggested that in the case of samples with mobile films such as condensed water and Z-15 films, alignment of liquid molecules (shear thinning) is responsible for the drop in friction force with an increase in scanning velocity. This could be another reason for the decrease in friction force for Si(100) and Z-15 film with velocity in this study.

To study the relative humidity effect on friction and adhesion, the variations of friction force, adhesive force,

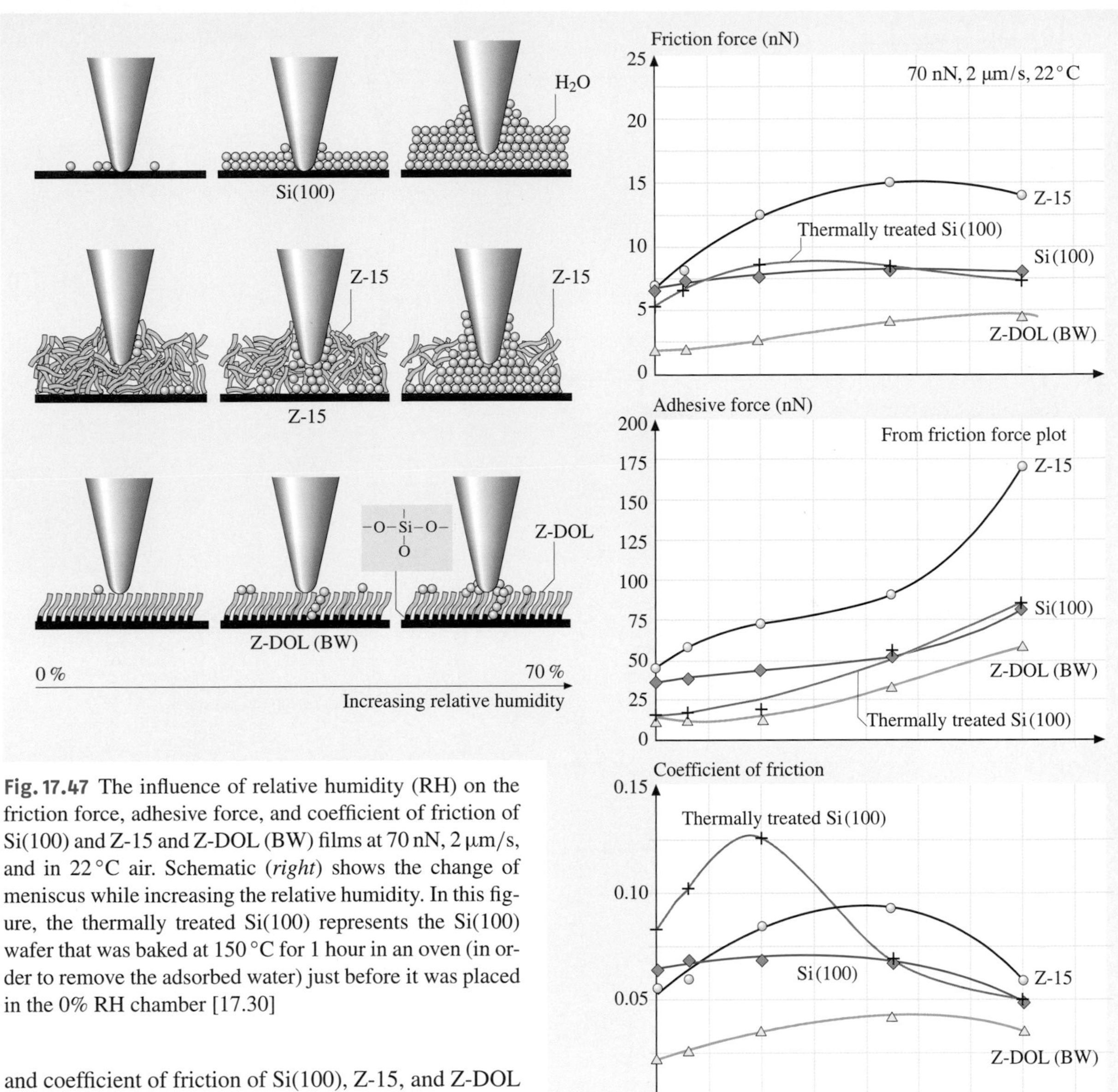

Fig. 17.47 The influence of relative humidity (RH) on the friction force, adhesive force, and coefficient of friction of Si(100) and Z-15 and Z-DOL (BW) films at 70 nN, 2 μm/s, and in 22 °C air. Schematic (*right*) shows the change of meniscus while increasing the relative humidity. In this figure, the thermally treated Si(100) represents the Si(100) wafer that was baked at 150 °C for 1 hour in an oven (in order to remove the adsorbed water) just before it was placed in the 0% RH chamber [17.30]

and coefficient of friction of Si(100), Z-15, and Z-DOL (BW) as a function of relative humidity are shown in Fig. 17.47. It shows that for Si(100) and Z-15 film, the friction force increases with a relative humidity increase of up to 45%, and then slightly decreases with a further increase in the relative humidity. Z-DOL (BW) has a smaller friction force than Si(100) and Z-15 in the whole testing range. And its friction force shows a relative apparent increase when the relative humidity is higher than 45%. For Si(100), Z-15, and Z-DOL (BW), their adhesive forces increase with relative humidity, and their coefficients of friction increase with relative humidity up to 45%, after which they decrease with further increasing of the relative humidity. It is also observed that the humidity effect on Si(100) really depends on the history of the Si(100) sample. As the surface of Si(100) wafer readily adsorbs water in air, without any pre-treatment the Si(100) used in our study almost

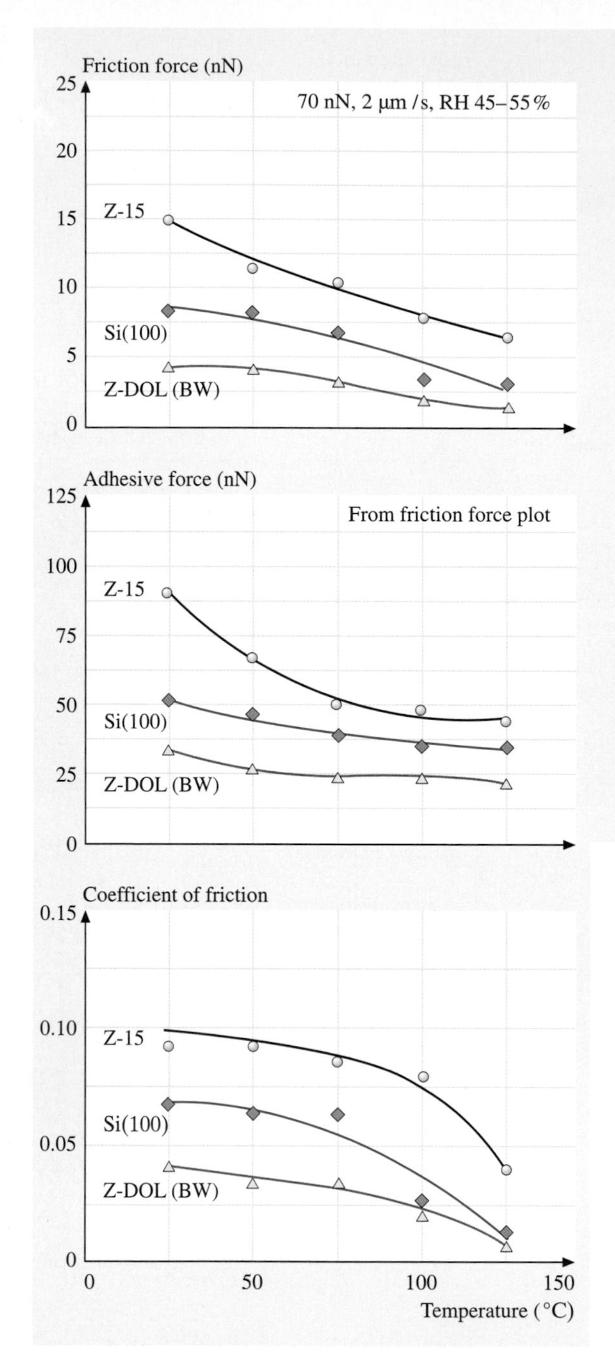

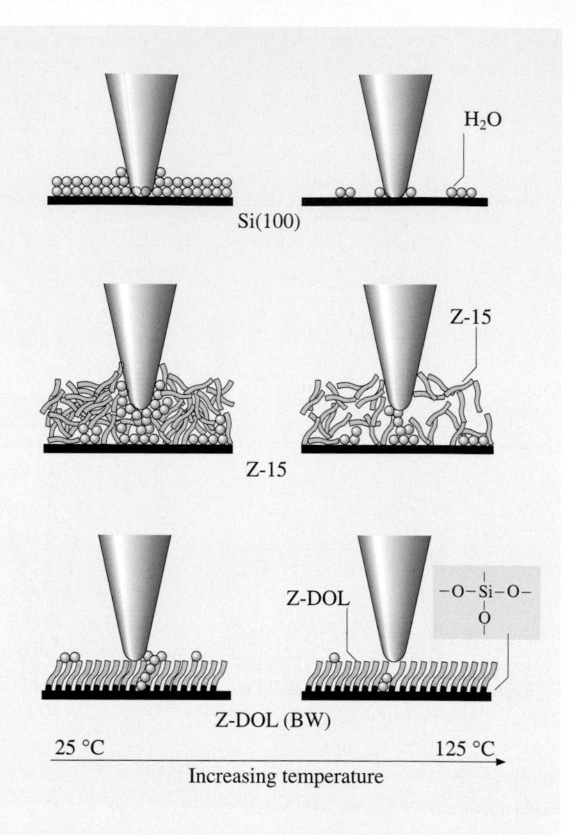

Fig. 17.48 The influence of temperature on the friction force, adhesive force, and coefficient of friction of Si(100) and Z-15 and Z-DOL (BW) films at 70 nN, at 2 μm/s, and in RH 40–50% air. The schematic (*right*) shows that at high temperature desorption of water decreases the adhesive forces. And the reduced viscosity of Z-15 leads to the decrease of the coefficient of friction. High temperature facilitates orientation of molecules in Z-DOL (BW) film, which results in lower coefficient of friction [17.30]

reaches to its saturate stage of adsorbed water and is responsible for less effect during increasing relative humidity. However, once the Si(100) wafer was thermally treated by baking at 150 °C for 1 hour, a bigger effect was observed.

The schematic in Fig. 17.47 (right) shows that Si(100), because its high free surface energy, can adsorb more water molecules with increasing relative humidity. As discussed earlier, for Z-15 film in the humid environment, the condensed water from the humid environment competes with the lubricant film present on the sample surface, interaction of the liquid lubricant film with the silicon substrate is weakened, and a boundary layer of the liquid lubricant forms puddles. This dewetting allows water molecules to be adsorbed on the Si(100) substrate mixed with Z-15 molecules [17.27, 28]. Obviously, more water molecules can be adsorbed on Z-15 surface while increasing relative humidity. The more adsorbed water molecules in the case of Si(100), along with lubricant molecules in Z-15 film case, form bigger water meniscus which leads to an increase of friction force, adhesive force, and coefficient of friction of Si(100) and Z-15 with humidity. But at a very high humidity of 70%, large quantities of adsorbed water can form a water layer that separates the tip and sample surface and act as a kind of lubricant, which causes a decrease in the friction force and coefficient of friction. For Z-DOL (BW) film, because of its hydrophobic surface properties, water molecules can be adsorbed at a humidity higher than 45% and causes an increase in the adhesive and friction forces.

To study the temperature effect on friction and adhesion, the variations of friction force, adhesive force, and coefficient of friction of Si(100), Z-15, and Z-DOL (BW) as a function of temperature are summarized in Fig. 17.48. It shows that the increasing temperature causes a decrease in friction force, adhesive force, and coefficient of friction of Si(100), Z-15, and Z-DOL (BW). The schematic (right) in Fig. 17.48 indicates that at high temperature, desorption of water leads to the decrease of friction force, adhesive forces, and coefficient of friction for all of the samples. For Z-15 film, the reduction of viscosity at high temperature also contributes to the decrease of friction force and coefficient of friction. In the case of Z-DOL (BW) film, molecules are more easily oriented at high temperature, which may be partly responsible for the low friction force and coefficient of friction.

To study the durability of lubricant films at the nanoscale, the friction of Si(100), Z-15, and Z-DOL (BW) as a function of the number of scanning cycles is shown in Fig. 17.49. As observed earlier, friction force and coefficient of friction of Z-15 is higher than that of Si(100) with the lowest values for Z-DOL (BW). During cycling, friction force and coefficient of friction of Si(100) show a slight decrease during the initial few cycles then remain constant. This is related to the removal of the top adsorbed layer. In the case of Z-15 film, the friction force and coefficient of friction show an increase during the initial few cycles, and then approach higher and stable values. This is believed to be caused by the attachment of the Z-15 molecules to the tip. The molecular interaction between these attached molecules to the tip and molecules on the film surface is responsible for an increase in friction. But after several scans, this molecular interaction reaches the equilibrium, and after that, friction force and coefficient of friction remain constant. In the case of Z-DOL (BW) film, the friction force and coefficient of friction start out low and remain

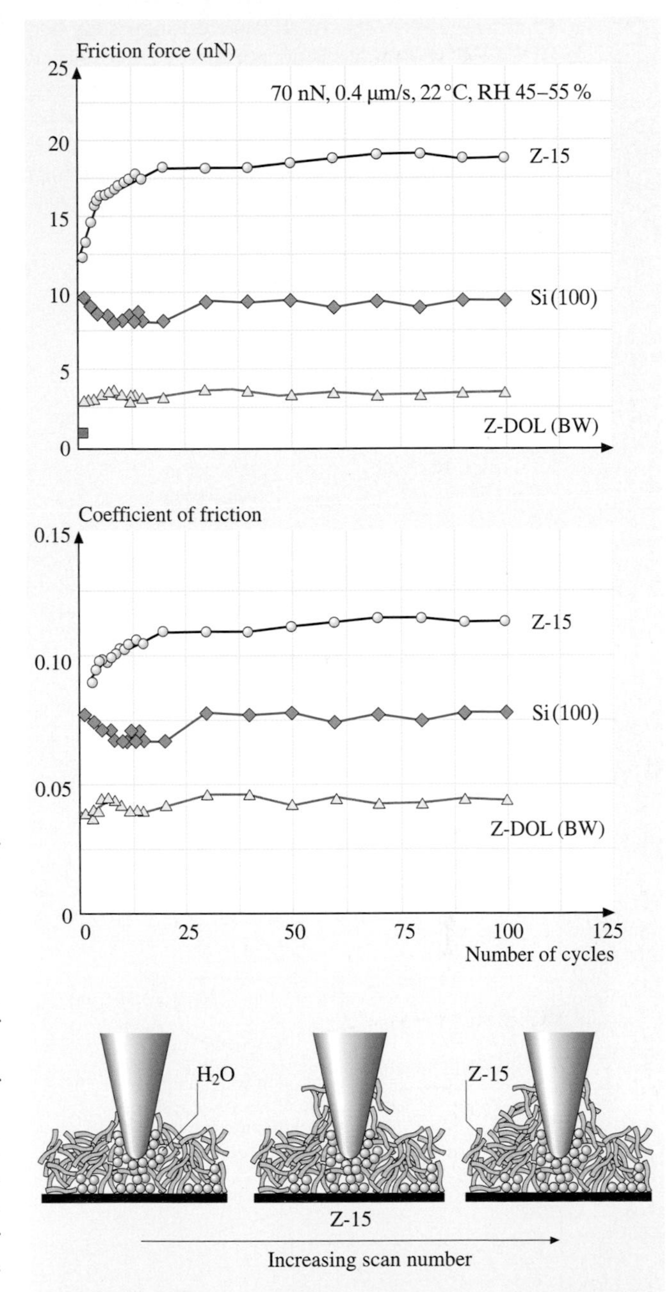

Fig. 17.49 Friction force and coefficient of friction versus number of sliding cycles for Si(100) and Z-15 and Z-DOL (BW) films at 70 nN, 0.8 μm/s, and in ambient air. Schematic (*bottom*) shows that some liquid Z-15 molecules can be attached to the tip. The molecular interaction between the attached molecules to the tip with the Z-15 molecules in the film results in an increase of the friction force with multiple scanning [17.30]

low during the entire test for 100 cycles, suggesting that Z-DOL (BW) molecules do not get attached or displaced as readily as Z-15 molecules.

As a brief summary, the influence of velocity, relative humidity, and temperature on the friction force of mobile Z-15 film is presented in Fig. 17.50. The changing trends are also addressed in this figure.

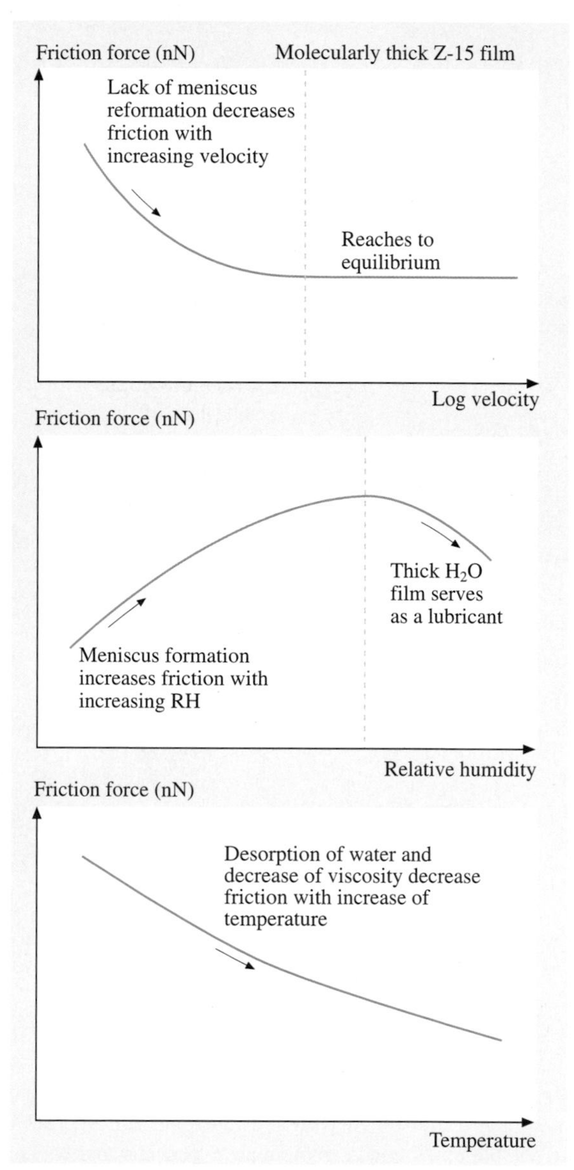

Fig. 17.50 Schematic shows the change of friction force of molecularly thick Z-15 films with log velocity, relative humidity, and temperature. The changing trends are also addressed in this figure [17.30]

17.5.2 Self-Assembled Monolayers

For lubrication of MEMS/NEMS, another effective approach involves the deposition of organized and dense molecular layers of long-chain molecules. Two common methods to produce monolayers and thin films are the Langmuir–Blodgett (L-B) deposition and self-assembled monolayers (SAMs) by chemical grafting of molecules. L-B films are physically bonded to the substrate by weak van der Waals attraction, while SAMs are chemically bonded via covalent bonds to the substrate. Because of the choice of chain length and terminal linking group that SAMs offer, they hold great promise for boundary lubrication of MEMS/NEMS. A number of studies have been conducted to study tribological properties of various SAMs [17.26, 29, 74–76]. It has been reported that SAMs with high-compliance long carbon chains exhibit low friction; chain compliance is desirable for low friction. Based on [17.29], the friction mechanism of SAMs is explained by a so-called "molecular spring" model (Fig. 17.51). According to this model, the chemically adsorbed self-assembled molecules on a substrate are just like assembled molecular springs anchored to the substrate. An asperity sliding on the surface of SAMs is like a tip sliding on the top of "molecular springs or brush". The molecular spring assembly has compliant features and can experience orientation and compression under load. The orientation of the molecular springs or brush under normal load reduces the shearing force at the interface, which, in turn, reduces the friction force.

The SAMs with high-compliance long carbon chains also exhibit the best wear resistance [17.29, 75]. In wear experiments, the wear depth as a function of normal load curves shows critical normal loads (Fig. 17.52). Below the critical normal load SAMs undergo orien-

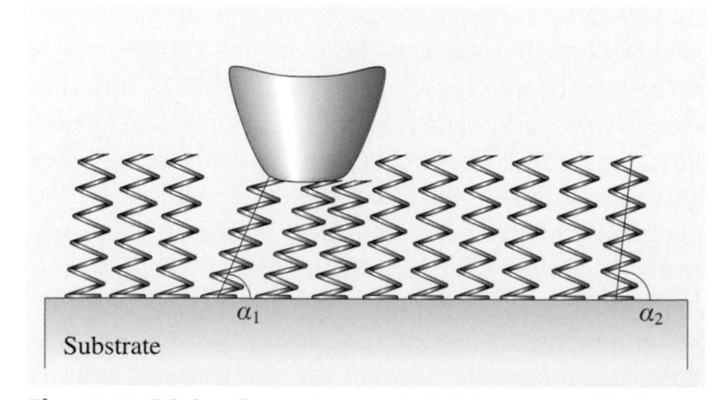

Fig. 17.51 Molecular spring model of SAMs. In this figure, $\alpha_1 < \alpha_2$, which is caused by the further orientation under the normal load applied by an asperity tip [17.29]

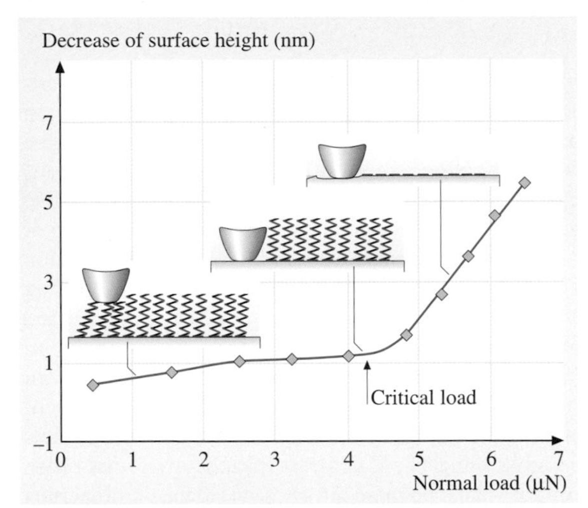

Fig. 17.52 Illustration of the wear mechanism of SAMs with increasing normal load [17.75]

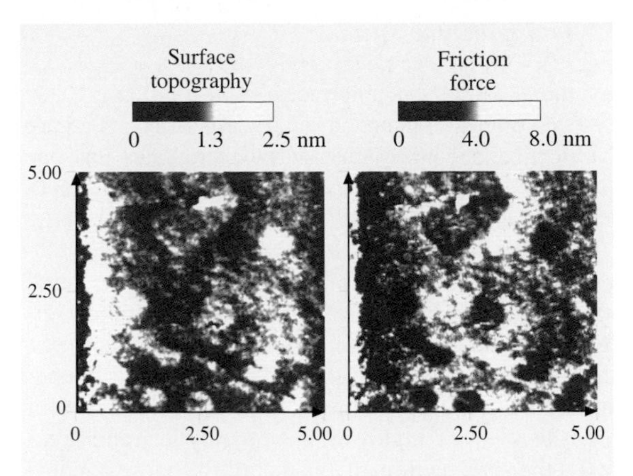

Fig. 17.53 Gray-scale plots of the surface topography and friction force obtained simultaneously for unbonded Demnum-type perfluoropolyether lubricant film on silicon [17.27]

tation, at the critical load SAMs wear away from the substrate due to weak interface bond strengths, while above the critical normal load severe wear takes place on the substrate.

17.5.3 Liquid Film Thickness Measurements

Liquid film thickness measurement of thin lubricant films (on the order of 10 nm or thicker) with nanometer lateral resolution can be made with the AFM [17.4, 59]. The lubricant thickness is obtained by measuring the force on the tip as it approaches, contacts, and pushes through the liquid film and ultimately contacts the substrate. The distance between the sharp snap-in (owing to the formation of a liquid meniscus between the film and the tip) at the liquid surface and the hard repulsion at the substrate surface is a measure of the liquid film thickness.

Lubricant film thickness mapping of ultrathin films (on the order of 2 nm) can be obtained using friction force microscopy [17.27] and adhesive force mapping [17.36]. Figure 17.53 shows gray-scale plots of the surface topography and friction force obtained simultaneously for unbonded Demnum-type PFPE lubricant film on silicon. The friction force plot shows well distinguished low and high friction regions roughly corresponding to high and low regions in surface topography (thick and thin lubricant regions). A uniformly lubricated sample does not show such a variation in the friction. Friction force imaging can thus be used to measure the lubricant uniformity on the sample surface, which cannot be identified by surface topography alone. Figure 17.54 shows the gray-scale plots of the adhesive force distribution for silicon samples coated uniformly and nonuniformly with Z-DOL type PFPE lubricant. It can be clearly seen that a region exists that has an adhesive force distinctly different from the other region for the nonuniformly coated sample. This implies that the liquid film thickness is nonuniform, giving rise to a difference in the meniscus forces.

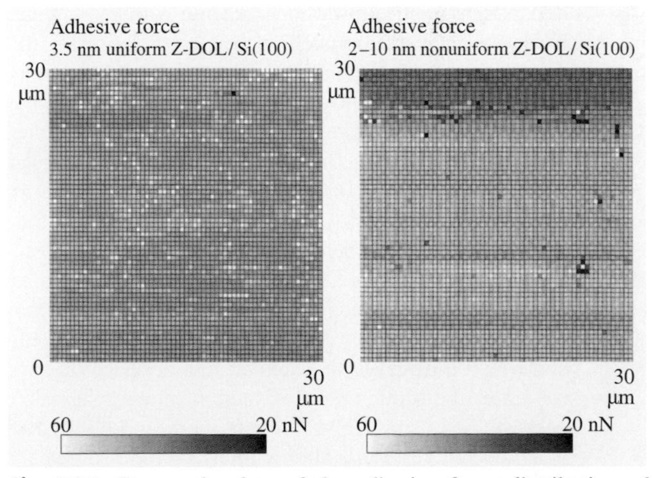

Fig. 17.54 Gray-scale plots of the adhesive force distribution of a uniformly coated, 3.5-nm thick unbonded Z-DOL film on silicon and 3- to 10-nm thick unbonded Z-DOL film on silicon that was deliberately coated nonuniformly by vibrating the sample during the coating process [17.36]

17.6 Closure

At most solid-solid interfaces of technological relevance, contact occurs at many asperities. A sharp AFM/FFM tip sliding on a surface simulates just one such contact. However, asperities come in all shapes and sizes. The effect of radius of a single asperity (tip) on the friction/adhesion performance can be studied using tips of different radii. AFM/FFM are used to study various tribological phenomena, including surface roughness, adhesion, friction, scratching, wear, indentation, detection of material transfer, and boundary lubrication. Measurement of atomic-scale friction of a freshly cleaved, highly oriented pyrolytic graphite exhibits the same periodicity as that of the corresponding topography. However, the peaks in friction and those in the corresponding topography are displaced relative to one another. Variations in atomic-scale friction and the observed displacement can be explained by the variation in interatomic forces in the normal and lateral directions. Local variations in microscale friction occur and are found to correspond to the local slopes, suggesting that a ratchet mechanism and collision effects are responsible for this variation. Directionality in the friction is observed on both micro- and macroscales, resulting from the surface roughness and surface preparation. Anisotropy in surface roughness accentuates this effect. Microscale friction is generally found to be smaller than the macrofriction, as there is less plowing contribution in microscale measurements. Microscale friction is load-dependent, and friction values increase with an increase in the normal load, approaching the macrofriction at contact stresses higher than the hardness of the softer material. The tip radius also affects adhesion and friction.

Mechanism of material removal on the microscale is studied. Wear rate for single-crystal silicon is negligible below 20 μN and is much higher and remains approximately constant at higher loads. Elastic deformation at low loads is responsible for negligible wear. Most of the wear debris is loose. SEM and TEM studies of the wear region suggest that the material on the microscale is removed by plastic deformation with a small contribution from elastic fracture; this observation corroborates the scratch data. Evolution of wear has also been studied using AFM. Wear is found to be initiated at nanoscratches. For a sliding interface requiring near-zero friction and wear, contact stresses should be below the hardness of the softer material to minimize plastic deformation, and surfaces should be free of nanoscratches. Furthermore, wear precursors can be detected at early stages of wear by using surface potential measurements. It is found that even in the case of zero wear (no measurable deformation of the surface using AFM), there can be a significant change in the surface potential inside the wear mark, which is useful for study of wear precursors. Detection of material transfer on a nanoscale is possible with AFM.

In situ surface characterization of local deformation of materials and thin coatings can be carried out using a tensile stage inside an AFM. An AFM can also be used for nanofabrication/nanomachining.

By using the force modulation technique, localized surface elasticity maps of composite materials with penetrating depths of less than 100 nm can be obtained. It is possible using phase contrast microscopy to get phase contrast maps or the contrast in viscoelastic properties of near surface regions. Scratching and indentation on nanoscales are powerful ways to screen for adhesion and resistance to deformation of ultrathin films. Modified AFM can be used to obtain load-displacement curves and for measurement of nanoindentation hardness and Young's modulus of elasticity, with depth of indentation as low as 1 nm. Hardness of ceramics on the nanoscale is found to be higher than that on the microscale. Ceramics exhibit significant plasticity and creep on a nanoscale.

Boundary lubrication studies and measurement of lubricant-film thickness with a lateral resolution on a nanoscale can be conducted using AFM. Chemically bonded lubricant films and self-assembled monolayers are superior in friction and wear resistance. For chemically bonded lubricant films, the adsorption of water, the formation of meniscus, and its change during sliding, viscosity, and surface properties play an important role on the friction, adhesion, and durability of these films. For SAMs, their friction mechanism is explained by a so-called "molecular spring" model. The films with high-compliance long carbon chains exhibit low friction and wear.

Investigations of wear, scratching, and indentation on the nanoscale using the AFM can provide insights into failure mechanisms of materials. Coefficients of friction, wear rates, and mechanical properties such as hardness have been found to be different on the nanoscale than on the macroscale; generally, coefficients of friction and wear rates on micro- and nanoscales are smaller, whereas hardness is greater. Therefore, micro/nanotribological studies may help define the regimes for ultralow friction and near-zero wear. These studies also provide insight into the atomic origins of adhesion, friction, wear, and lubrication mechanisms.

References

17.1 I. L. Singer, H. M. Pollock: *Fundamentals of Friction: Macroscopic and Microscopic Processes*, Nato Sci. Ser. E, Vol. 220 (Kluwer, Dordrecht 1992)

17.2 B. N. J. Persson, E. Tosatti: *Physics of Sliding Friction*, Nato Sci. Ser. E, Vol. 311 (Kluwer, Dordrecht 1996)

17.3 B. Bhushan: *Micro/Nanotribology and Its Applications*, Nato Sci. Ser. E, Vol. 330 (Kluwer, Dordrecht 1997)

17.4 B. Bhushan: *Handbook of Micro/Nanotribology*, 2nd edn. (CRC, Boca Raton 1999)

17.5 B. Bhushan: Nanoscale tribophysics and tribomechanics, Wear **225–229** (1999) 465–492

17.6 B. Bhushan: Wear and mechanical characterisation on micro- to picoscales using AFM, Int. Mat. Rev. **44** (1999) 105–117

17.7 B. Bhushan: *Modern Tribology Handbook, Vol. 1: Principles of Tribology* (CRC, Boca Raton 2001)

17.8 B. Bhushan: *Fundamentals of tribology and bridging the gap between the macro- and micro/nanoscales*, NATO Sci. Ser. E, Vol. 10 (Kluwer, Dordrecht 2001)

17.9 B. Bhushan: Nano- to microscale wear and mechanical characterization studies using scanning probe microscopy, Wear **251** (2001) 1105–1123

17.10 B. Bhushan: *Introduction to Tribology* (Wiley, New York 2002)

17.11 B. Bhushan, J. N. Israelachvili, U. Landman: Nanotribology: Friction, wear and lubrication at the atomic scale, Nature **374** (1995) 607–616

17.12 H. J. Guntherodt, D. Anselmetti, E. Meyer: *Forces in Scanning Probe Methods*, Nato Sci. Ser. E, Vol. 286 (Kluwer, Dordrecht 1995)

17.13 B. Bhushan: *Tribology and Mechanics of Magnetic Storage Devices*, 2nd edn. (Springer, New York 1996)

17.14 B. Bhushan: *Tribology Issues and Opportunities in MEMS* (Kluwer, Dordrecht 1998)

17.15 G. Binnig, C. F. Quate, Ch. Gerber: Atomic force microscopy, Phys. Rev. Lett. **56** (1986) 930–933

17.16 G. Binnig, Ch. Gerber, E. Stoll, T. R. Albrecht, C. F. Quate: Atomic resolution with atomic force microscope, Europhys. Lett. **3** (1987) 1281–1286

17.17 C. M. Mate, G. M. McClelland, R. Erlandsson, S. Chiang: Atomic-scale friction of a tungsten tip on a graphite surface, Phys. Rev. Lett. **59** (1987) 1942–1945

17.18 B. Bhushan, J. Ruan: Atomic-scale friction measurements using friction force microscopy: part II – application to magnetic media, ASME J. Tribol. **116** (1994) 389–396

17.19 J. Ruan, B. Bhushan: Atomic-scale friction measurements using friction force microscopy: part I – general principles and new measurement techniques, ASME J. Tribol. **116** (1994) 378–388

17.20 J. Ruan, B. Bhushan: Atomic-scale and microscale friction of graphite and diamond using friction force microscopy, J. Appl. Phys. **76** (1994) 5022–5035

17.21 J. Ruan, B. Bhushan: Frictional behavior of highly oriented pyrolytic graphite, J. Appl. Phys. **76** (1994) 8117–8120

17.22 B. Bhushan, V. N. Koinkar, J. Ruan: Microtribology of magnetic media, Proc. Inst. Mech. Eng., Part J: J. Eng. Tribol. **208** (1994) 17–29

17.23 B. Bhushan, S. Sundararajan: Micro/nanoscale friction and wear mechanisms of thin films using atomic force and friction force microscopy, Acta Mater. **46** (1998) 3793–3804

17.24 V. Scherer, W. Arnold, B. Bhushan: Active friction control using ultrasonic vibration. In: *Tribology Issues and Opportunities in MEMS*, ed. by B. Bhushan (Kluwer, Dordrecht 1998) pp. 463–469

17.25 V. Scherer, W. Arnold, B. Bhushan: Lateral force microscopy using acoustic friction force microscopy, Surf. Interface Anal. **27** (1999) 578–587

17.26 B. Bhushan, A. V. Kulkarni, V. N. Koinkar, M. Boehm, L. Odoni, C. Martelet, M. Belin: Microtribological characterization of self-assembled and Langmuir–Blodgett monolayers by atomic and friction force microscopy, Langmuir **11** (1995) 3189–3198

17.27 V. N. Koinkar, B. Bhushan: Micro/nanoscale studies of boundary layers of liquid lubricants for magnetic disks, J. Appl. Phys. **79** (1996) 8071–8075

17.28 V. N. Koinkar, B. Bhushan: Microtribological studies of unlubricated and lubricated surfaces using atomic force/friction force microscopy, J. Vac. Sci. Technol. A. **14** (1996) 2378–2391

17.29 B. Bhushan, H. Liu: Nanotribological properties and mechanisms of alkylthiol and biphenyl thiol self-assembled monolayers studied by AFM, Phys. Rev. B **63** (2001) 245412–1–245412–11

17.30 H. Liu, B. Bhushan: Nanotribological characterization of molecularly-thick lubricant films for applications to MEMS/NEMS by AFM, Ultramicroscopy **97** (2003) 321–340

17.31 B. Bhushan, V. N. Koinkar: Tribological studies of silicon for magnetic recording applications, J. Appl. Phys. **75** (1994) 5741–5746

17.32 V. N. Koinkar, B. Bhushan: Microtribological properties of hard amorphous carbon protective coatings for thin film magnetic disks and heads, Proc. Inst. Mech. Eng. Part J: J. Eng. Tribol. **211** (1997) 365–372

17.33 S. Sundararajan, B. Bhushan: Development of a continuous microscratch technique in an atomic force microscope and its application to study scratch resistance of ultra-thin hard amorphous carbon coatings, J. Mater. Res. **16** (2001) 75–84

17.34 B. Bhushan, V. N. Koinkar: Nanoindentation hardness measurements using atomic force microscopy, Appl. Phys. Lett. **64** (1994) 1653–1655
17.35 B. Bhushan, A. V. Kulkarni, W. Bonin, J. T. Wyrobek: Nano/picoindentation measurement using a capacitance transducer system in atomic force microscopy, Philos. Mag. **74** (1996) 1117–1128
17.36 B. Bhushan, C. Dandavate: Thin-film friction and adhesion studies using atomic force microscopy, J. Appl. Phys. **87** (2000) 1201–1210
17.37 B. Bhushan: Micro/nanotribology and its applications to magnetic storage devices and MEMS, Tribol. Int. **28** (1995) 85–95
17.38 D. DeVecchio, B. Bhushan: Use of a nanoscale Kelvin probe for detecting wear precursors, Rev. Sci. Instrum. **69** (1998) 3618–3624
17.39 B. Bhushan, A. V. Goldade: Measurements and analysis of surface potential change during wear of single crystal silicon (100) at ultralow loads using Kelvin probe microscopy, Appl. Surf. Sci **157** (2000) 373–381
17.40 B. Bhushan, A. V. Goldade: Kelvin probe microscopy measurements of surface potential change under wear at low loads, Wear **244** (2000) 104–117
17.41 M. S. Bobji, B. Bhushan: Atomic force microscopic study of the micro-cracking of magnetic thin films under tension, Scripta Mater. **44** (2001) 37–42
17.42 M. S. Bobji, B. Bhushan: In-situ microscopic surface characterization studies of polymeric thin films during tensile deformation using atomic force microscopy, J. Mater. Res. **16** (2001) 844–855
17.43 P. Maivald, H. J. Butt, S. A. C. Gould, C. B. Prater, B. Drake, J. A. Gurley, V. B. Elings, P. K. Hansma: Using force modulation to image surface elasticities with the atomic force microscope, Nanotechnol. **2** (1991) 103–106
17.44 D. DeVecchio, B. Bhushan: Localized surface elasticity measurements using an atomic force microscope, Rev. Sci. Instrum. **68** (1997) 4498–4505
17.45 V. Scherer, B. Bhushan, U. Rabe, W. Arnold: Local elasticity and lubrication measurements using atomic force and friction force microscopy at ultrasonic frequencies, IEEE Trans. Mag. **33** (1997) 4077–4079
17.46 H. U. Krotil, T. Stifter, H. Waschipky, K. Weishaupt, S. Hild, O. Marti: Pulse force mode: A new method for the investigation of surface properties, Surf. Interface Anal. **27** (1999) 336–340
17.47 S. Amelio, A. V. Goldade, U. Rabe, V. Scherer, B. Bhushan, W. Arnold: Measurements of elastic properties of ultra-thin diamond-like carbon coatings using atomic force acoustic microscopy, Thin Solid Films **392** (2001) 75–84
17.48 B. Anczykowski, D. Kruger, K. L. Babcock, H. Fuchs: Basic properties of dynamic force microscopy with the scanning force microscope in experiment and simulation, Ultramicroscopy **66** (1996) 251–259
17.49 J. Tamayo, R. Garcia: Deformation, contact time, and phase contrast in tapping mode scanning force microscopy, Longmuir **12** (1996) 4430–4435
17.50 R. Garcia, J. Tamayo, M. Calleja, F. Garcia: Phase contrast in tapping-mode scanning force microscopy, Appl. Phys. A **66** (1998) 309–312
17.51 W. W. Scott, B. Bhushan: Use of phase imaging in atomic force microscopy for measurement of viscoelastic contrast in polymer nanocomposites and molecularly-thick lubricant films, Ultramicroscopy **97** (2003) 151–169
17.52 B. Bhushan, J. Qi: Phase contrast imaging of nanocomposites and molecularly- thick lubricant films in magnetic media, Nanotechnol. **14** (2003) 886–895
17.53 V. N. Koinkar, B. Bhushan: Microtribological studies of Al_2O_3-TiC, polycrystalline and single-crystal Mn-Zn ferrite and SiC head slider materials, Wear **202** (1996) 110–122
17.54 E. Meyer, R. Overney, R. Luthi, D. Brodbeck, L. Howald, J. Frommer, H. J. Guntherodt, O. Wolter, M. Fujihira, T. Takano, Y. Gotoh: Friction force microscopy of mixed Langmuir–Blodgett films, Thin Solid Films **220** (1992) 132–137
17.55 C. D. Frisbie, L. F. Rozsnyai, A. Noy, M. S. Wrighton, C. M. Lieber: Functional group imaging by chemical force microscopy, Science **265** (1994) 2071–2074
17.56 V. N. Koinkar, B. Bhushan: Effect of scan size and surface roughness on microscale friction measurements, J. Appl. Phys. **81** (1997) 2472–2479
17.57 S. Sundararajan, B. Bhushan: Topography-induced contributions to friction forces measured using an atomic force/friction force microscope, J. Appl. Phys. **88** (2000) 4825–4831
17.58 B. Bhushan: *Principles and Applications of Tribology* (Wiley, New York 1999)
17.59 B. Bhushan, G. S. Blackman: Atomic force microscopy of magnetic rigid disks and sliders and its applications to tribology, ASME J. Tribol. **113** (1991) 452–458
17.60 M. Reinstaedtler, U. Rabe, V. Scherer, U. Hartmann, A. Goldade, B. Bhushan, W. Arnold: On the nanoscale measurement of friction using atomic-force microscope cantilever torsional resonances, Appl. Phys. Lett. **82** (2003) 2604–2606
17.61 T. Stifter, O. Marti, B. Bhushan: Theoretical investigation of the distance dependence of capillary and van der Waals forces in scanning probe microscopy, Phys. Rev. B **62** (2000) 13667–13673
17.62 U. D. Schwarz, O. Zwoerner, P. Koester, R. Wiesendanger: Friction force spectroscopy in the low-load regime with well-defined tips. In: *Micro/Nanotribology and Its Applications*, ed. by B. Bhushan (Kluwer, Dordrecht 1997) pp. 233–238

17.63 B. Bhushan, A. V. Kulkarni: Effect of normal load on microscale friction measurements, Thin Solid Films **278** (1996) 49–56; Errata: **293** 333

17.64 V. N. Koinkar, B. Bhushan: Scanning and transmission electron microscopies of single-crystal silicon microworn/machined using atomic force microscopy, J. Mater. Res. **12** (1997) 3219–3224

17.65 X. Zhao, B. Bhushan: Material removal mechanism of single-crystal silicon on nanoscale and at ultralow loads, Wear **223** (1998) 66–78

17.66 B. Bhushan, P. S. Mokashi, T. Ma: A new technique to measure Poisson's ratio of ultrathin polymeric films using atomic force microscopy, Rev. Sci. Instrumen. **74** (2003) 1043–1047

17.67 A. V. Kulkarni, B. Bhushan: Nanoscale mechanical property measurements using modified atomic force microscopy, Thin Solid Films **290-291** (1996) 206–210

17.68 A. V. Kulkarni, B. Bhushan: Nano/picoindentation measurements on single-crystal aluminum using modified atomic force microscopy, Mater. Lett. **29** (1996) 221–227

17.69 A. V. Kulkarni, B. Bhushan: Nanoindentation measurement of amorphous carbon coatings, J. Mater. Res. **12** (1997) 2707–2714

17.70 J. Ruan, B. Bhushan: Nanoindentation studies of fullerene films using atomic force microscopy, J. Mater. Res. **8** (1993) 3019–3022

17.71 N. A. Fleck, G. M. Muller, M. F. Ashby, J. W. Hutchinson: Strain gradient plasticity: Theory and experiment, Acta Metall. Mater. **42** (1994) 475–487

17.72 W. B. Li, J. L. Henshall, R. M. Hooper, K. E. Easterling: The mechanism of indentation creep, Acta Metall. Mater. **39** (1991) 3099–3110

17.73 F. P. Bowden, D. Tabor: *The Friction and Lubrication of Solids*, Vol. 1 (Clarendon, Oxford 1950)

17.74 H. Liu, B. Bhushan: Investigation of the adhesion, friction, and wear properties of biphenyl thiol self-assembled monolayers by atomic force microscopy, J. Vac. Sci. Technol. A **19** (2001) 1234–1240

17.75 H. Liu, B. Bhushan: Investigation of nanotribological properties of self-assembled monolayers with alkyl and biphenyl spacer chains, Ultramicroscopy **91** (2002) 185–202

17.76 B. Bhushan: Self-assembled monolayers for controlling hydrophobicity and/or friction and wear. In: *Modern Tribology Handbook, Vol. 2: Materials, Coatings, and Industrial Applications*, ed. by B. Bhushan (CRC, Boca Raton 2001) pp. 909–929

18. Surface Forces and Nanorheology of Molecularly Thin Films

In this chapter, we describe the static and dynamic normal forces that occur between surfaces in vacuum or liquids and the different modes of friction that can be observed between (i) bare surfaces in contact (dry or interfacial friction), (ii) surfaces separated by a thin liquid film (lubricated friction), and (iii) surfaces coated with organic monolayers (boundary friction).

Experimental methods suitable for measuring normal surface forces, adhesion and friction (lateral or shear) forces of different magnitude at the molecular level are described. We explain the molecular origin of van der Waals, electrostatic, solvation and polymer mediated interactions, and basic models for the contact mechanics of adhesive and nonadhesive elastically deforming bodies. The effects of interaction forces, molecular shape, surface structure and roughness on adhesion and friction are discussed.

Simple models for the contributions of the adhesion force and external load to interfacial friction are illustrated with experimental data on both unlubricated and lubricated systems, as measured with the surface forces apparatus. We discuss rate-dependent adhesion (adhesion hysteresis) and how this is related to friction. Some examples of the transition from wearless friction to friction with wear are shown.

Lubrication in different lubricant thickness regimes is described together with explanations of nanorheological concepts. The occurrence of and transitions between smooth and stick–slip sliding in various types of dry (unlubricated and solid boundary lubricated) and liquid lubricated systems are discussed based on recent experimental results and models for stick–slip involving memory distance and dilatency.

18.1 **Introduction: Types of Surface Forces** 544

18.2 **Methods Used to Study Surface Forces** 546
- 18.2.1 Force Laws 546
- 18.2.2 Adhesion Forces 547
- 18.2.3 The SFA and AFM 547
- 18.2.4 Some Other Force-Measuring Techniques 549

18.3 **Normal Forces Between Dry (Unlubricated) Surfaces** 550
- 18.3.1 Van der Waals Forces in Vacuum and Inert Vapors 550
- 18.3.2 Charge Exchange Interactions 552
- 18.3.3 Sintering and Cold Welding 553

18.4 **Normal Forces Between Surfaces in Liquids** 554
- 18.4.1 Van der Waals Forces in Liquids 554
- 18.4.2 Electrostatic and Ion Correlation Forces 554
- 18.4.3 Solvation and Structural Forces 557
- 18.4.4 Hydration and Hydrophobic Forces 559
- 18.4.5 Polymer-Mediated Forces 561
- 18.4.6 Thermal Fluctuation Forces 563

18.5 **Adhesion and Capillary Forces** 564
- 18.5.1 Capillary Forces 564
- 18.5.2 Adhesion Mechanics 566
- 18.5.3 Effects of Surface Structure, Roughness, and Lattice Mismatch 566
- 18.5.4 Nonequilibrium and Rate-Dependent Interactions: Adhesion Hysteresis 567

18.6 **Introduction: Different Modes of Friction and the Limits of Continuum Models** 569

18.7 **Relationship Between Adhesion and Friction Between Dry (Unlubricated and Solid Boundary Lubricated) Surfaces** 571
- 18.7.1 Amontons' Law and Deviations from It Due to Adhesion: The Cobblestone Model 571
- 18.7.2 Adhesion Force and Load Contribution to Interfacial Friction 572
- 18.7.3 Examples of Experimentally Observed Friction of Dry Surfaces... 576
- 18.7.4 Transition from Interfacial to Normal Friction with Wear 579

18.8 **Liquid Lubricated Surfaces** 580
18.8.1 Viscous Forces and Friction of Thick Films: Continuum Regime 580
18.8.2 Friction of Intermediate Thickness Films 582
18.8.3 Boundary Lubrication of Molecularly Thin Films: Nanorheology 584
18.9 **Role of Molecular Shape and Surface Structure in Friction** 591
References 594

18.1 Introduction: Types of Surface Forces

In this chapter, we discuss the most important types of surface forces and the relevant equations for the force and friction laws. Several different attractive and repulsive forces operate between surfaces and particles. Some forces occur in vacuum, for example, attractive van der Waals and repulsive hard-core interactions. Other types of forces can arise only when the interacting surfaces are separated by another condensed phase, which is usually a liquid. The most common types of surface forces and their main characteristics are listed in Table 18.1.

In *vacuum*, the two main long-range interactions are the attractive van der Waals and electrostatic (Coulomb) forces. At smaller surface separations (corresponding to molecular contact at surface separations of $D \approx 0.2$ nm), additional attractive interactions can be found such as covalent or metallic bonding forces. These attractive forces are stabilized by the hard-core repulsion. Together they determine the surface and interfacial energies of planar surfaces, as well as the strengths of materials and adhesive junctions. Adhesion forces are often strong enough to elastically or plastically deform bodies or particles when they come into contact.

In *vapors* (e.g., atmospheric air containing water and organic molecules), solid surfaces in or close to contact will generally have a surface layer of chemisorbed or physisorbed molecules, or a capillary condensed liquid bridge between them. A surface layer usually causes the adhesion to decrease, but in the case of capillary condensation, the additional Laplace pressure or attractive "capillary" force may make the adhesion between the surfaces stronger than in inert gas or vacuum.

When totally immersed in a *liquid*, the force between particles or surfaces is completely modified from that in vacuum or air (vapor). The van der Waals attraction is generally reduced, but other forces can now arise that can qualitatively change both the range and even the sign of the interaction. The attractive force in such a system can be either stronger or weaker than in the absence of the intervening liquid. For example, the overall attraction can be stronger in the case of two hydrophobic surfaces separated by water, but weaker for two hydrophilic surfaces. Depending on the different forces that may be operating simultaneously in solution, the overall force law is not generally monotonically attractive even at long range; it can be repulsive, or the force can change sign at some finite surface separation. In such cases, the potential energy minimum, which determines the adhesion force or energy, occurs not at true molecular contact between the surfaces, but at some small distance farther out.

The forces between surfaces in a liquid medium can be particularly complex at *short range*, i. e., at surface separations below a few nanometers or 4–10 molecular diameters. This is partly because with increasing confinement, a liquid ceases to behave as a structureless continuum with bulk properties; instead, the size and shape of its molecules begin to determine the overall interaction. In addition, the surfaces themselves can no longer be treated as inert and structureless walls (i. e., mathematically flat) and their physical and chemical properties at the atomic scale must now be taken into account. The force laws will then depend on whether the surfaces are amorphous or crystalline (and whether the lattices of crystalline surfaces are matched or not), rough or smooth, rigid or soft (fluid-like), and hydrophobic or hydrophilic.

It is also important to distinguish between *static* (i. e., equilibrium) interactions and *dynamic* (i. e., nonequilibrium) forces such as viscous and friction forces. For example, certain liquid films confined between two contacting surfaces may take a surprisingly long time to equilibrate, as may the surfaces themselves, so that the short-range and adhesion forces appear to be time-dependent, resulting in "aging" effects.

Table 18.1 Types of surface forces in vacuum vs. in liquid (colloidal forces)

Type of force	Subclasses or alternative names	Main characteristics
	Attractive forces	
van der Waals	Debye induced dipole force (v & s) London dispersion force (v & s) Casimir force (v & s)	Ubiquitous, occurs both in vacuum and in liquids
Electrostatic	Ionic bond (v) Coulombic force (v & s) Hydrogen bond (v) Charge-exchange interaction (v & s) Acid–base interaction (s) "Harpooning" interaction (v)	Strong, long-range, arises in polar solvents; requires surface charging or charge-separation mechanism
Ion correlation	van der Waals force of polarizable ions (s)	Requires mobile charges on surfaces in a polar solvent
Quantum mechanical	Covalent bond (v) Metallic bond (v) Exchange interaction (v)	Strong, short-range, responsible for contact binding of crystalline surfaces
Solvation	Oscillatory force (s) Depletion force (s)	Mainly entropic in origin, the oscillatory force alternates between attraction and repulsion
Hydrophobic	Attractive hydration force (s)	Strong, apparently long-range; origin not yet understood
Specific binding	"Lock-and-key" or complementary binding (v & s) Receptor–ligand interaction (s) Antibody–antigen interaction (s)	Subtle combination of different non-covalent forces giving rise to highly specific binding; main recognition mechanism of biological systems
	Repulsive forces	
van der Waals	van der Waals disjoining pressure (s)	Arises only between dissimilar bodies interacting in a medium
Electrostatic	Coulombic force (v & s)	Arises only for certain constrained surface charge distributions
Quantum mechanical	Hard-core or steric repulsion (v) Born repulsion (v)	Short-range, stabilizing attractive covalent and ionic binding forces, effectively determine molecular size and shape
Solvation	Oscillatory solvation force (s) Structural force (s) Hydration force (s)	Monotonically repulsive forces, believed to arise when solvent molecules bind strongly to surfaces
Entropic	Osmotic repulsion (s) Double-layer force (s) Thermal fluctuation force (s) Steric polymer repulsion (s) Undulation force (s) Protrusion force (s)	Due to confinement of molecular or ionic species; requires mechanism that keeps trapped species between the surfaces
	Dynamic interactions	
Non-equilibrium	Hydrodynamic forces (s) Viscous forces (s) Friction forces (v & s) Lubrication forces (s)	Energy-dissipating forces occurring during relative motion of surfaces or bodies

Note: (v) applies only to interactions in vacuum, (s) applies only to interactions in solution (or to surfaces separated by a liquid), and (v & s) applies to interactions occurring both in vacuum and in solution.

18.2 Methods Used to Study Surface Forces

18.2.1 Force Laws

The full force law $F(D)$ between two surfaces, i.e., the force F as a function of surface separation D, can be measured in a number of ways [18.2–6]. The simplest is to move the base of a spring by a known amount, ΔD_0. Figure 18.1 illustrates this method when applied to the interaction of two magnets. However, the method is applicable also at the microscopic or molecular level, and it forms the basis of all direct force-measuring apparatuses such as the surface forces apparatus (SFA; [18.3, 7]) and the atomic force microscope (AFM; [18.8–10]). If there is a detectable force between the surfaces, this will cause the force-measuring spring to deflect by ΔD_s, while the surface separation changes by ΔD. These three displacements are related by

$$\Delta D_s = \Delta D_0 - \Delta D\,. \tag{18.1}$$

The difference in force, ΔF, between the initial and final separations is given by

$$\Delta F = k_s \Delta D_s\,, \tag{18.2}$$

where k_s is the spring constant. The equations above provide the basis for measurements of the force difference between any two surface separations. For example, if a force-measuring apparatus with a known k_s can measure D (and thus ΔD), ΔD_0, and ΔD_s, the force difference ΔF can be measured between a large initial or reference separation D, where the force is zero ($F = 0$), and another separation $D - \Delta D$. By working one's way in increasing increments of $\Delta D = \Delta D_0 - \Delta D_s$, the full force law $F(D)$ can be constructed over any desired distance regime.

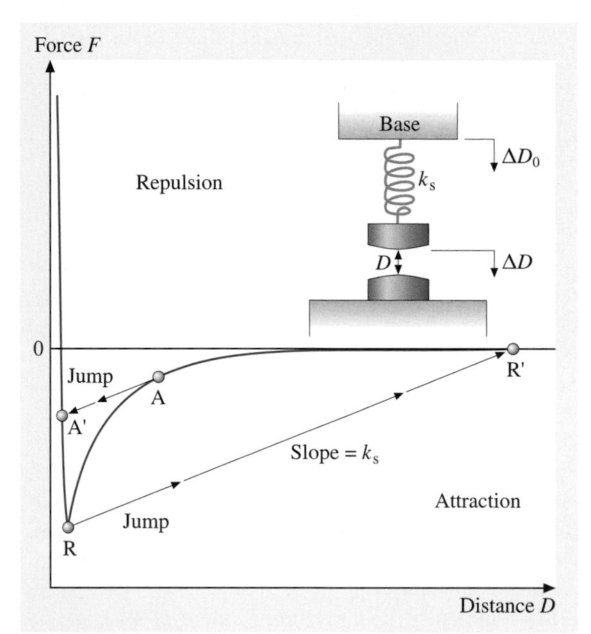

Fig. 18.1 Schematic attractive force law between two macroscopic objects such as two magnets, or between two microscopic objects such as the van der Waals force between a metal tip and a surface. On lowering the base supporting the spring, the latter will expand or contract such that at any equilibrium separation D the attractive force balances the elastic spring restoring force. If the gradient of the attractive force dF/dD exceeds the gradient of the spring's restoring force (defined by the spring constant k_s), the upper surface will jump from A into contact at A′ (A for "advancing"). On separating the surfaces by raising the base, the two surfaces will jump apart from R to R′ (R for "receding"). The distance R–R′ multiplied by k_s gives the adhesion force, i.e., the value of F at the point R (after [18.1] with permission)

In order to measure an equilibrium force law, it is essential to establish that the two surfaces have stopped moving before the displacements are measured. When displacements are measured while two surfaces are still in relative motion, one also measures a viscous or frictional contribution to the total force. Such dynamic force measurements have enabled the viscosities of liquids near surfaces and in thin films to be accurately determined [18.11–13].

In practice, it is difficult to measure the forces between two perfectly flat surfaces, because of the stringent requirement of perfect alignment for making reliable measurements at distances of a few tenths of a nanometer. It is far easier to measure the forces between curved surfaces, e.g., two spheres, a sphere and a flat surface, or two crossed cylinders. Furthermore, the force $F(D)$ measured between two curved surfaces can be directly related to the energy per unit area $E(D)$ between two flat surfaces at the same separation, D, by the so-called Derjaguin approximation [18.14]:

$$E(D) = \frac{F(D)}{2\pi R}\,, \tag{18.3}$$

where R is the radius of the sphere (for a sphere and a flat surface) or the radii of the cylinders (for two crossed cylinders).

18.2.2 Adhesion Forces

The most direct way to measure the adhesion of two solid surfaces (such as two spheres or a sphere on a flat) is to suspend one of them on a spring and measure the adhesion or "pull-off" force needed to separate the two bodies from its deflection. If k_s is the stiffness of the force-measuring spring and ΔD the distance the two surfaces jump apart when they separate, then the adhesion force F_s is given by

$$F_s = F_{max} = k_s \Delta D , \quad (18.4)$$

where we note that in liquids, the maximum or minimum in the force may occur at some nonzero surface separation (see Fig. 18.7). From F_s and a known surface geometry, and assuming that the surfaces were everywhere in molecular contact, one may also calculate the surface or interfacial energy γ. For an elastically deformable sphere of radius R on a flat surface, or for two crossed cylinders of radius R, we have [18.3, 15]

$$\gamma = \frac{F_s}{3\pi R} , \quad (18.5)$$

while for two spheres of radii R_1 and R_2

$$\gamma = \frac{F_s}{3\pi}\left(\frac{1}{R_1} + \frac{1}{R_2}\right) , \quad (18.6)$$

where γ is in units of $J\,m^{-2}$ (see Sect. 18.5.2).

18.2.3 The SFA and AFM

In a typical force-measuring experiment, at least two of the above displacement parameters – ΔD_0, ΔD, and ΔD_s – are directly or indirectly measured, and from these the third displacement and the resulting force law $F(D)$ are deduced using (18.1) and (18.2) together with a measured value of k_s. For example, in SFA experiments, ΔD_0 is changed by expanding or contracting a piezoelectric crystal by a known amount or by moving the base of the spring with sensitive motor-driven mechanical stages. The resulting change in surface separation ΔD is measured optically, and the spring deflection ΔD_s can then be obtained according to (18.1). In AFM experiments, ΔD_0 and ΔD_s are measured using a combination of piezoelectric, optical, capacitance or magnetic techniques, from which the change in surface separation ΔD is deduced. Once a force law is established, the geometry of the two surfaces (e.g., their radii) must also be known before the results can be compared with theory or with other experiments.

The SFA (Fig. 18.2) is used for measurements of adhesion and force laws between two curved molecularly smooth surfaces immersed in liquids or controlled vapors [18.3, 7, 16]. The surface separation is measured by multiple beam interferometry with an accuracy of ± 0.1 nm. From the shape of the interference fringes one also obtains the radius of the surfaces, R, and any surface deformation that arises during an interaction [18.17–19]. The resolution in the lateral direction is about 1 μm. The surface separation can be independently controlled to within 0.1 nm, and the force sensitivity is about 10^{-8} N. For a typical surface radius of $R \approx 1$ cm, γ values can be measured to an accuracy of about 10^{-3} mJ m^{-2}.

Several different materials have been used to form the surfaces in the SFA, including mica [18.20, 21], silica [18.22], sapphire [18.23], and polymer sheets [18.24]. These materials can also be used as supporting substrates in experiments on the forces between adsorbed or chemically bound polymer layers [18.13, 25–30], surfactant and lipid monolayers and bilayers [18.31–34], and metal and metal oxide layers [18.35–42]. The range of liquids and vapors that can be used is almost endless, and have thus far included aqueous solutions, organic liquids and solvents, polymer melts, various petroleum oils and lubricant liquids, and liquid crystals.

Friction attachments for the SFA [18.43–47] allow for the two surfaces to be sheared laterally past each other at varying sliding speeds or oscillating frequencies, while simultaneously measuring both the transverse (frictional or shear) force and the normal force (load) between them. The ranges of friction forces and sliding speeds that can be studied with such methods are currently 10^{-7}–10^{-1} N and 10^{-13}–10^{-2} m s^{-1}, respectively [18.48]. The externally applied load, L, can be varied continuously, and both positive and negative loads can be applied. The distance between the surfaces, D, their true molecular contact area, their elastic (or viscoelastic or elastohydrodynamic) deformation, and their lateral motion can all be monitored simultaneously by recording the moving interference fringe pattern.

In the atomic force microscope (Fig. 18.3), the force is measured by monitoring the deflection of a soft cantilever supporting a sub-microscopic tip ($R \approx 10$–200 nm) as this is interacting with a flat, macroscopic surface [18.8, 49, 50]. The measurements can be done in a vapor or liquid. The normal (bending) spring stiffness of the cantilever can be as small as 0.01 N m^{-1}, allowing measurements of normal forces as small as 1 pN (10^{-12} N), which corresponds to the bond strength

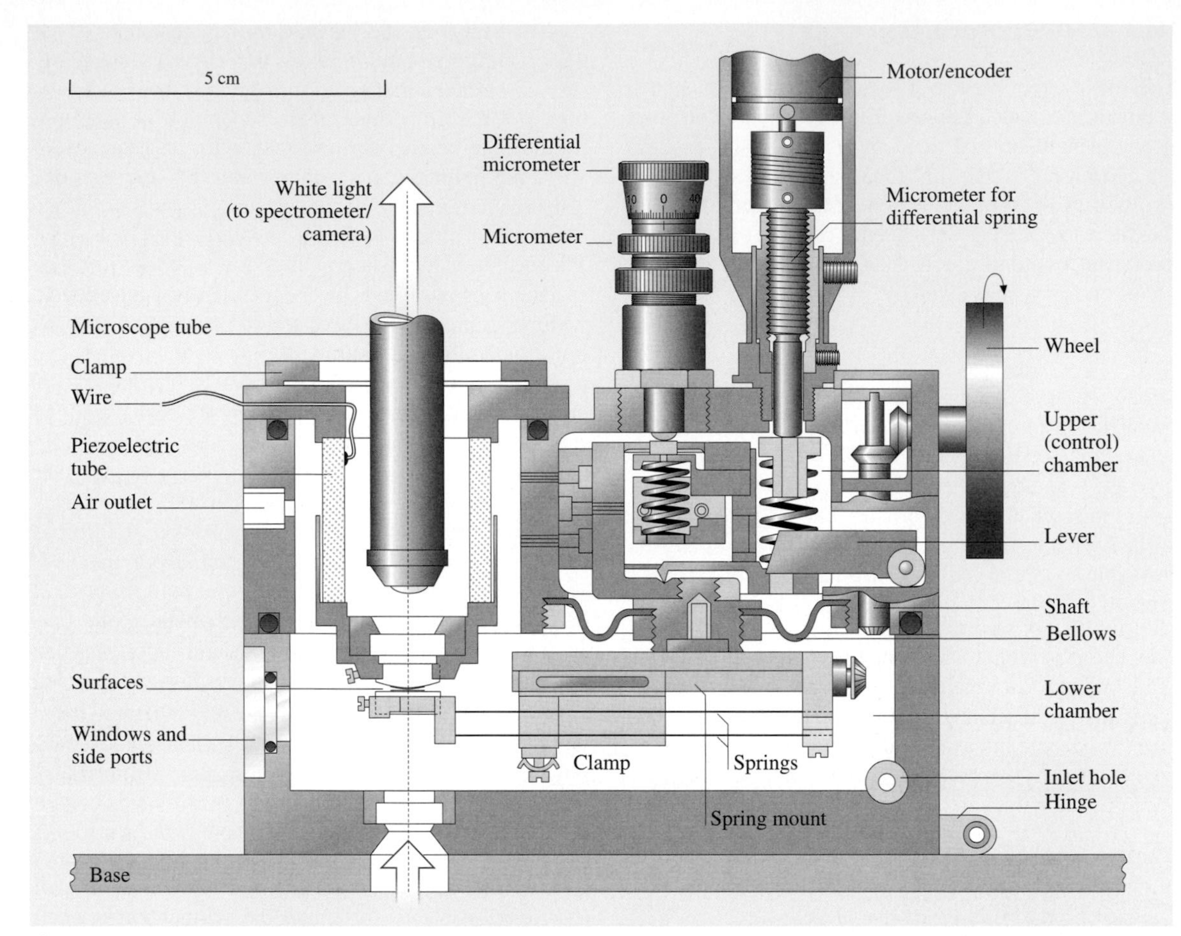

Fig. 18.2 A surface forces apparatus (SFA) where the intermolecular forces between two macroscopic, cylindrical surfaces of local radius R can be directly measured as a function of surface separation over a large distance regime from tenths of a nanometer to micrometers. Local or transient surface deformations can be detected optically. Various attachments for moving one surface laterally with respect to the other have been developed for friction measurements in different regimes of sliding velocity and sliding distance (after [18.16] with permission)

of single molecules [18.51, 52]. Distances can be inferred with an accuracy of about 1 nm, and changes in distance can be measured to about 0.1 nm. Since the contact area is small, different interaction regimes can be resolved on samples with a heterogeneous composition on lateral scales of about ten nanometers. Height differences and the roughness of the sample can be measured directly from the cantilever deflection or, alternatively, by using a feedback system to raise or lower the sample so that the deflection (the normal force) is kept constant during a scan over the area of interest. Weak interaction forces and larger (microscopic) interaction areas can be investigated by replacing the tip with a micrometer-sized sphere to form a "colloidal probe" [18.9].

The atomic force microscope can also be used for friction measurements (lateral force microscopy, LFM) by monitoring the torsion of the cantilever as the sample is scanned in the direction perpendicular to the long axis of the cantilever [18.10, 50, 53, 54]. Typically, the stiffness of the cantilever to lateral bending is much larger than to bending in the normal direction and to torsion, so that these signals are decoupled and height and friction can be detected simultaneously. The torsional spring constant can be as low as $0.1\,\mathrm{N\,m^{-1}}$, giving a lateral (friction) force sensitivity of 10^{-11} N.

Rapid technical developments have facilitated the calibrations of the normal [18.55, 56] and lateral spring constants [18.54, 57], as well as in situ measurements of

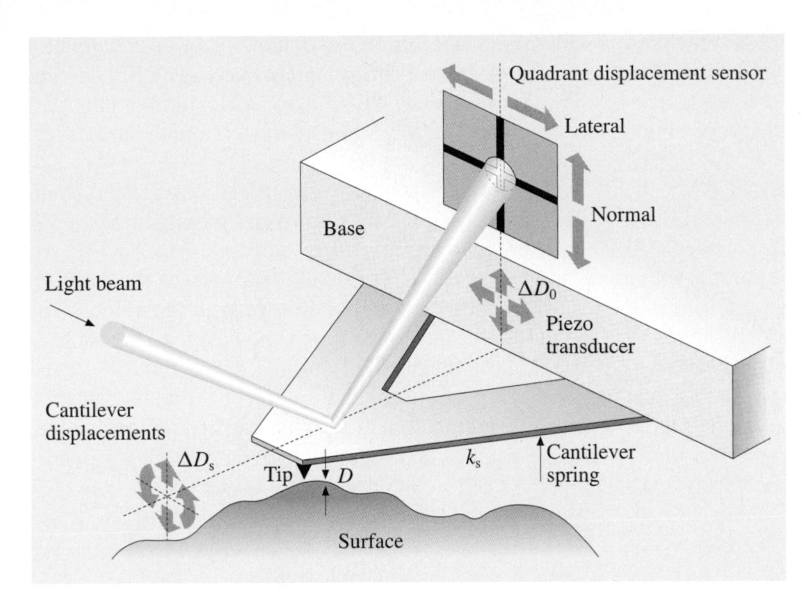

Fig. 18.3 Schematic drawing of an atomic force microscope (AFM) tip supported on a triangular cantilever interacting with an arbitrary solid surface. The normal force and topology are measured by monitoring the calibrated deflection of the cantilever as the tip is moved across the surface by means of a piezoelectric transducer. Various designs have been developed that move either the sample or the cantilever during the scan. Friction forces can be measured from the torsion of the cantilever when the scanning is in the direction perpendicular to its long axis (after [18.62] with permission)

the macroscopic tip radius [18.58]. Cantilevers of different shapes with a large range of spring constants, tip radii, and surface treatments (inorganic or organic coatings) are commercially available. The flat surface, and also the particle in the colloidal probe technique, can be any material of interest. However, remaining difficulties with this technique are that the distance between the tip and the substrate, D, and the deformations of the tip and sample are not directly measurable. Another important difference between the AFM/LFM and SFA techniques is the different size of the contact area, and the related observation that even when a cantilever with a very low spring constant is used in the AFM, the pressure in the contact zone is typically much higher than in the SFA. Hydrodynamic effects in liquids also affect the measurements of normal forces differently on certain time scales [18.59–61].

18.2.4 Some Other Force-Measuring Techniques

A large number of other techniques are available for the measurements of the normal forces between solid or fluid surfaces (see [18.5, 62]). The techniques discussed in this section are not used for lateral (friction) force measurements, but are commonly used to study normal forces, particularly in biological systems.

Micropipette aspiration is used to measure the forces between cells or vesicles, or between a cell or vesicle and another surface [18.63–65]. The cell or vesicle is held by suction at the tip of a glass micropipette and deforms elastically in response to the net interactions with another surface and to the applied suction. The shape of the deformed surface (cell membrane) is measured and used to deduce the force between the surfaces and the membrane tension [18.64]. The membrane tension, and thus the stiffness of the cell or vesicle, is regulated by applying different hydrostatic pressures. Forces can be measured in the range of 0.1 pN to 1 nN, and the distance resolution is a few nanometers. The interactions between a colloidal particle and another surface can be studied by attaching the particle to the cell membrane [18.66].

In the osmotic stress technique, pressures are measured between colloidal particles in aqueous solution, membranes or bilayers, or other ordered colloidal structures (viruses, DNA). The separation between the particle surfaces and the magnitude of membrane undulations are measured by X-ray or neutron scattering techniques. This is combined with a measurement of the osmotic pressure of the solution [18.67–70]. The technique has been used to measure repulsive forces, such as DLVO interactions, steric forces, and hydration forces [18.71]. The sensitivity in pressure is $0.1\,\mathrm{mN\,m^{-2}}$, and distances can be resolved to 0.1 nm.

The optical tweezers technique is based on the trapping of dielectric particles at the center of a focused laser beam by restoring forces arising from radiation pressure and light intensity gradients [18.72, 73]. The forces experienced by particles as they are moved toward or away from one another can be measured with a sensitivity in the pN range. Small biological molecules are typically attached to a larger bead of a material with suitable refractive properties. Recent development allows determinations of position with nanometer resolution [18.74], which makes this technique useful for studying the forces during extension of single molecules.

In total internal reflection microscopy (TIRM), the potential energy between a micrometer-sized colloidal particle and flat surface in aqueous solution is deduced from the average equilibrium height of the particle above the surface, measured from the intensity of scattered light. The average height ($D \approx 10-100$ nm) results from a balance of gravitational force, radiation pressure from a laser beam focused at the particle from below, and intermolecular forces [18.75]. The technique is particularly suitable for measuring weak forces (sensitivity ca. 10^{-14} N), but is more difficult to use for systems with strong interactions. A related technique is reflection interference contrast microscopy (RICM), where optical interference is used to also monitor changes in the shape of the approaching colloidal particle or vesicle [18.76].

An estimate of bond strengths can be obtained from the hydrodynamic shear force exerted by a fluid on particles or cells attached to a substrate [18.77, 78]. At a critical force, the bonds are broken and the particle or cell will be detached and move with the velocity of the fluid. This method requires knowledge of the contact area and the flow velocity profile of the fluid. Furthermore, a uniform stress distribution in the contact area is generally assumed. At low bond density, this technique can be used to determine the strength of single bonds (1 pN).

18.3 Normal Forces Between Dry (Unlubricated) Surfaces

18.3.1 Van der Waals Forces in Vacuum and Inert Vapors

Forces between macroscopic bodies (such as colloidal particles) across vacuum arise from interactions between the constituent atoms or molecules of each body across the gap separating them. These intermolecular interactions are electromagnetic forces between permanent or induced dipoles (van der Waals forces), and between ions (electrostatic forces). In this section, we describe the van der Waals forces, which occur between all atoms and molecules and between all macroscopic bodies (see [18.3]).

The interaction between two permanent dipoles with a fixed relative orientation can be attractive or repulsive. For the specific case of two freely rotating permanent dipoles in a liquid or vapor (orientational or Keesom interaction), and for a permanent dipole and an induced dipole in an atom or polar or nonpolar molecule (induction or Debye interaction), the interaction is on average always attractive. The third type of van der Waals interaction, the fluctuation or London dispersion interaction, arises from instantaneous polarization of one nonpolar or polar molecule due to fluctuations in the charge distribution of a neighboring nonpolar or polar molecule (Fig. 18.4a). Correlation between these fluctuating induced dipole moments gives an attraction that is present between any two molecules or surfaces across vacuum. At very small separations, the interaction will ultimately be repulsive as the electron clouds of atoms and molecules begin to overlap. The total interaction is thus a combination of a short-range repulsion and a relatively long-range attraction.

Except for in highly polar materials such as water, London dispersion interactions give the largest contribution (70–100 %) to the van der Waals attraction. The interaction energy of the van der Waals force between atoms or molecules depends on the separation r as

$$E(D) = \frac{-C_{\text{vdW}}}{r^6}\,, \tag{18.7}$$

where the constant C_{vdW} depends on the dipole moments and polarizabilities of the molecules. At large separations (> 10 nm), the London interaction is weakened by a randomizing effect caused by the rapid fluctuations. That is, the induced temporary dipole moment of one molecule may have changed during the time needed for the transmission of the electromagnetic wave (photon) generated by its fluctuating charge density to another molecule and the return of the photon generated by the induced fluctuation in this second molecule. This phenomenon is called retardation and causes the interaction energy to decay as r^{-7} at large separations [18.79].

Dispersion interactions are to a first approximation additive, and their contribution to the interaction energy between two macroscopic bodies (such as colloidal particles) across vacuum can be found by summing the pair-wise interactions [18.80]. The interaction is generally described in terms of the Hamaker constant, A_H. Another approach is to treat the interacting bodies and an intervening medium as continuous phases and determine the strength of the interaction from bulk dielectric properties of the materials [18.81, 82]. Unlike the pair-wise summation, this method takes into account the screening of the interactions between molecules inside the bodies by the molecules closer to the surfaces and the effects of the intervening medium. For the interaction between material 1 and material 3 across material 2, the non-retarded Hamaker constant given by the Lifshitz theory

Table 18.2 Van der Waals interaction energy and force between macroscopic bodies of different geometries

Geometry of bodies with surfaces D apart ($D \ll R$)			van der Waals interaction: Energy, E	Force, F
Two atoms or small molecules	r, σ	$r \gg \sigma$	$\frac{-C_{vdW}}{r^6}$	$\frac{-6C_{vdW}}{r^7}$
Two flat surfaces (per unit area)	r, $Area = \pi r^2$, D	$r \gg D$	$\frac{-A_H}{12\pi D^2}$	$\frac{-A_H}{6\pi D^3}$
Two spheres or macromolecules of radii R_1 and R_2	R_1, R_2, D	$R_1, R_2 \gg D$	$\frac{-A_H}{6D}\left(\frac{R_1 R_2}{R_1+R_2}\right)$	$\frac{-A_H}{6D^2}\left(\frac{R_1 R_2}{R_1+R_2}\right)$
Sphere or macromolecule of radius R near a flat surface	R, D	$R \gg D$	$\frac{-A_H R}{6D}$	$\frac{-A_H R}{6D^2}$
Two parallel cylinders or rods of radii R_1 and R_2 (per unit length)	R_1, R_2, Length	$R_1, R_2 \gg D$	$\frac{-A_H}{12\sqrt{2}\,D^{3/2}}\left(\frac{R_1 R_2}{R_1+R_2}\right)^{1/2}$	$\frac{-A_H}{8\sqrt{2}\,D^{5/2}}\left(\frac{R_1 R_2}{R_1+R_2}\right)^{1/2}$
Cylinder of radius R near a flat surface (per unit length)	R, D	$R \gg D$	$\frac{-A_H\sqrt{R}}{12\sqrt{2}\,D^{3/2}}$	$\frac{-A_H\sqrt{R}}{8\sqrt{2}\,D^{5/2}}$
Two cylinders or filaments of radii R_1 and R_2 crossed at 90°	R_1, R_2, D	$R_1, R_2 \gg D$	$\frac{-A_H\sqrt{R_1 R_2}}{6D}$	$\frac{-A_H\sqrt{R_1 R_2}}{6D^2}$

A negative force (A_H positive) implies attraction, a positive force means repulsion (A_H negative) (after [18.62] with permission)

is approximately [18.3]:

$$\begin{aligned} A_{\mathrm{H},123} &= A_{\mathrm{H},\nu=0} + A_{\mathrm{H},\nu>0} \\ &\approx \tfrac{3}{4} k_{\mathrm{B}} T \left(\tfrac{\varepsilon_1-\varepsilon_2}{\varepsilon_1+\varepsilon_2}\right)\left(\tfrac{\varepsilon_3-\varepsilon_2}{\varepsilon_3+\varepsilon_2}\right) \\ &+ \frac{3h\nu_{\mathrm{e}}}{8\sqrt{2}} \frac{(n_1^2-n_2^2)(n_3^2-n_2^2)}{\sqrt{(n_1^2+n_2^2)}\sqrt{(n_3^2+n_2^2)}\left(\sqrt{(n_1^2+n_2^2)}+\sqrt{(n_3^2+n_2^2)}\right)}, \end{aligned} \quad (18.8)$$

where the first term ($\nu = 0$) represents the permanent dipole and dipole–induced dipole interactions and the second ($\nu > 0$) the London (dispersion) interaction. ε_i and n_i are the static dielectric constants and refractive indexes of the materials, respectively. ν_{e} is the frequency of the lowest electron transition (around $3\times10^{15}\,\mathrm{s}^{-1}$). Either one of the materials 1, 2, or 3 in (18.8) can be vacuum or air ($\varepsilon = n = 1$). A_{H} is typically 10^{-20}–10^{-19} J (the higher values are found for metals) for interactions between solids and liquids across vacuum or air.

The interaction energy between two macroscopic bodies is dependent on the geometry and is always attractive between two bodies of the same material [A_{H} positive, see (18.8)]. The van der Waals interaction energy and force laws ($F = -\mathrm{d}E(D)/\mathrm{d}D$) for some common geometries are given in Table 18.2. Because of the retardation effect, the equations in Table 18.2 will lead to an overestimation of the dispersion force at large separations. It is, however, apparent that the interaction energy between macroscopic bodies decays more slowly with separation (i. e., has a longer range) than between two molecules.

For inert nonpolar surfaces, e.g., consisting of hydrocarbons or van der Waals solids and liquids, the Lifshitz theory has been found to apply even at molecular contact, where it can be used to predict the surface energies (surface tensions) of such solids and liquids. For example, for hydrocarbon surfaces, $A_{\mathrm{H}} = 5\times10^{-20}$ J. Inserting this value into the equation for two flat surfaces (Table 18.2) and using a "cut-off" distance of $D_0 \approx 0.165$ nm as an effective separation when the surfaces are in contact [18.3], we obtain for the surface energy γ (which is defined as half the interaction energy)

$$\gamma = \frac{E}{2} = \frac{A_{\mathrm{H}}}{24\pi D_0^2} \approx 24\,\mathrm{mJ\,m^{-2}}, \quad (18.9)$$

a value that is typical for hydrocarbon solids and liquids [18.83].

If the adhesion force is measured between a spherical surface of radius $R = 1$ cm and a flat surface using an SFA, we expect the adhesion force to be (see Table 18.2) $F = A_{\mathrm{H}}R/(6D_0^2) = 4\pi R\gamma \approx 3.0$ mN. Using a spring constant of $k_{\mathrm{s}} = 100\,\mathrm{N\,m^{-1}}$, such an adhesive force will cause the two surfaces to jump apart by $\Delta D = F/k_{\mathrm{s}} = 30\,\mu\mathrm{m}$, which can be accurately measured. (For elastic bodies that deform in adhesive contact, R changes during the interaction and the measured adhesion force is 25% lower, see Sect. 18.5.2). Surface energies of solids can thus be directly measured with the SFA and, in principle, with the AFM if the contact geometry can be quantified. The measured values are in good agreement with calculated values based on the known surface energies γ of the materials, and for nonpolar low-energy solids they are well accounted for by the Lifshitz theory [18.3].

18.3.2 Charge Exchange Interactions

Electrostatic interactions are present between ions (Coulomb interactions), between ions and permanent dipoles, and between ions and nonpolar molecules in which a charge induces a dipole moment. The interaction energy between ions or between a charge and a fixed permanent dipole can be attractive or repulsive. For an induced dipole or a freely rotating permanent dipole in vacuum or air, the interaction energy with a charge is always attractive.

Spontaneous charge transfer may occur between two dissimilar materials in contact [18.84]. The phenomenon is especially prominent in contacts between a metal, for example, mercury, and a material with low conductivity, but is also observed, for example, between two different polymer layers. During separation, rolling or sliding of one body over the other, the surfaces experience both charge transition from one surface to the other and charge transfer (conductance) along each surface (Fig. 18.4b). The latter process is typically slower, and, as a result, charges remain on the surfaces as they are separated in vacuum or dry nitrogen gas. The charging gives rise to a strong adhesion with adhesion energies of over 1,000 $\mathrm{mJ\,m^{-2}}$, similar to fracture or cohesion energies of the solid bodies themselves [18.84, 85]. Upon separating the surfaces farther apart, a strong, long-range electrostatic attraction is observed. The charging can be decreased through discharges across the gap between the surfaces (which requires a high charging) or through conducting in the solids. It has been suggested that charge exchange interactions are particularly important in rolling friction between dry surfaces (which can simplistically be thought of as an adhesion–separation process), where the distance dependence of forces acting normally to the surfaces plays a larger role than in sliding friction. Recent experiments on the sliding friction between metal–insulator surfaces indicate that stick–slip would be accompanied by charge transfer events [18.86].

Photo-induced charge transfer, or harpooning, involves the transfer of an electron between an atom in a molecular beam or at a solid surface (typically alkali or transition metal) to an atom or molecule in a gas (typically halides) to form a negatively charged molecular ion in a highly excited vibrational state. This transfer process can occur at atomic distances of 0.5–0.7 nm, which is far from molecular contact. The formed molecular ion is attracted to the surface and chemisorbs onto it. Photo-induced charge transfer processes also occur in the photosynthesis in green plants and in photoelectrochemical cells (solar cells) at the junction between two semiconductors or between a semiconductor and an electrolyte solution [18.87].

18.3.3 Sintering and Cold Welding

When macroscopic particles in a powder or in a suspension come into molecular contact, they can bond together to form a network or solid body with very different density and shear strength compared to the powder (a typical example is porcelain). The rate of bonding is dependent on the surface energy (causing a stress at the edge of the contact) and the atomic mobility (diffusion rate) of the contacting materials. To increase the diffusion rate, objects formed from powders are heated to about one-half of the melting temperature of the components in a process called sintering, which can be done in different atmospheres or in a liquid.

In the sintering process, the surface energy of the system is lowered due to the reduction of total surface area (Fig. 18.4c). In metal and ceramic systems, the most important mechanism is solid-state diffusion, initially surface diffusion. As the surface area decreases and the grain boundaries increase at the contacts, grain boundary diffusion and diffusion through the crystal lattice become more important. The grain boundaries will eventually migrate, so that larger particles are formed (coarsening). Mass can also be transferred through evaporation and condensation, and through viscous and plastic flow. In liquid-phase sintering, the materials can melt, which increases the mass transport. Amorphous materials like polymers and glasses do not have real grain boundaries and sinter by viscous flow [18.88].

Some of these mechanisms (surface diffusion and evaporation–condensation) reduce the surface area and increase the grain size (coarsening) without densification, in contrast to bulk transport mechanisms like grain boundary diffusion and plastic and viscous flow. As the material becomes denser, elongated pores collapse to form smaller, spherical pores with a lower surface energy. Models for sintering typically consider the size and growth rate of the grain boundary (the "neck") formed between two spherical particles. At a high stage of densification, the sintering stress σ at the curved neck between two particles is given by [18.88]

$$\sigma = \frac{2\gamma_{SS}}{G} + \frac{2\gamma_{SV}}{r_p} , \qquad (18.10)$$

where γ_{SS} is the solid–solid grain boundary energy, γ_{SV} is the solid–vapor surface energy, G is the grain size, and r_p is the radius of the pore.

A related phenomenon is cold welding, which is the spontaneous formation of strong junctions between clean (unoxidized) metal surfaces with a mutual solubility when they are brought in contact with or without an applied pressure. The plastic deformations accompanying the formation and breaking of such contacts on a molecular scale during motion of one surface normally (see Fig. 18.10c,d) or laterally (shearing) with respect to the other have been studied both experimentally [18.89] and theoretically [18.90–95]. The breaking of a cold welded contact is generally associated with damage or deformation of the surface structure.

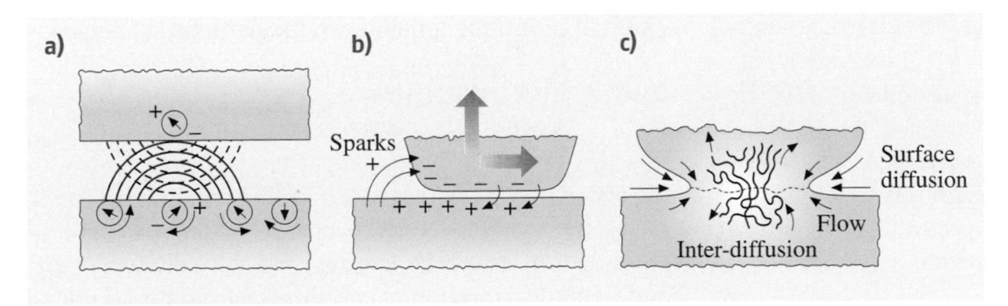

Fig. 18.4a–c Schematic representation of (**a**) van der Waals interaction (dipole–induced dipole interaction), (**b**) charge exchange, which acts to increase adhesion and friction forces, and (**c**) sintering between two surfaces

18.4 Normal Forces Between Surfaces in Liquids

18.4.1 Van der Waals Forces in Liquids

The dispersion interaction in a medium will be significantly lower than in vacuum, since the attractive interaction between two solute molecules in a medium (solvent) involves displacement and reorientation of the nearest-neighbor solvent molecules. Even though the surrounding medium may change the dipole moment and polarizability from that in vacuum, the interaction between two identical molecules remains attractive in a binary mixture. The extension of the interactions to the case of two macroscopic bodies is the same as described in Sect. 18.3.1. Typically, the Hamaker constants for interactions in a medium are an order of magnitude lower than in vacuum. Between macroscopic surfaces in liquids, van der Waals forces become important at distances below 10–15 nm and may at these distances start to dominate interactions of different origin that have been observed at larger separations.

Figure 18.5 shows the measured van der Waals forces between two crossed cylindrical mica surfaces in water and various salt solutions. Good agreement is obtained between experiment and theory. At larger surface separations, above about 5 nm, the measured forces fall off more rapidly than as D^{-2}. This retardation effect (see Sect. 18.3.1) is also predicted by Lifshitz theory and is due to the time needed for propagation of the induced dipole moments over large distances.

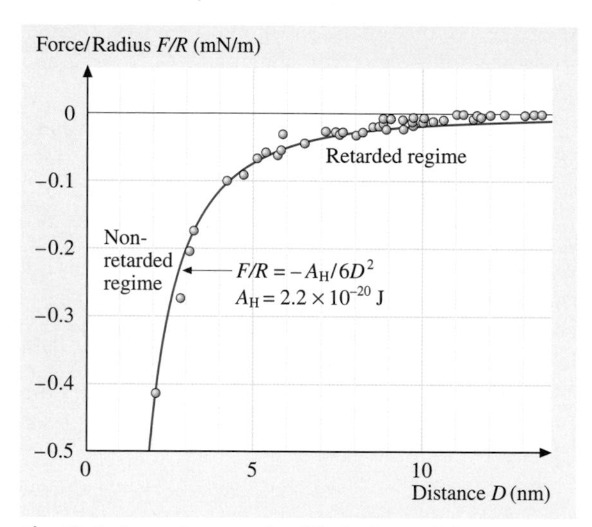

Fig. 18.5 Attractive van der Waals force F between two curved mica surfaces of radius $R \approx 1$ cm measured in water and various aqueous electrolyte solutions. The electrostatic interaction has been subtracted from the total measured force. The measured non-retarded Hamaker constant is $A_H = 2.2 \times 10^{-20}$ J. Retardation effects are apparent at distances larger than 5 nm, as expected theoretically. (After [18.3]. Copyright 1991, with permission from Elsevier Science)

From Fig. 18.5, we may conclude that at separations above about 2 nm, or 8 molecular diameters of water, the *continuum* Lifshitz theory is valid. This would mean that water films as thin as 2 nm may be expected to have bulk-like properties, at least as far as their interaction forces are concerned. Similar results have been obtained with other liquids, where in general continuum properties are manifested, both as regards their interactions and other properties such as viscosity, at a film thickness larger than 5 or 10 molecular diameters. In the absence of a solvent (in vacuum), the agreement of measured van der Waals forces with the continuum Lifshitz theory is generally good at all separations down to molecular contact ($D = D_0$).

Van der Waals interactions in a system of three or more different materials (see (18.8)) can be attractive or repulsive, depending on their dielectric properties. Numerous experimental studies show the attractive van der Waals forces in various systems [18.3], and also repulsive van der Waals forces have been measured directly [18.96]. A practical consequence of the repulsive interaction obtained across a medium with intermediate dielectric properties is that the van der Waals forces will give rise to preferential, nonspecific adsorption of molecules with an intermediate dielectric constant. This is commonly seen as adsorption of vapors or solutes to a solid surface. It is also possible to diminish the attractive interaction between dispersed colloidal particles by adsorption of a thin layer of material with dielectric properties close to those of the surrounding medium (matching of refractive index), or by adsorption of a polymer that gives a steric repulsive force that keeps the particles separated at a distance where the magnitude of the van der Waals attraction is negligible. Thermal motion will then keep the particles dispersed.

18.4.2 Electrostatic and Ion Correlation Forces

Most surfaces in contact with a highly polar liquid (such as water) acquire a surface charge, either by dissociation of ions from the surface into the solution or by preferential adsorption of certain ions from the solution.

The surface charge is balanced by a layer of oppositely charged ions (counterions) in the solution at some small distance from the surface (see [18.3]). In dilute solution, this distance is the Debye length, κ^{-1}, which is purely a property of the electrolyte *solution*. The Debye length falls with increasing ionic strength (i. e., with the molar concentration M_i and valency z_i) of the ions in solution:

$$\kappa^{-1} = \left(\frac{\varepsilon\varepsilon_0 k_B T}{e^2 N_A \sum_i z_i^2 M_i} \right)^{1/2} , \quad (18.11)$$

where e is the electronic charge. For example, for 1:1 electrolytes at 25 °C, $\kappa^{-1} = 0.304\,\text{nm}/\sqrt{M_{1:1}}$, where M_i is given in M (mol dm^{-3}). κ^{-1} is thus ca. 10 nm in a 1 mM NaCl solution and 0.3 nm in a 1 M solution. In totally pure water at pH 7, where $M_i = 10^{-7}$ M, κ^{-1} is 960 nm, or about 1 μm. The Debye length also relates the surface charge density σ of a surface to the electrostatic surface potential ψ_0 via the Grahame equation, which for 1:1 electrolytes can be expressed as:

$$\sigma = \sqrt{8\varepsilon\varepsilon_0 k_B T}\,\sinh\left(e\psi_0/2k_B T\right) \times \sqrt{M_{1:1}} . \quad (18.12)$$

Since the Debye length is a measure of the thickness of the diffuse atmosphere of counterions near a charged surface, it also determines the range of the electrostatic "double-layer" interaction between two charged surfaces. The electrostatic double-layer interaction is an entropic effect that arises upon decreasing the thickness of the liquid film containing the dissolved ions. Because of the attractive force between the dissolved ions and opposite charges on the surfaces, the ions stay between the surfaces, but an osmotic repulsion arises as their concentration increases. The long-range electrostatic interaction energy at large separations (weak overlap) between two similarly charged molecules or surfaces is typically repulsive and is roughly an exponentially decaying function of D:

$$E(D) \approx +C_{ES}\,e^{-\kappa D} , \quad (18.13)$$

where C_{ES} is a constant that depends on the geometry of the interacting surfaces, on their surface charge density, and the solution conditions (Table 18.3). We see that the Debye length is the decay length of the interaction energy between two surfaces (and of the mean potential away from one surface). C_{ES} can be determined by solving the so-called Poisson–Boltzmann equation or by using other theories [18.97, 98]. The equations in Table 18.3 are expressed in terms of a constant, Z, defined as

$$Z = 64\pi\varepsilon\varepsilon_0 (k_B T/e)^2 \tanh^2\left[ze\psi_0/(4k_B T)\right] , \quad (18.14)$$

which depends only on the properties of the *surfaces*.

The above approximate expressions are accurate only for surface separations larger than about one Debye length. At smaller separations one must use numerical solutions of the Poisson–Boltzmann equation to obtain the exact interaction potential, for which there are no simple expressions. In the limit of small D, it can be shown that the interaction energy depends on whether the surfaces remain at constant potential ψ_0 (as assumed in the above equations) or at constant charge σ (when the repulsion exceeds that predicted by the above equations), or somewhere between these limits. In the "constant charge limit" the total *number* of counterions in the compressed film does not change as D is decreased, whereas at constant potential, the *concentration* of counterions is constant. The limiting pressure (or force per unit area) at constant charge is the osmotic pressure of the confined ions:

$$\begin{aligned} F &= k_B T \times \text{ion number density} \\ &= 2\sigma k_B T/(zeD), \quad \text{for } D \ll \kappa^{-1} . \end{aligned} \quad (18.15)$$

That is, as $D \to 0$ the double-layer pressure at constant surface charge becomes infinitely repulsive and independent of the salt concentration (at constant potential the force instead becomes a constant at small D). However, at small separations, the van der Waals attraction (which goes as D^{-2} between two spheres or as D^{-3} between two planar surfaces, see Table 18.2) wins out over the double-layer repulsion, unless some other short-range interaction becomes dominant (see Sect. 18.4.4). This is the theoretical prediction that forms the basis of the so-called Derjaguin–Landau–Verwey–Overbeek (DLVO) theory [18.97, 99], illustrated in Fig. 18.6.

Because of the different distance dependence of the van der Waals and electrostatic interactions, the total force law, as described by the DLVO theory, can show several minima and maxima. Typically, the depth of the outer (secondary) minimum is a few $k_B T$, enough to cause reversible flocculation of particles from an aqueous dispersion. If the force barrier between the secondary and primary minimum is lowered, for example, by increasing the electrolyte concentration, particles can be irreversibly coagulated in the primary minimum. In practice, other forces (described

Table 18.3 Electrical "double-layer" interaction energy $E(D)$ and force ($F = -\mathrm{d}E/\mathrm{d}D$) between macroscopic bodies

Geometry of bodies with surfaces D apart ($D \ll R$)		Electric "double-layer" interaction Energy, E	Force, F
Two ions or small molecules	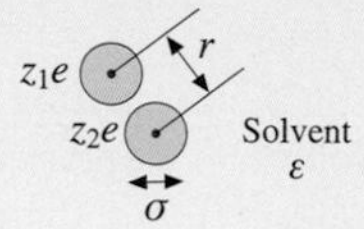 $r \gtrsim \sigma$	$\dfrac{+z_1 z_2 e^2}{4\pi\varepsilon\varepsilon_0 r}\dfrac{\mathrm{e}^{-\kappa(r-\sigma)}}{(1+\kappa\sigma)}$	$\dfrac{+z_1 z_2 e^2}{4\pi\varepsilon\varepsilon_0 r^2}\dfrac{(1+\kappa r)}{(1+\kappa\sigma)}\mathrm{e}^{-\kappa(r-\sigma)}$
Two flat surfaces (per unit area)	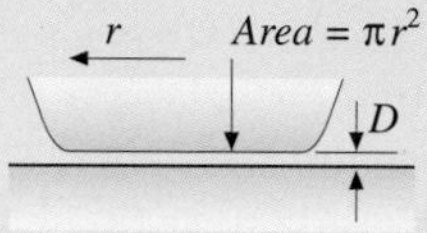 $r \gg D$	$(\kappa/2\pi)\, Z\mathrm{e}^{-\kappa D}$	$\left(\kappa^2/2\pi\right) Z\mathrm{e}^{-\kappa D}$
Two spheres or macromolecules of radii R_1 and R_2	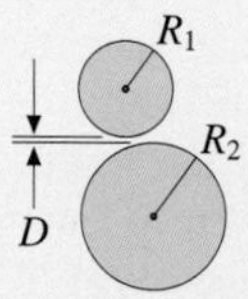 $R_1, R_2 \gg D$	$\left(\dfrac{R_1 R_2}{R_1+R_2}\right) Z\mathrm{e}^{-\kappa D}$	$\kappa\left(\dfrac{R_1 R_2}{R_1+R_2}\right) Z\mathrm{e}^{-\kappa D}$
Sphere or macromolecule of radius R near a flat surface	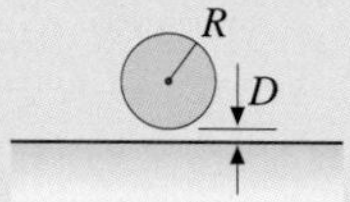 $R \gg D$	$RZ\mathrm{e}^{-\kappa D}$	$\kappa RZ\mathrm{e}^{-\kappa D}$
Two parallel cylinders or rods of radii R_1 and R_2 (per unit length)	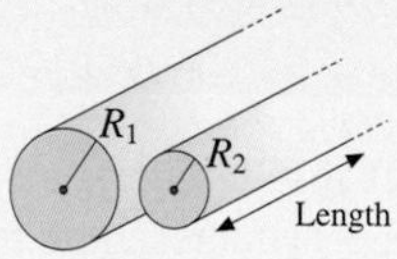 $R_1, R_2 \gg D$	$\dfrac{\kappa^{1/2}}{\sqrt{2\pi}}\left(\dfrac{R_1 R_2}{R_1+R_2}\right)^{1/2} Z\mathrm{e}^{-\kappa D}$	$\dfrac{\kappa^{3/2}}{\sqrt{2\pi}}\left(\dfrac{R_1 R_2}{R_1+R_2}\right)^{1/2} Z\mathrm{e}^{-\kappa D}$
Cylinder of radius R near a flat surface (per unit length)	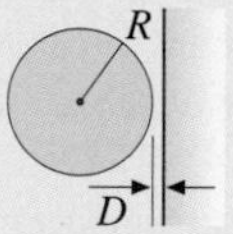 $R \gg D$	$\kappa^{1/2}\sqrt{\dfrac{R}{2\pi}}\, Z\mathrm{e}^{-\kappa D}$	$\kappa^{3/2}\sqrt{\dfrac{R}{2\pi}}\, Z\mathrm{e}^{-\kappa D}$
Two cylinders or filaments of radii R_1 and R_2 crossed at 90°	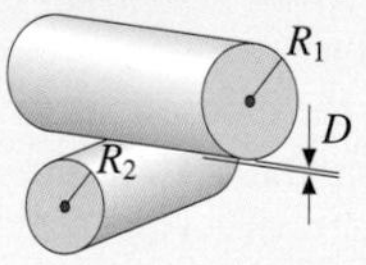 $R_1, R_2 \gg D$	$\sqrt{R_1 R_2}\, Z\mathrm{e}^{-\kappa D}$	$\kappa\sqrt{R_1 R_2}\, Z\mathrm{e}^{-\kappa D}$

The interaction energy and force for bodies of different geometries is based on the Poisson–Boltzmann equation (a continuum, mean-field theory). Equation (18.14) gives the interaction constant Z (in terms of surface potential ψ_0) for the interaction between similarly charged (ionized) surfaces in aqueous solutions of monovalent electrolyte. It can also be expressed in terms of the surface charge density σ by applying the Grahame equation (18.12) (after [18.62] with permission)

in the following sections) often appear at very small separations, so that the full force law between two surfaces or colloidal particles in solution can be more complex than might be expected from the DLVO theory.

There are situations when the double-layer interaction can be attractive at short range even between surfaces of similar charge, especially in systems with charge regulation due to dissociation of chargeable groups on the surfaces [18.100]; ion condensa-

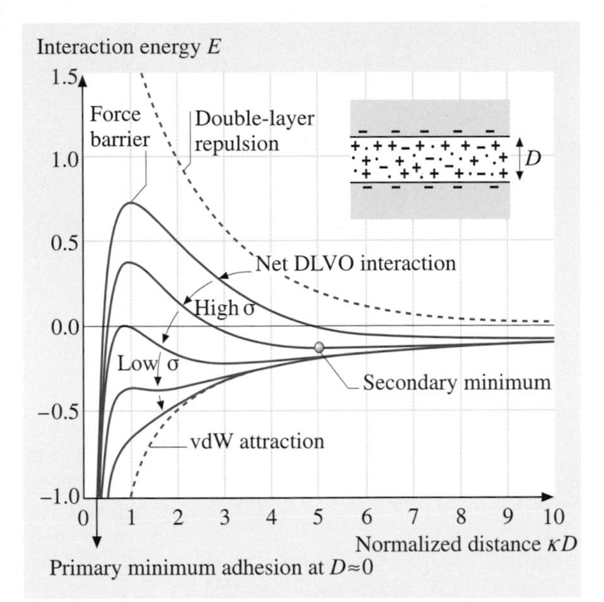

Fig. 18.6 Schematic plots of the DLVO interaction potential energy E between two flat, charged surfaces (or, according to the Derjaguin approximation, (18.3), the force F between two curved surfaces) as a function of the surface separation normalized by the Debye length, κ^{-1}. The van der Waals attraction (inverse power-law dependence on D) together with the repulsive electrostatic "double-layer" force (roughly exponential) at different surface charge σ (or potential, see (18.12)) determine the net interaction potential in aqueous electrolyte solution (after [18.62] with permission)

tion [18.101], which may lower the effective surface charge density in systems containing di-and trivalent counterions; or ion correlation, which is an additional van der Waals-like attraction due to mobile and therefore highly polarizable counterions located at the surface [18.102]. The ion correlation (or charge fluctuation) force becomes significant at separations below 4 nm and increases with the surface charge density σ and the valency z of the counterions. Computer simulations have shown that at high charge density and monovalent counterions, the ion correlation force can reduce the effective double-layer repulsion by 10–15 %. With divalent counterions, the ion correlation force was found to exceed the double-layer repulsion and the total force then became attractive at a separation below 2 nm even in dilute electrolyte solution [18.103]. Experimentally, such short-range attractive forces have been found in charged bilayer systems [18.104, 105].

18.4.3 Solvation and Structural Forces

When a liquid is confined within a restricted space, for example, a very thin film between two surfaces, it ceases to behave as a structureless continuum. At small surface separations (below about 10 molecular diameters), the van der Waals force between two surfaces or even two solute molecules in a liquid (solvent) is no longer a smoothly varying attraction. Instead, there arises an additional "solvation" force that generally oscillates between attraction and repulsion with distance, with a periodicity equal to some mean dimension σ of the liquid molecules [18.106]. Figure 18.7a shows the force law between two smooth mica surfaces across the hydrocarbon liquid tetradecane, whose inert, chain-like molecules have a width of $\sigma \approx 0.4$ nm.

The short-range oscillatory force law is related to the "density distribution function" and "potential of mean force" characteristic of intermolecular interactions in liquids. These forces arise from the confining effects two surfaces have on liquid molecules, forcing them to order into quasi-discrete layers. Such layers are energetically or entropically favored and correspond to the minima in the free energy, whereas fractional layers are disfavored (energy maxima). This effect is quite general and arises in all simple liquids when they are confined between two smooth, rigid surfaces, both flat and curved. Oscillatory forces do not require any attractive liquid–liquid or liquid–wall interaction, only two hard walls confining molecules whose shape is not too irregular and that are free to exchange with molecules in a bulk liquid reservoir. In the absence of any attractive pressure between the molecules, the bulk liquid density could be maintained by an external hydrostatic pressure – in real liquids attractive van der Waals forces play the role of such an external pressure.

Oscillatory forces are now well understood theoretically, at least for simple liquids, and a number of theoretical studies and computer simulations of various confined liquids (including water) that interact via some form of Lennard-Jones potential have invariably led to an oscillatory solvation force at surface separations below a few molecular diameters [18.107–114]. In a first approximation, the oscillatory force law may be described by an exponentially decaying cosine function of the form

$$E \approx E_0 \cos(2\pi D/\sigma) \mathrm{e}^{-D/\sigma} , \quad (18.16)$$

where both theory and experiments show that the oscillatory period and the characteristic decay length of the envelope are close to σ.

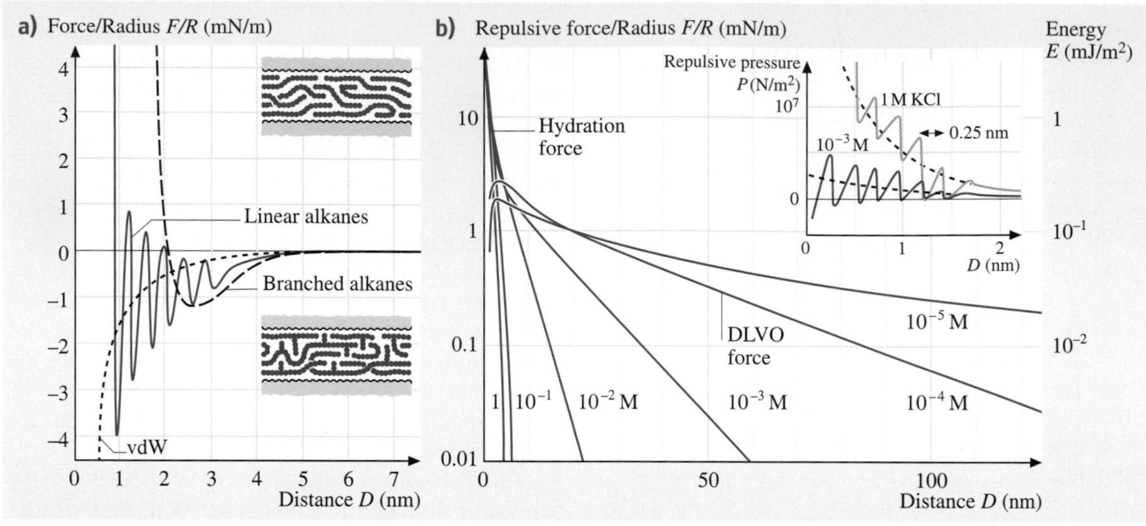

Fig. 18.7 **(a)** *Solid curve:* Forces measured between two mica surfaces across saturated linear chain alkanes such as n-tetradecane and n-hexadecane [18.115, 116]. The 0.4 nm periodicity of the oscillations indicates that the molecules are preferentially oriented parallel to the surfaces, as shown schematically in the *upper insert*. The theoretical continuum van der Waals attraction is shown as a *dotted curve*. *Dashed curve:* Smooth, non-oscillatory force law exhibited by irregularly shaped alkanes (such as 2-methyloctadecane) that cannot order into well-defined layers (*lower insert*) [18.116, 117]. Similar non-oscillatory forces are also observed between "rough" surfaces, even when these interact across a saturated linear chain liquid. This is because the irregularly shaped surfaces (rather than the liquid) now prevent the liquid molecules from ordering in the gap. **(b)** Forces measured between charged mica surfaces in KCl solutions of varying concentrations [18.20]. In dilute solutions (10^{-5} and 10^{-4} M), the measured forces are excellently described by the DLVO theory, based on exact solutions to the nonlinear Poisson–Boltzmann equation for the electrostatic forces and the Lifshitz theory for the van der Waals forces (using a Hamaker constant of $A_H = 2.2 \times 10^{-20}$ J). At higher concentrations, as more hydrated K^+ cations adsorb onto the negatively charged surfaces, an additional hydration force appears superimposed on the DLVO interaction at distances below 3–4 nm. This force has both an oscillatory and a monotonic component. *Insert:* Short-range hydration forces between mica surfaces shown as pressure versus distance. The lower and upper curves show surfaces 40% and 95% saturated with K^+ ions. At larger separations, the forces are in good agreement with the DLVO theory. (After [18.3]. Copyright 1991, with permission from Elsevier Science)

Once the solvation zones of the two surfaces overlap, the mean liquid density in the gap is no longer the same as in the bulk liquid. Since the van der Waals interaction depends on the optical properties of the liquid, which in turn depends on the density, the van der Waals and the oscillatory solvation forces are not strictly additive. It is more correct to think of the solvation force as *the* van der Waals force at small separations with the molecular properties and density variations of the medium taken into account. It is also important to appreciate that solvation forces do not arise simply because liquid molecules tend to structure into semi-ordered layers at surfaces. They arise because of the disruption or *change* of this ordering during the approach of a second surface. The two effects are related; the greater the tendency toward structuring at an isolated surface the greater the solvation force between two such surfaces, but there is a real distinction between the two phenomena that should be borne in mind.

Oscillatory forces lead to different adhesion values depending on the energy minimum from which two surfaces are being separated. For an interaction energy described by (18.16), "quantized" adhesion energies will be E_0 at $D=0$ (primary minimum), E_0/e at $D=\sigma$, E_0/e^2 at $D=2\sigma$, etc. E_0 can be thought of as a depletion force (see Sect. 18.4.5) that is approximately given by the osmotic limit $E_0 \approx -k_B T/\sigma^2$, which can exceed the contribution to the adhesion energy in contact from the van der Waals forces (at $D_0 \approx 0.15$–0.20 nm, as discussed in Sect. 18.3.1, keeping in mind that the

Lifshitz theory fails to describe the force law at *intermediate* distances). Such multivalued adhesion forces have been observed in a number of systems, including the interactions of fibers.

Measurements of oscillatory forces between different surfaces across both aqueous and nonaqueous liquids have revealed their richness of properties [18.118–121], for example, their great sensitivity to the shape and rigidity of the solvent molecules, to the presence of other components, and to the structure of the confining surfaces (see Sects. 18.5.3 and 18.9). In particular, the oscillations can be smeared out if the molecules are irregularly shaped (e.g., branched) and therefore unable to pack into ordered layers, or when the interacting surfaces are rough or fluid-like (see Sect. 18.4.6).

It is easy to understand how oscillatory forces arise between two flat, plane parallel surfaces. Between two curved surfaces, e.g., two spheres, one might imagine the molecular ordering and oscillatory forces to be smeared out in the same way that they are smeared out between two randomly rough surfaces (see Sect. 18.5.3); however, this is not the case. Ordering can occur so long as the curvature or roughness is itself regular or uniform, i. e., not random. This is due to the Derjaguin approximation (18.3). If the energy between two flat surfaces is given by a decaying oscillatory function (for example, a cosine function as in (18.16)), then the force (and energy) between two curved surfaces will also be an oscillatory function of distance with some phase shift. Likewise, two surfaces with regularly curved regions will also retain their oscillatory force profile, albeit modified, as long as the corrugations are truly regular, i. e., periodic. On the other hand, surface roughness, even on the nanometer scale, can smear out oscillations if the roughness is random and the confined molecules are smaller than the size of the surface asperities [18.122, 123]. If an organic liquid contains small amounts of water, the expected oscillatory force can be replaced by a strongly attractive capillary force (see Sect. 18.5.1).

18.4.4 Hydration and Hydrophobic Forces

The forces occurring in water and electrolyte solutions are more complex than those occurring in nonpolar liquids. According to continuum theories, the attractive van der Waals force is always expected to win over the repulsive electrostatic "double-layer" force at small surface separations (Fig. 18.6). However, certain surfaces (usually oxide or hydroxide surfaces such as clays or silica) swell spontaneously or repel each other in aqueous solution, even at high salt concentrations. Yet in all these systems one would expect the surfaces or particles to remain in strong adhesive contact or coagulate in a primary minimum if the only forces operating were DLVO forces.

There are many other aqueous systems in which the DLVO theory fails and where there is an additional short-range force that is not oscillatory but monotonic. Between hydrophilic surfaces this force is exponentially repulsive and is commonly referred to as the *hydration*, or *structural*, force. The origin and nature of this force has long been controversial, especially in the colloidal and biological literature. Repulsive hydration forces are believed to arise from strongly hydrogen-bonding surface groups, such as hydrated ions or hydroxyl ($-OH$) groups, which modify the hydrogen-bonding network of liquid water adjacent to them. Because this network is quite extensive in range [18.124], the resulting interaction force is also of relatively long range.

Repulsive hydration forces were first extensively studied between clay surfaces [18.125]. More recently, they have been measured in detail between mica and silica surfaces [18.20–22], where they have been found to decay exponentially with decay lengths of about 1 nm. Their effective range is about 3 to 5 nm, which is about twice the range of the oscillatory solvation force in water. Empirically, the hydration repulsion between two hydrophilic surfaces appears to follow the simple equation

$$E = E_0 \mathrm{e}^{-D/\lambda_0} , \quad (18.17)$$

where $\lambda_0 \approx 0.6{-}1.1$ nm for 1:1 electrolytes and $E_0 = 3{-}30\,\mathrm{mJ\,m^{-2}}$ depending on the hydration (hydrophilicity) of the surfaces, higher E_0 values generally being associated with lower λ_0 values.

The interactions between molecularly smooth mica surfaces in dilute electrolyte solutions obey the DLVO theory (Fig. 18.7b). However, at higher salt concentrations, specific to each electrolyte, hydrated cations bind to the negatively charged surfaces and give rise to a repulsive hydration force [18.20, 21]. This is believed to be due to the energy needed to dehydrate the bound cations, which presumably retain some of their water of hydration on binding. This conclusion was arrived at after noting that the strength and range of the hydration forces increase with the known hydration numbers of the electrolyte cations in the order: $Mg^{2+} > Ca^{2+} > Li^+ \sim Na^+ > K^+ > Cs^+$. Similar trends are observed with other negatively charged colloidal surfaces.

While the hydration force between two mica surfaces is overall repulsive below a distance of 4 nm, it is not always monotonic below about 1.5 nm but exhibits oscillations of mean periodicity of 0.25 ± 0.03 nm, roughly equal to the diameter of the water molecule. This is shown in the insert in Fig. 18.7b, where we may note that the first three minima at $D = 0$, 0.28, and 0.56 nm occur at negative energies, a result that rationalizes observations on certain colloidal systems. For example, clay platelets such as montmorillonite often repel each other increasingly strongly as they come closer together, but they are also known to stack into stable aggregates with water interlayers of typical thickness 0.25 and 0.55 nm between them [18.126, 127], suggestive of a turnabout in the force law from a monotonic repulsion to discretized attraction. In chemistry we would refer to such structures as stable hydrates of fixed stoichiometry, whereas in physics we may think of them as experiencing an oscillatory force.

Both surface force and clay swelling experiments have shown that hydration forces can be modified or "regulated" by exchanging ions of different hydration on surfaces, an effect that has important practical applications in controlling the stability of colloidal dispersions. It has long been known that colloidal particles can be precipitated (coagulated or flocculated) by increasing the electrolyte concentration, an effect that was traditionally attributed to the reduced screening of the electrostatic double-layer repulsion between the particles due to the reduced Debye length. However, there are many examples where colloids are stabilized at high salt concentrations, not at low concentrations. This effect is now recognized as being due to the increased hydration repulsion experienced by certain surfaces when they bind highly hydrated ions at higher salt concentrations. Hydration regulation of adhesion and interparticle forces is an important practical method for controlling various processes such as clay swelling [18.126, 127], ceramic processing and rheology [18.128, 129], material fracture [18.128], and colloidal particle and bubble coalescence [18.130].

Water appears to be unique in having a solvation (hydration) force that exhibits both a monotonic and an oscillatory component. Between hydrophilic surfaces the monotonic component is repulsive (Fig. 18.7b), but between hydrophobic surfaces it is attractive and the final adhesion is much greater than expected from the Lifshitz theory.

A hydrophobic surface is one that is inert to water in the sense that it cannot bind to water molecules via ionic or hydrogen bonds. Hydrocarbons and fluorocarbons are hydrophobic, as is air, and the strongly attractive hydrophobic force has many important manifestations and consequences such as the low solubility or miscibility of water and oil molecules, micellization, protein folding, strong adhesion and rapid coagulation of hydrophobic surfaces, non-wetting of water on hydrophobic surfaces, and hydrophobic particle attachment to rising air bubbles (the basic principle of froth flotation).

In recent years, there has been a steady accumulation of experimental data on the force laws between various hydrophobic surfaces in aqueous solution [18.131–143]. These studies have found that the hydrophobic force law between two macroscopic surfaces is of surprisingly long range, decaying exponentially with a characteristic decay length of 1 to 2 nm in the separation range of 0 to 10 nm, and then more gradually further out. The hydrophobic force can be far stronger than the van der Waals attraction, especially between hydrocarbon surfaces in water, for which the Hamaker constant is quite small. The magnitude of the hydrophobic attraction has been found to decrease with the decreasing hydrophobicity (increasing hydrophilicity) of lecithin lipid bilayer surfaces [18.31] and silanated surfaces [18.139], whereas examples of the opposite trend have been shown for some Langmuir–Blodgett-deposited monolayers [18.144].

For two surfaces in water the purely hydrophobic interaction energy (ignoring DLVO and oscillatory forces) in the range of 0 to 10 nm is given by

$$E = -2\gamma \, \mathrm{e}^{-D/\lambda_0} \,, \tag{18.18}$$

where typically $\lambda_0 = 1{-}2$ nm, and $\gamma = 10{-}50\,\mathrm{mJ\,m^{-2}}$. The higher value corresponds to the interfacial energy of a pure hydrocarbon–water interface.

At a separation below 10 nm, the hydrophobic force appears to be insensitive or only weakly sensitive to changes in the type and concentration of electrolyte ions in the solution. The absence of a "screening" effect by ions attests to the non-electrostatic origin of this interaction. In contrast, some experiments have shown that at separations greater than 10 nm, the attraction does depend on the intervening electrolyte, and that in dilute solutions, or solutions containing divalent ions, it can continue to exceed the van der Waals attraction out to separations of 80 nm [18.136, 145].

The long-range nature of the hydrophobic interaction has a number of important consequences. It accounts for the rapid coagulation of hydrophobic particles in water and may also account for the rapid folding of proteins. It also explains the ease with which water films rupture on hydrophobic surfaces. In this case, the van der Waals force across the water film is repulsive and therefore favors wetting, but this is more than offset by the attractive hydrophobic interaction acting between the two hydrophobic phases across water. Hydrophobic forces are increasingly being implicated in the adhesion and fusion of biological membranes and cells. It is known that both osmotic and electric field stresses enhance membrane fusion, an effect that may be due to the concomitant increase in the hydrophobic area exposed between two adjacent surfaces.

From the previous discussion we can infer that hydration and hydrophobic forces are not of a simple nature. These interactions are probably the most important, yet the least understood of all the forces in aqueous solutions. The unusual properties of water and the nature of the surfaces (including their homogeneity and stability) appear to be equally important. Some particle surfaces can have their hydration forces regulated, for example, by ion exchange. Others appear to be intrinsically hydrophilic (e.g., silica) and cannot be coagulated by changing the ionic condition, but can be rendered hydrophobic by chemically modifying their surface groups. For example, on heating silica to above 600 °C, two adjacent surface silanol (−OH) groups release a water molecule and form a hydrophobic siloxane (−O−) group, whence the repulsive hydration force changes into an attractive hydrophobic force.

How do these exponentially decaying repulsive or attractive forces arise? Theoretical work and computer simulations [18.109, 111, 146, 147] suggest that the solvation forces in water should be purely oscillatory, whereas other theoretical studies [18.148–153] suggest a monotonically exponential repulsion or attraction, possibly superimposed on an oscillatory force. The latter is consistent with experimental findings, as shown in the inset to Fig. 18.7b, where it appears that the oscillatory force is simply additive with the monotonic hydration and DLVO forces, suggesting that these arise from essentially different mechanisms.

It is probable that the short-range hydration force between all smooth, rigid, or crystalline surfaces (e.g., mineral surfaces such as mica) has an oscillatory component. This may or may not be superimposed on a monotonic force due to image interactions [18.150] and/or structural or hydrogen bonding interactions [18.148, 149].

Like the repulsive hydration force, the origin of the hydrophobic force is still unknown. *Luzar* et al. [18.152] carried out a Monte Carlo simulation of the interaction between two hydrophobic surfaces across water at separations below 1.5 nm. They obtained a decaying oscillatory force superimposed on a monotonically attractive curve. In more recent work [18.154, 155], it has been suggested that hydrophobic surfaces generate a depleted region of water around them, and that a long-range attractive force due to depletion arises between two such surfaces.

It is questionable whether the hydration or hydrophobic force should be viewed as an ordinary type of solvation or structural force that is reflecting the packing of water molecules. The energy (or entropy) associated with the hydrogen bonding network, which extends over a much larger region of space than the molecular correlations, is probably at the root of the long-range interactions of water. The situation in water appears to be governed by much more than the molecular packing effects that dominate the interactions in simpler liquids.

18.4.5 Polymer-Mediated Forces

Polymers or macromolecules are chain-like molecules consisting of many identical segments (monomers or repeating units) held together by covalent bonds. The size of a polymer coil in solution or in the melt is determined by a balance between van der Waals attraction (and hydrogen bonding, if present) between polymer segments, and the entropy of mixing, which causes the polymer coil to expand. In polymer melts above the glass transition temperature, and at certain conditions in solution, the attraction between polymer segments is exactly balanced by the entropy effect. The polymer solution will then behave virtually ideally, and the density distribution of segments in the coil is Gaussian. This is called the theta (θ) condition, and it occurs at the theta or Flory temperature for a particular combination of polymer and solvent or solvent mixture. At lower temperatures (in a poor or bad solvent), the polymer–polymer interactions dominate over the entropic, and the coil will shrink or precipitate. At higher temperatures (good solvent conditions), the polymer coil will be expanded.

High molecular weight polymers form large coils, which significantly affect the properties of a solution even when the total mass of polymer is very low. The radius of the polymer coil is proportional to the segment length, a, and the number of segments, n. At theta conditions, the hydrodynamic radius of the polymer coil (the root-mean-square separation of the ends of one polymer chain) is theoretically given by $R_\mathrm{h} = a n^{1/2}$, and the unperturbed radius of gyration (the average root-mean-square distance of a segment from the center of mass of the molecule) is $R_\mathrm{g} = a\,(n/6)^{1/2}$. In a good solvent the perturbed size of the polymer coil, the Flory radius R_F, is proportional to $n^{3/5}$.

Polymers interact with surfaces mainly through van der Waals and electrostatic interactions. The physisorption of polymers containing only one type of segment is reversible and highly dynamic, but the rate of exchange of adsorbed chains with free chains in the solution is low, since the polymer remains bound to the surface as long as one segment along the chain is adsorbed. The adsorption energy per segment is on the order of $k_B T$. In a good solvent, the conformation of a polymer on a surface is very different from the coil conformation in bulk solution. Polymers adsorb in "trains", separated by "loops" extending into solution and dangling "tails" (the ends of the chain). Compared to adsorption at lower temperatures, good solvent conditions favor more of the polymer chain being in the solvent, where it can attain its optimum conformation. As a result, the extension of the polymer is longer, even though the total amount of adsorbed polymer is lower. In a good solvent, the polymer chains can also be effectively repelled from a surface, if the loss in conformational entropy close to the surface is not compensated for by a gain in enthalpy from adsorption of segments. In this case, there will be a layer of solution (thickness $\approx R_g$) close to the surfaces that is depleted of polymer.

The interaction forces between two surfaces across a polymer solution will depend on whether the polymer is adsorbing onto the surfaces or is repelled from them, and also on whether the interaction occurs at "true" or "restricted" thermodynamic equilibrium. At true or full equilibrium, the polymer between the surfaces can equilibrate (exchange) with polymer in the bulk solution at all surface separations. Some theories [18.156, 157] predict that at full equilibrium, the polymer chains would move from the confined gap into the bulk solution where they could attain entropically more favorable conformations, and that a monotonic attraction at all distances would result from bridging and depletion interactions (which will be discussed below). Other theories suggest that the interaction at small separations would be ultimately repulsive, since some polymer chains would remain in the gap due to their attractive interactions with many sites on the surface (enthalpic) – more sites would be available to the remaining polymer chains if some others desorbed and diffused out from the gap [18.64, 158, 159].

At restricted equilibrium, the polymer is kinetically trapped, and the adsorbed amount is thus constant as the surfaces are brought toward each other, but the chains can still rearrange on the surfaces and in the gap. Experimentally, the true equilibrium situation is very difficult to attain, and most experiments are done at restricted equilibrium conditions. Even the equilibration of conformations assumed in theoretical models for restricted equilibrium conditions can be so slow that this condition is difficult to reach.

In systems of adsorbing polymer, bridging of chains from one surface to the other can give rise to a long-range attraction, since the bridging chains would gain conformational entropy if the surfaces were closer together. In poor solvents, both bridging and intersegment interactions contribute to an attraction [18.26]. However, regardless of solvent and equilibrium conditions, a strong repulsion due to the osmotic interactions is seen as small surface separations in systems of adsorbing polymers at restricted equilibrium.

In systems containing high concentrations of non-adsorbing polymer, the difference in solute concentration in the bulk and between the surfaces at separations smaller than the approximate polymer coil diameter ($2R_g$, i. e., when the polymer has been squeezed out from the gap between the surfaces) may give rise to an attractive osmotic force ("depletion attraction") [18.160–164]. In addition, if the polymer coils become initially compressed as the surfaces approach each other, this can give rise to a repulsion ("depletion stabilization") at large separations [18.163]. For a system of two cylindrical surfaces or radius R, the maximum depletion force, F_{dep}, is expected to occur when the surfaces are in contact and is given by multiplying the depletion (osmotic) pressure, $P_{dep} = \rho k_B T$, by the contact area πr^2, where r is given by the chord theorem: $r^2 = (2R - R_g)R_g \approx 2RR_g$ [18.3]:

$$F_{dep}/R = -2\pi R_g \rho k_B T \, , \tag{18.19}$$

where ρ is the number density of the polymer in the bulk solution.

If a part of the polymer (typically an end group) is different from the rest of the chain, this part may preferentially adsorb to the surface. End-adsorbed polymer is attached to the surface at only one point, and the extension of the chain is dependent on the grafting density, i. e., the average distance, s, between adsorbed end groups on the surface (Fig. 18.8). One distinguishes between different regions of increasing overlap of the chains (stretching) called pancake, mushroom, and brush regimes [18.165]. In the mushroom regime, where the coverage is sufficiently low so that there is no overlap between neighboring chains, the thickness of the adsorbed layer is proportional to $n^{1/2}$ (i. e., to R_g) at theta conditions and to $n^{3/5}$ in a good solvent.

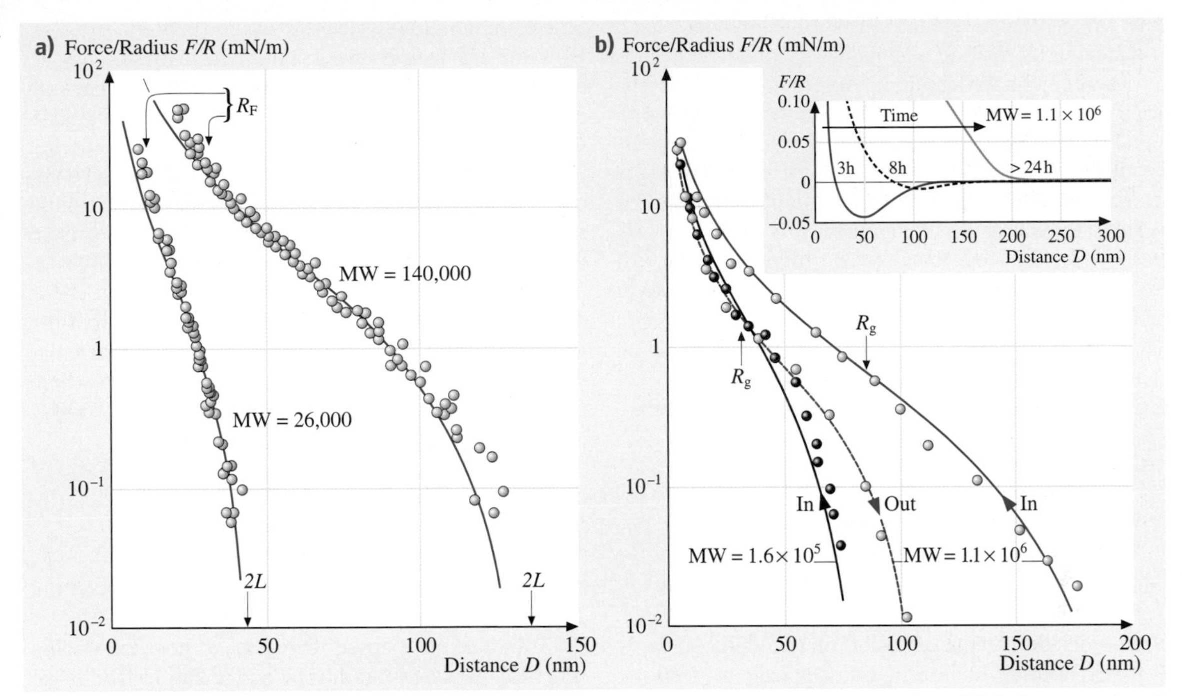

Fig. 18.8a,b Experimentally determined forces in systems of two interacting polymer brushes: (**a**) Polystyrene brush layers grafted via an adsorbing chain-end group onto mica surfaces in toluene (a good solvent for polystyrene). *Left curve:* MW = 26,000 g/mol, $R_F = 12$ nm. *Right curve*: MW = 140,000 g/mol, $R_F = 32$ nm. Both force curves were reversible on approach and separation. The *solid curves* are theoretical fits using the Alexander–de Gennes theory with the following measured parameters: spacing between attachments sites: $s = 8.5$ nm, brush thickness: $L = 22.5$ nm and 65 nm, respectively. (Adapted from [18.171]). (**b**) Polyethylene oxide layers physisorbed onto mica from 150 μg/ml solution in aqueous 0.1 M KNO_3 (a good solvent for polyethylene oxide). *Main figure:* Equilibrium forces at full coverage after ∼ 16 h adsorption time. *Left curve:* MW = 160,000 g/mol, $R_g = 32$ nm. *Right curve:* MW = 1,100,000 g/mol, $R_g = 86$ nm. Note the hysteresis (irreversibility) on approach and separation for this *physisorbed* layer, in contrast to the absence of hysteresis with *grafted* chains in case (**a**). The *solid curves* are based on a modified form of the Alexander–de Gennes theory. *Inset* in (**b**): Evolution of the forces with the time allowed for the higher MW polymer to adsorb from solution. Note the gradual reduction in the attractive bridging component. (Adapted from [18.172–174]. After [18.3]. Copyright 1991, with permission from Elsevier Science)

Several models [18.165–170] have been developed for the extension and interactions between two brushes (strongly stretched grafted chains). They are based on a balance between osmotic pressure within the brush layers (uncompressed and compressed) and elastic energy of the chains and differ mainly in the assumptions of the segment density profile, which can be a step function or parabolic. At high coverage (in the brush regime),where the chains will avoid overlapping each other, the thickness of the layer is proportional to n.

Experimental work on both monodisperse [18.27,28, 171,175] and polydisperse [18.30] systems at different solvent conditions has confirmed the expected range and magnitude of the repulsive interactions resulting from compression of densely packed grafted layers.

18.4.6 Thermal Fluctuation Forces

If a surface is not rigid but very soft or even fluid-like, this can act to smear out any oscillatory solvation force. This is because the thermal fluctuations of such interfaces make them dynamically "rough" at any instant, even though they may be perfectly smooth on a time average. The types of surfaces that fall into this category are fluid-like amphiphilic surfaces of micelles, bilayers, emulsions, soap films, etc., but also solid colloidal particle surfaces that are coated with surfactant monolayers, as occurs in lubricating oils, paints, toners, etc.

These thermal fluctuation forces (also called entropic or steric forces) are usually short-range and repulsive and are very effective at stabilizing the attractive van der Waals forces at some small but finite separation. This can reduce the adhesion energy or force by up to three orders of magnitude. It is mainly for this reason that fluid-like micelles and bilayers, biological membranes, emulsion droplets, or gas bubbles adhere to each other only very weakly.

Because of their short range it was, and still is, commonly believed that these forces arise from water ordering or "structuring" effects at surfaces, and that they reflect some unique or characteristic property of water. However, it is now known that these repulsive forces also exist in other liquids. Moreover, they appear to become stronger with increasing temperature, which is unlikely if the force would originate from molecular ordering effects at surfaces. Recent experiments, theory, and computer simulations [18.176–178] have shown that these repulsive forces have an entropic origin arising from the osmotic repulsion between exposed thermally mobile surface groups once these overlap in a liquid. These phenomena include undulating and peristaltic forces between membranes or bilayers, and, on the molecular scale, protrusion and head group overlap forces where the interactions also are influenced by hydration forces.

18.5 Adhesion and Capillary Forces

18.5.1 Capillary Forces

When considering the adhesion of two solid surfaces or particles in air or in a liquid, it is easy to overlook or underestimate the important role of capillary forces, i. e., forces arising from the Laplace pressure of curved menisci formed by condensation of a liquid between and around two adhering surfaces (Fig. 18.9).

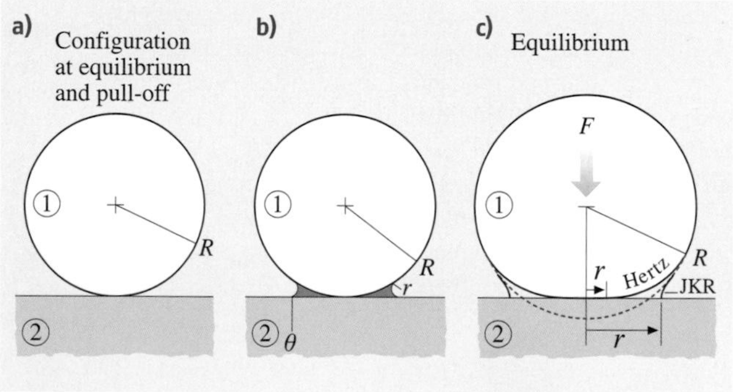

Fig. 18.9a–c Adhesion and capillary forces: (**a**) a non-deforming sphere on a rigid, flat surface in an inert atmosphere and (**b**) in a vapor that can "capillary condense" around the contact zone. At equilibrium, the concave radius, r, of the liquid meniscus is given by the Kelvin equation. For a concave meniscus to form, the contact angle θ has to be less than 90°. In the case of hydrophobic surfaces surrounded by water, a vapor cavity can form between the surfaces. As long as the surfaces are perfectly smooth, the contribution of the meniscus to the adhesion force is independent of r. (After [18.1] with permission.) (**c**) Elastically deformable sphere on a rigid flat surface in the absence (Hertz) and presence (JKR) of adhesion [(**a**) and (**c**) after [18.3]. Copyright 1991, with permission from Elsevier Science]

The adhesion force between a non-deformable spherical particle of radius R and a flat surface in an inert atmosphere (Fig. 18.9a) is

$$F_s = 4\pi R \gamma_{SV} . \qquad (18.20)$$

But in an atmosphere containing a condensable vapor, the expression above is replaced by

$$F_s = 4\pi R(\gamma_{LV} \cos\theta + \gamma_{SL}) , \qquad (18.21)$$

where the first term is due to the Laplace pressure of the meniscus and the second is due to the direct adhesion of the two contacting solids within the liquid. Note that the above equation does not contain the radius of curvature, r, of the liquid meniscus (see Fig. 18.9b). This is because for smaller r the Laplace pressure γ_{LV}/r increases, but the area over which it acts decreases by the same amount, so the two effects cancel out. Experiments with inert liquids, such as hydrocarbons, condensing between two mica surfaces indicate that (18.21) is valid for values of r as small as 1–2 nm, corresponding to vapor pressures as low as 40% of saturation [18.121, 179, 180]. Capillary condensation also occurs in binary liquid systems, e.g., when water dissolved in hydrocarbon liquids condense around two contacting hydrophilic surfaces or when a vapor cavity forms in water around two hydrophobic surfaces. In the case of water condensing from vapor or from oil, it also appears that the bulk value of γ_{LV} is applicable for meniscus radii as small as 2 nm.

The capillary condensation of liquids, especially water, from vapor can have additional effects on the

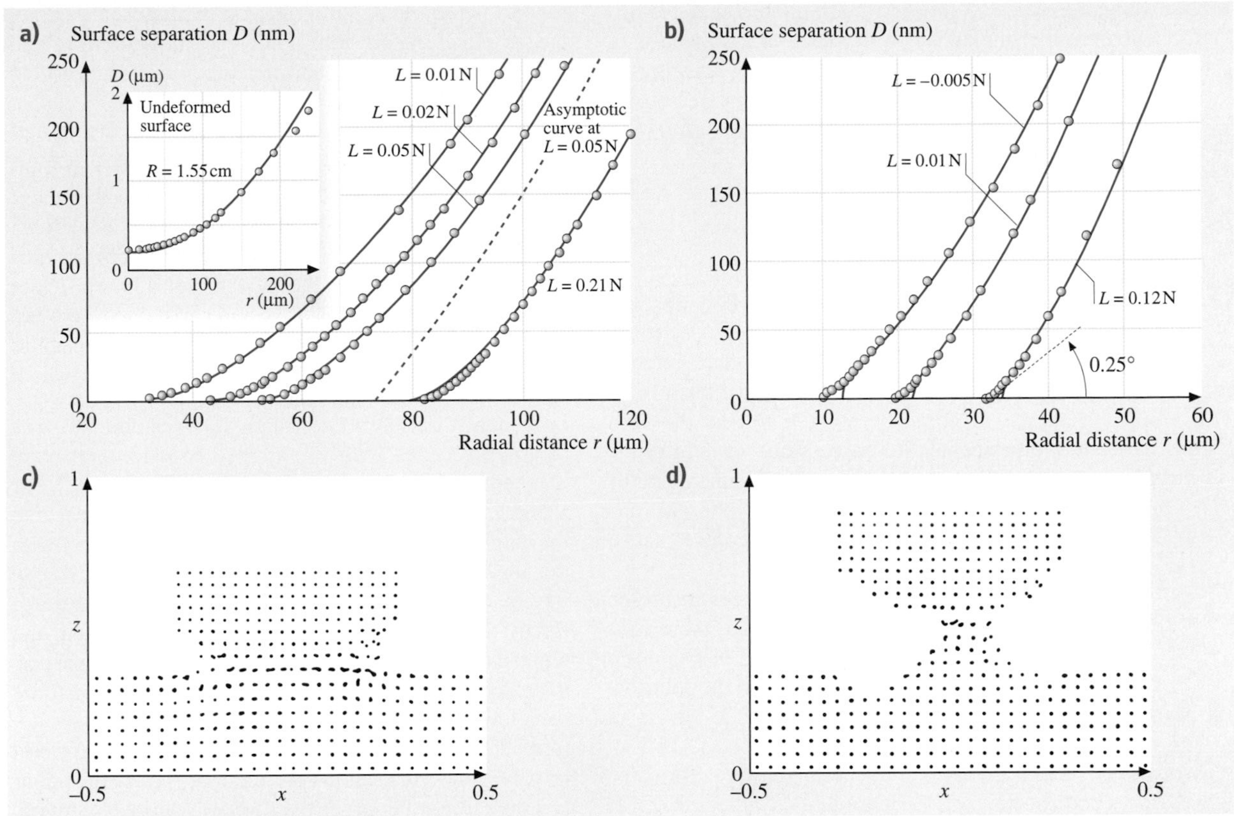

Fig. 18.10a–d Experimental and computer simulation data on contact mechanics for ideal Hertz and JKR contacts. **(a)** Measured profiles of surfaces in nonadhesive contact (*circles*) compared with Hertz profiles (*continuous curves*). The system was mica surfaces in a concentrated KCl solution in which they do not adhere. When not in contact, the surface shape is accurately described by a sphere of radius $R = 1.55$ cm (*inset*). The applied loads were 0.01, 0.02, 0.05, and 0.21 N. The last profile was measured in a different region of the surfaces where the local radius of curvature was 1.45 cm. The Hertz profiles correspond to central displacements of $\delta = 66.5$, 124, 173, and 441 nm. The *dashed line* shows the shape of the undeformed sphere corresponding to the curve at a load of 0.05 N; it fits the experimental points at larger distances (not shown). **(b)** Surface profiles measured with adhesive contact (mica surfaces adhering in dry nitrogen gas) at applied loads of -0.005, 0.01, and 0.12 N. The continuous lines are JKR profiles obtained by adjusting the central displacement in each case to get the best fit to points at larger distances. The values are $\delta = -4.2$, 75.6, and 256 nm. Note that the scales of this figure exaggerate the apparent angle at the junction of the surfaces. This angle, which is insensitive to load, is only about 0.25°. **(c)** and **(d)** Molecular dynamics simulation illustrating the formation of a connective neck between an Ni tip (topmost eight layers) and an Au substrate. The figures show the atomic configuration in a slice through the system at indentation **(c)** and during separation **(d)**. Note the crystalline structure of the neck. Distances are given in units of x and z, where $x = 1$ and $z = 1$ correspond to 6.12 nm. [**(a)** and **(b)** after [18.181]. Copyright 1987, with permission from Elsevier Science. **(c)** and **(d)** after [18.91], with kind permission from Kluwer Academic Publishers]

physical state of the contact zone. For example, if the surfaces contain ions, these will diffuse and build up within the liquid bridge, thereby changing the chemical composition of the contact zone, as well as influencing the adhesion. In the case of surfaces covered with surfactant or polymer molecules (amphiphilic surfaces), the molecules can overturn on exposure to humid air, so that the surface nonpolar groups become replaced by polar groups, which renders the surfaces hydrophilic. When two such surfaces come into contact, water will condense around the contact zone and the adhesion force will also be affected – generally increasing well above the value expected for inert hydrophobic surfaces. It is apparent that the adhesion in vapor or a solvent is often largely determined by capillary forces arising from the condensation of liquid that may be present only in very small quantities, e.g., 10–20 % of saturation in the vapor, or 20 ppm in the solvent.

18.5.2 Adhesion Mechanics

Two bodies in contact deform as a result of surface forces and/or applied normal forces. For the simplest case of two interacting elastic spheres (a model that is easily extended to an elastic sphere interacting with an undeformable surface (or vice versa)) and in the absence of attractive surface forces, the vertical central displacement (compression) was derived by *Hertz* [18.182]. In this model, the displacement and the contact area are equal to zero when no external force (load) is applied, i. e., at the points of contact and of separation.

In systems where attractive surface forces are present between the surfaces, the deformations are more complicated. Modern theories of the adhesion mechanics of two contacting solid surfaces are based on the Johnson–Kendall–Roberts (JKR) theory [18.15, 183], or on the Derjaguin–Muller–Toporov (DMT) theory [18.184–186]. The JKR theory is applicable to easily deformable, large bodies with high surface energy, whereas the DMT theory better describes very small and hard bodies with low surface energy [18.187].

In the JKR theory, two spheres of radii R_1 and R_2, bulk elastic modulus K, and surface energy γ will flatten due to attractive surface forces when in contact at no external load. The contact area will increase under an external load or normal force F, such that at mechanical equilibrium the radius of the contact area, r, is given by

$$r^3 = \frac{R}{K}\left[F + 6\pi R\gamma + \sqrt{12\pi R\gamma F + (6\pi R\gamma)^2}\right] , \tag{18.22}$$

where $R = R_1R_2/(R_1+R_2)$. In the absence of surface energy, γ, equation (18.22) is reduced to the expression for the radius of the contact area in the Hertz model. Another important result of the JKR theory gives the adhesion force or "pull-off" force:

$$F_S = -3\pi R\gamma_S , \tag{18.23}$$

where the surface energy, γ_S, is defined as $W = 2\gamma_S$, where W is the reversible work of adhesion. Note that according to the JKR theory a finite elastic modulus, K, while having an effect on the load–area curve, has no effect on the adhesion force, an interesting and unexpected result that has nevertheless been verified experimentally [18.15, 181, 188, 189].

Equation (18.22) and (18.23) provide the framework for analyzing results of adhesion measurements of contacting solids, known as contact mechanics [18.183, 190], and for studying the effects of surface conditions and time on adhesion energy hysteresis (see Sect. 18.5.4).

18.5.3 Effects of Surface Structure, Roughness, and Lattice Mismatch

In a contact between two rough surfaces, the real area of contact varies with the applied load in a different manner than between smooth surfaces [18.191, 192]. For a Hertzian contact, it has been shown that the contact area for rough surfaces increases approximately linearly with the applied normal force (load), F, instead of as $F^{2/3}$ for smooth surfaces. In systems with attractive surface forces, there is a competition between this attraction and repulsive forces arising from compression of high asperities. As a result, the adhesion in such systems can be very low, especially if the surfaces are not easily deformed [18.193–195]. The opposite is possible for soft (viscoelastic) surfaces where the real (molecular) contact area might be larger than for two perfectly smooth surfaces.

Adhesion forces may also vary depending on the commensurability of the crystallographic lattices of the interacting surfaces. *McGuiggan* and *Israelachvili* [18.196] measured the adhesion between two mica surfaces as a function of the orientation (twist angle) of their surface lattices. The forces were measured in air, water, and an aqueous salt solution where oscillatory structural forces were present. In humid air, the adhesion was found to be relatively independent of the twist angle θ due to the adsorption of a 0.4-nm-thick amorphous layer of organics and water at the interface. In contrast, in water, sharp adhesion peaks (energy minima) occurred at $\theta = 0°, \pm 60°, \pm 120°$ and $180°$, corresponding to the "coincidence" angles of the surface lattices (Fig. 18.11). As little as $\pm 1°$ away from these peaks, the energy decreased by 50%. In aqueous KCl solution, due to potassium ion adsorption the water between the surfaces becomes ordered, resulting in an oscillatory force profile where the adhesive minima occur at discrete separations of about 0.25 nm, corresponding to integral numbers of water layers. The whole interaction potential was now found to depend on the orientation of

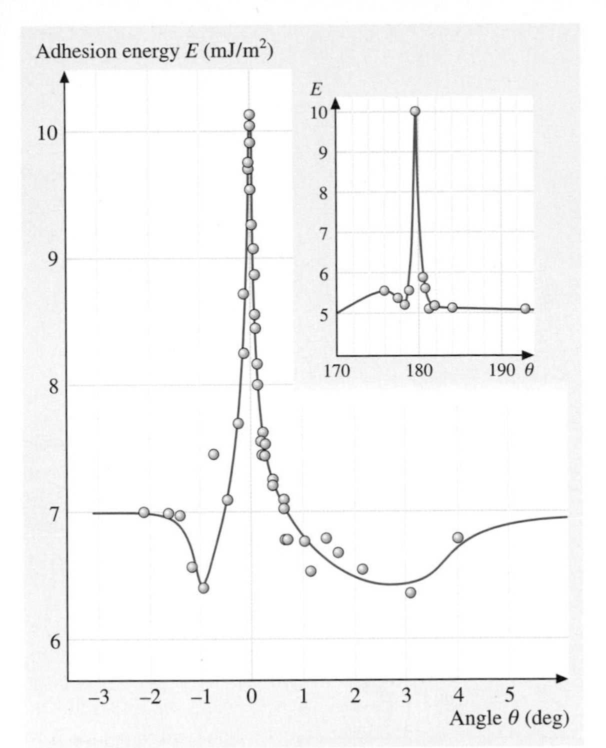

Fig. 18.11 Adhesion energy for two mica surfaces in contact in water (in the primary minimum of an oscillatory force curve) as a function of the mismatch angle θ about $\theta = 0°$ and 180° between the mica surface lattices. (After [18.197] with permission)

the surface lattices, and the effect extended at least four molecular layers.

It has also been appreciated that the structure of the confining surfaces is just as important as the nature of the liquid for determining the solvation forces [18.90, 122, 123, 198–202]. Between two surfaces that are completely flat but "unstructured", the liquid molecules will order into layers, but there will be no lateral ordering within the layers. In other words, there will be positional ordering normal but not parallel to the surfaces. If the surfaces have a crystalline (periodic) lattice, this may induce ordering parallel to the surfaces, as well, and the oscillatory force then also depends on the structure of the surface lattices. Further, if the two lattices have different dimensions ("mismatched" or "incommensurate" lattices), or if the lattices are similar but are not in register relative to each other, the oscillatory force law is further modified [18.196, 203] and the tribological properties of the film are also influenced, as discussed in Sect. 18.9 [18.203, 204].

As shown by the experiments, these effects can alter the magnitude of the adhesive minima found at a given separation within the last one or two nanometers of a thin film by a factor of two. The force barriers (maxima) may also depend on orientation. This could be even more important than the effects on the minima. A high barrier could prevent two surfaces from coming closer together into a much deeper adhesive well. Thus the maxima can effectively contribute to determining not only the final separation of two surfaces, but also their final adhesion. Such considerations should be particularly important for determining the thickness and strength of intergranular spaces in ceramics, the adhesion forces between colloidal particles in concentrated electrolyte solution, and the forces between two surfaces in a crack containing capillary condensed water.

For surfaces that are *randomly* rough, oscillatory forces become smoothed out and disappear altogether, to be replaced by a purely monotonic solvation force [18.116, 122, 123]. This occurs even if the liquid molecules themselves are perfectly capable of ordering into layers. The situation of *symmetric* liquid molecules confined between *rough* surfaces is therefore not unlike that of *asymmetric* molecules between *smooth* surfaces (see Sect. 18.4.3 and Fig. 18.7a). To summarize, for there to be an oscillatory solvation force, the liquid molecules must be able to be correlated over a reasonably long range. This requires that both the liquid molecules and the surfaces have a high degree of order or symmetry. If either is missing, so will the oscillations. A roughness of only a few tenths of a nanometer is often sufficient to eliminate any oscillatory component of the force law.

18.5.4 Nonequilibrium and Rate-Dependent Interactions: Adhesion Hysteresis

Under ideal conditions the adhesion energy is a well-defined thermodynamic quantity. It is normally denoted by E or W (the work of adhesion) or γ (the surface tension, where $W = 2\gamma$) and gives the reversible work done on bringing two surfaces together or the work needed to separate two surfaces from contact. Under ideal, equilibrium conditions these two quantities are the same, but under most realistic conditions they are not: The work needed to separate two surfaces is always greater than that originally gained by bringing them together. An understanding of the molecular mechanisms underlying this phenomenon is essential for understanding many adhesion phenomena, energy dissipation during loading–unloading cycles, contact angle hysteresis, and

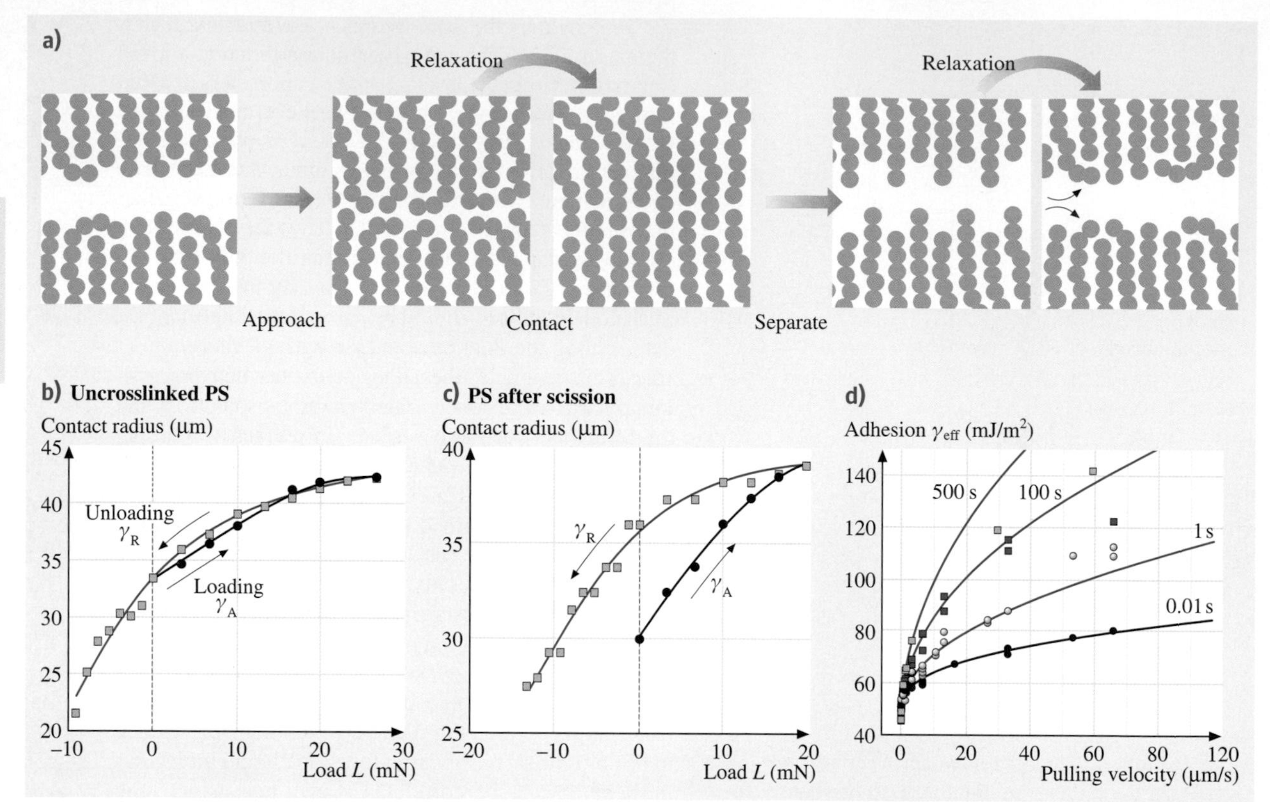

Fig. 18.12 (a) Schematic representation of interpenetrating chains. (b) and (c): JKR plots (contact radius r as a function of applied load L) showing small adhesion hysteresis for uncrosslinked polystyrene and larger adhesion hysteresis after chain scission at the surfaces after 18 h irradiation with ultraviolet light in an oxygen atmosphere. The adhesion hysteresis continues to increase with the irradiation time. (d) Rate-dependent adhesion of CTAB surfactant monolayers. The *solid curves* [18.205] are fits to experimental data on CTAB adhesion after different contact times [18.206] using an approximate analytical solution for a JKR model, including crack tip dissipation. Due to the limited range of validity of the approximation, the fits rely on the low effective adhesion energy part of the experimental data only. From the fits one can determine the thermodynamic adhesion energy, the characteristic dissipation velocity, and the intrinsic dissipation exponent of the model. [(a) after [18.207]. Copyright 1993 American Chemical Society. (b) and (c) after [18.208]. Copyright 2002 American Association for the Advancement of Science. (d) after [18.205]. Copyright 2000 American Chemical Society]

the molecular mechanisms associated with many frictional processes.

It is wrong to think that hysteresis arises because of some imperfection in the system such as rough or chemically heterogeneous surfaces, or because the supporting material is viscoelastic. Adhesion hysteresis can arise even between perfectly smooth and chemically homogenous surfaces supported by perfectly elastic materials. It can be responsible for such phenomena as rolling friction and elastoplastic adhesive contacts [18.190, 209–212] during loading–unloading and adhesion–decohesion cycles.

Adhesion hysteresis may be thought of as being due to mechanical effects such as instabilities, or chemical effects such as interdiffusion, interdigitation, molecular reorientations and exchange processes occurring at an interface after contact, as illustrated in Fig. 18.12. Such processes induce roughness and chemical heterogeneity even though initially (and after separation and re-equilibration) both surfaces are perfectly smooth and chemically homogeneous. In general, if the energy change, or work done, on separating two surfaces from adhesive contact is not fully recoverable on bringing the two surfaces back into contact again, the adhesion hysteresis may be expressed as

$$\underset{\text{Receding}}{W_R} > \underset{\text{Advancing}}{W_A}$$

or

$$\Delta W = (W_R - W_A) > 0\,, \tag{18.24}$$

where W_R and W_A are the adhesion or surface energies for receding (separating) and advancing (approaching) two solid surfaces, respectively.

Hysteresis effects are also commonly observed in wetting/dewetting phenomena [18.213]. For example, when a liquid spreads and then retracts from a surface the advancing contact angle θ_A is generally larger than the receding angle θ_R. Since the contact angle, θ, is related to the liquid–vapor surface tension, γ_L, and the solid–liquid adhesion energy, W, by the Dupré equation,

$$(1+\cos\theta)\gamma_L = W\,, \tag{18.25}$$

we see that *wetting hysteresis* or *contact angle hysteresis* ($\theta_A > \theta_R$) actually implies adhesion hysteresis, $W_R > W_A$, as given by (18.24).

Energy dissipating processes such as adhesion and contact angle hysteresis arise because of practical constraints of the *finite time* of measurements and the *finite elasticity* of materials. This prevents many loading–unloading or approach–separation cycles from being thermodynamically reversible, even though if carried out infinitely slowly they would be. By thermodynamically irreversible one simply means that one cannot go through the approach–separation cycle via a continuous series of equilibrium states, because some of these are connected via spontaneous – and therefore thermodynamically irreversible – instabilities or transitions where energy is liberated and therefore "lost" via heat or phonon release [18.214]. This is an area of much current interest and activity, especially regarding the fundamental molecular origins of adhesion and friction in polymer and surfactant systems, and the relationships between them [18.190, 205, 206, 208, 210, 215–218].

18.6 Introduction: Different Modes of Friction and the Limits of Continuum Models

Most frictional processes occur with the sliding surfaces becoming damaged in one form or another [18.209]. This may be referred to as "normal" friction. In the case of brittle materials, the damaged surfaces slide past each other while separated by relatively large, micron-sized wear particles. With more ductile surfaces, the damage remains localized to nanometer-sized, plastically deformed asperities. Some features of the friction between damaged surfaces will be described in Sect. 18.7.4.

There are also situations in which sliding can occur between two perfectly smooth, undamaged surfaces. This may be referred to as "interfacial" sliding or

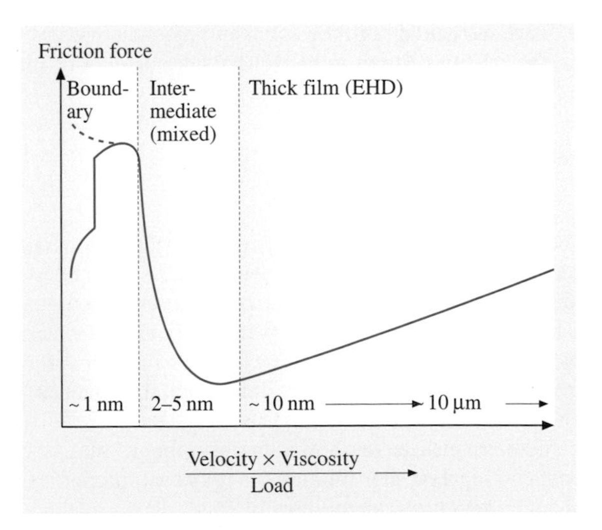

Fig. 18.13 Stribeck curve: an empirical curve giving the trend generally observed in the friction forces or friction coefficients as a function of sliding velocity, the bulk viscosity of the lubricating fluid, and the applied load. The three friction/lubrication regimes are known as the boundary lubrication regime (see Sect. 18.7), the intermediate or mixed lubrication regime (Sect. 18.8.2), and thick film or elastohydrodynamic (EHD) lubrication regime (Sect. 18.8.1). The film thicknesses believed to correspond to each of these regimes are also shown. For thick films, the "friction" force is purely viscous, e.g., Couette flow at low shear rates, but may become complicated at higher shear rates where EHD deformations of surfaces can occur during sliding. (After [18.1] with permission)

Table 18.4 The three main tribological regimes characterizing the changing properties of liquids subjected to an increasing confinement between two solid surfaces[a]

Regime	Conditions for getting into this regime	Static/equilibrium properties[b]	Dynamic properties[c]
Bulk	• Thick films (> 10 molecular diameters, $\gg R_g$ for polymers) • Low or zero loads • High shear rates	Bulk (continuum) properties: • Bulk liquid density • No long-range order	Bulk (continuum) properties: • Newtonian viscosity • Fast relaxation times • No glass temperature • No yield point • Elastohydrodynamic lubrication
Intermediate mixed	• Intermediately thick films (4–10 molecular diameters, $\sim R_g$ for polymers) • Low loads or pressure	Modified fluid properties include: • Modified positional and orientational order[a] • Medium to long-range molecular correlations • Highly entangled states	Modified rheological properties include: • Non-Newtonian flow • Glassy states • Long relaxation times • Mixed lubrication
Boundary	• Molecularly thin films (< 4 molecular diameters) • High loads or pressure • Low shear rates • Smooth surfaces or asperities	Onset of non-fluidlike properties: • Liquid-like to solid-like phase transitions • Appearance of new liquid-crystalline states • Epitaxially induced long-range ordering	Onset of tribological properties: • No flow until yield point or critical shear stress reached • Solid-like film behavior characterized by defect diffusion, dislocation motion, shear melting • Boundary lubrication

Based on work by *Granick* [18.219], *Hu* and *Granick* [18.220], and others [18.38, 207, 221] on the dynamic properties of short chain molecules such as alkanes and polymer melts confined between surfaces

[a] Confinement can lead to an increased or decreased order in a film, depending both on the surface lattice structure and the geometry of the confining cavity

[b] In each regime both the static and dynamic properties change. The static properties include the film density, the density distribution function, the potential of mean force, and various positional and orientational order parameters

[c] Dynamic properties include viscosity, viscoelastic constants, and tribological yield points such as the friction coefficient and critical shear stress

"boundary" friction and is the focus of the following sections. The term "boundary lubrication" is more commonly used to denote the friction of surfaces that contain a thin protective lubricating layer such as a surfactant monolayer, but here we shall use the term more broadly to include any molecularly thin solid, liquid, surfactant, or polymer film.

Experiments have shown that as a liquid film becomes progressively thinner, its physical properties change, at first quantitatively and then qualitatively [18.44, 47, 219, 220, 222, 223]. The quantitative changes are manifested by an increased viscosity, non-Newtonian flow behavior, and the replacement of normal melting by a glass transition, but the film remains recognizable as a liquid (Fig. 18.13). In tribology, this regime is commonly known as the "mixed lubrication" regime, where the rheological properties of a film are intermediate between the bulk and boundary properties. One may also refer to it as the "intermediate" regime (Table 18.4).

For even thinner films, the changes in behavior are more dramatic, resulting in a qualitative change in properties. Thus first-order phase transitions can now occur to solid or liquid-crystalline phases [18.46,201,207,221, 224–227], whose properties can no longer be characterized even qualitatively in terms of bulk or continuum liquid properties such as viscosity. These films now exhibit yield points (characteristic of fracture in solids) and their molecular diffusion and relaxation times can be ten orders of magnitude longer than in the bulk liquid or even in films that are just slightly thicker. The three friction regimes are summarized in Table 18.4.

18.7 Relationship Between Adhesion and Friction Between Dry (Unlubricated and Solid Boundary Lubricated) Surfaces

18.7.1 Amontons' Law and Deviations from It Due to Adhesion: The Cobblestone Model

Early theories and mechanisms for the dependence of friction on the applied normal force or load, L, were developed by *da Vinci*, *Amontons*, *Coulomb* and *Euler* [18.228]. For the macroscopic objects investigated, the friction was found to be directly proportional to the load, with no dependence on the contact area. This is described by the so-called Amontons' law:

$$F = \mu L\,, \tag{18.26}$$

where F is the shear or friction force and μ is a constant defined as the coefficient of friction. This friction law has a broad range of applicability and is still the principal means of quantitatively describing the friction between surfaces. However, particularly in the case of adhering surfaces, Amontons' law does not adequately describe the friction behavior with load, because of the finite friction force measured at zero and even negative applied loads.

When a lateral force, or shear stress, is applied to two surfaces in adhesive contact, the surfaces initially remain "pinned" to each other until some critical shear force is reached. At this point, the surfaces begin to slide past each other either smoothly or in jerks. The frictional force needed to initiate sliding from rest is known as the *static* friction force, denoted by F_s, while the force needed to maintain smooth sliding is referred to as the *kinetic* or *dynamic* friction force, denoted by F_k. In general, $F_s > F_k$. Two sliding surfaces may also move in regular jerks, known as stick–slip sliding, which is discussed in more detail in Sect. 18.8.3. Such friction forces cannot be described by models used for thick films that are viscous (see Sect. 18.8.1) and, therefore, shear as soon as the smallest shear force is applied.

Experimentally, it has been found that during both smooth and stick–slip sliding the local geometry of the contact zone remains largely unchanged from the static geometry. In an adhesive contact, the contact area as a function of load is thus generally well described by the JKR equation, (18.22). The friction force between two molecularly smooth surfaces sliding in adhesive contact is not simply proportional to the applied load, L, as might be expected from Amontons' law. There is an additional adhesion contribution that is proportional to the area of contact, A. Thus, in general, the interfacial friction force of dry, unlubricated surfaces sliding smoothly past each other in adhesive contact is given by

$$F = F_k = S_c A + \mu L\,, \tag{18.27}$$

where S_c is the "critical shear stress" (assumed to be constant), $A = \pi r^2$ is the contact area of radius r given by (18.22), and μ is the coefficient of friction. For low loads we have:

$$\begin{aligned} F &= S_c A = S_c \pi r^2 \\ &= S_c \pi \left[\frac{R}{K}\left(L + 6\pi R\gamma + \sqrt{12\pi R\gamma L + (6\pi R\gamma)^2}\right)\right]^{2/3}\,; \end{aligned} \tag{18.28}$$

whereas for high loads (or high μ), (18.27) reduces to Amontons' law: $F = \mu L$. Depending on whether the friction force in (18.27) is dominated by the first or second term, one may refer to the friction as *adhesion-controlled* or *load-controlled*, respectively.

The following friction model, first proposed by *Tabor* [18.229] and developed further by *Sutcliffe* et al. [18.230], *McClelland* [18.231], and *Homola* et al. [18.45], has been quite successful at explaining the interfacial and boundary friction of two solid crystalline surfaces sliding past each other in the absence of wear. The surfaces may be unlubricated, or they may be separated by a monolayer or more of some boundary lubricant or liquid molecules. In this model, the values of the critical shear stress S_c, and coefficient of friction μ, of (18.27) are calculated in terms of the energy needed to overcome the attractive intermolecular forces and compressive externally applied load as one surface is raised and then slid across the molecular-sized asperities of the other.

This model (variously referred to as the *interlocking asperity model*, *Coulomb friction*, or the *cobblestone model*) is similar to pushing a cart over a road of cobblestones where the cartwheels (which represent the molecules of the upper surface or film) must be made to roll over the cobblestones (representing the molecules of the lower surface) before the cart can move. In the case of the cart, the downward force of gravity replaces the attractive intermolecular forces between two material surfaces. When at rest, the cartwheels find grooves between the cobblestones where they sit in potential energy minima, and so the cart is at some stable mechanical equilibrium. A certain lateral force (the "push") is required to raise the cartwheels against the force of gravity in order to initiate motion. Motion will continue as long as the cart is pushed, and rapidly stops once it is no longer pushed. Energy is dissipated by the liberation of heat (phonons, acoustic waves, etc.) every time a wheel hits the next cobblestone. The cobblestone model is not unlike the *Coulomb* and *interlocking asperity* models of friction [18.228] except that it is being applied at the molecular level and for a situation where the external load is augmented by attractive intermolecular forces.

There are thus two contributions to the force pulling two surfaces together: the externally applied load or pressure, and the (internal) attractive intermolecular forces that determine the adhesion between the two surfaces. Each of these contributions affects the friction force in a different way, which we will discuss in more detail below.

18.7.2 Adhesion Force and Load Contribution to Interfacial Friction

Adhesion Force Contribution

Consider the case of two surfaces sliding past each other, as shown in Fig. 18.14. When the two surfaces are initially in adhesive contact, the surface molecules will adjust themselves to fit snugly together [18.233], in an analogous manner to the self-positioning of the cartwheels on the cobblestone road. A small tangential force applied to one surface will therefore not result in

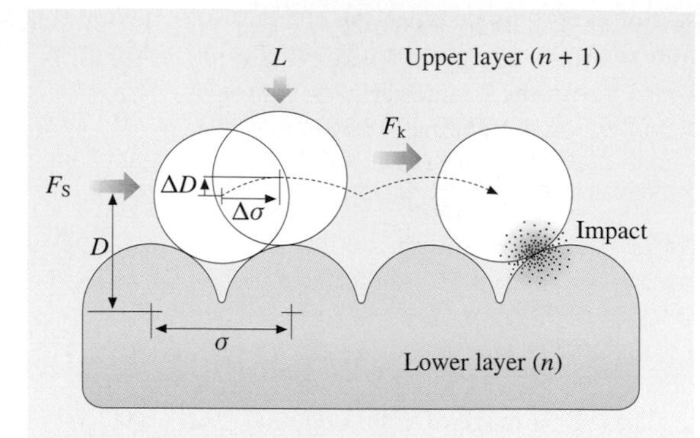

Fig. 18.14 Schematic illustration of how one molecularly smooth surface moves over another when a lateral force F is applied (the "cobblestone model"). As the upper surface moves laterally by some fraction of the lattice dimension $\Delta\sigma$, it must also move up by some fraction of an atomic or molecular dimension ΔD before it can slide across the lower surface. On impact, some fraction ε of the kinetic energy is "transmitted" to the lower surface, the rest being "reflected" back to the colliding molecule (upper surface). (After [18.232] with permission)

the sliding of that surface relative to the other. The attractive van der Waals forces between the surfaces must first be overcome by having the surfaces separate by a small amount. To initiate motion, let the separation between the two surfaces increase by a small amount ΔD, while the lateral distance moved is $\Delta\sigma$. These two values will be related via the geometry of the two surface lattices. The energy put into the system by the force F acting over a lateral distance $\Delta\sigma$ is

$$\text{Input energy: } F \times \Delta\sigma \,. \tag{18.29}$$

This energy may be equated with the change in interfacial or surface energy associated with separating the surfaces by ΔD, i. e., from the equilibrium separation $D = D_0$ to $D = (D_0 + \Delta D)$. Since $\gamma \propto D^{-2}$ for two flat surfaces, the energy cost may be approximated by:

Surface energy change × area:

$$2\gamma A\left[1 - D_0^2/(D_0 + \Delta D)^2\right] \approx 4\gamma A(\Delta D/D_0) \,, \tag{18.30}$$

where γ is the surface energy, A the contact area, and D_0 the surface separation at equilibrium. During steady state sliding (kinetic friction), not all of this energy will be "lost" or absorbed by the lattice every time the surface molecules move by one lattice spacing: Some fraction will be reflected during each impact of the "cartwheel" molecules [18.231]. Assuming that a fraction ε of the above surface energy is "lost" every time the surfaces move across the characteristic length $\Delta\sigma$ (Fig. 18.14), we obtain after equating (18.29) and (18.30)

$$S_c = \frac{F}{A} = \frac{4\gamma\varepsilon\Delta D}{D_0\Delta\sigma} . \tag{18.31}$$

For a typical hydrocarbon or a van der Waals surface, $\gamma \approx 25\,\mathrm{mJ\,m^{-2}}$. Other typical values would be: $\Delta D \approx 0.05\,\mathrm{nm}$, $D_0 \approx 0.2\,\mathrm{nm}$, $\Delta\sigma \approx 0.1\,\mathrm{nm}$, and $\varepsilon \approx 0.1$–0.5. Using the above parameters, (18.31) predicts $S_c \approx (2.5–12.5)\times 10^7\,\mathrm{N\,m^{-2}}$ for van der Waals surfaces. This range of values compares very well with typical experimental values of $2\times 10^7\,\mathrm{N\,m^{-2}}$ for hydrocarbon or mica surfaces sliding in air (see Fig. 18.16) or separated by one molecular layer of cyclohexane [18.45].

The above model suggests that all interfaces, whether dry or lubricated, dilate just before they shear or slip. This is a small but important effect: The dilation provides the crucial extra space needed for the molecules to slide across each other or flow. This dilation has been computed by *Thompson* and *Robbins* [18.201] and *Zaloj* et al. [18.234] and measured by *Dhinojwala* et al. [18.235].

This model may be extended, at least semi-quantitatively, to lubricated sliding, where a thin liquid film is present between the surfaces. With an increase in the number of liquid layers between the surfaces, D_0 increases while ΔD decreases, hence the lower the friction force. This is precisely what is observed, but with more than one liquid layer between two surfaces the situation becomes too complex to analyze analytically (actually, even with one or no interfacial layers, the calculation of the fraction of energy dissipated per molecular collision ε is not a simple matter). Furthermore, even in systems as simple as linear alkanes, interdigitation and interdiffusion have been found to contribute strongly to the properties of the system [18.114, 236]. Sophisticated modeling based on computer simulations is now required, as described in the following section.

Relation Between Boundary Friction and Adhesion Energy Hysteresis

While the above equations suggest that there is a direct correlation between friction and adhesion, this is not the case. The correlation is really between friction and adhesion hysteresis, described in Sect. 18.5.4. In the case of friction, this subtle point is hidden in the factor ε, which is a measure of the amount of energy absorbed (dissipated, transferred, or "lost") by the lower surface when it is impacted by a molecule from the upper surface. If $\varepsilon = 0$, all the energy is reflected, and there will be no kinetic friction force nor any adhesion hysteresis, but the absolute magnitude of the adhesion force or energy will remain finite and unchanged. This is illustrated in Figs. 18.17 and 18.19.

The following simple model shows how adhesion hysteresis and friction may be quantitatively related. Let $\Delta\gamma = \gamma_R - \gamma_A$ be the adhesion energy hysteresis per unit area, as measured during a typical loading–unloading cycle (see Figs. 18.17a and 18.19c,d). Now consider the same two surfaces sliding past each other and assume that frictional energy dissipation occurs through the same mechanism as adhesion energy dissipation, and that both occur over the same characteristic molecular length scale σ. Thus, when the two surfaces (of contact area $A = \pi r^2$) move a distance σ, equating the frictional energy $(F\times\sigma)$ to the dissipated adhesion energy $(A\times\Delta\gamma)$, we obtain

$$\text{Friction force: } F = \frac{A\times\Delta\gamma}{\sigma} = \frac{\pi r^2}{\sigma}(\gamma_R - \gamma_A) , \tag{18.32}$$

$$\text{or} \quad \text{Critical shear stress: } S_c = F/A = \Delta\gamma/\sigma , \tag{18.33}$$

which is the desired expression and has been found to give order of magnitude agreement between measured friction forces and adhesion energy hysteresis [18.207]. If we equate (18.33) with (18.31), since $4\Delta D/(D_0\Delta\sigma) \approx 1/\sigma$, we obtain the intuitive relation

$$\varepsilon = \frac{\Delta\gamma}{\gamma} . \tag{18.34}$$

External Load Contribution to Interfacial Friction

When there is no interfacial adhesion, S_c is zero. Thus, in the absence of any adhesive forces between two surfaces, the only "attractive" force that needs to be overcome for sliding to occur is the externally applied load or pressure.

For a preliminary discussion of this question, it is instructive to compare the magnitudes of the *externally* applied pressure to the *internal* van der Waals pressure between two smooth surfaces. The internal van der Waals pressure between two flat surfaces is given (see Table 18.2) by $P = A_H/6\pi D_0^3 \approx 1\,\mathrm{GPa}$ (10^4 atm), using a typical Hamaker constant of $A_H = 10^{-19}\,\mathrm{J}$, and assuming $D_0 \approx 2\,\mathrm{nm}$ for the equilibrium interatomic spacing. This implies that we should not expect the externally applied load to affect the interfacial friction force F, as defined by (18.27), until the externally applied pressure L/A begins to exceed ~ 100 MPa (10^3 atm). This is in agreement with experimental data [18.237] where the effect of load became dominant at pressures in excess of 10^3 atm.

For a more general semiquantitative analysis, again consider the cobblestone model used to derive (18.31), but now include an additional contribution to the surface energy change of (18.30) due to the work done against the external load or pressure, $L\Delta D = P_{\text{ext}} A \Delta D$ (this is equivalent to the work done against gravity in the case of a cart being pushed over cobblestones). Thus:

$$S_c = \frac{F}{A} = \frac{4\gamma\varepsilon\Delta D}{D_0\Delta\sigma} + \frac{P_{\text{ext}}\varepsilon\Delta D}{\Delta\sigma}\,, \qquad (18.35)$$

which gives the more general relation

$$S_c = F/A = C_1 + C_2 P_{\text{ext}}\,, \qquad (18.36)$$

where $P_{\text{ext}} = L/A$ and C_1 and C_2 are constants characteristic of the surfaces and sliding conditions. The constant $C_1 = 4\gamma\varepsilon\Delta D/(D_0\Delta\sigma)$ depends on the mutual adhesion of the two surfaces, while both C_1 and $C_2 = \varepsilon\Delta D/\Delta\sigma$ depend on the topography or atomic bumpiness of the surface groups (Fig. 18.14). The smoother the surface groups the smaller the ratio $\Delta D/\Delta\sigma$ and hence the lower the value of C_2. In addition, both C_1 and C_2 depend on ε (the fraction of energy dissipated per collision), which depends on the relative masses of the shearing molecules, the sliding velocity, the temperature, and the characteristic molecular relaxation processes of the surfaces. This is by far the most difficult parameter to compute, and yet it is the most important since it represents the energy transfer mechanism in any friction process, and since ε can vary between 1 and 0, it determines whether a particular friction force will be large or close to zero. Molecular simulations offer the best way to understand and predict the magnitude of ε, but the complex multi-body nature of the problem makes simple conclusions difficult to draw [18.239–241]. Some of the basic physics of the energy transfer and dissipation of the molecular collisions can be drawn from simplified models such as a 1-D three-body system [18.214].

Finally, the above equation may also be expressed in terms of the friction force F:

$$F = S_c A = C_1 A + C_2 L\,. \qquad (18.37)$$

Equations similar to (18.36) and (18.37) were previously derived by *Derjaguin* [18.242,243] and by *Briscoe*

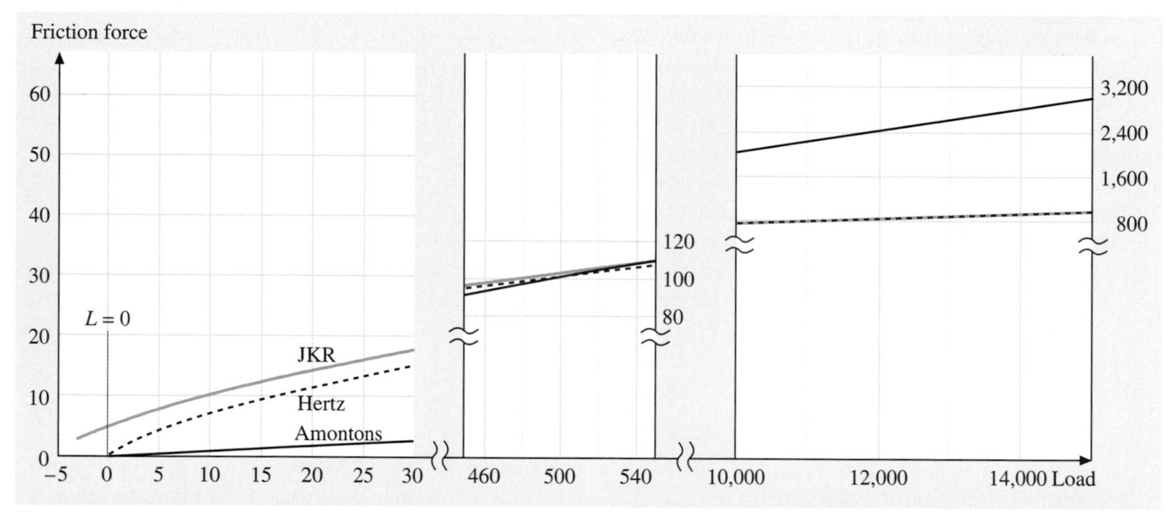

Fig. 18.15 Friction as a function of load for smooth surfaces. At low loads, the friction is dominated by the C_1A term of (18.37). The adhesion contribution (JKR curve) is most prominent near zero load where the Hertzian and Amontons' contributions to the friction are minimal. As the load increases, the adhesion contribution becomes smaller as the JKR and Hertz curves converge. In this range of loads, the linear C_2L contribution surpasses the area contribution to the friction. At much higher loads the explicit load dependence of the friction dominates the interactions, and the observed behavior approaches Amontons' law. It is interesting to note that for smooth surfaces the pressure over the contact area does not increase as rapidly as the load. This is because as the load is increased, the surfaces deform to increase the surface area and thus moderate the contact pressure. (After [18.238] with permission of Kluwer Academic Publishers)

and *Evans* [18.244], where the constants C_1 and C_2 were interpreted somewhat differently than in this model.

In the absence of any attractive interfacial force, we have $C_1 \approx 0$, and the second term in (18.36) and (18.37) should dominate (Fig. 18.15). Such situations typically arise when surfaces repel each other across the lubricating liquid film. In such cases, the total frictional force should be low and should increase *linearly* with the external load according to

$$F = C_2 L . \qquad (18.38)$$

An example of such lubricated sliding occurs when two mica surfaces slide in water or in salt solution (see Fig. 18.20), where the short-range "hydration" forces between the surfaces are repulsive. Thus, for sliding in 0.5 M KCl it was found that $C_2 = 0.015$ [18.245]. Another case where repulsive surfaces eliminate the adhesive contribution to friction is for polymer chains attached to surfaces at one end and swollen by a good solvent [18.175]. For this class of systems, $C_2 < 0.001$ for a finite range of polymer layer compressions (normal loads, L). The low friction between the surfaces in this regime is attributed to the entropic repulsion between the opposing brush layers with a minimum of entanglement between the two layers. However, with higher normal loads, the brush layers become compressed and begin to entangle, which results in higher friction (see [18.246]).

It is important to note that (18.38) has exactly the same form as Amontons' Law

$$F = \mu L , \qquad (18.39)$$

where μ is the coefficient of friction.

At the molecular level a thermodynamic analog of the Coulomb or cobblestone models (see Sect. 18.7.1) based on the "contact value theorem" [18.3, 238, 245] can explain why $F \propto L$ also holds at the microscopic or molecular level. In this analysis we consider the surface molecular groups as being momentarily compressed and decompressed as the surfaces move along. Under irreversible conditions, which always occur when a cycle is completed in a finite amount of time, the energy "lost" in the compression–decompression cycle is dissipated as heat. For two non-adhering surfaces, the stabilizing pressure P_i acting locally between any two elemental contact points i of the surfaces may be expressed by the contact value theorem [18.3]:

$$P_i = \rho_i k_B T = k_B T / V_i , \qquad (18.40)$$

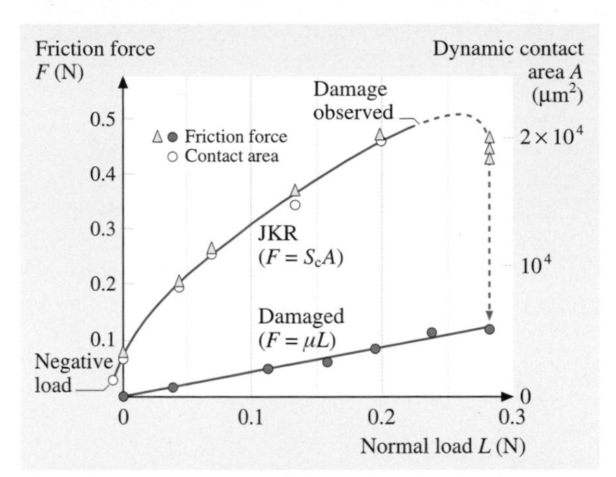

Fig. 18.16 Friction force F and contact area A vs. load L for two mica surfaces sliding in adhesive contact in dry air. The contact area is well described by the JKR theory, (18.22), even during sliding, and the friction force is found to be directly proportional to this area, (18.28). The *vertical dashed line* and *arrow* show the transition from interfacial to normal friction with the onset of wear (*lower curve*). Sliding velocity is 0.2 µm s^{-1}. (After [18.45] with permission, copyright 1989 American Society of Mechanical Engineers)

where $\rho_i = V_i^{-1}$ is the local number density (per unit volume) or activity of the interacting entities, be they molecules, atoms, ions or the electron clouds of atoms. This equation is essentially the osmotic or entropic pressure of a gas of confined molecules. As one surface moves across the other, local regions become compressed and decompressed by a volume ΔV_i. The work done per cycle can be written as $\varepsilon P_i \Delta V_i$, where ε ($\varepsilon \leq 1$) is the fraction of energy per cycle "lost" as heat, as defined earlier. The energy balance shows that for each compression–decompression cycle, the dissipated energy is related to the friction force by

$$F_i x_i = \varepsilon P_i \Delta V_i , \qquad (18.41)$$

where x_i is the lateral distance moved per cycle, which can be the distance between asperities or the distance between surface lattice sites. The pressure at each contact junction can be expressed in terms of the local normal load L_i and local area of contact A_i as $P_i = L_i / A_i$. The volume change over a cycle can thus be expressed as $\Delta V_i = A_i z_i$, where z_i is the vertical distance of confinement. Inserting these into (18.41), we get

$$F_i = \varepsilon L_i (z_i / x_i) , \qquad (18.42)$$

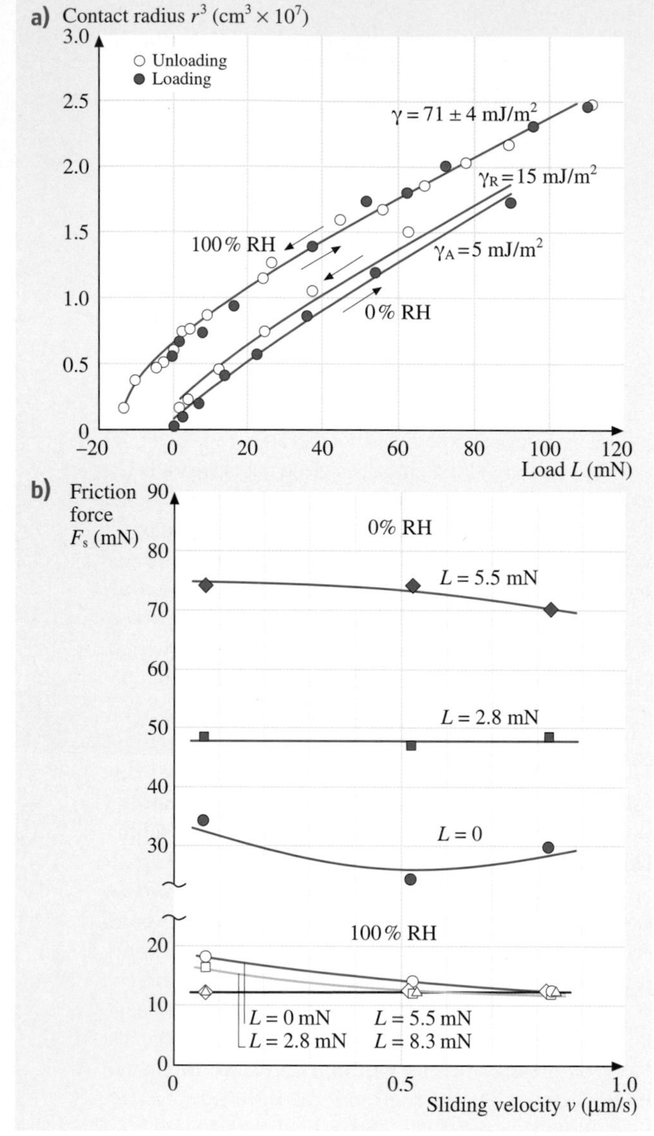

Fig. 18.17 (a) Contact radius r versus externally applied load L for loading and unloading of two hydrophilic silica surfaces exposed to dry and humid atmospheres. Note that while the adhesion is higher in humid air, the *hysteresis* in the adhesion is higher in dry air. (b) Effect of velocity on the static friction force F_s for hydrophobic (heat-treated electron beam evaporated) silica in dry and humid air. The effects of humidity, load, and sliding velocity on the friction forces, as well as the stick–slip friction of the hydrophobic surfaces, are qualitatively consistent with a "friction" phase diagram representation as in Fig. 18.28. (After [18.41]. Copyright 1994, with permission from Elsevier Science)

which is independent of the local contact area A_i. The total friction force is thus

$$\begin{aligned} F &= \sum F_i = \sum \varepsilon L_i (z_i/x_i) \\ &= \varepsilon \langle z_i/x_i \rangle \sum L_i = \mu L \,, \end{aligned} \tag{18.43}$$

where it is assumed that on average the local values of L_i and P_i are independent of the local "slope" z_i/x_i. Therefore, the friction coefficient μ is a function only of the average surface topography and the sliding velocity, but is independent of the local (real) or macroscopic (apparent) contact areas.

While this analysis explains non-adhering surfaces, there is still an additional explicit contact area contribution for the case of adhering surfaces, as in (18.37). The distinction between the two cases arises because the initial assumption of the contact value theorem, (18.40), is incomplete for adhering systems. A more appropriate starting equation would reflect the full intermolecular interaction potential, including the attractive interactions, in addition to the purely repulsive contributions of (18.40), much as the van der Waals equation of state modifies the ideal gas law.

18.7.3 Examples of Experimentally Observed Friction of Dry Surfaces

Numerous model systems have been studied with a surface forces apparatus (SFA) modified for friction experiments (see Sect. 18.2.3). The apparatus allows for control of load (normal force) and sliding speed, and simultaneous measurement of surface separation, surface shape, true (molecular) area of contact between smooth surfaces, and friction forces. A variety of both unlubricated and solid and liquid lubricated surfaces have been studied both as smooth single-asperity contacts and after they have been roughened by shear-induced damage.

Figure 18.16 shows the contact area, A, and friction force, F, both plotted against the applied load, L, in an experiment in which two molecularly smooth surfaces of mica in adhesive contact were slid past each other in an atmosphere of dry nitrogen gas. This is an example of the low load, adhesion controlled limit, which is excellently described by (18.28). In a number of different experiments, S_c was measured to be $2.5 \times 10^7\,\mathrm{N\,m^{-2}}$ and to be independent of the sliding velocity [18.45,247]. Note that there is a friction force even at negative loads, where the surfaces are still sliding in adhesive contact.

Figure 18.17 shows the correlation between adhesion hysteresis and friction for two surfaces consisting of silica films deposited on mica substrates [18.41].

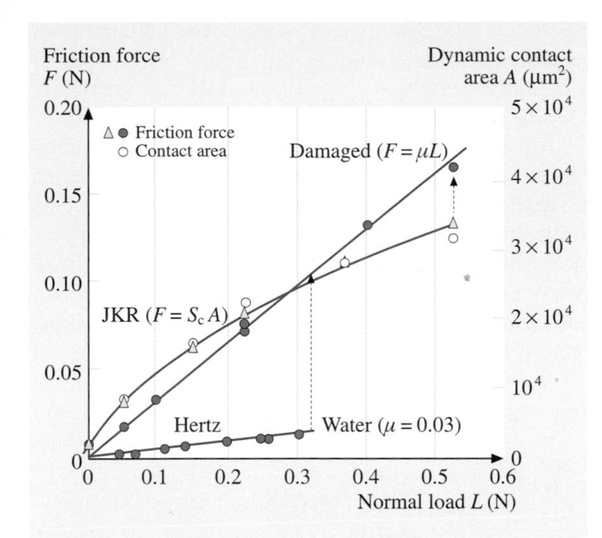

Fig. 18.18 Sliding of mica surfaces, each coated with a 2.5-nm-thick monolayer of calcium stearate surfactant, in the absence of damage (obeying JKR type boundary friction) and in the presence of damage (obeying Amontons' type normal friction). Note that both for this system and for the bare mica in Figs. 18.16 and 18.20, the friction force obeys Amontons' law with a friction coefficient of $\mu \approx 0.3$ after damage occurs. At much higher applied loads, the undamaged surfaces also follow Amontons' type sliding, but for a different reason: The dependence on adhesion becomes smaller. *Lower line:* Interfacial sliding with a monolayer of water between the mica surfaces, shown for comparison. (After [18.247]. Copyright 1990, with permission from Elsevier Science)

The friction between undamaged hydrophobic silica surfaces showed stick–slip both at dry conditions and at 100% relative humidity. Similar to the mica surfaces in Figs. 18.16, 18.18, and 18.20, the friction of damaged silica surfaces obeyed Amontons' law with a friction coefficient of 0.25 both at dry conditions and at 55% relative humidity.

The high friction force of unlubricated sliding can often be reduced by treating the solid surface with a boundary layer of some other solid material that exhibits lower friction such as a surfactant monolayer, or by ensuring that during sliding, a thin liquid film remains between the surfaces (as will be discussed in Sect. 18.8). The effectiveness of a solid boundary lubricant layer on reducing the forces of friction is illustrated in Fig. 18.18. Comparing this with the friction of the unlubricated/untreated surfaces (Fig. 18.16) shows that the critical shear stress has been reduced by a factor of about ten: from 2.5×10^7 to $3.5\times10^6\,\mathrm{N\,m^{-2}}$. At much higher applied loads or pressures, the friction force is proportional to the load, rather than the area of contact [18.237], as expected from (18.27).

Yamada and *Israelachvili* [18.248] studied the adhesion and friction of fluorocarbon surfaces (surfactant-coated boundary lubricant layers), which were compared to those of hydrocarbon surfaces. They concluded that well-ordered fluorocarbon surfaces have a high friction, in spite of their lower adhesion energy (in agreement with previous findings). The low friction coefficient of PTFE (Teflon) must, therefore, be due to some effect other than a low adhesion. For example, the softness of PTFE, which allows material to flow at the interface, which thus behaves like a fluid lubricant. On a related issue, *Luengo* et al. [18.249] found that C_{60} surfaces also exhibited low adhesion but high friction. In both cases the high friction appears to arise from the bulky surface groups – fluorocarbon compared to hydrocarbon groups in the former, large fullerene spheres in the latter. Apparently, the fact that C_{60} molecules rotate *in their lattice* does not make them a good lubricant: The molecules of the opposing surface must still climb over them in order to slide, and this requires energy that is independent of whether the surface molecules are fixed or freely rotating. Larger particles such as ~ 25 nm-sized nanoparticles (also known as "inorganic fullerenes") do appear to produce low friction by behaving like molecular ball bearings, but the potential of this promising new class of solid lubricant has still to be explored [18.250].

Figure 18.19 illustrates the relationship between adhesion hysteresis and friction for surfactant-coated surfaces under different conditions. This effect, however, is much more general and has been shown to hold for other surfaces as well [18.41, 232, 251].

Direct comparisons between absolute adhesion energies and friction forces show little correlation. In some cases, higher adhesion energies for the same system under different conditions correspond with lower friction forces. For example, for hydrophilic silica surfaces (Fig. 18.17) it was found that with increasing relative humidity the adhesion energy *increases*, but the adhesion energy hysteresis measured in a loading–unloading cycle *decreases*, as does the friction force [18.41]. For hydrophobic silica surfaces under dry conditions, the friction at load $L = 5.5$ mN was $F = 75$ mN. For the same sample, the adhesion energy hysteresis was $\Delta\gamma = 10\,\mathrm{mJ\,m^{-2}}$, with a contact area of $A \approx 10^{-8}\,\mathrm{m^2}$ at the same load. Assuming a value for the characteristic distance σ on the order of one lattice spacing, $\sigma \approx 1$ nm, and inserting these values into (18.32), the friction force is predicted to be $F \approx 100\,\mathrm{mN}$ for the kinetic friction force, which is close to the measured value of 75 mN. Alternatively, we may conclude that the dissipation factor is $\varepsilon = 0.75$, i. e., that almost all the energy is dissipated as heat at each molecular collision.

A liquid lubricant film (Sect. 18.8.3) is usually much more effective at lowering the friction of two surfaces than a solid boundary lubricant layer. However, to successfully use a liquid lubricant, it must "wet" the surfaces, that is, it should have a high affinity for the surfaces, so that not all the liquid molecules become squeezed out when the surfaces come close together, even under a large compressive load. Another important requirement is that the liquid film remains a liquid under tribological conditions, i. e., that it does not epitaxially solidify between the surfaces.

Effective lubrication usually requires that the lubricant be injected between the surfaces, but in some cases the liquid can be made to condense from the vapor. This is illustrated in Fig. 18.20 for two untreated mica surfaces sliding with a thin layer of water between them. A monomolecular film of water (of thickness 0.25 nm per surface) has reduced S_c from its value for dry surfaces (Fig. 18.16) by a factor of more than 30, which

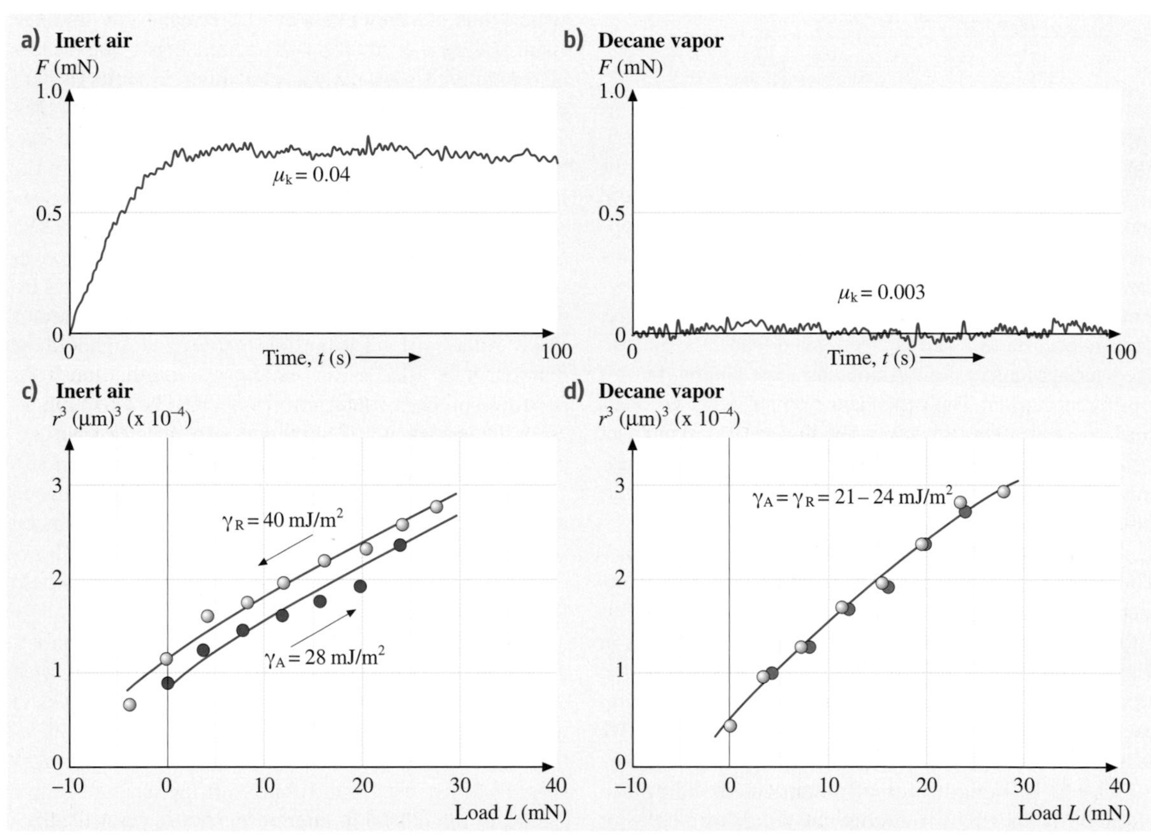

Fig. 18.19a–d *Top:* Friction traces for two fluid-like calcium alkylbenzene sulfonate monolayer-coated surfaces at 25 °C showing that the friction force is much higher between dry monolayers (**a**) than between monolayers whose fluidity has been enhanced by hydrocarbon penetration from vapor (**b**). *Bottom:* Contact radius vs. load (r^3 vs. L) data measured for the same two surfaces as above and fitted to the JKR equation (18.22), shown by the *solid curves*. For dry monolayers (**c**) the adhesion energy on unloading ($\gamma_R = 40\,\mathrm{mJ\,m^{-2}}$) is greater than that on loading ($\gamma_R = 28\,\mathrm{mJ\,m^{-2}}$), indicative of an adhesion energy hysteresis of $\Delta\gamma = \gamma_R - \gamma_A = 12\,\mathrm{mJ\,m^{-2}}$. For monolayers exposed to saturated decane vapor (**d**) their adhesion hysteresis is zero ($\gamma_A = \gamma_R$), and both the loading and unloading data are well fitted by the thermodynamic value of the surface energy of fluid hydrocarbon chains, $\gamma = 24\,\mathrm{mJ\,m^{-2}}$. (After [18.207] with permission. Copyright 1993 American Chemical Society)

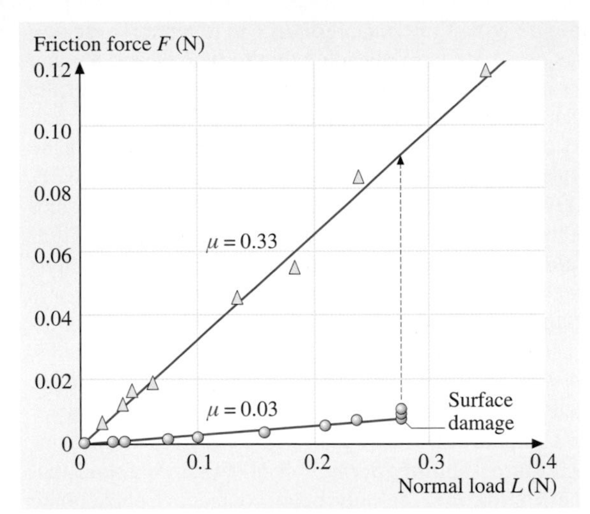

Fig. 18.20 Two mica surfaces sliding past each other while immersed in a 0.01 M KCl salt solution (nonadhesive conditions). The water film is molecularly thin, 0.25 to 0.5 nm thick, and the interfacial friction force is very low: $S_c \approx 5\times10^5\,\mathrm{N\,m^{-2}}$, $\mu \approx 0.03$ (before damage occurs). After the surfaces have become damaged, the friction coefficient is ca. 0.3. (After [18.247]. Copyright 1990, with permission from Elsevier Science)

may be compared with the factor of ten attained with the boundary lubricant layer (of thickness 2.5 nm per surface) in Fig. 18.18. Water appears to have unusual lubricating properties and usually gives wearless friction with no stick–slip.

The effectiveness of a water film only 0.25 nm thick to lower the friction force by more than an order of magnitude is attributed to the "hydrophilicity" of the mica surface (mica is "wetted" by water) and to the existence of a strongly repulsive short-range hydration force between such surfaces in aqueous solutions, which effectively removes the adhesion-controlled contribution to the friction force [18.245]. It is also interesting that a 0.25-nm-thick water film between two mica surfaces is sufficient to bring the coefficient of friction down to 0.01–0.02, a value that corresponds to the unusually low friction of ice. Clearly, a single monolayer of water can be a very good lubricant – much better than most other monomolecular liquid films – for reasons that will be discussed in Sect. 18.9.

Dry polymer layers (Fig. 18.21) typically show a high initial static friction ("stiction") as sliding commences from rest in adhesive contact. The development of the friction force after a change in sliding direction, a gradual transition from stick–slip to smooth sliding, is shown in Fig. 18.21. A correlation between adhesion hysteresis and friction similar to the one observed for silica surfaces in Fig. 18.17 can be seen for dry polymer layers below their glass transition temperature. As shown in Fig. 18.12b,c, the adhesion hysteresis for polystyrene surfaces can be increased by irradiation to induce scission of chains, and it has been found that the steady state friction force (kinetic friction) shows a similar increase with irradiation time [18.208].

Figure 18.22 shows an example of a computer simulation of the sliding of two unlubricated silicon surfaces (modeled as a tip sliding over a planar surface) [18.91]. The sliding proceeds through a series of stick–slip events, and information on the friction force and the local order of the initially crystalline surfaces can be obtained. Similar studies for cold welding systems [18.91] have demonstrated the occurrence of shear or friction damage within the sliding surface (tip) as the lowest layer of it adheres to the bottom surface.

18.7.4 Transition from Interfacial to Normal Friction with Wear

Frictional damage can have many causes such as adhesive tearing at high loads or overheating at high sliding speeds. Once damage occurs, there is a transition from "interfacial" to "normal" or load-controlled friction as the surfaces become forced apart by the torn-out as-

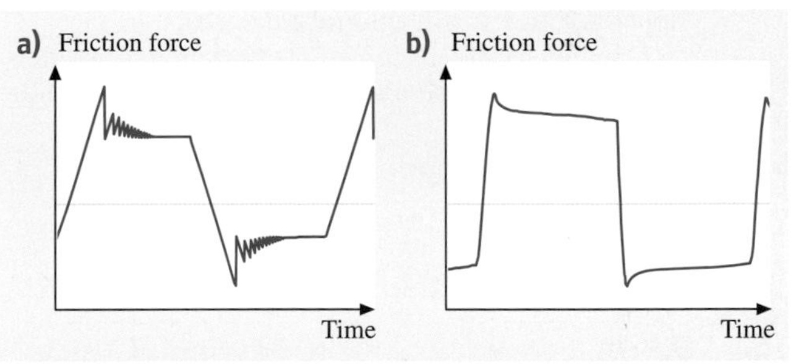

Fig. 18.21a,b Typical friction traces showing how the friction force varies with the sliding time for two symmetric, glassy polymer films at dry conditions. Qualitative features that are common to both polystyrene and polyvinyl benzyl chloride: **(a)** Decaying stick–slip motion is observed until smooth sliding is attained if the motion continues for a sufficiently long distance. **(b)** Smooth sliding observed at sufficiently high speeds. Similar observations have been made by *Berthoud* et al. [18.252] in measurements on polymethyl methacrylate. (After [18.208] with permission. Copyright 2002 American Association for the Advancement of Science)

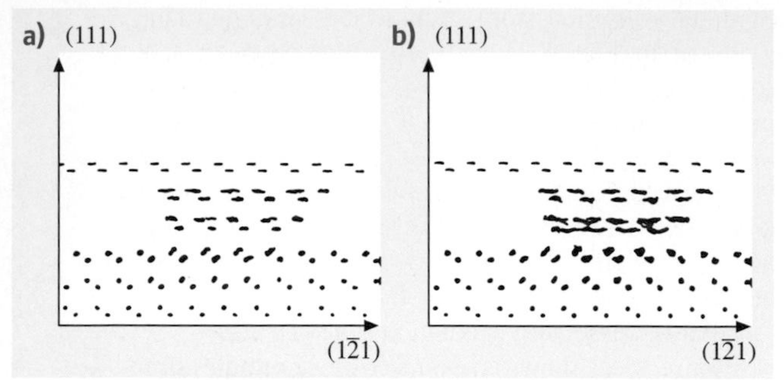

Fig. 18.22a,b Computer simulation of the sliding of two contacting Si surfaces (a tip and a flat surface). Shown are particle trajectories in a constant-force simulation, $F_{z,\text{external}} = -2.15 \times 10^{-8}$ N, viewed along the (10$\bar{1}$) direction just before (**a**) and after (**b**) a stick–slip event for a large, initally ordered, dynamic tip. (After [18.91] with permission of Kluwer Academic Publishers)

perities (wear particles). For low loads, the friction changes from obeying $F = S_c A$ to obeying Amontons' law, $F = \mu L$, as shown in Figs. 18.16 and 18.18, and sliding now proceeds smoothly with the surfaces separated by a 10–100 nm forest of wear debris (in this case, mica flakes). The wear particles keep the surfaces apart over an area that is much greater than their size, so that even one submicroscopic particle or asperity can cause a significant reduction in the area of contact and, therefore, in the friction [18.247]. For this type of frictional sliding, one can no longer talk of the molecular contact area of the two surfaces, although the macroscopic or "apparent" area is still a useful parameter.

One remarkable feature of the transition from interfacial to normal friction of brittle surfaces is that while the strength of interfacial friction, as reflected in the values of S_c, is very dependent on the type of surface and on the liquid film between the surfaces, this is not the case once the transition to normal friction has occurred. At the onset of damage, the material properties of the underlying substrates control the friction. In Figs. 18.16, 18.18, and 18.20 the friction for the damaged surfaces is that of any damaged mica–mica system, $\mu \approx 0.3$, *independent of the initial surface coatings or liquid films between the surfaces*. A similar friction coefficient was found for damaged silica surfaces [18.41].

In order to practically modify the frictional behavior of such brittle materials, it is important to use coatings that will both alter the interfacial tribological character and remain intact and protect the surfaces from damage during sliding. *Berman* et al. [18.253] found that the friction of a strongly bound octadecyl phosphonic acid monolayer on alumina surfaces was higher than for untreated, undamaged α-alumina surfaces, but the bare surfaces easily became damaged upon sliding, resulting in an ultimately higher friction system with greater wear rates than the more robust monolayer-coated surfaces.

Clearly, the mechanism and factors that determine *normal* friction are quite different from those that govern *interfacial* friction (Sects. 18.7.1–18.7.2). This effect is not general and may only apply to brittle materials. For example, the friction of ductile surfaces is totally different and involves the continuous plastic deformation of contacting surface asperities during sliding, rather than the rolling of two surfaces on hard wear particles [18.209]. Furthermore, in the case of ductile surfaces, water and other surface-active components do have an effect on the friction coefficients under "normal" sliding conditions.

18.8 Liquid Lubricated Surfaces

18.8.1 Viscous Forces and Friction of Thick Films: Continuum Regime

Experimentally, it is usually difficult to unambiguously establish which type of sliding mode is occurring, but an empirical criterion, based on the Stribeck curve (Fig. 18.13), is often used as an indicator. This curve shows how the friction force or the coefficient of friction is expected to vary with sliding speed, depending on which type of friction regime is operating. For thick liquid lubricant films whose behavior can be described by bulk (continuum) properties, the friction forces are essentially the hydrodynamic or viscous drag forces. For example, for two plane parallel surfaces of area A separated by a distance D and moving laterally relative to each other with velocity v, if the intervening liquid is *Newtonian*, i. e., if its viscosity η is independent of the shear rate, the frictional force experienced by the surfaces is given by

$$F = \frac{\eta A v}{D}, \tag{18.44}$$

where the shear rate $\dot{\gamma}$ is defined by

$$\dot{\gamma} = \frac{v}{D} \,. \tag{18.45}$$

At higher shear rates, two additional effects often come into play. First, certain properties of liquids may change at high $\dot{\gamma}$ values. In particular, the effective viscosity may become non-Newtonian, one form given by

$$\eta = \dot{\gamma}^n \,, \tag{18.46}$$

where $n = 0$ (i. e., $\eta_{\text{eff}} =$ constant) for Newtonian fluids, $n > 0$ for shear thickening (dilatent) fluids, and $n < 0$ for shear thinning (pseudoplastic) fluids (the latter become less viscous, i. e., flow more easily, with increasing shear rate). An additional effect on η can arise from the higher local stresses (pressures) experienced by the liquid film as $\dot{\gamma}$ increases. Since the viscosity is generally also sensitive to the pressure (usually increasing with P), this effect also acts to increase η_{eff} and thus the friction force.

A second effect that occurs at high shear rates is surface deformation, arising from the large hydrodynamic forces acting on the sliding surfaces. For example, Fig. 18.23 shows how two surfaces deform elastically when the sliding speed increases to a high value. These deformations alter the hydrodynamic friction forces, and this type of friction is often referred to as *elastohydrodynamic lubrication* (EHD or EHL), as mentioned in Table 18.4.

How thin can a liquid film be before its dynamic, e.g., viscous flow, behavior ceases to be described by bulk properties and continuum models? Concerning the static properties, we have already seen in Sect. 18.4.3 that films composed of simple liquids display continuum behavior down to thicknesses of 4–10 molecular diameters. Similar effects have been found to apply to the dynamic properties, such as the viscosity, of simple liquids in thin films. Concerning viscosity measurements, a number of dynamic techniques were recently developed [18.11–13, 43, 254, 255] for directly measuring the viscosity as a function of film thickness and shear rate across very thin liquid films between two surfaces. By comparing the results with theoretical predictions of fluid flow in thin films, one can determine the effective positions of the shear planes and the onset of non-Newtonian behavior in very thin films.

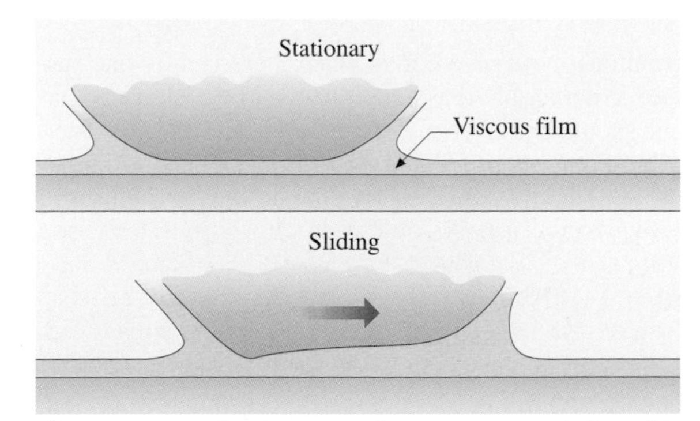

Fig. 18.23 *Top:* Stationary surfaces (one more deformable and one rigid) separated by a thick liquid film. *Bottom:* Elastohydrodynamic deformation of the upper surface during sliding. (After [18.1] with permission)

The results show that for simple liquids, including linear chain molecules such as alkanes, the viscosity in thin films is the same, within 10%, as the bulk even for films as thin as 10 molecular diameters (or segment widths) [18.11–13,254,255]. This implies that the shear plane is effectively located within one molecular diameter of the solid–liquid interface, and these conclusions were found to remain valid even at the highest shear rates studied (of $\sim 2\times10^5\,\text{s}^{-1}$). With water between two mica or silica surfaces [18.22, 254–256] this has been found to be the case (to within $\pm10\%$) down to surface separations as small as 2 nm, implying that the shear planes must also be within a few tenths of a nanometer of the solid–liquid interfaces. These results appear to be independent of the existence of electrostatic "double-layer" or "hydration" forces. For the case of the simple liquid toluene confined between surfaces with adsorbed layers of C_{60} molecules, this type of viscosity measurement has shown that the traditional no-slip assumption for flow at a solid interface does not always hold [18.257]. The C_{60} layer at the mica–toluene interface results in a "full slip" boundary, which dramatically lowers the viscous drag or effective viscosity for regular Couette or Poiseuille flow.

With polymeric liquids (polymer melts) such as polydimethylsiloxanes (PDMS) and polybutadienes (PBD), or with polystyrene (PS) adsorbed onto surfaces from solution, the far-field viscosity is again equal to the bulk value, but with the non-slip plane (hydrodynamic layer thickness) being located at $D = 1{-}2R_g$ away from each surface [18.11, 47], or at $D = L$ or less for polymer brush layers of thickness L per surface [18.13,258]. In contrast, the same technique was used to show that for non-adsorbing polymers in solution, there is actually a depletion layer of nearly pure solvent that exists at the surfaces that affects the confined solution flow properties [18.256]. These effects are observed from near contact to surface separations in excess of 200 nm.

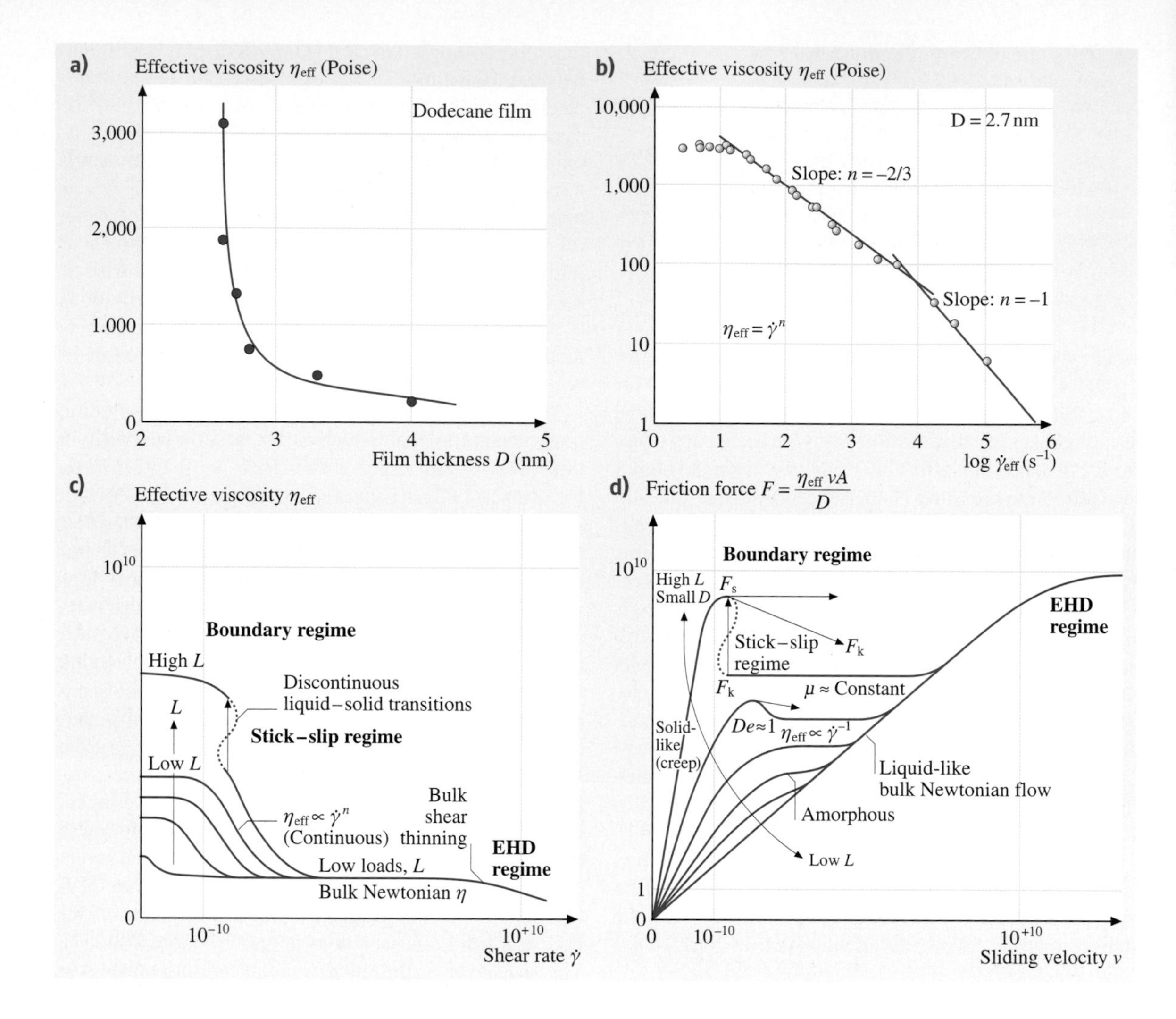

Further experiments with surfaces closer than a few molecular diameters ($D < 2$–4 nm for simple liquids, or $D < 2$–$4R_g$ for polymer fluids) indicate that large deviations occur for thinner films, described below. One important conclusion from these studies is, therefore, that the dynamic properties of simple liquids, including water, near an *isolated* surface are similar to those of the bulk liquid *already within the first layer of molecules adjacent to the surface*, only changing when another surface approaches the first. In other words, the viscosity and position of the shear plane near a surface are not simply a property of that surface, but of how far that surface is from another surface. The reason for this is because when two surfaces are close together, the constraining effects on the liquid molecules between them are much more severe than when there is only one surface. Another obvious consequence of the above is that one should not make measurements on a single, isolated solid–liquid interface and then draw conclusions about the state of the liquid or its interactions in a thin film *between* two surfaces.

18.8.2 Friction of Intermediate Thickness Films

For liquid films in the thickness range between 4 and 10 molecular diameters, the properties can be significantly different from those of bulk films. Still, the fluids remain recognizable as fluids; in other words, they do not undergo a phase transition into a solid

◂ **Fig. 18.24a–d** Typical rheological behavior of liquid films in the mixed lubrication regime. (**a**) Increase in effective viscosity of dodecane film between two mica surfaces with decreasing film thickness. At distances larger than 4–5 nm, the effective viscosity η_{eff} approaches the bulk value η_{bulk} and does not depend on the shear rate $\dot{\gamma}$. (After [18.219]. Copyright 1991 American Association for the Advancement of Science.) (**b**) Non-Newtonian variation of η_{eff} with shear rate of a 2.7-nm-thick dodecane film at a net normal pressure of 0.12 MPa and at 28 °C. The effective viscosity decays as a power law, as in (18.46). In this example, $n = 0$ at the lowest $\dot{\gamma}$ and changes to $n = -2/3$ and -1 at higher $\dot{\gamma}$. For films of bulk thickness, dodecane is a low viscosity Newtonian fluid ($n = 0$). (**c**) Proposed general friction map of effective viscosity η_{eff} (arbitrary units) as a function of effective shear rate $\dot{\gamma}$ (arbitrary units) on logarithmic scales. Three main classes of behavior emerge: (i) Thick films: elastohydrodynamic sliding. At $L = 0$, approximating bulk conditions, η_{eff} is independent of shear rate except when shear thinning might occur at sufficiently large $\dot{\gamma}$. (ii) Boundary layer films, intermediate regime. A Newtonian regime is again observed [$\eta_{eff} =$ constant, $n = 0$ in (18.46)] at low loads and low shear rates, but η_{eff} is much higher than the bulk value. As the shear rate $\dot{\gamma}$ increases beyond $\dot{\gamma}_{min}$, the effective viscosity starts to drop with a power-law dependence on the shear rate [see panel (**b**)], with n in the range $-1/2$ to -1 most commonly observed. As the shear rate $\dot{\gamma}$ increases still more, beyond $\dot{\gamma}_{max}$, a second Newtonian plateau is encountered. (iii) Boundary layer films, high load. The η_{eff} continues to grow with load and to be Newtonian provided that the shear rate is sufficiently low. Transition to sliding at high velocity is discontinuous ($n < -1$) and usually of the stick–slip variety. (**d**) Proposed friction map of friction force as a function of sliding velocity in various tribological regimes. With increasing load, Newtonian flow in the elastohydrodynamic (EHD) regimes crosses into the boundary regime of lubrication. Note that even EHD lubrication changes, at the highest velocities, to limiting shear stress response. At the highest loads (L) and smallest film thickness (D), the friction force goes through a maximum (the static friction, F_s), followed by a regime where the friction coefficient (μ) is roughly constant with increasing velocity (meaning that the kinetic friction, F_k, is roughly constant). Non-Newtonian shear thinning is observed at somewhat smaller load and larger film thickness; the friction force passes through a maximum at the point where $De = 1$. De, the Deborah number, is the point at which the applied shear rate exceeds the natural relaxation time of the boundary layer film. The velocity axis from 10^{-10} to 10^{10} (arbitrary units) indicates a large span. (Panels (**b**)–(**d**) after [18.259]. Copyright 1996, with permission from Elsevier Science)

or liquid-crystalline phase. This regime has recently been studied by *Granick* et al. [18.44, 219, 220, 222, 223], who used a friction attachment [18.43, 44] to the SFA where a sinusoidal input signal to a piezoelectric device makes the two surfaces slide back and forth laterally past each other at small amplitudes. This method provides information on the real and imaginary parts (elastic and dissipative components, respectively) of the shear modulus of thin films at different shear rates and film thickness. *Granick* [18.219] and *Hu* et al. [18.223] found that films of simple liquids become non-Newtonian in the 2.5–5 nm regime (about 10 molecular diameters, see Fig. 18.24). Polymer melts become non-Newtonian at much larger film thicknesses, depending on their molecular weight [18.47].

Klein and *Kumacheva* [18.46, 226, 260] studied the interaction forces and friction of small quasi-spherical liquid molecules such as cyclohexane between molecularly smooth mica surfaces. They concluded that surface epitaxial effects can cause the liquid film to already solidify at 6 molecular diameters, resulting in a sudden (discontinuous) jump to high friction at low shear rates. Such dynamic first-order transitions, however, may depend on the shear rate.

A generalized friction map (Fig. 18.24c,d) has been proposed by *Luengo* et al. [18.259] that illustrates the changes in η_{eff} from bulk Newtonian behavior ($n = 0$, $\eta_{eff} = \eta_{bulk}$) through the transition regime where n reaches a minimum of -1 with decreasing shear rate to the solid-like creep regime at very low $\dot{\gamma}$ where n returns to 0. A number of results from experimental, theoretical, and computer simulation work have shown values of n from $-1/2$ to -1 for this transition regime for a variety of systems and assumptions [18.219, 220, 239, 261–267].

The intermediate regime appears to extend over a narrow range of film thickness, from about 4 to 10 molecular diameters or polymer radii of gyration. Thinner films begin to adopt "boundary" or "interfacial" friction properties (described below, see also Table 18.5). Note that the intermediate regime is actually a very narrow one when defined in terms of film thickness, for example, varying from about $D = 2$ to 4 nm for hexadecane films [18.219].

A fluid's effective viscosity η_{eff} in the intermediate regime is usually higher than in the bulk, but η_{eff} usu-

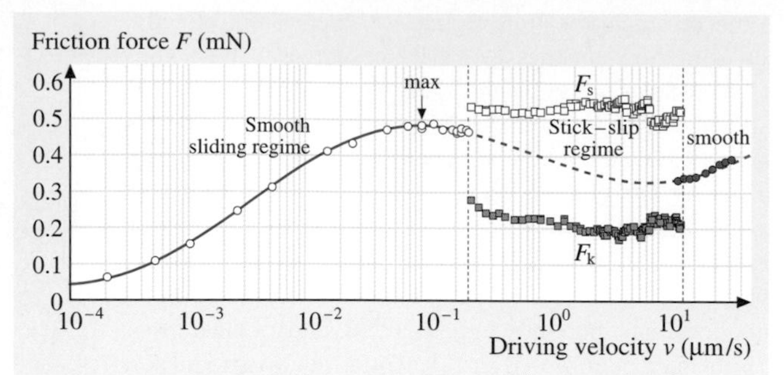

Fig. 18.25 Transition from smooth sliding to "inverted" stick–slip and to a second smooth sliding regime with increasing driving velocity during shear of two adsorbed surfactant monolayers in aqueous solution at a load of $L = 4.5$ mN and $T = 20\,°\mathrm{C}$. The smooth sliding (*open circles*) to inverted stick–slip (*squares*) transition occurs at $v_c \sim 0.3\,\mu\mathrm{m/s}$. Prior to the transition, the kinetic stress levels off at after a logarithmic dependence on velocity. The quasi-smooth regime persists up to the transition at v_c. At high driving velocities (*filled circles*), a new transition to a smooth sliding regime is observed between 14 and 17 μm/s. (After [18.268] with permission)

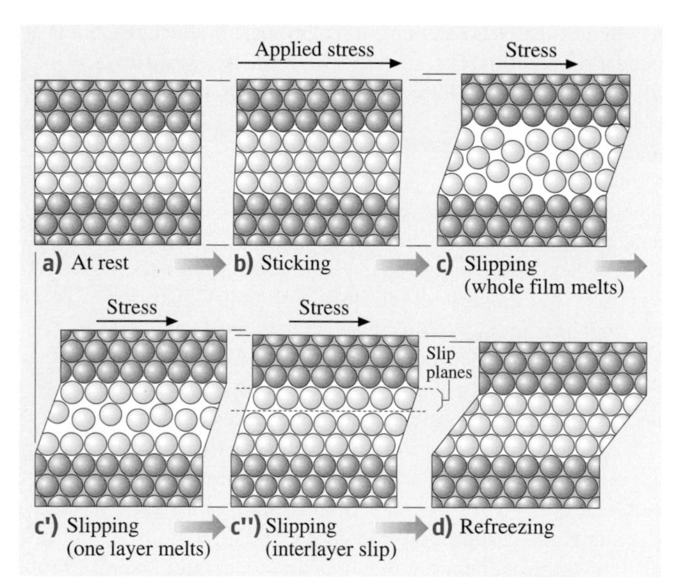

Fig. 18.26a–d Idealized schematic illustration of molecular rearrangements occurring in a molecularly thin film of spherical or simple chain molecules between two solid surfaces during shear. Depending on the system, a number of different molecular configurations within the film are possible during slipping and sliding, shown here as stages (c): total disorder as the whole film melts; (c′): partial disorder; and (c″): order persists even during sliding with slip occurring at a single slip plane either within the film or at the walls. A dilation is predicted in the direction normal to the surfaces. (After [18.224] with permission)

ally *decreases* with increasing sliding velocity, v (known as *shear thinning*). When two surfaces slide in the intermediate regime, the motion tends to thicken the film (dilatancy). This sends the system into the bulk EHL regime where, as indicated by (18.44), the friction force now *increases* with velocity. This initial decrease, followed by an increase, in the frictional forces of many lubricant systems is the basis for the empirical Stribeck curve of Fig. 18.13. In the transition from bulk to boundary behavior there is first a quantitative change in the material properties (viscosity and elasticity), which can be continuous, to discontinuous qualitative changes that result in yield stresses and non-liquidlike behavior.

The rest of this section is devoted to friction in the boundary regime. Boundary friction may be thought of as applying to the case where a lubricant film is present, but where this film is of molecular dimensions – a few molecular layers or less.

18.8.3 Boundary Lubrication of Molecularly Thin Films: Nanorheology

When a liquid is confined between two surfaces or within any narrow space whose dimensions are less than 4 to 10 molecular diameters, both the static (equilibrium) and dynamic properties of the liquid, such as its compressibility and viscosity, can no longer be described even qualitatively in terms of the bulk properties. The molecules confined within such molecularly thin films become ordered into layers ("out-of-plane" ordering), and within each layer they can also have lateral order ("in-plane" ordering). Such films may be thought of as behaving more like a liquid crystal or a solid than a liquid.

As described in Sect. 18.4.3, the measured *normal* forces between two solid surfaces across molecularly thin films exhibit exponentially decaying oscillations, varying between attraction and repulsion with a periodicity equal to some molecular dimension of the solvent molecules. Thus most liquid films can sustain a finite normal stress, and the adhesion force between two surfaces across such films is "quantized", depending on the thickness (or number of liquid layers) between the surfaces. The structuring of molecules in thin films and the oscillatory forces it gives rise to are now reasonably well understood, both experimentally and theoretically, at least for simple liquids.

Work has also recently been done on the dynamic, e.g., viscous or shear, forces associated with molecularly thin films. Both experiments [18.38, 46, 203, 221,

226,227,269,270] and theory [18.200,201,261,271] indicate that even when two surfaces are in steady state sliding they still prefer to remain in one of their stable potential energy minima, i. e., a sheared film of liquid can retain its basic layered structure. Thus even during motion the film does not become totally liquid-like. Indeed, if there is some "in-plane" ordering within a film, it will exhibit a yield-point before it begins to flow. Such films can therefore sustain a finite shear stress, in addition to a finite normal stress. The value of the yield stress depends on the number of layers comprising the film and represents another "quantized" property of molecularly thin films [18.200].

The dynamic properties of a liquid film undergoing shear are very complex. Depending on whether the film is more liquid-like or solid-like, the motion will be smooth or of the stick–slip type. During sliding, transitions can occur between n layers and $(n-1)$ or $(n+1)$ layers (see Fig. 18.27). The details of the motion depend critically on the externally applied load, the temperature, the sliding velocity, the twist angle between the two surface lattices, and the sliding direction relative to the lattices.

Smooth and Stick–Slip Sliding

Recent advances in friction-measuring techniques have enabled the interfacial friction of molecularly thin films to be measured with great accuracy. Some of these advances have involved the surface forces apparatus technique [18.38, 45, 47, 221, 247, 269, 270], while others have involved the atomic force microscope [18.10, 53, 231, 272]. In addition, computer simulations [18.90,200,201,271,273,274] have become sufficiently sophisticated to enable fairly complex tribological systems to be studied. All these advances are necessary if one is to probe such subtle effects as smooth or stick–slip friction, transient and memory effects, and ultralow friction mechanisms at the molecular level.

The theoretical models presented in this section will be concerned with a situation commonly observed experimentally: Stick–slip occurs between a static state with high friction and a low-friction kinetic state, and a transition from this sliding regime to smooth sliding can be induced by an increase in velocity. Experimental data on various systems showing this behavior are shown in Figs. 18.27, 18.30b, and 18.31a. Recent studies on adhesive systems have revealed the possibility of other dynamic responses such as "inverted" stick–slip between two kinetic states of higher and lower friction and with a transition from smooth sliding to stick–slip with increasing velocity, as shown in Fig. 18.25 [18.268].

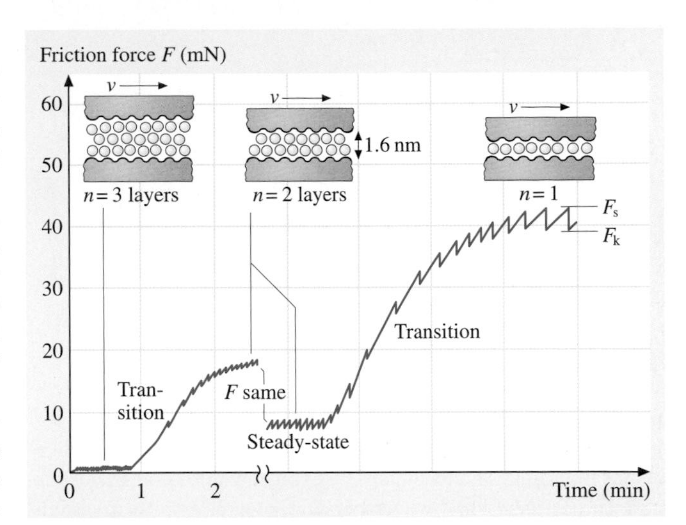

Fig. 18.27 Measured change in friction during interlayer transitions of the silicone liquid octamethylcyclotetrasiloxane (OMCTS), an inert liquid whose quasispherical molecules have a diameter of 0.8 nm. In this system, the shear stress $S_c = F/A$ was found to be constant as long as the number of layers, n, remained constant. Qualitatively similar results have been obtained with other quasi-spherical molecules such as cyclohexane [18.269]. The shear stresses are only weakly dependent on the sliding velocity v. However, for sliding velocities above some critical value, v_c, the stick–slip disappears and sliding proceeds smoothly at the kinetic value. (After [18.221] with permission)

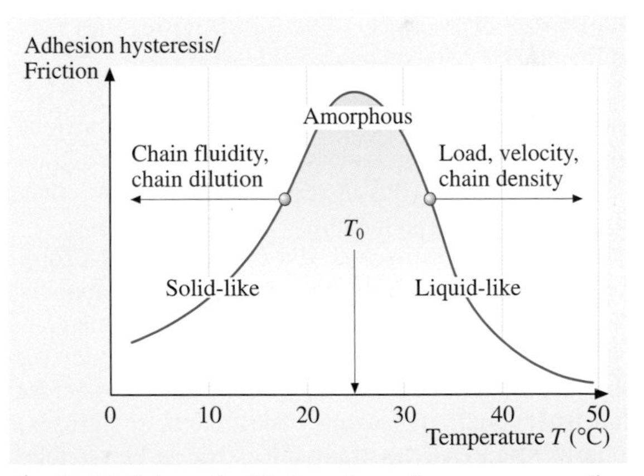

Fig. 18.28 Schematic friction phase diagram representing the trends observed in the boundary friction of a variety of different surfactant monolayers. The characteristic bell-shaped curve also correlates with the adhesion energy hysteresis of the monolayers. The *arrows* indicate the direction in which the whole curve is dragged when the load, velocity, etc., is increased. (After [18.232] with permission)

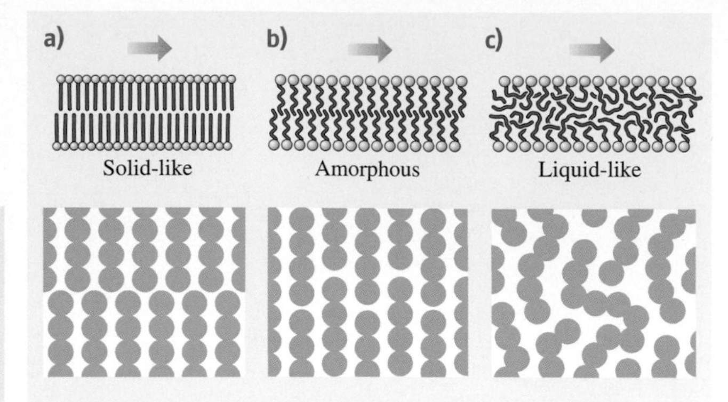

Fig. 18.29a–c Different dynamic phase states of boundary monolayers during adhesive contact and/or frictional sliding. Solid-like (a) and liquid-like monolayers (c) exhibit low adhesion hysteresis and friction. Increasing the temperature generally shifts a system from the left to the right. Changing the load, sliding velocity, or other experimental conditions can also change the dynamic phase state of surface layers, as shown in Fig. 18.28. (After [18.232] with permission)

With the added insights provided by computer simulations, a number of distinct molecular processes have been identified during smooth and stick–slip sliding in model systems for the more familiar static-to-kinetic stick–slip and transition from stick–slip to smooth sliding. These are shown schematically in Fig. 18.26 for the case of spherical liquid molecules between two solid crystalline surfaces. The following regimes may be identified:

Surfaces at rest (Fig. 18.26a): Even with no externally applied load, solvent–surface epitaxial interactions can cause the liquid molecules in the film to attain a solid-like state. Thus at rest the surfaces are stuck to each other through the film.

Sticking regime (frozen, solid-like film) (Fig. 18.26b): A progressively increasing lateral shear stress is applied. The solid-like film responds elastically with a small lateral displacement and a small increase or "dilatancy" in film thickness (less than a lattice spacing or molecular dimension, σ). In this regime the film retains its frozen, solid-like state: All the strains are elastic and reversible, and the surfaces remain effectively stuck to each other. However, slow creep may occur over long time periods.

Slipping and sliding regimes (molten, liquid-like film) (Fig. 18.26c,c′,c″): When the applied shear stress or force has reached a certain critical value, the *static* friction force, F_s, the film suddenly melts (known as "shear melting") or rearranges to allow for wall-slip or slip within the film to occur at which point the two surfaces begin to slip rapidly past each other. If the applied stress is kept at a high value, the upper surface will continue to slide indefinitely.

Refreezing regime (resolidification of film) (Fig. 18.26d): In many practical cases, the rapid slip of the upper surface relieves some of the applied force, which eventually falls below another critical value, the *kinetic* friction force F_k, at which point the film resolidifies and the whole stick–slip cycle is repeated. On the other hand, if the slip rate is smaller than the rate at which the external stress is applied, the surfaces will continue to slide smoothly in the kinetic state and there will be no more stick–slip. The critical velocity at which stick–slip disappears is discussed in more detail in Sect. 18.8.3.

Experiments with linear chain (alkane) molecules show that the film thickness remains quantized during sliding, so that the structure of such films is probably more like that of a nematic liquid crystal where the liquid molecules have become shear aligned in some direction, enabling shear motion to occur while retaining some order within the film. Experiments on the friction of two molecularly smooth mica surfaces separated by three molecular layers of the liquid OMCTS (Fig. 18.27) show how the friction increases to higher values in a quantized way when the number of layers falls from $n = 3$ to $n = 2$ and then to $n = 1$.

Computer simulations for simple spherical molecules [18.201] further indicate that during the slip the film thickness is roughly 15% higher than at rest (i. e., the film density falls), and that the order parameter within the film drops from 0.85 to about 0.25. Such dilatancy has been investigated both experimentally [18.235] and in further computer simulations [18.234]. The changes in thickness and in order parameter are consistent with a disorganized state for the whole film during the slip [18.271], as illustrated schematically in Fig. 18.26c. At this stage, we can only speculate on other possible configurations of molecules in the slipping and sliding regimes. This probably depends on the shapes of the molecules (e.g., whether spherical or linear or branched), on the atomic structure of the surfaces, on the sliding velocity, etc. [18.275]. Figure 18.26c,c′,c″ shows three possible sliding modes wherein the shearing film either totally melts, or where the molecules retain their layered structure and where slip occurs between two or more layers. Other sliding modes, for example, involving the movement of dislocations or disclinations are also possible, and it is unlikely that one single mechanism applies in all cases.

Both friction and adhesion hysteresis vary nonlinearly with temperature, often peaking at some

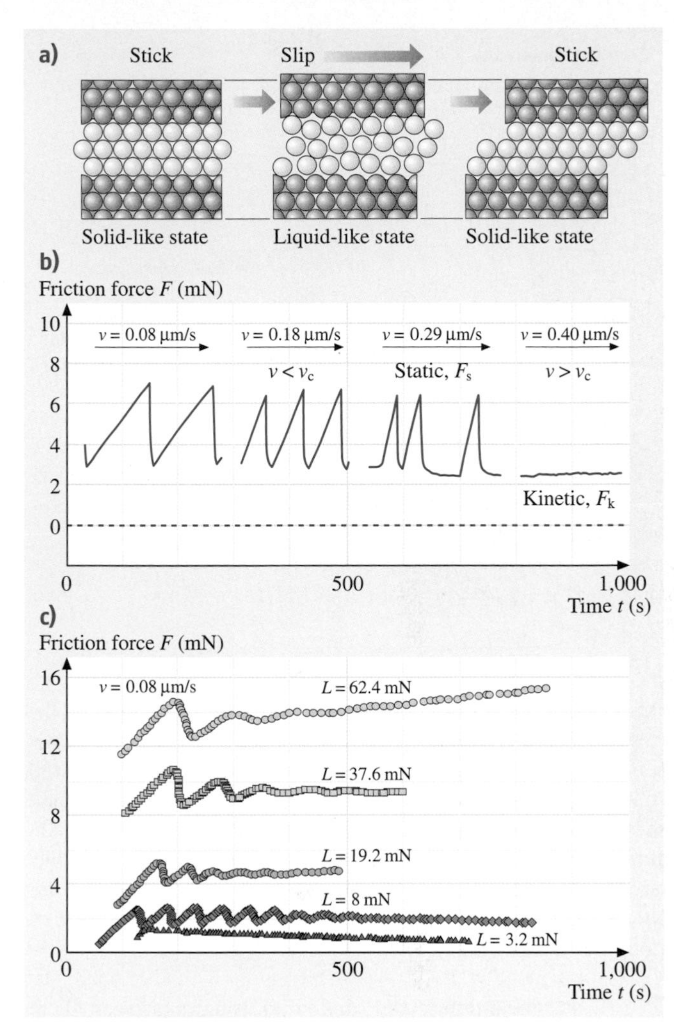

Fig. 18.30 (**a**) "Phase transitions" model of stick–slip where a thin liquid film alternately freezes and melts as it shears, shown here for 22 spherical molecules confined between two solid crystalline surfaces. In contrast to the velocity-dependent friction model, the intrinsic friction force is assumed to change abruptly (at the transitions), rather than smooth or continuously. The resulting stick–slip is also different, for example, the peaks are sharper and the stick–slip disappears above some critical velocity v_c. Note that while the slip displacement is here shown to be only two lattice spacings, in most practical situations it is much larger, and that freezing and melting transitions at surfaces or in thin films may not be the same as freezing or melting transitions between the bulk solid and liquid phases. (**b**) Exact reproduction of a chart-recorder trace of the friction force for hexadecane between two untreated mica surfaces at increasing sliding velocity v, plotted as a function of time. In general, with increasing sliding speed, the stick–slip spikes increase in frequency and decrease in magnitude. As the critical sliding velocity v_c is approached, the spikes become erratic, eventually disappearing altogether at $v = v_c$. At higher sliding velocities the sliding continues smoothly in the kinetic state. Such friction traces are fairly typical for simple liquid lubricants and dry boundary lubricant systems (see Fig. 18.31a) and may be referred to as the "conventional" type of static–kinetic friction (in contrast to Fig. 18.25). Experimental conditions: contact area $A = 4\times10^{-9}\,\mathrm{m}^2$, load $L = 10\,\mathrm{mN}$, film thickness $D = 0.4{-}0.8\,\mathrm{nm}$, $v = 0.08{-}0.4\,\mu\mathrm{m\,s}^{-1}$, $v_c \approx 0.3\,\mu\mathrm{m\,s}^{-1}$, atmosphere: dry N_2 gas, $T = 18\,^\circ\mathrm{C}$. [(**a**) and (**b**) after [18.276] with permission. Copyright 1993 American Chemical Society.] (**c**) Friction response of a thin squalane (a branched hydrocarbon) film at different loads and a constant sliding velocity $v = 0.08\,\mu\mathrm{m\,s}^{-1}$, slightly above the critical velocity for this system at low loads. Initally, with increasing load, the stick–slip amplitude and the mean friction force decrease with sliding time or sliding distance. However, at high loads or pressures, the mean friction force increases with time, and the stick–slip takes on a more symmetrical, sinusoidal shape. At all loads investigated, the stick–slip component gradually decayed as the friction proceeded towards smooth sliding. (*Gourdon* and *Israelachvili*, unpublished data)

particular temperature, T_0. The temperature dependence of these forces can, therefore, be represented on a friction phase diagram such as the one shown in Fig. 18.28. Experiments have shown that T_0, and the whole bell-shaped curve, are shifted along the temperature-axis (as well as in the vertical direction) in a systematic way when the load, sliding velocity, etc., are varied. These shifts also appear to be highly correlated with one another, for example, an increase in temperature producing effects that are similar to *decreasing* the sliding speed or load.

Such effects are also commonly observed in other energy dissipating phenomena such as polymer viscoelasticity [18.277], and it is likely that a similar physical mechanism is at the heart all of such phenomena. A possible molecular process underlying the energy dissipation of chain molecules during boundary layer sliding is illustrated in Fig. 18.29, which shows the three main dynamic phase states of boundary monolayers.

In contrast to the characteristic relaxation time associated with fluid lubricants, it has been established

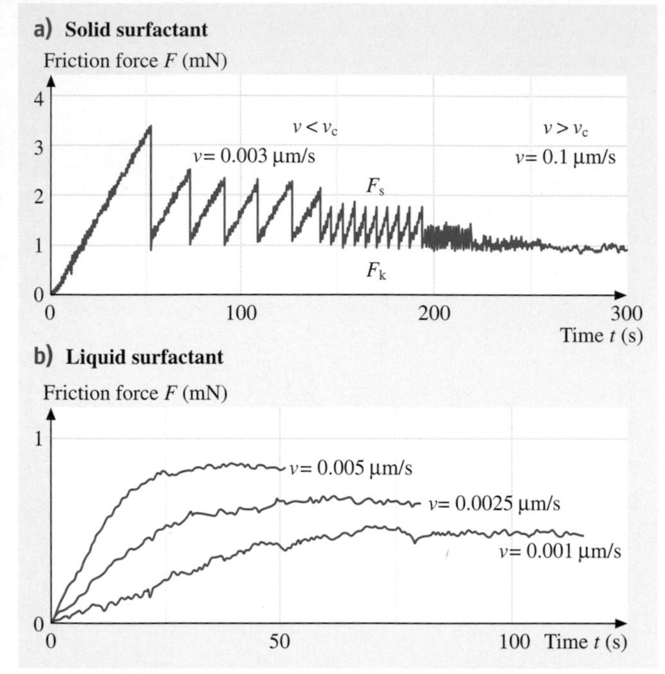

Fig. 18.31 (a) Exact reproduction of chart-recorder trace for the friction of closely packed surfactant monolayers (L-α-dimirystoyl-phosphatidyl-ethanolamine, DMPE) on mica (dry boundary friction) showing qualitatively similar behavior to that obtained with a liquid hexadecane film (Fig. 18.30b). In this case, $L = 0$, $v_c \approx 0.1\,\mu\mathrm{m\,s^{-1}}$, atmosphere: dry N_2 gas, $T = 25\,°\mathrm{C}$. (b) Sliding typical of liquid-like monolayers, here shown for calcium alkylbenzene sulfonate in dry N_2 gas at $T = 25\,°\mathrm{C}$ and $L = 0$. (After [18.207] with permission. Copyright 1993 American Chemical Society)

that for unlubricated (dry, solid, rough) surfaces, there is a characteristic memory distance that must be exceeded before the system looses all memory of its initial state (original surface topography). The underlying mechanism for a characteristic distance was first used to successfully explain rock mechanics and earthquake faults [18.278] and, more recently, the tribological behavior of unlubricated surfaces of ceramics, paper and elastomeric polymers [18.252, 279]. Recent experiments [18.275, 280, 281] suggest that fluid lubricants composed of complex branched-chained or polymer molecules may also have characteristic distances (in addition to characteristic relaxation times) associated with their tribological behavior – the characteristic distance being the total sliding distance that must be exceeded before the system reaches its steady-state tribological conditions (see Sect. 18.8.3).

Abrupt vs. Continuous Transitions Between Smooth and Stick–Slip Sliding

An understanding of stick–slip is of great practical importance in tribology [18.282], since these spikes are the major cause of damage and wear of moving parts. Stick–slip motion is a very common phenomenon and is also the cause of sound generation (the sound of a violin string, a squeaking door, or the chatter of machinery), sensory perception (taste texture and feel), earthquakes, granular flow, nonuniform fluid flow such as the spurting flow of polymeric liquids, etc. In the previous section, the stick–slip motion arising from freezing–melting transitions in thin interfacial films was described. There are other mechanisms that can give rise to stick–slip friction, which will now be considered. However, before proceeding with this, it is important to clarify exactly what one is measuring during a friction experiment.

Most tribological systems and experiments can be described in terms of an equivalent mechanical circuit with certain characteristics. The friction force F_0, which is generated at the surfaces, is generally measured as F at some other place in the setup. The mechanical coupling between the two may be described in terms of a simple elastic stiffness or compliance, K, or as more complex nonelastic coefficients, depending on the system. The distinction between F and F_0 is important because in almost all practical cases, the applied, measured, or detected force, F, is *not* the same as the "real" or "intrinsic" friction force, F_0, generated at the surfaces. F and F_0 are coupled in a way that depends on the mechanical construction of the system, for example, the axle of a car wheel that connects it to the engine. This coupling can be modeled as an elastic spring of stiffness K and mass m. This is the simplest type of mechanical coupling and is also the same as in SFA and AFM type experiments. More complicated real systems can be reduced to a system of springs and dashpots, as described by *Peachey* et al. [18.283] and *Luengo* et al. [18.47].

We now consider four different models of stick–slip friction, where the mechanical couplings are assumed to be of the simple elastic spring type. The first three mechanisms may be considered traditional or classical mechanisms or models [18.282], the fourth is essentially the same as the freezing–melting phase-transition model described in Sect. 18.8.3.

Rough Surfaces or Surface Topology Model. Rapid slips can occur whenever an asperity on one surface goes over the top of an asperity on the other surface. The extent of the slip will depend on asperity heights and slopes, on the speed of sliding, and on the elastic compliance of the surfaces and the moving stage. As in all cases of stick–slip motion, the driving velocity v may be constant, but the resulting motion at the surfaces v_0 will display large slips. This type of stick–slip has been described by *Rabinowicz* [18.282]. It will not be of much concern here since it is essentially a noise-type fluctuation, resulting from surface imperfections rather than from the intrinsic interaction between two surfaces. Actually, at the atomic level, the regular atomic-scale corrugations of surfaces can lead to periodic stick–slip motion of the type shown here. This is what is sometimes measured by AFM tips [18.10, 53, 231, 272].

Distance-Dependent or Creep Model. Another theory of stick–slip, observed in solid-on-solid sliding, is one that involves a characteristic *distance*, but also a characteristic time, τ_s, this being the characteristic time required for two asperities to increase their adhesion strength after coming into contact. Originally proposed by *Rabinowicz* [18.282, 284], this model suggests that two rough macroscopic surfaces adhere through their microscopic asperities of characteristic length. During shearing, each surface must first creep this distance (the size of the contacting junctions) after which the surfaces continue to slide, but with a lower (kinetic) friction force than the original (static) value. The reason for the decrease in the friction force is that even though, on average, new asperity junctions should form as rapidly as the old ones break, the time-dependent adhesion and friction of the new ones will be lower than the old ones.

The friction force, therefore, remains high during the creep stage of the slip. But once the surfaces have moved the characteristic distance, the friction rapidly drops to the kinetic value. For any system where the kinetic friction is less than the static force (or one that has a negative slope over some part of its F_0 versus v_0 curve) will exhibit regular stick–slip sliding motion for certain values of K, m, and driving velocity, v.

This type of friction has been observed in a variety of dry (unlubricated) systems such as paper-on-paper [18.285, 286] and steel-on-steel [18.284, 287, 288]. This model is also used extensively in geologic systems to analyze rock-on-rock sliding [18.289, 290]. While originally described for adhering macroscopic asperity junctions, the distance-dependent model may also apply to molecularly smooth surfaces. For example, for polymer lubricant films, the characteristic length would now be the chain–chain entanglement length, which could be much larger in a confined geometry than in the bulk.

Velocity-Dependent Friction Model. In contrast to the two friction models mentioned above, which apply mainly to unlubricated, solid-on-solid contacts, the stick–slip of surfaces with thin liquid films between them is better described by other mechanisms. The velocity-dependent friction model is the most studied mechanism of stick–slip and, until recently, was considered to be the only cause of intrinsic stick–slip. If a friction force decreases with increasing sliding velocity, as occurs with boundary films exhibiting shear thinning, the force (F_s) needed to initiate motion will be higher than the force (F_k) needed to maintain motion.

A decreasing intrinsic friction force F_0 with sliding velocity v_0 results in the sliding surface or stage moving in a periodic fashion, where during each cycle rapid acceleration is followed by rapid deceleration. So long as the drive continues to move at a fixed velocity v, the surfaces will continue to move in a periodic fashion punctuated by abrupt stops and starts whose frequency and amplitude depend not only on the function $F_0(v_0)$, but also on the stiffness K and mass m of the moving stage, and on the starting conditions at $t = 0$.

More precisely, the motion of the sliding surface or stage can be determined by solving the following differential equation:

$$m\ddot{x} = (F_0 - F) = F_0 - (x_0 - x)K$$
$$\text{or} \quad m\ddot{x} + (x_0 - x)K - F_0 = 0\,, \tag{18.47}$$

where $F_0 = F_0(x_0, v_0, t)$ is the intrinsic or "real" friction force at the shearing surfaces, F is the force on the spring (the externally applied or measured force), and $(F_0 - F)$ is the force on the stage. To fully solve (18.47), one must also know the initial (starting) conditions at $t = 0$, and the driving or steady-state conditions at finite t. Commonly, the driving condition is: $x = 0$ for $t < 0$ and $x = vt$ for $t > 0$, where $v =$ constant. In other systems, the appropriate driving condition may be $F =$ constant.

Various, mainly phenomenological, forms for $F_0 = F_0(x_0, v_0, t)$ have been proposed to explain various kinds of stick–slip phenomena. These models generally assume a particular functional form for the friction as a function of velocity only, $F_0 = F_0(v_0)$, and they may also contain a number of mechanically coupled elements comprising the stage [18.291, 292]. One version is a two-state model characterized by two friction forces, F_s and F_k, which is a simplified version of the phase transitions model described below. More complicated versions can have a rich F–v spectrum, as proposed by *Persson* [18.293]. Unless the experimental data is very detailed and extensive, these models cannot generally distinguish between different types of mechanisms.

Neither do they address the basic question of the *origin* of the friction force, since this is assumed to begin with.

Experimental data has been used to calculate the friction force as a function of velocity *within* an individual stick–slip cycle [18.294]. For a macroscopic granular material confined between solid surfaces, the data shows a velocity-weakening friction force during the first half of the slip. However, the data also shows a hysteresis loop in the friction–velocity plot, with a different behavior in the deceleration half of the slip phase. Similar results were observed for a 1–2 nm liquid lubricant film between mica surfaces [18.275]. These results indicate that a purely velocity-dependent friction law is insufficient to describe such systems, and an additional element such as the *state* of the confined material must be considered.

Phase Transitions Model. In recent molecular dynamics computer simulations it has been found that thin interfacial films undergo first-order phase transitions between solid-like and liquid-like states during sliding [18.201, 273]. It has been suggested that this is responsible for the observed stick–slip behavior of simple isotropic liquids between two solid crystalline surfaces. With this interpretation, stick–slip is seen to arise because of the abrupt change in the flow properties of a film at a transition [18.224,225,261], rather than the gradual or continuous change, as occurs in the previous example. Other computer simulations indicate that it is the stick–slip that induces a disorder ("shear-melting") in the film, not the other way around [18.295].

An interpretation of the well-known phenomenon of decreasing coefficient of friction with increasing sliding velocity has been proposed by *Thompson* and *Robbins* [18.201] based on their computer simulation. This postulates that it is not the friction that changes with sliding speed v, but rather the time various parts of the system spend in the sticking and sliding modes. In other words, at any instant during sliding, the friction at any local region is always F_s or F_k, corresponding to the "static" or "kinetic" values. The measured frictional force, however, is the sum of all these discrete values averaged over the whole contact area. Since as v increases each local region spends more time in the sliding regime (F_k) and less in the sticking regime (F_s), the overall friction coefficient falls. One may note that this interpretation reverses the traditional way that stick–slip has been explained. Rather than invoking a decreasing friction with velocity to explain stick–slip, it is now the more fundamental stick–slip phenomenon that is producing the apparent decrease in the friction force with increasing sliding velocity. This approach has been studied analytically by *Carlson* and *Batista* [18.296], with a comprehensive rate- and state- dependent friction force law. This model includes an analytic description of the freezing–melting transitions of a film, resulting in a friction force that is a function of sliding velocity in a natural way. This model predicts a full range of stick–slip behavior observed experimentally.

An example of the rate- and state-dependent model is observed when shearing thin films of OMCTS between mica surfaces [18.297, 298]. In this case, the static friction between the surfaces is dependent on the time that the surfaces are at rest with respect to each other, while the intrinsic kinetic friction $F_{k,0}$ is relatively constant over the range of velocities. At slow driving velocities, the system responds with stick–slip sliding with the surfaces reaching maximum static friction before each slip event, and the amplitude of the stick–slip, $F_s - F_k$, is relatively constant. As the driving velocity increases, the static friction decreases as the time at relative rest becomes shorter with respect to the characteristic time of the lubricant film. As the static friction decreases with increasing drive velocity, it eventually equals the intrinsic kinetic friction $F_{k,0}$, which defines the critical velocity v_c, above which the surfaces slide smoothly without the jerky stick–slip motion.

The above classifications of stick–slip are not exclusive, and molecular mechanisms of real systems may exhibit aspects of different models simultaneously. They do, however, provide a convenient classification of existing models and indicate which experimental parameters should be varied to test the different models.

Critical Velocity for Stick–Slip. For any given set of conditions, stick–slip disappears above some critical sliding velocity v_c, above which the motion continues smoothly in the liquid-like or kinetic state. The critical velocity is well described by two simple equations. Both are based on the phase transition model, and both include some parameter associated with the inertia of the measuring instrument. The first equation is based on both experiments and simple theoretical modeling [18.276]:

$$v_c \approx \frac{(F_s - F_k)}{5K\tau_0}, \qquad (18.48)$$

where τ_0 is the *characteristic nucleation time* or freezing time of the film. For example, inserting the following typically measured values for a ~1-nm-thick hexadecane film between mica: $(F_s - F_k) \approx 5\,\mathrm{mN}$, spring constant $K \approx 500\,\mathrm{N\,m^{-1}}$, and nucleation time [18.276]

$\tau_0 \approx 5$ s, we obtain $v_c \approx 0.4\,\mu\mathrm{m\,s^{-1}}$, which is close to typically measured values (Fig. 18.30b).

The second equation is based on computer simulations [18.273]:

$$v_c \approx 0.1\sqrt{\frac{F_s \sigma}{m}}\,, \qquad (18.49)$$

where σ is a molecular dimension and m is the mass of the stage. Again, inserting typical experimental values into this equation, viz., $m \approx 20$ g, $\sigma \approx 0.5$ nm, and $(F_s - F_k) \approx 5$ mN as before, we obtain $v_c \approx 0.3\,\mu\mathrm{m\,s^{-1}}$, which is also close to measured values.

Stick–slip also disappears above some critical temperature T_c, which is not the same as the melting temperature of the bulk fluid. Certain correlations have been found between v_c and T_c and between various other tribological parameters that appear to be consistent with the principle of time–temperature superposition (see Sect. 18.8.3), similar to that occurring in viscoelastic polymer fluids [18.277, 299, 300].

Recent work on the coupling between the mechanical resonances of the sliding system and molecular-scale relaxations [18.234, 301–303] has resulted in a better understanding of a phenomenon previously noted in various engineering applications: the vibrating of one of the sliding surfaces perpendicularly to the sliding direction can lead to a significant reduction of the friction. At certain oscillation amplitudes and a frequency higher than the molecular-scale relaxation frequency, stick–slip friction can be eliminated and replaced by an ultralow kinetic friction state.

18.9 Role of Molecular Shape and Surface Structure in Friction

The above scenario is already quite complicated, and yet this is the situation for the simplest type of experimental system. The factors that appear to determine the critical velocity v_c depend on the type of liquid between the surfaces (as well as on the surface lattice structure). Small spherical molecules such as cyclohexane and OMCTS have been found to have very high v_c, which indicates that these molecules can rearrange relatively quickly in thin films. Chain molecules and especially branched chain molecules have been found to have much lower v_c, which is to be expected, and such liquids tend to slide smoothly or with erratic stick–slip [18.275], rather than in a stick–slip fashion (see Table 18.5). With highly asymmetric molecules, such as multiply branched isoparaffins and polymer melts, no regular spikes or stick–slip behavior occurs at any speed, since these molecules can never order themselves sufficiently to solidify. Examples of such liquids are some perfluoropolyethers and polydimethylsiloxanes (PDMS).

Recent computer simulations [18.274, 304, 305] of the structure, interaction forces, and tribological behavior of chain molecules between two shearing surfaces indicate that both linear *and* singly or doubly branched chain molecules order between two flat surfaces by aligning into discrete layers parallel to the surfaces. However, in the case of the weakly branched molecules, the expected oscillatory forces do not appear because of a complex cancellation of entropic and enthalpic contributions to the interaction free energy, which results in a monotonically smooth interaction, exhibiting a weak energy minimum rather than the oscillatory force profile that is characteristic of linear molecules. During sliding, however, these molecules can be induced to further align, which can result in a transition from smooth to stick–slip sliding.

Table 18.5 shows the trends observed with some organic and polymeric liquid between smooth mica surfaces. Also listed are the bulk viscosities of the liquids. From the data of Table 18.5 it appears that there is a direct correlation between the shapes of molecules and their coefficient of friction or effectiveness as lubricants (at least at low shear rates). Small spherical or chain molecules have high friction with stick–slip, because they can pack into ordered solid-like layers. In contrast, longer chained and irregularly shaped molecules remain in an entangled, disordered, fluid-like state even in very thin films, and these give low friction and smoother sliding. It is probably for this reason that irregularly shaped branched chain molecules are usually better lubricants. It is interesting to note that the friction coefficient generally decreases as the bulk viscosity of the liquids *increases*. This unexpected trend occurs because the factors that are conducive to low friction are generally conducive to high viscosity. Thus molecules with side groups such as branched alkanes and polymer melts usually have higher bulk viscosities than their linear homologues for obvious reasons. However, in thin films the linear molecules have higher shear stresses, because of their ability to become ordered. The only exception to the

Table 18.5 Effect of molecular shape and short-range forces on tribological properties[a]

Liquid	Short-range force	Type of friction	Friction coefficient	Bulk liquid viscosity (cP)
Organic (water-free)				
Cyclohexane	Oscillatory	Quantized stick–slip	$\gg 1$	0.6
OMCTS[b]	Oscillatory	Quantized stick–slip	$\gg 1$	2.3
Octane	Oscillatory	Quantized stick–slip	1.5	0.5
Tetradecane	Oscillatory $\leftrightarrow$ smooth	stick–slip $\leftrightarrow$ smooth	1.0	2.3
Octadecane (branched)	Oscillatory $\leftrightarrow$ smooth	stick–slip $\leftrightarrow$ smooth	0.3	5.5
PDMS[b] ($M = 3{,}700\,\mathrm{g\,mol^{-1}}$, melt)	Oscillatory $\leftrightarrow$ smooth	Smooth	0.4	50
PBD[b] ($M = 3{,}500\,\mathrm{g\,mol^{-1}}$, branched)	Smooth	Smooth	0.03	800
Water				
Water (KCl solution)	Smooth	Smooth	0.01–0.03	1.0

[a] For molecularly thin liquid films between two shearing mica surfaces at 20 °C
[b] OMCTS: Octamethylcyclotetrasiloxane, PDMS: Polydimethylsiloxane, PBD: Polybutadiene

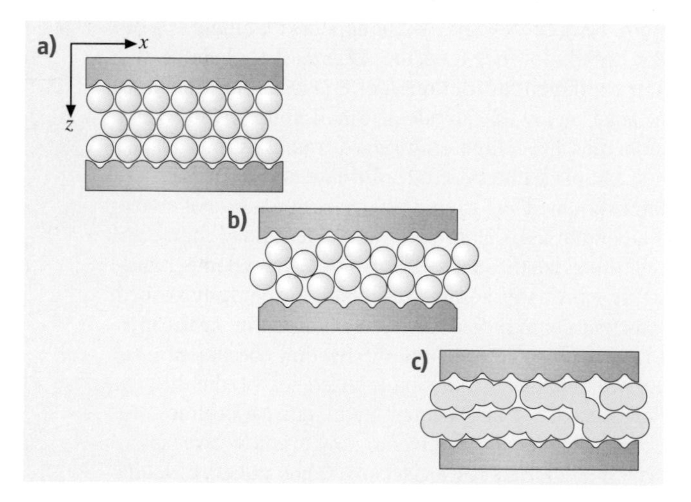

Fig. 18.32a–c Schematic view of interfacial film composed of spherical molecules under a compressive pressure between two solid crystalline surfaces. (**a**) If the two surface lattices are free to move in the x-y-z directions, so as to attain the lowest energy state, they could equilibrate at values of x, y, and z, which induce the trapped molecules to become "epitaxially" ordered into a "solid-like" film. (**b**) Similar view of trapped molecules between two solid surfaces that are not free to adjust their positions, for example, as occurs in capillary pores or in brittle cracks. (**c**) Similar to (**a**), but with chain molecules replacing the spherical molecules in the gap. These may not be able to order as easily as do spherical molecules even if x, y, and z can adjust, resulting in a situation that is more akin to (**b**). (After [18.276] with permission. Copyright 1993 American Chemical Society)

above correlations is water, which has been found to exhibit both low viscosity *and* low friction (see Fig. 18.20, and Sect. 18.7.3). In addition, the presence of water can drastically lower the friction and eliminate the stick–slip of hydrocarbon liquids when the sliding surfaces are hydrophilic.

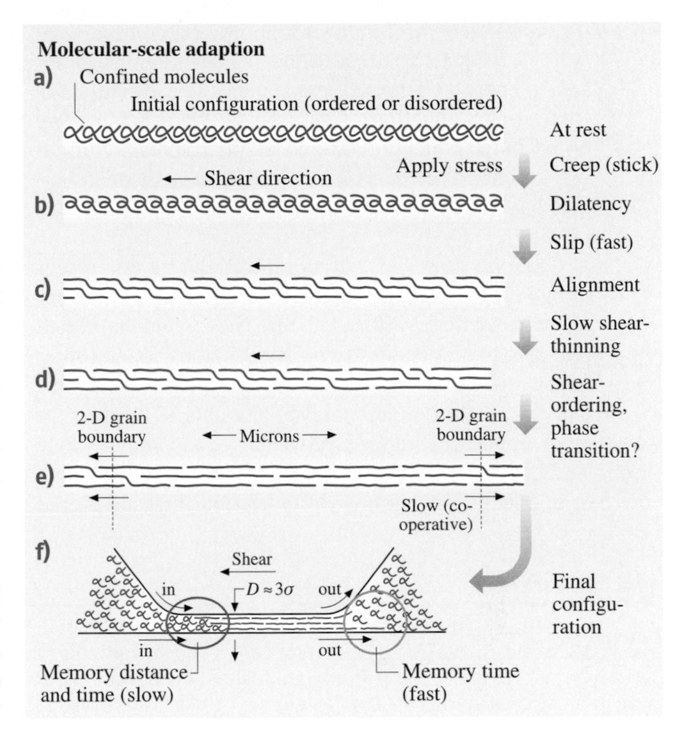

Fig. 18.33a–f Schematic representation of the film under shear. (**a**) The lubricant molecules are just confined, but not oriented in any particular direction. Because of the need to shear, the film dilates (**b**). The molecules disentangle (**c**) and get oriented in a certain direction related to the shear direction (**d**). (**e**) Slowly evolving domains grow inside the contact region. These macroscopic domains are responsible for the long relaxation times. (**f**) At the steady-state, a continuous gradient of confinement time and molecular order is established in the contact region, which is different for molecules adsorbed on the upper and lower surfaces. Molecules entering into the contact are not oriented or ordered. The required sliding distance to modify their state defines a characteristic distance. Molecules leaving the contact region need some (short) characteristic time to regain their bulk, unconfined configuration. (After [18.281] with permission. Copyright 2000 American Chemical Society)

If an "effective" viscosity, η_{eff}, were to be calculated for the liquids of Table 18.5, the values would be many orders of magnitude higher than those of the bulk liquids. This can be demonstrated by the following simple calculation based on the usual equation for Couette flow (see (18.44)):

$$\eta_{\mathrm{eff}} = F_{\mathrm{k}} D / A v \,, \tag{18.50}$$

where F_{k} is the kinetic friction force, D is the film thickness, A the contact area, and v the sliding velocity. Using typical values for experiments with hexadecane [18.276]: $F_{\mathrm{k}} = 5\,\mathrm{mN}$, $D = 1\,\mathrm{nm}$, $A = 3\times10^{-9}\,\mathrm{m}^2$, and $v = 1\,\mu\mathrm{m\,s}^{-1}$, one gets $\eta_{\mathrm{eff}} \approx 2{,}000\,\mathrm{Ns\,m}^{-2}$, or 20,000 Poise, which is $\sim 10^6$ times higher than the bulk viscosity, η_{bulk}, of the liquid. It is instructive to consider that this very high effective viscosity nevertheless still produces a low friction force or friction coefficient μ of about 0.25. It is interesting to speculate that if a 1 nm film were to exhibit bulk viscous behavior, the friction coefficient under the same sliding conditions would be as low as 0.000001. While such a low value has never been reported for any tribological system, one may consider it a theoretical lower limit that could conceivably be attained under certain experimental conditions.

Various studies [18.44, 219, 220, 222] have shown that confinement and load generally increase the effective viscosity and/or relaxation times of molecules, suggestive of an increased glassiness or solid-like behavior (Figs. 18.32 and 18.33). This is in marked contrast to studies of liquids in small confining capillaries where the opposite effects have been observed [18.306, 307]. The reason for this is probably because the two modes of confinement are different. In the former case (confinement of molecules between two structured solid surfaces), there is generally little opposition to any lateral or vertical displacement of the two surface lattices relative to each other. This means that the two lattices can shift in the x-y-z plane (Fig. 18.32a) to accommodate the trapped molecules in the most crystallographically commensurate or epitaxial way, which would favor an ordered, solid-like state. In contrast, the walls of capillaries are rigid and cannot easily move or adjust to accommodate the confined molecules (Fig. 18.32b), which will therefore be forced into a more disordered, liquid-like state (unless the capillary wall geometry and lattice are *exactly* commensurate with the liquid molecules, as occurs in certain zeolites [18.307]).

Experiments have demonstrated the effects of surface lattice mismatch on the friction between surfaces [18.203, 204, 308]. Similar to the effects of lattice mismatch on adhesion (Fig. 18.11), the static friction of a confined liquid film is maximum when the lattices of the confining surfaces are aligned. For OMCTS confined between mica surfaces [18.204] the static friction was found to vary by more than a factor of 4, while for bare mica surfaces the variation was by a factor of 3.5 [18.308]. In contrast to the sharp variations in adhesion energy over small twist angles, the variations in friction as a function of twist angle were much more broad both in magnitude and angular spread. Similar variations in friction as a function of twist or misfit angles have also been observed in computer simulations [18.309].

Robbins et al. [18.274] computed the friction forces of two clean crystalline surfaces as a function of the angle between their surface lattices. They found that for all non-zero angles (finite "twist" angles) the friction forces fell to zero due to incommensurability effects. They further found that submonolayer amounts of organic or other impurities trapped between two incommensurate surfaces can generate a finite friction force. They, therefore, concluded that any finite friction force measured between incommensurate surfaces is probably due to such "third body" effects.

With rough surfaces, i. e., those that have *random* protrusions rather than being periodically structured, we expect a smearing out of the correlated intermolecular interactions that are involved in film freezing and melting (and in phase transitions in general). This should effectively eliminate the highly regular stick–slip and may also affect the location of the slipping planes [18.123, 252, 279]. The stick–slip friction of "real" surfaces, which are generally rough, may, therefore, be quite different from those of perfectly smooth surfaces composed of the same material. We should note, however, that even between rough surfaces, most of the contacts occur between the tips of microscopic asperities, which may be smooth over their microscopic contact area.

References

18.1 J. N. Israelachvili: Surface Forces and Microrheology of Molecularly Thin Liquid Films. In: *Handbook of Micro/Nanotribology*, ed. by B. Bhushan (CRC Press, Boca Raton 1995) pp. 267–319

18.2 K. B. Lodge: Techniques for the measurement of forces between solids, Adv. Colloid Interface Sci. **19** (1983) 27–73

18.3 J. N. Israelachvili: *Intermolecular and Surface Forces*, 2nd edn. (Academic Press, London 1991)

18.4 P. F. Luckham, B. A. de L. Costello: Recent developments in the measurement of interparticle forces, Adv. Colloid Interface Sci. **44** (1993) 183–240

18.5 P. M. Claesson, T. Ederth, V. Bergeron, M. W. Rutland: Techniques for measuring surface forces, Adv. Colloid Interface Sci. **67** (1996) 119–183

18.6 V. S. J. Craig: An historical review of surface force measurement techniques, Colloids Surf. A **129–130** (1997) 75–94

18.7 J. N. Israelachvili, G. E. Adams: Measurements of forces between two mica surfaces in aqueous electrolyte solutions in the range 0–100 nm, J. Chem. Soc. Faraday Trans. I **74** (1978) 975–1001

18.8 G. Binnig, C. F. Quate, C. Gerber: Atomic force microscope, Phys. Rev. Lett. **56** (1986) 930–933

18.9 W. A. Ducker, T. J. Senden, R. M. Pashley: Direct measurement of colloidal forces using an atomic force microscope, Nature **353** (1991) 239–241

18.10 E. Meyer, R. M. Overney, K. Dransfeld, T. Gyalog: *Nanoscience: Friction and Rheology on the Nanometer Scale* (World Scientific, Singapore 1998)

18.11 J. N. Israelachvili: Measurements of the viscosity of thin fluid films between two surfaces with and without adsorbed polymers, Colloid Polym. Sci. **264** (1986) 1060–1065

18.12 J. P. Montfort, G. Hadziioannou: Equilibrium and dynamic behavior of thin films of a perfluorinated polyether, J. Chem. Phys. **88** (1988) 7187–7196

18.13 A. Dhinojwala, S. Granick: Surface forces in the tapping mode: Solvent permeability and hydrodynamic thickness of adsorbed polymer brushes, Macromolecules **30** (1997) 1079–1085

18.14 B. V. Derjaguin: Untersuchungen über die Reibung und Adhäsion, IV. Theorie des Anhaftens kleiner Teilchen, Kolloid Z. **69** (1934) 155–164

18.15 K. L. Johnson, K. Kendall, A. D. Roberts: Surface energy and the contact of elastic solids, Proc. R. Soc. London A **324** (1971) 301–313

18.16 J. N. Israelachvili, P. M. McGuiggan: Adhesion and short-range forces between surfaces. Part 1: New apparatus for surface force measurements, J. Mater. Res. **5** (1990) 2223–2231

18.17 J. N. Israelachvili: Thin film studies using multiple-beam interferometry, J. Colloid Interface Sci. **44** (1973) 259–272

18.18 Y. L. Chen, T. Kuhl, J. Israelachvili: Mechanism of cavitation damage in thin liquid films: Collapse damage vs. inception damage, Wear **153** (1992) 31–51

18.19 M. Heuberger, G. Luengo, J. Israelachvili: Topographic information from multiple beam interferometry in the surface forces apparatus, Langmuir **13** (1997) 3839–3848

18.20 R. M. Pashley: DLVO and hydration forces between mica surfaces in Li^+, Na^+, K^+, and Cs^+ electrolyte solutions: A correlation of double-layer and hydration forces with surface cation exchange properties, J. Colloid Interface Sci. **83** (1981) 531–546

18.21 R. M. Pashley: Hydration forces between mica surfaces in electrolyte solution, Adv. Colloid Interface Sci. **16** (1982) 57–62

18.22 R. G. Horn, D. T. Smith, W. Haller: Surface forces and viscosity of water measured between silica sheets, Chem. Phys. Lett. **162** (1989) 404–408

18.23 R. G. Horn, D. R. Clarke, M. T. Clarkson: Direct measurements of surface forces between sapphire crystals in aqueous solutions, J. Mater. Res. **3** (1988) 413–416

18.24 W. W. Merrill, A. V. Pocius, B. V. Thakker, M. Tirrell: Direct measurement of molecular level adhesion forces between biaxially oriented solid polymer films, Langmuir **7** (1991) 1975–1980

18.25 J. Klein: Forces between mica surfaces bearing adsorbed macromolecules in liquid media, J. Chem. Soc. Faraday Trans. I **79** (1983) 99–118

18.26 S. S. Patel, M. Tirrell: Measurement of forces between surfaces in polymer fluids, Annu. Rev. Phys. Chem. **40** (1989) 597–635

18.27 H. Watanabe, M. Tirrell: Measurements of forces in symmetric and asymmetric interactions between diblock copolymer layers adsorbed on mica, Macromolecules **26** (1993) 6455–6466

18.28 T. L. Kuhl, D. E. Leckband, D. D. Lasic, J. N. Israelachvili: Modulation and modeling of interaction forces between lipid bilayers exposing terminally grafted polymer chains. In: *Stealth Liposomes*, ed. by D. Lasic, F. Martin (CRC Press, Boca Raton 1995) pp. 73–91

18.29 J. Klein: Shear, friction, and lubrication forces between polymer-bearing surfaces, Annu. Rev. Mater. Sci. **26** (1996) 581–612

18.30 M. Ruths, D. Johannsmann, J. Rühe, W. Knoll: Repulsive forces and relaxation on compression of entangled, polydisperse polystyrene brushes, Macromolecules **33** (2000) 3860–3870

18.31 C. A. Helm, J. N. Israelachvili, P. M. McGuiggan: Molecular mechanisms and forces involved in the adhesion and fusion of amphiphilic bilayers, Science **246** (1989) 919–922

18.32 Y. L. Chen, C. A. Helm, J. N. Israelachvili: Molecular mechanisms associated with adhesion and contact angle hysteresis of monolayer surfaces, J. Phys. Chem. **95** (1991) 10736–10747

18.33 D. E. Leckband, J. N. Israelachvili, F.-J. Schmitt, W. Knoll: Long-range attraction and molecular rearrangements in receptor-ligand interactions, Science **255** (1992) 1419–1421

18.34 J. Peanasky, H. M. Schneider, S. Granick, C. R. Kessel: Self-assembled monolayers on mica for experiments utilizing the surface forces apparatus, Langmuir **11** (1995) 953–962

18.35 C. J. Coakley, D. Tabor: Direct measurement of van der Waals forces between solids in air, J. Phys. D **11** (1978) L77–L82

18.36 J. L. Parker, H. K. Christenson: Measurements of the forces between a metal surface and mica across liquids, J. Chem. Phys. **88** (1988) 8013–8014

18.37 C. P. Smith, M. Maeda, L. Atanasoska, H. S. White, D. J. McClure: Ultrathin platinum films on mica and the measurement of forces at the platinum/water interface, J. Phys. Chem. **92** (1988) 199–205

18.38 S. J. Hirz, A. M. Homola, G. Hadziioannou, C. W. Frank: Effect of substrate on shearing properties of ultrathin polymer films, Langmuir **8** (1992) 328–333

18.39 J. M. Levins, T. K. Vanderlick: Reduction of the roughness of silver films by the controlled application of surface forces, J. Phys. Chem. **96** (1992) 10405–10411

18.40 S. Steinberg, W. Ducker, G. Vigil, C. Hyukjin, C. Frank, M. Z. Tseng, D. R. Clarke, J. N. Israelachvili: Van der Waals epitaxial growth of α-alumina nanocrystals on mica, Science **260** (1993) 656–659

18.41 G. Vigil, Z. Xu, S. Steinberg, J. Israelachvili: Interactions of silica surfaces, J. Colloid Interface Sci. **165** (1994) 367–385

18.42 M. Ruths, M. Heuberger, V. Scheumann, J. Hu, W. Knoll: Confinement-induced film thickness transitions in liquid crystals between two alkanethiol monolayers on gold, Langmuir **17** (2001) 6213–6219

18.43 J. Van Alsten, S. Granick: Molecular tribometry of ultrathin liquid films, Phys. Rev. Lett. **61** (1988) 2570–2573

18.44 J. Van Alsten, S. Granick: Shear rheology in a confined geometry: Polysiloxane melts, Macromolecules **23** (1990) 4856–4862

18.45 A. M. Homola, J. N. Israelachvili, M. L. Gee, P. M. McGuiggan: Measurements of and relation between the adhesion and friction of two surfaces separated by molecularly thin liquid films, J. Tribol. **111** (1989) 675–682

18.46 J. Klein, E. Kumacheva: Simple liquids confined to molecularly thin layers. I. Confinement-induced liquid-to-solid phase transitions, J. Chem. Phys. **108** (1998) 6996–7009

18.47 G. Luengo, F.-J. Schmitt, R. Hill, J. Israelachvili: Thin film rheology and tribology of confined polymer melts: Contrasts with bulk properties, Macromolecules **30** (1997) 2482–2494

18.48 E. Kumacheva: Interfacial friction measurements in surface force apparatus, Prog. Surf. Sci. **58** (1998) 75–120

18.49 A. L. Weisenhorn, P. K. Hansma, T. R. Albrecht, C. F. Quate: Forces in atomic force microscopy in air and water, Appl. Phys. Lett. **54** (1989) 2651–2653

18.50 G. Meyer, N. M. Amer: Simultaneous measurement of lateral and normal forces with an optical-beam-deflection atomic force microscope, Appl. Phys. Lett. **57** (1990) 2089–2091

18.51 E. L. Florin, V. T. Moy, H. E. Gaub: Adhesion forces between individual ligand–receptor pairs, Science **264** (1994) 415–417

18.52 G. U. Lee, D. A. Kidwell, R. J. Colton: Sensing discrete streptavidin–biotin interactions with atomic force microscopy, Langmuir **10** (1994) 354–357

18.53 C. M. Mate, G. M. McClelland, R. Erlandsson, S. Chiang: Atomic-scale friction of a tungsten tip on a graphite surface, Phys. Rev. Lett. **59** (1987) 1942–1945

18.54 R. W. Carpick, M. Salmeron: Scratching the surface: Fundamental investigations of tribology with atomic force microscopy, Chem. Rev. **97** (1997) 1163–1194

18.55 J. P. Cleveland, S. Manne, D. Bocek, P. K. Hansma: A nondestructive method for determining the spring constant of cantilevers for scanning force microscopy, Rev. Sci. Instrum. **64** (1993) 403–405

18.56 J. E. Sader, J. W. M. Chon, P. Mulvaney: Calibration of rectangular atomic force microscope cantilevers, Rev. Sci. Instrum. **70** (1999) 3967–3969

18.57 Y. Liu, T. Wu, D. F. Evans: Lateral force microscopy study on the shear properties of self-assembled monolayers of dialkylammonium surfactant on mica, Langmuir **10** (1994) 2241–2245

18.58 C. Neto, V. S. J. Craig: Colloid probe characterization: Radius and roughness determination, Langmuir **17** (2001) 2097–2099

18.59 R. G. Horn, J. N. Israelachvili: Molecular organization and viscosity of a thin film of molten polymer between two surfaces as probed by force measurements, Macromolecules **21** (1988) 2836–2841

18.60 R. G. Horn, S. J. Hirz, G. Hadziioannou, C. W. Frank, J. M. Catala: A reevaluation of forces measured across thin polymer films: Nonequilibrium and pinning effects, J. Chem. Phys. **90** (1989) 6767–6774

18.61 O. I. Vinogradova, H.-J. Butt, G. E. Yakubov, F. Feuillebois: Dynamic effects on force measurements. I. Viscous drag on the atomic force microscope cantilever, Rev. Sci. Instrum. **72** (2001) 2330–2339

18.62 D. Leckband, J. Israelachvili: Intermolecular forces in biology, Quart. Rev. Biophys. **34** (2001) 105–267

18.63 E. Evans, D. Needham: Physical properties of surfactant bilayer membranes: Thermal transitions, elasticity, rigidity, cohesion, and colloidal interactions, J. Phys. Chem. **91** (1987) 4219–4228

18.64 E. Evans, D. Needham: Attraction between lipid bilayer membranes in concentrated solutions of nonadsorbing polymers: Comparison of mean-field theory with measurements of adhesion energy, Macromolecules **21** (1988) 1822–1831

18.65 S. E. Chesla, P. Selvaraj, C. Zhu: Measuring two-dimensional receptor–ligand binding kinetics by micropipette, Biophys. J. **75** (1998) 1553–1572

18.66 E. Evans, K. Ritchie, R. Merkel: Sensitive force technique to probe molecular adhesion and structural linkages at biological interfaces, Biophys. J. **68** (1995) 2580–2587

18.67 D. M. LeNeveu, R. P. Rand, V. A. Parsegian: Measurements of forces between lecithin bilayers, Nature **259** (1976) 601–603

18.68 A. Homola, A. A. Robertson: A compression method for measuring forces between colloidal particles, J. Colloid Interface Sci. **54** (1976) 286–297

18.69 V. A. Parsegian, N. Fuller, R. P. Rand: Measured work of deformation and repulsion of lecithin bilayers, Proc. Nat. Acad. Sci. USA **76** (1979) 2750–2754

18.70 R. P. Rand, V. A. Parsegian: Hydration forces between phospholipid bilayers, Biochim. Biophys. Acta **988** (1989) 351–376

18.71 S. Leikin, V. A. Parsegian, D. C. Rau, R. P. Rand: Hydration forces, Annu. Rev. Phys. Chem. **44** (1993) 369–395

18.72 S. Chu, J. E. Bjorkholm, A. Ashkin, A. Cable: Experimental observation of optically trapped atoms, Phys. Rev. Lett. **57** (1986) 314–317

18.73 A. Ashkin: Optical trapping and manipulation of neutral particles using lasers, Proc. Nat. Acad. Sci. USA **94** (1997) 4853–4860

18.74 K. Visscher, S. P. Gross, S. M. Block: Construction of multiple-beam optical traps with nanometer-resolution positioning sensing, IEEE J. Sel. Top. Quantum Electron. **2** (1996) 1066–1076

18.75 D. C. Prieve, N. A. Frej: Total internal reflection microscopy: A quantitative tool for the measurement of colloidal forces, Langmuir **6** (1990) 396–403

18.76 J. Rädler, E. Sackmann: On the measurement of weak repulsive and frictional colloidal forces by reflection interference contrast microscopy, Langmuir **8** (1992) 848–853

18.77 P. Bongrand, C. Capo, J.-L. Mege, A.-M. Benoliel: Use of hydrodynamic flows to study cell adhesion. In: *Physical Basis of Cell–Cell Adhesion*, ed. by P. Bongrand (CRC Press, Boca Raton 1988) pp. 125–156

18.78 G. Kaplanski, C. Farnarier, O. Tissot, A. Pierres, A.-M. Benoliel, A.-C. Alessi, S. Kaplanski, P. Bongrand: Granulocyte–endothelium initial adhesion: Analysis of transient binding events mediated by E-selectin in a laminar shear flow, Biophys. J. **64** (1993) 1922–1933

18.79 H. B. G. Casimir, D. Polder: The influence of retardation on the London–van der Waals forces, Phys. Rev. **73** (1948) 360–372

18.80 H. C. Hamaker: The London–van der Waals attraction between spherical particles, Physica **4** (1937) 1058–1072

18.81 E. M. Lifshitz: The theory of molecular attraction forces between solid bodies, Sov. Phys. JETP (English translation) **2** (1956) 73–83

18.82 I. E. Dzyaloshinskii, E. M. Lifshitz, L. P. Pitaevskii: The general theory of van der Waals forces, Adv. Phys. **10** (1961) 165–209

18.83 H. W. Fox, W. A. Zisman: The spreading of liquids on low-energy surfaces. III. Hydrocarbon surfaces, J. Colloid Sci. **7** (1952) 428–442

18.84 B.V. Derjaguin, V.P. Smilga: Electrostatic component of the rolling friction force moment, Wear **7** (1964) 270–281

18.85 R.G. Horn, D.T. Smith: Contact electrification and adhesion between dissimilar materials, Science **256** (1992) 362–364

18.86 R. Budakian, S.J. Putterman: Correlation between charge transfer and stick–slip friction at a metal–insulator interface, Phys. Rev. Lett. **85** (2000) 1000–1003

18.87 M. Grätzel: Photoelectrochemical cells, Nature **414** (2001) 338–344

18.88 R.M. German: *Sintering Theory and Practice* (Wiley, New York 1996)

18.89 R. Budakian, S.J. Putterman: Time scales for cold welding and the origins of stick–slip friction, Phys. Rev. B **65** (2002) 235429/1–5

18.90 U. Landman, W.D. Luedtke, N.A. Burnham, R.J. Colton: Atomistic mechanisms and dynamics of adhesion, nanoindentation, and fracture, Science **248** (1990) 454–461

18.91 U. Landman, W.D. Luedtke, E.M. Ringer: Molecular dynamics simulations of adhesive contact formation and friction, NATO Science Ser. E **220** (1992) 463–510

18.92 W.D. Luedtke, U. Landman: Solid and liquid junctions, Comput. Mater. Sci. **1** (1992) 1–24

18.93 U. Landman, W.D. Luedtke: Interfacial junctions and cavitation, MRS Bull. **18** (1993) 36–44

18.94 B. Bhushan, J.N. Israelachvili, U. Landman: Nanotribology: Friction, wear and lubrication at the atomic scale, Nature **374** (1995) 607–616

18.95 M.R. Sørensen, K.W. Jacobsen, P. Stoltze: Simulations of atomic-scale sliding friction, Phys. Rev. B **53** (1996) 2101–2113

18.96 A. Meurk, P.F. Luckham, L. Bergström: Direct measurement of repulsive and attractive van der Waals forces between inorganic materials, Langmuir **13** (1997) 3896–3899

18.97 E.J.W. Verwey, J.T.G. Overbeek: *Theory of the Stability of Lyophobic Colloids*, 1st edn. (Elsevier, Amsterdam 1948)

18.98 D.Y.C. Chan, R.M. Pashley, L.R. White: A simple algorithm for the calculation of the electrostatic repulsion between identical charged surfaces in electrolyte, J. Colloid Interface Sci. **77** (1980) 283–285

18.99 B. Derjaguin, L. Landau: Theory of the stability of strongly charged lyophobic sols and of the adhesion of strongly charged particles in solutions of electrolytes, Acta Physicochim. URSS **14** (1941) 633–662

18.100 D. Chan, T.W. Healy, L.R. White: Electrical double layer interactions under regulation by surface ionization equilibriums – dissimilar amphoteric surfaces, J. Chem. Soc. Faraday Trans. 1 **72** (1976) 2844–2865

18.101 G.S. Manning: Limiting laws and counterion condensation in polyelectrolyte solutions. I. Colligative properties, J. Chem. Phys. **51** (1969) 924–933

18.102 L. Guldbrand, V. Jönsson, H. Wennerström, P. Linse: Electrical double-layer forces: A Monte Carlo study, J. Chem. Phys. **80** (1984) 2221–2228

18.103 H. Wennerström, B. Jönsson, P. Linse: The cell model for polyelectrolyte systems. Exact statistical mechanical relations, Monte Carlo simulations, and the Poisson–Boltzmann approximation, J. Chem. Phys. **76** (1982) 4665–4670

18.104 J. Marra: Effects of counterion specificity on the interactions between quaternary ammonium surfactants in monolayers and bilayers, J. Phys. Chem. **90** (1986) 2145–2150

18.105 J. Marra: Direct measurement of the interaction between phosphatidylglycerol bilayers in aqueous-electrolyte solutions, Biophys. J. **50** (1986) 815–825

18.106 R.G. Horn, J.N. Israelachvili: Direct measurement of structural forces between two surfaces in a nonpolar liquid, J. Chem. Phys. **75** (1981) 1400–1411

18.107 I.K. Snook, W. van Megen: Solvation forces in simple dense fluids I, J. Chem. Phys. **72** (1980) 2907–2913

18.108 W.J. van Megen, I.K. Snook: Solvation forces in simple dense fluids II. Effect of chemical potential, J. Chem. Phys. **74** (1981) 1409–1411

18.109 R. Kjellander, S. Marcelja: Perturbation of hydrogen bonding in water near polar surfaces, Chem. Phys. Lett. **120** (1985) 393–396

18.110 P. Tarazona, L. Vicente: A model for the density oscillations in liquids between solid walls, Mol. Phys. **56** (1985) 557–572

18.111 D. Henderson, M. Lozada-Cassou: A simple theory for the force between spheres immersed in a fluid, J. Colloid Interface Sci. **114** (1986) 180–183

18.112 J.E. Curry, J.H. Cushman: Structure in confined fluids: Phase separation of binary simple liquid mixtures, Tribol. Lett. **4** (1998) 129–136

18.113 M. Schoen, T. Gruhn, D.J. Diestler: Solvation forces in thin films confined between macroscopically curved substrates, J. Chem. Phys. **109** (1998) 301–311

18.114 F. Porcheron, B. Rousseau, M. Schoen, A.H. Fuchs: Structure and solvation forces in confined alkane films, Phys. Chem. Chem. Phys. **3** (2001) 1155–1159

18.115 H.K. Christenson, D.W.R. Gruen, R.G. Horn, J.N. Israelachvili: Structuring in liquid alkanes between solid surfaces: Force measurements and mean-field theory, J. Chem. Phys. **87** (1987) 1834–1841

18.116 M.L. Gee, J.N. Israelachvili: Interactions of surfactant monolayers across hydrocarbon liquids, J. Chem. Soc. Faraday Trans. **86** (1990) 4049–4058

18.117 J.N. Israelachvili, S.J. Kott, M.L. Gee, T.A. Witten: Forces between mica surfaces across hydrocarbon liquids: Effects of branching and polydispersity, Macromolecules **22** (1989) 4247–4253

18.118 H. K. Christenson: Forces between solid surfaces in a binary mixture of non-polar liquids, Chem. Phys. Lett. **118** (1985) 455–458

18.119 H. K. Christenson, R. G. Horn: Solvation forces measured in non-aqueous liquids, Chem. Scr. **25** (1985) 37–41

18.120 J. Israelachvili: Solvation forces and liquid structure, as probed by direct force measurements, Acc. Chem. Res. **20** (1987) 415–421

18.121 H. K. Christenson: Non-DLVO forces between surfaces – solvation, hydration and capillary effects, J. Disp. Sci. Technol. **9** (1988) 171–206

18.122 L. J. D. Frink, F. van Swol: Solvation forces between rough surfaces, J. Chem. Phys. **108** (1998) 5588–5598

18.123 J. Gao, W. D. Luedtke, U. Landman: Structures, solvation forces and shear of molecular films in a rough nano-confinement, Tribol. Lett. **9** (2000) 3–13

18.124 H. E. Stanley, J. Teixeira: Interpretation of the unusual behavior of H_2O and D_2O at low temperatures: Tests of a percolation model, J. Chem. Phys. **73** (1980) 3404–3422

18.125 H. van Olphen: *An Introduction to Clay Colloid Chemistry*, 2nd edn. (Wiley, New York 1977) Chap. 10

18.126 U. Del Pennino, E. Mazzega, S. Valeri, A. Alietti, M. F. Brigatti, L. Poppi: Interlayer water and swelling properties of monoionic montmorillonites, J. Colloid Interface Sci. **84** (1981) 301–309

18.127 B. E. Viani, P. F. Low, C. B. Roth: Direct measurement of the relation between interlayer force and interlayer distance in the swelling of montmorillonite, J. Colloid Interface Sci. **96** (1983) 229–244

18.128 R. G. Horn: Surface forces and their action in ceramic materials, J. Am. Ceram. Soc. **73** (1990) 1117–1135

18.129 B.V. Velamakanni, J. C. Chang, F. F. Lange, D. S. Pearson: New method for efficient colloidal particle packing via modulation of repulsive lubricating hydration forces, Langmuir **6** (1990) 1323–1325

18.130 R. R. Lessard, S. A. Zieminski: Bubble coalescence and gas transfer in aqueous electrolytic solutions, Ind. Eng. Chem. Fundam. **10** (1971) 260–269

18.131 J. Israelachvili, R. Pashley: The hydrophobic interaction is long range, decaying exponentially with distance, Nature **300** (1982) 341–342

18.132 R. M. Pashley, P. M. McGuiggan, B. W. Ninham, D. F. Evans: Attractive forces between uncharged hydrophobic surfaces: Direct measurements in aqueous solutions, Science **229** (1985) 1088–1089

18.133 P. M. Claesson, C. E. Blom, P. C. Herder, B. W. Ninham: Interactions between water-stable hydrophobic Langmuir–Blodgett monolayers on mica, J. Colloid Interface Sci. **114** (1986) 234–242

18.134 Ya. I. Rabinovich, B.V. Derjaguin: Interaction of hydrophobized filaments in aqueous electrolyte solutions, Colloids Surf. **30** (1988) 243–251

18.135 J. L. Parker, D. L. Cho, P. M. Claesson: Plasma modification of mica: Forces between fluorocarbon surfaces in water and a nonpolar liquid, J. Phys. Chem. **93** (1989) 6121–6125

18.136 H. K. Christenson, J. Fang, B. W. Ninham, J. L. Parker: Effect of divalent electrolyte on the hydrophobic attraction, J. Phys. Chem. **94** (1990) 8004–8006

18.137 K. Kurihara, S. Kato, T. Kunitake: Very strong long range attractive forces between stable hydrophobic monolayers of a polymerized ammonium surfactant, Chem. Lett. (1990) 1555–1558

18.138 Y. H. Tsao, D. F. Evans, H. Wennerstrom: Long-range attractive force between hydrophobic surfaces observed by atomic force microscopy, Science **262** (1993) 547–550

18.139 Ya. I. Rabinovich, R.-H. Yoon: Use of atomic force microscope for the measurements of hydrophobic forces between silanated silica plate and glass sphere, Langmuir **10** (1994) 1903–1909

18.140 V. S. J. Craig, B. W. Ninham, R. M. Pashley: Study of the long-range hydrophobic attraction in concentrated salt solutions and its implications for electrostatic models, Langmuir **14** (1998) 3326–3332

18.141 P. Kékicheff, O. Spalla: Long-range electrostatic attraction between similar, charge-neutral walls, Phys. Rev. Lett. **75** (1995) 1851–1854

18.142 H. K. Christenson, P. M. Claesson: Direct measurements of the force between hydrophobic surfaces in water, Adv. Colloid Interface Sci. **91** (2001) 391–436

18.143 P. Attard: Nanobubbles and the hydrophobic attraction, Adv. Colloid Interface Sci. **104** (2003) 75–91

18.144 M. Hato: Attractive forces between surfaces of controlled "hydrophobicity" across water: A possible range of "hydrophobic interactions" between macroscopic hydrophobic surfaces across water, J. Phys. Chem. **100** (1996) 18530–18538

18.145 H. K. Christenson, P. M. Claesson, J. Berg, P. C. Herder: Forces between fluorocarbon surfactant monolayers: Salt effects on the hydrophobic interaction, J. Phys. Chem. **93** (1989) 1472–1478

18.146 N. I. Christou, J. S. Whitehouse, D. Nicholson, N. G. Parsonage: A Monte Carlo study of fluid water in contact with structureless walls, Faraday Symp. Chem. Soc. **16** (1981) 139–149

18.147 B. Jönsson: Monte Carlo simulations of liquid water between two rigid walls, Chem. Phys. Lett. **82** (1981) 520–525

18.148 S. Marcelja, D. J. Mitchell, B. W. Ninham, M. J. Sculley: Role of solvent structure in solution theory, J. Chem. Soc. Faraday Trans. II **73** (1977) 630–648

18.149 D. W. R. Gruen, S. Marcelja: Spatially varying polarization in water: A model for the electric double layer and the hydration force, J. Chem. Soc. Faraday Trans. 2 **79** (1983) 225–242

18.150 B. Jönsson, H. Wennerström: Image-charge forces in phospholipid bilayer systems, J. Chem. Soc. Faraday Trans. 2 **79** (1983) 19–35
18.151 D. Schiby, E. Ruckenstein: The role of the polarization layers in hydration forces, Chem. Phys. Lett. **95** (1983) 435–438
18.152 A. Luzar, D. Bratko, L. Blum: Monte Carlo simulation of hydrophobic interaction, J. Chem. Phys. **86** (1987) 2955–2959
18.153 P. Attard, M.T. Batchelor: A mechanism for the hydration force demonstrated in a model system, Chem. Phys. Lett. **149** (1988) 206–211
18.154 K. Leung, A. Luzar: Dynamics of capillary evaporation. II. Free energy barriers, J. Chem. Phys. **113** (2000) 5845–5852
18.155 D. Bratko, R.A. Curtis, H.W. Blanch, J.M. Prausnitz: Interaction between hydrophobic surfaces with metastable intervening liquid, J. Chem. Phys. **115** (2001) 3873–3877
18.156 P.G. de Gennes: Polymers at an interface. 2. Interaction between two plates carrying adsorbed polymer layers, Macromolecules **15** (1982) 492–500
18.157 J.M.H.M. Scheutjens, G.J. Fleer: Interaction between two adsorbed polymer layers, Macromolecules **18** (1985) 1882–1900
18.158 E.A. Evans: Force between surfaces that confine a polymer solution: Derivation from self-consistent field theories, Macromolecules **22** (1989) 2277–2286
18.159 H.J. Ploehn: Compression of polymer interphases, Macromolecules **27** (1994) 1627–1636
18.160 S. Asakura, F. Oosawa: Interaction between particles suspended in solutions of macromolecules, J. Polym. Sci. **33** (1958) 183–192
18.161 J.F. Joanny, L. Leibler, P.G. de Gennes: Effects of polymer solutions on colloid stability, J. Polym. Sci. Polym. Phys. **17** (1979) 1073–1084
18.162 B. Vincent, P.F. Luckham, F.A. Waite: The effect of free polymer on the stability of sterically stabilized dispersions, J. Colloid Interface Sci. **73** (1980) 508–521
18.163 R.I. Feigin, D.H. Napper: Stabilization of colloids by free polymer, J. Colloid Interface Sci. **74** (1980) 567–571
18.164 P.G. de Gennes: Polymer solutions near an interface. 1. Adsorption and depletion layers, Macromolecules **14** (1981) 1637–1644
18.165 P.G. de Gennes: Polymers at an interface; a simplified view, Adv. Colloid Interface Sci. **27** (1987) 189–209
18.166 S. Alexander: Adsorption of chain molecules with a polar head. A scaling description, J. Phys. (France) **38** (1977) 983–987
18.167 P.G. de Gennes: Conformations of polymers attached to an interface, Macromolecules **13** (1980) 1069–1075
18.168 S.T. Milner, T.A. Witten, M.E. Cates: Theory of the grafted polymer brush, Macromolecules **21** (1988) 2610–2619
18.169 S.T. Milner, T.A. Witten, M.E. Cates: Effects of polydispersity in the end-grafted polymer brush, Macromolecules **22** (1989) 853–861
18.170 E.B. Zhulina, O.V. Borisov, V.A. Priamitsyn: Theory of steric stabilization of colloid dispersions by grafted polymers, J. Colloid Interface Sci. **137** (1990) 495–511
18.171 H.J. Taunton, C. Toprakcioglu, L.J. Fetters, J. Klein: Interactions between surfaces bearing end-adsorbed chains in a good solvent, Macromolecules **23** (1990) 571–580
18.172 J. Klein, P. Luckham: Forces between two adsorbed poly(ethylene oxide) layers immersed in a good aqueous solvent, Nature **300** (1982) 429–431
18.173 J. Klein, P.F. Luckham: Long-range attractive forces between two mica surfaces in an aqueous polymer solution, Nature **308** (1984) 836–837
18.174 P.F. Luckham, J. Klein: Forces between mica surfaces bearing adsorbed homopolymers in good solvents, J. Chem. Soc. Faraday Trans. **86** (1990) 1363–1368
18.175 J. Klein, E. Kumacheva, D. Mahalu, D. Perahia, L.J. Fetters: Reduction of frictional forces between solid surfaces bearing polymer brushes, Nature **370** (1994) 634–636
18.176 J.N. Israelachvili, H. Wennerström: Hydration or steric forces between amphiphilic surfaces?, Langmuir **6** (1990) 873–876
18.177 J.N. Israelachvili, H. Wennerström: Entropic forces between amphiphilic surfaces in liquids, J. Phys. Chem. **96** (1992) 520–531
18.178 M.K. Granfeldt, S.J. Miklavic: A simulation study of flexible zwitterionic monolayers. Interlayer interaction and headgroup conformation, J. Phys. Chem. **95** (1991) 6351–6360
18.179 L.R. Fisher, J.N. Israelachvili: Direct measurements of the effect of meniscus forces on adhesion: A study of the applicability of macroscopic thermodynamics to microscopic liquid interfaces, Colloids Surf. **3** (1981) 303–319
18.180 H.K. Christenson: Adhesion between surfaces in unsaturated vapors – a reexamination of the influence of meniscus curvature and surface forces, J. Colloid Interface Sci. **121** (1988) 170–178
18.181 R.G. Horn, J.N. Israelachvili, F. Pribac: Measurement of the deformation and adhesion of solids in contact, J. Colloid Interface Sci. **115** (1987) 480–492
18.182 H. Hertz: Über die Berührung fester elastischer Körper, J. Reine Angew. Math. **92** (1881) 156–171
18.183 H.M. Pollock, D. Maugis, M. Barquins: The force of adhesion between solid surfaces in contact, Appl. Phys. Lett. **33** (1978) 798–799
18.184 B.V. Derjaguin, V.M. Muller, Yu.P. Toporov: Effect of contact deformations on the adhesion of particles, J. Colloid Interface Sci. **53** (1975) 314–326
18.185 V.M. Muller, V.S. Yushchenko, B.V. Derjaguin: On the influence of molecular forces on the deforma-

tion of an elastic sphere and its sticking to a rigid plane, J. Colloid Interface Sci. **77** (1980) 91–101

18.186 V. M. Muller, B. V. Derjaguin, Y. P. Toporov: On 2 methods of calculation of the force of sticking of an elastic sphere to a rigid plane, Colloids Surf. **7** (1983) 251–259

18.187 D. Maugis: Adhesion of spheres: The JKR–DMT transition using a Dugdale model, J. Colloid Interface Sci. **150** (1992) 243–269

18.188 V. Mangipudi, M. Tirrell, A. V. Pocius: Direct measurement of molecular level adhesion between poly(ethylene terephthalate) and polyethylene films: Determination of surface and interfacial energies, J. Adh. Sci. Technol. **8** (1994) 1251–1270

18.189 H. K. Christenson: Surface deformations in direct force measurements, Langmuir **12** (1996) 1404–1405

18.190 M. Barquins, D. Maugis: Fracture mechanics and the adherence of viscoelastic bodies, J. Phys. D: Appl Phys. **11** (1978) 1989–2023

18.191 J. A. Greenwood, J. B. P. Williamson: Contact of nominally flat surfaces, Proc. R. Soc. London A **295** (1966) 300–319

18.192 B. N. J. Persson: Elastoplastic contact between randomly rough surfaces, Phys. Rev. Lett. **87** (2001) 116101/1–4

18.193 K. N. G. Fuller, D. Tabor: The effect of surface roughness on the adhesion of elastic solids, Proc. R. Soc. London A **345** (1975) 327–342

18.194 D. Maugis: On the contact and adhesion of rough surfaces, J. Adh. Sci. Technol. **10** (1996) 161–175

18.195 B. N. J. Persson, E. Tosatti: The effect of surface roughness on the adhesion of elastic solids, J. Chem. Phys. **115** (2001) 5597–5610

18.196 P. M. McGuiggan, J. N. Israelachvili: Adhesion and short-range forces between surfaces. Part II: Effects of surface lattice mismatch, J. Mater. Res. **5** (1990) 2232–2243

18.197 P. M. McGuiggan, J. Israelachvili: Measurements of the effects of angular lattice mismatch on the adhesion energy between two mica surfaces in water, Mat. Res. Soc. Symp. Proc. **138** (1989) 349–360

18.198 C. L. Rhykerd, Jr., M. Schoen, D. J. Diestler, J. H. Cushman: Epitaxy in simple classical fluids in micropores and near-solid surfaces, Nature **330** (1987) 461–463

18.199 M. Schoen, D. J. Diestler, J. H. Cushman: Fluids in micropores. I. Structure of a simple classical fluid in a slit-pore, J. Chem. Phys. **87** (1987) 5464–5476

18.200 M. Schoen, C. L. Rhykerd, Jr., D. J. Diestler, J. H. Cushman: Shear forces in molecularly thin films, Science **245** (1989) 1223–1225

18.201 P. A. Thompson, M. O. Robbins: Origin of stick–slip motion in boundary lubrication, Science **250** (1990) 792–794

18.202 K. K. Han, J. H. Cushman, D. J. Diestler: Grand canonical Monte Carlo simulations of a Stockmayer fluid in a slit micropore, Mol. Phys. **79** (1993) 537–545

18.203 M. Ruths, S. Granick: Influence of alignment of crystalline confining surfaces on static forces and shear in a liquid crystal, 4′-*n*-pentyl-4-cyanobiphenyl (5CB), Langmuir **16** (2000) 8368–8376

18.204 A. D. Berman: Dynamics of molecules at surfaces. Ph.D. Thesis (Univ. of California, Santa Barbara 1996)

18.205 E. Barthel, S. Roux: Velocity dependent adherence: An analytical approach for the JKR and DMT models, Langmuir **16** (2000) 8134–8138

18.206 M. Ruths, S. Granick: Rate-dependent adhesion between polymer and surfactant monolayers on elastic substrates, Langmuir **14** (1998) 1804–1814

18.207 H. Yoshizawa, Y. L. Chen, J. Israelachvili: Fundamental mechanisms of interfacial friction. 1: Relation between adhesion and friction, J. Phys. Chem. **97** (1993) 4128–4140

18.208 N. Maeda, N. Chen, M. Tirrell, J. N. Israelachvili: Adhesion and friction mechanisms of polymer-on-polymer surfaces, Science **297** (2002) 379–382

18.209 F. P. Bowden, D. Tabor: *The Friction and Lubrication of Solids* (Clarendon, London 1971)

18.210 J. A. Greenwood, K. L. Johnson: The mechanics of adhesion of viscoelastic solids, Phil. Mag. A **43** (1981) 697–711

18.211 D. Maugis: Subcritical crack growth, surface energy, fracture toughness, stick–slip and embrittlement, J. Mater. Sci. **20** (1985) 3041–3073

18.212 F. Michel, M. E. R. Shanahan: Kinetics of the JKR experiment, C. R. Acad. Sci. II (Paris) **310** (1990) 17–20

18.213 C. A. Miller, P. Neogi: *Interfacial Phenomena: Equilibrium and Dynamic Effects* (Dekker, New York 1985)

18.214 J. Israelachvili, A. Berman: Irreversibility, energy dissipation, and time effects in intermolecular and surface interactions, Israel J. Chem. **35** (1995) 85–91

18.215 A. N. Gent, A. J. Kinloch: Adhesion of viscoelastic materials to rigid substrates. III. Energy criterion for failure, J. Polym. Sci. A-2 **9** (1971) 659–668

18.216 A. N. Gent: Adhesion and strength of viscoelastic solids. Is there a relationship between adhesion and bulk properties?, Langmuir **12** (1996) 4492–4496

18.217 H. R. Brown: The adhesion between polymers, Annu. Rev. Mater. Sci. **21** (1991) 463–489

18.218 M. Deruelle, M. Tirrell, Y. Marciano, H. Hervet, L. Léger: Adhesion energy between polymer networks and solid surfaces modified by polymer attachment, Faraday Discuss. **98** (1995) 55–65

18.219 S. Granick: Motions and relaxations of confined liquids, Science **253** (1991) 1374–1379

18.220 H. W. Hu, S. Granick: Viscoelastic dynamics of confined polymer melts, Science **258** (1992) 1339–1342

18.221 M. L. Gee, P. M. McGuiggan, J. N. Israelachvili, A. M. Homola: Liquid to solidlike transitions of molecularly thin films under shear, J. Chem. Phys. **93** (1990) 1895–1906

18.222 J. Van Alsten, S. Granick: The origin of static friction in ultrathin liquid films, Langmuir **6** (1990) 876–880

18.223 H.-W. Hu, G. A. Carson, S. Granick: Relaxation time of confined liquids under shear, Phys. Rev. Lett. **66** (1991) 2758–2761

18.224 J. Israelachvili, M. Gee, P. McGuiggan, P. Thompson, M. Robbins: Melting–freezing transitions in molecularly thin liquid films during shear, Fall Meeting of the Mater. Res. Soc., Boston, MA. 1990, ed. by J. M. Drake, J. Klafter, R. Kopelman (MRS Publications, Pittsburgh, PA1990) 3–6

18.225 J. Israelachvili, P. McGuiggan, M. Gee, A. Homola, M. Robbins, P. Thompson: Liquid dynamics in molecularly thin films, J. Phys.: Condens. Matter **2** (1990) SA89–SA98

18.226 J. Klein, E. Kumacheva: Confinement-induced phase transitions in simple liquids, Science **269** (1995) 816–819

18.227 A. L. Demirel, S. Granick: Glasslike transition of a confined simple fluid, Phys. Rev. Lett. **77** (1996) 2261–2264

18.228 D. Dowson: *History of Tribology*, 2nd edn. (Professional Engineering Publishing, London 1998)

18.229 D. Tabor: The role of surface and intermolecular forces in thin film lubrication, Tribol. Ser. **7** (1982) 651–682

18.230 M. J. Sutcliffe, S. R. Taylor, A. Cameron: Molecular asperity theory of boundary friction, Wear **51** (1978) 181–192

18.231 G. M. McClelland: Friction at weakly interacting interfaces. In: *Adhesion and Friction*, ed. by M. Grunze, H. J. Kreuzer (Springer, Berlin, Heidelberg 1989) pp. 1–16

18.232 J. N. Israelachvili, Y.-L. Chen, H. Yoshizawa: Relationship between adhesion and friction forces, J. Adh. Sci. Technol. **8** (1994) 1231–1249

18.233 D. H. Buckley: The metal-to-metal interface and its effect on adhesion and friction, J. Colloid Interface Sci. **58** (1977) 36–53

18.234 V. Zaloj, M. Urbakh, J. Klafter: Modifying friction by manipulating normal response to lateral motion, Phys. Rev. Lett. **82** (1999) 4823–4826

18.235 A. Dhinojwala, S. C. Bae, S. Granick: Shear-induced dilation of confined liquid films, Tribol. Lett. **9** (2000) 55–62

18.236 L. M. Qian, G. Luengo, E. Perez: Thermally activated lubrication with alkanes: The effect of chain length, Europhys. Lett. **61** (2003) 268–274

18.237 B. J. Briscoe, D. C. B. Evans, D. Tabor: The influence of contact pressure and saponification on the sliding behavior of steric acid monolayers, J. Colloid Interface Sci. **61** (1977) 9–13

18.238 A. Berman, J. Israelachvili: Control and minimization of friction via surface modification, NATO ASI Ser. E: Appl. Sci. **330** (1997) 317–329

18.239 M. Urbakh, L. Daikhin, J. Klafter: Dynamics of confined liquids under shear, Phys. Rev. E **51** (1995) 2137–2141

18.240 M. G. Rozman, M. Urbakh, J. Klafter: Origin of stick-slip motion in a driven two-wave potential, Phys. Rev. E **54** (1996) 6485–6494

18.241 M. G. Rozman, M. Urbakh, J. Klafter: Stick–slip dynamics as a probe of frictional forces, Europhys. Lett. **39** (1997) 183–188

18.242 B. V. Derjaguin: Molekulartheorie der äußeren Reibung, Z. Physik **88** (1934) 661–675

18.243 B. V. Derjaguin: Mechanical properties of the boundary lubrication layer, Wear **128** (1988) 19–27

18.244 B. J. Briscoe, D. C. B. Evans: The shear properties of Langmuir–Blodgett layers, Proc. R. Soc. London A **380** (1982) 389–407

18.245 A. Berman, C. Drummond, J. Israelachvili: Amontons' law at the molecular level, Tribol. Lett. **4** (1998) 95–101

18.246 P. F. Luckham, S. Manimaaran: Investigating adsorbed polymer layer behaviour using dynamic surface force apparatuses – a review, Adv. Colloid Interface Sci. **73** (1997) 1–46

18.247 A. M. Homola, J. N. Israelachvili, P. M. McGuiggan, M. L. Gee: Fundamental experimental studies in tribology: The transition from "interfacial" friction of undamaged molecularly smooth surfaces to "normal" friction with wear, Wear **136** (1990) 65–83

18.248 S. Yamada, J. Israelachvili: Friction and adhesion hysteresis of fluorocarbon surfactant monolayer-coated surfaces measured with the surface forces apparatus, J. Phys. Chem. B **102** (1998) 234–244

18.249 G. Luengo, S. E. Campbell, V. I. Srdanov, F. Wudl, J. N. Israelachvili: Direct measurement of the adhesion and friction of smooth C_{60} surfaces, Chem. Mater. **9** (1997) 1166–1171

18.250 L. Rapoport, Y. Bilik, Y. Feldman, M. Homyonfer, S. R. Cohen, R. Tenne: Hollow nanoparticles of WS_2 as potential solid-state lubricants, Nature **387** (1997) 791–793

18.251 J. Israelachvili, Y.-L. Chen, H. Yoshizawa: Relationship between adhesion and friction forces. In: *Fundamentals of Adhesion and Interfaces*, ed. by D. S. Rimai, L. P. DeMejo, K. L. Mittal (VSP, Utrecht, The Netherlands 1995) pp. 261–279

18.252 P. Berthoud, T. Baumberger, C. G'Sell, J. M. Hiver: Physical analysis of the state- and rate-dependent friction law: Static friction, Phys. Rev. B **59** (1999) 14313–14327

18.253 A. Berman, S. Steinberg, S. Campbell, A. Ulman, J. Israelachvili: Controlled microtribology of a metal oxide surface, Tribol. Lett. **4** (1998) 43–48

18.254 D.Y.C. Chan, R.G. Horn: The drainage of thin liquid films between solid surfaces, J. Chem. Phys. **83** (1985) 5311–5324

18.255 J.N. Israelachvili, S.J. Kott: Shear properties and structure of simple liquids in molecularly thin films: The transition from bulk (continuum) to molecular behavior with decreasing film thickness, J. Colloid Interface Sci. **129** (1989) 461–467

18.256 T.L. Kuhl, A.D. Berman, S.W. Hui, J.N. Israelachvili: Part 1: Direct measurement of depletion attraction and thin film viscosity between lipid bilayers in aqueous polyethylene glycol solutions, Macromolecules **31** (1998) 8250–8257

18.257 S.E. Campbell, G. Luengo, V.I. Srdanov, F. Wudl, J.N. Israelachvili: Very low viscosity at the solid–liquid interface induced by adsorbed C_{60} monolayers, Nature **382** (1996) 520–522

18.258 J. Klein, Y. Kamiyama, H. Yoshizawa, J.N. Israelachvili, G.H. Fredrickson, P. Pincus, L.J. Fetters: Lubrication forces between surfaces bearing polymer brushes, Macromolecules **26** (1993) 5552–5560

18.259 G. Luengo, J. Israelachvili, A. Dhinojwala, S. Granick: Generalized effects in confined fluids: New friction map for boundary lubrication, Wear **200** (1996) 328–335

18.260 E. Kumacheva, J. Klein: Simple liquids confined to molecularly thin layers. II. Shear and frictional behavior of solidified films, J. Chem. Phys. **108** (1998) 7010–7022

18.261 P.A. Thompson, G.S. Grest, M.O. Robbins: Phase transitions and universal dynamics in confined films, Phys. Rev. Lett. **68** (1992) 3448–3451

18.262 Y. Rabin, I. Hersht: Thin liquid layers in shear: Non-Newtonian effects, Physica A **200** (1993) 708–712

18.263 P.A. Thompson, M.O. Robbins, G.S. Grest: Structure and shear response in nanometer-thick films, Israel J. Chem. **35** (1995) 93–106

18.264 M. Urbakh, L. Daikhin, J. Klafter: Sheared liquids in the nanoscale range, J. Chem. Phys. **103** (1995) 10707–10713

18.265 A. Subbotin, A. Semenov, E. Manias, G. Hadziioannou, G. ten Brinke: Rheology of confined polymer melts under shear flow: Strong adsorption limit, Macromolecules **28** (1995) 1511–1515

18.266 A. Subbotin, A. Semenov, E. Manias, G. Hadziioannou, G. ten Brinke: Rheology of confined polymer melts under shear flow: Weak adsorption limit, Macromolecules **28** (1995) 3901–3903

18.267 J. Huh, A. Balazs: Behavior of confined telechelic chains under shear, J. Chem. Phys. **113** (2000) 2025–2031

18.268 C. Drummond, J. Elezgaray, P. Richetti: Behavior of adhesive boundary lubricated surfaces under shear: A new dynamic transition, Europhys. Lett. **58** (2002) 503–509

18.269 J.N. Israelachvili, P.M. McGuiggan, A.M. Homola: Dynamic properties of molecularly thin liquid films, Science **240** (1988) 189–191

18.270 A.M. Homola, H.V. Nguyen, G. Hadziioannou: Influence of monomer architecture on the shear properties of molecularly thin polymer melts, J. Chem. Phys. **94** (1991) 2346–2351

18.271 M. Schoen, S. Hess, D.J. Diestler: Rheological properties of confined thin films, Phys. Rev. E **52** (1995) 2587–2602

18.272 G.M. McClelland, S.R. Cohen: *Chemistry and Physics of Solid Surfaces VIII* (Springer, Berlin, Heidelberg 1990) pp. 419–445

18.273 M.O. Robbins, P.A. Thompson: Critical velocity of stick–slip motion, Science **253** (1991) 916

18.274 G. He, M. Müser, M. Robbins: Adsorbed layers and the origin of static friction, Science **284** (1999) 1650–1652

18.275 C. Drummond, J. Israelachvili: Dynamic phase transitions in confined lubricant fluids under shear, Phys. Rev. E. **63** (2001) 041506/1–11

18.276 H. Yoshizawa, J. Israelachvili: Fundamental mechanisms of interfacial friction. 2: Stick–slip friction of spherical and chain molecules, J. Phys. Chem. **97** (1993) 11300–11313

18.277 J.D. Ferry: *Viscoelastic Properties of Polymers*, 3rd edn. (Wiley, New York 1980)

18.278 A. Ruina: Slip instability and state variable friction laws, J. Geophys. Res. **88** (1983) 10359–10370

18.279 T. Baumberger, P. Berthoud, C. Caroli: Physical analysis of the state- and rate-dependent friction law. II. Dynamic friction, Phys. Rev. B **60** (1999) 3928–3939

18.280 J. Israelachvili, S. Giasson, T. Kuhl, C. Drummond, A. Berman, G. Luengo, J.-M. Pan, M. Heuberger, W. Ducker, N. Alcantar: Some fundamental differences in the adhesion and friction of rough versus smooth surfaces, Tribol. Ser. **38** (2000) 3–12

18.281 C. Drummond, J. Israelachvili: Dynamic behavior of confined branched hydrocarbon lubricant fluids under shear, Macromolecules **33** (2000) 4910–4920

18.282 E. Rabinowicz: *Friction and Wear of Materials*, 2nd edn. (Wiley, New York 1995) Chap. 4

18.283 J. Peachey, J. Van Alsten, S. Granick: Design of an apparatus to measure the shear response of ultrathin liquid films, Rev. Sci. Instrum. **62** (1991) 463–473

18.284 E. Rabinowicz: The intrinsic variables affecting the stick–slip process, Proc. Phys. Soc. **71** (1958) 668–675

18.285 T. Baumberger, F. Heslot, B. Perrin: Crossover from creep to inertial motion in friction dynamics, Nature **367** (1994) 544–546

18.286 F. Heslot, T. Baumberger, B. Perrin, B. Caroli, C. Caroli: Creep, stick–slip, and dry-friction dynamics: Experiments and a heuristic model, Phys. Rev. E **49** (1994) 4973–4988

18.287 J. Sampson, F. Morgan, D. Reed, M. Muskat: Friction behavior during the slip portion of the stick–slip process, J. Appl. Phys. **14** (1943) 689–700
18.288 F. Heymann, E. Rabinowicz, B. Rightmire: Friction apparatus for very low-speed sliding studies, Rev. Sci. Instrum. **26** (1954) 56–58
18.289 J. H. Dieterich: Time-dependent friction and the mechanics of stick–slip, Pure Appl. Geophys. **116** (1978) 790–806
18.290 J. H. Dieterich: Modeling of rock friction. 1. Experimental results and constitutive equations, J. Geophys. Res. **84** (1979) 2162–2168
18.291 G. A. Tomlinson: A molecular theory of friction, Phil. Mag. **7** (1929) 905–939
18.292 J. M. Carlson, J. S. Langer: Mechanical model of an earthquake fault, Phys. Rev. A **40** (1989) 6470–6484
18.293 B. N. J. Persson: Theory of friction: The role of elasticity in boundary lubrication, Phys. Rev. B **50** (1994) 4771–4786
18.294 S. Nasuno, A. Kudrolli, J. P. Gollub: Friction in granular layers: Hysteresis and precursors, Phys. Rev. Lett. **79** (1997) 949–952
18.295 P. Bordarier, M. Schoen, A. Fuchs: Stick–slip phase transitions in confined solidlike films from an equilibrium perspective, Phys. Rev. E **57** (1998) 1621–1635
18.296 J. M. Carlson, A. A. Batista: Constitutive relation for the friction between lubricated surfaces, Phys. Rev. E **53** (1996) 4153–4165
18.297 A. D. Berman, W. A. Ducker, J. N. Israelachvili: Origin and characterization of different stick–slip friction mechanisms, Langmuir **12** (1996) 4559–4563
18.298 A. D. Berman, W. A. Ducker, J. N. Israelachvili: Experimental and theoretical investigations of stick–slip friction mechanisms, NATO ASI Ser. E: Appl. Sci. **311** (1996) 51–67
18.299 K G. McLaren, D. Tabor: Viscoelastic properties and the friction of solids. Friction of polymers and influence of speed and temperature, Nature **197** (1963) 856–858
18.300 K. A. Grosch: Viscoelastic properties and friction of solids. Relation between friction and viscoelastic properties of rubber, Nature **197** (1963) 858–859
18.301 J. Gao, W. D. Luedtke, U. Landman: Friction control in thin-film lubrication, J. Phys. Chem. B **102** (1998) 5033–5037
18.302 M. Heuberger, C. Drummond, J. Israelachvili: Coupling of normal and transverse motions during frictional sliding, J. Phys. Chem. B **102** (1998) 5038–5041
18.303 L. Bureau, T. Baumberger, C. Caroli: Shear response of a frictional interface to a normal load modulation, Phys. Rev. E **62** (2000) 6810–6820
18.304 J. P. Gao, W. D. Luedtke, U. Landman: Structure and solvation forces in confined films: Linear and branched alkanes, J. Chem. Phys. **106** (1997) 4309–4318
18.305 J. Gao, W. D. Luedtke, U. Landman: Layering transitions and dynamics of confined liquid films, Phys. Rev. Lett. **79** (1997) 705–708
18.306 J. Warnock, D. D. Awschalom, M. W. Shafer: Orientational behavior of molecular liquids in restricted geometries, Phys. Rev. B **34** (1986) 475–478
18.307 D. D. Awschalom, J. Warnock: Supercooled liquids and solids in porous glass, Phys. Rev. B **35** (1987) 6779–6785
18.308 M. Hirano, K. Shinjo, R. Kaneko, Y. Murata: Anisotropy of frictional forces in muscovite mica, Phys. Rev. Lett. **67** (1991) 2642–2645
18.309 T. Gyalog, H. Thomas: Friction between atomically flat surfaces, Europhys. Lett. **37** (1997) 195–200

19. Scanning Probe Studies of Nanoscale Adhesion Between Solids in the Presence of Liquids and Monolayer Films

Adhesion between solids is a ubiquitous phenomenon whose importance is magnified at the micrometer and nanometer scales, where the surface-to-volume ratio diverges as we approach the single atom.

Numerous techniques for measuring adhesion at the atomic scale have been developed. Yet significant limitations exist. Instrumental improvements and reliable quantification are still needed.

Recent studies have highlighted the unique and important effect of liquid capillaries, particularly water, at the nanometer scale. The results demonstrate that macroscopic considerations of classic meniscus theory must be, at the very least, corrected to take into account new scaling and geometric relationships unique to the nanometer scale. More generally, a molecular-scale description of wetting and capillary condensation as it applies to nanometer-scale interfaces is clearly desirable, but remains an important challenge.

The measurement of adhesion between self-assembled monolayers has proven to be a reliable means for probing the influence of surface chemistry and local environment on adhesion. To date, however, few systems have been investigated quantitatively in detail. The molecular origins of adhesion down to the single bond level remain to be fully investigated. The most recent studies illustrate that although new information about adhesion in these systems has been revealed, further enhancements of current techniques, as well as the development of new methodologies, coupled with accurate theoretical modeling, are required to adequately tackle these complex measurements.

19.1 **The Importance of Adhesion at the Nanoscale** ... 605

19.2 **Techniques for Measuring Adhesion** ... 606

19.3 **Calibration of Forces, Displacements, and Tips** ... 610
- 19.3.1 Force Calibration ... 610
- 19.3.2 Probe Tip Characterization ... 611
- 19.3.3 Displacement Calibration ... 612

19.4 **The Effect of Liquid Capillaries on Adhesion** ... 612
- 19.4.1 Theoretical Background ... 612
- 19.4.2 Experimental and Theoretical Studies of Capillary Formation with Scanning Probes ... 614
- 19.4.3 Future Directions ... 618

19.5 **Self-Assembled Monolayers** ... 618
- 19.5.1 Adhesion at SAM Interfaces ... 618
- 19.5.2 Chemical Force Microscopy: General Methodology ... 619
- 19.5.3 Adhesion at SAM-Modified Surfaces in Liquids ... 620
- 19.5.4 Impact of Intra- and Inter-Chain Interactions on Adhesion ... 621
- 19.5.5 Adhesion at the Single-Bond Level ... 622
- 19.5.6 Future Directions ... 623

19.6 **Concluding Remarks** ... 624

References ... 624

19.1 The Importance of Adhesion at the Nanoscale

The mutual attraction and bonding of two surfaces, which can occur with or without an intervening medium, is a commonly observed phenomenon with far-reaching manifestations and applications in society. The adherence between a raindrop and a window pane, the climbing of a gecko up a vine, the sticking of multiple adhesive note pads to a professor's wall, the force required to separate hook-and-eye (Velcro^{TM}) strips, the building of a sand castle, and the book page turned by a wetted finger are all examples in which adhesion is important. Within the complexity of these examples and others lies a central theme: The mechanical forces between a pair of materials can be fundamentally affected not just by the macroscopic or microscopic structure of the surfaces, but also by the interatomic and intermolecular forces between them.

Adhesion and intermolecular forces have been studied for many years dating back to ancient times [19.1], and active research continues today of topics as broad as insect and reptile locomotion [19.2], interactions between cells in the body [19.3], and the design of self-healing composites [19.4], to name a few. While adhesion is clearly of interest for a wide range of macroscopic applications, its importance becomes dominant at the micrometer and nanometer scales. This is primarily due to the dramatic increase in the surface-to-volume ratio of materials at these scales, an effect that renders friction and interfacial wear critical phenomena [19.5]. For example, the dominating effect of adhesion at this scale has affected the development of microelectromechanical systems (MEMS), in which interfacial forces can prevent devices from functioning properly, since the small flexible parts often emerge from the fabrication line stuck together. Studies of adhesion in MEMS are ongoing, and commercially deployed MEMS devices rely upon sophisticated surface treatments to reduce and control adhesion [19.6–8].

Detailed control of adhesion at the molecular level will be essential for the proposed design of the smaller nanoelectromechanical systems (NEMS). Much has been written about the possibilities for nanoscale machines, sensors, and actuators, etc. It is crucial to understand that a molecule is *all* surface, and, therefore, molecular and nanoscale devices cannot be properly designed, implemented, or characterized unless an understanding of atomic-scale adhesion is thoroughly presented, particularly if these devices are to involve any moving parts that will come into and out of contact with each other. Studying adhesion at the nanoscale is important for other reasons. The protruding asperities in most MEMS materials often have nanoscale dimensions, and, therefore, a complete understanding of adhesion in MEMS requires the investigation of the adhesive properties of the individual nanoscale asperities. In addition, the experimental study of adhesion at the nanoscale is required for the development of detailed atomic-scale models of adhesion. Such progress requires close collaboration between experiment and theory.

There has been significant progress in the experimental study of adhesion at the nanometer scale using scanning probe methods, but numerous challenges exist. A discussion of solid-solid adhesion without an intervening medium is provided in part B and C of this book. This chapter focuses on how adhesion is affected by the ubiquitous presence of water, and how it can be controlled through the application of self-assembled monolayer (SAM) coatings, again in the presence of a liquid medium. In addition, we discuss specific instrumental challenges that are inherent to adhesion measurements. We do not delve into the realm of atomic-scale modeling of adhesion, nor do we discuss the role of more complex coatings such as polymer brushes and blends. Rather, we will focus on critically evaluating the relevant experimental techniques and critically reviewing recent results from studies of water and SAM films, which are perhaps the two most commonly encountered media in nanoscale adhesion applications.

19.2 Techniques for Measuring Adhesion

The experimental study of adhesion at the nanoscale has experienced two renaisances in this century. The first occurred with the development of the surface forces apparatus (SFA) [19.9, 10], and the second with the later development of the AFM [19.11] and other related scanning force techniques.

SFA experiments have contributed profoundly to our understanding of adhesion. The SFA consists of a pair of atomically smooth surfaces, usually mica sheets, mounted on crossed cylinders that can be pressed together to form a controlled circular contact. The applied load, normal displacement, surface separation, contact area, and shear force (if applied) can all be controlled and/or measured [19.9, 12–14]. The SFA can be operated in air, a controlled environment, or under liquid conditions. The surfaces are often treated to attach molecules whose behavior under confinement can be studied. Alternately, the behavior of a confined fluid layer can be observed. The surface separation can be measured and controlled in the Ångstrom (Å) regime. The lateral resolution is limited to the range of several tens of micrometers. However, the true contact area between the interfaces can be directly measured, which is a key advantage since it allows an adhesive force to be converted to a force (or energy) per unit area (or per molecule), thus separating geometrical contributions to adhesion from chemical contributions. In this chapter, we will refer to relevant results in the context of the scanning probe experiments. As a very brief summary, some of the most important results pertaining to adhesion include: the observation of capillary effects on adhesion [19.15, 16], the presence of hysteresis in adhesion due to pressure-induced restructuring of the interface [19.17], and numerous studies of how interfacial chemistry (hydrophilicity, surface charge, specific chemical groups, etc.) affect adhesive forces [19.1, 18, 19].

AFM instrumentation is described in part B of this book. Key ways in which the AFM differs from the SFA are: (1) the lateral contact size is nm, not μm, due to the fact that the tip is usually < 100 nm in radius, and the contact area at low applied loads will be a fraction of this radius [19.20]; (2) the force resolution is as good as 10^{-10} N or better, as opposed to $\sim 10^{-6}$ N with the SFA; (3) the contact area is not directly observable, although it may be derived through other means; (4) the actual separation between the tip and sample are not directly observable, which is a key disadvantage; (5) the relative separation between the sample and the *cantilever* (not the tip) is controlled in the 0.1 nm range or better; (6) the measurement bandwidth is typically in the kHz regime, but can extend into the MHz regime depending on the data acquisition technique; (7) the operating environment includes ultrahigh vacuum (UHV) and cryogenic to elevated temperatures; (8) there is virtually no limit to the range of materials that can be probed by the AFM; and (9) half of the interface (the tip) is essentially unknown or uncontrolled, whereas with the SFA, both surfaces are well-defined, which is another key challenge for AFM that has yet to be consistently addressed.

The general setup of the AFM is as follows. A small sharp tip (with a radius typically between 10–100 nm) is attached to the end of a compliant cantilever (see Fig. 19.1). The tip is brought into close proximity with a sample surface in a fashion that is identical to STM tip–sample approach mechanisms. Forces acting between the AFM tip and the sample will result in deflections of the cantilever (Fig. 19.2). The cantilever bends vertically (i. e., toward the sample) in response to attractive and/or repulsive forces acting on the tip. The deflection of the cantilever from its equilibrium position is proportional to the normal load applied to the tip by the cantilever. Lateral forces result in a twisting of the cantilever from

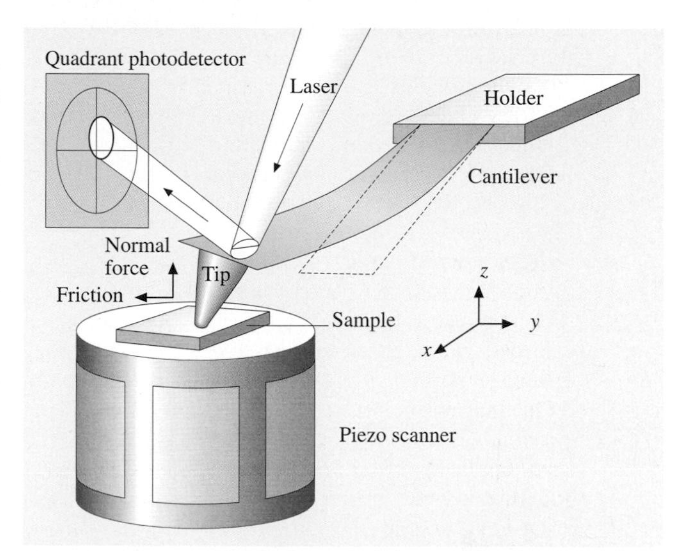

Fig. 19.1 Diagram of the AFM setup for the optical beam deflection method. The tip is in contact with a sample surface. A laser beam is focused on the back of the cantilever and reflects into a four-quadrant photodetector. Normal forces deflect the cantilever up or down, lateral forces twist the cantilever left and right. These deflections are simultaneously and independently measured by monitoring the deflection of the reflected laser beam

its equilibrium position. These measurements can be performed in a variety of environments: ambient air, controlled atmosphere, liquids [19.21], or UHV [19.22–24].

The simplest way (perhaps deceptively so) to measure adhesion with the AFM is through so-called force–displacement plots (also referred to in the literature as force curves and force–distance curves or plots). A force-displacement plot displays the vertical cantilever bending (i. e., the cantilever deflection, which can be calibrated to give the tip–sample interaction force) vs. lever–sample displacement. This displacement is measured between the sample and the rigidly held back end of the cantilever (as opposed to the front end with the tip, which will bend in response to interaction forces). This displacement is altered by varying the vertical position of the piezo tube, which, depending on the type of AFM, displaces the tip or the sample in the direction normal to the sample. Referring to Fig. 19.2a, the stages of acquisition of the force curve are as follows: (a) The lever and sample are initially far apart and no forces act. (b) As the lever is brought close to the sample, the tip senses attractive forces that cause the tip end of the lever to bend downward, thus signifying a negative (attractive) force. These forces may be of electrostatic, van der

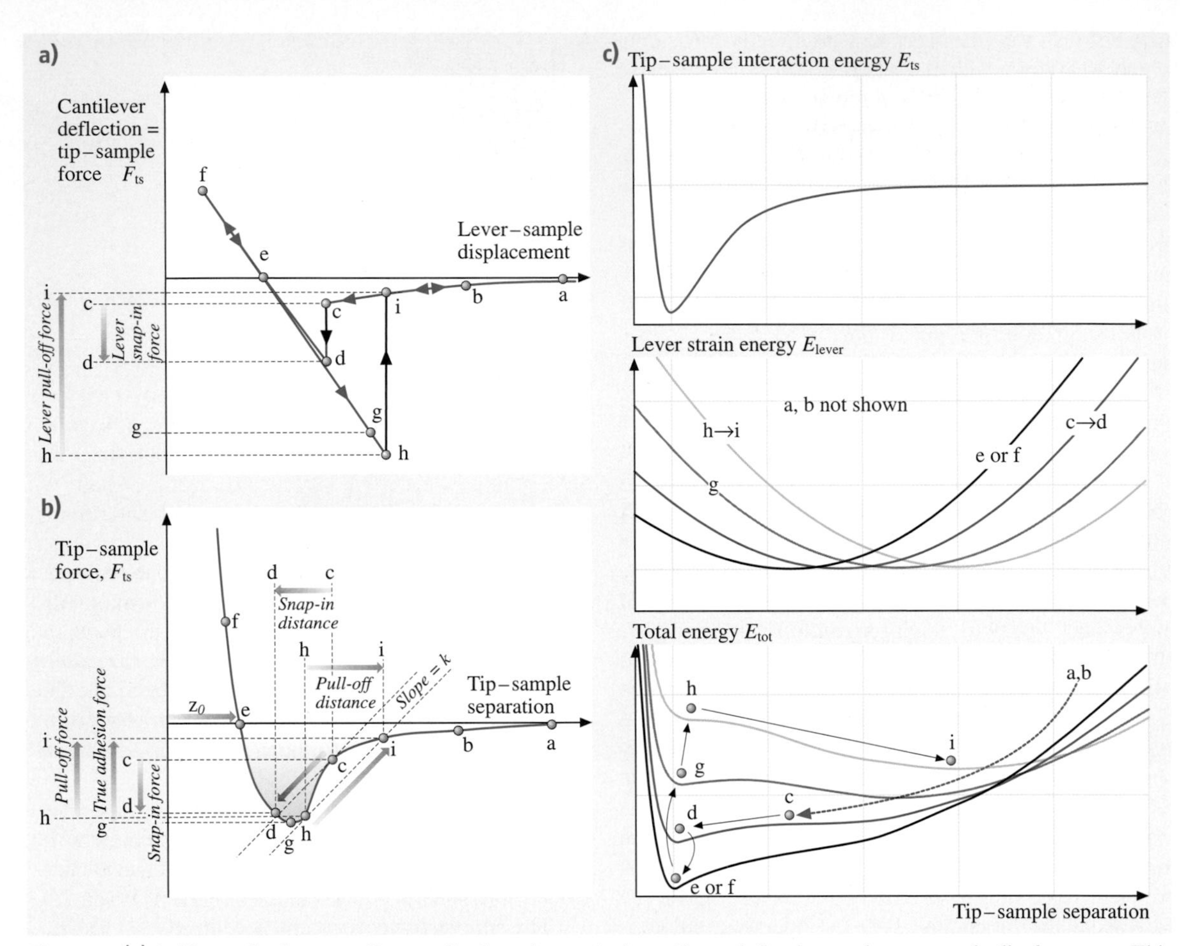

Fig. 19.2 **(a)** A "force-displacement" curve displays the vertical cantilever deflection vs. lever–sample displacement. This displacement is measured between the sample and the rigidly held back end of the cantilever (as opposed to the front end with the tip that will bend in response to interaction forces). **(b)** The true tip–sample interaction as a function of tip–sample separation. **(c)** Tip–sample interaction energy, lever strain energy, and total energy as a function of tip–sample separation

Waals, or other origin. (c) At this point, the attractive force gradient (slope) of the tip-sample interaction force exceeds the spring constant k of the lever, and this instability causes the tip to snap into contact (d) with the sample. The lever-sample displacement can continue to be reduced and eventually crosses the force axis (e), which corresponds to zero externally applied load. As this tip is in repulsive contact with the sample, the front end of the lever is pushed further and further upward, and the force corresponds to the externally applied load. (f) The displacement direction is reversed at a point chosen by the user. As the lever-sample distance is reduced, the force becomes negative. Adhesion between the tip and sample maintains the contact, although there is now a net negative (tensile) load. Eventually the tip passes through the point of maximum adhesion (g), which is not distinguishable on this plot (see Fig. 19.2b). Finally, an unstable point is reached at which the cantilever stiffness exceeds the force gradient of the tip-sample interaction, and the tip snaps out of contact with the sample (h–i). The resulting force relaxation is usually called the pull-off force. Note that the forces and distances are not drawn to scale, and, in particular, the attractive part of the interaction is exaggerated beyond that which often occurs for inert, neutral surfaces.

The force-displacement curve can be measured at a single point, or a series of measurements can be carried out over an area of interest. The so-called adhesion mapping techniques allow for spatially resolved adhesion measurements to be correlated with other sample properties such as friction, chemical termination, and other types of material heterogeneity.

In order to properly derive adhesion measurements from the AFM, it is critical to understand the nature of the mechanical instability of both the snap-in and pull-off processes. The fundamental point to comprehend is that *the pull-off force is* not *the adhesive bond strength.* In other words, *it is not a direct measure of the actual adhesive forces that were acting between the tip and sample in the absence of applied load.* This important point is often overlooked or misunderstood, and so we discuss it here in some detail.

A cartoon of the interaction force (F_{ts}) between the tip and sample is sketched (not to any particular scale) as a function of the true tip-sample separation in Fig. 19.2b. Superimposed on this is the force-distance relation of the cantilever: a straight line with slope k (N/m), the cantilever stiffness (diagonal dashed lines). Points of instability are shown as gray dots, labels, and arrows, as opposed to stable points, shown as black dots, labels, and arrows. Snap-in occurs when the attractive force gradient $\mathrm{d}F_{ts}/\mathrm{d}z$ just exceeds k, in other words, when the dashed line is tangent to the tip-sample force curve, as shown for point c. This instability is a direct consequence of Newton's 2nd law and is explained further below. Similarly, the pull-off occurs during retraction when k finally just exceeds $\mathrm{d}F_{ts}/\mathrm{d}z$, as shown for point h, where once again, the diagonal line is just tangent to the tip-sample force curve. By definition, since k is finite, this point *cannot* correspond to the point of maximum attractive force (or adhesive force), g. Thus, the force at which pull-off occurs is *not* precisely equal to the true adhesion force. The significance of this deviation is discussed below.

This can also be illustrated from an energetic perspective. In Fig. 19.2c, we show separately the tip-sample interaction energy, E_{ts}, as a function of tip-sample separation (top), followed by the quadratic elastic strain energy, E_{lever}, of the cantilever (middle) and the sum, E_{total}, of the two (bottom). As the lever-sample displacement is varied, the elastic energy curve is shifted to the appropriate position, which, for this example, is represented on the tip-sample separation axis. Far away from the sample, the system resides in a deep minimum, points a and b (not shown). As the tip-sample separation is reduced, this minimum becomes more and more shallow (evolving along the dashed arrow). At the snap-in point c, the minimum is eliminated, since the attractive energy of the tip-sample interaction has overwhelmed the energy minimum of the cantilever. Mathematically this is described as an inflection point, where:

$$\frac{\mathrm{d}^2 E_{tot}}{\mathrm{d}z^2} = 0 .$$

Now,

$$E_{tot} = E_{ts} + E_{lever} = E_{ts} + \tfrac{1}{2}k(z - z_0)^2 \tag{19.1}$$

$\therefore$ snap-in occurs when

$$\frac{\mathrm{d}^2 E_{ts}}{\mathrm{d}z^2} = -\frac{\mathrm{d}F_{ts}}{\mathrm{d}z} = -k .$$

This explains why the instability is described by the line of slope k being tangent to the tip-sample force curve, as shown in Fig. 19.2b. The system now rapidly finds the new minimum by following the gray arrow to point d, where a mechanical contact between the tip and sample is now formed.

Note that once the contact is formed, the tip and sample elastically deform, and so, strictly speaking, the tip-sample potential will be distinct from the potential shown before contact occurs. For simplicity, we have left out this change and instead have drawn the potential as a single-valued function. However, such changes in the energy landscape should be considered in a complete description of the problem [19.25].

Nevertheless, it will also hold true that upon retracting the tip from the surface, the stable minimum seen at e, f, and then g becomes more shallow, and eventually disappears as it is overwhelmed by the strain energy of the cantilever. Again, an inflection point is created when

$$\frac{\mathrm{d}^2 E_{attr}}{\mathrm{d}z^2} = -k , \tag{19.2}$$

in other words, when again

$$\frac{\mathrm{d}F_{ts}}{\mathrm{d}z} = k . \tag{19.3}$$

This second instability occurs at point h, the pull-off point. The system then follows the gray arrow to point i, whereby the tip is now out of contact with the sample.

If the spring is sufficiently compliant (low k), or the potential sufficiently stiff (large $\mathrm{d}F/\mathrm{d}z$ shortly past the minimum at g), then the pull-off force *does* nearly correspond to the attractive force minimum F_{min}. In other words, point h would occur very close to point g. But if the cantilever is somewhat stiff, or if the potential is rather compliant (which may be the case for organic, polymeric, biological, or liquid systems), then the pull-off force may differ substantially from F_{min}. For such cases, this distinction between pull-off force and true adhesion is important, and the limitations imposed by the AFM's intrinsic instabilities become apparent. Of course, if a cantilever is used whose stiffness exceeds the attractive force gradient at all points, one will avoid the instabilities. But since it is the deflection of the cantilever that is used to sense the force, one would have to trade force sensitivity for stability, which is usually an unsuitable compromise.

When a compliant spring or holder is used to manipulate a probe, as is the case for AFM, the technique is generally referred to as being "load-controlled", since the load can be prescribed, but the actual displacement of the probe with respect to the sample cannot (as illustrated by the jump in displacement that occurs during snap-in or pull-off).

In contrast, "displacement-controlled" techniques avoid this instability by effectively eliminating the compliance of the spring or holder. This has been carried out for decades in the mechanical testing community, and over the past 10 years, displacement-controlled scanning probe, have been developed. This is accomplished by displacing the tip by applying force to the tip itself. *Pethica* et al. [19.26,27] use a magnetic coating on the tip and external coils to apply forces to the tip. They refer to the instrument as a "force-controlled microscope". *Houston* et al. [19.28,29] control the force electrostatically and refer to the instrument as an "interfacial force microscope" (IFM). *Lieber* et al. use a variation on Pethica's method, whereby a magnetic coil is used to apply a force to the cantilever [19.30], and adapted it to work in solution.

An example of an adhesion measurement with the IFM is shown in [19.31]. which shows that the instrument is able to measure the entire interaction force curve without instabilities. Thus, this provides a *direct* measurement of the minimum interaction force (as well as forces at all other tip-sample separations) and is a more reliable measure of adhesion. The disadvantage with these techniques is primarily one of inconvenience: The probes require extra manufacturing steps and control electronics. However, given the importance of adhesion in nanoscale science and technology, the extra information gained makes these techniques clearly worth the effort.

Pull-off instabilities can occur even in a displacement-controlled experiment. In this case, the instability is an "intrinsic" instability, whereby the adhesive force gradient competes with the stiffness of the *contact itself* [19.32, 33]. Adhesive materials with low stiffness, such as polymers, may show this behavior.

Finally, the pull-off force may show a time dependence that arises from intrinsic viscoelasticity of the tip or sample materials [19.34], or kinetic effects due to adsorption or reaction of materials at the interface [19.35]. These regimes remain relatively unexplored, but certainly, to compare adhesion measurements between labs, the velocity of approach and retraction, as well as the time in contact ought to be reported for any published experimental results.

A quantitative and reliable examination of adhesion, therefore, requires careful consideration of the mechanics of the contact and the cantilever. For an AFM experiment in which instabilities occur, one can conclude that the pull-off force is a good measure of adhesion only if

1. the materials are fully elastic with little or no viscoelastic character,
2. the interface is chemically stable,
3. the cantilever stiffness is sufficiently low compared with the adhesive force gradient, and
4. the contact stiffness is sufficiently high compared with the adhesive force gradient. Otherwise, a more thorough investigation is required.

19.3 Calibration of Forces, Displacements, and Tips

Whether the forces are to be measured with load-controlled or displacement-controlled techniques, measurements cannot be compared between laboratories unless the forces are properly calibrated. Unfortunately, this can be a rather involved task, and adoption of standards has yet to become widespread and robust. Here we provide a summary of the pertinent issues with references to other work for further reading; these issues are also discussed in part B and C of this book.

19.3.1 Force Calibration

Commercially available AFM cantilevers often come in two forms: V-shaped and rectangular. Silicon and silicon nitride are the typical materials, and reflective coatings are often applied to enhance the reflectivity of the laser beam. The normal force constant of a monolithic rectangular cantilever beam is well-defined [19.36], but it relies on knowledge of all lever dimensions and its modulus. For a rectangular beam of length L, width w, thickness t, Young's modulus E, the stiffness k is given by:

$$k = \frac{Ewt^3}{4L^3} . \tag{19.4}$$

The cubic dependence on thickness is particularly problematic since in microfabrication processes t is usually determined by an etching process that is not precisely controlled, and is, therefore, difficult and cumbersome to measure experimentally. Variations in E can also occur depending on the type of cantilever, particularly for silicon nitride cantilevers, although the dependence on E is clearly not as critical. If a bulk value for E is used, the thickness can be determined by measuring the resonance frequency of the cantilever [19.37]. However, this method only works for uncoated monolithic cantilevers, since the metal coatings applied to enhance reflectivity, or for other purposes, will also alter substantially the spring constant [19.38]. In addition, formulae for V-shaped levers are substantially more complicated. Uncoated, single-crystal silicon cantilevers are perhaps the only ones for which the force constant can be reliably determined by using (19.4) and the resonance frequency [19.39]. Otherwise, an experimental, preferably in situ calibration method is strongly encouraged.

Experimental methods to calibrate the force constant of AFM cantilevers have been discussed extensively [19.37, 40–50]. It has been shown repeatedly that the manufacturer's quoted spring constants can be in error by large factors and should simply not be used in any quantitative research effort. We will not delve into the details of these calibration methods, but we do make note of one particularly recent method proposed by *Sader* et al. [19.51] that appears to be reliable and simple to carry out for rectangular levers. It relies on measuring the resonance frequency and the quality factor of the cantilever in air. Use of the hydrodynamic function relates the damping of air to the quality factor, and the dependence of E and t is eliminated from the resulting formula for the force constant.

19.3.2 Probe Tip Characterization

A problem of quite a different nature is that the geometry of the contact formed between the AFM tip and sample surface is not defined if the tip shape and composition is not known. This issue is of crucial importance since one is trying to understand the properties of an interface, and the tip is half of that interface.

The adhesion force between the tip and sample is a meaningless quantity in the absence of any knowledge of the tip shape [19.20]. The only use for such measurements is in cases where the *same* tip is used to compare different samples or different conditions, and verification by cyclic repetition of the experiments is carried out to ensure that the tip itself did not change during the experiment.

There are several in situ methods to characterize the tip shape. A topographic AFM image is actually a convolution of the tip and sample geometry [19.52]. Separation of the tip and sample contributions by contact imaging of known, or at least sharp, sample features allows determination of the tip shape to a significant extent [19.53–64]. Ex-situ tip imaging by transmission electron microscopy has also been performed [19.61, 65]. Some of these measurements have revealed that a *majority* of microfabricated cantilevers possess double tips and other unsuitable tip structures [19.56, 60, 64]. This convincingly demonstrates that tip characterization is absolutely necessary for useful nanotribological measurements with AFM. Thin film coatings applied to microfabricated levers can provide robust, smooth, and even conductive coatings [19.65–67]. Further work in this direction would be useful so as to provide a wider array of dependable tip structures and materials.

In addition to the shape of the tip, its chemical composition is equally important, but is also challenging to determine or control. *Xiao* et al. have shown that the AFM tip is readily chemically modified when scanned in contact with various materials [19.68], even tips that have been coated with self-assembled monolayers in order to control their chemistry. They recommend "running in" the tip with a standard sample to give reproducible results. The stresses that take place in a nano-contact can be very large [19.5], and so modification of both the chemistry and structure of the tip is important to consider.

One class of experiments for which the tip shape and chemistry is not as critical is when a molecule or nanostructure is tethered to the end of the tip and specific interactions are probed [19.69–73]. Another alternative method is to use colloid probes [19.74,75], whereby colloidal particles are attached to the cantilever on top of (or in place of) the tip. This method requires a unique calibration procedure [19.76], but provides particles whose structure and chemistry can be measured and perhaps controlled prior to attachment to the lever.

19.3.3 Displacement Calibration

Proper signal and spatial calibration also requires knowing the sensitivity of the piezoelectric scanning elements. This can involve complications due to instrumental drift [19.77] and inherent piezoelectric effects, namely nonlinearity, hysteresis, creep, and variations of sensitivity with applied voltage [19.78–82]. Caution must be exercised when determining and relying upon these parameters. Techniques such as laser interferometry [19.83], scanning sloped samples [19.45, 78], scanning known surface step heights [19.84], or the use of pre-calibrated piezoelectrics [19.85] can facilitate piezo calibration.

19.4 The Effect of Liquid Capillaries on Adhesion

The adsorption of water and other liquids onto surfaces, and their subsequent behavior at interfaces, continue to be a vibrant area of research. The importance of liquid-solid interface behavior is massive, encompassing topics as broad as paints, textiles, lubricants, geology, and environmental chemistry, and covering all corners of biology.

19.4.1 Theoretical Background

The ability to measure forces at the nanoscale using scanning probe microscopy has generated much interest in these fields. For any force measurement carried out in ambient laboratory conditions, one must consider the possibility of a capillary neck forming between the tip and sample. The study of such necks may, in turn, provide insight into the behavior of the liquid, which is discussed in other chapters in this book. In order to provide background for these emerging areas, we consider here the fundamental mechanical and chemical aspects of adhesion in the presence of a liquid meniscus. By way of introduction, *Israelachvili* [19.17] and *de Gennes* [19.87] provide rigorous coverage of the terminology, physics, and chemistry of liquid films and their wetting properties.

Water readily adsorbs at many surfaces. At a crack or sharp corner it can condense to form a meniscus if its contact angle is small enough. The small gap between an AFM tip and a surface is, therefore, an ideal occasion for such condensation. Early on, it was realized that liquid condensation plays a significant role in tip-sample interactions [19.88, 89].

The classic theory of capillary condensation starts by considering the thermodynamics of capillary formation. The general geometry of the capillary is shown in Fig. 19.3, adapted from [19.86], with hydrophobic surfaces sketched on the left half ($r_1 > 0$) and hydrophilic surfaces on the right half ($r_1 < 0$). The tip is idealized as a sphere of radius R. The water contact angles with the sample and tip respectively are θ_1 and θ_2. D represents the separation of the tip and sample. The angle ϕ is referred to as the "filling angle". The pressure difference, or Laplace pressure, across a curved interface is given by the Young–Laplace equation [19.17, 86]:

$$\Delta p = \gamma \left(\frac{1}{r_1} + \frac{1}{r_2} \right) , \tag{19.5}$$

where r_1 and r_2 are defined in Fig. 19.3 and γ is the surface tension of the liquid. The resulting force is attractive if $\Delta p < 0$. Note that $r_1 > 0$ and $r_2 < 0$. If the capillary formation is isothermal, then one can derive the Kelvin equation:

$$RT \ln \frac{p}{p_0} = \gamma V \left(\frac{1}{r_1} + \frac{1}{r_2} \right) , \tag{19.6}$$

where p_0 is the saturation pressure of the liquid, V is the molar volume of the liquid, T is the temperature, and R the molar gas constant. The ratio p/p_0 simply corresponds to the relative vapor pressure of the liquid,

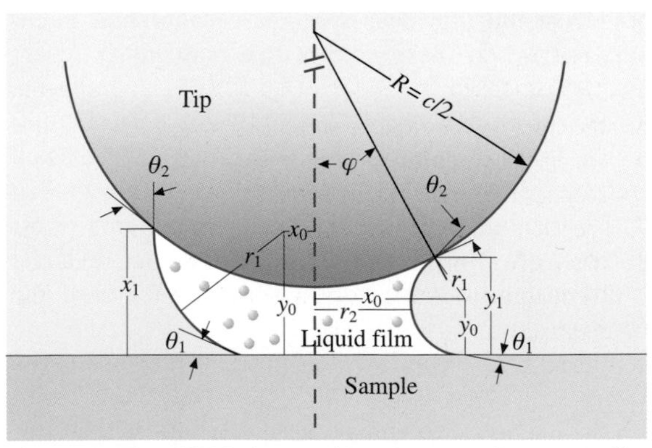

Fig. 19.3 The AFM tip considered as a sphere at a distance D from the sample. The liquid film in between may form a concave (*right*) or convex (*left*) shaped meniscus in the plane shown (adapted after [19.86])

which in the case of water is just the relative humidity (RH). This is often rewritten in terms of the Kelvin radius r_k:

$$r_k^{-1} = \left(\frac{1}{r_1} + \frac{1}{r_2}\right) = \frac{R_M T}{\gamma V} \ln \frac{p}{p_0} . \quad (19.7)$$

The Kelvin radius varies logarithmically with the partial pressure of the liquid and is by definition less than zero (negative values correspond to convex curvature). For water at 20 °C, $\gamma V/RT = 0.54$ nm. A graph of r_k vs. p/p_0 is shown in Fig. 19.4. Of particular note is the fact that starting from $p/p_0 = 0.35$ and higher, we find $|r_k| < 2$ nm. The application of this classic theory to capillaries with such small curvature radii is certainly a question to be considered carefully, and we will address it below.

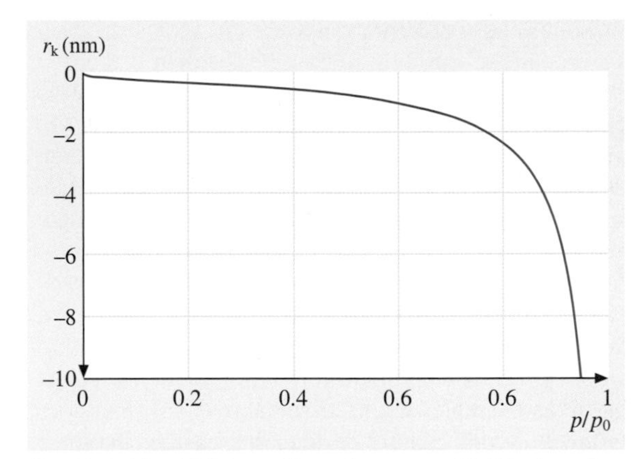

Fig. 19.4 The Kelvin radius of water at 20 °C plotted as a function of the relative humidity

A simple equation for the maximum attractive capillary force F_c between the tip and sample that is commonly used is:

$$F_c = -4\pi\gamma R \cos\theta . \quad (19.8)$$

This equation is calculated by considering the Laplace pressure only. The adhesion force is predicted to be independent of RH. Notice that an immediate problem arises, which is that it predicts a finite force even at 0% RH, where there can be no capillary formation. Derivations of (19.8) are presented in several other publications [19.17, 86, 90, 91]. Significantly, this equation and its use with AFM experiments contain several assumptions, *many of which may not be valid for an AFM experiment*:

1. $|r_1| \ll R$, which is equivalent to saying ϕ is small,
2. $|r_1| \ll |r_2|$,
3. $\theta_1 = \theta_2$,
4. $D \ll |r_1|, |r_2|$,
5. the tip is shaped like a perfect sphere,
6. the effect of solid-solid adhesion is negligible with respect to the meniscus force,
7. the meniscus cross sections are perfect, circular arcs,
8. the force from the Laplace pressure dominates the force due to the resolved surface tension of the meniscus,
9. the surface tension γ is independent of the meniscus size,
10. the meniscus volume remains constant as the tip is retracted,
11. the maximum force of attraction is equal to the pull-off force,
12. the tip and sample are perfectly rigid.

Assumption 1 may not be true, since the tip radius may indeed be small and comparable to the meniscus curvature radii. Assumption 2 may not be true since the small tip geometry may lead the two meniscus curvature radii to be similar. Assumption 3 is by no means true if the tip and sample are made of different materials. Assumption 4 may not be true since both the separation of the tip and sample, as well as the radii r_1 and r_2 may be in the nanometer range. Assumption 5 may be slightly or grossly in error and is a particularly dubious assumption in the absence of tip characterization. In addition, for large menisci, the capillary will grow beyond the end of the tip and start climbing up its shank, which may be pyramidal or conical in shape. Assumption 6 may also be inaccurate if van der Waals or other adhesive forces are significant (this is discussed further in the next section). Assumption 7 is not correct [19.90], and the regimes where it is a reasonable approximation require close scrutiny. Assumption 8 will be inaccurate at high relative vapor pressures [19.17, 91]. The nature of assumptions 1 to 8 and corrections to the theory to account for their violation are presented in the work of *Orr* and *Scriven* [19.90]. This theory remains within the bounds of the classical picture of capillary formation.

Assumption 9 concerns an important scientific question that has not been fully resolved and represents the possible violation of the classical framework of molecular effects at the nanoscale. SFA measurements by *Israelachvili* have indicated that for cyclohexane and other inert organic liquids, γ remains nearly equal to its macroscopic value even for Kelvin radii that, remarkably, correspond to one or two molecular diameters. However, for water, the adhesion force comes to within

10% of the bulk prediction only when $p/p_0 > 0.9$, which corresponds to a Kelvin radius greater than ~ 5 nm in magnitude [19.16]. This deviation from macroscopic thermodynamic predictions alone calls the use of (19.8) into serious question for AFM measurements. Rather, the exploration of this deviation at the molecular scale presents a unique opportunity for scanning probe measurements.

Assumptions 10 and 11 are not assumptions of (19.8) itself, but rather are assumptions that are often used when applying (19.8) to AFM measurements. Equation (19.8) simply gives the maximum force of attraction between the tip and sample. As discussed above, an AFM does *not* measure this quantity. Rather, it measures the force at which an instability occurs. If a capillary has formed between the tip and sample, then the force as a function of distance can be calculated. Calculating this force requires making one of two assumptions: Either the volume of the capillary is conserved (i. e., the rate of displacement is large with respect to the adsorption or desorption kinetics of the liquid), or the Kelvin radius is conserved (i. e., the rate of displacement is slow with respect to the adsorption or desorption kinetics of the liquid, and so the capillary remains in equilibrium). The constant volume assumption 10 has been used in every paper we have reviewed. *Israelachvili*, however, pointed out the difference between these two approaches in his book [19.17] and left the solution of the problem as an exercise to the reader! As with the problem of scale-dependent surface tension mentioned above, the kinetics of capillary formation and dissolution is a relatively unexplored problem and is therefore worthy of further investigation. A recent study of the humidity dependence of friction as a function of sliding speed is an example where this issue is raised [19.92].

Once an assumption about how the meniscus changes with displacement has been made, one still needs to consider the nature of the instability in order to relate the AFM pull-off measurement to the capillary's properties. As stated above and shown in Fig. 19.2, a low lever stiffness k or a strongly varying adhesive force will lead to a pull-off force that is nearly equal to the adhesive force, and so assumption 11 would be valid. However, if k is sufficiently large, or the capillary stiffness sufficiently weak, this assumption will fail. As we shall see below, experimental efforts to investigate this point, and theoretical work to predict it, have yet to be carried out.

Finally, assumption 12, if violated, requires a substantially more complex analysis. The question has been addressed independently by *Maugis* [19.93] and *Fogden* and *White* [19.94]. Both papers provide a nondimensional parameter that allows one to determine the severity of the effect. In the limit of small tips, stiff materials, large (in magnitude) Kelvin radii, and low surface tensions, the effect of elastic deformation is negligible. However, for relatively compliant materials, large tips, and small Kelvin radii, the meniscus can appreciably deform the contact in the immediate vicinity of the meniscus. This can substantially alter the mechanics of adhesion, as well as significantly affect the stresses. The dependence on the Kelvin radius is particularly critical. This effect may be of particular concern with soft materials like polymers or biological specimens. According to *Maugis*, the problem is analogous to the adhesive contact problem for solids, discussed by *Johnson* et al. [19.95] and studied further in many papers since [19.25, 96, 97].

19.4.2 Experimental and Theoretical Studies of Capillary Formation with Scanning Probes

There have been several experimental and theoretical investigations of how pull-off forces are affected by liquid capillaries. These studies have mostly focused on the effect of relative humidity. We do not present an exhaustive review here, but rather summarize a few key results that highlight the important trends observed and the outstanding questions that remain.

Early on, it was realized that water capillary formation occurred readily if at least one of the two surfaces in contact was hydrophilic. Higher adhesion will lead to higher contact forces and therefore larger elastic contact areas, and this can degrade the lateral spatial resolution of the AFM, as observed by *Thundat* et al. [19.98]. Furthermore, several observations have confirmed the expected result that capillary formation was readily prevented for hydrophobic surfaces. For example, *Binggeli* and *Mate* [19.88, 99] showed that tungsten tips in contact with clean silicon wafers with a hydrophilic native oxide exhibited strong adhesion and long pull-off lengths, whereas surfaces treated with a hydrophobic perfluoropolyether (such as Z-dol) showed pull-off forces reduced by a factor of 2 to 3. These results were confirmed by *Bhushan* and *Sundararajan* [19.100] who investigated the pull-off force between silicon nitride tips and Si(100) with and without Z-dol coatings for a range of relative humidities. As another example, MoO_3 films, likely hydrophilic, showed a sixfold increase in pull-off force measured with silicon tips from 0 to 50% RH [19.101].

While these examples illustrate some basic trends, the need for more finely resolved measurements as

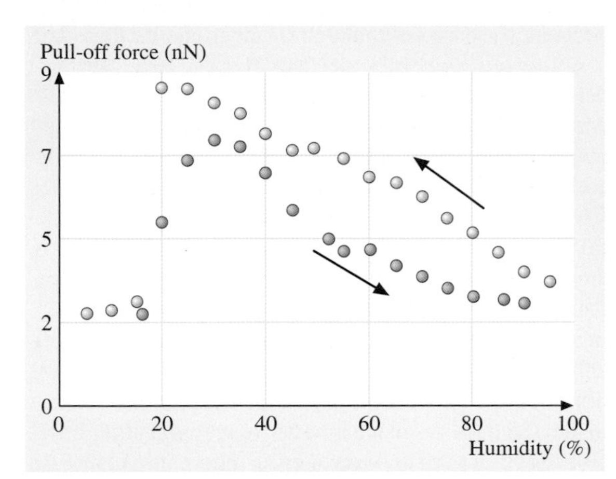

Fig. 19.5 Pull-off force between a silicon nitride tip and the muscovite mica surface as a function of RH. Two sets of data are shown: increasing (*grey circles*) and decreasing (*brown circles*) humidity (after [19.35])

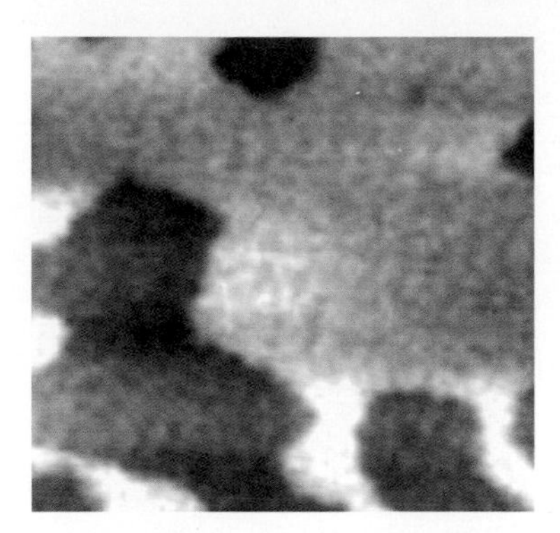

Fig. 19.6 Scanning polarization for microscopy image of water structures on the muscovite mica surface. A degree of polygonal shape to the boundaries is seen. The signal represents two distinct phases of water that are present as a molecular film below 45% RH (after [19.35])

a function of humidity has been addressed only quite recently. For example, *Xu* et al. measured adhesion between a silicon nitride tip and (hydrophilic) muscovite mica in $\sim 5\%$ RH increments. The result is shown in Fig. 19.5. There are three identifiable regions: constant adhesion at low RH ($< 20\%$), increasing adhesion, and then decreasing adhesion. Some hysteresis is seen between experiments conducted with increasing and decreasing RH, but the overall trend is preserved. *Xu* et al. correlated their measurements with detailed studies of the growth of molecular water films on the mica surface, which they could image directly using scanning polarization force microscopy (Fig. 19.6). They proposed that below 20%, capillary condensation does not occur, and, indeed, they saw no evidence of a water film at these low humidities. Above 20% RH, a strongly bound molecular water layer was formed on the bare mica surface (Fig. 19.6). In the presence of the tip, a capillary meniscus condensed. Above 40% RH, the pull-off force decreased. Recognizing one of the limitations of (19.8), *Xu* et al. attribute this to the violation of assumption 5, listed above. They argue that for a pyramidal AFM tip, r_1 and r_2 become comparable in magnitude (and remain opposite in sign) once the capillary reaches the shank of the tip, leading to a near cancellation of the Laplace pressure given in (19.5). This argument is certainly plausible, although a rigorous proof is not provided, and the other limitations of (19.8) are not discussed in relation to this issue. Nevertheless, the correlation between the onset of adhesion increase and the formation of the molecular water film, as seen directly in their dramatic images, is an extremely convincing case where the classical assumptions of (19.8) must be modified to account for the molecular structure of the water film.

Further considerations of the limitations to the (19.8) were measured, discussed, and modeled in a detailed paper by *Xiao* and *Qian* [19.91]. Adhesion measurements, collected in large numbers for good statistics, were carried out with the same silicon nitride tip on two different surfaces: hydrophilic SiO_2 and a hydrophobic layer of *n*-octadecyltrimethoxysilane (OTE) on SiO_2. Contact angle measurements to confirm the assertion of hydrophilicity were not presented, however, the OTE surface was confirmed to be hydrophobic with a water contact angle of 108°. The results are shown in Fig. 19.7. The hydrophobic surface shows no dependence on RH, whereas the hydrophilic bare SiO_2 surface

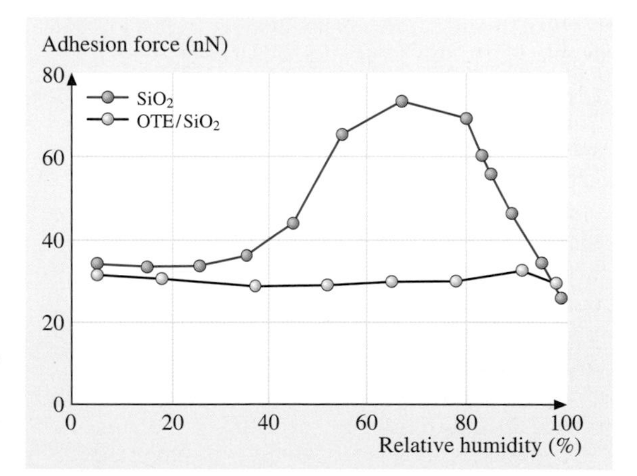

Fig. 19.7 Pull-off force as a function of RH for adhesion between a silicon nitride tip and SiO_2 (*brown circles*) and OTE/SiO_2 (*grey circles*) samples (after [19.91])

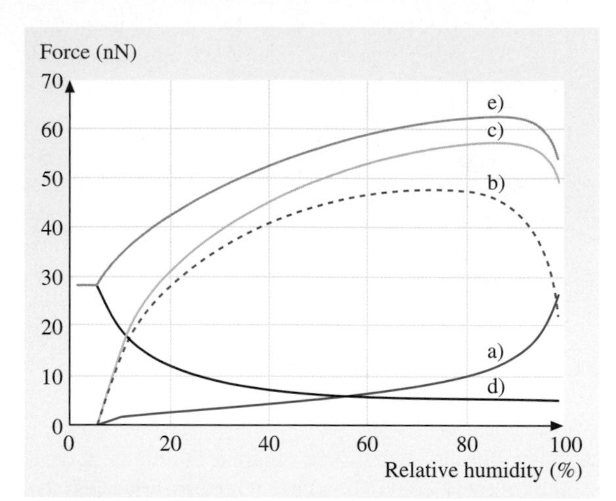

Fig. 19.8 Contributions to the adhesion force as a function of humidity: a) resolved surface tension force; b) Laplace pressure force; c) total capillary force (Laplace + surface tension); d) van der Waals force; e) total adhesion force. (After [19.91])

shows three regimes similar to the result of *Xu* et al.: constant pull-off force (>∼ 30% RH), increasing pull-off force (30%–70% RH), and then decreasing pull-off force. *Xiao* and *Qian* discuss the limitations of (19.8) in substantial detail. In particular, they consider the violation of assumptions 1 through 8 listed above. This includes a treatment of van der Waals adhesion, taking into consideration the effect of electrostatic screening of this force by the water itself. Equations for this more general case are presented. With all these aspects taken into account, they are only able to fit the model to their data qualitatively (Fig. 19.8). At low RH (< 10%), the van der Waals force dominates and the adhesion is initially constant. At intermediate values, the Laplace pressure contribution increases and then begins to saturate. The contribution from the resolved surface tension becomes significant at high RH (above ∼ 80%), and the contribution from the Laplace pressure begins to drop strongly around this same point. Qualitatively, this reproduces their results (and the aforementioned ones) by producing regimes of constant, increasing, and then decreasing adhesion. However, the humidities at which the transitions occur, and the relative changes in adhesion, do not match the data. Somewhat better agreement at high RH (> 70%) is found by considering alternate (blunt) tip shapes. Nevertheless, the most significant discrepancy occurs for low RH, where the extent of the constant adhesion force is underestimated by the classical theory. The authors attribute this to the failure of the classical continuum theory to properly describe the properties of a molecular-scale meniscus. Another interesting point of this study is that at low RH, the adhesion is very similar for both samples, a fact that the authors attribute to the dominance of the van der Waals force for both samples (which is largely determined by the substrate and not affected significantly by the OTE film).

Slightly more recently, *He* et al. have studied capillary forces for a variety of tip-sample pairs [19.102]. Hydrophilic tips (silicon and silicon nitride with no surface treatment) and hydrophobic tips (coated with *n*-octadecyltricholorsilane) were used. The hydrophobic character of the tip was asserted based on a water contact angle measurement of 105.5°, presumably taken on

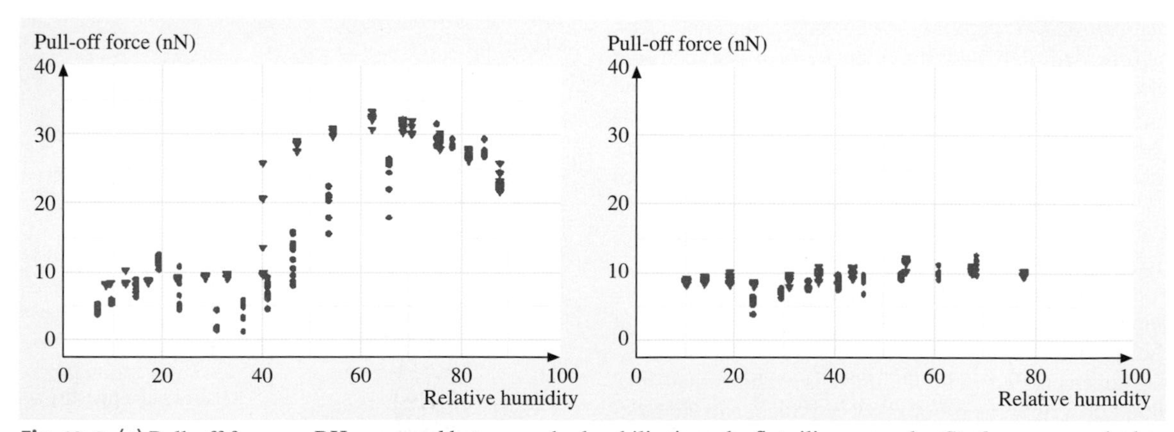

Fig. 19.9 (a) Pull-off force vs. RH measured between a hydrophilic tip and a flat silicon sample. *Circles*: measured when increasing RH. *Triangles*: measured when decreasing RH. (b) Pull-off force vs. RH measured between a sharp SFM tip coated with OTS and a flat silicon sample. The pull-off force is independent of humidity. (After [19.102])

a different region of the cantilever chip. Solvent-cleaned silicon samples and calcium fluoride films were used as hydrophilic substrates. As with the measurements of Xiao and Qian, contact angle measurements for the hydrophilic samples and tips were not presented. Results for hydrophobic and hydrophilic tips are shown in Fig. 19.9.

These results, which were carried out independently and without knowledge of Xiao and Qian's work, show impressive agreement. This is particularly interesting given that here the tip was varied from hydrophilic to hydrophobic (while the sample stayed hydrophilic), whereas in Xiao and Qian's work, the tip presumably remained hydrophilic and the sample was varied from hydrophilic to hydrophobic. For the hydrophobic tip, the pull-off force remains constant, indicating once again that capillary formation was suppressed. However, for the hydrophilic tip, three regimes of adhesion are found (Fig. 19.9). Similar results are found when a hydrophilic glass microsphere is used as a tip. The authors refer to these three regimes as the van der Waals regime, the mixed van der Waals–capillary regime, and the capillary regime. In agreement with Xiao and Qian's assessment, the authors propose that at low RH (<~ 35% in this case), the formation of the water meniscus is suppressed and the adhesion is dominated by solid-solid (presumably van der Waals) interactions. They propose, based on the work of *de Gennes* on the theory of spreading [19.87], that a minimum precursor film thickness is required to form the meniscus. The authors also present a calculation of the adhesion force when assumptions 1 and 2 are relaxed.

Note that the results discussed above consistently indicate that only one hydrophilic surface is required to form the water film. This means that should meniscus formation be undesirable in a nanoscale application (as is the case for MEMS), then hydrophobic treatment of *both* contacting surfaces is likely required.

Practical guidance for discerning the dependence of capillary and van der Waals force is provided by *Stifter* et al. [19.86], who have produced a detailed theoretical model of the meniscus as a function of contact angles, tip size, relative humidity, separation, and surface tension (to represent liquids other than water). An example is shown in Fig. 19.11. The model relaxes assumptions

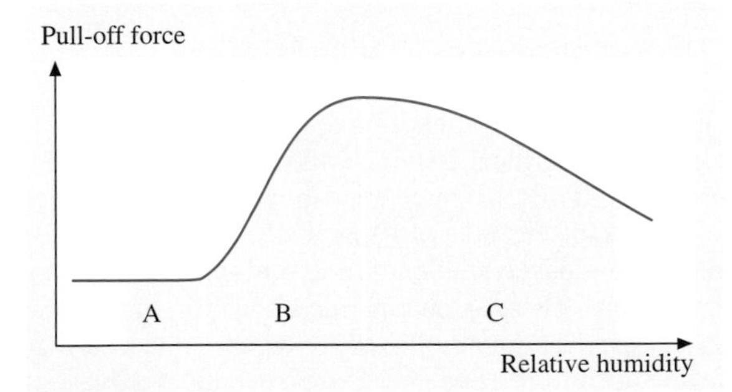

Fig. 19.10 Figure illustrating the distinct regimes of the pull-off force as a function of RH. Regimes A, B, and C are referred to as the van der Waals regime, mixed van der Waals–capillary regime, and capillary regime, respectively. (After [19.102])

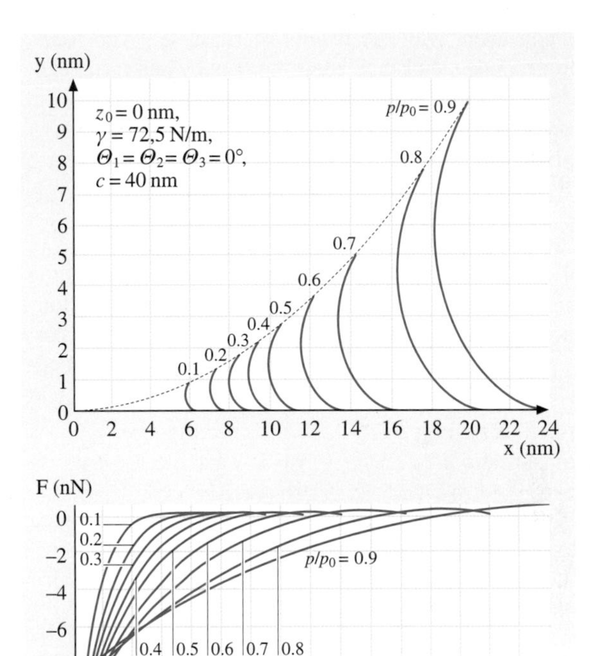

Fig. 19.11 *Top*: geometric shape of the meniscus for a range of RH values. *Bottom*: the meniscus force between the tip and the surface due to the Laplace pressure. The *dashed line* represents a cantilever with force constant 1 N/m. This relatively stiff value is still less stiff than the capillary force curve and suggests that the pull-off force would be very close to the true adhesion force. Of course, summation of the van der Waals and other contributions are necessary to make a full comparison of this sort. (After [19.86])

1 through 4 and 6. These studies show regimes where either van der Waals force or Laplace pressure forces dominate and provide guidance for actual values of the tip geometry and contact angles that minimize the capillary forces. It is not clear how much of an effect the resolved surface tension would have, as that is not included in this particular calculation. Another practical result is that for most of the plots of force vs. displacement, the "stiffness" (gradient) of the curve rarely drops below 1 N/m. This indicates that, within the assumptions of this paper, using cantilevers with force constants of 0.1 N/m will give pull-off forces that are reasonably accurate measurements of the adhesion force. This is consistent with the results of *He* et al., who found no dependence of their measurements when using cantilevers whose nominal force constants ranged from 0.032–0.5 N/m.

19.4.3 Future Directions

The results and modeling so far indicate the possibility of two trends. Hydrophobic surfaces will exhibit little dependence of adhesion on RH, whereas if one surface is hydrophilic, three regimes of behavior occur: constant solid-solid adhesion at low RH, increasing adhesion at intermediate RH, and decreasing adhesion at the highest RH. Some insight into the physical mechanisms behind these regimes has been presented, but it would be desirable to pursue further work in this area.

Perhaps most critically needed is an atomic-scale picture of the menisci and water films present under low partial pressures. This could address the question of why adhesion is initially independent of RH, and when the meniscus itself would start to form. Further theoretical developments that address the assumptions laid out above would help to clarify the picture. It is also important to extend these studies beyond simply the case of water, as the properties of other liquids are also of great interest and could be compared to previous experiments with the SFA and other tools.

Studies that clarify the kinetics of meniscus formation are also needed. There should be a noticeable transition in behavior once the rate of displacement becomes comparable to the appropriate kinetic rates, and this could provide valuable information about these kinetic processes at the nanometer scale.

Finally, as mentioned before, there continues to be a gap in reproducibility and comparability between laboratories that will only be bridged when standard techniques for tip characterization and force and displacement calibration are addressed. Efforts that take these considerations into account are worthy of support.

19.5 Self-Assembled Monolayers

When thinking about adhesion and the related phenomenon of friction, it is important to realize that the interfaces of real surfaces in contact are rarely atomically smooth. Surfaces that appear smooth on the macroscopic scale, upon closer inspection are found to consist of nanometer-scaled asperities (typically on the order of 10 nanometers) whose intentional or accidental interactions ultimately control adhesion, friction, and wear at contacts [19.6, 7, 103–105]. The size of these asperities becomes particularly important when one considers that the true contact area between interfaces for the load distribution is localized through these asperity-asperity interactions in wich extremely high pressures can be produced, sharply increasing local stress fields that can cause materials to yield and shear as they encounter each other during sliding and intermittent contact. In addition to load distribution at nanoscale asperity-asperity contacts, their size will influence surface wetting and adhesion due to capillary forces localized at the contacts [19.7, 104, 105]. The structure of applied lubricant films at such asperities will be highly dependent upon asperity curvature, and defects in lubricant film structure may form more readily here than on atomically flat surfaces.

19.5.1 Adhesion at SAM Interfaces

The minimization of adhesion at such asperity-asperity contacts is a critical issue in MEMS devices [19.6, 7, 104, 105]. In fact, the intentional introduction of surface roughness (on the order of 10 nm RMS) can be employed to lead to reduced stiction during post–processing feature release. These same asperities, however, must later resist wear during controlled or accidental contact during device operation. Thus, the specific details of adhesion and energy dissipation at such asperity-asperity contacts are required for the rational design of such systems.

To function as a protective lubricant layer in such systems, self-assembled monolayers (SAMs) of alkylsilane and fluorosilane compounds with chain lengths ranging from C_{10} to C_{18} have been useful in the reduction of friction and adhesion in MEMS [19.7, 104–106]. Such direct applications of SAMs as lubricant films, combined with the ease of sample preparation and the ability to generate model surfaces with well-defined film chemistry and structure, have made nanotribological and adhesion studies of SAMs a rich area of research [19.5]. Many of these nanotribological studies have used AFM to examine either SAMs of alkylsilane films on atomically smooth Si wafers, glass, or mica surfaces. Alternatively, many researchers have examined alkanethiol films on atomically smooth Au(111) surfaces. Using this approach, many molecular level details, such as the influence of film chemistry and molecular organization on friction, and the adhesion and wear of SAMs, can be obtained. Developing a clear understanding of the details of adhesion at SAM-modified surfaces allows for the complex links between surface chemistry, adhesion, and friction to be understood at the molecular level. In this section, we overview AFM studies of adhesion on SAM-modified surfaces.

19.5.2 Chemical Force Microscopy: General Methodology

In order to probe adhesion between chemically modified surfaces using AFM, the probe tips and sample surfaces are typically modified via self-assembly of monolayers using organosilanes on such surfaces as mica, glass, or oxidized Si or Si_3N_4 (the latter two being the typical materials of which AFM tips are made), or formed from thiols on Au-coated AFM tips and surfaces (Fig. 19.12) [19.107, 108]. While this has been shown to be a facile method for the modification of AFM tips for chemical force measurements, it should be noted that the details of the packing densities of the monolayers formed on the AFM tips are generally not known. This lack of detail regarding molecular overlayers on AFM probe tips can be a problem that requires careful consideration when using such chemically modified tips for the quantitative determination of adhesion forces and molecular interactions, as the number density of species in the tip sample contact is related to the measured adhesion.

Details of the environment in which the adhesion measurements are carried out are also an important consideration. Under ambient environmental lab conditions, surfaces are contaminated with organic compounds from the air, as well as with a layer of condensed water vapor, which varies with humidity. The condensed water layer can form a contact meniscus between the tip and sample, introducing a capillary force into the measured adhe-

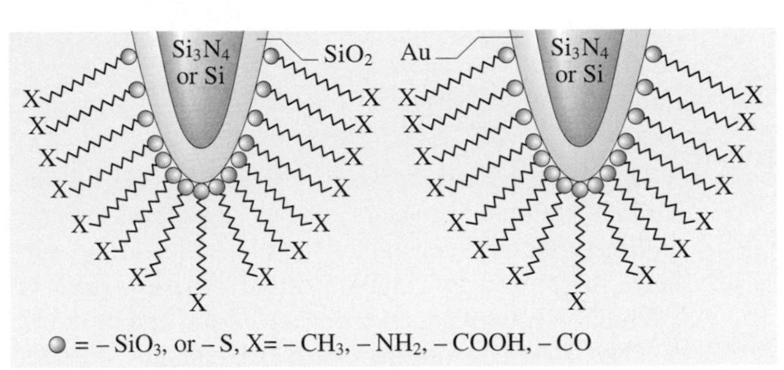

Fig. 19.12 The modification of AFM tips for chemical force microscopy is frequently carried out by chemical functionalization by alkylsilane monolayers on the oxidized surfaces of Si- or Si_3N_4-based tips, or by alkylthiol monolayers assembled on Au-coated (~ 50 nm) AFM tips. The Au tips also typically have a Cr binding layer (~ 5 nm) placed on before Au coating

sion [19.109]. The presence of this capillary force can overwhelm the details of the adhesion from the SAM-terminated surfaces to be probed. To eliminate capillary forces, many studies are performed under liquid environments or ultrahigh vacuum. Under liquids, the nature of the solvent will of course impact the measured adhesion for a given pair of interacting surfaces, as solvent exclusion plays an important role. Also, in the case of water, the pH and ionic strength of the water environment can influence the measured adhesion in the presence of any surface-bound charges.

In quantifying adhesion energies from AFM measurements, the contact mechanics model developed by Johnson, Kendall, and Roberts (JKR model) [19.20, 95] is often employed in the analysis of the adhesion data acquired by force-distance spectroscopy, whereby the number of interacting species (and consequently the average "unit" interaction force or energy) can be derived from the estimated contact area and the average molecular packing density. Using the JKR model, the force of adhesion (AFM pull-off force) is related to the work of adhesion, W_{adh}, and the reduced radius, R, of the tip-surface contact:

$$F_{adh} = -\frac{3}{2}\pi R W_{adh} . \tag{19.9}$$

The work of adhesion is a combination of the tip-surface (γ_{ts}), tip-solvent (γ_{tl}) and surface-solvent (γ_{sl}) interfacial energies ($W_{adh} = \gamma_{sl} + \gamma_{tl} - \gamma_{ts}$), and for tip-surface combinations that have the same chemical composition, the surface energy may be estimated directly from the adhesion measurement as W_{adh}. The effective contact radius at separation, r_s, from the JKR model is given as:

$$r_s = \left(\frac{3\pi W_{adh} R^2}{2K}\right)^{1/3} , \tag{19.10}$$

where K is the reduced elastic moduli of the tip and surface. Using the contact area at separation and the assumed packing density of the molecules at the surfaces in contact, an estimate of the adhesion force or interaction energy on a per molecule basis can be made.

The accuracy of the interfacial energies and per molecule values obtained using this approach, however, must be carefully considered, due to the accumulation of error carried by the imprecise knowledge of the contact, including the tip radius, molecular packing densities of the modified surfaces, as well as the associated elastic properties of the contact at the monolayer level. As the details of the elastic properties of self-assembled monolayers are generally not known, the elastic properties of the contacts are typically assumed to be dominated by those of the underlying substrate, and the bulk values of the surface and/or tip materials (Au, Si, mica, Si_3N_4, SiO_2) are often employed in these calculations. Moreover, as mentioned above, if the molecular packing densities of the monolayers being evaluated are not known (as is the case with a typical AFM tip), then estimations must be used. For contact areas at pull-off approaching 1 nm^2, if the error in packing density is as much as 1 molecule per nm^2, this can lead to an error as high as 25% for a typical alkanethiol, or more in the reported per molecule adhesion force. Such details should be taken into consideration when describing quantitative measurements.

An alternative approach to the measure of adhesive interactions based on Poisson statistics has been promoted by *Beebe* et al. for the statistical evaluation of single bond forces without a priori knowledge of the tip-surface contact details involved [19.119, 121, 122]. A principal limitation of this approach, however, is that a completely homogeneous chemical system is assumed, i. e., that there is only one type of discrete interaction present that gives rise to the observed adhesion. Unfortunately, for many solution phase systems a number of different interactions are typically operative, including energetic exchange with and reorganization of solvent molecules, depending on the solution conditions. These issues have never been thoroughly addressed in any molecular-level measurement of adhesion.

Table 19.1 List of various interactions evaluated by atomic force microscopy adhesion measurements

Chemical Contacts	References
$-CH_3/-CH_3$	[19.110–119]
$-CH_3/-COOH$	[19.110, 114, 116, 118]
$-CH_3/-CONH_2$	[19.118]
$-CH_3/-NH_2$	[19.118]
$-CH_3/-OH$	[19.118]
$-CF_3/-CF_3$	[19.113]
$-CF_3/-CH_3$	[19.113]
$-OH/-OH$	[19.30, 115, 117–121]
$-OH/NH_2$	[19.118]
$-OH/-COOH$	[19.117, 118]
$-OH/-CONH_2$	[19.118]
$-COOH/-COOH$	[19.30, 111, 114, 115, 117, 118, 122]
$-CONH_2/-CONH_2$	[19.118]
$-CONH_2/-COOH$	[19.118]
$-NH_2/-NH_2$	[19.116–118, 123]
$-CH_3/-NH_2$	[19.118, 123]
$-NH_2/-COOH$	[19.118]
$-NH_2/-CONH_2$	[19.118]
$-SO_3H/-SO_3H$	[19.123]
$-CH_3/-SO_3H$	[19.123]
$-NH_2/-SO_3H$	[19.123]
$-PO_3H_2/-PO_3H_2$	[19.116]
$-SH/-SH$	[19.119]

19.5.3 Adhesion at SAM-Modified Surfaces in Liquids

A number of researchers have used AFM to probe interfacial adhesion for a variety of different chemical systems. Most notably, *Lieber* et al. promoted the use of chemically modified substrates and AFM tips to study selective molecular interactions [19.73, 107, 110, 114, 117, 124]. A number of other researchers have adopted a similar methodology, leading to the measurement of adhesion forces and interfacial energies for the interactions of a variety of molecular functional groups in various environments, although to date few specific types of interactions have been thoroughly investigated (Table 19.1) [19.30, 107, 110–115, 120, 122, 125–129].

As described above, measurements in air suffer from difficulties of water vapor, thus, the adhesion of many SAM-terminated surfaces have been evaluated under liquids, and it is these systems that we shall focus on here. The value of the adhesion is of course modified depending on the solvent. For example, the interaction of methyl-terminated interfaces is much stronger in water than in nonpolar solvents, in agreement with the general concept that upon separation the generation of hydrophobic interfaces in a polar solvent is highly energetically unfavorable. The impact of solvent on adhesion can be addressed basically as a variation on the Hamaker constant.

When ionizable end groups are studied in water, the details of the adhesion measurements and results become more complicated, as the chemical nature of the surfaces are now dependent on the pH and ionic strength of the solution. In these circumstances, multiple interactions, including ionic, van der Waals and double-layer forces, come into play simultaneously. Under these conditions, the general form for the JKR adhesion force may be modified to include the following additional forces:

$$F_{\text{adh}} = -\frac{3}{2}\pi R W_{\text{adh}} - \frac{AR}{6D} + 6\pi R W_{\text{dl}} , \qquad (19.11)$$

where the second term now includes the attractive van der Waals component and the third term is the repulsive double layer.

Several groups have utilized the ability of AFM to function as a local probe of ionicity and to carry out local force titration measurements on several functional groups, including $-COOH$ [19.30, 114–118, 122], $-NH_2$ [19.116–118, 123], $-SO_3H$ [19.123] and $-PO_3H_2$ [19.116, 130] (see Table 19.1). The force-distance curves can give a general sense of the local chemical state based on whether the approach curve is attractive or repulsive. In force titrations, the adhesion is measured as a function of pH, and the change in force is dependent upon the equilibrium mixture of charged species within the tip-samples contact at the time of measurement. Peaks in the adhesion force vs. pH allow for the determination of the local pK values for the ionized species, which are shown to be dependent on ionic strength. A force titration of a diprotic acid (11-thioundecyl-1-phosphonic acid) is shown in Fig. 19.13. Shifts in the pK values from those measured in free solution have been ascribed to a variety of factors, including buildup of excess surface charge and solvation effects of the surface-bound ionic species. The localization of charge and change migration are effects that have yet to be fully explored in these systems, however.

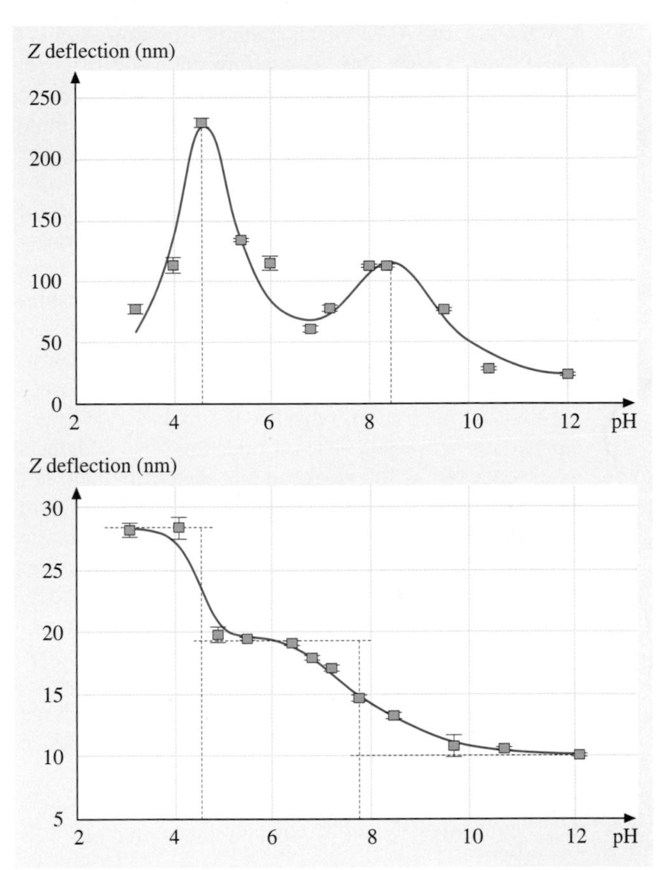

Fig. 19.13a,b Adhesion vs. pH for 11-thioundecyl-1-phosphonic acid in buffer illustrating the impact of pH and ionic strength, (**a**) 10^{-4} M and (**b**) 10^{-1} M on the measured adhesion. The peaks in adhesion provide local measures of the surface pK (adapted after [19.130])

19.5.4 Impact of Intra- and Inter-Chain Interactions on Adhesion

Several studies of the effects of chain length on friction and adhesion have found that, in general, adhesion decreases with increasing chain length, due to increased stability from lateral chain–chain van der Waals interactions within the film, which increases the overall film stiffness. The increased stiffness of the films acts to reduce the effective contact area that develops under compression, consequently reducing adhesion and friction [19.126, 127, 131]. In alkyl chain-based monolayers, the end-group orientation, however, is also dependent on chain length. This results in an "odd-even" effect on the measured adhesion/friction of the monolayer. With $-CH_3$-terminated films, the methyl group orientation differs for odd and even length molecules. This impacts the orientation of the methyl group net dipole and, hence, the local surface free energy [19.73]. In circumstances where interchain hydrogen bonding can occur, additional film stability can be introduced, also yielding an odd–even effect, as observed by *Houston* et al. [19.111, 125].

Studies by AFM of alkanethiol films on Au surfaces have shown that, depending upon the tip size, under varying loads, the tip can readily penetrate the SAM film, displacing the thiol layer from the tip-sample contact [19.132–134]. Upon reduction of the force, the tip again moves out of the SAM film and the surface structure is returned to its original condition. Of key importance in such studies is the mechanism by which the film is displaced. Recently, this has been modeled by *Harrison* using molecular dynamics simulations [19.135]. These studies have clearly demonstrated that in the initial stages of film compression and penetration by an asperity, gauche defects within the typically all-trans configurations of molecules in the SAM layers appear and propagate. The introduction of such defects is the catalyst for the weakening of the chain-chain lateral interactions that help stabilize and maintain film integrity. As this is lost, the asperity can rapidly penetrate the film, and alkanethiol displacement can occur either via chain collapse or bond scission from the surface. *Salmeron* et al. demonstrated that the prevalence of molecular displacement versus film compression depends heavily on the AFM tip size, with sharp tips readily penetrating and displacing surface-bound thiols, while large tips spread the load over more molecules and induce compression over displacement (Fig. 19.14).

The inherent stability of the film structure has also been confirmed by sum-frequency generation spectroscopic studies that have indicated that without the presence of lateral chain-chain interactions, gauche defects appear within the chain structure that reduce overall order and lateral interactions within films [19.136]. The ability to form these defects through poor film order will consequently increase adhesion, friction, and wear of the film. This same spectroscopic study further demonstrated that the appearance of gauche defects can be induced by controlling the local environment, i. e., with the presence of water, causing chains to collapse back upon themselves.

As molecular structure in SAMs moves away from ideally organized layers, chain-chain interactions between the contacting surfaces can also result in entanglements. A recent study of the adhesive interactions of Au(111) surfaces modified with dialkylsulfides [$CH_3(CH_2)_n-S-(CH_2)_9CH_3$; $n = 9, 11, 13, 15, 17$] with varying chain arm lengths probed the combined effect of chain length, solvent, and inter-surface chain entanglement on friction and adhesion using simultaneously modified surfaces and AFM tips [19.137]. This study found that chain-chain interpenetration produced the reverse dependence on chain length on the measured adhesion mentioned above. This work points to the need to examine the nature of inter-surface chain entanglements in nanoscaled systems. Such entanglements should be more prevalent in asperity-asperity contacts where ideal film structure will not be feasible.

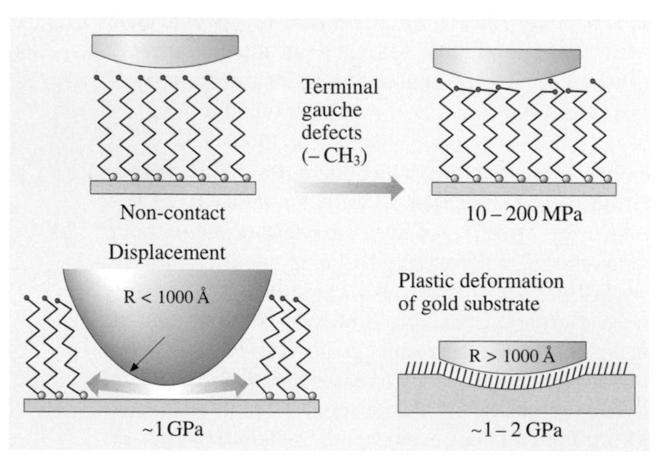

Fig. 19.14 Schematic illustrating the compression of a model lubricant layer under an AMF tip. At low pressures, the gauche defects can form at the tip–sample junction, but the molecules remain in place. Depending on tip size and load, displacement (for sharp tips) or trapping of monolayer molecules (for blunt tips) can occur (adapted after [19.5])

19.5.5 Adhesion at the Single-Bond Level

The ability to resolve the discrete components of interactions is highly desirable. There have been few reports of the direct observation of discrete force components observed with the separation of an AFM tip from a surface. *Beebe* et al., have utilized a statistical method (as described above) for the direct determination of single-bond forces for a variety of interactions [19.119, 121, 122], including biological systems such as biotin-avidin [19.138, 139]. The first report of quantized force measurements was described by *Hoh* et al., which lead to the estimation of single hydrogen bonding forces from studies of glass surfaces in water to be on the order of 10 pN [19.140]. The use of AFM for the study of the energetics of true single chemical bond cleavage has also received little attention. One previous report described discrete covalent bond scission using AFM. In that case, it was proposed that the jumps in the observed pull-off curves were due to sequential scission of chemical bonds contained in a large,

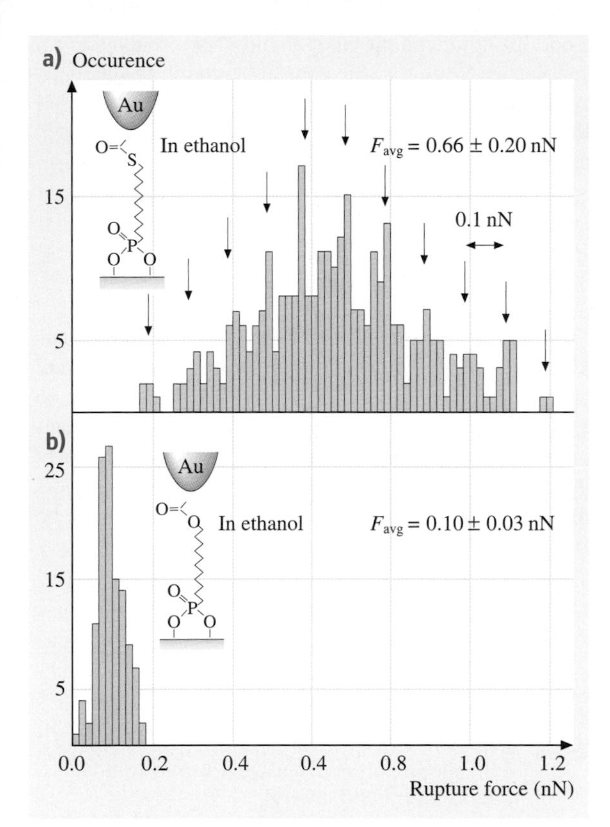

Fig. 19.15 Pull-off force distribution for Au atom abstraction in ethanol. The 100 pN peaks in the distribution represent quantized numbers of Au atoms being abstracted from the sample surface (adapted after [19.129])

multifunctional, polymeric species as it progressively detached from the substrate [19.141]. However, the identification of the relevant chemical bonds involved at each stage was largely based on the known (gas phase) bond strengths of the potentially active functional groups such that solvation effects on the bond energies were ignored – a simplification that profoundly affects the estimated energetics, as energetic exchange with the solvent must be included. More recently, the measurement of discrete bond scission was reported by *Frisbie* et al. for Au-thiol complexes (Fig. 19.15). Here the details of Au atom abstraction were reported with a quantized value of 100 pN (estimated at ~ 10 kJ/mol, based on an assumed bond rupture length of 1 Å) [19.129]. These studies have demonstrated the feasibility of probing local single bond energetics and have suggested some general requirements for the measurement of adhesion quantization in SAM layers, including the need for a significant negative tip-surface interfacial energy coupled with minimal solvent surface tension [19.129, 142].

19.5.6 Future Directions

Extending AFM adhesion measurements to reactive systems in which chemical bonds can form between the tip and surface affords an expansion of chemical reaction dynamics to solution-based chemistries, whereby the energetic details of single reaction events, previously only accessible for gas phase scattering experiments, may be obtained. Studies of such complex heterogeneous systems will open the door to the evaluation of the energetic pathways of solution-based chemistries for any system in which the appropriate functionalization of surfaces can be exploited. Adhesion has already been demonstrated as a reasonable local probe of surface reaction kinetics, whereby the local changes in the chemical forces may be followed as a function of time during surface chemical reactions [19.143, 144]. In addition to advances in measurements of reactive systems by AFM, complete insight into the operative molecular mechanisms can only be gained when combined with detailed theory that takes into account not only the specific types of interactions present between the surfaces, but also the requisite energetic exchange with local solvent molecules. Advances in computer technologies and in computational theory have made this realistic [19.128].

To advance the area of adhesion measurements at the molecular level, energetic barriers for specific interactions also need to be evaluated, with attention given to the nature of the molecular interactions being probed, *including* the details of energetic exchange with the solvent surrounding the interacting molecules. Studies by AFM of molecular interactions within sharply confined geometries (i. e., $\sim 1\,\text{nm}^2$ contact area) provide an opportunity to evaluate such contributions with molecular detail. Here again, *Lieber* et al. have been advancing the approach of chemically functionalizing carbon nanotubes to reduce both the type and number of specific interacting species [19.73]. This approach may hold some promise for probing well-defined, specific chemical interactions and/or reactions, as long as the nanotubes can be sufficiently stabilized against buckling [19.135, 145, 146].

In addition to the modification of probe geometries to improve localization of interactions for adhesion measurements, one of the principle difficulties in performing AFM measurements of adhesion at SAM surfaces is the unavoidable snap to contact. This makes the details of the long-range interaction potential almost completely inaccessible. To address this issue, *Lieber* et al. have developed a modified AFM system in which the cantilever is magnetized, providing an additional feedback mechanism to aid in avoiding snap-in [19.30]. When used, this approach provides a smooth approach and retract curve. Similarly, the capacitive coupling feedback mechanism of the interfacial force microscope (IFM) also affords measurements of this transition without snap-in [19.125].

19.6 Concluding Remarks

Scanning probes are powerful tools for determining the fundamental molecular basis of adhesion. Continuum models of adhesion and capillary condensations are useful, but careful attention must be paid to their limits and assumptions. Further progress in these areas requires detailed analysis of the structure and chemistry of both the tip and sample, as well as their environment (solvent, humidity, etc.).

The ability to determine the effects of molecular-scale water menisci or single bond energetics are truly spectacular accomplishments that continue to inspire researchers worldwide to pursue these measurements. With attention paid to previous literature and consideration of issues raised in this chapter, many more discoveries are sure to be made.

References

19.1 I. L. Singer, H. M. Pollock (Eds.): *Fundamentals of Friction* (Kluwer, Dordrecht 1992) p. 351

19.2 K. Autumn, Y. A. Liang, S. T. Hsieh, W. Zesch et al.: Adhesive force of a single gecko foot-hair, Nature **405** (2000) 681–685

19.3 A. L. Baldwin, G. Thurston: Mechanics of endothelial cell architecture and vascular permeability, Critical Rev. Biomed. Eng. **29** (2001) 247–278

19.4 S. R. White, N. R. Sottos, P. H. Guebelle, J. S. Moore et al.: Autonomic healing of polymer composites, Nature **409** (2001) 794–797

19.5 R. W. Carpick, M. Salmeron: Scratching the surface: Fundamental investigations of tribology with atomic force microscopy, Chem. Rev. **97** (1997) 1163–1194

19.6 R. Maboudian: Adhesion and friction issues associated with reliable operation of MEMS, MRS Bull. **23** (1998) 47–51

19.7 R. Maboudian, W. R. Ashurst, C. Caffaro: Tribological challenges in micromechanical systems, Tribol. Lett. **12** (2002) 95–100

19.8 R. Maboudian, R. T. Howe: Critical review: Adhesion in surface micromechanical structures, J. Vac. Sci. Technol. **15** (1997) 1

19.9 J. N. Israelachvili: Thin film studies using multiple-beam interferometry, J. Colloid Interface Sci. **44** (1973) 259–272

19.10 J. N. Israelachvili, D. Tabor: The measurement of van der Waals dispersion forces in the range 1.5 to 130 nm, Proc. R. Soc. London A **331** (1972) 19–38

19.11 G. Binnig, C. F. Quate, C. Gerber: Invention of AFM, Phys. Rev. Lett. **56** (1986) 930

19.12 P. Frantz, N. Agraït, M. Salmeron: SFA w/ capacitance, Langmuir **12** (1996) 3289–3294

19.13 J. N. Israelachvili, P. M. McGuiggan, A. M. Homola: Dynamic properties of molecularly thin liquid films, Science **240** (1988) 189–191

19.14 J. Peachey, J. Van Alsten, S. Granick: Design of an apparatus to measure the shear response of ultrathin liquid, Rev. Sci. Instrum. **62** (1991) 463–473

19.15 L. R. Fisher, J. N. Israelachvili: Direct measurement of the effect of meniscus forces on adhesion: A study of the applicability of macroscopic thermodynamics to microscopic liquid interfaces, Colloids Surf. **3** (1981) 303–319

19.16 L. R. Fisher, J. N. Israelachvili: Experimental studies on the applicability of the Kelvin equation to highly curved concave menisci, J. Colloid Interface Sci. **80** (1981) 528–541

19.17 J. N. Israelachvili: *Intermolecular and Surface Forces* (Academic Press, London 1992)

19.18 S. Granick: Molecular tribology of fluids. In: *Fundamentals of Friction*, ed. by I. L. Singer, H. M. Pollock (Kluwer, Dordrecht 1992) p. 387

19.19 G. Reiter, A. L. Demirel, J. Peanasky, L. L. Cai et al.: Stick to slip transition and adhesion of lubricated surfaces in moving contact, J. Chem. Phys. **101** (1994) 2606

19.20 K. L. Johnson: *Contact Mechanics* (Cambridge Univ. Press, Cambridge 1987)

19.21 O. Marti, B. Drake, P. K. Hansma: Atomic force microscopy of liquid-covered surfaces: Atomic resolution images, Appl. Phys. Lett. **51** (1987) 484–486

19.22 G. J. Germann, S. R. Cohen, G. Neubauer, G. M. McClelland et al.: Atomic scale friction of a diamond tip on diamond (100) and (111) surfaces, J. Appl. Phys. **73** (1993) 163

19.23 L. Howald, E. Meyer, R. Lüthi, H. Haefke et al.: UHV AFM design, Appl. Phys. Lett. **63** (1993) 117

19.24 M. Kageshima, H. Yamada, K. Nakayama, H. Sakama et al.: Development of an ultrahigh vacuum atomic force microscope for investigations of semiconductor surfaces, J. Vac. Sci. Technol. B **11** (1993) 1987–1991

19.25 J. A. Greenwood: Adhesion of elastic spheres, Proc. R. Soc. London A **453** (1997) 1277–1297

19.26 S. P. Jarvis, A. Oral, T. P. Weihs, J. B. Pethica: A novel force microscope and point contact probe, Rev. Sci. Instrum. **64** (1993) 3515

19.27 S. P. Jarvis, H. Yamada, S.-I. Yamamoto, H. Tokumoto: A new force controlled atomic force microscope for use in ultrahigh vacuum, Rev. Sci. Instrum. **67** (1996) 2281

19.28 S. A. Joyce, J. E. Houston: A new force sensor incorporating force-feedback control for interfacial force microscopy, Rev. Sci. Instrum. **62** (1991) 710–715

19.29 S. A. Joyce, J. E. Houston, T. A. Michalske: Differentiation of topographical and chemical structures using an interfacial force microscope, Appl. Phys. Lett. **60** (1992) 1175

19.30 P. D. Ashby, L. W. Chen, C. M. Lieber: Probing intermolecular forces and potentials with magnetic feedback chemical force microscopy, J. Am. Chem. Soc. **122** (2000) 9467–9472

19.31 H. I. Kim, V. Boiadjiev, J. E. Houston, X.-Y. Zhu, J. D. Kiely: Tribological properties of self-assembled monolayers on Au, SiO_x and Si surfaces, Tribol. Lett. **10**(1-2) (2001) 97–101

19.32 J. S. Nelson, B. W. Dodson, P. A. Taylor: Adhesive avalanche in covalently bonded materials, Phys. Rev. B **45** (1992) 4439–4444

19.33 P. A. Taylor, J. S. Nelson, B. W. Dodson: Adhesion between atomically flat metallic surfaces, Phys. Rev. B **44** (1991) 5834–5841

19.34 K. R. Shull: Contact mechanics and the adhesion of soft solids, Mater. Sci. Eng. R: Rep. **R 36** (2002) 1–45

19.35 L. Xu, A. Lio, J. Hu, D. F. Ogletree et al.: Wetting and capillary phenomena of water on mica, J. Phys. Chem. B **102** (1998) 540–548

19.36 S. P. Timoshenko, J. N. Goodier: *Theory of Elasticity* (McGraw Hill, New York 1987)

19.37 J. P. Cleveland, S. Manne, D. Bocek, P. K. Hansma: A nondestructive method for determining the spring constant of cantilevers for scanning force microscopy, Rev. Sci. Instrum. **64** (1993) 403

19.38 J. E. Sader: Parallel beam approximation for V-shaped atomic force microscope cantilevers, Rev. Sci. Instrum. **66** (1995) 4583

19.39 M. Tortonese, M. Kirk: Characterization of application specific probes for SPMs, Proc. SPIE **3009** (1997) 53–60

19.40 T. R. Albrecht, S. Akamine, T. E. Carver, C. F. Quate: Microfabrication of cantilever styli for the atomic force microscope, J. Vac. Sci. Technol. A **8** (1990) 3386–3396

19.41 H.-J. Butt, P. Siedle, K. Seifert, K. Fendler et al.: Scan speed limit in atomic force microscopy, J. Microsc. **169** (1993) 75–84

19.42 Y. Q. Li, N. J. Tao, J. Pan, A. A. Garcia et al.: Direct measurement of interaction forces between colloidal particles using the scanning force microscope, Langmuir **9** (1993) 637–641

19.43 R. Lüthi, E. Meyer, H. Haefke, L. Howald et al.: Nanotribology: An UHV-SFM study on thin films of C_{60} and AgBr, Surf. Sci. **338** (1995) 247–260

19.44 J. M. Neumeister, W. A. Ducker: Lateral, normal, and longitudinal spring constants of atomic force microscopy cantilevers, Rev. Sci. Instrum. **65** (1994) 2527–2531

19.45 D. F. Ogletree, R. W. Carpick, M. Salmeron: Calibration of frictional forces in atomic force microscopy, Rev. Sci. Instrum. **67** (1996) 3298–3306

19.46 J. A. Ruan, B. Bhushan: Atomic-scale friction measurements using friction force microscopy: Part I – general principles and new measurement techniques, Trans. ASME J. Tribol. **116** (1994) 378

19.47 J. E. Sader, I. Larson, P. Mulvaney, L. R. White: Method for the calibration of atomic force microscope cantilevers, Rev. Sci. Instrum. **66** (1995) 3789

19.48 U. D. Schwarz, P. Koster, R. Wiesendanger: Quantitative analysis of lateral force microscopy experiments, Rev. Sci. Instrum. **67** (1996) 2560

19.49 T. J. Senden, W. A. Ducker: Experimental determination of spring constants in atomic force microscopy, Langmuir **10** (1994) 1003

19.50 A. Torii, M. Sasaki, K. Hane, S. Okuma: A method for determining the spring constant of cantilevers for atomic force microscopy, Meas. Sci. Technol. **7** (1996) 179–184

19.51 J. E. Sader, J. W. M. Chon, P. Mulvaney: Calibration of rectangular atomic force microscope cantilevers, Rev. Sci. Instrum. **70** (1999) 3967–3969

19.52 J. S. Villarrubia: Morphological estimation of tip geometry for scanned probe microscopy, Surf. Sci. **321** (1994) 287–300

19.53 F. Atamny, A. Baiker: Direct imaging of the tip shape by AFM, Surf. Sci. **323** (1995) L314

19.54 R. W. Carpick, N. Agraït, D. F. Ogletree, M. Salmeron: Measurement of interfacial shear (friction) with an ultrahigh vacuum atomic force microscope, J. Vac. Sci. Technol. B **14** (1996) 1289–1295

19.55 R. Dixson, J. Schneir, T. McWaid, N. Sullivan et al.: Toward accurate linewidth metrology using atomic force microscopy and tip characterization, Proc. SPIE **2725** (1996) 589–607

19.56 L.S. Dongmo, J.S. Villarrubia, S.N. Jones, T.B. Renegar et al.: Experimental test of blind tip reconstruction for scanning probe microscopy, Ultramicroscopy **85** (2000) 141–153
19.57 K.F. Jarausch, T.J. Stark, P.E. Russell: Silicon structures for in situ characterization of atomic force microscope probe geometry, J. Vac. Sci. Technol. B **14** (1996) 3425–3430
19.58 P. Markiewicz, M.C. Goh: Atomic force microscope tip deconvolution using calibration arrays, Rev. Sci. Instrum. **66** (1995) 3186–3190
19.59 C. Odin, J.P. Aimé, Z. El Kaakour, T. Bouhacina: Tip's finite size effects on atomic force microscopy in the contact mode: Simple geometrical considerations for rapid estimation of apex radius and tip angle based on the study of polystyrene latex balls, Surf. Sci. **317** (1994) 321–340
19.60 S.S. Sheiko, M. Moller, E.M.C.M. Reuvekamp, H.W. Zandbergen: Evaluation of the probing profile of scanning force microscopy tips, Ultramicroscopy **53** (1994) 371–380
19.61 P. Siedle, H.-J. Butt, E. Bamberg, D.N. Wang et al.: Determining the form of atomic force microscope tips. In: *X-Ray Optics and Microanalysis*, ed. by P.B. Kenway, P.J. Duke, G.W. Lorimer et al. (Iop, Manchester 1992) pp. 361–364
19.62 J.S. Villarrubia: Algorithms for scanned probe microscope image simulation, surface reconstruction, and tip estimation, J. Res. Nat. Inst. Stand. Technol. **102** (1997) 425–454
19.63 K.L. Westra, D.J. Thomson: Atomic force microscope tip radius needed for accurate imaging of thin film surfaces, J. Vac. Sci. Technol. B **12** (1994) 3176–3181
19.64 S. Xu, M.F. Arnsdorf: Calibration of the scanning (atomic) force microscope with gold particles, J. Microsc. **3** (1994) 199–210
19.65 U.D. Schwarz, O. Zwörner, P. Köster, R. Wiesendanger: Friction force spectroscopy in the low-load regime with well-defined tips. In: *Micro/Nanotribology and Its Applications*, ed. by B. Bhushan (Kluwer, Dordrecht 1997)
19.66 P. Niedermann, W. Hanni, N. Blanc, R. Christoph et al.: Chemical vapor deposition diamond for tips in nanoprobe experiments, J. Vac. Sci. Technol. A **14** (1995) 1233–1236
19.67 S.J. O'Shea, R.N. Atta, M.E. Welland: Characterization of tips for conducting atomic force microscopy, Rev. Sci. Instrum. **66** (1995) 2508–2512
19.68 L.M. Qian, X.D. Xiao, S.Z. Wen: Tip in situ chemical modification and its effects on tribological measurements, Langmuir **16** (2000) 662–670
19.69 E.L. Florin, V.T. Moy, H.E. Gaub: Adhesion forces between individual ligand-receptor pairs, Science **264** (1994) 415–417
19.70 G.U. Lee, L.A. Chrisey, R.J. Colton: Direct measurement of the forces between complementary strands of DNA, Science **266** (1994) 771–773
19.71 V.T. Moy, E.L. Florin, H.E. Gaub: Intermolecular forces and energies between ligands and receptors, Science **266** (1994) 257–259
19.72 O.H. Willemsen, M.M.E. Snel, K.O. van der Werf, B.G. de Grooth et al.: Simultaneous height and adhesion imaging of antibody-antigen interactions by atomic force microscopy, Biophys. J. **75** (1998) 2220–2228
19.73 S.-S. Wong, H. Takano, M.D. Porter: Mapping orientation differences of terminal functional groups by friction force microscopy, Anal. Chem. **70** (1998) 5209–5212
19.74 H.J. Butt: Measuring electrostatic, van der Waals, and hydration forces in electrolyte solutions with an atomic force microscope, Biophys. J. **60** (1991) 1438–1444
19.75 W.A. Ducker, T.J. Senden, R.M. Pashley: Direct measurement of colloidal forces using an atomic force microscope, Nature **353** (1991) 239–241
19.76 V.S.J. Craig, C. Neto: In situ calibration of colloid probe cantilevers in force microscopy: Hydrodynamic drag on a sphere approaching a wall, Langmuir **17** (2001) 6018–6022
19.77 R. Staub, D. Alliata, C. Nicolini: Drift elimination in the calibration of scanning probe microscopes, Rev. Sci. Instrum. **66** (1995) 2513–2516
19.78 J. Fu: In situ testing and calibrating of z-piezo of an atomic force microscope, Rev. Sci. Instrum. **66** (1995) 3785–3788
19.79 J. Garnaes, L. Nielsen, K. Dirscherl, J.F. Jorgensen et al.: Two-dimensional nanometer-scale calibration based on one-dimensional gratings, Appl. Phys. A **66** (1998) 831–835
19.80 S.M. Hues, C.F. Draper, K.P. Lee, R.J. Colton: Effect of PZT and PMN actuator hysteresis and creep on nanoindentation measurements using force microscopy, Rev. Sci. Instrum. **65** (1994) 1561
19.81 J.F. Jorgensen, K. Carneiro, L.L. Madsen, K. Conradsen: Hysteresis correction of scanning tunneling microscope images, J. Vac. Sci. Technol. B **12** (1994) 1702–1704
19.82 J.F. Jorgensen, L.L. Madsen, J. Garnaes, K. Carneiro et al.: Calibration, drift elimination, and molecular structure analysis, J. Vac. Sci. Technol. B **12** (1994) 1698–1701
19.83 M. Jaschke, H.J. Butt: Height calibration of optical lever atomic force microscopes by simple laser interferometry, Rev. Sci. Instrum. **66** (1995) 1258
19.84 L.A. Nagahara, K. Hashimoto, A. Fujishima, D. Snowden-Ifft et al.: Mica etch pits as a height calibration source for atomic force microscopy, J. Vac. Sci. Technol. B **12** (1993) 1694–1697
19.85 H.M. Brodowsky, U.-C. Boehnke, F. Kremer: Wide range standard for scanning probe microscopy height calibration, Rev. Sci. Instrum. **67** (1996) 4198–4200
19.86 T. Stifter, O. Marti, B. Bhushan: Theoretical investigation of the distance dependence of capillary and

van der Waals forces in scanning force microscopy, Phys. Rev. B **62** (2000) 13667–13673
19.87 P. G. de Gennes: Wetting: Statistics and dynamics, Rev. Mod. Phys. **57** (1985) 827–863
19.88 M. Binggeli, C. M. Mate: Influence of capillary condensation of water on nanotribology studied by force microscopy, Appl. Phys. Lett. **65** (1994) 415
19.89 Y. Sugawara, M. Ohta, T. Konishi, S. Morita et al.: Effects of humidity and tip radius on the adhesive force measured with atomic force microscopy, Wear **168** (1993) 13–16
19.90 F. M. Orr, L. E. Scriven, A. P. Rivas: Pendular rings between solids: Meniscus properties and capillary force, J. Fluid Mech. **67** (1975) 723–742
19.91 X. Xiao, Q. Linmao: Investigation of humidity-dependent capillary force, Langmuir **16** (2000) 8153–8158
19.92 E. Riedo, F. Levy, H. Brune: Kinetics of capillary condensation in nanoscopic sliding friction, Phys. Rev. Lett. **88** (2002) 185505/1-4
19.93 D. Maugis, B. Gauthiermanuel: JKR-DMT transition in the presence of a liquid meniscus, J. Adhes. Sci. Technol. **8** (1994) 1311–1322
19.94 A. Fogden, L. R. White: Contact elasticity in the presence of capillary condensation. 1. The nonadhesive Hertz problem, J. Colloid Interface Sci. **138** (1990) 414
19.95 K. L. Johnson, K. Kendall, A. D. Roberts: Surface energy and the contact of elastic solids, Proc. R. Soc. London A **324** (1971) 301–313
19.96 K. Johnson, J. Greenwood: An adhesion map for the contact of elastic spheres, J. Colloid Interface Sci. **192** (1997) 326–333
19.97 K. L. Johnson: Adhesion and friction between a smooth elastic asperity and a plane surface, Proc. R. Soc. London A **453** (1997) 163
19.98 T. Thundat, X. Y. Zheng, G. Y. Chen, R. J. Warmack: Role of relative humidity in atomic force microscopy imaging, Surf. Sci. **294** (1993) 939–943
19.99 M. Binggeli, R. Christoph, H.-E. Hintermann: Observation of controlled, electrochemically induced friction force modulations in the nano-Newton range, Tribol. Lett. **1** (1995) 13
19.100 B. Bhushan, S. Sundararajan: Micro/nanoscale friction and wear mechanisms of thin films using atomic force and friction force microscopy, Acta Mater. **46** (1998) 3793–3804
19.101 W. Gulbinski, D. Pailharey, T. Suszko, Y. Mathey: Study of the influence of adsorbed water on AFM friction measurements on molybdenum trioxide thin films, Surf. Sci. **475** (2001) 149–158
19.102 M. He, A. S. Blum, D. E. Aston, C. Buenviaje et al.: Critical phenomena of water bridges in nanoasperity contacts, J. Chem. Phys. **114** (2001) 1355–1360
19.103 J. A. Greenwood, J. B. P. Williamson: Contact of nominally flat surfaces, Proc. R. Soc. London A **295** (1966) 300
19.104 K. Komvopoulous: Surface engineering and microtribology for microelectromechanical systems, Wear **200** (1996) 305–327
19.105 R. Maboudian, W. R. Ashurst, C. Carraro: Self-assembled monolayers as anti-stiction coatings for MEMS: Characteristics and recent developments, Sens. Actuators A **82** (2000) 219
19.106 R. Maboudian: Surface processes in MEMS technology, Surf. Sci. Rep. **30** (1998) 207–270
19.107 A. Noy, D. V. Vezenov, C. M. Lieber: Chemical force microscopy, Annu. Rev. Mater. Sci. **27** (1997) 381–421
19.108 H. Takano, J. R. Kenseth, S.-S. Wong, J. C. O'Brien et al.: Chemical and biochemical analysis using scanning force microscopy, Chem. Rev. **99** (1999) 2845
19.109 D. L. Sedin, K. L. Rowlen: Adhesion forces measured by atomic force microscopy in humid air, Anal. Chem. **72** (2000) 2183–2189
19.110 C. D. Frisbie, L. F. Rozsnyai, A. Noy, M. S. Wrighton et al.: Functional group imaging by chemical force microscopy, Science **265** (1994) 2071–2074
19.111 H. I. Kim, J. E. Houston: Separating mechanical and chemical contributions to molecular-level friction, J. Am. Chem. Soc. **122** (2000) 12045–12046
19.112 T. Nakagawa, K. Ogawa, T. Kurumizawa: Discriminating molecular length of chemically adsorbed molecules using an atomic force microscope having a tip covered with sensor molecules (an atomic force microscope having chemical sensing function), Jpn. J. Appl. Phys. **32** (1993) 294–296
19.113 T. Nakagawa, K. Ogawa, T. Kurumizawa: Atomic force microscope for chemical sensing, J. Vac. Sci. Techol. **B 12** (1994) 2215–2218
19.114 A. Noy, C. D. Frisbie, L. F. Rozsnyai, M. S. Wrighton et al.: Chemical force microscopy: Exploiting chemically-modified tips to quantify adhesion, friction, and functional group distributions in molecular assemblies, J. Am. Chem. Soc. **117** (1995) 7943–7951
19.115 S. K. Sinniah, A. B. Steel, C. J. Miller, J. E. Reutt-Robey: Solvent exclusion and chemical contrast in scanning force microscopy, J. Am. Chem. Soc. **118** (1996) 8925–8931
19.116 E. W. v.d. Vegte, G. Hadziioannou: Acid-base properties and the chemical imaging of surface-bound functional groups with scanning force microscopy, J. Phys. Chem. **B 101** (1997) 9563–9569
19.117 D. V. Vezenov, A. Noy, L. F. Rozsnyai, C. M. Lieber: Force titrations and ionization state sensitive imaging of functional group distributions in molecular assemblies, J. Am. Chem. Soc. **119** (1997) 2006–2015
19.118 E. W. v.d. Vegte, G. Hadziioannou: Scanning force microscopy with chemical specificity: An extensive study of chemically specific tip-surface interactions and the chemical imaging of surface functional groups, Langmuir **13** (1997) 4357–4368

19.119 L. A. Wenzler, G. L. Moyes, L. G. Olson, J. M. Harris et al.: Single-molecule bond rupture force analysis of interactions between AFM tips and substrates modified with organosilanes, Anal. Chem. **69** (1997) 2855–2861

19.120 T. Ito, M. Namba, P. Buhlmann, Y. Umezawa: Modification of silicon nitride tips with trichlorosilane self-assembled monolayers (SAMs) for chemical force microscopy, Langmuir **13** (1997) 4323–4332

19.121 L. A. Wenzler, G. L. Moyes, G. N. Raikar, R. L. Hansen et al.: Measurements of single-molecule bond rupture forces between self-assembled monolayers of organosilanes with the atomic force microscope, Langmuir **13** (1997) 3761–3768

19.122 T. Han, J. M. Williams, T. P. Beebe: Chemical bonds studied with functionalized atomic force microscopy tips, Anal. Chim. Acta **307** (1995) 365–376

19.123 V. Tsukruk, V. N. Bliznyuk: Adhesive and friction forces between chemically modified silicon and silicon nitride surfaces, Langmuir **14** (1998) 446–455

19.124 D.V. Vezenov, A.V. Zhuk, G.M. Whitesides, C. M. Likeber: Chemical force spectroscopy in heterogeneous systems: Intermolecular interactions involving epoxy polymer, mixed monolayers, and polar solvents, J. Am. Chem. Soc. **124** (2002) 10578–10588

19.125 J. E. Houston, H. I. Kim: Adhesion, friction, and mechanical properties of functionalized alkanethiol self-assembled monolayers, Accounts Chem. Res. **35** (2002) 547–553

19.126 H. I. Kim, M. Graupe, O. Oloba, T. Koini et al.: Molecularly specific studies of the frictional properties of monolayer films: A systematic comparison of CF_3-, $(CH3)_2(CH)$-, and CH_3- terminated films, Langmuir **15** (1999) 3179–3185

19.127 S. Lee, Y. S. Shon, R. Colorado, R. L. Guenard et al.: The influence of packing densities and surface order on the frictional properties of alkanethiol self-assembled monolayers (SAMs) on gold: A comparison of SAMs derived from normal and spiroalkanedithiols, Langmuir **16** (2000) 2220–2224

19.128 Y. Leng, S. Jiang: Dynamic simulations of adhesion and friction in chemical force microscopy, J. Am. Chem. Soc. **124** (2002) 11764–11770

19.129 H. Skulason, C. D. Frisbie: Detection of discrete interactiuons upon rupture of Au microcontacts to self-assembled monolayers terminated with $-S(CO)CH_3$ or $-SH$, J. Am. Chem. Soc. **122** (2000) 9750–9760

19.130 J. Zhang, J. Kirkham, C. Robinson, M. L. Wallwork et al.: Determination of the ionization state of 11-thioundecyl-1-phosphonic acid in self-assembled monolayers by chemical force microscopy, Anal. Chem. **72** (2000) 1973–1978

19.131 A. Lio, D. H. Charych, M. Salmeron: Comparative atomic force microscopy study of the chain length dependence of frictional properties of alkanethiols on gold and alkylsilanes on mica, J. Phys. Chem. **B 101** (1997) 3800

19.132 A. Lio, C. Morant, D. F. Ogletree, M. Salmeron: Atomic force microscopy study of the pressure-dependent structural and frictional properties of n-alkanethiols on gold, J. Phys. Chem. B **101** (1997) 4767–4773

19.133 G.-Y. Liu, M. Salmeron: Reversible displacement of chemisorbed n-alkane thiol molecules on Au(111) surface: An atomic force microscopy study, Langmuir **10** (1994) 367–370

19.134 X. Xiao, J. Hu, D. H. Charych, M. Salmeron: Chain length dependence of the frictional properties of alkylsilane molecules self-assembled on mica studied by atomic force microscopy, Langmuir **12** (1996) 235

19.135 A. B. Tutein, S. J. Stuart, J. A. Harrison: Indentation analysis of linear-chain hydrocarbon monolayers anchored to diamond, J. Phys. Chem. **B 103** (1999) 11357

19.136 R. L. Pizzolatto, Y. J. Yang, L. K. Wolf, M. C. Messmer: Conformational aspects of model chromatographic surfaces studied by sum-frequency generation, Anal. Chim. Acta **397** (1999) 81

19.137 E.W. v.d. Vegte, A. Subbotin, G. Hadziioannou: Nanotribological properties of unsymmetrical n-dialkyl sulfide monolayers on gold: Effect of chain length on adhesion, friction and imaging, Langmuir **16** (2000) 3249–3256

19.138 Y.-S. Lo, N. D. Huefner, W. S. Chan, F. Stevebs et al.: Specific interactions between biotin and avidin studies by atomic force microscopy using the poisson statistical analysis method, Langmuir **15** (1999) 1373–1382

19.139 Y.-S. Lo, J. Simons, T. P. Beebe, Jr.: Temperature dependence of the biotin–avidin bond rupture force studied by atomic force microscopy, J. Phys. Chem. **B 106** (2002) 9847–9857

19.140 J. H. Hoh, J. P. Cleavland, C. B. Prater, J.-P. Revel et al.: Quantitized adhesion detected with the atomic force microscope, J. Am. Chem. Soc. **114** (1992) 4917–4918

19.141 M. Grandbois, M. Beyer, M. Rief, H. Clausen-Schaumann et al.: How strong is a covalent bond?, Science **283** (1999) 1727–1730

19.142 H. Skulason, C. D. Frisbie: Contact mechanics modeling of pull-off measurements: Effect of solvent, probe radius, and chemical binding probability on the detection of single-bond rupture forces by atomic force microscopy, Anal. Chem. **74** (2002) 3096–3104

19.143 H. Schonherr, V. Chechik, C. J. M. Stirling, G. J. Vancso: Monitoring surface reactions at an AFM tip: An approach to following reaction kinetics in self-assembled monolayers on the nanometer scale, J. Am. Chem. Soc. **122** (2000) 3679–3687

19.144 M. P. L. Werts, E. W. v.d. Vegte, G. Hadziioannou: Surface chemical reactions probed with scanning force microscopy, Langmuir **13** (1997) 4939–4942
19.145 S. S. Wong, E. Joselevich, A. T. Woolley, C. Chin Li et al.: Covalently functionalized nanotubes as nanometre-sized probes in chemistry and biology, Nature **394** (1998) 52–55
19.146 S. S. Wong, A. T. Woolley, E. Joselevich, C. L. Cheung et al.: Covalently-functionalized single-walled carbon nanotube tips for chemical force microscopy, J. Am. Chem. Soc. **120** (1998) 8557–8558

20. Friction and Wear on the Atomic Scale

Friction is an old subject of research: the empirical da Vinci–Amontons laws are common knowledge. Macroscopic experiments systematically performed by the school of *Bowden* and *Tabor* have revealed that macroscopic friction can be related to the collective action of small asperities. During the last 15 years, experiments performed with the atomic force microscope gave new insight into the physics of single asperities sliding over surfaces. This development, together with complementary experiments by means of surface force apparatus and quartz microbalance, established the new field of nanotribology. At the same time, increasing computing power allowed for the simulation of the processes in sliding contacts consisting of several hundred atoms. It became clear that atomic processes cannot be neglected in the interpretation of nanotribology experiments. Experiments on even well-defined surfaces directly revealed atomic structures in friction forces. This chapter will describe friction force microscopy experiments that reveal, more or less directly, atomic processes in the sliding contact.

We will begin by introducing friction force microscopy, including the calibration of cantilever force sensors and special aspects of the ultra-high vacuum environment. The empirical Tomlinson model largely describes atomic stick-slip results and is therefore presented in detail. We review experimental results regarding atomic friction. These include thermal activation, velocity dependence, as well as temperature dependence. The geometry of the contact plays a crucial role in the interpretation of experimental results, as we will demonstrate, for example, for the calculation of the lateral contact stiffness. The onset of wear on atomic scale has recently come into the scope of experimental studies and is described here. In order to compare the respective results, we present molecular dynamics simulations that are directly related to atomic friction experiments. We close the chapter with a discussion of dissipation measurements performed in noncontact force microscopy, which may become an important complementary tool for the study of mechanical dissipation in nanoscopic devices.

20.1 **Friction Force Microscopy in Ultra-High Vacuum** ... 632
20.1.1 Friction Force Microscopy ... 632
20.1.2 Force Calibration ... 632
20.1.3 The Ultra-high Vacuum Environment ... 635
20.1.4 A Typical Microscope in UHV ... 635

20.2 **The Tomlinson Model** ... 636
20.2.1 One-dimensional Tomlinson Model 636
20.2.2 Two-dimensional Tomlinson Model 637
20.2.3 Friction Between Atomically Flat Surfaces ... 637

20.3 **Friction Experiments on Atomic Scale** ... 638
20.3.1 Anisotropy of Friction ... 642

20.4 **Thermal Effects on Atomic Friction** ... 642
20.4.1 The Tomlinson Model at Finite Temperature ... 642
20.4.2 Velocity Dependence of Friction ... 644
20.4.3 Temperature Dependence of Friction ... 645

20.5 **Geometry Effects in Nanocontacts** ... 646
20.5.1 Continuum Mechanics of Single Asperities ... 646
20.5.2 Load Dependence of Friction ... 647
20.5.3 Estimation of the Contact Area ... 647

20.6 **Wear on the Atomic Scale** ... 649
20.6.1 Abrasive Wear on the Atomic Scale . 649
20.6.2 Wear Contribution to Friction ... 650

20.7 **Molecular Dynamics Simulations of Atomic Friction and Wear** ... 651
20.7.1 Molecular Dynamics Simulation of Friction Processes ... 651
20.7.2 Molecular Dynamics Simulations of Abrasive Wear ... 652

20.8 **Energy Dissipation in Noncontact Atomic Force Microscopy** .. 654

20.9 **Conclusion** ... 656

References ... 657

20.1 Friction Force Microscopy in Ultra-High Vacuum

The *friction force microscope* (FFM, also called lateral force microscope, LFM) exploits the interaction of a sharp tip sliding on a surface to quantify dissipative processes down to the atomic scale (Fig. 20.1).

20.1.1 Friction Force Microscopy

The relative motion of tip and surface is realized by a *scanner* formed by piezoelectric elements, which moves the surface perpendicularly to the tip with a certain periodicity. The scanner can be also extended or retracted in order to vary the normal force, F_N, which is applied on the surface. This force is responsible for the deflection of the *cantilever* that supports the tip. If the normal force F_N increases while scanning due to the local slope of the surface, the scanner is retracted by a feedback loop. On the other hand, if F_N decreases, the surface is brought closer to the tip by extending the scanner. In such way, the surface topography can be determined line by line from the vertical displacement of the scanner. An accurate control of such vertical movement is made possible by a light beam, which is reflected from the rear of the lever into a photodetector. When the bending of the cantilever changes, the light spot on the detector moves up or down and causes a variation of the photocurrent that corresponds to the normal force F_N to be controlled.

Usually the relative sliding of tip and surface is also accompanied by *friction*. A lateral force, F_L, with the opposite direction of the scan velocity, v, hinders the motion of the tip. This force provokes the torsion of the cantilever, and it can be observed with the topography if the photodetector can reveal not only the normal deflection but also the lateral movement of the lever while scanning. In practice this is realized by four-quadrants photodetectors, as shown in Fig. 20.1. We should notice that friction forces also cause the lateral bending of the cantilever, but this effect is negligible if the thickness of the lever is much less than the width.

The FFM was first used by *Mate* et al. in 1987 to reveal friction with atomic features [20.1], i. e. just one year after *Binnig*, *Quate*, and *Gerber* introduced the atomic force microscope [20.2]. In their experiment, Mate used a tungsten wire and a slightly different technique to detect lateral forces (non-fiber interferometry). The optical beam deflection was introduced later by *Marti* et al., and *Meyer* et al. [20.3, 4]. Other methods to measure the forces between tip and surface are

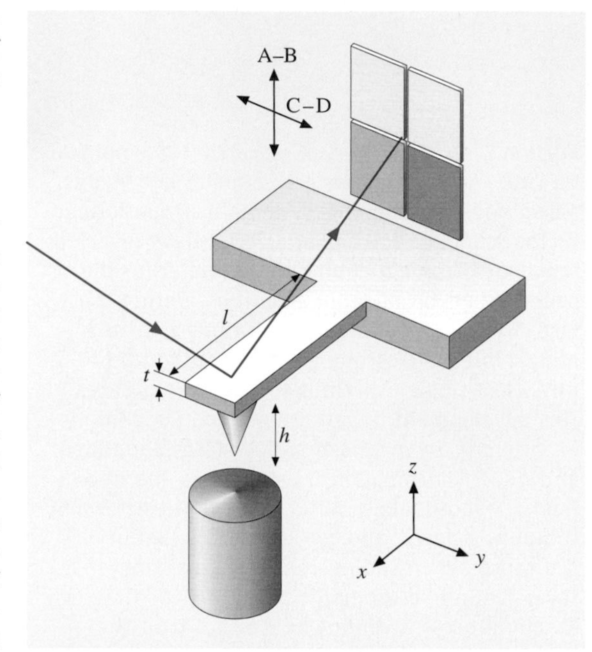

Fig. 20.1 Schematic diagram of a beam-deflection friction force microscope

given by capacitance detection [20.5], dual fiber interferometry [20.6], and piezolevers [20.7]. In the first method, two plates close to the cantilever reveal capacitance while scanning. The second technique uses two optical fibers to detect the cantilever deflection along two orthogonal directions angled 45° with respect to the surface normal. Finally, in the third method, cantilevers with two Wheatstone bridges at their base reveal normal and lateral forces, which are respectively proportional to the sum and the difference of both bridge signals.

20.1.2 Force Calibration

The force calibration is relatively simple if rectangular cantilevers are used. Due to possible discrepancies with the geometric values provided by manufacturers, one should use optical and electron microscopes to determine the width, thickness, and length of cantilever, w, t, l, the tip height, h, and the position of the tip with respect to the cantilever. The thickness of the cantilever can also be determined from the resonance frequency of the lever, f_0, using the relation [20.8]:

$$t = \frac{2\sqrt{12}\pi}{1.875^2}\sqrt{\frac{\rho}{E}} f_0 l^2 , \qquad (20.1)$$

In (20.1) ρ is the density of the cantilever and E is its Young's modulus. The normal spring constant, c_N, and the lateral spring constant, c_L, of the lever are given by

$$c_N = \frac{Ewt^3}{4l^3} , \quad c_L = \frac{Gwt^3}{3h^2 l} , \qquad (20.2)$$

where G is the shear modulus. Figure 20.2 shows some SEM images of rectangular silicon cantilevers used for FFM. In the case of silicon, $\rho = 2.33\times10^3\,\mathrm{kg/m^3}$, $E = 1.69\times10^{11}\,\mathrm{N/m^2}$ and $G = 0.5\times10^{11}\,\mathrm{N/m^2}$. Thus, for the cantilever in Fig. 20.2, $c_N = 1.9\,\mathrm{N/m}$ and $c_L = 675\,\mathrm{N/m}$.

The next step in force calibration consists of measuring the sensitivity of the photodetector, S_z (nm/V). For beam-deflection FFMs, the sensitivity S_z can be determined by force vs. distance curves measured on hard surfaces (e.g., Al_2O_3), where elastic deformations are negligible and the vertical movement of the scanner equals the deflection of the cantilever. A typical relation between the difference of the vertical signals on the four-quadrant detector, V_N, and the distance from the surface, z, is sketched in Fig. 20.3. When the tip is approached, no signal is revealed until the tip jumps into contact at $z = z_1$. Further extension or retraction of the scanner results in an elastic behavior until the tip jumps again out of contact at a distance $z_2 > z_1$. The slope of the elastic part of the curve gives the required sensitivity, S_z.

The normal and lateral forces are related to the voltage V_N, and the difference between the horizontal signals, V_L, as follows:

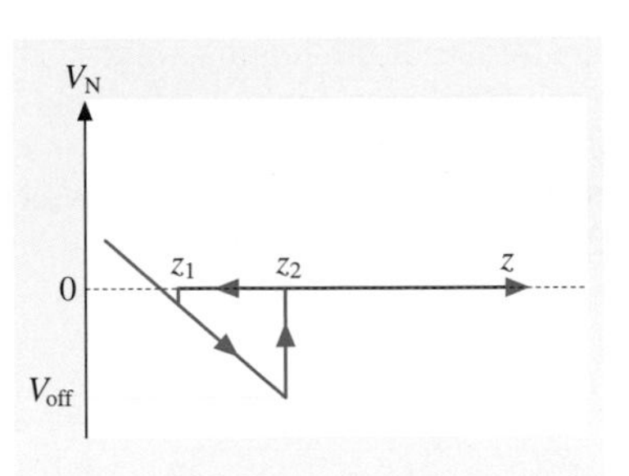

Fig. 20.3 Sketch of a typical force vs. distance curve

$$F_N = c_N S_z V_N , \quad F_L = \frac{3}{2} c_L \frac{h}{l} S_z V_L , \qquad (20.3)$$

in which it is assumed that the light beam is positioned above the probing tip.

The normal spring constant c_N can also be calibrated with alternative methods. *Cleveland* et al. [20.10] attached tungsten spheres to the tip, which changes the resonance frequency f_0 according to the formula

$$f_0 = \frac{1}{2\pi}\sqrt{\frac{c_N}{M + m^*}} . \qquad (20.4)$$

M is the mass of the added object, and m^* is an effective mass of the cantilever, which depends on its geometry [20.10]. The spring constant can be extrapolated from the frequency shifts corresponding to the different masses attached.

As an alternative, *Hutter* et al. observed that the spring constant c_N can be related to the area of the power spectrum of the thermal fluctuations of the cantilever, P [20.11]. The correct relation is $c_N = 4k_B T/3P$, where $k_B = 1.38\times10^{-3}\,\mathrm{J/K}$ is the Boltzmann's constant and T is the temperature [20.12].

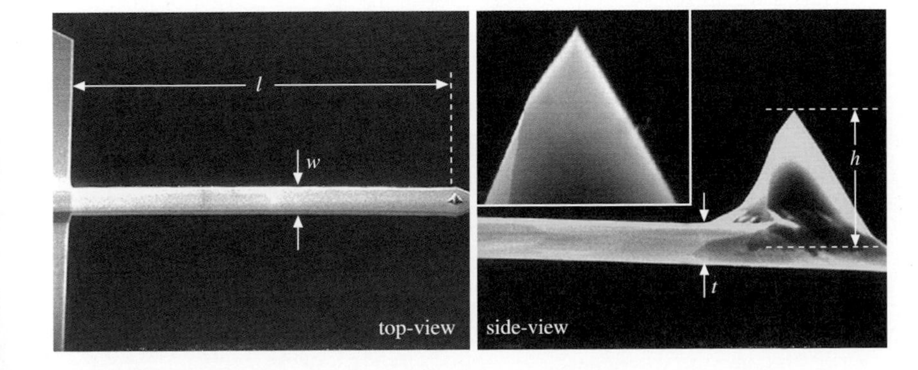

Fig. 20.2 SEM images of a rectangular cantilever. The relevant dimensions are $l = 445\,\mu\mathrm{m}$, $w = 43\,\mu\mathrm{m}$, $t = 4.5\,\mu\mathrm{m}$, $h = 14.75\,\mu\mathrm{m}$. Note that h is given by the sum of the tip height and half of the cantilever thickness. (After [20.9])

Cantilevers with different shape require finite element analysis, although in few cases analytical formulas can be derived. For V-shaped cantilevers, *Neumeister* et al. found the following approximation for the lateral spring constant c_L [20.13]:

$$c_L = \frac{Et^3}{3(1+\nu)h^2} \times \left(\frac{1}{\tan\alpha} \ln \frac{w}{d \sin\alpha} + \frac{L\cos\alpha}{w} - \frac{3\sin 2\alpha}{8} \right)^{-1} , \quad (20.5)$$

where the geometrical quantities L, w, α, d, t, and h are defined in Fig. 20.4. The expression for the normal constant is more complex and can be found in the cited reference.

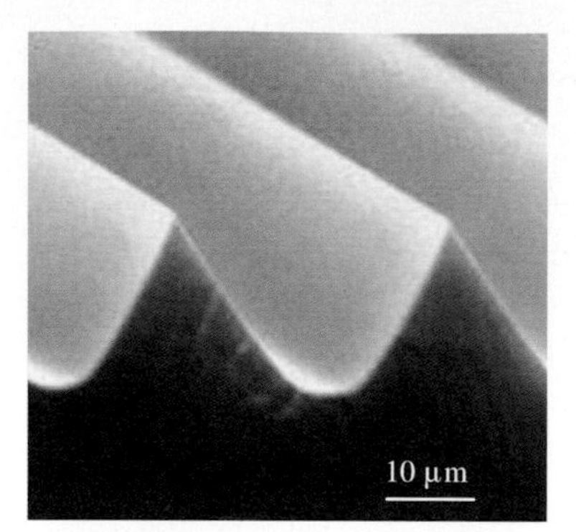

Fig. 20.5 Silicon grating formed by alternated faces angled ±55° (courtesy Silicon-MDT Ltd., POB 50, 103305, Moscow, Russia)

Surfaces with well-defined profiles provide an alternative in situ calibration of lateral forces [20.14]. We present a slightly modified version of the method [20.15]. Figure 20.5 shows a commercial grating formed by alternated faces with opposite inclinations with respect to the scan direction. When the tip slides on the inclined planes, the normal force, F_N, and the lateral force, F_L, with respect to the surface, are different from the two components $F_\perp$ and $F_\parallel$, which are separated by the photodiode (see Fig. 20.6a).

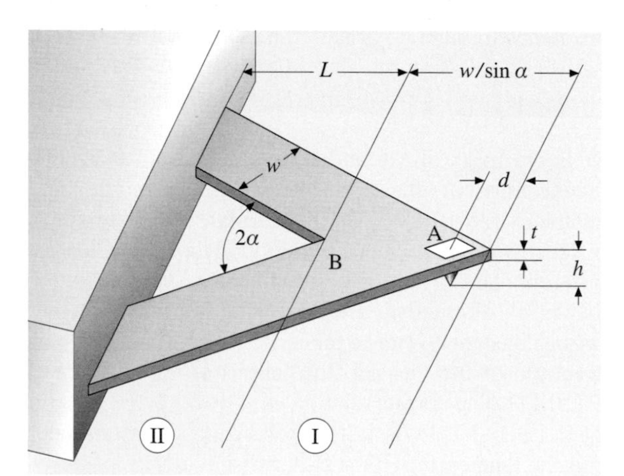

Fig. 20.4 Geometry of a V-shaped cantilever. (After [20.13])

If the linear relation $F_L = \mu F_N$ holds (see Sect. 20.5), the component $F_\parallel$ can be expressed in terms of $F_\perp$:

$$F_\parallel = \frac{\mu + \tan\theta}{1 - \mu\tan\theta} F_\perp . \quad (20.6)$$

The component $F_\perp$ is kept constant by the feedback loop. The sum and the difference of $F_\parallel$ on the two planes (1) and (2) are given by

$$F_+ \equiv F_\parallel^{(1)} + F_\parallel^{(2)} = \frac{2\mu\left(1+\tan^2\theta\right)}{1-\mu^2\tan^2\theta} F_\perp$$

$$F_- \equiv F_\parallel^{(1)} - F_\parallel^{(2)} = \frac{2\left(1+\mu^2\right)\tan\theta}{1-\mu^2\tan^2\theta} F_\perp . \quad (20.7)$$

The values of F_+ and F_- (in volts) can be measured by scanning the profile back and forth (Fig. 20.6b). If F_+ and F_- are recorded with different values of $F_\perp$, one can determine the conversion ratio between volts and nanonewtons as well as the coefficient of friction, μ.

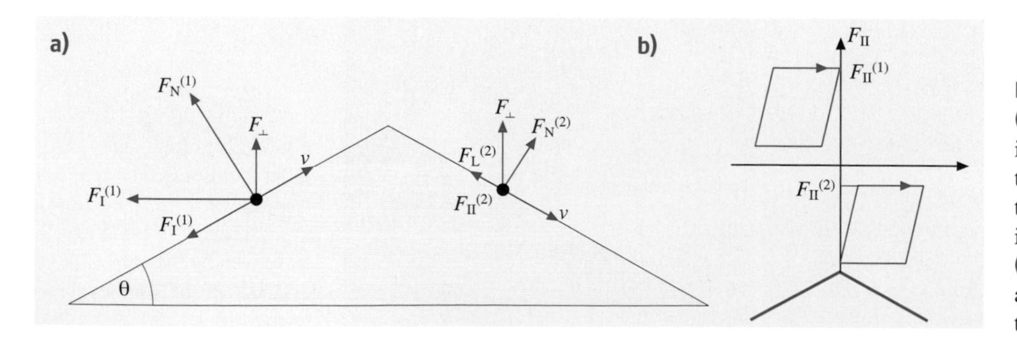

Fig. 20.6 **(a)** Forces acting on a FFM tip sliding on the grating in Fig. 20.5; **(b)** Friction loops acquired on the two faces

An accurate analysis of the errors in lateral force calibration was given by *Schwarz* et al., who revealed the important role of the cantilever's oscillations induced by the feedback loop and of the so-called pull-off force (Sect. 20.5) in friction measurements, aside from the geometrical setting of cantilevers and laser beams [20.16].

For some applications, an adequate estimation of the radius of curvature of the tip, R, is important (Sect. 20.5.2). This quantity can be evaluated with a scanning electron microscope. Otherwise, well-defined structures as step sites [20.17, 18] or whiskers [20.19] can be imaged. The image of these high-aspect ratio structures is a convolution with the tip structure. A deconvolution algorithm that allows for the extraction of the probing tip's radius of curvature was suggested by *Villarrubia* [20.21].

20.1.3 The Ultra-high Vacuum Environment

Atomic friction studies require well-defined surfaces and – whenever possible – tips. For the surfaces, the established methods of surface science can be employed working in ultra-high vacuum (UHV). Ionic crystals such as NaCl have become standard materials for friction force microscopy on atomic scale. Atomically clean and flat surfaces can be prepared by cleavage in UHV. The crystal has to be heated to about 150 °C for 1 h in order to remove charging after the cleavage process. Metal surfaces can be cleaned and flattened by cycles of sputtering with argon ions and annealing. Even surfaces prepared in air or liquids like self-assembled molecular monolayers can be transferred into the vacuum and studied after careful heating procedures, which remove water layers.

Tip preparation in UHV is a more difficult subject. Most force sensors for friction studies have silicon nitride or pure silicon tips. Tips can be cleaned and oxide layers removed by sputtering with argon ions. But the sharpness of the tip is normally reduced by sputtering. As an alternative, tips can be etched in fluoric acid directly before transfer to the UHV. The significance of tip preparation is limited by the fact that the chemical and geometrical structure of the tip can undergo significant changes when sliding over the surface.

The implementation of the friction force microscope into UHV requires some additional efforts. First of all, only materials with a low vapor pressure can be used, thereby excluding most plastics and lubricants. Beam-deflection type force microscopes employ either a light source in the vacuum chamber or an optical fiber guiding the light into the chamber. The positioning of the light beam onto the cantilever and of the reflected beam onto the position-sensitive detector is performed by motorized mirrors [20.20] or by moving light source or detector [20.22]. Furthermore, a motorized sample approach has to be realized.

The quality force sensor's electrical signal can seriously deteriorate on the way out of the vacuum chamber. Low noise and high bandwidth can be preserved by implementing a pre-amplifier into the vacuum. Again the choice of materials for print and devices is limited by the condition of low vapor pressure. Stronger heating of the electrical circuitry in vacuum, therefore, must be considered.

20.1.4 A Typical Microscope in UHV

A typical AFM for UHV application is shown in Fig. 20.7. The housing (1) contains the light source and a set of lenses that focus the light onto the cantilever. Alternatively, the light can be guided via an optical fiber into the vacuum. By using light emitting diodes with low coherency you avoid disturbing interference effects often found in instruments equipped with lasers as light source. A plane mirror fixed on the spherical rotor of a first stepping motor (2) can be rotated

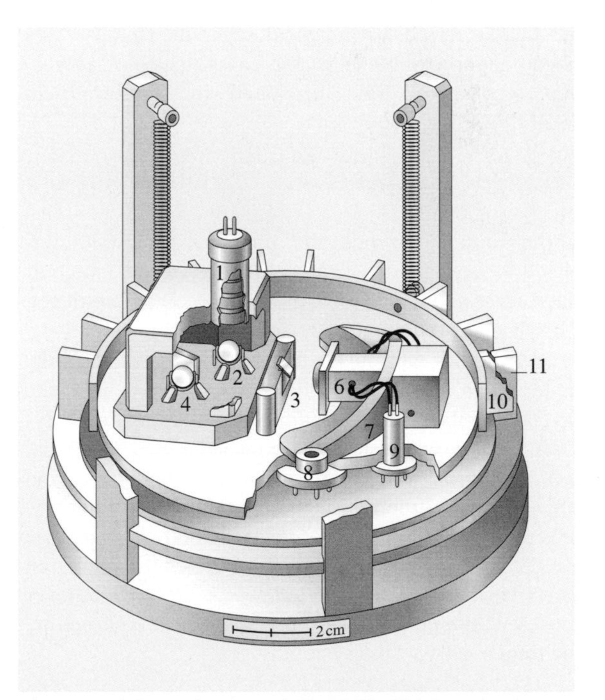

Fig. 20.7 Schematic view of the UHV-AVM realized at the University of Basel. (After [20.20])

around a vertical and horizontal axis to guide the light beam onto the rear of the cantilever, mounted on a removable carrier plate (3). The light is reflected off the cantilever toward a second motorized mirror (4) that guides the beam to the center of the quadrant photodiode (5), where the light is then converted into four photo-currents. Four pre-amplifiers in close vicinity to the photodiode allow low-noise measurements with 3 MHz bandwidth.

The two motors with spherical rotors, used to realign the light path after exchange of the cantilever, work as *inertial stepping motors*: the sphere is resting on three piezoelectric legs that can be moved by a small amount tangentially to the sphere. Each step of the motor consists of a slow forward motion of two legs followed by an abrupt jump backwards. During the slow forward motion the sphere follows the legs due to friction, whereas it cannot follow the sudden jump due to its inertia. A series of such tiny steps rotates the sphere macroscopically.

The sample, also on an exchangeable carrier plate, is mounted on the end of a tube scanner (6), which can move the sample in three dimensions over several micrometers. The whole scanning head (7) is the slider of a third inertial stepping motor for coarse positioning of the sample. It rests with its flat and polished bottom on three supports. Two of them are symmetrically placed piezoelectric legs (8), whereas the third central support is passive. The slider (7) can be translated in two dimensions and rotated about a vertical axis by several millimeters (rotation is achieved by antiparallel operation of the two legs). The slider is held down by two magnets, close to the active supports, and its travel is limited by two fixed posts (9) that also serve as cable attachments. The whole platform is suspended by four springs. A ring of radial copper lamellae (10), floating between a ring of permanent magnets (11) on the base flange, acts as efficient eddy current damping.

20.2 The Tomlinson Model

In Sect. 20.3, we will show that the FFM can reveal friction forces down to the atomic scale, which are characterized by a typical sawtooth pattern. This phenomenon can be seen as a consequence of a *stick-slip* mechanism, first discussed by *Tomlinson* in 1929 [20.23].

20.2.1 One-dimensional Tomlinson Model

In the Tomlinson model the motion of the tip is influenced by both the interaction with the atomic lattice of the surface and the elastic deformations of the cantilever. The shape of the tip-surface potential, $V(\boldsymbol{r})$, depends on several factors, such as the chemical composition of the materials in contact and the atomic arrangement at the tip end. For the sake of simplicity, we will start the analysis in the one-dimensional case considering a sinusoidal profile with the periodicity of the atomic lattice, a, and a peak-to-peak amplitude E_0. In Sect. 20.5, we will show how the elasticity of the cantilever and of the contact area are described in a unique framework by introducing an effective lateral spring constant, k_{eff}. If the cantilever moves with a constant velocity v along the x direction, the total energy of the system is

$$E_{\mathrm{tot}}(x,t) = -\frac{E_0}{2}\cos\frac{2\pi x}{a} + \frac{1}{2}k_{\mathrm{eff}}(vt-x)^2 \,. \quad (20.8)$$

Figure 20.8 shows the energy profile $E_{\mathrm{tot}}(x,t)$ at two different instants. When $t=0$ the tip is localized in the absolute minimum of E_{tot}. This minimum increases with time due to the cantilever motion, until the tip position becomes unstable when $t=t^*$.

At a given time t the position of the tip can be determined by equating to zero the first derivative of $E_{\mathrm{tot}}(x,t)$ with respect to x:

$$\frac{\partial E_{\mathrm{tot}}}{\partial x} = \frac{\pi E_0}{a}\sin\frac{2\pi x}{a} - k_{\mathrm{eff}}(vt-x) = 0 \,. \quad (20.9)$$

The critical position x^* corresponding to $t=t^*$ is determined by equating to zero the second derivative

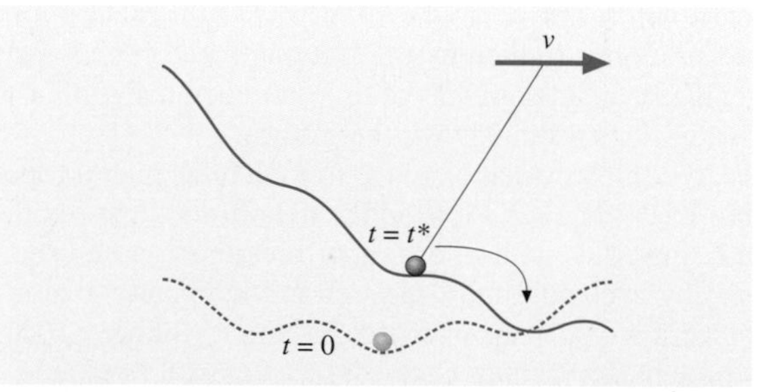

Fig. 20.8 Energy profile experienced by the FFM tip (*black circle*) at $t=0$ (*dotted line*) and $t=t^*$ (*continuous line*)

$\partial^2 E_{tot}(x,t)/\partial x^2$, which gives

$$x^* = \frac{a}{4}\arccos\left(-\frac{1}{\gamma}\right), \quad \gamma = \frac{2\pi^2 E_0}{k_{eff}a^2}. \tag{20.10}$$

The coefficient γ compares the strength of interaction between tip and surface with the stiffness of the system. When $t = t^*$ the tip suddenly *jumps* into the next minimum of the potential profile. The lateral force $F^* = k_{eff}(vt - x^*)$, which induces the jump, can be evaluated from (20.9) and (20.10):

$$F^* = \frac{k_{eff}\,a}{2\pi}\sqrt{\gamma^2 - 1}. \tag{20.11}$$

Thus the stick-slip is observed only if $\gamma > 1$, i. e. when the system is not too stiff or the tip–surface interaction is strong enough. Figure 20.9 shows the lateral force F_L as a function of the cantilever position, X. When the cantilever is moved rightward, the lower part of the curve in Fig. 20.9 is obtained. If, at a certain point, the cantilever's direction of motion is suddenly inverted, the force has the profile in the upper part of the curve. The area of the *friction loop* obtained by scanning back and forth gives the total energy dissipated.

20.2.2 Two-dimensional Tomlinson Model

In two dimensions, the energy of our system is given by

$$E_{tot}(\boldsymbol{r},t) = U(\boldsymbol{r}) + \frac{k_{eff}}{2}(\boldsymbol{v}t - \boldsymbol{r})^2, \tag{20.12}$$

where $\boldsymbol{r} \equiv (x, y)$ and $\boldsymbol{v}$ is arbitrarily oriented on the surface (note that $\boldsymbol{v} \neq d\boldsymbol{r}/dt$!). Figure 20.10 shows the total energy corresponding to a periodic potential of the form

$$U(x,y,t) = -\frac{E_0}{2}\left(\cos\frac{2\pi x}{a} + \cos\frac{2\pi y}{a}\right) + E_1\cos\frac{2\pi x}{a}\cos\frac{2\pi y}{a}. \tag{20.13}$$

The equilibrium condition becomes

$$\nabla E_{tot}(\boldsymbol{r},t) = \nabla U(\boldsymbol{r}) + k_{eff}(\boldsymbol{r} - \boldsymbol{v}t) = 0. \tag{20.14}$$

The stability of the equilibrium can be discussed introducing the Hessian matrix

$$H = \begin{pmatrix} \frac{\partial^2 U}{\partial x^2} + k_{eff} & \frac{\partial^2 U}{\partial x \partial y} \\ \frac{\partial^2 U}{\partial y \partial x} & \frac{\partial^2 U}{\partial y^2} + k_{eff} \end{pmatrix}. \tag{20.15}$$

When both eigenvalues $\lambda_{1,2}$ of the Hessian are positive, the position of the tip is stable. Figure 20.11 shows such regions for a potential of the form (20.13). The tip follows the cantilever adiabatically as long as it remains in the (++)-region. When the tip is dragged to the border of the region, it suddenly jumps into the next (++)-region. A comparison between a theoretical friction map deduced from the 2-D Tomlinson model and an experimental map acquired by UHV-FFM is given in the next section.

20.2.3 Friction Between Atomically Flat Surfaces

So far we have implicitly assumed that the tip is terminated by only one atom. It is also instructive to consider the case of a periodic surface sliding on another periodic surface. In the Frenkel–Kontorova–Tomlinson (FKT) model, the atoms of one surface are harmonically coupled with their nearest neighbors. We will restrict ourselves to the case of quadratic symmetries, with lattice constants a_1 and a_2 for the upper and lower surface, respectively (Fig. 20.12). In such context the role of *commensurability* is essential. It is well known that any real number z can be represented as a continued

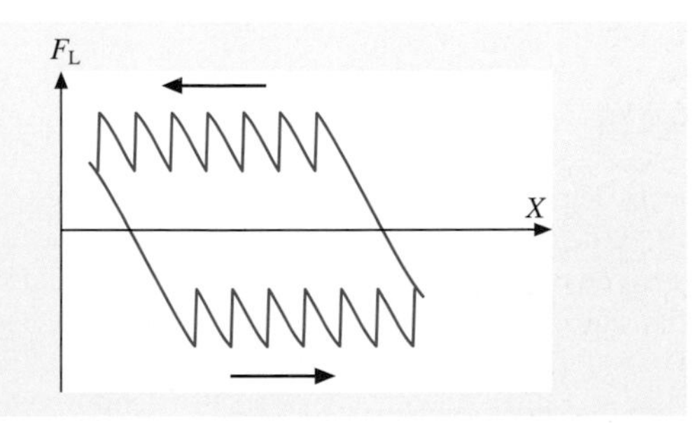

Fig. 20.9 Friction loop obtained by scanning back and forth in the 1-D Tomlinson model. The effective spring constant k_{eff} is the slope of the sticking part of the loop (if $\gamma \gg 1$)

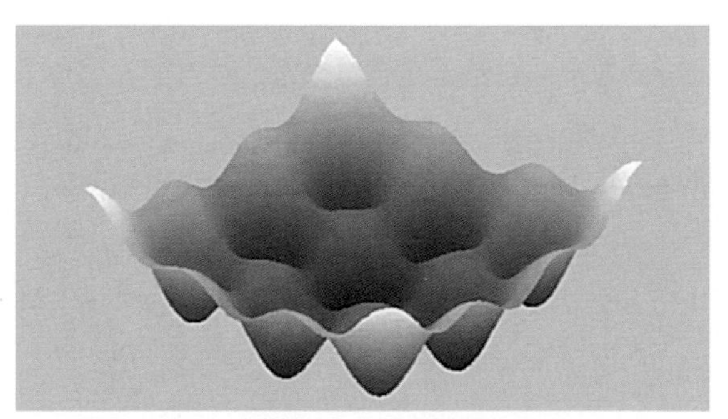

Fig. 20.10 Energy landscape experienced by the FFM tip in 2-D

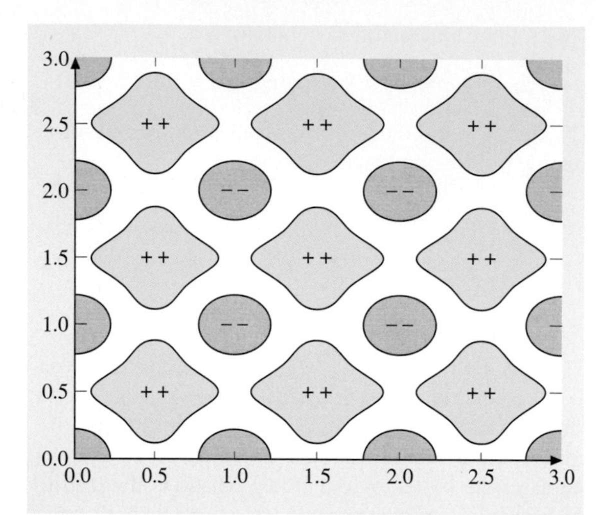

Fig. 20.11 Regions on the tip plane labeled according to the signs of the eigenvalues of the Hessian matrix. (After [20.24])

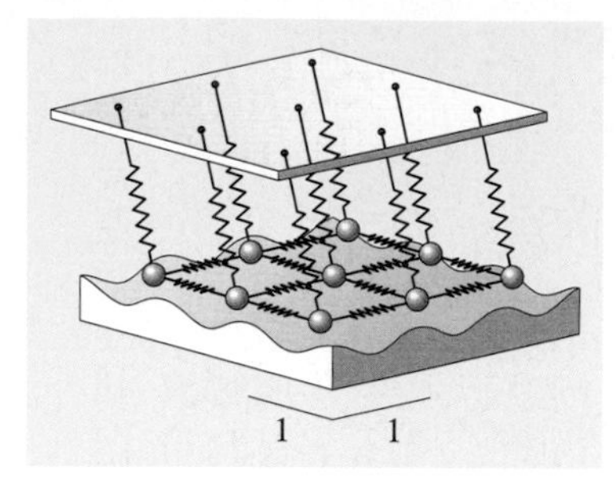

Fig. 20.12 The FKT model in 2-D. (After [20.25])

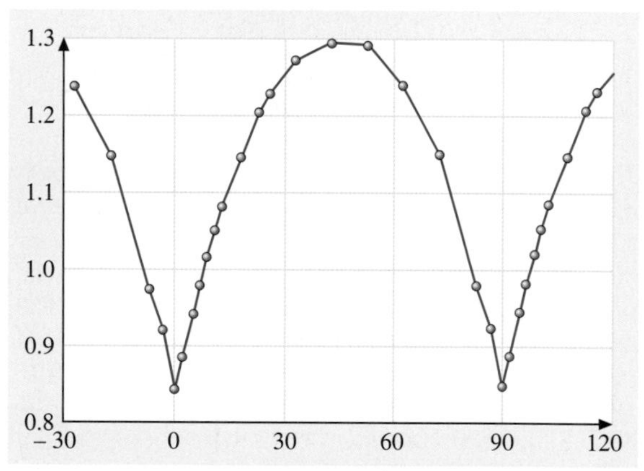

Fig. 20.13 Friction as a function of the sliding angle φ in the 2-D FKT model. (After [20.25])

fraction:

$$z = N_0 + \cfrac{1}{N_1 + \cfrac{1}{N_2 + \ldots}} . \quad (20.16)$$

The sequence that converges most slowly is obtained when all $N_i = 1$, which corresponds to the *golden mean* $\bar{z} = (\sqrt{5} - 1)/2$. In 1-D *Weiss* and *Elmer* predicted that friction should decrease with decreasing commensurability, the minimum of friction being reached when $a_1/a_2 = \bar{z}$ [20.26].

In 2-D *Gyalog* and *Thomas* studied the case $a_1 = a_2$, with a misalignment between the two lattices given by an angle θ [20.25]. When the sliding direction changes, friction also varies from a minimum value corresponding to the sliding angle $\varphi = \theta/2$ to a maximum value, which is reached when $\varphi = \theta/2 + \pi/4$ (Fig. 20.13). The misfit angle θ is related with commensurability. Since the misfit angles giving rise to commensurate structure form a dense subset, the dependence of friction on θ should be discontinuous. The numerical simulations performed by Gyalog are in agreement with this conclusion.

20.3 Friction Experiments on Atomic Scale

Figure 20.14 shows the first friction map observed by Mate on atomic scale. The periodicity of the lateral force is the same as of the atomic lattice of graphite. The series of friction loops in Fig. 20.15 reveals the stick-slip effect discussed in the previous section. The applied loads are in the range of tens of μN. According to the continuum models discussed in Sect. 20.5 these values correspond to contact diameters of 100 nm. A possible explanation for the atomic features observed at such high loads is that graphite flakes might have been detached from the surface and adhered on the tip [20.27]. Another explanation is that the contact between tip and surface consisted of few nm-scale asperities and the corrugation was not entirely averaged out while sliding. The load dependence of friction found by Mate is rather linear, with a small friction coefficient $\mu = 0.01$ (Fig. 20.16).

The UHV environment reduces the influence of contaminants on the surface and leads to more precise and reproducible results. *Meyer* et al. [20.28] obtained

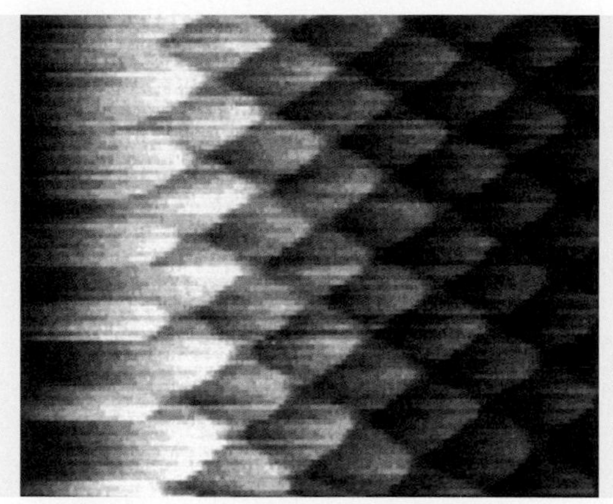

Fig. 20.14 First atomic friction map acquired on graphite with a normal force $F_N = 56\,\mu N$. Frame size: 2 nm. (After [20.1])

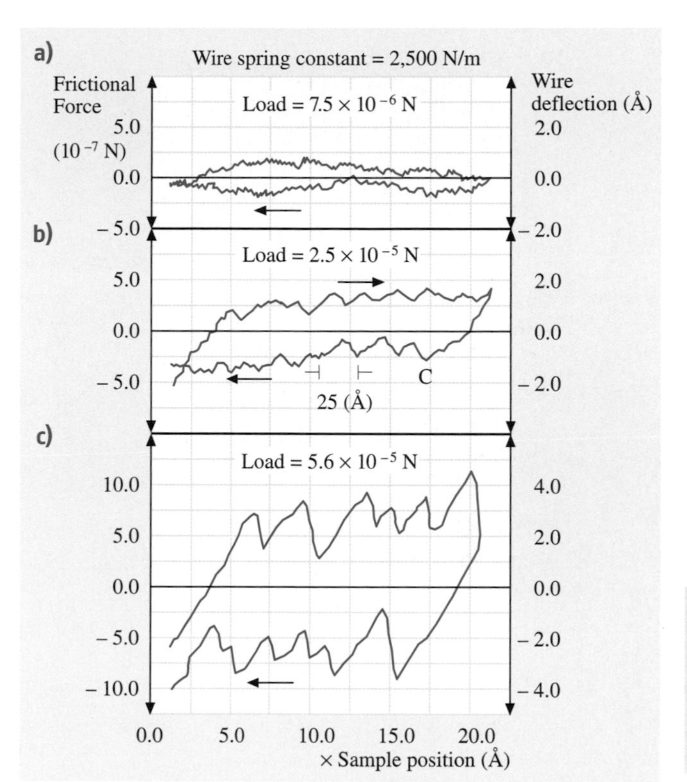

Fig. 20.15a–c Friction loops on graphite acquired with (a) $F_N = 7.5\,\mu N$, (b) $24\,\mu N$ and (c) $75\,\mu N$. (After [20.1])

a series of interesting results on ionic crystals using the UHV-FFM apparatus described in Sect. 20.1.4. In Fig. 20.17 a friction map recorded on KBr(100) is compared with a theoretical map, which was obtained with the 2-D-Tomlinson model. The periodicity $a = 0.47$ nm corresponds to the spacing between equally charged ions. No individual defects were observed. One possible reason is that the contact realized by the FFM tip is always formed by many atoms, which superimpose and average their effects. Molecular dynamics (MD) calculations (Sect. 20.7) show that even a single-atom contact may cause rather large stresses in the sample, which lead to the motion of defects far away from the contact area. In a picturesque frame, we can say that "defects behave like dolphins that swim away in front of an ocean cruiser" [20.28].

Lüthi et al. [20.29] detected atomic-scale friction even on the reconstructed Si(111)7×7 surface. But uncoated Si-tips and tips coated with Pt, Au, Ag, Cr, Pt/C damaged the sample irreversibly, and the observation of atomic features was achieved only after coating the tips with polytetrafluoroethylene (PTFE), which has lubricant properties and does not react with the dangling bonds of Si(111)7×7 (Fig. 20.18).

Recently friction could be resolved on atomic scale even on metallic surfaces in UHV [20.30]. In Fig. 20.19a a reproducible stick-slip process on Cu(111) is shown.

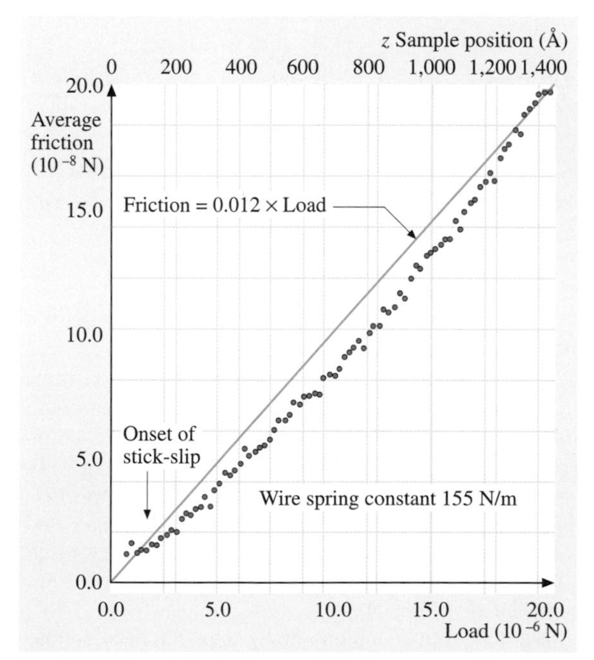

Fig. 20.16 Load dependence of friction on graphite. (After [20.1])

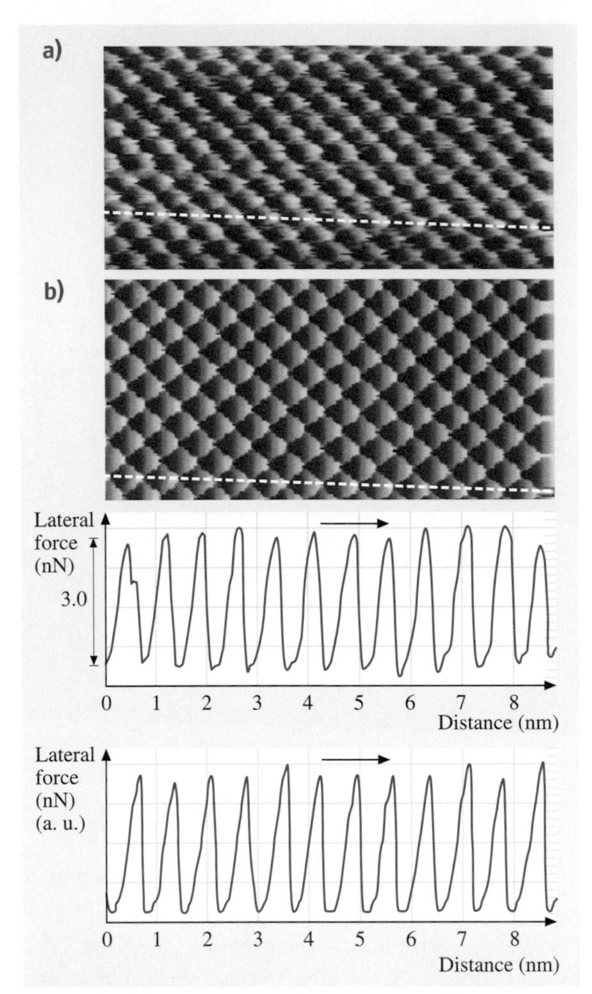

Fig. 20.17 (a) Measured and (b) theoretical friction map on KBr(100). (After [20.31])

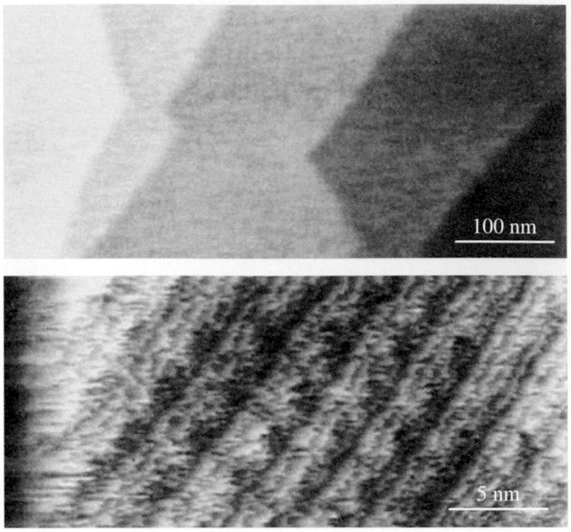

Fig. 20.18 (a) Topography and (b) friction image of Si(111)7×7 measured with a PTFE coated Si-tip. (After [20.29])

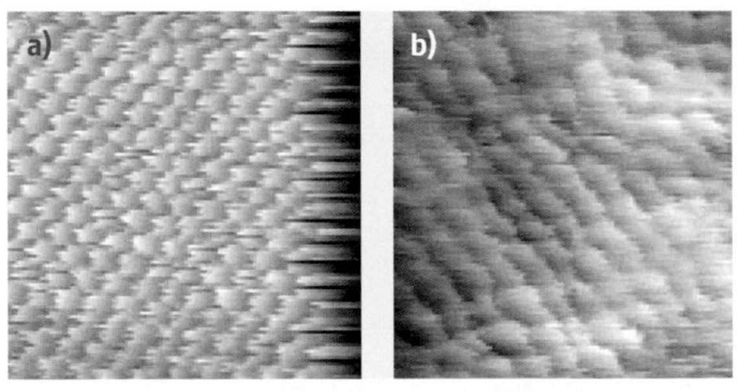

Fig. 20.19a,b Friction images of (a) Cu(111) and (b) Cu(100). Frame size: 3 nm. (After [20.34])

Sliding on the (100) surface of copper produced irregular patterns, although atomic features were recognized even in this case (Fig. 20.19b). Molecular dynamics suggests that wear should occur more easily on the Cu(100) surface than on the closed packed Cu(111) (Sect. 20.7). This conclusion was achieved by adopting copper tips in computer simulations. The assumption that the FFM tip used in the experiments was covered by copper is supported by current measurements performed at the same time.

Atomic stick-slip on diamond was observed by *Germann* et al. with an apposite diamond tip prepared by chemical vapor deposition [20.32] and, a few years later, by *van der Oetelaar* et al. [20.33] with standard silicon tips. The values of friction have huge discrepancies corresponding to the presence or absence of hydrogen on the surface.

Fujisawa et al. [20.35] measured friction on mica and on MoS_2 with a 2-D-FFM apparatus, which could also reveal forces perpendicular to the scan direction. The features in Fig. 20.20 correspond to a zig-zag walk of the tip, which is predicted by the 2-D-Tomlinson model [20.36]. The 2-D stick-slip on NaF was detected with normal forces below 14 nN, whereas loads up to 10 μN could be applied on layered materials. The contact between tip and NaF was thus formed by one or a few atoms. The zig-zag walk on mica was also observed by *Kawakatsu* et al. using an original 2-D-FFM with two laser beams and two quadrant photodetectors [20.37].

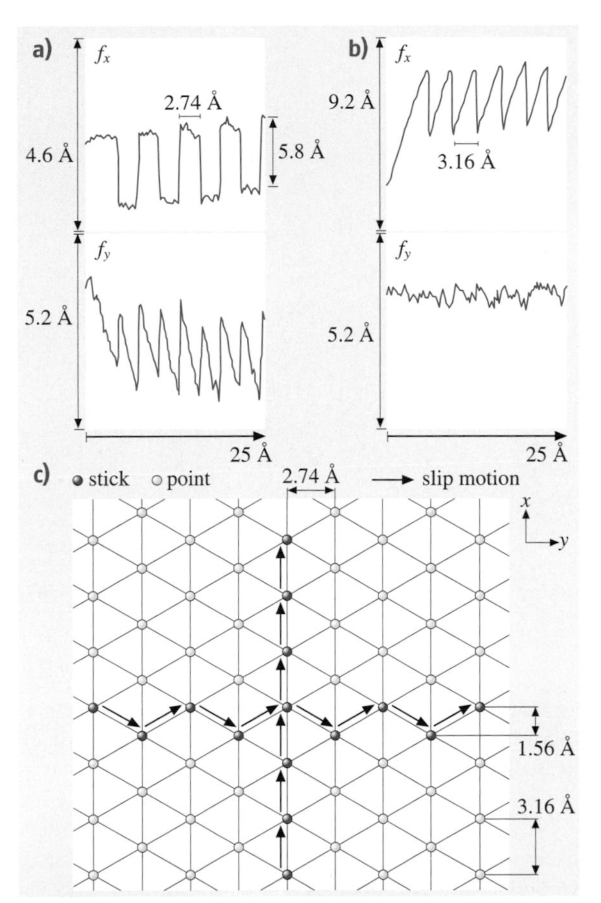

Fig. 20.20 (**a**) Friction force on MoS_2 acquired by scanning along the cantilever and (**b**) across the cantilever. (**c**) Motion of the tip on the sample. (After [20.35])

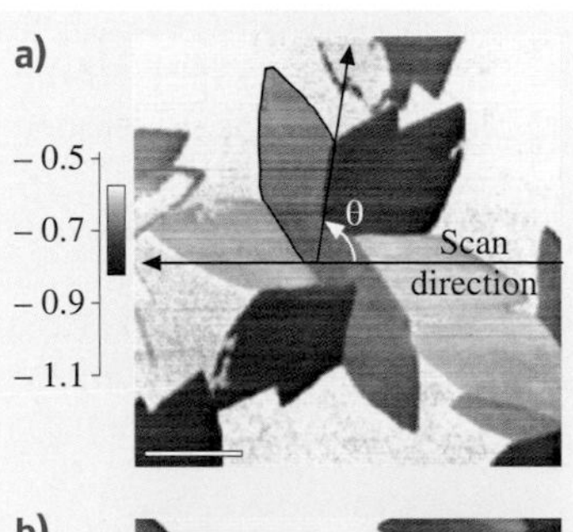

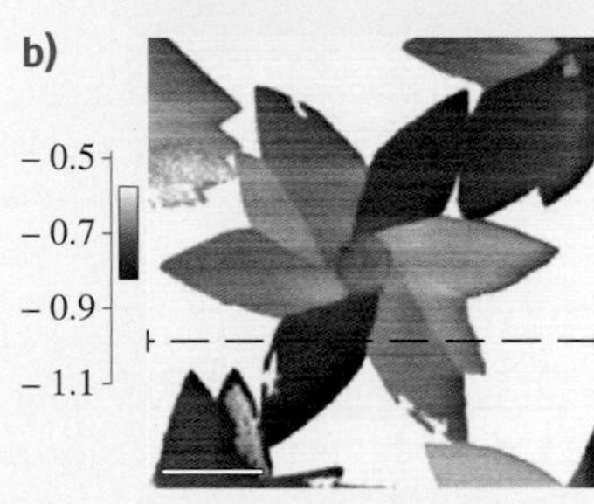

Fig. 20.21 Friction images of a thiolipid monolayer on a mica surface. (After [20.38])

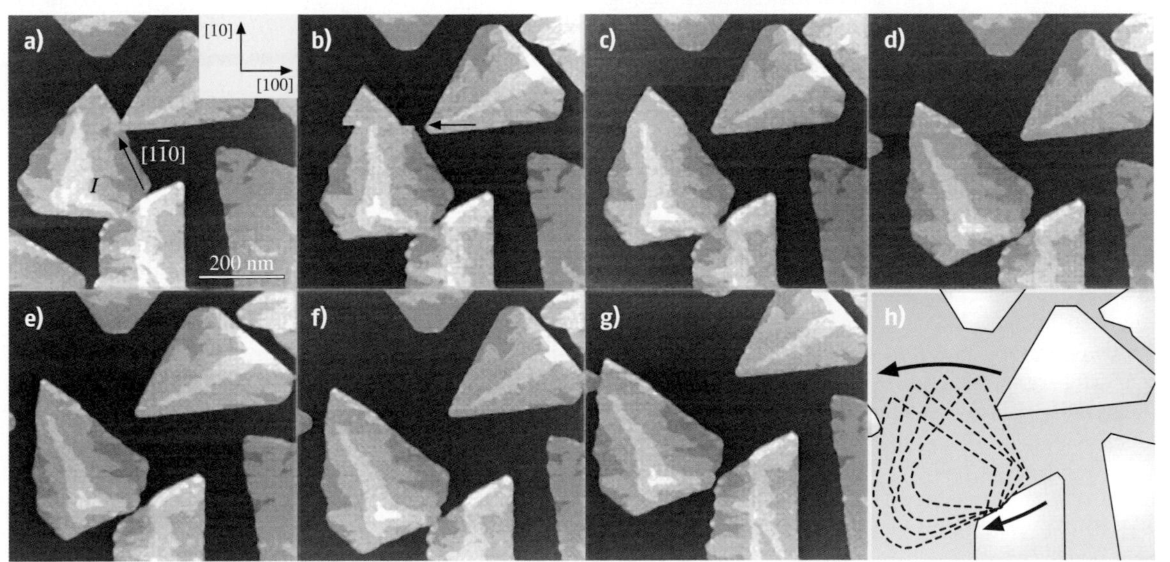

Fig. 20.22 Sequence of topography images of C_{60} islands on NaCl(100). (After [20.39])

20.3.1 Anisotropy of Friction

The importance of the misfit angle in the reciprocal sliding of two flat surfaces was first observed experimentally by *Hirano* et al. in the contact of two mica sheets [20.41]. Friction increased when the two surfaces formed commensurate structures, in agreement with the discussion in Sect. 20.2.3. In more recent measurements with a monocrystalline tungsten tip on Si(001), *Hirano* et al. observed *superlubricity* in the incommensurate case [20.42].

Overney et al. [20.43] studied the effects of friction anisotropy on a bilayer lipid film. In such case different molecular alignments resulted in significant variation of friction. Other measurements of friction anisotropy on stearic acid single crystals were reported by *Takano* and *Fujihira* [20.44].

Liley et al. [20.38] observed flower-shaped islands of a lipid monolayer on mica, which consisted of domains with different molecular orientation (Fig. 20.21). The angular dependence of friction reflects the tilt direction of the alkyl chains of the monolayer, which was revealed by other techniques.

Lüthi et al. [20.39] used the FFM tip to move C_{60} islands, which slided on sodium chloride in UHV without disruption (Fig. 20.22). In this experiment friction was found to be independent of the sliding direction. This was not the case in other experiments performed by *Sheehan* and *Lieber*, who observed that the misfit angle is relevant when MoO_3 islands are dragged on the MoS_2 surface [20.45]. In these experiments, sliding was possible only along low index directions. The weak orientation dependence found by *Lüthi* et al. [20.39] is probably due to the large mismatch of C_{60} on NaCl.

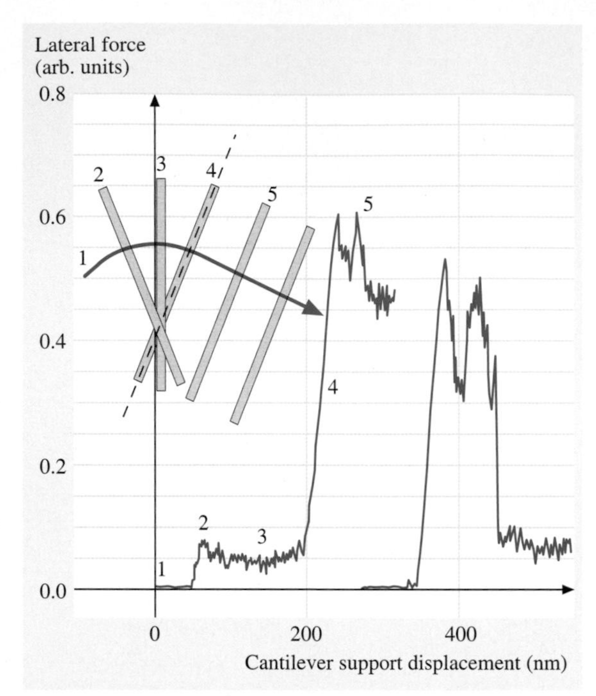

Fig. 20.23 Friction force experienced as a carbon nanotube is rotated into (*left trace*) and out of (*right trace*) commensurate contact. (After [20.40])

A recent example of friction anisotropy is related to carbon nanotubes. *Falvo* et al. [20.40] manipulated nanotubes on graphite using a FFM tip (Fig. 20.23). A dramatic increase of the lateral force was found in the directions corresponding to commensurate contact. At the same time, the nanotube motion changed from sliding/rotating to stick-roll.

20.4 Thermal Effects on Atomic Friction

Although the Tomlinson model gives a good interpretation of the basic mechanism of the atomic stick-slip discussed in Sect. 20.2, it cannot explain some minor features observed in the atomic friction. For example, Fig. 20.24 shows a friction loop acquired on NaCl(100). The peaks in the sawtooth profile have different heights, which is in contrast to the result in Fig. 20.9. Another effect is observed if the scan velocity v is varied: the mean friction force increases with the logarithm of v (Fig. 20.25). This effect cannot be interpreted within the mechanical approach in Sect. 20.2 without further assumptions.

20.4.1 The Tomlinson Model at Finite Temperature

Let us focus again on the energy profile discussed in Sect. 20.2.1. For sake of simplicity, we will assume that $\gamma \gg 1$. At a given time $t < t^*$, the tip jump is prevented by the energy barrier $\Delta E = E(x_{max}, t) - E(x_{min}, t)$, where x_{max} corresponds to the first maximum observed in the energy profile and x_{min} is the actual position of the tip (Fig. 20.26). The quantity ΔE decreases with time or, equivalently, with the frictional force F_L until it vanishes when $F_L = F^*$ (Fig. 20.27). Close to the critical

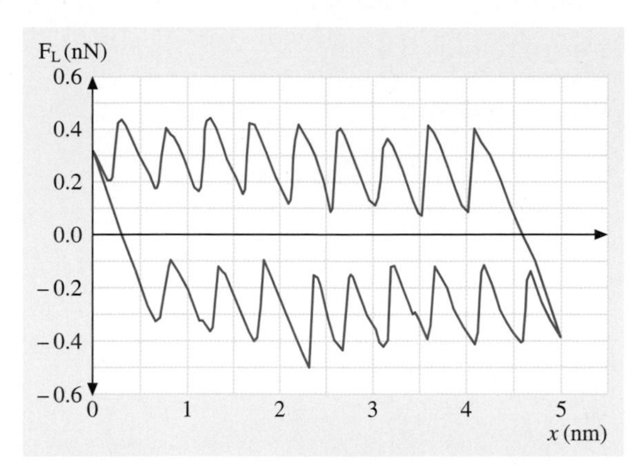

Fig. 20.24 Friction loop on NaCl(100). (After [20.46])

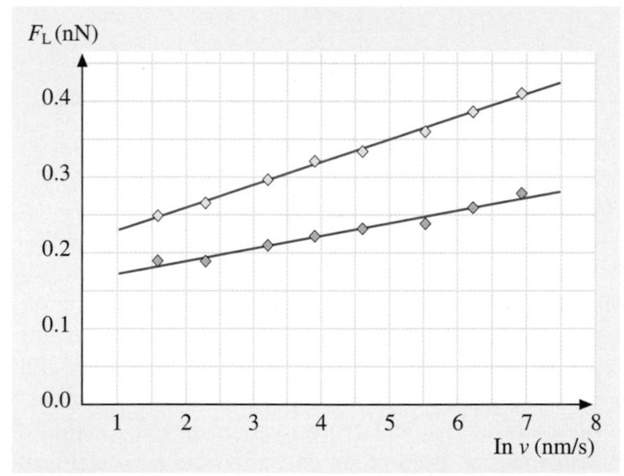

Fig. 20.25 Mean friction force vs. scanning velocity on NaCl(100) at $F_N = 0.44$nN (+) and $F_N = 0.65$nN (×). (After [20.46])

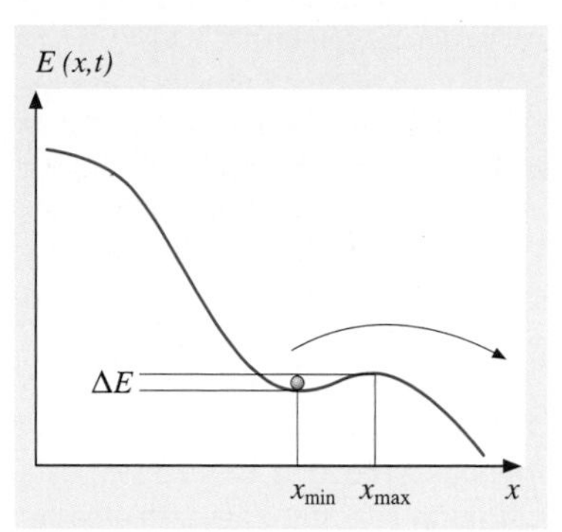

Fig. 20.26 Energy barrier that hinders the tip jump in the Tomlinson model

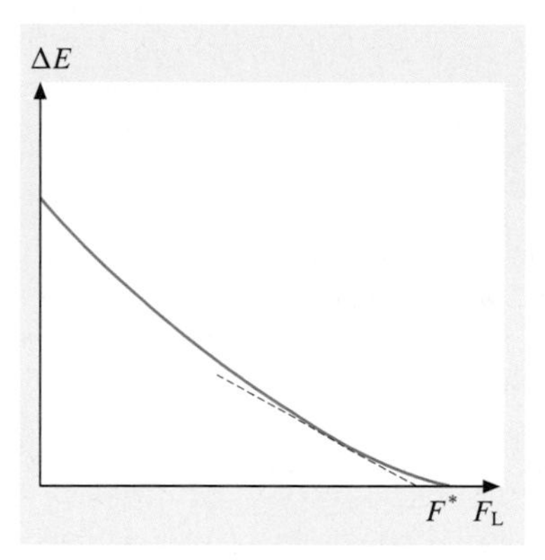

Fig. 20.27 Energy barrier ΔE as a function of the lateral force F_L. The *dashed line* close to the critical value corresponds to the linear approximation (20.17)

point the energy barrier can be written approximately as

$$\Delta E = \lambda(\tilde{F} - F_L) , \tag{20.17}$$

where $\tilde{F}$ is close to the critical value $F^* = \pi E_0/a$.

At finite temperature T, the lateral force required to induce a jump is lower than F^*. To estimate the most probable value of F_L at this point, we first consider the probability p that the tip does *not* jump. The probability p changes with time t according to the master equation

$$\frac{dp(t)}{dt} = -f_0 \exp\left(-\frac{\Delta E(t)}{k_B T}\right) p(t) , \tag{20.18}$$

where f_0 is a characteristic frequency of the system. The physical meaning of this frequency is discussed in Sect. 20.4.2. We should note that the probability of a reverse jump is neglected, since in this case the energy barrier to overcome is much higher than ΔE. If time is replaced by the corresponding lateral force, the master equation becomes

$$\frac{dp(F_L)}{dF_L} = -f_0 \exp\left(-\frac{\Delta E(F_L)}{k_B T}\right)\left(\frac{dF_L}{dt}\right)^{-1} p(F_L) . \tag{20.19}$$

At this point, we substitute

$$\frac{dF_L}{dt} = \frac{dF_L}{dX}\frac{dX}{dt} = k_{eff} v \tag{20.20}$$

and use the approximation (20.17). The maximum probability transition condition $d^2 p(F)/dF^2 = 0$ then yields

$$F_L(v) = F^* - \frac{k_B T}{\lambda} \ln \frac{v_c}{v} \tag{20.21}$$

with

$$v_c = \frac{f_0 k_B T}{k_{eff}\lambda} . \tag{20.22}$$

Thus the lateral force depends logarithmically on the sliding velocity, as observed experimentally. But the approximation (20.17) does not hold when the tip jump occurs very close to the critical point $x = x^*$, which is the case at high velocities. In this instance, the factor $(\mathrm{d}F_L/\mathrm{d}t)^{-1}$ in (20.19) is small and, consequently, the probability $p(t)$ does not change significantly until it suddenly approaches 1 when $t \to t^*$. Thus friction is constant at high velocities, in agreement with the classical Coulomb's law of friction [20.28].

Sang et al. [20.48] observed that the energy barrier close to the critical point is better approximated by a relation like

$$\Delta E = \mu(F^* - F_L)^{3/2} . \tag{20.23}$$

The same analysis performed using the approximation (20.23) instead of (20.17) leads to the expression [20.49]

$$\frac{\mu(F^* - F_L)^{3/2}}{k_B T} = \ln\frac{v_c}{v} - \ln\sqrt{1 - \frac{F^*}{F_L}} , \tag{20.24}$$

where the critical velocity v_c is now

$$v_c = \frac{\pi\sqrt{2}}{2}\frac{f_0 k_B T}{k_{\mathrm{eff}} a} . \tag{20.25}$$

The velocity v_c discriminates between two different regimes. If $v \ll v_c$ the second logarithm in (20.24) can be neglected, which leads to the logarithmic dependence

$$F_L(v) = F^* - \left(\frac{k_B T}{\mu}\right)^{2/3}\left(\ln\frac{v_c}{v}\right)^{2/3} . \tag{20.26}$$

In the opposite case, $v \gg v_c$, the term on the left in (20.23) is negligible and

$$F_L(v) = F^*\left(1 - \frac{v_c}{v}\right)^2 . \tag{20.27}$$

In such a case, the lateral force F_L tends to F^*, as expected.

20.4.2 Velocity Dependence of Friction

The velocity dependence of friction was studied by FFM only recently. *Zwörner* et al. observed that friction between silicon tips and diamond, graphite or amorphous carbon is constant with scan velocities of few μm/s [20.50]. Friction decreased when v is reduced below 1 μm/s. In their experiment on lipid films on mica (Sect. 20.3.1), *Gourdon* et al. [20.47] explored a range of velocities from 0.01 to 50 μm/s and found a critical velocity $v_c = 3.5$ μm/s, which discriminates between an increasing friction and a constant friction regime (Fig. 20.28). Although these results were not explained with thermal activation, we argue that the previous theoretical discussion gives the correct interpretative key. A clear observation of a logarithmic dependence of friction on the micrometer scale was reported by *Bouhacina* et al., who studied friction on triethoxysilane molecules and polymers grafted on silica with sliding velocity up to $v = 300$ μm/s [20.51]. The result was explained with a thermally activated Eyring model, which does not differ significantly from the model discussed in the previous subsection [20.52, 53].

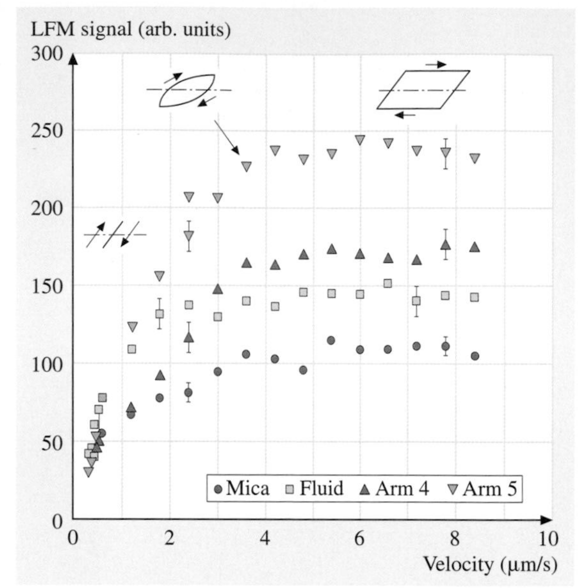

Fig. 20.28 Velocity dependence of friction on mica and on lipid films with different orientation (arms 4 and 5) and in a fluid phase. (After [20.47])

The first measurements on the atomic scale were performed by *Bennewitz* et al. on copper and sodium chloride [20.30, 46]; in both cases a logarithmic dependence of friction was revealed up to $v < 1$ μm/s (Fig. 20.25), in agreement with (20.21). Higher values of velocities were not explored, due to the limited range of the scan frequencies applicable by FFM on atomic scale. The same limitation does not allow a clear distinction between (20.21) and (20.26) in the interpretation of the experimental results.

At this point we would like to discuss the physical meaning of the characteristic frequency f_0. With lattice constants a of a few angstroms and effective spring

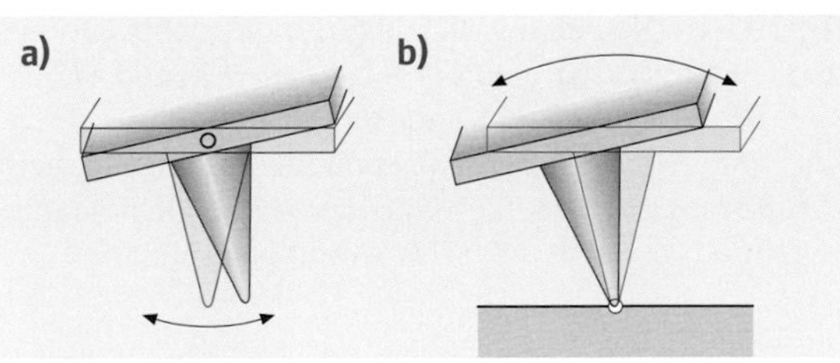

Fig. 20.29a,b Torsional modes of cantilever oscillations (**a**) when the tip is free and (**b**) when the tip is in contact with a surface. (After [20.54])

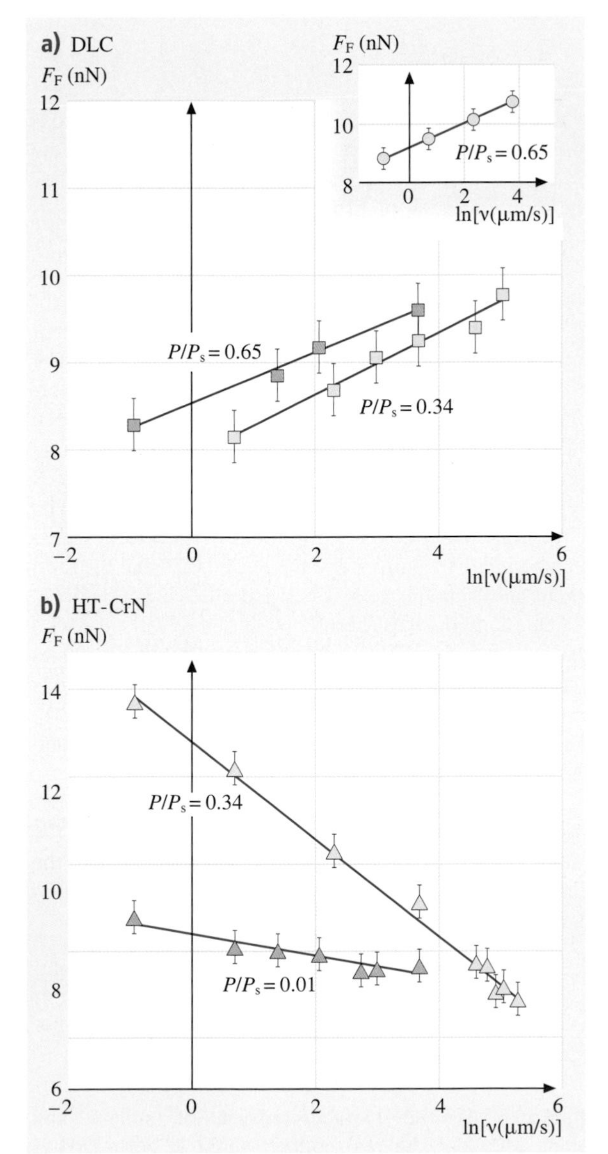

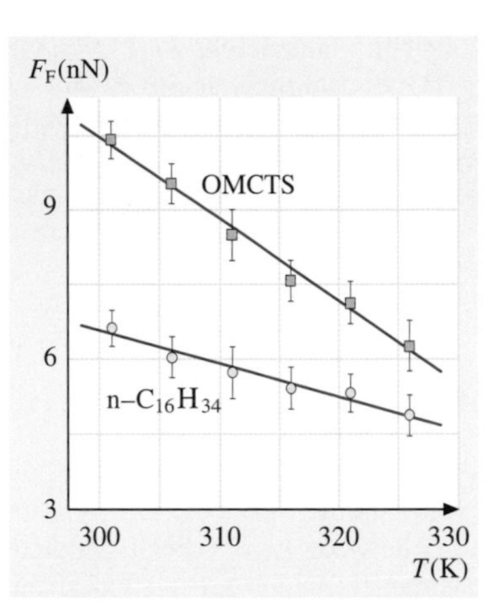

Fig. 20.31 Temperature dependence of friction on n-hexadecane and octamethylcyclotetrasiloxane. (After [20.56])

constants k_{eff} of about 1 N/m, which is typical in FFM experiments, (20.25) gives values of few hundreds of kHz for f_0. This is the characteristic range in which the torsional eigenfrequencies of the cantilevers are located in both contact and noncontact modes (Fig. 20.29). Future work might clarify whether or not f_0 has to be identified with these frequencies.

To conclude this paragraph, we should emphasize that the increase of friction with increasing velocity is ultimately related to the materials and the environment in which the measurements are realized. In humid environment, *Riedo* et al. observed that the surface wettability has an important role [20.55]. Friction *decreases* with increasing velocity on hydrophilic surfaces, and the rate of such decrease depends drastically on humidity. On partially hydrophobic surfaces, a logarithmic increase is found again (Fig. 20.30). These results were interpreted considering the thermally activated nucleation of water bridges between tip and sample asperities, as discussed in the cited reference.

20.4.3 Temperature Dependence of Friction

Thus far we have used thermal activation to explain the velocity dependence of friction. The same mechanism also predicts that friction should change with temperature. Master equation (20.18) shows that the probability of a tip jump is reduced at low temperatures T until it vanishes when $T = 0$. Within this limit case, thermal ac-

Fig. 20.30a,b Friction vs. sliding velocity (**a**) on hydrophobic surfaces and (**b**) on hydrophilic surfaces (After [20.55])

tivation is excluded, and the lateral force F_L is equal to F^*, independently of the scanning velocity v.

To our knowledge stick-slip processes at low temperatures have not been reported. A significant increase of the mean friction with decreasing temperature was recently measured by *He* et al. [20.56] by FFM (Fig. 20.31). Neglecting the logarithmic contributions (20.21) and (20.26) predict $(F^* - F_L) \sim T$ and $(F^* - F_L) \sim T^{2/3}$ respectively for the temperature dependence of friction. Although He et al. applied a linear fit to their data, the range of 30 K they considered is again not large enough to prove that a $T^{2/3}$ fit would be preferable.

20.5 Geometry Effects in Nanocontacts

Friction is ultimately related to the real shape of the contact between the sliding surfaces. On macroscopic scale the contact between two bodies is studied within continuum mechanics, which are all based on the elasticity theory developed by Hertz in 19th century. Various FFM experiments showed that continuum mechanics is still valid down to contact areas a few nanometers in size. Only when contact is reduced to few atoms does the continuum frame become unsuitable, and other approaches like molecular dynamics are necessary. This section will deal with continuum mechanics theory; molecular dynamics will be discussed in Sect. 20.7.

20.5.1 Continuum Mechanics of Single Asperities

The lateral force F_L between two surfaces in reciprocal motion depends on the size of the real area of contact, A, which can be a few orders of magnitude smaller than the apparent area of contact. The simplest assumption is that friction is proportional to A; the proportionality factor is called *shear strength* σ [20.57]:

$$F_L = \sigma A \,. \tag{20.28}$$

In case of plastic deformation, the asperities are compressed until the pressure, p, equals a certain yield value, p^*. The resulting contact area is thus $A = F_N/p^*$, and the well-known Amontons' law is obtained: $F_L = \mu F_N$, where $\mu = \sigma/p^*$ is the *coefficient of friction*. The same idea can be extended to contacts formed by many asperities, and it leads again to the Amontons' law. The simplicity of this analysis explains why most of the friction processes were related to plastic deformation for a long time. Such mechanism, however, should provoke a quick disruption of the surfaces, which is not observed in practice.

Elastic deformation can be easily studied in the case of a sphere of radius R pressed against a flat surface. In such case the contact area is

$$A(F_N) = \pi \left(\frac{R}{K}\right)^{2/3} F_N^{2/3} \,, \tag{20.29}$$

where $K = 3E^*/4$ and E^* is an effective Young's modulus, related to the Young's moduli, E_1 and E_2, and the Poisson's numbers, ν_1 and ν_2, of sphere and plane, by the following relation [20.58]:

$$\frac{1}{E^*} = \frac{1-\nu_1^2}{E_1} + \frac{1-\nu_2^2}{E_2} \,. \tag{20.30}$$

The result $A \propto F_N^{2/3}$ is in contrast with the Amontons' law. But a linear relation between F_L and F_N can be obtained for contacts formed by several asperities in particular cases. For example, the area of contact between a flat surface and a set of asperities with an exponential height distribution and the same radius of curvature R depends linearly on the normal force F_N [20.59]. The same conclusion holds approximately even for a Gaussian height distribution. But the hypothesis that the radii of curvature are the same for all asperities is not realistic. A general model was recently proposed by *Persson*, who derived analytically the proportionality between contact area and load for a large variety of elastoplastic contacts formed by surfaces with arbitrary roughness [20.60]. The discussion is not straightforward and goes beyond the purposes of this section.

Further effects are observed if adhesive forces between the asperities are taken into account. If the range of action of these forces is smaller than the elastic deformation, (20.29) is extended to the Johnson–Kendall–Roberts (JKR) relation

$$A(F_N) = \pi \left(\frac{R}{K}\right)^{2/3} \times \left(F_N + 3\pi\gamma R + \sqrt{6\pi\gamma R F_N + (3\pi\gamma R)^2}\right)^{2/3} \,, \tag{20.31}$$

where γ is the surface tension of sphere and plane [20.61]. The real contact area at zero load is

Part C | 20.5

finite and the sphere can be detached only by pulling it away with a certain force. This is also true in the opposite case, in which the range of action of adhesive forces is larger than the elastic deformation. In such a case, the relation between contact area and load has the simple form

$$A(F_N) = \pi \left(\frac{R}{K}\right)^{2/3} (F_N - F_{off})^{2/3} , \tag{20.32}$$

where F_{off} is the negative load required to break the contact. The Hertz-plus-offset relation (20.32) can be derived from the Derjaguin–Muller–Toporov (DMT) model [20.63]. To discriminate between the JKR or DMT models *Tabor* introduced a nondimensional parameter

$$\Phi = \left(\frac{9R\gamma^2}{4K^2 z_0^3}\right)^{1/3} , \tag{20.33}$$

where z_0 is the equilibrium distance in contact. The JKR model can be applied if $\Phi > 5$; the DMT model holds when $\Phi < 0.1$ [20.64]. For intermediate values of Φ, the Maugis–Dugdale model [20.65] could reasonably explain experimental results (Sect. 20.5.3).

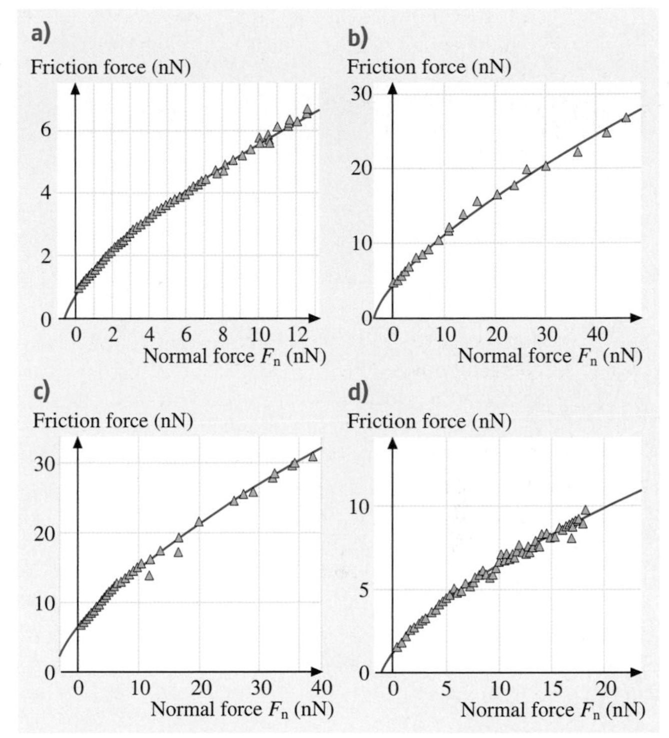

Fig. 20.32a–d Friction vs. load curve on amorphous carbon in argon atmosphere. Curves **(a)**–**(d)** refer to tips with different radii of curvature. (After [20.62])

20.5.2 Load Dependence of Friction

The FFM tip represents a single asperity sliding on a surface. The previous discussion suggests a nonlinear dependence of friction on the applied load, provided that continuum mechanics is applicable. Schwarz et al. observed the Hertz-plus-offset relation (20.32) on graphite, diamond, amorphous carbon and C_{60} in argon atmosphere (Fig. 20.32). In their measurements they used well-defined spherical tips with radii of curvature of tens of nanometers, obtained by contaminating silicon tips with amorphous carbon in a transmission electron microscope. In order to compare the tribological behavior of different materials, *Schwarz* et al. suggested the introduction of an effective coefficient of friction, $\tilde{C}$, which is independent of the tip curvature [20.62].

Meyer et al., *Carpick* et al., and *Polaczyc* et al. performed friction measurements in UHV in agreement with JKR theory [20.17, 66, 67]. In these experiments different materials were considered, i. e. ionic crystals, mica and metals. In order to correlate lateral and normal forces with improved statistics, Meyer et al. applied an original 2-D-histogram technique (Fig. 20.33). Carpick et al. extended the JKR relation (20.32) to include nonspherical tips. In the case of axisymmetric tip profile $z \propto r^{2n}$ $(n > 1)$, it can be proven analytically that the increase of friction becomes less pronounced with increasing n (Fig. 20.34).

20.5.3 Estimation of the Contact Area

In contrast to other tribological instruments, such as the surface force apparatus [20.68], the contact area cannot be directly measured by FFM. Indirect methods are provided by contact stiffness measurements. The contact between the FFM tip and the sample can be modeled by a series of two springs (Fig. 20.35). The effective constant k^z_{eff} of the series is given by

$$\frac{1}{k^z_{eff}} = \frac{1}{k^z_{contact}} + \frac{1}{c_N} , \tag{20.34}$$

where c_N is the normal spring constant of the cantilever and $k^z_{contact}$ is the normal stiffness of the contact. This quantity is related to the radius of the contact area a, by the simple relation

$$k^z_{contact} = 2aE^* , \tag{20.35}$$

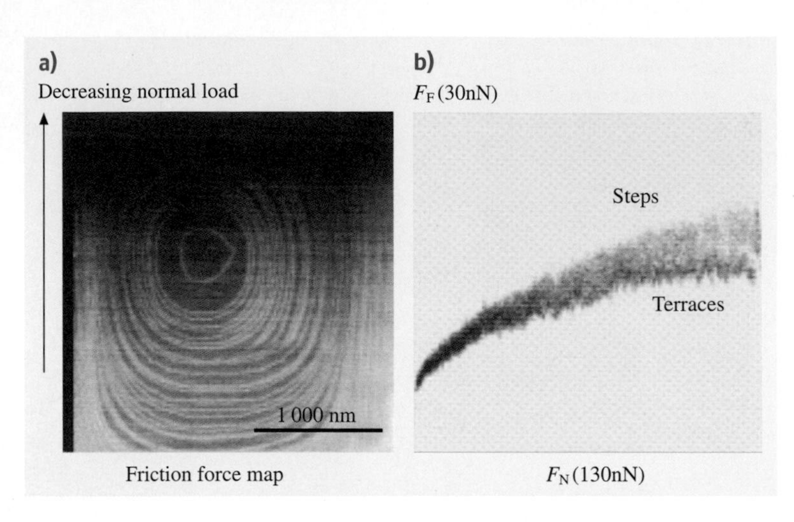

Fig. 20.33 (**a**) Friction force map on NaCl(100). The load is decreased from 140 to 0 nN (jump-off point) during imaging. (**b**) 2-D-histogram of (**a**). (After [20.17])

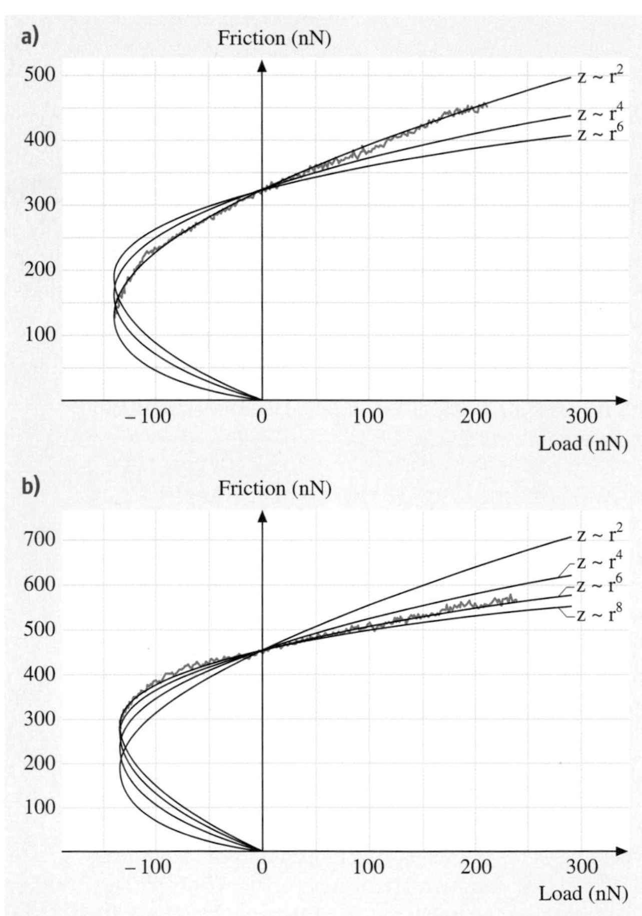

Fig. 20.34a,b Friction vs. load curves (**a**) for a spherical tip and (**b**) for a blunted tip. The *solid curves* are determined with the JKR theory. (After [20.66])

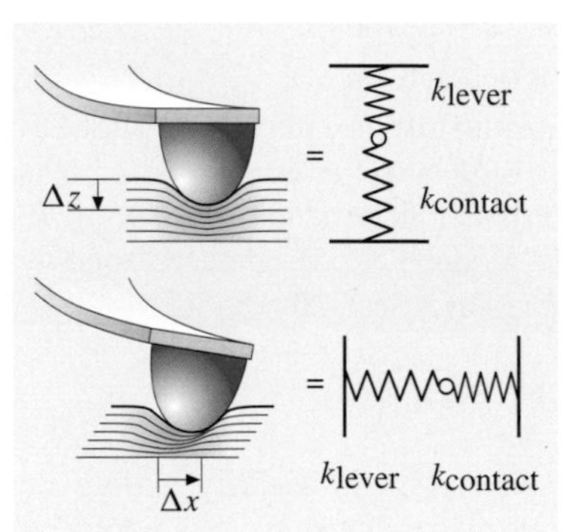

Fig. 20.35 Sketch of normal and lateral stiffness of the contact between tip and surface. (After [20.70])

where E^* is the effective Young's modulus previously introduced [20.69]. Typical values of $k^z_{contact}$ are order of magnitude larger than c_N, however, and a practical application of (20.34) is not possible.

Carpick et al. suggested an alternative method independently [20.70, 71]. According to various models, the *lateral* contact stiffness of the contact between a sphere and a flat surface is [20.72]

$$k^x_{contact} = 8aG^* , \tag{20.36}$$

where the effective shear stress G^* is defined by

$$\frac{1}{G^*} = \frac{2-\nu_1^2}{G_1} + \frac{2-\nu_2^2}{G_2} \tag{20.37}$$

(G_1, G_2 are the shear moduli of sphere and plane). The contact between the FFM tip and the sample can be modeled again by a series of springs (Fig. 20.35). The effective constant k^x_{eff} of the series is given by

$$\frac{1}{k^x_{eff}} = \frac{1}{k^x_{contact}} + \frac{1}{k^x_{tip}} + \frac{1}{c_L} , \tag{20.38}$$

where c_L is the lateral spring constant of the cantilever and $k^x_{contact}$ is the lateral stiffness of the contact. As suggested by Lantz, (20.38) includes also the lateral stiffness of the tip, k^x_{tip}, which can be comparable to the lateral spring constant. The effective spring constant k^x_{eff} is simply given by the slope dF_L/dx of the friction loops (Sect. 20.2.1). Once $k^x_{contact}$ is determined, the contact radius a is easily estimated by (20.36).

The lateral stiffness method was applied to contacts between silicon nitride and muscovite mica in air and between $NbSe_2$ and graphite in UHV. Both the spring constant k^x_{eff} and the lateral force F_L dependence on the load F_N were explained within the same models (JKR and Maugis–Dugdale, respectively), which confirms that friction is proportional to the contact area in the applied range of loads (up to $F_N = 40$ nN in both experiments).

Enachescu et al. estimated the contact area by measuring the contact conductance on diamond as a function of the applied load [20.73, 74]. Their experimental data were fitted with the DMT model, which was also used to explain the dependence of friction vs. load. Since the contact conductance is proportional to the contact area, the validity of the hypothesis (20.28) was confirmed again.

20.6 Wear on the Atomic Scale

If the normal force F_N applied with the FFM exceeds a critical value, which depends on the tip shape and on the material under investigation, the surface topography is permanently modified. In some cases wear is exploited to create patterns with well-defined shape. Here we will focus on the mechanisms acting on the nanometer scale, where recent experiments have proven the unique capability of the FFM in both scratching and imaging surfaces down to the atomic scale.

20.6.1 Abrasive Wear on the Atomic Scale

In the case of ionic crystals, *Lüthi* et al. observed the appearance of wear at very low loads, i. e. $F_N = 3$ nN [20.31]. Atomically resolved images of the damage produced by scratching the FFM tip areas on potassium bromide were obtained very recently by *Gnecco* et al. [20.75]. In Fig. 20.36 a small mound grown up at the end of a groove on KBr(100) is shown under different magnifications. The groove was created a few minutes before imaging by repeated scanning with the normal force $F_N = 21$ nN. The image shows a lateral force map acquired with a load of about 1 nN; no atomic features were observed in the corresponding topography signal. Figure 20.36a,b shows that the debris extracted from the groove recrystallized with the same atomic arrangement of the undamaged surface, which suggests that the wear process occurred like an epitaxial growth assisted by the microscope tip.

Although it is not straightforward to understand how wear is initiated and how the tip transports the debris, important indications are given by the profile of the lateral force F_L recorded while scratching. Figure 20.37 shows some friction loops acquired when the tip was scanned laterally on 5×5 nm^2 areas. The mean lateral force multiplied by the scanned length gives the total energy dissipated in the process. The result of the tip movement are the pits in Fig. 20.38a. Thanks to the pseudo-atomic

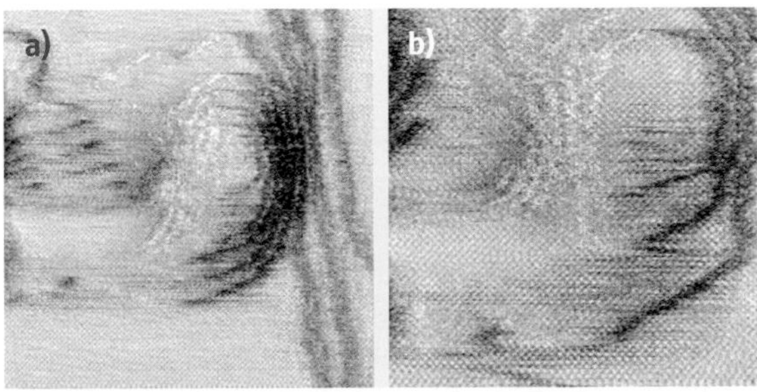

Fig. 20.36a,b Lateral force images acquired at the end of a groove scratched 256 times with a normal force $F_N = 21$nN. Frame sizes: **(a)** 39 nm, **(b)** 25 nm

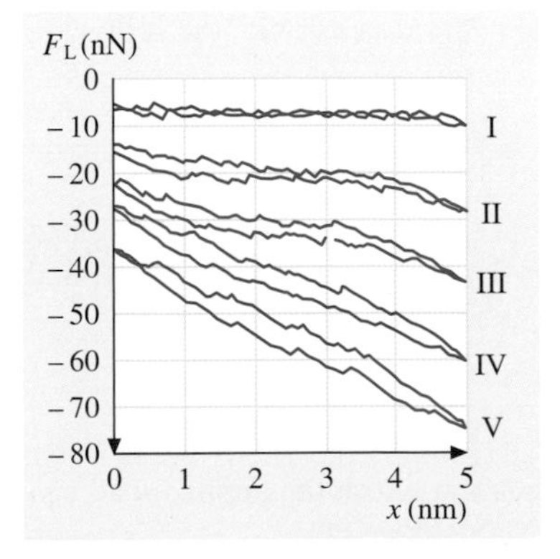

Fig. 20.37 Friction loops acquired while scratching the KBr surface on 5 nm long lines with different loads $F_N = 5.7$ to 22.8 nN. (After [20.75])

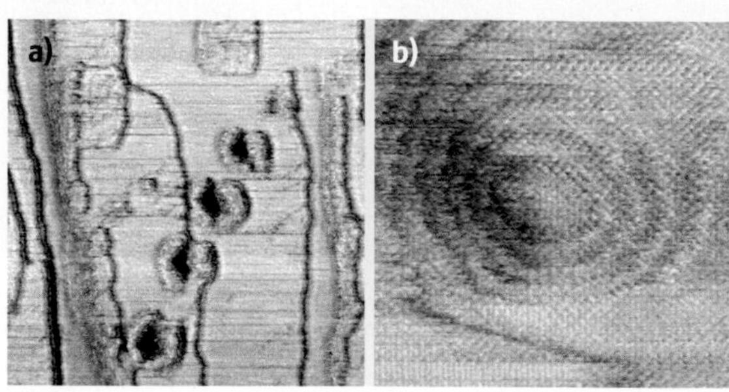

Fig. 20.38 (a) Lateral force images of the pits produced with $F_N = 5.7$ to 22.8 nN. Frame size: 150 nm; (b) Detailed image of the fourth pit from the top with psuedo-atomic resolution. Frame size: 20 nm

resolution obtained by FFM (Fig. 20.38b), the number of removed atoms could be determined from the lateral force images, which allowed estimating that 70% of the dissipated energy went into wear-less friction [20.75]. Figures 20.37 and 20.38 clearly show that the damage increases with increasing load. On the other hand, changing the scan velocity v between 25 and 100 nm/s did not produce any significant variation in the wear process.

A different kind of wear was observed on layered materials. *Kopta* et al. [20.76] could remove layers from a muscovite mica surface by scratching with the normal force $F_N = 230$ nN (Fig. 20.39a). Fourier filtered images acquired on very small areas revealed the different periodicities of the underlying layers, which reflect the complex structure of the muscovite mica (Fig. 20.39b,c).

20.6.2 Wear Contribution to Friction

In Fig. 20.40 the mean lateral force detected while scratching the KBr(100) surface with a fixed load $F_N = 11$ nN is shown. A rather continuous increase of "friction" with the number of scratches N is observed, which can be approximated with the following exponential law:

$$F_L = F_0\,\mathrm{e}^{-N/N_0} + F_\infty\left(1 - \mathrm{e}^{-N/N_0}\right) . \quad (20.39)$$

Equation (20.39) is easily interpreted assuming that friction is proportional to the contact area $A(N)$, and that time evolution of $A(N)$ can be described by

$$\frac{\mathrm{d}A}{\mathrm{d}N} = \frac{A_\infty - A(N)}{N_0} , \quad (20.40)$$

A_∞ being the limit area in which the applied load can be balanced without scratching.

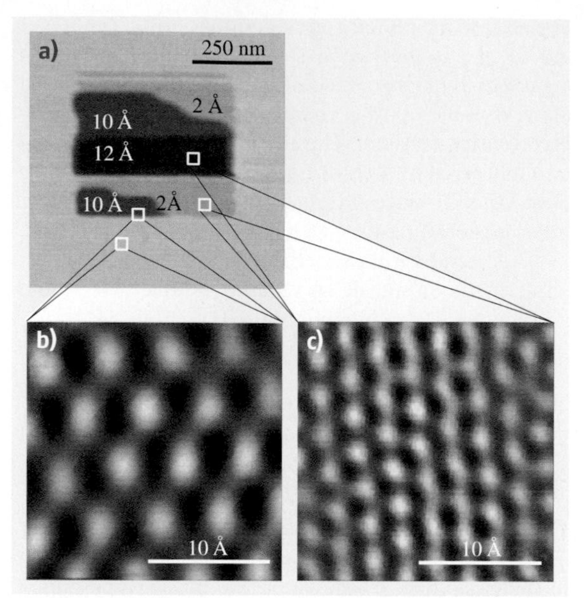

Fig. 20.39 (a) Topography image of an area scratched on muscovite mica with $F_N = 230$ nN; (b),(c) Fourier filtered images of different regions. (After [20.76])

To interpret their experiment on mica, Kopta et al. assumed that wear is initiated by atomic defects. When the defects accumulate beyond a critical concentration, they grow to form the scars shown in Fig. 20.39. Such process was once again related to thermal activation. The number of defects created in the contact area $A(F_N)$ is

$$N_{\mathrm{def}}(F_N) = t_{\mathrm{res}} n_0 A(F_N) f_0 \exp\left(-\frac{\Delta E}{k_B T}\right) , \quad (20.41)$$

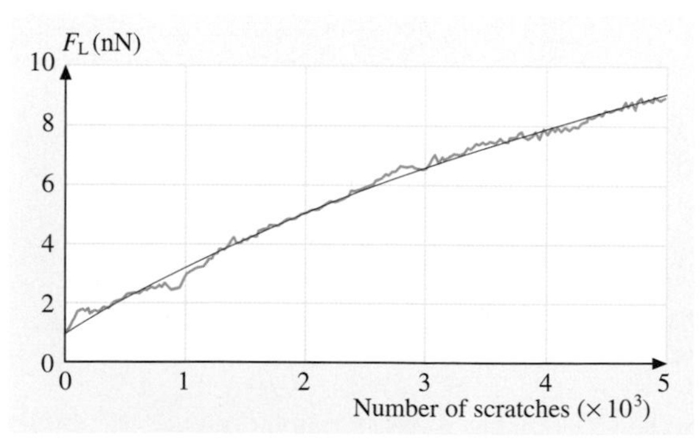

Fig. 20.40 Mean value of the lateral force during repeated scratching with $F_N = 11$ nN on a 500 nm line. (After [20.75])

where t_{res} is the residence time of the tip, n_0 is the surface density of atoms, and f_0 is the attempt frequency to overcome the energy barrier ΔE to break a Si−O bond, which depends on the applied load. When the defect density reaches a critical value, a hole is nucleated. The friction force during the creation of a hole could also be estimated with thermal activation by Kopta et al., who derived the formula

$$F_L = c(F_N - F_{off})^{2/3} + \gamma F_N^{2/3} \exp\left(B_0 F_N^{2/3}\right) . \tag{20.42}$$

The first term on the right gives the wearless dependence of friction in the Hertz-plus-offset model (Sect. 20.5.1); the second term is the contribution of the defect production. The agreement between (20.42) and the experiment can be observed in Fig. 20.41.

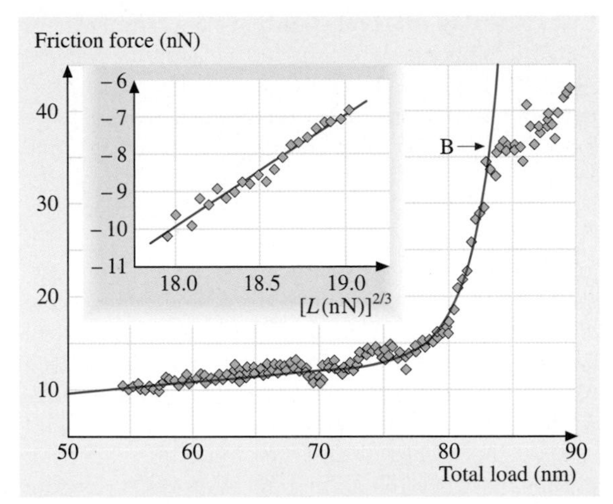

Fig. 20.41 Friction vs. load curve during the creation of a hole in the muscovite mica. (After [20.76])

20.7 Molecular Dynamics Simulations of Atomic Friction and Wear

Section 20.5 has introduced how small sliding contacts can be modeled by continuum mechanics. This modeling has several limitations. The first, most obvious, is that continuum mechanics cannot account for atomic-scale processes like atomic stick-slip. While this limit can be overcome by semiclassical descriptions like the Tomlinson model, a definite limit is the determination of contact stiffness for contacts with a radius of a few nanometers. Interpretation of experimental results with the methods introduced in Sect. 20.5.3 regularly yields contact radii of atomic or even smaller size, in clear contradiction to the minimal contact size given by adhesion forces. Such macroscopic quantities as shear modulus or pressure fail to describe the mechanical behavior of these contacts. A microscopic modeling including the atomic structure of the contact is therefore required. This is usually done by a *molecular dynamics* (MD) simulation of the contact. In such simulations, the sliding contact is set up by boundaries of fixed atoms in relative motion and the atoms of the contact, which are allowed to relax their positions according to interactions between each pair of atoms. The methods of computer simulations in tribology are discussed elsewhere in this book. In Sect. 20.7.1 we will discuss such simulations, which can be directly compared to experimental results showing atomic friction processes. The major outcome of the simulations beyond the inclusion of the atomic structure is the importance of displacement of atoms for a correct prediction of forces. In the following Sect. 20.7.2, we will present simulation studies including wear of tip or surface.

20.7.1 Molecular Dynamics Simulation of Friction Processes

The first experiments that exhibited atomic friction features were performed on layered materials, often graphite. A theoretical study of forces between an atomically sharp diamond tip and the graphite surface has been reported by *Tang* et al. [20.77]. The authors found a significant distance dependence of forces, in which the strongest contrast appeared at different distances for normal and lateral forces due to the strong displacement of surface atoms. The order of magnitude found in this study was one nanonewton, much less than in most experimental reports, which indicated that in such experiments a contact area of far larger dimension was realized. Tang et al. already determined that the distance dependence of forces could even change the symmetry appearance of the observed lateral forces. The experimental situation has also been studied in numerical simulations using simplified one-atom potential for the tip–surface interaction but including the spring potential of the probing force sensor [20.36]. The motivation for these studies was the observation of a hexagonal pattern in the friction force, while the surface atoms of graphite are ordered in a honeycomb structure. The simulations revealed how the jump path of the tip under lateral force is dependent on the force constant of the probing force sensor.

Surfaces of ionic crystals have become model systems for studies in atomic friction. Atomic stick-slip behavior has been observed by several groups with a lateral force modulation of the order of 1 nN. Pioneering work in atomistic simulation of sliding contacts has been done by *Landman* et al.. The first ionic system studied was a CaF_2 tip sliding over a $CaF_2(111)$ surface [20.78]. In MD calculation with controlled temperature, the tip was first approached toward the surface up to the point at which an attractive normal force of -3 nN acted on the tip. Then the tip was laterally moved, and the lateral force determined. An oscillation with the atomic periodicity of the surface and with an amplitude decreasing from 8 nN was found. Inspection of the atomic positions revealed a wear process by shear cleavage of the tip. Such transfer of atoms between tip and surface plays a crucial role in atomic friction studies as was shown by *Shluger* et al. [20.79]. These authors simulated a MgO tip scanning laterally over a LiF(100) surface. Initially an irregular oscillation of the system's energy is found together with transfer of atoms between surface and tip. After a while, the tip apex structure is changed by adsorption of Li and F ions in such a way that nondestructive sliding with perfectly regular oscillation of the energy with the periodicity of the surface could be observed. The authors called this effect self-lubrication and speculate that generally a dynamic self-organization of the surface material on the tip might promote the observation of periodic forces. In a less costly approach of a molecular mechanics study, in which the forces are calculated for each fixed tip–sample configuration, *Tang* et al. produced lateral and normal force maps for a diamond tip over a NaCl(100) surface including such defects as vacancies and a step [20.80]. In accordance with the studies mentioned before, they found that significant atomic force contrast can be expected for tip–sample distances below 0.35 nm, while distances below 0.15 nm result in destructive forces. For the idealized conditions of scanning at constant height in this regime, the authors predict that even atomic-sized defects could be imaged. Experimentally, these conditions cannot be stabilized in static modes, which so far have been used in lateral force measurements. But dynamic modes of force microscopy have proven atomic resolution of defects when entering the distance regime between 0.2 and 0.4 nm [20.81]. Recent experimental progress in atomic friction studies of surfaces of ionic crystals include the velocity dependence of lateral forces and atomic-scale wear processes. Such phenomena are not yet accessible by MD studies: the experimentally realized scanning time scale is too far from the atomic relaxation time scales that govern MD simulations. Furthermore, the number of freely transferable atoms that can be included in a simulation is simply limited by meaningful calculation time.

Landman et al. also simulated a system of high reactivity, namely a silicon tip sliding over a silicon surface [20.82]. A clear stick-slip variation of the lateral force was observed for this situation. Strong atom displacements created an interstitial atom under the influence of the tip, which was annealed as the tip moved on. Permanent damage was predicted, however, in the case that the tip enters the repulsive force regime. Although the simulated Si(111) surface is experimentally not accessible, it should be mentioned that on the Si(111)7×7 reconstructed surface the tip had to be passivated by a Teflon layer before nondestructive contact-mode measurements became possible (Sect. 20.3). It is worth noting that the simulations for the Cu(111) surface revealed a linear relation between contact area and mean lateral force, similar to classical macroscopic laws.

For metallic sliding over metallic surfaces, wear processes are predicted by several MD studies, which will be discussed in the following section. For a (111)-terminated copper tip over the Cu(111) surface, however, *Sørensen* et al. found that nondestructive sliding is possible while the lateral force exhibits the sawtooth-like shape characteristic of atomic stick-slip (Fig. 20.42). In contrast, a Cu(100) surface would be disordered by a sliding contact (Fig. 20.43). This difference between the (100) and (111) plane as well as the absolute lateral forces has been confirmed experimentally (Sect. 20.3).

20.7.2 Molecular Dynamics Simulations of Abrasive Wear

The long time scales characteristic of the wear processes and the large amount of material involved make any attempt to simulate these mechanisms on a computer a tremendous challenge. Despite these premises, MD can provide useful insight on the mechanism of removal and deposition of single atoms operated by the FFM tip, which is a kind of information not directly observable in the experiments. Complex processes like

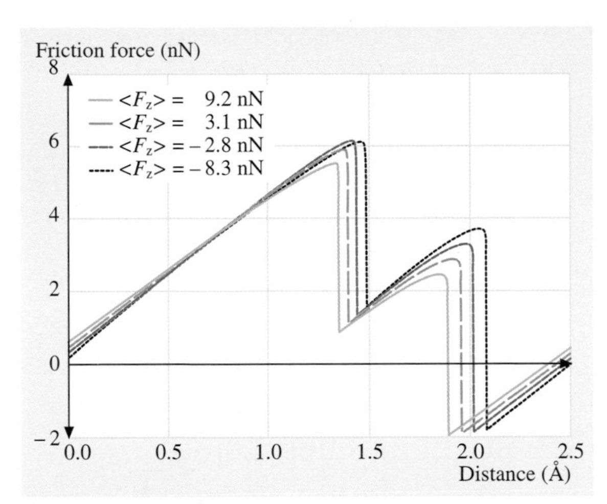

Fig. 20.42 Lateral force acting on a Cu(111) tip in matching contact with a Cu(111) substrate as a function of the sliding distance at different loads. (After [20.83])

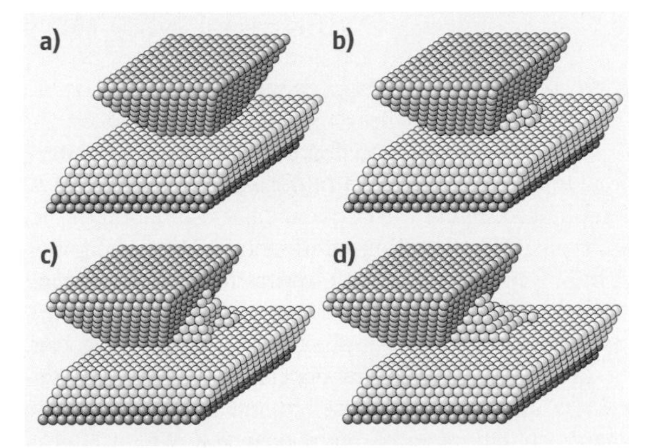

Fig. 20.43a–d Snapshot of a Cu(100) tip on a Cu(100) substrate during sliding. (**a**) Starting configuration; (**b**)–(**d**) snapshots after 2, 4, and 6 slips. (After [20.83])

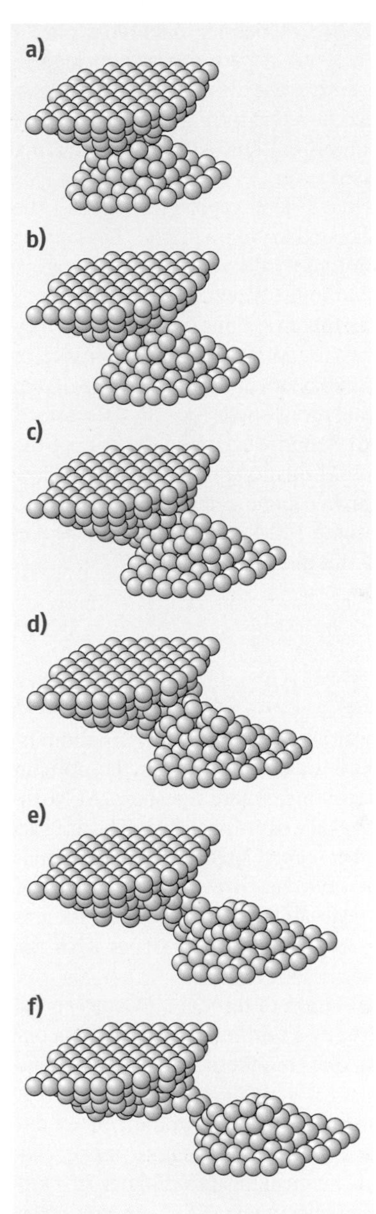

Fig. 20.44a–f Snapshot of a Cu(100) neck during shearing starting from the configuration (**a**). The upper substrate has been displaced 4.2 Å between subsequent pictures. (After [20.83])

abrasive wear and nanolithography can be investigated only within approximate classical mechanics.

The observation made by *Livshits* and *Shluger* that the FFM tip undergoes a process of self-lubrication when scanning ionic surfaces (Sect. 20.7.1) proves that friction and wear are strictly related phenomena. In their MD simulations on copper, *Sørensen* et al. considered not only ordered (111) and (100) terminated tips but even amorphous structures obtained by "heating" the tip at high temperatures [20.83]. The lateral motion of the neck thus formed revealed a stick-slip behavior due to combined sliding and stretching, and also ruptures, accompanied by deposition of debris on the surface (Fig. 20.44).

To our knowledge, only a few examples of abrasive wear simulations on the atomic scale have been reported. *Buldum* and *Ciraci*, for instance, studied nanoindentation and sliding of a sharp Ni(111) tip on Cu(110) and of a blunt Ni(001) tip on Cu(100) [20.84]. In the case of the sharp tip quasiperiodic variations of the lateral force were observed, due to stick-slip involving phase-transition. One layer of the asperity was deformed to match the substrate during the first slip and then two asperity layers merged into one through structural transition during the second slip. In the case of the blunt tip, the stick-slip was less regular.

Different results have been reported in which the tip is harder than the underlying sample. *Komanduri* et al. considered an infinitely hard Ni indenter scratching single crystal aluminum at extremely low depths (Fig. 20.45) [20.85]. In this case a linear relation between friction and load was found, with a high coefficient of friction $\mu = 0.6$, independent of the scratch depth. Nanolithography simulations were recently performed by *Fang* et al. [20.86], who investigated the role of the displacement of the FFM tip along the direction of slow motion between a scan line and the next one. They found a certain correlation with FFM experiments on silicon films coated with aluminum.

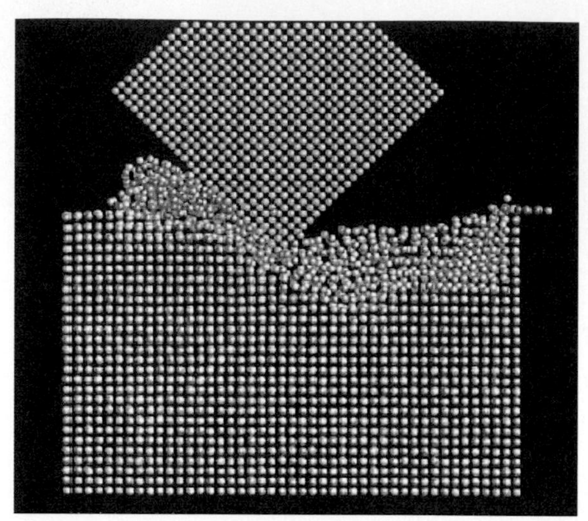

Fig. 20.45 MD simulation of a scratch realized with an infinitely hard tool. (After [20.85])

20.8 Energy Dissipation in Noncontact Atomic Force Microscopy

Historically, the measurement of energy dissipation induced by tip–sample interaction has been the domain of friction-force microscopy, where the sharp AFM-tip is sliding in gentle contact over a sample. The origins of dissipation in friction are related to phonon excitation, electronic excitation, and irreversible changes of the surface. In a typical stick-slip experiment, the energy dissipated in a single atomic slip event is of the order of 1 eV.

But the lateral resolution of force microscopy in the contact mode is limited by a minimum contact area containing several atoms due to adhesion between tip and sample.

This problem has been overcome in noncontact dynamic force microscopy. In the dynamic mode, the tip is oscillating with a constant amplitude A of typically 1–20 nm at the eigenfrequency f of the cantilever, which shifts by Δf due to interaction forces between tip and sample. This technique has been described in detail in Part B of this book.

Dissipation also occurs in noncontact mode of force microscopy, where the atomic structure of tip and sample are reliably preserved. In this dynamic mode, the damping of the cantilever oscillation can be deduced from the excitation amplitude A_{exc} required to maintain the constant tip oscillation amplitude on resonance.

Compared to friction force microscopy, the interpretation of noncontact-AFM (nc-AFM) experiments is complicated due to the perpendicular oscillation of the tip, typically with an amplitude that is large compared to the minimum tip–sample separation. Another problem is to relate the measured damping of the cantilever to the different origins of dissipation.

In all dynamic force microscopy measurements a power dissipation P_0 caused by internal friction of the freely oscillating cantilever is observed, which is proportional to the eigenfrequency ω_0 and to the square of the amplitude A and inverse proportional to the known Q-value of the cantilever. When the tip–sample distance is reduced, the tip interacts with the sample and therefore an additional damping of the oscillation is encountered. This extra dissipation P_{ts} caused by tip–sample interaction can be calculated from the excitation signal A_{exc} [20.87].

The observed energy losses per oscillation cycle (100 meV) [20.88] are roughly similar to the 1 eV energy loss in the contact slip-process. When estimating the contact area in the contact-mode to a few atoms, the energy dissipation per atom that can be associated with a bond being broken and reformed is also around 100 meV.

The idea to relate additional damping of the tip oscillation to dissipative tip–sample interaction has recently attracted much attention [20.89]. The origin of this additional dissipation are manifold: one may distinguish between apparent energy dissipation (for example from an inharmonic cantilever-motion, artifacts from the phase-controller, or slow fluctuations round the steady-state solution [20.89, 90]), velocity dependent dissipation (for example electric- and magnetic-field mediated Joule dissipation [20.91, 92]) and hysteresis-related dissipation (due to atomic instabilities [20.93, 94] or hysteresis due to adhesion [20.95]).

By recording Δf and A_{exc} simultaneously during a typical AFM-experiment, forces and dissipation can be measured. Many experiments show true atomic contrast in topography (controlled by Δf) and in the dissipation-signal A_{exc} [20.96]; however, the origin of the atomic energy dissipation process is still not completely resolved.

To prove that the observed atomic-scale variation of the damping is indeed due to atomic-scale energy dissipation and not an artefact of the distance feedback, *Loppacher* et al. [20.88] carried out an nc-AFM experiment on Si(111)-7×7 at constant height, i. e. with stopped distance feedback. Frequency-shift and dissipation exhibit an atomic-scale contrast, demonstrating a true atomic-scale variation of force and dissipation.

A strong atomic-scale dissipation contrast at step edges has been demonstrated in a few experiments (NaCl on Cu [20.81]; or measurements on KBr [20.97]). In Fig. 20.46 ultrathin NaCl islands grown on Cu(111) are shown. As it is shown in Fig. 20.46a, the island

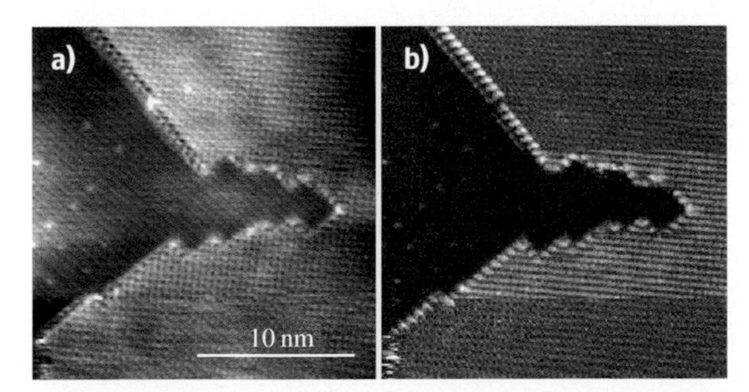

Fig. 20.46 (a) Topography and (b) A_{exc} images of a NaCl island on Cu(111). The tip changes after 1/4 of the scan, thereby changing the contrast in the topography and enhancing the contrast in A_{exc}. After 2/3 of the scan, the contrast from the *lower part* of the image is reproduced, indicating that the tip change was reversible. (After [20.81])

edges have a higher contrast compared to the NaCl-terrace and show an atomically resolved corrugation. The strongly enhanced contrast of the step edges and kink sites could be attributed to a slower decay of the electric field and to easier relaxation of theses positions of the ions at these locations. The dissipation image shown in Fig. 20.46b was recorded simultaneously. To establish a direct spatial correspondence between the excitation and the topography-signal, the match between topography and A_{exc} on many images has been studied. Sometimes topography and A_{exc} are in phase, sometimes they are shifted a little bit, and sometimes A_{exc} is at a minimum when topography is at a maximum. The local contrast formation thus depends strongly on the atomic tip structure. In fact, the strong dependence of the dissipation contrast on the atomic state of the tip apex is impressively confirmed by the tip change observed in the experimental images shown in Fig. 20.46b. The dissipation contrast is seriously enhanced, while the topography contrast remains nearly unchanged. Clearly, the dissipation depends strongly on the state of the tip and exhibits more short-range character than the frequency shift.

More directly related to friction measurements, where the tip is sliding in contact over the sample, are nc-AFM experiments, where the tip is oscillating parallel to the surface. *Stowe* et al. [20.98] oriented cantilever beams with in-plane tips perpendicular to the surface, so that the tip motion was approximately parallel to the surface. The noncontact damping of the lever was used to measure localized electrical Joule dissipation. They were able to image the dopant density for n- and p-type silicon samples with 150 nm spatial resolution. An U_{ts}^2 dependence on the tip–sample voltage for the dissipation was found, as proposed by *Denk* and *Pohl* [20.91] for electric-field Joule-dissipation.

Stipe et al. [20.99] measured the noncontact friction between a Au(111)-surface and a gold-coated tip with the same setup. They observed the same U_{ts}^2 dependence of the Bias-voltage and a distance dependence that follows a power law $1/d^n$, where n is between 1.3 and 3 [20.99, 100]. Even when the external Bias-voltage is zero, a substantial electric-field is present. The presence of inhomogeneous tip–sample electric fields is difficult to avoid, even under the best experimental conditions. Although this dissipation is electrical in origin, the detailed mechanism is not totally clear. The most straightforward mechanism is to assume that inhomogeneous fields emanating from the tip and the sample induce surface charges in the nearby metallic sample. When the tip moves, currents are induced, causing Ohmic dissipation [20.91, 98]. But in metals with good electrical conductivity, Ohmic dissipation is insufficient to account for the observed effect [20.101]. Thus the tip–sample electric field must have an additional effect, such as driving the motion of adsorbates and surface defects.

When exciting the torsional oscillation of commercial, rectangular AFM cantilevers, the tip is oscillating approximately parallel to the surface. In this mode, lateral forces acting on the tip at step edges and near impurities could be measured quantitatively [20.54]. An enhanced energy-dissipation was observed at the impurities as well. When the tip is further approached toward the sample, contact formation transforms the nearly free torsional oscillation of the cantilever into a different mode with the tip–sample contact acting as a hinge. When this contact is formed, a rapid increase of the power required to maintain a constant tip oscillation amplitude and a positive frequency shift are found. The onset of the simultaneously recorded damping and positive frequency shift are sharp and essentially coincide. It is assumed that these changes indicate the formation of a tip–sample contact. Two recent studies [20.102, 103] report on the use of the torsional eigenmode for measuring the elastic properties of the tip–sample contact, where the tip is in contact with the sample and the shear stiffness depends on the normal load.

Kawagishi et al. [20.104] scanned with lateral amplitudes in the order of 10 pm to 3 nm; their imaging technique gave contrast between graphite terraces, silicon and silicon dioxide, graphite and mica. Torsional self-excitation showed nanometric features of self-assembled monolayer islands due to different lateral dissipation.

Recently *Giessibl* et al. [20.105] established true atomic resolution of both conservative and dissipative forces by lateral force microscopy. The interaction between a single-tip atom oscillated parallel to an Si(111)-7×7 surface is measured. A dissipation energy of up to 4 eV per oscillation cycle is found, which is explained by a plucking action of one atom on to the other as described by *Tomlinson* in 1929 [20.23].

A detailed review of dissipation phenomena in noncontact force microscopy has been given by *Hug* [20.106].

20.9 Conclusion

Over the last 15 years, two instrumental developments have stimulated scientific activities in the field of nanotribology. On the one hand, the invention and development of friction force microscopy allows us to quantitatively study friction of a single asperity. As we have discussed in this chapter, atomic processes are observed by forces on the order of 1 nN, i. e. forces related to single chemical bonds. On the other hand, the enormous increase of achievable computing power provides the basis for molecular dynamics simulations of systems containing several hundred atoms. Therewithin, the development of the atomic structure in a sliding contact can be analyzed and forces predicted.

The most prominent observation of atomic friction is a stick-slip behavior with the periodicity of the surface atomic lattice. Semiclassical models can explain the experimental findings including the velocity dependence, which is a consequence of thermal activation of slip events. Classical continuum mechanics can also describe the load dependence of friction in contacts with an extension of several ten nanometers. But when trying to apply continuum mechanics to contacts formed at just ten atoms, obviously wrong numbers result, such as, for example, for the contact radius. Only a comparison with atomistic simulations can provide a full meaningful picture of the physical parameters of such sliding contacts. These simulations predict a close connection between wear and friction, in particular a transfer of atoms between surface and tip, which in some case can even lower friction in a process of self-lubrication.

First experiments have succeeded in studying the onset of wear with atomic resolution. The research in microscopic wear processes will certainly grow in importance as nanostructures are produced and their mechanical properties exploited. Simulation of such processes involving transfer of thousands of atoms will become feasible with further increase of computing power. Another perspective of nanotribology is the expansion of atomic friction experiments toward surfaces with a well-defined roughness. In general, the problem of bridging the gap between single asperity experiments on well-defined surfaces and macroscopic friction should be approached, both experimentally and in modeling.

References

20.1 C. M. Mate, G. M. McClelland, R. Erlandsson, S. Chiang: Atomic-scale friction of a tungsten tip on a graphite surface, Phys. Rev. Lett. **59** (1987) 1942–1945

20.2 G. Binnig, C. F. Quate, Ch. Gerber: Atomic force microscope, Phys. Rev. Lett. **56** (1986) 930–933

20.3 O. Marti, J. Colchero, J. Mlynek: Combined scanning force and friction microscopy of mica, Nanotechnology **1** (1990) 141–144

20.4 G. Meyer, N. Amer: Simultaneous measurement of lateral and normal forces with an optical-beam-deflection atomic force microscope, Appl. Phys. Lett. **57** (1990) 2089–2091

20.5 G. Neubauer, S. R. Cohen, G. M. McClelland, D. E. Horn, C. M. Mate: Force microscopy with a bidirectional capacitance sensor, Rev. Sci. Instrum. **61** (1990) 2296–2308

20.6 G. M. McClelland, J. N. Glosli: Friction at the atomic scale. In: *NATO ASI Series E*, Vol. 220, ed. by L. Singer, H. M. Pollock (Kluwer, Dordrecht 1992) pp. 405–425

20.7 R. Linnemann, T. Gotszalk, I. W. Rangelow, P. Dumania, E. Oesterschulze: Atomic force microscopy and lateral force microscopy using piezoresistive cantilevers, J. Vac. Sci. Technol. B **14** (1996) 856–860

20.8 M. Nonnenmacher, J. Greschner, O. Wolter, R. Kassing: Scanning force microscopy with micromachined silicon sensors, J. Vac. Sci. Technol. B **9** (1991) 1358–1362

20.9 R. Lüthi: Untersuchungen zur Nanotribologie und zur Auflösungsgrenze im Ultrahochvakuum mittels Rasterkraftmikroskopie. Ph.D. Thesis (Univ. of Basel, Basel 1996)

20.10 J. Cleveland, S. Manne, D. Bocek, P. K. Hansma: A nondestructive method for determining the spring constant of cantilevers for scanning force microscopy, Rev. Sci. Instrum. **64** (1993) 403–405

20.11 J. L. Hutter, J. Bechhoefer: Calibration of atomic-force microscope tips, Rev. Sci. Instrum. **64** (1993) 1868–1873

20.12 H. J. Butt, M. Jaschke: Calculation of thermal noise in atomic-force microscopy, Nanotechnology **6** (1995) 1–7

20.13 J. M. Neumeister, W. A. Ducker: Lateral, normal, and longitudinal spring constants of atomic-force microscopy cantilevers, Rev. Sci. Instrum. **65** (1994) 2527–2531

20.14 D. F. Ogletree, R. W. Carpick, M. Salmeron: Calibration of frictional forces in atomic force microscopy, Rev. Sci. Instrum. **67** (1996) 3298–3306

20.15 E. Gnecco: AFM study of friction phenomena on the nanometer scale. Ph.D. Thesis (Univ. of Genova, Genova 2001)

20.16 U. D. Schwarz, P. Köster, R. Wiesendanger: Quantitative analysis of lateral force microscopy experiments, Rev. Sci. Instrum. **67** (1996) 2560–2567

20.17 E. Meyer, R. Lüthi, L. Howald, M. Bammerlin, M. Guggisberg, H.-J. Güntherodt: Site-specific friction force spectroscopy, J. Vac. Sci. Technol. B **14** (1996) 1285–1288

20.18 S. S. Sheiko, M. Möller, E. M. C. M. Reuvekamp, H. W. Zandberger: Calibration and evaluation of scanning-force-microscopy probes, Phys. Rev B **48** (1993) 5675

20.19 F. Atamny, A. Baiker: Direct imaging of the tip shape by AFM, Surf. Sci. **323** (1995) L314

20.20 L. Howald, E. Meyer, R. Lüthi, H. Haefke, R. Overney, H. Rudin, H.-J. Güntherodt: Multifunctional probe microscope for facile operation in ultrahigh vacuum, Appl. Phys. Lett. **63** (1993) 117–119

20.21 J. S. Villarrubia: Algorithms for scanned probe microscope image simulation, surface reconstruction, and tip estimation, J. Res. Natl. Inst. Stand. Technol. **102** (1997) 425–454

20.22 Q. Dai, R. Vollmer, R. W. Carpick, D. F. Ogletree, M. Salmeron: A variable temperature ultrahigh vacuum atomic force microscope, Rev. Sci. Instrum. **66** (1995) 5266–5271

20.23 G. A. Tomlinson: A molecular theory of friction, Philos. Mag. Ser. **7** (1929) 905

20.24 T. Gyalog, M. Bammerlin, R. Lüthi, E. Meyer, H. Thomas: Mechanism of atomic friction, Europhys. Lett. **31** (1995) 269–274

20.25 T. Gyalog, H. Thomas: Friction between atomically flat surfaces, Europhys. Lett. **37** (1997) 195–200

20.26 M. Weiss, F. J. Elmer: Dry friction in the Frenkel–Kontorova–Tomlinson model: Static properties, Phys. Rev. B **53** (1996) 7539–7549

20.27 J. B. Pethica: Comment on "Interatomic forces in scanning tunneling microscopy: Giant corrugations of the graphite surface", Phys. Rev. Lett. **57** (1986) 3235

20.28 E. Meyer, R. M. Overney, K. Dransfeld, T. Gyalog: *Nanoscience, Friction and Rheology on the Nanometer Scale* (World Scientific, Singapore 1998)

20.29 L. Howald, R. Lüthi, E. Meyer, H.-J. Güntherodt: Atomic-force microscopy on the Si(111)7 × 7 surface, Phys. Rev. B **51** (1995) 5484–5487

20.30 R. Bennewitz, T. Gyalog, M. Guggisberg, M. Bammerlin, E. Meyer, H.-J. Güntherodt: Atomic-scale stick-slip processes on Cu(111), Phys. Rev. B **60** (1999) R11301–R11304

20.31 R. Lüthi, E. Meyer, M. Bammerlin, L. Howald, H. Haefke, T. Lehmann, C. Loppacher, H.-J. Güntherodt, T. Gyalog, H. Thomas: Friction on the atomic scale: An ultrahigh vacuum atomic force

microscopy study on ionic crystals, J. Vac. Sci. Technol. B **14** (1996) 1280–1284
20.32 G. J. Germann, S. R. Cohen, G. Neubauer, G. M. McClelland, H. Seki: Atomic scale friction of a diamond tip on diamond (100) and (111) surfaces, J. Appl. Phys. **73** (1993) 163–167
20.33 R. J. A.van den Oetelaar, C. F. J. Flipse: Atomic-scale friction on diamond(111) studied by ultra-high vacuum atomic force microscopy, Surf. Sci. **384** (1997) L828–L835
20.34 R. Bennewitz, E. Gnecco, T. Gyalog, E. Meyer: Atomic friction studies on well-defined surfaces, Tribol. Lett. **10** (2001) 51–56
20.35 S. Fujisawa, E. Kishi, Y. Sugawara, S. Morita: Atomic-scale friction observed with a two-dimensional frictional-force microscope, Phys. Rev. B **51** (1995) 7849–7857
20.36 N. Sasaki, M. Kobayashi, M. Tsukada: Atomic-scale friction image of graphite in atomic-force microscopy, Phys. Rev. B **54** (1996) 2138–2149
20.37 H. Kawakatsu, T. Saito: Scanning force microscopy with two optical levers for detection of deformations of the cantilever, J. Vac. Sci. Technol. B **14** (1996) 872–876
20.38 M. Liley, D. Gourdon, D. Stamou, U. Meseth, T. M. Fischer, C. Lautz, H. Stahlberg, H. Vogel, N. A. Burnham, C. Duschl: Friction anisotropy and asymmetry of a compliant monolayer induced by a small molecular tilt, Science **280** (1998) 273–275
20.39 R. Lüthi, E. Meyer, H. Haefke, L. Howald, W. Gutmannsbauer, H.-J. Güntherodt: Sled-type motion on the nanometer scale: Determination of dissipation and cohesive energies of C_{60}, Science **266** (1994) 1979–1981
20.40 M. R. Falvo, J. Steele, R. M. Taylor, R. Superfine: Evidence of commensurate contact and rolling motion: AFM manipulation studies of carbon nanotubes on HOPG, Tribol. Lett. **9** (2000) 73–76
20.41 M. Hirano, K. Shinjo, R. Kaneko, Y. Murata: Anisotropy of frictional forces in muscovite mica, Phys. Rev. Lett. **67** (1991) 2642–2645
20.42 M. Hirano, K. Shinjo, R. Kaneko, Y. Murata: Observation of superlubricity by scanning tunneling microscopy, Phys. Rev. Lett. **78** (1997) 1448–1451
20.43 R. M. Overney, H. Takano, M. Fujihira, W. Paulus, H. Ringsdorf: Anisotropy in friction and molecular stick-slip motion, Phys. Rev. Lett. **72** (1994) 3546–3549
20.44 H. Takano, M. Fujihira: Study of molecular scale friction on stearic acid crystals by friction force microscopy, J. Vac. Sci. Technol. B **14** (1996) 1272–1275
20.45 P. E. Sheehan, C. M. Lieber: Nanotribology and nanofabrication of MoO_3 structures by atomic force microscopy, Science **272** (1996) 1158–1161
20.46 E. Gnecco, R. Bennewitz, T. Gyalog, Ch. Loppacher, M. Bammerlin, E. Meyer, H.-J. Güntherodt: Velocity dependence of atomic friction, Phys. Rev. Lett. **84** (2000) 1172–1175
20.47 D. Gourdon, N. A. Burnham, A. Kulik, E. Dupas, F. Oulevey, G. Gremaud, D. Stamou, M. Liley, Z. Dienes, H. Vogel, C. Duschl: The dependence of friction anisotropies on the molecular organization of LB films as observed by AFM, Tribol. Lett. **3** (1997) 317–324
20.48 Y. Sang, M. Dubé, M. Grant: Thermal effects on atomic friction, Phys. Rev. Lett. **87** (2001) 174301
20.49 E. Riedo, E. Gnecco, R. Bennewitz, E. Meyer, H. Brune: Interaction potential and hopping dynamics governing sliding friction, Phys. Rev. Lett. **91** (2003) 084502
20.50 O. Zwörner, H. Hölscher, U. D. Schwarz, R. Wiesendanger: The velocity dependence of frictional forces in point-contact friction, Appl. Phys. A **66** (1998) 263–267
20.51 T. Bouhacina, J. P. Aimé, S. Gauthier, D. Michel, V. Heroguez: Tribological behavior of a polymer grafted on silanized silica probed with a nanotip, Phys. Rev. B **56** (1997) 7694–7703
20.52 H. J. Eyring: The activated complex in chemical reactions, J. Chem. Phys. **3** (1937) 107
20.53 J. N. Glosli, G. M. McClelland: Molecular dynamics study of sliding friction of ordered organic monolayers, Phys. Rev. Lett. **70** (1993) 1960–1963
20.54 O. Pfeiffer, R. Bennewitz, A. Baratoff, E. Meyer, P. Grütter: Lateral-force measurements in dynamic force microscopy, Phys. Rev. B **65** (2002) 161403
20.55 E. Riedo, F. Lévy, H. Brune: Kinetics of capillary condensation in nanoscopic sliding friction, Phys. Rev. Lett. **88** (2002) 185505
20.56 M. He, A. S. Blum, G. Overney, R. M. Overney: Effect of interfacial liquid structuring on the coherence length in nanolubrucation, Phys. Rev. Lett. **88** (2002) 154302
20.57 F. P. Bowden, F. P. Tabor: *The Friction and Lubrication of Solids* (Oxford Univ. Press, Oxford 1950)
20.58 L. D. Landau, E. M. Lifshitz: *Introduction to Theoretical Physics* (Nauka, Moscow 1998) Vol. 7
20.59 J. A. Greenwood, J. B. P. Williamson: Contact of nominally flat surfaces, Proc. R. Soc. Lond. A **295** (1966) 300
20.60 B. N. J. Persson: Elastoplastic contact between randomly rough surfaces, Phys. Rev. Lett. **87** (2001) 116101
20.61 K. L. Johnson, K. Kendall, A. D. Roberts: Surface energy and contact of elastic solids, Proc. R. Soc. Lond. A **324** (1971) 301
20.62 U. D. Schwarz, O. Zwörner, P. Köster, R. Wiesendanger: Quantitative analysis of the frictional properties of solid materials at low loads, Phys. Rev. B **56** (1997) 6987–6996
20.63 B. V. Derjaguin, V. M. Muller, Y. P. Toporov: Effect of contact deformations on adhesion of particles, J. Colloid Interface Sci. **53** (1975) 314–326
20.64 D. Tabor: Surface forces and surface interactions, J. Colloid Interface Sci. **58** (1977) 2–13

20.65 D. Maugis: Adhesion of spheres: the JKR-DMT transition using a Dugdale model, J. Colloid Interface Sci. **150** (1992) 243–269

20.66 R.W. Carpick, N. Agraït, D.F. Ogletree, M. Salmeron: Measurement of interfacial shear (friction) with an ultrahigh vacuum atomic force microscope, J. Vac. Sci. Technol. B **14** (1996) 1289–1295

20.67 C. Polaczyk, T. Schneider, J. Schöfer, E. Santner: Microtribological behavior of Au(001) studied by AFM/FFM, Surf. Sci. **402** (1998) 454–458

20.68 J.N. Israelachvili, D. Tabor: Measurement of van der Waals dispersion forces in range 1.5 to 130 nm, Proc. R. Soc. Lond. A **331** (1972) 19

20.69 S.P. Jarvis, A. Oral, T.P. Weihs, J.B. Pethica: A novel force microscope and point-contact probe, Rev. Sci. Instrum. **64** (1993) 3515–3520

20.70 R.W. Carpick, D.F. Ogletree, M. Salmeron: Lateral stiffness: A new nanomechanical measurement for the determination of shear strengths with friction force microscopy, Appl. Phys. Lett. **70** (1997) 1548–1550

20.71 M.A. Lantz, S.J. O'Shea, M.E. Welland, K.L. Johnson: Atomic-force-microscope study of contact area and friction on $NbSe_2$, Phys. Rev. B **55** (1997) 10776–10785

20.72 K.L. Johnson: *Contact Mechanics* (Cambridge Univ. Press, Cambridge 1985)

20.73 M. Enachescu, R.J.A. van den Oetelaar, R.W. Carpick, D.F. Ogletree, C.F.J. Flipse, M. Salmeron: Atomic force microscopy study of an ideally hard contact: the diamond(111)/tungsten carbide interface, Phys. Rev. Lett. **81** (1998) 1877–1880

20.74 M. Enachescu, R.J.A. van den Oetelaar, R.W. Carpick, D.F. Ogletree, C.F.J. Flipse, M. Salmeron: Observation of proportionality between friction and contact area at the nanometer scale, Tribol. Lett. **7** (1999) 73–78

20.75 E. Gnecco, R. Bennewitz, E. Meyer: Abrasive wear on the atomic scale, Phys. Rev. Lett. **88** (2002) 215501

20.76 S. Kopta, M. Salmeron: The atomic scale origin of wear on mica and its contribution to friction, J. Chem. Phys. **113** (2000) 8249–8252

20.77 H. Tang, C. Joachim, J. Devillers: Interpretation of AFM images – the graphite surface with a diamond tip, Surf. Sci. **291** (1993) 439–450

20.78 U. Landman, W.D. Luedtke, E.M. Ringer: Atomistic mechanisms of adhesive contact formation and interfacial processes, Wear **153** (1992) 3–30

20.79 A.I. Livshits, A.L. Shluger: Self-lubrication in scanning-force-microscope image formation on ionic surfaces, Phys. Rev. B **56** (1997) 12482–12489

20.80 H. Tang, X. Bouju, C. Joachim, C. Girard, J. Devillers: Theoretical study of the atomic-force-microscopy imaging process on the NaCl(100) surface, J. Chem. Phys. **108** (1998) 359–367

20.81 R. Bennewitz, A.S. Foster, L.N. Kantorovich, M. Bammerlin, Ch. Loppacher, S. Schär, M. Guggisberg, E. Meyer, A.L. Shluger: Atomically resolved edges and kinks of NaCl islands on Cu(111): Experiment and theory, Phys. Rev. B **62** (2000) 2074–2084

20.82 U. Landman, W.D. Luetke, M.W. Ribarsky: Structural and dynamical consequences of interactions in interfacial systems, J. Vac. Sci. Technol. A **7** (1989) 2829–2839

20.83 M.R. Sørensen, K.W. Jacobsen, P. Stoltze: Simulations of atomic-scale sliding friction, Phys. Rev. B **53** (1996) 2101–2113

20.84 A. Buldum, C. Ciraci: Contact, nanoindentation, and sliding friction, Phys. Rev. B **57** (1998) 2468–2476

20.85 R. Komanduri, N. Chandrasekaran, L.M. Raff: Molecular dynamics simulation of atomic-scale friction, Phys. Rev. B **61** (2000) 14007–14019

20.86 T.H. Fang, C.I. Weng, J.G. Chang: Molecular dynamics simulation of nanolithography process using atomic force microscopy, Surf. Sci. **501** (2002) 138–147

20.87 B. Gotsmann, C. Seidel, B. Anczykowski, H. Fuchs: Conservative and dissipative tip-sample interaction forces probed with dynamic AFM, Phys. Rev. B **60** (1999) 11051–11061

20.88 C. Loppacher, R. Bennewitz, O. Pfeiffer, M. Guggisberg, M. Bammerlin, S. Schär, V. Barwich, A. Baratoff, E. Meyer: Experimental aspects of dissipation force microscopy, Phys. Rev. B **62** (2000) 13674–13679

20.89 M. Gauthier, M. Tsukada: Theory of noncontact dissipation force microscopy, Phys. Rev. B **60** (1999) 11716–11722

20.90 J.P. Aimé, R. Boisgard, L. Nony, G. Couturier: Nonlinear dynamic behavior of an oscillating tip-microlever system and contrast at the atomic scale, Phys. Rev. Lett. **82** (1999) 3388–3391

20.91 W. Denk, D.W. Pohl: Local electrical dissipation imaged by scanning force microscopy, Appl. Phys. Lett. **59** (1991) 2171–2173

20.92 S. Hirsekorn, U. Rabe, A. Boub, W. Arnold: On the contrast in eddy current microscopy using atomic force microscopes, Surf. Interface Anal. **27** (1999) 474–481

20.93 U. Dürig: Forces in scanning probe methods. In: *NATO ASI*, Ser. E, Vol. 286, ed. by H.J. Güntherodt, D. Anselmetti, E. Meyer (Kluwer, Dordrecht 1995) pp. 353–366

20.94 N. Sasaki, M. Tsukada: Effect of microscopic nonconservative process on noncontact atomic force microscopy, Jpn. J. of Appl. Phys. **39** (2000) L1334–L1337

20.95 B. Gotsmann, H. Fuchs: The measurement of hysteretic forces by dynamic AFM, Appl. Phys. A **72** (2001) 55–58

20.96 M. Guggisberg, M. Bammerlin, A. Baratoff, R. Lüthi, C. Loppacher, F.M. Battiston, J. Lü, R. Bennewitz,

E. Meyer, H. J. Güntherodt: Dynamic force microscopy across steps on the Si(111)-(7 × 7) surface, Surf. Sci. **461** (2000) 255–265

20.97 R. Bennewitz, S. Schär, V. Barwich, O. Pfeiffer, E. Meyer, F. Krok, B. Such, J. Kolodzej, M. Szymonski: Atomic-resolution images of radiation damage in KBr, Surf. Sci. **474** (2001) 197–202

20.98 T. D. Stowe, T. W. Kenny, J. Thomson, D. Rugar: Silicon dopant imaging by dissipation force microscopy, Appl. Phys. Lett **75** (1999) 2785–2787

20.99 B. C. Stipe, H. J. Mamin, T. D. Stowe, T. W. Kenny, D. Rugar: Noncontact friction and force fluctuations between closely spaced bodies, Phys. Rev. Lett **87** (2001) 96801

20.100 B. Gotsmann, H. Fuchs: Dynamic force spectroscopy of conservative and dissipative forces in an Al-Au(111) tip-sample system, Phys. Rev. Lett. **86** (2001) 2597–2600

20.101 B. N. J. Persson, A. I. Volokitin: Comment on "Brownian motion of microscopic solids under the action of fluctuating electromagnetic fields", Phys. Rev. Lett. **84** (2000) 3504

20.102 K. Yamanaka, A. Noguchi, T. Tsuji, T. Koike, T. Goto: Quantitative material characterization by ultrasonic AFM, Surf. Interface Anal. **27** (1999) 600–606

20.103 T. Drobek, R. W. Stark, W. M. Heckl: Determination of shear stiffness based on thermal noise analysis in atomic force microscopy: Passive overtone microscopy, Phys. Rev. B **64** (2001) 045401

20.104 T. Kawagishi, A. Kato, Y. Hoshi, H. Kawakatsu: Mapping of lateral vibration of the tip in atomic force microscopy at the torsional resonance of the cantilever, Ultramicroscopy **91** (2002) 37–48

20.105 F. J. Giessibl, M. Herz, J. Mannhart: Friction traced to the single atom, PNAS **99** (2002) 12006–12010

20.106 H.-J. Hug, A. Baratoff: Measurement of dissipation induced by tip-sample interactions. In: *Noncontact Atomic Force Microscopy*, ed. by S. Morita, R. Wiesendanger, E. Meyer (Springer, Berlin, Heidelberg 2002) p. 395

21. Nanoscale Mechanical Properties – Measuring Techniques and Applications

The first part of this chapter describes local (at the scale of nanometers) measurements of mechanical properties. It includes detailed state-of-the-art presentation and in-depth analysis of experimental techniques, results, and interpretations.

After a short introduction, the second part describes local mechanical spectroscopy using coupled Atomic Force Microscopy and ultrasound. This technique allows us to map quickly not only spatial distribution of the elasticity but anelastic properties as well. At one point in the sample, semi-quantitative measurements can be made as a function of the temperature. On the nanometer scale, results have close similitudes to bulk measurements and interpretable differences. Local elasticity and damping were measured during phase transition of polymer samples and shape-memory alloys.

The third part describes the "nano-Swiss cheese" method of measuring the elastic properties of such tubular nanometer size objects as carbon nanotubes and microtubules. It is probably the only experiment in which properties of single-wall nanotube ropes were measured as a function of the rope diameter. We extended this idea to biological objects, microtubules, and successfully solved major experimental difficulties. We not only measured the temperature dependency of microtubule modulus in pseudo-physiological conditions but also estimated shear modulus using the same microtubule with several lengths of suspended segments.

The fourth section demonstrates the scanning nanoindentation technique as applied to human bone tissue. This instrument allows performing topography scans and indentation tests using the identical tip. The available surface scan allows a high positioning precision of the indenter tip on the structure of interest. For very inhomogeneous samples, such as bone tissue, this tool provides a probe to detect local variations of the mechanical properties. The indentation test supplies quantitative parameters like elastic modulus and hardness on the submicron level. Local mechanical properties of compact and trabecular bone lamellae were tested under both dry and pseudo-physiological conditions.

Finally, last part is given to a discussion of future prospects and conclusions.

21.1 **Local Mechanical Spectroscopy by Contact AFM** 662
21.1.1 The Variable-Temperature SLAM (T-SLAM) 663
21.1.2 Example One: Local Mechanical Spectroscopy of Polymers 664
21.1.3 Example Two: Local Mechanical Spectroscopy of NiTi 665

21.2 **Static Methods – Mesoscopic Samples** 667
21.2.1 Carbon Nanotubes – Introduction to Basic Morphologies and Production Methods 667
21.2.2 Measurements of the Mechanical Properties of Carbon Nanotubes by SPM 668
21.2.3 Microtubules and Their Elastic Properties 673

21.3 **Scanning Nanoindentation: An Application to Bone Tissue** 674
21.3.1 Scanning Nanoindentation 674
21.3.2 Application of Scanning Nanoindentation 674
21.3.3 Example: Study of Mechanical Properties of Bone Lamellae Using SN 675
21.3.4 Conclusion 681

21.4 **Conclusions and Perspectives** 682

References 682

Experiments measuring the mechanical properties at nanoscale are key to understanding mesoscopic materials and inhomogeneous materials.

The most prominent technique for such local measurements implements an Atomic Force Microscopy in static mode to obtain force–distance curves (f-d). Ideally this provides the force applied to the tip and the tip–sample distance allowing the local reduced Young's modulus to be determined. In practice, however, these values are not measured directly. Instead, the applied force is offset by adhesive and capillary forces. Furthermore, the tip–sample distance must be obtained by subtraction of the sample (Z-piezo) displacement and the deflection of the cantilever, neither of which are calibrated on most commercial instruments. Usually only the voltage applied to nonlinear and hysteretic Z-piezo and the uncalibrated cantilever deflection is available. These available data must be tediously calibrated and converted to real displacements and forces. The calibration procedure given by *Radmacher* [21.1] works well but only when applied to very compliant materials such as biological or polymer samples. An additional difficulty is that since one end of the cantilever is fixed, neither the orientation of the tip nor its position on the sample surface remains constant during f-d curve acquisition. Movement of the tip on the sample surface, therefore occurs not only vertically but also along the cantilever axis. As a result, applied contact mechanics may not be valid. Finally, uncertainty about the exact tip shape and radius complicate the procedure even further. Instrumental challenges are so significant that applying static AFM to obtain absolute values of the elastic modulus using commercial instruments is difficult at best, especially on stiff surfaces. And this, of course, is the motivation for developing of other techniques.

Although absolute values of mechanical properties are difficult to measure, relative measurements or maps (images of physical properties) are still very interesting. Using acoustic vibrations of the AFM sample surface one can access local elastic and anelastic properties. The first part of this chapter describes these techniques and applications.

Studies of mesoscopic specimens are even more challenging: it is difficult and time consuming to locate a sample, specially properly positioned. In the second part of the chapter, we used 'nano Swiss cheese' technique to get insight into the axial and shear modulus of several species of carbon nanotubes. The method was further extended to biological microtubules, where The experiment was performed in liquid and as a function of the temperature.

Reliable mechanical measurements can be obtained by combining the best of two worlds, AFM and nanoindenter. Classical nanoindenters use optical tip positioning, which are inadequate in the nanoworld. Scanning nanoindenter uses the same diamond tip for imaging (with microNewton forces) and measuring. Its application to human bones in liquid and 37 °C is described in the third part of this chapter.

One should underline the importance of AFM linearity and stability for such experiments: only good designs may limit frustration and experiment time of courageous PhD students, and lot of improvements are still lacking.

21.1 Local Mechanical Spectroscopy by Contact AFM

Mechanical properties of solids (elasticity, anelasticity, plasticity) are generally measured on macroscopic samples. But many phenomena in materials science require measurements of mechanical properties at the surface of a material or at the interfaces between thin layers deposited on a surface. High spatial resolution is also important, for example in the cases of multiphased materials or composite materials, phase transitions, lattice softening in shape-memory alloys, precipitation in light alloys, and glass transition of amorphous materials.

In the case of multiphased materials, such as nanomaterials, composites, alloys, or polymer blends, the location of the dissipative mechanism in one phase must be determined either through modeling or by separately studying each phase, when possible, without changing its behavior. The latter is only possible in a limited number of cases due to the interactions between the different phases within a material. To give an example, the global behavior of a composite is mostly driven by the stress transfer properties between the reinforcement and the matrix, which are controlled by the local mechanical properties in the interface region and in particular the dynamics of the structural defects in this area. It is obviously impossible to prepare a sample only composed of interface regions. Therefore a method for locally studying the dynamics of the structural defects will thus provide important steps in the understanding and the improvement of such materials.

Different techniques based on Scanning Probe Microscopies (SPM) have been developed to probe the elastic and anelastic properties of surfaces, interfaces, or phases of inhomogeneous materials at the micrometer and the nanometer scales. Scanning Acoustic Microscopy (SAM), first developped in the mid-1970s, allows one to study the materials properties at the micrometer scale.

Among the different ways explored to study local mechanical properties of materials, several groups have recently used techniques based on Scanning Force Microscopy (SFM) [21.2]. For most of these, the focus has been placed on "elasticity," using the so-called Force Modulation Mode (FMM) at low frequencies [21.3]. Force modulation mode generally uses a large amplitude (greater than 10 nm), low frequency (some kHz) vibration of the sample underneath the SFM tip. The component of the tip motion at the excitation frequency and the tip mean position are simultaneously recorded, giving several images of the sample surface. In particular, the in-phase and out-of-phase components of force modulation mode at room temperatures have been interpreted in terms of stiffness ("elasticity") and damping ("viscoelasticity") [21.4, 5]. But, it has been recently shown by *Mazeran* et al. [21.6] that the contrast of force modulation mode is dominated by friction properties, giving only little information on the elasticity. Consequently, some care has to be taken in the interpretation of these low-frequency studies. A way to suppress the influence of friction on the contrast is to use smaller amplitudes (some Å) at higher frequencies [21.7]. Scanning Local-Acceleration Microscopy (SLAM) implements this idea [21.8]: SLAM is a modification of contact-mode scanning force microscopy. Its principle is to vibrate the sample at a frequency just above the resonance of the tip–sample system. In this case, the inertia of the tip prevents it from completely following the imposed high frequency displacement, inducing nonnegligible forces and giving rise to elastic deformation of the sample. Contact stiffness is obtained from the measure of the residual displacement of the tip. Mapping the contact stiffness at different temperatures with SLAM [21.9] has allowed local mechanical spectroscopy. Some other techniques also use high frequencies but with different approaches to image elasticity at room temperature [21.7, 10, 11]. Each of these high-frequency techniques is capable of mapping properties such as stiffness or adhesion at constant temperature with a very high lateral resolution.

21.1.1 The Variable-Temperature SLAM (T-SLAM)

The technique described here combines the lateral resolution of scanning force microscopy with the physical information available from temperature ramps. It is the variable-temperature SLAM (T-SLAM) [21.12, 13]. Ramping the temperature of the sample during a SLAM measurement and acquiring both the amplitude and the phase of the tip's motion allows one to obtain local mechanical spectroscopy data. A simple model enables the interpretation of the measurement in terms of material properties.

T-SLAM is based on SLAM [21.8, 14], using a variable temperature sample holder [21.9, 15]. An ultrasonic transducer is placed beneath the sample in a commercial scanning force microscope (Fig. 21.1) and excited by means of a function generator. The ultrasound is

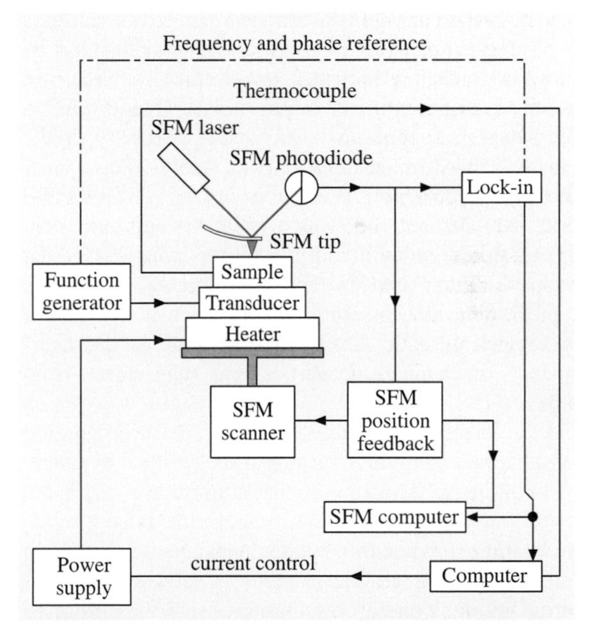

Fig. 21.1 *Schematic diagram of T-SLAM*. Ultrasound is generated with a transducer connected to a function generator and placed beneath the sample. The motion of the scanning force microscope tip is detected optically. The detection signal is fed to a lock-in amplifier, which extracts amplitude and phase relative to the transducer's motion. The temperature is controlled with a small resistive heater and measured with a thermocouple. The rest of the microscope head, to which the tip, the laser and the photodiode are attached, is not represented in this diagram

transmitted through the sample, forcing periodic local deformation of the sample's surface underneath the tip. The motion of the scanning force microscopy tip is detected optically by laser beam deflection from the backside of the cantilever. The detection signal is then fed to a lock-in amplifier, which extracts the tip's amplitude (related to elasticity) and phase (related to viscoelasticity) relative to the transducer's motion. The transducer's typical frequency is 825 kHz. This installation can operate in two ways, either by mapping the amplitude and phase of the signal as a function of position at fixed temperature or by recording the amplitude and phase as a function of temperature at a "fixed" location. The first method is known as "SLAM imaging," where the output signal of the lock-in is fed into an auxiliary data acquisition channel of the microscope. An extra computer is used to store local mechanical spectroscopy data as a function of the temperature at a fixed location and to control temperature. The heat is produced with a small resistive heater below the transducer, and the temperature is measured with a thermocouple. The sample must be prepared with a sufficiently low surface roughness in order to avoid artifacts in the measurements due to the sample's topography. Due to the geometry of the contact, SLAM measures only a small volume near the surface. Viewing SLAM as a very fast indentation measurement, the probed volume is approximately a half-sphere with a radius of $10a$, where a is the contact radius between tip and sample [21.16]. The typical value of a is some nm. So even if the sample is a thick film, T-SLAM gives access to the near-surface mechanical properties that may differ from bulk.

The mechanical properties of the deformed region are obtained from the measure of the residual displacement of the tip. The tip vibration amplitude d_1 is related to the contact stiffness, proportional to the dynamic elastic modulus, while the phase lag φ between the tip motion and the surface motions is related to the internal friction (energy dissipation inside the deformed volume). Mapping d_1 and φ at different temperatures with T-SLAM allows the study of the homogeneity of mechanical relaxations or of phase transitions. By recording d_1 and φ as a function of temperature at a fixed location, local mechanical spectroscopy can be performed.

When used just above the first resonance of the tip–sample system, the SLAM system can be described and analyzed using a point mass rheological model [21.9, 12], which allows one to obtain the equation relating the damping (loss factor η') to the measured parameters

$$\eta' = \left(\frac{1}{2}\frac{k_e}{k_c - m\omega^2}\right)\frac{\sin\varphi}{d_1/z_1}\,, \tag{21.1}$$

where ω is the measurement frequency, k_c is a parameter related to the elastic modulus of the AFM cantilever and k_e to the elastic modulus of the sample, z_1 is the transducer vibration amplitude, and $m = k_c/\omega_c^2$ is the equivalent point mass of the tip, where ω_c is the free resonant frequency of the cantilever.

The principal limitation of (21.1) comes from the model assumption that the cantilever is a point mass restricted to vertical displacements. Due to the possible lateral displacements of the tip [21.6] or the existence of other vibrational modes of the cantilever, which are not described by the point mass model, (21.1) is only reliable for the measurement of the damping just above the first resonance of the tip–sample system. In order to obtain a quantitative measurement of this quantity in a larger frequency spectrum, it is necessary to develop a more realistic model of the cantilever interacting with the sample surface such as, for example, the model presented by *Dupas* et al. [21.17], in which the cantilever is modeled as a beam.

21.1.2 Example One: Local Mechanical Spectroscopy of Polymers

Figure 21.2 shows local mechanical spectra [21.13, 18] obtained in bulk technical PVC (including plasticizer and pigments and taken off-the-shelf) as a function of temperature (Fig. 21.2a) and a Differential Scanning Calorimetry (DSC) measurement of the same sample (Fig. 21.2b). The mechanical measurement (Fig. 21.2a) displays the amplitude of vibration of the SLAM tip (thin line) and its phase lag (thick line) as a function of temperature. Four temperature domains can be identified. At lower temperatures (1), the first domain shows a phase lag peak associated with a decrease of amplitude. The second temperature (2) domain corresponds to a zone where the vibration amplitude increases, without any variation in the phase lag. The third temperature domain (3) is characterized by a large phase lag peak and a significant decrease of vibration amplitude. The last temperature domain (4) shows a slow increase in the phase lag without variation of the vibration amplitude.

The calorimetry curves (b) display the heat flow as a function of temperature for the same material but at a larger size scale. For clarity, the same temperature do-

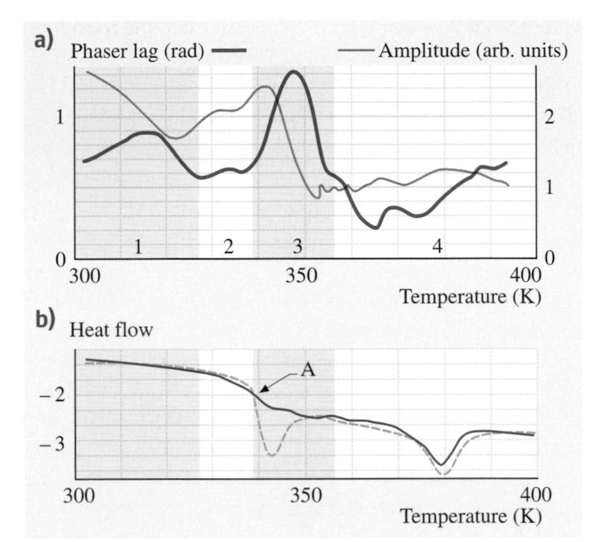

Fig. 21.2 **(a)** *Local mechanical spectroscopy spectra as a function of temperature of a technical PVC*: The vibration amplitude is displayed as a *thin line* and the phase lag as a *thick line*. Four temperature domains can be distinguished: From left to right, a small "vibration amplitude" decrease is associated with a first "phase lag" peak (1), in the next domain, "vibration amplitude" increases (2), then a large decrease of "vibration amplitude" is associated with a large "phase lag" peak (3) and finally "phase lag" increases slowly (4). **(b)** *DSC measurements of the same sample*. For clarity, the temperature domains observed on Fig. 21.5a have been reported. The graph displays the first (*dashed line*) and the second (*solid line*) heating. The glass transition can easily be recognized around 340 K (labeled *A*), slightly below the temperature range of domain 3. An irreversible endothermic relaxation takes place in the same temperature range (only visible on the first run). A reversible endothermic event occurs around 380 K

mains have been reported on the graph. The dashed line displays the first heating, where both reversible and irreversible events are present; the solid line shows the second heating, where only reversible events are still present. On the solid line, an endothermic event (labeled A) can be observed near the border between temperature domains 2 and 3. An irreversible endothermic relaxation is superimposed with (A) on the first heating. A reversible endothermic peak is observed around 380 K, in domain 4. The endothermic event (A) has the characteristics of a glass transition: the specific heat goes from one value to another without a peak [21.19]. The glass transition temperature of this PVC is, therefore, approximately 340 K. The irreversible relaxation occurring in the same temperature range is certainly associated with physical aging or structural relaxation. The reversible peak at higher temperature is associated with the melting of the small crystalline volume fraction.

These local mechanical spectroscopy results are in good qualitative agreement with macroscopic global measurements [21.13, 18]. They yield the same information as the macroscopic global measurements, but from a much smaller volume, and allow as a consequence the location of the different mechanisms. Both measurements (local and global) show a peak in phase for the primary and secondary relaxations, connected with a large decrease of the stiffness for the primary relaxation and a much smaller one for the secondary relaxation. Plasticity induces an increase of the phase lag in both cases. The temperatures observed for the relaxation related to the glass transition compare well with the calorimetric data. Based on all these examples, there is no doubt that the amplitude and phase lag of T-SLAM are functions of the stiffness and damping. In this regard, T-SLAM provides an extension of the global method, allowing location of dissipative phenomena and study of the spatial homogeneity of phase transitions or of relaxations. This will bring new insight into the field of inhomogeneous or composite materials. In addition, T-SLAM measures only a small volume near the surface. So even if the sample is a thick film, T-SLAM gives access to the near-surface mechanical properties, which may differ from bulk properties. This will bring interesting perspectives to the still debated field of surface dynamics.

21.1.3 Example Two: Local Mechanical Spectroscopy of NiTi

Near-stoichiometric NiTi alloys exhibit a martensitic phase transformation between a low-temperature monoclinic phase, called martensite, and a high temperature cubic phase with B2 structure, called austenite (1). This transformation is responsible for the shape memory and pseudo-elastic effects in deformed NiTi alloys. Upon transformation to the martensitic phase, an intermediate rhombohedral (R) phase can be formed [21.20, 21]. Although *Bataillard* [21.21] tends to demonstrate that the R phase is finely dispersed inside the material, a controversy remains over the spatial scale at which the decomposition of the transformation occurs. Another puzzling question concerns the transformation itself. Optical microscopy observation suggests that the transformation occurs very suddenly inside an austenite grain. This has led to the concept of "military transformation." The width of the globally measured transformation would then be a sum of different narrow contributions coming from different places inside the sample. This image is, however, not universally accepted.

A measurement inside one single grain of a polycrystal would be a way to address these questions. Both the spatial scale of the R phase distribution and the "military" character of the transformation will have an effect on the result of such a measurement. This is the reason for which local mechanical spectroscopy measurements of the martensitic phase transition of NiTi by T-SLAM have been performed [21.12, 22]. They are presented in Fig. 21.3. The global transformation behavior of NiTi is defined by calorimetry spectra (Fig. 21.3b, temperature scanning rate 10 K/min).

The vibration amplitude ("elasticity") and the phase lag ("internal friction") are shown in the top and bottom curves, respectively. The presented data is incomplete due to experimental limitations. A phase lag peak can be observed upon heating at approximately 370 K, associated with a change in the vibration amplitude level. This event can be correlated with the phase transformation from martensite to austenite observed by calorimetry (Fig. 21.3b). Upon cooling, no event is observed around 370 K: the vibration amplitude is stable and the phase lag curve does not exhibit any peak. But, an intense phase lag peak can be observed upon cooling near 330 K, associated with a restoration of the vibration amplitude level characteristic of the martensite. This event correlates with the phase transformation from austenite to martensite observed by calorimetry. A change in the vibration amplitude spectrum is correlated with each event on the phase lag spectrum. Two main features should be noted. First, the transformation peaks observed with the SLAM method are narrower than those observed with the (global) calorimetry measurement. Second, the measured temperatures for the local peak are located on the high temperature side of the peaks measured by calorimetry. Details of the two transformation peaks are displayed in Figs. 21.3c and 21.3d. The reverse transformation (Fig. 21.3c) is characterized by a change in the vibration amplitude level and a phase lag peak. This peak (A1) may have a shoulder on the low temperature

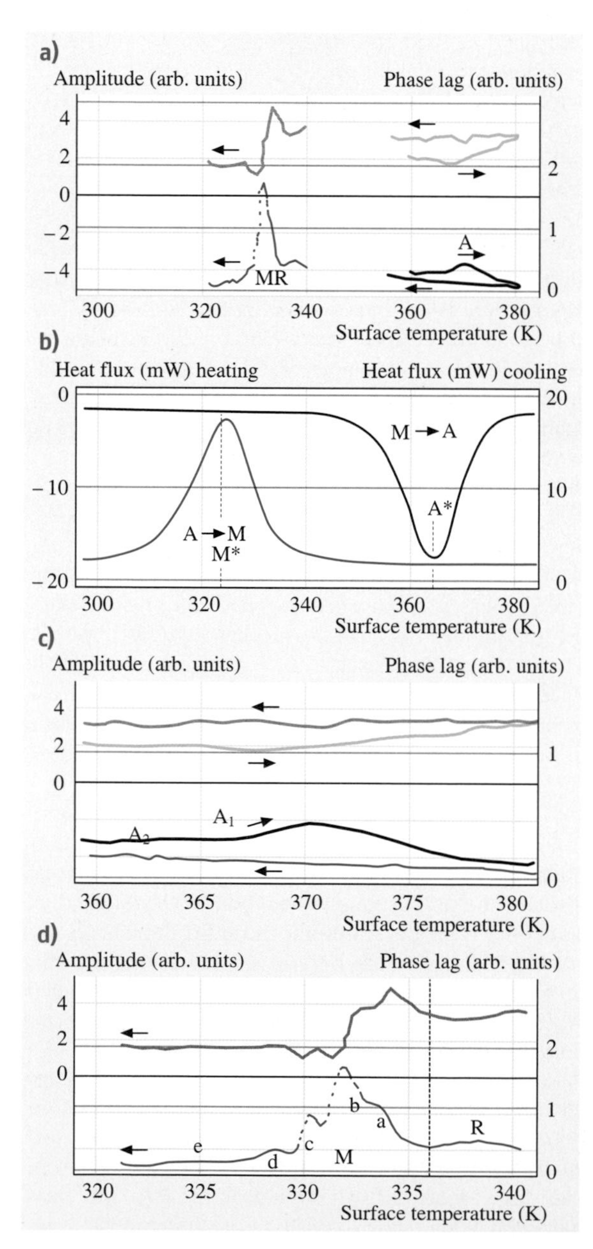

Fig. 21.3 **(a)–(b)** *Comparison of the local mechanical spectroscopy measurements of* NiTi *with calorimetric experiment.* **(a)** Both the reverse and direct transformation are associated with a phase lag peak and a modulus variation. **(b)** Calorimetric measurement of the same sample. **(c)–(d)** *Zoom on the reverse* **(c)** *and direct* **(d)** *transformation's temperature ranges.* The peak (A1) may exhibit a shoulder on the low temperature side (A2). The direct transformation is characterized by a recovery of the original vibration amplitude and a complex phase lag spectrum, formed of two main maxima, noted R and M. Peaks a,b,c,d,e are a substructure of the M peak

side (A2). A small decrease of the vibration amplitude may precede this increase. But, the magnitude of this softening and the very low intensity of peak A2 are too small to exclude experimental artifacts. The direct transformation (Fig. 21.3d) displays a recovery of the original vibration amplitude and a complicated phase lag maximum.

The correlation of the transition temperatures measured by calorimetry and the local mechanical spectroscopy lead to the conclusion that the peaks observed in phase lag and the change in vibration amplitude originate from the martensitic phase transition. At the scale of the observation, the martensite transforms into austenite around 370 K (A1 peak) and the austenite transforms into martensite close to 332 K (M peak), with the formation of the rhombohedral R phase at 337 K (R peak). The peaks A2 and e could be linked to the presence of two different types of martensite as already observed elsewhere [21.20, 21]. It is striking to note that all the events observed in macroscopic experiments seem to be reproducible at the scale of observation, namely less than $10^{-3}\ \mu m^3$. In particular, this would confirm that the R phase precipitates are very finely distributed in the austenite matrix as observed by Bataillard [21.21]. Otherwise the R peak could not be observed reproducibly using such a technique. The local measurement differs from global ones in at least two aspects: first, the width of the transformation temperature range is smaller; second, the M peak exhibits a substructure.

Global internal friction measurements on these materials have given spectra with peaks having a similar breadth as the calorimetry peaks [21.23], whereas the peaks measured locally are narrower. This is easy to understand if the martensitic transformation occurs inhomogeneously inside the sample. Grains tend to transform at different temperatures depending on their stress state. As the local measurement has sufficient spatial resolution to probe a single grain inside the material, it is logical that the transformation occurs over a narrower temperature range than in a global experiment probing a large number of grains.

The substructure in the M peak could have three origins: (i) the technique is sensitive to the mechanical relaxation inside newly formed plates, such as stress relaxation by twin motion; (ii) the probed volume contains several martensite plates that grow one by one, each with its own, distinct "elementary" peak (a, b, etc.); (iii) the analyzed region of the sample surface changes due to thermal drifts, therefore probing the transformation of several growing martensite plates, leading to multiple "elementary" peaks.

21.2 Static Methods – Mesoscopic Samples

21.2.1 Carbon Nanotubes – Introduction to Basic Morphologies and Production Methods

Carbon nanotubes (CNTs) are the newest form of carbon, found in 1991 by *Iijima* [21.24]. Because of their remarkable properties, this discovery has opened whole new fields of study in physics, chemistry, and material science. They possess a unique combination of small size (diameters ranging from ~ 1 to 50 nm with lengths up to $\sim 10\ \mu m$), low density (similar to that of graphite), high stiffness, high strength, and a broad range of electronic properties from metallic to p- and n-doped semiconducting. Their potential field of application is immense and includes reinforcing elements in high strength composites, electron sources in field emission displays and small X-ray sources, ultra-sharp and resistant AFM tips with high aspect ratios, gas sensors, and components of future, nanoscale electronics. In addition, they represent a widely used system for studying fundamental physical phenomena on the mesoscopic scale. Following advances in manufacturing and processing they are likely to be integral to many devices we use in our everyday life.

From the structural point of view, carbon nanotubes can be thought of as rolled up single sheets of graphite, *graphene*. They can be divided in two distinct groups. The first, multiwalled carbon nanotubes (MWNT), exhibit a Russian doll-like structure of nested, concentric tubes, Fig. 21.4a and b. They were also the first CNT to be discovered experimentally. The interlayer spacing can range from 0.342 to 0.375 nm, depending on the diameter and number of shells comprising the tube [21.25]. For comparison, the interlayer spacing in graphite is 0.335 nm, suggesting relatively weak interaction between individual shells. This fact has been corroborated by studies of mechanical and electronic properties of CNTs.

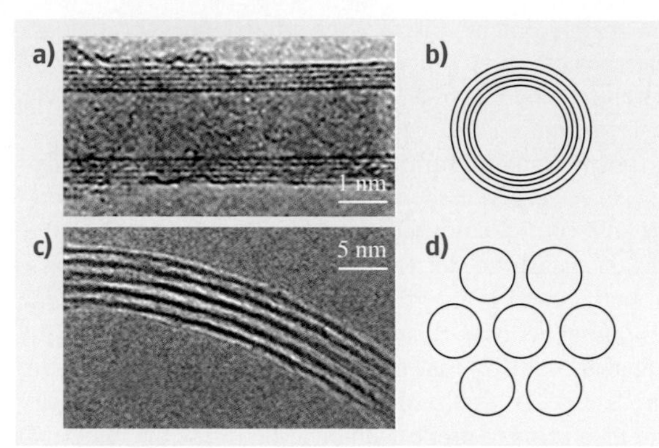

Fig. 21.4a–d TEM images and schematic drawings of cross sections for different morphologies of carbon nanotubes. (**a**) and (**b**) Multiwalled carbon nanotubes, consisting of concentric, nested tubes (MWNT). (**c**) and (**d**) ropes of single-walled carbon nanotubes, consisting of single carbon nanotubes bundled up in ropes and held together by van der Waals interaction

The second type of carbon nanotubes is in the basic form of a rolled-up graphitic sheet: a single-walled CNT (SWNT). During the production, their diameter distribution is relatively narrow so they often bundle in the form of crystalline "ropes" [21.26], Fig. 21.4c and d, in which the single tubes are held together by van der Waals interaction.

There are several distinct classes of production methods. The earliest is based on the cooling of carbon plasma. When voltage is applied between two graphitic electrodes in an inert atmosphere, they gradually evaporate and form plasma. On cooling, soot containing multiwalled nanotubes is formed [21.24]. If the anode is filled with catalysts such as, for example, cobalt of iron particles, SWNT ropes form. Another way of producing CNTs from carbon plasma is by laser ablation of a graphitic target [21.26]. It is considered that, in general, these methods produce CNTs of higher quality albeit in very small quantities and without the possibility of scale-up to industrial production. Other methods are based on chemical vapor deposition (CVD), a catalytic decomposition of various hydrocarbons, e.g. methane or acetylene mixed with nitrogen or hydrogen in the presence of catalysts [21.27]. This method offers the possibility of controlling the growth of nanotubes by patterning the catalyst [21.28] and is therefore more suitable for producing nanoscale structures with integrated CNTs. This method is also capable of producing CNTs in industrial quantities. Their main disadvantage is the higher concentration of defects, which as a consequence diminish their mechanical properties.

21.2.2 Measurements of the Mechanical Properties of Carbon Nanotubes by SPM

Mechanical measurements on CNTs performed with the AFM have confirmed theoretical expectations [21.29] of their superior mechanical properties. They involve measurements of deformations under controlled forces by bending immobilized carbon nanotubes either in the lateral [21.30] or normal direction [21.31] and also tensile stretching of CNTs fixed on their two ends to AFM tips observed in an electron microscope [21.32].

It is not obvious that continuum mechanics and its concepts like Young's, shear moduli, and tensile strength should work on the nanoscale. In order to apply them, one should also define the "thickness" of the nanotube's walls, a graphene sheet, in the frame of the continuous beam approximation. Most scientists working in this field are using the value 0.34 nm, close to the interlayer separation in graphite as the thickness of a nanotube. When comparing different results, however, one has to bear in mind that to convert relatively precise force–displacement measurements into macroscopic quantities like Young's or shear modulus, one has to introduce various geometrical factors, including diameter and length. Even a small imprecision in their determination is very unforgiving because the diameter d enters into equations for beam deformation as d^4 and length l as l^3, leading to large uncertainty in final results.

The first quantitative measurement of the Young's modulus of carbon nanotubes was reported by *Wong* et al. [21.30] in which they laterally bent MWNTs and SiC nanorods (similar in dimensions to MWNTs) deposited on flat surfaces. MWNTs were first randomly dispersed on a flat surface of MoS_2 single crystals that were used because of their low friction coefficient and exceedingly flat surface. Friction between the tubes and substrate was further reduced by performing the measurements in water. Tubes were then pinned on one side to this substrate by a deposition of an array of square pads through a shadow mask, Fig. 21.5a–c. AFM was used to locate and characterize the dimensions of protruding tubes. The beam was deformed laterally by the AFM tip, until at a certain deformation the tip would pass over the tube, allowing the tube to snap back to its relaxed position. During measurements, the force–distance curves were acquired at different positions along the chosen beam, Fig. 21.5d–e. Maximum

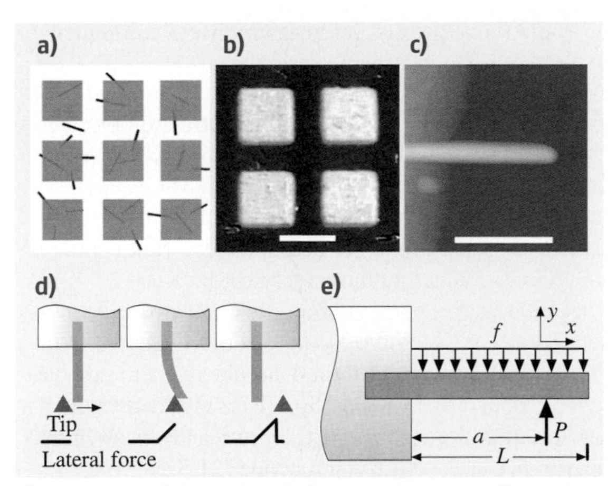

Fig. 21.5a–e Schematic of the experiment performed by *Wong* et al. [21.30] (a) nanotubes were dispersed on a substrate and pinned down by a deposition of SiO pads. (b) Optical micrograph showing the pads (*light*) and MoS_2 substrate (*dark*). The scale bar is 8 μm long. (c) AFM image of a SiC nanorod protruding from the pad. The scale bar is 500 nm long. The same experimental setup was used for elastically deforming MWNTs. (d) The tip shown as a triangle moves in the direction of the arrow. The lateral force is indicated by the trace at the bottom. Before the tip contacts the beam, the lateral force remains constant and equal to the friction force. During the bending a linear increase in the lateral force with deflection is measured. After the tip has passed over the beam the lateral force drops to its initial value and the beam snaps back to its equilibrium position. (e) Schematic of a pinned beam with a free end. The beam of length L is subjected to a point load P at $x = a$ and friction force f (abstracted with permission from [21.30] © American Association for the Advancement of Science)

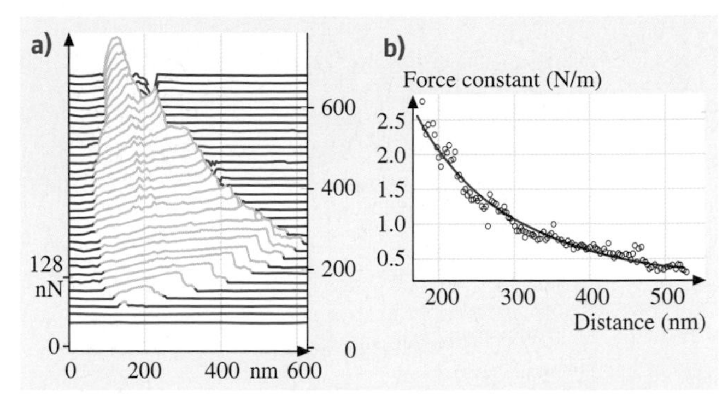

Fig. 21.6 (a) A series of lateral force distance curves acquired by *Wong* et al. [21.30] for different positions along a MWNT. (b) The lateral spring constant as a function of position on the beam. The curve is a fit to (21.3) (abstracted with permission from [21.30] © American Association for the Advancement of Science)

deflection of the nanobeam can be controlled to a certain degree by the applied normal load, and in this way tube breaking can be avoided or achieved in a controlled manner. The applied lateral load P in terms of lateral displacement y at the position x along the beam is given by the equation:

$$P(x, y) = 3EI\frac{y}{x^3} + \frac{f}{8}\left(x - 4L - 6\frac{L^2}{x}\right), \quad (21.2)$$

where E is the Young's modulus of the beam, I the second moment of the cross section, equal to $\pi d^4/64$ for a solid cylinder of diameter d and f the unknown friction force, presumably small due to the experimental design. The lateral force, Fig. 21.6a, is known only up to a factor of proportionality because the AFM lever's lateral force constant wasn't calibrated for these measurements. This uncertainty and the effect of friction were eliminated by calculating the nanobeam lateral force constant:

$$\frac{\mathrm{d}P}{\mathrm{d}y} \equiv k = \frac{3\pi d^4}{64x^3}E \quad (21.3)$$

shown in Fig. 21.6b.

The mean value for the Young's modulus of MWNTs was $E = 1.3 \pm 0.6$ TPa, similar to that of diamond ($E = 1.2$ TPa). For larger deformations, discontinuities in bending curves were also observed, attributed to elastic buckling of nanotubes [21.33].

In another series of experiments, *Salvetat* et al. measured the Young's modulus of isolated SWNTs and SWNT ropes [21.31], MWNTs produced using different methods [21.34], and the shear modulus of SWNT ropes [21.35]. The experimental setup that enabled them to perform measurements on such a wide range of CNT morphologies involved measuring the vertical deflection of nanotubes bridging holes in a porous membrane.

In their measurement method, they suspended CNTs in ethanol and deposited them on the surface of a well-polished alumina Al_2O_3 ultrafiltration membrane. Tubes adhered to the surface due to the van der Waals interaction occasionally spanning holes, Fig. 21.7a. After a suitable nanotube had been found, a series of AFM images was taken under different loads, in which every image would thus correspond to the surface (and the tube) under a given normal load. Extracted linescans across the tube revealed the vertical deformation, Fig. 21.7b. For the range of applied normal loads, the deflection of a thin, long nanotube at the midpoint δ as

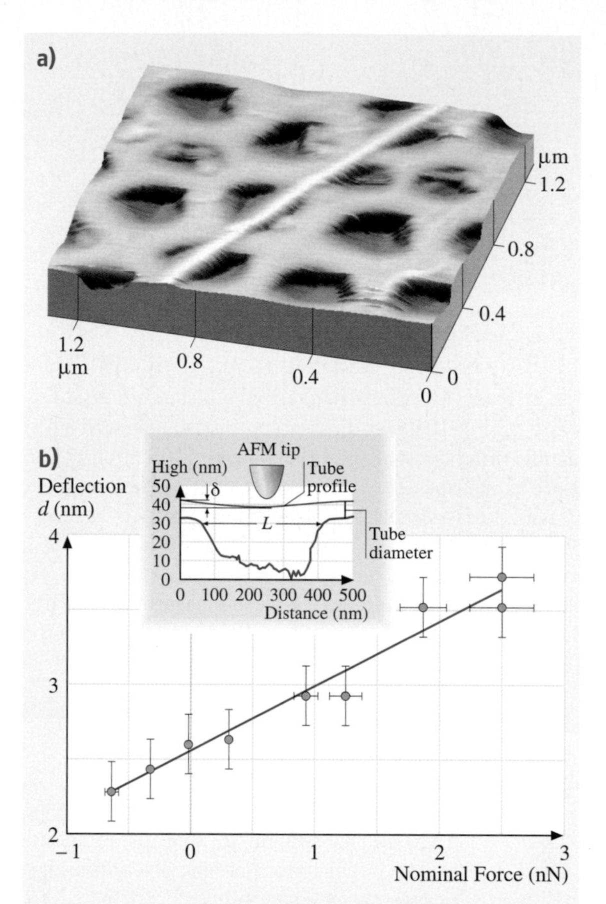

Fig. 21.7 (a) An AFM image of a CNT lying on a porous alumina [21.35] filter. (b) Measured dependence of vertical deflection on the applied nominal force. The inset shows a comparison between linescans taken on the tube and over a hole [21.34] (abstracted with permission (a) from [21.35] © 1999 American Physical Society, (b) from [21.34] © 1999 Wiley)

a function of the normal force F can be fitted using the clamped beam formula [21.36]:

$$\delta = \frac{FL^3}{192EI} , \tag{21.4}$$

where L is the suspended length.

The fitted line does not pass through the origin because the force acting on the nanotube is not equal to the nominal force alone; it contains a constant term arising from the attractive force between the AFM tip and the tube. The tube's deflection should also contain a term corresponding to the interaction between the tube and the substrate, but this is generally regarded as negligible.

This variable load imaging technique is advantageous for obtaining quantitative information as one is assured that the AFM tip is in the desired location when deforming the tube. Equation (21.4) is valid only if the tubes adhere well to the substrate, confirmed by the fact that the images reveal no displacement of the parts of the tube in contact with the membrane.

Using this technique, Young's modulus of 1 TPa was found for SWNTs. Values for the MWNTs show a strong dependence on the amount of disorder in the graphitic layers: an average value of $E = 870$ GPa was found for the arc-discharge grown tubes, while the catalytically grown MWNTs, known to include a high concentration of deffects, can have a Young's modulus as low as 12 GPa.

For the deflection of SWNT ropes, an additional term in the bending formula has to be taken into account because of the influence of shearing between the tubes comprising the rope. Single CNTs are held together in the tube only by weak, van der Waals interactions. Ropes, therefore, behave as an assembly of individual tubes rather than as a thick beam. The deflection can be modeled as a sum of deflections due to bending and shearing [21.36]:

$$\delta = \delta_{\text{bending}} + \delta_{\text{shearing}} = \frac{FL^3}{192EI} + f_s \frac{FL}{4GA} = \frac{FL^3}{192E_{\text{bending}} I} , \tag{21.5}$$

where f_s is a shape factor, equal to 10/9 for a cylinder, G the shear modulus and A the area of the beam's cross section. E_{bending} is the effective, bending modulus, equal to Young's modulus in the case in which the influence of shearing can be neglected (for thin, long ropes).

The Young's and the shear modulus can thus be extrapolated by measuring the E_{bending} of an ensemble of ropes with different diameter to length ratios: for thin ropes one obtains the value of the Young's modulus, while for the thick ones, the E_{bending} approaches the value of G on the order of 10 GPa (see Fig. 21.8).

Walters et al. pinned ropes of single-wall nanotubes beneath metal pads on an oxidized silicon surface, then released them by wet etching, Fig. 21.9a. The SWNT rope was deflected in the lateral direction using an AFM tip, Fig. 21.9b. As the suspended length is on the order of μm, the SWNT rope can be modeled as an elastic string stretched between the pads. Upon deformation all of the strain goes into stretching. In the simple case of a tube lying perpendicular to the trench and the AFM tip deforming the tube in the middle, the force F exerted on

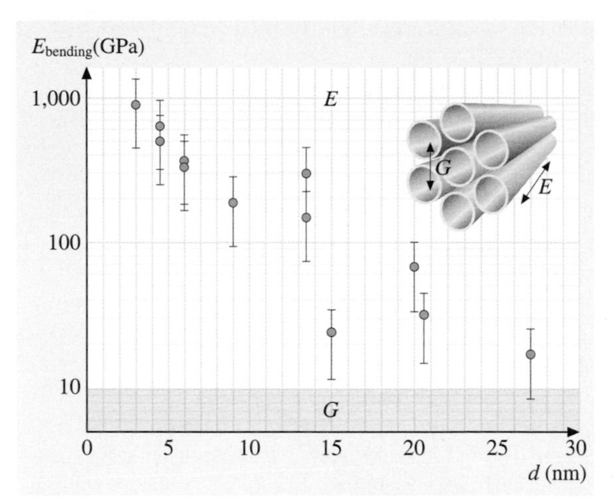

Fig. 21.8 Values of the shear modulus for 12 SWNT ropes of different diameters. The measured E_{bending} of thin ropes corresponds to E, while for thick ropes one obtains the value of shear modulus G (abstracted with permission from [21.35] © 1999 American Physical Society)

the tube by the AFM tip is given by the expression:

$$F = 2T\sin\theta = 2T\frac{2x}{L} \approx \frac{8kx^3}{L_0}\,, \tag{21.6}$$

where T is the tension in the string, L_0 its equilibrium length, k the spring constant and x the lateral deflection in the middle. Using this setup, they deformed SNWT ropes to the maximal strain of $5.8 \pm 0.9\,\%$ and determined a lower bound of 45 ± 7 GPa on the tensile strength, assuming a value of 1.2 TPa for the Young's modulus.

Kim et al. used a setup in which the SWNT rope was embedded in metallic electrodes deposited on a silicon substrate coated with poly(methylmethacrylate) (PMMA). In their experiment the tube can also be modeled as an elastic string. Using an AFM tip, they vertically deformed the rope and obtained an estimate of $E = 0.4$ TPa for the Young's modulus of an SWNT rope.

Finally, the first direct measurements of the elastic properties of CNTs that haven't relied on the beam or stretching string setup involved deforming MWNTs [21.38] and SWNT ropes [21.32] under axial strain. This was achieved by identifying and attaching opposite ends of MWNTs or SWNT ropes to two AFM tips, all inside a SEM. The AFM tips were integrated with different cantilevers, one rigid with a spring constant above 20 N/m and the other compliant with a spring constant of 0.1 N/m, Fig. 21.10. The rigid lever

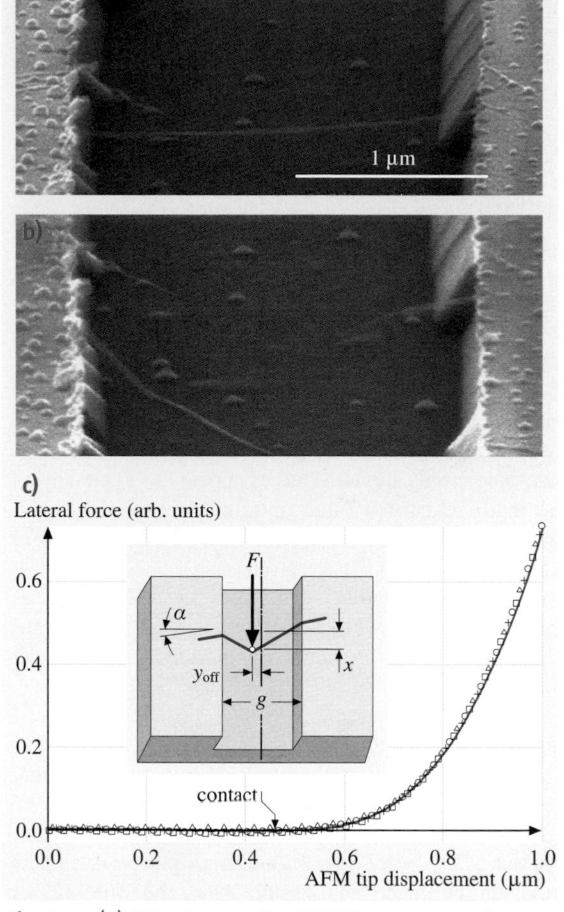

Fig. 21.9 (**a**) SEM image of a SWNT rope suspended over trench in silicon before and (**b**) after being deformed past its elastic limit. (**c**) Lateral force on a single-wall nanotube rope as a function of AFM tip displacement (abstracted with permission from [21.37] © American Institute of Physics)

was then driven using a linear piezomotor. On the other end, the compliant lever bent under the applied tensile load. The deflection of the compliant cantilever – corresponding to the force applied on the nanotube – and the strain of the nanotube were simultaneously measured. The force F is calculated as $F = kd$ where k is the spring constant of the flexible AFM lever and d its displacement in the vertical direction. The strain of the nanotube is $\delta L/L$, Fig. 21.11. From the stress–strain curves obtained in this fashion, Fig. 21.11b, Young's moduli ranging from 270–950 GPa were found. Exami-

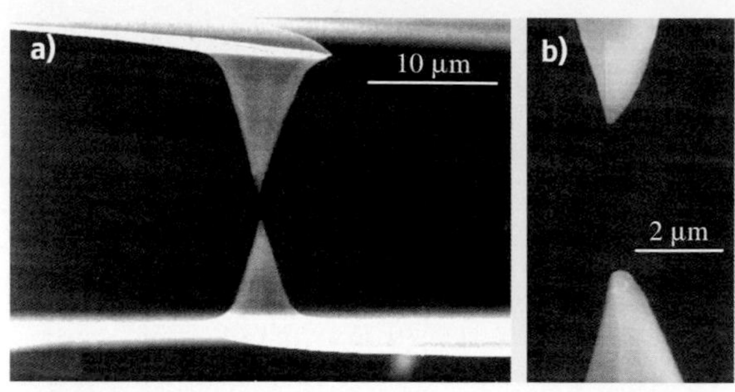

Fig. 21.10 (a) A SEM image of a MWNT mounted between two opposing AFM tips. (b) A close-up of the region indicated by a rectangle in (a) (abstracted with permission from [21.38] © 1999 American Association for the Advancement of Science)

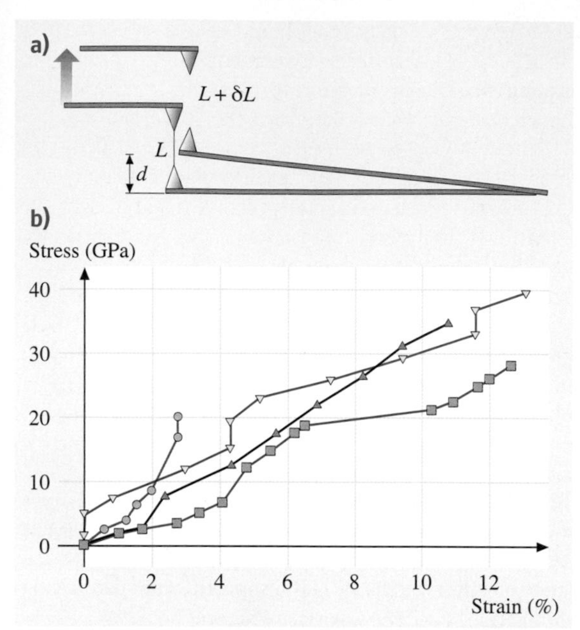

Fig. 21.11 (a) The principle of the experiment performed by *Yu* et al. As the rigid cantilever is driven upward, the lower, compliant cantilever bends by the amount d and the nanotube is stretched by δL. As a result, the nanotube is strained by $\delta L/L$ under the action of force $F = kd$, where k is the elastic constant of the lower AFM lever. (b) Plot of stress vs. strain curves for different individual MWNTs (abstracted with permission from [21.38] © 1999 American Association for the Advancement of Science)

nations of the same broken tubes inside a TEM revealed that nanotubes break with the "sword in sheath" mechanism, where only the outer layer appears to have carried the load. After it breaks, pullout of inner shells follows. An average bending strength of 14 GPa and axial strengths up to 63 GPa were found.

Firm attachment of nanotubes to AFM tips was ensured by a deposition of carbonaceous material induced by the electron beam concentrated in the contact area [21.40].

All these measurements of the elastic properties of carbon nanotubes are summarized in Table 21.1.

Before comparing them, it should be noted that the absolute values of these mechanical constants have relatively large uncertainties because of the huge influence of the precision of determining tube diameters and lengths on the final result. Also, the cited values represent mean values of results obtained on several tubes, with the exceptions of the lowest value given for catalytically grown MWNTs by Salvetat et al. [21.31], a single value for an individual SWNT rope from *Kim* et al. [21.39], and the lowest and highest values from *Yu* et al. [21.38]. One also has to bear in mind that concepts

Part C | 21.2

Table 21.1 Summary of the mechanical properties of carbon nanotubes measured using SPM methods

Young's modulus E (GPa)	Shear modulus G (GPa)	Tensile strength σ (GPa)	Nanotube type	Deformation method	Reference
1,300±600	–	–	MWNTs arc grown	Lateral bending	[21.30]
1,000±600	–	–	SWNTs	Normal bending	[21.31]
1,000±600	~ 1 GPa		SWNT ropes	Normal bending	[21.35]
870±400	–	–	MWNT arc-grown	Normal bending	[21.31]
12±6	–	–	MWNT catalytic	Normal bending	[21.31]
–	–	45±7	SWNT rope	Lateral bending	[21.37]
400	–	–	SWNT rope	Normal bending	[21.39]
1,020	–	30	SWNT ropes	Tensile loading	[21.32]
270–950	–	11–63	MWNTs arc-grown	Tensile loading	[21.38]

like Young's and shear moduli and tensile strength were introduced for describing macroscopic and continuous structures. Their application for describing mesoscopic objects like nanotubes, therefore, has its limitations.

The methods presented above are all in their nature "single-molecule" methods, in the sense that they measure properties of individual objects. A result of a single experiment, therefore, represents properties of a particular object and differs from case to case because of, for example, defects coming from production and purification, or for more prosaic reasons such as experimental errors. In order to perform comparisons, therefore, it is more practical to deal with values averaged for multiple tubes. The average values for the Young's modulus of high quality tubes are, within the experimental error, all on the order of 1 TPa, close to that of diamond (1.2 TPa), while the tensile strength can be 30 times higher than that of steel (1.9 GPa). Catalytically grown MWNTs with the Young's modulus that can be as low as 12±6 GPa are definitely disappointing and show that the production method plays an important role in the quality of carbon nanotubes from the point of view of their mechanical properties.

Future improvements in large scale production and processing are therefore necessary before applying them as the ultimate reinforcement fiber. Even so, the progress in measuring the mechanical properties of CNTs will continue to be closely related to, and often motivate, the progress in nanoscale manipulation in general.

21.2.3 Microtubules and Their Elastic Properties

Microtubules are a vital biological nanostructure, similar in dimension and shape to carbon nanotubes. In fact, the first name given to carbon nanotubes by their discoverer Iijima was "microtubules of graphitic carbon." From the structural point of view, they are much more complicated, as is common for biological structures. They self-assemble in buffers maintained at the physiological temperature of 37 °C out of protein subunits, α and β-tubulin, each having a molecular weight of 40 kDa. These protein subunits are bound laterally into protofilaments, which in turn are arranged in a shape of a hollow cylinder with an external diameter of 25 nm and an internal diameter of 15 nm. Microtubules are a remarkable material: inside living cells, their length incessantly fluctuates; they can even completely disassemble and consume energy in the form of GTP.

Together with actin and intermediate filaments, microtubules constitute the cellular cytoskeleton. In addition, they perform various unique vital functions: they act as tracks along which molecular motors move, they help pull apart chromosomes during cell division and form bundles that propel sperm and some bacteria. All these roles are determined by their structure and mechanical properties. Yet, after more than a decade, there is still a large discrepancy in the values of their Young's modulus reported in the literature. Several methods have been applied, such as bending or buckling microtubules using optical tweezers [21.42], hydrodynamic flow [21.43], thermally induced vibrations or shape fluctuations [21.44], buckling in vesicles [21.45], or squashing with an AFM tip [21.46]. These methods yielded results ranging from 1 MPa [21.46] to several GPa [21.42].

Since microtubules are geometrically similar to nanotubes, *Kis* et al. [21.41] used the suspended tube configuration. Microtubules were deposited on porous membranes, and AFM images were acquired under different nominal normal loads. All the measurements were performed in liquid and at controlled temperatures, in order to prevent the degradation of proteins, so the substrate had to be functionalized in order to ensure good adhesion between the tubes and the support. They used

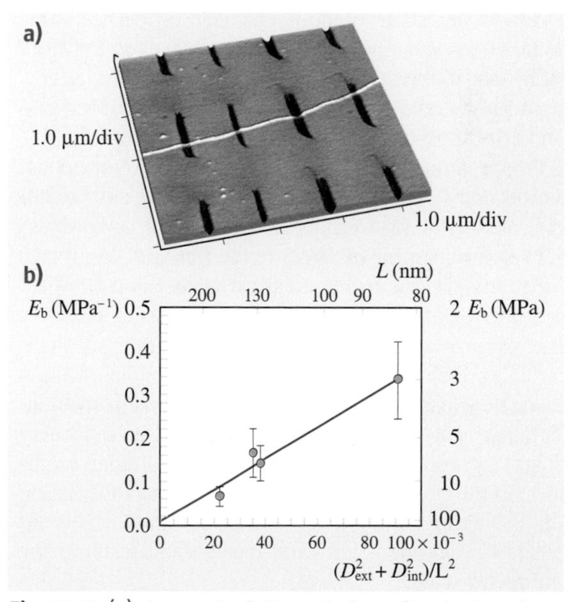

Fig. 21.12 (a) A pseudo 3-D rendering of a single microtubule suspended over four different-sized holes in PMMA. (b) From the variation of the $E_{bending}$ with varying length, the shear and the Young's modulus have been determined for the microtubule displayed in (a) (abtracted with permission from [21.41] © 2002 American Physical Society)

silicon with a layer of PMMA as a support. Slits were prepared in PMMA using electron-beam lithography, providing the possibility to measure the elastic response of individual microtubules lying over four different-sized holes ranging from 80–200 nm across and 400 nm deep, Fig. 21.12.

Clearly the results are dependent on the hole diameter, which could only result from a shear component within the microtubules. Simplifying (21.5), the deformation of microtubules can be modeled as:

$$\frac{1}{E_{\text{bending}}} = \frac{1}{E} + \frac{10}{3} \frac{D_{\text{ext}}^2 + D_{\text{int}}^2}{L^2} \frac{1}{G} , \tag{21.7}$$

where and D_{int} represent the external and internal MT's diameter. From the plot of $1/E_{\text{bending}}$ vs. the $\left(D_{\text{ext}}^2 + D_{\text{int}}^2\right)/L^2$ the shear modulus can be calculated from the slope of the fit, and the inverse Young's modulus will be equal to the intercept on the y-axis. In this way the shear modulus of $G = 1.4 \pm 0.3$ GPa and a lower limit of $E = 150$ MPa were simultaneously obtained from a measurement on an individual microtubule.

This anisotropy comes from the fact that microtubules are built of monomers that are strongly bound in the longitudinal direction, along single protofilaments. The link between neighboring protofilaments is much weaker. Microtubules, therefore, have to be considered as anisotropic beams, having an E_{bending} that depends on the scale on which they are deformed [21.41]. This is in all respects analogous to the situation in SWNT ropes that are built of stiff individual SWNTs, held together in bundles by a weak van der Waals interaction. This conclusion may explain the large discrepancy in the value of Young's moduli reported in the literature over the past ten years using inadequate modeling.

As in the case of carbon nanotubes, measuring the mechanical properties of microtubules can provide valuable insight into their structure and provide deeper understanding about the functioning of these remarkable structures.

21.3 Scanning Nanoindentation: An Application to Bone Tissue

In the following section we discuss nanoindentation as a tool to determine nanomechanical properties. This method was developed in the early 1980s evolving from traditional Vicker hardness testing devices. The latter is based on the concept that a pyramidal tip is loaded into a material applying a known force. After the test, the size of the remaining imprint is measured under an optical microscope. The ratio of the employed force and the imprint area after load removal was defined as hardness that represents a mean pressure the material can resist. Unfortunately this mechanical parameter is a complex combination of elastic and postyield properties and can't be easily explained on the level of continuum mechanics. This point raised important concerns for determining elastic constants such as the Young modulus from an indentation test. Further improvements on the transducer sensitivity were necessary to provide continuous acquisition of the employed load and the resulting indentation depth. Nanoindentation represents the state of the art of this development allowing mechanical tests on the nanometer scale.

21.3.1 Scanning Nanoindentation

Based on this continuous force–displacement data, an elastic modulus of volumes in the submicron regime can be quantified. To investigate features down to the micrometer regime, classical nanoindentation tools are combined with an optical microscope for positioning the indenter tip on the region of interest. But to defeat the limits of optical resolution, indentation and scanning probe microscopy were combined using the same tip to allow nanoscale control of the indenting tip position [21.47]. This instrument allows for scanning the material topography in an AFM mode and performing nanoindentation tests employing the same tip. The mechanical tests are restricted to the scanned area ($100\,\mu\text{m} \times 100\,\mu\text{m}$ maximum) where the indenter can be positioned with a precision of better than 100 nm. The surface roughness can be measured, which is helpful for choosing the area to be indented. Furthermore the available in situ scan of the indented region can provide information about the piling-up or sink-in behavior of the material. *Kulkarni* and *Bhushan* [21.48,49] introduced this device and demonstrated that measurements on aluminum and silicon were similar to results of non-scanning indentation systems.

21.3.2 Application of Scanning Nanoindentation

The development of the scanning nanoindentation (SN) technique has led to a variety of studies, primarily characterizing thin layers. Applications range from the mechanical characterization of corrosion-free film apposition on single-crystalline iron [21.50] to indentation and microscratch tests on Fe-N/Ti-N multilayers [21.51]. One author included nanoindentation to discuss two different electrochemical deposition techniques of thin Ni-P layers on pure Ni [21.52]. *Rar* et al. [21.53] reported studies on the growth of thin gold layer on native oxides of silicium while other investigations focussed on wear-resistant $TiB_2(N)$ coatings [21.54].

Studies of heterogeneous materials clearly demonstrate the advantage of the available surface scanning. *Malkow* and *Bull* [21.55] investigated the elastic/plastic behaviour of carbon-nitride films deposited on silicon. Therefore they determined the load-on and load-off hardness, the latter being accessible by scanning the remaining indent impression. *Shima* et al. [21.56] studied silicon oxynitride films on pure silicon and demonstrated hardness as a function of the employed deposition temperature. *Göken* et al. [21.57] used the high positioning precision to characterize individual lamellae of a TiAl alloy that consists of a two-phase structure and also to study the mechanical properties of nanometer-size precipitates of nickel-based super-alloys [21.58]. Performing in situ electrochemical treatments of iron single crystals, *Seo* et al. [21.59] used SN to study the variation of the remaining imprint shape with time.

In the field of biomechanics, work was done on bone tissue in wild-type and gene-mutated zebrafish. The characterization of the residual indentations in AFM mode supported the statement that gene mutation can change bone brittleness [21.60]. Other investigations focussed on the demineralising effects of soft drinks on tooth enamel by studying changes of elastic properties and surface topography [21.61]. *Habelitz* et al. [21.62] and *Marshall* et al. [21.63] characterized the junction between human tooth enamel and the mechanically different dentin.

Hengsberger et al. [21.64] took advantage of the nanometer positioning capability of the SN to investigate the elastic and plastic properties of individual human bone lamellae. Related work is presented in greater detail below to demonstrate the use and benefits of the SN technique at this example.

21.3.3 Example: Study of Mechanical Properties of Bone Lamellae Using SN

Before we discuss nanomechanical properties of bone lamellae, it is useful to know the structure on the macroscopic level. Figure 21.13 presents the hierarchy of human bone tissue at the example of the femoral neck. The outer shell is constituted of cortical bone while the porous trabecular bone structure gives inner support. On the next lower level, both bone types show structural units (BSU, after [21.65]) constituted of some tens of lamellae. For compact bone the lamellae have a concentric organization while for trabecular bone these lamellae are parallel. The nomenclature "BSU" was motivated by its underlying cellular process. The formation of such an individual structural unit occurs within a single cellular process. The optical contrast shows two alternating types of lamellae, thin lamellae that appear dark, and thick lamellae that appear bright. These bone lamellae show widths ranging from 1 to 3 μm for thin and 2 to 4 μm for thick lamellae [21.64]. On the next lower level the bone matrix is constituted of a complex collagen and mineral structure. It is still not entirely understood what type of variations in the collagen/mineral meshwork are responsible for the lamellar structure.

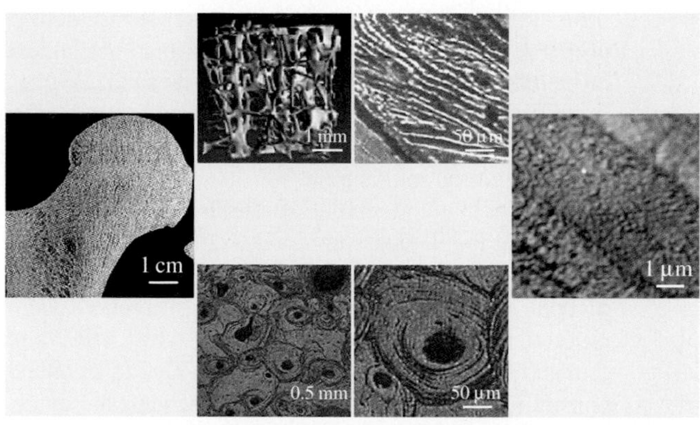

Fig. 21.13 Hierarchy of the human femoral neck (from left to the right) the far left figure shows a cut through a frontal plane of the femoral neck. The outer shell is constituted of compact bone while the inside is made of the spongier trabecular bone. The left pair of images shows the trabecular structure (*top*) and a transverse cut of the compact shell that shows vascular canals (*bottom*).
The right pair of images shows packets of trabecular bone lamellae (*top*) and a single osteon (*bottom*), a vascular canal surrounded by concentric lamellae. Packets of trabecular bone lamellae and osteons represent a structural unit (BSU). Note the alternation of bright (*thick*) and dark (*thin*) lamellae.
The figure far right shows three bone lamellae, structures that are similar for trabecular and compact bone. The bone matrix within the lamellae are mainly constituted of collagen fibers and hydroxyapatite crystals

The mechanical properties on these different levels of bone hierarchy are increasingly well understood. But, little is known about how the macroscopic mechanical properties of the whole bone relate to its nanomechanical properties. The possibility that factors such as fracture risk might be better understood with an analysis of small bone volumes motivates the application of nanoindentation in this field.

A set of nanoindentation studies is available (mainly employing the device combined with an optical microscope) that presents the intrinsic mechanical properties of bone tissue. Among recent work, differences were reported between donors, anatomical sites, BSU, and thin and thick lamellae [21.64, 66–69]. Due to technical constraints of this sensitive nanomechanical device, the majority of these studies present dehydrated or dried tissue properties measured at ambient temperature [21.70–73]. But, removal of the water content may lead to anisotropic shrinking of the matrix that creates microcracks and alters the mechanical properties of the bone constituents. For an accurate characterization of the in vivo properties, the nanomechanical tests should therefore be done under physiological conditions.

There have been a few attempts to characterize the in vivo bone properties [21.66, 68, 74]. This was achieved by studying fresh bone samples that are kept moist with a thin layer of liquid (less than a hundred microns) on the surface or with subsurface water irrigation. But, local evaporation of the thin liquid layer may have led to indents on areas that were partially dried during the test. One possible solution to such local drying might be to conduct measurements in which the indentation tip and tissue sample are both fully immersed in a liquid cell and simultaneously heated to body temperature.

Practical limitations on temperature stability and inaccuracy in contact force detection due to liquid on the surface make such measurements extraordinarily difficult, however. For statistically powerful studies it seems unavoidable to dry the samples.

The objective of this study was to use SN to determine the effect of drying on the stiffness of single bone lamellae. The goal was to determine a conversion factor that allows dry tissue properties to be recalculated into their in vivo properties.

For this purpose, we measured the identical set of lamellae selected from human trabecular and compact bone at first under physiological and then under dry conditions.

Experimental Setup and Technical Features: Figure 21.14 demonstrates an optical picture and a sketch of the combined AFM and nanoindenter device (Hysitron Inc. Minneapolis, MN). The transducer consists of a three-plate capacitor on whose central plate a Berkovich (three-sided pyramid) diamond tip is mounted. The transducer provides a contact force feedback between the tip and the sample surface. The sample is mounted on a piezoelectric scanner that allows moving in the x,y and z-directions. During x,y-surface scanning, the piezoelectric scanner keeps this feedback signal constant by correcting the z-height. The movement of the piezoelectric scanner, therefore, describes the negative surface of constant contact force, commonly called an AFM-scan. For the liquid cell tests, the sample was glued in a plexiglass cup for the addition of a several millimeter high liquid layer. A commercially available liquid cell tip (Hysitron Inc. Minneapolis, MN) was used, which contains an additional shaft of approximately 300 μm diameter and 5 mm length to protect the transducer from the fluid. A small layer of latex was placed between the sample holder and the magnetic stick of the piezoelectric scanner to protect the latter from liquid. The additional shaft of the indenter tip and the latex layer

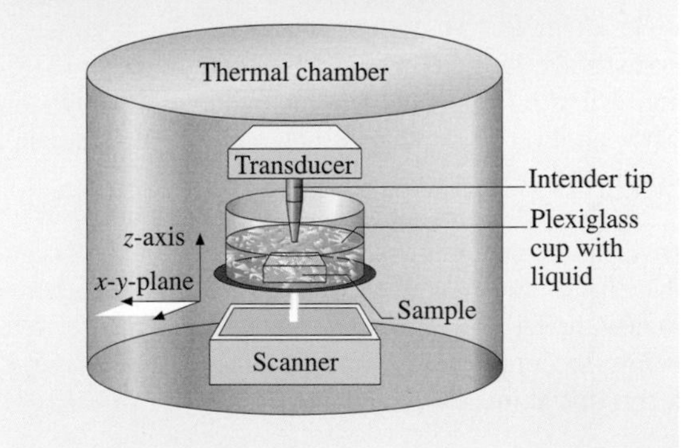

Fig. 21.14 Optical picture (*top*) and sketch (*bottom*) of the scanning nanoindentation device for measurements under physiological conditions. For this purpose the sample was installed in a plexiglass cup for addition of liquid. The entire instrument was heated in a custom-made thermal chamber

represent elastic components that increased the machine compliance ($C_m = 7$ nm/mN instead of 3 nm/mN). This variable corrects the deformation of all the machine components while indentation data are recorded and can be determined from the tip shape calibration curve (Hysitron, Minneapolis). The nanoindentation device was installed in a custom-made thermal chamber to allow sample heating to 37 °C. Note that the increased temperature and humidity also changed the value of the electrostatic force constant (EFC). The latter corrects the force due to the springs that support the central plate on which the indenter tip is mounted. The EFC can be calibrated doing out-of-contact indents and varying this value until a zero-line in the force–displacement curve is measured.

Based on an image acquired in AFM mode, the indentation area can be selected with a high spatial resolution ($< 0.1\,\mu$m). Figure 21.15 shows the force–displacement data of a typical nanoindentation curve. In the first step, the tip is pressed into the material, resulting in indistinguishable elastic and plastic deformation. Then the tip is held at maximum force resulting in creep of the material under the tip. In a third step, unloading is done that leads to elastic recovery of the material. Typically each indent requires between 15 seconds to several minutes, representing a compromise between a desired quasistatic strain rate and the thermal drift of the instrument (possibly below 0.1 nm/second). This device is load-controlled, and linearly increasing and decreasing loading protocols were therefore applied. The loading rate corresponds to

$$\frac{dP}{dt} = \frac{P_{max}}{T} \tag{21.8}$$

with P_{max} as the maximum load and T as the (un)loading time.

The elastic constants of the sample were determined using the unloading part (step 3) of the nanoindentation curve. Based on the analytical work by *Sneddon* [21.75], *Oliver* and *Pharr* [21.76] derived the following equation for an indenter of revolution pressed into an isotropic elastic material:

$$S(h_{max}) = \frac{dP}{dh}(h_{max}) = \frac{2}{\sqrt{\pi}} E_r \sqrt{A_c(h_{max})} \,. \tag{21.9}$$

Pharr et al. [21.77] showed that this equation is a good approximation for a Berkovich indenter tip. P represents the applied load, $S(h_{max})$ is the derivative of the unloading curve at the point of initial unloading h_{max}. This variable is determined by fitting typically 40% to 95% of the unloading curve to avoid the influence of viscoelastic effects at initial unloading when a singularity in the strain rate occurs. $A_c(h)$ is the contact area over which the material and the indenter are in instantaneous contact. The latter function has to be calibrated by performing indents with increasing depth in a standard material, typically fused silica. The latter has a known reduced modulus of $E_r = 69.9$ GPa that allows calibration of the contact area $A_c(h)$ using the measured contact stiffness $S(h)$ and (21.9). The reduced modulus E_r combines the deformation of the material and the diamond tip as follows [21.78]:

$$\frac{1}{E_r} = \frac{1-\nu_{specimen}^2}{E_{specimen}} + \frac{1-\nu_{tip}^2}{E_{tip}} \,. \tag{21.10}$$

We use the indentation modulus

$$E_{ind} = \left(\frac{1}{E_r} - \frac{1-\nu_{tip}^2}{E_{tip}}\right)^{-1} \tag{21.11}$$

that can be calculated with the reduced modulus and the elastic properties of the diamond indenter tip $\nu_{tip} = 0.07$ and $E_{tip} = 1{,}140$ GPa. This variable combines with

$$E_{ind} = \frac{E_{specimen}}{1-\nu_{specimen}^2} \tag{21.12}$$

the local Young's modulus $E_{specimen}$ and Poisson ratio $\nu_{specimen}$ of the specimen.

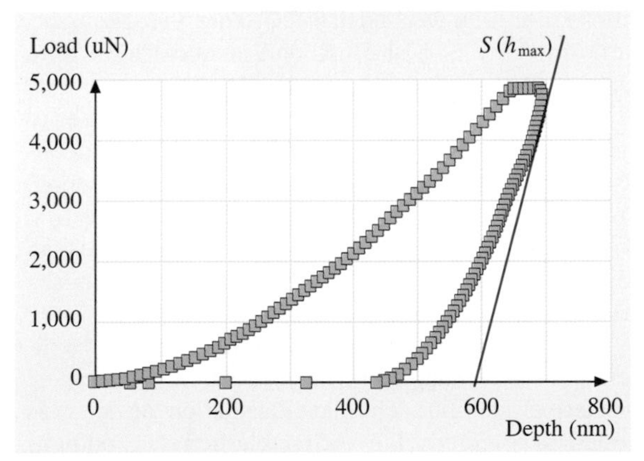

Fig. 21.15 Force–displacement curve obtained during a nanoindentation test that consists of three parts (see text). Hardness is determined at maximum load. The elastic indentation modulus is calculated with the slope at the point of initial unloading $S(h_{max})$

It is important to note that this theory assumes an isotropic material. For an anisotropic material the indentation modulus represents some average of the elastic properties in all directions. The latter strongly depends on the geometry of the indenter. For indents with a blunt indenter (such as Berkovich with an opening angle of 143°) nonnegligible deformations will occur in the plane perpendicular to the loading direction. In this context it is useful to present an approximation of the volume ganged by the indentation test.

The stress field generated by the indentation process is heterogeneous and leads to plastic deformation and damage in the vicinity of the tip. Using Hertz's theory [21.78], the spatial dependence of the stress components during indentation can be estimated by considering the elastic contact of a spherical indenter with a semi-infinite half space. In the direction of loading (z-axis), the stress component below the indenter decreases according to

$$\frac{\sigma_{zz}}{p_0} = -\left(1 + \frac{z^2}{a^2}\right) , \tag{21.13}$$

where p_0 indicates the maximum pressure below the indenter and a the contact radius.

In a horizontal plane at the surface ($z = 0$), the radial and circumferential components of the stress field next to the contact area obey

$$\frac{\sigma_{rr}}{p_0} = -\frac{\sigma_{\theta\theta}}{p_0} = \frac{(1-2\nu)\, a^2}{3r^2} , \tag{21.14}$$

where θ and r are the cylindrical coordinates of the periphery of the indenter and ν is the Poisson ratio. For $\nu = 0.3$ the stress field components reach their 10%-boundary defined by $\sigma_{zz} = -0.1 p_0$ and $\sigma_{rr} = -\sigma_{\theta\theta} = 0.1 p_0$ at a depth of $z \cong 3a$ or in lateral direction at $r \cong 2/\sqrt{3}a$. For a Berkovich tip, the ratio between maximum indentation depth and contact radius is approximately $a \cong 3h_{\max}$. The mechanical properties measured by nanoindentation, therefore, correspond to a semi-ellipsoidal volume extending to about nine times the employed indentation depth ($z \cong 3a \cong 9h_{\max}$) in the vertical direction (z) and about seven times this same depth ($2r \cong 4/\sqrt{3}a \cong 7h_{\max}$) in the radial direction (r).

The load-on hardness is determined with

$$H = \frac{P\,(h_{\max})}{A_c\,(h_{\max})} \tag{21.15}$$

as the ratio of the maximum load and the ("on load") contact area at maximum load. This is different from Vicker hardness in which the contact area is characterized by the remaining imprint after load removal. Differences between nanoindentation and Vicker, therefore, occur for materials with nonnegligible elastic recovery. After unloading, such materials may expose an imprint smaller than the area of contact at maximum load. Since the SN tool can be used to image the remaining imprint after unloading, hardness values using both definitions may thus be compared.

For fused silica the literature provides a hardness range between 8.3 to 9.5 GPa while an intercomparison of SN-users showed an average of 8.96 GPa (Surface, Hückelhoven). Conveniently, the hardness value also represents a possibility to check the area function of the tip based on the reduced modulus of the calibration material.

Tests of Bone Tissue Under Pseudo-physiological Conditions

Samples containing trabecular and compact bone lamellae were dissected from the medial part of the femoral neck of an 86-year-old female. After embedding in PMMA, the samples were polished with successive grades of carbide paper and finished with 0.05 μm alumina solution. Polishing represents an important preparation step since the here-employed theory assumes an infinitely flat surface. The mean surface roughness of the indented area should, therefore, be far below the employed indentation depth. Unfortunately no objective criteria that determines the maximum allowable surface roughness as a function of the indentation depth has been formulated so far.

Thin and thick lamellae of trabecular and compact bone were first characterized under physiological conditions, i. e. fully wet and at 37 °C. Then the specimens were dried for 24 h at 50 °C and identical tests were carried out again but under dry conditions.

In both cases 16 indentation tests were performed to maximum depths of 100 nm and 500 nm. Each test consisted of 10 s loading, 10 s holding, and 10 s unloading. The maximum allowable thermal drift was set 0.1 nm/s. To avoid proximity effects of neighbouring indents, an adjacent testing area in the identical lamellae was chosen after changing the testing conditions.

The tests under fully wet conditions at body temperature were found to be very sensitive with respect to thermal stability. The nanoindentation device was heated for several days to reach stable thermal conditions of the instrument components before the sample could be installed. The electrostatic force constant (EFC) was checked daily before beginning data acquisition. The approach of the indenter tip in the liquid environment was performed employing a contact force of 7 μN. This

Fig. 21.16 Surface topography of a trabecular bone structural unit that shows the lamellar structure. Thick (*bright*) lamellae correspond to the tops in the topography. Note, the two holes are not remaining indents but are ellipsoidal lacunae that embed bone cells

value is offset by approximately 1 μN per mm of water penetration, however, due to Archimedes force acting on the fluid cell tip. Such additional effects as water surface tension may also have a repulsive force on the tip.

AFM-scans as in Fig. 21.16, which shows the topography of trabecular bone lamellae under dry conditions, allowed the two lamella types to be identified. Thick lamellae correspond to the tops (bright contrast) and thin lamellae to the valleys (dark contrast) where the surface relief results from preferential polishing.

Indentation Modulus

Trabecular bone showed under wet conditions a mean indentation modulus of 12.5 ± 4.0 GPa when the data of both lamella types and indentation depths were combined. Under dry conditions the results showed a mean of 19.6 ± 2.6 GPa, an increase of 57%. Lamellae of compact bone increased their mean stiffness by 76.5% from 14.9 ± 4.5 GPa to 26.4 ± 3.8 GPa.

These determined conversion factors between dried and fully wet tissue properties should be compared with other studies. *Rho* [21.74] has reported an increase of indentation modulus of bovine compact bone by 15.8%. Our study demonstrates a change of +76.5% for human compact bone. This high discrepancy may be attributed to different preparation and testing protocols. Rho tested the bone samples at ambient temperature while kept moist by a thin film of deionized water. In our study the tissue properties were determined under fully wet conditions and at body temperature. Furthermore, Rho dried for 14 days at ambient temperature while in our study the drying process occurred during 24 h at 50 °C. These points may explain the different relative change of mechanical properties we detected.

Figure 21.17 presents the indentation moduli (combining both depths) normalized with respect to their initial wet values under physiological conditions. It is interesting that the increase of this elastic parameter after drying was significantly ($p < 0.00001$) higher for thin than for thick lamellae. The relative change of stiffness due to drying was +44% for thick lamellae and +109% for thin lamellae of compact bone. For trabecular bone the corresponding values were +37% and +78% for thick and thin lamellae, respectively.

Table 21.2 shows the results for both indentation depths. The differences are likely related to the volume sampled during the indentation, a semi-ellipsoid following the approximations of (21.13) and (21.14) with 0.7 μm diameter and 0.9 μm height for 100 nm indents (and 3.5 μm × 4.5 μm for 500 nm indents). Given the typical lamellae dimensions of 1–4 μm, the shallow indentation measurements, therefore, represent properties of single lamellae whereas the deeper indents include neighboring lamellae. It is also worth noting that thin lamellae showed a greater effect of drying when only the shallow indents were considered.

This important result should be discussed in the frame of published models that address the phenomenon of bone lamellation. *Marotti* [21.79] proposed that bone lamellae are the result of alternating changes of the collagen fiber density. According to his SEM-results the density of collagen is higher in thin lamellae than in thick lamellae. Collagen fibers are long chains of proteins and contain adsorption sites for polar water

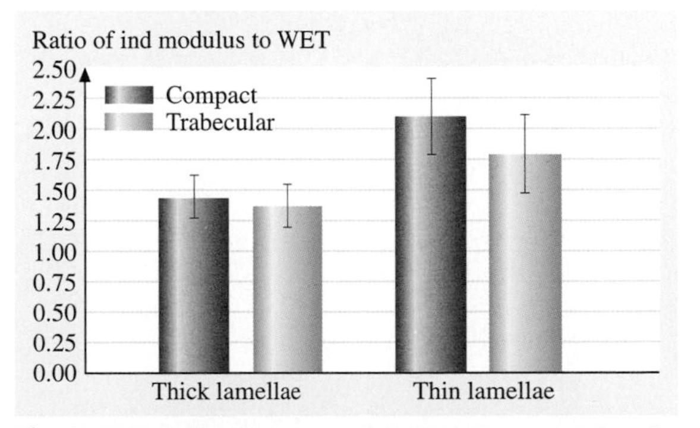

Fig. 21.17 Relative increase of indentation modulus for compact and trabecular bone lamellae after drying. The results are normalized with respect to their initial value under wet conditions. Note that thin lamellae are more affected by drying for both types of bone

Table 21.2 Absolute and relative changes of the indentation modulus with respect to the initial values under wet conditions. Note the greater increase for the thin lamellae after drying. Indents to 100 nm depth represent properties of single lamellae while 500 nm indents are also influenced by neighboring lamellae

Indentation modulus	Lamella	Ind. depth (nm)	Wet (GPa)	Dry (GPa)	Rel. change (%)
Trabecular bone	Thin	100	11.6±4.1	20.5±4.3	+76
		500	9.8±1.9	18.1±3.2	+84
	Thick	100	12.4±5	19.1±1.11	+54
		500	15.6±1.3	19.0±1.1	+22
Compact bone	Thin	100	10.4±0.2	23.4±3.6	+124
		500	13.9±2.0	27.1±3.1	+95
	Thick	100	19.9±4.8	27.9±4.6	+40
		500	15.5±2.6	23.0±0.8	+49

molecules. A higher density of the collagen fibers in the thinner lamellae results in a higher water binding capacity that may explain the higher relative influence of drying.

Giraud-Guille [21.80], on the other hand, proposed a nested-arc structure for bone lamellae with smooth orientation changes between adjacent collagen fibers. According to this model indents on thick lamellae load into the longitudinal direction of the fibers, whereas the load is perpendicular to the long axis for thin lamellae. Removal of the liquid phase leads to packaging of the collagen fibers, an effect that is intuitively anisotropic and that may explain why the effect of drying was different for both lamella types. In addition, drying leads to microfracture initiation within thick lamellae [21.64], possibly explaining the diminished effect of drying on these structure.

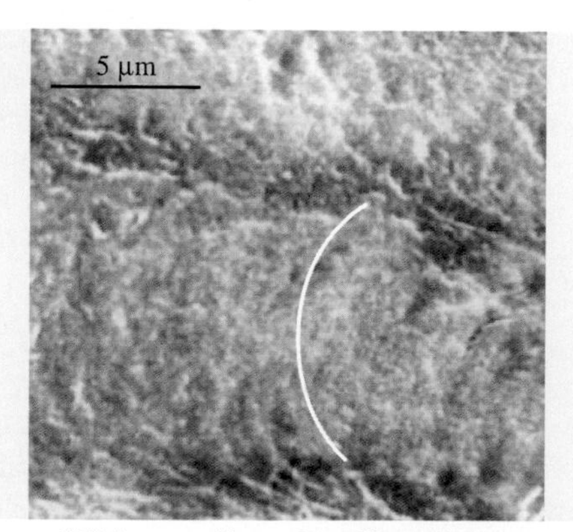

Fig. 21.18 SEM-scan of bone lamellae adjacent to the spot characterized by nanoindentation. This SEM-scan gives support for the bone lamellation model that is based on a smooth orientation change of collagen fibers [21.80]

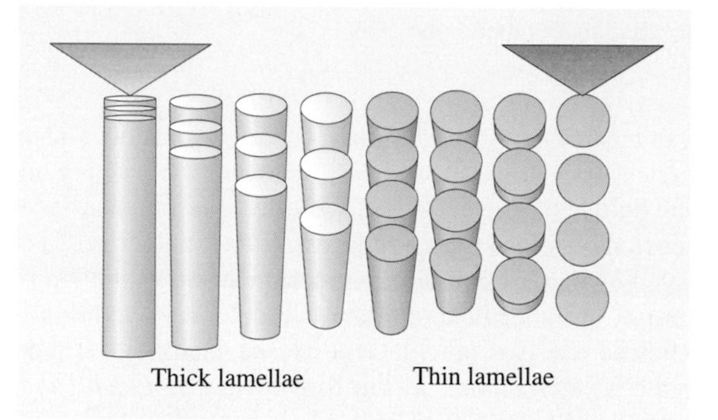

Fig. 21.19 Bone lamellation model proposed by this study. This model represents a combination of a smooth orientation change and density variations of the collagen fibers. Note also the spots where the indentations were done. For thin lamellae a greater change of mechanical properties was detected after drying

Figure 21.18 presents a SEM-image from a bone sample used in this study. This scan confirms the nested-arc structure Giraud-Guille observed with TEM.

Our results are, therefore, in agreement with both the Marotti and the Giraud-Guille models. This implies a model that combines smooth orientation changes of the collagen fibers with density changes. Such an architecture is sketched in Fig. 21.19. Note also the spots where the indentations were done.

Hardness

Compact bone revealed a mean hardness of 0.46 GPa under wet conditions that increased by 74% to 0.80 GPa after drying. Due to drying trabecular bone lamellae showed a mean hardness change of 76% from 0.41 up to 0.72 GPa.

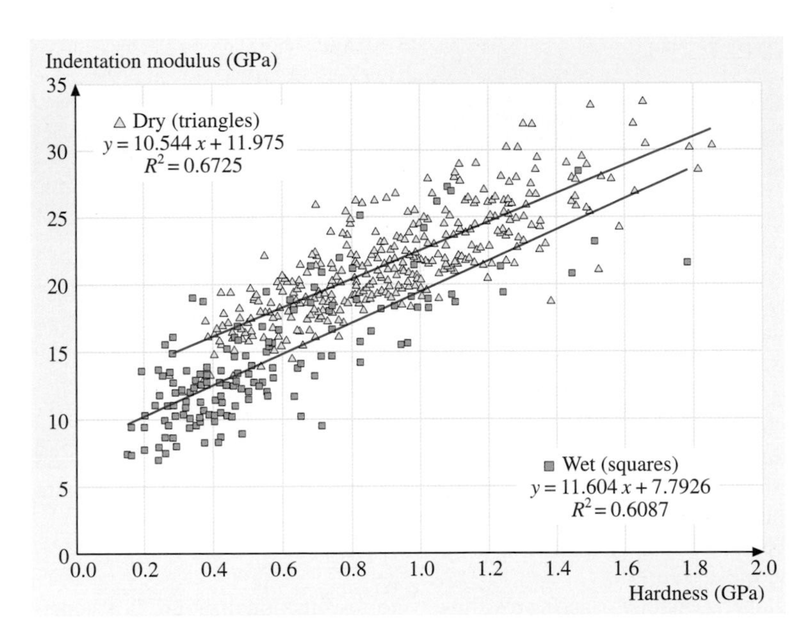

Fig. 21.20 Indentation modulus and hardness under dry and wet conditions using raw data from *Hengsberger* et al. [21.64]. Drying does not change the correlation coefficient and changes slightly but significantly the slope of the regression curve

For compact bone the effect of drying was again significantly ($p = 0.0006$) greater for thin lamellae (+108%) than for thick lamellae (+44%). Similar results were found for trabecular bone lamellae with a +99% increase for thin and a +56% increase for thick lamellae.

The results for hardness, therefore, appear to similar trends as the indentation modulus. This motivates us to discuss a further point, the correlation between indentation modulus and hardness. These mechanical parameters show an empirically proven relationship [21.81] that may change as a result of drying. For this purpose we related the data of the indentation modulus of Hengsberger et al. [21.64] with the corresponding unpublished hardness results. The correlation between hardness and indentation modulus is similar under wet ($R^2 = 0.61$) and under dry ($R^2 = 0.67$) conditions (Fig. 21.20). The slopes of the linear regression, however, show close but significantly different values ($p < 0.0001$) for the wet and dry samples. Hardness and indentation modulus show a similar but significantly different relative shift as a result of drying.

21.3.4 Conclusion

Modern scanning nanoindentation has clearly solved two problems: indentations in the nanometer scale can be performed with high force and displacement resolution, and nanometer lateral control of the indentation position allows the characterization of small structures of heterogeneous materials. Interfaces in composites, local density or composition variations of chemical products, or biological structures can then be investigated using the SN tool.

Based on recent theoretical progress [21.82], the nanoindentation technique can also be used to characterize anisotropic materials by indenting in different directions. Other parameters can also be determined. For instance, creep behavior is accessible when the indentation load is held constant at a maximum load. Hysteresis of the force–displacement curve represents the energy dissipated during an indentation providing information about post-elastic behavior of a sample. Such dynamical variables as viscoelastic properties are also accessible by including sinusoidal oscillations of different frequencies to the loading history.

Future interests of the nanotechnology community may direct the work towards tests on lower levels of structural organization like the molecular or atomic level. This would require strong improvements of the transducer sensitivity and of the indenter tip machining. New concepts to reduce the thermal drifting properties would also be necessary for the next generation of this device.

21.4 Conclusions and Perspectives

We have demonstrated that AFM is an ideal tool for investigating variations in local properties of bulk materials, like bone, and also for performing physical measurements on individual nanometer scale objects such as, for example, carbon nanotubes and protein polymers. In the latter case, the technique provides previously inaccessible quantities in living matter such as, for instance, the shear modulus. It should be emphasized that shear is omnipresent since biostructures are "composite" materials, with strong anisotropic interactions between their constituents. In the case of proteins, we have explored the mechanical properties as a function of temperature, as well. Interactions can vary remarkably even over a range of a few tens of degrees, providing deeper insight in the functioning of these structures. One task, which seems very difficult for the time being, is the frequency dependent response of nano and biostructures. In the latter case it is limited to a few kilohertz, mainly because of the surrounding liquid, but for a meaningful analysis one requires several decades in frequency.

Regarding the scanning probe tips themselves, improvements are necessary to provide better resolution, longer lifetimes, and easy functionalization for sensing different chemical environments. This might be achieved by carbon nanotube tips (Fig. 21.21), which have a Young's modulus of 1 TPa, a very well-defined and sharp tip structure, and pentagon "defects" at the apex providing a site for functionalization. It is generally agreed upon by the scanning probe community that carbon nanotubes will open new avenues for the study of living matter.

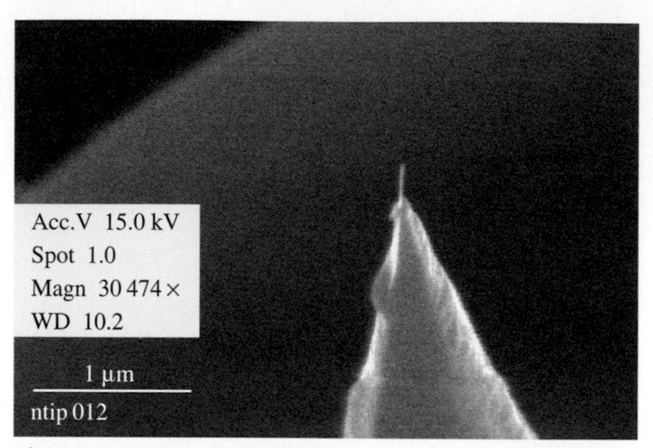

Fig. 21.21 SEM picture of an AFM tip with attached multiwall Carbon nanotube

Finally the development of the photonic force microscope in Heidelberg has allowed imaging of bio-structures with unprecedented three-dimensional resolution, even including features that are otherwise inaccessible to AFM tips. In the future, this instrument will certainly provide radical new insight in biological functioning.

References

21.1 M. Radmacher: Measuring the elastic properties of biological samples with the AFM, IEEE Eng. Med. Biol. Mag. **16**(2) (1997) 47–57

21.2 G. Binnig, C. F. Quate, C. Gerber: Atomic force microscope, Phys. Rev. Lett. **56** (1986) 930–933

21.3 P. Maivald, H. J. Butt, S. A. C. Gould, C. B. Prater, B. Drake, J. A. Gurley, V. B. Elings, P. K. Hansma: Using force modulation to image surface elasticities with the atomic force microscope, Nanotechnology **2** (1991) 103–106

21.4 T. Kajiyama, K. Takata, I. Ohki, S.-R. Ge, J.-S. Yoon, A. Takahara: Imaging of dynamic viscoelstic properties of a phase-separated polymer by forced oscillation atomic force microscopy, Macromolecules **27** (1994) 7932–7934

21.5 B. Nysten, R. Legras, J.-L. Costa: Atomic force microscopy imaging of viscoelastic properties in toughened polypropylene resins, J. Appl. Phys. **78** (1995) 5953–5958

21.6 P.-E. Mazeran, J.-L. Loubet: Normal and lateral modulation with a scanning force microscope, an analysis: Implication in quantitative elastic and friction imaging, Tribol. Lett. **3** (1997) 125–129

21.7 B. Cretin, F. Sthal: Scanning microdeformation microscopy, Appl. Phys. Lett. **62** (1993) 829–831

21.8 N. A. Burnham, A. J. Kulik, G. Gremaud, P.-J. Gallo, F. Oulevey: Scanning local-acceleration microscopy, J. Vac. Sci. Technol. B **14** (1996) 794–799

21.9 F. Oulevey, N. A. Burnham, A. J. Kulik, G. Gremaud, W. Benoit: Mechanical properties studied at the nanoscale using scanning local-acceleration microscopy (SLAM), J. Phys. **IV** (1996) C8-731-734

21.10 U. Rabe, W. Arnold: Acoustic microscopy by atomic force microscopy, Appl.Phys. Lett **64** (1994) 1493–1495

21.11 O. Kolosov, K. Yamanaka: Nonlinear detection of ultrasonic vibrations in an atomic force microscope, Jpn. J. Appl. Phys. **32** (1993) 22–25

21.12 F. Oulevey: Cartographie et spectrométrie des propriétés mécaniques à l'échelle nanométrique par microscopie acoustique en champ proche. Ph.D. Thesis (EPFL, Lausanne 1999)

21.13 F. Oulevey, G. Gremaud, A. Semoroz, A.J. Kulik, N.A. Burnham, E. Dupas, D. Gourdon: Local mechanical spectroscopy with nanometer-scale lateral resolution, Rev. Sci. Instrum. **69** (1998) 2085–2094

21.14 N.A. Burnham, G. Gremaud, A.J. Kulik, P.-J. Gallo, F. Oulevey: Materials' properties measurements: Choosing the optimal scanning probe microscope configuration, J. Vac. Sci. Technol. B **14** (1996) 1308–1312

21.15 F. Oulevey, G. Gremaud, A.J. Kulik, B. Guisolan: Simple low-drift heating stage for scanning probe microscopes, Rev. Sci. Instrum. **70** (1999) 1889–1890

21.16 H.M. Pollock: Nanoindentation. In: *Friction lubrication and wear technology*, AMS Handbook, Vol. 18 (AMS, Ohio 1992) p. 419

21.17 E. Dupas, G. Gremaud, A. Kulik, J.-L. Loubet: High-frequency mechanical spectroscopy with an atomic force microscope, Rev. Sci. Instrum. **72**(10) (2001) 3891–3897

21.18 F. Oulevey, N.A. Burnham, G. Gremaud, A.J. Kulik, H.M. Pollock, A. Hammiche, M. Reading, M. Song, D.J. Hourston: Dynamic mechanical analysis at the submicron scale, Polymer **41** (2000) 3087–3092

21.19 W.Wm. Wendlandt: *Thermal Analysis in Chemical Analysis*, Vol. 19, 3rd edn. (Wiley, New York 1996) p. 360

21.20 D. Mari, D.C. Dunand: NiTi and NiTi-TiC composites: Part I. Transformation and thermal cycling behavior, Metall. Mater. Trans. A **26A** (1995) 2833–2847

21.21 L. Bataillard: Transformation martensitique multiple dans un alliage à mémoire de forme Ni-Ti. Ph.D. Thesis (EPF, Lausanne 1996)

21.22 F. Oulevey, G. Gremaud, D. Mari, A.J. Kulik, N.A. Burnham, W. Benoit: Martensitic transformation of NiTi studied at the nanometer scale by local mechanical spectroscopy, Scripta Mater. **42** (2000) 31–36

21.23 D. Mari, L. Bataillard, D.C. Dunand, R. Gotthardt: Martensitic transformation of NiTi and NiTi-TiC composites, J. Phys. (France) **IV** (1995) C8-659–664

21.24 S. Iijima: Helical microtubules of graphitic carbon, Nature **354** (1991) 56–58

21.25 C.-H. Kiang, M. Endo, P.M. Ajayan, G. Dresselhaus, M.S. Dresselhaus: Size effects in carbon nanotubes, Phys. Rev. Lett. **81** (1998) 1869–1872

21.26 A. Thess, R. Lee, P. Nikolaev, H. Dai, P. Petit, J. Robert, C. Xu, Y.H. Lee, S.G. Kim, A.G. Rinzler, Daniel T. Colbert, G.U. Scuseria, D. Tománek, J.E. Fischer, R.E. Smalley: Crystalline ropes of metallic carbon nanotubes, Science **273** (1996) 483–487

21.27 W.Z. Li, S.S. Xie, L.X. Qian, B.H. Chang, B.S. Zou, W.Y. Zhou, R.A. Zhao, G. Wang: Large-scale synthesis of aligned carbon nanotubes, Science **274** (1996) 1701–1703

21.28 J. Kong, H.T. Soh, A.M. Cassell, C.F. Quate, H.J. Dai: Synthesis of individual single-walled carbon nanotubes on patterned silicon wafers, Nature **395** (1998) 878–881

21.29 J.P. Lu: Elastic properties of carbon nanotubes and nanoropes, Phys. Rev. Lett. **79** (1997) 1297–1300

21.30 E.W. Wong, P.E. Sheehan, C.M. Lieber: Nanobeam mechanics: Elasticity, strength and toughness of nanorods and nanotubes, Science **277** (1997) 1971–1975

21.31 J.-P. Salvetat, J.M. Bonard, N.H. Thomson, A.J. Kulik, L. Forró, W. Benoit, L. Zuppiroli: Mechanical properties of carbon nanotubes, Appl. Phys. A **69** (1999) 255–260

21.32 M.-F. Yu, B.S. Files, S. Arepalli, R.S. Ruoff: Tensile loading of ropes of single wall carbon nanotubes and their mechanical properties, Phys. Rev. Lett. **84** (2000) 5552–5555

21.33 S. Iijima, C. Brabec, A. Maiti, J. Bernholc: Structural flexibility of carbon nanotubes, J. Chem. Phys. **104** (1996) 2089–2092

21.34 J.-P. Salvetat, A.J. Kulik, J.-M. Bonard, G.A.D. Briggs, T. Stöckli, K. Méténier, S. Bonnamy, F. Béguin, N.A. Burnham, L. Forró: Elastic modulus of ordered and disordered multiwalled carbon nanotubes, Adv. Mater. **11** (1999) 161–165

21.35 J.-P. Salvetat, G.A.D. Briggs, J.-M. Bonard, R.R. Bacsa, A.J. Kulik, T. Stöckli, N. Burnham, L. Forró: Elastic and shear moduli of single-walled carbon nanotube ropes, Phys. Rev. Lett. **82** (1999) 944–947

21.36 J.M. Gere, S.P. Timoshenko: *Mechanics of Materials* (PWS-Kent, Boston 1990)

21.37 D.A. Walters, L.M. Ericson, M.J. Casavant, J. Liu, D.T. Colbert, K.A. Smith, R.E. Smalley: Elastic strain of freely suspended single-wall carbon nanotube ropes, Appl. Phys. Lett. **74** (1999) 3803–3805

21.38 M.-F. Yu, O. Lourie, M.J. Dyer, K. Moloni, T.F. Kelly, R.S. Ruoff: Strength and breaking mechanism of multiwalled carbon nanotubes under tensile load, Science **287** (2000) 637–640

21.39 G.-T. Kim, G. Gu, U. Waizmann, S. Roth: Simple method to prepare individual suspended nanofibers, Appl. Phys. Lett. **80** (2002) 1815–1817

21.40 T. Fujii, M. Suzuki, M. Miyashita, M. Yamaguchi, T. Onuki, H. Nakamura, T. Matsubara, H. Yamada, K. Nakayama: Micropattern measurement with an atomic force microscope, J. Vac. Sci. Technol. B **9** (1991) 666–669

21.41 A.S.K. Kis, B. Babic, A.J. Kulik, W. Benoît, G.A.D. Briggs, C. Schönenberger, S. Catsicas, L. Forró: Nanomechanics of microtubules, Phys. Rev. Lett. **89** (2002) 248101

21.42 M. Kurachi, M. Hoshi, H. Tashiro: Buckling of a single microtubule by optical trapping forces: Direct measurement of microtubule rigidity, Cell. Motil. Cyt. **30** (1995) 221–228

21.43 R. B. Dye, S. P. Fink, R. C. Williams: Taxol-induced flexibillity of microtubules and its reversal by Map-2 and Tau, J. Biol. Chem. **268** (1993) 6847–6850

21.44 F. Gittes, B. Mickey, J. Nettleton, J. Howard: Flexual rigidity of microtubules and actin filaments measured from thermal fluctuations in shape, J. Cell Biol. **120** (1993) 923–934

21.45 M. Elbaum, D. K. Fygenson, A. Libchaber: Buckling microtubules in vesicles, Phys. Rev. Lett. **76** (1996) 4078–4081

21.46 A. Vinckier, C. Dumortier, Y. Engelborghs, L. Hellemans: Dynamical and mechanical study of immobilized microtubules with atomic force microscopy, J. Vac. Sci. Technol. B **14** (1996) 1427–1431

21.47 Hysitron, Incorporated 2010 East Hennepin Avenue Minneapolis, MN 55413,

21.48 A. V. Kulkarni, B. Bhushan: Nanoscale mechanical property measurements using modified atomic force microscopy, Thin Solid Films **290-291** (1996) 206–210

21.49 A. V. Kulkarni, B. Bhushan: Nano/picoindentation measurements on single-crystal aluminum using modified atomic force microscopy, Mater. Lett. **29** (1996) 221–227

21.50 M. Chiba, M. Seo: Effects of dichromate treatment on mechanical properties of passivated single crystal iron (100) and (110) surfaces, Corros. Sci. **44** (2002) 2379–2391

21.51 X. C. Lu, B. Shi, L. K. Y. Li, J. Luo, X. Chang, Z. Tian, J. I. Mou: Nanoindentation and microtribological behavior of Fe-N/Ti-N multilayers with different thickness of Ti-N layers, Wear **251** (2001) 1144–1149

21.52 P. Peeters, Gvd. Horn, T. Daenen, A. Kurowski, G. Staikov: Properties of electroless and electroplated Ni-P and its application in microgalvanics, Electrochim. Acta **47** (2001) 161–169

21.53 A. Rar, J. N. Zhou, W. J. Liu, J. A. Barnard, A. Bennett, S. C. Street: Dendrimer-mediated growth of very flat ultrathin Au films, Appl. Surf. Sci. **175-176** (2001) 134–139

21.54 R. D. Ott, C. Ruby, F. Huang, M. L. Weaver, J. A. Barnard: Nanotribology and surface chemistry of reactively sputtered Ti-B-N hard coatings, Thin Solid Films **377-378** (2000) 602–606

21.55 T. Malkow, S. J. Bull: Hardness measurements on thin IBAD CN_x films – a comparative study, Surf. Coat. Technol. **137** (2001) 197–204

21.56 Y. Shima, H. Hasuyama, T. Kondoh, Y. Imaoka, T. Watari, K. Baba, R. Hatada: Mechanical properties of silicon oxynitride thin films prepared by low energy ion beam assisted deposition, Nucl. Instrum. Methods Phys. Res., Sect. B **148** (1999) 599–603

21.57 M. Göken, M. Kempf, W. D. Nix: Hardness and modulus of the lamellar microstructure in PST-TiAl studied by nanoindentations and AFM, Acta Mater. **49** (2001) 903–911

21.58 M. Göken, M. Kempf: Microstructural properties of superalloys investigated by nanoindentations in an atomic force microscope, Acta Mater. **47** (1999) 1043–1052

21.59 M. Seo, M. Chiba, K. Suzuki: Nano-mechano-electrochemistry of the iron (100) surface in solution, J. Electroanal. Chem. **473** (1999) 49–53

21.60 Y. Zhang, F. Z. Cui, X. M. Wang, Q. L. Feng, X. D. Zhu: Mechanical properties of skeletal bone in gene-mutated *stöpsel*^{dtl28d} and wild-type zebrafish (*Danio rerio*) measured by atomic force microscopy-based nanoindentation, Bone **30** (2002) 541–546

21.61 M. Finke, J. A. Hughes, D. M. Parker, K. D. Jandt: Mechanical properties of in situ demineralised human enamel measured by AFM nanoindentation, Surf. Sci. **491** (2001) 456–467

21.62 S. Habelitz, S. J. Marshall, G. W. Marshall, J. R. Balooch, M. Balooch: The functional width of the dentino-enamel junction determined by AFM-based nanoscratching, J. Struct. Biol. **135** (2001) 294–301

21.63 G. W. Marshall Jr., M. Balooch, R. R. Gallagher, S. A. Gansky, S. J. Marshall: Mechanical properties of the dentinoenamel junction: AFM studies of nanohardness, elastic modulus, and fracture, J. Biomed. Mater. Res. **54** (2001) 87–95

21.64 S. Hengsberger, A. Kulik: Nanoindentation discriminates the elastic properties of individual human bone lamellae under dry and physiological conditions, Bone **30** (2002) 178–184

21.65 E. F. Eriksen, D. W. Axelrod, F. M. Melsen: *Bone Histomorphometry*, 1st edn. (Raven, New York 1994)

21.66 C. E. Hoffler, K. E. Moore, K. Kozloff, P. K. Zysset, M. B. Brown, S. A. Goldstein: Heterogeneity of bone lamellar-level elastic moduli, Bone **26** (2000) 603–609

21.67 C. E. Hoffler, K. E. Moore, K. Kozloff, P. K. Zysset, S. A. Goldstein: Age, gender, and bone lamellae elastic moduli, J. Orth. Res. **18** (2000) 432–437

21.68 P. K. Zysset, X. E. Guo, C. E. Hoffler, K. E. Moore, S. A. Goldstein: Elastic modulus and hardness of cortical and trabecular bone lamellae measured by nanoindentation in the human femur, J. Biomech. **32** (1999) 1005–1012

21.69 J. Y. Rho, P. Zioupos, J. D. Currey, G. M. Pharr: Variations in the individual thick lamellar properties within osteons by nanoindentation, Bone **25** (1999) 295–300

21.70 J. Y. Rho, M. E. Roy, T. Y. Tsui, G. M. Pharr: Elastic properties of microstructural components of human bone tissue as measured by nanoindentation, J. Biomed. Mater. Res. **45** (1999) 48–54

21.71 S. Hengsberger, A. Kulik, P. Zysset: A combined atomic force microscopy and nanoindentation technique to investigate the elastic properties of bone structural units, Europ. Cells Mater. **1** (2001) 12–16

21.72 C. H. Turner, J. Y. Rho, Y. Takano, T. Y. Tsui, G. M. Pharr: The elastic properties of trabecular and cortical bone tissues are similar: Results from two microscopic measurement techniques, J. Biomech. **32** (1999) 437–441

21.73 M. E. Roy, J. Y. Rho, T. Y. Tsui, N. S. Evans, G. M. Pharr: Mechanical and morphological variation of the human lumbar vertebral cortical and trabecular bone, J. Biomed. Mater. Res. **44** (1999) 191–199

21.74 J. Y. Rho, G. M. Pharr: Effects of drying on the mechanical properties of bovine femur measured by nanoindentation, J. Mater. Sci.: Mater. Med. **10** (1999) 485–488

21.75 I. N. Sneddon: The relation between load and penetration in the axisymmetric Boussinesq problem for a punch of arbitrary profile, Int. J. Eng. Sci. **3** (1965) 47–57

21.76 W. C. Oliver, G. M. Pharr: An improved technique for determining hardness and elastic modulus using load and displacement sensing indentation experiments, Mat. Res. Soc. **7/6** (1992) 1564–1583

21.77 G. M. Pharr, W. C. Oliver, F. R. Brotzen: On the generality of the relationship among contact stiffness, contact area, and elastic modulus during indentation, J. Mater. Res. **7/3** (1992) 613–617

21.78 K. L. Johnson: *Contact Mechanics*, 1st edn. (Cambridge Univ. Press, Cambridge 1985) pp. 84–106

21.79 G. Marotti: A new theory of bone lamellation, Calcif Tissue Int. **53** (1993) 47–56

21.80 M. M. Giraud-Guille: Plywood structures in nature, Curr. Op. Solid State Mater. Sci. **3** (1998) 221–227

21.81 G. P. Evans, J. C. Behiri, J. D. Currey, W. Bonfield: Microhardness and Young's modulus in cortical bone exhibiting a wide range of mineral volume fractions, and in a bone analogue, J. Mater. Sci.: Mater. Med. **1** (1990) 38–43

21.82 J. G. Swadener, G. M. Pharr: Indentation of elastically anisotropic half-spaces by cones and parabolae of revolution, Philos. Mag. A **81** (2001) 447–466

22. Nanomechanical Properties of Solid Surfaces and Thin Films

Instrumentation for the testing of mechanical properties on the submicron scale has developed enormously in recent years. This has enabled the mechanical behavior of surfaces, thin films, and coatings to be studied with unprecedented accuracy. In this chapter, the various techniques available for studying nanomechanical properties are reviewed with particular emphasis on nanoindentation. The standard methods for analyzing the raw data obtained using these techniques are described, along with the main sources of error. These include residual stresses, environmental effects, elastic anisotropy, and substrate effects. The methods that have been developed for extracting thin-film mechanical properties from the often convoluted mix of film and substrate properties measured by nanoindentation are discussed. Interpreting the data is frequently difficult, as residual stresses can modify the contact geometry and, hence, invalidate the standard analysis routines. Work hardening in the deformed region can also result in variations in mechanical behavior with indentation depth. A further unavoidable complication stems from the ratio of film to substrate mechanical properties and the depth of indentation in comparison to film thickness. Even very shallow indentations may be influenced by substrate properties if the film is hard and very elastic but the substrate is compliant. Under these circumstances nonstandard methods of analysis must be used. For multilayered systems many different mechanisms affect the nanomechanical behavior, including Orowan strengthening, Hall–Petch behavior, image force effects, coherency and thermal stresses, and composition modulation.

The application of nanoindentation to the study of phase transformations in semiconductors, fracture in brittle materials, and mechanical properties in biological materials are described. Recent developments such as the testing of viscoelasticity using nanoindentation methods are likely to be particularly important in future studies of polymers and biological materials. The importance of using a range of complementary methods such as electron microscopy, in situ AFM imaging, acoustic monitoring, and electrical contact measurements is emphasized. These are especially important on the nanoscale because so many different physical and chemical processes can affect the measured mechanical properties.

22.1 **Instrumentation** ... 688
22.1.1 AFM and Scanning Probe Microscopy ... 688
22.1.2 Nanoindentation ... 689
22.1.3 Adaptations of Nanoindentation ... 690
22.1.4 Complimentary Techniques ... 691
22.1.5 Bulge Tests ... 691
22.1.6 Acoustic Methods ... 692
22.1.7 Imaging Methods ... 693

22.2 **Data Analysis** ... 694
22.2.1 Elastic Contacts ... 694
22.2.2 Indentation of Ideal Plastic Materials ... 694
22.2.3 Adhesive Contacts ... 695
22.2.4 Indenter Geometry ... 696
22.2.5 Analyzing Load/Displacement Curves ... 696
22.2.6 Modifications to the Analysis ... 699
22.2.7 Alternative Methods of Analysis ... 700
22.2.8 Measuring Contact Stiffness ... 701
22.2.9 Measuring Viscoelasticity ... 702

22.3 **Modes of Deformation** ... 702
22.3.1 Defect Nucleation ... 702
22.3.2 Variations with Depth ... 704
22.3.3 Anisotropic Materials ... 704
22.3.4 Fracture and Delamination ... 704
22.3.5 Phase Transformations ... 705

22.4 **Thin Films and Multilayers** ... 707
22.4.1 Thin Films ... 707
22.4.2 Multilayers ... 709

22.5 **Developing Areas** ... 711

References ... 712

When two bodies come into contact their surfaces experience the first and usually largest mechanical loads. Hence, characterizing and understanding the mechanical properties of surfaces is of paramount importance in a wide range of engineering applications. Obvious examples of where surface mechanical properties are important are in wear-resistant coatings on reciprocating surfaces and hard coatings for machine tool bits. This chapter details the current methods for measuring the mechanical properties of surfaces and highlights some of the key experimental results that have been obtained.

The experimental technique that is highlighted in this chapter is nanoindentation. This is for the simple reason that it is now recognized as the preferred method for testing thin film and surface mechanical properties. Despite this recognition, there are still many pitfalls for the unwary researcher when performing nanoindentation tests. The commercial instruments that are currently available all have attractive, user-friendly software, which makes the performance and analysis of nanoindentation tests easy. Hidden within the software, however, are a myriad of assumptions regarding the tests that are being performed and the material that is being examined. Unless the researcher is aware of these, there is a real danger that the results obtained will say more about the analysis routines than they do about the material being tested.

22.1 Instrumentation

The instruments used to examine nanomechanical properties of surfaces and thin films can be split into those based on point probes and those complimentary methods that can be used separately or in conjunction with point probes. The complimentary methods include a wide variety of techniques ranging from optical tests such as micro-Raman spectroscopy to high-energy diffraction studies using X-rays, neutrons, or electrons to mechanical tests such as bulge or blister testing.

Point-probe methods have developed from two historically different methodologies, namely, scanning probe microscopy [22.1] and microindentation [22.2]. The two converge at a length scale between 10–1,000 nm. Point-probe mechanical tests in this range are often referred to as nanoindentation.

22.1.1 AFM and Scanning Probe Microscopy

Atomic force microscopy (AFM) and other scanning probe microscopies are covered in detail elsewhere in this volume, but it is worth briefly highlighting the main features in order to demonstrate the similarities to nanoindentation. There are now a myriad of different variants on the basic scanning probe microscope. All use piezoelectric stacks to move either a probe tip or the sample with subnanometer precision in the lateral and vertical planes. The probe itself can be as simple as a tungsten wire electrochemically polished to give a single atom at the tip, or as complex as an AFM tip that is bio-active with, for instance, antigens attached. A range of scanning probes have been developed with the intention of measuring specific physical properties such as magnetism and heat capacity.

To measure mechanical properties with an AFM, the standard configuration is a hard probe tip (such as silicon nitride or diamond) mounted on a cantilever (see Fig. 22.1). The elastic deflection of the cantilever is monitored either directly or via a feedback mechanism

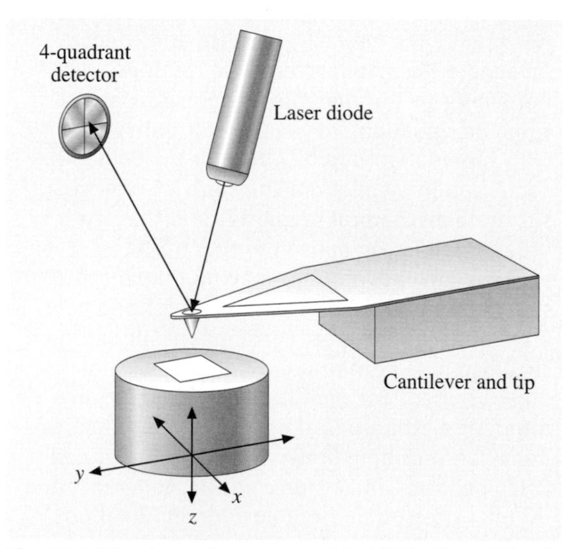

Fig. 22.1 Diagram of a commercial AFM. The AFM tip is mounted on a compliant cantilever, and a laser light is reflected off the back of the cantilever onto a position-sensitive detector (4-quadrant detector). Any movements of the cantilever beam cause a deflection of the laser light that the detector senses. The sample is moved using piezoelectric stack, and forces are calculated from the cantilever's stiffness and the measured deflection

to measure the forces acting on the probe. In general, the forces experienced by the probe tip split into attractive or repulsive forces. As the tip approaches the surface, it experiences intermolecular forces that are attractive, although they can be repulsive under certain circumstances [22.3]. Once in contact with the surface the tip usually experiences a combination of attractive intermolecular forces and repulsive elastic forces. Two schools of thought exist regarding the attractive forces when the tip is in contact with the surface. The first is often referred to as the DMT or Bradley model. It holds that attractive forces only act outside the region of contact [22.4–7]. The second theory, usually called the JKR model, assumes that all the forces experienced by the tip, whether attractive or repulsive, act in the region of contact [22.8]. Most real nanoscale contacts lie somewhere between these two theoretical extremes.

22.1.2 Nanoindentation

The fundamental difference between AFM and nanoindentation is that during a nanoindentation experiment an external load is applied to the indenter tip. This load enables the tip to be pushed into the sample, creating a nanoscale impression on the surface, otherwise referred to as a nanoindentation or nanoindent.

Conventional indentation or microindentation tests involve pushing a hard tip of known geometry into the sample surface using a fixed peak load. The area of indentation that is created is then measured, and the mechanical properties of the sample, in particular its hardness, is calculated from the peak load and the indentation area. Various types of indentation testing are used in measuring hardness, including Rockwell, Vickers, and Knoop tests. The geometries and definitions of hardness used in these tests are shown by Fig. 22.2.

When indentations are performed on the nanoscale there is a basic problem in measuring the size of the indents. Standard optical techniques cannot easily be used to image anything smaller than a micron, while electron microscopy is simply impractical due to the time involved in finding and imaging small indents. To overcome these difficulties, nanoindentation methods have been developed that continuously record the load, displacement, time, and contact stiffness throughout the indentation process. This type of continuously recording indentation testing was originally developed in the former Soviet Union [22.9–13] as an extension of microindentation tests. It was applied to nanoscale indentation testing in the early

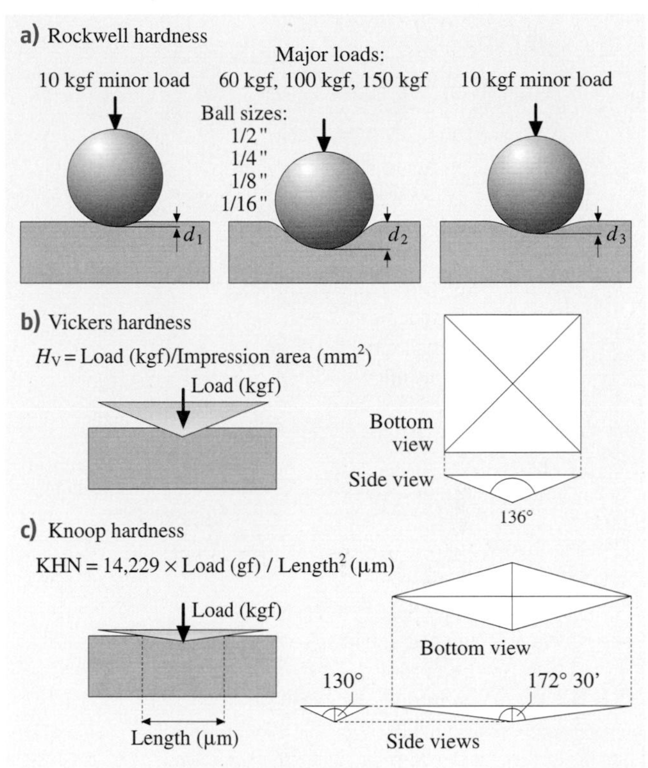

Fig. 22.2 (a) The standard Rockwell hardness test involves pushing a ball into the sample with a minor load, recording the depth, d_1, then applying a major load and recording the depth, d_2, then returning to the lower load and recording the depth, d_3. Using the depths, the hardness is calculated. (b) Vickers hardness testing uses a four-sided pyramid pushed into the sample with a known load. The area of the resulting indentation is measured optically and the hardness calculated as the load is divided by area. (c) Knoop indentation uses the same definition of hardness as the Vickers test, load divided by area, but the indenter geometry has one long diagonal and one short diagonal

1980s [22.14, 15], hence, giving rise to the field of nanoindentation testing.

In general, nanoindentation instruments include a loading system that may be electrostatic, electromagnetic, or mechanical, along with a displacement measuring system that may be capacitive or optical. Schematics of several commercial nanoindentation instruments are shown in Fig. 22.3a–c.

Among the many advantages of nanoindentation over conventional microindentation testing is the ability to measure the elastic, as well as the plastic properties of the test sample. The elastic modulus is obtained from the contact stiffness (S) using the following equation that

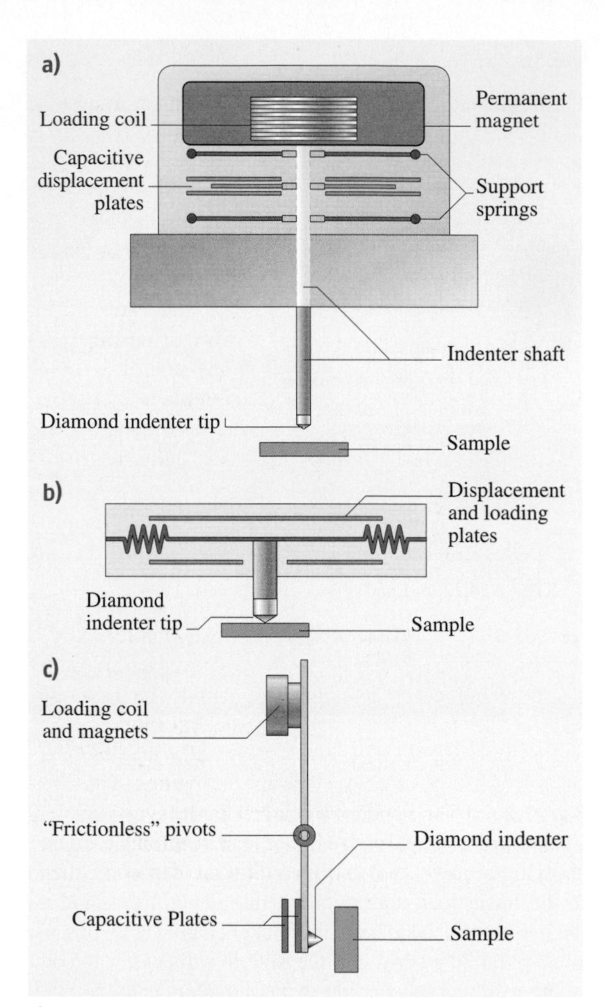

Fig. 22.3a–c Schematics of three commercial nanoindentation devices made by **(a)** MTS Nanoinstruments, Oak Ridge, Tennessee, **(b)** Hysitron Inc., Minneapolis, Minnesota, **(c)** Micro Materials Limited, Wrexham, UK. Instruments **(a)** and **(c)** use electromagnetic loading, while **(b)** uses electrostatic loads

appears to be valid for all elastic contacts [22.16, 17]:

$$S = \frac{2}{\sqrt{\pi}} E_r \sqrt{A} \,. \tag{22.1}$$

A is the contact area and E_r is the reduced modulus of the tip and sample as given by:

$$\frac{1}{E_r} = \frac{\left(1-\nu_t^2\right)}{E_t} + \frac{\left(1-\nu_s^2\right)}{E_s} \,, \tag{22.2}$$

where E_t, ν_t and E_s, ν_s are the elastic modulus and Poissons ratio of the tip and sample, respectively.

22.1.3 Adaptations of Nanoindentation

Several adaptations to the basic nanoindentation setup have been used to obtain additional information about the processes that occur during nanoindentation testing, for example, in situ measurements of acoustic emissions and contact resistance. Environmental control has also been used to examine the effects of temperature and surface chemistry on the mechanical behavior of nanocontacts. In general, it is fair to say that the more information that can be obtained and the greater the control over the experimental parameters the easier it will be to understand the nanoindentation results. Load-displacement curves provide a lot of information, but they are only part of the story.

During nanoindentation testing discontinuities are frequently seen in the load–displacement curve. These are often called "pop-ins" or "pop-outs", depending on their direction. These sudden changes in the indenter displacement, at a constant load (see Fig. 22.4), can be caused by a wide range of events, including fracture, delamination, dislocation multiplication, or nucleation and phase transformations. To help distinguish between the various sources of discontinuities, acoustic transducers have been placed either in contact with the sample or immediately behind the indenter tip. For example, the results of nanoindentation tests that monitor acoustic emissions have shown that the phase transformations seen in silicon during nanoindentation are not the sudden events that they would appear to be from the load–displacement curve. There is no acoustic emission associated with the pop-out seen in the unloading curve of silicon [22.18]. An acoustic emission would be expected if there were a very rapid phase transformation causing a sudden change in volume. Fracture and delamination of films, however, give very strong acoustic signals [22.19],

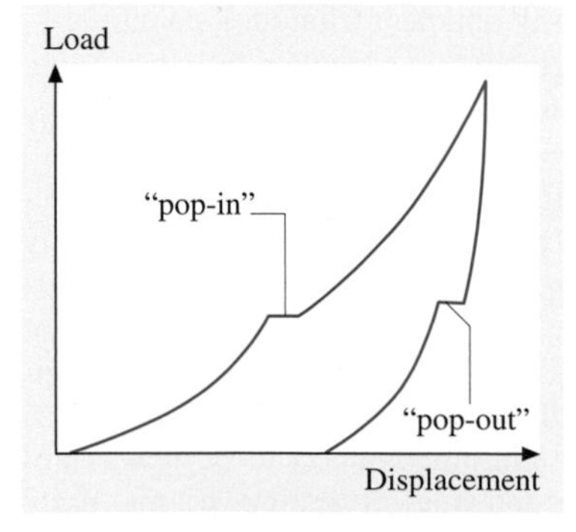

Fig. 22.4 Sketch of a load/displacement curve showing a pop-in and a pop-out

but the exact form of the signal appears to be more closely related to the sample geometry than to the event [22.20].

Additional information about the nature of the deformed region under the nanoindentation can be obtained by performing in situ measurements of contact resistance. The basic setup for this type of testing is shown in Fig. 22.5. An electrically conductive tip is needed to study contact resistance. Consequently, a conventional diamond tip is of limited use. Elastic, hard, and metallically conductive materials such as vanadium carbide can be used as substitutes for diamond [22.21, 22], or a thin conductive film (e.g., Ag) can be deposited on the diamond's surface (such a film is easily transferred to the indented surface so great care must be taken if multiple indents are performed). Measurements of contact resistance have been most useful for examining phase transformations in semiconductors [22.21, 22] and the dielectric breakdown of oxide films under mechanical loading [22.23].

One factor that is all too frequently neglected during nanoindentation testing is the effect of the experimental environment. Two obvious ways in which the environment can affect the results of nanoindentation tests are increases in temperature, which give elevated creep rates, and condensation of water vapor, which modifies the tip-sample interactions. Both of these environmental effects have been shown to significantly affect the measured mechanical properties and the modes of deformation that occur during nanoindentation [22.24–27]. Other environmental effects, for instance, those due to photoplasticity or hydrogen ion absorption, are also possible, but they are generally less troublesome than temperature fluctuations and variations in atmospheric humidity.

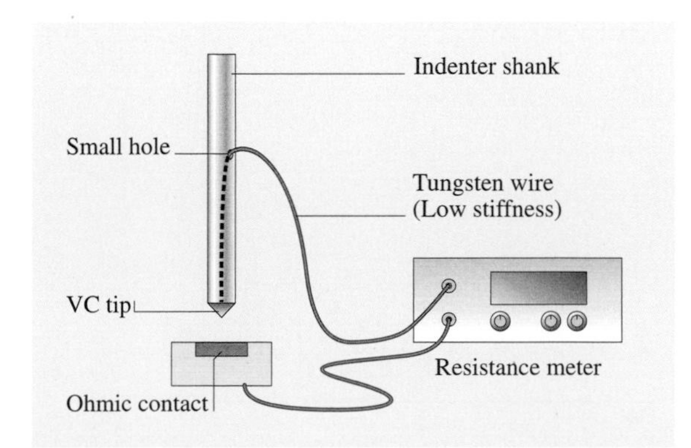

Fig. 22.5 Schematic of the basic setup for making contact resistance measurements during nanoindentation testing

22.1.4 Complimentary Techniques

Nanoindentation testing is probably the most important technique for characterizing the mechanical properties of thin films and surfaces, but there are many alternative or additional techniques that can be used. One of the most important alternative methods for measuring the mechanical properties of thin films uses bulge or blister testing [22.28]. Bulge tests are performed on thin films mounted on supporting substrates. A small area of the substrate is removed to give a window of unsupported film. A pressure is then applied to one side of the window causing it to bulge. By measuring the height of the bulge, the stress-strain curve and the residual stress are obtained. The basic configuration for bulge testing is shown in Fig. 22.6.

22.1.5 Bulge Tests

The original bulge tests used circular windows because they are easier to analyze mathematically, but now square and rectangular windows have become common [22.29]. These geometries tend to be easier to fabricate. Unfortunately, there are several sources of errors in bulge testing that can potentially lead to large errors in the measured mechanical properties. These errors at one time led to the belief that multilayer films can show a "super modulus" effect, where the elastic modulus of the multilayer is several times that of its constituent

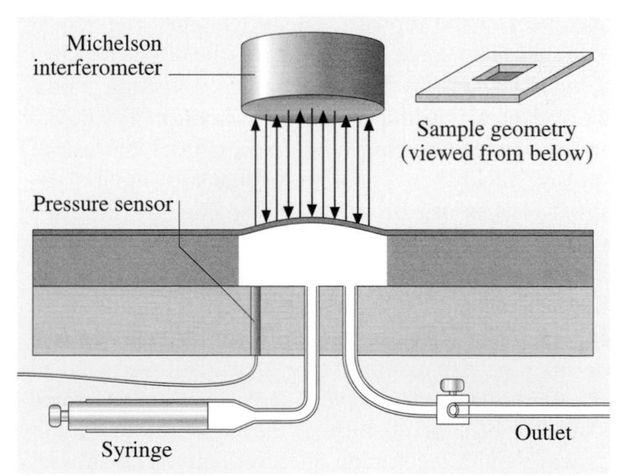

Fig. 22.6 Schematic of the basic setup for bulge testing. The sample is prepared so that it is a thin membrane, and then a pressure is applied to the back of the membrane to make it bulge upwards. The height of the bulge is measured using an interferometer

layers [22.30]. It is now accepted that any enhancement to the elastic modulus in multilayer films is small, on the order of 15% [22.31]. The main sources of error stem from compressive stresses in the film (tensile stress is not a problem), small variations in the dimensions of the window, and uncertainty in the exact height of the bulge. Despite these difficulties, one advantage of bulge testing over nanoindentation testing is that the stress state is biaxial, so that only properties in the plane of the film are measured. In contrast, nanoindentation testing measures a combination of in-plane and out-of-plane properties.

22.1.6 Acoustic Methods

Acoustic and ultrasonic techniques have been used for many years to study the elastic properties of materials. Essentially, these techniques take advantage of the fact that the velocity of sound in a material is dependent on the inter-atomic or inter-molecular forces in the material. These, of course, are directly related to the material's elastic constants. In fact, any nonlinearity of inter-atomic forces enables slight variations in acoustic signals to be used as a measure of residual stress.

An acoustic method ideally suited to studying surfaces is scanning acoustic microscopy (SAM) [22.32]. There are also several other techniques that have been used to study surface films and multilayers, but we will first consider SAM in detail. In a SAM, a lens made of sapphire is used to bring acoustic waves to focus via a coupling fluid on the surface. A small piezoelectric transducer at the top of the lens generates the acoustic signal. The same transducer can be used to detect the signal when the SAM is used in reflection mode. The use of a transducer as both generator and detector, a common imaging mode, necessitates the use of a pulsed rather than a continuous acoustic signal. Continuous waves can be used if phase changes are used to build up the image. The transducer lens generates two types of acoustic waves in the material: longitudinal and shear. The ability of a solid to sustain both types of wave (liquids can only sustain longitudinal waves) gives rise to a third type of acoustic wave called a Rayleigh, or surface, wave. These waves are generated as a result of superposition of the shear and longitudinal waves with a common phase velocity. The stresses and displacements associated with a Rayleigh wave are only of significance to a depth of ≈ 0.6 Rayleigh wavelengths below the solid surface. Hence, using SAM to examine Rayleigh waves in a material is a true surface characterization technique.

Using a SAM in reflection mode gives an image where the contrast is directly related to the Rayleigh wave velocity, which is in turn a function of the material's elastic constants. The resolution of the image depends on the frequency of the transducer used, i. e., for a 2 GHz signal a resolution better than 1 μm is achievable. The contrast in the image results from the interference of two different waves in the coupling fluid. Rayleigh waves that are excited in the surface "leak" into the coupling fluid and interfere with the acoustic signal that is directly reflected back from the surface. It is usually assumed that the properties of the coupling fluid are well characterized. The interference of the two waves gives a characteristic $V(z)$ curve, as illustrated by Fig. 22.7, where z is the separation between the lens and the surface. Analyzing the periodicity of the $V(z)$ curves provides information on the Rayleigh wave velocity. As with other acoustic waves, the Rayleigh velocity is related to the elastic constants of the material. When using the SAM for a material's characterization, the lens is usually held in a fixed position on the surface. By using a lens designed specifically to give a line-focus beam, rather than the standard spherical lens, it is possible to use SAM to look at anisotropy in the wave velocity [22.33] and hence in elastic properties by producing waves with a specific direction.

One advantage of using SAM in conjunction with nanoindentation to characterize a surface is that the measurements obtained with the two methods have a slightly different dependence on the test material's elastic properties, E_S and ν_S (the elastic modulus and Poisson's ratio). As a result, it is possible to use SAM

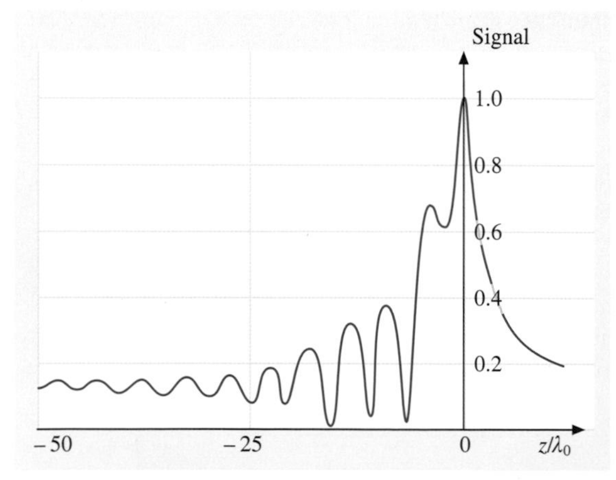

Fig. 22.7 A typical $V(z)$ curve obtained with a SAM when testing fused silica

and nanoindentation combined to find both E_S and ν_S, as illustrated by Fig. 22.8 [22.34]. This is not possible when using only one of the techniques alone.

In addition to measuring surface properties, SAM has been used to study thin films on a surface. However, the Rayleigh wave velocity can be dependent on a complex mix of the film and substrate properties. Other acoustic methods have been utilized to study freestanding films. A freestanding film can be regarded as a plate, and, therefore, it is possible to excite Lamb waves in the film. Using a pulsed laser to generate the waves and a heterodyne interferometer to detect the arrival of the Lamb wave, it is possible to measure the flexural modulus of the film [22.35]. This has been successfully demonstrated for multilayer films with a total thickness $< 10\,\mu m$. In the plate configuration, due to the nonlinearity of elastic properties, it is also possible to measure stress. This has been demonstrated for horizontally polarized shear waves in plates [22.36], but thin plates require very high frequency transducers or laser sources.

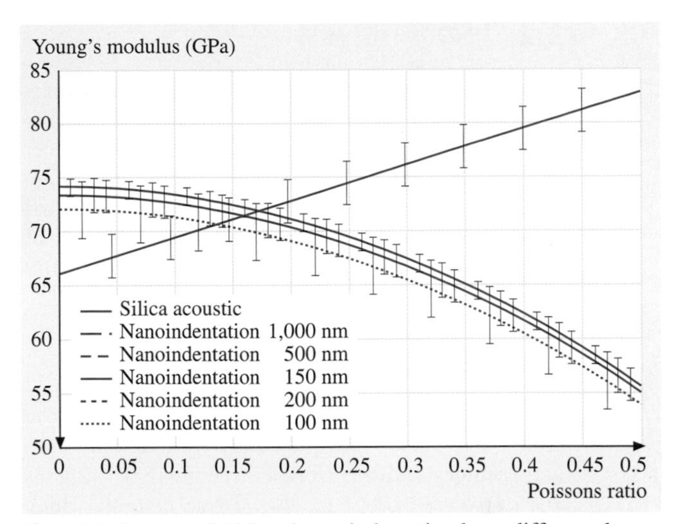

Fig. 22.8 Because SAM and nanoindentation have different dependencies on Young's modulus, E, and Poissons ratio, ν, it is possible to use the two techniques in combination to find E and ν [22.34]. On the graph, the intersection of the curves gives E and ν

22.1.7 Imaging Methods

When measuring the mechanical properties of a surface or thin film using nanoindentation, it is not always easy to visualize what is happening. In many instances there is a risk that the mechanical data can be completely misinterpreted if the geometry of the test is not as expected. To expedite the correct interpretation of the mechanical data, it is generally worthwhile to use optical, electron, or atomic force microscopy to image the nanoindentations. Obviously, optical techniques are only of use for larger indentations, but they will often reveal the presence of median or lateral cracks [22.37]. Electron microscopy and AFM, however, can be used to examine even the smallest nanoindentations. The principle problem with these microscopy techniques is the difficulty in finding the nanoindentations. It is usually necessary to make large, "marker" indentations in the vicinity of the nanoindentations to be examined in order to find them [22.38].

It is possible to see features such as extrusions with a scanning electron microscope (SEM) [22.39], as well as pile-up and sink-in around the nanoindents, though AFM is generally better for this. Transmission electron microscopy (TEM) is useful for examining what has happened subsurface, for instance, the indentation induced dislocations in a metal [22.40] or the phases present under a nanoindent in silicon [22.21]. However, with TEM there is the added difficulty of sample preparation and the associated risk of observing artifacts. Recently, there has been considerable interest in the use of focused ion beams to cut cross sections through nanoindents [22.41]. When used in conjunction with SEM or TEM this provides an excellent means to see what has happened in the subsurface region.

One other technique that has proved to be useful in studying nanoindents is micro-Raman spectroscopy. This involves using a microscope to focus a laser on the sample surface. The same microscope is also used to collect the scattered laser light, which is then fed into a spectroscope. The Raman peaks in the spectrum provide information on the bonding present in a material, while small shifts in the wave number of the peaks can be used as a measure of strain. Micro-Raman has proven to be particularly useful for examining the phases present around nanoindentations in silicon [22.42].

22.2 Data Analysis

The analysis of nanoindentation data is far from simple. This is mostly due to the lack of effective models that are able to combine elastic and plastic deformation under a contact. However, provided certain precautions are taken, the models for perfectly elastic deformation and ideal plastic materials can be used in the analysis of nanoindentation data. For this reason, it is worth briefly reviewing the models for perfect contacts.

22.2.1 Elastic Contacts

The theoretical modeling of elastic contacts can be traced back many years, at least to the late nineteenth century and the work of *Hertz* (1882) [22.43] and *Boussinesq* (1885) [22.44]. These models, which are still widely used today, consider two axisymmetric curved surfaces in contact over an elliptical region (see Fig. 22.9). The contact region is taken to be small in comparison to the radius of curvature of the contacting surfaces, which are treated as elastic half-spaces. For an elastic sphere, radius R, in contact with a flat, elastic half-space, the contact region will be circular and the Hertz model gives the following relationships:

$$a = \sqrt[3]{\frac{3PR}{4E_r}} , \tag{22.3}$$

$$\delta = \sqrt[3]{\frac{9P^2}{16RE_r^2}} , \tag{22.4}$$

$$P_0 = \sqrt[3]{\frac{6PE_r^2}{\pi^3 R^2}} , \tag{22.5}$$

where a is the radius of the contact region, E_r is given by (22.2), δ is the displacement of the sphere into the surface, P is the applied load and P_0 is the maximum pressure under the contact (in this case at the center of the contact).

The work of *Hertz* and *Boussinesq* was extended by *Love* [22.45, 46] and later by *Sneddon* [22.47], who simplified the analysis using Hankel transforms. *Love* showed how *Boussinesq*'s model could be used for a flat-ended cylinder and a conical indenter, while *Sneddon* produced a generalized relationship for any rigid axisymmetric punch pushed into an elastic half-space. *Sneddon* applied his new analysis to punches of various shapes and derived the following relationships between the applied load, P, and displacement, δ, into the elastic half-space for, respectively, a flat-ended cylinder, a cone of semi-vertical angle ϕ, and a parabola of revolution where $a^2 = 2k\delta$:

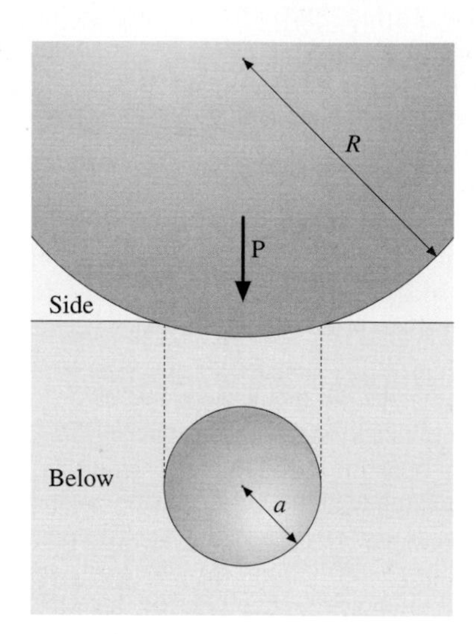

Fig. 22.9 Hertzian contact of a sphere, radius R, on a semi-infinite, flat surface. The contact in this case is a circular region of radius a

$$P = \frac{4\mu a\delta}{1-\nu} , \tag{22.6}$$

$$P = \frac{4\mu \cot\phi}{\pi(1-\nu)}\delta^2 , \tag{22.7}$$

$$P = \frac{8\mu}{3(1-\nu)}\left(2k\delta^3\right)^{1/2} , \tag{22.8}$$

where μ and ν are the shear modulus and Poisson's ratio of the elastic half-space, respectively.

The key point to note about (22.6), (22.7), and (22.8) is that they all have the same basic form, namely:

$$P = \alpha\delta^m , \tag{22.9}$$

where α and m are constants for each geometry.

Equation (22.9) and the relationships developed by Hertz and his successors, (22.3–22.8), form the foundation for much of the current nanoindentation data analysis routines.

22.2.2 Indentation of Ideal Plastic Materials

Plastic deformation during indentation testing is not easy to model. However, the indentation response of ideal plastic metals was considered by *Tabor* in his classic text, "The Hardness of Metals" [22.48]. An ideal plastic material (or more accurately an ideal elastic-plastic material) has a linear stress-strain curve until it reaches its elastic limit and then yields plastically at a yield stress, Y_0, that remains constant even after deformation has commenced. In a 2-D problem, the yielding occurs because the Huber-Mises [22.49] criterion has been reached. In other words, the maximum shear stress acting on the material is around $1.15Y_0/2$.

First, we consider a 2-D flat punch pushed into an ideal plastic material. By using the method of slip lines it is found that the mean pressure, P_m, across the end of the punch is related to the yield stress by:

$$P_m = 3Y_0 \,. \tag{22.10}$$

If the *Tresca* criterion [22.50] is used, then P_m is closer to $2.6Y_0$. In general, for both 2-D and three-dimensional punches pushed into ideal plastic materials, full plasticity across the entire contact region can be expected when $P_m = 2.6$ to $3.0Y_0$. However, significant deviations from this range can be seen if, for instance, the material undergoes work-hardening during indentation, or the material is a ceramic, or there is friction between the indenter and the surface.

The apparently straightforward relationship between P_m and Y_0 makes the mean pressure a very useful quantity to measure. In fact, P_m is very similar to the Vickers hardness, H_V, of a material:

$$H_V = 0.927 P_m \,. \tag{22.11}$$

During nanoindentation testing it is the convention to take the mean pressure as the nanohardness. Thus, the "nanohardness", H, is defined as the peak load, P, applied during a nanoindentation divided by the projected area, A, of the nanoindentation in the plane of the surface, hence:

$$H = \frac{P}{A} \,. \tag{22.12}$$

22.2.3 Adhesive Contacts

During microindentation testing and even most nanoindentation testing the effects of intermolecular and surface forces can be neglected. Very small nanoindentations, however, can be influenced by the effects of intermolecular forces between the sample and the tip. These adhesive effects are most readily seen when testing soft polymers, but there is some evidence that forces between the tip and sample may be important in even relatively strong materials [22.51, 52].

Contact adhesion is usually described by either the JKR or DMT model, as discussed earlier in this chapter. Both the models consider totally elastic spherical contacts under the influence of attractive surface forces. The JKR model considers the surface forces in terms of the associated surface energy, whereas the DMT model considers the effects of adding van der Waals forces to the Hertzian contact model. The differences between the two models are illustrated by Fig. 22.10.

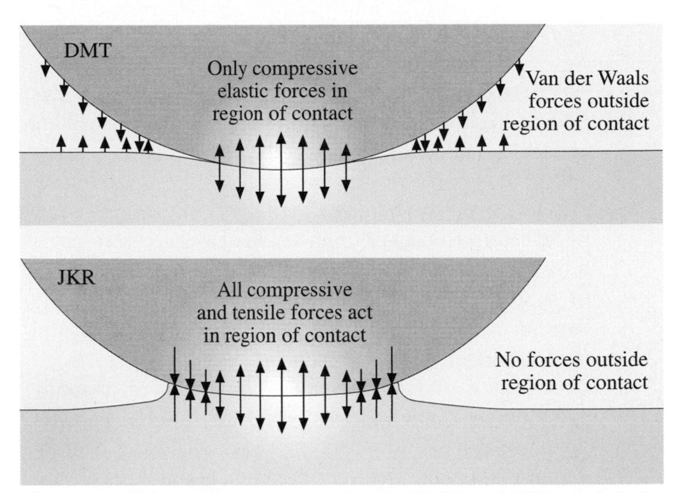

Fig. 22.10 The contact geometry for the DMT and JKR models for adhesive contact. Both models are based on the Hertzian model. In the DMT, model van der Waals forces outside the region of contact introduce an additional load in the Hertz model. But for the JKR model, it is assumed that tensile, as well as compressive stresses can be sustained within the region of contact

For nanoindentation tests conducted in air the condensation of water vapor at the tip-sample interface usually determines the size of the adhesive force acting during unloading. The effects of water vapor on a single nanoasperity contact have been studied using force-controlled AFM techniques [22.53] and, more recently, nanoindentation methods [22.26]. Unsurprisingly, it has also been found that water vapor can affect the deformation of surfaces during nanoindentation testing [22.27].

In addition to water vapor, other surface adsorbates can cause dramatic changes in the nanoscale mechanical behavior. For instance, oxygen on a clean metal surface can cause an increase in the apparent strength of the metal [22.54]. These effects are likely to be related to, firstly, changes in the surface and intermolecular forces acting between the tip and the sample and, secondly, changes in the mechanical stability of surface nanoasperities and ledges. Adsorbates can help stabilize atomic-scale variations in surface morphology, thereby making defect generation at the surface more difficult.

22.2.4 Indenter Geometry

All of the indenter geometries considered up to this point have been axisymmetric, largely because they are easier to deal with theoretically. Unfortunately, fabricating axisymmetric nanoindentation tips is extremely difficult, because shaping a hard tip on the scale of a few nanometers is virtually impossible. Despite these problems, there has been considerable effort put into the use of spherical nanoindentation tips [22.55]. This clearly demonstrates that the spherical geometry can be useful at larger indentation depths.

Because of the problems associated with creating axisymmetric nanoindentation tips, pyramidal indenter geometries have now become standard during nanoindentation testing. The most common geometries are the three-sided Berkovich pyramid and cube-corner (see Fig. 22.11). The Berkovich pyramid is based on the four-sided Vickers pyramid, the opposite sides of which make an 136° angle. For both the Vickers and Berkovich pyramids the cross-sectional area of the pyramid's base, A, is related to the pyramid's height, D, by:

$$A = 24.5D^2 . \quad (22.13)$$

The cube-corner geometry is now widely used for making very small nanoindentations, because it is much sharper than the Berkovich pyramid. This makes it easier to initiate plastic deformation at very light loads, but great care should be taken when using the cube-corner geometry. Sharp cube-corners can wear down quickly and become blunt, hence the cross-sectional area as a function of depth can change over the course of several indentations. There is also a potential problem with the standard analysis routines [22.56], which were developed for much blunter geometries and are based on the elastic contact models outlined earlier. The elastic contact models all assume the displacement into the surface is small compared to the tip radius. For the cube-corner geometry this is probably only the case for nanoindentations that are no more than a few nanometers deep.

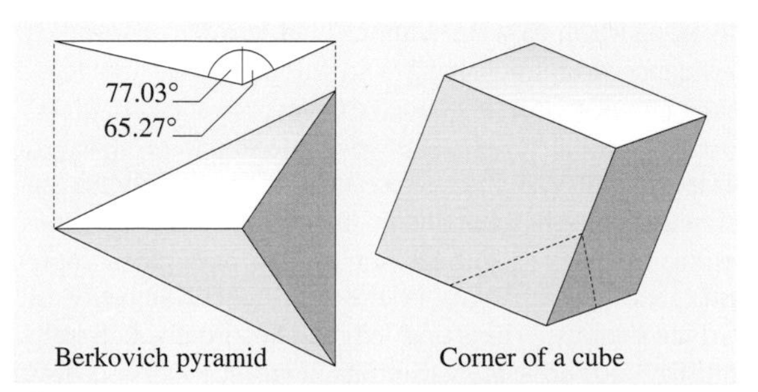

Fig. 22.11 The ideal geometry for the three-sided Berkovich pyramid and cube corner tips

22.2.5 Analyzing Load/Displacement Curves

The load/displacement curves obtained during nanoindentation testing are deceptively simple. Most newcomers to the area will see the curves as being somewhat akin to the stress/strain curves obtained during tensile testing. There is also a real temptation just to use the values of hardness, H, and elastic modulus, E, obtained from standard analysis software packages as the "true" values. This may be the case in many instances, but for very shallow nanoindents and tests on thin films the geometry of the contact can differ significantly from the geometry assumed in the analysis routines. Consequently, experimentalists should think very carefully about the test itself before concluding that the values of H and E are correct.

The basic shape of a load/displacement curve can reveal a great deal about the type of material being tested. Figure 22.12 shows some examples of ideal curves for materials with different elastic moduli and yield stresses. Discontinuities in the load/displacement curve can also provide information on such processes as fracture, dislocation nucleation, and phase transformations. Initially,

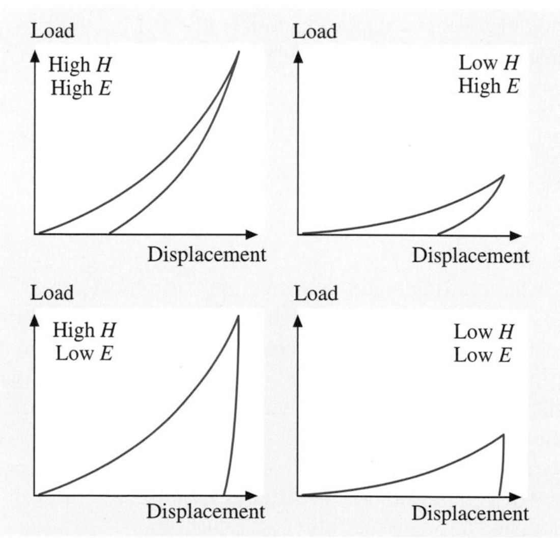

Fig. 22.12 Examples of load/displacement curves for idealized materials with a range of hardness and elastic properties

though, we will consider ideal situations such as those illustrated by Fig. 22.12.

The loading section of the load/displacement curve approximates a parabola [22.60] whose width depends on a combination of the material's elastic and plastic properties. The unloading curve, however, has been shown to follow a more general relationship [22.56] of the form:

$$P = \alpha\,(\delta - \delta_i)^m\,, \tag{22.14}$$

where δ is the total displacement and δ_i is the intercept of the unloading curve with the displacement axis shown in Fig. 22.13.

Equation (22.14) is essentially the same as (22.9) but with the origin displaced. Since (22.9) is obtained by considering purely elastic deformation, it follows that the unloading curve is exhibiting purely elastic behavior. Since the shape of the unloading curve is determined by

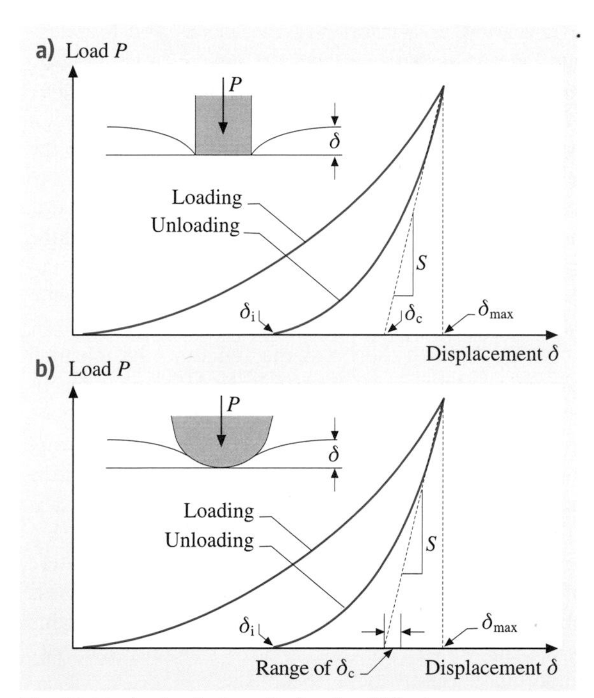

Fig. 22.13a,b Analysis of the load/displacement curve gives the contact stiffness, S, and the contact depth, δ_c. These can then be used to find the hardness, H, and elastic, or Young's modulus, E. **(a)** The first method of analysis [22.57–59] assumed the unloading curve could be approximated by a flat punch on an elastic half-space. **(b)** A more refined analysis [22.56] uses a paraboloid on an elastic half-space

the elastic recovery of the indented region, it is not entirely surprising that its shape resembles that found for purely elastic deformation. What is fortuitous is that the elastic analysis used for an elastic half-space seems to be valid for a surface where there is a plastically formed indentation crater present under the contact. However, the validity of this analysis may only hold when the crater is relatively shallow and the geometry of the surface does not differ significantly from that of a flat, elastic half-space. For nanoindentations with a Berkovich pyramid, this is generally the case.

Before *Oliver* and *Pharr* [22.56] proposed their now standard method for analyzing nanoindentation data, the analysis had been based on the observation that the initial part of the unloading curve is almost linear. A linear unloading curve, equivalent to $m = 1$ in (22.14), is expected when a flat punch is used on an elastic half-space. The flat punch approximation for the unloading curve was used in [22.57–59] to analyze nanoindentation data. When Oliver and Pharr looked at a range of materials they found m was typically larger than 1, and that $m = 1.5$, or a paraboloid, was a better approximation than a flat punch. Oliver and Pharr used (22.1) and (22.12) to obtain the values for a material's elastic modulus and hardness. Equation (22.1) relates the contact stiffness during the initial part of the unloading curve (see Fig. 22.13) to the reduced elastic modulus and the contact area at the peak load. Equation (22.12) gives the hardness as the peak load divided by the contact area. It is immediately obvious that the key to measuring the mechanical properties of a material is knowing the contact area at the peak load. This is the single most important factor in analyzing nanoindentation data. Most mistakes in the analysis come from incorrect assumptions about the contact area.

To find the contact area, a function relating the contact area, A_c, to the contact depth, δ_c, is needed. For a perfect Berkovich pyramid this would be the same as (22.13). But since making a perfect nanoindenter tip is impossible, an expanded equation is used:

$$A_c\,(\delta_c) = 24.5\delta_c^2 + \sum_{j=1}^{7} C_j \sqrt[2^j]{\delta_c}\,, \tag{22.15}$$

where C_j are calibration constants of the tip.

There is a crucial step in the analysis before A_c can be calculated, namely, finding δ_c. The contact depth is not the same as the indentation depth, because the surface around the indentation will be elastically deflected during loading, as illustrated by Fig. 22.14. Sneddon's analysis [22.47] provides a way to calculate the deflection of the surface at the edge of an axisymmetric

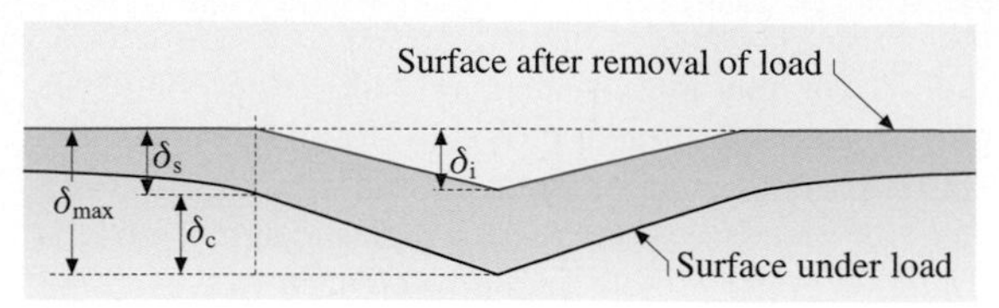

Fig. 22.14 Profile of surface under load and unloaded showing how δ_c compares to δ_i and δ_{max}

contact. Subtracting the deflection from the total indentation depth at peak load gives the contact depth. For a paraboloid, as used by *Oliver* and *Pharr* [22.56] in their analysis, the elastic deflection at the edge of the contact is given by:

$$\delta_s = \varepsilon \frac{P}{S} = 0.75 \frac{P}{S} , \qquad (22.16)$$

where S is the contact stiffness and P the peak load. The constant ε is 0.75 for a paraboloid, but ranges between 0.72 (conic indenter) and 1 (flat punch). Figure 22.15 shows how the contact depth depends on the value of ε. The contact depth at the peak load is, therefore:

$$\delta_c = \delta - \delta_s . \qquad (22.17)$$

Using the load/displacement data from the unloading curve and (22.1), (22.2), (22.12), (22.14–22.17), the hardness and reduced elastic modulus for the test sample can be calculated. To find the elastic modulus of the sample, E_s, it is also necessary to know Poisson's ratio, ν_s, for the sample, as well as the elastic modulus, E_t, and Poisson's ratio, ν_t, of the indenter tip. For diamond these are 1,141 GPa and 0.07, respectively.

There also remains the issue of calibrating the tip shape, or finding the values for C_j in (22.15). Knowing the exact expansion of $A_c(\delta_c)$ is vital if the values for E_s and H are to be accurate. Several methods for calibrating the tip shape have been used, including imaging the tip with an electron microscope, measuring the size of nanoindentations using SEM or TEM of negative replicas, and using scanning probes to examine either the tip itself or the nanoindentations made with the tip. There are strengths and weaknesses to each of these methods. In general, however, the accuracy and usefulness of the methods depends largely on how patient and rigorous the experimentalist is in performing the calibration.

Because of the experimental difficulties and time involved in calibrating the tip shape by these methods, *Oliver* and *Pharr* [22.56] developed a method for calibration based on standard specimens. With a standard specimen that is mechanically isotropic and has a known E and H that does not vary with indentation depth, it should be possible to perform nanoindentations to a range of depths, and then use the analysis routines in reverse to deduce the tip area function, $A_c(\delta_c)$. In other words, if you perform a nanoindentation test, you can find the contact stiffness, S, at the peak load, P, and the contact depth, δ_c, from the unloading curve. Then if you know E a priori, (22.1) can be used to calculate the contact area, A, and, hence, you have a value for A_c at a depth δ_c. Repeating this procedure for a range of depths will give a numerical version of the function $A_c(\delta_c)$. Then, it is simply a case of fitting (22.15) to the numerical data. If the hardness, H, is known and not a function of depth, and the calibration specimen was fully plastic during testing, then essentially the same approach could be used but based on (22.12). Situations where a constant H is used to calibrate the tip are extremely rare.

In addition to the tip shape function, the machine compliance must be calibrated. Basic Newtonian mechanics tells us that for the tip to be pushed into a surface the tip must be pushing off of another body. During nanoindentation testing the other body is the machine frame. As a result, during a nanoindentation test it is not just the sample, but the machine frame that is being loaded. Consequently, a very small elastic deformation of the machine frame contributes to the total stiffness obtained from the unloading curve. The machine frame is usually very stiff, $> 10^6$ N/m, so the effect is only important at relatively large loads.

To calibrate the machine frame stiffness or compliance, large nanoindentations are made in a soft material such as aluminum with a known, isotropic elastic modulus. For very deep nanoindentations made with a Berkovich pyramid, the contact area, $A_c(\delta_c)$, can be reasonably approximated to $24.5\delta_c^2$, thus (22.1) can be used to find the expected contact stiffness for the material. Any difference between the expected value of S

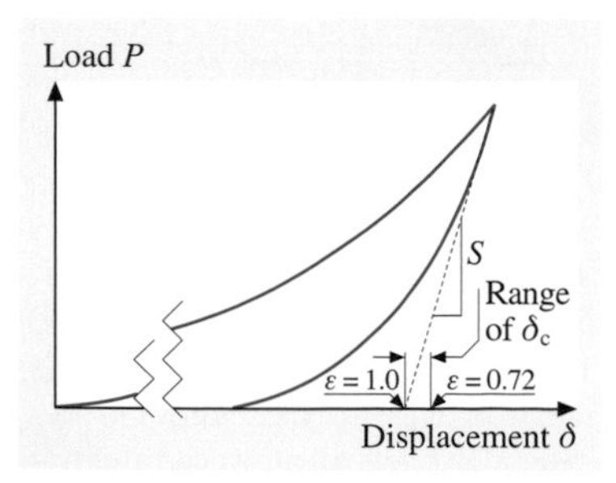

Fig. 22.15 Load-displacement curve showing how δ_c varies with ε

and the value measured from the unloading curve will be due to the compliance of the machine frame. Performing a number of deep nanoindentations enables an accurate value for the machine frame compliance to be obtained.

Currently, because of its ready availability and predictable mechanical properties, the most popular calibration material is fused silica ($E = 72\,\text{GPa}$, $\nu = 0.17$), though aluminum is still used occasionally.

22.2.6 Modifications to the Analysis

Since the development of the analysis routines in the early 1990s, it has become apparent that the standard analysis of nanoindentation data is not applicable in all situations, usually because errors occur in the calculated contact depth or contact area. *Pharr* et al. [22.61–64] have used finite element modeling (FEM) to help understand and overcome the limitations of the standard analysis. Two important sources of errors have been identified in this way. The first is residual stress at the sample surface. The second is the change in the shape of nanoindents after elastic recovery.

The effect of residual stresses at a surface on the indentation properties has been the subject of debate for many years [22.65–67]. The perceived effect was that compressive stresses increased hardness, while tensile stresses decreased hardness. Using FEM it is possible to model a pointed nanoindenter being pushed into a model material that is in residual tension or compression. An FEM model of nanoindentation into aluminum alloy 8009 [22.61] has confirmed earlier experimental observations [22.68] indicating that the contact area calculated from the unloading curve is incorrect if there are residual stresses. In the FEM model of an aluminum alloy the mechanical behavior of the material is modeled using a stress-strain curve, which resembles that of an elastic-perfectly-plastic metal with a flow stress of 425.6 MPa. Yielding starts at 353.1 MPa and includes a small amount of work hardening. The FEM model was used to find the contact area directly and using the simulated unloading curve in conjunction with Oliver and Pharr's method. The results as a function of residual stress are illustrated in Fig. 22.16. Note that the differences between the two measured contact areas lead to miscalculations of E and H.

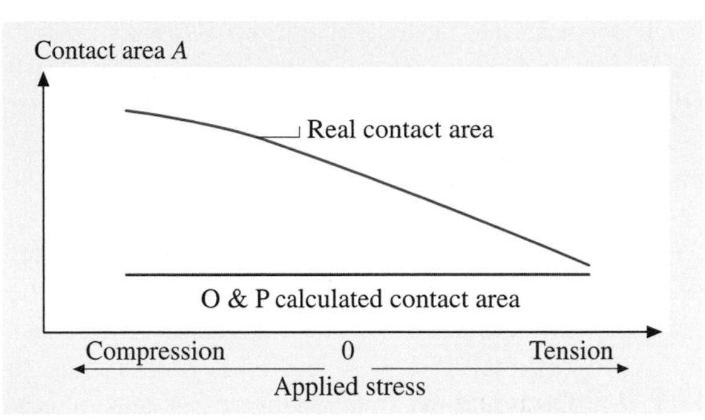

Fig. 22.16 When a surface is in a state of stress there is a significant difference between the contact area calculated using the Oliver and Pharr method and the actual contact area [22.61]. For an aluminum alloy this can lead to significant errors in the calculated hardness and elastic modulus

Errors in the calculated contact area stem from incorrect assumptions about the pile-up and sink-in at the edge of the contact, as illustrated by Fig. 22.17. The Oliver and Pharr analysis assumes the geometry of the sample surface is the same as that given by *Sneddon* [22.47] in his analytical model for the indentation of elastic surfaces. Clearly, for materials where there is significant plastic deformation, it is possible that there will be large deviations from the surface geometry found using Sneddon's elastic model. In reality, the error in the contact area depends on how much the geometry of the test sample surface differs from that of the calibration material (typically fused silica). It is possible that a test sample, even without a residual stress, will have a different surface geometry and, hence, contact area at a given depth, when compared to the calibration material. This is often seen for thin films on a substrate (e.g., *Tsui* et al. [22.69, 70]). Residual stresses increase the likelihood that the contact area calculated using Oliver and Pharr's method will be incorrect.

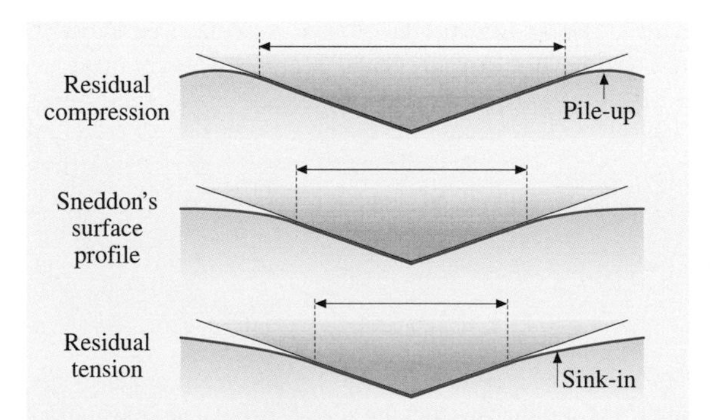

Fig. 22.17 Pile-up and sink-in are affected by residual stresses, and, hence, errors are introduced into standard Oliver and Pharr analysis

The issue of sink-in and pile-up is always a factor in nanoindentation testing. However, there is still no effective way to deal with these phenomena other than reverting to imaging of the indentations to identify the true contact area. Even this is difficult, as the edge of an indentation is not easy to identify using AFM or electron microscopy. One approach that has been used [22.71] with some success is measuring the ratio E_r^2/H, rather than E_r and H separately. Because E_r is proportional to $1/\sqrt{A}$ and H is proportional to $1/A$, E_r^2/H should be independent of A and, hence, unaffected by pile-up or sink-in. While this does not provide quantitative values for mechanical properties, it does provide a way to identify any variations in mechanical properties with indentation depth or between similar samples with different residual stresses.

Another source of error in the Oliver and Pharr analysis is due to incorrect assumptions about the nanoindentation geometry after unloading [22.63]. Once again, this is due to differences between the test sample and the calibration material. The exact shape of an unloaded nanoindentation on a material exhibiting elastic recovery is not simply an impression of the tip shape; rather, there is some elastic recovery of the nanoindentation sides giving them a slightly convex shape (see Fig. 22.18). The shape actually depends on Poisson's ratio, so the standard Oliver and Pharr analysis will only be valid for a material where $\nu = 0.17$, the value for fused silica, assuming it is used for the calibration.

To deal with the variations in the recovered nanoindentation shape, it has been suggested [22.63] that a modified nanoindenter geometry with a slightly concave side be used in the analysis (see Fig. 22.18). This

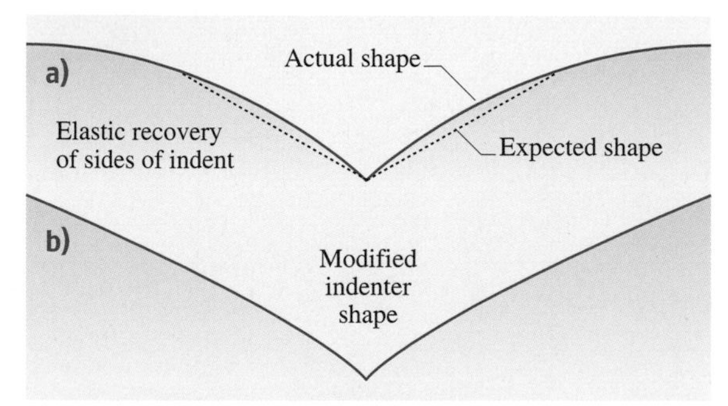

Fig. 22.18 (a) *Hay* et al. [22.63] found from experiments and FEM simulations that the actual shape of an indentation after unloading is not as expected. (b) They introduced a γ term to correct for this effect. This assumes the indenter has slightly concave sides

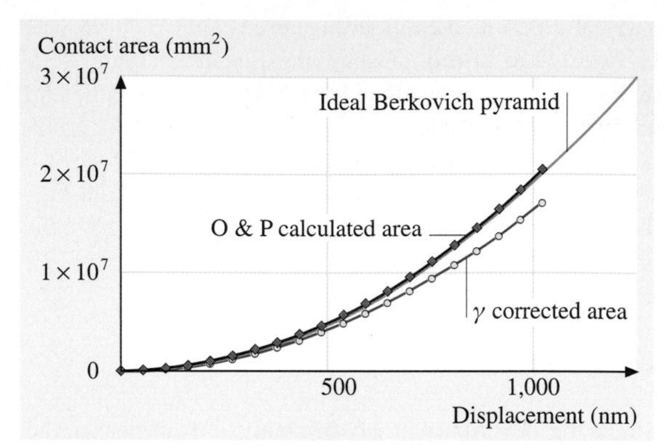

Fig. 22.19 For a real Berkovich tip the γ corrected area [22.63] is less at a given depth than the area calculated using the Oliver and Pharr method

requires a modification to (22.1):

$$S = \gamma 2 E_r \sqrt{\frac{A}{\pi}} , \qquad (22.18)$$

where γ is a correction term dependent on the tip geometry. For a Berkovich pyramid the best value is:

$$\gamma = \frac{\frac{\pi}{4} + 0.15483073 \cot \Phi \left(\frac{(1-2\nu_s)}{4(1-\nu_s)}\right)}{\left[\frac{\pi}{2} - 0.83119312 \cot \Phi \left(\frac{(1-2\nu_s)}{4(1-\nu_s)}\right)\right]^2} , \qquad (22.19)$$

where $\Phi = 70.32°$. For a cube corner the correction can be even larger and γ is given by:

$$\gamma = 1 + \left(\frac{(1-2\nu_s)}{4(1-\nu_s)\tan \Phi}\right) , \qquad (22.20)$$

where $\Phi = 42.28°$. Figure 22.19 shows how the modified contact area varies with depth for a real diamond Berkovich pyramid.

The validity of the γ-modified geometry is questionable from the perspective of contact mechanics since it relies on assuming an incorrect geometry for the nanoindenter tip to correct for an error in the geometry of the nanoindentation impression. The values for E and H obtained using the γ-modification are, however, good and can be significantly different to the values obtained with the standard Oliver and Pharr analysis.

22.2.7 Alternative Methods of Analysis

All of the preceding discussion on the analysis of nanoindentation curves has focused on the unloading curve, virtually ignoring the loading curve data. This is for the simple reason that the unloading curve can in many cases be regarded as purely elastic, whereas the shape of the loading curve is determined by a complex mix of elastic and plastic properties.

It is clear that there is substantially more data in the loading curve if it can be extracted. *Page* et al. [22.60, 72] have explored the possibility of curve fitting to the loading data using a combination of elastic and plastic properties. By a combination of analysis and empirical fitting to experimental data, it was suggested that the loading curve is of the following form:

$$P = E\left(\psi\sqrt{\frac{H}{E}} + \phi\sqrt{\frac{E}{H}}\right)^{-2}\delta^2\,, \tag{22.21}$$

where ψ and ϕ are determined experimentally to be 0.930 and 0.194, respectively. For homogenous samples this equation gives a linear relationship between P and δ^2. Coatings, thin film systems, and samples that strain-harden can give significant deviations from linearity. Analysis of the loading curve has yet to gain popularity as a standard method for examining nanoindentation data, but it should certainly be regarded as a prime area for further investigation.

Another alternative method of analysis is based on the work involved in making an indentation. In essence, the nanoindentation curve is a plot of force against distance indicating integration under the loading curve will give the total work of indentation, or the sum of the elastic strain energy and the plastic work of indentation. Integrating under the unloading curve should give only the elastic strain energy. Thus, the work involved in both elastic and plastic deformation during nanoindentation can be found. *Cheng* and *Cheng* [22.73] combined measurements of the work of indentation with a dimensional analysis that deals with the effects of scaling in a material that work-hardens to estimate H/E_r. They subsequently evaluated H and E using the Oliver and Pharr approach to find the contact area.

22.2.8 Measuring Contact Stiffness

As discussed earlier, it is possible to add a small AC load on top of the DC load used during nanoindentation testing, providing a way to measure the contact stiffness throughout the entire loading and unloading cycle [22.74, 75]. The AC load is typically at a frequency of ≈ 60 Hz and creates a dynamic system, with the sample acting as a spring with stiffness S (the contact stiffness), and the nanoindentation system acting as a series of springs and dampers. Figure 22.20 illustrates

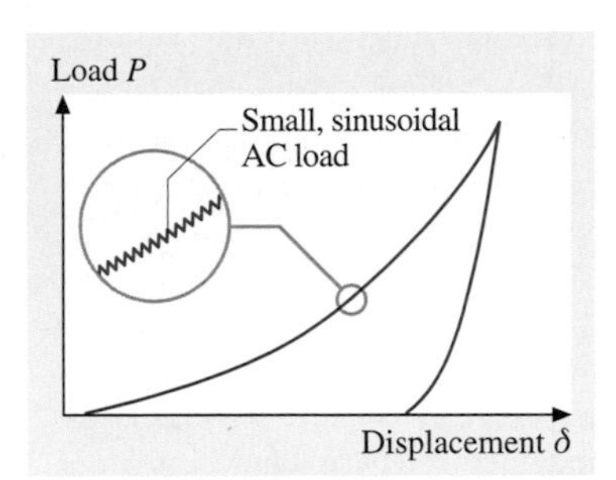

Fig. 22.20 A small AC load can be added to the DC load. This enables the contact stiffness, S, to be calculated throughout the indentation cycle

how the small AC load is added to the DC load. Figure 22.21 shows how the resulting dynamic system can be modeled. An analysis of the dynamic system gives the following relationships for S based on the amplitude of the AC displacement oscillation and the phase difference between the AC load and displacement signals:

$$\left|\frac{P_{\mathrm{os}}}{\delta(\omega)}\right| = \sqrt{\left[\left(S^{-1}+C_{\mathrm{f}}\right)^{-1}+K_{\mathrm{s}}-m\omega^2\right]^2+\omega^2D^2}\,, \tag{22.22}$$

$$\tan(\chi) = \frac{\omega D}{\left(S^{-1}+C_{\mathrm{f}}\right)^{-1}+K_{\mathrm{s}}-m\omega^2}\,, \tag{22.23}$$

where C_{f} is the load frame compliance (the reciprocal of the load frame stiffness), K_{s} is the stiffness of the support springs (typically in the region of 50–100 N/m), D is the damping coefficient, P_{os} is the magnitude of the load oscillation, $\delta(\omega)$ is the magnitude of the displacement oscillation, ω is the oscillation frequency, m is the mass of the indenter, and χ is the phase angle between the force and the displacement.

In order to find S using either (22.22) or (22.23), it is necessary to calibrate the dynamic response of the system when the tip is not in contact with a sample ($S^{-1} = 0$). This calibration combined with the standard

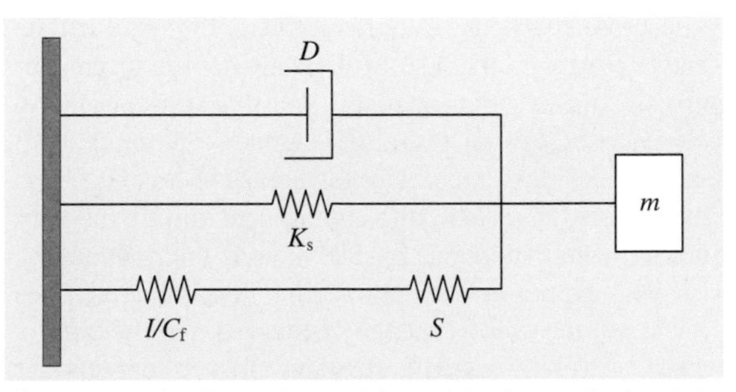

Fig. 22.21 The dynamic model used in the analysis of the AC response of a nanoindentation device

DC calibrations will provide the values for all of the constants in the two equations. All that needs to be measured in order to obtain S is either $\delta(\omega)$ or χ, both of which are measured by the lock-in amplifier used to generate the AC signal. Since the S obtained is the same as the S in (22.1), it follows that the Oliver and Pharr analysis can be applied to obtain E_r and H throughout the entire nanoindentation cycle.

The dynamic analysis detailed here was developed for the MTS Nanoindenter™ (Oakridge, Tennessee), but a similar analysis has been applied to other commercial instruments such as the Hysitron Triboscope™ (Minneapolis, Minnesota) [22.76]. For all instruments, an AC oscillation is used in addition to the DC voltage, and a dynamic model is used to analyze the response.

22.2.9 Measuring Viscoelasticity

Using an AC oscillation in addition to the DC load introduces the possibility of measuring viscoelastic properties during nanoindentation testing. This has recently been the subject of considerable interest with researchers looking at the loss modulus, storage modulus, and loss tangent of various polymeric materials [22.25, 77]. Recording the displacement response to the AC force oscillation enables the complex modulus (including the loss and storage modulus) to be found. If the modulus is complex, it is clear from (22.1) that the stiffness also becomes complex. In fact, the stiffness will have two components: S', the component in phase with the AC force and S'', the component out of phase with the AC force.

The dynamic model illustrated in Fig. 22.21 is no longer appropriate for this situation, as the contact on the test sample also includes a damping term, shown in Fig. 22.22. Equations (22.22) and (22.23) must also be revised. Neglecting the load frame compliance, C_f, which in most real situations is negligible, (22.22) and (22.23) when the sample damping, D_s, is included become:

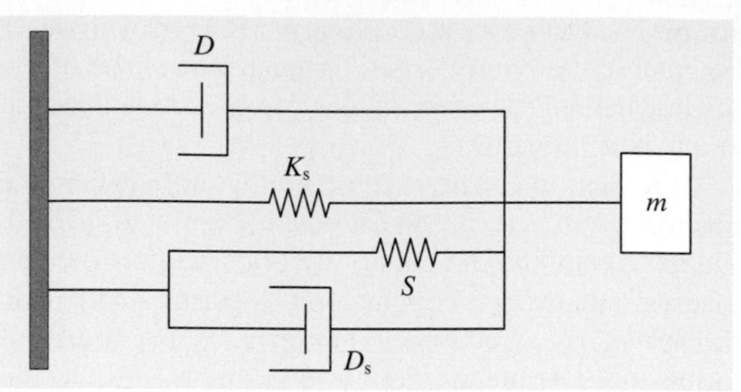

Fig. 22.22 The simplified dynamic model used when the sample is viscoelastic. It is assumed that the load frame compliance is negligible

$$\left|\frac{P_{os}}{\delta(\omega)}\right| = \sqrt{\left\{S+K_s-m\omega^2\right\}^2+\omega^2(D+D_s)^2}\,, \tag{22.24}$$

$$\tan(\chi) = \frac{\omega(D+D_s)}{S+K_s-m\omega^2}\,. \tag{22.25}$$

In order to find the loss modulus and storage modulus, (22.1) is used to relate S' (storage component) and S'' (loss component) to the complex modulus.

This method for measuring viscoelastic properties using nanoindentation has now been proven in principal, but has still only been applied to a very small range of polymers and remains an area of future growth.

22.3 Modes of Deformation

As described earlier, the analysis of nanoindentation data is based firmly on the results of elastic continuum mechanics. In reality, this idealized, purely elastic situation rarely occurs. For very shallow contacts on metals with thin surface films such as oxides, carbon layers, or organic layers [22.78, 79], the contact can initially be very similar to that modeled by Hertz and, later, Sneddon. It is very important to realize that this in itself does not constitute proof that the contact is purely elastic, because in many cases a small number of defects are present. These may be preexisting defects that move in the strain-field generated beneath the contact. Alternatively, defects can be generated either when the contact is first made or during the initial loading [22.52, 80]. When defects such as short lengths of dislocation are present the curves may still appear to be elastic even though inelastic processes like dislocation glide and cross-slip are taking place.

22.3.1 Defect Nucleation

Nucleation of defects during nanoindentation testing has been the subject of many experimental [22.81, 82] and theoretical studies [22.83, 84]. This is probably because nanoindentation is seen as a way to deform a small, defect-free volume of material to its elastic limit and beyond in a highly controlled geometry. There are, unfortunately, problems in comparing experimental results with theoretical predictions, largely because the kinetic processes involved in defect nucleation are difficult to model. Simulations conducted at 0 K do not permit kinetic processes, and molecular dynamics simulations are too fast (nanoseconds or picoseconds). Real nanoindentation experiments take place at ≈ 293 K and last for seconds or even minutes.

Kinetic effects appear in many forms, for instance, during the initial contact between the indenter tip and the surface when defects can be generated by the combined action of the impact velocity and surface forces [22.51]. A second example of a kinetic effect occurs during hold cycles at large loads when what appears to be an elastic contact can suddenly exhibit a large discontinuity in the displacement data [22.80]. Figure 22.23 shows how these kinetic effects can affect the nanoindentation data and the apparent yield point load.

During the initial formation of a contact, the deformation of surface asperities [22.51] and ledges [22.85] can create either point defects or short lengths of dislocation line. During the subsequent loading, the defects can help in the nucleation and multiplication of dislocations. The large strains present in the region surrounding the contact, coupled with the existence of defects generated on contact, can result in the extremely rapid multiplication of dislocations and, hence, pronounced discontinuities in the load-displacement curve. It is important to realize that the discontinuities are due to the rapid multiplication of dislocations, which may or may not occur at the same time that the first dislocation is nucleated. Dislocations may have been present for some time with the discontinuity only occurring when the existing defects are configured appropriately, as a Frank-Read source, for instance. Even under large strains, the time taken for a dislocation source to form from preexisting defects may be long. It is, therefore, not surprising that large discontinuities can be seen during hold cycles or unloading.

The generation of defects at the surface and the initiation of yielding is a complex process that is extremely dependent on surface asperities and surface forces. These, in turn, are closely related to the surface chemistry. It is not only the magnitude of surface forces, but also their range in comparison to the height of surface asperities that determines whether defects are generated on contact. Small changes in the surface chemistry or

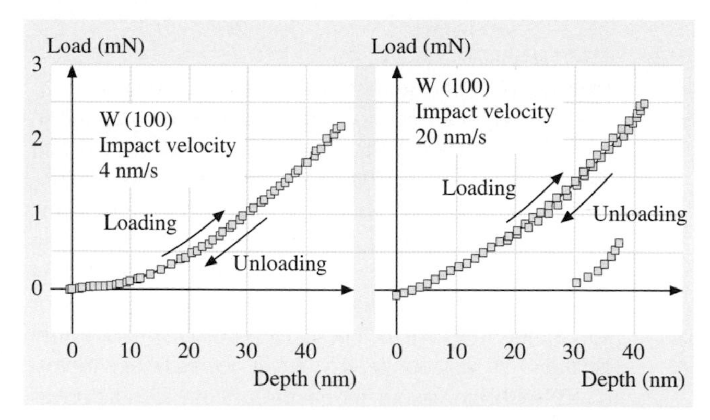

Fig. 22.23a,b Load-displacement curves for W(100) showing how changes in the impact velocity can cause a transition from perfectly elastic behavior to yielding during unloading

the velocity of the indenter tip when it first contacts the surface, can cause a transition from a situation in which defects are generated on contact to one where the contact is purely elastic [22.52].

When the generation of defects during the initial contact is avoided and the deformed region under the contact is truly defect free, then the yielding of the sample should occur at the yield stress of a perfect crystal lattice. The load at which plastic deformation commences under these circumstances becomes very reproducible [22.86]. Unfortunately, nanoindenter tips on the near-atomic scale are not perfectly smooth or axisymmetric. As a result, accurately measuring the yield stress is very difficult. In fact, a slight rotation in the plane of the surface of either the sample or the tip can give a substantial change in the observed yield point load. Coating the surface in a cushioning self-assembled monolayer [22.87] can alleviate some of these variations, but it also introduces a large uncertainty in the contact area. Surface oxide layers, which may be several nanometers thick, have also been found to enhance the elastic behavior seen for very shallow nanoindentations on metallic surfaces [22.78]. Removal of the oxide has been shown to alleviate the initial elastic response.

While nanoindentation testing is ideal for examining the mechanical properties of defect-free volumes and looking at the generation of defects in perfect crystal lattices, it should be clear from the preceding discussion that great care must be taken in examining how the surface properties and the loading rate affect the results, particularly when comparisons are being made to theoretical models for defect generation.

22.3.2 Variations with Depth

Ideal elastic-plastic behavior, as described by *Tabor* [22.48], can be seen during indentation testing, provided the sample has been work-hardened so that the flow stress is a constant. However, it is often the case that the mechanical properties appear to change as the load (or depth) is increased. This apparent change can be a result of several processes, including work-hardening during the test. This is a particularly important effect for soft metals like copper. These metals usually have a high hardness at shallow depths, but it decreases asymptotically with increasing indentation depth to a hardness value that may be less than half that observed at shallow depths. This type of behavior is due to the increasing density of geometrically necessary dislocations at shallow depths [22.88]. Hence the effects of work-hardening are most pronounced at shallow depths. For hard materials the effect is less obvious.

Work-hardening is one of the factors that contribute to the so-called indentation size effect (ISE), whereby at shallow indentation depths the material appears to be harder. The ISE has been widely observed during microindentation testing, with at least part of the effect appearing to result from the increased difficulty in optically measuring the area of an indentation when it is small. During nanoindentation testing the ISE can also be observed, but it is often due to the tip area function, $A_c(\delta_c)$, being incorrectly calibrated. However, there are physical reasons other than work-hardening for expecting an increase in mechanical strength in small volumes. As described in the previous section, small volumes of crystalline materials can have either no defects or only a small number of defects present, making plastic yielding more difficult. Also, because of dislocation line tension, the shear stress required to make a dislocation bow out increases as the radius of the bow decreases. Thus, the shear stress needed to make a dislocation bow out in a small volume is greater than it is in a large volume. These physical reasons for small volumes appearing stronger than large volumes are particularly important in thin film systems, as will be discussed later. Note, however, that these physical reasons for increased hardness do not apply for an amorphous material such as fused silica, which partially explains its value as a calibration material.

22.3.3 Anisotropic Materials

The analysis methods detailed earlier were concerned primarily with the interpretation of data from nanoindentations in isotropic materials where the elastic modulus is assumed to be either independent of direction or a polycrystalline average of a material's elastic constants. Many crystalline materials exhibit considerable anisotropy in their elastic constants, hence, these analysis techniques may not always be appropriate. The theoretical problem of a rigid indenter pressed into an elastic, anisotropic half-space has been considered by *Vlassak* and *Nix* [22.89]. Their aim was to identify the feasibility of interpreting data from a depth-sensing indentation apparatus for samples with elastic constants that are anisotropic. Nanoindentation experiments [22.90] have shown the validity of the elastic analysis for crystalline zinc, copper, and beta-brass. The observed indentation modulus for zinc, as predicted, varied by as much as a factor of 2 between different orientations. The variations in the observed hardness values for the same materials were smaller, with a maximum variation with orientation of 20% detected in zinc. While these variations are clearly detectable with nanoindentation techniques, the variations are small in comparison to the actual anisotropy of the test material's elastic properties. This is because the indentation modulus is a weighted average of the stiffness in all directions.

At this time the effects of anisotropy on the hardness measured using nanoindentation have not been fully explored. For materials with many active slip planes it is likely that the small anisotropy observed by *Vlassak* and *Nix* is correct once plastic flow has been initiated. It is possible, however, that for defect-free crystalline specimens with a limited number of active slip planes that very shallow nanoindentations may show a much larger anisotropy in the observed hardness and initial yield point load.

22.3.4 Fracture and Delamination

Indentation testing has been widely used to study fracture in brittle materials [22.91], but the lower loads and smaller deformation regions of nanoindentation tests make it harder to initiate cracks and, hence, less useful as a way to evaluate fracture toughness. To overcome these problems the cube corner geometry, which generates larger shear stresses than the Berkovich pyramid, has been used with nanoindentation testing to study fracture [22.92]. These studies have had mixed success, because the cube corner geometry blunts very quickly when used on hard materials. In many cases, brittle materials are very hard.

Depth sensing indentation is better suited to studying delamination of thin films. Recent work extends the research conducted by *Marshall* et al. [22.93, 94], who examined the deformation of residually stressed films by indentation. A schematic of their analysis is given by Fig. 22.24. Their indentations were several microns deep, but the basic analysis is valid for nanoindentations. The analysis has been extended to multilayers [22.95], which is important since it enables a quantitative assessment of adhesion energy when an additional stressed film has been deposited on top of the film and substrate of interest. The additional film limits the plastic deformation of the film of interest and also applies extra stress that aids in the delamination. After indentation, the area of the delaminated film is measured optically or with an AFM to assess the extent of the delamination. This measurement, coupled with the load-displacement data, enables quantitative assessment of the adhesion energy to be made for metals [22.96] and polymers [22.97].

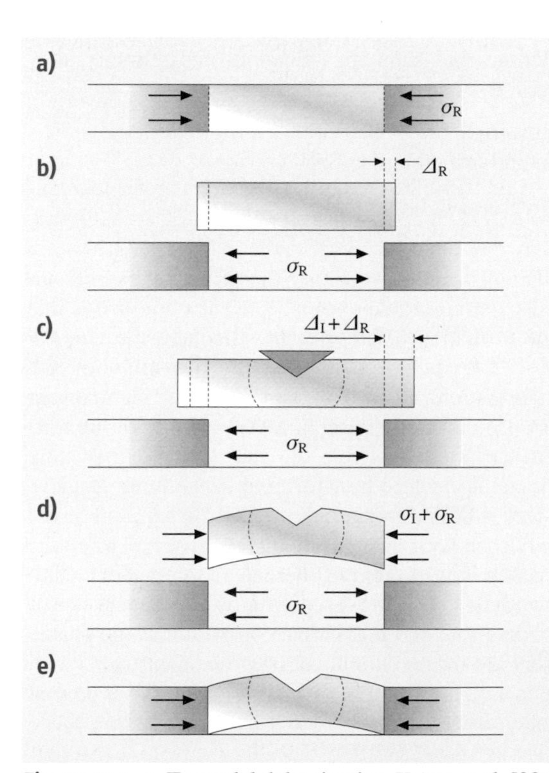

Fig. 22.24a–e To model delamination *Kriese* et al. [22.95] adapted the model developed by *Marshall* and *Evans* [22.93]. The model considers a segment of removed stressed film that is allowed to expand and then indented, thereby expanding it further. Replacing the segment in its original position requires an additional stress, and the segment bulges upwards

22.3.5 Phase Transformations

The pressure applied to the surface of a material during indentation testing can be very high. Equation (22.10) indicates that the pressure during plastic yielding is about three times the yield stress. For many materials, high hydrostatic pressures can cause phase transformations, and provided the transformation pressure is less than the pressure required to cause plastic yielding, it is possible during indentation testing to induce a phase transformation. This was first reported for silicon [22.98], but it has also been speculated [22.99] that many other materials may show the same effects. Most studies still focus on silicon because of its enormous technological importance, although there is some evidence that germanium also undergoes a phase transformation during the nanoindentation testing [22.100].

Recent results [22.21, 22, 41, 101, 102] indicate that there are actually multiple phase transformations during the nanoindentation of silicon. TEM of nanoindenta-

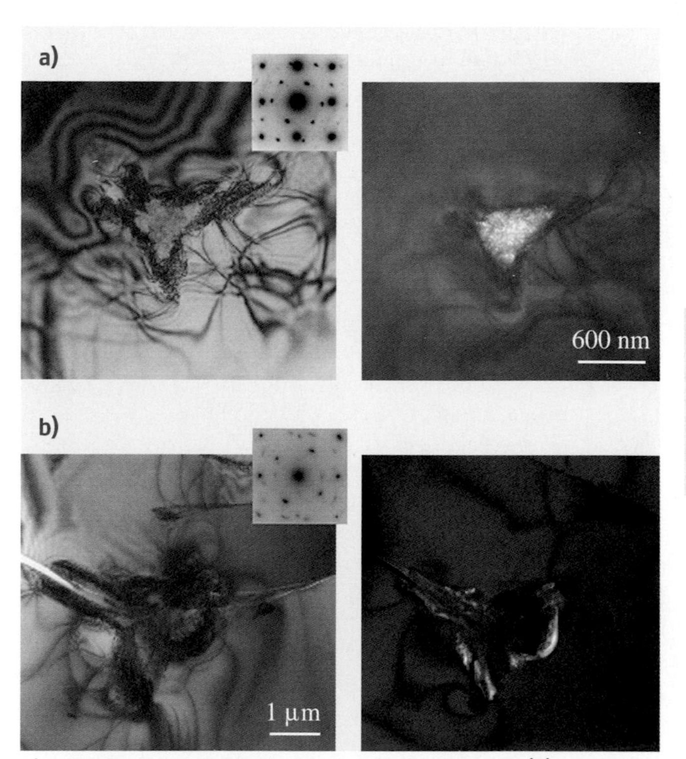

Fig. 22.25a,b Bright-field and dark-field TEM of **(a)** small and **(b)** large nanoindentations in Si. In small nanoindents the metastable phase BC-8 is seen in the center, but for large nanoindents BC-8 is confined to the edge of the indent, while the center is amorphous

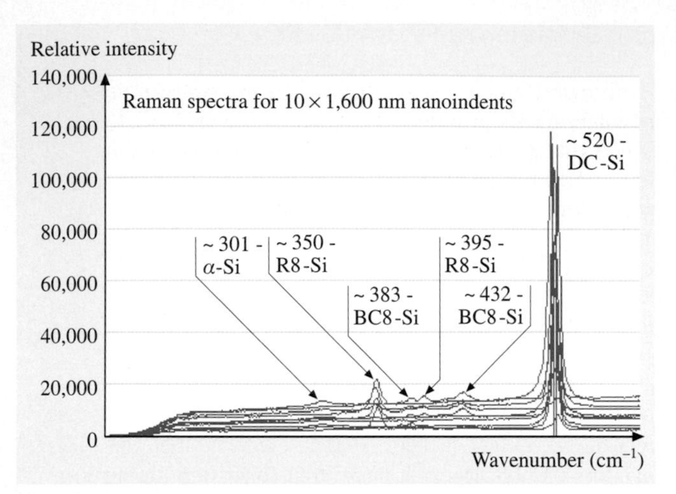

Fig. 22.26 Micro-Raman generally shows the BC-8 and R-8 Si phases that are at the edge of the nanoindents, but the amorphous phase in the center is not easy to detect, as it is often subsurface and the Raman peak is broad

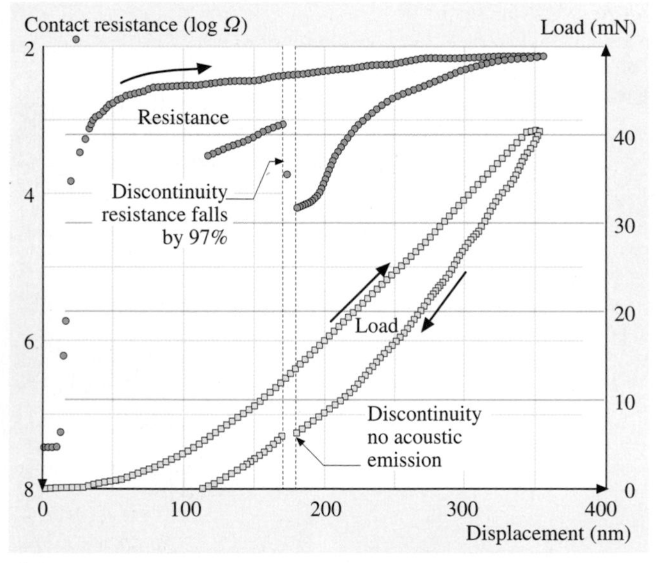

Fig. 22.27 Nanoindentation curves for deep indents on Si show a discontinuity during unloading and simultaneously a large drop in contact resistance

tions in diamond cubic silicon have shown the presence of amorphous-Si and the body-centered cubic BC-8 phase (see Fig. 22.25). Micro-Raman spectroscopy has indicated the presence of a further phase, the rhombhedral R-8 (see Fig. 22.26). For many nanoindentations on silicon there is a characteristic discontinuity in the unloading curve (see Fig. 22.27), which seems to correlate with a phase transformation. The exact sequence in which the phases form is still highly controversial with some [22.42], suggesting that the sequence during loading and unloading is:

Increasing load →
Diamond cubic Si → β-Sn Si

← Decreasing load
BC-8 Si and R-8 Si ← β-Sn Si

Other groups [22.21,22] suggest that the above sequence is only valid for shallow nanoindentations that do not exhibit an unloading discontinuity. For large nanoindentations that show an unloading discontinuity, they suggest the sequence will be:

Increasing load →
Diamond cubic Si → β-Sn Si

← Decreasing load
α-Si ← BC-8 Si and R-8 Si ← β-Sn Si

The disagreement is over the origin of the unloading discontinuity. *Mann* et al. [22.21] suggest it is due to the formation of amorphous silicon, while *Gogotsi* et al. [22.42] believe it is the β-Sn Si to BC-8 or R-8 transformation. Mann et al. argue that the high contact resistance before the discontinuity and the low contact resistance afterwards rule out the discontinuity being the metallic β-Sn Si transforming to the more resistive BC-8 or R-8. The counterargument is that amorphous Si is only seen with micro-Raman spectroscopy when the unloading is very rapid or there is a large nanoindentation with no unloading discontinuity. The importance of unloading rate and cracking in determining the phases present are further complications. The controversy will remain until in situ characterization of the phases present is undertaken.

22.4 Thin Films and Multilayers

In almost all real applications, surfaces are coated with thin films. These may be intentionally added such as hard carbide coatings on a tool bit, or they may simply be native films such as an oxide layer. It is also likely that there will be adsorbed films of water and organic contaminants that can range from a single molecule in thickness up to several nanometers. All of these films, whether native or intentionally placed on the surface, will affect the surface's mechanical behavior on the nanoscale. Adsorbates can have a significant impact on the surface forces [22.3] and, hence, the geometry and stability of asperity contacts. Oxide films can have dramatically different mechanical properties to the bulk and will also modify the surface forces. Some of the effects of native films have been detailed in the earlier sections on dislocation nucleation and adhesive contacts.

The importance of thin films in enhancing the mechanical behavior of surfaces is illustrated by the abundance of publications on thin film mechanical properties (see for instance *Nix* [22.88] or *Cammarata* [22.31] or *Was* and *Foecke* [22.103]). In the following sections, the mechanical properties of films intentionally deposited on the surface will be discussed.

22.4.1 Thin Films

Measuring the mechanical properties of a single thin surface film has always been difficult. Any measurement performed on the whole sample will inevitably be dominated by the bulk substrate. Nanoindentation, since it looks at the mechanical properties of a very small region close to the surface, offers a possible solution to the problem of measuring thin film mechanical properties. However, there are certain inherent problems in using nanoindentation testing to examine the properties of thin films. The problems stem in part from the presence of an interface between the film and substrate. The quality of the interface can be affected by many variables, resulting in a range of effects on the apparent elastic and plastic properties of the film. In particular, when the deformation region around the indent approaches the interface, the indentation curve may exhibit features due to the thin film, the bulk, the interface, or a combination of all three. As a direct consequence of these complications, models for thin-film behavior must attempt to take into account not only the properties of the film and substrate, but also the interface between them.

If, initially, the effect of the interface is neglected, it is possible to divide thin-coated systems into a number of categories that depend on the values of E (elastic modulus) and Y (the yield stress) of the film and substrate. These categories are typically [22.104, 105]:

1. coatings with high E and high Y, substrates with high E and high Y;
2. coatings with high E and high Y, substrates with high or low E and low Y;
3. coatings with high or low E and low Y, substrates with high E and high Y;
4. coatings with high or low E and low Y, substrates with high or low E and low Y.

The reasons for splitting thin film systems into these different categories have been amply demonstrated experimentally by *Whitehead* and *Page* [22.104, 105] and theoretically by *Fabes* et al. [22.106]. Essentially, hard, elastic materials (high E and Y) will possess smaller plastic zones than soft, inelastic (low E and Y) materials. Thus, when different combinations of materials are used as film and substrate, the overall plastic zone will differ significantly. In some cases, the plasticity is confined to the film, and in other cases, it is in both the film and substrate, as shown by Fig. 22.28. If the standard nanoindentation analysis routines are to be used, it is essential that the plastic zone and the elastic strain field are

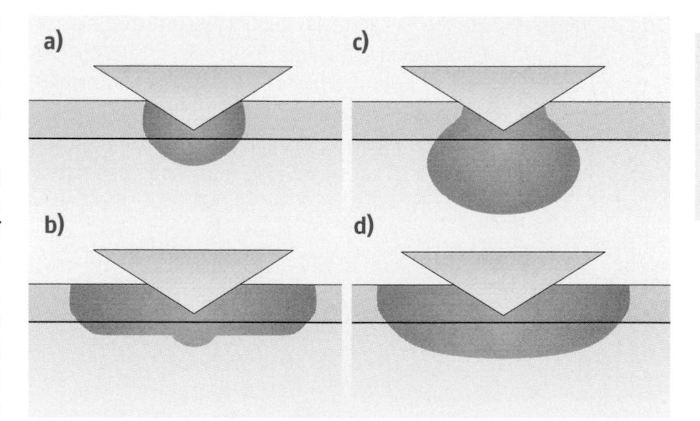

Fig. 22.28a–d Variations in the plastic zone for indents on films and substrates of different properties. **(a)** Film and substrate have high E and Y, **(b)** film has a high E and Y, substrate has a high or low E and low Y, **(c)** film has a high or low E and a low Y, and substrate has a high E and Y, **(d)** film has a high or low E and a low Y, and substrate has a high or low E and a low Y

both confined to the film and do not reach the substrate. Clearly, this is difficult to achieve unless extremely shallow nanoindentations are used. There is an often quoted 10% rule, that says nanoindents in a film must have a depth of less than 10% of the film's thickness if only the film properties are to be measured. This has no real validity [22.107]. There are film/substrate combinations for which 10% is very conservative, while for other combinations even 5% may be too deep. The effect of the substrate for different combinations of film and substrate properties has been studied using FEM [22.108], which has shown that the maximum nanoindentation depth to measure film only properties decreases in moving from soft on hard to hard on soft combinations. For a very soft film on a hard substrate, nanoindentations of 50% of the film thickness are alright, but this drops to $< 10\%$ for a hard film on a soft substrate. For a very strong film on a soft substrate, the surface film behaves like an elastic membrane or a bending plate.

Theoretical analysis of thin-film mechanical behavior is difficult. One theoretical approach that has been adopted uses the volumes of plastically deformed material in the film and substrate to predict the overall hardness of the system. However, it should be noted that this method is only really appropriate for soft coatings and indentation depths below the thickness of the coating (see cases c and d of Fig. 22.28), otherwise the behavior will be closer to that detailed later and shown by Fig. 22.29.

The technique of combining the mechanical properties of the film and substrate to evaluate the overall hardness of the system is generally referred to as the rule of mixtures. It stems from work by *Burnett* et al. [22.109–111] and *Sargent* [22.112], who derived a weighted average to relate the "composite" hardness (H) to the volumes of plastically deformed material in the film (V_f) and substrate (V_s) and their respective values of hardness, H_f and H_s. Thus,

$$H = \frac{H_f V_f + H_s V_s}{V_{total}} , \quad (22.26)$$

where V_{total} is $V_f + V_s$.

Equation (22.26) was further developed by *Burnett* and *Page* [22.109] to take into account the indentation size effect. They replaced H_s with $K\delta_c^{n-2}$, where K and n are experimentally determined constants dependent on the indenter and sample, and δ_c is the contact depth. This expression is derived directly from Meyer's law for spherical indentations, which gives the relationship $P = Kd^n$ between load, P, and the indentation dimension, d. Burnett and Page also employed a further refinement to enable the theory to fit experimental results from a specific sample, ion-implanted silicon. This particular modification essentially took into account the different sizes of the plastic zones in the two materials by multiplying H_s by a dimensionless factor (V_s/V_{total}). While this seems to be a sensible approach, it is mostly empirical, and the physical justification for using this particular factor is not entirely clear. Later, *Burnett* and *Rickerby* [22.110, 111] took this idea further and tried to generalize the equations to take into account all of the possible scenarios. Thus, the following equations were suggested:

$$H = \frac{H_f\left(\Omega^3\right) V_f + H_s V_s}{V_{total}} , \quad (22.27)$$

$$H = \frac{H_f V_f + H_s\left(\Omega^3\right) V_s}{V_{total}} . \quad (22.28)$$

The first of these, (22.27), deals with the case of a soft film on a hard substrate, and the second, (22.28), with a hard film on a soft substrate. The Ω term expresses the variation of the total plastic zone from the ideal hemispherical shape. This was taken still further by *Bull* and *Rickerby* [22.113], who derived an approximation for Ω based on the film and substrate zone radii being related to their respective hardness and elastic modulii [22.114, 115]. Hence:

$$\Omega = (E_f H_s / E_s H_f)^l , \quad (22.29)$$

where l is determined empirically. E_f and H_f and E_s and H_s are the elastic modulus and hardness of the film and substrate, respectively.

Experimental data [22.116] indicate that the effect of the substrate on the elastic modulus of the film can be quite different than the effect on hardness, due to

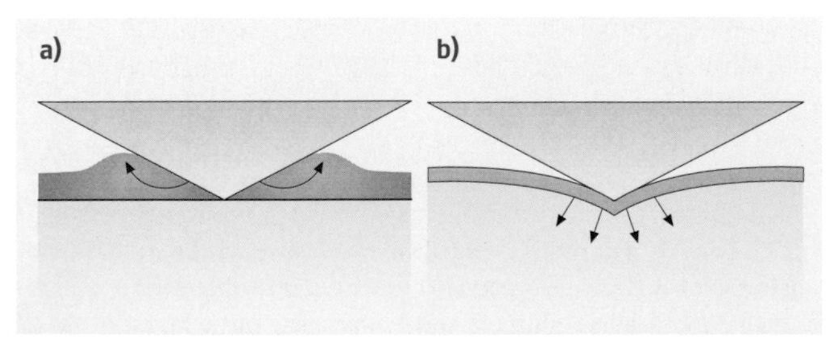

Fig. 22.29a,b Two different modes of deformation during nanoindentation of films. In **(a)** materials move upwards and outwards, while in **(b)** the film acts like a membrane and the substrate deforms

the zones of the elastic and plastic strain fields being different sizes.

Chechechin et al. [22.117] have recently studied the behavior of Al_2O_3 films of various thicknesses on different substrates. Their results indicate that many of the models correctly predict the transition between the properties of the film and those of the substrate, but do not always fit the observed hardness against depth curves. This group have also studied the pop-in behavior of Al_2O_3 films [22.118] and have attempted to model the range of loads and depths at which they occur via a Weibull-type distribution, as utilized in fracture analysis.

A point raised by Burnett and Rickerby should be emphasized. They state that there are two very distinct modes of deformation during nanoindentation testing. The first, referred to as *Tabor*'s [22.48] model for low Y/E materials, involves the buildup of material at the side of the indenter through movement of material at slip lines. The second, for materials with large Y/E does not result in surface pile-up. The displaced material is then accommodated by radial displacements [22.115]. The point is that a thin, strong, and well-bonded surface film can cause a substrate that would normally deform by Tabor's method to behave more like a material with high Y/E (see Fig. 22.29). It should be noted that this only applies as long as the film does not fail.

In recent theoretical and experimental studies the importance of material pile-up and sink-in has been investigated extensively. As discussed in an earlier section, pile-up can be increased by residual compressive stresses, but even in the absence of residual stresses pile-up can introduce a significant error in the calculated contact area. This is most pronounced in materials that do not work-harden [22.62]. For these materials using the Oliver and Pharr method fails to account for the pile-up and results in a large error in the values for E and H. For thin films *Tsui* et al. have used a focused ion beam to section through Knoop indentations in both soft films on hard substrates [22.69] and hard films on soft substrates [22.70]. The soft films, as expected, exhibit pile-up, while the hard film acts more like a membrane and the indentation exhibits sink-in with most of the plasticity in the substrate. Thus, there are three clearly identifiable factors affecting the pile-up and sink-in around nanoindents during testing of thin films:

1. Residual stresses
2. Degree of work-hardening
3. Ratio of film and substrate mechanical properties.

The bonding or adhesion between the film and substrate could also be added to this list. And it should not be forgotten that the depth of the nanoindentation relative to the film thickness also affects pile-up. For a very deep nanoindentation into a thin, soft film on a hard substrate pile-up is reduced, due to the combined constraints on the film of the tip and substrate [22.119]. Due to all of these complications, using nanoindentation to study thin film mechanical properties is fraught with danger. Many unprepared researchers have misguidedly taken the values of E and H obtained during nanoindentation testing to be absolute values only to find out later that the values contain significant errors.

Many of the problems associated with nanoindentation testing are related to incorrectly calculating the contact area, A. The *Joslin* and *Oliver* method [22.71] is one way that A can be removed from the calculations. This approach has been used with some success to look at strained epitaxial II/VI semiconductor films [22.120], but there is evidence that the lattice mismatch in these films can cause dramatic changes in the mechanical properties of the films [22.121]. This may be due to image forces and the film/substrate interface acting as a barrier to dislocation motion. Recently, it has been shown that using films and substrates with known matching elastic moduli, it is possible to use the assumption of constant elastic modulus with depth to evaluate H [22.122]. In effect, this is using (22.1) to evaluate A from the contact stiffness data, and then substituting the value for A into (22.12). The value of E is measured independently, for instance, using acoustic techniques.

22.4.2 Multilayers

Multilayered materials with individual layers that are a micron or less in thickness, sometimes referred to as superlattices, can exhibit substantial enhancements in hardness or strength. This should be distinguished from the super modulus effect discussed earlier, which has been shown to be largely an artifact. The enhancements in hardness can be as much as 100% when compared to the value expected from the rule of mixtures, which is essentially a weighted average of the hardness for the constituents of the two layers [22.123]. Table 22.1 shows how the properties of isostructural multilayers can show a substantial increase in hardness over that for fully interdiffused layers. The table also shows how there can be a substantial enhancement in hardness for non-isostructural multilayers compared to the values for the same materials when they are homogeneous.

Table 22.1 Results for some experimental studies of multilayer hardness

Study	Multilayer	Maximum hardness and multilayer repeat length	Reference hardness value	Range of hardness values for multilayers
Isostructural Knoop hardness [22.124]	Cu/Ni	524 at 11.6 nm	284 (interdiffused)	295–524
Non-isostructural Nanoindentation [22.125]	Mo/NbN	33 GPa at 2 nm	NbN – 17 GPa Mo – 2.7 GPa Wo – 7 GPa	12–33 GPa
	W/NbN	29 GPa at 3 nm	(individual layer materials)	23–29 GPa

There are many factors that contribute to enhanced hardness in multilayers. These can be summarized as [22.103]:

1. Hall-Petch behavior
2. Orowan strengthening
3. Image effects
4. Coherency and thermal stresses
5. Composition modulation

Hall-Petch behavior is related to dislocations piling-up at grain boundaries. (Note that pile-up is used to describe two distinct effects: One is material building up at the side of an indentation, the other is an accumulation of dislocations on a slip-plane.) The dislocation pile-up at grain boundaries impedes the motion of dislocations. For materials with a fine grain structure there are many grain boundaries, and, hence, dislocations find it hard to move. In polycrystalline multilayers, it is often the case that the size of the grains within a layer scales with the layer thickness so that reducing the layer thickness reduces the grain size. Thus, the Hall-Petch relationship (below) should be applicable to polycrystalline multilayer films with the grain size, d_g, replaced by the layer thickness.

$$Y = Y_0 + k_{HP}\, d_g^{-0.5} , \tag{22.30}$$

where Y is the enhanced yield stress, Y_0 is the yield stress for a single crystal, and k_{HP} is a constant.

There is an ongoing argument about whether Hall-Petch behavior really takes place in nanostructured multilayers. The basic model assumes many dislocations are present in the pile-up, but such large dislocation pile-ups are not seen in small grains [22.126] and are unlikely to be present in multilayers. As a direct consequence, studies have found a range of values, between 0 and -1, for the exponent in (22.30), rather than the -0.5 predicted for Hall-Petch behavior.

Orowan strengthening is due to dislocations in layered materials being effectively pinned at the interfaces. As a result, the dislocations are forced to bow out along the layers. In narrow films, dislocations are pinned at both the top and bottom interfaces of a layer and bow out parallel to the plane of the interface [22.127, 128]. Forcing a dislocation to bow out in a layered material requires an increase in the applied shear stress beyond that required to bow out a dislocation in a homogeneous sample. This additional shear stress would be expected to increase as the film thickness is reduced.

Image effects were suggested by *Koehler* [22.129] as a possible source of enhanced yield stress in multilayered materials. If two metals, A and B, are used to make a laminate and one of them, A, has a high dislocation line energy, but the other, B, has a low dislocation line energy, then there will be an increased resistance to dislocation motion due to image forces. However, if the individual layers are thick enough that there may be a dislocation source present within the layer, then dislocations could pile-up at the interface. This will create a local stress concentration point and the enhancement to the strength will be very limited. If the layers are thin enough that there will be no dislocation source present, the enhanced mechanical strength may be substantial. In Koehler's model only nearest neighbor layers were taken to contribute to the image forces. However, this was extended to include more layers [22.130] without substantial changes in the results. The consequence on image effects of reducing the thickness of the individual layers in a multilayer is that it prevents dislocation sources from being active within the layer.

For many multilayer systems there is an increase in strength as the bilayer repeat length is reduced, but there is often a critical repeat length (e.g., 3 nm for the W/NbN multilayer of Table 22.1) below which the strength falls. One explanation for the fall in strength involves the effects of coherency and thermal stresses on dislocation energy. Unlike image effects where the energy of dislocations are a maximum or minimum in the center of layers, the energy maxima and minima are at the interfaces for coherency stresses. Combining the effects of varying moduli and coherency stresses shows that the dependence of strength on layer thickness has a peak near the repeat period where coherency strains begin decreasing [22.131].

Another source of deviations in behavior at very small repeat periods is the imperfect nature of interfaces. With the exception of atomically perfect epitaxial films, interfaces are generally not atomically flat and there is some interdiffusion. For the Cu/Ni film of Table 22.1, the effects of interdiffusion on hardness were examined [22.124] by annealing the multilayers. The results were in agreement with a model by *Krzanowski* [22.132] that predicted the variations in hardness would be proportional to the amplitude of the composition modulation.

It is interesting to note that the explanations for enhanced mechanical properties in multilayered materials are all based on dislocation mechanisms. So it would seem natural to assume that multilayered materials that do not contain dislocations will show no enhanced hardness over their rule of mixtures values. This has been verified by studies on amorphous metal multilayers [22.133], which shows that the hardness of the multilayers, firstly, lies between that of the two individual materials and, secondly, has almost no variation with repeat period.

22.5 Developing Areas

Over the past 20 to 30 years, the driving force for studying nanomechanical behavior of surfaces and thin films has been largely, though not exclusively, the microelectronics industry. The importance of electronics to the modern world is only likely to grow in the foreseeable future, but other technological areas may overtake microelectronics as the driving force for research, including the broad fields of biomaterials and nanotechnology. In several places in this chapter, a number of developing areas have been mentioned. These include nanoscale measurements of viscoelasticity and the study of environmental effects (temperature and surface chemistry) on nanomechanical properties. Both of these topics will be vital in the study of biological systems and, as a result, will be increasingly important from a research point of view.

We still have a relatively rudimentary understanding of the nanomechanics of complex biological systems such as bone cells (osteoblasts and osteoclasts) and skin cells (fibroblasts), or even, for that matter, simpler biological structures such as dental enamel. For example, Fig. 22.30 shows how the mechanical properties of dental enamel can vary within a single tooth [22.134]. But this is still a relatively large-scale measure of mechanical behavior. The prismatic structure of enamel means that there are variations in mechanical properties on a range of scales from millimeters down to nanometers.

In terms of data analysis, there remains much to be done. If an analysis method can be developed that deals with the problems of pile-up and sink-in, the utility of nanoindentation testing will be greatly enhanced.

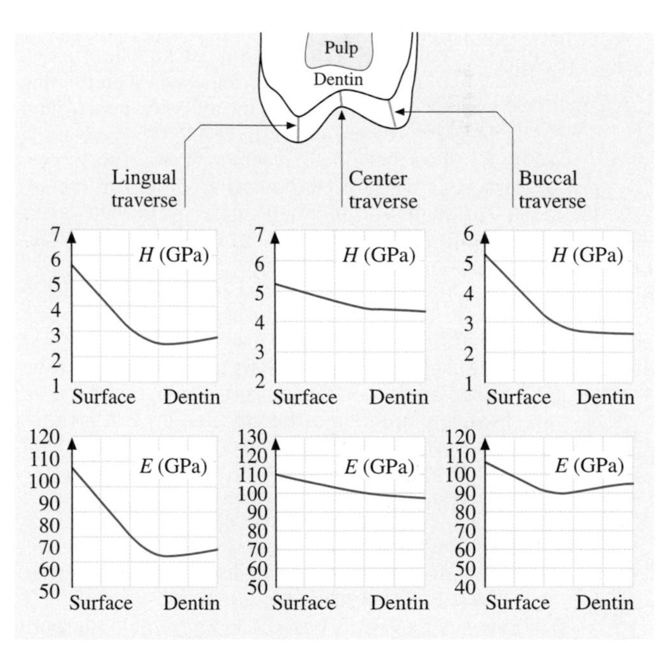

Fig. 22.30 Variations in E and H across human dental enamel. The sample is an upper 2nd molar cut in cross section from the lingual to the buccal side. Nanoindentations are performed across the surface to examine how the mechanical properties vary

References

22.1 G. Binnig, C. F. Quate, C. Gerber: Atomic force microscope, Phys. Rev. Lett. **56** (1986) 930–933

22.2 P. J. Blau, B. R. Lawn (Eds.): *Microindentation Techniques in Materials Science and Engineering* (ASTM, Pennsylvannia 1986)

22.3 J. N. Israelachvili: *Intermolecular and Surface Forces* (Academic, London 1992)

22.4 R. S. Bradley: The cohesive force between solid surfaces and the surface energy of solids, Philos. Mag. **13** (1932) 853–862

22.5 B.V. Derjaguin, V. M. Muller, Yu. P. Toporov: Effect of contact deformations on the adhesion of particles, J. Coll. Interface Sci. **53** (1975) 314–326

22.6 V. M. Muller, V. S. Yuschenko, B. V. Derjaguin: On the influence of molecular forces on the deformation of an elastic sphere and its sticking to a rigid plane, J. Coll. Interface Sci. **77** (1980) 91–101

22.7 V. M. Muller, B. V. Derjaguin, Yu. P. Toporov: On two methods of calculation of the force of sticking of an elastic sphere to a rigid plane, Coll. Surf. **7** (1983) 251–259

22.8 K. L. Johnson, K. Kendal, A. D. Roberts: Surface energy and the contact of elastic solids, Proc. R. Soc. A **324** (1971) 301–320

22.9 A. P. Ternovskii, V. P. Alekhin, M. Kh. Shorshorov, M. M. Khrushchov, V. N. Skvortsov: Zavod Lab. **39** (1973) 1242

22.10 S. I. Bulychev, V. P. Alekhin, M. Kh. Shorshorov, A. P. Ternovskii, G. D. Shnyrev: Determining Young's modulus from the indentor penetration diagram, Zavod Lab. **41** (1975) 1137

22.11 S. I. Bulychev, V. P. Alekhin, M. Kh. Shorshorov, A. P. Ternovskii: Mechanical properties of materials studied from kinetic diagrams of load versus depth of impression during microimpression, Prob. Prochn. **9** (1976) 79

22.12 S. I. Bulychev, V. P. Alekhin: Zavod Lab. **53** (1987) 76

22.13 M. Kh. Shorshorov, S. I. Bulychev, V. P. Alekhin: Sov. Phys. Doklady **26** (1982) 769

22.14 J. B. Pethica: Microhardness tests with penetration depths less than ion implanted layer thickness. In: *Ion Implantation into Metals*, ed. by V. Ashworth, W. Grant, R. Procter (Pergamon, Oxford 1982) p. 147

22.15 D. Newey, M. A. Wilkens, H. M. Pollock: An ultra-low-load penetration hardness tester, J. Phys. E: Sci. Instrum. **15** (1982) 119

22.16 D. Kendall, D. Tabor: An ultrasonic study of the area of contact between stationary and sliding surfaces, Proc. R. Soc. A **323** (1971) 321–340

22.17 G. M. Pharr, W. C. Oliver, F. R. Brotzen: On the generality of the relationship among contact stiffness, contact area and elastic-modulus during indentation, J. Mater. Res. **7** (1992) 613

22.18 T. P. Weihs, C. W. Lawrence, B. Derby, C. B. Scruby, J. B. Pethica: Acoustic emissions during indentation tests, MRS Symp. Proc. **239** (1992) 361–366

22.19 D. F. Bahr, J. W. Hoehn, N. R. Moody, W. W. Gerberich: Adhesion and acoustic emission analysis of failures in nitride films with 14 metal interlayer, Acta Mater. **45** (1997) 5163

22.20 D. F. Bahr, W. W. Gerberich: Relationships between acoustic emission signals and physical phemomena during indentation, J. Mat. Res. **13** (1998) 1065

22.21 A. B. Mann, D. van Heerden, J. B. Pethica, T. P. Weihs: Size-dependent phase transformations during point-loading of silicon, J. Mater. Res. **15** (2000) 1754

22.22 A. B. Mann, D. van Heerden, J. B. Pethica, P. Bowes, T. P. Weihs: Contact resistance and phase transformations during nanoindentation of silicon, Philos. Mag. A **82** (2002) 1921

22.23 S. Jeffery, C. J. Sofield, J. B. Pethica: The influence of mechanical stress on the dielectric breakdown field strength of SiO_2 films, Appl. Phys. Lett. **73** (1998) 172

22.24 B. N. Lucas, W. C. Oliver: Indentation power-law creep of high-purity indium, Metall. Trans. A **30** (1999) 601

22.25 S. A. Syed Asif: Time dependent micro deformation of materials. Ph.D. Thesis (Oxford Univ., Oxford 1997)

22.26 S. A. Syed Asif, R. J. Colton, K. J. Wahl: Nanoscale Surface Mechanical Property Measurements: Force Modulation Techniques Applied to Nanoindentation. In: *Interfacial Properties on the Submicron Scale*, ed. by J. Frommer, R. Overney (ACS Books, Whashington 2000)

22.27 A. B. Mann, J. B. Pethica: Nanoindentation studies in a liquid environment, Langmuir **12** (1996) 4583

22.28 J. W. Beams: Mechanical properties of thin films of gold and silver. In: *Structure and Properties of Thin Films*, ed. by C.A. Neugebauer, J. B. Newkirk, D. A. Vermilyea (Wiley, New York 1959) pp. 183–192

22.29 J. J. Vlassak, W. D. Nix: A new bulge test technique for the determination of Youngs modulus and Poissons ratio of thin-films, J. Mater. Res. **7** (1992) 3242

22.30 W. M. C. Yang, T. Tsakalakos, J. E. Hilliard: Enhanced elastic modulus in composition modulated gold-nickel and copper-palladium foils, J. Appl. Phys. **48** (1977) 876

22.31 R. C. Cammarata: Mechanical properties of nanocomposite thin-films, Thin Solid Films **240** (1994) 82

22.32 G. A. D. Briggs: *Acoustic Microscopy* (Clarendon, Oxford 1992)

22.33 J. Kushibiki, N. Chubachi: Material characterization by line-focus-beam acoustic microscope, IEEE Trans. Sonics Ultrasonics **32** (1985) 189–212

22.34 M. J. Bamber, K. E. Cooke, A. B. Mann, B. Derby: Accurate determination of Young's modulus and Poisson's ratio of thin films by a combination of acoustic microscopy and nanoindentation, Thin Solid Films **398** (2001) 299–305

22.35 S. E. Bobbin, R. C. Cammarata, J. W. Wagner: Determination of the flexural modulus of thin-films from measurement of the 1st arrival of the symmetrical Lamb wave, Appl. Phys. Lett. **59** (1991) 1544–1546

22.36 R. B. King, C. M. Fortunko: Determination of inplane residual-stress state in plates using horizontally polarized shear waves, J. Appl. Phys. **54** (1983) 3027–3035

22.37 R. F. Cook, G. M. Pharr: Direct observation and analysis of indentation cracking in glasses and ceramics, J. Am. Ceram. Soc. **73** (1990) 787–817

22.38 T. F. Page, W. C. Oliver, C. J. McHargue: The deformation-behavior of ceramic crystals subjected to very low load (nano)indentations, J. Mater. Res. **7** (1992) 450–473

22.39 G. M. Pharr, W. C. Oliver, D. S. Harding: New evidence for a pressure-induced phase-transformation during the indentation of silicon, J. Mater. Res. **6** (1991) 1129–1130

22.40 C. F. Robertson, M. C. Fivel: The study of submicron indent-induced plastic deformation, J. Mater. Res. **14** (1999) 2251–2258

22.41 J. E. Bradby, J. S. Williams, J. Wong-Leung, M. V. Swain, P. Munroe: Transmission electron microscopy observation of deformation microstructure under spherical indentation in silicon, Appl. Phys. Lett. **77** (2000) 3749–3751

22.42 Y. G. Gogotsi, V. Domnich, S. N. Dub, A. Kailer, K. G. Nickel: Cyclic nanoindentation and Raman microspectroscopy study of phase transformations in semiconductors, J. Mater. Res. **15** (2000) 871–879

22.43 H. Hertz: Über die Berührung fester elastischer Körper, J. reine angew. Math. **92** (1882) 156–171

22.44 J. Boussinesq: *Application des potentiels à l'étude de l'équilibre et du mouvement des solides élastiques* (Blanchard, Paris 1885) Reprint (1996)

22.45 A. E. H. Love: The stress produced in a semi-infinite solid by pressure on part of the boundary, Philos. Trans. R. Soc. **228** (1929) 377–420

22.46 A. E. H. Love: Boussinesq's problem for a rigid cone, Quarter. J. Math. **10** (1939) 161

22.47 I. N. Sneddon: The relationship between load and penetration in the axisymmetric Boussinesq problem for a punch of arbitrary profile, Int. J. Eng. Sci. **3** (1965) 47–57

22.48 D. Tabor: *Hardness of Metals* (Oxford Univ. Press, Oxford 1951)

22.49 R. von Mises: Mechanik der festen Körper in plastisch deformablen Zustand, Goettinger Nachr. Math.-Phys. **K1** (1913) 582–592

22.50 H. Tresca: Sur l'ecoulement des corps solids soumis s fortes pression, Compt. Rend. **59** (1864) 754

22.51 A. B. Mann, J. B. Pethica: The role of atomic-size asperities in the mechanical deformation of nanocontacts, Appl. Phys. Lett. **69** (1996) 907–909

22.52 A. B. Mann, J. B. Pethica: The effect of tip momentum on the contact stiffness and yielding during nanoindentation testing, Philos. Mag. A **79** (1999) 577–592

22.53 S. P. Jarvis: Atomic force microscopy and tip-surface interactions. Ph.D. Thesis (Oxford Univ., Oxford 1993)

22.54 J. B. Pethica, D. Tabor: Contact of characterised metal surfaces at very low loads: Deformation and adhesion, Surf. Sci. **89** (1979) 182

22.55 J. S. Field, M. V. Swain: Determining the mechanical-properties of small volumes of materials from submicrometer spherical indentations, J. Mater. Res. **10** (1995) 101–112

22.56 W. C. Oliver, G. M. Pharr: An improved technique for determining hardness and elastic-modulus using load and displacement sensing indentation experiments, J. Mater. Res. **7** (1992) 1564–1583

22.57 J. L. Loubet, J. M. Georges, O. Marchesini, G. Meille: Vickers indentation curves of magnesium oxide (MgO), Mech. Eng. **105** (1983) 91–92

22.58 J. L. Loubet, J. M. Georges, O. Marchesini, G. Meille: Vickers indentation curves of magnesium oxide (MgO), J. Tribol. Trans. ASME **106** (1984) 43–48

22.59 M. F. Doerner, W. D. Nix: A method for interpreting the data from depth sensing indentation experiments, J. Mater. Res. **1** (1986) 601–609

22.60 S. V. Hainsworth, H. W. Chandler, T. F. Page: Analysis of nanoindentation load-displacement loading curves, J. Mater. Res. **11** (1996) 1987–1995

22.61 A. Bolshakov, W. C. Oliver, G. M. Pharr: Influences of stress on the measurement of mechanical properties using nanoindentation. 2. Finite element simulations, J. Mater. Res. **11** (1996) 760–768

22.62 A. Bolshakov, G. M. Pharr: Influences of pileup on the measurement of mechanical properties by load and depth sensing instruments, J. Mater. Res. **13** (1998) 1049–1058

22.63 J. C. Hay, A. Bolshakov, G. M. Pharr: A critical examination of the fundamental relations used in the analysis of nanoindentation data, J. Mater. Res. **14** (1999) 2296–2305

22.64 G. M. Pharr, T. Y. Tsui, A. Bolshakov, W. C. Oliver: Effects of residual-stress on the measurement of hardness and elastic-modulus using nanoindentation, MRS Symp. Proc. **338** (1994) 127–134

22.65 T. R. Simes, S. G. Mellor, D. A. Hills: A note on the influence of residual-stress on measured hardness, J. Strain Anal. Eng. Des. **19** (1984) 135–137

22.66 W. R. Lafontaine, B. Yost, C. Y. Li: Effect of residual-stress and adhesion on the hardness of copper-films deposited on silicon, J. Mater. Res. **5** (1990) 776–783

22.67 W. R. Lafontaine, C. A. Paszkiet, M. A. Korhonen, C. Y. Li: Residual stress measurements of thin aluminum metallizations by continuous indentation and X-ray stress measurement techniques, J. Mater. Res. **6** (1991) 2084–2090

22.68 T. Y. Tsui, W. C. Oliver, G. M. Pharr: Influences of stress on the measurement of mechanical properties using nanoindentation. 1. Experimental studies in an aluminum alloy, J. Mater. Res. **11** (1996) 752–759

22.69 T. Y. Tsui, J. Vlassak, W. D. Nix: Indentation plastic displacement field: Part I. The case of soft films on hard substrates, J. Mater. Res. **14** (1999) 2196–2203

22.70 T. Y. Tsui, J. Vlassak, W. D. Nix: Indentation plastic displacement field: Part II. The case of hard films on soft substrates, J. Mater. Res. **14** (1999) 2204–2209

22.71 D. L. Joslin, W. C. Oliver: A new method for analyzing data from continuous depth-sensing microindentation tests, J. Mater. Res. **5** (1990) 123–126

22.72 M. R. McGurk, T. F. Page: Using the P-delta(2) analysis to deconvolute the nanoindentation response of hard-coated systems, J. Mater. Res. **14** (1999) 2283–2295

22.73 Y. T. Cheng, C. M. Cheng: Relationships between hardness, elastic modulus, and the work of indentation, Appl. Phys. Lett. **73** (1998) 614–616

22.74 J. B. Pethica, W. C. Oliver: Mechanical properties of nanometer volumes of material: Use of the elastic response of small area indentations, MRS Symp. Proc. **130** (1989) 13–23

22.75 W. C. Oliver, J. B. Pethica: Method for continuous determination of the elastic stiffness of contact between two bodies, United States Patent Number 4,848,141, (1989)

22.76 S. A. S. Asif, K. J. Wahl, R. J. Colton: Nanoindentation and contact stiffness measurement using force modulation with a capacitive load-displacement transducer, Rev. Sci. Instrum. **70** (1999) 2408–2413

22.77 J. L. Loubet, W. C. Oliver, B. N. Lucas: Measurement of the loss tangent of low-density polyethylene with a nanoindentation technique, J. Mater. Res. **15** (2000) 1195–1198

22.78 W. W. Gerberich, J. C. Nelson, E. T. Lilleodden, P. Anderson, J. T. Wyrobek: Indentation induced dislocation nucleation: The initial yield point, Acta Mater. **44** (1996) 3585–3598

22.79 J. D. Kiely, J. E. Houston: Nanomechanical properties of Au(111), (001), and (110) surfaces, Phys. Rev. B **57** (1998) 12588–12594

22.80 D. F. Bahr, D. E. Wilson, D. A. Crowson: Energy considerations regarding yield points during indentation, J. Mater. Res. **14** (1999) 2269–2275

22.81 D. E. Kramer, K. B. Yoder, W. W. Gerberich: Surface constrained plasticity: Oxide rupture and the yield point process, Philos. Mag. A **81** (2001) 2033–2058

22.82 S. G. Corcoran, R. J. Colton, E. T. Lilleodden, W. W. Gerberich: Anomalous plastic deformation at surfaces: Nanoindentation of gold single crystals, Phys. Rev. B **55** (1997) 16057–16060

22.83 E. B. Tadmor, R. Miller, R. Phillips, M. Ortiz: Nanoindentation and incipient plasticity, J. Mater. Res. **14** (1999) 2233–2250

22.84 J. A. Zimmerman, C. L. Kelchner, P. A. Klein, J. C. Hamilton, S. M. Foiles: Surface step effects on nanoindentation, Phys. Rev. Lett. **87** (2001) article 165507 (1–4)

22.85 J. D. Kiely, R. Q. Hwang, J. E. Houston: Effect of surface steps on the plastic threshold in nanoindentation, Phys. Rev. Lett. **81** (1998) 4424–4427

22.86 A. B. Mann, P. C. Searson, J. B. Pethica, T. P. Weihs: The relationship between near-surface mechanical properties, loading rate and surface chemistry, Mater. Res. Soc. Symp. Proc. **505** (1998) 307–318

22.87 R. C. Thomas, J. E. Houston, T. A. Michalske, R. M. Crooks: The mechanical response of gold substrates passivated by self-assembling monolayer films, Science **259** (1993) 1883–1885

22.88 W. D. Nix: Elastic and plastic properties of thin films on substrates: Nanoindentation techniques, Mater. Sci. Eng. A **234** (1997) 37–44

22.89 J. J. Vlassak, W. D. Nix: Indentation modulus of elastically anisotropic half-spaces, Philos. Mag. A **67** (1993) 1045–1056

22.90 J. J. Vlassak, W. D. Nix: Measuring the elastic properties of anisotropic materials by means of indentation experiments, J. Mech. Phys. Solids **42** (1994) 1223–1245

22.91 B. R. Lawn: *Fracture of Brittle Solids* (Cambridge Univ. Press, Cambridge 1993)

22.92 G. M. Pharr: Measurement of mechanical properties by ultra-low load indentation, Mater. Sci. Eng. A **253** (1998) 151–159

22.93 D. B. Marshall, A. G. Evans: Measurement of adherence of residually stressed thin-films by indentation. 1. Mechanics of interface delamination, J. Appl. Phys. **56** (1984) 2632–2638

22.94 C. Rossington, A. G. Evans, D. B. Marshall, B. T. Khuriyakub: Measurement of adherence of residually stressed thin-films by indentation. 2. Experiments with ZnO/Si, J. Appl. Phys. **56** (1984) 2639–2644

22.95 M. D. Kriese, W. W. Gerberich, N. R. Moody: Quantitative adhesion measures of multilayer films: Part I. Indentation mechanics, J. Mater. Res. **14** (1999) 3007–3018

22.96 M. D. Kriese, W. W. Gerberich, N. R. Moody: Quantitative adhesion measures of multilayer films: Part II. Indentation of W/Cu, W/W, Cr/W, J. Mater. Res. **14** (1999) 3019–3026

22.97 M. Li, C. B. Carter, M. A. Hillmyer, W. W. Gerberich: Adhesion of polymer-inorganic interfaces by nanoindentation, J. Mater. Res. **16** (2001) 3378–3388

22.98 D. R. Clarke, M. C. Kroll, P. D. Kirchner, R. F. Cook, B. J. Hockey: Amorphization and conductivity of

silicon and germanium induced by indentation, Phys. Rev. Lett. **60** (1988) 2156–2159
22.99 J.J. Gilman: Insulator-metal transitions at microindentation, J. Mater. Res. **7** (1992) 535–538
22.100 G.M. Pharr, W.C. Oliver, R.F. Cook, P.D. Kirchner, M.C. Kroll, T.R. Dinger, D.R. Clarke: Electrical-resistance of metallic contacts on silicon and germanium during indentation, J. Mater. Res. **7** (1992) 961–972
22.101 A. Kailer, Y.G. Gogotsi, K.G. Nickel: Phase transformations of silicon caused by contact loading, J. Appl. Phys. **81** (1997) 3057–3063
22.102 J.E. Bradby, J.S. Williams, J. Wong-Leung, M.V. Swain, P. Munroe: Mechanical deformation in silicon by micro-indentation, J. Mater. Res. **16** (2000) 1500–1507
22.103 G.S. Was, T. Foecke: Deformation and fracture in microlaminates, Thin Solid Films **286** (1996) 1–31
22.104 A.J. Whitehead, T.F. Page: Nanoindentation studies of thin-film coated systems, Thin Solid Films **220** (1992) 277–283
22.105 A.J. Whitehead, T.F. Page: Nanoindentation studies of thin-coated systems, NATO ASI Ser. E **233** (1993) 481–488
22.106 B.D. Fabes, W.C. Oliver, R.A. McKee, F.J. Walker: The determination of film hardness from the composite response of film and substrate to nanometer scale indentations, J. Mater. Res. **7** (1992) 3056–3064
22.107 T.F. Page, S.V. Hainsworth: Using nanoindentation techniques for the characterization of coated systems – a critique, Surface Coat. Technol. **61** (1993) 201–208
22.108 X. Chen, J.J. Vlassak: Numerical study on the measurement of thin film mechanical properties by means of nanoindentation, J. Mater. Res. **16** (2001) 2974–2982
22.109 P.J. Burnett, T.F. Page: Surface softening in silicon by ion-implantation, J. Mater. Sci. **19** (1984) 845–860
22.110 P.J. Burnett, D.S. Rickerby: The mechanical-properties of wear resistant coatings. 1. Modeling of hardness behavior, Thin Solid Films **148** (1987) 41–50
22.111 P.J. Burnett, D.S. Rickerby: The mechanical-properties of wear resistant coatings. 2. Experimental studies and interpretation of hardness, Thin Solid Films **148** (1987) 51–65
22.112 P.M. Sargent: A better way to present results from a least-squares fit to experimental-data – an example from microhardness testing, J. Test. Eval. **14** (1986) 122–127
22.113 S.J. Bull, D.S. Rickerby: Evaluation of coatings, Brit. Ceram. Trans. J. **88** (1989) 177–183
22.114 B.R. Lawn, A.G. Evans, D.B. Marshall: Elastic/plastic indentation damage in ceramics: The median/radial crack system, J. Am. Ceram. Soc. **63** (1980) 574–581
22.115 R. Hill: *The Mathematical Theory of Plasticity* (Clarendon, Oxford 1950)
22.116 W.C. Oliver, C.J. McHargue, S.J. Zinkle: Thin-film characterization using a mechanical-properties microprobe, Thin Solid Films **153** (1987) 185–196
22.117 N.G. Chechechin, J. Bottiger, J.P. Krog: Nanoindentation of amorphous aluminum oxide films. 1. Influence of the substrate on the plastic properties, Thin Solid Films **261** (1995) 219–227
22.118 N.G. Chechechin, J. Bottiger, J.P. Krog: Nanoindentation of amorphous aluminum oxide films. 2. Critical parameters for the breakthrough and a membrane effect in thin hard films on soft substrates, Thin Solid Films **261** (1995) 228–235
22.119 D.E. Kramer, A.A. Volinsky, N.R. Moody, W.W. Gerberich: Substrate effects on indentation plastic zone development in thin soft films, J. Mater. Res. **16** (2001) 3150–3157
22.120 A.B. Mann: Nanomechanical measurements: Surface and environmental effects. Ph.D. Thesis (Oxford Univ., Oxford 1995)
22.121 A.B. Mann, J.B. Pethica, W.D. Nix, S. Tomiya: Nanoindentation of epitaxial films: A study of pop-in events, Mater. Res. Soc. Symp. Proc. **356** (1995) 271–276
22.122 R. Saha, W.D. Nix: Effects of the substrate on the determination of thin film mechanical properties by nanoindentation, Acta Mater. **50** (2002) 23–38
22.123 S.A. Barnett: Deposition and mechanical properties of superlattice thin films. In: *Physics of Thin Films*, ed. by M.H. Francombe, J.L. Vossen (Academic, New York 1993)
22.124 R.R. Oberle, R.C. Cammarata: Dependence of hardness on modulation amplitude in electrodeposited Cu-Ni compositionally modulated thin-films, Scripta Metall. **32** (1995) 583–588
22.125 A. Madan, Y.Y. Wang, S.A. Barnett, C. Engstrom, H. Ljungcrantz, L. Hultman, M. Grimsditch: Enhanced mechanical hardness in epitaxial nonisostructural Mo/NbN and W/NbN superlattices, J. Appl. Phys. **84** (1998) 776–785
22.126 R. Venkatraman, J.C. Bravman: Separation of film thickness and grain-boundary strengthening effects in Al thin-films on Si, J. Mater. Res. **7** (1992) 2040–2048
22.127 J.D. Embury, J.P. Hirth: On dislocation storage and the mechanical response of fine-scale microstructures, Acta Mater. **42** (1994) 2051–2056
22.128 D.J. Srolovitz, S.M. Yalisove, J.C. Bilello: Design of multiscalar metallic multilayer composites for high-strength, high toughness, and low CTE

mismatch, Metall. Trans. A **26** (1995) 1805–1813
22.129 J. S. Koehler: Attempt to design a strong solid, Phys. Rev. B **2** (1970) 547–551
22.130 S. V. Kamat, J. P. Hirth, B. Carnahan: Image forces on screw dislocations in multilayer structures, Scripta Metall. **21** (1987) 1587–1592
22.131 M. Shinn, L. Hultman, S. A. Barnett: Growth, structure, and microhardness of epitaxial TiN/NbN superlattices, J. Mater. Res. **7** (1992) 901–911
22.132 J. E. Krzanowski: The effect of composition profile on the strength of metallic multilayer structures, Scripta Metall. **25** (1991) 1465–1470
22.133 J. B. Vella, R. C. Cammarata, T. P. Weihs, C. L. Chien, A. B. Mann, H. Kung: Nanoindentation study of amorphous metal multilayered thin films, MRS Symp. Proc. **594** (2000) 25–29
22.134 J. L. Cuy, A. B. Mann, K. J. Livi, M. F. Teaford, T. P. Weihs: Nanoindentation mapping of the mechanical properties of human molar tooth enamel, Arch. Oral Biol. **47** (2002) 281–291

23. Atomistic Computer Simulations of Nanotribology

Molecular dynamics (MD) and related simulation techniques are powerful tools for improving our understanding of nanotribology. In simulations, materials properties and boundary conditions can be varied at will, and the resulting changes in both macroscopic variables and the dynamics of individual atoms can be observed. This allows one to study systematically the effects of many different factors on friction and wear at the nano-scale. Some examples that are considered in this chapter are (i) the symmetry of contacting surfaces (disordered vs. crystalline and with or without common periods), (ii) surface elasticity, (iii) surface curvature or topology, (iv) interfacial adhesion, and (v) lubricant and/or contaminant molecules present at the interface. Results from simulations and experiments on isolated nano-scale contacts often contradict our experience from macroscopic systems. Kinetic friction coefficients can be orders of magnitude smaller than those observed in macroscopic experiments. Detailed calculations even suggest that there should be no static friction between most pairs of clean, chemically passivated surfaces unless the load is large enough to produce wear. Simulations that test a series of possible mechanisms for static friction are described. Geometrical interlocking can produce static friction in contacts containing only a few atoms, such as an atomic force microscope tip. Larger contacts only exhibit static friction when there is wear, or when the surfaces are separated by a glassy contaminant layer that locks them together. Most surfaces are coated by such films and they are shown to yield friction forces that agree with both nanoscale and macroscopic experiments.

23.1 **Molecular Dynamics** 718
23.1.1 Model Potentials 719
23.1.2 Maintaining Constant Temperature. 720
23.1.3 Imposing Load and Shear 721
23.1.4 The Time-Scale and Length-Scale Gaps 721
23.1.5 A Summary of Possible Traps 722

23.2 **Friction Mechanisms at the Atomic Scale**. 723
23.2.1 Geometric Interlocking 723
23.2.2 Elastic Instabilities 724
23.2.3 Role of Dimensionality and Disorder 727
23.2.4 Elastic Instabilities vs. Wear in Atomistic Models 727
23.2.5 Hydrodynamic Lubrication and Its Confinement-Induced Breakdown . 729
23.2.6 Submonolayer Films 731

23.3 **Stick-Slip Dynamics** 732

23.4 **Conclusions** 734

References 735

One of the main barriers to the design of nanodevices with moving parts is our limited understanding of nanotribology; friction, lubrication and wear at the nanometer scale. New experimental methods such as the atomic force microscope (AFM) [23.1,2], quartz-crystal microbalance (QCM) [23.3], and surface forces apparatus (SFA) [23.4, 5] have revolutionalized our ability to measure friction in systems with atomic scale dimensions. However, the results are often qualitatively different than in macroscopic systems. For example, solids may slide more easily than fluids [23.3, 6] and fluids may behave like solids [23.4, 7]. These results have also revealed that there are gaps in our understanding of the molecular underpinnings of macroscopic tribology. There is a growing recognition that the behavior of nanometer-thick interfacial layers is crucial to tribology in systems of all scales, which has led to a fruitful convergence of researchers from many disciplines.

Understanding friction is an interesting and challenging scientific task, owing to the many factors that come into play: elastic and plastic properties of the two solids in relative sliding motion, surface roughness in multi-asperity contacts and tip geometry in single asperity contacts, geometric correlation between the solids in a microcontact, the amount of lubricant between the two solids, the physical nature of the lubricants, chemical properties and reactions, adhesion, thermal fluctuations and aging, heat conductivity, etc.

Computer simulations have played an increasingly important role in improving our understanding of tribological phenomena. Unlike experiments, they allow us to perfectly control the geometry, the chemical composition, the sliding conditions, etc. and thus make it possible to explore the effect of each variable on friction, lubrication, and wear. Moreover, theorists have no other general approach to analyze processes like friction and wear. There is no known principle like minimization of free energy that determines the steady state of nonequilibrium systems. Even if there were, simulations would be needed to address the complex systems of interest, just as in many equilibrium problems.

Two different types of friction are commonly defined. Static friction, F_s, is the lateral force that must be applied to initiate sliding of one object over another, while kinetic friction, F_k, is the lateral force that must be applied to maintain sliding. At the macroscopic scale both forces usually obey Amontons's laws, which state that friction is proportional to the load pushing the surfaces together and is independent of area. However, there are obvious exceptions, such as tape, where friction occurs at zero load.

Despite significant experimental, theoretical, and computational efforts, no consensus has yet been reached on the relevant atomistic origins of friction between two solid bodies. Strictly speaking, static friction is not a force but a threshold. Its presence implies that the objects have locked together into a local energy minimum that must be overcome by the external force. This threshold force is observed between all the objects around us, yet its molecular origins have remained baffling. How does almost every pair of surfaces manage to lock together, and why does the friction usually obey Amontons's laws? Why does F_k not vanish linearly at small sliding velocities as it would if sliding occurred adiabatically? Different mechanisms have been suggested, such as geometric interlocking, elastic instabilities, interlocking mediated by so-called third bodies, plastic deformation, and crack propagation, to name a few. All these mechanisms may well be applicable, although each one under different circumstances.

In the following, we present a selected overview of the major results from the growing simulation literature. The emphasis is on providing a coherent picture of the key issues in the field, rather than a historical or a complete review. A much more detailed review of tribological computer simulations, including the sliding of adsorbed layers on surfaces, stick-slip motion, tribochemistry, cutting, high-velocity friction between metals, etc. has been given recently by *Robbins* et al. [23.8], and a review of the theoretical foundations of nanotribological issues is given in *Müser* et al. [23.9]. Reviews emphasizing the importance of surface chemistry and the role of self-assembled monolayers have been provided by *Harrison* et al. [23.10, 11].

23.1 Molecular Dynamics

The simulations described in this chapter all use an approach called classical molecular dynamics (MD), which is described extensively in a number of review articles and books [23.12, 13]. The basic outline of the method is straightforward. One begins by defining the interaction potentials. These produce forces on the individual particles whose dynamics will be followed, typically atoms or molecules. Next, the geometry and boundary conditions are specified, and initial coordinates and velocities are given to each particle. Then, the equations of motion for the particles are integrated numerically, stepping forward in time by discrete steps of size Δt. Quantities such as forces, velocities, work, heat flow, and correlation functions are calculated as a function of time to determine their steady-state values and dynamic fluctuations. The relation between changes in these quantities and the motion of individual molecules is also explored.

When designing a simulation, care must be taken to choose interaction potentials and ensembles that capture the essential physics that is to be addressed. The potentials may be as simple as ideal spring constants for studies of general phenomena, or as complex as electronic density-functional calculations in quantitative simulations. The ensemble can also be tailored to the problem of interest. Solving Newton's equations yields a constant energy and volume, or microcanonical, ensemble. By adding terms in the equation of motion that simulate heat baths or pistons, simulations can be done at constant temperature, pressure, lateral force, or velocity. Constant chemical potential can also be maintained by adding or removing particles using Monte Carlo methods or explicit particle baths.

The amount of computational effort typically grows linearly with both the number of particles, N, and the number of time-steps, M. The prefactor increases rapidly with the complexity of the interactions, and substantial ingenuity is required to achieve linear scaling with N for long-range interactions or density-functional approaches. Complex interactions also lead to a wide range of characteristic frequencies such as fast bond-stretching and slow bond-bending modes. Unfortunately, the time step Δt must be small compared to the period of the slowest mode ($\lessapprox 5\%$). This means that many time steps are needed before one obtains information about the slow modes.

The maximum feasible simulation size has increased continuously with advances in computer technology, but remains relatively limited. The product of N times M in the largest simulations described below is about 10^{12}. A cubic region of 10^6 atoms would have a side of about 50 nm. Such linear dimensions allow reasonable models of an atomic force microscope tip, the boundary lubricant in a surface force apparatus, or an individual asperity contact on a rough surface. However, 10^6 time steps is only about 10 ns, which is much shorter than experimental measurement times. This requires intelligent choices in the problems that are attacked and how results are extrapolated to experiments. It also limits sliding velocities to relatively high values, typically meters per second or above.

A number of efforts are underway to increase the accessible time scale, but the problem remains unsolved. Current algorithms attempt to follow the deterministic equations of motion, usually with the Verlet or predictor-corrector algorithms [23.12]. One alternative approach is to make stochastic steps. This would be a nonequilibrium generalization of the Monte Carlo approach that is commonly used in equilibrium systems. The difficulty is that there is no general principle for determining the appropriate probability distribution of steps in a nonequilibrium system.

In the following subsection, we describe some of the potentials that are commonly used and the situations where they are appropriate. The next two subsections describe methods for maintaining constant temperature and constant load. The section concludes with discussions of the challenges associated with the wide range of relevant time and length scales, and common pitfalls in simulations.

23.1.1 Model Potentials

A wide range of potentials has been employed in studies of tribology. Many of the studies described in the next section use simple ideal springs and sine-wave potentials. The Lennard-Jones potential gives a more realistic representation of typical interatomic interactions and is also commonly used in studies of general behavior. In order to model specific materials, more detail must be built into the potential. Simulations of metals frequently use the embedded atom method, while studies of hydrocarbons use potentials that include bond-stretching, bending, torsional forces, and even chemical reactivity. In this section, we give a brief definition of the most commonly used models. The reader may consult the original literature for more detail.

The Lennard-Jones (LJ) potential is a two-body potential that is commonly used for interactions between atoms or molecules with closed electron shells. It is applied not only to the interaction between noble gases, but also to the interaction between different segments on polymers. In the latter case, one LJ particle may reflect a single atom on the chain (explicit atom model), a CH_2 segment (united atom model), or even a Kuhn's segment consisting of several CH_2 units (coarse-grained model). United atom models [23.14] have been shown by *Paul* et al. [23.15] to successfully reproduce explicit atom results for polymer melts. Significant progress has been made recently to map back from the coarse-grained models to more detailed models [23.16, 17].

The 12-6 LJ potential has the form

$$U(r_{ij}) = 4\epsilon\left[\left(\frac{\sigma}{r_{ij}}\right)^{12} - \left(\frac{\sigma}{r_{ij}}\right)^{6}\right], \tag{23.1}$$

where r_{ij} is the distance between particles i and j, ϵ is the LJ interaction energy, and σ is the LJ diameter. The exponents 12 and 6 used above are very common, but other values may be chosen depending on the system of interest. Many of the simulation results discussed in subsequent sections are expressed in units derived from ϵ, σ, and a characteristic mass of the particles. For example, the standard LJ time unit is defined as $t_{LJ} = \sqrt{m\sigma^2/\epsilon}$ and would typically correspond to a few picoseconds. A convenient time step is $\Delta t = 0.005 t_{LJ}$ for a LJ liquid or solid at external pressures and temperatures that are not too large.

Most realistic atomic potentials cannot be expressed as two-body potentials. For example, bond angle potentials in a polymer are effectively three-body interactions and torsional potentials correspond to four-body interactions. Classical models of these interactions [23.18, 19] assume that a polymer is chemically inert and interactions between different molecules are modeled by two-body potentials. In the most sophisticated models, bond-stretching, bending, and torsional energies depend on the position of neighboring molecules and bonds are allowed to rearrange [23.20, 21]. Such generalized model potentials are needed to model friction-induced chemical interactions.

For the interaction between metals, a different approach has proven fruitful. The embedded atom method (EAM) developed by *Daw* et al. [23.22] includes a contribution in the potential energy associated with the cost of "embedding" an atom in the local electron density ρ_i produced by surrounding atoms. The total potential energy U is approximated by

$$U = \sum_i \tilde{F}_i(\rho_i) + \sum_i \sum_{j<i} \phi_{ij}(r_{ij}) \,, \tag{23.2}$$

where $\tilde{F}_i$ is the embedding energy, whose functional form depends on the particular metal. The pair potential $\phi_{ij}(r_{ij})$ is a doubly screened short-range potential reflecting core-core repulsion. The computational cost of the EAM is not substantially greater than pair potential calculations, because the density ρ_i is approximated by a sum of independent atomic densities. When compared to simple two-body potentials such as Lennard–Jones or Morse potentials, the EAM has been particularly successful in reproducing experimental vacancy formation energies and surface energies, even though the potential parameters were only adjusted to reproduce bulk properties. This feature makes the EAM an important tool in tribological applications, where surfaces and defects play a major role.

23.1.2 Maintaining Constant Temperature

An important issue for tribological simulations is temperature regulation. The work done when two walls slide past each other is ultimately converted into random thermal motion. The temperature of the system would increase indefinitely if there was no way for this heat to flow out of the system. In an experiment, heat flows away from the sliding interface into the surrounding solid. In simulations, the effect of the surrounding solid must be mimicked by coupling the particles to a heat bath.

Techniques for maintaining constant temperature T in equilibrium systems are well developed. Equipartition guarantees that the average kinetic energy of each particle along each Cartesian coordinate is $k_B T/2$, where k_B is Boltzmann's constant. To thermostat the system, the equations of motion are modified, so that the average kinetic energy stays at this equilibrium value. This approach assumes that T is not too small compared to the Debye temperature, so that quantum statistics are not important. The applicability of classical MD decreases at lower T.

One class of approaches removes or adds kinetic energy to the system by multiplying the velocities of all particles by the same global factor. In the simplest version, velocity rescaling, the factor is chosen to keep the kinetic energy exactly constant at each time step. However, in a true constant temperature ensemble there are fluctuations in the kinetic energy. Improvements such as the Berendsen and Nosé–Hoover methods [23.23] add equations of motion that gradually scale the velocities to maintain the correct average kinetic energy over a longer time scale.

Another approach is to couple each atom to its own local thermostat [23.24, 25]. The exchange of energy with the outside world is modeled by a Langevin equation that includes a damping coefficient γ and a random force $f_i(t)$ on each atom i. The equations of motion for the α component of the position $x_{i\alpha}$ become:

$$m_i \frac{\mathrm{d}^2 x_{i\alpha}}{\mathrm{d}t^2} = -\frac{\partial}{\partial x_{i\alpha}} U - m_i \gamma \frac{\mathrm{d}x_{i\alpha}}{\mathrm{d}t} + f_{i\alpha}(t) \,, \tag{23.3}$$

where U is the total potential energy and m_i is the mass of the atom. To produce the appropriate temperature, the forces must be completely random, have zero mean, and have a second moment given by

$$\langle \delta f_{i\alpha}(t)^2 \rangle = 2 k_B T m_i \gamma / \Delta t \,. \tag{23.4}$$

The damping coefficient γ must be large enough that energy can be removed from the atoms without substantial temperature increases. However, it should be small enough that the trajectories of individual particles are not perturbed too strongly.

The first issue in nonequilibrium simulations is what temperature means. Near equilibrium hydrodynamic theories define a local temperature in terms of the equilibrium equipartition formula and the kinetic energy relative to the local rest frame [23.26]. In d dimensions, the definition is

$$k_B T = \frac{1}{dN} \sum_i m_i \left[\frac{\mathrm{d}\boldsymbol{x}_i}{\mathrm{d}t} - \langle \boldsymbol{v}(\boldsymbol{x}) \rangle \right]^2 \,, \tag{23.5}$$

where the sum is over all N particles and $\langle \boldsymbol{v}(\boldsymbol{x}) \rangle$ is the mean velocity in a region around $\boldsymbol{x}$. As long as the change in mean velocity with position is sufficiently slow, $\langle \boldsymbol{v}(\boldsymbol{x}) \rangle$ is well-defined and this definition of temperature is on solid theoretical ground.

When the mean velocity difference between neighboring molecules becomes comparable to the random thermal velocities, temperature is not well-defined. An important consequence is that different strategies for defining and controlling temperature give very different structural order and shear forces [23.27–29]. In addition, the distribution of velocities may become non-Gaussian, and different directions, α, may have different effective temperatures. Care should be taken in drawing conclusions from simulations in this extreme regime. Fortunately, the above condition typically implies that the velocities of neighboring atoms differ by something approaching 10% of the speed of sound. This is generally higher than any experimental velocity and would certainly lead to heat buildup and melting at the interface.

In order to mimic experiments, the thermostat is often applied only to those atoms that are at the outer boundary of the simulation cell. This models the flow of heat into surrounding material that is not included explicitly in the simulation. The resulting temperature profile is peaked at the shearing interface (e.g., *Bowden* et al. [23.30], *Khare* et al. [23.31]). In some cases the temperature rise may lead to undesirable changes in the structure and dynamics, even at the lowest velocity, that can be treated in the available simulation time. In this case, a weak thermostat applied throughout the system may maintain the correct temperature and yield the dynamics that would be observed in longer simulations at lower velocities. The safest course is to couple the thermostat only to those velocity components that are perpendicular to the mean flow.

There may be a marginal advantage to local Langevin methods in nonequilibrium simulations, because they remove heat only from atoms that are too hot. Global methods like Nosé–Hoover remove heat everywhere. This can leave high temperatures in the region where heat is generated, while other regions are at an artificially low temperature.

23.1.3 Imposing Load and Shear

The magnitude of the friction that is measured in an experiment or simulation may be strongly influenced by the way in which the normal load and tangential motion are imposed [23.32]. Experiments almost always impose a constant normal load. The mechanical system applying shear can usually be described as a spring attached to a stage moving at controlled velocity. The effective spring constant includes the compliance of all elements of the loading device, including the material on either side of the interface. Very compliant springs apply a nearly constant force, while very stiff springs apply a nearly constant velocity.

In simulations, it is easiest to treat the boundary of the system as rigid and to translate atoms in this region at a constant height and tangential velocity. However, this does not allow atoms to move around or up and over atoms on the opposing surface. Even the atomic-scale roughness of a crystalline surface can lead to order of magnitude variations in normal and tangential force with lateral position when sliding is imposed in this way [23.33]. The difference between constant separation and pressure simulations of thin films can be arbitrarily large, since they produce different power law relations between viscosity and sliding velocity [23.34].

One way of minimizing the difference between constant separation and pressure ensembles is to include a large portion of the elastic solids that bound the interface. This extra compliance allows the surfaces to slide at a more uniform normal and lateral force. However, the extra atoms greatly increase the computational effort.

To simulate the usual experimental ensemble more directly, one can add equations of motion that describe the position of the boundary [23.34–36], much as equations are added to maintain constant temperature. The boundary is given an effective mass that moves in response to the net force from interactions with mobile atoms and from the external pressure and lateral spring. The mass of the wall should be large enough that its dynamics are slower than those of individual atoms, but not too much slower or the response time will be long compared to the simulation time. The spring constant should also be chosen to produce an appropriate response time and ensemble.

23.1.4 The Time-Scale and Length-Scale Gaps

One important motivation of tribological computer simulations is to find the relevant atomistic processes that occur in a single, well-defined sliding contact such as in the SFA or in the AFM. While it has become computationally feasible to study atomistically the core of a nanometer-scale AFM contact, the sliding speeds, v_{sl}, employed in the simulations are still about ten decades larger than in experiments. The use of clever new algorithms such as kinetic Monte Carlo might prove fruitful to overcome this time-scale gap in a brute-force way. However, such an approach is not necessarily required in order to make a meaningful comparison between experiment and simulation. Instead, one can try to find out what the relevant time scales are and what parameters define them, e.g., by studying a phenomenological description of an embedded system under shear. In such an analysis, *Porto* et al. [23.37] found that the motion of the top plate/tip is not sensitive to the characteristic atomic frequencies ω_{mic} – as long as there is a reasonable time-scale separation between ω_{mic} and the characteristic frequency of the tip Ω_{tip}. The experimentally relevant time scale Ω_{tip} is given by $\sqrt{K/M}$, where K is a restoring spring constant (defined by the tip's elasticity and/or the tip-substrate interactions) and M is the tip's inertia, eventually including the cantilever's mass. The experimental value of M will be about 20 decades larger than in the simulation, so values of the relevant parameter $v_{\text{sl}}\Omega_{\text{tip}}$ end up being similar in simulation and experiment. However, there is controvery over whether the tip can be treated as a mass point, or whether long-range elastic deformations in the tip are important [23.35], [23.38, Chap. 9.9].

Another way to overcome the gap in time scales is to alter interaction strengths, temperature, or other variables in order to accelerate the processes of interest. One such example is the cold welding study of *Müser* [23.40], where two dissimilar solids were placed on top of each other in vacuum. It would have been by no means desirable to base the calculations on interfacial interactions that correspond to typical experimental values, because the resulting time scales would be too long. By increasing the magnitude of the mixing energy, one drives the mixing process toward the same ground state on simulation time scales. This simple trick makes it possible to indirectly bridge several orders of magnitude in time scale between experiment and simulation.

Some open issues appear to require the explicit treatment of many different length scales. For example, the distribution of contacts and total contact area depend on both atomic-scale interactions at the surface and long-range elastic deformation of the bounding solid. Of particular interest is whether different contacting asperities can move independently, or whether those within an elastic coherence length l_{e} move in lock-step. Estimates of l_{e} range from the size of a single asperity [23.41] to many times the linear dimension of most objects of interest [23.42]. The correct value is of great importance to the question of whether clean, corrugated (macroscopic) surfaces have the potential to show elastic pinning and, hence, wearless friction. In order to address such issues, it will be necessary to do simulations that incorporate both atomistic and finite element methods (e.g., *Abraham* et al. [23.43]).

23.1.5 A Summary of Possible Traps

There are many possible traps that may render a well-done simulation useless or applicable only to highly idealized situations. In some aspects more care has to be taken than in equilibrium simulations, in other aspects one gets away with less care. We hope that this section helps to create some sensitivity as to what "ingredients" are relevant. For example, if we calculate the static friction force between two solids separated by a boundary lubricant in a MD simulation, the results will be fairly insensitive to the thermostat and to whether or not the thermostat is applied to the boundary lubricant. One should also avoid moves in which a "stuck" atom magically disappears from the contact through a grand canonical-type Monte Carlo move. The results depend more strongly on the thermostat if the lubricant is thick and the heating effects are of interest. In such a case, only the outermost atoms in the confining solids should be thermostated.

The typical situation in experiment is constant load, *not* constant separation. In some cases, there is something like a correspondence between these two "ensembles", but often there are *qualitative* differences between them, in particular, when strongly irreversible processes take place such as the production of wear particles [23.40]. Contacting experimental surfaces are also likely to have different geometry and chemistry – or at least different crystallographic orientations. However, most simulations consider the artificial case of contact

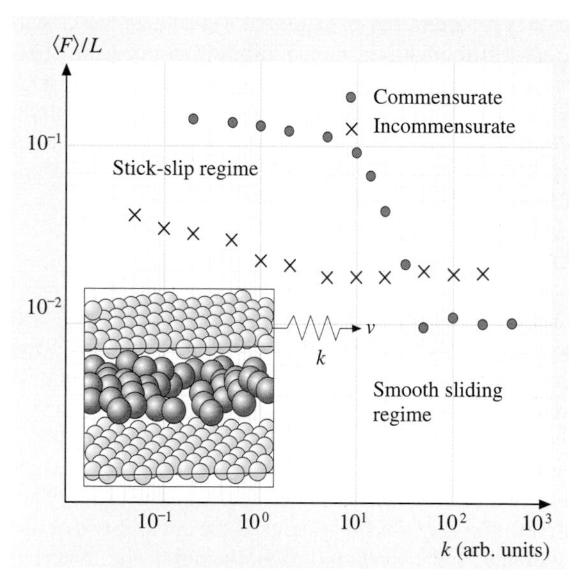

Fig. 23.1 Kinetic friction force F_{k} divided by load L as a function of spring constant k for commensurate and incommensurate walls lubricated by a quarter monolayer of LJ atoms. A schematic of the simulation is shown as well. The data shown was produced at a normal pressure of 0.4 GPa and a velocity of about 1 m/s. (After [23.39])

between identical, aligned crystals. This is the simplest example of a commensurate system, where the two surfaces share a common spatial period. As discussed below, the tribological behavior of commensurate systems is dramatically different than the more realistic case of incommensurate surfaces – even in the presence of a few layers of lubricant. One example of this is shown in Fig. 23.1 [23.39].

Another issue is the curvature of the surfaces. Real contacts typically take place between two curved surfaces or a curved and a flat surface. The pressure profiles differ from those in "flat" contacts, and that discrepancy can be relevant in certain circumstances. If we want to incorporate curvature effects, care should be taken that the resulting pressure profiles resemble a realistic profile, as was done by *Wenning* et al. [23.44].

23.2 Friction Mechanisms at the Atomic Scale

As noted in the introduction, static friction implies that the contacting surfaces have locked into a local free energy minimum. There have been many suggestions for how this can happen between nearly every pair of surfaces, whether macroscopic or nanoscale. The purpose of this section is to review some of these mechanisms and discuss their relevance to nanotribological experiments.

One obvious way that surfaces can interlock is through plastic deformation that brings the surfaces into conformity. A classic example is called plowing friction [23.30]. In this case a sharp feature on one surface penetrates into the opposing surface, causing permanent plastic deformation. The static friction represents the force needed to plow a groove through the surface. This type of motion has been observed in simulations [23.8, 9, 45]. and experiments (Sect. 20.6.1) using AFM probes with hard tips Because it leads to rapid wear, it is only desirable when one wants to remove material by abrasion or cutting.

Many nanotribological studies show friction without measurable wear, and this is the desirable mode of operation in most applications. For this reason, the following subsections focus on mechanisms that lock surfaces together without plastic deformation. We begin by discussing contact between rigid surfaces (Sect. 23.2.1), and then explore the ability of elastic deformations at the atomic scale (Sect. 23.2.2) or between contacts (Sect. 23.2.3) to lock surfaces together. We then describe one of the simulations that shows how unlikely these mechanisms are to yield static friction without wear unless the contact contains only a few tens of atoms (Sect. 23.2.4). The final two subsections consider the role of contaminants or "third bodies" that are present in the contacts between nearly all surfaces. First, changes in the behavior of lubricants with film thickness are discussed. Then molecularly thin films are shown to produce static and kinetic friction that is consistent with both nanoscale experiments and macroscopic observations.

23.2.1 Geometric Interlocking

Early theories of friction were based on the purely geometric argument that friction is caused by the interlocking of impenetrable and rigid surface asperities [23.30, 46]. The idea (Fig. 23.2) is that the top solid must be lifted up a typical slope, $\tan\alpha$, determined by roughness on the bottom surface. If there is no microscopic friction between the surfaces, then the minimum force to initiate sliding is

$$F_s = \mu_s L \qquad (23.6)$$

with $\mu_s = \tan\alpha$. This result satisfies Amontons's laws with a constant coefficient of friction $\mu_s \equiv F_s/L = \tan\alpha$. In 1737, Bélidor obtained a typical experimental value of $\mu_s \approx 0.35$ by modeling rough surfaces as spherical asperities arranged to form commensurate crystalline walls [23.46]. However, asperities on real surfaces do not match as well as envisioned in these models or as sketched in Fig. 23.2. In most cases the average force vanishes because for every asperity or atom going up a ramp, there is another going down.

The sum of forces between crystalline surfaces is considered explicitly in the next subsection, and shown

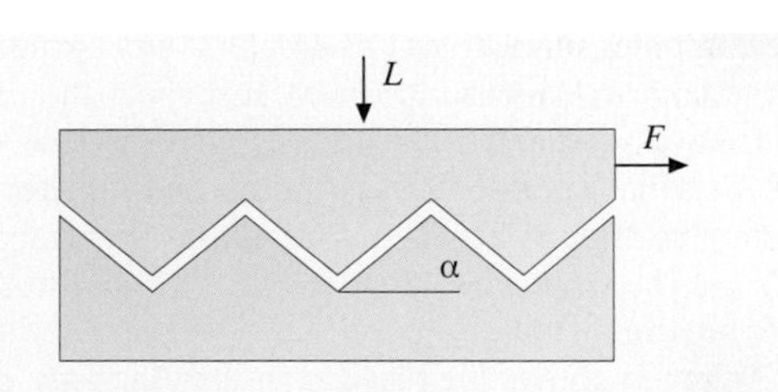

Fig. 23.2 Sketch of two surfaces with interlocking asperities. The top surface experiences a normal load L and a lateral force F, which attempts to pull the top surface up the slope $\tan\alpha$. The bottom wall is fixed

to vanish except in the unusual case where the surfaces have the same periodicity. In the case were one or both surfaces are random, simple analytic arguments show that μ_s will fall as $\mathcal{L}^{-d_{int}/2}$ when the linear size $\mathcal{L}$ of the contact goes to infinity [23.47], provided that the surface roughnesses are small and the solids are completely rigid. Here, d_{int} is the dimension of the contact and the pressure normal to the interface is assumed to be kept constant. This result is in contradiction with one of Amontons's laws, namely, the independence of F_s on the area of contact A. Nevertheless, the linear relation between F_s and L at fixed A is maintained if one assumes that the surfaces are not completely impenetrable but repel via an energy penalty that increases exponentially fast with increasing overlap of the two solids [23.47]. The resulting theoretical prediction that $F_s \propto L/\sqrt{A}$ between rigid, flat, disordered walls also holds reasonably well for more general interactions between the solids, as tested explicitly by atomistic computer simulations.

Most real surfaces are rough and disordered. The minimal model that includes the effect of roughness consists of a single asperity contact, for instance the curved tip of an AFM [23.2] on a flat substrate. The tip is usually assumed to be disordered. Moreover, the contact mechanics of a nanoscale tip and a substrate in vacuum can best be described by the so-called Derjaguin–Müller–Toporov (DMT) limit [23.48]. DMT is similar to Hertzian contact mechanics, but it includes an additional adhesive off-set load L_{adh} that is incorporated into the effective load $L_{eff} = L + L_{adh}$. The area of contact is then given by $A \propto L_{eff}^{2/3}$, Substituting this proportionality into the relation $F_s \propto L_{eff}/\sqrt{A}$ discussed above, *Wenning* et al. [23.44] found

$$F_s \propto L^{2/3} . \tag{23.7}$$

This relationship is indeed observed experimentally in ultrahigh vacuum (UHV) [23.49,50], as well as in atomistic computer simulations [23.44]. Of course, when the contacts are commensurate, we can expect the effective friction coefficient $\mu_{eff} = F_s/L_{eff}$ to be independent of L_{eff}, while for amorphous systems and in particular incommensurate systems, μ_{eff} decreases with increasing L_{eff}. The scaling with load in Fig. 23.3 confirms this expectation.

As noted above, the discussion thus far only deals with atomic/geometric interlocking. From Fig. 23.2 and the discussion above, one can understand how this interlocking leads to static friction, which is the threshold force to initiate sliding. However, kinetic friction $F_k(v)$ may nevertheless vanish in the limit of zero sliding velocity v, since in principle the energy that is gained when the system slides "downhill" can be stored and used later to make the slider go "uphill" again. Such a process would require that we can drive the system adiabatically. In the following, we will discuss atomistic processes (instabilities) that make it impossible to drive the system adiabatically. The presence of instabilities automatically leads to finite kinetic and static friction even in the absence of geometric interlocking.

23.2.2 Elastic Instabilities

In principle, elastic deflections of atoms can be large enough to prevent the interfacial forces from averaging to zero. Two simple ball and spring models are useful in illustrating how such motion can affect static and kinetic friction in principle as well as in understanding the results of detailed simulations or laboratory experiments. Both models consider two clean, flat, crystalline surfaces in direct contact (Fig. 23.4a). The bottom solid is assumed to be rigid, so that it can be treated as a fixed periodic substrate potential acting on the

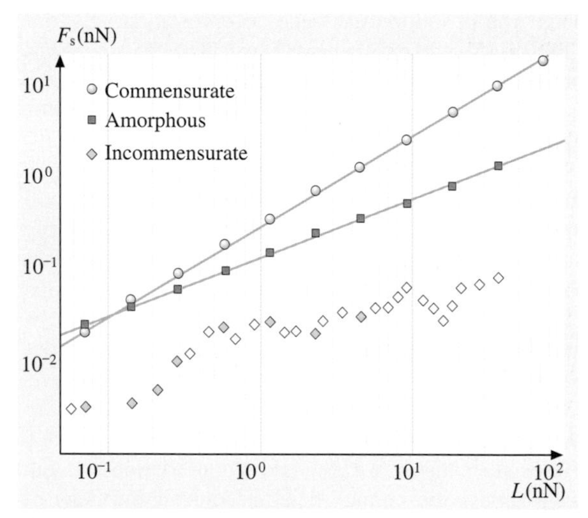

Fig. 23.3 Static friction force F_s vs. normal load L for a commensurate tip, an incommensurate tip, and an amorphous tip. In all three cases, the radius of curvature was $R_c = 70$ Å and contacts were nonadhesive. Straight lines are fits according to $F_s \propto L^{\beta}$ with the results $\beta = 0.97 \pm 0.005$ (commensurate) and $\beta = 0.63 \pm 0.01$ (amorphous). [After [23.44] plus new data from *L. Wenning (open diamonds)*]

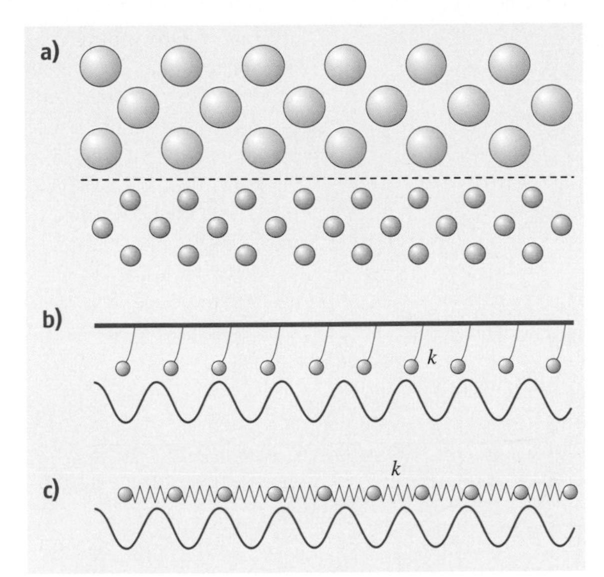

Fig. 23.4 (a) Two ideal, flat crystals making contact at the plane indicated by the *dashed line*. The nearest-neighbor spacings in the bottom and top walls are a and b, respectively. The Prandtl–Tomlinson model (b) and Frenkel–Kontorova model (c) replace the bottom surface by a periodic potential. The former model keeps elastic forces between atoms on the top surface and the center of mass of the top wall, and the latter includes springs of stiffness k between neighbors in the top wall. (After [23.56])

top solid. In order to make the problem analytically tractable, only the bottom layer of atoms from the top solid is retained, and the interactions within the top wall are simplified. In the Prandtl–Tomlinson model (Fig. 23.4b), the atoms are coupled to the center of mass of the top wall by springs of stiffness k, and coupling between neighboring atoms is ignored [23.51–53]. In the Frenkel–Kontorova model (Fig. 23.4c), the atoms are coupled to nearest-neighbors by springs, and the coupling to the atoms above is ignored [23.54, 55]. Due to their simplicity, these models arise in a number of different problems and a great deal is known about their properties.

Many features of the Prandtl–Tomlinson and Frenkel–Kontorova models can be understood from their 1-D versions. One important parameter is the ratio η between the lattice constants of the two surfaces $\eta \equiv b/a$. The other is the strength of the periodic potential from the substrate relative to the spring stiffness k that represents interactions within the top solid. If the substrate potential has a single Fourier component, then the periodic force can be written as

$$f(x) = -f_1 \sin\left(\frac{2\pi}{a}x\right) . \tag{23.8}$$

The relative strength of the potential and springs can be characterized by the dimensionless constant $\lambda \equiv 2\pi f_1/ka$.

In the limit of infinitely strong springs ($\lambda \to 0$), both models represent rigid solids. The atoms of the top wall are confined to lattice sites $x_l^0 = x_{\mathrm{CM}} + lb$, where the integer l labels successive atoms and x_{CM} represents a rigid translation of the entire wall. The total lateral or friction force is given by summing (23.8):

$$F = -f_1 \sum_{l=1}^{N} \sin\left[\frac{2\pi}{a}(lb + x_{\mathrm{CM}})\right] , \tag{23.9}$$

where N is the number of atoms in the top wall. In the special case of equal lattice constants ($\eta = b/a = 1$), the forces on all atoms add in phase and $F = -Nf_1 \sin(2\pi x_{\mathrm{CM}}/a)$. The maximum of this restraining force gives the static friction $F_s = Nf_1$.

As noted above, the case of equal lattice constants is artificial and for real surfaces η is most likely to be an irrational number. Such surfaces are called incommensurate, while surfaces with a rational value of η are commensurate. When η is irrational, atoms on the top surface sample all phases of the periodic force with equal probability and the net force (23.9) vanishes exactly. When η is a rational number, it can be expressed as p/q, where p and q are integers with no common factors. In this case, atoms only sample q different phases. The net force from (23.9) still vanishes, because the force is a pure sine wave and the phases are equally spaced. However, the static friction is finite if the potential has higher harmonics. A Fourier component with wave vector $q2\pi/a$ and magnitude f_q contributes Nf_q to F_s. Studies of surface potentials [23.57] show that f_q drops exponentially with increasing q and thus imply that F_s will only be significant for small q.

As the springs become weaker, the top wall deforms more easily into a configuration that lowers the potential energy. The Prandtl–Tomlinson model is the simplest case to consider, because each atom can be treated as an independent oscillator within the upper surface. The equations of motion for the position x_l of the lth atom can be written as

$$m\ddot{x}_l = -\gamma\dot{x}_l - f_1 \sin\left(\frac{2\pi}{a}x_l\right) - k(x_l - x_l^0) , \tag{23.10}$$

where m is the atomic mass and x_l^0 is the position of the lattice site. Here γ is a phenomenological damping coefficient, like that in a Langevin thermostat, that allows work done on the atom to be dissipated as heat. It represents the coupling to external degrees of freedom such as lattice vibrations in the solids.

In any steady state of the system, the total friction can be determined either from the sum of the forces exerted by the springs on the top wall, or from the sum of the periodic potentials acting on the atoms (23.9). If the time average of these forces differed, there would be a net force on the atoms and a steady acceleration. The static friction is related to the force in metastable states of the system where $\ddot{x}_l = \dot{x}_l = 0$. This requires that spring and substrate forces cancel for each l,

$$k(x_l - x_l^0) = -f_1 \sin\left(\frac{2\pi}{a} x_l\right) . \quad (23.11)$$

As shown graphically in Fig. 23.5a, there is only one solution for weak interfacial potentials and stiff solids ($\lambda < 1$). In this limit, the behavior is essentially the same as for infinitely rigid solids. There is static friction for $\eta = 1$, but not for incommensurate cases. Even though incommensurate potentials displace atoms from lattice sites, there are exactly as many displaced to the right as to the left, and the force sums to zero.

A new type of behavior sets in when λ exceeds unity. The interfacial potential is now strong enough compared to the springs that multiple metastable states are possible. These states must satisfy both (23.11) and the condition that the second derivative of the potential energy is positive: $1 + \lambda \cos(2\pi x_l / a) > 0$. The number of metastable solutions increases as λ increases. As illustrated in Fig. 23.5b, once an atom is in a given metastable minimum it is trapped there until the center of mass moves far enough away that the second derivative of the potential vanishes and the minimum becomes unstable. The atom then pops forward very rapidly to the nearest remaining metastable state. Such motion, where a system is stuck for a long time and then suddenly jumps forward by an atomic spacing, is called atomic-scale stick-slip motion. During the pops, atoms reach large velocities and hence dissipate a large amount of energy into the lattice vibrations. This makes it possible to have a finite kinetic friction that does not vanish linearly with sliding velocity even when the surfaces are incommensurate [23.8, 9].

A similar analysis can be done for the 1-D Frenkel–Kontorova model [23.55, 58, 59]. The main difference is that the static friction and ground state depend strongly on η. For any given irrational value of η, there is a threshold potential strength λ_c. For weaker potentials, the static friction vanishes. For stronger potentials, elastic metastability produces a finite static friction. The transition to the onset of static friction was termed a breaking of analyticity by *Aubry* [23.59] and is often called the Aubry transition. The metastable states for $\lambda > \lambda_c$ take the form of locally commensurate regions that are separated by domain walls where the two crystals are out of phase. Within the locally commensurate regions the ratio of the periods is a rational number p/q that is close to η. The range of η where locking occurs grows with increasing potential strength (λ) until it spans all values. At this point there is an infinite number of different metastable ground states that form a fascinating "Devil's staircase" as η varies [23.55, 59].

Although elastic instabilities can, in principle, induce finite static friction at the atomic scale, this appears to be rare. There have been a significant number of computer simulations showing that the *wearless* static friction between crystalline surfaces is extremely

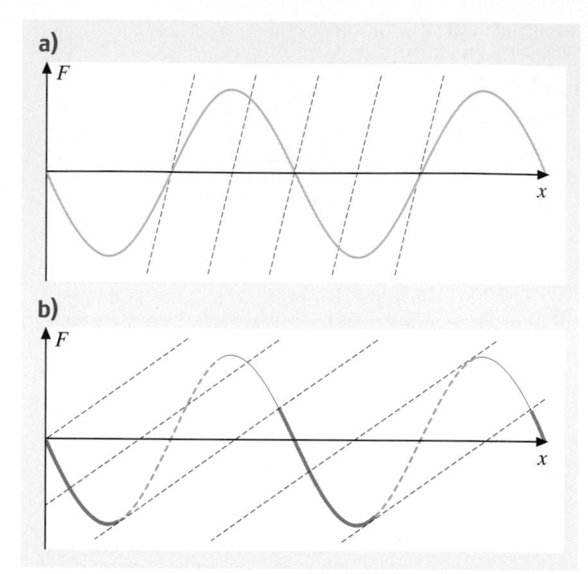

Fig. 23.5a,b Graphical solution for metastability of an atom in the Prandtl–Tomlinson model for **(a)** $\lambda = 0.5$ and **(b)** $\lambda = 3$. The *straight dashed lines* show the force from the spring, $k(x_l - x_l^0)$, at different x_l^0, and the *curved lines* show the periodic substrate potential. For $\lambda < 1$, there is a single intersection of the *dashed lines* with the substrate potential for each value of x_l^0, and thus a single metastable state. For $\lambda > 1$, there are multiple intersections with the substrate potential. *Dotted portions* of the potential curve indicate unstable maxima, and *solid regions* indicate metastable solutions. Fast pops with concurrent energy dissipation cannot be avoided for $\lambda > 1$ even when the slider moves slowly. (After [23.56])

small, except in the artificial case of identical and perfectly aligned walls [23.8, 10]. Studies of iron on iron [23.60], a blunt pyramidal diamond tip on a metal surface [23.61], a blunt pyramidal copper tip on incommensurate copper substrates [23.62], interlayer sliding in multiwalled carbon nanotubes [23.63], and Lennard-Jonesium on Lennard-Jonesium [23.40] all confirm the simple prediction that lateral forces cancel out. These simulations have in common that the solids were treated truly three-dimensionally and that the atoms (interacting via microscopic interaction potentials) were allowed to relax, thus making long-range elastic deformations possible. Significant lateral forces were only observed in combination with strongly irreversible processes such as production of wear debris, plastic deformation, material mixing, or cold welding and not due to geometric interlocking or elastic instabilities alone.

23.2.3 Role of Dimensionality and Disorder

The previous section considered whether local atomic displacements could lead to static friction. Long-range deformations between individual contacts could also allow surfaces to lock together [23.41, 42] Although we are concerned with the friction between two 3-D solid bodies, it is instructive to analyze the role of dimensionality in determining the effect of disorder. To do this, let us consider a d_{obj}-dimensional elastic solid, for example, a generalization of the elastic Frenkel–Kontorova chain to higher dimensions. We may safely assume the free elastic solid to be mechanically stable, meaning that the tensor of the elastic constants is positive definite. The dimension of the interface between the slider and disordered substrate is denoted by d_{int}.

In such a situation, there will be a competition between the random substrate-slider interactions and the elastic coupling within the solids. An important question to ask is how the interactions change when we change the scale of the system, for example, how strong are the random and the elastic interactions on a length scale $2\mathcal{L}$ if we know their respective strengths on a scale $\mathcal{L}$. Here $\mathcal{L}$ gives the linear dimension of the solids in all directions, that is to say, parallel and normal to the interface. As discussed above, the random forces between substrate and slider add incoherently and thus their sum scales as the square root of the interface area $\sim \mathcal{L}^{d_{\text{int}}/2}$. The elastic forces, on the other hand, scale as $\mathcal{L}^{d_{\text{obj}}-2}$. A linear chain can be more easily compressed if we replace one spring with two springs coupled in series. In two dimensions, springs are not only coupled in series but also in parallel, so that the elastic coupling remains invariant to a "block transformation". Each additional dimension strengthens the effect of "parallel" coupling.

In the thermodynamic limit $\mathcal{L} \to \infty$, the effect of disorder will always dominate the elastic interactions or vice versa, except in the special case where

$$\mathcal{L}^{d_{\text{int}}/2} \propto \mathcal{L}^{d_{\text{obj}}-2} . \tag{23.12}$$

For $\lim_{\mathcal{L}\to\infty} \mathcal{L}^{d_{\text{int}}/2}/\mathcal{L}^{d_{\text{obj}}-2} \gg 1$, the random interactions will dominate, and hence pinning via elastic instabilities cannot be avoided. This disorder-induced elastic pinning is similar to the pinning of compliant, ordered systems, as discussed above within the context of the Prandtl–Tomlinson and the Frenkel–Kontorova models. For $\lim_{\mathcal{L}\to\infty} \mathcal{L}^{d_{\text{int}}/2}/\mathcal{L}^{d_{\text{obj}}-2} \ll 1$, the long-range elastic forces dominate the long-range random forces. The slider's motion can only be opposed by elastic instabilities if the elastic coupling is sufficiently weak at finite $\mathcal{L}$ in order to make local pinning possible, again akin to the case of $\lambda > 1$ in the Prandtl–Tomlinson model.

The so-called *marginal* dimension, in which both contributions scale with the same exponent, $d_{\text{int}}/2 = d_{\text{obj}} - 2$, occurs in the important case of 3-D solid bodies with 2-D surfaces. In the marginal dimension, the friction force can stay finite, however, detailed analysis [23.42, 64] shows the friction force per unit area, and hence the friction coefficient, to be exponentially small. Even for objects with a dimensionality d_{mar} less than their marginal dimension $d_{\text{mar}} = 2 + d_{\text{int}}/2$ (see (23.12)), friction forces may turn out to be small. One example is an experimental quartz crystal microbalance study [23.65] of solid and liquid krypton films on disordered gold surfaces for which the pinning forces were undetectably small. For adsorbed layers, which satisfy $d_{\text{obj}} = d_{\text{int}}$, the marginal dimension would be $d_{\text{mar}} = 4$.

23.2.4 Elastic Instabilities vs. Wear in Atomistic Models

As mentioned in Sect. 23.2.2, there is no published computer simulation study known to us in which the solids were treated truly atomistically in three dimensions and where the interfacial interactions lead to elastic pinning according to Prandtl–Tomlinson or Frenkel–Kontorova. In this context, we want to discuss exemplarily computer simulations of a copper tip on copper by *Sørensen* et al. [23.62]. We consider their study to be rather generic and also well-done from a technical point of view for a variety of reasons: The effects of commensurability and finite contact area were examined in detail, the simulations were done at constant normal load instead of constant separation, the interaction potentials were realistic, and initial and boundary conditions were chosen such that the tip had the possibility to deform. Snapshots of one of their sliding simulations are shown in Fig. 23.6. Note that one must keep in mind that the simulation only addresses the tribological properties in UHV.

Flat, clean tips with (111) or (100) surfaces were brought into contact with corresponding crystalline substrates [23.62]. The two exterior layers of tip and surface were treated as rigid units, and the dynamics of the remaining mobile layers was followed. Interatomic forces and energies were calculated using semi-empirical potentials derived from effective medium theory [23.66]. At finite temperatures, the outer mobile layers of both tip and surface were coupled to a Langevin thermostat. Zero temperature simulations gave similar results.

In the commensurate Cu(111) case, *Sørensen* et al. [23.62] observed atomic-scale stick-slip motion of the tip like that described above for the Tomlinson model. The trajectory was of zigzag form, which could be related to jumps of the tip's surface between fcc and hcp positions. Detailed analysis of the slips showed that they occurred via a dislocation mechanism. Dislocations were nucleated at the corner of the interface, and then moved rapidly through the contact region.

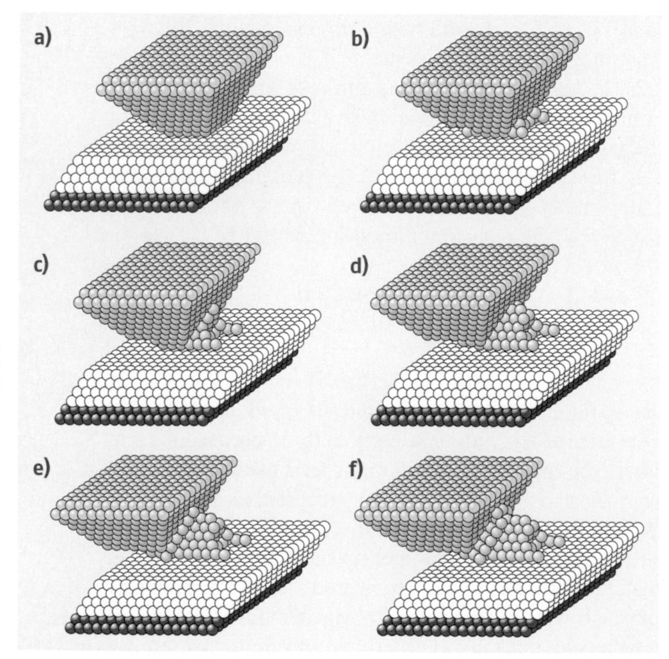

Fig. 23.6 Snapshots showing the evolution of a Cu(100) tip on a Cu(100) substrate during sliding to the left. (After [23.62])

Adhesion led to a large static friction at zero load: The static friction per unit area, or critical yield stress, dropped from 3.0 GPa to 2.3 GPa as T increased from 0 to 300 K. The kinetic friction increased linearly with load with a surprisingly small differential friction coefficient $\tilde{\mu}_k \equiv \partial F_k/\partial L \approx 0.03$. In the load regime investigated, $\tilde{\mu}_k$ was independent of temperature and load. No velocity dependence was detectable up to sliding velocities of $v = 5$ m/s. At higher velocities, the friction decreased. Even though the interactions between the surfaces are identical to those within the surfaces, no wear was observed. This was attributed to the fact that (111) surfaces are the preferred slip planes in fcc metals.

Adhesive wear was observed between a commensurate (100) tip and substrate (Fig. 23.6). Sliding in the (011) direction at either constant height or constant load led to inter-plane sliding between (111) planes inside the tip. As shown in Fig. 23.6, this plastic deformation led to wear of the tip, which left a trail of atoms in its wake. The total energy was an increasing function of sliding distance due to the extra surface area. The constant evolution of the tip kept the motion from being periodic, but the sawtoothed variation of force with displacement that is characteristic of atomic-scale stick-slip was still observed.

Sørensen et al. [23.62] also examined the effect of incommensurability. An incommensurate Cu(111) system was obtained by rotating the tip by 16.1° about the axis perpendicular to the substrate. For a small tip (5×5 atoms), they observed an Aubry transition from smooth sliding with no static friction at low loads to atomic-scale stick-slip motion at larger loads. Further increases in load led to sliding within the tip and plastic deformation. Finite systems are never truly incommensurate, and pinning was found to occur at the corners of the contact, suggesting it was a finite-size effect. Larger tips (19×19) slid without static friction at all loads. Similar behavior was observed for incommensurate Cu(100) systems.

These results confirm the conclusions of *Hirano* et al. [23.60] that even bare metal surfaces of the same material will not exhibit static friction if the surfaces are incommensurate. They also indicate that contact areas as small as a few hundred atoms are large enough to exhibit this effect. Similar results were obtained for Lennard Jonesium [23.6, 40, 67]: The interbulk interactions between unlike atoms (each solid is initially composed of one Lennard-Jones species) have to be distinctly larger than the intrabulk interactions between identical atoms in order to induce elastic instability. However, as soon as the interactions between unlike atoms are larger than those between like atoms, there is a mixing instability leading to a cold welded junction.

Lancon [23.69] reported that there is the possibility for elastic instabilities under carefully designed conditions between two incommensurate gold surfaces. The importance of the interfacial interactions with respect to the intrabulk interactions was increased through the application of a pressure normal to the interface. At 4 GPa, an elastic Aubry transition was observed concurrently with finite static friction. The normal pressures employed in those simulations were well above the yield strength of gold. However, plastic flow at the contact's circumference could be avoided by the use of flat surfaces that filled out the whole simulation cell parallel to the interface. Additional examples of elastic instabilities may be found for very compliant solids, but the majority of simple solids will not pin elastically on an atomic scale. Note, however, that cold welding may still occur on long time scales due to diffusion. On such time scales, the interface geometry may rearrange to produce locally commensurate regions that lower the interfacial free energy and allow geometrical pinning.

23.2.5 Hydrodynamic Lubrication and Its Confinement-Induced Breakdown

So far, we have been concerned with dry friction. However, most surfaces are covered by dirt and/or lubricants. Hydrodynamics and elasto-hydrodynamics have been very successful in describing lubrication by micron-thick films [23.70]. These continuum theories begin to break down as atomic structure becomes important. Experiments and simulations reveal a sequence of dramatic changes in the static and dynamic properties of fluid films as their thickness decreases from microns down to molecular scales. These changes have important implications for the function of boundary lubricants and the origins of static friction.

Let us consider two flat, lubricated walls separated by a distance d and sliding with a relative parallel velocity v. Let us further assume that the lubricant's viscosity μ in the center of the contact equals that of the bulk fluid phase. The shear stress must be uniform in steady state, because any imbalance would lead to accelerations. Since the velocity gradient times the viscosity μ gives the stress in the fluid, the kinetic friction per unit area is

$$F_{\mathrm{k}} = \frac{v\mu}{d + \mathcal{S}}\,, \tag{23.13}$$

where we introduce the so-called slip length $\mathcal{S}$. It represents the distance into the wall at which the velocity gradient would extrapolate to zero; see Fig. 23.7f for a graphical illustration of $\mathcal{S}$. The calculation of slip length from velocity profiles has some ambiguity, and we refer to work by *Bocquet* et al. [23.71] for the least ambiguous resolution to the definition of the slip length known to us.

Macroscopic experiments are generally well-described by a "no-slip" boundary condition (BC), where $\mathcal{S} = 0$ and the tangential component of the fluid velocity equals that of the solid at the surface. The one prominent exception is contact-line motion, where an interface between two fluids moves along a solid surface. This motion would require an infinite force in hydrodynamic theory, unless slip occurred near the contact line.

As d decreases, any deviation from $\mathcal{S} = 0$ becomes more evident in (23.13). Studies of confined films show

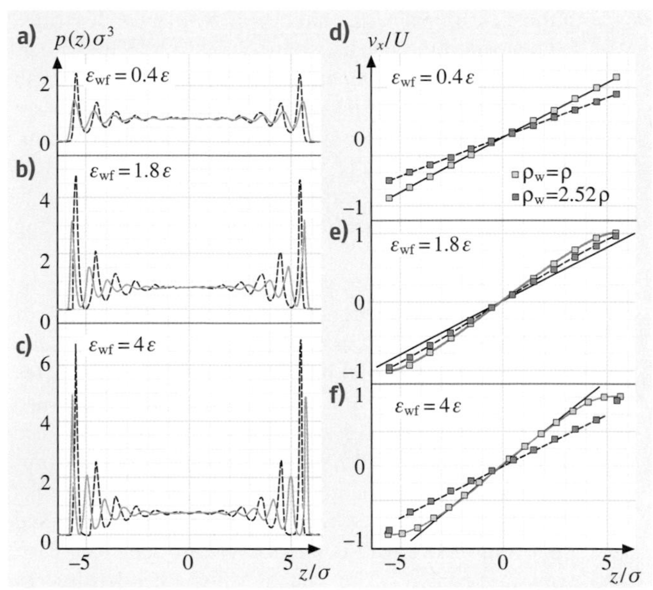

Fig. 23.7a–f Panels **(a)–(c)** show density as a function of position z relative to two walls whose atoms are centered at $z/\sigma = \pm 6.4$. Values of ϵ_{wf} are indicated. *Solid lines* are for equal wall and fluid densities ($\rho = 0.81\sigma^{-3}$), and *dotted lines* are for $\rho_w = 2.52\rho$. *Squares* in panels **(d)–(f)** show the average velocity in each layer as a function of z, and *solid* and *dotted lines* are fits through these values for low and high density walls, respectively. For this data, the walls were moved in opposite directions at speed $U = 1\sigma/t_{\mathrm{LJ}}$. The *dashed line* in **(e)** represents the flow expected from hydrodynamics with a no-slip BC ($\mathcal{S} = 0$). As shown in **(f)**, the slope (*dashed line*) far from the walls is used to define $\mathcal{S}$. (From [23.8] after [23.68])

that slip may occur ($\mathcal{S} > 0$) or layers may stick onto the solid ($\mathcal{S} < 0$) [23.68, 71–73]. These deviations from the no-slip BC have been correlated with structural changes in the fluid near the walls. The first observation was that layering occurs in planes parallel to the walls. It is induced by the sharp cutoff in fluid density at the wall and the pair correlation function $g(r)$ between fluid atoms. An initial fluid layer forms at the preferred wall-fluid spacing. Additional fluid molecules tend to lie in a second layer, at the preferred fluid-fluid spacing. This layer induces a third, and so on. Layering can alter the BC and the viscosity of the confined fluid. As suggested by the contrasting behavior of commensurate and incommensurate solids, the BC depends sensitively on the degree to which the walls imprint their in-plane density modulations into the lubricant.

Some of the trends that are observed in the degree and extent of layering are illustrated in Fig. 23.7 [23.68, 72]. The fluid density is plotted as a function of the distance between walls for a model that is also considered in almost all of the other studies of flow BCs described below. The fluid consists of spherical molecules interacting through a Lennard-Jones potential (23.1) with characteristic energy and length scales ϵ and σ. They are confined by crystalline walls containing discrete atoms. In this case, the walls were planar (001) surfaces of an fcc crystal. Wall and fluid atoms also interact with a Lennard-Jones potential, but with a different binding energy ϵ_{wf}.

The net adsorption potential from the walls can be increased by raising ϵ_{wf}, or by increasing the density of the walls ρ_w, so that more wall atoms attract the fluid. Figure 23.7 shows that both increases lead to increases in the height of the first density peak. The height also increases with the pressure in the fluid [23.73], since that forces atoms into steeper regions of the adsorption potential. The height of subsequent density peaks decreases smoothly with distance from the wall, and only four or five well-defined layers are seen near each wall in Fig. 23.7. The rate at which the density oscillations decay is determined by the decay length of structure in the bulk pair-correlation function of the fluid. Since all panels of Fig. 23.7 have the same conditions in the bulk, the decay rate is the same. The adsorption potential only determines the initial height of the peaks.

As mentioned above, solid surfaces also induce density modulations within the plane of the layers [23.7]. These modulations can be quantified by the 2-D static structure factor at the shortest reciprocal lattice vector Q of the substrate [23.35, 68]. When normalized by the number of atoms in the layer, N_l, this becomes an intensive variable that would correspond to the Debye–Waller factor in a crystal. The maximum possible value, $S(Q)/N_l = 1$, corresponds to fixing all atoms exactly at crystalline lattice sites. In a strongly ordered case such as $\rho_w = \rho$ in Fig. 23.7c, the small oscillations about lattice sites in the first layer only decrease $S(Q)/N_l$ to 0.71. This is well above the value of 0.6 that is typical of bulk 3-D solids at their melting point [23.74] and indicates that the first layer has crystallized onto the wall. The values of $S(Q)/N_l$ in the second and third layers are 0.31 and 0.07, respectively, and atoms in these layers exhibit typical fluid diffusion.

There is some correlation between the factors that produce strong layering and those that produce strong in-plane modulations. For example, chain molecules have several conflicting length scales that tend to frustrate both layering and in-plane order [23.36, 75, 76]. However, the dependence of in-plane order on the density of substrate atoms is more complicated than for layering. When $\rho_w = \rho$, the fluid atoms can naturally sit on the sites of a commensurate lattice, and $S(Q)$ is large. When the substrate density ρ_w is increased by a factor of 2.52, the fluid atoms no longer fit easily into the corrugation potential. The degree of induced in-plane order drops sharply, although the layering becomes stronger (Fig. 23.7). Sufficiently strong adsorption potentials may eventually lead to crystalline order in the first layers and stronger layering. However, this may actually increase slip.

Thompson et al. [23.68] found that all of their results for $\mathcal{S}$ collapsed onto a universal curve when plotted against the structure factor $S(Q)/N_l$. When one or more layers crystallized onto the wall, the same collapse could be applied as long as the effective wall position was shifted by a layer and the Q for the outer wall layer was used. The success of this collapse at small $S(Q)/N_l$ can be understood from a perturbation theory [23.6, 77] developed to quantitatively explain the kinetic friction on adsorbed monolayers [23.3].

One of the surprising features of Fig. 23.7 is that the viscosity remains the same even in regions near the wall where there is pronounced layering. Any change in viscosity would produce a change in the velocity gradient, since the stress is constant. However, the flow profiles in panel (d) remain linear throughout the cell. The profile for the dense walls in panel (f) is linear up to the last layer, which has crystallized onto the wall. From Fig. 23.7 it is apparent that local density variations by at least a factor of seven can be accommodated without a viscosity change.

Bitsanis et al. [23.78] examined the change in viscosity with film thickness. They found that results for film thicknesses $h > 4\sigma$ could be fit using the bulk viscosity for the average density. However, as h decreased below 4σ, the viscosity diverged much more rapidly than any model based on bulk viscosity could explain. These observations were consistent with surface force apparatus (SFA) experiments on nanometer thick films of a wide variety of small molecules [23.4, 5]. Layering in molecularly thin films gave rise to oscillations in the energy, normal force, and effective viscosity as the film thickness decreased. Most films of one to three molecular layers exhibited a yield stress characteristic of solid-like behavior, even though the molecules form a simple Newtonian fluid in the bulk. Deviations from the no-slip condition might cause μ_{eff} to differ from the bulk viscosity by an order of magnitude. However, they could not explain the observed changes of μ_{eff} by more than five orders of magnitude, or the even more dramatic changes by 10 to 12 orders of magnitude in the characteristic viscoelastic relaxation time determined from the shear rate dependence of μ_{eff} [23.5].

More recent experimental and theoretical studies confirm that most fluids exhibit a glass transition when confined to channels less than 3 to 10 molecular diameters [23.33, 79]. *Thompson* et al. [23.34, 36] considered changes in flow with film thickness in a wide range of model systems. As in experiment, they found dramatic divergences in the viscosity and relaxation time that suggested confinement induced a glass transition. Later work by *Baljon* et al. [23.33, 80] indicates that the same glass transition is produced by changes in thickness, pressure, or temperature. In particular, the shear-rate-dependent viscoelastic response of systems of all sizes, pressures, and temperatures can be collapsed onto the same universal curve above the glass transition [23.33].

23.2.6 Submonolayer Films

Physisorbed molecules such as short hydrocarbon chains can be expected to sit on any surface exposed to atmospheric conditions. Even in ultrahigh vacuum, special surface treatments are needed to remove strongly physisorbed species from surfaces. Given the generic observation of glassy behavior in thin films, *Robbins* et al. [23.81] suggested that these submonolayer contaminants could provide a general explanation for the universal observation of static friction. Subsequent simulations have confirmed that physisorbed molecules qualitatively alter tribological behavior. For example, they lead to static friction between two incommensurate walls [23.67, 82] or two disordered walls [23.47] that would slide without friction if they were clean. The theoretical explanation of this effect can be understood from a simple analytical model considered by *Müser* et al. [23.47]. In order to make two surfaces pin, a nonvanishing fraction of the density modulations must share the same period and lock into phase. This happens automatically for commensurate surfaces, but not for incommensurate surfaces unless they are unrealistically compliant. However, a submonolayer of atoms or molecules that does not form strong covalent bonds with the confining solids can simultaneously lock to the density modulations of two incommensurate walls and hence give rise to a finite friction coefficient.

Simulation results for the friction due to physisorbed layers are consistent with a large number of experimental observations. While Amontons's laws state that the friction is independent of the macroscopic area, it is well established that the friction is proportional to the real area of molecular contact A_{real} [23.30, 83, 84]. Experiments where both the friction and molecular area can be measured [23.4, 85–89] show that the local shear stress τ_{s} needed to initiate sliding rises linearly with the normal pressure P:

$$\tau_{\text{s}} = \tau_0 + \alpha P \,. \tag{23.14}$$

This gives a macroscopic friction force $F_{\text{s}} = \tau_0 A_{\text{real}} + \mu L$, where the total load is just $L = P A_{\text{real}}$. This friction law is actually more general than Amontons's laws, reducing to it at large loads and small areas, but explaining why the friction between adhesive surfaces (like tape) does not vanish at zero load.

For a wide range of parameters [23.82, 90], the friction calculated for physisorbed monolayers follows (23.14) up to the gigapascal pressures that are present in real contacts. Moreover, the results are remarkably insensitive to parameters that are not controlled in experiments such as the crystallographic orientation of the solids, the thickness of the adsorbed layer, and the direction of sliding. The only exceptions occur for the unlikely case of commensurate walls [23.90, 91], as discussed further below. Subsequent studies of kinetic friction [23.39, 92] explain both the close relation between experimental values of static friction and the slow variation with velocity. In particular, the logarithmic rate dependence seen in τ for many systems [23.83] can be explained through thermally activated hopping of physisorbed molecules [23.92].

Müser et al. [23.47] have constructed a simple analytical theory that captures many of the results and conclusions from the above simulations. At the pressures of interest, the atoms act like hard spheres. In order to slide, the bounding surfaces must lift up over these spheres against the normal pressure. The situation is analogous to the simple cartoon in Fig. 23.2, but with atomic-scale ramps, and the required stress increases linearly with pressure for the same reason. The offset, τ_0, is related to adhesive interactions between the two surfaces that increase the effective load [23.90]. Unlike the original model of geometrical interlocking, the atomic-scale ramps rearrange during sliding. The dissipation during these rearrangements is responsible for kinetic friction [23.47, 92], which, as we now discuss, is very different for commensurate and incommensurate walls.

The characteristic motion of individual boundary lubricant atoms depends on the relative displacement for commensurate walls, because the distribution of metastable states depends on the relative alignment of the two walls [23.90, 92]. For example, no "pops" of the lubricant atoms are observed when the two commensurate solids are aligned directly above each other. This behavior is in contrast with that for incommensurate surfaces for which the distribution of metastable states does not change as the slider is translated with respect to the substrate. Stressing the analogy to the Prandtl–Tomlinson model, static and kinetic friction are related for incommensurate, boundary-lubricated solids, while they are not necessarily related for commensurate, boundary-lubricated solids. Indeed, commensurate, boundary-lubricated surfaces exhibit a dramatic decrease in friction at the transition from stick-slip motion to smooth sliding, while this dramatic decrease is absent for incommensurate, boundary-lubricated walls [23.39].

It has been noted recently, that dimensionality not only plays an important role for elastic objects, but also for boundary lubricants [23.39]. It was predicted that kinetic friction would be Stokes-like (i. e., vanish linearly with velocity) between incommensurate surfaces, if it were possible to confine lubricant particles into trenches of (sub)nanometer depth and width, so that the motion of the lubricant embedded between the surfaces becomes 1-D. At the same time, however, static friction would remain finite.

The above results suggest that adsorbed molecules and other "third-bodies" may prove key to understanding macroscopic friction measurements. It will be interesting to extend these studies to more realistic molecular potentials and to rough surfaces. To date, realistic potentials have only been used for "between-sorbed" particles between commensurate surfaces. This work is reviewed elsewhere [23.8, 10, 11].

23.3 Stick-Slip Dynamics

The dynamics of sliding systems can be very complex and depend on many factors, including the types of metastable states in the system, the times needed to transform between states, and the mechanical properties of the device that imposes the stress. At high rates or stresses, systems usually slide smoothly. At low rates, the motion often becomes intermittent, with the system alternately sticking and slipping forward [23.30, 32]. Everyday examples of such stick-slip motion include the squeak of hinges and the music of violins.

The alternation between stuck and sliding states of the system reflects changes in the way energy is stored. While the system is stuck, elastic energy is pumped into the system by the driving device. When the system slips, this elastic energy is released into kinetic energy and eventually dissipated as heat. The system then sticks once more, begins to store elastic energy, and the process continues. Both elastic and kinetic energy can be stored in all the mechanical elements that drive the system. The whole coupled assembly must be included in any analysis of the dynamics.

The simplest type of intermittent motion is the atomic-scale stick-slip that occurs in the multistable regime ($\lambda > 1$) of the Prandtl–Tomlinson model. Energy is stored in the springs while atoms are trapped in a metastable state, and converted to kinetic energy as they pop to the next metastable state. This phenomenon is quite general and has been observed in many simulations of wearless friction [23.8, 10], as well as in the motion of atomic force microscope tips [23.1, 2]. In these cases, motion involves a simple ratcheting over the surface potential through a regular series of hops between neighboring metastable states. The slip distance is determined entirely by the periodicity of the surface potential.

Many examples of stick-slip involve a rather different type of motion that can lead to intermittency and chaos [23.93–97]. Instead of jumping between neighboring metastable states, the system slips for very long distances before sticking. For example, *Gee* et al. [23.4] and *Yoshizawa* et al. [23.98] observed slip distances of many microns in their studies of confined films with the

SFA. This distance is much larger than any characteristic periodicity in the potential and varied with velocity, load, and the mass and stiffness of the SFA. The fact that the SFA does not stick after moving by a lattice constant indicates that sliding has changed the state of the system in some manner, so that it can continue sliding even at forces less than the yield stress.

One simple property that depends on past history is the amount of stored kinetic energy. This can carry a system over potential energy barriers even when the stress is below the yield stress. The simplest example is the underdamped Prandtl–Tomlinson model, which has been thoroughly studied in the mathematically equivalent case of an underdamped Josephson junction [23.99]. One finds a hysteretic response function, where static and moving steady-states coexist over a range of forces between $F_{\rm min}$ and the static friction $F_{\rm s}$. There is a minimum stable steady-state velocity $v_{\rm min}$ corresponding to $F_{\rm min}$. At lower velocities, the only steady state is linearly unstable because $\partial v/\partial F < 0$ – pulling harder slows the system [23.30, 32]. If the top wall of the Tomlinson model is pulled at an average velocity less than $v_{\rm min}$ by a sufficiently compliant system, it will exhibit large-scale stick-slip motion.

Experimental systems typically have many other internal degrees of freedom that may depend on past history. Macroscopic models normally assume that a single unspecified state variable is relevant and add dynamical rules for this variable to create a phenomenological rate-state model [23.94, 100]. This approach has captured experimental behavior from a wide range of systems, including surface force apparatus experiments [23.38, 101], paper sliding on paper [23.93], and earthquakes [23.102]. However, in most cases, there is no microscopic model for the state variable and its time dependence.

One set of systems where microscopic pictures for the state variable have been considered are confined films in SFA experiments. Confined films have structural degrees of freedom that can change during sliding, and this provides an alternative mechanism for stick-slip motion [23.35]. Some of these structural changes are illustrated in Fig. 23.8, which shows stick-slip motion of a two-layer film of simple spherical molecules. The bounding walls were held together by a constant normal load. A lateral force was applied to the top wall through a spring k attached to a stage that moved with fixed velocity v in the x direction. The equilibrium configuration of the film at $v = 0$ is a commensurate crystal that resists shear. Thus at small times, the center of mass of the top wall remains pinned at $X = 0$. The force grows linearly with time, $F = kv$, as the stage advances ahead of the wall. When F exceeds $F_{\rm s}$, the wall slips forward. The force drops rapidly because the slip velocity $\dot{X}$ is much greater than v. When the force drops sufficiently, the

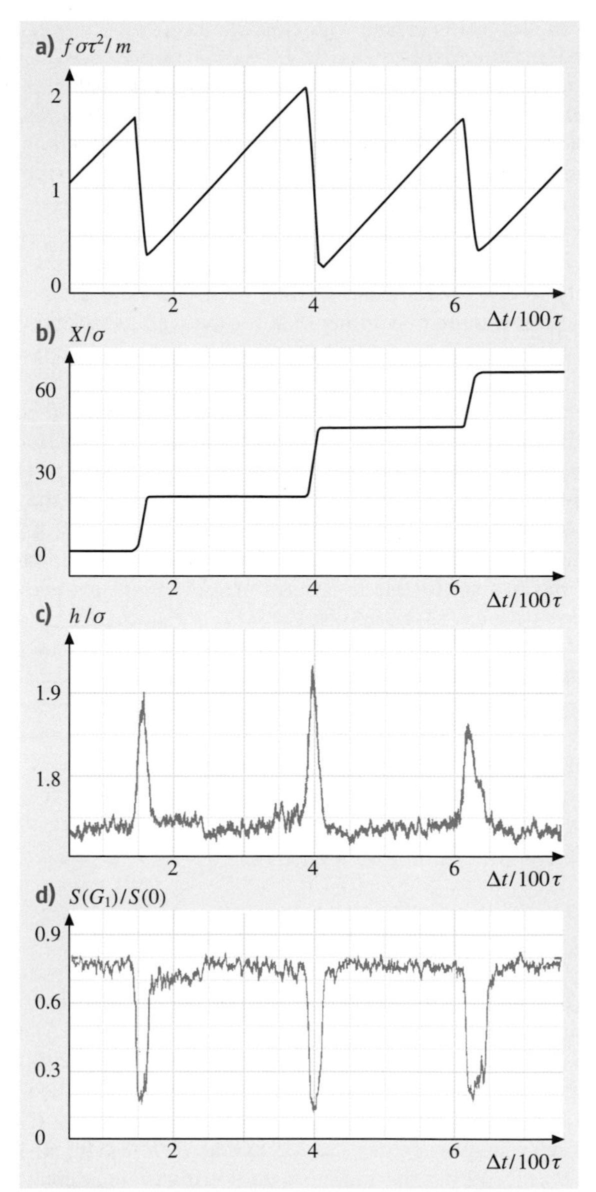

Fig. 23.8a–d Time dependence of (a) force per unit area f, (b) lateral position of the top wall X, (c) wall spacing h, and (d) degree of crystallinity $S(G_1)/S(0)$ as a function of time during stick-slip motion of walls confining two layers of spherical molecules. (After [23.35])

film recrystallizes, the wall stops, and the force begins to rise once more.

One structural change that occurs during each slip event is dilation by about 10% (Fig. 23.8c). *Dhinojwala* et al. [23.103] have recently confirmed that dilation occurs during slip in SFA experiments. The increased volume makes it easier for atoms to slide past each other and is part of the reason the sliding friction is lower than F_s. The system may be able to keep sliding in this dilated state as long as it takes more time for the volume to contract than for the wall to advance by a lattice constant. Dilation of this type plays a crucial role in the yield, flow, and stick-slip dynamics of granular media [23.104–106]. In some simulations, the dilation is localized at a single plane – either a wall/film interface or a plane within the film [23.33, 107]. In some cases, there is also in-plane ordering of the film to enable it to slide more easily over the wall. This ordering remains after sliding stops and provides a mechanism for the long-term memory seen in some experiments [23.4, 98, 108].

The degree of crystallinity throughout the film may also change during sliding. Deviations from an ideal crystalline structure can be quantified by the Debye–Waller factor $S(G_1)/S(0)$ (Fig. 23.8d), where G_1 is one of the shortest reciprocal lattice vectors and $S(0) = N$ is the total number of atoms in the film. When the system is stuck, $S(G_1)/S(0)$ has a large value that is characteristic of a 3-D crystal. During each slip event, $S(G_1)/S(0)$ drops dramatically. The minimum values are characteristic of simple fluids that would show a no-slip boundary condition (Sect. 23.2.5). The atoms also exhibit rapid diffusion that is characteristic of a fluid. The periodic melting and freezing transitions that occur during stick-slip are induced by shear, not by the negligible changes in temperature. Shear-melting transitions at constant temperature have been observed in both theoretical and experimental studies of bulk colloidal systems [23.29, 109]. While the above simulations of confined films used a fixed number of particles, *Lupowski* et al. [23.110] found equivalent results at fixed chemical potential.

Very similar behavior has been observed in simulations of sand [23.104], chain molecules [23.33], incommensurate or amorphous walls [23.35], and minimal models with single molecules between surfaces [23.95–97]. These systems transform between glassy and fluid states during stick-slip motion. As in equilibrium, the structural differences between glass and fluid states are small. However, there are strong changes in the self-diffusion and other dynamic properties when the film goes from the static glassy to sliding fluid state. *Buldum* and *Ciraci* [23.111] found stick-slip motion due to periodic structural transformations in the bottom layers of a pyramidal Ni(111) tip sliding on an incommensurate Cu(110) surface.

23.4 Conclusions

The discussion of computer simulations in this chapter aimed to build a general atomistic understanding of tribological processes rather than to elucidate a specific sliding system. The advantage of studying simplified models, as opposed to chemically realistic systems, is that system properties can be varied at will. This allows the role of adhesion, surface curvature, contaminant layers and other factors to be isolated and explored. For example, to explore the effects of adhesion one can compare results from a simulation that includes the long-range attractive interactions between atoms with those from a simulation where all attractive interactions between surfaces are eliminated, leaving all other parameters unchanged. Care should be taken to avoid simplifications that are known to yield atypical results. In particular, the use of commensurate, perfectly aligned, crystalline surfaces has been shown to produce qualitatively different results, even in the presence of a lubricating film! Surface curvature must also be included in a way that properly models stress distributions from the bounding solids.

One central question is how contacting surfaces lock together to produce a static friction force. This is particularly perplexing in the context of many nanotribological systems where friction occurs without significant plastic deformation or wear. Many theories for wearless solid friction assume that surfaces deform elastically to conform to each other, and that atoms advance in a series of elastic instabilities during sliding. However, the simulations discussed above show that simple solids should hardly ever show elastic instabilities on an atomic scale without concurrent plastic flow or material transfer. Despite extensive efforts to design an atomistic system whose static friction mechanism is entirely elastic, no such model could be found as long as the ingredients were: three-dimensional incommensurate solids, flat or smoothly curved contacts, and two-body interactions between all atoms. Under these conditions the relative lateral motion is only opposed by a viscous-like force that goes to zero for small sliding velocities. When interfacial interactions are increased in an attempt to produce elastic instabilities, plastic flow is also observed.

Wearless friction between clean solid surfaces is only found in simulations of very small contacts. The geometric interlocking in such contacts is particularly relevant to AFM experiments. When only a few atoms on the tip make contact, the potential from the substrate does not average to zero. Static friction can then occur without elastic instabilities within the tip. The Prandtl–Tomlinson model can be applied to these systems with the cantilever stiffness replacing the internal stiffness of the material in the tip. As the diameter of the contact increases, due to blunting of the tip or increased load, the friction coefficient produced by geometric interlocking decreases. In the specific case of non-adhesive contacts, the friction between an amorphous tip and a clean crystalline substrate scales as $F \propto L^{2/3}$ and μ vanishes as $L^{-1/3}$. Thus geometric interlocking is not expected to yield significant friction when the contact area contains more than $\sim 10^3$ atoms.

Most contacting solids are separated by contaminant layers, or third bodies, that introduce many additional structural degrees of freedom. The intervening atoms can rearrange into a local free energy minimum for any geometry of the bounding solids without paying the large elastic energy cost that limits rearrangements within these solids. Simulations that include these layers reproduce many details of SFA experiments. They also provide an explanation for Amontons's laws, and exceptions to them, in larger contacts.

Many important problems remain to be addressed by simulations. Of particular interest are the roles of realistic, random surface roughness and subsurface plastic deformation. Studies of these phenomena will require new multiscale approaches that treat long-range elastic and plastic deformations in the solids. Other important problems will require extensions in the time available to simulations. Examples include cold-welding through inter-surface diffusion, wear, and fatigue. Addressing these problems will be a continuing challenge to the computer modeling community.

References

23.1 C. M. Mate, G. M. McClelland, R. Erlandsson, S. Chiang: Atomic-scale friction of a tungsten tip on a graphite surface, Phys. Rev. Lett. **59** (1987) 1942–1945

23.2 R. W. Carpick, M. Salmeron: Scratching the surface: Fundamental investigations of tribology with atomic force microscopy, Chem. Rev. **97** (1997) 1163–1194

23.3 J. Krim: Friction at the atomic scale, Sci. Am. **275**(4) (1996) 74–80

23.4 M. L. Gee, P. M. McGuiggan, J. N. Israelachvili, A. M. Homola: Liquid to solid transitions of molecularly thin films under shear, J. Chem. Phys. **93** (1990) 1895–1906

23.5 S. Granick: Motions and relaxations of confined liquids, Science **253** (1992) 1374–1379

23.6 M. Cieplak, E. D. Smith, M. O. Robbins: Molecular origins of friction: The force on adsorbed layers, Science **265** (1994) 1209–1212

23.7 M. Schoen, C. L. Rhykerd, D. J. Diestler, J. H. Cushman: Fluids in micropores: I. Structure of a simple classical fluid in a slit-pore, J. Chem. Phys. **87** (1987) 5464–5476

23.8 M. O. Robbins, M. H. Müser: Computer simulations of friction, lubrication and wear. In: *Modern Tribology Handbook*, ed. by B. Bhushan (CRC, Boca Raton 2001) pp. 717–765

23.9 M. H. Müser, M. Urbakh, M. O. Robbins: Statistical mechanics of static and low-velocity kinetic friction, Adv. Chem. Phys. **126** (2003) 187–272

23.10 J. A. Harrison, S. J. Stuart, D. W. Brenner: Atomic-scale simulation of tribological and related phenomena. In: *Handbook of Micro/Nanotribology*, ed. by B. Bhushan (CRC, Boca Raton 1999) pp. 525–594

23.11 J. A. Harrison, G. T. Gao, R. J. Harrison, G. M. Chateauneuf, P. T. Mikulski: The friction of model self-assembled monolayers. In: *Encyclopedia of Nanoscience and Nanotechnology*, ed. by H. S. Nalwa (American Scientific, Los Angeles 2004)

23.12 M. P. Allen, D. J. Tildesley: *Computer Simulation of Liquids* (Clarendon, Oxford 1987)

23.13 D. Frenkel, B. Smit: *Understanding Molecular Simulation: From Algorithms to Applications* (Academic, San Diego 1996)

23.14 J. P. Ryckaert, A. Bellemans: Molecular dynamics of liquid alkanes, Faraday Discuss. **66** (1978) 95–106

23.15 W. Paul, D. Y. Yoon, G. D. Smith: An optimized united atom model for simulations of polymethylene melts, J. Chem. Phys. **103** (1995) 1702–1709

23.16 W. Tschöp, K. Kremer, J. Batoulis, T. Bürger, O. Hahn: Simulation of polymer melts. I. Coarse graining procedure for polycarbonates, Acta Polym. **49** (1998) 61–74

23.17 W. Tschöp, K. Kremer, J. Batoulis, T. Bürger, O. Hahn: Simulation of polymer melts. II. From coarse grained models back to atomistic description, Acta Polym. **49** (1998) 75–79

23.18 P. Flory: *Statistical Mechanics of Chain Molecules* (Hanser, München 1988)

23.19 K. Binder: *Monte Carlo and Molecular Dynamics Simulations in Polymer Science* (Oxford, New York 1995)

23.20 D.W. Brenner, O.A. Shenderova, J.A. Harrison, S.J. Stuart, B. Ni, S.B. Sinnot: Second generation reactive empirical bond order (REBO) potential energy expression for hydrocarbons, J. Phys. C **14** (2002) 783–802

23.21 S.J. Stuart, A.B. Tutein, J.A. Harrison: A reactive potential for hydrocarbons with intermolecular interactions, J. Chem. Phys. **112** (2000) 6472–6486

23.22 M.S. Daw, M.I. Baskes: Embedded-atom method: Derivation and application to impurities, surfaces, and other defects in metals, Phys. Rev. B **29** (1984) 6443–6453

23.23 S. Nosé: Constant temperature molecular dynamics methods, Prog. Theor. Phys. Supp. **103** (1991) 1–46

23.24 T. Schneider, E. Stoll: Molecular dynamics study of a three-dimensional one-component model for distortive phase transitions, Phys. Rev. B **17** (1978) 1302–1322

23.25 G.S. Grest, K. Kremer: Molecular dynamics simulations for polymers in the presence of a heat bath, Phys. Rev. A **33** (1986) 3628–3631

23.26 S.S. Sarman, D.J. Evans, P.T. Cummings: Recent developments in non-Newtonian molecular dynamics, Phys. Rep. **305** (1998) 1–92

23.27 D.J. Evans, G.P. Morriss: Shear thickening and turbulence in simple fluids, Phys. Rev. Lett. **56** (1986) 2172–2175

23.28 W. Loose, G. Ciccotti: Temperature and temperature control in nonequilibrium-molecular-dynamics simulations of the shear flow of dense liquids, Phys. Rev. A **45** (1992) 3859–3866

23.29 M.J. Stevens, M.O. Robbins: Simulations of shear-induced melting and ordering, Phys. Rev. E **48** (1993) 3778–3792

23.30 F.P. Bowden, D. Tabor: *The Friction and Lubrication of Solids* (Clarendon, Oxford 1986)

23.31 R. Khare, J.J. de Pablo, A. Yethiraj: Rheology of confined polymer melts, Macromolecules **29** (1996) 7910–7918

23.32 E. Rabinowicz: *Friction and Wear of Materials* (Wiley, New York 1965)

23.33 M.O. Robbins, A.R.C. Baljon: Response of thin oligomer films to steady and transient shear. In: *Microstructure and Microtribology of Polymer Surfaces*, ed. by V.V. Tsukruk, K.J. Wahl (American Chemical Society, Washington 2000) pp. 91–117

23.34 P.A. Thompson, G.S. Grest, M.O. Robbins: Phase transitions and universal dynamics in confined films, Phys. Rev. Lett. **68** (1992) 3448–3451

23.35 P.A. Thompson, M.O. Robbins: Origin of stick-slip motion in boundary lubrication, Science **250** (1990) 792–794

23.36 P.A. Thompson, M.O. Robbins, G.S. Grest: Structure and shear response in nanometer-thick films, Israel J. Chem. **35** (1995) 93–106

23.37 M. Porto, V. Zaloj, M. Urbakh, J. Klafter: Macroscopic versus microscopic description of friction: From Tomlinson model to shearons, Tribol. Lett. **9** (2000) 45–54

23.38 B.N.J. Persson: *Sliding Friction: Physical Principles and Applications* (Springer, Berlin 1998)

23.39 M.H. Müser: Nature of mechanical instabilities and their effect on kinetic friction, Phys. Rev. Lett. **89** (2002) 224301: 1–4

23.40 M.H. Müser: Dry friction between flat surfaces: Wearless multistable elasticity vs. material transfer and plastic deformation, Tribol. Lett. **10** (2001) 15–22

23.41 J.B. Sokoloff: Static friction between elastic solids due to random asperities, Phys. Rev. Lett. **86** (2001) 3312–3315

23.42 B.N.J. Persson, E. Tosatti: Theory of friction: Elastic coherence length and earthquake dynamics, Solid State Commun. **109** (1999) 739–744

23.43 F.F. Abraham, N. Bernstein, J.Q. Broughton, D. Hess: Dynamic fracture of silicon: Concurrent simulation of quantum electrons, classical atoms, and the continuum solid, MRS Bull. **25** (2000) 27–32

23.44 L. Wenning, M.H. Müser: Friction laws for elastic nano-scale contacts, Europhys. Lett. **54** (2001) 693–699

23.45 U. Landman, W.D. Luedtke, N.A. Burnham, R.J. Colton: Atomistic mechanisms and dynamics of adhesion, nanoindentation, and fracture, Science **248** (1990) 454–461

23.46 D. Dowson: *History of Tribology* (Longman, New York 1979)

23.47 M.H. Müser, L. Wenning, M.O. Robbins: Simple microscopic theory of Amontons's laws for static friction, Phys. Rev. Lett. **86** (2001) 1295–1298

23.48 M. Enachescu, R.J.A. van den Oetelaar, R.W. Carpick, D.F. Ogletree, C.J.F. Flipse, M. Salmeron: An AFM study of an ideally hard contact: The diamond (111)/tungsten carbide interface, Phys. Rev. Lett. **81** (1998) 1877–1880

23.49 C.A.J. Putman, M. Igarashi, R. Kaneko: Single-asperity friction in friction force microscope: The composite model, Appl. Phys. Lett. **66** (1995) 3221–3223

23.50 P. Schwarz, O. Zwörner, P. Köster, R. Wiesendanger: Quantitative analysis of the frictional properties of solid materials at low loads. II. Mica and germanium sulfide, Phys. Rev. B **56** (1997) 6987–7000

23.51 L. Prandtl: Ein Gedankenmodell zur kinetischen Theorie der festen Körper, Z. Angew. Math. Mech. **8** (1928) 85–106

23.52 G.A. Tomlinson: A molecular theory of friction, Philos. Mag. Series **7** (1929) 905–939

23.53 G.M. McClelland, S.R. Cohen: Tribology at the atomic scale. In: *Chemistry and Physics of Solid Surfaces*, Vol. VII, ed. by R. Vanselow, R. Rowe (Springer, Berlin, Heidelberg 1990) pp. 419–445

23.54 Y.I. Frenkel, T. Kontorova: On the theory of plastic deformation and twinning, Zh. Eksp. Teor. Fiz. **8** (1938) 1340

23.55 O.M. Braun, Y.S. Kivshar: Nonlinear dynamics of the Frenkel–Kontorova model, Phys. Rep. **306** (1998) 1–108

23.56 M.O. Robbins: Jamming, friction, and unsteady rheology. In: *Jamming and Rheology: Constrained Dynamics on Microscopic and Macroscopic Scales*, ed. by A.J. Liu, S.R. Nagel (Taylor and Francis, London 2000)

23.57 L.W. Bruch, M.W. Cole, E. Zaremba: *Physical Adsorption: Forces and Phenomena* (Oxford, New York 1997)

23.58 F.C. Frank, J.H. van der Merwe: One-dimensional dislocations. I. Static theory, Proc. R. Soc. A **198** (1949) 205–225

23.59 S. Aubry: The new concept of transitions by breaking of analyticity in a crystallographic model. In: *Solitons and Condensed Matter Physics*, ed. by A.R. Bishop, T. Schneider (Springer, Berlin, Heidelberg 1979) pp. 264–290

23.60 M. Hirano, K. Shinjo: Atomistic locking and friction, Phys. Rev. B **41** (1990) 11837–11851

23.61 J. Belak, I.F. Stowers: The indentation and scraping of a metal surface: A molecular dynamics study. In: *Fundamentals of Friction: Macroscopic and Microscopic Processes*, ed. by I.L. Singer, H.M. Pollock (Kluwer, Dordrecht 1992) pp. 511–520

23.62 M.R. Sørensen, K.W. Jacobsen, P. Stoltze: Simulations of atomic-scale sliding friction, Phys Rev. B **53** (1996) 2101–2113

23.63 A.N. Kolmogorov, V.H. Crespi: Smoothest bearings: Interlayer sliding in multiwalled carbon nanotubes, Phys. Rev. Lett. **85** (2000) 4727–4730

23.64 A. Volmer, T. Natterman: Towards a statistical theory of solid dry friction, Z. Phys. B **104** (1997) 363–371

23.65 C. Mak, J. Krim: Quartz-crystal microbalance studies of disorder induced lubrication, Faraday Discuss. **107** (1997) 389–397

23.66 K.W. Jacobsen, J.K. Norskov, M.J. Puska: Interatomic interactions in the effective-medium theory, Phys. Rev. B **35** (1987) 7423–7442

23.67 M.H. Müser, M.O. Robbins: Conditions for static friction between flat crystalline surfaces, Phys. Rev. B **61** (2000) 2335–2342

23.68 P.A. Thompson, M.O. Robbins: Shear flow near solids: Epitaxial order and flow boundary conditions, Phys. Rev. A **41** (1990) 6830–6837

23.69 F. Lancon: Aubry transition in a real material: Prediction for its existence in an incommensurate gold/gold interface, Eur. Phys. Lett. **57** (2002) 74–79

23.70 D. Dowson, G.R. Higginson: *Elastohydrodynamic Lubrication* (Pergamon, Oxford 1968)

23.71 L. Bocquet, J.-L. Barrat: Hydrodynamic boundary conditions and correlation functions of confined fluids, Phys. Rev. Lett. **70** (1993) 2726–2729

23.72 P.A. Thompson, M.O. Robbins: Simulations of contact-line motion: Slip and the dynamic contact angle, Phys. Rev. Lett. **63** (1989) 766–769

23.73 J.-L. Barrat, L. Bocquet: Large slip effect at a nonwetting fluid-solid interface, Phys. Rev. Lett. **82** (1999) 4671–4674

23.74 M.J. Stevens, M.O. Robbins: Melting of Yukawa systems: A test of phenomenological melting criteria, J. Chem. Phys. **98** (1993) 2319–2324

23.75 J. Gao, W.D. Luedtke, U. Landman: Structure and solvation forces in confined films: Linear and branched alkanes, J. Chem. Phys. **106** (1997) 4309–4318

23.76 A. Koike, M. Yoneya: Chain length effects on frictional behavior of confined ultrathin films of linear alkanes under shear, J. Phys. Chem. B **102** (1998) 3669–3675

23.77 E.D. Smith, M. Cieplak, M.O. Robbins: The friction on adsorbed monolayers, Phys. Rev. B **54** (1996) 8252–8260

23.78 I. Bitsanis, S.A. Somers, H.T. Davis, M. Tirrell: Microscopic dynamics of flow in molecularly narrow pores, J. Chem. Phys. **93** (1990) 3427–3431

23.79 A.L. Demirel, S. Granick: Glasslike transition of a confined simple fluid, Phys. Rev. Lett. **77** (1996) 2261–2264

23.80 A.R.C. Baljon, M.O. Robbins: Energy dissipation during rupture of adhesive bonds, Science **271** (1996) 482–484

23.81 M.O. Robbins, E.D. Smith: Connecting molecular-scale and macroscopic tribology, Langmuir **12** (1996) 4543–4547

23.82 G. He, M.H. Müser, M.O. Robbins: Adsorbed layers and the origin of static friction, Science **284** (1999) 1650–1652

23.83 J.H. Dieterich, B.D. Kilgore: Direct observation of frictional contacts: New insights for state-dependent properties, Pure Appl. Geophys. **143** (1994) 238–302

23.84 P. Berthoud, T. Baumberger: Shear stiffness of a solid-solid multicontact interface, Proc. R. Soc. Lond. A **454** (1998) 1615–1634

23.85 A.L. Demirel, S. Granick: Transition from static to kinetic friction in a model lubricated system, J. Chem. Phys. **109** (1998) 6889–6897

23.86 I.L. Singer: Solid lubrication processes. In: *Fundamentals of Friction: Macroscopic and Microscopic Processes* (Elsevier, Amsterdam 1992) pp. 237–261

23.87 B.J. Briscoe: Friction and wear of organic solids and the adhesion model of friction, Philos. Mag. A **43** (1981) 511–527

23.88 B. J. Briscoe, A. C. Smith: The interfacial shear strength of molybdenum disulphide and graphite films, Trans. ASLE **25** (1982) 349–354

23.89 A. Berman, C. Drummond, J. N. Israelachvili: Amontons's law at the molecular level, Tribol. Lett. **4** (1998) 95–101

23.90 G. He, M. O. Robbins: Simulations of the static friction due to adsorbed molecules, Phys. Rev. B **64** (2001) 035413

23.91 J. Gao, X. C. Zeng, D. J. Diestler: Nonlinear effects of physisorption on static friction, J. Chem. Phys. **113** (2000) 11293–11296

23.92 G. He, M. O. Robbins: Simulations of the kinetic friction due to adsorbed molecules, Tribol. Lett. **10** (2001) 7–14

23.93 F. Heslot, T. Baumberger, B. Perrin, B. Caroli, C. Caroli: Creep, stick-slip, and dry-friction dynamics: Experiments and a heuristic model, Phys. Rev. E **49** (1994) 4973–4988

23.94 A. Ruina: Slip instability and state variable friction laws, J. Geophys. Res. **88** (1983) 10359–10370

23.95 M. G. Rozman, M. Urbakh, J. Klafter: Stick-slip motion and force fluctuations in a driven two-wave potential, Phys. Rev. Lett. **77** (1996) 683–686

23.96 M. G. Rozman, M. Urbakh, J. Klafter: Stick-slip dynamics as a probe of frictional forces, Europhys. Lett. **39** (1997) 183–188

23.97 M. G. Rozman, M. Urbakh, J. Klafter, F. J. Elmer: Atomic scale friction and different phases of motion of embedded systems, J. Phys. Chem. B **102** (1998) 7924–7930

23.98 H. Yoshizawa, J. N. Israelachvili: Fundamental mechanisms of interfacial friction. 1. Stick-slip friction of spherical and chain molecules, J. Phys. Chem. **97** (1993) 11300–11313

23.99 D. E. McCumber: Effect of ac impedance on the dc voltage-current characteristics of superconductor weak-link junctions, J. App. Phys. **39**(7) (1968) 3113–3118

23.100 J. H. Dieterich: Modeling of rock friction. 2. Simulation of pre-seismic slip, J. Geophys. Res. **84** (1979) 2169–2175

23.101 A. A. Batista, J. M. Carlson: Bifurcations from steady sliding to stick slip in boundary lubrication, Phys. Rev. E **57** (1998) 4986–4996

23.102 J. H. Dieterich, B. D. Kilgore: Implications of fault constitutive properties for earthquake prediction, Proc. Natl. Acad. Sci. USA **93** (1996) 3787–3794

23.103 A. Dhinojwala, S. C. Bae, S. Granick: Shear-induced dilation of confined liquid films, Tribol. Lett. **9** (2000) 55–62

23.104 P. A. Thompson, G. S. Grest: Granular flow: Friction and the dilatancy transition, Phys. Rev. Lett. **67** (1991) 1751–1754

23.105 H. M. Jaeger, S. R. Nagel, R. P. Behringer: Granular solids, liquids, and gases, Rev. Mod. Phys. **68** (1996) 1259–1273

23.106 S. Nasuno, A. Kudrolli, J. P. Gollub: Friction in granular layers: Hysteresis and precursors, Phys. Rev. Lett. **79** (1997) 949–952

23.107 A. R. C. Baljon, M. O. Robbins: Stick-slip motion, transient behavior, and memory in confined films. In: *Micro/Nanotribology and Its Applications*, ed. by B. Bhushan (Kluwer, Dordrecht 1997) pp. 533–553

23.108 A. L. Demirel, S. Granick: Friction fluctuations and friction memory in stick-slip motion, Phys. Rev. Lett. **77** (1996) 4330–4333

23.109 B. J. Ackerson, J. B. Hayter, N. A. Clark, L. Cotter: Neutron scattering from charge stabilized suspensions undergoing shear, J. Chem. Phys. **84** (1986) 2344–2349

23.110 M. Lupowski, F. van Swol: Ultrathin films under shear, J. Chem. Phys. **95** (1991) 1995–1998

23.111 A. Buldum, S. Ciraci: Interplay between stick-slip motion and structural phase transitions in dry sliding friction, Phys. Rev. B **55** (1997) 12892–12895

24. Mechanics of Biological Nanotechnology

Mechanics of

One of the most compelling areas to be touched by nanotechnology is biological science. Indeed, we will argue that there is a fascinating interplay between these two subjects, with biology as a key beneficiary of advances in nanotechnology as a result of a new generation of single molecule experiments that complement traditional assays involving statistical assemblages of molecules. This interplay runs in both directions, with nanotechnology continually receiving inspiration from biology itself. The goal of this chapter is to highlight some representative examples of the exchange between biology and nanotechnology and to illustrate the role of nanomechanics in this field and how mechanical models have arisen in response to the emergence of this new field. Primary attention will be given to the particular example of the processes that attend the life cycle of bacterial viruses. Viruses feature many of the key lessons of biological nanotechnology, including self assembly, as evidenced in the spontaneous formation of the protein shell (capsid) within which the viral genome is packaged, and a motor-mediated biological process, namely, the packaging of DNA in this capsid by a molecular motor that pushes the DNA into the capsid. We argue that these processes in viruses are a compelling real-world example of nature's nanotechnology and reveal the nanomechanical challenges that will continue to be confronted at the nanotechnology-biology interface.

24.1 **Science at the Biology–Nanotechnology Interface** 740
24.1.1 Biological Nanotechnology 740
24.1.2 Self-Assembly as Biological Nanotechnology 740
24.1.3 Molecular Motors as Biological Nanotechnology 740
24.1.4 Molecular Channels and Pumps as Biological Nanotechnology 741
24.1.5 Biologically Inspired Nanotechnology........................... 742
24.1.6 Nanotechnology and Single Molecule Assays in Biology 743
24.1.7 The Challenge of Modeling the Bio-Nano Interface................. 744

24.2 **Scales at the Bio-Nano Interface**............ 746
24.2.1 Spatial Scales and Structures 747
24.2.2 Temporal Scales and Processes....... 749
24.2.3 Force and Energy Scales: The Interplay of Deterministic and Thermal Forces 750

24.3 **Modeling at the Nano-Bio Interface**....... 752
24.3.1 Tension Between Universality and Specificity 752
24.3.2 Atomic-Level Analysis of Biological Systems 753
24.3.3 Continuum Analysis of Biological Systems 753

24.4 **Nature's Nanotechnology Revealed: Viruses as a Case Study**.......................... 755

24.5 **Concluding Remarks** 760

References 761

Nanotechnology is the seat of a broad variety of interdisciplinary activity with applications as diverse as optoelectronics, microfluidics, and medicine. However, one of the richest interdisciplinary areas only now beginning to be harvested is the interface between the biological sciences and nanotechnology. The aim of the present chapter is threefold. First, we wish to illustrate the synergy that exists between biological nanotechnology (e.g., molecular motors, transmembrane pumps, etc.) and biologically inspired nanotechnology (e.g., synthetic proteins, artificial viruses aimed at delivering drugs, biofunctionalized cantilevers, etc.). Secondly, through a series of order-of-magnitude estimates and an associated examination of the units for describing spatial dimensions, temporal processes, forces at the nanoscale, and the energy budget in nanoscale systems, we aim to build an intuitive sense of the workings of nanotechnology in the biological setting. Finally, through several specific case studies, we hope to illustrate some of the challenges faced in modeling the mechanical processes of biological nanotechnology and show how such challenges have been met thus far and how they might be met in the future.

24.1 Science at the Biology–Nanotechnology Interface

24.1.1 Biological Nanotechnology

Though the innovations leading to the adoption of the expression "nanotechnology" are indeed impressive, the perspective of this chapter is that one need only look inward to the way that our muscles move, to how we digest and synthesize molecules of dazzling complexity, to the way in which we think the thoughts that permit us to fill the shelves of libraries with scientific journals to realize that the greatest nanotechnology of all is that which is revealed in the living world. That is, one of the central thrusts is the idea that the microscopic workings of life offer an inspiring vision of nanotechnology. Clearly, the diversity of the examples of "nanotechnology" seen in the living world can (and do) fill the pages of learned texts. A guided tour of the machinery of the cell can be found in *Alberts* et al. [24.1], while a vision of the cell as an assemblage of "protein machines" has been argued for by *Alberts* [24.2]. The ambition of the present section is to provide several cursory examples of the nanotechnological marvels that power the living realm.

24.1.2 Self-Assembly as Biological Nanotechnology

One of the most intriguing nanotechnological tricks of the living world is the central role played by self-assembly in such systems. Whether we consider the spontaneous assembly of viruses, either in test tubes or in the interior of an infected cell, or the fusion of one membrane-bound region to another through vesicle fusion, spontaneous formation of functional "materials" is a key part of the biological repertoire. To be more concrete, we note that self-assembly in biological systems takes place in a number of different guises. First, the self-assembly of linear assemblies is a key part of the cytoskeletal assembly process with G-actin associating to form actin filaments and, similarly, tubulin monomers joining to form microtubules. This is also the process that is used by certain bacteria such as *Listeria* for locomotion. Both of these examples are described more fully in [24.3]. This same type of process is taken to the next level of sophistication in simple viruses such as tobacco mosaic virus that involve not only the assembly of protein monomers, but also the genetic message in the form of long RNA molecules. A second broad class of self-assembly processes is associated with the formation of containers such as liposomes or viral capsids. In the case of liposomes, lipid molecules such as phosphatidylcholine spontaneously organize in a way that sequesters the hydrophobic tails from the surrounding water. Similarly, protein subunits spontaneously assemble to form viral capsids [24.4]. These structures play a variety of roles in the biological realm, from serving as containers for different macromolecules to providing for concentration gradients of small ions that are maintained and utilized by molecular motors such as ATPsynthase [24.1].

Beyond the simplistic description of self-assembly advanced above, a second key feature of biological self-assembly must also be considered. In particular, a variety of self-assembly processes in the biological realm are templated (or coded). As is well-known, proteins are a hugely versatile class of molecules all based upon the same fundamental building blocks. Interestingly, the enormous diversity of protein action is founded upon 20 distinct amino acid building blocks, and the template for assembling a given protein is carried in the form of messenger RNA, which is then read and used as the basis for protein synthesis by the ribosome. One of the most exciting developments in modern nanotechnology is the attempt to exploit templated self-assembly processes for the purposes of creating new materials. An example in the context of protein-based materials can be found in the article of *van Hest* and *Tirrell* [24.5], where the machinery of the cell has been tricked into incorporating artificial amino acids into synthetic proteins.

24.1.3 Molecular Motors as Biological Nanotechnology

One of the hallmarks of life is change and motion. At the cellular level, such motion is effected through a dizzying variety of mechanisms, most of which when viewed at the molecular level are seen to be the result of the action of molecular motors [24.3]. For example, muscle contraction reflects the coherent action of huge numbers of myosin motor molecules as they march along actin filaments. Similarly, the motion of certain bacteria can be traced to the rotation of a rotary motor embedded in the cell wall, which is attached to filaments known as flagella [24.6]. In a similar vein, just as our modern society is replete with examples of systems aimed at allowing for communication and transport between widely separated geographic locations, so, too, has the living world had to answer these same challenges. As will be shown in the section of this chapter aimed at making estimates of various scales and processes in nanotechnology, one of the key mechanisms for communication and transport is diffusion. That is, a chemical concentration at one place can make its presence known at a distance x with a characteristic time, $t_{\text{diffusive}} \approx x^2/2D$, where D is the diffusion constant. However, as will be shown later in the chapter, there are many cellular processes that cannot wait as long as $t_{\text{diffusion}}$. As a result, a host of molecular motors and an associated transport system (the elements of the cytoskeleton) permit active transport. For example, kinesin is a motor molecule that transports material along long, relatively rigid polymeric assemblies known as microtubules [24.1].

Even more incredible are rotary machines like those associated with ATPsynthase and bacterial flagella. ATPsynthase is a membrane-embedded machine that rotates and, in so doing, synthesizes new ATP molecules (adenosine triphosphate, energy currency of the cell) [24.1, 7–9]. Similarly, the bacterial flagellar motor rotates with the result that the attached bacterial flagellum rotates and thereby induces motion of the cell [24.3, 6]. The exquisite details of the construction of these devices are themselves breathtaking. The rotary motors in bacteria are constructed from several components that are much like a rotor and stator and perform periodic motions by deriving energy from the flow of protons across a membrane [24.6]. These rotary motors are very powerful, as evidenced by F_1-ATPsynthase, which can generate a torque large enough to rotate a molecule of actin 100 times its own length [24.10]. In addition, bacterial motors have another layer of sophistication in that stimuli from the external environment, which are sensed by the bacteria through the pores on its membrane, can change the direction of rotation of the motor in a process known as chemotaxis [24.3]. Feedback and signal transmission are implemented in engineering devices by means of complex circuitry, whereas in the living world these functions are often accomplished by means of chemistry and conformational changes in large molecules.

All of these molecular machines involve a rich interplay between chemistry, thermodynamics, and mechanics. From a structural perspective, most molecular motors are proteins with different subunits performing different functions [24.11]. Often there is some region within the motor where chemical energy is derived from the hydrolysis of adenosine triphosphate (ATP). This chemical energy is then converted to mechanical energy through a conformational change in the protein. These examples serve to call the reader's attention to the importance and variety of motor molecules found throughout the living world and which almost any sensible definition of nanotechnology would have to include as particulary sophisticated examples.

24.1.4 Molecular Channels and Pumps as Biological Nanotechnology

From a structural perspective, one of the most intriguing features of cellular systems is their division into a number of separate membrane-bound compartments. We have already touched upon this compartmentalization in the context of self-assembly, and note that the role of membranes and the proteins bound within them is described clearly in *Alberts* et al. [24.1]. The presence of such compartments, often marked by large concentration gradients with respect to the surrounding medium, hints at another nanotechnological wonder of the living world, namely, the presence of a wide range of transmembrane channels that mediate the exchange of material between these different compartments. Certain passive versions of these channels are gated by various mechanisms such as the arrival of signaling molecules or tension in the membrane within which they are found [24.12–14]. Once the channel is in the open state, ions pass through passively, by diffusion. Active versions of ion channels that transfer ions such as Na^+ and Ca^{2+} are similarly critical for the functioning of a cell, both in the case of unicellular and multicellular organisms. For example, in the case of Na^+ ions, typical concentrations within the cell can be as much as a factor of 10–20 less than those in the extracellular milieu. Such concentration gradients imply the need for sophisticated active "devices" which can do work against such gradients. One of the most remarkable of machines of this type is the Na^+–K^+ pump [24.1]. This machine is powered by hydrolysis of ATP (i. e., the consumption of ATP fuel), and it can pump ions up a potential gradient. In particular, this pump hydrolyzes ATP and pumps Na^+ ions *out* of the cell against a very steep concentration gradient while pumping K^+ ions *in* again against a steep gradient.

The point of this brief discussion has been to illustrate the first of several perspectives that we will bring to bear on the question of the biology-nanotechnology interface. Thus far, we have noted that nature is replete with examples of macromolecules and macromolecular assemblies that perform nanotechnological tasks and in this capacity serve as examples of biological nanotechnology. Next, we wish to examine the ways in which biological phenomena can inspire nanotechnology itself.

24.1.5 Biologically Inspired Nanotechnology

As noted in the beginning of this chapter, biological systems are nanotechnologically relevant for several reasons. First, as shown above, the living world is full of examples of nanotechnological devices. However, a second key point is that nanotechnology has been driven and inspired by the example of biological systems and the need (for example, in medicine) to influence biological systems at the scale of a single cell. In addition, preliminary steps have been taken to harness the nanotechnology of biological systems and use it to perform useful functions.

One compelling example of a proof-of-principle, biologically inspired device emerged from work aimed at exploring the function of ATPsynthase, the rotary device already described above. Fluorescently labelled filamentous proteins from the cytoskeleton, known as actin, were attached to the putative rotary component of the F_1 subunit. The point of this exercise is that such filaments are observable using light microscopy. It was then observed that the long actin filament rotated like a propeller when the ATPsynthase performed ATP hydrolysis [24.15].

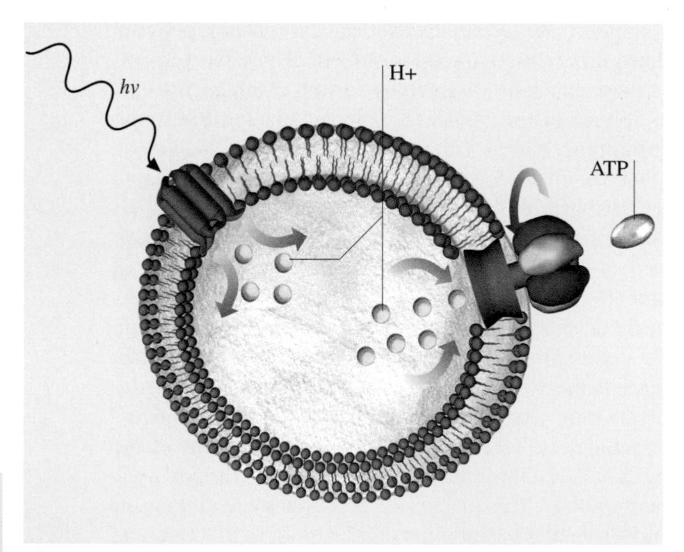

Fig. 24.1 Schematic of the experiment in which bacteriorhodopsin (*top of figure*) and ATPsynthase (*right of figure*) are artificially reconstituted in a liposome, and they act in unison to produce ATP molecules

A second example, also involving ATPsynthase, is suggested by a set of beautiful experiments done by *Racker* and others [24.16, 17]. The idea is illustrated schematically in Fig. 24.1. Two different protein machines are "reconstituted" in an artificial membrane-bound region known as a liposome. One of these devices is known as bacteriorhodopsin and has the capacity to pump hydrogen ions when it is exposed to light. Since both the bacteriorhodopsin and ATPsynthase are embedded in the same membrane, the ATPsynthase can then exploit the light-induced proton gradient to perform ATP synthesis. Again, these experiments were undertaken not for their role as possible devices, but rather to probe the nature of various molecular machines. Nevertheless, we view them as a provocative demonstration of both the manipulation and use of such machines in artificial environments. As such, they provide an inspiring vision of the possibilities for biologically inspired nanotechnology.

Another fascinating example of biologically inspired nanotechnology is that of biofunctionalized cantilevers (see Fig. 24.2). A typical example is provided by the experiments of *Wu* et al. [24.18], who have demonstrated the use of a biofunctionalized cantilever as a scheme for detecting small concentrations of biologically interesting molecules such as prostrate-specific antigen and single-stranded DNA. One surface of the cantilever is coated with an antibody, and then it is placed in environments containing different concentrations of the antigen. The key ideas from a mechanical perspective are: i) the difference in surface energy between the top and bottom surfaces of the cantilever induces spontaneous bending and ii) the binding of molecules of interest to target molecules initially present on the surface leads to surface energy differences and bending that can be detected by optical means. In this way the concentration of the molecules of interest can be measured. The specificity and sensitivity of the method makes it viable for use in the laboratory, as well as for commercial purposes. We return to this example in our discussion of the modeling challenges posed by problems at the interface between biology and nanotechnology.

Fig. 24.2 Schematic of the use of biofunctionalized cantilevers as a tool for detecting molecules of biological interest (figure courtesy of Arun Majumdar, Berkeley)

24.1.6 Nanotechnology and Single Molecule Assays in Biology

One of the key refrains of the nanotechnological era is Feynman's quip that "there is plenty of room at the bottom" [24.19]. The benefits of miniaturization are evident at every turn in applications ranging from our cars to our computers. Associated with the development of the inspiring new techniques made possible by nanotechnology has been the emergence of a host of scientific opportunities. One of the arenas to benefit from these new techniques is biology. As scientists and engineers have taken the plunge to the nanotechnological "bottom" foreseen by Feynman, opportunities have constantly arisen to manipulate biological systems in ways that were heretofore unimagined, culminating in a new era of single molecule biology. Single molecule experiments have, in fact, presented us with a view of Feynman's "room at the bottom" as being filled with very complicated machines whose functioning makes life possible.

Single molecule assays complement statistical/collective studies involving a large number of molecular actors by revealing the prominent role of fluctuations at the sub-cellular scale. For example, a photospectrometer measures the optical response of a huge collection of molecules, whereas optical tweezers pull on a single molecule of DNA and enable us to follow the changes in conformation or the breakage of bonds. In fact, experimental methods are so advanced that it is now possible to manipulate a single molecule even as we watch it on a screen as it jiggles around in different conformations. In what follows, we give several examples of how nanotechnology has reached out to help create single molecule biology and in the process has led to the advent of new quantitative opportunities to investigate biological systems.

Atomic-Force Microscopy

One of the tools that has revolutionized nanotechnology, in general, and single molecule biology, in particular, is the atomic-force microscope. The AFM has helped create the field of single molecule force spectroscopy [24.20]. We note that mechanics has a long tradition of using force-extension data (much like the electrical engineer uses current-voltage data) to probe the inner workings of various materials. It is now possible to apply forces of known magnitude on a macromolecule and study how it deforms under the force. This furnishes structural information and provides insights into the energy landscape the molecule needs to navigate as it undergoes force-induced conformational changes. The energy of deformation associated with such molecules is primarily determined by weak forces such as hydrogen bonds, van der Waals contacts, and hydration effects. On a more philosophical note, these experiments force us to think in terms of forces and not energy, complementing the traditional views held in molecular biology, and they can lead to many new insights about the relation between structure and function in proteins, polynucleotides, and other macromolecular entities.

The AFM has been used in a wide variety of single molecule experiments on many of the key classes of molecules found in the living world, including nucleic acids, proteins, and carbohydrates. One fascinating example is the use of atomic force microscopy to examine the mechanical properties of the muscle protein titin [24.21]. The experiment is illustrated schematically in Fig. 24.3, which shows that there is a series of force-extension signatures (increasing force and extension followed by a precipitous load drop) that correspond to the unbinding of the individual domains that make up this protein.

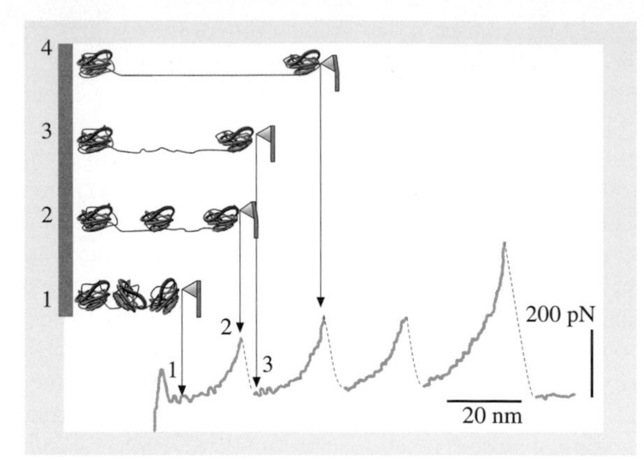

Fig. 24.3 Schematic illustrating the way that the AFM has been harnessed as a nanotechnological analogue of the Instron machine for the measurement of the force-extension properties of single molecules. Four snapshots in the life history of a globular protein subject to loading are shown, as well as the measured force-extension curve. Each sawtooth corresponds to the unfolding of a single protein domain. (Figure adapted from [24.22])

Optical Tweezers

Another instrument that has been used with great success in the realm of biophysics for the purposes of performing nanomechanical measurements on macromolecules and their assemblies is the optical tweezer [24.23,24]. While the AFM is relatively stiff and applies large forces (on the order of 100–1,000 pN), optical tweezers are compliant and can measure smaller forces (on the order of 0.1–10 pN). Some of the most interesting experiments performed with optical tweezers concern the functioning of molecular motors, which by themselves are marvels of nanotechnology. For example, *Svoboda* et al. [24.25] attached kinesin to an optically trapped bead and observed its movement along a microtubule. An example of the type of data to emerge from such experiments is shown in Fig. 24.4. One of the conclusions to emerge from such experiments is that kinesin can exert forces on the order of 5–7 pN before it stalls. Such experiments also permit an examination of the effect of changing the concentration of ATP on the functioning of kinesin [24.26]. Similar experiments have been performed on RNA polymerase as it advances along DNA to deduce not only the stall force, but also its velocity as a function of the constraining force [24.27]. Such measurements provide a mechano-chemical basis of biological function and go a long way in revealing connections between chemical kinetics and mechanical processes at the molecular level.

As noted in the abstract, one of the most fascinating examples of the biology-nanotechnology interface is that of bacterial viruses, known as bacteriophages. The life cycle of a large class of bacteriophages is characterized by self-assembly processes that lead to the formation of the protein shell of the virus followed by active packaging of the viral DNA within this shell by a motor. The structure of this so-called "viral portal motor" has recently been solved using X-ray crystallography [24.28]. In a recent experiment, *Smith* et al. used optical tweezers to study the characteristics of the DNA packaging process of the ϕ-29 bacteriophage [24.29]. One of the conclusions of this experiment is that the motor has to act against an increasing resistive force as more and more of the DNA is packed inside the capsid. From a quantitative perspective, this experiment yields the force and the rate of packing as a function of the fraction of the genome packed. It is also important to note that bacteriophages are not only an obscure subject of quiet enquiry, but are also the basis of a huge range of cloning products (see, for example, the lambda ZAP vectors of Stratagene) used for doing experiments with recombinant DNA, and more generally, viruses are being explored as the basis of gene therapy.

Our discussion thus far has been aimed at providing a rough overview of the vast landscape that sits at the interface between biology and nanotechnology. It is hoped that the few representative case studies set forth above suffice to illustrate our basic thesis, namely, that biological nanotechnology represents nanotechnology at its best.

24.1.7 The Challenge of Modeling the Bio-Nano Interface

As highlighted in the previous two subsections, there have been huge advances at the interface between nanotechnology and biology. We have argued that there are two distinct representations of the interface between biology and nanotechnology, and each has its own associated set of modeling challenges. The argument of the present discussion is that another key part of the infrastructure that must attend these developments is that associated with the modeling of these systems. One of the intriguing ways in which modeling at the biology-nanotechnology interface is assuming greater importance is that with increasing regularity, experimental data on biological systems is of a quantitative character. As a result, the models that are put forth to greet these experiments must similarly be of a quantitative character [24.30].

As an example of the type of modeling challenges that must be faced in contemplating the types of problems described above, we return to the example of biofunctionalized cantilevers as a problem in nanomechanics [24.18, 31]. The basic physics behind the use of biofunctionalized cantilevers as sensors is a competition between elastic bending energy and the surface free energy difference between the upper and lower faces of the cantilever. The face with the lower free energy per unit area tends to increase its area by bending the cantilever. The amount of bending, on the other hand, is limited by the elastic energy cost. The utility of this device derives from the fact that the difference in the surface free energy is affected by specific binding of target molecules to probe molecules that are initially deposited on one side of the cantilever.

To provide a quantitative model of the biofunctionalized cantilever, we construct an energy functional that takes into account the elastic energy of beam bending and the surface energy. Both contributions to the total energy can be written as functionals of $u(x)$, the deflection of the cantilever. Note that in the case in which the two surfaces have the same free energy per unit area, $\gamma_{\mathrm{up}} = \gamma_{\mathrm{down}}$, the equilibrium configuration corresponds to $u(x) = 0$. The case of interest here is that in which the two surfaces have different energies.

We recall that the energy associated with beam bending is of the form [24.32]

$$E_{\mathrm{bend}}[u(x)] = \frac{EI}{2}\int_0^L \left[u''(x)\right]^2 \mathrm{d}x\,, \tag{24.1}$$

where E is Young's modulus, L is the length of the beam, and $I = wt^3/12$ is the areal moment of inertia; t and w are the beam thickness and width. The main approximation we have made in writing (24.1) is that the cross section of the beam remains unchanged by the bending process. The surface contribution to the total energy is associated with the changes of the areas of the upper and lower surfaces by virtue of the beam deforming. In particular, we have

$$E_{\mathrm{surf}}[u(x)] = \gamma_{\mathrm{up}} w\left[L - \frac{t}{2}\int_0^L u''(x)\,\mathrm{d}x\right] + \gamma_{\mathrm{down}} w\left[L + \frac{t}{2}\int_0^L u''(x)\,\mathrm{d}x\right], \tag{24.2}$$

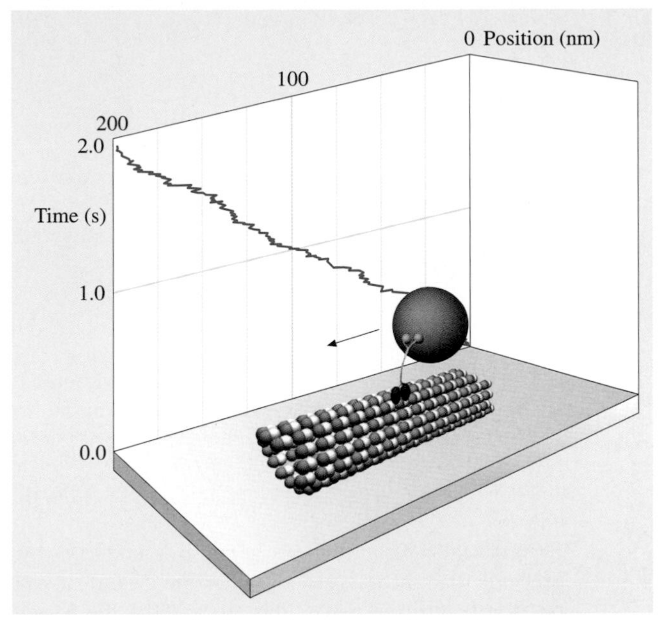

Fig. 24.4 Schematic illustrating the use of optical tweezers to measure the speed of the molecular motor kinesin in its journey along a microtubule

where the terms in parentheses are the arc lengths along the top and bottom surface of the beam, respectively. The physical content of this functional is the idea that if $u'' > 0$ (concave upward) then the upper surface area will shrink while the lower surface area will increase. Neglecting uninteresting constant terms, the total energy functional is

$$E_{\mathrm{tot}}[u(x)] = \frac{EI}{2}\int_0^L \left[u''(x)\right]^2 \mathrm{d}x + \frac{tw}{2}(\gamma_{\mathrm{down}} - \gamma_{\mathrm{up}})\int_0^L u''(x)\,\mathrm{d}x\,. \tag{24.3}$$

Our goal is to find the displacement profile $u(x)$ that yields the minimum value for the total energy. The general mathematical framework for effecting this minimization is the calculus of variations. For the functional in (24.3), a more direct route to the result can be obtained by completing the square. Namely, we note that the E_{tot} involves a term quadratic in $u''(x)$ and a second term that is linear in the same function. As a result, the total energy can be rewritten in the form

$$E_{\text{tot}}[u(x)] = \frac{EI}{2} \int_0^L \mathrm{d}x \left[u''(x) + \frac{tw\Delta\gamma}{2EI} \right]^2 , \quad (24.4)$$

where we have introduced $\Delta\gamma = \gamma_{\text{down}} - \gamma_{\text{up}}$, and we neglect an uninteresting constant term. Clearly the $u(x)$ that minimizes the energy is one for which the integrand vanishes everywhere, or

$$u''(x) + \frac{tw\Delta\gamma}{2EI} = 0 . \quad (24.5)$$

At this point, we are left with a standard differential equation whose solution, given the boundary conditions $u(0) = 0$ and $u'(0) = 0$, is

$$u(x) = -\frac{tw}{4EI} \Delta\gamma x^2 . \quad (24.6)$$

The physical meaning of this solution is hinted at by observing that if $\gamma_{\text{down}} > \gamma_{\text{up}}$ then the beam curves downward with the result that the area of the lower surface is reduced.

Measurements of *Wu* et al. [24.31] indicate an upward cantilever deflection when single stranded DNA (ssDNA), up to 20 nucleotides in length, hybridizes with complementary strands of ssDNA, which were deposited initially so as to functionalize the beam. This effect might be attributed to the difference in elastic properties of ssDNA and its double stranded counterpart. Namely, under physiological conditions, ssDNA has a persistence length equal to two nucleotides, while dsDNA is much stiffer (due to hydrogen bonding and base stacking interactions between the two strands) and has a persistence length of 150 nucleotides. Therefore, deposition of the flexible ssDNA molecules initially leads to bending of the cantilever downward due to entropic repulsion between the ssDNA chains. After hybridization, rigid dsDNA strands are formed, there is no longer any entropy to be gained by increasing the area of the top surface, and the beam bends back upwards. Remarkably, *Wu* et al. demonstrate that their biofunctionalized cantilever is sensitive to ssDNA differing in length by a single nucleotide!

To gain further insight into the physics of the biofunctionalized cantilever, it is instructive to examine some quantitative aspects of the experiment using (24.6). Namely, *Wu* et al. find a deflection of $u(L) = 12\,\text{nm}$ when a 20-nucleotide strand of ssDNA hybridizes with a 20-nucleotide-long complementary target. Using the quoted numbers for the cantilever, $E = 180\,\text{GPa}$, $L = 200\,\mu\text{m}$, $t = 0.5\,\mu\text{m}$, $w = 20\,\mu\text{m}$, leads to $\Delta\gamma = 4.5\,\text{fJ}/\mu\text{m}^2$. From this result we can estimate the change in surface free energy due to beam bending,

$$\Delta E_{\text{surf}} \approx wL\Delta\gamma . \quad (24.7)$$

We find $\Delta E_{\text{surf}} = -18\,\text{pJ}$, which, given the quoted areal chain density of $6 \times 10^{12}\,\text{cm}^{-2}$, leads to a decrease in free energy of 75 pN nm, or $18\,k_{\text{B}}T$, per chain. This is comparable to the entropy of a 20 nt ssDNA, which is on the order of $10\,k_{\text{B}}$, lending support to the idea that entropic repulsion between ssDNA strands is implicated in cantilever bending.

While this example gives a feel for the way in which quantitative models have been put forth to respond to biologically inspired nanotechnologies, the remainder of the chapter will emphasize attempts to construct nanomechanical models of biological nanotechnology itself. We turn first to a discussion of the various scales that arise in thinking about the biology-nanotechnology interface and then conclude with several modeling examples.

24.2 Scales at the Bio-Nano Interface

Every scientific discipline has a preferred set of units that lends itself to building intuition concerning the system at hand. For example, an astronomer thinks of distances between stars in light years, not kilometers. Though most of us have an intuitive sense of the meaning of a kilometer, by the time we add more than six zeros, all intuition is lost. At terrestrial scales, we talk of distances between cities in terms of the flying time or the driving time between them and usually not in terms of hundreds of miles. Similar choices must be faced in the biological setting. For example, a biologist might characterize the complexity of an organism by the size (in kilo basepairs) of the genome and not by the organism's physical size. The aim of the present section is to highlight some of the key scales and units that reveal themselves at the interface between biology and nanotechnology. Indeed, we go further and assert that as yet we are still in the process of searching for the most suitable units to characterize the biology-nanotechnology interface. Although our attempt to determine such units and scales might involve seemingly complicated interconversions, such as measuring distances in terms of time (via diffusion), or measuring concentrations in terms of distances, we hold that the approximate *numerical* characterization of the scales of interest is of crucial importance to the endeavor of considering nanomechanics at the biology-nanotechnology interface. In particular, the right choice of units can assist us in building intuition about these systems. Our goal in this section is to emphasize the scales in length, time, force, energy, and power that are relevant in contemplating the nanomechanics of biological systems.

24.2.1 Spatial Scales and Structures

We begin with a discussion of the length scales that arise in contemplating nanomechanics at the biology-nanotechnology interface. In this case, the prefix *nano* leads justifiably to a consideration of the nanometer as one possible choice as the fundamental unit of length. However, to prepare ourselves for the question of how best to describe the dimensions of the spatial structures of interest here, it is important to consider the *hierarchy* of length scales that arise in the nano-bio arena. After examining this hierarchy of structures, we reformulate these length scales in terms of the volumes of these structures measured in units of the volume of a typical bacterial cell, and conclude the present section with a discussion of the way that chemical concentrations can also be interpreted as determining a length scale.

As noted above, a first step in developing intuition concerning the spatial scales found at the biology-nanotechnology interface is through reference to the hierarchy of scales and structures that arise in this arena. The shortest distance in this hierarchy of scales that will interest us is that associated with the size of individual atoms. We recall that the size of a hydrogen atom is roughly 0.1 nm. The scale characterizing the linkage between atoms is that of typical bond lengths that range from roughly 0.1–0.3 nm. A step further in this hierarchy brings us to the basic building blocks of the biological world such as amino acids, nucleotides, individual sugars, lipid molecules, etc. A typical length scale that characterizes these building blocks is the nanometer itself. For example, we have already made reference to the importance of lipid-bilayer membranes in bounding different regions of the cell. Phosphatidylcholine is one of the molecular building blocks of many such membranes. It has a polar head group and a hydrocarbon tail with an overall dimension of 2–3 nm. Similarly, the dimensions of single amino acids and nucleotides are on the order of 1 nm as well.

As is well-known, individual nucleotides are assembled to form nucleic acids such as DNA, amino acids are assembled to form proteins, sugars combine to form polysaccharides, and lipids self-assemble to form membranes. One way of estimating the size of the resulting molecules is by taking the scale of the individual units and scaling up with the number of such units. The various molecular actors of relevance to the present discussion can also be characterized by a length, scale known as the persistence length, which gives a rough description of the length over which the molecule behaves as a stiff rod. For example, the persistence length of DNA is ≈ 50 nm, while that of the cytoskeletal filaments is in the range of 15 μm for actin and 6 mm for microtubules [24.11]. We note that in our later discussions of viruses as a profound example of biological nanotechnology, the ratio of the persistence length of DNA to the size of the viral capsid (the container within which the DNA is packaged) will serve as a measure of the energetic cost of packing the DNA within the capsid and will signal the need for molecular motors to take active part in the packaging process.

The next scale above that of the various macromolecular building blocks are assemblies of such molecules in the form of various molecular machines which are some of the most compelling examples of nature's nanotechnology. The ability to begin formulating mechanistic models of such machines is founded upon key advances in both X-ray crystallography and cryo-electron microscopy [24.33, 34]. Examples of such machines and their associated dimensions include: the machine responsible for making the message carried by DNA readable by the protein synthesis machinery, namely, RNA polymerase (≈ 15 nm) [24.35], the machine that produces ATP, the energy currency of the cell, namely, ATPsynthase (≈ 10 nm) [24.9], and the machine that carries out protein synthesis, namely, the ribosome (≈ 25 nm) [24.36]. A second class of assemblies of particular relevance to the present article are viruses, representative examples of which include lambda phage (≈ 27 nm), tobacco mosaic virus (≈ 250 nm in length), and the HIV virus (≈ 110–125 nm) [24.37].

From the standpoint of cellular function, the next level of structural organization is associated with the various organelles within the cell, structures such as the cell nucleus (3–10 μm), the mitochondria that serve as the power plant of the cell ($\approx 1{,}000$ nm), the Golgi apparatus wherein modifications are made to newly synthesized macromolecular components ($\approx 1{,}000$ nm), and so on [24.1]. These organelles should be thought of as factories in which many molecular motors of the type described earlier do their job simultaneously. At larger scales yet, and constituting a higher level of overall organization, life as nanotechnology is revealed as cells themselves, the fundamental unit of life that is self-replicating and self-sustaining. We will make special reference to one particular bacterial cell, namely, *E. coli* with typical dimensions of 1 μm. This should be contrasted with a typical eucaryotic cell such as yeast, which has linear dimensions roughly a factor of 10 larger than the *E. coli* cell.

We note that the various biological structures described above should be seen with reference to the sorts of man-made structures to which they are interfaced. For example, earlier in the chapter we mentioned the importance of optical tweezers as a means of communicating forces to macromolecules and their assemblies. Typical dimensions for the optical beads used in optical tweezers are on the scale of 500–1,000 nm. We note that though such dimensions are characteristic of organelles, they are much larger than the individual molecules they are used to study. A second way in which individual macromolecules are communicated with is through small tips such as those found on an AFM. In this case, the size of the tip can be understood through reference to its radius of curvature, which is typically of the order of 50 nm.

Our discussion thus far has centered on the use of a single characteristic length to describe the spatial extent of biological structures. In our quest to develop intuition about the typical spatial scales found at the biology-nanotechnology interface we note that a second way to evaluate such scales is through reference to the typical *volumes* of the various structures of interest. As noted in the introduction to this section, we claim that a key way to develop intuition is by making sure to use appropriate and revealing units. For the present purposes, we argue that one useful unit of volume is that of an *E. coli* bacterium, which if idealized as a cylinder with diameter 1 μm and height 2 μm, has a volume of $\approx 1.5 \times 10^9\ \text{nm}^3$. In particular, we will measure the sizes of the various entities described above in terms of how many of them can fit into a single *E. coli* bacterium. Our estimates of the comparative sizes of various structures of interest are given in Table 24.1. Note that we do not claim that this is literally how many of each of these entities are found in an *E. coli* bacterium, but rather we seek to give an impression of the relative volumes of some of the different structures that arise when contemplating biological nanotechnology.

Table 24.1 Sizes of entities in comparison to the size of *E. coli*

Entity	Size (nm)	# fitting in an *E. coli*
Amino acid	1	5.0×10^9
Nucleotide	1	5.0×10^9
Monosaccharide	1	5.0×10^9
Phosphatidyl choline	2–3	2.0×10^8
Proteins	5–6	3.0×10^7
Ribosome	20	1.0×10^6
Mitochondria	500–1,000	3–4
T4 phage	100	5,000

The idea of counting the number of molecules that fit into a given volume is actually quite standard. This is exactly what chemists do when they invoke the notion of concentration: measuring the strength of an acid or base using pH amounts to expressing the concentration of H^+ ions in solution. Similarly, when we refer to the molarity of a given solution, it is a statement of how many copies of a given molecule will be found per liter of solution. These ideas are pertinent for our examination of the scales that arise in contemplating biological nanotechnology. As an example, we consider the action of molecular pumps such as that which maintains the concentration gradient in Ca^{2+} ions between the cellular interior and the extracellular medium. Note that this is described in detail in chapter 12 of *Alberts* et al. [24.1]. The Ca^{2+} pump is responsible for insuring that the intercellular concentration of Ca^{2+} ions is 10^{-7} smaller than that in the cellular exterior. One explanation for this low a concentration of Ca^{2+} in the cell is the role played by these ions in signaling other activities. The low concentration might serve as a scheme for increasing the signal-to-noise ratio. The reported intracellular concentration of such ions is roughly 10^{-7} mM. It is interesting to ask how many Ca^{2+} ions this corresponds to in a eucaryotic cell. The volume of a eucaryotic cell measuring 20 μm in diameter is about 4×10^{-12} liters, which translates into ≈ 250 Ca^{2+} ions in the cytoplasm. For the case of *E. coli*, which has a volume approximately 1,000 smaller, this concentration would correspond to one ion for every four cells. This number gives us far greater insight than the standard molar (M = mol/liter) method of expressing concentrations. In a similar vein, we argue that distances between individual ions/molecules in solution are perhaps a better way of thinking about pH and molarity when using them in the context of biological nanosystems. To this end, Fig. 24.5 shows a plot of pH and molarity versus average distance d in nm between individual ions/molecules. Note that, in addition, we have also translated these concentrations into number of copies per *E. coli* cell and the time for an ion to diffuse over the average separation distance. For the time estimate we make use of the typical diffusion constant for small ions in water at 25 °C, $D = 2{,}000\,\mu\text{m}^2/\text{s}$.

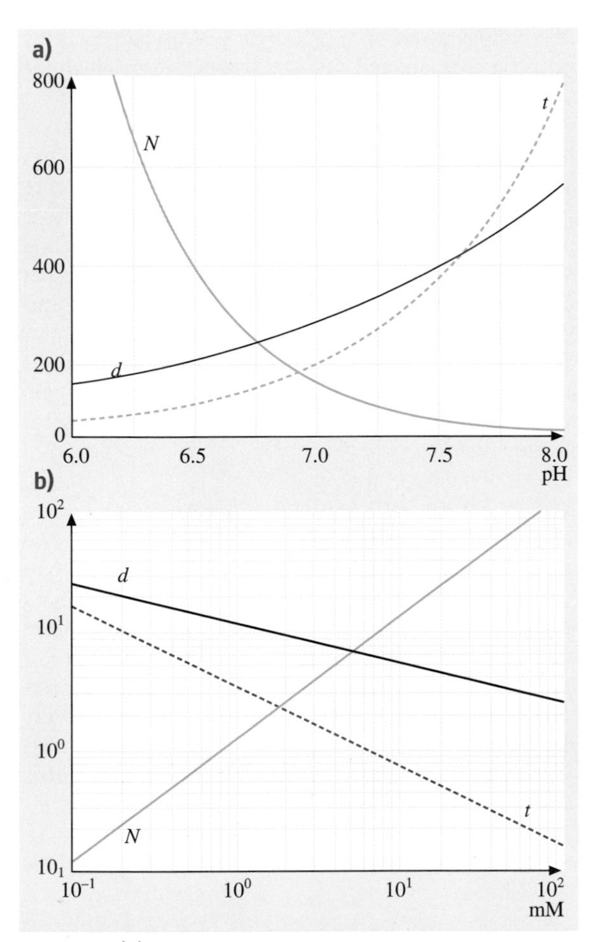

Fig. 24.5 **(a)** Representations of pH in terms of average distance between ions (d in nm), the number of ions per one *E. coli* cell volume (N), and the typical time for an ion to diffuse over a distance equal to their average separation (t in 100 ns). **(b)** Concentration unit millimolar (mM) represented in terms of d (nm), N (millions of molecules), and t (10 ns)

We have seen that the world of biological nanotechnology utilizes length scales that span a rather wide range, from fractions of a nanometer to tens of microns. This provides a challenge to modeling, whereby methods with atomic-scale resolution need to be combined in a consistent and seamless way with coarse-grained continuum descriptions to provide a complete picture. An example of such modeling will be provided in later sections when we take up the mechanical aspect of the viral life cycle.

24.2.2 Temporal Scales and Processes

We have seen that a study of spatial scales is a first step in our quest to understand the units that are suited to characterizing the biology-nanotechnology interface. We note that our understanding of spatial structures both in the living world, as well as those used in intriguing man-made technologies such as microelectronics have been built around important advances such as electron microscopy, X-ray crystallography, nuclear magnetic resonance, etc. that have revolutionized structural biology and materials science alike. On the other hand, one of the key challenges that remain in really appreciating the structure-properties paradigm, which is as central to biology as it is to materials science, is the need to acknowledge the dynamical evolution of these structures.

We note that just as there is a hierarchy of spatial scales that are important to consider in describing biological nanotechnology, so too is there a hierarchy of *temporal processes* that also demand a careful consideration of relevant units. An impressive representation of the temporal hierarchy that must be faced in contemplating biological systems is given by *Chan* and *Dill* [24.38] in their Fig. 5. They organize their temporal hierarchy according to factors of 1,000, starting at the femtosecond time scale and ending with time scales representative of the cell cycle itself. One of the elementary dynamical processes undergone by all of the molecular actors of the living world is thermal vibration. Vibrations of atoms occur at the time scales of 10^{-15} to 10^{-12} s. As will be noted later, from a modeling perspective, this depressing fact manifests itself in the necessity of using time steps of the same order when integrating the equations of motion describing molecular dynamics. The reason that this is unfortunate is that almost everything of dynamical interest occurs at time scales much longer than the femtosecond time scale characteristic of molecular jiggling, thus making molecular dynamics investigations computationally very expensive.

We follow *Chan* and *Dill* [24.38] in examining how successive thousandfold increases in our temporal resolution smears out features such as vibrational motion and brings into focus more interesting processes characteristic of macromolecular function. Indeed, the side chains of amino acids rotate with a characteristic period on the order of 10^{-9} s. Another thousandfold increase in time scale begins to bring key biological processes such as polymerization into view. Just as a length scale on the order of 10–100 nm is perhaps most characteristic of the length scales of biological nanotechnology, the time scale that is most relevant for considering the processes associated with biological nanotechnology is something between a microsecond and a millisecond. To drive home this point, we reconsider the machines described in the previous section from the length scale perspective, but now with an eye toward how fast they perform their function. Several of the molecular machines considered in the previous section mediate polymerization reactions. For example, RNA polymerase produces messenger RNA molecules by the repeated addition of nucleotides to a chain of ever increasing length. Such messenger RNA molecules serve in turn as the template for synthesis of new proteins by the ribosome, which reads the message contained in the RNA and adds individual amino acids onto the protein. From the temporal perspective of the present section, it is of interest to examine the rate at which new monomers are added to these polymers – roughly 10 per s in the case of RNA polymerase [24.39], and 2 per s in the case of the ribosome in eucaryotic cells [24.1] (bacterial ribosomes are 10 times faster). Once proteins are synthesized, in order to assume their full cellular responsibilities, they must fold into their native state, a process that occurs on time scales on the order of a millisecond.

In addition to translational machines like RNA polymerase and the ribosome, there are a host of intriguing rotary machines such as ATP synthase and the flagellar rotary motor. The flagellar motor, which is the means of locomotion for several bacteria, rotates at about 100 rpm [24.6]. ATP synthase makes use of a proton gradient across a lipid bilayer to provide rotation at a rate of roughly 6,000 rpm [24.41], approaching that of turbojet engines.

Thus far, our discussion has emphasized the rate of a variety of active processes of importance to biological nanotechnology. It is of interest to contrast these scales with those pertaining to diffusion. We note that in many instances in the biological setting the time scale of interest is that determined by the time it takes for diffusive communication of two spatially separated regions. In particular, as noted earlier in the chapter, the operative time scale is given by $t_{\text{diffusive}} = x^2/2D$, where D is the diffusion constant and x is the distance over which the diffusion has taken place.

To get a sense for all of the time scales described in this section, Table 24.2 presents a variety of results considered above in tabular form, but now with all times measured in units of the time it takes for ATPsynthase to make one rotation ($\approx$ 10 ms). Our reason for adopting ATPsynthase rotation as defining a unit of time is that it exhibits motions associated with one of the most important of life's processes, namely, the synthesis of new ATP molecules. In addition, our later discussion of units associated with both energy and power will once again appeal to the central role played by ATP in biological nanotechnology.

24.2.3 Force and Energy Scales: The Interplay of Deterministic and Thermal Forces

Another set of scales of great relevance and importance to the nano-bio interface concerns the nature of the forces that act in this setting. Thus far, we have examined the spatial extent and the cycle time of a variety of examples of biological nanotechnology. As a next step in our examination of scales, we consider the forces

Table 24.2 Time scales of various events in biological nanotechnology measured in units of the average time taken for ATPsynthase to make a single rotation

Process	Time scale (units of ATPase rotation)	Ref.
1/(frequency of amino acid addition) – eukaryotic ribosome	50	[24.1]
1/(frequency of monomer addition) – RNA polymerase	10	[24.39]
Time between motion reset of *E. coli*	10	[24.6]
Kinesin step	1	[24.26]
Diffusion time for protein (ion) to cross an *E. coli*	0.5 (0.02)	[24.11]
Vibrational period of nanocantilever	10^{-3}–10^{-5}	[24.40]

and energies associated with these structures. Two outstanding sources for developing a feel for the relevant numbers are the books of *Wrigglesworth* [24.42] and *Smil* [24.43]. Perhaps the most compelling feature when thinking about forces at these scales is the interplay between deterministic and thermal forces. To substantiate this claim, we note that at room temperature the fundamental energetic quantity is $k_B T \approx 4.1$ pN nm. The reason this number is of interest to the current endeavor is the realization that many of the molecular motors that have thus far been investigated act with forces on the piconewton scale over distances of nanometers.

The fact that the piconewton is the relevant unit of force can be gleaned from a simple estimate. Namely, consider a typical skeletal muscle in the human arm. The cross-sectional size of the muscle is of order 3 cm. The muscle consists of cylindrical rods of protein called myofibrils, which are roughly 2 μm in diameter, while the myofibrils themselves are made of strands of actin and myosin filaments, which total some 60 nm in diameter [24.1]. This gives 10^{12} myosin filaments per cross-sectional area of the muscle. As each myosin filament over the length of a single sarcomere (the contractile unit of myofibrils, some 2.5 μm long) contains some 300 myosin heads, lifting a 30 kg load corresponds to a force of 1 pN per myosin head. This is certainly an underestimate, since not all myosin heads are attached to the actin filament at the same time. Since our estimate leaves out many details, we might wonder how it compares to the measured forces exerted by molecular motors. Sophisticated experiments with optical tweezers have revealed that actin-myosin motors stall at a force of around 5 pN [24.44]; RNA polymerase stalls at a force of about 20 pN [24.39]; the portal motor of the ϕ-29 bacteriophage exerts forces up to 50 pN [24.29].

As noted earlier, RNA polymerase is a molecular motor that moves along DNA while transcribing genes into messenger RNA. The DNA itself is an elastic object that is deformed by forces exerted on it by various proteins. At forces less than 0.1 pN its response is that of an entropic spring with a stretch modulus of 0.1 pN, while at forces exceeding 10 pN its stretch modulus is determined by hydrogen bonding of the base pairs and is roughly 1,000 pN [24.45]. All the above mentioned data reinforces the argument that piconewton is the relevant unit of force in the nanomechanical world of the cell.

The concept of "stress" is closely related to that of force. Stress is a continuum mechanical concept of force per unit area, and it has been used with great success in solid and fluid mechanics. We extend the idea of stress to the nanomechanical level to see what numbers we arrive at. From the data above we can deduce that a single myosin fiber sustains a stress of about 10^{-2} pN/nm^2, which is the same as 10^{-2} MPa. Migratory animal cells such as fibroblasts, which are responsible for scavenging and destroying undesirable products in tissues, can generate a maximum stress on their substrate on the order of 3.0×10^{-2} MPa [24.46]. DNA can sustain stresses in the excess of 20 pN/nm^2, or 20 MPa, at which point an interesting structural transition accompanied by an overall increase of its contour length is observed [24.45]. Engineering materials such as steel and aluminium, on the other hand, can sustain stresses of about 100 MPa. This goes to show that nanotechnology in the context of biological systems is built from rather soft materials.

It is of interest to translate our intuition concerning piconewton forces into corresponding energetic terms. The kinesin motor advances 8 nm in each step against forces as high as 5 pN [24.25]. This translates to a work done on the order of 40 pN nm. A myosin motor suffers displacements of about 15 nm with forces in the piconewton range. ATP hydrolysis (to ADP) releases energy (at pH 7 and room temperature) of about 50 pN nm [24.1]. When a titin molecule is pulled, it unfolds under forces of 30–300 pN, causing discrete expansions of 10–30 nm [24.22], implying energies in the range of 300–9,000 pN nm. Experimental data provides a compelling argument in favor of thinking of the pN nm as a unit of energy for nanomechanics. However, the observation that $k_B T = 4.1$ pN nm at room temperature gives important insight, since it reveals that thermal forces and entropic effects play a competing role in biological nanomechanics. This provides nature with unique design challenges, whereby molecular motors that can perform useful work must do so in the presence of strong thermal fluctuations for the normal functioning of the cell. Operation of motors in such a noisy environment is governed by laws that are probabilistic in nature. This is manifest in single molecule experiments that observe motors stalling and sometimes reversing direction.

Energy conversion is crucial to any developmental or evolutionary process. The steam engine powered the industrial revolution. However, long before thinking beings with man-made machines founded the industrial revolution, power generation had already become a central part of life's nanotechnology. Indeed, the development of ATP-synthase is one of the cornerstones for the evolution of higher life-forms. An inevitable concomitant of evolution is the necessity for faster and more efficient operations. The industrial revolution led to the emergence of bigger and faster modes of transport; biological evolution led to the emergence of complex and intelligent organisms. Invariably, a machine or an organism is limited in its abilities by the speed at which it can convert one form of energy (usually chemical) to other forms of energy (usually mechanical). This is why a study of their "power plants" becomes important.

Power plants (or engines) are usually characterized by their force-velocity curves and compared using their power-to-weight (P/W) ratio. For example, the myosin motor has a P/W ratio of 2×10^4 W/kg, the bacterial flagellar motor stands at 100 W/kg, an internal combustion engine is at about 300 W/kg, and a turbojet engine stands at 3,000 W/kg [24.47]. These figures tell us that linear motors like myosin and kinesin are extremely powerful machines.

24.3 Modeling at the Nano-Bio Interface

We have already provided a number of different views of the biology-nanotechnology interface, all of which reveal the insights that can emerge from model building. Indeed, one of the key thrusts of this entire chapter is the view that as the type of data that emerges concerning biological systems becomes increasingly quantitative, it must be responded to with models that are also quantitative [24.30]. The plan of this section is to show how atomistic and continuum analyses each offer insights into problems of nanotechnological significance, but under some circumstances, both are found wanting and it is only through a synthesis of both types of models that certain problems will surrender. The plan of this section is to examine the advantages and difficulties associated with adopting both atomistic and continuum perspectives and then to hint at the benefits of seeking mixed representations.

24.3.1 Tension Between Universality and Specificity

One of the key insights concerning model building in nanomechanics, whether we are talking about the nanoscale tribological questions pertinent to magnetic recording or the operations of molecular motors, is that such questions live in the no-man's-land between traditional continuum analysis at one extreme and all-atom approaches such as molecular dynamics on the other. Indeed, there is much discussion about the breakdown of continuum mechanics in modeling the mechanics of systems at the nanoscale. This dichotomy between continuum theories, which treat matter as continuously distributed, and atomic-level models, which explicitly acknowledge the graininess of matter, can be restated in a different (and perhaps more enlightening) way. In particular, it is possible to see atomistic and continuum theories as offering complementary views of the same underlying physics. Continuum models are suitable for characterizing those features of a system that can be thought of as averages over the underlying microscopic fluctuations. By way of contrast, atomistic models reveal the details that a continuum model will never capture, and in particular, they shed light on the specificity of the problem at hand.

The perspective adopted here is that continuum models and atomistic models each reveal important features of a given problem. For example, in contemplating the competition between fracture and plasticity at crack tips, a continuum analysis provides critical insights into the nature of the elastic fields surrounding a defect such as a crack. These fields adopt a fundamental and universal form at large scales with all detailed material features buried in simple material parameters. By way of contrast, the precise details of the dissipative processes occurring at a crack tip (in particular, the competition between bond breaking with the creation of new free surface and dislocation nucleation) require detailed atomic-level descriptions of the energetics of bond stretching and breaking. In the biological setting, similar remarks can be made. In certain instances, the description of biological polymers as random coils suffices and yields insights into features such as the mean size of the polymer chain as a function of its length. On the other hand, if our objective is mechanistic understanding of processes such as how phosphorylation of a particular protein induces conformational change, this is an intrinsically atomic-level question. The language we invoke to describe this dichotomy is the use of the terms *universality* and *specificity*, where, as described above, insights of a universal character refer to those features of systems that are generic, while specificity refers to the features of systems that depend upon precise details such as whether or not a particular molecule is bound at a particular site. This fundamental tension between atomistic and continuum perspectives is elaborated in *Phillips* et al. [24.48].

24.3.2 Atomic-Level Analysis of Biological Systems

As already described in the introduction, one of the intriguing roles of nanotechnology in the biological setting is that it has brought the Instron technology of traditional solid mechanics to the nanoscale and has permitted the investigation of the force-extension characteristics of nanoscale systems (macromolecules and their assemblies, in particular). As noted earlier, mechanical force spectroscopy [24.20] is emerging as a profound tool for exploring the connection between structure, force, and chemistry, in much the same way that conventional stress-strain tests provided insights into the connection between structure and properties of conventional materials. Figure 24.3 gives one such example in the case of the muscle protein titin. Similar insights have been obtained through systematic examination of the force-extension properties of DNA [24.49].

The objective of this section is to call the reader's attention to the types of modeling that can be done from an all-atom perspective. What exactly is meant by a model in this setting? We begin by noting that for the purposes of the general discussion given here, the same basic ideas are present whether one is modeling tribological processes such as the sliding of adjacent surfaces, or attempting to examine the operation of a protein machine such as ATPsynthase. The set of degrees of freedom considered by the atomic-scale modeler is the full set of atomic positions $(\boldsymbol{R}_1, \boldsymbol{R}_2, \cdots \boldsymbol{R}_N)$, which we also refer to as $\{\boldsymbol{R}_i\}$. For most purposes, one proceeds through reference to a *classical* potential energy function, $E_{\text{tot}}(\{\boldsymbol{R}_i\})$, which is a rule that assigns an energy for every configuration $\{\boldsymbol{R}_i\}$. While it would be most appealing to be able to perform a full quantum mechanical analysis, such calculations are computationally prohibitive. Given the potential energy, the forces on each and every atom can be computed where, for example, the force in the αth Cartesian direction on the ith atom is given by

$$F_{i,\alpha} = -\frac{\partial E_{\text{tot}}}{\partial R_{i,\alpha}} . \tag{24.8}$$

For those interested in finding the energy minimizers, such forces can be used in conjunction with methods such as the conjugate gradient method or the Newton–Raphson method. Alternatively, many questions are of a dynamic character, and in these cases, Newton's equations of motion are integrated, thus permitting an investigation of the temporal evolution of the system of interest. We note that we have neglected to discuss subtleties of how one maintains the system at constant temperature, and we leave such subtleties to the curious reader, who can learn more about them in *Frenkel* and *Smit* [24.50].

To give a flavor of where such calculations can lead we note that the same force-extension characteristics already shown in Fig. 24.3 have been computed in a molecular dynamics simulation [24.51]. One of the insights to emerge from these calculations was the particular dynamical pathway, namely, the breaking of a particular collection of hydrogen bonds during the rupture process of each of the immunoglobulin domains. We further note that in the case of titin there has been an especially pleasing synergy between the atomic-scale calculations and the corresponding force-extension measurements. In particular, in response to the suggestion that it was a particular set of hydrogen bonds that were impugned in the rupture process, mutated versions of the titin protein were created in which the number of such hydrogen bonds was changed with the result that the rupture force was changed according to expectation [24.52]. This can be seen as a primitive example of the ultimate goal of tailoring new materials (both biological and otherwise) through appropriate computer modeling.

24.3.3 Continuum Analysis of Biological Systems

We have already noted that there are many appealing features to strictly continuum analyses of material systems. In particular, models based upon continuum mechanics result in a mathematical formulation that permits us to uncork the traditional tools associated with partial differential equations and functional analysis. In keeping with our argument that it is the role of models to serve our intuition, the ability to write down continuum models raises the possibility of obtaining analytic solutions to problems of interest, an eventuality that is nearly impossible once the all-atom framework has been adopted.

Macromolecules as Elastic Rods

Whether we contemplate the information carrying nucleic acids, the workhorse proteins, or energy storing sugars, ultimately, the molecular business of the living world is dominated by long chain molecules. For the model builder, such molecules suggest two complementary perspectives, each of which contains a part of the whole truth. On the one hand, polymer physicists have gained huge insights [24.53] by thinking of long chain molecules as random walks. Stated simply, the key virtue of the random-walk description of long chain molecules is that it reflects the overwhelming importance of entropy in governing the geometric conformations that can be adopted by macromolecules. The other side of the same coin considers long chain molecules as elastic rods with a stiffness that governs their propensity for bending. We will take up this perspective in great detail in the final section of this chapter when we consider the energetics of DNA packaging in viruses.

For our present purposes, we examine one case study in treating macromolecules as elastic rods. The example of interest here is chosen in part because it reflects another fascinating aspect of biological nanotechnology, namely, regulation and control. It is well-known that genes are switched on and off as they are needed. These topics are described beautifully in the work of *Ptashne* [24.54]. The basic idea can be elucidated through the example of a particular set of genes in *E. coli*: the *lac* operon. There are a set of enzymes exploited by this bacterial cell when it needs to digest the sugar lactose. The gene that codes for these enzymes is only turned on when lactose is present and certain other sugars are absent. The nanotechnological solution to the problem of regulating this gene has been solved by nature through the presence of a molecule known as the *lac* repressor, which binds onto the DNA in the region of this *lac* gene and prevents the gene from being expressed. More specifically, the *lac* repressor binds onto two different regions of the DNA molecule simultaneously, forming a loop in the intervening region, and thus rendering that region inoperative for transcription.

We have belabored the difficulties that attend the all-atom simulation of most problems of nanotechnological interest – system sizes are too small and simulated times are too short. On the other hand, clearly it is of great interest to probe the atomic-level dynamics of the way that molecules such as *lac* repressor interact with DNA and thereby serve as gene regulators. A compromise position adopted by *Balaeff* et al. [24.55] was to use elasticity theory to model the structure and energetics of DNA, which then served effectively as a boundary condition for their all-atom calculation of the properties of the *lac* repressor/DNA complex. The simulation cell is shown in Fig. 24.6, which also shows the looped region of DNA that serves as a boundary condition for the atomic-level model of the complex.

Membranes as Elastic Media

One way to classify biological structures is along the lines of their dimensionality. In the previous discussion, we examined the sense in which many of the key macromolecules of the living world can be thought of as one-dimensional rods. The next level in this dimensional hierarchy is to consider the various membranes that compartmentalize the cell and which can be examined from the perspective of two-dimensional elasticity. Just as there are a huge number of modeling questions to be posed concerning the structure and function of long chain molecules such as nucleic acids and proteins, there is a similar list of questions that attend the presence of a host of different membranes throughout the cell. To name a few, we first remind the reader that such membranes are full of various proteins that serve in a variety of different capacities, some of which depend upon the mechanical state of the membrane itself. In a different vein, there has been great interest in examining the factors that give rise to the equilibrium shapes of cells [24.56, 57]. In both instances, the logic associated

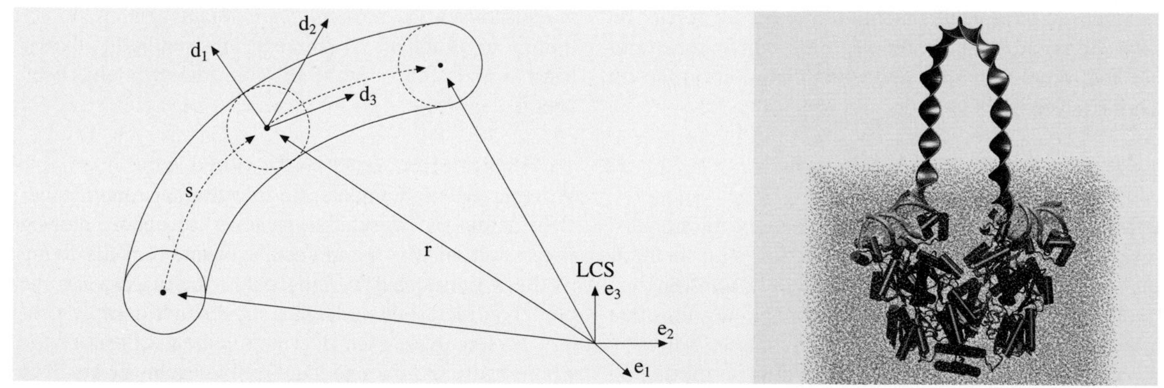

Fig. 24.6 Illustration of the way that the properties of a particular DNA-protein binding interaction were modeled by virtue of an elastic treatment of the looped region of the DNA [24.48]

with building models of these phenomena centers on the construction of an elastic Hamiltonian that captures the energetics of deformations expressed in terms of surface area, curvature, and variations in thickness. Also, just as there has been great progress in measuring the properties of single proteins and nucleic acids, there has been considerable progress in examining the force response of membranes as well [24.58].

The prospect of bringing the tools of traditional continuum theory to bear on problems of biological significance is indeed a daunting one. As noted earlier, to our way of thinking one of the biggest challenges posed by biological problems is the dependence of these systems on detailed molecular structures and dynamics – what we have earlier characterized as "specificity", one of the hallmarks of biological action.

24.4 Nature's Nanotechnology Revealed: Viruses as a Case Study

Over the course of this chapter, we have presented different facets of the relationship between biology and nanotechnology and the modeling challenges they encompass. For example, we have noted that two key aspects of nature's nanotechnologies are the exploitation of self-assembly processes for the construction of molecular machines and the role of active processes mediated by molecular motors. We have also argued for the synergistic role of nanotechnology in producing new methods for the experimental analysis of biological systems with examples ranging from new molecular dyes to the use of optical tweezers. Finally, we have described some of the modeling challenges posed by contemplating the biology-nanotechnology interface and the sorts of coarse-grained models that have arisen to meet these challenges. In this final section, we present a discussion of viruses as a case study at the confluence of these different themes.

Though the importance of viruses from a health perspective are well-known even to the casual observer, they are similarly important both from the technological perspective (as will be explained below) and as a compelling and profound example of nature's nanotechnology par excellence. To appreciate the sense in which viruses serve as a compelling example of biological nanotechnology, we begin by reviewing the nature of the viral life cycle with special reference to the case of bacterial viruses (i. e., viruses that infect bacteria), known as bacteriophages. For concreteness, we consider the life cycle of a bacteriophage such as the famed lambda phage that infects *E. coli*. The life cycle of such viruses is shown schematically in Fig. 24.7. Upon an encounter with the *E. coli* host, the virus attaches to a receptor (protein) embedded in the bacterial membrane and ejects its DNA into the host cell, leaving an empty capsid as refuse from the process. As an aside, it is worth noting that experiments such as the famed Hershey–Chase experiment used tagged DNA and tagged proteins on viruses to settle the question of whether proteins or nucleic acids are the carriers of genetic information. The outcome of these experiments was the conclusion that DNA is the genetic material. For the purposes of the present discussion, the other interesting outcome of the Hershey–Chase experiment is that it provides insights into the mechanistic process associated with delivery of the viral genome.

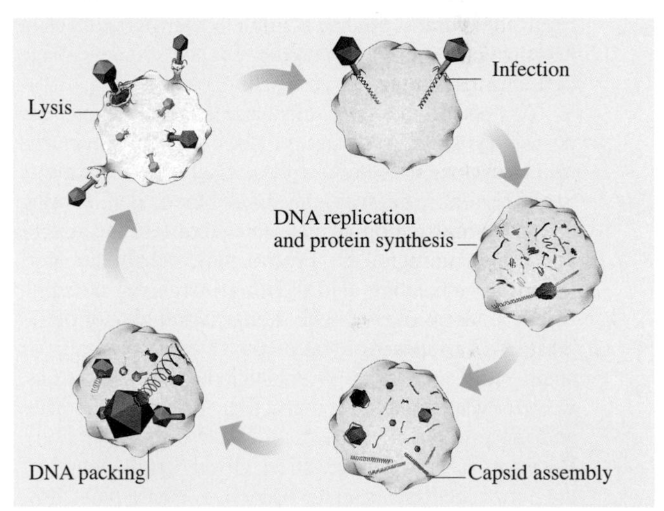

Fig. 24.7 Schematic representation of the viral life cycle illustrating the various nanotechnological actions in this cycle. During the infection stage, the viral genome is delivered to the victimized bacterial cell. During the process labeled DNA replication and protein synthesis, the machinery of the host cell is enlisted to produce renegade DNA and proteins to be used in the assembly of new viral particles. The capsid assembly process involves the attachment of the various protein building blocks that make up the capsid. DNA packing refers to the active packaging of DNA into the newly formed capsids by a molecular motor. Finally, once the assembly process is complete, the infectious phage particles are released from the victimized cell to repeat their infectious act elsewhere

Once the viral genome has been delivered to the host cell (we now oversimplify), the replication machinery of *E. coli* is hijacked to do the virus' bidding. In particular, the genes coded for in the viral DNA are expressed, and the proteins needed to make copies of the phage particle (i. e., the ingredients for an eventual self-assembly process) are created. Interestingly, part of the gene products associated with this process are the components of the molecular motor, which is responsible for packaging the replicated viral DNA into the new protein capsids that will eventually become the next generation of viruses. Indeed, once assembled, this motor takes the replicated viral DNA and packs it into the viral capsid. Recall our insistence that another of the important themes presented by biological examples of nanotechnology is the huge role played by active processes that reflect mechanochemical coupling. Once the packaging process is completed, and the remaining parts of the self-assembly process have been effected (such as the attachment of the viral sheath and legs whereby the phage attaches to host cells), enzymes are released that breakdown the cell wall of the infected cell with the ultimate result that what started as a single bacteriophage has in less than an hour become on the order of 100 new phage particles ready to infect new cells.

As noted above, a second sense in which viruses are deserving of case study status in the current chapter is the role viruses such as lambda phage play in biotechnology itself. For example, from the standpoint of both cloning and the construction of genomic libraries, the use of viruses is commonplace. From a more speculative perspective, viruses are also drawing increasing attention from the standpoint of gene therapy (as a way of delivering DNA to specific locations) and, more generally, as small-scale containers. To substantiate these assertions, we briefly consider the construction of genomic libraries and the role played by viruses in these manipulations. We pose the following question: Given that the length of the human genome is on the order of 10^9 base pairs, how can one organize and store all of this genetic information for the purposes of experiments such as the sequencing of the human genome? One answer to this challenge is the use of bacteriophage to deliver phage DNA to *E. coli* cells, but with the subtlety that the delivered DNA fragments have ligated within them fragments of the human genome whose lengths are on the order of 10 kb (kb = kilo base-pairs). The particulars of this procedure involve first cutting the DNA into fragments of roughly 10 kb in length using a class of enzymes known as restriction enzymes. The result of this operation is that the original genome is now separated into a random collection of fragments. These fragments are then mixed with the original lambda phage DNA that has also been cut at a single site such that the genomic fragments and the lambda fragments are complementary. Using a second enzyme known as ligase, the genomic DNA is joined to the original lambda phage fragments so that the resulting DNA resembles the original lambda DNA, but now with a 10 kb fragment inserted in the middle. These cloned lambda DNA molecules are then packaged into the lambda phage using a packaging reaction (see the website www.stratagene.com for an example of these products). The resulting lambda phage, now fully packed with cloned DNA, are used to infect *E. coli* cells, and the cloned DNA, once in the bacterial cell, circularizes into a DNA fragment known as a plasmid. The plasmid is then passed from one generation of *E. coli* to the next. This is the so-called prophage pathway in which, unlike the lytic pathway shown in Fig. 24.7, the bacteriophage is latent and does not destroy the cell [24.1]. Note that different *E. coli* cells are infected with viruses containing different cloned fragments. As a result, the collection of all such *E. coli* cells constitute a library of the original genomic DNA.

We round out our introductory discussion on viruses with a discussion of the compelling recent experiments that have been performed to investigate the problem of viral packing and, similarly, how ideas like those described in this chapter can be used to model these processes. *Smith* et al. [24.29] used optical tweezers to measure the force applied by the packaging motor of the ϕ-29 bacteriophage during the DNA packaging process. In particular, they measured the force and rate of packaging as a function of the amount of DNA packed into the viral head with the result that as more DNA is packed, the resistive force due to the packed DNA increases, and the packing rate is reduced.

As noted above, the viral problem is interesting not only because it exemplifies many of the features of biological nanotechnology introduced throughout this chapter, but also illustrates the way in which model building has arisen in response to experimental insights. The various competing energies that are implicated in the DNA packing process have been described by *Riemer* and *Bloomfield* [24.59]. The energetics of viral packing is characterized by a number of different factors, including: i) the entropic-spring effect that causes the DNA in solution to adopt a more spread out configuration than that in the viral capsid, ii) the energetics of elastic bending, which results from inducing curvature in the DNA on a scale that is considerably smaller than the persistence length of $\xi_p \approx 50$ nm, and iii) those factors related to the presence of charge both on the DNA itself and in the surrounding solution. As shown by *Riemer* and *Bloomfield* [24.59], the entropic contribution is smaller by a factor of 10 or more relative to the bending energies and those mediated by the charges on the DNA and the surrounding solution, and hence we make no further reference to it. As a result, just like in earlier work [24.60, 61], we examine the interplay of elastic and interaction forces, though we neglect surface terms originating from DNA-capsid and DNA-solvent interactions.

We note that the viral packing process involves DNA segments with lengths on the order of 10 μm and takes place on the time scale of minutes. As a result, from a modeling perspective, it is clear that such problems are clearly out of reach of conventional molecular dynamics. As a result, we exploit a continuum description of the DNA packing process with the proviso that such models will ultimately need to be refined to account for the sequence dependence of the elasticity of DNA.

A mathematical description of the energetics of viral packing must account for two competing factors, namely, the energy cost to bend the DNA and place it in the capsid and the repulsive interaction between adjacent DNA segments that are too close together. The structural picture of the packaged DNA is inspired by experiments that indicate that the DNA is packed in concentric rings from large radii inwards [24.62–64]. The bending energy cost of accumulating hoops within the capsid is given by

$$E_{\mathrm{el}} = \pi \xi_{\mathrm{p}} k_{\mathrm{B}} T \sum_i \frac{N(R_i)}{R_i} , \tag{24.9}$$

where $N(R_i)$ is the number of hoops that are packed at the radius R_i [24.59]. The basic idea of this expression is that we are adding up the elastic energy on a hoop by hoop basis, with each hoop penalized by the usual energy cost to bend an elastic beam into a circular arc. The presence of $N(R_i)$ reflects the fact that because of the shape of the capsid, as the radius gets smaller the DNA can pack higher up into the capsid, thus increasing the number of allowed hoops. To make analytic progress with the expression for the stored elastic energy given above, we convert it into an integral of the form

$$E_{\mathrm{el}} = \frac{\pi \xi_{\mathrm{p}} k_{\mathrm{B}} T}{\sqrt{3} d_{\mathrm{s}}/2} \int_R^{R_{\mathrm{out}}} \frac{N(R')}{R'} \, \mathrm{d}R' . \tag{24.10}$$

The summation $\sum_i$ has been replaced by an integral $\int_R^{R_{\mathrm{out}}} \mathrm{d}R'/(\sqrt{3} d_{\mathrm{s}}/2)$, where the integration bounds are the inner and outer radii of the inner spool and $\sqrt{3} d_{\mathrm{s}}/2$ is the horizontal spacing between adjacent strands of the DNA. The geometrical factor $\sqrt{3}/2$ owes its presence to the hexagonal packing of the DNA strands. R is the radius of the innermost stack of hoops, R_{out} is the radius of the capsid, $N(R')$ is the number of hoops at radius R', d_{s} is the spacing between adjacent hoops, and ξ_{p} is the persistence length of DNA. We note that this expression is a manifestation of the first term in the energy functional in (24.3), used earlier in the context of the biofunctionalized cantilever and now specialized to the geometry of a partially filled capsid. Just as the energy functional in the cantilever context reflected a competition between different contributions to the total energy (in that case, elastic bending energy and surface terms), the DNA packing problem reflects a similar competition, this time between the elastic bending and the interactions between nearby DNA segments by virtue of the charge along the DNA. In particular, the interaction energy between the DNA hoops packed in the viral capsid scales with the length and is given by

$$E_{\mathrm{int}} = F_0 L (c^2 + c d_{\mathrm{s}}) \exp(-d_{\mathrm{s}}/c) , \tag{24.11}$$

where c and F_0 are constants for a given solvent and L is the length of the packed DNA. This form of the interaction energy was shown to be very robust for a variety of solvents containing monovalent and divalent cations in a series of experiments by *Rau* et al. [24.65, 66] and *Parsegian* et al. [24.67]. The forces arising from this kind of an energy are purely repulsive in character. Interestingly, in this setting, rather than using temperature as a control parameter, it is much more common to change the circumstances in such experiments by tuning concentrations of various chemical constituents. In particular, when trivalent or tetravalent cations are present in solution there is an effective attractive interaction between DNA segments. In that case, the forces are repulsive only when d_{s} is smaller than a certain critical separation d_0 (and attractive otherwise) and the form of the interaction energy changes to

$$E_{\mathrm{int}} = F_0 L (c^2 + c d_{\mathrm{s}}) \exp\left(\frac{d_0 - d_{\mathrm{s}}}{c}\right) . \tag{24.12}$$

Once both the elastic and interaction contributions to the energy have been reckoned, we are in a position to compute the total energy. If we recognize that the length L of the packed DNA is given by

$$L = \frac{2}{\sqrt{3} d_{\mathrm{s}}} \int_R^{R_{\mathrm{out}}} 2\pi R' N(R') \, \mathrm{d}R' , \tag{24.13}$$

then the total energy for the packed DNA, in terms of L and d_s, is given by

$$E(L, d_s) = \frac{\pi \xi_p k_B T}{\sqrt{3} d_s/2} \int_{R}^{R_{out}} \frac{N(R')}{R'} \, dR' \qquad (24.14)$$

$$+F_0(c^2 + c d_s) \exp(-d_s/c) \frac{2}{\sqrt{3} d_s} \int_{R}^{R_{out}} 2\pi R' N(R') \, dR' \,.$$

The spacing at a given length L is determined by requiring that the packed DNA be in a minimum energy configuration, which is equivalent to asking that $\partial E/\partial d_s = 0$. Hence, for a given geometry and length packed, one obtains d_s and thereby calculates the energy. Thus the energy is known as a function of the length packed, and to compute the force as a function of length of DNA packed one need only differentiate this energy with respect to the length packed. The result of this procedure is an expression for the force of the form

$$F(L) = F_0(c^2 + c d_s) \exp\left(-\frac{d_s}{c}\right) + \frac{\xi_P k_B T}{2R^2} \,. \qquad (24.15)$$

This is a generic expression valid for a capsid of any shape. The effect of the geometry is captured in the variation of d_s as a function of the length packed, as well as through the inner packing radius R. The resulting force-packing curves are shown in Fig. 24.8. In particular, this figure shows the packing force as a function of a fraction of the genome packed for a number of different solutions, with the different curves revealing the large role played by positive counterions in dictating the overall energetics of these processes.

On the basis of calculations like those given above, *Kindt* et al. [24.61] estimate that the pressure inside the capsid of the phage is on the order of 35 atmospheres. A crude estimate by *Smith* et al. [24.29] gives a figure close to 60 atmospheres. These numbers are intriguing in their own right, but more importantly they demonstrate the promise that proteins and other biological materials hold as candidates for engineering materials. To further explore the structural integrity of viruses in their role as pressurized protein shells, we now turn to an examination of the rupture stress of viral capsids.

Throughout this chapter, one of our main arguments has been the idea that whether it is our ambition to model conventional materials or their biological counterparts, modeling at the nanoscale implies challenges that can not be met either by purely continuum ideas or by traditional all-atom thinking. Indeed, one set of powerful methods is built around the attempt to borrow those features of continuum and atomistic thinking that are most robust, while rejecting those that are inapplicable at the atomic scale. As an example of this type of thinking, we examine another question drawn from the viruses-as-nanotechnology setting. In particular, since viruses have been considered a means of transporting material other than the genetic material of the virus itself, it is of interest to consider the maximum internal pressure viral capsids can sustain without rupture, as this will determine how much material can be safely packaged. To that end, we use continuum mechanics to estimate the stresses within the capsid walls. These stresses are then mapped onto atomic-level forces by appealing to the details of the protein structure of the monomers making up the capsid and a knowledge of the forces that link them. By relating the continuum and atomistic calculations, we then determine the maximum sustainable internal pressure.

We imagine the capsid to be a hollow sphere loaded by a pressure p_i from inside and a pressure p_o from outside. The inner and outer radii are R_i and R_o, respectively. As a representative example, bacteriophage

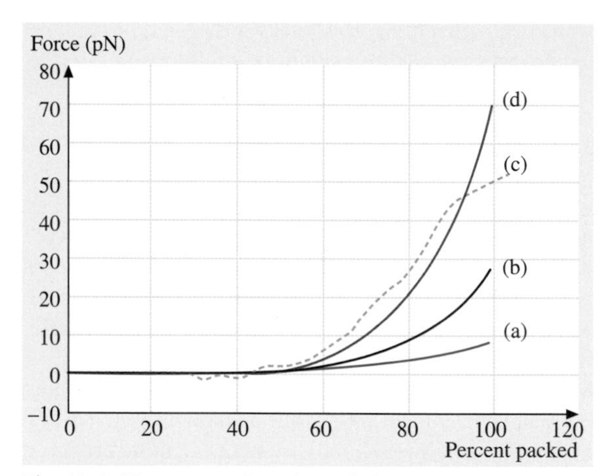

Fig. 24.8 Force as a function of the percent packed for a cylindrical capsid under purely repulsive solvent conditions. The dimensions of the capsid and the length of genome packed were chosen to represent the $\phi 29$ phage: $R_{out} = 210$ Å, $z = 540$ Å, and $L = 6.6\,\mu$m. Curve (c) shows the experimental results of *Smith* et al. [24.29], while theoretical curves (a), (b), and (d) are given by (24.15), with $F_0 = 10 \times 55{,}000$, $40 \times 55{,}000$, and $100 \times 55{,}000\,\text{pN/nm}^2$, respectively

GA is characterized geometrically by $R_\mathrm{i} = 123$ Å and $R_\mathrm{o} = 145$ Å. Evaluation of a number of different capsid types suggests that treating capsids as though they have a mean thickness of roughly 15 Å suffices for the level of modeling being considered here. For the purposes of computing the internal stresses within the capsid, we begin with a statement of equilibrium from continuum mechanics that requires that at every point in the capsid

$$\nabla \cdot \sigma = \mathbf{0}\,, \tag{24.16}$$

where σ is the stress tensor comprising three normal stresses and three shear stresses. For a problem with spherical symmetry, like that considered here, the stresses reduce to a radial stress σ_R and a circumferential stress σ_T. Solution of the equilibrium equations results in stresses of the form [24.32]

$$\sigma_R = \frac{C}{r^3} + D, \quad \sigma_T = -\frac{C}{2r^3} + D\,. \tag{24.17}$$

By using the boundary conditions $\sigma_R|_{r=R_\mathrm{i}} = -p_\mathrm{i}$ and $\sigma_R|_{r=R_\mathrm{o}} = -p_\mathrm{o}$, the constants C and D can be determined with the result

$$\sigma_R = \frac{p_\mathrm{o} R_\mathrm{o}^3 \left(r^3 - R_\mathrm{i}^3\right)}{r^3 \left(R_\mathrm{i}^3 - R_\mathrm{o}^3\right)} + \frac{p_\mathrm{i} R_\mathrm{i}^3 \left(R_\mathrm{o}^3 - r^3\right)}{r^3 \left(R_\mathrm{i}^3 - R_\mathrm{o}^3\right)}\,, \tag{24.18}$$

$$\sigma_T = \frac{p_\mathrm{o} R_\mathrm{o}^3 \left(2r^3 + R_\mathrm{i}^3\right)}{2r^3 \left(R_\mathrm{i}^3 - R_\mathrm{o}^3\right)} - \frac{p_\mathrm{i} R_\mathrm{i}^3 \left(2r^3 + R_\mathrm{o}^3\right)}{2r^3 \left(R_\mathrm{i}^3 - R_\mathrm{o}^3\right)}\,. \tag{24.19}$$

The stress σ_T is our primary concern, since it acts so as to tear the sphere apart. By looking at the expressions above, we can see that this stress is maximum at $r = R_\mathrm{i}$ and the maximum value is given by

$$\sigma_T^{\max} = \frac{3 p_\mathrm{o} R_\mathrm{o}^3 - p_\mathrm{i}\left(2R_\mathrm{i}^3 + R_\mathrm{o}^3\right)}{2\left(R_\mathrm{i}^3 - R_\mathrm{o}^3\right)}\,. \tag{24.20}$$

We note that elasticity theory in and of itself is unable to comment on $\sigma_T^{\max}$, since this is effectively a material parameter that characterizes the contacts between the various protein monomers that make up the capsid. As a result, we first examine how the rupture strength depends upon capsid dimensions in abstract terms and then turn to a concrete estimate of $\sigma_T^{\max}$ itself from several complementary perspectives. If (24.20) is rewritten using $p_\mathrm{i} = p_\mathrm{o} + \Delta p$ and $R_\mathrm{o} = R_\mathrm{i} + \Delta R$, and, further, it is realized that for typical capsid dimensions $\Delta R / R_\mathrm{i} \ll 1$, it can be shown by rearranging (24.20) and by considering the case in which $\sigma_T^{\max}$ is much larger than p_o, the maximum sustainable internal pressure is given by

$$p_\mathrm{i}^{\max} = \frac{2\sigma_T^{\max} \Delta R}{R_\mathrm{i}}\,, \tag{24.21}$$

where $p_\mathrm{i}^{\max}$ is really the quantity of interest, namely, the maximum sustainable internal pressure.

To make concrete progress to the point where we can actually estimate the rupture stress in atm, we need to consider capsids for which the structure is known and for which, at least approximately, the bonds between the monomers making up the capsid are understood. We note that in the language of fracture mechanics, what we seek is a cohesive surface model that provides a measure of the energy of interaction between two surfaces as a function of their separation [24.68]. There are a number of different ways to go about estimating the effective interaction between the monomers making up the capsid, one of which is by appealing to atomic-level calculations like those made by *Reddy* et al. [24.69]. The Viper website has systematized such information for a number of capsids, and one of the avenues we take to estimate $\sigma_T^{\max}$ is to appeal directly to their calculations. To that end, we assume that the energy of interaction per unit area as a function of separation x between adjacent monomers making up the capsid can be written in the form

$$E(x) = V_0 \left[\frac{1}{4}\left(\frac{x^*}{x}\right)^{12} - \frac{1}{2}\left(\frac{x^*}{x}\right)^{6}\right]. \tag{24.22}$$

The motivation for this functional form is the idea that the energy of interaction between adjacent monomeric units making up the capsid is the result of van der Waals contacts, and hence our cohesive surface law has inherited the properties of the underlying atomic force fields. To proceed to an estimate of $\sigma_T^{\max}$ itself, we must determine the parameters in the cohesive model described above. To that end, we note that *Reddy* et al. [24.69] have computed the association energies of various inequivalent contacts throughout a number of different icosahedral capsids. Their calculations result in a roughly constant value of ≈ -45 cal/mole Å^2 as the association energy, which in the language of our cohesive potential results in $V_0 \approx 12.5$ pN/Å. This may be seen by noting that the association energy is given by $E(x^*) = -V_0/4$.

Once we have fixed x^*, which amounts to choosing a particular value for the equilibrium separation between two monomers, the material parameter $\sigma_T^{\max}$ is obtained by evaluating $\partial E(x)/\partial x$ at a value of x corresponding to the point of inflection ($\partial^2 E(x)/\partial x^2 = 0$) in the cohesive surface function. For the cohesive surface function used above, this results in a maximum stress of the form

$$\sigma_T^{max} = \frac{7^{7/6} \times 18}{13^{13/6}} \frac{V_0}{x^*} . \quad (24.23)$$

With σ_T^{max} in hand, the maximum sustainable pressure is obtained from (24.21), and the results are shown in Fig. 24.9. These estimates suggest that the pressures within capsids as a result of packed DNA while large, are still smaller than our estimated rupture stresses. To more completely examine this question we have undertaken finite element elasticity calculations to examine the stresses in capsids exhibiting irregularities in both shape and thickness. In addition, further atomistic analysis of σ_T^{max} is needed with special reference to its dependence on distance between protein units. It would also be of interest to examine mutant versions of the monomeric units making up the capsid to see the implications of such mutations for σ_T^{max}.

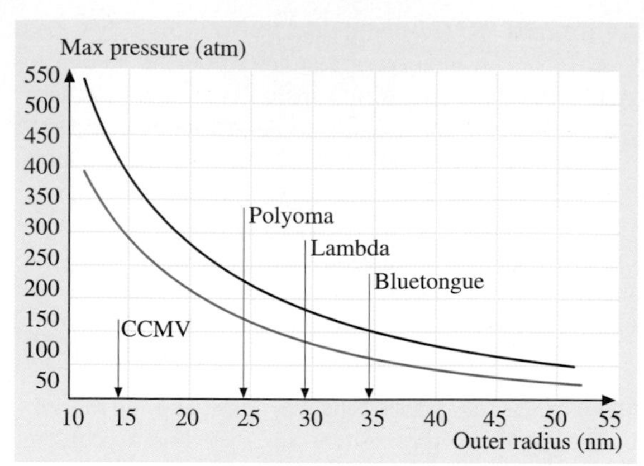

Fig. 24.9 Rupture pressure as a function of capsid radius for $x^* = 3$ Å (*upper curve*) and $x^* = 4$ Å (*lower curve*). The width of the capsid walls was set to $\Delta R = 1.5$ nm, while $V_0 = 180$ cal/molÅ^2

24.5 Concluding Remarks

One of the most compelling areas to be touched by nanotechnology is biological science. Indeed, we have argued that there is a fascinating interplay between these two subjects, with biology as a key beneficiary of advances in nanotechnology as a result of a new generation of single molecule experiments that complement traditional assays involving statistical assemblages of molecules. This interplay runs in both directions with nanotechnology continually receiving inspiration from biology itself. The goal of this chapter has been to highlight some representative examples of the interplay between biology and nanotechnology and to illustrate the role of nanomechanics in this field and how mechanical models have arisen in response to the emergence of this new field. Primary attention has been given to the particular example of the processes that attend the life cycle of bacterial viruses. Viruses feature many of the key lessons of biological nanotechnology, including self-assembly, as evidenced in the spontaneous formation of the protein shell (capsid) within which the viral genome is packaged and a motor-mediated biological process, namely, the packaging of DNA in this capsid by a molecular motor that pushes the DNA into the capsid. We argue that these processes in viruses are a compelling real-world example of nature's nanotechnology and reveal the nanomechanical challenges that will continue to be confronted at the nanotechnology-biology interface.

Though this chapter represents something of a departure from the rest of the chapters in this volume, it advances the view that biological nanotechnology serves as an inspiring vision of what nanotechnology can offer and reflects the authors' views that biologically inspired nanotechnology will play an ever increasing role as Feynman's view that "there is plenty of room at the bottom" continues to play out. Moreover, developing simple models of nature's nanotechnology can provide important insights into viable strategies for making machines at the nanoscale.

We have argued that the advance of single molecule technology presents both scientific and technological possibilities. Both classes of questions imply significant modeling demands, just as earlier advances in other settings, such as in materials for microelectronics applications, did. We claim that these modeling challenges are perhaps more acute as a result of the vicious chemical and structural nonhomogeneity of biological systems. As a result, we see an ever increasing role for modeling methods that aim to keep atomic-level specificity where needed, while rejecting such resolution elsewhere.

References

24.1 B. Alberts, D. Bray, A. Johnson, J. Lewis, M. Raff, K. Roberts, P. Walter: *Essential Cell Biology* (Garland, New York 1997) Chap. 12

24.2 B. Alberts: The cell as a collection of protein machines: Preparing the next generation of molecular biologists, Cell **92** (1998) 291

24.3 D. Bray: *Cell Movements From Molecules to Motility* (Garland, New York 2001)

24.4 S. J. Flint, L. W. Enquist, R. M. Krug, V. R. Racaniello, A. M. Skalka: *Principles of Virology* (ASM, Washington, DC 2000)

24.5 J. C. M. van Hest, D. A. Tirrell: Protein-based materials, toward a new level of structural control, Chem. Commun. **19** (2001) 1807–1904

24.6 H. Berg: Motile behavior of bacteria, Phys. Today **53** (2000) 24

24.7 J. E. Walker: ATP synthesis by rotary catalysis, Angew. Chem. Int. Edn. **37** (1998) 2308

24.8 P. D. Boyer: The ATP synthase – A splendid molecular machine, Annu. Rev. Biochem. **66** (1997) 717

24.9 M. Yoshida, E. Muneyuki, T. Hisabori: ATP synthase – A marvelous rotary engine of the cell, Nature Rev. Mol. Cell Bio. **2** (2001) 669

24.10 R. Yasuda, H. Noji, K. Kinosita, M. Yoshida: F_1-ATPase is a highly efficient molecular motor that rotates with discrete 120° steps, Cell **93** (1998) 1117

24.11 J. Howard: *Mechanics of Motor Proteins and the Cytoskeleton* (Sinauer, Sunderland 2001)

24.12 B. Hille: *Ion Channels of Excitable Membranes* (Sinauer, Sunderland 2001)

24.13 S. Sukharev, S. R. Durell, H. R. Guy: Structural Models of the MscL gating mechanism, Biophys. J. **81** (2001) 917

24.14 S. Sukharev, M. Betanzos, C.-S. Chiang, H. R. Guy: The gating mechanism of the large mechanosensitive channel MscL, Nature **409** (2001) 720

24.15 K. Kinosita, R. Yasuda, H. Noji: F_1-ATPase: A highly efficient rotary ATP machine, Essays Biochem. **35** (2000) 3

24.16 E. Racker, W. Stoeckenius: Reconstitution of purple membrane vesicles catalyzing light-driven proton uptake and adenosine triphosphate formation, J. Biol. Chem. **249** (1974) 662

24.17 G. Groth, J. E. Walker: ATP synthase from bovine heart mitochondria: reconstitution into unilamellar phospholipid vesicles of the pure enzyme in a functional state, Biochem. J. **318** (1996) 351

24.18 G. Wu, R. H. Datar, K. M. Hansen, T. Thundat, R. J. Cote, A. Majumdar: Bioassay of prostrate-specific antigen (PSA) using microcantilevers, Nature Biotech. **19** (2001) 856

24.19 R. P. Feynmann: There's plenty of room at the bottom, Eng. Sci. **23** (1960) 22–36, and www.zyvex.com/nanotech/feynman.html (1959)

24.20 E. Evans: Probing the relation between force-lifetime and chemistry in single molecular bonds, Annu. Rev. Biophys. Biomol. Struct. **30** (2001) 105

24.21 T. E. Fisher, P. E. Marszalek, J. M. Fernandez: Stretching single molecules into novel conformations using the atomic force microscope, Nature Struct. Bio. **7** (2000) 719

24.22 M. Rief, M. Gautel, F. Oesterhelt, J. M. Fernandez, H. Gaub: Reversible unfolding of individual titin immunoglobulin domains by AFM, Science **276** (1997) 1109

24.23 K. Svoboda, S. M. Block: Biological applications of optical forces, Annu. Rev. Biophys. Biomol. Struct. **23** (1994) 247

24.24 C. Bustamante, J. C. Macosko, G. J. L. Wuite: Grabbing the cat by the tail: Manipulating molecules one by one, Nature Rev. Mol. Cell Bio. **1** (2000) 130

24.25 K. Svoboda, S. M. Block: Force and velocity measured for single kinesin molecules, Cell **77** (1994) 773

24.26 M. J. Schnitzer, S. M. Block: Kinesin hydrolyses one ATP per 8-nm step, Nature **388** (1997) 386

24.27 B. Maier, T. R. Strick, V. Croquette, D. Bensimon: Study of DNA motors by single molecule micromanipulation, Single Mol. **1** (2000) 145

24.28 A. A. Simpson, Y. Tao, P. G. Leiman, M. O. Badasso, Y. He, P. J. Jardine, N. H. Olson, M. C. Morais, S. Grimes, D. L. Anderson, T. S. Baker, M. G. Rossmann: Structure of the bacteriophage ϕ-29 DNA packaging motor, Nature **408** (2000) 745

24.29 D. E. Smith, S. J. Tans, S. B. Smith, S. Grimes, D. L. Anderson, C. Bustamante: The bacteriophage ϕ-29 portal motor can package DNA against a large internal force, Nature **413** (2001) 748

24.30 D. Bray: Reasoning for results, Nature **412** (2001) 863

24.31 G. Wu, H. Ji, K. Hansen, T. Thundat, R. Datar, R. Cote, M. F. Hagan, A. K. Charkraborty, A. Majumdar: Origin of nanomechanical cantilever motion generated from biomolecular interactions, Proc. Nat. Acad. Sci. **98** (2001) 1560

24.32 L. D. Landau, E. M. Lifshitz: *Theory of Elasticity* (Pergamon, Oxford 1986)

24.33 D. H. Bamford, R. J. C. Gilbert, J. M. Grimes, D. I. Stuart: Macromolecular assemblies: greater than their parts, Curr. Op. Struc. Biology **11** (2001) 107

24.34 J. Frank: Single-particle imaging of macromolecules by cryo-electron microscopy, Annu. Rev. Biophys. Biomol. Struct. **31** (2002) 303

24.35 S. A. Darst: Bacterial RNA polymerase, Curr. Op. Struc. Biology **11** (2001) 155

24.36 N. Ban, P. Nissen, J. Hansen, P. B. Moore, T. A. Steitz: The complete atomic structure of the large ribosomal subunit at 2.4 Å resolution, Science **289** (2000) 905

24.37 T. S. Baker, N. H. Olson, S. D. Fuller: Adding the third dimension to virus life cycles: Three-dimensional reconstruction of icosahedral viruses from cryo-electron micrographs, Microbiol. Mol. Bio. Rev. **63** (1999) 862
24.38 H.-S. Chan, K. A. Dill: The protein folding problem, Phys. Today **46** (February 1993) 24
24.39 M. D. Wang, M. J. Schnitzer, H. Yin, R. Landick, J. Gelles, S. M. Block: Force and velocity measured for single molecules of RNA polymerase, Science **282** (1998) 902
24.40 M. L. Roukes: Nanoelectromechanical Systems, Technical Digest of the 2000 Solid-State Sensor and Actuator Workshop (2000)
24.41 K. Adachi, R. Yasuda, H. Noji, H. Itoh, Y. Harada, M. Yoshida, K. Kinosita: Stepping rotation of F_1-ATPase visualized through angle-resolved single-fluorophore imaging, Proc. Nat. Acad. Sci. **97** (2000) 7243
24.42 J. Wrigglesworth: *Energy and Life* (Taylor and Francis, London 1997)
24.43 V. Smil: *Energies* (MIT, Cambridge 1999)
24.44 J. T. Finer, R. M. Simmons, J. A. Spudich: Single myosin molecule mechanics: Piconewton forces and nanometre steps, Nature **368** (1994) 113
24.45 S. B. Smith, Y. Cui, C. Bustamante: Overstretching B-DNA: The elastic response of individual double-stranded and single-sranded DNA molecules, Science **271** (1996) 795
24.46 S. Munevar, Y. Wang, M. Dembo: Traction force microscopy of migrating normal and H-ras transformed 3T3 fibroblasts, Biophys. J. **80** (2001) 1744
24.47 L. Mahadevan, P. Matsudaira: Motility powered by supramolecular springs and ratchets, Science **288** (2000) 95
24.48 R. Phillips, M. Dittrich, K. Schulten: Quasicontinuum representations of atomic-scale mechanics: From proteins to dislocations, Ann. Rev. Mat. Sci. **32** (2002) 219
24.49 T. Strick, J.-F. Allemand, V. Croquette, D. Bensimon: Twisting and stretching single DNA molecules, Prog. Biophys. Mol. Bio. **74** (2000) 115
24.50 D. Frenkel, B. Smit: *Understanding Molecular Simulation* (Academic, San Diego 1996)
24.51 H. Lu, B. Isralewitz, A. Krammer, V. Vogel, K. Schulten: Unfolding of titin immunoglobulin domains by steered molecular dynamics simulation, Biophys. J. **75** (1998) 662
24.52 M. Carrion-Vazquez, A. F. Oberhauser, T. E. Fisher, P. E. Marszalek, H. Li, J. M. Fernandez: Mechanical design of proteins studies by single-molecule force spectroscopy and protein engineering, Prog. Biophys. Mol. Bio. **74** (2000) 63
24.53 A. Y. Grosberg, A. R. Khokhlov: *Giant Molecules* (Academic, San Diego 1997)
24.54 M. Ptashne, A. Gann: *Genes and Signals* (Cold Spring Harbor Laboratory Press, Cold Spring Harbor 2002)
24.55 A. Balaeff, L. Mahadevan, K. Schulten: Elastic rod model of a DNA loop in the Lac Operon, Phys. Rev. Lett. **83** (1999) 4900
24.56 U. Seifert: Configurations of fluid membranes and vesicles, Adv. Phys. **46** (1997) 13
24.57 D. Boal: *Mechanics of the Cell* (Cambridge Univ. Press, Cambridge 2002)
24.58 D. Leckband, J. Israelachvili: Intermolecular forces in biology, Q. Rev. Biophys. **34** (2001) 105
24.59 S. C. Riemer, V. A. Bloomfield: Packaging of DNA in bacteriophage heads: Some considerations on energetics, Biopolymers **17** (1978) 785
24.60 T. Odijk: Hexagonally packed DNA within bacteriophage T7 stabilized by curvature stress, Biophys. J. **75** (1998) 1223
24.61 J. Kindt, S. Tzlil, A. Ben-Shaul, W. Gelbart: DNA packaging and ejection forces in bacteriophage, Proc. Nat. Acad. Sci. **98** (2001) 13671
24.62 K. E. Richards, R. C. Williams, R. Calendar: Mode of DNA packing within bacteriophage heads, J. Mol. Bio. **78** (1973) 255
24.63 M. E. Cerritelli, N. Cheng, A. H. Rosenberg, C. E. McPherson, F. P. Booy, A. C. Steven: Encapsidated conformation of bacteriophage T7 DNA, Cell **91** (1997) 271
24.64 N. H. Olson, M. Gingery, F. A. Eiserling, T. S. Baker: The structure of isomeric capsids of bacteriophage T4, Virology **279** (2001) 385
24.65 D. C. Rau, B. Lee, V. A. Parsegian: Measurement of the repulsive force between polyelectrolyte molecules in ionic solution: Hydration forces between parallel DNA double helices, Proc. Natl. Acad. Sci. **81** (1984) 2621
24.66 D. C. Rau, V. A. Parsegian: Direct measurement of the intermolecular forces between counterion-condensed DNA double helices, Biophys. J. **61** (1992) 246
24.67 V. A. Parsegian, R. P. Rand, N. L. Fuller, D. C. Rau: Osmotic stress for the direct measurement of intermolecular forces, Meth. Enzymol. **127** (1986) 400
24.68 R. Phillips: *Crystals, Defects and Microstructures* (Cambridge Univ. Press, Cambridge 2001)
24.69 V. S. Reddy, H. A. Giesing, R. T. Morton, A. Kumar, C. B. Post, C. L. Brooks, J. E. Johnson: Energetics of quasiequivalence: Computational analysis of protein-protein interactions in icosahedral viruses, Biophys. J. **74** (1998) 546
The parameters can be found at http://www.scripps.edu/pub/olson-web/gmm/autodock/ad305/Using_AutoDock_305.a.html

25. Mechanical Properties of Nanostructures

Knowledge of the mechanical properties of nanostructures is necessary for designing realistic MEMS/NEMS devices. Microelectromechanical systems (MEMS) refer to microscopic devices that have a characteristic length of less than 1 mm but more than 1 μm and combine electrical and mechanical components. Nanoelectromechanical systems (NEMS) refer to nanoscopic devices that have a characteristic length of less than 1 μm and combine electrical and mechanical components.

Elastic and inelastic properties are needed to predict deformation from an applied load in the elastic and inelastic regimes, respectively. The strength property is needed to predict the allowable operating limit. Some of the properties of interest are hardness, elastic modulus, bending strength, fracture toughness, and fatigue strength.

Structural integrity is of paramount importance in all devices. Load applied during the use of devices can result in component failure. Cracks can develop and propagate under tensile stresses, leading to failure. Atomic force microscopy and nanoindenters can be used satisfactorily to evaluate the mechanical properties of micro/nanoscale structures for use in MEMS/NEMS.

The most commonly used materials are single-crystal silicon and silicon-based materials, e.g., SiO_2 and polysilicon films deposited by low-pressure chemical vapor deposition. An early study showed silicon to be a mechanically resilient material in addition to its favorable electronic properties. Single-crystal SiC deposited on large-area silicon substrates is used for high-temperature micro/nanosensors and actuators. Amorphous alloys can be formed on both metal and silicon substrates by sputtering and plating techniques, providing more flexibility in surface-integration. Electroless deposited Ni-P amorphous thin films have been used to construct microdevices, especially using the so-called LIGA techniques. Micro/nanodevices need conductors to provide power, as well as electrical/magnetic signals to make them functional. Electroplated gold films have found wide applications in electronic devices because of their ability to make thin films and process simply. Use of SiC, Ni-P, and Au films, together with silicon and silicon-based materials, opens up new design opportunities for MEMS/NEMS devices.

This chapter presents a review of mechanical property measurements on the nanoscale of various materials of interest and stress and deformation analyses of nanostructures.

25.1 **Experimental Techniques for Measurement of Mechanical Properties of Nanostructures** ... 765
25.1.1 Indentation and Scratch Tests Using Micro/Nanoindenters ... 765
25.1.2 Bending Tests of Nanostructures Using an AFM ... 765
25.1.3 Bending Tests Using a Nanoindenter ... 769

25.2 **Experimental Results and Discussion** ... 770
25.2.1 Indentation and Scratch Tests of Various Materials Using Micro/Nanoindenters ... 770
25.2.2 Bending Tests of Nanobeams Using an AFM ... 773
25.2.3 Bending Tests of Microbeams Using a Nanoindenter ... 777

25.3 **Finite Element Analysis of Nanostructures with Roughness and Scratches** ... 778
25.3.1 Stress Distribution in a Smooth Nanobeam ... 779
25.3.2 Effect of Roughness in the Longitudinal Direction ... 781
25.3.3 Effect of Roughness in the Transverse Direction and Scratches ... 781
25.3.4 Effect on Stresses and Displacements for Materials That Are Elastic, Elastic-Plastic, or Elastic-Perfectly Plastic ... 784

25.4 **Closure** ... 785

References ... 786

Microelectromechanical systems (MEMS) refer to microscopic devices that have a characteristic length of less than 1 mm but more than 1 μm and combine electrical and mechanical components. Nanoelectromechanical systems (NEMS) refer to nanoscopic devices that have a characteristic length of less than 1 μm and combine electrical and mechanical components. To put the dimensions in perspective, individual atoms are typically fraction of a nanometer in diameter, DNA molecules are about 2.5 nm wide, biological cells are in the range of thousands of nm in diameter, and human hair is about 75 μm in diameter. The mass of a micromachined silicon structure can be as low as 1 nN, and NEMS can be built with a mass as low as 10^{-20} N with cross sections of about 10 nm. In comparison, the mass of a drop of water is about 10 μN and the mass of an eyelash is about 100 nN. The acronym MEMS originated in the United States. The term commonly used in Europe and Japan is micro/nanodevices, which is used in a much broader sense. MEMS/NEMS terms are also now used in a broad sense. A micro/nanosystem, a term commonly used in Europe, is referred to as an intelligent miniaturized system comprising sensing, processing, and/or actuating functions.

A wide variety of MEMS, including Si-based devices, chemical and biological sensors and actuators, and miniature non-silicon structures (e.g., devices made from plastics or ceramics) have been fabricated with dimensions in the range of a couple to a few thousand microns (see e.g., [25.1–9]). A variety of NEMS have also been produced [25.10–15]. Two of the largest "killer" industrial applications of MEMS are accelerometers (about 85 million units in 2002) and digital micromirror devices (about $ 400 million in revenues in 2002). BIOMEMS and BIONEMS are increasingly used in commercial applications. The largest applications of BIOMEMS include silicon-based disposable blood pressure sensor chips for blood pressure monitoring (about 20 million units in 2002) and a variety of biosensors.

Structural integrity is of paramount importance in all devices. Load applied during the use of devices can result in component failure. Cracks can develop and propagate under tensile stresses, leading to failure [25.16, 17]. Friction/stiction and wear limit the lifetimes and compromise the performance and reliability of the devices involving relative motion [25.4, 18]. Most MEMS/NEMS applications demand extreme reliability. Stress and deformation analyses are carried out for an optimal design. MEMS/NEMS designers require mechanical properties on the nanoscale. Mechanical properties include elastic, inelastic (plastic, fracture, or viscoelastic), and strength. Elastic and inelastic properties are needed to predict deformation from an applied load in the elastic and inelastic regimes, respectively. The strength property is needed to predict the allowable operating limit. Some of the properties of interest are hardness, elastic modulus, bending strength (fracture stress), fracture toughness, and fatigue strength. Micro/nanostructures have some surface topography and local scratches dependent upon the manufacturing process. Surface roughness and local scratches may compromise the reliability of the devices and their effect needs to be studied.

Most mechanical properties are size-dependent [25.11, 16, 17]. Several researchers have measured mechanical properties of silicon and silicon-based milli- to microscale structures using tensile tests and bending tests [25.19–28], resonant structure tests for measurement of elastic properties [25.29], fracture toughness tests [25.20, 22, 30–34], and fatigue tests [25.32, 35, 36]. Most recently, a few researchers have measured mechanical properties of nanoscale structures using atomic force microscopy (AFM) [25.37, 38] and nanoindentation [25.39, 40]. For stress and deformation analyses of simple geometries and boundary condition, analytical models can be used. For analysis of complex geometries, numerical models are needed. The conventional finite element method (FEM) can be used down to a few tens of nanometers, although its applicability is questionable at the nanoscale. FEM has been used for simulation and prediction of residual stresses and strains induced in MEMS devices during fabrication [25.41], to perform fault analysis in order to study MEMS faulty behavior [25.42], to compute mechanical strain resulting from the doping of silicon [25.43], and to analyze micromechanical experimental data [25.22, 44, 45] and nanomechanical experimental data [25.38]. FEM analysis of nanostructures has been carried out to analyze the effect of types of surface roughness and scratches on stresses in nanostructures [25.46, 47].

The most commonly used materials for MEMS/NEMS are single-crystal silicon and silicon-based materials (e.g., SiO_2 and polysilicon films deposited by low pressure chemical vapor deposition (LPCVD) process) [25.11]. An early study showed silicon to be a mechanically resilient material in addition to its favorable electronic properties [25.48]. Single-crystal 3C-SiC (cubic or β-SiC) films deposited by atmospheric pressure chemical vapor deposition (APCVD) process on large-area silicon substrates are produced for high-temperature micro/nanosensor and actuator applications [25.49–51]. Amorphous alloys can be formed on both metal and silicon substrates by sputtering and plating techniques, providing more flexibility in surface-integration. Electroless deposited Ni-P amorphous thin films have been used to construct microdevices, especially using the so-called LIGA techniques [25.11,40]. Micro/nanodevices need conductors to provide power, as well as electrical/magnetic signals to make them functional. Electroplated gold films have found wide applications in electronic devices, because of their ability to make thin films and process simplicity [25.40]. Use of SiC, Ni-P, and Au films together with silicon and silicon-based materials open up new design opportunities for MEMS/NEMS devices.

This chapter presents a review of mechanical property measurements on the nanoscale of various materials of interest, and stress and deformation analyses of nanostructures.

25.1 Experimental Techniques for Measurement of Mechanical Properties of Nanostructures

25.1.1 Indentation and Scratch Tests Using Micro/Nanoindenters

A nanoindenter is commonly used to measure hardness, elastic modulus, and fracture toughness and to perform micro/nanoscratch studies to get a measure of scratch/wear resistance of materials [25.11,52].

Hardness and Elastic Modulus

The nanoindenter monitors and records the dynamic load and displacement of the three-sided pyramidal diamond (Berkovich) indenter during indentation with a force resolution of about 75 nN and displacement resolution of about 0.1 nm. Hardness and elastic modulus are calculated from the load-displacement data [25.11]. The peak indentation load depends on the mechanical properties of the specimen; a harder material requires higher load for a reasonable indentation depth.

Fracture Toughness

The indentation technique for fracture toughness measurement on the microscale is based on the measurement of the lengths of median-radial cracks produced by indentation. A Vickers indenter (a four-sided diamond pyramid) is used in a microhardness test. A load on the order of 0.5 N is typically used in making the Vickers indentations. The indentation impressions are examined using an optical microscope with Nomarski interference contrast to measure the length of median-radial cracks, c. The fracture toughness (K_{IC}) is calculated by the following relation [25.53]:

$$K_{\mathrm{IC}} = \alpha \left(\frac{E}{H}\right)^{1/2} \left(\frac{P}{c^{3/2}}\right) , \tag{25.1}$$

where α is an empirical constant depending on the geometry of the indenter, H and E are hardness and elastic moduli, and P is the peak indentation load. For Vickers indenters, α has been empirically found based on experimental data and is equal to 0.016 [25.11]. Both E and H values are obtained from the nanoindentation data. The crack length is measured from the center of the indent to the end of the crack using an optical microscope. For one indent, all crack lengths are measured. The crack length c is obtained from the average values of several indents.

Scratch Resistance

In micro/nanoscratch studies, a conical diamond indenter with a tip radius of about 1 μm and an included angle of 60° is drawn over the sample surface, and the load is ramped up until substantial damage occurs [25.11]. The coefficient of friction is monitored during scratching. In order to obtain scratch depths during scratching, the surface profile of the sample surface is first obtained by translating the sample at a low load of about 0.2 mN, which is insufficient to damage a hard sample surface. The 500-μm-long scratches are made by translating the sample while ramping the loads on the conical tip over different loads depending on the material hardness. The actual depth during scratching is obtained by subtracting the initial profile from the scratch depth measured during scratching. In order to measure the scratch depth after the scratch, the scratched surface is profiled at a low load of 0.2 mN and is subtracted from the actual surface profile before scratching.

25.1.2 Bending Tests of Nanostructures Using an AFM

Quasi-static bending tests of fixed nanobeam arrays are carried out using an AFM [25.37, 38, 54]. A three-sided pyramidal diamond tip (with a radius of about 200 nm)

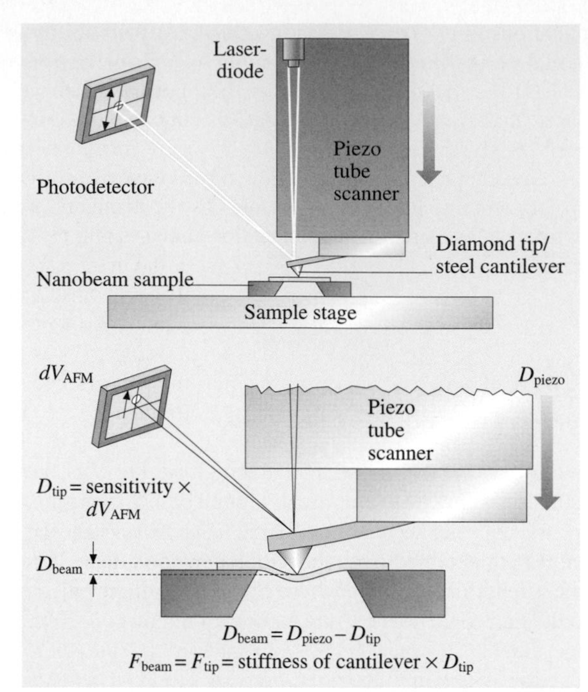

Fig. 25.1 Schematic showing the details of a nanoscale bending test. The AFM tip is brought to the center of the nanobeam and the piezo is extended over a known distance. By measuring the tip displacement, a load-displacement curve of the nanobeam can be obtained [25.38]

mounted on a rectangular stainless steel cantilever is used for the bending tests. The beam stiffness is selected based on the desired load range. The stiffness of the cantilever beams for application of normal load up to 100 μN is about 150–200 N/m.

The wafer with nanobeam array is fixed onto a flat sample chuck using double-stick tape [25.38]. For the bending test, the tip is brought over the nanobeam array with the help of the sample stage of the AFM and a built-in high magnification optical microscope (Fig. 25.1). For the fine positioning of the tip over a chosen beam, the array is scanned in contact mode at a contact load of about 2–4 μN, which results in negligible damage to the sample. After scanning, the tip is located at one end of a chosen beam. To position the tip at the center of the beam span, the tip is moved to the other end of the beam by giving the X-piezo an offset voltage. The value of this offset is determined after several such attempts have been made in order to minimize effects of piezo drift. Half of this offset is then applied to the X-piezo after the tip is positioned at one end of the beam, which usually results in the tip being moved to the center of the span. Once the tip is positioned over the center of the beam span, the tip is held stationary without scanning, and the Z-piezo is extended by a known distance, typically about 2.5 μm, at a rate of 10 nm/s, as shown in Fig. 25.1. During this time, the vertical deflection signal ($\mathrm{d}V_{\mathrm{AFM}}$), which is proportional to the deflection of the cantilever (D_{tip}), is monitored. The displacement of the piezo is equal to the sum of the displacements of the cantilever and nanobeam. Hence, the displacement of the nanobeam (D_{beam}) under the point of load can be determined as

$$D_{\mathrm{beam}} = D_{\mathrm{piezo}} - D_{\mathrm{tip}} . \tag{25.2}$$

The load (F_{beam}) on the nanobeam is the same as the load on the tip/cantilever (F_{tip}) and is given by

$$F_{\mathrm{beam}} = F_{\mathrm{tip}} = D_{\mathrm{tip}} \times k , \tag{25.3}$$

where k is the stiffness of the tip/cantilever. In this manner, a load displacement curve for each nanobeam can be obtained.

The photodetector sensitivity of the cantilever needs to be calibrated to obtain D_{tip} in nm. For this calibration, the tip is pushed against a smooth diamond sample by moving the Z-piezo over a known distance. For the hard diamond material, the actual deflection of the tip can be assumed to be the same as the Z-piezo travel (D_{piezo}), and the photodetector sensitivity (S) for the cantilever setup is determined as

$$S = D_{\mathrm{piezo}} / \mathrm{d}V_{\mathrm{AFM}} \ \mathrm{nm/V} . \tag{25.4}$$

In the measurements, D_{tip} is given as $\mathrm{d}V_{\mathrm{AFM}} \times S$.

Since a sharp tip would result in an undesirable large local indentation, *Sundararajan* and *Bhushan* [25.38] used a worn (blunt) diamond tip. Indentation experiments using this tip on a silicon substrate yielded a residual depth of less than 8 nm at a maximum load of 120 μN, which is negligible compared to displacements of the beams (several hundred nm). Hence, we can assume that negligible local indentation or damage is created during the bending process of the beams, and that the displacement calculated from (25.2) is entirely that of the beam structure.

Elastic Modulus and Bending Strength

Elastic modulus and bending strength (fracture stress) of the beams can be estimated by equations based on the assumption that the beams follow linear elastic theory of an isotropic material. This is probably valid since the beams have high length-to-width (ℓ/w) and length-to-thickness (ℓ/t) ratios and also since the length direction is along the principal stress direction during the test. For a fixed elastic beam loaded at the center of the span, the elastic modulus is expressed as

$$E = \frac{\ell^3}{192I} m \,, \quad (25.5)$$

where ℓ is the beam length, I is the area moment of inertia for the beam cross section, and m is the slope of the load-displacement curve during bending [25.55]. The area moment of inertia for a beam with a trapezoidal cross section is calculated from the following equation:

$$I = \frac{w_1^2 + 4w_1 w_2 + w_2^2}{36(w_1 + w_2)} t^3 \,, \quad (25.6)$$

where w_1 and w_2 are the upper and lower widths, respectively, and t is the thickness of the beam. According to linear elastic theory, for a centrally loaded beam, the moment diagram is shown in Fig. 25.2. The maximum moments are generated at the ends (negative moment) and under the loading point (positive moment), as shown in Fig. 25.2. The bending stresses generated in the beam are proportional to the moments and are compressive or tensile about the neutral axis (line of zero stress). The maximum tensile stress (σ_b, which is the bending strength or fracture stress) is produced on the top surface at both the ends and is given by [25.55]

$$\sigma_b = \frac{F_{max} \ell e_1}{8I} \,, \quad (25.7)$$

where F_{max} is the applied load at failure and e_1 is the distance of the top surface from the neutral plane of the beam cross section and is given by [25.55]

$$e_1 = \frac{t(w_1 + 2w_2)}{3(w_1 + w_2)} \,. \quad (25.8)$$

Although the moment value at the center of the beam is the same as at the ends, the tensile stresses at the center (generated on the bottom surface) are less than those generated at the ends, as per (25.7), because the distance from the neutral axis to the bottom surface is less than e_1. This is because of the trapezoidal cross section of the beam, which results in the neutral axis being closer to the bottom surface than the top (Fig. 25.2).

In the preceding analysis, the beams were assumed to have fixed ends. However, in the nanobeams used by *Sundararajan* and *Bhushan* [25.38], the underside of the beams was pinned over some distance on either side of the span. Hence, a finite element model of the beams was created to see if the difference in the boundary conditions affected the stresses and displacements of the beams. It was found that the difference in the stresses was less than

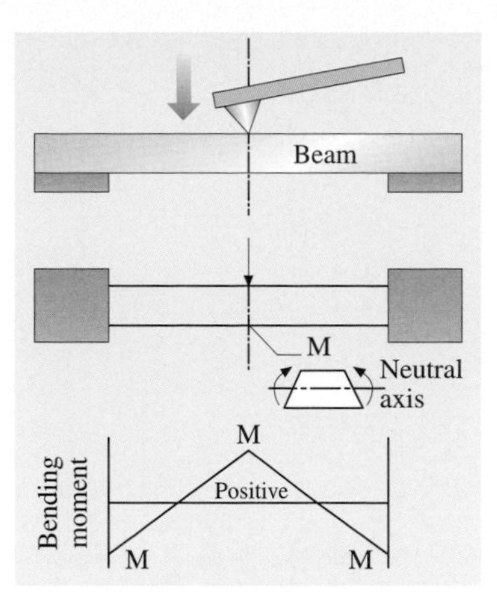

Fig. 25.2 A schematic of the bending moments generated in the beam during a quasi-static bending experiment with the load at the center of the span. The maximum moments occur under the load and at the fixed ends. Due to the trapezoidal cross section, the maximum tensile bending stresses occur at the top surfaces at the fixed ends

1%. This indicates that the boundary conditions near the ends of the actual beams are not that different from that of the fixed ends. Therefore, the bending strength values can be calculated from (25.7).

Fracture Toughness

Fracture toughness is another important parameter for brittle materials such as silicon. In the case of the nanobeam arrays, these are not best suited for fracture toughness measurements, because they do not possess regions of uniform stress during bending. *Sundararajan* and *Bhushan* [25.38] developed a methodology, and its steps are outlined schematically in Fig. 25.3a. First, a crack of known geometry is introduced in the region of maximum tensile bending stress, i. e., on the top surface near the ends of the beam. This is achieved by generating a scratch at high normal load across the width (w_1) of the beam using a sharp diamond tip (radius < 100 nm). A typical scratch thus generated is shown in Fig. 25.3b. By bending the beam as shown, a stress concentration will be formed under the scratch. This will lead to failure of the beam under the scratch once a critical load (fracture load) is attained. The fracture load and relevant dimensions of the scratch are input into the FEM model, which is used to generate the fracture stress plots. Figure 25.3c shows an FEM simulation of one such experiment, which reveals that the maximum stress does occur under the scratch.

If we assume that the scratch tip acts as a crack tip, a bending stress will tend to open the crack in Mode I. In this case, the stress field around the crack tip can be described by the stress intensity parameter K_I (for Mode I) for linear elastic materials [25.56]. In particular, the stresses corresponding to the bending stresses are

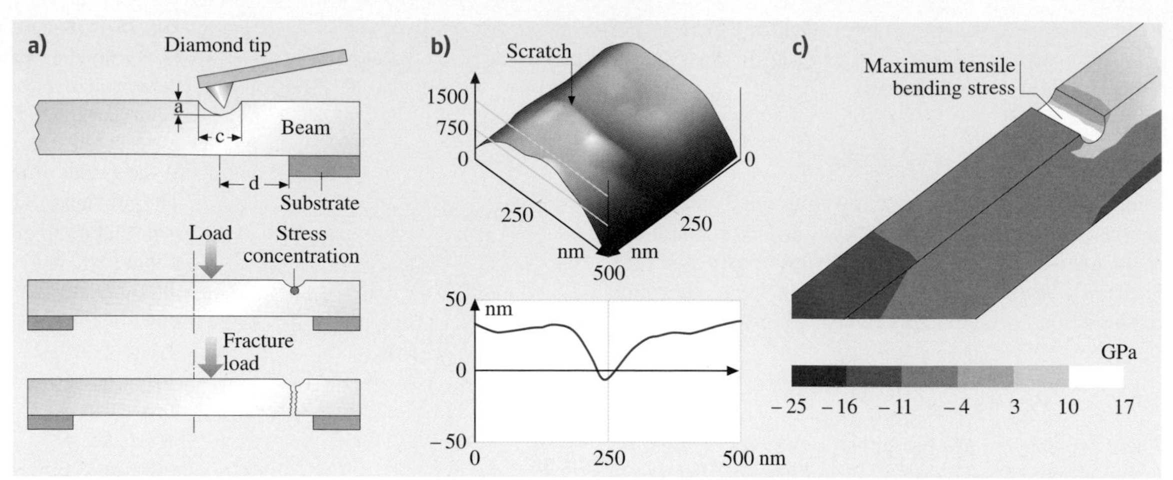

Fig. 25.3 (**a**) Schematic of technique to generate a defect (crack) of known dimensions in order to estimate fracture toughness. A diamond tip is used to generate a scratch across the width of the beam. When the beam is loaded as shown, a stress concentration is formed at the bottom of the scratch. The fracture load is then used to evaluate the stresses using FEM. (**b**) AFM 3-D image and 2-D profile of a typical scratch. (**c**) Finite element model results verifying that the maximum bending stress occurs at the bottom of the scratch [25.38]

described by

$$\sigma = \frac{K_\text{I}}{\sqrt{2\pi r}} \cos\left(\frac{\theta}{2}\right)\left[1+\sin\left(\frac{\theta}{2}\right)\sin\left(\frac{3\theta}{2}\right)\right] \tag{25.9}$$

for every point $p(r,\theta)$ around the crack tip, as shown in Fig. 25.4. If we substitute the fracture stress (σ_f) to the left hand side of (25.9), then the K_I value can be substituted with its critical value, which is the fracture toughness K_IC. Now, the fracture stress can be determined for the point ($r=0$, $\theta=0$), i.e., right under the crack tip as explained above. However, we cannot substitute $r=0$ in (25.9). The alternative is to substitute a value for r, which is as close to zero as possible. For silicon, a reasonable number is the distance between neighboring atoms in the (111) plane, the plane along which silicon exhibits the lowest fracture energy. This value was calculated from silicon unit cell dimensions of 0.5431 nm [25.57] to be 0.4 nm (half of the face diagonal). This assumes that Si displays no plastic zone around the crack tip, which is reasonable since in tension, silicon is not known to display much plastic deformation at room temperature. *Sundararajan* and *Bhushan* [25.38] used values $r=0.4{-}1.6\,\text{nm}$ (i.e., distances up to four times the distance between the nearest neighboring atoms) to estimate the fracture toughness for both Si and SiO_2 according to the following equation:

$$K_\text{IC} = \sigma_\text{f}\sqrt{2\pi r} \qquad r = 0.4{-}1.6\,\text{nm}\,. \tag{25.10}$$

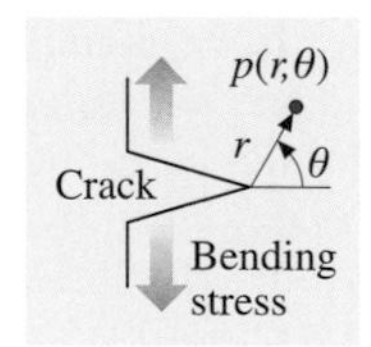

Fig. 25.4 Schematic of crack tip and coordinate systems used in (25.9) to describe a stress field around the crack tip in terms of the stress intensity parameter, K_I [25.38]

Fatigue Strength

In addition to the properties mentioned so far that can be evaluated from quasi-static bending tests, the fatigue properties of nanostructures are also of interest. This is especially true for MEMS/NEMS involving vibrating structures such as oscillators and comb drives [25.58] and hinges in digital micromirror devices [25.59]. To study the fatigue properties of the nanobeams, *Sundararajan* and *Bhushan* [25.38] applied monotonic cyclic stresses using an AFM, Fig. 25.5a. Similar to the bending test, the diamond tip is first positioned at the center of the beam span. In order to ensure that the tip is always in contact with the beam (as opposed to impacting it), the piezo is first extended by a distance D_1, which ensures a minimum stress on the beam. After this extension, a cyclic displacement of amplitude, D_2, is applied continuously until failure of the beam occurs.

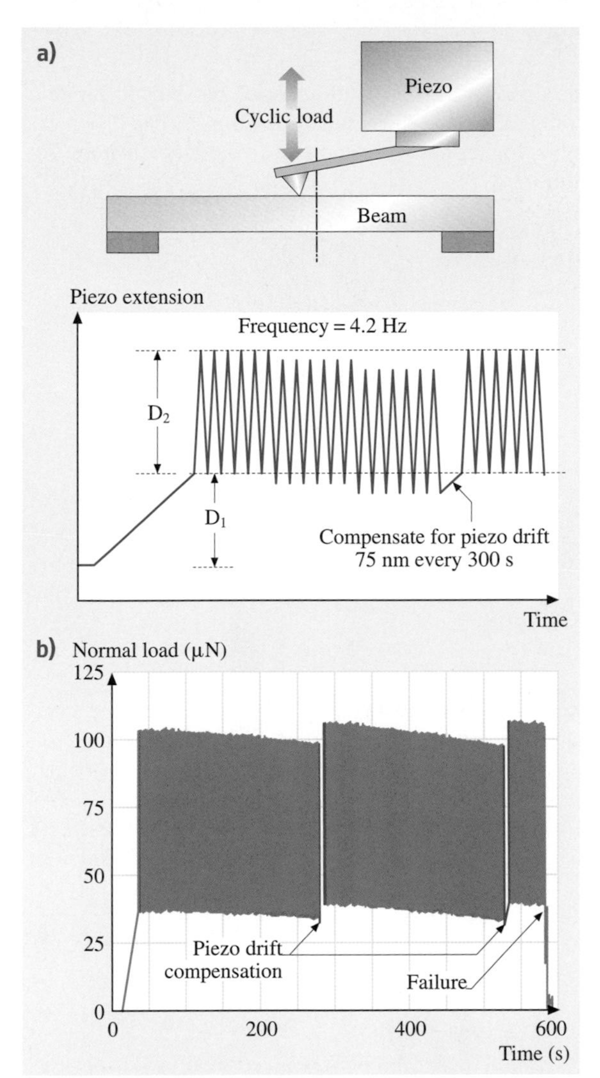

Fig. 25.5 **(a)** Schematic showing the details of the technique to study fatigue behavior of the nanobeams. The diamond tip is located at the middle of the span, and a cyclic load at 4.2 Hz is applied to the beam by forcing the piezo to move in the pattern shown. An extension is made every 300 s to compensate for the piezo drift to ensure that the load on the beam is kept fairly constant. **(b)** Data from a fatigue experiment on a nanobeam until failure. The normal load is computed from the raw vertical deflection signal. The compensations for piezo drift keep the load fairly constant [25.38]

This results in the application of a cyclic load to the beam. The maximum frequency of the cyclic load that could be attained using the AFM by *Sundararajan* and *Bhushan* [25.38] was 4.2 Hz. The vertical deflection signal of the tip is monitored throughout the experiment. The signal follows the pattern of the piezo input up to failure, which is indicated by a sudden drop in the signal. During initial runs, piezo drift was observed that caused the piezo to gradually move away from the beam (i. e., to retract), resulting in a continuous decrease in the applied normal load. In order to compensate for this, the piezo is given a finite extension of 75 nm every 300 s, as shown in Fig. 25.5a. This results in keeping the applied loads fairly constant. The normal load variation (calculated from the vertical deflection signal) from a fatigue test is shown in Fig. 25.5b. The cyclic stress amplitudes (corresponding to D_2) and fatigue lives are recorded for every sample tested. Values for D_1 are set such that minimum stress levels are about 20% of the bending strengths.

25.1.3 Bending Tests Using a Nanoindenter

Quasi-static bending tests of micro/nanostructures are also carried out using a nanoindenter [25.39, 40]. The advantage of the nanoindenter is that loads up to about 500 mN, higher than that in AFM (up to about 100 μN), can be used for structures requiring high loads for experiments. To avoid the indenter tip pushing into the specimen, a blunt tip is used in the bending and fatigue tests. *Li* et al. [25.40] used a diamond conical indenter with a radius of 1 μm and an included angle of 60°. The load position used was at the center of the span for the bridge beams and 10 μm off from the free end of the cantilever beams. An optical microscope with a magnification of 1,500× or an in situ AFM is used to locate the loading position. Then the specimen is moved under the indenter location with a resolution of about 200 nm in the longitudinal direction and less than 100 nm in the lateral direction.

Using the analysis presented earlier, elastic modulus and bending strength of the beams can be obtained from the load-displacement curves [25.40]. For fatigue tests, an oscillating load is applied and contact stiffness is measured during the tests. A significant drop in the contact stiffness during the test is a measure of the number of cycles to failure [25.39].

25.2 Experimental Results and Discussion

25.2.1 Indentation and Scratch Tests of Various Materials Using Micro/Nanoindenters

Studies have been conducted on five different materials: undoped single-crystal Si(100), undoped polysilicon film, SiO_2 film, SiC film, electroless deposited Ni-11.5 wt% P amorphous film, and electroplated Au film [25.40, 50, 51]. A 3-μm thick polysilicon film was deposited by a low pressure chemical vapor deposition (LPCVD) process on a Si(100) substrate. The 1-μm thick SiO_2 film was deposited by a plasma enhanced chemical vapor deposition (PECVD) process on a Si(111) substrate. A 3-μm thick 3C-SiC film was epitaxially grown using an atmospheric pressure chemical vapor deposition (APCVD) process on a Si(100) substrate. A 12-μm thick Ni-P film was electroless plated on a 0.8-mm-thick Al-4.5 wt% Mg alloy substrate. A 3-μm-thick Au film was electroplated on a Si(100) substrate.

Hardness and Elastic Modulus

Hardness and elastic modulus measurements are made using a nanoindenter [25.40]. The hardness and elastic modulus values of various materials at a peak indentation depth of 50 nm are summarized in Fig. 25.6 and Table 25.1. The SiC film exhibits the highest hardness of about 25 GPa and an elastic modulus of about 395 GPa among the samples examined, followed by the undoped Si(100), undoped polysilicon film, SiO_2 film, Ni-P film, and Au film. The hardness and elastic modulus data of the undoped Si(100) and undoped polysilicon film are comparable. For the metal alloy films, the Ni-P film exhibits higher hardness and elastic modulus than the Au film.

Table 25.1 Hardness, elastic modulus, fracture toughness, and critical load results of the bulk single-crystal Si(100) and thin films of undoped polysilicon, SiO_2, SiC, Ni-P, and Au

Samples	Hardness (GPa)	Elastic modulus (GPa)	Fracture toughness ($MPa\,m^{1/2}$)	Critical load (mN)
Undoped Si(100)	12	165	0.75	11
Undoped polysilicon film	12	167	1.11	11
SiO_2 film	9.5	144	0.58 (Bulk)	9.5
SiC film	24.5	395	0.78	14
Ni-P film	6.5	130		0.4 (Plowing)
Au film	4	72		0.4 (Plowing)

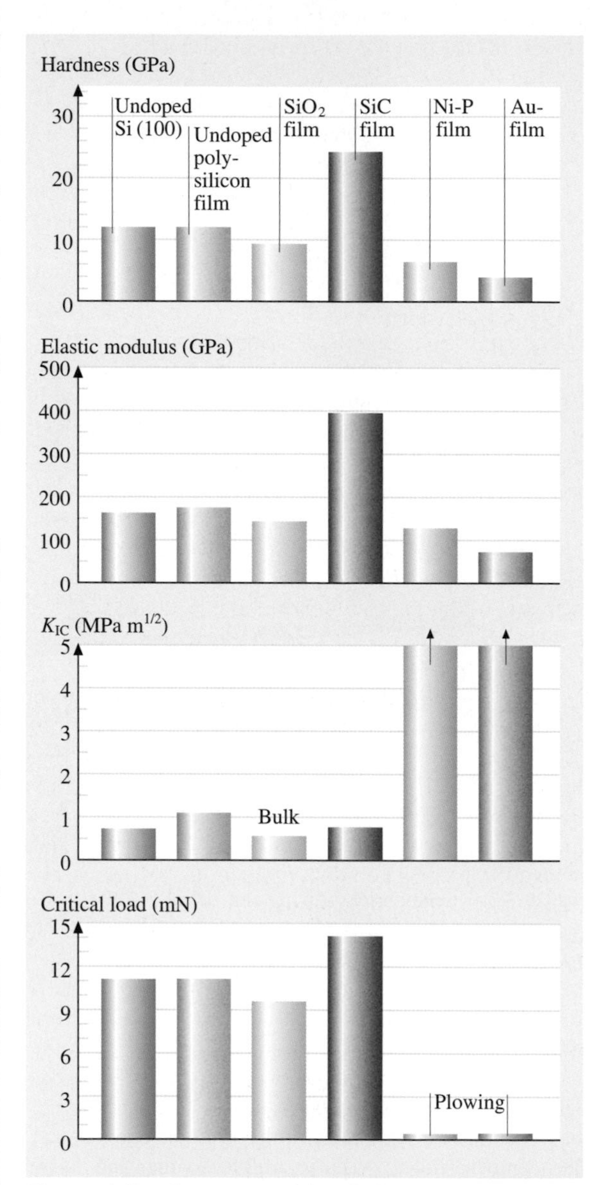

Fig. 25.6 Bar chart summarizing the hardness, elastic modulus, fracture toughness, and critical load (from scratch tests) results of the bulk undoped single-crystal Si(100) and thin films of undoped polysilicon, SiO_2, SiC, Ni-P, and Au [25.40]

Fracture Toughness

The optical images of Vickers indentations made using a microindenter at a normal load of 0.5 N held for 15 s on the undoped Si(100), undoped polysilicon film, and SiC film are shown in Fig. 25.7 [25.51]. The SiC film exhibits the smallest indentation mark, followed by the undoped polysilicon film and undoped Si(100). These Vickers indentation depths are smaller than one-third of the film thickness. Thus, the influence of the substrate on the fracture toughness of the films can be ignored. In addition to the indentation marks, radial cracks are observed emanating from the indentation corners. The SiC film shows the longest radial crack length, followed by the undoped Si(100) and undoped polysilicon film. The radial cracks for the undoped Si(100) are straight, whereas those for the SiC and undoped polysilicon film are not straight but go in a zigzag manner. The fracture toughness (K_{IC}) is calculated using (25.1).

The fracture toughness values of all samples are summarized in Fig. 25.6 and Table 25.1. The SiO_2 film used in this study is about 1 μm thick, which is not thick enough for a fracture toughness measurement. The fracture toughness values of bulk silica are listed instead for a reference. The Ni-P and Au films exhibit very high fracture toughness values that cannot be measured by indentation methods. For other samples, the undoped polysilicon film has the highest value, followed by the undoped Si(100), SiC film, and SiO_2 film. For the undoped polysilicon film, the grain boundaries can stop the radial cracks and change the propagation directions of the radial cracks, making the propagation of these cracks more difficult. Values of fracture toughness for the undoped Si(100) and SiC film are comparable. Since the undoped Si(100) and SiC film are single crystal, no grain boundaries are present to stop the radial cracks and change the propagation directions of the radial cracks. This is why the SiC film shows a lower fracture toughness value than the bulk polycrystal SiC materials of 3.6 MPa $m^{1/2}$ [25.60].

Scratch Resistance

The scratch resistance of various materials has been studied using a nanoindenter by *Li* et al. [25.40]. Figure 25.8 compares the coefficient of friction and scratch depth profiles as a function of increasing normal load and optical images of three regions over scratches: at the beginning of the scratch (indicated by A on the friction profile), at the point of initiation of damage at which the coefficient of friction increases to a high value or increases abruptly (indicated by B on the friction profile), and towards the end of the scratch (indicated by C on the friction profile) for all samples. Note that the ramp loads for the Ni-P and Au range are from 0.2 to 5 mN, whereas the ramp loads for other samples are from 0.2 to 20 mN. All samples exhibit a continuous increase in the coefficient of friction with increasing normal load from the beginning of the scratch. The continuous increase in the coefficient of friction during scratching is attributed to the increasing plowing of the sample by the tip with increasing normal load, as shown in the SEM images in Fig. 25.8. The abrupt increase in the coefficient of friction is associated with catastrophic failure, as well as significant plowing of the tip into the sample. Before the critical load, the coefficient of friction of the undoped polysilicon, SiC, and SiO_2 films increased at a slower rate and was smoother than that of the other samples. The undoped Si(100) exhibits some bursts in the friction profiles before the critical load. At the critical load, the SiC and undoped polysilicon films exhibit a small increase in the coefficient of friction, whereas the undoped Si(100) and undoped polysilicon film exhibit a sudden increase in the coefficient of friction. The Ni-P and Au films show a continuous increase in the coefficient of friction, indicating the behavior of a ductile metal. The bursts in the friction profile might result from the plastic deformation and material pile-up in front of the scratch tip. The Au film exhibits a higher coefficient of friction than the Ni-P film, because it has lower hardness and elastic modulus values than the Ni-P film.

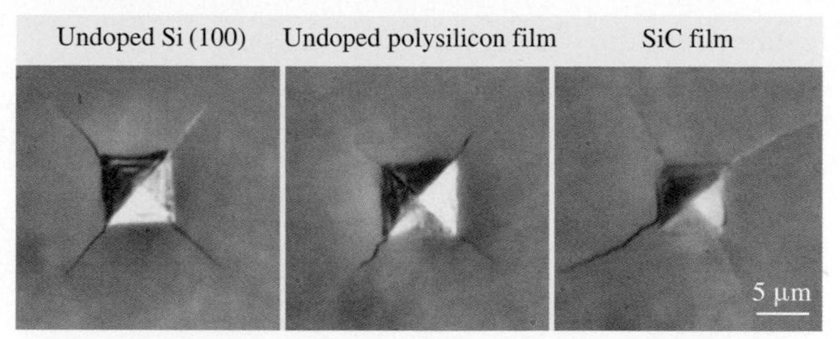

Fig. 25.7 Optical images of Vickers indentations made at a normal load of 0.5 N held for 15 s on the undoped Si(100), undoped polysilicon film, and SiC film [25.51]

The SEM images show that below the critical loads the undoped Si(100) and undoped polysilicon film were damaged by plowing, associated with the plastic flow of the material and formation of debris on the sides of the scratch. For the SiC and SiO_2 films, in region A, a plowing scratch track was found without any debris on the side of the scratch, which is probably responsible for the smoother curve and slower increase in the

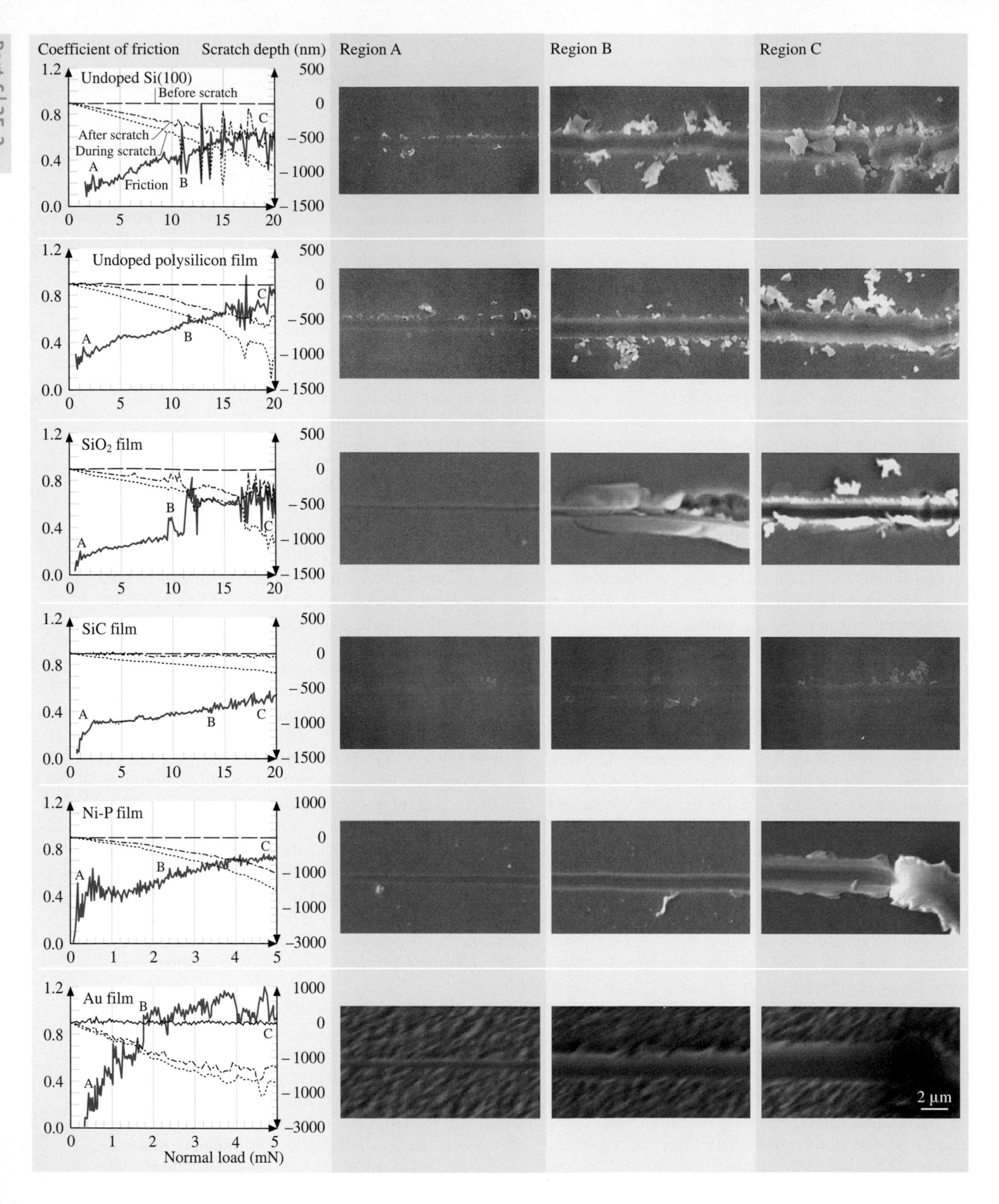

Coefficient of friction
Scratch depth (nm)
Region A
Region B
Region C
Undoped Si(100)
Before scratch
After scratch
During scratch
Friction
Undoped polysilicon film
SiO2 film
SiC film
Ni-P film
Au film
Normal load (mN)
2 µm

Fig. 25.8 Coefficient of friction and scratch depth profiles as a function of increasing normal load and optical images of three regions over scratches: at the beginning of the scratch (indicated by A on the friction profile), at the point of initiation of damage at which the coefficient of friction increases to a high value or increases abruptly (indicated by B on the friction profile), and towards the end of the scratch (indicated by C on the friction profile) for all samples [25.40]

coefficient of friction before the critical load. After the critical load, for the SiO_2 film, delamination of the film from the substrate occurred, followed by cracking along the scratch track. For the SiC film, only several small debris particles were found without any cracks on the side of the scratch, which is responsible for the small increase in the coefficient of friction at the critical load. For the undoped Si(100), cracks were found on the side of the scratch right from the critical load and up, which is probably responsible for the big bursts in the friction profile. For the undoped polysilicon film, no cracks were found on the side of the scratch at the critical load. This might result from grain boundaries, which can stop the propagation of cracks. At the end of the scratch, some of the surface material was torn away and cracks were found on the side of the scratch in the undoped Si(100). A couple of small cracks were found in the undoped polysilicon and SiO_2 films. No crack was found in the SiC film. Even at the end of the scratch, less debris was found in the SiC film. A curly chip was found at the end of the scratch in both Ni-P and Au films. This is a typical characteristic of ductile metal alloys. The Ni-P and Au films were damaged by plowing right from the beginning of the scratch with material pile-up at the side of the scratch.

The scratch depth profiles obtained during and after the scratch on all samples with respect to initial profile, after the cylindrical curvature is removed, are plotted in Fig. 25.8. Reduction in scratch depth is observed after scratching as opposed to during scratching. This reduction in scratch depth is attributed to an elastic recovery after removal of the normal load. The scratch depth after scratching indicates the final depth, which reflects the extent of permanent damage and plowing of the tip into the sample surface and is probably more relevant for visualizing the damage that can occur in real applications. For the undoped Si(100), undoped polysilicon film, and SiO_2 film, there is a large scatter in the scratch depth data after the critical loads, which is associated with the generation of cracks, material removal, and debris. The scratch depth profile is smooth for the SiC film. It is noted that the SiC film exhibits the lowest scratch depth among the samples examined. The scratch depths of the undoped Si(100), undoped polysilicon film, and SiO_2 film are comparable. The Ni-P and Au films exhibit much larger scratch depths than other samples. The scratch depth of the Ni-P film is smaller than that of the Au film.

The critical loads estimated from friction profiles for all samples are compared in Fig. 25.6 and Table 25.1. The SiC film exhibits the highest critical load of about 14 mN, as compared to other samples. The undoped Si(100) and undoped polysilicon film show comparable critical load of about 11 mN, whereas the SiO_2 film shows a low critical load of about 9.5 mN. The Ni-P and Au films were damaged by plowing right from the beginning of the scratch.

25.2.2 Bending Tests of Nanobeams Using an AFM

Bending tests have been performed on Si and SiO_2 nanobeam arrays [25.38, 54]. The single-crystal silicon bridge nanobeams were fabricated by bulk micromachining incorporating enhanced-field anodization using an AFM [25.37]. The Si nanobeams are oriented along the [110] direction in the (001) plane. Subsequent thermal oxidation of the beams results in the formation of SiO_2 beams. The cross section of the nanobeams is trapezoidal owing to the anisotropic wet etching process. SEM micrographs of Si and SiO_2 nanobeam arrays and a schematic of the shape of a typical nanobeam are shown in Fig. 25.9. The actual widths and thicknesses of nanobeams were measured using an AFM in tapping mode prior to tests using a standard Si tapping mode tip (tip radius < 10 nm). Surface roughness measurements of the nanobeam surfaces in tapping mode yielded a σ of 0.7 ± 0.2 nm and peak-to-valley (P–V) distance of 4 ± 1.2 nm for Si and a σ of 0.8 ± 0.3 nm and a P–V distance of 3.1 ± 0.8 nm for SiO_2. Prior to testing, the Si nanobeams were cleaned by immersing them in a "piranha etch" solution (3 : 1 solution by volume of 98% sulphuric acid and 30% hydrogen peroxide) for 600 s to remove any organic contaminants.

Bending Strength

Figure 25.10 shows typical load-displacement curves for Si and SiO_2 beams that were bent to failure [25.38, 54]. The upper width (w_1) of the beams is indicated in the figure. Also indicated are the elastic modulus values obtained from the slope of the load-displacement curve

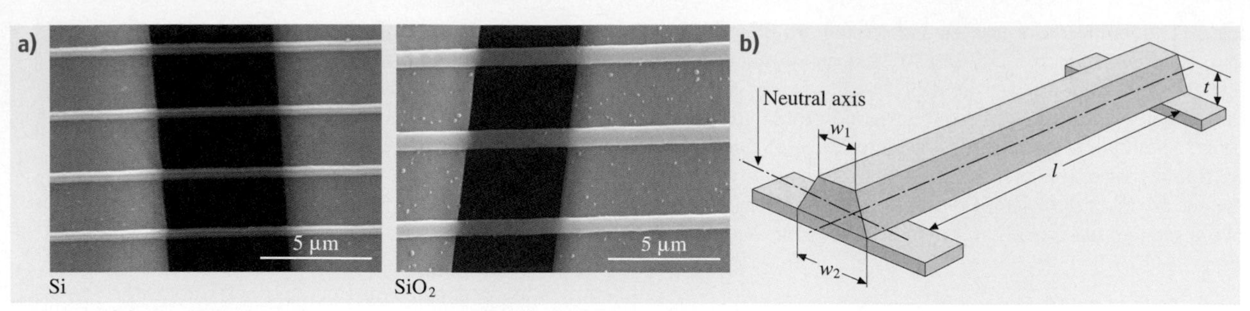

Fig. 25.9 **(a)** SEM micrographs of nanobeam arrays, and **(b)** a schematic of the shape of a typical nanobeam. The trapezoidal cross section is due to the anisotropic wet etching during the fabrication [25.54]

(25.5). All the beams tested showed linear elastic behavior followed by abrupt failure, which is suggestive of brittle fracture. Figure 25.11 shows the scatter in the values of elastic modulus obtained for both Si and SiO_2

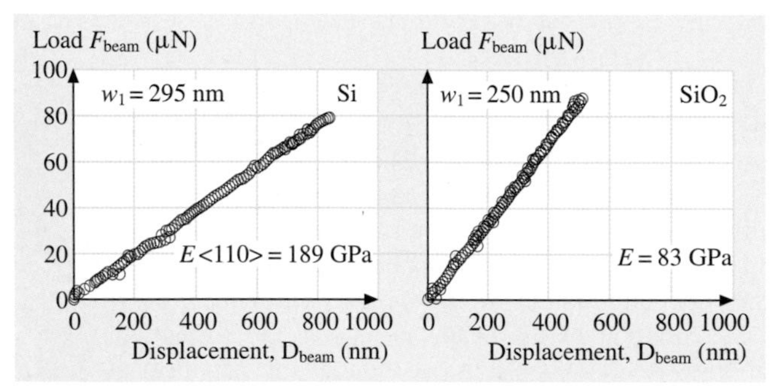

Fig. 25.10 Typical load-displacement curves of silicon and SiO_2 nanobeams. The curves are linear until sudden failure, indicative of brittle fracture of the beams. The elastic modulus (E) values calculated from the curves are shown. The dimensions of the Si beam were $w_1 = 295$ nm, $w_2 = 484$ nm, and $t = 255$ nm, while those of the SiO_2 beam were $w_1 = 250$ nm, $w_2 = 560$ nm, and $t = 425$ nm [25.54]

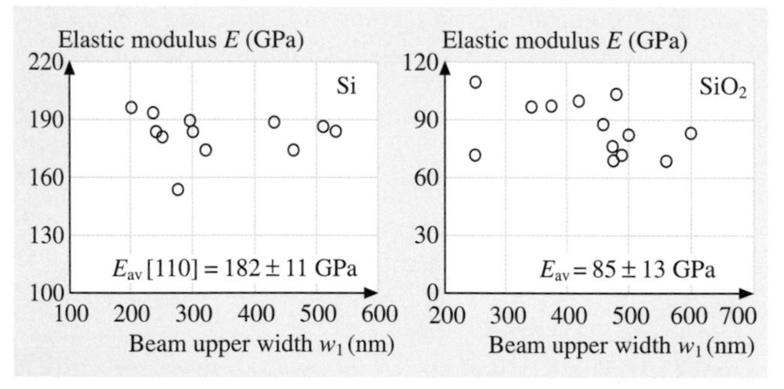

Fig. 25.11 Elastic modulus values measured for Si and SiO_2. The average values are shown. These are comparable to bulk values, which demonstrates that elastic modulus shows no specimen size dependence [25.38]

along with the average values (± standard deviation). The scatter in the values may be due to differences in the orientation of the beams with respect to the trench and the loading point being a little off center with respect to the beam span. The average values are a little higher than the bulk values (169 GPa for Si[110] and 73 GPa for SiO_2 in Table 25.2). However, the values of E obtained from (25.5) have an error of about 20% due to the uncertainties in beam dimensions and spring constant of the tip/cantilever (which affects the measured load). Hence the elastic modulus values on the nanoscale can be considered comparable to bulk values.

Most of the beams when loaded quasi-statically at the center of the span broke at the ends, as shown in Fig. 25.12a, which is consistent with the fact that maximum tensile stresses occur on the top surfaces near the ends. (See FEM stress distribution results in Fig. 25.12b.) Figure 25.13 shows the values of bending strength obtained for different beams. There appears to be no trend in bending strength with the upper width (w_1) of the beams. The large scatter is expected for the strength of brittle materials, since they are dependent on preexisting flaw population in the material and, hence, are statistical in nature. The Weibull distribution, a statistical analysis, can be used to describe the scatter in the bending strength values. The means of the Weibull distributions were found to be 17.9 GPa and 7.6 GPa for Si and SiO_2, respectively. Previously reported numbers of strengths range from 1 to 6 GPa for silicon [25.19, 20, 22–25, 27, 45, 61, 62] and about 1 GPa for SiO_2 [25.34] microscale specimens. This clearly indicates that bending strength shows a specimen size dependence. Strength of brittle materials is dependent on preexisting flaws in the material. Since for nanoscale specimens the volume is smaller than for micro- and macroscale specimens, the flaw population will be smaller as well, resulting in higher values of strength.

Table 25.2 Summary of measured parameters from quasi-static bending tests

Sample	Elastic modulus E (GPa)		Bending strength σ_b (GPa)		Fracture toughness K_{IC} (MPa $\sqrt{m}$)		
	Measured	Bulk value	Measured	Reported (microscale)	Estimated	Reported (microscale)	Bulk value
Si	182±11	169[1]	18±3	< 10[3]	1.67±0.4	0.6–1.65[5]	0.9[6]
SiO_2	85±13	73[2]	7.6±2	< 2[4]	0.60±0.2	0.5–0.9[4]	–

[1] Si(110), [25.63] [2] [25.64] [3] [25.19, 20, 22–25, 45, 61, 62] [4] [25.34] [5] [25.30–33] [6] [25.57]

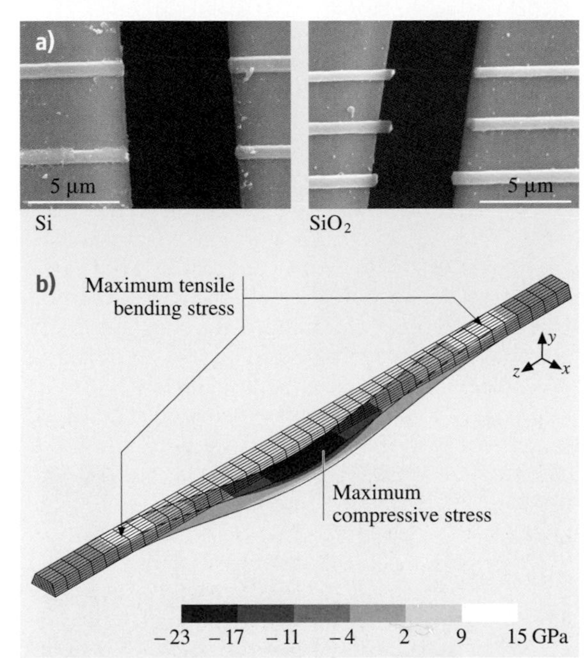

Fig. 25.12 (a) SEM micrographs of nanobeams that failed during quasi-static bending experiments. The beams failed at or near the ends, which is the location of maximum tensile bending stress [25.54], and (b) bending stress distribution for silicon nanobeam, indicating that the maximum tensile stresses occur on the top surfaces near the fixed ends

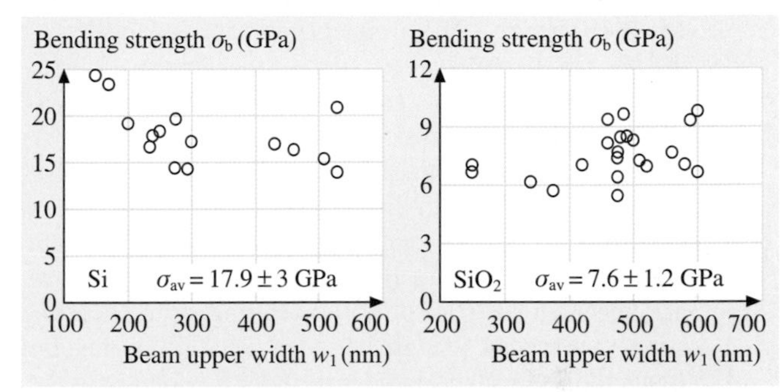

Fig. 25.13 Bending strength values obtained from bending experiments. Average values are indicated. These values are much higher than values reported for microscale specimens, indicating that bending strength shows a specimen size effect [25.38]

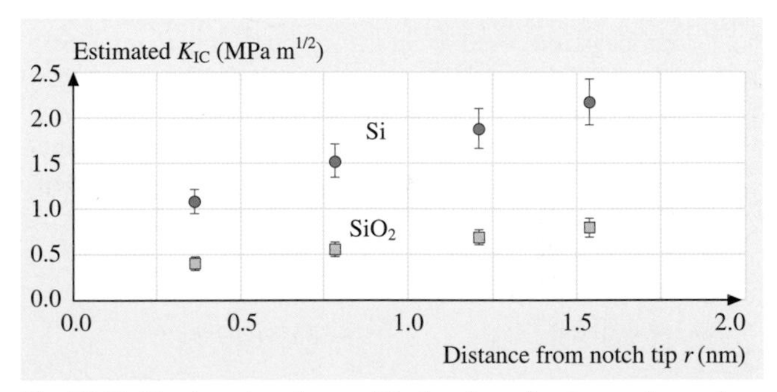

Fig. 25.14 Fracture toughness (K_{IC}) values for increasing values of r corresponding to distance between neighboring atoms in {111} planes of silicon (0.4 nm). Hence r values between 0.4 and 1.6 nm are chosen. The K_{IC} values thus estimated are comparable to values reported by others for both Si and SiO_2 [25.38]

Fracture Toughness

Estimates of fracture toughness calculated using (25.10) for Si and SiO_2 are shown in Fig. 25.14 [25.38]. The results show that the K_{IC} estimate for Si is about 1–2 MPa $\sqrt{m}$, whereas for SiO_2 the estimate is about 0.5–0.9 MPa $\sqrt{m}$. These values are comparable to values reported by others on larger specimens for Si [25.30–33] and SiO_2 [25.34]. The high values obtained for Si could be due to the fact that the scratches, despite being quite sharp, still have a finite radius of about 100 nm. The bulk value for silicon is about 0.9 MPa $\sqrt{m}$ (Table 25.2). Fracture toughness is considered a material property and is believed to be independent of specimen size. The values obtained in this study, given its limitations, appear to show that fracture toughness is comparable, if not a little higher on the nanoscale.

Fatigue Strength

Fatigue strength measurements of Si nanobeams have been carried out by *Sundararajan* and *Bhushan* [25.38] using an AFM and *Li* and *Bhushan* [25.39] using a nanoindenter. Various stress levels were applied to nanobeams by *Sundararajan* and *Bhushan* [25.38]. The minimum stress was 3.5 GPa for Si beams and 2.2 GPa for SiO_2 beams. The frequency of applied load was 4.2 Hz. In general, the fatigue life decreased with increasing mean stress as well as increasing stress amplitude. When the stress amplitude was less than 15% of the bending strength, the fatigue life was greater than 30,000 cycles for both Si and SiO_2. However, the mean stress had to be less than 30% of the bending strength for a life of greater than 30,000 for Si, whereas even at a mean stress of 43% of the bending strength, SiO_2 beams showed a life greater than 30,000. During fatigue, the beams broke under the loading point or at the ends when loaded at the center of the span. This was different from the quasi-static bending tests, where the beams broke at the ends almost every time. This could be due to the fact that the stress levels under the load and at the ends are not that different and fatigue crack propagation could occur at either location. Figure 25.15 shows a nanoscale *S-N* curve with bending stress (*S*) as a function of fatigue in cycles (*N*) with an apparent endurance life at lower stress. This study clearly demonstrates that fatigue properties of nanoscale specimens can be studied.

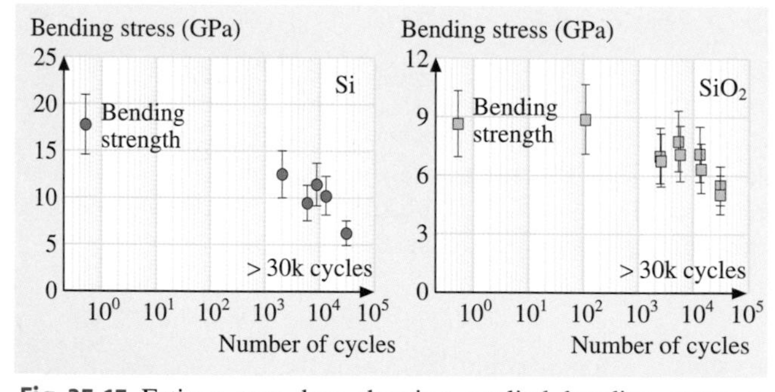

Fig. 25.15 Fatigue test data showing applied bending stress as a function of number of cycles. A single load-unload sequence is considered 1 cycle. The bending strength data points are, therefore, associated with $\frac{1}{2}$ cycle, since failure occurs upon loading [25.38]

SEM Observations of Fracture Surfaces

Figure 25.16 shows SEM images of the fracture surfaces of nanobeams broken during quasi-static bending, as well as fatigue [25.38]. In the quasi-static cases, the maximum tensile stresses occur on the top surface, so it is reasonable to assume that fracture initiated at or

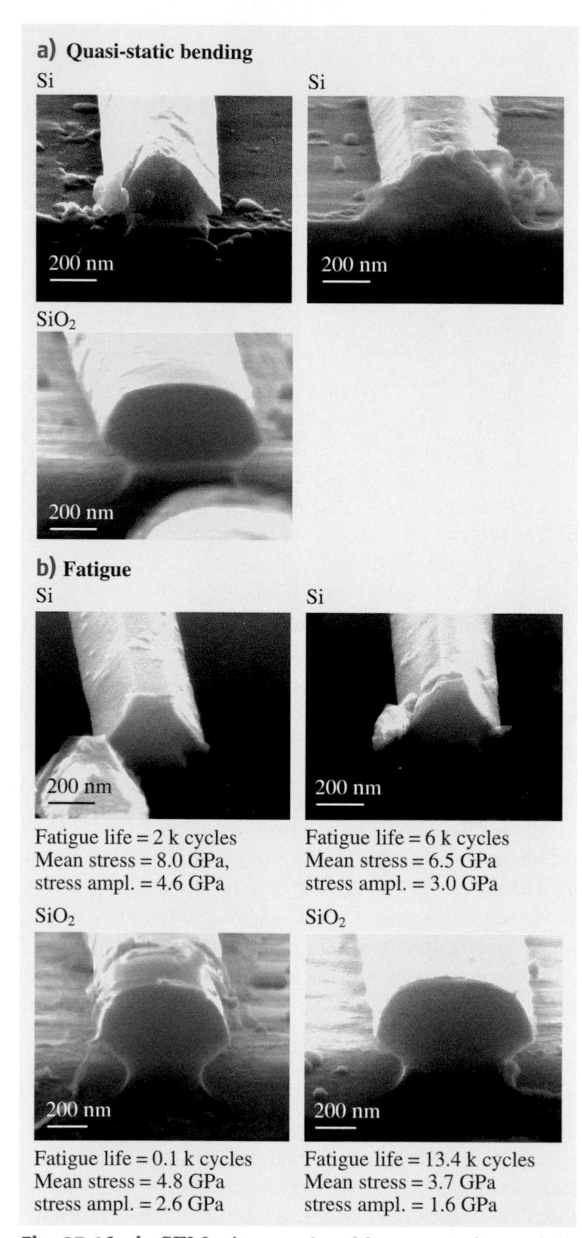

Fig. 25.16a,b SEM micrographs of fracture surfaces of silicon and SiO_2 beams subjected to (**a**) quasi-static bending and (**b**) fatigue [25.38]

near the top surface and propagated downward. The fracture surfaces of the beams suggest a cleavage type of fracture. Silicon beam surfaces show various ledges or facets, which is typical for crystalline brittle materials. Silicon usually fractures along the (111) plane, due to this plane having the lowest surface energy to overcome by a propagating crack. However, failure has also been known to occur along the (110) planes in microscale specimens, despite the higher energy required compared to the (111) planes [25.45]. The plane normal to the beam direction in these samples is the (110) plane, while (111) planes will be oriented at 35° from the (110) plane. The presence of facets and irregularities on the silicon surface in Fig. 25.16a suggest that it is a combination of these two types of fractures that has occurred. Since the stress levels are very high for these specimens, it is reasonable to assume that crack propagating forces will be high enough to result in (110) type failures.

In contrast, the silicon fracture surfaces under fatigue, shown in Fig. 25.16b, appear very smooth without facets or irregularities. This is suggestive of low energy fracture, i. e., of (111) type fracture. We do not see evidence of fatigue crack propagation in the form of steps or striations on the fracture surface. We believe that for the stress levels applied in these fatigue experiments, failure in silicon occurred via cleavage associated with "static fatigue" type of failures.

SiO_2 shows very smooth fracture surfaces for both quasi-static bending and fatigue. This is in contrast to the hackled surface one might expect for the brittle failure of an amorphous material on the macroscale. However, in larger scale fracture surfaces for such materials, the region near the crack initiation usually appears smooth or mirror-like. Since the fracture surface here is so small and very near the crack initiation site, it is not unreasonable to see such a smooth surface for SiO_2 on this scale. There appears to be no difference between the fracture surfaces obtained by quasi-static bending and fatigue for SiO_2.

Summary of Mechanical Properties Measured Using Quasi-Static Bending Tests

Table 25.2 summarizes the various properties measured via quasi-static bending in this study [25.38]. Also shown are bulk values of the parameters, along with values reported on larger scale specimens by other researchers. Elastic modulus and fracture toughness values appear to be comparable to bulk values and show no dependence on specimen size. However, bending strength shows a clear specimen size dependence with nanoscale numbers being twice as large as numbers reported for larger-scale specimens.

25.2.3 Bending Tests of Microbeams Using a Nanoindenter

Bending tests have been performed on Ni-P and Au microbeams [25.40]. The Ni-P cantilever microbeams were fabricated by focused ion beam machining technique. The dimensions were $10\times12\times50\,\mu m^3$. Notches with a depth of 3 μm and a tip radius of 0.25 μm were introduced in the microbeams to facilitate failure at a lower

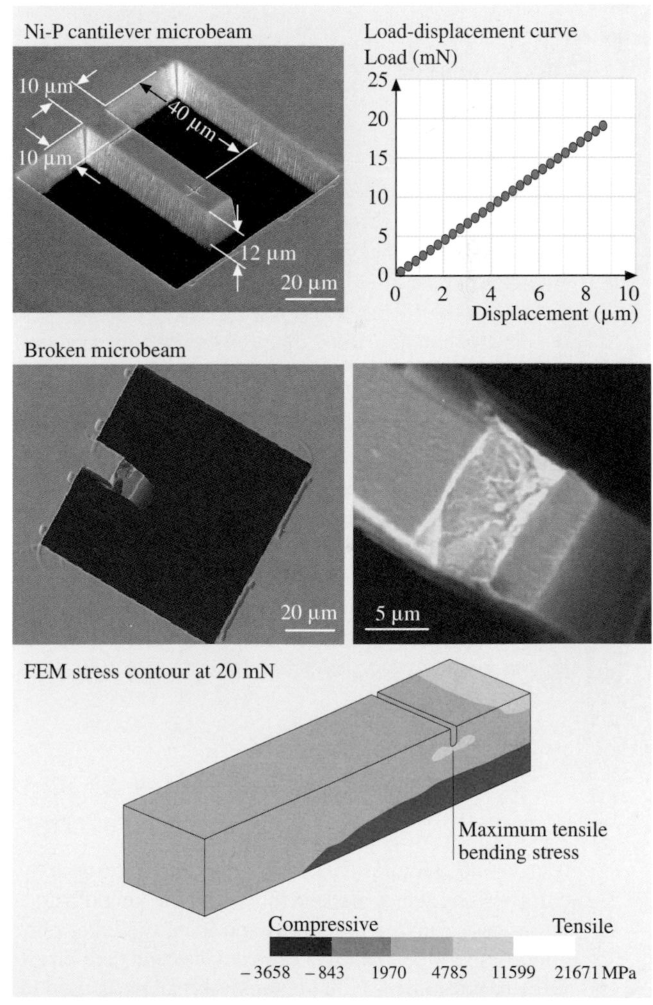

Fig. 25.17 SEM micrographs of the new and broken beams, load-displacement curve, and FEM stress contour for the notched Ni-P cantilever microbeam [25.40]

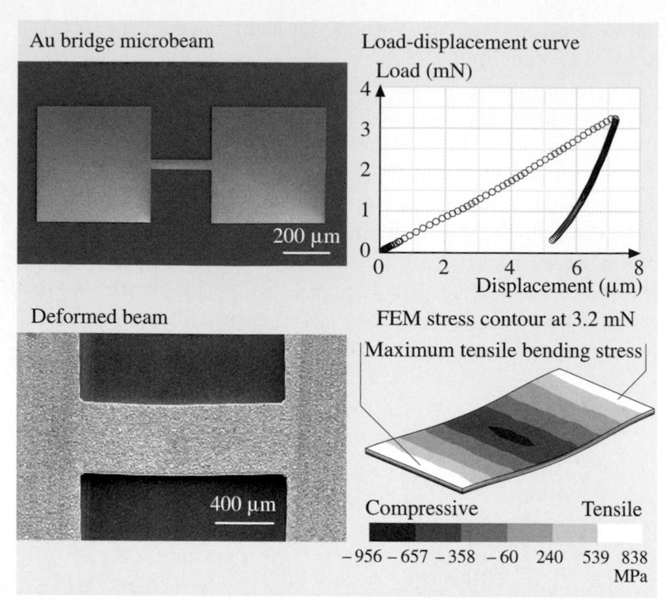

Fig. 25.18 SEM micrographs of the new and deformed beams, load-displacement curve, and FEM stress contour for the Au bridge microbeam [25.40]

load in the bending tests. The Au bridge microbeams were fabricated by electroplating technique.

Figure 25.17 shows the SEM images, load displacement curve, and FEM stress contour for the notched Ni-P cantilever microbeam that was bent to failure [25.40]. The distance between the loading position and the fixed end is 40 μm. The 3-μm-deep notch is 10 μm from the fixed end. The notched beam showed linear behavior followed by abrupt failure. The FEM stress contour shows that there is higher stress concentration at the notch tip. The maximum tensile stress σ_m at the notch tip can be analyzed using Griffith fracture theory as follows [25.53]:

$$\sigma_m \approx 2\sigma_o \left(\frac{c}{\rho}\right)^{1/2} , \qquad (25.11)$$

where σ_o is the average applied tensile stress on the beam, c is the crack length, and ρ is the crack tip radius. Therefore, elastic-plastic deformation will first occur locally at the end of the notch tip, followed by abrupt fracture failure after the σ_m reaches the ultimate tensile strength of Ni-P, even though the rest of the beam is still in the elastic regime. The SEM image of the fracture surface shows that the fracture started right from the notch tip with plastic deformation characteristic. This indicates that although local plastic deformation occurred at the notch tip area, the whole beam failed catastrophically. The present study shows that FEM simulation can predict well the stress concentration and helps in understanding the failure mechanism of the notched beams.

Figure 25.18 shows the SEM images, load-displacement curve and FEM stress contour for the Au bridge microbeam that was deformed by the indenter [25.40]. The recession gap between the beam and substrate is about 7 μm, which is not large enough to break the beam at the load applied. From the load-displacement curve, we note that the beam experienced elastic-plastic deformation. The FEM stress contour shows that the maximum tensile stress is located at the fixed ends, whereas the minimum compressive stress is located around the center of the beam. The SEM image shows that the beam has been permanently deformed. No crack was found on the beam surface. The present study shows a possibility for mechanically forming the Au film into the shape as needed. This may help in designing/fabricating functionally complex smart micro/nanodevices that need conductors for power supply and input/output signals.

25.3 Finite Element Analysis of Nanostructures with Roughness and Scratches

Micro/nanostructures have some surface topography and local scratches dependent upon the manufacturing process. Surface roughness and local scratches may compromise the reliability of the devices and their effect needs to be studied. Finite element modeling is used to perform parametric analysis to study the effect of surface roughness and scratches in different well-defined forms on tensile stresses that are responsible for crack propagation [25.46, 47]. The analysis has been carried out on trapezoidal beams supported at the bottom whose data (on Si and SiO_2 nanobeams) have been presented earlier.

The finite element analysis has been carried out using the static analysis of ANSYS 5.7, which calculates the deflections and stresses produced by the applied loading. The type of element selected for the present study was the SOLID95, which allows the use of different shapes without much loss of accuracy. This element is 3-D with 20 nodes, each node having three degrees of freedom that imply translation in the x, y, and z directions. Each nanobeam cross section is divided into six elements along the width and the thickness and 40 elements along the length. SOLID95 has plasticity, creep, stress stiffening, large deflection, and large strain capabilities. The large displacement analysis is used for large loads. The mesh is kept finer near the asperities and the scratches in order to take into account variation in the bending stresses. The beam materials studied are made of single-crystal silicon (110) and SiO_2 films whose data have been presented earlier. Based on bending experiments presented earlier, the beam materials can be assumed to be linearly elastic isotropic materials. Young's modulus of elasticity (E) and Poisson's ratio (ν) for Si and SiO_2 are 169 GPa [25.63] and 0.28 [25.57], and 73 GPa [25.64] and 0.17 [25.64], respectively. A sample nanobeam of silicon was chosen for performing most of the analysis, as silicon is the most widely used MEMS/NEMS material. The cross section of the fabricated beams used in the experiment is trapezoidal and supported at the bottom, Fig. 25.9. The following dimensions are used: $w_1 = 200$ nm, $w_2 = 370$ nm, $t = 255$ nm, and $\ell = 6\,\mu$m. In the boundary conditions, the displacements are constrained in all directions on the bottom surface for 1 μm from each end. A point load applied at the center of the beam is simulated with the load being applied at three closely located central nodes on the beam used. It has been observed from the experimental results that the Si nanobeam breaks at around 80 μN. Therefore, in this analysis, a nominal load of 70 μN is selected. At this load, deformations are large and a large displacement option is used.

To study the effect of surface roughness and scratches on the maximum bending stresses the following cases were studied. First, the semicircular and grooved asperities in the longitudinal direction with defined geometrical parameters are analyzed, Fig. 25.19a. Next, semicircular asperities and scratches placed along the transverse direction at a distance c from the end and separated by pitch p from each other are analyzed, Fig. 25.19b. Lastly, the beam material is assumed to be either purely elastic, elastic-plastic, or elastic-perfectly plastic. In the following section, we begin with the stress distribution in smooth nanobeams followed by the effect of surface roughness in the longitudinal and transverse directions and scratches in the transverse direction.

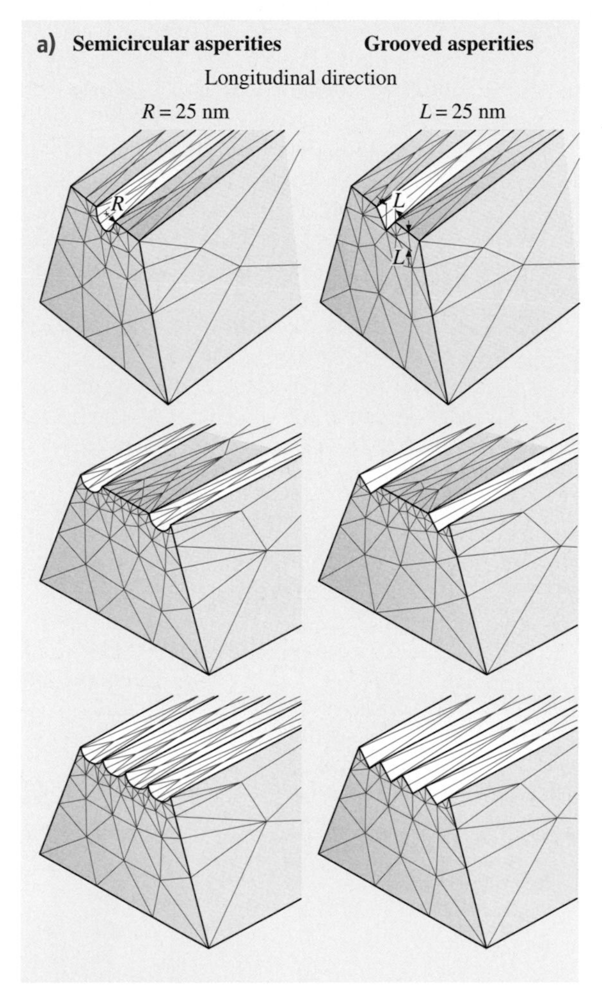

Fig. 25.19 **(a)** Plots showing the geometries of modeled roughness – semicircular and grooved asperities along the nanobeam length with defined geometrical parameters. **(b)** *(See on next page.)* Schematic showing semicircular asperities and scratches in the transverse direction, followed by the illustration of the mesh created on the beam with fine mesh near the asperities and scratches. Also shown are the semicircular asperities and scratches at different pitch values

25.3.1 Stress Distribution in a Smooth Nanobeam

Figure 25.20 shows the stress and vertical displacement contours for a nanobeam supported at the bottom and loaded at the center [25.46, 47]. As expected, the maximum tensile stress occurs at the ends, while the maximum compressive stress occurs under the load at

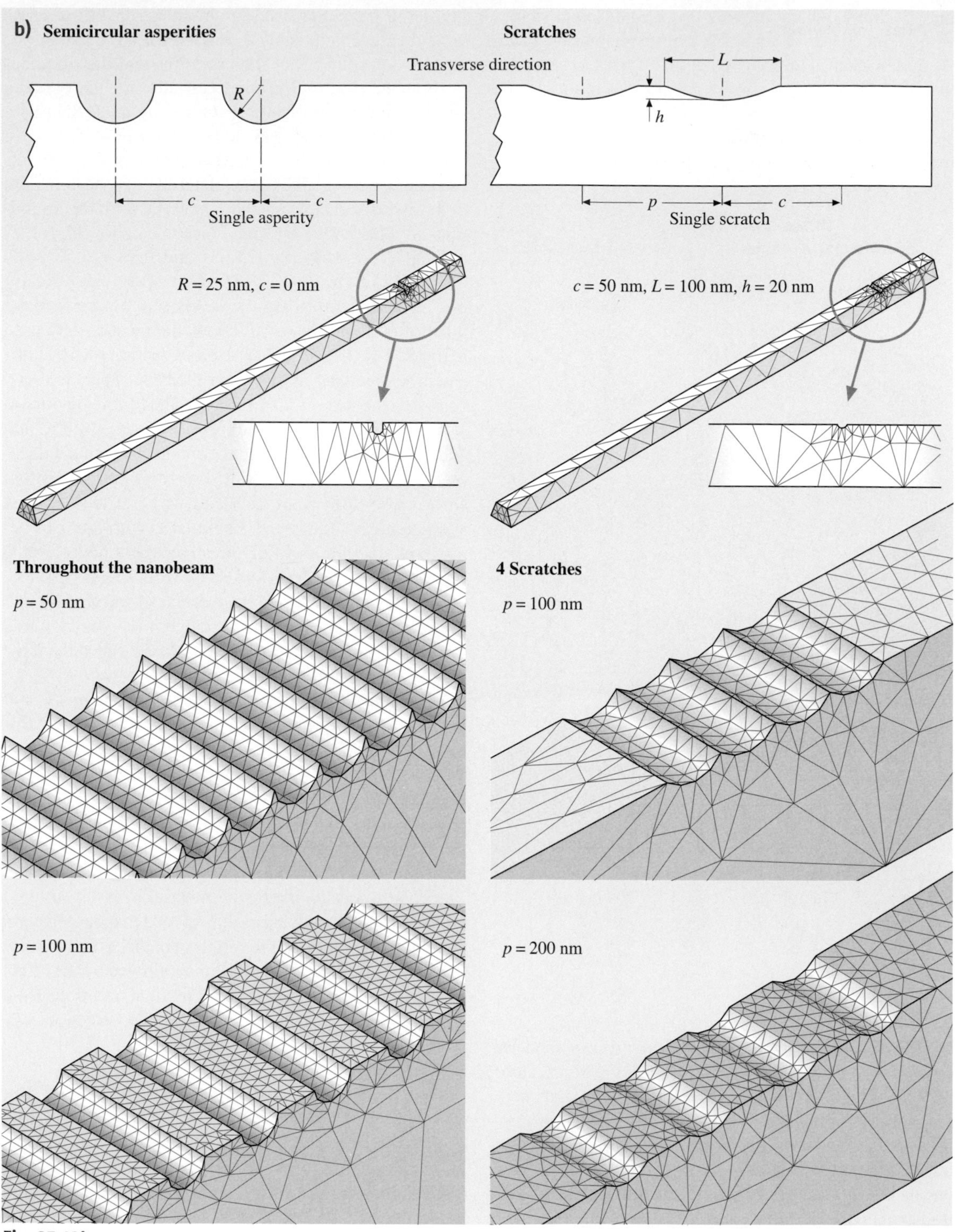
b) Semicircular asperities
Scratches
Transverse direction
R
L
h
c
c
Single asperity
p
c
Single scratch
R = 25 nm, c = 0 nm
c = 50 nm, L = 100 nm, h = 20 nm
Throughout the nanobeam
p = 50 nm
4 Scratches
p = 100 nm
p = 100 nm
p = 200 nm

Fig. 25.19b

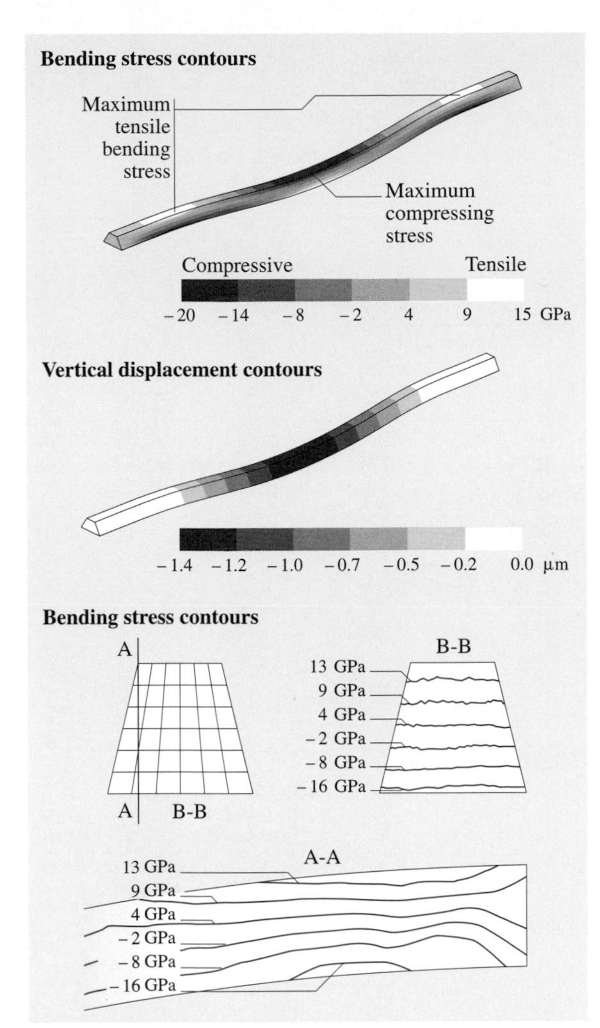

Fig. 25.20 Bending stress contours, vertical displacement contours, and bending stress contours after loading trapezoidal Si nanobeam ($w_1 = 200\,\text{nm}$, $w_2 = 370\,\text{nm}$, $t = 255\,\text{nm}$, $\ell = 6\,\mu\text{m}$, $E = 169\,\text{GPa}$, $\nu = 0.28$) at $70\,\mu\text{N}$ load [25.47]

the center. Stress contours obtained at a section of the beam from the front and side are also shown. In the beam cross section, the stresses remain constant at a given vertical distance from one side to another and change with a change in vertical location. This can be explained by the fact that the bending moment is constant at a particular cross section, so the stress is only dependent on the distance from the neutral axis. However, in cross section A-A the high tensile and compressive stresses are localized near the end of the beam at top and bottom, respectively, whereas the lower values are spread out away from the ends. High value of tensile stresses occurs near the ends because of high bending moment.

25.3.2 Effect of Roughness in the Longitudinal Direction

The roughness in the form of semicircular and grooved asperities in the longitudinal direction on the maximum bending stresses are analyzed [25.47]. The radius R and depth L are kept fixed at 25 nm, while the number of asperities is varied and their effect is observed on the maximum bending stresses. Figure 25.21 shows the variation of maximum bending stresses as a function of asperity shape and the number of asperities. The maximum bending stresses increase as the asperity number increases for both semicircular and grooved asperities. This can be attributed to the fact that as the asperity number increases, the moment of inertia decreases for that cross section. Also, the distance from the neutral axis increases because the neutral axis shifts downwards. Both these factors lead to the increase in the maximum bending stresses, and this effect is more pronounced in the case of semicircular asperity as it exhibits a higher value of maximum bending stress than that in grooved asperity. Figure 25.21 shows the stress contours obtained at a section of the beam from the front side for both cases when we have a single semicircular asperity and when four adjacent semicircular asperities are present. Trends are similar to those observed earlier for a smooth nanobeam (Fig. 25.20).

25.3.3 Effect of Roughness in the Transverse Direction and Scratches

We analyze semicircular asperities when placed along the transverse direction followed by the effect of scratches on the maximum bending stresses in varying numbers and pitch [25.47]. In the analysis of semicircular transverse asperities, three cases were considered, which included a single asperity and asperities throughout the nanobeam surface separated by pitch equal to 50 nm and 100 nm. In all of these cases, c value was kept equal to 0 nm. Figure 25.22 shows that the value of maximum tensile stress is 42 GPa, which is much larger than the maximum tensile stress value with no asperity of 16 GPa, or when the semicircular asperity is present in the longitudinal direction. It is also observed that the maximum tensile stress does not vary with the number of asperities or the pitch, while the maximum compres-

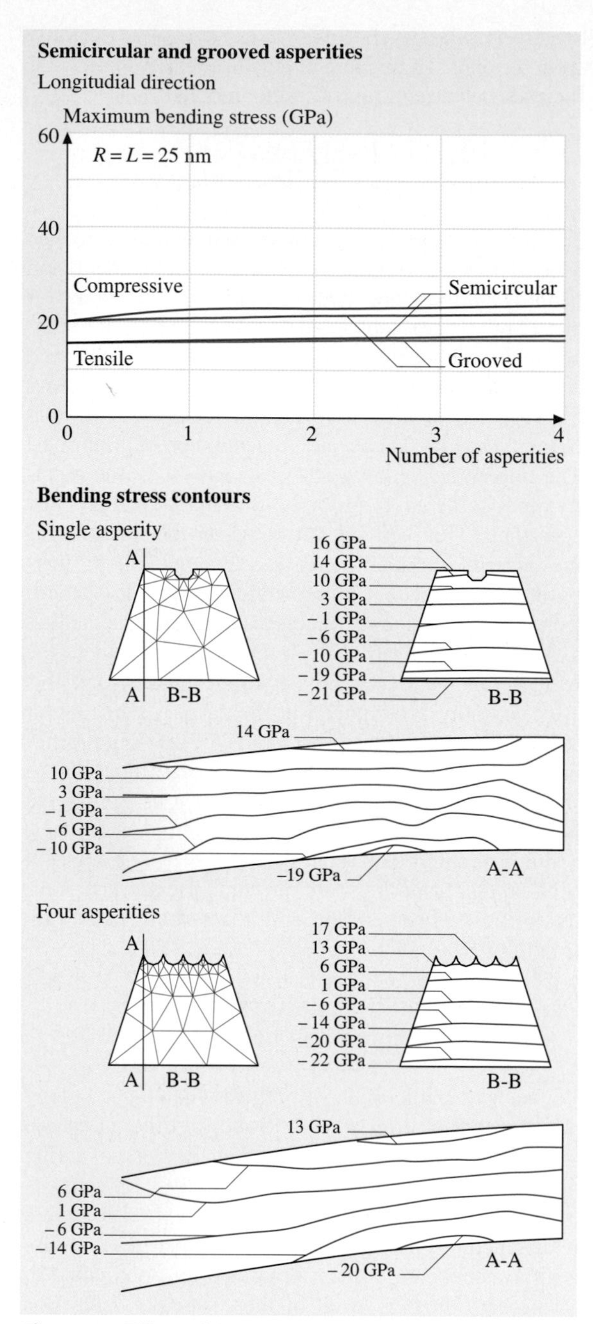

Fig. 25.21 Effect of longitudinal semicircular and grooved asperities in different numbers on maximum bending stresses after loading trapezoidal Si nanobeams ($w_1 = 200$ nm, $w_2 = 370$ nm, $t = 255$ nm, $\ell = 6$ μm, $E = 169$ GPa, $\nu = 0.28$, load = 70 μN). Bending stress contours obtained in the beam with semicircular single asperity and four adjacent asperities of $R = 25$ nm [25.47]

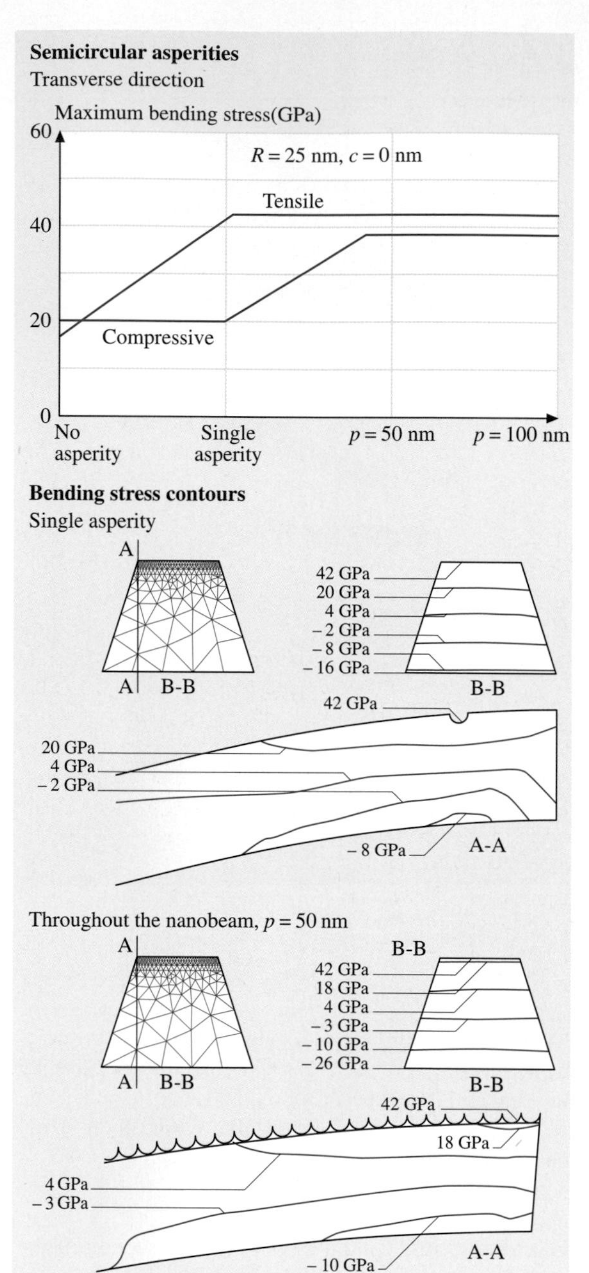

Fig. 25.22 Effect of transverse semicircular asperities located at different pitch values on the maximum bending stresses after loading trapezoidal Si nanobeams ($w_1 = 200$ nm, $w_2 = 370$ nm, $t = 255$ nm, $\ell = 6$ μm, $E = 169$ GPa, $\nu = 0.28$, load = 70 μN). Bending stress contours obtained in the beam with semicircular single asperity and semicircular asperities throughout the nanobeam surface at $p = 50$ nm [25.47]

sive stress does increase dramatically for the asperities present throughout the beam surface from its value when a single asperity is present. Maximum tensile stress occurs at the ends, and an increase in p does not add any asperities at the ends, whereas asperities are added in the central region where compressive stresses are maximum. The semicircular asperities present at the center cause the local perturbation in the stress distribution at the center of the asperity where load is being applied, leading to a high value of maximum compressive stress [25.65]. Figure 25.22 also shows the stress contours obtained at a section of the beam from the front and side for both cases when there is a single semicircular asperity and when asperities are present throughout the beam surface at a pitch equal to 50 nm. Trends are similar to those observed earlier for a smooth nanobeam (Fig. 25.20).

In the study pertaining to scratches, the number of scratches are varied along with the variation in the pitch. Furthermore, the load is applied at the center of the beam and at the center of the scratch near the end, as for all the cases. In all of these, c value was kept equal to 50 nm and L value was equal to 100 nm with h value being 20 nm. Figure 25.23 shows that the value of maximum tensile stress remains almost the same with the number of scratches for both types of loading – that is, when load is applied at the center of the beam and at the center of the scratch near the end. This is because the maximum tensile stress occurs at the beam ends no matter where the load gets applied. But the presence of scratch does increase the maximum tensile stress compared to its value for a smooth nanobeam, although the number of scratches no longer matter as the maximum tensile stress occurring at the nanobeam end is unaffected by the presence of more scratches beyond the first scratch in the direction toward the center. The value of the tensile stress is much lower when the load is applied at the center of scratch, and it can be explained as follows. The negative bending moment at the end near the applied load decreases with load offset after two-thirds of the length of the beam [25.66]. Since this negative bending moment is responsible for tensile stresses, their behavior with the load offset is the same as the negative bending moment. Also, the value of maximum compressive stress when load is applied at the center of the nanobeam remains almost the same as the center geometry is unchanged due to the number of scratches and, hence, the maximum compressive stress occurring below the load at the center is same. On the other hand, when the load is applied at the center of the scratch we observe that the maximum compressive stress in-

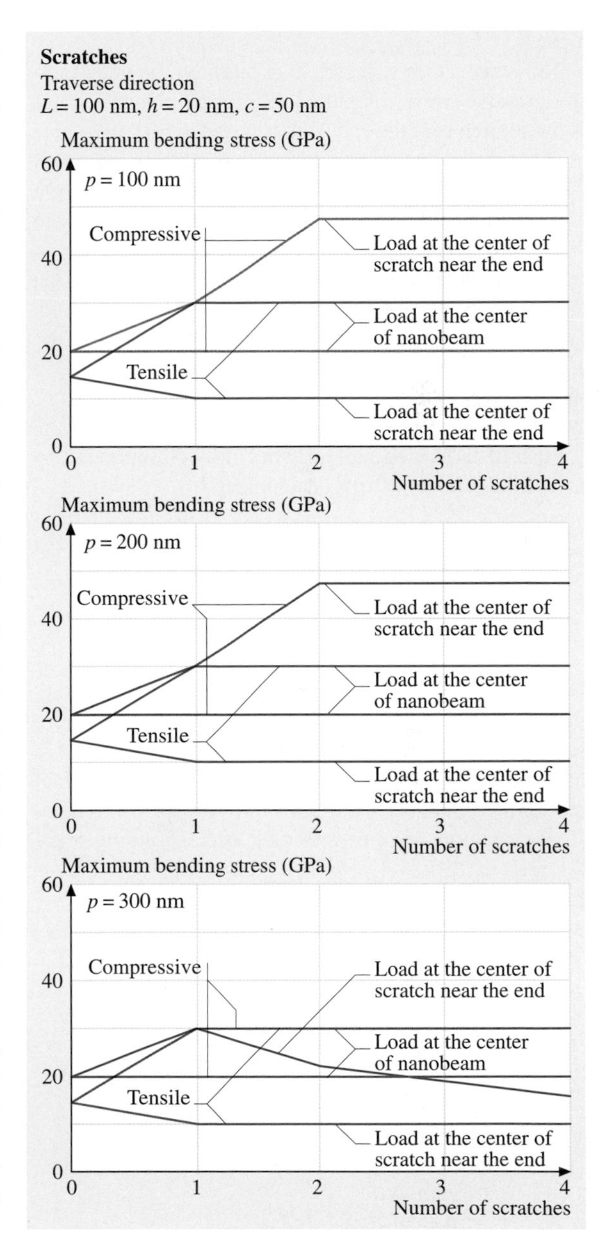

Fig. 25.23 Effect of number of scratches along with the variation in the pitch on the maximum bending stresses after loading trapezoidal Si nanobeams ($w_1 = 200$ nm, $w_2 = 370$ nm, $t = 255$ nm, $\ell = 6$ μm, $E = 169$ GPa, $\nu = 0.28$, load $= 70$ μN). Also shown is the effect of load when applied at the center of the beam and at the center of the scratch near the end [25.47]

creases dramatically because the local perturbation in the stress distribution at the center of the scratch where load is being applied leads to a high value of maximum compressive stress [25.65]. It increases further with the number of scratches and then levels off. This can be attributed to the fact that when there is another scratch present close to the scratch near the end, the stress concentration is more, as the effect of local perturbation in the stress distribution is more significant. However, this effect is insignificant when more than two scratches are present.

Now we address the effect of pitch on the maximum compressive stress when the load is applied at the center of the scratch near the end. When the pitch is up to a value of 200 nm the maximum compressive stress increases with the number of scratches, as discussed earlier. On the other hand, when the pitch value goes beyond 225 nm this effect is reversed. This is because the presence of another scratch no longer affects the local perturbation in the stress distribution at the scratch near the end. Instead, more scratches at a fair distance distribute the maximum compressive stress at the scratch near the end and the stress starts going down. Such observations of maximum bending stresses can help in identifying the number of asperities and scratches allowed separated by an optimum distance from each other.

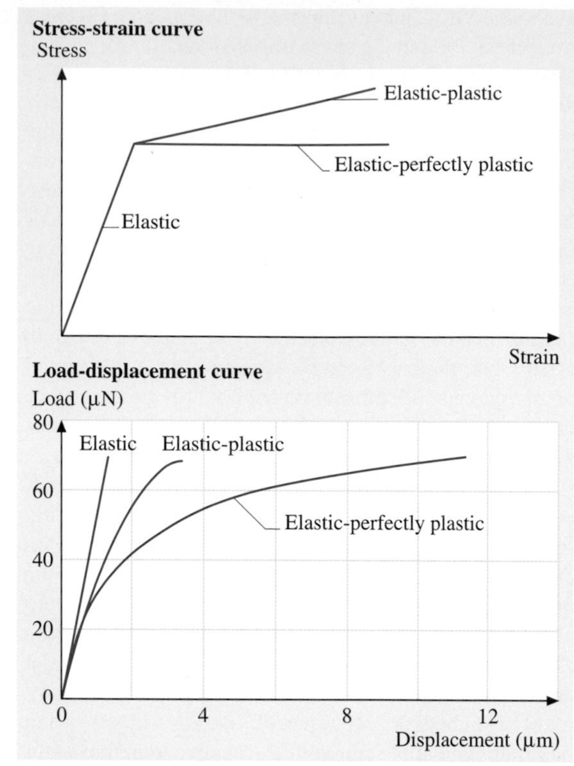

Fig. 25.24 Schematic representation of stress-strain curves and load-displacement curves for material when it is elastic, elastic-plastic, or elastic-perfectly plastic for a Si nanobeam ($w_1 = 200$ nm, $w_2 = 370$ nm, $t = 255$ nm, $\ell = 6$ μm, $E = 169$ GPa, tangent modulus in plastic range $= 0.5$ E, $\nu = 0.28$) [25.47]

25.3.4 Effect on Stresses and Displacements for Materials That Are Elastic, Elastic-Plastic, or Elastic-Perfectly Plastic

This section deals with the beam modeled as elastic, elastic-plastic, and elastic-perfectly plastic to observe the variation in the stresses and displacements from an elastic model used so far [25.47]. Figure 25.24 shows the typical stress-strain curves for the three types of deformation regimes and their corresponding load-displacement curves obtained from the model of an Si nanobeam that are found to exhibit the same trends.

Table 25.3 shows the comparison of maximum von Mises stress and maximum displacements for

Table 25.3 Stresses and displacements for materials that are elastic, elastic-plastic, or elastic-perfectly plastic (load $= 70$ μN, $w_1 = 200$ nm, $w_2 = 370$ nm, $t = 255$ nm, $\ell = 6$ μm, $R = 25$ nm, $E = 169$ GPa, tangent modulus in plastic range $= 0.5E$, $\nu = 0.28$)

	Elastic		Elastic-plastic		Elastic-perfectly plastic	
	Smooth nanobeam	Single semicircular longitudinal asperity	Smooth nanobeam	Single semicircular longitudinal asperity	Smooth nanobeam	Single semicircular longitudinal asperity
Max. von Mises stress (GPa)	18.2	19.3	13.5	15.2	7.8	9.1
Max. displacement (μm)	1.34	1.40	3.35	3.65	11.5	12.3

both smooth nanobeam and nanobeam with a defined roughness, which is single semicircular longitudinal asperity of R value equal to 25 nm for the three different models. It is observed that the maximum value of stress is obtained at a given load for elastic material, whereas the displacement is maximum for elastic-perfectly plastic material. Also, the pattern that the maximum bending stress value increases for a rough nanobeam still holds true in the other models as well.

25.4 Closure

Mechanical properties of nanostructures are necessary in designing realistic MEMS/NEMS devices. Most mechanical properties are scale-dependent. Micro/nanomechanical properties, hardness, elastic modulus, and scratch resistance of bulk materials of undoped single-crystal silicon (Si) and thin films of undoped polysilicon, SiO_2, SiC, Ni-P, and Au are presented. It is found that the SiC film exhibits higher hardness, elastic modulus, and scratch resistance as compared to other materials.

Bending tests have been performed on the Si and SiO_2 nanobeams and Ni-P and Au microbeams using an AFM and a depth-sensing nanoindenter, respectively. The bending tests were used to evaluate elastic modulus, bending strength (fracture stress), fracture toughness (K_{IC}), and fatigue strength of the beam materials. The Si and SiO_2 nanobeams exhibited elastic linear response with sudden brittle fracture. The notched Ni-P beam showed linear deformation behavior followed by abrupt failure. The Au beam showed elastic-plastic deformation behavior. Elastic modulus values of 182 ± 11 GPa for Si(110) and 85 ± 3 GPa for SiO_2 were obtained, which are comparable to bulk values. Bending strength values of 18 ± 3 GPa for Si and 7.6 ± 2 GPa for SiO_2 were obtained, which are twice as large as values reported on larger scale specimens. This indicates that bending strength shows a specimen size dependence. Fracture toughness value estimates obtained were $1.67\pm0.4\,\text{MPa}\sqrt{\text{m}}$ for Si and $0.60\pm0.2\,\text{MPa}\sqrt{\text{m}}$ for SiO_2, which are also comparable to values obtained on larger specimens. At stress amplitudes less than 15% of their bending strength and at mean stresses of less than 30% of the bending strength, Si and SiO_2 displayed an apparent endurance life of greater than 30,000 cycles. SEM observations of the fracture surfaces revealed a cleavage type of fracture for both materials when subjected to bending, as well as fatigue. The AFM and nanoindenters used in this study can be satisfactorily used to evaluate the mechanical properties of micro/nanoscale structures for use in MEMS/NEMS.

FEM simulations are used to predict the stress and deformation in nanostructures. The FEM has been used to analyze the effect of the type of surface roughness and scratches on stresses and deformation of nanostructures. We find that roughness affects the maximum bending stresses. The maximum bending stresses increase as the asperity number increases for both semicircular and grooved asperities in the longitudinal direction. When the semicircular asperity is present in the transverse direction the maximum tensile stress is much larger than the maximum tensile stress value with no asperity or when the semicircular asperity is present in the longitudinal direction. This observation suggests that the asperity in the transverse direction is more detrimental. The presence of scratches increases the maximum tensile stress. The maximum tensile stress remains almost the same with the number of scratches for two types of loading, that is, when load is applied at the center of the beam or at the center of the scratch near the end, although the value of the tensile stress is much lower when the load is applied at the center of the scratch. This means that the load applied at the ends is less damaging. This analysis shows that FEM simulations can be useful to designers to develop the most suitable geometry for nanostructures.

References

25.1 R. S. Muller, R. T. Howe, S. D. Senturia, R. L. Smith, R. M. White: *Microsensors* (IEEE, New York 1990)

25.2 I. Fujimasa: *Micromachines: A New Era in Mechanical Engineering* (Oxford Univ. Press, Oxford 1996)

25.3 W. S. Trimmer (Ed.): *Micromachines and MEMS, Classic and Seminal Papers to 1990* (IEEE, New York 1997)

25.4 B. Bhushan: *Tribology Issues and Opportunities in MEMS* (Kluwer, Dordrecht 1998)

25.5 G. T. A. Kovacs: *Micromachined Transducers Sourcebook* (WCB McGraw-Hill, Boston 1998)

25.6 S. D. Senturia: *Microsystem Design* (Kluwer, Boston 2001)

25.7 M. Gad-el-Hak: *The MEMS Handbook* (CRC, Boca Raton 2002)

25.8 T. R. Hsu: *MEMS and Microsystems* (McGraw-Hill, Boston 2002)

25.9 M. Madou: *Fundamentals of Microfabrication: The Science of Miniaturization*, 2nd edn. (CRC, Boca Raton 2002)

25.10 K. E. Drexler: *Nanosystems: Molecular Machinery, Manufacturing and Computation* (Wiley, New York 1992)

25.11 B. Bhushan: *Handbook of Micro/Nanotribology*, 2nd edn. (CRC, Boca Raton 1999)

25.12 G. Timp (Ed.): *Nanotechnology* (Springer, New York 1999)

25.13 E. A. Rietman: *Molecular Engineering of Nanosystems* (Springer, New York 2001)

25.14 H. S. Nalwa (Ed.): *Nanostructures Materials and Nanotechnology* (Academic, San Diego 2002)

25.15 W. A. Goddard, D. W. Brenner, S. E. Lyshevski, G. J. Iafrate: *Handbook of Nanoscience, Engineering, and Technology* (CRC, Boca Raton 2003)

25.16 B. Bhushan: *Principles and Applications of Tribology* (Wiley, New York 1999)

25.17 B. Bhushan: *Introduction to Tribology* (Wiley, New York 2002)

25.18 B. Bhushan: Macro- and microtribology of MEMS materials. In: *Modern Tribology Handbook*, ed. by B. Bhushan (CRC, Boca Raton 2001) pp. 1515–1548

25.19 S. Johansson, J. A. Schweitz, L. Tenerz, J. Tiren: Fracture testing of silicon microelements in-situ in a scanning electron microscope, J. Appl. Phys. **63** (1988) 4799–4803

25.20 F. Ericson, J. A. Schweitz: Micromechanical fracture strength of silicon, J. Appl. Phys. **68** (1990) 5840–5844

25.21 E. Obermeier: Mechanical and thermophysical properties of thin film materials for MEMS: Techniques and devices, Micromechan. Struct. Mater. Res. Symp. Proc. **444** (1996) 39–57

25.22 C. J. Wilson, A. Ormeggi, M. Narbutovskih: Fracture testing of silicon microcantilever beams, J. Appl. Phys. **79** (1995) 2386–2393

25.23 W. N. Sharpe, Jr., B. Yuan, R. L. Edwards: A new technique for measuring the mechanical properties of thin films, J. Microelectromech. Syst. **6** (1997) 193–199

25.24 K. Sato, T. Yoshioka, T. Anso, M. Shikida, T. Kawabata: Tensile testing of silicon film having different crystallographic orientations carried out on a silicon chip, Sens. Actuators A **70** (1998) 148–152

25.25 S. Greek, F. Ericson, S. Johansson, M. Furtsch, A. Rump: Mechanical characterization of thick polysilicon films: Young's modulus and fracture strength evaluated with microstructures, J. Micromech. Microeng. **9** (1999) 245–251

25.26 D. A. LaVan, T. E. Buchheit: Strength of polysilicon for MEMS devices, Proc. SPIE **3880** (1999) 40–44

25.27 E. Mazza, J. Dual: Mechanical behavior of a μm-sized single crystal silicon structure with sharp notches, J. Mech. Phys. Solids **47** (1999) 1795–1821

25.28 T. Yi, C. J. Kim: Measurement of mechanical properties for MEMS materials, Meas. Sci. Technol. **10** (1999) 706–716

25.29 H. Kahn, M. A. Huff, A. H. Heuer: Heating effects on the Young's modulus of films sputtered onto micromachined resonators, Microelectromech. Struct. Mater. Res. Symp. Proc. **518** (1998) 33–38

25.30 S. Johansson, F. Ericson, J. A. Schweitz: Influence of surface-coatings on elasticity, residual-stresses, and fracture properties of silicon microelements, J. Appl. Phys. **65** (1989) 122–128

25.31 R. Ballarini, R. L. Mullen, Y. Yin, H. Kahn, S. Stemmer, A. H. Heuer: The fracture toughness of polysilicon microdevices: A first report, J. Mater. Res. **12** (1997) 915–922

25.32 H. Kahn, R. Ballarini, R. L. Mullen, A. H. Heuer: Electrostatically actuated failure of microfabricated polysilicon fracture mechanics specimens, Proc. R. Soc. London A **455** (1999) 3807–3823

25.33 A. M. Fitzgerald, R. H. Dauskardt, T. W. Kenny: Fracture toughness and crack growth phenomena of plasma-etched single crystal silicon, Sens. Actuators A **83** (2000) 194–199

25.34 T. Tsuchiya, A. Inoue, J. Sakata: Tensile testing of insulating thin films: Humidity effect on tensile strength of SiO_2 films, Sens. Actuators A **82** (2000) 286–290

25.35 J. A. Connally, S. B. Brown: Micromechanical fatigue testing, Exp. Mech. **33** (1993) 81–90

25.36 K. Komai, K. Minoshima, S. Inoue: Fracture and fatigue behavior of single-crystal silicon microelements and nanoscopic AFM damage evaluation, Microsyst. Technol. **5** (1998) 30–37

25.37 T. Namazu, Y. Isono, T. Tanaka: Evaluation of size effect on mechanical properties of single-

crystal silicon by nanoscale bending test using AFM, J. Microelectromech. Syst. **9** (2000) 450–459
25.38 S. Sundararajan, B. Bhushan: Development of AFM-based techniques to measure mechanical properties of nanoscale structures, Sens. Actuators A **101** (2002) 338–351
25.39 X. Li, B. Bhushan: Fatigue studies of nanoscale structures for MEMS/NEMS applications using nanoindentation techniques, Surf. Coat. Technol. **163–164** (2003) 521–526
25.40 X. Li, B. Bhushan, K. Takashima, C.W. Baek, Y.K. Kim: Mechanical characterization of micro/nanoscale structures for MEMS/NEMS applications using nanoindentation techniques, Ultramicroscopy **97** (2003) 481–494
25.41 T. Hsu, N. Sun: Residual stresses/strains analysis of MEMS, Proc. of the Int. Conf. on Modeling and Simulation of Microsystems, Semiconductors, Sensors and Actuators, ed. by M. Laudon, B. Romanowicz (Computational Publications, Cambridge 1998) 82–87
25.42 A. Kolpekwar, C. Kellen, R.D. (Shawn) Blanton: Fault model generation for MEMS, Proc. of the Int. Conf. on Modeling and Simulation of Microsystems, Semiconductors, Sensors and Actuators, ed. by M. Laudon, B. Romanowicz (Computational Publications, Cambridge 1998) 111–116
25.43 H.A. Rueda, M.E. Law: Modeling of strain in boron-doped silicon cantilevers, Proc. of the Int. Conf. on Modeling and Simulation of Microsystems, Semiconductors, Sensors and Actuators, ed. by M. Laudon, B. Romanowicz (Computational Publications, Cambridge 1998) 94–99
25.44 M. Heinzelmann, M. Petzold: FEM analysis of microbeam bending experiments using ultramicro indentation, Comput. Mater. Sci. **3** (1994) 169–176
25.45 C.J. Wilson, P.A. Beck: Fracture testing of bulk silicon microcantilever beams subjected to a side load, J. Microelectromech. Syst. **5** (1996) 142–150
25.46 B. Bhushan, G.B. Agrawal: Stress analysis of nanostructures using a finite element method, Nanotechnology **13** (2002) 515–523
25.47 B. Bhushan, G.B. Agrawal: Finite element analysis of nanostructures with roughness and scratches, Ultramicroscopy **97** (2003) 495–501
25.48 K.E. Petersen: Silicon as a mechanical material, Proc. IEEE **70** (1982) 420–457
25.49 B. Bhushan, S. Sundararajan, X. Li, C.A. Zorman, M. Mehregany: Micro/nanotribological studies of single-crystal silicon and polysilicon and SiC films for use in MEMS devices. In: *Tribology Issues and Opportunities in MEMS*, ed. by B. Bhushan (Kluwer, Dordrecht 1998) pp. 407–430
25.50 S. Sundararajan, B. Bhushan: Micro/nanotribological studies of polysilicon and SiC films for MEMS applications, Wear **217** (1998) 251–261
25.51 X. Li, B. Bhushan: Micro/nanomechanical characterization of ceramic films for microdevices, Thin Solid Films **340** (1999) 210–217
25.52 B. Bhushan, X. Li: Nanomechanical characterization of solid surfaces and thin films, Int. Mater. Rev. **48** (2003) 125–164
25.53 B.R. Lawn, A.G. Evans, D.B. Marshall: Elastic/plastic indentation damage in ceramics: The median/radial system, J. Am. Ceram. Soc. **63** (1980) 574
25.54 S. Sundararajan, B. Bhushan, T. Namazu, Y. Isono: Mechanical property measurements of nanoscale structures using an atomic force microscope, Ultramicroscopy **91** (2002) 111–118
25.55 R.J. Roark: *Formulas for Stress and Strain*, 6th edn. (McGraw-Hill, New York 1989)
25.56 R.W. Hertzberg: *Deformation and Fracture Mechanics of Engineering Materials*, 3rd edn. (Wiley, New York 1989) pp. 277–278
25.57 Anonymous: Properties of silicon, EMIS Datareviews Series No. 4 (1988)
25.58 C.T.C. Nguyen, R.T. Howe: An integrated CMOS micromechanical resonator high-Q oscillator, IEEE J. Solid State Circ. **34** (1999) 440–455
25.59 L.J. Hornbeck: A digital light processing^{TM} update – status and future applications, Proc. SPIE **3634** (1999) 158–170
25.60 M. Tanaka: Fracture toughness and crack morphology in indentation fracture of brittle materials, J. Mater. Sci. **31** (1996) 749
25.61 T. Tsuchiya, O. Tabata, J. Sakata, Y. Taga: Specimen size effect on tensile strength of surface-micromachined polycrystalline silicon thin films, J. Microelectromech. Syst. **7** (1998) 106–113
25.62 T. Yi, L. Li, C.J. Kim: Microscale material testing of single crystalline silicon: Process effects on surface morphology and tensile strength, Sens. Actuators A **83** (2000) 172–178
25.63 B. Bhushan, S. Venkatesan: Mechanical and tribological properties of silicon for micromechanical applications: A review, Adv. Info. Storage Syst. **5** (1993) 211–239
25.64 B. Bhushan, B.K. Gupta: *Handbook of Tribology: Materials, Coatings, and Surface Treatments* (McGraw-Hill, New York 1991)
25.65 S.P. Timoshenko, J.N. Goodier: *Theory of Elasticity*, 3rd edn. (McGraw-Hill, New York 1970)
25.66 J.E. Shigley, L.D. Mitchell: *Mechanical Engineering Design*, 4th edn. (McGraw-Hill, New York 1993)

Part D

Molecular

Part D Molecularly Thick Films for Lubrication

26 Nanotribology of Ultrathin and Hard Amorphous Carbon Films
Bharat Bhushan, Columbus, USA

27 Self-Assembled Monolayers for Controlling Adhesion, Friction and Wear
Bharat Bhushan, Columbus, USA
Huiwen Liu, Columbus, USA

28 Nanoscale Boundary Lubrication Studies
Bharat Bhushan, Columbus, USA
Huiwen Liu, Columbus, USA

29 Kinetics and Energetics in Nanolubrication
Rene M. Overney, Seattle, USA
George W. Tyndall, San Jose, USA
Jane Frommer, San Jose, USA

26. Nanotribology of Ultrathin and Hard Amorphous Carbon Films

Diamond material and its smooth coatings are used for very low wear and relatively low friction. Major limitations of the true diamond coatings are that they need to be deposited at high temperatures, can only be deposited on selected substrates, and require surface finishing. Hard amorphous carbon, commonly known as diamond-like carbon or DLC coatings, exhibit mechanical, thermal, and optical properties close to that of diamond. These can be deposited with a large range of thicknesses by using a variety of deposition processes on a variety of substrates at or near room temperature. The coatings reproduce substrate topography, avoiding the need of post-finishing. Friction and wear properties of some DLC coatings can be very attractive for tribological applications. The largest industrial application of these coatings is in magnetic storage devices.

The prevailing atomic arrangement in the DLC coatings is amorphous or quasi-amorphous with small diamond, graphite, and other unidentifiable micro- or nanocrystallites. Most DLC coatings, except those produced by filtered cathodic arc, contain from a few to about 50 at% hydrogen. Sometimes hydrogen is deliberately incorporated in the sputtered and ion plated coatings to tailor their properties.

EELS and Raman spectroscopies can be successfully used for chemical characterization of amorphous carbon coatings. The prevailing atomic arrangement in the DLC coatings is amorphous or quasi-amorphous with small diamond (sp^3), graphite (sp^2) and other unidentifiable micro- or nanocrystallites. Most DLC coatings except those produced by filtered cathodic arc contain from a few to about 50 at% hydrogen. Sometimes hydrogen is deliberately incorporated in the sputtered and ion plated coatings to tailor their properties.

Amorphous carbon coatings deposited by various techniques exhibit different mechanical and tribological properties. The nanoindenter can be successfully used for measurement of hardness, elastic modulus, fracture toughness, and fatigue life. Microscratch and microwear experiments can be performed using either a nanoindenter or an AFM. Thin coatings deposited by filtered cathodic arc, ion beam, and ECR-CVD hold a promise for tribological applications. Coatings as thin as 5 nm or even thinner provide wear protection. Microscratch, microwear, and accelerated wear testing, if simulated properly, can be successfully used to screen coating candidates for industrial applications. In the examples shown in this chapter, trends observed in the microscratch, microwear, and accelerated macrofriction wear tests are similar to that found in functional tests.

In this chapter, the state-of-the-art of recent developments in the chemical, mechanical, and tribological characterization of ultrathin amorphous carbon coatings is presented.

26.1 **Description of Commonly Used Deposition Techniques** 795
26.1.1 Filtered Cathodic Arc Deposition Technique 798
26.1.2 Ion Beam Deposition Technique 798
26.1.3 Electron Cyclotron Resonance Chemical Vapor Deposition Technique 799
26.1.4 Sputtering Deposition Technique ... 799
26.1.5 Plasma-Enhanced Chemical Vapor Deposition Technique 799

26.2 **Chemical Characterization and Effect of Deposition Conditions on Chemical Characteristics and Physical Properties** ... 800
26.2.1 EELS and Raman Spectroscopy 800
26.2.2 Hydrogen Concentrations 804
26.2.3 Physical Properties 804
26.2.4 Summary 805

26.3 **Micromechanical and Tribological Characterizations of Coatings Deposited by Various Techniques** 805
26.3.1 Micromechanical Characterization . 805
26.3.2 Microscratch and Microwear Studies 813
26.3.3 Macroscale Tribological Characterization 822
26.3.4 Coating Continuity Analysis 826

References 827

Carbon exists in both crystalline and amorphous forms and exhibits both metallic and nonmetallic characteristics [26.1–3]. Crystalline carbon includes graphite, diamond, and a family of fullerenes, Fig. 26.1. The graphite and diamond are infinite periodic network solids with a planar structure, whereas the fullerenes are a molecular form of pure carbon with a finite network with a nonplanar structure. Graphite has a hexagonal, layered structure with weak interlayer bonding forces and exhibits excellent lubrication properties. The graphite crystal may be visualized as infinite parallel layers of hexagons stacked 0.34 nm apart with a 0.1415-nm interatomic distance between the carbon atoms in the basal plane. The atoms lying in the basal planes are trigonally coordinated and closely packed with strong σ (covalent) bonds to its three carbon neighbors using the hybrid sp^2 orbitals. The fourth electron lies in a p_z orbital lying normal to the σ bonding plane and forms a weak π bond by overlapping side to side with a p_z orbital of an adjacent atom to which carbon is attached by a σ bond. The layers (basal planes) themselves are relatively far apart and the forces that bond them are weak van der Waals forces. These layers can align themselves parallel to the direction of the relative motion and slide over one another with relative ease, thus providing low friction. Strong interatomic bonding and packing in each layer is thought to help reduce wear. The operating environment has a significant influence on lubrication, i. e., low friction and low wear, properties of graphite. It lubricates better in a humid environment than a dry one, resulting from adsorption of water vapor and other gases from the environment, which further weakens the interlayer bonding forces and results in easy shear and transfer of the crystallite platelets to the mating surface. Thus, transfer plays an important role in controlling friction and wear. Graphite oxidizes at high operating temperatures and can be used up to about 430 °C.

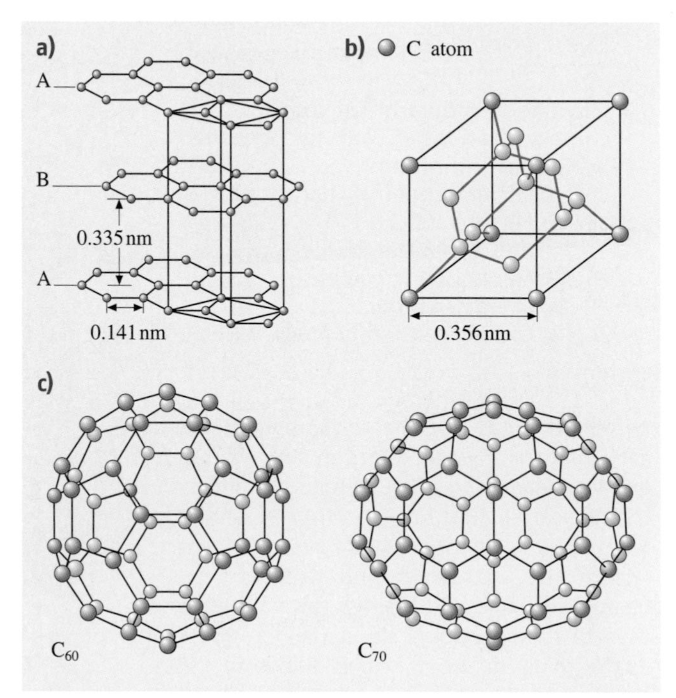

Fig. 26.1a–c The structure of three known forms of crystalline carbon (**a**) hexagonal structure of graphite, (**b**) modified face-centered cubic (fcc) structure, two interpenetrating fcc lattices displaced by one-quarter of the cube diagonal, of diamond (each atom is bonded to four others that form the corners of the pyramidal structure called tetrahedron), and (**c**) the structure of the two most common forms of fullerenes-soccer ball C_{60} and rugby ball C_{70} molecules

One of the fullerene molecules is C_{60}, commonly known as Buckyball. Since the C_{60} molecules are very stable and do not require additional atoms to satisfy chemical bonding requirements, they are expected to have low adhesion to the mating surface and low surface energy. Since C_{60} molecules with a perfect spherical symmetry are weakly bonded to other molecules, C_{60} clusters get detached readily, similar to other layered lattice structures, and either get transferred to the mating surface by mechanical compaction, or are present as loose wear particles that may roll like tiny ball bearings in a sliding contact, resulting in low friction and wear. The wear particles are expected to be harder than as-deposited C_{60} molecules, because of their phase transformation at high asperity contact pressures present in a sliding interface. The low surface energy, spherical shape of C_{60} molecules, weak intermolecular bonding, and high load bearing capacity offer potential for various mechanical and tribological applications. The sublimed C_{60} coatings and fullerene particles as an additive to mineral oils and greases have been reported to be good solid lubricants comparable to graphite and MoS_2 [26.4–6].

Diamond crystallizes in the modified face centered cubic (*fcc*) structure with an interactomic distance of 0.154 nm. The diamond cubic lattice consists of two interpenetrating fcc lattices displaced by one-quarter of the cube diagonal. Each carbon atom is tetrahedrally coordinated, making strong σ (covalent) bonds to its four carbon neighbors using the hybrid sp^3 atomic orbitals, which accounts for its highest hardness (80–104 GPa) and thermal conductivity (900–2,100 W/mK, on the order of five times that of copper) of any known solid, and a high electrical resistivity, optical transmission, and a large optical band gap. It is relatively chemically inert, and it exhibits poor adhesion with other solids with consequent low friction and wear. Its high thermal conductivity allows dissipation of frictional heat during sliding and protects the interface, and the dangling carbon bonds on the surface react with the environment to form hydrocarbons that act as good lubrication films. These are some of the reasons for low friction and wear of the diamond. Diamond and its coatings find many industrial applications: tribological applications (low friction and wear), optical applications (exceptional optical transmission, high abrasion resistance), and thermal management or heat sink applications (high thermal conductivity). The diamond can be used to high temperatures, and it starts to graphitize at about 1,000 °C in ambient air and at about 1,400 °C in vacuum. Diamond is an attractive material for cutting tools, as an abrasive for grinding wheels and lapping compounds, and other extreme wear applications.

The natural diamond, particularly in large sizes, is very expensive and its coatings, a low cost alternative, are attractive. The true diamond coatings are deposited by chemical vapor deposition (CVD) processes at high substrate temperatures (on the order of 800 °C). They adhere best on a silicon substrate and require an interlayer for other substrates. A major roadblock to the widespread use of true diamond films in tribological, optical, and thermal management applications is the surface roughness. Growth of the diamond phase on a non-diamond substrate is initiated by nucleation either at randomly seeded sites or at thermally favored sites, due to statistical thermal fluctuation at the substrate surface. Based on growth temperature and pressure conditions, favored crystal orientations dominate the competitive growth process. As a result, the grown films are polycrystalline in nature with relatively large grain size ($> 1\,\mu$m) and terminate in very rough surfaces with RMS roughnesses ranging from a few tenths of a micron to tens of microns. Techniques for polishing these films have been developed. It has been reported that the laser polished films exhibit friction and wear properties almost comparable to that of bulk polished diamond [26.7, 8].

Amorphous carbon has no long-range order, and the short-range order of carbon atoms can have one or more of three bonding configurations: – sp^3 (diamond), sp^2 (graphite), or sp^1 (with two electrons forming strong

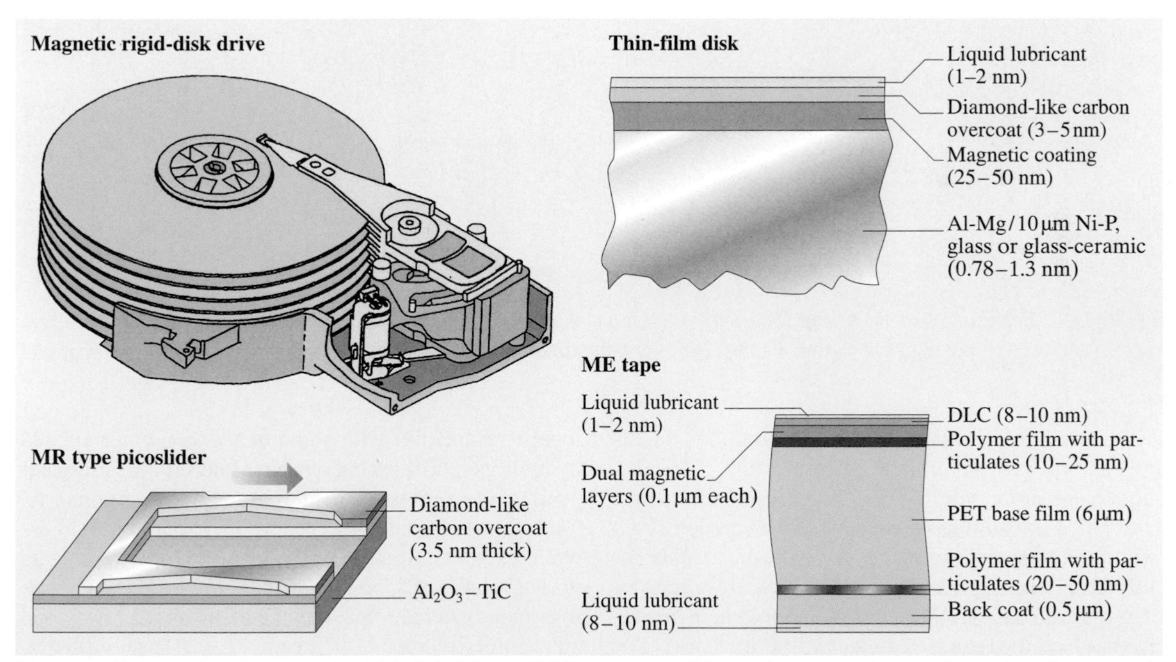

Fig. 26.2 Schematic of a magnetic rigid-disk drive and MR type picoslider and cross-sectional schematics of a magnetic thin-film rigid disk and a metal evaporated (ME) tape

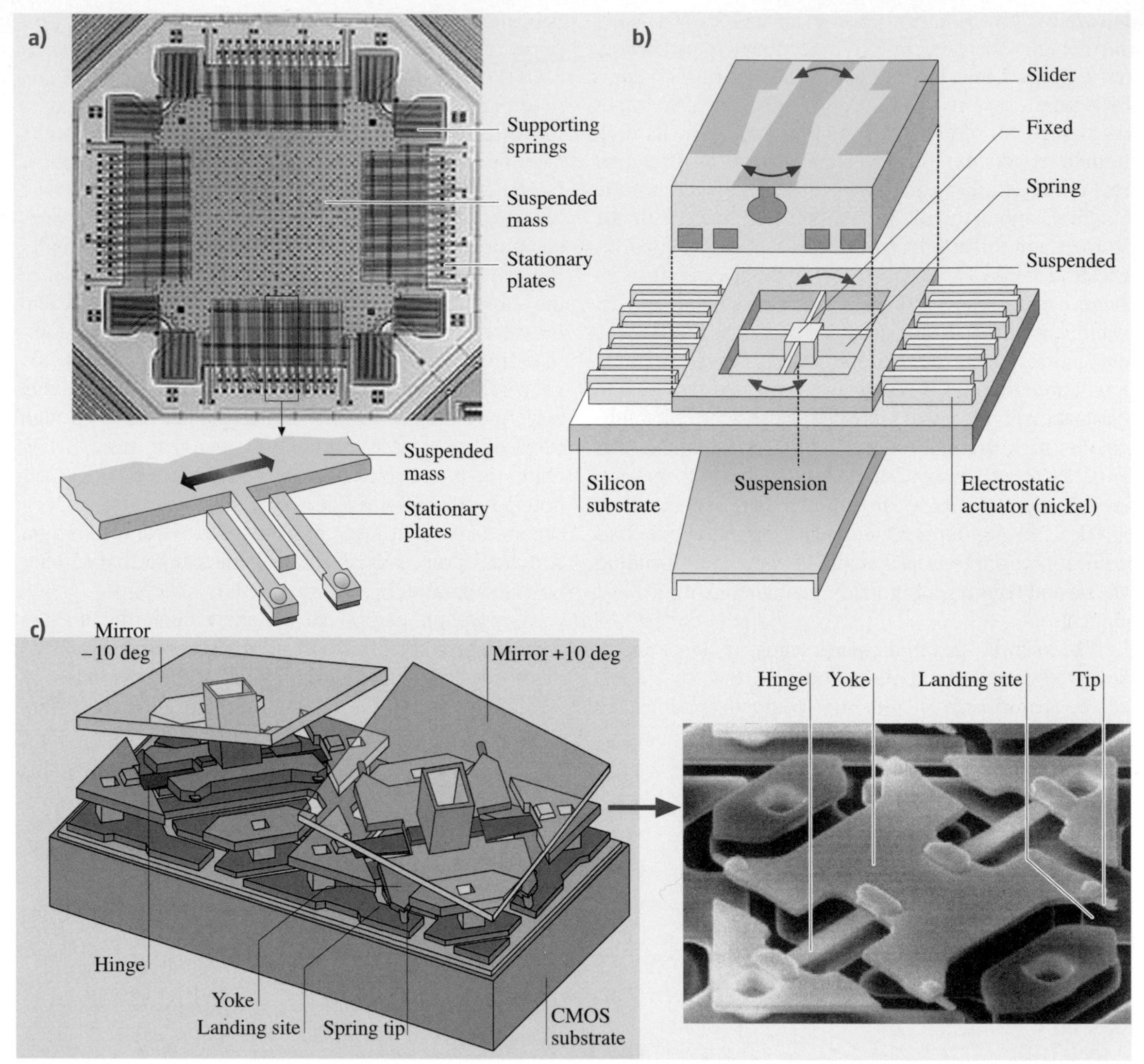

Fig. 26.3a–c Schematics of **(a)** a capacitive-type silicon accelerometer for automotive sensory applications, **(b)** digital micrometer devices for high-projection displays, and **(c)** polysilicon rotary microactuator for magnetic disk drives

σ bonds, and the remaining two electrons left in orthogonal p_y and p_z orbitals to form weak π bonds). Short-range order controls the properties of amorphous materials and coatings. Hard amorphous carbon (a-C) coatings, commonly known as diamond-like carbon or DLC (implying high hardness) coatings, are a class of coatings that are mostly metastable amorphous materials, but include a micro- or nanocrystalline phase. The coatings are random networks of covalently bonded carbon in hybridized tetragonal (sp^3) and trigonal (sp^2) local coordination with some of the bonds terminated by hydrogen. These coatings have been successfully deposited by a variety of vacuum deposition techniques on a variety of substrates at or near room temperature. These coatings generally reproduce substrate topography and do not require any post-finishing. However, these coatings mostly adhere best on silicon substrates. The best adhesion is obtained on substrates that form carbides, e.g., Si, Fe, and Ti. Based on depth profile analyses using Auger and XPS of DLC coatings deposited on silicon substrates, it has been reported that a substantial amount of silicon carbide (on the order of 5–10 nm in thickness) is present at the carbon-silicon interface for the coatings with good adhesion and high hardness (e.g., [26.9]). For good adhesion of DLC coatings to other substrates, in most cases, an interlayer of silicon is required except for cathodic arc deposited coatings.

There is significant interest in DLC coatings because of their unique combination of desirable properties. These properties include high hardness and wear resistance, chemical inertness to both acids and alkalis, lack of magnetic response, and an optical band gap ranging from zero to a few eV, depending upon the deposition conditions. These are used in a wide range of applications, including tribological, optical, electronic, and biomedical applications [26.1, 10, 11]. The high hardness, good friction and wear properties, versatility in deposition and substrates, and no requirements of post-finishing make them very attractive for tribological applications. Two primary examples include overcoats for magnetic media (thin-film disks and ME tapes) and MR type magnetic heads for magnetic storage devices, Fig. 26.2 [26.12–20] and the emerging field of microelectromechanical systems, Fig. 26.3 [26.21–24]. The largest industrial application of the family of amorphous carbon coatings, typically deposited by DC/RF magnetron sputtering, plasma-enhanced chemical vapor deposition, or ion beam deposition techniques, is in magnetic storage devices. These are employed to protect against wear and corrosion, magnetic coatings on thin-film rigid disks and metal evaporated tapes, and the thin-film head structure of a read/write disk head (Fig. 26.2). To maintain low physical spacing between the magnetic element of a read/write head and the magnetic layer of a media, thicknesses ranging from 3 to 10 nm are employed. Mechanical properties affect friction wear and therefore need to be optimized. In 1998, Gillette introduced Mach 3 razor blades with ultrathin DLC coatings, which has the potential of becoming a very large industrial application. DLC coatings are also used in other commercial applications such as glass windows of supermarket laser barcode scanners and sunglasses. These coatings are actively pursued in microelectromechanical systems (MEMS) components [26.23].

In this chapter, a state-of-the-art review of recent developments in the chemical, mechanical, and tribological characterization of ultrathin amorphous carbon coatings is presented. An overview of the most commonly used deposition techniques is presented, and followed by typical chemical and mechanical characterization data and typical tribological data both from coupon level testing and functional testing.

26.1 Description of Commonly Used Deposition Techniques

The first hard amorphous carbon coatings were deposited by a beam of carbon ions produced in an argon plasma on room-temperature substrates, as reported by *Aisenberg* and *Chabot* [26.25]. Subsequent confirmation by *Spencer* et al. [26.26] led to the explosive growth of this field. Following the first work, several alternative techniques have been developed. The amorphous carbon coatings have been prepared by a variety of deposition techniques and precursors, including evaporation, DC, RF or ion beam sputtering, RF or DC plasma-enhanced chemical vapor deposition (PECVD), electron cyclotron resonance chemical vapor deposition (ECR-CVD), direct ion beam deposition, pulsed laser vaporization and vacuum arc, from a variety of carbon-bearing solids or gaseous source materials [26.1, 27]. Coatings with both graphitic and diamond-like properties have been produced. Evaporation and ion plating techniques have been used to produce coatings with graphitic properties (low hardness, high electrical conductivity, very low friction, etc.), and all techniques have been used to produce coatings with diamond-like properties.

The structure and properties of a coating are dependent upon the deposition technique and parameters. High-energy surface bombardment has been used to produce harder and denser coatings. It is reported that sp^3/sp^2 fractions are in the decreasing order for cathodic arc deposition, pulsed laser vaporization, direct ion beam deposition, plasma-enhanced chemical vapor deposition, ion beam sputtering, and DC/RF sputtering [26.12,28,29]. A common feature to these techniques is that the deposition is energetic, i. e., carbon species arrive with an energy significantly greater than that represented by the substrate temperature. The resultant coatings are amorphous in structure, with hydrogen content up to 50%, and display a high degree of sp^3 character. From the results of previous investigations, it has been proposed that deposition of sp^3-bonded carbon requires that the depositing species have kinetic energies on the order of 100 eV or higher, well above

Table 26.1 Summary of the most commonly used deposition techniques and the kinetic energy of depositing species and deposition rates

Deposition technique	Process	Kinetic energy (eV)	Deposition rate (nm/s)
Filtered cathodic arc (FCA)	Energetic carbon ions produced by a vacuum arc discharge between a graphite cathode and grounded anode	100–2,500	0.1–1
Direct ion beam (IB)	Carbon ions produced from methane gas in an ion source and accelerated toward a substrate	50–500	0.1–1
Plasma-enhanced chemical vapor deposition (PECVD)	Hydrocarbon species, produced by plasma decomposition of hydrocarbon gases (e.g., acetylene), are accelerated toward a DC-biased substrate	1–30	1–10
Electron cyclotron resonance plasma chemical vapor deposition (ECR-CVD)	Hydrocarbon ions, produced by plasma decomposition of ethylene gas in the presence of a plasma in electron cyclotron resonance condition, are accelerated toward a RF-biased substrate	1–50	1–10
DC/RF sputtering	Sputtering of graphite target by argon ion plasma	1–10	1–10

those obtained in thermal processes like evaporation (0–0.1 eV). The species must then be quenched into the metastable configuration via rapid energy removal. Excess energy, such as that provided by substrate heating, is detrimental to the achievement of a high sp^3 fraction. In general, a high fraction of the sp^3-bonded carbon atoms in an amorphous network results in a higher hardness [26.29–36]. The mechanical and tribological properties of a carbon coating depend on the sp^3/sp^2-bonded carbon ratio, the amount of hydrogen in the coating, and adhesion of the coating to the substrate, which are influenced by the precursor material, kinetic energy of the carbon species prior to deposition, deposition rate, substrate temperature, substrate biasing, and the substrate itself [26.29, 33, 35, 37–46]. The kinetic energies and deposition rates involved in selected deposition processes used in the deposition of DLC coatings are compared in Table 26.1 [26.1, 28].

In the studies by *Gupta* and *Bhushan* [26.12, 47], *Li* and *Bhushan* [26.48, 49], and *Sundararajan* and *Bhushan* [26.50], DLC coatings ranging typically in thickness from 3.5 nm to 20 nm were deposited on single-crystal silicon, magnetic Ni-Zn ferrite, and Al_2O_3-TiC substrates (surface roughness $\approx 1–3$ nm RMS) by filtered cathodic arc (FCA) deposition, (direct) ion beam deposition (IBD), electron cyclotron resonance chemical vapor deposition (ECR-CVD), plasma-enhanced chemical vapor deposition (PECVD), and DC/RF planar, magnetron sputtering (SP) deposition techniques [26.51]. In this chapter, we will limit the presentation of data of coatings deposited by FCA, IBD, ECR-CVD, and SP deposition techniques.

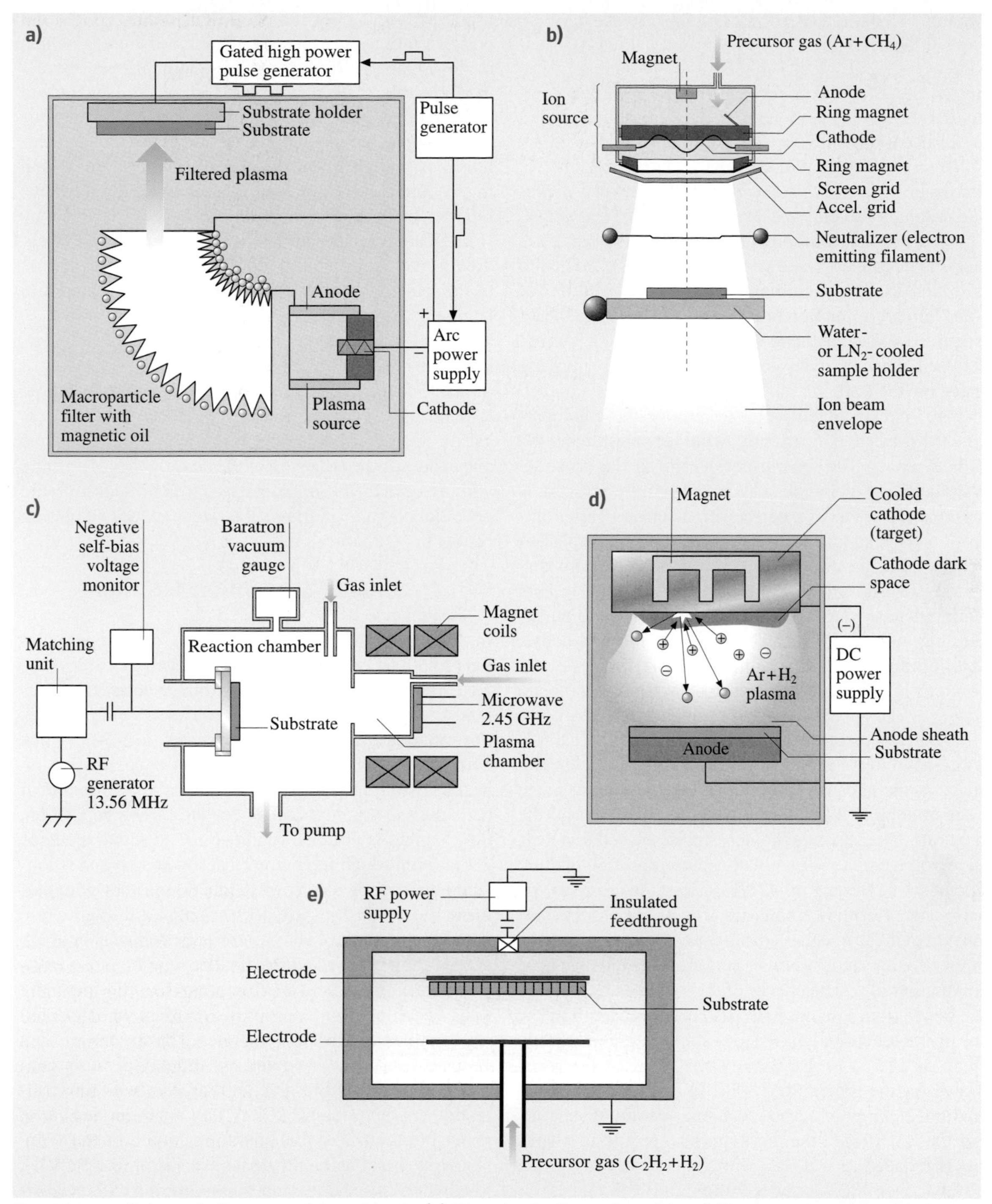

Fig. 26.4a–e Schematic diagrams of deposition by (**a**) filtered cathodic arc deposition, (**b**) ion beam deposition, (**c**) electron cyclotron resonance chemical vapor deposition (ECR-CVD), (**d**) DC planar magnetron sputtering, and (**e**) plasma-enhanced chemical vapor deposition (PECVD)

26.1.1 Filtered Cathodic Arc Deposition Technique

In the filtered cathodic arc deposition of carbon coating [26.29, 52–59], a vacuum arc plasma source is used to form carbon film. In the FCA technique used by *Bhushan* et al. (e.g., [26.12]), energetic carbon ions are produced by a vacuum arc discharge between a planar graphite cathode and grounded anode, Fig. 26.4a. The cathode is a 6-mm-diameter high-density graphite disk mounted on a water-cooled copper block. The arc is driven at an arc current of 200 A, arc duration of 5 ms, and arc repetition rate of 1 Hz. The plasma beam is guided by a magnetic field that transports current between the electrodes to form tiny, rapidly moving spots on the cathode surface. The source is coupled to a 90° bent magnetic filter to remove the macroparticles produced concurrently with the plasma in the cathode spots. The ion current density at the substrate is in the range of 10–50 mA/cm^2. The base pressure is less than 10^{-4} Pa. Compared with electron beam evaporation with auxiliary discharge, much higher plasma density is achieved with the aid of powerful arc discharge. In this process, the cathodic material suffers a complicated transition from the solid phase to an expanding, nonequilibrium plasma via liquid and dense, equilibrium non-ideal plasma phases [26.58]. The carbon ions in the vacuum arc plasma have a direct kinetic energy of 20–30 eV. The high voltage pulses are applied to the substrate mounted on a water-cooled sample holder, and ions are accelerated through the sheath and arrive at the substrate with an additional energy given by the potential difference between the plasma and the substrate. The substrate holder is pulse biased to a negative voltage up to −2 kV with a pulse duration of 1 μs. The negative biasing of −2 kV corresponds to 2 keV kinetic energy of the carbon ions. The use of a pulse bias instead of a DC bias has advantages of applying a much higher voltage and building a surface potential on a nonconducting film. The energy of the ions is varied during the deposition. For the first 10% of the deposition the substrates are pulsed biased to −2 keV with a pulse duty cycle of 25%, i. e., for 25% of the time the energy is 2 keV, for the remaining 75% it is 20 eV, which is the "natural" energy of carbon ions in a vacuum discharge. For the last 90% of the deposition the pulsed bias voltage is reduced to −200 eV with a pulse bias duty cycle of 25%, i. e., the energy is 200 eV for 25% and 20 eV for 75% of the deposition. The high energy at the beginning leads to a good intermixing and adhesion of the films, whereas the lower energy at the later stage leads to hard films. Under the conditions described, the deposition rate at the substrate is about 0.1 nm/s, which is slow. Compared with most gaseous plasma, the cathodic arc plasma is nearly fully ionized, and the ionized carbon atoms have high kinetic energy of carbon ions which help achieving a high fraction of sp^3-bonded carbon ions, which in turn result in a high hardness and higher interfacial adhesion. *Cuomo* et al. [26.42] have reported that based on electron energy loss spectroscopy (EELS) analysis, the sp^3-bonded carbon fraction of cathodic arc coating is 83% compared to 38% for the ion beam sputtered carbon. These coatings are reported to be *nonhydrogenated*.

This technique does not require an adhesion underlayer for non-silicon substrates. However, adhesion of the DLC coatings on the electrically insulating substrate is poor, as negative pulse biasing forms an electrical sheath that accelerates depositing ions to the substrate and enhances the adhesion of the coating to the substrate with associated ion implantation. It is difficult to build potential on an insulating substrate, and lack of biasing results in poor adhesion.

26.1.2 Ion Beam Deposition Technique

In the direct ion beam deposition of carbon coating [26.60–64], as used by *Bhushan* et al. (e.g., [26.12]), the carbon coating is deposited from an accelerated carbon ion beam. The sample is pre-cleaned by ion etching. For the case of non-silicon substrates, a 2–3-nm-thick amorphous silicon adhesion layer is deposited by ion beam sputtering using an ion beam of a mixture of methane and argon at 200 V. For the carbon deposition, the chamber is pumped to about 10^{-4} Pa, and methane gas is fed through the cylindrical ion source and is ionized by energetic electrons produced by a hot-wire filament, Fig. 26.4b. Ionized species then pass through a grid with a bias voltage of about 50 eV, where they gain a high acceleration energy and reach a hot-wire filament, emitting thermionic electrons that neutralize the incoming ions. The discharging of ions is important when insulating ceramics are used as substrates. The species are then deposited on a water-cooled substrate. Operating conditions are adjusted to give an ion beam with an acceleration energy of about 200 eV and a current density of about 1 mA/cm^2. At this operating condition, the deposition rate is about 0.1 nm/s, which is slow. Incidentally, tough and soft coatings are deposited at a high acceleration energy of about 400 eV and at a deposition rate of about 1 nm/s. The ion beam deposited carbon coatings are reported to be hydrogenated (30–40 at% hydrogen).

26.1.3 Electron Cyclotron Resonance Chemical Vapor Deposition Technique

ECR plasma's lack of electrodes and ability to create high densities of charged and excited species at low pressures ($\leq 10^{-4}$ Torr) make it an attractive processing discharge in the coating depositions [26.65]. In the ECR-CVD deposition process of carbon coating described by *Suzuki* and *Okada* [26.66] and used by *Li* and *Bhushan* [26.48, 49] and *Sundararajan* and *Bhushan* [26.50], microwave power is generated by a magnetron operating in continuous mode at a frequency of 2.45 GHz, Fig. 26.4c. The plasma chamber functions as a microwave cavity resonator. The magnetic coils arranged around the plasma chamber generate a magnetic field of 875 G, necessary for electron cyclotron resonance condition. The substrate is placed on a stage that is connected capacitively to a 13.56 MHz RF generator. The process gas is introduced into the plasma chamber and the hydrocarbon ions generated are accelerated by a negative self-bias voltage, which is generated by applying RF power to the substrate. Both the substrate stage and plasma chamber are water-cooled. The process gas used is 100% ethylene and its flow rate is held constant at 100 sccm. The microwave power is 100–900 W. The RF power is 30–120 W. The pressure during deposition is kept close to the optimum value of 5.5×10^{-3} Torr. Before the deposition, the substrates are cleaned using Ar ions generated in the ECR plasma chamber.

26.1.4 Sputtering Deposition Technique

In DC planar magnetron sputtered carbon coating [26.13, 33, 37, 40, 67–71], the carbon coating is deposited by the sputtering of graphite target with Ar ion plasma. In the glow discharge, positive ions from the plasma strike the target with sufficient energy to dislodge the atoms by momentum transfer, which are intercepted by the substrate. First, an about 5-nm-thick amorphous silicon adhesion layer is deposited by sputtering if the deposition is to be carried out on a non-silicon surface. In the process used by *Bhushan* et al. (e.g., [26.12]), the coating is deposited by the sputtering of a graphite target with Ar ion plasma at 300 W power for a 200-mm-diameter target and pressure of about 0.5 Pa (6 mTorr), Fig. 26.4d. Plasma is generated by applying a DC potential between the substrate and a target. *Bhushan* et al. [26.35] reported that sputtered carbon coating contains about 35 at% hydrogen. Hydrogen comes from the hydrocarbon contaminants present in the deposition chamber. To produce hydrogenated carbon coating with a larger concentration of hydrogen, deposition is carried out in Ar and hydrogen plasma.

26.1.5 Plasma-Enhanced Chemical Vapor Deposition Technique

In the RF-PECVD deposition of carbon coating, as used by *Bhushan* et al. (e.g., [26.12]), carbon coating is deposited by adsorption of most free radicals of hydrocarbon to the substrate and chemical bonding to other atoms on the surface. The hydrocarbon species are produced by the RF plasma decomposition of hydrocarbon precusors such as acetylene (C_2H_2), Fig. 26.4e [26.27, 69, 72–75]. Instead of requiring thermal energy in thermal CVD, the energetic electrons in the plasma (at pressures ranging from 1 to 5×10^2 Pa, typically less than 10 Pa) can activate almost any reaction among the gases in the glow discharge at relatively low substrate temperatures ranging from 100 to 600 °C (typically less than 300 °C). To deposit the coating on non-silicon substrates, an about 4-nm-thick amorphous silicon adhesion layer, used to improve adhesion, is first deposited under similar conditions from a gas mixture of 1% silane in argon [26.76]. In the process used by Bhushan and coworkers [26.12], the plasma is sustained in a parallel-plate geometry by a capacitive discharge at 13.56 MHz, at a surface power density on the order of 100 mW/cm^2. The deposition is performed at a flow rate on the order of 6 sccm and a pressure on the order of 4 Pa (30 mTorr) on a cathode-mounted substrate maintained at a substrate temperature of 180 °C. The cathode bias is held fixed at about −120 V with an external DC power supply attached to the substrate (powered electrode). The carbon coatings deposited by PECVD usually contain hydrogen up to 50% [26.35, 77].

26.2 Chemical Characterization and Effect of Deposition Conditions on Chemical Characteristics and Physical Properties

The chemical structure and properties of amorphous carbon coatings are a function of deposition conditions. It is important to understand the relationship of the chemical structure of amorphous carbon coatings to the properties in order to define useful deposition parameters. Amorphous carbon films are metastable phases formed when carbon particles are condensed on a substrate. The prevailing atomic arrangement in the DLC coatings is amorphous or quasi-amorphous with small diamond (sp^3), graphite (sp^2), and other unidentifiable micro- or nanocrystallites. The coating is dependent upon the deposition process and its conditions contain varying amounts of sp^3/sp^2 ratio and hydrogen. The sp^3/sp^2 ratio of DLC coatings ranges typically from 50% to close to 100% with an increase in hardness with the sp^3/sp^2 ratio. Most DLC coatings, except those produced by filtered cathodic arc, contain from a few to about 50 at% hydrogen. Sometimes hydrogen and nitrogen are deliberately added to produce hydrogenated (a-C:H) and nitrogenated amorphous carbon (a-C:N) coatings, respectively. Hydrogen helps to stabilize sp^3 sites (most of the carbon atoms attached to hydrogen have a tetrahedral structure); therefore, the sp^3/sp^2 ratio for hydrogenated carbon is higher [26.30]. Optimum sp^3/sp^2 in a random covalent network composed of sp^3 and sp^2 carbon sites (N_{sp^2} and N_{sp^3}) and hydrogen is [26.30]:

$$\frac{N_{sp^3}}{N_{sp^2}} = \frac{6X_H - 1}{8 - 13X_H}\,, \qquad (26.1)$$

where X_H is the atom fraction of hydrogen. The hydrogenated carbon has a larger optical band gap, higher electrical resistivity (semiconductor), and a lower optical absorption or high optical transmission. The hydrogenated coatings have a lower density, probably because of the reduction of cross-linking due to hydrogen incorporation. However, hardness decreases with an increase of the hydrogen, even though the proportion of sp^3 sites increases (that is, as the local bonding environment becomes more diamond-like) [26.78, 79]. It is speculated that the high hydrogen content introduces frequent terminations in the otherwise strong 3-D network, and hydrogen increases the soft polymeric component of the structure more than it enhances the cross-linking sp^3 fraction.

A number of investigations have been performed to identify the microstructure of amorphous carbon films using a variety of techniques such as Raman spectroscopy, EELS, nuclear magnetic resonance, optical measurements, transmission electron microscopy, and X-ray photoelectron spectroscopy [26.33]. The structure of diamond-like amorphous carbon is amorphous or quasi-amorphous with small graphitic (sp^2) and tetrahedrally coordinated (sp^3) and other unidentifiable nanocrystallites (typically on the order of a couple nm, randomly oriented) [26.33, 80, 81]. These studies indicate that the chemical structure and physical properties of the coatings are quite variable, depending on the deposition techniques and film growth conditions. It is clear that both sp^2 and sp^3-bonded atomic sites are incorporated in diamond-like amorphous carbon coatings and that the physical and chemical properties of the coatings depend strongly on their chemical bonding and microstructure. Systematic studies have been conducted to carry out chemical characterization and investigate how the physical and chemical properties of amorphous carbon coatings vary as a function of deposition parameters (e.g., [26.33, 35, 40]). EELS and Raman spectroscopy are commonly used to characterize the chemical bonding and microstructure. Hydrogen concentration of the coatings is obtained by means of forward recoil spectrometry (FRS). A variety of physical properties of the coatings relevant to tribological performance are measured.

To present the typical data obtained for characterization of typical amorphous carbon coatings and their relationships to physical properties, we present data on several sputtered coatings, RF-PECVD amorphous carbon and microwave-PECVD (MPECVD) diamond coatings [26.33,35,40]. The sputtered coatings were DC magnetron sputtered at a chamber pressure of 10 mTorr under sputtering power densities of 0.1 and 2.1 W/cm^2 in a pure Ar plasma, labeled as W1 and W2, respectively. They were prepared at a power density of 2.1 W/cm^2 with various hydrogen fractions of 0.5, 1, 3, 5, 7 and 10% of Ar/H, the gas mixtures labeled as H1, H2, H3, H4, H5, and H6, respectively.

26.2.1 EELS and Raman Spectroscopy

EELS and Raman spectra of four sputtered (W1, W2, H1, and H3) and one PECVD carbon samples were obtained. Figure 26.5 shows the EELS spectra of these carbon coatings. EELS spectra for bulk diamond and polycrystalline graphite in an energy range up to 50 eV are also shown in Fig. 26.5. One prominent peak is seen at 35 eV in diamond, while two peaks are seen at 27 eV and 6.5 eV in graphite, which are called ($\pi+\sigma$) and (π) peaks, respectively. These peaks are produced by the energy loss of transmitted electrons to the plasmon oscillation of the valence electrons. The $\pi+\sigma$ peak in each coating is positioned at a lower energy region than that of graphite. The π peaks in the W series and PECVD samples are also seen at a lower energy region than that of the graphite. However, the π peaks in the H-series are comparable to or higher than that of graphite (see Table 26.2). The plasmon oscillation frequency is proportional to the square root of the corresponding electron density to a first approximation. Therefore, the samples in the H-series most likely have a higher density of π electrons than the other samples.

Amorphous carbon coatings contain (mainly) a mixture of sp^2- and sp^3-bonds, even though there is some evidence for the presence of sp-bonds as well [26.82]. The PECVD coatings and the H-series coatings in this study have nearly the same mass density, as seen in Table 26.4, to be presented later, but the former have a lower concentration of hydrogen (18.1%) than the H-series (35–39%), as seen in Table 26.3, to be presented later. The relatively low energy position of π peaks of PECVD coatings, compared to those of the H-series, indicates that the PECVD coatings contain a higher fraction of sp^3-bonds than the sputtered hydrogenated carbon coatings (H-series).

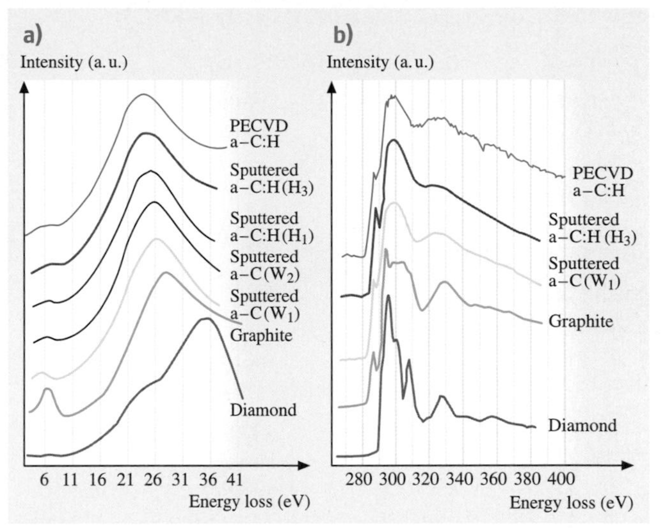

Fig. 26.5 (a) Low energy, and (b) high energy EELS of DLC coatings produced by DC magnetron sputtering and RF-PECVD techniques. Data for bulk diamond and polycrystalline graphite are included for comparison [26.35]

Figure 26.5b shows the EELS spectra associated with the inner-shell (K-shell) ionization. Again, the spectra for diamond and polycrystalline graphite are included for comparison. Sharp peaks are observed at

Table 26.2 Experimental results from EELS and Raman spectroscopy [26.35]

Sample	EELS peak position		Raman peak position		Raman FWHM[a]		I_D/I_G^d
	π (eV)	$\pi+\sigma$ (eV)	G-band[b] (cm^{-1})	D-band[c] (cm^{-1})	G-band (cm^{-1})	D-band (cm^{-1})	
Sputtered *a*-C coating (W1)	5.0	24.6	1541	1368	105	254	2.0
Sputtered *a*-C coating (W2)	6.1	24.7	1560	1379	147	394	5.3
Sputtered *a*-C:H coating (H1)	6.3	23.3	1542	1334	95	187	1.6
Sputtered *a*-C:H coating (H3)	6.7	22.4	e	e	e	e	e
PECVD *a*-C:H coating	5.8	24.0	1533	1341	157	427	1.5
Diamond coating	...	...	1525[f]	1333[g]	...	8[g]	...
Graphite (for reference)	6.4	27.0	1580	1358	37	47	0.7
Diamond (for reference)	...	37.0	...	1332[g]	...	2[g]	...

[a] Full width at half maximum width
[b] Peak associated with sp^2 "graphite" carbon
[c] Peak associated with sp^2 "disordered" carbon (not sp^3-bonded carbon)
[d] Intensity ratio of the D-band to the G-band
[e] Fluorescence
[f] Includes D and G band, signal too weak to analyze
[g] Peak position and width for diamond phonon

Table 26.3 Experimental results of FRS analysis [26.35]

Sample	Ar/H ratio	C (at% ±0.5)	H (at% ±0.5)	Ar (at% ±0.5)	O (at% ±0.5)
Sputtered *a*-C coating (W2)	100/0	90.5	9.3	0.2	...
Sputtered *a*-C:H coating (H2)	99/1	63.9	35.5	0.6	...
Sputtered *a*-C:H coating (H3)	97/3	56.1	36.5	...	7.4
Sputtered *a*-C:H coating (H4)	95/5	53.4	39.4	...	7.2
Sputtered *a*-C:H coating (H5)	93/7	58.2	35.4	0.2	6.2
Sputtered *a*-C:H coating (H6)	90/10	57.3	35.5	...	7.2
PECVD *a*-C:H coating	99.5% CH_4	81.9	18.1	...	...
Diamond coating	H_2-1 mole % CH_4	94.0	6.0	...	...

Table 26.4 Experimental results of physical properties [26.35]

Sample	Mass density (g/cm^3)	Nano-hardness (GPa)	Elastic modulus (GPa)	Electrical resistivity (Ohm-cm)	Compressive residual stress (GPa)
Sputtered *a*-C coating (W1)	2.1	15	141	1300	0.55
Sputtered *a*-C:H coating (W2)	1.8	14	136	0.61	0.57
Sputtered *a*-C:H coating (H1)	...	14	96	...	> 2
Sputtered *a*-C:H coating (H3)	1.7	7	35	$> 10^6$	0.3
PECVD *a*-C:H coating	1.6–1.8	33–35	~ 200	$> 10^6$	1.5–3.0
Diamond coating	...	40–75	370–430	...	...
Graphite (for reference)	2.267	Soft	9–15	5×10^{-5a}, 4×10^{-3b}	0
Diamond (for reference)	3.515	70–102	900–1050	10^7–10^{20}	0

[a] Parallel to layer planes

[b] Perpendicular to layer planes

285.5 eV and 292.5 eV in graphite, while no peak is seen at 285.5 eV in diamond. The general features of the K-shell EELS spectra for the sputtered and PECVD carbon samples resemble those of graphite, but with the higher energy features smeared. The observation of the peak at 285.5 eV in the sputtered and PECVD coatings also indicates the presence of sp^2-bonded atomic sites in the coatings. All these spectra peak at 292.5 eV, similar to the spectra of graphite, but the peak in graphite is sharper.

Raman spectra from samples W1, W2, H1, and PECVD are shown in Fig. 26.6. Raman spectra could not be observed in specimens H2 and H3 due to high flourescence signals. The Raman spectra of single crystal diamond and polycrstalline graphite are also shown for comparison in Fig. 26.6. The results of the spectral fits are summarized in Table 26.2. We will focus on the G-band position, which has been shown to be related to the fraction of sp^3-bonded sites. Increasing the power density in the amorphous carbon coatings (W1 and W2) results in a higher G-band frequency, implying a smaller fraction of sp^3-bonding in W2 than in W1. This is consistent with higher density of W1. H1 and PEVCD have still lower G-band positions than W1, implying an even higher fraction of sp^3-bonding, which is presumably caused by the incorporation of H atoms into the lattice. The high hardness of H3 might be attributed to efficient sp^3 cross-linking of small, sp^2-ordered domains.

The Raman spectrum of a MPECVD diamond coating is shown in Fig. 26.6. The diamond Raman peak is at 1,333 cm^{-1} with a line width of 7.9 cm^{-1}. There is a small broad peak around 1,525 cm^{-1}, which is attributed to a small amount of a-C:H. This impurity peak is not intense enough to fit to separate G- and D-bands. The diamond peak frequency is very close to that of natural diamond (1,332.5 cm^{-1}, e.g. Fig. 26.6), indicating that the coating is not under stress [26.83]. The large line width compared to that of natural diamond (2 cm^{-1})

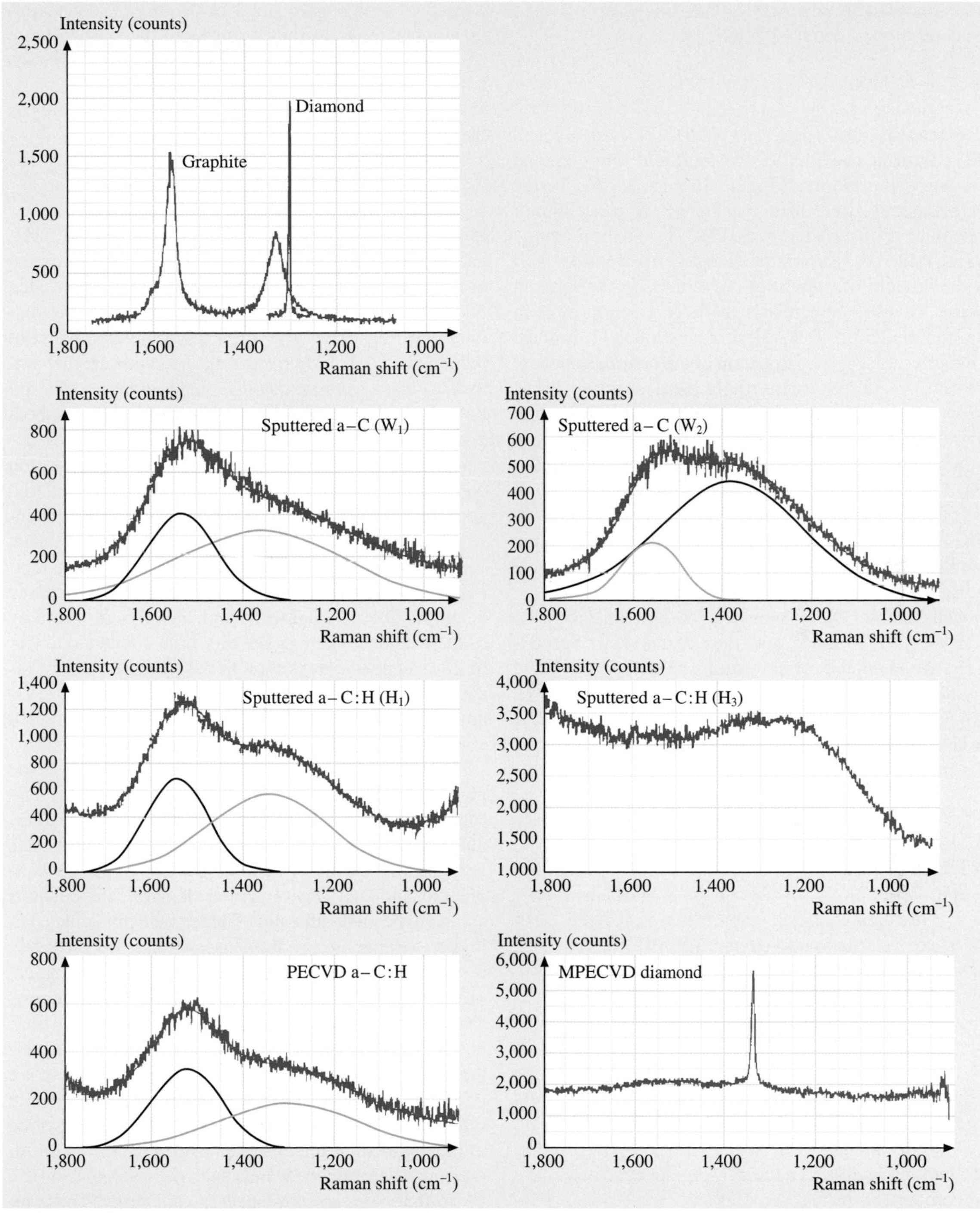

Fig. 26.6 Raman spectra of DLC coatings produced by DC magnetron sputtering and RF-PECVD techniques and a diamond film produced by MPECVD technique. Data for bulk diamond and microcrystalline graphite are included for comparison [26.35]

indicates that the microcrystallites likely have a high concentration of defects [26.84].

26.2.2 Hydrogen Concentrations

FRS analysis of six sputtered (W2, H2, H3, H4, H5, and H6) coatings, one PECVD coating, and one diamond coating was performed. Figure 26.7 shows an overlay of the spectra from the six sputtered samples. Similar spectra were obtained from the PECVD and the diamond films. Table 26.3 shows the H and C fractions, as well as the amount of impurities (Ar and O) in the films in atomic %. Most apparent is the large fraction of H in the sputtered films. Regardless of how much H_2 is in the Ar sputtering gas, the H content of the coatings is about the same, $\sim$ 35 at%. Interestingly, there is still $\sim$ 10% H present in the coating sputtered in pure Ar (W2). It is interesting to note that Ar is present only in coatings grown under low ($< 1\%$) H content in the Ar carrier gas. The presence of O in the coatings, combined with the fact that the coatings were prepared approximately nine months before the FRS analysis, caused suspicion that they had absorbed water vapor, and that this may be the cause for the H peak in specimen W2.

All samples were annealed for 24 h at 250 °C in a flowing He furnace and then reanalyzed. Surprisingly, the H content of all coatings measured increased slightly, even though the O content decreased, and W2 still had a substantial amount of H_2. This slight increase in H concentration is not understood. However, since H concentration did not decrease with the oxygen as a result of annealing, it suggests that high H concentration is not due to adsorbed water vapor. The PECVD film has more H ($\sim$ 18%) than the sputtered films initially, but after annealing it has the same fraction as specimen W2, the film sputtered in pure Ar. The diamond film has the smallest amounts of hydrogen, as seen in Table 26.3.

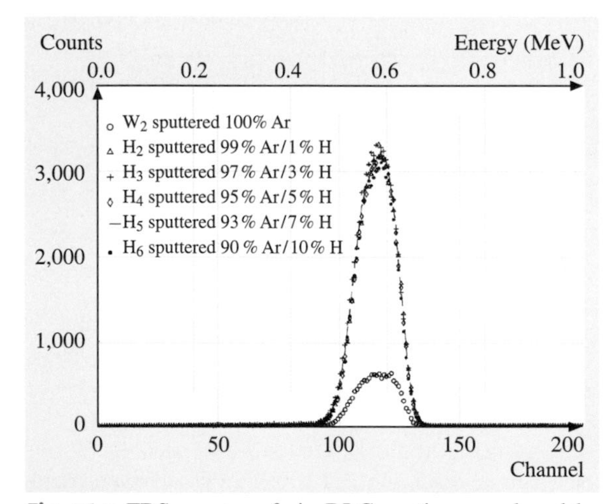

Fig. 26.7 FRS spectra of six DLC coatings produced by DC magnetron sputtering [26.35]

26.2.3 Physical Properties

Physical properties of the four sputtered (W1, W2, H1, and H3) coatings, one PECVD coating, one diamond coating, and bulk diamond and graphite are presented in Table 26.4. The hydrogenated carbon and the diamond coatings have very high resistivity compared to unhydrogenated carbon coatings. It appears that unhydrogenated carbon coatings have a higher density than the hydrogenated carbon coatings, although both groups are less dense than graphite. The density depends on the deposition technique and the deposition parameters. It appears that unhydrogenated sputtered coatings deposited at low power exhibit the highest density. Nanohardness of hydrogenated carbon is somewhat lower than that of the unhydrogenated carbon coatings. PECVD coatings are significantly harder than sputtered coatings. The nanohardness and modulus of elasticity of the diamond coatings are very high compared to that of DLC coatings, even though the hydrogen content is similar. The compressive residual stresses of the PECVD coatings are substantially higher than those of sputtered coatings, which is consistent with the hardness results.

Figure 26.8a shows the effect of hydrogen in the plasma on the residual stresses and the nanohardness for sputtered coatings W2 and H1 to H6. The coatings made with H_2 flow between 0.5 and 1.0% delaminate very quickly, even when only a few tens of nm thick. In pure Ar and at H_2 flows greater than 1%, the coatings appear to be more adhesive. The tendency of some coatings to delaminate can be caused by intrinsic stress in the coating, which is measured by substrate bending. All of the coatings in the figure are in compressive stress. The maximum stress occurs between 0 and 1% H_2 flow, but the stress cannot be quantified in this range because the coatings instantly delaminate upon exposure to air. At higher hydrogen concentrations the stress gradually diminishes. A generally decreasing trend is observed in the hardness of the coatings as hydrogen content increased. The hardness decreases slightly, going from 0% H_2 to 0.5% H_2, and then decreases sharply. These results are probably lower than the true values because of local delamination around the indentation point. This is especially likely for the 0.5% and 1.0% coatings where delamination is visually apparent, but may also be true to a lesser extent for the other coatings. Such an adjustment would bring the hardness profile into closer correlation with the stress profile. *Weissmantel* et al. [26.68] and *Scharff* et al. [26.85] observed a downturn in hardness for high bias and low pressure of hydrocarbon gas in ion plated carbon coating, and, therefore, presumably low hydrogen content in support of the above contention.

Figure 26.8b shows the effect of sputtering power (with no hydrogen added in the plasma) on the residual stresses and nanohardness for various sputtered coatings. As the power decreases, compressive stress does not seem to change while nanohardness slowly increases. The rate of change becomes more rapid at very low power levels.

The addition of H_2 during sputtering of carbon coatings increases H concentration in the coating. Hydrogen causes the character of the C-C bonds to shift from sp^2 to sp^3, and the rising number of C-H bonds, which ultimately relieves stress and produces a softer "polymer-like" material. Low power deposition, like the presence of hydrogen, appears to stabilize the formation of sp^3 C-C bonds, increasing hardness. These coatings have relieved stress and led to better adhesion. An increase in temperature during deposition at high power density results in graphitization of the coating material, responsible for a decrease in hardness with an increase in power density. Unfortunately, low power also means impractically low deposition rates.

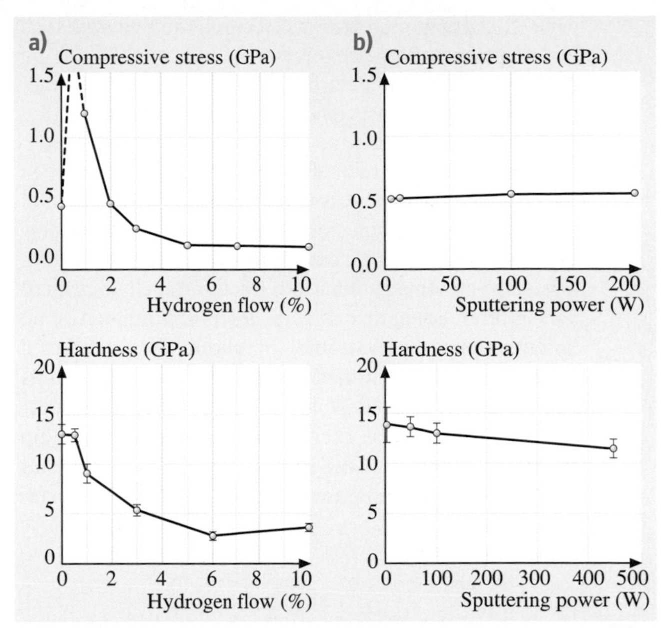

Fig. 26.8a,b Residual compressive stresses and nanohardness (**a**) as a function of hydrogen flow rate; sputtering power is 100 W and target diameter is 75 mm (power density = 2.1 W/cm^2), and (**b**) as a function of sputtering power over a 75-mm-diameter target with no hydrogen added in the plasma [26.40]

26.2.4 Summary

Based on the EELS and Raman data, all DLC coatings have both sp^2 and sp^3 bondings. The sp^2/sp^3 bonding ratio depends on the deposition techniques and parameters. The DLC coatings deposited by sputtering and PECVD contain significant concentrations of hydrogen, while the diamond coating contains only small amounts of hydrogen impurity. Sputtered coatings with no deliberate addition of hydrogen in the plasma contain a significant amount of hydrogen. Regardless of how much hydrogen is in the Ar sputtering gas, the hydrogen content of the coatings increases initially with no further increase.

Hydrogen flow and sputtering power density affect the mechanical properties of these coatings. Maximum compressive residual stress and hardness occur between 0 and 1% hydrogen flow, resulting in rapid delamination. Low sputtering power moderately increases hardness while relieving residual stress.

26.3 Micromechanical and Tribological Characterizations of Coatings Deposited by Various Techniques

26.3.1 Micromechanical Characterization

Common mechanical characterizations include measurement of hardness and elastic modulus, fracture toughness, fatigue life, and scratch and wear testing. Nanoindentation and atomic force microscopy (AFM) are used for mechanical characterization of ultrathin films.

Hardness and elastic modulus are calculated from the load displacement data obtained by nanoindentation at loads ranging typically from 0.2 to 10 mN using a commercially available nanoindenter [26.23,86].

This instrument monitors and records the dynamic load and displacement of the three-sided pyramidal diamond (Berkovich) indenter during indentation. For the fracture toughness measurement of ultrathin films ranging from 100 nm to a few μm, a nanoindentation-based technique is used in which through-thickness cracking in the coating is detected from a discontinuity observed in the load-displacement curve and energy released during the cracking is obtained from the curve [26.87–89]. Based on the energy released, fracture mechanics analysis is then used to calculate fracture toughness. An indenter with a cube-corner tip geometry is preferred because the through-thickness cracking of hard films can be accomplished at lower loads. In fatigue measurement, a conical diamond indenter having a tip radius of about one micron is used and load cycles of a sinusoidal shape are applied [26.90, 91]. The fatigue behavior of coatings is studied by monitoring the change in contact stiffness, which is sensitive to damage formation.

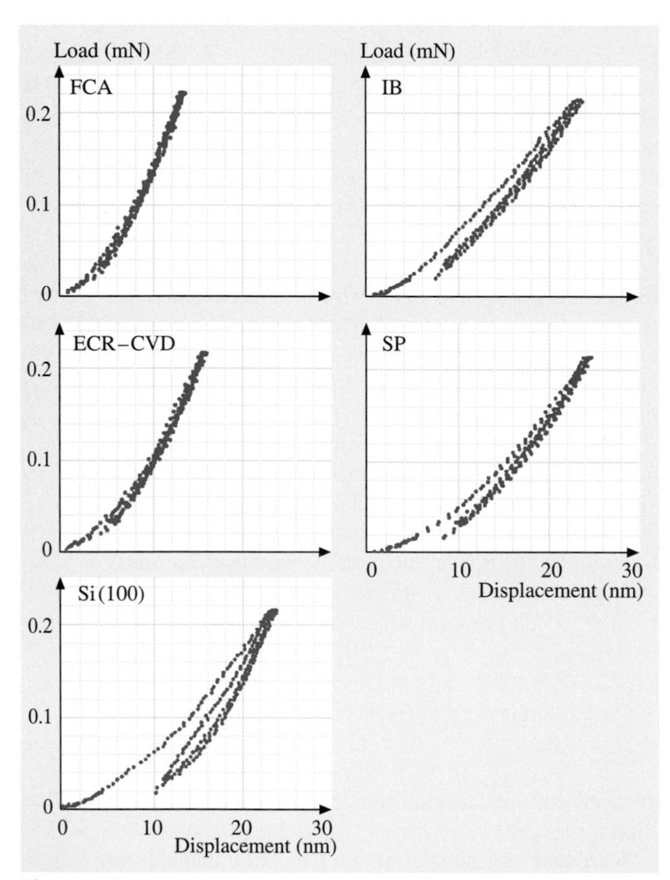

Fig. 26.9 Load versus displacement plots for various 100-nm-thick amorphous carbon coatings on single-crystal silicon substrate and bare substrate

Hardness and Elastic Modulus

For materials that undergo plastic deformation, high hardness and elastic modulus are generally needed for low friction and wear, whereas for brittle materials, high fracture toughness is needed [26.2, 3, 21]. DLC coatings used for many applications are hard and brittle, and values of hardness and fracture toughness need to be optimized.

Representative load-displacement plots of indentations made at 0.2 mN peak indentation load on 100-nm-thick DLC coatings deposited by the four deposition techniques on single-crystal silicon substrate are compared in Fig. 26.9. The indentation depths at the peak load range from about 18 to 26 nm, smaller than that of the coating thickness. Many of the coatings exhibit a discontinuity or pop-in marks in the loading curve, which indicate a sudden penetration of the tip into the sample. A nonuniform penetration of the tip into a thin coating possibly results from formation of cracks in the coating, formation of cracks at the coating-substrate interface, or debonding or delamination of the coating from the substrate.

The hardness and elastic modulus values at a peak load of 0.2 mN on the various coatings and single-crystal silicon substrate are summarized in Table 26.5 and Fig. 26.10 [26.47, 49, 89, 90]. Typical values for the peak and residual indentation depths ranged from 18 to 26 nm and 6 to 12 nm, respectively. The FCA coatings exhibit the highest hardness of 24 GPa and elastic modulus of 280 GPa of various coatings, followed by the ECR-CVD, IB, and SP coatings. Hardness and elastic modulus have been known to vary over a wide range with sp^3-to-sp^2 bonding ratio, which depends on the kinetic energy of the carbon species and amount of hydrogen [26.6, 30, 47, 92, 93]. The high hardness and elastic modulus of the FCA coatings are attributed to the high kinetic energy of carbon species involved in the FCA deposition [26.12, 47]. *Anders* et al. [26.57] also reported a high hardness, measured by nanoindentation, of about 45 GPa for cathodic arc carbon coatings. They observed a change in hardness from 25 to 45 GPa with pulsed bias voltage and bias duty cycle. The high hardness of cathodic arc carbon was attributed to the high percentage (more than 50%) of sp^3 bonding. *Savvides* and *Bell* [26.94] reported an increase in hardness from 12 to 30 GPa and an elastic modulus from 62 to 213 GPa with an increase of sp^3-to-sp^2 bonding ratio, from 3 to 6, for a C:H coating deposited by low-energy ion-assisted unbalanced magnetron sputtering of a graphite target in an Ar-H_2 mixture.

Bhushan et al. [26.35] reported hardnesses of about 15 and 35 GPa and elastic moduli of about 140 and 200 GPa, measured by nanoindentation, for a-C:H coatings deposited by DC magnetron sputtering and RF-plasma-enhanced chemical vapor deposition techniques, respectively. The high hardness of RF-PECVD a-C:H coatings is attributed to a higher concentration of sp^3 bonding than in a sputtered hydrogenated a-C:H coating. Hydrogen is believed to play a crucial role in the bonding configuration of carbon atoms by helping to stabilize tetrahedra-coordination (sp^3 bonding) of carbon species. *Jansen* et al. [26.78] suggested that the incorporation of hydrogen efficiently passivates the dangling bonds and saturates the graphitic bonding to some extent. However, a large concentration of hydrogen in the plasma in sputter deposition is undesirable. *Cho* et al. [26.33] and *Rubin* et al. [26.40] observed that hardness decreased from 15 to 3 GPa with increased hydrogen content. *Bhushan* and *Doerner* [26.95] reported a hardness of about 10–20 GPa and an elastic modulus of about 170 GPa, measured by nanoindentation, for 100-nm-thick DC-magnetron sputtered a-C:H on the silicon substrate.

Residual stresses measured using a well-known curvature measurement technique are also presented in Table 26.5. The DLC coatings are under significant compressive internal stresses. Very high compressive stresses in FCA coatings are believed to be partly responsible for their high hardness. However, high stresses result in coating delamination and buckling. For this reason, the coatings that are thicker than about 1 μm have a tendency to delaminate from the substrates.

Fracture Toughness

Representative load-displacement curves of indentations on the 400-nm-thick cathodic arc carbon coating on silicon at various peak loads are shown in Fig. 26.11. Steps are found in all curves, as shown by arrows in Fig. 26.11a. In the 30 mN SEM micrograph, in addition to several radial cracks, ring-like through-thickness cracking is observed with small lips of material overhanging the edge of indentation. The step at about 23 mN in the loading curves of indentations made at 30 and 100 mN peak indentation loads results from the ring-like through-thickness cracking. The step at 175 mN in the loading curve of the indentation made at 200 mN peak indentation load is caused by spalling and second ring-like through-thickness cracking.

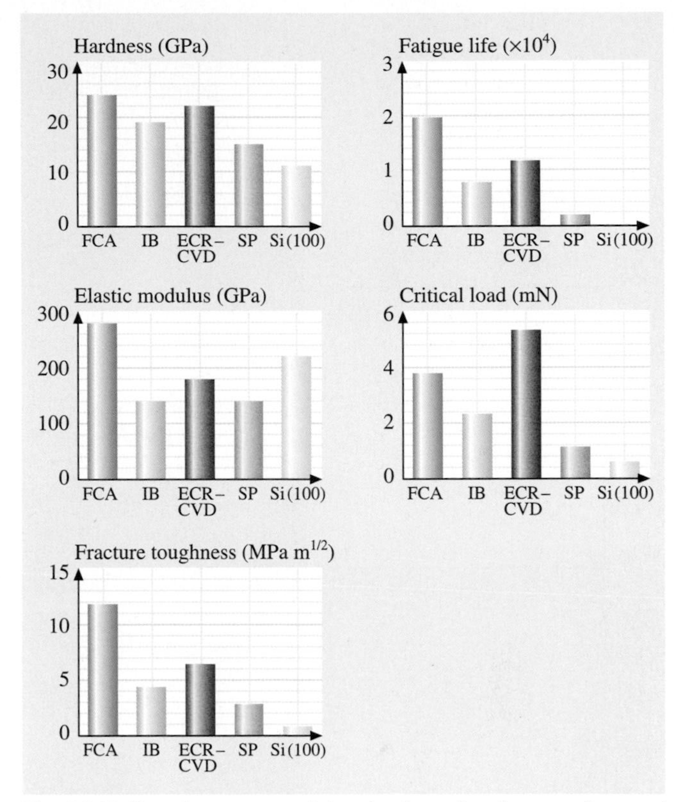

Fig. 26.10 Bar charts summarizing the data of various coatings and single-crystal silicon substrate. Hardness, elastic modulus, and fracture toughness were measured on 100-nm-thick coatings, and fatigue life and critical load during scratch were measured on 20-nm-thick coatings

Based on *Li* et al. [26.87], the fracture process progresses in three stages: (1) ring-like through-thickness cracks form around the indenter by high stresses in the contact area, (2) delamination and buckling occur around contact area at the coating/substrate interface by high lateral pressure, (3) second ring-like through-thickness cracks and spalling are generated by high bending stesses at the edges of the buckled coating, see Fig. 26.12a. In the first stage, if the coating under the indenter is separated from the bulk coating via the first ring-like through-thickness cracking, a corresponding step will be present in the loading curve. If discontinuous cracks form and the coating under the indenter is not separated from the remaining coating, no step appears in the loading curve, because the coating still supports the indenter and the indenter cannot suddenly advance into the material. In the second stage,

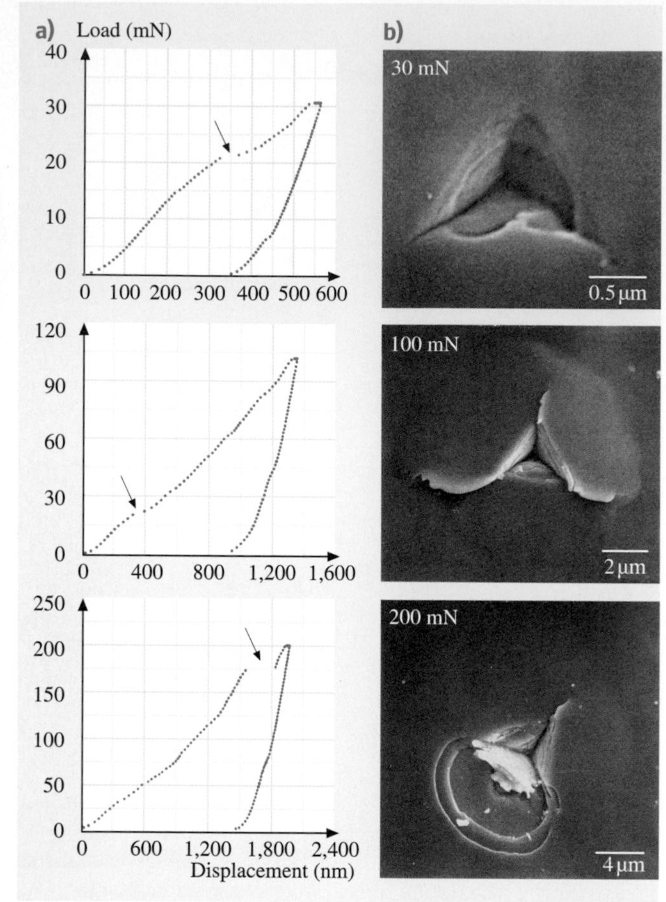

Fig. 26.11 (a) Load-displacement curves of indentations made at 30, 100, and 200 mN peak indentation loads using the cube corner indenter, and (b) the SEM micrographs of indentations on the 400-nm-thick cathodic arc carbon coating on silicon. *Arrows* indicate steps during loading portion of the load-displacement curve [26.87]

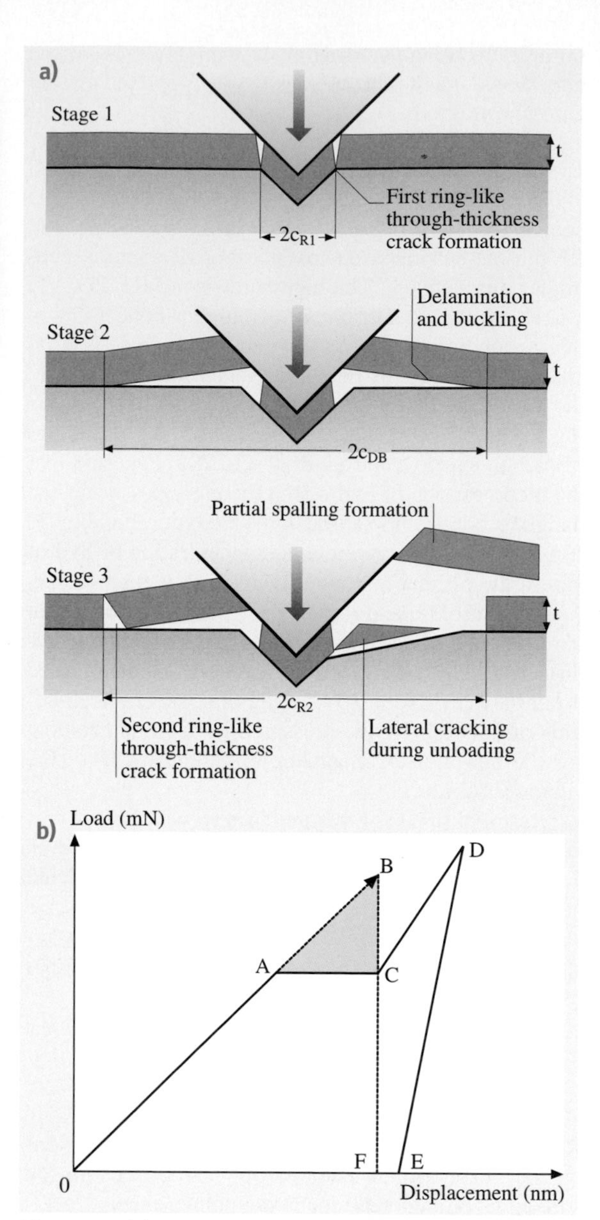

Fig. 26.12 (a) Schematic of various stages in nanoindentation fracture for the coatings/substrate system, and (b) schematic of a load-displacement curve showing a step during the loading cycle and associated energy release

for the coating used in the present study, the advances of the indenter during the radial cracking, delamination, and buckling are not big enough to form steps in the loading curve, because the coating around the indenter still supports the indenter, but generate discontinuities that change the slope of the loading curve with increasing indentation loads. In the third stage, the stress concentration at the end of the interfacial crack cannot be relaxed by the propagation of the interfacial crack. With an increase in indentation depth, the height of the bulged coating increases. When the height reaches a critical value, the bending stresses caused by the bulged coating around the indenter will result in the second ring-like through-thickness crack formation and spalling at the edge of the buckled coating, as shown in Fig. 26.12a, which leads to a step in the loading curve. This is a single event and results in the separation of the part of the coating around the indenter from the bulk

Table 26.5 Hardness, elastic modulus, fracture toughness, fatigue life, critical load during scratch, coefficient of friction during accelerated wear testing and residual stresses of various DLC coatings on single-crystal silicon substrate

Coating	Hardness[a] [26.48] (GPa)	Elastic modulus[a] [26.48] (GPa)	Fracture toughness[a] [26.89] (MPa $m^{1/2}$)	Fatigue life[b], N_f^d [26.90] $\times 10^4$	Critical load during scratch[b] [26.48] (mN)	Coefficient of friction during accelerated wear testing[b] [26.48]	Compressive residual stress[c] [26.47] (GPa)
Cathodic arc carbon coating (a-C)	24	280	11.8	2.0	3.8	0.18	12.5
Ion beam carbon coating (a-C:H)	19	140	4.3	0.8	2.3	0.18	1.5
ECR-CVD carbon coating (a-C:H)	22	180	6.4	1.2	5.3	0.22	0.6
DC sputtered carbon coating (a-C:H)	15	140	2.8	0.2	1.1	0.32	2.0
Bulk graphite (for comparison)	Very soft	9–15	–	–	–	–	–
Diamond (for comparison)	80–104	900–1050	–	–	–	–	–
Si(100) substrate	11	220	0.75	–	0.6	0.55	0.02

[a] Measured on 100-nm-thick coatings
[b] Measured on 20-nm-thick coatings
[c] Measured on 400-nm-thick coatings
[d] N_f was obtained at a mean load of 10 μN and a load amplitude of 8 μN

coating via cracking through coatings. The step in the loading curve results totally from the coating cracking and not from the interfacial cracking or the substrate cracking.

The area under the load-displacement curve is the work performed by the indenter during elastic-plastic deformation of the coating/substrate system. The strain energy release in the first/second ring-like cracking and spalling can be calculated from the corresponding steps in the loading curve. Fig. 26.12b shows a modeled load-displacement curve. OACD is the loading curve and DE is the unloading curve. The first ring-like through-thickness crack should be considered. It should be emphasized that the edge of the buckled coating is far from the indenter and, therefore, it does not matter if the indentation depth exceeds the coating thickness, or if deformation of the substrate occurs around the indenter when we measure fracture toughness of the coating from the released energy during the second ring-like through-thickness cracking (spalling). Suppose that the second ring-like through-thickness cracking occurs at AC. Now, let us consider the loading curve OAC. If the second ring-like through-thickness crack does not occur, it can be understood that OA will be extended to OB to reach the same displacement as OC. This means that the crack formation changes the loading curve OAB into OAC. For point B, the elastic-plastic energy stored in the coating/substrate system should be OBF. For point C, the elastic-plastic energy stored in the coating/substrate system should be OACF. Therefore, the energy difference before and after the crack generation is the area of ABC, i. e., this energy stored in ABC will be released as strain energy to create the ring-like through-thickness crack. According to the theoretical analysis by *Li* et al. [26.87], the fracture toughness of thin films can be written as

$$K_{Ic} = \left[\left(\frac{E}{(1-\nu^2)\, 2\pi C_R} \right) \left(\frac{U}{t} \right) \right]^{1/2} , \qquad (26.2)$$

where E is the elastic modulus, ν is the Poisson's ratio, $2\pi C_R$ is the crack length in the coating plane, t is the coating thickness, and U is the strain energy difference before and after cracking.

Using (26.2), the fracture toughness of the coatings is calculated. The loading curve is extrapolated along the tangential direction of the loading curve from the starting point of the step up to reach the same displacement as the step. The area between the extrapolated line and the step is the estimated strain energy difference before and after cracking. C_R is measured from SEM micrographs or AFM images of indentations. The second ring-like crack is where the spalling occurs. For example, for the 400-nm-thick cathodic arc carbon coating data presented in Fig. 26.11, U of 7.1 nNm is assessed from the steps in Fig. 26.11a at the peak indentation loads of 200 mN. For C_R of 7.0 μm from Fig. 26.11b, $E = 300$ GPa measured using a nanoindenter and an assumed value of 0.25 for ν, fracture toughness values are calculated as 10.9 MPa$\sqrt{m}$ [26.87, 88]. The fracture toughness and related data for various 100-nm-thick DLC coatings are presented in Fig. 26.10 and Table 26.5.

Nanofatigue

Delayed fracture resulting from extended service is called fatigue [26.96]. Fatigue fracturing progresses through a material via changes within the material at the tip of a crack, where there is a high stress intensity. There are several situations: cyclic fatigue, stress corrosion, and static fatigue. Cyclic fatigue results from cyclic loading of machine components. In a low-flying slider in a magnetic head-disk interface, isolated asperity contacts occur during use and the fatigue failure occurs in the multilayered thin-film structure of the magnetic disk [26.13]. In many MEMS components, impact occurs and the failure mode is cyclic fatigue. Asperity contacts can be simulated using a sharp diamond tip in an oscillating contact with the component.

Figure 26.13 shows the schematic of a fatigue test on a coating/substrate system using a continuous stiffness measurement (CSM) technique. Load cycles are applied to the coating, resulting in a cyclic stress. P is the cyclic load, P_{mean} is the mean load, P_o is the oscillation load amplitude, and ω is the oscillation frequency. The following results can be obtained: (1) endurance limit, i. e., the maximum load below which there is no coating failure for a preset number of cycles; (2) number of cycles at which the coating failure occurs; and (3) changes in contact stiffness measured by using the unloading slope of each cycle, which can be used to monitor the propagation of the interfacial cracks during cyclic fatigue process.

Figure 26.14a shows the contact stiffness as a function of the number of cycles for 20-nm-thick FCA coatings cyclically deformed by various oscillation load amplitudes with a mean load of 10 μN at a frequency of 45 Hz. At 4 μN load amplitude, no change in contact stiffness was found for all coatings. This indicates that 4 μN load amplitude is not high enough to damage the coatings. At 6 μN load amplitude, an abrupt decrease in contact stiffness was found at a certain number of cycles for each coating, indicating that fatigue damage had occurred. With increasing load amplitude, the number of cycles to failure, N_f, decreases for all coatings. Load amplitude versus N_f, a so-called S-N curve, is plotted in Fig. 26.14b. The critical load amplitude, below which no fatigue damage occurs (an endurance limit), was identified for each coating. This critical load amplitude, together with mean load, are of critical importance to the design of head-disk interfaces or MEMS/NEMS device interfaces.

To compare the fatigue lives of different coatings studied, the contact stiffness as a function of the number of cycles for 20-nm-thick FCA, IB, ECR-CVD and SP coatings cyclically deformed by an oscillation load amplitude of 8 μN with a mean load of 10 μN at a frequency of 45 Hz is shown in Fig. 26.14c. FCA coating has the longest N_f, followed by ECR-CVD, IB, and SP coatings. In addition, after the N_f, the contact stiffness of the FCA coating shows a slower decrease than the other coatings. This indicates that after the N_f, the FCA coating had less damage than the others. The fatigue behavior of FCA and ECR-CVD coatings of differ-

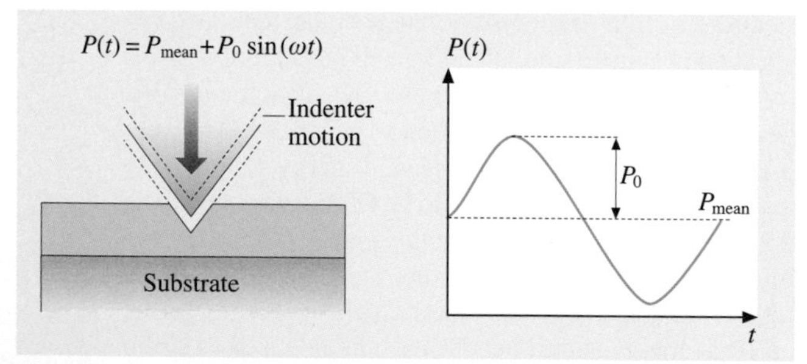

Fig. 26.13 Schematic of a fatigue test on a coating/substrate system using the continuous stiffness measurement technique

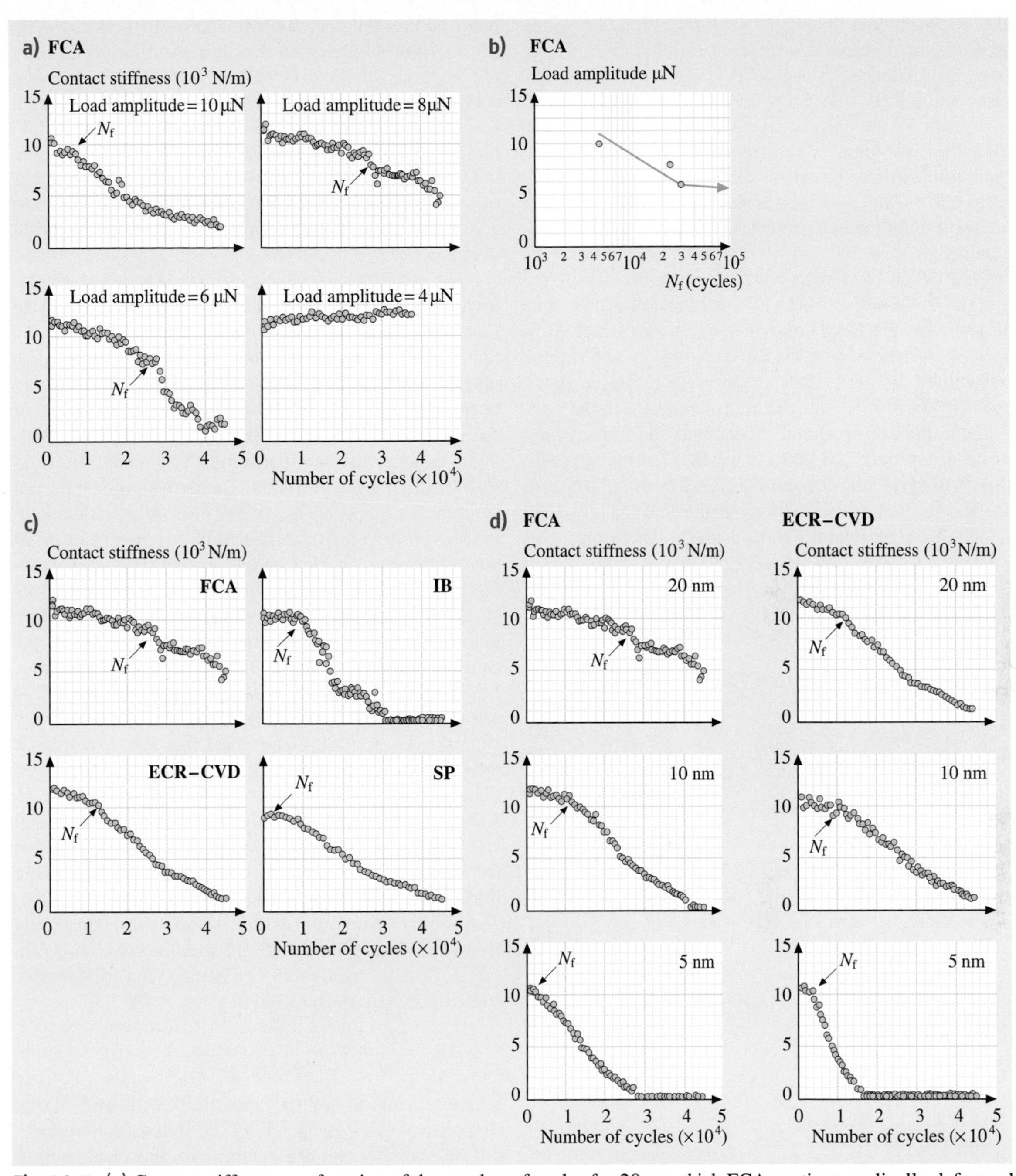

Fig. 26.14 (a) Contact stiffness as a function of the number of cycles for 20-nm-thick FCA coatings cyclically deformed by various oscillation load amplitudes with a mean load of 10 μN at a frequency of 45 Hz; (b) plot of load amplitude versus N_f; (c) contact stiffness as a function of the number of cycles for four different 20-nm-thick coatings with a mean load of 10 μN and load amplitude of 8 μN at a frequency of 45 Hz; and (d) contact stiffness as a function of the number of cycles for two coatings of different thicknesses at a mean load of 10 μN and load amplitude of 8 μN at a frequency of 45 Hz

ent thicknesses is compared in Fig. 26.14d. For both coatings, N_f decreases with decreasing coating thickness. At 10 nm, FCA and ECR-CVD have almost the same fatigue life. At 5 nm, ECR-CVD coating shows a slightly longer fatigue life than FCA coating. This indicates that even for nanometer-thick DLC coatings their microstructure and residual stresses are not uniform across the thickness direction. Thinner coatings are more influenced by interfacial stresses than thicker coating.

Figure 26.15a shows the high magnification SEM images of 20-nm-thick FCA coatings before, at, and after N_f. In the SEM images, the net-like structure is the gold film coated on the DLC coating, which should be ignored in analyzing the indentation fatigue damage. Before the N_f, no delamination or buckling was found except the residual indentation mark at magnifications up to 1,200,000× using SEM. This suggests that only plastic deformation occurred before the N_f. At the N_f, the coating around the indenter bulged upwards, indicating delamination and buckling. Therefore, it is believed that the decrease in contact stiffness at the N_f results from the delamination and buckling of the coating from the substrate. After the N_f, the buckled coating was broken down around the edge of the buckled area, forming a ring-like crack. The remaining coating overhung at the edge of the buckled area. It is noted that the indentation size increases with the increasing number of cycles. This indicates that deformation, delamination and buckling, and ring-like crack formation occurred over a period.

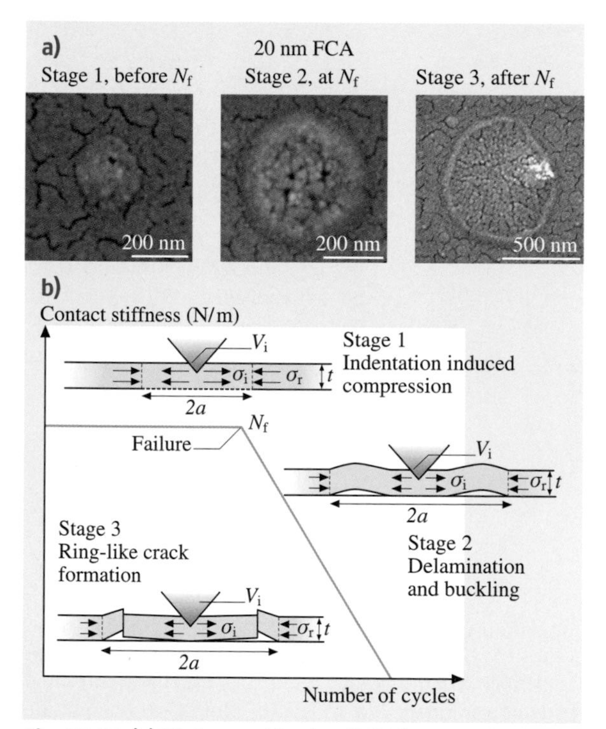

Fig. 26.15 (a) High magnification SEM images of a coating before, at, and after N_f, and (b) schematic of various stages in the indentation fatigue damage for a coating/substrate system [26.90]

The schematic in Fig. 26.15b shows various stages in the indentation fatigue damage for a coating/substrate system. Based on this study, three stages in the indentation fatigue damage appear to exist: (1) indentation-induced compression; (2) delamination and buckling; (3) ring-like crack formation at the edge of the buckled coating. The deposition process often induces residual stresses in coatings. The model shown in Fig. 26.15b considers a coating with the uniform biaxial residual compression σ_r. In the first stage, indentation induces elastic/plastic deformation, exerting an outward acting pressure on the coating around the indenter. Interfacial defects like voids and impurities act as original cracks. These cracks propagate and link up as the indentation compressive stress increases. At this stage, the coating, which is under the indenter and above the interfacial crack (with a crack length of $2a$), still maintains a solid contact with the substrate; the substrate still fully supports the coating. Therefore, this interfacial crack does not lead to an abrupt decrease in contact stiffness, but gives rise to a slight decrease in contact stiffness, as shown in Fig. 26.14. The coating above the interfacial crack is treated as a rigidly clamped disk. We assume that the crack radius, a, is large compared with the coating thickness t. Since the coating thickness ranges from 5 to 20 nm, this assumption is easily satisfied in this study (radius of the delaminated and buckled area, shown in Fig. 26.15a, is on the order of 100 nm). The compressive stress caused by indentation is given as [26.97]:

$$\sigma_i = \frac{E}{(1-\nu)}\varepsilon_i = \frac{EV_i}{2\pi t a^2(1-\nu)}\,, \tag{26.3}$$

where ν and E are the Poisson's ratio and elastic modulus of the coating, V_i is the indentation volume, t is the coating thickness, and a is the crack radius. With an increasing number of cycles, the indentation volume V_i increases. Therefore, the indentation compressive stress σ_i increases accordingly. In the second stage, buckling occurs during the unloading segment of the fatigue testing cycle when the sum of indentation compressive stress σ_i and the residual stress σ_r

exceed the critical buckling stress σ_b for the delaminated circular section as given by [26.98]

$$\sigma_b = \frac{\mu^2 E}{12\left(1-\nu^2\right)} \left(\frac{t}{a}\right)^2 , \quad (26.4)$$

where the constant μ equals 42.67 for a circular clamped plate with a constrained center point and 14.68 when the center is unconstrained. The buckled coating acts as a cantilever. In this case, the indenter indents a cantilever rather than a coating/substrate system. This ultrathin coating cantilever has much less contact stiffness than the coating/substrate system. Therefore, the contact stiffness shows an abrupt decrease at the N_f. In the third stage, with an increasing number of cycles, the delaminated and buckled size increases, resulting in a further decrease in contact stiffness since the cantilever beam length increases. On the other hand, a high bending stress acts at the edge of the buckled coating. The larger the buckled size, the higher the bending stress. The cyclically bending stress causes fatigue damage at the end of the buckled coating, forming a ring-like crack. The coating under the indenter is separated from the bulk coating (caused by the ring-like crack at the edge of the buckled coating) and the substrate (caused by the delamination and buckling in the second stage). Therefore, the coating under the indenter is not constrained, but is free to move with the indenter during fatigue testing. At this point, the sharp nature of the indenter is lost, because the coating under the indenter gets stuck on the indenter. The indentation fatigue experiment results in the contact of a relative huge blunt tip with the substrate. This results in a low contact stiffness value.

Compressive residual stresses result in delamination and buckling. A coating with a higher adhesion strength and a less compressive residual stress is required for a higher fatigue life. Interfacial defects should be avoided in the coating deposition process. We know that the ring-like crack formation occurs in the coating. Formation of fatigue cracks in the coating depends upon the hardness and fracture toughness. Cracks are more difficult to form and propagate in the coating with higher strength and fracture toughness.

It is now accepted that long fatigue life in a coating/substrate almost always involves "living with crack", that the threshold or limit condition is associated with the non-propagation of exiting cracks or defects, even though these cracks may be undetectable [26.96]. For all coatings studied, at 4 μN, contact stiffness does not change much. This indicates that delamination and buckling did not occur within the number of cycles tested in this study. This is probably because the indentation-induced compressive stress was not high enough to allow the cracks to propagate and link up under the indenter, or the sum of indentation compressive stress σ_i and the residual stress σ_r did not exceed the critical buckling stress σ_b.

Figure 26.10 and Table 26.5 summarize the hardness, elastic modulus, fracture toughness, and fatigue life of all coatings studied. A good correlation exists between fatigue life and other mechanical properties. Higher mechanical properties result in a longer fatigue life. The mechanical properties of DLC coatings are controlled by the sp^3-to-sp^2 ratio. The sp^3-bonded carbon exhibits the outstanding properties of diamond [26.51]. A higher deposition kinetic energy will result in a larger fraction of sp^3-bonded carbon in an amorphous network. Thus, the higher kinetic energy for the FCA could be responsible for its better carbon structure and higher mechanical properties [26.48–50, 99]. Higher adhesion strength between the FCA coating and substrate makes the FCA coating more difficult to delaminate from the substrate.

26.3.2 Microscratch and Microwear Studies

For microscratch studies, a conical diamond indenter (e.g., having a tip radius of about one micron and included angle of 60°) is drawn over the sample surface, and the load is ramped up (typically from 2 mN to 25 mN) until substantial damage occurs. The coefficient of friction is monitored during scratching. Scratch-induced damage of coating, specifically fracture or delamination, can be monitored by in situ friction force measurements and by optical and SEM imaging of the scratches after tests. A gradual increase in friction is associated with plowing, and an abrupt increase in friction is associated with fracture or catastrophic failure [26.100]. The load, corresponding to the abrupt increase in friction or an increase in friction above a certain value (typically 2× the initial value), provides a measure of scratch resistance or adhesive strength of a coating and is called "critical load". The depth of scratches with increasing scratch length or normal load is measured using an AFM, typically with an area of 10×10 μm [26.48, 49, 101].

The microscratch and microwear studies are also conducted using an AFM [26.23, 50, 99, 102, 103]. A square pyramidal diamond tip (tip radius ~ 100 nm) or a three-sided pyramidal diamond (Berkovich) tip with an apex angle of 60° and a tip radius of about 100 nm mounted on a platinum-coated, rectangular stainless steel cantilever of stiffness of about 40 N/m is scanned

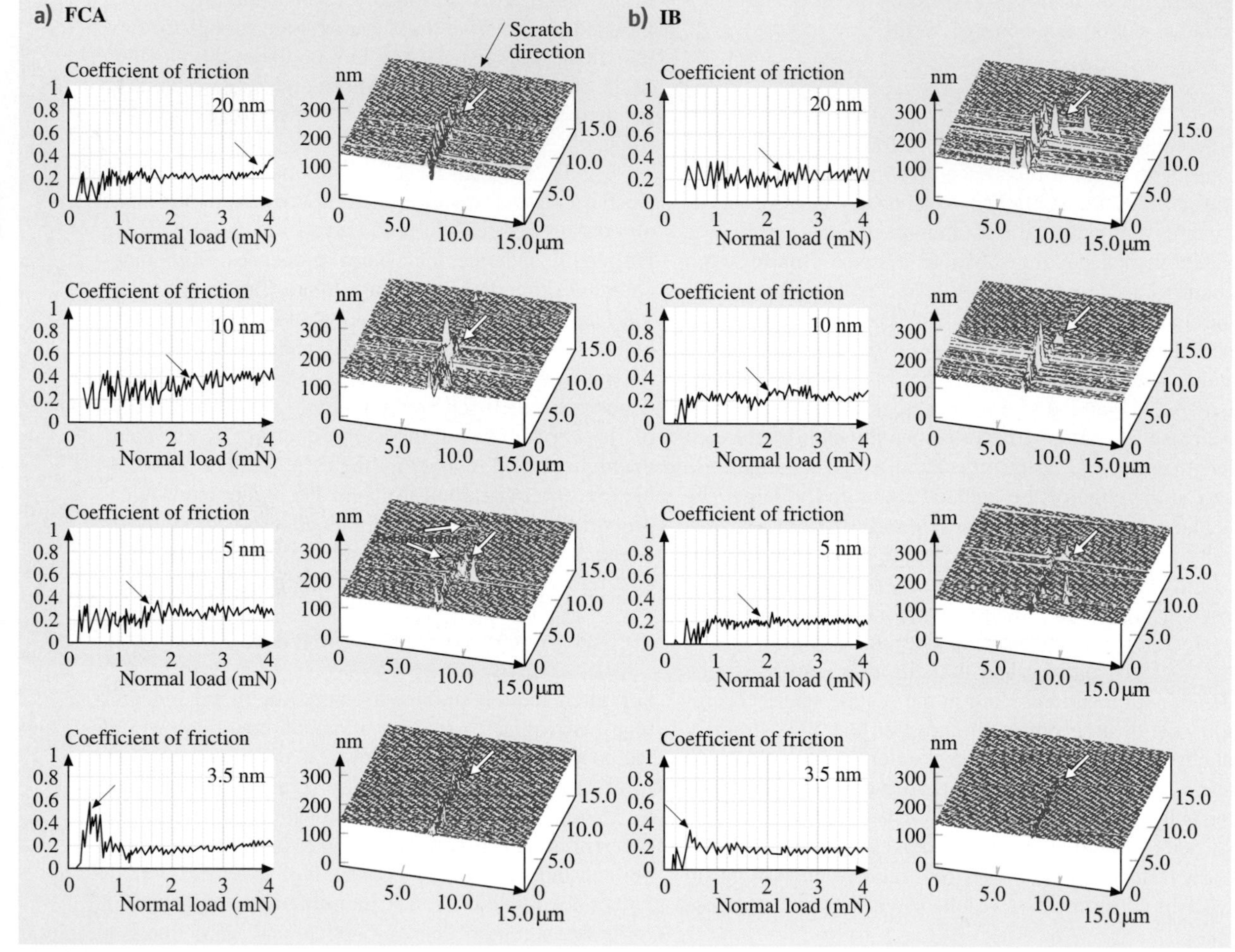

Fig. 26.16a,b

orthogonal to the long axis of the cantilever to generate scratch and wear marks. During the scratch test, the normal load is either kept constant or increased (typically from 0 to 100 µN) until damage occurs. Topography images of the scratch are obtained in situ with the AFM at a low load. By scanning the sample during scratching, wear experiments can be conducted. Wear at a constant load is monitored as a function of the number of cycles. Normal loads ranging from 10–80 µN are typically used.

Microscratch

Scratch tests conducted with a sharp diamond tip simulate a sharp asperity contact. In a scratch test, the cracking or delamination of a hard coating is signaled by a sudden increase in the coefficient of friction [26.23]. The load associated with this event is called the "critical load".

Wu [26.104], *Bhushan* et al. [26.70], *Gupta* and *Bhushan* [26.12,47], and *Li* and *Bhushan* [26.48,49,101] have used a nanoindenter to perform microscratch studies (mechanical durability) of various carbon coatings. The coefficient of friction profiles as a function of increasing normal load and AFM surface height maps of regions over scratches at the respective critical loads (indicated by the arrows in the friction profiles and AFM images) made on the various coatings of different thicknesses and single-crystal silicon substrate using a conical tip are compared in Figs. 26.16 and 26.17. *Bhushan* and *Koinkar* [26.102], *Koinkar* and

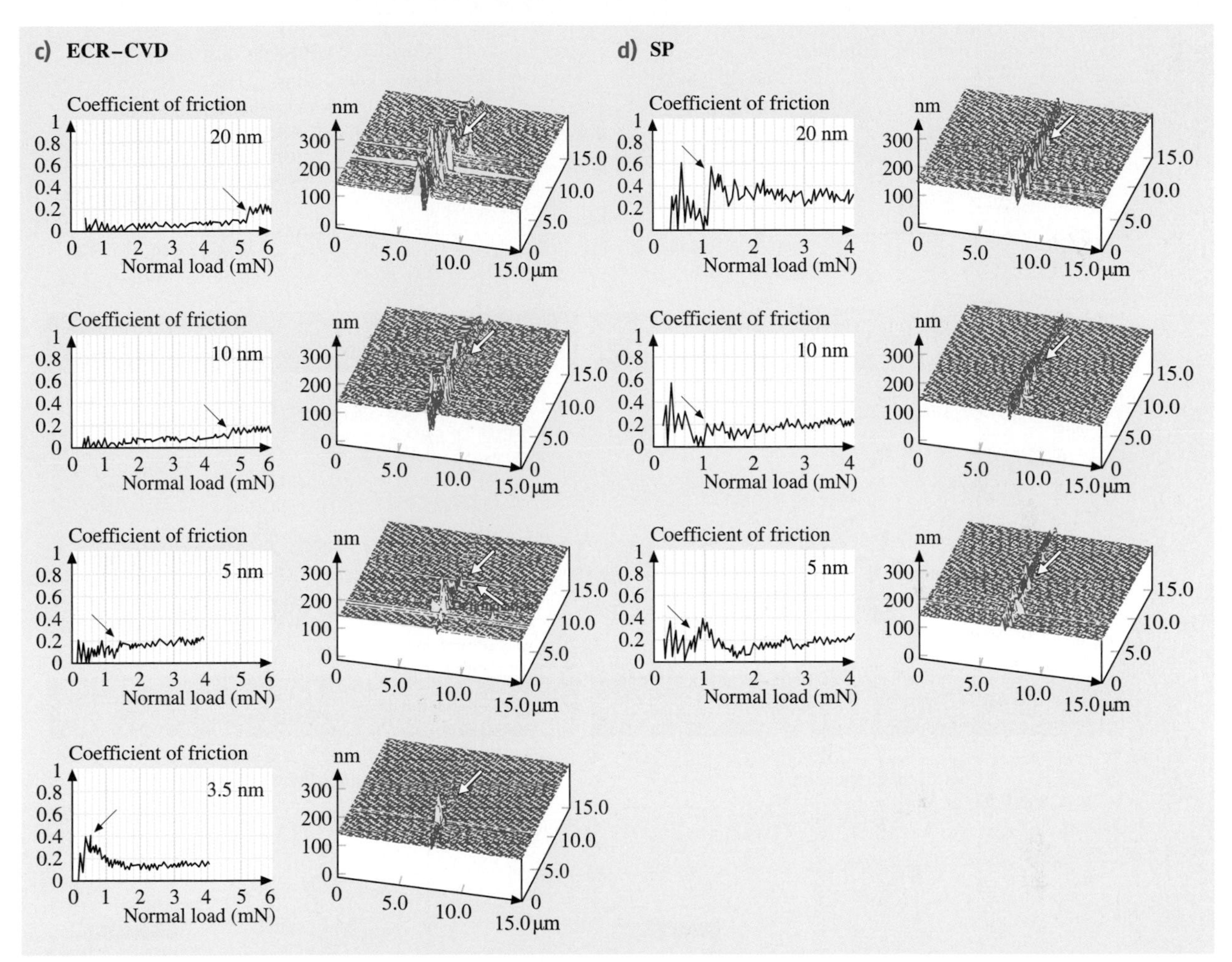

Fig. 26.16a–d Coefficient of friction profiles as a function of normal load and corresponding AFM surface height maps of regions over scratches at the respective critical loads (indicated by the arrows in the friction profiles and AFM images) for coatings of different thicknesses deposited by various deposition techniques: **(a)** FCA, **(b)** IB, **(c)** ECR-CVD, **(d)** SP

Bhushan [26.103], *Bhushan* [26.23], and *Sundararajan* and *Bhushan* [26.50, 99] have used an AFM to perform microscratch studies. Data for various coatings of different thicknesses and silicon substrate using a Berkovich tip are compared in Figs. 26.18 and 26.19. Critical loads for various coatings tested using a nanoindenter and AFM are summarized in Fig. 26.20. The selected data for 20-nm-thick coatings obtained using nanoindenter are also presented in Fig. 26.10 and Table 26.5.

It can be seen that a well-defined critical load exists for each coating. The AFM images clearly show that below the critical loads the coatings were plowed by the scratch tip, associated with the plastic flow of materials.

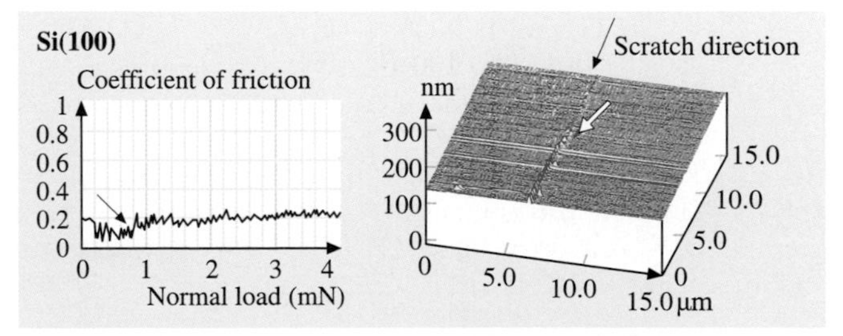

Fig. 26.17 Coefficient of friction profiles as a function of normal load and corresponding AFM surface height maps of regions over scratches at the respective critical loads (indicated by the arrows in the friction profiles and AFM images) for Si(100)

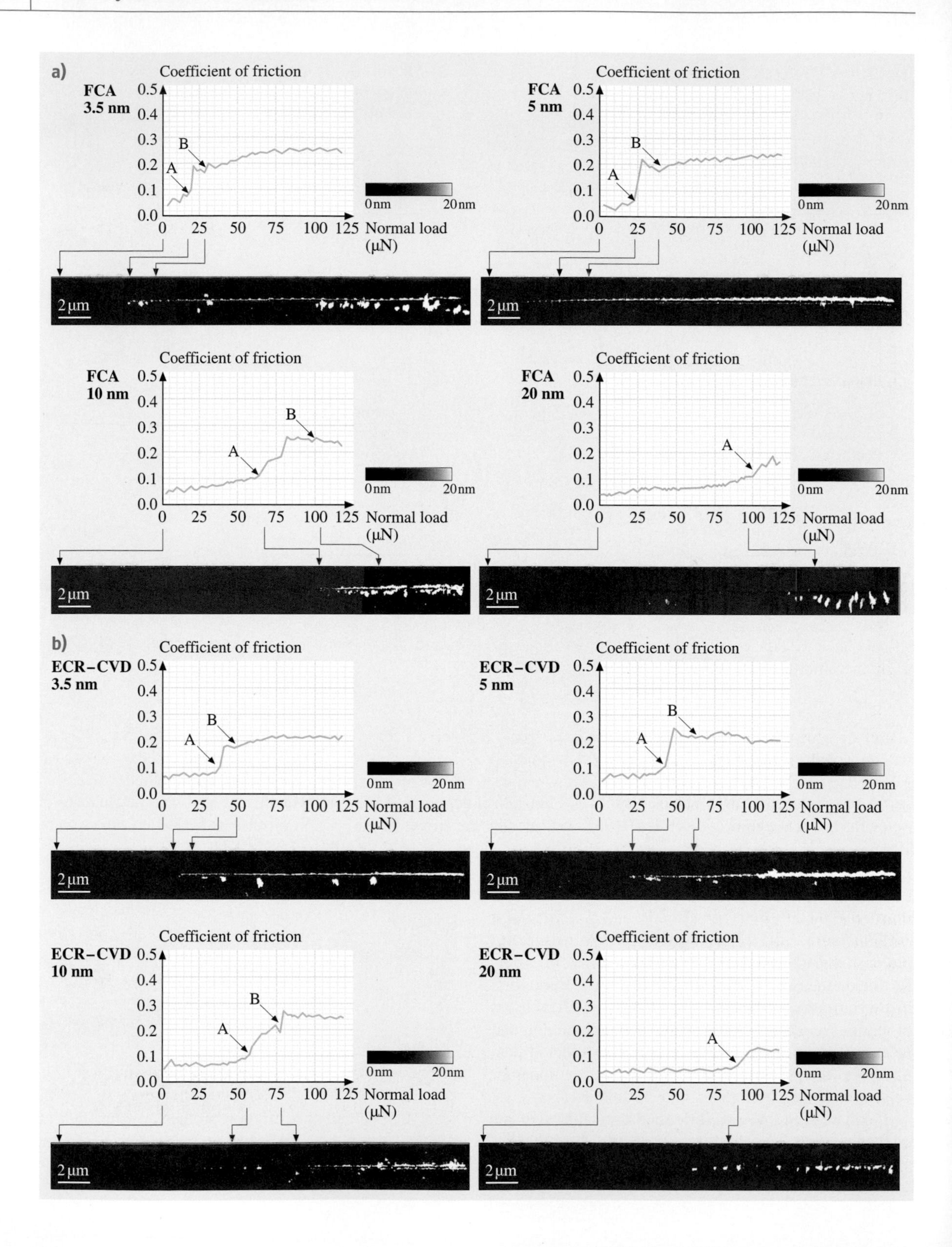

a)
Coefficient of friction
FCA 3.5 nm
FCA 5 nm
FCA 10 nm
FCA 20 nm
0.5
0.4
0.3
0.2
0.1
0.0
0
25
50
75
100
125
Normal load (μN)
A
B
0nm
20nm
2μm
b)
ECR-CVD 3.5 nm
ECR-CVD 5 nm
ECR-CVD 10 nm
ECR-CVD 20 nm

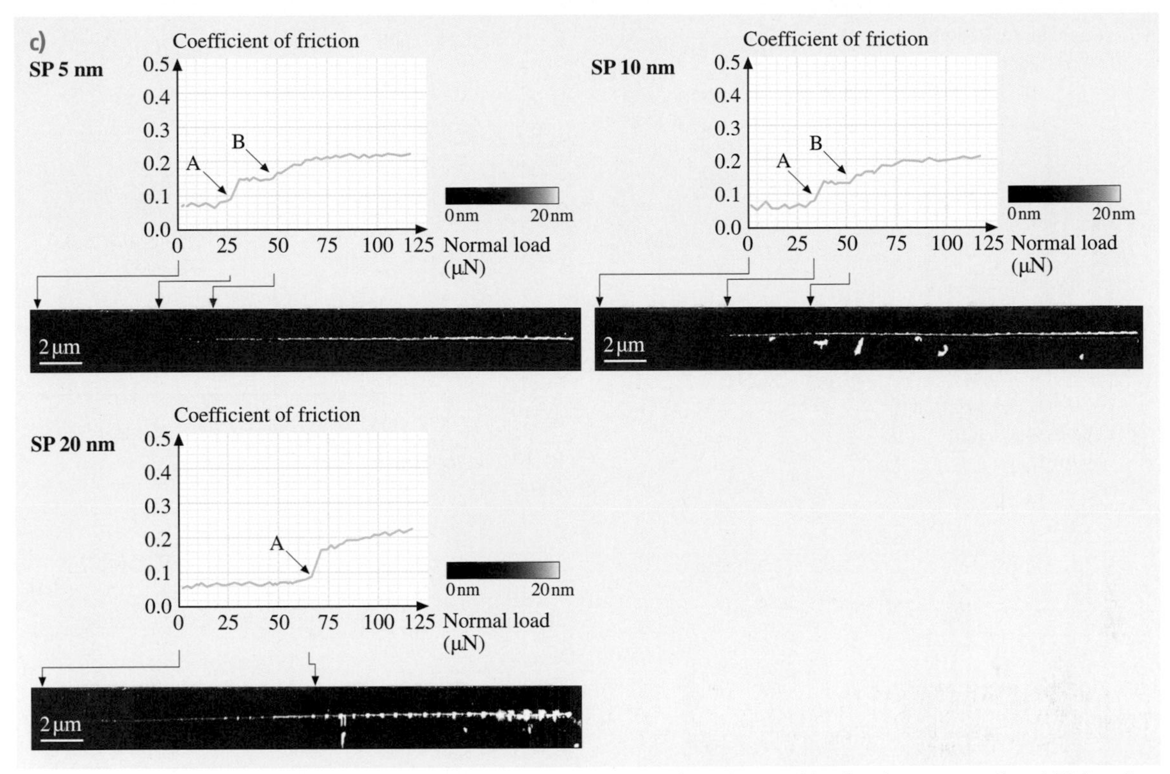

Fig. 26.18a–c Coefficient of friction profiles during scratch as a function of normal load and corresponding AFM surface height maps for (a) FCA, (b) ECR-CVD, and (c) SP coatings [26.99]

At and after the critical loads, debris (chips) or buckling was observed on the sides of scratches. Delamination or buckling can be observed around or after the critical loads, which suggests that the damage starts from delamination and buckling. For the 3.5- and 5-nm-thick FCA coatings, before the critical loads small debris is observed on the sides of scratches. This suggests that the thinner FCA coatings are not so durable. It is obvious that for a given deposition method, the critical loads increase with increasing coating thickness. This indicates that the critical load is determined not only by the adhesion strength to the substrate, but also by the coating thickness. We note that the debris generated on the thicker coatings is larger than that generated on the thinner coatings. For a thicker coating, it is more difficult to be broken; the broken coating chips (debris) for a thicker coating are larger than those for the thinner coatings. The difference in the residual stresses of the coatings of different thicknesses could also affect the size of debris. The AFM image shows that the silicon substrate was damaged by plowing, associated with the plastic flow of materials. At and after the critical load, small and uniform debris is observed and the amount of debris increases with increasing normal load.

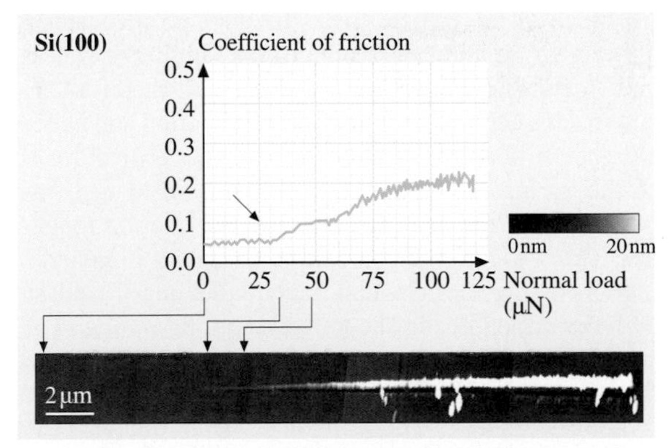

Fig. 26.19 Coefficient of friction profiles during scratch as a function of normal load and corresponding AFM surface height maps for Si(100) [26.99]

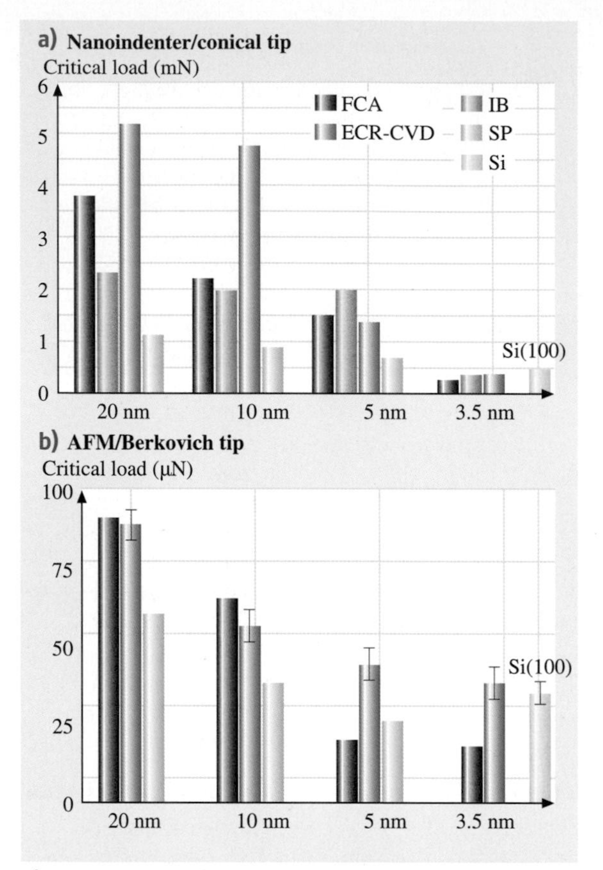

Fig. 26.20a,b Critical loads estimated from the coefficient of friction profiles from (a) nanoindenter and (b) AFM tests for various coatings of different thicknesses and Si(100) substrate

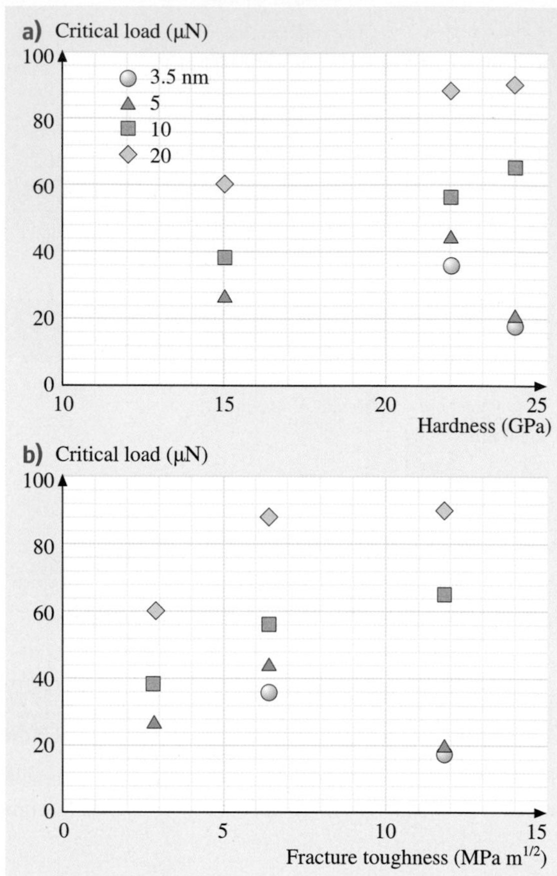

Fig. 26.21a,b Measured critical loads estimated from the coefficient of friction profiles from AFM tests as a function of (a) coating hardness and (b) fracture toughness. Coating hardness and fracture toughness values were obtained using a nanoindenter on 100-nm-thick coatings (Table 26.5)

Since at the critical load, the damage mechanism appears to be the onset of plowing, higher hardness and fracture toughness of a coating will therefore result in a higher load required for deformation and hence a higher critical load. Figure 26.21 shows critical loads of the various coatings, obtained with AFM tests, as a function of the coating hardness and fracture toughness (from Table 26.5). It can be seen that, in general, higher coating hardness and fracture toughness result in a higher critical load. The only exceptions are the FCA coatings at 5- and 3.5-nm coating thickness, which show the lowest critical loads despite their high hardness and fracture toughness. The brittleness of the thinner FCA coatings may be one reason for their low critical loads. The mechanical properties of coatings that are less than 10 nm thick are unknown. The FCA process may result in coatings with low hardness at such low thickness due to differences in coating stoichiometry and structure compared to the coatings of higher thickness. Also, at these thicknesses stresses at the coating-substrate interface may affect adhesion and load-carrying capacity of the coatings.

Based on the experimental results, a schematic of scratch damage mechanisms of the DLC coatings used in this study is shown in Fig. 26.22. Below the critical load, if a coating has a good combination of strength and fracture toughness, plowing associated with the plastic flow of materials is responsible for the damage of coating (Fig. 26.22a). Whereas if a coating has a lower fracture toughness, cracking could occur during plowing, asso-

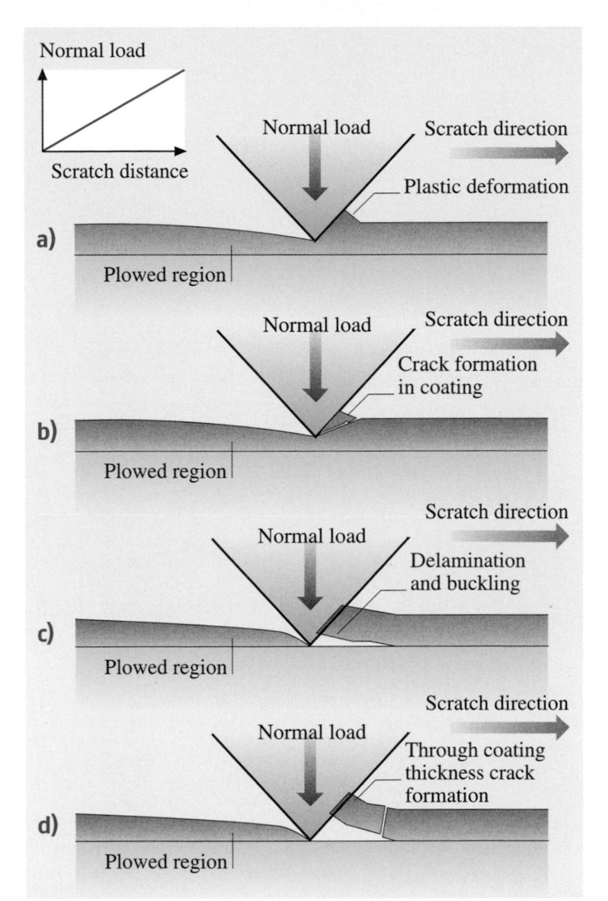

Fig. 26.22a–d Schematic of scratch damage mechanisms of the DLC coatings: (**a**) plowing associated with the plastic flow of materials, (**b**) plowing associated with the formation of small debris, (**c**) delamination and buckling at the critical load, and (**d**) breakdown via through coating thickness cracking at and after the critical load [26.48]

ciated with the formation of small debris (Fig. 26.22b). When normal load increases up to the critical load, delamination or buckling will occur at the coating/substrate interface (Fig. 26.22c). A further increase in normal load will result in the breakdown of coating via through coating thickness cracking, as shown in Fig. 26.22d. Therefore, adhesion strength plays a crucial role in the determination of critical load. If a coating has stronger adhesive strength with the substrate, the coating is more difficult to delaminate, which will result in a higher critical load. The interfacial and residual stresses of a coating could greatly affect the delamination and buckling [26.1]. The coating with higher interfacial and residual stresses is more easily delaminated and buckled, which will result in a low critical load. It has been reported earlier that the FCA coatings have higher residual stresses than the other coatings [26.47]. Interfacial stresses play a more important role when a coating gets thinner. A large mismatch in elastic modulus between the FCA coatings and the silicon substrate may cause large interfacial stresses. This may be why the thinner FCA coatings show relatively low critical loads compared with the thicker FCA coatings, even though the FCA coatings have a higher hardness and elastic modulus. The brittleness of the thinner FCA coatings may be another reason for the lower critical loads. The strength and fracture toughness of a coating also affect the critical load. Higher strength and fracture toughness will make the coating more difficult to be broken after delamination and buckling. The high scratch resistance/adhesion of FCA coatings is attributed to an atomic intermixing at the coating-substrate interface because of high kinetic energy (2 keV) plasma formed during the cathodic arc deposition process [26.57]. The atomic intermixing at the interface provides a graded compositional transition between the coating and the substrate materials. In all other coatings used in this study, the kinetic energy of the plasma was insufficient for atomic intermixing.

Gupta and *Bhushan* [26.12, 47] and *Li* and *Bhushan* [26.48,49] measured scratch resistance of DLC coatings deposited on Al_2O_3-TiC, Ni-Zn ferrite, and single-crystal silicon substrates. For good adhesion of DLC coating to other substrates, in most cases, an interlayer of silicon is required except for cathodic arc deposited coatings. The best adhesion with cathodic arc carbon coating is obtained on electrically conducting substrates such as Al_2O_3-TiC and silicon, as compared to Ni-Zn ferrite.

Microwear

Microwear studies can be conducted using an AFM [26.23]. For microwear studies, a three-sided pyramidal single-crystal natural diamond tip with an apex angle of about 80° and a tip radius of about 100 nm is used at relatively high loads of 1–150 μN. The diamond tip is mounted on a stainless steel cantilever beam with a normal stiffness of about 30 N/m. The sample is generally scanned in a direction orthogonal to the long axis of the cantilever beam (typically at a rate of 0.5 Hz). The tip is mounted on the beam such that one of its edges is orthogonal to the beam axis. In wear studies, typically an area of 2 μm × 2 μm is scanned for a selected number of cycles.

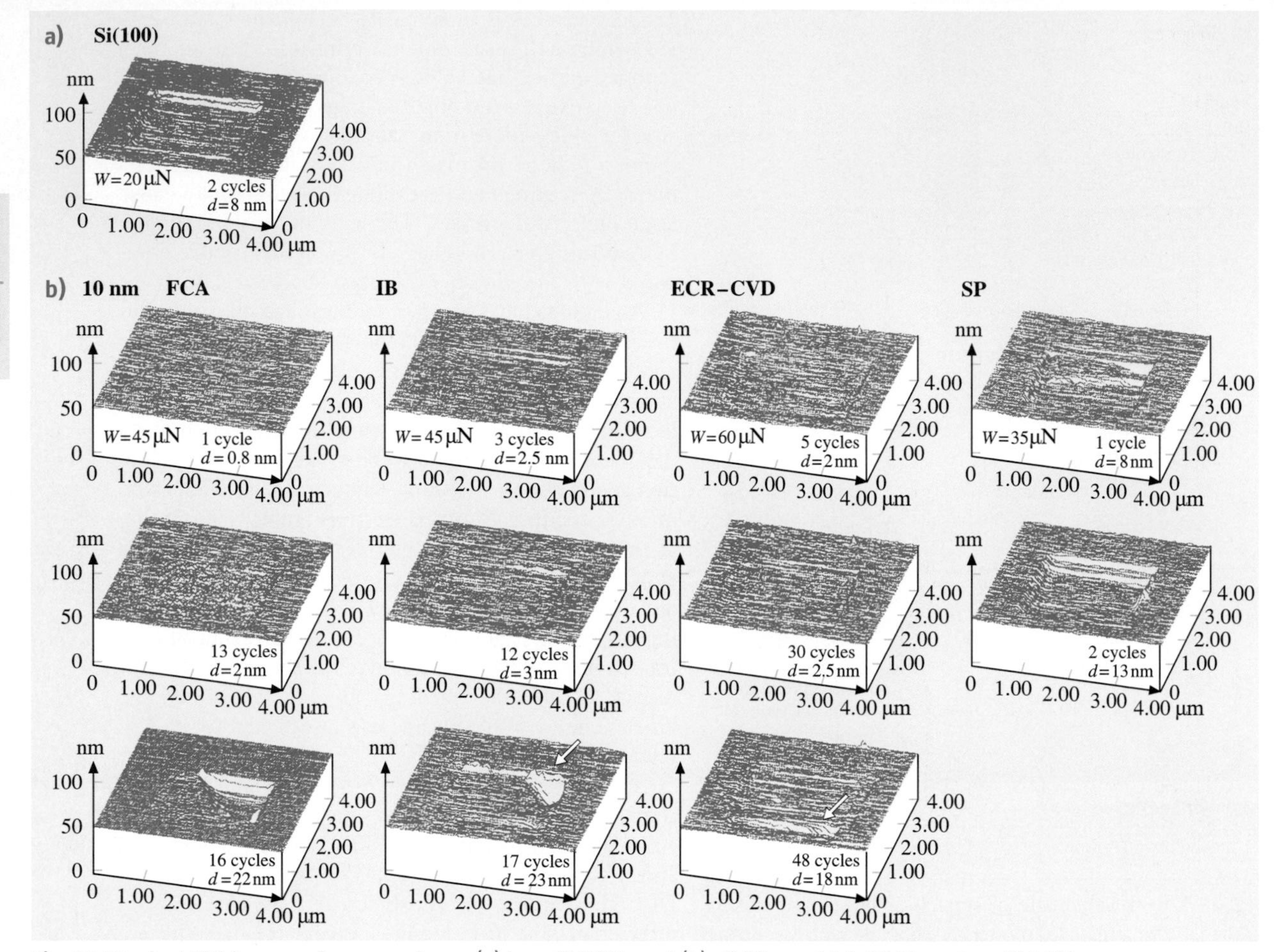

Fig. 26.23a,b AFM images of wear marks on (a) bare Si(100), and (b) all 10-nm-thick DLC coatings [26.50]

Microwear studies of various types of DLC coatings have been conducted [26.50, 102, 103]. Fig. 26.23a shows a wear mark on uncoated Si(100). Wear occurs uniformly and material is removed layer by layer via plowing from the first cycle, resulting in the constant friction force seen during the wear (Fig. 26.24a). Figure 26.23b shows AFM images of the wear marks on all 10 nm coatings. It is seen that coatings wear nonuniformly. Coating failure is sudden and accompanied by a sudden rise in the friction force (Fig. 26.24b). Figure 26.24 shows the wear depth of Si(100) substrate and various DLC coatings at two different loads. FCA and ECR-CVD, 20-nm-thick coatings show excellent wear resistance up to 80 μN, the load that is required for the IB 20 nm coating to fail. In these tests, "failure" of a coating results when the wear depth exceeds the quoted coating thickness. The SP 20 nm coating fails at the much lower load of 35 μN. At 60 μN, the coating hardly provides any protection. Moving on to the 10 nm coatings, ECR-CVD coating requires about 45 cycles at 60 μN to fail as compared to IB and FCA, which fail at 45 μN. The FCA coating exhibits slight roughening in the wear track after the first few cycles, which leads to an increase in the friction force. The SP coating continues to exhibit poor resistance, failing at 20 μN. For the 5 nm coatings, the load required to fail the coatings continues to decrease. But IB and ECR-CVD still provide adequate protection as compared to bare Si(100) in that order, failing at 35 μN compared to FCA at 25 μN and SP at 20 μN. Almost all the 20, 10, and 5 nm coatings provide better wear resistance than bare silicon. At 3.5 nm, FCA coating

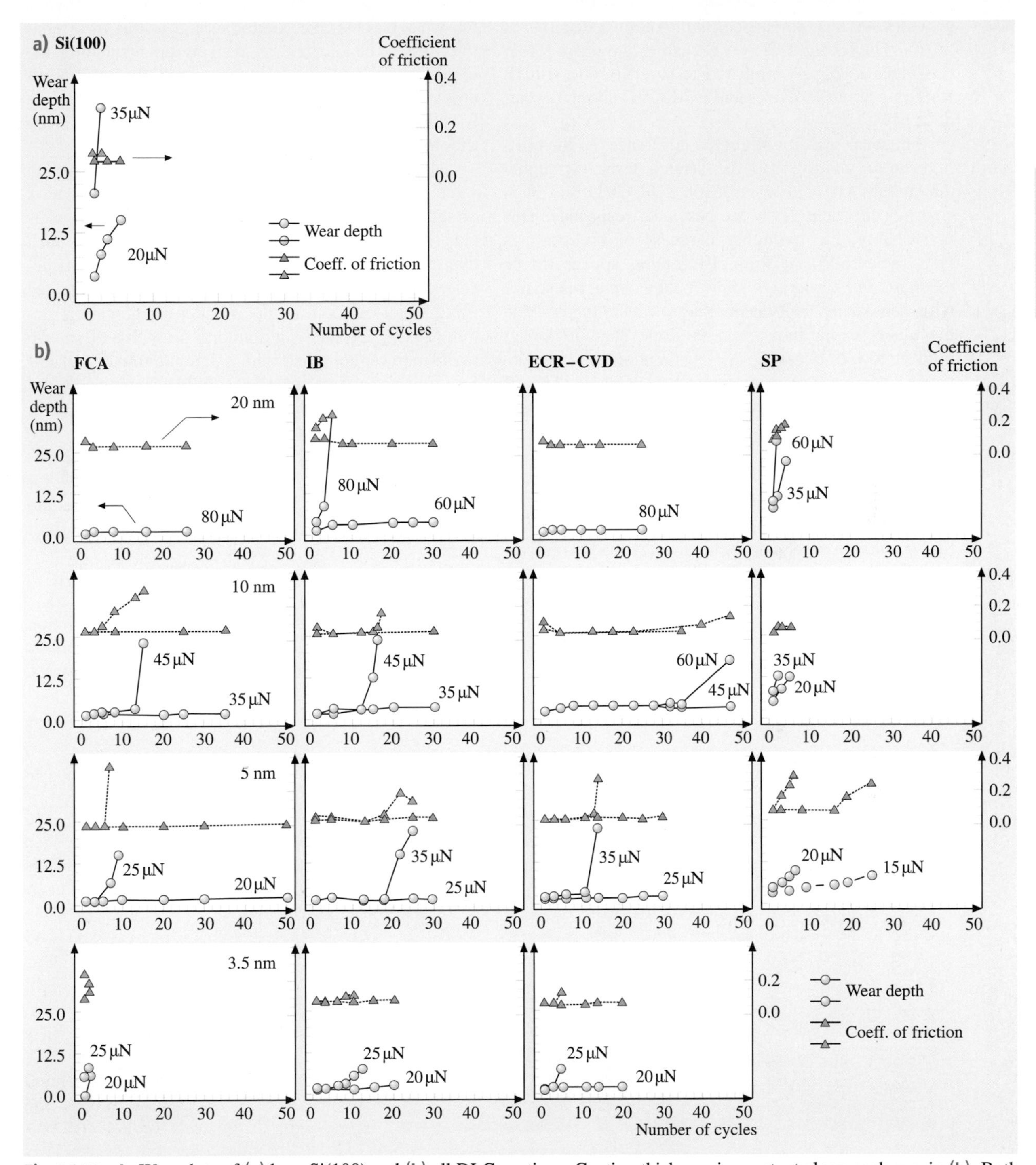

Fig. 26.24a,b Wear data of (a) bare Si(100) and (b) all DLC coatings. Coating thickness is constant along each row in (b). Both wear depth and coefficient of friction during wear for a given cycle are plotted [26.50]

provides no wear resistance, failing almost instantly at 20 μN. The IB and ECR-CVD coating show good wear resistance at 20 μN compared to bare Si(100). But IB lasts only about 10 cycles and ECR-CVD about 3 cycles at 25 μN.

The wear tests highlight the differences in the coatings more vividly than the scratch tests. At higher thicknesses (10 and 20 nm), the ECR-CVD and FCA coatings appear to show the best wear resistance. This is probably due to higher hardness of the coatings (see Table 26.5). At 5 nm, IB coating appears to be the best. FCA coatings show poorer wear resistance with decreasing coating thickness. This suggests that the trends in hardness seen in Table 26.5 no longer hold at low thicknesses. SP coatings showed consistently poor wear resistance at all thicknesses. The IB 3.5 nm coating does provide reasonable wear protection at low loads.

26.3.3 Macroscale Tribological Characterization

So far, data on mechanical characterization and microscratch and microwear studies using a nanoindenter and an AFM have been presented. Mechanical properties affect tribological performance of the coatings, and microwear studies simulate a single asperity contact, which helps in developing a fundamental understanding of the wear process. These studies are useful in screening various candidates, as well as understanding the relationships between deposition conditions and properties of various samples. As a next step, macroscale friction and wear tests need to be conducted to measure tribological performance of the coatings.

Macroscale accelerated friction and wear tests to screen a large number of candidates and functional tests on selected candidates have been conducted. An accelerated test is designed to accelerate the wear process such that it does not change the failure mechanism. The accelerated friction and wear tests are generally conducted using a ball-on-flat tribometer under reciprocating motion [26.70]. Typically, a diamond tip with a 20 μm tip radius or a sapphire ball with a 3 mm diameter and surface finish of about 2 nm RMS is slid against the coated substrates at selected loads. Coefficient of friction is monitored during the tests.

Functional tests are conducted using an actual machine under close-to-actual operating conditions for which coatings are developed. Generally, the tests are accelerated somewhat to fail the interface in a short time.

Accelerated Friction and Wear Tests

Li and *Bhushan* [26.48] conducted accelerated friction and wear tests on DLC coatings deposited by various deposition techniques using a ball-on-flat tribometer.

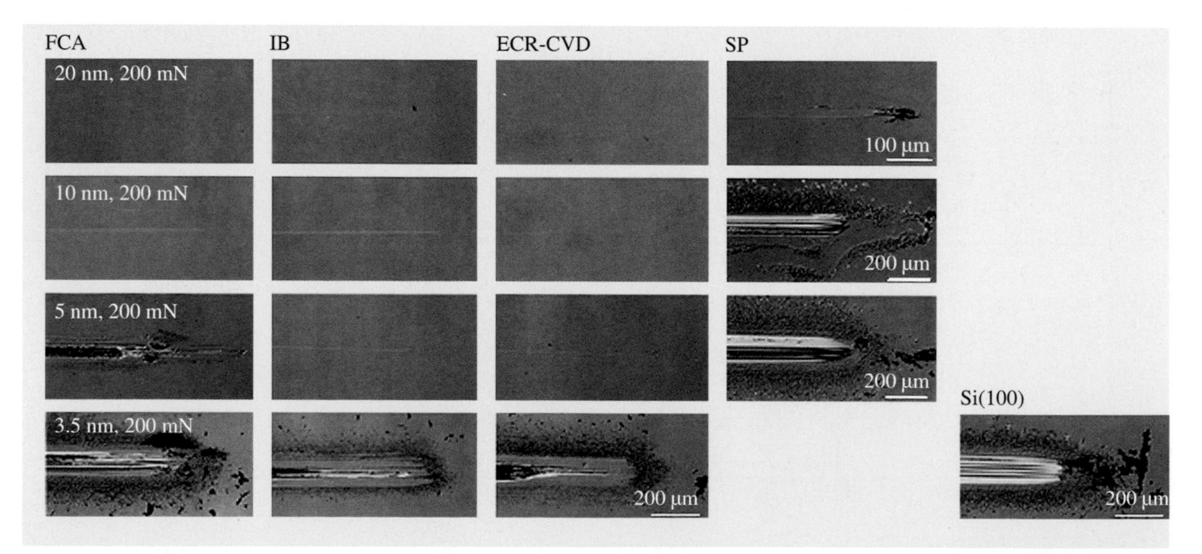

Fig. 26.25 Optical micrographs of wear tracks and debris formed on the various coatings of different thicknesses and silicon substrate when slid against a sapphire ball after sliding distance of 5 m. The end of the wear track is on the right-hand side of the image

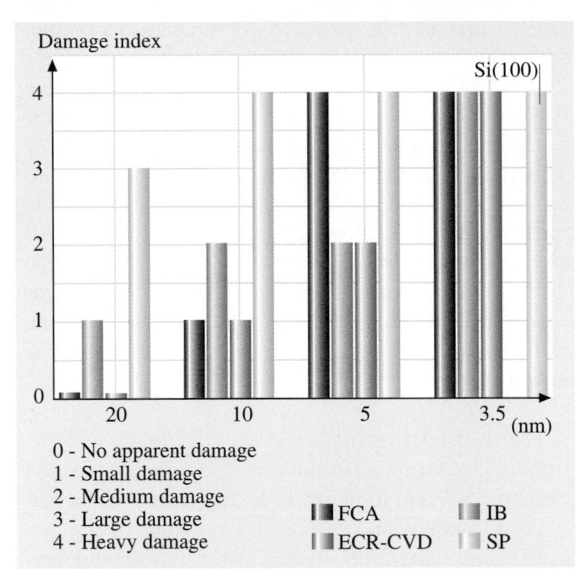

Fig. 26.26 Wear damage index bar chart of the various coatings of different thicknesses and Si(100) substrate based on optical examination of the wear tracks and debris

Average values of coefficient of friction are presented in Table 26.5. The optical micrographs of wear tracks and debris formed on all samples when slid against a sapphire ball after a sliding distance of 5 m are presented in Fig. 26.25. The normal load used for the 20- and 10-nm-thick coatings is 200 mN, and the normal load used for the 5- and 3.5-nm-thick coatings and silicon substrate is 150 mN.

Among the 20-nm-thick coatings, the SP coating exhibits a higher coefficient of friction of about 0.3 than the other coatings, which have comparable values of coefficient of friction of about 0.2. The optical micrographs show that the SP coating has a larger wear track and more debris than the IB coatings. No wear track and debris were found on the 20-nm-thick FCA and ECR-CVD coatings. The optical micrographs of 10-nm-thick coatings show that the SP coating was severely damaged, showing a large wear track with scratches and lots of debris. The FCA and ECR-CVD coatings show smaller wear tracks and less debris than the IB coatings.

For the 5-nm-thick coatings, the wear tracks and debris of the IB and ECR-CVD coatings are comparable. The bad wear resistance of the 5-nm-thick FCA coating is in good agreement with the low scratch critical load, which may be due to the higher interfacial and residual stresses as well as brittleness of the coating.

At 3.5 nm, all coatings exhibit wear. The FCA coating provides no wear resistance, failing instantly like the silicon substrate. Large block-like debris is observed on the sides of the wear track of the FCA coating. This indicates that large region delamination and buckling occurred during sliding, resulting in large block-like debris. This block-like debris, in turn, scratched the coating, making the coating damage even more severe. The IB and ECR-CVD coatings are able to provide some protection against wear at 3.5 nm.

In order to better evaluate the wear resistance of various coatings, based on the optical examination of the wear tracks and debris after tests, the wear damage index bar chart of the various coatings of different thicknesses and an uncoated silicon substrate is presented in Fig. 26.26. Among the 20- and 10-nm-thick coatings, the SP coatings show the worst damage, followed by FCA/ECR-CVD. At 5 nm, the FCA and SP coatings show the worst damage, followed by IB and ECR-CVD coatings. All 3.5-nm-thick coatings show the same heavy damage as the uncoated silicon substrate.

The wear damage mechanisms of thick and thin DLC coatings studied are believed to be as illustrated in Fig. 26.27. At the early stages of sliding, deformation zone and Hertzian and wear fatigue cracks formed beneath the surface extend within the coating on subsequent sliding [26.1]. Formation of fatigue cracks depend on the hardness and subsequent cycles. These are controlled by the sp^3-to-sp^2 ratio. For thicker coating, the cracks generally do not penetrate the coating. For a thinner coating, the cracks easily propagate down to the interface with the aid of the interfacial stresses and get diverted along the interface just enough to cause local delamination of the coating. When this happens, the coating experiences excessive plowing. At this point, the coating fails catastrophically, resulting in a sudden rise

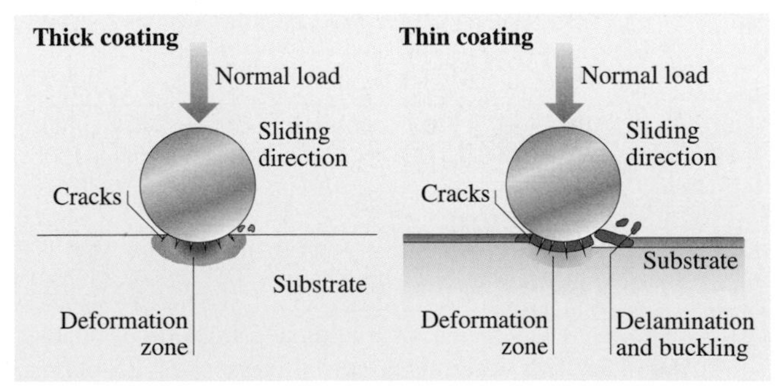

Fig. 26.27 Schematic of wear damage mechanisms of thick and thin DLC coatings [26.48]

in the coefficient of friction. All 3.5-nm-thick coatings failed much quicker than the thicker coatings. It appears that these thin coatings have very low load-carrying capacity and, therefore, the substrate undergoes deformation almost immediately. This generates stresses at the interface that weaken the coating adhesion and lead to delamination of the coating. Another reason may be that the thickness is insufficient to produce a coating comprised of the DLC structure. Instead, the bulk may be made up of a matrix characteristic of the interface region where atomic mixing occurs with the substrate and/or any interlayer used. This would also result in poor wear resistance and silicon-like behavior of the coating, especially in the case of FCA coatings, which show the worst performance at 3.5 nm. In contrast to the other coatings, all SP coatings show the worst wear performance at any thickness (Fig. 26.25). This may be due to their poor mechanical properties, such as lower hardness and scratch resistance, compared to the other coatings.

Comparison of Figs. 26.20 and 26.26 shows a very good correlation between the wear damage and scratch critical loads. Less wear damage corresponds to higher scratch critical load. Based on the data, thicker coatings do show better scratch and wear resistance than thinner coatings. This is probably due to better load-carrying capacity of the thick coatings compared to the thinner ones. For a given coating thickness, higher hardness and fracture toughness and better adhesion strength are believed to be responsible for the superior wear performance.

Effect of Environment

Friction and wear performance of amorphous carbon coatings are found to be strongly dependent on the water vapor content and partial gas pressure in the test environment. The friction data for an amorphous carbon film on the silicon substrate sliding against steel are presented as a function of the partial pressure of water vapor in Fig. 26.28 [26.1, 13, 69, 105, 106]. Friction increases dramatically above a relative humidity of about 40%. At high relative humidity, condensed water vapor forms meniscus bridges at the contacting asperities, and the meniscii result in an intrinsic attractive force that is responsible for an increase in friction. For completeness, the data for the coefficient of friction of bulk graphitic carbon are also presented in Fig. 26.28. Note that friction decreases with an increase in the relative humidity [26.107]. Graphitic carbon has a layered lattice crystal structure. Graphite absorbs po-

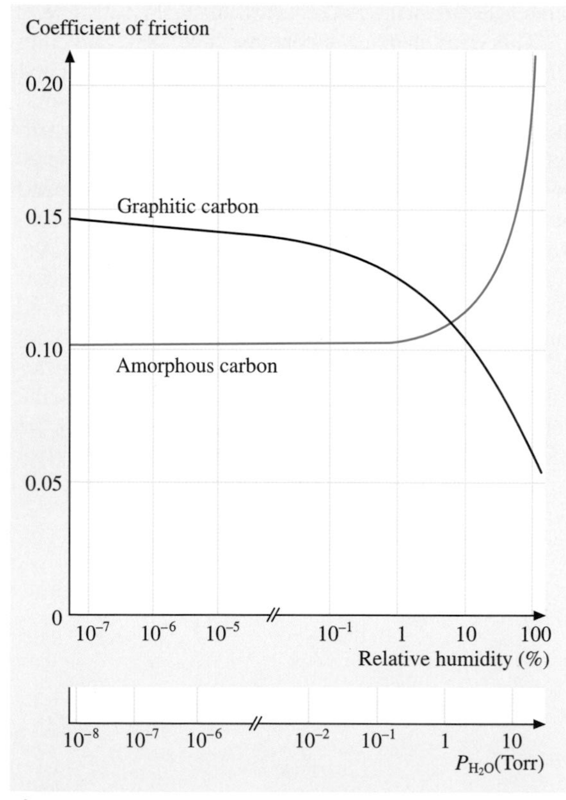

Fig. 26.28 Coefficient of friction as a function of relative humidity and water vapor partial pressure of RF-plasma deposited amorphous carbon coating and bulk graphitic carbon coating sliding against a steel ball

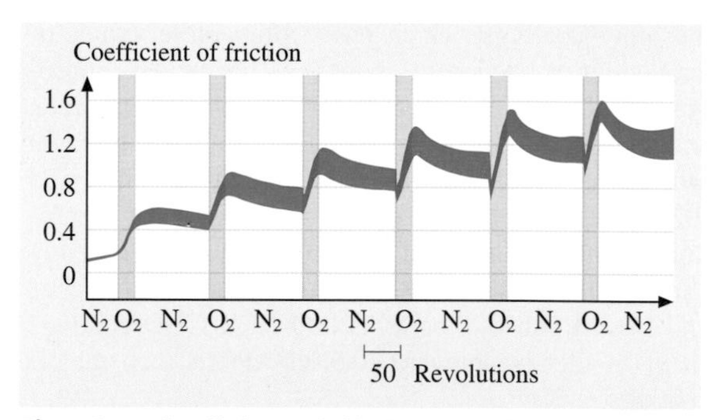

Fig. 26.29 Coefficient of friction as a function of sliding distance for a ceramic slider against a magnetic disk coated with 20-nm-thick DC magnetron sputtered DLC coating, measured at a speed of 0.06 m/s and 10 g load. The environment is alternated between oxygen and nitrogen gases [26.108]

lar gases (H_2O, O_2, CO_2, NH_3, etc.) at the edges of the crystallites, which weakens the interlayer bonding forces facilitating interlayer slip and results in lower friction [26.1].

To better study the effect of environment for carbon coated magnetic disks, a number of tests have been conducted in controlled environments. *Marchon* et al. [26.108] conducted tests in alternating environments of oxygen and nitrogen gases, Fig. 26.29. The coefficient of friction increases as soon as oxygen is added to the test environment, whereas in a nitrogen environment the coefficient of friction reduces slightly. Tribochemical oxidation of the DLC coating in the oxidizing environment is responsible for an increase in the coefficient of friction, implying wear. *Dugger* et al. [26.109], *Strom* et al. [26.110], *Bhushan* and *Ruan* [26.111], and *Bhushan* et al. [26.71] conducted tests with DLC coated magnetic disks (with about 2-nm-thick perfluoropolyether lubricant film) in contact with Al_2O_3-TiC sliders, in different gaseous environments including a high vacuum of 2×10^{-7} Torr, Fig. 26.30. The wear lives are shortest in high vacuum and longest in most atmospheres of nitrogen and argon with the following order (from best to worst): argon or nitrogen, $Ar+H_2O$, ambient, $Ar+O_2$, and $Ar+H_2O$, vacuum. From this sequence of wear performance, we can see that having oxygen and water in an operating environment worsens wear performance of the coatings, but having nothing in it (vacuum) is the worst of all. Indeed, failure mechanisms differ in various environments. In high vacuum, intimate contact between disk and slider surfaces results in significant wear. In ambient air, $Ar+O_2$, and $Ar+H_2O$, tribochemical oxidation of the carbon overcoat is responsible for interface failure. For experiments performed in pure argon and nitrogen, mechanical shearing of the asperities cause the formation of debris, which is responsible for the formation of scratch marks on the carbon surface as could be observed with an optical microscope [26.71].

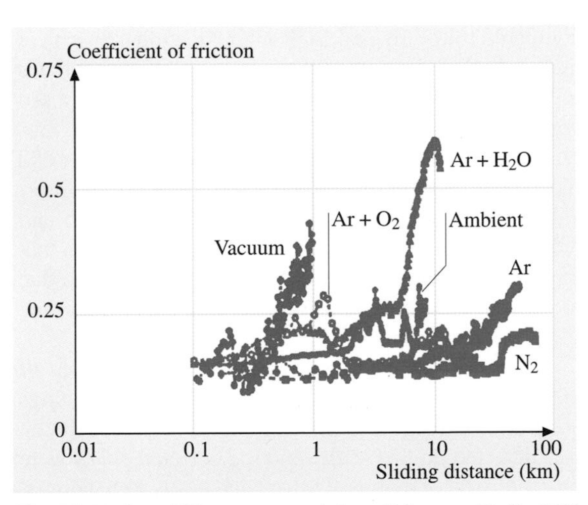

Fig. 26.30 Durability measured by sliding a Al_2O_3-TiC magnetic slider against a magnetic disk coated with 20-nm-thick DC sputtered amorphous carbon coating and 2-nm-thick perfluoropolyether film, measured at a speed of 0.75 m/s and 10 g load. Vacuum refers to 2×10^{-7} Torr [26.71]

Functional Tests

Magnetic thin-film heads made with Al_2O_3-TiC substrate are used in magnetic storage applications [26.13]. A multilayered thin-film pole-tip structure present on the head surface wears more rapidly than the much harder Al_2O_3-TiC substrate. Pole-tip recession (PTR) is a serious concern in magnetic storage [26.15–19, 112]. Two of the diamond-like carbon coatings superior in mechanical properties – ion beam and cathodic arc carbon – were deposited on the air bearing surfaces of Al_2O_3-TiC head sliders [26.15]. The functional tests were conducted by running a metal-particle (MP) tape in a computer tape drive. Average PTR as a function of sliding distance data are presented in Fig. 26.31. We note that PTR increases for the uncoated head, whereas for the coated heads there

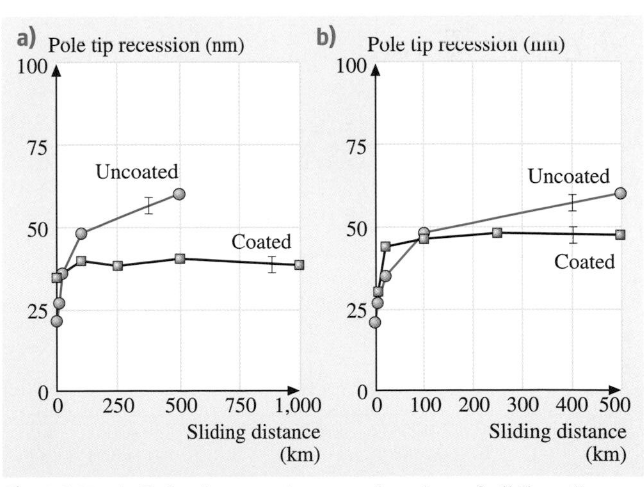

Fig. 26.31a,b Pole tip recession as a function of sliding distance measured with an AFM for (**a**) uncoated and 20-nm-thick ion beam carbon coated, and (**b**) uncoated and 20-nm-thick cathodic arc carbon coated Al_2O_3-TiC heads run against MP tapes [26.15]

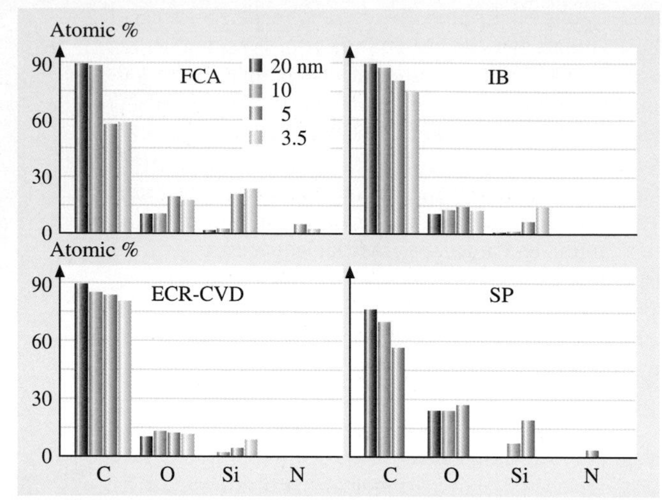

Fig. 26.32 Quantified XPS data for various DLC coatings on Si(100) substrate [26.50]. Atomic concentrations are shown

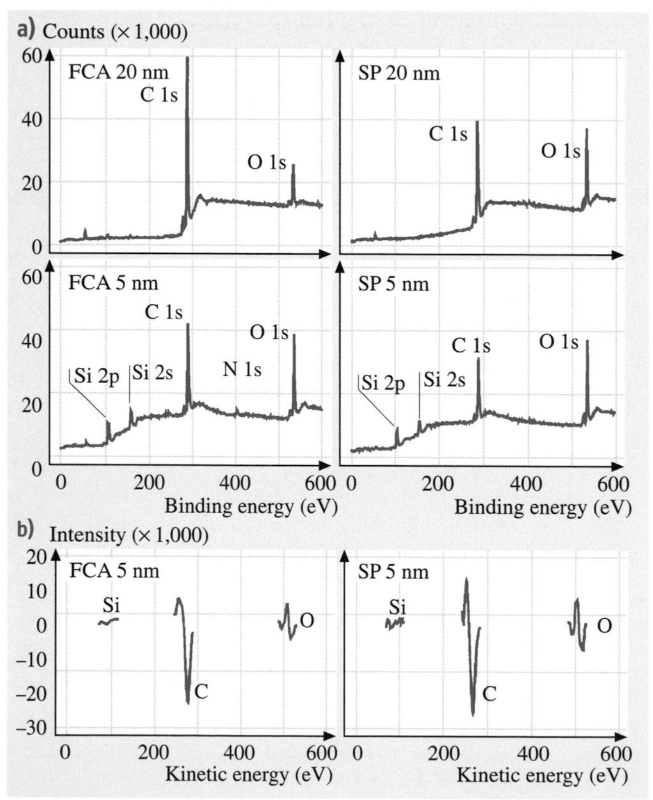

Fig. 26.33 (a) XPS spectra for FCA and SP coatings of 5 nm and 20 nm coating thicknesses on Si(100) substrate, and (b) AES spectra for FCA and SP coatings at 5 nm thickness on Si(100) substrate [26.50]

is a slight increase in PTR in early sliding followed by little change. Thus, coatings provide protection.

Micromechanical and accelerated and functional tribological data presented here clearly suggest that there is a good correlation between the scratch resistance and wear resistance measured using accelerated tests and functional tests. Thus, scratch tests can be successfully used to screen coatings for wear applications.

26.3.4 Coating Continuity Analysis

Ultrathin coatings less than 10 nm may not uniformly coat the sample surface. In other words, coating may be discontinuous and deposited in the form of islands on the microscale. A possible reason for poor wear protection and the nonuniform failure of the coatings may be due to poor coverage of the thin coatings on the substrate. Coating continuity can be studied by using surface analytical techniques such as Auger and/or XPS analyses. Any discontinuity with coating thickness less than the sampling depth of the instrument will detect locally the substrate species [26.49, 50, 102].

The results of XPS analysis on various coatings deposited on Si(100) substrates, over a 1.3 mm^2 region (single point measurement with spot diameter of 1,300 μm) are shown in Fig. 26.32. The sampling depth is about 2–3 nm. The poor SP coatings show much less carbon content ($< 75\%$ atomic concentration) as does the poor 5 nm and 3.5 nm FCA coatings ($< 60\%$) compared to the IB and ECR-CVD coatings. Silicon is detected in all 5 nm coatings. From the data it is hard to say if the Si is from the substrate or from exposed regions due to coating discontinuity. Based on the sampling depth any Si detected in 3.5 nm coatings would likely be from the substrate. The other interesting observation is that all poor coatings (all SP and FCA 5 and 3.5 nm) show almost twice the oxygen content than that of the other coatings. Any oxygen present may be due to leaks in the deposition chamber and is present in the form of silicon oxides.

AES measurements averaged over a scan area of 900 $μm^2$ were conducted on FCA and SP 5 nm coatings at six different regions on each sample. Very little silicon was detected on this scale, and the detected peaks were characteristic of oxides. The oxygen levels were comparable to that seen for the good coatings by XPS. These contrast the XPS measurements at a larger scale, suggesting that the coating possesses discontinuities only at isolated areas and that the 5 nm coatings are generally continuous on the microscale. Figure 26.33 shows representative XPS and AES spectra of selected samples.

References

26.1 B. Bhushan, B.K. Gupta: *Handbook of Tribology: Materials, Coatings, and Surface Treatments*, reprint edn. (Krieger, Malabar 1997)

26.2 B. Bhushan: *Principles and Applications of Tribology* (Wiley, New York 1999)

26.3 B. Bhushan: *Introduction to Tribology* (Wiley, New York 2002)

26.4 B. Bhushan, B.K. Gupta, G.W. VanCleef, C. Capp, J.V. Coe: Fullerene (C_{60}) films for solid lubrication, Tribol. Trans. **36** (1993) 573–580

26.5 B.K. Gupta, B. Bhushan, C. Capp, J.V. Coe: Material characterization and effect of purity and ion implantation on the friction and wear of sublimed fullerene films, J. Mater. Res. **9** (1994) 2823–2838

26.6 B.K. Gupta, B. Bhushan: Fullerene particles as an additive to liquid lubricants and greases for low friction and wear, Lubr. Eng. **50** (1994) 524–528

26.7 B. Bhushan, V.V. Subramaniam, A. Malshe, B.K. Gupta, J. Ruan: Tribological properties of polished diamond films, J. Appl. Phys. **74** (1993) 4174–4180

26.8 B. Bhushan, B.K. Gupta, V.V. Subramaniam: Polishing of diamond films, Diam. Films Technol. **4** (1994) 71–97

26.9 P. Sander, U. Kaiser, M. Altebockwinkel, L. Wiedmann, A. Benninghoven, R.E. Sah, P. Koidl: Depth profile analysis of hydrogenated carbon layers on silicon by X-ray photoelectron spectroscopy, auger electron spectroscopy, electron energy-loss spectroscopy, and secondary ion mass spectrometry, J. Vac. Sci. Technol. A **5** (1987) 1470–1473

26.10 A. Matthews, S.S. Eskildsen: Engineering applications for diamond-like carbon, Diam. Relat. Mater. **3** (1994) 902–911

26.11 A.H. Lettington: Applications of diamond-like carbon thin films, Carbon **36** (1998) 555–560

26.12 B.K. Gupta, B. Bhushan: Mechanical and tribological properties of hard carbon coatings for magnetic recording heads, Wear **190** (1995) 110–122

26.13 B. Bhushan: *Tribology and Mechanics of Magnetic Storage Devices*, 2nd edn. (Springer, New York 1996)

26.14 B. Bhushan: *Mechanics and Reliability of Flexible Magnetic Media*, 2nd edn. (Springer, New York 2000)

26.15 B. Bhushan, S.T. Patton, R. Sundaram, S. Dey: Pole tip recession studies of hard carbon-coated thin-film tape heads, J. Appl. Phys. **79** (1996) 5916–5918

26.16 J. Xu, B. Bhushan: Pole tip recession studies of thin-film rigid disk head sliders II: Effects of air bearing surface and pole tip region designs and carbon coating, Wear **219** (1998) 30–41

26.17 W.W. Scott, B. Bhushan: Corrosion and wear studies of uncoated and ultra-thin DLC coated magnetic tape-write heads and magnetic tapes, Wear **243** (2000) 31–42

26.18 W.W. Scott, B. Bhushan: Loose debris and head stain generation and pole tip recession in modern tape drives, J. Info. Storage Proc. Syst. **2** (2000) 221–254

26.19 W.W. Scott, B. Bhushan, A.V. Lakshmikumaran: Ultrathin diamond-like carbon coatings used for reduction of pole tip recession in magnetic tape heads, J. Appl Phys **87** (2000) 6182–6184

26.20 B. Bhushan: Macro- and microtribology of magnetic storage devices. In: *Modern Tribology Handbook*, ed. by B. Bhushan (CRC, Boca Raton 2001) pp. 1413–1513

26.21 B. Bhushan: Nanotribology and nanomechanics of MEMS devices, Proc. Ninth Annual Workshop on Micro Electro Mechanical Systems 1996 (IEEE, New York 91–98

26.22 B. Bhushan (Ed.): *Tribology Issues and Opportunities in MEMS* (Kluwer, Dordrecht 1998)

26.23 B. Bhushan: *Handbook of Micro/Nanotribology*, 2nd edn. (CRC, Boca Raton 1999)

26.24 B. Bhushan: Macro- and microtribology of MEMS materials. In: *Modern Tribology Handbook*, ed. by B. Bhushan (CRC, Boca Raton 2001) pp. 1515–1548

26.25 S. Aisenberg, R. Chabot: Ion beam deposition of thin films of diamond like carbon, J. Appl. Phys. **49** (1971) 2953–2958

26.26 E.G. Spencer, P.H. Schmidt, D.C. Joy, F.J. Sansalone: Ion beam deposited polycrystalline diamond-like films, Appl. Phys. Lett. **29** (1976) 118–120

26.27 A. Grill, B.S. Meyerson: Development and status of diamondlike carbon. In: *Synthetic Diamond: Emerging CVD Science and Technology*, ed. by K.E. Spear, J.P. Dismukes (Wiley, New York 1994) pp. 91–141

26.28 Y. Catherine: Preparation techniques for diamond-like carbon. In: *Diamond and Diamond-like Films and Coatings*, ed. by R.E. Clausing, L.L. Horton, J.C. Angus, P. Koidl (Plenum, New York 1991) pp. 193–227

26.29 J.J. Cuomo, D.L. Pappas, J. Bruley, J.P. Doyle, K.L. Seagner: Vapor deposition processes for amorphous carbon films with sp^3 fractions approaching diamond, J. Appl. Phys. **70** (1991) 1706–1711

26.30 J.C. Angus, C.C. Hayman: Low pressure metastable growth of diamond and diamondlike phase, Science **241** (1988) 913–921

26.31 J.C. Angus, F. Jensen: Dense diamondlike hydrocarbons as random covalent networks, J. Vac. Sci. Technol. A **6** (1988) 1778–1782

26.32 D.C. Green, D.R. McKenzie, P.B. Lukins: The microstructure of carbon thin films, Mater. Sci. Forum **52-53** (1989) 103–124

26.33 N.H. Cho, K.M. Krishnan, D.K. Veirs, M.D. Rubin, C.B. Hopper, B. Bhushan, D.B. Bogy: Chemical structure and physical properties of diamond-like

amorphous carbon films prepared by magnetron sputtering, J. Mater. Res. **5** (1990) 2543–2554
26.34 J. C. Angus: Diamond and diamondlike films, Thin Solid Films **216** (1992) 126–133
26.35 B. Bhushan, A. J. Kellock, N. H. Cho, J. W. Ager III: Characterization of chemical bonding and physical characteristic of diamond-like amorphous carbon and diamond films, J. Mater. Res. **7** (1992) 404–410
26.36 J. Robertson: Properties of diamond-like carbon, Surf. Coat. Technol. **50** (1992) 185–203
26.37 N. Savvides, B. Window: Diamondlike amorphous carbon films prepared by magnetron sputtering of graphite, J. Vac. Sci. Technol. A **3** (1985) 2386–2390
26.38 J. C. Angus, P. Koidl, S. Domitz: Carbon thin films. In: *Plasma Deposited Thin Films*, ed. by J. Mort, F. Jensen (CRC, Boca Raton 1986) pp. 89–127
26.39 J. Robertson: Amorphous carbon, Adv. Phys. **35** (1986) 317–374
26.40 M. Rubin, C. B. Hooper, N. H. Cho, B. Bhushan: Optical and mechanical properties of DC sputtered carbon films, J. Mater. Res. **5** (1990) 2538–2542
26.41 G. J. Vandentop, M. Kawasaki, R. M. Nix, I. G. Brown, M. Salmeron, G. A. Somorjai: Formation of hydrogenated amorphous carbon films of controlled hardness from a methane plasma, Phys. Rev. B **41** (1990) 3200–3210
26.42 J. J. Cuomo, D. L. Pappas, R. Lossy, J. P. Doyle, J. Bruley, G. W. Di Bello, W. Krakow: Energetic carbon deposition at oblique angles, J. Vac. Sci. Technol. A **10** (1992) 3414–3418
26.43 D. L. Pappas, K. L. Saegner, J. Bruley, W. Krakow, J. J. Cuomo: Pulsed laser deposition of diamondlike carbon films, J. Appl. Phys. **71** (1992) 5675–5684
26.44 H. J. Scheibe, B. Schultrich: DLC film deposition by laser-arc and study of properties, Thin Solid Films **246** (1994) 92–102
26.45 C. Donnet, A. Grill: Friction control of diamond-like carbon coatings, Surf. Coat. Technol. **94–95** (1997) 456
26.46 A. Grill: Tribological properties of diamondlike carbon and related materials, Surf. Coat. Technol. **94–95** (1997) 507
26.47 B. K. Gupta, B. Bhushan: Micromechanical properties of amorphous carbon coatings deposited by different deposition techniques, Thin Solid Films **270** (1995) 391–398
26.48 X. Li, B. Bhushan: Micro/nanomechanical and tribological characterization of ultra-thin amorphous carbon coatings, J. Mater. Res. **14** (1999) 2328–2337
26.49 X. Li, B. Bhushan: Mechanical and tribological studies of ultra-thin hard carbon overcoats for magnetic recording heads, Z. Metallkd. **90** (1999) 820–830
26.50 S. Sundararajan, B. Bhushan: Micro/nanotribology of ultra-thin hard amorphous carbon coatings using atomic force/friction force microscopy, Wear **225–229** (1999) 678–689
26.51 B. Bhushan: Chemical, mechanical, and tribological characterization of ultra-thin and hard amorphous carbon coatings as thin as 3.5 nm: Recent developments, Diam. Relat. Mater. **8** (1999) 1985–2015
26.52 I. I. Aksenov, V. E. Strel'Nitskii: Wear resistance of diamond-like carbon coatings, Surf. Coat. Technol. **47** (1991) 252–256
26.53 D. R. McKenzie, D. Muller, B. A. Pailthorpe, Z. H. Wang, E. Kravtchinskaia, D. Segal, P. B. Lukins, P. J. Martin, G. Amaratunga, P. H. Gaskell, A. Saeed: Properties of tetrahedral amorphous carbon prepared by vacuum arc deposition, Diam. Relat. Mater. **1** (1991) 51–59
26.54 R. Lossy, D. L. Pappas, R. A. Roy, J. J. Cuomo: Filtered arc deposition of amorphous diamond, Appl. Phys. Lett. **61** (1992) 171–173
26.55 I. G. Brown, A. Anders, S. Anders, M. R. Dickinson, I. C. Ivanov, R. A. MacGill, X. Y. Yao, K. M. Yu: Plasma synthesis of metallic and composite thin films with atomically mixed substrate bonding, Nucl. Instrum. Meth. Phys. Res. B **80–81** (1993) 1281–1287
26.56 P. J. Fallon, V. S. Veerasamy, C. A. Davis, J. Robertson, G. A. J. Amaratunga, W. I. Milne, J. Koskinen: Properties of filtered-ion-beam-deposited diamond-like carbon as a function of ion energy, Phys. Rev. B **48** (1993) 4777–4782
26.57 S. Anders, A. Anders, I. G. Brown, B. Wei, K. Komvopoulos, J. W. Ager III, K. M. Yu: Effect of vacuum arc deposition parameters on the properties of amorphous carbon thin films, Surf. Coat. Technol. **68–69** (1994) 388–393
26.58 S. Anders, A. Anders, I. G. Brown, M. R. Dickinson, R. A. MacGill: Metal plasma immersion ion implantation and deposition using arc plasma sources, J. Vac. Sci. Technol. B **12** (1994) 815–820
26.59 S. Anders, A. Anders, I. G. Brown: Transport of vacuum arc plasma through magnetic macroparticle filters, Plasma Sources Sci. **4** (1995) 1–12
26.60 D. M. Swec, M. J. Mirtich, B. A. Banks: Ion beam and plasma methods of producing diamondlike carbon films, Report No. NASATM102301, Cleveland 1988
26.61 A. Erdemir, M. Switala, R. Wei, P. Wilbur: A tribological investigation of the graphite-to-diamond-like behavior of amorphous carbon films ion beam deposited on ceramic substrates, Surf. Coat. Technol. **50** (1991) 17–23
26.62 A. Erdemir, F. A. Nicols, X. Z. Pan, R. Wei, P. J. Wilbur: Friction and wear performance of ion-beam deposited diamond-like carbon films on steel substrates, Diam. Relat. Mater. **3** (1993) 119–125
26.63 R. Wei, P. J. Wilbur, M. J. Liston: Effects of diamond-like hydrocarbon films on rolling contact fatigue of bearing steels, Diam. Relat. Mater. **2** (1993) 898–903

26.64 A. Erdemir, C. Donnet: Tribology of diamond, diamond-like carbon, and related films. In: *Modern Tribology Handbook*, ed. by B. Bhushan (CRC, Boca Raton 2001) pp. 871–908
26.65 J. Asmussen: Electron cyclotron resonance microwave discharges for etching and thin-film deposition, J. Vac. Sci. Technol. A **7** (1989) 883–893
26.66 J. Suzuki, S. Okada: Deposition of diamondlike carbon films using electron cyclotron resonance plasma chemical vapor deposition from ethylene gas, Jpn. J. Appl. Phys. **34** (1995) L1218–L1220
26.67 B.A. Banks, S.K. Rutledge: Ion beam sputter deposited diamond like films, J. Vac. Sci. Technol. **21** (1982) 807–814
26.68 C. Weissmantel, K. Bewilogua, K. Breuer, D. Dietrich, U. Ebersbach, H.J. Erler, B. Rau, G. Reisse: Preparation and properties of hard i-C and i-BN coatings, Thin Solid Films **96** (1982) 31–44
26.69 H. Dimigen, H. Hubsch: Applying low-friction wear-resistant thin solid films by physical vapor deposition, Philips Tech. Rev. **41** (1983/84) 186–197
26.70 B. Bhushan, B.K. Gupta, M.H. Azarian: Nanoindentation, microscratch, friction and wear studies for contact recording applications, Wear **181–183** (1995) 743–758
26.71 B. Bhushan, L. Yang, C. Gao, S. Suri, R.A. Miller, B. Marchon: Friction and wear studies of magnetic thin-film rigid disks with glass-ceramic, glass and aluminum-magnesium substrates, Wear **190** (1995) 44–59
26.72 L. Holland, S.M. Ojha: Deposition of hard and insulating carbonaceous films of an RF target in butane plasma, Thin Solid Films **38** (1976) L17–L19
26.73 L.P. Andersson: A review of recent work on hard i-C films, Thin Solid Films **86** (1981) 193–200
26.74 A. Bubenzer, B. Dischler, B. Brandt, P. Koidl: R.F. plasma deposited amorphous hydrogenated hard carbon thin films, preparation, properties and applications, J. Appl. Phys. **54** (1983) 4590–4594
26.75 A. Grill, B.S. Meyerson, V.V. Patel: Diamond-like carbon films by RF plasma-assisted chemical vapor deposition from acetylene, IBM J. Res. Develop. **34** (1990) 849–857
26.76 A. Grill, B.S. Meyerson, V.V. Patel: Interface modification for improving the adhesion of a-C:H to metals, J. Mater. Res. **3** (1988) 214
26.77 A. Grill, V.V. Patel, B.S. Meyerson: Optical and tribological properties of heat-treated diamond-like carbon, J. Mater. Res. **5** (1990) 2531–2537
26.78 F. Jansen, M. Machonkin, S. Kaplan, S. Hark: The effect of hydrogenation on the properties of ion beam sputter deposited amorphous carbon, J. Vac. Sci. Technol. A **3** (1985) 605–609
26.79 S. Kaplan, F. Jansen, M. Machonkin: Characterization of amorphous carbon-hydrogen films by solid-state nuclear magnetic resonance, Appl. Phys. Lett. **47** (1985) 750–753
26.80 H.C. Tsai, D.B. Bogy, M.K. Kundmann, D.K. Veirs, M.R. Hilton, S.T. Mayer: Structure and properties of sputtered carbon overcoats on rigid magnetic media disks, J. Vac. Sci. Technol. A **6** (1988) 2307–2315
26.81 B. Marchon, M. Salmeron, W. Siekhaus: Observation of graphitic and amorphous structures on the surface of hard carbon films by scanning tunneling microscopy, Phys. Rev. B **39** (1989) 12907–12910
26.82 B. Dischler, A. Bubenzer, P. Koidl: Hard carbon coatings with low optical-absorption, Appl. Phys. Lett. **42** (1983) 636–638
26.83 D.S. Knight, W.B. White: Characterization of diamond films by Raman spectroscopy, J. Mater. Res. **4** (1989) 385–393
26.84 J.W. Ager, D.K. Veirs, C.M. Rosenblatt: Spatially resolved Raman studies of diamond films grown by chemical vapor deposition, Phys. Rev. B **43** (1991) 6491–6499
26.85 W. Scharff, K. Hammer, O. Stenzel, J. Ullman, M. Vogel, T. Frauenheim, B. Eibisch, S. Roth, S. Schulze, I. Muhling: Preparation of amorphous i-C films by ion-assisted methods, Thin Solid Films **171** (1989) 157–169
26.86 B. Bhushan, X. Li: Nanomechanical characterization of solid surfaces and thin films, Int. Mater. Rev. **48** (2003) 125–164
26.87 X. Li, D. Diao, B. Bhushan: Fracture mechanisms of thin amorphous carbon films in nanoindentation, Acta Mater. **45** (1997) 4453–4461
26.88 X. Li, B. Bhushan: Measurement of fracture toughness of ultra-thin amorphous carbon films, Thin Solid Films **315** (1998) 214–221
26.89 X. Li, B. Bhushan: Evaluation of fracture toughness of ultra-thin amorphous carbon coatings deposited by different deposition techniques, Thin Solid Films **355-356** (1999) 330–336
26.90 X. Li, B. Bhushan: Development of a nanoscale fatigue measurement technique and its application to ultrathin amorphous carbon coatings, Scripta Mater. **47** (2002) 473–479
26.91 X. Li, B. Bhushan: Nanofatigue studies of ultrathin hard carbon overcoats used in magnetic storage devices, J. Appl. Phys. **91** (2002) 8334–8336
26.92 J. Robertson: Deposition of diamond-like carbon, Philos. Trans. R. Soc. London A **342** (1993) 277–286
26.93 S.J. Bull: Tribology of carbon coatings: DLC, diamond and beyond, Diam. Relat. Mater. **4** (1995) 827–836
26.94 N. Savvides, T.J. Bell: Microhardness and Young's modulus of diamond and diamondlike carbon films, J. Appl. Phys. **72** (1992) 2791–2796
26.95 B. Bhushan, M.F. Doerner: Role of mechanical properties and surface texture in the real area of contact of magnetic rigid disks, ASME J. Tribol. **111** (1989) 452–458
26.96 S. Suresh: *Fatigue of Materials* (Cambridge Univ. Press, Cambridge 1991)

26.97 D. B. Marshall, A. G. Evans: Measurement of adherence of residual stresses in thin films by indentation. I. Mechanics of interface delamination, J. Appl. Phys. **15** (1984) 2632–2638

26.98 A. G. Evans, J. W. Hutchinson: On the mechanics of delamination and spalling in compressed films, Int. J. Solids Struct. **20** (1984) 455–466

26.99 S. Sundararajan, B. Bhushan: Development of a continuous microscratch technique in an atomic force microscope and its application to study scratch resistance of ultrathin hard amorphous carbon coatings, J. Mater. Res. **16** (2001) 437–445

26.100 B. Bhushan, B. K. Gupta: Micromechanical characterization of Ni-P coated aluminum-magnesium, glass and glass-ceramic substrates and finished magnetic thin-film rigid disks, Adv. Info. Storage Syst. **6** (1995) 193–208

26.101 X. Li, B. Bhushan: Micromechanical and tribological characterization of hard amorphous carbon coatings as thin as 5 nm for magnetic recording heads, Wear **220** (1998) 51–58

26.102 B. Bhushan, V. N. Koinkar: Microscale mechanical and tribological characterization of hard amorphous coatings as thin as 5 nm for magnetic disks, Surf. Coat. Technol. **76-77** (1995) 655–669

26.103 V. N. Koinkar, B. Bhushan: Microtribological properties of hard amorphous carbon protective coatings for thin-film magnetic disks and heads, Proc. Inst. Mech. Eng., Part J **211** (1997) 365–372

26.104 T. W. Wu: Microscratch and load relaxation tests for ultra-thin films, J. Mater. Res. **6** (1991) 407–426

26.105 R. Memming, H. J. Tolle, P. E. Wierenga: Properties of polymeric layers of hydrogenated amorphous carbon produced by plasma-activated chemical vapor deposition: tribological and mechanical properties, Thin Solid Films **143** (1986) 31–41

26.106 C. Donnet, T. Le Mogne, L. Ponsonnet, M. Belin, A. Grill, V. Patel: The respective role of oxygen and water vapor on the tribology of hydrogenated diamond-like carbon coatings, Tribol. Lett. **4** (1998) 259

26.107 F. P. Bowden, J. E. Young: Friction of diamond, graphite and carbon and the influence of surface films, Proc. R. Soc. Lond. **208** (1951) 444–455

26.108 B. Marchon, N. Heiman, M. R. Khan: Evidence for tribochemical wear on amorphous carbon thin films, IEEE Trans. Magn. **26** (1990) 168–170

26.109 M. T. Dugger, Y. W. Chung, B. Bhushan, W. Rothschild: Friction, wear, and interfacial chemistry in thin film magnetic rigid disk files, ASME J. Tribol. **112** (1990) 238–245

26.110 B. D. Strom, D. B. Bogy, C. S. Bhatia, B. Bhushan: Tribochemical effects of various gases and water vapor on thin film magnetic disks with carbon overcoats, ASME J. Tribol. **113** (1991) 689–693

26.111 B. Bhushan, J. Ruan: Tribological performance of thin film amorphous carbon overcoats for magnetic recording rigid disks in various environments, Surf. Coat. Technol. **68/69** (1994) 644–650

26.112 B. Bhushan, G. S. A. M. Theunissen, X. Li: Tribological studies of chromium oxide films for magnetic recording applications, Thin Solid Films **311** (1997) 67–80

27. Self-Assembled Monolayers for Controlling Adhesion, Friction and Wear

Reliability of micro- and nanodevices, as well as magnetic storage devices require the use of lubricant films for the protection of sliding surfaces. To minimize high adhesion, friction, and because of small clearances in the devices, these films should be molecularly thick. Liquid films of low surface tension or certain hydrophobic solid films can be used. Ordered molecular assemblies with high hydrophobicity can be engineered using chemical grafting of various polymer molecules with suitable functional head groups, spacer chains and nonpolar surface terminal groups.

The classical approach to lubrication uses multi-molecular layers of liquid lubricants. Boundary lubricant films are formed by either physisorption, chemisorption, or chemical reaction. The physiosorbed films can be either monomolecularly or polymolecularly thick. The chemisorbed films are monomolecular, but stoichiometric films formed by chemical reaction can be multilayered. A good boundary lubricant should have a high degree of interaction between its molecules and the sliding surface. As a general rule, liquids are good lubricants when they are polar and thus able to grip on solid surfaces (or be adsorbed).

In this chapter, we focus on self-assembled monolayers (SAMs) for high hydrophobicity and/or low adhesion, friction, and wear. SAMs are produced by various organic precursors. We first present a primer to organic chemistry followed by an overview on suitable substrates, head groups, spacer chains, and end groups in the molecular chains and an overview of tribological properties of SAMs. The adhesion, friction, and wear properties of SAMs, having alkyl and biphenyl spacer chains with different surface terminal and head groups, are surveyed. The friction data are explained using a molecular spring model in which the local stiffness and intermolecular force govern its frictional performance. Based on the nanotribological studies of SAM films by AFM, they exhibit attractive hydrophobic and tribological properties.

27.1 **A Primer to Organic Chemistry** 834
27.1.1 Electronegativity/Polarity 834
27.1.2 Classification and Structure of Organic Compounds 835
27.1.3 Polar and Nonpolar Groups 838

27.2 **Self-Assembled Monolayers: Substrates, Head Groups, Spacer Chains, and End Groups** 839

27.3 **Tribological Properties of SAMs** 841
27.3.1 Surface Roughness and Friction Images of SAMs Films 844
27.3.2 Adhesion, Friction, and Work of Adhesion 844
27.3.3 Stiffness, Molecular Spring Model, and Micropatterned SAMs 848
27.3.4 Influence of Humidity, Temperature, and Velocity on Adhesion and Friction 850
27.3.5 Wear and Scratch Resistance of SAMs 853

27.4 **Closure** 856

References 857

Reliability of micro/nanodevices, also commonly referred to as micro/nanoelectromechanical systems (MEMS/NEMS), as well as magnetic storage devices (which include magnetic rigid disk drives, flexible disk drives, and tape drives) require the use of molecularly thick films for protection of sliding surfaces [27.1–8]. A solid or liquid film is generally necessary for acceptable tribological properties of sliding interfaces. However, a small quantity of high surface tension liquid present between smooth surfaces can substantially increase the adhesion, friction, and wear as a result of formation of menisci or adhesive bridges [27.9, 10]. It becomes a major concern in micro/nanoscale devices operating at ultralow loads, as the liquid mediated adhesive force may be on the same order as the external load.

The source of the liquid film can be either preexisting film of liquid and/or capillary condensates of water vapor from the environment. If the liquid wets the surface ($0 \leq \theta < 90°$, where θ is the contact angle between the liquid-vapor interface and the liquid-solid interface for a liquid droplet sitting on a solid surface, Fig. 27.1a), the liquid surface is thereby constrained to lie parallel to the surface [27.11], and the complete liquid surface must therefore be concave in shape, Fig. 27.1b. Surface tension results in a pressure difference across any meniscus surface, referred to as capillary pressure or Laplace pressure, and is negative for a concave meniscus [27.9, 10]. The negative Laplace pressure results in an intrinsic attractive (adhesive) force, which depends on the interface roughness (local geometry of interacting asperities and number of asperities), the surface tension, and the contact angle. During sliding, frictional effects need to be overcome, not only because of external load, but also because of intrinsic adhesive force.

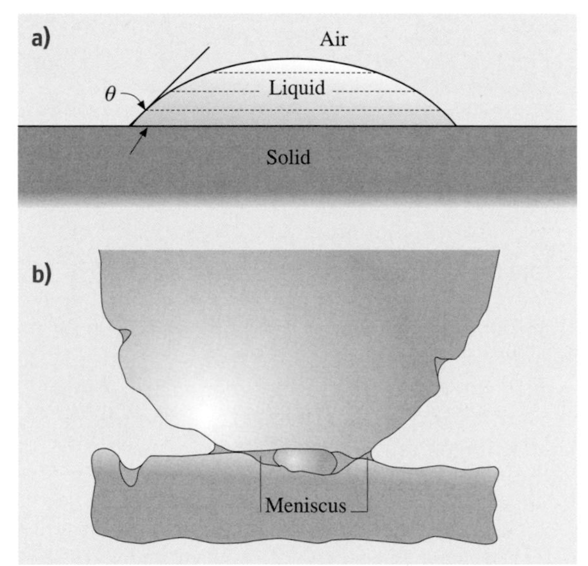

Fig. 27.1 (**a**) Schematic of a sessile-drop on a solid surface and the definition of contact angle, and (**b**) formation of meniscus bridges as a result of liquid present at an interface. The direct measurement of contact angle is most widely made from sessile-drops. The angle is generally measured by aligning a tangent with the drop profile at the point of contact with the solid surface using a telescope equipped with a goniometer eyepiece

Measured value of high static friction force contributed largely by liquid mediated adhesion (meniscus contribution) is generally referred to as "stiction". Basically, there are three ways to minimize the effect of liquid-mediated adhesion: increase in surface roughness, use of hydrophobic (water-fearing) rather than hydrophilic (water-loving) surfaces, and the use of a liquid with low surface tension [27.9, 10, 12, 13].

As an example, bulk silicon and polysilicon films used in the construction of micro/nanodevices can be dipped in hydrofluoric (HF) to make them hydrophobic (e.g., *Maboudian* [27.14]; *Scherge* and *Schaefer* [27.15]). In HF etching of silicon, hydrogen passivates the silicon surface by saturating the dangling bonds, and it results in a hydrogen-terminated silicon surface that is responsible for less adsorption of water. However, after exposure of the treated surface to the environment, it reoxidizes, which can adsorb water, and thus the surface again becomes hydrophilic.

The surfaces can also be treated or coated with a liquid with relatively low surface tension or a certain solid film to make them hydrophobic and/or to control adhesion, friction, and wear. The liquid lubricant film should be thin (about half of the composite roughness of the interface) to minimize liquid-mediated adhesion contribution [27.9, 10, 12, 13]. Thus, for ultra-smooth surfaces with RMS roughness on the order of a few nm, molecularly thick liquid films are required for liquid lubrication. The classical approach to lubrication uses freely supported multi-molecular layers of liquid lubricants [27.2, 4, 9, 10, 16–19]. Boundary lubricant films are formed by either physisorption, chemisorption, or chemical reaction. The physiosorbed films can be either monomolecularly or polymolecularly thick. The chemisorbed films are monomolecular, but stoichiometric films formed by chemical reaction can be multilayered. In general, stability and durability of surface films decrease in the following order: chemically reacted films, chemisorbed films, and physisorbed films. A good boundary lubricant should have a high degree of interaction between its molecules and the sliding surface. As a general rule, liquids are good lubricants when they are polar and thus able to grip on solid surfaces (or be adsorbed). Polar lubricants contain reactive functional end groups. Boundary lubrication properties are also dependent on the molecular conformation and lubricant spreading. It should be noted that the liquid films with a thickness on the order of a few nm, may be discontinuous and may deposit in an island form of nonuniform thickness with a lateral resolution on the nanometerscale.

Solid films are also commonly used for controlling hydrophobicity and/or adhesion, friction, and wear. Hydrophobic films have nonpolar surface terminal groups (to be described later) that repel water. These films have low surface energy (15–30 dyn/cm) and high contact angle ($\theta \geq 90°$), which minimize wetting (e.g., *Zisman* [27.20]; *Schrader* and *Loeb* [27.21]; *Neumann* and *Spelt* [27.22]). It should be noted that these films do not totally eliminate wetting. Multi-molecularly thick (few tenths of a nm) films of conventional solid-lubricants have been studied. *Hansma* et al. [27.23] reported the deposition of multi-molecularly thick, highly-oriented PTFE films from the melt or vapor phase or from solution by a mechanical deposition technique by dragging the polymer at controlled temperature, pressure, and speed against a smooth glass substrate. *Scandella* et al. [27.24] reported that the coefficient of nanoscale friction of MoS_2 platelets on mica, obtained by exfoliation of lithium intercalated MoS_2 in water, was a factor of 1.4 less than that of mica itself. However, MoS_2 is reactive to water and its friction and wear properties degrade with an increase in humidity [27.9, 10]. Amorphous diamond-like carbon (DLC) coatings can be produced with extremely high hardness and are commercially used as wear-resistant coatings [27.25, 26]. Their largest application is in magnetic storage devices [27.2]. Doping of the DLC matrix with elements like hydrogen, nitrogen, oxygen, silicon, and fluorine influences their hydrophobicity and tribological properties [27.25, 27, 28]. Nitrogen and oxygen reduce the contact angle (or increase the surface energy), due to the strong polarity that is formed when these elements are bonded to carbon. On the other hand, silicon and fluorine increase the contact angle ranging from 70–100° (or reduce the surface energy to 20–40 dyn/cm), making them hydrophobic [27.29, 30]. Nanocomposite coatings with a diamond-like carbon (a-C:H) network and a glass-like a-Si:O network are generally deposited using a plasma-enhanced chemical vapor deposition (PECVD) technique in which plasma is formed from a siloxane precursor using a hot filament. For a fluorinated DLC, CF_4 is added as the fluorocarbon source to an acetylene plasma. In addition, fluorination of DLC can be achieved by the post-deposition treatment of DLC coatings in a CF_4 plasma. Silicon- and fluorine-containing DLC coatings mainly reduce their polarity due to the loss of sp^2-bonded carbon (polarization potential of the involved π electrons) and dangling bonds of the DLC network. As silicon and fluorine are unable to form double bonds, they force carbon into a sp^3 bonding state [27.30]. Friction and wear properties of both silicon-containing and fluorinated DLC coatings have been reported to be superior to that of conventional DLC coatings [27.31,32]. However, DLC coatings require line of sight deposition process, which prevents deposition on complex geometries. Furthermore, it has been reported that some self-assembled monolayers (SAMs) are superior to DLC coatings in hydrophobicity and tribological performance [27.33].

Organized and dense molecular-scale layers of, preferably, long-chain organic molecules are known to be superior lubricants on both macro- and micro/nanoscales as compared with freely supported multi-molecular layers [27.4, 34–38]. Common techniques to produce molecular-scale organized layers are Langmuir–Blodgett (LB) deposition and chemical grafting of organic molecules to realize SAMs [27.39,40]. In the LB technique, organic molecules from suitable amphiphilic molecules are first organized at the air-water interface and then physisorbed on a solid surface to form mono- or multi-molecular layers [27.41]. In the case of SAMs, the functional groups of molecules chemisorb on a solid surface, which results in the spontaneous formation of robust, highly ordered, oriented, and dense monolayers [27.40]. In both cases, the organic molecules used have well distinguished amphiphilic properties (hydrophilic functional head and a hydrophobic aliphatic tail) so that an adsorption of such molecules on an active inorganic substrate leads to their firm attachment to the surface. Direct organization of SAMs on the solid surfaces allows coating in the tight areas such as the bearing and journal surfaces in an assembled bearing. The weak adhesion of classical LB films to the substrate surface restricts their lifetimes during sliding, whereas certain SAMs can be very durable [27.34]. As a result, SAMs are of great interest in tribological applications.

Much of the research in the application of SAMs has been carried out for the so-called soft lithographic technique [27.42, 43]. This is a non-photolithographic technique. Photolithography is based on a projection-printing system used for projection of an image from a mask to a thin-film photoresist and its resolution is limited by optical diffraction limits. In the soft lithography, an elastomeric stamp or mold is used to generate micropatterns of SAMs either by contact printing (known as microcontact printing or μCP [27.44]), by embossing (imprinting) [27.45], or by replica molding [27.46], which circumvents the diffraction limits of photolithography. The stamps are generally cast from photolithographically generated patterned masters, and the stamp material is generally polydimethylsiloxane (PDMS). In μCP, the ink is a SAM precursor to produce nm-thick resists with lines thinner than 100 nm. Although soft lithography requires little capital investment, it is doubtful that it would ever replace well established photolithography. However, μCP and embossing techniques may be used to produce microdevices that are a substantially cheaper and more flexible choice of material for construction than conventional photolithography (e.g., SAMs and non-SAM entities for μCP and elastomers for embossing).

Other applications for SAMs are in the areas of bio/chemical and optical sensors, for use as drug-delivery vehicles and in the construction of electronic components [27.39, 47]. Bio/chemical sensors require the use of highly sensitive organic layers with tailored biological properties that can be incorporated into electronic, optical, or electrochemical devices. Self-assembled microscopic vesicles are being developed to ferry potentially lifesaving drugs to cancer patients. By getting organic, metal, and phosphonate molecules (complexes of phosphorous and oxygen atoms) to assemble themselves into conductive materials, these can be produced as self-made sandwiches for use as electronic components. The application of interest here is an application requiring hydrophobicity and/or low friction and wear.

An overview of molecularly thick layers of liquid lubricants and conventional solid lubricant can be found in various references [27.2, 4, 9, 10, 18, 19, 25]. In this chapter, we focus on SAMs for high hydrophobicity and/or, low adhesion, friction, and wear. SAMs are produced by various organic precursors. We first present a primer to organic chemistry followed by an overview on suitable substrates, spacer chains, and end groups in the molecular chains, an overview of tribological properties of SAMs, and some concluding remarks.

27.1 A Primer to Organic Chemistry

All organic compounds contain the carbon (C) atom. Carbon, in combination with hydrogen, oxygen, nitrogen, and sulfur results in a large number of organic compounds. The atomic number of carbon is six, and its electron structure is $1s^2\,2s^2\,2p^2$. Two stable isotopes of carbon, ^{12}C and ^{13}C, exist. With four electrons in its outer shell, carbon forms four covalent bonds with each bond resulting from two atoms sharing a pair of electrons. The number of electron pairs that two atoms share determines whether or not the bond is single or multiple. In a single bond, only one pair of electrons is shared by the atoms. Carbon can also form multiple bonds by sharing two or three pairs of electrons between the atoms. For example, the double bond formed by sharing two electron pairs is stronger than a single bond, and it is shorter than a single bond. An organic compound is classified as saturated if it contains only the single bond and as unsaturated if the molecules possess one or more multiple carbon-carbon bonds.

27.1.1 Electronegativity/Polarity

When two different kinds of atoms share a pair of electrons a bond is formed in which electrons are shared unequally, one atom assumes a partial positive charge and the other a negative charge with respect to each other. This difference in charge occurs because the two atoms exert unequal attraction for the pair of shared electrons. The attractive force that an atom of an element has for shared electrons in a molecule or polyatomic ion is known as its electronegativity. Elements differ in their electronegativities. A scale of relative electronegatives, in which the most electronegative element, fluorine, is assigned a value of 4.0, was developed by Linus Pauling. The relative electronegativities of the elements in the periodic table can be found in most undergraduate chemistry textbooks (e.g., *Hein* et al. [27.48]). Relative electronegativity of the nonmetals is high compared to that of metals. Relative electronegativity of selected elements of interest with high values is presented in Table 27.1.

The polarity of a bond is determined by the difference in electronegativity values of the atoms forming the bond. If the electronegativities are the same, the bond is nonpolar and the electrons are shared equally. In this type of bond, there is no separation of positive and negative charge between atoms. If the atoms have greatly different electronegativities, the bond is very polar. A dipole is a molecule that is electrically asymmetrical, causing it to be oppositely charged at two points. As an example, both hydrogen and chlorine need one electron to form stable electron configurations. They share a pair of

Element	Relative electronegativity
F	4.0
O	3.5
N	3.0
Cl	3.0
C	2.5
S	2.5
P	2.1
H	2.1

Table 27.1 Relative electronegativity of selected elements

electrons in hydrogen chloride, HCl. Chlorine is more electronegative and, therefore, has a greater attraction for the shared electrons than hydrogen. As a result, the pair of electrons is displaced toward the chlorine atom, giving it a partial negative charge and leaving the hydrogen atom with a partial positive charge, Fig. 27.2. However, the entire molecule, HCl, is electrically neutral. The hydrogen atom with a partial positive charge (exposed proton on one end) can be easily attracted to the negative charge of other molecules, and this is responsible for the polarity of the molecule. A partial charge is usually indicated by δ, and the electronic structure of HClis given as:

$$\overset{\delta+}{\mathrm{H}}:\overset{\delta-}{\ddot{\underset{..}{\mathrm{Cl}}}}: \ .$$

Similar to the HCl molecule, HF is polar, and both behave as a small dipole. On the other hand, methane (CH_4), carbon tetrachloride (CCl_4), and carbon dioxide (CO_2) are nonpolar. In CH_4 and CCl_4, the four C−H and C−Cl polar bonds are identical, and because these bonds emanate from the center to the corners of a tetrahedron in the molecule, the effect of their polarities cancels out one another. CO_2 (O=C=O) is nonpolar, because carbon-oxygen dipoles cancel each other by acting in opposite directions. Symmetric molecules are generally nonpolar. Water (H−O−H) is a polar molecule. If the atoms in water were linear as in CO_2, two O−H dipoles would cancel each other, and the molecule would be nonpolar. However, water has a bent structure with an angle of 105° between the two bonds, which is responsible for the water being a polar molecule.

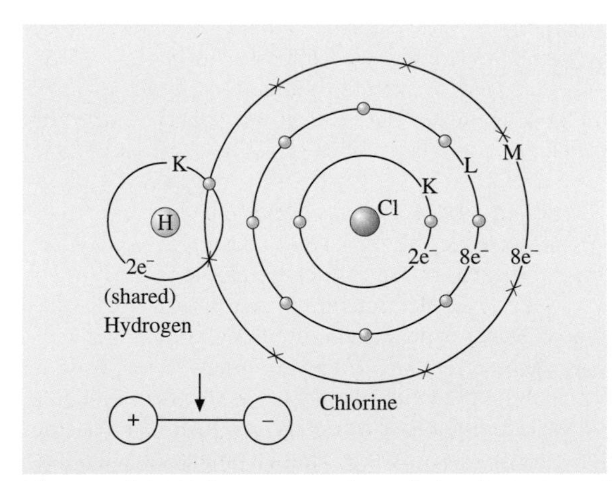

Fig. 27.2 Schematic representation of the formation of a polar HCl molecule

27.1.2 Classification and Structure of Organic Compounds

Tables 27.2–27.6 present selected organic compounds grouped into classes.

Hydrocarbons

Hydrocarbons are compounds that are composed entirely of carbon and hydrogen atoms bonded to each other by covalent bonds. Saturated hydrocarbons (alkanes) contain single bonds. Unsaturated hydrocarbons that contain a carbon-carbon double bond are called alkenes, and ones with a triple bond are called alkynes. Unsaturated hydrocarbons that contain an aromatic ring, e.g., benzene ring, are called aromatic hydrocarbons.

Saturated Hydrocarbons: Alkanes. The alkanes, also known as paraffins, are saturated hydrocarbons, straight- or branched-chain hydrocarbons with only single covalent bonds between the carbon atoms. The general molecular formula for alkanes is C_nH_{2n+2}, where n is the number of carbon atoms in the molecule. Each carbon atom is connected to four other atoms by four single covalent bonds. These bonds are separated by angles of 109.5° (the angle by lines drawn from the center of a regular tetrahedron to its corners). Alkane molecules contain only carbon-carbon and carbon-hydrogen bonds, which are symmetrically directed toward the corners of a tetrahedron. Therefore, alkane molecules are essentially nonpolar.

Common alkyl groups have the general formula C_nH_{2n+1} (one hydrogen atom less than the corresponding alkane). The missing H atom may be detached from any carbon in the alkane. The

Table 27.2 Names and formulas of selected hydrocarbons

Name	Formula
Saturated hydrocarbons	
Straight-chain alkanes	C_nH_{2n+2}
e.g., methane,	CH_4
ethane	C_2H_6 or CH_3CH_3
Alkyl groups	C_nH_{2n+1}
e.g., methyl,	$-CH_3$
ethyl	$-CH_2CH_3$
Unsaturated hydrocarbons	
Alkenes	$(CH_2)_n$
e.g., ethene,	C_2H_4 or $CH_2{=}CH_2$
propene	C_3H_6 or $CH_3CH{=}CH_2$
Alkynes	
e.g., acetylene	$HC{\equiv}CH$
Aromatic hydrocarbons	
e.g., benzene	C_6H_6 (structural formula: six-membered ring of C atoms with alternating double bonds, each C bearing one H)

Table 27.3 Names and formulas of selected alcohols, ethers, phenols, and thiols

Name	Formula
Alcohols	$R-OH$
e.g., methanol,	CH_3OH
ethanol	CH_3CH_2OH
Ethers	$R-O-R'$
e.g., dimethyl ether,	CH_3-O-CH_3
diethyl ether	$CH_3CH_2-O-CH_2CH_3$
Phenols	C_6H_5OH or (benzene ring with OH)
Thiols	$-SH$
e.g., methanethiol	CH_3SH

The letters R- and R′- represent an alkyl group. The R- groups in ethers can be the same or different and can be alkyl or aromatic (Ar) groups.

name of the group is formed from the name of the corresponding alkane by replacing -ane with -yl ending. Some examples are shown in Table 27.2.

Unsaturated Hydrocarbons. Unsaturated hydrocarbons consist of three families of compounds that contain fewer hydrogen atoms than the alkane with the corresponding number of carbon atoms and contain multiple bonds between carbon atoms. These include alkenes (with carbon-carbon double bonds), alkynes (with carbon-carbon triple bonds), and aromatic compounds (with benzene rings that are arranged in a six-member ring with one hydrogen atom bonded to each carbon atom and three carbon-carbon double bonds). Some examples are shown in Table 27.2.

Alcohols, Ethers, Phenols, and Thiols

Organic molecules with certain functional groups are synthesized for desirable properties. Alcohols, ethers, and phenols are derived from the structure of water by replacing the hydrogen atoms of water with alkyl groups (R) or aromatic (Ar) rings. For example, phenol is a class of compounds that have a hydroxy group attached to an aromatic ring (benzene ring). Organic compounds that contain the −SH group are analogs of alcohols and are known as thiols. Some examples are shown in Table 27.3.

Aldehydes and Ketones

Both aldehydes and ketones contain the carbonyl group, >C=O, a carbon-oxygen double bond. Aldehydes have at least one hydrogen atom bonded to the carbonyl group, whereas ketones have only an alkyl or aromatic group bonded to the carbonyl group. The general formula for the saturated homologous series of aldehydes and ketones is $C_nH_{2n}O$. Some examples are shown in Table 27.4.

Carboxyl Acids and Esters

The functional group of the carboxylic acids is known as a carboxyl group, represented as −COOH. Carboxylic acids can be either aliphatic (RCOOH) or aromatic (ArCOOH). The carboxylic acids with even numbers of carbon atoms, n, ranging from 4 to about 20 are called fatty acids (e.g., $n = 10, 12, 14, 16$ and 18 are called capric acid, lauric acid, myristic acid, palmitic acid, and stearic acid, respectively).

Esters are alcohol derivates of carboxylic acids. Their general formula is RCOOR′, where R may be a hydrogen, alkyl group, or aromatic group, and R′ may be an alkyl group or aromatic group but not a hydrogen. Esters are found in fats and oils. Some examples are shown in Table 27.5.

Table 27.4 Names and formulas of selected aldehydes and ketones

Name	Formula
Aldehydes	RCHO or R—C(=O)—H
	ArCHO or Ar—C(=O)—H
e.g., methanal or formaldehyde ethanal or acetaldehyde	HCHO CH_3CHO
Ketones	RCOR′ or R—C(=O)—R'
	RCOAr or R—C(=O)—Ar
	ArCOAr or Ar—C(=O)—Ar
e.g., butanone or methyl-ethyl ketone	$CH_3COCH_2CH_3$

The letters R and R′ represent alkyl groups, and Ar represents an aromatic group.

Name	Formula
Carboxylic acid*	RCOOH or R—C(=O)—OH
	ArCOOH or Ar—C(=O)—OH
e.g., methanoic acid (formic acid),	HCOOH
ethanoic acid (acetic acid),	CH_3COOH
octadecanoic acid (stearic acid)	$CH_3(CH_2)_{16}COOH$
Esters**	RCOOR′ or R—C(=O)—O—R′ (R—C(=O): acid; O—R′: alcohol)
e.g., methyl propanoate	$CH_3CH_2COOCH_3$

* The letter R represents an alkyl group and Ar represents an aromatic group.
** The letter R represents hydrogen, an alkyl group, or an aromatic group and R′ represents an alkyl group or aromatic group

Table 27.5 Names and formulas of selected carboxylic acids and esters

Name	Formula
Amides	$RCONH_2$ or R—C(=O)—NH_2
e.g., methanamide (formamide),	$HCONH_2$
ethanamide (acetamide)	CH_3CONH_2
Amines	RNH_2 or R—N(H)—H
	R_2NH
	R_3N
e.g., methylamine,	CH_3NH_2
ethylamine	$CH_3CH_2NH_2$

The letter R represents an alkyl group or aromatic group.

Table 27.6 Names and formulas of selected organic nitrogen compounds (amides and amines)

Amides and Amines

Amides and amines are organic compounds containing nitrogen. Amides are nitrogen derivates of carboxylic acids. The carbon atom of a carbonyl group is bonded directly to a nitrogen atom of a $-NH_2$, $-NHR$, or $-NR_2$ group. The characteristic structure of amide is $RCONH_2$.

An amine is a substituted ammonia molecule that has a general structure of RNH_2, R_2NH, or R_3N, where R is an alkyl or aromatic group. Some examples are shown in Table 27.6.

27.1.3 Polar and Nonpolar Groups

Table 27.7 summarizes polar and nonpolar groups commonly used in the construction of hydrophobic and hydrophilic molecules. Table 27.8 lists the relative polarity of selected polar groups [27.49]. Thiol, silane, carboxylic acid, and alcohol (hydroxyl) groups are the most commonly used polar anchor groups for their attachment to surfaces. Methyl and trifluoromethyl are commonly used end groups for hydrophobic film surfaces.

27.2 Self-Assembled Monolayers: Substrates, Head Groups, Spacer Chains, and End Groups

SAMs are formed as a result of spontaneous, self-organization of functionalized organic molecules onto the surfaces of appropriate substrates into stable, well-defined structures, Fig. 27.3. The final structure is close to or at thermodynamic equilibrium, and as a result, it tends to form spontaneously and rejects defects. SAMs consist of three building groups: a head group that binds strongly to a substrate, a surface terminal (tail) group that constitutes the outer surface of the film, and a spacer chain (backbone chain) that connects the head and surface terminal groups. The SAMs are named based on the surface terminal group followed by spacer chain and the head group (or type of compound formed at the surface). For a SAM to control hydrophobicity, adhesion, friction, and wear, it should be strongly adherent to the substrate, and the surface terminal group of the organic molecu-

Table 27.7 Some examples of polar (hydrophilic) and non-polar (hydrophobic) groups

Name	Formula
Polar	
Alcohol (hydroxyl)	–OH
Carboxyl acid	–COOH
Aldehyde	–COH
Ketone	R–C(=O)–R
Ester	–COO–
Carbonyl	>C=O
Ether	R–O–R
Amine	$-NH_2$
Amide	–C(=O)–NH_2
Phenol	benzene ring–OH
Thiol	–SH
Trichlorosilane	$SiCl_3$
Nonpolar	
Methyl	$-CH_3$
Trifluoromethyl	$-CF_3$
Aryl (benzene ring)	benzene ring–
The letter R represents an alkyl group.	

Table 27.8 Organic groups listed in the increasing order of polarity

Alkanes
Alkenes
Aromatic hydrocarbons
Ethers
Trichlorosilanes
Aldehydes, ketones, esters, carbonyls
Thiols
Amines
Alcohols, phenols
Amides
Carboxylic acids

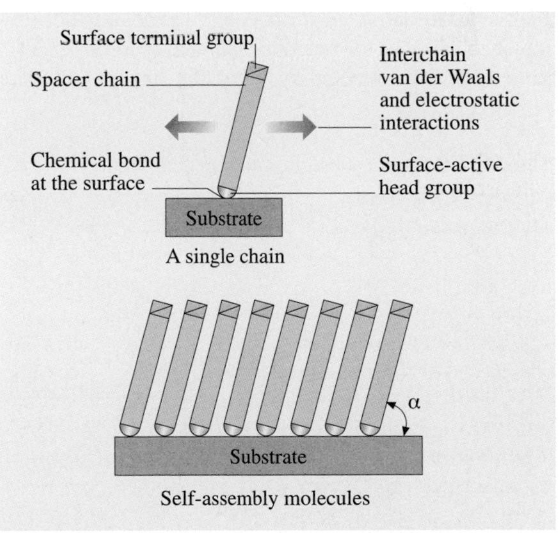

Fig. 27.3 Schematic of a self-assembled monolayer on a surface and the associated forces

lar chain should be nonpolar. For strong attachment of the organic molecules to the substrate, the head group of the molecular chain should contain a polar end group resulting in the exothermic process (energies on the order of tens of kcal/mol), i. e., it results in an apparent pinning of the head group to a specific site on the surface through a chemical bond. Furthermore, molecular structure and any crosslinking would have a significant effect on their friction and wear performance. The substrate surface should have a high surface energy (hydropolic), so that there will be a strong tendency for molecules to adsorb on the surface. The surface should be highly functional with polar groups and dangling bonds (generally unpaired electrons), so that they can react with organic molecules and provide a strong bond. Because of the exothermic head group–substrate interactions, molecules try to occupy every available binding site on the surface, and during this process they generally push together molecules that have already adsorbed. The process results in the formation of crystalline molecular assemblies. The interactions between molecular chains are van der Waals or electrostatic-type with energies on the order of a few (< 10) kcal/mol, exothermic. The molecular chains in SAMs are not perpendicular to the surface; the tilt angle depends on the anchor group, as well as on the substrate and the spacer group. For example, the tilt angle for alkanethiolate on Au is typically about 30–35° angle with respect to substrate normal.

SAMs are usually produced by immersing a substrate in a solution containing precursor (ligand) that is reactive to the substrate surface, or by exposing the substrate to a vapor of the reactive chemical species [27.39]. Table 27.9 lists selected systems that have been used for formation of SAMs [27.43]. The spacer chain of SAM is mostly an alkyl chain ($-C_nH_{2n+1}$), or made of a derivatized alkyl group. By attaching different terminal groups at the surface, the film surface can be made to attract or repel water. The commonly used surface terminal group of a hydrophobic film with low surface energy, in the case of a single alkyl chain, is a nonpolar methyl ($-CH_3$) or trifluoromethyl ($-CF_3$) group. For a hydrophilic film, the commonly used surface terminal groups are alcohol ($-OH$) or carboxylic acid ($-COOH$) groups. Surface active head groups most commonly used are thiol ($-SH$), silane (e.g., trichlorosilane or $-SiCl_3$), and carboxyl ($-COOH$) groups. The substrates most commonly used are gold, silver, platinum, copper, hydroxylated (activated) surfaces of SiO_2 on Si, Al_2O_3 on Al, glass, and hydrogen-terminated single-crystal silicon (H−Si). Hydroxylation of oxide surfaces is important to make them hydrophilic. Thermally grown silica can be activated through a sulfochromic treatment. The sample is dipped in a solution consisting of 100 mL of concentrated H_2SO_4, 5 mL of water, and 2 g of potassium

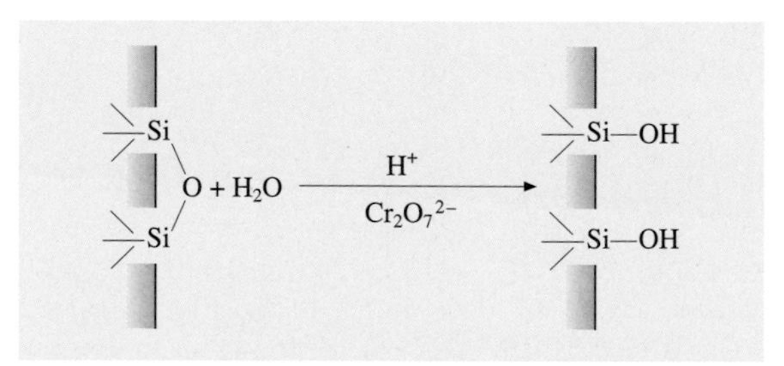

Fig. 27.4 Schematic showing the hydroxylation process occurring on a silica surface through a sulfochromic treatment

Table 27.9 Selected substrates and precursors that have commonly been used for formation of SAMs

Substrate	Precursor	Binding with substrate
Au	RSH (thiol)	RS−Au
Au	ArSH (thiol)	ArS−Au
Au	RSSR′ (disulfide)	RS−Au
Au	RSR′ (sulfide)	
Si/SiO_2, glass	$RSiCl_3$ (trichlorosilane)	Si−O−Si (Siloxane)
Si/Si−H	RCOOH (carboxyl)	R−Si
Metal oxides (e.g., Al_2O_3, SnO_2, TiO_2)	RCOOH (carboxyl)	RCOO− ... MO_n

R represents alkane (C_nH_{2n+2}) and Ar represents aromatic hydrocarbon. It consists of various surface active head groups and mostly with methyl terminal group

bichromate for 3–15 min [27.34]. The sample is then rinsed with flowing pure water. This results in a surface with silanol groups (−Si−OH) that is hydrophilic, Fig. 27.4. Bulk silicon, polysilicon film, or SiO_2 film surfaces can also be treated to produce activated hydrophilic silica surface by immersion in about three parts H_2SO_4 and one part H_2O_2 at temperatures ranging from ambient to 80 °C. For organic molecules to pack together and provide a better ordering, a substrate for given molecules should be selected such that the cross-sectional diameter of the spacer chains of the molecule is equal to or smaller than the distance between the anchor groups attached to the substrate. As an example, epitaxial Au film on glass, mica, or single-crystal silicon, produced by e-beam evaporation, is commonly used because it can be deposited on smooth surfaces as a film that is atomically flat and defect free. For the case of alkanethiolate film, the advantage of Au substrate over SiO_2 substrate is that it results in better ordering, because the cross-sectional diameter of the alkane molecule is slightly smaller than the distance between sulfur atoms attached to the Au substrate ($\sim$ 0.53 nm). The thickness of the film can be controlled by varying the length of the hydrocarbon chain, and the surface properties of the film can be modified by the terminal group.

Some of the SAMs have been widely reported. SAMs of long-chain fatty acids, $C_nH_{2n+1}COOH$ or $(CH_3)(CH_2)_nCOOH$ ($n = 10, 12, 14$ or 16), on glass or alumina have been used in films studies since the 1950s [27.16, 20]. Probably the most studied SAMs to date are n-alkanethiolate monolayers $CH_3(CH_2)_nS-$, prepared from adsorption of alkanethiol $-(CH_2)_nSH$ solution onto an Au film (n-alkyl and n-alkane are used interchangeably) [27.35–37, 43] and n-alkylsiloxane monolayers produced by adsorption of n-alkyltrichlorosilane $-(CH_2)_nSiCl_3$ solution onto hydroxylated Si/SiO_2 substrate [27.50] with siloxane (Si−O−Si) binding, Fig. 27.5. *Jung* et al. [27.51] have produced organosulfur monolayers – decanethiol $(CH_3)(CH_2)_9SH$, and didecyl $CH_3(CH_2)_9-S-(CH_2)_9CH_3$ on Au films. *Geyer* et al. [27.52], *Bhushan* and *Liu* [27.35], *Liu* et al. [27.36], and *Liu* and Bhushan [27.37] have produced monolayers of 1,1′-biphenyl-4-thiol (BPT) on Au surface, where the spacer chain of the film consists of two phenyl rings with a hydrogen end group. *Bhushan* and *Liu* [27.35] and *Liu* and *Bhushan* [27.37] have also reported monolayers of 4,4′-dihydroxybiphenyl on Si surface.

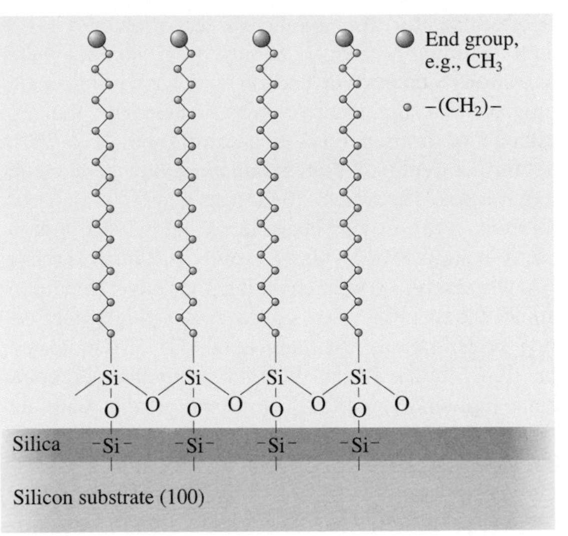

Fig. 27.5 Schematics of a methyl-terminated, n-alkylsiloxane monolayer on Si/SiO_2

27.3 Tribological Properties of SAMs

The basis for molecular design and tailoring of SAMs must start from a complete knowledge of the interrelationships between the molecular structure and tribological properties of SAMs, as well as a deep understanding of the adhesion, friction, and wear mechanisms of SAMs at the molecular level. Friction and wear properties of SAMs have been studied on macro- and nanoscales. Macroscale tests are conducted using a so-called pin-on-disk tribotester apparatus in which a ball specimen slides against a lubricated flat specimen [27.9, 10]. Nanoscale tests are conducted using an atomic force/friction force microscope (AFM/FFM) [27.4, 9, 10]. In the AFM/FFM experiments, a sharp tip of radius ranging from about 5–50 nm slides against a SAM specimen. A Si_3N_4 tip is commonly used for friction studies and a natural diamond tip is commonly used for scratch, wear, and indentation studies.

In early studies, the effect of chain length of the carbon atoms of fatty acid monolayers on the coefficient of friction and wear on the macroscale was studied by *Bowden* and *Tabor* [27.16] and *Zisman* [27.20]. *Zisman* reported that for the monolayers deposited on a glass surface sliding against a stainless steel surface, there is a steady decrease in friction with increasing chain length. At a significantly long chain length, the coefficient of friction reaches a lower limit, Fig. 27.6. He further reported that monolayers having a chain length below 12 carbon atoms behave as liquids (poor durability), those with chain length of 12–15 carbon atoms behave like a plastic solid (medium durability), whereas those with chain lengths above 15 carbon atoms behave like a crystalline solid (high durability). Investigations by *Ruhe* et al. [27.53] indicated that the lifetime of the alkylsilane monolayer coating on a silicon surface increases greatly with an increase in the chain length of the alkyl substituent. *DePalma* and *Tillman* [27.54] showed that a monolayer of n-octadecyltrichlorosilane (n-$C_{18}H_{37}SiCl_3$, OTS) is an effective lubricant on silicon, followed by n-undecyltrichlorosilane (n-$C_{11}H_{23}SiCl_3$, UTS) and (tridecafluoro-1,1,2,2-tetrahydrooct-1-yl) trichlorosilane (n-$C_6F_{13}CH_2CH_2SiCl_3$, FTS). *Ando* et al. [27.55] suggested that γ-(N,N-dioctadecylsuccinylamino) propyltriethoxysilane monolayer is a candidate lubricant. The film exhibited a low coefficient of kinetic friction of 0.1, without stick-slip. However, tests in all these investigations were carried out using a pin-on-disk tribotester under relatively large normal loads up to 0.15 N. Thus, the relevance of these tests is questionable for micro/nanodevice application.

With the development of AFM techniques, researchers have successfully characterized the nanotribological properties of self-assembled monolayers [27.1, 4, 34–38]. Studies by *Bhushan* et al. [27.34] showed that C_{18} alkylsiloxane films exhibit the lowest coefficient of friction and can withstand a much higher normal load during sliding compared to LB films, soft Au films, and hard SiO_2 coatings.

McDermott et al. [27.56] studied the effect of length of alkyl chains on the frictional properties of methyl-terminated n-alkylthiolate $CH_3(CH_2)_nS-$ films chemisorbed on Au(111) using an AFM. They reported that the longer chain monolayers exhibit a markedly lower friction and a reduced propensity to wear than shorter chain monolayers, Fig. 27.7. These results are in good agreement with the macroscale results by *Zisman* [27.20]. They also conducted infrared reflection spectroscopy to measure the bandwidth of the methylene stretching mode (ν_a (CH_2)), which exhibits a qualitative correlation with the packing density of the chains. They found that the chain structures of monolayers prepared with longer chain lengths are more ordered and more densely packed in comparison to those of monolayers prepared with shorter chain lengths. They further reported that the ability of the longer chain monolayers to retain molecular-scale order during shear leads to a lower observed friction. Monolayers with a chain length of more than 12 carbon atoms, preferably 18 or more, are desirable for tribological applications. (Incidentally, the monolayer with 18 carbon atoms, octadecanethiol films, are commonly studied.)

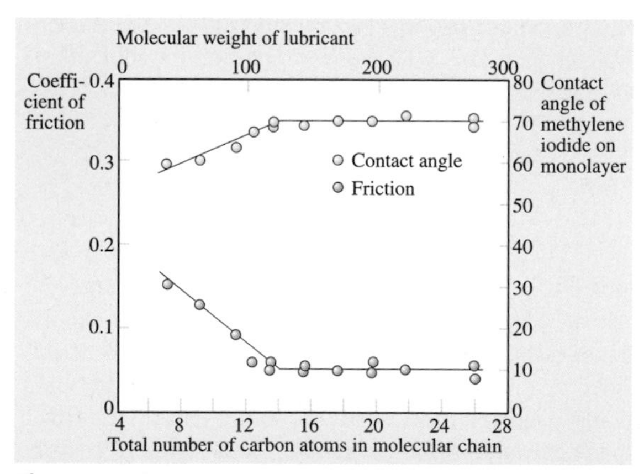

Fig. 27.6 Effect of chain length (or molecular weight) on coefficient of macroscale friction of stainless steel sliding on glass lubricated with a monolayer of fatty acid and contact angle of methyl iodide on condensed monolayers of fatty acid on glass [27.20]

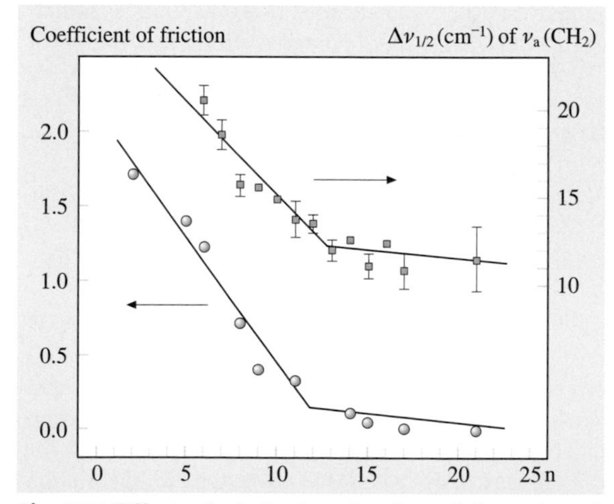

Fig. 27.7 Effect of chain length of methyl-terminated, n-alkanethiolate over Au film $AuS(CH_2)_nCH_3$ on the coefficient of microscale friction and peak bandwidth at half maximum ($\Delta\nu_{1/2}$) for the bandwidth of the methylene stretching mode [ν_a (CH_2)] [27.56]

Xiao et al. [27.57] and *Lio* et al. [27.58] also studied the effect of the length of the alkyl chains on the frictional properties of n-alkanethiolate films on gold and n-alkylsilane films on mica. Friction was found to be particularly high with short chains of less than eight carbon atoms. Thiols and silanes exhibit a similar friction force for the same n when $n > 11$; while for $n < 11$, silanes exhibit higher friction, larger than that for thiols by a factor of about 3 for $n = 6$. The increase in friction was attributed to the large number of dissipative modes in the less ordered chains that occurs when going from a thiol to a silane anchor or when decreasing n. Longer chains ($n > 11$), stabilized by van der Waals attraction, form more compact and rigid layers and act as better lubricants. *Schonherr* and *Vancso* [27.59] also correlated the magnitude of friction with the order of the alkane chains. The disorder of the short-chain hydrocarbon disulfides SAMs was found to result in a significant increase in the magnitude of friction.

Fluorinated carbon (fluorocarbon) molecules are commonly used for lubrication [27.2, 9, 10]. *Kim* et al. [27.60, 61] studied the influence of methyl-, isopropyl-, and trifluoromethyl-terminated alkanethiols on the friction properties of SAMs. They reported a factor of three increases in the friction in going from the methyl-terminated SAM to the trifluoromethyl-terminated SAM. They suggested that fluorinated monolayers exhibit higher frictional properties due to tighter packing at the interface, which arises from the larger van der Waals radii of the fluorine atoms. Subsequent steric and rotational factors between adjacent terminal groups give rise to long-range multi-molecular interactions in the plane of the $-CF_3$ groups. When these energetic barriers are overcome in the film structure, more energy is imparted to the film during sliding and results in higher friction for the fluorinated films.

Tsukruk et al. [27.62] and *Lee* et al. [27.63] studied the nanotribological properties of C_{60}-terminated alkyltrichlorosilanes and alkanedisulfides self-assembled monolayers. Both of their studies observed that the frictional forces of the C_{60}-terminated films are higher than those on normal methyl-terminated SAMs. However, *Tsukruk* et al. [27.62] reported that the C_{60}-terminated alkyltrichlorosilanes SAMs exhibit high wear resistance. *Tsukruk* and *Bliznyuk* [27.64] also studied the adhesion and friction between a Si sample and Si_3N_4 tip, in which both surfaces were modified by $-CH_3$, $-NH_2$, and $-SO_3H$ terminated silane-based SAMs. Various polymer molecules for the backbone were used. They reported that a very broad maximum adhesive force in the pH range of from 4 to 8 with a minimum adhesion at $pH > 9$ and $pH < 3$ were observed for all of the studied mating surfaces. This observation can be understood by considering a balance of electrostatic and van der Waals interaction between composite surfaces with multiple isoelectric points. The friction coefficient of NH_2/NH_2 and SO_3H/SO_3H mating SAMs is very high in aqueous solutions. Cappings of NH_2 modified surfaces (3-aminopropyltriethoxysilane) with rigid and soft polymer layers resulted in a significant reduction in adhesion to a level lower than that of untreated surfaces [27.65]. *Fujihira* et al. [27.66] studied the influence of surface terminal groups of SAMs and functional tip on adhesive force. It was found that the adhesive forces measured in air increase in the order of CH_3/CH_3, $CH_3/COOH$, and $COOH/COOH$.

Bhushan and *Liu* [27.35], *Liu* et al. [27.36], and *Liu* and *Bhushan* [27.37, 38] have studied the influence of spacer chains, surface terminal groups, and head groups on adhesion, friction, and wear properties of SAMs. They have explained the friction mechanisms using a molecular spring model in which local stiffness and the intermolecular forces govern the friction properties. They studied the influence of relative humidity, temperature, and velocity on adhesion and friction. They also investigated the wear mechanisms of SAMs by a continuous microscratch AFM technique.

To date, the nanotribological properties of alkyltrichlorosilanes, alkanethiols, and biphenyl thiol SAMs have been widely studied. In this chapter, we review, in some detail, the nanotribological properties of five kinds of SAMs, having alkyl and biphenyl spacer chains with different surface terminal groups ($-CH_3$, $-COOH$, and $-OH$), and head groups ($-SH$ and $-OH$), which have been investigated by AFM at various operating conditions, Fig. 27.8 [27.35–38]. Hexadecane thiol (HDT), 1,1′-biphenyl-4-thiol (BPT) and crosslinked BPT (BPTC), and 16-mercaptohexadecanoic acid thiol (MHA) were deposited on Au(111) substrates. A 4,4′-dihydroxybiphenyl (DHBp) film was deposited on a hydrogenated Si(111) substrate. Figure 27.8 shows that HDT and MHA have the same head groups and spacers, but different surface terminal groups; their surface terminals are $-CH_3$ and $-COOH$, respectively. BPT and DHBp have the same spacers, but the head and surface terminal groups of DHBp are $-OH$ instead of $-SH$ and $-CH$ in BPT. Crosslinked BPTC was produced by irradiation of BPT monolayers with low energy electrons.

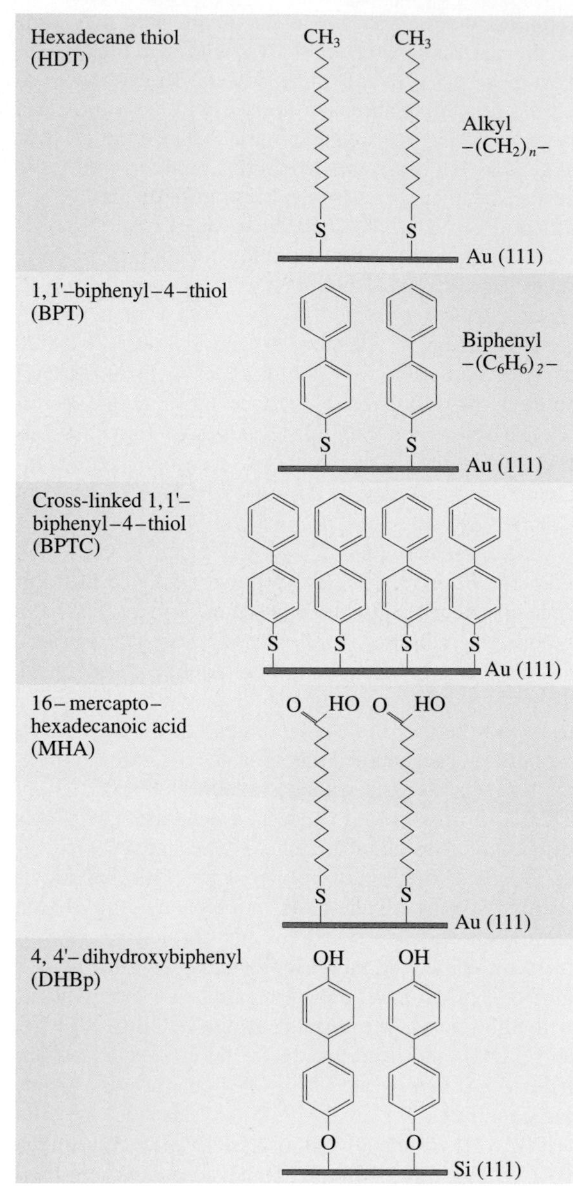

Fig. 27.8 Schematics of the structures of HDT, BPT, MHA, DHBp, and BPTC SAMs [27.35]

27.3.1 Surface Roughness and Friction Images of SAMs Films

Surface height and friction force images of SAMs were recorded simultaneously on an area of 1 μm × 1 μm by an AFM, Fig. 27.9 [27.35]. The topography of Au(111) film and SAMs deposited on Au(111) substrates appear to be granular. A good correlation between the surface height and the corresponding friction force images was observed. It was noticed that the friction force changed with surface height in the same direction (upwards and downwards) in both trace and retrace friction profiles of the friction loop. Thus, the change in friction force corresponds to transitions in surface slope [27.4, 35, 36, 67]. Figure 27.9 also indicates that the topography and friction images of Si(111) and DHBp/Si(111) are very similar; they exhibit featureless surfaces and their friction force images do not show any abrupt changes.

For further analysis presented later in this chapter, the roughness, thickness, tilt angles, and spacer chain lengths of Si(111), Au(111), and various SAMs are listed in Table 27.10 [27.35]. The roughness values of BPT and MHA are very close to that of Au(111). But the roughness of BPTC is lower than that of Au(111) and BPT; this is caused by electron irradiation. Table 27.10 indicates that the roughness values of HDT and DHBp are much higher than their substrate roughness of Au(111) and Si(111), respectively. This is caused by local aggregation of organic compounds on the substrates during SAMs deposition. Table 27.10 also indicates that the thickness of biphenyl thiol SAMs is generally thinner than the alkylthiol, which is responsible for shorter spacer chain in biphenyl thiol.

27.3.2 Adhesion, Friction, and Work of Adhesion

The average values and standard deviation of the adhesive force and coefficients of friction measured by contact mode AFM are presented in Fig. 27.10 [27.35]. Based on the data, the adhesive force and coefficient of friction of SAMs are less than their corresponding substrates. Among various films, HDT exhibits the lowest values. The ranking of adhesive forces F_a is in the following order: $F_{a\text{-Au}} > F_{a\text{-Si}} > F_{a\text{-DHBp}} \approx F_{a\text{-MHA}} > F_{a\text{-BPT}} > F_{a\text{-BPTC}} > F_{a\text{-HDT}}$. And the ranking of the coefficients of friction μ is in the following order: $\mu_{\text{Si}} > \mu_{\text{Au}} > \mu_{\text{DHBp}} > \mu_{\text{BPTC}} > \mu_{\text{BPT}} \approx \mu_{\text{MHA}} > \mu_{\text{HDT}}$. The ranking of various SAMs for adhesive force and coefficient of friction is similar. It suggests that alkylthiol and biphenyl SAMs can be used as effective molecular lubricants for micro/nanodevices fabricated from silicon.

In micro/nano scale contact, liquid capillary condensation is one of the factors that influence adhesion and friction. In the case of a sphere in contact with a flat surface, the attractive Laplace force caused by water

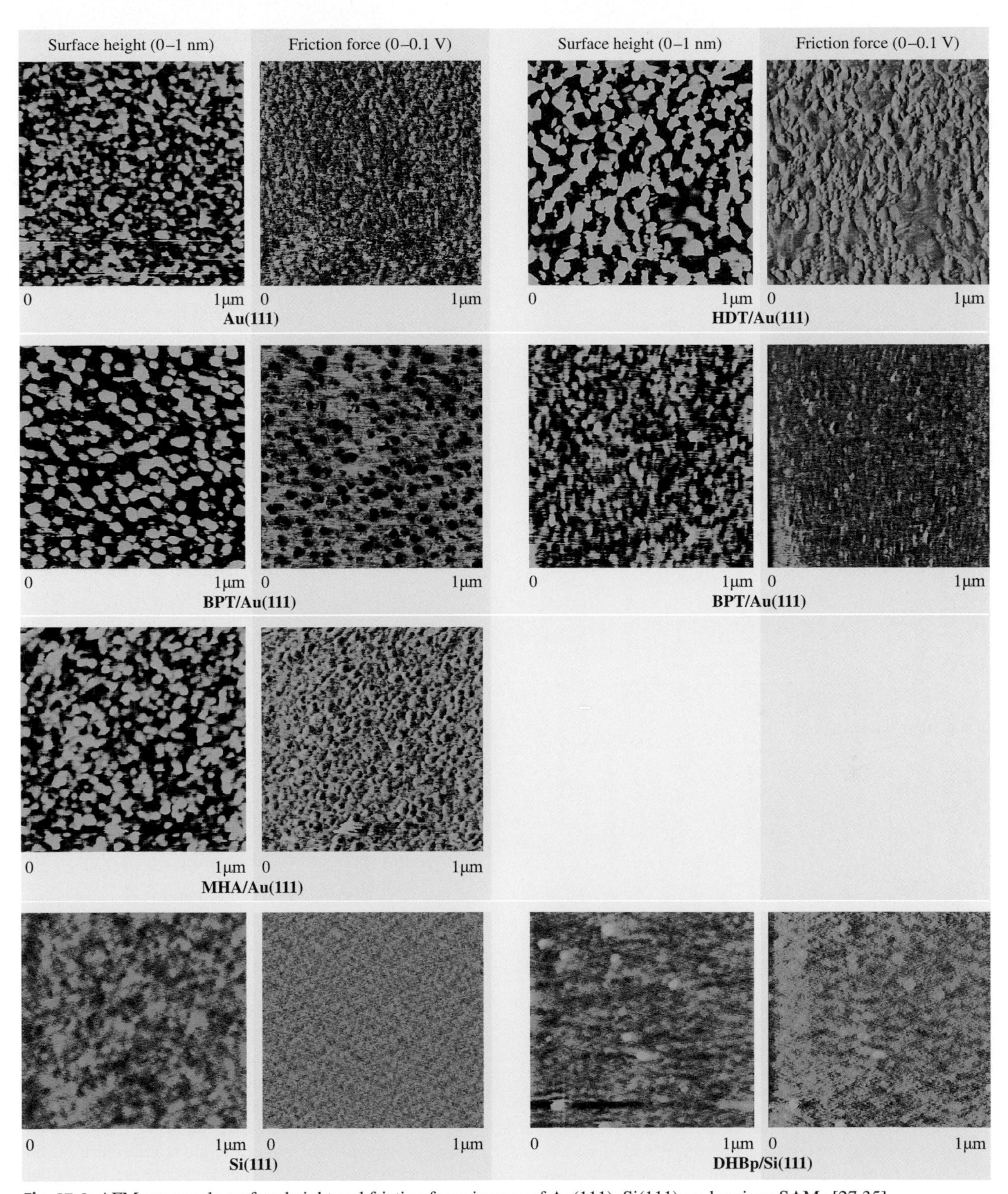

Fig. 27.9 AFM gray-scale surface height and friction force images of Au(111), Si(111), and various SAMs [27.35]

Table 27.10 The R_a roughness, thickness, tilt angles, and spacer chain lengths of SAMs

Samples	R_a roughness[1] (nm)	Thickness[2] (nm)	Tilt angle[2] (deg.)	Spacer length[3] (nm)
Si(111)	0.07			
Au(111)	0.37			
HDT	0.92	1.89	30	1.91
BPT	0.36	1.25	15	0.89
BPTC	0.14	1.14	25	0.89
MHA	0.37	2.01	30	1.91
DHBp	0.25	1.13	–	0.89

[1] Measured by an AFM with 1 μm × 1 μm scan size using a Si_3N_4 tip under 3.3 nN normal load.
[2] The thickness and tilt angles of BPT, BPTC, and DHBp are reported by *Geyer* et al. [27.52]. The thickness and tilt angles of HDT and MHA are reported by *Ulman* [27.40].
[3] The spacer chain lengths of alkylthiols were calculated by the method reported by *Miura* et al. [27.68]. The spacer chain lengths of biphenyl thiols were calculated by the data reported by *Ratajczak-Sitarz* et al. [27.69].

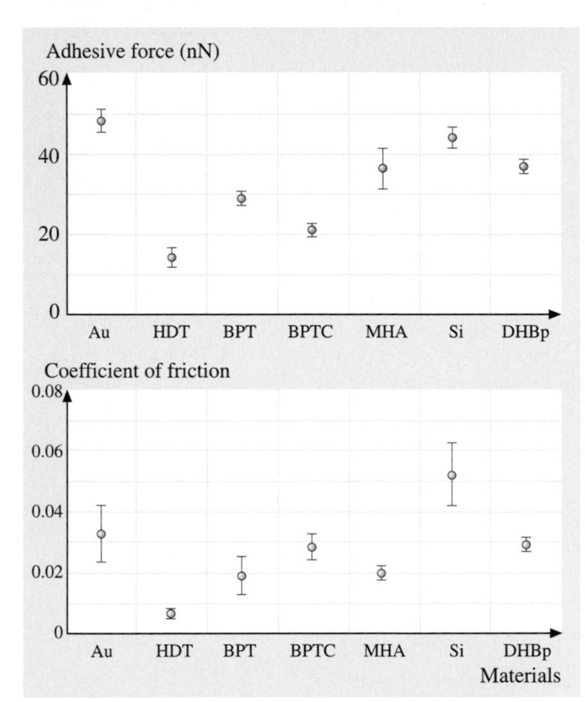

Fig. 27.10 Adhesive forces and coefficients of friction of Au(111), Si(111), and various SAMs [27.35]

capillary is:

$$F_L = 2\pi R\gamma_{la}(\cos\theta_1 + \cos\theta_2)\ , \tag{27.1}$$

where R is the radius of the sphere, γ_{la} is the surface tension of the liquid against air, θ_1 and θ_2 are the contact angles between liquid and flat and spherical surfaces, respectively [27.9, 10]. In an AFM adhesive study, the tip-flat sample contact is just like a sphere in contact with a flat surface and the liquid is water. Since a single tip was used in the adhesion measurements, $\cos\theta_2$ can be treated as a constant. Therefore,

$$\begin{aligned} F_L &= 2\pi R\gamma_{la}(1+\cos\theta_1) - 2\pi R\gamma_{la}(1-\cos\theta_2) \\ &= 2\pi R\gamma_{la}(1+\cos\theta_1) - C\ , \end{aligned} \tag{27.2}$$

where C is a constant.

Based on the following Young-Dupre equation, work of adhesion W_a (the work required to pull apart the unit area of the solid-liquid interface) can be written as [27.70]

$$W_a = \gamma_{la}(1+\cos\theta_1)\ . \tag{27.3}$$

It indicates that W_a is determined by the contact angle of SAMs, i. e., is influenced by the surface chemistry properties (polarization and hydrophobicity) of SAMs. By substituting (27.3) into (27.2), F_L can be expressed as

$$F_L = 2\pi R W_a - C\ . \tag{27.4}$$

When the influence on the adhesive force of other factors, such as van der Waals force, is very small, then adhesive force $F_a \approx F_L$. Thus the adhesive force F_a should be proportional to work of adhesion W_a.

Contact angle is a measure of the wettability of a solid by a liquid and determines the W_a value. The value of the contact angle changes with time. If the droplet expands, the contact angle is referred to as the advancing contact angle. The contact angles of distilled water on Si(111), Au(111), and SAMs were measured using a contact angle goniometer [27.35]. After dropping water on the sample surfaces, the dynamic advancing contact angle (DACA) was measured with a frequency of 60 s for spontaneous spreading for at least 420 s. The DACA as a function of time data is given in Fig. 27.11a. It indicates that there is a linear relationship between DACA and rest time. As stated by many authors, the static advancing contact angle (SACA) (at $t = 0$) should be used for surface characterization [27.71–73]. This SACA can be obtained by simple extrapolation of the data in Fig. 27.11a to $t = 0$. The SACA of Si(111), Au(111), and SAMs is summarized in Fig. 27.11b. For water, $\gamma_{la} = 72.6\,\mathrm{mJ/m^2}$ at 22 °C, and by using (27.3), the W_a data are obtained and presented in Fig. 27.11c. The W_a can be ranked in the following order: $W_{a\text{-Si}}(117.0) > W_{a\text{-DHBp}}(98.8) \approx W_{a\text{-MHA}}(101.8) \approx W_{a\text{-Au}}(97.1) > W_{a\text{-BPT}}(86.8) > W_{a\text{-BPTC}}(82.1) > W_{a\text{-HDT}}(61.4)$. Excluding $W_{a\text{-Au}}$, this order exactly matches the order of adhesion force in Fig. 27.10. The relationship between F_a and W_a is summarized in Fig. 27.12 [27.35]. It indicates that the adhesive force F_a (nN) increases with work of adhesion W_a (mJ/m^2) by the following linear relationship:

$$F_a = 0.57 W_a - 22\,. \qquad (27.5)$$

These experimental results agree well with the modeling prediction presented earlier in (27.4). It proves that on the nanoscale at ambient condition the adhesive force of SAMs is mainly influenced by the water capillary force. Comparing the results of Figure 27.12 with Figure 27.8, it is found that MHA and DHBp, which have polar surface terminal groups (−COOH and −OH), lead to larger W_a and eventually larger adhesive forces. Though both HDT and BPT do not have polar surface groups, the surface terminal of HDT has a symmetrical structure, which causes a smaller electrostatic attractive

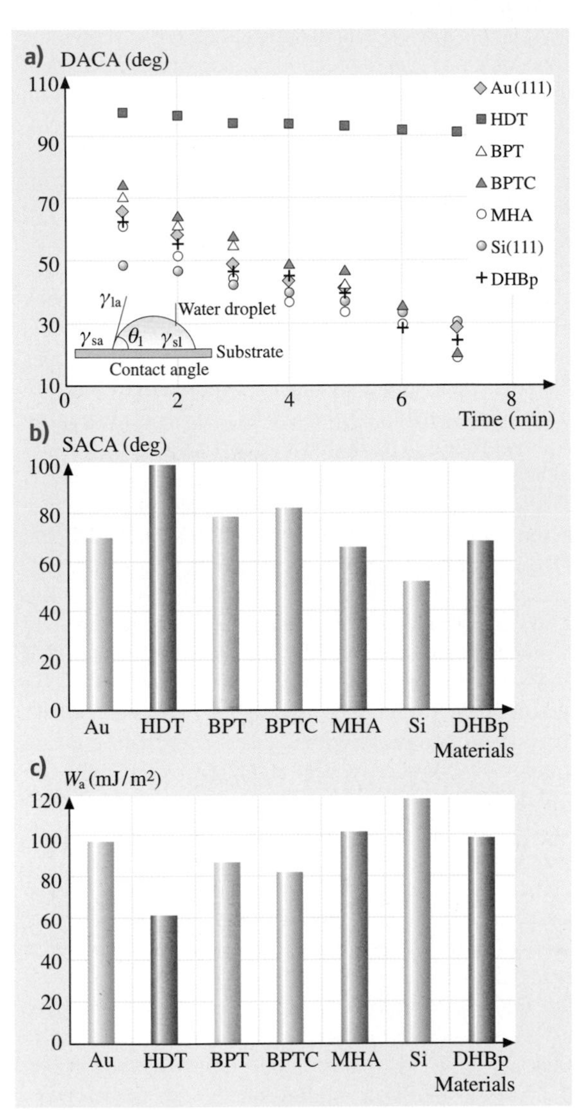

Fig. 27.11 (a) Variation of dynamic advancing contact angle (DACA) with time, (b) the static advancing contact angle (SACA), and (c) work of adhesion of Au(111), Si(111), and various SAMs. All of the points in this figure represent the mean value of six measurements. The uncertainty associated with the average contact angle is within ±2°. The insert schematic in (a) shows the contact angle of the water droplet on the sample surface [27.35]

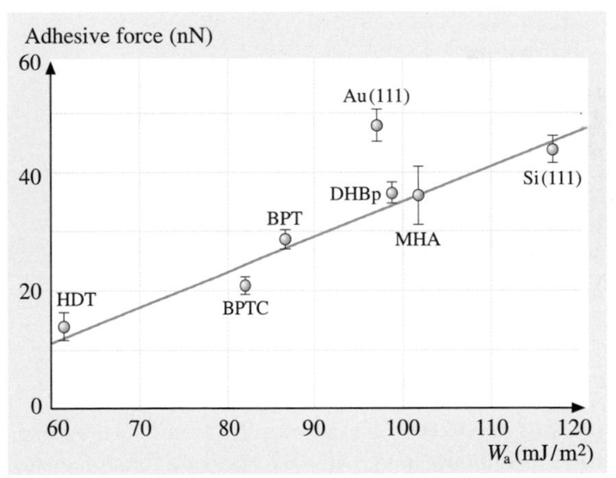

Fig. 27.12 Relationship between the adhesive forces and work of adhesion of different specimen [27.35]

force and yields smaller adhesive force than BPT. It is believed that the easy attachment of Au on the tip should be one of the reasons that cause the large adhesive force, which do not fit the linear relationship described by (27.5).

27.3.3 Stiffness, Molecular Spring Model, and Micropatterned SAMs

Next, the friction mechanisms of SAMs are examined. Monte Carlo simulation of the mechanical relaxation of $CH_3(CH_2)_{15}SH$ self-assembled monolayer performed by *Siepman* and *McDonald* [27.74] indicated that SAMs compress and respond nearly elastically to microindentation of an AFM tip under a critical normal load. Compression can lead to major changes in the mean molecular tilt (i. e., orientation), but the original structure is recovered as the normal load is removed.

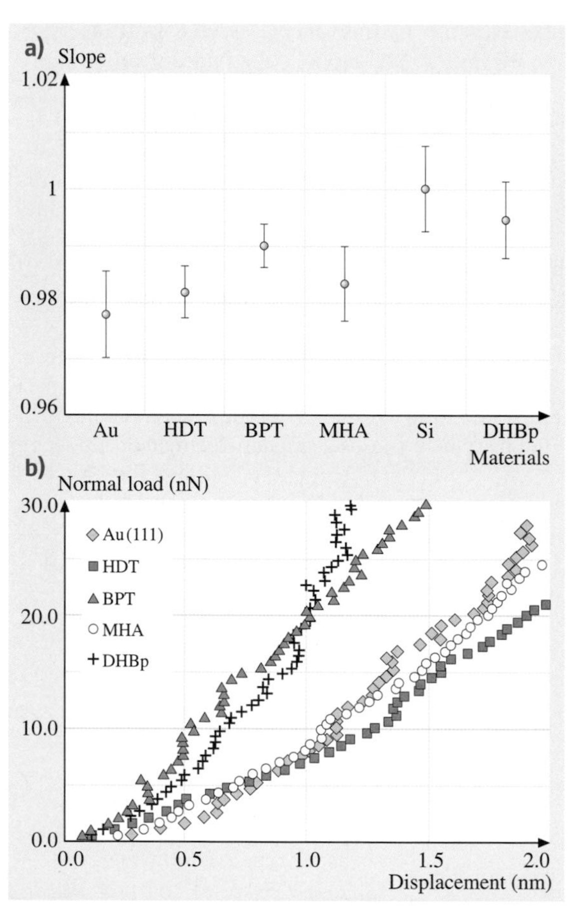

Fig. 27.13 **(a)** Slopes of cantilever deflection versus piezo displacement measured by force distance calibration mode AFM, and **(b)** normal load versus displacement curves of Au(111), Si(111), and various SAMs

To study the difference in the stiffness of various films, the stiffness properties were measured by an AFM in force-distance calibration mode, as well as in force modulation mode [27.4, 37]. Figure 27.13a shows the slope of cantilever deflection versus piezo movement obtained in the force-distance calibration mode. For an infinite hard material, since the surface can not be compressed, the cantilever deflection should equal the piezo movement distance, which means the slope of the cantilever deflection versus piezo movement equals 1. For a soft material, the surface can be compressed, causing a reduced cantilever deflection, and the slopes are smaller than 1. In this study the cantilever has been initially calibrated to slope $= 1$ against a clean, rigid Si(111) surface. Figure 27.13a indicates that the slopes of all of the SAMs are less than 1, which suggests that all of the SAMs can be compressed by a Si_3N_4 tip during the loading process. The slope value of SAMs can be ranked in the following order: $S_{DHBp} > S_{BPT} > S_{MHA} \approx S_{HDT}$. This order reflects the influence of molecular structure on the compression properties of SAMs. Since BPT and DHBp have rigid benzene structure, they are more difficult to compress than HDT and MHA. Figure 27.13b shows the variation of the displacement with normal load during indentation mode. It also clearly indicates that SAMs can be compressed. At a given normal load, long carbon chain structure SAMs such as HDT and MHA are easy to compress compared to rigid benzene-ring structure SAMs such as BPT and DHBp. *Garcia-Parajo* et al. [27.75] have also reported the compression and relaxation of octadecyltrichlorosilane (OTS) film in their loading and unloading tests.

In order to explain the frictional difference of SAMs, based on the friction and stiffness measurements by AFM and the Monte Carlo simulation, a molecular spring model is presented in Fig. 27.14. It is believed that the self-assembled molecules on a substrate are just like assembled molecular springs anchored to the substrates [27.35]. A Si_3N_4 tip sliding on the surface of SAMs is like a tip sliding on the top of "molecular springs or brush". The molecular spring assembly has compliant features and can experience compression and orientation under normal load. The orientation of the "molecular springs or brush" reduces the shearing force at the interface, which, in turn, reduces the friction force. The possibility of orientation is determined by the spring constant of a single molecule (local stiffness), as well as the interaction between the neighboring

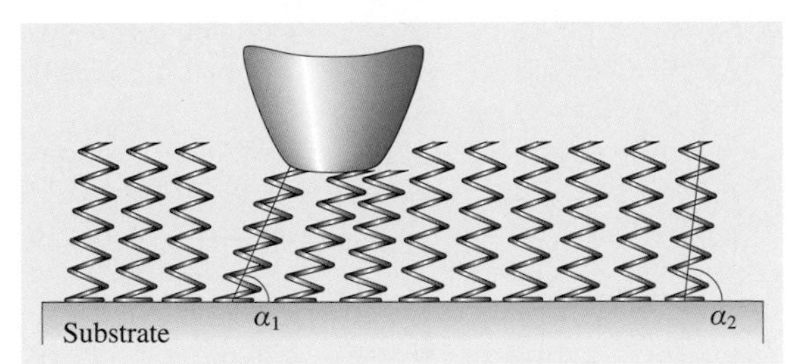

Fig. 27.14 Molecular spring model of SAMs. In this figure, $\alpha_1 < \alpha_2$, which is caused by the orientation under the normal load applied by an AFM tip. The orientation of the molecular springs reduces the shearing force at the interface, which, in turn, reduces the friction force. The molecular spring constant, as well as the inter-molecular forces can determine the magnitude of the coefficients of friction of SAMs. In this figure, the size of the tip and molecular springs are not drawn exactly to scale [27.35]

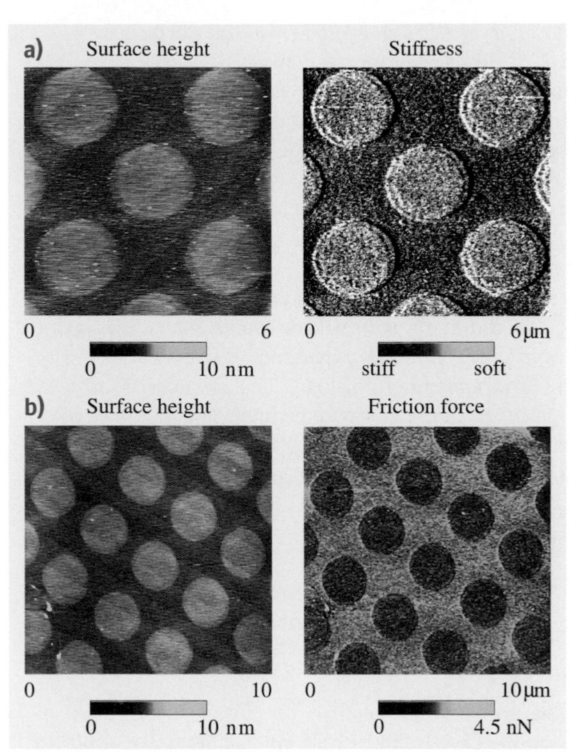

Fig. 27.15 (**a**) AFM gray-scale surface height and stiffness images, and (**b**) AFM gray-scale surface height and friction force images of micropatterned BDCS [27.37]

molecules, which can be reflected by packing density or packing energy. It should be noted that the orientation can lead to conformational defects along the molecular chains, which lead to energy dissipation. In the study of BPT by AFM, it was found that after the first several scans, the friction force is significantly reduced, but the surface height does not show any apparent change. This suggests that the molecular orientation can be facilitated by initial sliding and is reversible [27.38].

Based on the stiffness measurement results presented in Fig. 27.13 and the view of molecular structures in Fig. 27.14, biphenyl is a more rigid structure due to the contribution of two rigid benzene rings. Therefore, the spring constant of BPT is larger than that of HDT. The hydrogen (H^+) in a biphenyl chain has an electrostatic attractive force with the π electrons in the neighboring benzene ring. Thus, the intermolecular force between biphenyl chains is stronger than that for alkyl chains. The larger spring constant of BPT and stronger intermolecular force require a larger external force to allow it to orient, thus causing higher coefficient of friction. For MHA and DHBp, their basic chain structures are very close to HDT and BPT, respectively. But their surface terminals are different. The polar $-COOH$ and $-OH$ external functional groups in MHA and DHBp increase the adhesive force, thus leading to higher friction force than HDT and BPT, respectively. The crosslinking of BPT leads to a larger packing energy for BPTC. Therefore, it requires a larger external force to allow BPTC orientation, i. e., the coefficient of BPTC is higher than BPT.

An elegant way to demonstrate the influence of molecular stiffness on friction is to investigate SAMs with different structures on the same wafer. For this purpose, a micropatterned SAM was prepared. First, the biphenyldimethylchlorosilane (BDCS) was deposited on silicon by the typical self-assembly method [27.37]. Then the film was partially crosslinked using mask technique by low energy electron irradiation. Finally, the micropatterned BDCS films were realized that had the as-deposited and crosslinked coatings on the same wafer. The local stiffness properties of these micropatterned samples were investigated by force modulation AFM technique [27.76]. The variation in the deflection amplitude provides a measure of the relative local stiffness of the surface. Surface height, stiffness, and friction images of the micropatterned biphenyldimethylchlorosilane (BDCS) specimen are obtained and presented in Fig. 27.15 [27.37]. The circular areas correspond to the as-deposited film, and the remaining area to the crosslinked film. Figure 27.15a indicates that crosslinking caused by the low energy electron irradiation leads to about a 0.5 nm decrease of the surface height of BDCS films. The corresponding stiffness images indicate that the crosslinked area has a higher stiffness than the as-deposited area. Figure 27.15b indicates that the as-deposited area (higher surface height area) has lower friction force. Obviously, the data of the micropatterned sample prove that the local stiffness of SAMs has an influence on their friction performance. Higher stiffness leads to larger friction force. These results provide a strong proof of the suggested molecular spring model.

In summary, it has been found that SAMs exhibit compliance and can experience compression and orientation under normal load. The orientation of SAMs reduces the shear stress at the interface; therefore SAMs can serve as good lubricants. The molecular spring constant (local stiffness), as well as the intermolecular forces can influence the magnitude of the coefficients of friction of SAMs.

27.3.4 Influence of Humidity, Temperature, and Velocity on Adhesion and Friction

The influence of relative humidity on adhesion and friction was studied in an environmentally controlled chamber at 22 °C [27.35]. The results are presented in Fig. 27.16. It shows that for Si(111), Au(111), DHBp and MHA, the adhesive and friction forces increase with relative humidity. For BPT and BPTC, the adhesive force only slightly increases with relative humidity when the relative humidity is higher than 40%, but it is very interesting that their friction decreases in the same range. For HDT, in the testing range, the adhesive and friction force are not sensitive to changes in relative humidity.

It has already been shown that the adhesive force is linearly related to the W_a value of the surfaces in Fig. 27.12. DHBp and MHA have polar surface terminals and larger W_a value. This must therefore lead to the larger adhesive forces in higher relative humidity, which, in turn, increase the friction force. In comparison, HDT has a nonpolar $-CH_3$ surface terminal and a very small W_a value. Thus, the adhesive force and friction force of HDT are not sensitive to the change of humidity. The W_a values of BPT and BPTC are between Si(111) and HDT (see Fig. 27.11c); therefore their adhesive forces show a slight increase when RH > 40%. The fact that BPT and BPTC show lower friction force when RH > 40% suggests that a thin adsorbed water layer can act as a lubricant [27.35, 37]. The reason why the adsorbed water layer acts as a lubricant may be related to the thickness of the absorbed layer.

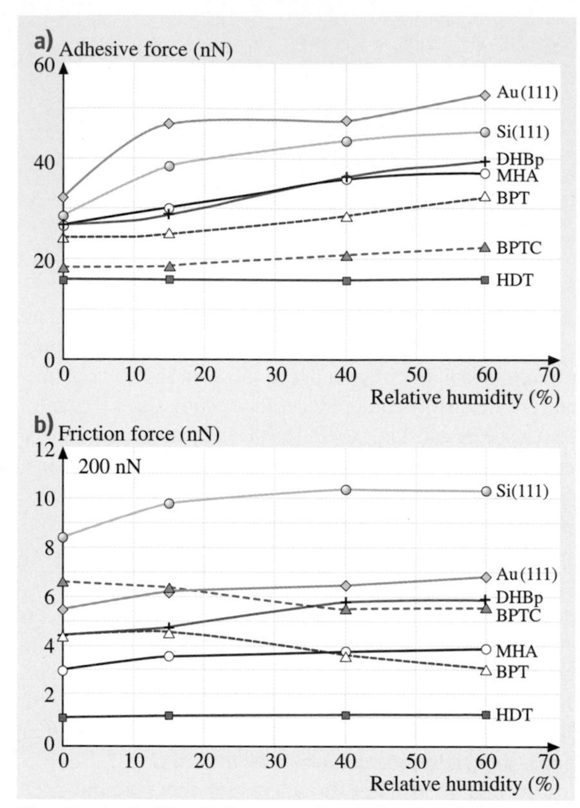

Fig. 27.16a,b The influence of relative humidity on the (a) adhesive force and (b) friction forces at 200 nN of Si(111), Au(111), and various SAMs [27.35]

Results of the influence of temperature on adhesion and friction of Si(111) and two selected typical SAMs – HDT and BPT – are presented in Fig. 27.17 [27.37]. The friction data are presented at various normal loads. In all cases, the normal load does not change the general trends. The data show that for Si(111), once the temperature is higher than 75 °C, the adhesive force decreases with temperature. The friction force of silicon decreases slightly with temperature from room temperature to 100 °C; once the temperature is higher than 100 °C, the friction force shows an apparent decrease. It is believed that desorption of the adsorbed water layer on Si(111) and the reduction of surface tension of water at high temperature are responsible for the decrease in adhesion and friction [27.77]. For HDT, in the testing range (22 °C–125 °C) its adhesion and friction show very slight change with temperature. For BPT, its adhesive force initially increases with temperature from 22 °C–75 °C, but when the temperature is higher than

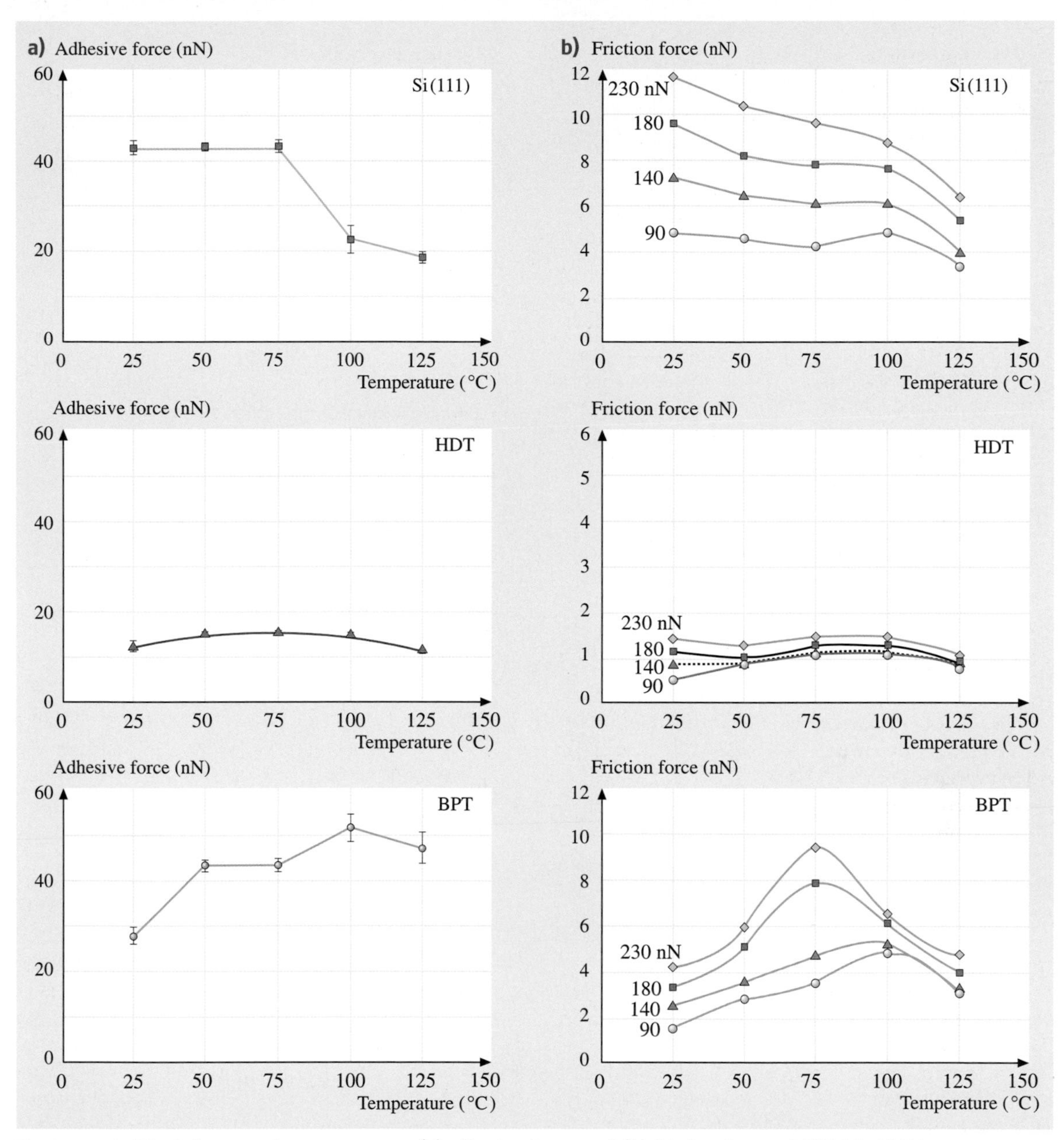

Fig. 27.17a,b The influence of temperature on (**a**) adhesive forces and (**b**) friction forces of Si(111), HDT, and BPT [27.37]

Compound type	Molecular formula	Melting points (°C) [27.77]
Linear carbon chain	$CH_3(CH_2)_{14}CH_2OH$	50
Benzene ring	3-HO−$C_6H_4C_6H_5$	78
	2-HO−$C_6H_4C_6H_5$	58–60

Table 27.11 Melting point of typical organic compounds similar to HDT and BPT SAMs

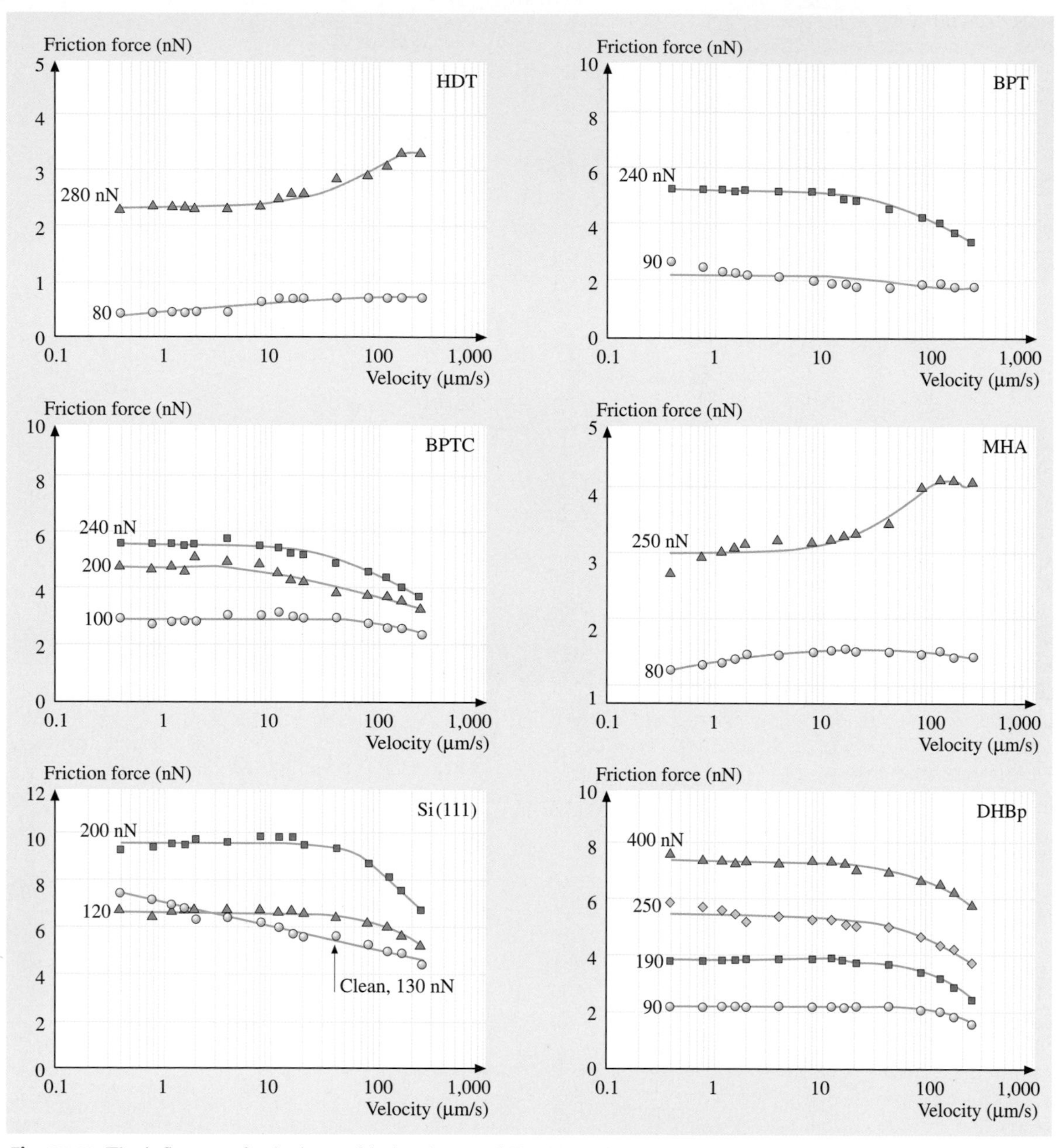

Fig. 27.18 The influence of velocity on friction forces of Si(111) and SAMs [27.37]

75 °C, the adhesive force remains steady within the experimental error. The variations in the friction force with temperature for BPT exhibit the maximum value, and the corresponding critical temperature is related to the normal load. The reason why BPT shows maximum friction force is believed to be caused by the melting of film at high temperature. When the temperature is below the melting point, an increase of temperature leads to the softening of SAM, which increases the real contact area and, consequently, the friction force. Once the temperature is higher than the melting point, the lubrication regime is changed from boundary lubrication in a solid

SAM to liquid lubrication in the melted SAM, and therefore the friction force is decreased. To date, the melting points of HDT and BPT SAMs are unknown. But from the literature [27.77], the melting points of typical linear carbon chain and benzene ring compounds are presented in Table 27.11. It is found that the reported melting of the benzene ring compounds is somewhat close to the frictional transition temperature of BPT measured by AFM. For HDT, their very compliant carbon chain can serve as an excellent lubricant even in solid state (i. e., at or near room temperature), so clear temperature effect is not observed as in the case of HDT.

Figure 27.18 shows the influence of velocity on the friction force of SAMs at different normal loads [27.37]. In all cases, normal load does not change the general trends. It is shown that for a silicon wafer, cleaned by Piranha solution just before the friction test, the friction force decreases linearly with increasing velocity. But for Si(111) without cleaning just before the AFM test, the friction force decreases only at high velocity. It is believed that this difference is caused by the adsorbed water and/or contamination on the surface during storage. Figure 27.18 also indicates that the velocity effects of SAMs depend on the molecular structures of SAMs. For SAMs that have compliant long carbon spacer chains, such as HDT and MHA, the friction force increases at high velocity, while for SAMs that have rigid biphenyl chains, such as BPT, BPTC, and DHBp, the friction force changes in the opposite way. The mechanisms responsible for the variation of the friction forces of SAMs with velocity are believed to be related to the viscoelastic properties of SAMs.

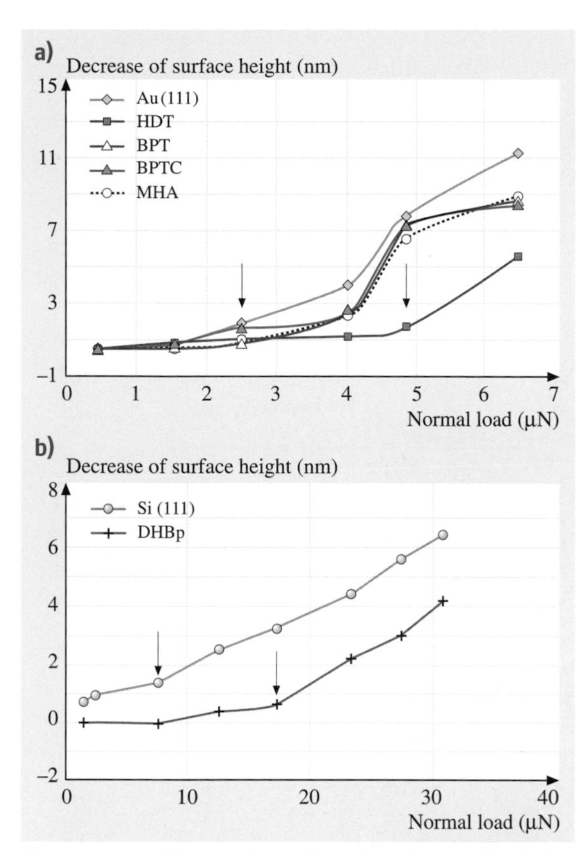

Fig. 27.19a,b Wear depth as a function of normal load after one scan cycle [27.35]. (**a**) on Au(111) and SAMs deposited on Au(111)/Si(111) (**b**) on Si(111) and DHPp deposited on Si(111)

27.3.5 Wear and Scratch Resistance of SAMs

Wear resistance was studied on an area of 1 μm × 1 μm using a diamond tip in an AFM. The variation of wear depth with normal loads is presented in Fig. 27.19 [27.35]. It clearly shows that in the whole testing range, DHBp on Si(111) exhibits much better wear resistance compared to Si(111), Au(111), and other SAMs that were deposited on the Au(111) substrate. For the SAMs deposited on Au(111), HDT exhibits the best wear resistance. For all of the tested SAMs, in the wear depth as a function of normal load curves, there appears a critical normal load, which is marked by arrows in Fig. 27.19. When the normal load is smaller than the critical normal load, the monolayer only shows slight height change in the scan areas. When the normal load is higher than the critical value, the height change of SAMs increase dramatically. Relocation and accumulation of BPT molecules have been observed during the initial several scans, which lead to the formation of a larger terrace. Wear studies of a single BPT terrace indicate that the wear life of BPT increases exponentially with terrace size [27.36, 37].

Scratch resistances of Si(111), Au(111), and SAMs were studied by a continuous AFM microscratch technique. Figure 27.20a shows coefficient of friction profiles as a function of increasing normal load and corresponding tapping mode AFM surface height images of the scratches captured on Si(111), Au(111), and SAMs [27.37]. Figure 27.20a indicates that there is an abrupt increase in the coefficients of friction for all of the tested samples. The normal load associated with this event is termed the critical load (indicated by the arrows labeled "A"). At the initial stages of the scratch, all the samples exhibit a low coefficient of friction, indicating that the friction force is dominated by the shear com-

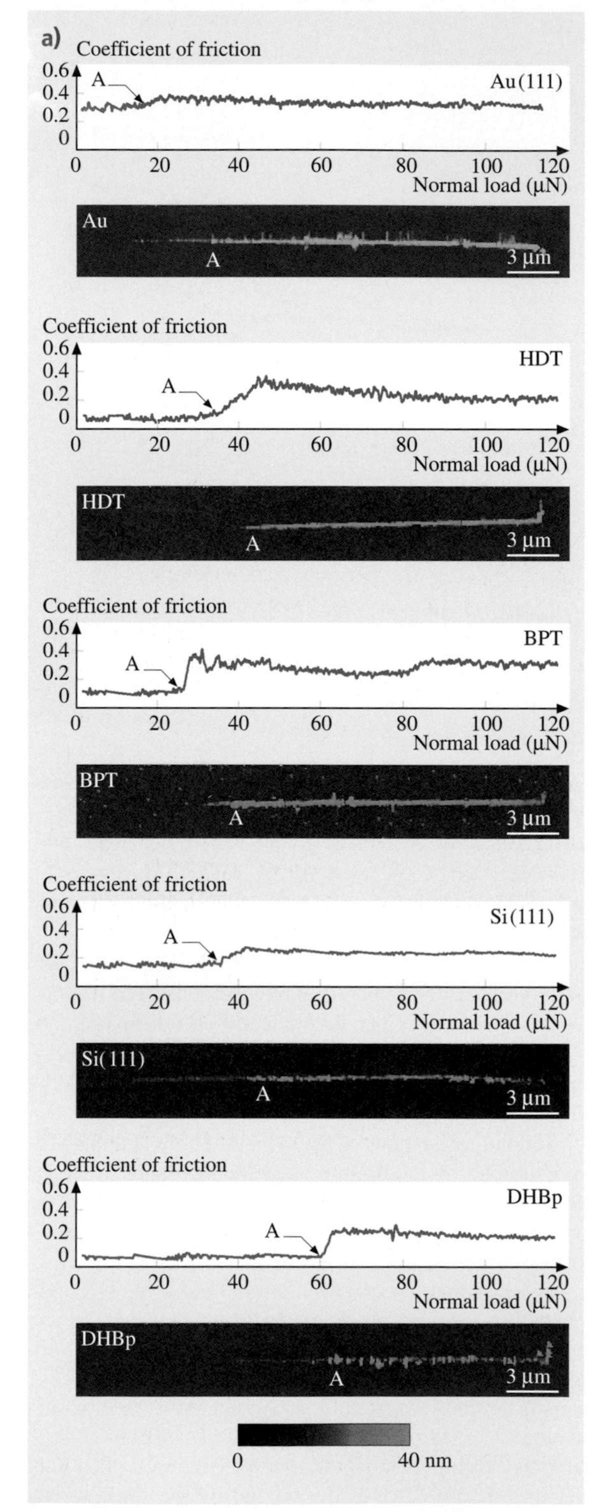

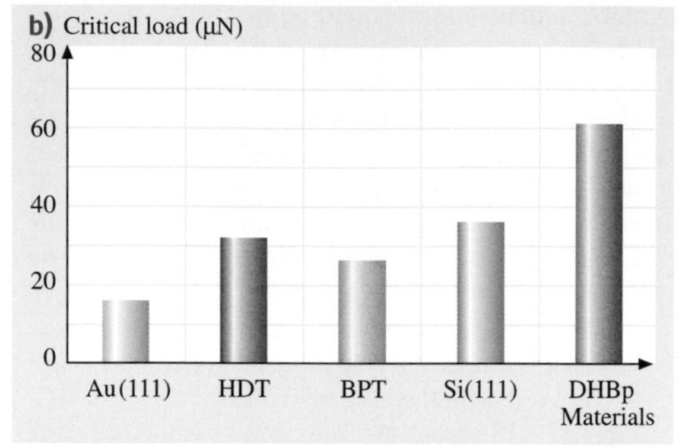

Fig. 27.20 (**a**) Coefficient of friction profiles during scratch as a function of normal load and corresponding AFM surface height images, and (**b**) critical loads estimated from the coefficient of friction profile and AFM images for Au(111), HDT/Au(111), BPT/Au(111), Si(111), and DHBp/Si(111) [27.37]

ponent. This is in agreement with analysis of the AFM images, which shows negligible damage on the surfaces prior to the critical load. At the critical load, a clear groove is formed, accompanied by the formation of material pileup at the sides of the scratch. This suggests that the initial damage that occurs at the critical load is due to ploughing associated with plastic deformation, which causes a sharp rise in the coefficient of friction. Beyond the critical load, debris can be seen in addition to material pile-up at the sides of the scratch. Figure 27.20b summarizes the critical loads for the various samples obtained in this study. It clearly indicates that all SAMs can increase the critical load of the corresponding substrate. DHBp, which is deposited on Si(111), shows superior scratch resistance in all of the tested samples.

Mechanisms responsible for a sudden drop in the decrease in surface height with an increase in load during wear and scratch tests need to be understood. *Barrena* et al. [27.78] observed that the height of self-assembled alkylsilanes decrease in discrete amounts with normal load. This step-like behavior is due to the discrete molecular tilts, which are dictated by the geometrical requirements of the close packing of molecules. Only certain angles are allowed due to the zigzag arrangement of the carbon atoms. The relative height of the monolayer under pressure can be calculated by the following equation:

$$\left(\frac{h}{L}\right) = \left[1+\left(\frac{na}{d}\right)^2\right]^{-\frac{1}{2}} , \qquad (27.6)$$

Table 27.12 Calculated $\left[1+\left(\frac{na}{d}\right)^2\right]^{-1/2}$ and measured $\left(\frac{h}{L}\right)$ relative heights of HDT and MHA self-assembled monolayers [27.35]

Steps (n)	Calculated[1] $\left[1+\left(\frac{na}{d}\right)^2\right]^{-1/2}$	Measured[2] $\left(\frac{h}{L}\right)$	
		HDT	MHA
1	0.883		
2	0.685	0.674[3]	0.643[3]
3	0.531	0.532[3]	0.552[3]
4	0.425	0.416[3]	
5	0.352	0.354[3]	
6	0.299		

[1] Calculations are based on the assumption that the molecules tilt in discrete steps (n) upon compression with a diamond AFM tip [27.78].

[2] Measured values are the mean values of three tests.

[3] These measured values correspond to the normal loads of 0.50 μN, 1.57 μN, 2.53 μN, and 4.03 μN, respectively.

where L is the total length of the molecule, h is the height of the SAMs in the tilt configuration (monolayer thickness), a is the distance between alternate carbon atoms in the molecule, d is the separation of the molecules, and n is the step number. For HDT and MHA, only the head groups are different in their work and this study. The spacer carbon chains in alkanethiol and alkylsilane are exactly the same. So the same a (0.25 nm) and d (0.47 nm) values are used in the calculation for HDT and MHA. The calculated and measured relative heights of HDT and MHA are listed in Table 27.12. When the normal loads are smaller than the critical values in Fig. 27.19, the measured relative height values of HDT and MHA are very close to the calculated values. This means that HDT and MHA underwent step tilting below critical normal loads.

Table 27.13 Calculated L_0 and measured residual film thickness for SAMs under critical load

	L_0^1 (nm)	Residual thickness[2] (nm)
HDT	0.24	0.25
BPT	0.39	0.42
BPTC	0.33	0.38
MHA	0.36	0.29
DHBp	–	0.52

[1] Calculated by the equation of $h = b\cos(\alpha)n + L_0$ [27.68]

[2] Measured by AFM using diamond tip under critical normal load. All of the data are the mean values of three tests

The residual SAMs thickness after wear under critical normal load was measured by profiling the worn film using AFM. The results are listed in Table 27.13. For an alkanethiol monolayer, the relationship between the monolayer thickness h and intercept length L_0 can be expressed as (see Fig. 27.21):

$$h = b\cos(\alpha)n + L_0 , \tag{27.7}$$

where b is the length of the projection of the C−C bond onto the main chain axis ($b = 0.127$ nm for alkanethiol), n is the chain length defined by $CH_3(CH_2)_nSH$, and α is the tilt angle [27.68]. For BPT and BPTC, based on the same principle and using the bond lengths reported in reference [27.69], the L_0 values are also calculated, Table 27.13. It indicates that the measured residual thickness values of SAMs under critical load are very close to the calculated intercept length L_0 values. It means that under the critical normal load, the Si_3N_4 tip approaches the interface and SAMs wear severely away from the substrate. This is due to the interface chemical adsorption bond strength (HS−Au and Si−O) being generally smaller than the other chemical bond strengths in SAMs

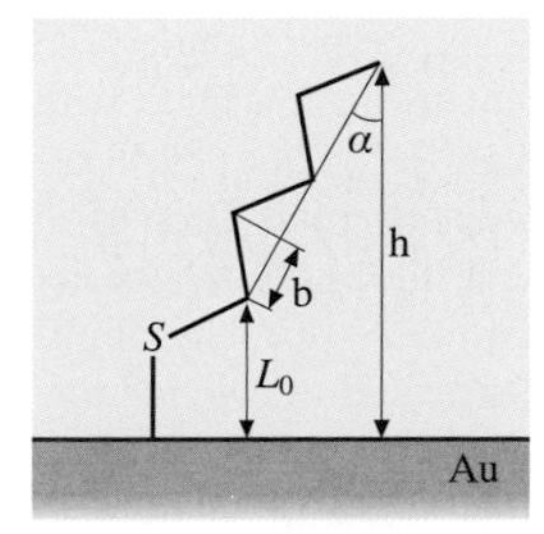

Fig. 27.21 Illustration of the relationship between the components of the equation $h = b\cos(\alpha)n + L_0$ [27.35]

Table 27.14 Bond strengths of the chemical bonds in SAMs [27.35]

SAMs	Bond	Bond strength[1] (kJ/mol)
Thiol on Au	S−Au	184 [27.58]
Hydroxyl on Si	O−Si	242.7 [27.79]
HDT	$H-CH_2$	464.8
	H−CH	421.7
	CH_3-t-C_4H_9	425.9±8
	t-C_4H_9−SH	286.2±6.3
MHA	O=CO	532.2±0.4
	$H-OCOC_2H_5$	445.2±8
	$HO-OCH_2C(CH_3)$	193.7±7.9
	$C_6H_5CH_2-COOH$	280
BPT	CH_3-CH_3	376.0±2.1
	$H_2C=CH_2$	733±8
	C_6H_5-SH	361.9±8
DHBp	C_6H_5-OH	361.9±8

[1] Most of the data are cited from [27.77], except where indicated. For MHA and DHBp, the bonds that are common as in HDT and DHBp are not repeated

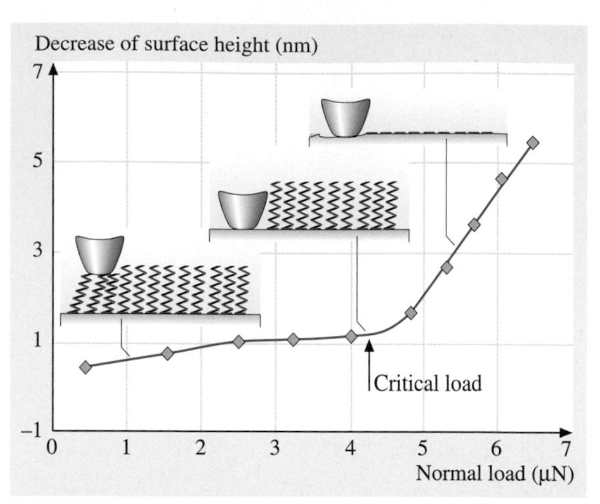

Fig. 27.22 Illustration of the wear mechanisms of SAMs with increasing normal load [27.37]

spacer chains (see Table 27.14). Based on this discussion, it is believed that the reason why DHBp has the best wear resistance is due to the rigid biphenyl ring structure (compared to linear carbon chain in alkylthiol), the hard Si(111) substrate (compared to Au(111) substrate), and the strong interface Si−O bond strength (compared to the weak S−Au bond strength in the other SAMs, see Table 27.14).

According to the wear and scratch results reported here and the above discussion, the transition of the wear mechanisms of SAMs with increasing normal load is illustrated in Fig. 27.22. Below the critical normal load, SAMs undergo step orientation, at the critical load SAMs wear away from the substrate due to the weak interface bond strengths, while above the critical normal load severe wear takes place on the substrate. In order to improve wear resistance, the interface bond must be enhanced; a rigid spacer chain and a hard substrate are also preferred.

27.4 Closure

Exposure of devices to humid environments results in condensates of water vapor from the environment. Condensed water, or a preexisting film of liquid, forms concave meniscus bridges between the mating surfaces. The negative Laplace pressure present in the meniscus results in an adhesive force that depends on the interface roughness, surface tension, and contact angle. The adhesive force can be significant in an interface with ultra-smooth surfaces, and it can be on the same order as the external load if the latter is small, such as in micro/nanodevices. Surfaces with high hydrophobicity can be produced by surface treatment. In many applications, hydrophobic films are expected to provide low adhesion, friction, and wear. To minimize high adhesion, friction, and/or because of small clearances in micro/nanodevices, these films should be molecularly thick. Liquid films of low surface tension or certain hydrophobic solid films can be used. Ordered molecular assemblies with high hydrophobicity can be engineered using chemical grafting of various polymer molecules with suitable functional head groups and nonpolar surface terminal groups.

The adhesion, friction, and wear properties of SAMs, having alkyl and biphenyl spacer chains with different surface terminal groups ($-CH_3$, −COOH, and −OH) and head groups (−SH and −OH), studied using an AFM, are reported in this chapter. It is found that the adhesive force varies linearly with the W_a value of SAMs, which indicates that capillary condensation of water plays an important role to the adhesion of SAMs on nanoscale at ambient conditions. It has been found that HDT exhibits the smallest adhesive force and friction force, because of low W_a of $-CH_3$ surface terminal groups, and high-compliance long carbon spacer chain. The friction data are explained using a molecular spring model, in which the local stiffness and intermolecular force governs its frictional performance. The results of the stiffness and friction characterization of the micropatterned sample with different structures support this model. The influence of relative humidity on adhesion and friction of SAMs is dominated by the thickness of the adsorbed water layer. At higher humidity, water can either increase friction through increased adhesion by meniscus effect in the contact zone or reduce friction through an enhanced water-lubricating effect. In the case of Si(111), the desorption of the adsorbed water layer and reduction of surface tension of water with increasing temperature reduce the adhesive force and friction force. The increase of temperature does not show an apparent influence on HDT, but strongly influences the adhesion and friction properties of BPT, which are believed to be related to its melting. The effect of velocity on friction properties of SAMs depends on their molecular structures. For alkylthiol SAMs, their friction is increased by increasing velocity, while for biphenyl SAMs it changes in the opposite way. The mechanism responsible for the variation of the friction forces of SAMs with velocity is believed to be related to their viscoelastic properties. Wear and continuous microscratch tests show that among the SAMs on Au(111), HDT exhibits the best wear resistance. DHBp on Si(111), due to its rigid biphenyl spacer chains, strong interface bonds, and a hard substrate, has the best wear resistance among all of the tested samples. For all of the SAMs, the wear depth as a function of normal load curves shows critical normal loads. Below the critical normal load SAMs undergo step orientation, at the critical load SAMs wear away from the substrate due to the weak interface bond strengths, while above the critical normal load severe wear take place on the substrate.

Based on the nanotribological studies of SAM films by AFM, they exhibit attractive hydrophobic and tribological properties. SAM films should find many tribological applications, including in micro/nanodevices.

References

27.1 B. Bhushan, J. N. Israelachvili, U. Landman: Nanotribology: Friction, wear and lubrication at the atomic scale, Nature **374** (1995) 607–616

27.2 B. Bhushan: *Tribology and Mechanics of Magnetic Storage Devices*, 2nd edn. (Springer, New York 1996)

27.3 B. Bhushan: *Tribology Issues and Opportunities in MEMS* (Kluwer, Dordrecht 1998)

27.4 B. Bhushan: *Handbook of Micro/Nanotribology*, 2nd edn. (CRC, Boca Raton 1999)

27.5 K. F. Man, B. H. Stark, R. Ramesham: *A Resource Handbook for MEMS Reliability* (California Institute of Technology, Pasadena 1998)

27.6 S. Kayali, R. Lawton, B. H. Stark: MEMS reliability assurance activities at JPL, EEE Links **5** (1999) 10–13

27.7 D. M. Tanner, N. F. Smith, L. W. Irwin et al.: *MEMS Reliability: Infrastructure, Test Structure, Experiments, and Failure Modes* (Sandia National Laboratories, Albuquerque 2000)

27.8 S. Arney: Designing for MEMS reliability, MRS Bull. **26** (2001) 296–299

27.9 B. Bhushan: *Principles and Applications of Tribology* (Wiley, New York 1999)

27.10 B. Bhushan: *Introduction to Tribology* (Wiley, New York 2002)

27.11 A. Ulman (Ed.): *Characterization of Organic Thin Films* (Butterworth-Heineman, Boston 1995)

27.12 B. Bhushan: Contact mechanics of rough surfaces in tribology: Multiple asperity contact, Tribol. Lett. **4** (1998) 1–35

27.13 B. Bhushan, W. Peng: Contact mechanics of multilayered rough surfaces, Appl. Mech. Rev. **55** (2002) 435–480

27.14 R. Maboudian: Surface processes in MEMS technology, Surf. Sci. Rep. **30** (1998) 209–269

27.15 M. Scherge, J. A. Schaefer: Surface modification and mechanical properties of bulk silicon. In: *Tribology Issues and Opportunities in MEMS*, ed. by B. Bhushan (Kluwer, Dordrecht 1998) pp. 529–537

27.16 F. P. Bowden, D. Tabor: *The Friction and Lubrication of Solids, Part I* (Clarendon, Oxford 1950)

27.17 V. N. Koinkar, B. Bhushan: Microtribological studies of unlubricated and lubricated surfaces using atomic force/friction force microscopy, J. Vac. Sci. Technol. A **14** (1996) 2378–2391

27.18 B. Bhushan, Z. Zhao: Macro- and microscale tribological studies of molecularly-thick boundary layers of perfluoropolyether lubricants for magnetic thin-film rigid disks, J. Info. Storage Proc. Syst. **1** (1999) 1–21

27.19 H. Liu, B. Bhushan: Nanotribological characterization of molecularly-thick lubricant films for applications to MEMS/NEMS by AFM, Ultramicroscopy **97** (2003) 321–340

27.20 W.A. Zisman: Friction, durability and wettability properties of monomolecular films on solids. In: *Friction and Wear*, ed. by R. Davies (Elsevier, Amsterdam 1959) pp. 110–148

27.21 M.E. Schrader, G.I. Loeb (Eds.): *Modern Approaches to Wettability* (Plenum, New York 1992)

27.22 A.W. Neumann, J.K. Spelt (Eds.): *Applied Surface Thermodynamics* (Dekker, New York 1996)

27.23 H. Hansma, F. Motamedi, P. Smith, P. Hansma, J.C. Wittman: Molecular resolution of thin, highly oriented poly(tetrafluoroethylene) films with the atomic force microscope, Polymer Commun. **33** (1992) 647–649

27.24 L. Scandella, A. Schumacher, N. Kruse, R. Prins, E. Meyer, R. Luethi, L. Howald, H.J. Guentherodt: Tribology of ultra-thin MoS_2 platelets on mica: Studies by scanning force microscopy, Thin Solid Films **240** (1994) 101–104

27.25 B. Bhushan: Chemical, mechanical and tribological characterization of ultra-thin and hard amorphous carbon coatings as thin as 3.5 nm: Recent developments, Diam. Relat. Mater. **8** (1999) 1985–2015

27.26 A. Erdemir, C. Donnet: Tribology of diamond, diamond-like carbon, and related films. In: *Modern Tribology Handbook Vol. 2: Materials, Coatings, and Industrial Applications*, ed. by B. Bhushan (CRC, Boca Raton 2001) pp. 871–908

27.27 V.F. Dorfman: Diamond-like nanocomposites (DLN), Thin Solid Films **212** (1992) 267–273

27.28 M. Grischke, K. Bewilogua, K. Trojan, H. Dimigan: Application-oriented modification of deposition process for diamond-like carbon based coatings, Surf. Coat. Technol. **74-75** (1995) 739–745

27.29 R.S. Butter, D.R. Waterman, A.H. Lettington, R.T. Ramos, E.J. Fordham: Production and wetting properties of fluorinated diamond-like carbon coatings, Thin Solid Films **311** (1997) 107–113

27.30 M. Grischke, A. Hieke, F. Morgenweck, H. Dimigan: Variation of the wettability of DLC coatings by network modification using silicon and oxygen, Diam. Relat. Mater. **7** (1998) 454–458

27.31 C. Donnet, J. Fontaine, A. Grill, V. Patel, C. Jahnes, M. Belin: Wear-resistant fluorinated diamondlike carbon films, Surf. Coat. Technol. **94-95** (1997) 531–536

27.32 D.J. Kester, C.L. Brodbeck, I.L. Singer, A. Kyriakopoulos: Sliding wear behavior of diamond-like nanocomposite coatings, Surf. Coat. Technol. **113** (1999) 268–273

27.33 H. Liu, B. Bhushan: Adhesion and friction studies of microelectromechanical systems/nanoelectromechanical systems materials using a novel microtriboapparatus, J. Vac. Sci. Technol. A **21** (2003) 1528–1538

27.34 B. Bhushan, A.V. Kulkarni, V.N. Koinkar, M. Boehm, L. Odoni, C. Martelet, M. Belin: Microtribological characterization of self-assembled and Langmuir–Blodgett monolayers by atomic and friction force microscopy, Langmuir **11** (1995) 3189–3198

27.35 B. Bhushan, H. Liu: Nanotribological properties and mechanisms of alkylthiol and biphenyl thiol self-assembled monolayers studied by atomic force microscopy, Phys. Rev. B **63** (2001) 245412-1–245412-11

27.36 H. Liu, B. Bhushan, W. Eck, V. Stadler: Investigation of the adhesion, friction, and wear properties of biphenyl thiol self-assembled monolayers by atomic force microscopy, J. Vac. Sci. Technol. A **19** (2001) 1234–1240

27.37 H. Liu, B. Bhushan: Investigation of nanotribological properties of alkylthiol and biphenyl thiol self-assembled monolayers, Ultramicroscopy **91** (2002) 185–202

27.38 H. Liu, B. Bhushan: Orientation and relocation of biphenyl thiol self-assembled monolayers, Ultramicroscopy **91** (2002) 177–183

27.39 A. Ulman: *An Introduction to Ultrathin Organic Films: From Langmuir–Blodgett to Self-Assembly* (Academic, San Diego 1991)

27.40 A. Ulman: Formation and structure of self-assembled monolayers, Chem. Rev. **96** (1996) 1533–1554

27.41 J.A. Zasadzinski, R. Viswanathan, L. Madsen, J. Garnaes, D.K. Schwartz: Langmuir–Blodgett films, Science **263** (1994) 1726–1733

27.42 J. Tian, Y. Xia, G.M. Whitesides: Microcontact printing of SAMs. In: *Thin Films – Self-Assembled Monolayers of Thiols*, Vol. 24, ed. by A. Ulman (Academic, San Diego 1998) pp. 227–254

27.43 Y. Xia, G.M. Whitesides: Soft lithography, Angew. Chem. Int. Edn. **37** (1998) 550–575

27.44 A. Kumar, G.M. Whitesides: Features of gold having micrometer to centimeter dimensions can be formed through a combination of stamping with an elastomeric stamp and an alkanethiol ink followed by chemical etching, Appl. Phys. Lett. **63** (1993) 2002–2004

27.45 S.Y. Chou, P.R. Krauss, P.J. Renstrom: Imprint lithography with 25-nanometer resolution, Science **272** (1996) 85–87

27.46 Y. Xia, E. Kim, X.M. Zhao, J.A. Rogers, M. Prentiss, G.M. Whitesides: Complex optical surfaces formed by replica molding against elastomeric masters, Science **273** (1996) 347–349

27.47 R.F. Service: Self-assembly comes together, Science **265** (1994) 316–318

27.48 M. Hein, L.R. Best, S. Pattison, S. Arena: *Introduction to General, Organic, and Biochemistry*, 6th edn. (Brooks/Cole, Pacific Grove 1997)

27.49 J.R. Mohrig, C.N. Hammond, T.C. Morrill, D.C. Neckers: *Experimental Organic Chemistry* (Freeman, New York 1998)

27.50 S.R. Wasserman, Y.T. Tao, G.M. Whitesides: Structure and reactivity of alkylsiloxane monolayers

formed by reaction of alkylchlorosilanes on silicon substrates, Langmuir **5** (1989) 1074–1089
27.51 C. Jung, O. Dannenberger, Y. Xu, M. Buck, M. Grunze: Self-assembled monolayers from organosulfur compounds: A comparison between sulfides, disulfides, and thiols, Langmuir **14** (1998) 1103–1107
27.52 W. Geyer, V. Stadler, W. Eck, M. Zharnikov, A. Golzhauser, M. Grunze: Electron-induced cross-linking of aromatic self-assembled monolayers: Negative resists for nanolithography, Appl. Phys. Lett. **75** (1999) 2401–2403
27.53 J. Ruhe, V. J. Novotny, K. K. Kanazawa, T. Clarke, G. B. Street: Structure and tribological properties of ultrathin alkylsilane films chemisorbed to solid surfaces, Langmuir **9** (1993) 2383–2388
27.54 V. DePalma, N. Tillman: Friction and wear of self-assembled tricholosilane monolayer films on silicon, Langmuir **5** (1989) 868–872
27.55 E. Ando, Y. Goto, K. Morimoto, K. Ariga, Y. Okahata: Frictional properties of monolayers of silane compounds, Thin Solid Films **180** (1989) 287–291
27.56 M. T. McDermott, J. B. D. Green, M. D. Porter: Scanning force microscopic exploration of the lubrication capabilities of n-alkanethiolate monolayers chemisorbed at gold: Structural basis of microscopic friction and wear, Langmuir **13** (1997) 2504–2510
27.57 X. Xiao, J. Hu, D. H. Charych, M. Salmeron: Chain length dependence of the frictional properties of alkylsilane molecules self-assembled on mica studied by atomic force microscopy, Langmuir **12** (1996) 235–237
27.58 A. Lio, D. H. Charych, M. Salmeron: Comparative atomic force microscopy study of the chain length dependence of frictional properties of alkanethiol on gold and alkylsilanes on mica, J. Phys. Chem. B **101** (1997) 3800–3805
27.59 H. Schonherr, G. J. Vancso: Tribological properties of self-assembled monolayers of fluorocarbon and hydrocarbon thiols and disulfides on Au(111) studied by scanning force microscopy, Mater. Sci. Eng. C **8-9** (1999) 243–249
27.60 H. I. Kim, T. Koini, T. R. Lee, S. S. Perry: Systematic studies of the frictional properties of fluorinated monolayers with atomic force microscopy: Comparison of CF_3- and CH_3-terminated groups, Langmuir **13** (1997) 7192–7196
27.61 H. I. Kim, M. Graupe, O. Oloba, T. Koini, S. Imaduddin, T. R. Lee, S. S. Perry: Molecularly specific studies of the frictional properties of monolayer films: A systematic comparison of CF_3-, $(CH_3)_2CH$-, and CH_3-terminated films, Langmuir **15** (1999) 3179–3185
27.62 V. V. Tsukruk, M. P. Everson, L. M. Lander, W. J. Brittain: Nanotribological properties of composite molecular films: C_{60} anchored to a self-assembled monolayer, Langmuir **12** (1996) 3905–3911
27.63 S. Lee, Y. S. Shon, T. R. Lee, S. S. Perry: Structural characterization and frictional properties of C_{60}-terminated self-assembled monolayers on Au(111), Thin Solid Films **358** (2000) 152–158
27.64 V. V. Tsukruk, V. N. Bliznyuk: Adhesive and friction forces between chemically modified silicon and silicon nitride surfaces, Langmuir **14** (1998) 446–455
27.65 V. V. Tsukruk, T. Nguyen, M. Lemieux, J. Hazel, W. H. Weber, V. V. Shevchenko, N. Klimenko, E. Sheludko: Tribological properties of modified MEMS surfaces. In: *Tribology Issues and Opportunities in MEMS*, ed. by B. Bhushan (Kluwer, Dordrecht 1998) pp. 607–614
27.66 M. Fujihira, Y. Tani, M. Furugori, U. Akiba, Y. Okabe: Chemical force microscopy of self-assembled monolayers on sputtered gold films patterned by phase separation, Ultramicroscopy **86** (2001) 63–73
27.67 S. Sundararajan, B. Bhushan: Topography-induced contribution to friction forces measured using an atomic force/friction force microscopy, J. Appl. Phys. **88** (2000) 4825–4831
27.68 Y. F. Miura, M. Takenga, T. Koini, M. Graupe, N. Garg, R. L. Graham, T. R. Lee: Wettability of self-assembled monolayers generated from CF_3-terminated alkanethiols on gold, Langmuir **14** (1998) 5821–5825
27.69 M. Ratajczak-Sitarz, A. Katrusiak, Z. Kaluski, J. Garbarczyk: 4,4′-biphenyldithiol, Acta Crystallogr. C **43** (1987) 2389–2391
27.70 J. N. Israelachvili: *Intermolecular and Surface Forces*, 2nd edn. (Academic, London 1992)
27.71 R. J. Good, C. J. V. Oss: *Modern Approaches to Wettability – Theory and Applications* (Plenum, New York 1992)
27.72 Y. Unno, H. Kawamura, H. Kita, S. Sekiyama: Friction characteristics of sintered Si_3N_4 in oil-lubricated medium, Proc. Int. Symp. Canadian Institute of Mining, Metallurgy and Petroleum, Montreal 1995 (Canadian Institute of Mining, Metallurgy and Petroleum, Montreal 1995) 275–284
27.73 M. H. V. C. Adao, B. J. V. Saramago, A. C. Fernandes: Estimation of the surface properties of styrene-acrylonitrile random copolymers from contact angle measurements, J. Coll. Interface Sci. **217** (1999) 94–106
27.74 J. I. Siepman, I. R. McDonald: Monte Carlo simulation of the mechanical relaxation of a self-assembled monolayer, Phys. Rev. Lett. **70** (1993) 453–456
27.75 M. Garcia-Parajo, C. Longo, J. Servat, P. Gorostiza, F. Sanz: Nanotribological properties of octadecyltrichlorosilane self-assembled ultrathin films studied by atomic force microscopy: Contact and tapping modes, Langmuir **13** (1997) 2333–2339

27.76 D. DeVecchio, B. Bhushan: Localized surface elasticity measurements using an atomic force microscope, Rev. Sci. Instrum. **68** (1997) 4498–4505
27.77 D. R. Lide: *CRC Handbook of Chemistry and Physics*, 75th edn. (CRC, Boca Raton 1994)
27.78 E. Barrena, S. Kopta, D. F. Ogletree, D. H. Charych, M. Salmeron: Relationship between friction and molecular structure: Alkysilane lubricant films under pressure, Phys. Rev. Lett. **82** (1999) 2880–2883
27.79 T. Hoshino, C. Yayoi, K. Inage: Adsorption of atomic and molecular oxygen and desorption of silicon monoxide on Si(111) surface, Phys. Rev. B **59** (1999) 2332–2340

28. Nanoscale Boundary Lubrication Studies

Boundary films are formed by physisorption, chemisorption, and chemical reaction. With physisorption, no exchange of electrons takes place between the molecules of the adsorbate and those of the adsorbant. The physisorption process typically involves van der Waals forces, which are relatively weak. In chemisorption, there is an actual sharing of electrons or electron interchange between the chemisorbed species and the solid surface. The solid surfaces bond very strongly to the adsorption species through covalent bonds. Chemically reacted films are formed by the chemical reaction of a solid surface with the environment. The physisorbed film can be either monomolecularly or polymolecularly thick. The chemisorbed films are monomolecular, but stoichiometric films formed by chemical reaction can have a large film thickness. In general, the stability and durability of surface films decrease in the following order: chemically reacted films, chemisorbed films, and physisorbed films. A good boundary lubricant should have a high degree of interaction between its molecules and the sliding surface. As a general rule, liquids are good lubricants when they are polar and, thus, able to grip solid surfaces (or be adsorbed). In this chapter, we focus on PFPEs. We first introduce details of the commonly used PFPE lubricants; then present a summary of nanodeformation, molecular conformation, and lubricant spreading studies; followed by an overview of nanotribological properties of polar and nonpolar PFPEs studied by atomic force microscopy (AFM) and some concluding remarks.

28.1 **Lubricants Details** 862

28.2 **Nanodeformation, Molecular Conformation, and Lubricant Spreading** . 864

28.3 **Boundary Lubrication Studies** 866
- 28.3.1 Friction and Adhesion 866
- 28.3.2 Rest Time Effect 869
- 28.3.3 Velocity Effect 871
- 28.3.4 Relative Humidity and Temperature Effect 873
- 28.3.5 Tip Radius Effect 876
- 28.3.6 Wear Study 879

28.4 **Closure** 880

References 881

Boundary films are formed by physisorption, chemisorption, and chemical reaction. With physisorption, no exchange of electrons takes place between the molecules of the adsorbate and those of the adsorbant. The physisorption process typically involves van der Waals forces, which are relatively weak. In chemisorption, there is an actual sharing of electrons or electron interchange between the chemisorbed species and the solid surface. The solid surfaces bond very strongly to the adsorption species through covalent bonds. Chemically reacted films are formed by the chemical reaction of a solid surface with the environment. The physisorbed film can be either monomolecularly or polymolecularly thick. The chemisorbed films are monomolecular, but stoichiometric films formed by chemical reaction can have a large film thickness. In general, the stability and durability of surface films decrease in the following order: chemically reacted films, chemisorbed films, and physisorbed films. A good boundary lubricant should have a high degree of interaction between its molecules and the sliding surface. As a general rule, liquids are good lubricants when they are polar and, thus, able to grip solid surfaces (or be adsorbed). Polar lubricants contain reactive functional groups with low ionization potential, or groups having high polarizability [28.1–3]. Boundary lubrication properties of lubricants are also dependent upon the molecular conformation and lubricant spreading [28.4–7].

Self-assembled monolayers (SAMs), Langmuir–Blodgett (LB) films, and perfluoropolyether (PFPE) films can be used as boundary lubricants [28.2, 3, 8–10].

PFPE films are commonly used for lubrication of magnetic rigid disks and metal evaporated magnetic tapes to reduce friction and wear of a head-medium interface [28.10]. PFPEs are well suited for this application because of the following properties: low surface tension and low contact angle, which allow easy spreading on surfaces and provide a hydrophobic property; chemical and thermal stability, which minimizes degradation under use; low vapor pressure,which provides low outgassing; high adhesion to substrate via organofunctional bonds; and good lubricity, which reduces the interfacial friction and wear [28.10–12]. While the structure of the lubricants employed at the head-medium interface has not changed substantially over the past decade, the thickness of the PFPE film used to lubricate the disk has steadily decreased from multilayer thicknesses to the sub-monolayer thickness regime [28.11, 13]. Molecularly thick PFPE films are also being considered for lubrication purposes of the evolving microelectromechanical systems (MEMS) industry [28.14]. It is well-known that the properties of molecularly thick liquid films confined to solid surfaces can be dramatically different from those of the corresponding bulk liquid. In order to efficiently develop lubrication systems that meet the requirements of the advanced rigid disk drive and MEMS industries, the impact of thinning the PFPE lubricants on the resulting of nanotribology should be fully understood [28.15, 16]. It is also important to understand lubricant-substrate interfacial interactions and the influence of the operating environment on the nanotribological performance of molecularly thick PFPEs.

An overview of nanotribological properties of SAMs and LB films can be found in many references such as [28.17]. In this chapter, we focus on PFPEs. We first introduce details of the commonly used PFPE lubricants; then present a summary of nanodeformation, molecular conformation, and lubricant spreading studies; followed by an overview of nanotribological properties of polar and nonpolar PFPEs studied by atomic force microscopy (AFM) and some concluding remarks.

28.1 Lubricants Details

Properties of two commonly used PFPE lubricants (Z-15 and Z-DOL) are reviewed here. Their molecular structures are shown schematically in Fig. 28.1. Z-15 has nonpolar $-CF_3$ end groups, whereas Z-DOL is a polar lubricant with hydroxyl ($-OH$) end groups. Their typical properties are summarized in Table 28.1. It shows that Z-15 and Z-DOL almost have the same density and surface tension. But Z-15 has larger molecular weight and higher viscosity. Both have low surface tension, low vapor pressure, low evaporation weight loss, and good oxidative stability [28.10, 12]. Generally, single-crystal Si(100) wafer with a native oxide layer was used as a substrate for deposition of molecularly thick lubricant films

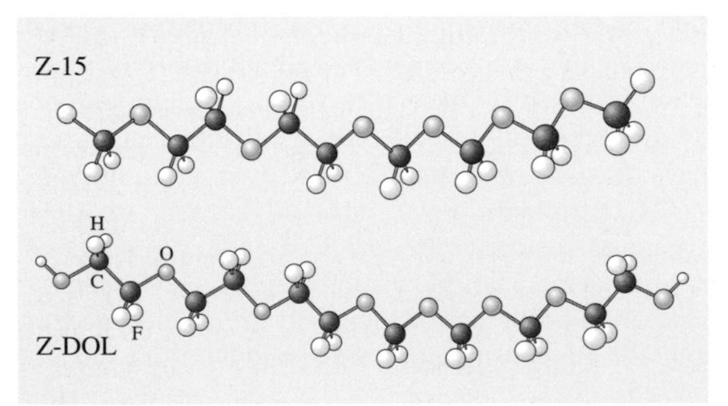

Fig. 28.1 Schematics of the molecular structures of Z-15 and Z-DOL. In this figure the m/n value, shown in Table 28.1, equals 2/3

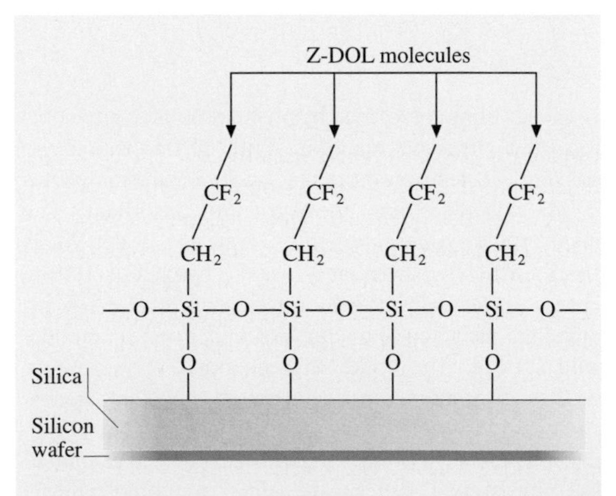

Fig. 28.2 Schematic of Z-DOL molecules that are chemically bonded on Si(100) substrate surface (which has native oxide) after thermal treatment at 150 °C for 30 min

Table 28.1 Typical properties of Z-15 and Z-DOL (data from Montefluous S.p.A., Milan, Italy)

	Z-15	Z-DOL (2000)
Formula	$CF_3{-}O{-}(CF_2{-}CF_2{-}O)_m{-}(CF_2{-}O)_n{-}CF_3$*	$HO{-}CH_2{-}CF_2{-}O{-}(CF_2{-}CF_2{-}O)_m{-}(CF_2{-}O)_n{-}CF_2{-}CH_2{-}OH$*
Molecular weight (Daltons)	9,100	2,000
Density *(ASTM D891)* 20 °C (g/cm^3)	1.84	1.81
Kinematic viscosity *(ASTM D445)* (cSt)		
20 °C	148	85
38 °C	90	34
99 °C	25	–
Viscosity index *(ASTM D2270)*	320	–
Surface tension *(ASTM D1331)* (dyn/cm) 20 °C	24	24
Vapor pressure (torr)		
20 °C	1.6×10^{-6}	2×10^{-5}
100 °C	1.7×10^{-5}	6×10^{-4}
Pour point *(ASTM D972)*		
°C	−80	–
Evaporation weight loss *(ASTM D972)*		
149 °C, 22 h (%)	0.7	–
Oxidative stability (°C)	–	320
Specific heat (cal/g °C)		
38 °C	0.21	–

* $m/n \sim 2/3$

for nanotribological characterization. Z-15 and Z-DOL films can be deposited directly on the Si(100) wafer by dip coating technique. The clean silicon wafer is vertically submerged into a dilute solution of lubricant in hydrocarbon solvent (HT-70) for a certain time. The silicon wafers are vertically pulled up from the solution with a motorized stage at a constant speed for deposition of desired thicknesses of Z-15 and Z-DOL lubricants. The lubricant film thickness obtained in dip coating is a function of the concentration and pulling-up speed, among other factors. The Z-DOL film is bonded to the silicon substrate by heating the as-deposited Z-DOL samples in an oven at 150 °C for about 30 s. The native oxide layer of Si(100) wafer reacts with the −OH groups of the lubricants during thermal treatment [28.18–21]. Subsequently, fluorocarbon solvent (FC-72) washing of the thermal treated specimen removes loosely absorbed species, leaving chemically bonded phase on the substrate. The chemical bonding between Z-DOL molecules and silicon substrate is illustrated in Fig. 28.2. The bonded and washed Z-DOL film is referred to as Z-DOL(BW) in this chapter. The as-deposited Z-15 and Z-DOL films are mobile phase lubricants (i. e., liquid-like lubricant), whereas the Z-DOL(BW) films are fully bonded soft solid phase (i. e., solid-like) lubricants. This will be further discussed in the next section.

28.2 Nanodeformation, Molecular Conformation, and Lubricant Spreading

Nanodeformation behavior of Z-DOL lubricants was studied using an AFM by *Blackman* et al. [28.22,23]. Before bringing a tungsten tip into contact with a molecular overlayer, it was brought into contact with a bare clean-silicon surface, Fig. 28.3. As the sample approaches the tip, the force initially is zero, but at point A the force suddenly becomes attractive (top curve), which increases until at point B, where the sample and tip come into intimate contact and the force becomes repulsive. As the sample is retracted, a pull-off force of 5×10^{-8} N (point D) is required to overcome adhesion between the tungsten tip and the silicon surface. When an AFM tip is brought into contact with an unbonded Z-DOL film, a sudden jump into adhesive contact is also observed. A much larger pull-off force is required to overcome the adhesion. The adhesion is initiated by the formation of a lubricant meniscus surrounding the tip. This suggests that the unbonded Z-DOL lubricant shows liquid-like behavior. However, when the tip was brought into contact with a lubricant film, which was firmly bonded to the surface, the liquid-like behavior disappears. The initial attractive force (point A) is no longer sudden, as with the liquid film, but, rather, gradually increases as the tip penetrates the film.

According to *Blackman* et al. [28.22, 23], if the substrate and tip were infinitely hard with no compliance and/or deformation in the tip and sample supports, the line for B to C would be vertical with an infinite slope. The tangent to the force-distance curve at a given point is referred to as the stiffness at that point and was determined by fitting a least-squares line through the nearby data points. For silicon, the deformation is reversible (elastic), since the retracting (outgoing) portion of the curve (C to D) follows the extending (ingoing) portion (B to C). For bonded lubricant film, at the point where slope of the force changes gradually from attractive to repulsive, the stiffness changes gradually, indicating compression of the molecular film. As the load is increased, the slope of the repulsive force eventually approaches that of the bare surface. The bonded film was found to respond elastically up to the highest loads of 5 μN that could be applied. Thus, bonded lubricant behaves as a soft polymer solid.

Figure 28.4 illustrates two extremes for the conformation on a surface of a linear liquid polymer without any reactive end groups and at submonolayer coverages [28.4, 6]. At one extreme, the molecules lie flat on the surface, reaching no more than their chain diameter δ above the surface. This would be the case if a strong attractive interaction exists between the molecules and the solid. On the other extreme, when a weak attraction exists between polymer segments and the solid, the molecules adopt conformation close to that of the molecules in the bulk, with the microscopic thickness equal to about the radius of gyration R_g. *Mate* and *Novotny* [28.6] used AFM to study conformation of 0.5–1.3 nm-thick Z-15 molecules on clean Si(100) surfaces. They found that the thickness measured by AFM is thicker than that measured by ellipsometry with the offset ranging from 3–5 nm. They found that the offset was the same for very thin submonolayer coverages. If the coverage is submonolayer and inadequate to make a liquid film, the relevant thickness is then the height (h_e) of the molecules extended above the solid surface. The offset should be equal to $2h_e$, assum-

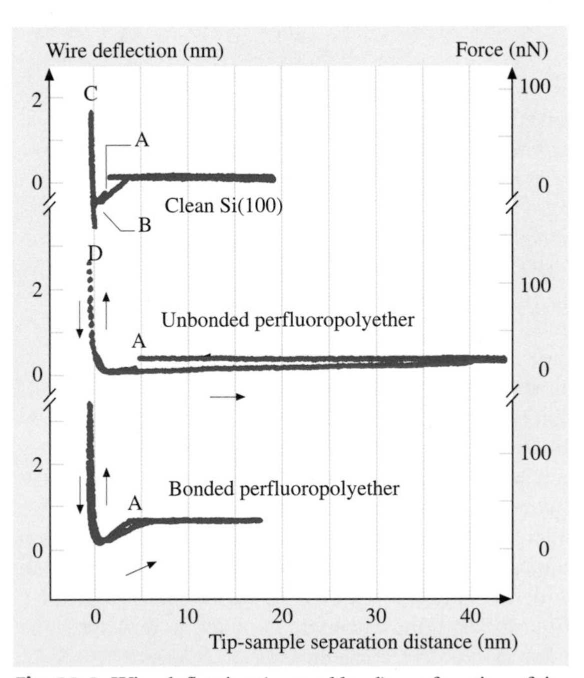

Fig. 28.3 Wire deflection (normal load) as a function of tip-sample separation distance curves comparing the behavior of clean Si(100) surface to a surface lubricated with free and unbonded PFPE lubricant, and a surface where the PFPE lubricant film was thermally bonded to the surface [28.22]

ing that the molecules extend the same height above both the tip and silicon surfaces. They therefore concluded that the molecules do not extend more than 1.5–2.5 nm above a solid or liquid surface, much smaller than the radius of gyration of the lubricants ranging between 3.2 and 7.3 nm, and to the approximate cross-sectional diameter of 0.6–0.7 nm for the linear polymer chain. Consequently, the height that the molecules extend above the surface is considerably less than the diameter of gyration of the molecules and only a few molecular diameters in height, implying that the physisorbed molecules on a solid surface have an extended, flat conformation. They also determined the disjoining pressure of these liquid films from AFM measurements of the distance needed to break the liquid meniscus that forms between the solid surface and the AFM tip. (Also see [28.7].) For a monolayer thickness of about 0.7 nm, the disjoining pressure is about 5 MPa, indicating strong attractive interaction between the liquid molecules and the solid surface. The disjoining pressure decreases with increasing film thickness in a manner consistent with a strong attractive van der Waals interaction between the liquid molecules and the solid surface.

Rheological characterization shows that the flow activation energy of PFPE lubricants is weakly dependent on chain length and is strongly dependent on the functional end groups [28.25]. PFPE lubricant films that contain polar end groups have lower mobility than those with nonpolar end groups of similar chain length [28.26]. The mobility of PFPE also depends on the surface chemical properties of the substrate. The spreading of Z-DOL on amorphous carbon surface has been studied as a function of hydrogen or nitrogen content in the carbon film, using scanning microellipsometry [28.24]. The diffusion coefficient data presented in Fig. 28.5 is thickness-dependent. It shows that the surface mobility of Z-DOL increased as the hydrogen content increased, but decreased as nitrogen content increased. The enhancement of Z-DOL surface mobility by hydrogenation may be understood from the fact that the interactions between Z-DOL molecules and the carbon surface can be significantly weakened, due to a reduction of the number of high-energy binding sites on the carbon surface. The stronger interactions between the Z-DOL molecules and carbon surface, as nitrogen content in the carbon coating increases, leads to the lowering of Z-DOL surface mobility.

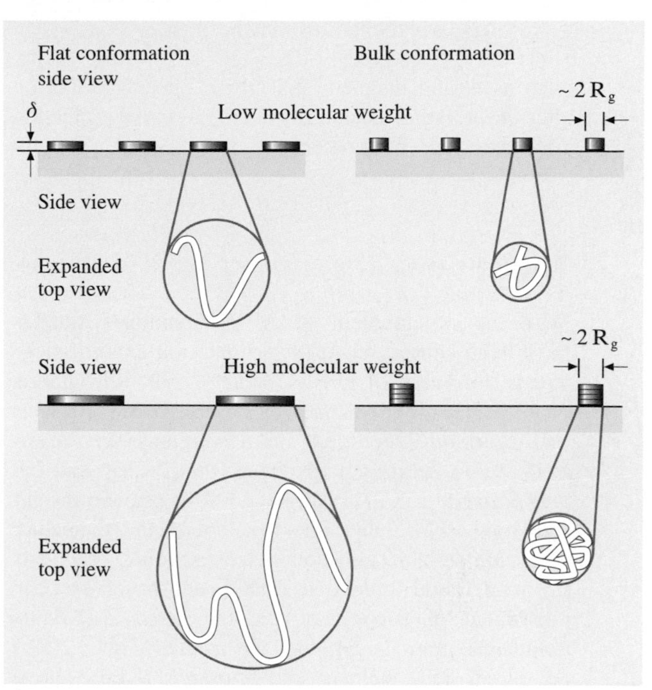

Fig. 28.4 Schematic representation of two extreme liquid conformations at the surface of the solid for low and high molecular weights at low surface coverage. δ is the cross-sectional diameter of the liquid chain, and R_g is the radius of gyration of the molecules in the bulk [28.6]

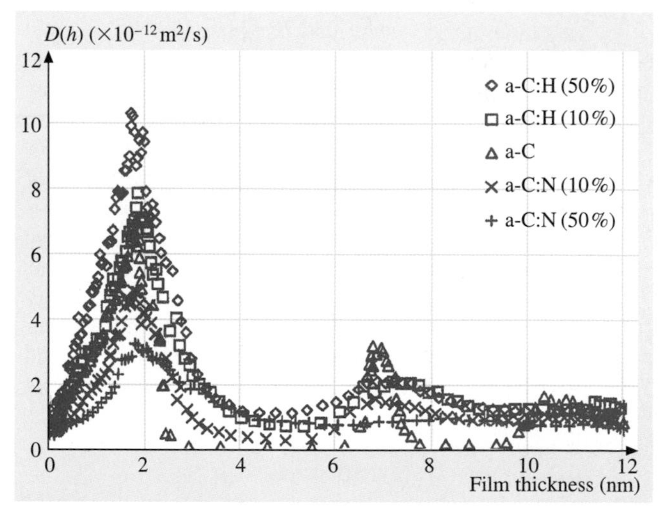

Fig. 28.5 Diffusion coefficient $D(h)$ as a function of lubricant film thickness for Z-DOL on different carbon films [28.24]

Molecularly thick films may be sheared at very high shear rates, on the order of $10^8-10^9\,s^{-1}$ during sliding, such as during magnetic disk drive operation. During such shear, lubricant stability is critical to the protection of the interface. For proper lubricant selection, viscosity at high shear rates and associated shear thinning need to be understood. Viscosity measurements of eight different types of PFPE films show that all eight lubricants display Newtonian behavior and their viscosity remains constant at a shear rate up to $10^7\,s^{-1}$ [28.27, 28].

28.3 Boundary Lubrication Studies

With the development of AFM techniques, studies have been carried out to investigate the nanotribological performance of PFPEs. *Mate* [28.29, 30], *O'Shea* et al. [28.31, 32], *Bhushan* et al. [28.15, 33], *Koinkar* and *Bhushan* [28.20, 34], *Bhushan* and *Sundararajan* [28.35], *Bhushan* and *Dandavate* [28.36], and *Liu* and *Bhushan* [28.21] used an AFM to provide insight into how PFPE lubricants function at the molecular level. *Mate* [28.29, 30] conducted friction experiments on bonded and unbonded Z-DOL and found that the coefficient of friction of the unbonded Z-DOL is about two times larger than the bonded Z-DOL (also see [28.31, 32]). *Koinkar* and *Bhushan* [28.20, 34] and *Liu* and *Bhushan* [28.21] studied the friction and wear performance of a Si(100) sample lubricated with Z-15, Z-DOL, and Z-DOL(BW) lubricants. They found that using Z-DOL(BW) could significantly improve the adhesion, friction, and wear performance of Si(100). They also discussed the lubrication mechanisms on the molecular level. *Bhushan* and *Sundararajan* [28.35] and *Bhushan* and *Dandavate* [28.36] studied the effect of tip radius and relative humidity on the adhesion and friction properties of Si(100) coated with Z-DOL(BW).

In this section, we review, in some detail, the adhesion, friction, and wear properties of two kinds of typical PFPE lubricants of Z-15 and Z-DOL at various operating conditions (rest time, velocity, relative humidity, temperature, and tip radius). The experiments were carried out using a commercial AFM system with pyramidal Si_3N_4 and diamond tips. An environmentally controlled chamber and a thermal stage were used to perform relative humidity and temperature effect studies.

28.3.1 Friction and Adhesion

To investigate the friction properties of Z-15 and Z-DOL(BW) films on Si(100), the friction force versus normal load curves were measured by making friction measurements at increasing normal loads [28.21]. The representative results of Si(100), Z-15, and Z-DOL(BW) are shown in Fig. 28.6. An approximately linear response of all three samples is observed in the load range of 5–130 nN. The friction force of solid-like Z-DOL(BW) is consistently smaller than that for Si(100), but the friction force of liquid-like Z-15 lubricant is higher than that of Si(100). *Sundararajan* and *Bhushan* [28.37] have studied the static friction force of silicon micromotors lubricated with Z-DOL by AFM. They also found that liquid-like lubricants of Z-DOL significantly increase the static friction force, whereas solid-like Z-DOL(BW) coating can dramatically reduce the static friction force. This is in good agreement with the results of *Liu* and *Bhushan* [28.21]. In Fig. 28.6, the nonzero value of the friction force signal at zero external load is due to the adhesive forces. It is well-known that the following relationship exists between the friction force F and external

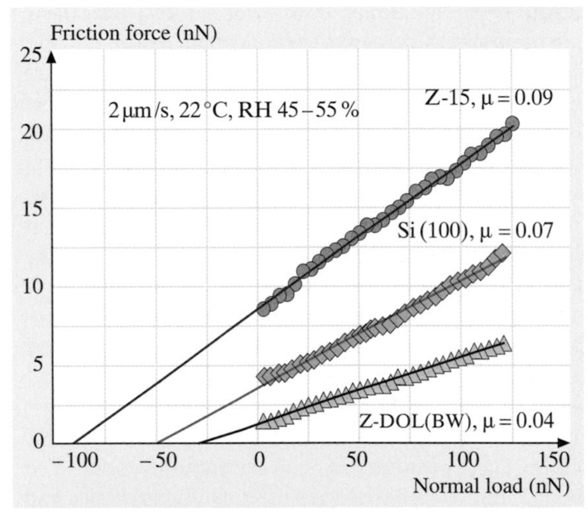

Fig. 28.6 Friction force versus normal load curves for Si(100), 2.8-nm-thick Z-15 film, and 2.3-nm-thick Z-DOL(BW) film at 2 μm/s, and in ambient air sliding against a Si_3N_4 tip. Based on these curves, coefficient of friction μ and adhesion force of W_a can be calculated [28.21]

normal load W [28.2, 3]

$$F = \mu(W + W_a) , \quad (28.1)$$

where μ is the coefficient of friction and W_a is the adhesive force. Based on this equation and the data in Fig. 28.6, we can calculate the μ and W_a values. The coefficients of friction of Si(100), Z-15, and Z-DOL are 0.07, 0.09, and 0.04, respectively. Based on (28.1), the adhesive force values are obtained from the horizontal intercepts of the friction force versus normal load curves at a zero value of friction force. Adhesive force values of Si(100), Z-15, and Z-DOL are 52 nN, 91 nN, and 34 nN, respectively.

The adhesive forces of these samples were also measured using a force calibration plot (FCP) technique. In this technique, the tip is brought into contact with the sample and the maximum force, needed to pull the tip and sample apart, is measured as the adhesive force. Figure 28.7 shows the typical FCP curves of Si(100), Z-15, and Z-DOL(BW) [28.21]. As the tip approaches the sample within a few nanometers (A), an attractive force exists between the tip and the sample surfaces. The tip is pulled toward the sample, and contact occurs at point B on the graph. The adsorption of water molecules and/or presence of liquid lubricant molecules on the sample surface can also accelerate this so-called "snap-in", due to the formation of meniscus of the water and/or liquid lubricant around the tip. From this point on, the tip is in contact with surface, and as the piezo extends further, the cantilever is further deflected. This is represented by the slope portion of the curve. As the piezo retracts, at point C the tip goes beyond the zero deflection (flat) line, because of the attractive forces, into the adhesive force regime. At point D, the tip snaps free of the adhesive forces and is again in free air. The adhesive force (pull-off force) is determined by multiplying the cantilever spring constant (0.58 N/m) by the horizontal distance between points C and D, which corresponds to the maximum cantilever deflection toward the samples before the tip is disengaged. Incidentally, the horizontal shift between the loading and unloading curves results from the hysteresis of the PZT tube.

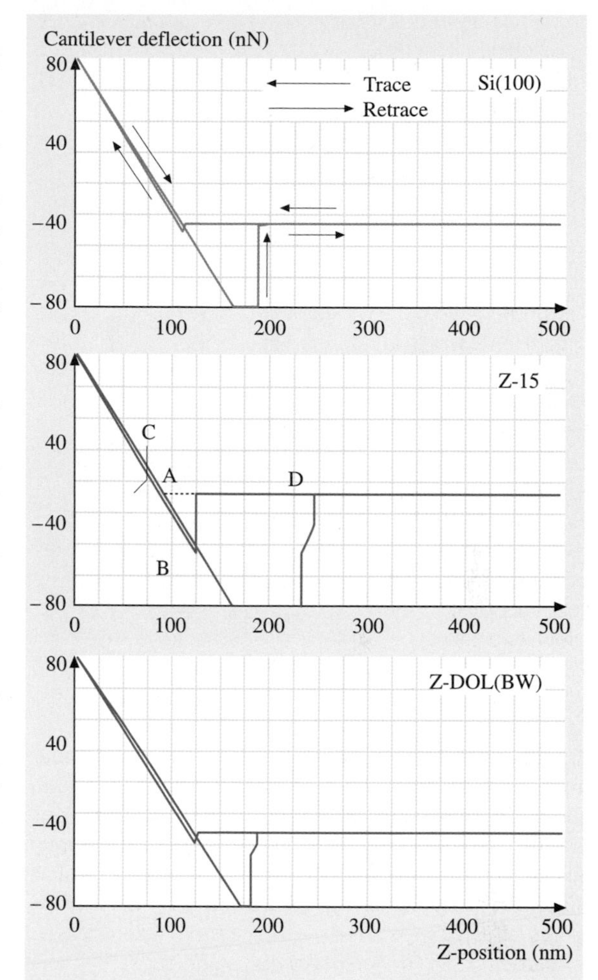

Fig. 28.7 Typical force calibration plots of Si(100), 2.8-nm-thick Z-15 film, and 2.3-nm-thick Z-DOL(BW) film in ambient air. The adhesive forces can be calculated from the horizontal distance between points C and D, and the cantilever spring constant of 0.58 N/m [28.21]

The adhesive forces of Si(100), Z-15, and Z-DOL(BW) measured by FCP and friction force versus normal load plot are summarized in Fig. 28.8 [28.21]. The results measured by these two methods are in good agreement. Figure 28.8 shows that the presence of mobile Z-15 lubricant film increases the adhesive force as compared to that of Si(100). In contrast, the presence of solid phase Z-DOL(BW) film reduces the adhesive force as compared to that of Si(100). This result is in good agreement with the results of *Blackman* et al. [28.22] and *Bhushan* and *Ruan* [28.38]. Sources of adhesive forces between the tip and the sample surfaces are van der Waals attraction and long-range meniscus force [28.2, 3, 16]. Relative magnitudes of the forces from the two sources are dependent on various factors, including the distance between the tip and the sample surface, their surface roughness, their hydrophobicity, and relative humidity [28.39]. For most surfaces with some roughness, meniscus contribution dominates at moderate to high humidities.

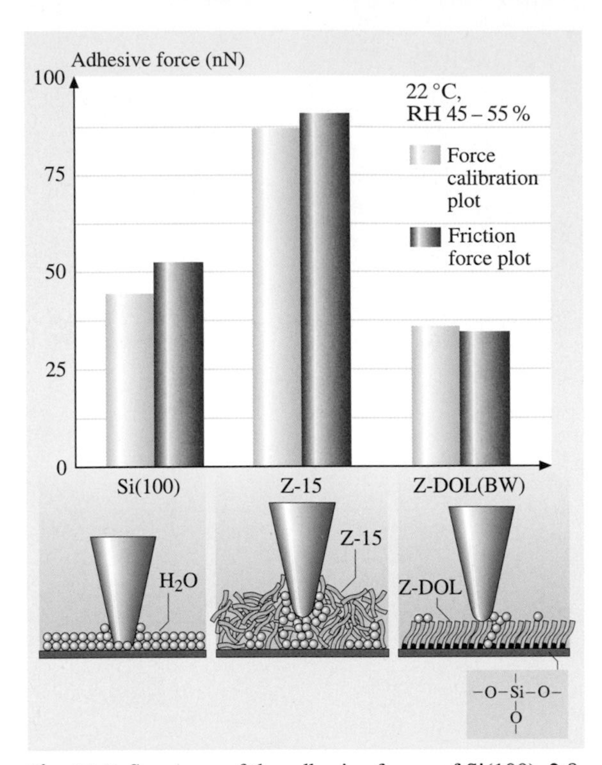

Fig. 28.8 Summary of the adhesive forces of Si(100), 2.8-nm-thick Z-15 film, and 2.3-nm-thick Z-DOL(BW) film measured by force calibration plots and friction force versus normal load plots in ambient air. The schematic (*bottom*) shows the effect of meniscus formed between the AFM tip and the sample surface on the adhesive and friction forces [28.21]

The schematic (bottom) in Fig. 28.8 shows relative size and sources of meniscus. The native oxide layer (SiO_2) on the top of Si(100) wafer exhibits hydrophilic properties, and some water molecules can be adsorbed on this surface. The condensed water will form meniscus as the tip approaches the sample surface. In the case of a sphere (such as a single-asperity AFM tip) in contact with a flat surface, the attractive Laplace force (F_L) caused by capillary is:

$$F_L = 2\pi R \gamma_{la}(\cos\theta_1 + \cos\theta_2) , \quad (28.2)$$

where R is the radius of the sphere, γ_{la} is the surface tension of the liquid against air, θ_1 and θ_2 are the contact angles between liquid and flat and spherical surfaces, respectively [28.2, 3, 40]. As the surface tension value of Z-15 (24 dyn/cm) is smaller than that of water (72 dyn/cm), the larger adhesive force in Z-15 cannot only be caused by the Z-15 meniscus. The nonpolarized Z-15 liquid does not have complete coverage and strong bonding with Si(100). In the ambient environment, the condensed water molecules will permeate through the liquid Z-15 lubricant film and compete with the lubricant molecules present on the substrate. The interaction of the liquid lubricant with the substrate is weakened, and a boundary layer of the liquid lubricant forms puddles [28.20, 34]. This dewetting allows water molecules to be adsorbed on the Si(100) surface as aggregates along with Z-15 molecules. And both of them can form meniscus while the tip approaches the surface. In addition, as the Z-15 film is pretty soft compared to the solid Si(100) surface, penetration of the tip in the film occurs while pushing the tip down. This leads to a large area of the tip involved to form the meniscus at the tip-liquid (water aggregates along with Z-15) interface. These two factors of the liquid-like Z-15 film result in higher adhesive force. It should also be noted that Z-15 has a higher viscosity compared to that of water. Therefore, Z-15 film provides higher resistance to sliding motion and results in a larger coefficient of friction. In the case of Z-DOL(BW) film, both of the active groups of Z-DOL molecules are strongly bonded on Si(100) substrate through the thermal and washing treatment. Thus, the Z-DOL(BW) film has relatively low free surface energy and cannot be displaced readily by water molecules or readily adsorb water molecules. Thus, the use of Z-DOL(BW) can reduce the adhesive force. We further note that the bonded Z-DOL molecules can be orientated under stress (behaves as a soft polymer solid), which facilitates sliding and reduces coefficient of friction.

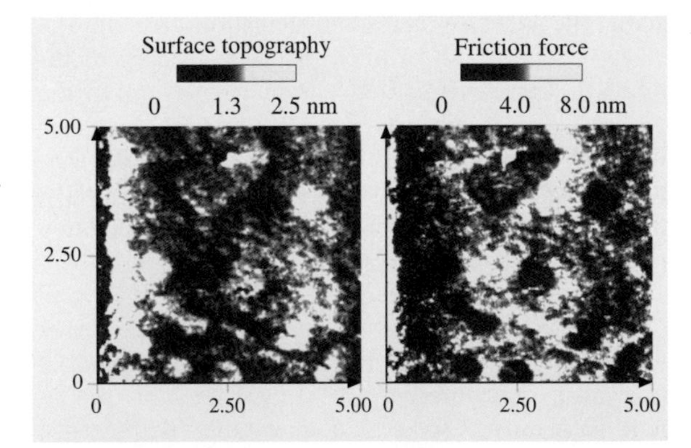

Fig. 28.9 Gray-scale plots of the surface topography and friction force obtained simultaneously for unbonded 2.3-nm-thick Demnum-type PFPE lubricant film on silicon [28.20]

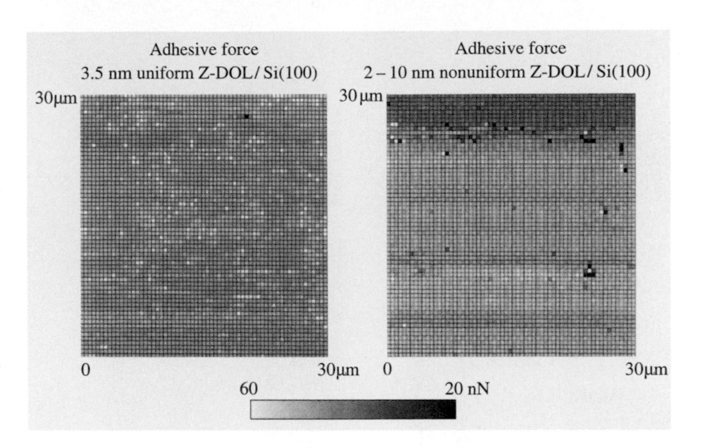

Fig. 28.10 Gray-scale plots of the adhesive force distribution of a uniformly coated, 3.5-nm-thick unbonded Z-DOL film on silicon and 3- to 10-nm-thick unbonded Z-DOL film on silicon that was deliberately coated nonuniformly by vibrating the sample during the coating process [28.36]

These studies suggest that if the lubricant films exist as liquid-like, such as Z-15 films, they easily form meniscus (by themselves and the adsorbed water molecules), and thus have higher adhesive force and higher friction force. Whereas if the lubricant film exists in solid-like phase, such as Z-DOL(BW) films, they are hydrophobic with low adhesion and friction.

In order to study the uniformity of lubricant film and its influence on friction and adhesion, friction force mapping and adhesive force mapping of PFPE have been carried out by *Koinkar* and *Bhushan* [28.34] and *Bhushan* and *Dandavate* [28.36], respectively. Figure 28.9 shows gray-scale plots of surface topography and friction force images obtained simultaneously for unbonded Demnum-type PFPE lubricant film on silicon [28.34]. The friction force plot shows well distinguished low and high friction regions corresponding roughly to high and low surface height regions in the topography image (thick and thin lubricant regions). A uniformly lubricated sample does not show such a variation in friction. Figure 28.10 shows the gray-scale plots of the adhesive force distribution for silicon samples coated uniformly and nonuniformly with Z-DOL lubricant. It can be clearly seen that there exists a region that has an adhesive force distinctly different from the other region for the nonuniformly coated sample. This implies that the liquid film thickness is nonuniform, giving rise to a difference in the meniscus forces.

28.3.2 Rest Time Effect

It is well-known that in the computer rigid disk drive, the stiction force increases rapidly with an increase in rest time between the head and magnetic medium disk [28.10, 11]. Considering that the stiction and friction are

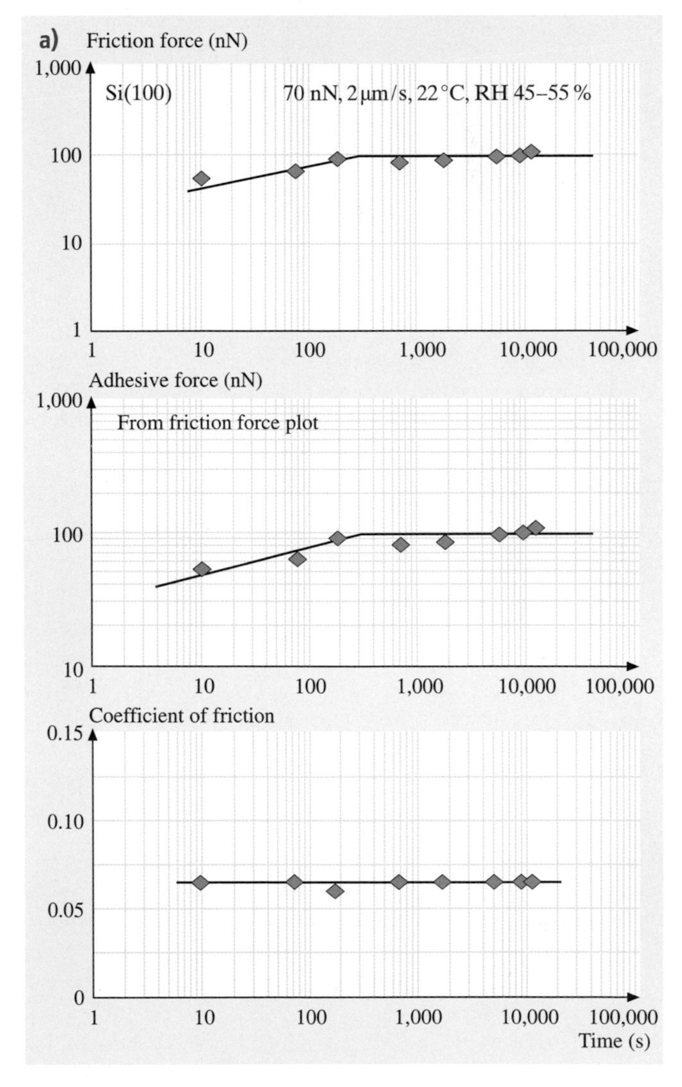

Fig. 28.11 (a) Rest time effect on friction force, adhesive force, and coefficient of friction of Si(100). **(b)** (*see next page*) Summary of the rest time effect on friction force, adhesive force, and coefficient of friction of Si(100), 2.8-nm-thick Z-15 film, and 2.3-nm-thick Z-DOL(BW) film. All of the measurements were carried out at 70 nN, 2 µm/s, and in ambient air [28.21]

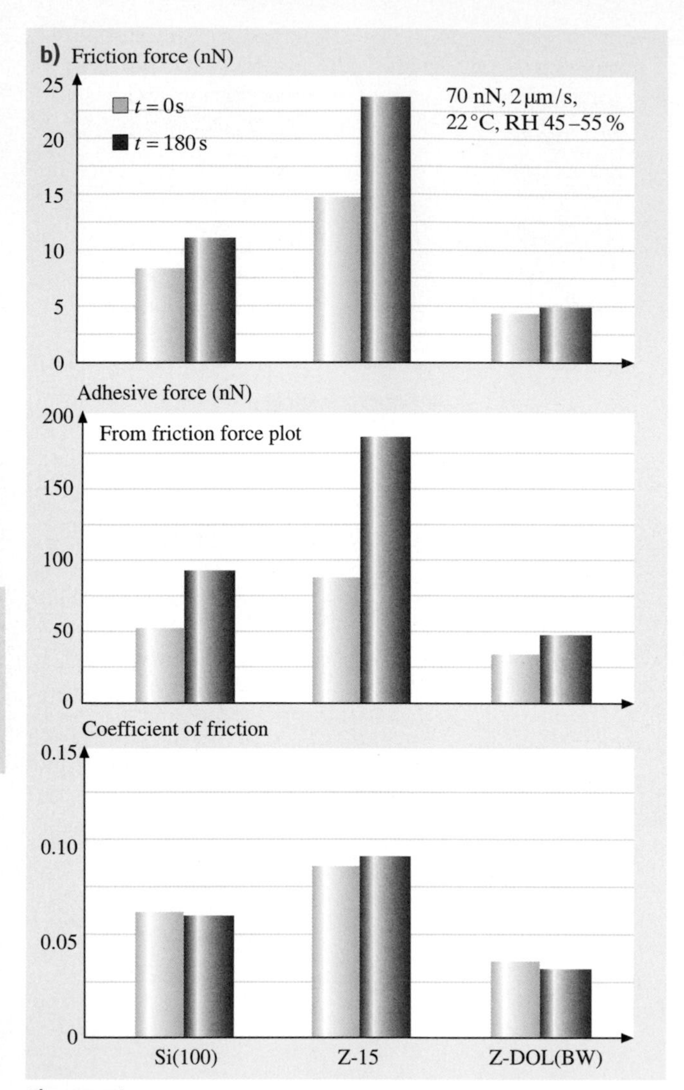

Fig. 28.11b

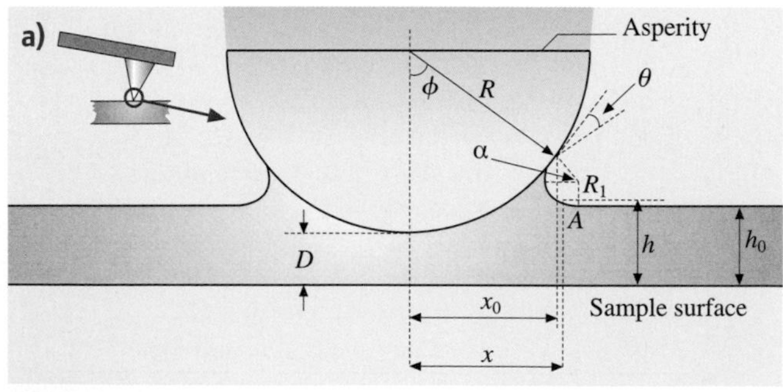

Fig. 28.12a

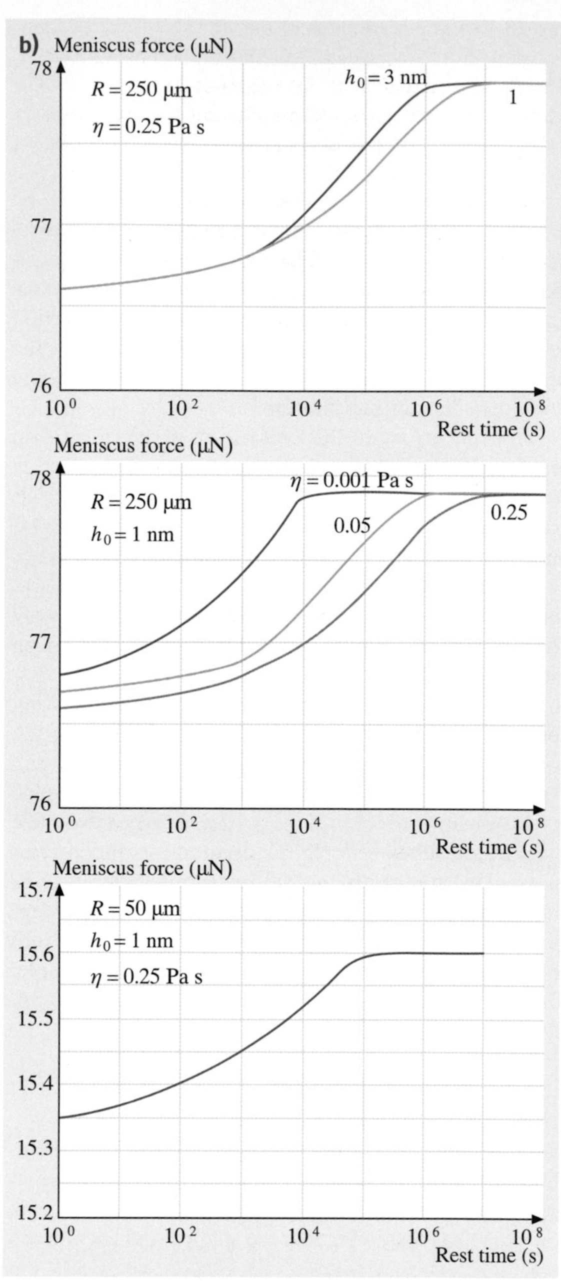

Fig. 28.12 (a) Schematic of a single asperity in contact with a smooth flat surface in the presence of a continuous liquid film when ϕ is large. (b) Results of the single asperity model. Effect of viscosity of the liquid, radius of the asperity, and film thickness is studied with respect to the time-dependent meniscus force [28.41]

two of the major issues that lead to the failure of computer rigid disk drives and MEMS, it is very important to find out if the rest time effect also exists on the nanoscale. First, the rest time effect on the friction force, adhesive force, and coefficient of Si(100) sliding against Si_3N_4 tip was studied, Fig. 28.11a [28.21]. It was found that the friction and adhesive forces logarithmically increase up to a certain equilibrium time after which they remain constant. Figure 28.11a also shows that the rest time does not affect the coefficient of friction. These results suggest that the rest time can result in growth of the meniscus, which causes a higher adhesive force, and in turn, a higher friction force. But in the whole testing range the friction mechanisms do not change with the rest time. Similar studies were also performed on Z-15 and Z-DOL(BW) films. The results are summarized in Fig. 28.11b [28.21]. It is seen that similar time effect has been observed on Z-15 film, but not on Z-DOL(BW) film.

AFM tip in contact with a flat sample surface is generally believed to represent a single asperity contact. Therefore, a Si_3N_4 tip in contact with Si(100) or Z-15/Si(100) can be modeled as a sphere in contact with a flat surface covered by a layer of liquid (adsorbed water and/or liquid lubricant), Fig. 28.12a. Meniscus forms around the contacting asperity and grows with time until equilibrium occurs [28.41]. The meniscus force, which is the product of meniscus pressure and meniscus area, depends on the flow of liquid phase toward the contact zone. The flow of the liquid toward the contact zone is governed by the capillary pressure P_c, which draws liquid into the meniscus, and the disjoining pressure Π, which tends to draw the liquid away from the meniscus. Based on the Young and Laplace equation, the capillary pressure, P_c, is:

$$P_c = 2K\gamma \,, \tag{28.3}$$

where $2K$ is the mean meniscus curvature ($= K_1 + K_2$, where K_1 and K_2 are the curvatures of the meniscus in the contact plane and perpendicular to the contact plane) and γ is the surface tension of the liquid. *Mate* and *Novotny* [28.6] have shown that the disjoining pressure decreases rapidly with increasing liquid film thickness in a manner consistent with a strong van der Waals attraction. The disjoining pressure, Π, for these liquid films can be expressed as:

$$\Pi = \frac{A}{6\pi h^3} \,, \tag{28.4}$$

where A is the Hamaker constant and h is the liquid film thickness. The driving forces that cause the lubricant flow that results in an increase in the meniscus force are the disjoining pressure gradient, due to a gradient in film thickness, and capillary pressure gradient, due to curved liquid-air interface. The driving pressure, P, can then be written as:

$$P = -2K\gamma - \Pi \,. \tag{28.5}$$

Based on these three basic relationships, the following differential equation has been derived by *Chilamakuri* and *Bhushan* [28.41], which can describe the meniscus at time t:

$$\begin{aligned} &2\pi x_0 \left(D + \frac{x_0^2}{2R} - h_0 \right) \frac{\mathrm{d}x_0}{\mathrm{d}t} \\ &= \frac{2\pi h_0^3 \gamma}{3\eta} \frac{(1+\cos\theta)}{D+a-h_0} - \frac{A x_0}{3\eta h} \cot\alpha \,, \end{aligned} \tag{28.6}$$

where η is the viscosity of the liquid and a is given as

$$a = R(1-\cos\phi) \sim \frac{R\phi^2}{2} \sim \frac{x_0^2}{2R} \,. \tag{28.7}$$

The differential equation (28.4) was solved numerically using Newton's iteration method. The meniscus force at any time t less than the equilibrium time is proportional to the meniscus area and meniscus pressure ($2K\gamma$), and it is given by

$$f_m(t) = 2\pi R\gamma(1+\cos\theta) \left(\frac{x_0}{(x_0)_{eq}} \right)^2 \left(\frac{K}{K_{eq}} \right) , \tag{28.8}$$

where $(x_0)_{eq}$ is the value of x_0 at the equilibrium time

$$\left[(x_0)_{eq}\right]^2 = 2R \left[\frac{-6\pi h_0^3 \gamma (1+\cos\theta)}{A} + (h_0 - D) \right] . \tag{28.9}$$

This modeling work (at the microscale) showed that the meniscus force initially increases logarithmically with the rest time up to a certain equilibrium time after which it remains constant. Equilibrium time decreases with an increase in liquid film thickness, a decrease in viscosity, and a decrease in the tip radius, Fig. 28.12b. This early numerical modeling work and the data at the nanoscale in Fig. 28.11a are in good agreement.

28.3.3 Velocity Effect

To investigate the velocity effect on friction and adhesion, the friction force versus normal load relationships of Si(100), Z-15, and Z-DOL(BW) at different velocities were measured, Fig. 28.13 [28.21]. Based on these data,

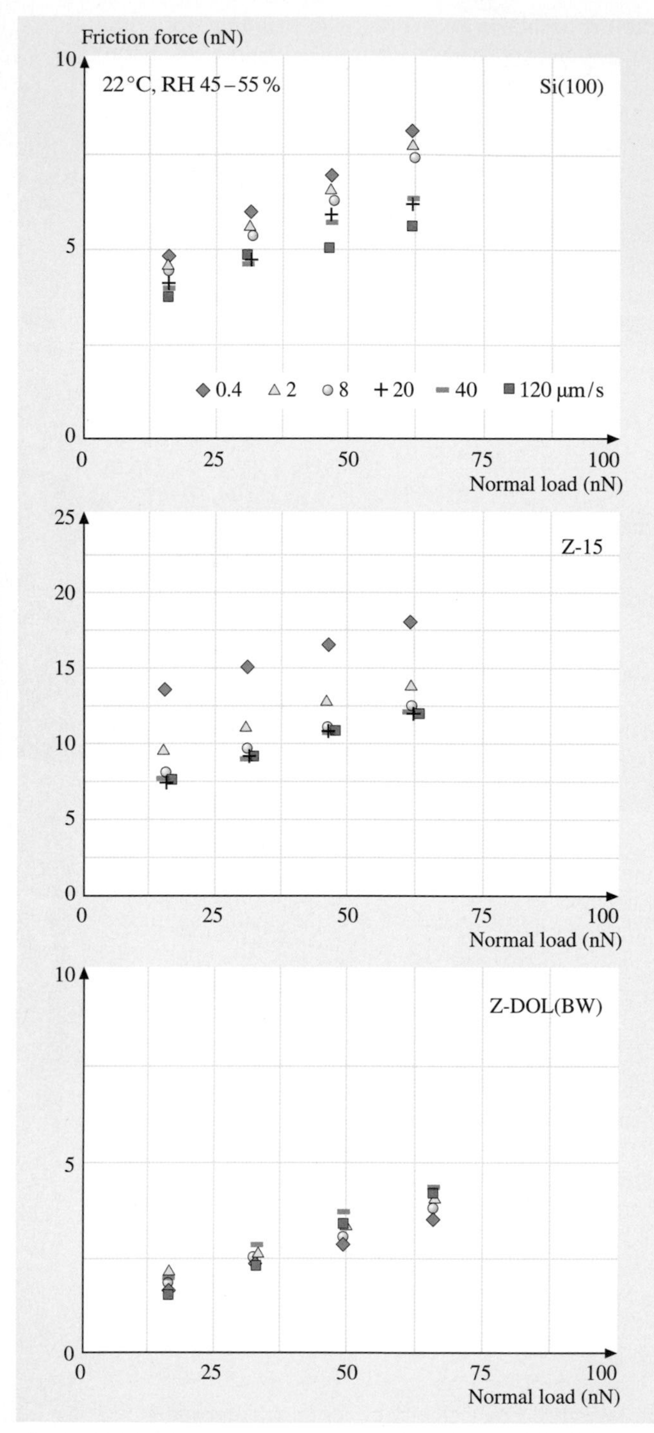

Fig. 28.13 Friction forces versus normal load data of Si(100), 2.8-nm-thick Z-15 film, and 2.3-nm-thick Z-DOL(BW) film at various velocities in ambient air [28.21]

the adhesive force and coefficient of friction values can be calculated by (28.1). The variation of friction force, adhesive force, and coefficient of friction of Si(100), Z-15, and Z-DOL(BW) as a function of velocity are summarized in Fig. 28.14. It indicates that for silicon wafer, the friction force decreases logarithmically with increasing velocity. For Z-15, the friction force decreases with increasing velocity up to 10 μm/s, after which it remains almost constant. The velocity has a much smaller effect on the friction force of Z-DOL(BW); it reduced slightly only at very high velocity. Figure 28.14 also indicates that the adhesive force of Si(100) is increased when the velocity is higher than 10 μm/s. The adhesive force of Z-15 is reduced dramatically with a velocity increase up to 20 μm/s, after which it is reduced slightly. And the adhesive force of Z-DOL(BW) also decreases at high velocity. In the testing range of velocity, only the coefficient of friction of Si(100) decreases with velocity, but the coefficients of friction of Z-15 and Z-DOL(BW) almost remain constant. This implies that the friction mechanisms of Z-15 and Z-DOL(BW) do not change with the variation of velocity.

The mechanisms of the effect of velocity on the adhesion and friction are explained based on the schematics shown in Fig. 28.14 (right). For Si(100), tribochemical reaction plays a major role. Although at high velocity the meniscus is broken and does not have enough time to rebuild, the contact stresses and high velocity lead to tribochemical reactions of Si(100) wafer and Si_3N_4 tip, which have native oxide (SiO_2) layers with water molecules. The following reactions occur:

$$SiO_2 + 2H_2O = Si(OH)_4 \quad (28.10)$$

$$Si_3N_4 + 16H_2O = 3Si(OH)_4 + 4NH_4OH . \quad (28.11)$$

The $Si(OH)_4$ is removed and continuously replenished during sliding. The $Si(OH)_4$ layer between the tip and Si(100) surface is known to be of low shear strength and causes a decrease in friction force and coefficient of friction in the lateral direction [28.42–46]. The chemical bonds of Si–OH between the tip and Si(100) surface induce large adhesive force in the normal direction. For Z-15 film, at high velocity the meniscus formed by condensed water and Z-15 molecules is broken and does not have enough time to rebuild. Therefore, the adhesive force and, consequently, friction force is reduced. For Z-DOL(BW) film, the surface can adsorb few water molecules in ambient condition, and at high velocity these molecules are displaced, which is responsible for a slight decrease in friction force and adhesive force. Even at high velocity range, the friction mechanisms for Z-15 and Z-DOL(BW) films still are shearing of the vis-

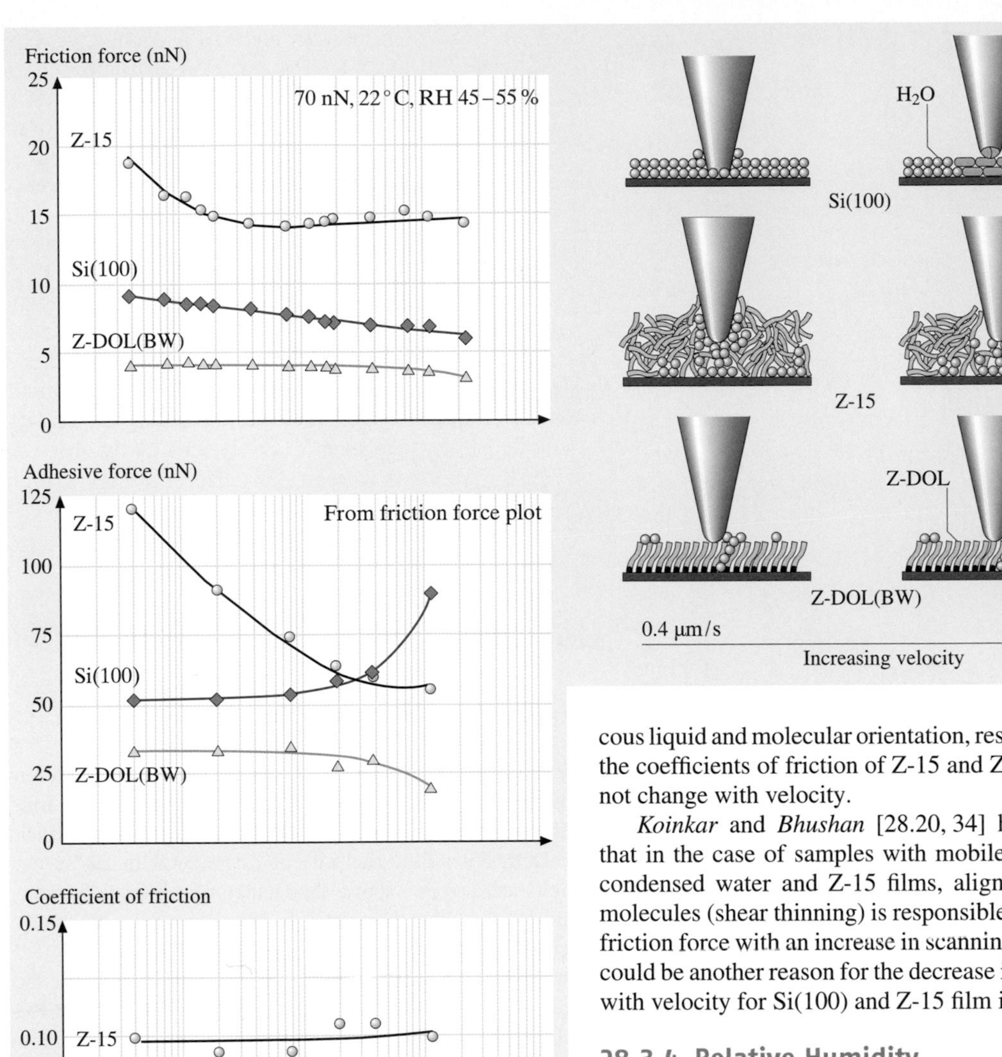

Fig. 28.14 The influence of velocity on the friction force, adhesive force, and coefficient of friction of Si(100), 2.8-nm-thick Z-15 film, and 2.3-nm-thick Z-DOL(BW) film at 70 nN, in ambient air. The schematic (*right*) shows the change of surface composition (by tribochemical reaction) and change of meniscus with increasing velocity [28.21]

cous liquid and molecular orientation, respectively. Thus the coefficients of friction of Z-15 and Z-DOL(BW) do not change with velocity.

Koinkar and *Bhushan* [28.20, 34] have suggested that in the case of samples with mobile films, such as condensed water and Z-15 films, alignment of liquid molecules (shear thinning) is responsible for the drop in friction force with an increase in scanning velocity. This could be another reason for the decrease in friction force with velocity for Si(100) and Z-15 film in this study.

28.3.4 Relative Humidity and Temperature Effect

The influence of relative humidity on friction and adhesion was studied in an environmentally controlled chamber. The friction force was measured by making measurements at increasing relative humidity, the results are presented in Fig. 28.15 [28.21]. It shows that for Si(100) and Z-15 film, the friction force increases with a relative humidity increase up to RH 45%, and then it shows a slight decrease with a further increase in relative humidity. Z-DOL(BW) has a smaller friction force than Si(100) and Z-15 in the whole testing range. And its friction force shows a relatively apparent increase when the relative humidity is higher than RH 45%. For Si(100), Z-15, and Z-DOL(BW), adhesive

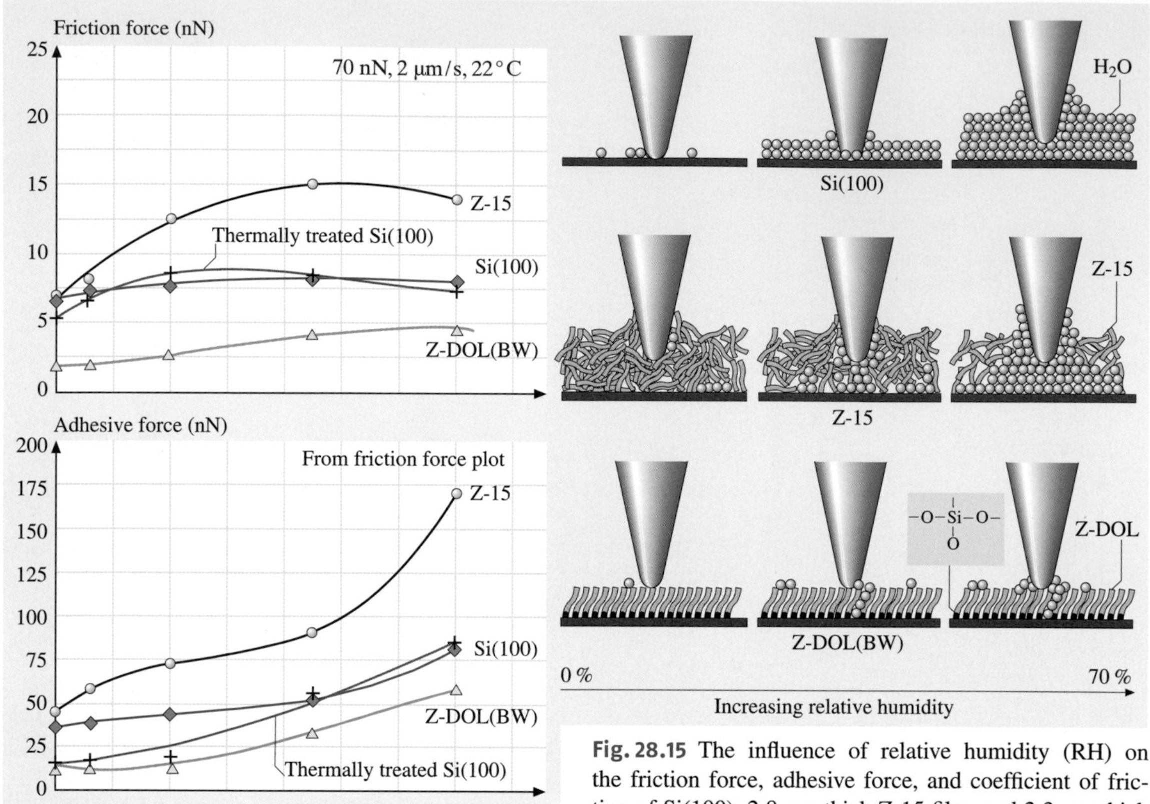

Fig. 28.15 The influence of relative humidity (RH) on the friction force, adhesive force, and coefficient of friction of Si(100), 2.8-nm-thick Z-15 film, and 2.3-nm-thick Z-DOL(BW) film at 70 nN, 2 μm/s, and in 22 °C air. Schematic (*right*) shows the change of meniscus with increasing relative humidity. In this figure, the thermally treated Si(100) represents the Si(100) wafer that was baked at 150 °C for 1 hour in an oven (in order to remove the adsorbed water) just before it was placed in the 0% RH chamber [28.21]

forces increase with relative humidity. And their coefficients of friction increase with a relative humidity up to RH 45%, after which they decrease with a further increase of the relative humidity. It is also observed that the humidity effect on Si(100) really depends on the history of the Si(100) sample. As the surface of Si(100) wafer readily adsorbs water in air, without any pre-treatment the Si(100) used in our study almost reaches its saturate stage of adsorbing water and is responsible for less effect during increasing relative humidity. However, once the Si(100) wafer was thermally treated by baking at 150 °C for 1 hour, a bigger effect was observed.

The schematic (right) in Fig. 28.15 shows that because its high free surface energy Si(100) can adsorb more water molecules with increasing relative humidity. As discussed earlier, for Z-15 film in a humid environment, condensed water competes with the lubricant film present on the sample surface. Obviously, more water molecules can also be adsorbed on a Z-15 surface with increasing relative humidity. The more adsorbed water

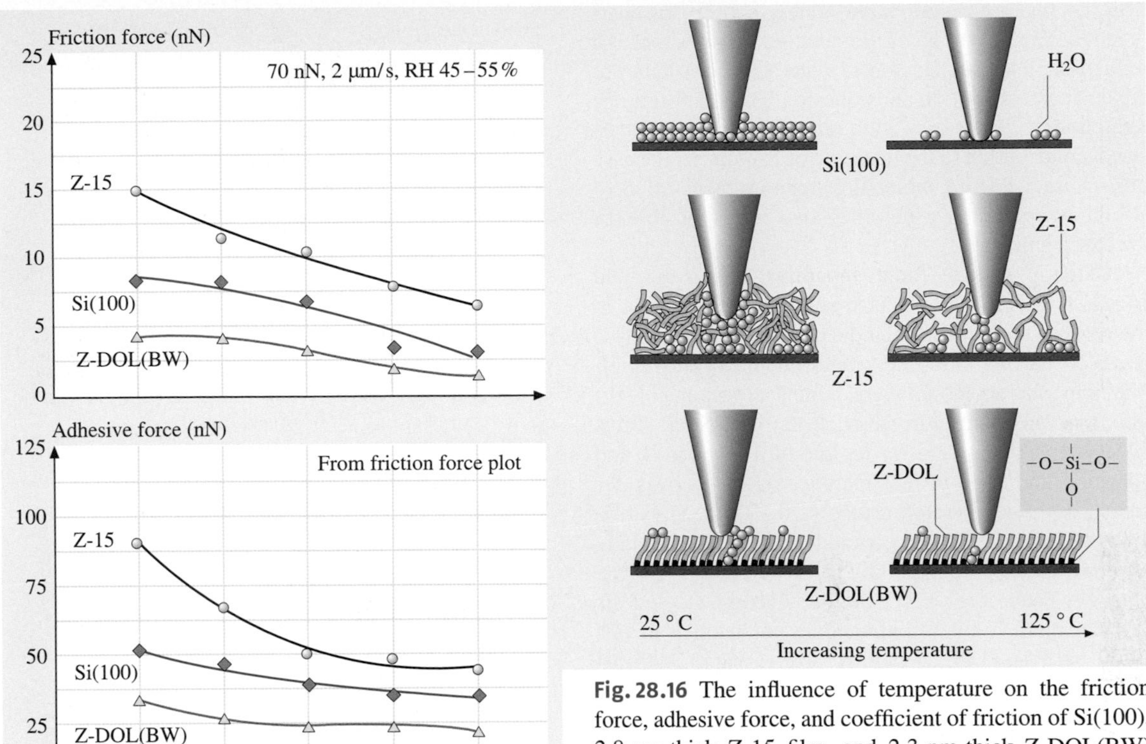

Fig. 28.16 The influence of temperature on the friction force, adhesive force, and coefficient of friction of Si(100), 2.8-nm-thick Z-15 film, and 2.3-nm-thick Z-DOL(BW) film at 70 nN, at 2 μm/s, and in RH 40–50% air. The schematic (*right*) shows that at high temperature, desorption of water decreases the adhesive forces. And the reduced viscosity of Z-15 leads to the decrease of coefficient of friction. High temperature facilitates orientation of molecules in Z-DOL(BW) film, which results in lower coefficient of friction [28.21]

molecules in the case of Si(100), along with lubricant molecules in Z-15 film, form a bigger water meniscus, which leads to an increase of friction force, adhesive force, and coefficient of friction of Si(100) and Z-15 with humidity. But at very high humidity of RH 70%, large quantities of adsorbed water can form a continuous water layer that separates the tip and sample surface, and acts as a kind of lubricant, which causes a decrease in the friction force and coefficient of friction. For Z-DOL(BW) film, because of its hydrophobic surface properties, water molecules can only be adsorbed at high humidity ($\geq$ RH 45%), which causes an increase in the adhesive force and friction force.

The effect of temperature on friction and adhesion was studied using a thermal stage attached to the AFM. The friction force was measured at increasing temperature from 22–125 °C. The results are presented in Fig. 28.16 [28.21]. It shows that the increasing temperature causes a decrease of friction force, adhesive force, and coefficient of friction of Si(100), Z-15, and Z-DOL(BW). The schematic (right) in Fig. 28.16 indicates that at high temperature, desorption of water leads to the decrease of friction force, adhesive force, and coefficient of friction for all of the samples. Besides that, the reduction of surface tension of water also contributes to the decrease of friction and adhesion. For Z-15 film, the reduction of viscosity at high temperature has an additional contribution to the decrease of friction. In the case of Z-DOL(BW) film, molecules are more easily oriented at high temperature, which may also be responsible for the low friction.

Using a surface force apparatus, *Yoshizawa* and *Israelachvili* [28.47, 48] have shown that a change in the velocity or temperature induces phase transformation (from crystalline solid-like to amorphous, then to liquid-like) in surfactant monolayers, which are responsible for the observed changes in the friction force. Stick-slip is observed in a low velocity regime of a few μm/s, and adhesion and friction first increase followed by a decrease in the temperature range of 0–50 °C. Stick-slip at low velocity and adhesion and friction curves peaking at some particular temperature (observed in their study), have not been observed in the AFM study. It suggests that the phase transformation may not happen in this study, because PFPEs generally have very good thermal stability [28.10, 12].

As a brief summary, the influence of velocity, relative humidity, and temperature on the friction force of Z-15 film is presented in Fig. 28.17. The changing trends are also addressed in this figure.

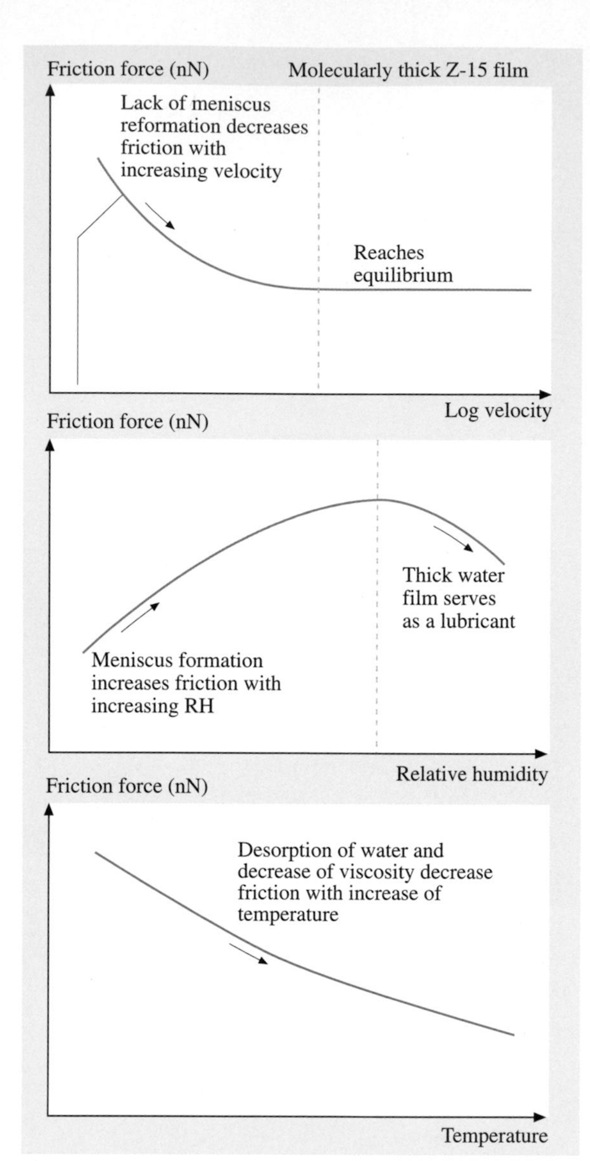

Fig. 28.17 Schematic shows the change of friction force of molecularly thick Z-15 films with log velocity, relative humidity, and temperature [28.21]

28.3.5 Tip Radius Effect

The tip radius and relative humidity affect adhesion and friction for dry and lubricated surfaces [28.35, 36]. Figure 28.18a shows the variation of single point adhesive force measurements as a function of tip radius on a Si(100) sample for several humidities. The adhesive force data are also plotted as a function of relative humidity for various tip radii. Figure 28.18a indicates that the tip radius has little effect on the adhesive forces at low humidities, but the adhesive force increases with tip radius at high humidity. Adhesive force also increases with an increase in humidity for all tips. The trend in adhesive forces as a function of tip radii and relative humidity, in Fig. 28.18a, can be explained by the presence of meniscus forces, which arise from capillary condensation of water vapor from the environment. If enough liquid is present to form a meniscus bridge, the meniscus force should increase with an increase in tip radius based on (28.2). This observation suggests that thickness of the liquid film at low humidities is insufficient to form continuous meniscus bridges and to affect adhesive forces in the case of all tips.

Figure 28.18a also shows the variation in coefficient of friction as a function of tip radius at a given humidity and as a function of relative humidity for a given tip radius on the Si(100) sample. It can be observed that for RH 0%, the coefficient of friction is about the same for the tip radii except for the largest tip, which shows

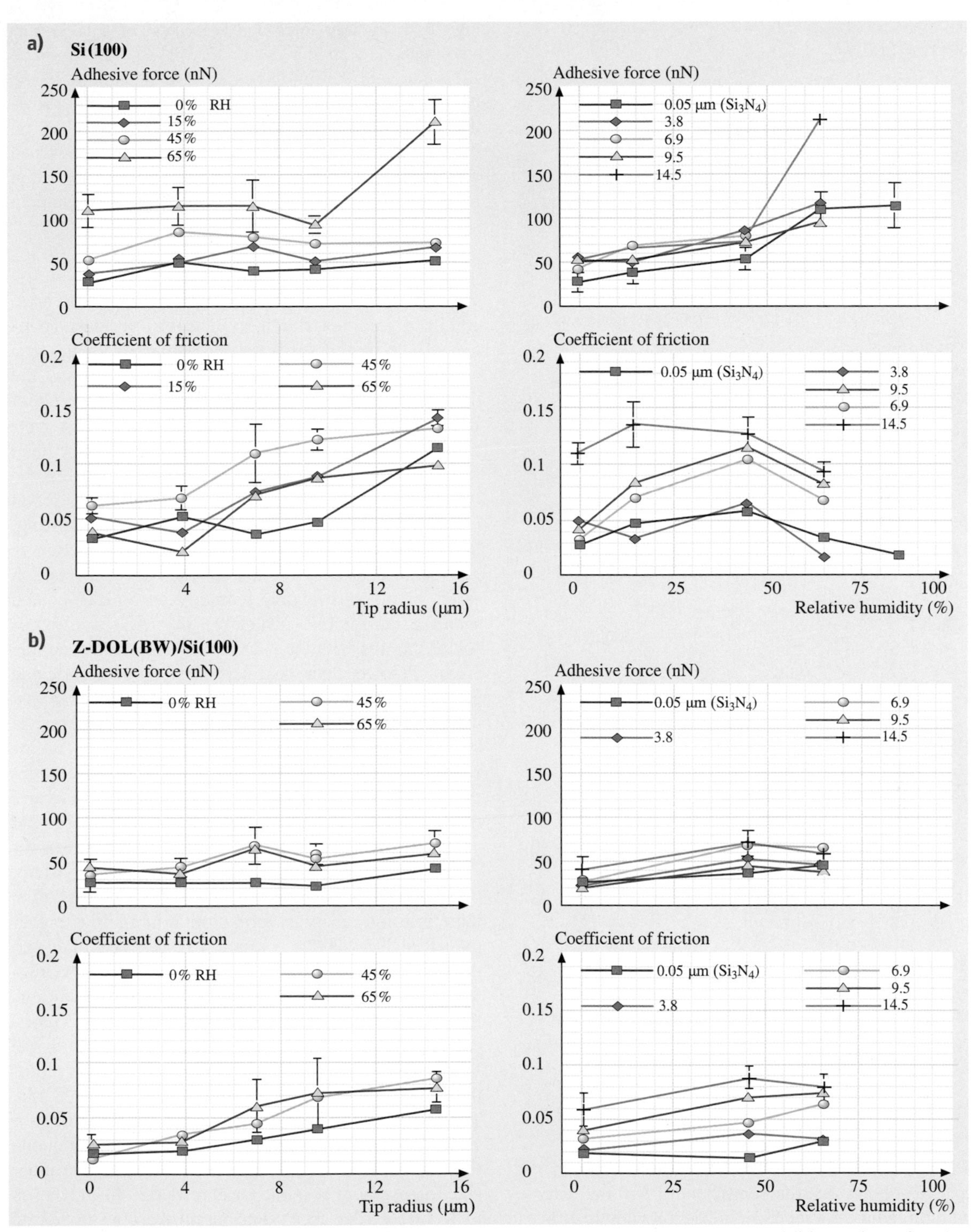

Fig. 28.18a,b Adhesive force and coefficient of friction as a function of tip radius at several humidities and as a function of relative humidity at several tip radii on (a) Si(100) and (b) 0.5-nm Z-DOL(BW) films [28.35]

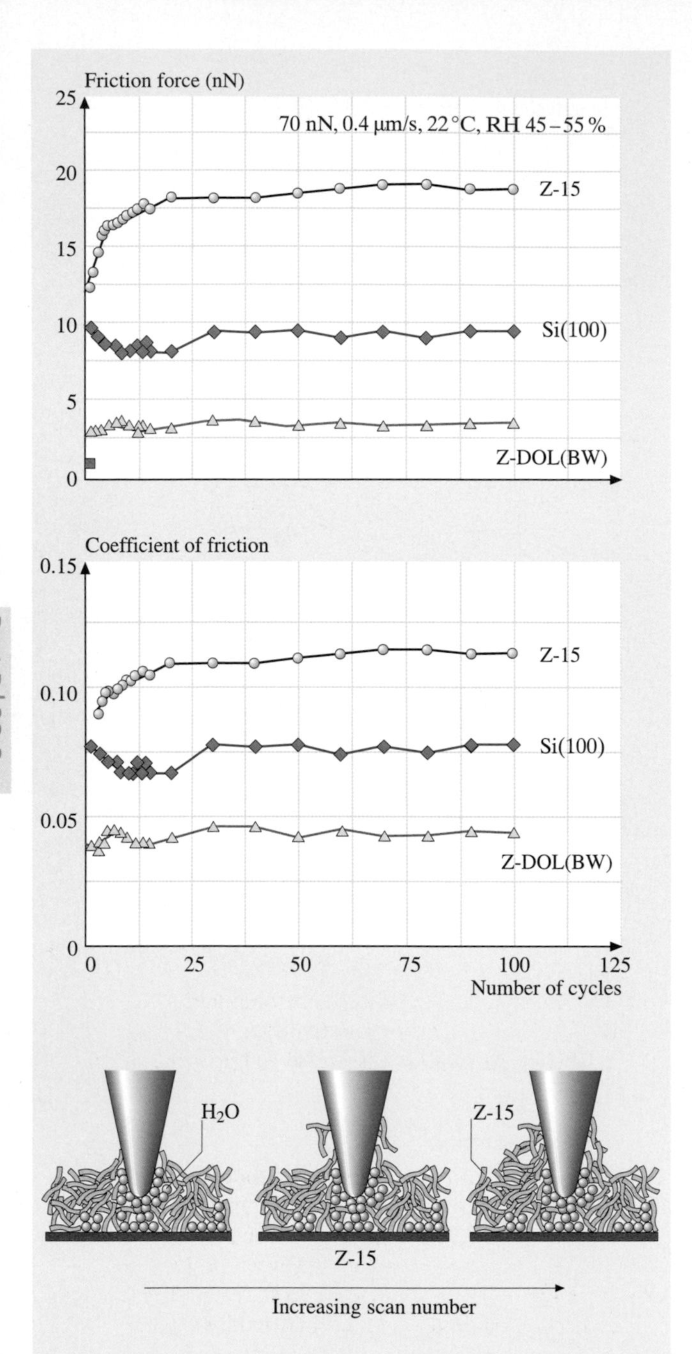

Fig. 28.19 Friction force and coefficient of friction versus number of sliding cycles for Si(100), 2.8-nm-thick Z-15 film, and 2.3-nm-thick Z-DOL(BW) film at 70 nN, 0.8 μm/s, and in ambient air. Schematic (*bottom*) shows that some liquid Z-15 molecules can be attached on the tip. The molecular interaction between the attached molecules on the tip with the Z-15 molecules in the film results in an increase of the friction force with multiple scanning [28.21]

a higher value. At all other humidities, the trend consistently shows that the coefficient of friction increases with tip radius. An increase in friction with tip radius at low to moderate humidities arises from increased contact area (i. e., higher van der Waals forces) and higher values of shear forces required for a larger contact area. At high humidities, similar to adhesive force data, an increase with tip radius occurs due to of both contact area and meniscus effects. It can be seen that for all tips, the coefficient of friction increases with humidity to about RH 45%, beyond which it starts to decrease. This is attributed to the fact that at higher humidities, the adsorbed water film on the surface acts as a lubricant between the two surfaces [28.21]. Thus the interface is changed at higher humidities, resulting in lower shear strength and, hence, lower friction force and coefficient of friction.

Figure 28.18b shows adhesive forces as a function of tip radius and relative humidity on Si(100) coated with 0.5-nm-thick Z-DOL(BW) film. Adhesive forces for all the tips with the Z-DOL(BW) lubricated sample are much lower than those experienced on unlubricated Si(100), shown in Fig. 28.18a. The data also show that even at a monolayer thickness of the lubricant there is very little variation in adhesive forces with tip radius at a given humidity. For a given tip radius, the variation in adhesive forces with relative humidity indicates that these forces slightly increase from RH 0% to RH 45%, but remain more or less the same with a further increase in humidity. This is seen even with the largest tip, which indicates that the lubricant is indeed hydrophobic; there is some meniscus formation at humidities higher than RH 0%, but the formation is very minimal and does not increase appreciably even up to RH 65%. Figure 28.18b also shows coefficient of friction for various tips at different humidities for the Z-DOL(BW) lubricated sample. Again, all the values obtained with the lubricated sample are much lower than the values obtained on unlubricated Si(100), shown in Fig. 28.18a. The coefficient of friction increases with tip radius for all humidities, as was seen on unlubricated Si(100), due to an increase in the contact area. Similar to the adhesive forces, there is an increase in friction from RH 0% to RH 45%, due to a contribution from an increased number of menisci bridges. But thereafter there is very little additional water film forming, due to the hydrophobicity of the Z-DOL(BW) layer, and, consequentially, the coefficient of friction does not change appreciably, even with the largest tip. These findings show that even a monolayer of Z-DOL(BW) offers good hydrophobic performance of the surface.

28.3.6 Wear Study

To study the durability of lubricant films at the nanoscale, the friction of Si(100), Z-15, and Z-DOL(BW) as a function of the number of scanning cycles was measured, Fig. 28.19 [28.21]. As observed earlier, friction force and coefficient of friction of Z-15 is higher than that of Si(100), and Z-DOL(BW) has the lowest values. During cycling, friction force and coefficient of friction of Si(100) show a slight variation during the initial few cycles then remain constant. This is related to the removal of the top adsorbed layer. In the case of Z-15 film, the friction force and coefficient of friction show an increase during the initial few cycles and then approach higher and stable values. This is believed to be caused by the attachment of the Z-15 molecules onto the tip. The molecular interaction between these attached molecules to the tip and molecules on the film surface is responsible for an increase in the friction. But after several scans, this molecular interaction reaches the equilibrium, and after that, friction force and coefficient of friction remain constant. In the case of Z-DOL(BW) film, the friction force and coefficient of friction start out low and remain low during the entire test for 100 cycles. It suggests that Z-DOL(BW) molecules do not get attached or displaced as readily as Z-15.

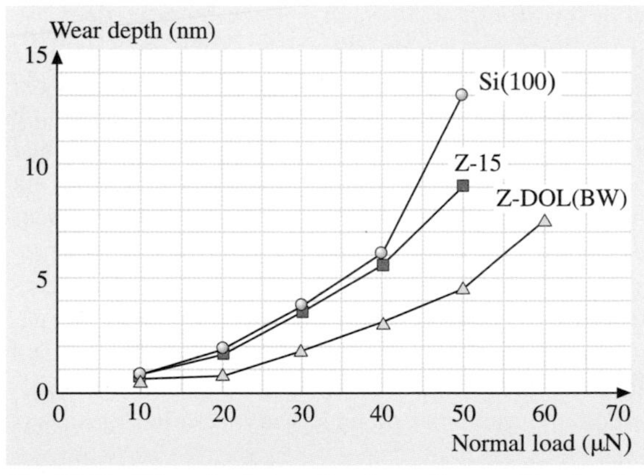

Fig. 28.20 Wear depth as a function of normal load using a diamond tip for Si(100), 2.9-nm-thick Z-15 film, and 2.3-nm-thick Z-DOL(BW) after one cycle [28.20]

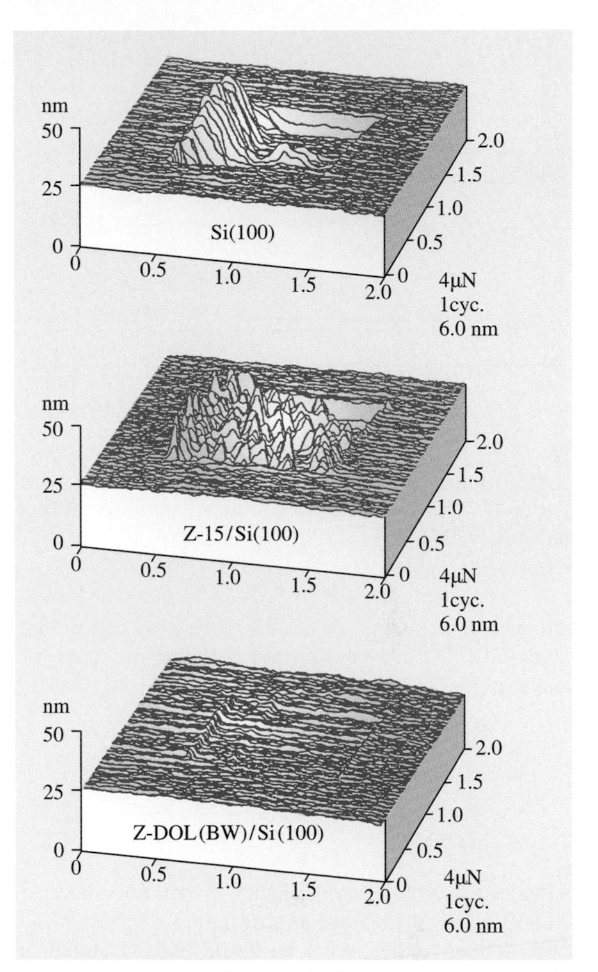

Fig. 28.21 Wear profiles for Si(100), 2.9-nm-thick Z-15 film, and 2.3-nm-thick Z-DOL(BW) film after wear studies using a diamond tip. Normal force used and wear depths are listed in the figure [28.20]

Koinkar and *Bhushan* [28.20, 34] conducted wear studies using a diamond tip at high loads. Figure 28.20 shows the plots of wear depth as a function of normal force, and Fig. 28.21 shows the wear profiles of the worn samples at 40 μN normal load. The 2.3-nm-thick Z-DOL(BW) lubricated sample exhibits better wear resistance than unlubricated and 2.9-nm-thick Z-15 lubricated silicon samples. Wear resistance of a Z-15 lubricated sample is little better than that of unlubricated sample. The Z-15 lubricated sample shows debris inside the wear track. Since the Z-15 is a liquid lubricant, debris generated is held by the lubricant and they become sticky. The debris moves inside the wear track

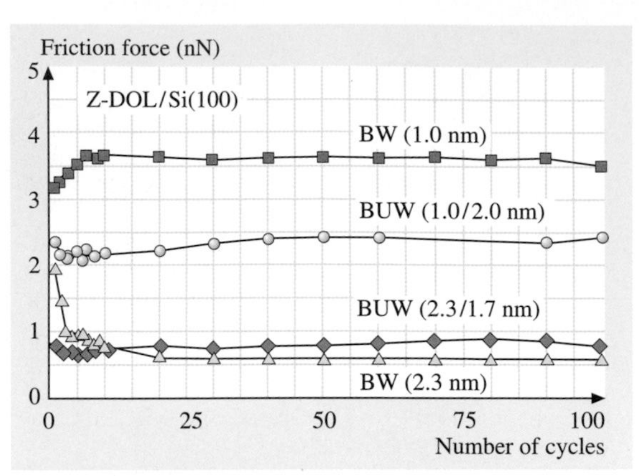

Fig. 28.22 Friction force as a function of number of cycles using a Si_3N_4 tip at a normal load of 300 nN for Z-DOL(BW) and Z-DOL(BUW) films with different film thicknesses [28.34]

and does damage, Fig. 28.20. These results suggest that using Z-DOL(BW) can improve the wear resistance of substrate.

To study the effect of the degree of chemical bonding, the durability tests were conducted on both fully bonded and partially bonded Z-DOL films. Durability results for Z-DOL(BW) and Z-DOL bonded and unwashed (Z-DOL(BUW), a partially bonded film that contains both bonded and mobile phase lubricants) with different film thicknesses are shown in Fig. 28.22 [28.34]. Thicker films, such as Z-DOL(BUW), with a thickness of 4.0 nm (bonded/mobile = 2.3 nm/1.7 nm) exhibit behavior similar to 2.3-nm-thick Z-DOL(BW) film. Figure 28.22 also indicates that Z-DOL(BW) and Z-DOL(BUW) films with a thinner film thickness exhibit a higher friction value. Comparing 1.0-nm-thick Z-DOL(BW) with 3.0-nm-thick Z-DOL(BUW) (bonded/mobile = 1.0 nm/2.0 nm), the Z-DOL(BUW) film exhibits a lower and stable friction value. This is because the mobile phase on a surface acts as a source of lubricant replenishment. A similar conclusion has also been reported by *Ruhe* et al. [28.19], *Bhushan* and *Zhao* [28.13], and *Eapen* et al. [28.49]. All of them indicate that using partially bonded Z-DOL films can dramatically reduce the friction and improve the wear life.

28.4 Closure

Nanodeformation study has shown that fully bonded Z-DOL lubricants behave as soft polymer solids, while the unbonded lubricants behave liquid-like. AFM studies have shown that the physisorbed nonpolar molecules on a solid surface have an extended, flat conformation. The spreading property of PFPE is strongly dependent on the molecular end groups and substrate chemistry.

Using solid-like Z-DOL(BW) film can reduce the friction and adhesion of Si(100), while using liquid-like lubricant of Z-15 shows a negative effect. Si(100) and Z-15 film show apparent time effect. The friction and adhesion forces increase as a result of growth of meniscus up to an equilibrium time, after which they remain constant. Using Z-DOL(BW) film can prevent time effects. High velocity leads to the rupture of meniscus and prevents its reformation, which leads to a decrease of friction and adhesive forces of Z-15 and Z-DOL(BW). The influence of relative humidity on the friction and adhesion is dominated by the amount of the adsorbed water molecules. Increasing humidity can either increase friction through increased adhesion by water meniscus, or reduce friction through an enhanced water-lubricating effect. Increasing temperature leads to desorption of water layer, decrease of water surface tension, decrease of viscosity, and easier orientation of the Z-DOL(BW) molecules. These changes cause a decrease of friction force and adhesion at high temperature. During cycling tests, the molecular interaction between the attached Z-15 molecules to the tip and the Z-15 molecules on the film surface causes the initial rise of friction. Wear tests show that Z-DOL(BW) can improve the wear resistance of silicon. Partially bonded PFPE film appears to be more durable than fully bonded films.

These results suggest that partially/fully bonded films are good lubricants for devices operating in different environments and under varying conditions.

References

28.1 B. Bhushan: Magnetic Recording Surfaces. In: *Characterization of Tribological Materials*, ed. by W.A. Glaeser (Butterworth-Heinemann, Boston 1993) pp. 116–133

28.2 B. Bhushan: *Principles and Applications of Tribology* (Wiley, New York 1999)

28.3 B. Bhushan: *Introduction to Tribology* (Wiley, New York 2002)

28.4 V.J. Novotny, I. Hussla, J.M. Turlet, M.R. Philpott: Liquid polymer conformation on solid surfaces, J. Chem. Phys. **90** (1989) 5861–5868

28.5 V.J. Novotny: Migration of liquid polymers on solid surfaces, J. Chem. Phys. **92** (1990) 3189–3196

28.6 C.M. Mate, V.J. Novotny: Molecular conformation and disjoining pressures of polymeric liquid films, J. Chem. Phys. **94** (1991) 8420–8427

28.7 C.M. Mate: Application of disjoining and capillary pressure to liquid lubricant films in magnetic recording, J. Appl. Phys. **72** (1992) 3084–3090

28.8 G.G. Roberts: *Langmuir–Blodgett Films* (Plenum, New York 1990)

28.9 A. Ulman: *An Introduction to Ultrathin Organic Films* (Academic, Boston 1991)

28.10 B. Bhushan: *Tribology and Mechanics of Magnetic Storage Devices*, 2nd edn. (Springer, New York 1996)

28.11 B. Bhushan: Macro- and microtribology of magnetic storage devices. In: *Modern Tribology Handbook Vol. 2: Materials, Coatings, and Industrial Applications*, ed. by B. Bhushan (CRC, Boca Raton 2001) pp. 1413–1513

28.12 Anonymous: Fomblin Z Perfluoropolyethers, Data sheet Montedism Group, Milan (2002)

28.13 B. Bhushan, Z. Zhao: Macroscale and microscale tribological studies of molecularly thick boundary layers of perfluoropolyether lubricants for magnetic thin-film rigid disks, J. Info. Storage Proc. Syst. **1** (1999) 1–21

28.14 B. Bhushan: *Tribology Issues and Opportunities in MEMS* (Kluwer, Dordrecht 1998)

28.15 B. Bhushan, J.N. Israelachvili, U. Landman: Nanotribology: Friction, wear and lubrication at the atomic scale, Nature **374** (1995) 607–616

28.16 B. Bhushan: *Handbook of Micro/Nanotribology*, 2nd edn. (CRC, Boca Raton 1999)

28.17 B. Bhushan: Self-assembled monolayers for controlling hydrophobicity and/or friction and wear. In: *Modern Tribology Handbook Vol. 2: Materials, Coatings, and Industrial Applications*, ed. by B. Bhushan (CRC, Boca Raton 2001) pp. 909–929

28.18 J. Ruhe, G. Blackman, V.J. Novotny, T. Clarke, G.B. Street, S. Kuan: Thermal attachment of perfluorinated polymers to solid surfaces, J. Appl. Polym. Sci. **53** (1994) 825–836

28.19 J. Ruhe, V. Novotny, T. Clarke, G.B. Street: Ultrathin perfluoropolyether films – influence of anchoring and mobility of polymers on the tribological properties, ASME J. Tribol. **118** (1996) 663–668

28.20 V.N. Koinkar, B. Bhushan: Microtribological studies of unlubricated and lubricated surfaces using atomic force/friction force microscopy, J. Vac. Sci. Technol. A **14** (1996) 2378–2391

28.21 H. Liu, B. Bhushan: Nanotribological characterization of molecularly-thick lubricant films for applications to MEMS/NEMS by AFM, Ultramicroscopy **97** (2003) 321–340

28.22 G.S. Blackman, C.M. Mate, M.R. Philpott: Interaction forces of a sharp tungsten tip with molecular films on silicon surface, Phys. Rev. Lett. **65** (1990) 2270–2273

28.23 G.S. Blackman, C.M. Mate, M.R. Philpott: Atomic force microscope studies of lubricant films on solid surfaces, Vacuum **41** (1990) 1283–1286

28.24 X. Ma, J. Gui, K.J. Grannen, L.A. Smoliar, B. Marchon, M.S. Jhon, C.L. Bauer: Spreading of PFPE lubricants on carbon surfaces: Effect of hydrogen and nitrogen content, Tribol. Lett. **6** (1999) 9–14

28.25 C.A. Kim, H.J. Choi, R.N. Kono, M.S. Jhon: Rheological characterization of perfluoropolyether lubricant, Polym. Prepr. **40** (1999) 647–649

28.26 M. Ruths, S. Granick: Rate-dependent adhesion between opposed perfluoropoly(alkylether) layers: Dependence on chain-end functionality and chain length, J. Phys. Chem. B **102** (1998) 6056–6063

28.27 U. Jonsson, B. Bhushan: Measurement of rheological properties of ultrathin lubricant films at very high shear rates and near-ambient pressure, J. Appl. Phys. **78** (1995) 3107–3109

28.28 C. Hahm, B. Bhushan: High shear rate viscosity measurement of perfluoropolyether lubricants for magnetic thin-film rigid disks, J. Appl. Phys. **81** (1997) 5384–5386

28.29 C.M. Mate: Atomic-force-microscope study of polymer lubricants on silicon surface, Phys. Rev. Lett. **68** (1992) 3323–3326

28.30 C.M. Mate: Nanotribology of lubricated and unlubricated carbon overcoats on magnetic disks studied by friction force microscopy, Surf. Coat. Technol. **62** (1993) 373–379

28.31 S.J. O'Shea, M.E. Welland, T. Rayment: Atomic force microscope study of boundary layer lubrication, Appl. Phys. Lett. **61** (1992) 2240–2242

28.32 S.J. O'Shea, M.E. Welland, J.B. Pethica: Atomic force microscopy of local compliance at solid–liquid interface, Chem. Phys. Lett. **223** (1994) 336–340

28.33 B. Bhushan, T. Miyamoto, V. N. Koinkar: Microscopic friction between a sharp diamond tip and thin-film magnetic rigid disks by friction force microscopy, Adv. Info. Storage Syst. **6** (1995) 151–161

28.34 V. N. Koinkar, B. Bhushan: Micro/nanoscale studies of boundary layers of liquid lubricants for magnetic disks, J. Appl. Phys. **79** (1996) 8071–8075

28.35 B. Bhushan, S. Sundararajan: Micro/nanoscale friction and wear mechanisms of thin films using atomic force and friction force microscopy, Acta Mater. **46** (1998) 3793–3804

28.36 B. Bhushan, C. Dandavate: Thin-film friction and adhesion studies using atomic force microscopy, J. Appl. Phys. **87** (2000) 1201–1210

28.37 S. Sundararajan, B. Bhushan: Static friction and surface roughness studies of surface micromachined electrostatic micromotors using an atomic force/friction force microscope, J. Vac. Sci. Technol. A **19** (2001) 1777–1785

28.38 B. Bhushan, J. Ruan: Atomic-scale friction measurements using friction force microscopy: Part II – application to magnetic media, ASME J. Tribol. **116** (1994) 389–396

28.39 T. Stifter, O. Marti, B. Bhushan: Theoretical investigation of the distance dependence of capillary and van der Waals forces in scanning probe microscopy, Phys. Rev. B **62** (2000) 13667–13673

28.40 J. N. Israelachvili: *Intermolecular and Surface Forces*, 2nd edn. (Academic, London 1992)

28.41 S. K. Chilamakuri, B. Bhushan: A comprehensive kinetic meniscus model for prediction of long-term static friction, J. Appl. Phys. **15** (1999) 4649–4656

28.42 H. Ishigaki, I. Kawaguchi, M. Iwasa, Y. Toibana: Friction and wear of hot pressed silicon nitride and other ceramics, ASME J. Tribol. **108** (1986) 514–521

28.43 T. E. Fischer: Tribochemistry, Annu. Rev. Mater. Sci. **18** (1988) 303–323

28.44 K. Mizuhara, S. M. Hsu: Tribochemical reaction of oxygen and water on silicon surfaces. In: *Wear Particles*, ed. by D. Dowson (Elsevier, New York 1992) pp. 323–328

28.45 S. Danyluk, M. McNallan, D. S. Park: Friction and wear of silicon nitride exposed to moisture at high temperatures. In: *Friction and Wear of Ceramics*, ed. by S. Jahanmir (Dekker, New York 1994) pp. 61–79

28.46 V. A. Muratov, T. E. Fischer: Tribochemical polishing, Annu. Rev. Mater. Sci. **30** (2000) 27–51

28.47 H. Yoshizawa, Y. L. Chen, J. N. Israelachvili: Fundamental mechanisms of interfacial friction I: Relationship between adhesion and friction, J. Phys. Chem. **97** (1993) 4128–4140

28.48 H. Yoshizawa, J. N. Israelachvili: Fundamental mechanisms of interfacial friction II: Stick slip friction of spherical and chain molecules, J. Phys. Chem. **97** (1993) 11300–11313

28.49 K. C. Eapen, S. T. Patton, J. S. Zabinski: Lubrication of microelectromechanical systems (MEMS) using bound and mobile phase of Fomblin Z-DOL, Tibol. Lett. **12** (2002) 35–41

29. Kinetics and Energetics in Nanolubrication

Lubrication, one of human kind's oldest engineering disciplines, in the 19th century gained from Reynolds' classical hydrodynamic description a theoretical base unmatched by most of the theories developed in tribology to date. In the 20th century, however, increasing demands on lubricants shifted the attention from bulk film to ultrathin film lubrication. Finite size limitations imposed constraints on the lubrication process that were not considered in bulk phenomenological treatments introduced by Reynolds. At this point, as is common in many engineering applications, empiricism took over. Functional relationships derived from the classical theories were tweaked to accommodate the new situation of reduced scales by introducing "effective" or "apparent" properties.

With the inception of nanorheological tools of complementary nature in the later decades of the 20th century (e.g., the surface forces apparatus and scanning force microscopy), tribology entered the realm of nanoscience. Through an increasing confidence in experimental findings on the nanoscale, kinetic and energetic theories incorporated interfacial and molecular constraints.

The very fundamentals have been challenged in recent years. Researchers have realized that bulk perceptions, such as "solid" and "liquid" are defied on the nanoscale. The reduction in dimensionality of the nanoscale imposes constraints that bring into question the use of classical statistical mechanics of decoupled events. The diffusive description of lubrication is failing in a system that is thermodynamically not well-equilibrated.

The challenge any nanotechnological endeavor encounters is the development of a theoretical

29.1 **Background: From Bulk to Molecular Lubrication** 885
29.1.1 Hydrodynamic Lubrication and Relaxation 885
29.1.2 Boundary Lubrication 885
29.1.3 Stick Slip and Collective Phenomena 885

29.2 **Thermal Activation Model of Lubricated Friction** 887

29.3 **Functional Behavior of Lubricated Friction** 888

29.4 **Thermodynamical Models Based on Small and Nonconforming Contacts** ... 890

29.5 **Limitation of the Gaussian Statistics – The Fractal Space** 891

29.6 **Fractal Mobility in Reactive Lubrication** .. 892

29.7 **Metastable Lubricant Systems in Large Conforming Contacts** 894

29.8 **Conclusion** 895

References 895

framework based on an appropriate statistics. In tribology this is met with spectral descriptions of the dynamic sliding process. Statistical kernels are being developed for probability density functions to explain anomalous transport processes that involve long-range spatial or temporal correlations. With such theoretical developments founded in nanorheological experiments, a more realistic foundation will be laid to describe the behavior of lubricants in the confined geometries of the nanometer length scale.

What is inaccessible today may become accessible tomorrow as has happened by the invention of the microscope. ... Coherent assumptions on what is still invisible may increase our understanding of the visible. ... Strong reasons have come to support a growing probability, and it can finally be said the certainty, in favor of the hypothesis of the atomists. (Jean Baptiste Perrin – Nobel Lecture, December 11, 1926)

Since technology is driving lubricant films to molecular thickness, kinetic friction and its dependence on the sliding parameters – especially the sliding velocity – have become of great interest. The complexity of the frictional resistance in lubricated sliding is illustrated in Fig. 29.1 with a *Stribeck Curve*. Various regimes of lubrication can be identified in the Stribeck curve. They express to what degree the hydrodynamic pressure is involved in the lubrication process. In the ultra-low speed regime, called the *boundary lubrication regime*, no hydrodynamic pressure is built up in the lubricant. Consequently the load is carried by contact asperities coated with adsorbed lubricant molecules. If the speed is raised, a hydrodynamic pressure builds that leads to a *mixed lubrication*, in which the load is carried by both asperities and hydrodynamic pressure. At even higher speeds, elastic contributions of the solid surfaces have to be considered paired with hydrodynamic pressure effects (elastohydrodynamic lubrication), until only *hydrodynamic lubrication* matters. Hence, the Stribeck curve combines various aspects of lubrication. The curve cannot be discussed without considering the lubricant thickness and the different models of asperity contact sliding.

In one of the first comprehensive physical models of "dry" friction, Bowden and Tabor introduced a plastic asperity model, in which the material's yield stress and adhesive properties play an important role [29.1]. Considering this model, which depends on surface energies and mechanical yield properties paired with all the properties that come along with a surface adsorbent lubricant, one can hardly grasp the difficulty level involved in describing the frictional kinetics in lubrication.

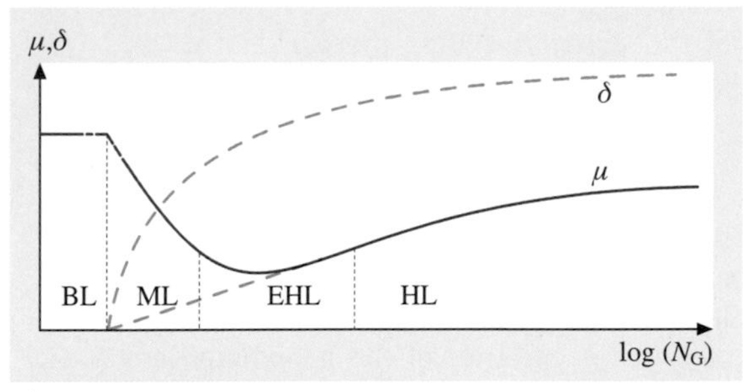

Fig. 29.1 Stribeck Curve (schematic) relates the fluid lubricant thickness, δ, and the friction coefficient μ to the Gumbel Number $N_G = \eta\omega P^{-1}$; i. e., the product of the liquid bulk viscosity, η, the sliding speed (or more precisely the shaft frequency), ω, and the inverse of the normal pressure, P. BL: Boundary Lubrication, ML: Mixed Lubrication, EHL: Elastohydrodynamic Lubrication, HL: Hydrodynamic Lubrication

Past and current engineering challenges in lubrication have been met with great and complex empiricism. The theoretical modeling of lubrication junctions generally involved only bulk property considerations with inadequately known adsorption mechanisms. The complexity of today's lubricants, most of them, such as motor oil, a product of empirical design over many years, increased exponentially, making it very difficult to meet future challenges. The problem of empiricism is that conventional laws and perceptions are unchallenged. *Effective* quantities are invented (e.g., effective viscosity), exponential fitting parameters are introduced (e.g., Kohlrausch relaxation parameter), and terminologies such as *solid* and *liquid* are taken as granted. Progress based on empiricism is only incremental and rarely revolutionary.

One of the reasons for empiricism is a lack of access to a system with fewer and better controlled parameters. In lubrication sliding that challenge has been addressed over the last two decades with the inception of the surface forces apparatus (SFA) by *Tabor* et al. [29.2] and scanning force microscopy (SFM) by *Binnig* et al. [29.3]. These two instrumental methods allow lubrication studies where roughness effects can be neglected, surface energies controlled, and wear from wearless friction distinguished. Lubricant properties can be studied at nearly mathematically described boundaries, atomistic friction events can be recorded, and fundamental models that have been considered to be mere Gedanken Experiments, such as the *Tomlinson model of friction*, can be verified. In the wake of these nanoscopic tools, exciting new theoretical lubrication and friction models have appeared.

This chapter considers these recent experimental and theoretical developments with a particular focus on sliding speed and real or apparent changes in the lubricant material properties. We will discuss kinetics and energetics in the "simplified" world of nanolubrication, in which our conventional perception is challenged. After a brief review of some of the classical lubrication concepts (hydrodynamic lubrication and boundary lubrication), we will turn our attention to a thermal activation model of friction, functional behavior of lubricated friction with velocity, and models based on small non-conforming contacts. We will critically discuss the limitation of the underlying Gaussian statistics, introduce fractal dynamics in lubrication, and will end our discussion with metastable lubricant systems.

29.1 Background: From Bulk to Molecular Lubrication

29.1.1 Hydrodynamic Lubrication and Relaxation

In the classical theories of tribology by da Vinci, Amonton, and Coulomb, not much attention was given to the dependence of kinetic friction on the sliding velocity. This clearly changed in the 19th century during the first industrial revolution, at which time lubricants became increasingly important, for instance, in ball and journal bearings. It was *Petrov* [29.4], *Tower* [29.5], and *Reynolds* [29.6] who established that the liquid viscous shear properties determine the frictional kinetics. *Reynolds* [29.6] combined the pressure-gradient determined *Poisseuille flow* with the bearing surface induced *Couette flow* assuming, based on Petrov's law [29.4], a no-slip condition at the interface between lubricant and solid. This led to the widely used linear relationship between friction and velocity. Reynolds' hydrodynamic theory of lubrication can be applied to steady state sliding at constant relative velocity and to transient decay sliding (sliding is stopped from an initial velocity v and a corresponding shear stress τ_0), which leads to the classical Debye exponential relaxation behavior, i. e.:

$$\tau = \tau_0 \exp\left(\frac{-D}{A\eta}t\right) \; ; \quad \tau_0 \propto \frac{v\eta}{D} \, . \tag{29.1}$$

D is the lubricant thickness, A the area of the slider, and η the viscosity of the fluid. We will later see that this classical exponential relaxation behavior, obtained in a thermodynamically well-mixed three dimensional medium, is distorted when the liquid film thickness is reduced to molecular dimensions.

29.1.2 Boundary Lubrication

Reynolds hydrodynamic description of lubrication was found to work well for thick lubricant films but to break down for thinner films. One manifestation is that for films on the order of ten molecular diameters, the stress in the film does not allow the tension to return to zero. It was also found that the motion in the steady state sliding regime was disrupted, exhibiting a stick–slip-like slider motion [29.7]). Consequently, this non-Newtonian behavior was treated with a modified viscosity parameter (effective viscosity), which was composed of the pressure, temperature, and rate of shear.

The term *Boundary Lubrication* is used to describe a lubricant that is reduced in thickness to molecular dimension and effectively reduces friction between two opposing solid surfaces. *Hardy* et al. [29.8] recognized that molecular properties such as molecular weight and molecular arrangement, are governing the frictional force. This confined concept of lubrication, often visualized by two highly ordered opposing films with shear taking place somewhere in between the two layers, contains many of the rate dependent manifestations of frictional sliding; e.g., stick–slip, ultra-low friction, transitions from high to low friction, phase transitions, dissipation due to dislocations (e.g., gauche and cis-transformations), and memory effects.

Boundary lubrication was found to be in many respects unique [29.9]. In macroscopic experiments, which involved rough surfaces, friction–velocity plots resembled logarithmic functions at moderate speeds. No static stiction force peaks were observed in boundary lubricants close to zero speed. On the contrary, retractive slips could be observed upon halting, constituting a static friction coefficient exceeded by the dynamic friction coefficient [29.9]. These unique manifestations of boundary lubrication were discussed in terms of a lubricated asperity–junction mechanism, which associated "an increase in the coefficient of friction with a decrease in the adsorptive coverage of the rubbing surfaces by the lubricant substance" [29.9]. It was argued that in the course of the sliding process of a macroscopic slider, more adsorbed lubricant is expected to exist within the interfacial area than outside the contact zone. This would lead upon halting to a relaxation process of the elastic restraints on the slider, causing the slider to a retractively slip.

29.1.3 Stick Slip and Collective Phenomena

Based on numerous friction experiments at the initiation of sliding with rough macroscopic contact, it was argued that the distinction between static and kinetic friction is not categorical but rather a manifestation of the apparatus [29.9]. This was a widely held opinion prior to *Briscoe*'s et al. [29.10] molecularly smooth monolayer SFA experiments of aliphatic carboxylic acids and their soaps. Briscoe found that the character of sliding motion (continuous vs. discontinuous), depends not only on the apparatus but also on the properties (chemistry) of the monolayer. As in the rough boundary layer experiments discussed above, Briscoe's molecularly smooth monolayer experiments exhibited logarithmic-like friction–velocity behaviors.

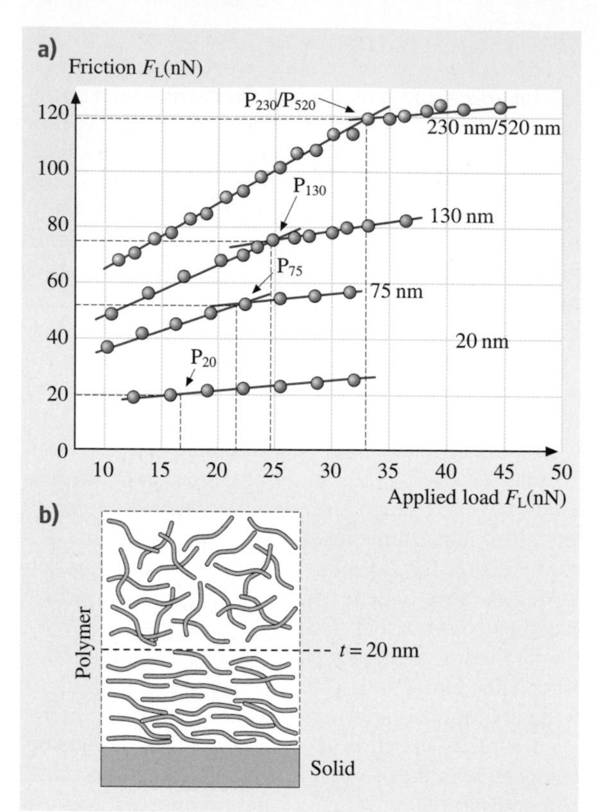

Fig. 29.2 (**a**) SFM friction measurements at a speed 1 μm/s: Cooperative molecular response of polyethylene co-propylene to frictional shear forces as a function of the applied load. P_t (t corresponds to the thickness of the polymer film) represents the critical activation load at which collective sliding is energetically more favorable than local plastic yielding. Adapted from [29.11]. (**b**) Sketch of the degree of disentanglement in the vicinity to the solid substrate surface

It was *Israelachvili* et al. [29.7] who, based on SFA experiments and computer simulations, provided a molecular picture of the stick–slip behavior caused by the lubricant material. The major achievement of this work was to draw our attention to the molecular structure of the lubricant, which is often different from the bulk and unstable during the sliding motion. It was recognized that bulk rheology failed to describe the lubrication process. Finally addressed in the Israelachvili study were in-plane structuring of simple liquids caused by compression forces and "freezing–melting" transitions due to shear.

The simple concept of a freezing–melting transition is based on a common perception of the two distinctive parts of a stick–slip occurrence: the solid (Hookian)-like sticking part and the liquid (Newtonian)-like slipping part. But a deformation of a solid can be both, coordinated or uncoordinated, and thus can exhibit both solid-like and liquid-like behavior. For instance, most of the plastic yielding processes are uncoordinated. On the other hand, slipping within a solid, along a crystal plane in a thermally activated strain-release process for instance, is a highly coordinated molecular process [29.12].

Similar arguments can be made for a liquid. For example, stick–slip behaviors were observed in more complex fluidic systems by *Reiter* et al. [29.13], who compared a molecularly "wet" lubricant film with a "dry" self-assembled monolayer lubricant. They concluded that sliding in liquid films is the result of slippage along an interface. In other words, the degree of molecular cooperation determined the frictional resistance.

The concept of local-*versus*-cooperative yield to shear is briefly illustrated here with a frictional-load study of a molecularly entangled polymer melt obtained in a SFM study of *Buenviaje* et al. [29.11]. Each of the curves presented in Fig. 29.2a represents a polymer film of polyethylene co-propylene of distinctly different degree of entanglement. Films of thickness above 230 nm exhibit the strongest entanglement strength. Films of 20 nm thickness or thinner are fully disentangled. The reason for the film thickness-dependent entanglement strength is given by the substrate distance-dependent shear strength during the spin coating process of the thin films. For entangled films SFM friction studies exhibit a critical applied load (identified by P_t, and the thickness t) that separates two friction regimes: One identified by a high friction coefficient and the other by a low friction coefficient. At loads below P_t the friction coefficients are high, indicating plastic yielding during sliding. In these plastic regimes of sliding, molecular cooperation is low, leading to high local shear stresses compensated by local yielding of the material. Above the critical load, the friction coefficient drops, independent of the film thickness, to a low value of 3.0, corresponding to the value obtained from the fully disentangled film. Note that the polymer molecules in the 20-nm-thick film experience high substrate tangential stresses during the spin coating process. Hence the disentangled polymer molecules can be considered to be aligned preferentially along the substrate surface as sketched in Fig. 29.2b. This leads to a decrease of the structural entropy the closer the material is to the solid substrate surface. Considering the matching friction coefficient of 0.3 above P_t for thicker films, we can assume that any entangled film above a critical load exhibits a similar molecular collective response toward shear as the 20 nm film during spin coating. The critical load and its related pressure represent a barrier that has to be overcome before a collective phenomenon is activated.

29.2 Thermal Activation Model of Lubricated Friction

With the discussion of shear in entangled polymer systems we have introduced structural entropy as one of the key players that affect frictional resistance in lubricants. We found that the structural entropy was affected by the load of the slider, which introduces an activation barrier in the form of a critical pressure. The terminology used here resembles the one of the Eyring theory of molecular liquid transport [29.14].

Eyring discussed a pure liquid at rest in terms of a thermal activation model. The individual liquid molecules experience a "cage-like" barrier that hinders molecular free motion, because of the close packing in liquids. To escape from the cage an activation barrier needs to be surmounted. In Eyring's model, two processes are considered in order to overcome the potential barrier: (i) shear stresses and (ii) thermal fluctuations. The potential barrier in the thermal activation model is depicted in Fig. 29.3 indicating the barrier modification by the applied pressure force P, and shear stress τ. *Briscoe* et al. [29.10] picked up on this idea to interpret the frictional behavior observed on molecularly smooth monolayer systems. Starting from the overall barrier height $E = Q + P\Omega - \tau\phi$ that is repeatedly overcome during a discontinuous sliding motion, using a Boltzmann distribution to determine the average time for single molecular barrier-hopping, and assuming a regular series of barriers and a high stress limit ($\tau\Phi/kT > 1$), the following shear strength versus velocity v relationship was derived [29.10]:

$$\tau = \frac{k_B T}{\phi} \ln\left(\frac{v}{v_0}\right) + \frac{1}{\phi}(Q + P\Omega) \;. \tag{29.2}$$

The barrier height, E, is composed of the process activation energy Q, the compression energy $P\Omega$, where P is the pressure acting on the volume of the junction Ω, and the shear energy $\tau\phi$, where τ is the shear strength acting on the stress activation volume ϕ. T represents the absolute temperature. The stress activation volume ϕ can be conceived as a process coherence volume and interpreted as the size of the moving segment in the unit shear process, whether it is a part of a molecule or a dislocation line. The most critical parameter in (29.2), v_0, is a characteristic velocity related to the frequency of the process and to a jump distance (discussed further below).

From (29.2) the following iso-relationships can be directly deduced [29.10]:

$$\tau = \tau_0 + \alpha P\;; \quad \tau_0 = \frac{1}{\phi}\left(k_B T \ln\left(\frac{v}{v_0}\right) + Q\right)\;;$$

$$\alpha = \frac{\Omega}{\phi}\;; \text{ at constant } v, T\;, \tag{29.3a}$$

$$\tau = \tau_1 - \beta T\;; \quad \tau_1 = \frac{1}{\phi}(Q + P\Omega)\;;$$

$$\beta = -\frac{k}{\phi}\ln\left(\frac{v}{v_0}\right)\;, \quad \text{at constant } P, v\;, \tag{29.3b}$$

$$\tau = \tau_2 + \theta \ln v\;; \quad \tau_2 = \frac{1}{\phi}(Q + P\Omega - kT\ln v_0)\;;$$

$$\theta = \frac{kt}{\phi}\;, \quad \text{at constant } P, T\;. \tag{29.3c}$$

Thus Eyring's model predicts a linear relationship of friction (the product of the shear strength and the active process area) in pressure and temperature and a logarithmic relationship in velocity.

Eyring's model has been verified in lubrication experiments of solid (soap-like) lubricants by Briscoe and liquid lubricants by *He* et al. [29.15] within

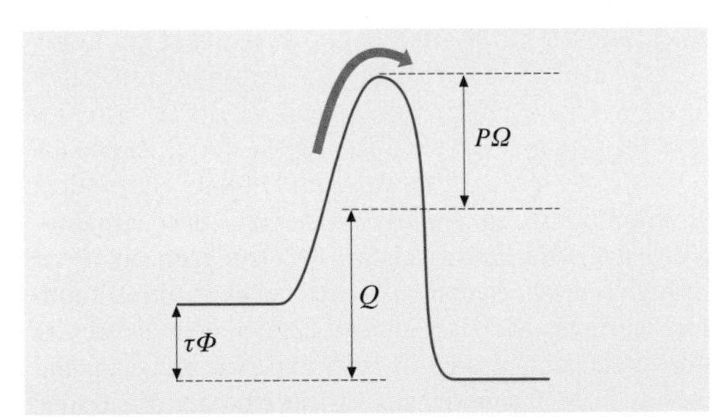

Fig. 29.3 Potential barrier in a lubricant based on Eyring's thermodynamic "cage-model." The normal pressure P and the shear stress τ are modifying the barrier height Q. Modified from [29.10]

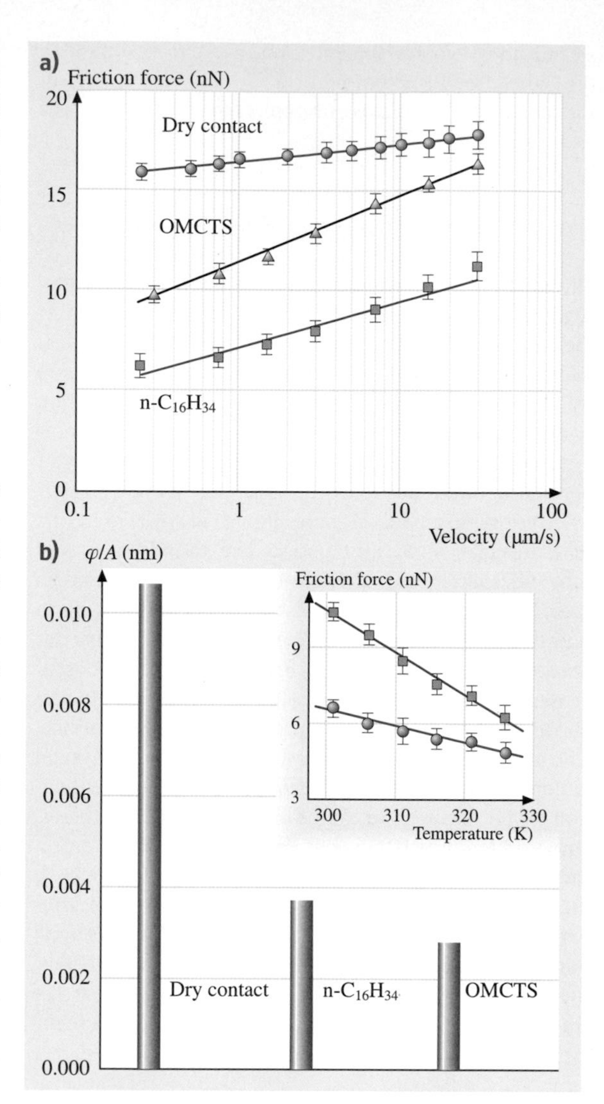

Fig. 29.4 **(a)** Logarithmic $F_F(v)$-plots. $F_F(v) = F_0 + \alpha \ln(v[\mu m/s])$: ● "dry" contact (18% relative humidity) with $F_0 = 16.4$ nN and $\alpha = 0.91$ nN, ▲ OMCTS lubricated with $F_0 = 11.3$ nN and $\alpha = 3.4$ nN, and ■ n-hexadecane (n-$C_{16}H_{34}$) lubricated with $F_0 = 7.1$ nN and $\alpha = 2.5$ nN. The measurements were obtained with rectangular SFM cantilevers (0.4–0.8 N/m) at 100 nN load and 21 °C, both feedback-controlled. **(b)** Stress activation length, ϕ/A. (A area of contact) for OMCTS, n-hexadecane and dry contact. The *inset* provides a linear relationship between friction and temperature at a velocity of 1 µm/s and a normal load of 100 nN. Adapted from [29.15]

three logarithmic decades of velocities. While *Briscoe* et al. [29.10] employed a SFA that confines and pressurizes the film over several square microns, He et al. used a SFM system in which the contact is on the order of the lubricant molecular dimension.

He et al. determined the degree of interfacial structuring and its effect on lubrication of n-hexadecane and octamethylcyclotetrasiloxane (OMCTS). For spherically shaped OMCTS molecules, only an interfacial "monolayer" was found; in contrast, a 2-nm-thick entropically cooled layer was detected for n-hexadecane in the boundary regime to an ultra-smooth silicon wafer. SFM measurements of the two lubricants (with similar chemical affinity to silicon) identified the molecular shape of n-hexadecane responsible for augmented interfacial structuring. Consequently, interfacial liquid structuring was found to reduce lubricated friction, Fig. 29.4. Again as reasoned above, these results can be discussed in terms of a collective phenomenon, i. e., in terms of increased molecular coordination in n-hexadecane versus OMCTS.

29.3 Functional Behavior of Lubricated Friction

Friction-rate experiments are well suited to evaluate the rheological nature of interfacial liquids. In classical theories, such as the Reynolds' hydrodynamic theory discussed above, drag forces in lubricated sliding over thick liquid films were found to depend linearly on the rate of sliding and on the viscosity of the bulk fluid. In high-pressure lubrication, described by the elastohydrodynamic lubrication theory [29.16, 17], it was found that the linear relationship between friction and velocity can be retained by adjusting the (apparent) viscosity by introducing an *apparent viscosity term*. Qualitatively, the same linear relationship has been observed for highly confined simple liquids between ultrasmooth mica surfaces such as alkanes [29.18]. Note that the lubricated contact area in SFA experiments is on the micron-scale. It significantly exceeds the size of the confined molecules. For small and unbranched molecules, such as simple alkanes, it is possible that the confined material undergoes a pressure-induced phase reconstruction, which leads to material properties that deviate significantly from the bulk. Larger and more complex (branched) molecules are less likely to exhibit pressure-induced phase reconstruction due to internal constraints and poor mixing within the contact area. This was shown by *Drummond* et al. [29.19] in SFA shear experiments. They found that the linear friction–velocity dependence does not apply for branched hydrocarbon lubricants. Also Drummond discussed "molecular lubrication" in terms of a logarithmic friction–velocity relationship, which is in accordance with the above-discussed thermal activation model, the solid lubricant SFA study by *Briscoe* et al. [29.10], and the liquid lubricant SFM study by *He* et al. [29.15].

Common to the three studies by *Briscoe*, *He*, and *Drummond* is that they operate on a single material phase that is disrupted or relaxed over a very specific lateral length scale. In the Eyring model, the length scale is deduced by assuming a regular series of barriers, separated by a *virtual jump distance*. The distance is embedded in v_0, the characteristic velocity, which is the product of the jump distance and the frequency of the process. *Briscoe* et al. [29.10] used the lattice constant of the highly oriented monolayers as the virtual jump distance. It was assumed that the process frequency was related to the vibrational frequency of the molecules ($10^{11}\,s^{-1}$), neglecting sliding velocity, temperature, and pressure effects. *He* et al. [29.15] assumed a jump distance of 0.2 nm and considered frequencies between a perfectly structured alkane layer (10^{11} Hz) and the bulk fluid ($10^{13}-10^{15}$ Hz, estimated from infrared absorption data for typical covalent bonds). With these assumptions *He* determined total "jump-energies" of $4-8\times10^{-20}$ J. *Briscoe* and *He* pointed out that a friction–velocity study alone provides only a qualitative measure of the microscopic origin of friction. Additional measurements have to be conducted that quantitatively address jump distances and frequencies.

The issue of the jump distance has been addressed by *Overney* et al. [29.20] in a SFM study on a highly ordered lubricant model system. This study avoided two levels of difficulties *Briscoe* et al. [29.10] and *He* et al. [29.15] encountered: (a) large contact areas of SFA studies, and (b) complex rheology with unknown structure parameters as in liquid lubricant studies. It involved contact dimensions on the order of 1 nm^2, and the crystalline form a bilayer model-lipid-lubricant with in-plane lattice spacings of 0.6 and 1.1 nm. The study mainly focused on the effect of the depth of the corrugation potential (barrier height) on the static and dynamic friction force. This is illustrated in Fig. 29.5 in the form of stick–slip amplitude plotted as a function of the drag direction (i. e., sliding with respect to the anisotropic row-like film structure). Relevant to our discussion about jump distance in lubrication events is Overney's discussion about the sliding speed and its effect on the slip distance. They demonstrated that within sliding speeds of 36 nm/s to 100 nm/s, the jump distance corresponded to the lattice spacing. At higher velocities, however, they could observe jumps over multiples of lattice distances and found the jump length distribution to become increasingly stochastic. They proposed molecular (or atomistic) friction as a white-noise driven system, which obeys a Gaussian fluctuation–dissipation relation. Hence, based on this finding one should consider discussing kinetic friction in terms of a statistical fluctuation model and understand the jump distance as a statistical quantity.

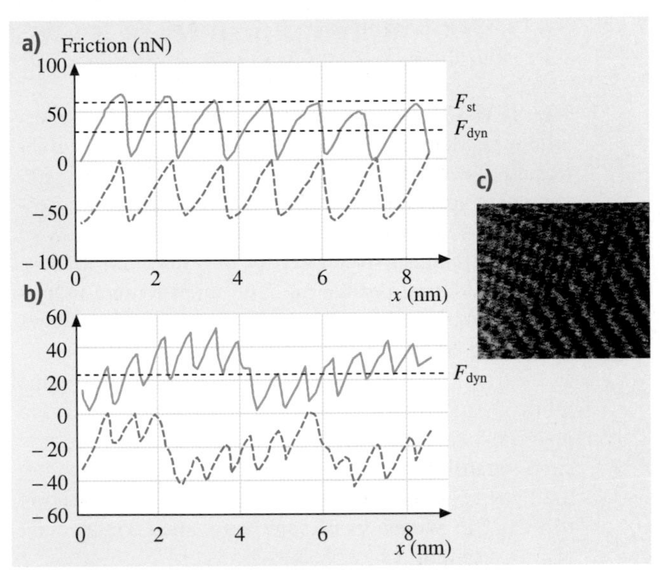

Fig. 29.5 SFM molecular stick-slip measurements of a bilayer lipid system (5-(4'-N,N-dihexadecylamino)benzylidene barbituric acid). **(a)** High amplitude frictional stick-slip behavior is observed for scans perpendicular to molecular rows as imaged in **(c)**. F_{st}, static friction, is assigned to the maximum force occurrence. The average value corresponds to the dynamic friction value, F_{dyn}, determined on large-scale micrometer scans. **(b)** A 30° out of row direction scan leads to decreased frictional stick-slip behavior due to smaller molecular corrugations. **(c)** $12\times12\,nm^2$ SFM lateral force image of a highly structured lipid bilayer. Two crystalline domains with a boundary are imaged. The anisotropic row-like structure is responsible for directional dependent friction forces. The molecular corrugation between the rows is larger than the molecular corrugation in between a single row. **(a)** and **(b)** are adapted from [29.20] and **(c)** from [29.21].

29.4 Thermodynamical Models Based on Small and Nonconforming Contacts

The SFM approach simulates a single asperity contact with a very high compliance, provided by a microfabricated and etched ultra-sharp tip and a typically soft cantilever spring. From a realistic, tribological perspective, the SFM approach is targeted toward the study of the intrinsic lubricant properties of a thin film in close vicinity to the solid substrate. The small contact area on the order of the lubricant's molecular dimension allows discussing SFM results in terms of a thermodynamic equilibrium. The area is insufficient in reorganizing the lubricant molecules coherently, to cause an apparent material phase-transition, or to generate a metastable situation as observed in SFA experiments (see below). SFM is therefore not appropriate to reflect on tribological issues involving large area confinement effects.

In our prior molecular discussion of friction above, we introduced for solid and liquid lubrication a thermal activation model, the *Eyring model*, which employed a regular series of potential barriers. Note that the concept applies for a solid lubricant of an inherent, highly ordered structure (e.g., [29.10]), but also for a liquid system in which the series of potentials is built up and overcome in the course of the shear process (e.g., [29.15]).

Gnecco et al. [29.22] showed in a ultrahigh vacuum study on sodium chloride that the concept of the Eyring model also applies for dry SFM friction studies. Thus a molecular theory of lubricated friction involving a molecular contact could be derived from a very simplistic model of an apparent sinusoidal-corrugated surface over which a cantilever tip is pulled. In a first attempt one could assume that the corresponding wave length of the shear process corresponds to the apparent lattice spacing of the corrugated surface. With such a simple attempt it is, however, assumed that there is no noise, such as thermal noise, existing in the system, and thus the driven tip leaves the total potential well when the barrier vanishes at the instability point. In the presence of noise, the transition to sliding can be expected to occur before the top of the barrier is reached. Such barrier-hopping fluctuations have been theoretically discussed by *Sang* et al. [29.23] and *Dudko* et al. [29.24].

The relationship between thermal fluctuations and velocity must be handled thoughtfully. *Sang* et al. [29.23] pointed out that in previous considerations of thermal fluctuations by *Heslot* et al. [29.25], the fluctuations were proportionally related to the velocity, which led to a friction force that is logarithmically dependent on the velocity. In Heslot's *linear creep model*, the barrier height is proportional to the frictional force. *Sang* argued that if one considered an absorbing boundary condition (i. e., an elastic deformation of the overall potential which is accomplished by shifting the x-axis) the barrier height becomes proportional to a 3/2-power law in the friction force. *Sang*'s extended linear creep model resembles a *ramped creep model* and leads analytically to a logarithmic distorted dynamic friction-*versus*-velocity relationship; i. e.,

$$F = F_c - \Delta F \left|\ln v^*\right|^{2/3} . \qquad (29.4)$$

In (29.4) v^* represents a dimensionless velocity, $\Delta F \propto T^{2/3}$, and F_c is an experimentally determined constant (by plotting F versus $T^{2/3}$ for a fixed ratio $T/v = 1\,\mathrm{K/(nm/sec)}$) that contains the critical position of the cantilever support. The same relationship of friction with velocity was also derived for the maximum spring force by *Dudko* et al. [29.24]. ΔF and v^* in (29.4) were derived as follows by *Sang*:

$$v^* = 2\left(\frac{v\beta\omega_0^2 U_0}{k_B T \lambda}\right)\frac{\Omega_k^2}{\left(1-\Omega_k^4\right)^{1/2}} ;$$

$$\Omega_k = \frac{\omega_0}{2\pi\omega_k} ;$$

$$\omega_0 = \sqrt{\frac{M\lambda^2}{U_0}} ; \quad \omega_k = \sqrt{\frac{M}{k}} , \qquad (29.5a)$$

$$\Delta F = \frac{\pi U_0}{\lambda}\left(\frac{3}{2}\frac{k_B T}{U_0}\right)^{\frac{2}{3}}\left(\frac{\left(1-\Omega_k^4\right)^{\frac{1}{6}}}{1+\Omega_k^2}\right) . \qquad (29.5b)$$

In (29.5a,b) v is the velocity of the cantilever stage, β is the microscopic friction coefficient or dissipation (damping) factor, ω_0 is the frequency of the small oscillations of the tip in the minima of the periodic potential, λ is the lattice constant, U_0 is the surface barrier potential height, M and k represent the mass and the spring constant of the cantilever, respectively, and $2\pi\Omega_k$ represents the ratio of ω_0 with the intrinsic cantilever resonance frequency ω_k.

Sang's and *Dudko*'s model was experimentally confirmed by *Sills* and *Overney* [29.26] on an unstructured amorphous surface of atactic polystyrene, Fig. 29.6. *Dudko* determined that the typically used weak spring constants in SFM measurements are responsible for

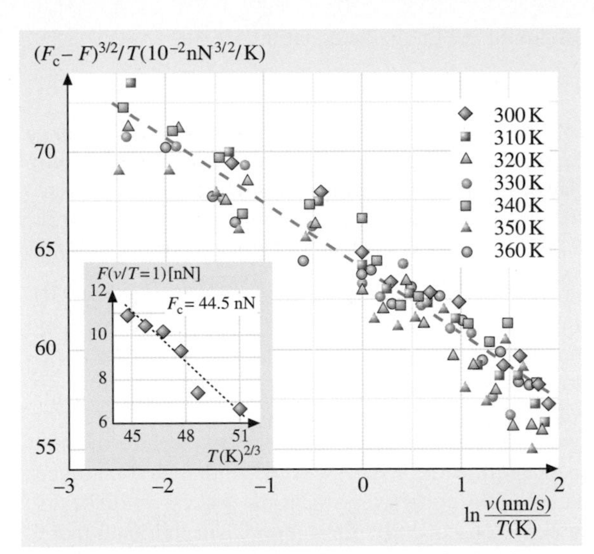

Fig. 29.6 Collapse of SFM friction data obtained on atactic polystyrene using the ramped creep scaling model. The regression parameters from the linear fit (*dashed line*) are $-2.158\times10^{-2}\,\text{N}^{3/2}\,\text{s/m}$ and $40.186\times10^{-2}\,\text{nN}^{-3/2}/\text{K}$ ($R^2 = 0.9124$). Lower *inset*: the constant F_c is determined from the intercept of the friction force F versus $T^{2/3}$ for a fixed ratio $T/v = 1\,\text{K/(nm/s)}$: $F_c = 44.5\,\text{nN}$. (Adapted from *Sills* and *Overney* [29.26])

the more pronounced logarithmic behavior of friction in velocity as found by *Gnecco* et al. [29.22] and *He* et al. [29.15]. The ramped creep model is also supported by numerical solutions of the Langevin equation.

The Langevin equation combines the equation of motion (including the sinusoidal potential and perfect cantilever oscillator in the total potential energy E) with the thermal noise in the form of the random force, $\xi(t)$, i.e.,

$$M\ddot{x} + M\beta\dot{x} + \frac{\partial E(x,t)}{\partial x} = \xi(t)\,, \tag{29.6}$$

$$\text{where } E(x,t) = \frac{k}{2}(R(t)-x)^2 - U_0\left(\frac{2\pi x}{\lambda}\right).$$

Equation (29.6) was solved numerically by both *Sang* et al. [29.23] and *Dudko* et al. [29.24] independently, assuming a Gaussian fluctuation–dissipation relation, $\langle\xi(t)\xi(t')\rangle = 2M\beta k_b T\delta(t-t')$ to express the random force. *Sang* confirmed the *ramped creep model*, and *Dudko* showed that a force reconstruction approach from the density of states (accumulated from the corresponding Fokker–Planck equation) is equivalent to the Langevin equation. From the dynamic spectral analysis, it could be concluded that the locked states (states within the potential wells) contribute mostly to the potential component of the friction force that dominates at low driving velocities, and sliding states contribute to viscous friction dominating at high driving velocities [29.24].

It should be noted that all of the above results considered an overdamped SFM system with respect to the driven spring (i.e., $\beta^2 > 4\text{kM}$), and an underdamped system with respect to the periodic potential (i.e., $\beta^2 < 4(M\omega)^2$). This aspect will be further addressed below in our discussion of metastable lubricant systems in large conforming contacts.

29.5 Limitation of the Gaussian Statistics – The Fractal Space

The spectral description of dynamic processes involving probability density functions has recently been the focus of numerous theoretical papers that treat various statistical kernels [29.27–30]. *Dudko*'s et al. [29.24] Fokker–Planck discussion of kinetic friction and *Luedtke*'s et al. [29.27] Lévy flight model of slip diffusion of adsorbed nanoclusters are two examples in which statistical methods are applied to describe diffusive properties relevant to the kinetics in tribology.

Currently most models used to describe tribological processes assume Gaussian statistics (e.g., [29.23], and [29.24]). One of the limitations of a Gaussian statistics is that there are no correlations between statistical incidences. In other words, the Gaussian dynamic system is without memory. This is important to remember as the Eyring model discussed above, with its equally spaced potential barriers, used a Gaussian statistics. Simple "inert" lubricants, such as short chain alkanes embedded between silicon wavers, are described satisfactorily with such a statistics; however, confined complex liquids are not, including branched molecules, polymers, and generally chemically interactive and entropically confined systems (e.g., perfluropolyether lubricants as discussed below).

Confined complex liquids, for instance, easily exhibit strongly interacting glass-like behavior. The dynamic and stress relaxation behaviors in glasses, frequently discussed only as a low interacting system with

Arrhenius laws (Gaussian statistics), are often distorted from processes described by independently occurring microscopic processes. For instance, deviations from the Debye exponential relaxation as introduced in (29.1) are expressed in the form of an *extended exponential* Kohlrausch relaxation function over time t; i. e.,

$$F(t) \equiv \left[\frac{X(t) - X(\infty)}{X(0) - X(\infty)}\right] = \mathrm{e}^{-\left(\frac{t}{\tau}\right)^b}; \quad 0 < b < 1 , \tag{29.7}$$

where X is the property that is relaxed. The exponent b, the Kohlrausch exponent, can theoretically be determined if one assumes that the process occurs in series, representing a well-determined microscopic origin that correlates the various degrees of freedom [29.31]. This approach is borrowed from magnetic spin models such as the *Ising spin model* [29.32]. The idea is that a given molecular motion is dependent on the availability of other degrees of freedom of mobile neighboring structural units. Finite relaxation times, $t_{\max}$, are gradually obtained with increasing spin levels (ergodic limit).

As mentioned above, the models by *Dudko* et al. [29.24] and *Sang* et al. [29.23] assumed Gaussian statistics. To illustrate how a diffusion process can deviate from Gaussian statistics, we introduce a simplified version of the Langevin equation, i. e.,

$$\ddot{x} = -\eta\dot{x} + \varsigma(t) \tag{29.8}$$

with the coordinate x, the dissipation (or dampening) parameter η, and the random acceleration $\zeta(t)$. Assuming Gaussian statistics, the mean squared displacement is

$$\left\langle x^2(t)\right\rangle = 2k_{\mathrm{B}}T\eta t = 2Dt , \tag{29.9}$$

where $D = \eta kT$ defines the diffusion constant [29.33]). It was already realized at the time of Smoluchovski at the beginning of the 20th century that a diffusive description of a dynamic process demands a thermodynamically well-equilibrated or mixed system. Especially in a confined tribological system that involves a third medium (e.g., a lubricant), it can be expected that the Markovian nature of the underlying stochastic process could be disturbed. Consequently, for a monolayer lubricant that is chemically interacting, a nonlinear relationship of the mean squared displacement in time can be expected. Manifestations of anomalous transport are long-range spatial or temporal correlations. Two extreme limits can be distinguished: (a) processes with strong temporal relations ("fractal time") [29.29], and (b) systems that exhibit long jumps ("Lévy flights") [29.30].

29.6 Fractal Mobility in Reactive Lubrication

The importance of the underlying kinetics is illustrated by ultra-thin wetting lubricants. The spreading of "completely wetting" polymer liquids on solid surfaces has revealed unexpected spatial and temporal features when examined at the molecular level. The spreading profile is typically characterized by the appearance of a precursor film of monomolecular thickness extending over macroscopic distances and, in many cases, a terracing (also on the order of molecular dimensions) of the fluid remaining in the reservoir [29.34]. These spatial features have been shown to be consistent with a Poiseuille-like flow in which the disjoining pressure gradients with film thickness drive the spreading process [29.35]. The temporal evolution of the spreading profile in this film thickness regime is, however, found to universally scale as $t^{1/2}$ even at short times [29.34]. That the spreading dynamics are reflective of a diffusive transport mechanism and not of a pressure driven "liquid" flow suggests that interfacial confinement substantially alters the mobility of molecularly thin polymer fluids [29.36].

The molecular mobility is of fundamental importance for monolayer lubrication purposes, such as in magnetic storage devices. It has, for instance, been shown that for low surface energy hydroxyl-terminated perfluoropolyether (PFPE-OH) films, the lubricant exhibits spatially terraced flow profiles indicative of film layering [29.35] and spreading dynamics that are diffusive in nature [29.37, 38]. In magnetic storage devices the hydroxylated chain ends of molecularly thin PFPE-OH films interact with the solid surface, an amorphous carbon surface, via the formation of hydrogen-bonds with the polar, carbon-oxygen functionalities located on the carbon surface. The bonding of the PFPE-OH polymer to carbon is predicated on the ability of the PFPE backbone to deliver spatially the hydroxyl end-group to within a sufficiently close distance to the surface active sites. Kinetic measurements probing the bonding of the PFPE-OH polymer to the carbon reveal two distinctive kinetic behaviors, as illustrated in Fig. 29.7 at two representative temperatures: 50 °C and 90 °C for the

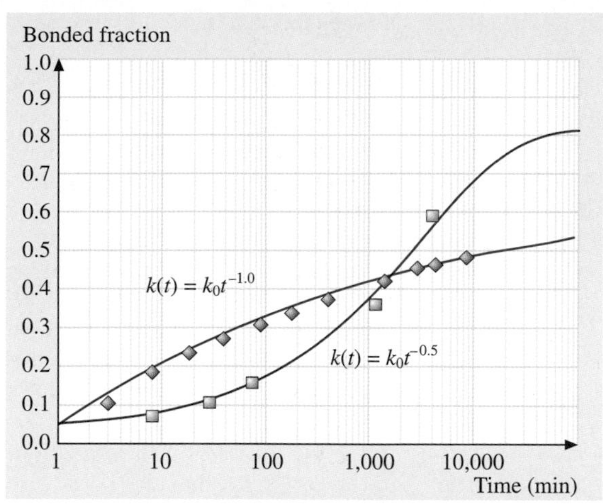

Fig. 29.7 Representative kinetic data for the bonding of PFPE-OH (tradename: Fomblin ZDOL) to amorphous carbon at $T = 50\,°\mathrm{C}$ and $T = 90\,°\mathrm{C}$. *Solid lines* represent fits using a rate coefficient of the form, $k(t) = k_0t^{-\alpha}$ with $\alpha = 0.5$ for $T = 50\,°\mathrm{C}$ and $\alpha = 1.0$ for $T = 90\,°\mathrm{C}$

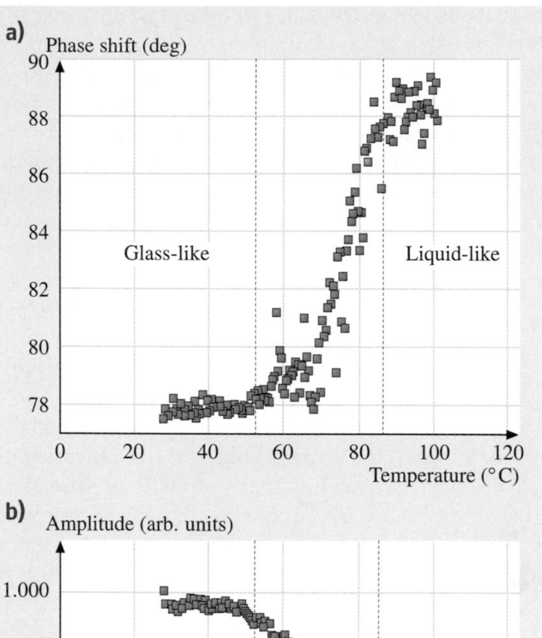

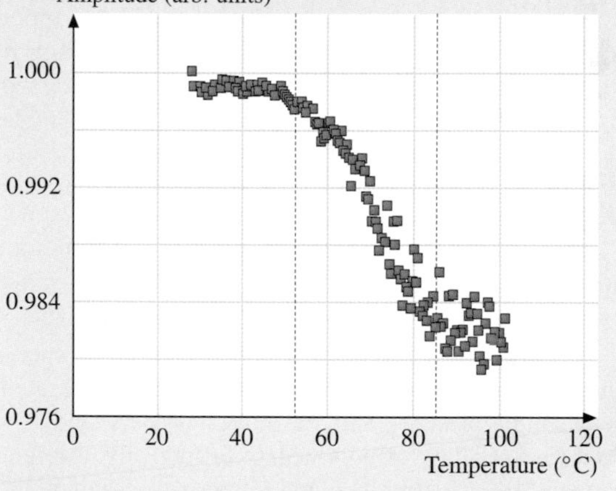

Fig. 29.8a,b Shear-modulated SFM experiments performed on a 10.7 ± 0.5 Å Fomblin Zdol film: (**a**) phase shift response between disturbance and response, and (**b**) contact stiffness response vs. temperature

two temperature regimes below 56 °C and above 85 °C. Below 56 °C the kinetics are described with a time-dependent (fractal time dependent) rate coefficient of the form

$$k(t) = k_0t^{-1/2} \tag{29.10}$$

and at temperatures above 85 °C with the form

$$k(t) = k_0t^{-1.0}\,. \tag{29.11}$$

The initial bonding rate constants, k_0, increased abruptly as the temperature rose above 50 °C.

The bonding kinetics in the low-temperature regime is characteristic of a diffusion-limited reaction occurring from a glass-like state of the molecularly thin PFPE-OH film [29.39]. The mobility of the PFPE chain in the glass-like state is limited by the propagation of holes or packets of free volume, which facilitate configurational rearrangements of the chain. The onset of changes in the bonding kinetics at nominally $T > 56°$ signifies a fundamental change in the mobility of the molecularly thin PFPE-OH film. Specifically, the transition in the fractal time dependence suggests that delivery of the hydroxyl moiety to the surface is no longer limited by hole diffusion, and the increase in the initial rate constant indicates an enhancement in the backbone flexibility. These results are consistent with a transition in the film from a glass-like to a liquid-like state in which the enhanced PFPE-OH segmental mobility results from rotations about the ether oxygen linkages in the chain that become increasingly facile. The time dependence observed in the high-temperature rate coefficients, $k(t) = k_0t^{-1.0}$ is characteristic of a process occurring from a confined liquid-like state in which the activation energy increases as the extent of the reaction increases.

The impact of this transition in the molecular mobility on tribology can be illustrated with sinusoidally modulated shear force experiments [29.40, 41]. In brief, a molecularly thin (10.7 ± 0.5 Å) PFPE-OH film is subjected to a local shear stress by means of a sinusoidal force applied *laterally* to a SFM probe (at constant load) where the modulation amplitude is initially set below that required to initiate sliding between the tip and the sample. The amplitude and phase-shift responses are measures of the contact stiffness and the effective viscous dampening, respectively. From Fig. 29.8 it can be inferred that the SFM-measured nanorheological properties of the PFPE-OH film exhibit the changes discussed above in the molecular mobility. Thus kinetic and rheological data suggest that the thermal transition observed is due to the formation of a two-dimensional (2-D) glass. The "glass transition" results from the preferential "freezing out" of the out-of-plane torsional motions of the energetically confined PFPE backbone. The confinement-induced solidification in the molecularly thin precursor film will significantly impact the lubrication properties and challenge thermodynamic, well-equilibrated models of lubrication as introduced above.

29.7 Metastable Lubricant Systems in Large Conforming Contacts

It is important to note that in an experiment of thermal activation, the critical time of the experiment t_{exp} decides the system response with its finite relaxation time t_{max}. If $t_{exp} > t_{max}$, the system behaves in an ergodic manner, and thermodynamic laws apply for interpreting lubrication results. On the contrary, if $t_{exp} < t_{max}$ the thermal evolution cannot be described by classical statistical thermodynamics. A metastable configuration is generated. The experimental time depends strongly on the contact area, the parameter that most differs between SFA and SFM measurements. SFA experiments involve large micron-scale contacts while SFM measurements are conducted with contacts on the nanoscale

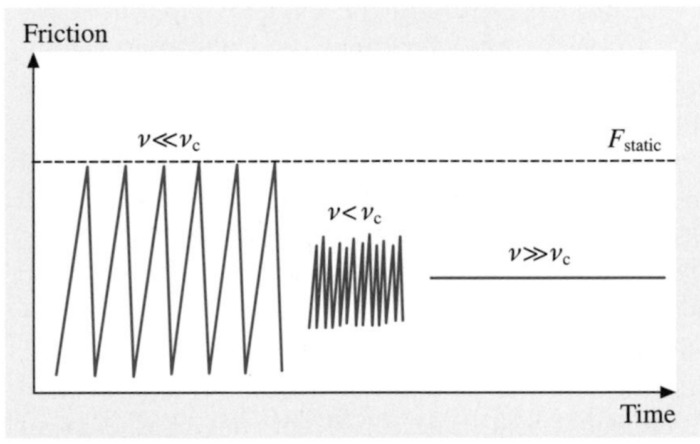

Fig. 29.9 Dynamic states of friction for an underdamped spring system

Because of the large contact area, SFA experiments are very susceptible to generating unequilibrated metastable lubricant configurations. The SFA study by *Yoshizawa* et al. [29.42], in which distinctively different *dynamic states of friction* were introduced, could be interpreted as such. To date there are three velocity regimes used to describe the dynamic state of friction for a system that is "underdamped" [29.43]. In an underdamped system, realized by a stiff spring compared to the friction constant, the characteristic slip time is comparable or smaller than the response time of the mechanical system. One distinguishes three velocity regimes, as depicted in Fig. 29.9. The three regimes distinguish themselves by a single discriminator, v_c, a material, pressure and temperature dependent critical velocity. The regimes are described as

1. highly regular with high amplitude stick–slip for $v \ll v_c$,
2. intermittent stochastic stick–slip for $v < v_c$, and
3. smooth low friction sliding for $v > v_c$.

Various models have been suggested to describe the different dynamic states, including melting freezing transition [29.44], chain adsorption on substrates [29.45], and embedded particle [29.43]. Until the embedded particle model by Rozman was introduced, the SFA approach seemed to be the tool of choice to investigate metastable lubricant configurations. Rozman's single particle model alerts us, however, to drawing unambiguous conclusions on the dynamical structure of a molecular system embedded between two plates and driven by an underdamped system. In Rozman's simple theoretical model, sketched in Fig. 29.10, a single particle is embedded between two corrugated

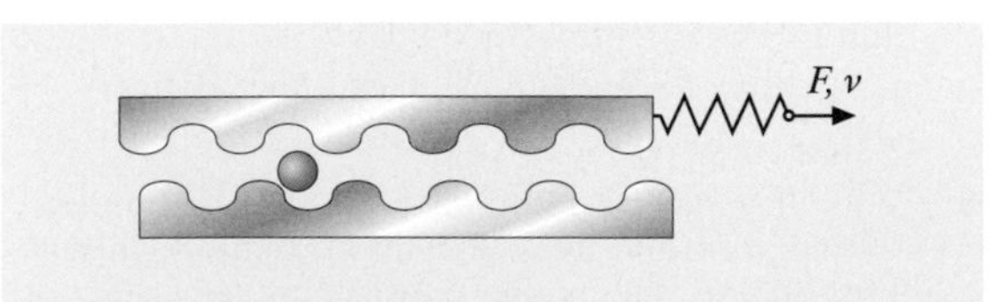

Fig. 29.10 Model of single particle embedded in a two-wave potential driven at constant force, F, and velocity v

surfaces (*two-wave potential*). The top plate is pulled at constant velocity by a linear spring, and the plate motion is monitored. Interestingly, the plate exhibits exactly the same dynamic state of frictional motion as introduced above with the three regimes, which were attributed to a rate-dependent configurational change of the lubricant.

Finally it shall be noted that the slip relaxation pattern depends condition under which the stick–slip motion is being studied. In an *overdamped* system, i. e., a system in which the spring constant is weak compared to the friction constant, *Rozman* et al. [29.46] found that one can control the experimentally observed relaxation pattern of the slip by controlling the spring constant.

29.8 Conclusion

Starting from a classical tribological Master Curve, the Stribeck curve, with its complex description of lubrication for thin lubricants, we launched a discussion of molecular lubrication pulling together disparate approaches to studying friction. We found that phenomenological descriptions of lubrication, such as the Reynolds' theory, were kept alive for ultra-thin lubricants by "adjusting" material properties, such as the viscosity. This is a common engineering approach to introduce "effective" or "apparent" properties, if tested models fail to describe new situations.

In the case of molecular lubrication, significant progress has been made in the last ten years. Two instrumental techniques have in particular contributed to this progress: the SFA and the SFM. The contributions of these two techniques have been complementary. While the SFA has tested lubricants under pressure constraints with large contacts in respect to the size of the trapped molecules, the SFM has probed the degree of collective mobility with disturbances in the size of the molecules themselves.

One feature common to interpreting lubrication results, and which was discussed here in detail, is the problem of finding the appropriate underlying statistics to describe the lubrication process. Most of the current molecular models that have been used to describe molecular friction and lubrication assumed a Gaussian statistics. Only recently has it been recognized important to consider statistics that embrace long-range spatial or temporal correlations. In the near future, it can be expected that the next leap in an improved fundamental understanding of kinetics and energetics in nanolubrication will come from interpretations that are challenging the Markovian nature of the underlying stochastic process.

References

29.1 F. P. Bowden, D. Tabor: *The Friction and Lubrication of Solids* (Clarendon, Oxford 1951)

29.2 D. Tabor, R. H. S. Winterton: The direct measurement of normal and retarded van der Waals forces, Proc. R. Soc. Lond. A **312** (1969) 435–450

29.3 G. Binnig, C. F. Quate, C. Gerber: Atomic Force Microscope, Phys. Rev. Lett. **56** (1986) 930–933

29.4 N. P. Petrov: *Friction in Machines and the Effect of the Lubricant* (St. Petersburg 1883)

29.5 B. Tower: First report on friction experiments (friction of lubricated bearings), Proc. Inst. Mech. Eng. (November 1883) 632–659

29.6 O. Reynolds: On the theory of lubrication and its application to Mr. Beauchamp tower's experiments, including an experimental determination of the viscosity of olive oil, Philos. Trans. Roy. Soc. Lond. **177** (1886) 157–234

29.7 J. P. Israelachvili, G. M. McGuiggan, M. Gee, A. Homola, M. Robbins, P. Thompson: Liquid dynamics in molecularly thin films, J. Phys. **2** (1990) 89–98

29.8 W. B. Hardy, I. Doubleday: Boundary lubrication – The paraffin series, Proc. Roy. Soc. Lond. A **100** (1922) 550–574

29.9 A. Dorinson, K. C. Ludema: *Mechanics and Chemistry in Lubrication* (Elsevier, Amsterdam 1985)

29.10 B. J. Briscoe, D. C. B. Evans: The shear properties of Langmuir–Blodgett layers, Proc. Roy. Soc. Lond. A **380** (1982) 389–407

29.11 C. Buenviaje, S. Ge, M. Rafailovich, J. Sokolov, J. M. Drake, R. M. Overney: Confined flow in polymer films at interfaces, Langmuir **19** (1999) 6446–6450

29.12 S. Blunier, H. Zogg, A. N. Tiwari, R. M. Overney, H. Haefke, P. Buffat, G. Kostorz: Lattice and thermal misfit dislocations in epitaxial CaF_2/Si(111) and

$BaF_2/CaF_2/Si(111)$ structures, Phys. Rev. Lett. **68** (1992) 3599–3602
29.13 G. Reiter, A. L. Demirel, J. Peanasky, L. L. Cai, S. Granick: Stick to slip transition and adhesion of lubricated surfaces in moving contact, J. Chem. Phys. **101** (1994) 2606–2615
29.14 S. Glasstone, K. J. Laidler, H. Eyring: *Theory of Rate Processes* (McGraw-Hill, New York 1941)
29.15 M. He, A. Szuchmacher Blum, G. Overney, R. M. Overney: Effect of interfacial liquid structuring on the coherence length in nanolubrication, Phys. Rev. Lett. **88**(15) (2002) 154302/1–4
29.16 K. L. Johnson: *Contact Mechanics* (Cambridge Univ. Press, Cambridge 1987)
29.17 E. Meyer, R. M. Overney, K. Dransfeld, T. Gyalog: *Nanoscience: Friction and Rheology on the Nanometer Scale* (World Scientific, Singapore 1998)
29.18 H. K. Christenson, D. W. R. Gruen, R. G. Horn, J. N. Israelachvili: Structuring in liquid alkanes between solid-surfaces – Force measurements and mean-field theory, J. Chem. Phys. **87**(3) (1987) 1834–1841
29.19 C. Drummond, J. Israelachvili: Dynamic behavior of confined branched hydrocarbon lubricant fluids under shear, Macromolecules **33**(13) (2000) 4910–4920
29.20 R. M. Overney, H. Takano, M. Fujihira, W. Paulus, H. Ringsdorf: Anisotropy in friction and molecular stick-slip motion, Phys. Rev. Lett. **72** (1994) 3546–3549
29.21 R. M. Overney, H. Takano, M. Fujihira: Elastic compliances measured by atomic force microscopy, Europhys. Lett. **26**(6) (1994) 443–447
29.22 E. Gnecco, R. Bennewitz, T. Gyalog, C. Loppacher, M. Bammerlin, E. Meyer, H.-J. Güntherodt: Velocity dependence of atomic friction, Phys. Rev. Lett. **84**(6) (2000) 1172–1175
29.23 Y. Sang, M. Dube, M. Grant: Thermal effects on atomic friction, Phys. Rev. Lett. **87**(17) (2001) 174301/1–4
29.24 O. K. Dudko, A. E. Filippov, J. Klafter, M. Urbakh: Dynamic force spectroscopy: A Fokker–Planck approach, Chem. Phys. Lett. **352** (2002) 499–504
29.25 F. Heslot, T. Baumberger, B. Perrin, B. Caroli, C. Caroli: Creep, stick-slip, and dry-friction dynamics: Experiments and a heuristic model, Phys. Rev. E **49** (1994) 4973–4988
29.26 S. Sills, R. M. Overney: Creeping friction dynamics and molecular dissipation mechanisms in glassy polymers, Phys. Rev. Lett. **91** (2003) 095501(1–4)
29.27 W. D. Luedtke, U. Landman: Slip diffusion and Levy flights of an adsorbed gold nanocluster, Phys. Rev. Lett. **82** (1999) 3835–3838
29.28 I. M. Sokolov: Levy flights from a continuous-time process, Phys. Rev. E **63** (2000) 011104/1–10
29.29 R. Metzler, J. Klafter: The random walks guide to anomalous diffusion: A fractional dynamics approach, Phys. Rep. **339** (2000) 1–77
29.30 R. Metzler, J. Klafter: Levy meets Boltzmann: Strange initial conditions for brownian and fractional Fokker–Planck equations, Physica A **302** (2001) 290–296
29.31 R. G. Palmer, D. L. Stein, E. Abrahams: Models of hierarchically constrained dynamics for glass relaxation, Phys. Rev. Lett. **53** (1984) 958–961
29.32 N. W. Ashcroft, N. D. Mermin: *Solid State Physics* (CBS Asia, Philadelphia 1976)
29.33 R. Becker: *Theorie der Wärme* (Springer, Berlin 1985)
29.34 F. Heslot, N. Fraysse, A. M. Cazabat: Molecular layering in the spreading of wetting liquid drops, Nature **338** (1989) 640–642
29.35 T. E. Karis, G. W. Tyndall: Calculation of spreading profiles for molecularly-thin films from surface energy gradients, J. Non-Newtonian Fluid Mech. **82** (1999) 287–302
29.36 S. F. Burlatsky, G. Oshanin, A. M. Cazabat, M. Moreau: Microscopic model of upward creep of an ultrathin wetting film, Phys. Rev. Lett. **76** (1996) 86–89
29.37 T. M. O'Connor, Y. R. Back, M. S. Jhon, B. G. Min, D. Y. Yoon, T. E. Karis: Surface diffusion of thin perfluoropolyalkylether films, J. Appl. Phys. **79** (1996) 5788–5790
29.38 X. Ma, J. Gui, L. Smoliar, K. Grannen, B. Marchon, C. L. Bauer, M. S. Jhon: Complex terraced spreading of perfluoropolyalkylether films on carbon surfaces, Phys. Rev. E **59** (1999) 722–727
29.39 A. Plonka, J. Bednarek, K. Pietrucha: Reaction dynamics in glass transition region: propagating radicals in ultraviolet-irradiated poly(methyl methacrylate), J. Chem. Phys. **104** (1996) 5279–5283
29.40 R. M. Overney, C. Buenviaje, R. Luginbuehl, F. Dinelli: Glass and structural transitions measured at polymer surfaces on the nanoscale, J. Therm. Anal. Calorimetry **59** (2000) 205–225
29.41 S. Ge, Y. Pu, W. Zhang, M. Rafailovich, J. Sokolov, C. Buenviaje, R. Buckmaster, R. M. Overney: Shear modulation force microscopy study of near surface glass transition temperature, Phys. Rev. Lett. **85**(11) (2000) 2340–2343
29.42 H. Yoshizawa, P. McGuiggan, J. N. Israelachvili: Identification of a second dynamic state during stick-slip motion, Science **259** (1993) 1305–1308
29.43 M. G. Rozman, M. Urbakh, J. Klafter: Stick-slip motion and force fluctuations in a driven two-wave potential, Phys. Rev. Lett. **77** (1996) 683–686
29.44 J. M. Carlson, A. A. Batista: Constitutive relation for the friction between lubricated surfaces, Phys. Rev. E **53** (1996) 4153–4164
29.45 Y. Braiman, F. Family, H. G. E. Hentschel: Array-enhanced friction in the periodic stick-slip motion of nonlinear oscillators, Phys. Rev. E **53** (1996) R3005–R3008
29.46 M. G. Rozman, M. Urbakh, J. Klafter: Controlling chaotic frictional forces, Phys. Rev. E **57** (1998) 7340–7343

Part E

Industrial

Part E Industrial Applications and Microdevice Reliability

30 Nanotechnology for Data Storage Applications
Dror Sarid, Tucson, USA
Brendan McCarthy, Tucson, USA
Ghassan E. Jabbour, Tucson, USA

31 The "Millipede" – A Nanotechnology-Based AFM Data-Storage System
Gerd K. Binnig, Rüschlikon, Switzerland
G. Cherubini, Rüschlikon, Switzerland
M. Despont, Rüschlikon, Switzerland
Urs T. Duerig, Rüschlikon, Switzerland
Evangelos Eleftheriou, Rüschlikon, Switzerland
H. Pozidis, Rüschlikon, Switzerland
Peter Vettiger, Rüschlikon, Switzerland

32 Microactuators for Dual-Stage Servo Systems in Magnetic Disk Files
Roberto Horowitz, Berkeley, USA
Tsung-Lin Chen, Shin Chu, Taiwan
Kenn Oldham, Berkeley, USA
Yunfeng Li, Berkeley, USA

33 Micro/Nanotribology of MEMS/NEMS Materials and Devices
Bharat Bhushan, Columbus, USA

34 Mechanical Properties of Micromachined Structures
Harold Kahn, Cleveland, USA

35 Thermo- and Electromechanics of Thin-Film Microstructures
Martin L. Dunn, Boulder, USA
Shawn J. Cunningham, Colorado Springs, USA

36 High Volume Manufacturing and Field Stability of MEMS Products
Jack Martin, Cambridge, USA

37 MEMS Packaging and Thermal Issues in Reliability
Yu-Ting Cheng, HsinChu, Taiwan
Liwei Lin, Berkeley, USA

30. Nanotechnology for Data Storage Applications

This chapter considers atomic force microscopy (AFM) as an enabling technology for data storage applications, considering already existing technologies such as hard disk drives (HDD), optical disk drives (ODD) and Flash Memories that currently dominate the nonvolatile data storage market, together with future devices based on magnetoresistive and phase change effects.
The issue at hand is the question of whether the novel AFM-based storage, dubbed "Probe Storage", can offer a competing approach to the currently available technologies by playing the role of a disruptive technology. Probe Storage will be contrasted to HDD and ODD who are purely mechanical, as they are based on a rotating disk that uses just a single probe to address billions of bits of data, and nonvolatile RAM that has no moving parts yet requires billions of interconnects. In particular, capacity, areal density, transfer rate, form factor and cost of various data storage devices will be discussed and the unique opportunity offered by Probe Storage in employing massive parallelism will be outlined. It will be shown that Probe Storage bridges the gap between HDD, ODD and other nonvolatile RAM, drawing from the strength of each one of these and adding a significant attribute neither of these has; namely, the possibility of addressing a very large number of nanoscale bits of data in parallel. This chapter differs from the other chapters in this book in that it addresses the important issue of whether a given scientific effort, namely, Probe Storage, is mature enough to evolve into a commercially viable technology. The answer seems to indicate that there indeed is a huge niche in the data storage arena that such a technology is uniquely qualified to fill, which is large enough to justify a major investment in research and development. Indeed, as other chapters indicate, such an effort is developing at a rapid pace, with hopes of having a viable product within a few years.

30.1 **Current Status of Commercial Data Storage Devices** 901
30.1.1 Non-Volatile Random Access Memory 904

30.2 **Opportunities Offered by Nanotechnology for Data Storage** 907
30.2.1 Motors 907
30.2.2 Sensors 909
30.2.3 Media and Experimental Results ... 913

30.3 **Conclusion** 918

References 919

This chapter will differ from the other chapters in this handbook in that it addresses the important issue of whether a given scientific effort is mature enough to evolve into a commercially viable technology. Specifically, we consider scanning probe microscopy (SPM), consisting of scanning tunneling microscopy (STM) [30.1–3] and atomic force microscopy (AFM) with all their variants, as a means for storing and retrieving nanoscale bits of data to and from a substrate. Indeed, many studies have already been published in which researchers demonstrated the feasibility of using SPM techniques that hold promise for realistic applications [30.4]. However, in approaching a topic from a practical point of view, that is, whether it has the ingredients that will spawn a commercial product, it is imperative to obtain a clear view of the status of competing technologies. The commercially available technologies that will concern us here entail two classes of devices, one consisting of hard disk drives (HDD) and optical disk drives (ODD), and the other consisting of Non-Volatile Random Access Memories (NVRAM) based on charge trapping (Flash Memory) that currently dominate the data storage market. The latter may soon be supplemented by magnetoresistive RAM (MRAM) and phase change RAM (PC-RAM). For brevity, we will refer to the first class as HDD and the one as NVRAM.

It is commonly thought that nanotechnology is a breakthrough technology that is a quantum leap beyond existing technologies in its capabilities. In reality, however, nanotechnology is in many cases just a limiting case to the already commercially available technologies. The latter may progress at such a high rate that the distinction between what is considered "conventional" and "nano" becomes blurred. A case in point is data storage in which distances and times are already specified by nanometers and nanoseconds. For example, the head of an HDD in a PC's hard disk flies across the storage medium (platter) at a speed of 10 m/s ($\sim$ 22 miles per h) at a height of $\sim$ 10 nm, reading and writing bits of data on a nanosecond time scale. Also, both giant magnetoresistance (GMR) devices that read magnetic data in an HDD and transistors belonging to NVRAM devices use nanoscale structures, one employing spintronic effects and the other tunneling junctions between pairs of transistors, where quantum laws of physics have to be used to describe their operation.

The issue at hand is, therefore, whether novel nanotechnologies can offer a competing approach to currently available technologies by playing the role of a "disruptive technology". This term has been coined in the high-tech industry to describe a situation where, figuratively speaking, a discovery developed in a basement can lead to a development of a new device that has the potential of becoming a commercial alternative to a commonly used one. Such a disruption of the success of a commercially established technology by a product developed by a new start-up, for example, has been witnessed more and more frequently in recent years. One would even venture to assume that the support companies give to research groups at universities can be attributed to their concern about a possible disruption of their own technology.

The important issue here is the question of what nanotechnology can offer to an already mature field such as data storage. As will be discussed later, HDD and NVRAM are worlds apart in technology. HDD is purely mechanical, as it is based on a rotating disk that uses just a single probe (head) to address billions of bits of data. NVRAM, on the other hand, has no moving parts yet requires billions of probes, i. e., interconnects, positioned at the intersection of horizontal and vertical interconnects with one probe for every bit of data. Another distinction between HDD and NVRAM is the time it takes to access a random bit of data. For an HDD the random access time is 10 ms (75 ms for ODD), while for NVRAM it is less than one μs. Clearly, NVRAM beats HDD (and ODD) in the random access arena, if that is the criteria for performance. On the other hand, the rate at which an HDD can read data sequentially is rather impressive. The current performance of an HDD stands at 436 Mb/s and 22 Mb/s for ODD, comparable to what NVRAM can offer, being of the order of 10 MByte/s. Another playing field in the battle between HDD and NVRAM performance concerns capacity and areal density of bits of data. Here, capacity measures the total number of bytes of data, while areal density measures the number of bits per square inch. The first addresses the ever growing need for storing larger and larger amounts of data, say, for high-definition TV applications, while the latter is related to what is commonly called the form factor, namely, the physical size of a storage device. The form factor is intimately connected to portability, as more and more hand-held devices have restrictions on the physical size of each of its components. Note also that a major consideration in comparing HDD and NVRAM is cost, a topic that will be addressed later.

As discussed before, both HDD and ODD use just a single probe that addresses all the bits of data on a platter where data is stored. NVRAM, on the other hand, has to resort to a grid of interconnects that contacts each bit of information individually. Such a grid, in which each bit is interfaced with one "word line" and one "bit line", occupies expensive real estate across a storage device, and, even more importantly, it requires that the interconnects be on the order of the bits of data they address. As long as these bits have dimensions on the order of, say, 1 μm, one is able to use conventional lithography for the fabrication of the interconnects. However, when the bits of data shrink to, say, 25 nm in size, interconnects are no longer a viable option due to lithographic limitations. HDD and ODD, in contrast, require no interconnects, so this limitation on the size of bits is not a concern. However, there are many other problems that arise as one tries to push the size of the stored bits to nanoscale dimensions, as will be discussed later.

By now we have touched on the heart of the challenge that nanotechnology-based storage faces if it is to compete with HDD and NVRAM in both performance and cost. Practically, one should restate this issue and ask a more modest question, namely, what niche nanotechnology will fill in consumer data-storage applications. To address this question, one notes that nanotechnology is concerned with the ability to miniaturize and characterize nanometer-sized structures. For data storage, the question is how to address nanoscale data bits rapidly without having to resort to the use of too many interconnects.

A possible solution to this challenge emerged from a brilliant idea developed at IBM and Stanford laboratories, as part of their scanning probe microscopy research [30.5–9].

A conventional AFM employs a single cantilever with a sharp tip at its end to raster scan a sample and its minute deflections, due to protrusion on the surface, generate a map of the surface topography. Under ideal conditions one can even obtain atomically resolved images. The IBM-Stanford group demonstrated that it is possible to employ a large number of such cantilevers and operate them in parallel. Such an operation provides a faster means of obtaining images of different parts of a sample simultaneously. These cantilever arrays are often referred to as types of microelectromechanical systems (MEMS), or if they are small enough, nanoelectromechanical systems (NEMS).

One can view this breakthrough concept as an enabling technology for data storage applications in that it bridges the gap between HDD and NVRAM technologies, drawing from the strength of each. To appreciate this concept, consider a medium whose area is sectioned into a thousand squares, each containing one million bits of data. Each square is addressed by its own interconnect whose width can now be as wide as a thousand bits of data. Consider now a square structure consisting of one thousand cantilevers whose tips are aligned such that each one addresses its own square. Such an arrangement, known as probe storage, and dubbed "millipede" by IBM, has some attributes belonging to an HDD, some attributes belonging to NVRAM, and potentially one significant attribute neither of these has: the possibility of addressing nanoscale bits of data. While random access is faster than that of a HDD, it is slower than that of NVRAM.

In this chapter, we will address the issues mentioned in the introduction with an emphasis on a comparison between existing and future development of conventional technologies and nanotechnologies. We will not address more futuristic approaches such as DNA and molecular electronics for data storage applications – fascinating topics on their own merit, yet deemed too far in the future for our more "technological" approach.

30.1 Current Status of Commercial Data Storage Devices

Before describing the current status of commercial storage devices, it is worthwhile to reiterate definitions and cover some new key features associated with HDD [30.10]. A schematic of a hard disk drive, shown in Fig. 30.1, is configured into sectors, each having radial tracks; there are typically 58,000 tracks per in. for HDD and 34,000 tracks per in. for ODD, 562 kbits per in. for HDD, and 100 kbits per in. for ODD. In 1993, a HDD head was flying above the storage media at a typical height of 100 nm, which dropped to 10 nm in 2002, and is expected to drop to 4 nm in 2004.

Hard Disk Drives and Optical Disk Drives

The time it takes the read-write head to get to a random bit along a track is called latency, while the time it takes the head to get to a random sector is called seek time. The time to get to any random bit is called access time. From 1990 to 2002 this time decreased linearly from 20 ms to 8 ms, and is expected to decrease to 4 ms by 2004.

Transfer time is the time it takes to transfer a given number of bits from the hard drive to a given destination in a computer. The total number of bits per unit area, usually given in bpsi (bits per square inch), is called areal density. The total number of bytes in a storage device is called capacity.

The physical size of a storage device and its price are paramount factors in its commercial viability. Disk drives are usually divided into magnetic and optical, although a combination of both is currently advanced by several companies and research groups around the world. A magnetic hard drive (HDD) has a higher capacity, lower cost, and higher performance than an optical hard drive (ODD), but its disk is not removable. The ODD, on the other hand, has a removable disk that can be mass produced by parallel replication (stamping), making it cheaper for large-scale distribution. Also, an ODD is configured to operate in such a way that the head reads data sequentially, while HDD is optimized for random access operation. Note that units of bits are usually denoted by a lower case "b" and used for areal density and transfer rate, while units of bytes are usually denoted by an upper case "B" and used for capacity. A note in passing: Areas are, oddly enough, expressed in square inches, and the holy grail for areal density is denoted by 1 tbpsi (one terabit per square inch).

Now, a crucial question when assessing the viability of a promised technology concerns the demand for the commercial product in question. Figure 30.2 estimates that a grand total of 2.12 million terabytes of information were produced in 1999 [30.11]. The breakdown is 83 terabytes for optical, 240 terabytes for paper, 427,000 terabytes for film, and 1.6 million terabytes for magnetic data storage. PC disk drives take the first place in demand: 700,000 terabytes of data!

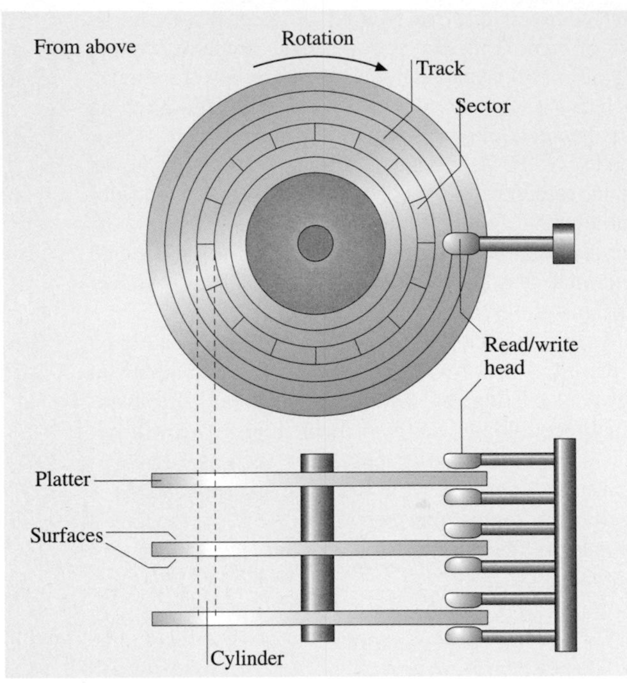

Fig. 30.1 A schematic diagram of a hard disk drive illustrating the concepts of cylinder, track, and sector

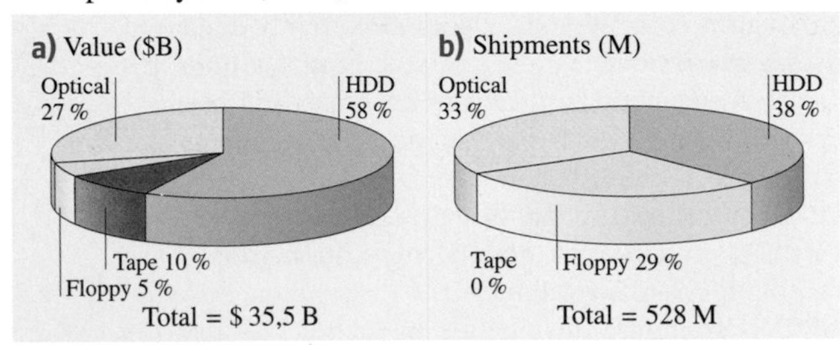

Fig. 30.3a,b Storage drive industry shipments in [30.10]. (a) Categorized by value and (b) units shipped

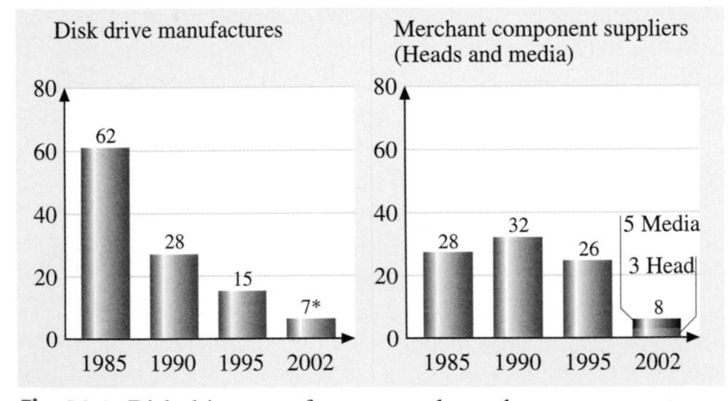

Fig. 30.4 Disk drive manufacturers and merchant component suppliers [30.12]. The number of disk drive manufacturers decreased from 62 in 1985 to a mere 7 by 2002

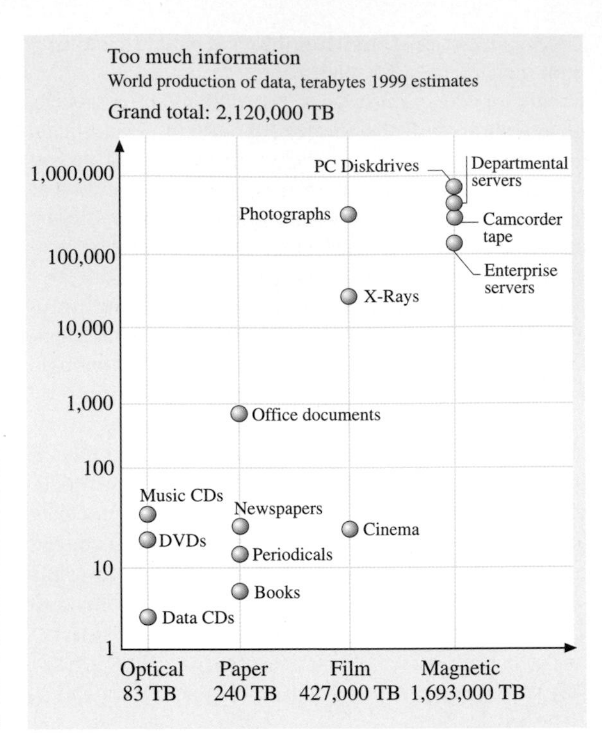

Fig. 30.2 World production of data in 1999 estimates

To appreciate the economic impact of the HDD storage drive industry (NVRAM will be discussed later), consider Fig. 30.3, which is current to 2002 [30.10]. Here, one observes that the industry shipped 528 million units with a value of $ 35.5 billion, 58% of the value consisting of HDD and 27% of ODD, while the numbers shipped are almost evenly divided between HDD, ODD, and floppies. Based on Fig. 30.3, one would think that such a huge market would involve a large number of companies competing with each other in price and performance.

However, as shown in Fig. 30.4, the cut-throat competition among various companies gave rise to the disappearance of many of them and mergers between the others [30.12]. Note that the number of disk drive manufacturers dropped from 62 in 1985 to only seven in 2002, while merchant components suppliers dropped from 28 to eight during that period! For example, Maxtor and Quantum merged, IBM sold its disk drive division to Hitachi, and Fujitsu exited desktop HDD altogether. In 2002, Seagate shipped 1,959,000 terabytes of storage, 10.2 million enterprise drives, 1.2 million 15,000 rpm drives, and 44.8 million personal storage devices. And they expect that there will be migration to smaller form factors and increased areal density in all segments of the industry and that mobility will be become increasingly important.

A projection of HDD areal density on a logarithmic scale is shown in Fig. 30.5 [30.13]. Whereas in 1990 the areal density was only 1 gigabit per sq. in., it reached 100 gigabits per square inch in 2002 and is projected to reach 1 Tbpsi in 2006. One wonders whether it will take nanotechnology to get to 50 Tbpsi in 2017, as the figure suggests. As quoted from a Seagate presentation, "in traditional applications, areal density has grown faster than consumption, moving to a single platter society in PC's that enables smaller form factors. In non-traditional markets, new applications drive capacity growth, especially with video content." [30.12].

Table 30.1 [30.10] presents several key parameters characterizing HDD and ODD that should serve as a reality check for the potential viability of any nanotechnology approach that aims to act as a disruptive technology. Thus any new concept should be judged against this current status of HDD and ODD and, as described later, NVRAM. Note that in contrast to these impressive parameters associated with HDD and ODD, seek time is limited to 10 ms and 75 ms, respectively.

A comparison of the perceived limits to HDD and ODD performance are presented in Table 30.2 [30.10]. For magnetic disk drives, the limit to areal density is the superparamagnetic effect in which too small a magnetic domain can be thermally excited into an opposite direction, losing the reliability of the stored data. Ways and means for approaching this limit and maintaining reliability will be discussed later. For optical data

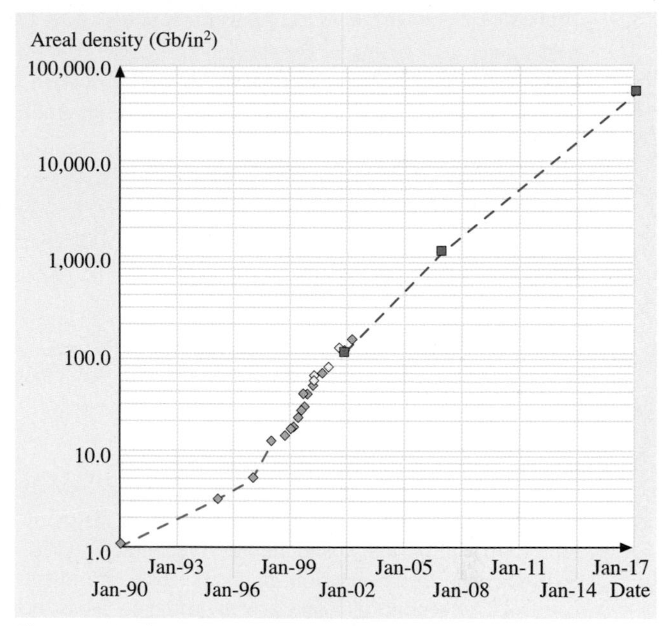

Fig. 30.5 Projected areal density for hard disk drives [30.13]

storage, the conventional limit is clearly the spot size of the focused laser beam. However, near-field methods are currently being developed to circumvent this limit. As far as speed goes, switching times for magnetic domains is around 1 ns, while for laser-induced thermal effects it is ten times slower.

The current and projected relationship between areal density and data rate for several technologies are

Table 30.1 A comparison of the parameters of HDD and ODD

Product comparison		
Parameter	HDD (80 Gb desktop)	ODD (4.7 Gb DVD-RAM)
Areal density	32.6 Gb/in.^{2}	3.3 Gb/in.^{2}
	Best = 50.0 Gb/in.^{2} (2.5″)	Best = 4.6 Gb/in.^{2} (50 mm)
TPI/BPI	58 K/562 K	34 K/100 K
Capacity/platter	40 Gb (2 side, 95 mm)	9.4 Gb (2 side, 120 mm)
Unit price (drive)	$80–$120	$365
Price/Gb (drive)	$1–$1.5	$39
Price/Gb (media)	–	~ $1.5–$2
Seek time	10 ms	75 ms
Transfer rate (writing)	436 Mb/s	22 Mb/s
Research laboratory comparison		
Parameter	HDD (longitudinal)	ODD (blue near-field)
Areal density	130 Gb/in.^{2}	45 Gb/in.^{2}
TPI/BPI	213 K/610 K	141 K/319 K

Table 30.2 Perceived technical limits for hard disk drive technology [30.10]

	Magnetic		Optical	
Areal density (Gb/in.^{2})	**Superparamagnetism**		**Focused spot size**	
	1994:	35 (scaling)	1994:	0.7 ($\lambda = 780$ nm)
	1997:	100 (lower bit aspect ratio)	1997:	3.3 ($\lambda = 650$ nm)
	2002:	500 (perpendicular)	2002:	19 (Blu-Ray)
		1,000 (patterned)		2× (ML)
		> 1,000 (HAMR)		2–3× (NF)
				? × (Volume)
Switching time (ns)	**Gyromagnetic ratio**		**Media crystallization**	
		~ 1		~ 10

shown in Fig. 30.6 [30.5]. These technologies consist of CD-ROM, magneto-optical, magnetic disks, magnetic tape, SIL, holography, and thermo-mechanical AFM. According to this projection, magnetic disks are in the forefront of technology and are expected to have the same areal density as that of the probe storage concept, yet have a much faster data rate. The latter technology, as predicted in this figure, will grow from a data rate of only 1 Mb/s to more than 10 Mb/s, accompanied by an increase in areal density from 10 gigabits per sq. in. to hundreds of gigabits per sq. in. More about the probe storage concept will be discussed later.

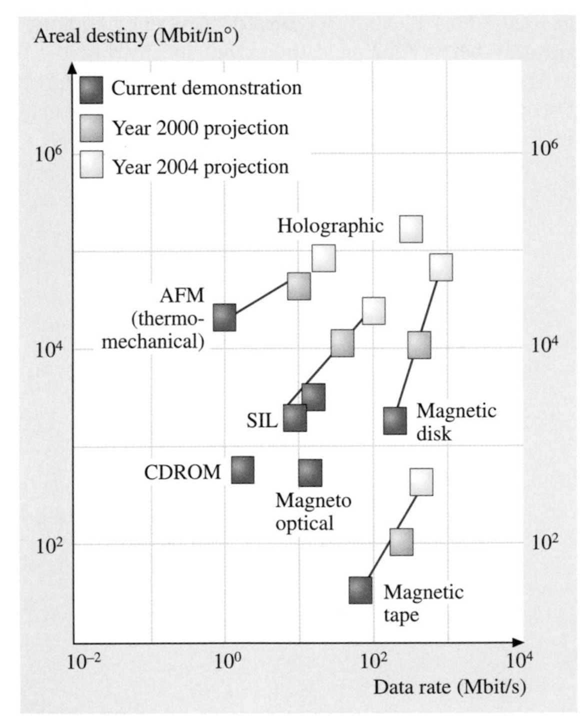

Fig. 30.6 Projected areal density vs. data rate for several technologies

30.1.1 Non-Volatile Random Access Memory

Demand for Flash Memory has been growing rapidly over the last few years, driven by the rapid expansion of several markets, including mobile communications and portable media devices (notably cameras and MP3 players). Flash Memory is by now the fastest growing segment of the memory market. Figure 30.7 [30.14] shows the total value of the Flash Memory market year by year. It can be seen that the market is increasing by nearly a factor of ten every five years and up to 100% every year. The market is projected to reach $ 23 billion by 2003, up from $ 2.5 billion in 1998. The reasons for this dramatic growth are depicted in Fig. 30.8 [30.14], which shows the demand by industry sector and the growth in these areas. By far the biggest driver of demand for Flash Memory is consumer electronics and, in particular, cellular phones, which account for nearly half the market and are growing at a compounded annual growth rate of 120%.

Flash Memory

Flash Memory, the most common non-volatile, programmable memory device, uses rows and columns of interconnects that address each data cell. At the intersection of these are two transistors, which comprise the Flash Memory cell, as shown in Fig. 30.9. Each cell is addressed by a network of interconnects, where the common connection to the transistor gates is called the word

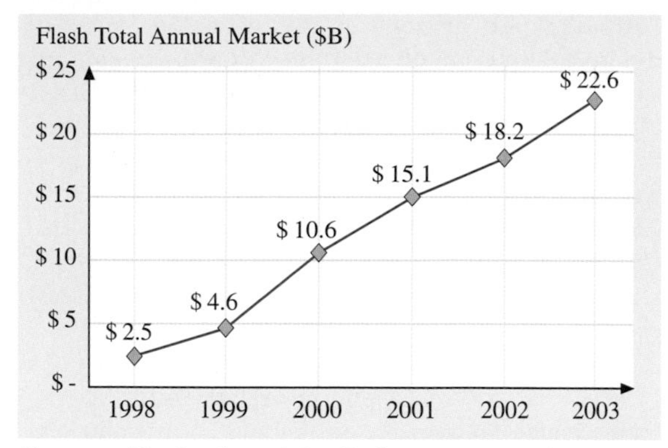

Fig. 30.7 Total value of the flash memory market, 1998–2003. The market has been growing rapidly over the last few years, and is now the fastest growing sector of the memory market

line and the common connection to the drain is called the bit line. Every cell in the memory array can be accessed by activating a unique combination of these lines. Here, the floating gate is linked to a word line via a second gate that serves as a control gate. The floating gate potential is altered by Fowler–Nordheim tunneling of electrons, an effect that takes place when the applied bias needed for switching is larger than the work function of the relevant electrode. The electrons arrive from the column, or bit line, and enter the floating gate from which they drain to the ground. A cell sensor monitors the flow of electrons through the gate. A value of 1 is assigned when the current is greater than a given threshold value, and 0 if it is below that value. Approximately 30,000 electrons are stored in the gate to make a 1 and 5,000 for a 0. The memory of the device derives from the very long storage time (tens of years) of the tunneling electrons in the gate capacitor. For the data to have a lifetime of ten years, the electrons can leak at a rate of no more than five a day. Flash memory can only be rewritten a limited number of times (10^{5-6}), as electrons get permanently trapped in the gate over time, impairing device efficiency. Memory devices using this technology are SmartMedia, CompactFlash cards, and Memory Sticks, for example, that read and write memory in 256- or 512-byte increments, enhancing their speed of operation. A Flash Memory device, in contrast to a hard disk drive, is noiseless, has no moving parts, is lighter and smaller, and, above all, has access times orders of magnitude faster than HDD. SmartMedia cards, which serve as solid-state floppy disk cards, were originally developed by Toshiba. They are currently available in a capacity of 128 Mbytes and higher and measure roughly 45 mm × 37 mm × 1 mm. In

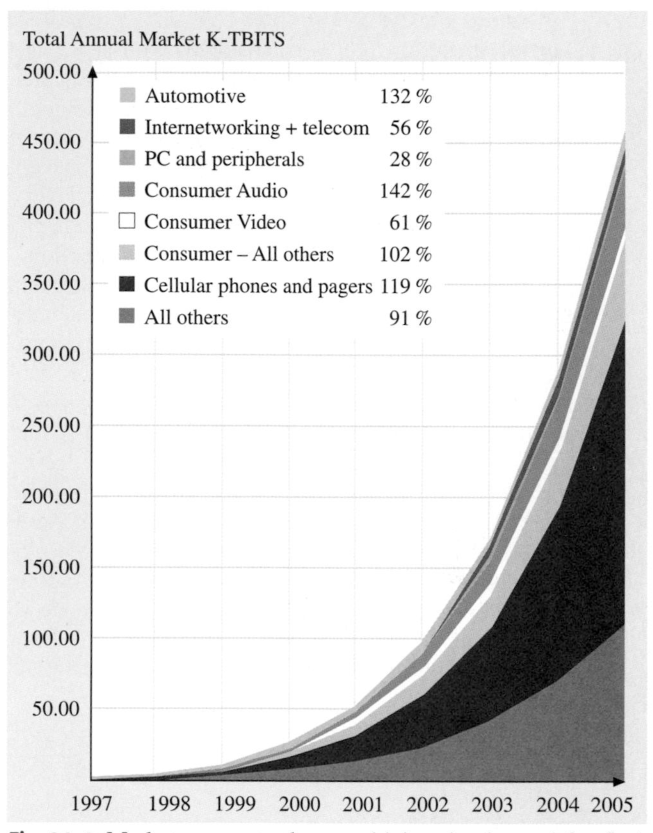

Fig. 30.8 Market segments that are driving the demand for flash memory where cellular phones are leading the demand

comparison to a HDD, however, SmartMedia cards are much more expensive. Note that a single storage cell occupies roughly 0.1 μm², a size limited by the complexity of its physical structure and width of the addressing interconnects.

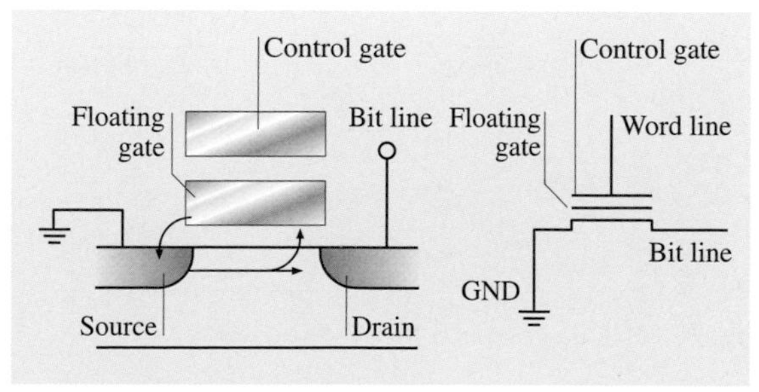

Fig. 30.9 Schematics diagram of a typical Flash Memory cell

Magnetoresistive Random-Access Memory

Figure 30.10 [30.15] is a schematic diagram of a magnetoresistive random-access memory (MRAM) that acts as a non-volatile storage device. As with Flash Memory, there are word lines and bit lines at the intersection of which is a cell with a switchable property. Whereas with Flash Memory the electric potential of a storage cell is modified by electron tunneling, here it is the magnetic domain orientation that is switched by the current flowing through the electrodes. Passage of electrons through the magnetic domain of the cell depends on the state of polarization of the spins of the injected electrodes, much the same as the operation of giant magnetoresistance elements used to read data in a hard disk drive. In contrast to HDD and Flash Memory, MRAM is still in a stage of development and, as such, is not yet commercially available. One of its main attributes is its high speed of programming, which is faster than that of Flash Memory. The main challenges confronting MRAM technology are: (i) the size of the cell is still too large, (ii) the power consumption required during the write mode is still too high, and (iii) there are manufacturing difficulties associated with high temperature steps that may damage the magnetic layers.

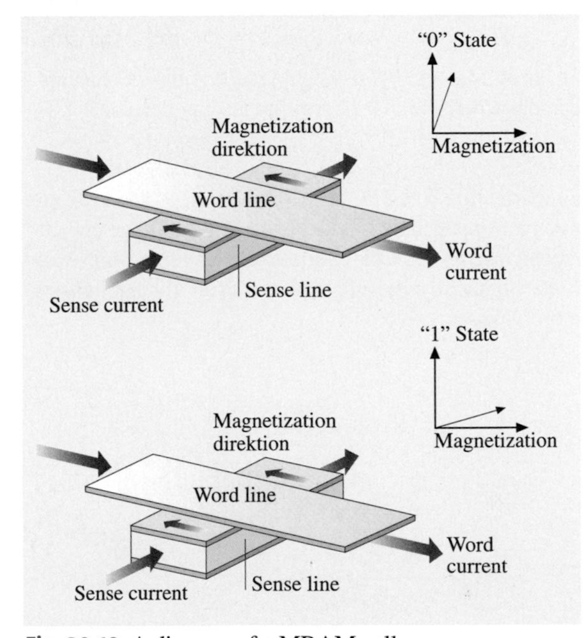

Fig. 30.10 A diagram of a MRAM cell

To exemplify the status of this technology, note that Motorola demonstrated a 1-Mbit MRAM test chip using a 0.6-μm process, yielding a cell size of 7.2 μm^2. Projections are that using a 0.18-μm process will shrink the cell size to about 0.7 μm^2, and thus be competitive with Flash Memory. Also, Sony demonstrated an MRAM test chip using a 0.35-μm process that yielded a cell size of 5.8 μm^2. Other companies like IBM, which is partnering with Infineon, are also active in MRAM research.

Phase-Change Random-Access Memory

Phase-change random-access memory is a technology that uses a chalcogenide as the data storage material in the memory cell. The medium is similar to the switchable material used in rewritable CDs. It is being pioneered by Ovonyx, in association with Intel. A schematic diagram of an Ovonyx Unified Memory (OUM) is shown in Fig. 30.11 [30.16]. The OUM is an example of a phase-change random-access memory that operates much as the Flash Memory and MRAM, as far as the electrodes are concerned, except that the memory bit cell consists of a chalcogenide material that undergoes a phase change. Upon passing current from the word line to the bit line, the chalcogenide, a GeTeSb alloy, changes its phase reversibly between amorphous and crystalline states, depending on the temperature it is heated to and the rate of cooling. This change in phase is accompanied by a several orders of magnitude change in conductivity, which makes it possible to obtain a high reading signal-to-noise ratio. Similar to Flash Memory and MRAM, the OUM requires a protective shield from the environment for reliable operation over a long period. We shall come back to this issue later.

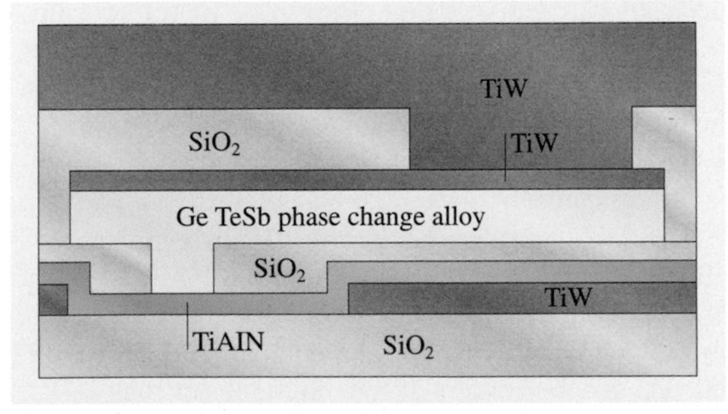

Fig. 30.11 A cross section of an OUM cell

30.2 Opportunities Offered by Nanotechnology for Data Storage

We have by now summarized the two mainstream commercial data storage devices, HDD and NVRAM, and the two near-term candidates for future commercial success, the MRAM and PC-RAM devices. As stated before, the technology utilized by these devices should be used as a measure against which novel nanotechnology approaches are judged. Note that even if a promising novel nanotechnology-based device comes into being, it will take time to bring it into maturity, and during that period, the state-of-the-art of HDD, Flash Memory, MRAM, and PC-RAM will steadily improve. One could state that we have here a competition on the "run" that will be decided based on the following four criteria: (i) capacity, (i) cell size, which translates to areal density or form factor, (iii) access time, and (iv) price, where each technology is looking for its own business niche.

Nanotechnology offers two unique opportunities to solve the problem of how to miniaturize storage devices. The first opportunity entails the use of a variety of chemical and physical processes to fabricate memory media capable of supporting nanoscale features that have reversible properties, be it magnetic, electric, or phase change. Such media should satisfy the first and second criteria, namely, a large data storage capacity and a small form factor. The second opportunity derives from the ability of scanning probe microscopy, the champion of nanotechnology, to characterize and modify structures down to atomic dimensions. Here, the use of a large number of probes operating in parallel should satisfy the third criterion, namely, fast access times. Also, because the fabrication of multiple cantilevers lends itself rather readily to mass production using currently available photolithographic processes, it should satisfy the fourth criterion: low cost.

In the next section, we describe how probe storage can be realized as a data storage technology that may find a commercially viable niche. It will be useful to divide this section into three parts, each one describing different aspects of what is known as probe storage. The first part, titled "Motors", will deal with the question of how one can access a large number of nanoscale memory cells using parallel and sequential processes. The second part, denoted by "Heads", will describe methods for reading and writing each individual memory cell using different SPM techniques. The third part, called "Media and Experimental Results", will describe the properties of different media applicable to data storage that have or can be used for probe storage.

30.2.1 Motors

This section considers two possible geometric configurations of data storage devices that operate in parallel. One consists of a rotating cylinder and an array of cantilevers positioned in an axial direction, as shown in Fig. 30.12, and the other is a 2-D, all-MEMS configuration, as shown in Fig. 30.13 [30.17]. Very little work has been done in the first, "mixed" configuration, and we will, therefore, present only a short survey of this work. Most of the section will be dedicated to the all-MEMS configuration, on which massive work has been proposed, tested, and published.

Cylinder

A cylindrical ROM media for optical data storage is currently being explored as an alternative to DVD devices, using an adhesive tape film as the recordable medium [30.18]. The purpose of the study is to find out whether this geometry will result in a smaller form factor

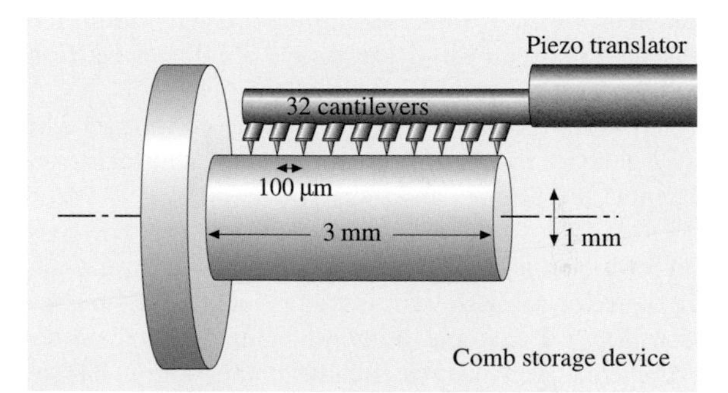

Fig. 30.12 A schematic diagram of a comb storage device

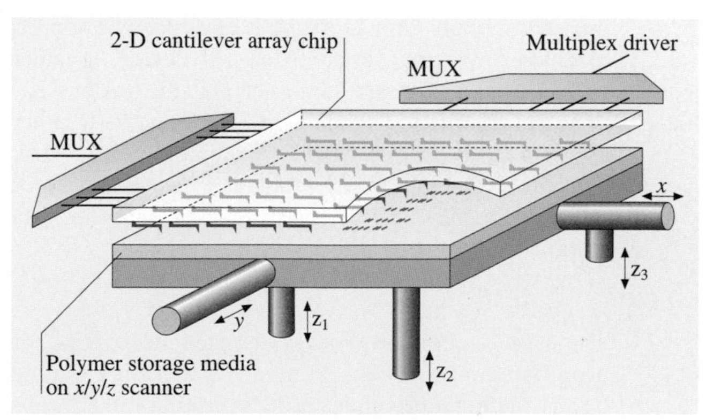

Fig. 30.13 A schematic diagram of a MEMS storage device

and reduced vibrations relative to a commercial DVD. These authors find that this geometry offers a larger writing area than that of a disk. For example, consider a disk with inner and outer radii of r and R, respectively, and a cylinder with the same radius and a height of h. For equal areas, the height of the cylinder is given by $h = \left(R^2 - r^2\right)/2R$. Therefore, for $r = 2.5$ mm and $R = 15$ mm, one obtains $h = 7.3$ mm. As a result, a cylinder with $R = h = 15$ mm offers more than twice the area of a disk. Such a device presents an interesting opportunity that the authors are currently pursuing. Figure 30.12 is an artist's concept of a possible geometry where a linear array of 32 cantilevers with their tips is mounted along the axis of a rotating cylinder. This array, which resembles a comb, is mounted on a piezoelectric actuator that can move the minute array very fast across a distance that equals the gap between adjacent cantilever tips. Let the cylinder circumference be divided into 32 sectors, each containing data bits arranged as circular tracks within each sector. As the cylinder rotates, each of the 32 tips reads independently a single track. Using the piezoelectric actuator, one can now move all the tips such that they read the next 32 tracks. Consider the particular geometry presented in Fig. 30.12, where the diameter of the cylinder is 1 mm, its length 3 mm, the distance between tips 100 μm, and a cylinder rotation rate of 10,020 rpm.

A crude assessment of the possible performance of such a device indicates promising values for the capacity, latency, access time, and transfer rate [30.19].

Note that while the rotational rates of these devices are expected to be similar to those of a HDD, one can conceive of MEMS-derived motors increasing the rotation rate. For example, a micro-engine with a diameter of 64 μm driven electrostatically at 300,000 rpm has recently been investigated [30.20]. The purpose here was to study the out-of-plane displacement and tilts about the rotating hub. Optical methods have been developed to characterize the operation of this device, together with a model that takes into account the various parameters relating to its operation. One wonders whether a combination of this MEMS technology with a disk or cylindrical recordable medium is a viable option for reducing latency and decreasing access time for data storage applications.

Probe Storage

The next section provides a detailed description of a probe storage device, so here we will only outline briefly the main principles of its operation [30.8, 9, 21–26]. Consider a typical schematic of a MEMS-based storage presented in Fig. 30.13, [30.17] where the cantilevers are laid out on a "sled" in both the x and y directions. The sled is translated, typically by electrostatic actuators, over the medium surface. This means that, unlike in conventional hard disks, the sectors are laid out adjacent to each other.

There are two possible architectures to this arrangement, one in which the media is moved relative to the tips and one in which the media is fixed and the tips are moved. The moving-tips model is very space inefficient, allowing only about 1% of the media area to be used for data storage, and, therefore, making it unviable in comparison to other storage technologies. In this configuration, the media is suspended by springs over the tips with actuators that control their positioning. In contrast, however, a typical design for a moving media model could have 10,000 tips in a 1 cm^2 area with a bit size of 50 nm^2. This geometry allows 30–50% of the area of the media to be utilized. Allowing for overheads such as error correction, this gives a storage capacity of 4 Gb/cm^2.

The disadvantage of this scheme is the greater mass that needs to be moved, resulting in large seek times. There is a settling time associated with each movement of the system, as the springs cause the sled to oscillate around its final position before coming to rest.

As shown in Fig. 30.14, the media is divided into $M \times N$ rectangular regions, each of which is accessible only by a single tip. The major advantage of this is that the data can be "striped" across multiple regions, dramatically increasing read times, as one seek operation is sufficient to read many bits.

The notation to address each bit is of the form $\langle x, y, tip\rangle$, where x and y are the horizontal and vertical coordinates of the bit within a rectangular region accessible by *tip*. A tip track is the data for which $\langle x, tip\rangle$ is the same, i. e., a vertical column within a rectangular region. By analogy with hard disk architecture, all data with identical x form a cylinder, accessible without moving the sled in the x direction. Cylinders are divided into tracks, which are accessible by the currently active tips, as not all tips can be active simultaneously for power considerations.

As mentioned, a benefit of this architecture is that data tracks are striped across tip tracks. This is depicted in Fig. 30.14 [30.23], where each data sector is mapped across several tip tracks. Sectors 1 and 2 can be read in parallel by activating tips 1–4 and moving them vertically downwards.

Simulated device performance for a prototype sled with parameters given in Table 30.3 is shown in Ta-

Table 30.3 A table of probe storage operating parameters

Sled mobility in x and y	100 μm
Bit cell width (area)	50 nm (0.0025 μm^2)
Number of tips	6,400
Simultaneously active tips	1,280
Tip sector length	80 bits (8 bytes)
Servo overhead	10 bits per tip sector
Device capacity (per sled)	2.1 GByte
Sled acceleration	114.8 m/s^2
Per-tip data rate	400 kbit/s
Settling time constants	1
Sled resonant frequency	220 Hz
Spring factor	75%

ble 30.4. Note that an important breakthrough in MEMS technology has recently been reported in which over one million cantilevers per sq. cm have been fabricated by anisotropic etching of silicon [30.27]. The cantilevers, which had tips at their end, measured a few micrometers in length, had spring constants of a few N/m, resonance frequencies around 10 MHz, and Q factors of 5 in air. Significantly, the resonance frequencies within the same row deviated only by 0.01%. Cantilevers as short as 100 nm and as thin as 20 nm were also fabricated, demonstrating the latitude that this technology offers. Also reported was the fabrication of a large number of cantilevers operating at frequencies up to 1 GHz with Q factors up to 8,000 [30.28]. These results will no doubt play a pivotal role in MEMS-related data storage technology for which mass parallelism combined with superfast sensors offer an important advantage.

Table 30.4 Probe storage projected performance

Average service time	1.49 ms (0.25)
Maximum service time	4.51 ms
Average seek time	1.27 ms (0.25)
Maximum seek time	1.66 ms
Average x seek time	1.24 ms (0.21)
Maximum x seek time	1.66 ms
Average y seek time	0.90 ms (0.31)
Maximum y seek time	1.62 ms
Settling time	0.72 ms
Average per request turnaround time	0.20 ms (0.20)
Maximum per request turnaround time	1.34 ms

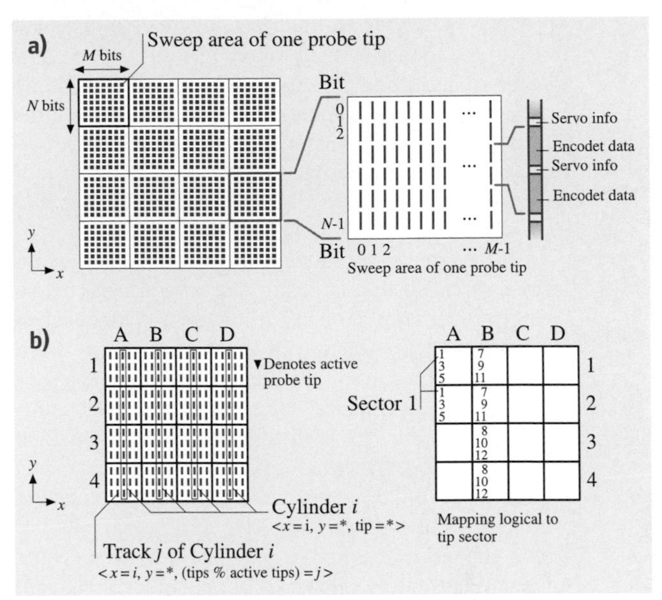

Fig. 30.14 A schematic diagram of a probe storage data organization

Table 30.4 [30.22] takes into account the physical performance of the device, including the acceleration of the actuators, settling time, etc. One observes that the results of the model compare favorably to current hard disk technology, having access times that are approximately six times faster. Note that recent results by the IBM group imply a potential bit areal densities of 400 Gbpsi [30.29].

30.2.2 Sensors

As stated before, the key problem associated with densely packed nanoscale data bits is how to address them, realizing that no interconnects of this small size exist. Clearly, the only known method is to utilize MEMS technology where sharp tips are used to read, write, and rewrite the data bits. Although such a technology is attractive, one has to realize that the number of physical phenomena available for effecting such an interaction is quite limited.

There are, in principle, three modes for interfacing, or sensing nanoscale data bits. In the first mode, denoted as contact mode, a sharp tip is constantly in contact with the medium on which the bits are written. Here, the apex of the tip has to be comparable in size to that of the bit. The second mode involves such a tip in close proximity to the bit without actually touching it. The third mode, dubbed the noncontact mode, is a combination of the other two modes. Here, the tip vibrates in close proximity to the media in such a way that on each cycle it touches or taps its surface for a short period. This mode is referred to as the tapping mode [30.30], which turns out to have several advantages relative to the other two modes. In either mode one can use an electric current within the tip to heat up a data bit, inject a current into it, or impose an electric field across it.

For the noncontact mode, one can use field emission from a sharp tip that is positioned several nanometers away from a sample, or use a tunneling current when the tip is within a fraction of a nanometer away from the sample. The current can then be used to heat a data bit and probe its conductivity as a means for data storage and retrieval. Yet both of these are accompanied with electronic noise when operating in ambient, thus requiring vacuum for a more steady operation. Note that a tunneling current is so sensitive to a tip-sample gap that a change of 0.1 nm in gap size will change the tunneling current by one order of magnitude. One can, therefore, rule out such a technique, as its restrictions on media flatness are too strenuous. In spite of some inherent difficulties, field emission maintained its position as a viable technology for data storage applications. A schematic diagram of the operation of a device based on this technology is shown in Fig. 30.15 [30.31].

Here, a multiplicity of stationary tips is positioned in close proximity to a conducting medium, such as a chalcogenide, and a controlled current emission for each tip is used to transform its phase reversibly from crystalline to amorphous. Either the tips or the media are raster scanned across each other such that, as with the probe storage device, one can address a large number of data bits. Both tips and media are placed under a mild vacuum that has to be maintained at all times.

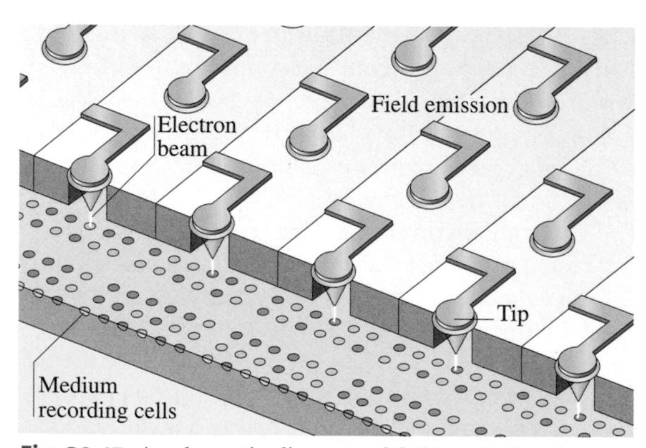

Fig. 30.15 A schematic diagram of field-emission tips

Although writing bits may be utilized efficiently, reading them back, namely, measuring their conductivity is a nontrivial task that requires further investigation. Along these lines, one may employ a MEMS-based approach and use conducting tips that are in direct contact with a conducting medium.

A different approach to locally heating a sample that does not require direct tip-sample contact, vacuum conditions, or conducting media, is to have a heating element embedded inside each tip. In spite of the fact that each tip has a complex structure, as shown in Fig. 30.16b, it was found possible to fabricate a 64×64 2-D array that raster scans a polymer medium. Heating a data bit and pressing a sharp tip into it generates a nanoscale pit that is estimated to persist for years. Sensing the pit for readout is accomplished by monitoring the heat

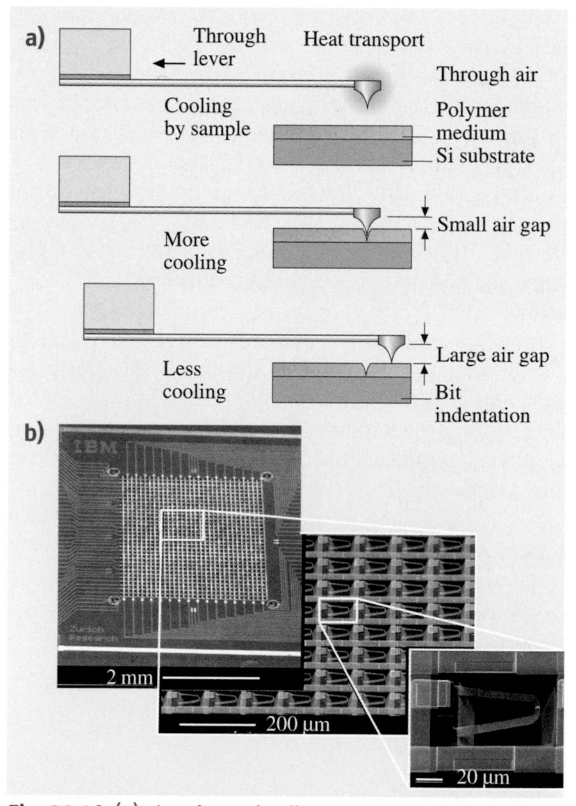

Fig. 30.16 **(a)** A schematic diagram of a cantilever with a heating and heat-sensing tip. Indentations are inscribed into a polymeric medium by means of a heated tip. The bits are read by sensing the increased heat loss in an indentation due to the greater surface area in contact. **(b)** Details of the tip structure

dissipation of the heated tip as a function of the media topography. When the tip is across a smooth surface, it is exposed mostly to air. When the tip is sunk into a pit, it is surrounded partially by the polymer material. Thus, the difference in heat conductivity between these two cases serves as a measure of the state of the polymer, whether it is smooth or dimpled. A key difficulty with this technology has been the erase mechanism, which is currently being addressed [30.29].

In the noncontact mode, the tip is vibrated at a distance such that the tip never touches the surface directly. However, the tip senses the surface of the sample indirectly through force gradients that modify the resonance frequency of the cantilever. This mode can be useful in magnetic recording if one attaches a small magnet to the tip and uses it to both modify and sense magnetic domains. This mode, however, is slow, as the tip has to vibrate many times during the sensing of a single data bit and, therefore, can be ruled out as a candidate for data storage.

We shall now consider the tapping mode that has three main advantages to the contact mode. First, the tip-sample contact time is reduced by almost two orders of magnitude, thus minimizing tip-media wear. The second advantage is that it touches the media in a much gentler way and, therefore, does not twist or bend. The third advantage is that the tip jumps between points of contacts, which constitute the data bits, so it is less prone to crash due to media roughness. If one uses the tapping mode with a conducting tip and taps on a conducting substrate, it is possible to follow the motion of the vibrating cantilever by monitoring the time the tip makes contact with the sample. This method makes it easy to control a feedback system that keeps the motion of the cantilever under control, provided the contact is good enough. The problem one encounters using this mode, however, is that such an operation is usually chaotic and, therefore, difficult to control. If one were to overcome this problem, this mode would be fast enough relative to the noncontact mode to be considered a viable method for data storage.

We will now consider the theory and experimental results of the tapping mode by first treating it as a grazing impact oscillator (GIO) for which one can obtain closed-form expressions connecting all of its parameters [30.32]. The GIO consists of a cantilever driven at a constant frequency and an amplitude by an external force. The vibrating cantilever is then brought into close proximity to the surface of a sample until its tip starts impacting the surface. At this point the amplitude of vibration of the GIO changes, together with the relative phase between the driver and the tip. Let the amplitude of vibration of the driver be denoted by a and the spring constant and quality factor of the cantilever be denoted by k_0 and Q, respectively. On impact, the GIO acquires amplitudes of vibration A and effective spring constant k_{eff}, establishing a new value for the input power and phase:

$$A = \frac{aQ}{\sqrt{Q^2\left(1-\frac{k}{k_0}\right)^2+\frac{k}{k_0}}}\,, \tag{30.1a}$$

$$\theta = \tan^{-1}\left(\frac{\sqrt{k/k_0}}{Q(1-k/k_0)}\right)\,, \tag{30.1b}$$

$$k = k_0\left[\sqrt{\left(\frac{a}{AA}\right)^2+\frac{1}{4}\left(\frac{1}{Q^4}-\frac{4}{Q^2}\right)} - \frac{1}{2Q^2}+1\right]\,. \tag{30.1c}$$

Equation (30.1) shows the relationship among the various parameters governing the operation of the GIO in a closed form, making it possible to model the effect that impacting has on the tapping cantilever. Figure 30.17a shows that on bringing the vibrating cantilever closer to the surface, its amplitude of vibration decreases and its effective spring constant increases. Figure 30.17b shows the phase of the vibrating cantilever as a function of the amplitude of vibration. Figure 30.17c and d depict the input power as a function of the amplitude of vibration of the cantilever and of its phase, respectively. The double-

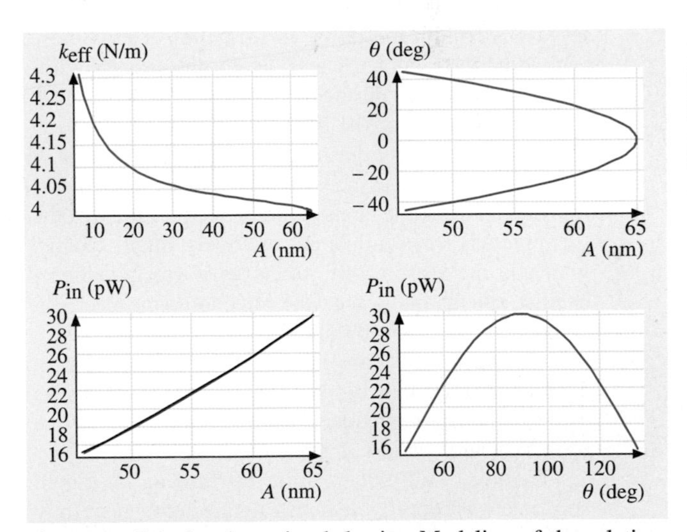

Fig. 30.17 Calculated tapping behavior. Modeling of the relationships among the amplitude A, effective spring constant k_{eff}, input power P_{in}, and phase ϕ

valued functions express the fact that one can operate either below or above the original resonance frequency of the cantilever.

This simplified model of a tapping mode neglects many effects such as indentation of the surface on impact and adhesion and other interface tip-sample forces. For our purpose, however, we will neglect these effects and explore the feasibility of using the tapping mode for data storage applications. The idea here is to have the tip jump from one pixel, or bit of stored data on the surface of a sample to another pixel without touching the surface in between. And while briefly touching the surface, the tip, which is conducting, will inject a pulse of current into the pixel and modify its conductivity [30.33]. Upon returning to the same pixel, the tip could monitor the conductivity of that particular pixel.

Considering the fact that the tapping mode involves a driven, damped oscillator, it is apparent that the motion of the cantilever traces a chaotic trajectory, necessitating an averaging process. For most imaging or nanofabrication purposes, such an averaging, combined with an appropriate image processing procedure, can indeed yield images with nanometer resolution. For data storage applications, however, it is desirable to have non-chaotic, reproducible interactions for every single tap because one could not afford an averaging procedure.

There are two key issues regarding the requirement that each tap produce a controlled interaction. The first concerns the identification of conditions under which (i) chaotic motion of the cantilever is minimized and (ii) good tip-sample contact is established. The second issue concerns the means by which one can observe these individual interactions. The difficulty in measuring tip-sample interactions during an individual tap stems from the fact that most of the time the tip is far away from the sample, and only during a small fraction of a cycle does it actually sense the presence of the sample. Also, it is the nature of the tapping mode of operation that the tip approaches the sample at a relatively small velocity, which is the reason why it acts as a grazing impact oscillator in the first place. One can choose the electrical contact between tip and sample as the measuring tool for the reproducibility of the interaction for individual taps. This choice is clearly compatible with storage modalities in which current injection into storage media gives rise to writing and reading of phase-change data.

The use of all-metal cantilevers tapping on a gold substrate was recently demonstrated, whereby carefully chosen parameters enable the generation and observation of individual current pulses [30.34]. These pulses were generated by applying a short bias pulse during the tip-sample contact time. Each current pulse, lasting up to 20 μs, could produce a current of ~ 10 μA in a nm^2 region yielding a current density of 10^{13} A/m^2. This density is large enough to be applicable for phase-change. The cantilever was fabricated from a rectangular PtIr wire whose length, width, and thickness were 1,200 μm, 100 μm, and 50 μm, respectively. One side of the cantilever was attached to a driving bimorph, and the other side etched to produce a tip whose apex radius was estimated to be 70 nm. The spring constant, resonance frequency, amplitude of vibration, and quality factor of the cantilever were 300 N/m, 13 kHz, 100 nm, and 100, respectively. The all-metal cantilever ensured good conductivity and could withstand a large number of

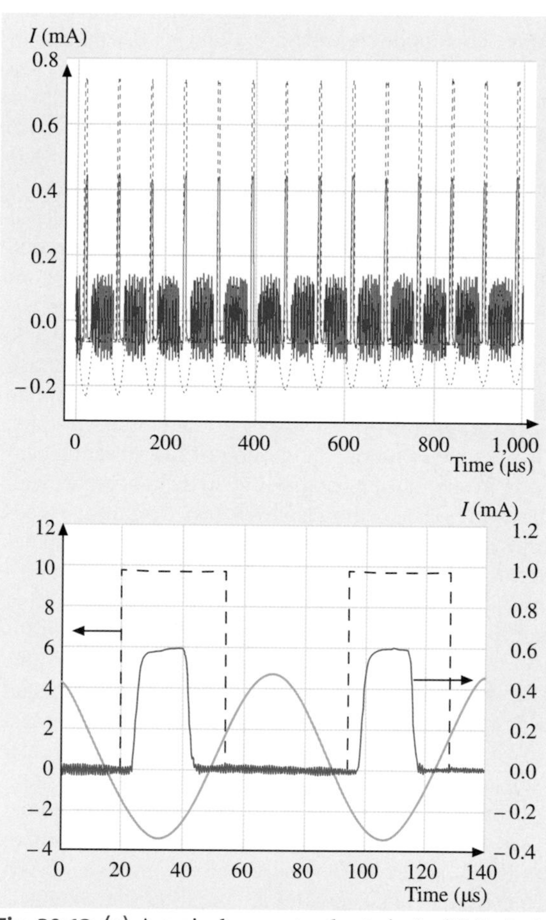

Fig. 30.18 **(a)** A typical current pulse train (*solid line*) obtained with pulsed voltage bias in phase with the cantilever's lowest position and the associated cantilever motion (*dotted line*), **(b)** two current pulses (*solid line*), bias voltage pulses (*dashed line*), and cantilever motion (*dotted line*)

tapping impacts without degrading and had a shiny surface that acted as a good mirror for the optical deflection technique.

The experimental results yielded long trains of individually injected current pulses. Figure 30.18 (left-hand side) shows 14 non-averaged experimental current pulses (solid line) and their associated cantilever motion (dashed line) with a background of instrumental noise [30.34]. The current pulses demonstrate the reproducibility of the conductance established during each tap. Figure 30.18 [30.34] (right-hand side) shows two averaged current pulses (solid line) with their input bias pulses (dashed line) and the associated cantilever motion (dotted line). Note that the tip-sample contact time is revealed by the duration of the current pulse, which is shorter than the duration of the applied bias pulse. Here, the amplitude of the bias pulse was 10 mV, its duration was 40 μs, and the timing of its application relative to the motion of the cantilever was provided by the output of the atomic force microscope photodiode. The current pulses, with a magnitude of 0.5 mA, are much larger than the tunneling current occurring while the tip is within a fraction of a nanometer from the surface. Also, the displacement current, due to the motion of the cantilever and the time varying nature of the applied bias, is expected to be much smaller than the observed current pulses. The contact conductance is shown in Fig. 30.19, where a linear relation between voltage and current implies a good ohmic contact. These results demonstrate that it is possible to minimize the chaotic behavior of a tapping cantilever and obtain reproducible conductance on each individual tap. The next step is to use the tapping mode in conjunction with current injection to write and read data bits on conducting media.

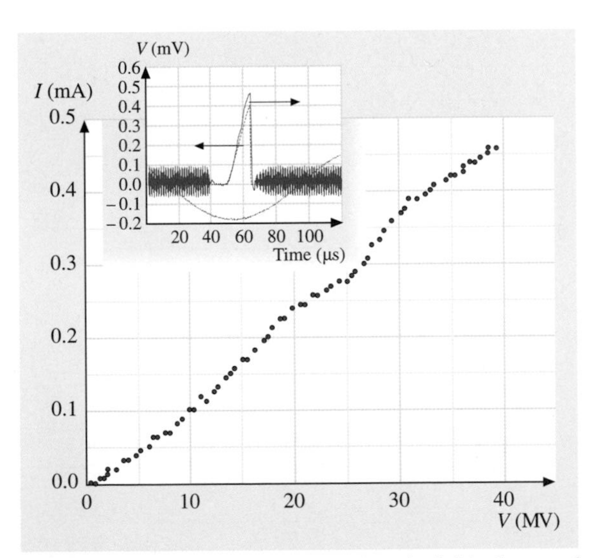

Fig. 30.19 An I–V curve taken with an individual ramped input voltage pulse demonstrating an ohmic contact. *Insert:* The current pulse (*solid line*), ramped input bias voltage (*dashed line*), and cantilever motion (*dotted line*)

30.2.3 Media and Experimental Results

This section summarizes several typical experimental results using different data storage media, electric charging, electric domain orientation, magnetic domain orientation, chemical modification, and phase change in chalcogenides and organic films.

Electric Charging

Quate's group at Stanford has studied charge trapping in a silicon nitride/silicon dioxide/silicon system using a scanning capacitance microscope [30.6]. The charge trapping in this system has been used for many years for non-volatilesemiconductor memory. Using chemical vapor deposition, 500 Å of silicon nitride was deposited onto a silicon wafer with a 50 Å oxide coating. By selectively storing charge into the substrate, they were able to write bits as small as 0.02 μm^2. Using this method, they were able to write the founding document of the Swiss Confederation onto a 120 μm^2 area, corresponding to 256 kilobits of information that yields a data density of 10 Gbpsi.

Electric Domains

Hidaka et al. have demonstrated the possibility of ultra-high density charge storage in a ferroelectric film, $PbZr_{1-x}Ti_xO_3$ [30.35]. They show that a change in the topography of the film can be introduced by domain reorientation induced by the application of a 10 V pulse with a duration of 100 μs. This is due to the piezoelectric effect caused by domain structures. Data bits as small as 40 nm were written this way. This result implies a data density of approximately 300 Gbpsi, suggesting that this medium is a good candidate for high-density data storage.

Thin Oxide Films

IBM Research in Switzerland has demonstrated reproducible switching in thin oxide perovskite films [30.36]. By constructing metal-insulator-metal capacitor-like devices, with $SrZr0_3$ as the insulator, they were able to observe an order of magnitude increase in the leakage current of the device at a threshold voltage of −0.5 V. This process is perfectly reversible, and the leakage current can be restored to its initial state upon application of a positive voltage. The switching occurs very rapidly, taking less than 100 ns to change from one conductivity state to another. Interestingly, the magnitude of the transition depends on the voltage applied, suggesting the possibility of storing multiple bits in one pixel.

Magnetic Domains

Sun et al. have demonstrated a novel way to overcome the limitations of media currently used for magnetic data storage [30.37]. By reducing iron and platinum-containing compounds, they demonstrated the formation of large-scale self-ordered arrays of magnetic particles. Initial studies suggest that an assembly of magnetic particles as small as 4 nm can support stable magnetization transitions at room temperature. By using the particles in these arrays as bits, the possibility of terabit-per-square-inch magnetic data storage could be realized.

Figure 30.20 shows transmission electron microscopy of these arrays that indicate a remarkable uniformity of particle size and large-scale periodicity. The figure also shows assemblies of these particles deposited onto a silicon oxide substrate under varying experimental conditions.

Chemical Modification

We have discussed the fact that the tapping mode is considerably less damaging to the tip and sample, as the tip spends less time contacting the sample and does not drag across it while raster scanning. For this purpose, diamond-coated cantilevers have been found useful for experiments in which wear is a problem. For sample oxidation purposes, the tip is first oscillated away from the sample with peak-to-peak amplitude of 10–15 nm, and then a feedback system reduces the amplitude to 7–10 nm by pressing the tip further into the sample. A 50 ms-long pulse with a 10–12 V magnitude is then applied to oxidize a thin Ti film. It was found that this technique produces much finer structures. Figure 30.21 [30.38] is a tapping mode AFM image of four dots created in such a way. The dots in this image have a diameter of less than 10 nm and a height of 3.5 nm above the surrounding Ti surface with a center separation of 20 nm. Around the dots there is what appears to be a lower ring of material with a channel in between. This experiment demonstrates that one can use the tapping mode to write fine pixels, albeit not to erase them.

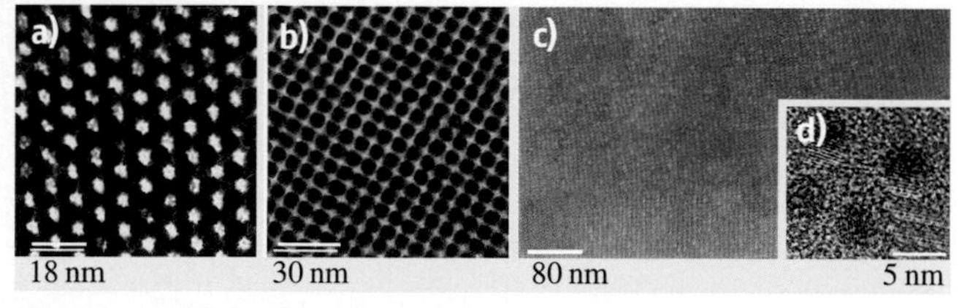

Fig. 30.20 (a) TEM micrograph of a 3-D assembly of 6-nm as-synthesized $Fe_{50}Pt_{50}$ particles deposited from a hexane/octane (v/v 1/1) dispersion onto a SiO-coated copper grid. (b) TEM micrograph of a 3-D assembly of a 6-nm $Fe_{50}Pt_{50}$ sample after replacing oleic acid/oleyl amine with hexanoic acid/hexylamine. (c) HRSEM image of a ∼180-nm-thick, 4-nm $F_{52}Pt_{48}$ nanocrystal assembly annealed at 560 °C for 30 min under 1 atm of N_2 gas. (d) High-resolution TEM image of 4-nm $Fe_{52}Pt_{48}$ nanocrystals annealed at 560 °C for 30 min on a SiO-coated copper grid reference

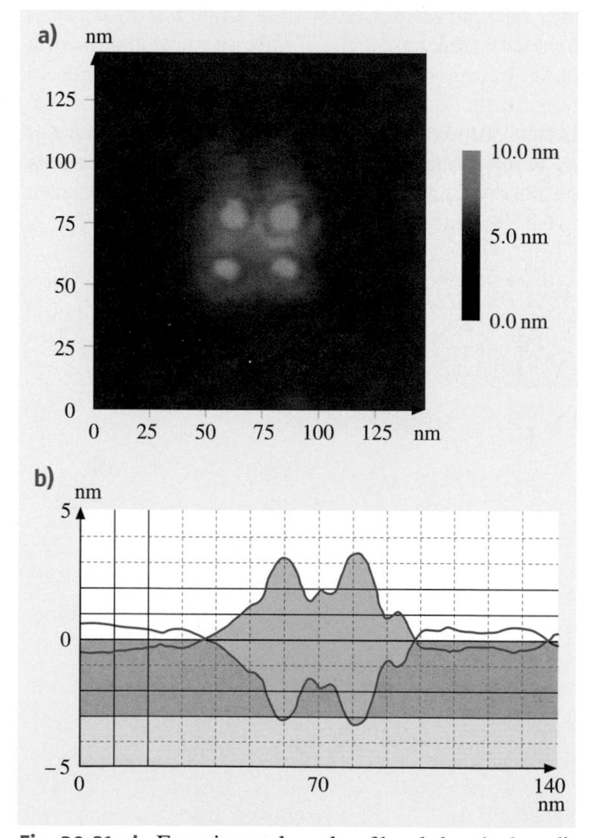

Fig. 30.21a,b Experimental results of local chemical modification using AFM tip: (a) 20-nm dots written on Ti film, (b) cross section of written bit

Phase Change in Chalcogenides

Chalcogenides, the active material in rewritable CDs, have the advantage of being a well-characterized and understood bistable phase change material that has already had successful commercial data storage applications, i. e., CDRWs. Currently, NVRAM using chalcogenides as the active material is being pursued by an alliance of Ovonyx and Intel. Chalcogenides are materials that have an easily reversible stable change between amorphous and crystalline phases. To switch from the crystalline to the amorphous phase, the material is heated sufficiently, so that the material subsequently cools rapidly into the amorphous phase. The reverse change is achieved by heating the material above the glass transition temperature, so that it recrystallizes upon cooling. For commercial CDRWs, writing is done by means of laser light heating and reading is done by measuring the difference in reflectivity of the laser light from each of these two phases.

Scanning probe microscopy can be used to switch the phase of a chalcogenide material by the application of a pulse of current that heats the material as it passes through it [30.39]. The state of the material can be read typically by measuring the resistance of the sample. Figure 30.22 shows the two different states, amorphous and crystalline of the chalcogenides. The reversibility of this transition is remarkable, and has been shown to be stable over 10^{12} repeated read-write cycles. *Gotoh* et al. demonstrated this effect using conducting AFM tips, where a transition from the high resistance amorphous state to the lower resistance crystalline state was observed both the in topography and the conductivity of the sample. The change from amorphous to crystalline phases was confirmed by X-ray diffraction. Figure 30.23a and a′ show a conductance AFM image and a profile of the surface after the application of varying voltage pulses. The topography of the substrate is shown in Fig. 30.23b. Figure 30.23c and c′ show the equivalent conductance AFM images after the substrate has been scanned at a negative voltage, erasing the lower power marks on the substrate.

These materials have been further characterized by *Gidon* et al. [30.26], who showed that the resistivity of the film in its amorphous phase was 100 Ω cm^{-1}, while that of the crystalline state was 0.1 Ω cm^{-1}. They also show that a data bit requires a 200 ns pulse with a power of 50 μW, which translates to a total energy of 10 pJ.

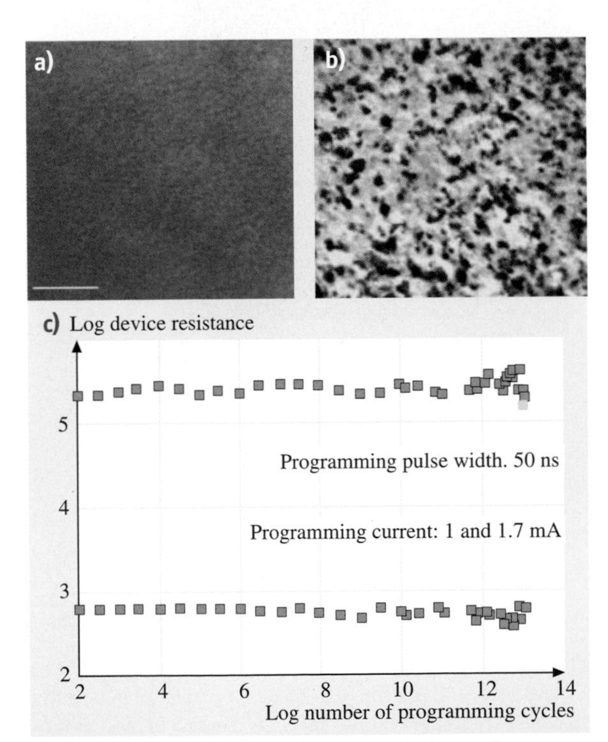

Fig. 30.22a–c Experimental results of phase change using interconnects. (**a**) Crystalline area, (**b**) amorphous area, (**c**) stability of bits over 1,012 rewrites (Ovonyx)

Phase Change in Organics

Organics have proved important in a variety of electronic and optoelectronic applications such as light-emitting diodes, thin-film displays, transistors, photovoltaics, and

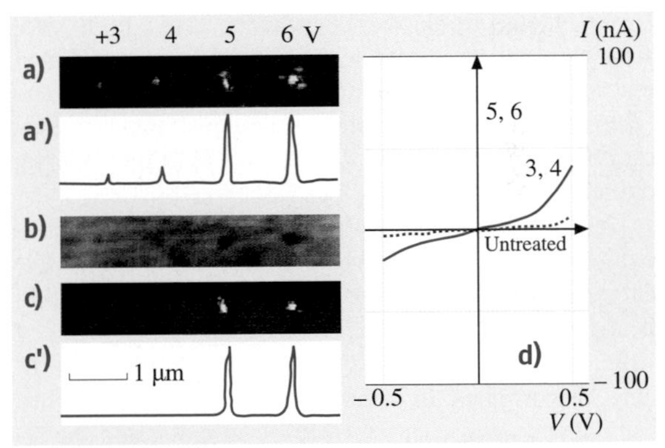

Fig. 30.23 Experimental results of phase change using an AFM tip (**a**), (**a′**), conducting AFM images after applying 100 ns pulses to the substrate, (**b**) topography AFM image of the substrate, (**c**), (**c′**) conducting AFM images showing erasure after scanning the substrate with an applied bias of −1 V, (**d**) *I–V* curves of untreated (*dotted*) and treated (*solid*) regions

lasers. There are several advantages to organics for economic and manufacturing reasons, and a large body of research already exists in this field. As such, there is significant interest and motivation in finding data storage applications based on organic technology.

There has been much research recently into organic materials and systems capable of reversible switching and thus capable of being exploited for data storage applications. There are two broad categories of organics data storage; one where the active material (which can be homogenous or heterogeneous) has inherent bistability, and one where the bistability is a property of the device structure.

There are many different materials that have been investigated for data storage. The fundamental property is that it should have two states (electronic, morphological, or some other easily measured physical characteristic) that the material can exist in, and it should be stable indefinitely in either state. Furthermore, the material should be able to be easily and reversibly switched between these states, and this should be reproducible over many millions of cycles for it to be useful as a data storage medium.

Several groups have been working on reversible transitions in organic complexes suitable for switching by scanning probe techniques. Many complexes have been demonstrated to exhibit reversible switching, e.g., 3-nitrobenzal malonitrile (NBMN) and 1,4-phenylenediamine (pDA) [30.40], tetrathiofulvane (TTF) and *m*-nitrobenzylidene propanedinitrile (*m*-NBP) [30.41], and many other variants.

Li et al. [30.41] have studied organic films consisting of tetrathiofulvane (TTF), an electron donor, and *m*-nitrobenzylidene propanedinitrile (*m*-NBP), an electron acceptor. This complex demonstrates electrical bistability, which can be written and read by scanning tunneling microscopy.

The nature of the measurement means that it is difficult to ascertain whether this is due to solely conductive changes in the sample, or whether topographic modifications might also contribute. By applying pulses of 1 ms duration, bits of 1.2 nm diameter could be written. This implies an areal density of $\sim 10^{18}$ bits/in^2. This change was characterized electrically, and the resistance was measured as changing from $10^8\,\Omega$ to $10^3\,\Omega$ with a threshold switching voltage of 3.1 V. The proposed switching mechanism is charge transfer between the two film constituents. This change was stable over time and is reversible. This change is reversible, but only by heating for prolonged periods, making it unsuitable for applications as of yet.

Gao et al. have demonstrated stable, local, reversible nanoscale switching [30.40] using a charge transfer complex of 3-nitrobenzal malonitrile (NBMN) and 1,4-phenylenediamine (pDA). A 200-nm-thick film of this complex was deposited onto highly ordered pyrolitic graphite. Macroscopic electrical measurements demonstrate the bistability and switching of the material. Figure 30.24a shows the current-voltage characteristics of the sample in these two conductivity states. Initially, the material is in a low conductivity state, but upon reaching a certain threshold voltage, in this case 3.2 V, the material abruptly changes to a conductive state. The transition time is very abrupt, 80 ns, as shown in Fig. 30.24b.

Of further interest for data storage applications is that, having demonstrated reversible switching in macroscopic films, Gao et al. also showed that it was possible to switch these materials on the nanoscale using an STM

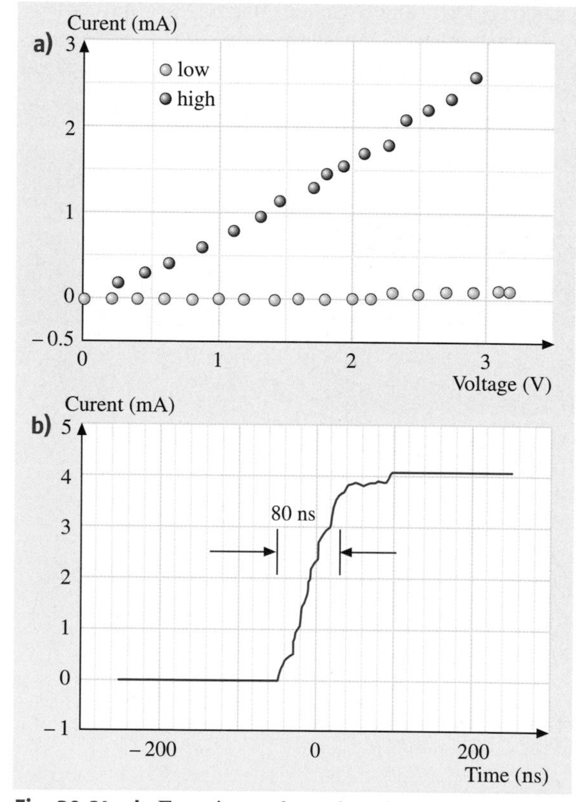

Fig. 30.24a,b Experimental results of bulk conductivity changes of organic films. **(a)** I–V curves of a 200-nm-thick film showing the two conductivity states. Switching voltage is 3.2 V. **(b)** Change in conductivity against time, showing a transition time of 80 ns

to apply pulses, i. e., to write bits, and also to measure conductivity, i. e., to read bits.

Figure 30.25a shows the local nanoscale film morphology of the sample. Figure 30.25b and c show patterns written on a sample by application of voltage pulses from the STM tip. Figure 30.25d demonstrates the erasure of a single bit from the pattern by application of a reverse-biased voltage pulse. Similar to the work of *Li* et al., the entire substrate could be erased by prolonged heating. Of much greater interest is the demonstration of individual bit erasure, achieved by applying reverse voltage pulses. The erasure is incomplete unless the pulse is reverse-biased, indicating that it is the combined effect of heating and electric field that erases the bits.

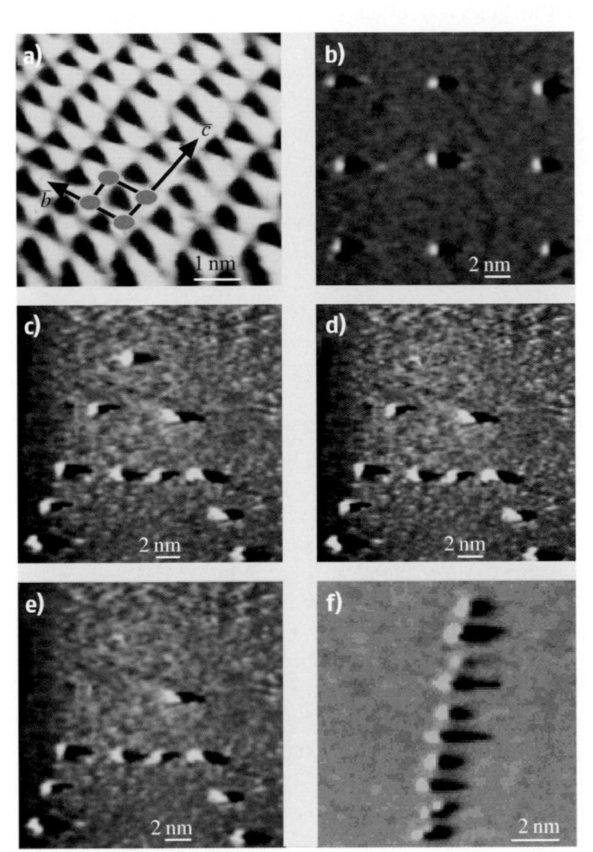

Fig. 30.25a–f STM of the NBMN-pDA film on HOPG. **(a)** 6 nm^2 image showing the crystalline order of the film; **(b)** a 3×3 array of bits (4 V, 1 μs); **(c)** an "A" pattern (3.5 V, 2 μs); **(d)**, **(e)**, **(f)** STM images after erasing marks with reverse-biased pulses (−4.5 V, 50 μs). Distance between neighboring marks is 1.7 nm

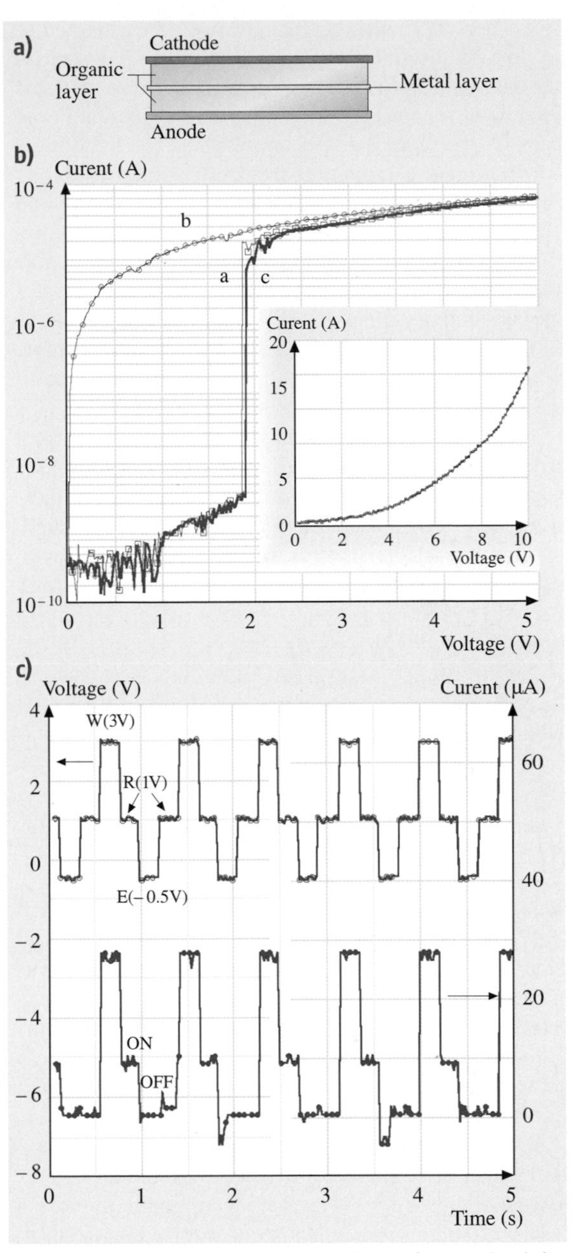

Fig. 30.26a–c Experimental results of conductivity changes of organic device structures: **(a)** the structure of the bistable device, **(b)** I–V characteristics. Voltage was scanned in steps of 0.1 V from 0 to 5 V. Curves a and b represent the I–V characteristics of the first and second bias scan, respectively. Curve c is the I–V curve of the third bias scan after the application of a reverse voltage pulse (−3 V). **(c)** Typical current responses of the device being written, read, erased, read over time

Several proposed mechanisms were investigated as the possible origin of this switching, including conformational switching and charge transfer. These could be discounted for reasons, including the stability and time scale of the switching. A more plausible hypothesis is a local phase transition from crystalline insulating regions to amorphous conducting regions. TEM and diffraction studies of the samples agree with this hypothesis. It should be pointed out that an alternative interpretation has been suggested by *Zhao* et al. [30.42]. The conductivity change could be due mainly to a change in the topography of the sample, reducing the tip-sample distance and, therefore, the tunneling distance, manifested as an apparent change in conductivity.

Ma et al. have developed a simple device structure that exhibits electrical bistability [30.43]. This device has a fairly simple structure consisting of alternating layers of aluminum and an organic compound, 2-amino-4,5-imidazoledicarbonitrile (AIDCN), as depicted in Fig. 30.26a. Several alternative organics (Alq3 and MEH-PPV) and metals (copper, silver, and gold) have been tried with varying degrees of success.

Figure 30.26b shows the current-voltage bistability of a device with structure Al/AIDCN (50 nm)/Al (20 nm)/AIDCN (50 nm)/Al. A first run shows an initially low current, followed by an abrupt and dramatic increase in the current of four orders of magnitude at 2 V. Subsequent repeated voltage runs (b) show that this transition is stable, being orders of magnitude higher than the initial measurement. The devices proved to be stable in this state over periods ranging from days to weeks.

Of further interest is that the device proved to be erasable by purely electrical means, the application of a -0.5 V pulse bringing the device back to its original state. This is denoted by c in Fig. 30.26b. Furthermore, the devices showed excellent stability over time, with over one million read-write-erase cycles being performed on the devices with excellent repeatability, as depicted in Fig. 30.26c. The origin of this bistability is unclear. It is clear that the central electrode is critical to the observed effect, as the bistability is only observed when the central electrode thickness is above a threshold value, 10 nm. Organics with high dielectric constants and low conductivity are believed to be important for device performance. The exact mechanism, however, is as yet unclear. Preliminary results on novel materials [30.44] have suggested the possibility of an all-organic variation of this concept, with an AFM tip switching between bistable states of the system.

30.3 Conclusion

Nanotechnology is perceived as a branch of science that deals with structures whose dimensions are measured in nanometers. Clearly, new phenomena become apparent when the number of atoms at the interface of such a structure becomes comparable to the number of atoms in its volume. As a result, nanotechnology evolved into a new kind of science that deals with the physics of small things. However, the suffix -technology here implies a far more reaching aspect than just a scientific curiosity; this branch of science should also be of some use. This poses a dilemma for researchers whose main interest is in understanding new phenomena, rather than in finding a use for them. Consider scanning probe microcopy, which is celebrating its twentieth birthday now with uncountable contributions to the science of measurement of surfaces and near-surface phenomena. Despite all the contributions scanning probe microscopy has made, it has always served as a tool for investigations, rather than as a practical device that can become a commercial commodity. And yet this is exactly what this chapter deals with. The challenge of converting scanning probe microscopy from a research tool into a probe storage device is daunting indeed and will require new ideas to make it happen. The danger associated with this challenge is that novel ideas developed by research groups are sometimes too immature to be measured against any existing technology, in general, and against data storage, in particular. One has to keep pursuing research opportunities whose outcomes are unknown and whose applications are unexpected with the hope that they will bring about good science and hopefully useful applications.

However, based on the technical breakthroughs delineated in this and the following chapter and judging from the history of data storage evolution, it seems plausible that probe storage will serve as a disruptive technology that replaces data storage devices in a particular market niche. These new devices should be non-volatile, have terabyte capacity with microsecond access times with gigabyte transfer rates, all for a cost that is highly competitive with current commercially available data storage devices.

References

30.1 D. Sarid: *Scanning Force Microscopy, with Applications to Electric, Magnetic, and Atomic Forces* (Oxford Univ. Press, Oxford 1991)

30.2 D. Sarid: *Scanning Force Microscopy, with Applications to Electric, Magnetic, and Atomic Forces*, revised edn. (Oxford Univ. Press, Oxford 1994)

30.3 D. Sarid: *Exploring Scanning Probe Microscopy with Mathematica* (Wiley, New York 1997)

30.4 B. G. Levi: Do atomic force microscope arrays have the write stuff?, Phys. Today **55**(10) (2002) 14–17

30.5 H. J. Mamin, B. D. Terris, L. S. Fan, S. Hoen, R. C. Barrett, D. Rugar: High-density data storage using proximal probe techniques, IBM J. Res. Dev. **39**(6) (1995) 681–699

30.6 R. C. Barrett, C. F. Quate: Large-scale charge storage by scanning capacitance microscopy, Ultramicroscopy **42–44** (1992) 262–267

30.7 H. J. Mamin, R. P. Ried, B. D. Terris, D. Rugar: High-density data storage based on the atomic force microscope, Proc. IEEE **87**(6) (1999) 1014–1027

30.8 P. Vettiger, J. Brugger, M. Despont, U. Drechsler, U. Durig, W. Haberle, H. Lutwyche, H. Rothuizen, R. Stutz, R. Widmer, G. Binnig: Ultrahigh density, high-data-rate NEMS-based AFM data storage system, Microelectron. Eng. **46** (1999) 11–17

30.9 P. Vettiger, M. Despont, U. Drechsler, U. Durig, W. Haberle, M. I. Lutwyche, H. E. Rothuizen, R. Stutz, R. Widmer, G. K. Binnig: The millipede-more than one thousand tips for future AFM data storage, IBM J. Res. Dev. **44**(3) (2000) 323–340

30.10 B. Schechtman: The future of optical disk storage, NSIC Annual Meeting Symposium, Monterey 2002, 1–14

30.11 P. Lyman, H. L. Varian: How much information (2000) retrieved from http://:www.sims.berkeley.edu/how-much-info

30.12 W. Perdue: Private Communication (September, 2002)

30.13 P. Frank: private communication, (2002)

30.14 H. De J. Ruiz, S. D. Morgan: , Witter Semiconductor and Systems Conference, Laguna Michael 2001

30.15 J. Daughton, NVE Inc., private communication

30.16 M. Gill, T. Lowrey, J. Park: Ovonics unified memory – A high performance nonvolatile memory technology for stand alone memory and embedded applications, 2002 IEEE Int. Solid State Circuits Conf., San Francisco 2002 (IEEE, Piscataway 2002)

30.17 M. Despont, J. Brugger, U. Drechsler, U. Dürig, W. Häberle, M. Lutwyche, H. Rothuizen, R. Stutz, R. Widmer, G. Binnig, H. Rohrer, P. Vettiger: VLSI-NEMS chip for AFM data storage, Technical Digest 12th IEEE Int. Conf. on Micro Electro Mechanical Systems (MEMS'99) (IEEE, Piscataway 1999) 564–569

30.18 K. Schulte-Wieking, S. Noehte, C. Dietrich, M. Mayer: A new cylindrically shaped optical data storage medium based on adhesive tape-concept and first results, IEEE Proc. Technical Digest of ISOM/ODS 2002, Joint Int. Symp. on Optical Memory and Optical Data Storage 2002 (IEEE, Piscataway 2002) 117–120

30.19 D. Sarid: unpublished results, (2002)

30.20 L. A. Romero, F. M. Dickey: A method for achieving constant rotation rates in a microorthogonal linkage system, J. Microelectromech. Syst. **9**(2) (2000) 236–244

30.21 L. R. Carley, G. R. Ganger, D. F. Nagle: MEMS-based integrated-circuit mass-storage systems, Commun. ACM **43**(11) (2000) 71–80

30.22 J. L. Griffin, S. W. Schlosser, G. R. Ganger, D. F. Nagle: Modeling and performance of MEMS-based storage devices, Int. Conf. on Measurement and Modeling of Computer Systems, Proc. of ACM Sigmetrics, Santa Clara 2000, **28**(1) (2000) 56–65, published as Performance Evaluation Review

30.23 S. W. Schlosser, J. L. Griffin, D. F. Nagle, G. R. Ganger: Designing computer systems with MEMS-based storage, Proc. of the 9th International Conf. on Architectural Support for Programming Languages and Operating Systems, Boston 2000, 1–12

30.24 L. R. Carley, G. Ganger, D. F. Guillou, D. Nagle: System design considerations for MEMS-actuated magnetic-probe-based mass storage, IEEE Trans. Magn. **37**(2) (2001) 657–662

30.25 R. T. El-Sayed, L. R. Carley: Performance analysis of beyond 100 Gb/in^2 MFM-based MEMS-actuated mass storage devices, IEEE Trans. Magn. **38**(5) (2002) 1–3

30.26 O. Bichet, S. Gidon, Y. Samson: Scanning probe based storage on phase change media, OSA Trends in Optics and Photonics Vol. 88 Optical Data Storage (OSA, Washington D.C. 2003)

30.27 H. Kawakatsu, D. Saya, A. Kato, K. Fukushima, H. Toshiyoshi, H. Fujita: Millions of cantilevers for atomic force microscopy, Rev. Sci. Instrum. **73**(3) (2002) 1188–1192

30.28 H. Kawakatsu, S. Kawai, D. Saya, M. Nagashio, D. Kobayashi, H. Toshiyoshi, H. Fujita: Towards atomic force microscopy up to 100 MHz, Rev. Sci. Instrum. **73**(6) (2002) 2317–2320

30.29 P. Vettiger, G. Cross, M. Despont, U. Drechsler, U. Dürig, B. Gotsman, W. Häberle, M. A. Lantz, H. E. Rothuizen, R. Stutz, G. K. Binnig: The 'millipede' – Nanotechnology entering data storage, IEEE Trans. Nanotechnol. **1** (2002) 39–55

30.30 Q. Zhong, D. Inniss, K. Kjoller, V. B. Elings: Fractured polymer/silica fiber surface studied by tapping mode atomic force microscopy, Surf. Sci. **290**(1-2) (1993) L688–L692

30.31 A. M. Hayashi: Punch cards of the future, Sci. Am. (May 2000)
30.32 J. P. Hunt, D. Sarid: Kinetics of lossy grazing impact oscillators, Appl. Phys. Lett. **72**(23) (1998) 2969–2971
30.33 D. Sarid: Tapping-mode scanning force microscopy: Metallic tips and samples, Comp. Mater. Sci. **5** (1996) 291–297
30.34 A. Fein, Y. Zhao, C. A. Peterson, G. E. Jabbour, D. Sarid: Individually injected current pulses with conducting-tip, tapping-mode atomic force microscopy, Appl. Phys. Lett. **79**(24) (2001) 3935–3937
30.35 T. Hidaka, T. Maruyama, M. Saitoh, N. Mikoshiba, M. Shimizu, T. Shiosaki, L. A. Willis, R. Hiskes, S. A. Dicarolis, J. Amano: Formation and observation of 50 nm polarized domains in $PbZr_{1-x}Ti_xO_3$ thin film using scanning probe microscope, Appl. Phys. Lett. **68**(17) (1996) 2358–2359
30.36 A. Beck, J. G. Bednorz, Ch. Gerber, C. Rossel, D. Widmer: Reproducible switching effect in thin oxide films for memory applications, Appl. Phys. Lett. **77**(1) (2000) 139–141
30.37 S. Sun, C. B. Murray, D. Weller, L. Folks, A. Moser: Monodisperse FePt nanoparticles and ferromagnetic FePt nanocrystal superlattices, Science **287** (2000) 1989–1992
30.38 C. Peterson: Characterization of semiconductor devices through scanned probe microscopy. Ph.D. Thesis (University of Arizona, Tucson 2002)
30.39 T. Gotoh, K. Sugawara, K. Tanaka: Nanoscale electrical phase-change in $GeSb_2Te_4$ films with scanning probe microscopes, J. Non-Cryst. Solids **299–302** (2002) 968–972
30.40 H. J. Gao, K. Sohlberg, Z. Q. Xue, H. Y. Chen, S. M. Hou, L. P. Ma, X. W. Fang, S. J. Pang, S. J. Pennycook: Reversible, nanometer-scale conductance transitions in an organic complex, Phys. Rev. Lett. **84**(8) (2000) 1780–1783
30.41 J. C. Li, Z. Q. Xue, W. M. Liu, S. M. Hou, X. L. Li, X. Y. Zhao: Study on a new organic-complex thin film with electrical bistable properties using a scanning tunneling microscope, Phys. Lett. A **266** (2000) 441–445
30.42 Y. Zhao, A. Fein, C. A. Peterson, D. Sarid: Comment on reversible, nanometer-scale conductance transitions in an organic complex, Phys. Rev. Lett. **87**(17) (2001) 179706-1
30.43 L. P. Ma, J. Liu, Y. Yang: Organic electrical bistable devices and rewritable memory cells, Appl. Phys. Lett. **80**(16) (2002) 2997–2999
30.44 B. McCarthy, K. Yamnitskiy, G. E. Jabbour, D. Sarid: unpublished results, (2002)

31. The "Millipede" – A Nanotechnology-Based AFM Data-Storage System

The "millipede" concept presented here is a new approach for storing data at high speed and ultrahigh density. The interesting part is that millipede stores digital information in a completely different way from magnetic hard disks, optical disks, and transistor-based memory chips. The ultimate locality is provided by a tip, and high data rates are a result of massive parallel operation of such tips. As storage medium, polymer films are being considered, although the use of other media, in particular, magnetic materials, has not been ruled out. The current effort is focused on demonstrating the millipede concept with areal densities of up to 0.5–1 Tb/in^2 and parallel operation of very large 2-D (up to 64 × 64) AFM cantilever arrays with integrated tips and write/read/erase functionality. The fabrication and integration of such a large number of mechanical devices (cantilever beams) will lead to what we envision as the VLSI age of micro- and nanomechanics.

In this chapter, the millipede concept for a MEMS-based storage device is described in detail. In particular, various aspects pertaining to AFM thermomechanical read/write/erase functions, 2-D array fabrication and characteristics, *x*/*y*/*z* microscanner design, polymer media properties, read channel modeling, servo control and synchronization, as well as modulation coding techniques suitable for probe-based data-storage devices are discussed.

31.1 **The Millipede Concept** ... 923
31.2 **Thermomechanical AFM Data Storage** ... 924
31.3 **Array Design, Technology, and Fabrication** ... 926
31.4 **Array Characterization** ... 927
31.5 ***x*/*y*/*z* Medium Microscanner** ... 929
31.6 **First Write/Read Results with the 32×32 Array Chip** ... 931
31.7 **Polymer Medium** ... 932
31.7.1 Writing Mechanism ... 932
31.7.2 Erasing Mechanism ... 935
31.7.3 Overwriting Mechanism ... 937
31.8 **Read Channel Model** ... 939
31.9 **System Aspects** ... 943
31.9.1 PES Generation for the Servo Loop ... 943
31.9.2 Timing Recovery ... 945
31.9.3 Considerations on Capacity and Data Rate ... 946
31.10 **Conclusions** ... 948
References ... 948

In the twenty-first century, the nanometer will probably play a role similar to the one played by the micrometer in the twentieth century. The nanometer scale will presumably also pervade the field of data storage, although there is so far no obvious way in conventional magnetic, optical, or transistor-based storage to achieve the nanometer scale in all three dimensions. After decades of spectacular progress, those mature technologies have entered the homestretch; imposing physical limitations loom before them.

One promising method involves the use of patterned magnetic media for which the ideal write/read concept still needs to be demonstrated. The biggest challenge, however, is to pattern the magnetic disk in a cost-effective way. If such approaches are successful, the basis for large-capacity storage in the twenty-first century might still be magnetism.

Other proposals call for totally different media and techniques such as local probes, near-field optics, magnetic super-resolution or holographic methods. In general, when an existing technology is about to reach its limits in the course of its evolution and alternatives are emerging in parallel, two things usually happen: First, the existing and well-established technology is explored further and every effort is made to push its limits to the utmost in order to get the maximum return on the considerable investments made. Then, when all possibilities for improvement have been exhausted, the technology may still survive for certain niche applications, but the emerging technology will take over, opening new perspectives and new directions.

In many fields today we are witnessing the transition of structures from the micrometer to the nanometer scale, a dimension that nature has long been using to build the finest devices with a high degree of local functionality. Many of the techniques we use today are not suitable for the nanometer age; some will require minor or major modifications, others will be partially or entirely replaced. It is certainly difficult to predict which techniques will fall into which category. In key information technology areas, such as nano-electronics and data storage, it is not yet obvious which technologies and materials will be used in the future.

In any case, for an emerging technology to be seriously considered as a candidate to replace an existing technology that is approaching its inherent limits, it must provide a long-term perspective. For instance, the silicon microelectronics and storage industries are huge and have exacted correspondingly enormous investments, which makes them long-term oriented by nature. The consequence for storage is that any novel technology with higher areal storage density than today's magnetic recording [31.1, 2] should have long-term potential for further scaling, preferably down to the nanometer or even atomic scale.

The only available tool known today that is simple and yet provides this very long-term perspective is the nanometer-sharp tip. The simple tip is a very reliable tool that concentrates on one functionality: the ultimate local confinement of interaction. Techniques that use nanometer-sharp tips for imaging and investigating the structure of materials down to the atomic scale, such as the atomic force microscope (AFM) and the scanning tunneling microscope (STM) [31.3, 4], are suitable for the development of ultrahigh-density storage devices.

In the early 1990s, Mamin and Rugar at the IBM Almaden Research Center pioneered the capability of using an AFM tip for writing and readback of topographic features for data storage. In one of their schemes [31.5], writing and reading were demonstrated with a single AFM tip in contact with a rotating polycarbonate substrate. The writing was performed thermomechanically by heating the tip. In this way, storage densities of up to $30\,\mathrm{Gb/in^2}$ were achieved, constituting a significant advance over the densities of that time. Later refinements included increasing readback speeds to attain a data rate of $10\,\mathrm{Mb/s}$ [31.6] and the implementation of track servoing [31.7].

When using single tips in AFM or STM operation for storage, one has to deal with their fundamental data-rate limitations. At present, the mechanical resonant frequencies of the AFM cantilevers limit the data rates of a single cantilever to a few Mb/s for AFM data storage [31.8, 9]. The feedback speed and low tunneling currents limit STM-based storage approaches to even lower data rates.

Currently a single AFM operates at best on the microsecond time scale. Conventional magnetic storage, however, operates at best on the nanosecond time scale, making it clear that AFM data rates have to be improved by at least three orders of magnitude to be competitive with current and future magnetic recording. One solution for substantially increasing the data rates achievable by tip-based storage devices is to employ micro-electro-mechanical system (MEMS) arrays of cantilevers operating in parallel, with each cantilever performing write/read/erase operations in an individual storage field. It is our conviction that very large-scale integrated (VLSI) micro/nanomechanics will greatly complement future micro- and nanoelectronics (integrated or hybrid) and may generate hitherto inconceivable applications of VLSI-MEMS.

Various efforts are under way to develop MEMS-based storage devices. For example, a MEMS-actuated magnetic-probe-based storage device capable of storing 2 GB of data on $2\,\mathrm{cm^2}$ of die area and whose fabrication is compatible with a standard integrated circuit manufacturing process is described by *Carley* et al. [31.10]. In their device, a magnetic storage medium is positioned in the x/y plane, and writing is achieved magnetically by using an array of probe tips, each tip being actuated in the z-direction. Another concept is the atomic resolution storage described by *Gibson* et al. [31.11], who employ electron field emitters to change the state of a phase-change medium in a bit-wise fashion from polycrystalline to amorphous or vice versa. Reading is done with lower currents by detecting either back-scattered electrons or changes in the semiconductor properties in the media.

The "millipede" concept presented here is a new approach for storing data at high speed and ultrahigh density. The interesting part is that millipede stores digital information in a completely different way from magnetic hard disks, optical disks, and transistor-based memory chips. The ultimate locality is provided by a tip, and high data rates are a result of massive parallel operation of such tips. As storage medium, polymer films are being considered, although the use of other media, in particular, magnetic materials, has not been ruled out. Our current effort is focused on demonstrating the millipede concept with areal densities of up to 0.5–1 Tb/in^2 and parallel operation of very large 2-D (up to 64×64) AFM cantilever arrays with integrated tips and write/read/erase functionality. The fabrication and integration of such a large number of mechanical devices (cantilever beams) will lead to what we envision as the VLSI age of micro- and nanomechanics.

In this chapter, the millipede concept for a MEMS-based storage device is described in detail. In particular, various aspects pertaining to AFM thermomechanical read/write/erase functions, 2-D array fabrication and characteristics, $x/y/z$ microscanner design, polymer media properties, read channel modeling, servo control and synchronization, as well as modulation coding techniques suitable for probe-based data-storage devices are discussed.

31.1 The Millipede Concept

A 2-D AFM cantilever array storage technique [31.13–15], internally called the millipede, is illustrated in Fig. 31.1. Information is stored as sequences of indentations and no indentations written on nanometer-thick polymer films using the array of AFM cantilevers. The presence and absence of indentations will also be referred to as logical marks. Each cantilever performs write/read/erase operations over an individual storage field with an area on the order of $100 \times 100\ \mu\text{m}^2$. Write/read operations depend on a mechanical parallel x/y scanning of either the entire cantilever array chip or the storage medium. The tip-medium contact is maintained and controlled globally, i. e., not on an individual cantilever basis, by using a feedback control in the z-direction for the entire chip, which greatly simplifies the system. This basic concept of the entire chip approach/leveling has been tested and demonstrated for the first time by parallel imaging with a 5×5 array chip [31.16, 17]. These parallel imaging results have shown that all 25 cantilever tips have approached the substrate within less than 1 μm of z-actuation, which indicates that overall chip tip-apex height control to within 500 nm is feasible. The stringent requirement for tip-apex uniformity over the entire chip is determined by the uniform force required to reduce tip and medium wear due to large force variations resulting from large tip-height nonuniformities [31.7]. Moreover, as the entire array is tracked without individual lateral cantilever positioning, thermal expansion of the array chip has to be small or well controlled. Thermal expansion considerations are a strong argument for a 2-D instead of a 1-D array arrangement.

Efficient parallel operations of large 2-D arrays can be achieved by a row/column time-multiplexed addressing scheme similar to that implemented in DRAMs. In the case of the millipede, the multiplexing scheme is used to address the array column by column with full parallel write/read operation within one column [31.18]. In particular, readback signal samples are obtained by applying an electrical read pulse to the cantilevers in a column of the array, low-pass filtering the cantilever response signals, and finally sampling the filter output signals. This process is repeated sequentially until all columns of the array have been addressed, and then restarted from the first column. The time between two pulses applied to the cantilevers of the same column corresponds to the time it takes for a cantilever to move from one logical mark position to the next. An alternative approach is to access all or a subset of the cantilevers simultaneously without resorting to the row/column multiplexing scheme. Clearly, the latter scheme yields higher data rates, whereas the former leads to lower implementation complexity of the channel electronics.

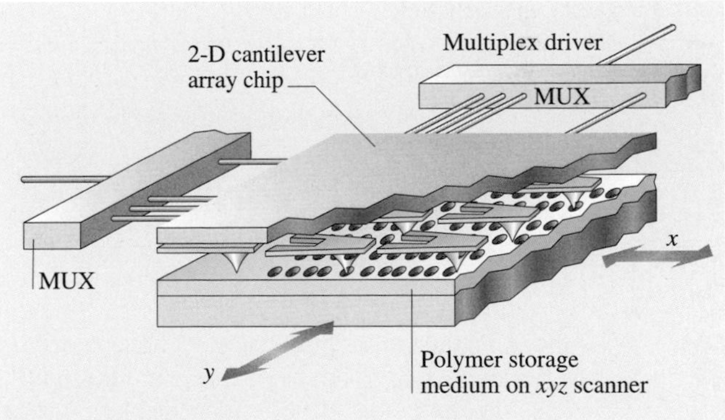

Fig. 31.1 The millipede concept (after [31.12])

31.2 Thermomechanical AFM Data Storage

In recent years, AFM thermomechanical recording in polymer storage media has undergone extensive modifications mainly with respect to the integration of sensors and heaters designed to enhance simplicity and to increase data rate and storage density. Thermomechanical writing in polycarbonate films and optical readback was first investigated and demonstrated by Mamin and Rugar (1992). Using heater cantilevers, thermomechanical recording at 30 Gb/in^2 storage density and data rates of a few Mb/s for reading and 100 kb/s for writing have been demonstrated [31.5, 6, 19]. Although the storage density of 30 Gb/in^2 obtained originally was not overwhelming, the results were encouraging enough to consider using polymer films to achieve density improvements. The current millipede storage approach is based on a new thermomechanical write/read process in nanometer-thick polymer films.

Thermomechanical writing is achieved by applying a local force through the cantilever/tip to the polymer layer and simultaneously softening the polymer layer by local heating. Initially, the heat transfer from the tip to the polymer through the small contact area is very poor, but it improves as the contact area increases. This means that the tip must be heated to a relatively high temperature of about 400 °C to initiate softening. Once softening has been initiated, the tip is pressed into the polymer, and hence the indentation size is increased. Rough estimates [31.20, 21] indicate that at the beginning of the writing process only about 0.2% of the heating power is used in the very small contact zone (10–40 nm^2) to soften the polymer locally, whereas about 80% is lost through the cantilever legs to the chip body and about 20% is radiated from the heater platform through the air gap to the medium/substrate. After softening has started and the contact area has increased, the heating power available for generating the indentations increases at least ten times to reach 2% or more of the total heating power.

With this highly nonlinear heat-transfer mechanism it is very difficult to achieve small tip penetration and hence small bit sizes, as well as to control and reproduce the thermomechanical writing process. This situation can be improved if the thermal conductivity of the substrate is increased and if the depth of tip penetration is limited. We have explored the use of very thin polymer layers deposited on Si substrates to improve these characteristics [31.22, 23], as illustrated in Fig. 31.2. The hard Si substrate prevents the tip from penetrating farther than the film thickness, and it enables more rapid transport of heat away from the heated region, as Si is a much better conductor of heat than the polymer. Using coated Si substrates with a 50-nm film of polymethylmethacrylate (PMMA), we have achieved indentation sizes of between 10 and 50 nm. However, increased tip wear has occurred, probably caused by contact between the Si tip and the Si substrate during writing. Therefore, a 70-nm layer of crosslinked photoresist (SU-8) was introduced between the Si substrate and the PMMA film to act as a softer penetration stop that avoids tip wear, but remains thermally stable.

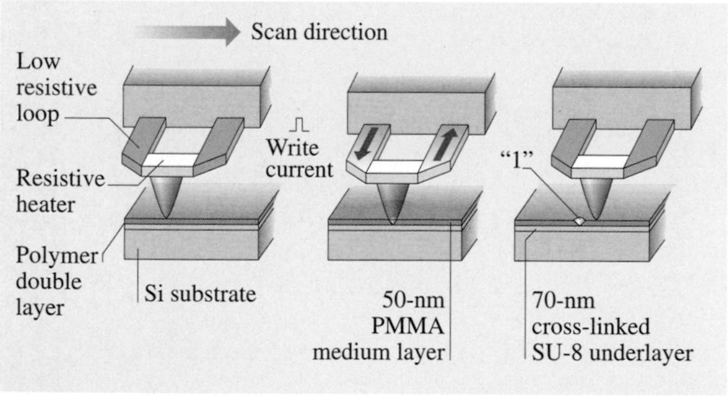

Scan direction
Low resistive loop
Resistive heater
Polymer double layer
Si substrate
Write current
"1"
50-nm PMMA medium layer
70-nm cross-linked SU-8 underlayer

Fig. 31.2 New storage medium used for writing small bits. A thin, writable PMMA layer is deposited on a Si substrate separated by a cross-linked film of epoxy photoresist (after [31.15])

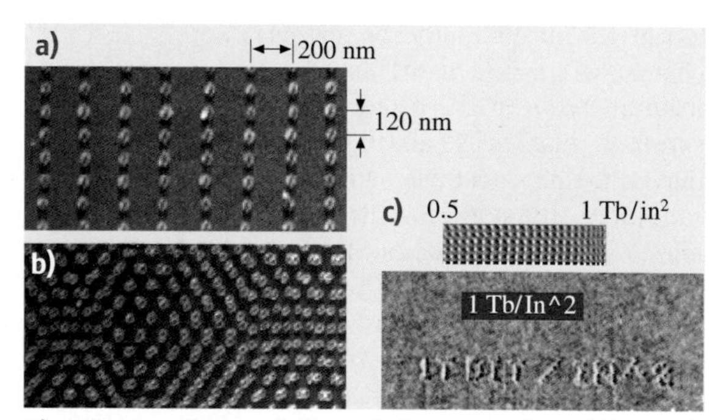

Fig. 31.3a–c Series of 40-nm indentations formed in a uniform array with **(a)** 120-nm pitch and **(b)** variable pitch ($\geq$ 40 nm), resulting in areal densities of up to 400 Gb/in^2. Images obtained with a thermal readback technique. **(c)** Ultra-high-density bit writing with areal densities approaching 1 Tb/in^2. The scale is the same for all three images (from [31.15] © 2002 IEEE)

Using this layered storage medium, indentations 40 nm in diameter have been written as shown in Fig. 31.3. These experiments were performed using a 1-μm-thick, 70-μm-long, two-legged Si cantilever [31.19]. The cantilever legs are made highly conducting by high-dose ion implantation, whereas the heater region remains low-doped. Electrical pulses 2 μs in duration were applied to the cantilever with a period of 50 μs. Figure 31.3a demonstrates that 40-nm bits can be written with 120-nm pitch or, as shown in Fig. 31.3b, very close to each other without merging, implying a potential areal density of 400 Gb/in^2. Figure 31.3c shows recent results from a single-lever experiment, where indentations are spaced as closely as 25 nm apart, resulting in areal densities of up to 1 Tb/in^2, although with a somewhat degraded write/read quality.

Imaging and reading are performed by a new thermomechanical sensing concept [31.24]. To read the written information, the heater cantilever originally used for writing is given the additional function of a thermal readback sensor by exploiting its temperature-dependent resistance. In general, the resistance increases nonlinearly with heating power/temperature from room temperature to a peak value at 500–700 °C. The peak temperature is determined by the doping concentration of the heater platform, which ranges from 1×10^{17} to 2×10^{18} at/cm^3. Above the peak temperature, the resistance drops as the number of intrinsic carriers increases due to thermal excitation [31.25]. For sensing, the resistor is operated at about 350 °C, a temperature that is not high enough to soften the polymer as in the case of writing. The principle of thermal sensing is based on the fact that the thermal conductance between the heater platform and the storage substrate changes according to the distance between them. The medium between the heater platform and the storage substrate, in our case air, transports heat from the cantilever to the substrate. When the distance between cantilever and storage substrate is reduced as the tip moves into an indentation, the heat transport through the air becomes more efficient. As a result, the evolution of the heater temperature differs in response to a pulse applied to the cantilever. In particular, the maximum value achieved by the heater temperature is higher in the absence of an indentation. As the value of the variable resistance depends on the temperature of the cantilever, the maximum value achieved by the resistance will be lower as the tip moves into an indentation. Therefore, during the read process, the cantilever resistance reaches different values depending on whether the tip moves into an indentation

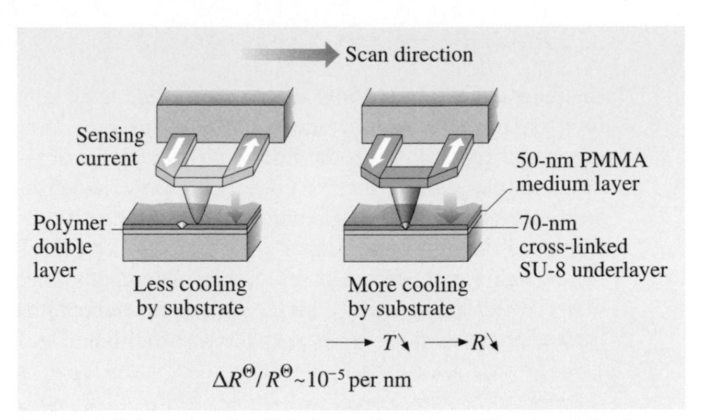

Fig. 31.4 Principle of AFM thermal sensing. The tip of the heater cantilever is continuously heated by a dc power supply while the cantilever is being scanned and the heater resistivity measured (after [31.15])

(logical bit "1") or over a region without an indentation (logical bit "0"). Figure 31.4 illustrates this concept.

Under typical operating conditions, the sensitivity of thermomechanical sensing is even greater than that of piezoresistive-strain sensing, which is not surprising because thermal effects in semiconductors are stronger than strain effects. The good sensitivity is demonstrated by the images of the 40-nm-sized indentations in Fig. 31.3, which were obtained using the described thermal-sensing technique.

The thermomechanical cantilever sensor, which transforms temperature into an electrical signal that carries information, is the electrical equivalent, to a first degree of approximation, of a variable resistance. A detection circuit must, therefore, sense a voltage that depends on the value of the cantilever resistance to decide whether a "1" or a "0" is written. The relative variation of thermal resistance is on the order of 10^{-5}/nm. Hence a written "1" typically produces a relative change of the cantilever thermal resistance $\Delta R^\Theta / R^\Theta$ of about 10^{-4} to 5×10^{-4}. Note that the relative change of the cantilever electrical resistance is of the same order of magnitude. Thus, one of the most critical issues in detecting the presence or absence of an indentation is the high resolution required to extract the signal that contains the information about the logical bit being "1" or "0". The signal carrying the information can be regarded as a small signal superimposed on a very large offset signal. The large offset problem can be mitigated by subtracting a suitable reference signal [31.12, 26, 27].

31.3 Array Design, Technology, and Fabrication

Encouraged by the results of the 5×5 cantilever array [31.16, 17], a 32×32 array chip was designed and fabricated [31.18]. With the findings from the fabrication and operation of the 5×5 array and the very dense thermomechanical writing/reading in thin polymers with single cantilevers, some important changes of the chip functionality and fabrication processes were made. The major differences are (1) surface micromachining to form cantilevers at the wafer surface, (2) all-silicon cantilevers, (3) thermal instead of piezoresistive sensing, and (4) first- and second-level wiring with an insulating layer for a multiplexed row/column-addressing scheme.

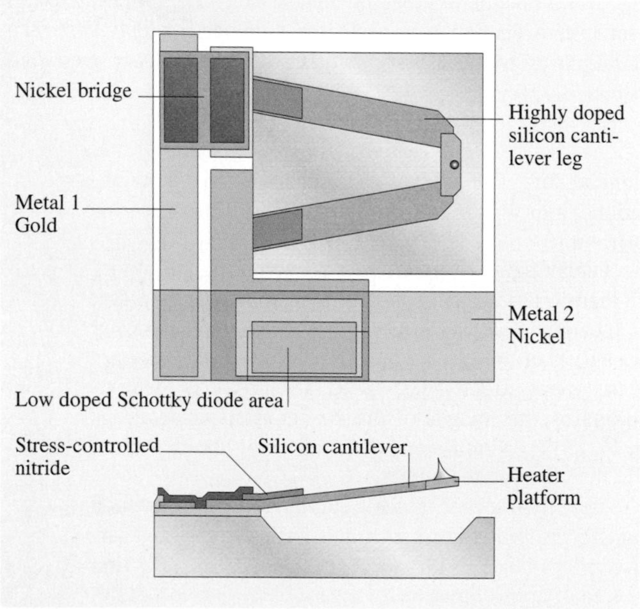

As the heater platform functions as a read/write element and no individual cantilever actuation is required, the basic array cantilever cell becomes a simple two-terminal device addressed by a multiplexed x/y wiring, as shown in Fig. 31.5. The cell area and x/y cantilever pitch are $92\times92\,\mu\mathrm{m}^2$, which results in a total array size of less than $3\times3\,\mathrm{mm}^2$ for the 1,024 cantilevers. The cantilevers are fabricated entirely of silicon for good thermal and mechanical stability. They consist of a heater platform with the tip on top, legs acting as soft mechanical springs, and electrical connections to the heater. They are highly doped to minimize interconnect resistance and to replace the metal wiring on the cantilever in order to eliminate electromigration and parasitic z-actuation of the cantilever due to a bimorph effect. The resistive ratio between the heater and the silicon interconnect sections should be as high as possible; currently the resistance of the highly doped interconnections is $\sim 400\,\Omega$ and that of the heater platform is $5\,\mathrm{k}\Omega$ (at 3 V reading bias).

Fig. 31.5 Layout and cross section of one cantilever cell (after [31.28])

Fig. 31.6 Photograph of fabricated chip ($14\times7\,\mathrm{mm}^2$). The 32×32 cantilever array is located at the center with bond pads distributed on either side (from [31.28] © 1999 IEEE)

The cantilever mass has to be minimized to obtain soft, high-resonant-frequency cantilevers. Soft cantilevers are required for a low loading force in order to eliminate or reduce tip and medium wear, whereas a high resonant frequency allows high-speed scanning. In addition, sufficiently wide cantilever legs are required for a small thermal time constant, which is partly determined by cooling via the cantilever legs [31.19]. These design considerations led to an array cantilever with 50-μm-long, 10-μm-wide and 0.5-μm-thick legs, and a 5-μm-wide, 10-μm-long and 0.5-μm-thick heater platform. Such a cantilever has a stiffness of $\sim 1\,\mathrm{N/m}$ and a resonant frequency of $\sim 200\,\mathrm{kHz}$. The heater time constant is a few microseconds, which should allow a multiplexing rate of up to 100 kHz.

The tip height should be as small as possible because the heater platform sensitivity depends strongly on the platform-to-medium distance. This contradicts the requirement of a large gap between the chip surface and the storage medium to ensure that only the tips, and not the chip surface, make contact with the medium. Instead of making the tips longer, we purposely bent the cantilevers a few micrometers out of the chip plane by depositing a stress-controlled plasma-enhanced chemical vapor deposition (PECVD) silicon-nitride layer at the base of the cantilever (see Fig. 31.5). This bending as well as the tip height must be well controlled in order to maintain an equal loading force for all cantilevers of an array.

Cantilevers are released from the crystalline Si substrate by surface micromachining using either plasma or wet chemical etching to form a cavity underneath the cantilever. Compared to a bulk-micromachined through-wafer cantilever-release process, as was done for our 5×5 array [31.16, 17], the surface micromachining technique allows an even higher array density and yields better mechanical chip stability and heat sinking. As mentioned above, the entire array is tracked without individual lateral cantilever positioning, therefore thermal expansion of the array chip has to be small or well controlled. For a 3×3-mm^2 silicon array area and 10-nm tip-position accuracy, the temperature difference between array chip and medium substrate has to be controlled to about 1 °C. This is ensured by four temperature sensors in the corners of the array and heater elements on each side of the array.

The photograph in Fig. 31.6 shows a fabricated chip with the 32×32 array located in the center (3×3 mm^2) and the electrical wiring interconnecting the array with the bonding pads at the chip periphery. Figure 31.7 shows the 32×32 array section of the chip with the independent approach/heat sensors in the four corners and the heaters on each side of the array, as well as zoomed scanning electron micrographs (SEMs) of an array section, a single cantilever, and a tip apex. The tip height is 1.7 μm and the apex radius is smaller than 20 nm, which is achieved by oxidation sharpening [31.29]. The cantilevers are connected to the column and row address lines using integrated Schottky diodes in series with the cantilevers. The diode is operated in reverse bias (high resistance) if the cantilever is not addressed, thereby greatly reducing cross talk between cantilevers. More details about the array fabrication are given in [31.28, 30].

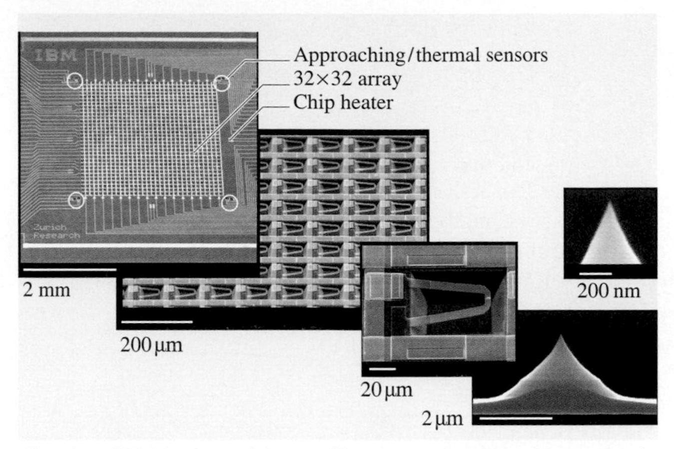

Fig. 31.7 SEM images of the cantilever array section with approaching and thermal sensors in the corners, array and single cantilever details, and tip apex (© 2000 International Business Machines Corporation; after [31.18])

31.4 Array Characterization

The array's independent cantilevers, which are located in the four corners of the array and used for approaching and leveling the chip and storage medium, serve to initially characterize the interconnected array cantilevers. Additional cantilever test structures are distributed over the wafer; they are equivalent to but independent of the array cantilevers. Figure 31.8 shows an I/V curve of such a cantilever; note the nonlinearity of the resistance. In the low-power part of the curve, the resistance increases as a function of heating power, whereas in the high-power regime, it decreases.

In the low-power, low-temperature regime, silicon mobility is affected by phonon scattering, which depends on temperature, whereas at higher power, the intrinsic temperature of the semiconductor is reached, which results in a resistivity drop owing to the increasing number of carriers [31.25]. Depending on the heater-platform doping concentration of 1×10^{17} to 2×10^{18} at/cm^3, our calculations estimate a resistance maximum at a temperature of 500 to 700 °C, respectively.

The cantilevers within the array are electrically isolated from one another by integrated Schottky diodes. As every parasitic path in the array to the cantilever addressed contains a reverse-biased diode, the cross talk current is drastically reduced, as shown in Fig. 31.9. Thus, the current response of an addressed cantilever in an array is nearly independent of the size of the array, as demonstrated by the I/V curves in Fig. 31.9. Hence, the power applied to address a cantilever is not shunted by other cantilevers, and the reading sensitivity is not degraded – not even for very large arrays (32×32). The introduction of the electrical isolation using integrated

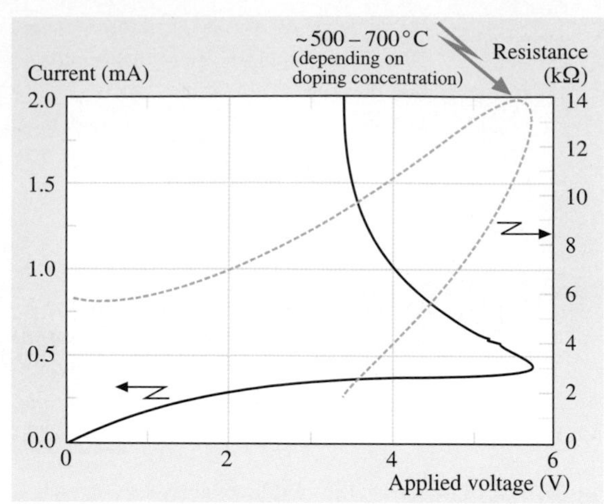

Fig. 31.8 I/V curve of one cantilever. The curve is nonlinear owing to the heating of the platform as the power and temperature are increased. For doping concentrations between 1×10^{17} and 2×10^{18} at/cm^3, the maximum temperature varies between 500 and 700 °C (after [31.30])

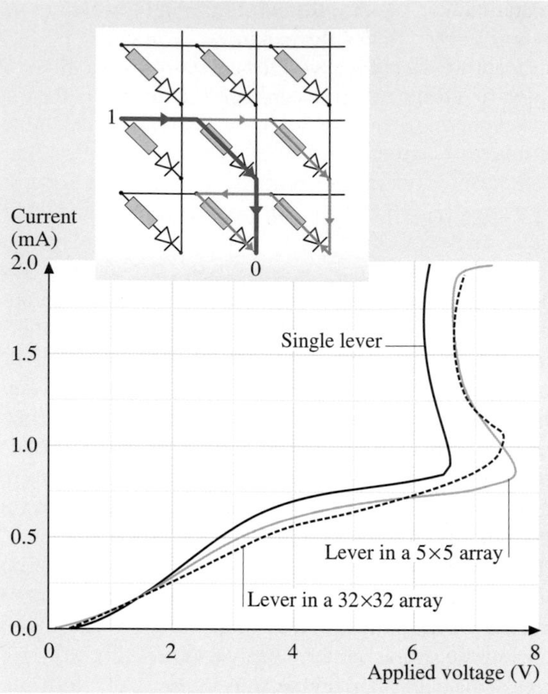

Fig. 31.9 Comparison of the I/V curve of an independent cantilever (*solid black line*) with the current response when addressing a cantilever in a 5×5 (*solid brown line*) or a 32×32 (*dashed line*) array with a Schottky diode serially connected to the cantilever. Little change is observed in the I/V curve between the different cases. Inset: sketch of the direct path (*bold line*) and a parasitic path (*thin line*) in a cantilever-diode array. In the parasitic path there is always one diode in reverse bias that reduces the parasitic current (after [31.30])

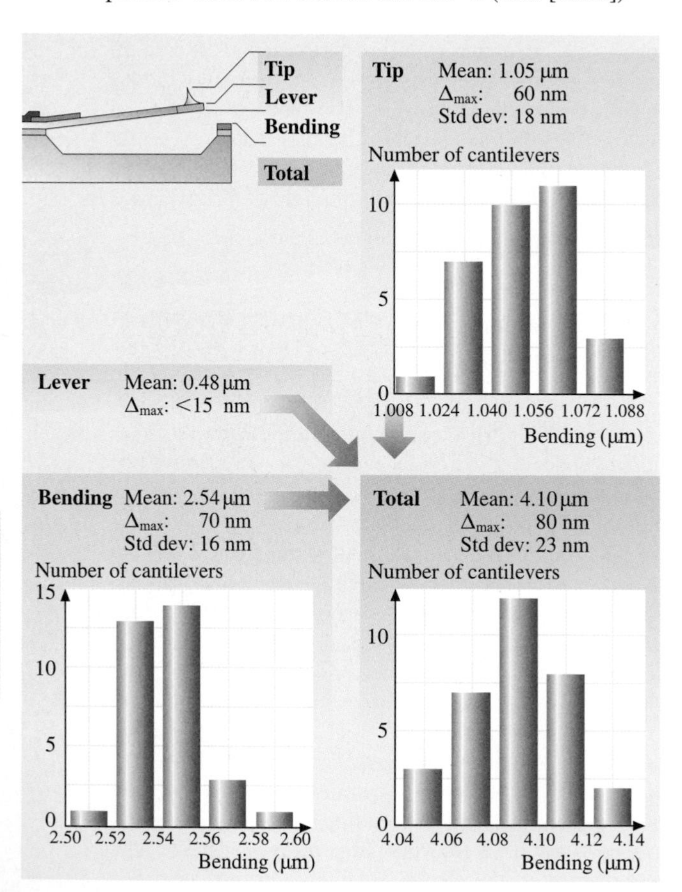

Fig. 31.10 Tip-apex height uniformity across one cantilever row of the array with individual contributions from the tip height and cantilever bending. (© 2000 International Business Machines Corporation; after [31.18])

Schottky diodes turned out to be crucial for the successful operation of interconnected cantilever arrays with a simple time-multiplexed addressing scheme.

The tip-apex height uniformity within an array is very important because it determines the force of each cantilever while in contact with the medium and hence influences write/read performance, as well as medium and tip wear. Wear investigations suggest that a tip-apex height uniformity across the chip of less than 500 nm is required [31.7], with the exact number depending on the spring constant of the cantilever. In the case of the millipede, the tip-apex height is determined by the tip height and the cantilever bending. Figure 31.10 shows the tip-apex height uniformity of one row of the array (32 tips) due to tip height and cantilever bending. It demonstrates that our uniformity is on the order of 100 nm, thus meeting the requirements.

31.5 *x*/*y*/*z* Medium Microscanner

A key issue for the millipede concept is the need for a low-cost, miniaturized scanner with $x/y/z$ motion capabilities and a lateral scanning range on the order of 100 μm. Multiple-probe systems arranged as 1-D or 2-D arrays [31.18] must also be able to control, by means of tilt capabilities, the parallelism between the probe array and the sample [31.18, 32].

We have developed a microscanner with these properties based on electromagnetic actuation. It consists of a mobile platform supported by springs and containing integrated planar coils positioned over a set of miniature permanent magnets [31.33]. A suitable arrangement of the coils and magnets allows us, by electrically addressing the various coils, to apply magnetically induced forces to the platform and drive it in the x, y, z, and tilt directions. Our first silicon/copper-based version of this device has proved the validity of the concept [31.34], and variations of it have since been used elsewhere [31.35]. However, the undamped copper spring system gave rise to excessive cross talk and ringing when driven in an open loop, and its layout limited the compactness of the overall device.

We investigate a modified microscanner that uses flexible rubber posts as a spring system and a copper-epoxy-based mobile platform, Fig. 31.11. The platform is made of a thick, epoxy-based SU-8 resist [31.36], in which the copper coils are embedded. The posts are made of polydimethylsiloxane (PDMS, Sylgard 184 silicon elastomer, Dow Corning, Midland, MI) and are fastened at the corners of the platform and at the ground plate, providing an optimally compact device by sharing the space below the platform with the magnets. The shape of the posts allows their lateral and longitudinal stiffness to be adjusted, and the dissipative rubber-like properties of PDMS provide damping to avoid platform ringing and to suppress nonlinearities.

Figure 31.12 shows the layout of the platform, which is scaled laterally, so that the long segments of the "racetrack" coils used for in-plane actuation coincide with commercially available 24 mm² SmCo magnets. The thickness of the device is determined by that of the magnets (1 mm), the clearance between magnet and platform (500 μm), and the thickness of the platform itself, which is 250 μm and determined mainly by the aspect ratio achievable in SU-8 resist during the exposure of the coil plating mold. The resulting device volume is approximately $15 \times 15 \times 1.6\,\text{mm}^3$.

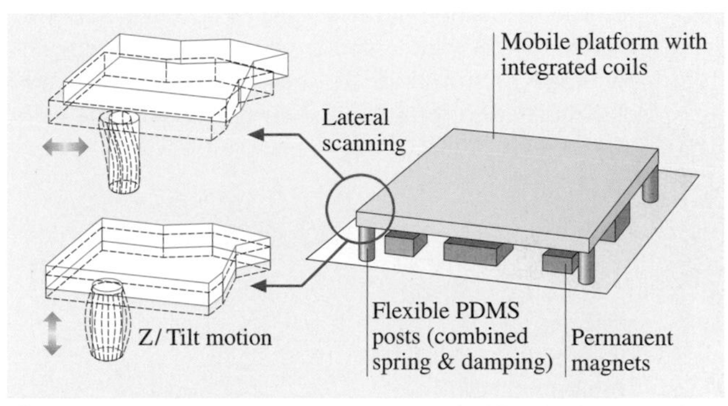

Fig. 31.11 Microscanner concept using a mobile platform and flexible posts (after [31.31])

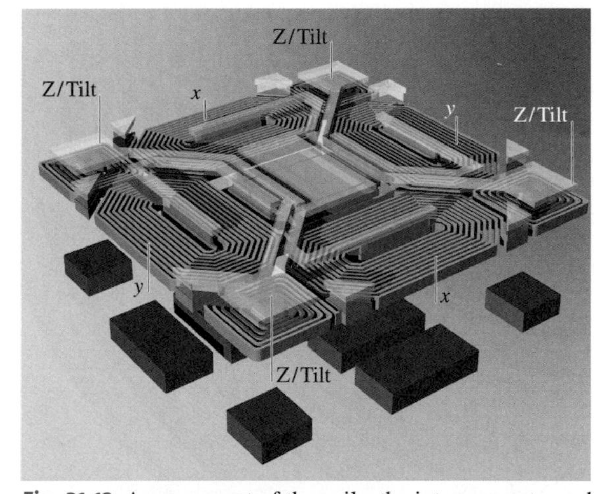

Fig. 31.12 Arrangement of the coils, the interconnects, and the permanent magnets, as well as the various motions addressed by the corresponding coils (from [31.15] © 2002 IEEE)

The SmCo magnets produce a measured magnetic field intensity of $\sim 0.14\,\mathrm{T}$ at the mid-thickness of the coils. The effective coil length is 320 mm, yielding an expected force $F_{x,y}$ of 45 μN per mA of drive current.

The principal design issue of the spring system is the ratio of its stiffnesses for in-plane and out-of-plane motion. Whereas for many scanning probe applications the required z-axis range need not be much larger than a few microns, it is necessary to ensure that the z-axis retraction of the platform due to the shortening of the posts as they take on an S-shape at large in-plane deflections can be compensated for at acceptable z-coil current levels. Various PDMS post shapes have been investigated to optimize and trade off the various requirements. Satisfactory performance was found for simple O-shapes [31.31].

The fabrication of the scanner, Fig. 31.13, starts on a silicon wafer with a seed layer and a lithographically patterned 200-μm-thick SU-8 layer, in which copper is electroplated to form the coils (Fig. 31.13a). The coils typically have 20 turns, with a pitch of 100 μm and a spacing of 20 μm. Special care was taken in the resist processing and platform design to achieve the necessary aspect ratio and to overcome adhesion and stress problems of SU-8. A second SU-8 layer, which serves as an insulator, is patterned with via holes, and another seed layer is then deposited (Fig. 31.13b). Next, an interconnect level is formed using a Novolac-type resist mask and a second copper-electroplating step (Fig. 31.13c). After stripping the resist, the silicon wafer is dissolved by a sequence of wet and dry etching, and the exposed seed layers are sputtered away to prevent shorts (Fig. 31.13d).

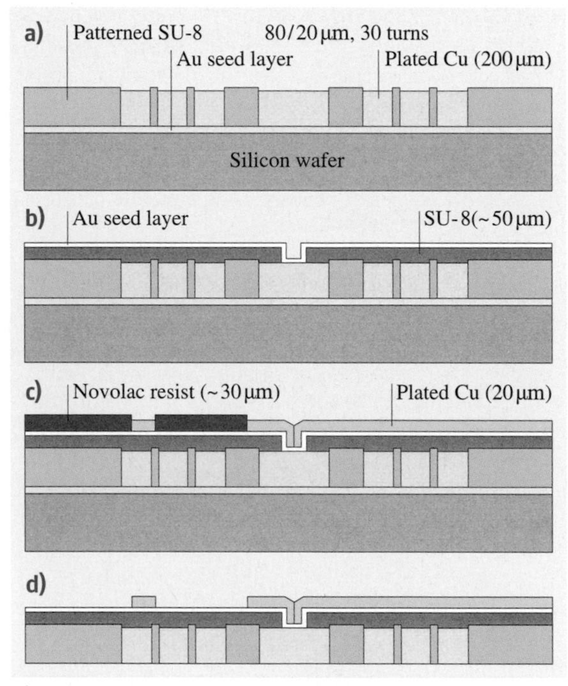

Fig. 31.13a–d Cross section of the platform fabrication process. **(a)** Coils are electroplated through an SU-8 resist mask, which is retained as the body of the platform; **(b)** an insulator layer is deposited; **(c)** interconnects are electroplated; **(d)** the platform is released from the silicon substrate (after [31.31])

The motion of the scanner was characterized using a microvision strobe technique [31.37]. The results presented below are based on O-type PDMS ports. Frequency response curves for in-plane motion (Fig. 31.14) show broad peaks (characteristic of a large degree of damping) at frequency values that are consistent with expectations based on the measured mass of the platform (0.253 mg). The amplitude response (Fig. 31.15) displays the excellent linearity of the spring system for displacement amplitudes up to 80 μm (160-μm displacement range). Based on these near-DC (10 Hz) responses (~ 1.4 μm/mA) and a measured circuit resistance of 1.9 Ω, the power necessary for a 50-μm displacement amplitude is approximately equal to 2.5 mW.

Owing to limitations of the measurement technique, it was not possible to measure out-of-plane displacements greater than 0.5 μm. However, the small-

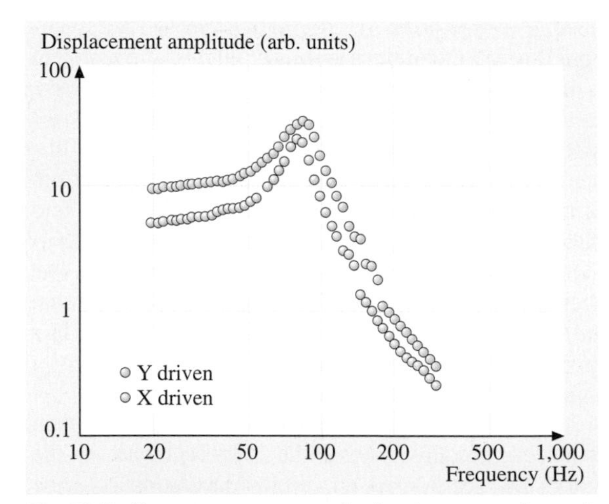

Fig. 31.14 Frequency response for in-plane x- and y-axis motion. The mechanical quality factors measured are between 3.3 and 4.6 (after [31.15])

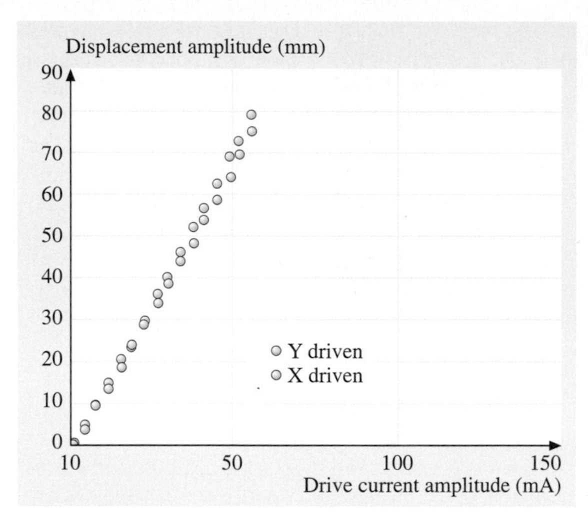

Fig. 31.15 In-plane displacement amplitude response for an ac drive current at 10 Hz (off resonance) (after [31.15])

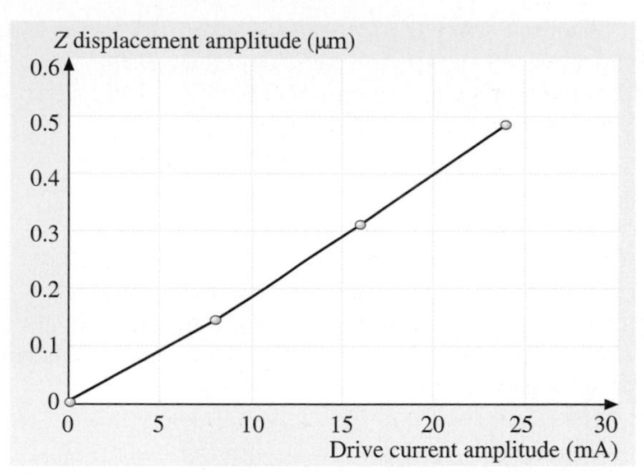

Fig. 31.16 Out-of-plane amplitude response for an ac drive current at 3 Hz. The drive current is the total for all four corner coils, which are driven in phase (after [31.31])

amplitude response for z-motion when all four corner coils are driven in-phase also displays good linearity over the range that can be measured (Fig. 31.16).

The electromagnetic scanner performs reliably and as predicted in terms of the scan range, device volume, and power requirements, achieving overall displacement ranges of 100 μm with approximately 3 mW of power. The potential access time in the 100-μm storage field is on the order of a few milliseconds. By being potentially cheap to manufacture, the integrated scanner presents a good alternative actuation system for many scanning probe applications.

31.6 First Write/Read Results with the 32×32 Array Chip

We have built a prototype that includes all the basic building blocks of the millipede concept (see Fig. 31.1) [31.38]. A 3×3-mm^2 silicon substrate is spin-coated with the SU-8/PMMA polymer medium, as described in Sect. 31.3. This storage medium is attached to the $x/y/z$ microscanner and approaching device. The magnetic z-approaching actuators bring the medium into contact with the tips of the array chip. The z-distance between the medium and the millipede chip is controlled by the approaching sensors in the corners of the array. The signals from these cantilevers are used to determine the forces on the z-actuators and, hence, the forces of the cantilever while it is in contact with the medium. This sensing/actuation feedback loop continues to operate during x/y scanning of the medium. The PC-controlled write/read scheme addresses the 32 cantilevers of one row in parallel. Writing is performed by connecting the addressed row for 20 μs to a high negative voltage and simultaneously applying data inputs ("0" or "1") to the 32 column lines. The data input is a high positive voltage for a "1" and ground for a "0". This row-enabling and column-addressing scheme supplies a heater current to all cantilevers, but only those cantilevers with high positive voltage generate an indentation ("1"). Those grounded are not hot enough to form an indentation and thus write a "0". When the scan stage has moved to the next logical mark position, the process is repeated, and this is continued until the line scan is finished. In the read process, the selected row line is connected to a moderate negative voltage, and the column lines are grounded via a protection resistor of about 10 kΩ, which keeps the cantilevers warm. During scanning, the voltages across the resistors are measured. Depending on the topography of the recording surface, the degree of cooling of each cantilever varies, thus changing the resistance and voltage across the series resistor and allowing written data to be read back.

The results of writing and reading in this fashion can be seen in Fig. 31.17, which shows 1,024 images corresponding to the 1,024 storage fields and associ-

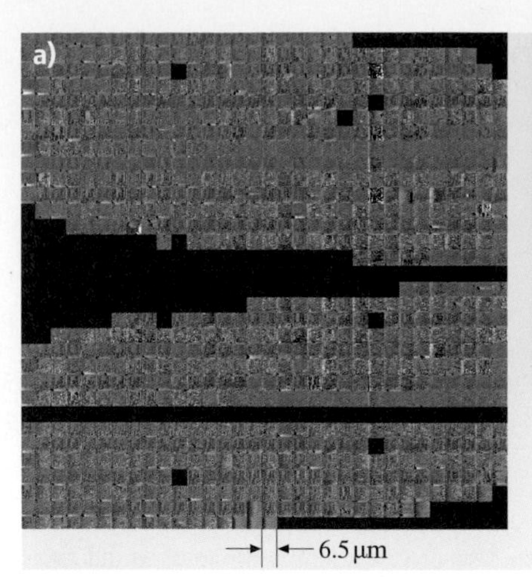

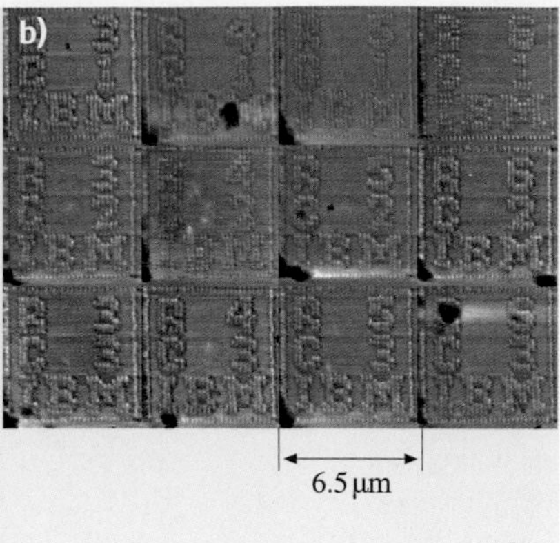

Fig. 31.17 (a) 1,024 images, each one obtained from a cantilever of the 2-D array. (b) Enlarged view of typical images from (a). The numbers in the images indicate the row and column of each lever (from [31.38])

ated cantilevers. Of the 1,024 levers, 834 were able to write and read data back, i. e., a success rate of more than 80%. The sequence is as follows. First, a bit pattern is written simultaneously by each of the levers in row 1, then read back simultaneously, followed by row 2, etc. through row 32. The data sent to the levers is different, each lever writing its own row and column number in the array. The bit pattern consists of 64×64 bits, where odd-numbered bits are always set to 0. In this case, the area used is $6.5 \times 6.5\,\mu m^2$. The readback image is a grey-scale bit map of 128×128 pixels. The distance between levers is 92 μm, so the images in Fig. 31.17 are also 92 μm apart. The areal density of the written information shown in Fig. 31.17 corresponds to $15-30\,Gb/in^2$, depending on the coding scheme adopted. More recently, an areal density of $150-200\,Gb/in^2$, at an array yield of about 60%, has been demonstrated.

Those levers that did not read back failed for one of four reasons: (i) a defective chip connector rendered an entire column unusable, (ii) a point defect occurred, meaning that a single lever or tip was broken, (iii) there was a nonuniformity of the tip contact due to tip/lever variability or storage substrate bowing due to mounting, (iv) there were thermal drifts. The latter two reasons were the most likely and major failure sources. At present, there is clearly a tradeoff between the number of working levers and the density, which will most likely be resolved by a better substrate/chip mounting technique and lower thermal drifts.

The writing and readback rates achieved with this system are 1 kb/s per lever, thus, the total data rate is about 32 kb/s. This rate is limited by the rate at which data can be transferred over the PC ISA bus, not by a fundamental time limitation of the read/write process.

31.7 Polymer Medium

The polymer storage medium plays a crucial role in millipede-like thermomechanical storage systems. The thin-film multilayer structure with PMMA as active layer (see Fig. 31.2) is not the only possible choice, considering the almost unlimited range of polymer materials available. The ideal medium should be easily deformable for writing, yet indentations should be stable against tip wear and thermal degradation. Finally, one would also like to be able to erase and rewrite data repeatedly. In order to be able to address all important aspects properly, some understanding of the basic physical mechanism of thermomechanical writing and erasing is required.

31.7.1 Writing Mechanism

In a *gedanken* experiment we visualize writing of an indentation as the motion of a rigid body (the tip) in a viscous medium (the polymer melt). Let us initially assume that the polymer, i. e., PMMA, behaves like a simple liquid after it has been heated above the glass-transition temperature in a small volume around the tip. As viscous drag forces must not exceed the loading force applied to the tip during indentation, we can estimate an upper bound for the viscosity ζ of the polymer melt using Stokes's equation:

$$F = 6\pi\zeta\varrho v\,. \tag{31.1}$$

In actual indentation formation, the tip loading force is on the order of $F = 50\,\mathrm{nN}$ and the radius of curvature at the apex of the tip is typically $\varrho = 20\,\mathrm{nm}$. Assuming a depth of the indentation of, say, $h = 50\,\mathrm{nm}$ and a heat pulse of $\tau_h = 10\,\mu\mathrm{s}$ duration, the mean velocity during indentation formation is on the order of $v = h/\tau_h = 5\,\mathrm{mm/s}$. Note that thermal relaxation times are on the order of microseconds [31.20, 21] and, hence, the heating time can be equated to the time it takes to form an indentation. With these parameters we obtain $\zeta < 25\,\mathrm{Pa\,s}$, whereas typical values for the shear viscosity of PMMA are at least seven orders of magnitude larger even at temperatures well above the glass-transition point [31.39].

This apparent contradiction can be resolved by considering that polymer properties are strongly dependent on the time scale of observation. At time scales on the order of 1 ms and below, entanglement motion is in effect frozen in and the PMMA molecules form a relatively static network. Deformation of the PMMA now proceeds by means of uncorrelated deformations of short molecular segments, rather than by a flow mechanism involving the coordinated motion of entire molecular chains. The price one has to pay is that elastic stress builds up in the molecular network as a result of the deformation (the polymer is in a so-called rubbery state). On the other hand, corresponding relaxation times are orders of magnitude smaller, giving rise to an effective viscosity at millipede time scales on the order of $10\,\mathrm{Pa\,s}$ [31.39], as required by our simple argument (see (31.1)). Note that, unlike normal viscosity, this high-frequency viscosity is basically independent of the detailed molecular structure of the PMMA, i. e., chain length, tacticity, polydispersity, etc. In fact, we can even expect that similar high-frequency viscous properties can be found in a large class of other polymer materials, which makes thermomechanical writing a rather robust process in terms of material selection.

We have argued above that elastic stress builds up in the polymer film during the formation of an indentation, creating a corresponding reaction force on the tip on the order of $F_r \sim 2\pi G\varrho^2$, where G denotes the elastic shear modulus of the polymer [31.40]. An important property for millipede operation is that the shear modulus drops by several orders of magnitude in the glass-transition regime, i. e., for PMMA from $\sim 1\,\mathrm{GPa}$ below Θ_g to $\sim 0.5{-}1\,\mathrm{MPa}$ above Θ_g, where Θ_g denotes the glass-transition temperature [31.39]. The bulk modulus, on the other hand, retains its low-temperature value of several GPa. Hence, in this elastic regime, formation of an indentation above Θ_g constitutes a volume-preserving deformation. For proper indentation formation, the tip load must be balanced between the extremes of the elastic reaction force F_r for temperatures below and above Θ_g, i. e., $F \ll 2.5\,\mu\mathrm{N}$ for PMMA to prevent indentation of the polymer in the cold state and $F \gg 2.5\,\mathrm{nN}$ to overcome the elastic reaction force in the hot state. Unlike the deformation of a simple liquid, the indentation represents a metastable state of the entire deformed volume, which is under elastic tension. Recovery of the unstressed initial state is prevented by rapid quenching of the indentation below the glass temperature with the tip in place. As a result, the deformation is frozen in, because below Θ_g motion of molecular-chain segments is, in effect, inhibited (see Fig. 31.18).

This mechanism also allows indentations to be erased locally – it suffices to heat the deformed volume locally above Θ_g, whereupon the indented volume reverts to its unstressed flat state driven by internal elastic stress. In addition, erasing is promoted by surface tension forces, which give rise to a restoring surface pressure on the order of $\gamma(\pi/\varrho)^2 h \approx 25\,\mathrm{MPa}$, where $\gamma \sim 0.02\,\mathrm{N/m}$ denotes the polymer-air surface tension.

One question immediately arises from these speculations: If the polymer behavior can be determined from the macroscopic characteristics of the shear modulus as a function of time, temperature, and pressure, can the time-temperature superposition principle also be applied in this case? The time-temperature superposition principle is a powerful concept of polymer physics [31.41]. It basically states that the time scale and the temperature are interdependent variables that determine the polymer behavior such as the shear modulus. A simple transformation can be used to translate time-dependent into temperature-dependent data and vice versa. It is not clear, however, whether this principle can be applied in our case, i. e., under such extreme conditions (high pressures, short time scales, and nanometer-sized volumes, which are clearly below the radius of gyration of individual polymer molecules).

To test this, we varied the heating time, the heating temperature, and the loading force in indentation-writing experiments on a standard PMMA sample. The results

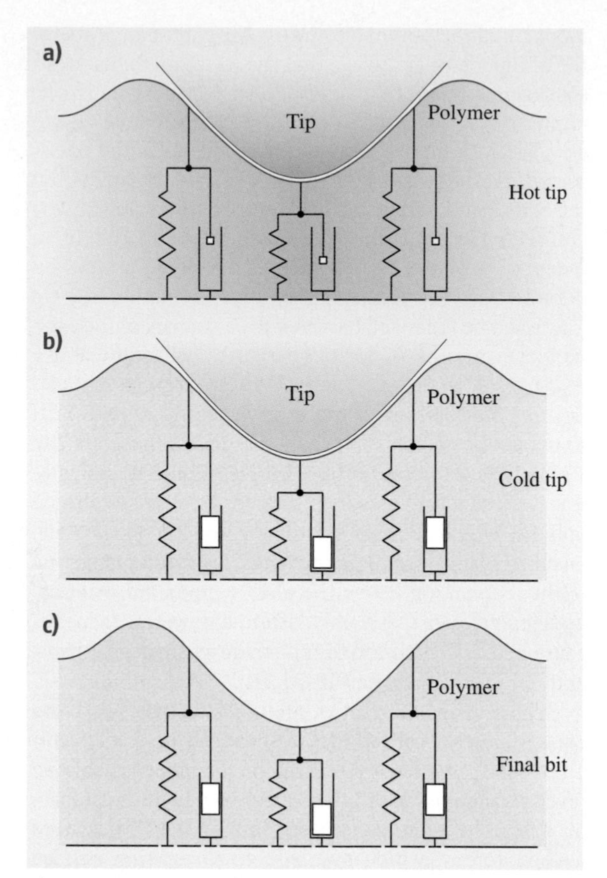

Fig. 31.18a–c Viscoelastic model of indentation writing. **(a)** The hot tip heats a small volume of polymer material to more than Θ_g. The shear modulus of the polymer drops drastically from GPa to MPa, which in turn allows the tip to indent the polymer. In response, elastic stress (represented as compression springs) builds up in the polymer. In addition, viscous forces (represented as pistons) associated with the relaxation time for the local deformation of molecular segments limit the indentation speed. **(b)** At the end of the writing process, the temperature is quenched on a microsecond time scale to room temperature: The stressed configuration of the polymer is frozen-in (represented by the locked pistons). **(c)** The final indentation corresponds to a metastable configuration. The original unstressed flat state of the polymer can be recovered by heating the indentation volume to more than Θ_g, which unlocks the compressed springs (after [31.15])

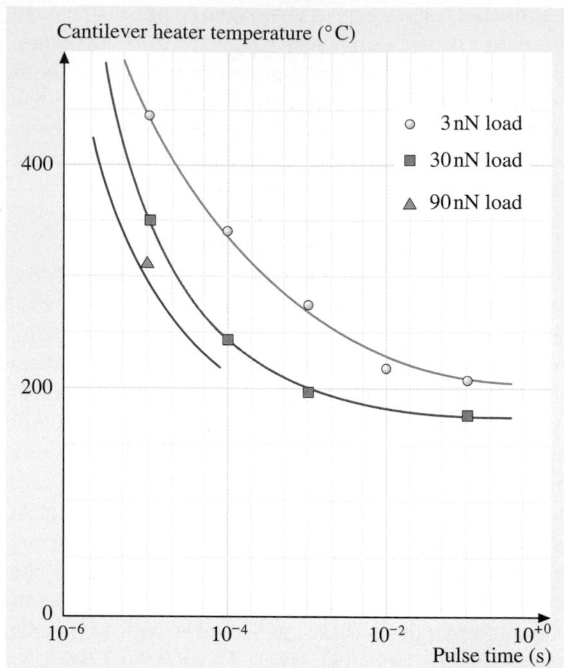

Fig. 31.19 Indentation-writing threshold measurements. The load was controlled by pushing the cantilever/tip into the sample with a controlled displacement and a known spring constant of the cantilever. When a certain threshold is reached, the indentations become visible in subsequent imaging scans (see also Fig. 31.21). The *solid lines* are guides to the eye. Curves of similar shape would be expected from the time-temperature superposition principle (after [31.15])

are summarized in Fig. 31.19. The minimum heater temperature at which the formation of an indentation starts for a given heating-pulse length and loading force was determined. This so-called threshold temperature is plotted against the heating-pulse length. A careful calibration of the heater temperature has to be done to allow a comparison of the data. The heater temperature was determined by assuming proportionality between temperature and electrical power dissipated in the heater resistor at the end of the heating pulse when the tip has reached its maximum temperature. An absolute temperature scale is established using two well-defined reference points. One is room temperature, corresponding to zero electrical power. The other is provided by the point of turnover from positive to negative differential resistance (see Fig. 31.8), which corresponds to a heater temperature of approximately 550 °C. The general shape of the measured threshold temperature versus heating time curves indeed shows the characteristics of time-temperature superposition. In particular, the curves are identical up to a load-dependent shift with respect to the time axis. Moreover, we observe that the time it takes to form an indentation at constant heater temperature is inversely proportional to the tip load. This property is exactly what one would expect if internal friction (owing to the high-frequency viscosity) is the rate-limiting step in forming an indentation (see (31.1)).

The time it takes to heat the indentation volume of polymer material higher than the glass-transition temperature is another potentially rate-limiting step. Here, the spreading resistance of the heat flow in the polymer and the thermal contact resistance are the most critical parameters. Simulations suggest [31.20, 21] that equilibration of temperature in the polymer occurs within less than 1 μs. Very little is known, however, about the thermal coupling efficiency across the tip-polymer interface. We have several indications that the heat transfer between tip and sample plays a crucial role, one of them being the asymptotic heater temperature for long writing times, which according to the graph in Fig. 31.19 is approximately 200 °C. The exact temperature of the polymer is unknown. However, the polymer temperature should approach the glass-transition temperature (around 120 °C for PMMA) asymptotically. Hence, the temperature drop between heater and polymer medium is substantial. Part of the temperature difference is due to a temperature drop along the tip, which according to heat-flow simulations [31.20, 21] is expected to be on the order of 30 °C at most. Therefore, a significant temperature gradient must exist in the tip-polymer contact zone. Further experiments on the heat transfer from tip to surface are needed to clarify this point.

We also find that the heat transfer for a nonspherical tip is anisotropic. As shown in Fig. 31.20, in the case of a pyramid-shaped tip, the indentation not only exhibits sharp edges, but also the region around the indentation, where polymer material is piled up, is anisotropic. The pile-up characteristics will be discussed in detail below. At this point we take it as an indication of the relevance of the heat transfer to the measurements.

One of the most striking conclusions of our model of the indentation-writing process is that it should in principle work for most polymer materials. The general behavior of the mechanical properties as a function of temperature and frequency is similar for all polymers [31.41]. The glass-transition temperature Θ_g would then be one of the main parameters that determines the threshold writing temperature.

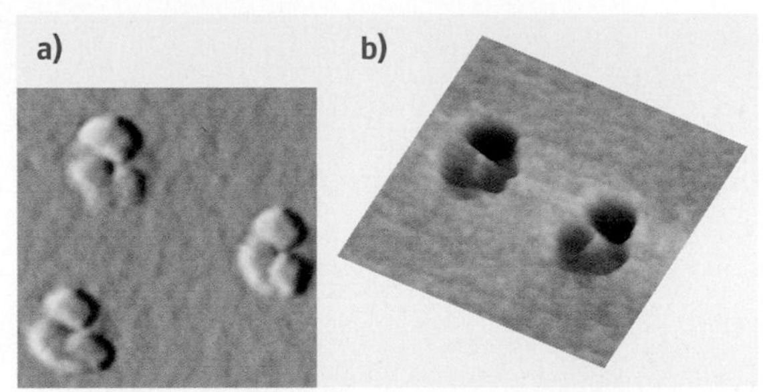

Fig. 31.20a,b Topographic image of individual indentations. **(a)** The region around the actual indentations clearly shows the threefold symmetry of the tip, here a three-sided pyramid. **(b)** The indentations themselves exhibit sharp edges, as can be seen from the inverted 3-D image. Image size is $2\times2\,\mu m^2$ (from [31.15] © 2002 IEEE)

A verification of this was found experimentally by comparing various polymer films. The samples were prepared in the same way as the PMMA samples discussed above [31.18]: by spin-casting thin films (10–30 nm) onto a silicon wafer with a photo-resist buffer. Then, threshold measurements were done by applying heat pulses with increasing current (or temperature) to the tip while the load and the heating time were held constant (load about 10 nN and heating time 10 μs). Examples of such measurements are shown in Fig. 31.21, where the increasing size and depth of indentations can be seen for different heater temperatures. A threshold can be defined based on such data and compared with the glass-transition temperature of these materials. The results show a clear correlation between the threshold heater temperature and the glass-transition temperature (see Fig. 31.22).

With our simple viscoelastic model of writing we are able to formulate a set of requirements that potential candidate materials for millipede data storage have to fulfill. First, the material should ideally exhibit a well-defined glass-transition point with a large drop of the shear modulus at Θ_g. Second, a rather high value of Θ_g (on the order of 150 °C) is preferred to facilitate thermal read back of the data without destroying the information. We have investigated a number of materials to explore the Θ_g parameter space. The fact that all polymer types tested are suitable for forming small indentations leaves us free to choose which polymer type to optimize in terms of the technical requirements for a device such as lifetime of indentations, polymer endurance of the read and write process, power consumption, etc. These are subjects of ongoing research.

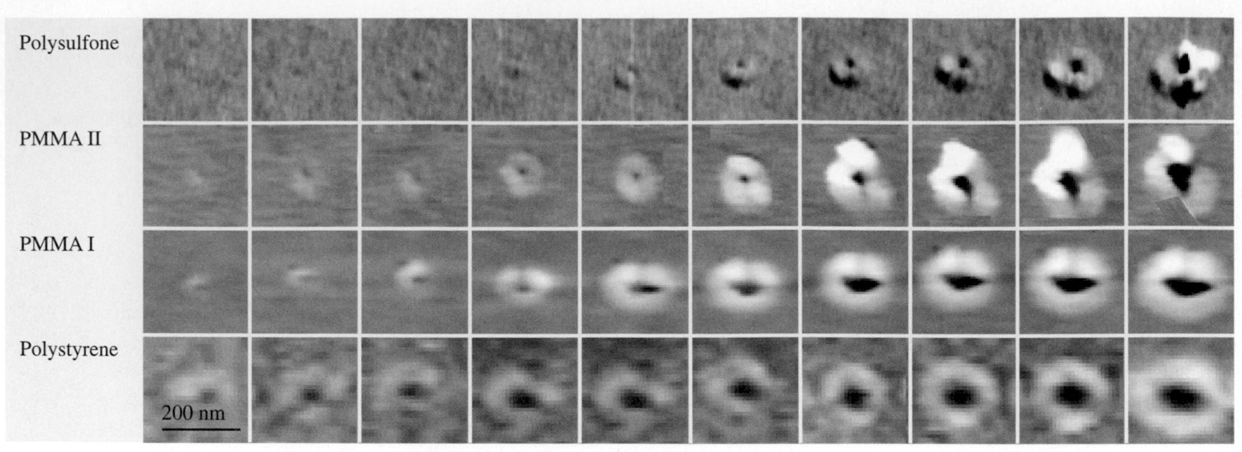

Fig. 31.21 Written indentations for different polymer materials. The heating pulse length was 10 μs, the load about 10 nN. The grey scale is the same for all images. The heater temperatures for the indentation on the left-hand side are 445, 400, 365, and 275 °C for the polymers Polysulfone, PMMA II (anionically polymerized PMMA, $M \approx 26\,\mathrm{k}$), PMMA I (Polymer Standard Service, Germany, $M \approx 500\,\mathrm{k}$), and Polystyrene, respectively. The temperature increase between events on the horizontal axis is 14, 22, 20, and 9 °C, respectively (from [31.15] © 2002 IEEE)

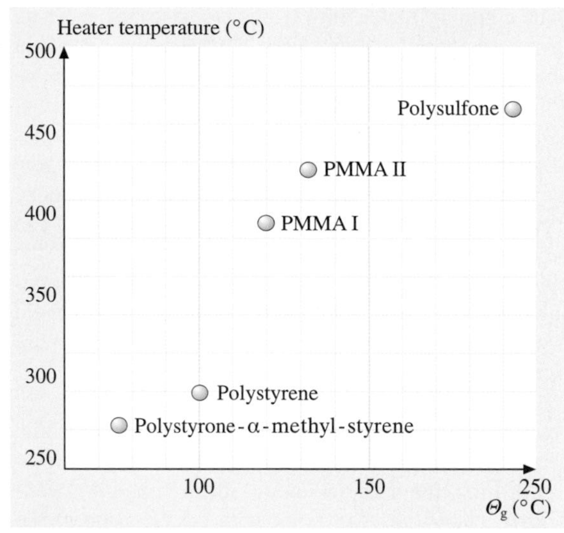

Fig. 31.22 The heater temperature threshold for writing indentations with the same parameters as in Fig. 31.21 is plotted against the glass-transition temperature for these polymers, including poly-α-methyl-styrene (after [31.15])

31.7.2 Erasing Mechanism

It is worthwhile to look at the detailed shapes of the written indentations. The polymer material around an indentation appears piled up as can be seen, for example, in Fig. 31.20. This is not only material that was pushed aside during indentation formation as a result of volume conservation. Rather, the flash heating by the tip and subsequent rapid cooling result in an increase of the specific volume of the polymer. This phenomenon, that the specific volume of a polymer can be increased by rapidly cooling a sample through the glass transition [31.41], is well-known. Our system allows a cooling time on the order of microseconds, which is much faster than the highest rates that can be achieved with standard polymer-analysis tools. However, a quantitative measurement of the specific volume change cannot be easily performed with our type of experiments. On the other hand, the pile-up effect serves as a convenient threshold thermometer. The outer perimeter of the pile-ups surrounding the indentations corresponds to the Θ_g isotherm, and the temperature in the enclosed area has certainly reached values greater than Θ_g during the indentation process. Based on our viscoelastic model, one would thus conclude that previously written indentations that overlap with the piled up region of a subsequently written indentation should be erased.

That this pile-up effect actually works against the formation of an indentation can clearly be seen in the line scans of a series of indentations written in Polysulfone (Fig. 31.23). Here, the heating of the tip was accompanied by a rather high normal force. The force was high enough to create a small indentation, even if the tip was too cold to modify the polymer (Fig. 31.23a).

Part E | 31.7

Then, with increasing tip heating, the indentations initially fill up in the piled up region (Fig. 31.23b) before they finally become deeper (Fig. 31.23c).

The pile-up phenomenon turns out to be particularly beneficial for data-storage applications. The following example demonstrates the effect. If we look at the sequence of images in Fig. 31.24, taken on a standard PMMA sample, we find that the piled up regions can overlap each other without disturbing the indentations. However, if the piled up region of an individual writing event extends over the indented area of a previously written "1", the depth of the corresponding indentation decreases markedly (Fig. 31.24d). This can be used for erasing written data. On the other hand, if the pitch between two successive indentations is decreased even further, this erasing process will no longer work. Instead, a broader indentation is formed, as shown in Fig. 31.24e. Hence, to exclude mutual interference, the minimum pitch between successive indentations, which we denote by minimum-indentation pitch (BP_{min}), must be larger than the radius of the piled up area around an indentation.

In the example shown in Fig. 31.24, the temperature chosen was so high that the ring around the indentations was very large, whereas the depth of the indentation was limited by the stop layer underneath the PMMA material. Clearly, the temperature was too high here to form small indentations, the minimum pitch of which is around 250 nm. However, by carefully optimizing all parameters it is possible to achieve areal densities of up to 1 Tb/in^2, as demonstrated in Fig. 31.3c.

The new erasing scheme based on this volume effect switches from writing to erasing merely by decreasing the pitch of writing indentations. This can be done in a very controlled fashion, as shown in Fig. 31.25, where individual lines or predefined subareas are erased. Hence, this new erasing scheme can be made to work in a way that is controlled on the scale of individual indentations. Compared with earlier global erasing schemes [31.23], this simplifies erasing significantly.

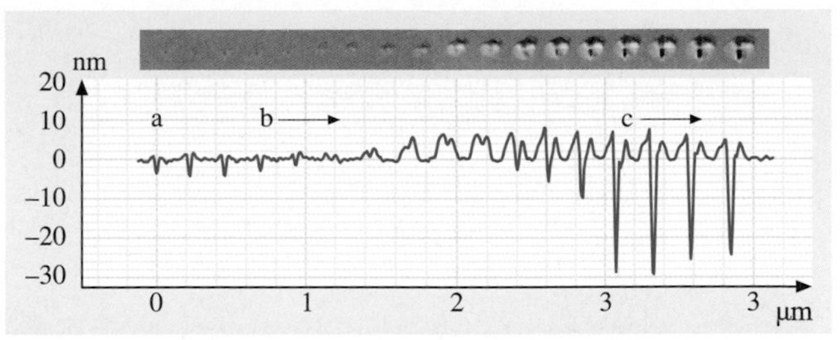

Fig. 31.23 Section through a series of indentations similar to Fig. 31.21. Here, a load of about 200 nN was applied before a heating pulse of 10-µs length was fired. The temperature of the heater at the end of the pulse has been increased from 430 to 610 °C in steps of about 10.6 °C. (a) The load was sufficient to form a plastic indentation even if the polymer was not heated enough to come near the glass transition. (b) With increasing heater temperature, the polymer swells. This eliminates the indentation, thus erasing previously written "cold" marks. (c) As this process continues, the thermomechanical formation of indentations begins to dominate until, finally, normal thermomechanical indentation writing occurs (after [31.15])

31.7.3 Overwriting Mechanism

Overwriting data on some part of the storage medium can be achieved by first erasing the entire area and then writing the desired data on the erased surface. Although this process works well, it is time-consuming and dissipates a significant amount of power. In a millipede-based storage device, where data rate and power consumption are at a premium, such a two-step overwriting mech-

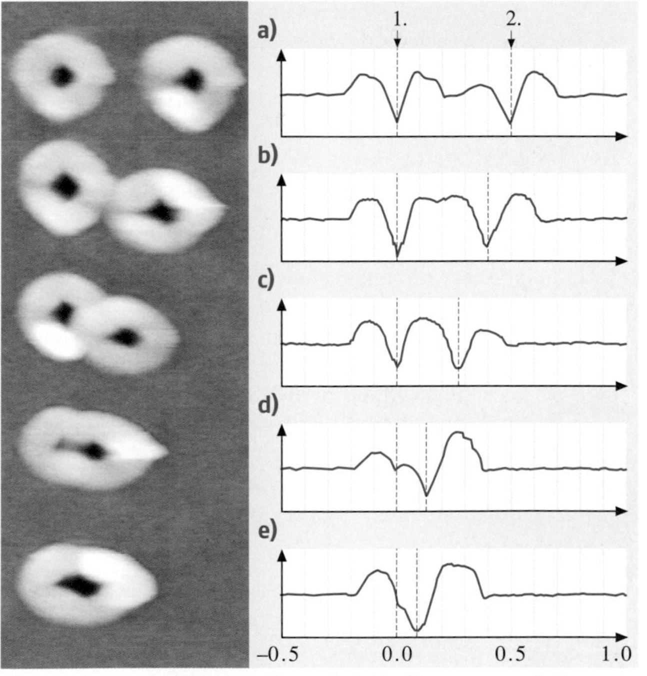

Fig. 31.24a–e Indentations in a PMMA film at several distances. The depth of the indentations is ~15 nm, roughly the same as the thickness of the PMMA layer. The indentations on the left-hand side were written first, then a second series of indentations was made with decreasing distance from the first series going from (a) to (e) (after [31.15])

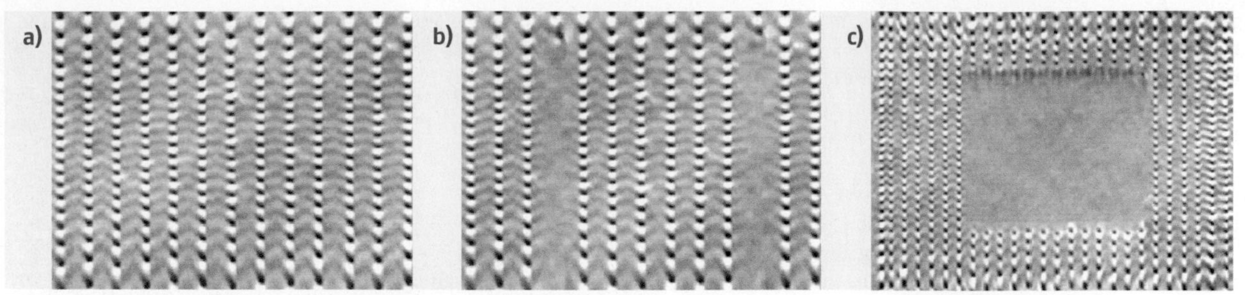

Fig. 31.25a–c Demonstration of the new erasing scheme: (a) A bit pattern recorded with variable pitch in the vertical axis (fast scan axis) and constant pitch in the horizontal direction (slow scan axis) was prepared. (b) Then two of the lines were erased by decreasing the pitch in the vertical direction by a factor of three, showing that the erasing scheme works for individual lines. One can also erase entire fields of indentations without destroying indentations at the edges of the fields. This is demonstrated in (c), where a field has been erased from an indentation field similar to the one shown in (a). The distance between the lines is 70 nm (from [31.15] © 2002 IEEE)

anism may be impractical. Instead, a one-step, direct overwriting process similar to those applied in magnetic hard-disk drives and rewritable optical drives is desired.

As discussed above, switching from writing to erasing may be achieved by decreasing the pitch of writing. It has been found experimentally that erasing can be performed effectively by halving the pitch of writing successive indentations, which is denoted as BP, provided the condition $\text{BP} \approx \text{BP}_{\text{min}}$ is satisfied. This suggests that the basic distance unit for combined write-erase operations should be BP/2. Written indentations are spaced n units apart, where $n \geq 2$. Let us recall that the presence of an indentation corresponds to a logical bit "1" and the absence of an indentation to a logical bit "0". Logical bits are then stored in the medium at the points of a regular lattice with minimum distance between points equal to BP/2 in the on-track direction, with successive "1"s separated by at least one "0". This condition is necessary in order to avoid mutual interference between successive "1"s. It is also the basis for an important category of codes known as (d, k) codes, which are described in Sect. 31.9.3. Coding can thus be used to enable direct data overwriting in an elegant way. Direct overwriting requires the simultaneous realization of two conditions: If previously written "1"s exist where "0"s are to be written, then these "1"s have to be *erased*. On the other hand, if "0"s exist where "1"s are to be written, then these "1"s have to be *written*. Writing an indentation is performed thermomechanically as described above. Erasing an existing indentation is done by writing another indentation next to it, at a distance of BP/2 units. However, as this operation creates a new indentation shifted by BP/2 with respect to the one erased, the erasing process must be performed repeatedly until the newly created indentation lies at a position corresponding to a "1" in the new data pattern. The basic principle of erasing is illustrated in Fig. 31.26a and b. The figures show how the four bit strings 001, 010, 100, and 101 are modified into the string 010. Figure 31.27 depicts the results of a rewriting experiment; the top track shows a prestored sequence, which is to be overwritten by another sequence, shown on the bottom track for comparison. The result of direct

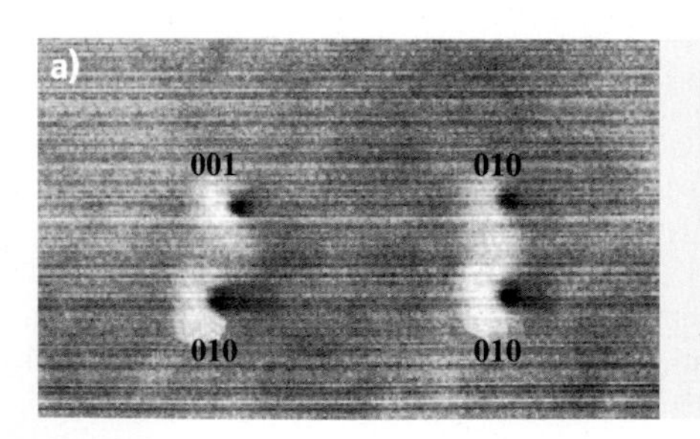

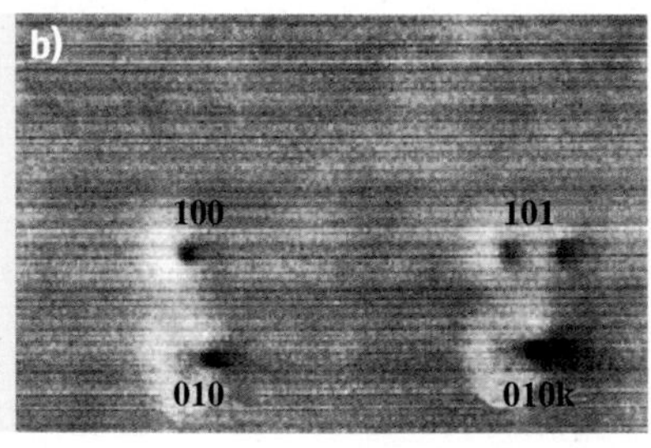

Fig. 31.26a,b Bit strings (a) 001 and 010, and (b) 100 and 101, overwritten to 010

overwriting of the prestored sequence is shown on the middle track.

Comparison with the sequence on the bottom track, which is written on a clean surface, illustrates the effectiveness of the proposed procedure. Although the write/read quality of overwritten data is somewhat inferior to that of data written on a clean storage surface, detection of the newly written sequence is not affected. However, repeated overwriting may further degrade the quality of stored data. As the extent and rate of degradation are important characteristics of a storage system, this remains an area of ongoing investigation.

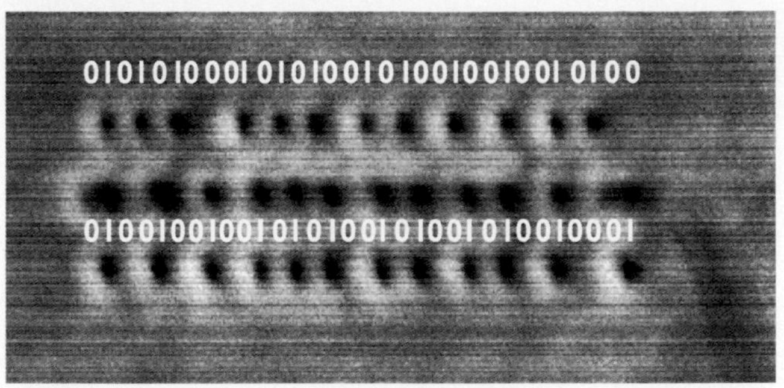

Fig. 31.27 Experimental result of overwriting a bit sequence

31.8 Read Channel Model

Let us now consider the readback channel for a single cantilever that is scanning a storage field where bits are written as indentations or absence of indentations in the storage medium. As discussed above, a cantilever is modeled as a variable resistor that depends on the temperature at the cantilever tip. The model of the read channel, used for the design and analysis of the detection system, is illustrated in Fig. 31.28 [31.12, 26, 27].

To evaluate the evolution of the temperature of a heated cantilever during the read process, we resort to a simple RC-equivalent thermal circuit, illustrated in Fig. 31.29, where $(1+\eta_x)R^{\Theta}$ and C^{Θ} denote the thermal resistance and capacitance, respectively. The parameter $\eta_x = \Delta R^{\Theta}(x)/R^{\Theta}$ indicates the relative variation of thermal resistance that results from the small change in air-gap width between the cantilever and the storage medium. The subscript x indicates the distance in the direction of scanning from the initial point. Therefore, the parameter η_x will assume the largest absolute value when the tip of the cantilever is located at the center of an indentation. The heating power that is dissipated in the cantilever heater region is expressed as

$$P^e[t, \Theta(t,x)] = \frac{V_C^2(t)}{R^e[\Theta(t,x)]} , \qquad (31.2)$$

where $V_C(t)$ is the voltage across the cantilever, $\Theta(t,x)$ is the cantilever temperature, and $R^e[\Theta(t,x)]$ is the temperature-dependent cantilever resistance.

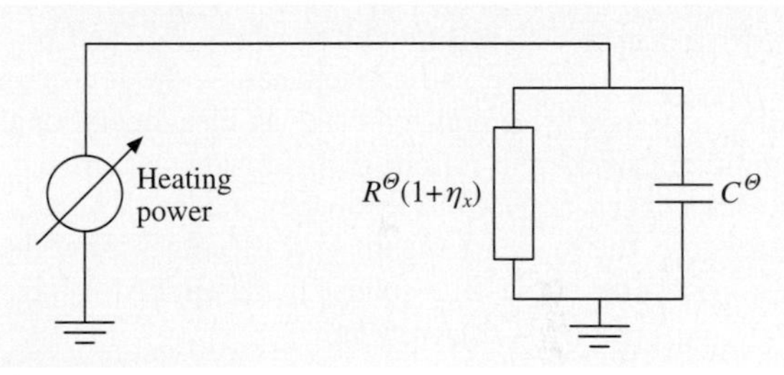

Fig. 31.29 RC-equivalent thermal model of the heat transfer process (after [31.12])

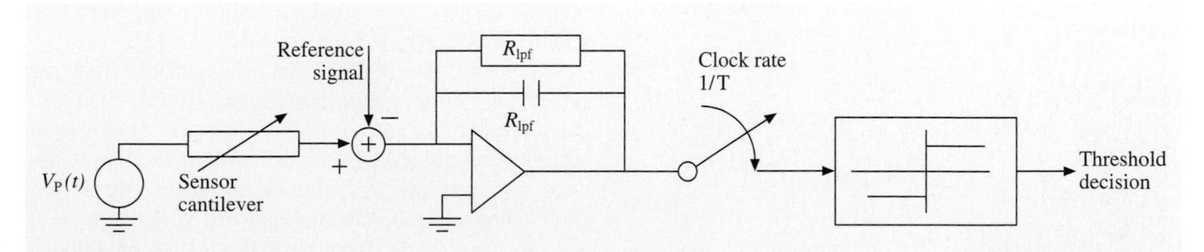

Fig. 31.28 Block diagram of the detection circuit

As the heat-transfer process depends on the value of thermal resistance and on the read pulse waveform, the cantilever temperature $\Theta(t, x)$ depends on time t and distance x. However, because the time it takes for the cantilever to move from the center of one logical mark to the next is greater than the duration of a read pulse, we assume that $\Theta(t, x)$ does not vary significantly as a function of x while a read pulse is being applied, and that it decays to the ambient temperature Θ_0 before the next pulse is applied. Therefore, the evolution of the cantilever temperature in response to a pulse applied at time $t_0 = x_0/v$ at a certain distance x_0 from the initial point of scanning and for a certain constant velocity v of the scanner obeys a differential equation expressed as

$$\Theta'(t, x_0) + \frac{1}{(1+\eta_{x_0})R^{\Theta}C^{\Theta}}[\Theta(t, x_0) - \Theta_0] = \frac{1}{C^{\Theta}} \frac{V_C^2(t)}{R^e[\Theta(t, x_0)]}, \tag{31.3}$$

where $\Theta'(t, x_0)$ denotes the derivative of $\Theta(t, x_0)$ with respect to time.

With reference to the block diagram of the read channel illustrated in Fig. 31.28, the source generates the read pulse $V_P(t)$ applied to the cantilever. Clearly, because of the virtual ground at the operational amplifier input, the voltage $V_C(t)$ across the cantilever variable resistance is equal to $V_P(t)$. Furthermore, the active low-pass RC detector filter, where R_{lpf} and C_{lpf} denote the resistance and capacitance of the low-pass filter, respectively, is realized using an ideal operational amplifier that exhibits infinite input impedance, zero output impedance, and infinite frequency-independent gain. Therefore, the readback signal $V_o(t, x_0)$ obtained at the low-pass filter output in response to the applied voltage $V_P(t) = A\,\mathrm{rect}\,[(t - T_0)/\tau]$, where

$$\mathrm{rect}\left(\frac{t}{\tau}\right) = \begin{cases} 1 & \text{if } 0 \le t \le \tau \\ 0 & \text{otherwise} \end{cases}, \tag{31.4}$$

and A denotes the pulse amplitude, obeys the differential equation

$$V_o'(t, x_0) = \frac{1}{R_{lpf}C_{lpf}}\left[-V_o(t, x_0) + \frac{R_{lpf}}{R^e[\Theta(t, x_0)]}V_P(t)\right]. \tag{31.5}$$

As the voltage at the output of the low-pass filter depends on the value of the variable resistance $R^e[\Theta(t, x_0)]$, the readback signal is determined by solving jointly the differential equations (31.3) and (31.5), with initial conditions $\Theta(t_0, x_0) = \Theta_0$ and $V_o(t_0, x_0) = 0$. For example, a comparison between experimental and synthetic readback signals is shown in Figs. 31.30 and 31.31 for a time constant of the low-pass filter $\tau_{lpf} = 1.18\,\mu\mathrm{s}$ and two values of the duration of the applied rectangular pulse. For a given cantilever design the function $R^e(\Theta)$ is determined experimentally. Finally, the parameters R^{Θ} and C^{Θ} used in the simple readback channel model are obtained via simulated annealing, where the cost function is given by the mean-square error between experimental and synthetic signals at the low-pass filter output.

Assuming that ideal control of the scanner is performed such that the time of application of a read pulse corresponds either to the cantilever located at the center of an indentation for detecting a "1" bit, or away from an indentation for detecting a "0" bit, two possible responses are obtained at the output of the low-pass filter as solutions of (31.3) and (31.5), which we denote as $V_{o,1}(t, x_0)$ and $V_{o,0}(t, x_0)$, respectively. By sampling the readback signal at the instant $t_s = t_0 + \tau$, simple threshold detection may in principle be applied to detect a written bit, where the value of the threshold is given by

$$V_{Th} = \frac{1}{2}\left[V_{o,1}(t_s, x_0) + V_{o,0}(t_s, x_0)\right]. \tag{31.6}$$

As mentioned in Sect. 31.2, one of the most critical issues in detecting the presence or absence of an indentation is the high resolution required to extract the small signal $V_{o,1}(t_s, x_0) - V_{o,0}(t_s, x_0)$ that contains the information about the bit being "1" or "0" superimposed on the offset signal $V_{o,0}(t, x_0)$. As illustrated in Fig. 31.28, this problem can be solved by subtracting a suitable reference signal $V_{o,ref}(t, x_0)$ from the readback signal. The readback signal is thus given by

$$\tilde{V}_o(t, x_0) = V_o(t, x_0) - V_{o,ref}(t, x_0), \tag{31.7}$$

and the threshold is set at $\tilde{V}_{Th} = \frac{1}{2}\left[V_{o,1}(t_s, x_0) - V_{o,0}(t_s, x_0)\right]$. A VLSI implementation of the detection scheme analyzed here is presented in [31.42].

Now consider read pulses of duration τ that are periodically applied at instants $t_n = nT$, where $1/T$ denotes the symbol rate. Assuming that the response of the previous pulse has vanished as a new pulse is applied and that the temperature of the cantilever has approached the ambient temperature, i.e., $V_o(t_n, x_n) = 0$ and $\Theta(t_n, x_n) = \Theta_0$, then the analysis presented above still holds. In particular, the readback signal samples ob-

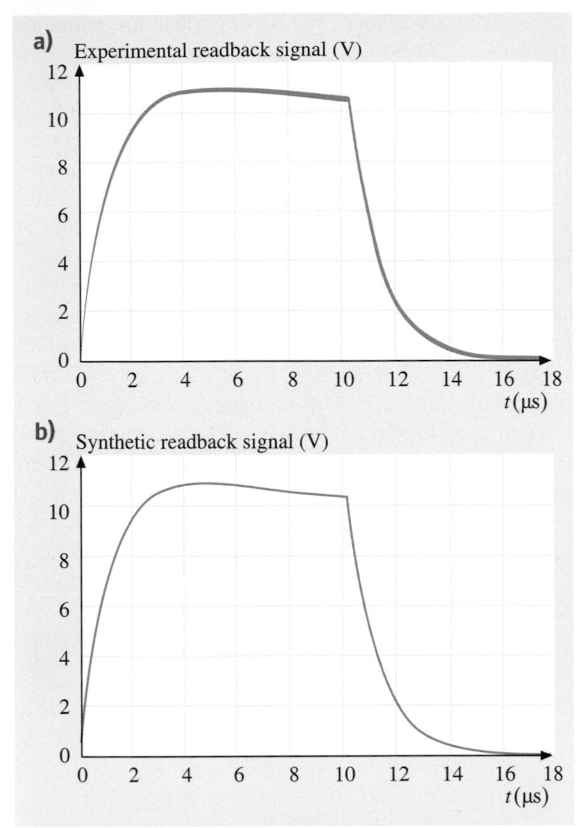

Fig. 31.30 (**a**) Experimental and (**b**) synthetic readback signal for $\tau = 10.25\,\mu\text{s}$ (after [31.12])

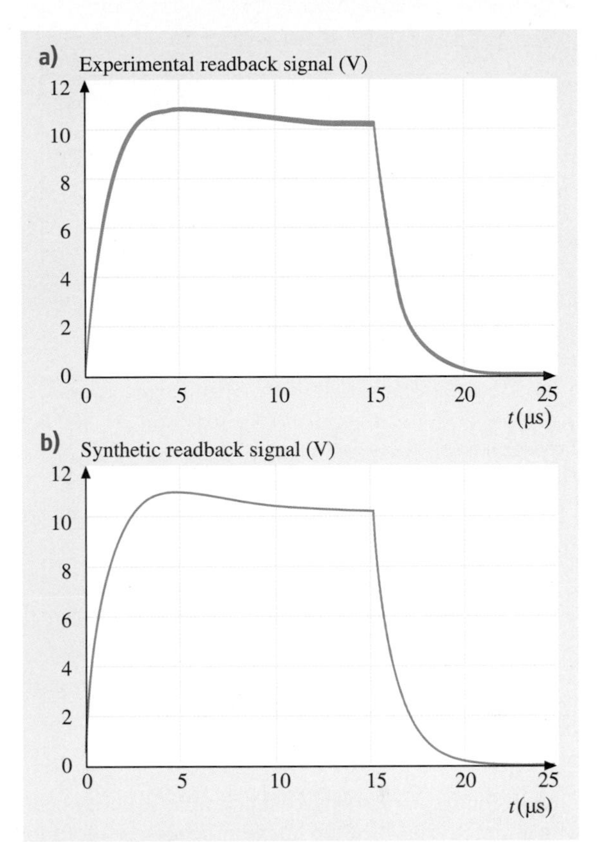

Fig. 31.31 (**a**) Experimental and (**b**) synthetic readback signal for $\tau = 15.25\,\mu\text{s}$ (after [31.12])

tained in response to N pulses applied to the cantilever for detecting a sequence of N binary written symbols are expressed as

$$s(t_{s,n}) = \tilde{V}_o(t_{s,n}, x_n), \quad t_{s,n} = nT + \tau, \quad n = 0, \dots, N-1\,, \tag{31.8}$$

where $\tilde{V}_o(t, x_n)$ is given by (31.7) for pulses applied at time t_n and at distance $x_n = nTv$, $n = 0, \dots, N-1$ from the initial point of scanning. Note that the functions $V_o(t, x_n)$ and $V_{o,\text{ref}}(t, x_n)$ in (31.7) are ideally given by the solution of the differential equations (31.3) and (31.5) for $\eta_x = \Delta R^\Theta(x_n)/R^\Theta$ and $\eta_x = 0$, respectively.

The readback signal (31.7) at the output of the low-pass filter is observed in the presence of additive noise. Therefore, the readback signal for the detection of the n-th binary symbol is given by

$$r(t_{s,n}) = s(t_{s,n}) + w(t_{s,n})\,, \tag{31.9}$$

where $w(t)$ denotes the noise signal. The components of the noise signal that must be taken into account are thermal noise (Johnson's noise) from the sensor, which reaches a temperature of about $\Theta_1 = 350\,°\text{C}$ during the read process, as well as from front-end analog circuitry. However, note that besides thermal noise, medium-related noise also affects the overall system performance.

Based on the above analysis, the response to a pulse applied to the cantilever at a distance x from some initial point can be calculated given the parameter $\eta_x = \Delta R^\Theta(x)/R^\Theta$. Recall that the value of η_x is proportional to the distance of the cantilever from the storage medium at the current location of the tip. Therefore, during tip displacement due to scanner motion, η_x is modulated from the topographical features of the storage surface such as written indentations, rings, and dust particles. This indicates that the modeling of the readback signal is a two-step process. First,

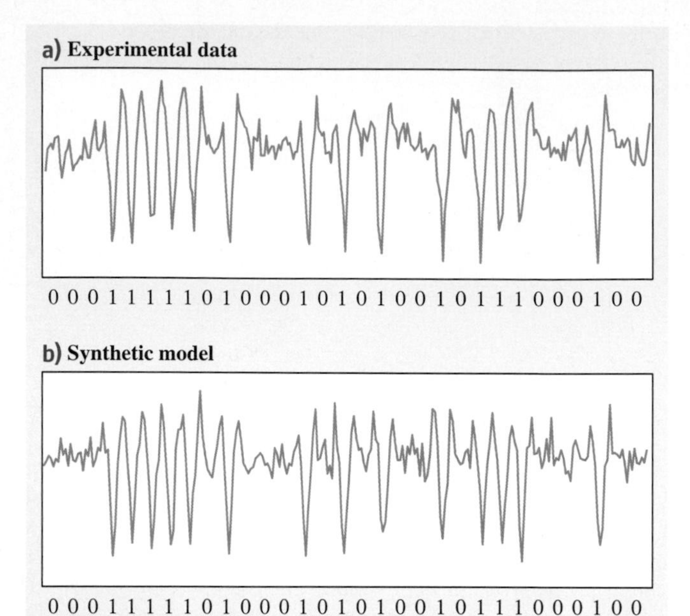

Fig. 31.32a,b Comparison between (**a**) the readback signal obtained experimentally along a data track and (**b**) the readback signal obtained by the synthetic model (after [31.12])

a model for the storage surface topography is developed, which directlyand then the above procedure is used to calculate the readback signal samples in response to pulses applied at selected points in the particular storage area.

In the absence of any imperfections during the manufacturing and the writing process, the storage surface would consist of completely flat regions interrupted by uniformly shaped indentations, possibly surrounded by polymer rings. A 1-D cross section of an indentation along the scanning direction is modeled by a function with one main lobe and two side lobes, one on each side of the main lobe, the magnitude and the extent of which can be varied independently. The side lobes are of opposite sign than the main lobe and simulate the polymer rings around written indentations. By varying their magnitude and extent while keeping the total extent of the pulse fixed, one can simulate indentations/rings of varying width and asymmetric ring formation, phenomena that are caused by different recording conditions. In practice, however, no polymer surface is entirely flat and indentation shapes are far from uniform. The deviation of indentations from uniformity is simulated by scaling the amplitude of each pulse shape by a random number drawn from a Gaussian distribution with unit mean and adjustable variance. The surface roughness is in turn simulated by adding white Gaussian noise to the height of every point in the area of interest. Note that surface roughness is a medium-related effect and manifests itself in the readback signal as a noise process, which is, however, of a very different nature than thermal noise. The advantage of the adopted two-step model is that it naturally decouples these unrelated noise sources, as well as the write and the read processes.

Figure 31.32 illustrates the experimental and synthetic readback signals obtained along a data track. The waveforms shown in Fig. 31.32 have been obtained by applying pulses at the oversampling rate of q/T, where q denotes the oversampling factor. For a more detailed comparison between model and actual signals, Figs. 31.33 and 31.34 illustrate 3-D views of isolated indentations from experimental and synthetic readback

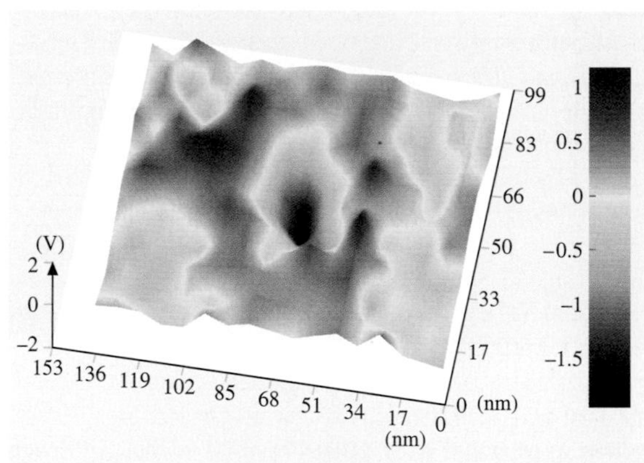

Fig. 31.33 Three-dimensional view of an isolated indentation obtained experimentally

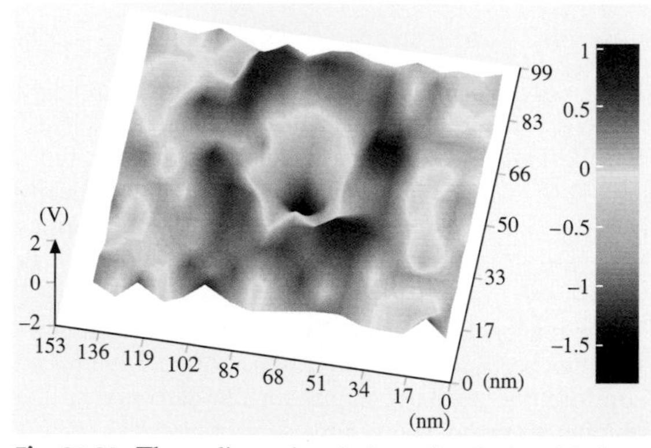

Fig. 31.34 Three-dimensional view of an isolated indentation obtained by the synthetic model

signals, respectively. The dark regions in the center of both figures correspond to the indentation centers, whereas dark regions around them are due to rings. Note also the irregular height of the surrounding surface, which is attributed to the roughness of the storage medium.

31.9 System Aspects

In this section, we describe various aspects of a storage system based on our millipede concept. Each cantilever can write data to and read data from a dedicated area of the polymer surface, called a *storage field*. As mentioned above, in each storage field the presence (absence) of an indentation corresponds to a logical "1" ("0"). All indentations are nominally of equal depth and size. The logical marks are placed at a fixed horizontal distance from each other along a data track. We refer to this distance, measured from one logical mark center to the next, as the *bit pitch* (BP). The vertical (cross-track) distance between logical mark centers, the *track pitch* (TP), is also fixed. To read and write data the polymer medium is moved under the (stationary) cantilever array at a constant velocity.

A robust way to achieve synchronization and servo control in an x/y-actuated large 2-D array is by reserving a small number of storage fields exclusively for timing recovery and servo-control purposes, as illustrated in Fig. 31.35. Because of the large number of levers in the millipede, this solution is advantageous in terms of overhead compared with the alternative of timing and servo information being embedded in all data fields.

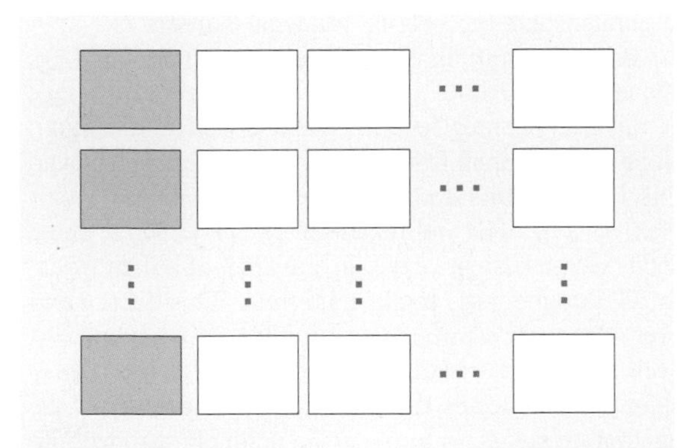

Fig. 31.35 Layout of data and servo/timing fields. *Dark boxes* represent dedicated servo/timing fields, *white boxes* represent data fields

31.9.1 PES Generation for the Servo Loop

With logical marks as densely spaced as in the millipede, accurate track following becomes a critical issue. Track following means controlling the position of each tip such that the tip is always positioned over the center of a desired track during reading. During writing, the tip position should be such that the written marks are aligned in a predefined way. In electromechanical systems, track following is performed in a servo loop, which is driven by an appropriate error signal called position-error signal (PES). Ideally, its magnitude is a direct estimate of the vertical (cross-track) distance of the tip from the closest track centerline, and its polarity indicates the direction of this offset.

Several approaches exist to generate a PES for AFM-based storage devices [31.9]. However, based on the results reported, none of these methods can achieve the track-following accuracy required for the millipede system. The quality of the PES directly affects the stability and robustness of the associated tracking servo loop [31.43].

We describe a method for generating a uniquely decodable PES for the millipede system [31.12, 27]. The method is based on the concept of *bursts* that are vertically displaced with respect to each other, arranged in such a way as to produce two signals in quadrature, which can be combined to provide a robust PES. This concept is borrowed from magnetic recording [31.43]. However, servo marks, as opposed to magnetic transitions, are placed in bursts labeled A and B for the in-phase signal and C and D for the quadrature signal. The centers of servo marks in burst B are vertically offset from mark centers in burst A by d' units of length. This amount of vertical spacing is related to the diameter of the written marks. The same principle applies to marks in the quadrature bursts C and D, with the additional condition that mark centers in burst C are offset by $d'/2$ units from mark centers in A in the cross-track direction. The latter condition is required in order to generate a quadrature signal. The configuration of servo bursts is illustrated in Fig. 31.36 for a case where $\mathrm{TP} = 3d'/2$. Al-

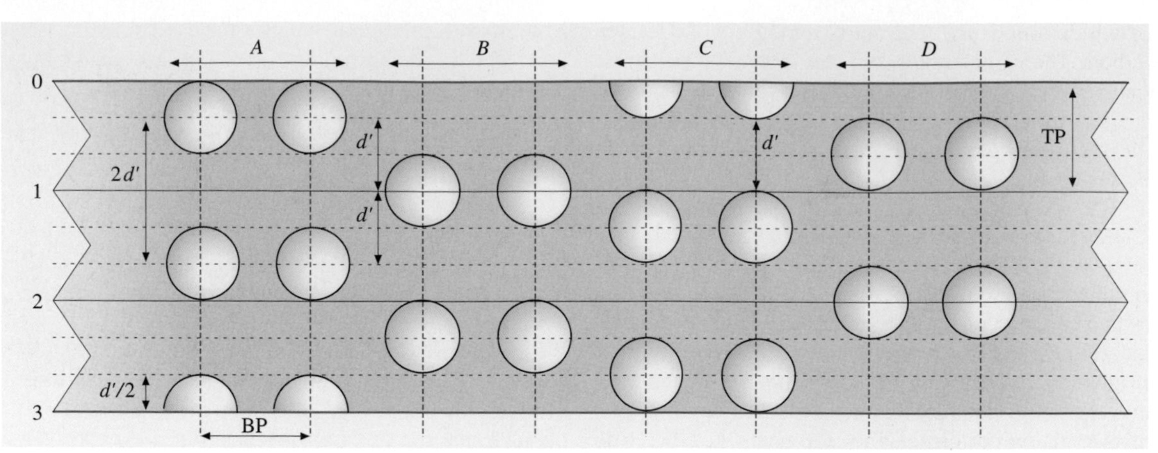

Fig. 31.36 Servo burst configuration (after [31.12])

though each burst typically consists of many marks to enable averaging of the corresponding readout signals, only two marks per burst are shown here to simplify the presentation. The solid horizontal lines depict track centerlines, and circles represent written marks, which are modeled here as perfect conical indentations on the polymer storage surface.

To illustrate the principle of PES generation, let us assume that marks in all bursts are spaced BP units apart in the longitudinal direction, and that sampling occurs exactly at mark centers, so that timing is perfect. The assumption of perfect timing is made only for the purpose of illustration. In actual operation, sampling is performed with the aid of a timing recovery loop, as described in Sect. 31.9.2. Referring to Fig. 31.36, let us further assume that the cantilever/tip is located on the line labeled "0" and moves vertically toward line "3" in a line crossing the centers of the left-most marks in burst A (shown as a brown dashed line). The tip moves from the edge of the top mark toward its center, then toward its bottom edge, then to a blank space, again to a mark, and so on. The readout signal magnitude decreases linearly with the distance from the mark center and reaches a constant, background level value at a distance greater than the mark radius from the mark center, according to the adopted (conical mark) model. To synthesize the in-phase signal, the readout signal is also captured as the tip (conceptually) moves in a vertical line crossing the mark centers of burst B (brown dashed line in Fig. 31.36). The in-phase signal is then formed as the difference $\bar{A} - \bar{B}$, where $\bar{A}$ and $\bar{B}$ stand for the measured signal amplitudes in bursts A and B, respectively. This signal is represented by the line labeled "I" in Fig. 31.37. It has zero-crossings at integer multiples of d', which do not generally correspond to track centers because we set $\mathrm{TP} = 3d'/2$ in this example. Therefore, the I-signal is not a valid PES in itself. This is why the quadrature (Q) signal becomes necessary in this case. The Q-signal is generated from the servo readback signals of bursts C and D as $\bar{C} - \bar{D}$ and is also shown in Fig. 31.37 (Q-curve). Note that it exhibits zero-crossings at points where the I-signal has local extrema.

A certain combination of the two signals (I and Q), shown as solid lines in Fig. 31.37, has zero-crossings at all track center locations and constant (absolute) slope, which qualifies it as a valid PES. However, this PES exhibits zero-crossings at all integer multiples of $d'/2$. For our example of $\mathrm{TP} = 3d'/2$, three such zero-crossings exist in an area of width equal to TP around any track centerline. This fact, however, does not hamper unique position decoding. At even-numbered tracks, it is the zero of the *in-phase* signal that indicates the track center. The zeros of the quadrature signal, in turn, can be uniquely mapped into a position estimate by examining the polarity of the in-phase signal at the corresponding positions. This holds for any value of the combined PES within an area

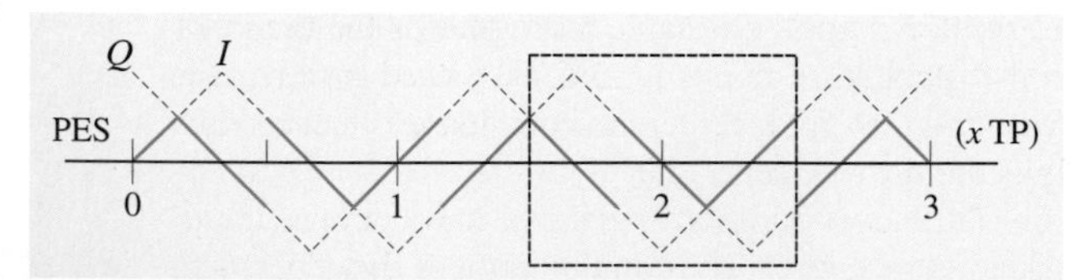

Fig. 31.37 Ideal position-error signal (after [31.12])

of width equal to TP around each current track centerline. The signals exchange roles for odd-numbered tracks. The current track number, which is known a priori from the seek operation, is used to determine the mode of operation for the position demodulation procedure.

The principle of PES generation based on servo marks has been verified experimentally. For this purpose, A, B, C, and D bursts were written by an AFM cantilever/tip on an appropriate polymer medium consisting of a polymer coating on a silicon substrate. The bit pitch was set to 42 nm, and the track pitch was taken to be approximately equal to d', the cross-track distance between A (C) and B (D) bursts. An image created by reading the written pattern with the same cantilever is shown in Fig. 31.38. Shaded areas indicate indentations. The readout signal from the cantilever was also used for servo demodulation, as described above. The resulting in-phase and quadrature signals are shown in Fig. 31.39. The track centerlines are indicated by vertical dotted lines in the graph.

It can be observed that the zero-crossings of the in-phase signal are closely aligned with the track centerlines, as well as with the minima and maxima of the quadrature signal, as required for unique position decoding across all possible cross-track positions, at least in cases where TP $\neq d'$. Moreover, the PES slope is nearly linear along a cross-track width of one track pitch around each track center, as TP $\approx d'$ in this case, although deviations from the ideal signal shape exist. These deviations occur mainly because written indentations do not have perfect conical shapes, and also because of medium noise due to the roughness of the recording medium. Nevertheless, the experimentally generated error signals indicate that the proposed concept is valid and promising. Specifically, the results indicate that servo self-writing is feasible, that servo demodulation is almost identical to data readout and can be performed by any cantilever without special provisions, and that the PES generated closely approximates the desirable features described above.

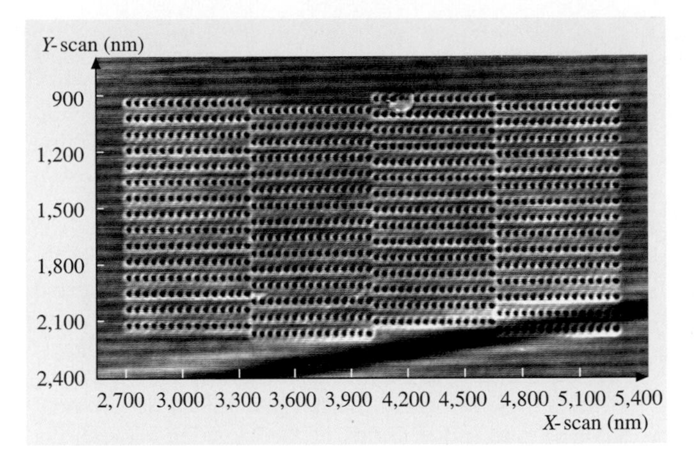

Fig. 31.38 Experimental A, B, C, and D servo bursts (BP = 42 nm) (after [31.12] © 2003 IEEE)

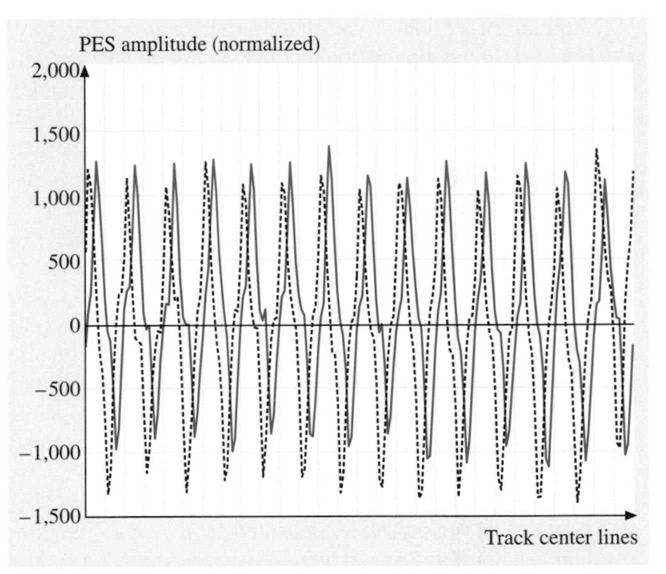

Fig. 31.39 Demodulated in-phase (*solid line*) and quadrature (*dashed line*) PES based on the servo burst of Fig. 31.38 (after [31.12])

31.9.2 Timing Recovery

Similar to obtaining servo information based on using dedicated servo fields, we employ separate dedicated clock fields for recovery of timing information [31.12, 26, 27]. The concept is to have continuous access to a pilot signal for synchronization after initial phase acquisition and gain estimation. The recovered clock is then distributed to all remaining storage fields to allow reliable detection of random data. Initial phase acquisition is obtained by a robust correlation algorithm, gain estimation is based on averaging of the readback signal obtained from a predefined stored pattern, and tracking of the optimum sampling phase is achieved by a second-order digital loop.

At the beginning of the read process, several signal parameters have to be estimated prior to data detection. Besides the clock phase and frequency, it is necessary to estimate the gain of the overall read channel. To solve the problem of initial estimation of signal parameters prior to data detection, the sequence written in the clock field consists of a preamble, followed by a pattern of all "1"s for tracking the optimum sampling phase during the detection of random data. The transition between the preamble and the pattern of all "1"s must be detected reliably, as it indicates the start of data records to the remaining storage fields. Assuming that the initial frequency offset is within a small, predetermined range, usually 1,000 parts-per-million (PPM), we distinguish the tasks needed for timing recovery as follows: (i) acquisition of the optimum sampling phase; (ii) estimation of the overall channel gain needed for threshold detection; (iii) detection of the transition between the preamble and the pattern of all "1"s; and (iv) tracking of the optimum sampling phase.

At the beginning of the acquisition process, an estimate of the optimum sampling phase is obtained by resorting to a correlation method. We rely on the knowledge of the preamble and of an ideal reference-channel impulse response, which closely resembles the actual impulse response (see Sect. 31.9). The channel output samples obtained at the oversampling rate q/T are first processed by removing the dc-offset, then averaging is performed to reduce the noise level, and finally the resulting sequence is correlated with the reference impulse response to determine the phase estimate.

After determining the estimate of the optimum sampling phase, an estimate of the overall channel gain is obtained by averaging the amplitude of the channel output samples at the optimum sampling instants. The gain estimate is obtained from an initial segment of the preamble corresponding to an "all 1" binary pattern. As mentioned above, it is necessary that the end of the preamble is indicated by a "sync" pattern, which marks the transition between acquisition mode and tracking mode. Detection of the sync pattern is also based on a robust correlation method. After the sync pattern, an "all 1" pattern, as in the case of robust phase acquisition and gain estimation, is employed for tracking. The "all 1" pattern corresponds to regularly spaced indentations, which convey reliable timing information.

Tracking of the optimum sampling phase is achieved by the second-order loop configuration shown in Fig. 31.40. Assuming data detection is performed at instants that correspond to integer multiples of the oversampling factor q, the deviation of the sampling phase from the optimum sampling phase is estimated as

$$\Delta\tau_n = r(t_{s,nq+1}) - r(t_{s,nq-1}) . \qquad (31.10)$$

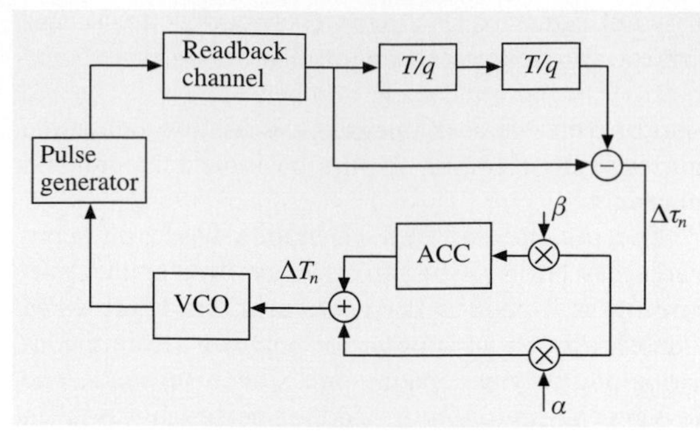

Fig. 31.40 Second-order loop for tracking the optimum sampling phase (after [31.12])

This estimate of the phase deviation is input to a second-order loop filter, which provides an output given by

$$\Delta T_n = u_n + \alpha\,\Delta\tau_n , \qquad (31.11)$$

where the discrete-time integrator is recursively updated as

$$u_{n+1} = u_n + \beta\,\Delta\tau_n . \qquad (31.12)$$

The loop-filter output then determines the control signal for a voltage-controlled oscillator (VCO).

Note that a similar concept for timing recovery can also be applied if no separate clock field is available. In this case, the timing information is extracted from the random user data on each storage field.

31.9.3 Considerations on Capacity and Data Rate

The ultimate locality provided by nanometer-sharp tips represents the pathway to the high areal densities that will be needed in the foreseeable future. The intrinsic nonlinear interactions between closely spaced indentations, however, may limit the minimum distance between successive indentations and, hence, the areal density. The storage capacity of a millipede-based storage device can be further increased by applying modulation or constrained codes [31.12].

With modulation coding, a desired constraint is imposed on the data-input sequence, so that the encoded data stream satisfies certain properties in the time or frequency domain. These codes are very important in digital recording devices and have become ubiquitous in all data-storage applications. The particular class

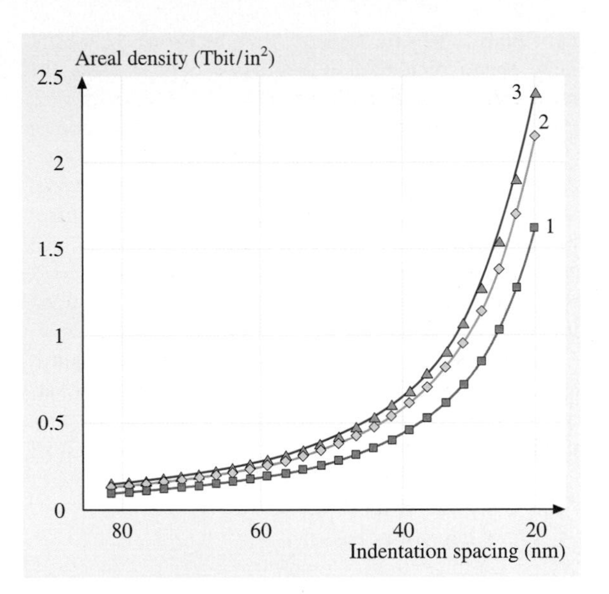

Fig. 31.41 Areal density versus indentation spacing. Curve 1: $d=0$; curve 2: $d=1$; and curve 3: $d=2$

of codes that imposes restrictions on the number of consecutive "1"s and "0"s in the encoded data sequence, generally known as run-length-limited (RLL) (d,k) codes [31.44], can be used to facilitate overwriting and also increase the effective areal density of a millipede-based storage device. The code parameters d and k are nonnegative integers with $k>d$, where d and k indicate the minimum and maximum number of "0"s between two successive "1"s, respectively. In the past, the precoded (RLL) (d,k) codes were mainly used for spreading the magnetic transitions further apart via the d-constraint, thereby minimizing intersymbol interference and nonlinear distortion, and for preventing loss of clock synchronization via the k-constraint. In optical recording, precoded RLL codes are primarily used for increasing the shortest pit length in order to improve the reliability of bit detection, as well as for limiting the number of identical symbols, so that useful timing information can be extracted from the readback signal.

For the millipede application, where dedicated clock fields are used, the k-constraint does not really play an important role and, therefore, can, in principle, be set to infinity, thereby facilitating the code design process. In a precoderless RLL code design, where the presence or absence of an indentation represents a "1" or "0", respectively, the d-constraint is instrumental in limiting the interference between successive indentations, as well as in increasing the effective areal density of the storage device. In particular, the quantity $(d+1)R$, where R denotes the rate of the (d,k) code, is a direct measure of the increase in linear recording density. Clearly, the packing density can be made larger by increasing d. On the other hand, large values of d lead to codes with very low rate, which implies high recording symbol rates, thus rendering these codes impractical for storage systems that are limited by the clock speed. The choice of $d=1$ and $k\geq 6$ guarantees the existence of a code with rate $R=2/3$. Use of $(d=1, k\geq 6)$ modulation coding reduces the bit distance by half while maintaining the pitch between "1"s constant, thereby

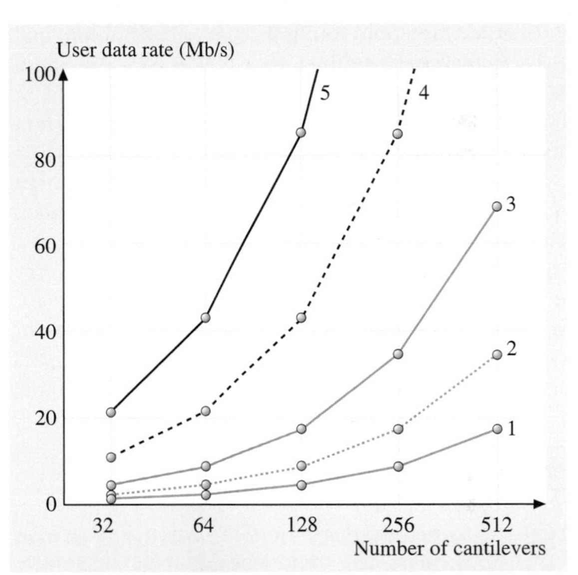

Fig. 31.42 User data rate versus number of active cantilevers for the $(d=1,\ k\geq 6)$ coding scheme. Curve 1: $T=20\,\mu s$; curve 2: $T=10\,\mu s$; curve 3: $T=5\,\mu s$; curve 4: $T=2\,\mu s$; and curve 5: $T=1\,\mu s$ (after [31.12])

Table 31.1 Areal density and storage capacity (after [31.12] with permission © 2003 IEEE)

Coding	Linear density (kb/in)	Track density (kt/in)	Areal density (Gb/in^2)	Capacity (Gb)
Uncoded	847	847	717	1.21
$(d=1, k\geq 6)$	1,129	847	956	1.61
$(d=2, k>6)$	1,269	847	1,075	1.81

increasing the linear density by a factor of $4/3$. Similarly, the choice of $d = 2$ and $k > 6$ guarantees the existence of a code with rate $R = 1/2$. Use of ($d = 2$, $k > 6$) modulation coding reduces the bit distance to a third while maintaining the pitch between "1"s constant, thereby increasing the linear density by a factor of $3/2$. Figure 31.41 shows the areal density as a function of the indentation spacing for an uncoded system, as well as for systems coded with ($d = 1$, $k \geq 6$) and ($d = 2$, $k > 6$), where coding is applied only in the on-track direction.

Table 31.1 shows the achievable areal densities and storage capacities for a 32×32 cantilever array with 1,024 storage fields, each having an area of $100 \times 100\,\mu m^2$, resulting in a total storage area of $3.2 \times 3.2\,mm^2$. The indentation pitch and the track pitch are set equal to 30 nm. Finally, for the computation of the storage capacity an overall efficiency of 85% has been assumed, taking into account the redundancy of the outer error-correction coding, as well as the presence of dedicated servo and clock fields.

Figure 31.42 shows the user data rate as a function of the total number of cantilevers accessed simultaneously, for various symbol rates and a ($d = 1$, $k \geq 6$) modulation coding scheme. For example, for a 32×32 cantilever array, a system designed to access a maximum of 256 cantilevers every $T = 5\,\mu s$ provides a user data rate of 34.1 Mb/s. Alternatively, by resorting to the row/column multiplexing scheme with $T = 80\,\mu s$ a data rate of 8.5 Mb/s is achieved.

31.10 Conclusions

A very large 2-D array of AFM probes has been operated for the first time in a multiplexed/parallel fashion, and write/read/erase operations in a thin polymer medium have been successfully demonstrated at densities of or significantly higher than those achieved with current magnetic storage systems.

The millipede has the potential to achieve ultrahigh storage areal densities on the order of $1\,Tb/in^2$ or higher. The high areal storage density, small form factor, and low power consumption render the millipede concept a very attractive candidate as a future storage technology for mobile applications, as it offers several Gigabytes of capacity at data rates of several Megabytes per second. Dedicated servo and timing fields allow reliable system operation with a very small overhead. The read channel model provides the methodology for analyzing system performance and assessing various aspects of the detection and servo/timing algorithms that are key to achieving the system reliability required by the applications envisaged.

Although the first high-density storage operations with the largest 2-D AFM array chip ever built have been demonstrated, there are a number of issues that need further investigation such as overall system reliability, including long-term stability of written indentations, tip and medium wear, limits of data rates, array and cantilever size, as well as trade-offs between data rate and power consumption.

References

31.1 E. Grochowski, R. F. Hoyt: Future trends in hard disk drives, IEEE Trans. Magn. **32** (1996) 1850–1854
31.2 D. A. Thompson, J. S. Best: The future of magnetic data storage technology, IBM J. Res. Dev. **44** (2000) 311–322
31.3 G. Binnig, H. Rohrer, C. Gerber, E. Weibel: 7×7 reconstruction on Si(111) resolved in real space, Phys. Rev. Lett. **50** (1983) 120–123
31.4 G. Binnig, C. F. Quate, C. Gerber: Atomic force microscope, Phys. Rev. Lett. **56** (1986) 930–933
31.5 H. J. Mamin, D. Rugar: Thermomechanical writing with an atomic force microscope tip, Appl. Phys. Lett. **61** (1992) 1003–1005
31.6 R. P. Ried, H. J. Mamin, B. D. Terris, L. S. Fan, D. Rugar: 6-MHz 2-N/m piezoresistive atomic-force-microscope cantilevers with incisive tips, J. Microelectromech. Syst. **6** (1997) 294–302
31.7 B. D. Terris, S. A. Rishton, H. J. Mamin, R. P. Ried, D. Rugar: Atomic force microscope-based data storage: Track servo and wear study, Appl. Phys. A **66** (1998) S809–S813
31.8 H. J. Mamin, B. D. Terris, L. S. Fan, S. Hoen, R. C. Barrett, D. Rugar: High-density data storage using proximal probe techniques, IBM J. Res. Dev. **39** (1995) 681–699
31.9 H. J. Mamin, R. P. Ried, B. D. Terris, D. Rugar: High-density data storage based on the atomic force microscope, Proc. IEEE **87** (1999) 1014–1027
31.10 L. R. Carley, J. A. Bain, G. K. Fedder, D. W. Greve, D. F. Guillou, M. S. C. Lu, T. Mukherjee, S. San-

thanam, L. Abelmann, S. Min: Single-chip computers with microelectromechanical systems-based magnetic memory, J. Appl. Phys. **87** (2000) 6680–6685

31.11 G. Gibson, T. I. Kamins, M. S. Keshner, S. L. Neberhuis, C. M. Perlov, C. C. Yang: Ultra-high density storage device, (1996) US Patent 5,557,596

31.12 E. Eleftheriou, T. Antonakopoulos, G. K. Binnig, G. Cherubini, M. Despont, A. Dholakia, U. Dürig, M. A. Lantz, H. Pozidis, H. E. Rothuizen, P. Vettiger: millipede – A MEMS-based scanning-probe data-storage system, IEEE Trans. Magn. **39** (2003) 938–945

31.13 G. K. Binnig, H. Rohrer, P. Vettiger: Mass-storage applications of local probe arrays, (1998) US Patent 5,835,477

31.14 P. Vettiger, J. Brugger, M. Despont, U. Drechsler, U. Dürig, W. Häberle, M. Lutwyche, H. Rothuizen, R. Stutz, R. Widmer, G. Binnig: Ultrahigh density, high-data-rate NEMS-based AFM data storage system, J. Microelectron. Eng. **46** (1999) 11–17

31.15 P. Vettiger, G. Cross, M. Despont, U. Drechsler, U. Dürig, B. Gotsmann, W. Häberle, M. A. Lantz, H. E. Rothuizen, R. Stutz, G. K. Binnig: The "millipede" – Nanotechnology entering data storage, IEEE Trans. Nanotechnol. **1** (2002) 39–55

31.16 M. Lutwyche, C. Andreoli, G. Binnig, J. Brugger, U. Drechsler, W. Häberle, H. Rohrer, H. Rothuizen, P. Vettiger: Microfabrication and parallel operation of 5 × 5 2D AFM cantilever array for data storage and imaging, Proc. IEEE 11th Int. Workshop MEMS, Heidelberg 1998 (IEEE, Piscataway 1998) 8–11

31.17 M. Lutwyche, C. Andreoli, G. Binnig, J. Brugger, U. Drechsler, W. Häberle, H. Rohrer, H. Rothuizen, P. Vettiger, G. Yaralioglu, C. Quate: 5 × 5 2D AFM cantilever arrays: A first step towards a terabit storage device, Sens. Actuators A **73** (1999) 89–94

31.18 P. Vettiger, M. Despont, U. Drechsler, U. Dürig, W. Häberle, M. I. Lutwyche, H. E. Rothuizen, R. Stutz, R. Widmer, G. K. Binnig: The "millipede" – More than one thousand tips for future AFM data storage, IBM J. Res. Dev. **44** (2000) 323–340

31.19 B. W. Chui, H. J. Mamin, B. D. Terris, D. Rugar, K. E. Goodson, T. W. Kenny: Micromachined heaters with 1-μs thermal time constants for AFM thermomechanical data storage, Proc. IEEE Transducers, Chicago 1997 (IEEE, Piscataway 1997) 1085–1088

31.20 W. P. King, J. G. Santiago, T. W. Kenny, K. E. Goodson: Modelling and prediction of sub-micrometer heat transfer during thermomechanical data storage, 1999 Microelectromechanical Systems (MEMS). Proc. ASME Intl. Mechanical Engineering Congress and Exposition, ed. by A. P. Lee, L. Lin, F. K. Forster, Y. C. Young, K. Goodson, R. S. Keynton (ASME, New York 1999) 583–588

31.21 W. P. King, T. W. Kenny, K. E. Goodson, G. L. W. Cross, M. Despont, U. Dürig, H. Rothuizen, G. Binnig, P. Vettiger: Design of atomic force microscope cantilevers for combined thermomechanical writing and thermal reading in array operation, J. Microelectromech. Syst. **11** (2002) 765–774

31.22 G. K. Binnig, M. Despont, W. Häberle, P. Vettiger: Method of forming ultrasmall structures and apparatus therefore, (March 1999) US Patent Office, Application No. 147865

31.23 G. Binnig, M. Despont, U. Drechsler, W. Häberle, M. Lutwyche, P. Vettiger, H. J. Mamin, B. W. Chui, T. W. Kenny: Ultra high-density AFM data storage with erase capability, Appl. Phys. Lett. **74** (1999) 1329–1331

31.24 G. K. Binnig, J. Brugger, W. Häberle, P. Vettiger: Investigation and/or manipulation device, (March 1999) US Patent Office, Application No. 147867

31.25 S. M. Sze: *Physics of Semiconductors Devices* (Wiley, New York 1981)

31.26 G. Cherubini, T. Antonakopoulos, P. Bächtold, G. K. Binnig, M. Despont, U. Drechsler, A. Dholakia, U. Dürig, E. Eleftheriou, B. Gotsmann, W. Häberle, M. A. Lantz, T. Loeliger, H. Pozidis, H. E. Rothuizen, R. Stutz, P. Vettiger: The millipede, a very dense, highly parallel scanning-probe data-storage system, ESSCIRC – Proceedings 28th European Solid-State Circuits Conference, ed. by A. Baschirotto, P. Malcovati (Univ. Bologna, Bologna 2002) 121–125

31.27 E. Eleftheriou, T. Antonakopoulos, G. K. Binnig, G. Cherubini, M. Despont, A. Dholakia, U. Dürig, M. A. Lantz, H. Pozidis, H. E. Rothuizen, P. Vettiger: "millipede": A MEMS-based scanning-probe data-storage system, Digest of the Asia-Pacific Magnetic Recording Conference 2002, APMRC '02 (IEEE, Piscataway 2002) CE-2-1–CE2-2

31.28 M. Despont, J. Brugger, U. Drechsler, U. Dürig, W. Häberle, M. Lutwyche, H. Rothuizen, R. Stutz, R. Widmer, G. Binnig, H. Rohrer, P. Vettiger: VLSI-NEMS chip for AFM data storage, Technical Digest 12th IEEE Int. Micro Electro Mechanical Systems Conf. "MEMS '99" (IEEE, Piscataway 1999) 564–569

31.29 T. S. Ravi, R. B. Marcus: Oxidation sharpening of silicon tips, J. Vac. Sci. Technol. B **9** (1991) 2733–2737

31.30 M. Despont, J. Brugger, U. Drechsler, U. Dürig, W. Häberle, M. Lutwyche, H. Rothuizen, R. Stutz, R. Widmer, G. Binnig, H. Rohrer, P. Vettiger: VLSI-NEMS chip for parallel AFM data storage, Sens. Actuators A **80** (2000) 100–107

31.31 H. Rothuizen, M. Despont, U. Drechsler, G. Genolet, W. Häberle, M. Lutwyche, R. Stutz, P. Vettiger: Compact copper/epoxy-based micromachined electromagnetic scanner for scanning probe applications, Technical Digest, 15th IEEE Int. Conf. on Micro Electro Mechanical Systems "MEMS 2002" (IEEE, Piscataway 2002) 582–585

31.32 S. C. Minne, G. Yaralioglu, S. R. Manalis, J. D. Adams, A. Atalar, C. F. Quate: Automated parallel high-speed atomic force microscopy, Appl. Phys. Lett. **72** (1998) 2340–2342

31.33 M. Lutwyche, U. Drechsler, W. Häberle, R. Widmer, H. Rothuizen, P. Vettiger, J. Thaysen: Planar micromagnetic x/y/z scanner with five degrees of freedom. In: *Magnetic Materials, Processes, and Devices: Applications to Storage and Micromechanical Systems (MEMS)*, Vol. 98-20, ed. by L. Romankiw, S. Krongelb, C. H. Ahn (Electrochemical Society, Pennington 1999) pp. 423–433

31.34 H. Rothuizen, U. Drechsler, G. Genolet, W. Häberle, M. Lutwyche, R. Stutz, R. Widmer, P. Vettiger: Fabrication of a micromachined magnetic x/y/z scanner for parallel scanning probe applications, Microelectron. Eng. **53** (2000) 509–512

31.35 J.-J. Choi, H. Park, K. Y. Kim, J. U. Jeon: Electromagnetic micro x-y stage for probe-based data storage, J. Semicond. Technol. Sci. **1** (2001) 84–93

31.36 H. Lorenz, M. Despont, N. Fahrni, J. Brugger, P. Vettiger, P. Renaud: High-aspect-ratio, ultrathick, negative-tone near-UV photoresist and its applications for MEMS, Sens. Actuators A **64** (1998) 33–39

31.37 C. Q. Davis, D. Freeman: Using a light microscope to measure motions with nanometer accuracy, Opt. Eng. **37** (1998) 1299–1304

31.38 M. I. Lutwyche, M. Despont, U. Drechsler, U. Dürig, W. Häberle, H. Rothuizen, R. Stutz, R. Widmer, G. K. Binnig, P. Vettiger: Highly parallel data storage system based on scanning probe arrays, Appl. Phys. Lett. **77** (2000) 3299–3301

31.39 K. Fuchs, C. Friedrich, J. Weese: Viscoelastic properties of narrow-distribution poly(methyl metacrylates), Macromolecules **29** (1996) 5893–5901

31.40 U. Dürig, B. Gotsman: This estimate is based on a fluid dynamic deformation model of a thin film, private communication

31.41 J. D. Ferry: *Viscoelastic Properties of Polymers, 3rd edition* (Wiley, New York 1980)

31.42 T. Loeliger, P. Bächtold, G. K. Binnig, G. Cherubini, U. Dürig, E. Eleftheriou, P. Vettiger, M. Uster, H. Jäckel: CMOS sensor array with cell-level analog-to-digital conversion for local probe data storage, ESSCIRC – Proceedings 28th European Solid-State Circuits Conference, ed. by A. Baschirotto, P. Malcovati (Univ. Bologna, Bologna 2002) 623–626

31.43 A. H. Sacks: Position signal generation in magnetic disk drives. Ph.D. Thesis (Carnegie Mellon University, Pittsburgh 1995)

31.44 K. A. S. Immink: *Coding Techniques for Digital Recorders* (Prentice Hall, Hemel 1991)

32. Microactuators for Dual-Stage Servo Systems in Magnetic Disk Files

This chapter discusses the design and fabrication of electrostatic MEMS microactuators and the design of dual-stage servo systems in disk drives. It introduces fundamental requirements of disk drive servo systems, along with challenges posed by storage densities increases. It describes three potential dual-stage configurations and focuses on actuated slider assemblies using electrostatic MEMS microactuators. The authors discuss major electrostatic actuator design issues, such as linear versus rotary motion, electrostatic array configuration, and differential operation. Capacitive and piezoresistive elements may be used to sense relative slider position, while integrated gimbals and structural isolation may prove useful in improving performance. A detailed design example based on a translational, micromolded actuator illustrates several of these concepts and is accompanied by theoretical and experimental results.

The chapter continues to discuss MEMS microactuator fabrication. It describes several processes for obtaining appropriate electrostatic devices including micromolding, deep-reactive ion etching, and electroplating. The primary goal of these processes is to obtain very high-aspect ratio structures, which improve both actuation force and structural robustness. Other fabrication issues, such as electrical interconnect formation, material selection, and processing cost, are also considered. Actuated slider fabrication is compared to that of actuated suspension and actuated head assemblies; this includes instrumentation of a suspension with strain sensors to aid in vibration detection.

The section on controller design reviews dual-stage servo control design architectures and methodologies. The considerations for controller design of MEMS microactuator dual-stage servo systems are discussed. The details of control designs using a decoupled SISO design method and the robust multivariable design method of μ-synthesis are also presented. In the decoupled design, a self-tuning control algorithm has been developed to compensate for the variations in the microactuator's resonance mode. In the μ-synthesis design, robust controllers can be synthesized by using additive and parametric uncertainties to characterize the un-modeled dynamics of the VCM and the variations in the microactuator's resonance mode. Finally, the chapter also introduces a dual-stage short-span seek control scheme based on decoupled feed forward reference trajectories generations.

32.1 **Design of the Electrostatic Microactuator** ... 952
32.1.1 Disk Drive Structural Requirements ... 952
32.1.2 Dual-Stage Servo Configurations ... 953
32.1.3 Electrostatic Microactuators: Comb-Drives vs. Parallel-Plates ... 954
32.1.4 Position Sensing ... 956
32.1.5 Electrostatic Microactuator Designs for Disk Drives ... 958

32.2 **Fabrication** ... 962
32.2.1 Basic Requirements ... 962
32.2.2 Electrostatic Microactuator Fabrication Example ... 962
32.2.3 Electrostatic Microactuator Example Two ... 963
32.2.4 Other Fabrication Processes ... 966
32.2.5 Suspension-Level Fabrication Processes ... 967
32.2.6 Actuated Head Fabrication ... 968

32.3 **Servo Control Design of MEMS Microactuator Dual-Stage Servo Systems** ... 968
32.3.1 Introduction to Disk Drive Servo Control ... 969
32.3.2 Overview of Dual-Stage Servo Control Design Methodologies ... 969
32.3.3 Track-Following Controller Design for a MEMS Microactuator Dual-Stage Servo System ... 971
32.3.4 Dual-Stage Seek Control Design ... 976

32.4 **Conclusions and Outlook** ... 978

References ... 979

This chapter discusses the design and fabrication of electrostatic MEMS microactuators, and the design of dual-stage servo systems, in disk drives. The focus of the chapter is an actuated slider assembly using an electrostatic MEMS microactuator. We discuss major design issues, including linear versus rotary actuation, electrostatic array configuration, and integrated sensing capability. We describe several fabrication processes for obtaining the necessary devices, such as micromolding, deep-reactive ion etching, and electroplating. Dual-stage servo control design architectures and methodologies are then reviewed . We present in detail track-following controller designs based on a sensitivity function decoupling single-input-single-output design methodology and the robust μ-synthesis design methodology. Finally, we introduce a 2-DOF short span seek control design using a dual-stage actuator.

Since the first hard disk drive (HDD) was invented in the 1950s by IBM, disk drives' storage density has been following Moore's law, doubling roughly every 18 months. The current storage density is 10 million times larger than that of the first HDD [32.1]. Historically, increases in storage density have been achieved by almost equal increases in track density, the number of tracks encircling the disk, and bit density, the number of bits in each track. However, because of superparamagnetism limitations, future areal storage density increases in HDDs are predicted to be achieved mainly through an increase in track density [32.2].

Research in the HDD industry is now targeting an areal density of one terabit per square inch. For a predicted bit aspect ratio of 4 : 1, this translates to a linear bit density of 2 M bits per inch (BPI) and a radial track density of 500 K tracks per inch (TPI), which in turn implies a track width of 50 nm. A simple rule of thumb for servo design in HDDs is that three times the statistical standard deviation of the position error between the head and the center of the data track should be less than 1/10 of the track width. Thus, to achieve such a storage density, nanometer-level precision of the servo system will be required.

A disk drive stores data as magnetic patterns, forming bits, on one or more disks. The polarity of each bit is detected (read) or set (written) by an electromagnetic device known as the read/write head. The job of a disk drive's servo system is to position the read/write head over the bits to be read or written as they spin by on the disk. In a conventional disk drive, this is done by sweeping over the disk or disks a long arm consisting of a voice coil motor (VCM), an E-block, suspensions, and sliders, as shown in Figs. 32.1 and 32.2. A read/write head is fabricated on the edge of each slider (one for each disk surface). Each slider is supported by a suspension and flys over the surface of a disk on an air-bearing. The VCM actuates the suspensions and sliders about a pivot in the center of the E-block. We describe this operation in more detail in the following section.

The key to increasing HDD servo precision is to increase servo control bandwidth. However, the bandwidth of a traditional single-stage servo system as shown in Fig. 32.2 is limited by the multiple mechanical resonance modes of the pivot, the E-block, and the suspension between the VCM and the head. Nonlinear friction of the pivot bearing also limits achievable servo precision. Dual-stage actuation, with a second stage actuator placed between the VCM and the head, has been proposed as a solution that would increase servo bandwidth and precision.

Several different secondary actuation forces and configurations have been proposed, each having strengths and weaknesses given the requirements of HDDs. The dual-stage configurations can be categorized into three groups: actuated suspension, actuated slider, and actuated head. Within these, actuation forces include piezoelectric, electrostatic, and electromagnetic. In this chapter, we discuss design, fabrication and control of an electrostatic MEMS microactuator (MA) for actuated slider dual-stage positioning.

32.1 Design of the Electrostatic Microactuator

The servo system of a hard disk drive is the mechatronic device that locates and reads data on the disk. In essence, it is a large arm that sweeps across the surface of the disk. At the end of the arm is the read-write head, containing the magnetic reading and writing elements that transfer information to and from the disk.

32.1.1 Disk Drive Structural Requirements

This read-write head is contained in a box-like structure known as a slider. The slider has a contoured lower surface that acts as an air bearing between the head and the disk. The high-velocity airflow generated by the spinning disk pushes up on the air-bearing surface, maintaining the slider and read-write head at a constant distance from the disk, despite unevenness of the disk, permitting reliable data reading and writing.

The arm over an HDD's disk has three primary stages: the voice-coil motor (VCM), the E-block, and the suspension. In a conventional disk drive, the VCM performs all positioning of the head, swinging it back and forth across the disk. The E-block lies between the VCM and the suspension and contains the pivot point. The suspension projects from the E-block over the disk as a thin flexible structure, generally narrowing to a point at the location of the slider.

For a disk drive servo to operate effectively, it must maintain the read-write head at a precise height above the disk surface and within a narrow range between the disk tracks arranged in concentric circles around the disk. It must also be able to seek from one track on the disk to another. Information about the track that the head is following is encoded in sectors radiating from center of the disk, allowing the head to identify its position and distance from the center of the track. To maintain a correct flying height, the suspension must be designed with an appropriate stiffness in the vertical direction to balance an air-bearing force corresponding to the slider design in use in the drive. Meanwhile, the suspension must be flexible to roll and pitch at the slider location to permit adaptation to unevenness of the disk surface. This is accomplished using a gimbal structure. The suspension as a whole, however, should not bend or twist during the operation, as this would misdirect the head away from the track it is following.

As data densities in HDDs increase and track widths diminish, single-stage, conventional servo systems become less able to position the head precisely. Because the VCM/E-block/suspension assembly is large and massive as a unit, the speed at which the head can be controlled is limited. Furthermore, the assembly tends to have a low natural frequency, which can accentuate vibration in the disk drive and cause off-track errors. At track densities in the future approaching one Terabit per square inch, the vibration induced in a disk drive by airflow alone is enough to force the head off-track.

A solution to these problems is to complement the VCM with a smaller, secondary actuator to form a dual-stage servo system. The VCM continues to provide rough positioning, while the second stage actuator does fine positioning and damps out vibration and other disturbances. The smaller second stage actuator can typically be designed to have a much higher natural frequency and less susceptibility to vibration than the VCM. Any actuator used in a dual-stage system should be inexpensive to build, require little power to operate, and preserve the stiffness properties described above necessary to preserve the flying height.

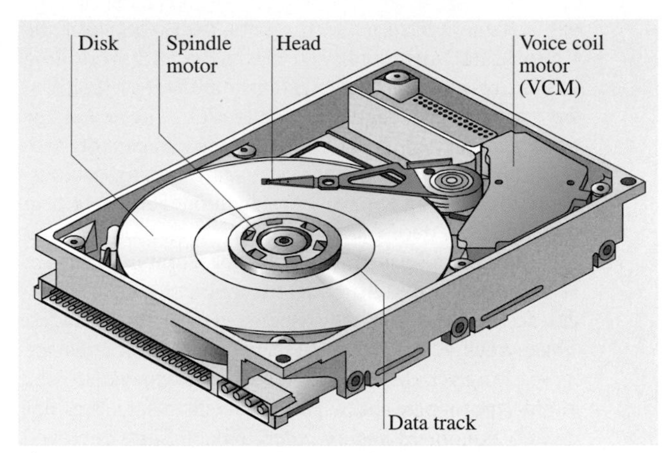

Fig. 32.1 A schematic diagram of an HDD

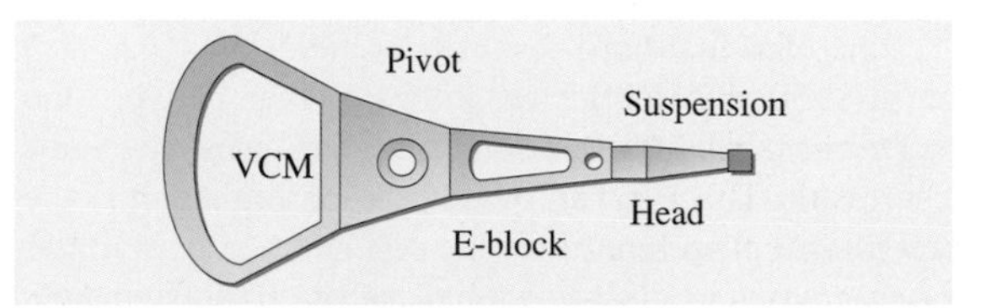

Fig. 32.2 VCM actuator in an HDD

32.1.2 Dual-Stage Servo Configurations

In the past six years, much research work has been dedicated to the exploration of suitable secondary actuators for constructing dual-stage servo systems for HDDs. These dual-stage configurations can be categorized into three groups: "actuated suspension," "actuated slider," and "actuated head."

Actuated Suspension

In this approach, the suspension is redesigned to accommodate an active component, typically a piezoelectric material. This piezoelectric material stretches or flexes the suspension to position the slider and magnetic head. Piezoelectric material is an active actuation element that produces a large actuation force but small actuation stroke. In the "actuated-suspension" configuration, therefore, the piezoelectric actuators are usually implemented in a leverage mechanism that can convert small actuation displacements into large head displacements. Typically this is done by placing the piezoelectric actuators away from the magnetic head (between the E-block and suspension) so that they can have a long leverage arm to gain mechanical amplification and produce a sufficient magnetic-head motion. The advantage of this approach is that the suspension can be fabricated by a conventional suspension-making process, and its dual-stage servo configuration is effective in attaining low frequency runout attenuation in the positioning servo loop. The major drawback of this approach is that the system is still susceptible to instabilities due to the excitation of suspension resonance modes. Thus, track-per-inch (TPI) servo performance can be increased but remains limited when compared to the alternative approaches. The actuated suspension approach, nonetheless is expected to be the first deployed in commercial HDDs [32.3, 4].

Actuated Slider

In this approach, a microactuator is placed between the slider and gimbal to position the slider/magnetic heads. The resulting servo bandwidth can be higher than the previous approach because the secondary actuation bypasses the mechanical resonances of the suspension. This approach uses existing sliders and microactuators that can be batch fabricated, and thus could be cost effective. However, the size and mass of the microactuator are significant relative to those of current sliders and may interfere with the slider flying stability. Current suspensions, therefore, need to be redesigned to adopt this secondary actuator. Suitable driving forces in this approach include electrostatic, electromagnetic, and piezoelectric [32.5–8]. To further reduce the assembly task of placing the microactuator in between gimbal structure and slider, some researchers have proposed microactuators that are either integrated with the gimbal structure [32.6] or the slider [32.9].

Actuated Head

In this approach, the slider is redesigned so that the microactuators can be placed inside the slider block and actuate the magnetic heads with respect to the rest of the slider body. As these microactuators are very small, they only slightly increase the slider weight and are thus capable of working with the current suspension assembly. Researchers have successfully demonstrated the integrated fabrication process for fabricating the electrostatic microactuators and magnetic heads within one piece of ceramic block (slider). The embedded electrostatic microactuator has its resonance close to 30 kHz and was able to position the magnetic heads relative to the rest of the slider body by 0.5 μm [32.10, 11]. Full-fledged integration of slider, actuator, and read/write head remains a challenge.

In this chapter, we focus on actuated slider configurations, as they involve a great deal of interesting microscale engineering. In particular, we will discuss electrostatic actuation, probably the most common method of implementing microactuation in microelectromechanical devices. Any such microactuator will exhibit certain features:

- a fixed base, which attaches to the suspension,
- a movable platform, upon which the slider rests,
- springs between the base and platform, flexible in the direction of desired motion, and stiff in all other directions, and
- an electrostatic actuation array that generates the force used to move the platform and slider.

Microactuators must also include a wiring scheme for transferring signals to and from the slider and often incorporate a structure for sensing the motion of the slider relative to the suspension. Electrostatic microactuators to be discussed in this chapter include HexSil and DRIE fabricated actuators from the University of California, Berkeley, and electroplating-formed actuators by IBM and the University of Tokyo.

Electromagnetic or piezoelectric force are alternatives to electrostatic actuation in the actuated-slider configuration. Electromagnetic microactuators use ferromagnetic films to produce force perpendicular to an applied electric field. This type of actuation has potentially low voltage requirements but requires special fabrication techniques to integrate the magnetic components into the assembly. A microactuator of this type for hard disk drives is under development at Seagate, with results as yet unpublished. Piezoelectric microactuators use a piezoelectric material, which expands or contracts in response to applied voltage, to move the slider. These actuators have simple fabrication, the patterning a piece of piezoelectric material to sit between the suspension and slider, but it is difficult to obtain an adequate range of motion. A short stroke from the piezoelectric piece must be leveraged into a much larger motion at the read/write head. A piezoelectric microactuator has been produced by TDK corporation with a 0.5 μm stroke length at 10 V, with a 10 V bias [32.12].

32.1.3 Electrostatic Microactuators: Comb-Drives vs. Parallel-Plates

Electrostatic microactuators have been studied as the secondary actuators in HDDs for their relative ease of fabrication, particularly in the configurations of "actuated slider" and "actuated head," since the structural material needs only to be conductive rather than ferromagnetic or piezoelectric. Electrostatic force is generated by applying a voltage difference between the moving shuttle and a fixed stator element. Depending on the designated motion for the shuttle, electrostatic actuators are often categorized into two groups: comb-drives and parallel-plates, as illustrated in Fig. 32.3.

The magnitude of the electrostatic force generated equals the rate of change of energy that is retained within the finger-like structure and varied by shuttle motion. Therefore, the electrostatic force for comb-drive actuators, in which the designated shuttle motion moves along the x-direction, as shown in Fig. 32.3, equals

$$F_{\text{comb}} = \frac{\partial E}{\partial x} = \frac{\epsilon h}{2d} V^2 , \tag{32.1}$$

where ϵ is the permittivity of air, x is the overlap between two adjacent plates, h is plate thickness, and d is the gap between two parallel plates. Similarly, the electrostatic force for parallel-plates actuators is

$$F_{\text{parallel}} = \frac{\partial E}{\partial y} = \frac{\epsilon x h}{2d^2} V^2 , \tag{32.2}$$

where x is the finger overlap.

As indicated in (32.1) and (32.2), the electrostatic force for comb-drives actuators does not depend on the displacement of the moving shuttle and thus allows a long stroke while maintaining a constant electrostatic force. The electrostatic force for parallel-plates actuators, in contrast, is a nonlinear function of its shuttle motion ($\propto 1/d^2$), and the maximum stroke is limited by the nominal gap between shuttle and stator. A longer stroke is achieved with a larger gap, at the expense of lower electrostatic force. For applications that require small stroke but large force output, parallel-plates actuators are preferred since the output force from parallel-plates can be x/d times larger than the force from comb-drives. The following equation (32.3) is easily derived from (32.1) and (32.2).

$$\frac{F_{\text{parallel}}}{F_{\text{comb}}} = \frac{x}{d} . \tag{32.3}$$

A simplified second-order differential equation is often utilized to describe the dynamic response of an electrostatic microactuator

$$m\ddot{x}(t) + b\dot{x}(t) + K_{\text{m}}x(t) = F\,[V, x(t)] , \tag{32.4}$$

where m is the mass of the moving shuttle, b is the damping coefficient of the microactuator, K_{m} is the spring constant of the mechanical spring that connects the moving shuttle to an anchor point, and F is the electrostatic force that can be obtained from (32.1) and (32.2).

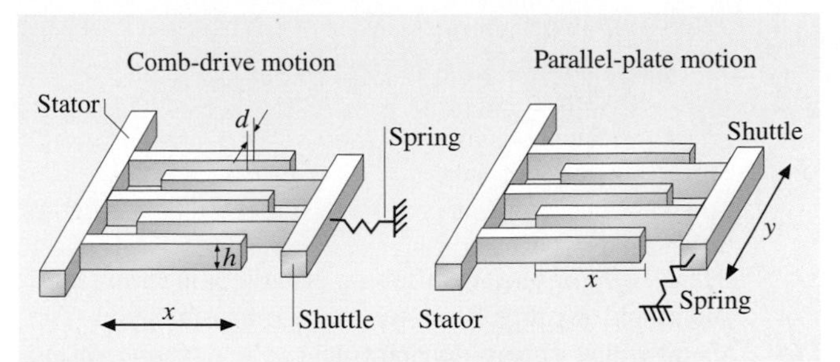

Fig. 32.3 Electrostatic microactuators. Comb-drives versus parallel-plates

Differential Drives

Because electrostatic force is always attractive, electrostatic microactuators need other features to actively control the direction of shuttle motion in servo applications, as opposed to relying on the restoring force from a mechanical spring. For this reason, the differential-drives approach, as shown in Fig. 32.4, is frequently adopted in electrostatic microactuator designs. Based on the differential-drive configuration, the simplified second-order differential equation (32.4) is rewritten as

$$\begin{aligned} m\ddot{x}(t)+b\dot{x}(t)+K_{\text{m}}x(t) = {}& F[V_{\text{bias}}+V_{\text{dr}}, x_o-x(t)]- \\ & F[V_{\text{bias}}-V_{\text{dr}}, x_o+x(t)] , \end{aligned} \tag{32.5}$$

where x_0 is the nominal position of the moving shuttle. If the differential drive is operated at the bias voltage (V_{bias}) with a small perturbation voltage (V_{dr}), the nonlinear force input in (32.5) can be linearized with a first-order

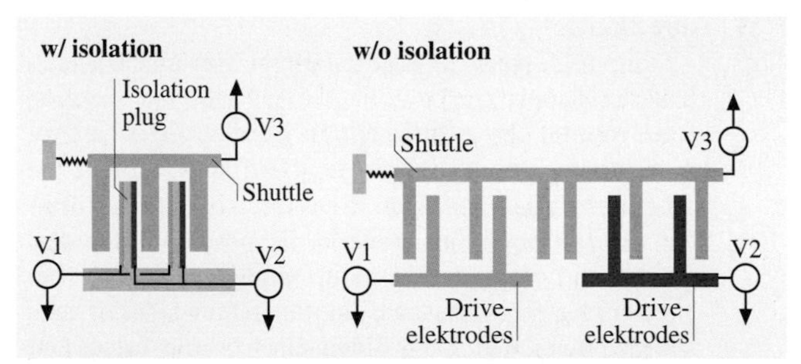

Fig. 32.4 Differential parallel-plates actuators and electrical-isolation features

approximation.

$$m\ddot{x}(t)+b\dot{x}(t)+(K_m-K_e)x(t)=K_v V_{dr}\,,$$
$$K_e=2\frac{\partial F}{\partial x}\,|V_{bias}\,,$$
$$K_v=2\frac{\partial F}{\partial V_{dr}}\,|V_{bias}\,. \qquad (32.6)$$

Here K_e represents a softening electrostatic spring constant, and K_v represents the voltage-to-force gain. The electrostatic spring constant acts as a negative spring during the electrostatic microactuator operation, and its value varies with the bias voltage (V_{bias}). When the electrostatic spring constant K_e exceeds the spring constant K_m from the mechanical spring, the microactuator becomes unstable; this is often described as pull-in instability. As shown in (32.5) and (32.6), the differential configuration cancels the even order harmonics in voltage and thus linearizes the voltage–force relation to some extent. Furthermore, in parallel-plates actuators, the differential configuration reduces the nonlinearity in actuation voltage as well as in shuttle displacement.

Electrical Isolation

Because electrostatic actuation requires multiple voltage levels for actuation force and position sensing (as discussed in the following section), electrical isolation is another challenge for designing an electrostatic microactuator. Generally speaking, when multiple voltage levels are needed in MEMS devices, electrical isolation is achieved by breaking up the parts that need to be on different voltage level and anchoring them separately to a nonconductive substrate. This approach has many drawbacks, not only because it requires a substrate in a device but also because structures have to be mechanically separated to be electrically isolated. The electrical isolation problem is far more severe in parallel-plates microactuators than comb-drive microactuators since parallel-plates actuation generally requires different voltage levels for stator fingers pulling in opposite directions.

Figure 32.4 shows an example of how an electrical-isolation feature can be utilized to increase the actuation force output in a differential parallel-plates microactuator design. As shown in the figure, without the proper electrical isolation, drive-electrodes with different voltage potentials have to be placed in separate groups and result in the same voltage difference between drive-electrodes and shuttle on both sides of each shuttle finger [32.5, 6]. Since electrostatic forces are always attractive, gaps on two sides of the interlaced structure cannot be made equal, otherwise the forces on two sides of a shuttle finger will be equal and the shuttle's movement direction will be uncontrollable. With such electrical-isolation features as the "isolation plug," shown on the left in Fig. 32.4, gaps of the interlaced structure can be the same width, since different voltages can be applied on the two sides of the shuttle fingers. As shown in Fig. 32.4, the design with integrated electrical-isolation features is more compact than without isolation features. Consequently, more finger structures can fit in the same amount of space, and the actuation voltage can be reduced.

32.1.4 Position Sensing

Most proposed HDD dual-stage servo controllers utilize only the position of the magnetic head relative to the center of data track, known in the industry as the position error signal, or PES, for closed-loop track following control. These systems have a single-input-multi-output (SIMO) control architecture. In some instances, however, it is also possible to measure the relative position error signal (RPES) of the magnetic head relative to the VCM. In this case, the control architecture is multi-input-multi-output (MIMO). As shown in [32.13], RPES can be used in a MIMO controller to damp out the second stage actuator's resonance mode and enhance the overall robustness of the servo system.

Capacitive position sensing and piezoresistive position sensing are two popular sensing mechanisms among electrostatic microactuator designs. Each of these sensing mechanisms is discussed in more detail in the following sections.

Capacitive Position Sensing

Capacitive position sensing is based on shuttle movement causing a capacitance change between the moving shuttle and fixed stators. By measuring the change in capacitance, it is possible to determine the shuttle location relative to the fixed stator. The output voltage (V_o) for both differential drives in Fig. 32.5 equals $2\delta C/C_i V_s$. The capacitance change due to shuttle movement (dC/dx) can be derived and the output voltage (V_o) for the comb-drives and parallel-plates can be formulated as a function of shuttle displacement.

$$V_{comb}=2\frac{C_s}{C_i}\frac{\delta x}{x_0}V_s\,,$$
$$V_{parallel}=2\frac{C_s}{C_i}\frac{y_0\delta y}{y_0^2-\delta y^2}V_s\,,$$
$$\approx 2\frac{C_s}{C_i}\frac{\delta y}{y_0}V_s\,, \qquad (32.7)$$

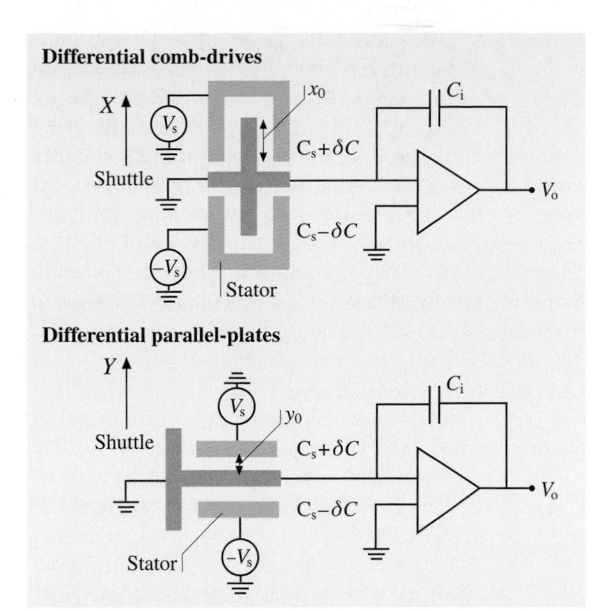

Fig. 32.5 Capacitive position sensing. Comb-drives versus parallel-plates motion

where x_0 is the nominal overlap for interlaced fingers and y_0 is the nominal gap between overlapped fingers, as shown in Fig. 32.5.

As indicated by (32.7), the voltage output for the comb-drives sensing structure, V_{comb}, is linear with shuttle displacement. On the other hand, the sensing configuration that makes use of parallel-plates motion has better sensitivity for detecting shuttle motion since y_0 is usually smaller than x_0. Although the nonlinearity in parallel-plates sensing can be linearized by the differential drive configuration to some extent, in a design example of 4 μm gap with 1 μm stroke, the linear model shown in (32.7) still produces 6% deviation from the nonlinear model.

Among electrostatic microactuators that use capacitive position sensing, the capacitance variation due to shuttle motion ($\mathrm{d}C/\mathrm{d}x$) is typically at the level of 100 fF/μm. In order to obtain 10 nm position sensing resolution, the capacitance sensing circuit must be able to detect capacitance variation of 1 fF in the presence of parasitic capacitance and offset/mismatches from op-amps, which can easily result in an output voltage orders of magnitude larger than the output voltage from the designated capacitance variation. In most capacitive position sensing, the limiting factor for the sensing resolution is not the thermal noise but the sensing circuit's design.

Here we introduce two basic capacitance sensing circuits suitable for high-resolution position sensing [32.14]. The concept of the "synchronous scheme," as shown in Fig. 32.6 I, is to reduce the impedance of sense capacitors as well as the offset and 1/f noises from op-amps by applying modulation techniques on the sense voltage (V_s). The R_{dc} resistor on the feedback loop sets the DC voltage level at the input nodes of the "charge integrator." The effect of the presence of parasitic capacitance (C_p) is nullified by the virtual ground condition from the op-amps. The major drawback of the synchronous scheme is that the DC-setting resistor (R_{dc}) has to be large to ensure the proper gain for the capacitance sensing [32.14], which introduces excessive thermal noise into the sensing circuit. In addition, a large resistor usually consumes a large die space in implementation.

A switched-capacitance scheme, as shown in Fig. 32.6 II, is one alternative that avoids the use of DC-setting resistor. The capacitance sensing period is broken into two phases: reset phase and sense phase. During the reset phase, input/output nodes of capacitors

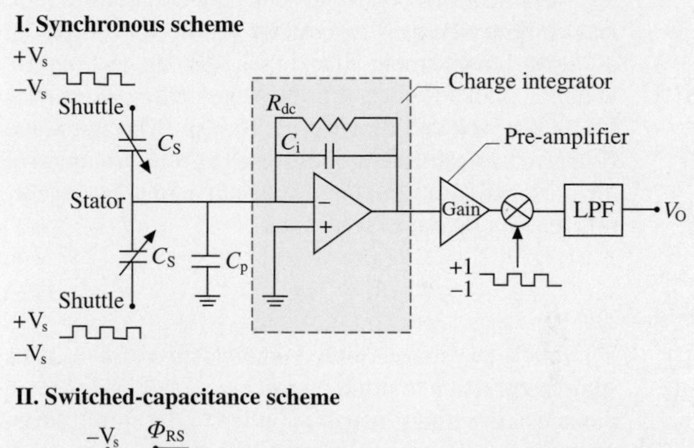

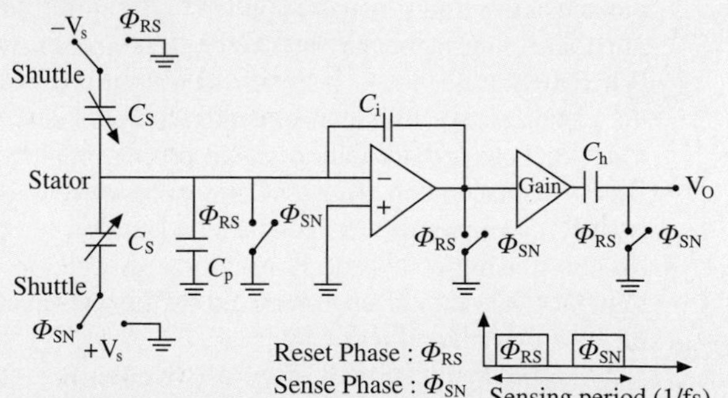

Fig. 32.6 Capacitive position sensing circuits. Synchronous scheme versus switched-capacitance scheme

and input voltage to op-amps are set to ground or reference level to ensure proper DC voltage for the charge integrator. During the sense phase, sense voltage $\pm V_s$ is applied to the sense capacitors, and the amount of charge proportional to the mismatch in the sense capacitors is integrated on the capacitor C_i, thus producing an output voltage proportional to the capacitance mismatch from the sense capacitors. This approach replaces the large DC-setting resistor by capacitors and switches and results in a much smaller die compared to the synchronous scheme. Furthermore, the switching technique allows more design flexibility for system integration and performance improvement because of the ability to allocate separate phases for various operations. The major drawback of this design is that it draws noise into sensing circuits from switches and sampling capacitor C_h. However, these noises can be compensated by dividing the sense phase into $2 \sim 3$ sub-sense phases [32.14], at the expense of a complicated circuit design.

Piezoresistive Sensing

Piezoresistive films have been widely used as strain-sensitive components in a variety of MEMS devices, including pressure sensors and vibration sensors. Generally speaking, piezoresistive sensing techniques require less complicated sensing circuits and perform better in a severe environment than other sensing techniques. When a piezoresistive film is subjected to stress, the film resistivity and dimensions change. The fractional change of resistance is proportional to the deformation of the piezoresistive film. For a small change of resistance, this relation can be expressed as,

$$\frac{\Delta R}{R} = K \cdot \epsilon \,, \tag{32.8}$$

where R is resistance of the piezoresistive film, K is its gage factor and ϵ is strain. In microactuator designs, the piezoresistive film is usually applied to the spring structure that connects the moving shuttle to an anchor point. When the shuttle moves, it stretches the spring as well as the piezoresistive film, consequently, the piezoresistive film produces a deformation signal proportional to the shuttle displacements. As a result, piezoresistive sensing is easier to implement than capacitive position sensing, but the sensing resolution is usually less accurate due to higher thermal noises introduced by the resistance of the piezoresistive film.

Another application of piezoresistive sensing, aside from measuring relative slider position, is to detect vibration in the suspension itself. The idea is to sense airflow-induced vibration of the suspension and feed that information forward to an actuated slider to damp out motion at the head. Piezoresistors used for this purpose can be made from metal or semiconductor materials, arranged as a strip or series of strips oriented along the direction of vibration strain. It is important that these sensors observe all vibration modes that contribute to off-track error, so a number of optimization schemes for locating the sensors have been developed [32.15]. One method is to maximize the minimum eigenvalue of the observability matrix of the sensor or sensors. This ensures that all relevant modes are observed. Another method is to minimize a linear quadratic gaussian control problem over potential sensor locations [32.16]. This serves to determine an optimal placement from the perspective of a linear controller.

32.1.5 Electrostatic Microactuator Designs for Disk Drives

Various electrostatic microactuators have been designed for secondary actuation in HDDs. To incorporate an electrostatic microactuator into a HDD without altering much of current suspension configuration, many design constrains are imposed. In this section, we will first discuss some design issues and then present one specific design example.

Translational Microactuators versus Rotary Microactuators

Depending on the motion of the magnetic head actuated, microactuator designs are categorized into two groups: translational actuators and rotary actuators. Either type can be implemented by comb-drives [32.8, 17] or parallel-plates [32.5, 6, 18] actuation.

When employing a translational microactuator in a dual-stage HDD servo, previous research [32.13] has shown that a force coupling between the suspension and the translational microactuator exists, consisting of transmitted actuation force from the VCM and suspension vibration induced by windage. The force coupling from the VCM not only complicates the dual-stage servo controller but also imposes a design constraint on a translational microactuator design, in that the translational microactuator has to provide a large force output to counterbalance the coupling force. When the VCM makes a large movement, as in seeking a new data track, the microactuator may be overpowered. One solution is to pull the actuator to one side and lock it momentarily in place. Even then, the use of the two actuation-stages must be carefully coordinated to moderate the influence of the VCM on the microactuator. On the other hand, the linear springs in the translational microactuator can also aid in damping out motion of the suspension. The portion of suspension vibration induced by windage mostly consists of high frequency excitation, so the resulting magnetic head's position error can be passively attenuated by low resonant frequency translational microactuators.

Generally speaking, rotary actuators are more difficult to design/analyze than translational actuators because of their nonuniform gap between shuttle and stator. Still, their different operating properties have both strengths and weaknesses. Unlike a translational actuator, no obvious force coupling is transmitted from the suspension to a rotary actuator, as the microactuator is nearly always attached to the end of suspension at the microactuator's center of rotation, which acts as a pivot point. With no mechanical coupling, the dual-stage servo system using a rotary microactuator does not suffer from the force coupling between VCM and microactuator seen in translational designs. However, the rotary microactuator has to compensate for the magnetic head's position error induced by suspension vibration without any passive attenuation of the vibration. Overall, a rotary acuator is likely to behave better than a translational microactuator during track seeking and worse during track following.

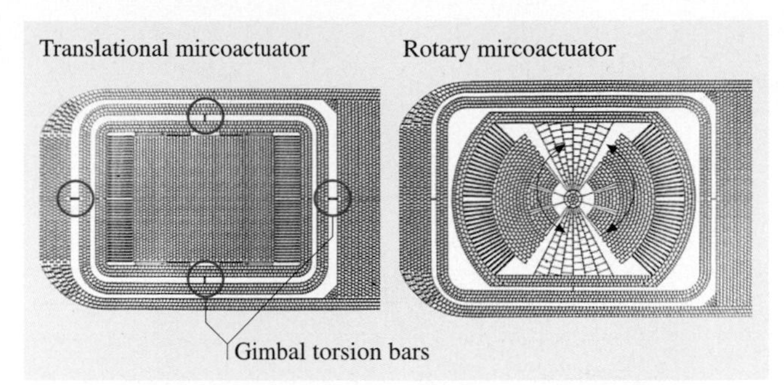

Fig. 32.7 Schematics of translational microactuator versus rotary microactuator. Courtesy from Lilac Muller

Gimballed Microactuator Design

Proper flying height and orientation of a slider and read-write head over a hard disk is maintained by the interaction of the suspension, the air bearing of the slider, and a gimbal structure. The gimbal structure is located at the tip of the suspension and holds the slider/microactuator in its center coupon. The dynamic characteristic requirements of gimbal structures are that they be flexible in pitch and roll motion but stiff in in-plane and out-of-plane bending motion. To meet all these requirements using only one piece of metal is a highly challenging task. For this reason, most commercially available suspension/gimbal designs consist of two to four pieces of steel, each with different thickness.

The goal of a gimballed microactuator design is to seamlessly integrate an actuator and gimbal into a one-piece structure, as shown in Fig. 32.7, to simplify both suspension design and HDD assembly. A full integrated suspension that includes suspension, gimbal, and microactuators in one part has also been proposed [32.19]. The dimple structure, existing in most current gimbal structures, is excluded from the gimballed microactuator design and electrical interconnects are in situ fabricated on the gimballed microactuator, replacing the flexible cable in current HDD suspension assembly.

The dimple structure in current suspension assemblies provides out-of-plane stiffness while preserving the necessary torsional compliance in the gimbal structure. Without a dimple structure in the suspension assembly, the gimbal itself must provide high out-of-plane bending stiffness. Otherwise, it would unbalance the suspension pre-load, which is an overbend of the suspension that balances the upward air-bearing force on the suspension during operation. The electrical interconnects are implemented to transmit data between magnetic heads, located at the center coupon of the gimbal structure, and IC circuits located at the end of the suspension. The in situ fabricated electrical interconnects are inevitably passed through torsion bars of the gimbal and thus set a design constraint for the minimum width of torsional bars. Furthermore, both the gimbal structure and microactuator should be the same thickness to simplify the MEMS fabrication process.

To summarize the design constrains discussed above, the integrated gimbal structure has to meet performance requirements with a single, uniform piece of material that would previously have been achieved by two to four metal pieces with different thicknesses, while the minimum width of any torsion bars that may be used pre-determined. To solve this problem, *Muller* [32.6] proposed a T-shaped structure (a beam structure with overhang surface sheet) for the torsion bars, and *Chen* [32.19] proposed "double-flexured" torsion bars. Additionally, many suspension manufacturers have developed new gimbal structures for their suspensions designed specifically for use with MEMS microactuators, moving the gimbal location back to the suspension from the microactuator.

Fig. 32.8 Pico-slider mounted on a translational microactuator. Courtesy from Horsely 1998

An Electrostatic Microactuator Design Example

Figure 32.8 shows a translational microactuator design suitable for the HDD dual-stage actuation by *Horsley* in 1998 [32.5]. The translational electrostatic microactuator dimensions are 2.2 mm × 2.0 mm × 0.045 mm and weight 67 μg. The dimensions of the pico-slider on the top are 1.2 mm × 1.0 mm × 0.3 mm and weight 1.6 mg. This microactuator design does not include "electrical-isolation" features, and thus the electrical isolation and electrical-interconnects were fabricated on a separate substrate and subsequently bonded to the microactuator. This microactuator uses parallel-plates for actuation force but does not have dedicated position sensing structures due to fabrication process limitations. Table 32.1 summarizes key parameters of this microactuator design.

Based on these parameters, the characteristics of this electrostatic microactuator can be estimated by the linear differential equation shown in (32.6).

Figure 32.9 shows the schematics of a circuit design by *Wongkomet* in 1998 [32.14], in which actuation driving voltage and capacitive position sensing were implemented for the electrostatic microactuator designed by Horsley. As mentioned before, the electrostatic microactuator design doesn't have dedicated structure for a position sensing; as a consequence, the input nodes for actuation and output nodes for capacitive position sensing have to share the same electrodes. Capacitors C_c and C_{c0}, shown in Fig. 32.9, are carefully designed to shield the high voltage presented in actuation circuit from sensing circuit, which is mostly low voltage, and thus enable driving/sensing circuit integration.

The driving circuit, shown in the left in Fig. 32.9, demonstrates how to generate the bias voltage (V_{bias}, ±40 V) and drive voltage (V_{dr}, −40–+40 V) from 0–5 V CMOS compatible circuits. The switches at the output of the charge pumps were synchronized with the switching period ϕ_{RS}. During the sense phases ϕ_{SN1} and ϕ_{SN2}, therefore, the switches are left open and thus no voltage fluctuation is seen by the sensing circuits. This arrangement was utilized to reduce feedthrough from the driving circuit to position sensing circuit.

The design target for the capacitive position sensing circuit was to achieve position sensing resolution of 10 nm, and this goal was approached by two main techniques implemented in the circuit: differential sensing and Correlated-Double-Sampling (CDS). The main

Table 32.1 Parameters of the electrostatic microactuator design by *Horsley* 1998 [32.5]

Parameters	Source*	Value
Nominal gap	D	10 μm
Structure thickness	D	45 μm
Rotor mass, m	I	44 μg
dC/dx	C	68 fF/μm
Actuation voltage, bias voltage	D	40 V
Actuation voltage, maximum driving voltage	D	±40 V
Voltage-to-force gain K_v	I	50 nN/V
Mechanical spring constant K_m	I	29 N/m
Electrostatic spring constant K_e	I	9.6 N/m
Damping coefficient, b	I	1.03×10^{-4} N/(m/s)
Voltage-to-position DC gain	M	0.05 μm/V
Resonance frequency, w_r	M	550 Hz

*: D = Design value, C = Calculation, M = Measurement, I = Inferred from measurements

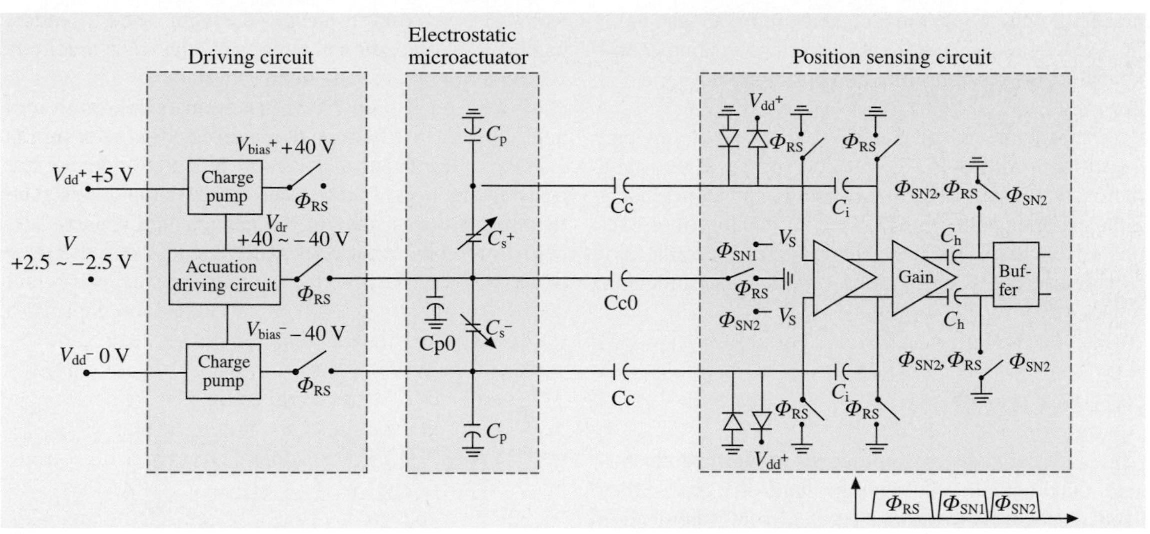

Fig. 32.9 A simplified schematic including driving and sensing circuits for the electrostatic microactuator. Courtesy from Wongkomet 1998

benefits of the differential sensing scheme are reduced noise coupling and feedthrough, elimination of even-order harmonics, and improvement of dynamic range by doubling the output swing. To adapt this differential sensing scheme in a differential parallel-plates electrostatic microactuator, a bias voltage (V_{bias}) was applied to the stator and the drive voltage (V_{dr}) was applied to the shuttle. The CDS technique, a modified capacitance sensing technique based on the switched-capacitance scheme, was implemented along with the differential sensing scheme to compensate for sensing noises including 1/f, KT/C, switch charge injection and offset from op-amps [32.14]. The concepts of CDS can be briefly described as follows: the sense period is broken into three phases: one reset phase, ϕ_{RS}, and two sense phases, ϕ_{SN1} and ϕ_{SN2}. During the ϕ_{RS} sense phase, the voltages for capacitors and input nodes to op-amps are set to the reference level of the DC-voltage setting for the charge integrator, same as for the switched-capacitance scheme discussed in Sect. 32.1.4. During the ϕ_{SN1} sense phase, a sensing voltage $-V_{\text{s}}$ is applied to the shuttle and results in a voltage difference, $\alpha(-V_{\text{s}}) + V_{\text{error}}$, across the sampling capacitor C_{h}, where α is the transfer function from sense voltage to the output voltage at the pre-amplifier, shown as *Gain* in the plot, and V_{error} is the voltage at the output node of pre-amplifier resulting from noise, leakage charge, and offset of op-amps. Lastly, during the ϕ_{SN2} sense phase, the sensing voltage is switched from $-V_{\text{s}}$ to V_{s} and the switch next to C_{h} is switched open. This results in a voltage output $\alpha(V_{\text{s}}) + V_{\text{error}}$ at the output node of the pre-amplifier and a voltage, $2\alpha(V_{\text{s}})$, at the input node of the buffer. As a consequence, the voltage resulting from sensing error (V_{error}), which appears at the output node of pre-amplifier, disappears from the voltage-input node to the buffer. Switches utilized in

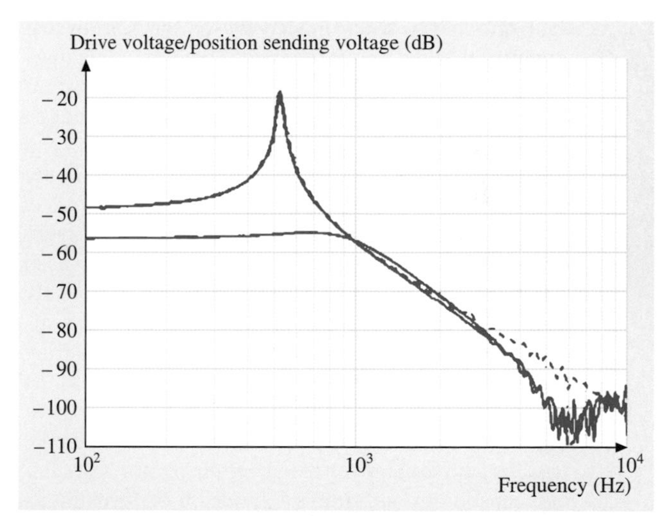

Fig. 32.10 Open-loop and closed-loop frequency response of the prototype microactuator with a pico-slider. The capacitive position measurement (*solid line*) is compared to the measurements from LDV (*dashed line*). Courtesy from Wongkomet

the circuit and their correspondent timing are shown in Fig. 32.9. Be aware that for presentation simplicity components in the pre-amplifier and buffer are not shown in detail in Fig. 32.9.

The frequency response, both open-loop and closed-loop with a PD controller, of the electrostatic microactuator measured by a capacitive position sensing circuit is shown in Fig. 32.10. The position measurements from an LDV are also shown in the same plot for comparison. The deviation between measurements from the different position sensing devices appears in the high frequency region of the plot and has been identified as the feedthrough from the capacitance sensing circuit. The effect from feedthrough was negligible at low frequency region but becomes significant and results in a deviation in magnitude for the transfer function of $-80\,\mathrm{dB} \sim -90\,\mathrm{dB}$ after 2 kHz. The feedthrough presented in the capacitance sensing circuits limited the position sensing resolution to the level of 10 nm.

32.2 Fabrication

While there are several approaches to building electrostatic microactuators suitable for hard disk drives, they all exhibit certain common features from a fabrication standpoint.

32.2.1 Basic Requirements

As we discussed earlier, nearly all electrostatic microactuators rely on a system of interlaced fingers or plates to provide actuation force. As a result, a method for producing arrays of these fingers or plates with narrow gaps between them is usually the central concern in developing a fabrication process. The resulting structure must then be strong enough to support both the slider on the microactuator and the microactuator on the suspension, particularly when loaded by the air-bearing that supports the slider above the hard drive's spinning disk. In addition, the design and fabrication process must include a way to perform electrical interconnection on the microactuator. This involves transferring signals to and from the actuator and slider and isolating the parts of the microactuator requiring different voltage levels.

Meanwhile, the microfabrication process is subject to certain basic constraints. The materials used in fabrication must either be thermally and chemically compatible with any processing steps that take place after their deposition or must somehow be protected during steps that would damage them. This often constrains the choice of materials, deposition techniques, and processing order for microdevices. Another concern is that the surface of the structure be planar within photoresist spinning capabilities and lithography depth-of-focus limits if patterning is to be performed. This can be a major challenge for disk drive microactuators, which are large in size and feature high-aspect ratio trenches compared to other MEMS devices.

32.2.2 Electrostatic Microactuator Fabrication Example

Section 32.1.5 described the design and operation of a translational electrostatic microactuator. This section examines the fabrication process by which that microactuator was built [32.5]. The process is a variation on a micro-molding process known as HexSil [32.20]. In a micro-molding process, a mold wafer defines the structure of the microactuator and may be reused many times like dies in macro-scale molding; in the HexSil version of the procedure, the mechanical parts of the microactuator are formed in the mold by polysilicon. A second wafer is used to create metallized, patterned target dies. Upon extraction from the mold, the HexSil structure is bonded to this target substrate, which is patterned to determine which sections of the HexSil structure are electrically connected.

Naturally, the mold wafer is the first item to be processed and is quite simple. It is a negative image of the desired structure, etched down into the wafer's surface, as shown in Fig. 32.11a. A deep but very straight etch is critical for successful fabrication and subsequent operation of the devices. If the trench is too badly bowed or is undercut, the finished devices will be stuck in the mold and difficult to release. On the other hand, a tall device will have a larger electrostatic array area, will generate more force, and be better able to support a slider while in a disk drive.

Fabrication of the HexSil structure forms the majority of the processing sequence Fig. 32.11b–g. HexSil fabrication begins by coating the surface of the mold wafer with a sacrificial silicon oxide. This layer, deposited by low-pressure chemical vapor deposition (LPCVD), must coat all surfaces of the mold, so that removal of the oxide at the end of the process will

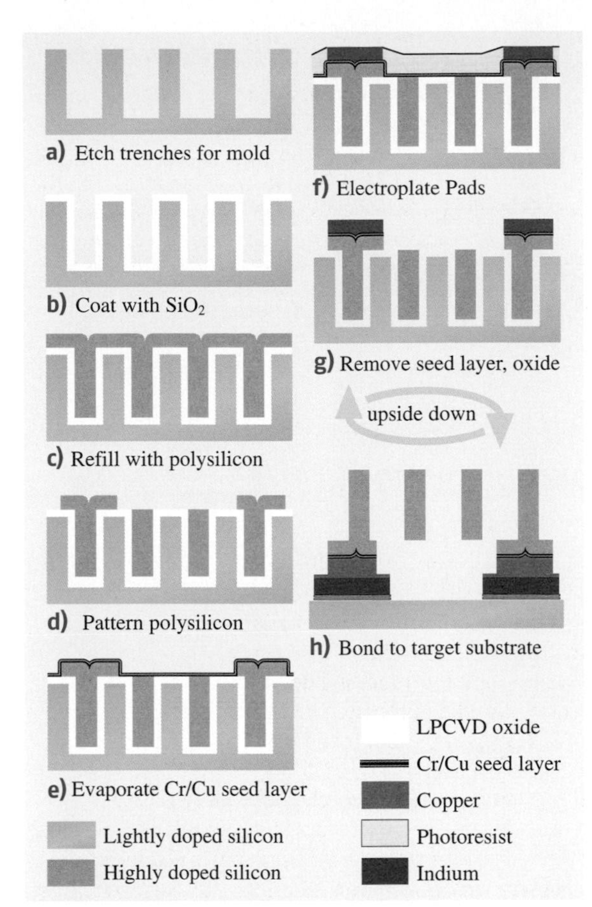

Fig. 32.11 Fabrication example 1: HexSil

leave the device completely free of the mold. This microactuator uses a 3 to 4 μm thick oxide layer to ensure clearance between the polysilicon structure and the trench walls during release. The mold is then refilled completely with LPCVD polysilicon, which will form the desired mechanical structure. Polysilicon is chosen for its conductivity and good conformality in refilling trenches. This also leaves a planar surface ready for photolithography to cover bonding locations. After the lithography step, the polysilicon deposited on top of wafer is removed, except where the bonding points were defined.

The structure is then prepared for bonding by forming a soldering surface. First a chrome/copper seed layer is evaporated on the surface of the wafer. The chrome promotes copper adhesion to the silicon, while the copper forms the starting point for electroplating. The seed layer is covered with photoresist, which is cleared only in the locations where contacts are desired. There, electroplating will produce a metal film, forming soldering points for the device to the target substrate that will form its base. In this case, copper is used as the plating material, thanks to its high conductivity and solderability. The device is released from the mold by ion milling away the thin seed layer, then dissolving the oxide lining with a hydroflouric (HF) acid wet etch. The free standing portion of the microactuator is thus ready for bonding to the target. The mold, meanwhile, may be used again after a simple cleaning step.

The target substrate may be prepared in several ways. The target shown above consists simply of indium solder bumps and interconnects on a patterned seed layer; this is known as a "plating bus" arrangement. The seed layer is used for the same reasons as on the HexSil structure, but in this case the seed layer is patterned immediately, with the unwanted portions removed by sputter etching. This provides isolation where desired without having to etch the seed layer after the indium is in place. Photoresist is spun again and patterned to uncover the remaining seed layer. A layer of indium is then electroplated everywhere the seed layer is visible. Indium acts as the solder when the two parts of the microactuator are pressed together, forming a cold weld to the copper at 200–300 MPa as shown in Fig. 32.11h, that was found suitable for hard disk requirements.

32.2.3 Electrostatic Microactuator Example Two

The high-aspect ratio microactuator described in this section is in development at the University of California, Berkeley, and demonstrates another way to create the features required of an actuated slider. The microactuator is translational, with parallel-plates actuation, and includes two of the design options described previously: integrated isolation plugs, as described in Sect. 32.1.3, and a capacative sensing array, as described in Sect. 32.1.4. A picture of the basic design is shown in Fig. 32.12; a central shuttle holding the slider is supported by four folded-flexure springs and driven by parallel-plates arrays.

Basic Process: Silicon-on-Insulator

Successful processing of the design is centered around the ability to etch very straight, narrow trenches using deep reactive ion etching (DRIE). DRIE is a special plasma etching sequence that uses polymerization of the sidewalls of a trench to keep the walls straight even at extreme aspect ratios. Deep, narrow trenches

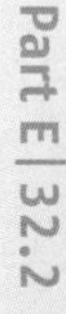

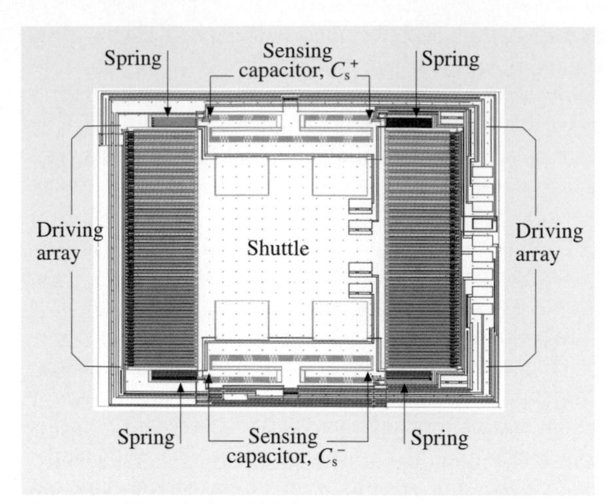

Fig. 32.12 Translational high-aspect ratio microactuator

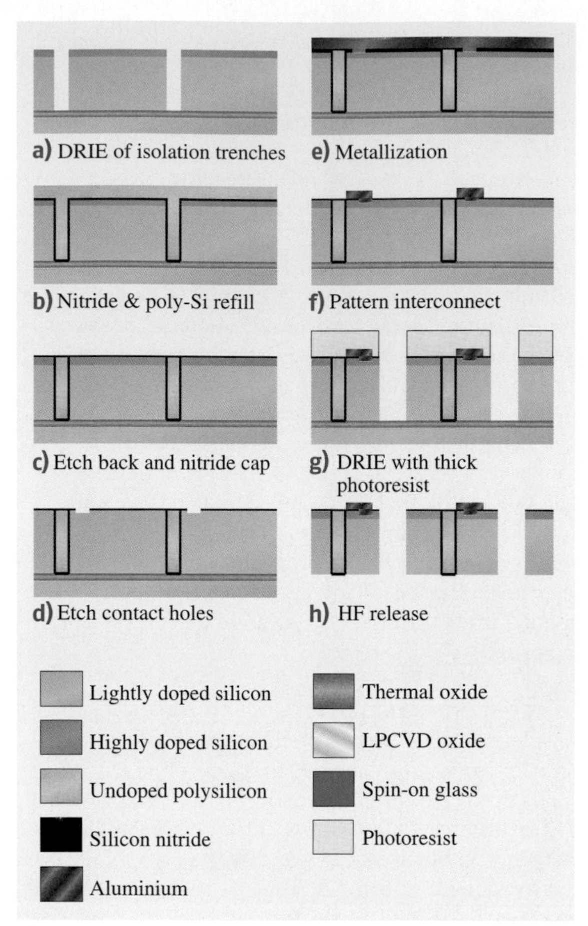

Fig. 32.13 Basic silicon-on-insulator fabrication process

make possible a microactuator with both closely packed electrostatic arrays and good mechanical strength. Conventional surface micromaching is used to produce the majority of the electrical interconnects.

The basic fabrication process uses a silicon-on-insulator (SOI) wafer to control DRIE trench depth. SOI wafers are very useful in microfabrication as they provide a layer of single-crystal silicon (the device layer) separated from the bulk of the wafer (the handle layer) by a thin layer of buried silicon oxide. This gives a very well controlled thickness to finished device but is very expensive. The microactuators described here are fabricated from an SOI wafer with a 100 μm device layer. A variation on the processing sequence and layout eliminates the need for SOI wafers and will be discussed in the following section.

The first stage of fabrication is creation of deep isolation trenches (see Fig. 32.13a–c). This procedure has been adapted to MEMS from integrated circuit processing for isolating thick MEMS structures [32.21]. The isolation pattern is formed by photolithography and etched by DRIE down to the buried oxide. With trenches only 2 μm wide, this corresponds to an aspect ratio of 50 : 1. The wafer surface and trenches are then coated with LPCVD silicon nitride, which acts as the electrical insulator. The trench is then refilled in its entirety with LPCVD polysilicon, which is a more conformal material better fills the trench than silicon nitride would alone; polysilicon also has much lower residual stresses than silicon nitride. The refill leaves a layer of polysilicon on the surface of the substrate, which must be etched or polished back to the silicon nitride. After etch-back, a second layer of LPCVD nitride completes isolation between regions.

Next, electrical interconnects are formed, as shown in Fig. 32.13d–f. Contact holes to the substrate are defined by photolithography and etched by reactive ion etching of the silicon nitride. Metal lines may then be patterned directly by photolithography or formed by lift-off. In a lift-off process, photoresist is deposited and patterned before the metal. The metal is then deposited vertically, so that the sidewalls of the resist are uncovered. Dissolving the resist allows the metal on top to float away, leaving behind the interconnect pattern. In the figures shown here, the interconnects are formed by evaporated aluminum, eliminating the need for electroplating and a seed layer. The ability to isolate portions of the substrate is useful for simplifying the interconnect layout, as one interconnect can cross another by passing underneath it through an isolated portion of the substrate.

Finally, structural lithography is performed and a second etch by DRIE is done, as shown in Fig. 32.13g. These trenches will define the shape of the microactuator, and are larger, 4 μm wide, to allow sufficient rotor travel. After this etch step, the device layer on the SOI buried oxide consists of fully defined microactuators. When the buried oxide is removed by wet etching, the microactuator is released from the substrate Fig. 32.13h and, after removal of the photoresist that protected the device surface from HF during release, ready for operation. Thanks to the high-aspect ratios of DRIE and the efficient integrated isolation scheme, the microactuator can theoretically be operated with a DC bias voltage of 15 V on the rotor, and a dynamic stator voltages below 5 V. In practice, higher voltages are required due to the spreading of trench walls over the course of DRIE. A picture of the electrostatic structure, with the rotor fingers on the left and the stator fingers on the right, divided down the center by isolation trenches, is shown in Fig. 32.14.

Variation: Anisotropic Backside Release

An alternative fabrication approach is to build the microactuator from a regular silicon wafer (~ 550 μm thick), but then, since the microactuator shouldn't be more than 100 μm thick to fit in a HDD, etch away the backside of the wafer to obtain devices of that size. The main reason for using a silicon-only wafer is to reduce the need for expensive SOI wafers, which cost approximately ten times as much as single-material wafers. Another benefit is the elimination of the need for etch holes to reach the buried oxide layer, which increase vulnerability of microactuators to particle or moisture contamination. However, the thickness of the devices cannot be controlled as accurately as with the SOI process, and the processing sequence is more complicated.

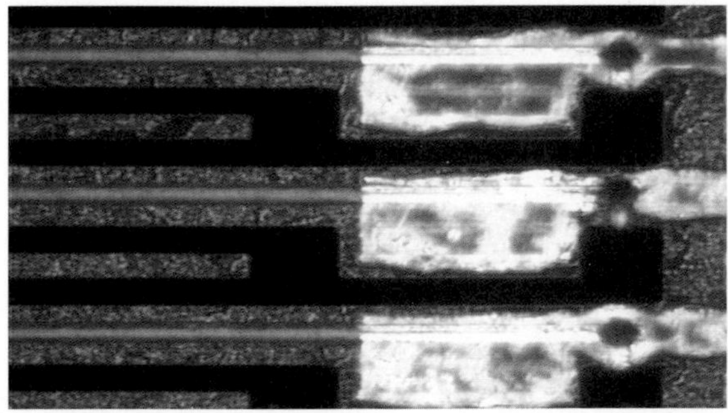

Fig. 32.14 Electrostatic driving array of microactuator from SOI

Control of backside etching is difficult, making it hard to obtain devices with uniform thickness across a wafer. A proposed solution is to use an anisotropic wet etchant during the backside etch step. Anisotropic etchants work very slowly on certain crystal planes of a single-crystal silicon wafer; a coating on trench sidewalls can protect fast-etching planes when the etchant reaches the trenches from the back, causing slow-etching crystal planes to begin coming together slowing the etch. The result is a nearly self-stopping etch with a V-shaped profile that prevents overetching even in the presence of nonuniformities in etch rate and trench depth across the wafer. The most common anisotropic etchants are potassium hydroxide (KOH) and tetramethyl ammonium hydroxide (TMAH). This release process is referred to here as an anisotropic backside release (ABR).

The key to a good release by this method is to achieve excellent protection of the top of the wafer and the deep trenches from the etchant, and this neccessitates changes in the process flow from that for an SOI wafer. Without a good protective film, the interconnects on the surface of the wafer could be destroyed, or the fast-etching planes might be attacked along the trench sidewalls. Of the etchants mentioned above, TMAH is preferred, due to high selectivity to both silicon oxide and nitride and, especially, thermally grown oxide, which can be used as protective films. However, growing thermal oxide and/or depositing high-quality nitride are high temperature processes not compatible with most metals, so the structural trenches must be etched and oxidized before forming the electrical interconnects. As shown in Fig. 32.15c–d, the isolation trenches are formed as before, but the structural trenches are patterned and etched in the very next step. A thermal oxide is then grown in the trenches to produce the protective oxide layer Fig. 32.15e.

To do lithography for the interconnects at this point, then, the surface of the wafer must be planarized. This is done by refilling the trenches with spin-on-glass. The spin-on-glass plugs the trenches nearly to the top and can be sacrificed along with the thermal oxide at the end of the process. Extra glass on the wafer surface can be removed by a quick HF dip or chemical mechanical polishing. The resulting surface is uneven, as shown in Fig. 32.15f, but a thick photoresist layer can be applied over the trenches evenly enough to accomplish contact and metal patterning lithographies. After metallization much the same as for the SOI wafer Fig. 32.15g, a final coating of etch resistant material is required to protect the metal during backside release, as shown in

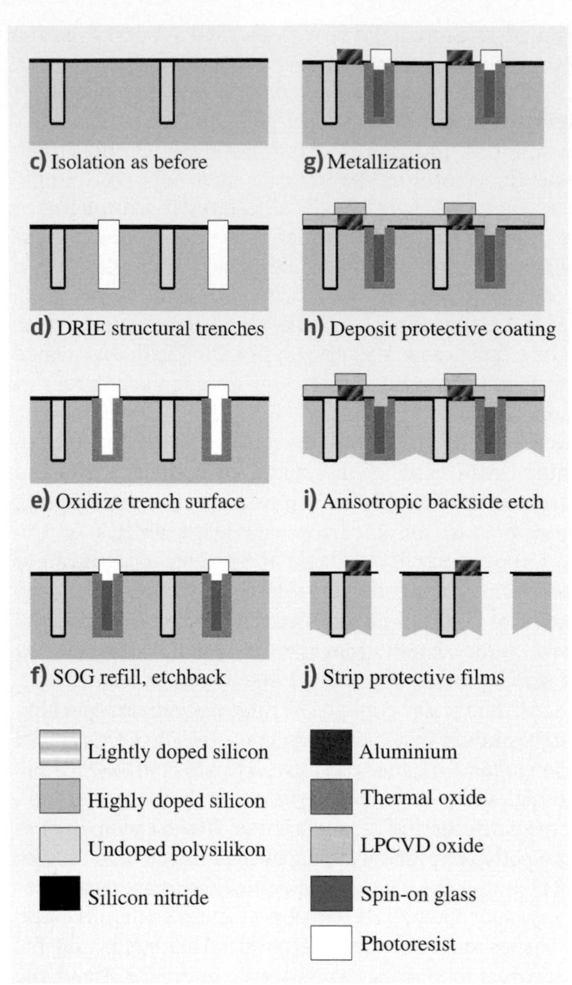

Fig. 32.15 Deep trench fabrication process adapted for backside release

Fig. 32.15h–j. After ABR, the backside of the device will be shaped by crystal planes, but the thickness of the device at the edge of DRIE features will be fixed. While a combination of silicon oxide and silicon nitride has been effective over a "metallization" of undoped polysilicon, an effective combination of true metal interconnects and protective films has yet to be found and is an ongoing area of research for improving the ABR process sequence.

A prototype microactuator with integrated gimbal fabricated using ABR is shown in Fig. 32.16. This prototype was operated with a 30 V DC bias and ±8 V AC driving voltage for ±1 μm displacement, while a larger version (shown in Fig. 32.12) was operated at 15 V bias and ±3 V driving. This layout and fabrication concept is currently being adjusted to fit a MEMS-ready suspension for more advanced testing.

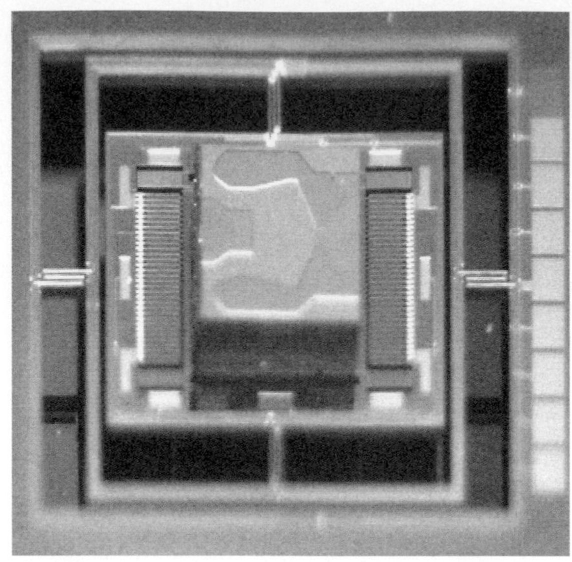

Fig. 32.16 Microactuator with integrated gimbal fabricated by ABR

32.2.4 Other Fabrication Processes

Where the DRIE-based process described above "digs" the microactuator out of a wafer, a procedure developed by IBM "grows" a microactuator with high-aspect ratio trenches on top of a wafer [32.22].

IBM Electroplated Microactuator

The microactuator is fabricated by a clever sequence of electroplating steps and sacrificial depositions. An oxide sacrificial film is first patterned then covered with a metal seed layer. When later removed, the sacrificial will have left a gap between the microactuator and substrate portions of the microactuator needed to move freely. A polymer, 40 μm thick, is spun onto the wafer and patterned with a reverse image of the main structural layer. Trenches are etched down to the sacrificial layer by plasma etching then refilled by electroplating, much like micro-molding. Another sacrificial polymer, photoresist this time, separates the top of the fixed portions of the structure from the platform upon which the slider will sit, while conveniently planarizing the surface for further lithography. The platform is created by two more electroplating steps on top of the parts of the structure that will move freely. Removal of the sacrificial layers (photoresist, polymer, and oxide) releases the microactuator for use.

This process has been used to produce both rotary and linear microactuators and is comparatively advanced from a commercial standpoint. The resulting devices are thinner than those described in the previous example, but a similar aspect ratio (20 : 1) can be achieved. This procedure also has the benefits of being a low temperature process, which helps decrease processing cost and is compatible with the thin film magnetic head manufacturing process. Moreover, it includes an upper structure that covers the fingers, which improves device reliability by shielding the fingers from particles.

Additional fabrication details and dynamic testing results for the IBM design may be found in references [32.22] and [32.23]. Certain control results are described in Sect. 32.3.2.

Electroplated Microactuators with Fine Gaps

Finally, fabrication processes developed at the University of Tokyo demonstrate additional techniques for obtaining very small gaps between electrostatic fingers [32.24]. Similar to the IBM process, electrostatic fingers are electroplated, in this case by nickel, within a polymer pattern. The gap between closely spaced fingers, however, is then formed by a sacrificial metal. Photoresist is used to cover the fingers except for the surfaces where a small gap is desired; these are electroplated with sacrificial copper, making narrow, well-defined gaps possible, in perhaps the most reliable of the processes described here. Continuing with a second nickel electroplating step creates the interlaced fingers, which will also support the slider; the wafer is polished down to level the structure. Finally the copper, photoresist, and seed layer are etched away, leaving the free standing structure.

In another version of process, photoresist alone separates the stator and rotor fingers. In this case, a thin layer of photoresist is left where small gaps are desired, instead of growing a layer of copper. This method is dependent on excellent alignment but can eliminate the need for polishing after the second electroplating step.

32.2.5 Suspension-Level Fabrication Processes

The long, highly integrated fabrication processes required by actuated slider microactuators contrast greatly with the fabrication of actuated suspension microactuators.

PZT Actuated Suspensions

In most cases, actuated suspensions use piezoelectric (PZT) drivers cut or etched as single, homogeneous pieces and attached to the steel suspension. Nevertheless, microfabrication techniques can still be useful in suspension processing. For example, some recent suspensions have used thin film PZT deposited on a substrate by sputtering or spinning-on of a sol-gel for incorporation in a suspension. This is intended to produce a higher quality PZT films and begins to introduce MEMS-style processing techniques even into suspension-scale manufacturing [32.25, 26].

Integrated Silicon Suspension

Another area where silicon processing has been suggested for use on a suspension is the gimbal region, or even the entire suspension, as described in [32.19]. An integrated silicon suspension would incorporate aspects of both actuated suspension and actuated slider fabrication. Force is provided by four piezes of bulk PZT cut down to size, as in other actuated suspensions, but the rest of the suspension is fabricated from single-crystal silicon by bulk micromachining.

The fabrication process for the integrated suspension is the precursor to the ABR process used to fabricate the microactuator in Sect. 32.2.3. Layers of LPCVD silicon nitride and polysilicon are deposited on a bare silicon wafers. The polysilicon is doped by ion implantation and covered with a second layer of nitride. The outline of the suspension and any spaces within it are then lithographically patterned. After a reactive-ion nitride etch to reach the substrate, trenches are etched by DRIE. The trenches are then plugged with spin-on-polymer for further lithography. First, the top layer of polysilicon is patterned into piezoresisistive strips for sensing vibration. Second, a layer of photoresist is patterned for copper lift-off, to create metal interconnects. Last, the planarizing polymer is removed, and the entire surface is coated with silicon nitride. This final coating protects the top surface during an anisotropic backside wet etch, in this case by potassium hydroxide.

Instrumented Suspensions

Finally, microfabrication techniques may be useful for installing sensors on conventional steel suspensions. The concept is to deposit a dielectric and a piezoresistive material on the stainless steel sheet that will be formed into the central piece of the suspension and then to pattern piezoresistive layer directly by photolithography and etching to form strain gages. A second lithography may be used to add metal lines back to the E-block. Vibration sensors on the suspension are very desirable for controlling slider position, and the use of thin-film deposition and photolithography permits the sensors to be located at the points most effective for detecting vibration. An example of a sensor formed by this method is shown in Fig. 32.17.

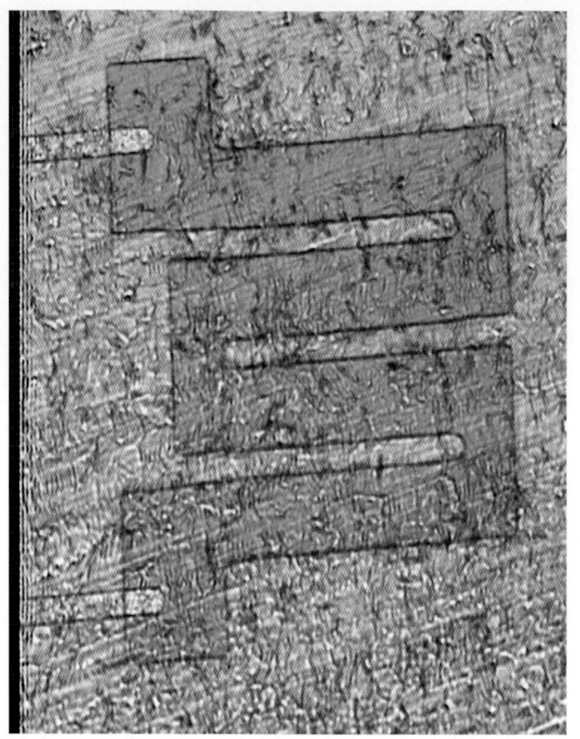

Fig. 32.17 Piezoresistive strain sensor on steel substrate

32.2.6 Actuated Head Fabrication

Several interesting fabrication processes have been proposed for another approach to creating a dual-stage disk drive servo: the actuated head. In an actuated head, the actuator is built into the slider and moves just the read/write head. This could potentially eliminate many of the mechanical limits of actuated suspensions or actuated sliders. Actuation techniques are typically similar to those of actuated sliders, with the key complication being the need to integrate an actuator fabrication process with a slider fabrication process. We discuss ways in which this might be done below.

One fabrication approach to actuating a read/write head is to enclose an electrostatic driving array inside a slider. The slider is built on a glass or silicon substrate, beginning with the air-bearing surface (ABS), followed by a microactuator, and then surrounded by the remainder of the slider. The ABS is formed by silicon oxide deposited over aluminum and tapered photoresist layers that give it a contoured shape. The majority of the ABS is then covered with photoresist, while a block in the center is plated with nickel and etched to form the electrostatic actuator. The electrostatic microactuator, in turn, is covered while the remainder of the slider body is electroplated. The slider may then be bonded to a suspension and the ABS surface released from the substrate by etching away the original aluminum and photoresist surface [32.9].

The sacrificial metal technique described in Sect. 32.2.4 is another possibility for building an electrostatic array onto a slider. It can be used to form small array with very narrow gaps, and the process is compatible with slider and head materials [32.27].

The third actuated-head process makes use of an SOI wafer and silicon based materials. For this structure, a 20 μm device layer is coated with silicon nitride. The nitride is patterned with contact holes for connections to the substrate and used as insulation over the rest of the device. A layer of molybdenum is sputter deposited to perform interconnections to both the head and the electrostatic array. This array is formed by DRIE to the buried oxide, using a lithographically patterned hard mask of tetraethylorthosilicate. The result is a simple, tiny, actuator that must be bonded to the edge of a slider after release from the SOI wafer [32.28].

32.3 Servo Control Design of MEMS Microactuator Dual-Stage Servo Systems

The objective of disk drive servo control is to move the read/write head to the desired track as quickly as possible, referred to as track seek control, and, once on-track, position the head on the center of the track as precisely as possible, referred to as track-following control, so that data can be read/written quickly and reliably. The implementation of the servo controller relies on the position error signal (PES), which is obtained by reading the position information encoded on the disk's data tracks.

32.3.1 Introduction to Disk Drive Servo Control

The position error is also called track mis-registration (TMR) in the disk drive industry. Major TMR sources in the track-following mode include spindle runout, disk fluttering, bias force, external vibration/shock disturbance, arm and suspension vibrations due to air turbulence, PES noise, written-in repeatable runout, and residual vibration due to seek/settling [32.29]. These sources can be categorized as runout, r, input disturbance, d, and measurement noise, n, by the locations they are injected into the control system, as shown in Fig. 32.18. In Fig. 32.18, $G_P(s)$ and $G_C(s)$ represent the disk drive actuator and the controller, respectively; r and x_p represent track runout and head position, respectively.

From Fig. 32.18, the PES can be written as

$$\mathrm{PES} = S(s)r + S(s)G_P(s)d - S(s)n\,, \tag{32.9}$$

where $S(s)$ is the closed loop sensitivity function defined by

$$S(s) = \frac{1}{1+G_C(s)G_P(s)}\,. \tag{32.10}$$

The higher the bandwidth of the control system, the higher the attenuation of the sensitivity function $S(s)$ below the bandwidth. Thus, one of the most effective methods to reduce PES and increase servo precision is to increase the control system bandwidth.

Traditional disk drive servo systems utilize a single voice coil motor (VCM) to move the head. Multiple structural resonance modes of the E-block arm and the suspension located between the pivot and the head impose a major limitation on the achievable control bandwidth. A dual-stage servo system using a MEMS actuated-slider microactuator can achieve high bandwidth because the microactuator is located between the suspension and the slider, thus by-passing the pivot, E-block arm, and suspension resonance modes.

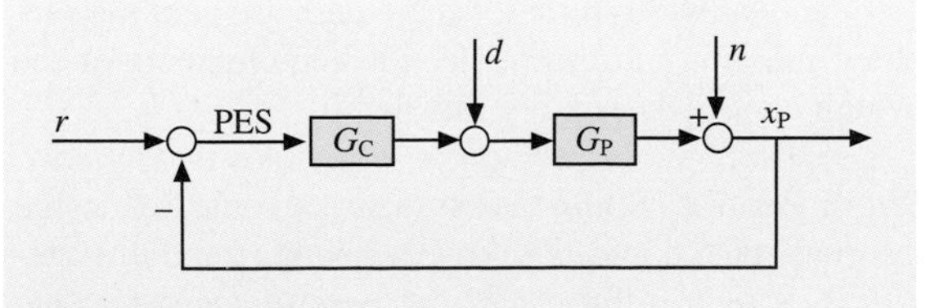

Fig. 32.18 Disk drive servo control

One basic task of dual-stage servo control design is to increase control bandwidth using the second stage actuator. Many design methodologies have been developed to accomplish this objective. In this section, we first review major dual-stage servo control design methodologies, and then we discuss design considerations for controller design of a MEMS microactuator dual-stage servo system. As design examples, the details of track-following control designs using a sensitivity function decoupling design method and the μ-synthesis design method are presented in Sect. 32.3.3. In Sect. 32.3.4, we introduce a short-span seek control scheme using a dual-stage actuator based on decoupled feed forward reference trajectories generations.

32.3.2 Overview of Dual-Stage Servo Control Design Methodologies

Various control design architectures and methodologies have been developed for dual-stage servo control design. They can be largely classified into two categories: those based on decoupled or sequential single-input-single-output (SISO) designs, and those based on modern optimal design methodologies, such as LQG, LQG/LTR, H_∞, and μ-synthesis, in which the dual-stage controllers are obtained simultaneously.

Two constraints must be considered in dual-stage servo control design. First, the contribution from each actuator must be properly allocated. Usually the first stage actuator, or the coarse actuator, has a large moving range but a low bandwidth while the second stage actuator, or the fine actuator, has a high bandwidth but small moving range. Second, the destructive effect, in which the two actuators fight each other by moving in opposite directions, must be avoided.

Classical SISO Design Methodologies

Several architectures and design methodologies have been proposed to transform the dual-stage control design problem into decoupled or sequential multiple SISO compensators design problems, for example, master–slave design, decoupled design [32.30], PQ method [32.31], and direct parallel design [32.32]. Figures 32.19 to 32.22 shows the block diagrams of dual-stage controller designs using these methods. In these figures, G_1 and G_2 represent the coarse actuator (VCM) and fine actuator (microactuator) respectively; x_1 is the position of the coarse actuator; x_r is the position of the fine actuator relative to the coarse actuator; x_p is the total position output; r is the reference input (runout), and PES is the position error, $e = r - x_p$.

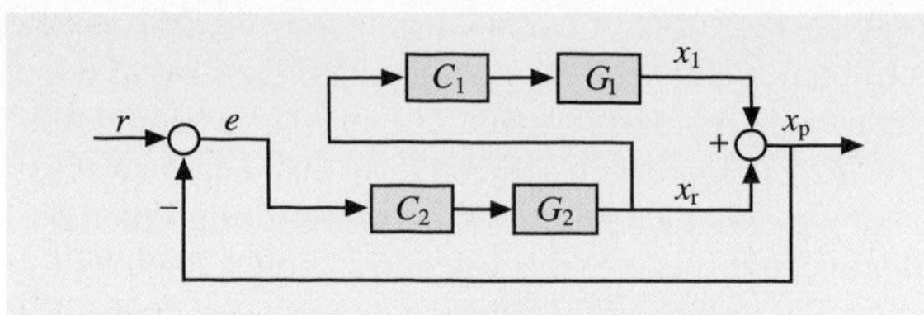

Fig. 32.19 Master–slave design structure

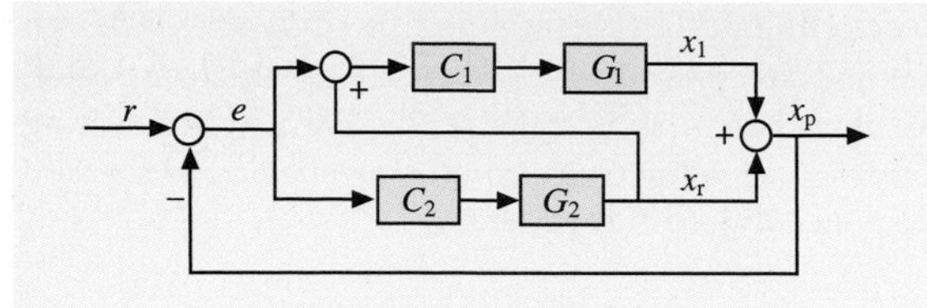

Fig. 32.20 Decoupled control design structure

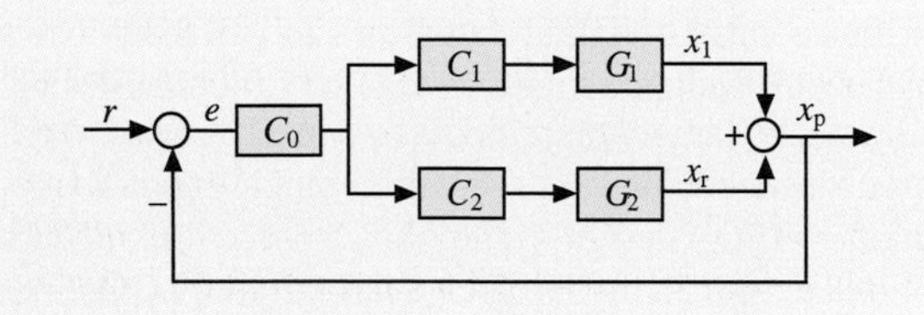

Fig. 32.21 PQ control design structure

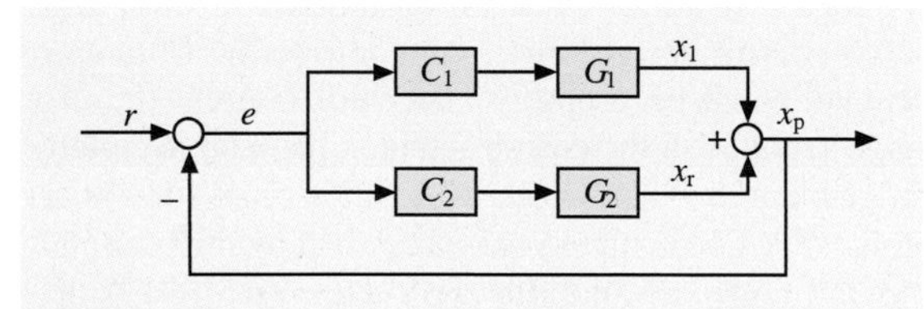

Fig. 32.22 Parallel control design structure

Master–slave Design. In a traditional master–slave structure, the absolute position error is fed to the fine actuator, and output of the fine actuator is fed to the coarse actuator, as shown in Fig. 32.19. The position error will be compensated by the high bandwidth fine actuator. The coarse actuator will follow the fine actuator to prevent its saturation.

Decoupled Design. Figure 32.20 shows the decoupled design structure [32.30], which is similar to the master–slave structure. A summation of the fine actuator output and the position of the coarse actuator, is fed the coarse actuator. A very nice feature of this structure is that the control system is decoupled into two independent control loops, and the total sensitivity function is the product of the sensitivity functions of each of the control loops [32.13]. Thus the two compensators C_1 and C_2 can be designed independently. Decoupled design is also referred to as decoupled master–slave design, or sensitivity function decoupling design. In Sect. 32.3.3, we will discuss details of a track-following controller for MEMS microactuator dual-stage servo system designed using this method.

Both the master–slave and the decoupled designs use the relative position of the fine actuator, x_r. If a relative position sensor is unavailable, x_r can be estimated using a model of the fine actuator.

PQ Design Method. The PQ method is another innovative design technique for control design of dual-input, single-output system [32.31]. A block diagram of a dual-stage control design using this method is shown in Fig. 32.21.

In PQ design, P is defined by

$$P = \frac{G_1}{G_2}\,, \tag{32.11}$$

and a dual-stage controller can be designed in two steps. The first involves the design of an auxiliary compensator Q for plant P, which is defined by

$$Q = \frac{C_1}{C_2}\,. \tag{32.12}$$

Q is designed to parameterize the relative contribution of the coarse and fine actuators. The 0 dB crossover frequency and phase margin of the open loop transfer function PQ are the design parameters in the design of Q. At frequencies below the 0 dB crossover frequency of PQ, the output is dominated by the coarse actuator, while at frequencies above the 0 dB crossover frequency, the output is dominated by the fine actuator. At the 0 dB crossover frequency, the contributions from the two actuators are equal. A large phase margin of PQ will ensure that the two actuators will not fight each other when their outputs are close in magnitude, thus avoiding any destructive effects.

The second step in the PQ design methodology is to design a compensator C_0 for SISO plant PQ such that the bandwidth (crossover frequency), gain margin, phase margin, and error rejection requirements of the overall control system are satisfied.

Direct Parallel Design. It is also possible to design the dual-stage controller directly using a parallel structure, as shown in Fig. 32.22, by imposing some design constraints and by sequential loop closing [32.32].

The two constraints for parallel design in terms of the PES open loop transfer functions are [32.32]:

$$C_1(s)G_1(s)+C_2(s)G_2(s) \to C_2(s)G_2(s)\ , \quad (32.13)$$

at high frequencies and

$$|C_1(s)G_1(s)+C_2(s)G_2(s)| \gg |C_2(s)G_2(s)|\ , \quad (32.14)$$

at low frequencies. The first constraint implies that open loop frequency response of the dual-stage control system at high frequencies approximately equals that of the fine actuator control loop. Thus the compensator $C_2(s)$ can be first designed independently as a SISO design problem to satisfy the bandwidth, gain margin, and phase margin requirements of the dual-stage control system. The second constraint ensures that the fine actuator will not be saturated. Compensator $C_1(s)$ can then be designed for the SISO plant model with the fine actuator control loop closed, such that the low frequency constraint and overall stability requirement are satisfied. This model is defined by:

$$G(s)=\frac{G_1(s)}{1+C_2(s)G_2(s)}\ . \quad (32.15)$$

A dual-stage controller for a MEMS dual-stage actuator servo system has been designed using this method and implemented at IBM. An open loop gain crossover frequency of 2.39 kHz with gain margin 5.6 dB and phase margin 33° was obtained. In experimental testing, a 1-σ TMR of 0.024 μm has been achieved [32.32].

Modern MIMO Design Methodologies

Since the dual-stage actuator servo system is a MIMO system, it is natural to utilize modern MIMO optimal design methodologies, such as LQG, LQG/LTR, H_∞, and μ-synthesis, to design the dual-stage controller. Usually MIMO optimal designs are based on the parallel structure shown in Fig. 32.22, augmented with noise/disturbances models and other weighting functions to specify the control design performance objectives.

Linear quadratic Gaussian (LQG) control combines a Kalman filter and optimal state feedback control based on the separation principle. However, the Kalman filter weakens the desirable robustness properties of the optimal state feedback control. Linear quadratic Gaussian/loop-transfer recovery (LQG/LTR) control recovers robustness by a Kalman filter redesign process. Examples of dual-stage control designs using LQG and LQG/LTR have been reported in [32.33–35].

The LQG design methodology minimizes the H_2 norm of the control system, while the H_∞ design methodology minimizes the H_∞ norm of the control system. μ-synthesis design methodology is based on the H_∞ design and accounts for plant model uncertainties during the controller synthesis process with guaranteed robustness. Examples of dual-stage control designs based on the H_∞ and μ-synthesis design methodologies have been reported in [32.33, 36, 37]. A control design using μ-synthesis for the MEMS microactuator dual-stage servo system will be presented in Sect. 32.3.3

Other advanced control theories also have been applied to dual-stage servo control designs, such as sliding mode control [32.38] and neural networks [32.39].

32.3.3 Track-Following Controller Design for a MEMS Microactuator Dual-Stage Servo System

MEMS microactuators for dual-stage servo application are usually designed to have a single flexure resonance mode between 1–2 kHz [32.8,40]. This resonance mode is usually very lightly damped and can have a ±15% variations due to the differences in the fabrication processes. Since the uncertain resonance frequency is relatively low and close to the open loop gain crossover frequency of the control system, robustness to these variations must be considered in the controller design.

Controller Design Considerations for MEMS Microactuator Dual-stage Servo System

Capacitive sensing can be utilized in MEMS microactuators to measure the position of the microactuator relative to the VCM actuator [32.14, 40]. However, this requires additional sensing electronics and wires to and from the head gimbal assembly (HGA), which result in additional fabrication and assembly costs. Thus, whether or not a relative position sensor will be used in MEMS dual-stage servo systems is still an open question. If the microactuator's resonance mode is lightly damped, it is susceptible to airflow turbulence and external disturbances. The relative position signal can be utilized to damp the microactuator's resonance mode. Dual-stage control can be classified as MIMO or SIMO designs according to the availability of the relative position signal.

Given the fact that the inertia of the MEMS microactuator is very small compared to that of the VCM, the effect of the motion of the microactuator on the VCM can be neglected. By feeding the control input of the VCM to the input of the microactuator with a proper gain, we can cancel the coupling effect of the VCM on the microactuator [32.13]. Thus the dual-stage control model can be decoupled as the sum of the outputs of the VCM and the microactuator, as shown in Fig. 32.23, in which G_V, G_M are the VCM and microactuator models respectively, and x_p is the absolute position of the read/write head, which equals the absolute position of the VCM, x_v, and the position of the microactuator relative to the VCM, RPES.

Decoupled Track-Following Control Design and Self-Tuning Control

Figure 32.24 shows a block diagram used for decoupled control design of a MEMS microactuator dual-stage servo systems [32.13].

Sensitivity Function Decoupling Design Methodology. The part enclosed in the dashed box on the upper-right corner is the dual-stage plant model shown in Fig. 32.23. In the figure, G_V and G_M are the VCM and microactuator models, respectively; x_P is the absolute position of the read/write head; x_v is the absolute position of the VCM actuator; and RPES is the position of the microactuator relative to the VCM; r represents the track runout, and PES is the position error signal of the head relative to the data track.

The decoupling control approach, originally introduced in [32.30], utilizes the PES and RPES signals to generate the position error of the VCM relative to the data track center, labeled as $VPES$,

$$VPES = \text{PES} + \text{RPES} = r - x_v , \quad (32.16)$$

and this signal is fed to the VCM loop compensator.

As shown in the block diagram, three compensators need to be designed: the VCM loop compensator K_V; the microactuator PES loop compensator K_M; and the microactuator RPES minor loop compensator $K_{MM} = K_{RR}/K_{RS}$ is used to damp the microactuator's flexure resonance mode and place the closed loop poles of the microactuator RPES loop at appropriate locations. The damped microactuator closed loop transfer function G_R, shown in the lower-middle dashed box, is defined by

$$G_R = \frac{G_M}{K_{RS} + G_M K_{RR}} . \quad (32.17)$$

The total dual-stage open loop transfer function from r to x_P, G_T, is

$$G_T = K_V G_V + K_M G_R + K_M G_R K_V G_V . \quad (32.18)$$

It can be shown that the block diagram in Fig. 32.24 is equivalent to the sensitivity block diagram shown in Fig. 32.25, and the total closed loop sensitivity function from r to PES equals the product of the VCM and microactuator loop sensitivities, S_V and S_M, respectively:

$$S_T = \frac{1}{1 + G_T} = S_V S_M , \quad (32.19)$$

where

$$S_V = \frac{1}{1 + K_V G_V} , \; S_M = \frac{1}{1 + K_M G_R} . \quad (32.20)$$

Thus the dual-stage servo control design can be decoupled into two independent designs: the VCM loop and the microactuator loop designs. The VCM loop sensitivity can be designed using traditional single-stage servo design methodologies. The microactuator loop sensitivity is designed to expand the bandwidth and increase the attenuation of low frequency runout and disturbances.

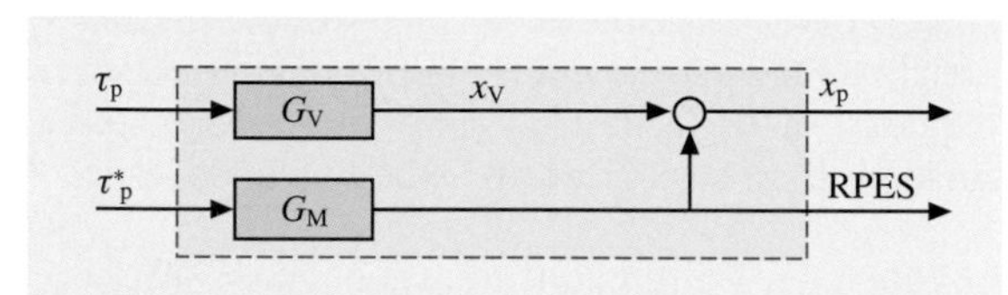

Fig. 32.23 Dual-stage control plant

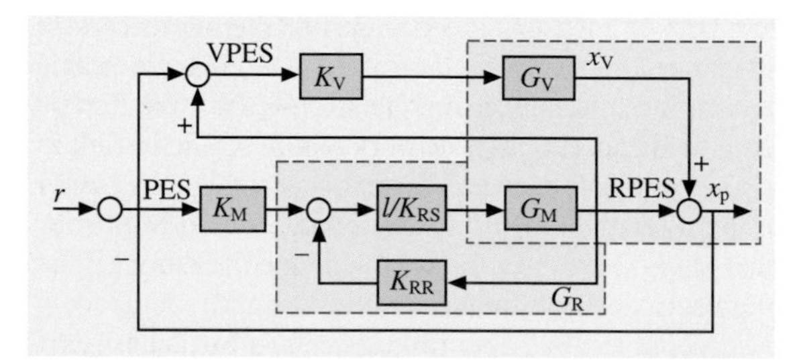

Fig. 32.24 Decoupled dual-stage control design block diagram

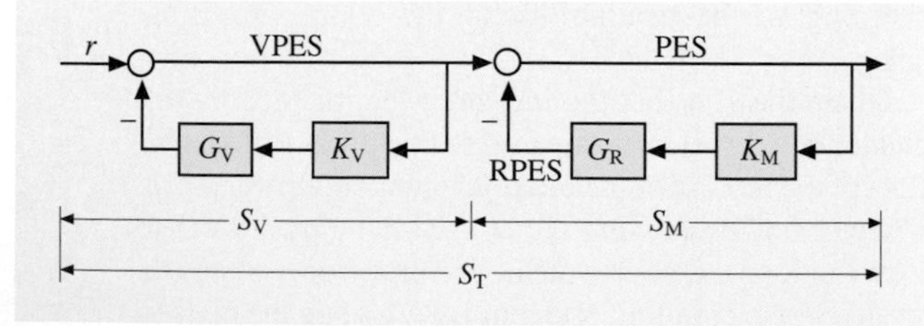

Fig. 32.25 The sensitivity block diagram

Closed Loop Sensitivity Design by Pole Placement. For a typical sampled second order system with the discrete time transfer function

$$G\left(q^{-1}\right)=\frac{q^{-1}\,B_o\left(q^{-1}\right)}{A_o\left(q^{-1}\right)}\,, \tag{32.21}$$

where q^{-1} is one step delay operator; $B_o\left(q^{-1}\right)$ is the plant zero polynomial, and $A_o\left(q^{-1}\right)=1+a_1q^{-1}+a_2q^{-2}$ is the plant pole polynomial. The closed loop sensitivity function can be designed by pole placement, solving the following Diophantine equation [32.41]:

$$\begin{aligned}A_c\left(q^{-1}\right)&=A_o\left(q^{-1}\right)S\left(q^{-1}\right)\\&\quad+q^{-1}B_o\left(q^{-1}\right)R\left(q^{-1}\right)\,,\end{aligned} \tag{32.22}$$

where $A_c\left(q^{-1}\right)$ is the desired closed loop characteristic polynomial. The closed loop sensitivity function, G_S, is

$$G_S\left(q^{-1}\right)=\frac{A_o\left(q^{-1}\right)\,S\left(q^{-1}\right)}{A_c\left(q^{-1}\right)}\,, \tag{32.23}$$

and the discrete time controller $C\left(q^{-1}\right)$ is

$$C\left(q^{-1}\right)=\frac{R\left(q^{-1}\right)}{S\left(q^{-1}\right)}\,. \tag{32.24}$$

$A_c\left(q^{-1}\right)$ can be chosen such that the required bandwidth and system response are satisfied. Usually it is more intuitive to describe it with its continuous time equivalent parameters: the damping ratio, ζ, and the natural frequency, ω_n. The damping ratio is directly related to the phase margin of the open loop transfer function. For a typical design, it can be chosen to be equal to or greater than one, in order to ensure adequate phase margin. ω_n is related to the control system's bandwidth, which is limited by sampling frequency and time delay, or the high frequency structural dynamics of the system.

Dual-Stage Sensitivity Function Design. The compensators in the dual-stage servo system depicted in Fig. 32.24 can be designed by a two-step design process, as illustrated in Fig. 32.26.

First, the VCM loop compensator K_V is designed to attain a desired VCM closed loop sensitivity S_V, as shown in the top part of Fig. 32.26. Its bandwidth, ω_{VC} in Fig. 32.26, is generally limited by the E-block and suspension resonance modes. The design of this compensator can be accomplished using conventional SISO frequency shaping techniques.

The second step of the design process involves the design of the microactuator loop compensators to attain

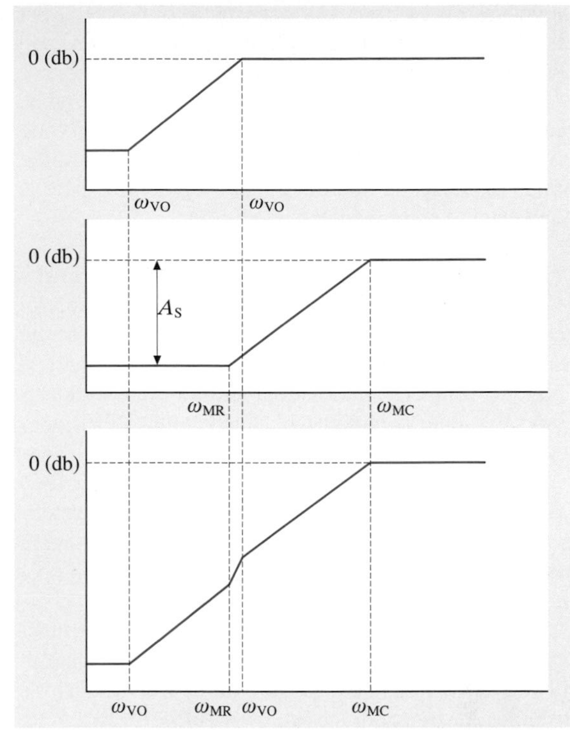

Fig. 32.26 Illustration of dual-stage sensitivity S_T design

additional attenuation S_M, as shown in the middle part of Fig. 32.26. This step is itself accomplished in two steps. First, the minor RPES loop compensator, K_{RR}/K_{RS}, is determined in order to damp the microactuator resonance mode and place the poles of G_R, or equivalently ω_{MR} in Fig. 32.26, at desired locations. The compensator can be obtained by solving a Diophantine equation by pole placement as discussed in the previous subsection. The poles of G_R will become the zeros of the microactuator loop sensitivity function, S_M.

Finally, the PES loop compensator K_M is designed to place the poles, or equivalently. ω_{MC} in Fig. 32.26, of the microactuator loop closed loop sensitivity S_M. ω_{MC} is limited by the PES sampling frequency and computational time delay [32.42].

The total dual-stage sensitivity is shown in the bottom part of Fig. 32.26. For a given ω_{MC}, the additional attenuation A_S, provided by the microactuator loop, will be determined by ω_{MR}. In our proposed procedure, the initial value of ω_{MR} can be chosen to be the same as ω_{VC}. It is then adjusted so the desired attenuation and phase margin requirements of the overall dual-stage system are satisfied. Decreasing ω_{MR} increases the low frequency attenuation of the closed loop sensitivity function S_T but also generally reduces the phase margin of open loop transfer function G_T.

For SIMO design when the RPES is not available, the microactuator model can be used as an open loop observer to estimate the RPES. The combination of the open loop observer and the minor loop compensator is equivalent to a notch filter.

Self-Tuning Control to Compensate Variations in the Microactuator's Resonance Mode. An adaptive control scheme can be combined with the decoupled discrete time pole placement design methodology described above to compensate variations in the microactuator's resonance mode by tuning the microactuator RPES inner loop compensator.

A block diagram for the microactuator inner loop self-tuning control is shown in Fig. 32.27. The parameter adaptation algorithm (PAA) is a direct self-tuning algorithm, based on the microactuator inner loop pole placement design.

Consider the microactuator open loop transfer function in (32.21). Since the microactuator's resonance mode is lightly damped, the zero of the microactuator's discrete time transfer function is very close to 1 ($z_o \approx 1$). Thus it is possible to factor out the "known" term $(1+z_o q^{-1})$ from the Diophantine equation (32.22). The resulting RPES minor-loop closed loop dynamics is given by

$$A_c\left(q^{-1}\right) y(k) = q^{-1} b_0 \left(1+z_o q^{-1}\right) \times\left[K_{RS}\left(q^{-1}\right) u_M(k) + K_{RR}\left(q^{-1}\right) y(k)\right],$$

where u_M is the control input to the microactuator; y denotes the position of the microactuator relative to the VCM, i.e. $y = \text{RPES}$. Defining:

$$S\left(q^{-1}\right) = b_0 K_{RS}\left(q^{-1}\right) = s_0 + s_1 q^{-1}, \quad (32.25)$$

$$R\left(q^{-1}\right) = b_0 K_{RR}\left(q^{-1}\right) = r_0 + r_1 q^{-1}, \quad (32.26)$$

the regressor vector, $\phi(k)$ and filtered regressor vector, $\phi_f(k)$,

$$\phi(k) = \left[u_M(k)\ u_M(k-1)\ y(k)\ y(k-1)\right], \quad (32.27)$$

$$A_c\left(q^{-1}\right)\phi_f(k) = \left(1+z_o q^{-1}\right)\phi(k), \quad (32.28)$$

and the controller parameter vector $\theta = [s_0\ s_1\ r_0\ r_1]^T$, the closed loop RPES dynamics (32.25) can be rewritten as

$$y(k) = \theta^T \phi_f(k-1). \quad (32.29)$$

From (32.29), the controller parameter vector estimate $\hat{\theta}(k) = [\hat{s}_0(k)\ \hat{s}_1(k)\ \hat{r}_0(k)\ \hat{r}_1(k)]^T$ can be updated using a standard recursive least square algorithm (RLS) [32.41]:

$$\hat{\theta}(k) = \hat{\theta}(k-1) + P(k)\phi_f(k)e^o(k), \quad (32.30)$$

$$e^o(k) = y(k) - \hat{\theta}^T(k-1)\phi_f(k-1), \quad (32.31)$$

$$P(k) = \left[P(k-1) - \frac{P(k-1)\phi_f(k-1)\phi_f^T(k-1)P(k-1)}{1+\phi_f^T(k-1)P(k-1)\phi_f(k-1)}\right], \quad (32.32)$$

The control law is

$$\hat{S}\left(k,q^{-1}\right) u_M(k) = \hat{S}_o(k) u_M(k) - \hat{R}\left(k,q^{-1}\right) y(k), \quad (32.33)$$

with

$$\hat{S}\left(k,q^{-1}\right) = \hat{s}_0(k) + \hat{s}_1(k) q^{-1}, \quad (32.34)$$

$$\hat{R}\left(k,q^{-1}\right) = \hat{r}_0(k) + \hat{r}_1(k) q^{-1}, \quad (32.35)$$

and $u_M(k)$ being the output of the microactuator fixed outer loop compensator K_M, $u_M(k)$ being the control input to the microactuator.

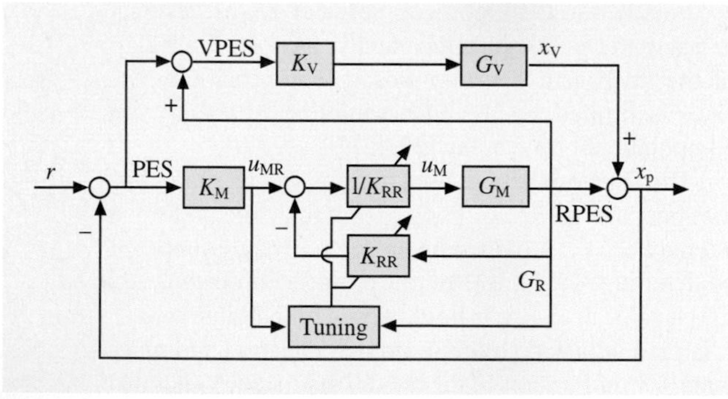

Fig. 32.27 Self-tuning control of the microactuator

Dual-Stage Track-Following Control Design Using μ-Synthesis

The structured singular value, μ, is a measure of how big a perturbation to a system needs to be to make the sys-

Part E | 32.3

tem unstable. By incorporating a fictitious uncertainty block, robust performance in terms of the H_∞ norm of the closed loop transfer function can be related to the value of μ. μ-synthesis is a robust optimal controller design technique that attempts to minimize μ through an iterative process [32.43].

Design Methodology. For controller design using μ-synthesis, the model uncertainties are represented using linear fractional transformations (LFT). Disturbances and outputs are weighted to characterize the real plant environment and the performance requirements. A block diagram used for dual-stage μ-synthesis controller design is shown in Fig. 32.28.

In the block diagram, two model uncertainties, δ_1 and δ_2, are considered. δ_1 is an additive uncertainty used to describe the VCM unmodeled dynamics. δ_2 represents the parametric uncertainty in the stiffness of the microactuator.

Disturbance signals accounted for in the model include the track runout r, input disturbances to the VCM d_{VCM}, input disturbances to the microactuator d_{MA}, PES sensor noise n_{PES}, and the $RPES$ sensor noise n_{RPES}. These disturbance signals are generated by passing normalized signals $\overline{r}$, $\overline{d_{\mathrm{VCM}}}$, $\overline{d_{\mathrm{MA}}}$, $\overline{n_{PES}}$, and $\overline{n_{RPES}}$ through weighting functions W_r, $W_{d\mathrm{VCM}}$, $W_{d\mathrm{MA}}$, W_{nPES}, and W_{nRPES}, respectively, which can be either constants or frequency shaping filters. These weights are selected by the designer so that disturbances are modeled with sufficient fidelity.

The output signals in the synthesis model are head position error signal, PES, the microactuator relative position signal, $RPES$, the VCM control input u_{VCM}, and the microactuator control input u_{MA}. These signals are multiplied by scaling factors, W_{PES}, W_{RPES}, $W_{u\mathrm{VCM}}$, and $W_{u\mathrm{MA}}$, respectively, to produce the weighted performance signals $\overline{PES}$, $\overline{RPES}$, $\overline{u_{\mathrm{VCM}}}$, $\overline{u_{\mathrm{MA}}}$. The scaling factors are selected to characterize the performance requirements.

Given a set of input and output weights and plant uncertainties, if the synthesized controller achieves

$$\mu \le \beta\,, \tag{32.36}$$

the closed loop transfer function, $\overline{T}$, from the normalized disturbances

$$\overline{d} = \left[\overline{r}\ \overline{d_{\mathrm{VCM}}}\ \overline{d_{\mathrm{MA}}}\ \overline{n_{\mathrm{PES}}}\ \overline{n_{\mathrm{RPES}}}\right], \tag{32.37}$$

to the weighted performance signals

$$\overline{e} = \left[\overline{\mathrm{PES}}\ \overline{\mathrm{RPES}}\ \overline{u_{\mathrm{VCM}}}\ \overline{u_{\mathrm{MA}}}\right], \tag{32.38}$$

will have infinity norm

$$\|\overline{T}\|_\infty \le \beta\,, \tag{32.39}$$

for perturbations

$$\|\Delta\|_\infty = \left\|\begin{bmatrix}\delta_1 & 0\\ 0 & \delta_2\end{bmatrix}\right\|_\infty \le \frac{1}{\beta}\,. \tag{32.40}$$

The interpretation of H_∞ as the RMS gain of the sinusoidal signals to help understand the design is as follows. Assume that each element $\overline{d_i}$ of the disturbance

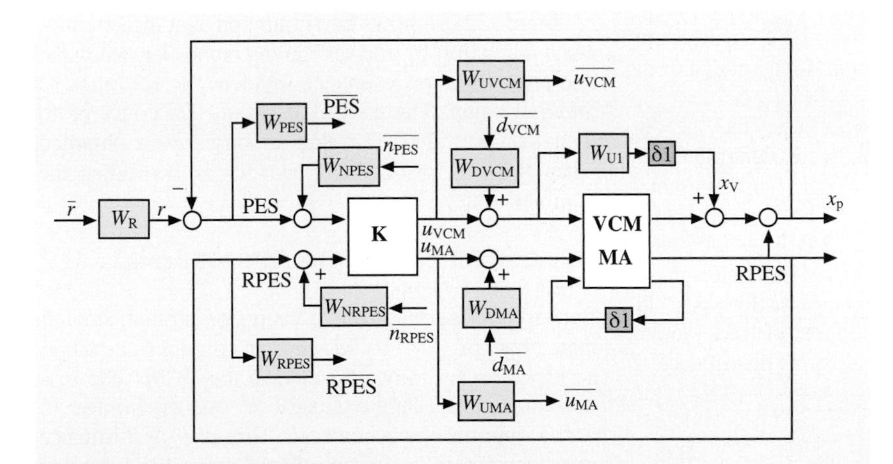

Fig. 32.28 μ-synthesis design block diagram

input vector in (32.37) is a sinusoid of the form

$$\overline{d_i}(t) = D_i \sin(\omega_i t + \psi_i)\,, \tag{32.41}$$

such that

$$\sum_{i=1}^{5} D_i^2 \leq 1\,. \tag{32.42}$$

Then the steady state response of output in (32.38) will also be a sinusoid of the form

$$\overline{e_i}(t) = E_i \sin(\omega_i t + \phi_i)\,, \tag{32.43}$$

and

$$\sum_{i=1}^{4} E_i^2 \leq \beta\,, \tag{32.44}$$

under the perturbations defined in (32.40).

Weighting Functions for μ-Synthesis Design. Performance weights are chosen based on limits that the error signals should not exceed. The limit on the PES is based on a rule of thumb commonly used in the disk drive industry which states that the position error of the head should not exceed $1/10$ of a track width during track following. RPES is limited by the stroke limit of the microactuator. Thus,

$$W_{\text{PES}} = \frac{10}{\text{track width}}\,; \quad W_{\text{RPES}} = \frac{1}{\text{stroke limit}}\,. \tag{32.45}$$

Weights on control inputs are based on the control input saturations.

$$W_{U\text{VCM}} = \frac{1}{(i_{\text{V}})_{\max}}\,; \quad W_{U\text{MA}} = \frac{1}{(v_{\text{M}})_{\max}}\,, \tag{32.46}$$

where $(i_{\text{V}})_{\max}$ and $(v_{\text{M}})_{\max}$ are the maximum drive current of the VCM and the maximum drive voltage of the microactuator, respectively.

The weights for PES noise, n_{PES}, and the disturbance to the VCM, $w_{d\text{VCM}}$, were determined by extrapolating the PES decomposition results presented in [32.29]. The weight of the RPES sensing noise, n_{RPES}, is assumed to be 8 nm [32.14]. The disturbance weight, $w_{d\text{MA}}$, is assumed to be 0.1 V in the design.

The runout weighting function is a low pass filter that captures the combined effects of track runout, mechanical vibrations, and bias force disturbance to PES,

$$W_{\text{r}} = \frac{2\times 10^{-8}(s+2\pi\times 10)^2}{(s+2\pi\times 1000)^2}\,. \tag{32.47}$$

The combination of W_{PES} and W_{r} determines the closed loop sensitivity function from r to PES, $S(s)$, of the dual-stage system. If the designed control system achieves a peak μ value of β, we have

$$\left\| W_{\text{PES}} S(s) W_{\text{r}} \right\|_{\infty} \leq \beta\,. \tag{32.48}$$

If $\beta = 1$, the magnitude of bode plot of $S(s)$ lies below that of $1/W_{\text{PES}}W_{\text{r}}$.

For SIMO design, the controller to be synthesized has only one input. In the synthesis model, RPES is not fed to the controller K, and RPES noise n_{rpes} and its weights W_{nRPES} are not used.

Design and Simulation Results

Figure 32.29 shows the Bode plots of the open loop transfer functions from r to $x_{\text{P}}p$ and the closed loop sensitivity functions of the two designs, respectively. Sampling frequency was chosen to be 20 kHz in both designs.

For the decoupled MIMO design, the gain crossover frequency, gain margin, phase margin of the open loop system are 2,201 Hz, 9.1 dB, and 48.8°, respectively. For the decoupled SIMO design, they are 2,381 Hz, 8.7 dB, and 30.6°, respectively.

For μ-synthesis MIMO design, the gain crossover frequency, gain margin, phase margin of the open loop system are 2,050 Hz, 9.5 dB, 66.0°, respectively. For μ-synthesis SIMO design, they are 2,223 Hz, 2.9 dB, 32.0°, respectively.

The above controller designs have not been implemented on a MEMS microactuator dual-stage servo system. Controllers designed using both methods have been tested on a PZT actuated suspension dual-stage servo system, however, validating their effectiveness [32.44].

Figure 32.30 shows the simulation of control parameters adaptation for the self-tuning controller, when the real microactuator resonance frequency is 1.2 times its nominal value. The controller parameters converge to their desired values. Similar responses were obtained when the real resonance frequency is 0.8 times the nominal value.

32.3.4 Dual-Stage Seek Control Design

Because the inertia of a microactuator is much smaller than that of the VCM, it can produce a larger acceleration and move faster than the VCM. The motion range of a microactuator is usually limited to a few micrometers, however. Thus the performance improvement in seek control made possible by using

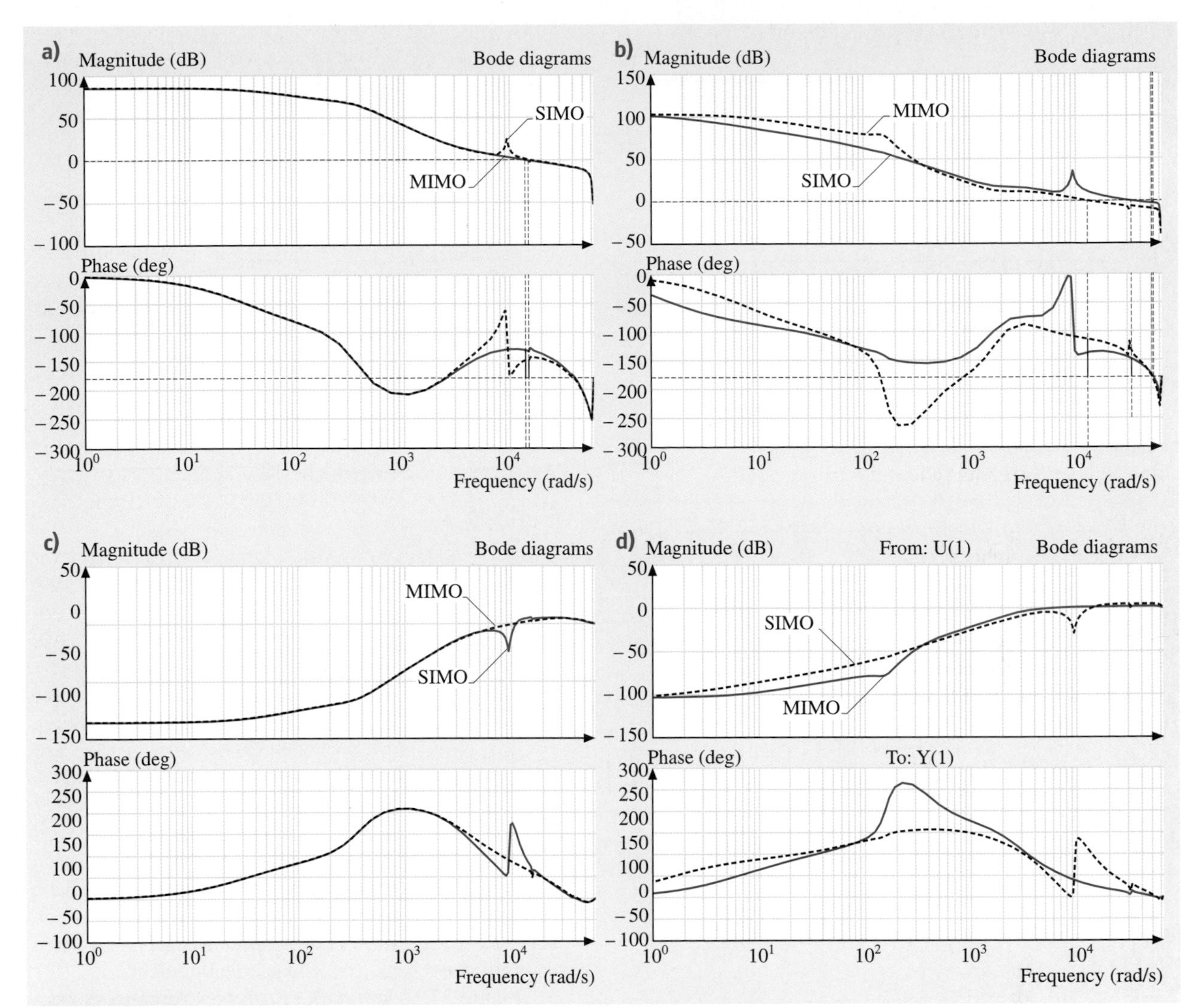

Fig. 32.29 (a) Open loop Bode plot of the decoupled design; (b) open loop Bode plot of the μ-synthesis design; (c) closed loop sensitivity Bode plot of the decoupled design; (d) closed loop sensitivity Bode plot of the μ-synthesis design

dual-stage actuation will mainly be in short distance seeks.

Two degree-of-freedom (DOF) position control has been a very popular control technique in short distance seeks. A 2-DOF control technique utilizing decoupled feed forward reference trajectories has been developed for short span seek control using a PZT actuated suspension dual-stage servo system [32.45]. The same technique can also be applied to seek control of MEMS microactuator dual-stage servo system.

Figure 32.31 shows a block diagram for dual-stage short span seek control design using this method. In the figure, x_{V}^{d} and x_{R}^{d} are the desired seek trajectories of the VCM and the microactuator respectively. K_{VF} and K_{RF} are Zero Phase Error Tracking Feed Forward Controllers (ZPETFFC) generated using with the VCM model G_{V} and the damped microactuator model G_{R} [32.46].

To minimize the residual vibration after seek operation, minimum jerk seek trajectories were applied to both the VCM and the microactuator. These can be generated

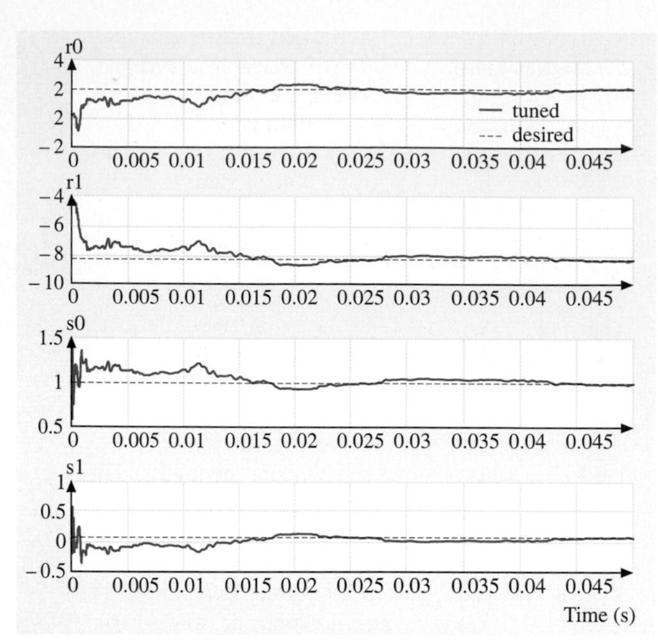

Fig. 32.30 Control parameters adaptation response

by [32.47]

$$x_{\mathrm{V}}^{d}(t)=60\,d_s\left[\frac{1}{10}\left(\frac{t}{T_{\mathrm{V}}}\right)^5-\frac{1}{4}\left(\frac{t}{T_{\mathrm{V}}}\right)^4+\frac{1}{6}\left(\frac{t}{T_{\mathrm{V}}}\right)^3\right], \tag{32.49}$$

$$x_{\mathrm{R}}^{d}=\begin{cases}60\,\left(d_s-x_{\mathrm{V}}^{d}(T)\right) & \\ \times\left[\frac{1}{10}\left(\frac{t}{T}\right)^5-\frac{1}{4}\left(\frac{t}{T}\right)^4+\frac{1}{6}\left(\frac{t}{T}\right)^3\right] & t\le T\,,\\ d_s-x_{\mathrm{V}}^{d}(t) & t>T\end{cases} \tag{32.50}$$

where d_s is the distance of the head from the target track; T is the time when the head reaches the target track if dual-stage actuator is used, while T_{V} is the time when the head reaches the target track if only VCM is used. T_{V} and T can be chosen based on the control force saturation and seek performance requirements.

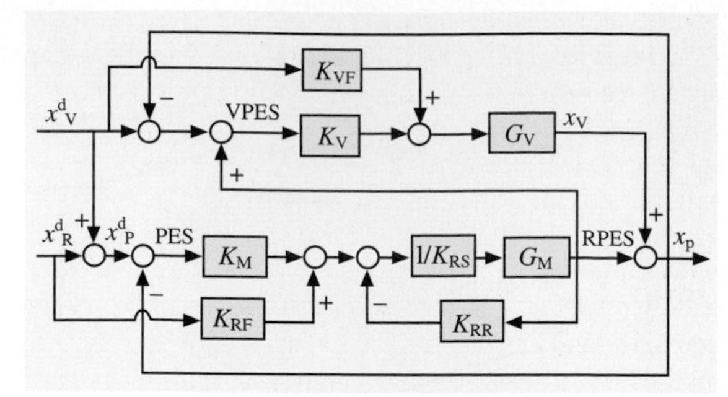

Fig. 32.31 Dual-stage seek control design

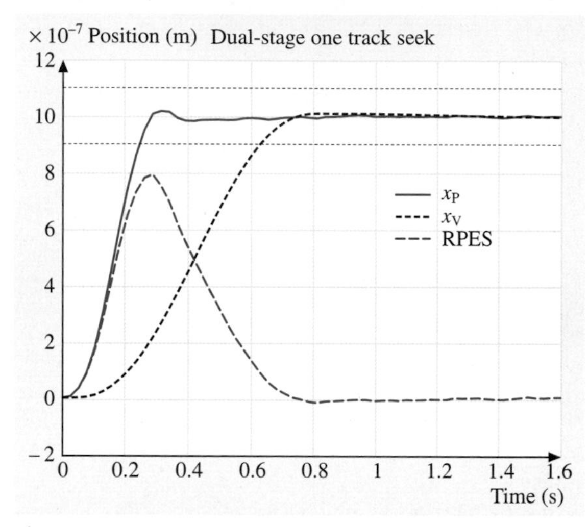

Fig. 32.32 Dual-stage short span seek response

Figure 32.32 shows the 1 μm seek responses of the dual-stage actuator for the MIMO design. Overshoot was eliminated, and no obvious residual vibrations occured. The seek time using the dual-stage actuator is about 0.25 ms, compared to a seek time of about 0.7 ms if only the VCM is used.

32.4 Conclusions and Outlook

In order to increase hard disk drive storage density, high bandwidth dual-stage servo systems are necessary to suppress disturbances and increase servo precision. Various prototype MEMS microactuators have been designed and fabricated to provide a dual-stage actuation system. The most common approach for these prototypes is to use electrostatic driving arrays produced by MEMS fabrication methods.

MEMS microactuators provide a potentially high performance and low cost solution for achieving the servo requirements for extremely high HDD storage density. Simulations and experiments show that many

MEMS microactuator features, such as integrated silicon gimbals and capacitive relative position sensing arrays, can meet important requirements for operation of HDDs. The use of structural electrical-isolation in the microactuator design could considerably reduce microactuator driving voltages, while processes that avoid the use of silicon-on-insulator wafers or high-temperature processing steps could reduce manufacturing costs. High bandwidth dual-stage track-following controllers have been designed using both a decoupled SISO design method and robust μ-synthesis design method, among others. In addition, short-span seeking control using a 2-DOF control structure with decoupled feed forward trajectory generation can greatly reduce the short span seek time.

Research remains to be done in the areas of system integration, reliability, and performance enhancement for this technology to be utilized in commercial products. Cost and reliability are probably the most important obstacles to commercial implementation of actuated slider dual-stage servo systems at this time. Streamlining of microactuator fabrication processes or development of a process compatible with that of the slider and head fabrication could reduce manufacturing costs, which are not quite yet economical. Meanwhile, dynamic behavior and reliability of the microactuator under disturbances from airflow, head-disk interaction, and particle presence are being experimentally studied by industry.

Further in the future, to achieve nanometer servo precision, third-generation dual-stage servo systems will have to employ an actuated head approach. Research in this area has just started, with the key problem being how to combine microactuator fabrication with read/write head fabrication. Another area of research is the use of MEMS technology to incorporate additional sensors, such as accelerometers and strain gauge vibration sensors, to suppress the TMR due to airflow and external disturbances excited structural vibrations. New robust, adaptive MIMO control architectures and algorithms must be developed for such a multisensing and multiactuation servo system.

Besides the application of MEMS microactuators in dual-stage servo control of traditional rotating media magnetic HDDs, recent advances in MEMS and nanotechnology have made possible the development of a new class of miniaturized ultra-high density storage devices that use probe arrays for recording and reading data. This new architecture abandons the traditional rotating media and flying slider paradigm in favor of parallel x-y scanning of entire probe arrays over a storage medium. The probe arrays and/or the media are moved with electrostatic or electromagnetic microactuators that generate x-y in-plane relative motion, as well as z-axis out-of-plane relative motion. One example of this new generation of storage devices is IBM's "millipede," which utilizes atomic force microscopy (AFM) recording technology [32.48]. Data is recorded on a polymer media by tiny depressions melted by a thermomechanical process by the AFM tips. $400\,\mathrm{Gb/in.}^2$ bit patterns have been demonstrated using thermal probes.

By any of the paths above, MEMS and nanotechnology will become a basic enabling technology for the development of future data storage devices, offering the capabilities of small size, low power consumption, ultra-high densities, low cost, and high performance.

References

32.1 T. Yamaguchi: Modelling and control of a disk file head-positioning system, Proc. Inst. Mech. Eng. I (J. Syst. Cont. Eng.) **215** (2001) 549–567

32.2 T. Howell, R. Ehrlich, M. Lippman: TPI growth is key to delaying superparamagnetism's arrival, Data Stor. (1999) 21–30

32.3 R. B. Evans, J. S. Griesbach, W. C. Messner: Piezoelectric microactuator for dual-stage control, IEEE Trans. Magn. **35** (1999) 977–981

32.4 I. Naniwa, S. Nakamura, S. Saegusa, K. Sato: Low voltage driven piggy-back actuator of hard disk drives, IEEE International MEMS 99 Conference (IEEE, Piscataway 1999) 49–52

32.5 D. Horsley: Microfabricated electrostatic actuators for magnetic disk drives. Ph.D. Thesis (University of California, Berkeley 1998)

32.6 L. Muller: Gimballed electrostatic microactuators with embedded interconnects. Ph.D. Thesis (University of California, Berkeley 2000)

32.7 S. Koganezawa, K. Takaishi, Y. Mizoshita, Y. Uematsu, T. Yamada: Development of integrated piggyback milli-actuator for high density magnetic recording, International Conference on Micromechatronics for Information and Precision Equipment 1997 (IEEE, New York 1997) 20–23

32.8 L.-S. Fan, T. Hirano, J. Hong, P. R. Webb, W. H. Juan, W. Y. Lee, S. Chan, T. Semba, W. Imaino,

T. S. Pan, S. Pattanaik, F. C. Lee, I. McFadyen, S. Arya, R. Wood: Electrostatic microactuator and design considerations for hdd application, IEEE Trans. Magn. **35** (1999) 1000–1005
32.9 T. Imamura, M. Katayama, Y. Ikegawa, T. Ohwe, R. Koishi, T. Koshikawa: MEMS-based integrated head/actuator/slider for hard disk drives, IEEE/ASME Trans. Mechatron. **3** (1998) 166–174
32.10 T. Imamura, T. Koshikawa, M. Katayama: Transverse mode electrostatic microactuator for MEM-based HDD slider, Proc. IEEE MEMS Workshop, San Diego 1996 (IEEE, New York 1996) 216–221
32.11 S. Nakamura, K. Suzuki, M. Ataka, H. Fujita: An electrostatic micro actuator for a magnetic head tracking system of hard disk drives, Transducers '97 (IEEE, Piscataway 1997) 1081–1084
32.12 Y. Soeno, S. Ichikawa, T. Tsuna, Y. Sato, I. Sato: Piezoelectric piggy-back microactuator for hard disk drive, IEEE Trans. Magn. **35** (1999) 983–987
32.13 Y. Li, R. Horowitz: Mechatronics of electrostatic microactuator for computer disk drive dual-stage servo systems, IEEE/ASME Trans. Mechatron. **6** (2001) 111–121
32.14 N. Wongkomet: Position sensing for electrostatic micropositioners. Ph.D. Thesis (University of California, Berkeley 1998)
32.15 A. Hac, L. Liu: Sensor and actuator location in motion control of flexible structures, J. Sound Vib. **167** (1993) 239–261
32.16 K. Hiramoto, H. Doki, G. Obinata: Optimal sensor/actuator placement for active vibration control using explicit solution of algebraic Riccati equation, J. Sound Vib. **229** (2000) 1057–1075
32.17 P. Cheung, R. Horowitz, R. Howe: Design, fabrication and control of an electrostatically driven polysilicon microactuator, IEEE Trans. Magn. **32** (1996) 122–128
32.18 T. Chen, Y. Li, K. Oldham, R. Horowitz: MEMS application in computer disk drive drive dual-stage servo systems, J. Soc. Instrum. Control Eng. **41** (2002) 412–420
32.19 T.-L. Chen: Design and fabrication of PZT-actuated silicon suspensions for hard disk drives. Ph.D. Thesis (University of California, Berkeley 2001)
32.20 C. G. Keller, R. T. Howe: HexSil tweezers for teleoperated micro-assembly, 10th Int'l Workshop on Micro Electro Mechaninical Systems (MEMS'97), Nagoya 1997 (IEEE, New York 1997) 72–77
32.21 T. J. Brosnihan, J. M. Bustillo, A. P. Pisano, R. T. Howe: Embedded interconnect and electrical isolation for high-aspect-ratio, SOI inertial instruments, International Conference on Solid-State Sensors and Actuators, New York 1997 (IEEE, New York 1997) 637–640
32.22 T. Hirano, L.-S. Fan, T. Semba, W. Y. Lee, J. Hong, S. Pattanaik, P. Webb, W.-H. Juan, S. Chan: Microactuator for tera-storage, IEEE Int'l MEMS 1999 Conference, Orlando 1999 (IEEE, Piscataway 1999) 6441–6446
32.23 T. Hirano, L.-S. Fan, T. Semba, W. Lee, J. Hong, S. Pattanaik, P. Webb, W.-H. Juan, S. Chan: High-bandwidth HDD tracking servo by a moving-slider micro-actuator, IEEE Trans. Magn. **35** (1999) 3670–3672
32.24 T. Iizuka, T. Oba, H. Fugita: Electrostatic micro actuators with high-aspect-ratio driving gap for hard disk drive applications, Int'l Symposium on micromechatronics and human science (IEEE, Piscataway 2000) 229–236
32.25 H. Kuwajima, K. Matsuoka: Thin film piezoelectric dual-stage actuator for HDD, InterMag Europe, Session BS04 (IEEE, Piscaraway 2000) BS4
32.26 Y. Lou, P. Gao, B. Qin, G. Guo, E.-H. Ong, A. Takada, K. Okada: Dual-stage servo with on-slider PZT microactuator for hard disk drives, InterMag Europe, Session BS03, Amsterdam 2002 (IEEE, Piscataway 2002)
32.27 H. Fujita, K. Suzuki, M. Ataka, S. Nakamura: A microactuator for head positioning system of hard disk drives, IEEE Trans. Magn. **35** (1999) 1006–1010
32.28 B.-H. Kim, K. Chun: Fabrication of an electrostatic track-following micro actuator for hard disk drives using SOI wafer, J. Micromech. Microeng. **11** (2001) 1–6
32.29 R. Ehrlich, D. Curran: Major HDD TMR sources, and projected scaling with TPI, IEEE Trans. Magn. **35** (1999) 885–891
32.30 K. Mori, T. Munemoto, H. Otsuki, Y. Yamaguchi, K. Akagi: A dual-stage magnetic disk drive actuator using a piezoelectric device for a high track density, IEEE Trans. Magn. **27** (1991) 5298–5300
32.31 S. J. Schroeck, W. C. Messner, R. J. McNab: On compensator design for linear time-invariant dual-input single-output systems, IEEE/ASME Trans. Mechatron. **6** (2001) 50–57
32.32 T. Semba, T. Hirano, L.-S. Fan: Dual-stage servo controller for HDD using MEMS actuator, IEEE Trans. Magn. **35** (1999) 2271–2273
32.33 T. Suzuki, T. Usui, M. Sasaki, F. Fujisawa, T. Yoshida, H. Hirai: Comparison of robust track-following control systems for a dual stage hard disk drive, Proc. of International Conference on Micromechatronics for Information and Precision Equipment, ed. by B. Bhushan, K. Ohno (Word Scientific, Singapore 1997) 101–118
32.34 X. Hu, W. Guo, T. Huang, B. M. Chen: Discrete time LQG/LTR dual-stage controller design and implementation for high track density HDDs, Proc. of American Automatic Control Conference (IEEE, Piscataway 1999) 4111–4115
32.35 S.-M. Suh, C. C. Chung, S.-H. Lee: Design and analysis of dual-stage servo system for high track density HDDs, Microsyst. Technol. **8** (2002) 161–168

32.36 D. Hernandez, S.-S. Park, R. Horowitz, A. K. Packard: Dual-stage track-following servo design for hard disk drives, Proc. of American Automatic Control Conference (IEEE, Piscataway 1999) 4116–4121
32.37 M. Rotunno, R. A. de Callafon: Fixed order H_∞ control design for dual-stage hard disk drives, Proc. of the 39th IEEE Conference on Decision and Control (IEEE, Piscataway 2000) 3118–3119
32.38 S.-H. Lee, S.-E. Baek, Y.-H. Kim: Design of a dual-stage actuator control system with discrete-time sliding mode for hard disk drives, Proc. of the 39th IEEE Conference on Decision and Control (IEEE, Piscataway 2000) 3120–3125
32.39 M. Sasaki, T. Suzuki, E. Ida, F. Fujisawa, M. Kobayashi, H. Hirai: Track-following control of a dual-stage hard disk drive using a neuro-control system, Eng. Appl. Artif. Intell. **11** (1998) 707–716
32.40 D. Horsley, N. Wongkomet, R. Horowitz, A. Pisano: Precision positioning using a microfabricated electrostatic actuator, IEEE Trans. Magn. **35** (1999) 993–999
32.41 K. J. Åström, B. Wittenmark: *Adaptive Control*, 2nd edn. (Addison-Wesley, Reading 1995)
32.42 M. T. White, W.-M. Lu: Hard disk drive bandwidth limitations due to sampling frequency and computational delay, Proc. of the 1999 IEEE/ASME International Conference on Intelligent Mechatronics (IEEE, Piscataway 1999) 120–125
32.43 G. J. Balas, J. C. Doyle, K. Glover, A. Packard, R. Smith: *μ-Analysis and Synthesis ToolBox* (MUSYN Inc. and The MathWorks, Natick 1995)
32.44 Y. Li, R. Horowitz: Design and testing of track-following controllers for dual-stage servo systems with PZT actuated suspensions [in HDD], Microsyst. Technol. **8** (2002) 194–205
32.45 M. Kobayashi, R. Horowitz: Track seek control for hard disk dual-stage servo systems, IEEE Trans. Magn. **37** (2001) 949–954
32.46 M. Tomizuka: Zero phase error tracking algorithm for digital control, Trans. ASME J. Dynam. Syst. Meas. Control **109** (1987) 65–68
32.47 Y. Mizoshita, S. Hasegawa, K. Takaishi: Vibration minimized access control for disk drive, IEEE Trans. Magn. **32** (1996) 1793–1798
32.48 P. Vettiger, G. Cross, M. Despont, U. Drechsler, U. Durig, B. Gotsmann, W. Haberle, M. A. Lantz, H. W. Rothuizen, R. Stutz, G. K. Binnig: The "millipede" – nanotechnology entering data storage, IEEE Trans. Nanotechnol. **1** (2002) 39–55

33. Micro/Nanotribology of MEMS/NEMS Materials and Devices

The field of MEMS/NEMS has expanded considerably over the last decade. The length scale and large surface-to-volume ratio of the devices result in very high retarding forces such as adhesion and friction that seriously undermine the performance and reliability of the devices. These tribological phenomena need to be studied and understood at the micro- to nanoscales. In addition, materials for MEMS/NEMS must exhibit good microscale tribological properties. There is a need to develop lubricants and identify lubrication methods that are suitable for MEMS/NEMS. Using AFM-based techniques, researchers have conducted micro/nanotribological studies of materials and lubricants for use in MEMS/NEMS. In addition, component level testing has also been carried out to aid in better understanding the observed tribological phenomena in MEMS/NEMS.

Macroscale and microscale tribological studies of silicon and polysilicon films have been performed. The effects of doping and oxide films and environment on the tribological properties of these popular MEMS/NEMS materials have also been studied. SiC film is found to be a good tribological material for use in high-temperature MEMS/NEMS devices. Hexadecane thiol self-assembled monolayers and bonded perfluoropolyether lubricants appear to be well suited for lubrication of microdevices under a range of environmental conditions. DLC coatings can also be used for low friction and wear. Surface roughness measurements of micromachined polysilicon surfaces have been made using an AFM. The roughness distribution on surfaces is strongly dependent on the fabrication process. Roughness should be optimized for low adhesion, friction, and wear. Adhesion and friction of microstructures can be measured using novel apparatuses. Adhesion and friction measurements on silicon-on-silicon confirm AFM measurements that hexadecane thiol and bonded perfluoropolyether films exhibit superior adhesion and friction properties. Static friction force measurements of micromotors have been performed using an AFM. The forces are found to vary considerably with humidity. A bonded layer of perfluoropolyether lubricant is found to satisfactorily reduce the friction forces in the micromotor.

AFM/FFM-based techniques can be satisfactorily used to study and evaluate micro/nanoscale tribological phenomena related to MEMS/NEMS devices.

This chapter presents a review of macro- and micro/nanoscale tribological studies of materials and lubrication studies for MEMS/NEMS and component-level studies of stiction phenomena in MEMS/NEMS devices.

33.1 **Introduction to MEMS** 985

33.2 **Introduction to NEMS** 988

33.3 **Tribological Issues in MEMS/NEMS** 989
33.3.1 MEMS 989
33.3.2 NEMS 994
33.3.3 Tribological Needs 995

33.4 **Tribological Studies of Silicon and Related Materials** 995
33.4.1 Tribological Properties of Silicon and the Effect of Ion Implantation 996
33.4.2 Effect of Oxide Films on Tribological Properties of Silicon 998
33.4.3 Tribological Properties of Polysilicon Films and SiC Film.... 1000

33.5 **Lubrication Studies for MEMS/NEMS** 1003
33.5.1 Perfluoropolyether Lubricants 1003
33.5.2 Self-Assembled Monolayers (SAMs) 1004
33.5.3 Hard Diamond-like Carbon (DLC) Coatings 1008

33.6 **Component-Level Studies** 1009
33.6.1 Surface Roughness Studies of Micromotor Components 1009
33.6.2 Adhesion Measurements 1011
33.6.3 Static Friction Force (Stiction) Measurements in MEMS 1014
33.6.4 Mechanisms Associated with Observed Stiction Phenomena in Micromotors 1016

References 1017

Microelectromechanical systems (MEMS) refer to microscopic devices that have a characteristic length of less than 1 mm but more than 1 μm and combine electrical and mechanical components. Nanoelectromechanical systems (NEMS) refer to nanoscopic devices that have a characteristic length of less than 1 μm and combine electrical and mechanical components. In mesoscale devices, if the functional components are on the micro- or nanoscale, they may be referred to as MEMS or NEMS, respectively. To put the dimensions and masses in perspective, see Fig. 33.1 and Table 33.1. The acronym MEMS originated in the United States. The term commonly used in Europe and Japan is micro/nanodevices, which is used in a much broader sense. MEMS/NEMS terms are also now used in a broad sense. A micro/nanosystem, a term commonly used in Europe, is referred to as an intelligent miniaturized system comprising sensing, processing, and/or actuating functions.

Fabrication techniques include top-down methods, in which one builds down from the large to the small, and the bottom-up methods, in which one builds up from the small to the large. Top-down methods include micro/nanomachining methods and methods based on lithography, as well as nonlithographic miniaturization for MEMS and NEMS fabrication. In the bottom-up methods, also referred to as nanochemistry, the devices and systems are assembled from their elemental constituents for NEMS fabrication, much like the way nature uses proteins and other macromolecules to

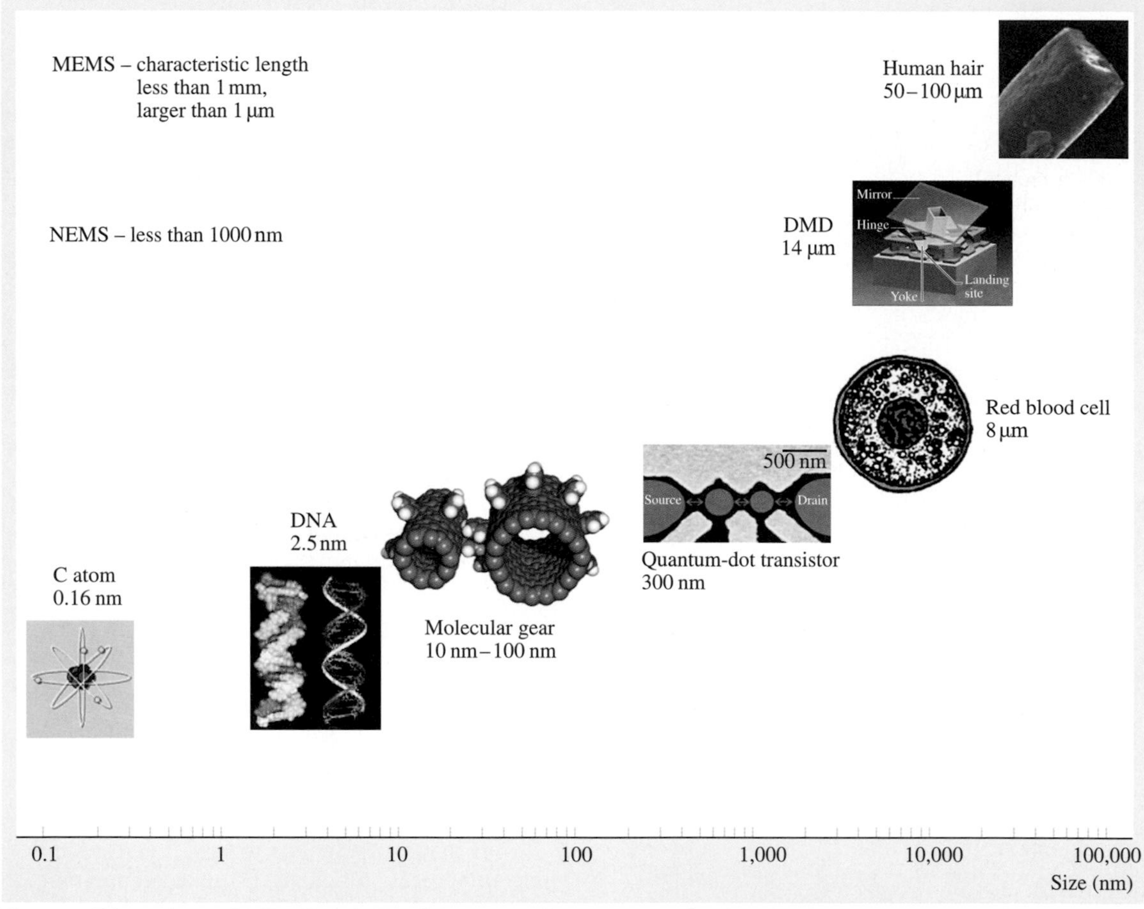

Fig. 33.1 Dimensions of MEMS and NEMS in perspective. An example of molecular dynamic simulations of carbon-nanotube-based gear is obtained from [33.1], quantum-dot transistor is obtained from [33.2], and DMD is obtained from [33.3]

Table 33.1 Dimensions and masses in perspective

A Dimensions in Perspective	
NEMS characteristic length	< 1,000 nm
MEMS characteristic length	< 1 mm and > 1 μm
Individual atoms	typically fraction of a nm in diameter
DNA molecules	~ 2.5 nm wide
Molecular gear	~ 50 nm
Biological cells	in the range of thousands of nm in diameter
Human hair	~ 75,000 nm in diameter
B Masses in Perspective	
NEMS built with cross sections of about 10 nm	as low as 10^{-20} N
Micromachine silicon structure	as low as 1 nN
Water droplet	~ 10 μN
Eyelash	~ 100 nN

construct complex biological systems. The bottom-up approach has the potential to go far beyond the limits of top-down technology by producing nanoscale features through synthesis and subsequent assembly. Furthermore, the bottom-up approach offers the potential to produce structures with enhanced and/or completely new functions. It allows the combination of materials with distinct chemical composition, structure, and morphology.

MEMS and emerging NEMS are expected to have a major impact on our lives, comparable to that of semiconductor technology, information technology, or cellular and molecular biology [33.4, 5]. MEMS/NEMS are used in mechanical, information/communication, chemical, and biological applications. The MEMS industry in 2000 was worth about $15 billion and, with a projected 10–20% annual growth rate, is expected to be more than $100 billion by the end of this decade [33.6]. Growth of Si MEMS may slow down and non-silicon MEMS may pick up during this decade. The NEMS industry is expected to expand in this decade, mostly in biomedical applications, as well as in nanoelectronics or molecular electronics. The NEMS industry was worth about $100 million in 2002 and integrated NEMS are expected to be more than $25 billion by the end of this decade. Due to the enabling nature of these systems, and because of the significant impact they can have on both commercial and defense applications, industry, as well as the federal government have taken a special interest in seeing growth nurtured in this field. Micro- and nanosystems are the next logical step in the "silicon revolution."

33.1 Introduction to MEMS

The advances in silicon photolithographic process technology beginning in the 1960s led to the development of MEMS in the early 1980s. More recently, lithographic processes have been developed to process nonsilicon materials. The lithographic processes are being complemented with non-lithographic processes for fabrication of components or devices made from plastics or ceramics. Using these fabrication processes, researchers have fabricated a wide variety of miniaturized devices, including Si-based devices, chemical and biological sensors and actuators, and miniature non-silicon structures (e.g., devices made from plastics or ceramics) with dimensions in the range of a couple to a few thousand microns (see e.g., [33.7–15]). MEMS for mechanical applications include acceleration, pressure, flow, gas sensors, linear and rotary actuators, and other microstructures or microcomponents such as electric motors, gear trains, gas turbine engines, nozzles, fluid pumps, fluid valves, switches, grippers, and tweezers. MEMS for chemical applications include chemical sensors and various analytical instruments. Microoptoelectromechanical systems, or MOEMS, include micromirror arrays and fiber optic connectors. BIOMEMS include biochips. Radio frequency MEMS or RF-MEMS include inductors, capacitors, and antennas. High-aspect-ratio MEMS (HARMEMS) have also been introduced.

The fabrication techniques for MEMS devices include lithographic and non-lithographic techniques. The lithographic techniques fall into three basic categories: bulk micromachining, surface micromachining, and LIGA (a German acronym for Lithographie Galvanoformung Abformung), a German term for lithography, electroforming, and plastic molding. The first two approaches, bulk and surface micromachining, use planar photolithographic fabrication processes developed for semiconductor devices in producing two-dimensional (2-D) structures [33.10, 15–17]. The various steps involved in these two fabrication processes are shown schematically in Fig. 33.2. Bulk micromachining employs anisotropic etching to remove sections through the thickness of a single-crystal silicon wafer, typically 250 to 500 μm thick. Bulk micromachining is a proven high-volume production process and is routinely used to fabricate microstructures such as accelerometers, pressure sensors, and flow sensors. In surface micromachining, structural and sacrificial films are alternatively deposited, patterned and etched to produce a freestanding structure. These films are typically made of low-pressure chemical vapor deposition (LPCVD) polysilicon film with 2 to 20 μm thickness. Surface micromachining is used to produce sensors, actuators, and complex microdevices such as micromirror arrays, motors, gears, and grippers.

The LIGA process is based on the combined use of X-ray lithography, electroforming, and molding processes. The steps involved in the LIGA process are shown schematically in Fig. 33.3. LIGA is used to produce high-aspect-ratio MEMS (HARMEMS) devices that are up to 1 mm in height and only a few microns in width or length [33.18]. The LIGA process yields very sturdy 3-D structures due to their increased thickness. One of the limitations of silicon microfabrication processes originally used for fabrication of MEMS devices is the lack of suitable materials that can be processed. With LIGA, a variety of non-silicon materials such as metals, ceramics, and polymers can be processed. Nonlithographic micromachining processes, primarily in Europe and Japan, are also being used for fabrication of millimeter-scale devices using direct material microcutting or micromechanical machining (such as microturning, micromilling, and microdrilling), or removal by energy beams (such as microspark erosion, focused ion beam, laser ablation, and laser polymerization) [33.19, 20]. Hybrid technologies, including LIGA and high-precision micromachining techniques, have been used to produce miniaturized motors, gears, actuators, and connectors [33.21–24]. These millimeter-scale devices may find more immediate applications.

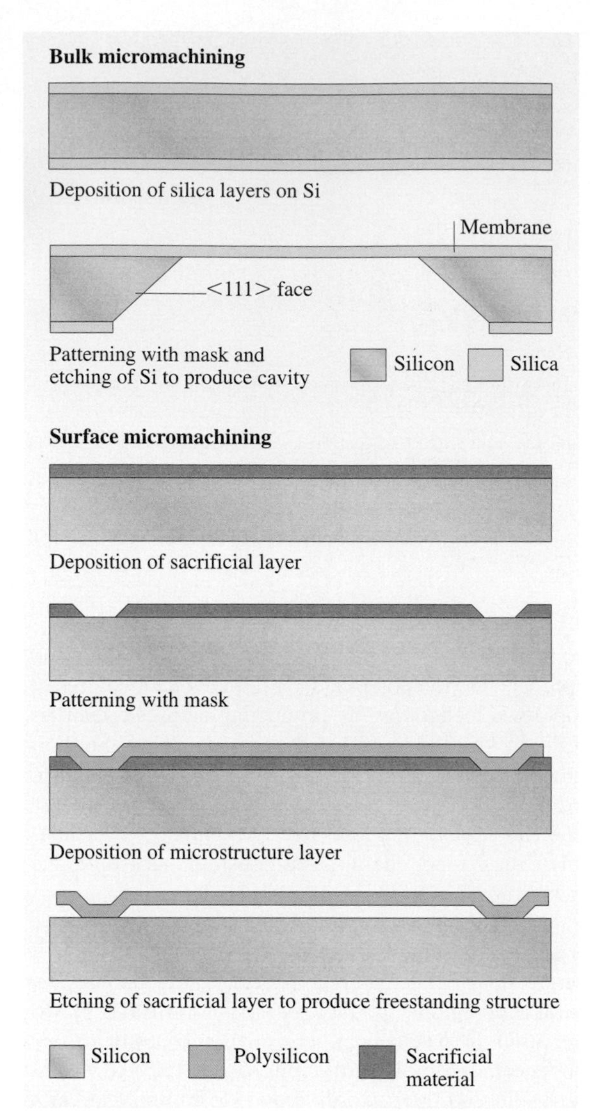

Fig. 33.2 Schematics of process steps involved in bulk micromachining and surface micromachining fabrication of MEMS

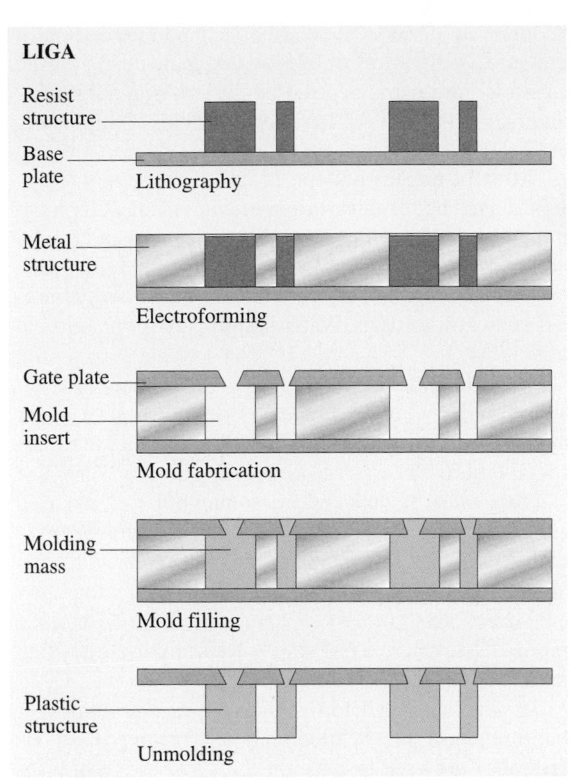

Fig. 33.3 Schematic of process steps involved in LIGA fabrication of MEMS

A microfabrication technique, so-called "soft lithography," is a nonlithographic technique [33.25, 26] in which an elastomeric stamp or mold is used to generate micropatterns by replica molding [33.27], embossing (imprinting) [33.28], or by contact printing (known as microcontact printing or μCP) [33.29]. Replica molding is commonly used for mass-produced, disposable plastic microcomponents, for example, microfluidic chips, generally made of poly(dimethylsiloxane) (PDMS) and polymethyl methacrylate (PMMA). The elastomeric stamps are generally cast from photolithographically generated micropatterned masters. This technique is substantially cheaper and more flexible in choice of materials for construction than conventional photolithography.

To assemble microsystems, microrobots are used. Microrobotics include building blocks such as steering links, microgrippers, conveyor system, and locomotive robots [33.13].

A variety of MEMS devices have been produced and some are commercially used [33.8, 10–12, 14, 15]. A variety of sensors are used in industrial, consumer, defense, and biomedical applications. Various microstructures or microcomponents are used in micro-instruments and other industrial applications such as micromirror arrays. Two of the largest "killer" industrial applications are accelerometers (about 85 million units in 2002) and digital micromirror devices (about $400 million in sales in 2001). Integrated capacitive-type, silicon accelerometers have been used in airbag deployment in automobiles since 1991 [33.30, 31]. Accelerometer technology was about a billion-dollar-a-year industry in 2001 dominated by Analog Devices, followed by Motorola and Bosch. Commercial digital light processing (DLP) equipments, using digital micromirror devices (DMD), were launched in 1996 by Texas Instruments for digital projection displays in portable and home theater projectors, as well as table top and projection TVs [33.32–34]. More than 1.5 million projectors were sold before 2002. Other major industrial applications include pressure sensors, inkjet printer heads and optical switches. Silicon-based piezoresistive pressure sensors for manifold absolute pressure sensing for engines were launched in 1991 by NovaSensor, and their annual sales were about 25 million units in 2002. Capacitive pressure sensors for tire pressure measurements were launched by Motorola. Annual sales of inkjet printer heads with microscale functional components were about 400 million units in 2002. Other applications of MEMS devices include chemical sensors, gas sensors, infrared detectors and focal plane arrays for earth observation, space science, and missile defense applications, pico-satellites for space applications, and many hydraulic, pneumatic, and other consumer products. MEMS devices are also being pursued in magnetic storage systems [33.35], where they are being developed for super-compact and ultrahigh-recording-density magnetic disk drives. Several integrated head/suspension microdevices have been fabricated for contact recording applications [33.36, 37]. High-bandwidth servo-controlled microactuators have been fabricated for ultrahigh-track-density applications that serve as the fine-position control element of a two-stage, coarse/fine servo system, coupled with a conventional actuator [33.38–40].

BIOMEMS are used increasingly in commercial and defense applications (e.g., [33.8, 41–44]). Applications of BIOMEMS include biofluidic chips (microfluidic chips or bioflips or simply biochips) for chemical and biochemical analyses (biosensors) in medical diagnostics (e.g., DNA, RNA, proteins, cells, blood pressure and assays, and toxin identification) and implantable pharmaceutical drug delivery. The biosensors, also referred to as lab-on-a-chip, integrate sample handling, separation, detection, and data analysis onto one platform. Biosensors are designed to either detect a single or class of (bio)chemicals or system-level analytical capabilities for a broad range of (bio)chemical species known as micro total analysis systems (μTAS). The chips rely on microfluidics and involve manipulation of tiny amounts of fluids in microchannels using microvalves. The chips consist of several basic components, including microfluidic channels and reservoirs, microvalves, micropumps, flow sensors, and biosensors. The test fluid is injected into the chip generally using an external pump or a syringe for analyses. Some chips have been designed with an integrated electrostatically actuated diaphragm-type micropump. Micropumps both with and without valves are used. The sample, which can have volume measured in nanoliters, flows through microfluidic channels via an electric potential and capillary action using microvalves (having various designs, including membrane type) for various analyses. The fluid is preprocessed and then analyzed using a biosensor. For a review on micropumps and microvalves, see [33.11, 45, 46]. Silicon-based disposable blood-pressure sensor chips were introduced in the early 1990s by NovaSensor for blood pressure monitoring (about 20 million units in 2002). A blood sugar monitor, referred to as GlucoWatch, was introduced in 2002. It automatically checks blood sugar every 10 min by detecting glucose through the skin, without having to draw blood. If glucose is out of the acceptable range, it sounds an alarm so the diabetic can address the problem quickly. A variety of biosensors, many using plastic substrates, are manufactured by various companies, including ACLARA, Agilent Technologies, Calipertech, and I-STAT.

After the tragedy of Sept. 11, 2001, concern over biological and chemical warfare has led to the development of handheld units with bio- and chemical sensors for detection of biological germs, chemical or nerve agents, and mustard agents, and chemical precursors to protect subways, airports, water supply, and population at large [33.47].

Other BIOMEMS applications include minimal invasive surgery, including endoscopic surgery, laser angioplasty, and microscopic surgery. Implantable artificial organs can also be produced.

Micro-instruments and micromanipulators are used to move, position, probe, pattern, and characterize nanoscale objects and nanoscale features. Miniaturized analytical equipment include gas chromatography and mass spectrometry. Other instruments include micro-STM, where STM stands for scanning tunneling microscope.

In some cases, MEMS devices are used primarily for their miniature size. While in others, as in the case of air bags, they are used because of their low-cost manufacturing techniques, since semiconductor-processing costs have reduced drastically over the last decade, allowing the use of MEMS in many fields.

33.2 Introduction to NEMS

NEMS are produced by nanomachining in a typical top-down approach (from large to small) and bottom-up approach (from small to large) largely relying on nanochemistry [33.48–53]. The top-down approach relies on fabrication methods, including advanced integrated-circuit (IC) lithographic methods – electron-beam lithography and STM writing by removing material atom by atom. The bottom-up approach includes chemical synthesis, the spontaneous "self-assembly" of molecular clusters (molecular self-assembly) from simple reagents in solution, or biological molecules (e.g., DNA) as building blocks to produce 3-D nanostructures, quantum dots (nanocrystals) of arbitrary diameter (about $10–10^5$ atoms), molecular beam epitaxy (MBE) and organometallic vapor phase epitaxy (OMVPE) to create specialized crystals one atomic or molecular layer at a time, and manipulation of individual atoms by an atomic force microscope or atom optics. The self-assembly must be encoded. That is, one must be able to precisely assemble one object next to another to form a designed pattern. A variety of nonequilibrium plasma chemistry techniques are also used to produce layered nanocomposites, nanotubes, and nanoparticles. The NEMS field, in addition to fabrication of nanosystems, has provided the impetus to develop experimental and computation tools.

Examples of NEMS include nanocomponents, nanodevices, nanosystems, and nanomaterials such as microcantilevers with integrated sharp nanotips for STM and atomic force microscopy (AFM), AFM array (millipede) for data storage, AFM tips for nanolithography, dip-pen nanolithography for printing molecules, biological (DNA) motors, molecular gears, molecularly thick films (e.g., in giant magnetoresistive or GMR heads, and magnetic media), nanoparticles (e.g., nanomagnetic particles in magnetic media), nanowires, carbon nanotubes, quantum wires (QWRs), quantum boxes (QBs), and quantum transistors. BIONEMS include nanobiosensors – a microarray of silicon nanowires, roughly a few nm in size, to selectively bind and detect even a single biological molecule such as DNA or protein by using nanoelectronics to detect the slight electrical charge caused by such binding, or a microarray of carbon nanotubes to electrically detect glucose, implantable drug-delivery devices – e.g., micro/nanoparticles with drug molecules encapsulated in functionized shells for a site-specific targeting applications and a silicon capsule with a nanoporous membrane filled with drugs for long-term delivery – , nanodevices for sequencing single molecules of DNA in the Human Genome Project, cellular growth using carbon nanotubes for spinal cord repair, nanotubes for nanostructured materials for various applications such as spinal fusion devices, organ growth, and growth of artificial tissues using nanofibers.

Nanoelectronics can be used to build computer memory using individual molecules or nanotubes to store bits of information, molecular switches, molecular or nanotube transistors, nanotube flat-panel displays, nanotube integrated circuits, fast logic gates, switches, nanoscopic lasers, and nanotubes as electrodes in fuel cells.

33.3 Tribological Issues in MEMS/NEMS

Tribological issues are important in MEMS/NEMS requiring intended and/or unintended relative motion. In MEMS/NEMS, various forces associated with the device scale down with size. When the length of the machine decreases from 1 mm to 1 μm, the area decreases by a factor of a million and the volume decreases by a factor of a billion. As a result, surface forces such as adhesion, friction, meniscus forces, viscous drag forces, and surface tension that are proportional to area become a thousand times larger than the forces proportional to the volume such as inertial and electromagnetic forces. In addition to the consequence of a large surface-to-volume ratio, since MEMS/NEMS are designed for small tolerances, physical contact becomes more likely, which makes them particularly vulnerable to adhesion between adjacent components. Since the start-up forces and torques involved in MEMS/NEMS operation available to overcome retarding forces are small, the increase in resistive forces such as adhesion and friction become a serious tribological concern that limits the life and reliability of MEMS/NEMS [33.10]. A large lateral force required to initiate relative motion between two surfaces, large static friction, is referred to as "stiction," which has been studied extensively in tribology of magnetic storage systems [33.35, 54, 55]. A large normal force required to separate two surfaces is also referred to as stiction. Adhesion, friction/stiction (static friction), wear, and surface contamination affect MEMS/NEMS performance and in some cases, can even prevent devices from working. Some examples of devices that experience tribological problems follow.

33.3.1 MEMS

Figure 33.4a shows examples of several microcomponents that can encounter the aforementioned tribological problems. The polysilicon electrostatic micromotor has 12 stators and a four-pole rotor and is produced by surface micromachining. The rotor diameter is 120 μm and the air gap between the rotor and stator is 2 μm [33.56]. It is capable of continuous rotation up to speeds of 100,000 RPM. The intermittent contact at the rotor-stator interface and physical contact at the rotor-hub flange interface result in wear issues, and high stiction between the contacting surfaces limits the repeatability of operation, or may even prevent the operation altogether. Next, a bulk micromachined silicon stator/rotor pair is shown with bladed rotor and nozzle guide vanes on the stator with dimensions less than a mm [33.57]. These are being developed for high-temperature micro-gas turbine generators with an operating speed up to 1 million RPM. Erosion of blades and vanes and wear of bearings used in the turbine generators are some of the concerns. Next, is an SEM micrograph of a surface micromachined polysilicon six-gear chain from Sandia National Lab. (For more examples of previous versions, see [33.58, 59].) As an example of non-silicon components, a milligear system produced using the LIGA process for a DC brushless permanent magnet millimotor (diameter = 1.9 mm, length = 5.5 mm) with an integrated milligear box [33.21–23] is also shown. The gears are made of metal (electroplated Ni-Fe), but can also be made from injected polymer materials (e.g.,

Electrostatic micromoter (*Tai* et al., [33.56])

Microturbine bladed rotor and nozzle guide vanes on the stator (*Spearing* and *Chen*, [33.57])

Six gear chain (www.sandia.gov)

Ni-Fe Wolfrom-type gear system by LIGA (*Lehr* et al., [33.21])

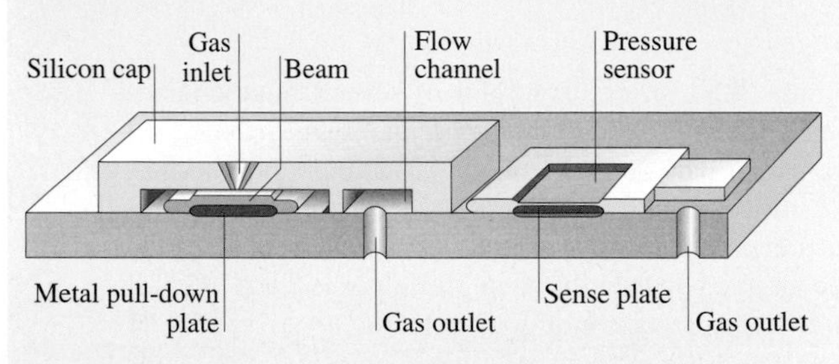

a) Low-pressure flow modulator with electrostatically actuated microvalves (*Robertson* and *Wise*, [33.60])

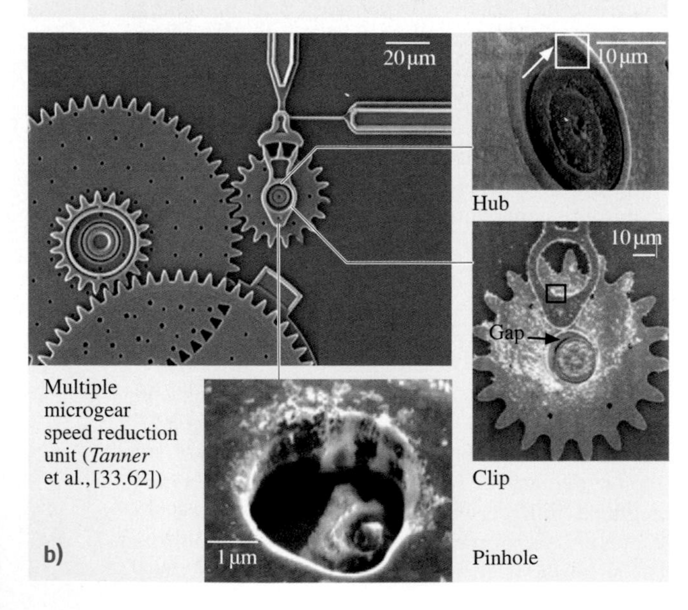

Multiple microgear speed reduction unit (*Tanner* et al., [33.62])

b)

Fig. 33.4a,b Examples of MEMS devices and components that experience tribological problems: (**a**) several micro-component and (**b**) a polysilicon, multiple microgear speed reduction unit

Polyoxy-methylene, or POM) using the LIGA process. Even though the torque transmitted at the gear teeth is small, on the order of a fraction of a nNm, because of small dimensions of gear teeth, the bending stresses are large where the teeth mesh. Tooth breakage and wear at the contact of gear teeth are concerns. Next, in a micromachined flow modulator, several micromachined flow channels are integrated in series with electrostatically actuated microvalves [33.60]. The flow channels lead to a central gas outlet hole drilled in the glass substrate. Gas enters the device through a bulk micromachined gas inlet hole in the silicon cap. After passing through an open microvalve, the gas flows parallel to the glass substrate through flow channels and exits the device through an outlet. The normally open valve structure consists of a freestanding double-end-clamped beam, which is positioned beneath the gas inlet orifice. When deflected electrostatically upwards, the beam seals against the inlet orifice, and the valve is closed. In these microvalves used for flow control, the mating valve surfaces should be smooth enough to seal while maintaining a minimum roughness to ensure low adhesion [33.54, 55, 61]. High adhesion (stiction) is a major issue.

Figure 33.4b shows a polysilicon, multiple microgear speed reduction unit and its components after laboratory wear tests conducted in air [33.62]. These units have been developed for an electrostatically driven microactuator (microengine) developed at Sandia National Lab for operation in the kHz frequency range. Wear of various components is clearly observed in the figure. Humidity was shown to be a strong factor in the wear of rubbing surfaces. In order to improve the wear characteristics of rubbing surfaces, 20-nm-thick tungsten (W) coating using chemical vapor deposition (CVD) technique was used [33.63]. Tungsten-coated microengines tested for reliability showed improved wear characteristics with longer lifetimes than polysilicon microengines.

Commercially available MEMS devices also exhibit tribological problems. Figure 33.5 shows an integrated capacitive-type silicon accelerometer fabricated using surface micromachining by Analog Devices, a couple of mm in dimension, which is used for the deployment of airbags in automotives and more recently for

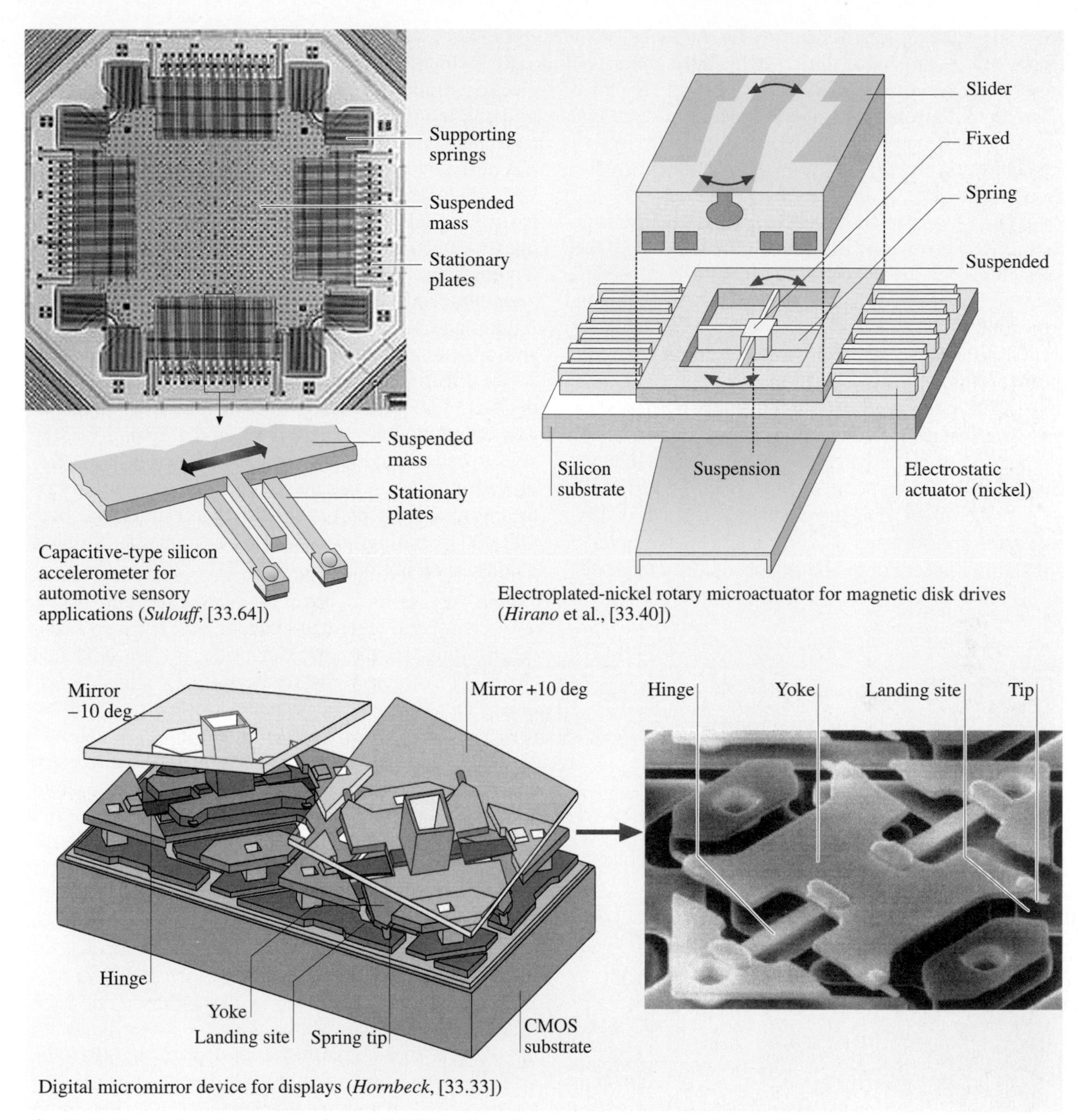

Fig. 33.5 Examples of commercial use MEMS devices that experience tribological problems

the consumer electronics market [33.30, 64, 65]. These accelerometers are now being developed for many other applications such as vehicle stability, rollover control, and gyro sensor applications. The central suspended beam mass (about 0.7 μg) is supported on the four corners by spring structures. The central beam has interdigitated, cantilevered electrode fingers (about 125 μm long and 3 μm thick) on all four sides that alternate with those of the stationary electrode fingers, as shown, with about a 1.3 μm gap. Lateral motion of the central beam causes a change in the capacitance between these electrodes, which is used to measure the acceleration. Here stiction between the adjacent electrodes, as well as stiction of the beam structure with the underlying substrate are detrimental to the operation of the sensor [33.30, 64]. Wear during unintended contacts of these polysilicon fingers is also a problem. A molecularly thick diphenyl siloxane lubricant film with high resistance to temperature and oxidation, applied by a vapor deposition process is used on the electrodes to reduce stiction and wear.

Figure 33.5 also shows two digital micromirror device (DMD) pixels used in digital light processing (DLP) technology for digital projection displays in portable and home theater projectors, as well as table-top and projection TVs [33.32–34]. The entire array (chip set) consists of a large number of rotatable aluminum micromirrors (digital light switches) that are fabricated on top of a CMOS static random access memory integrated circuit. The surface micromachined array consists of half of a million to more than two million of these independently controlled, reflective micromirrors (mirror size on the order of 14 μm square and 15 μm pitch) that flip backward and forward at a frequency on the order of 5,000 times per s. For the binary operation, a micromirror/yoke structure mounted on torsional hinges is rotated ±10° (with respect to the horizontal plane) as a result of electrostatic attraction between the micromirror structure and the underlying memory cell and is limited by a mechanical stop. Contact between cantilevered spring tips at the end of the yoke (four present on each yoke) with the underlying stationary landing sites is required for true digital (binary) operation. Stiction and wear during a contact between aluminum alloy spring tips and landing sites, hinge memory (metal creep at high operating temperatures), hinge fatigue, shock and vibration failure, and sensitivity to particles in the chip package and operating environment are some of the important issues affecting the reliable operation of a micromirror device [33.66–68]. Perfluorodecanoic acid (PFDA) self-assembled monolayers are used on the tip and landing sites to reduce stiction and wear. The spring tip is employed in order to use the spring stored energy to pop up the tip during pull-off. A lifetime estimate of over 100,000 operating hours with no degradation in image quality is the norm.

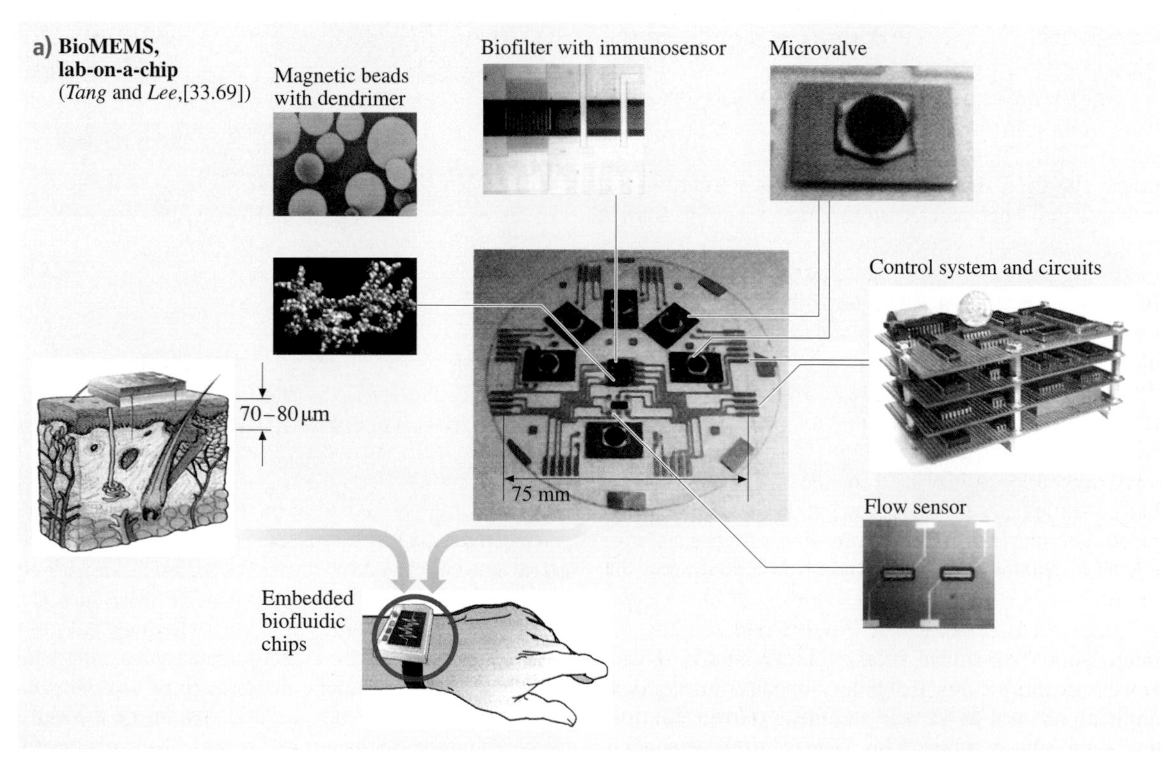

Fig. 33.6a,b Schematics of (a) a MEMS-based biofluidic chip, commonly known as disposable lab-on-a-chip, and (b) (*see next page*) a NEMS-based submicroscopic drug-delivery device, and an intravascular micro/nanoparticles for search-and-destroy diseased blood cells

A third MEMS device, shown in Fig. 33.5, is an electrostatically driven rotary microactuator for a magnetic disk drive surface-micromachined by a multilayer electroplating method [33.40]. This high-bandwidth servo-controlled microactuator, located between a slider and a suspension, is being developed for ultrahigh-track-density applications, which serves as the fine-position and high-bandwidth control element of a two-stage, coarse/fine servo system when coupled with a conventional actuator [33.38–40]. A slider is placed on top of the central block of a microactuator, which gives rotational motion to the slider. The bottom of the silicon substrate is attached to the suspension. The radial flexure beams in the central block give the rotational freedom of motion to the suspended mass (slider), and the electrostatic actuator drives the suspended mass. Actuation is accomplished via interdigitated, cantilevered electrode fingers, which are alternatingly attached to the central body of the moving part and to the stationary substrate to form pairs. A voltage applied across these electrodes results in an electrostatic force, which rotates the central block. The inter-electrode gap width is about 2 μm. Any unintended contacts between the moving and stationary electroplated-nickel electrodes may result in wear and stiction.

An example of disposable plastic lab-on-a-chip using microfluidics and BIOMEMS technologies is shown in Fig. 33.6a [33.69]. This biofluidic chip integrates multiple fluidic components on a microfluidic motherboard that can be worn like a wristwatch. It enables the unobstructive assessment of a human's medical condition through continuous blood sampling. It can be used for the analysis of a number of different compounds, including lactate, glucose, carbon dioxide, and oxygen levels in the blood and for the detection of infectious diseases. The test fluid is injected into the microchannels using a micropump, and the flow is regulated using microvalves. A magnetic bead approach is used for identification of target biomolecules based on sandwich immunoassay and electrochemical detection. Adhesion in micropumps and microvalves involving moving parts and adhesion of fluid in microchannels are some of the tribological issues. In both active and passive microvalves, the main wear mechanism appears to be erosion, corrosion, and fatigue. As the magnetic bead particulates are pumped through the microfluidic system, they could erode the valve material. Also, several different reagents are used to perform the necessary analyses. The interaction of these reagents with the valve material could cause corrosive wear. Finally, because the valves are continuously being activated, fatigue failure becomes a concern.

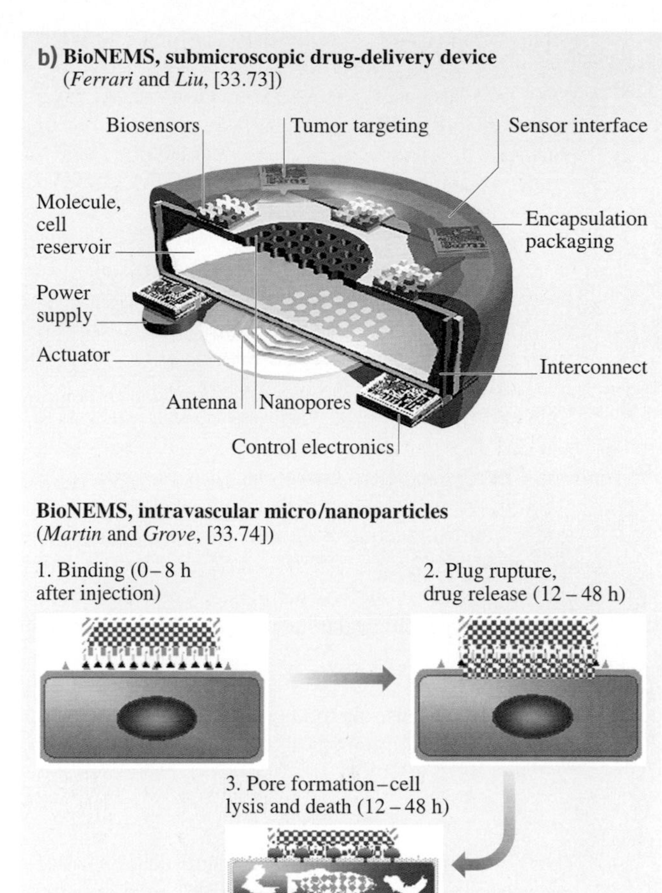

Fig. 33.6b

Stiction/friction and wear clearly limit the lifetimes and compromise the performance and reliability of microdevices. Figure 33.7 summarizes tribological problems encountered in some of the MEMS devices and components just discussed. In addition, tribological issues are present in the processes used for fabrication of MEMS/NEMS. For example, in surface micromachining, the suspended structures can sometimes collapse and permanently adhere to the underlying substrate due to meniscus effects during the final rinse and dry process, as shown in Fig. 33.7 [33.70]. Adhesion is caused by water molecules adsorbed on the adhering surfaces and/or because of formation of adhesive bonds by silica residues that remain on the surfaces after the water has evaporated. This so called release stiction is overcome

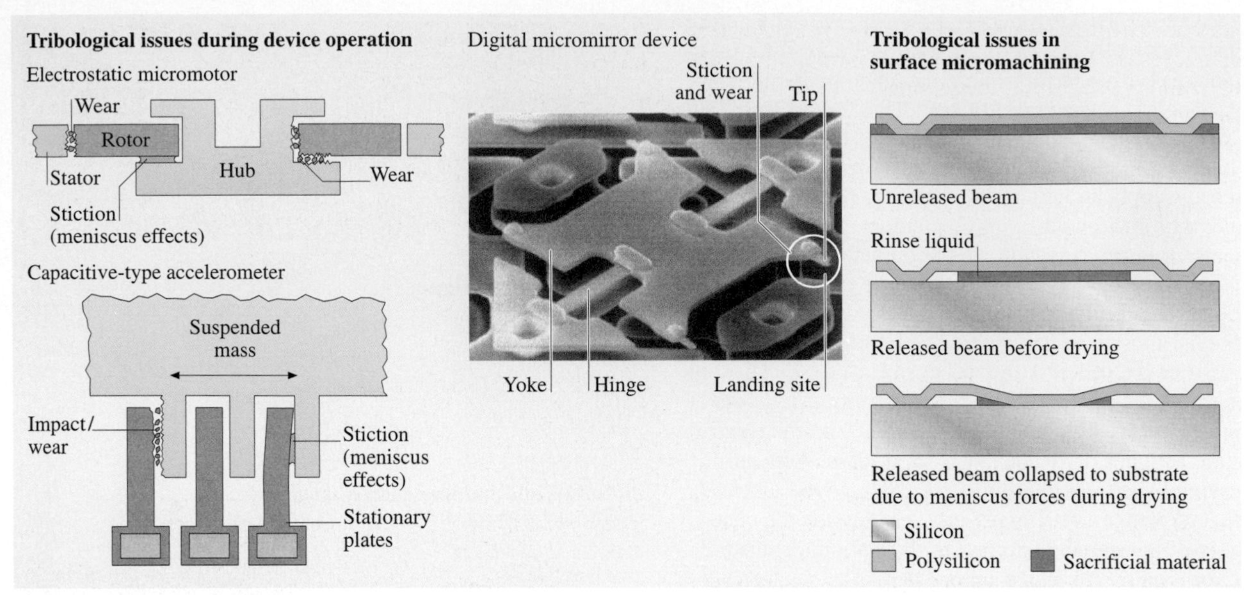

Fig. 33.7 Summary of tribological issues in MEMS device operation and fabrication via surface micromachining

by using dry release methods (e.g., CO_2 critical point drying or sublimation methods [33.71]).

33.3.2 NEMS

Figure 33.8 shows an AFM-based nanoscale data storage system for ultrahigh density magnetic recording that experiences tribological problems [33.72]. The system uses arrays of several thousand silicon microcantilevers ("millipede") for thermomechanical recording and playback on an about 40-nm-thick polymer (PMMA) medium with a harder Si substrate. The cantilevers are integrated with integrated tip heaters that have tips of nanoscale dimensions. Thermomechanical recording is a combination of applying a local force to the polymer layer, and softening it by local heating. The tip heated to about 400 °C is brought in contact with the polymer for recording. Imaging and reading are done using the heater cantilever, originally used for recording, as a thermal read-back sensor by exploiting its temperature-dependent resistance. The principle of thermal sensing is based on the fact that the thermal conductivity between the heater and the storage substrate changes according to the spacing between them. When the spacing between the heater and sample is reduced as the tip moves into a bit, the heater's temperature and hence its resistance will decrease. Thus, changes in temperature of the continuously heated resistor are monitored while the cantilever is scanned over data bits, providing a means of detecting the bits. Erasing for subsequent rewriting is carried out by thermal reflow of the storage field by heating the medium to 150 °C for a few seconds. The smoothness of the reflown medium allows multiple rewriting of the same storage field. Bit sizes ranging between 10 and 50 nm have been achieved by using a 32 × 32 (1024) array write/read chip (3 mm × 3 mm). It has been reported that tip wear occurs through the contact between tip and Si substrate during writing. Tip wear is considered a major factor in device reliability.

Figure 33.6b shows two examples of BIONEMS [33.73, 74]. The figure shows a conceptual model of a submicroscopic drug delivery device with the ability to localize in the areas of need. It is a silicon capsule with a nanoporous membrane filled with drugs for long-term delivery. It uses a nanomembrane with pores as small as 6 nm that are used as flux regulators for the long-term release of drugs. The nanomembrane also protects therapeutic substances from attack by the body's immune system. The figure also shows a conceptual model of an intravascular drug delivery device – micro/nanoparticle used for search-and-destroy disease cells. With the lateral dimensions of 1 μm or less, the particle is smaller than any blood cells. These particles can be injected into the blood stream and travel freely through the circulatory system. In order to direct these drug-delivery micro/nanoparticles to cancer sites, their external surfaces are chemically modified to carry molecules that have lock-and-key binding specificity with molecules that support a growing cancer mass. As soon as the particles dock on the cells, a compound is released that forms a pore on the membrane of the cells, which leads to cell death and ultimately to that of the cancer mass that was nourished by the blood vessel. Adhesion between micro/nanodevices and disease cells is required.

33.3.3 Tribological Needs

The MEMS/NEMS need to be designed to perform expected functions typically in the ms to ps range. Expected life of the devices for high speed contacts can vary from a few hundred thousand to many billions of cycles, e.g., over a hundred billion cycles for DMDs, which puts serious requirements on materials [33.10, 62, 75–78]. Most mechanical properties are known to be scale-dependent [33.79]. The properties of nanoscale structures need to be measured [33.80]. Tribology is an important factor affecting the performance and reliability of MEMS/NEMS [33.10, 49, 81]. There is a need for developing a fundamental understanding of adhesion, friction/stiction, wear, and the roles of surface contamination and environment [33.10]. MEMS/NEMS materials need to exhibit good mechanical and tribological properties on the micro/nanoscale. There is a need to develop lubricants and identify lubrication methods that are suitable for MEMS/NEMS. Component-level studies are required to provide a better understanding of the tribological phenomena occurring in MEMS/NEMS. The emergence of micro/nanotribology and atomic force microscopy-based techniques has provided researchers with a viable approach to address these problems [33.49, 82]. This chapter presents a review of macro- and micro/nanoscale tribological studies of materials and lubrication studies for MEMS/NEMS and component-level studies of stiction phenomena in MEMS/NEMS devices.

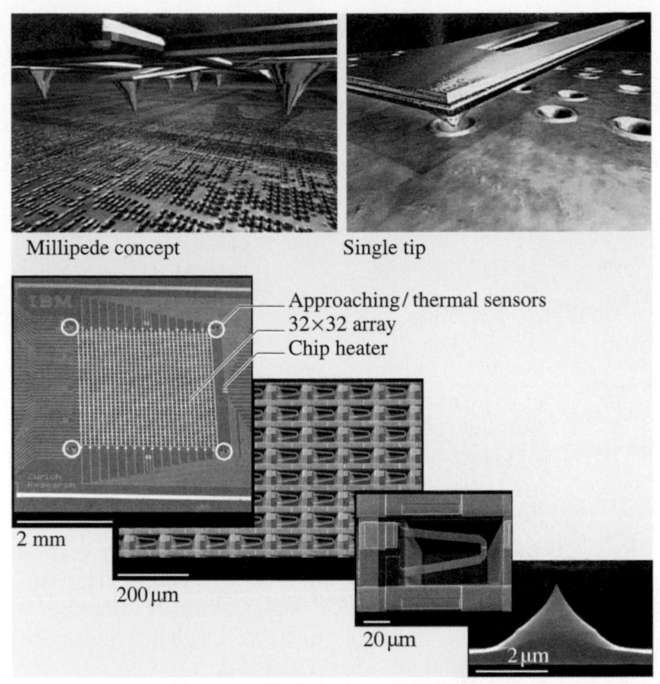

Fig. 33.8 Example of a NEMS device – AFM-based nanoscale data storage system with 32×32 tip array – that experiences tribological problems [33.72]

33.4 Tribological Studies of Silicon and Related Materials

Materials of most interest for planar fabrication processes using silicon as the structural material are undoped and boron-doped (p^+-type) single-crystal silicon for bulk micromachining and phosphorus (n^+-type) doped and undoped LPCVD polysilicon films for surface micromachining. Silicon-based devices lack high-temperature capabilities with respect to both mechanical and electrical properties. Researchers have been pursuing SiC as a material for high-temperature microsensor and microactuator applications, for some time [33.83, 84]. SiC is a likely candidate for such applications, since it has long been used in high temperature electronics, high frequency and high power devices. SiC can also be desirable for high-frequency micromechanical resonators, in the GHz range, because of its high modulus of elasticity and, consequently, high resonant frequency. Table 33.2 compares selected bulk properties of SiC and Si(100). Researchers have found low cost techniques of producing single-crystal 3C-SiC (cubic or β-SiC) films via epitaxial growth on large-area silicon substrates for bulk micromachining [33.85] and polycrystalline 3C-SiC films on polysilicon and silicon dioxide layers for surface micromachining of SiC [33.86].

Table 33.2 Selected bulk properties[a] of 3C (β- or cubic) SiC and Si(100)

Sample	Density (kg/m^3)	Hardness (GPa)	Elastic modulus (GPa)	Fracture toughness ($MPa\,m^{1/2}$)	Thermal conductivity[b] (W/m K)	Coeff. of thermal expansion[b] ($\times10^{-6}/°C$)	Melting point (°C)	Band gap (eV)
β-SiC	3,210	23.5–26.5	440	4.6	85–260	4.5–6	2,830	2.3
Si(100)	2,330	9 –10	130	0.95	155	2 –4.5	1,410	1.1

[a] Unless stated otherwise, data shown were obtained from [33.87]
[b] Obtained from [33.88]

As will be shown, bare silicon exhibits inadequate tribological performance and needs to be coated with a solid and/or liquid overcoat or surface treated (e.g., oxidation and ion implantation, commonly used in semiconductor manufacturing), which exhibit lower friction and wear. SiC films exhibit good tribological performance. Both macroscale and microscale tribological properties of virgin and treated/coated silicon, polysilicon films, and SiC are presented next.

33.4.1 Tribological Properties of Silicon and the Effect of Ion Implantation

Friction and wear of single-crystalline and polycrystalline silicon samples, as well as the effect of ion implantation with various doses of C^+, B^+, N_2^+, and Ar^+ ion species at 200 keV energy to improve their friction and wear properties have been studied [33.89–91]. The coefficient of macroscale friction and wear factor of virgin single-crystal silicon and C^+-implanted silicon samples as a function of ion dose are presented in Fig. 33.9 [33.89]. The macroscale friction and wear tests were conducted using a ball-on-flat tribometer. Each data bar represents the average value of four to six measurements. The coefficient of friction and wear factor for bare silicon are very high and decrease drastically with ion dose. Silicon samples bombarded above the ion dose of 10^{17} C^+ cm^{-2} exhibit extremely low values of coefficients of friction (typically 0.03 to 0.06 in air) and wear factor (reduced by as much as four orders of magnitude). *Gupta* et al. [33.89] reported that a decrease in coefficient of friction and wear factor of silicon as a result of C^+ ion bombardment occurred because of formation of silicon carbide, rather than amorphization of silicon. *Gupta* et al. [33.90] also reported an improvement in friction and wear with B^+ ion implantation.

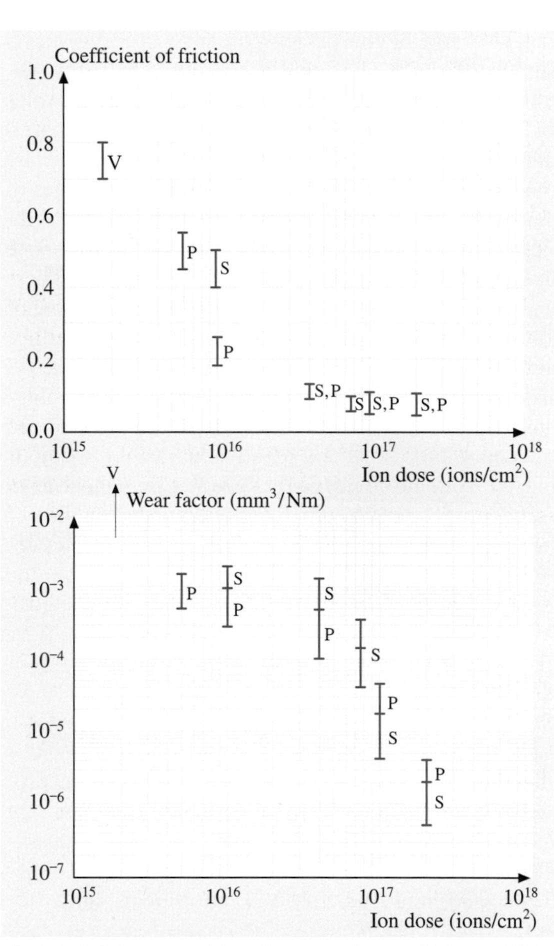

Fig. 33.9 Influence of ion doses on the coefficient of friction and wear factor on C^+ ion bombarded single-crystal and polycrystalline silicon slid against an alumina ball. V corresponds to virgin single-crystal silicon, while S and P denote tests that correspond to doped single- and polycrystalline silicon, respectively [33.89]

Table 33.3 Surface roughness and micro- and macroscale coefficients of friction of selected samples

Material	RMS roughness (nm)	Coefficient of microscale friction[a]	Coefficient of macroscale friction[b]
Si(111)	0.11	0.03	0.33
C^+-implanted Si(111)	0.33	0.02	0.18

[a] Versus Si_3N_4 tip, tip radius of 50 nm in the load range of 10–150 nN (2.5–6.1 GPa) at a scanning speed of 5 μm/s over a scan area of 1 μm × 1 μm in an AFM

[b] Versus Si_3N_4 ball, ball radius of 3 mm at a normal load of 0.1 N (0.3 GPa) at an average sliding speed of 0.8 mm/s using a tribometer

Microscale friction measurements were performed using an atomic force/friction force microscope (AFM/FFM) [33.49]. Table 33.3 shows values of surface roughness and coefficients of macroscale and microscale friction for virgin and doped silicon. There is a decrease in coefficients of microscale and macroscale friction values as a result of ion implantation. When measured for small contact areas and very low loads used in microscale studies, the indentation hardness and elastic modulus are higher than at the macroscale. This, added to the effect of the small apparent area of contact reducing the number of trapped particles on the interface, results in less plowing contribution and lower friction in the case of microscale friction measurements. Results of microscale wear resistance studies of ion-implanted silicon samples studied using a diamond tip in an AFM [33.92] are shown in Fig. 33.10a, b. For tests conducted at various loads on Si(111) and C^+-implanted Si(111), it is noted that wear resistance of an implanted sample is slightly poorer than that of virgin silicon up to about 80 μN. Above 80 μN, the wear resistance of implanted Si improves. As one continues to run tests at 40 μN for a larger number of cycles, the implanted sample exhibits higher wear resistance than the unimplanted sample. Damage from the implantation in the top layer results in poorer wear resistance. However, the implanted zone at the subsurface is more wear resistant than the virgin silicon.

Hardness values of virgin and C^+-implanted Si(111) at various indentation depths (normal loads) are presented in Fig. 33.10c [33.92]. The hardness at a small indentation depth of 2.5 nm is 16.6 GPa and drops

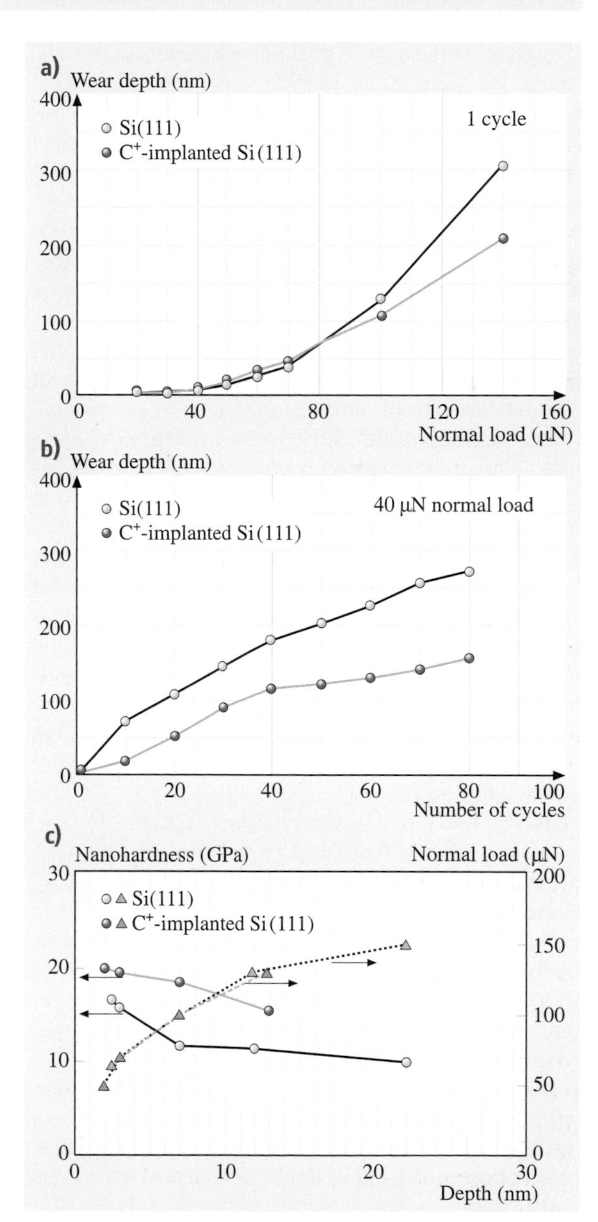

Fig. 33.10a–c Wear depth as a function of (**a**) load (after one cycle), and (**b**) cycles (normal load = 40 mN) for Si(111) and C^+-implanted Si(111). (**c**) Nanohardness and normal load as a function of indentation depth for virgin and C^+-implanted Si(111) [33.92]

to a value of 11.7 GPa at a depth of 7 nm and a normal load of 100 μN. Higher hardness values obtained in low-load indentation may arise from the observed pressure-induced phase transformation during the nanoindentation [33.93, 94]. An additional increase in the hardness at an even lower indentation depth of 2.5 nm reported here may arise from the contribution of complex chemical films (not from native oxide films) present on the silicon surface. At small volumes there is a lower probability of encountering material defects (dislocations, etc.). Furthermore, according to the strain gradient plasticity theory advanced by *Fleck* et al. [33.95], large strain gradients inherent in small indentations lead to the accumulation of geometrically necessary dislocations that cause enhanced hardening. These are some of the plausible explanations for an increase in hardness at smaller volumes. If the silicon material were to be used at very light loads such as in microsystems, the high hardness of surface films would protect the surface until it is worn.

From Fig. 33.10c, hardness values of C^+-implanted Si(111) at a normal load of 50 μN is 20.0 GPa with an indentation depth of about 2 nm, which is comparable to the hardness value of 19.5 GPa at 70 μN, whereas measured hardness value for virgin silicon at an indentation depth of about 7 nm (normal load of 100 μN) is only about 11.7 GPa. Thus, ion implantation with C^+ results in an increase in hardness in silicon. Note that the surface layer of the implanted zone is much harder compared to the subsurface and may be brittle, leading to higher wear on the surface. The subsurface of the implanted zone is harder than the virgin silicon, resulting in higher wear resistance, which is also observed in the results of the macroscale tests conducted at high loads.

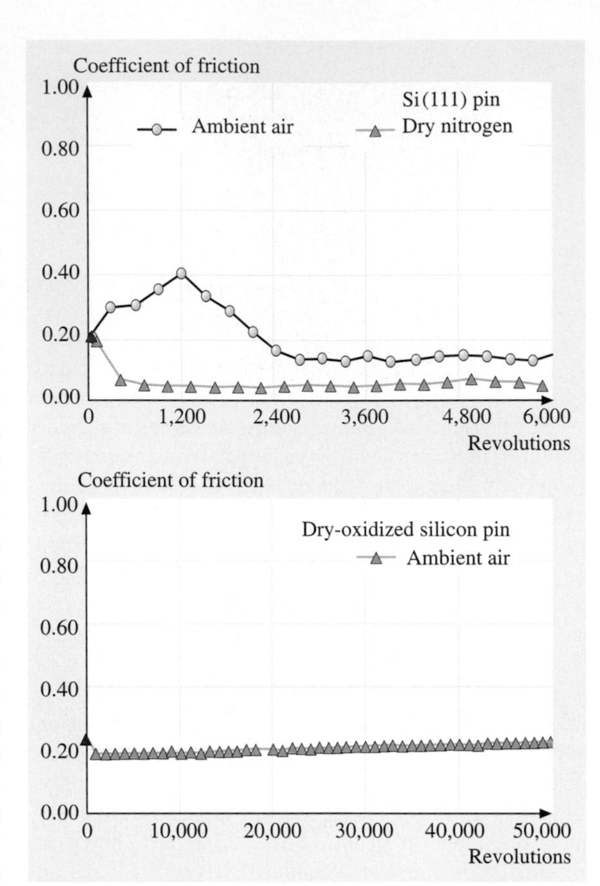

Fig. 33.11 Coefficient of friction as a function of number of sliding revolutions in ambient air for Si(111) pin in ambient air and dry nitrogen and dry-oxidized silicon pin in ambient air [33.96]

33.4.2 Effect of Oxide Films on Tribological Properties of Silicon

Macroscale friction and wear experiments have been performed using a magnetic disk drive with bare, oxidized, and implanted pins sliding against amorphous-carbon coated magnetic disks lubricated with a thin layer of perfluoropolyether lubricant [33.96–99]. Representative profiles of the variation of the coefficient of friction with a number of sliding cycles for Al_2O_3-TiC slider and bare and dry-oxidized silicon pins are shown in Fig. 33.11. For bare Si(111), after an initial increase, the coefficient of friction drops to a steady state value of 0.1 following the increase, as seen in Fig. 33.11. The rise in the coefficient of friction for the Si(111) pin is associated with the transfer of amorphous carbon from the disk to the pin and the oxidation-enhanced fracture of pin material followed by tribochemical oxidation of the transfer film, while the drop is associated with the formation of a transfer coating on the pin, as shown in Fig. 33.12. Dry-oxidized Si(111) exhibits excellent characteristics, and no significant increase was observed over 50,000 cycles (Fig. 33.11). This behavior has been attributed to the chemical passivity of the oxide and lack of transfer of DLC from the disk to the pin. The behavior of PECVD-oxide (data are not presented here) was comparable to that of dry oxide, but for the wet oxide there was some variation in the coefficient of friction (0.26 to 0.4). The difference between dry and wet oxide was attributed to increased

Ambient air, 6,000 cycles

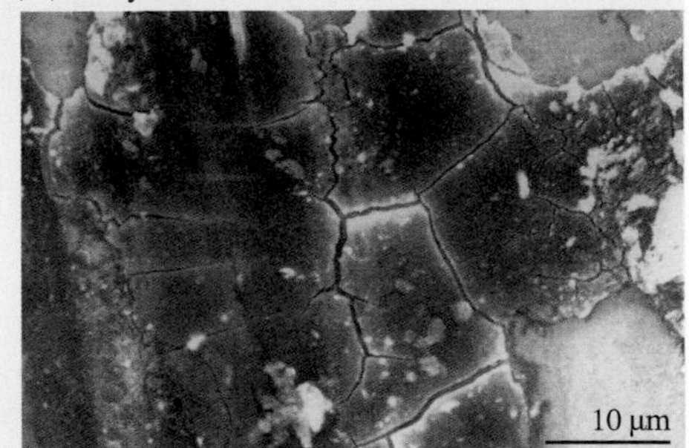

Dry nitrogen, 15,000 cycles

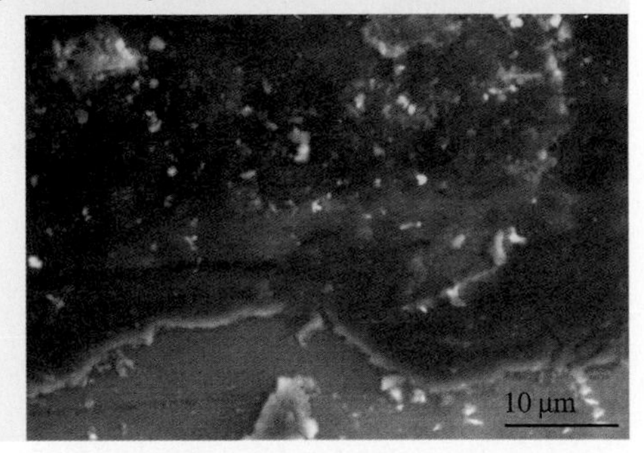

Fig. 33.12 Scanning electron micrographs of Si(111) after sliding against a magnetic disk in a rigid disk drive in ambient air for 6,000 cycles and in dry nitrogen after 15,000 cycles

porosity of the wet oxide [33.96]. Since tribochemical oxidation was determined to be a significant factor, experiments were conducted in dry nitrogen [33.97, 98]. The variation of the coefficient of friction for a silicon pin sliding against a thin-film disk in dry nitrogen is shown in Fig. 33.11. It is seen that in a dry nitrogen environment, the coefficient of friction of Si(111) sliding against a disk decreased from an initial value of about 0.2 to 0.05 with continued sliding. Based on SEM and chemical analysis, this behavior has been attributed to the formation of a smooth amorphous-carbon/lubricant transfer patch and suppression of oxidation in a dry nitrogen environment. Based on macroscale tests using disk drives, it was found that the friction and wear performance of bare silicon is not adequate. With dry-oxidized or PECVD SiO_2-coated silicon, no significant friction increase or interfacial degradation was observed in ambient air.

Table 33.4 and Fig. 33.13 show surface roughness, microscale friction and scratch data, and nanoindentation hardness for various silicon samples [33.92]. Scratch experiments were performed using a diamond tip in an AFM. Results on polysilicon samples are also shown for comparison. Coefficients of microscale friction values for all the samples are about the same. These samples could be scratched at 10 μN load. Scratch depth increased with normal load. Crystalline orientation of silicon has little influence on

Table 33.4 RMS, microfriction, microscratching/microwear, and nanoindentation hardness data for various virgin, coated, and treated silicon samples

Material	RMS roughness[a] (nm)	Coefficient of microscale friction[b]	Scratch depth[c] at 40 μN (nm)	Wear depth[c] at 40 μN (nm)	Nanohardness[c] at 100 μN (GPa)
Si(111)	0.11	0.03	20	27	11.7
Si(110)	0.09	0.04	20		
Si(100)	0.12	0.03	25		
Polysilicon	1.07	0.04	18		
Polysilicon (lapped)	0.16	0.05	18	25	12.5
PECVD-oxide coated Si(111)	1.50	0.01	8	5	18.0
Dry-oxidized Si(111)	0.11	0.04	16	14	17.0
Wet-oxidized Si(111)	0.25	0.04	17	18	14.4
C^+-implanted Si(111)	0.33	0.02	20	23	18.6

[a] Scan size of 500 nm × 500 nm using AFM.

[b] Versus Si_3N_4 tip in AFM/FFM, radius 50 nm; at 1 μm × 1 μm scan size.

[c] Measured using an AFM with a diamond tip of radius 100 nm.

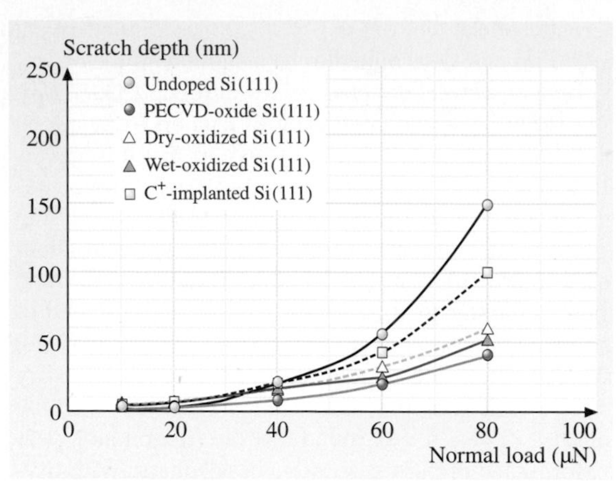

Fig. 33.13 Scratch depth as a function of normal load after 10 cycles for various silicon samples: virgin, treated, and coated [33.92]

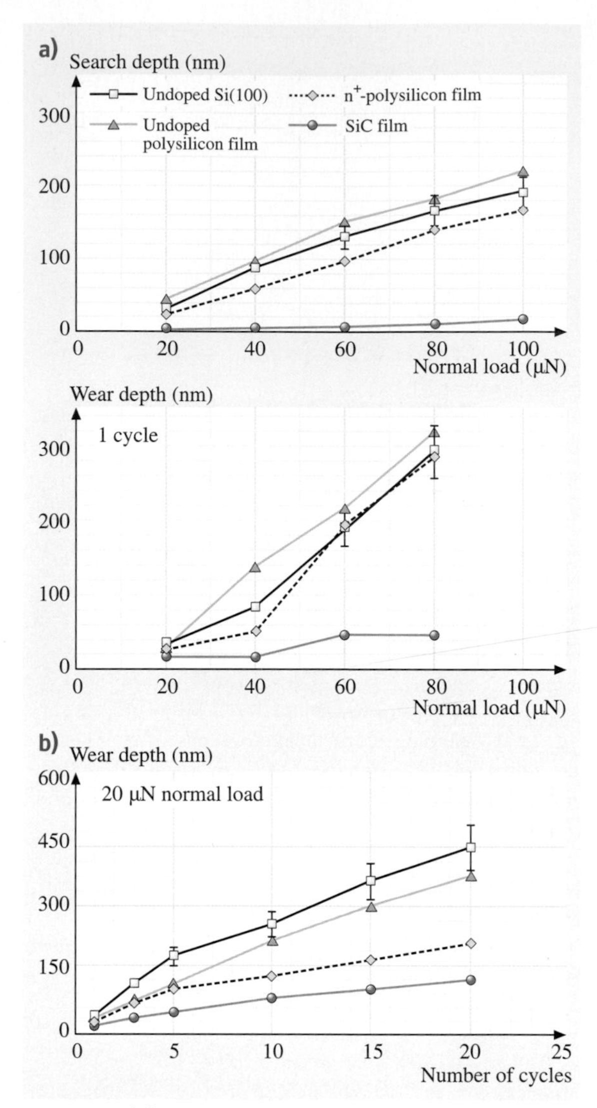

Fig. 33.14 (a) Scratch depths for 10 cycles as a function of normal load and (b) wear depths as a function of normal load and as a function of number of cycles for various samples [33.10]

scratch resistance, because natural oxidation of silicon in ambient masks the expected effect of crystallographic orientation. PECVD-oxide samples showed the best scratch resistance, followed by dry-oxidized, wet-oxidized, and ion-implanted samples. Ion implantation with C^+ does not appear to improve scratch resistance.

Wear data on the silicon samples are also presented in Table 33.1 [33.92]. PECVD-oxide samples showed superior wear resistance followed by dry-oxidized, wet-oxidized, and ion-implanted samples. This agrees with the trends seen in scratch resistance. In PECVD, ion bombardment during deposition improves the coating properties such as suppression of columnar growth, freedom from pinhole, decrease in crystalline size, and increase in density, hardness, and substrate-coating adhesion. These effects may help in improving the mechanical integrity of the sample surface. Coatings and treatments improved nanohardness of silicon. Note that dry-oxidized and PECVD films are harder than wet-oxidized films, as these films may be porous. High hardness of oxidized films may be responsible for measured high scratch/wear resistance.

33.4.3 Tribological Properties of Polysilicon Films and SiC Film

Studies have also been conducted on undoped polysilicon film, heavily doped (n^+-type) polysilicon film, heavily doped (p^+-type) single-crystal Si(100), and 3C-SiC (cubic or β-SiC) film [33.10, 100, 101]. The polysilicon films studied here are different from the ones discussed previously.

Table 33.5 presents a summary of the tribological studies conducted on polysilicon and SiC films. Values for single-crystal silicon are shown for comparison. Polishing of the as-deposited polysilicon and SiC films drastically affects the roughness, as the values reduce by two orders of magnitude. Si(100) appears to be

Table 33.5 Summary of micro/nanotribological properties of the sample materials

Sample	RMS roughness[a] (nm)	P–V distance[a] (nm)	Coefficient of friction Micro[b]	Coefficient of friction Macro[c]	Scratch depth[d] (nm)	Wear depth[e] (nm)	Nano-hardness[f] (GPa)	Young's modulus[f] (GPa)	Fracture toughness[g], K_{IC} MPa $m^{1/2}$
Undoped Si(100)	0.09	0.9	0.06	0.33	89	84	12	168	0.75
Undoped polysilicon film (as deposited)	46	340	0.05						
Undoped polysilicon film (polished)	0.86	6	0.04	0.46	99	140	12	175	1.11
n^+-Type polysilicon film (as deposited)	12	91	0.07						
n^+-Type polysilicon film (polished)	1.0	7	0.02	0.23	61	51	9	95	0.89
SiC film (as deposited)	25	150	0.03						
SiC film (polished)	0.89	6	0.02	0.20	6	16	25	395	0.78

[a] Measured using AFM over a scan size of 10 μm × 10 μm

[b] Measured using AFM/FFM over a scan size of 10 μm × 10 μm

[c] Obtained using a 3-mm-diameter sapphire ball in a reciprocating mode at a normal load of 10 mN and average sliding speed of 1 mm/s after 4 m sliding distance

[d] Measured using AFM at a normal load of 40 μN for 10 cycles, scan length of 5 μm

[e] Measured using AFM at normal load of 40 μN for 1 cycle, wear area of 2 μm × 2 μm

[f] Measured using nanoindenter at a peak indentation depth of 20 nm

[g] Measured using microindenter with Vickers indenter at a normal load of 0.5 N

the smoothest followed by polished undoped polysilicon and SiC films, which have comparable roughness. The doped polysilicon film shows higher roughness than the undoped sample, which is attributed to the doping process. Polished SiC film shows the lowest friction followed by polished and undoped polysilicon film, which strongly supports the candidacy of SiC films for use in MEMS/NEMS devices. Macroscale friction measurements indicate that SiC film exhibits one of the lowest friction values as compared to the other samples. The doped polysilicon sample shows low friction on the macroscale as compared to the undoped polysilicon sample, possibly due to the doping effect.

Figure 33.14a shows a plot of scratch depth vs. normal load for various samples [33.10, 100]. Scratch depth increases with increasing normal load. Figure 33.15 shows AFM 3-D maps and averaged 2-D profiles of the scratch marks on the various samples. It is observed that scratch depth increases almost linearly with the normal load. Si(100) and the doped and undoped polysilicon films show similar scratch resistance. It is clear from the data that the SiC film is much more scratch resistant than the other samples. Figure 33.14b shows results from microscale wear tests on the various films. For all the materials, the wear depth increases almost linearly with increasing number of cycles. This suggests that the material is removed layer by layer in all the materials. Here also SiC film exhibits lower wear depths than the other samples. Doped polysilicon film wears less than the undoped film. Higher fracture toughness and hardness of SiC compared to Si(100) are responsible for its lower wear. Also the higher thermal conductivity of SiC (see Table 33.2) compared to the other materials leads to lower interface temperatures, which generally results in less degradation of the surface [33.35, 54, 55]. Doping of the polysilicon does not affect the scratch/wear resistance and hardness much. The measurements made on the doped sample are affected by the presence of grain boundaries. These studies indicate that SiC film exhibits desirable tribological properties for use in MEMS devices. Recently, researchers have fabricated SiC micromotors and have reported satisfactory operation at high temperatures [33.102].

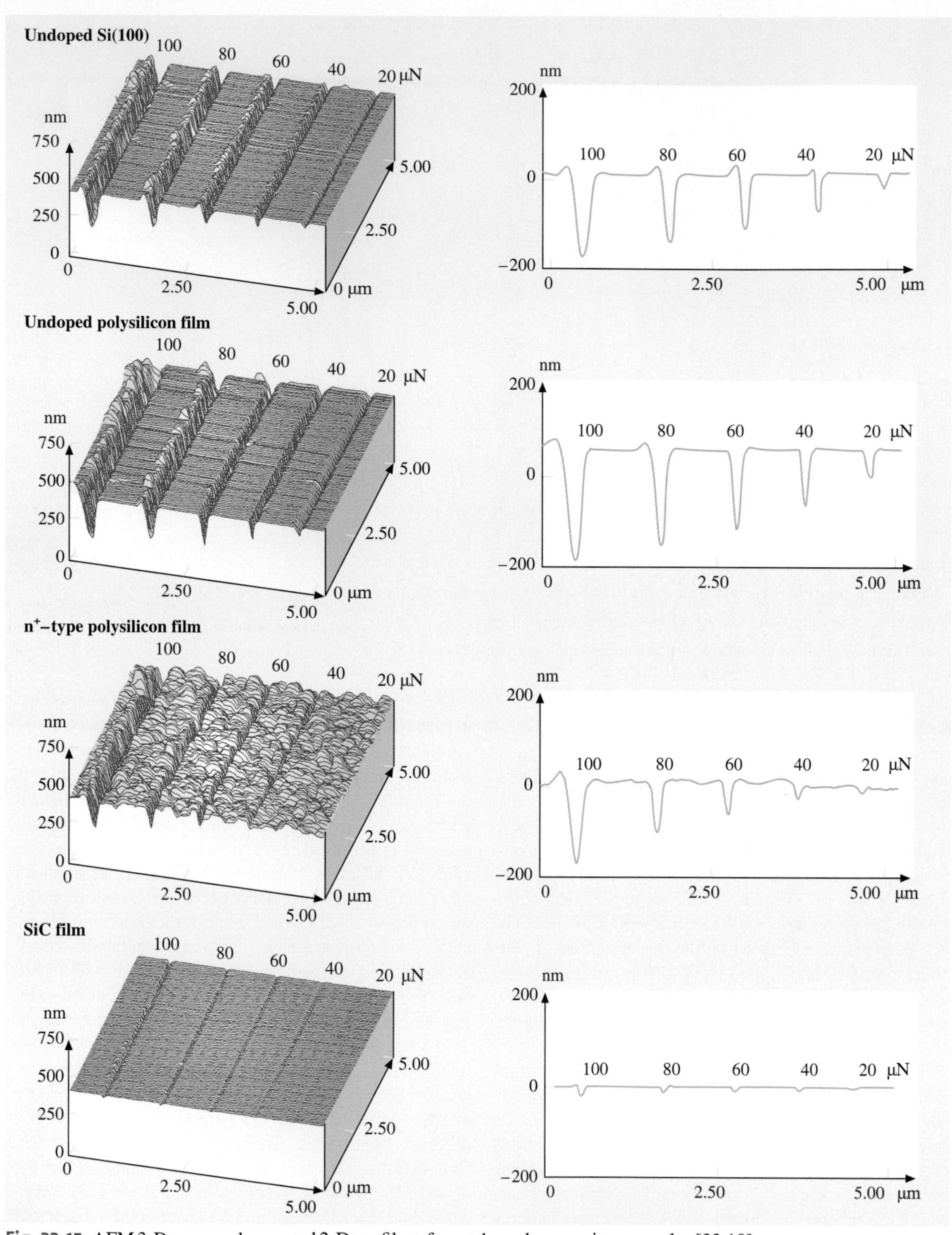

Fig. 33.15 AFM 3-D maps and averaged 2-D profiles of scratch marks on various samples [33.10]

33.5 Lubrication Studies for MEMS/NEMS

Several studies of liquid perfluoropolyether (PFPE) lubricant films, self-assembled monolayers (SAMs), and hard diamond-like carbon (DLC) coatings have been carried out for the purpose of minimizing adhesion, friction, and wear [33.49, 66, 99, 103–108]. Many variations of these films are hydrophobic (low surface tension and high contact angle) and have low shear strength, which provide low adhesion, friction, and wear. Relevant details are presented here.

33.5.1 Perfluoropolyether Lubricants

The classical approach to lubrication uses freely supported multimolecular layers of liquid lubricants [33.49, 54, 55]. The liquid lubricants are sometimes chemically bonded to improve their wear resistance. Partially chemically bonded, molecularly thick perfluoropolyether (PFPE) lubricants are widely used for lubrication of magnetic storage media [33.35] and are suitable for MEMS/NEMS devices.

Adhesion, friction, and durability experiments have been performed on virgin Si(100) surfaces and silicon surfaces lubricated with two commonly used PFPE lubricants – Z-15 (with $-CF_3$ nonpolar end groups) and Z-DOL (with -OH polar end groups) [33.49, 103, 105, 106]. Z-DOL film was thermally bonded at 150 °C for 30 min and an unbonded fraction was removed by a solvent (BW) [33.35]. The thicknesses of Z-15 and Z-DOL (BW) films were 2.8 nm and 2.3 nm, respectively. Nanoscale measurements were made using an AFM. The adhesive forces of Si(100), Z-15 and Z-DOL (BW) measured by force calibration plot and friction force versus normal load plot are summarized in Fig. 33.16. The results measured by these two methods are in good agreement. Figure 33.16 shows that the presence of mobile Z-15 lubricant film increases the adhesive force compared to that of Si(100) by meniscus formation [33.54, 55, 109]. The presence of solid phase Z-DOL (BW) film reduces the adhesive force compared to that of Si(100) due to the absence of mobile liquid. The schematic (bottom) in Fig. 33.16 shows the relative size and sources of meniscus. It is well-known that the native oxide layer (SiO_2) on the top of Si(100) wafer exhibits hydrophilic properties, and some water molecules can be adsorbed on this surface. The condensed water will form meniscus as the tip approaches the sample surface. The larger adhesive force in Z-15 is not only caused by the Z-15 meniscus. The nonpolarized Z-15 liquid does not have good wettability and strong bonding with Si(100). In the ambient environment, the condensed water molecules from the environment permeate through the liquid Z-15 lubricant film and compete with the lubricant molecules presented on the substrate. The interaction of the liquid lubricant with the substrate is weakened, and a boundary layer of the liquid lubricant forms puddles [33.105, 106]. This dewetting allows water molecules to be adsorbed on the Si(100) surface as aggregates, along with Z-15 molecules. And both of them can form meniscus while the tip approaches the surface. Thus, the dewetting of liquid Z-15 film results in higher adhesive force and poorer lubrication performance. In addition, as the Z-15 film is pretty soft compared to the solid Si(100) surface, penetration of the tip in the film occurs while pushing the tip downwards. This leads to the large area of the tip involved

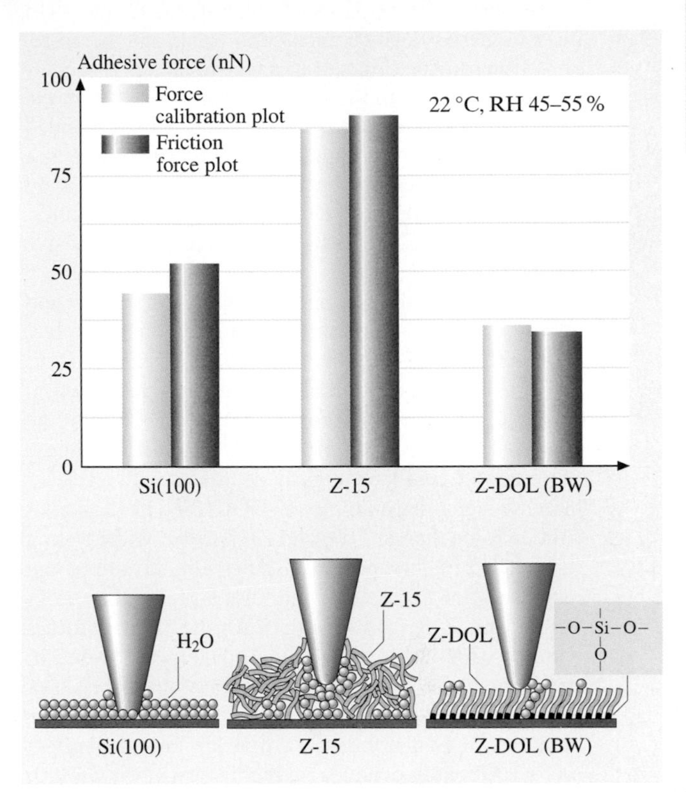

Fig. 33.16 Summary of the adhesive forces of Si(100) and Z-15 and Z-DOL (BW) films measured by force calibration plots and friction force versus normal load plots in ambient air. The schematic (*bottom*) showing the effect of meniscus, formed between the AFM tip and the surface sample, on the adhesive and friction forces [33.103]

to form the meniscus at the tip-liquid (mixture of Z-15 and water) interface. It should also be noted that Z-15 has a higher viscosity compared to water, and therefore, Z-15 film provides higher resistance to motion and coefficient of friction. In the case of Z-DOL (BW) film, the active groups of Z-DOL molecules are mostly bonded on Si(100) substrate, thus the Z-DOL (BW) film has low free surface energy and cannot be displaced readily by water molecules, or readily adsorb water molecules. Thus, the use of Z-DOL (BW) can reduce the adhesive force.

To study the relative humidity effect on friction and adhesion, the variations of friction force, adhesive force, and coefficient of friction of Si(100), Z-15, and Z-DOL (BW) as a function of relative humidity are shown in Fig. 33.17. It shows that for Si(100) and Z-15 film, the friction force increases with a relative humidity increase of up to 45% and then slightly decreases with a further increase in relative humidity. Z-DOL (BW) has a smaller friction force than Si(100) and Z-15 in the whole testing range. And its friction force shows a relative apparent increase when the relative humidity is higher than 45%. For Si(100), Z-15, and Z-DOL (BW), the adhesive forces increase with relative humidity. And their coefficients of friction increase with a relative humidity up to 45%, after which they decrease with a further increase in relative humidity. It is also observed that the humidity effect on Si(100) really depends on the history of the Si(100) sample. As the surface of Si(100) wafer readily adsorbs water in air, without any pre-treatment the Si(100) used in our study almost reaches its saturated stage of adsorbed water and is responsible for less effect during increasing relative humidity. However, once the Si(100) wafer was thermally treated by baking at 150 °C for 1 h, a bigger effect was observed.

The schematic (right) in Fig. 33.17 shows that Si(100), because its high free-surface energy, can adsorb more water molecules with increasing relative humidity. As discussed earlier, for the Z-15 film in the humid environment, the condensed water from the humid environment competes with the lubricant film present on the sample surface, and interaction of the liquid lubricant film with the silicon substrate is weakened, and a boundary layer of the liquid lubricant forms puddles. This dewetting allows water molecules to be adsorbed on the Si(100) substrate mixed with Z-15 molecules [33.105, 106]. Obviously, more water molecules can be adsorbed on Z-15 surface while increasing relative humidity. The more adsorbed water molecules in the case of Si(100), along with lubricant molecules in Z-15 film case, form bigger water meniscus, which leads to an increase in friction force, adhesive force, and coefficient of friction of Si(100) and Z-15 with humidity. But at a very high humidity of 70%, large quantities of adsorbed water can form a continuous water layer that separates the tip and sample surface and acts as a kind of lubricant, which causes a decrease in the friction force and coefficient of friction. For Z-DOL (BW) film, because of its hydrophobic surface properties, water molecules can be adsorbed at a humidity higher than 45% and cause an increase in the adhesive and friction forces.

To study the durability of lubricant films at the nanoscale, the friction of Si(100), Z-15, and Z-DOL (BW) as a function of the number of scanning cycles are shown in Fig. 33.18. As observed earlier, friction force and coefficient of friction of Z-15 is higher than that of Si(100) with the lowest values for Z-DOL (BW). During cycling, friction force and coefficient of friction of Si(100) show a slight decrease during the initial few cycles, then remain constant. This is related to the removal of the top adsorbed layer. In the case of Z-15 film, the friction force and coefficient of friction show an increase during the initial few cycles and then approach higher and stable values. This is believed to be caused by the attachment of the Z-15 molecules onto the tip. The molecular interaction between these attached molecules and molecules on the film surface is responsible for an increase in friction. But after several scans, this molecular interaction reaches the equilibrium, and after that, friction force and coefficient of friction remain constant. In the case of Z-DOL (BW) film, the friction force and coefficient of friction start out low and remain low during the entire test for 100 cycles, suggesting that Z-DOL (BW) molecules do not get attached or displaced as readily as Z-15.

33.5.2 Self-Assembled Monolayers (SAMs)

A preferred method of lubrication of MEMS/NEMS is by the deposition of organized and dense molecular-scale layers of long-chain molecules, as they have been shown to be superior lubricants [33.54, 55, 104, 107, 108, 110, 111]. Two common methods of producing monolayers are the Langmuir–Blodgett (L-B) deposition and self-assembled monolayers (SAMs) by chemical grafting of molecules. L-B films are physically bonded to the substrate by weak van der Waals forces, while SAMs are bonded covalently to the substrate and provide high durability.

SAMs can be formed spontaneously by the immersion of an appropriate substrate into a solution of an

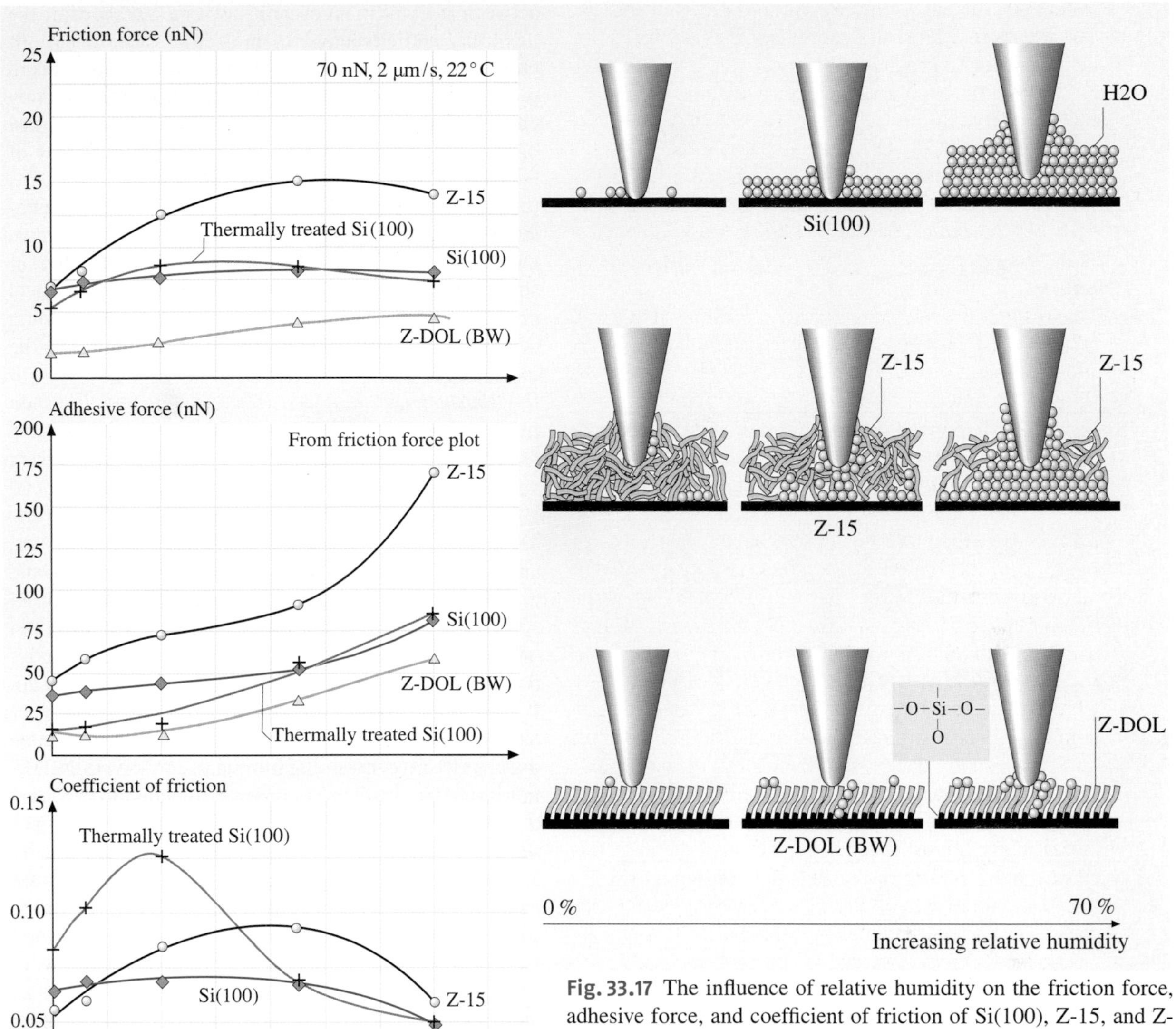

Fig. 33.17 The influence of relative humidity on the friction force, adhesive force, and coefficient of friction of Si(100), Z-15, and Z-DOL (BW) films at 70 nN, 2 μm/s, and in 22 °C air. Schematic (*right*) shows the change of meniscus with increasing relative humidity. In this figure, the thermally treated Si(100) represents the Si(100) wafer that was baked at 150 °C for 1 h in an oven (in order to remove the adsorbed water) just before it was placed in the 0% RH chamber [33.103]

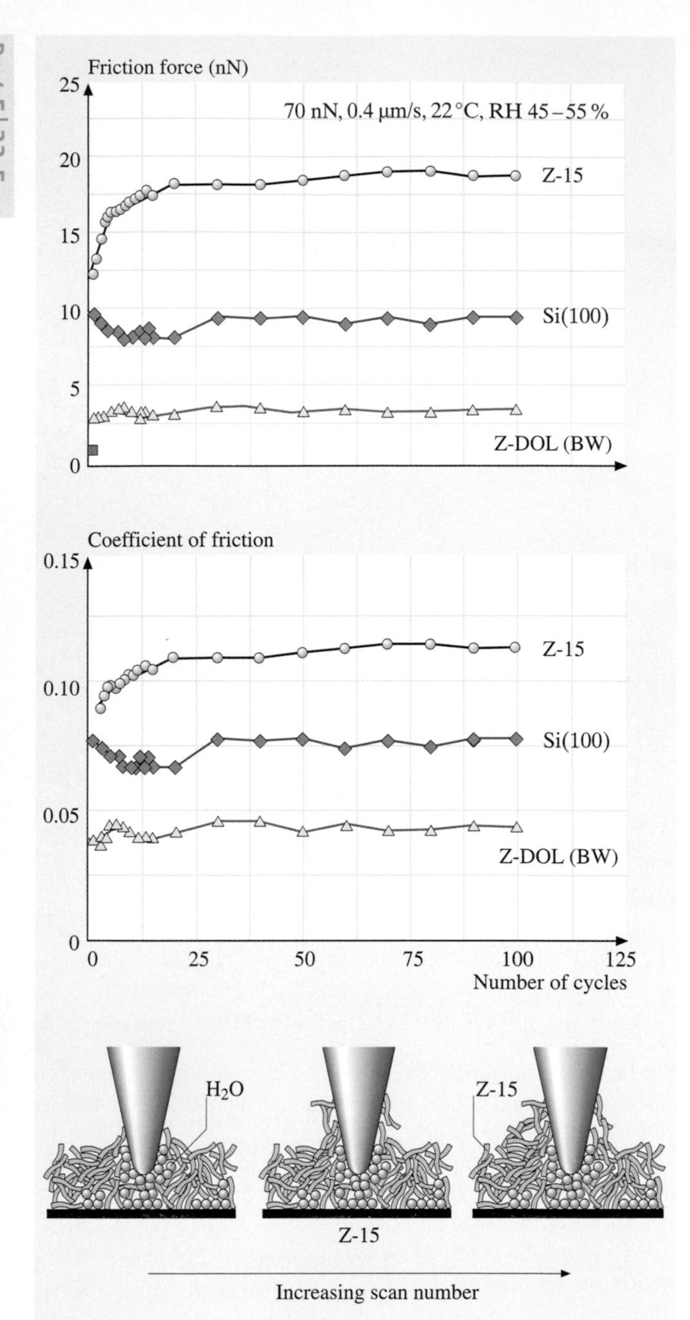

Fig. 33.18 Friction force and coefficient of friction versus number of sliding cycles for Si(100) and Z-15 and Z-DOL (BW) films at 70 nN, 0.4 μm/s, and in ambient air. Schematic (*bottom*) shows that some liquid Z-15 molecules can be attached onto the tip. The molecular interaction between the attached molecules with the Z-15 molecules in the film results in an increase of the friction force with multi-scanning [33.103]

active surfactant in an organic solvent. SAMs offer the flexibility and advantage of molecular tailoring to obtain a variety of different tribological and mechanical properties. For example, researchers have shown that by changing the head groups, tail groups, chain lengths, or types of bonds within a chain, varying degrees of friction, adhesion and/or compliance can be obtained (e.g., [33.107, 108]). These studies indicate that the basis for molecular design and tailoring of SAMs must include a complete understanding of the interrelationships between the molecular structure and tribological properties of SAMs, as well as a deep understanding of the friction and wear mechanisms of SAMs at the molecular level.

Bhushan and *Liu* [33.107] and *Liu* and *Bhushan* [33.108] studied nanotribological properties of four different kinds of alkylthiol and biphenyl thiol monolayers with different surface terminals, spacer chains, and head groups using AFM/FFM techniques. These monolayers, along with a schematic of their structures and substrates, are listed in Fig. 33.19. Surface roughness, adhesion, and microscale friction studies were carried out using an AFM/FFM. The average values and standard deviation of the adhesive force and coefficient of microscale friction for the various SAMs are summarized in Fig. 33.20. It shows that SAMs can reduce the adhesive and friction forces. Alkylthiol and biphenyl thiol SAMs can be used as effective molecular lubricants for MEMS/NEMS fabricated from silicon. In order to explain the frictional differences in SAMs, a molecular spring model is presented in Fig. 33.21. It is suggested that the chemically adsorbed self-assembled molecules on a substrate are just like assembled molecular springs anchored to the substrate. An AFM tip sliding on the surface of a SAM is like a tip sliding on the top of "molecular springs or brush." The molecular assembly has compliant features and can experience orientation and compression under load. The orientation of the molecular springs under normal load reduces the shearing force at the interface, which, in turn, reduces the friction force. The possibility of orientation is determined by the spring constant of a single molecule (local stiffness), as well as the interaction between the neighboring molecules, which is reflected by the packing density or packing energy. It is noted that HDT exhibits the lowest adhesive and friction forces. Based on local stiffness measurements [33.108] and the view of molecular structures, biphenyl is a more rigid structure due to the contribution by two benzene rings. Therefore, the spring constant of BPT is larger than that of HDT. A more compliant film (HDT) exhibits lower friction force than a rigid film (BPT/BPTC).

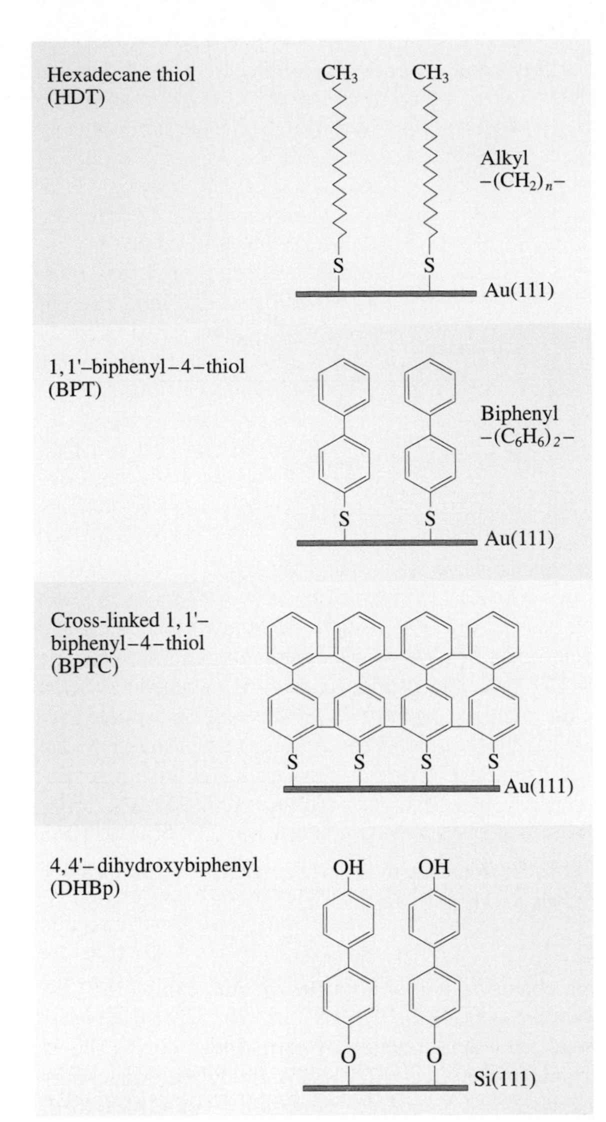

Fig. 33.19 Structure of SAMs studied using AFM/FFM techniques

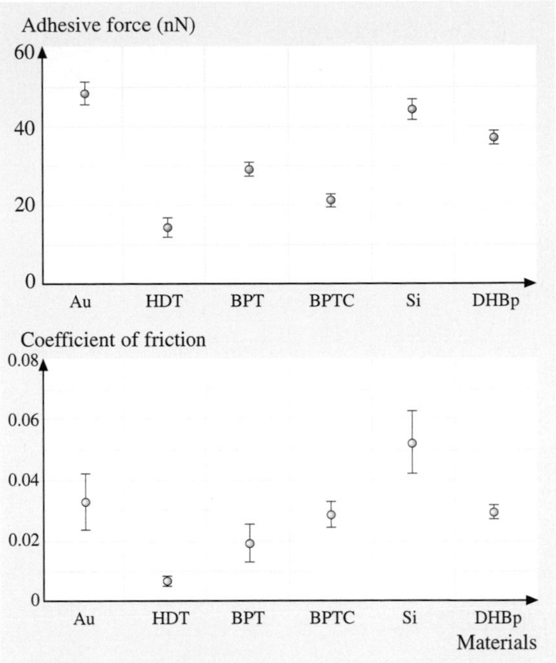

Fig. 33.20 Adhesive force and coefficient of friction values for the various SAMs studied

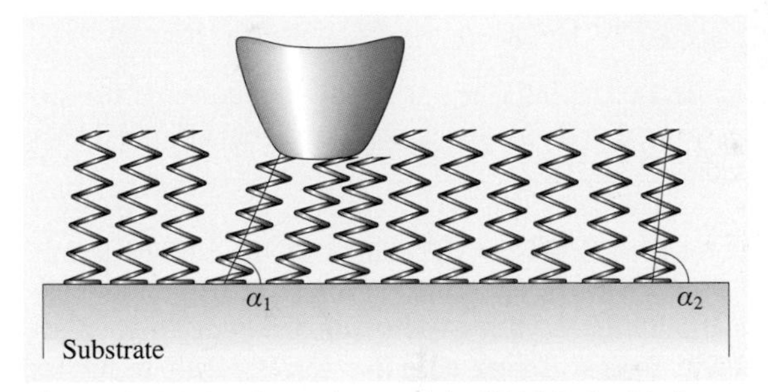

Fig. 33.21 Molecular spring model of SAMs. In this figure, $\alpha_1 < \alpha_2$, which is caused by the further orientation under the normal load applied by an asperity tip [33.107]

Crosslinking of chains in BPTC resulted in higher coefficient of friction. It should be noted that the orientation can lead to conformational defects along the molecular chains that lead to energy dissipation.

The influence of relative humidity on adhesion and friction was studied in an environmentally controlled chamber. The results are given in Fig. 33.22, which show that for Si(111), Au(111), and DHBp, the adhesive and frictional forces increase with relative humidity. For BPT and BPTC (cross-linked BPT), the adhesive forces slightly increase with relative humidity when the relative humidity is higher than 40%, but it is very interesting that their coefficients of friction decrease slightly in the same range. For HDT, over the testing range, the adhesive and friction force do not seem to be sensitive to the change in relative humidity. The influence of relative humidity on adhesive and frictional forces can be mainly understood by comparing their surface terminal polarization properties and work of adhesion. Meniscus forces are related linearly to the works of adhesion of a surface, which in

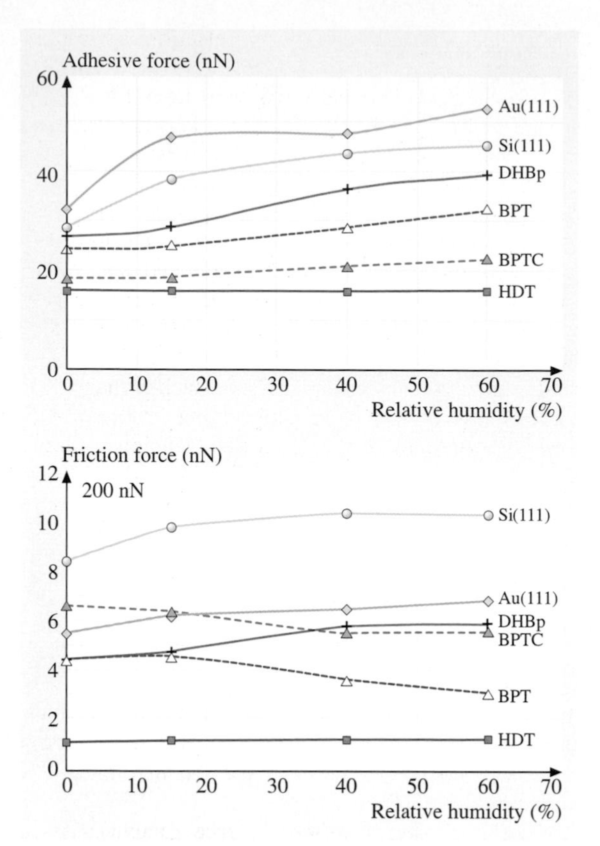

Fig. 33.22 The influence of relative humidity on the adhesive and friction forces at 200 nN of various SAMs studied [33.107]

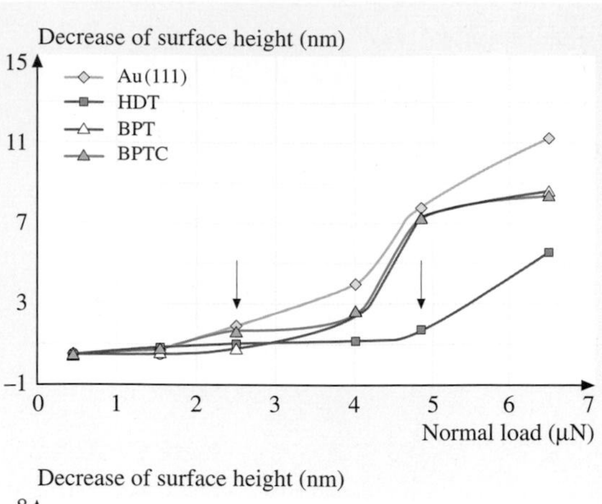

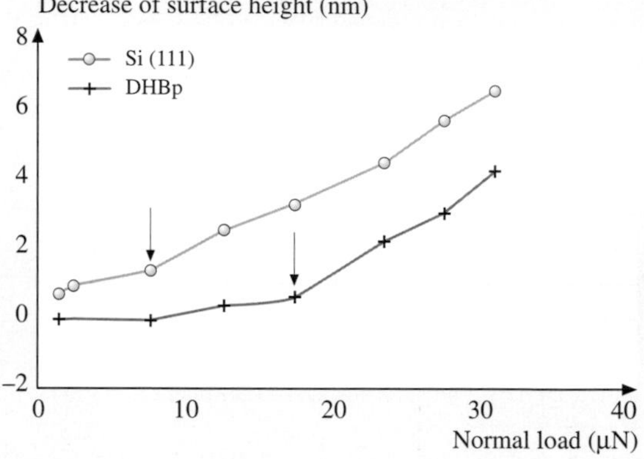

Fig. 33.23 Wear depth as a function of normal load for various SAMs [33.107]

the case of SAMs, is dependent on the surface terminals. Polar surface terminals such as DHBp on Si(111) result in higher work of adhesion and hence higher adhesive forces. Larger adhesive forces result in higher friction forces. Nonpolar surface terminals (HDT) have very small work of adhesion and hence low adhesive and friction forces. In higher humidity, water capillary condensation can either increase friction through increased adhesion in the contact zone, or reduce friction through an enhanced water-lubricating effect.

Figure 33.23 shows results of microscale wear tests on SAMs using a diamond tip in an AFM. DHBp on Si(111) shows the best wear resistance. For the SAMs deposited on gold substrate, HDT exhibits the best wear resistance. Scratch tests corroborated the wear results. The wear resistance of SAMs is influenced by the interfacial bond strength, the molecular structure of the spacers, and the substrate hardness. Based on the wear data, the SAMs with high-compliance long carbon chains, in addition to low friction, exhibit the best wear resistance [33.107, 108]. In wear experiments, the wear depth as a function of normal load curves shows critical normal load (Fig. 33.23). Below the critical normal load, SAMs undergo orientation. At the critical load, SAMs wear away from the substrate due to weak interfacial bond strength, while above the critical normal load, severe wear takes place on the substrate, Fig. 33.24.

According to these results, it is suggested that a dual SAM with a compliant layer on a stiff layer, deposited on hydrogenated silicon, may have optimized tribological performance for MEMS/NEMS applications.

33.5.3 Hard Diamond-like Carbon (DLC) Coatings

Hard amorphous carbon (a-C), commonly known as DLC (implying high hardness), coatings are deposited by a variety of deposition techniques including

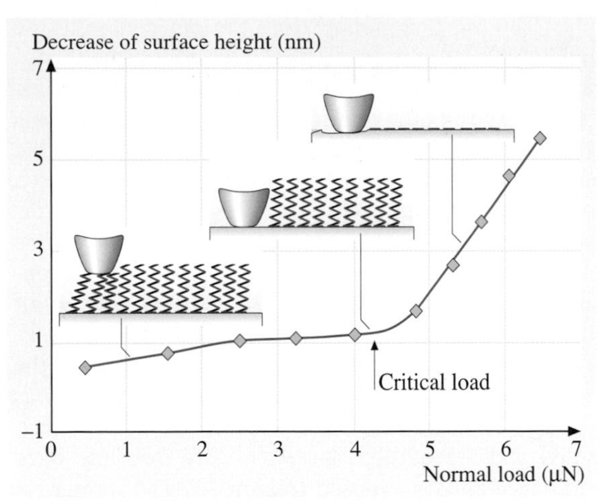

Fig. 33.24 Illustration of the wear mechanism of SAMs with increasing normal load [33.108]

filtered cathodic arc, ion beam, electron cyclotron resonance chemical vapor deposition (ECR-CVD), plasma-enhanced chemical vapor deposition (PECVD), and sputtering [33.83, 99]. These coatings are used in a wide range of applications, including tribological, optical, electronic, and biomedical applications. Ultrathin coatings (3.5 to 10 nm thick) are employed to protect against wear and corrosion in magnetic storage applications (thin-film rigid disks, metal evaporated tapes) and thin-film read/write head (Gillette Mach 3 razor blades, glass windows, and sunglasses). The coatings exhibit low friction, high hardness and wear resistance, chemical inertness to both acids and alkalis, lack of magnetic response, and optical band gap ranging from zero to a few eV, depending on the deposition technique and its conditions. Selected data on DLC coatings relevant for MEMS/NEMS applications are presented in a following section on adhesion measurements.

33.6 Component-Level Studies

33.6.1 Surface Roughness Studies of Micromotor Components

Most of the friction forces resisting motion in the micromotor are concentrated near the rotor-hub interface, where continuous physical contact occurs. Surface roughness of the surfaces usually has a strong influence on the friction characteristics on the micro/nanoscale. A catalog of roughness measurements on various components of a MEMS device does not exist in the literature. Using an AFM, measurements on various component surfaces was made for the first time by *Sundararajan* and *Bhushan* [33.112].

Table 33.6 shows various surface roughness parameters obtained from $5\times5\,\mu m$ scans of the various component surfaces of several unlubricated micromotors using the AFM in tapping mode. A surface with a Gaussian height distribution should have a skewness of zero and kurtosis of three. Although the rotor and stator top surfaces exhibit comparable roughness pa-

Table 33.6 Surface roughness parameters and microscale coefficient of friction for various micromotor component surfaces measured using an AFM. Mean and $\pm1\sigma$ values are given

	RMS roughness[a] (nm)	Peak-to-valley distance[a] (nm)	Skewness[a], Sk	Kurtosis[a], K	Coefficient of microscale friction[b] μ
Rotor topside	21 ± 0.6	225 ± 23	1.4 ± 0.30	6.1 ± 1.7	0.07 ± 0.02
Rotor underside	14 ± 2.4	80 ± 11	-1.0 ± 0.22	3.5 ± 0.50	0.11 ± 0.03
Stator topside	19 ± 1	246 ± 21	1.4 ± 0.50	6.6 ± 1.5	0.08 ± 0.01

[a] Measured from a tapping mode AFM scan of size $5\,\mu m\times5\,\mu m$ using a standard Si tip scanning at $5\,\mu m/s$ in a direction orthogonal to the long axis of the cantilever

[b] Measured using an AFM in contact mode at $5\,\mu m\times5\,\mu m$ scan size using a standard Si_3N_4 tip scanning at $10\,\mu m/s$ in a direction parallel to the long axis of the cantilever

rameters, the underside of the rotors exhibits lower RMS roughness and peak-to-valley distance values. More importantly, the rotor underside shows negative skewness and lower kurtosis than the topsides both of which are conducive to high real area of contact and hence high friction [33.54, 55]. The rotor underside also exhibits higher coefficient of microscale friction than the rotor topside and stator, as shown in Table 33.6. Figure 33.25 shows representative surface height maps of the various surfaces of a micromotor measured using the AFM in tapping mode. The rotor underside exhibits varying topography from the outer edge to the middle and inner edge. At the outer edges, the topography shows smaller circular asperities, similar to the topside. The middle and inner regions show deep pits with fine edges that may have been created by the etchants used for etching the sacrificial layer. It is known that etching can affect the surface roughness of surfaces in surface micromachining. The residence time of the etchant near the inner region is high, responsible for larger pits. Figure 33.26 shows roughness of the surface directly beneath the rotors (the base polysilicon layer). There appears to be a difference in the roughness between the portion of this surface that was initially underneath the rotor (region B) during fabrication and the portion that was away from the rotor and hence always exposed (region A). The former region shows lower roughness than the latter region. This suggests that the surfaces at the rotor-hub interface that come into contact at the end of the fabrication process exhibit large real areas of contact that result in high friction.

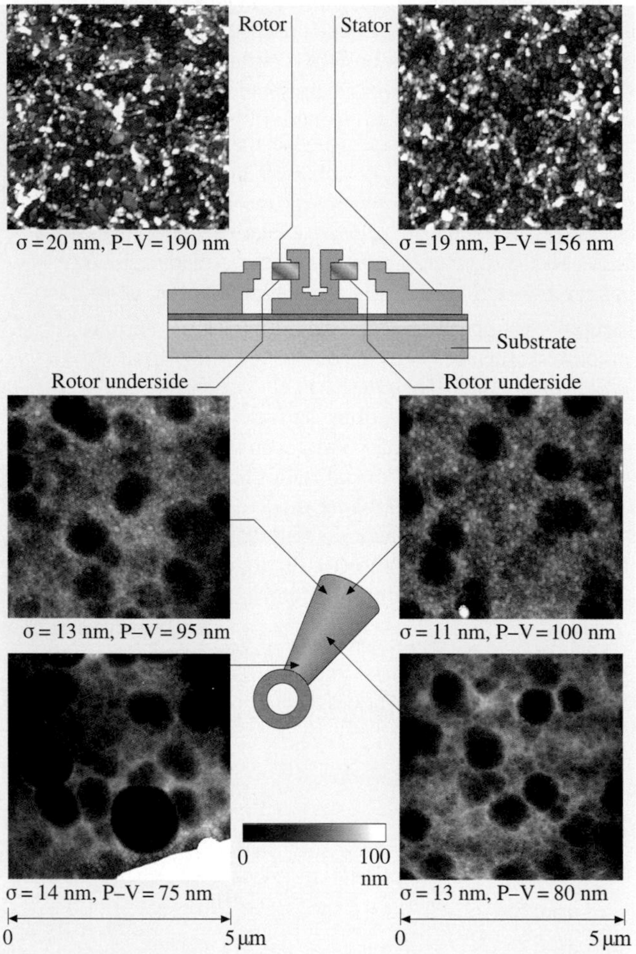

Fig. 33.25 Representative AFM surface height images obtained in tapping mode (5 μm × 5 μm scan size) of various component surfaces of a micromotor. RMS roughness and peak-to-valley (P-V) values of the surfaces are given. The underside of the rotor exhibits drastically different topography from the topside [33.112]

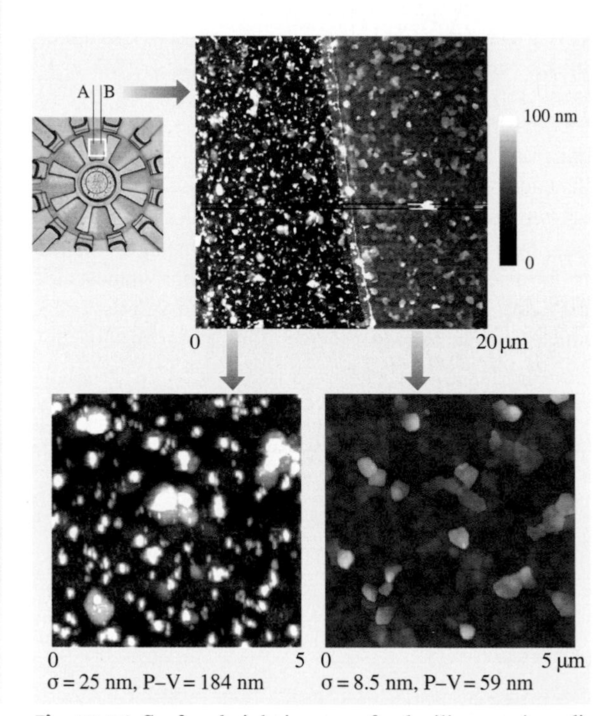

Fig. 33.26 Surface height images of polysilicon regions directly below the rotor. Region A is away from the rotor, while region B was initially covered by the rotor prior to the release etch of the rotor. During this step, slight movement of the rotor caused region B to be exposed [33.112]

33.6.2 Adhesion Measurements

Surface force apparatus (SFA) and AFMs are used to measure adhesion between two surfaces on micro- to nanoscales. In the SFA, the adhesion of liquid films sandwiched between two curved and smooth surfaces is measured. In an AFM, as discussed earlier, adhesion between a sharp tip and the surface of interest is measured. To measure adhesion between two beams in the mesoscopic length scale, a cantilever beam array (CBA) technique is used [33.113, 115–117]. The technique utilizes an array of micromachined polysilicon beams (for Si MEMS applications) anchored to the substrate at one end and with different lengths parallel to the surface. It relies on the peeling and detachment of cantilever beams. Change in free energy, or reversible work done to separate unit areas of two surfaces from contact is called work of adhesion. To measure the work of adhesion, electrostatic actuation is used to bring all beams in contact with the substrate, Fig. 33.27 [33.113, 117]. Once the actuation force is removed, the beams begin to peel themselves off the substrate, which can be observed with an optical interference microscope. For beams shorter than a characteristic length, the so-called detachment length, their stiffness is sufficient to completely free them from the substrate underneath. Beams larger than the detachment length remain adhered. The beams at the transition region start to detach and remain attached to the substrate just at the tips. For this case, by equating the elastic energy stored within the beam and the beam-substrate interfacial energy, the work of adhesion, W_{ad}, can be calculated by the following equation [33.117]:

$$W_{ad} = \frac{3Ed^2t^3}{8l_d^4} , \qquad (33.1)$$

where E is the Young's modulus of the beam, d is the spacing between the undeflected beam and the substrate, t is the beam thickness, and l_d is the detachment length. The technique has been used to screen methods for adhesion reduction in polysilicon microstructures.

To measure adhesion, friction, and wear between two microcomponents, a microtriboapparatus has been used. Figure 33.28 shows a schematic of a microtriboapparatus capable of adopting MEMS components [33.114]. In this apparatus, an upper specimen mounted on a soft cantilever beam comes in contact with a lower specimen mounted on a lower specimen holder. The apparatus consists of two piezos (x- and z-piezos), and four fiber optic sensors (x- and z-displacement sensors, and x- and z-force sensors). For adhesion and friction studies, z- and x-piezos are used to bring the upper and lower specimens in contact and to apply a relative motion in the lateral direction, respectively. The x- and z-displacement sensors are used to measure the lateral position of the

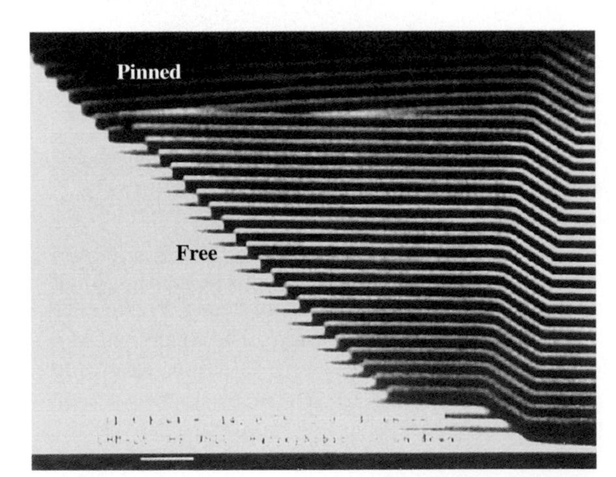

Fig. 33.27 SEM micrograph of a micromachined array of polysilicon cantilever beams of increasing length. The micrograph shows the onset of pinning for beams longer than 34 μm [33.113]

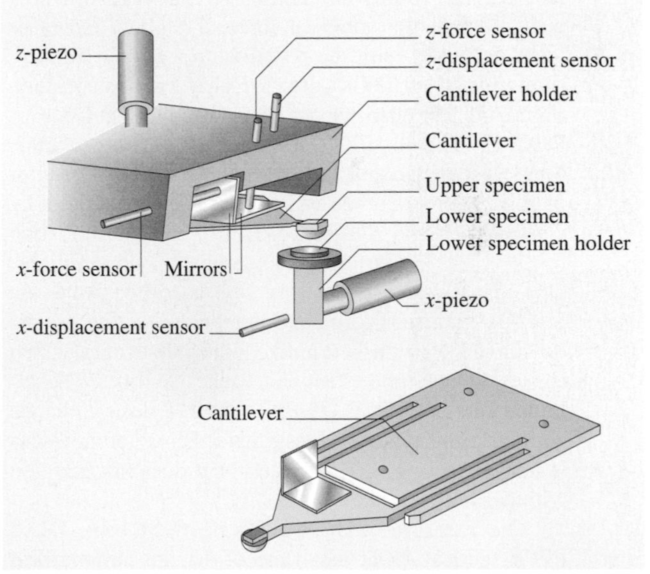

Fig. 33.28 Schematic of the microtriboapparatus, including specially designed cantilever (with two perpendicular mirrors attached on the end), lower specimen holder, two piezos (x- and z-piezos), and four fiber optic sensors (x- and z-displacement sensors and x- and z-force sensors) [33.114]

lower specimen and vertical position of the upper specimen, respectively. The x- and z-force sensors are used to measure friction force and normal load/adhesive force between these two specimens, respectively, by monitoring the deflection of the cantilever.

As most of the MEMS/NEMS devices are fabricated from silicon, the study of silicon-on-silicon contacts is important. This contact was simulated by a flat single-crystal Si(100) wafer (phosphorus doped) specimen sliding against a single crystal Si(100) ball (1 mm in diameter, 5×10^{17} atoms/cm^3 boron doped) mounted on a stainless steel cantilever [33.114, 118]. Both of them have a native oxide layer on their surfaces. The other materials studied were 10-nm-thick DLC deposited by filtered cathodic arc deposition on Si(100), 2.3-nm-thick chemically bonded PFPE (Z-DOL, BW) on Si(100), and hexadecane thiol (HDT) monolayer on evaporated Au(111) film to investigate their anti-adhesion performance.

It is well-known that in the computer rigid disk drives, the adhesive force increases rapidly with an increase in rest time between a magnetic head and a magnetic disk [33.35]. Considering that adhesion and friction are the major issues that lead to the failure of MEMS/NEMS devices, the rest time effect on microscale on Si(100), DLC, PFPE, and HDT was studied, and the results are summarized in Fig. 33.29a. It is found that the adhesive force of Si(100) increases logarithmically with the rest time to a certain equilibrium time ($t = 1{,}000$ s), after which it remains constant. Figure 33.29a also shows that the adhesive force of DLC, PFPE, and HDT does not change with rest time. Single-asperity contact modeling of the dependence of meniscus force on the rest time has been carried out by *Chilamakuri* and *Bhushan* [33.119], and the modeling results (Fig. 33.29b) verify experimental observations. Due to the presence of thin film adsorbed water on Si(100), meniscus forms around the contacting asperities and grows with time until equilibrium occurs, which causes the rest time effect on its adhesive force. The adhesive forces of DLC, PFPE, and HDT do not change with rest time, which suggests that either water meniscus is not present on their surfaces, or it does not increase with time.

The measured adhesive forces of Si(100), DLC, PFPE, and HDT at rest time of 1 s are summarized in Fig. 33.30. It shows that the presence of solid films of DLC, PFPE, and HDT greatly reduces the adhesive force of Si(100), whereas HDT film has the lowest adhesive force. It is well-known that the native oxide layer (SiO_2) on the top of Si(100) wafer exhibits hydrophilic

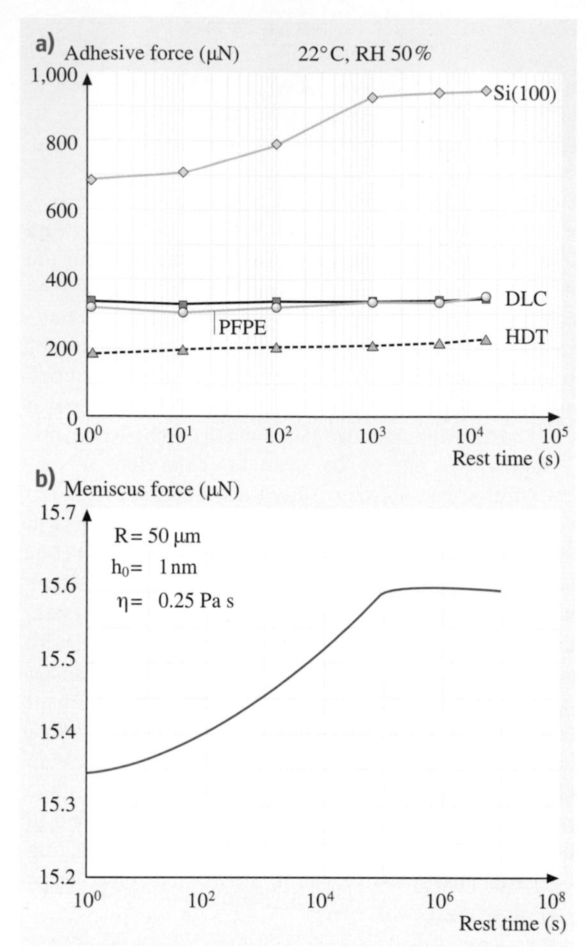

Fig. 33.29 (a) The influence of rest time on the adhesive force of Si(100), DLC, chemically bonded PFPE, and HDT, and (b) single asperity contact modeling results of the rest time effect on the meniscus force for an asperity of R in contact with a flat surface with a water-film thickness of h_0 and absolute viscosity of η_0 [33.119]

properties, and water molecules produced by capillary condensation of water vapor from the environment can be adsorbed easily on this surface. The condensed water will form meniscus as the upper specimen approaches the lower specimen surface. The meniscus force is a major contributor to the adhesive force. In the case of DLC, PFPE, and HDT, the films are found to be hydrophobic based on contact angle measurements, and the amount of condensed water vapor is low compared to that on Si(100). It should be noted that the measured adhesive force is generally higher than that measured in AFM,

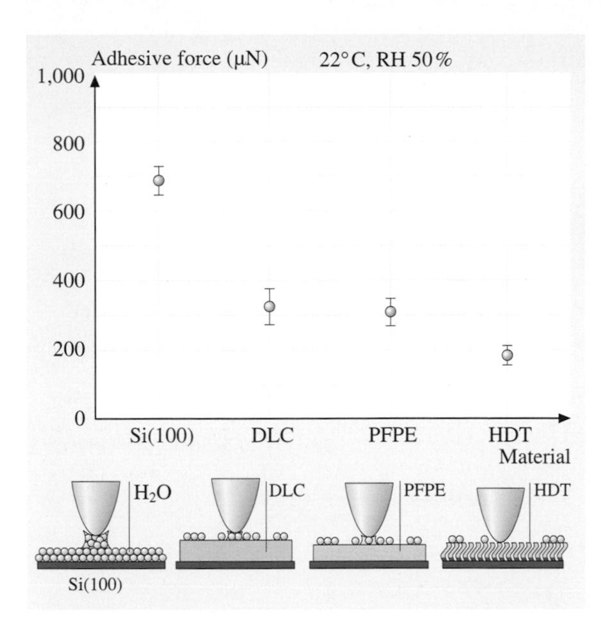

Fig. 33.30 Adhesive forces of Si(100), DLC, chemically bonded PFPE, and HDT at ambient conditions and a schematic showing the relative size of water meniscus on different specimens

because the larger radius of the Si(100) ball compared to that of an AFM tip induces larger meniscus and van der Waals forces.

To investigate the velocity effect on friction, the friction force as a function of velocity was measured and is summarized in Fig. 33.31a. It indicates that for Si(100) the friction force initially decreases with increasing velocity until equilibrium occurs. Figure 33.31a also indicates that the velocity almost has no effect on the friction properties of DLC, PFPE, and HDT. This implies that the friction mechanisms of DLC, PFPE, and HDT do not change with the variation in velocity. For Si(100), at high velocity, the meniscus is broken and does not have enough time to rebuild. In addition, it is believed that tribochemical reaction plays an important role. The high velocity leads to tribochemical reactions of Si(100), which has native oxide SiO_2, with water molecules to form a $Si(OH)_4$ film. This film is removed and continuously replenished during sliding. The $Si(OH)_4$ layer at the sliding surface is known to be of low shear strength. The breaking of the water meniscus and the formation of a $Si(OH)_4$ layer result in a decrease in friction force of Si(100). For DLC, PFPE, and HDT, surfaces exhibit hydrophobic properties and can only adsorb a few water molecules in ambient conditions. The above mentioned meniscus breaking and tribochemical reaction mechanisms do not exist for these films. Therefore, their friction force does not change with velocity.

The influence of relative humidity was studied in an environmentally controlled chamber. The adhesive and friction forces were measured at increasing relative humidity, and the results are summarized in Fig. 33.31b. The figure shows that for Si(100), the adhesive force increases with relative humidity, but the adhesive forces of DLC and PFPE only show slight increases when humidity is higher than 45%, while the adhesive force of HDT does not change with humidity. Figure 33.31b also shows that for Si(100), the friction force increases with a relative humidity increase up to 45%, and then it shows a slight decrease with a further increase in the relative humidity. For PFPE, there is an increase in the friction force when humidity is higher than 45%. In the whole testing range, relative humidity does not have any apparent influence on the friction properties of DLC and HDT. In the case of Si(100), the initial increase of relative humidity up to 45% causes more adsorbed water molecules and forms bigger water meniscus, which leads to an increase of friction force. But at a very high humidity of 65%, large quantities of adsorbed water can form a continuous water layer that separates the tip and sample surfaces and acts as a kind of lubricant, which causes a decrease in the friction force. For PFPE, dewetting of lubricant film at humidity larger than 45% results in an increase in the adhesive and friction forces. DLC and HDT surfaces show hydrophobic properties, and increasing relative humidity does not play much of a role in their friction forces.

The influence of temperature was studied using a heated stage. The adhesive force and friction force were measured at increasing temperatures from 22 °C to 125 °C. The results are presented in Fig. 33.31c. It shows that once the temperature is higher than 50 °C, increasing temperature causes a significant decrease in adhesive and friction forces of Si(100) and a slight decrease in the cases of DLC and PFPE. But the adhesion and friction forces of HDT do not show any apparent change with test temperature. At high temperature, desorption of water, and reduction of surface tension of water lead to the decrease of adhesive and friction forces of Si(100), DLC, and PFPE. However, in the case of HDT film, as only a few water molecules are adsorbed on the surface, the above mentioned mechanisms do not play a big role. Therefore, the adhesive and friction forces of HDT do not show any apparent change with temperature. Figure 33.31 shows that in the whole ve-

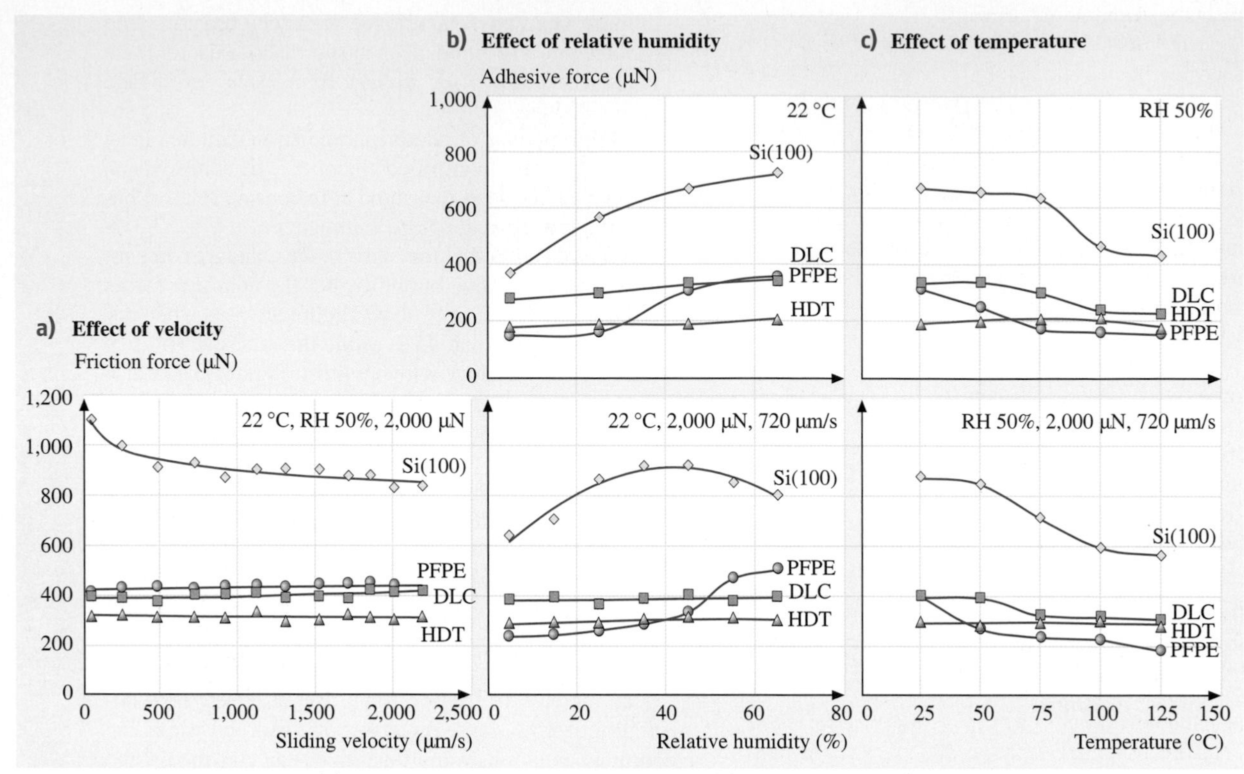

Fig. 33.31a–c The influence of (a) sliding velocity on the friction forces, (b) relative humidity on the adhesive and friction forces, and (c) temperature on the adhesive and friction forces of Si(100), DLC, chemically bonded PFPE, and HDT

locity, relative humidity, and temperature test range, the adhesive forces and friction forces of DLC, PFPE, and HDT are always smaller than that of Si(100), whereas HDT has the smallest value.

To summarize, several methods can be used to reduce adhesion in microstructures. MEMS surfaces can be coated with hydrophobic coatings such as PFPEs, SAMs, and passivated DLC coatings. It should be noted that other methods to reduce adhesion include the formation of dimples on the contact surfaces to reduce contact area [33.10, 54, 55, 116, 117]. Furthermore, an increase in hydrophobicity of the solid surfaces (high contact angle approaching 180°) can be achieved by using surfaces with suitable roughness, in addition to lowering their surface energy [33.120, 121]. The hydrophobicity of surfaces is dependent on a subtle interplay between surface chemistry and mesoscopic topography. The self-cleaning mechanism, or so-called "Lotus-effect", is closely related to the ultra-hydrophobic properties of these biological surfaces, which usually show micro-sculptures in specific dimensions.

33.6.3 Static Friction Force (Stiction) Measurements in MEMS

In MEMS devices involving parts in relative motion to each other, such as micromotors, large friction forces become the limiting factor to their successful operation and reliability. It is generally known that most micromotors cannot be rotated as manufactured and require some form of lubrication. It is therefore critical to determine the friction forces present in such MEMS devices. To measure in situ the static friction of a rotor-bearing interface in a micromotor, *Tai* and *Muller* [33.122] measured the starting torque (voltage) and pausing position for different starting positions under a constant-bias voltage. A friction-torque model was used to obtain the coefficient of static friction. To measure the in situ kinetic friction of the turbine and gear structures, *Gabriel* et al. [33.123] used a laser-based measurement system to monitor the steady-state spins and decelerations. *Lim* et al. [33.124] designed and fabricated a polysilicon microstructure to in situ measure the static friction of various films. The microstructure consisted of a shuttle suspended above the underlying electrode by a folded beam suspension. A known normal force was applied, and lateral force was measured to obtain the coefficient of static friction. *Beerschwinger* et al. developed a cantilever-deflection rig to measure friction of LIGA-processed micromotors [33.125, 126]. Table 33.7 presents static friction coefficients of various MEMS devices evaluated by various researchers. Most of these techniques employ indirect methods to determine the friction forces, or involve fabrication of complex structures.

A novel technique to measure the static friction force (stiction) encountered in surface micromachined polysilicon micromotors using an AFM has been developed by *Sundararajan* and *Bhushan* [33.112]. Continuous physical contact occurs during rotor movement (rotation) in the micromotors between the rotor and lower hub flange. In addition, contact occurs at other locations between the rotor and the hub surfaces and between the rotor and the stator. Friction forces will be present at these contact regions during motor operation. Although the actual distribution of these forces is not known, they can be expected to be concentrated near the hub, where there is continuous contact. If we, therefore, represent the static friction force of the micromotor as a single force F_s acting at point P_1 (as shown in Fig. 33.32a), then the magnitude of the frictional torque about the center of the motor (O) that must be overcome before rotor movement can be initiated is:

$$T_s = F_s l_1 , \tag{33.2}$$

where l_1 is the distance OP_1, which is assumed to be the average distance from the center at which the friction force F_s occurs. Now consider an AFM tip moving against a rotor arm in a direction perpendicular to the long axis of the cantilever beam (the rotor arm edge closest to the tip is parallel to the long axis of the cantilever beam), as shown in Fig. 33.32a. When the tip encounters the rotor at point P_2, the tip will twist, generating a lateral force between the tip and the rotor (event A in Fig. 33.32b). This reaction force will generate a torque about the center of the motor. Since the tip is trying to move farther in the direction shown, the tip will continue to twist to a maximum value at which the lateral force between the tip and the rotor becomes high enough such that the resultant torque T_f about the center of the motor equals the static friction torque T_s. At this point, the ro-

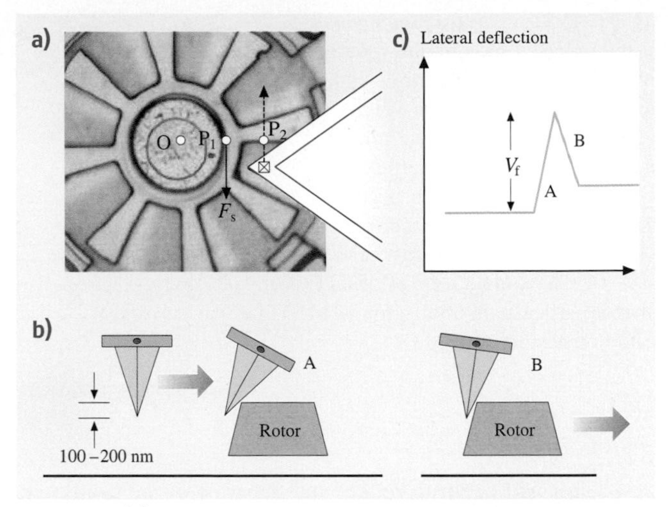

Fig. 33.32 (a) Schematic of the technique used to measure the force, F_s, required to initiate rotor movement using an AFM/FFM. (b) As the tip is pushed against the rotor, the lateral deflection experienced by the rotor due to the twisting of the tip prior to rotor movement is a measure of static friction force, F_s, of the rotors. (c) Schematic of lateral deflection expected from the above mentioned experiment. The peak V_f is related to the state of the rotor [33.112]

Table 33.7 Published data on coefficient of static friction measurements of MEMS devices and structures

Reference	Test method	Device/structure	Material pairs	Environment	Coefficient of static friction
[33.122]	Starting voltage	IC-processed micromotor	PolySi/Si_3N_4	Air	0.20–0.40
[33.124]	Electrostatic loading	Comb-drive microstructure	PolySi/PolySi PolySi/Si_3N_4	Air	4.9 ± 1.0 2.5 ± 0.5
[33.127]	Pull-off force	Silicon microbeams	SiO_2/SiO_2	Air	2.1 ± 0.8
[33.126]	Cantilever/fiber deflection rig	LIGA micromotors	Ni/Alumina	Air	0.6–1.2

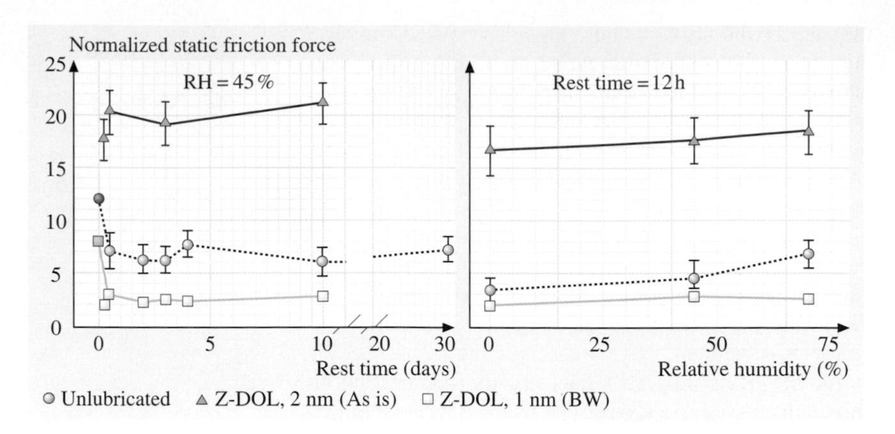

Fig. 33.33 Static friction force values of unlubricated motors and motors lubricated using PFPE lubricants, normalized over the rotor weight, as a function of rest time and relative humidity. Rest time is defined as the time elapsed between a given experiment and the first experiment in which motor movement was recorded (time 0). The motors were allowed to sit at a particular humidity for 12 h prior to measurements [33.112]

tor will begin to rotate, and the twist of the cantilever decreases sharply (event B in Fig. 33.32b). The twist of the cantilever is measured in the AFM as a change in the lateral deflection signal (in volts), which is the underlying concept of friction force microscopy (FFM). The change in the lateral deflection signal corresponding to the above mentioned events as the tip approaches the rotor is shown schematically in Fig. 33.32b. The value of the peak V_f is a measure of the force exerted on the rotor by the tip just before the static friction torque is matched and the rotor begins to rotate.

Using this technique, the viability of PFPE lubricants for micromotors has been investigated and the effect of humidity on the friction forces of unlubricated and lubricated devices was studied as well. Figure 33.33 shows static friction forces normalized over the weight of the rotor of unlubricated and lubricated micromotors as a function of rest time and relative humidity. Rest time here is defined as the time elapsed between the first experiment conducted on a given motor (solid symbol at time zero) and subsequent experiments (open symbols). Each open symbol data point is an average of six measurements. It can be seen that for the unlubricated motor and the motor lubricated with a bonded layer of Z-DOL (BW), the static friction force is highest for the first experiment and then drops to an almost constant level. In the case of the motor with an as-is mobile layer of Z-DOL, the values remain very high up to 10 days after lubrication. In all cases, there is negligible difference in the static friction force at 0% and 45% RH. At 70% RH, the unlubricated motor exhibits a substantial increase in the static friction force, while the motor with bonded Z-DOL shows no increase in static friction force due to the hydrophobicity of the lubricant layer. The motor with an as-is mobile layer of the lubricant shows consistently high values of static friction force that vary little with humidity.

33.6.4 Mechanisms Associated with Observed Stiction Phenomena in Micromotors

Figure 33.34 summarizes static friction force data for two motors, M1 and M2, along with schematics of the meniscus effects for the unlubricated and lubricated surfaces. Capillary condensation of water vapor from the environment results in the formation of meniscus bridges between contacting and near-contacting asperities of two surfaces in close proximity to each other, as shown in Fig. 33.34. For unlubricated surfaces, more menisci are formed at higher humidity, resulting in a higher friction force between the surfaces. The formation of meniscus bridges is supported by the fact that the static friction force for unlubricated motors increases at high humidity (Fig. 33.34). Solid bridging may occur near the rotor-hub interface due to silica residues after the first etching process. In addition, the drying process after the final etch can result in liquid bridging formed by the drying liquid due to meniscus force at these areas [33.54,55,113,116]. The initial static friction force therefore will be quite high, as evidenced by the solid data points in Fig. 33.34. Once the first movement of the rotor permanently breaks these solid and liquid bridges, the static friction force of the motors will drop (as seen in Fig. 33.34) to a value dictated predominantly by the adhesive energies of rotor and hub surfaces, the real area of contact between these surfaces and meniscus forces due to water vapor in the air, at which point, the effect of lubricant films can be observed. Lubrication with a mobile layer, even a thin one, results in very high static friction forces

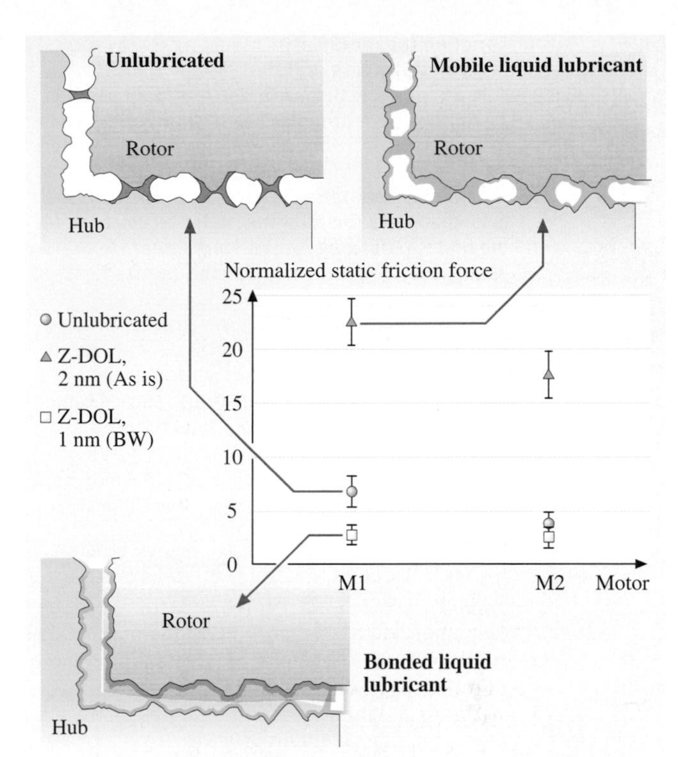

Fig. 33.34 Summary of effect of liquid and solid lubricants on static friction force of micromotors. Despite the hydrophobicity of the lubricant used (Z-DOL), a mobile liquid lubricant (Z-DOL as is) leads to very high static friction force due to increased meniscus forces, whereas a solid-like lubricant (bonded Z-DOL, BW) appears to provide some amount of reduction in static friction force

due to meniscus effects of the lubricant liquid itself at and near the contact regions. It should be noted that a motor submerged in a liquid lubricant would result in a fully flooded lubrication regime. In this case, there is no meniscus contribution and only the viscous contribution to the friction forces would be relevant. However, submerging the device in a lubricant may not be a practical method. A solid-like hydrophobic lubricant layer (such as bonded Z-DOL) results in favorable friction characteristics of the motor. The hydrophobic nature of the lubricant inhibits meniscus formation between the contact surfaces and maintains low friction even at high humidity (Fig. 33.34). This suggests that solid-like hydrophobic lubricants are ideal for lubrication of MEMS, while mobile lubricants result in increased values of static friction force.

References

33.1 NASA Ames Research Center, Mottett Field, CA, USA, http://www.ipt.arc.nasa.gov/gallery.html

33.2 W. G. van der Wiel, S. De Franceschi, J. M. Elzerman, T. Fujisawa, S. Tarucha, L. P. Kouwenhoven: Electron transport through double quantum dots, Rev. Mod. Phys. **75** (2003) 1–22

33.3 Texas Instruments DLP Products, Plano, TX, USA, http://www.dlp.com

33.4 Anonymous: *Microelectromechanical Systems: Advanced Materials and Fabrication Methods* (National Academy Press, Washington, D.C. 1997)

33.5 M. Roukes: Nanoelectromechanical systems face the future, Phys. World (February 2001) 25–31

33.6 M. A. Huff: A distributed MEMS processing environment, http://www.mems-exchange.org/ (2002)

33.7 R. S. Muller, R. T. Howe, S. D. Senturia, R. L. Smith, R. M. White: *Microsensors* (IEEE, New York 1990)

33.8 I. Fujimasa: *Micromachines: A New Era in Mechanical Engineering* (Oxford Univ. Press, Oxford 1996)

33.9 W. S. Trimmer (Ed.): *Micromachines and MEMS: Classic and Seminal Papers to 1990* (IEEE, New York 1997)

33.10 B. Bhushan (Ed.): *Tribology Issues and Opportunities in MEMS* (Kluwer, Dordrecht 1998)

33.11 G. T. A. Kovacs: *Micromachined Transducers Sourcebook* (WCB McGraw-Hill, Boston 1998)

33.12 S. D. Senturia: *Microsystem Design* (Kluwer, Boston 2001)

33.13 M. Gad-el-Hak: *The MEMS Handbook* (CRC, Boca Raton 2002)

33.14 T. R. Hsu: *MEMS and Microsystems* (McGraw-Hill, Boston 2002)

33.15 M. Madou: *Fundamentals of Microfabrication: The Science of Miniaturization*, 2nd edn. (CRC, Boca Raton 2002)

33.16 R. C. Jaeger: *Introduction to Microelectronic Fabrication*, 5th edn. (Addison-Wesley, Reading 1988)

33.17 J. W. Judy: Microelectromechanical systems (MEMS): Fabrication, design, and applications, Smart Mater. Struct. **10** (2001) 1115–1134

33.18 E. W. Becker, W. Ehrfeld, P. Hagmann, A. Maner, D. Munchmeyer: Fabrication of microstructures with high aspect ratios and great structural heights by synchrotron radiation lithography, galvanoforming, and plastic moulding (LIGA process), Microelectron. Eng. **4** (1986) 35–56

33.19 C. R. Friedrich, R. O. Warrington: Surface characterization of non-lithographic micromachining. In: *Tribology Issues and Opportunities in MEMS*, ed. by B. Bhushan (Kluwer, Dordrecht 1998) pp. 73–84

33.20 M. Madou: Facilitating choices of machining tools and materials for miniaturization science: A review. In: *Tribology Issues and Opportunities in MEMS*, ed. by B. Bhushan (Kluwer, Dordrecht 1998) pp. 31–51

33.21 H. Lehr, S. Abel, J. Doppler, W. Ehrfeld, B. Hagemann, K. P. Kamper, F. Michel, Ch. Schulz, Ch. Thurigen: Microactuators as driving units for microrobotic systems, SPIE Proc. Microrobotics: Components and Applications **2906** (1996) 202–210

33.22 H. Lehr, W. Ehrfeld, B. Hagemann, K. P. Kamper, F. Michel, Ch. Schulz, Ch. Thurigen: Development of micro-millimotors, Min. Invas. Ther. Allied Technol. **6** (1997) 191–194

33.23 F. Michel, W. Ehrfeld: Microfabrication technologies for high performance microactuators. In: *Tribology Issues and Opportunities in MEMS*, ed. by B. Bhushan (Kluwer, Dordrecht 1998) pp. 53–72

33.24 M. Tanaka: Development of desktop machining microfactory, Riken Rev. **34** (2001) 46–49

33.25 Y. Xia, G. M. Whitesides: Soft lithography, Angew. Chem. Int. Edn. **37** (1998) 550–575

33.26 H. Becker, C. Gaertner: Polymer microfabrication methods for microfluidic analytical applications, Electrophoresis **21** (2000) 12–26

33.27 Y. Xia, E. Kim, X. M. Zhao, J. A. Rogers, M. Prentiss, G. M. Whitesides: Complex optical surfaces formed by replica molding against elastomeric masters, Science **273** (1996) 347–349

33.28 S. Y. Chou, P. R. Krauss, P. J. Renstrom: Imprint lithography with 25-nanometer resolution, Science **272** (1996) 85–87

33.29 A. Kumar, G. M. Whitesides: Features of gold having micrometer to centimeter dimensions can be formed through a combination of stamping with an elastomeric stamp and an alkanethiol ink followed by chemical etching, Appl. Phys. Lett. **63** (1993) 2002–2004

33.30 T. A. Core, W. K. Tsang, S. J. Sherman: Fabrication technology for an integrated surface-micromachined sensor, Solid State Technol. **36** (1993) 39–47

33.31 J. Bryzek, K. Peterson, W. McCulley: Micromachines on the march, IEEE Spectrum (May 1994) 20–31

33.32 L. J. Hornbeck, W. E. Nelson: Bistable deformable mirror device. In: *OSA Technical Digest Series, Vol. 8, Spatial Light Modulators and Applications* (OSA, Washington 1988) pp. 107–110

33.33 L. J. Hornbeck: A digital light processing(tm) update – Status and future applications, Proc. SPIE **3634** (1999) 158–170

33.34 L. J. Hornbeck: The DMD^{TM} projection display chip: A MEMS-based technology, MRS Bull. **26** (2001) 325–328

33.35 B. Bhushan: *Tribology and Mechanics of Magnetic Storage Devices*, 2nd edn. (Springer, New York 1996)

33.36 H. Hamilton: Contact recording on perpendicular rigid media, J. Mag. Soc. Jpn. **15** (1991) (Suppl. S2) 483–481

33.37 T. Ohwe, Y. Mizoshita, S. Yonoeka: Development of integrated suspension system for a nanoslider with an MR head transducer, IEEE Trans. Magn. **29** (1993) 3924–3926

33.38 L. S. Fan, S. Woodman: Batch fabrication of mechanical platforms for high-density data storage, 8th Int. Conf. Solid State Sensors and Actuators (Transducers '95)/Eurosensors IX, Stockholm 1995, 434–437

33.39 D. A. Horsley, M. B. Cohn, A. Singh, R. Horowitz, A. P. Pisano: Design and fabrication of an angular microactuator for magnetic disk drives, J. Microelectromech. Syst. **7** (1998) 141–148

33.40 T. Hirano, L. S. Fan, D. Kercher, S. Pattanaik, T. S. Pan: HDD tracking microactuator and its integration issues, Proc. ASME **2** 2000, ed. by A. P. Lee, J. Simon, F. K. Foster, R. S. Keynton (ASME, New York 2000) 449–452

33.41 P. Gravesen, J. Branebjerg, O. S. Jensen: Microfluidics – A review, J. Micromech. Microeng. **3** (1993) 168–182

33.42 C. Lai Poh San, E. P. H. Yap (Eds.): *Frontiers in Human Genetics* (World Scientific, Singapore 2001)

33.43 C. H. Mastrangelo, H. Becker (Eds.): Microfluidics and BioMEMS, Proc. SPIE **4560** (2001)

33.44 H. Becker, L. E. Locascio: Polymer microfluidic devices, Talanta **56** (2002) 267–287

33.45 S. Shoji, M. Esashi: Microflow devices and systems, J. Micromech. Microeng. **4** (1994) 157–171

33.46 P. Woias: Micropumps – Summarizing the first two decades, *Proc. SPIE – Microfluidics and BioMEMS*, ed. by C. H. Mastrangelo, H. Becker, Proc. SPIE **4560** (2001) 39–52

33.47 M. Scott: MEMS and MOEMS for national security applications, *Reliability, Testing, and Characterization of MEMS/MOEMS II*, Proc. SPIE **4980** (2003) 37–44

33.48 K. E. Drexler: *Nanosystems: Molecular Machinery, Manufacturing and Computation* (Wiley, New York 1992)

33.49 B. Bhushan: *Handbook of Micro/Nanotribology*, 2nd edn. (CRC, Boca Raton 1999)

33.50 G. Timp (Ed.): *Nanotechnology* (Springer, New York 1999)

33.51 E. A. Rietman: *Molecular Engineering of Nanosystems* (Springer, New York 2001)

33.52 H. S. Nalwa (Ed.): *Nanostructured Materials and Nanotechnology* (Academic, San Diego 2002)

33.53 W.A. Goddard, D.W. Brenner, S.E. Lyshevski, G.J. Iafrate: *Handbook of Nanoscience, Engineering, and Technology* (CRC, Boca Raton 2003)

33.54 B. Bhushan: *Principles and Applications of Tribology* (Wiley, New York 1999)

33.55 B. Bhushan: *Introduction to Tribology* (Wiley, New York 2002)

33.56 Y.C. Tai, L.S. Fan, R.S. Muller: IC-processed micromotors: Design, technology and testing, Proc. IEEE Micro Electro Mechanical Systems 1989, 1–6

33.57 S.M. Spearing, K.S. Chen: Micro-gas turbine engine materials and structures, Ceramic Eng. Sci. Proc. **18** (2001) 11–18

33.58 M. Mehregany, K.J. Gabriel, W.S.N. Trimmer: Integrated fabrication of polysilicon mechanisms, IEEE Trans. Electron. Dev. **35** (1988) 719–723

33.59 E.J. Garcia, J.J. Sniegowski: Surface micromachined microengine, Sens. Actuators A **48** (1995) 203–214

33.60 J.K. Robertson, K.D. Wise: An electrostatically actuated integrated microflow controller, Sens. Actuators A **71** (1998) 98–106

33.61 B. Bhushan: Nanotribology and nanomechanics of MEMS devices, Proc. Ninth Annual Workshop on Micro Electro Mechanical Systems 1996 (IEEE, New York 1996) 91–98

33.62 D.M. Tanner, N.F. Smith, L.W. Irwin et al.: *MEMS Reliability: Infrastructure, Test Structures, Experiments, and Failure Modes* (Sandia National Laboratories, Albuquerque 2000) 2000–2091

33.63 S.S. Mani, J.G. Fleming, J.A. Walraven, J.J. Sniegowski et al.: Effect of W coating on microengine performance, Proc. 38th Annual Inter. Reliability Phys. Symp. 2000 (IEEE, New York 2000) 146–151

33.64 R.E. Sulouff: MEMS opportunities in accelerometers and gyros and the microtribology problems limiting commercialization. In: *Tribology Issues and Opportunities in MEMS*, ed. by B. Bhushan (Kluwer, Dordrecht 1998) pp. 109–120

33.65 Analog Devices Inc., Berkeley, CA, USA, http://www.analog.com

33.66 S.A. Henck: Lubrication of digital micromirror devices, Tribol. Lett. **3** (1997) 239–247

33.67 M.R. Douglass: Lifetime estimates and unique failure mechanisms of the digital micromirror devices (DMD), Proc. 36th Annual Inter. Reliability Phys. Symp. 1998 (IEEE, New York 1998) 9–16

33.68 M.R. Douglass: DMD reliability: A MEMS success story, Reliability, Testing, and Characterization of MEMS/MOEMS II, Proc. SPIE **4980** (2003) 1–11

33.69 W.C. Tang, A.P. Lee: Defense applications of MEMS, MRS Bull. **26** (2001) 318–319. Also see www.darpa.mil/mto/mems

33.70 H. Guckel, D.W. Burns: Fabrication of micromechanical devices from polysilicon films with smooth surfaces, Sens. Actuators **20** (1989) 117–122

33.71 G.T. Mulhern, D.S. Soane, R.T. Howe: Supercritical carbon dioxide drying of microstructures, Proc. Int. Conf. on Solid-State Sensors and Actuators 1993 (IEEE, New York 1993) 269–299

33.72 P. Vettiger, J. Brugger, M. Despont, U. Drechsler, U. Duerig, W. Haeberle: Ultrahigh density, high data-rate NEMS based AFM data storage system, Microelectron. Eng. **46** (1999) 11–27, also see http://www.ibm.com

33.73 M. Ferrari, J. Liu: The engineered course of treatment, Mech. Eng. (December 2001) 44–47

33.74 F.J. Martin, C. Grove: Microfabricated drug delivery systems: Concepts to improve clinical benefits, Biomed. Microdev. **3** (2001) 97–108

33.75 K.F. Man, B.H. Stark, R. Ramesham: *A Resource Handbook for MEMS Reliability*, Rev. A (California Institute of Technology, Pasadena 1998)

33.76 S. Kayali, R. Lawton, B.H. Stark: MEMS reliability assurance activities at JPL, EEE Links **5** (1999) 10–13

33.77 S. Arney: Designing for MEMS reliability, MRS Bull. **26** (2001) 296–299

33.78 K.F. Man: MEMS reliability for space applications by elimination of potential failure modes through testing and analysis (2001), http://www.rel.jpl.nasa.gov/Org/5053/atop/products/Prod-map.html

33.79 B. Bhushan, A.V. Kulkarni, W. Bonin, J.T. Wyrobek: Nano/picoindentation measurement using a capacitance transducer system in atomic force microscopy, Philos. Mag. **74** (1996) 1117–1128

33.80 S. Sundararajan, B. Bhushan: Development of AFM-based techniques to measure mechanical properties of nanoscale structures, Sens. Actuators A **101** (2002) 338–351

33.81 B. Bhushan (Ed.): *Modern Tribology Handbook* (CRC, Boca Raton 2001)

33.82 B. Bhushan, J.N. Israelachvili, U. Landman: Nanotribology: Friction, wear and lubrication at the atomic scale, Nature **374** (1995) 607–616

33.83 J.S. Shor, D. Goldstein, A.D. Kurtz: Characterization of n-type β-SiC as a piezoresistor, IEEE Trans. Electron. Dev. **40** (1993) 1093–1099

33.84 M. Mehregany, C.A. Zorman, N. Rajan, C.H. Wu: Silicon carbide MEMS for harsh environments, Proc. IEEE **86** (1998) 1594–1610

33.85 C.A. Zorman, A.J. Fleischmann, A.S. Dewa, M. Mehregany, C. Jacob, S. Nishino, P. Pirouz: Epitaxial growth of 3C-SiC films on 4 in. diam Si(100) silicon wafers by atmospheric pressure chemical vapor deposition, J. Appl. Phys. **78** (1995) 5136–5138

33.86 C.A. Zorman, S. Roy, C.H. Wu, A.J. Fleischman, M. Mehregany: Characterization of polycrystalline silicon carbide films grown by atmospheric pressure chemical vapor deposition on polycrystalline silicon, J. Mater. Res. **13** (1998) 406–412

33.87 B. Bhushan, B.K. Gupta: *Handbook of Tribology: Materials, Coatings and Surface Treatments* (Krieger, Malabar 1997) Reprint

33.88 J. F. Shackelford, W. Alexander, J. S. Park (Eds.): *CRC Material Science and Engineering Handbook*, 2nd edn. (CRC, Boca Raton 1994)

33.89 B. K. Gupta, J. Chevallier, B. Bhushan: Tribology of ion bombarded silicon for micromechanical applications, ASME J. Tribol. **115** (1993) 392–399

33.90 B. K. Gupta, B. Bhushan, J. Chevallier: Modification of tribological properties of silicon by boron ion implantation, Tribol. Trans. **37** (1994) 601–607

33.91 B. K. Gupta, B. Bhushan: Nanoindentation studies of ion implanted silicon, Surf. Coat. Technol. **68-69** (1994) 564–570

33.92 B. Bhushan, V. N. Koinkar: Tribological studies of silicon for magnetic recording applications, J. Appl. Phys. **75** (1994) 5741–5746

33.93 G. M. Pharr: The anomalous behavior of silicon during nanoindentation. In: *Thin Films: Stresses and Mechanical Properties III*, Vol. 239, ed. by W. D. Nix, J. C. Bravman, E. Arzt, L. B. Freund (Materials Research Soc., Pittsburgh 1991) pp. 301–312

33.94 D. L. Callahan, J. C. Morris: The extent of phase transformation in silicon hardness indentation, J. Mater. Res. **7** (1992) 1612–1617

33.95 N. A. Fleck, G. M. Muller, M. F. Ashby, J. W. Hutchinson: Strain gradient plasticity: Theory and experiment, Acta Metall. Mater. **42** (1994) 475–487

33.96 B. Bhushan, S. Venkatesan: Friction and wear studies of silicon in sliding contact with thin-film magnetic rigid disks, J. Mater. Res. **8** (1993) 1611–1628

33.97 S. Venkatesan, B. Bhushan: The role of environment in the friction and wear of single-crystal silicon in sliding contact with thin-film magnetic rigid disks, Adv. Info Storage Syst. **5** (1993) 241–257

33.98 S. Venkatesan, B. Bhushan: The sliding friction and wear behavior of single-crystal, polycrystalline and oxidized silicon, Wear **171** (1994) 25–32

33.99 B. Bhushan: Chemical, mechanical, and tribological characterization of ultra-thin and hard amorphous carbon coatings as thin as 3.5 nm: Recent developments, Diam. Relat. Mater. **8** (1999) 1985–2015

33.100 S. Sundararajan, B. Bhushan: Micro/nanotribological studies of polysilicon and SiC films for MEMS applications, Wear **217** (1998) 251–261

33.101 X. Li, B. Bhushan: Micro/nanomechanical characterization of ceramic films for microdevices, Thin Solid Films **340** (1999) 210–217

33.102 A. A. Yasseen, C. H. Wu, C. A. Zorman, M. Mehregany: Fabrication and testing of surface micromachined polycrystalline SiC micromotors, IEEE Electron. Dev. Lett. **21** (2000) 164–166

33.103 H. Liu, B. Bhushan: Nanotribological characterization of molecularly-thick lubricant films for applications to MEMS/NEMS by AFM, Ultramicroscopy **97** (2003) 321–340

33.104 B. Bhushan, A. V. Kulkarni, V. N. Koinkar, M. Boehm, L. Odoni, C. Martelet, M. Belin: Microtribological characterization of self-assembled and Langmuir–Blodgett monolayers by atomic force and friction force microscopy, Langmuir **11** (1995) 3189–3198

33.105 V. N. Koinkar, B. Bhushan: Micro/nanoscale studies of boundary layers of liquid lubricants for magnetic disks, J. Appl. Phys. **79** (1996) 8071–8075

33.106 V. N. Koinkar, B. Bhushan: Microtribological studies of unlubricated and lubricated surfaces using atomic force/friction force microscopy, J. Vac. Sci. Technol. A **14** (1996) 2378–2391

33.107 B. Bhushan, H. Liu: Nanotribological properties and mechanisms of alkylthiol and biphenyl thiol self-assembled monolayers studied by AFM, Phys. Rev. B **63** (2001) 245412:1–11

33.108 H. Liu, B. Bhushan: Investigation of nanotribological properties of self-assembled monolayers with alkyl and biphenyl spacer chains, Ultramicroscopy **91** (2002) 185–202

33.109 T. Stifter, O. Marti, B. Bhushan: Theoretical investigation of the distance dependence of capillary and van der Waals forces in scanning force microscopy, Phys. Rev. B **62** (2000) 13667–13673

33.110 B. Bhushan: Self-assembled monolayers for controlling hydrophobicity and/or friction and wear. In: *Modern Tribology Handbook*, ed. by B. Bhushan (CRC, Boca Raton 2001) pp. 909–929

33.111 H. Liu, B. Bhushan, W. Eck, V. Stadler: Investigation of the adhesion, friction, and wear properties of biphenyl thiol self-assembled monolayers by atomic force microscopy, J. Vac. Sci. Technol. A **19** (2001) 1234–1240

33.112 S. Sundararajan, B. Bhushan: Static friction and surface roughness studies of surface micromachined electrostatic micromotors using an atomic force/friction force microscope, J. Vac. Sci. Technol. A **19** (2001) 1777–1785

33.113 C. H. Mastrangelo, C. H. Hsu: Mechanical stability and adhesion of microstructures under capillary forces – Part II: Experiments, J. Microelectromech. Syst. **2** (1993) 44–55

33.114 H. Liu, B. Bhushan: Adhesion and friction studies of microelectromechanical systems/nanoelectromechanical systems materials using a novel microtriboapparatus, J. Vac. Sci. Technol. A **21** (2003) 1528–1538

33.115 M. P. De Boer, T. A. Michalske: Accurate method for determining adhesion of cantilever beams, J. Appl. Phys. **86** (1999) 817

33.116 R. Maboudian, R. T. Howe: Critical review: Adhesion in surface micromechanical structures, J. Vac. Sci. Technol. B **15** (1997) 1–20

33.117 C. H. Mastrangelo: Surface force induced failures in microelectromechanical systems. In: *Tribology Issues and Opportunities in MEMS*, ed. by B. Bhushan (Kluwer, Dordrecht 1998) pp. 367–395

33.118 B. Bhushan, H. Liu, S. M. Hsu: Adhesion and friction studies of silicon and hydrophobic and low fric-

tion films and investigation of scale effects, ASME J. Tribol. in press
33.119 S.K. Chilamakuri, B. Bhushan: A comprehensive kinetic meniscus model for prediction of long-term static friction, J. Appl. Phys. **15** (1999) 4649–4656
33.120 J. Kijlstra, K. Reihs, A. Klamt: Roughness and topology of ultra-hydrophobic surfaces, Colloids Surf. A **206** (2002) 521–529
33.121 D. Quere, P. Aussillous: Non-stick droplets, Chem. Eng. Technol. **25** (2002) 925–928
33.122 Y.C. Tai, R.S. Muller: Frictional study of IC processed micromotors, Sens. Actuators A **21-23** (1990) 180–183
33.123 K.J. Gabriel, F. Behi, R. Mahadevan, M. Mehregany: In situ friction and wear measurement in integrated polysilicon mechanisms, Sens. Actuators A **21-23** (1990) 184–188
33.124 M.G. Lim, J.C. Chang, D.P. Schultz, R.T. Howe, R.M. White: Polysilicon microstructures to characterize static friction, Proc. IEEE Micro Electro Mechanical Systems (IEEE, New York 1990) 82–88
33.125 U. Beerschwinger, S.J. Yang, R.L. Reuben, M.R. Taghizadeh, U. Wallrabe: Friction measurements on LIGA-processed microstructures, J. Micromech. Microeng. **4** (1994) 14–24
33.126 D. Matheison, U. Beerschwinger, S.J. Young, R.L. Rueben, M. Taghizadeh, S. Eckert, U. Wallrabe: Effect of progressive wear on the friction characteristics of nickel LIGA processed rotors, Wear **192** (1996) 199–207
33.127 R. Maboudian: Adhesion and friction issues associated with reliable operation of MEMS, MRS Bull. (June 1998) 47–51

34. Mechanical Properties of Micromachined Structures

To be able to accurately design structures and make reliability predictions, in any field it is necessary first to know the mechanical properties of the materials that make up the structural components. In the fields of microelectromechanical systems (MEMS) and nanoelectromechanical systems (NEMS), the devices are necessarily very small. The processing techniques and microstructures of the materials in these devices may differ significantly from bulk structures. Also, the surface-area-to-volume ratio in these structures is much higher than in bulk samples, and so the surface properties become much more important. In short, it cannot be assumed that mechanical properties measured using bulk specimens will apply to the same materials when used in MEMS and NEMS. This chapter will review the techniques that have been used to determine the mechanical properties of micromachined structures, especially residual stress, strength, and Young's modulus. The experimental measurements that have been performed will then be summarized, in particular the values obtained for polycrystalline silicon (polysilicon).

34.1 **Measuring Mechanical Properties of Films on Substrates** 1023
34.1.1 Residual Stress Measurements 1023
34.1.2 Mechanical Measurements Using Nanoindentation 1024

34.2 **Micromachined Structures for Measuring Mechanical Properties** 1024
34.2.1 Passive Structures 1025
34.2.2 Active Structures 1028

34.3 **Measurements of Mechanical Properties** . 1034
34.3.1 Mechanical Properties of Polysilicon 1034
34.3.2 Mechanical Properties of Other Materials 1036

References 1037

34.1 Measuring Mechanical Properties of Films on Substrates

In order to determine accurately the mechanical properties of very small structures, it is necessary to test specimens made from the same materials, processed in the same way, and of the same approximate size. Not surprisingly it is often difficult to handle specimens this small. One solution is to test the properties of films remaining on substrates. Micro- and nanomachined structures are typically fabricated from films that are initially deposited onto a substrate, are subsequently patterned and etched into the appropriate shapes, and then finally released from the substrate. If the testing is performed on the continuous film, before patterning and release, the substrate can be used as an effective "handle" for the specimen (in this case, the film). Of course since the films are adhered to the substrate, the types of tests possible are severely limited.

34.1.1 Residual Stress Measurements

One common measurement easily performed on films attached to substrates is residual film stress. The curvature of the substrate is measured before and after film deposition. Curvature can be measured in a number of ways. The most common technique is to scan a laser across the surface (or scan the substrate beneath the laser) and detect the angle of the reflected signal. Alternatively, profilometry, optical interferometry, or even atomic force microscopy can be used. As expected, tools that map a surface or perform multiple linear scans can give more accurate readings than tools that measure only a single scan.

Assuming that the film is thin compared to the substrate, the average residual stress in the film, σ_f, is given by the Stoney equation,

$$\sigma_f = \frac{1}{6}\frac{E_s}{(1-\nu_s)}\frac{t_s^2}{t_f}\left(\frac{1}{R_1}-\frac{1}{R_2}\right), \quad (34.1)$$

where the subscripts f and s refer to the film and substrate, respectively; t is thickness, E is Young's modulus, ν is Poisson's ratio, and R is radius of curvature before (R_1) and after (R_2) film deposition [34.2]. For the typical (100)-oriented silicon substrate, $E/(1-\nu)$ (also known as the biaxial modulus) is equal to 180.5 GPa, independent of in-plane rotation [34.3]. This investigation can be performed on the as-deposited film or after any subsequent annealing step, provided no changes occur to the substrate.

This measurement will reveal the average residual stress of the film. Typically, however, the residual stresses of deposited films will vary through the film thickness. One way to detect this, using substrate curvature techniques, is to etch away a fraction of the film and repeat the curvature measurement. This can be iterated any number of times to obtain a residual stress profile for the film [34.4]. Alternatively, tools have been designed to measure the substrate curvature during the deposition process itself, to obtain information on how the stresses evolve [34.5].

An additional feature of some of these tools is the ability to heat the substrates while performing the stress measurement. An example of the results obtained in such an experiment is shown in Fig. 34.1 [34.1], for an aluminum film on a silicon substrate. The slope of the heating curve gives the difference in thermal expansion between the film and the substrate. When the heating curve changes slope and becomes nearly horizontal, the yield strength of the film has been reached.

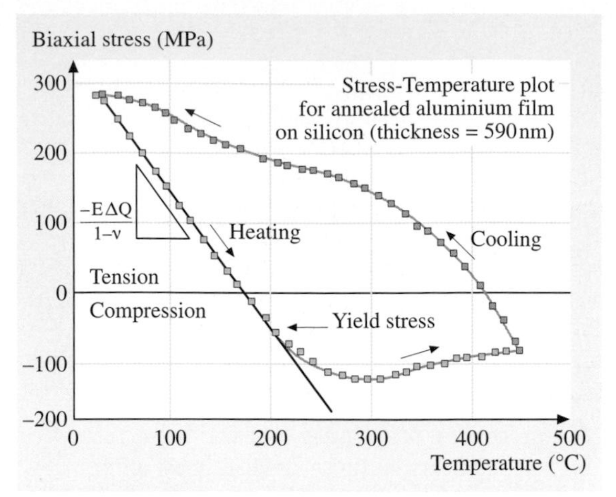

Fig. 34.1 Typical results for residual stress as a function of temperature for an aluminum film on a silicon substrate [34.1]. The stresses were determined by measuring the curvature of the substrate before and after film deposition, using the reflected signal of a laser scanned across the substrate surface

34.1.2 Mechanical Measurements Using Nanoindentation

Aside from residual stress, it is difficult to measure mechanical properties of films adhered to substrates without the measurement being affected by the presence of the substrate. Recent developments in nanoindentation equipment have allowed this technique to be used in some cases. With specially designed tools, indentation can be performed using very low loads. If the films being investigated are thick and rigid enough, measurements can be made that are not influenced by the presence of the substrate. Of course, this can be verified by depositing the same film onto different substrates. By continuously monitoring the displacement as well as the load during indentation, a variety of properties can be measured, including hardness and Young's modulus [34.6]. This area is covered in more detail in a separate chapter.

For brittle materials, cracks can be generated by indentation, and strength information can be gathered. But the exact stress fields created during the indentation process are not exactly known, and, therefore, quantitative values for strength are difficult to determine. Anisotropic etching of single crystal silicon has been done to create 30-μm-tall structures that were then indented to examine fracture toughness [34.7], however this is not possible with most materials.

34.2 Micromachined Structures for Measuring Mechanical Properties

Certainly the most direct way to measure the mechanical properties of small structures is to fabricate structures that would be conducive to such tests. Fabrication techniques are sufficiently advanced that virtually any design can be realized, at least in two dimensions. Two basic types of devices are used for mechanical property testing: "passive" structures and "active" structures.

34.2.1 Passive Structures

As mentioned previously, the main difficulty in testing very small specimens is in handling. One way to circumvent this problem is by using passive structures. These structures are designed to act as soon as they are released from the substrate and to provide whatever information they are designed to supply without further manipulation. For all of these passive structures, the forces acting on them come from the structural materials' own internal residual stresses. For devices on the micron scale or smaller, gravitational forces can be neglected, and therefore internal stresses are the only source of actuation force.

Stress Measurements

Since the internal residual stresses act upon the passive devices when they are released, it is natural to design a device that can be used to measure residual stresses. One such device, a rotating microstrain gauge, is shown in Fig. 34.2. There are many different microstrain gauge designs, but all operate under the same principle. In Fig. 34.2a, the large pads, labeled A, will remain anchored to the substrate when the rest of the device is released. Upon release, the device will expand or contract in order to relieve its internal residual stresses. A structure under tension will contract, and a structure under compression will expand. For the structure in Fig. 34.2, a compressive stress will cause the legs to lengthen. Since the two opposing legs are not attached to the central beam at the same point but are offset, they will cause the central beam to rotate when they expand. The device in Fig. 34.2 contains two independent gauges that point to one another. At the ends of the two central beams are two parts of a Vernier scale. By observing this scale, one can measure the rotation of the beams.

If the connections between the legs and the central beams were simple pin connections, the strain, ε, of the legs (the fraction of expansion or contraction) could be determined simply by the measured rotation and the geometry of the device, namely

$$\varepsilon = \frac{d_{\text{beam}} d_{\text{offset}}}{2 L_{\text{central}} L_{\text{leg}}} , \qquad (34.2)$$

where d_{beam} is the lateral deflection of the end of one central beam, d_{offset} is the distance between the connections of the opposing legs, L_{central} is the length of the central beam (measured to the center point between the leg connections), and L_{leg} is the length of the leg. But since the entire device was fabricated from a single polysilicon film, this cannot be the case; there must be some bending occurring at the connections. As a result, to get an accurate determination of the strain relieved upon release, finite element analysis (FEA) of the struc-

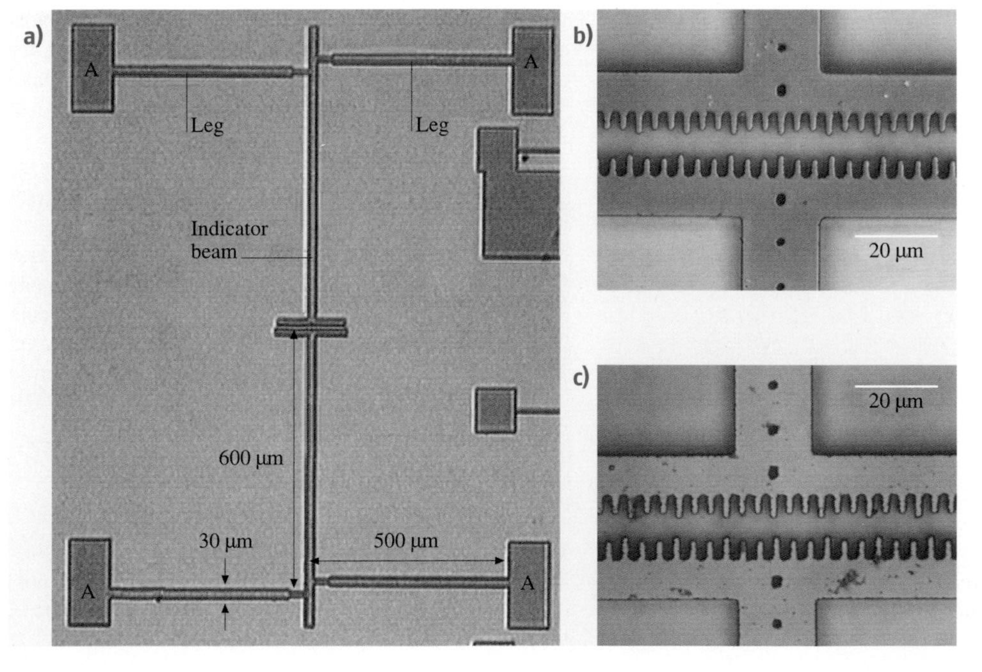

Fig. 34.2 (a) Microstrain gauge fabricated from polysilicon. (b) shows a close-up of the Vernier scale before release, and (c) shows the same area after release

ture must be performed. This is a common situation for microdevices. FEA is a powerful tool in determining the displacements and stresses of nonideal geometries. One drawback is that Young's modulus of the material must be known in order to do the FEA as well as to convert the measured strain into a stress value. But Young's moduli for many micromachined materials are known or can be measured using other techniques.

Other devices, besides rotating strain gauges, are designed to measure residual stresses. One of the most simple is a doubly clamped beam, a long, narrow beam of constant width and thickness that is anchored to the substrate at both ends. If the beam contains a tensile stress, it will remain straight. But if the beam contains a compressive stress, it will buckle if its length exceeds a critical value, l_{cr}, according to the Euler buckling criterion [34.8],

$$\varepsilon_r = -\frac{\pi^2}{3}\left(\frac{h}{l_{cr}}\right)^2 , \quad (34.3)$$

where ε_r is the residual strain in the beam, and h is the width or thickness of the beam, whichever is less. To determine the residual strain, a series of doubly clamped beams are fabricated of varying lengths. In this way, the critical length, l_{cr}, for buckling can be deduced after release. One problem with this technique is that during the release process, any turbulence in the solution will lead to enhanced buckling of beams, and a low value for l_{cr} will be obtained.

For films with tensile stresses, a similar analysis can be done using ring-and-beam structures, also called Guckel rings after their inventor, Henry Guckel. A schematic of this design is shown in Fig. 34.3 [34.8]. A tensile stress in the outer ring will cause it to contract. This will lead to a compressive stress in the central beam, even though the material was originally tensile before release. The amount of compression in the central beam can be determined analytically from the geometry of the device and the residual strain of the material. Again, by changing the length of the central beam, the l_{cr} will be determined, and then the residual strain can be deduced.

Stress Gradient Measurements

For structures fabricated from thin deposited films, the stress gradient can be just as important as the stress itself. Figure 34.4 shows a portion of a silicon microactuator. The device is designed to be completely planar; however, stress gradients in the film cause the structures to bend. This figure illustrates the importance of characterizing and controlling stress gradients, and it also demonstrates that the easiest structure for measuring stress gradients is a simple cantilever beam. By measuring the end deflection δ, of a cantilever beam of length l, and thickness t, the stress gradient, $\mathrm{d}\sigma/\mathrm{d}t$ is determined by [34.9]

$$\frac{\mathrm{d}\sigma}{\mathrm{d}t} = \frac{2\delta}{l^2}\frac{E}{1-\nu} . \quad (34.4)$$

The magnitude of the end deflection can be measured by microscopy, optical interferometry, or any other technique.

Another useful structure for measuring stress gradients is a spiral. For this structure, the end of the spiral not anchored to the substrate will move out-of-plane. Also

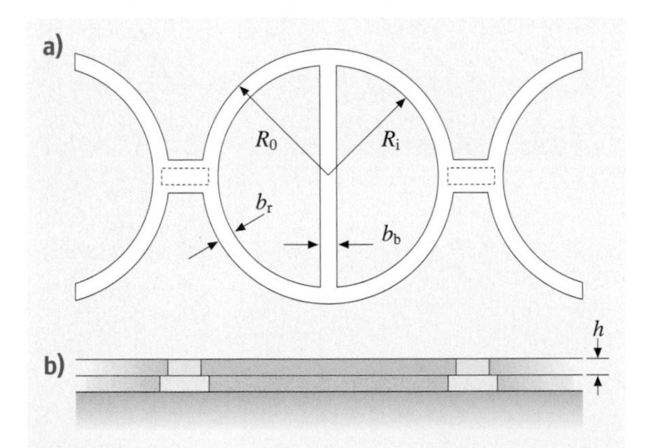

Fig. 34.3a,b Schematic (a) top view and (b) side view of Guckel ring structures [34.8]. The *dashed lines* in (a) indicate the anchors

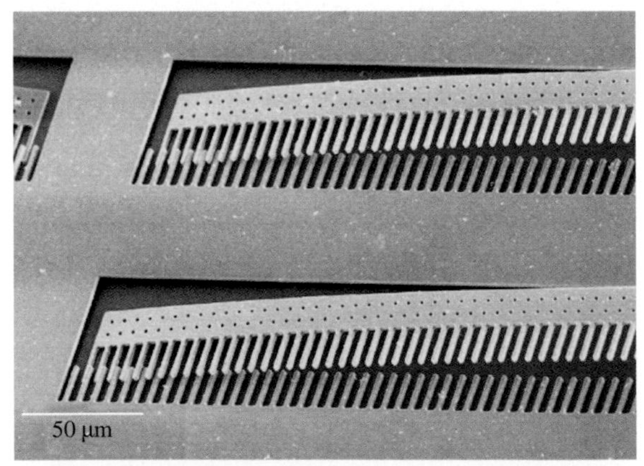

Fig. 34.4 Scanning electron micrograph (SEM) of a portion of a silicon microactuator. Residual stress gradients in the silicon cause the structure to bend

the diameter of the spiral will contract, and the free end of the spiral will rotate when released [34.10].

Strength and Fracture Toughness Measurements

As mentioned above, if a doubly clamped beam contains a tensile stress, it will remain taut when released because it cannot relieve any of its stress by contracting. This tensile stress can be thought of as a tensile load being applied at the ends of the beam. If this tensile load exceeds the tensile strength of the material, the beam will break. Since the tensile stress can be measured, as discussed in the Sect. 34.2.1, this technique can be used to gather information on the strength of materials. Figure 34.5 shows two different beam designs that have been used to measure strength. The device shown in Fig. 34.5a was fabricated from a tensile polysilicon film [34.11]. Different beams were designed with varying lengths of the wider regions (marked l_1 in the figure). In this manner, the load applied to the narrow center beam was varied, even though the entire film contained a uniform residual tensile stress. For l_1 greater than a critical value, the narrow center beam fractured, giving a measurement for the tensile strength of polysilicon.

The design shown in Fig. 34.5b was fabricated from a tensile Si_xN_y film [34.12]. As seen in the figure, a stress concentration was designed into the beam, to ensure the fracture strength would be exceeded. In this case, a notch was etched into one side of the beam. Since the stress concentration is not symmetric with regard to the beam axis, this results in a large bending moment at that position, and the test measures the bend strength of the material. Again, like the beams shown in Fig. 34.5a, the geometry of various beams fabricated from the same film were varied, to vary the maximum stress seen at the notch. By seeing which beams fracture at the stress concentration after release, the strength can be determined.

Through a similar technique but by using an atomically sharp pre-crack instead of a stress concentration, the fracture toughness of a material can be determined. Sharp pre-cracks can be introduced into micromachined structures before release by placing a Vickers indent on the substrate, near the device; the radial crack formed by the indent will propagate into the overlying structure [34.14]. Accordingly, the beam with a sharp pre-crack, shown in Fig. 34.6, was fabricated using polysilicon [34.13]. Due to the stochastic nature of indentation, the initial pre-crack length will vary from beam to beam. Because of this, even though the geometry of the beams remains identical, the stress intensity, K, at the pre-crack tip will vary. Upon release, only those pre-cracks whose K exceeds the fracture toughness of the material, K_{Ic}, will propagate, and in this way upper and lower bounds for K_{Ic} for the material can be determined.

For all of the beams discussed in this section, finite element analysis is required to determine the stress concentrations and stress intensities. Even though approximate analytical solutions may exist for these designs, the actual fabricated structure will not have idealized geometries. For example, corners will never be perfectly sharp, and cracks will never be perfectly straight. This

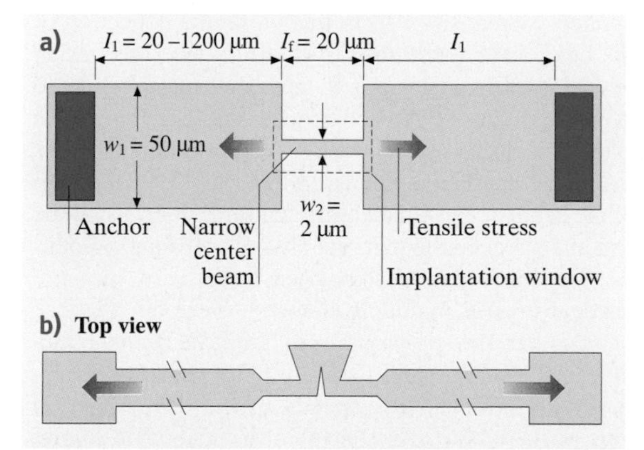

Fig. 34.5a,b Schematic designs of doubly clamped beams with stress concentrations for measuring strength. (a) was fabricated from polysilicon [34.11], and (b) was fabricated from Si_xN_y [34.12]

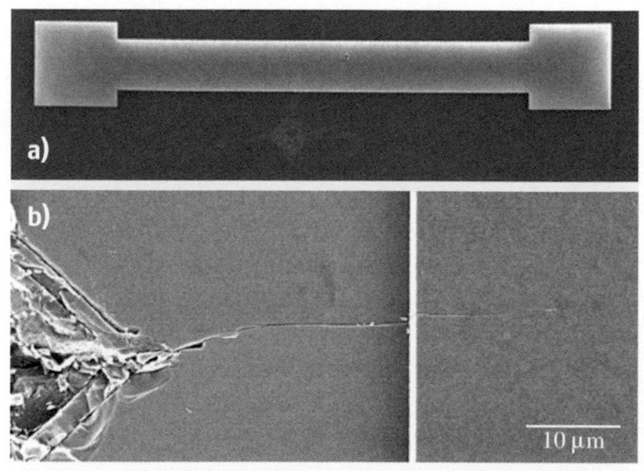

Fig. 34.6 (a) SEM of a 500 μm long polysilicon beam with a Vickers indent placed near its center; (b) higher magnification SEM of the area near the indent showing the pre-crack traveling from the substrate into the beam [34.13]

reinforces the idea that FEA is a powerful tool in determining mechanical properties of very small structures.

34.2.2 Active Structures

As discussed above, it is very convenient to design structures that act upon release to provide information on the mechanical properties of the structural materials. This is not always possible, however. For example, those passive devices just discussed rely on residual stresses to create the changes (rotation or fracture) that occur upon release, but many materials do not contain high residual stresses as deposited, or the processing scheme of the device precludes the generation of residual stresses. Also, some mechanical properties, such as fatigue resistance, require motion in order to be studied. Active devices are therefore used. These are acted upon by a force (the source of this force can be integrated into the device itself or can be external to the device) in order to create a change, and by the response to the force, the mechanical properties are studied.

Young's Modulus Measurements

Young's modulus, E, is a material property critical to any structural device design. It describes the elastic response of a material and relates stress, σ, and strain, ε, by

$$\sigma = E\varepsilon \,. \tag{34.5}$$

In bulk samples, E is often measured by loading a specimen in tension and measuring displacement as a function of stress for a given length. While this is considerably more difficult for small structures, such as those fabricated from thin deposited films, it can be achieved with careful experimental techniques. Figure 34.7 shows the schematic of one such measurement system [34.15]. The fringe detectors in the figure detect the reflected laser signal from two gold lines deposited onto the polysilicon specimen, which act as gage markers. In this manner, the strain in the specimen during loading can be monitored. Besides gold lines, Vickers indents placed in a nickel specimen can also serve as gage markers [34.16], or a speckle interferometry technique [34.17] can be used to determine strain in the specimen. Once the stress-versus-strain behavior is measured, the slope of the curve is equal to E.

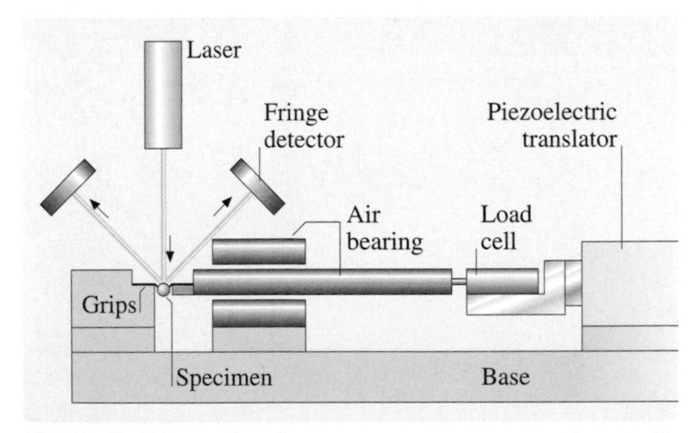

Fig. 34.7 Schematic of a measurement system for tensile loading of micromachined specimens [34.15]

In addition to the tensile test, Young's modulus can be determined by other measures of stress-strain behavior. As seen in Fig. 34.8, a cantilever beam can be bent by pushing on the free end with a nanoindenter [34.18]. The nanoindenter can monitor the force applied and the displacement, and simple beam theory can convert the displacement into strain to obtain E. A similar technique, shown schematically in Fig. 34.9 [34.19], involves pulling downward on a cantilever beam by means of an electrostatic force. An electrode is fabricated into the substrate beneath the cantilever beam, and a voltage is applied between the beam and the bottom electrode. The force acting on the beam is equal to the electrostatic force corrected to include the effects of fringing fields acting on the sides of the beam, namely

$$F(x) = \frac{\varepsilon_0}{2}\left(\frac{V}{g+z(x)}\right)^2\left(1+\frac{0.65[g+z(x)]}{w}\right) , \tag{34.6}$$

where $F(x)$ is the electrostatic force at x, ε_0 is the dielectric constant of air, g is the gap between the beam and the bottom electrode, $z(x)$ is the out-of-plane deflection of the beam, w is the beam width, and V is the applied voltage [34.19]. In this work, the deflection of the beam as a function of position is measured using optical interferometry. These measurements combine to give stress-strain behavior for the cantilever beam. An extension of this technique uses doubly clamped beams instead of cantilever beams. In this case, the deflection of the beam at a given electrostatic force depends on the residual stress in the material as well as Young's modulus. This method can also be used, therefore, to measure residual stresses in doubly clamped beams.

Another device possible to fabricate from a thin film and also possible to use for investigating stress-strain behavior is a suspended membrane, seen in Fig. 34.10 [34.20]. As shown in the schematic figure, the membrane is exposed to an elevated pressure on one side, causing it to bulge in the opposite direction. The deflection of the membrane is measured by optical or other techniques and related to the strain in the mem-

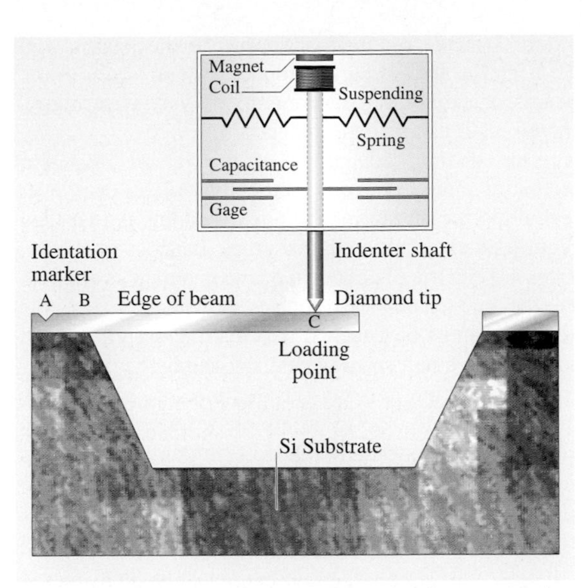

Fig. 34.8 Schematic of a nanoindenter loading mechanism pushing on the end of a cantilever beam [34.18]

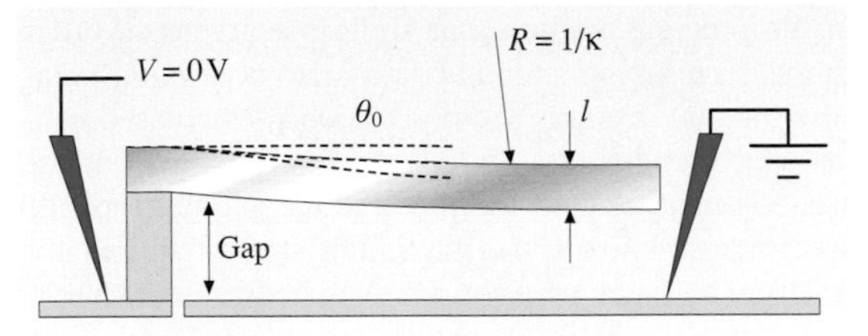

Fig. 34.9 Schematic of a cantilever beam bending test using an electrostatic voltage to pull the beam toward the substrate [34.19]

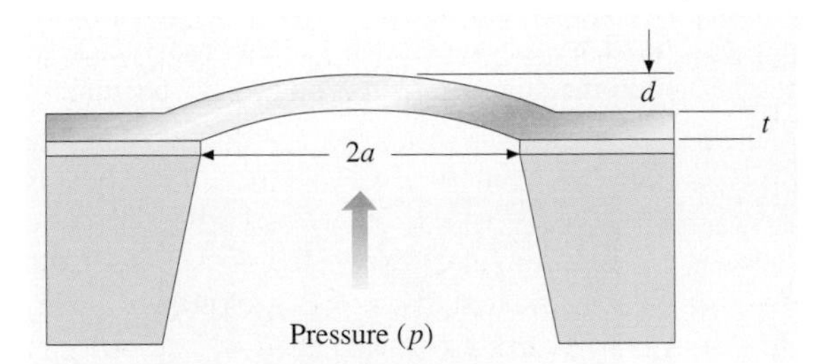

Fig. 34.10 Schematic cross section of a microfabricated membrane [34.20]

brane. These membranes can be fabricated in any shape, typically square or circular. Both analytical solutions and finite element analyses have been done to relate the deflection to the strain. Like the doubly clamped beams, both Young's modulus and residual stress play a role in the deflected shape. Both of these mechanical properties can therefore be determined by the pressure-versus-deflection performance of the membrane.

Another measurement besides stress-strain behavior that can reveal Young's modulus of a material is the determination of the natural resonance frequency. For a cantilever, the resonance frequency, f_r, for free undamped vibration is given by

$$f_r = \frac{\lambda_i^2 t}{4\pi l^2}\left(\frac{E}{3\rho}\right)^{1/2} , \quad (34.7)$$

where ρ, l, and t are density, length, and thickness of the cantilever; λ_i is the eigenvalue, where i is an integer that describes the resonance mode number; for the first mode $\lambda_1 = 1.875$ [34.22]. By measuring f_r and knowing the geometry and density, therefore, E can be determined. The cantilever can be vibrated by a number of techniques, including a laser, loudspeaker, or piezoelectric shaker. The frequency that produces the highest amplitude of vibration is the resonance frequency.

A micromachined device that uses an electrostatic comb drive and an AC signal to generate the vibration of the structure is known as a lateral resonator [34.23]. One example is shown in Fig. 34.11 [34.21]. When a voltage is applied across either set of interdigitated comb fingers shown in Fig. 34.11, an electrostatic attraction will be generated due to the increase in capacitance as the overlap between the comb fingers increases. The force,

Fig. 34.11 SEM of a polysilicon lateral resonator [34.21]

F, generated by the comb drive is given by

$$F = \frac{1}{2}\frac{\partial C}{\partial x}V^2 = n\varepsilon\frac{h}{g}V^2\,, \quad (34.8)$$

where C is capacitance, x is distance traveled by one comb-drive toward the other, n is the number of pairs of comb fingers in one drive, ε is the permittivity of the fluid between the fingers, h is the height of the fingers, g is the gap spacing between the fingers, and V is the applied voltage [34.23]. When an AC voltage at the resonance frequency is applied across either of the two comb drives, the central portion of the device will vibrate. In fact, since force depends on the square of the voltage for electrostatic actuation, for a time, t, dependent drive voltage, $v_D(t)$, of

$$v_D(t) = V_P + v_d \sin(\omega t)\,, \quad (34.9)$$

where V_P is the DC bias and v_d is the AC drive amplitude, the time dependent portion of the force will scale with

$$2\omega V_P v_d \cos(\omega t) + \omega v_d^2 \sin(2\omega t) \quad (34.10)$$

[34.23]. Therefore, if an AC drive signal is used with no DC bias, at resonance, the frequency of the AC drive signal will be one half the resonance frequency. For this device, the resonance frequency, f_r, will be

$$f_r = \frac{1}{2\pi}\left(\frac{k_{sys}}{M}\right)^{1/2}\,, \quad (34.11)$$

where k_{sys} is the spring constant of the support beams and M is the mass of the portion of the device that vibrates. The spring constant is given by

$$k_{sys} = 24EI/L^3\,, \quad (34.12)$$

$$I = \frac{hw^3}{12}\,, \quad (34.13)$$

where I is the moment of inertia of the beams, and L, h, and w are the length, thickness, and width of the beams. By combining these equations and measuring f_r, therefore, E can be determined.

One distinct advantage of the lateral resonator technique and the electrostatically pulled cantilever technique for measuring Young's modulus is that they require no external loading sources. Portions of the devices are electrically contacted, and a voltage is applied. For the pure tension tests, such as shown in Fig. 34.7, the specimen must be attached to a loading system, which for the very small specimens discussed here, can be extremely difficult, and any misalignment or eccentricity in the test could lead to unreliable results. But the advantage of the externally loaded technique is that there are no limitations on the type of materials that can be tested. Conductivity is not a requirement, nor is any compatibility with electrical actuation.

Strength and Fracture Toughness Measurements

As one might expect, any of the techniques discussed in the previous section that strain specimens in order to measure Young's modulus can also be used to measure fracture strength. Simply, the load is increased until the specimen breaks. As long as either the load or the strain is measured at fracture, and the geometry of the specimen is known, the maximum stress required for fracture, σ_{crit}, can be determined, either through analytical analysis or FEA. Depending on the geometry of the test, σ_{crit} will represent the tensile or bend strength of the material.

If the available force is limited, or if it is desired to have a localized fracture site, stress concentrations can be added to the specimens. These are typically notches micromachined into the edges of specimens.

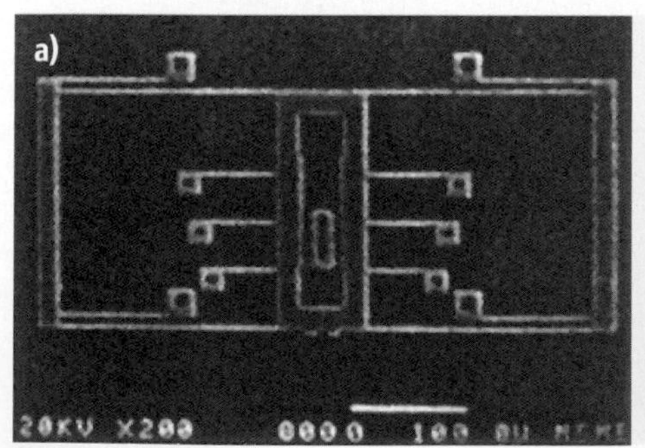

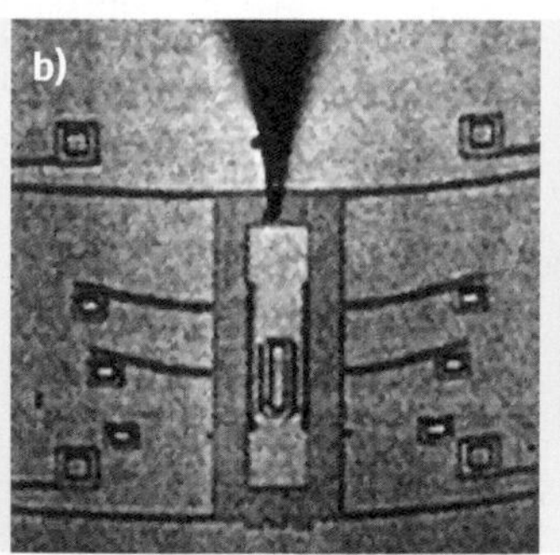

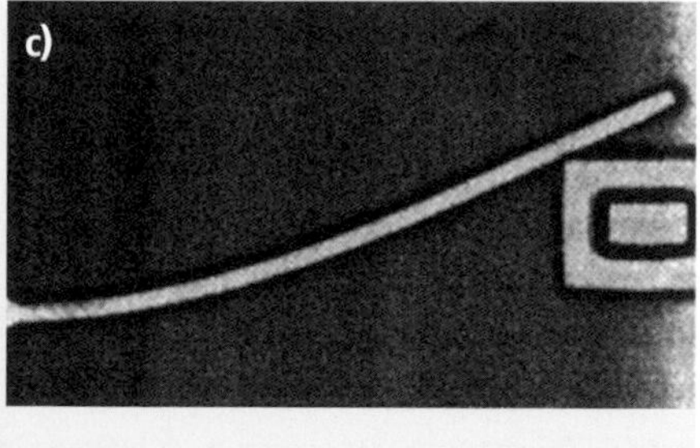

Fig. 34.12 (a) SEM of a device for measuring bend strength of polysilicon beams; (b) image of a test in process; (c) higher magnification view of one beam shortly before breaking [34.24]

Focused ion beams have also been used to carve stress concentrations into fracture specimens.

All of the external loading schemes, such as those shown in Figs. 34.7 and 34.8, have been used to measure fracture strength. Also, the electrostatically loaded doubly clamped beams can be pulled until they fracture. In this case, there is one complication. The electrostatic force is inversely proportional to the distance between the electrodes, and at a certain voltage, called the "pull-in voltage," the attraction between the beam and the substrate will become so great that the beam will immediately be pulled into contact with the bottom electrode. As long as the fracture takes place before the pull-in voltage is reached, the experiment will give valid results.

Other loading techniques have been used to generate fracture of microspecimens. Figure 34.12 [34.24] shows one device designed to be pushed by the end of a micromanipulated needle. The long beams that extend from the sides of the central shuttle come into contact with anchored posts, and, at a critical degree of bending, the beams will break off. Since the applied force cannot be measured in this technique, the experiment is optically monitored continuously during the test, and the image of the beams just before fracture is analyzed to determine σ_{crit}.

Another loading scheme that has been demonstrated for micromachined specimens utilizes scratch drive actuators to load the specimens [34.25]. These types of actuators work like inchworms, traveling across a substrate in discrete advances as an electrostatic force is repeatedly applied between the actuator and the substrate. The stepping motion can be made in the nm scale, depending on the frequency of the applied voltage, and so it can be an acceptable approximation to continuous loading. One advantage of this scheme is that very large forces can be generated by relatively small devices. The exact forces generated cannot be measured, so like the technique that used micromanipulated pushing, the test is continuously observed to determine the strain at fracture. Another advantage of this technique is that, like the lateral resonator and the electrically pulled cantilever, the loading takes place on-chip, and, therefore, the difficulties associated with attaching and aligning an external loading source are eliminated.

Another on-chip actuator used to load microspecimens is shown in Fig. 34.13, along with three different microspecimens [34.14]. Devices have been fabricated with each of the three microspecimens integrated with the same electrostatic comb-drive actuator. In all three cases, when a DC voltage is applied to the actuator, it moves downward, as oriented in Fig. 34.13. This pulls down on the left end of each of the three microspecimens, which are anchored on the right. The actuator contains 1,486 pairs of comb fingers. The maximum voltage that can be applied is limited by the breakdown

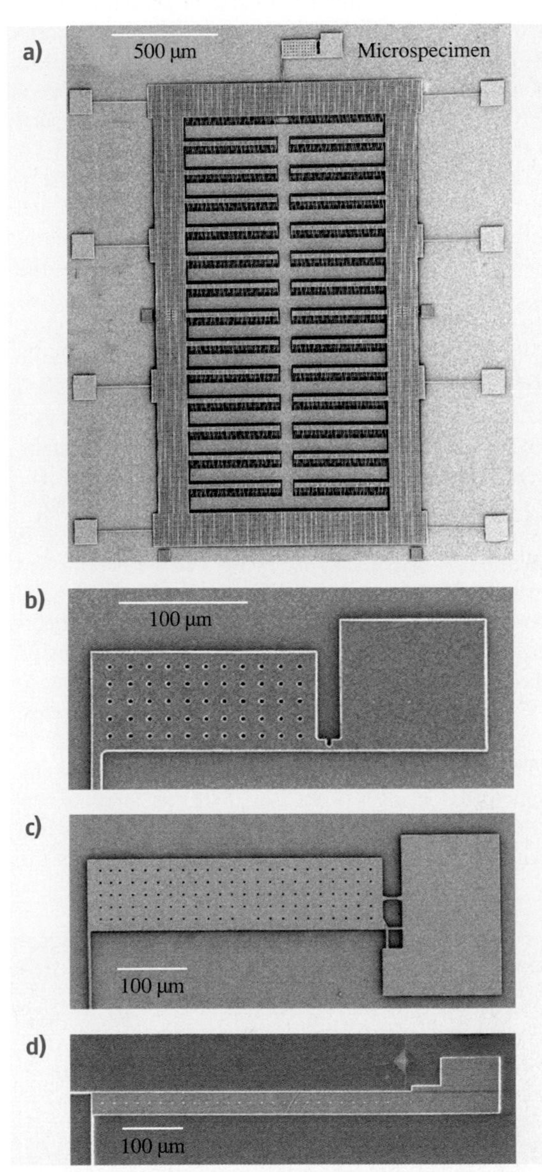

Fig. 34.13 (a) SEM of a micromachined device for conducting strength tests; the device consists of a large comb drive electrostatic actuator integrated with a microspecimen; (b)–(d) SEMs of various microspecimens for testing bend strength, tensile strength, and fracture toughness, respectively [34.14]

voltage of the medium in which the test takes place. In air, that limits the voltage to less than 200 V. As a result, given a finger height of 4 μm and a gap of 2 μm, and using (34.8), the maximum force generated by this actuator is limited to about 1 mN. Standard optical photolithography has a minimum feature size of about 2 μm. As a result, the electrostatic actuator cannot generate sufficient force to perform a standard tensile test on MEMS structural materials such as polysilicon. The microspecimens shown in Fig. 34.13 are therefore designed such that the stress is amplified.

The specimen shown in Fig. 34.13b is designed to measure bend strength. It contains a micromachined notch with a root radius of 1 μm. When the actuator pulls downward on the left end of this specimen, the notch serve as a stress concentration, and when the stress at the notch root exceeds σ_{crit}, the specimen fractures. The specimen in Fig. 34.13c is designed to test tensile strength. When the left end of this specimen is pulled downward, a tensile stress is generated in the upper thin horizontal beam near the right end of the specimen. As the actuator continues to move downward, the tensile stress in this beam will exceed the tensile strength, causing fracture. Finally, the specimen in Fig. 34.13d is similar to that in Fig. 34.13b, except that the notch is replaced by a sharp pre-crack that was produced by the Vickers indent placed on the substrate near the specimen. When this specimen is loaded, a stress intensity, K, is generated at the crack tip. When the stress intensity exceeds a critical value, K_{Ic}, the crack propagates. K_{Ic} is also referred to as the fracture toughness.

The force generated by the electrostatic actuator can be calculated using (34.8). But (34.8) assumes a perfectly planar, two-dimensional device. In fact, when actuated, the electric fields extend out of the plane of the device, and so (34.8) is just an approximation. Instead, like many of the techniques discussed in this section, the test is continuously monitored, and the actuator displacement at the time of fracture is recorded. Then FEA is used to determine the magnitude of the stress or stress intensity seen by the specimen at the point of fracture.

Fatigue Measurements

A benefit of the electrostatic actuator shown in Fig. 34.13 is that, besides monotonic loading, it can generate cyclic loading. This allows the fatigue resistance of materials to be studied. Simply by using an AC signal instead of a DC voltage, the device can be driven at its resonant frequency. The amplitude of the resonance depends on the magnitude of the AC signal. This amplitude can be increased until the specimen breaks; this will investigate the low-cycle fatigue resistance. Otherwise the amplitude of resonance can be left constant at a level below that is required for fast fracture, and the device will resonate indefinitely until the specimen breaks; this will investigate high-cycle fatigue. It should be noted that the resonance frequency of such a device is about 10 kHz. Therefore, it is possible to stress a specimen over 10^9 cycles in less than a day. In addition to simple cyclic loading, if a DC bias is added to the AC signal, a mean stress can be superimposed on the cyclic load. In this way, nonsymmetric cyclic loading (either with a large tensile stress alternating with a small compressive stress, or vice versa) can be studied.

Another device that can be used to investigate fatigue resistance in MEMS materials is shown in Fig. 34.14 [34.26]. In this case, a large mass is attached to the end of a notched cantilever beam. The mass contains two comb drives on opposite ends. When an AC signal is applied to one comb drive, the device will res-

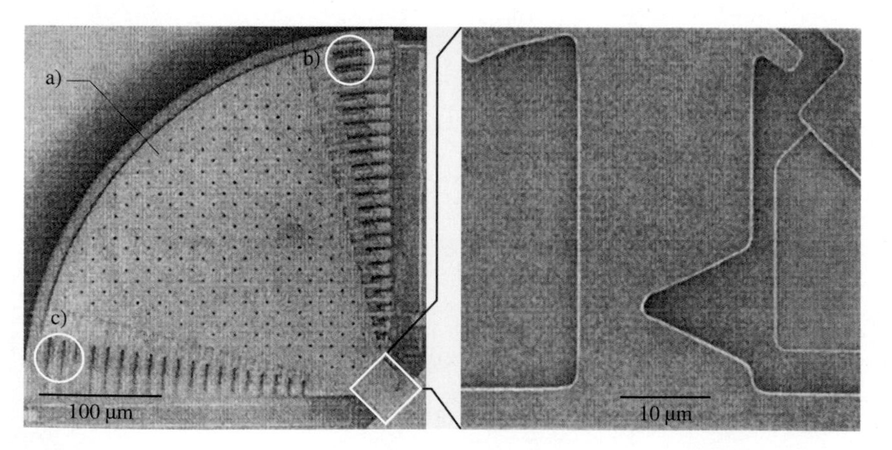

Fig. 34.14 SEM of a device for investigating fatigue; the image on the right is a higher magnification view of the notch near the base of the moving part of the structure [34.26] (a) mass, b) comb drive actuator, c) capacitive displacement sensor)

onate, cyclically loading the notch. The comb drive on the opposite side is used as a capacitive displacement sensor. This device contains many fewer comb fingers than the device shown in Fig. 34.13. As a result, it can apply cyclic loads by exploiting the resonance frequency of the device but cannot supply sufficient force to achieve monotonic loading.

Fatigue loading has also been studied using the same external loading techniques shown in Fig. 34.7. In this case, the frequency of the cyclic load is considerably lower, since the resonance frequency of the device is not being utilized. This leads to longer high-cycle testing times. Since the force is essentially unlimited, however, this technique allows a variety of frequencies to be studied to determine their effect on the fatigue behavior.

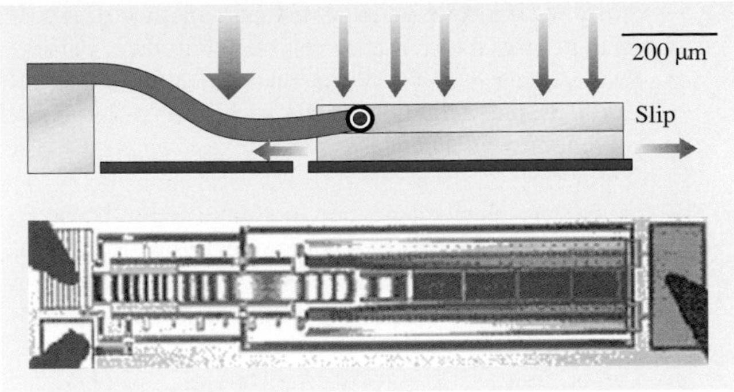

Fig. 34.16 Schematic cross section and top-view optical micrograph of a hinged-cantilever test structure for measuring friction in micromachined devices [34.28]

Friction Measurements

Friction is another property that has been studied in micromachined structures. To study friction, of course, two surfaces must be brought into contact with each other. This is usually avoided at all costs for these devices because of the risk of stiction. (Stiction is the term used when two surfaces that come into contact adhere so strongly that they cannot be separated.) Even so, a few devices have been designed to investigate friction. One of these is shown schematically in Fig. 34.15 [34.27]. It consists of a movable structure with a comb drive on one end and a cantilever beam on the other. Beneath the cantilever, on the substrate, is planar electrode. The device is moved to one side using the comb drive. Then a voltage is applied between the cantilever beam and the substrate electrode. The voltage on the comb drive is then released. The device would normally return to its original position, to relax the deflection in the truss suspensions, but the friction between the cantilever and the substrate electrode holds it in place. The voltage to the substrate electrode is slowly decreased until the device starts to slide. Knowing the electrostatic force generated by the

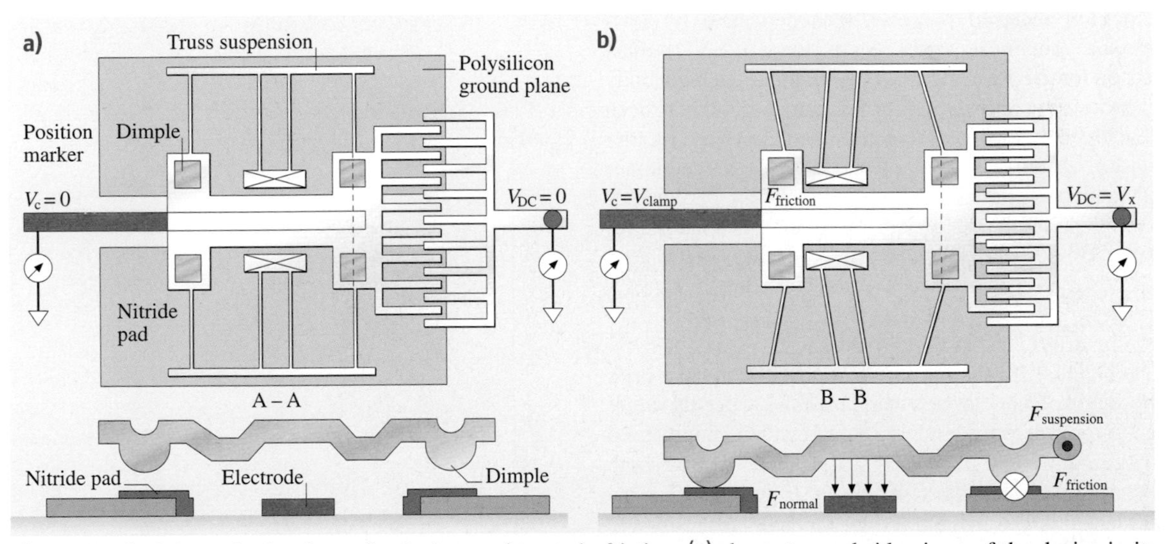

Fig. 34.15a,b Schematic drawings of a device used to study friction. **(a)** shows top and side views of the device in its original position, and **(b)** shows views of the device after it has been displaced using the comb drive and clamped using the substrate electrode [34.27]

substrate electrode and the stiffness of the truss suspensions, the static friction can be determined. For this device, bumps were fabricated on the bottoms of the cantilever beams. This limited the surface area that came into contact with the substrate and so lowered the risk of stiction.

Another device designed to study friction is shown in Fig. 34.16 [34.28]. This technique uses a hinged cantilever. The portion near the free end acts as the friction test structure, and the portion near the anchored end acts as the driver. The friction test structure is attracted to the substrate by means of electrostatic actuation, and when a second electrostatic actuator pulls down the driver, the friction test structure slips forward by a length proportional to the forces involved, including the frictional force. This distance, however, has a maximum of 30 nm, so all measurements must be exceedingly accurate in order to investigate a range of forces. This test structure can be used to determine the friction coefficients for surfaces with and without lubricating coatings.

34.3 Measurements of Mechanical Properties

All of the techniques discussed in Sects. 34.1 and 34.2 have been used to measure the mechanical properties of MEMS and NEMS materials. As a general rule, the results from the various techniques have agreed well with each other, and the argument becomes which of the measurement techniques is easiest and most reliable to perform. It is crucial to keep in mind, however, that certain properties, such as strength, are process dependent, and so the results taken at one laboratory will not necessarily match those taken from another. This will be discussed in more detail in Sect. 34.3.1.

34.3.1 Mechanical Properties of Polysilicon

In current MEMS technology, the most widely used structural material is polysilicon deposited by low-pressure chemical vapor deposition (LPCVD). One reason for the prevalence of polysilicon is the large body of processing knowledge for this material that has been developed by the integrated circuit community. Another reason, of course, is that polysilicon possesses a number of qualities beneficial for MEMS devices, particularly its high strength and Young's modulus. As a natural result, most of the mechanical properties investigations on MEMS materials have focused on polysilicon.

Residual Stresses in Polysilicon

The residual stresses of LPCVD polysilicon have been thoroughly characterized using both the wafer curvature technique, discussed in Sect. 34.1.1, and the microstrain gauges, discussed in Sect. 34.2.1. The results from both techniques give consistent values. Figure 34.17 summarizes the residual stress measurements as a function of deposition temperature taken from five different investigations at five different laboratories [34.29]. All five sets of data show the same trend. The stresses change from compressive at the lowest deposition temperatures to tensile at intermediate temperatures and back to compressive at the highest temperatures. The exact transition temperatures vary somewhat among the different investigations, probably due to differences in the deposition conditions: the silane or dichlorosilane pressure, the gas flow rate, the geometry of the deposition system, and the temperature uniformity. But in each data set the transitions are easily discernible. The origin of these residual stress changes lies with the microstructure of the LPCVD polysilicon films.

As with all deposited films, the microstructure of LPCVD polysilicon films is dependent on the deposition conditions. In general, the films are amorphous at

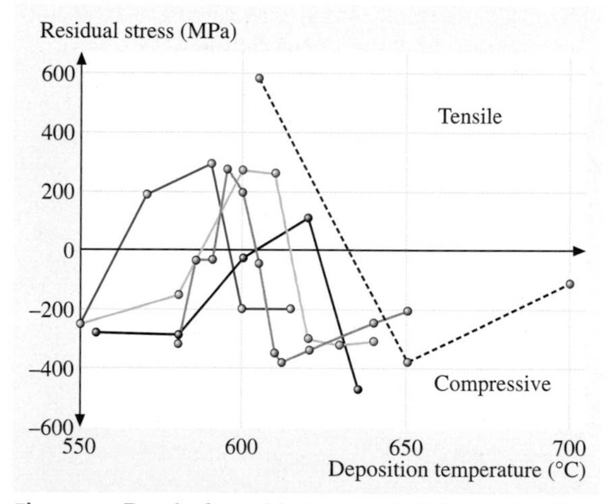

Fig. 34.17 Results for residual stress of LPCVD polysilicon films taken from five different investigations [34.29]. The data from each investigation are connected by a line

the lowest growth temperatures (lower than $\sim 570\,°C$), display fine ($\sim 0.1\,\mu m$ diameter) grains at intermediate temperatures ($\sim 570\,°C$ to $\sim 610\,°C$), and contain columnar (110)-textured grains with a thin fine-grained nucleation layer at the substrate interface at higher temperatures ($\sim 610\,°C$ to $\sim 700\,°C$) [34.29]. The fine-grained microstructure results from the homogeneous nucleation and growth of silicon grains within an as-deposited amorphous silicon film. In this regime, the deposition rate is just slightly faster than the crystallization rate. The as-deposited films will be crystalline near the substrate interface and amorphous at the free surface. (The amorphous fraction can be quickly crystallized by annealing above 610 °C.) The columnar microstructure seen at the higher growth temperatures results from the formation of crystalline silicon films as deposited, with growth being fastest in the $\langle 110 \rangle$ directions.

The origin of the tensile stress in the fine-grained polysilicon arises from the volume decrease that accompanies the crystallization of the as-deposited amorphous material. The origins of the compressive stresses in the amorphous and columnar films are less well understood. One proposed explanation for compressive stress generation during thin film growth postulates that an increase in the surface chemical potential is caused by the deposition of atoms from the vapor; the increase in surface chemical potential induces atoms to flow into newly formed grain boundaries, creating a compressive stress in the film [34.30].

Stress gradients are also typical of LPCVD polysilicon films. The partially amorphous films contain large stress gradients since they are essentially bilayers of compressive amorphous silicon on top of tensile fine-grained polysilicon. The fully crystalline films also exhibit stress gradients. The columnar compressive films are most highly stressed at the film-substrate interface, with the compressive stresses decreasing as the film thickness increases; the fine-grained films are less tensile at the film-substrate interface, with the tensile stresses increasing as the film thickness increases [34.29]. Both stress gradients are associated with microstructural variations. For the columnar films, the initial nucleation layer corresponds to a very high compressive stress, which decreases as the columnar morphology develops. For the fine-grained films, the region near the film-substrate interface has a slightly smaller average grain size, due to heterogeneous nucleation at the interface. This region displays a slightly lower tensile stress than the rest of the film, since the increased grain boundary area reduces the local density.

Young's Modulus of Polysilicon

The Young's modulus of polysilicon films has been measured using all the techniques discussed in Sect. 34.2.2. A good review of the experimental results taken from bulge testing, tensile testing, beam bending, and lateral resonators are contained in [34.31]. All of the reported results are in reasonable agreement, varying from 130 to 175 GPa, though many values are reported with a relatively high experimental scatter. The main origin of the error in these results is the uncertainties involving the geometries of the small specimens used to make the measurements. For example, from (34.13), the Young's modulus determined by the lateral resonators depends on the cube of the tether beam width, typically about $2\,\mu m$. In general, the beam width and other dimensions can be measured with scanning electron microscopy to within about $0.1\,\mu m$; however, the width of the beams is not perfectly constant along the entire length or even through the thickness. These uncertainties in geometry lead to uncertainties in modulus.

In addition, the various experimental measurements lie close to the Voigt and Reuss bounds for Young's modulus calculated using the elastic stiffnesses and compliances for single crystal silicon [34.31]. This strongly implies that Young's modulus of micro- and nanomachined polysilicon structures will be the same as for bulk samples made from polysilicon. This is not unexpected, since Young's modulus is a material property. It is related to the interatomic interactions and should have no dependence on the geometry of the sample. It should be noted that polysilicon can display a preferred crystallographic orientation depending on the deposition conditions, and that this could affect the Young's modulus of the material, since the Young's modulus of silicon is not isotropic. But the anisotropy is fairly small for the cubic silicon.

A more recent investigation that utilized electrostatically actuated cantilevers and interferometric deflection detection yielded a Young's modulus of 164 GPa [34.19]. They found the grains in their polysilicon films to be randomly oriented and calculated the Voigt and Reuss bounds to be 163.4–164.4 GPa. This appears to be a very reliable value for randomly oriented polysilicon.

Fracture Toughness and Strength of Polysilicon

Using the device shown in Fig. 34.13a and the specimen shown in Fig. 34.13d, the fracture toughness, K_{Ic}, of polysilicon has been shown to be $1.0 \pm 0.1\,MPa\,m^{1/2}$ [34.32]. Several different polysilicon microstructures were tested, including fine-grained, columnar, and multilayered. Amorphous silicon was also investigated. All of the microstructures displayed the same K_{Ic}. This indicates that, like Young's modulus, fracture toughness is a material property, independent of the material microstructure or the geometry of the sample.

A tensile test, such as shown in Fig. 34.7 but using a sample with indentation-induced sharp pre-cracks, reveals a K_{Ic} of $0.86\,MPa\,m^{1/2}$ [34.33]. The passive, residual stress loaded beams with sharp pre-cracks, shown in Fig. 34.6, gave a K_{Ic} of $0.81\,MPa\,m^{1/2}$ [34.13].

Given that K_{Ic} is a material property for polysilicon, measured fracture strength, σ_{crit}, is related to K_{Ic} by

$$K_{Ic} = c\sigma_{crit}(\pi a)^{1/2} , \quad (34.14)$$

where a is the crack-initiating flaw size, and c is a constant of order unity. The value for c will depend on the exact size, shape, and orientation of the flaw; for a semicircular flaw, c is equal to 0.71 [34.34]. Therefore, any differences in the reported fracture strength of polysilicon will be the result of changes in a.

A good review of the experimental results available in the literature for polysilicon strength is contained in [34.35]. The tensile strength data vary from about 0.5 to 5 GPa. Like many brittle materials, the measured strength of polysilicon is found to obey Weibull statistics. This implies that the polysilicon samples contain a random distribution of flaws of various sizes, and that the failure of any particular specimen will occur at the largest flaw that experiences the highest stress. One consequence of this behavior is that, since larger specimens have a greater probability of containing larger flaws, they will exhibit decreased strengths. More specifically, it was found that the most important geometrical parameter is the surface area of the sidewalls of a polysilicon specimen [34.35]. The sidewalls, as opposed to the top and bottom surfaces, are those surfaces created by etching the polysilicon film. This is not surprising since LPCVD polysilicon films contain essentially no flaws within the bulk, and the top and bottom surfaces are typically very smooth.

As a result, the etching techniques used to create the structures will have a strong impact on the fracture strength of the material. For single-crystal silicon specimens it was found that the choice of etchant could change the observed tensile strength by a factor of two [34.36]. In addition, the bend strength of amorphous silicon was measured to be twice that of polysilicon for specimens processed identically [34.32]. It was found that the reactive ion etching used to fabricate the specimens produced much rougher sidewalls on the polysilicon than on the amorphous silicon.

Fatigue of Polysilicon

Fatigue failure involves fracture after a number of load cycles, when each individual load is not sufficient by itself to generate catastrophic cracking in the material. For ductile materials, such as metals, fatigue occurs due to accumulated damage at the site of maximum stress and involves local plasticity. As a brittle material, polysilicon would not be expected to be susceptible to cyclic fatigue. But fatigue has been observed for polysilicon tensile samples [34.33], polysilicon bend specimens with notches [34.26, 37], and polysilicon bend specimens with sharp cracks [34.38]. The exact origins of the fatigue behavior are still subject to debate. But some aspects of the experimental data are that the fatigue lifetime does not depend on the loading frequency [34.33], the fatigue behavior is affected by the ambient [34.13, 38], and the fatigue depends on the ratio of compressive to tensile stresses seen in the load cycle [34.13].

Friction of Polysilicon

The friction of polysilicon structures has been measured using the techniques described in Sect. 34.2.2. The measured coefficient of friction was found to vary from 4.9 [34.27] to 7.8 [34.28].

34.3.2 Mechanical Properties of Other Materials

As discussed above, of all the materials used for MEMS and NEMS, polysilicon has generated the most interest as well as the most research in mechanical properties characterization. But measurements have been taken on other materials, and these are summarized in this section.

As discussed in Sect. 34.2.2, one advantage of the externally loaded tension test, as shown in Fig. 34.7, is that essentially any material can be tested using this technique. As such, tensile strengths have been measured to be 0.6 to 1.9 GPa for SiO_2 [34.39] and 0.7 to 1.1 GPa for titanium [34.40]. The yield strength for electrodeposited nickel was found to vary from 370 to 900 MPa, depending on annealing temperature [34.16]. In addition, the yield strength was strongly affected by the current density during the electrodeposition process. Both the annealing and current density effects were correlated to changes in the microstructure of the material. Young's moduli were determined to be 100 GPa for titanium [34.40] and 215 GPa for electrodeposited nickel [34.16].

The technique of bending cantilever beams, shown in Fig. 34.8, can also be performed on a variety of materials. The yield strength and Young's modulus of gold were found to be 260 MPa and 57 GPa, respectively, using this method [34.18]. Another technique that can be used with a number of materials is the membrane deflection method, shown in Fig. 34.10. A polyimide membrane gave a residual stress of 32 MPa, a Young's modulus of 3.0 GPa, and an ultimate strain of about four percent [34.20]. Membranes were also fabricated from polycrystalline SiC films with two different grain structures [34.41]. The film with (110)-texture columnar grains had a residual stress of 434 MPa and a Young's modulus of 349 GPa. The film with equiaxed (110)- and (111)-textured grains had a residual stress of 446 MPa and a Young's modulus of 456 GPa.

Other devices used for measuring mechanical properties require more complicated micromachining, namely patterning, etching, and release, in order to operate. These devices are more difficult to fabricate with materials not commonly used as MEMS structural materials. The following examples, however, demonstrate work in this area. The structure shown in Fig. 34.5b was fabricated from Si_xN_y and revealed a nominal fracture toughness of 1.8 MPa m$^{1/2}$ [34.12]. Lateral resonators of the type shown in Fig. 34.11 were processed using polycrystalline SiC, and Young's modulus was determined to be 426 GPa [34.42]. The device shown in Fig. 34.12 was fabricated from polycrystalline germanium and used to measure a bend strength of 1.5 GPa for unannealed Ge and 2.2 GPa for annealed Ge [34.43].

References

34.1 W. Nix: Mechanical properties of thin films, Metall. Trans. A **20** (1989) 2217–2245

34.2 G. G. Stoney: The tension of metallic films deposited by electrolysis, Proc. R. Soc. Lond. A **82** (1909) 172–175

34.3 W. Brantley: Calculated elastic constants for stress problems associated with semiconductor devices, J. Appl. Phys. **44** (1973) 534–535

34.4 A. Ni, D. Sherman, R. Ballarini, H. Kahn, B. Mi, S. M. Phillips, A. H. Heuer: Optimal design of multilayered polysilicon films for prescribed curvature, J. Mater. Sci., in press

34.5 J. A. Floro, E. Chason, S. R. Lee, R. D. Twesten, R. Q. Hwang, L. B. Freund: Real-time stress evolution during $Si_{1-x}Ge_x$ heteroepitaxy: Dislocations, islanding, and segregation, J. Electron. Mater. **26** (1997) 969–979

34.6 X. Li, B. Bhusan: Micro/nanomechanical characterization of ceramic films for microdevices, Thin Solid Films **340** (1999) 210–217

34.7 M. P. de Boer, H. Huang, J. C. Nelson, Z. P. Jiang, W. W. Gerberich: Fracture toughness of silicon and thin film micro-structures by wedge indentation, Mater. Res. Soc. Symp. Proc. **308** (1993) 647–652

34.8 H. Guckel, D. Burns, C. Rutigliano, E. Lovell, B. Choi: Diagnostic microstructures for the measurement of intrinsic strain in thin films, J. Micromech. Microeng. **2** (1992) 86–95

34.9 F. Ericson, S. Greek, J. Soderkvist, J.-A. Schweitz: High sensitivity surface micromachined structures for internal stress and stress gradient evaluation, J. Micromech. Microeng. **7** (1997) 30–36

34.10 L. S. Fan, R. S. Muller, W. Yun, R. T. Howe, J. Huang: Spiral microstructures for the measurement of average strain gradients in thin films, Proc. IEEE Micro Electro Mechanical Systems Workshop, Napa Valley 1990 (IEEE, New York 1990) 177–182

34.11 M. Biebl, H. von Philipsborn: Fracture strength of doped and undoped polysilicon, Proc. Intl. Conf. Solid-State Sensors and Actuators, Stockholm 1995, ed. by S. Middelhoek, K. Cammann (Royal Swedish Academy of Engineering Sciences, Stockholm 1995) 72–75

34.12 L. S. Fan, R. T. Howe, R. S. Muller: Fracture toughness characterization of brittle films, Sens. Actuators A **21–23** (1990) 872–874

34.13 H. Kahn, R. Ballarini, J. J. Bellante, A. H. Heuer: Fatigue failure in polysilicon is not due to simple stress corrosion cracking, Science **298** (2002) 1215–1218

34.14 H. Kahn, N. Tayebi, R. Ballarini, R. L. Mullen, A. H. Heuer: Wafer-level strength and fracture toughness testing of surface-micromachined MEMS devices, Mater. Res. Soc. Symp. Proc. **605** (2000) 25–30

34.15 W. N. Sharpe Jr., B. Yuan, R. L. Edwards: A new technique for measuring the mechanical properties of thin films, J. Microelectromech. Syst. **6** (1997) 193–199

34.16 H. S. Cho, W. G. Babcock, H. Last, K. J. Hemker: Annealing effects on the microstructure and mechanical properties of LIGA nickel for MEMS, Mater. Res. Soc. Symp. Proc. **657** (2001) EE5.23.1–EE5.23.6

34.17 W. Suwito, M. L. Dunn, S. J. Cunningham, D. T. Read: Elastic moduli, strength, and fracture initiation at sharp notches in etched single crystal silicon microstructures, J. Appl. Phys. **85** (1999) 3519–3534

34.18 T. P. Weihs, S. Hong, J. C. Bravman, W. D. Nix: Mechanical deflection of cantilever microbeams: A new technique for testing the mechanical properties of thin films, J. Mater. Res. **3** (1988) 931–942

34.19 B. D. Jensen, M. P. de Boer, N. D. Masters, F. Bitsie, D. A. La Van: Interferometry of actuated microcantilevers to determine material properties and test structure nonidealities in MEMS, J. Microelectromech. Syst. **10** (2001) 336–346

34.20 M. G. Allen, M. Mehregany, R. T. Howe, S. D. Senturia: Microfabricated structures for the in situ measurement of residual stress, Young's modulus, and ultimate strain of thin films, Appl. Phys. Lett. **51** (1987) 241–243

34.21 H. Kahn, S. Stemmer, K. Nandakumar, A. H. Heuer, R. L. Mullen, R. Ballarini, M. A. Huff: Mechanical properties of thick, surface micromachined polysilicon films, Proc. IEEE Micro Electro Mechanical Systems Workshop, San Diego 1996, ed. by M. G. Allen, M. L. Redd (IEEE, New York 1996) 343–348

34.22 L. Kiesewetter, J.-M. Zhang, D. Houdeau, A. Steckenborn: Determination of Young's moduli of micromechanical thin films using the resonance method, Sens. Actuators A **35** (1992) 153–159

34.23 W. C. Tang, T.-C. H. Nguyen, R. T. Howe: Laterally driven polysilicon resonant microstructures, Sens. Actuators A **20** (1989) 25–32

34.24 P. T. Jones, G. C. Johnson, R. T. Howe: Fracture strength of polycrystalline silicon, Mater. Res. Soc. Symp. Proc. **518** (1998) 197–202

34.25 P. Minotti, R. Le Moal, E. Joseph, G. Bourbon: Toward standard method for microelectromechanical systems material measurement through on-chip electrostatic probing of micrometer size polysilicon tensile specimens, Jpn. J. Appl. Phys. **40** (2001) L120–L122

34.26 C. L. Muhlstein, E. A. Stach, R. O. Ritchie: A reaction-layer mechanism for the delayed failure of micron-scale polycrystalline silicon structural films subjected to high-cycle fatigue loading, Acta Mater. **50** (2002) 3579–3595

34.27 M. G. Lim, J. C. Chang, D. P. Schultz, R. T. Howe, R. M. White: Polysilicon microstructures to characterize static friction, Proc. IEEE Micro Electro Mechanical Systems Workshop, Napa Valley 1990 (IEEE, New York 1990) 82–88

34.28 B. T. Crozier, M. P. de Boer, J. M. Redmond, D. F. Bahr, T. A. Michalske: Friction measurement in MEMS using a new test structure, Mater. Res. Soc. Symp. Proc. **605** (2000) 129–134

34.29 J. Yang, H. Kahn, A. Q. He, S. M. Phillips, A. H. Heuer: A new technique for producing large-area as-deposited zero-stress LPCVD polysilicon films: The MultiPoly process, J. Microelectromech. Syst. **9** (2000) 485–494

34.30 E. Chason, B. W. Sheldon, L. B. Freund, J. A. Floro, S. J. Hearne: Origin of compressive residual stress in polycrystalline thin films, Phys. Rev. Lett. **88** (2002) 156103-1–156103-4

34.31 S. Jayaraman, R. L. Edwards, K. J. Hemker: Relating mechanical testing and microstructural features of polysilicon thin films, J. Mater. Res. **14** (1999) 688–697

34.32 R. Ballarini, H. Kahn, N. Tayebi, A. H. Heuer: Effects of microstructure on the strength and fracture toughness of polysilicon: A wafer level testing approach, ASTM STP **1413** (2001) 37–51

34.33 J. Bagdahn, J. Schischka, M. Petzold, W. N. Sharpe Jr.: Fracture toughness and fatigue investigations of polycrystalline silicon, Proc. SPIE **4558** (2001) 159–168

34.34 I. S. Raju, J. C. Newman Jr.: Stress intensity factors for a wide range of semi-elliptical surface cracks in finite-thickness plates, Eng. Fract. Mech. **11** (1979) 817–829

34.35 J. Bagdahn, W. N. Sharpe Jr., O. Jadaan: Fracture strength of polysilicon at stress concentrations, J. Microelectromech. Syst. (2003) 302–312

34.36 T. Yi, L. Li, C.-J. Kim: Microscale material testing of single crystalline silicon: Process effects on surface morphology an dtensile strength, Sens. Actuators A **83** (2000) 172–178

34.37 H. Kahn, R. Ballarini, R. L. Mullen, A. H. Heuer: Electrostatically actuated failure of microfabricated polysilicon fracture mechanics specimens, Proc. R. Soc. Lond. A **455** (1999) 3807–3923

34.38 W. W. Van Arsdell, S. B. Brown: Subcritical crack growth in silicon MEMS, J. Microelectromech. Syst. **8** (1999) 319–327

34.39 T. Tsuchiya, A. Inoue, J. Sakata: Tensile testing of insulating thin films; humidity effect on tensile strength of SiO_2 films, Sens. Actuators A **82** (2000) 286–290

34.40 H. Ogawa, K. Suzuki, S. Kaneko, Y. Nakano, Y. Ishikawa, T. Kitahara: Measurements of mechanical properties of microfabricated thin films, Proc. IEEE Micro Electro Mechanical Systems Workshop (IEEE, New York 1997) 430–435

34.41 S. Roy, C. A. Zorman, M. Mehregany: The mechanical properties of polycrystalline silicon carbide films determined using bulk micromachined diaphragms, Mater. Res. Soc. Symp. Proc. **657** (2001) EE9.5.1–EE9.5.6

34.42 A. J. Fleischman, X. Wei, C. A. Zorman, M. Mehregany: Surface micromachining of polycrystalline SiC deposited on SiO_2 by APCVD, Mater. Sci. Forum **264–268** (1998) 885–888

34.43 A. E. Franke, E. Bilic, D. T. Chang, P. T. Jones, T.-J. King, R. T. Howe, G. C. Johnson: Post-CMOS integration of germanium microstructures, Proc. Intl. Conf. Solid-State Sensors and Actuators (IEE Jpn., Tokyo 1999) 630–637

35. Thermo- and Electromechanics of Thin-Film Microstructures

Applications using thin-film micromechanical structures for actuation and sensing require the coupling of energy between various physical domains. This chapter focuses on two important couplings: thermomechanics and electromechanics. Thermomechanical phenomena is considered in Sect. 35.1 where we describe broad aspects of the deformation characteristics and stress states that arise when dealing with a large class of thin-film microstructures. These include the origin of stresses in multiplayer films and their qualitative evolution through processing and release from the substrate. A basic framework is described for the analysis of the thermo mechanics of multiplayer films, emphasizing linear response. Issues of geometric and material no linearity are then taken up, and equal emphasis is put on the generality of the analysis approach and specific applications. As much as possible, we show comparisons between theoretical predictions and companion experimental results.

A common use of electro mechanics in microsystems involves the application of an electric potential between two electrodes where one is fixed and the other is connected to a deformable elastic structure. The electric potential produces an electric field and an associated electrostatic force that deforms the structure, and in turn alters the electrostatic force, resulting in fully-coupled nonlinear behavior. At some point an instability can occur where the deformable structure snaps into contact with the fixed electrode. This phenomena, called pull-in is often used for switching applications. In Sect. 35.2 we describe the basic electromechanical phenomena using a parallel-plate electrostatic actuator as a reference. We discuss many important phenomena including pull-in, external forcing, stabilization, time response, the effects of dielectric charging, and breakdown of gases in small gaps. We address these phenomena for a wide range of micro mechanical structures including cantilevered beams and plates, torsion ally suspended plates, and zipper actuators with curved electrodes.

35.1 **Thermomechanics of Multilayer Thin-Film Microstructures** .. 1041
- 35.1.1 Basic Phenomena ... 1041
- 35.1.2 A General Framework for the Thermomechanics of Multilayer Films ... 1046
- 35.1.3 Nonlinear Geometry ... 1054
- 35.1.4 Nonlinear Material Behavior ... 1058
- 35.1.5 Other Issues ... 1061

35.2 **Electromechanics of Thin-Film Microstructures** ... 1061
- 35.2.1 Applications of Electromechanics . 1061
- 35.2.2 Electromechanics Analysis ... 1063
- 35.2.3 Electromechanics – Parallel-Plate Capacitor ... 1064
- 35.2.4 Electromechanics of Beams and Plates ... 1066
- 35.2.5 Electromechanics of Torsional Plates ... 1068
- 35.2.6 Leveraged Bending ... 1069
- 35.2.7 Electromechanics of Zipper Actuators ... 1070
- 35.2.8 Electromechanics for Test Structures ... 1072
- 35.2.9 Electromechanical Dynamics: Switching Time ... 1073
- 35.2.10 Electromechanics Issues: Dielectric Charging ... 1074
- 35.2.11 Electromechanics Issues: Gas Discharge ... 1075

35.3 **Summary and Mention of Topics not Covered** ... 1078

References ... 1078

Microsystems rely heavily on thin-film technology: the deposition, patterning, etching, and so on, of multiple film layers to yield micromechanical structures. We do not discuss fabrication techniques here but instead refer the interested reader to many excellent references that describe the fabrication of micromechanical structures by means of thin-film technology [35.1, 2] and, in particular, surface micromachining [35.3]. Regardless of the fabrication process flow, the end result is a thin-film microstructure that is connected to a substrate through one or more *sacrificial* film layers, which must be released to render it freestanding and thus useful. Figure 35.1a illustrates a simple, yet typical in terms of the microstructure and the materials involved, case: a cantilevered beam made of polycrystalline silicon (polysilicon) anchored to a silicon substrate and encased in an SiO_2 sacrificial layer. Also shown is a metal electrode used for electrostatic actuation of the cantilever after release. This is the state of affairs after the thin-film fabrication steps; we refer to it as the as-processed but unreleased condition. To render the beam freestanding, although anchored to the substrate at one end, the sacrificial material must be removed (Fig. 35.1b). For SiO_2, this is usually accomplished by etching in an HF solution, followed by drying. The latter step is important because during wet etching, strong (at the microscale) capillary forces can pull the beam into contact with the substrate. If the microstructure is too compliant, adhesive forces between the microstructure and the substrate can pin the beam to the substrate, rendering it useless. This phenomena is often referred to as stiction, and many recent references describe the phenomena, its impact regarding reliability, and practical ways to overcome it [35.4, 5]. For our purposes, we assume that the microstructure can be successfully released so that it is free from the substrate where desired and anchored where desired.

High stresses can develop in the film layers during processing, and these can vary significantly from layer to layer and even within a single layer. When the sacrificial layers are removed, the stresses redistribute, but at the expense of deformation. This can be detrimental in applications in which planarity is essential, or it can be used advantageously for actuation purposes. Indeed many applications are based on the use of thermally induced bending of bilayer beams and plates that arises due to the difference in thermal expansion coefficients between the film layers. An example is shown in Fig. 35.2 where a micromirror is suspended above the substrate by 2.0-μm-thick gold/polysilicon bilayer beams that position the mirror (or any desired device) and can then be electrostatically actuated to control the position of the micromirror. Upon release, microstructures are typically used for actuation and/or sensing functions. This requires the coupling of the mechanical behavior of the microstructure with various energy domains. These couplings include, among others, thermomechanics, electromechanics, electrothermomechanics, and magnetomechanics. The most common, and perhaps the most easily realizable, is electromechanics, specifically the coupling of electrostatics and mechanics. This coupling has been successfully demonstrated by many applications as both sensors and actuators.

This chapter focuses on the behavior of thin-film microstructures under the action of thermomechanical and electromechanical loadings. Section 35.1 is devoted to the development of stresses and deformation in multilayer thin-film microstructures when subjected to thermomechanical loading. It focuses heavily on linear thermoelastic behavior but describes important issues regarding geometric and material nonlinearity. Section 35.2 is devoted to the electromechanics of microstructures, emphasizing general approaches to modeling and then focusing on the behavior of typ-

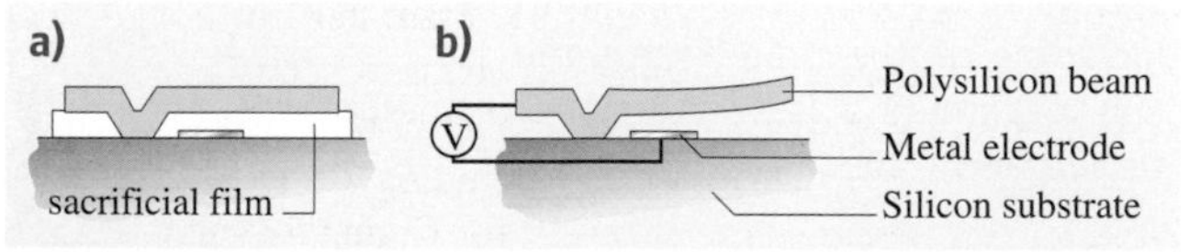

Fig. 35.1a,b Schematic of thin-film microstructures (**a**) attached to a substrate layer, and (**b**) freestanding after release from the substrate by etching of the sacrificial film

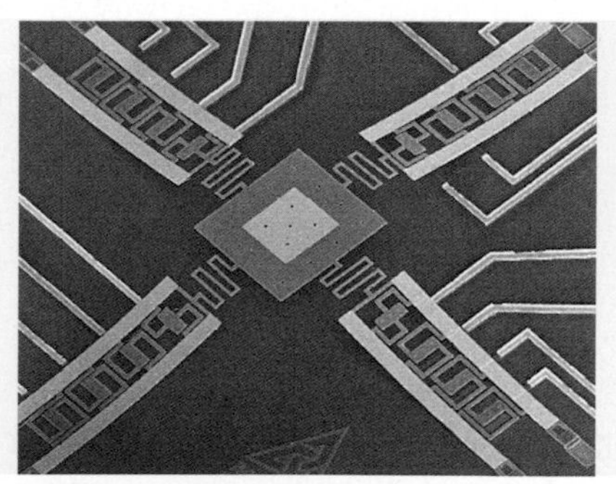

Fig. 35.2 Scanning electron microscope image of a micromirror (300 μm square) supported above the substrate by gold/polysilicon bilayer beams that can be individually actuated electrostatically to control the position of the micromirror

ical classes of electromechanical microstructures. The treatment is certainly not exhaustive, but we hope it is detailed enough to provide the interested reader with an entry-level understanding of the basic phenomena and some guidance regarding where to turn to obtain more in-depth information.

35.1 Thermomechanics of Multilayer Thin-Film Microstructures

In this section we describe aspects of the deformation characteristics and stress states in multilayer thin-film microstructures. Deformation and stress states will be considered for microstructures that are both attached to a substrate and freestanding after release from the substrate. In the modeling formalism that we will describe, the substrate simply serves as another layer. We will refer to microstructures as beams and/or plates, depending on the relevant dimensions. At times we will use the terms loosely, but we will describe the difference of their deformation characteristics in detail. The remainder of this section will be organized as follows: Section 35.1.1 provides an overview of the basic phenomena including the origin of stresses in multilayer films and their qualitative evolution through processing and then release from the substrate. Section 35.1.2 outlines a general framework for the analysis of the thermomechanical behavior of multilayer thin films, emphasizing *linear* response. Issues of *geometric* and *material* nonlinearity are then briefly taken up in Sects. 35.1.3 and 35.1.4, respectively. Section 35.1.5 touches on a number of related and important issues that were not covered in the other sections.

The objectives of this section are to describe the thermomechanical phenomena that arise when dealing with a large class of thin-film microstructures. A basic framework is described for analysis, and equal emphasis is put on the generality of the approach and specific applications. When possible, we show comparisons between theoretical predictions and companion experimental results. To this end, a specific material and geometric system must be chosen, and we use a gold/polysilicon multilayer for most of these cases. We note, though, that the phenomena discussed are not restricted to the gold/polysilicon system but arise in many multilayer film systems.

35.1.1 Basic Phenomena

During the multiple thin-film deposition and etching processes, many complicated mechanisms occur that result in straining of the film layers, and these generally differ from layer to layer. The requirement that the film layers maintain coherency at the interfaces between them means that these strains cannot occur freely, and thus they are constrained. This constraint leads to stresses in the layers and curvature of the multilayer film system; both can have significant practical implications. The mechanisms that lead to internal stresses include chemical reactions, lattice mismatch, and grain growth, among others, and are often called *intrinsic* stresses (see, for example [35.6]). The details of intrinsic stress development in thin films have been the subject of extensive study for both scientific and technological reasons, but they are beyond the scope of our study. In addition, stresses may arise due to thermal expansion mismatch between layers when subjected to a temperature change. Such stresses are often called *extrinsic* stresses. Now, in the unreleased configuration of Fig. 35.1a, a complicated stress distribution exists as a result of the intrinsic stresses. The release process to yield the configuration of Fig. 35.1b obviously alters the stress distribution as well as the state of deformation. In order to gain physical insight, the remainder of this section discusses rather qualitatively the nature of the stress states in the unreleased and released configurations. In subsequent sections details are given regarding analysis of these thermomechanical phenomena.

Transformation Strains, Misfit Strains, and Thin-Film Stresses

The important concepts here are most easily described in terms of a bilayer film system in which both layers are isotropic. Furthermore, they are easier to grasp in the context of a specific example, of which many would suffice. Once this understanding is in place, the generalization of the concepts to multilayer film systems follows naturally. To begin, we define a couple of terms. A *transformation strain* ε^* is a strain that occurs in a solid, but with no accompanying stress. It is thus an inelastic strain. In the literature a number of other terms are used to describe this concept including *stress-free strain* and *eigenstrain*. It can result, for example, from thermal expansion, a crystallographic phase transformation, or a variety of sources. When thermal expansion is the source, ε^* can be specified as $\varepsilon^* = \alpha T$ where

α is the thermal expansion coefficient of the material and T is the temperature change from a reference (usually stress-free) configuration. In general ε^* and α are second-order tensors. Stresses and deformation develop in bilayer film/substrate systems when, due to external or internal sources, each layer wants to undergo a *different* transformation strain, but the requirement of bonding of the films at the interface constrains each film to some degree. The source of the stress and deformation is then the difference in the transformation strains between the layers, i. e., $\Delta\varepsilon = \varepsilon_1^* - \varepsilon_2^*$ where ε_1^* and ε_2^* are the transformation strains of the two layers. For example, if a bilayer is subjected to a uniform temperature change T, then $\Delta\varepsilon = (\alpha_1 - \alpha_2)\,T = \Delta\alpha T$ where α_1 and α_2 are the thermal expansion coefficients of the two layers. $\Delta\varepsilon = \varepsilon_1^* - \varepsilon_2^*$ is termed the *misfit strain*. Its role as the direct source of deformation is obvious and clear for two-layer systems. The concept of misfit strain is not as direct for multilayer systems, and so we simply characterize the behavior in terms of the individual transformation strains of each layer.

It is important to understand the character of the stress distribution and deformation. To make the ideas concrete, consider the bilayer film system with planar dimensions (normal to the film thickness direction) that far exceed the total thickness. The two layers, denoted by 1 and 2, are perfectly bonded along the interface and have different isotropic thermal expansion coefficients and elastic moduli. The bilayer is initially flat, and each layer is stress free. Now assume the bilayer is subjected to a uniform temperature change T, which would lead to equibiaxial ($\varepsilon^* = \varepsilon_{xx}^* = \varepsilon_{yy}^*$, $\gamma_{xy}^* = 0$) transformation strains $\varepsilon_1^* = \alpha_1 T$ and $\varepsilon_2^* = \alpha_2 T$ in the two layers if they were not bonded. Because they are bonded, though, this results in the development of equibiaxial stresses ($\sigma = \sigma_{xx} = \sigma_{yy}$, $\sigma_{xy} = 0$) in the layers, a change of length in the planar directions (in-plane straining) and bending of the bilayer (curvature). This can be understood by the following thought experiment, which is often used in the calculation of film stresses. We will only use it, though, to explain the basics of the resulting deformation and will leave the actual calculations to the next section.

Consider the bilayer system shown schematically in Fig. 35.3a. The film layers are stress free and perfectly bonded. Our discussion will be in the context of biaxial transformation strains and stresses, but the ideas can be immediately applied to situations of uniaxial stress/strain behavior or plane stress or strain situations. We are interested in obtaining the deformation and stress state when the bilayer is subject to a uniform temperature change T. Since the films are stress free in Fig. 35.3a, one can cut the bonds that connect the two layers and separate the films (Fig. 35.3b) without generating any stresses or deforming either layer. Now we (Fig. 35.3c) subject the layers to a uniform temperature change T; each layer will undergo its stress-free biaxial strain, $\varepsilon_1^* = \alpha_1 T$ and $\varepsilon_2^* = \alpha_2 T$ (for simplicity, Fig. 35.3c illustrates the deformation for the case where $\varepsilon_2^* = 0$, $T < 0$, and $\alpha_1 > 0$, but the ideas are the same for more general situations). Now the interfaces, which were originally in perfect registry, are no longer. To bring them into registry, a uniform stress must be applied to the film (Fig. 35.3d) by means of the application of forces along the film edge. At this stage the film is subjected to the uniform biaxial stress $\sigma_1 = M_1\varepsilon_1^*$, where M_1 is the biaxial modulus of the film (see Sect. 35.1.2), the substrate is still stress free, $\sigma_2 = 0$, and the layers are in perfect registry. They can then be reconnected (imagine reattaching the original bonds that were cut in Fig. 35.3b), without generating additional stresses or deforming the layers (Fig. 35.3e). The stress

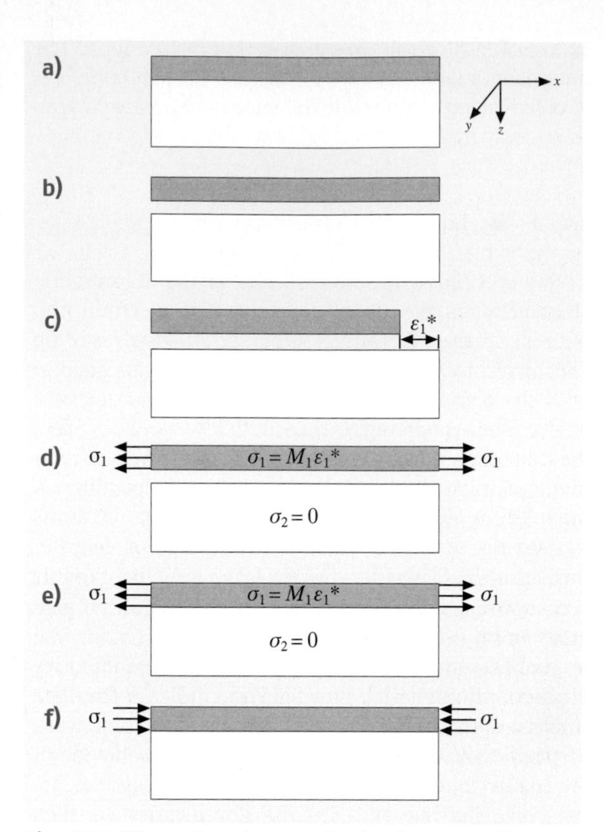

Fig. 35.3 Illustration showing the development of stresses and deformation in a bilayer film/substrate system

state in Fig. 35.3e is the same as that in Fig. 35.3d, the only difference is that the films are now bonded. Now, to recover the solution for the bilayer subjected to a uniform temperature change T, we only have to remove the edge forces. This can be accomplished by applying forces of equal magnitude but opposite sign to the edges (Fig. 35.3f). The solution is then obtained as the superposition of the problems in Fig. 35.3e, which is trivial, and Fig. 35.3f, which is difficult in general. Instead of now trying to solve the problem in Fig. 35.3f, we will simply describe the character of the resulting deformation and stress state. Two aspects are important: the behavior near the free edge where the stress in Fig. 35.3f is applied, and the behavior far away from the free edge. One can appeal to St. Venant's principle to simplify the latter, but more involved analysis is required for the former (see for example [35.7]).

At a distance of ten or so film thicknesses away from the edges, the solution to the problem in Fig. 35.3f can be decomposed, as shown in Fig. 35.4. The uniform force distribution on the edges of the film (Fig. 35.4a) can be replaced by a concentrated force acting at the middle z-coordinate of the film thickness (Fig. 35.4b). The solution to this problem can then be obtained by the superposition of the two problems in Fig. 35.4c and 35.4d, the loadings of which are statically equivalent to that in Fig. 35.4b. That in Fig. 35.4c is a concentrated force applied at the as-yet-unknown neutral surface of the bilayer; it will lead to a uniform in-plane (x-y plane) straining of the bilayer. That in Fig. 35.4d is a moment of magnitude Pd, where d is the distance between the z-coordinates of the middle of the film and the neutral surface of the bilayer; it will lead to bending of the bilayer. If there were also a film on the bottom of the substrate so that the layered film was symmetric in material properties and geometry (film thicknesses) in the z-direction, then no moment would result, and the layered system would not bend but would only strain in-plane.

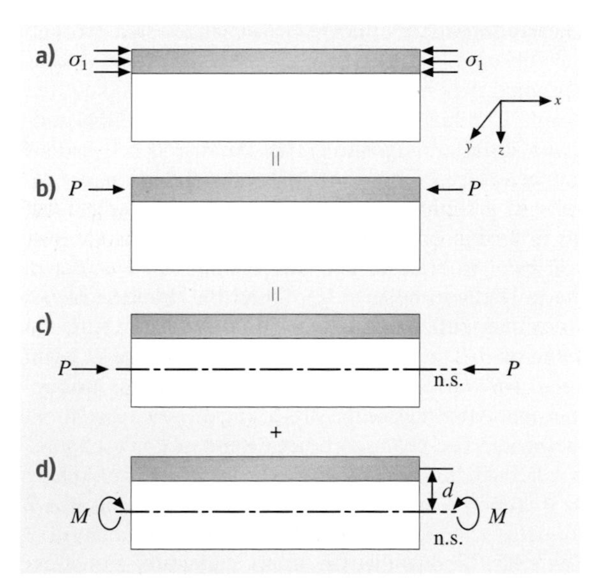

Fig. 35.4 Decomposition of the edge-loading problem into the superposition of a force and moment applied to the multilayer. Away from the edge, the force gives rise to uniform straining, and the moment gives rise to bending

It is probably obvious that very near the edge, the stress state is quite complicated. In fact, in the context of linear elasticity, a stress singularity generally exists at the intersection of the free edge and the interface between the two layers. This singular stress state is similar to that which exists at a crack tip, but the strength of the singularity differs depending on the mismatch in elastic constants of the materials [35.8, 9]. The region within which this stress state exists depends on the film thickness, and for very thin films it is often so small that it is insignificant. It has recently been shown, however, that in some cases it can be appreciable and in fact can be the primary source of failure [35.10]. As described in detail by *Hui* et al. [35.7], aside from the singular stresses, the stress state near the edges contains highly localized shear and normal stress (perpendicular to the interface) components. By St. Venant's principle these stress components decay over a rather short distance from the edge, but they can play a significant role in the durability of the film/substrate system as they can lead to delamination of the interface or cracking into one of the layers. The character of these stress components at the interface is shown qualitatively in Fig. 35.5. Note that in order to satisfy equilibrium, the normal stress (often called the peel stress) changes sign from tensile to compressive. Furthermore, since these stress components are highly localized at the edge, in the interior of the system the interface is free of normal or shear stresses. In this chapter we will not further consider these localized stresses but instead will concern ourselves with the stresses away from the edges and the companion deformation.

Our discussion has been based on the assumption that transformation strains are constant throughout the cross section of each film. In microsystems applications, though, films will often have transformation strains that vary through the thickness. This will lead to bending of the film after it is released from the substrate. Transformation strains that vary through the thickness, i. e.,

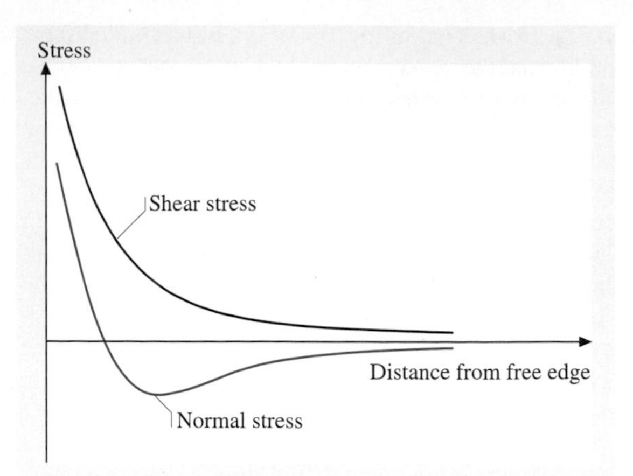

Fig. 35.5 Character of the normal and shear stress distribution at the film/substrate interface near the free edge

$\varepsilon^*(z)$, can be easily accommodated in the description of Fig. 35.3. It is altered in that now in Fig. 35.3c an in-plane strain results, the magnitude of which is given by the through-thickness average of $\varepsilon^*(z)$, but in addition, the film bends with constant curvature. Then, both a force (uniform stress) and moment (linearly varying stress) must be applied in Fig. 35.3d to flatten the film and lengthen it so that it is again in perfect registry with the substrate and can then be rebonded [35.11].

This description provides the starting point for understanding the effects of transformation strains on stresses and deformation in a multilayer thin-film system. It is an accurate picture of the state of affairs before release of a film from the substrate to yield a freestanding microstructure. In the next sections we will discuss the changes in the deformation that occur upon release as the constraint of the substrate is removed.

Single-Layer Beams

To keep things simple, in this and the following section we focus on *beams*, which have one nonzero normal stress component, that along the beam axis. In subsequent sections we will discuss the limitations of approximating such structures as beams and proceed with a more appropriate development in terms of *plates*. Consider the beam shown in Fig. 35.1. For simplicity we assume that the sacrificial material is much thinner than the substrate, has the same elastic moduli as the substrate, and so before release is mechanically equivalent to the substrate. In this way we can describe the stress distribution before release as that of a two-layer beam. To make the discussion more tangible, consider a specific but realistic case of a polysilicon beam on a silicon substrate. The top layer of thickness 1.5 μm is the polysilicon film, and the bottom layer of thickness 500 μm is the silicon substrate (including the very thin sacrificial layer that will be removed). Assume that the polysilicon film has an intrinsic mean stress S and a linear stress gradient β (we can easily consider a nonlinear stress gradient, but the simple linear gradient suffices to demonstrate the concepts involved). This intrinsic stress is interpreted as the stress state in the film if it were free, i. e., a transformation stress. It could be expressed in terms of a transformation strain $\varepsilon^*(z)$, but since reference is often made to thin-film stresses we will carry out this and the following example using this transformation stress rather than a transformation strain. We consider three cases to isolate the effects of the mean stress and the stress gradient: (a) $S = -10\,\text{MPa}$, $\beta = 0$, (b) $S = 0$, $\beta = 0.8\,\text{MPa/μm}$, and (c) $S = -10\,\text{MPa}$, $\beta = 0.8\,\text{MPa/μm}$. In all cases, the intrinsic stress in the substrate is zero.

In Table 35.1 we give calculated results using a simple beam theory (to be discussed in the following sections) for the three scenarios. Case (a) $S = -10\,\text{MPa}$, $\beta = 0$ results in a slight contraction and a negative curvature (bent downward) of the film/substrate system prior to release. The compressive stress in the polysilicon beam is uniform but slightly lower than the value of S. The stress induced in the substrate is small and varies linearly through its thickness. It changes sign through the substrate, a necessary requirement to satisfy equilibrium. After release the stress is zero throughout the beam. To relax the stress, the beam extends (the mid-plane strain ε^0 is positive), but because $\beta = 0$ it does not bend. Case (b) $S = 0\,\text{MPa}$, $\beta = 0.8\,\text{MPa/μm}$ results in a slight positive curvature (bent upward) but no extension or contraction of the film/substrate system prior to release. The mean compressive stress in the polysilicon beam is zero, and the gradient of the stress distribution through the film is decreased from the value of β. The stress induced in the substrate is again small and varies linearly through its thickness, changing sign. After release the stress is again zero throughout the beam. The beam does not extend or contract since $S = 0$, but it bends, assuming a curvature $\kappa = \beta/E$ where E is Young's modulus of polysilicon (see Sect. 35.1.2 for details regarding this result). Careful examination of Table 35.1 shows that the strain, curvature, and stress for case (c) $S = -10\,\text{MPa}$, $\beta = 0.8\,\text{MPa/μm}$ can be obtained as the superposition of the results of cases (a) and (b). Specifically, after release the beam extends and bends.

Table 35.1 Stresses, curvature, and strains in a film/substrate system consisting of a 1.5-μm-thick polysilicon film and a 500-μm-thick silicon film. The elastic moduli of the film and the substrate are assumed equal, and the film has an intrinsic mean stress and stress gradient. The sign convention for curvature is that (−) indicates bending downward and (+) indicates bending upward

		(a) $S=-10$ MPa $\beta=0$ MPa/μm		(b) $S=0$ MPa $\beta=0.8$ MPa/μm		(c) $S=-10$ MPa $\beta=0.8$ MPa/μm	
		Unreleased	**Released**	**Unreleased**	**Released**	**Unreleased**	**Released**
	ε^0 (με)	0.0624	62.5	0	0	0.0624	62.5
	κ (m^{-1})	-7.5×10^{-4}	0	5.0×10^{-9}	5.0	-7.5×10^{-4}	5.0
σ (MPa)	Top	−9.96	0	0.2	0	−9.76	0
Polysilicon Film	Bottom	−9.96	0	−0.2	0	−10.16	0
σ (MPa)	Top	0.04	0	-2.0×10^{-7}	0	0.04	0
Silicon Substrate	Bottom	−0.02	0	2.0×10^{-7}	0	−0.02	0

These results illustrate the behavior of beams before and after release when they contain a mean stress and/or a linear stress gradient. The results also suggest natural means to attempt to measure the mean stress and the gradient. Test structures that are sensitive to length changes of beams after release can be used to characterize mean stress. For example, fracture of arrays of beams of varying width upon release can be used to estimate tensile stress, and fixed-fixed beams that buckle can be used to estimate compressive stress. The stress gradient in a beam can be determined by measuring its curvature after release. A survey of test structures suitable for these purposes can be found in [35.12].

Bilayer Beams

While stresses in single-layer microstructures are important for many applications, multilayer film microstructures abound in microsystems technology, and the development of stresses and deformation in them is also important. In order to convey the basic ideas, we consider an example of a two-layer beam fixed to a substrate in the same manner as that in the previous section. Specifically we consider the layered structure shown in the inset of Fig. 35.6: a polysilicon beam of thickness 1.5 μm attached to a silicon substrate of thickness 500 μm, covered with a gold film of thickness 0.5 μm. Photos of actual beams of this form are shown in Fig. 35.7 where the length of the longest beam shown is 600 μm, and the beams are 50 μm wide. Again we assume that the sacrificial material is very thin and has the same elastic moduli as the substrate, and so before release it is mechanically equivalent to the substrate. We assume that the polysilicon film has an intrinsic mean stress S and a linear stress gradient β as in the previous example, and that the gold film has an intrinsic mean stress but no stress gradient. We consider the reasonably typical stress states: $S=-10$ MPa and $\beta=0.8$ MPa/μm in the polysilicon beam and $S=50$ MPa in the gold film. The resulting curvature and stress distribution before and after release are shown in Fig. 35.6. They show that before release the stresses in the polysilicon and gold layer are nearly uniform, compared to the range of stresses throughout the layers. The multilayer experiences a positive curvature; recall that the system consisting of just the polysilicon

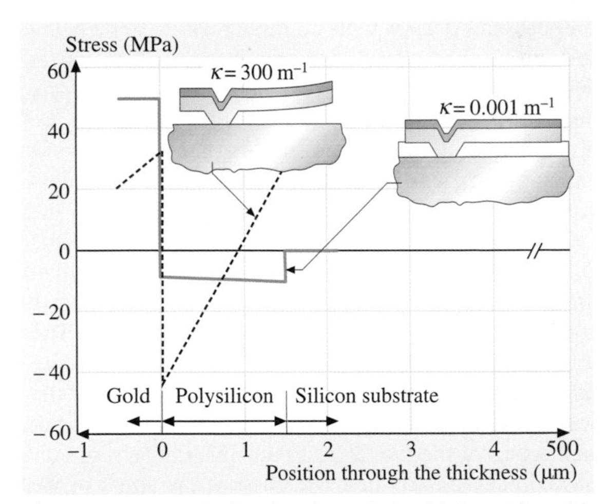

Fig. 35.6 Stress distribution through the thickness for a bilayer film consisting of a 0.5-μm-thick gold film on a 1.5-μm-thick polysilicon film subjected to an internal stresses and stress gradients. Stress distributions are shown before and after release from a 500-μm-thick silicon

Fig. 35.7 Scanning electron microscope image of gold/polysilicon bilayer cantilever beams as released from the substrate. All beams have a 0.5-μm-thick gold film, but the beams on the *left* have a 1.5-μm-thick polysilicon film, and those on the *right* a 3.5-μm-thick polysilicon layer

film and the substrate experienced a negative curvature of about the same magnitude. Thus the tensile stress in the gold is significant enough to drive the curvature from negative (bending down) to positive (bending up).

After release from the substrate, the gold/polysilicon beam contracts and bends substantially to a curvature of about $300\,m^{-1}$. This should be compared with $5\,m^{-1}$ for the polysilicon beam with a stress gradient. For a beam 300 μm long cantilevered from one end as shown in Fig. 35.6, this amounts to a tip deflection of about 13 μm. The stress states in both the gold and polysilicon are significantly altered upon release. Perhaps most significantly, large stress gradients exist through the thickness of each layer. Starting at the top of the gold film, the tensile stress increases through the gold thickness. Due to the discontinuity in elastic modulus and the residual stress and stress gradients, there is a jump in the stress of about −77 MPa at the gold/polysilicon interface. The stress is compressive at the top of the polysilicon film layer and increases linearly through the thickness of the polysilicon film, taking on its maximum tensile value at the bottom of the two-layer beam. In summary, the release process results in a reduction of the stress in the gold, an increase of the stress in the polysilicon, stress gradients in both layers, a small contraction, and a large curvature change.

These examples are meant to convey the basic phenomena and give a feel for the magnitudes of the quantities involved. The actual numerical results are based on calculations performed using a relatively simple multilayer beam theory and are reasonable estimates. Details regarding the analysis will be given in the following section, including a discussion of the suitability of the simple beam theory.

35.1.2 A General Framework for the Thermomechanics of Multilayer Films

Numerous studies have elucidated the basic thermomechanical response of multilayer material systems when subjected to temperature changes or other sources of transformation strains between the layers. These have come in the context of many technological applications, the most common being structural composite materials [35.13, 14] and thin film/substrate systems for microelectronics [35.15–22], which are directly applicable to the analysis of multilayer films.

The General Multilayer Film

When a layered film is subjected to transformation strains, two aspects of deformation play a primary role: straining of the midplane and bending. When the transverse deflections due to bending are of prime importance, as is often the case, one way to broadly characterize the deformation response, especially for plates with relatively large in-plane dimensions as compared to their thickness, is in terms of the average curvature developed as a function of the transformation strains. Formally the curvature is a second-rank tensor, and for the type of layered film problems considered here it can be wholly described by the two principal curvature components, e.g., in the x- and y- directions, κ_x and κ_y. The curvature is a pointwise quantity, meaning it varies from point to point over the in-plane dimensions of the plate. In terms of the average curvature variation as a function of misfit strain, three deformation regimes have been identified [35.16–19, 22, 23]. The first regime is a linear relation between the average curvature and temperature change where $\kappa_x = \kappa_y$, i. e., the average curvature is spherically symmetric (when the materials are isotropic). This symmetric deformation would not exist if the material properties were anisotropic. This deformation regime is characterized by both small transverse displacements and rotations, and so conventional thin-plate theory adequately describes the deformation. It is the subject of the analysis in this section. The second and third regimes, which arise due to geometric nonlinearity, will be taken up in Section 35.1.3.

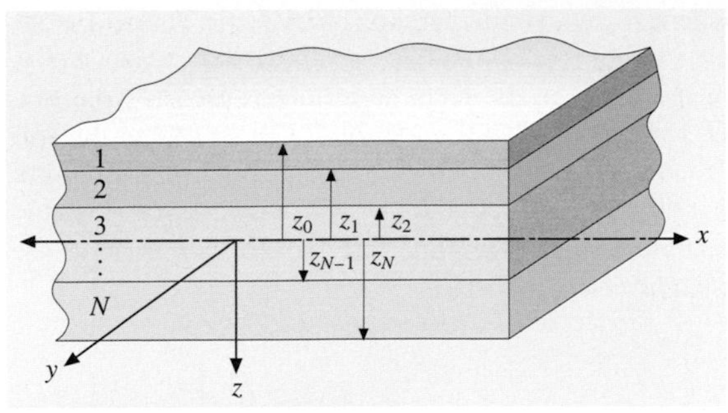

Fig. 35.8 Multilayer film system with a definition of the parameters involved in the analysis

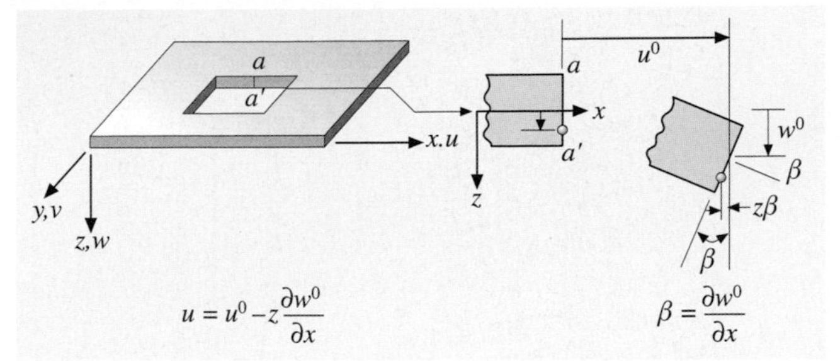

Fig. 35.9 Kinematics of the deformation of the thin-film multilayer

Figure 35.8 shows a schematic of the geometry considered in this study; a multilayer film microstructure with lateral dimensions that far exceed the total thickness. As such we ignore the effects of the free edges and focus on a representative point in the layered film that lies at least a distance of a few film thicknesses from the edge. The layers are numbered $1, 2, \dots, N$ where 1 is the top layer and N is the bottom layer. Each layer has a thickness t_i $(i = 1 \rightarrow N)$, and the total thickness of the microstructure is t. Each of the film layers consists of an isotropic material with a Young's modulus E_i, Poisson's ratio, ν_i, and may undergo a transformation strain $\varepsilon_i^*(z)$, which may vary with position z through the film thickness. The source of $\varepsilon_i^*(z)$ may be arbitrary, and the effect of multiple sources can be considered as the sum of the individual ones. A common source, and one that we will focus on, is thermal expansion upon a uniform temperature change. In this case $\varepsilon_i^*(z) = \varepsilon_i^* = \alpha_i T$ where α_i is the thermal expansion coefficient and T is the temperature change from a reference state. In addition, each layer may contain an internal stress $\hat{\sigma}(z)$, which also may vary with position z through the thickness. For simplicity we focus our results on isotropic layers, but the framework is applicable to anisotropic behavior of each layer. As a result of the difference in stress-free strains and/or internal stresses between each layer, stresses are developed in the layers and the structure deforms; our intent is to compute these.

Analysis

The analysis presented here is somewhat standard in the treatment of laminated composite plates but has not been often used in the analysis of multilayer thin film problems for microelectronics and/or microelectromechanical systems. Detailed accounts of parts of the theory can be found in texts on composite materials (see for example [35.24,25]), and so only a brief account is presented here.

The displacements u and v at any point z through the thickness of the laminate are assumed to be described by the classical Kirchhoff hypothesis. As illustrated in Fig. 35.9, the Kirchhoff hypothesis states that after loads are applied to the layered film, the line a–a' which is initially straight and normal to the geometric midplane remains straight, normal to the deformed geometric midplane, and its length is unchanged. In other words, the normal does not deform but only translates and rotates. As a consequence of this assumption, the displacements of any point in the layered film are given by:

$$\begin{aligned} u &= u^0 - z\frac{\partial w^0}{\partial x} \\ v &= v^0 - z\frac{\partial w^0}{\partial y} \\ w &= w^0 , \end{aligned} \tag{35.1}$$

where u^0, v^0, and w^0 are the displacements of the geometric midplane at $z = 0$. Note that in addition to this explicit dependence of the displacements on z, they are generally functions of x and y.

The nonzero normal and shear strains are then:

$$\begin{aligned} \varepsilon_x &= \frac{\partial u}{\partial x} = \frac{\partial u^0}{\partial x} - z\frac{\partial^2 w^0}{\partial x^2} = \varepsilon_x^0 + z\kappa_x \\ \varepsilon_y &= \frac{\partial v}{\partial y} = \frac{\partial v^0}{\partial x} - z\frac{\partial^2 w^0}{\partial x^2} = \varepsilon_y^0 + z\kappa_y \\ \gamma_{xy} &= \frac{\partial u}{\partial y} + \frac{\partial v}{\partial x} = \frac{\partial u^0}{\partial y} + \frac{\partial v^0}{\partial x} - 2z\frac{\partial^2 w^0}{\partial x \partial y} \\ &= \gamma_{xy}^0 + z\kappa_{xy} \end{aligned} \tag{35.2a}$$

or in matrix form:

$$\{\varepsilon\} = \left\{\varepsilon^0\right\} + z\{\kappa\} , \tag{35.2b}$$

where:

$$\{\varepsilon\} = \begin{Bmatrix} \varepsilon_x \\ \varepsilon_y \\ \gamma_{xy} \end{Bmatrix} \left\{\varepsilon^0\right\} = \begin{Bmatrix} \varepsilon_x^0 \\ \varepsilon_y^0 \\ \gamma_{xy}^0 \end{Bmatrix} = \begin{Bmatrix} \dfrac{\partial u^0}{\partial x} \\ \dfrac{\partial v^0}{\partial y} \\ \dfrac{\partial u^0}{\partial y} + \dfrac{\partial v^0}{\partial x} \end{Bmatrix}$$

$$\{\kappa\} = \begin{Bmatrix} \kappa_x \\ \kappa_y \\ \kappa_{xy} \end{Bmatrix} = \begin{Bmatrix} -z\dfrac{\partial^2 w^0}{\partial x^2} \\ -z\dfrac{\partial^2 w^0}{\partial y^2} \\ -2z\dfrac{\partial^2 w^0}{\partial x \partial y} \end{Bmatrix} . \tag{35.3}$$

In (35.3), $\left\{\varepsilon^0\right\}$ is the midplane strain and $\{\kappa\}$ is the midplane curvature. As with the displacements, $\left\{\varepsilon^0\right\}$ and $\{\kappa\}$ are generally functions of x and y.

The stress-strain relations for each layer are:

$$\begin{Bmatrix} \sigma_x(z) \\ \sigma_y(z) \\ \sigma_{xy}(z) \end{Bmatrix}_k = \begin{bmatrix} Q_{11} & Q_{12} & Q_{16} \\ Q_{12} & Q_{22} & Q_{26} \\ Q_{16} & Q_{26} & Q_{66} \end{bmatrix}_k \times \left[\begin{Bmatrix} \varepsilon_x \\ \varepsilon_y \\ \gamma_{xy} \end{Bmatrix} - \begin{Bmatrix} \varepsilon_x^*(z) \\ \varepsilon_y^*(z) \\ \gamma_{xy}^*(z) \end{Bmatrix}_k \right] + \begin{Bmatrix} \hat{\sigma}_x(z) \\ \hat{\sigma}_y(z) \\ \hat{\sigma}_{xy}(z) \end{Bmatrix}_k \tag{35.4a}$$

or in compact form:

$$\{\sigma(z)\}_k = \left\{\overline{Q}\right\}_k \left[\{\varepsilon\} - \{\varepsilon(z)\}_k^*\right] + \left\{\hat{\sigma}(z)\right\}_k , \tag{35.4b}$$

where $k = 1 \rightarrow N$ denotes the layer number. Equation (35.4) is written in a global coordinate system for the layered structure, (x, y, z). As such, $\left[\bar{Q}\right]_k$ are the (generally anisotropic) elastic stiffness coefficients of the kth layer in the global coordinate system. They are obtained by transformation (see [35.24, 25], for example) of the stiffnesses, $[Q]_k$, from a natural coordinate system (x_1, x_2, x_3). For orthotropic materials the nonzero components of $[Q]$ for each layer are,

$$[Q] = \begin{bmatrix} \dfrac{E_1}{1-\nu_{12}\nu_{21}} & \dfrac{\nu_{12}E_2}{1-\nu_{12}\nu_{21}} & 0 \\ \dfrac{\nu_{21}E_1}{1-\nu_{12}\nu_{21}} & \dfrac{E_2}{1-\nu_{12}\nu_{21}} & 0 \\ 0 & 0 & G_{12} \end{bmatrix} , \tag{35.5a}$$

where E_i, ν_{ij}, and G_{ij} are the orthotropic Young's moduli, Poisson's ratios, and the shear modulus, respectively, of the layer, in the local coordinate system. Note that $Q_{12} = Q_{21}$, and so the equality of these expressions in (35.5a) shows the reciprocal relationship between E_i and ν_{ij}; only three of the four are independent. For isotropic materials $\left[\bar{Q}\right] = [Q]$, and the nonzero components for each layer are:

$$\left[\bar{Q}\right] = [Q] = \begin{bmatrix} \dfrac{E}{1-\nu^2} & \dfrac{\nu E}{1-\nu^2} & 0 \\ \dfrac{\nu E}{1-\nu^2} & \dfrac{E}{1-\nu^2} & 0 \\ 0 & 0 & G \end{bmatrix} \tag{35.5b}$$

In (35.4) $\left\{\hat{\sigma}(z)\right\}_k$ is the residual internal stress in the layer and may be an arbitrary function of z, i. e., a stress gradient may exist within each layer; $\{\varepsilon\}$ is the total strain, $\{\varepsilon^*\}_k$ is the inelastic transformation strain, and the term in rectangular brackets is the elastic strain. The stresses and strains are all defined in the (x, y, z) coordinate system. Substituting (35.2) into (35.4) yields the stress distribution through the thickness:

$$\{\sigma(z)\}_k = \left\{\overline{Q}\right\}_k \left[\left\{\varepsilon^0\right\} + z\{\kappa\} - \left\{\varepsilon^*(z)\right\}_k\right] + \left\{\hat{\sigma}(z)\right\}_k . \tag{35.6}$$

For the layered film system we define a resultant force $\{N\}$ and moment $\{M\}$ per unit length as:

$$\begin{Bmatrix} N_x \\ N_y \\ N_{xy} \end{Bmatrix} = \{N\} = \int_{-t/2}^{t/2} \{\sigma(z)\}_k \, \mathrm{d}z$$
$$\begin{Bmatrix} M_x \\ M_y \\ M_{xy} \end{Bmatrix} = \{M\} = \int_{-t/2}^{t/2} \{\sigma(z)\}_k \, z \, \mathrm{d}z . \tag{35.7}$$

For example, N_x is the force in the x-direction per unit length in the y-direction. $\{N\}$ and $\{M\}$ have units of $[F/L]$, $[F \cdot L/L]$, respectively.

Substituting (35.6) into (35.7) and breaking the integrals through the thickness into sums of integrals over each layer yields:

$$\{N\} = [A]\left\{\varepsilon^0\right\} + [B]\{\kappa\} - \left\{N^\varepsilon\right\} + \left\{N^\sigma\right\}$$
$$\{M\} = [B]\left\{\varepsilon^0\right\} + [D]\{\kappa\} - \left\{M^\varepsilon\right\} + \left\{M^\sigma\right\} , \tag{35.8}$$

where:

$$[A] = \int_{-t/2}^{t/2} \{\overline{Q}\}_k \, \mathrm{d}z = \sum_{k=1}^{N} \{\overline{Q}\}_k (z_k - z_{k-1})$$

$$[B] = \int_{-t/2}^{t/2} \{\overline{Q}\}_k z \, \mathrm{d}z = \frac{1}{2} \sum_{k=1}^{N} \{\overline{Q}\}_k \left(z_k^2 - z_{k-1}^2\right)$$

$$[D] = \int_{-t/2}^{t/2} \{\overline{Q}\}_k z^2 \, \mathrm{d}z = \frac{1}{3} \sum_{k=1}^{N} \{\overline{Q}\}_k \left(z_k^3 - z_{k-1}^3\right)$$

$$\{N^\varepsilon\} = \int_{-t/2}^{t/2} \{\overline{Q}\}_k \{\varepsilon^*(z)\}_k \, \mathrm{d}z$$

$$\{N^\sigma\} = \int_{-t/2}^{t/2} \{\hat{\sigma}(z)\}_k \, \mathrm{d}z$$

$$\{M^\varepsilon\} = \int_{-t/2}^{t/2} \{\overline{Q}\}_k \{\varepsilon^*(z)\}_k z \, \mathrm{d}z$$

$$\{M^\sigma\} = \int_{-t/2}^{t/2} \{\hat{\sigma}(z)\}_k z \, \mathrm{d}z \,. \qquad (35.9)$$

In the terminology of laminated composite materials, $[A]$, $[B]$, and $[D]$ are the extensional, coupling, and bending constants and are functions of the elastic moduli of each layer and the arrangement of the layers through the thickness. $[A]$, $[B]$, and $[D]$ have units of $[F/L]$, $[F]$, and $[F \cdot L]$, respectively. Physically, $[A]$ describes the connection between in-plane forces and straining of the midplane, $[D]$ describes the connection between moments and curvature, and $[B]$ connects moments to midplane strain and forces to curvature. If both the geometry and material properties, including $\varepsilon^*(z)$ and $\hat{\sigma}(z)$, of the layered film system are symmetric about $z = 0$, then $[B] = [0]$ and there is no coupling between in-plane straining and bending. Note that $[A]$, $[B]$, and $[D]$ are 3×3 matrices, as are $[\bar{Q}]_k$, while all other terms a 3×1 column vectors of the form of (35.3) and (35.4).

Equation (35.8) can be written in compact form as:

$$\begin{Bmatrix} \{N\} \\ \{M\} \end{Bmatrix} = \begin{bmatrix} [A] & [B] \\ [B] & [D] \end{bmatrix} \begin{Bmatrix} \{\varepsilon^0\} \\ \{\kappa\} \end{Bmatrix} - \begin{Bmatrix} \{N^\varepsilon\} \\ \{M^\varepsilon\} \end{Bmatrix} + \begin{Bmatrix} \{N^\sigma\} \\ \{M^\sigma\} \end{Bmatrix} . \qquad (35.10)$$

Equation (35.10) is most easily implemented in matrix form as a 6×6 system of equations that hold at each (x, y) coordinate. These are the constitutive equations for the layered film, incorporating the Kirchoff hypothesis and the strain-displacement relations. In order to completely describe the response of a layered film due to external or internal loads, (35.10) must be supplemented with the stress equilibrium equations and appropriate boundary conditions. In general the solution of these equations for a specified multilayer film geometry and loading is a complex undertaking and recourse is often taken to numerical methods such as the finite element method (FEM). Alternatively, the equations can be formulated using an energy approach that will be advantageous when we consider geometric nonlinearity in Section 35.1.3.

Fortunately, using this formulation we can solve a number of important problems simply using (35.10). For example, if there are no externally applied loads then $\{N\} = \{M\} = \{0\}$ and (35.10) can be inverted to yield:

$$\begin{Bmatrix} \{\varepsilon^0\} \\ \{\kappa\} \end{Bmatrix} = \begin{bmatrix} [A] & [B] \\ [B] & [D] \end{bmatrix}^{-1} \times \left\{ \begin{Bmatrix} \{N^\varepsilon\} \\ \{M^\varepsilon\} \end{Bmatrix} - \begin{Bmatrix} \{N^\sigma\} \\ \{M^\sigma\} \end{Bmatrix} \right\} . \qquad (35.11)$$

Recall that all terms on the right-hand side are functions of the elastic moduli, the known transformation strains and stresses, and the geometrical arrangement of the layers. Once the midplane strain and curvature are computed using (35.11), the stress distribution in each layer can be computed using (35.6).

Special Cases

Here we apply (35.11) to yield some simple, yet important, results. Before doing so, it is useful to explicitly express the form of the transformation stress distribution $\{\hat{\sigma}(z)\}_k$. To this end we restrict $\{\hat{\sigma}(z)\}_k$ to be a linearly varying function through each layer. This can be written as:

$$\{\hat{\sigma}(z)\}_k = \{S\}_k - \{\beta\}_k \left(z - \frac{z_{k-1} + z_k}{2}\right) , \qquad (35.12)$$

where $\{S\}_k$ is the average stress in layer k and $\{\beta\}_k$ is the stress gradient as shown in Fig. 35.10. For this form of the stress, $\{N^\sigma\}$ and $\{M^\sigma\}$ can be expressed

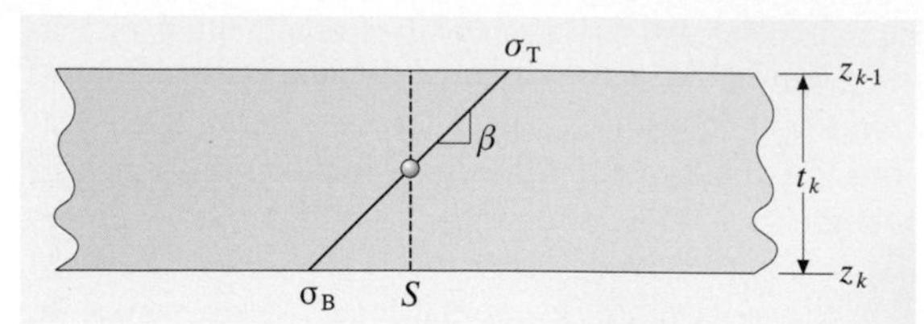

Fig. 35.10 Schematic showing a linear stress distribution characterized by a mean stress S and linear through-thickness stress gradient β

as:

$$\{N^\sigma\} = \sum_{k=1}^{N} \{S\}_k (z_k - z_{k-1})$$

$$\{M^\sigma\} = \frac{1}{2}\sum_{k=1}^{N} \{S\}_k \left(z_k^2 - z_{k-1}^2\right) - \frac{1}{12}\sum_{k=1}^{N} \{\beta\}_k \left(z_k^3 - z_{k-1}^3\right) . \tag{35.13}$$

Furthermore, it is useful to specify the form of $\{\varepsilon^* (z)\}$ when its source is thermal expansion during a uniform temperature change T from a reference temperature. In this case:

$$\left\{\varepsilon^* (z)\right\}_k = \left\{\varepsilon^*\right\}_k = \{\alpha\}_k T . \tag{35.14}$$

For this form of the stress-free strain, $\{N^\varepsilon\}$ and $\{M^\varepsilon\}$ can be expressed as:

$$N^\varepsilon = \int_{-t/2}^{t/2} \left\{\overline{Q}\right\}_k \{\alpha\}_k T \,\mathrm{d}z = \Delta T \sum_{k=1}^{N} \left\{\overline{Q}\right\}_k \{\alpha\}_k (z_k - z_{k-1})$$

$$M^\varepsilon = \int_{-t/2}^{t/2} \left\{\overline{Q}\right\}_k \{\alpha\}_k Tz \,\mathrm{d}z = \frac{\Delta T}{2} \sum_{k=1}^{N} \left\{\overline{Q}\right\}_k \{\alpha\}_k \left(z_k^2 - z_{k-1}^2\right) . \tag{35.15}$$

In all of the examples described in this section we will consider only cases in which all layers have elastic constants, thermal expansion properties, and misfit strains that are isotropic. As a result, $A_{16} = A_{26} = B_{16} = B_{26} = D_{16} = D_{26} = 0$. The ramifications of this are that the midplane shear strains $\gamma_{xy}^0 = 0$ and the twist curvatures $\kappa_{xy} = 0$. In addition, the midplane strains and curvatures are equibiaxial, i. e., $\varepsilon_x^0 = \varepsilon_y^0 = \varepsilon^0$ and $\kappa_x = \kappa_y = \kappa$. So in the remainder of this section this equibiaxial deformation state is understood, and we refer to it simply by ε^0 and κ. While this is the case in the examples presented in this section, it is not the case when material anisotropy exists, the film is patterned in lines instead of as a blanket, or large deflections occur which result in geometric nonlinearity. These issues will be discussed briefly in later sections.

Single-Layer Plate. Consider a single-layer film with an internal biaxial mean stress $S = S_{11} = S_{22}$ and stress gradient $\beta = \beta_{11} = \beta_{22}$ that is subjected to a uniform temperature change T. This scenario models the deformation of a film after release from a substrate when subjected to a temperature change where before release the film had an internal stress S, and stress gradient β. In this case, after significant manipulation (35.11), with (35.12) and (35.13), yields:

$$\kappa = \frac{\beta}{M} \tag{35.16}$$

$$\varepsilon^0 = -\frac{S}{M} + \alpha T , \tag{35.17}$$

where $M = \frac{E}{1-\upsilon}$ is the biaxial modulus. As previously discussed, after release there is straining due to S and $\varepsilon^* = \alpha T$, and the plate bends with a biaxial curvature κ. The result of (35.16), and especially the simplification that results for beams (simply obtained by setting $\nu = 0$ and interpreting κ and ε^0 as the uniaxial curvature and midplane strain in the direction of the beam length) is commonly used in conjunction with measurements of the curvature after release to determine the stress gradient in a film.

Bilayer Plate. Another simple yet practically important special case is that of two isotropic layers of thicknesses t_1 and t_2, subject only to a uniform temperature change T. In this case, (35.11), with (35.12) and (35.13), yields:

$$\kappa = \frac{6\Delta\alpha T}{t_2} hm \left(\frac{1+h}{1+2hm\left(2+3h+2h^2\right)+h^4m^2}\right) \tag{35.18}$$

$$\varepsilon^0 = T\left(\frac{\alpha_1 mh\left(1+3h+3h^2+mh^3\right)}{1+2hm\left(2+3h+2h^2\right)+h^4m^2} + \frac{\alpha_2\left[1+mh\left(3+3h+h^2\right)\right]}{1+2hm\left(2+3h+2h^2\right)+h^4m^2}\right) , \tag{35.19}$$

where $h = \frac{t_1}{t_2}$, $m = \frac{M_1}{M_2}$, and $M_k = \frac{E_k}{1-\upsilon_k}$ (here $k = 1, 2$ corresponds to layers 1 and 2) and $\Delta\alpha = \alpha_2 - \alpha_1$.

It is often convenient to define a nondimensional curvature $\hat{\kappa}$:

$$\hat{\kappa} = \frac{\kappa t_2}{6\Delta\alpha T} \tag{35.20}$$

Equation (35.18) can then be written as

$$\hat{\kappa} = \frac{hm(1+h)}{1+2hm(2+3h+2h^2)+h^4m^2} . \tag{35.21}$$

Two limiting values of (35.21) are of interest. The first is the case where there is no elastic mismatch between the layers, i. e., $m = 1$. In this case (35.21) simplifies to:

$$\hat{\kappa} = \frac{h}{(1+h)^3} . \tag{35.22}$$

The other is the *thin-film limit* where $t_1 \ll t_2$. Expanding (35.18) in powers of h and retaining only the lowest order term recovers *Stoney's* [35.26] well-known result:

$$\kappa = \frac{6\Delta\alpha T}{t_2} hm$$
$$\hat{\kappa} = hm . \tag{35.23}$$

In the thin film limit, (35.19) reduces to $\varepsilon^0 = \alpha_2 T$.

Figure 35.11 shows the normalized curvature $\hat{\kappa}$ as a function of the layer thickness ratio h for various ratios of the biaxial moduli, m. The curves from top to bottom are for $m = 10$, 0.66, and 0.1. $m = 10$ and 0.1 represent a stiff film on a compliant substrate and a compliant film on a stiff substrate, respectively. $m = 0.65$ represents a gold film on a polysilicon substrate. The solid curves are the exact results from (35.21) and the dashed curves are the Stoney's result for the thin-film limit, (35.23). In the latter case, $\hat{\kappa}$ varies linearly with h, with slope m. The results in Fig. 35.10 show how the range of applicability of Stoney's result depends on both h and m; the range decreases for increasing h and decreasing m. For most microstructure applications where the layer thicknesses are comparable ($h > 0.1$), Stoney's result is not accurate unless the film is much more compliant than the substrate. Thus physically, Stoney's result is accurate when the substrate stiffness controls the deformation, either because it is much thicker or has much higher elastic moduli than the film. Over the years many papers have discussed *corrections* to Stoney's result to increase the region of validity in terms of layer thicknesses and modulus mismatch. We emphasize, though, that (35.18) is valid for arbitrary layer thicknesses and modulus mismatch and is quite simple to use itself. We refer the reader interested in more details to [35.7, 18, 22].

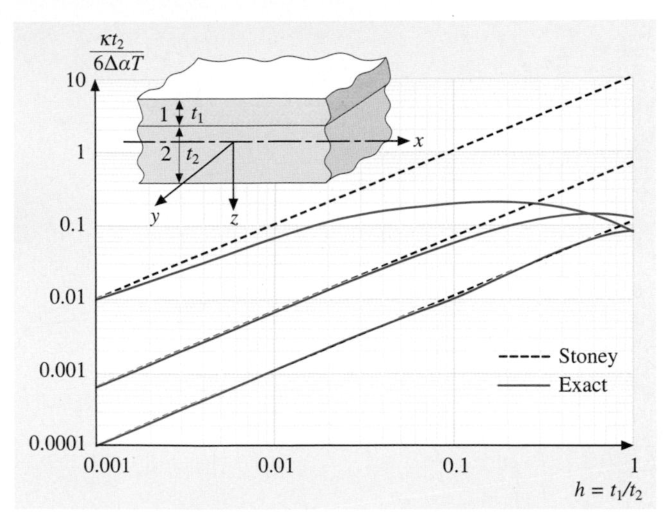

Fig. 35.11 Normalized curvature $\hat{\kappa}$ as a function of the film thickness ratio $h = t_1/t_2$ for three values of the biaxial modulus ratio, $m = M_1/M_2$. From *top* to *bottom*, the curves are for $m = 10$, 0.65 (gold/polysilicon), and 0.1. The *solid lines* are the exact solution, (35.21) and the *dashed lines* are Stoney's result in the thin-film limit, (35.23)

Equally important from a technological viewpoint is the equibiaxial stress distribution through the thickness $\sigma_x(z) = \sigma_y(z) = \sigma(z)$ obtained from (35.6):

$$\sigma(z)_k = M_k\left(\varepsilon^0 + z\kappa - \alpha_k T\right) , \tag{35.24}$$

where $k = 1, 2$ denotes the layer. In (35.24), κ and ε^0 are obtained from (35.18) and (35.19), respectively, and M_k and α_k are the values for the kth layer. In the thin-film limit the equibiaxial stress varies so little through the film thickness that it can be taken as constant, i. e., $\sigma_x = \sigma_y = \sigma$. This is the reason one often refers to the *stress* in a thin-film by a single number. In the thin film limit, the substrate (layer 2) thermal expansion dominates the deformation process, and so the film strains by an amount $\Delta\alpha T$, and the biaxial stress in the film is $\sigma_1 = M_1\Delta\alpha T$, or in terms of the curvature:

$$\sigma_1 = \frac{M_2}{6}\frac{t_2^2}{t_1}\kappa . \tag{35.25}$$

Note that, as written, the thermoelastic properties of the film do not enter (35.25). We will discuss the practical ramifications of this shortly.

Figure 35.12 shows the stress distribution through the thickness of both layers for a negative unit temperature change, (35.24), as a function of $h = t_1/t_2$ for

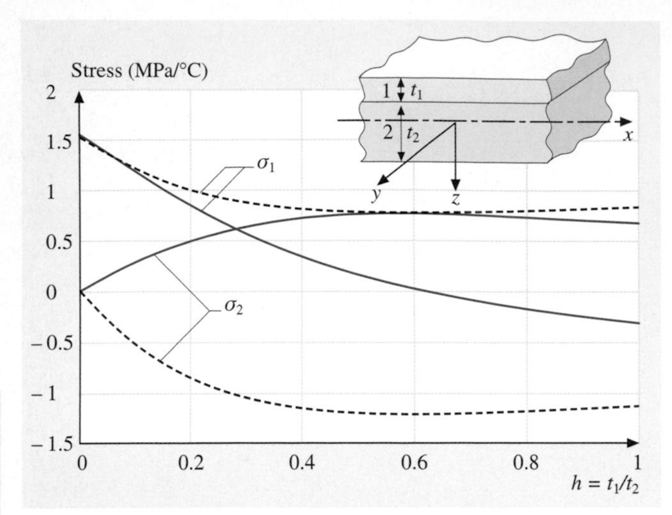

Fig. 35.12 Biaxial stress per unit temperature change, (35.24), for a gold/polysilicon bilayer ($m = 0.67$) for a 0.5-µm-thick gold film as a function of the polysilicon thickness $h = t_1/t_2$. Biaxial stresses are shown at four locations: the interface on the gold and polysilicon sides (*dashed lines*), and the surfaces of the gold and polysilicon (*solid lines*)

$m = 0.67$, and $t_1 = 0.5\,\mu m$, representative of a gold film on a silicon substrate. Equation (35.24) shows, and we have previously discussed, that the biaxial stress varies linearly through each material. Thus we can completely specify the stress distribution by plotting it at four points: the top of material 1, the interface in material 1, the interface in material 2, and the bottom of material 2. Referring to Fig. 35.8, these positions are $z = z_0$, z_1, and z_2. Because of the perfect bonding between the two films, the total strain must be continuous across the interface. The elastic moduli are different for the two films, though, and so the stresses will be discontinuous at the interface. Hence we interpret $z = z_1$ as a position just above, or just below, the interface, and use M_k and α_k in (35.24) as appropriate. As $h \to 0$ the stresses at the top and at the interface in the film (layer 1) converge and take on a maximum value given by $\sigma = M_1 \Delta\alpha T$. The stress in the substrate vanishes in this limit. As the substrate thickness increases (increasing h) the stress in the film decreases with the maximum value occurring at the interface. The stress in the substrate increases at the interface but decreases on the bottom surface. In general this results in a stress distribution in the substrate that is compressive on the bottom surface and tensile at the interface. Depending on h, the stresses in the film are either tensile throughout or tensile on the interface and compressive on the surface. The difference in stress between the interface and the surface is a measure of the stress gradient in each layer; it increases with increasing h. For different values of m, the behavior with varying h is similar to that in Fig. 35.12, with the stress in the film increasing with increasing m.

Equation (35.25) is the theoretical basis of what is perhaps the most common technique to measure thin-film stresses – the wafer curvature method [35.6]. It involves measuring the curvature of a thick (compared to the film thickness) wafer with and without a film, and then using the curvature difference between the two in (35.25) to determine the film stress. It is particularly convenient because the film stress is obtained from the measured curvature regardless of the source of the stress and the thermoelastic moduli of the film. It is necessary, though, to know accurately the thickness of the film and substrate, as well as the elastic properties (biaxial modulus, $E/(1-\nu)$) of the substrate. An appealing feature of the method is that the elastic properties of the film, typically difficult to measure themselves, do not need to be known. This is because the much thicker substrate controls the deformation response.

Three-Layer Symmetric Plate. Considerable simplification also results for the technologically important case of a thin-film multilayer subject to a uniform temperature change T that consists of three layers, but the top and bottom layer are identical in terms of thickness ($t_3 = t_1$) and material properties ($M_3 = M_1, \alpha_3 = \alpha_1$). In this case, from (35.9) $[B] = [0]$, and thus the plate does not bend, i. e., $\kappa = 0$. Equation (35.11), with (35.12) and (35.13), then yields:

$$\varepsilon^0 = \frac{(2\alpha_1 mh + \alpha_2)\,T}{2mh+1}\,. \qquad (35.26)$$

In the thin-film limit $h \to 0$, $\varepsilon^0 = \alpha_2 T$.

The equibiaxial stress distribution is such that it does not vary through the thickness within each layer, i. e., $\sigma(z)_3 = \sigma(z)_1 = \sigma_1$, and $\sigma(z)_2 = \sigma_2$. From (35.6) with $\kappa = 0$ and ε^0 from (35.26):

$$\sigma_1 = \frac{M_1 \Delta\alpha T}{2mh+1}$$
$$\sigma_2 = \frac{-2mhM_2 \Delta\alpha T}{2mh+1}\,. \qquad (35.27)$$

In the thin-film limit, the stresses of (35.27) reduce to $\sigma_1 = M_1 \Delta\alpha T$ and $\sigma_2 = -2M_1 h \Delta\alpha T$.

Figure 35.13 shows the uniform biaxial stress in each material for a negative unit temperature change as a function of $h = t_1/t_2$ for $m = 0.65$, and $t_1 = 0.5\,\mu m$, again representative of a gold film on a silicon substrate. As $h \to 0$, the stresses in the films (layers 1 and 3) are given by $\sigma_1 = M_1 \Delta\alpha T$, and that in the substrate (layer 2) approaches zero. The stress in the films is tensile and decreases with increasing h while that in the substrate is compressive, increasing in magnitude with increasing h.

Beams. Our analysis thus far has focused on multilayer plates where both in-plane dimensions far exceed the thickness and where the transformation strains, and thus the stresses, are equibiaxial. When one in-plane dimension, say x, is much larger than the other, the structure is often referred to as a beam. In the classical treatment of the flexure of beams, it is assumed that only one nonzero normal stress component exists, σ_x. This greatly simplifies the analysis for homogeneous beams, but the situation is not as straightforward for multilayers with biaxial transformation strains. From the results in the previous section, one can obtain results for a beam theory by simply setting $\nu_1 = \nu_2 = 0$ in the formulae developed for biaxial stresses, but these results should be used with caution as the transverse component of the moment that results from the equibiaxial transformation strains cannot be neglected as shown by three-dimensional finite element calculations [35.27, 28]. *Swanson* [35.29] describes how to apply the results from a multilayer plate theory to beams and specifically focuses on the effect of the beam width, giving easily usable results for narrow and wide beams. The main results are an effective structural stiffness $(EI)_{\text{eff}}$ that can be used to replace EI (E is Young's modulus and I is the cross-sectional moment of inertia) in existing beam theory results, such as those that will be presented in Sect. 35.2 for electromechanics analysis. Insight into the behavior can be seen in the results in Table 35.2, which show measurements (using interferometry as described by *Dunn* et al. [35.30]) and predictions based on (35.18) for plates and the simplified version for beams obtained by setting $\nu_1 = \nu_2 = 0$. In all calculations both the gold and polysilicon are modeled as linear thermoelastic with isotropic material properties. Input parameters to the finite element calculations are $E_2 = 163\,\text{GPa}$, $\nu_2 = 0.22$, $E_1 = 78\,\text{GPa}$, $\nu_1 = 0.42$ [35.31]. The thermal expansion coefficients of the materials were assumed to vary linearly with temperature, and values at 100(24) °C used are $\alpha_2 = 3.1(2.6) \times 10^{-6}$ /°C, and $\alpha_1 = 14.6(14.2) \times 10^{-6}$ /°C [35.31]. Although some uncertainty exists in the values of these material properties for the gold and polysilicon films, we think they are sufficiently accurate for the purpose of modeling the observed phenomena. Young's modulus and Poisson's ratio of the polysilicon are in line with many measurements over many MUMPS runs [35.32] and agree adequately with bulk polycrystal averages of single-crystal elastic constants. Good agreement exists between the predictions using plate theory and the measurements, while predictions based on beam theory are significantly less accurate, demonstrating the need to use plate, rather than beam, theory.

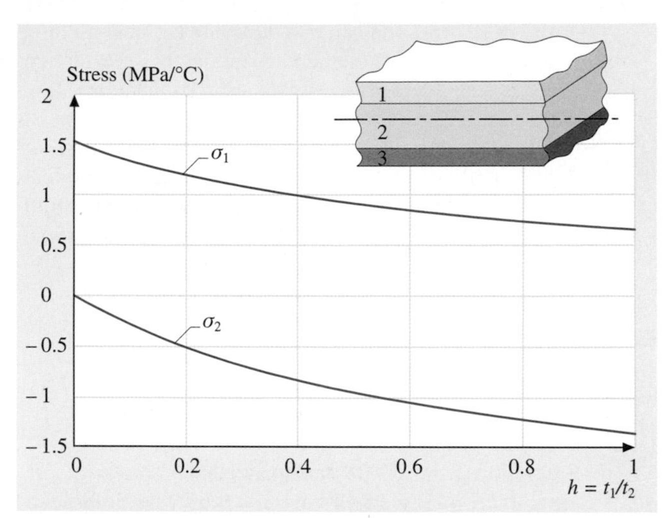

Fig. 35.13 Biaxial stress per unit temperature change for a gold/polysilicon/gold ($m = M_1/M_2 = 0.65$) three-layer film system. The calculations, (35.27), are for a 0.5-μm-thick gold film as a function of the polysilicon thickness $h = t_1/t_2$. Biaxial stresses are shown in the gold (*top curve*) and polysilicon (*bottom curve*)

Patterned Films. In microsystems applications, additional design freedom can be obtained by using multilayer films with patterned, rather than blanketed,

Table 35.2 Predicted and measured curvature per unit negative temperature change for gold/polysilicon microstructures as a function of the polysilicon thickness. In all cases the gold thickness is 0.5 μm. The microstructures are 300 μm long and 50 μm wide and cantilevered from one end

Polysilicon thickness (μm)	$d\kappa/dT$ (m^{-1}/°C) Measured	Predicted (plate)	Predicted (beam)
1.5	−5.36	−5.56	−4.86
3.5	−1.42	−1.44	−1.16

films. In general one can use the finite element method with multilayer plate elements to model the deformation and stresses in a multilayer with arbitrary patterned layers. A useful practical example is a beam of width w_2 that is covered with a film in the pattern of a stripe of width w_1 as shown in Fig. 35.14. A simple modification of (35.18) yields the curvature along the length of the patterned beam as a function of the nondimensional linewidth, $c = w_1/w_2$:

$$\kappa_{\text{patterned}} = \frac{6\Delta\alpha T}{t_2} hcm \times \left(\frac{1+h}{1+2hcm\left(2+3h+2h^2\right)+h^4c^2m^2} \right) . \tag{35.28}$$

Predictions from (35.28), normalized as $\eta = \kappa_{\text{patterned}}/T$, are shown in Fig. 35.14 for $m = 0.65$ (the gold/polysilicon bilayer) along with predictions from detailed finite element calculations and measurements over the entire range of $c = w_1/w_2$ [35.33]. Equation (35.28) accurately describes the effect of linewidth on curvature development and can be used to design bilayer beams.

Deflections. Thus far our discussion of deformation has been cast in terms of curvature. An important quantity for many applications where control of the deformation is important is the deflection of the multilayer. Formally this can be obtained by integrating the curvature-displacement relationship and applying the relevant boundary conditions. For example, assuming constant curvature, the tip deflection $w(x = L)$ for a cantilever beam of length L, is:

$$w(x = L) = \frac{\kappa L^2}{2} . \tag{35.29}$$

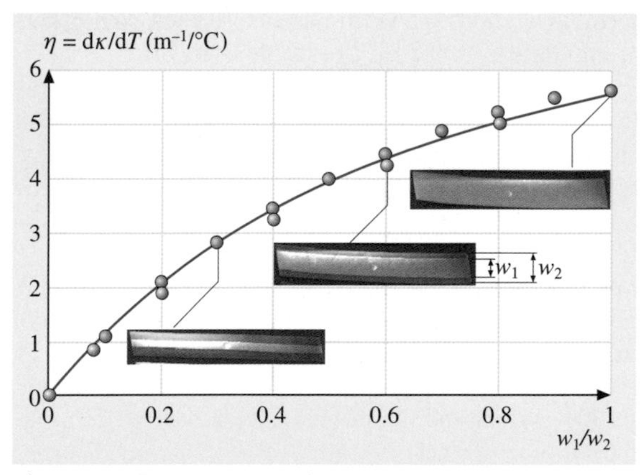

Fig. 35.14 Curvature per unit temperature change as a function of the ratio of gold to polysilicon film width w_1/w_2 for gold (0.5 μm thick)/polysilicon (1.5 μm thick) strip-patterned beams. *Grey circles* are finite element simulation results, the *solid line* is the analytical result of (35.28), and the *brown circles* are measurements

35.1.3 Nonlinear Geometry

In the previous section we noted that in general the deformation of a layered film microstructure subjected to transformation strains consists of three regimes but focused on the first, linear regime. Here we focus on the second and third regimes. To illustrate, consider the seemingly simple case of a plate with total thickness much less than the in-plane dimensions, composed of two isotropic layers with different material properties (elastic moduli and thermal expansion coefficients) subjected to a temperature change. Although we use a temperature change to demonstrate, the results hold for any type of transformation strains. In terms of the average curvature variation as a function of temperature change, three deformation regimes have been identified as illustrated in Fig. 35.15 [35.16–19, 22, 23]. The first regime, I, consists of a linear relation between the average curvature and temperature change where $\kappa_x = \kappa_y$, i. e., the average curvature is spherically symmetric. This deformation regime is characterized by both small transverse displacements and rotations, and so conventional thin-plate theory adequately describes the deformation. In Sect. 35.1.2 we outlined the analysis in this regime. The second regime, II, consists of a nonlinear relation between the average curvature and temperature, but again $\kappa_x = \kappa_y$. The behavior is due to *geometric nonlinearity* that results when the deflections become excessively large relative to the plate thickness, and they contribute significantly to the in-plane strains. It has been shown [35.15–19, 22, 23] that in these two regimes the symmetric deformation modes are stable. The second regime ends at the point when the deformation response bifurcates from a spherical to ellipsoidal deformation, i. e., $\kappa_x \neq \kappa_y$. At this point, the beginning of regime III, it becomes energetically favorable for the plate to assume the ellipsoidal shape because to retain the spherical deformation under an increasing temperature change requires increased mid-plane straining. After the bifurcation, the curvature in one direction increases while that perpendicular to it decreases; the plate tends toward a state of cylindrical curvature. This observation helps to explain the energetic argument, as unlimited cylindrical curvature can be obtained with no midplane straining, while spherical curvature cannot. This discussion has been cast in the context of linear material behavior. Additional deformation regimes result if material nonlinearity is present, for example, yielding [35.18].

Analysis

The understanding described above derives from a number of studies with different technological motivations, primarily structural laminated composites and thin films for microelectronics. Most of these studies are analytical [35.15–19, 22, 23, 34] and build upon the original work of Hyer [35.13, 14]. To illustrate the computational approach in a reasonably simple setting, here we consider a two-layer plate with layer thicknesses t_1 and t_2. Each layer is isotropic and characterized by the Young's modulus E_i, Poisson's ratio ν_i, and thermal expansion coefficient α_i $(i = 1, 2)$. We consider a square plate with side length L; a similar analysis can be carried out for circular [35.22] or other plate shapes. The plate is subject to a uniform temperature change, T. In other words, the stress-free strains $\{\varepsilon^*\}_k$ are due to thermal expansion as given by (35.14).

In order to compute the deformed shape when the plate is subject to a temperature change we use the approach of *Hyer* [35.14] as applied by *Masters* and *Salamon* [35.16]. The basic idea is to assume an admissible displacement field $w(x, y)$ in terms of unknown parameters (d_i) that are suitably chosen to be consistent with observed deformation modes. Values of the parameters d_i are then determined via a Ritz procedure so as to minimize the total potential energy of the system. Different choices of the assumed displacement field have been considered, and details of the procedures are given in the above references. Such analyses are sufficient to qualitatively, and in many cases quantitatively, explain the three regimes of deformation shown in Fig. 35.15. In fact, quite simple closed-form expressions result for special cases that provide illuminating descriptions of observed phenomena (see for example [35.22]). A disadvantage of the analytical approaches is that for simplicity a displacement field consistent with a spatially constant curvature deformation mode is usually chosen. Additionally, these formulations are useful for only simple plate shapes. While this may be adequate for the structures demonstrated here, it is not for more complex in-plane shapes, of either or both layers, that arise in microsystems applications. In this more general case the best approach is probably to use the finite element method for the approximate analysis.

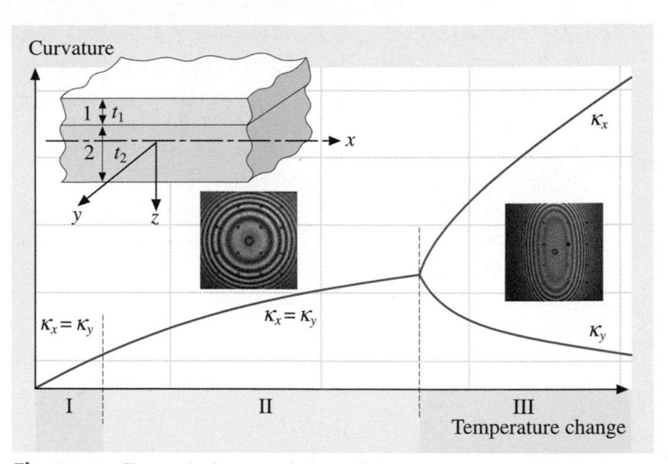

Fig. 35.15 General characteristics of the average curvature-versus-temperature change of a two-layer plate microstructure showing the three deformation regimes. The interferograms in the *inset* show measured displacement contours for gold/polysilicon plates in Regimes II and III

To proceed, we assume the transverse midplane displacement is of the form:

$$w^0(x, y) = d_1 x^2 + d_2 y^2 . \tag{35.30}$$

d_1 and d_2 are immediately recognized as one-half the curvature in the x- and y-directions, and the deformation is seen to be a constant curvature mode. The midplane displacements u^0 and v^0 are assumed to be described by third order polynomials in x and y, also with unknown constant coefficients d_3–d_8 to be determined:

$$\begin{aligned} u^0(x, y) &= d_3 x + \frac{d_5}{3} x^3 + d_7 x y^2 - \frac{d_1^2}{6} x^3 \\ v^0(x, y) &= d_4 y + \frac{d_6}{3} y^3 + d_8 x^2 y - \frac{d_2^2}{6} y^3 . \end{aligned} \tag{35.31}$$

The midplane strains are then computed from the nonlinear strain-displacement relations of the von Karman plate theory:

$$\begin{aligned} \varepsilon_x^0 &= \frac{\partial u^0}{\partial x} + \frac{1}{2}\left(\frac{\partial w^0}{\partial x}\right)^2 \\ \varepsilon_y^0 &= \frac{\partial v^0}{\partial y} + \frac{1}{2}\left(\frac{\partial w^0}{\partial y}\right)^2 \\ \gamma_{xy}^0 &= \frac{\partial u^0}{\partial y} + \frac{\partial v^0}{\partial x} + \left(\frac{\partial w^0}{\partial x}\right)\left(\frac{\partial w^0}{\partial y}\right) . \end{aligned} \tag{35.32}$$

The strains at any point through the thickness are then computed using the standard kinematic relations for thin plates, (35.2), and the stresses are computed from the strains using the conventional linear thermoelastic constitutive relations for each layer, (35.6) with $\{\varepsilon^*\}_k = \{\alpha\}_k T$. The first term in (35.32) is recognized as the conventional small deformation, and the second term arises from the large transverse deflections w^0, which result in straining of the midplane. This is shown schematically in Fig. 35.16 for the strain component ε_x^0; similar results are obtained for ε_y^0 and γ_{xy}^0.

The potential energy density of each layer is computed from the stress and strain in each layer, and the total potential energy of the plate is then computed by integrating it over the volume of the plate. This yields an expression for the potential energy of the plate in terms of the unknown coefficients d_i:

$$U(d_i) = \int_{-L/2}^{L/2} \int_{-L/2}^{L/2} \int_{-t/2}^{t/2} \{\varepsilon\}^T \{\sigma\}_k \, dz \, dx \, dy \,, \tag{35.33}$$

where the superscript T denotes the transpose of the 3×1 column vector. The coefficients d_i are determined by application of the principle of minimum potential energy:

$$\frac{\partial U(d_i)}{\partial d_i} = 0 \,. \tag{35.34}$$

This yields eight equations for the eight unknown d_i and thus solutions for the deformation response of Fig. 35.15. Complete details regarding this analysis can be found elsewhere [35.14, 16]; here we present only the pertinent results in what we hope is an accessible form. When the displacements are small, the solution of (35.34) and (35.35) recovers (35.18) and (35.19).

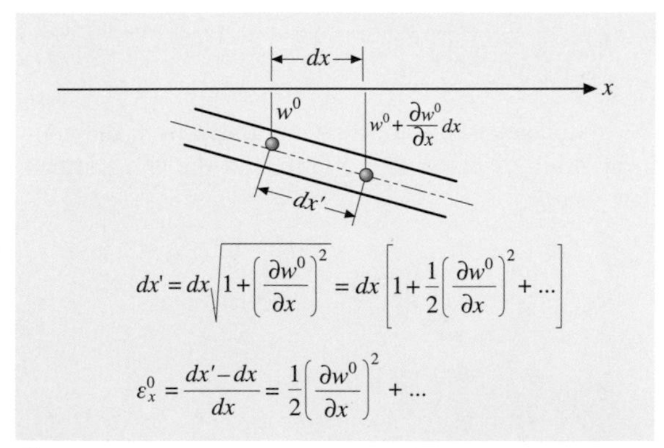

Fig. 35.16 Illustration of the development of midplane strain due to bending when nonlinear geometry effects arise

In the nonlinear but symmetric deformation regime II, the relationship between both the curvature and midplane strain, and the thermal expansion mismatch is too complex to present here. This is also the case in the nonlinear regime III after bifurcation. An important difference between the linear and nonlinear response, though, is that the nonlinear response depends on the plate size. The critical curvature at which bifurcation occurs, though, can be obtained explicitly and is given by:

$$\kappa_{cr} = \frac{12\sqrt{2}}{L^2}\sqrt{\frac{6+A_{12}/A_{11}}{1+A_{12}/A_{11}}}\,\frac{\sqrt{A_{66}D_{66}-B_{66}^2}}{A_{66}^2} \tag{35.35}$$

for a square plate of side length t, where A_{ij}, B_{ij}, and D_{ij} are the composite moduli of (35.8). In the simplified case where there is no elastic mismatch between the layers and the elastic, response can be expressed in terms of the Young's modulus E and Poisson's ratio ν, (35.35) reduces to:

$$\kappa_{cr} = 2\sqrt{6}\sqrt{\frac{6+\nu}{1+\nu}}\frac{t}{L^2} \,, \tag{35.36}$$

where $t = t_1 + t_2$. These results have also been obtained by *Finot* and *Suresh* [35.18]; similar results exist for a circular plate [35.22, 30].

The analysis just discussed is made tractable by assuming a displacement field consistent with a constant curvature. In the nonlinear case, this assumption is questionable. While the assumed displacement field used in the Ritz procedure could be modified to incorporate the dependence of curvature on position, perhaps the simplest approach to tackle these more general problems is to use the finite element method to solve the geometrically nonlinear equations over an arbitrary spatial domain. This is also the most viable approach for more complicated in-plane geometries, including patterned films. Complete details regarding the use of the finite element method to carry out such calculations can be found in *Dunn* et al. [35.30]. Here we will simply show the result of some of these calculations. The input parameters used are the same as those given in Sect. 35.1.2.

Deformation Behavior

In order to illustrate the nature of the deformation phenomena in the nonlinear regimes, Fig. 35.17 shows contour plots of measured (using an interferometric microscope as described by *Dunn* et al. [35.30]) and predicted displacement fields $w(x, y)$ for square plates of four plate sizes that have been cooled from about 100 °C where they are flat to room temperature. Due to the thermal expansion mismatch between the polysilicon and gold, the $L = 150\,\mu m$ samples deform in a spherically symmetric manner; contours of constant transverse displacement $w(x, y)$ are nearly circles. This is also the case as the size increases to $L = 200$ and $250\,\mu m$, although the displacements increase as the plate size increases. At $L = 300\,\mu m$, though, the transverse displacement contours are not circular but elliptical, indicating that the deformation is no longer spherically symmetric. Thus when subjected to the same temperature change, both the magnitude and deformation mode depend on the plate size. As the in-plane dimension of the plate increases with the thickness held constant, the deformation mode changes from one of spherical symmetry to one more like cylindrical symmetry.

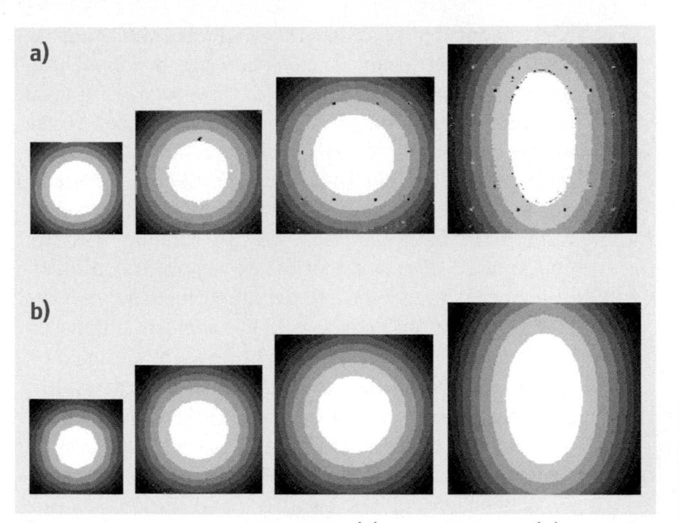

Fig. 35.17a,b Contour plots of the (**a**) measured, and (**b**) predicted transverse displacements $w(x, y)$ at room temperature following cooling from 100 °C for the four gold/polysilicon square plates: $L = 150$, 200, 250, and $300\,\mu m$ from *left* to *right*. Each contour band represents a displacement of 0.23, 0.35, 0.45, and $0.6\,\mu m$ for the $L = 150$, 200, 250, and $300\,\mu m$ plates, respectively

A complete picture of the deformation as a function of plate size and temperature change could be obtained by acquiring full-field displacements like those in Fig. 35.17 as a function of temperature. But a reasonable picture of the deformation behavior can be obtained in much simpler form by considering the *average* curvature in the x- and y-directions as a function of temperature. This is shown in Fig. 35.18, in which the average curvature in the x- and y-directions is plotted as a function of the magnitude of the temperature change during cooling. The temperature change is actually negative according to our convention, but its magnitude is plotted for convenience. The average curvatures are determined from the measured and computed $w(x, y)$ by averaging $\kappa_x = -\partial^2 w(x, 0)/\partial x^2$ and $\kappa_y = -\partial^2 w(0, y)/\partial y^2$ along the paths $y = 0$ and $x = 0$, respectively [35.30]. The x- and y-directions are taken to be aligned with the principal curvatures after bifurcation. The use of the average curvature as a measure of the plate deformation seems appropriate if the curvature is, or is close to, spatially uniform. This aspect will not be taken up in detail, but we refer the interested reader to [35.22,30] for details. It is apparent from both the measurements and predictions that in regime I, the curvature-temperature response is

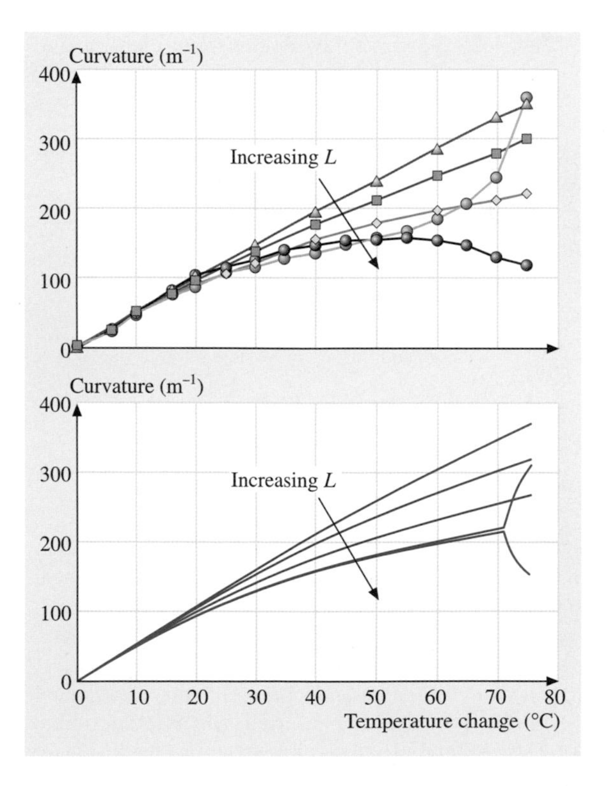

Fig. 35.18 Average measured (*top*) and predicted (*bottom*) curvature as a function of temperature change upon cooling from 100 °C to room temperature. The curves from *top* to *bottom* are for the $L = 150$, 200, 250, and $300\,\mu m$ structures, respectively

independent of plate size and shape. In regime II, though, there is a strong dependence on plate size, and this is also the case in regime III. The major discrepancy between the measurements and predictions is the sharpness of the bifurcation for the $L = 300\,\mu m$ plate; it is quite sharp in the predictions but much more gradual in the measurements. To understand this we note that the source of the bifurcation is an *imperfection* of some sort that breaks the ideal symmetry, and in general it is difficult to accurately model the detailed imperfection.

Figure 35.19 demonstrates the connection between the thermomechanical loading (the temperature change), the geometry (plate size, L), and the boundaries between the three deformation regimes for gold/polysilicon plate microstructures. Specifically, it shows the temperature change necessary to initiate nonlinear effects (the transition between regions I and II), and bifurcation (the transition between regions II and III) as a function of polysilicon thickness when the gold film thickness is kept constant at $0.5\,\mu m$. Despite the fact that the constant curvature approximation becomes questionable for larger plate sizes, (35.35) is a good approximation as seen by the agreement with the finite element calculations, at least for the elastic mismatch and plate sizes considered here. In fact, although not shown, the simplified result of (35.36) for no elastic mismatch is in reasonable agreement with the finite element and

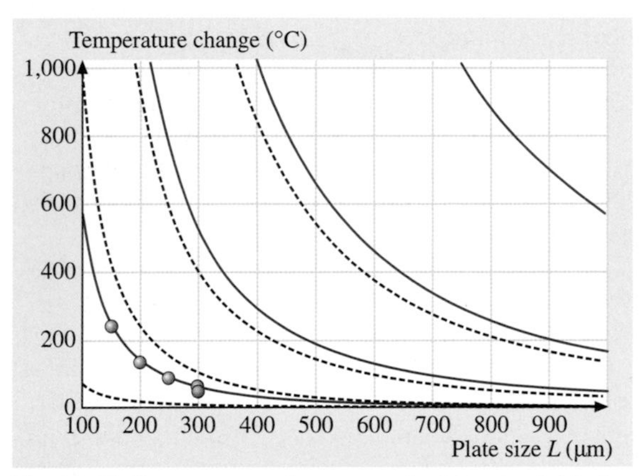

Fig. 35.19 Temperature change required for the initiation of nonlinear geometry effects (*dotted lines*) and critical temperature change for bifurcation (*solid lines*) as a function of square plate size and the thickness of polysilicon for the gold/polysilicon microstructures. Lines in each set from *left* to *right* represent $1.5\,\mu m$, $3.5\,\mu m$, $5.5\,\mu m$, and $8.5\,\mu m$ thick polysilicon, respectively. *Grey circles* are finite element calculations, and the *brown circle* is a measurement

the complete analytical results. The measurement for $L = 300\,\mu m$ is accurately described by both the analytical and finite element results.

In the analytical treatments discussed previously, it is assumed that the curvature is spatially uniform. The power of finite element calculations is that this requirement is relaxed and the spatial variation of the curvature can be studied theoretically. Full-field interferometry measurements allow one to study this experimentally as well. Briefly we discuss the resulting behavior for gold/polysilicon plate microstructures but refer the interested reader to [35.22, 30] for more details. In the linear regime, measurements and predictions both show that curvature is essentially uniform across the plate. In the nonlinear regime, though, the curvature varies appreciably with position, increasing by about a factor of two from the center to the periphery of the plate. The spatial nonuniformity of the curvature increases as the plate size increases. The spatial variation of the curvature raises concern regarding the suitableness of an analysis based on constant curvature. As mentioned previously, such an analysis may be adequate to describe the general deformation behavior but not for finer details.

35.1.4 Nonlinear Material Behavior

So far we have discussed in some detail the linear thermoelastic response of multilayer films and to a lesser degree, the effects of geometric nonlinearity. An additional complication is material nonlinearity. It can arise in numerous forms including plasticity, creep, stress relaxation, and evolution of the material microstructure (densification, grain growth, defect annihilation, etc.) during thermomechanical loading. In this section we briefly discuss nonlinear material behavior of multilayer films, focusing on the phenomena most relevant for the realization of reliable devices. The discussion in this section is concerned primarily with multilayer films where one of the films is a metal.

Nonlinear Material Behavior During Initial Thermal Cycles

As deposited, the material microstructure of a metal film is generally not stable. As the film is heated, microstructural evolution leads to the development of internal stress, which then can lead to changes in the curvature of the multilayer film. The general behavior is common to films made of many different metals, such as aluminum, copper, nickel, and gold. It has been studied extensively in the context of microelectronics applications (see for example [35.6, 35, 36]) where the substrate

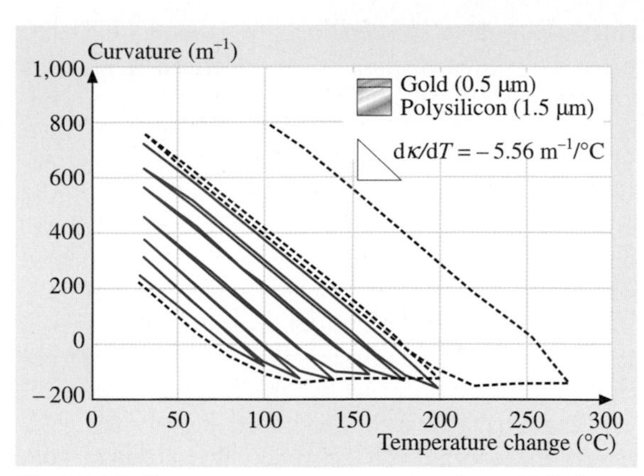

Fig. 35.20 Curvature versus temperature for gold (0.5-µm-thick)/polysilicon (1.5-µm-thick) 300 µm × 50 µm beam microstructures. The *solid lines* are tests with six cycles from room temperature to 200 °C with the maximum temperature in each successive cycle increasing by 20 °C. The *dashed lines* are tests with two cycles: the first from room temperature to 200 °C and then back to 30 °C, and the second to 275 °C and then back to 30 °C

is much thicker than the film and the thin-film limit results, which simplifies the interpretation of the results.

Figure 35.20 shows the general behavior of the curvature vs. temperature for a bilayer film consisting of a 0.5 µm gold film that was evaporated onto a 1.5 µm polysilicon film. During the initial heating the deformation is thermoelastic, characterized by a decrease in curvature with a constant $\mathrm{d}\kappa/\mathrm{d}T$, until the temperature reaches about 70 °C (the theoretical thermoelastic slope from (35.18) is shown on the figure). Between 70 °C and 100 °C, the curvature decreases at a much lower rate; in fact the curve flattens, i. e., $\mathrm{d}\kappa/\mathrm{d}T \approx 0$. Interestingly, between room temperature and 100 °C, the curvature changes from positive to negative, our convention being that a positive curvature means a beam with the gold on top is curved upward. The strong departure of the curvature from the thermoelastic behavior results because the microstructure of the evaporated film is not stable in the as-deposited condition. The temperature increase during the first cycle promotes microstructural changes in the film. These result in tensile straining of the film, which competes with the thermoelastic deformation (see for example [35.6, 36]). Upon cooling from 100 °C, the response is again thermoelastic, this time throughout the entire cooling process. Although it does not occur here, if the temperature change is large enough, yielding of the metal film can occur upon cooling. The curvature upon return to room temperature is greater than that initially at room temperature. It is important to understand this behavior because many post-fabrication and packaging processes expose a microstructure to elevated temperature excursions. During these, the film is susceptible to this nonlinear deformation. In the tests in Fig. 35.20, the material has not been taken to a temperature high enough to stabilize fully the gold microstructure in the first cycle. As a result, subsequent cycles going to successively higher temperatures appear to have the effect of continuing to stabilize the microstructure. The response during the second cycle is similar to the first with the exception that the temperature change required to initiate nonlinear behavior is increased. Indeed, over the range of temperature cycles studied, each subsequent cycle is similar, exhibiting thermoelastic response upon heating until a point where nonlinear κ-T behavior commences. The cooling process is again thermoelastic with the subsequent room temperature curvature increasing with each cycle. Also shown in Fig. 35.20 is a result for a second test on a nominally identical microstructure with only two cycles: the first from room temperature to 200 °C and back to 30 °C, and the second to 275 °C and back to 30 °C. The results of the first cycle to 200 °C neatly envelope the results of the set of tests with cycles up to 200 °C. This shows that although there is a path dependence of the deformation behavior, it is controlled by the maximum temperature reached during cycling, independent of how many increments are carried out to reach that temperature. The second cycle to 275 °C shows the same behavior observed in all cycles. The results in Fig. 35.20 have been obtained at a constant heating/cooling rate. In terms of the stabilization of the material microstructure that results in nonlinear material behavior, the rate, along with the time held at the elevated temperature, plays a significant role.

Figure 35.20 suggests that a simple description of the evolution of curvature with temperature can be obtained with parameters as shown in Fig. 35.21: the as-released curvature κ_r (and associated temperature, T_r), a critical temperature at which inelastic mechanisms are activated, T_cr (and the associated curvature, κ_cr), the slope of the inelastic κ-T behavior (assuming simple linear behavior) for $T > T_\mathrm{cr}$, ξ, and the thermoelastic slope $\eta = \mathrm{d}\kappa/\mathrm{d}T$. The as-released curvature is determined primarily by the intrinsic stress development upon film deposition, cooling to room temperature, and release. These also influence T_cr. The parameter ξ is physically related to the curvature per unit temperature change due to inelastic mechanism(s) activated beyond T_cr. The situation in Fig. 35.20 is much simplified in that $\xi = 0$.

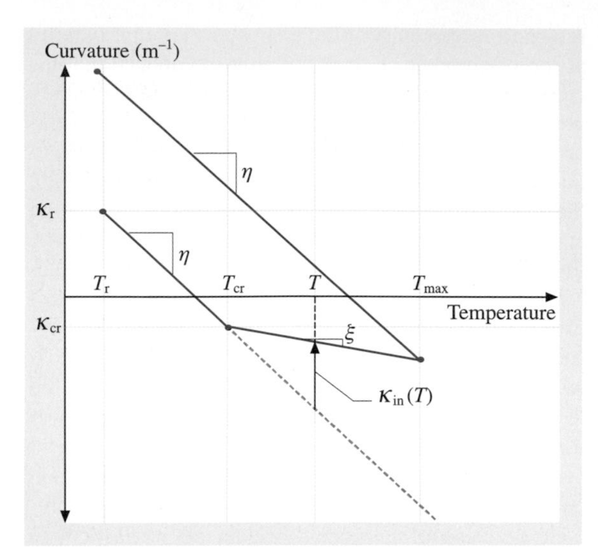

Fig. 35.21 Schematic diagram showing the parameters in the simple model that describes the curvature-versus-temperature behavior of the bilayer beam during thermal cycling after release

With knowledge of these parameters, one can compute the curvature following a defined thermal loading history. The qualitative behavior in Fig. 35.21 also results for a substrate with patterned lines (Sect. 35.1.2), and the same formalism can be used with κ_r, η, and ξ depending on linewidth [35.33]. For example, (35.28) gives the dependence of η on linewidth.

Plasticity

Plasticity in thin-film multilayers has been studied in depth for thin metal films on thick substrates (see, for example [35.6,37,38]), but little work has been directed toward freestanding multilayers. This is probably because, at least when subject to thermal loading, the stresses are typically smaller in freestanding multilayer films than in systems with thin films on thick substrates (see Fig. 35.12). Many interesting open issues exist with regard to the general issue of plasticity in microstructures including the understanding of size effects and strain gradient effects (see for example [35.39,40]).

Creep and Stress Relaxation

For many microsystems applications, it is important to control accurately the deformation of thin-film microstructures over a significant period of time in order to meet performance and reliability requirements. This is especially important for microstructures subjected to thermal loading and/or operated at elevated temperatures. When one or more of the layers consists of a metal or polymer film, creep and stress relaxation in the film can significantly influence deformation and compromise device performance, and so their effects must be carefully considered. For example, *Miller* et al. [35.41] designed and fabricated microrelay switch arrays for RF communications applications using prestressed gold/polysilicon bimaterial beams as electrostatically actuatable switches. They observed a change in the switch shape and position over time and attributed it to stress relaxation in the gold. *Vickers-Kirby* et al. [35.42] report that creep in gold and nickel cantilever beams leads to voltage drops in micromachined tunneling accelerometers over time.

Creep and stress relaxation phenomena have been investigated in some detail for thin film on thick substrate systems, motivated primarily by microelectronics applications (see for example [35.11,35,43–46]). These studies have focused on measuring, understanding, and modeling the stress-versus-temperature behavior that occurs when a metal film-substrate system is heated from room temperature, then cooled back to room temperature, over one or a few cycles. The stress-temperature curves in the experimental works are typically obtained using the wafer curvature method that is now widely used to measure stress in thin films. During a typical test thermoelastic and inelastic mechanisms contribute to produce a complex nonlinear stress-temperature curve like that in Fig. 35.20. In most of these studies, the stress-temperature response is studied using wafer curvature measurements at a fixed heating or cooling rate so that rate dependent and independent phenomena are coupled in the response. Stress relaxation during an isothermal hold has been studied by [35.35,43,45,47], among others. These studies all show that significant stress relaxation can occur over periods of only a few hours at modest temperatures of only about 100 °C for many metal thin-film systems. Due to the thin-film/thick-substrate system, the stress in the metal films is quite high, on the order of hundreds of MPa. Models incorporating power law creep of the metal film were successfully able to describe the observed response qualitatively and, to a large degree, quantitatively. The latter, however, sometimes required modification of the power-law exponent. *Shen* and *Suresh* [35.35] showed experimentally that a thin passivation layer on a metal film on thick silicon substrates can significantly reduce creep and stress relaxation. *Thouless* et al. [35.45] also studied the effect of a thin passivation layer on the stress relaxation of thin films with a focus on the deformation mechanisms. Their experimental results suggest that the presence of a passivation layer on the surface of a film can have a substantial effect on relaxation rates likely by suppressing mechanisms associated with diffusion and dislocation motion.

Thermomechanical Fatigue

It is also important to understand the development of damage and its effect on deformation during cyclic thermomechanical loading. To date, only very limited attention has been directed toward this issue. *Zhang* and *Dunn* [35.33] showed that if gold/polysilicon multilayers were cycled (Fig. 35.20) to an elevated temperature and then cooled to room temperature, they followed the thermoelastic path on subsequent cycling to a temperature below the maximum reached during the initial thermal cycle. Presumably over this range of temperature and time, the gold microstructure has been stabilized by the first cycle, and the polysilicon microstructure is not changing. This was confirmed for only a few, perhaps ten, cycles. *Gall* et al. [35.48] have shown that for thousands of cycles, though, a gradual shift of the thermoelastic curve downward is observed, possibly due to creep and stress relaxation in the gold, although the thermoelastic slope is maintained.

35.1.5 Other Issues

There are a number of additional issues regarding the thermomechanical behavior of thin-film microstructures that warrant discussion. Due to space limitations, however, we will only mention them and refer the reader to appropriate references.

In addition to the deformation itself, the stiffness characteristics of a microstructure, such as for example a beam or plate, can depend significantly on the stress in the beam. *Senturia* [35.49] describes the phenomena and discusses implications for microstructure design. This includes tensile stresses that tend to stiffen beams and plates as well as compressive stresses that can lead to buckling.

An important strategy has emerged to characterize the effects of processing on the residual stress state and thermomechanical behavior of films: the use of on-chip test structures to extract material properties including film stresses. Many test structures, both passive and active, have been developed to measure residual stresses (tensile and compressive), stress gradients, elastic moduli, strength, toughness, thermal expansion, and many others. Nice surveys of many test structures are given by Masters et al. [35.12] and *de Boer* et al. [35.50] where further references to more extensive details can be found.

35.2 Electromechanics of Thin-Film Microstructures

Electromechanics, the coupling of electrostatics and mechanics, is perhaps the most common of the energy couplings used in microsystems technology for sensing and actuation. In this section we summarize the contemporary application of electromechanics, the analysis of fundamental electromechanics problems, and the practical issues that arise in the design of an electromechanical microsystem.

35.2.1 Applications of Electromechanics

Our reference to electromechanics will be limited to the coupling of electrostatics and the mechanics of constrained elastic media such as a cantilever beam. In this case, the electrical energy is transformed into strain energy of the elastic media but with typically negligible energy dissipated due to material damping. An alternative application of electromechanics is the coupling of electrostatics with rigid body mechanics. In this case the mechanical structure is not constrained by a compliant support to the substrate but by a kinematic constraint, which is a pin joint between multiple structural layers or anchor between the substrate and a structural element. An example of these rigid body constraints is the anchor in the electrostatic micromotors. In this case, the electrostatic energy is dissipated by friction (Coulomb type friction between structural elements and pin joints, other structural elements, and/or the substrate) and is stored in the kinetic energy of the rigid body. In this case, the energy stored as strain energy in the structural element is by design negligible. In both electromechanics cases, the energy is dissipated by viscous forces under dynamic conditions. Finally the coupling of electrostatics and mechanics can more broadly; consider the coupling of electrostatics and fluids or electro-thermo-mechanics.

Microsystems have enabled many different commercial applications of electromechanics. These applications have included pressure sensors, accelerometers, gyros, resonators, micropumps, optical mirrors, optical shutters/VOAs, and DC and RF switches. In these examples, electrostatic forces are produced between a stationary electrode and a movable electrode, which is often the proof mass for an accelerometer or the diaphragm for a pump. The electrostatic forces are imparted by a voltage difference between the movable and stationary electrode. The electromechanics is typically implemented in two ways, with the first characterized by a stable, analog behavior, and the second characterized by an unstable, digital behavior. The stable behavior is implemented as a self-test capability for accelerometers such that a specific, stable displacement is produced with the application of the voltage. The unstable behavior is implemented with a switch so that when the instability point (snap-in or pull-in) is reached, the switch will instantaneously close.

Electrostatic actuation has been used for switches and microrelays [35.51–53], accelerometers [35.54,55], micropumps [35.56, 57], and micromirror [35.58–65]. The specific case of switches and relays involves multiple regions of electromechanical behavior and represents one of the richer electromechanics design problems. The general behavior is shown in Fig. 35.22. The first region is the stable actuation region. In this region, the switch is actuated to an instability point in the voltage displacement curve called the pull-in or snap-in voltage. Up to the instability, the displacement increases monotonically until the pull-in voltage is reached. At the pull-in voltage, the switch instantaneously closes the gap because of the instability and establishes contact with the substrate. In the case of the switch, this means the closure of a contact. The electromechanics problem continues with a new set of boundary conditions in the next region because the switch is now "simply" constrained at the contact and fixed at its anchors to the substrate. In Fig. 35.22, the switches tip displacement is shown, which is the reason the displacement is unchanged as the voltage increases. Region 2 is then characterized by a constant displacement for increasing voltage. The shape of the beam continues to change as the voltage is increased to the point that the electrodes could short if necessary precautions are not taken.

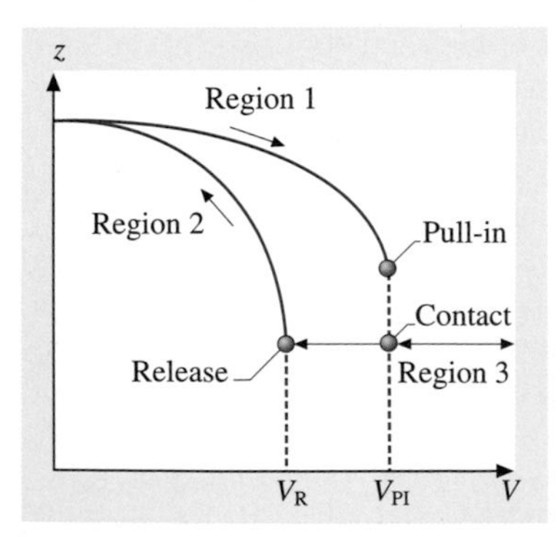

Fig. 35.22 Characteristic electromechanical hysteresis curve showing pull-in, contact, and release voltages

If the first consideration of the switch design is pull-in voltage, the second consideration of the switch design is the development of contact force, which is directly related to the contact resistance. The electromechanical analysis is continued beyond the pull-in voltage once contact is established to develop an increasing contact force. The electrostatic force is distributed between the anchor and the contacts, assuming other portions of the switch are not in contact. The electrostatic force increases as the square of the voltage difference, so the force acting on the contact will increase in proportion to this. Of course other physical limitations must be considered such as gas breakdown or a secondary pull-in that cause shorting between electrodes. The contact areas of the switch that are not the contacts should be minimized, so this does not detract from the contact force.

A third design consideration is the overdrive capability, defined as the maximum voltage the switch control electrodes can support before they touch and short. In some cases, the electrodes may be isolated from each other by a thin dielectric film, whose implications will be discussed in a later section. As suggested in the later section, the insulation material can be patterned to reduce the contacting area between the electrodes and the insulating material, in order to reduce charge accumulation. With the patterned isolation, the switch could be driven by such a large voltage to short the uninsulated regions. This failure mode could be evaluated by the same electromechanics analysis.

A fourth design issue would be the switch's release voltage, which is the voltage when the contact force reaches zero and the switch opens. This is shown as the beginning of region 3, where the deformation is reduced as the voltage is decreased. The release voltage is lower than the pull-in voltage because the electrostatic forces are much greater for the reduced gap than compared to the larger gap just before the pull-in instability. A fifth design consideration is the switch's self-actuation voltage, which is that which occurs across open contact and unintentionally causes the switch to close. By design, the switch's self-actuation voltage should be maximized in the electromechanical analysis. A final design issue is the determination of breakdown limits across the switch's control electrodes. This is not limited to the voltage supported across the equilibrium gap of the control electrodes and needs to be considered for different cases. The breakdown will be described by the Paschen curve or by vacuum breakdown proportional to the gap.

35.2.2 Electromechanics Analysis

An electromechanics analysis requires modeling of the structural domain, the electrostatic domain, and the coupling between the electrostatic fields and the structural elements. It has been approached at three levels of complexity. The simplest approach is the development of lumped parameter models, which have provided the basic understanding of instability, the impact of nonlinear springs, and the pull-in voltage-gap relationship. The lumped parameter model is comprised of the parallel plate capacitor and a spring (linear or nonlinear). One plate is constrained in space, by a substrate for example. The second plate is elastically constrained to one degree-of-freedom, typically, of motion, which is usually closing the "air" gap. The spring element represents the effective spring stiffness K of the structural element that could be a beam, a plate, a membrane. The parallel plate capacitor represents the electrostatic coupling between the fixed and moving elements.

Increasing in complexity, the second level of analysis considers continuum structural elements with a distributed electrostatic load. For example, the solution for an isotropic homogeneous beam would satisfy the Euler–Bernoulli beam equation:

$$EI\frac{\mathrm{d}^4 w}{\mathrm{d}x^4} = q(x)\,, \qquad (35.37)$$

where E is Young's modulus, I is the structure's moment of inertia, $w(x)$ is the transverse deflection of the beam, x is the coordinate along the length of the beam, and $q(x)$ represents the distributed electrostatic load. For a multilayer film, an effective EI can be used, as mentioned in Sect. 35.1.2. The third level of analysis is to model the fully coupled problem as continuum solids. The approach has used different implementations of finite difference (FD), finite element (FE), and boundary element (BE). A typical example would be to implement a finite element code to solve the continuum mechanics problem and a boundary element code to solve the electrostatics problem. Another implementation would use finite element methods to solve both the mechanics and the electrostatics problems. In each case, a similar algorithm is followed for developing a self-consistent solution to the coupled electromechanics analysis.

The coupled electromechanics problem can be described simply as a pair of conductors, where one conductor has a fixed constraint and the second conductor is attached to an elastic structure, such as a beam, a plate, a membrane, or an elastic substrate. The conductors support a voltage difference between them that causes charge to be induced on the surface of the conductors. This charge produces an electrostatic force that attracts the two conductors together. As the elastically supported conductor deforms toward the fixed conductor, the charge redistributes and modifies the field, thereby the force distribution. The process continues until a redistribution of charge is no longer necessary to maintain equilibrium. The typical algorithm begins by calculating the displacements of the undeformed geometry, which is updated before proceeding to the electrostatic analysis. In the electrostatic analysis, the surface charge density distribution is calculated and then used to calculate the electrostatic forces. The electrostatic force distribution is the important parameter because it is passed to the mechanical analysis as the loading condition on the conductors in the undeformed state to deform the structure until a state of equilibrium is reached.

Electromechanical Systems Energy Balance

A general analytical methodology can be applied to many different structures including parallel plate actuators with linear or nonlinear springs and plates supported by torsional suspensions [35.66]. They consider a single input system with charge control or voltage control. To begin, a generalized actuator is considered that consists of two conductors. One conductor is fixed. The second conductor is constrained by an elastic support.

In the charge controlled actuator, the total energy for the generalized actuator is written in terms of the mechanical energy as a function of the generalized coordinate χ and of the electrical energy stored in the capacitance of the actuator. The total energy is:

$$U_\mathrm{T}(\chi, Q) = U_\mathrm{M}(\chi) + \frac{Q^2}{2C(\chi)}\,, \qquad (35.38)$$

where U_T is the total energy, U_M is the mechanical energy, and the electrical energy is determined by the charge Q, and the capacitance $C(\chi)$. The first derivative of the total energy with respect to the generalized coordinate set equal to zero determines the equilibrium. The second derivative of the total energy with respect to the generalized coordinate set equal to zero determines the stability. The first and second derivatives are determined and then combined to determine what is referred to as the charge-controlled pull-in equation:

$$\frac{\partial U_M(\chi_{PI})}{\partial \chi}\frac{\partial^2}{\partial \chi^2}\left(\frac{1}{C(\chi_{PI})}\right) - \frac{\partial^2 U_M(\chi_{PI})}{\partial \chi^2}\frac{\partial}{\partial \chi}\left(\frac{1}{C(\chi_{PI})}\right) = 0 \qquad (35.39)$$

and the pull-in charge Q_{PI}:

$$Q_{PI} = \left(\frac{-2\dfrac{\partial U_M(\chi_{PI})}{\partial \chi}}{\dfrac{\partial C(\chi_{PI})}{\partial \chi}}\right)^{1/2} . \qquad (35.40)$$

In the voltage controlled case, the total co-energy for the generalized actuator is used because the capacitor is nonlinear, and the actuator is voltage controlled. The co-energy is written in terms of the capacitance between the generalized conductors, the voltage across the conductors, and the mechanical energy. The total co-energy is:

$$U_T^*(\chi, Q) = \frac{1}{2}C(\chi)V^2 - U_M(\chi) . \qquad (35.41)$$

The relationship between energy and co-energy is shown in Fig. 35.23.

In a similar process to the charge-controlled system, the voltage-controlled pull-in equation is determined by the first derivative of the total co-energy with respect to the generalized coordinate and by the second derivative of the total co-energy with respect to the generalized coordinate. The equations for the first and second derivative are combined to determine the voltage-controlled pull-in equation

$$\frac{\partial U_M(\chi_{PI})}{\partial \chi}\frac{\partial^2 C(\chi_{PI})}{\partial \chi^2} - \frac{\partial^2 U_M(\chi_{PI})}{\partial \chi^2}\frac{\partial C(\chi_{PI})}{\partial \chi} = 0 \qquad (35.42)$$

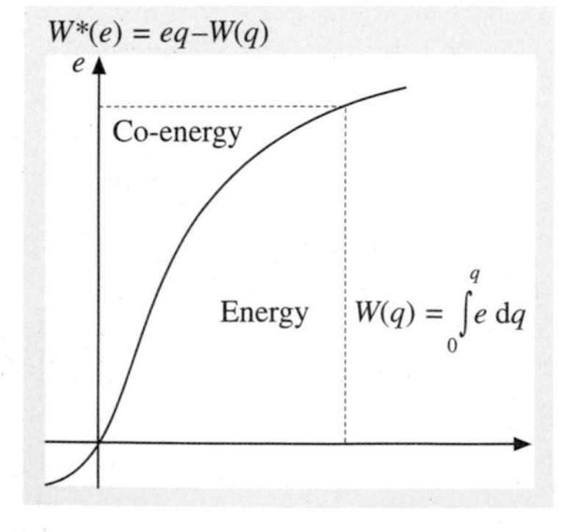

Fig. 35.23 Effort-Displacement curve showing the relationship between energy and co-energy

The pull-in voltage V_{PI} is:

$$V_{PI} = \left(\frac{2\dfrac{\partial U_M(\chi_{PI})}{\partial \chi}}{\dfrac{\partial C(\chi_{PI})}{\partial \chi}}\right)^{1/2} . \qquad (35.43)$$

The application of this approach, to either the voltage-controlled or the charge-controlled case involves determining the mechanical energy and the capacitance. From the mechanical energy and the capacitance, the pull-in parameters and voltage are determined.

Lagrangian Approach to Electromechanics
Li and *Aluru* [35.67] and *Aluru* and *White* [35.68] have proposed a Lagrangian approach for both the electrostatic analysis and the mechanical analysis. Their algorithm begins by calculating the structural displacements for the undeformed geometry. They proceed to the electrostatic analysis to calculate the charge distribution on the conductors in the undeformed state. The charge distribution is used to calculate the electrostatic forces on the undeformed geometry until the system reaches equilibrium. The first advantage of their approach is that the algorithm does not require the structural geometry to be updated. A second advantage is the elimination of the integration error, wich occurs because flat panels are used in other approaches to approximate curved surfaces. The overall algorithm is shortened because the interpolation functions do not have to be updated as the structure shape changes.

35.2.3 Electromechanics – Parallel-Plate Capacitor

The classic element for understanding electromechanics is the simple parallel-plate capacitor model with one plate fixed and the second, moveable, plate, suspended by a spring. This basic electromechanics element is shown in Fig. 35.24. The capacitor is described by an area, A; an initial gap, g_0; and a dielectric constant of ε in the air gap (where ε is the product of the relative dielectric of the media filling the gap, ε_r, and the permittivity of free space, $\varepsilon_0 = 8.854\times10^{-12}$ F/m). The moveable plate is shown suspended by the spring, with spring constant K, and a single degree-of-freedom represented by the coordinate z. In this case, the equilibrium position is represented by having no charge on the plates (i. e. no electrostatic force) and no displacement of the spring (i. e. no spring force). It should be noted that as the spring length grows by increasing z, the gap decreases. The behavior of this simple model is described in detail in many sources including [35.49, 66].

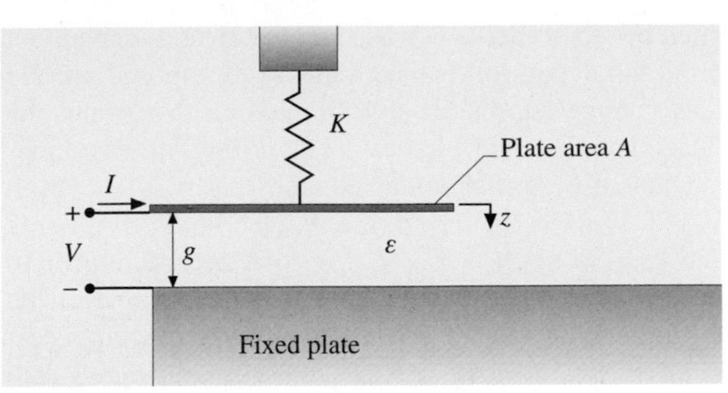

Fig. 35.24 Lumped parameter parallel plate capacitor electromechanical model

The parallel plate capacitor can be analyzed with either a voltage source or a current source. With a voltage source, the charge on the capacitor plates is

$$Q = \frac{\varepsilon A}{g} V , \tag{35.44}$$

where Q is the charge on the capacitor plates, ε is the dielectric constant in the gap, A is the capacitor plate area, g is the instantaneous gap between plates, and V is the voltage across the plates.

The electrostatic force on the capacitor plates is determined by the dielectric constant, the capacitor plate area, the instantaneous gap, and the voltage across the capacitor:

$$F = \frac{\varepsilon A V^2}{2g^2} . \tag{35.45}$$

The electrostatic force scales as the voltage squared and inversely diminishes as the instantaneous gap squared. This is the force that acts on the movable plate; hence the spring stretches (or is relaxed if the voltage is lowered). The amount of stretch in the spring is:

$$z = \frac{F}{K} . \tag{35.46}$$

The instantaneous gap is

$$g = g_0 - z = g_0 - \frac{\varepsilon A V^2}{2Kg^2} . \tag{35.47}$$

Electromechanics – Parallel Plate Capacitor with External Forcing

Nemirovsky et al. [35.66] examined the case of constant applied force acting on the suspended capacitor plate of a parallel-plate electrostatic actuator. This situation represents a force-balanced system, in which the electrostatic force is used to balance a pressure load acting on a membrane or an inertial load (e.g. acceleration induced) on a suspended proof mass. With an applied mechanical force, the mechanical energy is:

$$U_M(z) = \frac{1}{2} K z^2 - F_{ext} z , \tag{35.48}$$

where F_{ext} is the externally applied force. The solution to the pull-in equation is

$$\zeta_{PI} = \frac{z}{z_{max}} = \frac{1}{3}\left(1 + 2\frac{F_{ext}/K}{z_{max}}\right) , \tag{35.49}$$

where ζ_{PI} is the ratio z/z_{max}. It can be seen when the applied force is zero. Equation (35.49) reduces to the familiar result of $z = z_{max}/3$. The pull-in voltage is

$$V_{PI} = \left(\frac{8K\left(z_{max} - \frac{F_{ext}}{K}\right)^3}{27\varepsilon_0 A} \right)^{1/2} . \tag{35.50}$$

Electromechanics – Parallel Plate Stabilization

An important practical consideration is the stabilization of the parallel plate electrostatic actuator. *Seeger* et al. [35.69] and Nemirovsky et al. [35.66] showed that the addition of a series capacitance (feedback capacitance) stabilizes the actuator such that it does not pull-in to the substrate. The series capacitance and the variable actuator capacitance combine to form a voltage divider that provides passive control of the voltage across the actuator. In this case, Nemirovsky et al. [35.66] identified three regimes, defined in terms of δ, which is the ratio of the initial actuator capacitance to the series capacitance (Nemirovsky et al. [35.66] had defined the series capacitance in terms of a dielectric layer on one capacitor plate). The first regime is defined for $0 < \delta < 0.5$, when the series capacitance is less than half the nominal actuator capacitance. In this case, the actuator is stable so the pull-in effect is eliminated. The second regime is defined by $\delta > 0.5$ and the pull-in occurs between $1/3$ of the gap and the full distance of the gap. In the third regime $\delta \gg 1$, which means the series capacitance is much greater than the nominal capacitance. The pull-in parameter is given by:

$$\zeta_{PI} = \frac{1}{3}\left(1 + \frac{1}{\delta}\right) , \tag{35.51}$$

which reduces to $\zeta_{PI} = \zeta_{max}/3$ when the series capacitance is eliminated. The pull-in voltage is:

$$V_{PI} = \left[\frac{8Kz_{max}^3}{27\varepsilon_0 A}\left(1+\frac{1}{\delta}\right)^3\right]^{1/2} . \qquad (35.52)$$

Electromechanics – Extending the Range of Motion

Electromechanical systems have two obvious characteristics: nonlinearity and instability. If one desires digital operation or fast switching times, the inherent electromechanical instability is a significant benefit. On the other hand if analog control is desired, the control is faced with both the nonlinearity and the instability. In both cases, the range of motion of the electromechanical device is limited. The range of motion to the instability was extended by the addition of a series capacitor, but this did not address the linearity. *Toshiyoshi* et al. [35.59] addressed the linearity of electrostatically actuated mirrors in an optical scanner. Their's was a two degree-of-freedom scanner, so the increased linearity was necessary to improve the crosstalk between the two different scanner modes. The scanner in question is a square plate suspended by torsion beams within a frame that is itself suspended by another pair of torsion beams. The plate and frame are suspended above the substrate, which has four stationary electrodes defined in the four quadrants of the plate as shown in Fig. 35.25. To actuate the mirror with increased linearity, a bias voltage V_{bias} was applied between each electrode and the plate. Independent differential control voltages are superimposed on the bias voltage supplied between each electrode and the plate. The driving voltages are described by two independent differential voltages such that $V_{diff1} = (V_x + V_y)/2$, $V_{diff2} = (-V_x + V_y)/2$, $V_{diff3} = (-V_x + -V_y)/2$ and $V_{diff4} = (V_x + -V_y)/2$. A second approach taken to improve the linearity was to consider the shape of the mirror, because of the geometric relationship that determines the net electrostatic torque. They considered such simple geometries as square, diamond (square rotated by 45), and a circular plate, as shown in Fig. 35.26, but more complex shapes could be considered. The square plate is presumed to have worse linearity because as the plate rotates the corners deflect toward the control electrodes, which increases the applied torque and further distorts the shape of the mirror. This suggested the removal of the corners could contribute to increased linearity and reduced distortion of the mirror surface. They found the circular plates reduced the distortion by 50% over the rectangular and diamond shaped plates, which is significant for the optical performance. The authors defined distortion as the normalized angle deviation from the ideal value. It is normalized relative to the maximum scan angle. *Nadal-Guardia* et al. [35.70] have described extending the range of travel by using current drive methods instead of voltage drive.

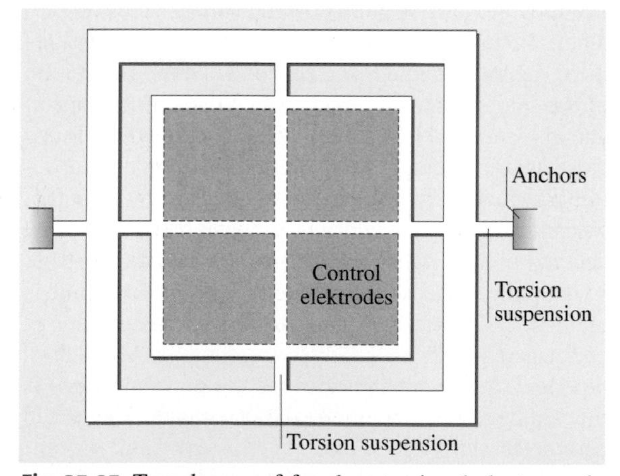

Fig. 35.25 Two degree-of-freedom torsional electromechanical actuator with a rectangular plate and a stationary electrode divided into four sections

35.2.4 Electromechanics of Beams and Plates

The basic understanding of electromechanics and pull-in phenomena has been developed around the simple parallel plate capacitor model developed above. Many MEMS designers innately understand that the electrostatic force is proportional to the square of the voltage, is inversely proportional to the square of the gap, and

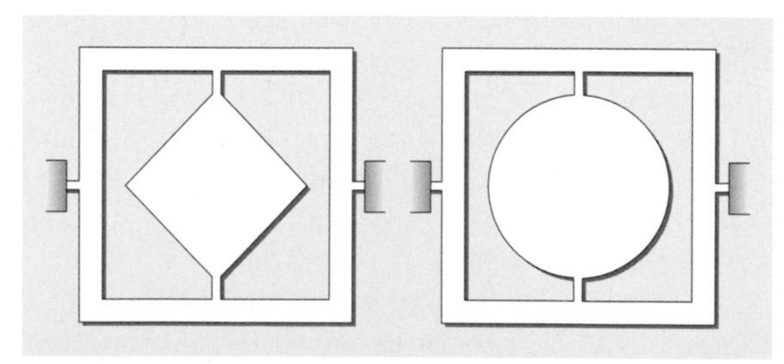

Fig. 35.26 Alternate two degree-of-freedom torsional, electromechanical actuators with diamond and circular plates

that the plate will pull in after displacing 1/3 of the gap. While this is good to first order, it is not adequate for the electromechanical design of other structures such as cantilevered beams/plates, beams/plates with multiple supports, plates with torsional supports, and other structures.

O'Brien et al. [35.71] and *Choi* and *Lovell* [35.72] have studied the electromechanics of cantilever beams by presenting two improved models for the cantilever beam electrostatic actuators. In both cases, the cantilever beam is a homogeneous material anchored to the substrate at one end and otherwise freely suspended along its length, which is supported above a stationary electrode. The new models are compared to the parallel plate capacitance model and a simple extension of the parallel plate model to a cantilever beam. The cantilever beam model uses a simple cantilever of width, W, length, L, thickness, T, and initial air gap of Z_0. The parallel plate capacitance model is based on a capacitance determined by the length and width of the beam. The beam deflection was represented by the displacement function of a beam with a force applied at the beam tip. The beam spring constant was determined by the ratio of the force at the beam tip to the displacement at the beam tip (maximum displacement). The capacitance was determined by integrating the parallel plate capacitance for a differential length element along the length of the beam, where the beam gap is defined by the initial gap in the mechanical displacement function. This showed the beam tip displaced 45% of the initial gap before pull-in occurred. In this case the pull-in voltage can be expressed as a function of the beam's Young's modulus E, cross-sectional moment of inertia I, length L, and width W, the initial gap Z_0, and the dielectric constant in the air gap as

$$V_1 = \sqrt{\frac{18EIZ_0^3}{5\varepsilon_0 L^4 W}} = 0.5477\sqrt{\frac{ET^3 Z_0^3}{\varepsilon_0 L^4}} \quad \text{(Volts)} . \tag{35.53}$$

In the first improved model, the cantilever beam is defined by the same characteristics, but the electrostatic force acting along the length of the beam is replaced by an equivalent bending moment M applied at the beam tip. The beam displacement was determined for a uniform, homogeneous beam with a bending moment applied at its end. The beam's spring constant was determined by the ratio of the force at the beam tip and the displacement at the beam tip. The capacitance was determined by integrating the parallel plate capacitance for a differential length element along the length of the beam, where the beam gap is defined by the initial gap in the mechanical displacement function. This model showed the beam tip displacement was approximately 46% of the initial gap before pull-in occurred. In this case the pull-in voltage is

$$V_2 = \frac{43}{50}\sqrt{\frac{18EIZ_0^3}{5\varepsilon_0 L^4 W}} = 0.471\sqrt{\frac{ET^3 Z_0^3}{\varepsilon_0 L^4}} \quad \text{(Volts)} . \tag{35.54}$$

In the second improved model, the cantilever beam was defined by a rigid beam, whose anchor was replaced by a simple hinge and continuum elasticity replaced by an equivalent spring at the beam's free end. Three new variables were introduced to describe the behavior in terms of radial coordinates centered at the hinge. The two radial coordinates describe the beginning of the electrode at r_0 and the end of the beam/electrode at r_1. The angle θ was introduced to describe the angular displacement of the beam. In this model, the tip of the beam displaced 44% of the initial gap before pulling into the substrate. The pull-in voltage is estimated by (35.55) or (35.56) where r_0 and r_1 have been replaced by 0 and L.

$$V_3 = \frac{11}{25r_1^2}\sqrt{\frac{77ET^3 Z_0^3}{25\varepsilon_0\left[\frac{11(r_1-r_0)}{r_1-\frac{11}{25}r_0} - 14\ln\left(\frac{25r_1-11r_0}{14r_1}\right)\right]}} , \tag{35.55}$$

$$V_3 = \frac{11}{25L^2}\sqrt{\frac{77ET^3 Z_0^3}{25\varepsilon_0}} = 0.4548\sqrt{\frac{ET^3 Z_0^3}{\varepsilon_0 L^4}} \quad \text{(Volts)} . \tag{35.56}$$

As expected, the dependence on parameters, such as Young's modulus, beam thickness, beam length, initial air gap, and the air gap dielectric constant is the same for all cases presented. The difference in the three models is the coefficient of the pull-in voltage function and the tip deflection at pull-in. The coefficients of the pull-in voltage predictions are 0.5479, 0.471 and 0.4548, respectively. The beam tip deflected 45%, 44%, and 46% of the initial gap before pull-in occurred. This is in contrast to the parallel plate model that predicts the plate deflection will be 33% of the initial gap before pull-in. By comparison to experimental test results, the models presented in (35.54) and (35.56) were more accurate at predicting the pull-in voltage for cantilever beams. The parallel plate electromechanics model had greater error and under-predicted the pull-in voltage. The model represented by (35.53) over-predicted the pull-in voltage with greater error than either of (35.54) or (35.55). The beam models predicted more accurately for wider beams, which means they are better applied to beams with dimensions representative of a parallel plate model used as a basis of the capacitance estimate. In all three cases, the beam models presented in (35.53)–(35.55) predicted poorly the pull-in voltage for narrow beams, which are dominated by the effects of nonuniform fringing fields and not the uniform parallel plate fields. The typical application of the parallel plate capacitor model is to geometries with a width W and length L that are much greater than the capacitor gap Z_0 ($W \gg Z_0$, $L \gg Z_0$; where greater than a factor of 10 is a common rule of thumb). In Fig. 35.27, two capacitors are shown such that in one case the plate dimensions are much greater than the gap, and in the second case the plate dimensions are much smaller than the gap. When the plate dimensions are much larger than the gap, the uniform fields in the gap dominate the capacitive coupling. A small contribution to the capacitive coupling comes from the *fringing* fields at the edge of the plate (or the perimeter of the plate in three dimensions). As shown in Fig. 35.6, the fringing fields could depend on the thickness of the plate. When the plate dimensions are of the same order as the gap, the fringing fields may dominate or be of the same order as the contribution of the uniform field in the gap. This is the nature of the narrow cantilever beams that were not predicted well by a parallel plate capacitor model. A correction can be introduced to the parallel plate model to approximate the contribution of the fringing fields. This approximation is often referred to as the Love approximation.

Segmented Control of Electromechanical Membrane Deformation

Wang and *Hadaegh* [35.62] describe the electrostatic actuation of a deformable membrane mirror with segmented electrodes on the substrate. The membrane mirror, in this case silicon nitride with a conductive film, was suspended above the substrate containing the segmented electrode. The segmented electrodes are defined along angular and radial pattern. A specified voltage can be applied between the membrane and each electrode segment to develop a specific deformation for the mirror. This increases the complexity of the control system but also the flexibility of controlling the shape of the mirror surface. The optical properties could be optimized by individual characterization of membranes or by feedback from the optical output. This basic concept was also applied by *Zhang* and *Dunn* [35.73] who were able to linearize the voltage–deflection relationship by tailoring the shape of the electrodes and the beam structure used to support them.

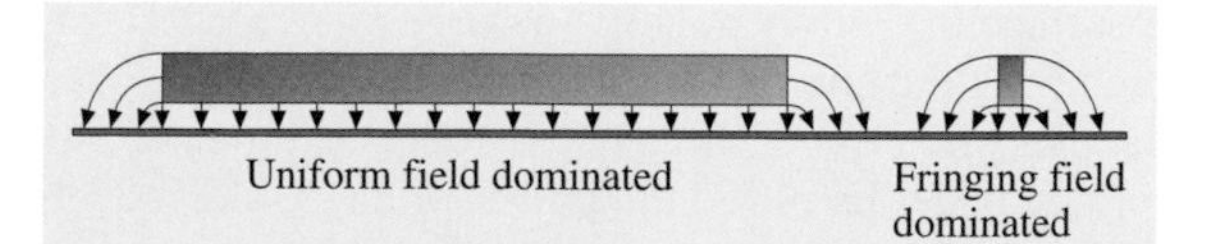

Fig. 35.27 Cross-sections of wide and narrow electrodes showing uniform electric fields in the gap and fringing fields at the edges (perimeter)

35.2.5 Electromechanics of Torsional Plates

The plate suspended by a torsional suspension instead of bending is the final configuration to be considered in terms of electromechanics. The electrostatic torsional plate has many applications, such as optical mirrors [35.58,63] and accelerometers [35.55]. The plate will define the optical surface of the mirror, the moveable plate of a varactor, or the proof mass of an accelerometer. It will be conductive or have a conductive film so that it can couple electrostatically to a stationary plate on the substrate. The moveable plate will be suspended by a torsional suspension such that the bending modes are suppressed and the primary kinematic mode is rotation of the plate. The variable capacitive coupling is used in two ways. First, it is used for actuation by applying a voltage between the moveable plate and a stationary plate. Second, it is used for sensing or use of the variable capacitance. Both actuation function and the variable capacitance function can be present in a device. The key elements of interest are the voltage-displacement function and the pull-in voltage.

Small Angular Deflection Approximation

Degani et al. [35.74], *Xiao* et al. [35.75], and Nemirovsky et al. [35.66] have considered the torsional electrostatic actuator from an analytical perspective with comparison to experimental results and other modeling results. The effective torsional spring constant is determined from the torque-twist angle relationship, where the torque is the electrostatic torque. The electrostatic torque is developed for both small and large angle cases. The pull-in angle and pull-in voltage are determined from the pull-in equations for voltage control (35.42, 43) and charge control (35.39, 40). Their method is an improved approach for modeling torsional electrostatic actuators over approximations involving parallel plate models and effective spring constants as they have shown by comparison to experimental measurements. The torsional plate considered is suspended by a torsional suspension, whose axis of rotation defines a gap of d. It is assumed the plates extend into the page (z-axis) by a depth b and have a total length a. The torsional plate electrostatic actuator is shown in Fig. 35.28. The solution assumes the tilted plates are semi-infinite, ignoring the effects of fringing fields hence their corrections, and are tilted by angles that satisfy the small angle approximation ($\tan\alpha \approx \sin\alpha \approx \alpha$ and $\cos\alpha \approx 1$). The capacitance is:

$$C(\Theta) = \frac{\varepsilon_0 b}{\alpha_{\max}} \frac{1}{\Theta} \ln\left(\frac{1}{1-\Theta}\right) , \tag{35.57}$$

where $\alpha_{\max} = d/a$ and $\Theta = \alpha/\alpha_{\max}$.

The capacitance function and the linear relationship for the elastic restoring force ($F_M = K\,a$) are used to develop the electrostatic and mechanical energy function to substitute in the pull-in equations. For voltage controlled actuation, the pull-in parameter is developed from (35.42) to yield:

$$4 + \frac{5\Theta_{PI} - 4}{(1-\Theta_{PI})^2} - 3\ln(1-\Theta_{PI}) = 0 . \tag{35.58}$$

Equation (35.58) is solved numerically to show the voltage-controlled pull-in angle is $\Theta_{PI} \approx 0.44$. The pull-in angle is substituted in (35.43) to determine the voltage controlled pull-in voltage:

$$V_{PI} = \left(\frac{0.827 K \alpha_{\max}^3}{\varepsilon_0 b}\right)^{1/2} . \tag{35.59}$$

The charge-controlled pull-in parameter relationship is determined by substituting the electrostatic energy and mechanical energy functions in the charge-controlled pull-in parameter (35.39). This provides the nonlinear relationship for the charge controlled pull-in angle:

$$\ln(1-\Theta_{PI}) - \frac{\Theta_{PI}}{(1-\Theta_{PI})^2} - \frac{2\Theta_{PI}^2}{(1-\Theta_{PI})^2 \ln(1-\Theta_{PI})} = 0 . \tag{35.60}$$

The charge-controlled pull-in angle is given by $\Theta_{PI} \approx 0.71$. The pull-in angle is used in the charge-controlled pull-in equation (35.40) to determine the pull-in charge relationship:

$$Q_{PI} = (1.798 K \varepsilon_0 b \alpha_{\max})^{1/2} . \tag{35.61}$$

Large Angular Deflection and Plate Geometry

Nemirovsky and *Bochobza-Degani* [35.66] have extended the work to the case of large angle deflections, which included using the trigonometric relationship in the capacitance function and a nonlinear relationship for the mechanical spring constant. These nonlinear relationships were used in the voltage-controlled pull-in equation. In addition, their methodology was extended to planar plates with different shapes (Fig. 35.29). They specifically considered triangular plates (the apex of the triangle is near the axis of rotation), a reversed triangular plate (the triangle apex is at the distal end of the plate), a square plate (with the rotational axis on the edge of the plate), and a circular plate (with the axis of rotation passing through a point on the perimeter and the opposite point on the perimeter defining the most distal point of the plate). Only simple geometries were considered, and their methodology can be applied to other more general geometry. The rectangular plate torsional, electrostatic actuator will provide the lowest actuation voltage relative to a fixed plate length. The reversed triangular plate has proved beneficial with regard to pixel area in mirror application and has been able to provide a higher torque for a specific plate length.

35.2.6 Leveraged Bending

As discussed, when the actuation electrode spans the entire length of the actuator, the stable controllable travel distance of the actuator plate is severely limited due to the pull-in instability. One can get around the problem of

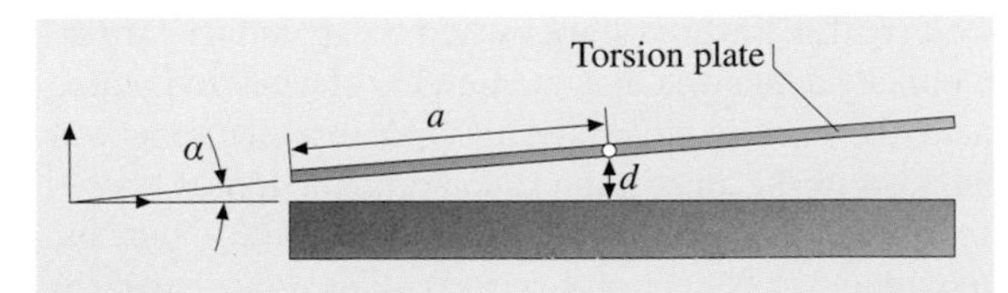

Fig. 35.28 Cross-section of a torsional actuator with a uniform cross section into the page showing the angular degree-of-freedom, α; plate length, a; and initial gap, d

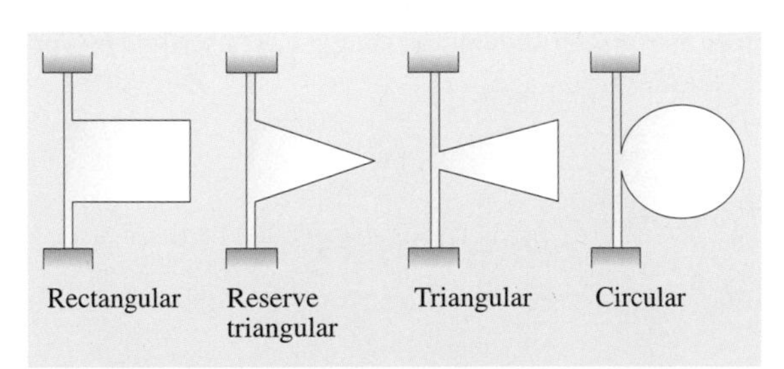

Fig. 35.29 Rectangular, triangular, reverse triangular, and circular plates are shown suspended by torsion beams for application to electromechanical mirrors or other actuators

pull-in by applying an electrostatic force to only part of the structure and using the rest of the structure as a lever to move the structure through a larger range of motion. This concept, referred to as leveraged bending [35.76], is shown schematically in Fig. 35.30 for a cantilever beam. The key idea is that the portions of the beam above the electrodes do not deflect far enough to violate the pull-in limit. The concept can be applied to other microstructure shapes and support conditions, but for fixed-fixed beams, the effect can be diminished by tensile residual stresses [35.76].

35.2.7 Electromechanics of Zipper Actuators

The design requirements for electromechanical actuators usually require minimizing the actuation voltage but maximizing the largest displacement. With the parallel plate actuator, be it a plate, beam, or torsional plate, the pull-in voltage is reduced by minimizing the air gap, but this limits the maximum displacement. Actuators with a zippering behavior have been implemented to achieve greater deflections for a lower actuation voltage. In one example (see [35.77, 78]), a plate is suspended by a torsional suspension. A long slender electrode extends from the plate and terminates with plate-like area. This electrode is supported above the fixed substrate electrode by a largely uniform gap. A voltage is applied between the substrate electrode and the extended electrode, such that the plate-like area is first pulled into the substrate. As the actuation voltage is increased, the long-slender electrode continues to collapse against the substrate electrode until a desired maximum deflection is achieved or until the drive voltage becomes impractical. This is the so-called zippering effect, which has been described for a motion perpendicular to the substrate. This uniform gap scenario is not the best suited for minimizing the pull-in voltage, so other configurations such as the curved electrode actuator have been proposed and applied to such devices as a switch and optical shutters as described by *Jin* et al. [35.79].

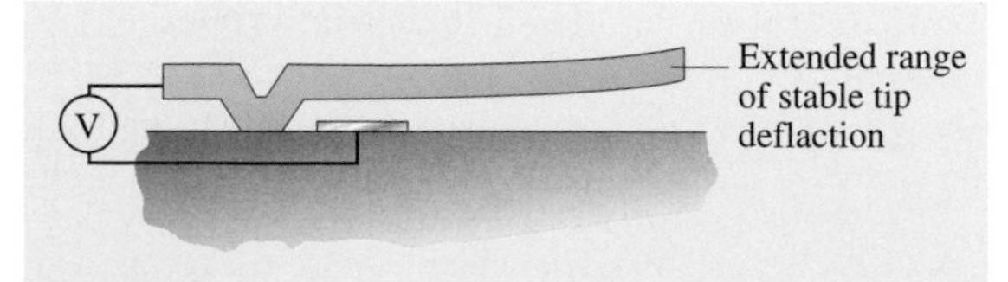

Fig. 35.30 Illustration of the concept of leveraged bending to increase the range of stable motion

Legtenberg et al. [35.80, 81] have described the basic operating features of the curved-electrode actuator. These actuators will provide large deflections in the plane of the substrate. The curved electrode actuator is composed of a stationary electrode and a compliant, moving electrode. The stationary electrode has a shape described by the function $s(x)$ shown in (35.62) and Fig. 35.31, which is written in terms of the maximum gap between electrodes δ_{max}, the length of the electrode L, the distance along the length of the electrode x, and the polynomial order of the curve n.

$$s(x) = \delta_{max}\left(\frac{x}{L}\right)^n . \tag{35.62}$$

A necessary feature of the curved electrode actuators and zipper actuators, in general, is the contact between the moving and stationary electrodes. Without due consideration, these electrodes will short upon contact and damage the actuator. Two isolation solutions are implemented to prevent the shorting scenario. First, a continuous insulation film can be added between the electrodes to prevent them from shorting. In this case, the simplest solution is to add the insulation layer to the fixed, curved electrode. The alternative solutions include adding the insulating film to only the moving electrode or to both the moving and stationary electrodes. If the insulation is added to the moving electrode, it will become part of the stiffness of the beam. A second solution would be to add discrete, isolated contact bumpers that are made from the same conducting material as the moving electrode. As it closes the gap, the moving electrode makes contact with the isolated bumpers but not the actuation electrode. These have their operational differences as described by *Legtenberg* et al. [35.81]. Three other issues that must be considered during the design of a curved electrode actuator or zipper actuators are dielectric charging, stiction, and dielectric breakdown.

Legtenberg et al. [35.81] found that as the polynomial order increased for the electrode curve, the pull-in voltage will decrease for about the same maximum displacement. This means larger displacements can be achieved for lower voltages when compared to the parallel plate actuator. Once the moving electrode is pulled in to the curved electrode, the moving electrode will collapse to the maximum displacement. It was shown for polynomial curves of order above $n = 2$ that the maximum tip deflection becomes stable because of the geometric constraints provided by the curved electrode. This means to achieve the maximum displacement with the lowest pull-in voltage, a polynomial curve of order $n = 2$ should be specified. For polynomial curves with $n > 2$, stable continuous deflection along the curved electrode can be implemented with the maximum deflection occurring at lower voltages for polynomial orders just above two. If an approximate linear range of operation is desired (i. e. the tip deflection versus voltage is approximately linear), the higher order curves ($n > 2$) are desired. *Legtenberg* et al. [35.81] analyzed the curved electrode problem by application of analytical energy methods, specifically the Rayleigh–Ritz method and of three-dimensional self-consistent electromechanical numerical simulations using CoSolve-EM in MEMCAD (now integrated into CoventorWare). From an energy perspective, the total potential energy U_T is determined in terms of the mechanical energy associated with the strain energy of bending and the electrostatic energy stored in the electric field. For the curved electrode of Fig. 35.31, the total potential energy is:

$$U_\mathrm{T} = U_\mathrm{M} + U_\mathrm{E} = \frac{1}{2}\int_0^L EI\left(\frac{\mathrm{d}^2 w(x)}{\mathrm{d}x^2}\right)^2 \mathrm{d}x - \frac{1}{2}\int_0^L \frac{\varepsilon_0 h V^2}{\frac{d}{\varepsilon_r} + s(x) - w(x)}\,\mathrm{d}x\,, \tag{35.63}$$

where L is the beam length, E is the beam's Young's modulus, I is the beam's cross-sectional moment of inertia, h is the thickness of the beam, V is the voltage applied between the beam and electrode, d is the initial gap at the start of the electrode, ε_0 is the permittivity of free space, and $w(x)$ is the beam's deflection curve. The next step in the solution of this problem is to determine an admissible trial function that satisfies the beam boundary conditions and can adequately approximate the solution. In this case the deflection curve for a uniformly loaded cantilever beam is chosen as the trial function:

$$\omega(x) = cx^2\left(6L^2 - 4Lx + x^2\right) = c f(x)\,, \tag{35.64}$$

where c is a constant to be determined for specific cases.

The approach of Nemirovsky et al. [35.66] can be taken by substituting the potential energy functions in the voltage-controlled pull-in equations. The simultaneous solution of the pull-in equations will result in the pull-in voltage and an implicit equation in terms of the constant c at pull-in. The pull-in voltage relationship is

$$V_\mathrm{PI}^2 = \frac{EI}{\varepsilon_0 h}\frac{\int_0^L \left(\frac{\mathrm{d}^2 f(x)}{\mathrm{d}x^2}\right)^2 \mathrm{d}x}{\int_0^L \frac{f^2(x)}{[d/\varepsilon_r + s(x) - c_\mathrm{PI} f(x)]^3}\,\mathrm{d}x}\,. \tag{35.65}$$

The implicit equation to determine the coefficient of the beam deformation function is

$$\int_0^L \frac{c_\mathrm{PI} f^2(x)\,\mathrm{d}x}{[d/\varepsilon_\mathrm{r} + s(x) - c_\mathrm{PI} f(x)]^3} = \int_0^L \frac{f(x)\,\mathrm{d}x}{[d/\varepsilon_\mathrm{r} + s(x) - c_\mathrm{PI} f(x)]^2}\,. \tag{35.66}$$

DeReus [35.82] designed the curved electrode actuator to convert the curvilinear motion into a translational motion at the tip of the moving electrode. The transformation is accomplished by a folded spring that connects the electrode to the body that will be translating. A solid model of this actuator design is shown in Fig. 35.32. A pull-in analysis was performed with the results before pull-in and after pull-in presented in Fig. 35.33. This shows that a 60 μm translation was accomplished

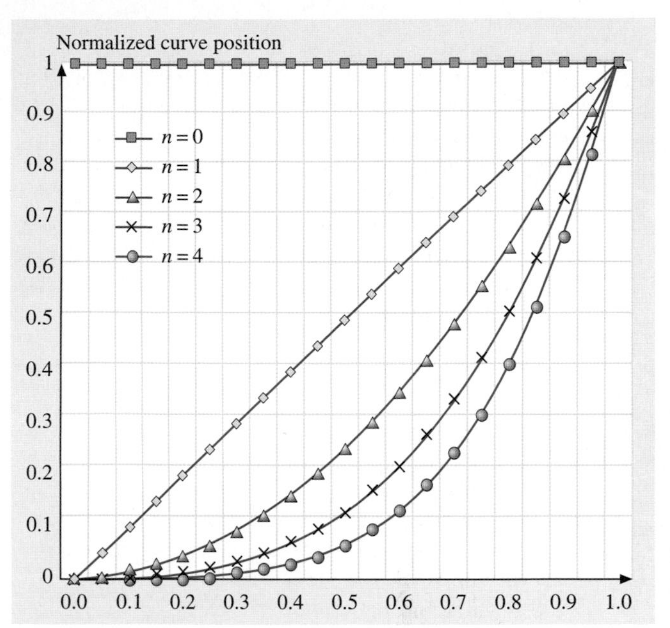

Fig. 35.31 Curved electrode shape curve shown as a function of order of the curve n

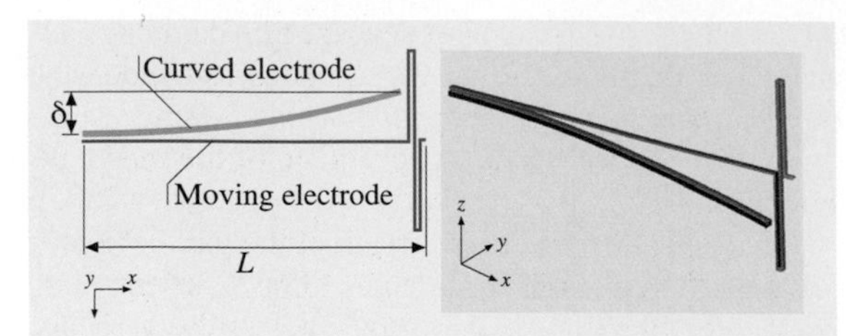

Fig. 35.32 Solid model of a curved electrode actuator with a folded suspension showing the curved electrode δ and the actuator beam length L

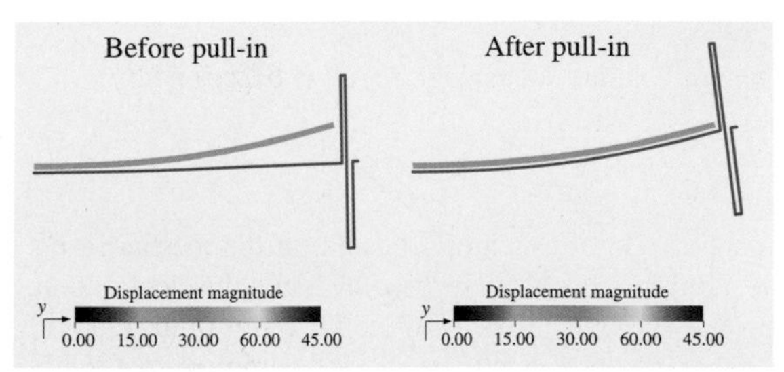

Fig. 35.33 Results of electromechanical pull-in analysis showing curved electrode actuator before and after pull-in has occurred

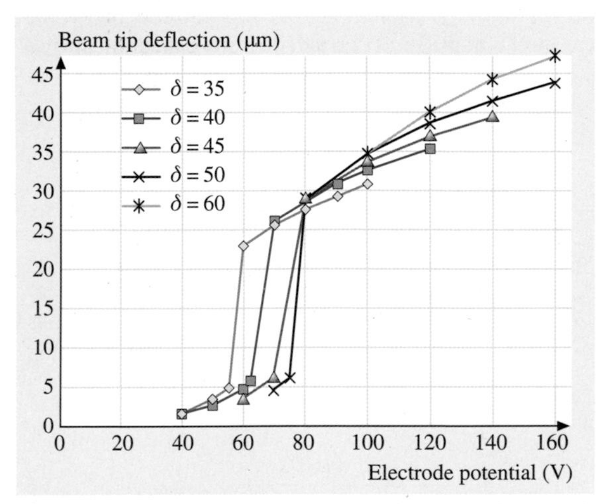

Fig. 35.34 Beam tip deformation as a function of electrode potential and δ for a suspension with a single fold

with this particular design, which has a twofold flexure and could have application as a optical shutter or VOA. In Fig. 35.34, electrode tip displacement is shown as a function of electrode voltage and electrode curve δ for a onefold flexure design. The displacement as a function of electrode potential and curve δ is shown for a twofold flexure in Fig. 35.35.

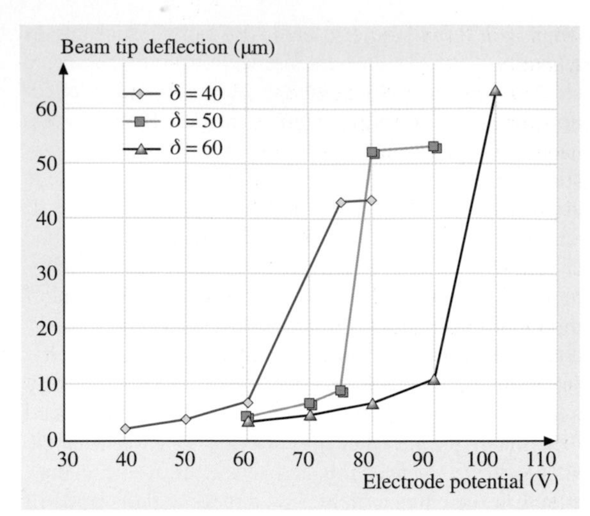

Fig. 35.35 Beam tip deformation as a function of electrode potential and δ for a suspension with a doubly folded suspension

35.2.8 Electromechanics for Test Structures

Electromechanics test structures are an important contributor to the development of microsystems products as they are used to extract information about material properties of thin films for specific process parameters. They are also used to monitor the material property variation during the fabrication process. In this case, they are distributed around the wafer on specific test die or they may be included in the dicing streets of more mature processes. These distributed test structures enable the manufacturer to track the variation in material properties with its associated within-wafer variation, wafer-to-wafer variation, and lot-to-lot variation. Another application of test structures is to characterize the impact of the first level packaging on the performance of the product. With all the critical information being extracted from the test structures, it is important for the information to be easy to interpret. Test structures have been developed that involve electrical, optical, and mechanical interrogation of the structure, but these are not easily instituted at all levels of characterization. The purely electrical test structures offer simple evaluation at the wafer, die, and package level so manufacturers often prefer them. The most common electrical test structures used for extracting information on mechanical properties involve electromechanics, of course. This section has provided a basis for interpreting and modeling the behavior of specific test structures, such as a homogeneous cantilever beam, and can be extended to other structures.

Osterberg and *Senturia* [35.83] describe in detail the application of electromechanics to the extraction of mechanical properties for microsystems design. They labeled their test structures as M-TEST for mechanical test structure. Their test structures comprised arrays of cantilever beams, fixed-fixed beams, and clamped, circular plates of various lengths, widths, or diameters. With these test structures, they demonstrated the extraction of important mechanical properties such as Young's modulus, plate modulus, and the residual film stress. Why choose an electromechanics test structure unless for the breadth and ease of application? An additional reason is the inherent instability that has been described as the pull-in of an electromechanical structure. The pull-in voltage is very distinct and can be measured in terms of a single parameter, the pull-in voltage. Given a control range of pull-in voltages for different structural layers in the manufacturing process, the pull-in parameter can be used as a simple control parameter.

Osterberg and *Senturia* [35.83] describe the methodology for extracting the mechanical properties. Their methodology includes the measurement of plan-view dimensions (i. e., width and length), out-of-plane curvature, pull-in voltage, film thickness, and air gap. With these measured parameters, the mechanical properties can be extracted. Their extraction methodology includes general closed form expressions for pull-in voltage that depend on the material properties and geometry of a homogeneous, isotropic material.

35.2.9 Electromechanical Dynamics: Switching Time

Switching time is an important factor in the design of such electromechanical devices as electrical or optical switches. The switching time analysis requires consideration of the dynamic equations representing the device and the electrical source. *Castaner* et al. [35.84–86] have considered the optimization of the speed-energy product. In this case, a parallel plate electrostatic actuator is described by an equivalent mechanical circuit that includes the mass m, the damping b, and the spring k, and a variable gap g. The variable gap depends on the initial gap g_0 and the displacement of the moving electrode x, such that $g = g_0 - x$. In the parallel plate, variable gap model the damping is described by squeeze film damping. (Other electrostatic actuators that move parallel to the substrate with variable gaps defined by groups of moving and stationary fingers.) In these cases, the damping model will need to include squeeze film damping and shear flow damping (Couette or Stokes Flow models). The electrical source is modeled as a voltage source and a source series resistance. A schematic of the model is shown in Fig. 35.36.

The equivalent electrical circuit of the parallel plate electrostatic actuator is modeled by (35.67), where V_s is the source voltage, V is the voltage across the variable capacitance, R is the series source resistance, and Q is the charge on the capacitor plates

$$\frac{V_s}{R} = \frac{V}{R} + \frac{dQ}{dT} \,. \tag{35.67}$$

The equivalent mechanical circuit is described by:

$$m\frac{d^2X}{dT^2} + b\frac{dX}{dT} + kX = \frac{Q^2}{2C_0G_0} \,. \tag{35.68}$$

Equation (35.68) includes the sum of the inertial force, the damping force, the spring force, and the electrostatic force. Equation (35.68) is written in terms of the mass of the moving capacitor plate, m; the damping coefficient due to squeeze film damping, b; the elastic constant of the spring by which the capacitor plate is suspended, k; the charge on the capacitor plate, Q; the initial gap, G_0; and the initial capacitance, C_0. The initial capacitance is written in terms of the area of the capacitor A, the dielectric constant of the capacitor air gap ε, and the initial gap:

$$C_0 = \frac{\varepsilon A}{G_0} \,. \tag{35.69}$$

The dynamics of the parallel plate electrostatic actuation were divided into four events. The first occurrence is a short charging time for the capacitor. The second phase is the acceleration phase, which defines the rapidly increasing velocity of the mass, but the

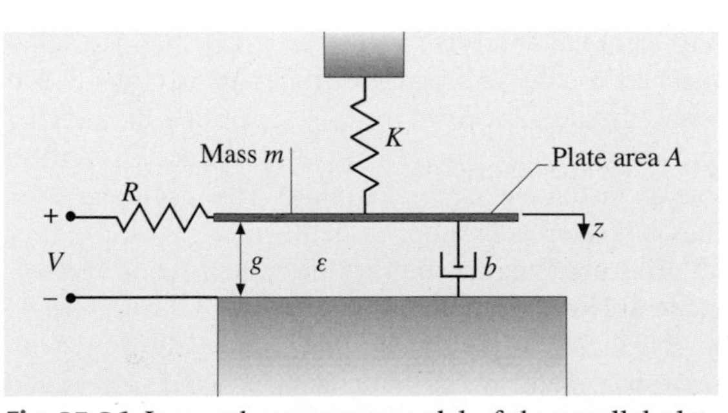

Fig. 35.36 Lumped parameter model of the parallel plate electromechanical actuator including damping and series resistance for dynamic analysis

position is still close to the initial equilibrium position. The third phase is the damping phase, defined by the velocity reaching a maximum, then decreasing due to the increased damping, and then finally increasing due to the increasing electrostatic forces as the gap closes. The fourth and final phase is the pull-in phase, characterized by a rapidly increasing velocity. *Hung* and *Senturia* [35.76] have described generating efficient dynamical models from a minimal set of finite element simulations. *Pons-Nin* et al. [35.87] have described the behavior of pull-in voltage and pull-in time for electromechanical actuators with current drives.

35.2.10 Electromechanics Issues: Dielectric Charging

Electrostatic actuators are operated in three different regimes depending on the application. In the first regime, the electrostatic actuator is operated over the voltage range that does not include the pull-in voltage. In this case, the actuated element does not contact the opposing substrate. In the second regime, the electrostatic actuator is operated through a voltage range that does include the pull-in voltage ($V_{act} \geq V_{PI}$). In this case, the desired response is to establish contact with or without some overdrive to achieve greater contact force or other response parameter. In the third regime, the electrostatic actuator is operated through a voltage range beyond the pull-in voltage. This is characteristic of electrostatic zipper actuators. In this case, the operating voltage is programmed to quickly establish pull-in, thereby contact, then the actuator operates in a range above the pull-in voltage ($V_{PI} < V_{act}$). In all three regimes, dielectric isolation layers are considered to prevent shorting between actuation electrodes. For the first and second case, the use of the insulating material is a preventative measure to protect against electrode shorting during an unintentional overdrive of the actuator. It is preventative because the actuation electrodes are not intended to touch. In the third case, the gap separating the moving and stationary actuation electrodes is reduced to zero so that an insulating layer is required. The insulating layer may be patterned or unpatterned.

The third regime defines the touch mode electrostatic actuators. *Cabuz* et al. [35.88] have described many of the failure modes that can be observed with the touch mode electrostatic actuators and that should be addressed during the design of the actuator and process. The observations and failure modes included: 1) a significant difference between the experimental actuation voltage (DC) and the theoretical actuation voltage, 2) a closed state electrostatic pressure lower than predicted, 3) an actuation voltage that increases with each subsequent cycle until electrical breakdown occurs, 4) actuator vibration observed between an open and a closed state, 5) actual release voltage is higher than theoretical, but the electrostatic forces should be much higher in the closed state such that the release voltage is always lower than the pull-in voltage, 6) the actuator remains closed temporarily after shorting, 7) permanent stiction will occur when in the closed state for a long period of time, 8) permanent stiction occurs after many actuation cycles.

Wibbeler et al. [35.89] have analyzed this problem in terms of one of the contributing factors, the accumulation and storage of localized charge in dielectrics. The common dielectrics in microsystems devices are SiO_2 and Si_3N_4, which show low mobility for surface charges and provide trap sites for positive or negative charge within their volume and at interfaces. *Kubo* et al. [35.90] showed this charge can be produced by the fabrication process by contact electrification, and by gas discharge in the air gap (the case focused on that by *Wibbeler* et al. [35.89]), and static electrification during handling (ESD). The contact electrification created electrical charge through the mechanical contact of two materials with different work functions and the separation of the two materials.

To understand the impact of the accumulated charge, *Wibbler* et al. [35.89] considered the impact of stored surface charge on the electrostatic force acting on a simple parallel plate capacitor. In Fig. 35.37, a simple capacitor is considered with a dielectric (ε_d) filling part of the air gap (ε). The dielectric has a surface charge of $\sigma_s(x, y)$. The capacitor plates carry charge shown as σ_1 and σ_2, which are provided by the battery, of voltage V, connected across the capacitor plates. The thicknesses of the dielectric and air gap are d_d and d, respectively. The force was estimated from the applied voltage and from the offset voltage due to the accumulated surface charge. The force acting on the capacitor plate as a function of the applied voltage, V; the offset

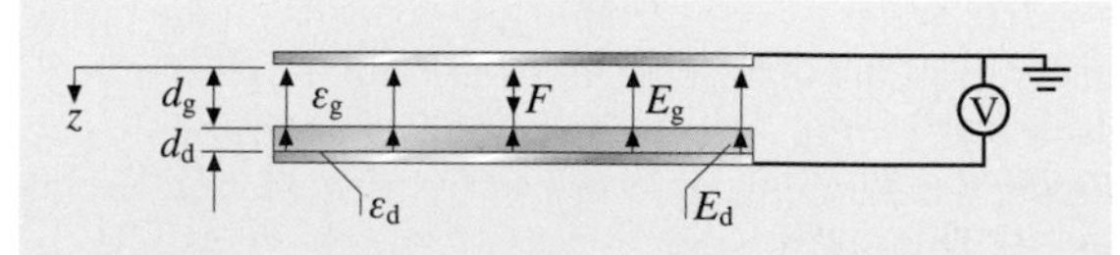

Fig. 35.37 Parallel plate capacitor with dielectric and parasitic charges

voltage, V_s; and the capacitance, C, is:

$$F = \frac{dC}{dz} \int_A \frac{[V - V_s(x, y)]^2}{2} dA \,. \tag{35.70}$$

The offset voltage is shown in (35.71) as a function of the surface charge; the thickness of the dielectric, d_d; the dielectric constant, ε_d; or the dielectric's contribution to the total capacitance, C_d. The capacitance of the dielectric layer, per unit area, being defined as $C_d = \varepsilon_d / d_d$.

$$V_s = -\sigma_s(x, y) \frac{d_d}{\varepsilon_d} = -\frac{\sigma_s(x, y)}{C_d} \,. \tag{35.71}$$

Electromechanical Pull-In with Dielectric Charging

Nemirovsky et al. [35.66] and *Chan* et al. [35.91] studied the effect of residual charge at the interface of a dielectric and the air in the capacitor gap. With the dielectric in the capacitor gap and the residual charge, the total energy for the actuator is:

$$U_T = \frac{1}{2} K z^2 + \frac{Q_{res}^2}{2(C(z) + C_d)} - \frac{C(z) C_d}{2(C(z) + C_d)} V^2 \,. \tag{35.72}$$

In (35.72), Q_{res} is the residual charge, $C(z)$ is the air gap capacitance, and C_d is the dielectric capacitance. From the total energy, the pull-in parameter and pull-in equations (35.42) and (35.43) can be solved for the pull-in parameter ζ_{PI} and pull-in voltage V_{PI}:

$$\zeta_{PI} = \frac{1}{3}\left(1 + \frac{1}{\delta}\right) , \tag{35.73}$$

$$V_{PI} = \left[\frac{8 K z_{max}^3}{27 \varepsilon_0 A} \left(1 + \frac{1}{\delta}\right)^3 - \frac{Q_{res}^2}{C_d^2} \right]^{1/2} \,. \tag{35.74}$$

In (35.73), ζ_{PI} is the ratio z/z_{max} and δ is the ratio $\varepsilon_d z_{max}/z_d$. Equation (35.74) shows that the dielectric layer will increase the pull-in voltage and the residual charge will lower the pull-in voltage. It can be seen that the residual charge does not impact the normalized pull-in coordinate shown in (35.73).

Test structures were fabricated to verify the shift predicted by (35.70) that would result with a dielectric in the air gap. Test structures with a dielectric did show a shift in the voltage–deflection curve, whereas a shift was not observed when a dielectric was not present. A shift of −25 V was observed. Following electrostatic discharge events, the offset voltages of −210 V and 145 V were observed during different phases of a charge-discharge cycle. This demonstrates the significant bias the accumulated surface charge can create. As a result, *Wibbler* et al. [35.89] did not recommend trying the fix shorting by adding dielectric films over the electrodes. They did recommend estimating the influence of accumulated charge on the response of the device being designed. Instead of complete dielectric film coverage of an electrode, they recommended patterning the dielectric such that it was limited to the edges of the electrodes or patterned at locations that will touch first. By patterning small areas of insulation, the surface area available to store charge is smaller and the distance to the electrode is shorter for more timely dissipation of stored charge.

Cabuz et al. [35.88] identified the effects of humidity, DC or AC driving voltage, and SAM coatings on the performance of electromechanical actuators. With a DC operating voltage, the devices operated as expected below 35% RH except in dry air or dry N_2. In the dry environment the actuator would temporarily remain closed, would remain closed after high fields are applied, or would remain closed after a high number of cycles. When operated above 35% RH, the DC voltage was higher than expected, the closed state electrostatic pressure was lower, the actuator did not release when expected, the actuator vibrated, and the voltage increased with subsequent cycles. When an AC actuation voltage is used the actuators operated up to 55% RH. At higher RH, the actuation voltage increased. The dry condition operation was improved with AC driving compared to DC operation. AC voltage control is recommended for eliminating RC effects but cannot overcome the effects of humidity. To avoid the effects of humidity, two SAM coatings (octadecyltrichlorosilane (OTS) and perfluorodecyltrichlorosilane (PFTS)) were tried to influence the hydration of the surfaces. The OTS eliminated the effects observed for uncoated samples up to a 95% RH level for both DC and AC voltages. The PTFS improved operation up to 95% RH for AC voltages and 70% RH for DC control voltages. The residual in-use stiction was determined to be associated with accumulated and trapped charge on the surfaces or at interfaces.

35.2.11 Electromechanics Issues: Gas Discharge

As described in previous sections, the electromechanical actuation of a microstructure is performed by applying a voltage across two plates. The two plates can be defined by two conductors (e.g. Al to Al), by semiconductors (e.g. polysilicon), or by a combination of a conductor and a semiconductor (e.g. Al and polysilicon). When these two plates (or other similarly described structures) are fabricated in proximity with each other, they are usually fabricated with an "air" gap less than 10 μm. With the electrostatic actuators, a small gap is desirable to minimize the voltage required for actuation.

The presence of small gaps in microsystems devices has prompted the further investigation of electrical breakdown in air and other ambient conditions at small gaps (i.e. less than 50 μm). *Slade* and *Taylor* [35.92] have demonstrated that breakdown in small gaps is defined by three regimes: less than 4 μm, 4 to 6 μm, and greater than 6 μm. In the first case, the breakdown event follows the behavior for a vacuum breakdown process at a larger gap. In the last case, the breakdown follows the behavior of Paschen's law for breakdown in air. The middle case of gaps between 4 and 6 μm, the breakdown process is in a transition from following the vacuum breakdown process to following the Paschen curve. In this case, the breakdown voltage is lower than predicted in either case, which indicates significant contributions from factors in both cases. The deviation from Paschen's curve was observed by *Torres* and *Dhariwal* [35.93] for gaps less than 4 μm.

The goal of this section is to describe some basic breakdown principles that can be used to determine whether or not a particular design will break down during an actuation event. Different processes will be considered for different gap regimes so that breakdown estimates can be made.

The vacuum breakdown region is characterized by a breakdown voltage that is a function of the gap separating the two electrodes. This contrasts with the breakdown behavior that follows the Paschen curve, which shows electrode gap-pressure product dependence. The breakdown voltage in vacuum is

$$V_B = K_V t_g \,. \tag{35.75}$$

Equation (35.75) shows the breakdown voltage V_B is proportional to the gap t_g where the proportionality constant is $K_v = 97\,\mathrm{V/\mu m}$ for data from many sources as indicated by *Slade* and *Taylor* [35.92]. The breakdown data for vacuum was determined for electrode gaps between 35 and 200 μm. For electrode gaps between 0.2 and 40 μm in air, the proportionality constant was between 65 and 110 V/μm.

The electric field at the surface of the electrode is a critical parameter for determining breakdown in vacuum. This is estimated by the ratio of the voltage between the electrodes V and the gap between the electrodes t_g with a modification factors β_g. β_g is a geometrical enhancement factor that captures the field enhancement due to the macroscopic geometry:

$$E_g = \beta_g V/t_g \,. \tag{35.76}$$

The geometrical enhancement factor can be determined by performing finite element analysis to determine the macroscopic electric field as a function of the electrode gap for the electrode geometry under consideration. The geometric enhancement factor is determined from the ratio of the macroscopic electric field E_g, from finite element simulations, to the electric field estimated by V/t_g [35.92, 94]. The estimated electric field, V/t_g, can be used as a reasonable approximation when the electrode gap is much smaller than that radius of curvature of the cathode electrode. If this is not the case, the geometric enhancement factor needs to be estimated from numerical analysis such as finite element or boundary element methods. The geometric enhancement factor is shown, by *Slade* and *Taylor* [35.92], to be between 1.01 and 1.57 for gaps ranging from 0.2 to 40 μm, respectively, for the geometry of a needle cathode and a plate anode.

A second enhancement factor is attributed to the microscopic field enhancement due to the surface roughness of the electrodes. The electric field containing the effects of the geometric enhancement and the microscopic enhancement is:

$$E = \beta_m \beta_g V/t_g \,. \tag{35.77}$$

In (35.77) β_m is the microscopic enhancement factor. The total enhancement factor is $\beta_g \beta_m$, which can have a range of 100 to 250 for polished electrodes [35.95].

Following the estimate of the electric field at the surface of the electrodes, the current density for the field emission of electrons is calculated. This current density j_{FE} is described by the Fowler–Nordheim equation, which depends on the electric field E, the work function of the electrode material ϕ, and the dimensionless parameters of $t(y)$ and $v(y)$, where y is a parameter defined as $3.79\times10^{-5}\sqrt{E}/\phi$. Typically, $t(y)$ is one. $v(y)$ is given by $0.956 - 1.06y^2$. The Fowler–Nordheim equation is presented in (35.78) with j_{FE} in $\mathrm{A/m^2}$, E in V/m, and ϕ in eV.

$$j_{FE} = \frac{1.54\times10^{-6}E^2}{\phi t(y)^2} \exp\left(\frac{-6.83\times10^9\phi^{3/2}v(y)}{E}\right) . \tag{35.78}$$

Equation (35.78) can be rewritten with E substituted from (35.77) and the current density replaced by an emission current I_e and the area A_e of a microprojection where the emission occurs. Equation (35.78) is shown rewritten in (35.79), such that the slope from the Fowler–Nordheim plot will provide the field enhancement factor as shown in (35.80) where m is the slope. The Fowler–Nordheim plot is the plot of I_e/V^2 versus $1/V$, which is constructed by measuring the voltage across the electrode gap and measuring I_e.

$$\mathrm{Log}_{10}\left(\frac{I_e}{V^2}\right) = \mathrm{Log}_{10}\left(\frac{1.54\times 10^{-6} A_e \beta^2}{\phi\, t(y)^2 t_g^2}\right) - \frac{1}{2.303}\left(\frac{6.83\times 10^9 \phi^{3/2} t_g v(y)}{\beta}\right)\frac{1}{V}, \quad (35.79)$$

$$\beta = -\frac{2.303m}{6.83\times 10^9 \phi^{3/2} t_g v(y)}. \quad (35.80)$$

For gaps larger than 6 μm, the electrical breakdown as a function of gap does follow the Paschen curve for air and does not follow the vacuum breakdown curve for these larger gaps. The electrode gap is now larger than the mean free path of the air molecules so it is expected that emitted electrons will collide with the air molecules in the gap. For these larger electrode gaps, it is assumed the electrical breakdown will follow the Townsend electron avalanche theory. A similar process is followed to estimate the breakdown. First, the electric field at the surface of the cathode electrode is estimated including the enhancement factors for geometry and surface microstructure. An estimate of the total field enhancement factor is required for this. The electric field is substituted into the Fowler–Nordheim equation (35.78) to estimate the field emission current. In this case, the emitted electrons will interact with the background gas molecules causing an avalanche in the current. The current avalanche process is described by (35.79) and (35.80), where t_g is the electrode gap, α is the first Townsend coefficient, and γ is the second Townsend coefficient. The first Townsend coefficient represents the number of electrons produced per unit distance along the direction of the electric field. The second Townsend coefficient is the number of electrons generated by secondary processes per each primary avalanche. Breakdown will occur when the denominator of the expression for current approaches zero: $\gamma \exp(\alpha t_g)$ approaches unity.

In the transition region, the breakdown voltage can be lower than predicted by the Paschen curve or vacuum breakdown. It is suggested that in this regime, the partial electron avalanche process enhances the vacuum breakdown process to lead to a lower breakdown voltage [35.92].

The choice of materials is another consideration for the design of an electrostatic microactuator. As mentioned, the electrode pairs could comprise the following combinations: conductor-conductor, conductor-semiconductor, or semiconductor-semiconductor. In *Torres* and *Dhariwal* [35.93], the breakdown voltage was measured for electrode pairs comprised of nickel, brass, or aluminum. A significant difference was not observed between the breakdown voltages of the different conductors or for the size of the electrode. At larger gaps, the breakdown voltage was different for different electrode shapes (cylindrical versus spherical) but this can be explained by the relative difference in the geometric enhancement factor for the electric field for the different electrodes. *Ono* et al. [35.96] made a comparison between electrode combinations where both a semiconductor (silicon) and a silicon–metal combination were used. In the case of a silicon electrode with a metal electrode, the breakdown voltage deviated from the Paschen curve at approximately 6 μm. It was lower than the breakdown voltage predicted by the Paschen curve and trailed off with an approximate linear dependence on the electrode gap. With the silicon as one electrode, Au/Cr or Pt/Cr was used as the second electrode but little difference was found for the different metals. With the silicon-silicon electrode combination, the breakdown voltage followed the Paschen curve for air. For similar gaps, the breakdown voltage was significantly higher for the silicon-silicon electrodes than for the silicon-metal electrode. This was true for a thin silicon film over a Cr/Pt/Cr electrode.

Another option for reducing the electric breakdown in the air gap is to introduce an insulating material over the electrode. But *Wibbeler* et al. [35.89] have demonstrated this does not solve the problem of breakdown by itself and can lead to parasitic charging of the insulator. An electric breakdown will produce a large current, as described previously, that is destructive to the electrodes without insulation. The dielectric limits the current so that the electrical breakdown event is not so destructive. *Wibbeler* et al. [35.89] demonstrated through highly chaotic behavior in the voltage–displacement response of an electrostatic actuator. It is reasoned that the local field strength at the electrode reached a magnitude to cause an avalanche of free electrons and ions, as described here. These free charges travel the length of the air gap, but they are stopped and neutralized at the dielectric surface. This reduces the local electric field at the location of the surface charge so that the force on the cantilever is reduced and the process repeats itself. Throughout this process, a large amount of charge can be introduced onto the dielectric surface, which will typically have a very long dissipation time constant.

35.3 Summary and Mention of Topics not Covered

This chapter has focused on the thermomechanics and electromechanics of thin-film microstructures. Emphasis has been placed on the basic physical phenomena, approaches to analyze the phenomena, and the presentation of accessible results for special cases of practical importance in a variety of applications. We have not discussed many other energy domain couplings that play major roles in microsystems technology such as electrothermomechanics and magnetomechanics. We have also not discussed the role and behavior of materials such as piezoelectrics or shape memory alloys, which play important roles in electromechanics and thermomechanics, respectively. We refer the interested reader to *Senturia* [35.49] for an introduction and further references.

References

35.1 M. Madou: *Fundamentals of Microfabrication* (CRC, Boca Raton 1997)

35.2 M. Gad-el-Hak: *The MEMS Handbook* (CRC, Boca Raton 2001)

35.3 J.J. Sniegowski, M.P. de Boer: IC-compatible polysilicon surface micromachining, Annu. Rev. Mater. Sci. **30** (2000) 299–333

35.4 C.H. Mastrangelo: Adhesion-related failure mechanisms in microelectromechanical devices, Tribol. Lett. **3** (1997) 223–238

35.5 M.P. de Boer, T.M. Mayer: Tribology of MEMS, MRS Bull. **26** (2001) 302–304

35.6 W.D. Nix: Mechanical properties of thin films, Metall. Trans. A **20A** (1989) 2217–2245

35.7 C.Y. Hui, H.D. Conway, Y.Y. Lin: A reexamination of residual stresses in thin films and of the validity of Stoney's estimate, J. Electron. Packaging **122** (2000) 267–273

35.8 D.B. Bogy, K.C. Wang: Stress singularities at interface corners in bonded dissimilar materials, Int. J. Solids Struct. **7** (1971) 993–1005

35.9 V.L. Hein, F. Erdogan: Stress singularities in a two material wedge, Int. J. Fract. Mech. **7** (1971) 317–330

35.10 E.D. Reedy, T.R. Guess: Nucleation and propagation of an edge crack in a uniformly cooled epoxy/glass bimaterial, Int. J. Solids Struct. **39** (2002) 325–340

35.11 M.D. Thouless: Modeling the development and relaxation of stresses in films, Annu. Rev. Mater. Sci. **25** (1995) 69–96

35.12 N.D. Masters, M.P. de Boer, B.D. Jensen, M.S. Baker, D. Koester: Side-by-side comparison of passive MEMS residual strain test structures under residual compression. In: *Mechanical Properties of Structural Films*, ASTM STP **1413**, ed. by C.L. Muhlstein, S.B. Brown (2001) 168–200

35.13 M.W. Hyer: Calculation of the room-temperature shapes of unsymmetric laminates, J. Compos. Mater. **15** (1981) 296–310

35.14 M.W. Hyer: The room-temperature shape of four-layer unsymmetric cross-ply laminates, J. Compos. Mater. **16** (1982) 318–340

35.15 D.E. Fahnline, C.B. Masters, N.J. Salamon: Thin film stress from nonspherical substrate bending measurements, J. Vac. Sci. Technol. A **9** (1991) 2483–2487

35.16 C.B. Masters, N.J. Salamon: Geometrically nonlinear stress-deflection relations for thin film/substrate systems, Int. J. Eng. Sci. **31** (1993) 915–925

35.17 C.B. Masters, N.J. Salamon: Geometrically nonlinear stress-deflection relations for thin film/substrate systems with a finite element comparison, J. Appl. Mech. **61** (1994) 872–878

35.18 M. Finot, S. Suresh: Small and large deformation of thick and thin film multilayers: effects of layer geometry, plasticity and compositional gradients, J. Mech. Phys. Solids **44** (1996) 683–721

35.19 M. Finot, I.A. Blech, S. Suresh, H. Fujimoto: Large deformation and geometric instability of substrates with thin film deposits, J. Appl. Phys. **81** (1997) 3457–3464

35.20 L.B. Freund: The stress distribution and curvature of a general compositionally graded semiconductor layer, J. Cryst. Growth **132** (1993) 341–344

35.21 L.B. Freund: Some elementary connections between curvature and mismatch strain in compositionally graded thin films, J. Mech. Phys. Solids **44** (1996) 723–736

35.22 L. B. Freund: Substrate curvature due to thin film mismatch strain in the nonlinear deformation range, J. Mech. Phys. Solids **48** (2000) 1159–1174
35.23 N. J. Salamon, C. B. Masters: Bifurcation in isotropic thin film/substrate plates, Int. J. Solids Struct. **32** (1995) 473–481
35.24 R. M. Jones: *Mechanics of Composite Materials*, 2nd edn. (Taylor & Francis, London 1999)
35.25 M. W. Hyer: *Stress Analysis of Fiber-reinforced Composite Materials* (McGraw-Hill, Boston 1998)
35.26 G. G. Stoney: The tension of metallic films deposited by electrolysis, Proc. R. Soc. Lond. A **82** (1909) 172–175
35.27 C. D. Pionke, G. Wempner: The various approximations of the bimetallic thermostatic strip, J. Appl. Mech. **58** (1991) 1015–1020
35.28 P. Krulevitch, G. C. Johnson: Curvature of a cantilever beam subjected to an equibiaxial bending moment, Mater. Res. Symp. Proc. **518** (1998) 67–72
35.29 S. R. Swanson: *Introduction to Design and Analysis with Advanced Composite Materials* (Prentice-Hall, Upper Siddel River 1997)
35.30 M. L. Dunn, Y. Zhang, V. Bright: Deformation and structural stability of layered plate microstructures subjected to thermal loading, J. Microelectromech. Syst. **11** (2002) 372–383
35.31 J. A. King: *Materials Handbook for Hybrid Microelectronics* (Teledyne Microelectronics, Los Angeles 1988)
35.32 W. N. Sharpe: Mechanical properties of MEMS materials. In: *The CRC Handbook of MEMS*, ed. by M. Gad el Hak (CRC, Boca Raton 2001) Chap. 3
35.33 Y. Zhang, M. L. Dunn: Deformation of blanketed and patterned bilayer thin film microstructures during post-release thermal and cyclic thermal loading, J. Microelectromech. Syst. (2003) in press
35.34 B. D. Harper, C.-P. Wu: A geometrically nonlinear model for predicting the intrinsic film stress by the bending plate method, Int. J. Solids Struct. **26** (1990) 511–525
35.35 Y. L. Shen, S. Suresh: Thermal cycling and stress relaxation response of Si-Al and Si-Al-SiO_2 layered thin films, Acta. Metall. Mater. **43** (1995) 3915–3926
35.36 S. P. Baker, A. Kretschmann, E. Arzt: Thermomechanical behavior of different texture components in Cu thin films, Acta Mater. **49** (2001) 2145–2160
35.37 W. D. Nix: Elastic and plastic properties of thin films on substrates, Mater. Sci. Eng. A **234-236** (1997) 37–44
35.38 C. Thompson: The yield stress of polycrystalline thin films, J. Mater. Res. **8** (1993) 237–238
35.39 H. Gao, Y. Huang, W. D. Nix, J. W. Hutchinson: Mechanism-based strain gradient plasticity – I. Theory, J. Mech. Phys. Solids **47** (1999) 1239–1263
35.40 H. D. Espinosa, B. C. Prorok, M. Fischer: A novel method for measuring elasticity, plasticity, and fracture of thin films and MEMS materials, J. Mech. Phys. Solids **51** (2003) 47–67
35.41 D. C. Miller, M. L. Dunn, V. M. Bright: Thermally induced change in deformation of multimorph MEMS structures, Proc. SPIE **4558** (2001) 32–44
35.42 D. J. Vickers-Kirby, R. L. Kubena, F. P. Stratton, R. J. Joyce, D. T. Chang, J. Kim: Anelastic creep phenomena in thin film metal plated cantilevers for MEMS, Mater. Res. Soc. Symp. **657** (2001) EE2.5.1–EE2.5.6
35.43 R. M. Keller, S. P. Baker, E. Arzt: Stress-temperature behavior of unpassivated thin copper films, Acta Mater. **47** (1999) 415–426
35.44 M. D. Thouless, J. Cupta, J. M. E. Harper: Stress development and relaxation in copper films during thermal cycling, J. Mater. Res. **8** (1993) 1845–1852
35.45 M. D. Thouless, K. P. Rodbell, C. Cabral Jr.: Effect of a surface layer on the stress relaxation of thin films, J. Vac. Sci. Technol. **14** (1996) 2454–2461
35.46 R. P. Vinci, E. M. Zielinski, J. C. Bravman: Thermal stress and strain in copper thin films, Thin Solid Films **262** (1995) 142–153
35.47 J. Koike, S. Utsunomiya, Y. Shimoyama, K. Maruyama, H. Oikawa: Thermal cycling fatigue and deformation mechanism in aluminum alloy thin films on silicon, J. Mater. Res. **13** (1998) 3256–3264
35.48 K. Gall, M. L. Dunn, Y. Zhang, B. Corff: Thermal cycling response of layered gold/polysilicon MEMS structures, Mech. Mater. (2003) in press
35.49 S. D. Senturia: *Microsystems Design* (Kluwer, Dordrecht 2001)
35.50 M. P. de Boer, N. F. Smith, N. D. Masters, M. B. Sinclair, E. J. Pryputniewicz: Integrated platform for testing MEMS mechanical properties at the wafer scale by the IMaP methodology. In: *Mechanical Properties of Structural Films*, ASTM STP **1413** (2001) 85–95
35.51 Z. J. Yao, S. Chen, S. Eshelman, D. Denniston, C. Goldsmith: Micromachined low-loss microwave switches, J. Microelectromech. Syst. **8** (1999) 129–134
35.52 B. McCarthy, G. G. Adams, N. E. McGruer, D. Potter: A dynamic model, including contact bounce of an electrostatically actuated microswitch, J. Microelectromech. Syst. **11** (2002) 276–283
35.53 H.-S. Lee, C. H. Leung, J. Shih, S.-C. Chang, S. Lorincz, I. Nedelescu: Integrated microrelays: Concept and initial results, J. Microelectromech. Syst. **11** (2002) 147–153
35.54 S. J. Cunningham, S. Tatic-Lucic, J. Carper, J. Lindsey, L. Spangler: A high aspect ratio accelerometer fabricated using anodic bonding, dissolved wafer, and deep RIE processes, Proc. Transducers '99, the 10th Intl. Conf. Solid-State Sensors and Actuators, Sendai 1999, ed. by M. Esashi (IEE Jpn. 1999) 1522–1525
35.55 L. Spangler, C. Kemp: ISAAC: integrated silicon automotive acceleromter, Sens. Actuators A **54** (1996) 523–529

35.56 M. T. A. Saif, B. E. Alaca, H. Sehitoglu: Analytical modeling of electrostatic membrane actuator for micropumps, J. Microelectromech. Syst. **8** (1999) 335–345

35.57 C. Huang, C. Christophorou, K. Najafi, A. Naguib, H. M. Naguib: An electrostatic microactuator system for application in high-speed jets, J. Microelectromech. Syst. **11** (2002) 222–235

35.58 H. Toshiyoshi, H. Fujita: Electrostatic micro torsion mirrors for optical switch matrix, J. Microelectromech. Syst. **5** (1996) 231–237

35.59 H. Toshiyoshi, W. Piyawattanametha, C.-T. Chan, M. C. Wu: Linearization of electrostatically actuated surface micromachined 2-D optical scanner, J. Microelectromech. Syst. **10** (2001) 205–214

35.60 M. Fischer, M. Giousouf, J. Schaepperle, D. Eichner, M. Weinmann, W. von Münch, F. Assmus: Electrostatically deflectable polysilicon micromirrors-dynamic behaviour and comparison with the results from FEM modelling with Ansys, Sens. Actuators A **67** (1998) 89–95

35.61 P. K. C. Wang, R. C. Gutierrez, R. K. Bartman: A method for designing electrostatic-actuator electrode pattern in micromachined deformable mirrors, Sens. Actuators A **55** (1996) 211–217

35.62 P. K. C. Wang, F. Y. Hadaegh: Computation of static shapes and voltages for micromachined deformable mirrors with nonlinear electrostatic actuators, J. Microelectromech. Syst. **5** (1996) 205–220

35.63 J. Bühler, J. Funk, J. G. Korvink, F.-P. Steiner, P. M. Sarro, H. Baltes: Electrostatic aluminum micromirrors using double-pass metallization, J. Microelectromech. Syst. **6** (1997) 126–135

35.64 M. Fischer, H. Graef, W. von Münch: Electrostatically deflectable polysilicon torsional mirrors, Sens. Actuators A **44** (1994) 83–89

35.65 H. Schenk, P. Dürr, D. Kunze, H. Lakner, H. Kück: An electrostatically excited 2D-micro-scanning-mirror with an in-plane configuration of the driving electrodes, Proc. IEEE MEMS 2000, The 13th Ann. Intl. Conf. Micro Electro Mechanical Systems, Miyazaki 2000, ed. by I. Shimoyama, H. Kuwano (IEEE, Piscataway 2000)

35.66 Y. Nemirovsky, O. Bochobza-Degani: Methodology and model for the pull-in parameters of electrostatic actuators, J. Microelectromech. Syst. **10** (2001) 601–615

35.67 G. Li, N. R. Aluru: A Lagrangian approach for electrostatics analysis of deformable conductors, J. Microelectromech. Syst. **11** (2002) 245–254

35.68 N. R. Aluru, J. White: A multilevel Newton method for mixed-energy domain simulation of MEMS, J. Microelectromech. Syst. **8** (1999) 299–308

35.69 J. I. Seeger, S. B. Crary: Stabilization of electrostatically actuated mechanical devices, Proc. Transducers '97, 1997 Intl. Conf. Solid-State Sensors and Actuators, Chicago 1997, ed. by K. Wise (IEEE, Piscataway 1997) 1133–1136

35.70 R. Nadal-Guardia, A. Dehe, R. Aigner, L. M. Castaner: Current drive methods to extend the range of travel of electrostatic microactuators beyond the voltage pull-in point, J. Microelectromech. Syst. **11** (2002) 255–263

35.71 G. J. O'Brien, D. J. Monk, L. Lin: Electrostatic latch and release; a theoretical and empirical study, Proc. Micro-Electro-Mechanical Syst. (MEMS) 2000, MEMS-Vol. 2, The 2000 ASME Intl. Mech. Eng. Cong. and Expo., Orlando 2000, ed. by A. J. Malshe, Q. Tan, A.¶. Lee, F. R. Forster, R. S. Kenten (ASME, New York 2000) 19–26

35.72 B. Choi, E. G. Lovell: Improved analysis of microbeams under mechanical and electrostatic loads, J. Micromech. Microeng. **7** (1997) 24–29

35.73 Y. Zhang, M. L. Dunn: A vertical electrostatic actuator with extended digital range via tailored topology, Proc. SPIE **4700** (2002) 147–156

35.74 O. Degani, E. Socher, A. Lipson, T. Leitner, D. J. Setter, S. Kaldor, Y. Nemirovsky: Pull-in study of an electrostatic torsion microactuator, J. Microelectromech. Syst. **7** (1998) 373–379

35.75 Z. Xiao, X. Wu, W. Peng, K. R. Farmer: An angled-based design approach for rectangular electrostatic actuators, J. Microelectromech. Syst. **10** (2001) 561–568

35.76 E. S. Hung, S. D. Senturia: Generating efficient dynamical models for microelectromechanical systems from a few finite-element simulation runs, J. Microelectromech. Syst. **8** (1999) 280–289

35.77 J. R. Gilbert, R. Legtenberg, S. D. Senturia: 3D coupled electro-mechanics for MEMS: Application of CoSolve-EM, Proc. IEEE MEMS Conference, Amsterdam 1995, ed. by M. Elwenspoek, N. de Rooij (IEEE, Piscataway 1995) 122–127

35.78 J. R. Gilbert, G. K. Ananthasuresh: 3D modeling of contact problems and hysteresis in coupled electro-mechanics, Proc. IEEE MEMS 1996, The ninth Ann. Intl. Workshop on Micro Electro Mechanical Syst., San Diego 1996, ed. by M. G. Allen, M. L. Reed (IEEE, Piscataway 1996) 127–132

35.79 Y.-H. Jin, K.-S. Seo, Y.-H. Cho, S.-S. Lee, K.-C. Song, J.-U. Bu: An integrated SOI optical microswitch using electrostatic micromirror actuators with insulated touch-down beams and curved electrodes, Proc. Micro-Electro-Mechanical Syst. (MEMS) 2000, MEMS-Vol. 2. The 2000 ASME Intl. Mech. Eng. Cong. and Expo., Orlando 2000, ed. by A. J. Malshe, Q. Tan, A. P. Lee, F. R. Forster, R. S. Kenten (ASME, New York 2000) 177–181

35.80 R. Legtenberg, E. Berenschot, M. Elwenspoeke, J. Fluitman: Electrostatic curved electrode actuators, Proc. IEEE MEMS Conference, Amsterdam 1995 (IEEE, Piscataway 1995) 37–42

35.81 R. Legtenberg, J. Gilbert, S. D. Senturia, M. Elwenspoek: Electrostatic curved electrode actuators, J. Microelectromech. Syst. **6** (1997) 257–265

35.82 D. DeReus: Personal communications and internal reports (2002)

35.83 P. M. Osterberg, S. D. Senturia: M-TEST: A test chip for MEMS material property measurement using electrostatically actuated test structures, J. Microelectromech. Syst. **6** (1997) 107–118

35.84 L. M. Castaner, S. D. Senturia: Speed-energy optimization of electrostatic actuators based on pull-in, IEEE J. Microelectromech. Syst. **8** (1999) 290–298

35.85 L. Castaner, A. Rodriguez, J. Pons, S. D. Senturia: Measurement of power-speed product of electrostatic actuators, Proc. Transducers '99, the 10th Intl. Conf. Solid-State Sensors and Actuators, Sendai 1999, ed. by M. Esashi (IEE Jpn., Tokyo 1999) 1772–1775

35.86 L. Castaner, A. Rodriguez, J. Pons, S. D. Senturia: Pull-in time-energy product of electrostatic actuators: Comparison of experiments and simulation, Sens. Actuators **83** (2000) 263–269

35.87 J. Pons-Nin, A. Rodriguez, L. M. Castaner: Voltage and pull-in time in current drive of electrostatic actuators, J. Microelectromech. Syst. **11** (2002) 196–205

35.88 C. Cabuz, E. I. Cabuz, T. R. Ohnstein, J. Neus, R. Maboudian: Factors enhancing the reliability of touch-mode electrostatic actuators, Sens. Actuators **79** (2000) 245–250

35.89 J. Wibbeler, G. Pfeifer, M. Hietschold: Parasitic charging of dielectric surfaces in capacitive microelectromechanical systems (MEMS), Sens. Actuators A **71** (1998) 74–80

35.90 H. Kubo, T. Namura, K. Yoneda, H. Ohishi, Y. Todokoro: Evaluation of charge build-up in wafer processing by using MOS capacitors with charge collecting electrodes, ICMTS 1995 Proc. IEEE Intl. Conf. Microelectronic Test Structures, Vol. 8, Nara 1995, ed. by T. Sugano, K. Asada (IEEE, Piscataway 1995) 5–9

35.91 E. K. Chan, K. Garikipati, R. W. Dutton: Characterization of contact electromechanics through capacitance-voltage measurement and simulations, IEEE J. Microelectromech. Syst. **8** (1999) 208–217

35.92 P. G. Slade, E. D. Taylor: Electrical breakdown in atmospheric air between closely spaced (0.2 μm–40 μm) electrical contacts, Proc. 47th IEEE Holm Conf. Electrical Contacts, Montreal 2001, ed. by K. Leung (IEEE, Piscataway 2001) 245–250

35.93 J-M. Torres, R. S. Dhariwal: Electric field breakdown at micrometer separations, Nanotechnology **10** (1999) 102–107

35.94 R. Longwitz, H. van Lintel, R. Carr, C. Hollenstein, P. Renaud: Study of gas ionization schemes for microdevices, Proc. Transducers '01, Eurosensors XV, The 11th Intl. Conf. Solid-State Sensors and Actuators, Munich 2001, ed. by O. Obermeier (Springer, Berlin, Heidelberg 2001) 1258–1261

35.95 D. K. Davies, M. F. Biondi: Vacuum breakdown between plane-parallel copper plates, J. Appl. Phys. **37** (1966) 2969–2977

35.96 T. Ono, D. Y. Sim, M. Esashi: Micro-discharge and electric breakdown in a micro-gap, J. Micromech. Microeng. **10** (2000) 445–451

36. High Volume Manufacturing and Field Stability of MEMS Products

Low volume MEMS/NEMS production is practical when an attractive concept is implemented with business, manufacturing, packaging, and test support. Moving beyond this to high volume production adds requirements on design, process control, quality, product stability, market size, market maturity, capital investment, and business systems. In a broad sense, this chapter uses a case study approach: It describes and compares the silicon-based MEMS accelerometers, pressure sensors, image projection systems, and gyroscopes that are in high volume production. Although they serve several markets, these businesses have common characteristics. For example, the manufacturing lines use automated semiconductor equipment and standard material sets to make consistent products in large quantities. Standard, well controlled processes are sometimes modified for a MEMS product. However, novel processes that cannot run with standard equipment and material sets are avoided when possible. This reliance on semiconductor tools, as well as the organizational practices required to manufacture clean, particle-free products partially explains why the MEMS market leaders are integrated circuit manufacturers. There are other factors. MEMS and NEMS are enabling technologies, so it can take several years for high volume applications to develop. Indeed, market size is usually a strong function of price. This becomes a vicious circle, because low price requires low cost – a result that is normally achieved only after a product is in high volume production. During the early years, IC companies reduced cost and financial risk by using existing facilities for low volume MEMS production. As a result, product architectures are partially determined by capabilities developed for previous products. This chapter includes a discussion of MEMS product architecture with particular attention to the impact of electronic integration, packaging, and surfaces. Packaging and testing are critical, because they are significant factors in MEMS product cost. These devices have extremely high surface/volume ratios, so performance and stability may depend on the control of surface characteristics after packaging. Looking into the future, the competitive advantage of IC suppliers will decrease as small companies learn to integrate MEMS/NEMS devices on CMOS foundry wafers. Packaging challenges still remain, because most MEMS/NEMS products must interact with the environment without degrading stability or reliability. Generic packaging solutions are unlikely. However, packaging subcontractors recognize that MEMS/NEMS is a growth opportunity. They will spread the overhead burden of high-capital-cost-facilities by developing flexible processes in order to package several types of moderate volume integrated MEMS/NEMS products on the same equipment.

36.1 **Manufacturing Strategy** 1086
36.1.1 Volume 1086
36.1.2 Standardization 1086
36.1.3 Production Facilities 1086
36.1.4 Quality 1087
36.1.5 Environmental Shield 1087

36.2 **Robust Manufacturing** 1087
36.2.1 Design for Manufacturability 1087
36.2.2 Process Flow and Its Interaction with Product Architecture 1088
36.2.3 Microstructure Release 1095
36.2.4 Wafer Bonding 1095
36.2.5 Wafer Singulation 1097
36.2.6 Particles 1098
36.2.7 Electrostatic Discharge and Static Charges 1098
36.2.8 Package and Test 1099
36.2.9 Quality Systems 1101

36.3 **Stable Field Performance** 1102
36.3.1 Surface Passivation 1102
36.3.2 System Interface 1105

References 1106

Solid-state pressure sensors were first reported in the 1960s and commercialized by a number of start-up companies in the 1970s. By 1983, a *Scientific American* cover story [36.1] described pressure sensors, accelerometers, and ink-jet print heads and featured a gas chromatograph that combined an injection valve, a detector, and a chromatographic column on a silicon wafer. The future of the budding micromechanical device industry looked bright, but commercial reality has seriously lagged the rosy market projections. The reason was simple: Market forecasters failed to recognize the difference between lab prototypes and high volume production of stable products. Many companies have demonstrated that small quantities can be produced for niche markets. However, routine high volume production of reliable packaged devices is much more challenging. It requires the production prototype skills plus the manufacturing disciplines that are critical to long-term stable production.

Successful companies understand and apply these lessons. Indeed, over a hundred million MEMS pressure sensor, accelerometer, gyro, gas flow, and optical projection devices are shipped to customers annually. There is no universally accepted definition of MEMS. This chapter focuses on silicon-based products with movable micromechanical elements. It excludes disk drive heads, ink-jet print heads, hearing aids, microscale plastic, ceramic, quartz and metal components, and test strips for in vitro diagnostics that are sometimes included within the definition of MEMS. Many MEMS products do not integrate support electronics with the MEMS element on the chip even though they are produced with IC equipment, materials, and processes. However, integrated MEMS products are becoming ubiquitous. By 2002 Analog Devices had shipped over a hundred million surface micromachined MEMS accelerometers with integrated electronics to customers.

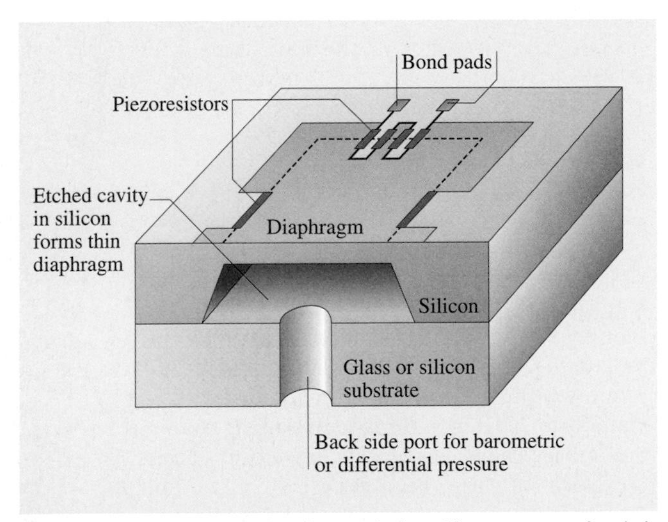

Fig. 36.1 Cross section of a piezoresistive silicon sensor after it is bonded to a silicon or glass substrate. When pressure is applied, the thin silicon diaphragm deforms. This causes changes in four implanted resistors that form a Wheatstone bridge. The resistors are located and oriented to cause resistance to increase in two resistors and decrease in the other two

The literature related to high volume production of MEMS products is quite uneven. Some suppliers have published detailed process and design descriptions, while others limit disclosures to market-focused publications. The two major suppliers of integrated MEMS products have discussed their designs and processes in detail. *Core* et al. [36.2], *Kuehnel* et al. [36.3], *Chau* et al. [36.4], and *Sulouff* [36.5] describe the Analog Devices *i*MEMS accelerometers and gyros. *Mignardi* et al. [36.6] give an in-depth review of the Texas Instruments DLP (Digital Light Processing) products.

MEMS devices are components in larger systems, so every MEMS product must define its system interface in a way that adds value from the perspective of the system designer. Some MEMS pressure sensor products are simply Wheatstone bridge elements in a circuit (Fig. 36.1). The image projection systems that use the Texas Instruments Digital Light Processing Technology^{TM} could never be designed in this manner, because up to 1.3 million mirrors are driven on each chip (Fig. 36.2). Such a system would be impractical unless electronics are integrated with the MEMS mirrors on the chip. Air bag sensors also illustrate how integration adds value. Non-MEMS ball-and-spring air bag sensors were often used in the early 1990s. Upon impact, a ball would be released and detected as it passed through the sensor tube. Unfortunately, these binary sensors could not always distinguish a frontal collision from the jolt of a pothole or a side collision. Therefore, multiple ball-and-spring sensors were placed at different locations in a vehicle and wired together in order to reliably identify accidents that should result in air bag deployment. High cost limited the use of these distributed systems. Analog Devices changed the air bag market when it introduced the ADXL50 in 1993 (Fig. 36.3). The ADXL50 integrated electronics and an accelerometer on one chip to provide an analog output of deceleration versus time when an impact event occurred. This allowed automobile manufacturers to design air bag modules that compared the ADXL50 output to the known signature of a frontal collision. The result was greater reliability

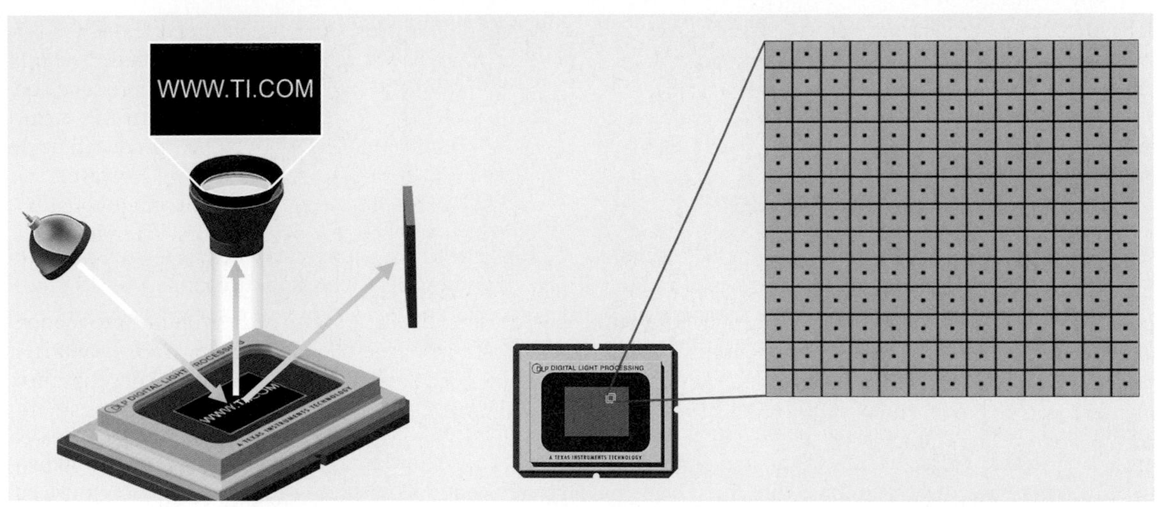

Fig. 36.2 Digital Light Processing technology™ is based on Digital Micromirror Device (DMD) chips that address and drive a "sea of mirrors." The DMD is the key element in small computer-based projection systems. Each of the 14 or 17 μm mirrors represents one pixel. The diagram on the *left* represents light reflecting from a mirror to the system lens. The illustration on the *right* is an exploded view of a small region of mirrors in a packaged DMD. (Courtesy of Digital Light Processing™, Texas Instruments, Inc.)

and lower cost, which led to the installation of air bags in almost every new automobile.

This chapter will not recite detailed descriptions of particular commercial products. Instead, brief descriptions will be used to illustrate basic principles and concepts. Part of the chapter is organized according to MEMS-specific issues like wafer singulation. This is useful because some designs and unit processes are well suited for lab scale or low volume production, but impractical in a high volume manufacturing environment. Complete reliance on such an organization would be artificial, because partitioning omits the greatest challenge – combining these individual topics into a manufacturing flow that routinely produces products that meet cost, performance, and reliability expectations. The importance of process integration cannot be

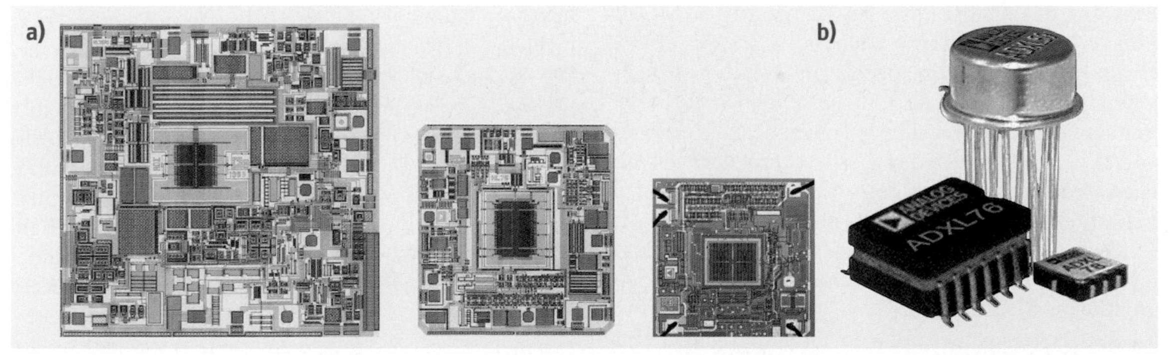

Fig. 36.3a,b Three generations of integrated accelerometers designed for single-axis air bag applications. **(a)** Die size is reduced by a factor of two about every four years. **(b)** Package technology also evolved to meet size, cost, and automated handling requirements. The ADXL50 was packaged in TO-100 seam-sealed metal headers. ADXL76 cerpacs (*bottom left*) are being replaced by the ADXL78 in 5 mm × 5 mm LCC packages (*bottom right*). (Courtesy of Micromachined Products Division, Analog Devices, Inc.)

overemphasized, because changes in one step almost always affect other steps. Solutions may involve design modifications, so design-for-manufacturability is an important topic. Manufacturing and business skills are included to the extent that they apply to stable production of reliable MEMS products.

36.1 Manufacturing Strategy

MEMS manufacturing is based on existing IC technology. However, there are only a few examples of routine MEMS and IC production intermixed on the same equipment (or of MEMS devices with electronics integrated on the chip). Even when a MEMS process step is nominally the same as a standard IC process, the specific conditions or end point or some quality such as film stress will differ. Technologists from Analog Devices and Motorola have discussed these thin film issues in depth [36.7–10].

Manufacturers prefer to use dry etch processes when possible in order to achieve better control of features, minimize waste disposal, and utilize automated equipment developed for the IC industry. However, anisotropic wet etching provides unique value in the production of some MEMS products like pressure, mass flow, and yaw rate sensors [36.11].

Although MEMS production economics is closely tied to the semiconductor model, useful lessons can be drawn from many industries. Some of the key characteristics that affect success are as follows:

36.1.1 Volume

Semiconductor economics is rooted in the principle that expensive processes are cost-effective when many devices are made simultaneously. Obviously, a MEMS product that is sold in low quantities loses this critical advantage. MEMS components are "disruptive technologies" – they are the key element in higher-level products that were previously impractical. It can take several years to develop a higher-level product and build its customer base. Therefore, it is prudent to include an incubation phase in the business plan for a novel MEMS device. Low sales volume during this phase causes high unit costs – batch process cost benefits are achieved only after the market develops.

Start-up companies often do not understand the implications of design-for-volume. It is a major challenge to transfer products made on lab tools to manufacturing equipment. The challenge is sufficient to justify the use of manufacturing people and equipment for product development.

Realistic price-volume estimates are also essential to a good business plan. The highest volume MEMS products (accelerometers and pressure sensors) all represent a small part of the total system cost, even though the functions they provide are absolutely critical to system performance.

Market-pull business models that respond to the needs of a specific large application in a particular industry are usually more successful than technology-driven models that seek applications for new devices. Given this reality, it is prudent to thoroughly understand the target industry and its quirks. Biotech, wireless, image projection, optical communications, and consumer markets offer tremendous growth potential. However, the past and present growth in high volume MEMS products have been driven primarily by the automotive industry, with pressure sensors also benefiting from medical applications. *Sulouff* [36.12], *Weinberg* [36.13], *Marek* et al. [36.14], *Eddy* et al. [36.15], and *Verma* et al. [36.16] discuss automotive applications and the challenges a successful supplier must master.

36.1.2 Standardization

Mature industries develop standards in response to equipment vendors, material suppliers, and customer interface requirements. The MEMS industry is fragmented, so even after 30 years it is not mature. Deviations from semiconductor industry standards are inevitable because MEMS products impose unique requirements on wafer processing, packaging, handling, and testing. For example, KOH wet etching raises mobile ion concerns. Gold used in some optical products is incompatible with CMOS. Any deviation from standard process and material sets is serious, because the price-volume relationship is a steep curve in high volume applications.

36.1.3 Production Facilities

The semiconductor industry is divided into front-end (wafer fabrication) and back-end (product assembly and test) operations. IC design rules are based on prequalified fab, assembly, and test capabilities at specific sites. Consider how each of these functions matches the needs of MEMS production:

1. An IC product might be produced at an internal wafer fab, or designed to run on a standard foundry process. In contrast, production challenges and "know-how" have caused MEMS suppliers to retain wafer fab production within their internal facilities. High volume production in MEMS foundries was limited to pressure sensors. However, Motorola is now outsourcing accelerometer production. This suggests that the MEMS industry will follow the IC model, with foundry services becoming more substantial in the future.
2. IC suppliers have shifted assembly operations to subcontractor facilities located in Southeast Asia. Some MEMS suppliers have followed this model. It is common to assemble low volume products in North America, Europe, or Japan. However, semiconductor assembly and package technology are evolving rapidly and are driven by subcontractors based in Southeast Asia. MEMS companies that do not utilize these resources pay a significant cost penalty. Obviously, the break point at which offshore assembly is attractive differs from one product to the next. Products that have unique assembly requirements and support high market prices may never reach this point.
3. IC testing is often carried out at the assembly subcontractor site. MEMS package and testing comprise a substantial portion of the total product cost, so a high volume manufacturing strategy should consider Southeast Asia packaging and testing. This is complicated by the fact that MEMS testing requires unique stimuli that are not in the standard IC portfolio.

36.1.4 Quality

The goal in lab and low volume production environments is to make functioning devices. As quantities increase, other factors become more important: lot-to-lot repeatability, yield, device performance that meets well-defined specifications, process stability to ensure predictable on-time delivery, documentation and procedures that enable traceability and corrective actions when problems arise, etc. Methodologies that promote stable, high-quality manufacturing practices have been established in several industries. These standard procedures such as ISO9000, QS9000, and cGMP overlap to a large extent. QS9000 was developed to help automotive suppliers implement effective, well controlled processes to produce reliable products. It has several unique features, so it is used as the model in this chapter.

36.1.5 Environmental Shield

IC wafers are completely passivated before packaging in order to meet product stability and reliability requirements. Aside from ink-jet print heads, the active regions of all MEMS devices currently in high volume production replicate this hermetic barrier in some manner. This passivation requirement is noteworthy because it has been a primary limitation on the growth of the MEMS industry. For example, it limits electrochemical sensors to benign or "throw-away" applications.

36.2 Robust Manufacturing

Stable production requires a manufacturing flow that is well controlled when measured against the product performance specifications. This has several implications.

36.2.1 Design for Manufacturability

Low maintenance products and processes must be designed to run on standard equipment. The semiconductor industry has invested billions of dollars to develop equipment that is automated, maintainable, reliable, and capable of supporting well controlled processes. Processes that are implemented on custom-designed equipment put this experience base aside and invariably have a long learning curve.

Equipment

Microstructure release illustrates this issue and how it changes over time. Virtually all semiconductor processes operate in vacuum or at atmospheric pressure. In a brief, unpublished 1992 study related to ADXL50 development, the author demonstrated the use of supercritical CO_2 mixtures for microstructure release and particle cleaning. Unfortunately, supercritical processes were incompatible with the industry infrastructure, because they require pressures in excess of 7.38 MPa (72.8 atm; 1,070 psi). Implementation would have required process and equipment development, so a release process that ran on existing equipment was selected.

Supercritical equipment designed for MEMS release was introduced a few years later. However, maintenance and cycle time concerns, as well as single wafer capabilities limited its adoption to university and low volume manufacturing.

The intrinsic limitations of non-standard equipment have caused high volume MEMS suppliers to avoid supercritical processes. However supercritical equipment will continue to evolve and become competitive, particularly in applications where the design requirements make alternative processes unusable.

Materials

Process control can be no better than the materials used in the process. SAM (*s*elf-*a*ssembled *m*onolayer) coatings illustrate this limitation. These materials suppress stiction following aqueous release of microstructures by treating the wafers in a SAM solution before drying. Although straightforward in principle, high volume manufacturers have not (yet) adopted SAM processes (Sect. 36.2.3 reviews the processes that are used commercially). There are several reasons for this reticence. Classical SAM materials are chlorosilanes that have at least one organic substituent. Organosilanes have been used to treat the surface of inorganic materials for many years. Applications include coupling agents on the surface of fillers and reinforcing agents in polymers, as well as agglomeration control of particles. The chlorine sites hydrolyze when dissolved in solutions that contain a small amount of water. The resulting hydroxyl groups react with the microstructure surface oxide to produce a chemically bonded layer that reduces stiction due to the low surface energy and hydrophobic nature of the organic-rich surface. The concerns that cause manufacturers to use alternative stiction solutions include:

1. High variability due to the SAM chemical reactivity
2. Particles are common by-products of the reactivity
3. Many SAM process flows require organic baths. This raises health, safety, and waste disposal concerns
4. Possibility of chloride corrosion on aluminum interconnects

Maboudian et al. [36.17], *Srinivasan* et al. [36.18], *Kim* et al. [36.19], and *Pamidighantam* et al. [36.20] describe the use of SAM solutions to suppress stiction. There is also ongoing research focused on solving the manufacturing concerns [36.21–25].

Note that stiction also arises in packaged MEMS products. This yield and reliability problem is discussed in Sect. 36.3.1.

36.2.2 Process Flow and Its Interaction with Product Architecture

MEMS products have structures and functions that do not exist in standard IC devices, so it is unrealistic to expect that every fab, assembly, and test step will reapply a previously qualified IC process. The challenge is to maximize utilization of available IC technology within the constraints of the product function and cost requirements. This leads to fundamental choices in product architecture, product design, and process flow, as discussed below.

Integration of MEMS and Circuits

The integration of MEMS and electronics onto one chip has proven difficult. It adds little value in some MEMS products. However, as noted earlier, integration is essential to products based on the Texas Instruments (TI) DLP (Digital Light Processing^{TM}). The third (and largest) category consists of applications where integrated and non-integrated products compete for market share. Pressure sensors and accelerometers are examples of this group.

Image Projection. The first projectors based on the DLP technology were shipped in 1996. DLP chips are produced on CMOS wafers in a mature TI process. The circuitry addresses and drives aluminum mirror arrays that function as on-off pixels – 1.3 million mirrors in the latest SXGA products (Fig. 36.2). *Mignardi* et al. [36.6] describe the DMD (Digital Mirror Device) fabrication process and illustrate some of the factors that must be considered when manufacturing flows are changed. The manufacturing flow patterns three aluminum depositions over SRAM cells that are positioned with 14 or 17 μm center-to-center spacing (Figs. 36.4 and 36.5). The first of these metal films form electrostatic drive electrodes and landing pads. The yoke and hinge layer has spring tips that land on the first metal layer when mirrors are rotated (Fig. 36.6). Mirrors are mechanically connected to the yokes through center pedestals. Sacrificial organic films separate the metal layers. These organics, along with a protective organic cover film, remain on the device until after the wafer is sawn and the chips are mounted in a ceramic package. They are removed by dry etching.

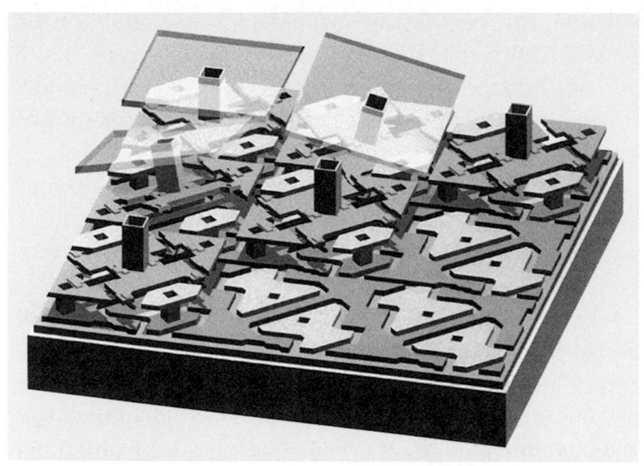

Fig. 36.4 DMD pixel array with tilted and non-tilted mirrors. (Courtesy of Digital Light Processing™, Texas Instruments, Inc.)

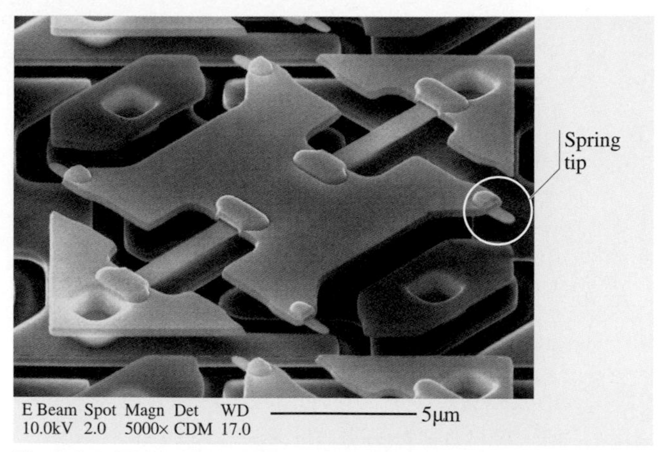

Fig. 36.6 DMD yoke and hinge layer. The spring tip touches landing pads when the drive electrode tilts the mirror. Elastic energy stored in the deformed spring is part of the restoring force that overcomes stiction when the drive electrode voltage is removed. (Courtesy of Digital Light Processing™, Texas Instruments, Inc.)

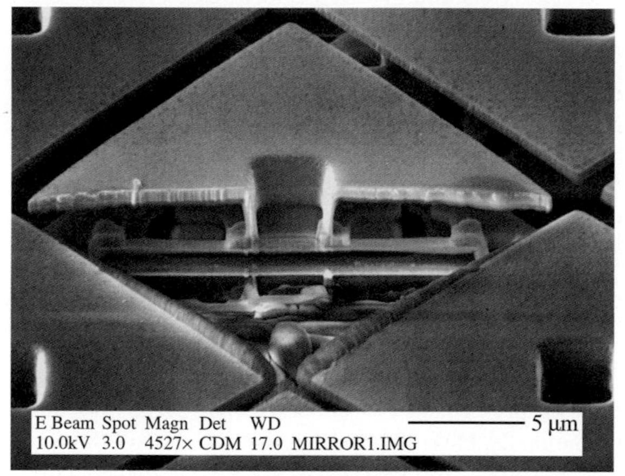

Fig. 36.5 Ion milled section of DMD pixel showing one mirror on its center support and its relation to the underlying layers. Note the close mirror spacing. (Courtesy of Digital Light Processing™, Texas Instruments, Inc.)

Following surface passivation, inspection, and testing, an optical glass sub-assembly is hermetically sealed to the ceramic package (Fig. 36.7). At this point, the device is ready for burn-in and final test – a significant task because each mirror is actuated on chips with as many as 1.3 million mirrors.

In operation, a (nominal) 26 V electrode bias generates an electrostatic force that causes the mirror/yoke assembly to rotate either plus or minus 10 degrees around the hinge. Over one million DMD devices had been shipped by 2002 [36.26].

Unlike the pressure sensors and accelerometers discussed below, TI had no choice but to integrate. It would be difficult, if not impossible, to devise a cost-effective multi-chip solution to control and drive a matrix of over a million mirrors. Successful integration of circuits and mirrors was only one step in the commercialization process. To remain competitive against lower cost LCD products, TI continues to develop their package and test technologies, because package and test comprise a major part of the total product cost. This is discussed further in Sect. 36.2.8.

Pressure Sensors. Early (1970s-era) bulk micromachined piezoresistive pressure sensors from Honeywell, ICT (later Foxboro-ICT), and Kulite were not integrated. However, in the 1980s, Motorola commercialized the

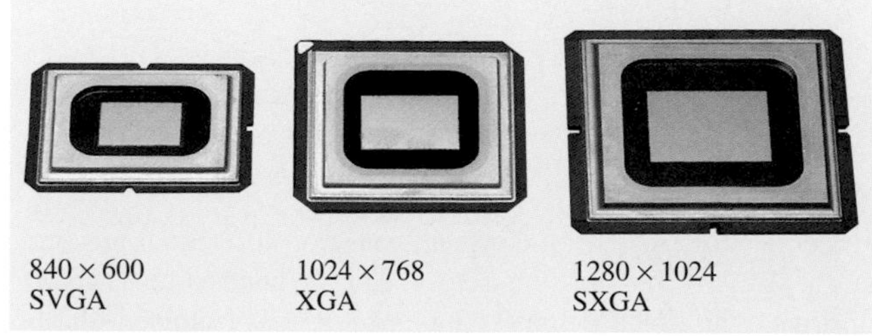

Fig. 36.7 DMD products in hermetic packages. (Courtesy of Digital Light Processing™, Texas Instruments, Inc.)

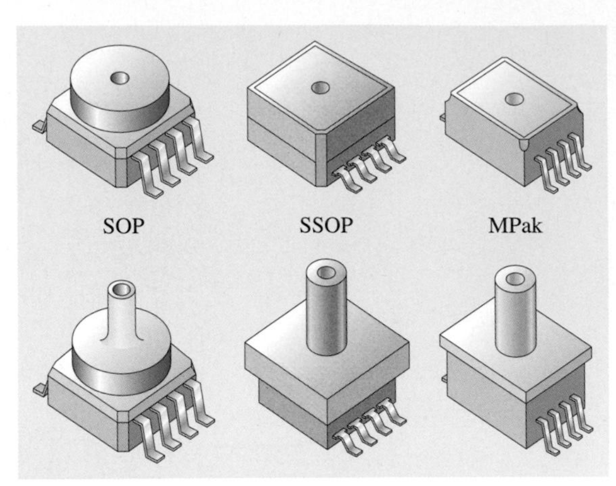

Fig. 36.8 Pressure sensor surface mount packages. (Courtesy of Semiconductor Products Sector, Motorola, Inc.)

first of its MPX5100 series piezoresistive pressure sensor products. These products had bipolar signal conditioning electronics and temperature compensation on the sensor chip. Current MPX products are offered in various plastic packages (Figs. 36.8 and 36.9) with maximum pressure ranges between 10 and 300 kPa (1.45–44 psi). Medical and tire pressure products are also available.

Integrated Motorola piezoresistive pressure sensor products are cited here for illustrative purposes, but capacitive and piezoresistive silicon pressure sensors are available from many companies. Capacitive designs are less temperature sensitive and require less power than piezoresistive designs. However, die size tends to be larger. The fabrication process and interface electronics are also more involved. Piezoresistive products (sensor plus signal processing) are less expensive, smaller, easier to manufacture and suitable for most applications. They rely on the fact that the resistance of silicon changes in response to strain. This effect is very temperature sensitive. However, configuring four piezoresistors as a Wheatstone bridge, judicious selection of the doping level and junction depth and the use of a temperature compensating resistor network minimizes temperature error.

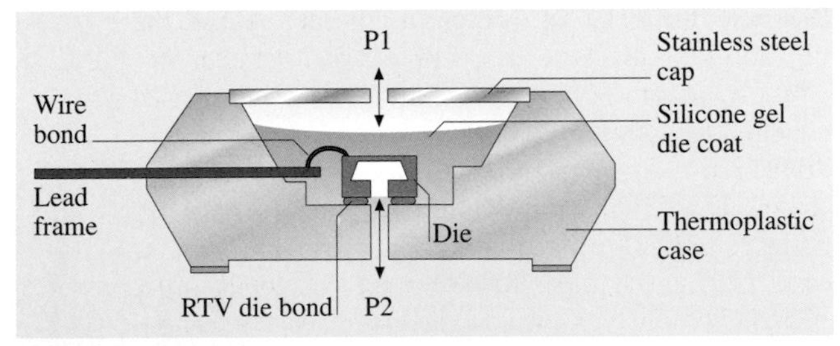

Fig. 36.9 Cross section of a typical gauge or differential pressure sensor. The silicon pressure sensor chip is bonded to a silicon substrate. This bonded unit is mounted in a pre-molded plastic package and protected from the ambient environment by silicone gel. (Courtesy of Semiconductor Products Sector, Motorola, Inc.)

Piezoresistive pressure sensors (Fig. 36.1) require double-sided polished wafers and front-to-back alignment. The piezoresistors are implanted into the front side of (100) wafers and precisely located with respect to the edges of thin diaphragms. The diaphragms are produced by anisotropically wet etching cavities into the back surface of the wafers. KOH etch solutions are usually used to avoid the safety, toxicity, and waste disposal concerns that arise when large quantities of organic etchants like EDP (Ethylenediamine-pyrocatechol-water) are used. Front-to-back side wafer alignment is critical in order to achieve precise resistor location with respect to the cavity edges. Alignment to the wafer crystal axis is also important, because the slow-etching (111) planes determine the cavity wall, and thus the location of the diaphragm edges. The piezoresistors are commonly placed near these edges to maximize the signal (the effect of pressure on diaphragm stress is greatest at the edges). An etch stop is sometimes used at additional cost because diaphragm thickness variations also have a large effect on sensitivity.

Control of mount stress is critical to achieving predictable performance. After wafer fab, some suppliers hermetically bond pressure-sensor wafers to a backup silicon wafer using glass frit. Others anodically bond them to a borosilicate glass wafer that has a thermal expansion coefficient close to silicon (Pyrex 7740 and Schott 8330 are two glasses used for this purpose). Absolute pressure sensors are produced when this bonding step is done in vacuum. As shown in Fig. 36.1 and Fig. 36.9, differential pressure and gauge pressure products incorporate through holes in the backup wafers. These ports allow fluid pressure to be applied through the cavities to the back side of the diaphragms.

After wafer singulation, the chips are mounted and sealed in cavity packages. Soft die attach materials, often silicone-based, are used to decouple the sensor from package and substrate stresses. In some products, the sensor chip is directly mounted to a substrate with a soft die attach without the intermediate backup wafer. Many package variations of MEMS pressure sensors are available, as summarized in Sect. 36.2.8.

The MEMS pressure sensor market is believed to exceed a hundred million pressure sensors annually [36.27]. Automotive applications like manifold absolute pressure (MAP) sensors and a wide range of gas and liquid pressure sensors form the largest market segment. Health care and medical uses such as disposable blood pressure transducers and sensors to measure pressure in angioplasty catheters, infusion pumps, and intrauterine products are the second largest market. Industrial products such as process control pressure and differential pressure transmitters, household appliance, and aeronautical products are also significant. *Maudie* et al. [36.28] review the performance and reliability issues related to appliance applications like household washing machines.

Accelerometers. Most MEMS accelerometer designs apply some variant of Newton's Law, $F = ma$, to sense the response of a proof mass. Many sensing principles, including piezoresistive, capacitive, piezoelectric, and resonant, have been examined [36.29]. Figure 36.10 has a conceptual view of the capacitive design used in Analog Devices ADXL accelerometers.

SensorNor was an early pioneer in silicon accelerometer manufacturing. Their product had a mass bonded to the end of a silicon piezoresistive element in an oil-filled package. However, the first MEMS accelerometer product to achieve large-scale market acceptance was the integrated ADXL50 air bag sensor from Analog Devices. When it was fully qualified in 1993, the ADXL50 was the first integrated surface-micromachined MEMS device of any type in production. The MEMS element in the ADXL50 was part of a closed loop differential capacitance circuit, but was designed to allow routine self-testing after the air bag module was installed in vehicles. This self-test capability eased concerns related to the adoption of a new technology in a critical safety application. The analog output feature also allowed the industry to design single point sensors – a significant system cost savings. *Core* et al. [36.2] and *Sulouff* [36.5] describe the ADXL wafer fabrication process. In essence, circuits are fabricated with a well-established BiCMOS process. The sensor polysilicon is deposited after these high temperature steps, followed by the lower temperature metal and passivation processes. The in situ doped polysilicon is connected to the circuit through N+ doped runners. Following wafer fabrication, thin film resistors are laser trimmed in an automated trim system to meet the specific end user's requirements. After wafer singulation (Sect. 36.2.5), the chips are assembled in hermetic cavity packages and screened in an automated test system to ensure compliance to the performance specification (Sect. 36.2.8).

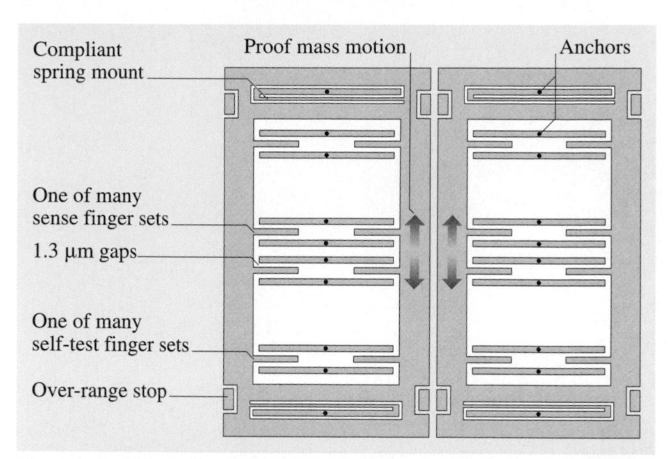

Fig. 36.10 Block diagram of the MEMS element in an ADXL78 single-axis accelerometer. The detailed design applies design-for-manufacturability principles to ensure close control of critical dimensions and residue-free removal of the sacrificial layer. (Courtesy of Micromachined Products Division, Analog Devices, Inc.)

Front air bag sensors have an output of 2 V at a full scale acceleration that is specified by the user (typically 35–50 g). Satellite air bag sensors are located in door pillars and near the front of the automobile and typically have higher ranges (250 g). Regardless of range, all air bag sensors must meet stringent cross-axis specifications, i. e., a front air bag sensor must not react to a side collision or to potholes in the road. This cross-axis requirement also applies to products that are designed to sense in two axes (Fig. 36.11).

Progressive reductions in chip and package sizes (Fig. 36.3) led to cost reductions and penetration into new markets. One example is the ADXL202 (Fig. 36.11). This two-axis, low-g accelerometer has a full scale of 2 g in each axis. It is widely used in industrial, consumer, and automotive applications. Production of integrated MEMS accelerometers at Analog Devices has grown rapidly in recent years (several million devices per month by 2001). It is worth noting that these products are developed using an organizational philosophy that emphasizes design-for-manufacturability. For example, development is conducted on production equipment and directly involves production personnel. This may appear to be expensive and unwieldy, however, it ensures that product designs remain within the bounds of practical man-

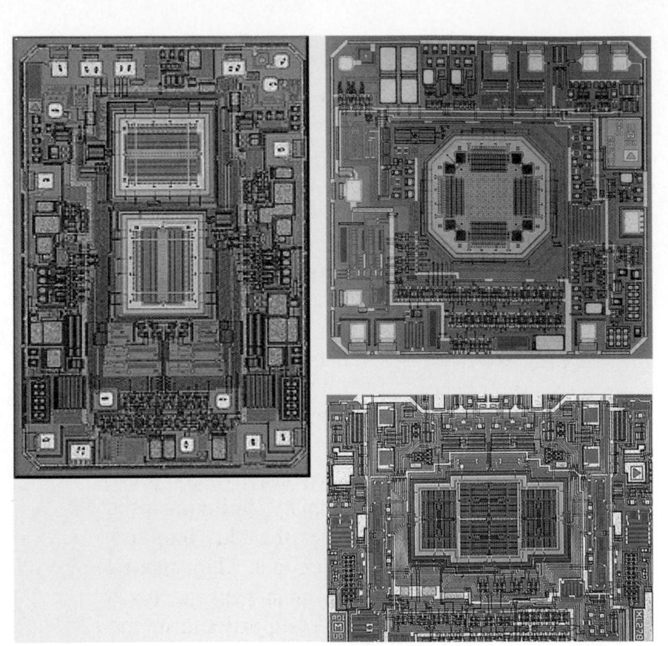

Fig. 36.11 Three two-axis accelerometers. *Top left:* ADXL276 was the first two-axis air bag accelerometer. The ADXL278 family (*bottom right*) is replacing it. *Top right:* 2-g ADXL202. The ADXL202 and ADXL278 are packaged in the LCC shown in Fig. 36.3. (Courtesy of Micromachined Products Division, Analog Devices, Inc.)

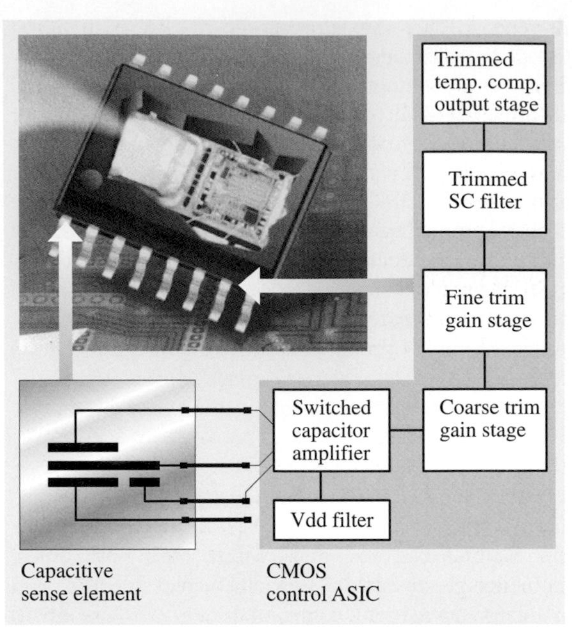

Fig. 36.12 Two-chip accelerometer in 16 lead plastic SOIC package. (Courtesy of Semiconductor Products Sector, Motorola, Inc.)

ufacturing. It also minimizes the difficulties that are normally encountered when new products are transferred to production.

Prior to the introduction of the ADXL50, most of the MEMS air bag candidates were nonintegrated, bulk-micromachined piezoresistive sensors [36.3]. However, Motorola and Bosch soon followed with an integrate-in-the-package strategy. Denso, Delphi, and VTI Hamlin are also active in some market sectors. The Motorola and Bosch products are nonintegrated capacitive MEMS products that are hermetically sealed at the wafer level (Sect. 36.2.4). By interconnecting a separate circuit chip and a sealed accelerometer chip in one package, they are able to use low cost CMOS for signal processing and near-standard plastic packaging (Sect. 36.2.8).

In contrast to the ADXL sensors, Motorola's MMA air bag products measure acceleration normal to the plane of the chip (z-axis). These differential capacitance sensors are formed from three layers of polysilicon. The middle layer is a "proof mass" that is suspended with ligaments and is free to move in response to an accelerating force. Ranges and outputs are similar to the ADXL products. Recent Motorola expansions add x-axis air bag products and a new line of low-g z-axis products (Fig. 36.12).

The Bosch x-axis sensor has interdigitated fingers that are patterned in a thick (11 μm) polysilicon film that is deposited in an epi reactor. Tight control of stress and stress gradients in these films is essential in order to meet sensor performance requirements [36.30, 31]. After patterning, the polysilicon is dry etched using fluorine-based chemistry [36.32] to form the differential capacitance sensor. An HF vapor process is used to remove the sacrificial oxide and release the microstructures. These thick structures have working capacitances near 1 pF [36.33]. *Laermer* et al. [36.34] explain how Bosch applies deep silicon etch processing to several production and experimental inertial sensors. Better known as the "Bosch etch," this deep reactive ion etching technique has pioneered the field of high-aspect-ratio silicon-based MEMS.

The Future of Integration. The high growth rate of integrated MEMS products demonstrates the value of integration in MEMS applications. To date, integration has been limited to large semiconductor companies that have in-house signal processing expertise and the willingness to invest substantial resources over a number of years. Each company identified a large volume market that could support a profitable integrated product. This "large semiconductor company" barrier-to-entry for integrated MEMS products will diminish in the future due to new SOI wafer technology (Sect. 36.2.4), new flip-chip equipment, and the development of low temperature MEMS processes. With these capabilities, MEMS-only companies will be able to integrate microstructures on circuit wafers that are preprocessed in CMOS foundries. To be successful, they will also have to develop cost-effective manufacturing and quality expertise (Sect. 36.2.9).

Risk

High risk is normal in research programs. However, the tremendous financial investments required to manufacture new products make the risk of failure a key consideration. This section discusses how the perception of manufacturing risk changes with time and is influenced by available knowledge and expertise.

Every development program makes fundamental product architecture decisions. These decisions can lead competing companies down different paths. For example, Analog Devices pursued chip-level integration, while Bosch and Motorola chose package-level integration. Why did these companies choose different paths? The answer is related to available resources, cost, time-to-market, and technical maturity. Analog Devices chose to integrate because of its expertise in process innovation and signal processing. This path did have risks. When the ADXL50 was designed, the fatigue life of polysilicon MEMS devices was unknown. Therefore, a closed loop circuit architecture was chosen in order to keep the proof mass motion under 100 Å. Polysilicon actually has excellent fatigue life. Once this was established, Analog changed their ADXL products to an open loop architecture that can detect capacitance changes as low as 20 zF (10^{-21} F). Most companies do not have the expertise to work in this regime, but the decision to integrate on one chip allowed Analog to make full use of its signal processing knowledge.

The two-chip Motorola and Bosch designs allow the circuit chip to be fabricated on any IC process. This product concept also removed the concern that integrating MEMS and circuits onto one chip might reduce overall product yield. However, it does require higher output signals from the MEMS chip in order to overcome parasitics that arise in the bond wires between the circuit and sensor chips. This requirement led Motorola toward large-area z-axis designs, while Bosch utilized its deep-etch process capability. To minimize cost and ensure reliability, both companies seal the sensors with glass at the wafer level. Sensor and circuit chips are then molded together in one plastic package. Each company built on expertise that it already had in order to minimize risk and cost. Motorola had in-house manufacturing expertise in wafer-level glass frit sealing of pressure sensors that are subsequently molded in plastic packages. Bosch had deep-etch process capability and the world's largest hybrid manufacturing plant, so screen printing and glass sealing were well-established. These considerations plus the concern that MEMS yield loss would undermine product yield contributed to their two-chip decision.

Semiconductor companies continually drive down cost and serve evolving markets by shrinking product size. Analog Devices believes that this long-term roadmap requires an integrated product. They chose to deal with one-chip risks, rather than package uncertainties. ADXL50 chips had been capped and molded in plastic long before the first qualified products were shipped in 1993 [36.35]. However, production-worthy capping equipment did not exist at that time. The effects of plastic package stress on yield and long-term device parametrics were also uncertain. By introducing the initial product in TO-100 hermetic metal packages, the risks associated with custom equipment and package stress were removed.

Reusable Engineering and Facilities to Achieve Economies of Scale

The IC industry produces many products on each process flow. Foundries and packaging subcontractors routinely apply this "economy-of-scale" principle. Only basic items like mask sets and bond pad diagrams are changed. The MEMS industry has been less successful in this area, although techniques like wafer-level laser trim and blowing poly fuses allow accelerometer suppliers to offer "specialty" products without losing the advantages of high volume production.

If only a few devices are produced in a capital-intensive facility, unit costs are high. This issue is particularly critical during the first years of product life when sales volume is normally low. One solution is to build the market by manufacturing the new product in an existing facility. The two-axis low-g ADXL202 (Fig. 36.11) was built using a new mask set with an existing air bag sensor process. Many ADXL202 consumer, industrial, and automotive applications are too small to justify the capital investment risk required to develop and produce a two-axis accelerometer, but use of an existing design and manufacturing base removed this constraint.

Market Dynamics Drive New Product Opportunities

In the 1980s, low cost and high reliability allowed MEMS pressure sensors to penetrate automotive applications like the measurement of manifold vacuum (MAP sensors). Once a technology is accepted, new opportunities usually develop. For example, tire pressure monitors with RF communication capability are now available in some car models. In the future, this type of monitor will be a safety feature expected by all consumers. Indeed, wireless communication capabilities will also be a common MEMS product requirement.

This example illustrates the fact that MEMS products are merely system components. They must support seamless integration into the larger system. MEMS opportunities in automobile safety systems will be affected by both communication protocols and interconnect technology (copper, fiber optic, and wireless). The growth spurt of the 1990s started with front-air-bag crash sensors located in the center module of automobiles. It has evolved to include satellite crash sensors behind bumpers, side-impact sensors in the B pillar (the vertical post between doors), impact sensors inside doors, rollover sensors, and vehicle dynamic control sensors to improve ride and handling. Each of these applications has unique characteristics that can favor one technology over another. For example, side-impact sensing systems must activate the air bags faster than front-impact systems. This response time characteristic can favor the use of pressure sensors that monitor air pressure inside door cavities – a good solution as long as the cavity remains sealed. An alternative is to place an accelerometer in the B pillar. These thin structural members have very little space, so z-axis accelerometers are sometimes easier to implement. Delphi and Motorola products have this characteristic. Analog's small x-axis accelerometers also meet this space requirement.

Gyros – New Products Produced with Preexisting Manufacturing Capabilities

Gyroscopes are transducers for rotational motion. They have a wide range of uses in platform stabilization and robotics, as well as automotive applications (dead reckoning backup for GPS navigation systems when satellite communication links are lost, skid control systems, and rollover detection). Unfortunately, the non-MEMS and resonant quartz gyros used in military and aerospace applications are rather expensive for many high volume applications. BEI Systron Donner tailored their high performance quartz gyro to meet automotive needs and has shipped several million gyros. Other lower cost gyro and angular rate products, like the Murata vibrating ceramic bimorph products, are used in cameras. However, as a group, these low cost products do not meet the performance required for many applications.

Analog Devices and Bosch reapplied their accelerometer capabilities to develop and manufacture low cost gyroscopes. *Geen* et al. [36.36] describe the Analog Devices product. *Funk* et al. [36.37] and *Lutz* et al. [36.38] describe the Bosch product. These three publications reference previous work and discuss the challenges of producing gyros that meet performance, reliability, and cost targets.

The Analog Devices ADXRS150 gyro (Fig. 36.13) illustrates the technology. It has a resonant microstructure that is integrated with two electronic systems on one chip. One of the electronic systems drives the MEMS resonator. When the ADXRS150 is rotated, Coriolis ac-

Fig. 36.13 **(a)** ADXRS150 gyro with integrated electronics in 7 mm × 7 mm BGA package. **(b)** ADXRS150 chip. **(c)** MEMS portion of the ADXRS150 chip. (Courtesy of Micromachined Products Division, Analog Devices, Inc.)

celeration is generated in a direction that is perpendicular to the vibration axis of the resonator. This motion is detected by the second accelerometer system. Full scale is rated at 150 deg/s, although it can sustain overloads up to 10,000 deg/s. Integration of electronics on the chip, microstructure modeling, and predictable manufacturing were essential to success, because the full-scale Coriolis motion is only about 1 Å. The combination of differential capacitance detection, area averaging, and correlation techniques allow this motion to be resolved from thermal noise down to about 0.00016 Å [36.36].

Resonant gyros are not new. However, as noted above, their high cost restricts them to specialty applications. Many are non-MEMS, or utilize the piezoelectric properties of quartz, so chip-level integration is not an option. Most require expensive vacuum packaging to achieve high mechanical Q at resonance. Unfortunately, vacuum packaging also makes them susceptible to mechanical damage. Viscous damping in gas-filled devices like the ADXRS150 suppresses the destructive impact events that occur in vacuum-packaged MEMS products during shipping and handling and in-use conditions. Equally important are ADXRS150 design features that allow the measurement to reject mechanical shock, vibration, and other environmental noise sources. As a result, under-the-hood location is practical in automotive applications, an unmatched characteristic that significantly reduces system cost.

Analog Devices and Bosch both needed several years to develop their gyro products. Use of existing accelerometer infrastructure during the prototype and early production phases significantly reduced development costs and manufacturing risks. Early ADXRS150 gyros were sold as engineering grade products in socketable side braze packages. This allowed the development team and customers to understand the design and process parametrics that affect performance, reliability, and yield. Side braze packages are unacceptable in high volume applications, so the ADXRS150 is sold in a hermetic, ceramic, surface mount package.

36.2.3 Microstructure Release

Release stiction occurs when microstructures stick together after the sacrificial material is removed in a wet etch. It can cause considerable yield loss and must be considered in devices like capacitive pressure sensors and accelerometers that have closely spaced, mechanically compliant microstructures. Piezoresistive pressure sensors are not susceptible to release stiction, because these sensing diaphragms are not near other surfaces.

Release stiction is most frequently observed in silicon MEMS devices that use a sacrificial oxide. The oxide is removed by etching in wet HF, rinsed in deionized water, and then dried. Water promotes growth of a hydrophilic surface oxide. It also has a high surface tension. To minimize energy, the high surface tension in the shrinking water droplets causes the microstructures to be pulled together. When they touch, clean oxide surfaces stick, thus destroying device functionality. *Maboudian* et al. [36.17] review release stiction mechanisms, as well as the solutions that have been reported in the literature.

Some manufacturers do not publicly discuss their fab process. Those that do avoid release stiction by using gas phase processes to remove the sacrificial material. Texas Instruments uses photoresist, rather than silicon dioxide as the sacrificial material in their DMD manufacturing process. Two photoresist layers are under the aluminum mirrors. The wafer is also covered with a third organic layer to provide mechanical damage protection and allow for particle cleaning after the wafer is singulated into individual devices. By removing these organics in a dry etch process, the stiction problems that characterize wet etch processes are avoided.

Analog Devices avoids release stiction by dividing its accelerometer process flow into several steps, each of which is well controlled with standard manufacturing equipment. After the microstructure is formed, a few small channels are etched in the sacrificial oxide and filled with photoresist. These photoresist pedestals hold the microstructures in place when the remaining sacrificial oxide is etched, rinsed, and dried. The pedestals are then removed in a dry etch process. *Core* et al. [36.39] and *Sulouff* [36.5] describe this technique in more detail.

Bosch avoids release stiction by holding accelerometer wafers above an HF solution. HF vapors from the solution react with the sacrificial oxide, converting it to gaseous SiF_4 and water. The wafers are heated to prevent water from condensing. This process is noted in *Offenberg* et al. [36.40]. *Anguita* et al. [36.41] describe a similar process.

36.2.4 Wafer Bonding

To date, the growth of MEMS wafer bonding has been driven by three factors:

- Sealing microstructures by bonding wafers together addresses two sources of yield loss in MEMS manufacturing: wafer singulation and particles.

- Wafer-level mounting can be a cost-effective way to control stress in products that are sensitive to variations in mount stress.
- Fabrication of 3-D microstructures from elements that are formed on multiple wafers.

Piezoresistive pressure sensor wafers have been anodically bonded to borosilicate glass wafers for about 25 years. The earliest products were manufactured using little more than a hot plate and a high-voltage power supply. The era of custom built manual equipment has passed, because both Electronic Visions and Karl Suss now offer automated production-worthy tools for anodic, glass frit, organic, and silicon-silicon wafer bonding. *Schmidt* [36.42] reviews wafer-wafer bonding processes, while *Mirza* [36.43,44] addresses MEMS wafer-level bonding applications and equipment.

Anodic Bonding

Anodic bonding is commonly used to seal glass wafers to the back side of bulk micromachined pressure sensor wafers. The process requires that a silicon wafer be placed in intimate contact with a glass wafer that contains a mobile ion at elevated temperature. An applied electric field causes mobile ions (usually sodium) in the glass to move away from the silicon interface toward the cathode at the other side of the glass. Bound negative charges remain in the glass near the silicon interface. These charges produce an electric field that pulls the wafers together and drives oxygen across the interface to anodically oxidize the silicon surface. Hermetic seals are routinely produced between flat wafers when particles are rigorously excluded. Process conditions depend on the glass composition and thickness, but 500–1,000 V at 400 °C with 10 min cycle times is typical. Borosilicate glasses with thermal expansion coefficients close to silicon are used to minimize stress. Anodic bonding promotes surface conformation so hermetic glass-silicon seals can be formed with surface grooves as deep as 50 nm [36.45]. This process has been demonstrated to work with many material combinations [36.46].

Glass Frit Bonding

Glass frit bonding reapplies techniques and materials developed for hybrid processes and cerdip sealing. It is often used to seal pressure sensor wafers to silicon substrate wafers. A second common application is to glass-frit seal bulk-micromachined silicon cap wafers over the microstructures in accelerometer wafers. In this process, fine glass powder is dispersed in an organic binder to form a paste. This paste is screen printed or stenciled in the desired pattern on a wafer. The wafer is then passed through a furnace (typically 430–500 °C). This causes the organics to burn off and the individual glass particles to coalesce into the desired pattern.

The wafer bonding process starts with a wafer prepared as described in the preceding paragraph. This wafer is aligned to a second wafer, where they are held in intimate contact and heated until the glass softens. Liquid glass readily wets the surface of the adjacent wafer when at least trace amounts of oxygen are present. The liquid-wetting mechanism produces hermetic seals even when the wafer surfaces are relatively rough. However, seal widths less than 100 μm are difficult to achieve. Fixtures that maintain wafer alignment through the bonding process are also required. Low-melting lead oxide and boron oxide glasses are commonly used in processes that typically run at 450 °C. Glass wet-out occurs quickly, but process cycle times are limited by the cooling rate (rapid cooling can cause stress gradients and cracking in the glass if proper design practices are not followed). Glass frit bonding to semiconductor wafers requires a good understanding of glass composition effects and process conditions [36.47].

Silicon–Silicon Wafer Bonding

Silicon-silicon bonding has been demonstrated in many IC and MEMS applications over the last 30 years. These processes bring highly polished wafers into intimate contact and are often promoted by a thin hydrophilic oxide that is left on the surfaces by the aqueous cleaning process. Clean, well polished wafers will bond at room temperature. However, heating (800–1,200 °C) increases bond strength by an order of magnitude. A fundamental difference between MEMS silicon–silicon bonding and either anodic or glass frit bonding is that it is used to produce complex microstructures that cannot be made in one wafer. In contrast, most anodic and glass frit bonding applications address assembly and packaging issues.

Recent silicon-silicon advances have solved technical factors like stress control that had limited MEMS applications. For example, Analog Devices balances the stresses at oxide interfaces and the handle wafer to produce flat, three- and four-layer SOI wafers [36.48, 49]. The thickness of each layer is precisely controlled using grinding and CMP (chemical-mechanical polishing). Intermediate layers can be patterned, electrically isolated, electrically interconnected and locally etched to produce single-crystal silicon microstructures that are hundreds of microns thick on CMOS compatible wafers. With this new capability, several organizations are fab-

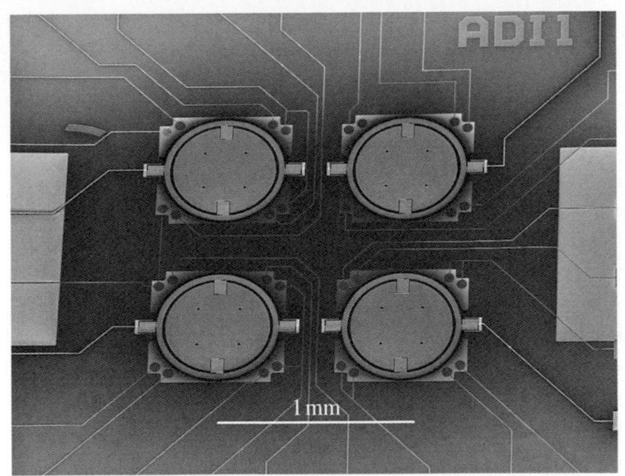

Fig. 36.14 Array of steerable mirrors designed for fiber optic network switching applications. The gimbal-mounted mirrors in this experimental product are electrostatically driven. Closed loop control based on capacitive sense electrodes ensures that the proper angle is maintained. High and low voltage circuits are integrated with mirrors on one chip. (Courtesy of Micromachined Products Division, Analog Devices, Inc.)

ricating MEMS devices for biotech, automotive, and communications applications. One example is the array of gimbal-mounted mirrors shown in Figs. 36.14 and 36.15. These mirrors can be tilted in any axis to steer light between fibers in optical fiber bundles. The mirrors are driven by electrostatic force applied from electrodes that are patterned under the mirrors. Capacitance measurements based on a second set of electrodes are also part of a closed loop control system that maintains position control. Integration of high voltage drive electronics, low voltage control electronics, and MEMS mirrors would be impractical without this SOI capability and related MEMS processes.

36.2.5 Wafer Singulation

The cooling water and the particles generated in a standard IC diamond saw will destroy MEMS wafers unless some form of protection is used. Manufacturers solve this problem in several ways. Bosch and Motorola bulk micromachine cap wafers that are glass frit bonded to surface micromachined wafers. The caps protect the microstructures from water, particles, and mechanical damage so the wafers can be sawn with standard equipment.

The optical function of DMD chips required Texas Instruments to find a different solution for DMD wafers. Originally, they used a partial saw process on wafers that were protected with an organic film. This allowed wafer-level testing, but caused particle contamination and die loss when wafers were broken along the partial saw cuts. They now use a standard saw process, but do it before the mirrors are released. The wafers have the protective organic film over the mirrors, so normal cleaning processes are used. The singulated chips are

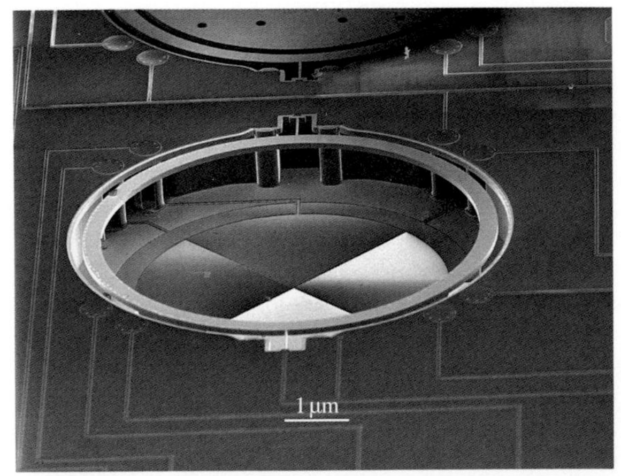

Fig. 36.15 View of the deep etched trenches, interlayer connections, and electrodes under a mirror of the SOI MEMS device illustrated in Fig. 36.14. (Courtesy of Micromachined Products Division, Analog Devices, Inc.)

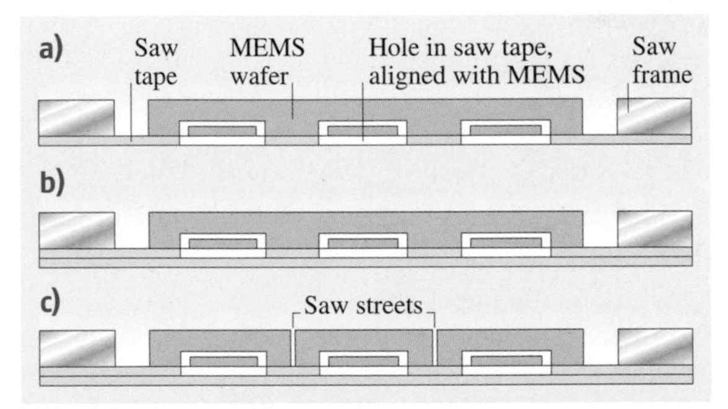

Fig. 36.16a–c Analog Devices "upside-down" saw process flow. **(a)** Saw tape is mounted on a saw frame. Holes are punched in the tape to match the microstructures on the wafer. The wafer is mounted-upside down on the saw tape and aligned so that the microstructures are in the holes. **(b)** A second layer of tape is placed over the first layer to form watertight pockets. **(c)** Wafer is aligned and sawn using standard equipment (adapted from [36.50])

mounted in ceramic packages before the organic layers are dry etched. This flow eliminates a primary source of particle contamination and problems associated with handling partially sawn wafers [36.6].

Analog Devices singulates uncapped MEMS wafers using an upside-down saw process and standard equipment and fixtures that are slightly modified (Fig. 36.16) (*Roberts* et al. [36.50]). A major attraction is that it allows wafer-level test and trim, so only chips that meet the product specification are assembled. It also avoids the added cost and yield loss associated with wafer capping. Although originally developed for accelerometers, this technology was extended to gyro wafers and to wafers that contain optical mirror arrays (see Figs. 36.13 and 36.14).

36.2.6 Particles

Particles cause yield loss in every product that has closely spaced movable microstructures. They also cause large area defects in anodic and silicon-silicon wafer bonded processes. Commercial MEMS suppliers use several solutions:

1. Cap the MEMS wafers before they leave the wafer fab clean area. Bosch and Motorola use this approach to produce accelerometer products (Fig. 36.12). It solves the wafer saw dilemma faced by most MEMS suppliers and allows the use of plastic packaging. Disadvantages include yield loss due to cap misalignment and stress, as well as the cost of the added capping steps. Hermeticity testing is also difficult because capped devices require sensitivity levels beyond the capabilities of standard analytical test equipment.
   Capping protects MEMS from particle contamination during assembly, but does not prevent wafer fab contamination. A greater concern is that glass-frit seal processes and equipment do not usually meet fab cleanliness standards. Once capped, inspection is essentially impossible.
2. Analog Devices and Texas Instruments assemble their MEMS products in clean rooms. This option increases capital costs. TI invested in environmental and process control, rather than develop optical quality capping for the DMD. Analog Devices had solved the wafer saw dilemma, so environmental and process control was the low risk path for particle control.

Particle control is central to high yield and reliability. The first control level is to minimize particle contamination through proper handling procedures, equipment maintenance, environment controls, and process design. The DMD protective organic film illustrates how process design can be used to control particles. This film allows the wafer to be cleaned after wafer saw and die attach.

Even with the best of controls, some particle contamination occurs in every clean room. Electrical tests in the automated wafer- and package-level test programs are designed to detect particles in Analog Devices products. Texas Instruments also finds particles in DMD products by driving each mirror in an array to ensure proper operation. Such screens are good but imperfect. Attempts have been made to replace visual inspection with automated particle inspection based on pattern recognition software. To date, however, machine vision systems have not been able to match the human eye and mind aided by a high quality microscope.

36.2.7 Electrostatic Discharge and Static Charges

Susceptibility to electrostatic discharges (ESD) led semiconductor manufacturers to make ESD avoidance a central criterion in equipment and fixture design, as well as handling procedures. Control of ESD events on the manufacturing floor is only the first step. Analog Devices and Texas Instruments incorporate ESD protective circuitry into their MEMS products as a standard practice to avoid failures in their customer assembly lines, as well as in the final system.

Most MEMS products use electrostatic force to actuate or control suspended microstructures. This design characteristic introduces performance parasitics and failure mechanisms that are not encountered in normal IC devices. For example, charges that build up on a dielectric surface can deflect unshielded microstructures, even when they are several millimeters apart. Such effects are insidious, because closed loop control systems do not always correct for electrostatic-induced errors.

Section 36.3.1 discusses surface treatments designed to suppress stiction caused by surface forces. These treatments often produce a dielectric surface. If the dielectric properties of the treatment and the device design cause the surface to hold a charge, that anti-stiction coating may actually promote electrostatic stiction! Thus, any "solution" must be critically evaluated in order to identify and remove undesirable side effects.

36.2.8 Package and Test

IC packages are environmental (mechanical, chemical, electromagnetic, optical) barriers that protect the chips from surrounding media. Only power and electrical signals and inertial forces pass unimpeded, so most MEMS sensors and actuators require that this barrier be selectively penetrated. Package and test functions are unique to each company's design and MEMS product type. Standard semiconductor packaging equipment and processes are modified when possible. However, even the line of demarcation that separates wafer fab from packaging is unique to each product.

High volume MEMS testing is even more challenging than MEMS packaging. In essence, the task is to measure and trim the response of products mounted in non-standard packages to calibrated stimuli over a range of temperatures. Automated IC test equipment does not have the calibrated pressure, acceleration, or optical test functionality required for this task. *Maudie* et al. [36.51] detail the test system elements that are required to support volume production of MEMS products.

Pressure Sensors

Pressure sensors are extremely susceptible to stress, so die-attach stress variations broaden performance distributions. The uniformity produced by wafer-level bonding is a primary reason why these processes are used to make the first-level package in many pressure sensor products. Soft silicone and fluorosilicone rubber products are often used to attach the sensor chip to the package to maximize isolation from mechanical stresses transmitted through the mount.

Most semiconductor products are assembled in plastic packages. The automated plastic presses transfer viscous epoxy compounds into multi-cavity molds at 175 °C and pressures near 6,900 kPa (1,000 psi). The hot liquid plastic creates high shear stress as it flows around devices in the mold cavities. Further stresses develop as the plastic hardens and cools. Even if this technology were adapted to create pressure ports in packages, the mold stresses would substantially affect pressure sensor performance.

Pre-molded plastic packages avoid the stress problem and incorporate pressure ports (Figs. 36.8 and 36.9). Polyphenylene sulfide (a high temperature thermoplastic) and medical-grade polysulfone are commonly used to make pre-molded pressure sensor packages.

The liquid or gas that is being measured can cause corrosion of interconnects and chip metallurgy, as well as parasitic leakage paths in the sensor. For that reason, a barrier such as silicone or fluorosilicone gel is often applied over the sensor. The soft gel transmits pressure with high fidelity over a wide temperature range and passivates the chip against many types of chemical attack. Parylene is also used for this purpose. This vapor-deposited organic coating is much stiffer than the gels, so thickness is typically controlled near 1 μm.

Petrovic et al. [36.52] tested the ability of fluorosilicone gels and several Parylene C thicknesses to protect powered pressure sensors against automotive and white goods benchmark liquids. One purpose of their study was to propose and demonstrate a formal media compatibility test protocol for pressure sensors similar to the IC industry standard tests. They found that the lifetime of sensors coated with both gel and parylene was considerably longer than sensors coated with either gel or parylene alone. *Petrovic* [36.53] later summarized the advantages and limitations of gels, parylene, and other techniques.

Figure 36.8 has examples of commercial plastic packaged pressure sensors. Pressure calibration equipment and software must be custom designed and built for these products. However, most pressure products are dimensioned to match IC packages in order to maximize compatibility with standard IC test handlers.

The harsh environment of some under-the-hood applications has been used to justify the high cost of metal cavity packages. For example, Bosch introduced a piezoresistive MAP sensor with signal conditioning that was hermetically sealed in an evacuated T08 header [36.54]. In-package trim was based on planar thyristors that were zapped as needed to bring the sensor within spec. The design avoided chip metallurgy corrosion and surface electrical parasitics by using the back surface of the silicon diaphragm as the fluid interface.

Much more expensive packaging is used in the pressure and differential pressure transmitters that are designed for industrial process control. These products seal the sensor in silicone oil behind thin metal diaphragms. Such products can measure pressure differences of 20 kPa superimposed on a pressure of 20,000 kPa (3,000 psi) over wide temperature ranges with high accuracy and stability. Soft die attach materials decouple the sensor from thermal stresses that arise when they are mounted in metal housings. Note, however, that silicone die attach materials cannot be used if the transmitter is filled with silicone oil. Fluorosilicones are one alternative. Piezoresistive sensors have been used in pressure and differential pressure transmitters since the 1980s. *Fung* et al. [36.55] describe a relatively new version. This product uses a piezoresistive polysilicon sensor to measure both absolute and differential pressure on the same chip. See *Chau* et al. [36.56] for early work on a multirange version of this technology.

Image Projection

The Texas Instruments DMD poses unusual challenges because the package lid must be an optical quality glass with anti-reflection coatings to improve optical performance and reduce heat load. The glass lid has opaque borders around the image area to minimize stray light effects and create a sharp edge on the projected image. It is fused to Kovar frames to produce lid sub-assemblies. The package base is a multilayer alumina substrate with co-fired tungsten to provide the electrical interconnects. The package sidewalls are formed by brazing a Kovar seal ring to the substrate. Heat dissipation through this substrate is a significant consideration because the service life of DMD mirror hinges (Fig. 36.6) is largely determined by operating temperature [36.57]. Excessive high temperature creep results in "hinge memory" and would cause a gradual drift in mirror orientation.

The DMD chips are die attached in the package cavity before the mirrors are released. A dry etch process is used to remove the organic sacrificial layers. The die are then passivated and tested. Getter strips are attached on the inside surface of the glass at the sides to control vapor composition in the package cavity before the lid is aligned and seam sealed to the package seal ring. Image quality requirements place stringent requirements on handling, alignment, and spacing tolerances. Thermal stresses also arise when dissimilar materials are joined in high temperature processes to produce the final hermetic package. The DMD packages (Fig. 36.7) and the related processes are discussed further in *O'Connor* [36.58], *Mignardi* et al. [36.6], *Bang* et al. [36.59], and *Poradish* et al. [36.60].

DMD mirror release does not occur until after die attach. This has advantages with respect to wafer saw and particle suppression, but it sacrifices the economic attraction of wafer-level testing. There is no commercial test system capable of combining CMOS electrical testing with 100% testing of optical mirrors. Therefore, a custom electro-optic test system was built around an x-y-θ translation stage with a CCD camera, light source, control hardware and software, and test programs [36.6]. A characteristic of the papers published by the DMD group is their effective use of this test system to examine problems and statistically validate the solutions.

Assembly of the DMD on the projector electronics board requires alignment with the system optics, in addition to electrical connections. Initially, the DMD was held with a plastic clamp and electrically connected through elastomer pads that had alternating layers of conductive and nonconductive material. This allowed easy replacement if a DMD was damaged during assembly. However, the impedance of this connector system was too high for new, higher speed products. Electrical intermittents were also observed. Therefore, it was replaced [36.61] by a grid of c-shaped springs (cLGA^{TM}, Intercon Systems, Inc., Harrisburg, PA).

The DMD has about 40% of the business and entertainment image projection market [36.62]. Products based on liquid crystal technology serve the balance of the market. Market share is largely driven by price, so considerable effort has been placed on reducing the cost of DMD package and board-level assembly. *Migl* [36.61] reports on the assembly benefits achieved by replacing the epoxied heat sink with a mechanically attached heat sink. *Jacobs* et al. [36.62] describes the effort to replace the seam sealed window mount with a lower cost epoxy-bonded design. In theory, such a bond is not hermetic. However, the team realized that proper material selection, design, analysis, and use of moisture getters would produce a low humidity package through the product life. This change was not released to production. If it is implemented, success will require that adhesive bond integrity also be maintained, because the adhesive joint sustains thermal expansion and mechanical clamp stress cycling each time the projector is used.

Accelerometers and Gyros

The Bosch and Motorola capping processes are a first-level package that is applied at the end of wafer fab. After capping, the wafers can be sawn using standard equipment and do not require clean room assembly conditions. In principle, capped devices are compatible with standard plastic packaging. However, molding stresses can be a serious problem. To minimize stress, capped sensor chips are mounted on lead frames with a soft elastomer and often coated with a silicone gel to isolate the sensors from the package stress. Motorola products are calibrated by burning EPROMS after molding. Their early air bag accelerometers were molded in DIP and SIP packages for assembly on through-hole circuit boards. The shift to smaller surface mount SOIC packages reduces the quantity of plastic, so package stresses are reduced. To maintain compatibility with standard tooling in small thin plastic packages, either the cap or the MEMS wafers must be backlapped.

Analog Devices uses a near-standard saw process (Sect. 36.2.5 and Fig. 36.16), so capping is not required if hermetic cavity packages are used. Such packages are standard, but more expensive than molded plastic. A major attraction of cavity packages is that they eliminate plastic package stress (see page 1099), because the only mechanical connection between the chip and the package is through the die attach and bond wires. As a result, ADXL accelerometers and ADXRS gyro products are fully tested and trimmed on automated systems before the wafers are sawn. A lot-tracking system transfers this data to automated die attach systems. These systems are programmed to pick only chips that meet performance specifications from the wafer saw film frames. Thus, reject die are not assembled into packages. After packaging, devices are tested to ensure conformance to specification. However, package-level trim is not required, because the assembly process and cavity packages do not appreciably shift device parametrics.

The evolution of ADXL cavity packages is illustrated in Fig. 36.3. Initial products were packaged in TO-100 metal packages. This seam-sealed package is useful for development purposes. However, it is expensive and incompatible with the automated equipment used to assemble electronic circuit boards. Therefore, the early air bag sensors were soon switched to cerdips. Cerdips are made from low cost, molded, ceramic bases that have glass seal surfaces. The high process temperature requires use of a silver-glass die attach product. The full assembly flow includes two or three furnace passes near 450 °C to produce hermetic cavity packages.

IC products have been packaged in cerdips for decades, but through-hole circuit boards are seldom used today. A simple change in lead forming allows cerdips to be fully compatible with standard surface mount boards ("cerpacs"), so cerdips and cerpacs are assembled on the same equipment. Most of the ADXL products shipped in the late 1990s were packaged in cerpacs. However, small LCCs have become the package of choice for one- and two-axis accelerometers. The LCC (Leadless Chip Carrier) uses solder-sealed ceramic bases and metal lids. Organic die attach materials are practical, because the furnace gas is nitrogen, rather than air, and temperatures are about a hundred degrees lower than cerpac furnace temperatures. ADXRS gyros are assembled in a solder-sealed 7 mm × 7 mm ceramic ball grid array package (Fig. 36.13).

The fundamental message of this section is that IC package technology – both plastic and hermetic – is evolving very rapidly. Cost-competitive suppliers must remain cognizant of these trends in order to use them to best advantage. MEMS products that are packaged in a way that is not compatible with standard IC equipment and interfacing standards bear a significant cost premium. High package cost seriously limits market size because the price-volume curve is steep for most products. The package size trend is also critical to new market penetration. In summary, customer interface and new application requirements will continue to drive down both the cost and the size of most MEMS products.

The ADXL and ADXRS automated testers and handlers measure multiaxial linear and rotational acceleration. Analog Devices has a division that designs and manufactures automated test systems (many of the corporation's IC products have unique test requirements), so it avoided some of the problems associated with the procurement of custom-automated test equipment. MEMS suppliers have relied on custom-designed test equipment, but this situation is gradually changing. For example, Multitest GmbH now makes handlers with integrated shakers that are specifically designed for accelerometer testing.

36.2.9 Quality Systems

Quality systems are an intrinsic part of stable MEMS production. Management must drive a systemic approach to quality and set continuous improvement as a high priority goal.

Several quality systems define continuous improvement methodologies. In general, they formalize the process used to minimize defects and variations in products and in the processes used to manufacture them. Automotive supply companies must implement quality systems that meet QS-9000 [36.63] before product volume ramps up and continue them through the product life cycle. Four elements that are worthy of note:

1. *Failure Mode Effect Analysis* (*FMEA*). Early in the development phase, a team with representatives from several disciplines reviews the design (or process) in order to identify possible causes of failure. Each potential failure mode is given three numerical rankings. One ranking represents the likelihood of an occurrence, while the others rank the severity of that result and detectivity. These numerical scores are combined for each potential failure mode in order to identify the issues that merit the most attention before they become problems. FMEA spreadsheets are periodically updated as the product or process moves into production. More targeted versions are used in products or processes that are in stable production.

2. *Process Control*. Every significant process must have a "short loop" monitor to ensure stability. This measurement is tracked and statistically analyzed with respect to the control limits to identify changes before the process strays beyond the control limits.
3. *Review Boards*. Every change, unusual occurrence, and proposed solution is assessed by boards that meet on a regular basis to ensure that it does not put product quality at risk. Affected production material is put aside until the appropriate board approves its release. This may appear to be expensive and bureaucratic. However, the cost of a scraped wafer lot is insignificant when compared to the cost of a field replacement program.
4. *Procedures and Specifications*. Each step must be fully specified and identified in the process flow, along with the appropriate metrics for each lot. The compilation of this data in a retrievable form is an essential part of every continuous improvement program. By combining these product and process databases with lot tracking software, Pareto charts linking yield loss and test failures to process variations can be identified and eliminated. Experience has shown that unexpected second-order effects are present in every production line – but data-driven decisions on indirect effects cannot be made on small test populations.

Implementation of a continuous improvement program requires well-informed failure analysis teams, methodologies, and programs. Such analyses start with gathering facts and making relatively simple tests. Often this is sufficient. However, MEMS products are susceptible to uncommon failure modes, so *Walraven* et al. [36.64] gathered examples of more powerful analytical techniques and showed how they are applied to MEMS devices.

Each supplier uses the information and insights gathered from product performance, as well as control and yield data at different points in the process flow to refine their operations. *Douglass* [36.65], for example, outlined the yield loss and failure mechanisms observed in early Texas Instruments DMD products. Concerns like hinge memory, hinge fatigue, particles, stiction, and environmental robustness were each addressed and mitigated by focused teams. For example, *Mignardi* et al. [36.6] describe the partial wafer saw process used in the original DMD manufacturing flow process. Breaking these delicate wafers into individual product chips generated particles that caused yield loss and were potential sources of field failures. A new flow based on a standard full saw process eliminates this particle source. The result was an increase in both yield and reliability. Hinge fatigue characterization showed that bulk metal fatigue models do not properly describe thin film behavior. The hinge memory effort required an understanding of how thin film metal creep is affected by alloy composition and the environment [36.57]. Stiction control involves surface passivation, spring design, mirror dynamics, and moisture level [36.26,65]. These publications give an insight into the quantities of data and the time required to bring robust products to market. They also illustrate how data-based evaluations can uncover unexpected effects like the acceleration of hinge creep by adsorbed moisture [36.26].

The second continuous improvement example applies to MEMS integration. High yield loss is perceived to be a serious risk when electronics and MEMS are integrated on one chip. Indeed, the initial yields on the Analog Devices integrated MEMS accelerometers were not impressive. However, it increased each year due to the work of many mission-oriented teams. Yield has reached defect density limited levels because the teams eliminated all significant failure mechanisms.

Coincident with the annual yield increases were reductions in customer failure rates. Failure rates in Analog's accelerometer products are in the low single digit ppm range. Quality does pay.

36.3 Stable Field Performance

Some topics discussed in Sect. 36.2 like particles and ESD are equally relevant to long-term stability. Most microstructure products have elements that are in close proximity, so stability is also affected by surface characteristics and mechanical shock.

36.3.1 Surface Passivation

Early, 1970s-era piezoresistive pressure sensors were not passivated. Like early ICs, these wafers only had oxide over the piezoresistors. Performance was inconsistent and drifted over time, because surface interactions with moisture and other atmospheric gases created a variety of shunts, parasitics, and charging issues.

Electrical Surface Passivation

The introduction of silicon nitride passivation and conductive field plates led to stable products that were hitherto unobtainable. The ability to achieve measurement stability was absolutely critical to growth of the MEMS pressure sensor industry.

Aside from pressure sensors, few MEMS products are passivated. Indeed, standard plasma nitride processes are usually impractical, because they produce dielectric coatings that support static charges. Such charges can cause electrical drift, or be the source of electrostatic forces that cause stiction.

There have been published reports attributing instability to the lack of passivation on microstructures. For example, Analog Devices implemented a special process in ADXL50 accelerometers before the product was released in order to suppress high temperature electrical drift [36.66]. This was not a moisture effect, because it occurred in dry nitrogen packages. Later ADXL designs eliminated the root cause.

Scientists at Lucent Technologies [36.67] observed anodic oxidation in polysilicon electrodes used to electrostatically drive mirrors in optical cross-connect products. If allowed to occur, this corrosion mechanism would cause the mirror position to drift and the product to eventually fail. The study concluded that, within the limits of the test, maintaining low moisture in these optical packages eliminates anodic oxidation.

The preceding paragraphs suggest that lack of MEMS passivation can affect long-term product stability. This issue is very design-related and can be driven by factors other than humidity. Even when low package humidity is determined to be adequate, design reviews should consider how normal manufacturing variations, outgassing and diffusion over the product life affect the moisture level in the gas adjacent to the microstructures.

Mechanical – Stiction and Wear

In-use stiction is difficult to predict and may not become an issue until manufacturing volumes increase, or when endusers handle the part in ways that are not anticipated [36.68]. The resulting product liability and field replacement programs carry substantial financial costs and have caused MEMS suppliers to withdraw from the market. End users have little tolerance for failure. Since in-use stiction is a failure mechanism in MEMS devices, a company that solves this problem creates an effective market barrier against competitors.

The scale of this problem varies between product areas. For example, the Texas Instruments DMD has moving mirrors that are designed to touch down on a substrate and later release. This product must address both stiction and wear. In contrast, the accelerometer products from Analog Devices, Bosch, and Motorola have proof masses that are suspended on compliant springs. By design, an automobile collision only displaces the proof mass by a few percent of the gap that separates it from adjacent surfaces. It is difficult to generate a shock wave in an air bag module that is sufficient to cause these proof masses to contact adjacent surfaces (acceleration levels of several thousand g are required). However, handling of the discrete packaged parts during test, shipment, and module assembly frequently cause shock events that bring MEMS surfaces into contact. For example, *Li* et al. [36.69] used both math modeling and tests to show that dropping a packaged MEMS accelerometer from the height of a table top generates several tens of thousands of g in the device when it lands on the floor. Thus, noncontacting MEMS products must be designed to withstand at least transient stiction.

Design. When shock, vibration, or functional operation causes MEMS elements to touch, the mounting springs are designed to pull them apart (Fig. 36.10). High stiffness springs are desirable to suppress stiction, but low stiffness springs increase measurement sensitivity and signal-to-noise ratio. Thus stiction is a fundamental MEMS design constraint that often limits product performance. Each manufacturer has proprietary design practices to address this performance versus reliability trade-off. In general, designers suppress stiction by minimizing the sources and applying supplementary techniques to ensure recovery when contact occurs:

1. Thorough analysis and testing to move harmful resonances beyond the range where they might be excited. Note that MEMS component-level analysis is insufficient – this is a system issue based on the packaged part as it is handled through manufacturing, system assembly, and end use.
2. Surface modification (discussed in the next section).
3. Minimizing contact area by integrating bumps and "stoppers" into the design [36.70].
4. Elimination of dielectric surfaces that may accumulate surface charges and result in electrostatic attraction.

5. Release of DMD mirrors from the touchdown position is assisted by pulsing the reset voltage to excite a mirror resonant frequency [36.65, 71].
6. Use of gas-filled packages to reduce contact velocity, as discussed later in this chapter.
7. ESD-protected designs that prevent voltage transients that may cause electrostatic attraction.

The only product in significant production that contains MEMS elements that are designed to touch is the Texas Instruments DMD. The proprietary aluminum alloy springs used in the mirror contact points touch and release millions of times during the product life. Considerable development was required to overcome metal creep that occurred in early designs.

Surface Modification. In addition to the touchdown springs, early DMD die were treated with perfluorodecanoic acid vapor before the glass lid was sealed onto the package. This treatment [36.72] created a low energy monolayer coating that suppressed initial stiction. However, improvements were required to achieve longer wear life, higher production rates, and lower contamination levels. A new process [36.73] places capsules inside DMD packages. After cleaning, the packages are immediately sealed. An oven bake releases anti-stiction vapor from the capsules to create a monolayer organic coating on surfaces inside these sealed packages. Excess anti-stiction material in the capsules maintains a low vapor pressure in the package cavities throughout the product life. Therefore, wear damage at the mirror contact springs is continuously repaired to prevent stiction failures before they occur.

Maintaining a low level of organic vapor inside the DMD reapplies a concept used with other organics to prevent corrosion of electronic metallurgy and contact fretting of separable connectors (see, for example, [36.74, 75]). These treatments have been shown to add years to the useful life of electronic systems that are in enclosures but also exposed to aggressive environments.

Stiction can be minimized by design and process spin-offs – stiff springs, control of contact area, etch residues, etc. Some suppliers design within these limits; others seek to modify the MEMS surfaces. The requirements of any surface modification technology are quite substantial. In addition to normal manufacturing requirements (stable, scalable process and materials), the process must be conformal and must uniformly treat all areas, including microstructure bottom and side surfaces. If a conductive material is used, it must somehow be patterned in order to avoid electrical shorting. If a dielectric surface is formed, it must be extremely thin because thick dielectrics support surface charges. Such charges cause performance shifts and induce electrostatic stiction.

Since devices with movable microstructures are particularly susceptible to stiction, most suppliers who develop production-worthy surface modification processes consider this information a trade secret. Analog Devices chose to patent portions of its technology. The following paragraphs give an overview of the published Analog Devices surface treatment technologies. *Martin* [36.66] describes the evolution of these technologies from 1992 to 1999. Further information is contained in *Martin* et al. [36.76] and *Martin* [36.77].

The earliest of the published Analog surface treatment technologies was implemented after stiction was observed during automated testing and module assembly of air bag sensors that were packaged in cerdips. Cerdip assembly typically includes at least two furnace processes in air at 440–500 °C, so they seldom contain organic materials. The high temperature furnace air thermo-oxidizes adsorbed materials and removes them from the MEMS surface. The resulting surfaces are extremely clean. Unfortunately, clean inorganics have high surface energies. Therefore, adjacent microstructures readily stick together if shock and vibration during handling cause them to touch. Cerdips (and the surface mount cerpacs) have very low moisture levels (Mil specs allow up to 5,000 ppm moisture, but < 100 ppm moisture is common in a well controlled manufacturing line). Deliberately raising the moisture level was found to be a potential solution because, within limits, moisture adsorption reduces surface energy. However, the solution adopted by Analog Devices in the mid-1990s was the anti-stiction agent (ASA) process. This process involved dispensing a controlled amount of a pure siloxane liquid (ASA) into each package immediately before sealing. ASA selection was based on thermo-oxidative stability, low volatility, purity, and the requirement that it be a liquid at the dispensing temperature. As packages heat in the furnace, adsorbed material on the microstructures is removed, leaving chemically active, high energy surfaces. Further heating volatilizes the ASA. Before escaping into the furnace, the ASA vapor surrounds the reactive MEMS surfaces and chemically bonds to them. The result is infusion of organic groups into the native oxide surface. Organic-rich surfaces are unreactive and have low surface energy. Thus the treatment is self-limiting after the first vapor molecules react. Obviously, ASA volatilization must be complete before the glass softens and seals the package cavity. Some

ASA degrades in this process. However, silicone thermo-oxidation simply adds a few angstroms of silicon oxide to the native oxide, so the ASA process, by design, is non-contaminating.

Although variations and refinements of the ASA process were developed, this technology required that each individual package be treated. Automated dispense equipment made ASA acceptable when product volumes were moderate. However, as production approached several million per month, it became evident that a wafer-level anti-stiction process was desirable.

Assessment of candidate processes and materials led to a vapor treatment based on a custom-synthesized solid polymeric siloxane. This polymer is volatilized in a standard CVD furnace that treats up to a hundred wafers at a time. Repeatability is extraordinary because thickness variation within a furnace run, or from run to run, is only about one angstrom.

Package Environment Effects. The viscosity and density of the gas that contacts MEMS resonators limit the device Q, so resonant devices are often designed to operate in vacuum. Unfortunately, this makes them susceptible to handling damage. Viscous fill gases reduce destructive excursions of high-Q microstructures when they are mechanically shocked. They also provide squeeze film damping to cushion the contact event as surfaces approach each other. None of the high volume MEMS products on the market are evacuated. Presumably, this is due to the high cost of vacuum packaging and the susceptibility to handling damage.

Many investigators believe that suppression of in-use stiction requires low humidity in the package. This view is not surprising because most MEMS research groups have observed stiction caused by aqueous surface tension during microstructure release. However, in-use stiction and wear test results published by several investigators lead to a different conclusion [36.76, 78–83]. In general, they find high stiction in both dry and moist environments. However, stiction is substantially suppressed at intermediate humidity levels (approximately 15–40 % relative humidity at room temperature). This is equivalent to a moisture level of 3,000–10,000 ppm. Note that the limits cited by different authors vary considerably due to the use of different materials and test conditions. Stiction at low humidity is probably caused by high surface energy on inorganic oxides that do not have adsorbed surface films. The low stiction and wear rates observed at intermediate humidity are attributed to the passivating and lubricating effects of adsorbed water. Capillary forces caused by the high surface tension of water become dominant at high humidity.

36.3.2 System Interface

MEMS products are mounted in systems that have much greater mass than the device itself. For this reason, it is difficult to transmit a mechanical shock wave to a MEMS sensor that is sufficient to cause failure after it is installed in a module. However, failures are possible when the devices are handled during module assembly. Improper functional response is also possible if the module mount system has a mechanical resonance that amplifies or reduces mechanical transmission under different conditions. This is a particular concern in automotive safety applications because every automobile platform has a different mechanical signature. Proper module mount design is required to ensure that the impact signal from an automobile crash is properly transmitted to air bag modules.

Suppliers of large IC die closely examine board-level stresses in order to minimize solder joint fatigue. Such stresses are more serious in MEMS products because the device, as well as the package are susceptible to mounting conditions. Furthermore, customer processes impose these stresses, so a supplier may have little information with respect to their origin or existence. One solution is to define and qualify a product-specific mount system, as Texas Instruments has done with the DMD. The ideal, however, is to devise products that are compatible with standard board footprint and assembly processes. This challenge became more difficult with the shift from through-hole to surface mount technology, because the mechanical isolation provided by the package pins is no longer present. Small package size, in-package, and within-chip isolation techniques, as well as thorough package-board stress analysis are used to address this concern.

References

36.1 J. B. Angell, S. C. Terry, P. W. Barth: Silicon micromechanical devices, Sci. Am. **248** (1983) 44–50

36.2 T. A. Core, W. K. Tsang, S. J. Sherman: Fabrication technology for an integrated surface-micromachined sensor, Solid State Technol. **36**(10) (1993) 39–47

36.3 W. Kuehnel, S. Sherman: A surface micromachined silicon accelerometer with on-chip detection circuitry, Sens. Actuators A **45** (1994) 7–16

36.4 K. H.-L. Chau, R. E. Sulouff Jr.: Technology for the high-volume manufacturing of integrated surface-micromachined accelerometer products, Microelectron. J. **29** (1998) 579–586

36.5 B. Sulouff: Integrated surface micromachined technology. In: *Sensors for Automotive Technology*, Sensors Applications, Vol. 4, ed. by J. Marek, H.-P. Trah, Y. Suzuki, I. Yokomori (Wiley-VCH, Weinheim 2003) Chap. 5.2

36.6 M. A. Mignardi, R. O. Gale, D. J. Dawson, J. C. Smith: The digital micromirror device – A micro-optical electromechanical device of display applications. In: *MEMS and MOEMS Technology and Applications*, ed. by P. Rai-Choudhury (SPIE, Bellingham 2000) Chap. 4

36.7 K. Nunan, G. Ready, J. Sledziewski: LPCVD & PECVD operations designed for iMEMS sensor devices, Vac. Technol. Coat. **2**(1) (2001) 26–37

36.8 M. Williams, J. Smith, J. Mark, G. Matamis, B. Gogoi: Development of low stress, silicon-rich nitride film for micromachined sensor applications, *Micromachining and Microfabrication Process Technology VI*, Proc. SPIE **4174**, ed. by J. Karam, J. Yasaitis, Proc. SPIE **4174** (2000) 436–442

36.9 Z. Zhang, K. Eskes: Elimination of wafer edge die yield loss for accelerometers, *Micromachining and Microfabrication Process Technology VI*, ed. by J. Karam, J. Yasaitis, Proc. SPIE **4174** (2000) 477–484

36.10 G. Bitko, A. C. McNeil, D. J. Monk: Effect of inorganic thin film material processing and properties on stress in silicon piezoresistive pressure sensors, Proc. Mat. Res. Soc. Symp. **444** (1997) 221–226

36.11 A. Hein, S. Finkbeiner, J. Marek, E. Obermeier: Material related effects on wet chemical micromachining of smart MEMS devices, *Micromachined Devices and Components V*, ed. by P. French, E. Peeters, Proc. SPIE **3876** (1999) 29–36

36.12 B. Sulouff: Commercialization of MEMS automotive accelerometers, 7th International Conference on the Commercialization of Micro and Nano Systems (COMS), Ypsilanti 2002 (MANCEF, Albuquerque 2002) 267–270

36.13 H. Weinberg: MEMS sensors are driving the automotive industry, Sensors **19**(2) (2002) 36–41

36.14 J. Marek, M. Illing: Microsystems for the automotive industry, Electron Devices Meeting, IEDM Technical Digest International, San Francisco 2000 (IEEE, New York 2000) 3–8

36.15 D. S. Eddy, D. R. Sparks: Application of MEMS technology in automotive sensors and actuators, Proc. IEEE **86**(8) (1998) 1747–1755

36.16 R. Verma, I. Baskett, B. Loggins: Micromachined electromechanical sensors for automotive applications, SAE Special Pub. **1312** (1998) 55–59

36.17 R. Maboudian, R. T. Howe: Critical review: Adhesion in surface micromechanical structures, J. Vac. Sci. Technol. B **15**(1) (1997) 1–20

36.18 U. Srinivasan, M. R. Houston, R. T. Howe, R. Maboudian: Alkyltrichlorosilane-based self-assembled monolayer films for stiction reduction in silicon micromachines, J. Microelectromech. Syst. **7**(2) (1998) 252–260

36.19 B. H. Kim, T. D. Chung, C. H. Oh, K. Chun: A new organic modifier for anti-stiction, J. Microelectromech. Syst. **10**(1) (2001) 33–40

36.20 S. Pamidighantam, W. Laureyn, A. Salah, A. Verbist, H. Tilmans: A novel process for fabricating slender and compliant suspended poly-Si micromechanical structures with sub-micron gap spacing, 15th IEEE 2002 Micro Electro Mechanical Systems (MEMS) Conf. (IEEE, New York 2002) 661–664

36.21 B. C. Bunker, R. W. Carpick, R. A. Assink, M. L. Thomas, M. G. Hankins, J. A. Voigt, D. Sipola, M. P. de Boer, G. L. Gulley: Impact of solution agglomeration on the deposition of self-assembled monolayers, Langmuir **16** (2000) 7742–7751

36.22 Y. Jun, V. Boiadjiev, R. Major, X.-Y. Zhu: Novel chemistry for surface engineering in MEMS, *Materials and Devices Characterization in Micromachining III*, ed. by Y. Vladimirsky, P. Coane, Proc. SPIE **4175** (2000) 113–120

36.23 R. Maboudian, W. R. Ashurst, C. Carraro: Self-assembled monolayers as anti-stiction coatings for MEMS: Characteristics and recent developments, Sens. Actuators A **82**(1) (2000) 219–223

36.24 W. R. Ashurst, C. Yau, C. Carraro, R. Maboudian, M. T. Dugger: Dichlorodimethylsilane as an anti-stiction monolayer for MEMS: A comparison to the octadecyltrichlorosilane self-assembled monolayer, J. Microelectromech. Syst. **10** (2001) 41–49

36.25 W. R. Ashurst, C. Yau, C. Carraro, C. Lee, G. J. Kluth, R. T. Howe, R. Maboudian: Alkene based monolayer films as anti-stiction coatings for polysilicon MEMS, Sens. Actuators A **91**(3) (2001) 239–248

36.26 S. J. Jacobs, S. A. Miller, J. J. Malone, W. C. McDonald, V. C. Lopes, L. K. Magel: Hermeticity and stiction in MEMS packaging, 40th Annual IEEE International Reliability Physics Symp., Dallas 2002 (IEEE, New York 2002) 136–139

36.27 Nexus Task Force Report: *Market Analysis for Microsystems 1996-2002* (NEXUS, Grenoble 1998) www.nexus-mems.com

36.28 T. Maudie, J. Wertz: Pressure sensor performance and reliability, IEEE Industry Appl. Mag. **3**(3) (1997) 37–43

36.29 Lj. Ristic, R. Gutteridge, B. Dunn, D. Mietus, P. Bennett: Surface micromachined polysilicon accelerometer, IEEE 1992 Solid State Sensor and Actuator Workshop, IEEE 5th Technical Digest, Hilton Head 1992 (IEEE, New York 1992) 118–121

36.30 M. Furtsch, M. Offenberg, H. Munzel, J. R. Morante: Influence of anneals in oxygen ambient on stress of thick polysilicon layers, Sens. Actuators **76** (1999) 335–342

36.31 P. Lange, M. Kirsten, W. Riethmuller, B. Wenk, G. Zwicker, J. R. Morante, F. Ericson, J. A. Schweitz: Thick polycristalline silicon for surface micromechanical applications: Deposition, structuring and mechanical characterization, Proc. 8th International Conf. on Solid State Sensors and Actuators and Eurosensors IX, Transducers '95, Vol. 1 (IEEE, New York 1995) 202–205

36.32 F. Laermer, A. Schilp: Method of anisotropically etching silicon, (1996) U.S. Patent 5,501,893

36.33 M. Offenberg, H. Munzel, D. Schubert, O. Schatz, F. Laermer, E. Muller, B. Maihofer, J. Marek: Acceleration sensor in surface micromachining for airbag applications with high signal/noise ratio, SAE Special Pub. **1133** (1996) 35–41

36.34 F. Laermer, A. Schilp, K. Funk, M. Offenberg: Bosch deep silicon etching: Improving uniformity and etch rate for advanced MEMS applications, Proc 12th International Conf. on Micro Electro Mechanical Systems MEMS (IEEE, New York 1999) 211–216

36.35 J. R. Martin, C. M. Roberts Jr.: Package for sealing an integrated circuit die, (2001) U.S. Patent 6,323,550

36.36 J. A. Geen, S. J. Sherman, J. F. Chang, S. R. Lewis: Single chip surface micromachined integrated gyroscope with 50 deg/hour allan deviation, IEEE J. Solid-State Circuits **37**(12) (2002) 1860–1866

36.37 K. Funk, H. Emmerich, A. Schilp, M. Offenberg, R. Neul, F. Larmer: A surface micromachined silicon gyroscope using a thick polysilicon layer, Proc. 12th International Conf. on Micro Electro Mechanical Systems MEMS (IEEE, New York 1999) 57–60

36.38 M. Lutz, W. Golderer, J. Gerstenmeier, J. Marek, B. Maihofer, S. Mahler, H. Munzel, U. Bischof: A precision yaw rate sensor in silicon micromachining, 11th International Conf. on Solid State Sensors and Actuators, Transducers '97 (IEEE, New York 1997) 847–850

36.39 T. A. Core, R. T. Howe: Method for fabricating microstructures, (1994) U.S. Patent 5,314,572

36.40 M. Offenberg, F. Laermer, B. Elsner, H. Munzel, W. Reithmuller: Novel process for a monolithic integrated accelerometer, Proc. 8th International Conf. on Solid State Sensors and Actuators and Eurosensors IX, Transducers '95 (IEEE, New York 1995) 589–592

36.41 J. Anguita, F. Briones: HF/H_2O vapor etching of SiO_2 sacrificial layer for large-area surface-micromachined membranes, Sens. Actuators A **64** (1998) 247–251

36.42 M. A. Schmidt: Wafer-to-wafer bonding for microstructure formation, Proc. IEEE **86**(8) (1998) 1575–1585

36.43 A. R. Mirza: Wafer-level bonding technology for MEMS, Proc. 7th Intersociety Conf. on Thermal and Thermomechanical Phenomena in Electronic Systems, ITHERM 2000, Vol. 1 (IEEE, New York 2000) 113–119

36.44 A. R. Mirza: One micron precision, wafer-level aligned bonding for interconnect, MEMS and packaging applications, Proc. 50th Electronic Components & Technology Conf. (IEEE, New York 2000) 676–680

36.45 S. Mack, H. Baumann, U. Goesele: Gas tightness of cavities sealed by silicon wafer bonding, Proc. 10th Annual International Workshop on Micro Electro Mechanical Systems, MEMS; IEEE Micro Electro Mechanical Systems (MEMS) (IEEE, New York 1997) 488–493

36.46 G. Wallis: Field assisted glass sealing, SAE Automotive Engineering Congress, Detroit 1971 (Society of Automotive Engineers, New York 1971) Paper 71023

36.47 S. A. Audet, K. M. Edenfeld: Integrated sensor wafer-level packaging, Proc. International Conf. on Solid State Sensors and Actuators, Transducers '97, Vol. 1 (IEEE, New York 1997) 287–289

36.48 C. Gormley, A. Boyle, V. Srigengan, S. Blackstone: HARM processing techniques for MEMS and MOEMS devices using bonded SOI substrates and DRIE, *Micromachining and Microfabrication Process Technology VI*, ed. by J. Karam, J. Yasaitis, Proc. SPIE **4174** (2000) 98–110

36.49 K. Somasundram, D. Cole, C. McNamara, A. Boyle, P. McCann, C. Devine, A. Nevin: Fusion-bonded multilayered SOI for MEMS applications, Smart Sensors, Actuators and MEMS, Proc. SPIE **5116**, ed. by J.-C. Chiao, V. Varadan, C. Cané (2003) 12–19

36.50 C. M. Roberts Jr., L. H. Long, P. A. Ruggerio: Method for separating circuit dies from a wafer, (1994) U.S. Patent 5,362,681

36.51 T. Maudie, T. Miller, R. Nielsen, D. Wallace, T. Ruehs, D. Zehrbach: Challenges of MEMS device characterization in engineering development and final manufacturing, Proc. 1998 IEEE AUTOTESTCON (IEEE, New York 1998) 164–170

36.52 S. Petrovic, A. Ramirez, T. Maudie, D. Stanerson, J. Wertz, G. Bitko, J. Matkin, D. J. Monk: Reliability test methods for media-compatible pressure sensors, IEEE Trans. Industrial Electron. **45**(6) (1998) 877–885

36.53 S. Petrovic: Progress in media compatible pressure sensors, Proc. of InterPACK'01, the Pacific Rim/International Intersociety Electronic Packaging Technical/Business Conf. & Exhibition (ASME, New York 2001) IPACK2001-15517

36.54 H.-J. Kress, J. Marek, M. Mast, O. Schatz, J. Muchow: Integrated pressure sensors with electronic trimming, Automotive Eng. **103**(4) (1995) 65–68

36.55 C. Fung, R. Harris, T. Zhu: Multifunction polysilicon pressure sensors for process control, Sensors **16**(10) (1999) 75–79

36.56 K. H.-L. Chau, C. D. Fung, P. R. Harris, J. G. Panagou: High-stress and overrange behavior of sealed cavity polysilicon pressure sensors, IEEE 4th Technical Digest, Solid State Sensor and Actuator Workshop (IEEE, New York 1990) 181–183

36.57 A. B. Sontheimer: Digital micromirror device (DMD) hinge memory lifetime reliability modeling, Proc. 40th Annual IEEE International Reliability Physics Symp. (IEEE, New York 2002) 118–121

36.58 J. P. O'Connor: Packaging design considerations and guidelines for the digital micromirror device^{TM}, Proc. of InterPACK'01, the Pacific Rim/International Intersociety Electronic Packaging Technical/Business Conf. & Exhibition (ASME, New York 2001) IPACK2001-15526

36.59 C. Bang, V. Bright, M. A. Mignardi, D. J. Monk: Assembly and test for MEMS and optical MEMS. In: *MEMS and MOEMS Technology and Applications*, ed. by P. Rai-Choudhury (SPIE, Bellingham 2000) Chap. 7

36.60 F. Poradish, J. T. McKinley: Package for a semiconductor device, (1994) U.S. Patent 5,293,511

36.61 T. W. Migl: Interfacing to the digital micromirror device for home entertainment applications, Proc. of InterPACK'01, the Pacific Rim/International Intersociety Electronic Packaging Technical/Business Conf. & Exhibition (ASME, New York 2001) IPACK2001-15712

36.62 S. J. Jacobs, J. J. Malone, S. A. Miller, A. Gonzalez, R. Robbins, V. C. Lopes, D. Doane: Challenges in DMD^{TM} assembly and test, Proc. Mater. Res. Soc. **657** (2001) EE6.1.1-EE6.1.12

36.63 Automotive Industry Action Group: *Quality Systems Requirements QS-9000*, 3rd edn. (Automotive Industry Action Group of the American Society for Quality, Milwaukee 1998)

36.64 J. A. Walraven, B. A. Waterson, I. De Wolf: Failure analysis of micromechanical systems (MEMS). In: *Microelectronic Failure Analysis*, 4th edn. (ASM, Materials Park 2002) 2002 Suppl.

36.65 M. R. Douglass: Lifetime estimates and unique failure mechanisms of the digital micromirror device (DMD), Proc. 1998 36th Annual IEEE International Reliability Physics Symp. (IEEE, New York 1998) 9–16

36.66 J. R. Martin: Surface characteristics of integrated MEMS in high volume production. In: *Nanotribology: Critical Assessment and Research Needs*, ed. by S. M. Hsu, Z. C. Ying (Kluwer, Dordrecht 2002) Chap. 14

36.67 H. R. Shea, A. Gasparyan, C. D. White, R. B. Comizzoli, D. Abusch-Magder, S. Arney: Anodic oxidation and reliability of MEMS poly-silicon electrodes at high relative humidity and high voltages, *MEMS Reliability for Critical Applications*; Proc. SPIE **4180** (2000) 117–122

36.68 J. Martin: Stiction suppression in high volume MEMS products, Proc. 2003 STLE/ASME Joint International Tribology Conf. (ASME, New York 2003) 2003TRIB-266

36.69 G. X. Li, F. A. Shemansky Jr.: Drop Test and analysis on micro-machined structures, Sens. Actuators **85** (2000) 280–286

36.70 R. T. Howe, H. J. Barber, M. Judy: Apparatus to minimize stiction in micromachined structures, (1996) U.S. patent 5,542,295

36.71 L. J. Hornbeck, W. E. Nelson: Spatial light modulator and method, (1992) U.S. Patent 5,096,279

36.72 L. J. Hornbeck: Low reset voltage process for DMD, (1994) U.S. Patent 5,331,454

36.73 E. C. Fisher, R. Jascott, R. O. Gale: Method of passivating a micromechanical device within a hermetic package, (1999) U.S. Patent 5,936,758

36.74 B. A. Miksic: Use of vapor phase inhibitors for corrosion protection of metal products, Proc. 1983 NACE Annual Conf., Corrosion 83 (National Assoc. Corrosion Engineers, Houston 1983) Paper 308

36.75 D. Vanderpool, S. Akin, P. Hassett: Corrosion inhibitors in the electronics industry: Organic copper corrosion inhibitors, Proc. 1986 NACE Annual Conf., Corrosion 86 (National Assoc. Corrosion Engineers, Houston 1986) Paper 1

36.76 J. R. Martin, Y. Zhao: Micromachined device packaged to reduce stiction, (1997) U.S. Patent 5,694,740

36.77 J. R. Martin: Process for wafer level treatment to reduce stiction and passivate micromachined surfaces and compounds used therefor, (2001) International Patent Application WO 01/57920 A1

36.78 S. T. Patton, J. S. Zabinski: Failure mechanisms of a MEMS actuator in very high vacuum, Tribol. Int. **35**(6) (2002) 373–379

36.79 S. T. Patton, K. C. Eapen, J. S. Zabinski: Effects of adsorbed water and sample aging in air on the μN level adhesion force between Si(100) and silicon nitride, Tribol. Int. **34**(7) (2001) 481–491

36.80 S. T. Patton, W. D. Cowan, K. C. Eapen, J. S. Zabinski: Effect of surface chemistry on the tribological performance of a MEMS electrostatic lateral output motor, Tribol. Lett. **9** (2000) 199–209

36.81 S.T. Patton, W.D. Cowan, J.S. Zabinski: Performance and reliability of a new MEMS electrostatic lateral output motor, Proc. 1999 37th Annual IEEE International Reliability Physics Symp. (IEEE, New York 1999) 179–188
36.82 D.M. Tanner, J.A. Walraven, L.W. Irwin, M.T. Dugger, N.F. Smith, W.P. Eaton, W.M. Miller, S.L. Miller: The effect of humidity on the reliability of a surface micromachined microengine, Proc. 1999 37th Annual IEEE International Reliability Physics Symp. (IEEE, New York 1999) 189–197
36.83 M.P. de Boer, P.J. Clews, B.K. Smith, T.A. Michalske: Adhesion of polysilicon microbeams in controlled humidity ambients, Proc. Mater. Res. Soc. Symp. **518** (1998) 131–136

37. MEMS Packaging and Thermal Issues in Reliability

The potential of MEMS/NEMS technologies has been viewed as a comparable or even bigger revolution than that of microelectronics. These scientific and engineering advancements in MEMS/NEMS could bring applications to reality previously unthinkable, from space systems, environmental instruments, to daily life appliances. As presented in previous chapters, the development of core MEMS/NEMS processes has already demonstrated a lot of commercial applications as well as future potentials with elaborated functionalities. However, a low cost and reliable package for the protection of these MEMS/NEMS products is still a very difficult task. Without addressing the packaging and reliability issues, no commercial products can be sold on the market. Packaging design and modeling, packaging material selection, packaging process integration, and packaging cost are main issues to be considered when developing a new MEMS packaging process. In this chapter, we will present the fundamentals of MEMS/NEMS packaging technology, including packaging processes, hermetic and vacuum encapsulations, thermal issues, packaging reliability, and future packaging trends. The future development of MEMS packaging will rely on the success of the implementation of several unique echniques, such as packaging design kits for system and circuit designer, low cost and high yield wafer level, chip-scale packaging techniques, effective testing techniques at the wafer-level to reduce overall testing costs; and reliable fabrication of an interposer [37.1] with vertical through-interconnects for device integrations.

37.1 **MEMS Packaging** 1111
37.1.1 MEMS Packaging Fundamentals 1112
37.1.2 Contemporary MEMS Packaging Approaches 1113

37.2 **Hermetic and Vacuum Packaging and Applications** 1116
37.2.1 Integrated Micromachining Processes 1117
37.2.2 Post-Packaging Processes 1118
37.2.3 Localized Heating and Bonding 1119

37.3 **Thermal Issues and Packaging Reliability** 1122
37.3.1 Thermal Issues in Packaging 1122
37.3.2 Packaging Reliability 1124
37.3.3 Long-Term and Accelerated MEMS Packaging Tests 1125

37.4 **Future Trends and Summary** 1128

References 1129

37.1 MEMS Packaging

MEMS are miniaturized systems in a size of micrometer to millimeter that may have mechanical, chemical, or biomedical features integrated with IC circuitry for sensor or actuator applications [37.2]. For example, pressure [37.3], temperature, flow [37.4], accelerometers [37.5], gyroscopes [37.6], and chemical sensors [37.7] can be fabricated by MEMS technologies for sensing applications. Fluidic valves [37.8], pumps [37.9], and inkjet printer heads are examples of actuation devices for medical, environmental, office, and industrial applications. Silicon is typically used as the primary substrate material for MEMS fabrication because it can provide unique electrical, thermal, and mechanical properties however also can be easily micromachined in a form of batch-processing and be incorporated with a microelectronic circuits by using most of the conventional semiconductor manufacturing processes and tools. As a result, smaller size, lighter weight, lesser power consumption and cheaper fabrication cost become the advantages of MEMS devices as compared with the existing macroscale systems with similar functionalities. With the advances of MEMS fabrication technology in the past decades, the MEMS market at the component-level is currently in excess of

$ 5 billion and is driving the end-product markets of large than $ 100 billion [37.10].

Nevertheless, the road to the commercialization of MEMS doesn't look as promising as expected. Many industrial companies took the advantages of MEMS technology of high production volumes and high added value created by product integrations. Therefore, cost-efficiency becomes the major factor driving MEMS toward commercialization. Several MEMS devices have been developed for and applied in automotive industry and information technology, and they dominate the MEMS market due to high production volumes. Most of custom-designed MEMS products are still very diverse, aiming for different applications, and their initial costs in small-to-medium-scale production are still much higher than market-acceptable levels. Consequently, high packaging and testing costs have slowed the commercialization of MEMS. Furthermore, based on the past experience of IC industry, the cost of packaging processes is about 30%, and sometimes can be more than 70%, of the total production expenses. MEMS packaging process is expected to be even more costly because of the challenging and stringent packaging issues with regard to the additional MEMS components, in addition to the microelectronic circuitry, in a typical MEMS product [37.11].

Part E | 37.1

37.1.1 MEMS Packaging Fundamentals

The functions of conventional IC packaging are to protect, power, and cool the microelectronic chips or components and provide electrical and mechanical connection between the microelectronic part and the outside world. With the increase of the needs of high performance and multifunctional consumer electronic products, IC packaging processes have incorporated more complex designs and advanced fabrication technologies, such as Cu interconnects [37.12], flip-chip bonding [37.13], ball grid array [37.14], wafer-level chip scale packaging [37.15], and 3-D packaging [37.16], etc., in order to satisfy the need for high I/O density, large die area, and high clock frequency. In addition to the requirements of electrical interconnects, MEMS components may need to interface outside environment (for example, fluidic interconnectors [37.17]), some other components may need to be hermetically sealed in vacuum (for example, accelerometers [37.5]). Therefore, MEMS packaging processes have to provide more functionalities including better mechanical protection, thermal management, hermetic sealing, and complex electricity and signal distribution.

It has been suggested that MEMS packaging should be incorporated in the device fabrication stage as part of the micromachining process. Although this approach solves the need for some specific devices, it does not solve the packaging need for general microsystems. Many MEMS devices are now fabricated by various foundry services [37.18, 19], and there is a tremendous need for a uniform packaging process. Figure 37.1 shows a typical MEMS device being encapsulated by a packaging cap. The most fragile part of this device is the suspended mechanical sensor, which is a freestanding mechanical, mass-spring microstructure. This mechanical part must be protected during the packaging and handling process. Moreover, vacuum encapsulation may be required for these microstructures in applications such as resonant accelerometers [37.5] or gyroscopes [37.6, 20]. A "packaging cap" with properly designed micro cavity is to be fabricated to encapsulate and protect the fragile MEMS structure as the first-level MEMS post-packaging process. The wafer can be diced afterward, and the well-established packaging technology in IC industry can follow and finish the final packaging step. Unlike the packaging requirements for ICs, however, the common MEMS packaging requirement is hermetic seal and sometimes vacuum encapsulations. Hermetic seal is important to ensure that no moisture or contaminant can enter the package

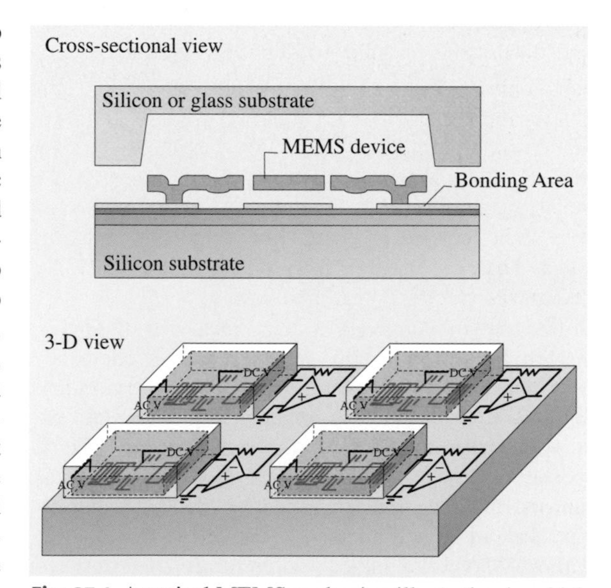

Fig. 37.1 A typical MEMS packaging illustration in which a MEMS structure is encapsulated and protected by the packaging cap

and affect the functionality of microstructures. This increases the difficulty of common IC packaging processes tremendously. Although most single function MEMS chips can employ typical IC packaging techniques, such as die-attached processes, wiring interconnects, molded plastic, ceramic, and metal for packaging [37.21], with the increase of complexity MEMS devices, more advanced packaging techniques, especially wafer-level packaging, are required for the integration of multi-chips for multifunctional applications.

Before the state-of-the-art MEMS packaging processes are discussed, several primary micro-fabrication processes for packaging applications are described. These processes may include Flip-Chip (FC) technique, Ball Grid Arrays (BGA), through-wafer etching, and plating. Other silicon-based processes, such as thin-film deposition, wet and dry chemical etching, lithography, lift-off, and wiring bonding processes can be found in many textbooks [37.22].

Flip-Chip Technique (FC)

This technique is commonly used in the assembly process between a chip with microelectronics and a package substrate [37.13]. The microelectronic chip is "flipped joined" with the packaging substrate, and metal solder bumps are used as both the bonding agents and electrical paths between bond pads on the microelectronic chip and metal pads on the package substrate. Because the vertical bonding space can be very small as controlled by the heights of the solder bumps, and the lateral distributions of bond pads can be on the whole chip instead of being only on the edge, this technique can provide high density Input/Output (I/O) connections. In the FC technique, solder bumps are generally fabricated by means of electroplating. Before the bumping process, multiple metal layers, such as TiW-Cu, Cr-Cu, Cr-Ni, TaN/Ta/Ni, have to be deposited as a seed layer for electroplating and as a barrier to prevent the diffusion of solder into its underneath electrical interconnect.

Ball Grid Arrays (BGA)

This technology is very similar to the FC technique. An area array of solder balls on a single chip module or multi-chip module are used in the packaging process as electrical, thermal, and mechanical connects to join the module with the next level package, usually a printed circuit board [37.14]. The major difference between a typical BGA and FC is the size of solder bump. In BGA chip, the bumps are in the order of 750 μm in diameter which is ten times larger than those commonly used in FC.

Through-Wafer Etching

This is a chemical etching process to make through-wafer channels on a silicon substrate for the fabrication of vertical through-wafer interconnects. The chemical etching process can be either wet or dry. Anisotropic or isotropic etching solutions can be used in the wet etching process. The dry etch process is based on plasma and ion-assisted chemical reactions that can be either isotropic or anisotropic. In order to create high density and high aspect ratio through-wafer vias, Deep Reactive Ion Etching (DRIE) is typically used. Two popular DRIE approaches, Bosch and Cyro, are well described in literature [37.23].

Electroplating

Electroplating is another common micro-fabrication process. It can be conducted for the deposition of an adherent metallic layer onto a conductive or nonconductive substrate. The process on a conductive substrate is called electrolytic plating, which utilizes a seed layer as the anode to transfer metal ions onto the cathode surface when a DC current is passed through the plating solution. The plating process without applying electrical current is called electroless plating, which can happen in both conducive and nonconductive surfaces. Electroless plating process requires a layer of noble metal such as Pd, Pt, or Ru on the substrate as the catalysis to trigger the self-decomposition reaction in the plating solution. The electroplating processes are very important for electrical interconnect and solder bump fabrications for packaging applications because of low process temperature and cost.

These processes are generally developed for providing electrical and thermal paths for various IC/MEMS packaging approaches.

37.1.2 Contemporary MEMS Packaging Approaches

Several MEMS packaging issues and approaches before 1985 were discussed in [37.24], and researchers have been working on MEMS packaging approaches continuously. For example, *Senturia* and *Smith* [37.25] discussed the packaging and partitioning issues for microsystems. *Smith* and *Collins* [37.26] used epoxy to bond glass and silicon for chemical sensors. Several MCM methods have been proposed. *Butler* et al. [37.27] proposed adapting multi-chip module foundries using chip-on-flex (COF) process. *Schuenemann* et al. [37.28] introduced a 3-D stackable packaging concept for the Top-Bottom Ball Grid Array (TB-BGA) that includes electric, fluidic, optic, and communication interfaces. *Lee* et al. [37.29] and *Ok* et al. [37.10] presented a direct-chip-attach MEMS packaging using through-wafer electrical interconnects. *Laskar* and *Blythe* [37.30] developed a multi-chip modules (MCM) type packaging process by using epoxy. *Reichl* [37.31] discussed different materials for bonding and interconnection. *Grisel* et al. [37.32] designed a special process to package micro-chemical sensors. Special processes have also been developed for MEMS packaging, such as packaging for microelectrodes [37.33], packaging for biomedical systems [37.34], and packaging for space systems [37.35]. These specially designed, device-oriented packaging methods are meant for individual systems. There is no reliable method yet that would qualify as a versatile post-packaging process for MEMS with the rigorous process requirements of low temperature, hermetic sealing and long-term stability.

An integrated process by using surface-micromachined microshells has previously been developed [37.36]. This process applies the concepts of sacrificial layer and LPCVD sealing to achieve wafer-level post-packaging. Similar processes have been demonstrated. For example, *Guckel* et al. [37.37] and *Sniegowski* et al. [37.38] developed a reactive sealing method to seal vibratory micromachined beams. *Ikeda* et al. [37.39] adopted epitaxial silicon to seal microstructures. *Mastrangelo* et al. [37.40] used silicon nitride to seal mechanical beams as light sources. *Smith* et al. [37.41] accomplished a new fabrication technology by embedding microstructures and CMOS circuitry. All of these methods have integrated the MEMS process together with the post-packaging such that no extra bonding is required. These schemes are highly process dependent, however, and not suitable for prefabricated circuitry.

Recently several new efforts for MEMS post-packaging processes have been reported. *Butler* et al. [37.27] demonstrated an advanced MCM packaging scheme. It adopts the high density interconnect (HDI) process consisting of embedding bare die into premilled substrates. Because the MEMS structures have to be released after the packaging process, it is undesirable for general microsystems. *Van der Groen* et al. [37.42] reported a transfer technique for CMOS circuits based on epoxy bonding. This process overcomes the surface roughness problem, but epoxy is not a good material for hermetic sealing. In 1996, *Cohn* et al. [37.43] demonstrated a wafer-to-wafer to-wafer vacuum packaging process by using Silicon–Gold eutectic bonding with a 2-μm-thick polysilicon microcap. But experimental results show substantial leakage after a period of 50 days. *Cheng* et al. [37.44] developed a vacuum packaging technology using localized aluminum/silicon-to-glass bonding. In 2002, *Chiao* and *Lin* [37.45] demonstrated vacuum packaging of microresonators by rapid thermal processing. These recent and ongoing research efforts indicate the strong need for a versatile MEMS post-packaging process.

Bonding Processes for MEMS Packaging Applications

Previously, silicon-bonding technologies have been used in many MEMS fabrication and packaging applications, including epoxy bonding, eutectic bonding, anodic bonding, fusion bonding, and solder bonding. For example, such devices as pressure sensors, micro-pumps, biomedical sensors or chemical sensors require mechanical interconnectors to be bonded on the substrate (see for example [37.7, 17, 46]). Glass has been commonly used as the bonding material by anodic bonding at a temperature of about $300 \sim 450\,^{\circ}C$ (see for example [37.47, 48]). *Klaassen* et al. [37.49] and *Hsu* et al. [37.50] have demonstrated different types of silicon fusion bonding and Si–SiO_2 bonding processes at very high temperatures of over $1,000\,^{\circ}C$. *Ko* et al. [37.24], *Tiensuu* et al. [37.51], *Lee* et al. [37.52] and *Cohn* et al. [37.43] have used eutectic bonding for different applications. All of these bonding techniques have different mechanisms that determine the individual bonding characteristics and process parameters. This section discusses the details of these processes.

Fusion Bonding for MEMS Packaging

Silicon fusion bonding is an important fabrication technique of the SOI (Silicon On Insulator). The bonding is based on Si−O, Si−N, or Si−Si strong covalent bond. But very high bonding temperature (higher than $1,000\,^{\circ}C$) and flat bonding surfaces (less than 6 nm) are the two basic requirements for strong, uniform, and hermetic bonding. Although hydrophilic surface treatment can lower the bonding temperature, an annealing step higher than $800\,^{\circ}C$ is still needed to remove possible bubble formation at the bonding interface. *Bower* et al. [37.53] proposed that low temperature Si_3N_4 fusion bonding could be achieved at the temperature under $300\,^{\circ}C$. *Takagi* et al. [37.54] proposed that silicon fusion bonding could be made at room temperature by using Ar^+ beam treatment on the wafer surface, and the bond strength is comparable to the conventional fusion bonding. In summary, fusion bonding has been a popular fabrication technique in MEMS fabrication and packaging.

Anodic Bonding for MEMS Packaging Applications

The invention of anodic bonding dates back to 1969 when *Wallis* and *Pomerantz* [37.55] found that glass and metal can be bonded together at about $200 \sim 400\,°C$ below the melting point of glass with the aid of a high electrical field. This technology has been widely used for protecting on-board electronics in biosensors (see for example [37.56–58]) and sealing cavities in pressure sensors (see for example [37.59]). Many reports have also discussed the possibility of lowering the bonding temperature by different mechanisms [37.60, 61]. Anodic bonding forms Si−O or Si−Si covalent bond and is one of the strongest chemical bonds available for silicon-based systems. The bonding process can be accomplished on a hot plate with a temperature between 180 °C and 500 °C in atmosphere or vacuum environment. When a static electrical field is built within the Pyrex glass (7740 from Dow Corning) and silicon, the electrostatic force can pull two surfaces close for a strong bond. In order to create high electrical field, a flat bonding surface with less than 50 nm roughness is required. In addition, the electrical field required for bonding is larger than 3×10^6 V/cm [37.24]. Such a high electrical field is generated by a power supply of 200 to 1,000 V, which may cause damages on integrated circuits. *Hanneborg* et al. [37.62] have successfully bonded silicon with other thin solid films, such as silicon dioxide, nitride, and polysilicon, together with an intermediate glass layer using anodic bonding technique. In practice, electrostatic bonding has become widely accepted in MEMS fabrications and packaging applications. Unfortunately, the possible contamination due to excessive alkali metal in the glass; possible damage to microelectronics due to the high electrical field; and the requirement of flat surface for bonding limit the application of anodic bonding to MEMS post-packaging applications [37.63].

Epoxy Bonding (Adhesive Bonding)

Four major components comprise epoxy : epoxy resin, filler – like silver slake, solvent or reactive epoxy diluent, and additives like hardener and catalyst [37.64, 65]. The bonding mechanism of epoxy is very complicated and depends on the type of epoxy. In general, the main source of bonding strength is the van der Waals force. Because epoxy is a soft polymer material and its curing temperature for bonding is only around 150 °C, low residual stress and process temperature are the major advantages of epoxy bonding. But the properties of epoxy can be easily changed with environmental humidity and temperature so the bonding strength decays over time. In addition, epoxy bonding has low moisture resistance and, due to its additives, is a dirty process. These disadvantages have made epoxy unfavorable for the special MEMS packaging need for hermetic or vacuum sealing.

Eutectic Bonding

In many binary systems, a eutectic point corresponds to the alloy composition with the lowest melting temperature. If the environmental temperature is kept higher than the eutectic point, two contacted surfaces containing two elements with the eutectic composition can form liquid phase alloy. The solidification of the eutectic alloy forms "eutectic bonding" at a temperature lower than the melting temperature of either element in the alloy. Eutectic bonding can be a strong bond. For example, in the case of Au/Si alloy system, eutectic temperature is only 363 °C when the composition is at the atomic ratio of 81.4% Au to 18.6% Si and bonding strength is higher than 5.5 GPa [37.66]. Because other alloy systems may have lower eutectic temperatures than Al/Si system, they present great potential for MEMS packaging applications. In addition to Au/Si, Al/Ge/Si, Au/SnSi, and Au/Ge/Si systems have been applied for MEMS packaging.

Solder Bonding

Solder bonding has been widely applied in microelectronic packaging [37.67]. Both low bonding temperature and high bonding strength are good characteristics for packaging. Furthermore, there are a variety of choices of solder material for specific applications. *Singh* et al. [37.68] have successfully applied solder bump bonding in the integration of electronic components and mechanical devices for MEMS fabrication [37.69]. In this case, indium metal was used for bonding two separated silicon surfaces together by applying 350 MPa pressure and the bonding strength was as strong as 10 MPa. Glass frits can also be treated as a solder material and have been extensively used for vacuum encapsulation in MEMS industry. Glass frits are ceramic materials that can provide strong bonding strength with silicon with good hermeticity. Its bonding temperature is lower than 400 °C and is suitable for electronic components. But a bonding area more than 200 μm wide is required to achieve good results, which may become a drawback because area is the measure of manufacturing cost in IC industry. Nevertheless, glass frit is the

Table 37.1 Summary of bonding mechanisms

Bonding methods	Temperature	Roughness	Hermeticity	Post-packaging	Reliability
Fusion bonding	very high	highly sensitive	yes	yes by LH	good
Anodic bonding	medium	highly sensitive	yes	difficult	good
Epoxy bonding	low	low	no	yes	???
Integrated process	high	medium	yes	no	good
Low Temp. bonding	low	highly sensitive	???	no	???
Eutectic bonding	medium	low	yes	yes by LH	???
Brazing	very high	low	yes	yes by LH	good
???: no conlusive data		LH: Localized Heating			

most popular bonding process used in the current MEMS products.

Localized Heating and Bonding

Low bonding temperature and short process time are desirable process parameters in MEMS packaging fabrication to provide less thermal budget and high throughput. Most chemical bonding reactions, however, require a minimum and sufficient thermal energy to overcome the reaction energy barrier, or activation energy, to start the reaction and to form a strong bond. As a result, high bonding temperature generally results in shorter processing time to reach the same bonding quality at a lower bonding temperature [37.70]. The common limitations for the above techniques are their individual bonding characteristics and temperature requirements. In general, MEMS packaging requires a good bonding for hermetic sealing while the processing temperature must be kept low at the wafer-level to have less thermal effects on the existing devices. For example, a MEMS device may have prefabricated circuitry, biomaterial, or other temperature-sensitive materials such as organic polymer, magnetic metal alloy, or pizeo-ceramic. Since the packaging step comes after the MEMS device fabrication processes, bonding temperature should be kept low to avoid high temperature effects on the system. Possible temperature effects include residual stress due to the mismatch of thermal expansion coefficient of bonding materials and substrates, electrical contact failure due to atom inter-diffusion at the interface, and contamination due to the outgas or evaporation of materials. In addition to the control of bonding temperature, Other factors to consider are the magnitude of applied force to create intimate contact for bonding and atmospherics environment control. Based on the heat transfer simulation study [37.71], it is possible to confine high temperature area in a small region by localized heating without heating the whole substrate. Assembly steps can therefore always be processed after device fabrication without having detrimental effects. As such, localized heating and bonding technique is introduced and implemented for the fabrication of MEMS packaging for post-processing approaches [37.70, 72].

Table 37.1 summarizes these MEMS packaging technologies and their limitations, including the localized heating and bonding approach. The localized heating approach introduces several new opportunities. First, better and faster temperature control can be achieved. Second, higher temperature can be applied to improve the bonding quality. Third, new bonding mechanisms that require high temperature such as brazing [37.73] may now be explored in MEMS applications. As such, it has potential applications for a widerange of MEMS devices and is expected to advance the field of MEMS packaging.

37.2 Hermetic and Vacuum Packaging and Applications

Hermetic packaging is desirable for MEMS packaging requirements in most cases because it provides a moisture free environment to avoid charge separation in capacitive devices, corrosion in metallization, or electrolytic conduction in order to prolong the lifetime of the electronic circuitry. In several device applications, vacuum encapsulation is necessary but can be costly. Many surface micromachined resonant devices need vacuum to improve the performance including comb-shape μ-resonators and ring-type μ-gyroscopes that have very large surface-to-volume ratios and vibrate in a very tight space [37.20,39]. Two major approaches of MEMS hermetic and vacuum packaging have been demonstrated: (1) the integrated encapsulation approach and (2) the post-process packaging approach. Both are discussed in this section. Vacuum encapsulation by means of localized heating and bonding is discussed separately as another example for issues related to hermetic and vacuum packaging.

37.2.1 Integrated Micromachining Processes

Several MEMS hermetic and vacuum packaging processes have been demonstrated before based on the integrated micromachining processes. An integrated vacuum sealing process by LPCVD is presented here as the illustration example. This integrated process can encapsulate comb-shape microresonators [37.74] in vacuum at the wafer level. Figure 37.2 illustrates the cross-sectional view of the manufacturing process. First, standard surface-micromachining process [37.75] is conducted by using four masks to define the first polysilicon layer, anchors to the substrate, dimples and, the second polysilicon layer as shown in Fig. 37.2a. The process so far is similar to the MCNC MUMPs process [37.19], and comb-shape microstructures are fabricated at the end of these steps. In the standard surface micromachining process, the sacrificial layer (oxide) is etched away to release the microstructures. In the MEMS post-packaging process, a thick PSG (phosphorus-doped glass) of 7 μm is deposited to cover the microstructure and patterned by using 5:1 BHF (buffered HF) to define the microshell area as shown in Fig. 37.2b. A thin PSG layer of 1 μm is then deposited and defined to form etch channels as illustrated in Fig. 37.2c. The microshell material, low-stress silicon nitride, is now deposited with a thickness of 1 μm. Etch holes are defined and opened on the silicon nitride layer by using a plasma etcher. Silicon dioxide inside the packaging shell is now etched away by concentrated HF, and the wafer is dried using the supercritical CO_2 drying process [37.76]. After these steps, Fig. 37.2d applies. A 2-μm-thick LPCVD low-stress nitride is now deposited at a deposition pressure of 300 mtorr to seal the shell in the vacuum condition. Finally, the contact pads are opened as shown in Fig. 37.2e. Figure 37.3 is the SEM (Scanning Electron Microscope) microphoto of a finished device with protected microshell on top. The total packaging area

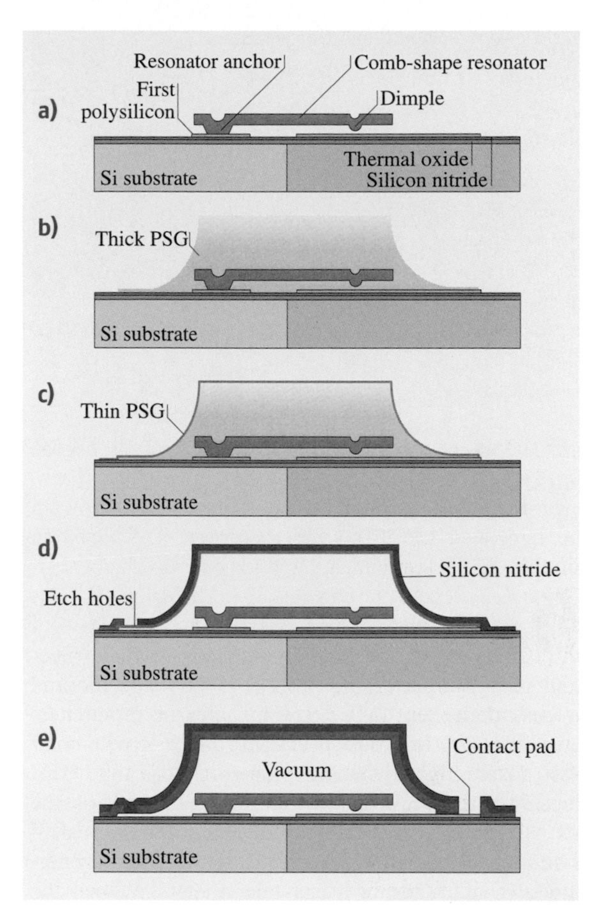

Fig. 37.2 An integrated vacuum encapsulation process using LPCVD nitride sealing to package micro-mechanical resonators [37.71]

(microshell) is about $400\times400\,\mu m^2$. A contact pad is shown with the covering nitride layer removed. The shape of the microresonator, with beams of 150-μm-long and 2-μm-wide is reflected on the surface of the microshell due to the integrated packaging process. The total height of the nitride shell is 12 μm, as seen standing above the substrate. Spectrum measurement of the comb resonator inside the packaging reveals that a vacuum level of about 200 mtorr has been accomplished [37.77].

Although the above vacuum sealing process successfully achieves MEMS hermetic and vacuum packaging, it has several drawbacks. First, several high temperature steps were used after the standard surface-micromachining process. As such, no circuitry (such as those shown in Fig. 37.1) or temperature-sensitive materials

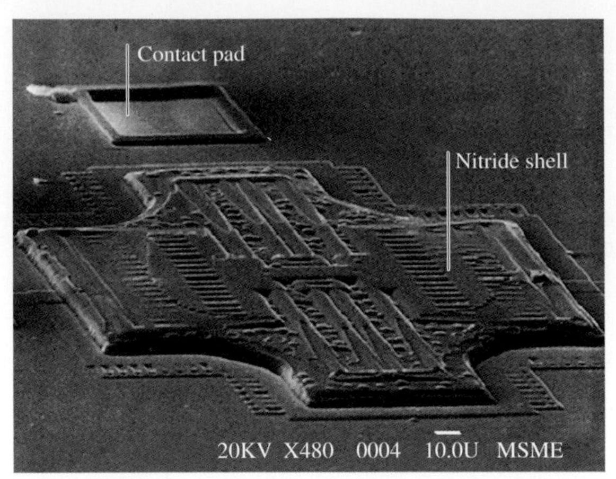

Fig. 37.3 SEM microphoto showing a vacuum packaged MEMS mechanical comb-shape resonator by an the integrated LPCVD sealing process as depicted in Fig. 37.2 [37.71]

will survive due to the global heating effect. Second, this post-packaging process is very specific and process dependent. MEMS companies or researchers have to adopt this post-packaging process with their own device manufacturing process. It can't be conducted in the multiuser MEMS process such as the MCNC MUMPs. Third, the thickness of the microshell is limited by the thin film deposition step that generally is in the range of a few micrometers. Wether the thin microshell can survive the high-pressure plastic molding process afterward in some device packaging design is a big concern. Finally, although integrated encapsulation can achieve low pressure by wafer-level fabrication and provide lower manufacturing cost, it does not provide the controllability of the cavity pressure.

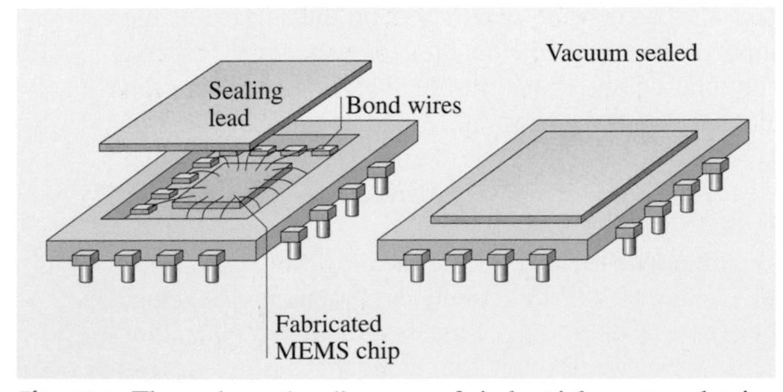

Fig. 37.4 The schematic diagram of industrial post-packaging (DIPS) using a ceramic holder to be covered by a sealing lid

37.2.2 Post-Packaging Processes

The second approach is defined as post-packaging process. The packaging process starts when the device fabrication processes are completed, so this approach has high flexibility for various microsystems. For example, Fig. 37.4 shows a common industrial post-hermetic packaging called Dual-in-line packaging (DIP) [37.78, 79]. A die is placed inside a ceramic holder covered by a sealing lid. Solder or ceramic joining is generally used for assembling lid and holder under a pressure-controlled environment. High cost is the major drawback of this method because of expensive ceramic holder and low fabrication throughput. Another example of post-packaging method is based on wafer bonding techniques combined with a microshell encapsulation. Devices are sealed by stacking another micromachined silicon or glass substrate as illustrated in Fig. 37.1. Integrated microsystems and protection shells are fabricated on different wafers, either silicon or glass, at the same time. After the two substrates are assembled together using silicon fusion, anodic, or low temperature solder bonding to achieve the final encapsulation, these micro-shells will provide mechanical support, thermal path, or electrical contact for the MEMS devices. Low packaging cost can be expected due to wafer-level processing.

The related packaging issues are discussed by using a specific example of hermetic and vacuum post-packaging by rapid thermal processing (RTP) that provides the advantages of wafer-level processing and low thermal budget. Figure 37.5 shows the

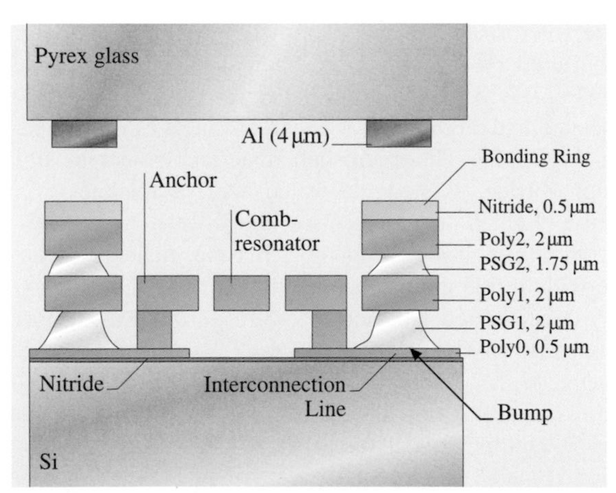

Fig. 37.5 The schematic diagram of the RTP (Rapid Thermal Processing) aluminum-to-nitride bonding setup [37.45]

schematic illustration of the RTP MEMS vacuum packaging process. Comb-shape microresonators are chosen as the vacuum packaging examples in this chapter, and a standard surface-micromachining process is used to fabricate these microresonators [37.19]. One major addition to a regular microstructure is that an integrated sealing ring using silicon nitride as the topmost layer is incorporated in the manufacturing process for the purpose of aluminum-to-nitride bonding [37.45]. A glass cap wafer is deposited and patterned with 4-μm-thick aluminum sealing-rings with width ranging from 100 to 250 μm and bonding area ranging from 450×450 to $1,000\times1,000\,\mu m^2$. Before the vacuum packaging process, both the device and cap wafers are baked in vacuum at 300 °C for over four hours to dry water and gas species adhere at the surface. Afterward, the device and cap wafers are flip-chip assembled immediately and loaded into a quartz chamber and put into a RTP chamber. The base vacuum estimated at 10 mtorr inside the quartz chamber is achieved by using a mechanical pump. After heating for 10 seconds at 750 °C, the aluminum-to-nitride bond is formed. Figure 37.6 shows the measured spectrum of a vacuum-packaged, double-folded beam comb-drive resonator by using a micro-stroboscope [37.80]. The central resonant frequency is about 18,625 Hz and the quality factor is extracted as $1,800\pm200$ corresponding to a pressure level about 200 mtorr inside the package [37.81].

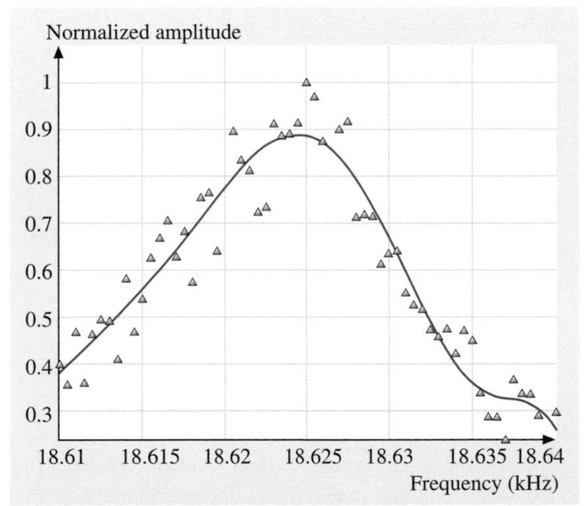

Fig. 37.6 Spectrum measurement results of a vacuum encapsulated comb-shape resonator by using the RTP aluminum-to-nitride bonding method [37.45]

This type of post-packaging process at wafer level has become the favorite approach to fabricate a hermetic encapsulation because it can provide lower cost and more process flexibility. But the packaging process relies on "good" bonding techniques. A strong and reliable bonding between two substrates should be provided, and this procedure should be compatible with the other microsystem fabrication processes.

37.2.3 Localized Heating and Bonding

The approach of MEMS post-packaging by localized heating and bonding is proposed to address the problems of global heating effects. In this chapter, resistive micro-heaters are used as the examples to provide localized heating, although several other means of localized heating have been demonstrated recently, including laser welding [37.82], inductive heating [37.83], and ultrasonic bonding [37.84]. The principle of localized heating is to achieve high temperature for bonding while maintaining low temperature globally at the wafer level. Resistive heating by using micro-heaters on top of device substrate is applied to form strong bond with silicon or glass cap. According to the results of a 2-D heat conduction finite element analysis as shown in Fig. 37.7, the steady-state heating region of a 5-μm-wide polysilicon micro-heater capped with a Pyrex glass substrate

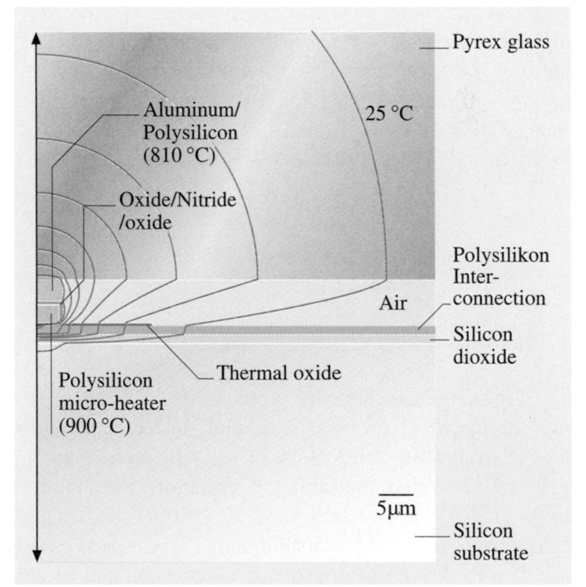

Fig. 37.7 The schematic diagram of 2-D heat transfer model, geometry and boundary conditions

can be confined locally as long as the bottom of the silicon substrate is constrained to the ambient temperature. The physics of localized heating behind this design can be understood by solving the governing heat conduction equations of the device structure without a cap [37.81]. As long as the width of micro-heater and the thickness of silicon substrate are much smaller than the die size and a good heat sink is placed underneath the silicon substrate, the heating can be confined locally. The temperature of silicon substrate can be kept low or close to room temperature. Several localized resistive heating and bonding techniques have been successfully developed for packaging applications including localized silicon-to-glass fusion bonding, gold-to-silicon eutectic bonding, and localized solder bonding. Several solder materials have been successfully tested, including PSG, indium, and aluminum alloy [37.85].

The vacuum packaging example presented in this chapter is based on localized aluminum/silicon-to-glass solder bonding technique. Built-in folded-beam comb drive μ-resonators are used to monitor the pressure of the package. Figure 37.8 shows the fabrication process of the package and resonators. Thermal oxide (2 μm) and LPCVD Si_3N_4 (3,000 Å) are first deposited on a silicon substrate for electrical insulation followed by the deposition of 3,000 Å LPCVD polysilicon. This polysilicon is used as both the ground plane and the electrical interconnect to the μ-resonators as shown in Fig. 37.8a. Figure 37.8b shows a 2 μm LPCVD SiO_2 layer that is deposited and patterned as a sacrificial layer for the fabrication of polysilicon μ-resonators using a standard surface micromachining process. A 2-μm-thick phosphorus doped polysilicon is used for both the structural layer of microresonators and the on-chip micro-heaters. This

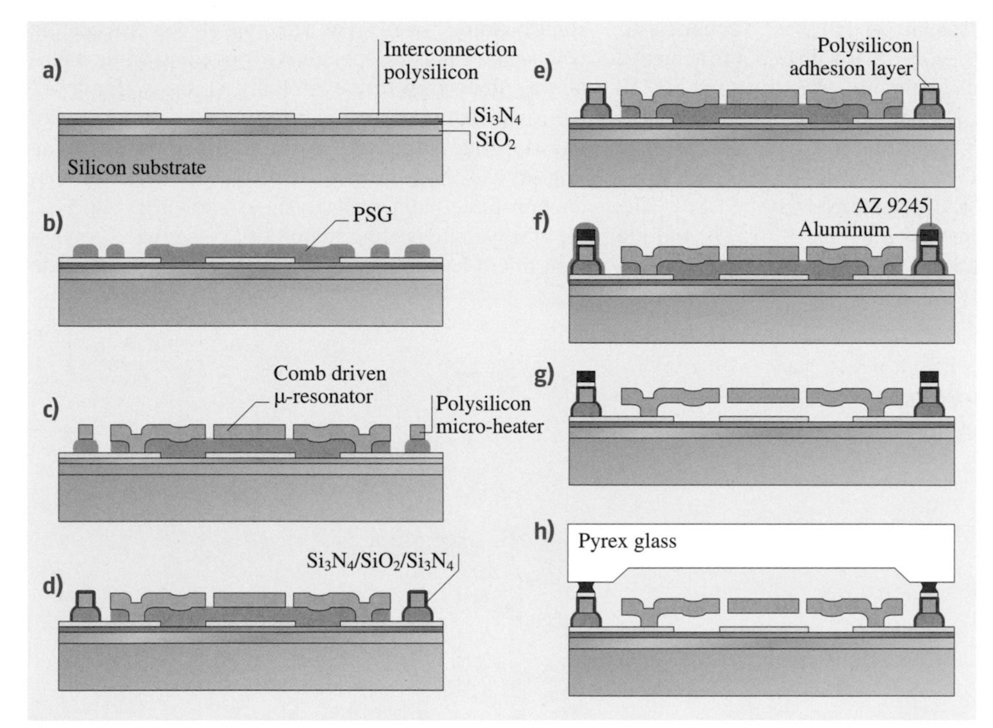

Fig. 37.8a–h The schematic process flow of vacuum encapsulation using localized aluminum/silicon-to-glass bonding, **(a)** electrical ground plane and device interconnects fabrication **(b)** a 2 μm LPCVD SiO_2 sacrificial layer deposition for the fabrication of polysilicon μ-resonator **(c)** a 2 μm-thick phosphorus doped polysilicon deposition and patterning for both the structural layer of micro resonators and the on-chip microheaters **(d)** a LPCVD Si_3N_4 (750 Å)/SiO_2 (1,000 Å)/Si_3N_4 (750 Å) sandwich layer deposition and patterning on top of the microheater for the electrical isolation **(e)** an aluminum (2.5 μm) and polysilicon (5,000 Å) bonding materials are deposition and patterning. **(f)** a thick AZ 9245 photoresist deposition and patterning for the protection of aluminum/silicon-to-glass bonding layers **(g)** 20 minutes sacrificial release in concentrated HF **(h)** vacuum encapsulation with 0.2 MPa contact pressure and 3.4 W input power under a 25 mtorr vacuum environment

layer is formed over the sacrificial oxide in two steps to achieve a uniform doping profile. Lower input power and better process compatibility are two major advantages in using the on-chip micro-heater in the glass package. The resonators are separated from the heater by a short distance, 30 μm, to effectively prevent their exposure to the high heater temperature as shown in Fig. 37.8c. This concludes the fabrication of μ-resonators.

In order to prevent the current supplied to the micro-heater from leaking into the aluminum solder during bonding, a LPCVD Si_3N_4 (750 Å)/SiO_2 (1,000 Å)/Si_3N_4 (750 Å) sandwich layer is grown and patterned on top of the micro-heater as shown in Fig. 37.8d. Figure 37.8e and f show that aluminum (2.5 μm) and polysilicon (5,000 Å) bonding materials are deposited and patterned. The sacrificial release is the final step to form freestanding μ-resonators. Figure 37.8f shows a thick AZ 9245 photoresist is applied to cover aluminum/silicon-to-glass bonding system to ensure that the system withstands the attack from concentrated hydrofluoric acid. After 20 minutes sacrificial release in concentrated HF, the system as shown in Fig. 37.8g is ready for vacuum packaging. A Pyrex glass cap with a 10-μm deep recess is then placed on top with an applied pressure of ∼ 0.2 MPa under a 25 mtorr vacuum, and the heater is warmed using 3.4 watt input power (exact amount depends on the design of the micro-heaters) for 10 minutes to complete the vacuum packaging process as shown in Fig. 37.8h.

To evaluate the integrity of the resonators packaged using localized aluminum/silicon-to-glass solder bonding, the glass cap is forcefully broken and removed from the substrate. No damage is found on the μ-resonator and a part of the micro-heater is stripped away as shown in Fig. 37.9, demonstrating that a strong and uniform bond can be achieved without detrimental effects on the encapsulated device. Figure 37.10 shows a vacuum encapsulated unannealed μ-resonator (∼ 57 kHz) after

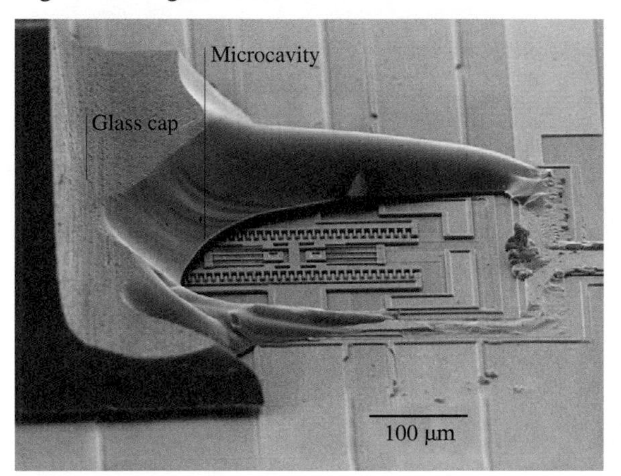

Fig. 37.9 The SEM microphoto of encapsulated micro-resonators after the glass cap is forcefully broken

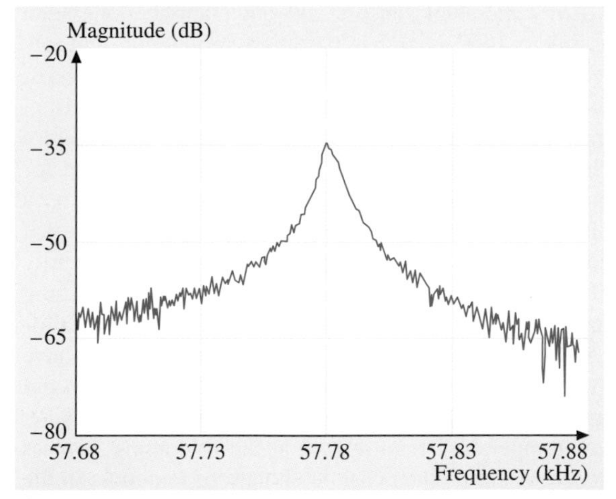

Fig. 37.10 The transmission spectrum of a glass-encapsulated μ-resonator after 120 minute pump down time in vacuum environment ($Q = 9,600$)

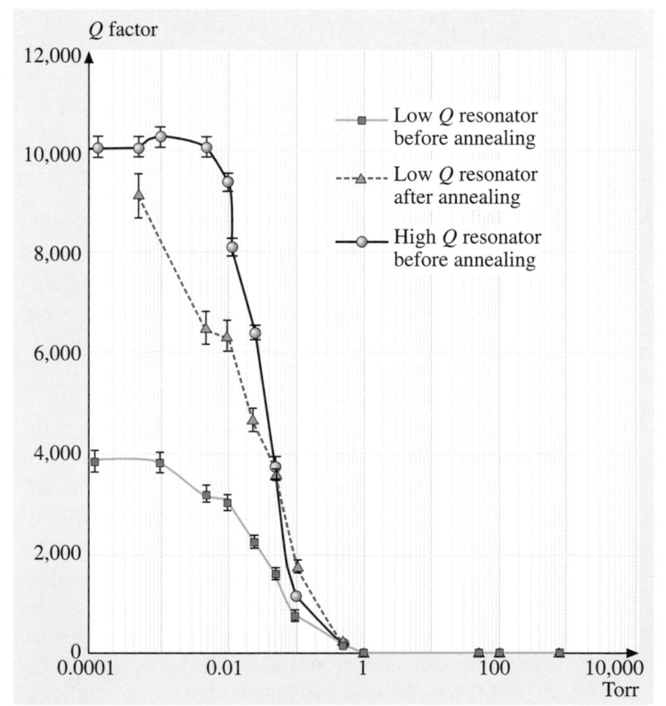

Fig. 37.11 Measured Q factor vs. pressure of unpackaged μ-resonators

120 minutes of wait time. The measured Q-factor after packaging is 9,600. Based on the measurement of Q vs. pressure of a high Q unpackaged μ-resonator as shown in Fig. 37.11, it is demonstrated that the pressure inside the packaging is comparable to the vacuum level of packaging chamber.

The post-process packaging using localized heating and bonding technique includes four basic components: (1) an electrical and thermal insulation layer such as silicon dioxide or silicon nitride should be used for localized heating, (2) resistive micro-heaters will be fabricated to provide the heating source for localized bonding, (3) materials, including metal and polysilicon that can provide good bonding and hermeticity with silicon or glass substrates are considered as the bonding materials, and (4) a good heat sink under the device substrate for localized heating is provided during the bonding experiments. MEMS devices will be fabricated on the device chip and hermetically sealed in the cavity formed by the device chip, resistive micro-heaters, and protection cap. The process can be either die level or wafer level. The schematic design of wafer-level packaging process is shown in Fig. 37.12. The resistive micro-heaters are parallel to each other and connected in order to ensure that identical current density is applied for individual packages at the same time. These heaters can be fabricated either on the chip or protection cap and can be built in a larger wafer for current inputs. The interconnections for these packaging cavities can be built in the dicing area, so no extra space is required for the packaging process.

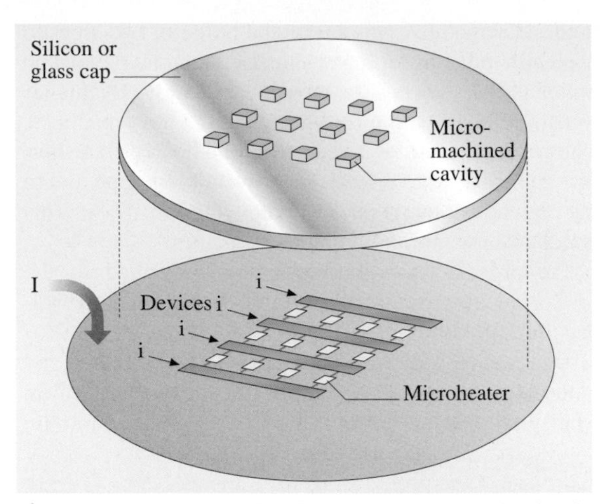

Fig. 37.12 Illustration of wafer-level vacuum packaging in the wafer level using localized heating and bonding technique

37.3 Thermal Issues and Packaging Reliability

37.3.1 Thermal Issues in Packaging

The two key thermal issues related to MEMS packaging are: 1) heat dissipation from actuators and integrated circuitry components and 2) thermal stress generated during the packaging process. These two topics are discussed separately.

Heat Dissipation Issues

In IC, heat dissipation is a serious problem when the size of transistor continues to shrink and the density of transistor on a chip continue to increase with the advance of IC fabrication technology. The trend of power packing into smaller packages has created increasing thermal management challenges [37.86]. Since the electrical characteristics of transistors change with working temperature, inefficient power dissipation could raise the working temperature and affect device performance. Present MEMS devices don't need high power, high performance microprocessors so power dissipation is not a problem. Nevertheless, some functional components in packaged MEMS are very sensitive to temperature variation, such as biomaterials or laser diodes. Several MEMS chemical sensors and other applications such as micro PCR chamber for DNA replication actually require elevated temperature for operation and microthermal platforms are built for these devices. Thermal management to maintain working temperature on these chips for stable operation is still an essential issue for packaging considerations. The geometrical complexity of MEMS resulting from packing various functional components in the tight space increases the difficulty of the thermal management. As the MEMS packaging integration process becomes more complex, the fabrication constraint in the packaging process will have great impact on the heterogeneous integration process in the front-end MEMS and IC processes. For example, low temperature requirement at the packaging process generally limits the possible choices of materials in the back-end process. In general, conventional IC packaging employs a heat sink attached to the chip to remove heat. The heat sink is generally made of a copper or stainless steel bar with an array of fin structures on one side for better natural or forced heat convection to dissipate heat into the environment. In addition to heat sinks, thermal vias, heat pipe cooling, immersion cooling, and thermoelectric cooling can also be used for effective heat removal. Because most MEMS packages still follow the typical IC packaging architecture, one promising thermal management method, heat pipe, is discussed for possible MEMS packaging applications.

A heat pipe is a sealed slender tube containing a wick structure and a working fluid, typically water in electronics cooling. It is composed of three sections, the evaporator section at one end, condenser section at the other end, and adiabatic section in the middle. In the evaporator section, heat is absorbed by the working fluid via phase transformation from liquid to vapor. In the section of condenser, heat dissipates into the outside environment. Thus the fluid goes back to liquid phase. Vapor phase is at a high pressure and temperature state that forces the vapor to flow into the condenser section at a lower temperature. Once the vapor condenses and gives up its latent heat, the condensed fluid is pumped back to the evaporator section by the capillary force developed in the wick structure. Therefore the middle adiabatic section contains two phases, the vapor phase in the core region and the liquid phase in the wick, flowing opposite to each other with no significant heat transfer between the fluid and the surrounding medium. Silicon has good thermal conductivity (1.41 W/cm/°C) and is easily micro-machined to fabricate the heat pipe. The implementation of the silicon micro heat pipe in IC and MEM packaging has therefore great potential and several approaches have been proposed on this topic [37.87–89].

Packaging Induced Thermal Stresses

Thermal based bonding processes have been used in MEMS packaging applications for many years, as introduced previously in this chapter. Thermal management is extremely important during the bonding process in order to avoid fracture in the substrate or MEMS device itself. Extremely high temperatures or rapid cooling conditions can cause damage and should be carefully evaluated both analytically and experimentally. There are many ways to provide heating energy including electrical resistive heating, oven heating or induction heating [37.83]. These bonding processes may be put into two categories. The first is localized bonding in which the heat is directly applied only to the adhesive material used to bond the package to the MEMS device. The second category is global heating where the entire system (MEMS device, adhesive, and packaging material) is heated to produce bonding of the materials. The latter is the common approach for all MEMS packaging processes. This section focuses therefore on the thermal stress effects during the heating and cooling procedures in MEMS packaging. The RTP aluminum-to-glass bonding process is used as the specific example for the discussion of thermal stresses [37.90]. The bonding process heats the packaging system to 750 °C for 10 seconds, and then cools it back to room temperature. To simulate this process, an ANSYS program [37.91] is established to examine the shear stress due to CTE (Coefficient of Thermal Expansion) variations in the bonding system as the result of temperature changes. The shear stress was recorded from the ANSYS analysis on the aluminum/Pyrex glass interface and the aluminum/silicon interface.

Two different models were analyzed. The first was the quartz-aluminum-silicon bonding system, and the second was the Pyrex glass-aluminum-silicon bonding system [37.92]. The results of the ANSYS analysis were then analyzed and compared with experimental observations. Figure 37.13 shows the ANSYS results for a Pyrex glass bonding system: the width of the aluminum solder is 100 μm and the maximum residual stress is be 60 MPa in glass, slightly lower than the fracture strength of Pyrex glass at 70 MPa. It was discovered that increasing the aluminum width leads to lower residual stresses. This likely occurred because the length of the Pyrex glass, quartz, and silicon remains constant. As a result, it will always want to contract the same amount with a constant temperature change independent of the aluminum width. When increasing the aluminum width, however, the stress is not in such a concentrated area and therefore decreases. For example, the maximum residual stress an-

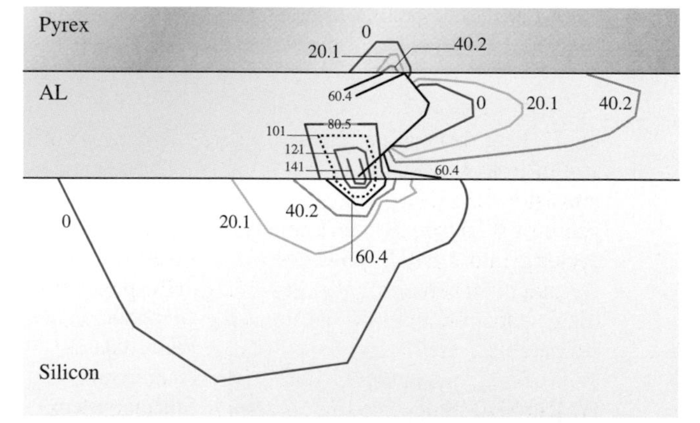

Fig. 37.13 Residual Stress (GPa) for an aluminum solder width of 100 μm under the RTP silicon-aluminum-glass bonding system

alyzed from ANSYS in the Pyrex glass bonding systems is 74.5 Gpa, 58 Gpa, and 60 GPa for aluminum widths of 30, 50, and 100 μm, respectively [37.92]. Pyrex glass has a documented strength of around 69 GPa [37.93]. Fracture should always occur with an aluminum width of 30 μm or less, according to the ANSYS analysis. Fracture may occur sporadically at widths of 50 or 100 μm, depending on the amount and magnitude of the flaws in the Pyrex glass. Experiments were done on the Pyrex glass bonding systems with a width of 100 μm. The samples were heated to 750 °C and then cooled by taking them out of the oven. In all four experimental cases, small cracks were observed in Pyrex glass as shown in Fig. 37.14. These cracks may have occurred consistently for several reasons. First, it may be a result of handling of the Pyrex glass before bonding. The Pyrex glass samples were kept in containers with each other, which may have resulted in abrasive contact and possibly caused flaws in the materials. These flaws can cause a reduced strength from the 69 GPa and therefore the predictions from the ANSYS analysis could be correct. Second, the observation is made that the cracks are small, only occurring tens of microns away from the aluminum and not propagating completely through the Pyrex glass. These cracks could be caused by the high stress applied, but the crack has not reached a critical size and therefore has not propagated completely through the Pyrex glass. Therefore, the strength remains at the theoretical strength of 69 GPa, and the Pyrex glass is only partially cracked. Experimental analysis done by *Chaio* and *Lin* [37.90] shows that fracture was not observed when using aluminum widths greater than 150 μm. This is consistent with the results of the ANSYS analysis which show that as the width of the aluminum is increased, the residual stress decreases.

The quartz bonding system ANSYS stress predictions are conducted, and the maximum stress predicted

Fig. 37.14 Microphoto of the experimental result on Pyrex Glass-Aluminum-Silicon system. Small cracks can be observed

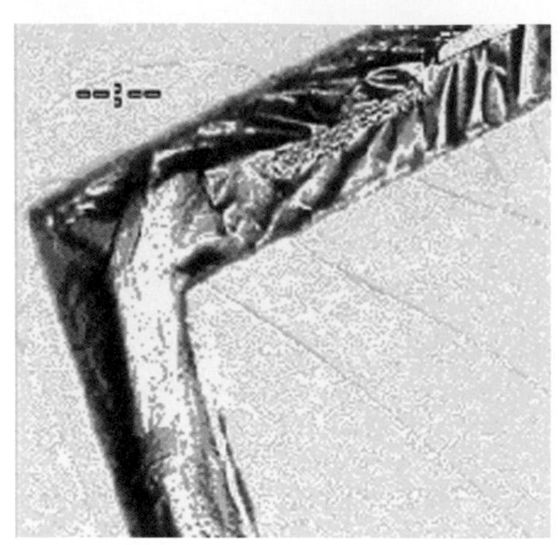

Fig. 37.15 Bonding result of Quartz-Aluminum-Silicon system, fracture can be observed

in the analysis is 207 GPa, 117 GPa, and 100 GPa, for aluminum widths of 30 μm, 50 μm, and 100 μm, respectively. All three of these stresses are much larger than the theoretical strength of Quartz at 48 GPa, and therefore fracture should always occur. Quartz has a much larger coefficient of thermal expansion (CTE) than silicon, which is why this was predicted. Experimentally, a quartz substrate was tested for the silicon-aluminum-quartz bonding test and the result is shown in Fig. 37.15.Cracks happen all over the place and cause serious damages on the quartz wafer. These cracks could be the failure mechanism of the hermetic package. Pyrex glass is therefore identified as a better bonding substrate than quartz.

The thermal stresses generated in the packaging process in quartz are much larger than the stresses in Pyrex glass because of the difference between the CTE in the two systems. Quartz has a low CTE ($0.54\times10^{-6}\,K^{-1}$) compared to the aluminum CTE ($23\times10^{-6}\,K^{-1}$) and silicon CTE ($3.5\times10^{-6}\,K^{-1}$). On the other hand Pyrex glass has a much closer CTE ($3.2\times10^{-6}\,K^{-1}$) to silicon and aluminum, resulting in smaller stresses. The practical implications of the ANSYS results and the preceding information are that materials must be chosen carefully when carrying out bonding. Material should not be used with a much higher or lower CTE than silicon to ensure fracture will not occur. This finding is a valuable component of this particular packaging system and should benefit the other packaging processes involving bonding MEMS packaging because it supports the prediction that Pyrex glass is an excellent material to use when bonding with silicon in MEMS packaging, as long as sufficiently wide adhesive material is used.

37.3.2 Packaging Reliability

The thermal stress induced by the CTE mismatch is one of the main factors that affect the packaging reliability. As a matter of fact, the formation of the stress can happen not only during packaging process but also during the operation of devices. In particular, during device operation, the package will go through various temperature cycles because of environment change. Such temperature variation causes the expansion of packaging materials when they are constrained by the packaged assembly. As a result of such thermal mismatch, significant stresses are induced in the package and finally cause the device to fail. In addition to thermal mismatch, corrosion creep, fracture fatigue, crack initiation and propagation, and delamination of thin films are all possible factors to cause failure of packaged devices [37.94]. These failure mechanisms could be prevented or deferred under proper packaging designs. For instance, the thermally induced strain inside the packaging material is generally below the tolerance of the material and can't cause immediate catastrophic damage. But cyclic loading can generate and accumulate stresses, eventually causing failure. Several common designs in IC packaging have been used to prolong the lifetime of devices. For example, the strain in solder interconnects of BGA or flip-chip packaging can be effectively reduced by introducing a polymer underfill material between the chip and the substrate to distribute thermal stress induced by CTE mismatch [37.95]. The strain can further be reduced if excellent thermal paths are built around interconnects to diminish the thermal stress originated from the temperature gradient between ambient and operation temperatures. Delamination phenomenon is another source to cause reliability problems, and it happens in the interface of adjacent material layers. In MEMS, components made of dissimilar materials are commonly bonded together to provide specific functions. Delamination can result in electrical or mechanical failures of devices in the package such as mechanically cracking through the electrical via wall to make an electrical open because of the propagation of the delamination of metal line from the dielectric layer or over heating of the die because of the delamination of die from its underneath to make an opening on the heat dissipation path. Because the stress and thermal loading, the geometry, and the material properties are complex in MEMS, the development of the packaging designs to increase the reliability is very important and requires more extensive investigations.

Reliability testing is required before a new device can be delivered to the market. The test results can provide the information for the following improvement of packaging design and fabrication. How to analyze the failure data, called the "reliability metrology", is thus very important in the packaging industry. The analysis method is to use the mathematical tools of probability and statistical distributions to evaluate data to understand the patterns and identify the sources of failure. For example, a failure density function is defined as the time derivative of the cumulative failure function

$$f(t) = \frac{\mathrm{d}F(t)}{\mathrm{d}t} , \tag{37.1}$$

$$F(t) = \int_0^t f(s)\,\mathrm{d}s . \tag{37.2}$$

The cumulative failure function, $F(t)$, is the fraction of a group of original devices that has failed at time t. The Weibull distribution function is one of the analytic mathematic models commonly used in the packaging reliability evaluation to represent the failure density function [37.21].

$$f(t) = \frac{\beta}{\lambda}\left(\frac{t}{\lambda}\right)^{\beta-1} \exp\left[-\left(\frac{t}{\lambda}\right)^{\beta}\right] , \tag{37.3}$$

where β and λ are the Weibull parameters. The parameters β, called a shape factor, measures how the failure frequency is distributed around the average lifetime. The parameter λ is called the lifetime parameter and indicates the time at which 63.2% of the devices faile. By integrating both sides of the equation (37.3). $F(t)$ becomes:

$$F(t) = 1 - \exp\left[-\left(\frac{t}{\lambda}\right)^{\beta}\right] . \tag{37.4}$$

Using the Weibull distribution function with the two parameters extrapolated by experimental data, one can estimate the number of failures at any time during the test. Moreover, by knowing the meaning and values of the parameters, one can compare two sets of test data. For example, the λ with greater number means this set of samples has longer lifetime. Because all of the mathematical models are statistical approximations based on the real experimental data, more testing samples can provide more accurate estimation.

37.3.3 Long-Term and Accelerated MEMS Packaging Tests

The reliability of MEMS packages is best characterized by means of long-term tests with statistical data analyses. But it is very difficult and time-consuming to obtain these experimental data so accelerated tests that put samples in extreme environment for accelerated failure are commonly used to predict the lifetime of the devices. Unfortunately, few research publications deal with these two issues. In this section, two MEMS packaging examples that aim to address these two tests we discuss.

Figure 37.16 shows long-term measurements of the Q factor of a vacuum packaged μ-resonators using localized aluminum/silicon-to-glass bonding [37.81] vacuum encapsulation process is described in details as shown in Fig. 37.8. It is found that the vacuum package by means of localized heating and bonding provides stable vacuum environments for the μ-resonator and a quality factor of 9,600 has been achieved with no degradation for at least one year. Since the performance of high Q μ-resonator is very sensitive to environmental pressure as shown in Fig. 37.11, any leakage can be easily detected. The fact that this high Q value can hold for one year indicates the packaging process is well performed and both aluminum and Pyrex glass are suitable materials for vacuum packaging applications. According to a previous study of hermeticity in different materials, metal has lower permeability to moisture than other materials such as glass, epoxy, and silicon. With a width of 1 μm, metal can effectively block moisture for more than ten years [37.21]. In this vacuum package system, the bonding width is 30 μm so it can sufficiently block the diffusion process of moisture. On the other hand, the diffusion effects of air molecules into these tiny cavities have not been studied extensively, and the design guidelines for vacuum encapsulations are not clearly defined. Further investigations will be needed in this area, and the example presented here serves as a good starting point.

On the other hand, accelerated testing puts a large amount of samples in harsh environments, such as elevated temperature, elevated pressure, and 100% humidity, to accelerate the corrosion process. The statistical failure data are gathered and analyzed to predict the lifetime of packages under normal usage environment. As a result, the long-term reliability of the package can be predicted without going through the true long-term tests. Unfortunately, accelerated tests have not been addressed in MEMS research papers. Although the MEMS industry must have done some extensive reliability tests, they do not publish their results, probably due to liability concerns. Among the very limited publications, this section uses as the illustration example a specific MEMS packaging system that has gone through accelerated tests [37.45].

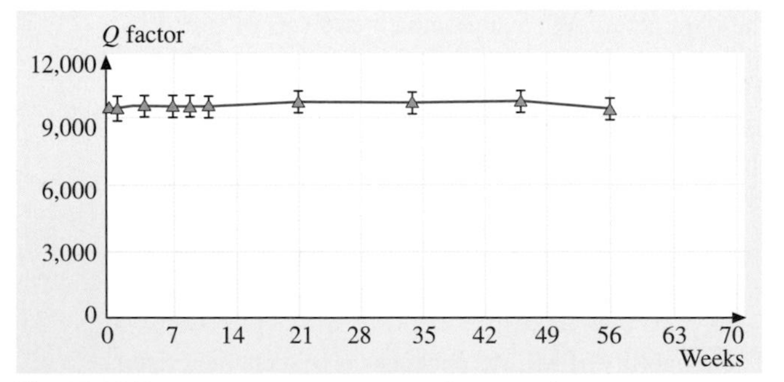

Fig. 37.16 Long-term measurement of encapsulated μ-resonators. No degradation of Q factors is found after 56 weeks

The MEMS package is accomplished by means of RTP (Rapid Thermal Processing) bonding as described previously in this chapter. The goal of the accelerated test is to examine the failure rate at the bonding interface. The accelerated tests start by putting the packaged samples into the autoclave chamber filled with high temperature and pressurized steam at 130 °C, 2.7 atm, and 100% relative humidity for accelerated testing. The pressurized steam can penetrate small cracks if any defect exists at the bonding interface [37.96]. Elevated temperature and humid environment speed up the corrosion process. A package is considered a failure if water is condensed or diffused into the package. Statistical failure data are gathered every 24 h under optical examination for a period of 864 h, during which time new failure is seldom observed. Because of the robustness of the samples, it is difficult to conduct the tests to the point at which all the packages may fail. The cumulative failure function is recorded, and it is found that most failures occur in the first 96 h and such high early failure rate reflects the yielding issue of the bonding process. Weibull and Lognormal models are compared to predict the lifetime of the packages, and it is found that Lognormal model better describes the statistical data. Figure 37.17 shows the inverse standard normal distribution function vs. ln(time) and the Maximum Likelihood Estimator (MLE) is then used to predict the mean, standard deviation, and the MTTF (mean time to failure). Table 37.2 shows the MLE calculation results of MTTF. The wide interval of confidence level comes from the fact that only a small number of samples failed at the end of the test. It is also observed that packages with larger bonding width and smaller bonding areas have larger MTTF values. The lower bound of the MTTF provides the worst-case scenario. For example, only 4 our of 31 samples failed when tests stopped in the case of ring width of 200 μm and sealing area of $450\times450\,\mu m^2$. The MTTF predicts, in the worst-case scenario, that there

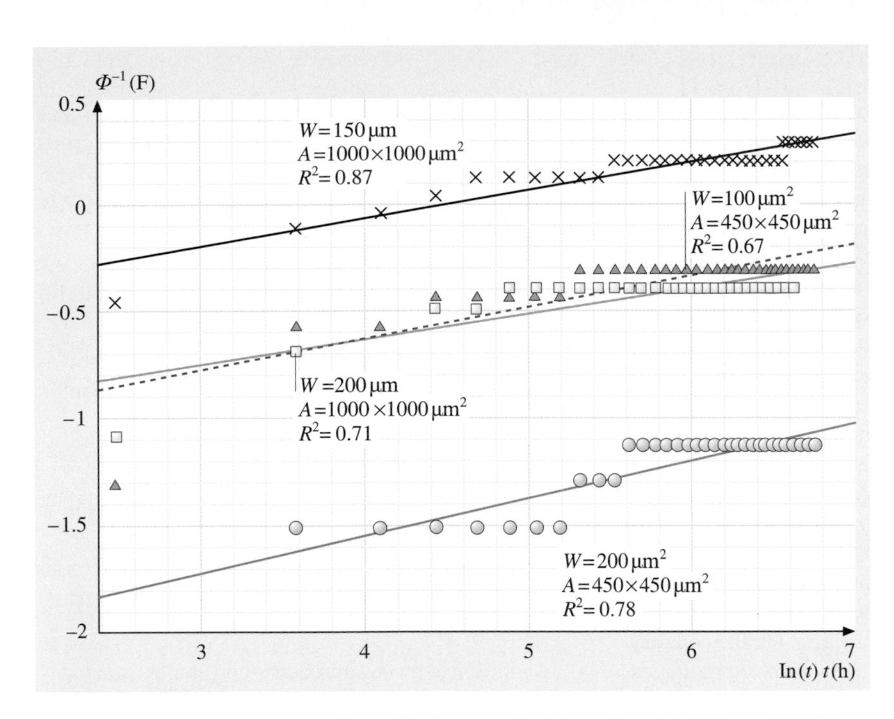

Fig. 37.17 Life data fitted by Lognormal distribution. R^2 is the coefficient of determination

Table 37.2 The Maximum Likelihood Estimation for Mean Time To Failure (MTTF)

MLE calculation resuls of MMTF[a]				
Bonding width, W (µm)	Area, A (µm^2)	MTTF		Worst cases in jungle condition (years)
		UB (years)	LB (years)	
200	450× 450	1.8×10^7	0.57	1,700
100	450× 450	5.3	0.10	300
200	1,000×1,000	6.5×10^3	0.09	270
150	1,000×1,000	0.50	0.017	50

[a] UB is the upper bound, and LB is the lower bound of the 90% confidence interval, respectively. The MTTF LB times AF is the worst case MTTF used in jungle condition.

is 90% chance a package will fail in 0.57 years in the autoclave environment.

It is widely accepted that the acceleration factor (AF) for autoclave tests follows the Arrhenius equation [37.21] and can be modeled as:

$$AF = \frac{(RH^{-n}\,e^{\Delta E_a/kT})_{normal}}{(RH^{-n}\,e^{\Delta E_a/kT})_{accelerated}}\,, \tag{37.5}$$

where RH is relative humidity (85%, RH = 85), k the Boltzmann constant and T the absolute temperature. The recommended value for n, an empirical constant, is 3.0 [37.97], and ΔE_a, the activation energy, is 0.9 eV for plastic dip package and 0.997 eV for anodically bonded glass-to-silicon package [37.58]. If $\Delta E_a = 0.9$ is used to estimate the AF for the accelerated testing condition as compared with the jungle condition (35 °C, 1 atm and 95% RH), AF is about 3,000, and the worst-case lifetime values in jungle condition are also listed in Table 37.2. The high values of estimated MTTF in jungle condition could be a result of overestimation of AF because plastic dip package may have smaller AFs than those of glass packages. Nevertheless, these data and analyses provide important guidelines in the area of accelerated tests for MEMS packages.

37.4 Future Trends and Summary

In the past, the development of MEMS packaging is mainly originated from IC packaging advancement because existing packaging techniques can significantly reduce the development cost of MEMS. It is expected that the situation will be changed very soon, hoewever, in a way the MEMS packaging approaches will assist the IC packaging development. Recent advances in IC packaging aim to provide high I/O density and more chip integration capability for the needs of high speed and high data communication rate. In order to satisfy those requirements, several packaging concepts and techniques are developed, including 3-D packaging, wafer-level packaging, BGA, and flip-chip technique. Although all of the concepts and methods can provide a package with more I/O density, flexibility in chip integration, and lower manufacturing cost for IC fabrication, they are still insufficient for providing solutions for the future applications because of the increase of complexity and requirements of MEMS packaging. On the contrary, with the progress of MEMS fabrication technologies, several key processes such as DRIE (Deep Reactive Ion Etching), wafer bonding, and thick photoresist processes [37.98] have been utilized for IC packaging fabrication. Technologies developed in MEMS fabrication can therefore also assist the development of new IC packaging approaches.

In order to address the future needs of process integration, adaptive multi-chip module (MCM) [37.28] or 3-D packaging combined with vertical through substrate interconnects [37.10, 99] are promising approaches for the development of future MEMS packaging processes. Based on low temperature flip-chip solder bonding technique, these packaging methods can provide more flexibility in device fabrication and packaging. Devices can be fabricated before they are integrated to form microsystems to dramatically reduce the packaging cost. Vertical through substrate interconnects can have higher I/O density, smaller resistance, parasitic capacitance, and mutual inductance. Although this approach provides many possible advantages, technical challenges exist. For instance, metal is commonly used as the fill material inside the vertical vias as electrical interconnects, and it may introduce large thermal mismatch with respect to silicon substrate and generate huge thermal stress to cause packaging reliability problems. Moreover, it will be an interesting engineering challenge to fill the materials into those high aspect ratio vias.

The future development of MEMS packaging depends on the successful implementation of unique techniques as described in the following:

1. Development of mechanical, thermal, and electrical models for packaging designs and fabrication processes;
2. Wafer-level, chip-scale packaging with low packaging cost and high yield;
3. Effective testing techniques at the wafer level to reduce the testing costs;
4. Device integrations by vertical through-interconnects as an interposer [37.1] to avoid thermal mismatch problems.

In addition to these approaches and challenges, many other possibilities have not been listed but also require dedicated investigations. For example, several key nanotechnologies have been introduced in the previous chapters, but the packaging solutions for the NEMS devices have not been addressed. Because it is feasible to use MEMS as the platform for NEMS fabrication, all of the packaging issues discussed in the chapter can be directly applied to NEMS devices. On the other hand, nanotechnology may introduce new opportunities for MEMS/NEMS packaging applications by providing superior electrical, mechanical, and thermal properties [37.100–103]. For example, carbon nanotubes have very high thermal conductivity [37.102] and may be suitable to increase the thermal cooling effects for better IC/MEMS/NEMS packaging applications.

In summary, MEMS packaging issues have been introduced in the areas of fabrication, application, reliability, and future development. Packaging design and modeling, packaging material selection, packaging process integration, and packaging cost are the main issues to be considered when developing a new MEMS packaging process.

References

37.1 M. Matsuo, N. Hayasaka, K. Okumura, E. Hosomi, C. Takubo: Silicon interposer technology for high-density package, IEEE ECTC (2000) 1455–1459

37.2 K. E. Peterson: Silicon as a mechanical material, Proc. IEEE **70** (1982) 420–457

37.3 L. Lin, H.-C. Chu, Y.-W. Lu: A simulation program for the sensitivity and linearity of piezoresistive pressure sensors, IEEE/ASME J. Microelectromech. Syst. **8** (1999) 514–522

37.4 Y. C. Tai, R. S. Muller: Lightly-doped polysilicon bridge as a flow meter, Sens. Actuators **15** (1988) 63–75

37.5 D. Hicks, S.-C. Chang, M. W. Putty, D. S. Eddy: Piezoelectrically activated resonant bridge micro-acceleromenter, 1994 Solid-State Sensors and Actuators Workshop (1994) 225–228

37.6 J. Berstein, S. Cho, A. King, A. Kourepenis, P. Maciel, M. Weinberg: A micromachined comb-drive tuning fork rate gyroscope, 6th IEEE Int. Conf. on MEMS (1993) 143–148

37.7 M. Madou: Compatibility and incompatibility of chemical sensors and analytical equipment with micromachining, Solid-State Sensor and Actuator Workshop, Hilton Head (1994) 164–171

37.8 M. J. Zdeblick, J. B. Angell: A microminiature electric-to-fluidic valve, Proceedings of Transducers'87, the 4th International Conference on Solid-State Transducers and Actuators (1987) 827–829

37.9 J. H. Tsai, L. Lin: A thermal bubble actuated micro nozzle-diffuser pump, IEEE/ASME J. Microelectromech. Syst. **11** (2002)

37.10 S. J. Ok, D. Baldwin: High density, aspect ratio through-wafer electrical interconnect vias for low cost, generic modular MEMS packaging, 8th IEEE Int. Symp. On Adv. Pack. Mater. (2002) 8–11

37.11 H. Reichl, V. Grosser: Overview and development trends in the field of MEMS packaging, 14th IEEE Int. Conf. on MEMS (2001) 1–5

37.12 P. Kapur, P. M. McVittie, K. Saraswat: Technology and reliability constrained future copper interconnects=Part II: Performance implication, Trans. Electron. Dev. **49** (2002) 598–604

37.13 J. Lau: *Flip Chip Technologies* (McGraw-Hill, New York 1996)

37.14 R. Prasad: *Surface Mount Technology: Principles and Practice*, 2nd edn. (Chapman, New York 1989)

37.15 M. Töpper, J. Auersperg, V. Glaw, K. Kaskoun, E. Prack, B. Keser, P. Coskina, D. Jäger, D. Petter, O. Ehrmann, K. Samulewicz, C. Meinherz, S. Fehlberg, C. Karduck, H. Reichl: Fab integrated packaging (FIP) a new concept for high reliability wafer-level chip size packaging, IEEE, ECTC (2000) 74–81

37.16 S. Savastiouk, O. Siniaguine, Ed. Korczynski: 3D wafer level packaging, IEEE, ECTC. (2000) 26–31

37.17 J. H. Tsai, L. Lin: Micro-to-macro fluidic interconnectors with an integrated polymer sealant, J. Micromech. Microeng. **11** (2001) 577–581

37.18 T. A. Core, W. K. Tsang, S. Sherman: Fabrication technology for an integrated surface-micromachined sensor, Solid State Technol. (October 1993) 39–47

37.19 K. Koester, R. Majedevan, A. Shishkoff, K. Marcus: *Multi-Sser MEMS Processes (MUMPS) Introduction and Design Rules* (MCNC MEMS Technology Applications Center, Research Triangle Park, NC 27709 1996)

37.20 R. Lengtenberg, H. A. C. Tilmans: Electrically driven vacuum-encapsulated polysilicon resonantor, Part I: Design and fabrication, Sens. Actuators A **45** (1994) 57–66

37.21 R. R. Tummala, E. J. Rymaszewski, A. G. Klopfenstein: *Microelectronics Packaging Handbook, Semiconductor Packaging* (Chapman, New York 1997)

37.22 S. Wolf: *Silicon Processing for the VLSI era, Vol. I: Process Technology* (Lattice Press, Sunset Beach 1995)

37.23 J. K. Bhardway, H. Ashraf: Advanced silicon etching using high density plasmas, SPIE Micromach. Fabr. Technol. **2639** (1995) 224–233

37.24 W. H. Ko, J. T. Suminto, G. J. Yeh: *Bonding Techniques for Microsensors, Micromachining and Micropackaging for Transducers* (Elsevier, Amsterdam 1985)

37.25 S. D. Senturia, R. L. Smith: Microsensor packaging and system partitioning, Sens. Actuators **15** (1988) 221–234

37.26 R. L. Smith, S. D. Collins: Micromachined packaging for chemical microsensors, IEEE Trans. Electron Dev. **35** (1988) 787–792

37.27 J. T. Butler, V. M. Bright, P. B. Chu, R. J. Saia: Adapting multichip module foundries for MEMS packaging, Int. Conf. on Multi. Mod. High Den. Pack. (1998) 106–111

37.28 M. Schuenemann, A. J. Kourosh, V. Grosser, R. Leutenbauer, G. Bauer, W. Schaefer, H. Reichl: MEM modular packaging and interfaces, IEEE ECTC (2000) 681–688

37.29 D. W. Lee, T. Ono, T. Abe, M. Esashi: Fabrication of microprobe array with sub-100nm nano-heater for nanometric thermal imaging and data storage, Proc. IEEE Micro Electro Mechanical Systems Conf., IEEE Int. Conf. on MEMS (January 2001) 204–207

37.30 A. S. Laskar, B. Blythe: Epoxy multichip modules, a solution to the problem of packaging and interconnection of sensors and signal-processing chips, Sens. Actuators A **36** (1993) 1–27

37.31 R. Reichl: Packaging and interconnection of sensors, Sens. Actuators A **25–27** (1991) 63–71
37.32 A. Grisel, C. Francis, E. Verney, G. Mondin: Packaging technologies for integrated electrochemical sensors, Sens. Actuators **17** (1989) 285–295
37.33 J.L. Lund, K.D. Wise: Chip-level encapsulation of implantable CMOS microelectrode arrays, 1994 Solid-State Sensor and Actuator Workshop, Hilton Head (1994) 29–32
37.34 T. Akin, B. Siaie, K. Najafi: Modular micromachined high-density connector for implantable biomedical systems, IEEE Micro Electro Mechanical Systems Workshop (1996) 497–502
37.35 L. Muller, M.H. Hecht, L.M. Miller et al.: Packaging qualification for MEMS-based space systems, IEEE Micro Electro Mechanical Systems Workshop 503–508
37.36 L. Lin, K. McNair, R.T. Howe, A.P. Pisano: Vacuum encapsulated lateral microresoantors, 7th Int. Conference on Solid State Sensors and Actuators (June 1993) 270–273
37.37 H. Guckel: Surface micromachined pressure transducers, Sens. Actuators A **28** (1991) 133–146
37.38 J.J. Sniegowski, H. Guckel, R.T. Christenson: Performance characteristics of second generation polysilicon resonating beam force transducers, IEEE Solid-State Sensor and Actuator Workshop, Hilton Head Island (1990) 9–12
37.39 K. Ikeda, H. Kuwayama, T. Kobayashi, T. Watanabe, T. Nishikawa, T. Oshida, K. Harada: Three dimensional micromachining of silicon pressure sensor integrating resonant strain gauge on diaphragm, Sens. Actuators A **21–23** (1990) 1001–1010
37.40 C.H. Mastrangelo, R.S. Muller: Vacuum-sealed silicon micromachined incandescent light source, IEEE IEDM (1989) 503–506
37.41 J. Smith, S. Montague, J. Sniegowski, R. Manginell, P. McWhorter, R. Huber: Characterization of the embedded micromechanical device approach to the monolithic integration of MEMS with CMOS, SPIE 2879
37.42 S. Van der Groen, M. Rosmeulen, P. Jansen, K. Baert, L. Deferm: CMOS compatible wafer scale adhesive bonding for circuit transfer, 1997 International Conference on Solid-State Sensors and Actuators, Transducers'97 (1997) 629–632
37.43 M.B. Cohn, Y. Liang, R. Howe, A.P. Pisano: Wafer to wafer transfer of microstructures for vacuum package, Solid State Sensor and Actuator Workshop, Hilton Head (1996) 32–35
37.44 Y.T. Cheng, Y.T. Hsu, L. Lin, C.T. Nguyen, K. Najafi: Vacuum packaging using localized aluminum/silicon-to-glass bonding, Int. Conf. on MEMS (2001) 18–21
37.45 M. Chiao, L. Lin: Vacuum packaging of microresonators by rapid thermal processing, Proc. of SPIE on Smart Electronics, MEMS, and Nanotechnology (2002) 17–21
37.46 M. Esashi, S. Shoji, A. Nakano: Normally closed microvalve and micropump fabricated on a silicon wafer, Sens. Actuators **20** (1989) 163–169
37.47 M.E. Poplawski, R.W. Hower, R.B. Brown: A simple packaging process for chemical sensors, Solid-State Sensor and Actuator Workshop, Hilton Head (1994) 25–28
37.48 S.F. Trautweiler, O. Paul, J. Stahl, H. Baltes: Anodically bonded silicon membranes for sealed and flush mounted microsensors, Micro Electro Mechanical Systems Workshop (1996) 61–66
37.49 E.H. Klaassen, K. Petersen, J.M. Noworolski, J. Logan, N.I. Malfu, J. Brown, C. Storment, W. McCulley, G.T.A. Kovac: Silicon fusion bonding and deep reactive ion etching: A new technology for microstructures, Sens. Actuators A **52** (1996) 132–139
37.50 C.H. Hsu, M.A. Schmidt: Micromachined structures fabricated using a wafer-bonded sealed cavity process, Solid State Sensor and Actuator Workshop, Hilton Head (1994) 151–155
37.51 A.L. Tiensuu, J.A. Schweitz, S. Johansson: In situ investigation of precise high strength micro assembly using Au-Si eutectic bonding, Int. Conf. Solid-State Sens. Actuators, Eurosens. IX (1995) 236–239
37.52 A.P. Lee, D.R. Ciarlo, P.A. Krulevitch, S. Lehew, J. Trevino, M.A. Northrup: Practical microgripper by fine alignment, eutectic bonding and SMA actuation, Int. Conf. Solid-State Sens. Actuators, Eurosens. IX (1995) 368–371
37.53 R.W. Bower, M.S. Ismail, B.E. Roberds: Low temperature Si_3N_4 direct bonding, Appl. Phys. Lett. **62** (1993) 3485–3487
37.54 H. Takagi, R. Maeda, T.R. Chung, T. Suga: Low temperature direct bonding of silicon and silicon dioxide by the surface activation method, Int. Conf. on Solid-State Sensors and Actuators, Transducer 97 (1997) 657–660
37.55 G. Wallis, D. Pomerantz: Filed assisted glass-metal sealing, J. Appl. Phys. **40** (1969) 3946–3949
37.56 L. Bowman, J. Meindl: The packaging of implantable integrated sensors, IEEE Trans. Biomed. Engin. **33** (1986) 248–255
37.57 M. Esashi: Encapsulated micro mechanical sensors, Microsyst. Technol. **1** (1994) 2–9
37.58 B. Ziaie, J. Von Arx, M. Dokmeci, K. Najafi: A hermetic glass-silicon micropackages with high-density on-Chip feedthroughs for sensors and actuators, J. Microelectromech. Syst. **5** (1996) 166–179
37.59 Y. Lee, K. Wise: A batch-fabricated silicon capacitive pressure transducer with low temperature sensitivity, IEEE Trans. Electron Dev. **29** (1982) 42–48
37.60 S. Shoji, H. Kicuchi, H. Torigoe: Anodic bonding below 180 degree C for packaging and assembling of MEMS using lithium aluminosilicate-beta-quartz glass-ceramic, Proc. 10th Annual Int. Workshop Micro Electro Mechanical Systems (1997) 482–487

37.61 M. Esashi, N. Akira, S. Shoji, H. Hebiguchi: Low-temperature silicon-to-silicon anodic bonding with intermediate low melting point glass, Sens. Actuators A **23** (1990) 931–934
37.62 A. Hanneborg, M. Nese, H. Jakobsen, R. Holm: Silicon-to-thin film anodic bonding, J. Micromech. Microeng. **2** (1992) 117–121
37.63 S.A. Audet, K.M. Edenfeld: Integrated sensor wafer-level packaging, International Conference on Solid-State Sensors and Actuators, Transducers'97 (1997) 287–289
37.64 R.C. Benson, N. deHaas, P. Goodwin, T.E. Phillips: Epoxy adhesives in microelectronic hybrid applications, Johns Hopkins APL Tech. Dig. **13** (1992) 400–406
37.65 M. Shimbo, J. Yoshikawa: New silicon bonding method, J. Electrochem. Soc. **143** (1996) 2371–2377
37.66 P.M. Zavracky, B. Vu: Patterned eutectic bonding with Al/Ge thin film for MEMS, SPIE **2639** (1995) 46–52
37.67 G. Humpston, D.M. Jacobson: Principles of soldering and brazing, ASM Int. (1993) 241–244
37.68 A. Singh, D. Horsely, M.B. Cohn, R. Howe: Batch transfer of microstructures using flip-chip solder bump bonding, International Conference on Solid State Sensors and Actuators, Transducer 97 (1997) 265–268
37.69 M.M. Maharbiz, M.B. Cohn, R.T. Howe, R. Horowitz, A.P. Pisano: Batch micropackaging by compression-bonded wafer-wafer transfer, 12th Int. Conf. MEMS (1999) 482–489
37.70 Y.T. Cheng, L. Lin, K. Najafi: Localized silicon fusion and eutectic bonding for MEMS fabrication and packaging, IEEE/ASME J. Microelectromech. Syst. **9** (2000) 3–8
37.71 L. Lin: Selective encapsulations of MEMS: micro channels, needles, resonators, and electromechanical filters. Ph.D. Thesis (University of California, Berkeley 1993)
37.72 Y.C. Su, L. Lin: Localized plastic bonding for micro assembly, packaging and liquid encapsulation, Proceedings of IEEE Micro Electro Mechanical Systems Conference, Interlaken (2001) 50–53
37.73 M. Schwartz: *Brazing* (Chapman, London 1995)
37.74 L. Lin, R.T. Howe, A.P. Pisano: Microelectromechanical Filters for Signal Processing, IEEE/ASME J. Microelectromech. Syst. **7** (1998) 286–294
37.75 W.C. Tang, C.T.-C. Nguyen, R.T. Howe: Laterally driven polysilicon resonant microstructures, Sens. Actuators A **20** (1989) 25–32
37.76 G.T. Mulhern, D.S. Soane, R.T. Howe: Supercritical carbon dioxide drying of microstructures, 7th International Conference on Solid State Sensors and Actuators, Yokohama (1993) 296–299
37.77 M. Judy: Micromechanisms using sidewall beams. Ph.D. Thesis (University of California, Berkeley 1994)
37.78 A.M. Leung, J. Jones, E. Czyzewska, J. Chen, B. Woods: Micromachined accelerometer based on convection heat transfer, International Conference on MEMS (1998) 627–630
37.79 D.R. Spark, L. Jordan, J.H. Frazee: Flexible vacuum-packaging method for resonating micromachines, Sens. Actuators A **55** (1996) 179–183
37.80 SensArray Corporation, 47451 Fremont Blvd. Fremont, CA 94538
37.81 Y.T. Cheng, W.T. Hsu, K. Najafi, C.T. Nguyen, L. Lin: Vacuum packaging technology using localized aluminum/silicon-to-glass bonding, IEEE/ASME J. Microelectromech. Syst. **11** (2002) 556–565
37.82 C. Luo, L. Lin: The application of nanosecond-pulsed laser welding technology in MEMS packaging with a shadow mask, Sens. Actuators A **97-98** (2002) 398–404
37.83 A. Cao, M. Chiao, Liwei. Lin: Selective and localized wafer bonding using induction heating, Technical Digest of Solid-State Sensors and Actuators Workshop (2002) 153–156
37.84 J.B. Kim, M. Chiao, L. Lin: Ultrasonic Bonding of In/Au and Al/Al for Hermetic Sealing of MEMS Packaging, Proceedings of IEEE Micro Electro Mechanical Systems Conference (2002) 415–418
37.85 Y.T. Cheng, L. Lin, K. Najafi: Localized bonding with PSG or indium solder as intermediate layer, 12th International Conference on MEMS (1999) 285–289
37.86 G. Thyrum, E. Cruse: Heat pipe simulation, a simplified technique for modeling heat pipe assisted heat sinks, Adv. Packaging (2001) 23–27
37.87 G.P. Peterson, A.B. Duncan, M.H. Weichold: Experimental investigation of micro heat pipes fabricated in silicon wafers, J. Heat Trans. (1993) 751–756
37.88 L. Jiang, M. Wong, Y. Zohar: Forced convection boiling in a microchannel heat sink, J. Microelectromech. Syst. **10** (2001) 80–87
37.89 F. Arias, S.R.J. Oliver, B. Xu, E. Holmlin, G.M. Whitesides: Fabrication of metallic exchangers using sacrificial polymer mandrils, IEEE/ASME J. Microelectromech. Syst. **10** (2001) 107–112
37.90 M. Chiao, L. Lin: Hermetic wafer bonding based on rapid thermal processing, Sens. Actuators A **91** (2001) 404–408
37.91 *ANSYS Modeling and Meshing Guide*, 3rd edn. (SAS IP, Inc., Canousburg 2002)
37.92 D. Bystrom, L. Lin: Residual stress analysis of silicon-aluminum-glass bonding processes, ASME International Mechanical Engineering Congress and Exposition, MEMS Symposium (November 2002)
37.93 H. Scholze: *Glass: Nature, Structure, and Properties*, 1st edn. (Springer, Berlin, Heidelberg 1991)
37.94 R.R. Tummala: *Fundamentals of Microsystems packaging* (McGraw-Hill, New York 2001)
37.95 S.J. Adamson: BGA, CSP, and flip chip, Adv. Packaging (2002) 21–24
37.96 D. Steoehle: On the penetration of water vapor into packages with cavities and on maximum allow-

able leak rate, 15th Annual Proceeding Reliability Physics Symposium (1977) 101–106

37.97 W. D. Brown: *Advanced Electronic Packaging* (IEEE Press, 1999)

37.98 F. Niklaus, P. Znoksson, E. Käluesten, G. Stemme: Void free full wafer adhesion bonding, Proceedings of IEEE Micro Electro Mechanical Systems Conf. 2000) 241–252

37.99 C. H. Cheng, A. S. Ergun, B. T. Khuri-Yakub: Electrical through-wafer interconnects with subpicofarad parasitic capacitance, IEEE ECTC (2002) 18–21

37.100 D. Routkevitch, A. A. Tager, J. Haruyama, D. Almawlawi, M. Moskovits, J. M. Xu: Nonlithographic nano-array arrays: fabrication, physics, and device applications, IEEE Trans. Electron. Dev. **43** (1996) 1646–1658

37.101 J. Gou, M. Lundstrom, S. Datta: Performance projections for ballistic carbon nanotube fieldeffect transistors, Appl. Phys. Lett. **80** (2002) 3192–3194

37.102 S. U. S. Choi, Z. G. Zhang, W. Yu, F. E. Lockwood, E. A. Grulke: Anomalous thermal conductivity enhancement in nanotube suspensions, Appl. Phys. Lett. **79** (2001) 2252–2254

37.103 K. Velikov, A. Moroz, A. Blaaderen: Photonic crystals of core-shell colloidal particles, Appl. Phys. Lett. **80** (2002) 49–51

Part F

Social and

Part F Social and Ethical Implication

38 Social and Ethical Implications of Nanotechnology
William S. Bainbridge, Arlington, USA

38. Social and Ethical Implications of Nanotechnology

Nanotechnology will have very broad applications across all fields of engineering, so it will be an amplifier of the social effects of other technologies. There is an especially great potential for it to combine with three other powerful trends – biotechnology, information technology, and cognitive science – based on the material unity of nature at the nanoscale and on technology integration from that scale. Technological convergence highlights such existing issues as the treatment of the disabled, communication breakdowns, economic stagnation, and threats to national security. Nanotechnology itself may possibly raise distinctive ethical and social issues in the future, but much of the public discussion to this point has been misdirected and misinformed, lacking a firm social scientific basis. Thus it will be important to integrate social and ethical studies into nanotechnology developments from their very beginning.

Social scientific and economic research can help manufacturers and governments make the right decisions when deploying a new technology, maximizing its benefit for human beings. In addition, technically competent research on the societal implications of nanotechnology will help give policymakers and the general public a realistic picture free of unreasonable hopes or fears. The costs of premature or excessive regulation would be extremely high, harming the very people it was intended to protect, and failure to develop beneficial nanotechnology applications would be unethical.

The significance of nanotechnology depends largely on how its development relates to wider trends going on in the world such as the impending population declines of most advanced industrial nations, the apparent diminishing returns to increased medical research and health care investment, and the threatened deceleration of progress in microelectronics. Well established social-scientific explanations for unethical behavior – such as learning, strain, control, and subculture theories – could help us understand possible future cases in nanotechnology industries. Ethics and social implications are largely matters of social perception, and the public conception of nanotechnology is still in the early stages of developing. Social science can now begin to examine its unfolding impacts in all sectors of the economy, in most spheres of life, and both short-term and long-term time scales.

38.1 **Applications and Societal Impacts** 1136

38.2 **Technological Convergence** 1139

38.3 **Major Socio-technical Trends** 1141

38.4 **Sources of Ethical Behavior** 1143

38.5 **Public Opinion** 1145

38.6 **A Research Agenda** 1148

References 1149

The views expressed in this essay do not necessarily represent the views of the National Science Foundation or the United States.

Science and engineering have only begun to explore the potential for discovery and creativity at the nanoscale, but already some intemperate voices call for government regulation or outright banning of nanotechnology [38.1–4]. Such action would be extremely premature, because we have just started on a very long road of research that we must traverse before we will know what is technically practical, what real-world applications nanotechnology might actually have, and what the appropriate societal responses would be to any hazards that might be directly or indirectly related to these applications. Some writers in the popular press seem to treat nanotechnology as a new kind of Frankenstein's Monster, but it is worth remembering that biological science is still not able to create such a monster, nearly two centuries after Mary Shelly imagined one.

A good starting point for consideration of social and ethical issues is a formal definition of the topic, but unfortunately a somewhat different definition has lodged in the public mind from that employed by professionals working in the field. For example, *Merriam-Webster's Collegiate Dictionary* defines *nanotechnology* as: "the art of manipulating materials on an atomic or molecular scale especially to build microscopic devices (as robots)." Similarly, the online reference About.com defines it thus: "The development and use of devices that have a size of only a few nanometres." Later, we will examine public conceptions more closely, but it is worth noting here that these definitions focus on robot-like devices constructed at the nanoscale and ignore the topics of most chapters in this volume.

A more authoritative if long-winded definition of *nanotechnology* has been provided by the U.S. Government's National Nanotechnology Initiative: "Research and technology development at the atomic, molecular or macromolecular levels, in the length scale of approximately 1–100 nm range, to provide a fundamental understanding of phenomena and materials at the nanoscale and to create and use structures, devices and systems that have novel properties and functions because of their small and/or intermediate size. The novel and differentiating properties and functions are developed at a critical length scale of matter typically under 100 nm. Nanotechnology research and development includes manipulation under control of the nanoscale structures and their integration into larger material components, systems and architectures."

While this definition includes the word "devices," it also covers materials, structures, and systems. Significantly, it envisions that many nanotechnology advances will be integrated into "larger material components, systems and architectures," which means that we would need to understand the social and ethical implications of the larger-scale systems of which nanoscale structures are only components.

There is good reason to believe that engineering at the nanoscale will have very broad applications across all fields of technology and most fields of science [38.5–7]. Therefore, it will be an enabler or amplifier of the effects of other technologies. This means that it is foolish to ask, "Is nanotechnology harmful?" Nanotechnology is not one thing, either ethical or unethical. Rather, it is a myriad of different things – many nanotechnologies – each of which has a range of potential uses and misuses in conjunction with other technologies and applied to different goals.

This line of reasoning could lead some people to conclude that we did not need to worry at all about the social and ethical implications of nanotechnology, because they will be subsumed by the implications of larger systems. Others might conclude that it is far too early in the history of nanoscale engineering to consider societal implications, because we cannot study the impact of a technology that does not yet exist.

However, the very fact that some writers are urging regulation or banning of nanotechnology suggests that the debate about its value to humanity has begun, and we need to bring the best possible thinking and information to that debate [38.8–10]. In addition, a few relatively simple but significant applications have begun to appear, so in those cases at least an impact can be observed. Social scientists, natural scientists, philosophers, and engineers have indeed begun considering the future of nanotechnology, so we can draw upon the very first serious intellectual efforts in this important field.

38.1 Applications and Societal Impacts

In the year 2000, the National Science and Technology Council sponsored a major workshop at the National Science Foundation, which led to a published report, *Societal Implications of Nanoscience and Nanotechnology* [38.11]. Sixty-four representatives of academic science, government laboratories, and corporate research listed a large number of potential nanotechnology applications, a few of which had already entered production, but many of which could be expected only after decades of research. Following are diverse examples of possible applications suggested by workshop participants concerning a variety of sectors of the economy and spheres of life:

- More efficient components for the semiconductor industry such as integrated circuits containing transistors constructed from carbon nanotubes.

- Nanostructured catalysts for the chemical industry and for more effective converters to handle pollution from automobile exhausts and other combustion.
- Lighter and stronger nanomaterials in bulk quantities to enable safer and more efficient transportation vehicles, including automobiles, aircraft, and train systems.
- Improved pharmaceuticals with features such as programmed delivery to desired targets like tumors and the ability to employ substances that are not soluble in water as medications.
- Cost-effective and reliable filters for water decontamination, desalination, containment of industrial pollutants, and air purification.
- More efficient solar energy conversion, thereby reducing current reliance on oil and offering an alternative to nuclear power for future electricity needs.
- Efficient fuel cells and hydrogen storage systems, leading to nonpolluting automobiles, trucks, and buses.
- More durable composite materials, such as nanoparticle-reinforced polymers, designed for optimal performance in specific uses with reduced waste and greatly increased lifetime.
- Molecularly engineered biodegradable fertilizers and insecticides designed for precision agriculture with efficient delivery to where they are needed and prevention of unwanted side effects.
- Revolutionary launch vehicles with high mass-ratios, nonpolluting but powerful engines, and low electric power needs in order to finally realize the promise of space exploration.
- Nanoscale components in sensor systems that can quickly detect and identify pollutants and disease organisms, as well as chemical or biological warfare agents to allow quick and appropriate medical treatment or security responses.
- Coatings that would give ordinary materials extraordinary properties like self-cleaning window glass and heat-reflecting exterior architectural surfaces.

Knowledgeable scientists and engineers reported that these and many other similar applications were plausible benefits of nanotechnology research, but they recognized that progress has costs. For example, the creation of a new industry often results in the decline or complete destruction of an older industry, as in the proverbial case of buggy-whip manufacturers who were rendered obsolete by the automobile. The potential benefit from a new technology must be calculated in a sophisticated manner, taking into account the full life-cycle costs [38.12]. For example, an electric car may reduce some forms of pollution while increasing others such as the release of lead or other dangerous substances into the environment during the manufacture of the electric battery and the disposal of it when the car is eventually junked.

Many ethical issues of fairness will arise within nanotechnology industries themselves. For example, it will be important to invest in appropriate safeguards for workers engaged in hazardous production processes. In general, scientific discoveries cannot be patented, whereas new products and production methods can, so there will be serious issues of intellectual property protection in nanoscience, a nontraditional field where the legal line between discovery and invention has not yet been clearly drawn. Often, it will be necessary to bring together information and expertise from several different sources in order to create an economically profitable application, drawing upon government-funded university research projects, as well as several specialized industrial corporations, which will require improved models of collaboration across organizations.

As the *Societal Implications* report observed, the general public has a significant stake in the National Nanotechnology Initiative (NNI) that must be managed both directly through public participation and indirectly through the involvement of social scientists:

> It is important to include a wide range of interests, values, and perspectives in the overall decision process that charts the future development of nanotechnology. Involvement of members of the public or their representatives has the added benefit of respecting their interests and enlisting their support. ... The inclusion of social scientists and humanistic scholars, such as philosophers of ethics, in the social process of setting visions for nanotechnology is an important step for the NNI. As scientists or dedicated scholars in their own right, they can respect the professional integrity of nanoscientists and nanotechnologists, while contributing a fresh perspective. Given appropriate support, they could inform themselves deeply enough about a particular nanotechnology to have a well-grounded evaluation. At the same time, they are professionally trained representatives of the public interest and capable of functioning as communicators between nanotechnologists and the public or government officials. Their input may help maximize the societal benefits of the technology while reducing the possibility of debilitating public controversies [38.11].

Social scientific and economic research can help manufacturers and the government make the right decisions when deploying a new technology, maximizing its benefit for human beings. In addition, technically competent research on the societal implications of nanotechnology will help give policymakers and the general public a realistic picture free from unreasonable hopes or fears. As the report notes, the National Nanotechnology Initiative is just now commencing, so "there is a rare opportunity to integrate the societal studies and dialogues from the very beginning and to include societal studies as a core part of the NNI investment strategy" [38.11].

The workshop developed a number of recommendations for government, academia, and industry suggesting how the social and ethical implications of nanotechnology could best be addressed [38.11]. Here are some of the most important principles in education, social science, and ethics:

- Nanoscale concepts should be introduced into science and engineering education at all levels, thereby giving the widest possible range of students a fundamental understanding of the field while intellectually linking nanotechnology to many other fields.
- The training of nanotechnologists should include societal implications and ethical sensitivity, so their future work will be guided by principles that will maximize human benefit.
- A sufficient number and variety of social and economic scientists should receive effective multidisciplinary training, so they will be well prepared to work in the nanotechnology area.
- Formal measurement methods, such as social and economic indicators, should be developed and consistently employed to chart the actual widespread changes in industry, education, and public welfare as they occur.
- Government agencies, private foundations, and industry should support a wide range of high-quality, theory-based social and economic research studies on nanotechnology, examining both the decision processes that shape the emerging technology and itsspecific societal impacts once it is deployed.
- A knowledge base and institutional infrastructure must be created to evaluate the probable future intellectual and societal impacts of nanoscience and nanotechnology over the short, medium, and long term.
- Government and the private sector should establish effective channels for informing the public promptly about new concepts, projects, potential applications, and ethical issues as they emerge.
- There should be formal organizations and mechanisms to ensure participation by diverse societal institutions and the general public in setting priorities for research and development and in providing timely input to decision makers.
- Government and industry should develop management plans and policies that can effectively incorporate all relevant information and legitimate stakeholder interests to ensure that we can respond flexibly to social, ethical, legal, and economic implications as they appear.

It will take time and effort to implement these recommendations, but some progress has already been made. The National Science Foundation has recently funded Rosalyn Berne at the University of Virginia to study the developing ethics of nanotechnology through a narrative analysis of discourse from engineers and scientists in the field, and a team headed by Michael Gorman at the same institution was funded to explore the potential for multidisciplinary research on social and ethical dimensions of nanotechnology. The Environmental Protection Agency, which has supported several research projects on how nanotechnology might remedy conventional forms of pollution, has funded Darrell Velegol and Kristen Fichthorn at Pennsylvania State University to study "Green Engineering of Dispersed Nanoparticles," exploring how valuable nanoparticles might be produced without the use of polluting additives. In its review of the National Nanotechnology Initiative, the National Research Council has recommended the development of "a new funding strategy to ensure that the societal implications of nanoscale science and technology become an integral and vital component of the NNI" [38.13].

38.2 Technological Convergence

In the year following the *Societal Implications* workshop, NSF and the U.S. Department of Commerce staged an even larger meeting to examine the potential for *nanotechnology* to combine with three other powerful scientific and engineering trends: *biotechnology*, *information technology*, and *cognitive science* – four "NBIC" fields. The resulting report, *Converging Technologies for Improving Human Performance* [38.14], examined benefits of technological convergence that would specifically enhance the ability of humans to satisfy their needs and achieve their legitimate goals.

Convergence of diverse NBIC technologies will be based on the material unity of nature at the nanoscale and on technology integration from that scale. This is the scale at which much future biotechnology and the hardware for information technology will operate, and it is the scale from which fundamental research on the human nervous system will take place. Key transforming tools, including research instruments and production methods, will arise in the conjunction of previously separated fields of science. Drawing upon advances in mathematics and computation, it will be possible to model phenomena from the nanoscale to the cosmic scale as complex, coupled, hierarchical systems.

The *Converging Technologies* scientists, engineers, and policymakers sketched six general areas in which the human benefits and social issues are likely to result:

- Overall potential of converging technologies
- Expanding human cognition and communication
- Improving human health and physical capabilities
- Enhancing group and societal outcomes
- National security
- Unifying science and education

Numerous valuable applications were sketched in greater or lesser detail that could reasonably be expected to result from a decade or two of vigorous research and development, including the following ideas that depend on partnerships between nanotechnology and one or more of the other NBIC fields:

- Comfortable, wearable sensors and computers will enhance every person's awareness of his or her health condition, the environment, chemical pollutants, potential hazards, and information of interest about local businesses, natural resources, and the like.
- The human body will be more durable, healthier, more energetic, easier to repair, and more resistant to many kinds of stress, biological threats, and aging processes.
- Machines and structures of all kinds, from homes to aircraft, will be constructed of materials that have exactly the desired properties, including the ability to adapt to changing situations, high energy efficiency, and environmental friendliness.
- A combination of technologies and treatments will compensate for many physical and mental disabilities and will eradicate altogether some handicaps that have plagued the lives of millions of people.
- National security will be greatly strengthened by lightweight, information-rich war fighting systems, capable uninhabited combat vehicles, adaptable smart materials, invulnerable data networks, superior intelligence-gathering systems, and effective measures against biological, chemical, radiological, and nuclear attacks.
- The ability to control the genetics of humans, animals, and agricultural plants will greatly benefit human welfare; widespread consensus about ethical, legal, and moral issues will be built in the process.
- Agriculture and the food industry will greatly increase yields and reduce spoilage through networks of cheap, smart sensors that constantly monitor the condition and needs of plants, animals, and farm products.
- Transportation will be safe, cheap, and fast due to ubiquitous real-time information systems, extremely high-efficiency vehicle designs, and the use of synthetic materials and machines fabricated from the nanoscale for optimum performance.
- The work of scientists will be revolutionized by importing approaches pioneered in other sciences, for example, genetic research employing principles from natural language processing and cultural research employing principles from genetics.
- Formal education will be transformed by a unified but diverse curriculum based on a comprehensive, hierarchical intellectual paradigm for understanding the architecture of the physical world from the nanoscale through the cosmic scale.

Several of these application areas depend upon the development of nanoscale components for sensor, computation, and communication networks. Others involve a synergy between inorganic nanoscale engineering and organic chemistry or biology. Others extrapolate that it will be possible to build large-scale structures that are engineered at the nanoscale to enhance the qualities of architecture, transportation vehicles, or even the human body. Fundamental to all of the application areas is a conceptual revolution that unifies science and provides humanity with a comprehensive view of nature and technology.

Contributors to the report showed how nanoscale science and engineering were essential for progress in biology, information science, and cognitive science. The result of NBIC convergence will be the strengthening of science and the empowerment of humanity, as we will find solutions to many of humanity's problems in health, natural resources, environment, security, and even communication with fellow human beings. Often, the technologies based in the sister sciences will raise ethical issues. In such cases, nanotechnology may become implicated in controversies without being central to them. The correct policy response would then be to ignore the superficial nanotechnology aspects and focus directly on the core problem that lies in a different field. *Converging Technologies* contributors examined how technological convergence highlighted such existing issues as the treatment of the disabled, communication breakdowns, economic stagnation, and threats to national security.

For greater clarity, we can consider here the very specific example of the proliferation of nuclear weapons. Nuclear physics and atomic technology were specifically not included in the constellation of NBIC converging technologies considered by the workshop, so the following discussion supplements its findings. As scientists already understood in the early 1940s, there are fundamentally two ways that a rogue nation can build up its own supply of weapons-grade nuclear materials. One is breeding plutonium in a nuclear reactor fueled by uranium, transmuting the common uranium isotope of atomic weight 238. But this is hard to conceal from other nations, especially if reactors are open to international inspection. The other way is separation of the relatively rare uranium isotope ^{235}U, which, like plutonium, is suitable for use in weapons. The original separation method was gas diffusion, but then the centrifuge method was developed, and other approaches are possible such as laser or electromagnetic separation.

In the gas diffusion method, a chemical containing uranium (uranium hexaflouride) is passed through a porous barrier such as a ceramic wall that has tiny holes in it. Molecules containing ^{235}U travel through the barrier slightly faster than molecules containing ^{238}U. This is a very inefficient process, because the isotopes are very similar to each other, so many passes through a barrier are required to enrich the proportion of ^{235}U sufficiently for use in a nuclear weapon. The other approach uses centrifuges, rapidly spinning rotors containing the uranium hexaflouride to separate the two isotopes on the basis of their slight difference in mass. Both approaches are very difficult and costly, however. Conceivably, nanotechnology could slightly reduce the difficulty, by producing porous barriers with exactly the optimum sized holes and by providing lighter and stronger centrifuge rotors that could operate at higher speeds.

Therefore, without a more careful analysis someone might leap to the conclusion that nanotechnology is a danger because it facilitates proliferation of nuclear weapons. This ignores several crucial facts. First, the world already faces a very severe crisis of nuclear proliferation without nanotechnology. Second, the fundamental requirement for production of nuclear weapons is the availability of uranium, so the most logical preventive would be international monitoring and control over its mining and distribution. Third, there are many other technical requirements for production and delivery of nuclear weapons, and it is the whole collection of them that presents a threat, not any one alone. Fourth, nanotechnology is inherently nonnuclear, because nuclear reactions occur at a smaller scale than the nanoscale, so its connections to proliferation of nuclear weapons will always be highly indirect if they exist at all. Finally, the fundamental problem is not nuclear proliferation, but weapons of mass destruction in general, and nanotechnology has a major role to play in the defense against chemical and biological warfare agents in sensor systems that can detect them before they have caused harm.

Perhaps a very few nano products will need to be "controlled substances," analogous to narcotics, insecticides, asbestos, and antibiotics [38.15]. Conceivably, porous barriers with exactly the right pore size for uranium isotope separation would be controlled in some way. Some products could be produced and used under strict regulations, whereas a few others might be banned altogether. Most would not require restrictions of any kind.

Such decisions would need to be based on the careful analysis of facts related to the specific issue at hand. For example, it may be that no improvements in gas diffusion separation are possible through nanotechnology, or they are so slight as to be negligible. Perhaps technological developments in laser or electromagnetic separation, completely unrelated to nanotechnology, may render both the gas diffusion and centrifuge methods obsolete. Therefore, the mere fact that we can imagine a way in which nanotechnology might possibly encourage the proliferation of nuclear weapons is far too weak a justification for international regulation. We would need to go beyond vague fears to precise analysis based on rigorous scientific research to distinguish real dangers from imaginary ones and design appropriate safeguards in those rare cases when they are necessary.

The costs of excessive regulation would be extremely high, harming the very human beings it was intended to protect. Porous barriers precisely designed at the nanoscale are among the high benefit applications we can anticipate early in the history of nanotechnology. They would have a wide range of applications, from water purification and desalination to separation of valuable substances in biotechnology production processes. Similarly, new high-performance materials with great strength but low weight will have a myriad of applications throughout the civilian sectors of the economy. Among the products that would benefit millions of ordinary people are light but indestructible eyeglass frames, safety helmets, and hand tools.

Nanotechnology has the potential to increase the effectiveness of biotechnology, information technology, and technologies based on cognitive science through NBIC convergence. Therefore, it could become entangled in ethical issues that already involve those other technologies such as the conflict over genetically modified foods, the dispute over information privacy, and the debate over treatment of mental illness. These are not fundamentally nanotechnology issues, however, and it will be important to keep a proper perspective on their real nature.

38.3 Major Socio-technical Trends

The significance of nanotechnology depends not only on its own accomplishments and on convergence with other technologies, but also on how these purely technical developments relate to wider trends going on in the world [38.16, 17]. At present, many of these trends are confused and are the subject of great controversy in the social science and public policy arenas. It is always hazardous to extrapolate trends from any particular moment in history, but the present time appears so chaotic that the events of a single day might send humanity down a very different road.

A few years ago, there was a broad consensus that the rapid growth in the number of people on the planet, the so-called *population explosion*, was the dominant quantitative trend that needed to be factored into any projections [38.18]. But by the mid-1980s, it became obvious that population growth had been replaced by the threat of population decline in most advanced industrial nations, with the possible exception of the United States [38.19, 20]. Even before collapsing fertility begins to reduce the number of people in the affected nations, their average age and pension costs increase, social dynamism and creativity decrease, and governments become gridlocked trying to deal with ultimately insoluble problems of funding demanded services. This is one of the explicit reasons why the United States has begun to discount the influence of its European allies and Japan [38.21, 22].

Although nanotechnology has little obvious direct connection with fertility, there are two potentially counterbalancing, indirect connections. Rapid progress in nanotechnology could revitalize economic expansion and technical creativity, thereby to some extent offsetting the negative consequences of fertility collapse. However, the aversion to change that marks societies having a high proportion of elderly citizens may dampen the intellectual inquisitiveness and investment risk-taking required to develop nanotechnology and technological convergence.

A related trend is the constant improvement in health resulting from progress in medical science and the economic growth needed to fund increasingly costly health care. Continued progress in health is by no means assured, however. During the nineteenth century, American health may actually have declined significantly for a number of decades, despite economic growth [38.23]. We tend to attribute the undeniable improvement in health over the twentieth century to medical progress, but health education and public sanitation may have been more important [38.24]. The U.S. Centers for Disease Control argue that substantial improvements in health and longevity could be achieved by lifestyle changes, notably, exercising more, reducing the fats in our diet, and avoiding smoking [38.25, 26]. The introduction of antibiotics helped increase longevity, and modern cardiology saves the lives of many people who might otherwise die prematurely of heart attacks, but the progress against cancer, AIDS, and most aging-related illnesses has been agonizingly slow. On balance, economists find that the increasing investments in heath care are paying off, but not necessarily in all areas, or with very great benefit [38.27–29].

There are several reasons to be pessimistic about the future benefits of medicine, unless there are very major scientific breakthroughs. We have yet to find a cure for any chronic viral infections, of which AIDS is merely the most publicized example. Bacteria are rapidly evolving resistance to antibiotics, yet the rate of development of new prescription medicines is declining.

The average white American male born in 1900 could expect to live 48 years, but in 2000 this life expectancy had increased to 74 years. For white females, the average life expectancy increased from 51 to 80. Projecting these figures forward at the same rate of increase suggests that life expectancy in 2100 might be 114 for males and 125 for females. However, using more realistic assumptions, the U.S. Census Bureau has projected that life expectancy for Americans born in the year 2100 might be only around 88 for males and 92 for females [38.30]. This does represent progress, but at a steeply declining rate, and it depends upon continued economic growth.

Perhaps the most economically significant technical trend of recent years has been *Moore's Law*, Gordon Moore's observation that the density of transistors on the most advanced microchip doubles about every 18 months. The rapid improvement of computer and communications hardware has been fundamental to the implementation of new software applications, networking, and information technology in general. These, in turn, have been responsible for much economic growth. However, the semiconductor industry is approaching physical limits in the traditional methods of making integrated circuits, including computer chips, so Moore's Law will stall without breakthroughs in nanoscale technologies such as molecular logic gates and carbon nanotube transistors [38.31–33].

Arguably, several of the great thrusts of twentieth-century technology have already stalled. Except for the improved capacity of scientific instruments to collect and process astronomical data, the space program has hardly advanced since the mid-1970s. Civilian aviation and ground transportation have improved only in relatively minor design details since the 1950s. Despite great efforts, little progress has been made in controlling nuclear fusion for power generation. Nuclear fission power generation has largely been halted by public controversies, and no technological breakthroughs in safe operation of reactors or disposal of radioactive wastes have assuaged the public's concerns. The social sciences seem no better able to define good public policies than they were decades ago. Outside of nanotechnology itself, the two main areas where rapid progress obviously continues are information technology and genetic engineering. Convergence with nanotechnology is essential for further progress in both of these areas.

What would happen to the world economy if technological progress slowed or even halted? First, the advanced industrial nations would lose the advantage they currently hold over developing countries. For example, the American semiconductor industry would probably collapse if it lost its technological superiority, because nations with lower wage rates could produce the same things more cheaply. Technological products, from computer chips to pharmaceuticals, would become commodities, manufactured wherever it was cheapest to do so. With their heavy pension burdens and costly social services, governments of advanced nations would face fiscal crises of unprecedented magnitude. Indeed, the economies of these nations could fall rapidly down to the average of all nations, unleashing social discontent that could lead to unpredictable violence.

At the same time, some developing nations would advance economically, which would mean increased net world industrial production with concomitant increase in environmental pollution and resource depletion. Without technological progress to solve these problems, the average human welfare could decline, rather than remaining stable. In order to improve human welfare, technological progress must continue. Perhaps the best way to achieve that is through nanoscience and nanotechnology in convergence with the other NBIC fields, gaining understanding and control at the physical scale that is the basis for most science and technology.

It is impossible to predict how rapidly the conditions in the richest nations would deteriorate, or whether humanity could navigate through these crises to achieve a stable world. But clearly a halt in technological progress would be a shock to the world economy, global political institutions, and human welfare – which it would be wise to avoid.

38.4 Sources of Ethical Behavior

Ethical questions related to nanotechnology are not limited to the ways people might use it to harm others intentionally, but also include obligations to avoid potentially harmful unintended consequences [38.34]. An example of the former might be a weapon and of the latter, a kind of environmental pollution. In either case, the harm might be morally justified, as when a nation employs a weapon to defend against attack, or when limited pollution of the environment is offset by substantial benefits to humanity. Other ethical questions concern harm to the owners of nanotechnology, for example, if a competing company violates a nanotech patent, or if a government bans a nano-related product without careful examination of scientific evidence about its value. News reporters or popular writers who spread false information about nanotechnology may also be acting unethically, whether they are willfully lying or merely failing to be professionally diligent in checking their facts.

It is important to recognize that there are sins of omission as well as commission. To fail to develop a beneficial nanotechnology application could also be unethical. Imagine, for instance, that a laboratory had discovered how nanotechnology could enable a much more effective form of chemotherapy to cure cancer reliably and cheaply, but the company owning the patents prevented it from being developed, because its less effective traditional chemotherapy products were more profitable. This could be considered just as unethical as doing aggressive harm.

For over a century, social scientists have been studying the origins of ethical behavior – or its opposite, unethical behavior – in fields like the social psychology of groups, the sociology of deviance, and criminology. Although it is impossible to predict when a particular person will commit a specific ethical or unethical act, we can identify with good confidence the general factors that steer human behavior in one direction or another. An individual's behavior is determined by a complex of factors, few of which can be measured accurately, and chance seems to play a significant role in human affairs. Thus, the findings of social science tend to be statistical or conditional in nature. This does not render them useless, because one can develop rational laws, government policies, and investment strategies based on probabilities rather than certainties.

The theoretical and empirical scientific literature on the topic is vast, including literally thousands of publications dating back to the 1840s, expressing the views of countless schools of thought, often employing idiosyncratic terminology. For present purposes, we must distill that huge and incoherent intellectual heritage down to its most meaningful essentials. A good start is the following set of four themes, each of which can be expressed as a theory about why people might violate an ethical rule:

1. Learning – A person violates a rule because the person has learned it is rewarding to do so.
2. Strain – A person violates a rule because the person is frustrated in the attempt to fulfill one of the expectations of society.
3. Control – A person violates a rule because the person lacks stable social bonds that might prevent it.
4. Subculture – A person violates a rule because the person belongs to a group or network of other people who encourage violation.

When each of these ideas first appeared in the social-scientific literature, the theory's proponents tended to present it as the complete explanation for deviant behavior, but now we understand that each of these four has merit and the correct explanation is usually a mixture of two or more. We will illustrate each of the theories with hypothetical illustrations, because actual examples may not exist yet.

Learning Theory is sometimes interpreted in economic or "rational choice" terms, as meaning that a person will do whatever he or she perceives it is profitable to do. But human life is not spent entirely in a marketplace where one can compare prices and shop for a good deal. Rather, there are many quite different contexts in which people make decisions. Regardless of whether options can be quantified in dollars, people tend to rely upon their own past experiences in selecting between familiar courses of action [38.35–37].

If the managers of an industrial corporation have learned they can achieve higher profits by ignoring regulations against polluting the environment, and they have not suffered heavy fines from the government for doing so, they will be more likely to pollute in the future. Therefore, a corporation guilty of polluting with conventional technologies is apt to continue with this history of pollution when it adopts new nanotechnologies. This suggests that one way to reduce the likelihood that industrial companies will pollute when they adopt new nanotechnologies is to enact and enforce strong regulations against pollution in general, to make them learn to avoid polluting.

In Strain Theory, a person violates a norm because he or she is frustrated in the attempt to achieve one of the values endorsed by society [38.38, 39]. This idea has been used to explain why many members of poor communities or groups that have been subject to discrimination turn to crime to gain wealth and status, but it can also explain why a major corporation might begin to engage in dubious accounting practices when its growth in real business stalls. The central idea is that an individual or group becomes committed to a particular kind of success that society encourages – whether it is wealth, political power, widespread fame, or honor within a particular sector of society – then experiences frustration in attaining success, which motivates ethical violations.

Conceivably, a nanotechnology corporation could be affected by strain if it has committed itself to a new and untested line of products that turns out to be less effective, more harmful, or more costly than originally estimated. The corporation has committed not only its money and effort, but also its public prestige to success, but success may elude it. It may then unethically advertise its products as more effective than they really are, conceal knowledge of their danger to human health or the environment, and cook its financial records to deceive investors. Of course, many ordinary companies have done this in the past, and, ironically, they sometimes solve their problems before the public becomes aware of them, thereby becoming ethical and successful again (e.g., [38.40]).

When Merton proposed the most influential version of Strain Theory in 1938, he identified three modes of adaptation to three increasing levels of strain. If a person was unable to achieve society's values by following conventional norms, frustration might drive him or her to violate the norms in pursuit of the value, what Merton called *innovation*. If this did not work, he or she might substitute both new values and new norms for those of society, in what he called *rebellion*. But if this failed as well, the person might abandon values, as well as norms and fall into a demoralized state of inaction called *retreatism*. Retreatism is inconsequential, and successful rebellion tends to radicalize society, but innovation (in Merton's sense of the term) can promote scientific and technical progress. In other work, Merton [38.41, 42] was one of the founders of the sociology of science and technology, but he never followed through on the implications of Strain Theory for science, and his distinctive concept of innovation was chiefly used to explain criminal behavior by poor people who violated the law in order to the get the values (i. e., money) endorsed by capitalist society.

There are many examples in the history of science in which Strain Theory can be used to explain scientific and technological innovations that turned out to be beneficial, rather than unethical, and others where the result was a mixture of benefit and harm that presumably could have been managed to emphasize the benefit. In the seventeenth century, England led the world in scientific progress, perhaps in a nationalistic reaction to its inferiority in population and other resources relative to France and Spain [38.43]. In the late nineteenth and early twentieth centuries, Germany lacked access to as abundant a supply of natural resources as had the great colonial powers, Britain and France. This inferiority may have stimulated many advances in German chemistry such as Adolf Bayer's synthesis of indigo dye and the nitrogen fixation method developed by Fritz Haber and Karl Bosch, which allowed Germany to continue producing explosives for World War I after it had lost access to foreign sources of nitrates. When the Treaty of Versailles concluding World War I limited German long-range artillery, the Germans developed the V-2 rocket largely to get around this restriction, thereby ushering in the age of space flight [38.44, 45].

In Control Theory, a person violates norms because he or she lacks stable social bonds that might prevent it [38.46–48]. This is a very successful scientific explanation of many common crimes, such as ordinary theft, and many kinds of criminals tend to lack stable friendships or family relationships that might restrain their deviant behavior. Control Theory might seem irrelevant to nanotechnology ethics because no isolated individual is in a position to make major decisions about new applications. The theory perfectly fits the Hollywood image of Dr. Frankenstein living like a hermit in a remote castle, where he is free to conduct any ghastly experiment he wishes, but no real scientist lives like this.

However, a variant of Control Theory explains that a community or organization can suffer from a partial but pervasive form of isolation, called *social disorganization*, in which communication channels are poor and many people deviate to a moderate degree because they are somewhat detached from each other. In a socially disorganized research laboratory studying the properties of a new nanostructured material, the quality of the science could suffer, pieces of information could fail to get to scientists who need them, and management would get incorrect impressions about the material's properties. This could lead to a chain reaction of unethical behavior in which no one individual was primarily to blame. For instance, the result could be a corporate decision to put into production a new kind of aircraft component that was subject to unpredictable catastrophic failure, causing crashes and loss of life. In a case such as this, the unethical behavior is scattered in many small pieces, committed by many individuals, any time one of them presents a dubious observation as a scientific fact. The best way to reassert the truth-oriented professional norms of science would be to rebuild good channels of communication and cooperation, reattaching the researchers to each other and to the scientific community.

Subculture Theory is similar to Control Theory, but operates on the level of groups rather than individuals. A person violates a rule because he or she belongs to a subculture that rejects the rule and even encourages violation of it [38.49, 50]. A subculture is a group or network having a set of norms, roles, and values that differ from those of the surrounding culture. At the extreme, it can have a radical ideology or dogma, but practically every social group is a subculture to at least a very mild extent. Social psychologist *Janis* [38.51] has shown that even very powerful groups can become deviant subcultures, if they isolate themselves from the wider community, as the Nixon White House did during the Watergate episode.

Even if a research laboratory is functioning properly, the organization of which it is a part can make harmful decisions if the top management is a subculture practicing what Janis calls "groupthink," the refusal to accept information that conflicts with the group's beliefs and commitments. Sometimes this may result in harm to others, for instance, if a new nanotechnology-enabled treatment for cancer were put into production before safe methods for administering it to patients in hospital settings were worked out. Or the harm could fall on the organization itself, for example, if a semiconductor company decided to invest its future in molecular logic gates without fully assessing the potential of carbon nanotube transistors instead. In that case, we might say the company deserves what it gets, but there is the ethical issue of its responsibility to stockholders and employees, including to the scientists in the research laboratory who were ready to do the studies needed for a correct decision.

As the example of Strain Theory illustrates, sometimes strain has positive consequences, rather than negative ones. Similarly, Learning Theory explains why an individual or organization might develop habits of ethical innovation, and both Control Theory and Subculture Theory could show that a person or group may innovate best when somewhat detached from conventional thinking. Thus, the issue is not one of avoiding these four factors, but of finding the right levels of them to encourage healthy innovation without unleashing serious unethical behavior. This is the fundamental trade-off in any free society: How to liberate human creativity while restraining greed, immorality, and inhumanity.

Strain Theory has been used to explain the emergence of collective behavior and social movements [38.52], including reform movements such as the environmental movement or crusades to establish professional codes of ethics. Concern about the social and ethical implications of nanotechnology can, therefore, energize the development of appropriate institutions to manage it wisely. Learning Theory suggests that a rewarding market for beneficial nanotechnology products could teach scientists, engineers, and managers to invest their energy in projects that would serve the needs of humanity. Both Control Theory and Subculture Theory stress the importance of strong networks of communication in restraining unethical behavior. Convergence of nanotechnology with the other NBIC fields would support healthy communication by unifying science, providing a universal technical language for interaction across fields, and creating a network of collaborative relationships across sciences, institutions, and sectors of society.

38.5 Public Opinion

Ethics and social implications are largely matters of social perception, and the public conception of nanotechnology is still in the early stages of developing. Nanoscale science and engineering are evolving from conventional work in materials science, physics, chemistry, and related fields, as the gradual emergence of new methodologies permits many kinds of observation and experimentation with inorganic structures having at least one dimension less than about 100 nm. This progress in *real nanotechnology* is solidly based in existing scientific knowledge, even as it achieves new discoveries and inventions.

At the same time, something quite different has been emerging that also calls itself nanotechnology: a social movement based on metaphors, approximations, and hopes. *Tolles* [38.53] calls it an "irrational vision." In social science, popular enthusiasm for a loosely defined set of unreasonable hopes is often called a *mania* [38.54], so I will call this movement *nanomania*.

The key moment in the emergence of nanomania, but not of nanoscience and nanotechnology themselves, was the 1986 publication of *Engines of Creation* by Drexler. This is a popular book, inspirational rather than technical, filled with verbal metaphors about what might possibly be accomplished if nanoscale assemblers could be created, rather like industrial robot arms and assembly lines. The plausibility of Drexler's argument, such as it is, comes from his constant reminder that biological systems include "machines" that assemble themselves and manufacture or manipulate other nanoscale structures: proteins, RNA, and DNA.

In a rambling discourse that reminds us that Leonardo da Vinci could not have predicted the details of late twentieth century technology, Drexler asserts that some time in the future it will be possible to build nanoscale assemblers that are able to make *anything that can be designed*. At this point he has outstripped his biological analogy, because the "nanomachines" of biology cannot do this; they can make only those specific things that evolution has programmed them to make [38.55]. Drexler says that suitably programmed assemblers will be able to make copies of themselves, as living organisms do.

The specific image he offers is of tiny robot arms grabbing atoms from the surrounding raw material and putting them together in the correct configurations. These imaginary assemblers can make anything, he suggests, so they can quickly replace all the factories and construction companies on Earth, giving humankind perfect abundance of anything we might want. The book then scans quickly through other dreams of science fiction, including artificial intelligence, cheap space flight, near immortality, and the danger to human life of uncontrolled self-replicating intelligent machines.

In some ways the book is laudable, inspiring nontechnical readers with many conceivable applications of nanotechnology, some of which might actually turn out to be practical in the long run, discussing their social implications, and offering potential solutions to problems that might arise. The problem is that many people have come to the unfounded conclusion that Drexler's self-replicating universal assemblers are not only theoretically possible, but likely to be created in the near future. This is the fundamental error of nanomania, and it has called forth a reaction in the form of a *nanophobia* – the fear that all kinds of horrible evils will soon emerge from nanotechnology.

At the 2000 *Societal Implications* workshop, Drexler's assemblers were debunked by several participants, including Nobel prize-winner Smalley, discoverer of fullerenes and a pioneer of carbon nanotubes. *Smalley* observes that technical details of chemistry and a correct understanding of the properties of atoms and molecules are largely absent from nanomania [38.56]. In particular, the universal nanoscale assemblers envisioned by Drexler are physically impossible: "Because the fingers of a manipulator arm must themselves be made out of atoms, they have a certain irreducible size. There just isn't enough room in the nanometer-size reaction region to accommodate all the fingers of all the manipulators necessary to have complete control of the chemistry.... Manipulator fingers on the hypothetical self-replicating nanobot are not only too fat; they are also too sticky: the atoms of the manipulator hands will adhere to the atom that is being moved. So it will often be impossible to release this miniscule building block in precisely the right spot" [38.57].

The difficulty of developing reasonable public policies about nanotechnology when popular thinking is dominated by fanciful notions is illustrated by Chen's paper, "The Ethics of Nanotechnology", distributed online both by the Markkula Center for Applied Ethics at Santa Clara University and by BioScience Productions, Inc., which describes itself as "a non-partisan organization whose goal is the promotion of public literacy in the biosciences". Following Drexler's approach, *Chen* takes assumes that nanotechnology means the creation of molecular machines, "which could be used as molecular assemblers and disassemblers to build, repair, or tear down any physical or biological objects" [38.58]. Recognizing that such "nanites" may not be practical in the near future, Chen nonetheless suggests that we consider taking three possible actions soon:

1. Nanotechnology R&D should be banned
2. A nongovernmental regulatory or advisory commission should be set up
3. Adopt design guidelines
   - Nanomachines should only be specialized, not general purpose
   - Nanomachines should not be self-replicating

- Nanomachines should not be made to use an abundant natural compound as fuel
- Nanomachines should be tagged so they can be tracked

Chen notes that the first choice would undesirably prevent beneficial applications of molecular machines, but might not prevent "rogue researchers" from developing nanites anyway. However, readers of his article who are unfamiliar with real nanotechnology may not realize that most research in the field has nothing to do with these hypothetical nanites. It would take considerable wisdom to figure out how a regulatory commission might be set up, empowered, and managed effectively in such a way that it could ensure nanotechnology will be used for human benefit without stifling worthwhile innovations. The design guidelines flow from Chen's assumptions about what nanotechnology is, but any formal regulations embodying them could have serious negative unintended consequences, given that they start with a flawed definition of nanomachines.

Chen's rule that nanomachines be "specialized, not general purpose" may mean little, because a fully "general purpose" machine is as mythical as a "universal solvent" or "perpetual motion machine." However, a regulatory commission could do great harm by adopting poorly conceived definitions of *nanomachine* and *general purpose*. Suppose a very useful new abrasive were developed consisting of nanoparticles that changed shape, perhaps in response to temperature or force fluctuations, in some complex way that helped them do their intended job. *Webster's II New Riverside University Dictionary* defines *machine* thus: "a system, usually of rigid bodies, constructed and connected to change, transmit, and direct applied forces in a predetermined way to accomplish a particular objective, as performance." By this definition, the nanoscale abrasive particles would be machines. If they could be used to shape a wide range of industrial materials, they could be called general purpose. Yet this abrasive could be no more dangerous to humanity than ordinary sandpaper.

Strictly speaking, no entities on Earth are "self-replicating." Living organisms cannot replicate themselves without the help afforded them by the particular environment in which they dwell. All life is part of an ecology. Any regulation prohibiting self-replication would have to define very clearly what *self-replication* meant in the context of particular environments. Today, factory robots build parts of robots, but nobody worries about the ethical implications, perhaps because anybody can see that the manufacture process is entirely under the control of human beings. Because nanites are imaginary, and only vaguely defined, there are no facts to give nontechnical people comparable confidence.

Chen's rule against nanomachines that use "an abundant natural compound as fuel" presumably would prevent a nanite from rampaging uncontrolled in the natural environment. But this could outlaw the development of non-replicating nanoscale devices that can function only in very restricted laboratory or industrial systems, because most chemicals used in motors or batteries are abundant natural compounds.

Requiring that all nanoscale machines be "tagged so that they can be tracked" is reminiscent of the British Red Flag Act of 1865, which required that each horseless carriage be preceded by someone on foot waving a red flag. In the context of Chen's recommendations, *tracked* would seem to mean located, not merely identified once found. This could add an extremely heavy technical burden such as radioactive tagging, addition of a distinctive adulterating chemical, or even complex design features above the nanoscale to allow the nanomachine to be detected by microwave radio locators. If the machines are incapable of escaping into the natural environment, this burden would be not only ill-advised but ludicrous.

Has nanomania distorted the reception real nanotechnology is receiving? A survey done in 1999 found that the American business community was rather unaware of nanotechnology, only 2% saying they knew what it was and a further 2% reporting they had heard of it, but did not know what it was [38.59]. Two years later, a Web-based survey sponsored by the National Geographic Society found that science-attentive members of the general public are very enthusiastic about nanotechnology, and a rather large number of ideas about its benefits have already entered popular culture [38.60]. Statistical analysis of responses from 3,909 people revealed that they associate nanotechnology with other science-based technology, but do not connect it with pseudoscience. Significantly, when 598 people wrote paragraphs about nanotechnology, not a single respondent expressed concern about hypothetical dangers from self-reproducing nanites. Future surveys can examine the deepening popular awareness of nanotechnology, as the field itself progresses, and assess any public concerns that develop.

38.6 A Research Agenda

Sociologist *Etzkowitz* [38.61] has noted that the social sciences can play three different but mutually supportive roles in the development of nanotechnology:

1. Analyzing and contributing to the improvement of the processes of scientific discoveries that increasingly involve organizational issues where the social sciences have a long-term research and knowledge base.
2. Analyzing the effects of nanotechnology, whether positive or negative, expected or unintended, hypothetically and proactively and as they occur in real-time.
3. Evaluation of public and private programs to promote nanoscience and nanotechnology.

There are several ways social scientists can study the nanotechnology innovation process as it occurs [38.11, 62]. Cultural anthropologists and participant observation sociologists can enter nanoscience research teams, ethnographically documenting the behavior of the researchers over time as they frame their scientific problems, seek solutions, and labor collectively to understand the meaning of research results [38.63]. Fruitful sites for research on the innovation process include academe, industry, government laboratories, federal agencies, and professional societies.

Historical methods for collecting and analyzing written documents are also useful, even in cases where the documents are only days or weeks old, rather than the decades-old materials employed by most professional historians. Formal interviews, content analysis of communications such as e-mail messages, and questionnaires can provide more systematic kinds of data.

Application of a scientific idea to a technical problem, a technology transfer, and the introduction of products into the marketplace can be tracked through economic statistics on research and development investments, patent applications, and advertisements for new products and services. Some geographic areas, strata of society, and kinds of individuals and institutions experience technological change earlier than others, notably, the first adopters who try out an innovation before influencing other people to try it as well [38.64, 65]. Start-up companies are in a sense early adopters, even though they are also innovators, and thus they may be an especially fruitful if logistically difficult field for research.

There is a long tradition of systematic research on scientific publication, tracking the introduction of new concepts, charting the changing patterns of citation in scientific literature, and measuring the bulk of publications in different areas [38.66]. The progress of nanotechnology in modern culture could also be studied by tracking the introduction of new topics and courses in school and university curricula, mentions of nanoscale phenomena in the popular press, and the proliferation of both commercial and noncommercial Websites dedicated to nanotechnology. Professional societies and entrepreneurs are already creating series of new forums, symposia, journals, and job fairs oriented toward nanoscience and nanotechnology.

Nanotechnology may have significant impacts in all sectors of the economy, in most spheres of life, and on both short-term and long-term time scales. The intended consequences of any particular innovation can be determined simply by interviewing its investors or promoters and by inspecting their publications about it. Unintended consequences are much harder to study, because they require not only extensive research on the impacts on diverse areas of society, but also good scientific intuition about where to look for them. Consequences have consequences, in a continuing cascade, so the second-order consequences will also have to be studied [38.67]. Ultimately, this would require an extensive, vigorous interdisciplinary scientific community that is dedicated to research on social and technological change in general.

Technology is an elaborate system, embedded in the even more elaborate system that is global society. Through feedback among complex social subsystems, major phenomena can produce chaotic and unpredictable effects. Nanotechnology will have substantial impacts on many aspects of society in often different ways, so predictions of its influence will be difficult to make and evaluate scientifically. This means that researchers will have to invest considerable sophistication and effort in their work.

The social and ethical implications of nanotechnology are the results not merely of what the technology itself does, but also of how people react to it. Therefore, it will be important to understand the social acceptance, resistance, or rejection of nanotechnology at different times and in different places and human contexts. Interviews, focus groups, and questionnaire surveys can measure emotional, cognitive, and psychosocial parameters [38.68]. The combined methodologies of political science, sociology, and socio-legal studies will be required to chart the process of regulatory review and approval, court decisions that actively permit or prohibit the use of the technology, mobilization of political support and opposition, and the activities of relevant social movements.

Ultimately, the test of the various nanotechnologies will be their benefit for human beings, as measured by economic growth, improved health and longevity, environmental protection, strengthened security, social vitality, and enhanced human capabilities. Convergence with other NBIC technologies will complicate the challenge for researchers and policymakers who wish to understand the social and ethical implications of nanotechnology, but it will also magnify the potential benefits to humanity many times over. Success will be most complete and most probable if scientific research on these implications is an integral part of the effort from the very beginning.

References

38.1 B. Joy: Why the future doesn't need us, Wired **8**(4) (April 2000) 238–262

38.2 B.J. Feder: Opposition to nanotechnology, New York Times (August 19 2002) (http://www.nytimes.com/)

38.3 G. H. Reynolds: *Forward to the Future: Nanotechnology and Regulatory Policy* (Pacific Research Institute, San Francisco 2002)

38.4 J. S. Brown, P. Duguid: Don't count society out: A response to Bill Joy. In: *Societal Implications of Nanoscience and Nanotechnology*, ed. by M. Roco, W. S. Bainbridge (Kluwer, Dordrecht 2001) pp. 37–45

38.5 R. W. Siegel, E. Hu, M. C. Roco (Eds.): *Nanostructure Science and Technology* (Kluwer, Dordrecht 1999)

38.6 M. C. Roco, R. S. Williams, P. Alivisatos (Eds.): *Nanotechnology Research Directions* (Kluwer, Dordrecht 2000)

38.7 M. Roco, R. Tomellini: *Nanotechnology: Revolutionary Opportunities and Societal Implications* (Office for Official Publications of the European Communities, Luxembourg 2002)

38.8 M. S. Meyer: Socio-economic research on nanoscale science and technology: A european overview and illustration. In: *Societal Implications of Nanoscience and Nanotechnology*, ed. by M. Roco, W. S. Bainbridge (Kluwer, Dordrecht 2001) pp. 278–311

38.9 R. H. Smith: Social, ethical, and legal implications of nanotechnology. In: *Societal Implications of Nanoscience and Nanotechnology*, ed. by M. C. Roco, W. S. Bainbridge (Kluwer, Dordrecht 2001) pp. 257–271

38.10 V. Weil: Ethical issues in nanotechnology. In: *Societal Implications of Nanoscience and Nanotechnology*, ed. by M. C. Roco, W. S. Bainbridge (Kluwer, Dordrecht 2001) pp. 244–251

38.11 M. C. Roco, W. S. Bainbridge (Eds.): *Societal Implications of Nanoscience and Nanotechnology* (Kluwer, Dordrecht 2001)

38.12 L. B. Lave: Lifecycle/sustainability implications of nanotechnology. In: *Societal Implications of Nanoscience and Nanotechnology*, ed. by M. Roco, W. S. Bainbridge (Kluwer, Dordrecht 2001) pp. 205–212

38.13 National Research Council: *Small Wonders, Endless Frontiers: A Review of the National Nanotechnology Initiative* (National Academy Press, Washington D.C. 2002)

38.14 M. C. Roco, W. S. Bainbridge (Eds.): *Converging Technologies to Improve Human Performance* (Kluwer, Dordrecht 2003)

38.15 M. C. Suchman: Envisioning life on the nanofrontier. In: *Societal Implications of Nanoscience and Nanotechnology*, ed. by M. C. Roco, W. S. Bainbridge (Kluwer, Dordrecht 2001) pp. 271–276

38.16 N. Gingrich: The age of transition. In: *Societal Implications of Nanoscience and Nanotechnology*, ed. by M. Roco, W. S. Bainbridge (Kluwer, Dordrecht 2001) pp. 29–35

38.17 G. Yonas, S.T. Picraux: National needs drivers for nanotechnology. In: *Societal Implications of Nanoscience and Nanotechnology*, ed. by M. C. Roco, W. S. Bainbridge (Kluwer, Dordrecht 2001) pp. 46–55

38.18 P. R. Ehrlich: *The Population Bomb* (Ballantine, New York 1968)

38.19 K. Davis, M. S. Bernstam, R. Ricardo-Campbell (Eds.): *Below-Replacement Fertility in Industrial Societies: Causes, Consequences, Policies (A supplement to Population and Development Review)* 1986)

38.20 B. J. Wattenberg: *The Birth Dearth* (Ballantine, New York 1987)

38.21 Central Intelligence Agency: *Long-Term Global Demographic Trends: Reshaping the Geopolitical Landscape* (Central Intelligence Agency, Washington D.C. 2001)

38.22 Demography and the west, Economist **364**(8287) (August 2002) 20–22

38.23 D. L. Costa, R. H. Steckel: Long-term trends in health, welfare, and economic growth in the United States. In: *Health and Welfare During Industrialization*, ed. by R. H. Steckel, R. Floud (Univ. Chicago Press, Chicago 1997) pp. 47–89

38.24 S. H. Preston: *American Longevity: Past, Present, and Future* (Center for Policy Research, Syracuse University, Syracuse 1996)
38.25 HHS (U.S. Department of Health and Human Services): *Chronic Diseases and Their Risk Factors: The Nation's Leading Causes of Death* (Centers for Disease Control, Atlanta 1999)
38.26 HHS (U.S. Department of Health and Human Services): *Reducing Tobacco Use: A Report of the Surgeon General* (Centers for Disease Control, Atlanta 2000)
38.27 D. M. Cutler, M. McClellan, J. P. Newhouse, D. Remler: Are medical prices declining, Quarterly J. Economics **113** (1998) 991–1024
38.28 D. M. Cutler, E. Richardson: Your money and your life: The value of health care and what affects it. In: *Frontiers in Health Policy Research*, ed. by A. M. Garber (MIT, Cambridge 1999) pp. 99–132
38.29 V. R. Fuchs: The future of health economics, J. Health Economics **19** (2000) 141–157
38.30 F. W. Hollmann, T. J. Mulder, J. E. Kallan: *Methodology and Assumptions for the Population Projections of the United States: 1999 to 2100* (United States Census Bureau, Washington D.C. 2000)
38.31 R. Doering: Social implications of scaling to nanoelectronics. In: *Social Implications of Nanoscience and Nanotechnology*, ed. by M. C. Roco, W. S. Bainbridge (Kluwer, Dordrecht 2001) pp. 84–93
38.32 R. S. Williams, P. J. Kuekes: W've only just began. In: *Social Implications of Nanoscience and Nanotechnology*, ed. by M. C. Roco, W. S. Bainbridge (Kluwer, Dordrecht 2001) pp. 103–108
38.33 Central Intelligence Agency: *Experts See Risks to US IT Superiority Beyond Moore's Law*, report OTI IA 2002-151 (Central Intelligence Agency, Directorate of Intelligence, Washington D.C. 2002) pp. 2002–151
38.34 E. Tenner: Nanotechnology and unintended consequences. In: *Societal Implications of Nanoscience and Nanotechnology*, ed. by M. C. Roco, W. S. Bainbridge (Kluwer, Dordrecht 2001) pp. 311–318
38.35 G. C. Homans: *Social Behavior: Its Elementary Forms* (Harcourt Brace Jovanovich, New York 1974)
38.36 R. L. Akers: *Deviant Behavior: A Social Learning Approach* (Wadsworth, Belmont 1985)
38.37 R. Stark, W. S. Bainbridge: *A Theory of Religion* (Rutgers University Press, New Brunswick 1987)
38.38 R. K. Merton: *Social structure and anomie* (Free Press, New York 1968) pp. 185–214, first published 1938
38.39 R. A. Cloward, L. E. Ohlin: *Delinquency and Opportunity* (Free Press, New York 1960)
38.40 A. P. Sloan: *My Years with General Motors* (Doubleday, Garden City 1964)
38.41 R. K. Merton: *Science, Technology and Society in Seventeenth Century England* (H. Fertig, New York 1970)
38.42 R. K. Merton: *The Sociology of Science: Theoretical and Empirical Investigations* (University of Chicago Press, Chicago 1973)
38.43 L. Greenfeld: *Nationalism: Five Roads to Modernity* (Harvard Univ. Press, Cambridge 1992)
38.44 W. S. Bainbridge: *The Spaceflight Revolution* (Wiley, New York 1976)
38.45 W. S. Bainbridge: *Goals in Space: American Values and the Future of Technology* (State Univ. New York Press, Albany 1991)
38.46 T. Hirschi: *Causes of Delinquency* (University of California Press, Berkeley 1969)
38.47 C. Shaw, H. D. McKay: *Delinquency Areas* (University of Chicago Press, Chicago 1929)
38.48 M. R. Gottfredson, T. Hirschi: *A General Theory of Crime* (Stanford Univ. Press, Stanford 1990)
38.49 E. H. Sutherland: *Principles of Criminology* (Lippincott, Philadelphia 1947)
38.50 G. Sykes, D. Matza: Techniques of neutralization: A theory of delinquency, Am. Sociol. Rev. **22** (1957) 664–670
38.51 I. L. Janis: *Groupthink: Psychological Studies of Policy Decisions and Fiascoes* (Houghton Mifflin, Boston 1982)
38.52 N. J. Smelser: *Theory of Collective Behavior* (Free Press, New York 1962)
38.53 W. M. Tolles: National security aspects of nanotechnology. In: *Societal Implications of Nanoscience and Nanotechnology*, ed. by M. C. Roco, W. S. Bainbridge (Kluwer, Dordrecht 2001) pp. 218–237
38.54 W. S. Bainbridge: Collective behavior and social movements. In: *Sociology*, ed. by R. Stark (Wadsworth, Belmont 1985) pp. 493–523
38.55 V. Vogel: Societal impacts of nanotechnology in education and medicine. In: *Societal Implications of Nanoscience and Nanotechnology*, ed. by M. C. Roco, W. S. Bainbridge (Kluwer, Dordrecht 2001) pp. 180–186
38.56 R. F. Service: Is nanotechnology dangerous?, Science **290** (2000) 1526–1527
38.57 R. E. Smalley: Of chemistry, love and nanobots, Sci. Am. **285** (2001) 76–77
38.58 A. Chen: *The Ethics of Nanotechnology* (BioScience Productions and Markkula Center for Applied Ethics, Santa Clara University, 2002) www.actionbioscience.org/newfrontiers/chen.html and www.scu.edu/ethics/publications/submitted/chen/nanotechnology.html
38.59 J. Canton: The strategic impact of nanotechnology on the future of business and economics. In: *Societal Implications of Nanoscience and Nanotechnology*, ed. by M. Roco, W. S. Bainbridge (Kluwer, Dordrecht 2001) pp. 114–121
38.60 W. S. Bainbridge: Public attitudes toward nanotechnology, J. Nanopart. Res. **4** (2002) 561–570

38.61 H. Etzkowitz: Nano-science and society: Finding a social basis for science policy. In: *Societal Implications of Nanoscience and Nanotechnology*, ed. by M. C. Roco, W. S. Bainbridge (Kluwer, Dordrecht 2001) pp. 121–128

38.62 J. S. Carroll: Social science research methods for assessing societal implications of nanotechnology. In: *Societal Implications of Nanoscience and Nanotechnology*, ed. by M. C. Roco, W. S. Bainbridge (Kluwer, Dordrecht 2001) pp. 238–244

38.63 B. A. Nardi: A cultural ecology of nanotechnology. In: *Societal Implications of Nanoscience and Nanotechnology*, ed. by M. C. Roco, W. S. Bainbridge (Kluwer, Dordrecht 2001) pp. 318–324

38.64 E. Katz, P. F. Lazarsfeld: *Personal Influence* (Free Press, Glencoe 1955)

38.65 E. M. Rogers: *Social Change in Rural Society* (Appleton-Century-Crofts, New York 1960)

38.66 D. Crane: *Invisible Colleges: Diffusion of Knowledge in Scientific Communities* (University of Chicago Press, Chicago 1972)

38.67 R. A. Bauer: *Second-Order Consequences: A Methodological Essay on the Impact of Technology* (MIT, Cambridge 1969)

38.68 P. B. Thompson: Social acceptance of nanotechnology. In: *Societal Implications of Nanoscience and Nanotechnology*, ed. by M. C. Roco, W. S. Bainbridge (Kluwer, Dordrecht 2001) pp. 251–256

Acknowledgements

4 Nanowires

by Mildred S. Dresselhaus, Yu-Ming Lin, Oded Rabin, Marcie R. Black, Gene Dresselhaus

The authors gratefully acknowledge the stimulating discussions with Professors Charles Lieber, Gang Chen, S. T. Lee, Arun Majumdar, Peidong Yang, and Jean-Paul Issi, Dr. Joseph Heremans, and Ted Harman. The authors are grateful for support for this work by the ONR Grant #000140-21-0865, the MURI program subcontract PO #0205-G-7A114-01 through UCLA, and DARPA contract #N66001-00-1-8603.

6 Stamping Techniques for Micro and Nanofabrication: Methods and Applications

by John A. Rogers

The author extends his deepest thanks to all of the collaborators who contributed the work described here.

8 MEMS/NEMS Devices and Applications

by Darrin J. Young, Christian A. Zorman, Mehran Mehregany

The authors wish to thank Wen H. Ko for the helpful discussions and suggestions and Michael Suster and Joseph Seeger for preparing the figures.

10 Therapeutic Nanodevices

by Stephen C. Lee, Mark Ruegsegger, Philip D. Barnes, Bryan R. Smith, Mauro Ferrari

The authors gratefully acknowledge many friends and colleagues whose input helped shape this article. Particularly, we acknowledge Beth S. Lee for many, many helpful discussions and for unflagging support. We also acknowledge Phil Streeter for many of the same services, performed in the office of friend rather than spouse. We acknowledge the support services of Anita Bratcher in manuscript preparation and the artistic stylings of Vladimir Marukhlenko for the figures incorporated in the manuscript. We also acknowledge Carol Bozarth for her kind and spontaneous support of the physical process of manuscript editing. This work is dedicated to the memories of Mildred A. Lee, Antonio Ferrari and Marialuisa Ferrari and the multitude of others whose lives have been tragically shortened by cancer, with the determination that therapeutic nanotechnology be used to help abolish the terrible power of the disease over human life.

13 Noncontact Atomic Force Microscopy and Its Related Topics

by Seizo Morita, Franz J. Giessibl, Yasuhiro Sugawara, Hirotaka Hosoi, Koichi Mukasa, Akira Sasahara, Hiroshi Onishi

Thanks to Tom Albrecht, Alexis Baratoff, Hartmut Bielefeldt, Gerd Binnig, Dominik Brändlin, Peter van Dongen, Urs Dürig, Christoph Gerber, Stefan Hembacher, Markus Herz, Lukas Howald, Christian Laschinger, Ulrich Mair, Jochen Mannhart, Thomas Ottenthal, Calvin Quate, Marco Tortonese, and the BMBF for funding under project no. 13N6918.

18 Surface Forces and Nanorheology of Molecularly Thin Films

by Marina Ruths, Alan D. Berman, Jacob N. Israelachvili

This work was supported by ONR grant N00014-00-1–0214. M. R. Luths thanks the Academy of Finland for financial support.

19 Scanning Probe Studies of Nano-scale Adhesion Between Solids in the Presence of Liquids and Monolayer Films

by Robert W. Carpick, James D. Batteas

We gratefully acknowledge the help of Ms. Erin Flater, who provided valuable assistance and insights into the literature on capillary formation. RWC acknowledges support of a career award from the National Science Foundation, grant #CMS-0134571.

24 Mechanics of Biological Nanotechnology

by Rob Phillips, Prashant K. Purohit, Jané Kondev

We happily acknowledge useful discussions with Kai Zinn, Jon Widom, Bill Gelbart, Andy Spakowitz, Zhen-Gang Wang, Ken Dill, Carlos Bustamante, Tom Powers, Larry Friedman, Jack Johnson, Pamela Bjorkman, Paul Wiggins, Steve Williams, Wayne Falk, Adrian Parsegian, Alasdair Steven and Steve Quake. RP and PP acknowledge support of the NSF through grant number 9971922, the NSF supported CIMMS center, and the support of the Keck Foundation. JK is supported by the NSF under grant number DMR-9984471 and is a Cottrell Scholar of Research Corporation.

30 Nanotechnology for Data Storage Applications

by Dror Sarid, Brendan McCarthy, Ghassan E. Jabbour

The authors would like to thank Digital Instruments (Veeco) for contributing their Multi-Mode Nanoscope III, the Department of Energy (DE-FG-3-02ER46013/A001) for a generous grant, EMC for their generous gift, and the Vice President for Research, University of Arizona, for equipment support.

31 The "Millipede" – A Nanotechnology-Based AFM Data-Storage System

by Gerd K. Binnig, G. Cherubini, M. Despont, Urs T. Duerig, Evangelos Eleftheriou, H. Pozidis, Peter Vettiger

It is our pleasure to acknowledge our colleagues T. Albrecht, T. Antonakopoulos, P. Bächtold, A. Dholakia, U. Drechsler, B. Gotsmann, W. Häberle, D. Jubin, M. A. Lantz, T. Loeliger, H. E. Rothuizen, R. Stutz, and D. Wiesmann for their invaluable contributions to the millipede project.

In addition, thanks and appreciation go to H. Rohrer for his contribution to the initial millipede vision and concept and to our former collaborators, J. Brugger, now at the Swiss Federal Institute of Technology, Lausanne (Switzerland), M. I. Lutwyche, now at Seagate, Pittsburg, IL, and W. P. King, now at Georgia Tech, Atlanta, GA, as well as to K. Goodson, T. W. Kenny, and C. F. Quate of Stanford University, CA.

We are also pleased to acknowledge stimulating discussions with and encouraging support from our colleagues W. Bux and P. F. Seidler of the IBM Zurich Research Laboratory, J. Mamin, D. Rugar, and B. D. Terris of the IBM Almaden Research Center, San Jose, CA, and G. Hefferon of IBM, East Fishkill, NY.

Special thanks go to J. Frommer, C. Hawker, J. Mamin, and R. Miller of the IBM Almaden Research Center for their enthusiastic support in identifying and synthesizing alternative polymer media materials, and to H. Dang, A. Sharma, and S. Sri-Jayantha of the IBM T. J. Watson Research Center, Yorktown Heights, NY, for their contributions to the work on servo control.

35 Thermo- and Electromechanics of Thin-Film Microstructures

by Martin L. Dunn, Shawn J. Cunningham

We are most grateful for the assistance of Ms. Yanhang Zhang, a Ph.D. student at the University of Colorado, for her help with the preparation of figures. Many of the calculations and measurements also draw from her work. MLD acknowledges support from DARPA, Sandia National Laboratories, and the AFOSR for support of aspects of his research that appear in this work.

About the Authors

Chong H. Ahn

University of Cincinnati
Department of Electrical
and Computer Engineering
and Computer Science
Cincinnati, OH, USA
chong.ahn@uc.edu
http://www.BioMEMS.uc.edu/

Chapter 9

Dr. Chong Ahn is a Professor of Electrical and Computer Engineering at the University of Cincinnati. He obtained his Ph.D. degree in Electrical Engineering from the Georgia Institute of Technology in 1993 and then worked as a postdoctoral fellow at IBM T.J. Watson Research Center. His research interests include all aspects of design, fabrication, and characterization of magnetic MEMS devices, microfluidic devices, protein chips, lab-on-a-chips, nano biosensors, point-of-care testing and BioMEMS systems. He is an associate editor of the IEEE Sensors Journal.

Boris Anczykowski

nanoAnalytics GmbH
Münster, Germany
anczykowski@nanoanalytics.com
http://www.nanoanalytics.com

Chapter 15

Dr. Boris Anczykowski is a physicist with an extensive research background in the field of dynamic Scanning Force Microscopy. He co-invented the Q-Control technique and received the Innovation Award Münsterland for Science and Economy in 2001 for this achievement. He is a managing director and co-founder of nanoAnalytics GmbH, a company specialized in the characterization of surfaces and interfaces on the micro- and nanometer scale.

Massood Z. Atashbar

Western Michigan University
Department of Electrical
and Computer Engineering
Kalamazoo, MI, USA
massood.atashbar@wmich.edu
http://homepages.wmich.edu/~zandim

Chapter 5

Professor Atashbar's main area of research is nanotechnology and wireless sensors. He has been working in the field of micro/nanosensors for chemical and physical sensing specifically integrated smart wireless surface acoustic wave sensors and systems. He is a senior member of IEEE and an associate editor of the International Journal of Modelling and Simulation

Wolfgang Bacsa

Université Paul Sabatier
Laboratoire de Physique des Solides (LPST)
Toulouse, France
bacsa@lpst.ups-tlse.fr
http://www.lpst.ups-tlse.fr/users/wolfgang/

Chapter 3

Wolfgang Bacsa is an expert in the emerging field of Nano-Optics and Carbon Nanotubes. He has a Ph.D. from the Swiss Federal Institute of Technology (ETH) Zurich in Physics and has extensive experience in condensed matter physics, optics, microscopy, synthesis of ultra-thin films and nanostructured carbon. Professor Bacsa worked at the ETH Zürich, PennState Universiy and EPFL Lausanne.

William Sims Bainbridge

National Science Foundation
Division of Information
and Intelligent Systems
Arlington, VA, USA
wbainbri@nsf.gov
http://mysite.verizon.net/william.bainbridge

Chapter 38

William Sims Bainbridge earned his Ph.D. in Sociology from Harvard University in 1975 and is the author of 10 books, 4 textbook-software packages, and about 150 shorter publications in information science, social science of technology, and the sociology of culture. He has represented the social and behavioral sciences on five US government advanced technology initiatives: High Performance Computing and Communications, Knowledge and Distributed Intelligence, Digital Libraries, Information Technology Research, and Nanotechnology.

Antonio Baldi

Institut de Microelectronica
de Barcelona (IMB)
Centro National Microelectrónica
(CNM-CSIC)
Barcelona, Spain
Antoni.baldi@cnm.es

Chapter 5

Antonio Baldi received his PhD from the Universitat Autonoma de Barcelona, Spain, in 2001. He is a postdoctoral fellow at the University of Minnesota. His current research is focused on the fabrication and testing of bioMEMS, on the development of new sensors and microfluidic devices incorporating stimuli-sensitive hydrogels and on the development of inductively coupled wireless MEMS.

Philip D. Barnes

Ohio State University
Biomedical Engineering Center
Columbus, OH, USA
d_skill@yahoo.com

Chapter 10

Phillip D. Barnes is a graduate student at the Biomedical Engineering Center at the Ohio State University. His background is in electrical engineering and his current interests are in sensors and imaging devices.

James D. Batteas

National Institute
of Standards and Technology
Surface and Microanalysis Science
Division
Gaithersburg, MD, USA
james.batteas@nist.gov
http://www.cstl.nist.gov/div837/837.03/

Chapter 19

Dr. James D. Batteas is a research chemist in the Surface and Interface Research Group of the National Institute of Standards and Technology. The group's current research involves developing and charcterizing the organization and properties of nanoscale materials and devices on surfaces, this includes nanoparticles for use in optoelectric and optomagnetic gating, molecular electronics, and sensors. His group is also investigating friction, adhesion and wear mechanisms on the nanoscale.

Roland Bennewitz

McGill University
Physics Department
Montreal, QC, Canada
roland@physics.mcgill.ca

Chapter 20

Roland Bennewitz studied physics in Freiburg and Berlin where he received his PhD for work on defects at surfaces of insulators. He is now assistant at the University of Basel where his research activities focus on high-resolution force microscopy as a tool in nanotribology and surface science.

Alan D. Berman

Monitor Venture Enterprises
Los Angeles, CA, USA
alan.berman.2001@anderson.ucla.edu

Chapter 18

Alan Berman received his Ph.D. from the Chemical Engineering Department at the University of California, Santa Barbara. He is involved in technology commercialisation, assisting large technology companies to maximize return on R&D investments by finding and developing applications for their intellectual property outside of their core business. Value is realized through technology licensing, joint ventures and venture capital funded new entities. Technical interests are in fields of tribology, MEMS, and nanotechnology.

Bharat Bhushan

The Ohio State University
Nanotribology Laboratory
for Information Storage
and MEMS/NEMS
Columbus, OH, USA
bhushan.2@osu.edu

Chapters 1, 11, 17, 25, 26, 27, 28, 33

Dr. Bharat Bhushan is an Ohio Eminent Scholar and The Howard D. Winbigler Professor in the Department of Mechanical Engineering, a Graduate Research Faculty Advisor in the Department of Materials Science & Engineering, and the Director of the Nanotribology Laboratory for Information Storage & MEMS/NEMS (NLIM) at the Ohio State University, Columbus, Ohio. He holds two M.S., a Ph.D. in mechanical engineering/mechanics, an MBA, and three semi-honorary and honorary doctorates. His research interests are in micro/nanotribology and its applications to magnetic storage devices and MEMS/NEMS (Nanotechnology). He has authored 5 technical books, 45 handbook chapters, more than 450 technical papers in referred journals, and more than 60 technical reports, edited more than 25 books, and holds 14 U.S. patents.

Gerd K. Binnig

IBM Zurich Research Laboratory
Micro-/Nanomechanics
Rüschlikon, Switzerland
gbi@zurich.ibm.com

Chapter 31

Gerd Binnig obtained his Ph.D. from the Johann Wolfgang Goethe University, Frankfurt, Germany, and joined IBM Research in 1978. He was corecipient of the 1986 Nobel Prize in Physics for the invention of the scanning tunnelling microscope, and he also invented the atomic force microscope. His current research interests are micro- and nanosystem techniques and "Fractal Darwinism", a theory he developed to describe complex systems.

Marcie R. Black

Massachusetts Institute of Technology
Department of Electrical Engineering and Computer Science
Cambridge, MA, USA
marcie@alum.mit.edu
http://web.mit.edu/mrb/www/prof.html

Chapter 4

Marcie Black recently received her Ph.D. from Prof. Dresselhaus's research group at MIT studying the optical properties of nanowires. In particular, she identified the dominant optical absorption mechanism in the IR of bismuth nanowires as an indirect interband transition that is enhanced over bulk bismuth. Currently she is studying organic opto-electronics with an emphasis on photovoltaics.

Jean-Marc Broto

University Toulouse III
Laboratoire National des Champs Magnétiques Pulsés (LNCMP)
Toulouse, France
broto@insa-tlse.fr
http://www.insa-tlse.fr/

Chapter 3

Jean-Marc Broto is Professor at the Universite Toulouse III, France. He is a specialist in electronic transport and magnetization properties under high magnetic fields and contributed to the discovery of the Giant Magnetoresistance in 1988.

Robert W. Carpick

University of Wisconsin-Madison
Department of Engineering Physics
Madison, WI, USA
carpick@engr.wisc.edu
http://www.engr.wisc.edu/ep/faculty/carpick_robert.html

Chapter 19

Robert Carpick has a Ph.D. in Physics (1997) from the University of California at Berkeley. He has been an Assistant Professor at the University of Wisconsin-Madison since 2000. He carries out research and publishes in the areas of nanotribology, nanomechanics, nanostructured materials, and scanning probe development. He serves on the editorial board of Review of Scientific Instruments. In 2002 he received a Faculty Early Career Development award from the U.S. National Science Foundation.

Tsung-Lin Chen

National Chiao Tung University
Department
of Mechanical Engineering
Shin Chu, Taiwan
tsunglin@mail.nctu.edu.tw

Chapter 32

Tsung-Lin Chen received his B.S. and M.A. degrees in mechanical engineering from the National Tsing-Hua University, Taiwan in 1990 and 1992, respectively. He received Ph.D. degree in mechanical engineering from University of California at Berkeley, USA in 2001. He is currently an assistant professor in National Chiao Tung University in Taiwan. His research interests include MEMS devices design and MEMS fabrication process development.

Yu-Ting Cheng

National Chiao Tung University
Department
of Electronics Engineering
& Institute of Electronics
HsinChu, Taiwan
ytcheng@faculty.nctu.edu.tw

Chapter 37

Yu-Ting Cheng received his Ph.D. degree in Electrical Engineering at the University of Michigan, Ann Arbor, in 2000. After his graduation, he worked for IBM Watson Research Center, Yorktown Heights, as a research staff member. He is a member of IEEE, IOP, and Phi Tau Phi. Currently he is Assistant Professor at the National Chiao Tung University, Taiwan. His research interests include the development of novel materials and fabrication technologies for MEMS/NEMS applications and microsystems integration.

Giovanni Cherubini

IBM Zurich Research Laboratory
Storage Technologies
Rüschlikon, Switzerland
cbi@zurich.ibm.com

Chapter 31

Dr. Giovanni Cherubini received a Ph.D. degree in electrical engineering from the University of California, San Diego, in 1986, and joined IBM Research in 1987. His interests include high-speed data transmission and data storage systems. He is Editor for CDMA systems, for IEEE Transactions on Communications, and served as Guest Editor for the IEEE Journal on Selected Areas in Communications issues on access technologies (1995) and on multiuser detection techniques (2001–2002). He is co-author of the book Algorithms for Communications Systems and their Applications.

Jin-Woo Choi

Louisiana State University
Department of Electrical
and Computer Engineering
Baton Rouge, LA, USA
choi@ece.lsu.edu

Chapter 9

Jin-Woo Choi received his B.S. and M.S. degree in Electrical Engineering from Seoul National University in Korea in 1994 and 1996, respectively. He received his Ph.D. degree in Electrical Engineering from the University of Cincinnati in 2000. Now he is an Assistant Professor at Louisiana State University, Baton Rouge, Louisiana. His current research activities include magnetic particle separators, microfluidic systems for biochemical detection, micro total analysis systems (μ-TAS), bioelectronics, and BioMEMS components and systems.

Shawn J. Cunningham

WiSpry, Inc.
Colorado Springs Design Center
Colorado Springs, CO, USA
shawn.cunningham@wispry.com

Chapter 35

Shawn Cunningham is working on the development of RF MEMS switch and associated processes with wiSpry, Inc. His interests include materials characterization, reliability, and Design for MEMS Manufacturability. Prior to joining wiSpry, Shawn pursued MEMS research and product development at Coventor, Ford Microelectronics, and the University of Utah's Center for Engineering Design and in collaboration with the Univerisity of Colorado.

Michel Despont

IBM Zurich Research Laboratory
Micro-/Nanomechanics
Rüschlikon, Switzerland
dpt@zurich.ibm.com

Chapter 31

Michel Despont received his Ph.D. in physics from the University of Neuchâtel, Switzerland, in 1996. After a postdoctoral fellowship at IBM's Zurich Research Laboratory, he was visiting scientist at the Seiko Instrument Research Laboratory in Japan in 1997. His current research at IBM focuses on the development of micro- and nanomechanical devices and of processes to fabricate so-called system-on-chip.

Gene Dresselhaus

Massachusetts Institute
of Technology
Francis Bitter Magnet Laboratory
Cambridge, MA, USA
gene@mgm.mit.edu
http://web.mit.edu/fbml/cmr/

Chapter 4

Gene Dresselhaus received his Ph.D. in physics from the University of California in 1955. He was a faculty member at the University of Chicago, and assistant professor at Cornell before joining MIT Lincoln Laboratory in 1960 as a staff member. In 1976 he assumed his current position at the MIT Francis Bitter Magnet Laboratory. His area of interest is the electronic structure of nanomaterials and he has co-authored with M.S. Dresselhaus several books on fullerenes, nanowires, and nanotubes.

Mildred S. Dresselhaus

Massachusetts Institute
of Technology
Department
of Electrical Engineering
and Computer Science
and Department of Physics
Cambridge, MA, USA
millie@mgm.mit.edu
http://www.eecs.mit.edu/faculty/index.html

Chapter 4

Mildred Dresselhaus received her Ph.D. in physics from the University of Chicago in 1958. She joined the MIT faculty in 1967. She has been active in research across broad areas of solid state physics, especially in carbon science. Her present research activities focus on carbon nanotubes, bismuth nanowires, low dimensional thermoelectricity, and novel forms of carbon. She is the recipient of the National Medal of Science and 17 honorary degrees.

Martin L. Dunn

University of Colorado at Boulder
Department of Mechanical Engineering
Boulder, CO, USA
martin.dunn@colorado.edu

Chapter 35

Martin L. Dunn received the Ph.D. in Mechanical Engineering from the University of Washington and was a postdoctoral appointee at Sandia National Laboratories. His research focuses on the micromechanical behavior of materials and structures. He has published over 75 articles in archival journals, and his research has been sponsored by NSF, DOE, NIST, DARPA, AFOSR, and Sandia National Laboratories.

Urs T. Dürig

IBM Zurich Research Laboratory
Micro-/Nanomechanics
Rüschlikon, Switzerland
drg@zurich.ibm.com

Chapter 31

Urs Dürig received a Ph.D. degree from the Swiss Federal Institute of Technology, Zurich, in 1984. He joined IBM as a post-doc working on near-field optical microscopy. He is Research Staff Member since 1986: He worked in the field of scanning tunnelling and dynamic force microscopy. In 1997, he joined the Micro/Nanomechanics group focusing on polymer material issues and thermal modelling.

Evangelos Eleftheriou

IBM Zurich Research Laboratory
Storage Technologies
Rüschlikon, Switzerland
ele@zurich.ibm.com

Chapter 31

He received a Ph.D. in Electrical Engineering from Carleton University, Ottawa, Canada, in 1985, and joined IBM Research in 1986. His research focuses on signal processing and coding for recording and transmission systems. His current research interests include nanotechnology, in particular probe-storage techniques.
Dr. Eleftheriou was elected IEEE Fellow in 2002.

Mauro Ferrari

Ohio State University
Biomedical Engineering Center
Columbus, OH, USA
Ferrari.5@osu.edu
http://bmew.bme.ohio- state.edu/bmeweb3/bme_faculty.htm

Chapter 10

Mauro Ferrari is the Edgar Hendrickson Professor of Biomedical Engineering of the Ohio State University. A mechanical engineer by training, he also holds appointments in Internal Medicine, Mechanical Engineering and Materials Science. Professor Ferrari is a globally recognized figure in biomedical nanotechnology and micro and nanofabrication. His current interests are application of novel fabrication technologies to innovative drug delivery devices for treatment of cardiovascular disease, palliative care and oncology.

Emmanuel Flahaut

Université Paul Sabatier
CIRIMAT (Centre Interuniversitaire
de Recherche et d'Ingénierie
des Matériaux)
Toulouse, France
flahaut@chimie.ups-tlse.fr
http://eflahaut.nano.free.fr

Chapter 3

Emmanuel Flahaut obtained his Ph.D in Materials Science in Toulouse working on CCVD synthesis of carbon nanotubes (CNTs) and dense ceramic-based composites including CNTs. He spent then more than one year as a post-doctoral researcher in Malcolm Green's Group in Oxford to work mainly on the filling of CNTs. He is now a permanent CNRS researcher at the University of Toulouse.

Lásló Forró

Swiss Federal Institute
of Technology (EPFL)
Institute of Physics
of Complex Matter
Lausanne, Switzerland
laszlo.forro@epfl.ch
http://nanotubes.epfl.ch

Chapter 21

Professor László Forró is working on the synthesis, physical properties and manipulation of carbon nanostuctures and nanostuctured arrays, as well as, mechanical properties of carbon nanotubes, carbon onions, biological tubular systems. Transport and electron spin resonance studies of molecular materials, quasi-one-dimensional organic metals, organic superconductors, cuprates, manganates and fullerenes – up to high pressures. Tunneling spectroscopy in cuprate and fullerene superconductors. Optical properties of strongly correlated systems and biomaterials.

Jane Frommer

IBM Almaden Research Center
Department of Science
and Technology
San Jose, CA, USA
frommer@Almaden.ibm.com

Chapter 29

Following a Ph.D. in Organometallic Chemistry from Caltech, Dr. Jane Frommer has been involved in a diverse set of research areas, including electronically conducting polymers and scanning probe microscopy. Her present AFM lab at IBM collaborates with a wide variety of laboratories involved in materials research, including lithography, chromatography, and storage. Common to all these studies is Frommer's interest in the properties and structure of molecules in confined geometries.

Harald Fuchs

Universität Münster
Physikalisches Institut
Münster, Germany
fuchsh@uni-muenster.de
http://www.uni-muenster.de/Physik/PI/Fuchs

Chapter 15

1984 Ph.D. Universität des Saarlandes with Prof. H. Prof Gleiter (nano crystalline Systems), 1984–1985, Post doc with IBM Research Lab. Zurich in the group of G. Binnig and H. Rohrer, 1985–1993 Project manager 'Ultrathin Organic Films' with BASF AG, Ludwigshafen, Germany. Since 1993 he is full Professor and Director at the Physical Institute of the University of Münster, 2000: Cofounder and Scientific Director of the Center for Nanotechnology (CeNTech).

Franz J. Giessibl

Universität Augsburg
Lehrstuhl für Experimentalphysik VI
Augsburg, Germany
franz.giessibl@physik.uni-augsburg.de
http://www.physik.uni-augsburg.de/exp6/

Chapter 13

Dr. Giessibl is working in atomic force microscopy and scanning tunnelling microscopy in ultrahigh vacuum at room temperature and low temperatures. He is a Steering Committee Member of the International Conference on Noncontact Atomic Force Microscopy. He received the R&D 100 Award Chicago 1994, the German Nanoscience Award 2000, and the Rudolf Kaiser Price in 2001.

Enrico Gnecco

University of Basel
Department of Physics
Basel, Switzerland
Enrico.Gnecco@unibas.ch
http://monet.physik.unibas.ch/~gnecco/

Chapter 20

Enrico Gnecco studied physics in Genoa, where he received his PhD for work on the mechanism of growth of nanostructured carbon films. He is now assistant at the University of Basel, where his research activities focus on friction force microscopy and molecular machinery.

Gérard Gremaud

Swiss Federal Institute
of Technology (EPFL)
Institute of Physics
of Complex Matter
Lausanne, Switzerland
gremaud@epfl.ch

Chapter 21

Dr. Gérard Gremaud is a physicist and senior lecturer at the Ecole Polytechnique Fédérale de Lausanne (EPFL). He is active in the research fields of dislocation dynamic, acoustic and atomic force microscopy and granular physics. He is also responsible for the teaching of metrology and practical works to the physics students.

Jason H. Hafner

Rice University
Department of Physics
& Astronomy
Houston, TX, USA
hafner@rice.edu
http://www.ruf.rice.edu/~hafner

Chapter 12

Jason Hafner earned his Ph.D. in physics from Rice University in 1998. He then held an NIH postdoctoral fellowship at Harvard University working on nanotube probes for high resolution biological atomic force microscopy. In 2001 he returned to Rice as an assistant professor where his group is pursuing various biophysical applications of scanned probe microscopy and nanomaterials. Dr. Hafner received a Beckman Young Investigator award in 2002.

Stefan Hengsberger

University of Applied Science
of Fribourg
Fribourg, Switzerland
stefan.hengsberger@eif.ch
http://www.anwalt-homburg.de/
Stefan-index.htm

Chapter 21

Stefan Hengsberger received his diploma in physics from the University of Saarbrücken in 1997. He started as a research scientist at Fraunhofer/ Miami (Florida) where he worked until mai 1998. In July 1998 he joined the swiss federal institute of Technology (EPFL) where he earned his PhD in biomechanics in spring 2002. He stayed for another year at EPFL as a postdoc in biomechanics and physics until summer 2003. Since October 2003 he is Professor of Physics at the University of Applied Science in Fribourg, Switzerland.

Peter Hinterdorfer

Johannes Kepler University of Linz
Institute for Biophysics
Linz, Austria
peter.hinterdorfer@jku.at
http://at22.bphys.uni-linz.ac.at/bioph/
staf/hipe/hipe.htm

Chapter 16

Peter Hinterdorfer earned a Dr. tech. from the University of Linz, Austria, Institute for Biophysics in 1992. He was a postdoctoral fellow at the University of Virginia, Department of Molecular Physiology and Biological Physics (1992/1993). Since then he is at the University of Linz, Institute for Biophysics, where he holds a position as Associate Professor. His current and ongoing research includes single molecule force spectroscopy and high resolution topography and recognition imaging of biological samples.

Roberto Horowitz

University of California at Berkeley
Department
of Mechanical Engineering
Berkeley, CA, USA
horowitz@me.berkeley.edu
http://www.me.berkeley.edu/~horowitz

Chapter 32

Roberto Horowitz joined the Department of Mechanical Engineering at the University of California at Berkeley in 1982, where he is currently a Professor. Dr. Horowitz teaches and conducts research in the areas of adaptive, learning, nonlinear and optimal control, with applications to Micro-Electromechanical Systems (MEMS), computer disk file systems, robotics, mechatronics and intelligent vehicle and highway systems (IVHS).

Hirotaka Hosoi

Japan Science and Technology
Corporation
Sapporo, Japan
hosoi@sapporo.jst-plaza.jp

Chapter 13

Hirotaka Hosoi received the D.E. degree in electronic enginnering from Hokkaido University in 1999. Since 2002 he is at Innovation Plaza Hokkaido, Japan Science and Technology Corporation (JST). His main research focus is in high-resolution magnetic imaging of magnetic materials surfaces using a scanning force microscope. His current research interests includes magnetism on metal-oxide surfaces.

Jacob N. Israelachvili

University of California
Department of Chemical Engineering and Materials Department
Santa Barbara, CA, USA
Jacob@engineering.ucsb.edu
http://www.chemengr.ucsb.edu/people/faculty/israelachvili

Chapter 18

Jacob Israelachvili earned his Ph.D. 1971 at the Cavendish Laboratory, University of Cambridge, UK. He held various positions at the Department of Applied Mathematics at the Australian National University (1974–1986), including those of Professional Fellow and Head of Department. In 1986 he joined the faculty at University of California, Santa Barbara as Professor in the Department of Chemical Engineering and Materials Department. In 1988 he was elected a Fellow of the Royal Society of London, and in 1991 he was awarded the Alpha Chi Sigma Award for Chemical Engineering Research by the AIChE. He was elected as a foreign associate of the US National Academy of Engineering in 1996.

Ghassan E. Jabbour

University of Arizona
Optical Sciences Center
Tucson, AZ, USA
gej@optics.arizona.edu

Chapter 30

Ghassan E. Jabbour is the head of the research group Organic Optoelectronic Materials and Devices at the Optical Sciences Center working on organic and hybrid materials and their applications to light emitting dvices, solar cells, memory storage, solid state lighting, and other areas. Professor Jabbour is a SPIE Fellow, Track Chair of the Nanotechnology Program for SPIE, and Associate Editor of the Journal of the Society for Information Display (JSID). He has over 200 publications, invited talks, and conference proceedings.

Harold Kahn

Case Western Reserve University
Department of Materials Science and Engineering
Cleveland, OH, USA
kahn@cwru.edu

Chapter 34

Harold Kahn is Researcher Associate Professor of Materials Science and Engineering at Case Western Reserve University, Cleveland, Ohio. His research is focused on MEMS device processing and testing, particularly wafer-level mechanical testing and shape-memory actuated microfluidics. He received a B.S. in metallurgical engineering from Lafayette College and a Ph.D. in electronic materials from the Massachusetts Institute of Technology.

András Kis

Swiss Federal Institute of Technology (EPFL)
Institute of Physics of Complex Matter
Lausanne, Switzerland
andras@igahpse.epfl.ch
http://www.andras.kis.name

Chapter 21

András Kis is a Ph.D. student of Prof. László Forró at the Swiss Federal Institute of Technology (EPFL) where he works together with Andrzej Kulik. His main subject is the measurement of the mechanical properties of nanoscale objects, including carbon nanotubes and microtubules with the aim of understanding the interplay between their nanoscale structure and mechanical properties.

Jané Kondev

Brandeis University
Physics Department
Waltham, MA, USA
kondev@brandeis.edu
http://mattter.cc.brandeis.edu

Chapter 24

Professor Kondev graduated with a PhD from Cornell University and did postdoctoral work at Brown University and Princeton University before joining the faculty at Brandeis University. He is a recipient of the CAREER award from the National Science Foundation and is a Cotrell Scholar of the Research Corporation.

Andrzej J. Kulik

Swiss Federal Institute of Technology (EPFL)
Institute of Physics of Complex Matter
Lausanne, Switzerland
andrzej.kulik@epfl.ch

Chapter 21

Andrzej Kulik is the Head of the Biostructures and Nanomechanics Laboratory of the Swiss Federal Institute of Technology (EPFL) in Lausanne, Switzerland. His research concentrates on quantitative nanoscale materials properties, nanotribology. scanning probe microscopy, contact mechanics, nanoindentation, near-field ultrasonics, optical tweezers, nanolithography, and nanomanipulation.

Christophe Laurent

Université Paul Sabatier
CIRIMAT (Centre Interuniversitaire de Recherche et d'Ingénierie des Matériaux)
Toulouse, France
laurent@chimie.ups-tlse.fr
http://ncn.f2g.net/

Chapter 3

Dr. Ch. Laurent is Professor of Materials Chemistry at University Paul Sabatier and is the head of the Nanocomposites and Carbon Nanotubes group of CIRIMAT. His research include in the synthesis, characterization and mechanical properties of ceramic-matrix nanocomposites, and since 1994, carbon nanotubes (synthesis of single- and double-walled CNTs, formation mechanisms, characterization, localized growth, hydrogen storage, ceramic-matrix composites).

Stephen C. Lee

Ohio State University
Biomedical Engineering Center
Columbus, OH, USA
Lee@bme.ohio-state.edu
http://medicine.osu.edu/mcbiochem/

Chapter 10

Stephen C. Lee is a pioneer in the field of semi-biological nanodevices (nanobiological devices), having published the first monograph devoted to the topic in 1998. His interests are in enabling technologies for the incorporation of functional proteins and nucleic acids into nanodevices, particularly for application in oncology and cardiovascular disease. He is currently Associate Professor of Cellular and Molecular Biochemistry, Chemical Engineering and Biomedical Engineering at the Ohio State University.

Yunfeng Li

University of California at Berkeley
Department of Mechanical Engineering
Berkeley, CA, USA
yunfeng@me.Berkeley.edu

Chapter 32

Yunfeng Li received the B.S. and M.S. degrees from Beijing University of Aeronautics and Astronautics, Beijing, China, in 1992 and 1995, respectively. He is currently working towards the Ph.D. degree in the Department of Mechanical Engineering, University of California, Berkeley, CA, USA. His research interests include motion control, vibration control, control of MEMS devices, and mechatronics.

Liwei Lin

University of California at Berkeley
Mechanical Engineering Department
Berkeley, CA, USA
lwlin@me.berkeley.edu
http:// me.berkeley.edu/faculty/lin/index.html

Chapter 37

Liwei Lin received his Ph.D. from the University of California, Berkeley, in 1993. He was an Associate Professor in the Institute of Applied Mechanics, National Taiwan University, Taiwan (1994–1996) and an Assistant Professor in Mechanical Engineering Department, University of Michigan (1996–1999). He joined UC-Berkeley in 1999 and is now an Associate Professor at Mechanical Engineering Department and Co-Director at Berkeley Sensor and Actuator Center.

Yu-Ming Lin

Massachusetts Institute of Technology
Department of Electrical Engineering and Computer Science
Cambridge, MA, USA
yming@mgm.mit.edu

Chapter 4

Lin Yu-Ming performed experimental and theoretical studies on Bi-based nanowires for next-generation thermoelectric materials. The electronic transport properties of these systems are studied to investigate quantum size effects. He received the Masterworks Award of the Department of EECS, MIT (2000) and the Gold Medal, MRS Graduate Student Award (2002)

Huiwen Liu

Ohio State University
Nanotribology Laboratory for Information Storage and MEMS/NEMS
Columbus, OH, USA
liu.403@osu.edu

Chapters 27, 28

Dr. Huiwen Liu is the associate director of the Nanotribology Laboratory for Information Storage and MEMS/NEMS at The Ohio State University. His research interests are study of mechanical, tribological, and physical properties of advanced materials and MEMS/NEMS devices on micro/nano scale. He obtained an Alexander von Humboldt fellowship and the Japanese Science and Technology Agency Fellowship in 1997 and 1998, respectively

Adrian B. Mann

Rutgers University
Department of Ceramics and Materials Engineering
Piscataway, NJ, USA
abmann@rci.rutgers.edu

Chapter 22

Dr. Mann's research focuses on the nanomechanics of materials and the fabrication of nanostructured materials. His research is predominantly on biomedical materials, but also includes ceramics, polymers and metals. He is currently an Assistant Professor at Rutgers University, New Jersey. Prior to this he was a lecturer at the University of Manchester, England and a Fulbright scholar at The Johns Hopkins University, Maryland.

Othmar Marti

University of Ulm
Department of Experimental Physics
Ulm, Germany
Othmar.Marti@physik.uni-ulm.de
http://servex.physik.uni-ulm.de/marti

Chapter 11

Profesor Othmar Marti is Head of the Department of Experimental Physics at the University of Ulm. His main research topics are polymers, scanning force microscopy, friction, near-field optics, and the optics of nanoparticles. Studies: diploma from ETH Zürich, Dr. sc. nat. ETH Zürich, Habilitation University Konstanz. He worked at IBM Research Zurich, Switzerland, University of California, USA, ETH Zurich, Switzerland and University of Konstanz, Germany.

Jack Martin

Analog Devices, Inc.
Micromachined Products Division
Cambridge, MA, USA
jack.martin@analog.com

Chapter 36

Jack Martin has been a technologist and manager in the design, development and manufacture of industrial MEMS products for 25 years. His accomplishments include development of wafer fab and packaging processes that are used by Analog Devices to produce iMEMS integrated accelerometers and gyroscopes. He has a Ph.D. in Materials Science, a BS and MS in Chemical Engineering, and is a Licensed Professional Engineer.

Brendan McCarthy

University of Arizona
Optical Sciences Center
Tucson, AZ, USA
bmccarthy@optics.arizona.edu

Chapter 30

Brendan Mc Carthy is a postdoctoral researcher in the Optical Sciences Center, University of Arizona. He received his PhD in 2001, from Trinity College, University of Dublin, Ireland. His thesis topic was the characterisation of composite materials based on carbon nanotubes and conjugated polymers, using optical and vibrational spectroscopy, scanning tunnelling microscopy, and transmission electron microscopy.

Mehran Mehregany

Case Western Reserve University
Department of Electrical Engineering and Computer Science
Cleveland, OH, USA
mxm31@cwru.edu
http://mems.cwru.edu/

Chapters 7, 8

Mehran Mehregany is currently the Chairman of the Electrical Engineering and Computer Science Department at Case and the Goodrich Professor of Engineering Innovation. His research interests include: silicon and silicon carbide micro/nano systems technology (including MEMS/NEMS); micromachining and microfabrication technologies; materials and modeling issues related to MEMS/NEMS and (in some cases) integrated circuits technologies; and MEMS packaging.

Ernst Meyer

University of Basel
Institute of Physics
Basel, Switzerland
Ernst.Meyer@unibas.ch
http://monet.unibas.ch/cgi-bin/people/meyer/

Chapter 20

Ernst Meyer is professor of physics at the University of Basel. He is interested in friction force and dynamic force microscopy with true atomic resolution. He is also active in the field of sensors based upon micromechanics and magnetic spin resonance detection with force microscopy. Awarded from the Swiss Physical Society, he is member of the Swiss and American Physical Society, of the Editorial Board of Tribology Letters, and co-editor of books on atomic force microscopy.

Marc Monthioux

UPR A-8011 CNRS
Centre d'Elaboration des Matériaux et d'Etudes Structurales (CEMES)
Toulouse, France
monthiou@cemes.fr

Chapter 3

Marc Monthioux has been working on carbon materials for more than 20 years. He is involved in research on carbon nanotubes since 1998, discovered the ability of single-wall nanotubes in being filled by foreign molecules the same year, associated with B. Smith and Prof. D.E. Luzzi from University of Pennsylvania. He is currently Director of Research at the French National Center for Scientific Research and European Associate Editor of CARBON Journal.

Markus Morgenstern

University of Hamburg
Institute of Applied Physics
Hamburg, Germany
mmorgens@physnet.uni-hamburg.de
http://www.nanoscience.de/group_r/stm-sts/

Chapter 14

Markus Morgenstern earned his Ph.D. from the Institute of Interface Research and Vacuum Physics of the Forschungszentrum Jülich, Germany in 1996. After one year of research at the University of Paris VII he joined the Group of Prof. Dr. R. Wiesendanger in 1997 as a Senior Scientist. In 2002 he completed his Habilitation at the University of Hamburg with the subject Scanning tunnelling spectroscopy on semiconductor systems and nanostructures

Seizo Morita

Osaka University
Department of Electronic Engineering
Suita-Citiy, Osaka, Japan
smorita@ele.eng.osaka-u.ac.jp
http://www-e2.ele.eng.osaka-u.ac.jp/

Chapter 13

Prof. Seizo Morita works in the atomic force microscopy (AFM). He has discovered two-dimensional friction with a lattice periodicity, and two-dimensional solid phase of densely contact-electrified electrons on SiO_2 thin films under ambient conditions. He has achieved mapping, discrimination and control of atomic force and atom with atomic resolution, and also atom manipulation based on a mechanical method using the noncontact AFM apparatus.

Koichi Mukasa

Hokkaido University
Nanoelectronics Laboratory
Sapporo, Japan
mukasa@nano.eng.hokudai.ac.jp
http://www.nano.eng.hokudai.ac.jp/nano/

Chapter 13

K. Mukasa is a Professor of electronics at Hokkaido University, Sapporo, Japan. In 1980 he joined Alps Electric Co.Ltd, where he worked on the magnetic thin film heads and materials. In 1987 he moved to the university. His research interests include spin-polarized STM, exchange force microscopy, magnetic force microscopy, Mott spin detectors, nanostructure concerning electron spin and molecular/biological materials and devices.

Martin H. Müser

University of Western Ontario
Department of Applied Mathematics
London, Ontario, Canada
mmuser@uwo.ca
http://publish.uwo.ca/~mmuser/

Chapter 23

Prof. Martin Müser received his Ph.D. (1995) and habilitation (2002) from the University of Mainz, Germany. He was a post doctoral researcher at Columbia University (1996/97) as Feodor Lynen fellow and at Johns Hopkins University (1998). He joined the faculty of the University of Western Ontario (U.W.O.) in 2002. His research interests are statistical physics with emphasis on computational materials science, tribology, and quantum solids.

Kenn Oldham

University of California at Berkeley
Department
of Mechanical Engineering
Berkeley, CA, USA
oldham@newton.berkeley.edu

Chapter 32

Kenn Oldham is a graduate student in Mechanical Engineering at the University of California at Berkeley, pursuing M.S. and Ph.D. through National Science Foundation Graduate Fellowship. He received his B.S. from Carnegie Mellon University, Pittsburgh, PA. He is currently researching microdevices for hard disk drives. Research interests include MEMS design for control and reliability, optimal and robust control, and materials for microdevices.

Hiroshi Onishi

Kanagawa Academy
of Science and Technology
Surface Chemistry Laboratory
Kanagawa, Japan
oni@net.ksp.or.jp
http://home.ksp.or.jp/onishipro/home.eng/

Chapter 13

Dr. Onishi Hiroshi is an experimental chemist at the Kanagawa Academy of Science and Technology interested in molecule-scale reaction kinetics at interfaces. He likes to observe molecules moving and reacting over metal oxide surfaces by time-lapse imaging with scanning probe microscopes. Domestic societies encouraged him with awards to further develop his research towards nano-scale chemistry.

René M. Overney

University of Washington
Department of Chemical Engineering
Seattle, WA, USA
roverney@u.Washington.edu
http://depts.washington.edu/nanolab/

Chapter 29

René Overney received his Ph.D. in Physics at the University of Basel in 1992. After his postdoctoral years in Japan and at Exxon, CR, Annandale (NJ) he joined the University of Washington in 1996. His research interests are in mesoscale sciences involving nanorheological interpretations of processes that are as diverse as lubrication, membrane transport, or quantum yield in optoelectronic devices.

Alain Peigney

Université Paul Sabatier
CIRIMAT (Centre Inter-universitaire de Recherches et d'Ingénierie
des Matériaux) – UMR CNRS 5085
Toulouse, France
peigney@chimie.ups-tlse.fr
http://ncn.f2g.net/

Chapter 3

Ceramic Engineer and Doctor in Physical-Chemistry – Associate Professor of Materials Chemistry at the Paul Sabatier University – Researches in the synthesis, sintering and microstructural characterization of ceramics and ceramic matrix nanocomposites, and from 1994 on the synthesis on single- and double-walled carbon nanotubes and preparation of nanocomposites containing carbon nanotubes.

Oliver Pfeiffer

University of Basel
Institute of Physics
Basel, Switzerland
Oliver.Pfeiffer@stud.unibas.ch

Chapter 20

Oliver Pfeiffer is PhD student at the University of Basel. The main topic of his work is the energy dissipation of oscillating cantilevers in non-contact AFM. Related to this research field is the examination of damping of torsional oscillations of cantilevers when approaching the sample.

Rob Phillips

California Institute of Technology
Mechanical Engineering
and Applied Physics
Pasadena, CA, USA
phillips@aero.caltech.edu
http://www.rpgroup.caltech.edu/people/phillips.html

Chapter 24

Rob Phillips is Professor of Mechanical Engineering and Applied Physics at CALTECH, Pasadena. Phillips' work aims to examine nanomechanics of both crystalline solids and biological molecules and their assemblies. Recent efforts have been aimed at investigating mechanics of DNA packing in viruses, the tension-induced gating of ion channels and the mechanical response of bio-functionalized cantilevers.

Haralampos Pozidis

IBM Zurich Research Laboratory
Storage Technologies
Rüschlikon, Switzerland
hap@zurich.ibm.com

Chapter 31

He received the Ph.D. in electrical engineering from Drexel University, Philadelphia, PA, in 1998. After working with Philips Research, Eindhoven, the Netherlands, on signal processing and coding for optical storage technologies, with focus on DVD and Blue-Ray-Disc, Dr. Pozidis joined IBM Research in 2001. His research focuses on receiver design for alternative storage technologies, particularly scanning probe microscopy-based techniques.

Prashant K. Purohit

California Institute of Technology
Mechanical Engineering
Pasadena, CA, USA
prashant@caltech.edu
http://www.rpgroup.caltech.edu/~prashant/

Chapter 24

Purohit Prashant is currently a Postdoctoral Scholar in Applied Mechanics in the Mechanical Engineering department at Caltech. The overall theme of his research is to develop systematic methods to understand the physics operative at the nanometer scales. His research falls broadly into two categories. The first concerns the subject of coarse-graining methods for crystalline solids and the second mechanics problems in biology. His group hopes to extend the coarse-graining methods in solids to the arena of biological macromolecules.

Oded Rabin

Massachusetts Institute of Technology
Department of Chemistry
Cambridge, MA, USA
oded@mgm.mit.edu

Chapter 4

Oded Rabin is a Ph.D. candidate in Prof. Mildred S. Dresselhaus' group working on the thermoelectric properties of bismuth-antimony nanowire systems, and on electrochemistry-based nanowire synthesis methods. He earned a B.A. degree in Chemistry from the Technion – Israel Institute of Technology, Haifa, Israel and an M.A. degree in Chemistry from the Weizmann Institute of Science, Rehovot, Israel.

Françisco M. Raymo

University of Miami
Department of Chemistry
Coral Gables, FL, USA
fraymo@miami.edu
http://www.as.miami.edu/chemistry/FMRaymo.html

Chapter 2

Françisco M. Raymo is Assistant Professor in the Department of Chemistry at the University of Miami. His research interests lie at the interface of chemistry and materials science. In particular, he is exploring innovative strategies to process optical signals with molecular switches, design fluorescent probes for chemical sensing and assemble nanostructured films from electroactive building blocks.

Manitra Razafinimanana

Université Paul Sabatier
Centre de Physique des Plasmas et leurs Applications (CPPAT)
Toulouse, France
razafinimanana@cpat.ups-tlse.fr
http://cpat.ups-tlse.fr

Chapter 3

Manitra Razafinimanana was born in Analalava, Madagascar, on June 28, 1951. He received the 3rd cycle degree, and Doctorat d'Etat ès Sciences Physiques degree from the Université Paul Sabatier, Toulouse, France, in 1982 and 1986 respectively. Since 2000, he was held the position of Professor. He has worked on plasma diagnostics, arc-electrode interaction, transport coefficients and thermodynamical properties calculation.

Mark O. Robbins

Johns Hopkins University
Department of Physics and Astronomy
Baltimore, MD, USA
mr@jhu.edu
http://www.pha.jhu.edu/~mr

Chapter 23

Mark Robbins is a Professor in the Department of Physics and Astronomy of the Johns Hopkins University. His research focuses on non-equilibrium processes like friction, adhesion, spreading, fluid invasion, and shear-induced phase transitions. A common goal is to understand the atomic origins of macroscopic behaviour. He is a fellow of the American Physical Society and was a Sloan Fellow and Presidential Young Investigator.

John A. Rogers

University of Illinois
Department of Materials Science and Engineering
Urbana, IL, USA
jrogers@uiuc.edu
http://www.mse.uiuc.edu/faculty/Rogers.html

Chapter 6

John A. Rogers earned the Ph.D. degree in physical chemistry from M.I. T. in 1995. Until 2002, he was at Bell Labs serving first as a Member of Technical Staff and subsequently as research Director. John is currently Founder Professor of Engineering at University of Illinois at Urbana/Champaign. His research interests include methods for micro/nanofabrication, plastic and molecular electronics, photonics

Mark Ruegsegger

Ohio State University
Biomedical Engineering Center
Columbus, OH, USA
mark@bme.ohio-state.edu

Chapter 10

Mark Ruegsegger received his PhD in biomedical engineering at Case Western Reserve Unviersity and is currently an Assistant Professor of Biomedical Engineering at The Ohio State University. His research focus is the development of superparamagnetic particles that can target specific cell or tissue types, in particular atherosclerotic plaque. Other projects include biomimetic surface coatings on biomaterials and characterization of biomolecular flow in nanochannels.

Marina Ruths

Åbo Akademi University
Department of Physical Chemistry
Åbo, Finland
mruths@abo.fi

Chapter 18

Marina Ruths received her Ph.D. from the University of California, Santa Barbara in 1996 followed by postdoctoral research at the University of Illinois at Urbana-Champaign, at the Max-Planck-Institute for Polymer Research, an at Åbo Akademi University. Her current research includes adhesion, friction and nanorheology of surfactant, polymer, and liquid crystal systems. She received an ASLA-Fulbright grant in 1991 and an Alexander von Humboldt fellowship in 1998.

Dror Sarid

University of Arizona
Optical Sciences Center
Tucson, AZ, USA
sarid@optics.arizona.edu
http://www.optics.arizona.edu/spm

Chapter 30

Dror Sarid is Professor of Optical Sciences and Director of the Optical Data Storage Center. He is conducting research in scanning tunneling microcopy, atomic force microscopy and related systems and in particular probe storage, nano-optics and nano-technology. He published three books and more than 150 papers, mostly on optics and nanotechnology related topics.

Akira Sasahara

Kanagawa Academy
of Science and Technology
Surface Chemistry Laboratory
Kanagawa, Japan
ryo@net.ksp.or.jp
http://home.ksp.or.jp/onishipro/home.eng/

Chapter 13

Dr. Sasahara Akira is interested in local structures formed on solid surfaces. His current research focuses on elucidation of chemical and physical properties of nano-scale structures on metal oxide surfaces and their effect on chemical reaction.

André Schirmeisen

University of Münster
Institute of Physics
Münster, Germany
schira@uni-muenster.de
http://www.andre-schirmeisen.de

Chapter 15

Dr. André Schirmeisen is currently working with his research group on nanoscale mechanical phenomena at Münster University. First to combine field ion microscopy with force microscopy to investigate atomically defined nanocontacts. He spent several years in Canada at McGill University to earn his PhD degree in physics. Before he worked as a strategic business consultant and is always interested in connecting nanoscience research with business applications.

Alexander Schwarz

University of Hamburg
Institute of Applied Physics
Hamburg, Germany
aschwarz@physnet.uni-hamburg.de
http://www.nanoscience.de/group_r/

Chapter 14

Dr. Alexander Schwarz belongs to the scientific staff of the Center of Microstructure Research at the Institute of Applied Physics at the University of Hamburg, Germany. The group has 10 years (since 1993) experience in scientific research in the field of force microscopy and spectroscopy at cryogenic temperatures in ultrahigh vacuum and high magnetic fields. He is a Senior Scientist and works on Magnetic Force Microscopy (MFM) at low temperatures

Udo D. Schwarz

Yale University
Department
of Mechanical Engineering
New Haven, CT, USA
udo.schwarz@yale.edu
http://www.eng.yale.edu/nanomechanics/

Chapter 14

Udo D. Schwarz received his Ph.D. from the University of Basel in 1993 already using scanning force microscopy. Subsequently he moved to the University of Hamburg where he specialised on low-temperature scanning force microscopy and nanotribology. After spending a year at the Lawrence Berkeley National Laboratory he accepted a position as Associate Professor of Mechanical Engineering at Yale University in 2002.

Philippe Serp

Ecole Nationale Supèrieure d'Ingénieurs
en Arts Chimiques
et Technologiques
Laboratoire de Catalyse,
Chimie Fine et Polymères
Toulouse, France
Philippe.Serp@ensiacet.fr
http://www.ensiacet.fr/

Chapter 3

Philippe Serp is associate professor in the LCCFP at ENSIACET. After receiving is PhD from Paul Sabatier University, Toulouse, in 1994 where he worked on the preparation of supported catalyst, he moved to Universidade do Porto to carry out post-doctoral research on catalytic CVD to prepare carbon fibers. His current research interests include CVD preparation of nanostructured materials and catalysis.

Bryan R. Smith

Ohio State University
Biomedical Engineering Center
Columbus, OH, USA
bryan@bme.ohio-state.edu

Chapter 10

Bryan Smith is a Ph.D. candidate in biomedical engineering at The Ohio State University, specializing in the microfabrication and bioconjugation of micro- and nanoparticles for proteomics as well as imaging and drug delivery in breast cancer and atherosclerosis.

Anisoara Socoliuc

University of Basel
Institute of Physics
Basel, Switzerland
A.Socoliuc@unibas.ch
http://monet.physik.unibas.ch/gue/uhvafm/

Chapter 20

Anisoara Socoliuc studied physics in Iasi, Romania. She is now PhD student at the University of Basel. The main topic of her work is the study of wear processes on the nanometer scale by friction force microscopy.

Yasuhiro Sugawara

Osaka University
Department of Applied Physics
Suita, Japan
sugawara@ap.eng.osaka-u.ac.jp

Chapter 13

Yasuhiro Sugawara received his Ph.D. in 1988 from Tohoku University and is Professor in the Department of Applied Physics of the Graduate School of Engineering at Osaka University since 2002. His research focuses on the further development of scanning probe microscopes and their applications, especially the noncontact atomic force microscope for the observation of solid surfaces at the atomic and molecular level. His aim is also to develop new nanomaterials and nanodevices by manipulation of single atoms and molecules using the atomic force microscope.

George W. Tyndall

IBM Almaden Research Center
Science and Technology
San Jose, CA, USA
tyndallgw@netscape.net

Chapter 29

Dr. George Tyndall's current research focuses on the friction and adhesion in boundary lubricants especially as it pertains to the tribology of the magnetic recording head-disk interface in hard-disk drive applications.

Peter Vettiger

IBM Zurich Research Laboratory
Manager Micro-/Nanomechanics
Rüschlikon, Switzerland
pv@zurich.ibm.com

Chapter 31

In 1963 Peter Vettiger joined IBM Research (Rüschlikon). He established and headed micro/nanoscale fabrication activities for superconducting, electronic and opto-electronic devices. Together with G.K. Binnig, he initiated Millipede probe-storage activities in 1995. His research interests are micro/nanomechanical devices and systems for probe-storage and biological applications. He is an IEEE Fellow and received a Doctor honoris causa from the University of Basel, Switzerland.

Darrin J. Young

Case Western Reserve University
Electrical Engineering
and Computer Science
Cleveland, OH, USA
djy@po.cwru.edu
http://home.cwru.edu/~djy

Chapter 8

Darrin J. Young received his BS, MS, and PhD degrees from the EECS Department at University of California at Berkeley in 1991, 1993, and 1999, respectively. He pioneered RF MEMS high-Q tunable passive devices for wireless communications. He joined the EECS Department at Case Western Reserve University as an assistant professor in 1999. His main research interests include MEMS device design and fabrication.

Babak Ziaie

University of Minnesota
Department of Electrical
and Computer Engineering
Minneapolis, MN, USA
ziaie@ece.umn.edu

Chapter 5

Dr. Babak Ziaie is an assistant professor in the Electrical and Computer Engineering Department of the University of Minnesota. His research interests are located at the boundaries between engineering and biology. He collaborates with physicians and biologists of all kinds and his lab is actively involved in the design and fabrication of biomedical micro and nanoelectromechanical systems.

Christian A. Zorman

Case Western Reserve University
Department of Electrical Engineering
and Computer Science
Cleveland, OH, USA
caz@po.cwru.edu
http://mems.cwru.edu/

Chapters 7, 8

Christian A. Zorman received a B.S. cum laude in physics and a B.A. cum laude in economics from The Ohio State University in 1988. He received M.S. and Ph.D. degrees in physics from Case Western Reserve University (CWRU) in 1991 and 1994, respectively. Dr. Zorman served in several research capacities at CWRU between 1994 and 2002, before being appointed an Associate Professor in 2002. His research interests involve materials and processes for MEMS.

Philippe K. Zysset

Technische Universität Wien
Institut für Leichtbau
und Flugzeugbau (ILFB)
Wien, Austria
philippe.zysset@epfl.ch
http://ilfb.tuwien.ac.at/

Chapter 21

Philippe Zysset graduated in physics in 1987 and obtained a Ph.D. in biomechanics in 1994 at Ecole Polytechnique Fédérale de Lausanne (EPFL) in Lausanne, Switzerland. He completed his postdoctoral training at the Orthopaedic Research Laboratories of the University of Michigan and returned to EPFL as an assistant professor in 1997. His current research interests are in bone biomechanics.

Detailed Contents

List of Tables XXIX
List of Abbreviations XXXIII

1 **Introduction to Nanotechnology** 1
1.1 Background and Definition of Nanotechnology 1
1.2 Why Nano? 2
1.3 Lessons from Nature 2
1.4 Applications in Different Fields 3
1.5 Reliability Issues of MEMS/NEMS 4
1.6 Organization of the Handbook 5
References 5

Part A Nanostructures, Micro/Nanofabrication, and Micro/Nanodevices

2 **Nanomaterials Synthesis and Applications: Molecule-Based Devices** 9
2.1 Chemical Approaches to Nanostructured Materials 10
2.1.1 From Molecular Building Blocks to Nanostructures 10
2.1.2 Nanoscaled Biomolecules: Nucleic Acids and Proteins 10
2.1.3 Chemical Synthesis of Artificial Nanostructures 12
2.1.4 From Structural Control to Designed Properties and Functions 12
2.2 Molecular Switches and Logic Gates 14
2.2.1 From Macroscopic to Molecular Switches 15
2.2.2 Digital Processing and Molecular Logic Gates 15
2.2.3 Molecular AND, NOT, and OR Gates 16
2.2.4 Combinational Logic at the Molecular Level 17
2.2.5 Intermolecular Communication 18
2.3 Solid State Devices 22
2.3.1 From Functional Solutions to Electroactive and Photoactive Solids 22
2.3.2 Langmuir–Blodgett Films 23
2.3.3 Self-Assembled Monolayers 27
2.3.4 Nanogaps and Nanowires 31
2.4 Conclusions and Outlook 35
References 36

3 **Introduction to Carbon Nanotubes** 39
3.1 Structure of Carbon Nanotubes 40
3.1.1 Single-Wall Nanotubes 40
3.1.2 Multiwall Nanotubes 43

3.2 Synthesis of Carbon Nanotubes ... 45
3.2.1 Solid Carbon Source-Based Production Techniques for Carbon Nanotubes ... 45
3.2.2 Gaseous Carbon Source-Based Production Techniques for Carbon Nanotubes ... 52
3.2.3 Miscellaneous Techniques ... 57
3.2.4 Synthesis of Aligned Carbon Nanotubes ... 58
3.3 Growth Mechanisms of Carbon Nanotubes ... 59
3.3.1 Catalyst-Free Growth ... 59
3.3.2 Catalytically Activated Growth ... 60
3.4 Properties of Carbon Nanotubes ... 63
3.4.1 Variability of Carbon Nanotube Properties ... 63
3.4.2 General Properties ... 63
3.4.3 SWNT Adsorption Properties ... 63
3.4.4 Transport Properties ... 65
3.4.5 Mechanical Properties ... 67
3.4.6 Reactivity ... 67
3.5 Carbon Nanotube-Based Nano-Objects ... 68
3.5.1 Hetero-Nanotubes ... 68
3.5.2 Hybrid Carbon Nanotubes ... 68
3.5.3 Functionalized Nanotubes ... 71
3.6 Applications of Carbon Nanotubes ... 73
3.6.1 Current Applications ... 73
3.6.2 Expected Applications Related to Adsorption ... 76
References ... 86

4 **Nanowires** ... 99
4.1 Synthesis ... 100
4.1.1 Template-Assisted Synthesis ... 100
4.1.2 VLS Method for Nanowire Synthesis ... 105
4.1.3 Other Synthesis Methods ... 107
4.1.4 Hierarchical Arrangement and Superstructures of Nanowires ... 108
4.2 Characterization and Physical Properties of Nanowires ... 110
4.2.1 Structural Characterization ... 110
4.2.2 Transport Properties ... 115
4.2.3 Optical Properties ... 126
4.3 Applications ... 131
4.3.1 Electrical Applications ... 131
4.3.2 Thermoelectric Applications ... 133
4.3.3 Optical Applications ... 134
4.3.4 Chemical and Biochemical Sensing Devices ... 137
4.3.5 Magnetic Applications ... 137
4.4 Concluding Remarks ... 138
References ... 138

5 **Introduction to Micro/Nanofabrication** 147
5.1 Basic Microfabrication Techniques 148
5.1.1 Lithography 148
5.1.2 Thin Film Deposition and Doping 149
5.1.3 Etching and Substrate Removal 153
5.1.4 Substrate Bonding 157
5.2 MEMS Fabrication Techniques 159
5.2.1 Bulk Micromachining 159
5.2.2 Surface Micromachining 163
5.2.3 High-Aspect-Ratio Micromachining 166
5.3 Nanofabrication Techniques 170
5.3.1 E-Beam and Nano-Imprint Fabrication 171
5.3.2 Epitaxy and Strain Engineering 172
5.3.3 Scanned Probe Techniques 173
5.3.4 Self-Assembly and Template Manufacturing 176
References 180

6 **Stamping Techniques for Micro and Nanofabrication: Methods and Applications** 185
6.1 High Resolution Stamps 186
6.2 Microcontact Printing 187
6.3 Nanotransfer Printing 190
6.4 Applications 193
6.4.1 Unconventional Electronic Systems 193
6.4.2 Lasers and Waveguide Structures 198
6.5 Conclusions 200
References 200

7 **Materials Aspects of Micro- and Nanoelectromechanical Systems** . 203
7.1 Silicon 203
7.1.1 Single Crystal Silicon 204
7.1.2 Polysilicon 205
7.1.3 Porous Silicon 208
7.1.4 Silicon Dioxide 208
7.1.5 Silicon Nitride 209
7.2 Germanium-Based Materials 210
7.2.1 Polycrystalline Ge 210
7.2.2 Polycrystalline SiGe 210
7.3 Metals 211
7.4 Harsh Environment Semiconductors 212
7.4.1 Silicon Carbide 212
7.4.2 Diamond 215
7.5 GaAs, InP, and Related III-V Materials 217
7.6 Ferroelectric Materials 218
7.7 Polymer Materials 219
7.7.1 Polyimide 219
7.7.2 SU-8 220

7.7.3 Parylene 220
7.8 Future Trends 220
References 221

8 **MEMS/NEMS Devices and Applications** 225
8.1 MEMS Devices and Applications 227
8.1.1 Pressure Sensor 227
8.1.2 Inertial Sensor 229
8.1.3 Optical MEMS 233
8.1.4 RF MEMS 239
8.2 NEMS Devices and Applications 246
8.3 Current Challenges and Future Trends 249
References 250

9 **Microfluidics and Their Applications to Lab-on-a-Chip** 253
9.1 Materials for Microfluidic Devices and Micro/Nano Fabrication Techniques 254
9.1.1 Silicon 254
9.1.2 Glass 254
9.1.3 Polymer 255
9.2 Active Microfluidic Devices 257
9.2.1 Microvalves 258
9.2.2 Micropumps 260
9.3 Smart Passive Microfluidic Devices 262
9.3.1 Passive Microvalves 262
9.3.2 Passive Micromixers 265
9.3.3 Passive Microdispensers 266
9.3.4 Microfluidic Multiplexer Integrated with Passive Microdispenser 267
9.3.5 Passive Micropumps 269
9.3.6 Advantages and Disadvantages of the Passive Microfluidic Approach 269
9.4 Lab-on-a-Chip for Biochemical Analysis 270
9.4.1 Magnetic Micro/Nano Bead-Based Biochemical Detection System 270
9.4.2 Disposable Smart Lab-on-a-Chip for Blood Analysis 273
References 276

10 **Therapeutic Nanodevices** 279
10.1 Definitions and Scope of Discussion 280
10.1.1 Design Issues 281
10.1.2 Utility and Scope of Therapeutic Nanodevices 285
10.2 Synthetic Approaches: "top-down" versus "bottom-up" Approaches for Nanotherapeutic Device Components 285
10.2.1 Production of Nanoporous Membranes by Microfabrication Methods: A top-down Approach 285

10.2.2 Synthesis of Poly(amido) Amine (PAMAM) Dendrimers: A bottom-up Approach 286
10.2.3 The Limits of top-down and bottom-up Distinctions with Respect to Nanomaterials and Nanodevices 287
10.3 Technological and Biological Opportunities 288
10.3.1 Assembly Approaches 288
10.3.2 Targeting: Delimiting Nanotherapeutic Action in Three-Dimensional Space 296
10.3.3 Triggering: Delimiting Nanotherapeutic Action in Space and Time 298
10.3.4 Sensing Modalities 302
10.3.5 Imaging Using Nanotherapeutic Contrast Agents 304
10.4 Applications for Nanotherapeutic Devices 307
10.4.1 Nanotherapeutic Devices in Oncology 307
10.4.2 Cardiovascular Applications of Nanotherapeutics 310
10.4.3 Nanotherapeutics and Specific Host Immune Responses 311
10.5 Concluding Remarks: Barriers to Practice and Prospects 315
10.5.1 Complexity in Biology 315
10.5.2 Dissemination of Biological Information 315
10.5.3 Cultural Differences Between Technologists and Biologists 316
References 317

Part B Scanning Probe Microscopy

11 **Scanning Probe Microscopy – Principle of Operation, Instrumentation, and Probes** 325
11.1 Scanning Tunneling Microscope 327
11.1.1 Binnig et al.'s Design 327
11.1.2 Commercial STMs 328
11.1.3 STM Probe Construction 330
11.2 Atomic Force Microscope 331
11.2.1 Binnig et al.'s Design 333
11.2.2 Commercial AFM 333
11.2.3 AFM Probe Construction 338
11.2.4 Friction Measurement Methods 342
11.2.5 Normal Force and Friction Force Calibrations of Cantilever Beams 346
11.3 AFM Instrumentation and Analyses 347
11.3.1 The Mechanics of Cantilevers 347
11.3.2 Instrumentation and Analyses of Detection Systems for Cantilever Deflections 350
11.3.3 Combinations for 3-D-Force Measurements 358
11.3.4 Scanning and Control Systems 359
References 364

12 **Probes in Scanning Microscopies** 371
12.1 Atomic Force Microscopy 372
12.1.1 Principles of Operation 372
12.1.2 Standard Probe Tips 373
12.1.3 Probe Tip Performance 374
12.1.4 Oxide-Sharpened Tips 375
12.1.5 FIB tips 376
12.1.6 EBD tips 376
12.1.7 Carbon Nanotube Tips 376
12.2 Scanning Tunneling Microscopy 382
12.2.1 Mechanically Cut STM Tips 382
12.2.2 Electrochemically Etched STM Tips 383
References 383

13 **Noncontact Atomic Force Microscopy and Its Related Topics** 385
13.1 Principles of Noncontact Atomic Force Microscope (NC-AFM) 386
13.1.1 Imaging Signal in AFM 386
13.1.2 Experimental Measurement and Noise 387
13.1.3 Static AFM Operating Mode 387
13.1.4 Dynamic AFM Operating Mode 388
13.1.5 The Four Additional Challenges Faced by AFM 388
13.1.6 Frequency-Modulation AFM (FM-AFM) 389
13.1.7 Relation Between Frequency Shift and Forces 390
13.1.8 Noise in Frequency-Modulation AFM – Generic Calculation . 391
13.1.9 Conclusion 391
13.2 Applications to Semiconductors 391
13.2.1 Si(111)7×7 Surface 392
13.2.2 Si(100)2×1 and Si(100)2×1:H Monohydride Surfaces 393
13.2.3 Metal-Deposited Si Surface 395
13.3 Applications to Insulators 397
13.3.1 Alkali Halides, Fluorides, and Metal Oxides 397
13.3.2 Atomically Resolved Imaging of a NiO(001) Surface 402
13.3.3 Atomically Resolved Imaging Using Noncoated and Fe-Coated Si Tips 402
13.4 Applications to Molecules 404
13.4.1 Why Molecules and What Molecules? 404
13.4.2 Mechanism of Molecular Imaging 404
13.4.3 Perspectives 407
References 407

14 **Low Temperature Scanning Probe Microscopy** 413
14.1 Microscope Operation at Low Temperatures 414
14.1.1 Drift 414
14.1.2 Noise 415
14.1.3 Stability 415
14.1.4 Piezo Relaxation and Hysteresis 415
14.2 Instrumentation 415

14.2.1 A Simple Design for a Variable Temperature STM 416
14.2.2 A Low Temperature SFM Based on a Bath Cryostat 417
14.3 Scanning Tunneling Microscopy and Spectroscopy 419
14.3.1 Atomic Manipulation 419
14.3.2 Imaging Atomic Motion 420
14.3.3 Detecting Light from Single Atoms and Molecules 421
14.3.4 High Resolution Spectroscopy 422
14.3.5 Imaging Electronic Wave Functions 427
14.3.6 Imaging Spin Polarization: Nanomagnetism 431
14.4 Scanning Force Microscopy and Spectroscopy 433
14.4.1 Atomic-Scale Imaging 434
14.4.2 Force Spectroscopy 436
14.4.3 Electrostatic Force Microscopy 438
14.4.4 Magnetic Force Microscopy 439
References 442

15 **Dynamic Force Microscopy** 449
15.1 Motivation: Measurement of a Single Atomic Bond 450
15.2 Harmonic Oscillator: A Model System for Dynamic AFM 454
15.3 Dynamic AFM Operational Modes 455
15.3.1 Amplitude-Modulation/ Tapping-Mode AFMs 456
15.3.2 Self-Excitation Modes 461
15.4 *Q*-Control 464
15.5 Dissipation Processes Measured with Dynamic AFM 468
15.6 Conclusion 471
References 471

16 **Molecular Recognition Force Microscopy** 475
16.1 Ligand Tip Chemistry 476
16.2 Fixation of Receptors to Probe Surfaces 478
16.3 Single-Molecule Recognition Force Detection 479
16.4 Principles of Molecular Recognition Force Spectroscopy 482
16.5 Recognition Force Spectroscopy: From Isolated Molecules to Biological Membranes 484
16.5.1 Forces, Energies, and Kinetic Rates 484
16.5.2 Complex Bonds and Energy Landscapes 486
16.5.3 Live Cells and Membranes 489
16.6 Recognition Imaging 490
16.7 Concluding Remarks 492
References 493

Part C Nanotribology and Nanomechanics

17 **Micro/Nanotribology and Materials Characterization Studies Using Scanning Probe Microscopy** 497
17.1 Description of AFM/FFM and Various Measurement Techniques 499

17.1.1 Surface Roughness and Friction Force Measurements 500
17.1.2 Adhesion Measurements 502
17.1.3 Scratching, Wear and Fabrication/Machining 503
17.1.4 Surface Potential Measurements 503
17.1.5 In Situ Characterization of Local Deformation Studies 504
17.1.6 Nanoindentation Measurements 504
17.1.7 Localized Surface Elasticity and Viscoelasticity Mapping 505
17.1.8 Boundary Lubrication Measurements 507
17.2 Friction and Adhesion 507
17.2.1 Atomic-Scale Friction 507
17.2.2 Microscale Friction 507
17.2.3 Directionality Effect on Microfriction 511
17.2.4 Velocity Dependence on Microfriction 513
17.2.5 Effect of Tip Radii and Humidity on Adhesion and Friction .. 515
17.2.6 Scale Dependence on Friction 518
17.3 Scratching, Wear, Local Deformation, and Fabrication/Machining ... 518
17.3.1 Nanoscale Wear 518
17.3.2 Microscale Scratching 519
17.3.3 Microscale Wear 520
17.3.4 In Situ Characterization of Local Deformation 524
17.3.5 Nanofabrication/Nanomachining 526
17.4 Indentation 526
17.4.1 Picoindentation 526
17.4.2 Nanoscale Indentation 527
17.4.3 Localized Surface Elasticity and Viscoelasticity Mapping 528
17.5 Boundary Lubrication 530
17.5.1 Perfluoropolyether Lubricants 530
17.5.2 Self-Assembled Monolayers 536
17.5.3 Liquid Film Thickness Measurements 537
17.6 Closure 538
References 539

18 **Surface Forces and Nanorheology of Molecularly Thin Films** 543
18.1 Introduction: Types of Surface Forces 544
18.2 Methods Used to Study Surface Forces 546
18.2.1 Force Laws 546
18.2.2 Adhesion Forces 547
18.2.3 The SFA and AFM 547
18.2.4 Some Other Force-Measuring Techniques 549
18.3 Normal Forces Between Dry (Unlubricated) Surfaces 550
18.3.1 Van der Waals Forces in Vacuum and Inert Vapors 550
18.3.2 Charge Exchange Interactions 552
18.3.3 Sintering and Cold Welding 553
18.4 Normal Forces Between Surfaces in Liquids 554
18.4.1 Van der Waals Forces in Liquids 554
18.4.2 Electrostatic and Ion Correlation Forces 554
18.4.3 Solvation and Structural Forces 557

18.4.4 Hydration and Hydrophobic Forces 559
18.4.5 Polymer-Mediated Forces 561
18.4.6 Thermal Fluctuation Forces 563
18.5 Adhesion and Capillary Forces 564
18.5.1 Capillary Forces 564
18.5.2 Adhesion Mechanics 566
18.5.3 Effects of Surface Structure, Roughness, and Lattice Mismatch 566
18.5.4 Nonequilibrium and Rate-Dependent Interactions: Adhesion Hysteresis 567
18.6 Introduction: Different Modes of Friction and the Limits of Continuum Models 569
18.7 Relationship Between Adhesion and Friction Between Dry (Unlubricated and Solid Boundary Lubricated) Surfaces 571
18.7.1 Amontons' Law and Deviations from It Due to Adhesion: The Cobblestone Model 571
18.7.2 Adhesion Force and Load Contribution to Interfacial Friction 572
18.7.3 Examples of Experimentally Observed Friction of Dry Surfaces 576
18.7.4 Transition from Interfacial to Normal Friction with Wear 579
18.8 Liquid Lubricated Surfaces 580
18.8.1 Viscous Forces and Friction of Thick Films: Continuum Regime 580
18.8.2 Friction of Intermediate Thickness Films 582
18.8.3 Boundary Lubrication of Molecularly Thin Films: Nanorheology 584
18.9 Role of Molecular Shape and Surface Structure in Friction 591
References 594

19 Scanning Probe Studies of Nanoscale Adhesion Between Solids in the Presence of Liquids and Monolayer Films 605
19.1 The Importance of Adhesion at the Nanoscale 605
19.2 Techniques for Measuring Adhesion 606
19.3 Calibration of Forces, Displacements, and Tips 610
19.3.1 Force Calibration 610
19.3.2 Probe Tip Characterization 611
19.3.3 Displacement Calibration 612
19.4 The Effect of Liquid Capillaries on Adhesion 612
19.4.1 Theoretical Background 612
19.4.2 Experimental and Theoretical Studies of Capillary Formation with Scanning Probes 614
19.4.3 Future Directions 618
19.5 Self-Assembled Monolayers 618
19.5.1 Adhesion at SAM Interfaces 618
19.5.2 Chemical Force Microscopy: General Methodology 619
19.5.3 Adhesion at SAM-Modified Surfaces in Liquids 620
19.5.4 Impact of Intra- and Inter-Chain Interactions on Adhesion 621

19.5.5 Adhesion at the Single-Bond Level ... 622
19.5.6 Future Directions ... 623
19.6 Concluding Remarks ... 624
References ... 624

20 **Friction and Wear on the Atomic Scale** ... 631
20.1 Friction Force Microscopy in Ultra-High Vacuum ... 632
20.1.1 Friction Force Microscopy ... 632
20.1.2 Force Calibration ... 632
20.1.3 The Ultra-high Vacuum Environment ... 635
20.1.4 A Typical Microscope in UHV ... 635
20.2 The Tomlinson Model ... 636
20.2.1 One-dimensional Tomlinson Model ... 636
20.2.2 Two-dimensional Tomlinson Model ... 637
20.2.3 Friction Between Atomically Flat Surfaces ... 637
20.3 Friction Experiments on Atomic Scale ... 638
20.3.1 Anisotropy of Friction ... 642
20.4 Thermal Effects on Atomic Friction ... 642
20.4.1 The Tomlinson Model at Finite Temperature ... 642
20.4.2 Velocity Dependence of Friction ... 644
20.4.3 Temperature Dependence of Friction ... 645
20.5 Geometry Effects in Nanocontacts ... 646
20.5.1 Continuum Mechanics of Single Asperities ... 646
20.5.2 Load Dependence of Friction ... 647
20.5.3 Estimation of the Contact Area ... 647
20.6 Wear on the Atomic Scale ... 649
20.6.1 Abrasive Wear on the Atomic Scale ... 649
20.6.2 Wear Contribution to Friction ... 650
20.7 Molecular Dynamics Simulations of Atomic Friction and Wear ... 651
20.7.1 Molecular Dynamics Simulation of Friction Processes ... 651
20.7.2 Molecular Dynamics Simulations of Abrasive Wear ... 652
20.8 Energy Dissipation in Noncontact Atomic Force Microscopy ... 654
20.9 Conclusion ... 656
References ... 657

21 **Nanoscale Mechanical Properties – Measuring Techniques and Applications** ... 661
21.1 Local Mechanical Spectroscopy by Contact AFM ... 662
21.1.1 The Variable-Temperature SLAM (T-SLAM) ... 663
21.1.2 Example One: Local Mechanical Spectroscopy of Polymers ... 664
21.1.3 Example Two: Local Mechanical Spectroscopy of NiTi ... 665
21.2 Static Methods – Mesoscopic Samples ... 667
21.2.1 Carbon Nanotubes – Introduction to Basic Morphologies and Production Methods ... 667
21.2.2 Measurements of the Mechanical Properties of Carbon Nanotubes by SPM ... 668
21.2.3 Microtubules and Their Elastic Properties ... 673

21.3 Scanning Nanoindentation: An Application to Bone Tissue 674
21.3.1 Scanning Nanoindentation 674
21.3.2 Application of Scanning Nanoindentation 674
21.3.3 Example: Study of Mechanical Properties of Bone Lamellae Using SN 675
21.3.4 Conclusion 681
21.4 Conclusions and Perspectives 682
References 682

22 **Nanomechanical Properties of Solid Surfaces and Thin Films** 687
22.1 Instrumentation 688
22.1.1 AFM and Scanning Probe Microscopy 688
22.1.2 Nanoindentation 689
22.1.3 Adaptations of Nanoindentation 690
22.1.4 Complimentary Techniques 691
22.1.5 Bulge Tests 691
22.1.6 Acoustic Methods 692
22.1.7 Imaging Methods 693
22.2 Data Analysis 694
22.2.1 Elastic Contacts 694
22.2.2 Indentation of Ideal Plastic Materials 694
22.2.3 Adhesive Contacts 695
22.2.4 Indenter Geometry 696
22.2.5 Analyzing Load/Displacement Curves 696
22.2.6 Modifications to the Analysis 699
22.2.7 Alternative Methods of Analysis 700
22.2.8 Measuring Contact Stiffness 701
22.2.9 Measuring Viscoelasticity 702
22.3 Modes of Deformation 702
22.3.1 Defect Nucleation 702
22.3.2 Variations with Depth 704
22.3.3 Anisotropic Materials 704
22.3.4 Fracture and Delamination 704
22.3.5 Phase Transformations 705
22.4 Thin Films and Multilayers 707
22.4.1 Thin Films 707
22.4.2 Multilayers 709
22.5 Developing Areas 711
References 712

23 **Atomistic Computer Simulations of Nanotribology** 717
23.1 Molecular Dynamics 718
23.1.1 Model Potentials 719
23.1.2 Maintaining Constant Temperature 720
23.1.3 Imposing Load and Shear 721
23.1.4 The Time-Scale and Length-Scale Gaps 721
23.1.5 A Summary of Possible Traps 722

23.2 Friction Mechanisms at the Atomic Scale ... 723
23.2.1 Geometric Interlocking ... 723
23.2.2 Elastic Instabilities ... 724
23.2.3 Role of Dimensionality and Disorder ... 727
23.2.4 Elastic Instabilities vs. Wear in Atomistic Models ... 727
23.2.5 Hydrodynamic Lubrication and Its Confinement-Induced Breakdown ... 729
23.2.6 Submonolayer Films ... 731
23.3 Stick-Slip Dynamics ... 732
23.4 Conclusions ... 734
References ... 735

24 **Mechanics of Biological Nanotechnology** ... 739
24.1 Science at the Biology–Nanotechnology Interface ... 740
24.1.1 Biological Nanotechnology ... 740
24.1.2 Self-Assembly as Biological Nanotechnology ... 740
24.1.3 Molecular Motors as Biological Nanotechnology ... 740
24.1.4 Molecular Channels and Pumps as Biological Nanotechnology ... 741
24.1.5 Biologically Inspired Nanotechnology ... 742
24.1.6 Nanotechnology and Single Molecule Assays in Biology ... 743
24.1.7 The Challenge of Modeling the Bio-Nano Interface ... 744
24.2 Scales at the Bio-Nano Interface ... 746
24.2.1 Spatial Scales and Structures ... 747
24.2.2 Temporal Scales and Processes ... 749
24.2.3 Force and Energy Scales: The Interplay of Deterministic and Thermal Forces ... 750
24.3 Modeling at the Nano-Bio Interface ... 752
24.3.1 Tension Between Universality and Specificity ... 752
24.3.2 Atomic-Level Analysis of Biological Systems ... 753
24.3.3 Continuum Analysis of Biological Systems ... 753
24.4 Nature's Nanotechnology Revealed: Viruses as a Case Study ... 755
24.5 Concluding Remarks ... 760
References ... 761

25 **Mechanical Properties of Nanostructures** ... 763
25.1 Experimental Techniques for Measurement of Mechanical Properties of Nanostructures ... 765
25.1.1 Indentation and Scratch Tests Using Micro/Nanoindenters ... 765
25.1.2 Bending Tests of Nanostructures Using an AFM ... 765
25.1.3 Bending Tests Using a Nanoindenter ... 769
25.2 Experimental Results and Discussion ... 770
25.2.1 Indentation and Scratch Tests of Various Materials Using Micro/Nanoindenters ... 770
25.2.2 Bending Tests of Nanobeams Using an AFM ... 773
25.2.3 Bending Tests of Microbeams Using a Nanoindenter ... 777

25.3 Finite Element Analysis of Nanostructures with Roughness and Scratches 778
25.3.1 Stress Distribution in a Smooth Nanobeam 779
25.3.2 Effect of Roughness in the Longitudinal Direction 781
25.3.3 Effect of Roughness in the Transverse Direction and Scratches 781
25.3.4 Effect on Stresses and Displacements for Materials That Are Elastic, Elastic-Plastic, or Elastic-Perfectly Plastic 784
25.4 Closure 785
References 786

Part D Molecularly Thick Films for Lubrication

26 **Nanotribology of Ultrathin and Hard Amorphous Carbon Films** 791
26.1 Description of Commonly Used Deposition Techniques 795
26.1.1 Filtered Cathodic Arc Deposition Technique 798
26.1.2 Ion Beam Deposition Technique 798
26.1.3 Electron Cyclotron Resonance Chemical Vapor Deposition Technique 799
26.1.4 Sputtering Deposition Technique 799
26.1.5 Plasma-Enhanced Chemical Vapor Deposition Technique 799
26.2 Chemical Characterization and Effect of Deposition Conditions on Chemical Characteristics and Physical Properties 800
26.2.1 EELS and Raman Spectroscopy 800
26.2.2 Hydrogen Concentrations 804
26.2.3 Physical Properties 804
26.2.4 Summary 805
26.3 Micromechanical and Tribological Characterizations of Coatings Deposited by Various Techniques 805
26.3.1 Micromechanical Characterization 805
26.3.2 Microscratch and Microwear Studies 813
26.3.3 Macroscale Tribological Characterization 822
26.3.4 Coating Continuity Analysis 826
References 827

27 **Self-Assembled Monolayers for Controlling Adhesion, Friction and Wear** 831
27.1 A Primer to Organic Chemistry 834
27.1.1 Electronegativity/Polarity 834
27.1.2 Classification and Structure of Organic Compounds 835
27.1.3 Polar and Nonpolar Groups 838
27.2 Self-Assembled Monolayers: Substrates, Head Groups, Spacer Chains, and End Groups 839
27.3 Tribological Properties of SAMs 841
27.3.1 Surface Roughness and Friction Images of SAMs Films 844
27.3.2 Adhesion, Friction, and Work of Adhesion 844

27.3.3 Stiffness, Molecular Spring Model, and Micropatterned SAMs 848
27.3.4 Influence of Humidity, Temperature, and Velocity on Adhesion and Friction 850
27.3.5 Wear and Scratch Resistance of SAMs 853
27.4 Closure 856
References 857

28 **Nanoscale Boundary Lubrication Studies** 861
28.1 Lubricants Details 862
28.2 Nanodeformation, Molecular Conformation, and Lubricant Spreading 864
28.3 Boundary Lubrication Studies 866
28.3.1 Friction and Adhesion 866
28.3.2 Rest Time Effect 869
28.3.3 Velocity Effect 871
28.3.4 Relative Humidity and Temperature Effect 873
28.3.5 Tip Radius Effect 876
28.3.6 Wear Study 879
28.4 Closure 880
References 881

29 **Kinetics and Energetics in Nanolubrication** 883
29.1 Background: From Bulk to Molecular Lubrication 885
29.1.1 Hydrodynamic Lubrication and Relaxation 885
29.1.2 Boundary Lubrication 885
29.1.3 Stick Slip and Collective Phenomena 885
29.2 Thermal Activation Model of Lubricated Friction 887
29.3 Functional Behavior of Lubricated Friction 888
29.4 Thermodynamical Models Based on Small and Nonconforming Contacts 890
29.5 Limitation of the Gaussian Statistics – The Fractal Space 891
29.6 Fractal Mobility in Reactive Lubrication 892
29.7 Metastable Lubricant Systems in Large Conforming Contacts 894
29.8 Conclusion 895
References 895

Part E Industrial Applications and Microdevice Reliability

30 **Nanotechnology for Data Storage Applications** 899
30.1 Current Status of Commercial Data Storage Devices 901
30.1.1 Non-Volatile Random Access Memory 904
30.2 Opportunities Offered by Nanotechnology for Data Storage 907
30.2.1 Motors 907
30.2.2 Sensors 909
30.2.3 Media and Experimental Results 913
30.3 Conclusion 918

References 919

31 **The "Millipede" –
A Nanotechnology-Based AFM Data-Storage System** 921
31.1 The Millipede Concept 923
31.2 Thermomechanical AFM Data Storage 924
31.3 Array Design, Technology, and Fabrication 926
31.4 Array Characterization 927
31.5 *x*/*y*/*z* Medium Microscanner 929
31.6 First Write/Read Results with the 32×32 Array Chip 931
31.7 Polymer Medium 932
31.7.1 Writing Mechanism 932
31.7.2 Erasing Mechanism 935
31.7.3 Overwriting Mechanism 937
31.8 Read Channel Model 939
31.9 System Aspects 943
31.9.1 [peserror]PES Generation for the Servo Loop 943
31.9.2 Timing Recovery 945
31.9.3 Considerations on Capacity and Data Rate 946
31.10Conclusions 948
References 948

32 **Microactuators for Dual-Stage Servo Systems
in Magnetic Disk Files** 951
32.1 Design of the Electrostatic Microactuator 952
32.1.1 Disk Drive Structural Requirements 952
32.1.2 Dual-Stage Servo Configurations 953
32.1.3 Electrostatic Microactuators:
Comb-Drives vs. Parallel-Plates 954
32.1.4 Position Sensing 956
32.1.5 Electrostatic Microactuator Designs for Disk Drives 958
32.2 Fabrication 962
32.2.1 Basic Requirements 962
32.2.2 Electrostatic Microactuator Fabrication Example 962
32.2.3 Electrostatic Microactuator Example Two 963
32.2.4 Other Fabrication Processes 966
32.2.5 Suspension-Level Fabrication Processes 967
32.2.6 Actuated Head Fabrication 968
32.3 Servo Control Design
of MEMS Microactuator Dual-Stage Servo Systems 968
32.3.1 Introduction to Disk Drive Servo Control 969
32.3.2 Overview of Dual-Stage Servo Control Design Methodologies 969
32.3.3 Track-Following Controller Design
for a MEMS Microactuator Dual-Stage Servo System 971
32.3.4 Dual-Stage Seek Control Design 976
32.4 Conclusions and Outlook 978
References 979

33 **Micro/Nanotribology of MEMS/NEMS Materials and Devices** 983
33.1 Introduction to MEMS 985
33.2 Introduction to NEMS 988
33.3 Tribological Issues in MEMS/NEMS 989
33.3.1 MEMS 989
33.3.2 NEMS 994
33.3.3 Tribological Needs 995
33.4 Tribological Studies of Silicon and Related Materials 995
33.4.1 Tribological Properties of Silicon and the Effect of Ion Implantation 996
33.4.2 Effect of Oxide Films on Tribological Properties of Silicon 998
33.4.3 Tribological Properties of Polysilicon Films and SiC Film 1000
33.5 Lubrication Studies for MEMS/NEMS 1003
33.5.1 Perfluoropolyether Lubricants 1003
33.5.2 Self-Assembled Monolayers (SAMs) 1004
33.5.3 Hard Diamond-like Carbon (DLC) Coatings 1008
33.6 Component-Level Studies 1009
33.6.1 Surface Roughness Studies of Micromotor Components 1009
33.6.2 Adhesion Measurements 1011
33.6.3 Static Friction Force (Stiction) Measurements in MEMS 1014
33.6.4 Mechanisms Associated with Observed Stiction Phenomena in Micromotors 1016
References 1017

34 **Mechanical Properties of Micromachined Structures** 1023
34.1 Measuring Mechanical Properties of Films on Substrates 1023
34.1.1 Residual Stress Measurements 1023
34.1.2 Mechanical Measurements Using Nanoindentation 1024
34.2 Micromachined Structures for Measuring Mechanical Properties 1024
34.2.1 Passive Structures 1025
34.2.2 Active Structures 1028
34.3 Measurements of Mechanical Properties 1034
34.3.1 Mechanical Properties of Polysilicon 1034
34.3.2 Mechanical Properties of Other Materials 1036
References 1037

35 **Thermo- and Electromechanics of Thin-Film Microstructures** 1039
35.1 Thermomechanics of Multilayer Thin-Film Microstructures 1041
35.1.1 Basic Phenomena 1041
35.1.2 A General Framework for the Thermomechanics of Multilayer Films 1046
35.1.3 Nonlinear Geometry 1054
35.1.4 Nonlinear Material Behavior 1058
35.1.5 Other Issues 1061
35.2 Electromechanics of Thin-Film Microstructures 1061
35.2.1 Applications of Electromechanics 1061
35.2.2 Electromechanics Analysis 1063

35.2.3 Electromechanics – Parallel-Plate Capacitor 1064
35.2.4 Electromechanics of Beams and Plates 1066
35.2.5 Electromechanics of Torsional Plates 1068
35.2.6 Leveraged Bending 1069
35.2.7 Electromechanics of Zipper Actuators 1070
35.2.8 Electromechanics for Test Structures 1072
35.2.9 Electromechanical Dynamics: Switching Time 1073
35.2.10 Electromechanics Issues: Dielectric Charging 1074
35.2.11 Electromechanics Issues: Gas Discharge 1075
35.3 Summary and Mention of Topics not Covered 1078
References 1078

36 **High Volume Manufacturing and Field Stability of MEMS Products** 1083
36.1 Manufacturing Strategy 1086
36.1.1 Volume 1086
36.1.2 Standardization 1086
36.1.3 Production Facilities 1086
36.1.4 Quality 1087
36.1.5 Environmental Shield 1087
36.2 Robust Manufacturing 1087
36.2.1 Design for Manufacturability 1087
36.2.2 Process Flow and Its Interaction with Product Architecture . 1088
36.2.3 Microstructure Release 1095
36.2.4 Wafer Bonding 1095
36.2.5 Wafer Singulation 1097
36.2.6 Particles 1098
36.2.7 Electrostatic Discharge and Static Charges 1098
36.2.8 Package and Test 1099
36.2.9 Quality Systems 1101
36.3 Stable Field Performance 1102
36.3.1 Surface Passivation 1102
36.3.2 System Interface 1105
References 1106

37 **MEMS Packaging and Thermal Issues in Reliability** 1111
37.1 MEMS Packaging 1111
37.1.1 MEMS Packaging Fundamentals 1112
37.1.2 Contemporary MEMS Packaging Approaches 1113
37.2 Hermetic and Vacuum Packaging and Applications 1116
37.2.1 Integrated Micromachining Processes 1117
37.2.2 Post-Packaging Processes 1118
37.2.3 Localized Heating and Bonding 1119
37.3 Thermal Issues and Packaging Reliability 1122
37.3.1 Thermal Issues in Packaging 1122
37.3.2 Packaging Reliability 1124
37.3.3 Long-Term and Accelerated MEMS Packaging Tests 1125

37.4 Future Trends and Summary 1128
References 1129

Part F Social and Ethical Implication

38 **Social and Ethical Implications of Nanotechnology** 1135
38.1 Applications and Societal Impacts 1136
38.2 Technological Convergence 1139
38.3 Major Socio-technical Trends 1141
38.4 Sources of Ethical Behavior 1143
38.5 Public Opinion 1145
38.6 A Research Agenda 1148
References 1149

Acknowledgements 1153
About the Authors 1155
Detailed Contents 1171
Subject Index 1189

Subject Index

1, 1′-biphenyl-4-thiol (BPT) 843
4, 4′-dihydroxybiphenyl (DHBp) 843
μCP (microcontact printing) 187, 197, 834
μTAS (micro-total analysis systems) 257
γ-modified geometry 700
1,4-phenylenediamine (pDA) 916
1/f noise 387, 957
16-mercaptohexadecanoic acid thiol (MHA) 843
1-D localization
– effects 120
– theory 120
1-D to 3-D transition in magnetic field 122
1-ethyl-3-(3-diamethylaminopropyl) carbodiimide (EDC) 80, 296
2-D FKT model 638
2-DEG
– two-dimensional electron gas 438
2-D-histogram technique 647
2-amino-4,5-imidazoledicarbonitrile (AIDCN) 918
2-mercaptoethylamine HCl 478
2-pyridyldithiopropionyl (PDP) 478
3,3′-dimethyl bipyridinium 14
3-D bulk state 428
3-D coil inductor 242
3-D switching architecture 238
3-D-force measurements 358
3-nitrobenzal malonitrile (NBMN) 916

A

abrasive wear 649, 652
Abrikosov lattice 440
ABS (air-bearing surface) 968
ac electrochemical deposition 104
ac electrodeposition 103
accelerated friction 822
acceleration energy 798
accelerometer 230
– lateral 231
– packaging 1100
– test 1101
– three-axis 233
– vertical 232
– wafer fabrication 1091
acetonitrile 14, 19
acetylene (C_2H_2) 799
acid-treated SWNT 64
acoustic emission 690
actin-myosin motor 751
activation energy barrier 482
active linearization 361
active matrix electronic paper 195
active microfluidic devices 257
active microvalves 258
active structures 1028
actuated head 954
– fabrication 968
actuated slider 954, 968
– configuration 954
actuated suspension 953
actuation
– pneumatic 258
actuator
– miniaturized 986
actuator failure modes 1074
adatoms 392
addressing nanoscale bits of data 901
adenine 10
adenosine triphosphate (ATP) 741
adhesion 325, 498, 507, 515, 605, 799, 844, 930, 983
– at SAM interfaces 618
– control 832
– energy 619
– force 452, 546, 547, 552, 564, 566, 567, 619, 621
– force, quantized 558, 584
– hysteresis 567, 568, 573
– hysteresis, relation to friction 573, 576–579, 585
– influence of humidity on 850
– measurement 1011
– measurement techniques 606
– mechanics 564–566
– performance 866
– primary minimum 544, 555, 559
– promoter 188
– quantization 623
– rate-dependent 567, 568
adhesion force
– total 616
adhesion-controlled friction 571, 572, 574
adhesive
– coating 479
– interaction 732
– material 610
– surfaces 731
– wear 728
adhesive force 335, 646, 662, 867, 1004
– increase 876
– intrinsic 832
– mapping 537, 869
– measurement 503
adhesive tape film 907
adiabatic limit 482
adsorbate 404, 695
adsorbed
– insulator substrate 400
– water 875
– water film 878
adsorption 64, 79, 476
– capacity 80
– sites in MWNT 65
AES (Auger electron spectroscopy) measurement 826
AFAM (atomic force acoustic microscopy) 326
affinity maturation 316
affinity selection 297
affinity-based targeting 301
AFM (atomic force microscope) 173–175, 743, 765
– adhesion measurement 620
– *Binnig* design 333
– calculated sensitivity 355
– cantilever 332, 350, 610
– cantilever array 923
– commercial 333
– construction 288
– contact mode 331
– control electronics 362
– design optimization 348
– designs 376
– feedback loop 362
– for UHV application 635
– image 392
– imaging signal 386
– instrumentation 347
– manufacturers 333
– microscratch technique 853
– probe construction 338
– probes 378
– pull-off force 619

– resolution 375
– set-up 450, 453
– spectroscopy 458
– surface height map 817
– test 818
– thermomechanical recording 924
– tip 288, 379, 386, 450, 452, 477, 480, 1013
– tip containing antibodies 489
– tip radius 518
– tip sensor design 476
– tip size 622
AFM image
– tapping mode 914
Ag 395, 397
– on Nb(110) 429
– trimer 395
air bag sensor 1084, 1091
air damping 470
air gap capacitance 1075
air induced oscillations 348
air/water interface 479
air-bearing surface (ABS) 968
Al_2O_3-TiC head 825
Al_2O_3 54, 462, 633
– grains 508
– ultrafiltration membrane 669
Al_2O_3-TiC composite 508
Al_2O_3-TiC slider 998
$Al_xGa_{1-x}As$ 248
AlAs 248
alcohol 837
aldehydes 837
aligned carbon nanotubes 58
alignment accuracy 195
alignment of nanowires 109
alkali halides 397
alkaline phosphatase (AP) 271
alkane 835, 888, 889
alkanethiolate 187
– film 841
– on gold 187
– on palladium 187
– on silver 187
alkanethiols 404
alkylsilane film 619
all-fiber interferometer 418
all-MEMS configuration 907
all-metal cantilever 912
all-optical logic gates 21
all-optical switching network 237
alloy 662
all-silicon cantilever 926
alumina template 127
aluminum
– gates 34
– mirror array 1088
aluminum oxide 54, 462, 633
aluminum-based surface micromachining 241
AM (amplitude modulation)
– AFM 388
– mode 455
amide 838
amine 838
Amontons law 454, 517, 571, 574, 577, 580, 646, 718
amorphous
– oxide layer 106
– surfaces 544
amorphous carbon 49, 793
– chemical structure 800
– coatings 800, 806
– phase 48
amperometric time-based detection method 272
amphiphilic molecular building blocks 23
amplitude feedback 465
amplitude modulation (AM) 388, 456
– mode 331
– SFM (scanning force microscope) 433
amplitude of vibration 911
anchor group 840
AND operator 15
angle of helicity 41
angle of twist formula 355
anion-terminated tip 399
anisotropy 704
– of friction 642
annealing effect 1036
anodic alumina (Al_2O_3) 100
– oxide films 102
– templates 103
anodic bonding 1096, 1115
anodization 100
ANSYS
– analysis 1123
– stress prediction 1124
anthracene 20
– channel 21
antibiotic-resistant pathogens 293
antibody 296
– directed enzyme-prodrug therapy (ADEPTS) 301
– production 312
antibody–antigen 475
– binding 283
– complex 484
– recognition 491
antibody-coated beads 271
antiferromagnetic spin ordering 403
antigen presenting cell (APC) 312
antigen recognition 282
antimicrobial chemotherapy 293
anti-stiction
– agent (ASA) 1104
– material 1104
AP (alkaline phosphatase) 271
APC (antigen presenting cell) 312
APCVD (atmospheric pressure chemical vapor deposition) 770
apparent viscosity 888
applications
– electrical 131
arc discharge 798
areal density 900, 901, 907, 925, 932, 947, 952
Arg-Gly-Asp (RGD) 489
armchair-type SWNT 42
array
– cantilever 927
– characterization 927
– chip 923
– of aligned nanowires 115
– of nanotubes 53
Arrhenius equation 1127
artifacts 415
artificial double helix 12
as-deposited film 1035
aspect ratio 127
asperity-asperity interaction 618
as-released curvature 1059
assembled nanotube probes 378
assembly approaches 288
assembly of nanostructures 294
association process 476
atmospheric pressure chemical vapor deposition (APCVD) 770
atomic
– interaction force 452
– manipulation 419
– motion imaging 420
atomic force acoustic microscopy (AFAM) 326
atomic force microscope (AFM) 114, 288, 289, 303, 325, 331, 371, 385, 404, 476, 498, 546, 547, 549, 585, 588, 688, 717, 743, 841, 861, 862, 899, 922
atomic resolution 330, 382, 388, 389, 427, 476, 480
– imaging 391, 394–397, 462
atomic-scale
– dissipation 655
– force measurement 347

– friction 507
– hysteresis 468
– image 327, 342, 434
– roughness 721
atomistic computer simulation 398, 724
ATP hydrolysis 742
ATP(adenosintriphosphate) synthase 741
attraction
– long-range 550, 562
attractive force–distance profile 480
attractive interaction 864
Au film 771
Au microbeam 777, 778, 785
Au tip 619
Au(111) 423, 848, 850
– surface 619
Aubry transition 728
Au-coated AFM tip 619
austenite 665
average
– distance 749
– lifetime 482
– stress 1049
axial strength 672
azopyridine 19

B

backbone chain 839
backflow 261
bacmid 283
bacteriophage 744, 755
bacteriorhodopsin 742
BaF_2 398
ball and spring model 724
ball grid arrays (BGA) 1113
ballistic phonon transport 126
ballistic transport 115, 121
balloon angioplasty 189
bamboo texture 52, 61
band gap tunability 128
band structure
– electronic 128
barcode reading 233
barcode tags 136
barrier-hopping 887
– fluctuations 890
basic switching operations 15
batch fabrication 249
– techniques 373
batch nanotube tip fabrication 380
B-cell epitopes 311
bc-MWNT (bamboo-concentric) 44
BDCS (biphenyldimethylchlorosilane) 849
beam
– two-layer 1045
beam bending energy 745
beam deflection 455
beam failure 775
beam-deflection FFM 633
beam-steering mirror 239
behavior
– unethical 1143
Bell's formula 483
bending
– moment 767, 783
– stiffness 349
– strength 767, 774, 775, 785
– stress 767, 781
– test 765, 766, 769
Berkovich
– indenter 765
– indenter tip 677
– pyramid 696
– tip 815
BGA (ball grid arrays) 1113
bh-MWNT (bamboo-herringbone) 44
$Bi_2Sr_2CaCu_2O_{8+\delta}$ (BSCCO) 431
$Bi_{1-x}Sb_x$ nanowires 123
bias voltage 328
biaxial strain 1042
biaxial stress 1053
BiCMOS 229
bidirectional micropump 261
bifurcation 1056
bilayer
– beam 1054
– cantilever 1046
– film 1059
– model-lipid-lubricant 889
– plate 1050
– system 1042
bimetallic catalyst 51, 62
binary compound 397
bio force probe (BFP) 480
biochemical fluids 254
biochemical reaction 262, 270
biocompatibility 263, 311
bioconjugate chemistry 295, 296
bioconjugation method 288
biofilter 271
– surface-mounted 272
biofluidic chip 273, 987
biofunctionalized cantilever 742, 743, 745
biogenic amorphous silicon shells 314
biological
– activity 316
– affinity 297
– affinity reagents 303
– device 742
– device components 304
– evolution 751
– macromolecules 299
– nanotechnology 739, 760
– self-assembly 293
biomaterials 711
biomechanics 675
biomedical device 189
BioMEMS (Biological or Biomedical Microelectromechanical Systems) 253, 262, 764, 987
– actuation 269
biomimetic valves 262
biomolecules bound to carbon nanotubes 303
biomolecules in therapeutic nanodevices 293
bio–nano interface 744
BioNEMS (Biological or Biomedical Nanoelectromechanical Systems) 764, 989
bio-sampling
– magnetic bead-based 271
biosensors 80, 302
bio-surfaces 489
biotechnology 1139
– limits of 281
biotin 80, 476
biotin-avidin 622
– spectrum 487
biotin-directed IgG 476
biotinylated AFM tip 489
biphenyl 419
bipyridine
– building block 12, 31
– centered LUMOs 26
birefringent crystal 352
bismuth nanowire 118, 134
bistability 459, 916
bit pitch (BP) 943
bit strings 938
bits per inch (BPI) 952
blister test 691
Bloch states 428
Bloch wave 427
block co-polymer 291
block-like debris 823
blood cell analysis 310
blood flow measurement 304
blood-pool agents 306
BN nanotubes 68

Bode plot 976
Bohr radius of excitons 135
Boltzmann ansatz 482
Boltzmann distribution 887
bond
– angle 67
– breakage 483
– lifetime 482
– rupture 487
– scission 623
– strength 550
bonded lubricant film 864
bonded PFPE 1012
bonding
– energy 386
– kinetics 893
– techniques 1114
bone
– lamellae 675
– lamellation 679
– tissue 675
boron
– diffusion 227
– doped Si 205
– ion implantation 330
bottom-up
– approach 99, 107, 285
– chemical strategies 10
– materials 281
– scheme 133
boundary
– element (BE) 1063
– film formation 861
– lubricant 729, 861
– lubricant film 832
– lubrication 530, 569–571, 580, 583, 585, 852, 866, 885
– lubrication measurement 507
– lubrication regime 884
– scattering effects 125
bovine serum albumin (BSA) 476
bpsi (bits per square inch) 901
BPT (1, 1′-biphenyl-4-thiol) 854, 1007
BPTC (cross-linked BPT) 1007
branched hydrocarbon lubricants 889
breakdown
– electrical 1076
bridging of polymer chains 562, 563
broken beams 777
broken coating chip 817
brush *see* polymer brush
brush-like structures 109
buckling 808, 812, 1026
– force 377
– stress 813
bulge test 691
bulk
– addressing 22
– atoms 387
– conduction band 424
– diamond 793, 801, 803
– etched silicon wafer 263
– fluid transport 261
– graphitic carbon 824
– micromachining 229
– Si wafers 205
– state 428
– viscosity 731
– xenon 436

C

C_{70} 792
C_2 moieties 51
C_{60} 32, 41, 49, 404, 647, 792
– film 330
– fullerenes 289
– island 641, 642
– multilayered film 404
– SWNT (single wall nanotube) 70
C_{60}-terminated film 843
Ca ion 399
cadherin-mediated adhesion 488
CaF_2 398
CaF_2 tip 652
$CaF_2(111)$ 399
$CaF_2(111)$ surface 399, 652
calculated tapping behavior 911
calculus of variations 745
calibration 610, 698, 699
calorimetric experiments 482
cantilever 387, 390, 457, 481
– all-metal 912
– all-silicon 926
– array 926, 927
– axis 418
– base 469
– beam 1015, 1061
– beam array (CBA) 1011
– beam model 1067
– biofunctionalized 742, 743, 745
– biosensors 303
– cell 926
– deflection 332, 372, 387, 389, 418, 465, 480, 632, 745, 848
– deflection calculation 344
– diode array 928
– driven 455
– effective mass 633
– eigenfrequency 415
– elasticity 636
– fabrication 909
– flexible 332
– foil 373
– heater 925
– material 339, 387
– motion 331, 913
– mount 336
– oscillation 457, 469
– *Q*-factor 433
– resonance 451
– resonance behavior 350
– resonance frequency 481, 611, 890, 1029
– sensor 925
– spring 454
– spring constant 485
– stainless steel 813
– stiffness 338, 505, 610, 926
– temperature 940
– thickness 632, 633
– tip 303, 454
– triangular 339, 349, 354
– untwisted 335
cantilever-based probes 433
capacitance detection 356
capacitive
– accelerometers 230
– detection 361
– detector 331
– displacement sensor 1033
– forces 465
– position sensing 956
– sensor 227, 230
– transducer 245
capacitor-like device 913
capacity growth 903
capillarity-driven stop valve 265
capillary
– effect on adhesion 607
– electrophoresis (CE) 255
– force 544, 559, 564, 619, 662
– force curve 617
– formation 614
– pressure 265
– stiffness 614
capsid wall 760
carbon 792
– crystalline 792
– film 865
– magnetron sputtered 799
– shells 52
– source 57
– spacer chain 853
– superactivated 77

carbon coating
– unhydrogenated 804
carbon nanofiber
– vapor grown 56
carbon nanotube (CNT) 55, 105, 287, 425, 667
– adsorption properties 76
– application 73
– catalyst-free growth 59
– catalyst-supporting materials 75
– chemical reactivity 67
– diameters 63
– FETS 303
– field emission 74
– formation 46
– growth mechanisms 59
– heterogeneity 63
– in situ filling 69
– maximum current density 74
– mechanical properties 376
– molten state filling 69
– oxidation 72
– production 46
– properties 63
– quantum wires 66
– sublimation filling 70
– synthesis conditions 61
– tip 342, 375, 376, 501
– wet chemistry filling 69
carbon–carbon distance 436
carboxyl acid 837
carboxylates ($RCOO^-$) 404
cardiovascular
– sensors 310
– system 310
– tissue engineering 310
carrier
– density 118, 122
– gas 804
– mean free path 116, 118, 121
– mobility 118
Casimir force 545
casting 186
catalysis 397
catalysis-enhanced disproportionation 54
catalysis-enhanced thermal cracking 54
catalyst 49, 106, 668
– nanoparticles 58
– preparation 54
catalyst-based SWNT 61
catalyst-free 57
catalytic chemical vapor deposition (CCVD) 52, 55, 58, 60, 61
catalytic decomposition 54
catalytically grown MWNT 673
cathode deposit 49, 51
cathodic arc carbon 806
CCVD *see* catalytic chemical vapor deposition
CCVD method 61
CDS (correlated double sampling) technique 961
CdSe nanorods 136
cell adhesion 489
cellular immune responses 311
cellular phone 904
$CeO_2(111)$ 400
ceramic
– matrix composites 81
– slider 824
– tip 74
cerpacs 1101
CFM (chemical force microscopy) 73, 489, 619
chain adsorption 894
chalcogenide 906, 910, 915
change of meniscus 1005
changing of surface conditions 117
channel etching process 255
channel-liquid pair 261
chaotic mixer 266
characteristic distance 588
characteristic slip time 894
characteristic velocity 889
charge density wave (CDW) 428
charge exchange interactions 545, 552, 553
charge fluctuation forces *see* ion correlation forces
charge separation 1116
charge transfer interactions *see* charge exchange interactions
charge trapping 913
charge-controlled pull-in equation 1064
check valve design 260
chelate metal cations 12
chemical
– bond 386
– bonding 388, 863
– bonding force 331
– characterization 800
– composition 115
– detection devices 137
– force 386
– force microscopy (CFM) 73, 489, 619
– heterogeneity 568
– input 16
– interaction force 452
– sensors 74
– signal 19, 20
– synthesis 12
– vapor deposition (CVD) 104, 793
– vapor infiltration 57
chemisorption 861
chemistry route 57
chemoselective conjugation 295, 296
chemotherapy 307
chip level integration 1093
chip-on-flex (COF) 1113
clamped-clamped mechanical beams 247
classical finite size 116
classical size effect 120
clock field 946
clogging 56
closed loop sensitivity 973
clustered acetate 406
CMOS electronics 230
CMOS static random access memory (SRAM) 235
CMOS sustaining electronics 245
c-MWNT (concentric multiwall nanotube) 43, 48, 60, 67
Co clusters 425
CO disproportionation 54
CO on Cu(110) 419
coated particles 306
coating
– continuity 826
– damage 823
– failure 820
– hardness 818
– hydrophobic 1014
– mass density 801
– microstructure 800
– thickness 810, 812
coating–substrate interface 819
coefficient
– effective 647
coefficient of friction 501, 510, 516, 535, 579, 634, 771, 773, 809, 813, 823, 824, 844, 868, 872, 996, 1005
– Si(100) 531
– Z-15 531
– Z-DOL (BW) 531
coefficient of friction *see* friction coefficient
coefficient of friction relationship 343
coefficient of thermal expansion (CTE) 1124
co-energy 1064
coercivity 137

COF (chip-on-flex) 1113
cognate receptors 298
cognitive ligands 476
cognitive science 1139
coherence length 423
cohesive
– potential 759
– surface model 759
coil inductor 242
cold welding 553, 579
collagen fibers 679
collective phenomenon 885, 888
colloidal
– forces 545
– probe 548
colossal magneto resistive effect 439
comb storage device 907
comb-drive 955, 1030
– actuator 955
– resonator 245
combined AFM nanoindenter device 676
commensurate Cu(111) 728
commensurate system 723
commercial data-storage device 907
commercial MEMS devices 991
communicating
– between compatible molecular components 19
– force 748
– molecular switches 19
compact bone
– lamellae 678
– mean hardness 680
complex
– bonds 486
– dielectric function 127
– fluidic systems 886
– logic functions with molecular switches 19
component
– failure 764
– supplier 902
– surface 1009
composite 662
composite material 81
compositionally modulated nanowires 106
compression energy 887
compressive stress 805, 812, 1035, 1044
computer simulation 718
– forces 553, 557, 561, 564, 565, 585
– friction 553, 573, 574, 579, 580, 583, 585, 586, 590, 591, 594
concentration
– critical 650
concentric
– graphenes 43
– texture 61
– type (c-MWNT) 43
conditional release of therapeutics 288
conductance 117
conductance quantization in metallic nanowires 115
conducting polymers 103, 304
conductive polymeric materials 304
conductivity 136
– electrical 133
confined
– complex liquids 891
– nanostructure 68
– simple liquids 888
confinement 544, 545, 570, 584, 593
conformation 864
conformational defect 849
conforming contacts 894
connector
– miniaturized 986
constant
– amplitude (CA) 464
– amplitude FM mode 468
– current mode 328
– excitation mode (CE) 464
– force mode 362
– height mode 328, 399
constrained elastic media 1061
consumer electronics 904
consumer nanotherapeutic device 288
contact
– AFM (atomic force microscope) 455
– AFM dynamic method 662
– angle 616, 833
– angle goniometer 847
– angle of SAM 846
– conductance 649, 913
– elastic 703
– mechanics *see* adhesion mechanics
– meniscus 619
– mode 373, 909
– mode photolithography 195
– printing 186, 200, 833
– radius 678
– resistance 690
– stiffness 677, 689, 698, 701, 810, 894
– stress 872
– value theorem 575
contact area 548, 650, 651, 677, 699, 709
– apparent macroscopic 571, 576, 580
– true molecular 547, 549, 566, 571, 574–576, 594
contamination 332
continuous micropump 269
continuous stiffness measurement (CSM) 810
continuum
– analysis 753
– model 638
– theory 554, 556, 558, 559, 569, 580
contour plot 1057
contour-mode disk resonator 246
contraction 1044
contrast 435
– agent design 306
– agents 304
– enhancing agents for medical imaging 304
– enhancing nanoparticle 306
– formation 397
control
– of nanotherapeutic action 302
– of position 109
– over nanowire diameter distribution 106
– system 359, 362
– theory 1144
controlled desorption 419
controlled evaporation 269
controlled geometry (CG) 330
controlled nanoscale architecture 294
controlled substances 1140
controlled triggering of therapeutic action 298
controller design 969
CoO 402
Co-O film on PET 524
Cooper pairs 429
coordinated molecular process 886
copper adhesion 963
copper hexadecafluorophthalocyanine 196
copper tip 652, 727
corannulene 61
cores of carbon nanotubes 105
core-sheath nanowire 114

core-shell structure 132
Coriolis acceleration 232, 1095
correlated-double-sampling (CDS) 960
corrosion 1116, 1125
cost of goods (COGs) 285, 307
co-transporter 489
Couette flow 569, 593, 885
Coulomb
– force 396
– interaction 66
– law of friction 644
coupled electromechanics 1063
covalent
– bonds 10
– scaffolds 12
Cr binding layer 619
Cr coating 432
Cr(001) 427
crack spacing 524
cracks 807
creativity at the nanoscale 1135
creep 415, 691, 1125
– effect 528
– relaxation 1060
creep model
– linear 890
critical
– concentration 650
– curvature 1056
– degree of bending 1031
– load 809
– magnetic field strength 122
– normal load 853
– position 636
– shear stress 570–572, 577, 585, 586
– temperature 423, 434, 1059
– time 894
– velocity 585, 587, 590, 591
crosshair alignment 195
crosslinked BPT (BPTC) 843
crosstalk 363, 927, 1066
cross-track distance 945
cryostat 415
crystal
– growth direction 103
– orientation 206
– structure 333
– surfaces 375
crystal growth orientation, preferred 104
crystalline
– carbon 792
– silicon 254
– surfaces 544
CSM (continuous stiffness measurement) 810
CTE (coefficient of thermal expansion) 1124
CTL (cytotoxic T-cell) responses 311
Cu(001) 397
Cu(100) 640
Cu(100) substrate 728
Cu(100) tip 728
Cu(111) 423, 639, 640, 655
Cu(111) surface states 427
Cu(111) tip 653
cube corner 696, 704
culture of engineering 315
cuprates 431
current density effect 1036
current rectification 116
current-voltage characteristics 116, 916
curvature 1044
– critical 1056
– effects 723
– measured 1053
– variation 1046
curved electrode
– actuator 1070
– shape 1071
custom-designed MEMS product 1112
cut-off distance 552, 573
CVD (chemical vapor deposition) 104, 793
cyclic
– fatigue 810, 1036
– olefin copolymers (COC) 257
– thermomechanical load 1061
– voltammograms 29
cytoplasmic surface 478
cytosine 10
cytoskeletal filament 747
cytotoxic therapeutics 307
cytotoxic T-lymphocytes (CTLs) 311
cytotoxin deposition 309

D

(d, k) codes 947
D_2O 425
damage 553, 569–588
damage mechanism 818
damped harmonic oscillator 454
damping 664
– effective 470
– effects 451
– pneumatic 418
data
– lifetime 905
– rate limitation 922
– rates 948
– storage 900
– storage application 912
– storage device 922
– storage system 994
dc electrochemical deposition 103, 104
de Broglie wavelength 116
Deborah number 583
debris 823
Debye exponential relaxation 885, 892
Debye length 555, 557
decoupled design 970
decoupled track-following control 972
deep reactive ion etching (DRIE) 156, 162, 254, 963
defect motion 415
defect nucleation 702
defect production 651
defects in channel 255
deflection 457, 1054
– curve 1071
– maximum 1071
– measurement 387
– noise 433
deformation 547, 549, 1041, 1047
– behavior 1058
– elastic 450, 566, 581, 647
– of microtubule 674
– plastic 553, 580
– regime 1046
deformed beam 778
degradability 311
degrees of freedom 753
delamination 704, 804, 808, 812, 1125
delivery of therapeutics 296
Demnum-type PFPE 537
– lubricant film 869
dendrimer-based therapeutic 284
dendritic
– cells (DCs) 312
– polymers 310
density
– functional theory (DFT) 78
– modulation 730
– of defects 193, 195
density of states (DOS) 66
– electronic 133
dental enamel 711

depletion
– attraction 545, 562
– interaction 562
– stabilization 545, 562
deployment of functional biomolecules 288
deposition
– conditions 206
– rate 798
– techniques 795, 796
Derjaguin approximation 546, 557, 559
Derjaguin–Landau–Verwey–Overbeek (DLVO) theory 555
Derjaguin–Muller–Toporov (DMT)
– model 647
– theory 566
design rule 348
designed molecules 12
design-for-manufacturability 1086
design-for-volume 1086
detection 939
– circuit 939
– systems 332, 350
device
– architectures 288
– components communication 304
– design 280
– electronic 133
– fluidic 253
– molecule-based 24, 28
– scaling 989
dewetting 868, 1003, 1013
DFM (dynamic force microscopy) 433, 471, 490, 501
DFS (dynamic force spectroscopy) 484
DHBp (4, 4′-dihydroxybiphenyl) 843, 854, 1007
diameter-dependent 116
diamond 67, 373, 379, 673, 698, 792
– coated cantilever 914
– coating 804, 833
– film 793
– like amorphous carbon coating 800
– like carbon (DLC) 524, 1003
– powder 51
– tip 504, 651, 662, 676, 814, 819, 841, 866, 999
diatoms 314
diblock copolymers 102, 103, 138
dicationic BIPY 17
dielectric
– breakdown 691
– capacitance 1075
– charging 1074
– function 127, 129
differential scanning calorimetry (DSC) 664
differential-drive configuration 955
diffuser micropumps 261
diffusion 1061
– based extractor 266
– coefficient 265, 865
– flame synthesis 58
– limited reaction 893
– parameters 420
– thermally activated 415
diffusive
– communication 750
– relaxation 482
– transport 115, 116
digital
– feedback 362
– light processing (DLP) 987
– micromirror device (DMD) 235, 987, 992
– signal processor (DSP) 328
– transmission between molecules 20
dilation 573, 584, 586, 593
dimension 387
dimensions of spatial structures 747
dimer structure 394
dimer-adatom-stacking (DAS) 392
Diophantine equation 973
DIP (dual in-line packaging) 1118
dip coating technique 863
diphtheria toxin (DT) 300
dipole molecule 834
dip-pen nanolithography (DPN) 175, 288
diprotic acid (11-thioundecyl-1-phosphonic acid) 621
direct overwriting 938
direct parallel design 970
direct write electron beam/focused ion beam lithography 186
directly growing nanotubes 379
discontinuous sliding motion 887
disjoining pressure 871
disk drive
– manufacturer 902
– servo control 968
dislocation
– line tension 704
– motion 570
– nucleation 696
dispersion
– force 406
– mixing 266
displacement 1047, 1057
– amplitude 930
– calibration 612
– controlled scanning probe 610
– field 1055
– maximum 1070
– resolution 681
– vertical 779
display
– paper-like 193
disposable biosensors 274
disposable plastic lab-on-a-chip 993
dissemination of biological information 315
dissipated power 939
dissipation 468
– force 461
– measurement 631
dissipative tip-sample interaction 468
dissociation 476
distortion 360
distributed
– Bragg reflector (DBR) 199
– feedback (DFB) 199
– laser resonators 199
dithio-bis(succinimidylundecanoate) 476
dithio-phospholipids 479
D-L amino acids (aas) 291
DLC 1012
– coating 524, 794, 796, 809, 819, 826, 1008
– coating microstructure 812
DLP technology 1088
DLVO interactions 549, 555–558
DMD (digital mirror device)
– fabrication 1088
– manufacturing 1095
– packaging 1100
DMT (Derjaguin–Muller–Toporov)
– model 695
– theory 724
DNA 404, 464, 467, 476, 747
– analysis 272
– analysis system 265
– nanostructure 294
– nanowire 33
– packing 756
domain
– pattern 439
– re-orientation 913
donor impurity concentration 127

doped
– anode 49
– polysilicon film 1001
– silicon 996
– silicon wafer 31
DOS structure 423
double barrier resonant tunneling 124
double-layer
– force 621
– interaction 545, 555–557
double-wall nanotubes (DWNTs) 51
doubly clamped beam 1028
doxorubicin 308
DRAM 923
Drexler 1146
DRIE (deep reactive ion etching) 156, 162, 166, 168–170, 254, 954, 963
dried tissue properties 676
drive amplitude 470
drive current
– maximum 976
driven cantilever 455
driving frequency 351
driving/sensing circuit integration 960
drug
– delivery 285
– resistant organism 293
– targeting 297
dry
– etching 205
– nitrogen environment 999
– (plasma) etching 254
– thermal oxidation 286
dry surfaces 571, 587
– forces 550, 552
– friction 571, 576, 588, 589
dual-axis gyroscope 233
dual-in-line packaging (DIP) 1118
dual-stage
– configuration 952
– sensitivity 973
– servo system 953
– short span seek 977
Dupré equation 569
DWNT (double-wall nanotubes) 51
dynamic
– interactions 544–546, 570, 581, 584
– mode 331
– operation mode 388
– oscillation force 456
– spectral analysis 891
– states of friction 894
– viscosity 264
dynamic AFM 387, 453–455, 461, 468, 469
dynamic force microscopy (DFM) 457, 471, 490, 501
dynamic force spectroscopy (DFS) 484
dynamic friction *see* kinetic friction

E

E. coli
– bacterium 748
– cells 756
early failure 1126
early tumor detection 310
EBD (electron beam deposited) tips 376
ECR (electron cyclotron resonance) CVD 795, 815, 1009
– coating 822
EDP (ethylene diamine pyrocatechol) 254
EELS (electron resonance loss spectroscopy) 800
EFC (electrostatic force constant) 678
effect of surface roughness 779, 781
effective
– coefficient of friction 647
– damping constant 470
– force gradient 463
– ligand concentration 485
– mass 433
– medium 127
– medium theory 127
– shear stress 648
– spring constant 637, 645, 721
– structural stiffness 1053
– tether length 485
– viscosity 583, 593, 885
effects of doping and annealing 120
EHD
– induction type 262
– injection type 262
eigenstrain 1041
elastic
– contact 703
– energy 732
– force 727, 757
– Frenkel–Kontorova chain 727
– Hamiltonian 755
– instabilities 727
– mismatch 1051, 1056
– modulus 664, 698, 767, 770, 774, 805, 806, 809
– properties 673
– properties of ssDNA 746
– stiffness coefficient 1048
elastic deformation 450, 647
– long-range 722
elasticity 433, 663
– of DNA 757
– two-dimensional 754
elastic-plastic deformation 785
elastohydrodynamic lubrication 569, 570, 581, 583
elastomer precursor 256
elastomeric stamp 190
electric charging 913
electric field gradient microscopy (EFM) 114
electric force gradient 338
electrical
– applications 131
– breakdown 1076
– conductivity 133
– isolation 956
– surface passivation 1103
electric-arc
– method 61
– reactor 48
electroactive
– fragments 14
– layer 24
– solids 22
electrochemical
– AFM 338
– cell 24
– degradation 304
– deposition 103, 675
– detection 270
– etch 383
– etching 373
– immunosensor 272
– STM 330
electrochemically etched tips 383
electrode
– arrays 26
– geometry 1076
– pair 1077
electrodeposition 103, 153, 180
electrodes with molecular layers 27
electrohydrodynamic (EHD) pumping 261
electroless deposition 188
electro-luminescence (EL) 132, 135
electrolytic
– conduction 1116

– method 57
electromagnetic
– actuators 259
– forces 550
– microvalves 259
electromechanical
– coupling 245
– load 1040
– mirror 1069
electromechanics
– analysis 1063
– application 1061
– test structure 1072
electromigration 419
electron
– beam deposition (EBD) 376
– cyclotron resonance chemical vapor deposition (ECR-CVD) 795, 799
– energy loss spectrometer (EELS) 113
– energy loss spectroscopy (EELS) 798
– Fermi wavelength 116
– field emission 133
– interactions 428
– mean free path 115
– scattering 120
– tunneling 327
electron-beam lithography 674
electronegativity 834
electron-electron interaction 423
electronic
– band structure 128
– density of states 133
– device 133
– ink 195
– newspaper 196
– noise 333
electron-phonon interaction 428
electro-optical modulation 136
electro-osmosis (EO) 261
electro-osmotic
– flow (EOF) 255
– pumping 260
electroplated microactuator 966
electroplating 153, 1113
electro-rheological fluid 260
electrostatic
– actuation 236, 244, 248
– binding 478
– coupling 1063
– discharge (ESD) 1098
– interaction 393, 402, 464
– microactuator 952
– micromotor 1061
– potential 398, 438
– short-range interaction 398
– spring constant 956
electrostatic force 386, 437, 544–558, 1062, 1103
– constant (EFC) 677
– interaction 393
– microscopy 438
electrostatically actuated devices 258
electrostatic-induced errors 1098
embedded atom method (EAM) 720
embedded system 721
embossing 833
– technique 191
end caps 47
end deflection 1026
end group 864
endofullerene 71
endothelial cell surfaces 488
energy
– barrier 482
– conservation 469
– dissipation 468, 567–569, 572–575, 578, 587, 654, 849
– elastic 732
– electrical 1063
– mechanical 1063
– resolution 422
– scales 750
– separation between subband 115
– total 1063
energy dispersive X-ray spectrometer (EDS) 113
engineered nanostructures 13
engineering materials 758
enhanced permeability and retention effect (EPR) 307
entangled states 570
entanglement 622, 886
– strength 886
entropic force 564
entropically
– confined systems 891
– cooled layer 888
entropy
– structural 886, 887
environmental shield 1087
enzymatic cleavage 287
enzymatic prodrug activator 301
enzymes 482
epitaxial silicon seal 1114
epitaxy 152, 170, 172
epitope mapping 491
epoxy bonding 1115
EPR tumor targeting 310
equibiaxial
– deformation state 1050
– stress distribution 1051
equilibrium
– equipartition 720
– interactions *see* static interactions
– true (full) or restricted 562
equivalent mechanical circuit 1073
erasing mechanism 936
ESD protective circuitry 1098
ester 837
etching 186, 1041
ethanolamine 477
ether 837
ethical rule violation 1143
ethics of nanotechnology 1138
ethylene diamine 287
ethylene diamine pyrocatechol (EDP) 254
Euler
– buckling criterion 1026
– equation 349
Euler–Bernoulli beam equation 1063
eutectic bonding 1115
evolution of curvature 1059
examples of NEMS 988
exchange
– force 402
– force interaction 402
– interaction energy 402
exchangeable carrier plate 636
excitation
– external 456
excitation amplitude 461
exciton 127, 136
– binding energy 135
– confinement 135
– laser action 135
extended exponential Kohlrausch relaxation function 892
external
– excitation 456
– excitation frequency 466
– noise 415
– normal load 867
– triggering 284
– vibrations 329
extraction of mechanical properties 1073
extravasation 297, 307
extrinsic stress 1041
Eyring
– model 644, 887, 890
– theory 887

F

Fab molecule 476
fabrication
– of membrane 314
– of microfluidic systems 254
– of mold masters 255
– techniques 254
– technology 225
face-centered cubic (fcc) 792
faceted 109, 111, 135
failure
– density function 1125
– mechanism 825, 1125
– mode 1101
– mode effect analysis (FMEA) 1101
– of MEMS/NEMS device 1012
– rate 1126
Fano resonance 425
fatigue 810
– behavior 769
– crack 776, 823, 1125
– damage 810
– failure 1036
– life 809
– measurement 1032
– properties 768, 776
– resistance 1028
– strength 776
– test 776, 813
fatty acid monolayer 841
FC (flip chip) technique 1113
FCA (filtered cathodic arc) 815
– coating 806, 811
$Fe(NO_3)_3$ 381
feasible simulation
– maximum 719
Fe-coated tip 403
feedback
– architecture 228
– circuit 333, 464, 465
– loop 363, 373, 450, 456, 461, 465, 485, 634, 957
– network 327
– signal 388
Fe-N/Ti-N multilayer 675
FeO 402
Feridex 306
Fermi
– energy 122
– level 33, 422
– points 428
ferrocene 59
– hydrophobic 24
ferroelectric film 913
ferromagnetic
– particles 305
– probe 439
– tip 441
FE-SEM image 136
FFM (friction force microscopy) 632
– dynamic mode 652
– on atomic scale 644
– tip 507, 636, 639
FIB (focused ion beam) tips 376
fiber
– optic telecommunication 233
– optical interferometer 352
– optics 299
FIB-milled probe 342
FID
– free induction decay 305
field
– emission 910
– emission tip 378, 910
– emission-based display 58
– emission-based screen 74
– ion microscope (FIM) 452
field-effect transistor (FET) 85, 116, 132, 137, 289, 290, 302
figure of merit 133
filled nanotube 69
filling efficiency 71
film/substrate system 1045
films on substrate 1023
film–substrate interface 1035
film–substrate system 1060
filtered cathodic arc (FCA) deposition 796–811
filtering 246
FIM (field ion microscope) 452
– technique 464
finite
– difference (FD) 1063
– element (FE) 1063
– element analysis 778
– element method (FEM) 764, 1049
– element modeling (FEM) 699
– size effect 131
first generation NEMS devices 247
first principle calculation 396
first principle simulation 434
FKT (Frenkel–Kontorova–Tomlinson)
– model 637
flagellar motor rotation rate 750
flap type valves 261
flash memory 899, 904–907
flat punch 697
flavin adenine dinucleotide (FAD) 303
flavoenyzme glucose oxidase (Gox) 303
flexible cantilever 332
flexible electronic devices 193
flip chip technique (FC) 1113
floating catalyst method 55
flocculation 555
Flory temperature 561
flow
– activation energy 865
– laminar 264
– rate 269
– resistance 266
fluid mosaic membranes 293
fluidic
– devices 253
– motion 270
– sampling 253
fluorescence
– quantum yield 17
– spectroscopy 127
fluoride 397, 398
fluoride (111) surface 400
fluorinated DLC 833
fluorinated silane monolayer 190
FM (frequency modulation)
– AFM 464
– AFM images 435
– mode 461
FMEA (failure mode effect analysis) 1101
focused ion beam (FIB) 330, 342, 376, 693, 709
foil cantilever 373
Fokker–Planck 891
folded protein structures 293
force
– advection 266
– and displacement calibration 618
– between macroscopic bodies 551, 556
– between surfaces in liquids 554
– calibration 610, 632
– calibration mode 526
– calibration plot (FCP) 867, 1003
– cantilever-based 433
– constant 618
– curve 382
– detection 348
– distribution 1043
– effective gradient 463
– elastic 727, 757
– extension curve 744
– induced unbinding 487

– long-range 389, 398, 437
– mapping 490
– measurement 387
– measuring techniques 546, 547, 549
– modulation mode (FMM) 663
– modulation technique 505
– packing curves 758
– repulsive 480
– resolution 433, 481
– scales 750
– sensing tip 358
– sensitivity 481
– sensor 332, 389, 635
– spectroscopy (FS) 362, 434, 436, 476
– titration 621
– undulation 545, 564
– velocity curve 752
force field spectroscopy
– three-dimensional 437
force-displacement
– curve 608, 677, 681
– plot 607
force-distance
– calibration 848
– curve 338, 451, 456, 621, 662, 668
– cycle 480
– diagram 469
form factor 900, 907
formate
– covered surface 405
– $(HCOO^-)$ 404
formation of SAM 840
four helical bundle proteins 316
four-quadrants photodetectors 632
Fowler–Nordheim
– equation 1076
– plot 1077
– tunneling 905
fractal
– mobility 892
– space 891
– time 892
– time dependence 893
fracture 704, 1124, 1125
– strength 1036
– stress 767
– surfaces 777
– toughness 765, 767, 771, 775, 806, 807, 809, 810, 818
– toughness measurement 1027, 1030
Frank-Read source 703
free carrier density 127
free induction decay (FID) 305
free surface energy 874
freezing-melting transitions 588, 590, 886
Frenkel–Kontorova–Tomlinson (FKT) 637
– elastic chain 727
– model 725
frequency
– measurement precision 358
– modulation (FM) 388
– modulation (FM) mode 331
– modulation AFM (FM-AFM) 389
– modulation SFM (FM-SFM) 415
– shift calculation 390
– shift curve 463, 464
friction 325, 498, 507, 515, 606, 618, 631, 717, 793, 844, 983
– anisotropy 642
– characteristics 1009
– coefficient 569–572, 576, 577, 579, 580, 583, 590, 592, 593, 886
load dependence 518
– control 832
– directionality effect 512
– effect of humidity on 515
– effect of tip on 515
– experiments on atomic scale 638
– global internal 667
– image 640
– kinetic 571–573, 583, 585–587, 589, 590, 593, 718, 728, 729, 884, 885
– lateral 453
– loop 637, 638, 649
– macroscale 842, 998
– macroscopic 732
– map 583, 640
– measurement 1033
– measurement methods 342
– measuring techniques 547, 548, 576, 583, 585
– mechanism 510, 723, 843, 848, 871, 1013
– molecular dynamics simulation 651
– performance 866
– rate experiments 888
– scale dependent 518
– torque 1015
– torque model 1014
– total 726
friction force 331, 335, 501, 516, 642, 813, 843, 850, 866, 1004, 1005
– calibration 335, 346
– curve 363
– decrease of 875
– image 845
– kinetic 722
– magnitude 345
– map 511, 648
– mapping 869
– microscope (FFM) 325, 498, 499, 841
– microscopy (FFM) 537, 631, 632
– Z-15 531
– Z-DOL (BW) 531
friction models
– cobblestone model 572, 574
– Coulomb model 572
– creep model 589
– distance-dependent model 589
– interlocking asperity model 572
– phase transitions model 587, 589, 590
– rate-and-state dependent 590
– rough surfaces model 588
– surface topology model 588
– velocity-dependent model 587, 589
fringe detector 1028
fringing field 1068
frustules 314
fullerene 41, 792
fullerene-like structure 49
fullerite 42
functionalized nanotube 71, 72
fundamental resonant frequency 350
fused silica 699, 700, 704
– hardness 678
fusion bonding 1114

G

GaAs 248
– stamp 191
– wafers 192
GaAs/AlGaAs
– heterostructures 438
– quantum well heterostructures 248
gage marker 1028
gain 465
gain control circuit 389
GaN nanowires 135
gap stability 327
GaP(110) 435
gas separation 79
gate potential
– floating 905

gauche defect 622
Gaussian
– fluctuation–dissipation relation 889, 891
– statistics 884, 891, 892
Gd on Nb(110) 429
Ge membranes 210
gear
– miniaturized 986
generalized actuator 1064
genetic engineering 1142
genomic libraries 756
genuine graphite 40
geometric
– interlocking 724
– nonlinearity 1054
– parameters 228
– restrictions 249
geometry effects in nanocontacts 646
germanium 705
g-factor 424
giant magnetoresistance (GMR) 900
gimballed microactuator design 959
gimbal-mounted mirror 1097
GIO (grazing impact oscillator) 911
glass 619
– bulk etching 255
– frit bonding 1096
– frit sealing 1093
– transition 894
– transition temperature 256, 570, 579, 665
glassiness 593
glass-like behavior 891
glass-to-glass direct bonding 271
glass-to-silicon package 1127
global internal friction 667
glucose oxidase (GOD) 274
GMR
– giant magnetoresistance 900
gold 330
– coated tip 476
– electrodes 27
– film on polysilicon substrate 1051
– nanowires 136
– to-silicon eutectic bonding 1120
Grahame equation 555
grain
– boundary 553, 1035
– boundary scattering 120, 121
– size 189, 793
graphene 40, 63, 667
– defect 70
graphite 415, 417, 507, 638, 651, 792
– cathode 798
– pellet 47
– surface 463
graphite(0001) 427
gratings 200
grazing impact oscillator (GIO) 911
Griffith fracture theory 778
grinding 382
growth
– by pressure injection 102
– of nanotubes 668
guanine 10
gyro
– packaging 1100
– test 1101
gyroscope 230, 232, 1094

H

H_2O 425
Hagen–Poiseuille equation 264
half-generation precursor 287
Hall resistance 438
Hall–Petch behavior 710
Hamaker constant 386, 464, 551–577, 621, 871
Hamilton–Jacobi method 390
hand-held biochip analyzer 274
handling damage 1105
haptenization 313
haptenized nanostructure 313
hard amorphous carbon coatings 795
hard disk drive (HDD) 899–907, 952
– head 901
– performance 903
hard disk drive technology
– limits for 904
hard-core or steric repulsion 545
hardness 504, 770, 800, 802, 805, 809, 1024
harmonic oscillator 454
harpooning interaction 545, 553
HARPSS 166, 169, 170
HDT (hexadecanethiol) 188, 843, 854, 1006, 1012
head
– displacement 953
– gimbal assembly (HGA) 971
– group 839
– position 969
health care 1141
heat
– curable prepolymer 186
– dissipation 1122
– pipe cooling 1123
– transfer model 1119
heater
– cantilever 925
– temperature 934
heavy ion bombardment 441
heavy metal toxicity 306
helicity 41
helper T-cell epitopes 311
hermetic
– packaging 1116
– seal 1112
herringbone
– MWNT 44, 60, 76
– texture 61
Hershey–Chase experiment 755
Hertz model 694
Hertzian contact model 695
Hertz-plus-offset relation 647
heterodyne interferometer 351
heterogeneous CCVD 53
hetero-nanotube 68
hexadecanethiol (HDT) 188, 843, 854, 1006, 1012
HexSil 166, 168, 169, 954
– micro-molding process 962
hierarchical arrangement 108
hierarchy
– of structures 747
– of temporal processes 749
high modulus elastomer 188
high quality manufacturing 1087
high reflectivity 136
high speed scanning 467
high temperature conditions 61
high temperature superconductivity (HTCS) 419, 431, 440
high volume production 1093
high-aspect-ratio
– MEMS (HARMEMS) 986
– tips 376, 377
high-definition television 235
higher orbital tip states 422
highest resolution images 382
highly ordered monolayers 187
highly oriented pyrolytic graphite (HOPG) 337, 507
high-performance material 1141
high-resolution
– electrodes 196
– FM-AFM 436
– image 305
– imaging 407

– MR images 305
– patterning 193
– printing 186
– spectroscopy 422
– stamps 186
– tips 379
high-temperature operation STM 417
Hill coefficient 488
hinged cantilever 1034
HiPCo process 56
HiPCo technique 48
h-MWNT (herringbone) 44, 60, 76
homodyne interferometer 351
honeycomb
– chained trimer (HCT) 395
– lattice 68
– pattern 399
Hooke's law 433, 450, 454, 469, 480
horizontal coupling 329
host immune responses 311
hot embossing 257
HtBDC (hexa-tert-butyl-decacyclene) on Cu(110) 420
HTCS (high temperature superconductivity) 419, 431, 440
Huber-Mises 695
human bone tissue 675
humanity 1136
humidity 645
– effect 1105
hybrid
– carbon nanotube 68
– manufacturing 1093
– nanotube tip fabrication 380
hybridization 41, 397
hydration forces 545–575
hydration regulation 560
hydrocarbon 53, 57, 835
– chains 731
– precusors 799
– unsaturated 836
hydrodynamic
– forces 545, 549
– lubrication 729, 884
– radius 561
hydrofluoric (HF) 832
hydrogen 394
– bonding 10, 561
– bonding force 622
– concentrations 804
– content 804
– end group 841
– flow rate 805
– in carbon nanotube 77
– sensor 137
– storage 77
– storage capacity 78
– termination 394
hydrogenated carbon 800
hydrogenated coating 800
hydrophilic
– control 262
– surfaces 544, 560
– tip 616
hydrophobic
– coating 1014
– control 262
– ferrocene 24
– force 559, 560
– surfaces 544, 560
– tip 616
hydroxyl polymethacrylamide (HPMA) 308
hydroxylated surfaces 29
hydroxylation 840
hydroxyl-terminated perfluoropolyether 892
Hysitron 676
hysteresis 360, 382, 388, 415, 439, 459, 469, 681
– loop 333, 363

I

IB (ion beam)
– coating 822
IBD (ion beam deposition) 796, 798, 815
IC (integrated circuit) 226, 988
– industry 1112
– package technology 1101
– packaging 1112
ideal nanobiological devices 282
IFM (interfacial force microscope) 610
IgG
– repetitive 484
image
– effects 710
– processing software 364
– topography 331, 406
image projection 1100
– system 1084
imaging
– bandwidth 339
– electronic wave functions 427
– signal noise 389, 391
– tools 111
immunoassay
– magnetic bead-based 271
immunogenic synthetic nanomaterials 313
immunoisolation 286, 313
immunosensor 271
immunosuppression 313
immunosurveillance 313
impregnation method 54
imprinted polymers 200
in situ
– environmental TEM chamber 112
– screening 137
– sharpening of the tips 328
in vivo properties 676
InAs 417
InAs(110) 423, 435
incipient wetness impregnation 76
incommensurate surface 725, 731
indentation 498, 526, 923, 932, 935, 936
– creep process 528
– depth 527, 678, 806, 809, 998
– fatigue damage 812
– hardness 325, 527
– induced compression 812
– modulus 679
– size 527, 924
– size effect (ISE) 704
– technique 765
independent molecular operators 19
indirect sensor-device coupling 304
indirect transition in bismuth 129
indium tin oxide (ITO) 193
induction type EHD 262
inductive heating 1119
inelastic tunneling 419
inertial sensor 229
infected cell 756
influence of humidity
– on adhesion 850
information technology 1139, 1142
initial contact 703
injection type EHD 262
ink
– electronic 195
ink jet printing 193
inorganic-organic solar cells 136
InP(110) 392
in-phase signal 944
in-plane
– actuation 929
– mechanical strain 188
instability position 461
instability, jump-to-position 388, 610

instrumented suspension 967
insulator 397
insuline pump 269
integrated
– circuit (IC) 226, 988
– MEMS 259, 1084, 1092
– MEMS accelerometer 1091
– micromachining process 1117
– silicon suspension 967
– suspension 967
– tip 341, 373
– vacuum sealing 1117
Intel 915
intellectual property protection 1137
interaction
– energy 757
– force 434, 757
– potential 718
interatomic
– attractive force 344
– bonding 393
– force 332
– force constants 387
– interaction 1035
– spring constant 332
interbulk interactions 728
intercalated MoS_2 833
intercellular adhesion molecule-1 (ICAM-1) 489
intercellular concentration 748
intercept length 855
interchain hydrogen bonding 621
interchain interactions 484
interconnect 900
interconnect technology 1094
interconnected operators 18
interdiffusion 568
interdigitation 568
interface sensor components 302
interfacial
– chemistry 607
– defects 813
– energy (tension) *see* surface energy (tension)
– force microscope (IFM) 610
– friction *see* also boundary lubrication, 571–574, 579, 580, 585, 862
– liquid structuring 888
– potential 726
– stress 819
– structuring 888
– wear 606
interferometeric detection sensitivity 353
interlocking macrocycles 26
intermediate or mixed lubrication 569, 570, 582, 583
intermittent contact mode 373, 460
intermolecular communication 18
intermolecular force 289, 481, 484, 606
internal friction 935
internal stress 1025, 1041
intersymbol interference 947
intertube distance 43
intraband transitions 423
intramolecular
– forces 484
– self-assembly 293
intrinsic
– adhesive force 832
– damping 469
– mean stress 1045
– stress 804
iodine 419
iodobenzene 419
ion
– correlation forces 554, 557
– displacement 398
– implantation 205, 528, 925, 996
– implanted silicon 997
– plating techniques 795
– source 798
ion beam
– deposition (IBD) 796, 798
– etching 248
– sputtered carbon 798
ion correlation forces 545
ionic bond 545
ionic strength 619
Ising spin model 892
isolated nanotubes 381
isotropic etching 189
isotropic layer 1050
itinerant nanotube levels 425
ITO (indium-tin-oxide) 195
I–V characterization 116, 124

J

JKR (Johnson–Kendall–Roberts)
– model 619, 621, 695
– relation 646
– theory 566
Joule
– dissipation 655
– heating 439
jump length distribution 889
jump-to-contact 388, 436, 452
junction area 132, 135

K

KBr 397
KBr(100) 640, 650
$KCl_{0.6}Br_{0.4}(001)$ surface 398
Kelvin
– equation 564, 612
– radius 613, 614
Kelvin probe force microscopy (KPFM) 407, 503
ketones 837
key scale at bio-nano interface 746
kidney exclusion limit 296
kinesin 741, 745
kinetic
– friction 571–573, 583, 585–587, 589, 590, 593, 718, 728, 729, 884, 885
– friction force 722
– friction ultralow 591
– measurements 892
– off-rate 487, 488
– processes 703
– rate constant 482, 486
kinetics 883
– of capillary formation 614
Kirchoff hypothesis 1047
knife-edge blocking 361
Koanda effect 266
KOH 330
Kohlrausch
– exponent 892
– relaxation 884
Kondo
– effect 414, 425
– temperature 425
K-shell EELS spectra 802

L

L amino acid 291
lab-on-a-chip 270, 988
– application 270
– concept 253
– disposable plastic 993
– systems 265
– technology 255
lac repressor 754
Lagrangian approach 1064
Lamb waves 693
laminar
– flow 264
– flow characteristics 266
laminated n-channel transistor 196
Landau
– levels 122, 415, 438

– quantization 424
Langevin equation 891, 892
Langmuir–Blodgett (LB) 861
– deposition 536, 833
– film (ethyl-2,3-dihydroxyoctadecanoate) 467
– methodology 35
Laplace
– force 844, 868
– pressure 544, 564, 613–618, 832
– pressure force 616
large angular deflection 1069
large array of nanowires 109
larger-scale systems 1136
large-scale stick-slip motion 733
Larmor frequency 305
laser
– ablation method 61
– beam deflection method 453
– deflection sensing 453
– deflection technique 333
– devices 46
– interference lithography 186
– vaporization method 61
– welding 1119
laser-to-fiber coupler 236
lateral
– accelerometer 231
– contact stiffness 631
– cracks 693
– deflection 671
– deflection signal 1016
– displacement 669
– force 632, 643, 669, 718, 1015
– force calculation 354
– force microscope (LFM) 325, 499, 632
– force microscopy (LFM) 187, 404, 548
– friction 453
– resolution 328, 331, 477, 500, 606, 663
– resonator 1037
– scanning range 929
– spring constant 338, 636
– stiffness 362, 376, 649
lattice
– constant 725
– imaging 332
– vibration 726
layer of resist 186
layered structure 1045
LCC (Leadless Chip Carrier) 1101
leakage 1126
– rate 263
learning theory 1143
length scale 722, 747
Lennard-Jones potential 436, 457, 719, 730
Lennard-Jonesium 727
leukocyte function-associated antigen-1 (LFA-1) 489
leveraged bending 1070
levers 932
lever-sample displacement 608
levitated copper inductor 242
Lévy flight model 891
LiF 397
LiF(100) surface 652
life-cycle costs 1137
lifetime broadening 419, 423
lifetime-force relation 485
Lifshitz theory 551, 552, 554
lift mode 338, 503
LIGA 166–168, 986
– fabrication of MEMS 987
– technique 765
ligand concentration
– effective 485
ligand-receptor affinity 288, 316
ligand-receptor interaction 476
light beam deflection galvanometer 353
light emission 135
light-emitting diode (LED) 132, 135
limits for hard disk drive technology 904
limits of biotechnology 281
linear
– creep model 890
– fractional transformations (LFT) 975
– friction–velocity dependence 889
– quadratic Gaussian (LQG) 971
– recording density 947
– stress gradient 1045
– variable differential transformers (LVDT) 361
lipid 747
lipid film 644
liposome 742
liquid
– capillaries 612
– capillary condensation 844
– film meniscus 612
– film thickness 537, 885
– helium 415
– helium operation STM 417
– lubricant 530, 832
– lubricated surfaces 572, 576, 578, 587
– mediated adhesion 832
– nitrogen (LN) 416
– perfluoropolyether (PFPE) 1003
– solid interface 832
– vapor interface 832
load
– critical 809
load contribution to friction 572–574
load dependence of friction 647
load-carrying capacity 824
load-controlled 610
load-controlled friction 571, 579
load-displacement 528
load-displacement curve 774, 778, 808, 809
loading curve 701
loading rate 483
local deformation 504, 518
local deformation of material 524
local density of states (LDOS) 422
local mechanical spectroscopy 664
local pinning 727
local shear stress 731
local stiffness 362, 1006
localization
– effect 122–125, 134
– theory 120
localized heat bonding 1116–1120
logarithmic-like friction–velocity behavior 885
logic
– devices 132
– gates 14, 15, 132
– protocols 15
Lognormal distribution 1127
London
– dispersion interaction 545, 550, 552
– penetration depth 431, 440
longitudinal
– magnetoresistance 120, 121
– piezo-resistive effect 356
– relaxation time (T_1) 305
long-range
– attraction 550, 562
– elastic deformation 722
– force 389, 398, 437
– tip-molecule force 406
long-term
– measurements 414
– memory 734
– reliability 1126
– stability 414, 948, 1102

– test 1125
loop-transfer recovery (LTR) 971
loss modulus 702
lot-to-lot variation 1072
low cost gyroscope 1094
low pressure chemical vapor deposition (LPCVD) 770
low temperature
– AFM/STM 332
– condition 60
– NC-AFM 402
– SFM (LTSFM) 417
low-cycle fatigue resistance 1032
lower lasing threshold 135
low-noise measurement 636
low-temperature microscope operation 414
low-temperature scanning tunneling spectroscopy (LTSTM) 419
low-temperature SPM (LTSPM) 414
LPCVD (low pressure chemical vapor deposition) 770, 962, 1034
– nitride sealing 1117
LTR
– loop-transfer recovery 971
LTSFM 417
LTSTM 419
lubricant 723
– atoms 732
– film 618
– meniscus 864
– spreading 864
– thickness 885
lubricated sliding 884
lubrication 717
– elastohydrodynamic 569, 581, 583
– intermediate or mixed 569, 570, 582, 583
– method 995
lumped
– parameter 1065
– parameter model 1063, 1073
Luttinger liquid 66
LVDT
– linear variable differential transformers 361

M

macrocyclic polyether 13, 25
macromolecular
– building block 747
– interaction 475
macromolecules as elastic rods 753
macroscale friction 842, 998
macroscopic
– building blocks 10
– friction 732
magnetic
– actuators 259
– applications 137
– dipole interaction 403
– disk 824
– disk drive 794
– information 138
– Ni 431
– ordering 129
– particles
self-ordered arrays 914
– quantum flux 440
– recording 911, 922
– storage device 795, 831
– tape 466, 513, 524, 862
– thin-film head 825
– tip 465
magnetic field
– critical strength 122
– length 122
– microscopy (MFM) 114
– triggering 299
magnetic force 386
– gradient 338
– microscope (MFM) 338
– microscopy (MFM) 326, 439
magnetic resonance
– force microscopy (MRFM) 441
– imaging (MRI) 304
magnetically coated tip 490
magnetohydrodynamic (MHD) pumping 261
magnetomotive transduction 248
magneto-optics 128
magnetoresistance 120
– longitudinal 120, 121
– transverse 121
magnetoresistive RAM (MRAM) 899, 906
magnetostatic interaction 439
magnetron sputtered carbon 799
major histocompatibility complex (MHC) 311
manganites 431
manifold absolute pressure (MAP) 1091
manipulation of individual atoms 325
manually assembled MWNT tips 378
manufacturability 1087
manufacturing “risk” change 1093
MAP sensor 1094
marginal dimension 727
martensite 665
mask, high resolution stamp 186
mass
– effective 433
mass of cantilever 331
master–slave design 970
material
– for microfluidic devices 254
– hardness 765
– molecule-based 35
– property variation 1072
– structural 1028
Matthiessen rule 120
maximum likelihood estimator (MLE) 1126
MBE (molecular beam epitaxy) 152, 172
MBI (multiple beam interferometry) 547
mean time to failure (MTTF) 1127
mean-field theory 556
measurement of hardness 505
mechanical
– coupling 588
– dissipation in nanoscopic device 631
– instability 452, 609
– micropumps 260
– protection 1112
– relaxation 457, 848
– resonance 591
– spectroscopy 664
– stability 927
– surface relaxations 468
– wear 263
mechanical properties 764, 1023
– characterization 1036
– of bone 675
– of carbon nanotubes 668, 672
– of DLC coating 813
mechanically cut tips 382
mechanics of cantilevers 347
mechatronic device 952
media flatness 910
mediator molecules 304
medium
– effective 127
medium theory
– effective 127
melting freezing transition 894
melting point of SAM 853
membrane
– deflection method 1037
– fabrication of 314

– proteins 375
membrane-embedded machine 741
membranes as elastic media 754
memory
– cells and switches 116
– distance 588
– long-term 734
MEMS 147–171, 225, 270, 376, 606, 617, 764, 795, 831, 862, 901, 983, 1023, 1084
– accelerometer manufacturing 1091
– accelerometers 230
– applications 985
– capacitive switch 244
– components 1011
– device 247, 987, 990
– device operation 994
– fabrication 1111
– fiber optic switching 238
– industry 1093
– integration 1102
– manufacturing 1086
– microactuator (MA) 952, 971
– mounting condition 1105
– packaging 1111, 1116, 1128
– packaging requirement 1112
– passivation 1103
– post-packaging 1114, 1117
– pressure sensor production 1090
– production economics 1086
– resonators 244
– sensor 229
– standardization 1086
– storage device 907
– structure 1112
– surface 1104
– switch, metal-to-metal 244
– technology 233, 237, 249, 1034
– testing 1099
– tribological problems 989
– tunable capacitors 241
– vacuum packaging 1117
– wafer bonding 1095
– wafer protection 1097
MEMS/NEMS 530
MEMS-based
– microfluidic system 270
– storage concept 904
– storage device 922
meniscus
– bridge 831, 876
– force 516, 871
– of liquid 564
mercaptopropyltrimethoxysilane (MPTMS) 191
merocyanine 18
mesoporous molecular sieves 102
mesoscale osmotic actuator 269
messenger RNA 750
metabolic facility 282
metal 103
– catalyst 379
– cavity packaging 1099
– cluster agents 306
– electroplating 255
– evaporated (ME) tape 793
– matrix composites 81
– nanowires 69
– oxide 397, 400
– particle catalyst 54
– porphyrin (Cu-TBPP) 404
– seed layer 966
metal/insulator/metal (MIM) 197
metal-catalyzed
– chemical vapor deposition (CVD) 379
– polymerization method 287
metal-deposited Si surface 395
metallic nanotube 34
metallic SWNT 66
metal-organic compound 56
metal-particle (MP) tape 825
metal-to-metal MEMS switch 244
metastable lubricant systems 894
methylene stretching mode 842
methyl-terminated SAM 843
Meyer's law 708
MFM sensitivity 338
MgO tip 398
MgO(001) surface 403
Mg-terminated tip 398
MHD (magnetohydrodynamic) pumping 261
mica 102, 337, 615, 619, 640, 888
– films 100
– muscovite 615, 650
– surface 641
micelles 291, 292
micro/nanofabrication 147–150, 154
microactuator 990
– design 960
– fabrication requirements 962
– loop 972
– shape 965
microbially-derived antibiotics 293
microcantilever 994
microchannel 260
microcomponent 989
microcontact printing (μCP) 187, 834
microcrystalline graphite 803
microdevice 1026
microdispenser 266
microelectromechanical
– motor 226
– resonators 244
– switch 244
microelectronics 397, 749
micro-electro-rheological valve 260
micro-engine 908
microfabricated
– cantilever 372, 453
– silicon cantilevers 387
microfabrication 187, 226, 249
– methods 285
– techniques 285, 372
microfluidic 253
– applications 255
– channel 254, 269
– channels on silicon substrates 254
– control 260
– line 258
– mixers 265
– motherboards 270
– multiplexer 268
– network 268
– passive 262
– pumping 260
– structures 256
microfluidic device 269
– disadvantages 270
– materials for 254
– passive 257, 269
microfluidic dispenser
– passive 273
microfriction 511
micro-hinge technology 236
microindentation 689
micro-injection molding 273
micromachined
– high-Q resonators 245
– inductors 242
– notch 1032
– silicon 310
micromachining 147, 150, 151, 154, 155, 159–163, 166–168, 170
micromanipulator 988
micromechanical
– cantilevers 303
– device 1084
– structures 204
micromirror 992, 1040
– array 237
micromixer category 265

micromotor 1016
micron-sized channels 265
micropatterned SAM 848
micropipette aspiration 549
micropump 260
– passive 269
micro-Raman spectroscopy 693
microrelay 1062
micro-resonators 244
microscale
– diffusion 265
– friction 1006, 1010
– material removal 521
– scratching 503
– silicon gate 34
– wear 503, 520
– wear test 1008
microscanner 929
microscope eigenfrequency 348
microscopic friction 723
microscopic origin of friction 889
microscratch 813
– test 675
microscratching measurement 519
microsensors 229
microstrain gauge 1025, 1034
microstructure 1040
microsystem 249
micro-total analysis systems (μTAS) 257
microtransformer 197
microtriboapparatus 1011
microtubule 740
– bending 673
– buckling 673
microvalve 258
– design 260
microwear 813, 819
midplane 1046
– curvature 1048
– strain 1048, 1049
milligear system 989
millipede 901, 994
– chip 931
– system concept 921, 923, 943
MIM capacitors 197
MIMO design 971
miniaturized
– actuator 986
– connector 986
– gear 986
– motor 986
minimal detectable depression 374
minimally invasive screening 285
minimum-indentation pitch 937
mirror/yoke assembly 1089
misfit
– angle 638
– strain 1042
mismatch
– elastic 1051, 1056
mismatch of crystalline surfaces 566, 567, 593
mitochondria 747
mixed C-N nanotube 56
mixed lubrication *see* intermediate or mixed lubrication, 884
mixing performance 265
MMA air bag sensor 1092
Mn on Nb(110) 429
Mn on W(110) 432
m-nitrobenzylidene propanedinitrile (*m*-NBP) 916
MOCVD (metallorganic chemical vapor deposition) 152, 172
mode coupling gratings 199
model potentials 719
modeling
– at the nanoscale 752
– mechanics of systems 752
– tribological process 753
modular multifunctionality 280
modulation codes 946
modulus
– elastic 664, 698, 767, 770, 774, 805, 806, 809
modulus of elasticity 325, 516
molarity 749
molding 186
molecular
– AND gate 17
– assembly 26
– beam epitaxy (MBE) 988
– building blocks 12, 35, 747
– chain 839
– conduction band 33
– conformation 864
– cooperation 886
– dynamics (MD) 476, 718, 749
– dynamics (MD) calculation 639
– dynamics simulation (MDS) 488, 622, 631, 651, 753
– engineering 281
– interaction 879
– interconverting state 15
– logic gates 15
– lubrication 885
– machines 13
– mobility 894
– motor 13, 673, 740
– NOT gate 16
– placement by force microscopy 285
– precursor 58
– pump 741, 748
– recognition force microscopy (MRFM) 476
– recognition force spectroscopy principles 482
– resolution 421
– reversible transformation 15
– scale layer 833
– shape 558, 559, 586, 591, 592
– shuttle 14
– spring 536
– spring model 843, 848
– stiffness 849
– switch 14, 35
– theory of lubricated friction 890
– wires 289
molecule-based
– device 24, 28
– materials 35
– switches 15
moment of inertia 1030
monodispersity 286
monolayer 885
– switches 28
– thickness 855
monolithic accelerometers 231
monolithic high-Q MEMS 245
Monte Carlo simulation 848
MoO_3 film 614
Moore's law 1142
Morse potential 464
MoS_2 friction 640, 641, 668
motion
– unperturbed 390
motor
– actin-myosin 751
– flagellar 750
– micro 1016
– microelectromechanical 226
– miniaturized 986
– molecular 13, 752
– myosin 741, 745
– oil 884
– voice coil 952, 969
movement-constrained macromolecules 305
moving media model 908
moving-tips model 908
MRAM (magnetoresistive RAM) 899–907
– cell 906
MRFM

– magnetic resonance force microscopy 441
MRI contrast agents 305
MTTF (mean time to failure) 1126
multicomponent nanodevice 288
multi-input-multi-output (MIMO) 956
multilayer 709
– devices 197
– Fe-N/Ti-N 675
– structures 104
– thin-film 825
multimode AFM 338
multiple beam interferometry (MBI) 547
multiple cantilever 907
multiple therapeutic nanodevice 294
multiple-asperity contact 517
multiplexed sensor arrays 304
multiplication of dislocations 703
multivalent targeting strategy 298
multiwall carbon nanotubes (MWCNT) 43–49, 289, 342, 376, 667, 727
muscle contraction 740
muscovite mica 615, 650
μ-synthesis 974
μ-synthesis controller design 975
MWCNT
– multiwall carbon nanotubes 43–49, 289, 342, 376, 667, 727
MWNT (multiwall nanotube) 64
– based catalyst-support 76
– bc (bamboo-concentric) 44
– bh (bamboo-herringbone) 44
– bunches 58
– catalytically grown 673
– composites
  MWNT-Al composite 81
– flexural modulus 67
– purified 378
– surface area 64
– tips
  manually assembled 378
– Young's modulus 669
myosin motor 751

N

N acceptors 10
Na on Cu (111) 421
NaCl 397
– island on Cu(111) 655
– islands 468
– thin film on Cu(111) 398
NaCl(001) 397
NaCl(100) 641–643, 648
NaF 397, 640
NAND gate 15
nanites 1146
nanoasperity 515, 695
nanobeam 669
– array 766, 774
nanobiological
– applications 313
– design 312
– devices 281
– vaccine design 312
nanobiotechnological device 300
nanobiotechnological therapeutic device 313
nanochannel glass 100, 102
nanochemistry 984
nanocluster 400
nanocomposite
– coating 833
– materials 28
nanocrystallites 800
nanodeformation 864, 880
nanodevice 85, 281
– multiple therapeutic 294
nanoelectrodes 32
nanoelectromechanical systems (NEMS) 226, 246, 606, 764, 901, 984
nanoelectronic device 32, 34
nano-electronics 84
nanofabrication 147, 170, 175, 187, 326, 526
– techniques 253
nanofatigue 810
nanofiber 44, 53
nanofilament 44, 60
nanogaps 31
nanohardness 695, 804, 805, 1000
nanohelix 12
nanoimprint 171
nanoindentation 674, 688, 697, 805, 1024
– curve 677
– measurement 504
nanoindenter 765, 769, 818, 1028
nanolubrication 883
nanomachine 1147
nanomachining 325, 326
– techniques 205
nanomagnetism 431
nanomania 1146
nanomaterial 662
nanomechanical Si beams 247
nanometer resolution 186
nanometer servo precision 979
nano-objects 68
nanoparticle 49, 374
– contrast agents 305, 306
nanoparticulate carriers 310
nanophobia 1146
nanoporous membranes 285, 314
– microfabrication 286
nano-positioning 359
nanorheology 584
nanoscale
– antimicrobials 293
– architectural property 283
– assemblers 1146
– biomolecules 10, 12
– control of molecular events 310
– data bit 909
– drug delivery 291
– electrodes 23, 27
– FET 117
– film morphology 917
– interventions 285
– mechanical properties 661, 662
– molecular therapeutic device 308
– molecules 35
– Schottky barrier 34
– supramolecular assembly 407
– therapeutic device 281, 315
– therapeutic platforms 284
– wear 518
Nanoscope I 328
nanoscratch 522
– studies 765
nanostructures 147, 170, 171, 179
– containing proteins 307
– mechanical properties 764, 1023
nanoswitches 15
nanotechnological
– actions 755
– therapy 307
nanotechnology 280, 918, 1139
– definition 1136
– ethical issues 1136
– in the living world 740
– possible applications 1136
– recommendations 1138
– role of social sciences 1148
– social issues 1136
– unintended consequences 1143
nanotherapeutic 280, 288, 303
– contrast agents 304, 310
– delivery device 298
– design paradigms 282
– device application 307
– device components 285
– devices in oncology 307

– drug delivery 311
– triggering external stimuli 299
– vaccines 312
nanotransfer printing (nTP) 190
nanotransistor 32
nanotribology 717, 862
nanotube 12, 105, 133
– AFM tips 379
– buckling 377
– chemistry 67
– defects 47
– functionalization of wall 72
– functionalized 71, 72
– growth 45
– length 382
– morphology 49
– nucleation 45
– oxidation 72
– resistance 34
– strain 671
– surface energy 380
– tip fabrication 380
– tips, surface growth 380
– transistor 34
– volume fraction 83
– yield 49, 59
nanotube production 56
– efficiency 49
– electric-arc method 48
– laser ablation 46
– solar furnace 52
– techniques 57
nanotube-based
– emitter 74
– hybrid material 71
– sensor 75
– SPM tip 73
nanotube-Co-MgO 81
nanotube-containing
– composite 82
– electrical device 289
nanotube-Fe/Co-$MgAl_2O_4$ 81
nanotube-Fe-Al_2O_3 81
nanotube-polymer composite 83
nano-tweezer 85
nanowire 31
– applications 131, 138
– metal 69
– nucleation 110
– photodetectors 135
– semimetallic 118
– stress-induced crystalline 107
– superlattice 132
– Zn 120, 123
– ZnO 135
nanowire device
– two-terminal 132
naphthalene channel 21
National Nanotechnology Initiative (NNI) 1137
native oxide layer 868
Nb superconductor 423
NBIC convergence 1140
$NbSe_2$ 431, 649
NC-AFM (noncontact atomic force microscopy) 397, 433
near-field
– method 903
– scanning optical microscopy (NSOM) 128, 372
– technique 434
near-surface mechanical properties 665
necrotic domains 307
negative logic convention 26
negative magnetoresistance 121
neighbouring indents 678
NEMS 831, 983, 988, 1023
– devices 246
– nanoelectromechanical systems 226, 246, 606, 764, 901, 984
neocarzinostatin (NCS) 309
nested-arc structure 680
net electrostatic torque 1066
net fluid transport 261
newspaper
– electronic 196
Newton's first law 454
Newtonian flow 570, 580, 581, 583
Newton–Raphson method 753
n-hexadecane 888
N-hydroxysuccinimidyl (NHS) 478
Ni(001) tip on Cu(100) 653
Ni(111) tip on Cu(110) 653
Ni, Co catalyst 47
Ni/Y catalyst 49
Ni^{2+}-chelating 485
nickel-iron permalloy 259
Ni-Fe valve membrane 259
NIH3T3 fibroblast cell 489
NiO 402, 417
NiO(001) 397, 400, 402
NiO(001) surface 403
Ni-P
– beam 785
– cantilever 778
– film 770, 771
– microbeam 777, 785
NiTi 665
– transformation behavior 666
nitrogenated carbon 800
NMP (no moving part) valves 265
noble metal surfaces 424
n-octadecyltrichlorosilane ($n\text{-}C_{18}H_{37}SiCl_3$, OTS) 616, 842
n-octadecyltrimethoxysilane (OTE) 615
noise 387, 391, 453
– 1/f 387, 957
– electronic 333
– external 415
– performance 391
– signal 941
– source 355
– vertical 391
Nomarski-interferometer 352
nonconducting film 798
nonconductive
– materials 404
– sample 329
– surface 462
nonconforming contacts 890
noncontact
– dynamic force microscopy 654
– friction 438, 655
– imaging 331
– mode 373, 388, 909
noncontact atomic force microscopy (NC-AFM) 374, 385, 391, 393, 397, 404, 654
nondestructive contact-mode measurement 652
nondimensional curvature 1051
nonequilibrium interactions *see* dynamic interactions
nonequilibrium simulations 720
noninvasive device 283
nonlinear
– force 460
– geometry effect 1058
– I–V behavior 124
– material behavior 1058, 1059
– optical property 129
– spring 491
– strain-displacement relation 1055
nonliquidlike behavior 584
nonmagnetic Zn 431
non-Newtonian behavior 885
non-Newtonian flow 570, 581, 583
nonpolar group 838
nonspherical tip 647
nonsymmetrical passive valves 268
nontherapeutic nanobiological device 283
nonvolatile random access memories (NVRAM) 899–913
nonwetting 560
NOR gate 15

normal friction 569, 577, 579, 580
normal load 721, 866
– external 867
normalized frequency shift 463
no-slip boundary condition 729
NOT gate 15
N-succinnimidyl-3-(S-acethylthio)propionate (SATP) 478
NTA (nitrilotriacetate) - His_6 478
nTP (nanotransfer printing) 197
nuclear weapon 1140
nucleation of nanowires 110
nucleic
– acid (NA) 294, 299
– acid hybridization 294, 300
nucleotides 747
n-undecyltrichlorosilane ($n\text{-}C_{11}H_{23}SiCl_3$, UTS) 842

O

O acceptors 10
$(OCN)^-$ 398
octamethylcyclotetrasiloxane 888
ODD performance 903
odd-even effect 621
off-track error 958
OH-terminated tip 398
on-chip
– actuator 1031
– electronic detection 231
– integration 210
– microvalves 260
– spiral inductors 242
oncological nanotherapeutics 291
one-dimensional quantum effect 115
one-dimensional system 100
open loop architecture 1093
operation 387
opsonized particle 296
optical
– absorption 129
– applications 134
– beam deflection 607
– cavity 135
– data switching 237
– deflection systems 372
– detector 331
– disk drives (ODD) 899, 900
– gap behavior 127
– head 336
– head mount 337
– integration 236
– MEMS 233
– output 16
– property 126, 127
– reflection 127
– switch 135
– transmission 127
– trap 480
– tweezers (OT) 480, 549, 744
optical lever 353
– angular sensitivity 355
– deflection method 362
– optimal sensitivity 355
optimal beam waist 355
optimum sampling phase 946
optoelectronic
– components 198
– module 236
– switch 136
OR
– operation 23
– operator 15
order
– in-plane 567, 584, 585, 593
– long-range 570
– out-of-plane 557–559, 567, 570, 584, 585, 591
– parameter 570, 586
ordering nanowires 108
organelles 281
organic
– compounds 835
– device structure 917
– film 916
– inverter circuit 196
– technology 916
– transistor 193, 196
organics 915
– data storage 916
organized molecular arrays 23
organometallic vapor phase epitaxy (OMVPE) 988
Orowan strengthening 710
orthogonal conjugation 295, 296
orthotropic Young's modulus 1048
oscillating tip 334, 469
oscillation
– amplitude 389, 453, 454, 457, 461
– loop 462
oscillatory force 544, 545, 557–561, 567
osmosis response time 269
osmotic
– interactions 545, 555, 562, 564
– pressure 555, 562, 575
– stress technique 549
OTE (octadecyltrimethoxysilane) 615
OTS (octadecyltrichlorsilane) 848, 1075
OUM cell 906
out-of-plane stiffness 959
overall barrier height 887
overdamped SFM system 891
overdrive capability 1062
overwriting mechanism 937
Ovonyx 915
Ovonyx Unified Memory (OUM) 906
oxide layer 703
– amorphous 106
oxide-enhanced nanowire growth mechanism 107
oxide-sharpened tips 375
oxidized nanotube 72
oxygen
– content 826
– sensor 274

P

package level integration 1093
packaging 248, 249
– cap 1112
– equipment 1099
– induced thermal stress 1123
– reliability 1125
– uniform 1112
packing density
– chains 842
palladium nanowires 137
PAMAM (polyamido amine) dendrimer 286, 287
p-aminophenyl phosphate (PAPP) 271
paper-like displays 193
paraboloid load displacement 697
paraffins 835
parallel-plate
– actuator 955
– capacitor model 1064
– sensing 957
paramagnetic particles 305
parasitic charging 1077
particle
– contamination 1098
– control 1098
– opsonized 296
– track-etched mica films 103
particle model
– embedded 894
Paschen curve 1076
passive
– check valves 262

– linearization 361
– microfluidic device 257, 269
– microfluidic dispenser 273
– microfluidics 262
– micromixers 265
– micropump 269
– microvalves 257, 262
– structure 1025
– valve geometry 264
patient-monitoring systems 274
pattern of ink 186
pattern transfer 191
patterned
– conductive substrate 110
– film 1053
– self-assembled monolayer (SAM) 188
– silicon chip 25
patterning 1075
Pb on Ge(111) 428
PbI_2 filled SWNT 70
PDP (2-pyridyldithiopropionyl) 478
peak indentation load 810
peapods 71, 289
PECVD (plasma enhanced chemical vapor deposition) 833, 926, 1009
– carbon sample 800
PEG (polyethylene glycol)
– crosslinkers 479
pegylated molecules 313
Peierls instability 428
per molecule adhesion force 620
perfluorodecanoic acid (PFDA) 992
perfluoropolyether (PFPE) 530, 861
perfluoropolyether Z-DOL 614
performance of electromechanical actuator 1075
performance parasitics 1098
periodic
– force 725
– potential 637, 891
permanent dipole moment 407
perpendicular scan 344
persistence length 488
perturbation approach 390
perylene 404
PES (photoemission spectroscopy) 956
– generation 944
– loop compensator 973
PFPE (perfluoropolyether) 864, 1012
– lubricant 862, 868, 1003, 1016
– OH 892
pH of water environment 619, 621
pH sensor 137
phage display 297, 298
pharmacoeconomics 285
phase
– angle image 530
– breaking length 120
– curve 458
– images 470
– lag spectrum 666
– measurement 471
– signal change 458
– transformation 690, 705
phase change
– material 915
– medium 922
– RAM (PC-RAM) 899, 906
phenol 837
phenoxy-centered HOMOs 26
phonon
– confinement effects 125, 127, 131
– excitation 468
– scattering 120
phosphorus-doped polysilicon 206
photoacid 19
photoactive
– fragments 14
– solids 22
– stopper 14
photo-conductance 128
photo-crosslinker 478
photodetector sensitivity 766
photoemission spectroscopy (PES) 423
photogenerated hole 30
photoinduced electron transfer 16, 17
photoinduced proton transfer 19
photolithography 285
– near-field conformal 187
– proximity mode 186
photolithography patterns 186
photo-luminescence (PL) 127, 135
– optical imaging 128
photoresist 188, 190, 1095, 1121
photothermal effect 465
physical properties of nanowires 110
physical vapor deposition 104
physisorbed protein layer 476
physisorption 861
pick-up
– SWNT tips 382
– tip method 381
pico-slider 960
piezo
– ceramic material 360
– effect 359
– element 461
– excitation 454
– hysteresis 415
– relaxation 415
– stacks 361
– tube 359
piezoelectric
– actuation 259
– actuator 908
– drive 327
– leg 636
– positioning elements 414
– scanner 372, 415, 676
piezoresistive
– cantilever 355
– coefficients 356
– detection 355, 387
– pressure sensor 1089
– sensing 958
– sensor 227
piezoscanner 363, 382
piezotranslator 363
piezotube calibration 329
pigment particles 195
pile-up, nanoindent 693, 699, 700, 709, 711
– characteristic 935
piling-up behavior 674
pilot signal 945
pinning 441
– forces 727
pin-on-disk tribotester 841
PL (photoluminescence) NSOM imaging 135
plane-to-plane separation 12
plasma
– etch 286
– frequency 127
– resonance 127
plasma enhanced chemical vapor deposition (PECVD) 197, 339, 770, 795, 799
plasminogen activator (PA) 302
plasmon mode 421
plastic
– asperity model 884
– circuit 193
– contact regime 517
– deformation 522, 728, 812, 854
– electronics 197
– fluidic chips 273
– large area circuit 194
– packaging 1099
plasticity 1060
platelet nanofiber 60

platform 929
– fabrication process 930
platinum-iridium 330
– tip 29
PMMA (poly(methyl methacrylate)) 247, 257, 671, 924, 933, 936, 994
p-n
– diodes 116
– junction diode 132
– junctions 132, 135
pneumatic
– actuation 258
– damping 418
p-nitrophenolate 19
point probes 688
Poisseuille flow 885
Poisson statistics 620
Poisson's ratio 698, 1047
polar
– group 838
– lubricant 832, 861
polarization dependence 129
polishing 793, 1000
poly(amido) amine (PAMAM) dendrimer 286
poly(ethylene glycol) (PEG) 477
poly(ethyleneterephthalate) (PET) 192
poly(methyl methacrylate) (PMMA) 247, 257, 671, 924, 933, 936, 994
polyaromatic shells 63
polycarbonate (PC) 257
polycrystalline
– films 210
– graphite 801
– Si 226
– SiGe 210
polydimethylsiloxane (PDMS) 187, 834
polyethylene (PE) 257
polyethylene co-propylene 886
polyethylene terephthalate (PET) 524
poly-Ge
– mechanical properties 210
polymer 255, 610, 664, 910
– blend 471, 662
– brush 563, 575, 581
– chemistry 286, 291
– conjugation (pegylation) 296
– fluids 892
– intermediates 287
– matrix composites 82
– medium 932
– melt 886
– membranes 103
– microfabrication techniques 255
– microstructures 257
– mushroom 562
– nanojunction 32
– properties 263
– therapeutics 307
– transparent microcapsules 195
polymeric
– liquids (melts) 561, 581, 583
– magnetic tape 526
polypeptides 10
polypropylene 404, 471
polysilicon 963, 967, 1010, 1040
– beam 1044
– fatigue 1036
– film 229, 995, 1027
– fracture strength 1036
– fracture toughness 1035
– friction 1036
– layer 263
– mechanical properties 1034
– MEMS 206
– microstructure 1035
– residual stress 1034
– resonators 245
– strip 245
– surface micromachining 210
– Young's modulus 1035
polystyrene (PS) 257, 936
polysulfone 936
polytetrafluoroethylene (PTFE) 639
polyurethane 471
poor mixing 889
pop-ins 690
pop-outs 690
population explosion 1141
porous
– silica or alumina 120
– vycor glass 102
position
– accuracy 363
– error 972
– error signal (PES) 943, 968
– sensing resolution 957
positioning 109
positive logic convention 18, 26
post-packaging process 1118
potential barrier 887
potential energy
– total 720
power dissipation 237, 470, 654
power to weight ratio 752
PQ design method 970
Prandtl–Tomlinson model 725–727
pre-crack length 1027
precursor 56, 69, 70
– gases 210
preferential flow direction 261
preparation improvements 54
pressure 871
– injection 104
pressure sensor 227
– fabrication process 1090
– packaging 1099
pressure-induced phase reconstruction 888
primary minimum adhesion 544, 555, 559
printed
– coil 197
– DFB resonator 200
probability density functions 891
probe
– FIB-milled 342
probe storage 901, 907, 908
– data organization 909
– operating parameters 909
probe tip 372
– characterization 611
– performance 374
probe-sample distance 457, 462
probe-surface distance 455
process
– control 1088
– flow 1088
– gas 799
process coherence volume 887
processing conditions 44
prodrugs 298
product architecture 1088
production efficiency 47
programmable microfluidic systems (sPROMs) 266
programmed assembler 1146
projection mode photolithography 186
propagation of cracks 524
properties of a coating 795
protection of sliding surface 831
protein 477
– bioactivity 294
– coupled to nanomaterials 294
– folding 293
– primary structure 293
– secondary structure 293
– synthetic chemistry 294
– tertiary structure 293
protofilament 673
prototyping of biological nanodevices 282
protrusion force 545, 564

Subject Index

P-selectin glycoprotein ligand-1 (PSGL-1) 488
PSG (phosphorus-doped glass) 1117
Pt alloy tip 330
PTCDA 404
PTFE (polytetrafluoroethylene)
– coated Si-tip 640
– film 833
Pt-Ir
– tip 330
– wire cantilever 912
pull-hold-release cycle 485
pull-in 1072
– angle 1068
– charge 1064
– instability 956
– voltage 1062, 1064, 1067, 1068, 1071, 1075
pull-off force *see* adhesion force, 608, 614, 615, 617, 635, 864
pulse-etching 379
pumping chamber 261
purified MWNT 378
pyramidal
– AFM tip 615
– etch 373
pyrazoline 16
pyrolysis of hydrocarbons 59
pyrolytic method 58
PZT (led zirconate titanate)
– actuated suspensions 967
– scanner 500
– tube scanner 328, 334

Q

Q-Control 464, 465
Q-factor 388, 436, 455, 1126
quad photodetector 334
quadrant detector 354
quadrature signal 943–945
quality factor *Q* 339, 348, 389, 454, 911
quality system 1101
quantization energy 117
quantum
– corrals 427
– dot 134, 172, 173
– effects 126, 134
– Hall regime 438
– limit for the thermal conductance 125
– subbands 116
– well 421
– wire superlattices 123
– yield 16
quantum confinement 99, 126, 127, 129
– effect 116, 135, 138
quantum size
– effect 118, 134
– regimes 116
quantum-confined structures 133
quartz-crystal microbalance (QCM) 717
quasi-optical experiment 421
quasi-static
– bending test 775, 777
– mode 388
– strain rate 677

R

radial cracking 808
radio-frequency (RF) MEMS 239
radiotherapy 307
radius of gyration 561
Raleigh's method 348
Raman
– spectra 131, 802
– spectroscopy 800
ramped creep model 890
random access memory (RAM) 28
random access time 900
random force 891
randomly oriented polysilicon 1035
rapid thermal processing (RTP) 1118
raster scan 901
Rayleigh wave 692
Rayleigh–Ritz method 1071
RbBr 397
RC-oscillators 358
reaction kinetics 483
reactive
– lubrication 892
– sealing 1114
– spreading 189
reactive ion etching (RIE) 205, 247, 254
read
– channel 939
– time 908
readback
– rates 932
– signal 940, 941
readout
– electronics 354
– signal 944
receptor binding 295
receptor-ligand
– bond 482
– interaction 484, 545
– unbinding 483
– unbinding force 476
receptor-mediated uptake of synthetic nanomaterial 296
recognition force microscopy (RFM) 476
recognition force spectroscopy (RFS) 484
recognition imaging 489
recombinant protein therapeutics 285
rectangular cantilever 339, 610, 656
rectifying properties 132
recursive least square algorithm (RLS) 974
reduced modulus 690
reflection interference contrast microscopy (RICM) 550
reflectivity 129
relative humidity (RH) 613
– effects of 533, 873, 1004
relative position error signal (RPES) 956
relaxation
– mechanical 457, 848
relaxation time 570–574, 583, 587, 588, 593, 894
relaxation time (T_1)
– longitudinal 305
relaxation time (T_2)
– transverse 305
release
– long-term of drugs 994
– stiction 1095
– voltage 1062
reliability 831
– long-term 1126
– of MEMS 995
– testing 1125
relief on a surface 186
remanence 137
remote detection system 354
removable disk 901
repetitive IgG 484
replica molding 833
repulsive force 480
residual
– film stress 1023
– stress 692, 699, 709, 804, 809, 812
– thickness 855
resistive micro-heater 1119
resolution 233, 374
– vertical 331

resonance
– curve detection 339
– frequency 454
– mechanical 591
resonant
– enhancement effects 131
– frequency 373
Resovist 306
response time 467
rest time 869
retardation effect 550, 552
reticuloendothelial system (RES) 296
retractive slips 885
Reynolds hydrodynamic theory of lubrication 885
RF magnetron sputtering 795
rheological
– characterization 865
– model 469, 664
rhombhedral R-8 706
rigid
– biphenyl chain 853
– disk drive 862, 869
– surface asperities 723
Ritz procedure 1055
RLL (run-length-limited) code 947
RMS roughness 1010
RNA polymerase 747
robust manufacturing 1087
rolling friction 552, 568
root mean square 433
ropes (SWNT) 66
rotary microactuator 958, 993
rotating cylinder 907
rotating strain gauge 1026
rotaxane 13
rough macroscopic contact 885
roughness 558, 559, 566–568, 594
RTP (rapid thermal processing) MEMS vacuum packaging 1119
Ru(II)-trisbipyridine 13
rule of mixtures 708, 709
run-length-limited (RLL) 947
rupture forces 488

S

sacrificial film layers 1040
sales volume 1093
SAM (self-assembled monolayer) 165, 176, 178, 622, 1088
– coating 1075
– melting point of 853
– modified surface 620
sample holder 416
sandwich immunoassays 270
sandwiched molecules 26
saturated calomel electrode 24
saturated hydrocarbons 835
scale-dependent surface tension 614
scaling factor 975
scan
– area 335
– direction 345
– frequency 335
– head 417
– range 359, 360
– rate 329
– size 337
– speed 360
scanner, piezo 436
scanning
– acoustic microscopy (SAM) 663, 692
– capacitance microscopy (SCM) 326
– chemical potential microscopy (SCPM) 326
– electrochemical microscopy (SEcM) 326
– electron microscope (SEM) 110, 373, 521
– electrostatic force microscopy (SEFM) 326
– head 328
– ion conductance microscopy (SICM) 326
– Kelvin probe microscopy (SKPM) 326
– lateral range 929
– local-acceleration microscopy (SLAM) 663
– magnetic microscopy (SMM) 326
– nanoindentation (SN) 674
– nanoindenter 662
– near field optical microscopy (SNOM) 326
– polarization force microscopy (SPFM) 615
– speed 335
– system 359
– thermal microscopy (SThM) 114, 326
– tunneling microscope (STM) 385, 396, 450, 498, 922
– tunneling microscopy (STM) 114, 325, 371, 382, 397, 899
– tunneling probe 114
– velocity 519
scanning force
– acoustic microscopy (SFAM) 326
– microscopy (SFM) 326, 433, 663, 884
– spectroscopy (SFS) 433
scanning probe
– lithography 186
– methods 606
– microscope (SPM) 326
– microscopy (SPM) 371, 385, 612, 663, 899
schematic diagram of T-SLAM 663
Schottky diode 132
scission 622
scope of therapeutic nanodevices 285
scratch
– critical load 824
– damage mechanism 818, 819
– depth 771, 773, 1000
– depth profile 773
– drive actuator 1031
– profile 519
– resistance 765, 771, 853, 1001
– test 520, 814
scratch-induced damage 813
scratching measurement 325
SCREAM 161
screen printing 193
sealing 1112
second anodization 100
secondary signaling drug delivery 301
Seebeck coefficient 123, 133
seek
– control 976
– time 901
selected area electron diffraction (SAED) 111
selective electrodeposition 108
self-actuation voltage 1062
self-aligning optical system 236
self-assembled
– growth 420
– microscopic vesicles 834
– monolayers (SAMs) 27, 165, 176, 187, 476, 536, 606, 618, 703, 833, 839, 861, 1003, 1004
– nanoparticles 312
– nanotube bunches 58
– structures 291, 310
self-assembled monolayer (SAM)
– patterned 188
self-assembly 131, 147, 170, 173, 176, 178, 179, 288
– in biological systems 740
– of nanostructures 289
– of organic layers 23

self-enhancing instability 458
self-excitation
– modes 461
– scheme 464
self-lubrication 652, 653
self-regulation of a therapeutic device 302
self-replicating nanobot 1146
self-tuning control 974, 976
semibiological nanodevices 281
semiconductor 103, 391
– economics 1086
– quantum dots 428
– surface 392
– SWNT 74, 85
semimetallic nanowire 118
semimetal–semiconductor transition 118, 123
sensing
– channel 303
– element 227
– modalities of therapeutic nanotechnology 302
sensing devices
– biological 137
– chemical 137
sensitivity 356, 357, 390, 467, 479, 633, 972
sensitizer 30
sensor 233
– and effector communication 283
– applications 117
– noise 975
– systems 302
– transducer interface 304
sequential logic circuits 20
sequential microfluidic manipulation 268
servo
– demodulation 945
– field 943
– loop 943
– self-writing 945
servo control design 968
sessile-drop 832
SFA (surface forces apparatus) 884, 894, 1011
– measurement 613
SFM (scanning force microscope) 884, 888, 890, 894
– shear force experiments 893
shallow indentation measurement 679
shape memory alloys (SMA) 260
sharpened metal wire tip 382
shear
– energy 887
– force *see* kinetic friction, static friction
– melting 570, 586
– modulus 633, 933
– rate 866
– strength 646
– stress 850, 887
– thinning 581, 583, 584, 589
shear flow
– damping 1073
– detachment (SFD) 479
shear stress
– critical 571, 572, 577, 585, 586
– effective 648
shearing force 848
shearing interface 721
shear-melting transitions 734
SH-group 476
short-cut carbon nanotubes 428
short-range
– chemical force 437, 464
– chemical interaction 402
– contribution 386
– electrostatic attraction 399
– electrostatic interaction 398
– energy dissipation 468
– interatomic force 455
– magnetic interaction 397, 402, 403
shot noise 355
Shubnikov–de Haas quantum oscillatory effect 122
shuttle
– displacement 957
– finger 956
Si 373, 375, 619
– adatom 393
– cantilever 373, 458, 465
– MEMS 247, 985
– nanobeam 773, 781, 785
– stamp 190
– tip 374, 392
– trimer 395
– wafer 373, 459, 619
Si(001)2×1 394
Si(001)2×1-H surface 394
Si(100) 515, 520, 614, 818, 820, 862, 867–869, 875, 877, 878, 995, 1012
– wafer 863
Si(100)2×1:H monohydride 393
Si(111) 519, 527, 848, 850, 999
Si(111) surface 404
Si(111)$\sqrt{3}\times\sqrt{3}$-Ag 395, 396
Si(111)(7×7) 434, 468, 640
Si(111)(7×7) surface 389, 397, 462, 639
$Si(OH)_4$ film 1013
Si_3N_4 375, 501, 619
– based tip 619
– cantilever 373
– layer 373
– tip 341, 374, 531, 841, 848, 866, 871, 880
Si-Ag covalent bond 397
Si-based devices 764
SiC 995
– bulk properties 995
– film 770
signal
– distribution 1112
– transduction 20
silane (SiH_4) 206
silica
– thermally grown 840
silicon 332, 356, 388, 610, 693, 705, 1111
– AFM tip 478
– cantilever 387, 633
– chemistry 478
– crystalline 254
– dioxide (SiO_2) 254
– dioxide layer 31, 33
– FET 303
– fractures 777
– fusion 1118
– grain 1035
– membrane 263
– MEMS 1095
– micromachining 227
– micromotor 866
– oxide 106, 477
– surface 477
– tip 392, 434, 463, 635
– wafer 25, 27, 186, 189, 228, 286, 453, 614, 930
silicon nitride (Si_3N_4) 254, 332, 477, 610, 615, 913, 964, 968
– cantilever 372
– layer 1117
– tip 375, 635
silicon-based MEMS sensor 229
silicon-based mold masters 255
silicon-on-insulator (SOI) 205, 241, 964
– substrates 247
silicon-only wafer 965
silicon-on-silicon contact 1012
silicon-silicon bonding 1096
silicon-terminated tip 435

silicon-to-glass fusion bonding 1120
siloxane 833
simultaneous imaging 397
– of a $TiO_2(110)$ surface 401
single asperity 646
– AFM tip 868
– contact 517, 1012
– size of 722
single bond 622
single crystal
– aluminum (100) 522
– nanowires 104
– Si(100) 770, 862
– silicon 504, 964, 995, 1024
– silicon cantilever 341, 611
– wafers 204
single crystalline iron 675
single domain magnetic 138
single magnetic flux quantum 122
single molecular motor 480
single molecule
– assays 743
– biology 743
– detection 477
– experiment 760
– studies 485
single particle model 894
single point sensor 1091
single protein therapeutics 281
single receptor-ligand pair 480
single stranded DNA (ssDNA) 746
single-input-multi-output (SIMO) 956
single-input-single-output (SISO) 969
single-layer
– beam 1044
– film 1050
single-particle wave function 422
single-stranded nucleic acid molecule (SSNA) 302
single-wall carbon nanotube (SWCNT) 34, 40, 289, 315, 342, 376, 668
single-wall nanocapsule 49
singularity in joint density of states 134
sink-in 693, 699, 700, 709, 711
– behavior 674
sintering 553
SiO_2 54, 373, 375, 531, 615
– film 770
– nanobeam 773, 774, 785
– thin films 205
SiOH groups 478
Si-SiO_2 stress formation 375
SISO 973
– design 969
site-specific drug delivery 298
Si-wafer 460
SLAM 663
slider 952
sliding
– contact 651
– direction 585
– distance 587
– friction 734
– of tip 632
– speed 889
– velocity 574, 585, 644, 884
SLIGA 168
s-like tip states 422
slip distance 732, 889
slipping and sliding regimes 586
SMA (shape memory alloys) 260
– actuation schemes 260
– driven micropumps 261
small angular deflection 1068
small specimens handling 1025
SMANCS (S-Methacryl-neocarzinostatin) 309
– albumin complexes 309
smart media card 905
smart plastic biochip 273
smooth sliding 585
snap-in process 609
$SnO_2(110)$ 400
soap-like lubricants 887
social
– perception 1145
– services 1142
societal impacts 1136
socio-technical trends 1141
soft
– cantilever 453
– coatings 708
– lithography 833
– polymer stamps 190
SOI wafer 1096
solar furnace
– devices 52
– method 61
solder bonding 1115, 1118
solid boundary lubricated surfaces 571, 572, 576–578, 587, 588
solid solution catalyst 55
solid xenon 436
SOLID95 779
solidification 578, 582
solid-like
– behavior 593
– Z-DOL(BW) 866
solution-phase synthesis 107
solvation
– effect 621, 623
– forces 545, 557, 558, 561, 567
sonolubrication 513
soot 46
SP 815
spacer chain 839–841
– length 844
spark plasma sintering (SPS) 81
spatial variation curvature 1058
specific accumulation of nanoparticles 311
specificity 752
spectroscopic resolution in STS 419
speed-energy product 1073
spin
– casting 192
– coating 887
– density waves 431
– quantization 423
spin-on-glass 965
spiropyran 18
SPM (scanning probe microscopy) 73, 372
– methods 672
– tip 452
spreading
– dynamics 892
– profile 892
spring
– sheet cantilever 341
– system 327, 929
spring constant 377, 387, 390, 393, 433, 450, 451, 469, 480, 483, 611, 633, 671, 911, 956, 1030, 1067
– calculation 347
– changes 415
– effective 637, 645, 721
– lateral 338, 636
– measurement 346
– torsional 1068
– vertical 338
sputtered coatings
– physical properties 804
sputtering
– deposition 799
– power 805
sp^3 bonding 806
sp^3-bonded carbon 813
squeeze film damping 248, 1073
SRAM
– cells 1088

– CMOS static random access memory 235
SrF_2 398, 399
$SrTiO_3$ 402
$SrTiO_3(100)$ 400, 401
stability
– mechanical 927
stabilization 1065
stable field performance 1102
stainless steel cantilever 813
stamp
– Au coated PDMS 192
– Au/Ti coated 191
– composite 188
– depth 190
– fabrication 186
– mechanical properties 187
– positioning 195
standardization 249
static
– advancing contact angle (SACA) 847
– AFM 388, 469
– charge 1103
– deflection AFM 450
– friction 571, 576, 579, 583, 585–587, 589, 590, 593, 718, 731, 1034
– friction force 722, 866, 1014
– interactions 544, 584
– mode 387, 433, 453
– mode AFM 354
statistical fluctuation model 889
statistical kernels 891
steady-state sliding *see* kinetic friction
steady-state velocity 733
step tilting 855
stepping motor 636
stereoelectronic properties 12
stick slip 592
sticking regime 586
stick–slip 571, 576, 579, 583–585, 587, 590, 876, 885, 886, 889, 894
– mechanism 636
– motion 732
stiction 579, 832, 885, 991, 1088, 1103
– phenomena in micromotors 1016
– suppression 1103
stiffness 390, 663, 819
– torsional 349
stiffness coefficient
– elastic 1048
stiffness measurement
– continuous (CSM) 810
STM (scanning tunneling microscope) 173–176, 325, 397
– cantilever 330
– cantilever material 339
– principle 327
– probe construction 330
– tip 325, 417
stoichiometry of compound nanowires 115
Stokes's equation 933
Stone–Wales defect 62, 68
Stoney equation 1024
storage
– capacity 946
– drive industry 902
– field 943
– modulus 702
– surface topography 942
strain
– energy 609
– energy difference 810
– engineering 147, 170
– theory 1144
– thermally induced 1125
– transformation 1041
strain–displacement relation 1049
stray capacitance 357
strength 1023
streptavidin 478
– mutants 484
stress 751, 930, 1023
– activation volume 887
– component 1043
– distribution 779, 1049
– field 1024
– gradient 1026
– maximum 1027
– measurement 1025
– relaxation 1060
– singularity 1043
– tensor 759
– transformation 1044
– uniaxial 1042
stresses within the capsid walls 758
stress–strain
– behavior 1028
– curve 671, 784
– relation 1048
stress–temperature response 1060
stretch modulus 751
Stribeck curve 569, 584, 884
strip domain 439
structural
– characterization 110
– entropy 886, 887
– forces 545, 557, 559, 566
– integrity 764
– material 1028
structurally programmable microfluidic system (sPROMs) 264
structure factor 730
structuring of simple liquids 886
SU-8 resist 929
sub-Angstrom deflections 372
subcontractor 1087
subculture theory 1145
sublimation 52
submicron
– electron-beam lithography 246
– ICs 247
– lithographic techniques 246
submonolayer
– coverage 864
– lubrication 731
subnanometer precision 10
sub-nanonewton precision 452
substrate 839, 840
substrate curvature technique 1024
substrate–slider interactions 727
subsurface water irrigation 676
super modulus 691
supercapacitor 85
superconducting
– gap 431
– magnetic levitation 327
– matrix 440
superconductivity 419
superconductors 103, 429
– type-I 431
– type-II 431
superficial nanotechnology aspects 1140
superlattice
– nanowire 103, 114
– structure 108
superlubricity 642
superparamagnetic materials 305
superparamagnetism 138
superstructures of nanowires 108
supramolecular
– assembly 17
– forces 35
surface
– adsorption 263
– amorphous 544
– asperity effect 513
– atom layer 388
– band 423
– barrier potential height 890
– charge 557, 621, 1074
– charge density 555–557

– chemistry 190, 296
– crystalline 544
– diffusion 190
– effects 129
– elasticity 529
– energy 619, 792, 1088
– energy (tension) 547, 552, 566, 567, 572, 574, 578
– forces 544
– forces apparatus (SFA) 479, 498, 546–548, 552, 576, 585, 588, 606, 717, 731, 884
– free energy 745, 746
– hydrophobic 544, 560
– micromachined accelerometers 231
– micromachining 163, 226, 229
– micromechanical properties 498
– microtribological properties 498
– mobility 865
– nanomechanical properties 498
– nanotribological properties 498
– passivation 1102
– potential 419, 555–557, 798
– potential measurement 503
– slope 844
– state lifetime 423
– stiffness 506
– structure 111, 400, 544, 558, 559, 566, 567, 570, 586, 591
– tension 265, 516, 618, 832, 846, 862
– terminal group 839
– topography 632, 778, 901
– transfer chemistry 190
– treatment 1104
surface roughness 383, 674, 678, 724, 773, 778, 844, 942, 997, 1009
– map 510
– measurement 500
surface-mounted biofilter 272
surfactant monolayers 547, 564, 568–570, 577, 580, 584, 585, 588
surfactants 103, 107
surgical resection 307
suspended membrane 1028
suspension 967
– vibration 959
suspension-level fabrication 967
sustainable internal pressure
– maximum 759
sustained drug delivery 269
switch
– two-state 19
switch design 1062
switches
– molecule-based 15
– monolayer 28
switching
– architecture 238
– devices 15
– energy 135
– speed 135
– time 1073
SWNT (single wall nanotube) 40, 52
– adsorption properties 63
– based FET 85
– based materials 64
– catalyst-based 61
– conductance 75
– electronic structure 66
– epoxy composite 82
– flexural modulus 67
– magneto-resistance 67
– matrix interaction 82
– production techniques 45
– rope 668, 670
– structure 41
– surface area 64
– tensile strength 67
– thermal conductivity 66
SWNT-PMMA composite 83
symmetric deformation mode 1054
sync pattern 946
synchronization 945
synthesis 100
– reactor 47
– yield 47
synthetic
– devices 281
– nanoporous membrane 313
– readback signal 941
system
– underdamped 894

T

tailoring of SAMs 841, 1006
tapping amplitude 529
tapping-mode 332, 373, 456, 501, 910
– AFM 456, 459
– etched silicon probe (TESP) 341
– resistance measurement 914
– tip assembly 507
targeting of nanomaterials 296
Taylor dispersion pattern 266
technical PVC 664
technological convergence 1139
Teflon layer 652
temperature
– annealing 227
– critical 423, 434, 1059
– dependence of friction 645
– domain 664
– effect 873
– regulation 720
– sensitive microcantilever 303
temperature-dependent resistance 118
template
– manufacturing 170, 176, 179
– synthesis 100
template-assisted synthesis 100
templates 103
templating technique 57
temporal scales 749
tensile
– load 1027
– maximum stress 783
– straining 1059
– stress 767, 1026, 1035, 1046
– test 1036
terabit 903
terabyte 901
test environment 824
tethering
– flexible 491
tetraarylmethane 13
tetracene 20
– channel 21
tetramethyl-aluminum hydroxide (TMAH) 204, 254
tetrapods 108
tetrathiofulvane (TTF) 916
T-filter 266
therapeutic
– bioactivity 312
– nanodevices 280
– nanomaterials 281
– nanotechnology 285
– photo-oxidation 299
thermal
– activation 894
– activation model of friction 884
– activation model of lubricated friction 887
– conductivity 125, 133
– drift 414, 417
– effect 642
– energy 481
– expansion coefficient 388, 1042
– expansion mismatch 1056
– fluctuation 633
– fluctuation forces 564
– fluctuations and velocity 890
– force 751

– frequency noise 415
– management 1112, 1122
– mismatch 1125
– noise 481, 891
– sensing 925
– vibration amplitude 377
thermocouples 123
thermodynamic forces 293
thermoelastic behavior 1040
thermoelectric
– applications 125, 133
– figure of merit 125, 133
– properties 122, 123
thermo-gravimetric analysis (TGA) 72
thermomechanical
– load 1040, 1058
– noise 433
– write/read 924
thermomechanics of multilayer film 1046
thermoplastics 257
thermopneumatic actuation 258
theta condition 561
thick film lubrication *see* elastohydrodynamic lubrication
thickness 1047
thickness of nanotubes 668
thin film 701, 707
thin oxide film 913
thin-film
– limit 1051
– lubrication 498
– microstructure 1040, 1041
thin-plate theory 1046, 1054
thiol 464, 837
three-axis accelerometer 233
three-digit input/output strings 20
three-dimensional force field spectroscopy 437
three-dimensional micromixers 265
three-input NOR gate 22
three-layer symmetric plate 1052
three-state molecular switch 18
three-terminal devices 31
three-way microvalve system 258
threshold potential 726
through wafer etching 1113
through-thickness
– average 1044
– cracking 807
thymine 10
Ti atom 405
TiAl alloy 675
TiC grains 508
tight-binding approximation 402
tilt angle of SAMs 855
time scales 721
timing field 943
timing recovery 945
TiO_2 400, 402, 425
– substrate 405
$TiO_2(100)$ 400
$TiO_2(110)$ 405
TiO_2-terminated layer 401
tip
– anion-terminated 399
– apex 435, 477, 655, 927
– artifact 392
– atom(s) 415
– atomic structure 383
– characterization 618
– deflection 503
– displacement 941
– Fe-coated 403
– ferromagnetic 441
– geometry 377
– hydrophilic 616
– load 933
– material 383
– mount 329
– oscillation 373
– oscillation amplitude 654
– performance 382
– preparation in UHV 635
– preparation method 330
– properties 382
– radius effect 339, 515, 813, 876, 878
– shape 611
– surface 477
– vibration 460
– vibration amplitude 664
tip-bound
– antigens 484
– biomolecule 303
tip-broadened image 374
tip–cantilever assembly 376
tip-induced atomic relaxation 435
tip-induced quantum dot 424
tip–liquid interface 531, 868
tip–molecule
– distance 419
– gap 406
– long-range force 406
tip–particle distance 29
tip–polymer interface 935
tip–sample 451
– contact 467
– dissipative interaction 468
– distance 388, 396, 399, 401, 435, 451, 459, 507
– electric field 656
– energy dissipation 389
– force 373, 388, 433, 436, 450, 453, 455, 460, 461
– force gradient 391
– gap 910
– interaction 391, 433, 441, 455, 456, 462, 654
– interaction force 608
– interaction potential 464
– interface 372
– junction 468
– potential 390
– separation 467, 608, 609
– system 664
tip–surface
– distance 406, 468
– interaction 490, 651
– potential 636
– separation 480
tip-terminated atom 400
tire pressure monitor 1094
tissue dendritic cells (DCs) 296
tissue engineering scaffolds 310
titin 485
TMR sources 969
Tomlinson model 631, 636
– finite temperature 642
– of friction 884
– one-dimensional 636
– two-dimensional 637
tooth enamel 675
top-down
– approach 10, 133
– materials 281
topographic
– AFM image 611
– images 331
topographical asymmetry 403
topography measurement 334
torsion mirror
– vertical 238
torsional plate 1068
torus model calculation 360
total internal reflection microscopy (TIRM) 550
touch mode architecture 228
Townsend electron avalanche theory 1077
toxicity monitoring 274
trabecular bone 679
track
– centerline 943, 945
– following 943
– mis-registration (TMR) 969
– pitch (TP) 943

track-etched
– polycarbonate membranes 102
– polymers 100
track-following controller 971
tracking servo loop 943
tracks per inch (TPI) 952
transducer component 302
transfer time 901
transformation
– strain 1041
– stress 1044
transistors 132
transition metal complexes 31
transition metal oxides 402
transitions between smooth and stick–slip sliding 585
translational microactuator 958
transmembrane channel 741
transmission 129
transmission electron microscopy (TEM) 111, 374, 521, 693
transplant rejection 311
transport properties of nanowires 115, 118
transversal piezo-resistive effect 356
transverse
– magnetoresistance 121
– midplane displacement 1055
– relaxation time (T_2) 305
traveling direction of the sample 346
Tresca criterion 695
triangular cantilever 339, 349, 354
triaryl 13
tribochemical
– oxidation 999
– reaction 531, 872
tribological
– characterization of coating 805
– computer simulations 721
tribological issues
– in MEMS 989
– in NEMS 989
tribological performance 996
– of coatings 822
tribological properties
– of SAM 841
– of silicon 996
tribometer 996
(tridecafluoro-1,1,2,2-tetrahydro-oct-1-yl) trichlorosilane (n-$C_6F_{13}CH_2CH_2SiCl_3$, FTS) 842
triethanolamine 14
triethoxysilane 644
trifluoromethyl-terminated SAM 843
triggering 288
– external 284
– strategy 298
trimer tip apex 452
true atomic resolution 389, 392, 434
true tip-sample interaction 608
truth table 17, 21
T-SLAM 663
tubulin monomer 740
tumor
– architecture 307
– associated antigens (TAAs) 307
– deposition 309
– properties 307
– targeting 307
– vasculature 307
tungsten 330
– sphere 633
– tip 330, 468, 614
tunneling
– current 327, 388, 389, 424
– detector 331
– junctions 25
– tip 431
turn-on voltage 117
two-body potentials 719
two-digit input strings 20
two-dimensional
– elasticity 754
– electron gas (2-DEG) 438
– electron system (2-DES) 428
two-layer beam 1045
two-state switch 19
two-terminal nanowire device 132

U

UHV environment 638
UHV-AVM 635
ultrahigh vacuum (UHV) 391, 418, 461, 607, 635
ultrasonic
– bonding 1119
– lubrication 513
– transducer 663
ultrasound triggering 299
ultrathin DLC coatings 795
umklapp processes 125
unbinding
– force 476, 481
– force distribution 485
– pathway 488
uncapped MEMS wafer 1097, 1098
uncoated monolithic cantilever 611
uncommon failure mode 1102
unconstrained binding 477
unidirectional electron transfer 28
unimolecular level 22
universality 752
unlubricated sample 879
unlubricated surfaces *see* dry surfaces
upward cantilever deflection 746
UV-LIGA lithography 267

V

vacuum encapsulation 1116, 1117, 1126
vacuum packaged MEMS 1118
vacuum packaging 231, 1116, 1121
van der Waals 607, 617, 621
– adhesion 616
– attraction 515
– contact 759
– force 372, 386, 398, 404, 406, 450, 544, 545, 550–554, 557, 558, 618
– interaction 58, 289, 402, 457, 843, 865
– surfaces 435
vanadium carbide 691
vapor deposition 104, 120
vapor grown
– carbon fiber 53
– carbon nanofiber 56
vapor-liquid-solid (VLS) mechanism 105
varactor diodes 241
variable capacitor 240
variable force mode 362
variable temperature STM
– setup 416
variable-temperature SLAM (T-SLAM) 663
vascular
– address system mapping 297, 298
– prosthesis 310
VCM loop compensator 972
vehicle stability 232
velocity
– critical 585, 587, 590, 591
velocity dependence of friction 644
velocity effect 871
Verlet algorithm 457
vertical coupling 329
vertical rms-noise 434
vertically aligned MWNT 59
Verwey transition temperature 439

very large-scale integration (VLSI) 922, 940
vibration 749
– external 329
vibratory rate gyroscope 232
vibromotor 236
Vickers
– hardness 674, 678, 695
– indentations 771
– indenter 765
vinylidene fluoride 404
viral life cycle 755
virgin single-crystal silicon 996
virtual jump distance 889
viruses 747
viscoelastic
– effect 468
– model 934
– properties 681
– relaxation time 731
– response 731
viscoelasticity 663, 702, 711
– mapping 505, 528
viscosity 730, 885
viscous
– damping 469
– force 569, 580, 584, 1061
VLS (Vapor-Liquid-Solid) 60
– growth method 116
– mechanism 60
– method 105, 108, 137
VLSI (very large scale integration) 922, 940
voice coil motor (VCM) 952
voltage bias 383
voltage-controlled oscillator (VCO) 239, 946
voltage-displacement function 1068
voltage-to-force gain 956
von Karman plate theory 1055
vortex in superconductor 440
V-shaped cantilever 341, 610, 634

W

W tip 330
wafer
– curvature technique 1034
– fabrication 1087
– saw dilemma 1097, 1098
wafer-level vacuum packaging 1122
wafer-to-wafer
– vacuum packaging 1114
– variation 1072
waste disposal concern 1090
water 613
– capillary force 847
– meniscus 461
water vapor 691, 695
– content 824
Watson-Crick base pairing 484
wear 498, *see* damage, 618, 793, 809
– contribution to friction 650
– control 832
– damage 824
– damage mechanism 823
– debris 521
– depths 1000
– measurement 325
– mechanical 263
– mechanism 522
– of tip 728
– performance 866
– process 822
– region 523
– resistance 822, 823, 853, 856, 879, 997, 1008
– study 879
– test 822
wearless
– friction 884
– static friction 726
Weibull
– distribution 774
– distribution function 1125
– statistics 1036
weight function 390
weighting functions for μ-synthesis 976
well-aligned nanowires 111
wet (chemical) etching 254
wet etching in HF 247
wettability 846
wetting 569
– lubricants 892
white-noise driven system 889
wire boundary scattering 121
wire cantilever 339
wireless communication 1094
work
– hardening 704, 709
– of adhesion 844, 847, 1011
– of indentation 701
write/read scheme 931
writing mechanism 932

X

xenon 417
XNOR 17
X-ray
– analysis 115
– diffraction (XRD) patterns 103
– lithographic techniques 286

Y

yield point 570, 571, 585
– load 703
yield stress 703
yielding 1055
Young's modulus 81, 346, 362, 377, 415, 611, 646, 668, 677, 779, 1023, 1024, 1044, 1071
– measurement 1028
Young–Dupre equation 846
Young–Laplace equation 612

Z

Z-15 530, 862, 874–876, 878, 1006
– film 531, 879
– lubricant 1003
– lubricant film 867
– properties 863
z-axis accelerometer 230
z-axis gyroscope 233
Z-DOL 530, 614, 862, 869
– partially bonded film 880
– properties 863
Z-DOL(BW) 868, 874, 877–879, 1006
– film 530, 867, 1003
zeolites 54
zeta potential 261
zig-zag axis 42
zig-zag-type SWNT 42
zinc 704
zipper actuator 1070
Zn nanowire 120, 123
ZnO nanowire 135

Applied Physics A

Materials Science & Processing

Editor-in-Chief: M. Stuke
Max-Planck-Institut für biophysikalische Chemie,
P.O. Box 2841,
37018 Göttingen, Germany
Fax: +49 (551) 201-1330

ISSN (print) 0947-8396
ISSN (electronic) 1432-0630
Title No. 339

→ For detailed information
springeronline.com

Applied Physics A publishes regular articles, rapid communications, and invited papers about new results, focussing on the condensed phase including nanostructured materials and their applications. In addition to those about surfaces and thin films, papers on advanced processing and characterization techniques are of interest for the readers of this journal. The high print quality ensures excellent reproduction of photographs obtained in fields such as scanning probe microscopy.

Be the first to know with *Online First™*

- Read journal articles online before they appear in print
- Papers are published in their final version online shortly after acceptance
- Each online first article is fully retrievable, searchable and citable by its DOI (Digital Object Identifier), which is linked to the article.

Instructions how to prepare and submit manuscripts are to be found on an extra page of the journal.
Once a manuscript has been accepted, it is published quickly and free of page charges, if page restrictions are observed. Fifty offprints are delivered at no cost.
Applied Physics A - Materials Science & Processing is included in the [SpringerLink] Information Service. Access to the full articles is provided at no extra charge to those institutions that have subscribed to the paper version of **Applied Physics A**. The table of contents and the abstracts are accessible free of charge to all readers via the World Wide Web several weeks before shipping of the printed issue. Details can be obtained from SpringerLink Information Service.

Abstracted / Indexed in:
CAS, Current Contents, FIZ, INIS, Inspec, Techn. Inf. Center